Air Pollution Engineering Manual

Air Pollution Engineering Manual

Edited by
Anthony J. Buonicore
Wayne T. Davis

VNR VAN NOSTRAND REINHOLD
New York

This project has been funded in part by the United States Environmental Protection Agency under assistance agreement T901763 to the Air and Waste Management Association. The contents of this document do not necessarily reflect the views and policies of the Environmental Protection Agency, nor does mention of trade names or commercial products constitute endorsement or recommendation for use.

Library of Congress Catalog Card Number 91-46007
ISBN 0-442-00843-0

Manufactured in the United States of America

Published by Van Nostrand Reinhold
115 Fifth Avenue
New York, New York 10003

Chapman and Hall
2-6 Boundary Row
London, SE1 8HN, England

Thomas Nelson Australia
102 Dodds Street
South Melbourne 3205
Victoria, Australia

Nelson Canada
1120 Birchmount Road
Scarborough, Ontario M1K 5G4, Canada

16 15 14 13 12 11 10 9 8 7 6 5 4 3 2 1

Library of Congress Cataloging-in-Publication Data

Air pollution engineering manual / Air & Waste Management Association ; edited by Anthony J. Buonicore, Wayne Davis.
p. cm.
Includes bibliographical references and index.
ISBN 0-442-00843-0
1. Air—Pollution—Equipment and supplies. 2. Gases, Asphyxiating and poisonous—Environmental aspects. 3. Particles—Environmental aspects. I. Buonicore, Anthony J. II. Davis, Wayne T. III. Air & Waste Management Association.
TD889.A39 1992
628.5′3—dc20 91-46007
CIP

Contents

Foreword

The *Air Pollution Engineering Manual,* since its inception in the early sixties, has been recognized by practitioners, engineers, academics, and students as a fundamental and practical source of information on the control of air pollution. The manual was originally published by the Los Angeles County Air Pollution Control District under the editorship of John A. Danielson with assistance from personnel of the agency.

In 1973 a revised second edition was published by the U.S. Environmental Protection Agency as AP-40. The editorial and technical content of the second edition was developed exclusively by the District under the direction of Mr. Robert G. Lunche, Mr. Eric Lemke, and Mr. John A. Danielson (as editor).

The manual, with its many flowcharts, photographs, and practical descriptions of various sources and control of air pollution has continued to provide valuable, if not indispensable, insight to practitioners even though it has been two decades since its last revision. In recent years, the development of the PM-10 standard for ambient air quality has increased the need to characterize emissions from sources by particle size. Further, the Clean Air Act Amendments of 1990 mandate an aggressive review of a broad range of hazardous air pollutants and an increased understanding of the processes and the control technologies for both particulate and gaseous pollutants. In addition, it requires further reductions in acidic emissions, such as sulfur dioxide and nitrogen oxides, and in ozone precursors such as volatile organic compounds.

In light of the historical significance of the manual and the increasing need for an updated manual due to the rapidly changing technology of air pollution control, the Air & Waste Management Association proposed in 1989 to undertake the task of preparing this latest *Air Pollution Engineering Manual.* With partial funding from the U.S. Environmental Protection Agency, the Association has overseen the development of the manual, with co-editors, Mr. Anthony J. Buonicore and Dr. Wayne T. Davis, who represented the Technical and Education Councils of Air and Waste Management Association, respectively.

The editors and numerous authors from industry, federal, state, and local government agencies, trade associations, consulting firms, universities, and research organizations volunteered their efforts to produce the manual. We can only hope that this latest edition, inspired by its predecessor, will continue to be a practical and valuable source of information on air pollution control.

Acknowledgment

The editors would like to express their appreciation to the many authors who volunteered their time to make this manual possible, and to the U.S. Environmental Protection Agency for providing partial funding to initiate the project.

The editors would also like to express their appreciation to Mr. Paul Kueser, formerly the director of programs and planning at Air and Waste Management Association, for his role in initiating the preparation of the manual. A special thanks is expressed to Ms. Kate Rau, monographs editor at Air and Waste Management Association for her persistence and many long hours of work coordinating communication between the editors and the authors, and for her assistance in keeping the manual on schedule.

Anthony J. Buonicore
Wayne T. Davis

Contributor List

1
Air Pollution Control Engineering

Anthony J. Buonicore, P.E.
President
Environmental Data Resources, Inc.
Southport, CT

Louis Theodore
Professor
Manhattan College
Bronx, NY

Wayne T. Davis
Professor
Civil Engineering Department
University of Tennessee
Knoxville, TN

2
Control of Gaseous Pollutants

Absorption
Anthony J. Buonicore, P.E.
President
Environmental Data Resources, Inc.
Southport, CT

Adsorption
Anthony J. Buonicore, P.E.
President
Environmental Data Resources, Inc.
Southport, CT

Condensation
Anthony J. Buonicore, P.E.
President
Environmental Data Resources, Inc.
Southport, CT

Incineration
Anthony J. Buonicore, P.E.
President
Environmental Data Resources, Inc.
Southport, CT

3
Control of Particulates

Cyclones and Inertial Separators
Elizabeth Ashbee
Civil Engineering Department
University of Tennessee
Knoxville, TN

Wayne T. Davis
Professor
Civil Engineering Department
University of Tennessee
Knoxville, TN

Wet Scrubbers
Kenneth Schifftner
Technical Sales/Marketing
Advanced Concepts, Inc.
San Diego, CA

Dr. Howard Hesketh
Professor
Mechanical Engineering and Energy Processes
Southern Illinois University
Carbondale, IL

Electrostatic Precipitators
James H. Turner
Senior Research Chemical Engineer
Research Triangle Institute
Research Triangle Park, NC

Phil A. Lawless
Senior Research Physicist
Research Triangle Institute
Research Triangle Park, NC

Toshiaki Yamamoto
Research Mehanical Engineer
Research Triangle Institute
Research Triangle Park, NC

David W. Coy
Manager, Environmental Engineering
Research Triangle Institute
Research Triangle Park, NC

Gary P. Greiner
President
ETS, Inc.
Roanoke, VA

John D. McKenna
President
ETS International, Inc.
Roanoke, VA

William M. Vatavuk
Senior Chemical Engineer
U.S. Environmental Protection Agency
Research Triangle Park, NC

Fabric Filters

John D. McKenna
President
ETS International, Inc.
Roanoke, VA

Dale A. Furlong
President
Clean Air Technologies
El Toro, CA

4
Fugitive Emissions

John S. Kinsey
Principal Environmental Scientist
Midwest Research Institute
Kansas City, MO

Chatten Cowherd, Jr.
Principal Advisor
Midwest Research Institute
Kansas City, MO

5
Odors

William H. Prokop, P.E.
President
PROKOP Enviro Consulting
Deerfield, IL

6
Ancillary Equipment for Local Exhaust Ventilation Systems

H. D. Goodfellow
President
Goodfellow Consultants, Inc.
Missiauga, Ontario
Canada
Adjunct Professor
Department of Chemical Engineering and Applied Chemistry
University of Toronto
Toronto, Ontario
Canada

7
Combustion Sources

Coal

Wayne T. Davis
Professor
Civil Engineering Department
University of Tennessee
Knoxville, TN

Arijit Pakrasi
Civil Engineering Department
University of Tennessee
Knoxville, TN

Fuel Oil

Wayne T. Davis
Professor
Civil Engineering Department
University of Tennessee
Knoxville, TN

Arijit Pakrasi
Civil Engineering Department
University of Tennessee
Knoxville, TN

Natural Gas

Wayne T. Davis
Professor
Civil Engineering Department
University of Tennessee
Knoxville, TN

Arijit Pakrasi
Civil Engineering Department
University of Tennessee
Knoxville, TN

Wood Waste

Anthony J. Buonicore, P.E.
President
Environmental Data Resources, Inc.
Southport, CT

8
Waste Incineration Sources

Refuse
James R. Donnelly
Director
Environmental Services and Technology
Davy Environmental
San Ramon, CA

Hazardous Waste
Anthony J. Buonicore, P.E.
President
Environmental Data Resources, Inc.
Southport, CT

Medical Waste Incineration
Anthony J. Buonicore, P.E.
President
Environmental Data Resources, Inc.
Southport, CT

Sewage Sludge
Calvin R. Brunner
President
Incinerator Consultants Inc.
Reston, VA

Drum Reconditioning
Stanley M. Krinov
S.M. Krinov, Inc.
Cincinnati, OH

9
Evaporative Loss Sources

Dry Cleaning
Stephen V. Capone
Manager, Air Planning and Engineering Programs
Alliance Technologies Corp.
Lowell, MA

Petroleum Storage
Randy J. McDonald
U.S. Environmental Protection Agency
Research Triangle Park, NC

Gasoline Marketing
Stephen A. Shedd
U.S. Environmental Protection Agency
Research Triangle Park, NC

Organic Solvent Cleaning (Degreasing)
Mark B. Turner
Senior Chemical Engineer
Midwest Research Institute
Cary, NC

Richard V. Crume
Head, Air Quality Engineering Section
Midwest Research Institute
Cary, NC

10
Surface Coating

Mark B. Turner
Senior Chemical Engineer
Midwest Research Institute
Cary, NC

11
Graphic Arts Industry

Nonheatset Web Printing
H. Wilson Cunningham
Director
American Newspaper Publishers Association
Reston, VA

Heatset Web Offset
Robert M. Birkland
Vice President
Lebanon Valley Offset
Annville, PA

Gerald Bender
Vice President, Environmental Affairs
R.R. Donnelley & Sons Company
Lisle, IL

William Schaeffer
Retired
c/o Graphic Arts Technical Foundation
Pittsburgh, PA

Flexography
Warren J. Weaver
Manager
Motter Press
York, PA

Sheetfed Offset Printing
Jerry Bender
Vice President, Environmental Affairs
R.R. Donnelley & Sons Company
Lisle, IL

Nelson Ho
Director
Graphic Arts Technical Foundation
Pittsburgh, PA

David Johnson
CEO
Princeton Polychrome
Princeton, NJ

Frank Little
Manager

NIES/Artcraft
St. Louis, MO

William Schaeffer
Retired
c/o Graphic Arts Technical Foundation
Pittsburgh, PA

Gravure Printing

Jerry Bender
Vice President, Environmental Affairs
R.R. Donnelley & Sons Company
Lisle, IL

Robert Oppenheimer
Retired
Glen Cove, NY

Screen Printing

Marcia Y. Kinter
Director
Screen Printing Association International
Fairfax, VA

12
Chemical Process Industry

Acrylonitrile

Ronald D. Bell
Senior Program Manager
Radian Corporation
Austin, TX

Alan J. Mechtenberg
Staff Engineer
Radian Corporation
Austin, TX

Carbon Black

Barry R. Taylor
Cabot Corporation
Billerica, MA

Charcoal/Activated Carbon

William D. Byers
Director, Process Engineering
CH2M HILL
Corvallis, OR

Chlor-Alkali

Kenneth S. Walborn
Manager, Environmental Control
PPG Industries, Inc.
Natrium, WV

Ethylene Oxide

Robert F. Dye
Principal Engineer
Dye Engineering and Technology Company
Sugar Land, TX

Explosives

Institute of Makers of Explosives
Washington, DC

Phosphoric Acid Manufacturing

Gordon F. Palm
Gordon F. Palm & Associates
Lakeland, FL

Phthalic Anhydride

Dr. Herbert P. Dengler
Senior Engineering Associate
Exxon Chemical Company
Baton Rouge, LA

Printing Inks

Stephen W. Paine
Thiele-Engdahl, a Business Unit of ICI Americas, Inc.
Wilmington, DE

Stephen K. Harvey
ICI Specialty Chemicals
Wilmington, DE

Soaps and Detergents

Richard C. Scherr
Vice President, Southern Region
ENSR Consulting and Engineering
Houston, TX

Sodium Carbonate

Michael J. Barboza, P.E.
Chief Air Quality Engineer
Paulus, Sokolowski and Sartor, Inc.
Warren, NJ

Neil M. Haymes, MSPH
Senior Project Manager
Paulus, Sokolowski and Sartor, Inc.
Warren, NJ

Sulfuric Acid

Thomas L. Muller
Engineering Associate
E.I. du Pont de Nemours & Company, Inc.
Deepwater, NJ

Sulfur

Bruce Scott
President
Bruce Scott, Inc.
San Rafael, CA

Natural Fiber Textile Industry

Anthony J. Buonicore, P.E.
President
Environmental Data Resources, Inc.
Southport, CT

Terephthalic Acid

Ronald D. Bell
Senior Program Manager
Radian Corporation
Austin, TX

Alan J. Mechtenberg
Staff Engineer
Radian Corporation
Austin, TX

Kevin McQuigg
Senior Engineer
Formerly with Radian Corp., now with IT McGill
Tulsa, OK

Thermoplastic Resins

John A. Fey
Manager, Environmental Affairs
Union Carbide Chemicals and Plastics Company, Inc.
South Charleston, WV

J. Mitchell Jenkins III
Corporate Engineering Fellow
Union Carbide Chemicals and Plastics Company, Inc.
South Charleston, WV

Thermosetting Resins

John A. Gannon
Consultant
Formerly Senior Staff Chemist
Ciba-Geigy Corporation
Ardsley, NY

13
Food and Agricultural Industry

Bread Baking

Patrick J. Cafferty
Tuttle & Taylor
San Francisco, CA

Coffee Processing

Ronald G. Ostendorf, Editor
Section Head
Procter & Gamble Company
Cincinnati, OH

Grain Handling and Processing

Dennis Wallace
Senior Environmental Scientist
Midwest Research Institute
Cary, NC

Fermentation

Joseph A. Mulloney, Jr., P.E.
Senior Technical Specialist
EA Engineering, Science, and Technology, Inc.
Baltimore, MD

Fish, Meat, and Poultry Processing

William H. Prokop, P.E.
President
PROKOP Enviro Consulting
Deerfield, IL

John M. Sweeten, Ph.D., P.E.
Associate Department Head, Professor and Extension Specialist
Texas A & M University
College Station, TX

John R. Blandford
(Deceased)
Oscar Mayer Foods Corporation
Madison, WI

Nitrate Fertilizers

Horace C. Mann
National Fertilizer and Environmental Research Center
Tennessee Valley Authority
Muscle Shoals, AL

Ammonium Phosphates

Horace C. Mann
National Fertilizer and Environmental Research Center
Tennessee Valley Authority
Muscle Shoals, AL

Normal Superphosphate

Horace C. Mann
National Fertilizer and Environmental Research Center
Tennessee Valley Authority
Muscle Shoals, AL

Triple Superphosphate

Horace C. Mann
National Fertilizer and Environmental Research Center
Tennessee Valley Authority
Muscle Shoals, AL

Urea

Horace C. Mann
National Fertilizer and Environmental Research Center
Tennessee Valley Authority
Muscle Shoals, AL

14
Metallurgical Industry

Primary Aluminum Industry

Maurice W. Wei
Environmental Engineering Consultant
Aluminum Company of America
Pittsburgh, PA

Metallurgical Coke
Thomas W. Easterly, P.E.
Senior Environmental Engineer
Bethlehem Steel Corporation
Bethlehem, PA

Stefan P. Shoup
Advisor, Environmental Affairs
Inland Steel Company
East Chicago, IN

Dennis P. Kaegi
Section Manager, Process Research
Inland Steel Company
East Chicago, IN

Copper Smelting
Jacques Moulins, Editor
Noranda, Inc.
Rouyn-Noranda, Quebec
Canada

Ferroalloy Industry Particulate Emissions
Steve Stasko, Editor
Manager, Technical Programs
Air & Waste Management Association
Pittsburgh, PA

Steel Industry
Bruce A. Steiner
Vice President, Environment & Energy
American Iron and Steel Institute
Washington, DC

Miscellaneous Fugitive Emission Sources
Robert E. Sistek
Environmental Management Engineer
LTV Steel Company
Cleveland, OH

Frank Pendleton
Environmental Scientist
Midwest Research Institute
Kansas City, MO

Blast Furnace
Marek S. Klag, B. Eng.
Environmental Engineer
Dofasco Inc.
Hamilton, Ontario
Canada

Basic Oxygen Furnace Shops
S. S. Felton
Senior Staff Engineer, Environmental Affairs
Armco Steel Company, L.P.
Middletown, OH

E. Cocchiarella, P.Eng.
Environmental Engineer
Dofasco Inc.
Hamilton, Ontario
Canada

M.S. Greenfield
Director, Environmental Control
Dofasco Inc.
Hamilton, Ontario
Canada

Casters
Manfred Bender
President
Bender Corporation, Inc. Consulting Engineers
Beverly Hills, CA

Michael S. Peters
Manager, Environment
Structural Metals, Inc.
Sequin, TX

Electric Arc Furnaces/Argon–Oxygen Decarburization Process
Manfred Bender
President
Bender Corporation, Inc. Consulting Engineers
Beverly Hills, CA

Michael S. Peters
Manager, Environmental
Structural Metals, Inc.
Sequin, TX

Ladle Metallurgy Vacuum Degassing
Edward Cocchiarella, P.Eng.
Environmental Engineer
Dofasco Inc.
Hamilton, Ontario
Canada

Fluxed Iron Ore Pellet Production
Gus R. Josephson
Safety and Environmental Engineer
Inland Steel Mining Company
East Chicago, IN

Pickling
R.M. Hudson
Retired
Formerly Senior Research Consultant
USS Group of USX Corporation
Pittsburgh, PA

Direct Reduction
Murray S. Greenfield, P.E.
Director, Environmental Control
Dofasco Inc.
Hamilton, Ontario, Canada

Rolling
Michael T. Unger
Inland Steel Company
East Chicago, IN

Scarfing
S.J. Manganello
Research Consultant
Heavy Products Division
USS Technical Center
Monroeville, PA

Sinter Plants
S.S. Felton
Senior Staff Engineer, Environmental Affairs
Armco Steel Company, L.P.
Middletown, OH

L.M. Stuart
Supervisor, Environmental Processes and Environmental Control
Bethlehem Steel Corporation
Bethlehem, PA

Primary Lead Smelting
Paul Deveau
Director of Environment
Brunswick Mining and Smelting Corporation
Belledune, New Brunswick
Canada

Zinc Smelting
Philippe Krick
Technical Superintendent
Canadian Electrolytic Zinc Ltd.
Valleyfield, Quebec
Canada

Secondary Aluminum
Charles A. Licht, P.E.
President
Charles Licht Engineering Associates
Olympia Fields, IL

Secondary Brass and Bronze Melting Processes
Charles A. Licht, P.E.
President
Charles Licht Engineering Associates
Olympia Fields, IL

Iron Foundries
American Foundrymen's Society Air Quality Committee (10-E)
Des Plaines, IL

Secondary Lead Smelting
Charles A. Licht, P.E.
President
Charles Licht Engineering Associates
Olympia Fields, IL

Secondary Zinc
Charles A. Licht, P.E.
President
Charles Licht Engineering Associates
Olympia Fields, IL

Steel Foundries
Ezra L. Kotzin
American Foundrymen's Association
Des Plaines, IL

15
Mineral Products Industry

Hot Mix Asphalt Mixing Facilities
Kathryn O'C. Gunkel
Project Engineer
Wildwood Environmental Engineering Consultants, Inc.
Baltimore, MD

Portland Cement
Walter L. Greer
Vice President, Production and Environment
Ash Grove Cement Company
Overland Park, KS

Michael D. Johnson
Sales Representative
BHA Group, Inc.
Kansas City, MO

Edward L. Morton
Director of Engineering
Patchem Inc.
Middlesex, NJ

Errol C. Raught
Project Sales Manager
Fuller Company
Lehigh Valley, PA

Hans E. Steuch
Director of Engineering
Ash Grove Cement Company
Portland, OR

Claude B. Trusty, Jr.
Vice President, Technical Services
CBR Cement Corporation
San Mateo, CA

Fiberglass Manufacturing Operations
Aaron J. Teller
Vice President
R-C Environmental Services & Technologies
A Research Cottrell Company
Branchburg, NJ

Joseph Y. Hsieh
Manager
R-C Environmental Services & Technologies
A Research Cottrell Company
Branchburg, NJ

Glass Manufacturing
Aaron J. Teller
Vice President

R-C Environmental Services & Technologies
A Research Cottrell Company
Branchburg, NJ

Joseph Y. Hsieh
Manager
R-C Environmental Services & Technologies
A Research Cottrell Company
Branchburg, NJ

Sand and Gravel Processing

John E. Yocom, P.E., C.I.H.
Environmental Consultant
West Simsbury, CT

Stone and Quarrying Processing

John E. Yocom, P.E., C.I.H.
Environmental Consultant
West Simsbury, CT

Lime Manufacturing

National Lime Association, Editor
Arlington, VA

Coal Processing

Larry L. Simmons
Principal
E^2M, Inc.
Pittsburgh, PA

Lisa E. Lambert
E^2M, Inc.
Pittsburgh, PA

16
Pharmaceutical Industry

Richard V. Crume
Head, Air Quality Engineering Section
Midwest Research Institute
Cary, NC

Jeffrey W. Portzer
Research Chemical Engineer
Research Triangle Institute
Cary, NC

17
The Petroleum Industry

J. Eldon Rucker
Deputy Director, Health & Environment
American Petroleum Institute
Washington, DC

Robert P. Strieter
Senior Regulatory Analyst
American Petroleum Institute
Washington, DC

18
Wood Processing Industry

Dr. Arun V. Someshwar
Senior Research Engineer
National Council of the Paper Industry
for Air and Stream Improvement, Inc.
New York, NY

Dr. John E. Pinkerton
Director, Air Quality Program
National Council of the Paper Industry
for Air and Stream Improvement, Inc.
New York, NY

Chemical Wood Pulping

Dr. Arun V. Someshwar
Senior Research Engineer
National Council of the Paper Industry
for Air and Stream Improvement, Inc.
New York, NY

Dr. John E. Pinkerton
Director, Air Quality Program
National Council of the Paper Industry
for Air and Stream Improvement, Inc.
New York, NY

Mechanical Pulping

Dr. Arun V. Someshwar
Senior Research Engineer
National Council of the Paper Industry
for Air and Stream Improvement, Inc.
New York, NY

Dr. John E. Pinkerton
Director, Air Quality Program
National Council of the Paper Industry
for Air and Stream Improvement, Inc.
New York, NY

Paper and Paperboard Manufacture

Dr. Arun V. Someshwar
Senior Research Engineer
National Council of the Paper Industry
for Air and Stream Improvement, Inc.
New York, NY

Dr. John E. Pinkerton
Director, Air Quality Program
National Council of the Paper Industry
for Air and Stream Improvement, Inc.
New York, NY

19
Treatment and Land Disposal

Treatment and Land Disposal

T.T. Shen
Senior Research Scientist

New York State Department of Environmental Conservation
Albany, NY

C.E. Schmidt, Ph.D.
Environmental Consultant
Red Bluff, CA

T.P. Nelson
Program Manager
Radian Corporation
Austin, TX

Municipal Solid Waste Landfill Gas Emissions

Michael J. Barboza, P.E.
Chief Air Quality Engineer
Paulus, Sokolowski and Sartor, Inc.
Warren, NJ

20
Groundwater and Soil Treatment Processes

Lori P. Andrews
Program Director
Center for Labor Education and Research
University of Alabama at Birmingham
Birmingham, AL

Reviewers

Acott, S. Mike
National Asphalt Paving Association
Lanham, MD

Allie, Gary
Inland Steel Industries, Inc.
East Chicago, IN

Anderson, Dr. David M.
Bethlehem Steel
Bethlehem, PA

Atkins, Dr. P.R.
Aluminum Company of America
Pittsburgh, PA

Bimbo, Anthony P.
Zapata Haynie Corp.
Reedsville, VA

Brumagin, Thomas E.
National Asphalt Paving Association
Lanham, MD

Buonicore, Anthony J.
Environmental Data Resources, Inc.
Southport, CT

Collins, Liam E.
Cabot Corp.
Waltham, MA

Curtis, Daniel L.
Cabot Corp.
Waltham, MA

Davis, Wayne T.
University of Tennessee
Knoxville, TN

Deveau, Paul
Brunswick Mining & Smelting
Belledune, New Brunswick, Canada

Dorfman, Ira H.
American Bakers Association
Washington, D.C.

Dungan, A.E.
The Chlorine Institute
Washington, D.C.

EE-6 Odor Committee
Air & Waste Management Association
Pittsburgh, PA

Fisher, Perry
Campbell Taggart
Dallas, TX

Frank, W.B.
Alcoa Technical Center
Pittsburgh, PA

Gjersvik, Charles B.
Continental Baking Co.
St. Louis, MO

Grantz, James A.
Armco Steel Co., L.P.
Middletown, OH

Gray, Charles A.
Cabot Corp.
Waltham, MA

Gunderson, Edward L.
Stepan Co.
Elwood, IL

Gwinnell, Harry J.
Cabot Corp.
Waltham, MA

Hackmann, Frank H.
Sonnenschein Nath & Rosenthal
St. Louis, MO

Harmon, Mary Lou
LTV Steel Co.
Cleveland, OH

Hurst, Ronald C.
Cabot Corp.
Waltham, MA

Kerch, Richard L.
Consolidation Coal Co.
Pittsburgh, PA

Grendel, Robert W.
Monsanto Enviro-Chem Systems, Inc.

King, P.M.
PPG Industries, Inc.
Pittsburgh, PA

Krick, Philippe
Canadian Electrolytic Zinc, Inc.
Valleyfield, Quebec, Canada

Laurie, Keith
Unocal Corp.
Kenai, AK

Licht, Charles A.
Charles Licht Engineering Associates
Olympia Fields, IL

List, Stephen J.
Cabot Corp.
Waltham, MA

Lockwood, J. Douglas
Cabot Corp.
Waltham, MA

Loos, K.R.
Shell Development Co.
Houston, TX

Madrazo, Roger
Anchor Glass Co.
Tampa, FL

Majetich, J.C.
USX Research
Monroeville, PA

Meijer, Jon
International Fabricare Institute
Silver Spring, MD

Moulins, Jacques
Noranda, Inc.
Rouyn-Noranda, Quebec, Canada

Mullins, J.A.
Shell Oil Co.
Houston, TX

Newsom, Michael
Anchor Glass Co.
Tampa, FL

Ostendorf, Ronald
The Procter & Gamble Co.
Cincinnati, OH

Peters, Jim
Sterling Chemical
Texas City, TX

Pinkerton, Dr. John E.
National Council of the Paper Industry for Air and Stream Improvement
New York, NY

Rau, Robert B.
PPG Industries, Inc.
Pittsburgh, PA

Samelson, R.J.
PPG Industries, Inc.
Pittsburgh, PA

Sedlak, Richard I.
The Soap and Detergent Association
New York, NY

Shoup, Stefan P.
Inland Steel Co.
East Chicago, IN

Sistek, Robert E.
LTV Steel Corp.
Cleveland, OH

Stasko, Steve
Air & Waste Management Association
Pittsburgh, PA

Steppy, W.C.
Alcoa Technical Center
Pittsburgh, PA

Symons, Carl R.
Bethlehem Steel
Bethlehem, PA

Turner, Ronald V.
Cabot Corp.
Waltham, MA

Vanderzwaag, Dirk
Stelco, Inc.
Hamilton, Ontario, Canada

Watkins, E. Charles
PPG Industries, Inc.
Lexington, NC

Weiskircher, Roy J.
USX Corp.
Monroeville, PA

Weiss, Kenneth N.
ERM Inc.
Exton, PA

West, William L.
LTV Steel Co.
Cleveland, OH

Trade Associations

American Bakers Association
1111 14th Street N.W.
Suite 300
Washington, DC 20005
202 296-5800

American Foundrymen's Society
505 State Street
Des Plaines, IL 60016-8399
708 824-0181

American Newspaper and Publishers Association
The Newspaper Center
11600 Sunrise Valley Drive
Reston, VA 22091
703 648-1000

American Petroleum Institute
1220 L Street N.W.
Washington, DC 20005
202 682-8000

American Textile Manufacturers Institute
1801 K Street N.W.
Suite 900
Washington, DC 20006
202 862-0500

Chemical Manufacturers Association
2501 M Street N.W.
Washington, DC 20037
202 887-1100

Environmental Conservation Board of the Graphic Communications Industries
1899 Preston White Drive
Reston, VA 22091
703 264-7200

Flexible Packaging Association
1090 Vermont Avenue N.W.
Suite 500
Washington, DC 20005
202 842-3880

Flexographic Technical Association
900 Marconi Avenue
Ronkonkoma, NY 11779
516 737-6813

Graphic Arts Technical Foundation
4615 Forbes Avenue
Pittsburgh, PA 15213
412 621-6941

Gravure Association of America
90 Fifth Avenue
4th Floor
New York, NY 10011-7601
212 255-0070

Institute of the Makers of Explosives
1120 Nineteenth Street N.W.
Suite 310
Washington, DC 20036-3605
202 429-9280

International Fabricare Institute
12251 Tech Road
Silver Spring, MD 20904
301 622-1900

National Association of Printers and Lithographers
780 Palisade Avenue
Teaneck, NJ 07666
201 342-0700

National Association of Printing Ink Manufacturers
47 Halstead Avenue
Harrison, NY 10528
914 835-5650

National Printing Equipment and Supply Association
1899 Preston White Drive
Reston, VA 22091-4326
703 264-7200

National Council of the Paper Industry for Air and Stream Improvement
260 Madison Avenue
New York, NY 10016
212 532-9000

Printing Industries of America
1730 N Lynn Street
Arlington, VA 22209
703 841-8100

Screen Printing Association International
10015 Main Street
Fairfax, VA 22031
703 385-1335

The Chlorine Institute
2001 L Street Common N.W.
Washington, DC 20036
202 775-2790

The Soap and Detergent Association
475 Park Avenue S
New York, NY 10016
212 725-1262

The Fertilizer Institute
501 2nd Street N.E.
Washington, DC 20002
202 675-8250

1
Air Pollution Control Engineering

Anthony J. Buonicore, Louis Theodore, and Wayne T. Davis

INTRODUCTION

In the past two decades, the engineering profession has been heavily influenced by its responsibility to society. This responsibility has been directed toward the protection of public health and welfare and is guided by a host of environmental regulations. The extent to which the engineering profession responds to this challenge depends largely on the limits imposed by three principal considerations:

1. Legal limitations imposed for the protection of public health and welfare.
2. Social limitations imposed by the community in which the pollution source is or will be located.
3. Economic limitations imposed by marketplace constraints.

Careful evaluation within the framework of all three limitations is now essential and often integral to corporate strategic planning processes.

The control strategy for environmental impact assessment often focuses on five alternatives whose purpose would be the reduction and/or elimination of pollutant emissions:

1. Elimination of the operation entirely or in part
2. Modification of the operation
3. Relocation of the operation
4. Application of appropriate control technology
5. Combinations thereof

In view of the relatively high costs often associated with pollution control systems, engineers today are directing considerable effort toward process modification to eliminate as much of the pollution problem as possible *at the source*. This includes evaluation of alternative manufacturing and production techniques, substitution of raw materials, and improved process control methods. Unfortunately, if there is no alternative, the application of pollution control equipment must be considered. Considering the relatively high costs, proper selection of this equipment is essential. The equipment must be designed to comply with regulatory emission limitations on a continual basis, with interruptions subject to severe penalty depending on the circumstances. The requirement for design performance on a continual basis places very heavy emphasis on operation and maintenance practices. In fact, it is not unusual that favorable operation and maintenance requirements associated with a particular piece of equipment can strongly influence its selection, despite the fact that its capital cost may be higher. The rapidly escalating costs of energy, labor, and materials can make operation and maintenance considerations even more important than original cost.

FACTORS IN CONTROL EQUIPMENT SELECTION

There are a number of factors to be considered prior to selecting a particular piece of air pollution control hardware.[1] In general, they can be grouped into three categories: environmental, engineering, and economic.

Environmental

1. Equipment location
2. Available space
3. Ambient conditions
4. Availability of adequate utilities (i.e., power, water, etc.) and ancillary system facilities (i.e., waste treatment and disposal, etc.)
5. Maximum allowable emissions (air pollution regulations)
6. Aesthetic considerations (i.e., visible steam or water vapor plume, impact on scenic vistas, etc.)

7. Contribution of air pollution control system to wastewater and solid waste
8. Contribution of air pollution control system to plant noise levels

Engineering

1. Contaminant characteristics (i.e., physical and chemical properties, concentration, particulate shape and size distribution—in the case of particulates, chemical reactivity, corrosivity, abrasiveness, toxicity, etc.)
2. Gas stream characteristics (i.e., volume flow rate, temperature, pressure, humidity, composition, viscosity, density, reactivity, combustibility, corrosivity, toxicity, etc.)
3. Design and performance characteristics of the particular control system (i.e., size and weight, fractional efficiency curves—in the case of particulates, mass transfer and/or contaminant destruction capability—in the case of gases or vapors, pressure drop, reliability and dependability, turndown capability, power requirements, utility requirements, temperature limitations, maintenance requirements, flexibility with regard to complying with more stringent air pollution regulations, etc.)

Economic

1. Capital cost (equipment, installation, engineering, etc.)
2. Operating cost (utilities, maintenance, etc.)
3. Expected equipment lifetime and salvage value

Prior to the purchase of control equipment, experience has shown that the following points should be emphasized:

1. Control equipment should not be purchased without reviewing *certified independent test data* on its performance in a similar application. The manufacturer should be asked to provide performance information and design specifications.
2. In the event that sufficient performance data are unavailable, the equipment supplier can often provide a small pilot model for evaluation under existing conditions.
3. Participation of the local control authorities in the decision-making process is strongly recommended.
4. A good set of specifications is essential. A *strong performance guarantee* from the manufacturer should be obtained to ensure that the control equipment will meet all applicable local, state, and federal codes at specific process conditions.
5. Process and economic fundamentals should be closely reviewed. The possibility for emission trade-offs (offsets) and/or applying the "bubble concept" should be assessed. The bubble concept permits a plant to find the most efficient way to control its emissions as a whole, rather than having the U.S. Environmental Protection Agency (EPA) regulate the emissions from individual sources. Reductions at a source where emissions can be lessened for the least cost can offset emissions of the same pollutant from another source in the plant.
6. A careful material balance study should be made before authorizing an emission test or purchasing control equipment.
7. Equipment should not be purchased until *firm* installation cost estimates have been added to the equipment cost. *Escalating installation costs are the rule rather than the exception.*
8. Operation and maintenance costs should be given high priority on the list of equipment selection factors.
9. Equipment should not be purchased until a solid commitment from the utility supplier(s) is obtained. Every effort should be made to ensure that the new system will utilize fuel, controllers, filters, motors, and so on, that are compatible with those already available at the plant.
10. The specification should include written assurance of *prompt* technical assistance from the equipment supplier. This, together with a complete operating manual (with parts list and full schematics), is essential and is too often forgotten in the rush to get the equipment operating.
11. Schedules, particularly for projects being completed under a court order or consent judgment, can be critical. In such cases, delivery guarantees should be obtained from the manufacturers and penalties identified.
12. The air pollution equipment should be of fail-safe design with built-in indicators to show when performance is deteriorating.
13. A portion of the purchase price (10–15%) should be withheld until compliance is clearly demonstrated.

The usual design–procurement–construction–start-up problems can be further compounded by any one or combination of the following:

1. Unfamiliarity of process engineers with air pollution engineering
2. New and changing air pollution regulations
3. New suppliers with unproven equipment
4. Lack of industry standards in some key areas
5. Inaccurate interpretations by control agency field personnel
6. Compliance schedules that are too tight
7. Vague specifications
8. Weak guarantees for the new control equipment
9. Unreliable delivery schedules
10. Variability; unreliable process operation

Proper selection of a particular system for a specific application can be extremely difficult and complicated. In view of the multitude of complex and often ambiguous pollution control regulations, it is in the best interest of the prospective user to work closely with regulatory officials as early in the process as possible. Finally, previous experience on a similar application cannot be overemphasized.

GENERALIZED DESIGN REVIEW PROCEDURE

Design reviews for air pollution control equipment are performed for a variety of reasons, including:

1. To anticipate compliance with applicable air pollution codes
2. To estimate performance of existing control equipment
3. To evaluate the feasibility of a proposed equipment design
4. To assess the effect on control equipment of process modification

A typical generalized design review approach is presented in Figure 1. The design review investigation is an activity performed early in the evaluation process. Other activities that must be accomplished before final compliance is achieved are presented in Figure 2.

COMPARING CONTROL EQUIPMENT ALTERNATIVES

The final choice in equipment selection is usually dictated by that equipment capable of achieving compliance with regulatory codes at the lowest uniform annual cost (amortized capital investment plus operation and maintenance costs). In order to compare specific control equipment alternatives, knowledge of the particular application and site is essential. A preliminary screening, however, may be performed by reviewing the advantages and disadvantages of each type of air pollution control equipment. For example, if water or a waste treatment system is not available at the site, this may preclude use of a wet scrubber system and instead focus particulate removal on dry systems, such as cyclones, baghouses, and/or electrostatic precipitators. If auxiliary fuel is unavailable on a continuous basis, it may not be possible to combust organic pollutant vapors in an incineration system. If the particulate-size distribution in the gas stream is relatively fine, cyclone collectors most probably would not be considered. If the pollutant vapors can be reused in the process, control efforts may be directed to adsorption systems. There are many more situations in which knowledge of the capabilities of the various control options, combined with common sense, will simplify the

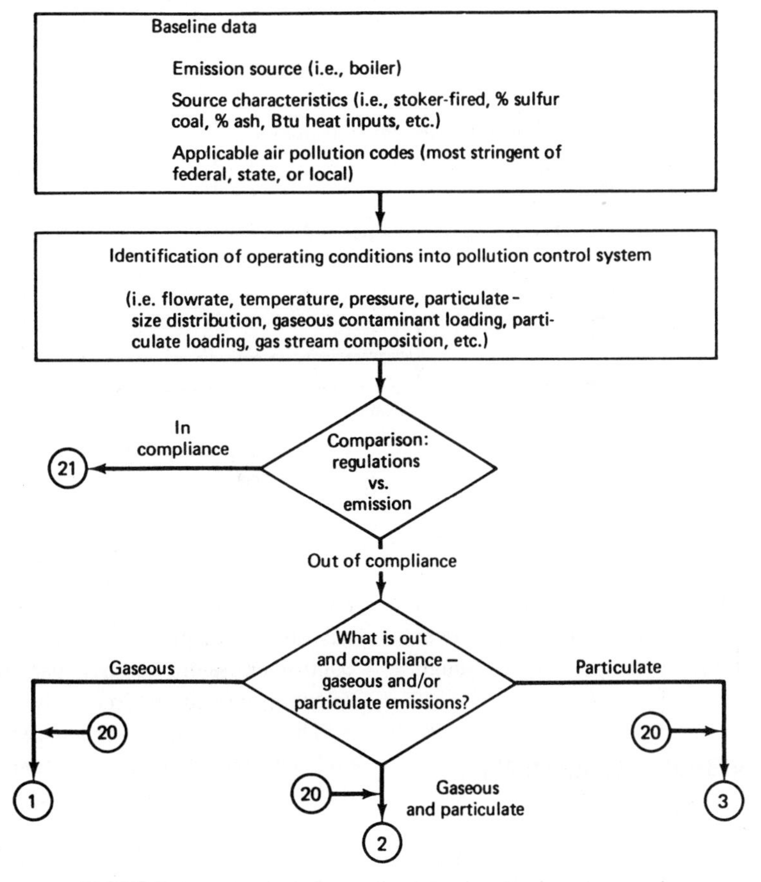

FIGURE 1. Typical Generalized Design Review Approach

FIGURE 1. (*Continued*)

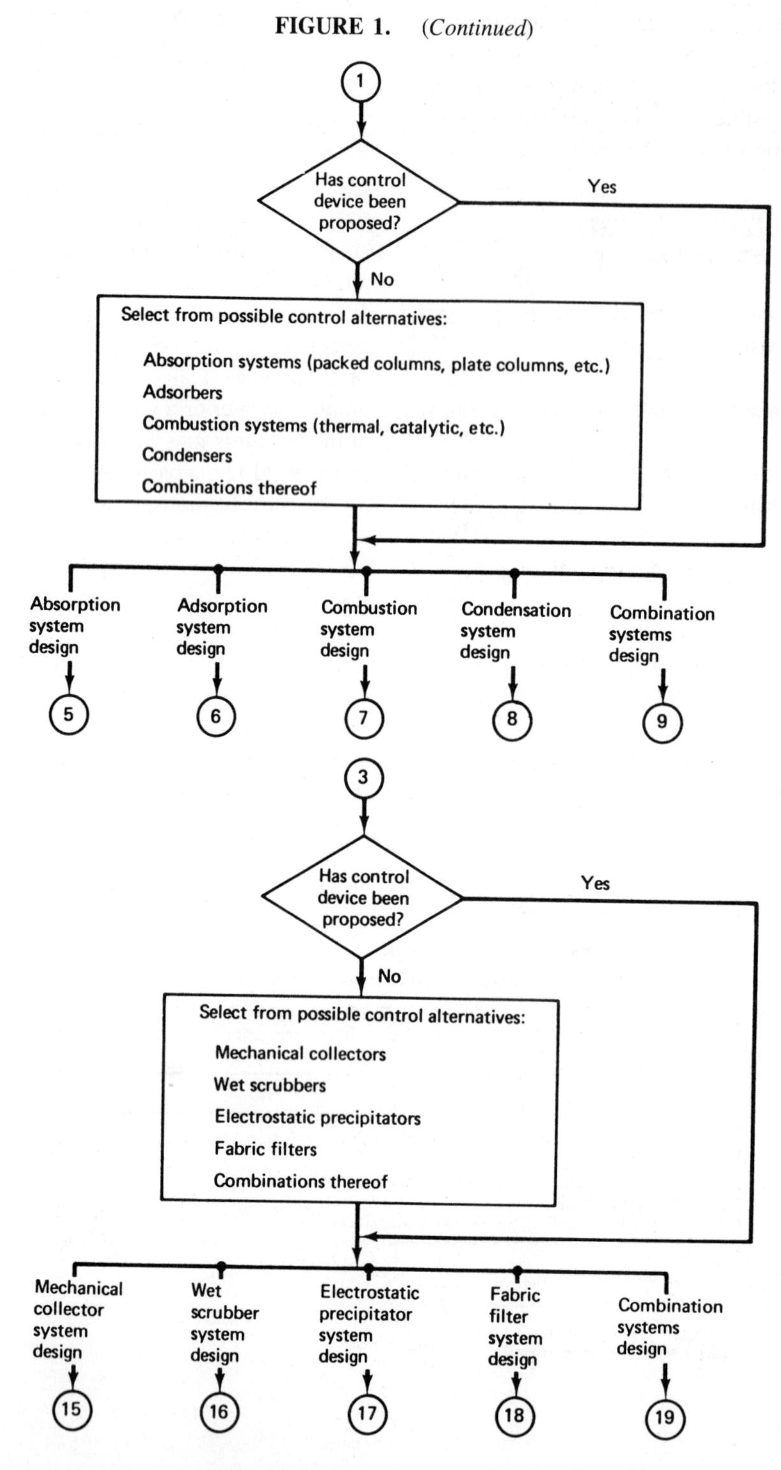

selection procedure. General advantages and disadvantages of the most popular types of air pollution control equipment for gases and particulates are presented in Tables 1 through 9.

SELECTING CONTROL EQUIPMENT FOR SPECIFIC INDUSTRIES

The basic types of emission control devices are mechanical collectors, wet scrubbers, baghouses, electrostatic precipitators, combustion systems, condensers, absorbers, and adsorbers. All of these have been used to some extent to control emissions from a variety of processes, with the selection procedure almost always dictated by experience.

The chapters in this manual provide detailed information on (1) the basic principles of operation of each of the control devices, (2) a variety of industries affected by air pollution control regulations, (3) the general process descriptions of the industries, and (4) the specific control systems used in each industry.

FIGURE 1. *(Continued)*

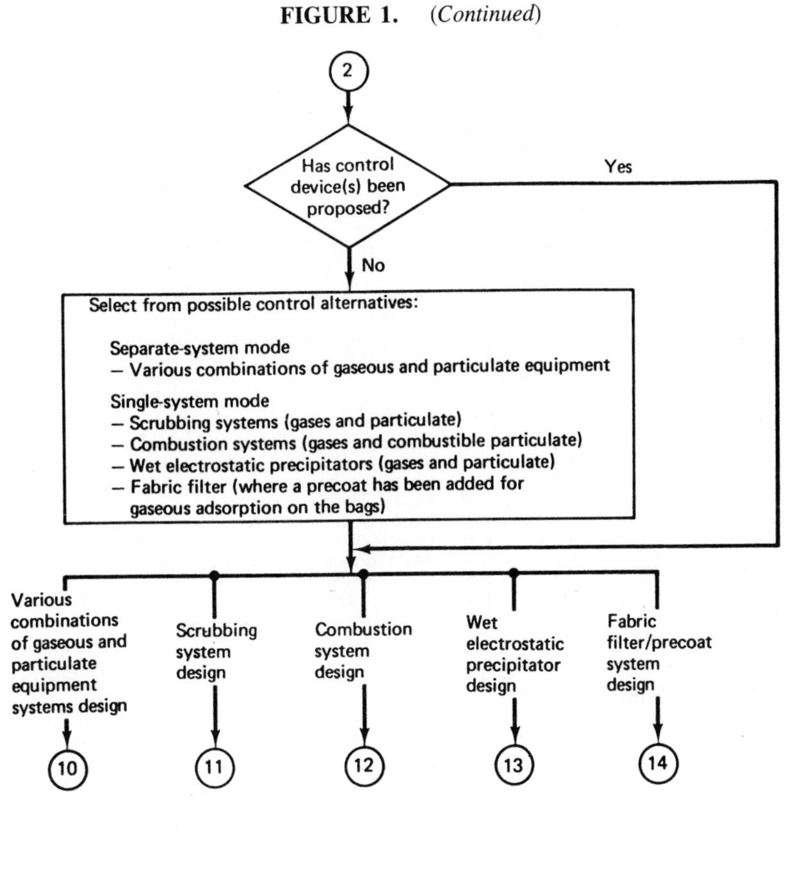

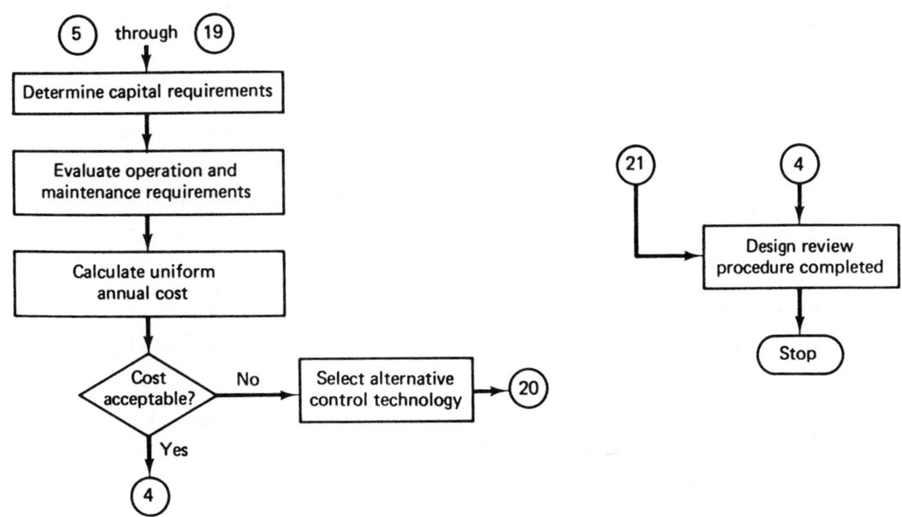

REGULATORY CONSIDERATIONS

The federal government's involvement in air pollution control began in 1955 with Public Law 159. This law authorized funding for the U.S. Public Health Service to initiate research into the nature and extent of the nation's air pollution problem. With the passage of the Clean Air Act of 1963, grants were authorized to state and local agencies to assist them in their own control programs. It also provided some limited authority to the federal government to take action to relieve interstate pollution problems. The basic federal control authority was expanded and strengthened by the Air Quality Act of 1967. One of the more significant measures gave citizens, for the first time, a statutory right to participate in the control process (through public hearings). However, it was not until the Clean Air Act Amendments of 1970 that the regulatory effort first had a significant impact.

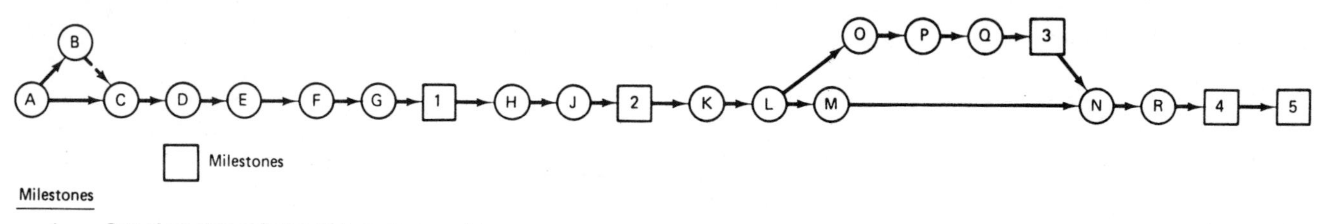

Milestones

1	Date of submittal of final control plan to appropriate agency.
2	Date of award of control device contract.
3	Date of initiation of onsite construction or installation of emission control equipment.
4	Date by which onsite construction or installation of emission control equipment is completed.
5	Date by which final compliance is achieved.

Activities

Designation	
A-C	Preliminary investigation
A-B	Source tests, if necessary
C-D	Evaluate control alternatives
D-E	Commit funds for total program
E-F	Prepare preliminary control plan and compliance schedule for agency
F-G	Agency review and approval
G-1	Finalize plans and specifications
1-H	Procure control device bids
H-J	Evaluate control device bids
J-2	Award control device contract
2-K	Vendor prepares assembly drawings

Designation	
K-L	Review and approval of assembly drawings
L-M	Vendor prepares fabrication drawings
M-N	Fabricate control device
L-O	Prepare engineering drawings
O-P	Procure construction bids
P-Q	Evaluate construction bids
Q-3	Award construction contract
3-N	Onsite construction
N-R	Install control device
R-4	Complete construction (system tie-in)
4-5	Startup, shakedown, source test

FIGURE 2. Compliance Activity and Schedule Chart

The Clean Air Act Amendments of 1970

The historic Clean Air Act Amendments of 1970 broadened and accelerated the nation's earlier air pollution control programs and are the basis for most of the present efforts toward abating air pollution. The amendments reaffirmed that state and local governments had the primary responsibility to control air pollution, but strengthened the federal government's role in air pollution control. This legislation empowered the EPA to establish National Ambient Air Quality Standards (NAAQS) to protect the public health and welfare and then ensure that they were enforced. The EPA was given authority to grant permits to construct or modify an emission source in any of the categories of significant stationary sources. New Source Performance Standards (NSPS) were promulgated, together with specific sampling methodologies and reporting requirements. An ongoing procedure to update the list of significant source categories was included. As of 1991, there were over 60 NSPSs established under Section 111, including sources ranging from coal-fired electric utilities, municipal solid-waste combustors, and other major industrial sources to residential wood heaters and dry cleaners. Sources with applicable NSPSs are summarized in Table 10.

TABLE 1. Advantages and Disadvantages of Cyclone Collectors

Advantages

1. Low cost of construction.
2. Relatively simple equipment with few maintenance problems.
3. Relatively low operating pressure drop (for degree of particulate removal obtained) in the range of approximately 2 to 6 inches water column.
4. Temperature and pressure limitations imposed only by the materials of construction used.
5. Dry collection and disposal.
6. Relatively small space requirements.

Disadvantages

1. Relatively low overall particulate collection efficiencies, especially on particulates below 10 μm in size.
2. Inability to handle tacky materials.

TABLE 2. Advantages and Disadvantages of Wet Scrubbers

Advantages

1. No secondary dust sources.
2. Relatively small space requirements.
3. Ability to collect gases as well as particulates (especially "sticky" ones).
4. Ability to handle high-temperature, high-humidity gas streams.
5. Low capital cost (if wastewater treatment system is not required).
6. For some processes, the gas stream is already at high pressures (so pressure drop considerations may not be significant).
7. Ability to achieve high collection efficiencies on fine particulates (however, at the expense of pressure drop).

Disadvantages

1. May create water disposal problem.
2. Product is collected wet.
3. Corrosion problems are more severe than with dry systems.
4. Steam plume opacity and/or droplet entrainment may be objectionable.
5. Pressure drop and horsepower requirements may be high.
6. Solids buildup at the wet–dry interface may be a problem.
7. Relatively high maintenance costs.

TABLE 3. Advantages and Disadvantages of Electrostatic Precipitators

Advantages
1. Extremely high particulate (coarse and fine) collection efficiencies can be attained (at a relatively low expenditure of energy).
2. Dry collection and disposal.
3. Low pressure drop (typically less than 0.5 inch water column).
4. Designed for continuous operation with minimum maintenance requirements.
5. Relatively low operating costs.
6. Capable of operation under high pressure (to 150 psi) or vacuum conditions.
7. Capable of operation at high temperatures (to 1300°F).
8. Relatively large gas flow rates can be effectively handled.

Disadvantages
1. High capital cost.
2. Very sensitive to fluctuations in gas stream conditions (in particular, flow rates, temperatures, particulate and gas composition, and particulate loadings).
3. Certain particulates are difficult to collect due to extremely high or low resistivity characteristics.
4. Relatively large space requirements required for installation.
5. Explosion hazard when treating combustible gases and/or collecting combustible particulates.
6. Special precautions required to safeguard personnel from the high voltage.
7. Ozone is produced by the negatively charged electrode during gas ionization.
8. Relatively sophisticated maintenance personnel required.

Development of emission standards for existing stationary sources (except sources of hazardous pollutants) was required of the states. The states were also responsible for enforcing the NSPSs, as well as attaining and maintaining the NAAQSs within their respective boundaries. The State Implementation Plan (SIP) was the means whereby the states were to comply with the Clean Air Act.

Criteria Pollutants

Criteria pollutants were identified pursuant to Sections 108 and 109 of the 1970 Clean Air Act Amendments (CAAA). Total suspended particulates (TSP), sulfur dioxide, nitrogen dioxide, hydrocarbons, carbon monoxide, and photochemical oxidants were defined as the first six criteria pollutants under the CAAA of 1970. In 1976, lead (Pb) was added to the list of criteria pollutants. In 1979, the photochemical oxidants standard was revised and restated as ozone (O_3), and the hydrocarbon standard was reviewed and withdrawn in 1983. In 1987, the TSP standard was reviewed and revised to include only particles with an aerodynamic particle diameter less than or equal to 10 μm, referred to as PM_{10}.

Each state was given the responsibility for promulgating regulations to control these pollutants such that the NAAQSs would be achieved. Although major progress was made in many cities and metropolitan areas in attaining and/or maintaining the standards, many larger cities have

TABLE 4. Advantages and Disadvantages of Fabric Filter Systems

Advantages
1. Extremely high collection efficiency on both coarse and fine (submicron) particulates.
2. Relatively insensitive to gas stream fluctuation. Efficiency and pressure drop are relatively unaffected by large changes in inlet dust loadings for continuously cleaned filters.
3. Filter outlet air may be recirculated within the plant in many cases (for energy conservation).
4. Collected material is recovered dry for subsequent processing or disposal.
5. No problems with liquid waste disposal, water pollution, or liquid freezing.
6. Corrosion and rusting of components are usually not problems.
7. There is no hazard of high voltage, simplifying maintenance and repair and permitting collection of flammable dusts.
8. Use of selected fibrous or granular filter aids (precoating) permits the high-efficiency collection of submicron smokes and gaseous contaminants.
9. Filter collectors are available in a large number of configurations, resulting in a range of dimensions and inlet and outlet flange locations to suit installation requirements.
10. Relatively simple operation.

Disadvantages
1. Temperatures much in excess of 550°F require special refractory mineral or metallic fabrics that are still in the developmental stage and can be very expensive.
2. Certain dusts may require fabric treatments to reduce dust seeping or, in other cases, assist in the removal of the collected dust.
3. Concentrations of some dusts in the collector (~50 g/m^3) may represent a fire or explosion hazard if a spark or flame is admitted by accident. Fabrics can burn if readily oxidizable dust is being collected.
4. Relatively high maintenance requirements (bag replacement, etc.).
5. Fabric life may be shortened at elevated temperatures and in the presence of acid or alkaline particulate or gas constituents.
6. Hygroscopic materials, condensation of moisture, or tarry adhesive components may cause crusty caking or plugging of the fabric or require special additives.
7. Replacement of fabric may require respiratory protection for maintenance personnel.
8. Medium pressure-drop requirements, typically in the range of 4 to 10 inches water column.

TABLE 5. Advantages and Disadvantages of Absorption Systems (Packed and Plate Columns)

Advantages
1. Relatively low pressure drop.
2. Standardization in fiberglass-reinforced plastic (FRP) construction permits operation in highly corrosive atmospheres.
3. Capable of achieving relatively high mass-transfer efficiencies.
4. Increasing the height and/or type of packing or number of plates can improve mass transfer without purchasing a new piece of equipment.
5. Relatively low capital cost.
6. Relatively small space requirements.
7. Ability to collect particulates as well as gases.

Disadvantages
1. May create water (or liquid) disposal problem.
2. Product collected wet.
3. Particulates deposition may cause plugging of the bed or plates.
4. When FRP construction used, it is sensitive to temperature.
5. Relatively high maintenance costs.

TABLE 6. Comparison of Plate and Packed Columns

Packed column
1. Lower pressure drop.
2. Simpler and cheaper to construct.
3. Preferable for liquids with high foaming tendencies.

Plate column
1. Less susceptible to plugging.
2. Less weight.
3. Less of a problem with channeling.
4. Temperature surge will result in less damage.

TABLE 7. Advantages and Disadvantages of Adsorption Systems

Advantages
1. Product recovery may be possible.
2. Excellent control and response to process changes.
3. No chemical disposal problem when pollutant (product) is recovered and returned to process.
4. Capability of systems to provide fully automatic, unattended operation.
5. Capability to remove gaseous or vapor contaminants from process streams to extremely low levels.

Disadvantages
1. Product recovery may require an exotic, expensive distillation (or extraction) scheme.
2. Adsorbent progressively deteriorates in capacity as the number of cycles increases.
3. Adsorbent regeneration requires a steam or vacuum source.
4. Relatively high capital cost.
5. Prefiltering of gas stream may be required to remove any particulate capable of plugging the adsorbent bed.
6. Cooling of the gas stream may be required to get to the usual range of operation (less than 120°F).
7. Relatively high steam requirements to desorb high-molecular-weight hydrocarbons.

TABLE 8. Advantages and Disadvantages of Combustion Systems

Advantages
1. Simplicity of operation.
2. Capability to provide steam generation or heat recovery in other forms.
3. Capability for high destruction efficiency of organic contaminants.

Disadvantages
1. Relatively high operating costs (particularly associated with fuel requirements).
2. Potential for flashback and subsequent explosion hazard.
3. Catalyst poisoning (in the case of catalytic incineration).
4. Incomplete combustion can create potentially worse pollution problems.

failed to do so to date. In 1991, there were 96 cities, 41 cities, and 70 cities or areas that were still nonattainment for ozone, carbon monoxide, and particulate matter (PM_{10}), respectively.

TABLE 9. Advantages and Disadvantages of Condensers

Advantages
1. Pure product recovery (in the case of indirect-contact condensers).
2. Water used as the coolant in an indirect contact condenser (i.e., shell-and-tube heat exchanger) does not contact the contaminated gas stream and can be reused after cooling.

Disadvantages
1. Relatively low removal efficiency for gaseous contaminants (at concentrations typical of pollution control applications).
2. Coolant requirements may be extremely expensive.

Hazardous Pollutants

The CAAA of 1970 differentiated between nonhazardous and hazardous pollutants. The EPA Administrator was charged with promulgation of National Emission Standards for Hazardous Air Pollutants (NESHAPs) from both new and existing sources. Under Section 112, only eight pollutants were designated as hazardous air pollutants between 1970 and 1989: asbestos, beryllium, mercury, vinyl chloride, arsenic, benzene, radionuclides, and coke oven emissions. Enforcement of these standards was the responsibility of the EPA; at the states' option, implementation plans submitted to the EPA could include regulatory control procedures for these pollutants. After EPA approval of the SIPs, the states were authorized to enforce the hazardous air pollutant standards.

State Permit Systems

Since the early 1970s, and in many jurisdictions well before that, operating permits have been required for various processes that are vented to the atmosphere. A thorough review of regulations applicable to each plant site is required to determine precisely the processes for which permits are required.

The law requires that permits be obtained before a new source is constructed or an existing source is modified. Permit forms for these operations allow the control agency to evaluate the planned emission control equipment and assess potential compliance with applicable regulations. If the agency judges that the source as planned will not operate in compliance with regulations, agency officials may require changes in the design of the process or installation. Only processes believed to operate in compliance with the applicable regulations can receive permits.

Prevention of Significant Deterioration Areas

Guidelines for the prevention of significant deterioration (PSD) were issued in 1974 by the EPA to prevent significant deterioration of the air quality in areas that were already in attainment of the NAAQSs. The requirements

TABLE 10. Source Categories for Which New Source Performance Standards Have Been Set as of 1991

- Fossil-Fuel–Fired Steam Generators for Which Construction Commenced After August 17, 1971
- Electric Utility Steam Generating Units for Which Construction Commenced After September 18, 1978
- Industrial–Commercial–Institutional Steam Generating Units
- Incinerators
- Portland Cement Plants
- Nitric Acid Plants
- Sulfuric Acid Plants
- Asphalt Concrete Plants
- Petroleum Refineries
- Storage Vessels for Petroleum Liquids for Which Construction, Reconstruction, or Modification Commenced After June 11, 1973, and Prior to May 19, 1978
- Storage Vessels for Petroleum Liquids for Which Construction, Reconstruction, or Modification Commenced After May 18, 1978, and Prior to July 23, 1984
- Volatile Organic Liquid Storage Vessels (Including Petroleum Liquid Storage Vessels) for Which Construction, Reconstruction, or Modification Commenced After July 23, 1984
- Secondary Lead Smelters
- Secondary Brass and Bronze Production Plants
- Primary Emissions from Basic Oxygen Process Furnaces for Which Construction Commenced After June 11, 1973
- Secondary Emissions from Basic Oxygen Process Steelmaking Facilities for Which Construction Commenced After January 20, 1983
- Sewage Treatment Plants
- Primary Copper Smelters
- Primary Zinc Smelters
- Primary Lead Smelters
- Primary Aluminum Reduction Plants
- Phosphate Fertilizer Industry: Wet-Process Phosphoric Acid Plants
- Phosphate Fertilizer Industry: Superphosphoric Acid Plants
- Phosphate Fertilizer Industry: Diammonium Phosphate Plants
- Phosphate Fertilizer Industry: Triple Superphosphate Plants
- Phosphate Fertilizer Industry: Granular Triple Superphosphate Storage Facilities
- Coal Preparation Plants
- Ferroalloy Production Facilities
- Steel Plants: Electric Arc Furnaces Constructed After October 21, 1974, and on or Before August 17, 1983
- Steel Plants: Electric Arc Furnaces and Argon-Oxygen Decarburization Vessels Constructed After August 7, 1983
- Kraft Pulp Mills
- Glass Manufacturing Plants
- Grain Elevators
- Surface Coating of Metal Furniture
- Stationary Gas Turbines
- Lime Manufacturing Plants
- Lead-Acid Battery Manufacturing Plants
- Metallic Mineral Processing Plants
- Automobile and Light-Duty Truck Surface Coating Operations
- Phosphate Rock Plants
- Ammonium Sulfate Manufacture
- Graphic Arts Industry: Publication Rotogravure Printing
- Pressure-Sentitive Tape and Label Surface Coating Operations
- Metal Coil Surface Coating
- Asphalt Processing and Asphalt Roofing Manufacture
- Equipment Leaks of VOC in the Synthetic Organic Chemicals Manufacturing Industry
- Beverage Can Surface Coating Industry
- Bulk Gasoline Terminals
- New Residential Wood Heaters
- Rubber Tire Manufacturing Industry
- Flexible Vinyl and Urethane Coating and Printing
- Equipment Leaks of VOC in Petroleum Refineries
- Synthetic Fiber Production Facilities
- Petroleum Dry Cleaners
- Equipment Leaks of VOC from Onshore Natural Gas Processing Plants
- Onshore Natural Gas Processing; SO_2 Emissions
- Nonmetallic Mineral Processing Plants
- Wool Fiberglass Insulation Manufacturing Plants
- VOC Emissions from Petroleum Refinery Wastewater Systems
- Magnetic Tape Coating Facilities
- Industrial Surface Coating: Surface Coating of Plastic Parts for Business Machines
- Volatile Organic Compound Emissions from Synthetic Organic Chemical Manufacturing Industry (SOCMI) Air Oxidation Unit Processes
- Volatile Organic Compound Emissions from Synthetic Organic Chemical Manufacturing Industry Distillation Operations
- Polymeric Coating of Supporting Substrates Facilities

made use of an area classification scheme whose basic premise was that a moderate amount of industrial development should be routinely permitted in all areas, but that industrialization should not be allowed to degrade air quality to the point where it barely complies with air quality standards.

Three classes were subsequently designated. The Class I category was to include the pristine areas subject to tightest control. Class II covered areas of moderate growth and Class III was for areas of major industrialization. The EPA regulations also established another critical concept known as the increment. This would be the numerical definition of the amount of additional pollution that would be allowed through the combined effects of all new growth in a particular area.

The EPA did impose one major additional requirement for PSD areas to assure that the increments would not be used hastily. It specified that each major new plant must install the best available control technology (BACT) to limit its emissions. Major sources subject to PSD review include 28 categories with the potential to emit (after application of control technology) 100 tons per year or more of any regulated pollutant under the CAAA. All sources emitting greater than 250 tons per year are subject to PSD review. For new sources regulated under NSPS, the NSPS generally defines the requirements for BACT. For new non-NSPS sources, a BACT review document is prepared in the preconstruction review.

Nonattainment Areas

Nonattainment is designated to those areas of the country not yet in compliance with the National Ambient Air Quality Standards. In any nonattainment area, no major new source can be constructed without a permit. The permit imposes stringent control requirements and requires sufficient "offsets" to assure progress toward compliance. Approval for a new source in a nonattainment area requires that the new source be equipped with pollution controls to assure the lowest achievable emission rate (LAER); that all existing sources owned by an applicant in the same region be in compliance or under an approved schedule to achieve compliance, the applicant to come up with sufficient "offsets" to more than make up for the emissions to be generated by the new source (after application of LAER); and that the emission offset provide a positive net air quality benefit in the affected area.

The PSD and nonattainment guidelines gave birth to the concept of emissions trading. In nonattainment areas, offsets are required to construct or modify a major source; in PSD areas, offsets have been used on occasion to allow the source to "net" out of the PSD review process.[2] Voluntary offsets in PSD areas have provided the impetus for the development of emissions trading policies in many states, whereby reductions in emissions generated emission reduction credits (ERCs) that can be "banked" for later use or transferred to a third party. In 1985, more than 2000 offset transactions had occurred.

Clean Air Act Amendments of 1977

The 1977 amendments further strengthened air pollution control standards and enforcement.[3] The basic control strategy remained the same; however, additional enforcement power was given to the EPA. The significance of the PSD program established by the EPA in 1974 and the Offsets Policy Interpretive Ruling (for nonattainment areas) issued by the EPA in 1976 was that, when Congress in the 1977 amendments provided the statutory foundations for PSD and nonattainment areas, it adopted in toto the basic concepts of the EPA programs.

Clean Air Act Amendments of 1990

In November, 1990 Congress adopted the Clean Air Act Amendments of 1990, providing substantial changes to many aspects of the existing CAAA. The concepts of NAAQS, NSPS, and PSD remain virtually unchanged, except in a few areas to be discussed below. However, significant changes have occurred in several areas that directly affect industrial facilities and electric utilities and air pollution control at these facilities. These include changes and additions in the following major areas:

Title I	Nonattainment areas
Title III	Hazardous air pollutants
Title IV	Acid deposition control

Title I: Nonattainment Areas

The existing regulations for nonattainment areas have been made more stringent in several areas. The CAAA of 1990 require the development of comprehensive emission inventory tracking for all nonattainment areas and establishes a classification scheme defining nonattainment areas into levels of severity. For example, ozone nonattainment areas are designated as marginal, moderate, serious, severe (two levels), and extreme, with compliance deadlines of 3, 6, 9, 15–17, and 20 years, respectively, with each classification having more stringent requirements regarding strategies for compliance. (See Table 11). Volatile organic compound (VOC) emissions reductions of 15% are required in moderate and above areas by 1996 and 3% a year thereafter for severe and above areas until compliance is achieved. In addition, the definition of a major source of ozone precursors (previously 100 tons per year of NO_x, CO, or VOC emissions) was redefined to as little as 10 tons per year in the "extreme" classification, with increased offset requirements of 1.5 to 1 for new and modified sources. These

TABLE 11. Ozone Nonattainment Area Classifications and Associated Requirements

Nonattainment Area Classification	One-Hour Ozone Concentration Design Value ppm	Attainment Date	Major Source Threshold Level tons VOCs/yr	Offset Ratio for New/Modified Sources
Marginal	0.121–0.138	Nov. 15, 1993	100	1.1 to 1
Moderate	0.138–0.160	Nov. 15, 1996	100	1.15 to 1
Serious	0.160–0.180	Nov. 15, 1999	50	1.2 to 1
Severe	0.180–0.190	Nov. 15, 2005	25	1.3 to 1
	0.190–0.280	Nov. 15, 2007	25	1.3 to 1
Extreme	0.280 and up	Nov. 15, 2010	10	1.5 to 1

Source: Reference 4

requirements place major contraints on the affected industries in these nonattainment areas. A similar approach is being taken in PM_{10} and CO nonattainment areas.

Title III: Hazardous Air Pollutants

The Title III provisions on hazardous air pollutants (HAPs) represent a major departure from the previous approach of developing NESHAPs. While only eight HAPs were designated in the 20 years since enactment of the CAAA of 1970, the new CAAA of 1990 designated 189 pollutants as HAPs requiring regulation. These are summarized in Table 12 and will affect over 300 listed major source categories. Major sources are defined as any source (new or existing) that emits (after control) 10 tons a year or more of any regulated HAP or 25 tons a year or more of any combination of HAPs. The deadlines for promulgation of the source categories and appropriate emission standards are as follows:

First 40 source categories	November 15, 1992
Coke oven batteries	December 31, 1992
25% of all listed categories	November 15, 1994
Publicly owned treatment works	November 15, 1995
50% of all listed categories	November 15, 1997
100% of all categories	November 15, 2000

Each source will be required to meet maximum achievable control technology (MACT) requirements. For existing sources, MACT is defined as a stringency equivalent to the average of the best 12% of the sources in the category. For new sources, MACT is defined as the best controlled system. New sources are required to meet MACT immediately, while existing sources have three years from the date of promulgation of the appropriate MACT standard. As an early incentive, existing sources that undergo at least a 90% reduction in emissions of an HAP (or 95% for a hazardous particulate) prior to the promulgation of the MACT standard will be issued a six-year extension on the deadline for final compliance.

Title IV: Acid Deposition Control

The Acid Deposition Control Program is designed to reduce emissions of SO_2 in the United States by 10 million tons per year, resulting in a net yearly emission of 8.9 million tons by the year 2000. Phase I of the program requires 111 existing uncontrolled coal-fired power plants ($\geq$ 100 MW) to reduce emissions to 2.5 pounds of SO_2 per 10^6 Btu by 1995 (1997 if scrubbers are used to reduce emissions by at least 90%). The reduction is to be accomplished by issuing all affected units emission "allowances" equivalent to what their annual average SO_2 emissions would have been in the years 1985–1987 based on 2.5 pounds SO_2 per 10^6 Btu coal. The regulations represent a significant departure from previous regulations where specified SO_2 removal efficiencies were mandated; rather, the utilities will be allowed the flexibility of choosing which strategies will be used (i.e., coal washing, low-sulfur coal, flue gas desulfurization) and which units will be controlled, as long as the overall "allowances" are not exceeded. Any excess reduction in SO_2 by a utility will create "banked" emissions that can be sold or used at another unit.

Phase II of Title IV limits the majority of plants $\geq$20 MW and all plants $\geq$75 MW to maximum emissions of 1.2 pounds of SO_2 per 10^6 Btu after the year 2000. In general, new plants would have to acquire banked emission allowances in order to be built. Emission allowances will be traded through a combination of sell/purchase with other utilities, EPA auctions and direct sales. The details of these procedures are to be developed.

Control of NO_x under the CAAA of 1990 will be accomplished through the issuance of a revised NSPS by 1994, with the objective of reducing emissions by 2 million tons a year from 1980 emission levels. The technology being considered is the use of low-NO_x burners (LNBs). The new emission standards will not apply to cyclone and wet-bottom boilers, unless alternative technologies are found, as these cannot be retrofitted with existing LNB technologies.

REGULATORY DIRECTION

The current direction of regulations and air pollution control efforts is clearly toward significantly reducing the emissions to the environment of a broad range of compounds, including:

1. Volatile organic compounds and other ozone precursors (CO and NO_x)
2. Hazardous air pollutants, including carcinogenic organic emission and heavy metals emissions
3. Acid rain precursors, including SO_x and NO_x

In addition, the recently developed PM_{10} NAAQS will continue to place emphasis on quantifying and reducing particulate emissions in the less than 10-μm particle-size range. Particle size-specific emission factors have been developed for many sources and size-specific emission standards have been developed in some states. These standards are addressing concerns related to HAP emissions of heavy metals, which are generally associated with the submicron particles.

Although it is not possible to predict the future, it is possible to prepare for it and to influence it. It is highly recommended that maximum flexibility be designed into new air pollution control systems to allow for the increasingly stringent emissions standards for both particulates and gases that are on the horizon. Further, it is everyone's responsibility to provide a thorough review of existing and proposed new processes and to make every attempt to identify economical process modifications and/or material substitutions that reduce or, in some cases, eliminate both the emissions to the environment and the overdependency on retrofitted or new end-of-pipe control systems.

TABLE 12. List of Hazardous Air Pollutants Regulated by The Clean Air Act Amendments of 1990

CAS[a] Number	Chemical Name	CAS[a] Number	Chemical Name
75070	Acetaldehyde	111422	Diethanolamine
60355	Acetamide	121697	*N,N*-Diethylaniline
75058	Acetonitrile		(*N,N*-Dimethylaniline)
98862	Acetophenone	64675	Diethyl sulfate
53963	2-Acetylaminofluorene	119904	3,3-Dimethoxybenzidine
107028	Acrolein	60117	Dimethyl aminoazobenzene
79061	Acrylamide	119937	$3,3^{1}$ -Dimethyl benzidine
79107	Acrylic acid	79447	Dimethyl carbamoyl chloride
107131	Acrylonitrile	68122	Dimethyl formamide
107051	Allyl chloride	57147	1,1-Dimethyl hydrazine
92671	4-Aminobiphenyl	131113	Dimethyl phthalate
62533	Aniline	77781	Dimethyl sulfate
90040	*o*-Anisidine	534521	4,6-Dinitro-o-cresol, and
1332214	Asbestos		salts
71432	Benzene (including benzene	51285	2,4-Dinitrophenol
	from gasoline)	121142	2,4-Dinitrotoluene
92875	Benzidine	123911	1,4-Dioxane
98077	Benzotrichloride		(1,4-Diethyleneoxide)
100447	Benzyl chloride	122667	1,2-Diphenylhydrazine
92524	Biphenyl	106898	Epichlorohydrin
117817	Bis(2-ethylhexyl)phthalate		(1-Chloro-2,3-epoxypropane)
	(DEHP)	106887	1,2-Epoxybutane
542881	Bis(chloromethyl)ether	140885	Ethyl acrylate
75252	Bromoform	100414	Ethyl benzene
106990	1,3-Butadiene	51796	Ethyl carbamate (Urethane)
156627	Calcium cyanamide	75003	Ethyl chloride (Chloroethane)
105602	Caprolactam	106934	Ethylene dibromide
133062	Captan		(Dibromoethane)
63252	Carbaryl	107062	Ethylene dichloride
75150	Carbon disulfide		(1,2-Dichloroethane)
56235	Carbon tetrachloride	107211	Ethylene glycol
463581	Carbonyl sulfide	151564	Ethylene imine (Aziridine)
120809	Catechol	75218	Ethylene oxide
133904	Chloramben	96457	Ethylene thiourea
57749	Chlordane	75343	Ethylidene dichloride
7782505	Chlorine		(1,1-Dichloroethane)
79118	Chloroacetic acid	50000	Formaldehyde
532274	2-Chloroacetophenone	76448	Heptachlor
108907	Chlorobenzene	118741	Hexachlorobenzene
510156	Chlorobenzilate	87683	Hexachlorobutadiene
67663	Chloroform	77474	Hexachlorocyclopentadiene
107302	Chloromethyl methyl ether	67721	Hexachloroethane
126998	Chloroprene	822060	Hexamethylene-1,6-diisocyanate
1319773	Cresols/Cresylic acid	680319	Hexamethylphosphoramide
	(isomers and mixture)	110543	Hexane
95487	*o*-Cresol	302012	Hydrazine
108394	*m*-Cresol	7647010	Hydrochloric acid
106445	*p*-Cresol	7664393	Hydrogen fluoride
98828	Cumene		(Hydrofluoric acid)
94757	2,4-D, salts and esters	123319	Hydroquinone
3547044	DDE	78591	Isophorone
334883	Diazomethane	58899	Lindane (all isomers)
132649	Dibenzofurans	108316	Maleic anhydride
96128	1,2-Dibromo-3-chloropropane	67561	Methanol
84742	Dibutylphthalate	72435	Methoxychlor
106467	1,4-Dichlorobenzene(p)	74839	Methyl bromide
91941	3,3-Dichlorobenzidene		(Bromomethane)
111444	Dichloroethyl ether	74873	Methyl chloride
	(Bis(2-chloroethyl)ether)		(Chloromethane)
542756	1,3-Dichloropropene	71556	Methyl chloroform
62737	Dichlorvos		(1,1,1-Trichloroethane)

TABLE 12. *(Continued)*

CAS[a] Number	Chemical Name	CAS[a] Number	Chemical Name
78933	Methyl ethyl ketone (2-Butanone)	1746016	2,3,7,8-Tetrachlorodibenzo-*p*-dioxin
60344	Methyl hydrazine	79345	1,1,2,2-Tetrachloroethane
74884	Methyl iodide (Iodomethane)	127184	Tetrachloroethylene (Perchloroethylene)
108101	Methyl isobutyl ketone (Hexone)		
624839	Methyl isocyanate	7550450	Titanium tetrachloride
80626	Methyl methacrylate	108883	Toluene
1634044	Methyl tert butyl ether	95807	2,4-Toluene diamine
101144	4,4-Methylene bis(2-chloroaniline)	584849	2,4-Toluene diisocyanate
		95534	*o*-Toluidine
75092	Methylene chloride (Dichloromethane)	8001352	Toxaphene (chlorinated camphene)
101688	Methylene diphenyl diisocyanate (MDI)	120821	1,2,4-Trichlorobenzene
		79005	1,1,2-Trichloroethane
101779	4,4'-Methylenedianiline	79016	Trichloroethylene
91203	Naphthalene	95954	2,4,5-Trichlorophenol
98953	Nitrobenzene	88062	2,4,6-Trichlorophenol
92933	4-Nitrobiphenyl	121448	Triethylamine
100027	4-Nitrophenol	1582098	Trifluralin
79469	2-Nitropropane	540841	2,2,4-Trimethylpentane
684935	*N*-Nitroso-N-methylurea	108054	Vinyl acetate
62759	*N*-Nitrosodimethylamine	593602	Vinyl bromide
59892	*N*-Nitrosomorpholine	75014	Vinyl chloride
56382	Parathion	75354	Vinylidene chloride (1,1-Dichloroethylene)
82688	Pentachloronitrobenzene (Quintobenzene)	1330207	Xylenes (isomers and mixture)
87865	Pentachlorophenol	95476	*o*-Xylenes
108952	Phenol	108383	*m*-Xylenes
106503	*p*-Phenylenediamine	106423	*p*-Xylenes
75445	Phosgene	0	Antimony compounds
7803512	Phosphine	0	Arsenic compounds (inorganic including arsine)
7723140	Phosphorus		
85449	Phthalic anhydride	0	Beryllium compounds
1336363	Polychlorinated biphenyls (Aroclors)	0	Cadmium compounds
		0	Chromium compounds
1120714	1,3-Propane sultone	0	Cobalt compounds
57578	*β*-Propiolactone	0	Coke oven emissions
123386	Propionaldehyde	0	Cyanide compounds
114261	Propoxur (Baygon)	0	Glycol ethers
78875	Propylene dichloride (1,2-Dichloropropane)	0	Lead compounds
		0	Manganese compounds
75569	Propylene oxide	0	Mercury compounds
75558	1,2-Propylenimine (2-Methyl aziridine)	0	Fine mineral fibers
		0	Nickel compounds
91225	Quinoline	0	Polycylic organic matter
106514	Quinone	0	Radionuclides (including radon)
100425	Styrene		
96093	Styrene oxide	0	Selenium compounds

[a]CAS = Chemical Abstract Service.

Source: Title II, CAAA, 1990.

References

1. L. Theodore and A. J. Buonicore, *Industrial Air Pollution Control Equipment for Particulates,* Chap. 6., CRC Press, West Palm Beach, FL, 1976.
2. M. Weiss and J. Palmisano, "Emissions trading gives flexibility in meeting clean-air laws." *Power,* 55–58 (March 1985).
3. A. J. Buonicore, "Air pollution control," *Chem. Eng.*, 87 (13): 81–101 (June 30, 1980).
4. Clean Air Act Amendments, Public Law 101 549, Titles I–XI, November 15, 1990.

2
Control of Gaseous Pollutants

Absorption
Adsorption
Condensation
Incineration
Anthony J. Buonicore

There are four commonly used control technologies for gaseous pollutants: absorption, adsorption, condensation, and incineration (combustion).

The choice of control technology depends on the pollutant(s) that must be removed, the removal efficiency required, pollutant and gas stream characteristics, and specific characteristics of the site. Experience has led to numerous generalities that should only be considered as such.

1. At low gaseous pollutant concentrations (less than 100 ppmv), incineration can generally achieve 90–95% efficiency with thermal incineration slightly better than catalytic at the lowest concentrations.
2. At higher gaseous pollutant concentrations (greater than 100 ppmv), incineration typically can achieve 95–99% efficiency.
3. Incineration can produce products of incomplete combustion or otherwise undesirable by-products that may require additional controls.
4. Carbon adsorption typically can achieve 90+% efficiency at gaseous pollutant concentrations greater than a few hundred ppmv. At higher concentrations (greater than 1000 ppmv), efficiencies can exceed 95%.
5. Adsorption becomes a less favored technology option if the mixture of recovered organics cannot be returned to the process with minimum additional treatment. In such cases, it may be more cost-effective to incinerate the recovered organics that have been significantly concentrated by the adsorption process.
6. Absorption removal efficiency is a function of inlet concentration. In general, lower inlet concentrations (below a few hundred ppmv) will result in efficiency levels in the low 90s. Higher inlet concentrations will result in efficiencies in the upper 90s.
7. Absorption becomes a less favored technology option if a liquid blowdown stream cannot be accommodated at the facility.
8. Condensation generally needs relatively high inlet concentrations (greater than a few thousand ppmv) to achieve efficiencies in the 80+% range.
9. Condensation generally cannot meet high efficiency requirements without the use of very low temperatures (e.g., use of liquid nitrogen) or high pressure.
10. Typically, only incineration and absorption technologies are able to achieve greater than 99% gaseous pollutant removal efficiency on a consistent basis.

ABSORPTION

Anthony J. Buonicore

The removal of one or more selected components from a gas mixture by absorption is probably the most important operation in the control of gaseous pollutant emissions. The process of absorption conventionally refers to the intimate contacting of a mixture of gases with a liquid so that part of one or more of the constituents of the gas will dissolve in the liquid. Such action takes place in all types of wet scrubbers; however, this discussion will deal exclusively with packed or plate absorbers since these are the most commonly used. Only equipment and design procedures are emphasized; a detailed presentation of the theory, including the diffusional process, mass transfer coefficients, equilibrium (lines), operating lines, and so on, is beyond the scope and may be found elsewhere.[1–6]

Gas absorption as applied to the control of air pollution is concerned with the removal of one or more pollutants from a contaminated gas stream by treatment with a liquid. The necessary condition is the solubility of these pollutants in the absorbing liquid. The rate of transfer of the soluble constituents from the gas to the liquid phase is determined by diffusional processes occurring on each side of the gas–liquid interface. Consider, for example, the process taking place when a mixture of air and gaseous hydrogen chloride (HCl) is brought into contact with water. The HCl is soluble in water, and those molecules that come into contact with the water surface dissolve immediately. However, the HCl molecules are initially dispersed throughout the gas phase, and they can only reach the water surface by diffusion through the air, which is substantially insoluble in the water. When the HCl at the water surface has dissolved, it is distributed throughout the water phase by a second diffusional process. Consequently, the rate of absorption is determined by the rates of diffusion in both the gas and liquid phases.

Equilibrium is another important factor to be considered in controlling the operation of absorption systems. The rate at which the pollutant will diffuse into an absorbent liquid will depend on the departure from equilibrium that is maintained. The rate at which equilibrium is established is then essentially dependent on the rate of diffusion of the pollutant through the nonabsorbed gas and through the absorbing liquid. The rate at which the pollutant mass is transferred from one phase to another depends also on a so-called mass transfer, or rate coefficient, which equates the quantity of mass being transferred with the driving force. As can be expected, this transfer process ceases upon the attainment of equilibrium.

In gas absorption operations the equilibrium of interest is that between a nonvolatile absorbing liquid (solvent) and a solute gas (usually the pollutant). The solute is ordinarily removed from its mixture in a relatively large amount of a carrier gas that does not dissolve in the absorbing liquid. Temperature, pressure, and the concentration of solute in one phase are independently variable. The equilibrium relationship of importance is a plot of x, the mole fraction of solute in the liquid, against y^*, the mole fraction in the vapor in equilibrium with x. If Henry's law applies:

$$y^* = mx \tag{1}$$

where m is the Henry's law constant.

The engineering design of gas absorption equipment must be based on a sound application of the principles of diffusion, equilibrium, and mass transfer. The main requirement in equipment design is to bring the gas into intimate contact with the liquid; that is, to provide a large interfacial area and a high intensity of interface renewal and to minimize resistance and maximize driving force. This contacting of the phases can be achieved in many different types of equipment, the most important being either packed or plate columns. The final choice rests with the various criteria that must be met. For example, if the pressure drop through the column is large enough that horsepower costs become significant, a packed column may be preferable to a plate-type column because of the lower pressure drop.

In most processes involving the absorption of a gaseous pollutant from an effluent gas stream, the gas stream is the processed fluid; hence its inlet conditions (flow rate, composition, and temperature) are usually known. The temperature and composition of the inlet liquid and the composition of the outlet gas are usually specified. The main objectives, then, in the design of an absorption column are the determination of the solvent flow rate and the calculation of the principal dimensions of the equipment (column diameter and height to accomplish the operation). These objectives can be attained by evaluating, for a selected solvent at a given flow rate, the number of theoretical separation units (stages or plates) and converting them into practical units of column height or number of actual plates by means of existing correlations.

The general design procedure consists of a number of steps that have to be taken into consideration. These include the following:

1. Solvent selection.
2. Equilibrium data evaluation.
3. Estimation of operating data (usually consisting of a mass and energy balance, where the energy balance decides whether the absorption process can be considered as isothermal or adiabatic).
4. Column selection (Should the column selection not be obvious or specified, calculations must be carried out for the different types of columns and the final selection based on economic considerations.)
5. Calculation of column diameter (For packed columns, this is usually based on flooding conditions, and for plate columns, on the optimum gas velocity or the liquid-handling capacity of the plate.)
6. Estimation of column height or the number of plates (For packed columns, the column height is obtained by multiplying the number of transfer units, obtained from a knowledge of equilibrium and operating data, by the height of a transfer unit; for plate columns, the number of theoretical plates determined from the plot of equilibrium and operating lines is divided by the estimated overall plate efficiency to give the number of actual plates, which, in turn, allows the column height to be estimated from the plate spacing.)
7. Determination of pressure drop through the column (For packed columns, correlations dependent on packing type, column operating data, and physical properties of the constituents involved are available to estimate the pressure drop through the packing; for plate columns, the pressure drop per plate is obtained and multiplied by the number of plates.)

The operating data to be determined or estimated are the flow rates, terminal concentrations, and terminal temperature of the phases. The flow rates and concentrations fix the operating line, while the terminal temperatures give an indication as to what extent the operation can be considered isothermal; that is, whether or not the equilibrium line needs to be corrected for changes in liquid temperature. The operating line is obtained by a mass balance, and the outlet liquid temperature is evaluated from an energy balance on the column. In air pollution control, where relatively small quantities of gaseous pollutants are typically being absorbed, temperature effects are usually negligible.

In gas-absorption operations, the choice of a particular solvent is also important. Frequently, water is used, as it is very inexpensive and plentiful, but the following properties must also be considered:

1. Gas solubility. A high gas solubility is desired since this increases the absorption rate and minimizes the quantity of solvent necessary; generally, solvent of a chemical nature similar to that of the solute to be absorbed will provide good solubility.
2. Volatility. A low solvent vapor pressure is desired since the gas leaving an absorption unit is ordinarily saturated with the solvent and much of it may thereby be lost.
3. Corrosiveness.
4. Cost (for solvents other than water).
5. Viscosity. Low viscosity is preferred for reasons of rapid absorption rates, improved flooding characteristics, lower pressure drops, and good heat transfer characteristics.
6. Chemical stability. The solvent should be chemically stable and, if possible, nonflammable.
7. Toxicity.
8. Low freezing point. If possible, a low freezing point is favored since any solidification of the solvent in the column could prove disastrous.

Once the solvent is specified, the choice (and design) of the absorption system may be determined.

DESCRIPTION OF EQUIPMENT

The principal types of gas absorption equipment may be classified as follows:

1. Packed columns (continuous operating)
2. Plate columns (stage operating)
3. Miscellaneous

Of the three categories, the packed column is by far the most commonly used for the absorption of gaseous pollutants.

Packed Columns

For the case of continuous-contact operation, the gas and liquid phases flow through the equipment in a continuous manner with intimate contact throughout. Equilibrium between two phases at any position in the equipment is never established. (Should equilibrium occur anywhere in the system, the end result would be the equivalent of an infinitely tall column.)

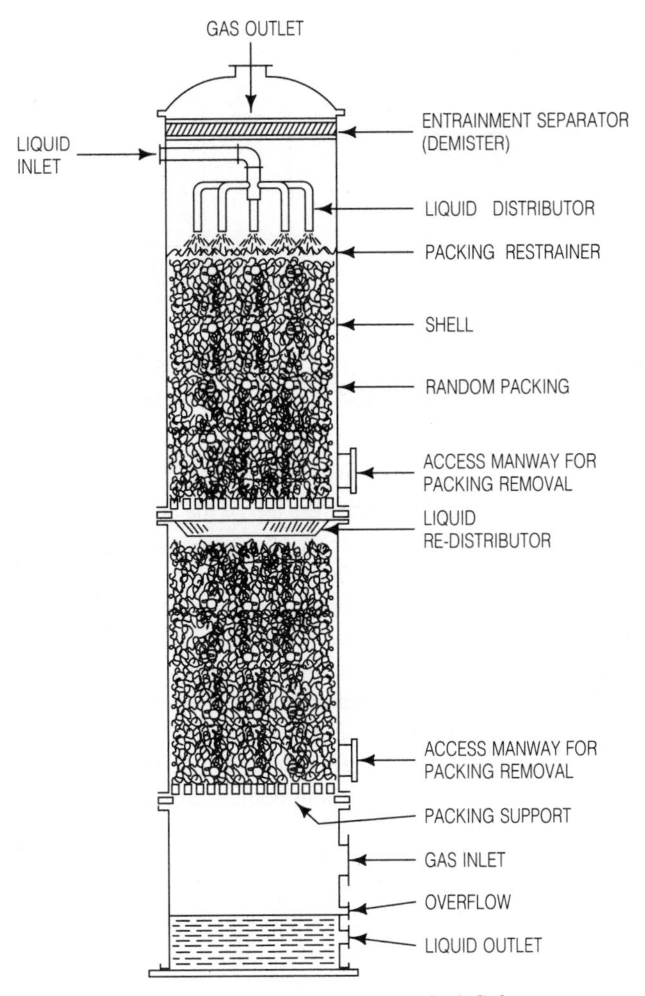

FIGURE 1. Typical Countercurrent Packed Column

Packed columns, used for the continuous contact of liquid and gas, are usually vertical columns that have been filled with packing or devices of large surface area. The liquid is distributed over and trickles down through the packed bed, thus exposing a large surface area to contact the gas. The countercurrent packed column (see Figure 1) is the most common type of unit encountered in gaseous pollutant control for the removal of the undesirable gas, vapor, or odor.

The gas stream (containing the pollutant) moves upward through the packed bed against an absorbing or reacting liquor (solvent-scrubbing solution), which is injected at the top of the packing. This results in the highest possible efficiency. Since the solute concentration in the gas stream decreases as it rises through the column, there is constantly fresher solvent available for contact. This provides the maximum average driving force for the diffusion process throughout the packed bed.

Occasionally, cocurrent flow may be used, where the gas

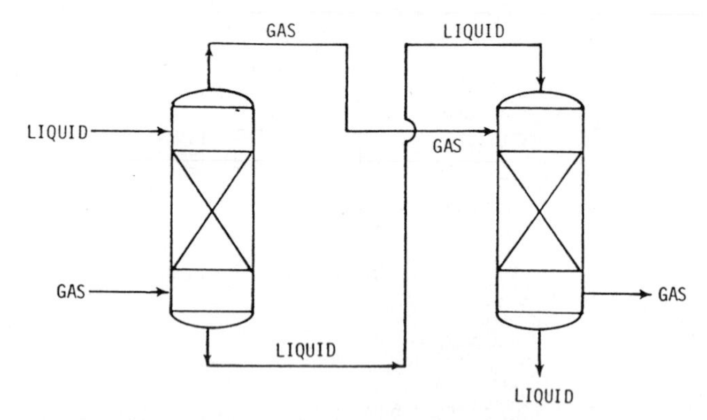

FIGURE 2. Countercurrent–Cocurrent Arrangement for Very Tall Columns

stream and solvent both enter the top of the column. Initially, there is a very high rate of absorption that constantly decreases until, with an infinitely tall column, the gas and liquid would leave in equilibrium. In this case, high gas and liquid rates are possible since the pressure drop tends to be rather low. However, these columns are efficient only when large driving forces are available, for example, with very soluble gases or acid scrubbing in caustic media. The design for this case utilizes minimum column diameters because of the low pressure drop and nonflooding characteristics. In general, cocurrent flow is not used too often except in the case of a very tall column built in two sections, both located on the ground as shown in Figure 2, with the second section operating in cocurrent flow merely as an economy measure to save on the large diameter gas pipe connecting the two. For an operation requiring an exceptionally high solvent flow rate, however, cocurrent flow might be used to prevent flooding that could occur in countercurrent operation.

Packed columns may also operate in cross flow (see Figure 3) where the airstream moves horizontally through the packed bed and is irrigated by the scrubbing liquid, which flows vertically down through the packing. Cross-flow designs are characterized by low water consumption and fairly high airflow capacity at low pressure drop. Where highly soluble gases are to be removed, the cross-flow packed scrubber has several advantages over the countercurrent scrubber. Using the same liquid and gas mass flow rates, a cross-flow scrubber has a lower pressure drop. Besides reducing water consumption drastically, the cross-flow principle also reduces pump and fan motor sizes. Other advantages include less plugging from solids dropout at the packing support plate and the possible use of higher gas and liquid rates because of extremely low pressure drop. On the other hand, liquid entrainment from these systems is rather high and mist eliminators are usually required downstream.

Packed columns are characterized by a number of features to which their widespread popularity may be attributed.

1. Minimum structure. The packed column usually needs only a packing support and liquid distributor about every 10 feet along its height.
2. Versatility. The packing material can be changed by simply dumping it and replacing it with a type giving better efficiency, lower pressure drop, or higher capacity; the depth of packing can also be easily changed if efficiency turns out to be less than anticipated or if feed or product specifications change.
3. Corrosive-fluids handling. Ceramic packing is common and often preferable to metal or plastic because of its corrosion resistance. When such packing does deteriorate, it is quickly and easily replaced; it is also preferred when handling hot combustion gases.
4. Low pressure drop. Unless operated at very high liquid

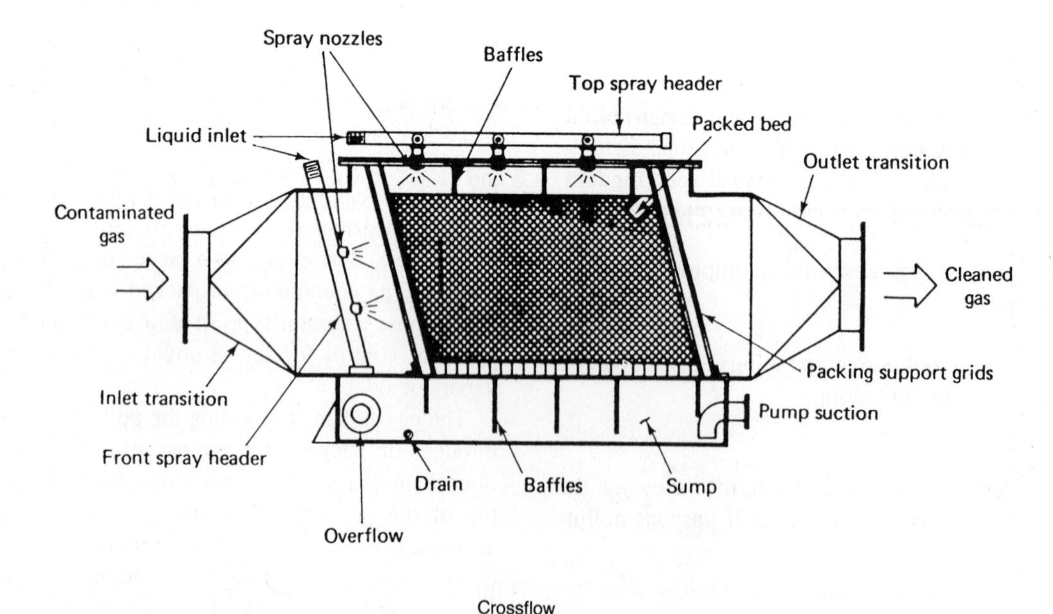

FIGURE 3. Cross-Flow Operation in a Packed Column

rates, where the liquid becomes the continuous phase as its films thicken and merge, the pressure drop per lineal foot of packed height is relatively low.

5. Range of operation. Although efficiency varies with gas and liquid feed rates, the range of operation is relatively broad.
6. Low investment. When plastic packings are satisfactory or when the columns are less than about 3 to 4 feet in diameter, cost is relatively low.

The packing is the heart of the performance of this type of equipment. Its proper selection entails an understanding of packing operational characteristics and the effect of performance of the points of significant physical difference among the various types. The main points to be considered in choosing the column packing include:

1. Durability and corrosion resistance. (The packing should be chemically inert to the fluids being processed.)
2. Free space per cubic foot of packed space. (This controls the liquor holdup in the column, as well as the pressure drop across it. Ordinarily, the fractional void volume, or fraction of free space, in the packed bed should be large.)
3. Wetted surface area per unit volume of packed space. (This is very important since it determines the interfacial surface between liquid and gas. It is rarely equal to the actual geometric surface since the packing is usually not completely wetted by the fluid.)
4. Frictional resistance to the flow of gas. (This affects the pressure drop over the column.)
5. Packing stability and structural strength to permit easy handling and installation.
6. Weight per cubic foot of packed space.
7. Cost per square foot of effective surface.

Table 1 illustrates some of the various types and applications of the different column packings available. Table 2 contains data on common packings. One additional distinction should also be made—the difference between random and stacked packings. Random packings are those that are simply dumped into the column during installation and allowed to fall at random. This is the most common method of packing installation. During installation, before pouring the packing into the column, the column is filled with water. This prevents breakage of the more fragile packing by reducing the velocity of the fall.

The fall should be as gentle as possible since broken packing tightens the bed and increases pressure drop. Stacked packing, on the other hand, is specially laid out and stacked by hand, making it a tedious operation and rather costly. It is avoided where possible except for the initial layers on supports. Liquid distributed in this way usually flows straight down through the packing immediately adjacent to the point of contact.

Liquid distribution plays an important role in the efficient operation of a packed column. A good packing from the process viewpoint can be reduced in effectiveness by poor liquid distribution across the top of its upper surface. Poor distribution reduces the effective wetted packing area and promotes liquid channeling. The final selection of the mechanism of distributing the liquid across the packing depends on the size of the column, the type of packing, the tendency of packing to divert liquid to column walls, and the materials of construction for distribution. For stacked packing, the liquid usually has little tendency to cross-distribute and thus moves down the column in the cross-sectional area that it enters. In the dumped condition, most packings follow a conical distribution down the column, with the apex of the cone at the liquid impingement point. For uniform liquid flow and reduced channeling of gas and liquid with as efficient use of the packed bed as possible, the impingement of the liquid onto the bed must be as uniform as possible. The liquid coming down through the packing and on the wall of the column should be redistributed after a bed depth of approximately three column diameters for Raschig rings and 5 to 10 column diameters for saddle packings. As a guide, Raschig rings usually have a maximum 10 to 15 feet of packing per section, while saddle packing can use a maximum 12 to 20 feet. As a general rule of thumb, however, the liquid should be redistributed every 10 feet of packed height. The redistribution brings the liquid off the wall and outer portions of the column and directs it toward the center area of the column for a new start as distribution and contact in the next lower section. Redistribution is usually not necessary for stacked bed packings, as the liquid flows essentially in vertical streams.

Plate Columns

Plate columns are essentially vertical cylinders in which the liquid and gas are contacted in stepwise fashion (staged operation) on plates or trays, in a manner shown schematically for one type in Figure 4. The liquid enters at the top and flows downward via gravity. On the way, it flows across each plate and through a downspout to the plate below. The gas passes upward through openings of one sort or another in the plate, then bubbles through the liquid to form a froth, disengages from the froth, and passes on to the next plate above. The overall effect is a multiple countercurrent contact of gas and liquid. Each plate of the column is a stage since on the plate on which the fluids are brought into intimate contact, interphase diffusion occurs and the fluids are separated. The number of theoretical plates (or stages) depends on the difficulty of the separation to be carried out and is determined solely from material balances and equilibrium considerations. The diameter of the column, on the other hand, depends on the quantities of liquid and gas flowing through the column per unit time. The actual number of plates required for a given separation is

TABLE 1. Some Typical Packings and Applications

Packing	Application Features
Raschig rings 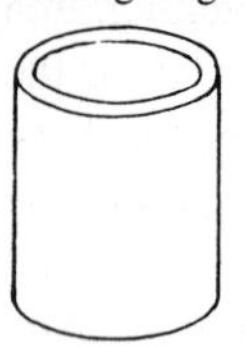	Most popular type, usually cheaper per unit cost, but sometimes less efficient than others; available in widest variety of materials to fit service; very sound structurally; usually packed by dumping wet or dry, with larger 4- to 6-inch sizes sometimes hand stacked; wall thickness varies among manufacturers, also some dimensions; available surface changes with wall thickness; produces considerable side thrust on tower; usually has more internal liquid channeling and directs more liquid to walls of column.
Berl saddles 	More efficient than Raschig rings in most applications, but more costly; packing nests together and creates "tight" spots in bed, which promotes channeling, but not as much as Raschig rings; do not produce much side thrust and have lower unit pressure drops with higher flooding point than Raschig rings; easier to break in bed than Raschig rings.
Intalox saddles 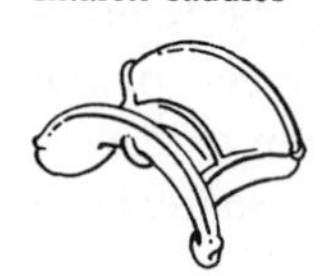	One of the most efficient packings, but more costly; very little tendency or ability to nest and block areas of bed; gives fairly uniform bed; higher flooding limits and lower pressure drop than Raschig rings or Berl saddles; easier to break in bed than Raschig rings.
Pall rings 	Lower pressure drop (less than half) than Raschig rings; higher flooding limit; good liquid distribution, high capacity; considerable side thrust on column wall; available in metal, plastic, and ceramic.
Spiral rings 	Usually installed as stacked, taking advantage of internal whirl of gas–liquid and offering extra contact surface over Raschig rings, Lessing rings, or cross-partition rings, available in single, double, and triple internal spiral designs, higher pressure drop, wide variety of performance data not available.
Teller rosette (Tellerette) 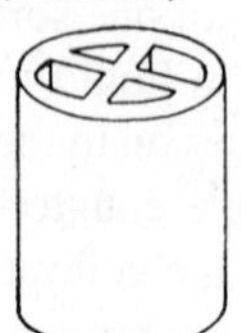	Available in plastic; lower pressure drops, higher flooding limits than Raschig rings or Berl saddles; very low unit weight, low side thrust; relatively expensive.
Cross-partition rings 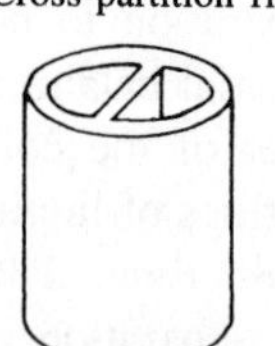	Usually used stacked, and as first layers on support grids for smaller packing above; pressure drop relatively low, channeling reduced for comparative stacked packings; no side-wall thrust.

TABLE 1. (*Continued*)

Packing	Application Features
Lessing rings 	Not many performance data available, but in general slightly better than Raschig rings; pressure drop slightly higher; high side-wall thrust.
Ceramic balls 	Tend to fluidize in certain operating ranges, self-cleaning, uniform bed structure, higher pressure drop, and better contact efficiency than Raschig rings; high side thrust; not many commercial data.
Goodloe packing and wire mesh packing 	Available in metal only, used in large and small columns for distillation, absorption, scrubbing, liquid extraction; high efficiency, low pressure drop.

greater than the theoretical number owing to plate inefficiency. To achieve high plate efficiencies, the contact time between gas and liquid on each plate should be long so as to permit the diffusion to occur, the interfacial surface between phases must be as large as possible, and a relatively high degree of turbulence is required to obtain high mass-transfer coefficients. In order to increase contact time, the liquid pool on each plate should be deep so that the gas bubbles will require a relatively long time to rise through the liquid. Relatively high gas velocities are also preferred for high plate efficiencies. The result is that the gas is very thoroughly dispersed into the liquid and causes froth formation, which provides large interfacial surface areas. On the other hand, great depths of liquid on the plates, although leading to high plate efficiencies, result in higher pressure drop per plate. High gas velocities, although providing

TABLE 2. Packing Factors—Dumped Packing

		Nominal Packing Size (inches)										
Packing Type	Material	¼	⅜	½	⅝	¾	1	1¼	1½	2	3	3½
Hy-Pak™	Metal						43			18	15	
Super Intalox® saddles	Ceramic						60			30		
Super Intalox saddles	Plastic						33			21	16	
Pall rings	Plastic				97		52		40	25		16
Pall rings	Metal				70		48		28	20		16
Intalox® saddles	Ceramic	725	330	200		145	98		52	40	36	
Raschig rings	Ceramic	1600	1000	640	380	255	160	95	65	65		
Raschig rings	1/32″ metal	700	390	300	170	185	115					
Raschig rings	1/16″ metal			410	290	230	137	110	83	57	32	
Berl saddles	Ceramic	900		240		170	110		65	45		
Tri-packs	Plastic						28			15		14
Tri-packs	Metal									18	14	

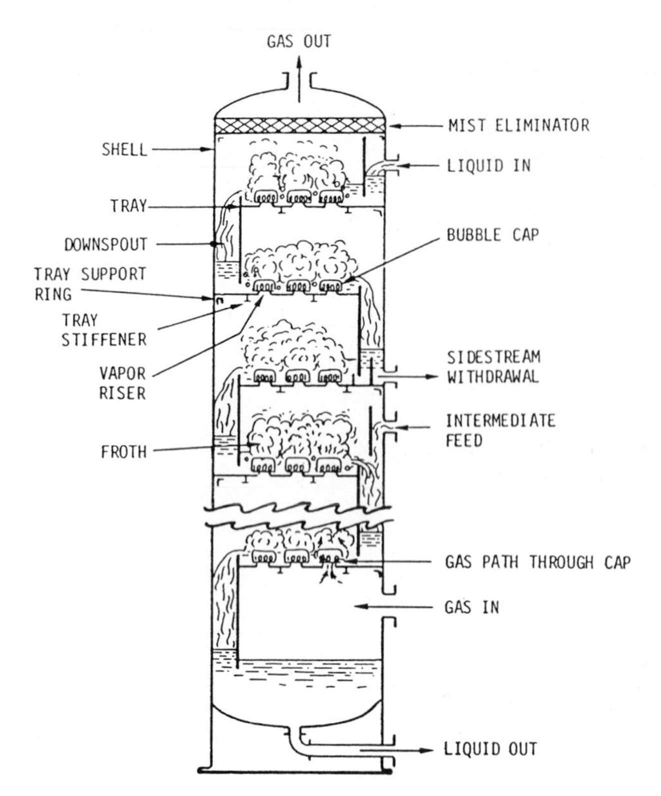

FIGURE 4. Typical Bubble-Cap Plate Column

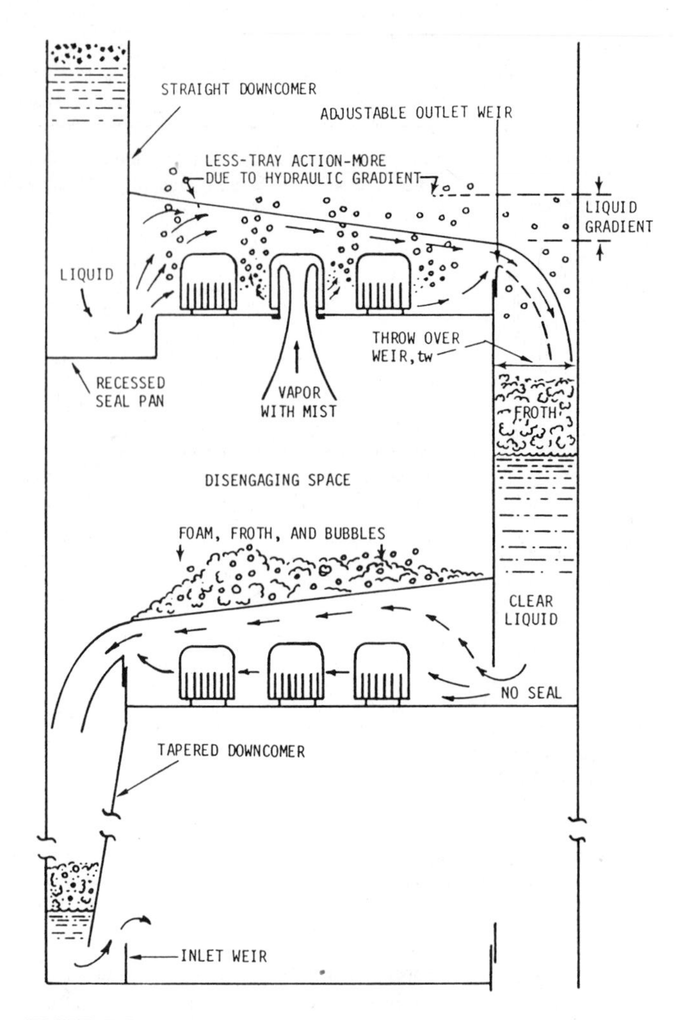

FIGURE 5. Bubble-Cap Plate Schematic—Dynamic Operation

good vapor–liquid contact, may lead to excessive entrainment accompanied by high pressure drop. Hence the various arrangements and dimensions chosen for the design of plate columns are usually those that experience has proven to be reasonably good compromises. Detailed information on the design of plate columns can be found in any standard distillation text.

The particular plate selection and its design can materially affect the performance of a given absorption operation. Each plate should be designed so as to give as efficient a contact between the vapor and liquid as possible, within reasonable economic limits. The principal types of plates encountered are discussed below.

In bubble-cap plates, the vapor rises up through "risers" into the bubble cap, out through the slots as bubbles, and into the surrounding liquid on the plates. Figure 5 demonstrates vapor–liquid action for a bubble-cap plate. The bubble-cap plate design is the most flexible of plate designs for high and low vapor and liquid rates. On the average, plates are usually spaced approximately 24 inches apart.

In sieve or perforated plates, the vapor rises through small holes (usually ⅛ to 1 inch in diameter) in the plate floor and bubbles through the liquid in a fairly uniform manner. The perforated plate is made with or without the downcomer. With the downcomer, the liquid flows across the plate floor and over a weir (if used), and then through the downcomer to the plate below. Figure 6 shows the operation schematically. These plates are not generally suitable for columns operating under variable load. Plate spacing in this case usually averages about 15 inches. In perforated plates without the downcomer, at the same time the vapor rises through the holes, the liquid head forces liquid countercurrent through these holes and onto the plate below. For this case, 12 inches is usually the average plate spacing. In general, perforated plates are used in systems where high-capacity near-design rates are to be maintained.

DESIGN AND PERFORMANCE EQUATIONS

Once all the streams entering and leaving the column and their constituents are identified, flow rates calculated, and operating conditions determined, the physical dimensions of the column can be calculated. The column must be of sufficient diameter to accommodate the gas and liquid and of sufficient height to insure that the required amount of mass is transferred with the existing driving force.

Consider a packed column that is operating at a given liquid rate and the gas rate is gradually increased. After a certain point, the gas rate is so high that the drag on the liquid is sufficient to keep the liquid from flowing freely down the column. Liquid begins to accumulate and tends to

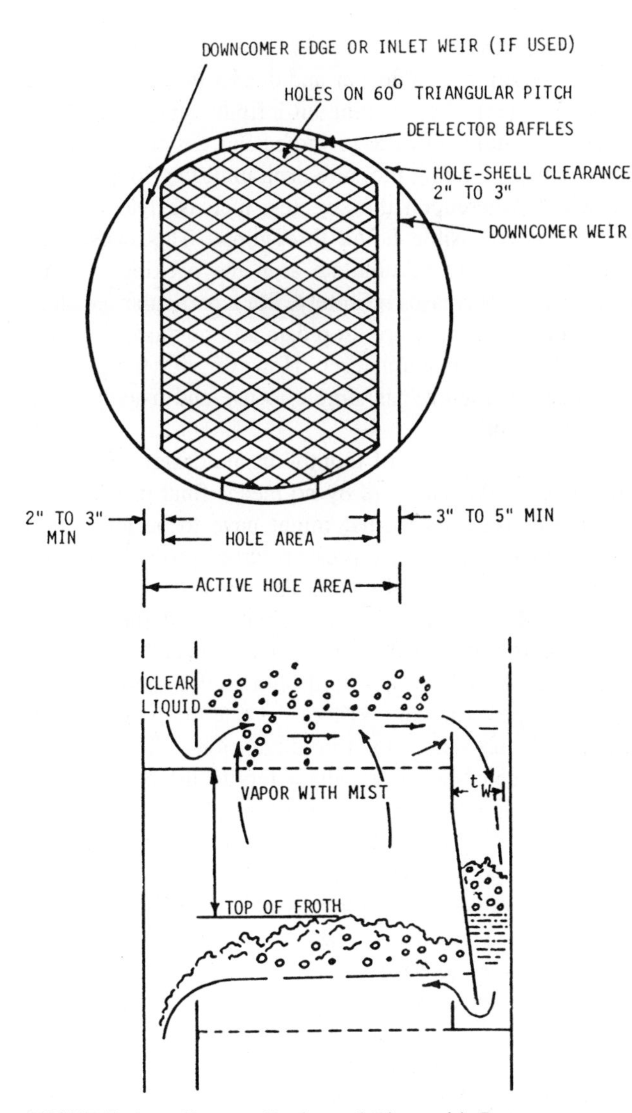

FIGURE 6. Sieve or Perforated Plate with Downcomers

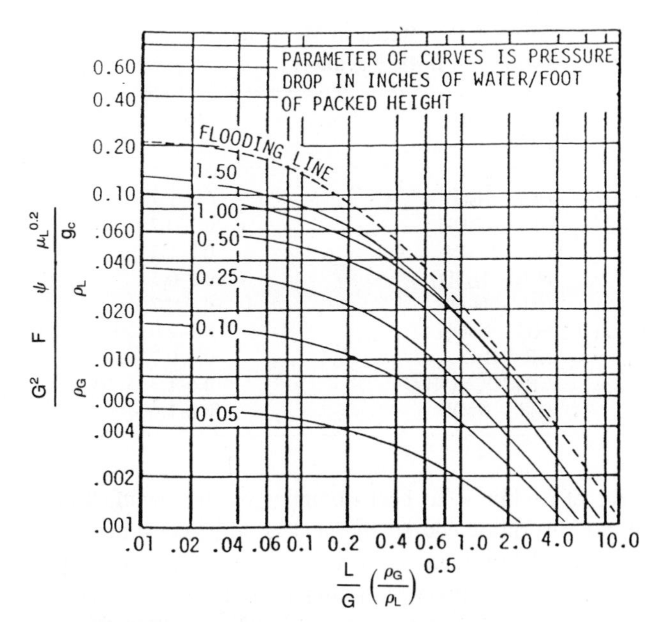

FIGURE 7. Generalized Pressure Drop Correlation to Estimate Column Diameter (G = gas flow rate, lb/sec ft^2; L = liquid flow rate, lb/sec ft^2; F = packing factor; Ψ = ratio, density of water/density of liquid; μ_L = liquid viscosity, cP$_a$; ρ_G = gas density, lb/ft^3; ρ_L = liquid density, lb/ft^3; g_c = 32.2).

block the entire cross section to flow (so-called loading). This, of course, both increases the pressure drop and prevents the packing from mixing the gas and liquid effectively, and ultimately some liquid is even carried back up the column. This undesirable condition, known as flooding, occurs fairly abruptly, and the superficial gas velocity at which it occurs is called the flooding velocity. The calculation of column diameter is based on flooding considerations, the usual operating range being taken as 50–75% of the flooding rate.

One of the more commonly used correlations is U.S. Stoneware's[6] generalized pressure drop correlation, as presented in Figure 7. The procedure to determine the column is as follows:

1. Calculate the abscissa, $(L/G)\ (\rho_G/\rho_L)^{0.5}$.
2. Proceed to the flooding line and read the ordinate (design parameter).
3. Solve the ordinate equation for G at flooding.
4. Calculate the column cross-sectional area, S, for the fraction of flooding velocity chosen for operation, f, by the equation:

$$S = W/fG \tag{2}$$

where: W = pounds per second of gas
S = area, square feet

5. The diameter of the column is then determined by

$$D = 1.13(S)^{0.5}$$

Note that the proper units as designated in the correlation must be used as the plot is not dimensionless. The flooding rate should also be evaluated using total flows of the phases at the bottom of the column, where they are at their highest value. The pressure drop may be evaluated directly from Figure 4 using a revised ordinate that contains the actual, not flooding, value of G. The column height is given by

$$Z = N_{OG}H_{OG} \tag{3}$$

where: N_{OG} = number of overall transfer units
H_{OG} = height of an overall transfer unit, feet
Z = height of packing, feet.

In most air pollution design practice, the number of transfer units (N_{OG}) is obtained experimentally or calculated using any of the methods to be explained later in this section. The height of a transfer unit (H_{OG}) is also usually determined experimentally for the system under consideration. Information on many different systems using various

TABLE 3. Range of Manufacturers' H_{OG} Data for Plastic Packings

Chemical System (in air)	H_{OG}, (feet)
$HCl - H_2O$	0.6–1.1
$HCl - NaOH$	0.5–0.7
$Cl_2 - NaOH$	0.8–1.2
$NH_3 - H_2SO_4$	0.3–0.5
$NH_3 - H_2O$	0.3–0.7
$SO_2 - NaOH$	0.7–2.0
$HF - H_2O$	0.4–0.7
$CH_3COCH_3–H_2O$	0.8–1.3
$H_2S - NaOH$	0.8–1.6

types of packings has been compiled by the manufacturers of gas absorption equipment and should be consulted prior to design. The data are usually in the form of graphs depicting, for a specific system and packing, the H_{OG} versus the gas rate (lb/hr-ft^2) with the liquid rate (lb/hr-ft^2) as a parameter. The packing height (Z) is then simply the product of the H_{OG} and the N_{OG}. Although there are many different approaches to determine the column height, the H_{OG}-N_{OG} approach is the most commonly used at the present time, with the H_{OG} usually being obtained from the manufacturer. Manufacturers' H_{OG} data are summarized in Table 3 for plastic packings.

In most air pollution applications, the pollutant to be absorbed is in the very dilute range. If the equilibrium and operating lines are both parallel and straight (Henry's law applies), and the number of transfer units is given by:

$$N_{OG} = \frac{\ln\left[\frac{(y_1 - mx_2)}{(y_2 - mx_2)}\left(1 - \frac{1}{A}\right) + \frac{1}{A}\right]}{1 - (1/A)} \tag{4}$$

where: $A = L/mG$

m = slope of the equilibrium curve (Henry's law constant)

Qualitatively, the height of a transfer unit is a measure of the height of a contactor required to effect a standard separation, and it is a function of the gas flow rate, the liquid flow rate, the type of packing, and the chemistry of the system. As indicated above, experimental values for H_{OG} are generally available in the literature or from vendors.

The pressure drop through a packed column for any combination of liquid and gas flows in the operable range is an important economic consideration in the design of such columns. For most random packings, the pressure drop suffered by the gas is influenced by the gas and liquid-flow rates. At constant gas rate, an increase in liquid throughput—which takes up more room in the packing (increased holdup) and, therefore, leaves less room for the gas (greater restriction)—is accompanied by an increase in pressure drop until the liquid flooding rate is reached. At this point, any slight excess that cannot pass through remains atop the packing, building up a deeper and deeper head (or pressure drop), hypothetically reaching an infinite value. Similarly, at constant liquid downflow, increasing gas flow is again accompanied by rising pressure drop until the flooding rate is reached, whereupon the slightest increase will cause a decline in permissible liquid throughput. This causes the remainder again to accumulate atop the packing, so that pressure again increases infinitely. For a particular packing, the most accurate pressure drop data will be those available directly from the manufacturer. However, for the purpose of estimation, Figure 7 is simple to use and usually gives reasonable results.

Some general "rules of thumb" in the design of packed columns do exist. They are by no means final in that there are other considerations that might have to be taken into account, that is, allowable pressure drop, possible column height restrictions, and the like. The rules, therefore, must be applied discriminately. The normal size of the packing typically does not exceed about 1/20th of the column diameter. It is also usual practice to design so that the operating gas rate is approximately 60% of the rate that would cause flooding. If possible, column dimensions should be in readily available sizes (i.e., diameters to the nearest one-half foot and heights to the nearest foot). If the column can be purchased "off-the-shelf" as opposed to being specially made, a substantial savings can be realized.

The most important design considerations for plate columns include calculation of the column diameter, type and number of plates to be used (for example, bubble cap or sieve plates), actual plate layout and physical design, and plate spacing (which, in turn, determines the column height). To consider each of these to any great extent is beyond the scope here. Details are available in any standard chemical engineering unit operations (i.e., distillation) or mass transfer text.[1–3] The discussion that follows, therefore, will be a relatively concise presentation of some of the general design techniques that will provide satisfactory results for the purpose of estimation.

The column diameter, and consequently its cross section, must be sufficiently large to handle the gas and liquid at velocities that will not cause flooding or excessive entrainment. The superficial gas velocity for a given type of plate at flooding is given by the relation

$$V_F = C_F[(\rho_L - \rho_G)/\rho_G]^{0.5} \tag{5}$$

where: V_F = gas velocity through the net column cross-sectional area for gas flow, ft^3/s-ft^2

ρ_L, ρ_G = liquid and gas densities, respectively, lb/ft^3

C_F = an empirical coefficient that depends on the type of plate and operating conditions

The net cross section is the difference between the column cross section and the area taken up by downcomers. In actual design, some percent of V_F is usually used: for nonfoaming liquids 80 to 85% of V_F, and 75% or less for foaming liquids. Of course, the value is subject to a check of entrainment and pressure drop characteristics. The calculation of column diameter based on equation 5 assumes that the gas flow rate is the controlling factor in its determination. After a plate layout has been assumed, it is, therefore, necessary to check the plate for its liquid-handling capacity. If the liquid-to-gas ratio is high and the column diameter large, the check will indicate whether the column will show a tendency toward flooding or gas maldistribution on the plate. If this is the case, then the liquid rate is the controlling factor in estimating column diameter and a satisfactory assumption for design purposes is a plate-handling capacity of 30 gal/min of liquid per foot of diameter.[4] However, a well-designed single-pass, cross-flow plate can ordinarily be expected to handle up to 60 gal/min of liquid per foot of diameter without excessive liquid gradient.

The column height is determined from the product of the number of actual plates (theoretical plates divided by the overall plate efficiency) and the plate spacing chosen. The theoretical plate (or stage, as it is sometimes called) is the theoretical unit of separation in plate column calculations. It is defined as a plate in which two dissimilar phases are brought into intimate contact with each other and then are mechanically separated. During the contact, various diffusing components of the mixture redistribute themselves between the phases. In an equilibrium stage, the two phases are well mixed for a time sufficient to allow establishment of equilibrium between the phases leaving the stage. At equilibrium, no further net change of composition of the phases is possible for a given set of operating conditions. The number of theoretical plates is usually determined graphically from the operating diagram composed of an operating line and equilibrium curve.

In the above discussion of equilibrium stages, it was assumed that the phases leaving the stage were in equilibrium. In actual countercurrent multistage equipment, it is not practical to provide the combination of residence time and intimacy of contact required to accomplish equilibrium. Hence the concentration change for a given stage is less than that predicted by equilibrium considerations. Stage efficiencies are employed to characterize this condition. The efficiency term frequently used is the overall stage (plate) efficiency given by the ratio of the theoretical contacts required for a given separation to the actual number of contacts required for the same operation. While reliable information on such an efficiency is most desirable and convenient to use, so many variables come into play that really reliable values for the overall stage efficiency are difficult to come by. This value is generally obtained by experiment or field test data or may be specified by the vendor.

For cases in which both the operating line and the equilibrium curve may be considered straight (dilute solutions), the number of theoretical plates may be determined directly without recourse to graphical techniques. This will frequently be the case for relatively dilute gas (as usually encountered in air pollution control) and liquid solutions where, more often than not, Henry's law is applicable. Since the quantity of gas absorbed is small, the total flows of liquid and gas entering and leaving the column remain essentially constant. Hence the operating line will be substantially straight. For such cases, the Kremser–Brown–Souders[7,8] equation applies for determining the number of theoretical plates, N_P.

$$N_p = \log\left\{\left[\left(\frac{Y_{Np+1} - mX_o}{Y_1 - mX_o}\right)\left(1 - \frac{1}{A}\right) + \frac{1}{A}\right](\log A)^{-1}\right\} \tag{6}$$

Here, mX_o is the gas composition in equilibrium with the entering liquid (m is Henry's law constant = slope of equilibrium curve). If the entering liquid contains no solute gas, then $X_o = 0$ and equation 6 can be simplified further. The solute concentrations in the gas stream, $Y_{N_{p+1}}$ and Y_1 represent inlet and outlet conditions, and L and V the total pound mole rates of liquid and gas flow per unit time per unit column cross-sectional area. Small variations in L and V may be roughly compensated for by using the geometric average value of each taken at the top and bottom of the column.

The general procedure to follow in sizing a plate tower is as follows:

1. Calculate the number of theoretical stages, N, using equation 6.
2. Estimate the efficiency of separation, E. This may be determined at the local (across plate), plate (between plates), or overall (across column) level. The overall efficiency, E_o, is generally employed.
3. Calculate the actual number of plates:

$$N_{\text{act}} = N/E_o$$

4. Obtain the height between plates, h. This is usually in the 12- to 36-inch range. Most towers use a 24-inch plate spacing.
5. The tower height, Z, is then

$$Z = (N_{\text{act}})(h)$$

6. The diameter may be calculated directly from equation 5.
7. The plate or overall pressure drop is difficult to quantify. It is usually in the 2- to 6-inches water per plate range for most columns.

Of the various types of gas absorption devices available, packed columns and plate columns are the most commonly used in practice. Although packed columns are used more often in air pollution control, both have their special area of usefulness, and the relative advantages and disadvantages of each are worth considering. In general:

1. The pressure drop of the gas passing through the packed column is smaller.
2. The plate column can stand an arbitrarily low liquid feed and permits a higher gas feed than the packed column. It can also be designed to handle liquid rates that would ordinarily flood the packed column.
3. If the liquid deposits a sediment, the plate column is more advisable. By fitting the column with manholes, the plate column can be cleaned of accumulated sediment that would clog many packing materials and warrant costly removal and refilling of the column. Packed columns are also susceptible to plugging if the gas contains particulate contaminate.
4. In mass transfer processes accompanied by considerable heat effects, cooling or heating the liquid is much easier in the plate column. A system of pipes immersed in the liquid can be placed on the plates between the caps, and heat can be removed or supplied through the pipe wall directly to the area in which the process is taking place. The solution of the same problem for a packed column leads to the division of this process into a number of sections, the cooling or heating of the liquid taking place between these sections.
5. The total weight of the plate column is usually less than that of the packed column designed for the same capacity.
6. A well-installed plate column avoids serious channeling difficulties, insuring good, continuous contact between the gas and liquid throughout the column.
7. In highly corrosive atmospheres, the packed column is simpler and cheaper to construct.
8. The liquid holdup in the packed column is considerably less than in the plate column.
9. Temperature changes are apt to do more damage to the packed column than to the plate column.
10. Plate columns are advantageous for absorption processes with an accompanying chemical reaction (particularly when it is not very rapid). The process is favored by a long residence time of the liquid in the column and by easier control of the reaction.
11. Packed columns are preferred for liquids with high foaming tendencies.
12. The relative merits of the plate column and packed column for a specified purpose are properly determined only by comparison of the actual cost figures resulting from a detailed design analysis for each type. Most conditions being equal, packed columns in the smaller sizes (diameters up to approximately 2 to 3) feet are on the average less expensive. In the larger sizes, plate columns tend to be the more economical.

OPERATION AND MAINTENANCE[10]

Since packed towers are primarily used for gaseous pollutant control, this device will be emphasized in the presentation below, although much of this material will also be applicable to plate and other towers.

After start-up, the system liquid- and gas-flow rates should be checked to ensure that they are operating within the design parameters of the system. After approximately two weeks of operation, the system should be shut down and inspected for any possible nozzle pluggage or settling of the packing, which frequently occurs during these first two weeks. If the packing has settled, the packing should be topped off to the appropriate design level. During normal operation, daily checks should be made of the recycle liquid flow to the liquid distributor and the liquid makeup rate. Also, if a chemical feed system is employed, daily checks of the chemical solution should be made to ensure an adequate supply of chemical. Daily logs should also be kept of any instrumentation readings, such as pressure indicators, temperature indicators, flow indicators, and all other operations pertinent to the specific system. Should any drastic variance be noted in these readings, a close investigation should be conducted to indicate the cause of these abnormal readings. The following list is a guide for any abnormalities that may be encountered during operation of the equipment.

1. If the static pressure drop across the scrubber continually increases over a long period, this could indicate one of the following.
   a. The liquid flow rate to the liquid distributor has increased and should be checked.
   b. The packing in the irrigated bed could be partially plugged as a result of solids deposition and may require cleaning.
   c. The entrainment separator could be partially plugged and may require cleaning.
   d. The packing support plate at the bottom of the packed section could be blinding and causing increased pressure drop and so require cleaning.
   e. The packing could be settling due to corrosion or to solids deposition, again requiring cleaning or additional packing.
   f. The airflow rate through the absorber could have been increased by a change in the damper setting, which may need readjustment.
2. The pressure drop across the absorber begins to decrease either slowly or rapidly. This could be caused by the following possibilities.
   a. The liquid flow rate to the distributor has decreased and should be adjusted accordingly.
   b. The air flow rate to the scrubber has decreased because of a change in the fan characteristics or a change in the system damper settings.

c. Partial plugging of the spray or liquid distributor, causing channeling through the scrubber, could be occurring. The liquid distributor should be inspected to ensure that it is totally operable.
d. The packing support plate could have been damaged and have fallen into the bottom of the absorber, allowing the packing to fall to the bottom and produce a lower pressure drop. This should be checked.

3. A pressure or flow change in the recycle liquid, indicating a lower liquid flow rate, may be caused by the following.
   a. A plugged strainer or filter in the recycle piping, which may require cleaning.
   b. Plugged spray nozzles, which may require cleaning.
   c. The piping may be becoming partially plugged with solids and need cleaning.
   d. The liquid level in the sump could have decreased, causing pump cavitation.
   e. The pump impeller could have been worn excessively.
   f. A valve in either the suction or discharge side of the pump could have been closed inadvertently.
4. A high liquid flow indicates the following.
   a. A break in the internal distributor piping.
   b. A spray nozzle that has been inadvertently "uninstalled."
   c. A spray nozzle that may have come loose or eroded away, creating a low pressure drop.
   d. A change in the throttling valve setting on the discharge side of the pump, allowing larger liquid flow; reset to the proper conditions.
5. Excessive entrainment carryover from the outlet of the absorber could be caused by the following.
   a. A partially plugged entrainment separator, causing channeling and reentrainment of the collected liquid droplets.
   b. The air flow rate to the absorber could have increased above the design capability, causing reentrainment.
   c. If a packed-type entrainment separator is used, the packing may not be level, causing channeling and reentrainment of moisture.
   d. If a packed entrainment separator is used, and a sudden surge of air through the absorber has occurred, this could have caused the packing to be carried out of the absorber or the packing being blown aside, creating an open area "hole" through which the air passes.
   e. The entrainment support plate could be damaged and have dropped, causing channeling through the separator.
   f. The velocity through the absorber has decreased to a point that absorption does not effectively take place and low removal is achieved.
   g. Conceivably during wintertime operation, some moisture resulting from condensation may be observed and considered a problem with the entrainment separator. However, it should be recognized that this is not uncommon, since the air off an absorber is saturated with water vapor and any difference or lowering of the temperature will cause condensation to occur.
6. If the reading indicates a low air flow or no air flow across the absorber, the following may be the cause.
   a. The packing in the absorber may be plugged, causing a restriction to the air flow.
   b. The liquid flow rate to the absorber could have been increased inadvertently, again causing greater restriction and pressure drop, creating a lower gas flow rate.
   c. The fan belts have worn or loosened, reducing the air flow to the equipment.
   d. The fan impeller could be partially corroded, reducing fan efficiency.
   e. The ductwork to or from the absorber could be partially plugged with solids and may need cleaning.
   f. A damper in the system has been inadvertently closed or the setting changed.
   g. A break or a leak in the duct could have occurred due to corrosion.
7. Should the airflow through the absorber be increasing or has suddenly increased, it may be due to the following.
   a. A sudden opening of a damper in the system.
   b. Low liquid flow rate to the absorber.
   c. The packing has suddenly been damaged and has fallen to the bottom of the absorber.
8. A sudden decrease in the absorber efficiency could indicate the following.
   a. The liquid makeup rate to the absorber has been inadvertently shut off or throttled to a low level, decreasing the absorber efficiency.
   b. If a chemical feed system is employed, the system may have run out of the chemical required.
   c. The foregoing could indicate a malfunction of pH probes if employed, requiring replacement.
   d. The set point on the pH control may have to be adjusted to allow more chemical feed.
   e. A problem may exist with the chemical metering pump, control valve, or line pluggage.
   f. The liquid flow rate to the scrubber may be too low for effective removal.
   g. Pluggage or solids buildup in the absorber liquid distributor or packing may have caused channeling.

This list indicates only some of the common operational problems encountered with absorber systems. There are a number of related problems. Excessive moisture may be blown from either the fan or the blower, if it is on the suction side of the system. Excessive moisture droplets being emitted could be caused by condensation in the stack or in the fan. Improper condensate drainage and trapping or line pluggage from the fan could be the problem. Scrubber liquid levels can also be too high. This can occur by having too large an inlet continuous freshwater makeup, with the overflow piping being too restrictive or with the overflow and drain piping trap or barometric seal being broken. Generally, an absorber system is overdesigned for the capacity it is to handle. On occasion, lower volumes are introduced through the system below the design condition. These lower volumes usually increase gas absorption. However, if too low a gas input is used, creating very low static pressure drops through the absorber, channeling can occur and can lower the performance of the absorber and entrainment separator performance. Caution should be taken to use the manufacturer's minimum gas-flow-rate recommendation to prevent this problem from occurring.

Normal preventative maintenance requires only periodic checks of the fan, pumps, chemical feed system, piping, duct, and absorber liquid distributor. The normally irrigated packing may never have to be cleaned during the life of the absorber as long as the absorber has been properly designed and operated. However, should the absorber be run dry or the entering airstream contain unexpectedly high solids loadings, a heavy formation of solids, crystallized salts, or other foreign matter may accumulate on the packing, and this must be removed. In most cases, removal can be accomplished by recirculating, for a short time, a chemical solution into which the solids will dissolve or react. Sometimes, chemical removal will not remove the buildup and it may be necessary to use high-pressure water, hot water, or atmospheric steam. Prior to using any chemical, hot water, or steam, the absorber manufacturer should be consulted to verify the resistance of the internals and shell. In a very few instances, such as calcium fluoride deposition or extremely heavy solids deposition, it may be necessary to remove the packing medium from the absorber for cleaning.

The normally unirrigated entrainment separator must be periodically flushed with sprays to prevent buildup and eventual plugging. The intervals between routine washings must be determined by experience, as the collection of solid materials is a function of the specific operating conditions. To ensure proper operation and prevention of unexpected problems with the absorber system, the following is a suggested maintenance checklist.

1. Pump maintenance
   a. Check the bearing lubrication to make sure that the proper amount of oil is present (weekly).
   b. If a packed stuffing box is provided, make sure that the drip rate is not excessive or that the packing is lubricated (weekly).
   c. Check pump and motor bearings for unusual noise or heat. If this is noticed, it could indicate excessive wear on the bearings, which may result in excessive shaft runout, requiring frequent repacking of the stuffing box (weekly).
   d. Inspect gasketing at the suction and discharge connections for leaks (weekly).
   e. Check shaft alignment and levelness of the base plate (monthly).
   f. Look for crystallization of solution or solids in the packing, which could cause scoring of the shaft (monthly).
   g. Check for water flush to the mechanical seal (if provided) and adjust if necessary (weekly).
2. Fan maintenance
   a. Check the fan motor and fan bearings for unusual vibration, noise or heat (weekly).
   b. Check the bearing lubrication to ensure that the proper amount of oil is present (weekly).
   c. Inspect the belts for wear and replace if necessary. Also check the belts for proper tension. Adjust if necessary (monthly).
   d. Check all fan bolts for looseness and tighten if required (monthly).
   e. Inspect the fan impeller and blades for any solids buildup or erosion. Clean or replace the impeller if excessive buildup or erosion has occurred. Such buildup can unbalance the impeller and cause premature failure (monthly).
   f. Check the fan housing drain to make sure it is not plugged so that it will permit proper condensate drainage (monthly).
3. Absorber maintenance
   a. Check the pressure drop across the absorber. Increased pressure drop will usually indicate that plugging of the packing is occurring (weekly).
   b. Open the absorber access door provided for inspection of the liquid distributor. (Access doors should never be opened while the fan is in operation.) Observe the liquid distribution. Clean if required (monthly to quarterly).
   c. Inspect the absorber sump for solids buildup and flush if necessary (semiannually).
   d. During the inspection of the liquid distributor, check the packing level for possible settling. Add packing to the proper level if necessary (quarterly to semiannually).
   e. Check the entrainment separator for solids or crystallization buildup and clean with sprays if required (quarterly to semiannually).
   f. Inspect the absorber packing for solids buildup and clean if required (semiannually).
   g. Open all access openings and inspect the tower internals for corrosion or breakage (semiannually).
   h. During the inspection of the tower internals, also inspect the absorber shell for any corrosion and repair if necessary (semiannually).
4. System maintenance
   a. Check and clean all strainers or filters in the piping system (weekly to monthly).
   b. Inspect the recycle piping, duct and absorber

flanged connections, access doors, and flexible connections for leaks and tighten or replace gaskets if necessary (monthly to quarterly).

c. Check all system bypass or control dampers and cycle each one a few times to ensure proper operation (weekly).
d. Check all piping to and from the absorber, including the recycle piping, for erosion or plugging and clean or replace if required (quarterly).
e. If a chemical feed system is employed, check the metering pump or valves and the pH probes weekly for proper operation. It is not unusual to replace the pH probe elements frequently if solids buildup occurs (weekly).
f. Inspect the duct for solids buildup due to settling. This may require periodic flushing to prevent excessive restriction to air flow or weight (semiannually).
g. Check the duct, fan, and absorber for any external corrosive attack or obvious deterioration (semiannually).
h. Inspect structural supports, duct, and pipe hangers for any deterioration caused by corrosion (semiannually).
i. Frequent checks of any heat tracing or freeze protection during the winter season, if applicable, should be conducted on a weekly basis.
j. All instrumentation should be inspected, checked, and calibrated to ensure proper function and control (monthly).

The periodic time intervals indicated for maintenance will vary depending on the specific system's operating conditions and the equipment manufacturer's recommendation.

MAINTAINING AND IMPROVING EQUIPMENT PERFORMANCE[10]

The earlier sections discussed the basic design parameters associated with absorbers, primarily packed absorbers, and the operation and maintenance of this equipment. Much of the information presented is also applicable to other absorber types, such as spray towers, baffle towers, tray towers, and wet cyclones.

There are options available that should be considered when purchasing a packed absorber, or any absorber, to reduce maintenance costs and allow the equipment to be more operationally reliable between scheduled maintenance shutdowns. Any options generally increase the initial cost of the equipment, and for this reason are often not considered. However, some of these items can pay for themselves in a short time by reducing maintenance costs.

One such item is the provision of adequate access openings to all internal portions of the scrubber. The areas that should be considered are the sump section of the absorber; the lower packed section for removal of the packing and internal packing support plate, if required; the liquid distributor for spray nozzle cleaning or removal of the internal liquid distributor; and access to the entrainment separator for inspection and removal of the packing and support plates, if necessary. The access openings should be large enough for accessibility and total removal of the internal components of the absorber. To reduce the initial cost of equipment, some manufacturers provide only two or three access doors. In many cases, it may be advantageous to consider clear, see-through access openings, especially in the liquid distributor region, for external observation of the absorber of the liquid distributor to ensure its proper operation without equipment shutdown.

The liquid distributor, especially spray-type distributors, is one area where maintenance is required on a periodic basis. Where spray-type distributors are used, many manufacturers provide these as an integral part of the absorber shell construction. Should the spray nozzles need cleaning or removal with this type of spray distributor, a worker must physically enter the absorber to remove the spray nozzles. This can be a continual source of irritation. Since most absorbers are handling corrosive-type gases, the absorber must be shut down and totally purged with air to allow entry. In addition, in most plants, it is necessary that a worker wear a rubber suit and mask for safety when entering a vessel. The need for entry into the absorber for maintenance of a spray-type distributor can be eliminated by designing the distributor so that it can be totally removed from the absorber.

The entrainment separator is another source requiring continuous maintenance. Because it is generally an unirrigated section, solids and/or salt deposition, which is not uncommon, needs to be eliminated periodically by washing. To clean the entrainment separator it is necessary to shut down the absorber, open the equipment, and physically flush down the separator media with liquid. Equipment shutdown for this maintenance can be eliminated by installation of a spray wash-down flush header above the entrainment separator, which can provide intermittent flushing of the separator media. This eliminates equipment shutdown and maintenance cost. If this is provided, it should be recognized that if the fan is operating, liquid will be carried over through the outlet of the system during the flush cycle. To prevent this, the fan should be shut down during this flushing cycle.

The absorber can be designed to provide future additional packing-height requirements should greater future gas absorption capability be required. This can be accomplished in two ways: the shell of the absorber can be flanged to allow a future stub section to be inserted, or the tower can be initially designed with a void space above the packing between the liquid distributors so that future packing heights can be added without shell modification. Although the initial cost of this may seem high, it could preclude the need to field-modify the absorber or to purchase a new absorber for greater efficiency, when and if the air pollution laws are tightened.

Another source of periodic maintenance relates to the strainers or filters in the recycle piping and the recycle pumps. To prevent total system shutdown during cleaning or maintenance of these items, dual strainers or pumps may be considered, valved separately, so that the system can be maintained in a fully operational mode during strainer cleaning or pump maintenance checks.

Proper instrumentation used for the purpose of providing operational data is useful in determining areas where problems may arise or maintenance may be required. The type of instrumentation will depend totally on the system requirements. Areas of specific interest would be pressure drop across the absorber, both the wetted bed and entrainment separator; low liquid recycle flow and/or pressure; liquid makeup rates to the system; and liquid temperatures.

ILLUSTRATIVE EXAMPLES

Example 1

A packed column is designed to absorb ammonia from a gas stream. The unit operates at 60% of the flooding gas mass velocity, the actual liquid flow rate is 25% more than the minimum, and 90% ammonia is to be collected based on state regulations. Given the operating conditions and type of packing below, calculate the height of packing and column diameter. Gas mass flow rate = 5000 lb/h; NH_3 concentration in inlet gas stream = 2.0 mol%; scrubbing liquid = pure water; packing type = 1 inch Raschig rings; H_{OG} of the column = 2.5 feet; Henry's law constant, m = 1.20; density of gas (air) = 0.075 lb/ft^3; density of water = 62.4 lb/ft^3; viscosity of water = 1.8 cP.

Solution First calculate the equilibrium outlet liquid composition and the outlet gas composition for 90% removal.

$$x^*_1 = y_1/m = (0.02)/(1.20) = 0.0167$$

$$y_2 = (0.1)y_1/[(1 - y_1) + (0.1)y_1] = (0.1)(0.02)/[(1 - 0.02) + (0.1)(0.02)] = 0.00204$$

The minimum liquid to gas ratio (molar basis) is obtained by a material balance.

$$(L_m/G_m)_{\min} = (y_1 - y_2)/(x^*_1 - x_2) = (0.02 - 0.00204)/(0.0167 - 0) = 1.08$$

The actual ratio is 25% above the minimum. Thus,

$$L_m/G_m = (1.25)(L_m/G_m)_{\min} = (1.25)(1.08) = 1.35$$

To evaluate N_{OG}

$$(y_1 - mx_2)/(y_2 - mx_2) = [(0.02) - (1.2)(0)]/[(0.00204) - (1.2)(0)] = 9.80$$

$$(mG_m)/L_m = (1.2)/(1.35) = 0.889$$

From equation 4:

$$N_{OG} = 6.2$$

The packing height is then

$$Z = (N_{OG})(H_{OG}) = (6.2)(2.5) = 15.5 \text{ feet}$$

Figure 7 is employed to calculate the tower diameter and packing pressure drop.

$$(L/G)(\rho/\rho_1)^{0.5} = (L_m/G_m)(18/29)(\rho_G/\rho_L)^{0.5} = (1.35)(18/29)(0.075/62.4)^{0.5} = 0.0291$$

From Figure 7

$$(G^2F\Psi\mu_L^{0.2})/(\rho_G\rho_L) = 0.19$$

Thus, the flooding velocity is

$$G_f = (0.19)(\rho_G\rho_L g_c)/(F\Psi\mu_L^{0.2})^{0.5} = [(0.19)(62.4)(0.075)(32.2)/(160)(1)(1.8)^{0.2}]^{0.5} = 0.409 \text{ lb/ft}^2\text{/s}$$

The actual velocity is

$$G_{act} = (0.6)(0.409) = 0.245 \text{ lb/ft}^2\text{/s} = 884 \text{ lb/ft}^2\text{/h}$$

The tower diameter may now be calculated directly from equation 2.

$$D_T = [(1.13)(4000)/(884)]^{0.5} = 2.67 \text{ feet}$$

Example 2

Qualitatively outline how one could size (diameter and height) a packed tower to achieve a given degree of separation without any information on the physical and chemical properties of the gas to be cleaned.

Solution In order to calculate the height, one needs both the height of a gas transfer unit, H_{OG}, and the number of gas transfer units, N_{OG}. Since equilibrium data are not available, assume that m (slope of equilibrium curve) approaches zero. This is not an unreasonable assumption for most solvents that preferentially absorb the pollutant. For this condition:

$$N_{OG} = \ln(y_1/y_2)$$

Since the scrubbing medium is usually water or a solvent that effectively has the physical and chemical properties of water, H_{OG} can be assumed to take the values usually encountered for water systems. These are given below.

Packing diameter (inches)	Plastic packing H_{OG} (feet)	Ceramic packing H_{OG} (feet)
1	1.0	2.0
1.5	1.25	2.5
2	1.5	3.0
3	2.25	4.5
3.5	2.75	5.5

For plastic packing, the liquid and gas flow rates are both typically in the range of 1500 to 2000 lb-h-ft^2. For ceramic packing, the range of flow rates is 500 to 1000 lb-h-ft^2. For difficult to absorb gases, the gas flow rate is usually lower and the liquid flow rate higher. The height, Z, is then calculated from:

$$Z = (H_{OG})(N_{OG})(SF)$$

where SF is a safety factor whose value can range from 1.25 to 1.50. One may use Table 4 to estimate packing height requirements without the safety factor included.

Typical superficial throughput velocities for packed columns are in the range of 3 to 6 fps, with lower and higher values applying to ceramic packing and plastic packing, respectively. A value of 4 fps is typical. Dividing the volumetric flow rate by this value provides a reasonable estimate of the tower cross-sectional area, S.

$$S = q(\text{acfs})/4;\ \text{ft}^2$$

Tower pressure drop in air pollution control for ceramic packing is in the neighborhood of 0.25 in. H_2O per foot of packing; the value for plastic packing is approximately 0.2 in. H_2O per foot of packing.

For tower diameters in the 3-foot range, packing diameters of 1 inch are recommended. For larger-diameter towers, one should use larger packing; for smaller towers, smaller packing is usually employed. Regarding the calculation of N_{OG}, if vapor pressure data for the pollutant are available, one can calculate Henry's constant (assuming ideal behavior) from $m = P'/P$; P' is the vapor pressure at operating temperature and P is the total pressure. If the solvent is something other than water—as in many chemical process and/or petroleum applications—techniques are available for estimating H_{OG}. This usually requires information on the pollutant–air–solvent diffusivities. This calculational procedure is beyond the scope here; however, it can be found in many chemical engineering mass-transfer books. For chemical reacting systems, the suggested procedure is to assume that $m = 0$. This is a very reasonable assumption since the mole fraction in the solvent is zero immediately following reaction with the solvent. This simple procedure eliminates the need for either obtaining or calculating equilibrium data for a reacting system, particularly when the normality of an acidic or basic solvent is variable. The results (procedures of this problem) can be extended to plate towers in a manner similar to that described above.

TABLE 4. Typical Packed Height (feet) to Obtain Specified Efficiency

Removal Efficiency, (%)	Plastic Packing Size (inch): 1	1½	2	3	3½
63.2	1	1.25	1.5	2.25	3
77.7	1.5	2	2.25	3.5	4.25
86.5	2	2.5	3.0	4.5	5.75
90	2.3	2.88	3.45	5.18	6.5
95	3	3.75	4.5	6.75	8.5
98	4	5	6	9	11.25
99	4.6	5.75	7	10.25	13
99.5	5.25	6.5	8	12	14.75
99.9	7	8.75	10.5	15.75	19.75
99.99	9.25	11.5	14	21	26

References

1. R. E. Treybal, *Mass Transfer Operations,* 2nd ed., McGraw-Hill Book Co., New York, 1968.
2. E. J. Henley and H. K. Staffin, *Stage Process Design,* John Wiley & Sons, New York, 1963.
3. M. Van Winkle, *Distillation,* McGraw-Hill Book Co., New York, 1967.
4. H. Sawistowski and W. Smith, *Mass Transfer Process Calculations,* Interscience, New York, 1963, p. 54.
5. L. Theodore and A. J. Buonicore (eds.), *Selection, Design, Operation and Maintenance: Air Pollution Control Equipment,* Prentice-Hall, Englewood Cliffs, NJ, 1982.
6. A. Buonicore and L. Theodore, *Industrial Air Pollution Control Equipment for Gaseous Pollutants* (Vol. II), CRC Press, Boca Raton, FL, 1975.
7. A. Kremser, *Nat'l Petrol. News,* 22(21): 42 (1930).
8. M. Souders, and G. G. Brown, *Ind. Eng. Chem.,* 24:519 (1932).
9. Ceilcote Co., *Installation, Operation and Maintenance Manual,* Berea, OH, 1975.
10. J. W. Macdonald, in *Selection, Design, Operation and Maintenance: Air Pollution Control Equipment,* L. Theodore and A. J. Buonicore, eds., Prentice-Hall, Englewood Cliffs, NJ, 1982.

ADSORPTION

Anthony J. Buonicore

It has already been well established that the molecular forces at the surface of a liquid are in a state of unbalance or unsaturation. The same is true of the surface of a solid, where the molecules or ions on the surface may not have all their forces satisfied by union with other particles. As a result of this unsaturation, solid and liquid surfaces tend to satisfy their residual forces by attracting to and retaining on their surface gases or dissolved substances with which they come in contact. This phenomenon of concentration of a substance on the surface of a solid (or liquid) is called adsorption. The substance thus attracted to a surface is said to be the adsorbed phase or adsorbate, while the substance to which it is attached is the adsorbent. Adsorption should be carefully distinguished from absorption, the latter process being characterized by a substance's not only being retained on a surface, but also passing through the surface to become distributed throughout the phase. Where doubt exists as to whether a process is true adsorption or absorption, the noncommittal term "sorption" is sometimes employed.

Study of adsorption of various gases (or vapors) on solid surfaces has revealed that the forces operative in adsorption are not the same in all cases. Two types of adsorption are generally recognized, namely, "physical" or van der Waals adsorption and "chemical" or activated adsorption.

Physical adsorption is the result of the intermolecular forces of attraction between molecules of the solid and the substance adsorbed. When, for example, the intermolecular attractive forces between a solid and a gas (or vapor) are greater than those existing between molecules of the gas itself, the gas will condense on the surface of the solid even though its pressure may be lower than the vapor pressure corresponding to the prevailing temperature. The adsorbed substance does not penetrate within the crystal lattice of the solid and does not dissolve in it, but remains entirely upon the surface. Should the solid, however, be highly porous, containing many fine capillaries, the adsorbed substance will penetrate these interstices if it "wets" the solid. At equilibrium, the partial pressure of the adsorbed substance equals that of the contacting gas phase, and by lowering the pressure of the gas phase or raising the temperature, the adsorbed gas is readily removed or desorbed in unchanged form. Physical adsorption is characterized by low heats of adsorption (of the order of 40 Btu or less per pound per mole of adsorbate) and by the fact that the adsorption equilibrium is reversible and is established rapidly.

Chemisorption, or activated adsorption, on the other hand, is the result of chemical interaction between the solid and the adsorbed substance. The strength of the chemical bond may vary considerably and identifiable chemical compounds in the usual sense may not form. Nevertheless, the adhesive force is generally much greater than that found in physical adsorption. Chemisorption is also accompanied by much higher heat changes (ranging from 80 to as high as 400 Btu/lb-mol) with the heat liberated being on the order of the heat of chemical reaction. The process is frequently irreversible, and, on desorption, the original substance will often be found to have undergone a chemical change.

Although it is probable that all solids adsorb gases (or vapors) to some extent, adsorption as a rule is not very pronounced unless an adsorbent possesses a large surface for a given mass. For this reason, such adsorbents as silica gel, charcoals, and molecular sieves are particularly effective as adsorbing agents. These substances have a very porous structure, and with their large exposed surface, they can take up appreciable volumes of various gases. The extent of adsorption can be increased further by "activating" the adsorbents in various ways. For example, wood charcoal can be "activated" by heating between 350° and 1000°C in a vacuum or in air, steam, and certain other gases to a point where the adsorption of carbon tetrachloride, for example, at 24°C can be increased from 0.011 g/g of charcoal to 1.48. The activation apparently involves a distilling out of hydrocarbon impurities from a charcoal and leads thereby to exposure of a larger free surface for possible adsorption.

The amount of gas (or vapor) adsorbed by a solid depends on the natures of the adsorbent and gas being adsorbed, the surface area of the adsorbent, the temperature, and the pressure of the gas. As may be expected, an increase in the surface area of adsorbent increases the total amount of gas adsorbed. Since this adsorbent surface area cannot always be readily determined, common practice is to employ the adsorbent mass as a measure of the surface available and to express the amount of adsorption per unit mass of adsorbing agent used.

A concept that becomes especially important in determining adsorbent capacity is that of "available" surface, that is, surface area accessible to the adsorbate molecule. It is apparent from pore size distribution data that the major contribution to surface area is located in pores of molecular dimensions. It seems logical to assume that a molecule, because of steric effects, will not readily penetrate into a pore smaller than a certain minimum diameter—hence the concept that molecules are "screened out." This minimum diameter is the so-called critical diameter and is characteristic of the adsorbate and related to molecular size. Thus for any molecule, the effective surface area for adsorption can exist only in pores that the molecule can enter. Figure 1 attempts to illustrate this concept for the case in which two adsorbate molecules in a solvent (not shown) compete with each other for adsorbent surface. Because of the irregular shapes of both pores and molecules, and also by virtue of constant molecular motion, the fine pores are not blocked by the large molecules, but are still free for entry by small molecules. As a contributing factor, the greater mobility of the smaller molecule should permit it to diffuse ahead of the larger molecule and penetrate the fine pores first.

Four important adsorbents widely used industrially will be considered briefly, namely, activated carbon, activated alumina, silica gel, and molecular sieves. The first three of these are amorphous adsorbents with a nonuniform internal structure. Molecular sieves, however, are crystalline and have, therefore, an internal structure of regularly spaced cavities with interconnecting pores of definite size. Details of the properties peculiar to the various materials are best obtained directly from the manufacturer. The following is a brief description of these principal adsorbents.

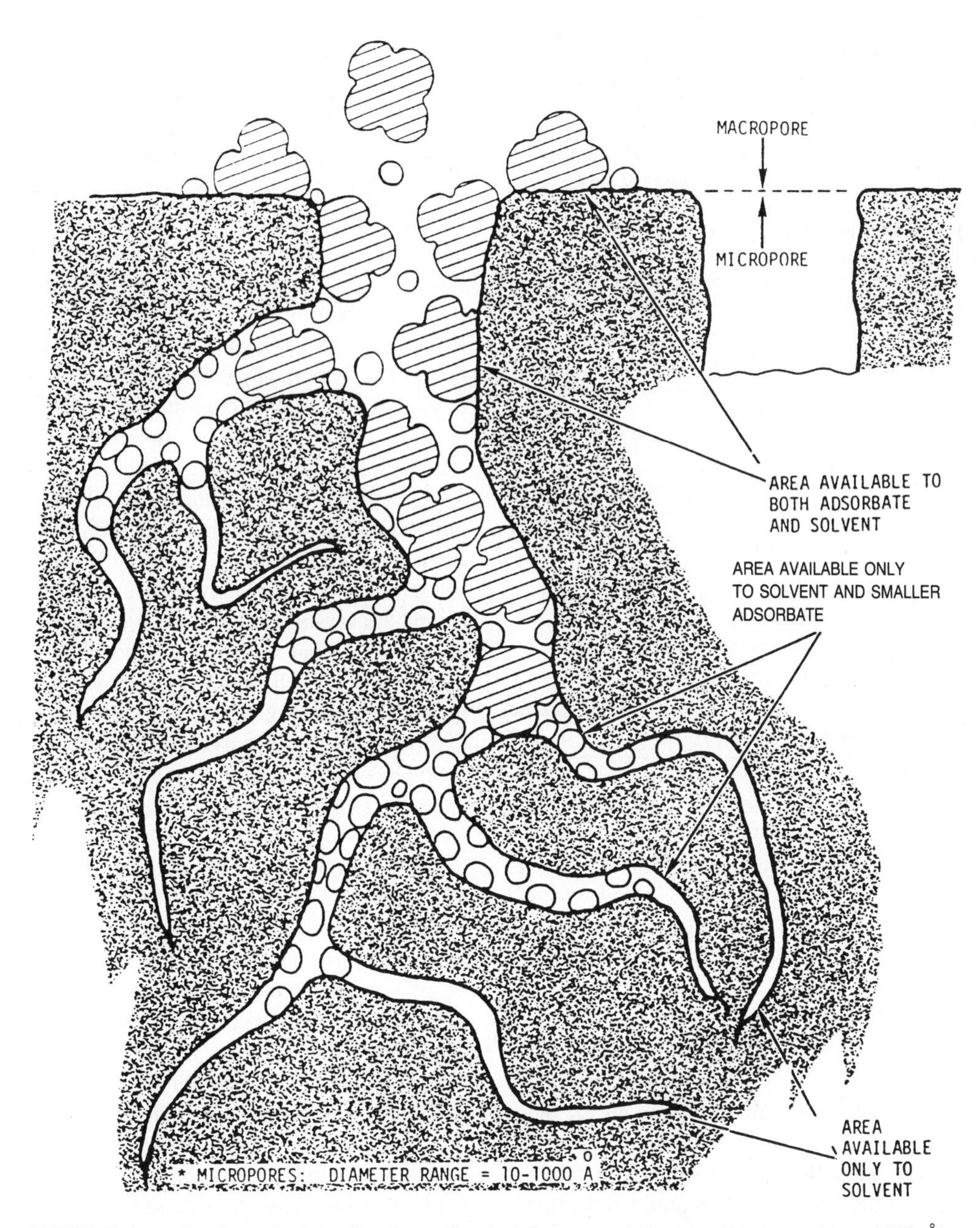

FIGURE 1. Concept of Molecular Screening in Micropores (Diameter Range = 10 to 1000 Å)

THE PRINCIPAL ADSORBENTS

Activated Carbon

Charcoal, an inefficient form of activated carbon, is obtained by the carbonization of wood. Various raw materials have been used in the preparation of adsorbent chars, resulting in the development of active carbon, a much more adsorbent form of charcoal. Industrial manufacture of activated carbon is today largely based on nut shells and coal, which are subjected to heat treatment in the absence of air; in the case of coal, this is followed by steam activation at high temperatures. Other substances of a carbonaceous nature also used in the manufacture of active carbons include wood, coconut shells, peat, and fruit pits. Zinc chloride, magnesium chloride, calcium chloride, and phosphoric acid have also been used in place of steam as activating agents. Some approximate properties of typical granular gas (or vapor) adsorbent carbons are given in Table 1. Gas (or vapor) adsorbent carbons find primary application in pollution control, solvent recovery (hydrocarbon vapor emissions), odor elimination, and gas purification.

Activated Alumina

Activated alumina (hydrated aluminum oxide) is produced by special heat treatment of precipitated or native aluminas or bauxite. It is available in either granule or pellet form with the typical properties given in Table 2. Activated alumina is mainly used for the drying of gases, and it is particularly useful for the drying of gases under pressure.

TABLE 1. Properties of Activated Carbon

Bulk density	22–34 lb/ft^3
Heat capacity	0.27–0.36 Btu-lb°F
Pore volume	0.56–1.20 cm^3/g
Surface area	600–1600 m^2/g
Average pore diameter	15–25 Å
Regeneration temperature (steaming)	100–140°C
Maximum allowable temperature	150°C

TABLE 2. Properties of Activated Alumina

Density in bulk	
Granules	38–42 lb/ft^3
Pellets	54–58 lb/ft^3
Specific heat	0.21–0.25 Btu/lb-°F
Pore volume	0.29–0.37 cm^3/g
Surface area	210–360 m^2/g
Average pore diameter	18–48 Å
Regeneration temperature	200–250°C
Stable up to	500°C

TABLE 3. Properties of Silica Gel

Bulk density	44–46 lb/ft^3
Heat capacity	0.22–0.26 Btu/lb-°F
Pore volume	0.37 cm^3/g
Surface area	750 m^2/g
Average pore diameter	22 Å
Regeneration temperature	120–250°C
Stable up to	400°C

Silica Gel

The manufacture of silica gel consists of the neutralization of sodium silicate by mixing it with dilute mineral acid and washing the gel formed free from salts produced during the neutralization reaction, followed by drying, roasting, and grading processes. The name "gel" arises from the jellylike form of the material during one stage of its production. It is generally used in granular form, although bead forms are available. The material has the typical physical properties shown in Table 3. Silica gel also finds primary use in gas drying, although it also finds application in gas desulfurization and purification.

Molecular Sieves

Unlike the amorphous adsorbents, that is, activated carbon, activated alumina and silica gel, molecular sieves are crystalline, being essentially dehydrated zeolites—that is, aluminosilicates in which atoms are arranged in a definite pattern. The complex structural units of molecular sieves have cavities at their centers to which access is by pores or windows. For certain types of crystalline zeolites, these pores are precisely uniform in diameter. Due to the crystalline porous structure and precise uniformity of the small pores, adsorption phenomena only take place with molecules that are small enough in size and of suitable shape to enter the cavities through the pores. The fundamental building block is a tetrahedron of four oxygen anions surrounding a smaller silicon or aluminum cation. The sodium ions or other cations serve to make up the positive charge deficit in the alumina tetrahedra. Each of the four oxygen anions is shared, in turn, with another silica or alumina tetrahedron to extend the crystal lattice in three dimensions. The resulting crystal is unusual in that it is honeycombed with relatively large cavities, each cavity connected with six adjacent ones through apertures or pores. The sieves are manufactured by hydrothermal crystal growth from aluminosilicate gels followed by specific heat treatment to effect dehydration. They have the typical properties shown in Table 4.

The adsorption process involves three necessary steps. The fluid must first come into contact with the adsorbent, at which time the adsorbate is preferentially, or selectively, adsorbed on the adsorbent. Next, the unadsorbed fluid must be separated from the adsorbent-adsorbate, and, finally, the adsorbent must be regenerated by removing the adsorbate or discarding used adsorbent and replacing it with fresh material. Regeneration is performed in a variety of ways, depending on the nature of the adsorbate. Gases or vapors are usually desorbed by either raising the temperature (thermal cycle) or reducing the pressure (pressure cycle) of the adsorbent-adsorbate. The more popular thermal cycle is accomplished by passing hot gas through the adsorption bed in the direction opposite to that of the flow during the adsorption cycle. This ensures that the gas passing through the unit during the adsorption cycle always meets the most active adsorbent last and that the adsorbate concentration in the

TABLE 4. Properties of Molecular Sieves

	Anhydrous Sodium Aluminosilicate	Anhydrous Calcium Aluminosilicate	Anhydrous Aluminosilicate
Type	4A	5A	13X
Density in bulk (lb/ft^3)	44	44	38
Specific heat (Btu/lb-°F)	0.19	0.19	—
Effective diameter of pores (Å)	4	5	13
Regeneration temperature	200–300°C	200–300°C	200–300°C
Stable up to (short period)	600°C	600°C	600°C

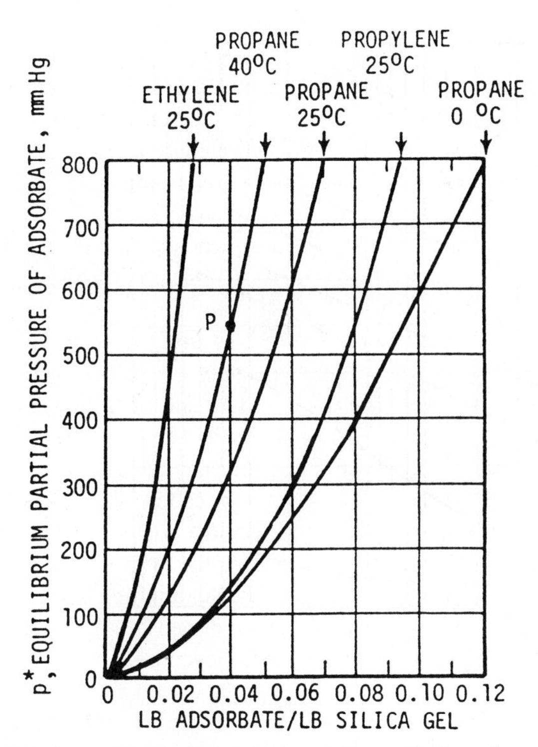

FIGURE 2. Adsorption Isotherm of Some Hydrocarbons on Silica Gel

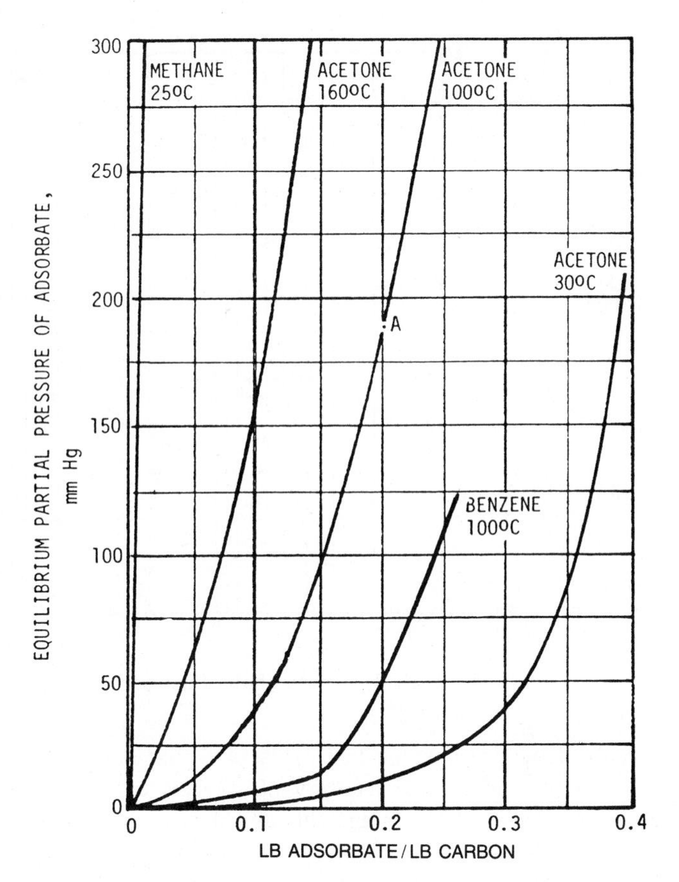

FIGURE 3. Vapor–Solid Equilibrium

adsorbent at the outlet end of the unit is always maintained at a minimum.

In the first step mentioned above, where the molecules of the fluid come into contact with the adsorbent, an equilibrium is established between the adsorbed fluid and that remaining in the fluid phase. Figures 2 through 4 show experimental equilibrium adsorption isotherms.

Consider Figure 2, where the concentration of adsorbed gas on the solid is plotted against the equilibrium partial pressure, p^*, of the vapor or gas at constant temperature. At 40°C, for example, pure propane vapor at a pressure of 550 mmHg is in equilibrium with an adsorbate concentration at point P of 0.04 pound of adsorbed propane per pound silica gel. Increasing the pressure of the propane will cause more to be adsorbed, while decreasing the pressure of the system at P will cause propane to be desorbed from the carbon.

Adsorption is an exothermic process; hence the concentration of adsorbed gas (or vapor) decreases with increased temperature at a given equilibrium pressure. This is evident from the behavior of the isotherm curves.

The process of a gas or vapor being brought into contact with an evacuated porous solid, and part of it being taken up by the solid, is always accompanied by the liberation of heat. The extent to which the process is exothermic depends on the type of sorption and the particular system. For physical adsorption, the amount of heat liberated is usually equal to the latent heat of condensation of the adsorbate plus the heat of wetting of the solid by the adsorbate. The heat of wetting is usually only a small fraction of the heat of adsorption. On the other hand, in chemisorption, the heat involved approximates the heat of chemical reaction.

DESCRIPTION OF EQUIPMENT

Because of the high cost of maintenance of air purification systems in applications of high concentrations of organic vapors, scientists and engineers have been forced into the research and design of systems for solvent recovery. The result has been the development of three types of systems, differentiated by the manner in which the adsorbent bed is maintained or handled during both phases of the adsorption-regeneration cycle: (1) fixed or stationary bed, (2) moving bed and (3) fluidized bed.[1]

Figure 5 presents a flow diagram of a dual stationary-bed solvent-recovery system with auxiliaries for collecting the vapor–air mixture from various point sources, then transporting through the particulate filter and into the on-stream carbon adsorber, in this case, bed 1. The effluent air, which is virtually free of vapors, is usually vented outdoors. The lower carbon adsorber (number 2) is regenerated during the service time of bed 1. A steam generator or other source of steam is required. The effluent steam–solvent mixture from the adsorber is directed through the condenser and the liquified mixture, then run into the decanter and/or distillation column for separation of the solvent from the steam condensate.

Figures 6 and 7 show two designs of stationary-bed solvent-recovery systems. The type shown employs vertical cylindrical beds wherein the solvent-laden air flows

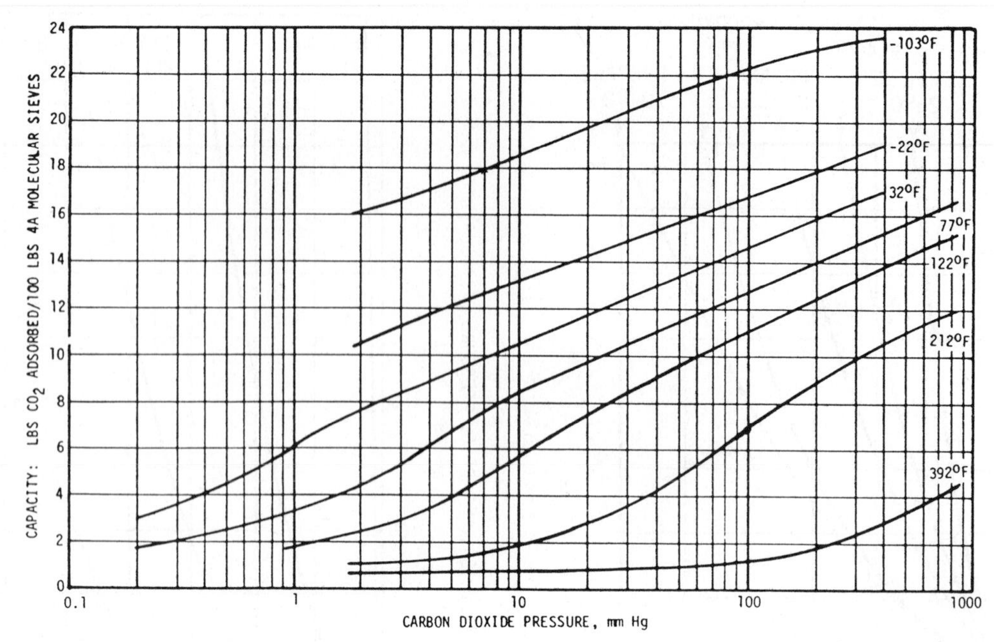

FIGURE 4. Vapor–Solid Equilibrium

axially down through the bed. This particular design is commonly used for the recovery of solvents emitted in degreasing and dry cleaning, although it is equally well suited for the recovery of solvents from other industries. Regeneration is accomplished with steam flowing upward through the bed and, if solvents are immiscible in water, decantation is used to separate the condensed solvent from the steam condensate. The valves are of the disk type, opened and closed by air-driven pistons. Water is used as a coolant in the condenser. Steam, electric power to drive the blower, and cooling water are three operating-cost items. The cost of steam and cooling water increases with frequency of regeneration.

Figure 6 shows the features and arrangement of the component parts of a two-adsorber system. This particular unit has two adsorber beds used alternately (i.e., while one unit is on-stream adsorbing, the other is regenerating).[2]

Dual-adsorber systems can also be operated with both beds on-stream simultaneously, especially when solvent concentrations are low. In this situation, regeneration is less frequent than a full work shift and may be accomplished during off-work hours. Operation in parallel almost doubles the air-handling capacity of the adsorption unit and may be an advantage in terms of operating cost.

The concept of the moving-bed system is illustrated in Figure 8. The rotary component of the system consists of coaxial cylinders. The outer cylinder is impervious to gas flow, except through the slots near at one end. The carbon bed is retained between two cylinders made of screen or perforated metal and also segmented by partitions placed radially between the two cylinders. The inner cylinder is again impervious to gas flow, except at the slots near the end of the rotary. These slots serve as outlet ports for the purified air and as inlet ports for the regenerating steam. On each rotation of the rotary, each segment of the bed un-

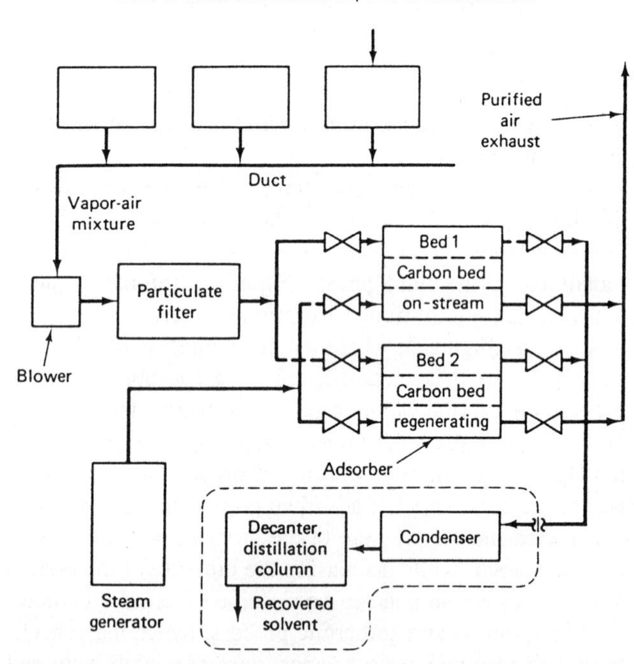

FIGURE 5. Stationary-bed Carbon System with Auxiliaries for Vapor Collection and Solvent Separation from Steam Condensate

dergoes adsorption and regeneration. The desorbed solvent can then be separated from the steam by decantation or distillation.

Because of the continuous regeneration capability of the rotary bed, more efficient utilization of the carbon is possible than with stationary-bed systems. In most solvent-recovery operations, the adsorption zone and the saturated bed behind it are idle, but add to the bulk of the system and increase airflow resistance through the bed. In deep beds of

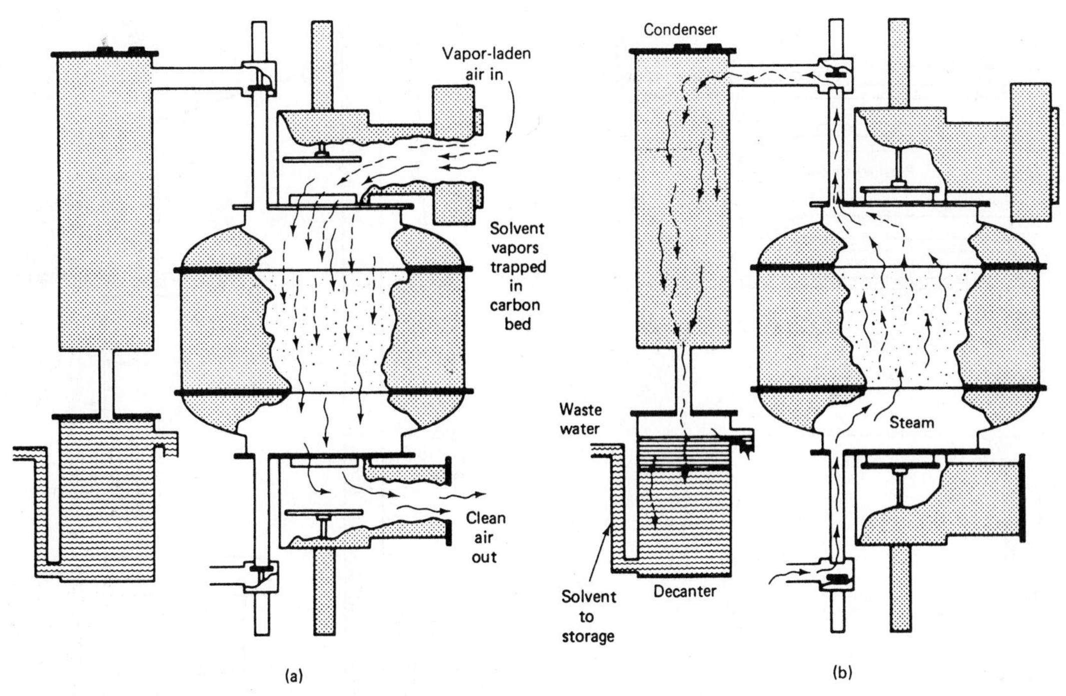

FIGURE 6. Adsorption Cycle (a) and Desorption Cycle (b)

12 to 36 inches, as required in stationary-bed systems, a large portion of the bed is idle at any one time. By continuous regeneration, the regeneration time for each segment of the bed is shortened, and thus shorter bed lengths can be used. This leads to two advantages: a more compact system and reduced pressure drop. The disadvantages are those associated with wear on moving parts and maintaining seals in contact with moving parts. The use of shorter beds also decreases the steam utilization efficiency.

Figure 9 shows a flow diagram for a fluidized-bed solvent-recovery system. The carbon is recirculated continuously through the adsorption–regeneration cycle. The spent carbon, saturated with solvent, is elevated to the surge bin and then passed down into the regeneration bed, where it is contacted with an upward flow of steam. The regenerated carbon is then metered into the adsorber, where the carbon traverses nine beds while fluidized with the upward flow of the vapor-laden air. An air velocity of approximately 240 fpm is required to cause fluidization of the bed. In both the regeneration and adsorption phases, the carbon is moved countercurrent to the gas or vapor. The countercurrent movement increases the efficiency of regeneration, that

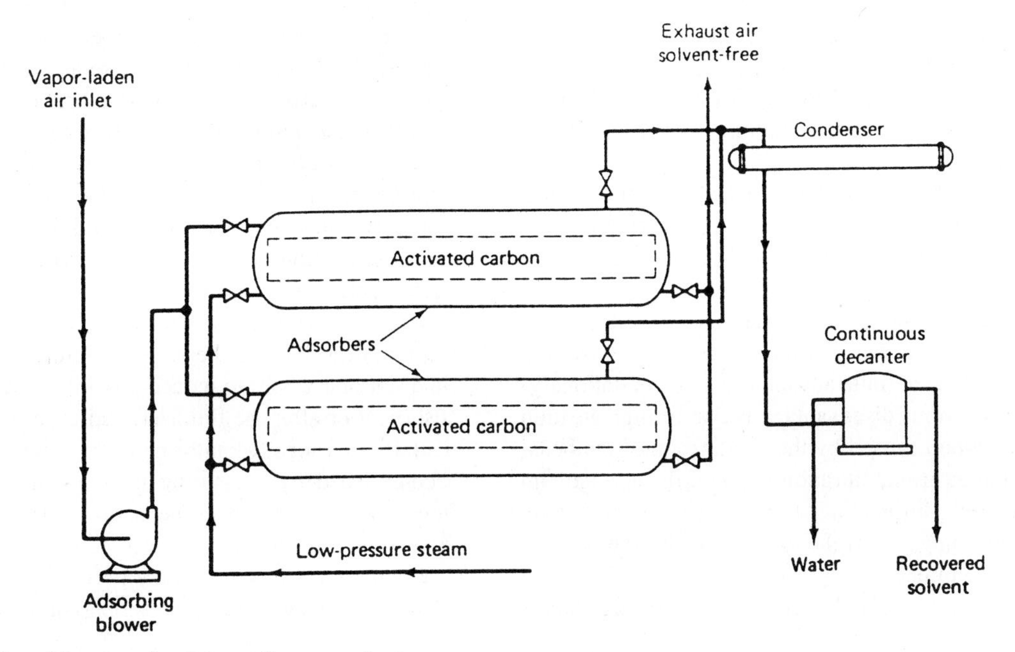

FIGURE 7. Flow Diagram of a Solvent Recovery System

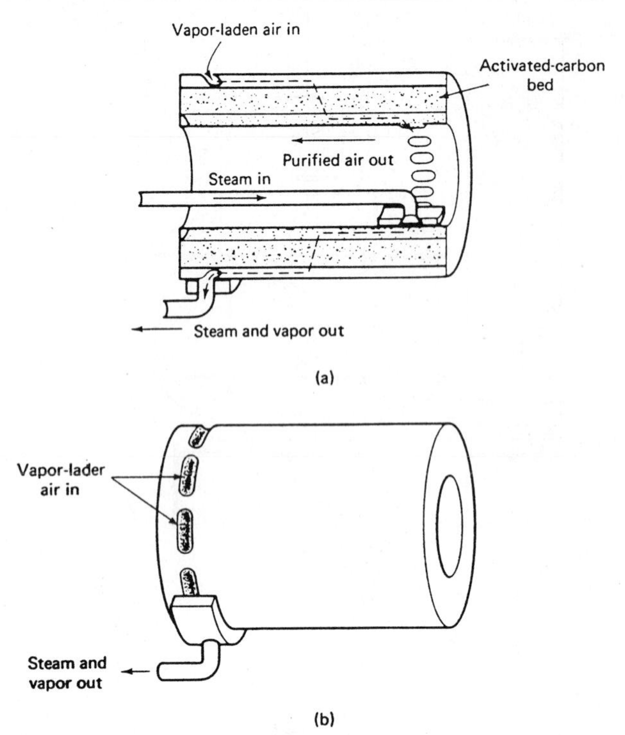

FIGURE 8. Continuous Rotary Bed: Cross-Sectional View (a) and Horizontal Exterior View (b)

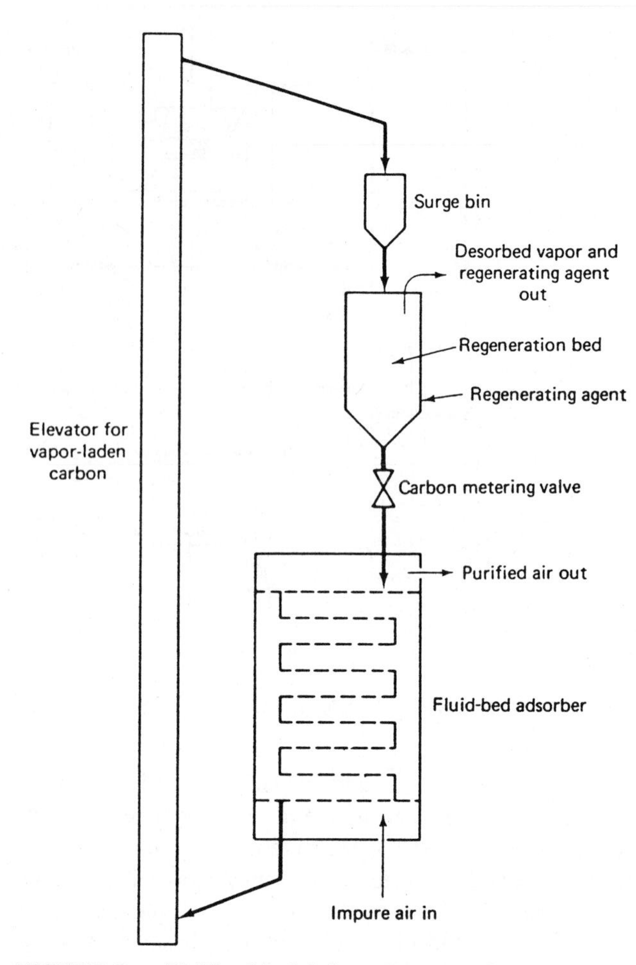

FIGURE 9. Fluidized-bed Solvent Recovery System

is, the steam-solvent ratio is less than for a stationary bed under otherwise comparable conditions. In addition to the beneficial effects of the countercurrent movement, the bed length can be increased to improve steam utilization further.

The countercurrent movement also increases the effective use of the carbon; more solvent can be recovered with less carbon than with stationary- or rotary-bed systems. By adjustments or balance of the carbon and solvent input rates, the total carbon in the adsorber can be made part of the adsorption zone reaching saturation in the lowest bed just before it is discharged into the elevator. Very little of the carbon is then idle; hence maximum utilization is made of the carbon.

The fact that the carbon has reached saturation when delivered to the regeneration phase is another factor in the reduced steam requirement; in this respect, the fluidized-bed system is attractive.

When large air volumes are treated and available space for the installation is at a premium, the smaller size and lower initial cost are definite advantages over the stationary-bed system. A serious disadvantage is that of high attrition losses of the carbon caused by the fluidization of the beds. Because of the attrition, filtration of the effluent airstream may be required. Influent airstream need not be highly filtered because plugging of the fluidized beds with particulate matter is minimal.

No adsorber system could operate without enough auxiliary equipment and components to collect, transport, and filter vapor-laden airstreams being delivered to the adsorber. Proper design of these components is necessary to provide proper service to the adsorber. The ducts and piping need to be sized properly for required air velocities to optimize the efficiency of the adsorber. If the air velocity is too high, the stationary-bed adsorber may become a fluidized-bed adsorber, or low flows may create severe channeling through the beds. The fan is the catalyst for forcing the gas stream into and out of the unit, so it is important that careful attention to design and sizing be given to this equipment.[3]

Because there are several adsorber configurations, the location of a filter can be varied—before the inlet airstream (in a stationary bed) to reduce possible contamination to the adsorbent or after the fluidized-bed adsorber to reduce particulate emissions. Monitoring of the filter efficiency can be accomplished by measuring the pressure drop across the filter using either a manometer or pressure gauge that will read from 1 to 20 inches water column.

Compressed-air systems are necessary in some adsorber systems for valve and damper operation. For best results,

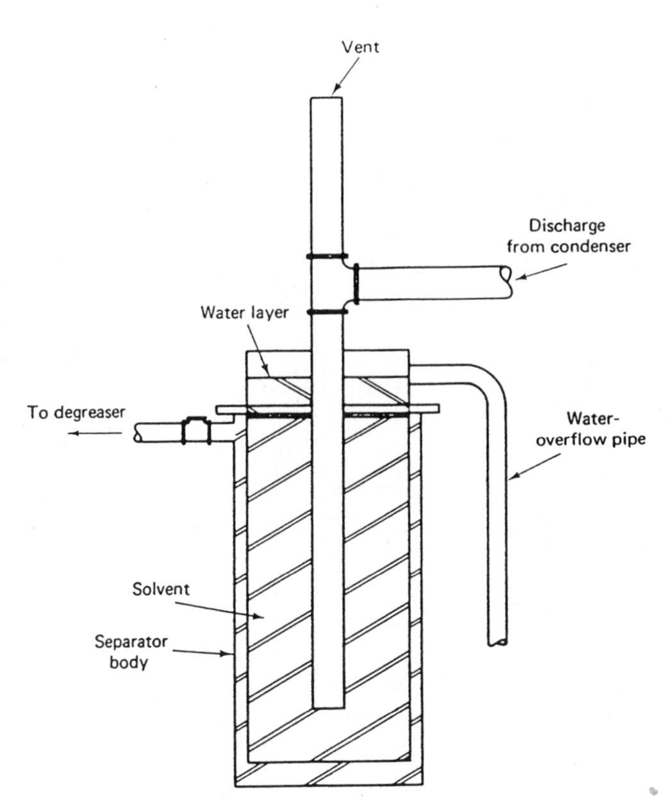

FIGURE 10. Water Separator

the air supply should be kept contaminant-free through the use of a filter installed close to the adsorber. The compressed-air supply should be equipped with an in-line filter, pressure regulator, and lubricator.

Several devices are installed in series following adsorption for recovery of the contaminant after regeneration. Condensers and separators are examples of recovery devices. The condenser is installed just after the system for removal of the heat from the vapors. There are two basic types of condensers: surface condensers and contact condensers. In a surface condenser the coolant does not contact the vapors or condensate. Most surface condensers are of shell-and-tube configurations. Water flows through the tubes and vapors condense on the shell side. In contact condensers the coolant vapors and condensate are intimately mixed. These condensers are more flexible, simpler, and considerably less expensive to install. Sizing of this condenser is also more straightforward than the design for surface condensers.

Separators (decanters) are installed following the condenser to separate the contaminant from the water. Figure 10 shows a typical separator used with activated-carbon bed adsorbers. Separators work on the principle of gravitational forces where the heavier material to be separated is removed from the bottom of the canister and the lighter material is removed through a line located at the top of the canister. Water separators are more effective with single solvent applications and only when the solvent is immiscible in water.

DESIGN AND PERFORMANCE EQUATIONS

Fixed-bed adsorbers are the usual choice for the control of gaseous pollutants when adsorption is the desired method of control. Consider Figure 11 for a binary solution containing a strongly adsorbed solute (gaseous pollutant) at concentration C_0. The gas stream containing the pollutant is to be passed continuously down through a relatively deep bed of adsorbent, which is initially free of adsorbate. The top layer of adsorbent, in contact with the contaminated gas entering, at first adsorbs the pollutant rapidly and effectively, and what little pollutant is left in the gas is substantially removed by the layers of adsorbent in the lower part of the bed. At this point, the effluent from the bottom of the bed is practically pollutant-free as at C_1. The top layer of the bed is practically saturated, and the bulk of the adsorption takes place over a relatively narrow adsorption zone in which there is a rapid change in concentration. The saturated bed length is L_{St}. As the gas stream continues to flow, the adsorption zone of length L_Z travels downward, similar to a wave, at a rate usually much less than the linear velocity of the gas stream through the bed. At some later time, roughly half of the bed is saturated with the pollutant, but the efficient concentration C_2 is still substantially zero. Finally, at C_3, the lower portion of the adsorption zone has reached the bottom of the bed and the concentration of pollutant in the effluent has suddenly risen to an appreciable value for the first time. The system is said to have reached the "break point." The pollutant concentration in the effluent gas stream now rises rapidly as the adsorption zone passes through the bottom of the bed and at C_4 has just about reached the initial value C_0. At this point, the bed is just about fully saturated with pollutant. The portion of the curve between C_3 and C_4 is termed the "breakthrough" curve. If the gas stream continues to flow, little additional adsorption takes place, since the bed is, for all practical purposes, entirely in equilibrium with the gas stream. Theoretically, zero concentration is attained only at infinite bed length, but for practical purposes, concentrations of $0.01C_0$ to $0.001C_0$ are considered essentially zero, and hence the finite bed lengths.

The time at which the breakthrough curve appears and its shape greatly influences the method of operating a fixed-bed adsorber. The curves usually have an S shape, but they may be steep or relatively flat and, in some cases, considerably distorted. The actual rate and mechanism of the adsorption process, the nature of the adsorption equilibrium, the fluid velocity, the pollutant concentration in the entering gas, and the adsorber bed length all contribute to the shape of the curve produced for any system. The break point is very sharply defined in some cases, while in others it is not. The break-point time generally decreases with decreased bed height, increased adsorbent particle size, increased gas flow rate through the bed, and increased initial pollutant concentration in the entering gas stream. It is usual industrial practice to determine experimentally for a

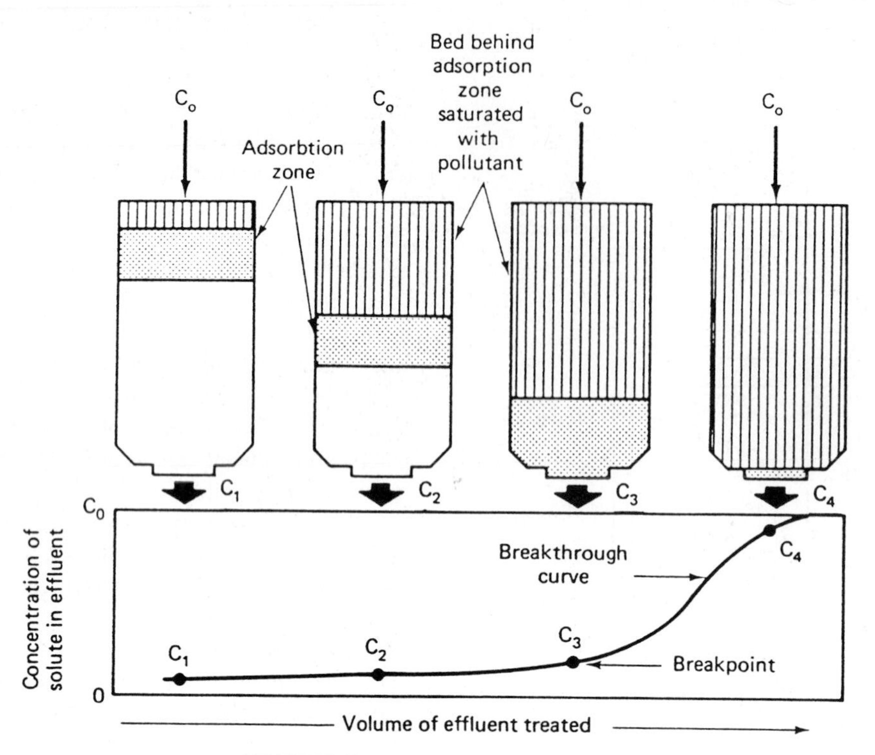

FIGURE 11. Adsorption Wave Front

particular system the break point and breakthrough curve under conditions as close as possible to those expected in the process.

The service time is one of the prime factors to be considered in sizing adsorption equipment. When deep or long adsorbent beds are used, as in solvent recovery (with activated-carbon adsorbent material), L_{St} is large relative compared to L_Z and essentially determines the service time. With usual solvent-recovery practices, the solvent air velocity through the bed is maintained at near 100 fpm. In these cases, L_Z is of the order of 3 inches and Z, the adsorbent bed length or depth, can range from 16 to over 36 inches. The capacity of the saturated bed length, L_{St}, can be calculated with a considerable degree of accuracy if an adsorption isotherm has been determined.

The capacity of the adsorption zone, besides varying with C_0 and the temperature, T, also varies with the adsorbent mesh size and flow velocity. The service time, t_B, to penetration concentration, C_B, is dependent on the adsorptive capacities of the saturated bed and adsorption zone as expressed by the following equation.

$$t_B = (A/VC_0)(X_{St}L_{St} + X_ZL_Z) \quad (1)$$

where: A = area of adsorbent bed, ft^2 or cm^2

V = total gas volumetric flow rate, ft^3/min or L/min

X_{St} = adsorptive capacity per unit adsorbent bed volume in the saturated zone, lb/ft^3 or g/cm^3

X_Z = mean adsorptive capacity per unit bed volume in the adsorption zone, lb/ft^3 or g/cm^3

In equation 1, X_ZL_Z is a fixed amount wherein the L_Z (often referred to as the mass transfer zone) is defined as the bed length in which C decreases from C_0 to essentially zero. L_{St} varies with the C_B/C_0 ratio; at $C_B = C_0$, $L_{St} = Z$, while at $C_B \approx O$, $L_{St} = Z - L_Z$.

In thin beds of 2 inches and less, flow velocities near 40 fpm are usually employed, which shortens L_Z to the 2-inch level. In these applications, the service time is essentially

$$t_B = (A/VC_0)X_ZL_Z \quad (2)$$

The usual procedure in practice is to work with a term defined as the working charge (or working capacity) rather than the X terms presented above. It provides a numerical value for the actual adsorbing capacity of the bed under operating conditions. If experimental data are available, the working charge, WC, may be estimated from

$$WC = CAP\left(\frac{Z\text{-}MTZ}{Z}\right) + 0.5\left(\frac{MTZ}{Z}\right) - HEEL \quad (3)$$

where CAP has replaced (symbolically) X and HEEL is the residual adsorbate present in the bed following regeneration. Since many of these data are rarely available, or just simply ignored, the working charge is taken to be some fraction of the saturated (equilibrium) capacity of the adsorbent, that is,

$$\mathrm{WC} = (f)(\mathrm{CAP});\ 0 \leq f \leq 1.0 \quad (4)$$

For multicomponent adsorption, the working charge may be calculated from the following:

$$WC = \frac{1.0}{\sum_{i=1}^{n} (w_i/\mathrm{CAP}_i)} \quad (5)$$

where: n = number of components
w_i = mass fraction of i in n components (not including carrier gas)
CAP_i = equilibrium capacity of component i

For a two-component (A,B) system, the above equation reduces to

$$\mathrm{WC} = (\mathrm{CAP}_A)(\mathrm{CAP}_B)/[(w_A)(\mathrm{CAP}_B) + (w_B)(\mathrm{CAP}_A)]$$

After determining the service time and/or working charge necessary for a particular application, there are several other factors to be considered:

1. The adsorbent particle size.
2. The physical adsorbent bed depth.
3. The gas velocity.
4. The temperature of the inlet gas stream and the adsorbent.
5. The contaminant concentration to be adsorbed.
6. The contaminants not to be adsorbed, including moisture.
7. The removal efficiency.
8. Possible contaminant polymerization on the adsorbent.
9. The frequency of operation.
10. Regeneration conditions.
11. The system pressure.

These factors are considered next in more detail.

1. The shape and size (d) of the adsorbent particle affects several parameters—in particular, the pressure drop and the diffusion rate. The pressure drop is lowest when the particles are uniform in size and spherical. The pressure decrease will vary with the Reynolds number. The mass transfer rate increases inversely with $d^{3/2}$ and the internal adsorption rate inversely with d^2. With all other parameters considered equal, the pressure drop increases with smaller particle size, and yet provides a more efficient adsorptive environment.
2. The actual adsorbent bed depth affects the adsorption mass transfer in two ways: (1) the bed depth is sized to be deeper than the transfer zone, which is unsaturated; and (2) any multiple of minimum bed depth yields a greater than proportional increase in capacity. The mass transfer zone (MTZ) can be determined as follows[4]:

$$MTZ = \frac{\text{Total bed depth}}{[t_{st}/(t_{st} - t_B)] - X} \quad (6)$$

where: t_{st} = time required to saturation
t_B = time required until break point
X = degree of saturation in the mass transfer zone

3. The velocity of gas stream through the adsorption bed is determined by the "crushing velocity" of the adsorbent. Crushing-velocity data are available from manufacturers of adsorbents. The length of the MTZ is proportional to the velocity; usually, the higher the velocity, the longer is the mass transfer zone.
4. Temperature yields an inversely proportional effect on adsorption capacity, such as when temperature increases, the adsorption capacity decreases. The adsorption process is exothermic. As the adsorption activity moves through the bed, a temperature front follows and heat is transferred to the gas stream. When the gas leaves the area of adsorption activity, the heat exchange reverses (gas will transfer heat to the bed). The temperature differential during the adiabatic operation of the adsorber can be estimated as follows:

$$\Delta T = \frac{6.1}{(C_p/C_1) \times 10^5 + 0.51(C_A/\mathrm{CAP})} \quad (7)$$

where: ΔT = temperature rise, °F
C_p = heat capacity of air, Btu/ft^3-°F
C_1 = inlet concentration of pollutant, ppm
C_A = heat capacity of adsorbent, Btu/ft^3-°F (see Table 5)
CAP = equilibrium capacity of adsorbent at bed average temperatures ($T + \Delta T$, °F), lb/100 lb

5. There is a direct correlation between adsorbate concentration and adsorption capacity. The MTZ length is inversely proportional to the adsorbate concentration. Considering these two observations only, a deeper or longer bed will remove a lower-concentration contaminant and a higher concentration of the same contaminant with equal efficiency. In considering recovery, the value of the adsorbate and the concentration are primary factors. In handling combustible gas, it is mandatory that the concentration of the adsorbate be well below the lower explosive limit (LEL). All gases present will be

TABLE 5. Specific Adsorbent Heat Capacity Values (Ambient Conditions, Btu/ft^3-°F)

Activated carbon	0.25
Alumina	0.21
Molecular sieve	0.25

adsorbed on the adsorbent surface to some degree. The competition between the gases for available surface area lowers the adsorption capacity for the preferred adsorbate to be removed. Under ambient conditions, air is adsorbed only slightly on commercial adsorbents, but moisture and carbon dioxide adsorb more significantly. Even though activated carbon is less sensitive than other adsorbents to moisture, its adsorption capacity is limited considerably when compared with adsorption from a dry airstream.

6. Varying percentages of removal are required with different applications. Of course, deeper beds are required to achieve complete removal than for a partial removal efficiency of 60% to 80%.
7. Decomposition and polymerization of solvents can occur when in contact with adsorbents. The decomposed product takes on properties different from those of the original substance. These different properties include corrosivity and the ability to be adsorbed at a lower capacity. Polymerization can significantly reduce the adsorption capacity and render it nonregenerable by conventional methods. When adsorption of acetylene on activated carbon is accomplished at higher temperatures, polymerization occurs.
8. Even with the most precise of design calculations and selection of equipment, the frequency of operation greatly affects the performance of an adsorption system. If operation is continuous, the best conditions normally result; if operation is intermittent, the performance of the system will be impaired. Short periods of intermittent operation do not affect the overall performance significantly, but continued over long periods, particularly in overloaded systems, it can cause a serious reduction of capacity.
9. Adsorption processes in practice use various techniques to accomplish regeneration or desorption. The adsorption–desorption cycles are usually classified into four types, used separately or in combination. This important topic is now considered in more detail.
   a. *Thermal swing cycles*, using either direct heat transfer by contacting the bed with a hot fluid or indirect transfer through a surface, reactivate the adsorbent by raising the temperature. Usually a temperature between 300°F and 600°F is reached, the bed is flushed with a dry purge gas or reduced in pressure, and then it is returned to adsorption conditions. High design loadings on the adsorbent can usually be obtained, but a cooling step is needed.
   b. *Pressure swing cycles* use either a lower pressure or vacuum to desorb the bed. The cycle can be operated at nearly isothermal conditions with no heating or cooling steps. The advantages include fast cycling with reduced adsorber dimensions and adsorbent inventory, direct production of a high-purity product, and the ability to utilize gas compression as the main source of energy.
   c. *Purge gas stripping cycles* use an essentially nonadsorbent purge gas to desorb the bed by reducing the partial pressure of the adsorbed component. Such stripping is more efficient the higher the operating temperature and the lower the operating pressure. The use of a condensible purge gas has the advantages of reduced power requirements, gained by using a liquid pump instead of a blower, and an effluent stream that may be condensed to separate the desorbed material by simple distillation.
   d. *Displacement cycles* use an adsorbable purge to displace the previously adsorbed material on the bed. The stronger the adsorption of the purge, the more completely the bed is desorbed using lesser amounts of purge, but the more difficult it becomes subsequently to remove the adsorbed purge itself from the bed.

When deciding whether to employ a regenerative system, several factors should be considered. The principal one is that of economics. It is important to establish if recovery of the adsorbate will be cost effective or whether or not regeneration of the adsorbent is the prime consideration. If solvent recovery is the main objective, the design should be based on past experimental data to establish the ratio of sorbent fluid to recoverable adsorbent at different working capacities. Most systems today employ steam as the regenerating medium, but some of the new systems use a hot inert gas, such as nitrogen. With regard to steam systems, a well-designed system will have steam consumption in the range of 1 to 4 pounds of steam per pound of recovered solvent. The steam entering the bed not only introduces heat, but creates adsorption and capillary action of the moisture, which supplies additional heat for the desorption process. Certain parameters should be considered in the design of a stripping process:

1. The time required for the regeneration should be minimized. If continuous adsorption and recovery are required, multiple systems have to be installed.
2. A short regeneration time requires a higher steaming rate, thus increasing the heat duty of the condenser system.
3. The steaming direction should be opposite that of adsorption to prevent the possible accumulation of polymerizable substances.
4. To enable fast stripping and efficient heat transfer, it is necessary to sweep out the carrier gas from the adsorber and condenser system as fast as possible.
5. A larger fraction of the heat content of the steam is used to heat the adsorber vessel and the adsorbent; thus it is essential that the steam condense quickly in the bed. The steam should contain only slight superheat to allow condensation.
6. It is advantageous to use a low-retentivity carbon to enable the adsorbate to be stripped out easily. When

empirical data are not available, the following heat requirements have to be taken into consideration: heat to the adsorbent and vessel; heat of adsorption and specific heat of adsorbate leaving the adsorbent; latent and specific heat of water vapor accompanying the adsorbate; heat in condensed, indirect steam; and radiation and convection heat loss.

During the desorb cycle, condensation and adsorption will take place in the adsorbent bed, increasing the moisture content of the adsorbent. Also, a certain portion of the contaminant will remain; this is referred to as the "heel." To achieve a minimum efficiency drop from successive adsorption cycles, a drying and cooling cycle should occur before returning to the adsorb mode. When using high adsorbate concentrations, it may be desirable to leave some moisture; with other contaminants, a moisture-free bed is desired.

In air pollution control applications where pollutant concentrations are generally low and have little no recovery value, steam may not be the best regenerating agent. Because of the adsorption at low concentrations, the vapors may be held tightly by the adsorbent, and, relative to the amount of pollutant adsorbed, the amount of steam required may be large. Under these conditions, even slightly soluble organic compounds may be completely dissolved in the steam condensate. Under these circumstances, the more economical approach may be to regenerate with a noncondensible gas such as air, or if there is danger of explosion, regenerate with an inert noncondensible gas such as nitrogen or flue gas. For example, miscible pollutants such as the solvents 4-methyl-2-pentanone and propanone would completely dissolve in the large amounts of steam condensate. Their recoveries by distillation usually are not economical, thereby creating a disposal problem if water pollution is to be avoided. In these cases, the solution to the problem may be to regenerate the adsorbent with a noncondensible gas and burn the released vapors immediately in a small thermal incinerator.

Generally, the adsorption capacity of an adsorbent increases with increasing pressure if the partial pressure of the contaminant increases. However, at high pressures (over 500 psig), a decrease in capacity will be observed. The design of fixed-bed adsorption systems also requires the capability of estimating pressure drop through the bed. Ergun[5] derived a correlation to estimate the pressure drop for the flow of a fluid through a bed of packed solids when it alone fills the voids in the bed. This correlation is given by the relationship

$$\frac{\Delta P\, g_c d_p \epsilon^3}{(Z)(2)\rho u^2(1-\epsilon)} = \frac{75(1-\epsilon) + 0.875}{\text{Re}} \tag{8}$$

where: ΔP = pressure drop of gas, lb_f/ft^2
Z = depth of packing, feet
g_c = conversion constant, $4.18(10^8)$ ft-lb/lb_f-h^2
d_p = effective particle diameter (the diameter of a sphere of the same surface/volume ratio as the packing in place), feet = $6(1-\epsilon)/a_p$
ϵ = fractional void volume in dry-packed bed, ft^3 voids/ft^3 packed volume
a_p = surface of solid particles, ft^2/ft^3 of packed volume
ρ = gas density, lb/ft^3
u = superficial velocity of gas through bed, fph
Re = Reynolds number, dimensionless = $d_p\rho u/\mu$
μ = gas viscosity, lb/ft-h

Information on different mesh carbon sizes is presented in Figure 12. These data are often used in pressure-drop calculations, There is a simpler form of the Ergun equation provided by Union Carbide for molecular sieves.[6]

$$\frac{\Delta P}{Z} = \frac{f C_T G^2}{\rho d_p} \tag{9}$$

where: C_T = pressure-drop coefficient, ft-h^2/in.^{2}
f = friction factor
G = superficial mass velocity, lb/h-ft^2 = $\rho\mu$
ΔP = pressure drop, psi

The friction factor, f, is determined from Figure 13 as a function of the modified Reynolds number. The pressure drop coefficient, C_T, is determined from the same figure, which has C_T plotted as a function of ϵ. For molecular-sieve pellets, the effective particle diameter can be obtained from

$$d_p = \frac{d_c}{\frac{2}{3} + \frac{1}{3}(d_c/l_c)} \tag{10}$$

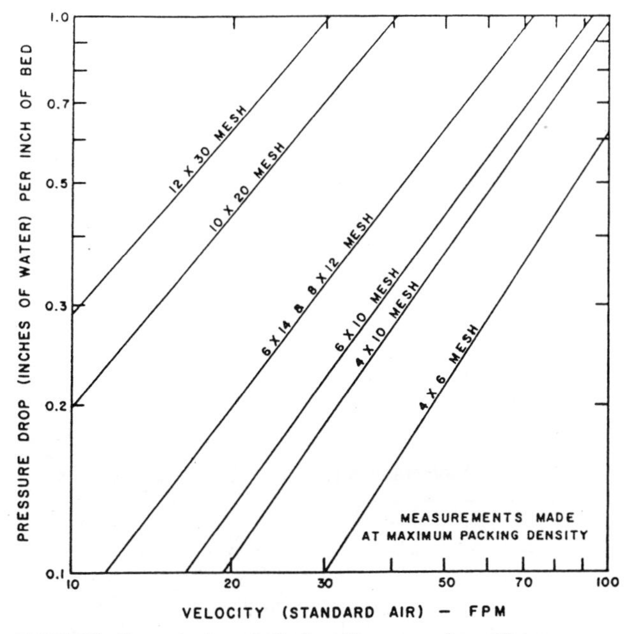

FIGURE 12. Activated Carbon Pressure along Curves

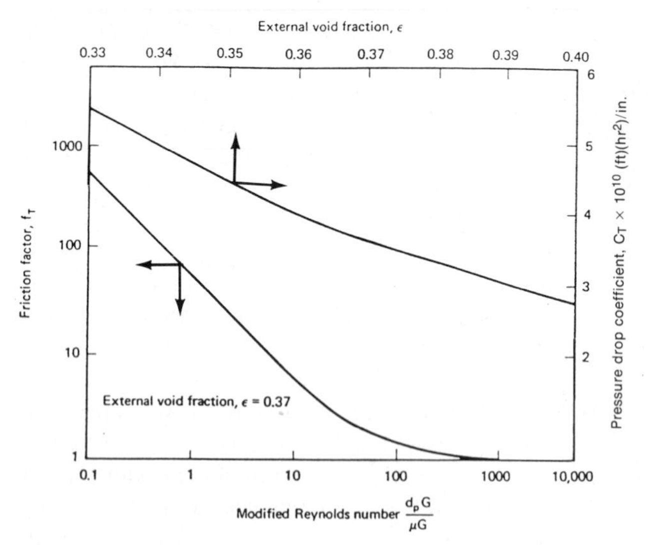

FIGURE 13. Friction Factor as a Function of Reynolds Number

where: d_c = particle diameter, feet
l_c = particle length, feet

The suggested values for ϵ and d_p for various sizes of molecular sieve are given in Table 6.

This section is concluded with a rather simplified overall design procedure for a system adsorbing an organic. The system consists of two horizontal units (one on/one off) that are regenerated with steam.[7]

1. Select adsorbent type and size.
2. Select cycle time; estimate regeneration time; set adsorption time equal to regeneration time; set cycle time to twice the regeneration time; generally, try to minimize regeneration time.
3. Set velocity; v is usually 80 fpm; can increase to 100 fpm.
4. Set steam/solvent ratio.
5. Calculate (or obtain) WC for above.
6. Calculate amount of solvent adsorbed during ½ cycle time $(t_{ads})M_s$.

$$M_s = Qc_s t_{ads};\ c_s = \text{inlet solvent concentration}$$

7. Calculate adsorbent required, M_{AC}.

$$M_{AC} = M_s/\text{WC}$$

TABLE 6. Molecular Sieve Pressure Drop Coefficients

	ϵ	d_p
⅛-inch pellets	0.37	0.0122
1/16-inch pellets	0.37	0.0061
14 × 30 mesh granules	0.37	0.0033

8. Calculate adsorbent volume requirement.

$$V_{AC} = M_{AC}/\rho_B;\ \rho_B = \text{adsorbent bulk density}$$

9. Calculate the face area of the bed.

$$A_{AC} = Q/V$$

10. Calculate bed height.

$$H = (V/A)_{AC}$$

11. Set L/D (length-to-diameter) ratio.
12. Calculate L/D from $A = LD$ constraints: $L < 30$ feet, $D < 10$ feet; L/D of 3 to 4 acceptable if $v < 30$ fpm
13. Design (structurally) to handle if filled with water.
14. Design vertically if $Q < 2500$ acfm.

OPERATION AND MAINTENANCE[8]

Utilizing adsorption as an air pollution control measure on sources emitting volatile organic compounds has proved to be extremely effective if proper design is applied and rigid operating procedures are established and followed. It is always desirable to check over the process being controlled to determine that normal operation is being experienced. Of course, the blower must be running with the fan turning in the correct direction. Three-phase motors that are running backward may be corrected by changing two of the phase wires to the motor. Initial start-ups will probably require a system balance of the exhaust ducts. If multiple processes are being exhausted into the same system, adjustments of individual slide dampers may be required to obtain correct airflow in each branch duct. Airflow switches are inexpensive and should be placed in each branch to sense airflow and allow the process to operate when the airflow is adequate. Airflows below 100 fpm through the carbon beds will provide adequate retention time for solvents in the airstream to be adsorbed on the charcoal. Excessive flows will reduce carbon efficiencies and allow volatiles to escape into the atmosphere. Excessive airflows are also detrimental to process operations where unnecessary solvents are evaporated and lost from process tanks and delivered to the carbon beds, posing additional loads on the system.

Prior to start-up, the prefilter, ahead of the carbon tanks, should be checked to verify that it is properly secured to the housing. This filter should be closely monitored during the first few days of operation. It may tend to accumulate dirt or lint from the construction and installation activities, causing the filter to become ineffective and restricting solvent entry into the carbon beds. After about three days of operation, replace or clean the dirty filter bag and observe the manometer gauge that registers air pressure between the blower and the carbon bed. There is no exact pressure differential that is proper for all machines, as the duct restrictions of specific installations will yield different gauge readings. Each system should stabilize and provide consistent manometer readings. However, during the first two or three days of operation, the carbon beds will be settling, so the readings will not reflect final operating conditions. With both carbon beds set on the adsorption cycle, place a mark on the manometer dial face to indicate a normal reading. Any changes in the reading can be used to help maintain the correct airflow to the adsorber. A change of 1 inch of water pressure on this gauge is significant, and the problem should be corrected. If the reading drops by 1 inch or more, check the following points ahead of the blower: (1) dirty filter bag, (2) improper slide damper setting, and (3) obstruction in the duct. If the reading increases by 1 inch or more, check the following: (1) appropriate beds should be in the adsorption cycle when the reading is noted; (2) the adsorber tank dampers must be open; (3) there must be no restriction in the exhaust duct; and (4) the filter bag may be loose or missing.

A black discharge may be noticed coming from the adsorber exhaust upon start-up. This is due to the carbon dust contained in all new charcoal and should subside after a couple of cycles. The duration of the adsorption cycle may be preset by manually set timers within the control panel of the unit. The precise time from start-up to saturation depends on a number of factors. No simple, universal formula can be given to cover every situation. The type of solvent, airstream temperature, airflow rates, variations in concentration, and total amount of solvent evaporated from the process per unit of time must be considered. The initial adsorb cycle should be sufficiently long to saturate the carbon beds completely with solvent. All subsequent cycles, then, should be shortened because total regeneration cannot be effectively achieved. On start-up, initial cycle time and timer settings can be estimated by using the experimental data for carbon efficiencies contained in Table 7. During actual operation, the inlet and outlet solvent concentrations may be continuously monitored with an organic vapor analyzer (OVA) to determine the break point on the adsorption cycle. Timers can then be adjusted to discontinue adsorption and begin regeneration. A desirable feature of activated carbon in the control of solvent emissions is its ability to recover the adsorbed solvents on steam regeneration. As discussed earlier, regeneration may be accomplished by passing hot gas through the carbon bed. Low-pressure steam at 3 to 15 psig is the usual source of heat and is sufficient to remove most solvents. Normally, the flow of steam passes in a direction opposite to the flow of gases during adsorption. With this arrangement, the steam passes upward through the carbon. The steam flow through the bed is only one fifth to one tenth of the air velocity and is too low to initiate boiling or cratering of the bed. This countercurrent flow is an advantage in regeneration because a solvent gradient exists across the adsorbent bed and, de-

TABLE 7. Physical Properties of Common VOCs

	Boiling Point (°F)	Molecular Weight	Soluble in Water	Flammable	Lower Explosive Limit[a] (vol%)	Carbon Adsorption Efficiency[b]
Acetone	133	58.1	Yes	Yes	2.15	8
Benzene	176	78.1	No	Yes	1.4	6
Butyl acetate	259	116.2	No	Yes	1.7	8
Butyl alcohol	241	74.1	Yes	Yes	1.7	8
Carbon tetrachloride	170	153.8	No	No	—	10
Ethyl acetate	171	88.1	Yes	Yes	2.2	8
Ethyl alcohol	165	46.1	Yes	Yes	3.3	8
Heptane	209	100.2	No	Yes	1	6
Hexane	156	86.2	No	Yes	1.36	6
Isobutyl alcohol	241	74.1	Yes	Yes	1.68	8
Isopropyl alcohol	205	60.1	Yes	Yes	2.5	8
Methyl alcohol	153	32	Yes	Yes	6	7
Methylene chloride	104	84.9	Yes	No	—	10
Methylethyl ketone	174	72.1	Yes	Yes	1.81	8
Methylisobutyl ketone	237	100.2	Yes	Yes	1.4	7
Perchloroethylene	250	165.8	No	No	—	20
Toluene	231	92.1	No	Yes	1.27	7
Trichloroethane	189	131.4	No	No	—	15
Trichlorotrifluoroethane	117.6 (113)	186.3	No	No	—	8
Naphtha	208	—	No	No	0.81	7
Xylene	292	106.2	No	Yes	1	10

[a]Lower explosive limit: the lowest concentration value of a vapor that will support propagation of a flame upward through a cylindrical tube.
[b]Carbon adsorption efficiency, lbs adsorbed/100 lbs carbon; efficiencies are based on 200 cfm of 100°F solvent-laden air per 100 pounds of carbon per hour at concentrations above 15 ppm.

pending on the concentration of adsorbate and bed depth, the inlet side of the bed may be saturated before the outlet reaches the break point. Thus, with countercurrent regeneration, the solvent, driven out of the adsorbent from the outlet side by the incoming steam, will, in turn, start to remove vapor at the inlet before it becomes heated because it is already saturated. This results in lower steam consumption and more efficient operation.

Steam consumption per pound of solvent recovered varies with strip time and with the particular solvent adsorbed. As the steam strip time is extended, more steam per pound of solvent recovered is required and a point is reached at which expended steam cost exceeds recovery benefits. Therefore, it is more economical to operate the strip cycle to retrieve only part of the adsorbed solvent—that part recovered generally in less than 90 minutes, leaving a "heel" of solvent within the bed. All carbon systems, therefore, should be operated with the heel left in the beds in order to maximize solvent recovery versus steam usage and/or energy consumption.

After the solvent is steam-stripped, the carbon beds not only are hot, but are saturated with water. Cooling and drying are accomplished by opening the bed to the incoming airstream, allowing the entrapped water to evaporate and adsorb the heat of the carbon and to be swept away to the atmosphere by the airstream.

The solvent vapors and steam that emerge from the bed during stripping are then condensed in a water-cooled condenser. The cooling required for condensation may typically be 5 gal/min at an inlet temperature of 60°F. Temperature rise in the condenser will be approximately 40°F, making the outlet water temperature about 100°F. High cooling-water temperatures or low water-flow rates will slow the condensation process, thereby increasing the steam-strip cycle time and adding unnecessary cost to the process. The final step in the regeneration cycle is the separation of the steam and solvent condensate by using a gravity-type decanter that is (1) adequately vented to eliminate "air locks," (2) properly leveled and drained into an open container or tank to reduce back pressure, and (3) precharged with a sufficient amount of solvent to prevent initial steam condensate from entering the solvent discharge pipe and contaminating the solvent storage tank.

It is important to sample both the wastewater discharge and the recovered solvent periodically to maintain the necessary effluent quality. Any additional solvent discharges into municipal sewer systems or to local streams should be eliminated. The integrity of the recovered solvent can be determined through tests for moisture content and acidity. Moisture content should be less than the solubility of water in solvent. Acid acceptance tests will reflect losses of solvent inhibitors that have been added to reduce breakdown of the solvent.

The adsorption phenomenon becomes somewhat more complex if the gases or vapors to be adsorbed consist of several compounds. Carbon adsorption of the various components in a mixture is not uniform, and generally, these components are adsorbed in an approximately inverse relationship to their relative volatilities. Hence when air containing a mixture of organic vapors is passed through an activated-carbon bed, the vapors are equally adsorbed at the beginning, but as the amount of the higher boiling constituent retained in the bed increases, the more volatile vapor revaporizes. After the breakpoint is reached, the exit vapor consists largely of the more volatile material. At this stage, the higher boiling component has displaced the lower boiling component, and this is repeated for each additional component. Two or more volatile organic compounds (VOCs) in the airstream, as a general rule, will have the following effects:

1. The adsorption of organic compounds having higher molecular weights will tend to displace those having lower molecular weights. Lighter compounds will tend to be separated or partitioned from the heavier compounds and will pass through the bed at a faster rate. This will increase the mass transfer zone and may require additional carbon bed depth or shorter operating cycles.
2. Carbon retentivity may be reduced.
3. Efficiencies of any given system will tend to be lower with multiple organic compound applications.
4. The LEL of the mixture will vary directly with the LEL of the individual components. Safety considerations may dictate more or less dilution air to reduce the flammability potential.

Decanter operation will also be affected by multiple organics in the system. Where multiple solvents are involved, the organic layer will consist of a mixture of the compounds, roughly paralleling their feed-stream concentrations. Some compound mixtures of immiscible organics may possess a variety of densities, resulting in multilayer decanter separations requiring multiple decantation for recovery. The recovered organics may also be a compound mixture and their value as a reclaimed solvent will depend upon the following questions:

1. Can the mixture be reused, as decanted?
2. Is it possible to reconstitute the mixture for reuse by solvent additions?
3. If the mixture cannot be reused, is it possible to sell it to a central refiner who would fractionate it and resell it as used solvent?
4. If the mixture cannot be made suitable for reuse, can it be used as a fuel for generation of thermal energy?
5. Is the recovered mixture of high enough value to consider on-site distillation equipment?

Water-soluble organics present a special problem to steam regeneration systems. Because of water contact in the decanter, the water-soluble materials will be carried away in the wastewater stream. If losses of these organics are detrimental to the process or to water quality, alternative stripping mechanisms should be considered. Complex distillation columns may also be employed to recover the water solubles and return them to the source.

The solvent corrosion factor should be considered in selection of the materials of construction for the adsorption equipment. Corrosion should be prevented or retarded if long-term operation of the equipment is expected. Many VOCs are not particularly corrosive and may be handled in conventional carbon-steel tanks. Some solvents are subject to hydrolysis and/or other chemical reactions and to the formation of corrosive by-products. These must be dealt with by the use of coatings or base metals such as stainless steel or other materials that have a high resistance to oxidation. Most carbon-steel carbon adsorption systems will employ some type of protective coating to isolate the activated carbon from direct contact with the interior of the adsorption vessel. This preventive measure is necessary to reduce galvanic action between the carbon and the tank wall that will result in excessive pitting and corrosion in the tank. The cycle time and/or manner in which the system is to be utilized will also be a factor. Systems that are to be regenerated only once or twice a day will tend to be drier than those that cycle hourly and may be satisfactorily designed using internal coatings rather than expensive base materials. More expensive materials of construction should be provided only in those inaccessible areas such as vapor lines, small valves, drain connections, and damper housings, where the satisfactory applications of coatings would be difficult. Where corrosive compounds are utilized and where systems are to be cycled on a frequent basis, the more resistant types of construction materials should be considered.

Many hydrocarbon materials introduced into the adsorption system can be characterized as either aliphatic or aromatic. Aromatic compounds exhibit a special type of unsaturation having to do with resonance and the stable nucleus provided by the six-carbon benzene-ring configuration. Aromatics of the simpler type will resist the formation of by-products and may be used with mild steel. Many of the aliphatic compounds are saturated alkanes or paraffins, and do not usually react with aqueous solutions of acids, bases, or oxidizing agents and so are compatible with mild steel constructions.

The major families of corrosive hydrocarbons include halogenated ketones or aldehydes and esters. Esters are the product of reaction between an acid and an alcohol. Such reactions are reversible, and the hydrolysis of an ester will yield the acid and alcohol from which it was produced. In use, most esters involve the formation of acetic acid and will require processing in stainless-steel equipment.

The halogenated compounds represent a saturated hydrocarbon in which a hydrogen atom is replaced by a halide. The halogen atom is easily displaced from its associated carbon atom and the formation of other compounds—often acidic—can be expected. Many halogenated materials are stabilized by the addition of corrosion retardants and inhibitors and can generally be processed in carbon-steel equipment with an impervious surface coating. Monel, hastelloy, or titanium are sometimes used in connection with difficult halogenated compounds.

Ketones and aldehydes are both characterized by the presence of a carbonyl group. This carbon/oxygen group is subject to chemical reactions that may produce a number of corrosive by-products, requiring handling in equipment of stainless-steel construction. The ketones form the major category of "reactive organics," and it is the presence of the carbonyl group, with its ability to undergo chemical reaction, that causes the problem in carbon adsorption systems. Often these reactions are exothermic, and these reactive organic operations will require additional hardware and modified control sequences to control maximum temperature within the beds. Live steam injection or permanently installed cooling coils are effective in controlling temperatures that may approach the LEL of the solvent.

Undesirable contaminants in the airstream may also produce detrimental effects if not removed prior to their introduction into carbon beds. Contaminants may be broken down into three categories: (1) particulates, (2) entrained liquids, and (3) high boilers. Almost all industrial applications will require solid-medium-type filtration systems for the removal of airborne dust, lint, and general dirt in a particle size down to 3 to 5 μm. Solid-medium filters of the type generally available are made of cloth or fiberglass and are usually satisfactory for this application. Automatically operated filters of the moving-medium type, controlled by pressure drop across the filter material, usually offer a satisfactory solution for excessive particulate contamination.

In other applications, where fine particles of resin or other solids used in the coating process have diameters of 1 μm or less, special filtration systems will be required. Electrostatic precipitators, for instance, may find application in this area.

Entrained liquid is also a form of contaminant. There are a number of mist eliminators available, most of which are designed to be mounted in the solvent airstream.

Effective air pollution control utilizing carbon adsorption must be accompanied by a routine maintenance program. The program should provide for scheduled inspections of all equipment components, as well as necessary monitoring of operating parameters to ensure correct operation and optimum performance of the control equipment. Monitoring of the inlet and outlet gas streams, quality of reclaimed solvent, wastewater quality, and so on, will provide valuable information about the performance of the carbon adsorber.

Routine maintenance of air pollution control equipment can also be important because equipment failure can be expensive in terms of lost production, lost solvents, degradation of air resources, and potential effects on employee health. Each component must operate properly to ensure steady, efficient output and desired results from the system.

Establishing an equipment maintenance program need not be elaborate or complicated. The actual work in routine inspection and servicing may be performed, in large part, by shop personnel operating the control equipment. Of course, extensive repairs or rebuilding should be accomplished by skilled and trained maintenance personnel. System components of the carbon adsorber that require routine maintenance fall into four major categories: (1) air handling, (2) adsorbing, (3) stripping, and (4) reclaiming.

The function of the air-handling apparatus is to collect, transport, and deliver particulate-free, solvent-laden air to the adsorber. Any leaks in the ductworks on the suction side of the fan will introduce excessive ambient air into the system, resulting in a reduction of solvent concentration and poor adsorber efficiency. Air-duct leakage on the pressure side of the fan will discharge unwanted solvent vapors into the workplace. Leakage checks should be performed periodically, especially at flexible connections, at joints in the ductwork, on the fan and filter housing, and around the adsorber bed dampers. Accurate collection velocity data should be established by routinely checking the capture velocity at the source using a vane-type velocity meter or a thermal anemometer. Correct operation of flow-indicating devices within the ducts can be verified by mechanically stopping flow to the duct or by turning off the fan motor. Ventilation system imbalance may also occur from time to time and may require periodic adjustments to dampers to rebalance the system. The particulate filter bag installed in-line ahead of the adsorber beds should be equipped with a differential pressure gauge to indicate dirty or stopped-up filter media. The bag should be changed or cleaned when the differential pressure increases by 1 inch (water) or more. Observance of the operation of the disk dampers within the adsorber tanks should occur periodically to determine correct seating and opening of the valves.

Maintenance inspections on the adsorption-cycle equipment are somewhat more complex than for the air-handling apparatus. The integrity of the activated carbon must be maintained to ensure efficient removal of solvent vapors from the airstream. As the carbon particles erode with time and the capillaries become plugged with contaminants and polymers, the granules gradually lose their ability to adsorb and retain solvent molecules. Carbon adsorbability and retentivity should be tested regularly by opening up the bed and extracting carbon samples from the top, center, and bottom layers. Laboratory analysis will reveal the effectiveness of the carbon bed. Most manufacturers of activated charcoal will perform adsorbability and retentivity tests for their customers. If the carbon fails the tests, all the adsorbent should be removed from the system and be regenerated or replaced.

Maintenance requirements on the desorption-cycle equipment involves primarily the steam supply, valving, and timer controls. Steam pressure on the carbon tanks should be regulated to minimize steam stripping pressure (3 to 10 psi on perchloroethylene [PCE]). Lower steam pressure will require too much run time to strip the beds adequately, while pressure that is too high will tend to fluidize the bed and create excessive erosion of the carbon granules. Steam traps must be operative or water will be carried into the carbon beds and retard proper stripping action. Periodically, the steam pressure relief valves located in the supply line and in the main carbon tanks should be checked for correct pressure settings by increasing steam pressure until the valves "pop off." Steam leaks around gaskets and operating dampers should be corrected by replacing the gaskets and seals. Gasket materials that are in contact with the solvent vapors must withstand the chemical properties of that particular solvent. Often, the gasket material supplied with the system may not be suitable for the solvent presented to the adsorber and a substitute material may be required. Leaks around the carbon tanks may create additional problems of corrosion around the leak. The appearance of corrosion throughout the system may indicate a loss of inhibitors in the solvent that have been lost through numerous steam-stripping cycles. Makeup inhibitors or makeup solvent may alleviate this condition. Boiler feedwater treatment may require some modification if "carryover" chemicals are introduced into the carbon beds and create corrosion problems.

The apparatus utilized to reclaim the solvent requires little maintenance. The automatic cooling-water valve should be checked for proper opening and closing operation. Automatic mechanical valve shafts and other mechanisms should be lightly oiled. The condenser, in time, may become inefficient because of excessive buildup of solubles from the cooling water. Acidizing of the water jacket or tubes may be required to renew condenser efficiency. Inadequate separation of water and solvent in the decanter may indicate a plugged vent line. All vent lines and drain lines must be unrestricted for correct operation of the system.

In general, normal maintenance procedures should be followed in routine cleaning of electrical contacts; lubrication of all bearings, compressed-air components, and air cylinder shafts; replacement of obviously broken or worn parts; and housekeeping practices around the adsorber. In the final analysis, common sense is the best maintenance tool available in view of the fact that a large percentage of carbon adsorber equipment failures can be traced to neglect, improper operation, or just plain abuse.

MAINTAINING AND IMPROVING EQUIPMENT PERFORMANCE[8]

Optimizing performance of air pollution control equipment such as carbon adsorbers involves consideration and monitoring of the following aspects: (1) operation of manufacturing process controls to minimize solvent emissions; (2) quality of the solvent/air inlet stream; (3) characteristics of the inlet stream, for example, concentration, temperature, and flow; (4) duration of the adsorption cycle; saturation, and working bed capacities; (5) quality and quantity of available steam for regeneration; (6) duration of steam-strip cycle; (7) saturation and retentivity of the carbon; (8) quantity and quality of cooling water; (9) effectiveness of the water–solvent separator; (10) quality of reclaimed solvent; (11) quality of wastewater; and (12) quality of exhaust stream from the adsorber bed.

Operation of the Manufacturing Process

Prior to applying end-of-the-pipe air pollution control equipment such as the carbon adsorber, consideration must be given to process controls that will minimize solvent losses to the airstream and/or workplace. Listed below are some important considerations.

1. Consider substitution of the solvent by one of lower volatility or environmental impact.
2. Minimize local exhaust ventilation from the process.
3. Utilize adequate freeboard height to reduce losses.
4. Provide canopies, hoods, and enclosures over tanks if possible, and minimize all openings to process.
5. Provide a parts-drying chamber within the process if possible.
6. Cover tanks when not in use.
7. Minimize the travel rate of parts into and out of the cleaning tank vapor zone.
8. Perform solvent spraying in the vapor zone, preferably with a gentle flush rather than an atomized spray.
9. Allow parts to drain properly prior to exiting the cleaning tank.
10. Maintain a cold-air blanket above tanks with either chilled-water coils or direct-expansion refrigeration coils.
11. Do not clean porous or adsorbent materials.
12. Do not vigorously boil solvent sump.
13. Do not agitate the solvent bath with compressed air.
14. Use compressed air blow-off as a last resort.
15. Do not direct ventilating fans on process.
16. Establish and maintain filling, draining, start-up, shutdown, operation, and maintenance procedures.

After the process is optimized to minimize emissions, consideration can be given to end-of-the-pipe controls.

Quality of Solvent–Air-Inlet Stream

The large surface areas initially offered by the activated-carbon granules will gradually be lost in the adsorption–desorption process. The effective life of the carbon depends on the quality of the incoming airstream presented to the beds. Airborne contaminants must be removed from the vapor-laden stream prior to introduction to the adsorbent to eliminate "plugging" of capillaries within the granules. Periodic and routine maintenance on the bag filter will ensure longer life to the carbon beds. Some solvents offered to the carbon beds may decompose or polymerize when contacting the adsorbent. Polymerization will significantly lower the adsorption capacity of the adsorbent and also reduce its regenerating capabilities by conventional means. These polymerizing solvents must be avoided.

Characteristics of Inlet Stream

In many instances, exhaust ventilation design is concerned primarily with contaminant control at the point of emission to protect worker health and reduce employee exposures to hazardous chemicals. As a result, many exhaust systems have been designed to satisfy only these requirements, with little regard for system efficiencies. Earlier discussions of process design characteristics indicated a desire to minimize exhausted air losses and excessive carryout of solvent vapors. Lower airflow not only will improve the cleaning process efficiency, but will reduce fan cost, ventilation system cost, and power consumption, and will reduce previously conditioned room air losses to the atmosphere.

Duration of the Adsorption Cycle

Working bed capacities vary considerably, depending on the particular solvent being reclaimed and its regeneration characteristics. To maximize performance of the carbon adsorber, the duration of the adsorption cycle should be extended to just below the break point of the beds. Breakthrough can be determined using organic vapor analyzers simultaneously on the inlet and outlet streams of the adsorber bed. Breakthrough history can then be determined on the particular process being controlled, and regeneration can be initiated only when absolutely necessary.

Steam Quality

Low steam pressure and wet steam will reduce the effectiveness of the stripping operation. Steam traps must be continually checked for correct operation. Because of the intermittent steam demand, condensate in the supply lines must be removed prior to stripping. Automatic steam traps should ensure the availability of dry steam for the desorption of the beds.

Duration of Steam-Strip Cycle

To provide efficient recovery of solvents, steam stripping must not be excessively long, to try to recover too much of the "heel," or too short, requiring more frequent regeneration. Desirable strip-cycle times can be assured by continuous monitoring of reclaimed solvent quantities that flow from the decanter.

Saturation and Retentivity of Carbon

Ensure good operation and performance of the carbon beds by periodically testing the charcoal for its ability to adsorb solvent and to retain the solvent after the manufacturing process flow is discontinued. Adsorbent manufacturers will assist in this test. The adsorbability and retentivity of the carbon will be affected by airborne contaminants that enter the beds and plug the capillaries in the granule.

Cooling-Water Quantity

Monitor cooling-water inlet and outlet temperatures. Usually, the steam-strip cycle is operated on a timer mode and cooling water is required to condense all the vapors presented to the heat exchanger in the same given amount of time. Inadequate cooling water will result in solvent losses to the airstream. The inlet and outlet temperatures will vary depending on the particular solvent being reclaimed. The cooling-water supply must be sufficient to condense vapors of steam as well as solvent vapors that are presented to the condenser for the duration of the steam-strip cycle.

Effectiveness of the Water–Solvent Separator

Vent properly, provide unrestricted flow of solvent and water discharging from the decanter, and check the quality of both solvent and wastewater.

Quality of Reclaimed Solvent

Laboratory analysis is required to determine losses of inhibitors, decomposition of solvent, acidity, or contamination.

Quality of Wastewater

Wastewater analysis can also detect the presence of pollutants in the stream. Because of the rapid separation of water and solvent, gross amounts of nonsoluble solvents can be detected by observation. Dissolved contaminants must be detected through laboratory analysis.

Quality of Exhaust Stream

The final analysis required to determine efficient operation of the adsorber is in the exhaust stream. By using an organic vapor analyzer, the concentration of organic solvent in the airstream can be determined. The total quantity of solvent losses in pounds per unit time should be recorded to establish historical records on the system. Deviations from the file data can then be noticed, rectified, and carbon adsorber efficiencies maintained.

Improvements in the operation and performance of carbon adsorption systems can be obtained through a well-defined program of operation and maintenance, a program that encompasses the total manufacturing process, beginning with the solvent used and the design and operation of the process and its exhaust system. To ensure proper operation and maintenance, a formal monitoring program should be established. Implementation of this program will provide an accurate data record of such parameters as inlet concentrations, outlet concentrations, solvent-removal efficiency, cycle times, and solvent quantities reclaimed.

Monitoring of the organic vapors in the airstream is done for several reasons: to demonstrate compliance, to establish organic vapor removal efficiencies, and to determine optimum time intervals for adsorption and regeneration cycles.

Two methods recognized for the monitoring of organic emissions are the source test screening method, which specifies the organic vapor analyzer instrument, and the source testing verification method, which specifies an integrated bag as the sampling device with a gas chromatograph or infrared spectrophotometer analysis.

Solvent concentration monitoring must be accompanied by adequate airflow determination to calculate solvent losses and to perform the required material balance on the system. Airflow may be determined using the standard pitot tube and inclined water-filled manometer. In cool, dry gas streams, a thermal anemometer may be used.

Monitoring of wastewater discharge and recovered solvent is also important to maintain the necessary effluent quality. Discharges to municipal sewers or to local streams should be minimized to reduce solvent losses. The integrity of the solvent to be reused depends on individual cleanliness requirements; however, samples of the solvent should be tested periodically for moisture content and acidity, as well as for cleanliness. Often, visual inspection will provide needed information pertaining to either solvent or wastewater quality. Discolorations in solvents that have been returned to the process may signal a need for further investigation. When water and immiscible solvents are present together, a clear boundary line may be visible.

Sight glasses on holding tanks can be misleading if two immiscible solvents are present in the tank, such as water and PCE. Because of the density difference, the PCE will assume the bottom position in the tank and, in conventional sight-glass installations, will fill from the bottom of the tank. The presence of water may not be reflected at all in sight-glass readings. Because of this density difference, the sight-glass levels may also be misleading, not reflecting the true level of liquid within the tank.

Certainly, a viable monitoring program is mandatory in "tracking" the performance of any air pollution control equipment. The costs to implement such a program will be returned many times over through improvements in source operations, in the recovery of organic solvents, and in the protection of air resources *through reduced emissions.*

ILLUSTRATIVE EXAMPLES

Example 1

1. Estimate the pounds of carbon dioxide (CO_2) that can be adsorbed by 100 pounds of Davison 4A Molecular Sieve from a discharge gas mixture at 77°F and 40 psia containing 10,000 ppm (by volume) CO_2.
2. What percentage of this adsorbed vapor would be recovered by passing superheated steam at a temperature of 392°F through the adsorbent until the partial pressure of the CO_2 in the stream leaving is reduced to 1.0 mmHg?
3. What is the residual CO_2 partial pressure in a gas mixture at 77°F in contact with the freshly stripped sieve in the second part?

Solution Refer to Figure 4.

1. The CO_2 partial pressure is $10{,}000 \times 10^{-6}$ (40 psia) = 0.4 psia = 20.7 mmHg. At 77°F and this partial pressure, the adsorbent capacity is 9.8 pounds CO_2 per 100 pounds sieve.
2. The capacity (extrapolated) at 392°F and 1.0 mmHg is approximately 0.8 pound CO_2 per 100 pounds sieve. The CO_2 recovered is then 9.8 – 0.8 = 9.0 pounds CO_2 per 100 pounds sieve, and the percent recovery: (9.0/9.8) 100 = 91.8%.
3. The partial pressure of CO_2 in equilibrium at 77°F with sieve containing 0.8 pound CO_2 per 100 pounds sieve is approximately (by extrapolation) 0.044 mmHg. The corresponding concentration (residual CO_2 partial pressure) at a pressure of 40 psia is $[(0.044/40)/(760/14.7)]10^6 = 21$ ppm CO_2.

Example 2

You are asked to determine the required height of adsorbent for an adsorber that treats a degreaser ventilation stream contaminated with trichloroethylene (TCE) given the following operating and design data. The adsorption column cycle is set at four hours in the adsorption mode, two hours in heating and desorbing, one hour in cooling, and one hour in standby. The adsorber recovers 99.5% by weight of TCE. A horizontal unit with an inside diameter of 6 feet and length of 15 feet is to be used. Volumetric flow rate of contaminated air stream is 10,000 scfm; standard conditions are 60°F, 1 atm; operating temperature is 70°F; operating pressure is 20 psia; adsorbent is activated carbon; bulk density of activated carbon is 28 pounds TCE per 100 pounds carbon; inlet concentration of TCE is 2000 ppm (by volume); and molecular weight of TCE is 131.5.

Solution The total and TCE actual volumetric flow rates are first calculated.

$$\begin{aligned} Q_{act} &= 10{,}000[(70 + 460)/(60 + 460)][(14.7)/(20)] \\ &= 7491 \text{ acfm} \\ &= 4.5 \times 10 \text{ acfh} \end{aligned}$$

$$\begin{aligned} Q_{\text{TCE}} &= (Y_{\text{TCE}})(Q_{\text{act}}) \\ &= (2000 \times 10^{-6})(4.5 \times 10^5) \\ &= 900 \text{ acfh} \end{aligned}$$

The mass flow rate of the TCE, W, is then

$$\begin{aligned} W = (Q)(\rho) &= Q(PM/RT) \\ &= (20)(131.5)(900)/(10.73)(70 + 460) \\ &= 416.2 \text{ lb/h} \end{aligned}$$

$$\begin{aligned} \text{TCE adsorbed} &= (W)(0.995)(4) \\ &= (416.2)(0.995)(4) \\ &= 1656.6 \text{ pounds} \end{aligned}$$

Activated carbon required = (TCE to be adsorbed)(100 pounds carbon/28 pounds TCE adsorbed)/(bulk density)
= (1656.6)(100/28)/(36)
= 164 ft^3

Height of adsorbent = (activated-carbon volume)/(cross-sectional area)
= (164)/(6)(15)
= 1.83 feet

Example 3

Design an adsorber system to collect 500 ppm benzene from an air–benzene mixture at 1.1 atm, 50°F, flowing at 500 acfm. Use 4 × 6 mesh activated carbon. Steam is available twice per day at the plant and a steam solvent ratio of 3.0 is to be employed. Thus each unit is to be operated for 24 hours per total cycle (one on/one off). What are the pressure-drop and horsepower requirements for this process? Also, calculate the steam flow per regeneration for this system. Pertinent data are as follows:

ρ_B, carbon = 30 lb/ft^3
CAP = 0.45 pound benzene per pound carbon
HEEL = 0.03 pound benzene per pound carbon
Velocity = 75 fpm
Fan efficiency = 58% = 0.58

Solution Key calculations are as follows:

V = (500 ft^3/min) (60) (12 hours) = 360,000 ft^3 per 12 hours
V_B = 360,000 (500/10) = 180 ft^3
ρ_B = (78) (1.1)/(0.73) (510) = 0.2304 lb/ft^3
M_B = (180) (0.2304) = 41.47 pounds benzene
WC = CAP – HEEL = 0.45 – 0.03 = 0.42 pound benzene per pound carbon
M_{AC} = 41.47/0.42 = 98.74 pounds activated carbon
V_{AC} = 98.74/30 = 3.291 ft^3
A = 500/75 = 6.666 ft^2

If vertical, D = 2.91 feet

H = 3.291/6.667 = 0.4937 feet
= 5.9245 inches ≈ 6 inches

From Figure 12:

ΔP = 0.4 in. water in. bed
ΔP_{Total} = (0.4)(6.0) = 2.4 in.
= (2.4)(5.2) = 12.3 lb_f/ft^2
Horsepower = (12.3)(500)/(60)(550)(0.58)
= 0.3219 hp; choose commercial size 1/3 hp
M_{Steam} = (3.0)(41.5)
= 124.5 pounds steam during regeneration

References

1. MSA Research Corp., *Package Sorption Device System Study*. EPA, April 1973.
2. VIC Manufacturing Co., *Installation, Operation and Maintenance for VIC Air Pollution Control System*.
3. F. Cross, and H. E. Hesketh, *Handbook for the Operation and Maintenance of Air Pollution Control Equipment*, Technomic Publishing, Westport, CT, 1975.
4. E. W. Brothers, Electron Scanning Microscope Photographs, Western Electric Co., June 1980.
5. S. Ergun, *Chem. Eng. Progr.*, 48:89 (1952); *Ind. Eng. Chem.*, 41:1179 (1949).
6. Union Carbide Corp., Linde Division, Molecular Sieve Department, New York, Bulletins F-34, F-34-1, and F-34-2.
7. L. Theodore and A. J. Buonicore, *Air Pollution Control Equipment: Selection, Design, Operation and Maintenance*, Prentice-Hall, Englewood Cliffs, NJ, 1982.
8. R. G. Wynne and L. P. Spencer, in *Air Pollution Control Equipment: Selection, Design, Operation, and Maintenance*, L. Theodore and A. J. Buonicore (Eds.), Prentice-Hall, Englewood Cliffs, NJ, 1982.

CONDENSATION

Anthony J. Buonicore

Condensation is the process of converting a gas or vapor to liquid. Any gas can be reduced to a liquid by sufficiently lowering its temperature and/or increasing its pressure. The most common approach is to reduce the temperature of the gas stream, since increasing the pressure of a gas can be expensive.

Condensers are simple, relatively inexpensive devices that normally use water or air to cool and condense a vapor stream. Since these devices are usually not required or capable of reaching low temperatures (below 100°F), high removal efficiencies of most gaseous pollutants are not obtained unless the vapors will condense at high temperatures. Condensers are typically used as pretreatment devices. They can be used ahead of adsorbers, absorbers, and incinerators to reduce the total gas volume to be treated by more expensive control equipment. Employed in this manner, this device may help reduce the overall cost of the control system.

When a hot vapor stream contacts a cooler medium, heat is transferred from the hot gases to the cooler medium. As the temperature of the vapor stream is cooled, the average kinetic energy of the gas is reduced. Ultimately, the gas molecules are slowed down and crowded so closely together that the attractive forces (van der Waals forces) between the molecules cause them to condense to a liquid. The two conditions that aid condensation are low temperatures, so that the kinetic energy of the gas molecules are low, and high pressures, so that the molecules are brought close together. The actual conditions at which a particular gas molecule will condense depends on its physical and chemical properties. Condensation occurs when the partial pressure of the pollutant in the gas stream equals its vapor pressure as a pure substance at the operating temperature.

As described above, condensation of a gas can occur in three ways: (1) at a given temperature, the system pressure is increased (compressing the gas volume) until the partial pressure of the gas equals its vapor pressure; (2) at a fixed pressure, the gas is cooled until the partial pressure equals its vapor pressure; or (3) in a combined technique, the gas is compressed and cooled until its partial pressure equals its vapor pressure.

DESCRIPTION OF EQUIPMENT

As indicated in the previous section, condensation can be accomplished by increasing pressure or decreasing temperature (removing heat). In practice, air pollution control condensers operate by extracting heat. Condensers differ in the means of removing heat and the type of device used. The two different means of condensing are direct contact, where the cooling medium with vapors and the condensate are intimately mixed and combined, and indirect (or surface), where the cooling medium and vapor/condensate are separated by a surface area of some type.

Contact condensers are simpler, less expensive to install, and require less auxiliary equipment and maintenance. The condensate/coolant from a contact condenser has a volume 10 to 20 times that of a surface condenser. This condensate often cannot be reused and may pose a waste disposal problem unless the dilution of the pollutant is sufficient to meet regulatory requirements. Some typical contact condensers are shown in Figure 1.

Surface condensers form the bulk of the condensers used for air pollution control. Among the applicable types of surface condensers are shell and tube, double pipe, spiral plate, flat plate, air-cooled, and various extended-surface tubular units. This discussion focuses on shell-and-tube condensers, because they are so widely used in industry and

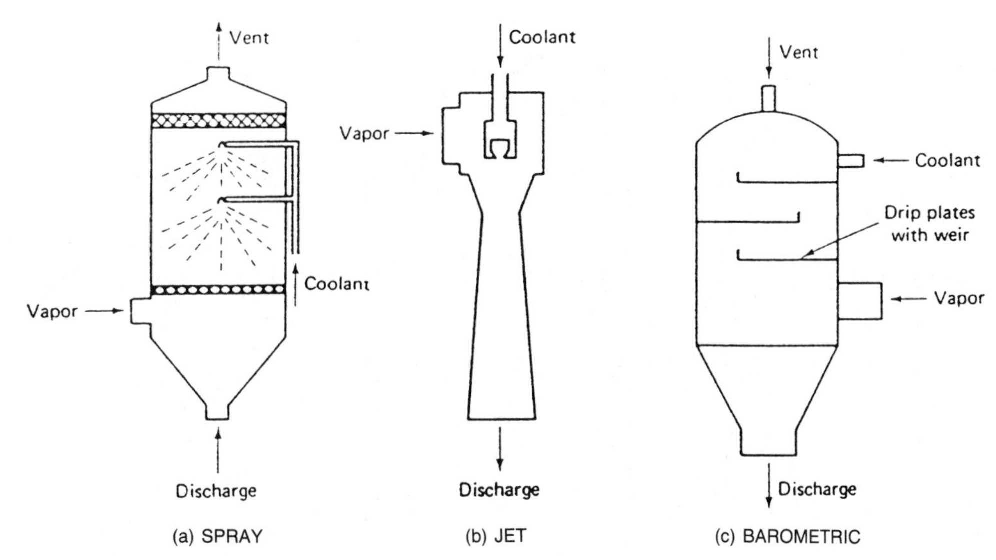

FIGURE 1. Contact Condensers: (a) Spray; (b) Jet; (c) Barometric

have been standardized by the Tubular Exchanger Manufacturers Association (TEMA).

Condensing can be accomplished in either the shell or the tubes. The economics, maintenance, and operational ramifications of the allocation of fluids are extremely important, especially if extended-surface tubing is being considered. The designer should be given as much latitude as possible in specifying the condenser. Details of TEMA designations for shell-and-tube heat exchangers are shown in Figure 2.

Air-cooled condensers consist of a rectangular bank of high finned tubes, a fan, a plenum for even distribution of air to the rectangular face of the tube bank, a header for vapor inlet and condensate outlet, and a steel supporting structure. Fins are usually aluminum, 0.5 to 0.625 inch high, applied to a bare tube by tension winding or soldering, or cold-extruded from the tube itself. Bimetallic tubes can be used to provide process corrosion resistance on the inside and aluminum extruded fins on the outside. Headers for the tube side frequently contain removable gasketed plugs corresponding to each tube end and are used for access to the individual tube-to-tube sheet joint for maintenance.

Nearly all tubular exchangers employ roller expanding of the tube ends into tube holes drilled into tube sheets as a means of providing a leakproof seal. According to TEMA requirements, tube holes must be of close tolerance and contain two concentric grooves. These grooves give the tube joint greater strength, but do not increase sealing ability and may actually decrease it. Research is currently being done to clarify what the best procedure is to follow for a given set of materials, pressures, temperatures, and loadings. Welding or soldering of tubes to tube sheets may be performed for additional leak tightness or in lieu of roller expanding.

DESIGN AND PERFORMANCE EQUATIONS

Design of contact condensers involves calculating the quantity of coolant required to condense and subcool the vapor and proper sizing of discharge piping and hotwell. Calculation of coolant flow rate is as follows:

$$M_c = \frac{M_v H_v + M_l C_{pl}(T_i - T_d)}{C_{pc}(T_d - t_i)} \quad (1)$$

where: M_v, M_1, M_c = flow rate of vapor, liquid condensate, and coolant, respectively, lb/h,

C_{p1}, C_{pc} = heat capacity of condensate and coolant, Btu/lb-°F

T_i, T_d = inlet temperature of vapor and discharge temperature, °F

t_i = inlet temperature of coolant, °F

H_v = latent enthalpy of vaporization of condensate, Btu/lb

Use of a contact condenser requires consideration of coolant availability and liquid waste disposal or treatment facilities. Contact condensers are relatively efficient scrubbers as well as condensers and have the lowest equipment capital cost. Contaminants that are not ordinarily condensable or even particulates can be removed from the vapor stream. Reputable manufacturers of the various types of contact condensers should be consulted for sizing and layout recommendations.

The heat transfer for a surface condenser is governed by the following relationship:

$$Q = U_o A T_m \quad (2)$$

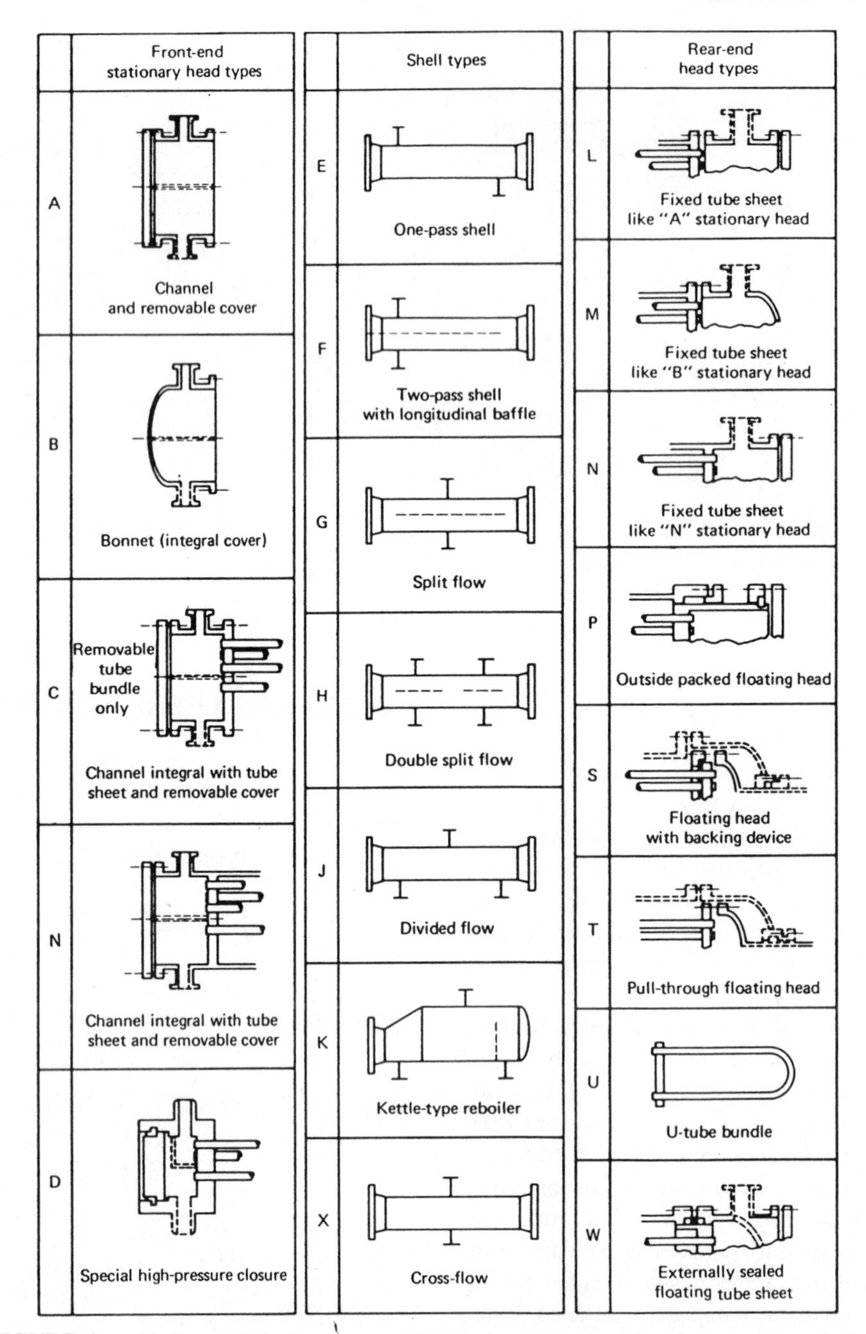

FIGURE 2. Shell-and-Tube Condensers (© Tubular Exchange Manufacturers Association)

$$A = \frac{Q}{U_o T_m} \tag{3}$$

where: Q = total heat load, Btu/h
U_o = overall heat-transfer coefficient, Btu/h-°F-ft^2
T_m = mean temperature difference driving force, °F
A = surface area, ft^2

The first determination is the heat load. In the simplest case, it is the latent heat to condense plus the subcooling load. The next step is specifying the coolant flow rate and temperatures to balance the vapor heat load. Temperature crosses where the outlet coolant temperature is higher than the vent and/or condensate outlet temperature should be avoided.

After the terminal temperatures of the vapor and coolant have been determined, a mean temperature difference can be established. For isothermal or linear condensing and liquid subcooling, the logarithmic mean temperature difference (LMTD) is applicable. The LMTD is calculated as follows:

$$\text{LMTD} = \frac{t_g - t_l}{\ln(t_g/t_l)} \quad (4)$$

where: t_g = greater temperature difference between the hot and cold streams
t_l = lesser temperature difference between the hot and cold streams
ln = the natural log

Fluids are usually in counterflow, but in some cases, as with a fluid near its freezing point, parallel flow is used to afford a greater degree of safety from freezing the condenser. Frequently, there is a combination of parallel flow and counterflow as in a two-pass tube side, single-pass shell tubular condenser. The procedure is to apply a correction to the LMTD. The correction factors have been reduced to a series of graphs available in most heat-transfer texts.[1] Strictly, the correction factor applies only to the subcooling or desuperheating zones, but it is sometimes applied to condensers.

If condensation is nonlinear and has a range exceeding 20°F, the LMTD is not applicable and a condensing curve or calculations at several points over the temperature range is required to determine a correct T_m. A mixture of miscible vapors or a vapor and noncondensables are examples of where the LMTD is not applicable.

The remaining unknown to calculate the required surface is the overall heat-transfer coefficient, U_o. The overall heat-transfer coefficient is the inverse of the sum of the resistances to heat transfer and is given by the following relationship.

$$U_o = \frac{1}{1/h_o + 1/h_{io} + r_w + f_o + f_{io}} \quad (5)$$

where: h_o = outside or shell-side coefficient, Btu/h-°F-ft^2
h_{io} = inside or tube-side coefficient corrected to the outside area
r_w = resistance of the surface area or tube wall, h-°F-ft^2/Btu
f_o = fouling resistance on outside or shell side of surface area
f_{io} = fouling resistance on inside corrected to outside of surface area

Some typical industrial values of U are given in Table 1.

If there is noncondensable gas present or a vapor mixture of miscible and immiscible components with a large condensing range, diffusion of the condensing vapor through another vapor or gas must be considered. The presence of a noncondensable gas in a vapor stream reduces the film coefficient; the amount of reduction is related to the size of the gas cooling load relative to the total load. The greater the ratio of sensible heat to total, the closer the heat-transfer coefficient approaches that of cooling the gas only. The calculation procedure is essentially that established by Colburn and Hougen.[2] It is a trial-and-error method performed for at least five points along the condensing curve. Typical condensing coefficients (Btu/°F-ft^2-h) are given in Table 2.

TABLE 1. Typical Overall Heat Transfer Coefficients in Tubular Heat Exchangers

Condensing Vapor (Shell Side)	Cooling Liquid (Tube Side)	U (Btu/°F-ft^2-h)
Alcohol vapor	Water	100–200
High-boiling hydrocarbons (vacuum)	Water	20–50
Low-boiling hydrocarbons	Water	80–200
Organic solvents	Water	100–200
Organic solvents/high percent of noncondensables present	Water or brine	20–80
Naphtha	Water	50–75
Stabilizer reflux vapors	Water	80–120
Sulfur dioxide	Water	150–200
Tall oil derivatives, vegetable oil vapors	Water	20–50
Steam	Feedwater	400–1000

TABLE 2. Typical Film Coefficients (Btu/h-°F-ft)

Fluid	h
Steam	1500
Steam/10% gas	600
Steam/20% gas	400
Steam/40% gas	220
Pure light hydrocarbons	250
Mixed light hydrocarbons	175
Medium hydrocarbons	100
Medium hydrocarbons with steam	125
Pure organic solvents	250

The choice of a coolant will depend on the particular plant and the efficiency required of the condenser. The most common coolant is the primary plant coolant, usually a cooling tower or river water. It has been shown[3] that the vapor outlet temperature is critical to the efficiency of a condenser. In some instances, use of a chilled brine or a boiling refrigerant can achieve a collection efficiency that will be sufficient without additional control devices. It is not unusual to specify multiple-stage condensing, where condensers, usually two, connected in series use cooling mediums with successively lower temperatures. For example, a condenser using cooling-tower water can be used prior to a unit using chilled water or brine—thereby achieving maximum efficiency while minimizing the use of chilled water.

Careful consideration should be given to the fouling resistances f_o and f_{io} applied to the condenser design. The purpose of adding a fouling resistance is to provide consistent performance without excessive cleaning. Theoretically, increased fouling factors will provide greater safety allowance. In practice, however, they can actually decrease the safety allowance. Consider, for example, a typical application where a maximum pressure drop and overall length are specified. The most efficient design will use the length and pressure drop to the limit specified. If the fouling resistance specified is so large that this is not possible, velocities of the respective streams will be lower than desirable and fouling will be greater than in a design using a lower value of fouling resistance. Fouling resistances used indiscriminately can actually be "self-fulfilling." The most effective design feature to minimize fouling is high velocity and efficient use of pressure drop.[4]

Limited information is available in the literature to calculate the (vapor) collection efficiency of the condenser. The following calculational procedure is recommended. If F, V, and L represent the molar flow rates entering, vented, and (condensate) discharged, respectively, then:

$$F = V + L \tag{6}$$

If z_i, y_i, and x_i are the mole fraction of organic i in the feed, vent, and condensate streams, respectively, then:

$$Fz_i = Vy_i \quad \text{and} \quad Lx_i \tag{7}$$

Assuming the vent and condensate streams leave the unit in equilibrium:

$$K_i = y_i/x_i \tag{8}$$

The simultaneous solution of equations 6 through 8 can provide information on the amount of organics (condensables) remaining in the gas stream.

$$\Sigma Fz_i/[V + (L/K_i)] = 1.0 \tag{9}$$

OPERATION AND MAINTENANCE[5]

The maximum allowable working pressures and temperatures are indicated on the nameplate of the condenser. These values must not be exceeded. Special precautions should be taken if any individual part of the condenser is designed for a maximum temperature lower than that of the condenser as a whole. The most common example is some copper-alloy tubing with a maximum allowable temperature lower than the actual inlet gas temperature. This is done to compensate for the low strength levels of some brasses or other copper alloys at elevated temperatures. An adequate flow of the cooling medium must be maintained at all times.

Condensers are designed for a particular fluid throughput. Generally, a reasonable overload can be tolerated without causing damage. If operated at excessive flow rates, erosion or destructive vibration could result. Erosion could occur at normally acceptable flow rates if other conditions, such as entrained liquids or particulates in a gas stream or abrasive solids in a liquid stream, are present. Evidence of erosion should be investigated to determine the cause. Vibration can be propagated by other than flow overloads (e.g., improper design, fluid maldistribution, or corrosion/erosion of internal flow-directing devices such as baffles). Considerable study and research has been undertaken in recent years to develop a reliable vibration analysis procedure to predict or correct damaging vibration. At this point, the developed correlations are considered "state of the art," yet most manufacturers have the capability of applying some type of vibration check when designing a condenser. Vibration can produce severe mechanical damage, and operation should not be continued when an audible vibration disturbance is evident.

Condensers should be warmed up slowly and uniformly; the higher the temperature ranges, the slower the warm-up should be. This is generally accomplished by introducing the coolant and bringing the flow rate to design level and gradually adding the vapor. For fixed-tube-sheet units with different shell and tube material, consideration should be given to differential expansion of shell and tubes. As fluids are added, the respective areas should be vented to ensure complete distribution. A procedure other than this could cause large differences in temperature between adjacent parts of the condenser and result in leaks or other damage. It is recommended that gasketed joints be inspected after continuous full-flow operation has been established. Handling, temperature fluctuations, and yielding of gaskets or bolting may necessitate retightening of the bolting.

Cooling down is generally accomplished by shutting off the vapor stream first and then the cooling stream. Again, fixed-tube-sheet condensers require consideration of differential expansion of the shell and tubes. Condensers containing flammable, corrosive, or high-freezing-point fluids should be thoroughly drained during prolonged outages.

The recommended maintenance of condensers requires regular inspection to ensure mechanical soundness of the unit and a level of performance consistent with the original design criteria. A brief general inspection should be performed on a regular basis while the unit is operating. Vibratory disturbance, leaking gasketed joints, excessive pressure drop, decreased efficiency indicated by higher gas outlet temperature or lower condensate rates, and intermixing of fluids are all signs that thorough inspection and maintenance procedures are required.

Complete inspection requires shutdown of the condenser for access to internals and pressure testing and cleaning. Scheduling can only be determined from experience and general inspections. Tube internals and exteriors, where accessible, should be visually inspected for fouling, corrosion, or damage. The nature of any metal deterioration should be investigated to determine properly the anticipated life of equipment or possible corrective action. Possible causes of deterioration include general corrosion, intergranular corrosion, stress cracking, galvanic corrosion, impingement, and erosion attack.

Fouling of condensers is the deposition of foreign material on the interior or exterior of tubes. Evidence of fouling in operation is increased pressure drop and general decreasing performance. Fouling can be so severe that tubes are completely plugged, resulting in thermal stresses and subsequent mechanical damage of equipment.

The nature of the deposited foul determines the method of cleaning. Soft deposits can be removed by steam, hot water, various chemical solvents, or brushing. Plant experience can determine which method to use. Chemical cleaning should be performed by contractors specialized in the field who will consider the deposit to be removed and the materials of construction. If the cleaning method involves elevated temperature, consideration should be given to thermal stresses induced in the tubes; steaming out individual tubes can loosen the tube-to-tube sheet joints.

Mechanical methods of cleaning are useful for soft and hard deposits. There are numerous tools for cleaning tube interiors: brushes, scrapers, and various rotating cutter-type devices. The condenser manufacturer or suppliers of tube tools can be consulted in the selection of the correct tool for the particular deposit. When cutting or scraping deposits, care should be exercised to avoid damaging tubes.

Cleaning of tube exteriors is generally performed using chemicals, steam, or other suitable fluids. Mechanical cleaning is performed, but it requires that the tubes be exposed, as in a typical air-cooled condenser, or capable of being exposed, as in a removable bundle shell-and-tube condenser. The layout pattern of the tubes must provide sufficient intersecting empty lanes between the tubes, as in square pitch. Mechanical cleaning of tube bundles, if necessary, requires the utmost care to avoid damaging tubes or fins.

Proper maintenance requires testing of a condenser to check the integrity of the tubes, tube-to-tube sheet joints, welds, and gasketed joints. The normal procedure consists of pressuring the shell with water or air at the nameplate-specified test pressure and viewing the shell welds and the face of the tube sheet for leaks in the tube sheet joints or tubes. Water should be at ambient temperature to avoid false indications due to condensation. Pneumatic testing requires extra care because of the destructive nature of a rupture or of explosion or fire hazards when residual flammable materials are present. Condensers of straight-tube, floating-head construction require a test gland to perform the test. Tube bundles without shells are tested by pressurizing the tubes and viewing the length of the tubes and back face of the tube sheets.

Corrective action for leaking tube-to-tube sheet joints requires expanding the tube end with a suitable roller-type tube expander. Good practice calls for an approximate 8% reduction in wall thickness after metal-to-metal contact between the tube and tube hole. Tube expanding should not extend beyond 1/8 inch of the inner tube-sheet face to avoid cutting the tube. Care should be exercised to avoid overrolling the tube, which can cause work hardening of the material, an insecure seal, and/or stress-corrosion cracking of the tube.

Defective tubes can be either replaced or plugged. Replacing tubes requires special tools and equipment. The user should contact the manufacturer or a contractor qualified in repair. Plugging of tubes, although a temporary solution, is acceptable provided that the percentage of the total number of tubes per tube pass to be plugged is not excessive. The type of plug to be used is a tapered one-piece or two-piece metal plug suitable for the tube material and inside diameter. Care should be exercised in seating plugs to avoid damaging the tube sheets. If a significant number of tube or tube joint failures are clustered in a given area of the tube layout, their location should be noted and reported to the manufacturer. A concentration of failures is usually caused by other than corrosion (e.g., impingement, erosion, or vibration).

MAINTAINING AND IMPROVING EQUIPMENT PERFORMANCE[5]

Within the constraints of the existing system, improving operation and performance refers to maintaining operation and original or consistent performance. There are several factors previously mentioned that are critical to the design and performance of a condenser: operating pressure, amount of noncondensable gases in the vapor stream, coolant temperature and flow rate, fouling resistance, and mechanical soundness. Any pressure drop in the vapor line upstream of the condenser should be minimized. Deaerators or similar devices should be operational where necessary to remove gases in solution with liquids. Proper and regular venting of equipment and leakproof gasketed joints in vacuum systems is necessary to prevent gas binding and alteration of condensing equilibrium. Coolant flow rate and temperatures should be checked regularly to ensure that they are in accordance with the original design criteria. The importance of this can be illustrated merely by comparing the winter and summer performance of a condenser using cooling-tower or river water. Decreased performance due to fouling will generally be exhibited by a gradual decrease in efficiency and should be corrected as soon as possible. Mechanical malfunctions can also be gradual, but will eventually be evidenced by near total lack of performance.

Fouling and mechanical soundness can only be controlled by regular and complete maintenance. In some cases, fouling is much worse than predicted and requires frequent cleaning regardless of the precautions taken in the original design. These cases require special designs to alleviate the problems associated with fouling. A leading manufacturer of polyvinyl chloride found that carryover of polymer reduced the efficiency of its monomer condenser and caused frequent downtime. The solution was providing polished internals and high condensate loading in a vertical downflow shell-and-tube condenser. A major pharmaceutical intermediate manufacturer had catalyst carryover to a vertical downflow shell-and-tube condenser that accumulated on the tube internals. The solution was to recirculate condensate to the top of the unit and spray it over the tube-sheet face to create a film descending down the tubes to rinse the tubes clean.

Most condenser manufacturers will provide designs for alternative conditions as a guide to estimating the cost of improving efficiency via alternate coolant flow rates and temperatures, as well as alternative configurations (i.e., vertical, horizontal, shell side, or tube side).

ILLUSTRATIVE EXAMPLES

Example 1

Steam at atmospheric pressure is being condensed in a heat exchanger with water that enters and exits the unit at 80°F and 115°F, respectively. Assuming no subcooling of the condensed steam, calculate the log mean temperature driving force for the exchanger.

Solution Saturated steam at 1.0 atm is at 212°F. The log mean temperature difference, LMTD, is given by equation 4.

$$\text{LMTD} = (t_g - t_l)/\ln(t_g/t_l)$$

Substituting temperatures gives

$$\begin{aligned} t_g &= 212 - 80 = 132 \\ t_l &= 212 - 115 = 97 \\ \text{LMTD} &= (132 - 97)/\ln(132/97) \\ &= 113.6°\text{F} \end{aligned}$$

Example 2

The discharge gases from a meat-rendering plant contain essentially all steam with a small fraction of odor-carrying gases. This steam is typically condensed (to remove the water) prior to treatment with an adsorber or incinerator. Estimate the size of a condenser to treat 60,000 lb/h of this exhaust gas. Assume condensing occurs in a heat exchanger operating with the temperature driving force calculated in Example 1 and an overall heat-transfer coefficient of 135 Btu/h-ft^2-°F.

Solution Assuming that the discharge gas is all steam with an enthalpy of vaporization of approximately 1000 Btu/lb, the condenser heat load becomes

$$\begin{aligned} Q &= (\text{m})(\Delta H) \\ &= (60{,}000)(1{,}000) \\ &= 6.0 \times 10^7 \text{ Btu/h} \end{aligned}$$

The condenser area is as follows:

$$\begin{aligned} A &= Q/(U)(\text{LMTD}) \\ &= 6.0 \times 10^7/(135)(113.6) \\ &= 3912 \text{ ft}^2 \end{aligned}$$

One should consult a vendor if information on the number of tubes and the tube diameter and length is required.

References

1. D. Q. Kern, *Process Heat Transfer*, McGraw-Hill Book Co., New York, 1950.
2. A. P. Colburn and O. A. Hougen, *Ind. Eng. Chem.*, 26(11): 1178 (1934).
3. W. F., Connery et al., Energy and the environment, *Proclamation Third National Conference AIChE*, 1975, p. 276.
4. L. Theodore and A. J. Buonicore (Eds.), *Air Pollution Control Equipment: Selection, Design, Operation and Maintenance*, Prentice-Hall, Englewood Cliffs, NJ, 1982.
5. W. F. Connery, in *Air Pollution Control Equipment: Selection, Design, Operation and Maintenance*, L. Theodore and A. J. Buonicore, Eds., Prentice-Hall, Englewood Cliffs, NJ, 1982.

INCINERATION

Anthony J. Buonicore

Any presentation on incineration must focus first on the principles of combustion. The process of combustion is most often used to control the emissions of organic compounds from process industries. At a sufficiently high temperature and adequate residence time, any hydrocarbon can be oxidized to carbon dioxide and water by the combustion process. Combustion systems are often relatively simple devices capable of achieving very high destruction efficiencies. They consist of burners, which ignite the fuel and organic, and a chamber, which provides appropriate residence time for the oxidation process. Because of the high cost and decreasing supply of fuels, combustion systems may be designed to include some type of heat recovery. Combustion is also used for the more serious emission problems that require high destruction efficiencies, such as emission of toxic or hazardous gases. There are, however, some problems that may occur when using combustion. Incomplete combustion of many organic compounds results in the formation of aldehydes and organic acids that may create an additional pollution problem. Oxidizing organic compounds containing sulfur or halogens produce unwanted pollutants such as sulfur dioxide, hydrochloric acid, hydrofluoric acid, and phosgene. If present, these pollutants may require some type of scrubber to remove them prior to release into the atmosphere. Several basic combustion systems can be used. Although these devices are physically similar, the conditions under which they operate may be different. Choosing the proper device depends on many factors, including type of hazardous contaminants in the waste stream, concentration of combustibles in the stream, process flow rate, control requirements, and an economic evaluation.

Combustion is a chemical process arising from the rapid combination of oxygen with various elements or chemical compounds resulting in the release of heat. The process of combustion has also been referred to as oxidation or incineration. Most fuels used for combustion along with the waste are composed essentially of carbon and hydrogen, but can include other elements, such as sulfur, nitrogen, and chlorine. Although combustion seems to be a very simple process that is well understood, in reality it is not. The exact manner in which a fuel or waste is oxidized occurs in a series of complex, free radical chain reactions. The precise set of reactions by which combustion occurs is termed the mechanism of combustion. By analyzing the mechanism of combustion, the rate at which the reaction proceeds and the variables affecting the rate can be predicted. For most combustion devices, the rate of reaction proceeds extremely fast compared with the mechanical operation of the device. Maintaining efficient and complete combustion is somewhat of an art rather than a science. Therefore, this discussion will focus on the factors that influence the completeness of combustion, rather than on analyzing the mechanisms involved.

To achieve complete combustion once the air (oxygen), waste, and fuel have been brought into contact, the following conditions must be provided: a temperature high enough to ignite the waste–fuel mixture, turbulent mixing of the air and waste–fuel, and sufficient residence time for the reaction to occur. These three conditions are referred to as the "three T's of combustion." Time, temperature, and turbulence govern the speed and completeness of reaction. They are not independent variables since changing one affects the other two.

The rate at which a combustible compound is oxidized is greatly affected by temperature. The higher the temperature, the faster the oxidation reaction will proceed. The chemical reactions involved in the combination of a fuel and oxygen can occur even at room temperature, but very slowly. For this reason, a pile of oily rags can be a fire hazard. Small amounts of heat are liberated by the slow oxidation of the oils. This, in turn, raises the temperature of the rags and increases the oxidation rate, liberating more heat. Eventually, a full-fledged fire can break out. For combustion processes, ignition is accomplished by adding heat to speed up the oxidation process. Heat is needed to combust any mixture of air and fuel until the ignition temperature of the mixture is reached. By gradually heating a mixture of fuel and air, the rate of reaction and the energy released will gradually increase until the reaction no longer depends on the outside heat source. More heat is being generated than is lost to the surroundings. The ignition temperature must be reached or exceeded to ensure complete combustion. To maintain combustion of a waste, the amount of energy released by the combusted waste must be sufficient to heat the incoming waste (and air) up to its ignition temperature; otherwise, a fuel must be added. The ignition temperature of various fuels and compounds can be found in combustion handbooks. These temperatures are dependent on combustion conditions and, therefore, should be used only as a guide. Ignition depends on:

1. Concentration of combustibles in the waste stream
2. Inlet temperature of the waste stream
3. Rate of heat loss from the combustion chamber
4. Residence time and flow pattern of the waste stream
5. Combustion-chamber geometry and materials of construction

Most incinerators operate at higher temperature than the ignition temperature, which is a minimum temperature. Thermal destruction of most organic compounds occurs between 590°C and 650°C (1100°F and 1200°F). However, most hazardous waste incinerators are operated at 1800–2200°F to ensure nearly complete destruction of the organics in the waste.

Time and temperature affect combustion in much the same manner as temperature and pressure affect the volume of a gas. When one variable is increased, the other may be decreased with the same end result. With a higher temperature, a shorter residence time can achieve the same degree of oxidation. The reverse is also true; a higher residence time allows the use of a lower temperature. In describing incinerator operation, these two terms are always mentioned together. This effect is demonstrated in graphical form in Figure 1. The choice between higher temperature or longer residence time is based on economic considerations. Increasing residence time involves using a larger combustion chamber, resulting in a higher capital cost. Raising the operating temperature increases fuel usage, which also adds to the operating cost. Fuel costs are the major operating expense for most incinerators. Within certain limits, lowering the temperature and adding volume to increase residence time can be a cost-effective alternative method of operation. The residence time of gases in the combustion chamber may be calculated from:

$$t = V/Q \tag{1}$$

where: t = residence time, seconds
V = chamber volume, ft^3
Q = gas volumetric flow rate at combustion conditions, ft^3/s

Q is the total flow of hot (flue) gases in the combustion chamber. Adjustments to the flow rate must include outside air added for combustion.

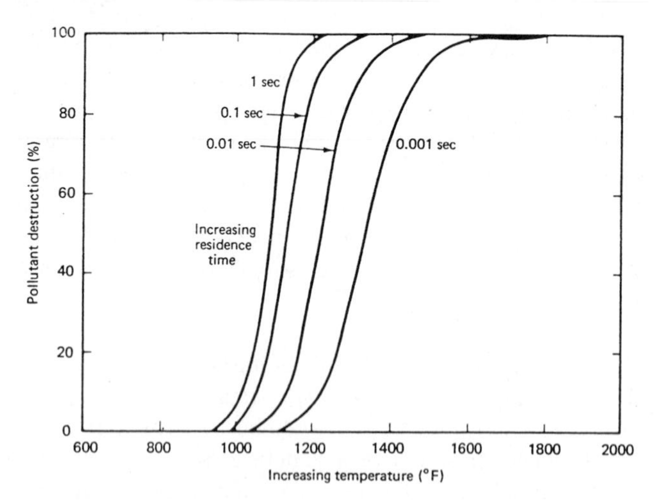

FIGURE 1. Coupled Effects of Temperature and Time on Rate of Pollutant Oxidation

Proper mixing is important in combustion processes for two reasons. First, for complete combustion to occur, every particle of waste and fuel must come in contact with air (oxygen). If this does not happen, unreacted waste and fuel will be exhausted from the stack. Second, not all of the fuel or waste stream is able to be in direct contact with the burner flame. In most incinerators, a portion of the waste stream may bypass the flame and be mixed at some point downstream of the burner with the hot products of combustion. If the two streams are not completely mixed, a portion of the waste stream will not react at the required temperature and incomplete combustion will occur. A number of methods are available to improve mixing the air and waste (combustion) streams, including the use of refractory baffles, swirl-fired burners, and baffle plates. The problem of obtaining complete mixing is not easily solved. Unless properly designed, many of these mixing devices may create "dead spots" and reduce operating temperatures. Merely inserting obstructions to increase turbulence is not the answer. The process of mixing flame and waste stream to obtain a uniform temperature for the decomposition of wastes is the most difficult part in the design of the incinerator.

Oxygen is necessary for combustion to occur. To achieve complete combustion of a compound, a sufficient supply of oxygen must be present to convert all of the carbon to carbon dioxide. This quantity of oxygen is referred to as the stoichiometric or theoretical amount. The stoichiometric amount of oxygen is determined from a balanced chemical equation summarizing the oxidation reactions as described earlier. If an insufficient amount of oxygen is supplied, the mixture is referred to as rich. There is not enough oxygen to combine with all the fuel and waste so that incomplete combustion occurs. If more than the stoichiometric amount of oxygen is supplied, the mixture is referred to as lean. The added oxygen plays no part in the oxidation reaction and passes through the incinerator. Oxygen for the combustion process is supplied by using air. Since air is essentially 79% nitrogen and 21% oxygen, a larger volume of air is required than if pure oxygen were used. A listing of theoretical air requirements is given in Table 1.

In most applications, more than the stoichiometric amount of air is used to ensure complete combustion. This extra volume is referred to as excess air. If ideal mixing were achievable, no excess air would be necessary. However, most combustion devices are not capable of achieving ideal mixing of the fuel and airstreams. The amount of excess air is sometimes held to a minimum in order to reduce heat losses. Excess air takes no part in the reaction, but it does absorb some of the heat produced.

To raise the excess air to the combustion temperature, additional fuel must be used to make up for this loss of heat. Operating at a high volume of excess air can be very costly in terms of the added fuel required.

In addition to the theoretical air required, Table 1 also lists the volume of combustion products produced from oxidizing a substance. This is an important term used to determine the size of the combustion chamber. The values in Table 1 are given in volume percent and weight percent. Volume percent is the more important term for combustion of substances since gas flows are measured in cubic feet per second instead of weight units. For example, from Table 1, when 1 ft^3 of methane is combusted with the theoretical amount of air, 10.53 ft^3 of flue gas is produced. Natural gas is not listed in Table 1 since its chemical composition can vary. When 1 ft^3 of natural gas is burned with a stoichiometric amount of air, it produces approximately 11.4 ft^3 (average value) of flue gas. This is an important number to remember in incinerator calculations.

Afterburning is another term sometimes used to describe the combustion process for control of gaseous emissions. The term afterburner is generally appropriate only to describe a thermal oxidizer used to control gases coming from a process where combustion was not complete. Incinerators are used to combust solid, liquid, and gaseous materials. When used here, the term incinerator will refer to combusting waste streams.

HYDROCARBON INCINERATION KINETICS

Although the actual mechanism of hydrocarbon (HC) combustion is undoubtedly quite complex, in the presence of excess oxygen (a dilute fume in air), the rate equation can be represented by:

$$\frac{d[\mathrm{HC}]}{dt} = -k[\mathrm{HC}] \tag{2}$$

where k = pseudo–first-order rate constant, s^{-1}, which includes the oxygen concentration.

TABLE 1. Combustion Constants

			Heat of Combustion				For 100% Total Air (mol/mol of combustible) (ft^3/ft^3 of combustible)						For 100% Total Air (lb/lb of combustible)						Flammability Limits (% by volume)	
			(Btu/ft^3)		(Btu/lb)		Required for Combustion			Flue Products			Required for Combustion			Flue Products				
Substance	lb/ft^3	ft^3/lb	Gross (high)	Net (low)	Gross (high)	Net (low)	O_2	N_2	Air	CO_2	H_2O	N_2	O_2	N_2	Air	CO_2	H_2O	N_2	Lower	Upper
Carbon, C[a]	—	—	—	—	14,093	14,093	1.0	3.76	4.76	1.0	—	3.76	2.66	8.86	11.53	3.66	—	8.86	—	—
Hydrogen, H_2	0.0053	187.723	325	275	61,100	51,623	0.5	1.88	2.38	—	1.0	1.88	7.94	26.41	34.34	—	8.94	26.41	4.00	74.20
Oxygen, O_2	0.0846	11.819	—	—	—	—	—	—	—	—	—	—	—	—	—	—	—	—	—	—
Nitrogen (atm), N_2	0.0744	13.443	—	—	—	—	—	—	—	—	—	—	—	—	—	—	—	—	—	—
Carbon monoxide, CO	0.0740	13.506	322	322	4,347	4,347	0.5	1.88	2.38	1.0	—	1.88	0.57	1.90	2.47	1.57	—	1.90	12.50	74.20
Carbon dioxide, CO_2	0.1170	8.548	—	—	—	—	—	—	—	—	—	—	—	—	—	—	—	—	—	—
Paraffin series																				
Methane, CH_4	0.0424	23.565	1013	913	23,879	21,520	2.0	7.53	9.53	1.0	2.0	7.53	3.99	13.28	17.27	2.74	2.25	13.28	5.00	15.00
Ethane, C_2H_6	0.0803	12.455	1792	1641	22,320	20,432	3.5	13.18	16.68	2.0	3.0	13.18	3.73	12.39	16.12	2.93	1.80	12.39	3.00	12.50
Propane, C_3H_8	0.1196	8.365	2590	2385	21,661	19,944	5.0	18.82	23.82	3.0	4.0	18.82	3.63	12.07	15.70	2.99	1.68	12.07	2.12	9.35
n-Butane, C_4H_{10}	0.1582	6.321	3370	3113	21,308	19,680	6.5	24.47	30.97	4.0	5.0	24.47	3.58	11.91	15.49	3.03	1.55	11.91	1.86	8.41
Isobutane, C_4H_{10}	0.1582	6.321	3363	3105	21,257	19,629	6.5	24.47	30.97	4.0	5.0	24.47	3.58	11.91	15.49	3.03	1.55	11.91	1.80	8.44
n-Pentane, C_5H_{12}	0.1904	5.252	4016	3709	21,091	19,517	8.0	30.11	38.11	5.0	6.0	30.11	3.55	11.81	15.35	3.05	1.50	11.81	—	—
Isopentane, C_5H_{12}	0.1904	5.252	4008	3716	21,052	19,478	8.0	30.11	38.11	5.0	6.0	30.11	3.55	11.81	15.35	3.05	1.50	11.81	—	—
Neopentane, C_5H_{12}	0.1904	5.252	3993	3693	20,970	19,396	8.0	30.11	38.11	5.0	6.0	30.11	3.55	11.81	15.35	3.05	1.50	11.81	—	—
n-Hexane, C_6H_{14}	0.2274	4.398	4762	4412	20,940	19,403	9.5	35.76	45.26	6.0	7.0	35.76	3.53	11.74	15.27	3.06	1.46	11.74	1.18	7.40
Olefin series																				
Ethylene, C_2H_4	0.0746	13.412	1614	1513	21,644	20,295	3.0	11.29	14.29	2.0	2.0	11.29	3.42	11.39	14.81	3.14	1.29	11.39	2.75	28.60
Propylene, C_3H_6	0.1110	9.007	2336	2186	21,041	19,691	4.5	16.94	21.44	3.0	3.0	16.94	3.42	11.39	14.81	3.14	1.29	11.39	2.00	11.10
n-Butene, C_4H_8	0.1480	6.756	3084	2885	20,840	19,496	6.0	22.59	28.59	4.0	4.0	22.59	3.42	11.39	14.81	3.14	1.29	11.39	1.75	9.70
Isobutene, C_4H_8	0.1480	6.756	3068	2869	20,730	19,382	6.0	22.59	28.59	4.0	4.0	22.59	3.42	11.39	14.81	3.14	1.29	11.39	—	—
n-Pentene, C_5H_{10}	0.1852	5.400	3836	3586	20,712	19,363	7.5	28.23	35.73	5.0	5.0	28.23	3.42	11.39	14.81	3.14	1.29	11.39	—	—
Aromatic series																				
Benzene, C_6H_6	0.2060	4.852	3751	3601	18,210	17,480	7.5	28.23	35.73	6.0	3.0	28.23	3.07	10.22	13.30	3.38	0.69	10.22	1.40	7.10
Toluene, C_7H_8	0.2431	4.113	4484	4284	18,440	17,620	9.0	33.88	42.88	7.0	4.0	33.88	3.13	10.40	13.53	3.34	0.78	10.40	1.27	6.75
Xylene, C_8H_{10}	0.2803	3.567	5230	4980	18,650	17,760	10.5	39.52	50.02	8.0	5.0	39.52	3.17	10.53	13.70	3.32	0.85	10.53	1.00	6.00
Miscellaneous gas																				
Acetylene, C_2H_2	0.0697	14.344	1499	1448	21,500	20,776	2.5	9.41	11.91	2.0	1.0	9.41	3.07	10.22	13.30	3.38	0.69	10.22	—	—
Napthalene, $C_{10}H_8$	0.3384	2.955	5854	5654	17,298	16,708	12.0	45.17	57.17	10.0	4.0	45.17	3.00	9.97	12.96	3.43	0.56	9.97	—	—
Methyl alcohol, CH_3OH	0.0846	11.820	868	768	10,259	9,078	1.5	5.65	7.15	1.0	2.0	5.65	1.50	4.98	6.48	1.37	1.13	4.98	6.72	36.50
Ethyl alcohol, C_2H_5OH	0.1216	8.221	1600	1451	13,161	11,929	3.0	11.29	14.29	2.0	3.0	11.29	2.08	6.93	9.02	1.92	1.17	6.93	3.28	18.95
Ammonia, NH_3	0.0456	21.914	441	365	9,668	8,001	0.75	2.82	3.57	—	1.5	3.32	1.41	4.69	6.10	—	1.59	5.51	15.50	27.00
Sulfur, S[a]	—	—	—	—	3,983	3,983	1.0	3.76	4.76	SO_2 1.0	—	3.76	1.00	3.29	4.29	SO_2 2.00	—	3.29	—	—
Hydrogen sulfide, H_2S	0.0911	10.979	647	596	7,100	6,545	1.5	5.65	7.15	1.0	1.0	5.65	1.41	4.69	6.10	1.88	0.53	4.69	4.30	45.50
Sulfur dioxide, SO_2	0.1733	5.770	—	—	—	—	—	—	—	—	—	—	—	—	—	—	—	—	—	—
Water vapor, H_2O	0.0476	21.017	—	—	—	—	—	—	—	—	—	—	—	—	—	—	—	—	—	—
Air	0.0766	13.063	—	—	—	—	—	—	—	—	—	—	—	—	—	—	—	—	—	—
Gasoline	—	—	—	—	—	—	—	—	—	—	—	—	—	—	—	—	—	—	1.40	7.60

[a]Carbon and sulfur are considered as gases for molal calculations only.

Source: Adapted from *Fuel Flue Gases*, American Gas Association, *Combustion Flame and Explosions of Gases*, 1951.

TABLE 2. Thermal Oxidation Parameters[1]

Compound	A	E, cal/g-mol
Acrolein	3.30×10^{10}	35900
Acrylonitrile	2.13×10^{12}	52100
Allyl alcohol	1.75×10^{6}	21400
Allyl chloride	3.89×10^{7}	29100
Benzene	7.43×10^{21}	95900
Butene-1	3.74×10^{14}	58200
Chlorobenzene	1.34×10^{17}	76600
Cyclohexane	5.13×10^{12}	47600
1-2 Dichloroethane	4.82×10^{11}	45600
Ethane	5.65×10^{14}	63600
Ethanol	5.37×10^{11}	48100
Ethyl acrylate	2.19×10^{12}	46000
Ethylene	1.37×10^{12}	50800
Ethyl formate	4.39×10^{11}	44700
Ethyl mercaptan	5.20×10^{5}	14700
Hexane	6.02×10^{8}	34200
Methane	1.68×10^{11}	52100
Methyl chloride	7.34×10^{8}	40900
Methyl ethyl ketone	1.45×10^{14}	58400
Natural gas	1.65×10^{12}	49300
Propane	5.25×10^{19}	85200
Propylene	4.63×10^{8}	34200
Toluene	2.28×10^{13}	56500
Triethylamine	8.10×10^{11}	43200
Vinyl acetate	2.54×10^{9}	35900
Vinyl chloride	3.57×10^{14}	63300

If the initial concentration is $C_{A,o}$ the solution to equation 2 is as follows:

$$\ln(C_A/C_{A,o}) = -kt \tag{3}$$

Thus, for a typical afterburner with mole fractions in the range of 0.15 for oxygen and 0.001 for HC, equation 3 is often used to model the kinetics. In this model, the rate constant is usually presumed to be of Arrhenius form, that is,

$$k = Ae^{-E/RT} \tag{4}$$

where: A = pre-exponential factor, s^{-1}
E = activation energy, cal/g-mol
R = universal gas constant, 1.987 cal/g-mol °K
T = absolute temperature, °K

Although this pseudo–first-order model is often used to describe HC combustion reactions (especially in afterburner applications), it is by no means universal. Some kinetic data that have been reported in the literature are summarized in Table 2.[1] Only models that are first order in the HC are reported in this table.

DESCRIPTION OF EQUIPMENT

Equipment used to control waste gases by combustion can be divided into three categories: direct combustion or flaring, thermal oxidation, and catalytic oxidation. A direct combustor or flare is a device in which air and all the combustible waste gases react at the burner (see Figure 2). Complete combustion must occur instantaneously since there is no residence chamber. Therefore, the flame temperature is the most important variable in flaring waste gases. In contrast, for thermal oxidation, the combustible waste gases pass over or around a burner flame into a residence chamber where oxidation of the waste gases is completed (see Figure 3). Catalytic oxidation is very similar to thermal oxidation. The main difference is that after passing through the flame area, the gases pass over a catalyst bed, which promotes oxidation at a lower temperature than does thermal oxidation (see Figure 4). Details on these three control devices are given below.

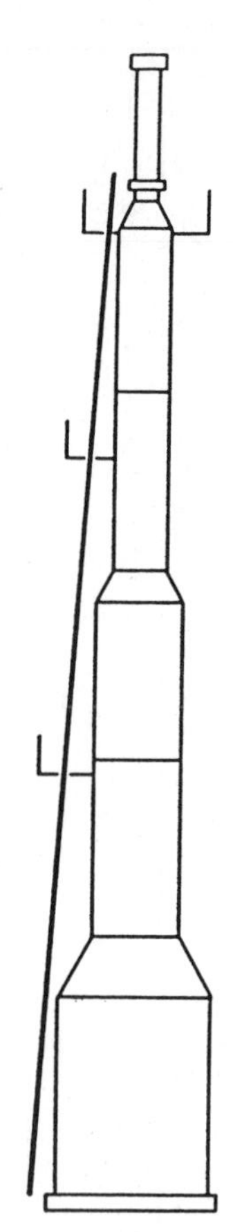

FIGURE 2. Elevated Flare Schematic

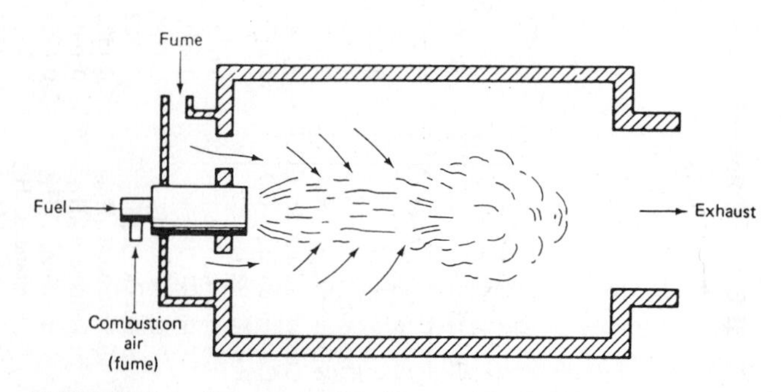

FIGURE 3. Schematic of a Thermal Incinerator

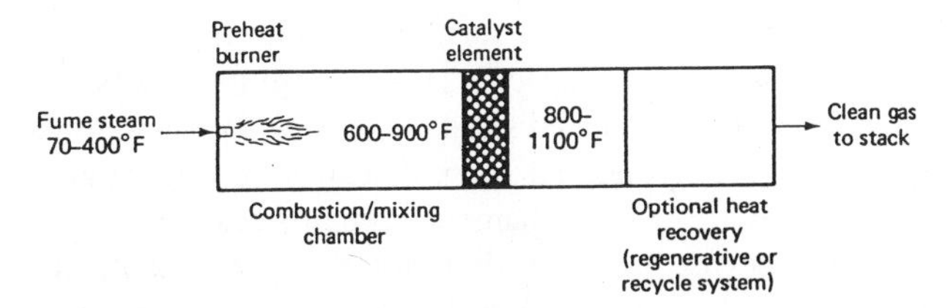

FIGURE 4. Schematic of a Catalytic Incinerator

Thermal oxidizers or afterburners can be used over a fairly wide, but low, range of organic vapor concentrations. The concentration of the organics in air must be substantially below the lower flammable level (lower explosive limit). As a rule, a factor of four is employed for safety precaution. Reactions are conducted at elevated temperatures in order to ensure high chemical reaction rates for the organics. To achieve this temperature, it is necessary to preheat the feed stream with auxiliary energy. Along with the contaminant-laden gas stream, air and fuel are continuously delivered to the reactor, where the fuel is combusted with air in the firing unit (burner). The burner may utilize the air in the process waste stream as the combustion air for auxiliary fuel, or it may use a separate source of outside air for this purpose. The products of combustion and the unreacted feed stream are intensely mixed and enter the reaction zone of the unit. The pollutants in the process gas stream are then reacted at the elevated temperature. The unit requires operating temperatures in the 1200–2000°F range for combustion of most pollutants. A residence time of 0.2 to 2.0 seconds is recommended in the literature, but this factor is primarily dictated by kinetic considerations. A length-to-diameter ratio of 2.0 to 3.0 is usually employed. The end products are continuously discharged at the outlet of the reactor. The average gas velocity can range from as low as 10 fps to as high as 50 fps. (The velocity increases from inlet to outlet due to the increase in the number of moles of the reacting fluid and the increase in temperature due to reaction.) These high velocities are required to prevent settling of particulates (if present) and to minimize the dangers of flashback and fire hazards. The fuel is usually natural gas. The energy liberated by the reaction may be directly recovered in the process or indirectly recovered by suitable external heat exchange. This should be included in a design analysis since energy is the only commodity of value that is usually derived from the combustion process.

Because of the high operating temperatures, the unit must be constructed of metals capable of withstanding this condition. Combustion devices are usually constructed with an outer shell that is lined with refractory material. However, refractory material is heavy, with densities as low as 50 lb/ft^3 for lightweight insulating firebrick and as high as 175 lb/ft^3 for castable refractories. Refractory wall thickness is in the 3- to 9-inch range. This weight adds considerably to the cost. Because of its light weight, firebrick wall construction is being used in some units. All-metal construction has found limited application. These combustion reactors also vary in shape from tanks to tubular pipes. The latter type is usually used since it has a high surface-to-volume ratio, which is advantageous for energy recovery and suitable for continuous operation.

The incineration process may be thought of as occurring in two separate stages:

1. Combustion of fuel
2. Combustion of pollutants

The combustion process in the first stage is an extremely rapid and highly irreversible chemical reaction. The oxygen supplied by the (primary) air may be in excess or obtained directly from the process gas stream (secondary air). The carbon content of the fuel burns almost quantitatively to carbon dioxide and the hydrogen to water. Combustion reactors should be operated with the least amount of excess air compatible with the need for complete combustion and/or fuel requirements. The mixing of the air and the fuel in the burner section of the reactor also determines the completeness of combustion. One can show that any fuel completely burned without excess air produces a discharge gas containing the maximum CO_2 possible for that fuel. Thus with no excess air and perfect mixing, combustion will be complete, with the resulting gas containing maximum CO_2 and no O_2, CO, and H_2. Since the reaction is extremely rapid, the combustion of the fuel occurs in a rather narrow zone in the reactor, perhaps within a foot of reactor length at entrance conditions. The average residence time approaches zero. The energy liberated on combustion of the fuel is used to heat the combustion gases. Under these conditions, the rate of the combustion reaction is dictated by the fuel rate; kinetics effects need not be considered. The calculations reduce to one of stoichiometry and overall mass and energy balances.

However, mixing of the fuel combustion gases and the process gas stream does not occur instantaneously. A finite residence time (or reactor length) is required to achieve "complete" system mixing that will yield a uniform temperature profile through the cross section of the reactor. This is a critical design and operational consideration, as failure to achieve the above can result in incomplete combustion, high(er) fuel requirements, or both. In fact, variations in performance, in many cases, can be either directly or indirectly attributed to this mixing process. Care must be exercised in this part of the operation since flame quenching can result if the mixing process is too severe. Once mixing is complete and the temperature uniform, the so-called residence requirement is applicable to the combustion of the pollutant in the process gas stream. Thus the residence time specified in the codes and regulations applies to both the mixing process, where the process gas stream and fuel combustion products achieve a uniform temperature, and the combustion process for the pollutant(s), which is assumed to occur at this elevated temperature.

In the second stage of the process, the heated gases from the burner pass through the reactor, where the pollutants (organics) are reacted (oxidized) to harmless end products.

Although the pollutants also serve as a source of fuel, enthalpy of reaction effects is often small enough to be neglected so that the calculation of conversion reduces to a kinetic problem. However, the reaction mechanism is quite complex. It is difficult to obtain and interpret experimental data for combustion reactions. In fact, it is common practice to treat a mixture of hydrocarbons in terms of a single hydrocarbon component. Since many chemical (combustion) reacting systems of interest to the environmental engineer can be closely approximated by first-order or pseudo–first-order reactions,[2,3] industry usually employs a simple first-order, irreversible reaction mechanism in the design calculation. (It is well known that even zero- or second-order reactions can often be satisfactorily represented by first-order reactions.) The reaction velocity constant is an empirically determined parameter that is applicable over the desired range of operating conditions. Although this approach is displaced from earlier presentations, the procedure has merit and is thoroughly justifiable if the resulting equation works.

Catalytic reactors are an alternative to thermal reactors or afterburners. If a solid catalyst is added to the reactor, the reaction is said to be heterogeneous. For simple reactions, the effect of the presence of a catalyst is to:

1. Increase the rate of reaction.
2. Permit the reaction to occur at a lower temperature.
3. Permit the reaction to occur at a more favorable pressure.
4. Reduce the reactor volume.
5. Increase the yield of a reactant(s) relative to the other(s).

In a typical catalytic reactor, the gas stream is delivered to the reactor continuously by a fan at a velocity in the 10- to 30-fps range but at a lower temperature—usually in the 650°F to 800°F range—than a thermal unit. A length-to-diameter ratio less than 0.5 is usually employed. To achieve destruction efficiencies in the 90–95% range, approximately 1.5–2.0 ft^3 of catalyst is required per 1000 scfm (exhaust stream plus supplementary fuel combustion products). The gases, which may or may not be preheated, pass through the catalyst bed where the reaction occurs. The combustion products, which are again made up of water vapor, carbon dioxide, inerts, and unreacted vapors, are continuously discharged from the outlet at a higher temperature. Energy savings can again be achieved with heat recovery from the exit stream. Overall heat transfer coefficients in the heat-recovery exchangers typically range between 2 and 8 Btu/h-ft^2-°F.

Metals in the platinum family are recognized for their ability to promote combustion at low temperatures. Other catalysts include various oxides of copper, chromium, vanadium, nickel, and cobalt. These catalysts are subject to poisoning, particularly from halogens, halogen and sulfur compounds, zinc, arsenic, lead, mercury, and particulates. High temperatures can reduce catalyst activity. It is, therefore, important that catalyst surfaces be clean and active to ensure optimum performance. Catalysts can usually be regenerated with superheated steam. Catalysts may be porous pellets, usually cylindrical or spherical in shape, ranging from 1/16 to 1/2 inch in diameter. Small sizes are recommended, but the pressure drop through the reactor increases. Other shapes include honeycombs, ribbons, and wire mesh. Since catalysis is a surface phenomenon, a physical property of these particles is that the internal pore surface is infinitely greater than the outside surface. The reader is referred to the literature[4] for more information on catalyst preparation, properties, comparisons, costs, and impurities.

From a macroscopic point of view, the following sequence of steps is involved in the catalytic conversion of reactants to products:

1. Transfer of reactants to and products from the outer catalyst surface.
2. Diffusion of reactants and products within the pore of the catalyst.
3. Active adsorption of reactants and the desorption of the products on the active centers of the catalyst.
4. Reaction(s) on active centers on the catalyst surface.

At the same time, energy effects arising due to chemical reaction can result in the following:

1. Heat transfer to or from active centers to the catalyst particle surface.
2. Heat transfer to and from reactants and products within the catalyst particle.
3. Heat transfer to and from moving streams in the reactor.
4. Heat transfer from one catalyst particle to another within the reactor.
5. Heat transfer to or from the walls of the reactor.

Some of the advantages of catalytic combustion reactors over afterburners include the following:

1. Lower fuel requirements
2. Lower operating temperatures
3. Little or no insulation requirements
4. Reduced fire hazards
5. Reduced flashback problems

The disadvantages include the following:

1. Initial cost is high.
2. Catalyst poisoning is possible.
3. Particulate often must first be removed.
4. There may be a spent catalyst disposal problem.

The basic problem in the design of a heterogeneous reactor, as in the case of a catalytic combustion unit, is to determine the quantity of catalyst and/or reactor size required for a given conversion and flow rate. In order to obtain this, information on the rate equation(s) and their parameter(s) must be made available. A rigorous approach to the evaluation of these reaction velocity constants, and so forth, has yet to be accomplished. At this time, industry still relies on the procedures set forth earlier.

Although not treated in any detail in this chapter, flares provide another means of control. If the concentration of the organics in air equals or exceeds the lower flammability level, a flare unit may be employed. Elevated, ground-level, and open-pit flares may be used. However, ground-level flares have been outlawed in most states and open-pit burning is used only under special conditions. Elevated flares are usually employed to insure sufficient dilution and subsequent dispersion from adjacent structures and potential receptors of the energy and end products of combustion. Flares find their primary application in the petroleum and petrochemical industries. The suggested design procedures are included in the equations. In operation, the gas containing the organics is continuously fed to and discharged from a stack, with the combustion occurring near the top of the stack and characterized by a flame at the end of the stack. The discharge temperature is in the 1500–3000°F range. The combustion flame is located at a point where the velocity of flow equals the velocity of flame propagation, provided the mixture is at or above the ignition temperature. Good mixing and a H/C ratio in excess of 0.3 in the process gas stream can help insure proper combustion. A blue flame, appearing colorless against a blue-sky background, indicates good operation; a yellow-orange flame with a trail of black smoke indicates poor operation. The addition of steam (jets) at the top of the stack can help remove operational problems resulting from incomplete combustion. Steam rates in the range of 0.1 to 0.5 pound of steam per pound of process gas are often employed. Operating stack velocities are in excess of the flame propagation rate, and usually exceed 200 fps. The design of this unit also includes flame arresters in the stack to help remove some of the flashback and explosion possibilities. The diameter of the stack is obtained from flow (velocity) considerations and the height from atmospheric dispersion calculations.

DESIGN AND PERFORMANCE EQUATIONS

There are two key calculations associated with combustion devices. These include determining:

1. The fuel requirements
2. The physical dimensions of the unit

Both of these calculations are interrelated. The general procedure to follow, with pertinent equations, is given below. It is assumed that the process gas stream flow rate, inlet temperature, and combustion temperature are known and that the required residence time has been specified.

1. Calculate the heat load required to raise the process stream from its inlet temperature to the operating temperature of the combustion device.

$$Q = \Delta H \qquad (5)$$

2. Correct the heat load term for any radiant losses, *RL*.

$$Q = (1 + RL)(\Delta H); \; RL = \text{fractional basis} \qquad (6)$$

3. Assuming natural gas of known heating value HV_G as fuel, calculate the available heat at the operating temperature. For engineering purposes, one may use a short-cut method that bypasses these detailed calculations.

$$HA_T = (HV_G)(HA_T/HV_G)_{ref} \qquad (7)$$

where the subscript ref refers to a reference fuel. For natural gas with a reference HV_G of 1059 Btu/scf, the available heat (assuming stoichiometric air) is given by the following:[5]

$$(HA_T)_{ref} = -0.237(T) + 981; \; T = °F \qquad (8)$$

4. Calculate the natural gas required, *NG*.

$$NG = Q/HA; \text{ consistent units} \qquad (9)$$

5. Determine the volumetric flowrate of *both* the process gas stream, *p*, and the flue products of combustion of the natural gas, *q*, at *the operating temperature*.

$$q_T = q_p + q_c \qquad (10)$$

A good estimate for q_c is

$$q_c = (11.5)(NG) \qquad (11)$$

6. The diameter of the combustion device is given by:

$$S = q_T/v_t \qquad (12)$$

where v_t is the velocity of the hot gases in the incinerator.

OPERATION AND MAINTENANCE[6]

There are three distinct categories of incinerator operation:

1. Automatic
2. Semiautomatic
3. Manual

The categories refer specifically to the method of start-up and then flame supervision, and have been adopted by the various insurance agencies (such as Factory Mutual) and by the National Fire Protection Association (NFPA). Most incinerator burners and fuel systems are designed in accordance with one of these insurance agencies' guidelines, whether or not insurance is actually obtained from that agency. Therefore, these publications should be consulted before attempting to operate the equipment.

Typically, the following procedure must be followed to start up any thermal incinerator. Each function, or group of functions, is interlocked electrically so that each must be accomplished before the next step can occur:

1. Prepurge
   a. Power on to control system
   b. Fan motor on
   c. Airflow verified
   d. Fuel valves closed
2. Purge
   a. Airflow to maximum rate
   b. Purge timer starts (length of time set to allow furnace to be purged of any combustible fumes; eight complete air changes considered sufficient)
3. Ignitor sequence
   a. Airflow to light-off rate
   b. Auxiliary fuel pressure verified as sufficient; temperature verified, if required
   c. Ignitor electric spark mechanism starts
   d. Ignitor timer starts
   e. Ignitor fuel valve opens, fuel is ignited in furnace by sparking mechanism
   f. Ignitor flame sensed by flame scanner
   g. Ignitor timer ends time cycle, shuts off sparking mechanism (if no flame sensed by flame scanner, ignitor fuel valve immediately closes, and postpurge cycle starts)
4. Main burner sequence
   a. Main burner fuel-control valve verified in light-off position
   b. Atomizing fluid (air or stream) valve opens (needed only on certain oil-fired units)
   c. Burner timer starts cycle
   d. Main burner fuel shut off valves open—allowing fuel to main burner
   e. Main flame ignites from ignitor flame
   f. Burner timer ends time cycle, ignitor fuel valve closes (if main flame not sensed by flame scanner, fuel shutoff valves immediately close, and postpurge cycle starts)
   g. Interlocks verifying light-off positions on fan and control valves are now by-passed, allowing fuel and air rates to be adjusted
   h. Main burner is now in service subject to the following conditions:
      (1) Fan on
      (2) Airflow proven
      (3) Flame detected
      (4) Fuel pressure adequate
      (5) Furnace temperature not high
5. Normal operation
   a. Fuel and airflow are now adjusted to provide optimum furnace temperature for incinerator; this may be done manually or automatically; on refractory-lined
      units, warm-up must be done gradually to prevent thermal shock to the refractory
   b. Once the incinerator reaches its operating temperature, fumes can be introduced into the furnace
   c. Proper incinerator operation is maintained by controlling incinerator outlet temperature and fume, auxiliary fuel, and airflow rates
6. Normal shutdown
   a. Normal shutdown is accomplished by closing the fume flow valves, then the auxiliary fuel valves; the systems then proceed to the postpurge cycle
7. Postpurge
   a. Fan adjusted to maximum flow rate
   b. Postpurge timer starts time cycle
   c. Keeps air flowing until five complete air changes have occurred in the furnace
   d. Postpurge timer ends cycle, fan turned off, incinerator is shut down

An automatic system requires only that an operator throw a switch to start operation. The control mechanism then starts the incinerator, including the fans and other related equipment, in a presequenced order; brings it up to temperature; and maintains proper incinerator operation for as long as required. Shutdown is usually initiated by manually throwing another switch; the control system will then carry out all the operations necessary to shut down the incinerator.

A distinguishing feature of the "automatic" system is its ability to "recycle." This includes recycling to reattempt light-off if the previous attempt failed, and shutting down or starting up again as fume flow dictates. This is not a feature of the semiautomatic system.

Semiautomatic operation is the most common method. It involves considerably more human input than does the completely automatic operation. In this type of operation, the operator must perform certain operations when called for by the control system. This input is especially required during the start-up sequence. Semiautomatic control is a broadly applied term and may refer to a system that requires the operator to operate valves manually or merely to push certain buttons. However, in all cases, proper sequencing is enforced by the control systems. Emergency shutdown, however, is completely automatic.

In the "manual" type of control system, all the actions required by the start-up sequence are performed manually. Proper sequencing must be followed, but the control logic allows the operator sufficient time to accomplish the tasks. Again, emergency shutdown is automatic.

The length of time required to bring the incinerator from "cold" to its normal operating temperature varies with the type and thickness of the lining and with the actual operating temperature. Typically, the warm-up period may require several hours. For a thick lining (6 to 12 inches of refractory plus any insulating block), the manufacturer's recommendation is usually to start initially with an outlet temperature of 200°F and hold for one hour, and thereafter, to increase the outlet temperature at the rate of 100°F per hour until the outlet temperature is 600°F. Then the outlet temperature may be increased at the rate of 200–400°F per hour until the final operating temperature is reached. This process may take an entire shift and wastes considerable amounts of fuel; it is one of the principal disadvantages of using a refractory-lined furnace.

In actual practice, however, the manufacturer's recommended warm-up rate schedule is seldom followed, especially for relatively thin linings and for incinerators that operate only on an 8- to 16-hour shift. Warm-up periods of an hour or less are certainly not uncommon, regardless of lining thickness. The penalty for this is increased inspection and maintenance (patching with plastic) of the lining; usually twice a week is sufficient. Certainly, reduced refractory life can also be expected from such a procedure, but this must be balanced against the wasted fuel usage required for longer warm-up periods. Using short warm-up periods, one can expect to replace the lining approximately once a year compared with a 10- to 15-year interval using the recommended procedure. A shortened warm-up period should *never* be used for the initial curing period or if the refractory has been wetted or soaked with water. A quick warm-up period under these circumstances will result in the immediate loss of the lining.

No special cool-down procedure is usually followed. Usually, the incinerator and fans are simply turned off and allowed to cool naturally. Under these conditions, it will take from 8 to 24 hours for the refractory to reach ambient temperature; however, the lining will usually be cool enough for inspection after a 4- to 12-hour period. Leaving the fan on during the cool-down period usually does not cause a serious problem, but should be avoided, as it does reduce refractory life. Water should *never* be sprayed on the refractory to cool it down. Although this may seem obvious, it nevertheless happens quite frequently if the maintenance people have not been properly trained.

Normal operation of a fume incinerator should be quite simple. A controller should be incorporated into the design to maintain the outlet temperature at a fixed value by varying the auxiliary fuel input. Combustion air (assuming limited air in the fume stream) is usually controlled by the average amount of fumes to be incinerated. This adjustment is normally set manually if the fume flow rate and fume heat content are fairly constant. If the fume heat content or flow rate is likely to be highly variable, a more sophisticated control system may be appropriate—one that analyzes combustion efficiency at the outlet as well as fume flow and outlet temperature, and varies both auxiliary fuel input and combustion air accordingly. But normal operation of even a complicated control system usually requires nothing more than occasional monitoring of the instruments.

Most incinerators are custom-designed within certain basic parameters. Therefore, they are likely to be accompanied by a very complete instruction manual, which should include the manufacturer's basic maintenance instructions from all subcontractors. That is, the incinerator manufacturer will have bought equipment, such as the refractory, valves, and controls, from other suppliers. The operating and maintenance instructions for this equipment will be quite extensive and complete because it has been written by the original manufacturer, who is concerned only with particular items. The instruction manual is, therefore, a very useful document from a maintenance viewpoint and should be followed carefully. (The instruction manual may not be so useful from a *system* operations viewpoint because incinerator systems are usually custom-designed and so *system* problems cannot always be anticipated, which often results in some field modifications to the operating instructions.)

There are, however, some general maintenance guidelines that can be discussed. The refractory should be inspected on a regular basis. Cracks that may develop, especially in brick joints, and thermal shock damage (spalling) should be repaired with a suitable plastic of the same thermal properties. The burner should be inspected at regular intervals for signs of warpage or corrosion. Moving parts should be lubricated with graphite or a similar high-temperature lubricant. Lubricants that carbonize under any circumstances should not be used. Also, any dirt, mortar, carbon, or other foreign matter should be cleared from the burner area. The pressure seals around any parts projecting through the burner or incinerator shell should be inspected. Usually, these are asbestos rope packing glands and should be fairly tight after the adjusting/retaining screws have been loosened. These seals should be lubricated only with flake or powder graphite. This should never be mixed with oil, as the oil will carbonize. On burners using gas as an auxiliary fuel, the gas jets should be free of corrosion and should be cleared of any deposits.

The outer shell of the incinerator should be inspected, especially when new or when a new lining has been installed, for signs of thermal shock. That is, welds, especially at the outlet, should be checked for hairline cracks, which are the first signs of poor thermal design.

The auxiliary fuel piping train should be inspected in accordance with the manufacturer's instructions. Electrically operated valves and interlock switches should be inspected frequently for conditions that might cause "shorting" (e.g., dirty contacts, moisture leaks, deteriorating insulation). Air-supply lines and filters (to air-operated valves) should be inspected for dirt or blockages. The valves themselves are usually provided with air-signal and air-supply pressure gauges, and these should be checked occasionally for accuracy.

If there are shutoff dampers in the ductwork to or from the incinerator, their seals should be checked frequently.

Maintenance procedures for catalytic incinerators should include catalyst cleaning every three months to a year. Cleaning is usually accomplished by blowing clean compressed air through the catalyst element, by vacuuming, or by washing with water or a mild detergent not containing phosphates. Iron oxide deposits can be removed by soaking with a mild organic acid, followed by a water rinse.

MAINTAINING AND IMPROVING EQUIPMENT PERFORMANCE[6]

Improving the performance of a thermal incinerator basically involves the optimization of fume combustion. Ideally, no more combustion air should be used than is required for complete combustion of the fumes and the auxiliary fuel. The auxiliary fuel should be used only in amounts required to maintain the design furnace temperature.

An incinerator operating efficiently may have only 1 to 2% oxygen and 0 to 1% combustibles in the outlet gases. Monitors are available that can indicate these parameters to the operator, as well as provide automatic control of the incinerator when required.

ILLUSTRATIVE EXAMPLES

Example 1

Determine the volumetric flow rate (in standard cubic feet per minute) of natural gas required to heat 20,000 scfm (60°F) of an exhaust process gas stream from 200°F to 1800°F. Assume that there are no heat losses and that the natural gas fuel has an available heat, HA, of 950 Btu/scf. The average heat capacity of air over the temperature range may be assumed to be equal to 7.5 Btu/lb mol-°F.

Solution Calculate the molar flow rate, N, of the exhaust process gas.

$$\begin{aligned} N &= (20{,}000 \text{ scfm})(1 \text{ lb mol}/379 \text{ scf}) \\ &= 52.8 \text{ lb mol/min} \end{aligned}$$

Calculate the heat required to incinerate the exhaust gas.

$$\begin{aligned} Q &= N\bar{C}_p\Delta T \\ &= (52.8)(7.5)(1800 - 200) \\ &= 6.34 \times 10^5 \text{ Btu/min} \end{aligned}$$

Finally, calculate the amount of natural gas required, NG. For engineering purposes, this may be simply calculated by dividing the above result by the available heat.

$$\begin{aligned} NG &= Q/HA = 6.34 \times 10^5/950 \\ &= 667 \text{ scfm} \end{aligned}$$

For most afterburner systems, approximately 150 scfm of natural gas is required to heat 5000 scfm process gas stream. This is a good number to remember for purposes of (estimating) engineering calculations.

Example 2

Consider a 1000 scfm (60°F) effluent at 150°F that needs to be treated to control a very low concentration of a gaseous pollutant at the parts-per-million level. This is to be accomplished in an afterburner with an incineration at 1200°F for at least 0.3 second. The intake combustion air is at 60°F. Assuming a length-to-diameter ratio of 2.0 and a throughput velocity of 12 fps, calculate:

1. The natural gas (HV_G = 1059 Btu/scf) required for preheating the contaminated effluent to 1200°F using all fresh combustion air intake (primary air).
2. The afterburner throat diameter to give 20 fps throat velocity for good mixing. Approximately 11.0 scf of combustion gases are produced from 1.0 scf of natural gas combusted.
3. The length of the afterburner.

Solution

1. Calculate the waste effluent mass flow rate. Assume the waste effluent to have the physical properties of air.

$$\begin{aligned} W &= (q_w)(\rho) \\ &= (1000)(0.0766 \text{ lb/scf}) \\ &= 76.6 \text{ lb/min} \\ &= 4596 \text{ lb/hr} \end{aligned}$$

Obtain the enthalpy of air at 1200 and 150°F:

$$H \text{ at } 1200°\text{F}, H_1 = 294.22 \text{ Btu/lb}$$

$$H \text{ at } 150°\text{F}, H_2 = 28.4 \text{ Btu/lb}$$

Calculate the heat required to increase the effluent waste stream temperature from 150°F to 1200°F, allowing for 10% loss.

$$\begin{aligned} Q &= (1.10)(W)(H_1 - H_2) \\ &= (1.1)(4596)(294.2 - 28.4) \\ &= 1.34 \times 10^6 \text{ Btu/h} \end{aligned}$$

Calculate the available heat of the natural gas at 1200°F. For purposes of engineering calculation, the available heat for a natural gas with a HHV of 1059 Btu/scf may be calculated using the following equation.[5]

$$\begin{aligned} HA_T &= (-0.237)(\text{T}) + 981; \; T = °\text{F} \\ &= (-0.237)(1200) + 981 \\ &= 697 \text{ Btu/scf} \end{aligned}$$

Calculate the natural gas required, NG.

$$NG = Q/HA_T$$
$$= 1.929 \times 10^3 \text{ scf of natural gas per hour}$$

2. Calculate the volumetric flow rate of combustion products at 1200°F, q.

$$q_c = (NG)(11.0 \text{ scf of combustion gas/scf of natural gas})(460 + 1200)/(460 + 60)$$
$$= 67{,}740 \text{ ft}^3/\text{h}$$
$$= 18.8 \text{ ft}^3/\text{s}$$

Calculate the volumetric flow rate of waste effluent at 1200°F.

$$q_{w,1200°F} = (1000 \text{ scfm})(460 + 1200)/(460 + 60)$$
$$= 3192 \text{ ft}^3/\text{min}$$
$$= 53.2 \text{ ft}^3/\text{s}$$

Obtain the total volumetric flow rate of gases to the afterburner.

$$q_T = q_c + q_{w,1200°F}$$
$$= 72.0 \text{ ft}^3/\text{s}$$

Determine the afterburner throat diameter.

$$S = \pi D_t^2/4 = q_T/v_t$$
$$D_t = (4q_T/\pi v_t)^{0.5}$$
$$= 2.14 \text{ feet}$$

3. Calculate the length of the afterburner chamber.

$$L = 2D_t$$
$$= 4.3 \text{ feet}$$

Determine whether the residence time is greater than the required minimum residence time of 0.3 second.

$$= L/V_{\text{throughput}}$$
$$= 4.3/12$$
$$= 0.36 \text{ second}$$

The design is therefore satisfactory.

Note that since the pollutant concentration in the gas stream is very low (in the parts-per-million range), there is no contribution to the heating value associated with the combustion process. Also, note that the effluent gas stream has been assumed to have the same physical and chemical properties as air. Finally, combustion of the fuel was accomplished with fresh (primary) air intake. Energy savings could be realized by employing secondary air—air from the waste effluent stream—for combustion. This would reduce fuel requirements.

Example 3

You must review plans for a permit to construct a direct-flame afterburner serving a lithographer. Review is for the purpose of judging whether the proposed system, when operating as it is designed to operate, will meet emission standards. The permit application provides operating and design data. Agency experience has established design criteria that if met in an operating system, typically ensure compliance with standards. Operating and design data from the permit application are given in the following.

Effluent exhaust volumetric flow rate = 7000 scfm (60°F, 1 atm)
Exhaust temperature = 300°F
Hydrocarbons in effluent air to afterburner (assume hydrocarbons to be toluene) = 30 lb/h
Afterburner entry temperature of effluent = 738°F
Afterburner heat loss = 10%
Afterburner dimensions = 4.2 feet in diameter, 14 feet in length

Agency Design Criteria
Afterburner temperature = 1300°F to 1500°F
Residence time = 0.3 to 0.5 second
Afterburner velocity = 20 to 40 fps

Standard Data
Gross-heating value of natural gas = 1059 Btu/scf of natural gas
Combustion products per cubic foot of natural gas burned = 11.5 scf
Available heat of natural gas at 1400°F = 600 Btu/scf of natural gas
Molecular weight of toluene = 92
Specific heat of effluent gases at 738°F (above 0°F) = 7.12 Btu/lb/mol-°F
Specific heat of effluent gases at 1400°F (above 0°F) = 7.39 Btu/lb/mol-°F
Volume of primary air required to combust 1 ft^3 of natural gas = 10.33 scf air/scf natural gas

Solution The permit application design temperature is already within agency criteria. Therefore, you need be concerned only with whether, under the conditions given, the residence time and the afterburner velocity will be within agency design criteria. Once you determine the volumetric flow rate, you can calculate the afterburner velocity and residence time since the dimensions of the afterburner are given.

First, calculate the molar flow rate of effluent, N.

$$N = (7000 \text{ scfm})/(379 \text{ scf/lb mol})$$
$$= 18.47 \text{ lb mol/min}$$

The heat load can now be calculated.

$$q = N[C_{P2}(T_2 - T_b) - C_{P1}(T_1 - T_b)]$$
$$= 18.47[(7.39)(14 - 0) - (7.12)(738 - 0)]$$
$$= 93{,}740 \text{ Btu/min}$$

Account for a 10% heat loss in the heat load.

$$\text{Actual heat load} = (1.1)(q)$$
$$= (1.1)(93{,}740)$$
$$= 103{,}114 \text{ Btu/min}$$

$$\text{Rate of natural gas required} = \text{heat load/available heat}$$
$$= (103{,}114)/(600)$$
$$= 171.9 \text{ scfm}$$

The total volumetric flow rate is the sum of the process effluent from the lithographer, *plus* the combustion products (flue) of the natural gas.

$$Q_1 = \text{(rate of natural gas)(11.5 scf of combustion products/scf of natural gas)}$$
$$= (171.9)(11.5)$$
$$= 1976 \text{ scfm of combustion products (flue)}$$

Volumetric flow rate of the effluent = 7000 scfm

$$Q_2 = \text{(rate of natural gas)(10.33 scf of air/scf of natural gas)}$$
$$= (171.9)(10.33)$$
$$= 1776 \text{ scfm of air required to combust natural gas}$$

Since primary air is employed in the combustion of the natural gas, Q_2 is not subtracted from Q_T.

$$Q_T = 7000 + Q_1$$
$$= 7000 + 1976$$
$$= 8976 \text{ scfm}$$

The total actual flow is then

$$Q_T = (8976)(1400 + 460)/(60 + 460)$$
$$= 32{,}106 \text{ acfm}$$

The area is

$$S = \pi D^2/4$$
$$= (\pi)(4.2)^2/4$$
$$= 13.85 \text{ ft}^2$$

and the velocity

$$V_t = (32{,}106)/13.85$$
$$= 2318.12 \text{ fpm}$$
$$= 38.6 \text{ fps}$$

The gas resistance time in the incenerator is

$$14/38.6 = 0.36 \text{ sec}$$

The resistance time and afterburner velocity comply with the agency designer criteria.

References

1. K. C. Lee, et al., "Revised model for the prediction of the time-temperature requirements for thermal destruction of dilute organic vapors and its usage for predicting compound destructability," Paper No. 82-5.3, 75th Annual Meeting of the Air Pollution Control Association, New Orleans, June 20–25, 1982.
2. J. Wei and C. D. Prater, "A new approach to first chemical reaction systems", *AIChE J.*, 9:77 (1963).
3. J. Wei and C. D. Prater, *Advances in Catalysis*, Vol. 13, Academic Press, New York, 1962.
4. J. M. Smith, *Chemical Engineering Kinetics*, 2nd ed., McGraw-Hill Book Co., New York, 1972.
5. L. Theodore and J. Reynolds, *Introduction to Hazardous Waste Incineration*, Wiley-Interscience, New York, 1987.
6. J. P. Bilotti and T. K. Sutherland, in *Air Pollution Control Equipment: Selection, Design, Operation and Maintenance*, L. Theodore and A. J. Buonicore Eds.; Prentice-Hall, Englewood Cliffs, NJ, 1982.

3
Control of Particulates

Cyclones and Inertial Separators
Elizabeth Ashbee and Wayne T. Davis
Wet Scrubbers
Kenneth Schifftner and Howard Hesketh
Electrostatic Precipitators
James H. Turner, Phil A. Lawless, Toshiaki Yamamoto, David W. Coy, Gary P. Greiner, John D. McKenna, and William M. Vatavuk
Fabric Filters
John D. McKenna and Dale A. Furlong

CYCLONES AND INERTIAL SEPARATORS

Elizabeth Ashbee and Wayne T. Davis

Inertial separators are widely used for the collection of medium-sized and coarse particles. Their relatively simple construction and absence of moving parts mean that both the capital and the maintenance costs are lower than for baghouses and electrostatic precipitators. However, the efficiency is not as high, and inertial separators are usually used as precleaners upstream of these other devices, to reduce the dust loading and to remove larger, abrasive particles.

The general principle of inertial separation is that the particulate-laden gas is forced to change direction. As the gas changes direction, the inertia of the particles causes them to continue in the original direction and be separated from the gas stream.

Cyclones, where the gas is forced to spin in a vortex through a tube, are the most common type of inertial separators (see Figure 1). They can be used singly, or in a multiple-tube arrangement for large volumes. Centrifugal separators use the same circular motion, but have a rotating vane in the collector itself to both move the air and separate the dust. Baffle chambers (vertical baffles in a gravity settling chamber) also use the inertial principle, as do baffles for mist eliminators.

CYCLONES

Cyclones provide a low-cost, low-maintenance method of removing larger particulates from gas streams. On their

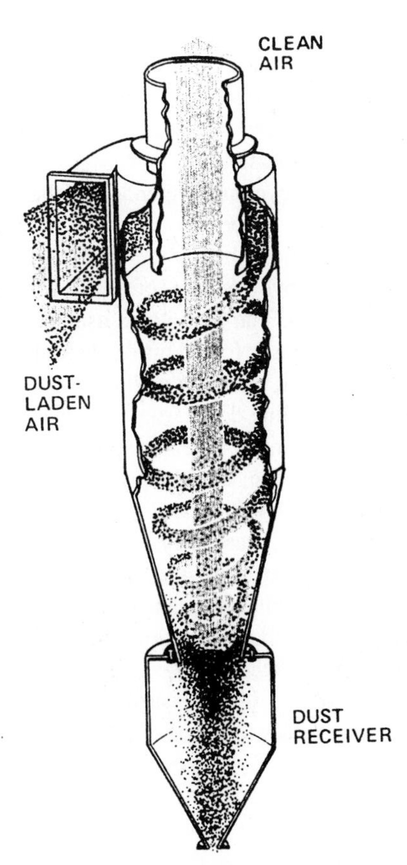

FIGURE 1. Operation of Cyclone (From Reference 10)

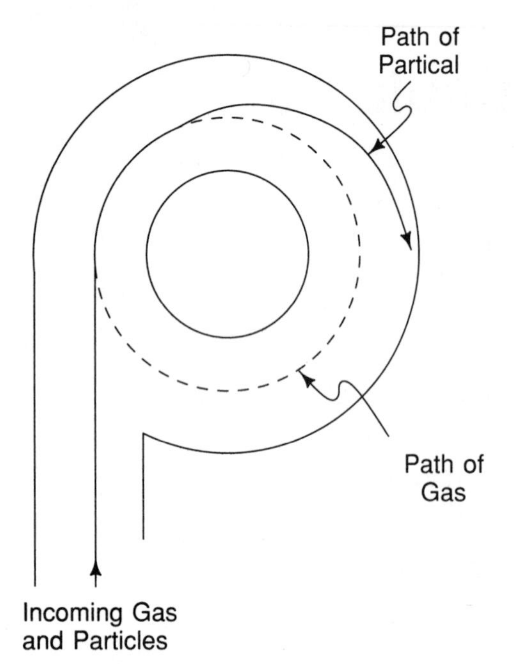

FIGURE 2. Movement of Particles Across Gas Streamlines

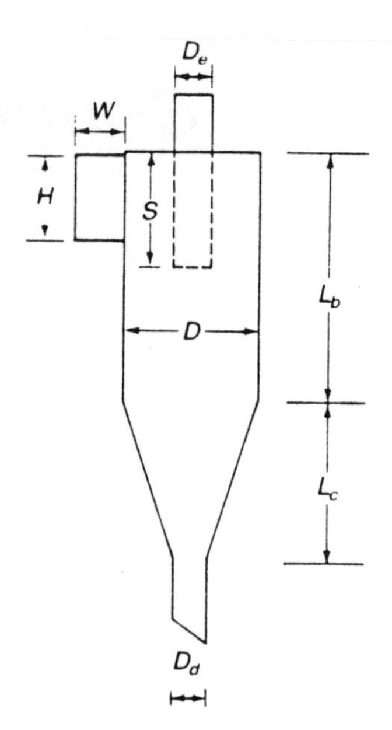

FIGURE 3. Standard Cyclone Dimensions (From Reference 2)

own, they are not usually sufficient to meet stringent emissions standards, but they serve well as precleaners for other, more expensive control devices and for dry-product recovery. Common uses are collection of grinding and machining dust in woodworking plants and tool rooms and in industrial materials handling. A cyclone reduces dust loading and removes larger abrasive particles, in this way extending the life of the fabric filter, which is usually used as the final collector. Multiple-tube cyclone arrangements are used on fossil fuel boilers, where they handle the large volumes effectively as precleaners.

Particles enter the cyclone suspended in the gas stream, which is forced into a vortex by the shape of the cyclone. The inertia of the particles resists the change in direction of the gas and they move radially outward across the gas streamlines (Figure 2). The gas inlet is usually at the top of the cyclone, and the gas is forced into a cyclonic path by the shape of the inlet itself, the outer wall of the device and the cylindrical tube extending down from the top to the cyclone known as the vortex finder. The gas spirals in the cylindrical section of the cyclone, and the particles move outward under the influence of centrifugal force until they strike the body of the cyclone. There they are caught in the thin laminar layer of air next to the wall and are carried downward by gravity to be collected in the dust hopper. When the gas reaches the cone-shaped section, increased rotational gas velocity helps to keep the dust against the wall. The direction of the vortex reverses near the bottom of the cone and the cleaned gas rises to exit through the central tube of the vortex finder.

Cyclones are usually designed with geometric similarity such that the ratios of the dimensions remain constant at different diameters, and these dimensions can be expressed in terms of the body diameter, D_0 (Figure 3). The values of the ratios determine whether the cyclone has conventional proportions or is a high-efficiency or high-throughput type. See Table 1.

The number of turns of the vortex, N_e, can be estimated from the cyclone dimensions, since it depends on the height of one turn of the vortex (i.e., the height of the inlet) and the length of the cyclone:

$$N_e \cong \frac{1}{H}\left[L_b + \frac{L_c}{2}\right] \tag{1}$$

Cyclones can be optimized for high collection efficiency by using small diameters, long cylinders, and high inlet velocities, or for high throughput, with converse dimensions. Smaller diameters, however, increase the pressure drop and hence the cost of operation.

Different shapes of the inlet are used, all with the aim of merging the incoming gas with that already in the cyclone. Deflector vanes may also be used to improve the gas flow and to minimize turbulence (Figure 4). The inlet may be near the base of the cyclone, with the clean gas exiting from the top, Figure 5. This arrangement is often used for cyclones after wet scrubbers to collect the particulate-laden droplets from the wet collector.

A scroll outlet may be used to reduce pressure loss resulting from turbulence when operating in the suction mode or in limited space. Dust reentrainment from the hopper into the exit vortex reduces collection efficiency, and an airtight dust receiver with a large dead space is needed to minimize this.

Multiple-Tube Cyclones

When high efficiency (which requires small cyclone diameter) and large throughput are both desired, a number of

TABLE 1. Characteristics of Cyclones

	Cyclone Type					
	High Efficiency		Conventional		High Throughput	
	(1)	(2)	(3)	(4)	(5)	(6)
Body diameter D_0/D_0	1.0	1.0	1.0	1.0	1.0	1.0
Height of inlet H/D_0	0.5	0.44	0.5	0.5	0.75	0.8
Width of inlet W/D	0.2	0.21	0.25	0.25	0.375	0.35
Diameter of gas exit, D_e/D_0	0.5	0.4	0.5	0.5	0.75	0.75
Length of vortex finder, S/D_0	0.5	0.5	0.625	0.6	0.875	0.85
Length of body L_b/D_0	1.5	1.4	2.0	1.75	1.5	1.7
Length of cone L_c/D_0	2.5	2.5	2.0	2.0	2.5	2.0
Diameter of dust outlet, D_d/D_0	0.375	0.4	0.25	0.4	0.375	0.4

[a]After Cooper and Alley.[7] Columns 1 and 5 from reference 8; columns 2, 4, and 6 from reference 9; and column 3 from reference 2.

cyclones can be operated in parallel. In a multiple-cyclone separator, the housing contains a large number of tubes that have a common gas inlet and outlet in the chamber. The gas enters the tubes through axial inlet vanes that impart a circular motion (Figure 6).

The arrangement of the tubes, as well as their diameter, affects the overall collection efficiency of a multitube cyclone. There is a small pressure drop in the incoming gas stream for each row of tubes it must pass; a decreasing length of outlet tube may be used to compensate for this, producing an organ-pipe arrangement. Alternatively, tubes can be arranged in groups, each with its own outlet plenum, and space in between the groups, which allows even and unimpeded gas flow to all tubes (Figure 7).

Collection efficiency can be improved by hopper evacuation or slip streaming, in which a small portion (about 15%) of the total gas flow is drawn off through the hopper. This suction considerably reduces dust reentrainment into the cyclone tubes. An increase of 40–50% in collection efficiency is possible.[1] The dust may be reinjected into the combustion zone in the case of fossil fuel boilers or cleaned by a fabric filter and returned to the exit with the cleaned air from the cyclone tubes. Such systems have been used as a retrofit when higher efficiencies were necessary (Figure 8).

An alternative solution to the problem of dust reentrainment is to use a straight-through type of cyclone, rather than the reverse-flow type already discussed. Multiple-tube

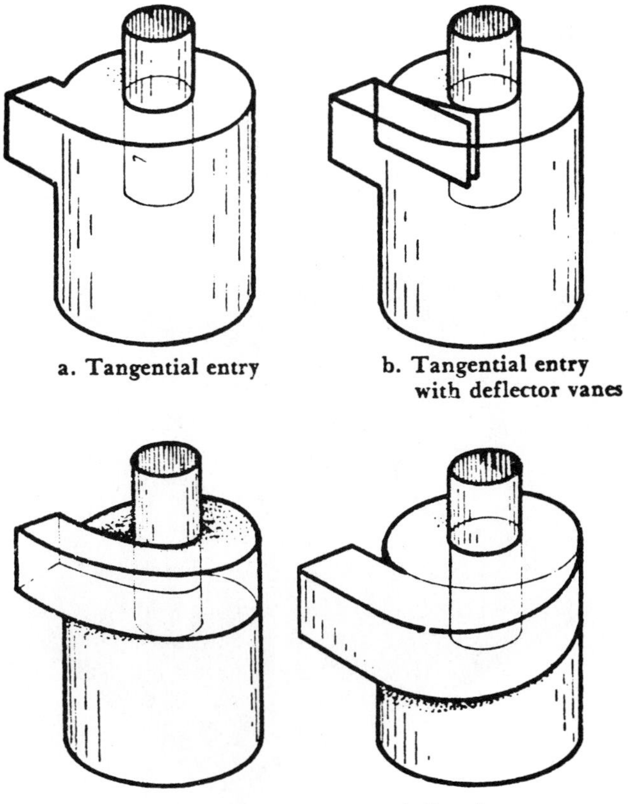

a. Tangential entry

b. Tangential entry with deflector vanes

c. Helical entry

d. Involute entry

FIGURE 4. Types of Cyclone Inlets (From reference 11)

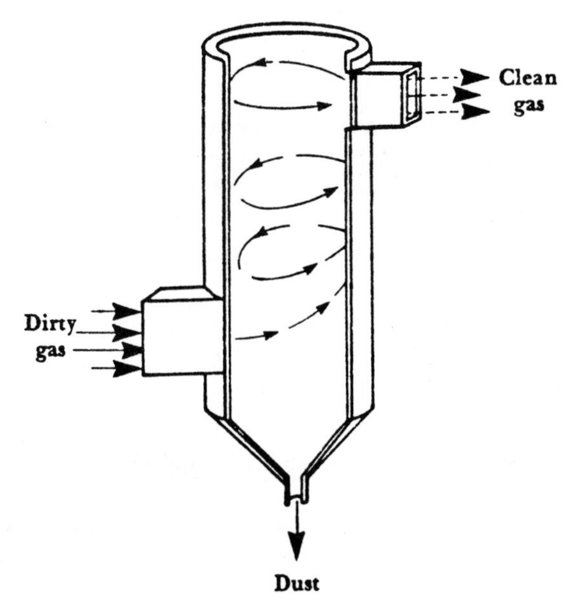

FIGURE 5. Bottom Inlet Cyclone Separator (From Reference 11)

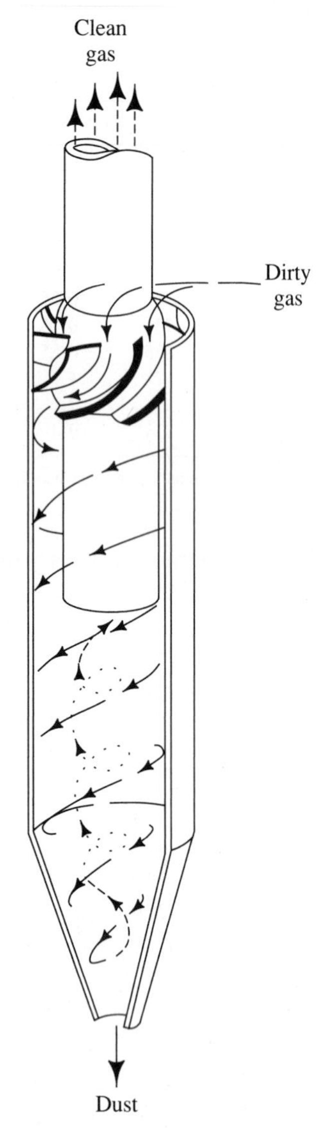

FIGURE 6. Vane Axial Entry Cyclone (From Reference 11)

FIGURE 7a. Organ-Pipe Arrangement of Multitube Cyclones. (From Reference 12)

arrangements of these are common in older coal-fired boilers (Figure 9). The tubes are arranged horizontally and the air is drawn axially into the tube. The fixed vanes impart a cyclonic motion to the gas, causing the particles to move to the edges of the gas stream. In this case, the peripheral dust-laden gas stream is removed through an annular channel and the cleaned gas passes out axially through the tube. The concentrated dust stream is then recirculated or further separated into a hopper.

PREDICTION OF COLLECTION EFFICIENCY

Collection efficiency is a strong function of particle size and increases with increasing particle size. The collection efficiency of a single particle size can be determined by either a semi-empirical approach, developed by Lapple in 1951,[2] or by one of several theoretical equations that have since been developed. Determination of the overall collection efficiency requires knowledge of the particle size distribution of the dust particles. Collection efficiencies for each size range are then weighted accordingly and the values summed.

Theoretical Approach (Laminar Flow)

The basis of the simple theoretical derivation of efficiency is as follows: Particles enter with the gas stream, but tend to move outward under the influence of centrifugal force. This is resisted by the drag of the particles moving radially

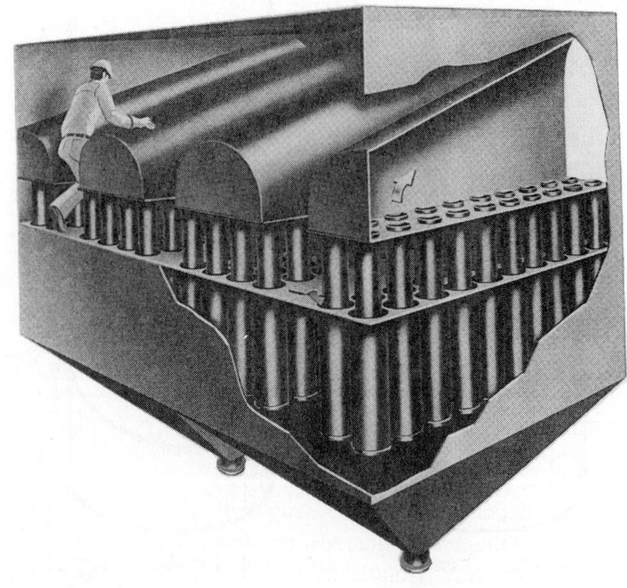

FIGURE 7b. Grouped Tube Arrangement of Multitube Cyclone (From Reference 12)

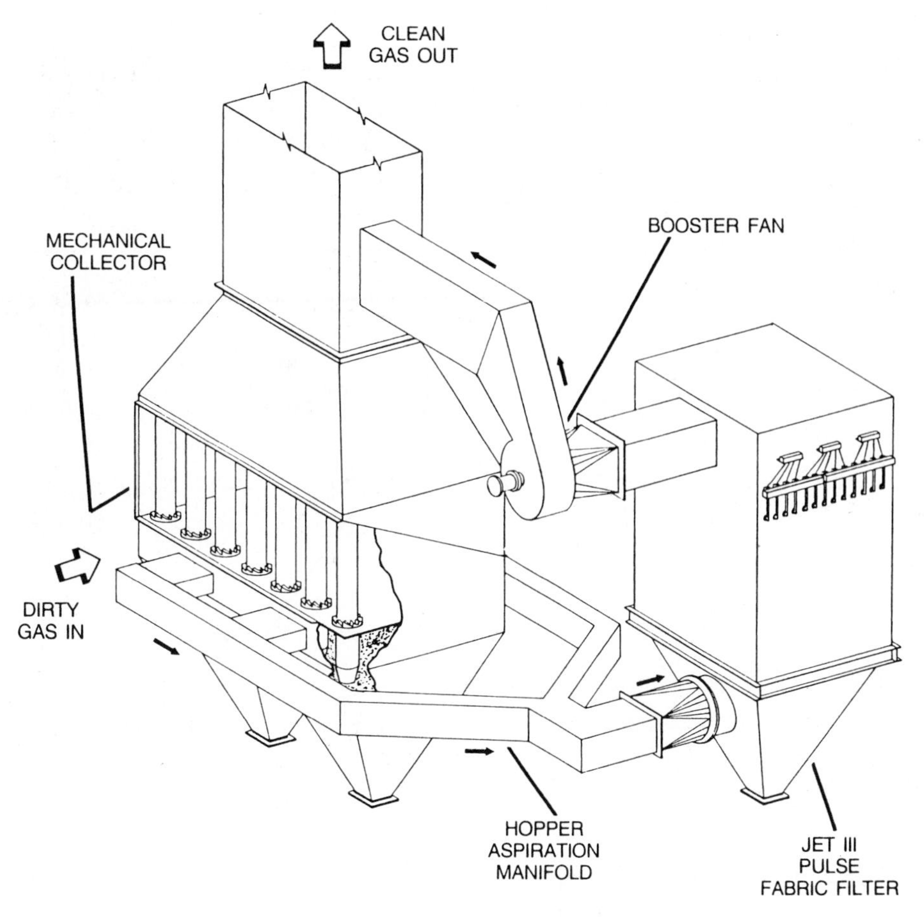

FIGURE 8. Hopper Aspiration System (From Reference 1)

through the gas, and the resultant terminal or radial velocity of the particles is found by equating the centrifugal and drag forces. To be collected, particles must reach the outer wall before the gas leaves the outer vortex. The time and distance are both known: The time is the gas residence time, which depends on gas inlet velocity, radius of the cyclone, and number of turns in the vortex, and the maximum value of the distance to be traveled is the length from the inner edge of the inlet to the outer wall. Assuming laminar flow, an expression is derived that relates the collection efficiency to the cyclone parameters and operating conditions:

$$\eta = \frac{\pi N_e \rho_p d_p^2 V_g}{9 \mu W} \tag{2}$$

where: η = efficiency
N_e = effective number of turns
ρ_p = particle density
d_p = particle diameter
V_g = gas velocity
μ = gas viscosity

This model indicates the strong dependence of efficiency on particle diameter (squared), the dependence on the number of vortex turns (related to the length of the cyclone) and inlet velocity, and the inverse dependence on cyclone inlet width, which is proportional to body diameter.

However, the model predicts a finite value of particle diameter above which collection efficiency is 100% ("critical size"), whereas experimental evidence shows that efficiency approaches 100% asymptotically with increasing particle diameter.

$$d_{pcrit} = \left[\frac{9 \mu W}{\pi N_e \rho_p V_g}\right]^{1/2} \tag{3}$$

Cut Diameter Approach

The semiempirical approach developed by Lapple[2] used this laminar flow treatment, but introduced the concept of a cut size, d_{p50}, defined as the size of particle that is collected with 50% efficiency. The value is a characteristic of the control device and operating conditions, not of the size range of the dust particles. Using experimental particle collection efficiency versus particle-size data, correlated for cyclones of standard proportions, he plotted a generalized curve, as shown in Figure 10.

A particle whose collection efficiency is to be found is characterized by the ratio of its diameter, d_p, to the cut size, d_{p50}, which gives the value of the abscissa of the graph, and the collection efficiency is read from the ordinate. The value of the cut size is calculated from equation 2 by setting efficiency equal to 0.5 and solving for d_p, which is now d_{p50} by definition:

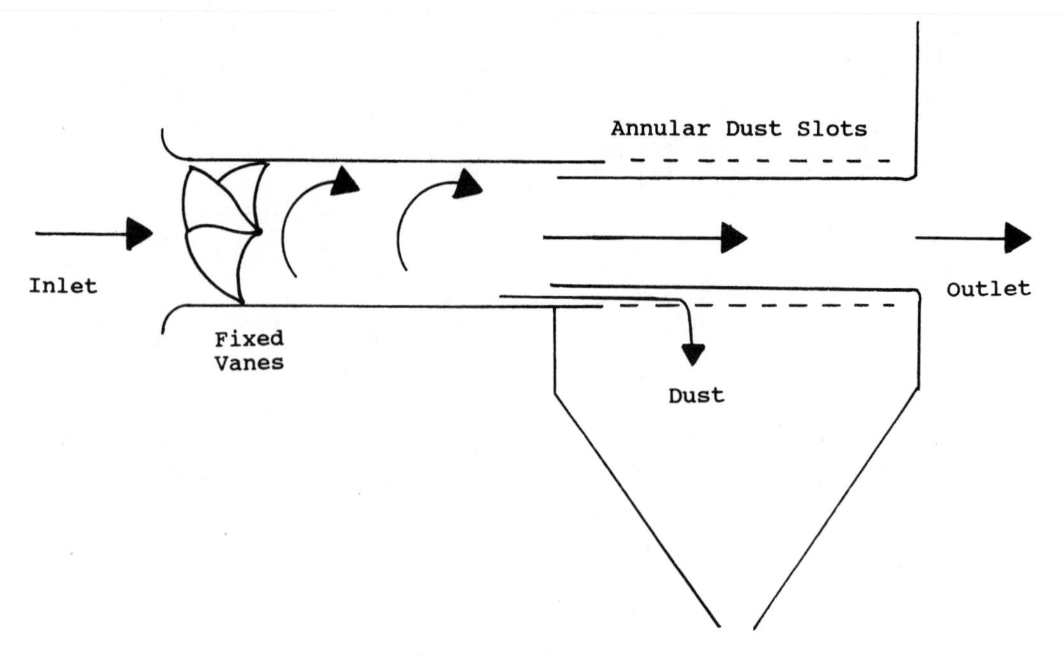

FIGURE 9. Fixed-Impeller Straight-Through Cyclone

$$d_{p50} = \left[\frac{9\mu W}{2\pi N_e \rho_p V_g}\right]^{1/2} \quad (4)$$

Lapple's graph has been fitted to an algebraic equation by Theodore and De Paola,[3] which makes it more convenient for computer applications:

$$\eta_j = 1/(1 + (d_{p50}/d_{pj})^2) \quad (5)$$

where: d_{p50} = particle cut diameter
d_{pj} = particle diameter in range j
η_j = fractional efficiency in range j

More Recent Approaches

The laminar flow model has limitations, as gas flow in a cyclone is not simply laminar (nor is it fully turbulent, because the boundary layer has a significant depth).

Leith and Licht[4] developed an equation for collection efficiency from theoretical considerations that takes into account the back-mixing of uncollected particles and determines an appropriate average residence time for the gas in the cyclone. The equation is of the form

$$\eta = 1 - \exp[-Ad_p^{\,B}] \quad (6)$$

which is a common form for efficiency expressions, and indicates the asymptotic approach of efficiency to 100%. It uses a parameter C, which is a function of the cyclone's dimensional ratios, and Ψ, which is a modified inertial impaction parameter.

$$\eta = 1 - \exp\left[-2\,(C\Psi)^{1/2n+2}\right] \quad (7)$$

where:

$$n = 1 - \frac{(D)^{0.14}}{2.5}\left(\frac{T}{283}\right)^{0.3}$$

with cyclone diameter D in inches and T in °Kelvin.

$$C = \frac{\pi D^2}{WH}\left[2\left(1 - \left(\frac{D_e}{D}\right)^2\right)\left(\frac{S}{D} - \frac{H}{2D}\right) + \frac{1}{3}\left(\frac{S + k - L_b}{D}\right)\left(1 + \frac{d}{D} + \left(\frac{d}{D}\right)^2\right) + \frac{L_b}{D} - \left(\frac{D_e}{D}\right)^2\frac{k}{D} - \frac{S}{D}\right] \quad (8)$$

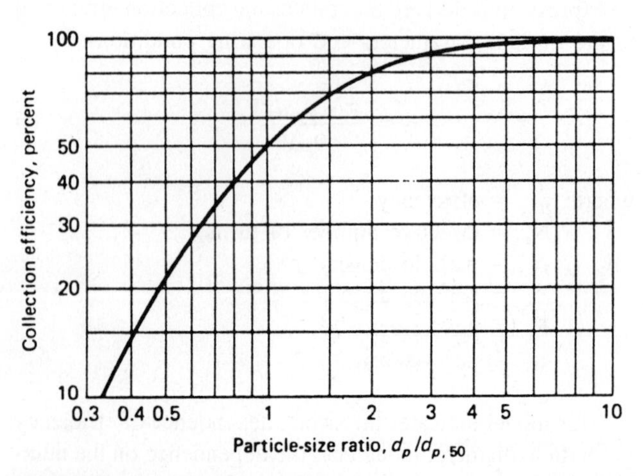

FIGURE 10. Cyclone Efficiency versus Particle-Size Ratio (From Reference 2)

where k is the furthest distance that the vortex extends below the gas exit duct,

$$k = 2.3\, D_e \left(\frac{D^2}{WH}\right)^{1/3}$$

and d is the diameter of the conical section at k

$$d = D - (D - D_d)\,(S + k - L_b)/L_c$$

In addition,

$$\Psi = \frac{\rho_p d_p^2 V_g}{18 \mu_g D}(n + 1) \quad (9)$$

The efficiency expression in equation 7 produces a curve of collection efficiency versus single-particle diameter, which, like the experimental data, is everywhere concave downwards, in contrast to the S-shaped curve of some of the many other theoretical models, as shown in Figure 11. The expression was shown to fit the experimental data to within 4%.

As an example of the application of the Leith and Licht expression, overall particle-collection efficiency has been calculated, using equation 7, for cyclones with diameters from 6 inches to 36 inches, with an inlet velocity of 60 fps at 20°C for a Lapple conventional cyclone. The particle size distribution was assumed to be log normal with a geometric standard deviation of 2.0, and the dust was assumed to have a density of 2.5 g/cm^3 (typical of fly ash and many other dusts). Figure 12 shows the theoretical relationship between overall collection efficiency and mass mean diameter. Efficiency is seen to increase with increasing particle mass mean diameter and with decreasing cyclone diameter. The observation that the typical efficiency ranges from 70% to 90% further illustrates that cyclones are generally not used as final primary control devices, but as precleaners.

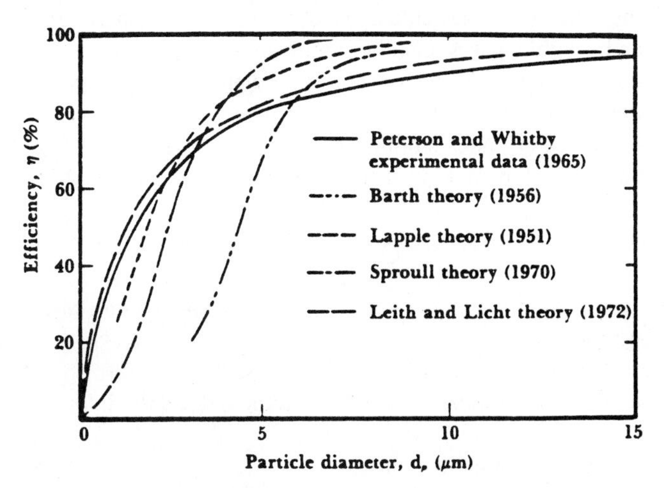

Source: Leith and Mehta, 1973.

FIGURE 11. Comparison of Experimental Data with Theoretical Predictions. (From Reference 6)

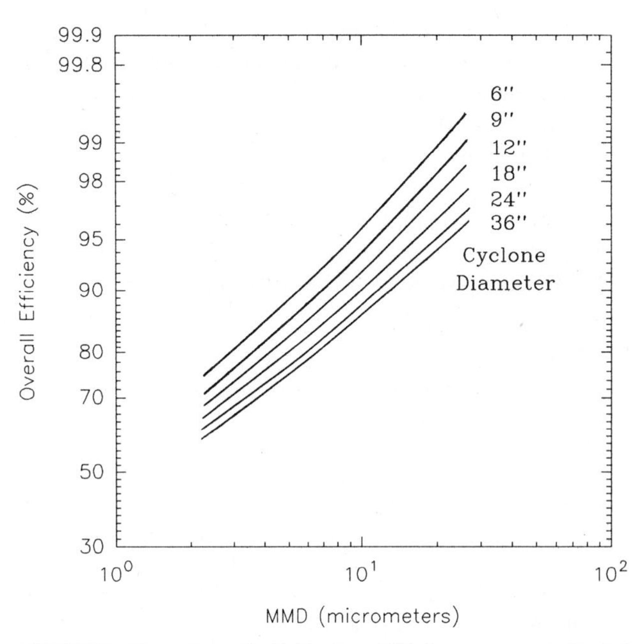

FIGURE 12. Overall Collection Efficiency versus Particle Mass Mean Diameter for a Range of Cyclone Sizes at 70°F

PREDICTION OF PRESSURE DROP

Pressure drop is an important parameter because it relates directly to operating costs. Higher efficiencies for a given cyclone can be obtained by higher inlet velocities, but this also increases the pressure drop, and a trade-off must be made. Shepherd and Lapple[5] found empirically that pressure drop (expressed as the number of inlet velocity heads) depends inversely on exit diameter squared:

$$H_v = K \frac{HW}{D_e^2} \quad (10)$$

where H_v is the number of velocity heads and K is an empirical constant with a value of 16 for a tangential inlet cyclone and 7.5 for one with an inlet vane. The pressure drop is as follows:

$$\Delta P = \frac{1}{2} \rho_g V_g^2 H_v \quad (11)$$

where: ΔP = pressure drop, Pa (N/m^2)
ρ = gas density, kg/m^3
V_g = gas velocity, m/s

Although other equations have been derived from theoretical considerations, they have generally not been found to be more accurate than the equation developed by Lapple, which stands as the most useful one for calculating pressure drop.[6]

Pressure drop is a function of the square of inlet velocity, so too high a velocity will cause excessive pressure drop. On the other hand, too low a velocity would cause a low efficiency. A very high velocity would also actually decrease efficiency because of increased turbulence and saltation/reentrainment of particles; generally it is found that the best operating velocity is around 60 fps. Common ranges of pressure drops are as follows:

Low-efficiency cyclones	2–4 inches water
Medium-efficiency cyclones	4–6 inches water
High-efficiency cyclones	8–10 inches water

References

1. W. I. Olson, "Upgrading mechanical collector performance with hopper aspiration," Wheelabrator Air Pollution Control Technical Paper TP88-123, 1988.
2. C. E. Lapple, "Processes use many collector types," *Chem. Eng.*, 58:144–151 (1951).
3. L. Theodore and V. DePaola, "Predicting cyclone efficiency," *J. Air Pollution Cont. Assoc.*, 30(10) (1980).
4. D. Leith and W. Licht, "The collection efficiency of cyclone type particle collectors—a new theoretical approach," American Institute of Chemical Engineers, *Symposium Series 126*, 68:196–206 (1972).
5. C. B. Shepherd and C. E. Lapple, "Flow pattern and pressure drop in cyclone dust collectors," *Ind. Eng. Chem.*, 31(8) (1939); 32(9) (1940).
6. D. Leith and D. Mehta "Cyclone performance and design," *Atmospheric Environ.* 7:527–549 (1973).
7. C. D. Cooper and F. C. Alley, *Air Pollution Control: A Design Approach,* PWS Publishers, 1986.
8. C. J. Stairmand, "The design and performance of cyclone separators," *Trans. Ind. Chem. Eng.*, 29 (1951).
9. P. Swift "Dust control in industry," *Steam Heating Eng.*, 38 (1969).
10. M. Heumann, "Understanding cyclone dust collectors, *Plant Eng.* (May 26, 1983).
11. *Control of Particulate Emissions* Course 413 Student Manual, EPA Air Pollution Training Institute.
12. Bulletin MC 581, Zurn Industries, Inc.

Bibliography

Air Pollution, Vol. IV—Engineering Control of Air Pollution, 3rd ed., A. C. Stern, Ed.; Academic Press, New York, 1977.

L. W. Briggs, *Trans. AIChE,* 42(3):511–526 (1946).

W. Strauss, *Industrial Gas Cleaning,* International Series of Monographs in Chemical Engineering, Vol. 8, Pergamon Press, New York, 1966.

L. Theodore and A. J. Buonicore, *Air Pollution Control Equipment, Volume I—Particulates,* CRC Press, 1988.

K. Wark and C. F. Warner, *Air Pollution—Its Origin and Control,* 2nd ed., Harper & Row, New York, 1981.

WET SCRUBBERS

Kenneth Schifftner and Howard Hesketh

When it comes to separating particulate and gases from other gas streams, few techniques are as commonly used as wet scrubbing. Most industrial processes utilize wet scrubbing technology at some point either to recover valuable components or to prevent harmful compounds from escaping into the atmosphere. When car bumpers are plated, wet scrubbers prevent the escape of corrosive gases. When paper is bleached white, the residual gases from the process are scrubbed clean. When hazardous waste is incinerated, wet scrubbing technology acts as a protective barrier between the 2000°F combustion process and the environment.

Few technologies are as well developed, as "mature," as wet scrubbing. There are literally hundreds of thousands of wet scrubbers in operation worldwide, some of which have been operating for over five decades.

Modern wet scrubbing technology can be traced to the beginning of the industrial revolution. Replacing manual labor with machine labor created the need to control airborne emissions. Early "smut collectors" ("smut" in this case meaning particulate pollutants) were usually simple baffle or spray chambers applied to combustion processes. Near the turn of the century, lime calcining kilns had baffle chambers, sometimes made of wood, where cascading water or sprays helped to knock down and recover valuable lime product. During World War I, a special fan-driven device was used to aspirate trench warfare chemicals, spraying the liquids and gases out over enemy troops. That device in its modern form, the venturi scrubber, is now used to remove thousands of tons per year of pollutants, thus protecting the people who live nearby.

Though the basic principles have changed little over the years, refinement of the application and design of wet scrubbing technology has produced systems with high reliability and high efficiency that are relatively low in cost.

One of the keys to the successful solution of the problems of air pollution control involves a knowledge of *where* to apply *which* technology. Elsewhere in this book, specific applications are defined and described for the reader. This chapter is intended to provide background information on wet gas scrubbing, its basic tenets, and how it is applied to the removal of particulate. With wet scrubbers, particulate and soluble gases can be removed simultaneously. Our comments will center on the particulate-removal capabilities of wet scrubbers.

A clearer understanding of wet scrubber technology can be obtained by looking at some general areas of application.

AREAS OF APPLICATION

Wet scrubbing technology is used under circumstances where:

1. The contaminant cannot be removed easily in a dry form.
2. Soluble gases are present.
3. Soluble or wettable particulates are present.
4. The contaminant will undergo some subsequent wet process (such as recovery, wet separation or settling, or neutralization).

5. The pollution control system must be compact.
6. The contaminants are most safely handled wet rather than dry (where the dry particulate may ignite or explode).

As a result, wet scrubbing typically is used to control sticky emissions that would otherwise plug filter-type collectors, to control both particulate and gaseous emissions simultaneously, to control acid gases, to recover soluble dusts and powders, to scrub particulate from incinerator exhausts, and to control metallic powders such as aluminum dust that tend to explode if handled dry. Compact wet scrubbing systems are used on oil tankers to help provide inert gases for use in the ship's hold to reduce the chances of explosion. Some portable hazardous-waste incinerators use wet scrubbing technology, trailer mounted, to provide an easily moved system that allows the incinerator to be brought to the site of contamination.

WET SCRUBBING BASICS

Literally hundreds of types of wet scrubbers currently are available. The application of certain basic design parameters varies from vendor to vendor, resulting in similar yet significantly different designs. Some systems are designed for highest removal efficiency, others for ease of maintenance, and still others simply for low cost.

Although engineers are not unanimous as to the exact mechanisms, the following popular explanations of the capture mechanisms utilized in wet scrubbers should be helpful.

How Wet Scrubbers Remove Particles

Wet scrubbers remove particles from gas by capturing the particles in liquid (usually water) droplets and separating the droplets from the gas stream. The droplets act as conveyors of the particulate out of the gas stream.

Wet scrubbers are believed to capture particulate through three mechanisms:

1. *Impaction* of the particle directly into a target droplet (Figure 1).
2. *Interception* of the particle by a target droplet as the particle comes near the droplet (Figure 2).
3. *Diffusion* of the particle through the gas surrounding the target droplet until the particle is close enough to be captured (Figure 3).

Note the term "target droplet." Nearly all high-efficiency wet scrubbers are adept at creating target droplets. The wet-scrubber designer labors to create a closely packed fine dispersion of droplets to act as targets for particle capture. Some do this by accelerating gases to high velocity, injecting liquid, then pneumatically shearing the liquid into a fine spray. Others use mechanical energy (such as spinning disks) or high-pressure sprays to create the target droplet dispersion.

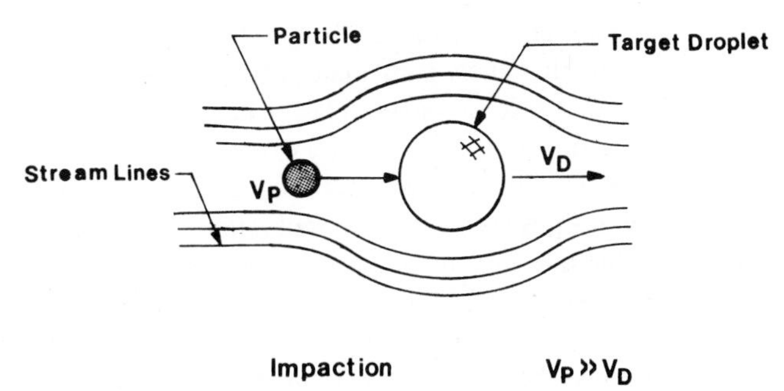

FIGURE 1. Most Particles Are Removed by Direct Impaction into a Droplet

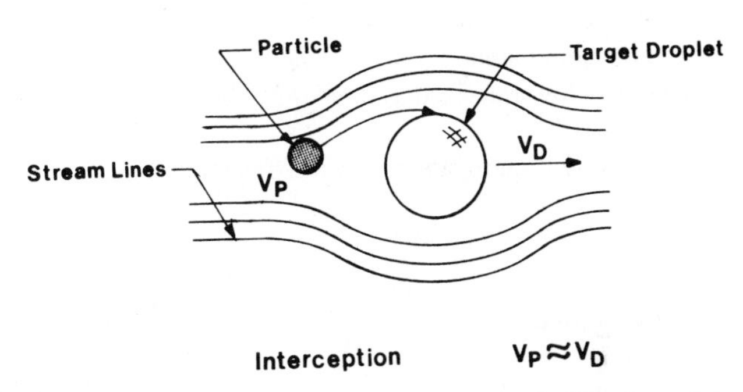

FIGURE 2. Other Particles Come Close to the Droplet and Are Intercepted

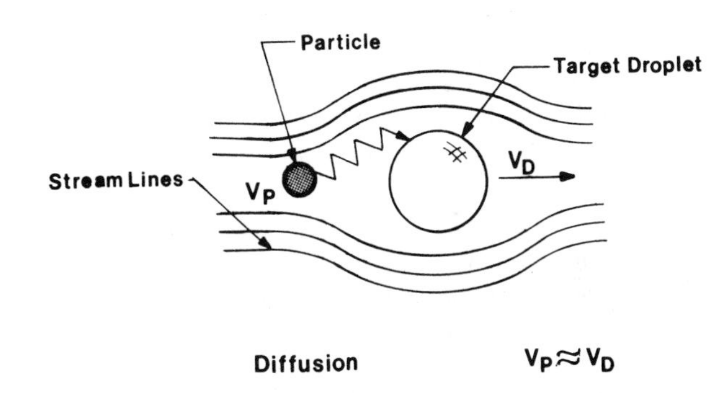

FIGURE 3. Smaller Particles Are Captured by Diffusion

The goal is to cause the tiny pollutant particle to be lodged inside the collecting droplet and then to remove the larger droplet from the gas stream. In each case, the designer's goal is to create a dense dispersion of fine target droplets. In general, it is believed that the smaller the target droplet, the smaller the size of particulate that can be captured. The more densely the droplets are packed, the greater is the probability of capture. This sounds simple, but it is not.

As we will see later, the method by which this droplet dispersion is created can vary widely from design to design, but most, if not all, wet scrubbers share this common design parameter. Successful scrubbers create and control the droplet dispersion effectively.

We will look at one scrubber design, the venturi scrubber, as an example of how particulates are captured. Venturi scrubbers (described in more detail later) generate fine droplet dispersion by pneumatically atomizing the scrubbing liquid in a high-velocity zone called the venturi throat.

Scrubber designs start with a desired removal rate for the scrubber. This rate may be dictated by the process, by the regulated emissions limitations, or by a combination of the two.

To determine the pressure drop, we first need to know:

1. The amount of particulate entering the system.
2. The size distribution of the particulate (how much is below 0.5 μm, below 1 μm, etc.).
3. Code requirements.
4. The gas flow rate in dry standard cubic measure.

The required aggregate removal efficiency is simple to calculate:

$$\text{Required efficiency} = \frac{(\text{Mass of particulate in}) - (\text{out})}{\text{Mass of particulate in}}$$

The mass out is determined by the code requirements. If the code is given in concentration per unit dry standard volume of gas, knowing the dry gas volume will yield the mass out if one simply multiplies the allowable concentration by the dry gas volume. If the code is process weight related, one would have to know the process production rate and then multiply that rate by the allowable emissions factor mentioned in the code.

Next we need to know the removal efficiency of the particular wet scrubber in question for various particle sizes in order to determine the overall efficiency required of the collector. From pilot testing and field analysis of existing scrubbers, scrubber designers develop curves that show the ability of the wet scrubber to remove particulates of a given size. If the curve is plotted as particle size versus removal efficiency, it is usually called simply an "efficiency curve." If the particle size is plotted versus the quantity of particulate of a given size that penetrates through the collector, it is usually called a "grade penetration curve." The penetration is 100% minus the efficiency (in percent). Since most emissions codes are based on limiting what comes out of a system, some argue that grade penetration curves are better suited for determining scrubber pressure drop requirements. Figure 4 shows a typical grade penetration curve for various scrubber pressure drops. This series of curves is characteristic of a venturi scrubber, but other wet scrubbers would have similar curves.

Note that the efficiency of removal (or the penetration) is needed for each particle size range. We cannot simply use an "overall" efficiency without knowing how it was determined. Many scrubber advertisements claim "99% removal of particulate," but this is meaningless unless we know what *size* of particle is involved. Some scrubbers operating at only 1 w.c. can remove 99% of particles above 5 μm, but will remove less than 5% of particles below 1 μm. The removal efficiency for a given size must be known in order to predict true performance.

To create the efficiency curves, designers accumulate empirical data on the collection efficiency of their particular scrubber pressure drop. The performance curves are then used to predict the removal efficiency of the scrubber in question for a given distribution of particle sizes. Though these curves vary from design to design, they share a similar trend: As the particle becomes smaller, the energy required to collect it becomes greater. This is why the scrubbing system will require more energy to collect submicron fumes than it would to collect large (above 10 μm) powders or dusts. Figure 4 was generated from such an analysis.

For particular sizing, many designers rely on the aerodynamic mean diameter of the particle, not its physical diameter. Wet scrubbers are aerodynamic devices. The aerodynamic characteristics of the particulate are more important than the physical dimensions. In addition, many particles are not round, or nearly round, but are oblong, agglomerated, sintered together, or otherwise nonuniform dimensionally. To determine the aerodynamic diameter of the particles, a device called a cascade impactor is typically used (Figure 5). This precision device inertially separates particulate by aerodynamic diameter. The collection stages are weighed and the results are usually presented in percent by mass below a given size range.

Given the particle-size distribution of the source in question and the known removal characteristics of the wet scrubber in question, we can determine what pressure drop is needed to meet our control requirements. The mass of particles in a given size range is multiplied by the capture capability of the scrubber for that size range. The results for each size range are totaled. The amount of particulate our scrubber is predicted not to collect must be less than the amount we are allowed to emit. An iteration procedure is followed for various pressure drops until a pressure drop is found that predicts emissions will be safely below the allowable limit.

Pressure drops across scrubbing towers can be predicted by flow behaviors and rates, accounting for system geometry. For common wet scrubbers, such as the venturi type, pressure drop equations have been developed to simplify this task. One of the simplest equations for predicting venturi scrubber pressure drops, and one of the few to incorporate the venturi throat size, is given by Hesketh[1] in metric and English units:

$$P \cong \frac{v^2 d_g A^{0.133} L^{0.78}}{3870}$$

where: P = venturi scrubber pressure drop, centimeters H_2O

v = throat velocity of the gas and particles, cm/s

A = throat cross-sectional area, cm^2

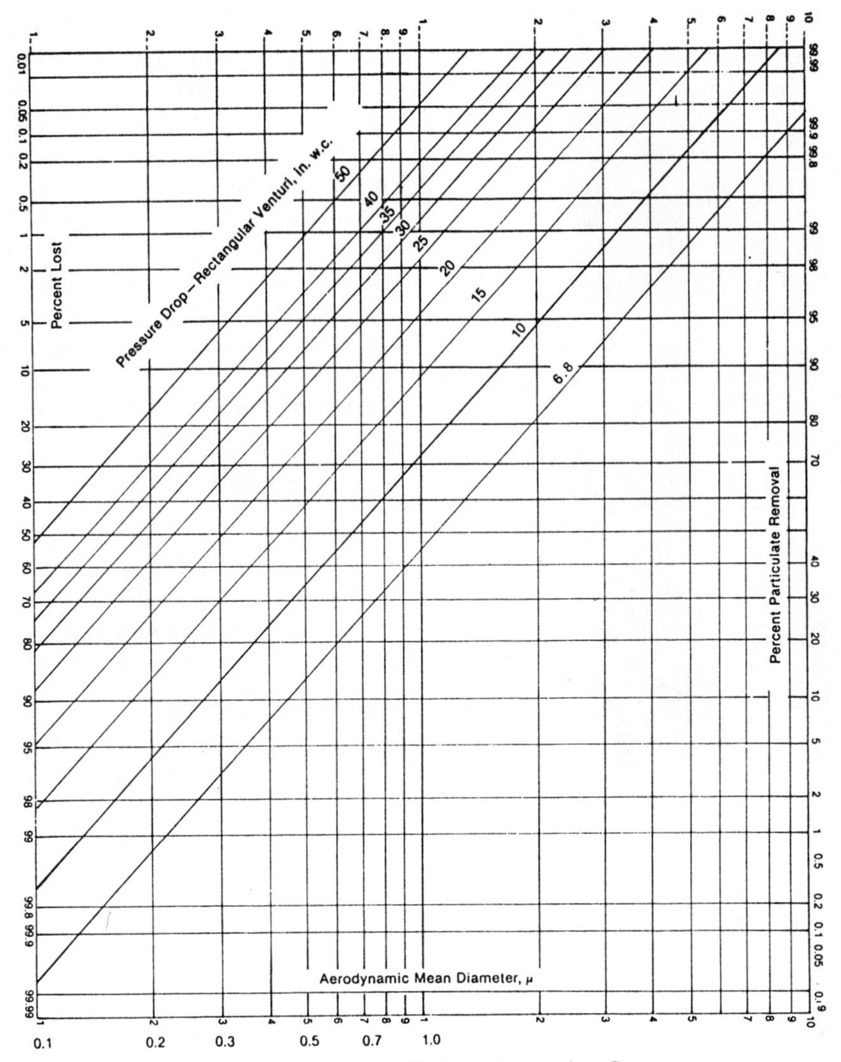

FIGURE 4. Grade Efficiency/Penetration Curve

L = liquid/gas ratio, l/m^3
d_g = gas density, g/cm^3

$$P \cong \frac{v^2 d_g A^{0.133} L^{0.78}}{1270}$$

where: P = venturi scrubber pressure drop, inches H_2O
v = throat velocity of the gas and particles, ft/s
A = throat cross-sectional area, ft^2
L = liquid/gas ratio, gallons per 1000 actual cubic feet gas
d_g = gas density, lb/ft^3

Note that venturi scrubber pressure drop is the pressure drop across the working portion of the scrubber. This is measured from the venturi inlet duct, where a uniform free stream velocity has been established, to the outlet duct, where the free stream velocity resumes. These inlet–outlet static pressure measurements should be made in sections where gas velocities are relatively similar and care must be taken to assure that the pressure taps and lines are not plugged or contain water.

Others, such as Yung et al.[2] and Boll,[3] also have developed equations for predicting venturi scrubber pressure drop. Additional researchers have determined key design parameters for scrubbers like the venturi type, such as throat length, entry, and diverging angles, that serve to optimize the design. This information accumulates as the "know-how" from which the scrubber designer earns his or her living.

References 4–7 describe particulate (and gas) separation techniques in more detail and are recommended reading for those requiring further information regarding pressure drop.

Applied to collection of particles by wet scrubbers, this theory states that the particle-collection efficiency is directly related to the energy expended in the gas–liquid contacting. This varies from one wet scrubbing device to another as there is some dependence on the condition of the gas (temperature and degree of saturation), type of particles (wettability), and scrubber configuration (how the contacting occurs). The contacting power usually comes mainly from the gravity or mechanical energy input.

A correlation of gas-phase contacting power was presented by Young et al.[2]; these data were based on gas stream energy only and assume 60% motor fan efficiency.

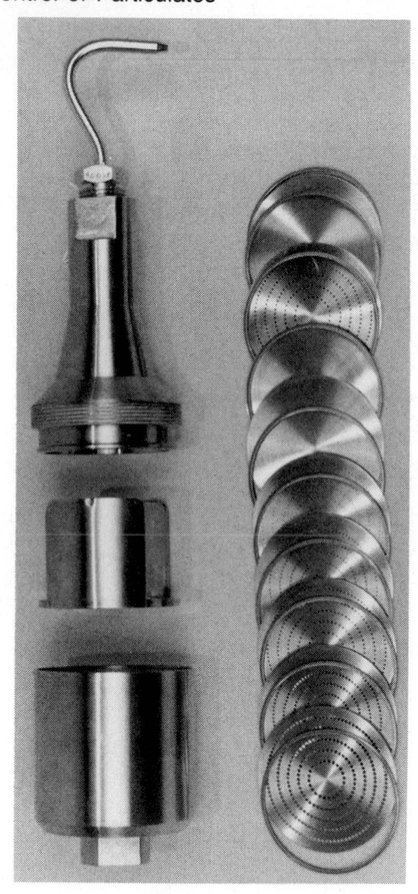

FIGURE 5. Particle-Size Cascade Impactor (Anderson Samplers)

Figure 6 shows this correlation for several wet scrubbers plus some data from personal experience. The ordinate of this figure is cut diameter, which is the size particle captured with 50% efficiency. Therefore, the lower the cut diameter, the more efficient will be the wet scrubber.

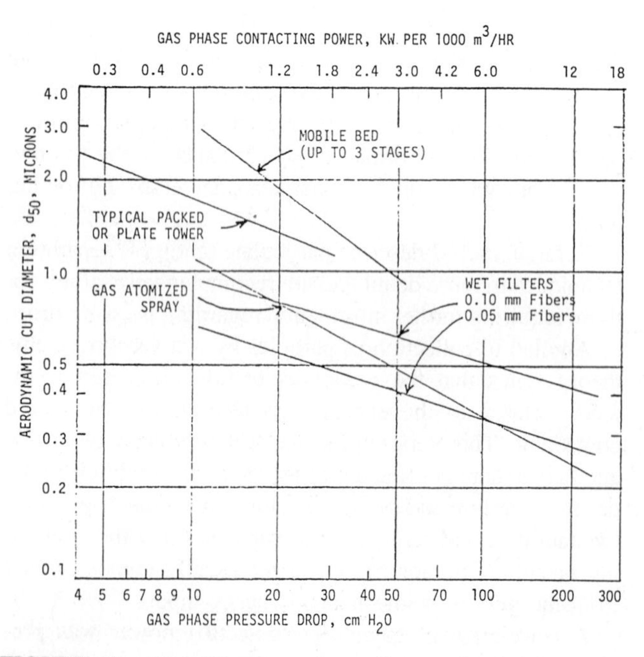

FIGURE 6. Cut Diameter versus Contact Power for Wet Scrubbers

Particle wettability was noted as one factor that can affect scrubber particle-collection efficiency. Basically, the easier the particle wets, the easier it is to capture. Many studies have been made to evaluate this. We maintain (and discuss later) that particles should be cooled and gas saturated before entering the working portion of a wet scrubber. This gas conditioning results in a positive flux force condensation, that is, water vapor that condenses productively serves to remove particulate.

Types of Wet-Particulate Collectors

Wet-particulate collectors can be grouped into the following major categories:

1. Venturi scrubbers
2. Mechanically aided scrubbers
3. Pump-aided scrubbers
4. Wetted-filter-type scrubbers
5. Tray- or sieve-type scrubbers

Venturi Scrubbers

Figure 7 shows the basic components of a venturi scrubber. The entrance zone is called the "converging section," the

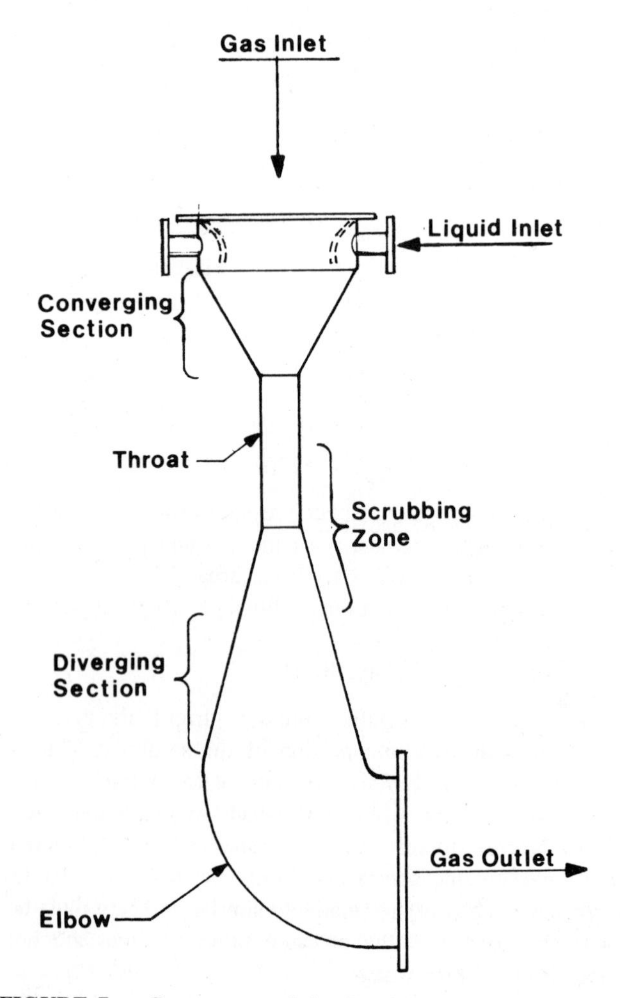

FIGURE 7. Components of the Common Venturi Scrubber

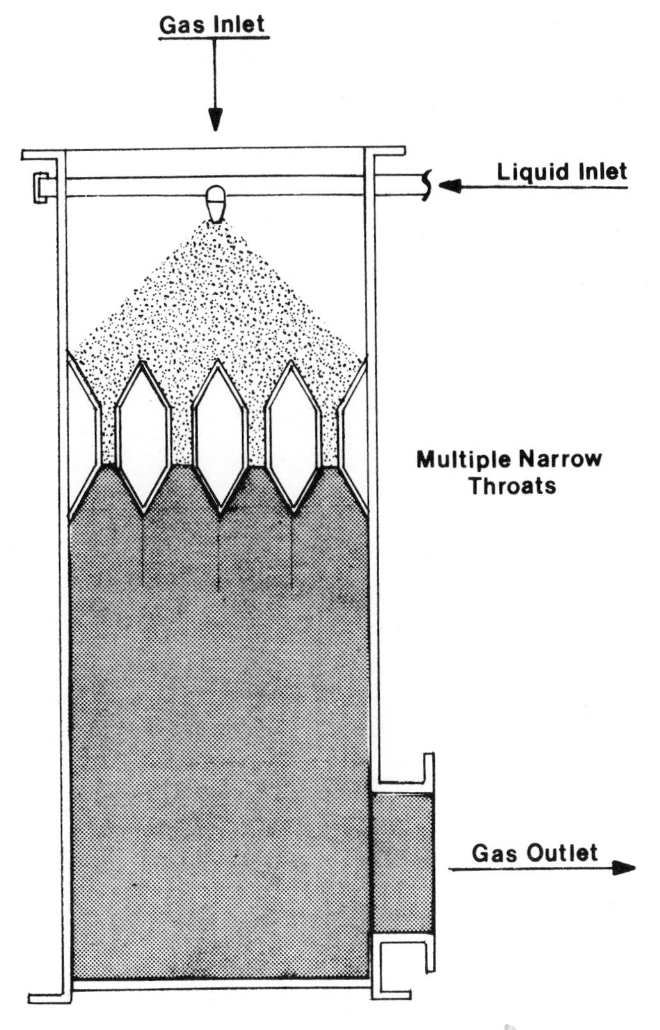

FIGURE 8. Multiple-Throat Venturi

high-velocity center zone is called the "throat," and the expanding section after the throat is called the "diverging section" or "velocity pressure recovery zone." In the venturi, the static pressure of the gases is converted to velocity pressure (i.e., kinetic energy) as the gases move through the constriction, or throat. The required dispersion of target droplets is created by accelerating the gas stream to a high velocity and then using this kinetic energy to shear the scrubbing liquid into fine droplets. The motive force comes primarily from gas-stream kinetic energy, usually injected into the system by a fan. The energy imparted to the gas stream acts on the high-velocity center of the throat. Most venturi scrubbers have throat widths of less than about 15 cm. Some, as shown in Figure 8, have throat widths of less than 3 cm, thus effectively eliminating sneak-by.

To avoid sneak-by and to tolerate larger gas flows, annular-type venturi scrubbers are sometimes used. These devices have round throat zones in which a center body is smaller in diameter than the inside diameter of the scrubber. An annular gap is created through which the gas passes.

Scrubbing liquid is injected into venturi scrubbers in a variety of ways. Some designs inject at the throat zone, others at the gas inlet, and still others upwards, against the gas flow in the throat. Some designs use supplemental hydraulically or pneumatically atomized sprays to augment target droplet creation.

Venturi scrubbers are typically considered high-energy particulate-control devices. Pressure drop equations for these devices were noted earlier. Studies made on venturi scrubbers have provided data that enable particle-collection efficiencies to be predicted for venturi scrubbers.[2] This is shown in Figure 9, which plots fraction overall penetration (i.e., one minus fraction overall efficiency) against the ratio of aerodynamic cut diameter (d_{50}) to aerodynamic mass mean diameter (d_a). The parameters on this plot are the particle standard geometric deviation (σ_g). This fully accounts for particle size and distribution and is only applicable to gas-atomizing-type wet scrubbers, such as the venturi scrubbers.

Mechanically-Aided Scrubbers

Figure 10 depicts a mechanically-aided scrubber. In this device, the droplet dispersion is created by a whirling mechanical device (usually a fan wheel or disk). Liquid is injected into or onto the disk and mechanical energy is added to the system to break the liquid into fine droplets. The mechanically driven device acts on the liquid and gas, providing more than pneumatic shearing, as in the venturi scrubber.

Other mechanically-aided designs involve spraying a rotating fan wheel or similar design, which helps shear the liquid stream into droplets and compress the scrubbing zone.

Mechanically-aided devices typically use lower fan energy than other devices, but must be evaluated on a total energy-input basis, since the collection energy comes from supplemental, driven equipment.

Pump-Aided Scrubber

In Figure 11, a pump-aided scrubber is shown. This particular unit is an eductor-type venturi scrubber that uses high-velocity liquid spray to entrain the gas and pull it through the unit. Other pump-aided devices spray the scrubbing liquid into the gas stream counter to the direction of the gas, in the same direction, or at an angle. Still others use air-atomized sprays. In each case, the predominant energy input comes from the pressurized liquid stream. The energy injected into the liquid by a pump acts upon the gas. Since pumps are mechanically more efficient than fans, the claim is made that pump-aided scrubbers are more efficient than fan-aided scrubbers. The liquid circuit, however, may rely on nozzles with high-velocity orifices. The applications engineer must chose the type of unit that best serves his or her purposes.

Wetted Filter Scrubber

Wetted-filter-type scrubbers place a tortuous path in the way of the gas stream. Energy is imparted into the gas and sometimes into the liquid. The liquid and gas must pass

Pt = exp (-A d_a^2)
Overall Penetration, Pt_0
0.5
0.1
0.01
0.001
0.001
0.01
0.1
1.0
σ_g = 6
5
4
3
2.5
2
1.5
1
Ratio of $d_{50}/\bar{d}_a$

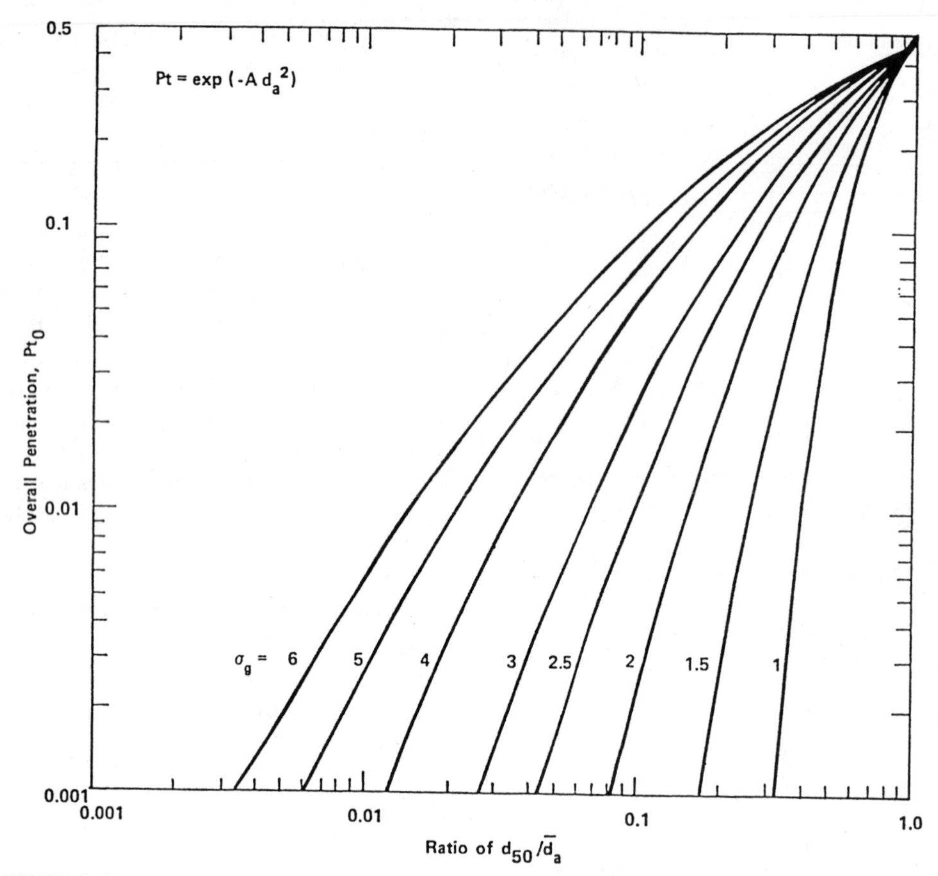

FIGURE 9. Overall Penetration versus Ratio of Aerodynamic Cut Diameter to Aerodynamic Mass Mean Diameter

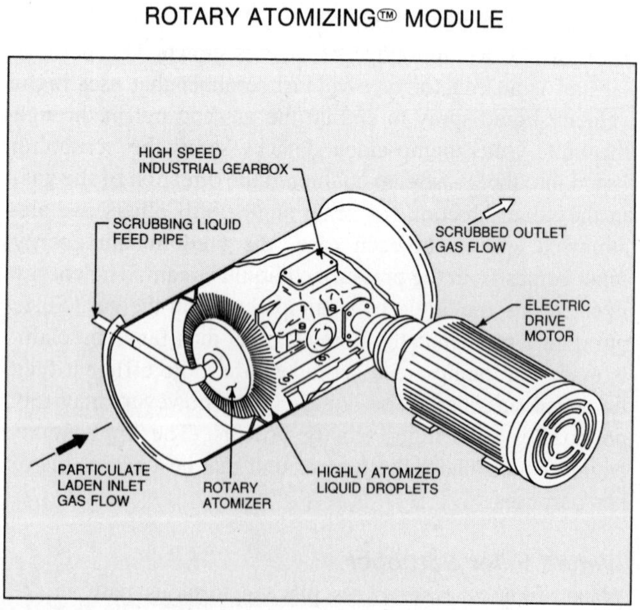

ROTARY ATOMIZING® MODULE
HIGH SPEED INDUSTRIAL GEARBOX
SCRUBBING LIQUID FEED PIPE
SCRUBBED OUTLET GAS FLOW
ELECTRIC DRIVE MOTOR
PARTICULATE LADEN INLET GAS FLOW
ROTARY ATOMIZER
HIGHLY ATOMIZED LIQUID DROPLETS

FIGURE 10. Mechanically-Aided Scrubber

GAS INPUT
CHEM INPUT
VENTURI
TO TOWER
FLOW METER
PUMP
TANK
GAS OUT

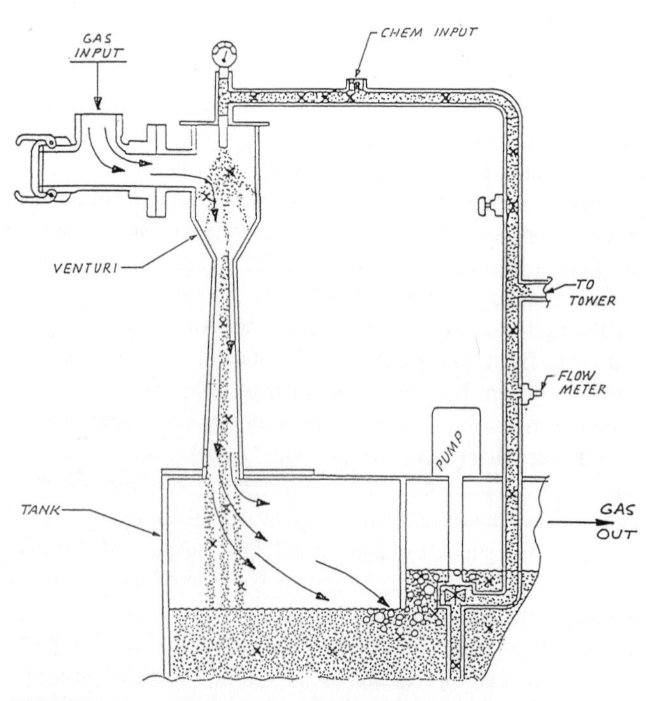

FIGURE 11. Pump-Aided Scrubber (Eductor Type)

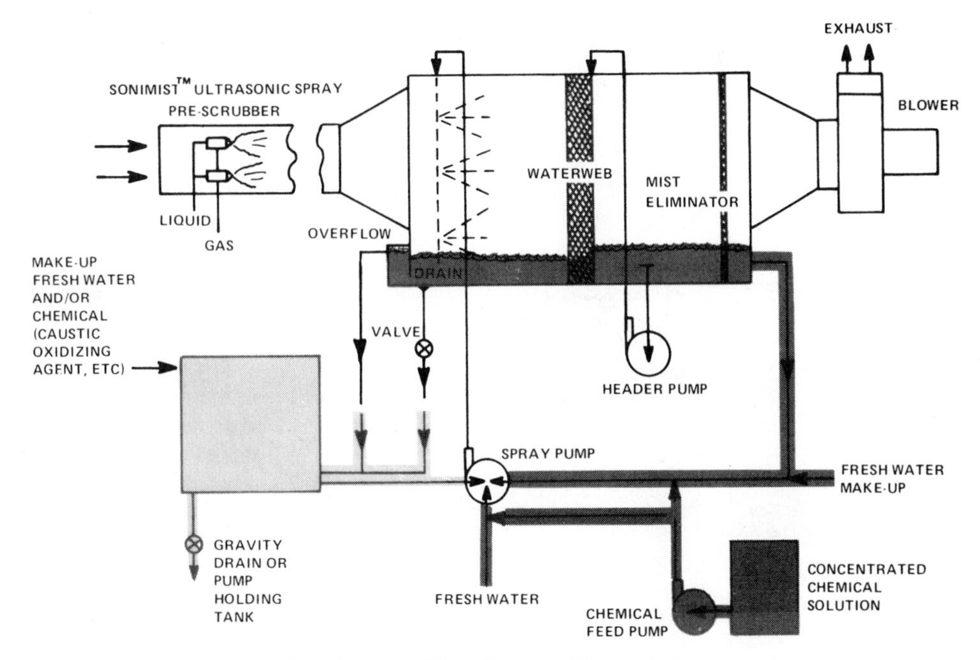

FIGURE 12. Wetted Filter Scrubber (Heat Systems Ultrasonics)

through a medium with small openings. A filtration-like process occurs, with the particulate temporarily sticking to the filter. This type of device is shown in Figure 12. These devices are sometimes used in series and are usually found on applications with low loadings of particulate.

Tray or Sieve Scrubber

Tray- or sieve-type wet scrubbers have small orifices that accelerate the gas stream. The kinetic energy of the gas is applied to a liquid surface, creating a froth. The particulate is injected into the liquid stream, using the energy of the gas. Figure 13 shows a tray-type scrubber of this variety. Scrubbers that use perforated plates are typically called "sieve tray towers"; those that place rigid baffles opposite each perforation are called "impingement plate" scrubbers. Another version of a tray- or sieve-type scrubber is shown in Figure 14. In this unit, a curved wire mesh grid allows the utilization of the velocity pressure of the gas. A turbulent "fluidized" bed of liquid and gas is created above the grid. The energy of the gas is utilized, along with the energy of the descending liquid.

Hybrid (Combination) Scrubbers

Combination scrubbers, or "hybrids," also exist. Some use eductor scrubbers followed by packed tower or tray units. Others are tray towers with different types of trays. The combinations are created to meet certain design criteria, sometimes unique to a specific application. Applications engineers can combine various technologies to achieve the desired results. Particulate that would plug a packed tower can be first removed using a low-energy venturi or pump-aided scrubber, for example, and then the gases can proceed to a packed tower.

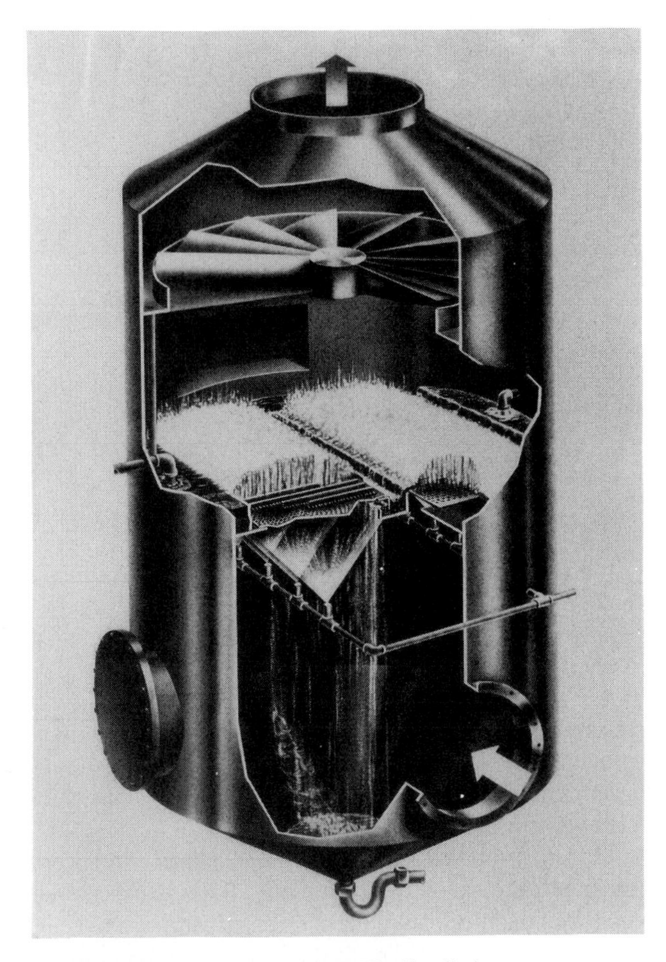

FIGURE 13. Tray Tower (W. W. Sly Co.)

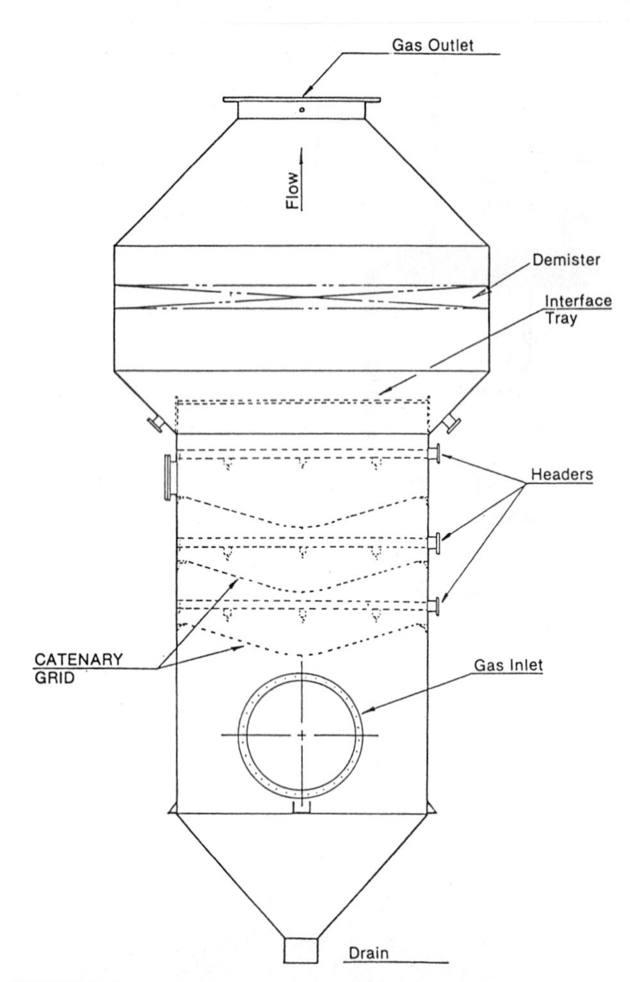

FIGURE 14. Catenary Grid Scrubber™ (Otto H. York Co.)

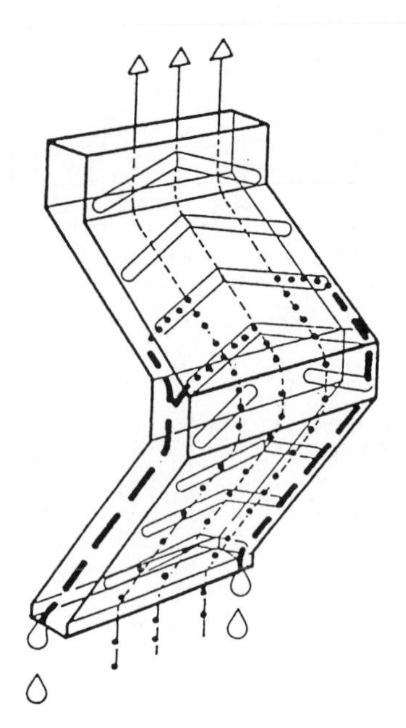

FIGURE 15. Vertical Gas Flow Chevron (Munters Corp.)

DROPLET REMOVAL

We mentioned earlier that wet scrubbers remove particulate by impacting the particulate into target droplets and absorb gases by extending the liquid surface area. A critical component of effective wet scrubbing is the efficient removal of the residual droplets or mists that are created in the scrubber.

Many wet scrubbers remove droplets through centrifugal force. Separators of this type are usually termed "cyclonic" separators, because a whirling cyclone of gas is created inside the scrubber by imparting a tangential velocity to the gases. This velocity is imparted through the use of vanes, rotating elements, or simply a tangential gas inlet in a cylindrical vessel. The cyclonic action throws the liquid against the vessel wall, where it drains by gravity or is otherwise trapped.

Modern scrubbers also use "chevron" droplet eliminators for either vertical or horizontal gas flows. Shaped like curved yet parallel blades, the chevron provides a surface in the path of the droplet. The droplet's inertia tends to carry it directly forward, however, the chevron blade's profile causes the droplet to zigzag from side to side. The droplet impacts onto the blade surface, accumulates, and drains. Chevrons have "active" surfaces in the direct path of the gas flow, which must be cleaned periodically when solids are present. This cleaning is usually accomplished using a timed spray of scrubbing liquid or water. The "passive" or lee side of the chevron can also build up with solids. This surface can similarly be flushed, but only when the gas flow is halted. Otherwise the cleaning spray would pass out of the scrubber. Two popular types of chevron are shown in Figures 15 and 16.

Another droplet-removal device, called a mist eliminator or mesh pad, is used to coalesce fine liquid droplets until they enlarge enough to fall, by gravity or capillary action, out of the pad. Used most often where little or no particulate is present, these pads are made in a fine mesh from a variety of materials, both metallic and nonmetallic. The mesh is

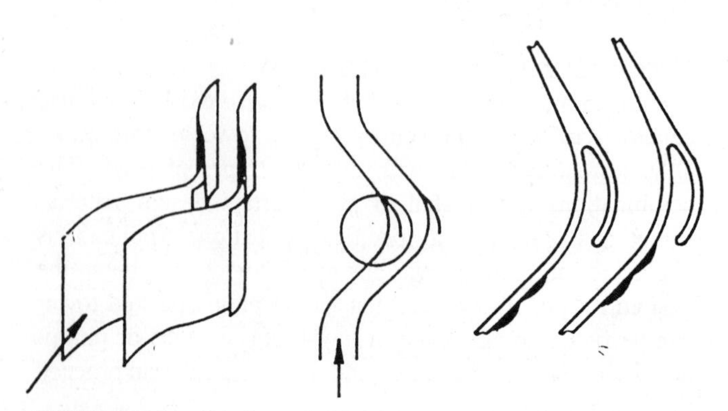

FIGURE 16. Horizontal Gas Flow Chevron (Munters Corp.)

FIGURE 17. Mesh-Type Mist Eliminator

woven and then layered to form the pad. Figure 17 shows a mesh type of mist eliminator.

The mesh-pad material, usually consisting of a filament or wire made from metallic and/or nonmetallic compounds, provides surfaces that permit droplets to coalesce, enlarge, and be removed from the gas stream. The pad material also takes up space in the scrubber vessel. The net open area of the vessel is, therefore, reduced by the presence of the mesh pad. Designers balance the pad type, diameter, and density, while providing acceptable levels of gas throughput. This results in a wide variety of styles of mesh pads from which to choose.

Since the pad occupies tower area and offers flow resistance, mesh pads can usually only be back-flushed when the gas flow is low enough to allow drainage; otherwise flooding will occur. Horizontally oriented pads are back-flushed from above, using liquid uniformly distributed across the surface of the pad.

APPLICATION ENGINEERING BASICS

The key to successful air-pollution control using wet scrubbers lies in the proper selection of the scrubbing method for a given application. This area of engineering is called "application engineering" and is usually performed by a consultant, vendor, or trained specialist.

Some application engineering basics are given in the following.

1. For hot gas streams:
   a. Avoid spraying a liquid containing a high dissolved-solids stream into a hot, nonsaturated gas stream. The liquid will evaporate, leaving fine particulates that are difficult to remove. Use only clean liquid to cool hot (above 250°C) gas streams.
   b. The order of gas cleaning is important. Saturate the gas stream first; then remove particulates. Then remove contaminant gases, if at all possible. If you do not saturate or come near the saturation (within 10°C) temperature, evaporation will occur in the scrubber, driving particulates away from the target droplets, not toward them. Particulates are removed next because gas absorption usually requires media or spray nozzles, which do not perform well with high solids. Gas absorption is best performed with minimal interference from solids or adverse temperature conditions.
   c. Condense the flue gas moisture, if possible. Condensing moisture helps sweep particles out of the gas stream, and encapsulates submicron droplets, making them easier targets for collection.[4]
   d. Allow for thermal expansion and contraction.
   e. Inject chemicals, if used, so that they proceed through the scrubber at least once before being bled away to water treatment or disposal. To save chemicals, bleed solids away first so that the solids do not interfere with the chemicals.
2. For particulate control:
   a. Make certain that the scrubber can handle the peak expected dust loading. The higher the loading, the more you need a scrubber that resists solids fouling. For low loadings, scrubbers with fine perforations, nozzles, or other orifices can be used.
   b. If the process varies in flow, make certain that the scrubber can vary as well.
   c. For high particulate flows, allow for high bleed rates from the scrubber.
   d. In general, bleed solids from the system at the highest solids concentration and highest temperature. If possible, inject clean liquid where the gas is cleanest (the exception being the hot gas stream, mentioned earlier). Avoid running high solids zones where the cleanest gas is desired, that is, clean droplet eliminators with clean liquid. Avoid adding clean liquid to a dirty sump or scrubber tank.
3. For gas absorption:
   a. Place the cleanest, coolest scrubbing liquid nearest the cleanest gas.
   b. Make certain that any scrubbing chemicals (for neutralization, etc.) pass at least once through the system.
   c. If particulate is present at the absorption stage or is created through chemical reaction, make certain that the scrubber design can tolerate these solids.
   d. Allow adequate absorption surface area to remove the peak expected contaminant flow.

In Table 1, common scrubber pressure drops are listed for a variety of applications. The actual pressure loss can vary, so individual vendors should be contacted regarding the pressure loss for their equipment on each specific application.

In subsequent parts of this book, specific applications of

TABLE 1. Typical Applications

Source	Type Scrubber (See Notes)	Flow—ACFM	Pressure Drop Inches W.C.
Incinerators			
Hazardous Waste	HEV/PT	1,000–100,000	45–50
	F/C MV	1,000–100,000	35–45
Hospital Waste	HEV	500–50,000	45–50
	F/C MV	500–50,000	35–40
Liquid Waste	HEV/PT	500–40,000	40–45
Gaseous Waste	PT	100–100,000	5–15
Pulp/Paper Applications			
Lime Sludge Kiln	MEV or	2,000–100,000	18–30
	PAV	2,000–60,000	10–15
Lime Slaker	LEV or	500–20,000	6–12
	FBS or	500–20,000	6–8
	ST	500–20,000	6–10
Dissolving Tank	LEV or	500–30,000	8–16
	MP or	500–30,000	4–8
	FBS or	500–30,000	6–8
	ST	500–30,000	4–8
	LEV/PT	500–30,000	10–20
Bleach Plant	PT or	1,000–60,000	6–10
	FBS	1,000–60,000	6–8
BRN. Stock Washer	PT or	1,000–60,000	6–8
	CS or	1,000–60,000	8–10
	FBS	1,000–60,000	6–8
Chemical Industry			
Soluble Gas	PT or	50–60,000	3–20
	FBS or	50–60,000	6–12
	WF	50–60,000	3–10
Dust	LEV or PAV	50–60,000	6–20
	FBS	50–60,000	6–12

Notes: CS = cyclonic spray scrubber; F/C MV = flue-gas condensation, multiple-throat venturi; FBS = fluidized-bed scrubber (tray or catenary grid); HEV = high-energy venturi; LEV = low-energy venturi; MEV = medium-energy venturi; MP = mesh pad; PAV = pump-aided venturi; PT = packed tower; ST = spray tower; WF = wetted filter.

wet scrubbers will be discussed in detail. We hope this description of wet scrubbing techniques will contribute to the understanding of those specific applications and lead to the successful application of wet scrubbing technology.

References

1. H. E. Hesketh, *Air Pollution Control for Traditional and Hazardous Pollutants,* Technomic Publishers, Lancaster, PA, 1991.
2. S. C. Yung, H. F. Barbarika, and S. Calvert, "Pressure losses in venturi scrubbers," *JAPCA* 27(4):348 (1977).
3. R. H. Boll, "Particulate collection and pressure drop in venturi scrubbers," *IEC Fund* 12(40) (1973).
4. S. Calvert, *Study of Flux Force Condensation Scrubbing of Fine Particles,* EPA-600/2-75-018, U.S. Environmental Protection Agency, Research Triangle Park, NC, 1975.
5. K. Schifftner and H. Hesketh, *Wet Scrubbers,* Lewis Publishers, Chelsea, MI, 1980, pp 44–50.
6. P. A. Schweitzer, *Handbook of Separation Techniques for Chemical Engineers,* 2nd ed., McGraw-Hill Book Co., New York, 1988, Section 3.2.
7. R. Perry, C. H. Chilton, Eds., *Chemical Engineer's Handbook,* 5th ed., McGraw-Hill Book Co., New York, 1974.

ELECTROSTATIC PRECIPITATORS

James H. Turner, Phil A. Lawless, Toshiaki Yamamoto, David W. Coy, Gary P. Greiner, John D. McKenna, and William M. Vatavuk

An electrostatic precipitator (ESP) is a particle control device that uses electrical forces to move the particles out of the flowing gas stream and onto collector plates. The particles are given an electric charge by forcing them to pass through a corona, a region in which gaseous ions flow. The electrical field that forces the charged particles to the walls comes from electrodes maintained at high voltage in the center of the flow lane.

Once the particles are collected on the plates, they must be removed from the plates without reentraining them into the gas stream. This is usually accomplished by knocking them loose from the plates, allowing the collected layer of particles to slide down into a hopper, from which they are evacuated. Some precipitators remove the particles by intermittent or continuous washing with water.

Figure 1 shows an ESP with its various components identified. Figure 2 shows two variations of charging electrode/collector electrode arrangements used in ESPs.

TYPES OF ESPs

Electrostatic precipitators are configured in several ways. Some of these configurations have been developed for special control action, and others have evolved for economic reasons. The types that will be described here are (1) the plate-wire precipitator, the most common variety; (2) the flat-plate precipitator; (3) the tubular precipitator; (4) the wet precipitator, which may have any of the previous mechanical configurations; and (5) the two-stage precipitator.

Plate-Wire Precipitator

Plate-wire ESPs are used in a wide variety of industrial applications, including coal-fired boilers, cement kilns, solid waste incinerators, paper mill recovery boilers, petroleum refining catalytic cracking units, sinter plants, basic oxygen furnaces, open hearth furnaces, electric arc furnaces, coke oven batteries, and glass furnaces.

In a plate-wire ESP, gas flows between parallel plates of sheet metal and high-voltage electrodes. These electrodes are long wires weighted and hanging between the plates or are supported there by mastlike structures (rigid frames). Within each flow path, gas flow must pass each wire in sequence as it flows through the unit.

The plate-wire ESP allows many flow lanes to operate in parallel, and each lane can be quite tall. As a result, this type of precipitator is well-suited for handling large volumes of gas. The need for rapping the plates to dislodge the collected material has caused the plate to be divided into sections, often three or four in series with one another, which can be rapped independently. The power supplies are often sectionalized in the same way to obtain higher operating voltages, and further electrical sectionalization may be used for increased reliability. Dust also deposits on the discharge electrode wires and must be periodically removed similarly to the collector plate.

The power supplies for the ESP convert the industrial ac voltage (220–480 volts) to pulsating dc voltage in the range of 20,000–100,000 volts as needed. The supply consists of a step-up transformer, high-voltage rectifiers, and sometimes filter capacitors. The unit may supply either half-wave or full-wave rectified dc voltage. There are auxiliary components and controls to allow the voltage to be adjusted to the highest level possible without excessive sparking and to protect the supply and electrodes in the event a heavy arc or short circuit occurs.

The voltage applied to the electrodes causes the gas between the electrodes to break down electrically, an action known as a "corona." The electrodes usually are given a negative polarity because a negative corona supports a higher voltage than does a positive corona before sparking occurs. The ions generated in the corona follow electric field lines from the wires to the collecting plates. Therefore, each wire establishes a charging zone through which the particles must pass.

Particles passing through the charging zone intercept some of the ions, which become attached. Small aerosol particles (<1 μm diameter) can absorb tens of ions before their total charge becomes large enough to repel further ions, and large particles (>10 μm diameter) can absorb tens of thousands. The electrical forces are therefore much stronger on the large particles.

As the particles pass each successive wire, they are driven closer and closer to the collecting walls. The turbulence in the gas, however, tends to keep them uniformly mixed with the gas. The collection process is therefore a competition between the electrical and dispersive forces. Eventually, the particles approach close enough to the walls so that the turbulence drops to low levels and the particles are collected.

If the collected particles could be dislodged into the hopper without losses, the ESP would be extremely efficient. The rapping that dislodges the accumulated layer also projects some of the particles (typically 12% for coal fly ash) back into the gas stream. These reentrained particles are then processed again by later sections, but the particles reentrained in the last section of the ESP have no chance to be recaptured and so escape the unit.

Practical considerations of passing the high voltage into

Used with permission from the *Journal of the Air and Waste Management Association.*

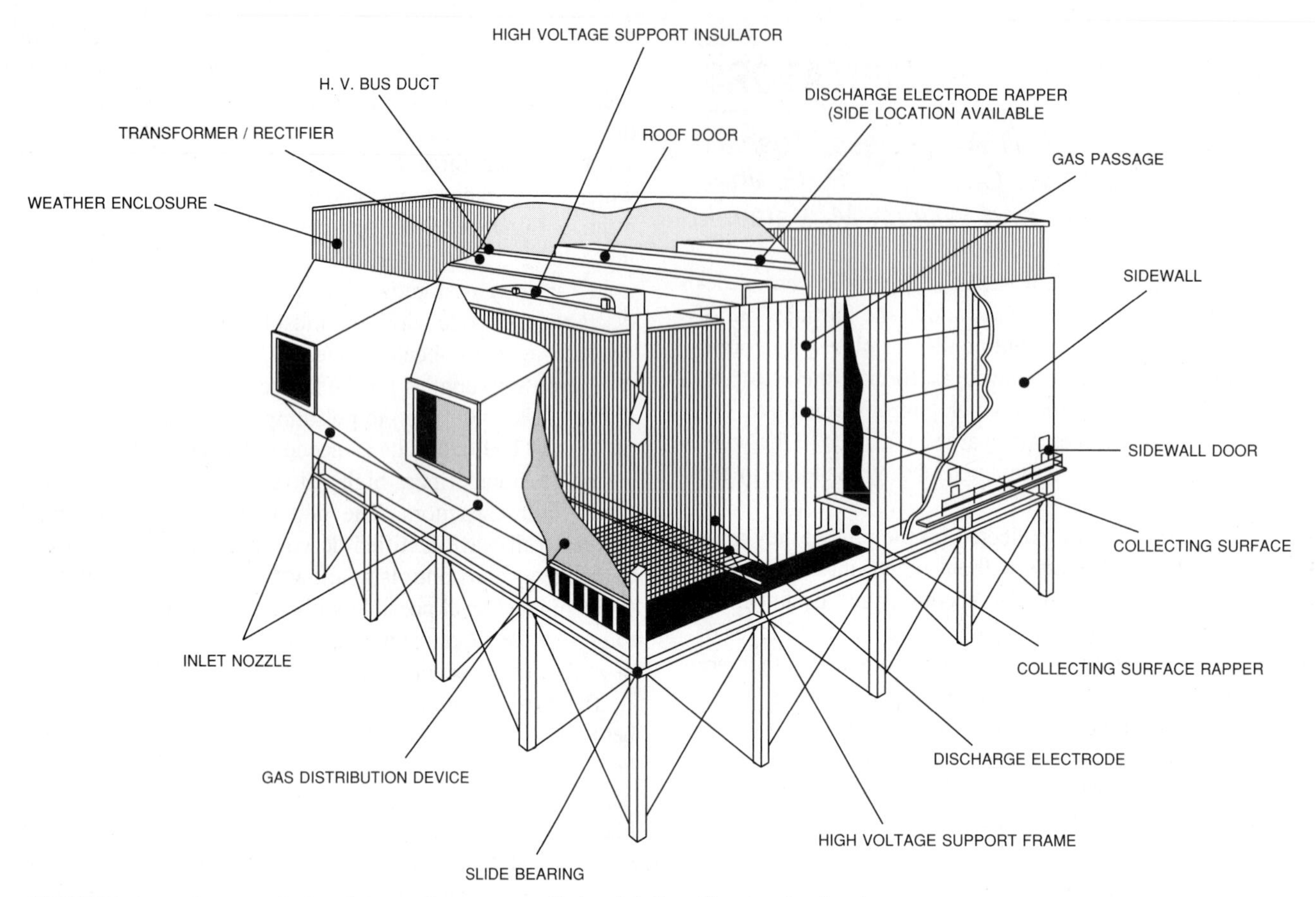

FIGURE 1. Electrostatic Precipitator Components (Industrial Gas Cleaning Institute)

the space between the lanes and allowing for some clearance above the hoppers to support and align electrodes leave room for part of the gas to flow around the charging zones. This is called "sneakage" and amounts to 5–10% of the total flow. Antisneakage baffles usually are placed to force the sneakage flow to mix with the main gas stream for collection in later sections. But, again, the sneakage flow around the last section has no opportunity to be collected.

These losses play a significant role in the overall performance of an ESP. Another major factor is the resistivity of the collected material. Because the particles form a continuous layer on the ESP plates, all the ion current must pass through the layer to reach the ground plates. This current creates an electric field in the layer, and it can become large enough to cause local electrical breakdown. When this occurs, new ions of the wrong polarity are injected into the wire-plate gap where they reduce the charge on the particles and may cause sparking. This breakdown condition is called "back corona."

Back corona is prevalent when the resistivity of the layer is high, usually above 2×10^{11} ohm-cm. For lower resistivities, the operation of the ESP is not impaired by back corona, but resistivities much higher than 2×10^{11} ohm-cm considerably reduce the collection ability of the unit because the severe back corona causes difficulties in charging the particles. At resistivities below 10^8 ohm-cm, the particles are held on the plates so loosely that rapping and nonrapping reentrainment become much more severe. Care must be taken in measuring or estimating resistivity because it is strongly affected by such variables as temperature, moisture, gas composition, particle composition, and surface characteristics.

Flat-Plate Precipitators

A significant number of smaller precipitators (100,000–200,000 acfm) use flat plates instead of wires for the high-voltage electrodes. The flat plates (United McGill Corp. patents) increase the average electric field that can be used to collect the particles, and they provide an increased surface area for the collection of particles. Corona cannot be generated on flat plates by themselves, so corona-generating electrodes are placed ahead of and sometimes behind the flat-plate collecting zones. These electrodes may be sharp-pointed needles attached to the edges of the plates or independent corona wires. Unlike plate-wire or tubular ESPs, this design operates equally well with either negative or positive polarity. The manufacturer has chosen to use positive polarity to reduce ozone generation.

A flat-plate ESP operates with little or no corona current

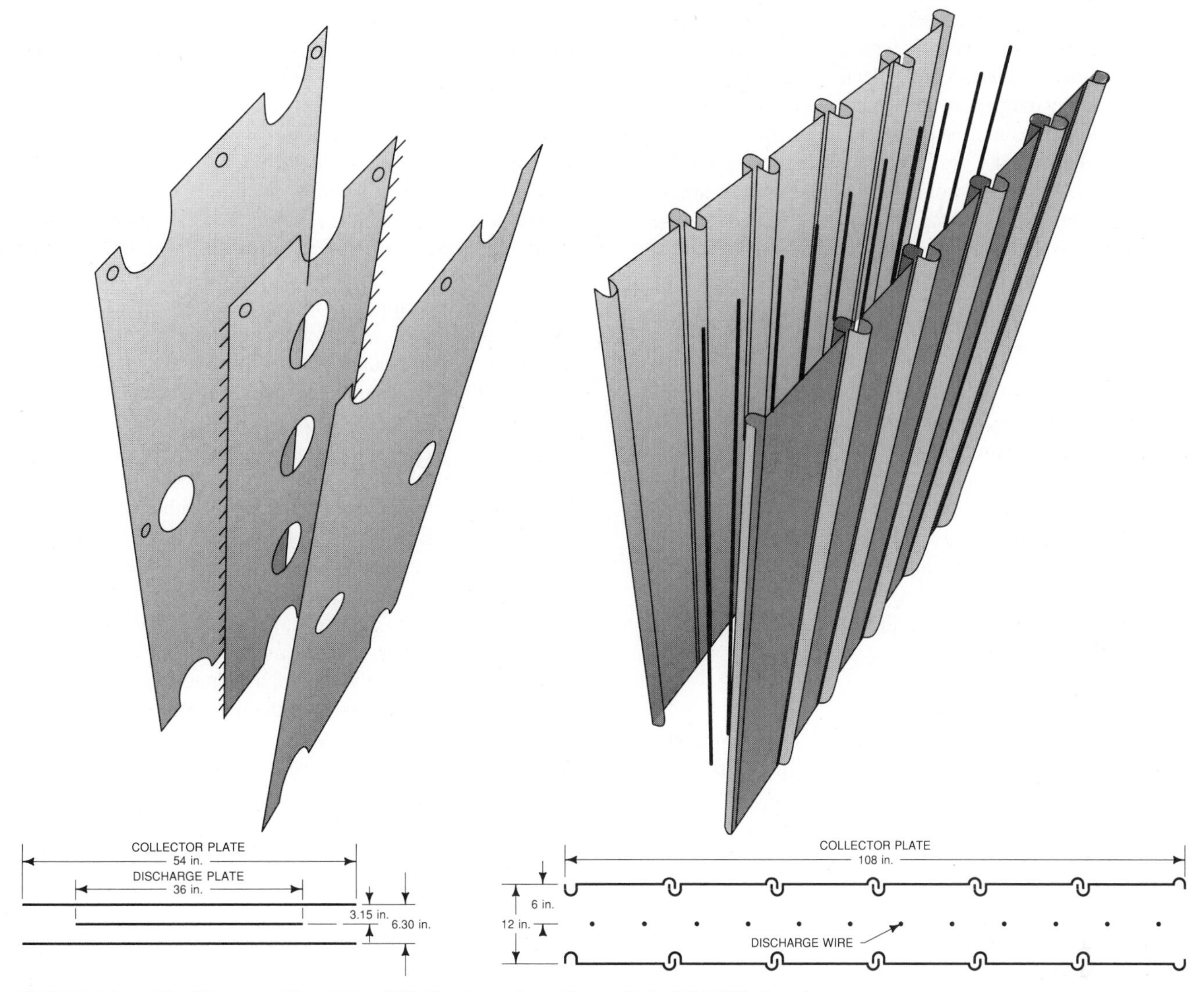

FIGURE 2. Flat-Plate and Plate-Wire ESP Configurations (From United McGill Corp.)

flowing through the collected dust, except directly under the corona needles or wires. This has two consequences. The first is that the unit is somewhat less susceptible to back corona than conventional units are because no back corona is generated in the collected dust, and particles charged with both polarities of ions have large collection surfaces available. The second consequence is that the lack of current in the collected layer causes an electrical force that tends to remove the layer from the collecting surface; this can lead to high rapping losses.

Flat-plate ESPs seem to have wide application for high-resistivity particles with small (1–2 μm) mass median diameters (MMDs). These applications especially emphasize the strengths of the design because the electrical dislodging forces are weaker for small particles than for large ones. Fly ash has been successfully collected with this type of ESP, but low-flow velocity appears to be critical for avoiding high rapping losses.

Tubular Precipitators

The original ESPs were tubular, like the smokestacks on which they were placed, with the high-voltage electrode running along the axis of the tube. Tubular precipitators have typical applications in sulfuric acid plants, coke oven by-product gas cleaning (tar removal), and, recently, iron and steel sinter plants. Such tubular units are still used for some applications, with many tubes operating in parallel to handle increased gas flows. The tubes may be formed as a circular, square, or hexagonal honeycomb with gas flowing upward or downward. The length of the tubes can be selected to fit conditions. A tubular ESP can be tightly sealed to prevent leaks of material, especially valuable or hazardous material.

A tubular ESP is essentially a one-stage unit and is unique in having all the gas pass through the electrode region. The high-voltage electrode operates at one voltage for the entire length of the tube, and the current varies along the length as the particles are removed from the system. No sneakage paths are around the collecting region, but corona nonuniformities may allow some particles to avoid charging for a considerable fraction of the tube length.

Tubular ESPs make up only a small portion of the ESP population and are most commonly applied where the particulate is either wet or sticky. These ESPs, usually cleaned with water, have reentrainment losses of a lower magnitude than do the dry particulate precipitators.

Wet Precipitators

Any of the precipitator configurations discussed above may be operated with wet walls instead of dry. The water flow may be applied intermittently or continuously to wash the collected particles into a sump for disposal. The advantage of the wet wall precipitator is that it has no problems with rapping reentrainment or with back corona. The disadvantage is the increased complexity of the wash and the fact that the collected slurry must be handled more carefully than a dry product, adding to the expense of disposal.

Two-Stage Precipitators

The previously described precipitators are all parallel in nature; that is, the discharge and collecting electrodes are side by side. The two-stage precipitator invented by Penney is a series device with the discharge electrode, or ionizer, preceding the collector electrodes. For indoor applications, the unit is operated with positive polarity to limit ozone generation.

Advantages of this configuration include more time for particle charging, less propensity for back corona, and economical construction for small sizes. This type of precipitator is generally used for gas-flow volumes of 50,000 acfm and less and is applied to submicrometer sources emitting oil mists, smokes, fumes, or other sticky particulates because there is little electrical force to hold the collected particulates on the plates. Modules consisting of a mechanical prefilter, ionizer, collecting-plate cell, afterfilter, and power pack may be placed in parallel or series-parallel arrangements. Preconditioning of gases is normally part of the system. Cleaning may be by water wash of modules removed from the system up to automatic, in-place detergent spraying of the collector followed by air-blow drying.

Two-stage precipitators are considered to be a separate and distinct type of device as compared with large, high-gas-volume, single-stage ESPs. The smaller devices are usually sold as preengineered package systems.

AUXILIARY EQUIPMENT

Typical auxiliary equipment associated with an ESP system is shown schematically in Figure 3. Along with the ESP itself, a control system usually includes the following auxiliary equipment: a capture device (i.e., hood or direct exhaust connection); ductwork; dust removal equipment (screw conveyor, etc.); fans, motors, and starters; and a stack. In addition, spray coolers and mechanical collectors may be needed to precondition the gas before it reaches the ESP. Capture devices are usually hoods that exhaust pollutants into the ductwork or are direct exhaust couplings attached to a combustor or process equipment. These devices are usually refractory lined, water cooled, or simply fabricated from carbon steel, depending on the gas-stream temperatures.

Refractory or water-cooled capture devices are used where the wall temperatures exceed 800°F; carbon steel is used for lower temperatures. The ducting, like the capture device, should be water cooled, refractory, or stainless steel for hot processes and carbon steel for gas temperatures below approximately 1150°F (duct wall temperatures

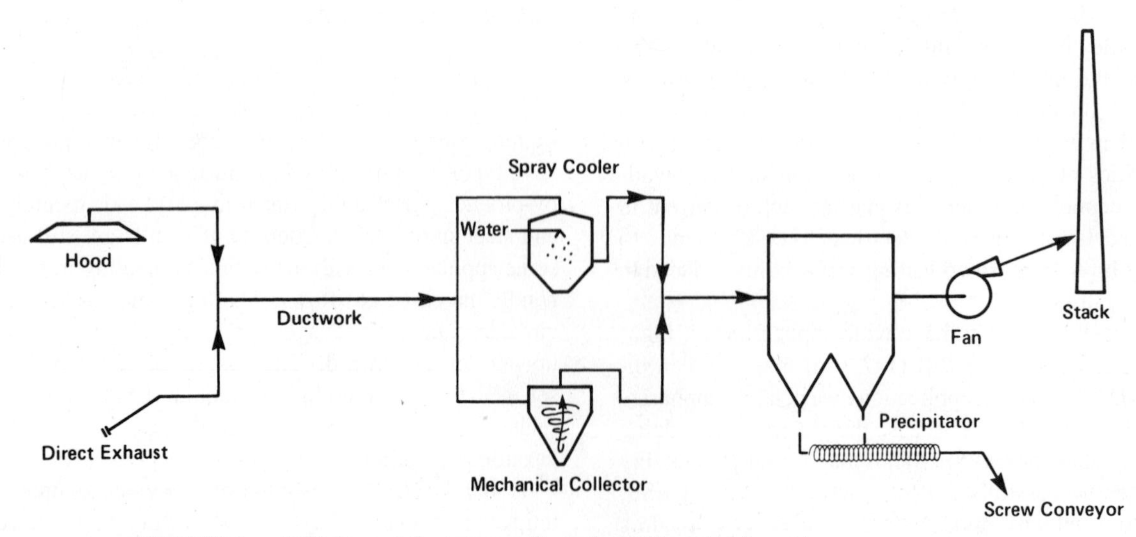

FIGURE 3. ESP Control System and Typical Auxiliary Equipment

<800°F). The ducts should be sized for a gas velocity of approximately 4000 fpm for the average case to prevent particle deposition in the ducts. Large or dense particles might require higher velocities, but rarely would lower velocities be used. Spray chambers may be required for processes where the addition of moisture, or decreased temperature or gas volume, will improve precipitation or protect the ESP from warpage. For combustion processes with exhaust gas temperatures below approximately 700°F, cooling would not be required, and the exhaust gases can be delivered directly to the precipitator.

When much of the pollutant loading consists of relatively large particles, mechanical collectors, such as cyclones, may be used to reduce the load on the ESP, especially at high inlet concentrations. The fans provide the motive power for air movement and can be mounted before or after the ESP. A stack, normally used, vents the cleaned stream to the atmosphere. Screw conveyors or pneumatic systems are often used to remove captured dust from the bottom of the hoppers.

Wet ESPs require a source of wash water to be injected or sprayed near the top of the collector plates either continuously or at timed intervals. The water flows with the collected particles into a sump from which the fluid is pumped. A portion of the fluid may be recycled to reduce the total amount of water required. The remainder is pumped directly to a settling pond or passed through a dewatering stage, with subsequent disposal of the sludge.

Gas conditioning equipment to improve ESP performance by changing dust resistivity is occasionally used as part of the original design, but more frequently it is used to upgrade existing ESPs. The equipment injects an agent into the gas stream ahead of the ESP. Usually, the agent mixes with the particles and alters their resistivity to promote higher migration velocity and thus higher collection efficiency. However, electrical properties of the gas may change, rather than dust resistivity. For instance, cooling the gas will allow more voltage to be applied before sparking occurs. Significant conditioning agents that are used include SO_3, H_2SO_4, sodium compounds, ammonia, and water, but the major conditioning agent by usage is SO_3. A typical dose rate for any of the gaseous agents is 10–30 ppm by volume.

The equipment required for conditioning depends on the agent being used. A typical SO_3 conditioner requires a supply of molten sulfur. It is stored in a heated vessel and supplied to a burner, where it is oxidized to SO_2. The SO_2 gas is passed over a catalyst for further oxidation to SO_3. The SO_3 gas is then injected into the flue gas stream through a multioutlet set of probes that breaches a duct. In place of a sulfur burner to provide SO_2, liquid SO_2 may be vaporized from a storage tank. Although more expensive in annualized cost, such systems have lower capital cost and are easier to operate.

Water or ammonia injection requires a set of spray nozzles in the duct, along with pumping and control equipment.

Sodium conditioning is often done by coating the coal on a conveyor with a powder compound or a water solution of the desired compound. A hopper or storage tank is often positioned over the conveyor for this purpose.

ELECTROSTATIC PRECIPITATION THEORY

The theory of ESP operation requires many scientific disciplines to describe it thoroughly. The ESP is basically an electric machine. The principal actions are the charging of particles and forcing them to the collector plates. The amount of charged particulate matter affects the electrical operating point of the ESP. The transport of the particles is affected by the level of turbulence in the gas. The losses mentioned earlier, sneakage and rapping reentrainment, are major influences on the total performance of the system. The particle properties also have a major effect on the operation of the unit.

The following subsections will explain the theory behind (1) electrical operating points in the ESP, (2) particle charging, (3) particle collection, and (4) sneakage and rapping reentrainment. General references for these topics are White[1] or Lawless and Sparks.[2]

Electrical Operating Point

The electrical operating point of an ESP section is the value of voltage and current at which the section operates. As will become apparent, the best collection occurs when the highest electric field is present, which roughly corresponds to the highest voltage on the electrodes.

In this work, the term "section" represents one set of plates and electrodes having a common power source. In the direction of flow, this unit is commonly called a "field," and a section or "bus section" represents a subdivision of a field perpendicular to the direction of flow. In an ESP model, and in sizing applications, the two terms section and field are used equivalently because the subdivision into bus sections should have no effect on the model. This terminology has probably arisen because of the frequent use of the word field to refer to the electric field.

The lowest acceptable voltage is the voltage required for the formation of a corona, the electrical discharge that produces ions for charging particles. The (negative) corona is produced when an occasional free electron near the high-voltage electrode, produced by a cosmic ray, gains enough energy from the electric field to ionize the gas and produce more free electrons. The electric field for which this process is self-sustained has been determined experimentally. For round wires, the field at the surface of the wire is given by:

$$E_c = 3.126 \times 10^6 [d_r + 0.0301 (d_r / r_w)] \tag{1}$$

where: E_c = corona onset field *at the wire surface* V/m

d_r = relative gas density, referred to 1 atm pressure and 20°C

r_w = radius of the wire, meters

This is the field required to produce "glow" corona, the form usually seen in the laboratory on smooth, clean wires. The glow appears as a uniform, rapidly moving diffuse light around the electrode. After a period of operation, the movement concentrates into small spots on the wire surface, and the corona assumes a tuftlike appearance. The field required to produce "tuft" corona, the form found in full-scale ESPs, is 0.6 times the value of E_c.

The voltage that must be applied to the wire to obtain this value of field, V_c, is found by integrating the electric field from the wire to the plate. The field follows a simple $1/r$ dependence in cylindrical geometry. This leads to a logarithmic dependence of voltage on electrode dimensions. In the plate-wire geometry, the field dependence is somewhat more complex, but the voltage still shows the logarithmic dependence. V_c is given by:

$$V_c = E_c r_w \ln(d/r_w) \tag{2}$$

where: V_c = corona onset voltage, volts
d = outer cylinder radius in a tubular ESP, meters, or
d = $4/\pi \cdot$ wire-plate separation for plate-wire ESP, meters

No current will flow until the voltage reaches this value, but the amount of current will increase steeply for voltages above this value. The maximum current density (A/m^2) on the plate or cylinder directly under the wire is given by:

$$j = \mu \epsilon V^2 / L^3 \tag{3}$$

where: j = maximum current density, A/m^2
μ = ion mobility, m^2/V s
ϵ = free-space permittivity, 8.845×10^{-12} F/m
V = applied voltage, volts
L = shortest distance from wire to collecting surface, meters

For tuft corona, the current density is zero until the corona onset voltage is reached, when it jumps almost to this value of j within a few hundred volts.

The region near the wire is strongly influenced by the presence of ions there, and the corona onset voltage magnitude shows strong spatial variations. Outside the corona region, it is quite uniform.

The electric field is strongest along the line from wire to plate and is approximated very well, except near the wire, by:

$$E_{max} = V/L \tag{4}$$

where: E_{max} = maximum field strength, V/m

When the electric field throughout the gap between the wire and the plate becomes strong enough, a spark will occur, and the voltage cannot be increased without severe sparking occurring. The field at which sparking occurs is not sharply defined, but a reasonable value is given by:

$$E_s = 6.3 \times 10^5 \times (273/T \times P)^{1.65} \tag{5}$$

where: E_s = sparking field strength, V/m
T = absolute temperature, K
P = gas pressure, atm

This field would be reached at a voltage of, for example, 35,000 volts for a wire-plate spacing of 11.4 cm (4.5 inches) at a temperature of 149°C (300°F). The ESP will generally operate near this voltage in the absence of back corona. E_{max} will be equal to or less than E_s.

Instead of sparking, back corona may occur if the electric field in the dust layer, resulting from the current flow in the layer, reaches a critical value of about 1×10^6 V/m. Depending on conditions, the back corona may enhance sparking or may generate so much current that the voltage cannot be raised any higher. The field in the layer is given by:

$$E_l = j \times \rho \tag{6}$$

where: E_l = electric field in dust layer, V/m
ρ = resistivity of the collected material, ohm-m

Particle Charging

Charging of particles takes place when ions bombard the surface of a particle. Once an ion is close to the particle, it is tightly bound because of the image charge within the particle. The "image charge" is a representation of the charge distortion that occurs when a real charge approaches a conducting surface. The distortion is equivalent to a charge of opposite magnitude to the real charge, located as far below the surface as the real charge is above it. The motion of the fictitious charge is similar to the motion of an image in a mirror, hence the name. As more ions accumulate on a particle, the total charge tends to prevent further ionic bombardment.

There are two principal charging mechanisms: diffusion charging and field charging. Diffusion charging results from the thermal kinetic energy of the ions overcoming the repulsion of the ions already on the particle. Field charging occurs when ions follow electric field lines until they terminate on a particle. In general, both mechanisms are operative for all sizes of particles. Field charging, however, adds a larger percentage of charge on particles greater than about 2 μm in diameter, and diffusion charging adds a greater percentage on particles smaller than about 0.5 μm.

Diffusion charging, as derived by White,[1] produces a logarithmically increasing level of charge on particles, given by:

$$q(t) = rkT/e \ln(1 + \tau) \quad (7)$$

where: $q(t)$ = particle charge (C) as function of time, seconds
r = particle radius, meters
k = Boltzmann's constant, j/K
T = absolute temperature, K
e = electron charge, 1.67×10^{-19} C
τ = dimensionless time given by:

$$\tau = \pi r v N e^2 \theta/(kT) \quad (8)$$

where: v = mean thermal speed of the ions, m/s
N = ion number concentration near the particle, No./m^3
θ = real time (exposure time in the charging zone), seconds

Diffusion charging never reaches a limit, but it becomes very slow after about three dimensionless time units. For fixed exposure times, the charge on a particle is proportional to its radius.

Field charging also exhibits a characteristic time dependence, given by:

$$q(t) = q_s\theta/(\theta + \tau') \quad (9)$$

where: q_s = saturation charge (charge at infinite time), C
θ = real time, seconds
τ' = another dimensionless time unit

The saturation charge is given by:

$$q_s = 12\pi\epsilon r^2 E \quad (10)$$

where: ϵ = free-space permittivity, F/m
E = external electric field applied to the particle, V/m

The saturation charge is proportional to the square of the radius, which explains why field charging is the dominant mechanism for larger particles. The field charging time constant is given by:

$$\tau' = 4\epsilon/Ne\mu \quad (11)$$

where: μ = ion mobility (all other terms are as defined previously)

Strictly speaking, both diffusion and field charging mechanisms operate at the same time on all particles, and neither mechanism is sufficient to explain the charges measured on the particles. It has been found empirically that a very good approximation to the measured charge is given by the sum of the charges predicted by equations 7 and 9 independently of one another:

$$q_{tot}(t) = q_d(t) + q_f(t) \quad (12)$$

where: q_{tot} = particle charge due to both mechanisms
q_d = particle charge due to diffusion charging
q_f = particle charge due to field charging

Particle Collection

The electric field in the collecting zone produces a force on a particle proportional to the magnitude of the field and to the charge:

$$F_e = qE \quad (13)$$

where: F_e = force due to electric field, N
q = charge on particle, C
E = electric field, V/m

Because the field charging mechanism gives an ultimate charge proportional to the electric field, the force on large particles is proportional to the square of the field, which shows the advantage of maintaining as high a field as possible.

The motion of the particles under the influence of the electric field is opposed by the viscous drag of the gas. By equating the electric force and the drag force component due to the electric field (according to Stokes' law), we can obtain the particle velocity:

$$v(q,E,r) = q(E,r)EC(r)/(6\pi\eta r) \quad (14)$$

where: v = particle velocity, m/s
q = particle charge, C
C = Cunningham correction to Stokes' law (dimensionless)
η = gas viscosity, kg/ms

The particle velocity is the rate at which the particle moves along the electric field lines, that is, toward the walls.

For a given electric field, this velocity is usually at a minimum for particles of about 0.5 μm diameter. Smaller particles move faster because the charge does not decrease very much, but the Cunningham factor increases rapidly as radius decreases. Larger particles have a charge increasing as r^2 and a viscous drag only increasing as r^1; the velocity then increases as r.

Equation 14 gives the particle velocity with respect to still air. In the ESP, the flow is usually very turbulent, with instantaneous gas velocities of the same magnitude as the particles' velocities, but in random directions. The motion of particles toward the collecting plates is therefore a statistical process, with an average component imparted by the electric field and a fluctuating component from the gas turbulence.

This statistical motion leads to an exponential collection equation, given by:

$$N(r) = N_0(r) \exp(-v(r)/v_0) \quad (15)$$

where: N = particle concentration of size r at the exit of the collecting zone, No./m^3
N_0 = particle concentration at the entrance of the zone, No./m^3
v = size-dependent particle velocity, m/s
v_0 = characteristic velocity of the ESP, m/s, given by:

$$v_0 = Q/A = 1/SCA \quad (16)$$

where: Q = volume flow rate of the gas, m^3/s
A = plate area for the ESP collecting zone, m^2
SCA = specific collection area (area per unit volume), s/m

When this collection equation is averaged over all the particle sizes and weighted according to the concentration of each size, the Deutsch equation results, with the penetration (fraction of particles escaping) given by:

$$p = \exp(-w_e\ SCA) \quad (17)$$

where: p = penetration (fraction)
w_e = effective migration velocity for the particle ensemble, m/s

The efficiency is given by:

$$Eff = 100 \times (1 - p) \quad (18)$$

This is the number most often used to describe the performance of an ESP.

Sneakage and Rapping Reentrainment

Sneakage and rapping reentrainment are best considered on the basis of the sections within an ESP. Sneakage occurs when a part of the gas flow bypasses the collection zone of a section. Generally, the portion of gas that bypasses the zone is thoroughly mixed with the gas that passes through the zone before all the gas enters the next section. This mixing cannot always be assumed, and when sneakage paths exist around several sections, the performance of the whole ESP is seriously affected. To describe the effects of sneakage and rapping reentrainment mathematically, we first consider sneakage by itself and then consider the effects of rapping as an average over many rapping cycles.

On the assumption that the gas is well mixed between sections, the penetration for each section can be expressed as:

$$p_s = S_N + (1 - S_N)p_c(Q') \quad (19)$$

where: p_s = section's fractional penetration
S_N = fraction of gas bypassing the section (sneakage)
$p_c(Q')$ = fraction of particles penetrating the collection zone, which is functionally dependent on Q', the gas volume flow in the collection zone, reduced by the sneakage, m^3/s

The penetration of the entire ESP is the product of the section penetrations. The sneakage sets a lower limit on the penetration of particles through the section.

To calculate the effects of rapping, we first calculate the amount of material captured on the plates of the section. The fraction of material that was caught is given by:

$$m/m_0 = 1 - p_s = 1 - S_N - (1 - S_N)p_c(Q') \quad (20)$$

where: m/m_0 = mass fraction collected from the gas stream

This material accumulates until the plates are rapped, whereupon most of the material falls into the hopper for disposal, but a fraction of it is reentrained and leaves the section. Experimental measurements have been conducted on fly ash ESPs to evaluate the fraction reentrained, which averages about 12%.

The average penetration for a section, including sneakage and rapping reentrainments, is:

$$p_s = S_N + (1 - S_N)p_c(Q') + RR\{(1 - S_N)[1 - p_c(Q')]\} \quad (21)$$

where: RR = fraction reentrained

This can be written in a more compact form as:

$$p_s = LF + (1 - LF)\ p_c(Q') \quad (22)$$

by substituting LF (loss factor) for $S_N + RR - S_N \times RR$. These formulas can allow for variable amounts of sneakage and rapping reentrainment for each section, but there is no experimental evidence to suggest that it is necessary.

Fly ash precipitators analyzed in this way have an average S_N of 0.07 and an RR of 0.12. These values are the best available at this time, but some wet ESPs, which presumably have no rapping losses, have shown S_N values of 0.05 or less. These values offer a means for estimating the performance of ESPs whose actual characteristics are not known, but about which general statements can be made. For instance, wet ESPs would be expected to have $RR = 0$, as would ESPs collecting wet or sticky particles. Particulate materials with a much smaller MMD than fly ash would be expected to have a lower RR factor because they are held more tightly to the plates and each other. Sneakage factors are harder to account for; unless special efforts have been made in the design to control sneakage, the 0.07 value should be used.

ESP DESIGN PROCEDURE

Specific Collecting Area

Specific collecting area (SCA) is a parameter used to compare ESPs and roughly estimate their collection efficiency. The SCA is the total collector plate area divided by gas volume flow rate and has the units of s/m or s/ft. It is often expressed as $m^2/(m^3/s)$ or ft^2/kacfm (thousand acfm). It is also one of the most important factors in determining the capital and several of the annual costs (for example, maintenance and dust disposal costs) of the ESP because it determines the size of the unit. Because of the various ways in which SCA can be expressed, Table 1 gives equivalent SCAs in the different units for what would be considered a small, medium, and large SCA.

TABLE 1. Small, Medium, and Large SCAs as Expressed by Various Units

Units[a]	Small	Medium	Large
ft^2/kacfm	100	400	900
s/m	19.7	78.8	177
s/ft	6	24	54

[a] ft^2/kacfm = (s/m) × 5.079.

The design procedure is based on the loss factor approach of Lawless and Sparks[2] and considers a number of process parameters. It can be calculated by hand, but it is most conveniently used with a spreadsheet program. For many uses, tables of effective migration velocities can be used to obtain the SCA required for a given efficiency. In the following subsection, tables have been calculated using the design procedure for a number of different particle sources and for differing levels of efficiency. If a situation is encountered that is not covered in these tables, then the full procedure that appears at the end of the subsection should be used.

TABLE 2. Plate-Wire ESP Migration Velocities (cm/s [fps])[a]

Particle Source	Design Efficiency, %			
	95	99	99.5	99.9
Bituminous coal fly ash[b]				
(no BC)	12.6 (0.41)	10.1 (0.33)	9.3 (0.31)	8.2 (0.27)
(BC)	3.1 (0.10)	2.5 (0.082)	2.4 (0.079)	2.1 (0.069)
Subbituminous coal fly ash in tangential-fired boiler[b]				
(no BC)	17.0 (0.56)	11.8 (0.39)	10.3 (0.34)	8.8 (0.29)
(BC)	4.9 (0.16)	3.1 (0.10)	2.6 (0.085)	2.2 (0.072)
Other coal[b] (no BC)	9.7 (0.32)	7.9 (0.26)	7.9 (0.26)	7.2 (0.24)
(BC)	2.9 (0.095)	2.2 (0.072)	2.1 (0.069)	1.9 (0.062)
Cement kiln[c] (no BC)	1.5 (0.049)	1.5 (0.049)	1.8 (0.059)	1.8 (0.059)
(BC)	0.6 (0.020)	0.6 (0.020)	0.5 (0.016)	0.5 (0.016)
Glass plant[d] (no BC)	1.6 (0.052)	1.6 (0.052)	1.5 (0.049)	1.5 (0.049)
(BC)	0.5 (0.016)	0.5 (0.016)	0.5 (0.016)	0.5 (0.016)
Iron/steel sinter plant dust with mechanical precollector[b]				
(no BC)	6.8 (0.20)	6.2 (0.20)	6.6 (0.22)	6.3 (0.21)
(BC)	2.2 (0.072)	1.8 (0.059)	1.8 (0.059)	1.7 (0.056)
Kraft-paper-recovery boiler[b]				
(no BC)	2.6 (0.085)	2.5 (0.082)	3.1 (0.10)	2.9 (0.095)
Incinerator fly ash[e] (no BC)	15.3 (0.50)	11.4 (0.37)	10.6 (0.35)	9.4 (0.31)
Copper reverberatory furnace[f]				
(no BC)	6.2 (0.20)	4.2 (0.14)	3.7 (0.12)	2.9 (0.095)
Copper converter[g] (no BC)	5.5 (0.18)	4.4 (0.14)	4.1 (0.13)	3.6 (0.12)
Copper roaster[h] (no BC)	6.2 (0.20)	5.5 (0.18)	5.3 (0.17)	4.8 (0.16)
Coke plant combustion stack[c] (no BC)	1.2[j] (0.039)	—	—	—

[a]To convert cm/s to fps, multiply cm/s by 0.0328, but computational procedure uses SI units. To convert cm/s to m/s, multiply cm/s by 0.01. Assumes same particle size as given in full computational procedure.
[b]At 300°F. Depending on individual furnace/boiler conditions, chemical nature of the fly ash, and availability of naturally occurring conditioning agents (e.g., moisture in the gas stream), migration velocities may vary considerably from these values. Likely values are in the range from back corona to no back corona. BC = back corona.
[c]At 600°F.
[d]At 500°F.
[e]At 250°F.
[f]450°F to 570°F.
[g]500°F to 700°F.
[h]600°F to 660°F.
[i]360°F to 450°F.
[j]Data available only for inlet concentrations in the range of 0.02 to 0.2 g/s m^3 and for efficiencies less than 90%.

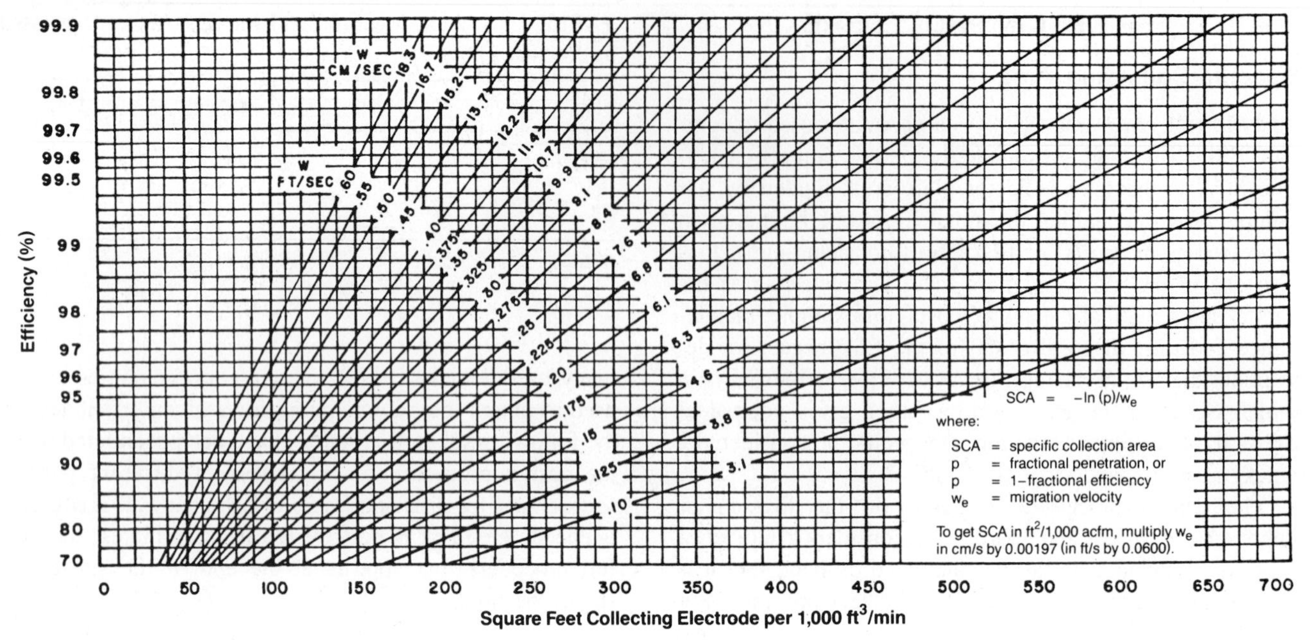

FIGURE 4. Chart for Finding Specific Collection Area (SCA)

Migration Velocity

If the migration velocity is known, then equation 17 can be rearranged to give the SCA:

$$SCA = -\ln(p)/w_e \tag{23}$$

A graphical solution to equation 23 is given in Figure 4. The migration velocities have been calculated for three main precipitator types: plate-wire, flat-plate, and wet wall ESPs of the plate-wire type. The following three tables, keyed to design efficiency as an easily quantified variable, summarize the velocities under various conditions. In Table 2, the migration velocities are given for conditions of no back corona and severe back corona; temperatures appropriate for each process have been assumed.

In Table 3, the migration velocities calculated for a wet wall ESP of the plate-wire type assume no back corona and no rapping reentrainment.

In Table 4, the flat-plate ESP migration velocities are given only for no back corona conditions because they appear to be less affected by high-resistivity dusts than the plate-wire types.

It is generally expected from experience that the migration velocity will decrease with increasing efficiency. In Tables 2 through 4, however, the migration velocities show some fluctuations. This is so because the number of sections must be increased as the efficiency increases, and the changing sectionalization affects the overall migration velocity. This effect is particularly noticeable, for example, in Table 4 for glass plants. When the migration velocities in the tables are used to obtain SCAs for the different efficiencies in the tables, the SCAs will increase as the efficiency increases.

The full procedure for determining the SCA for large ESPs is given here. The smaller, two-stage precipitators are generally sold on the basis of a nominal ft³/min rating for packaged modules.

TABLE 3. Wet Wall Plate-Wire ESP Migration Velocities (cm/s [fps])[a]

	Design Efficiency, %			
Particle Source[b]	95	99	99.5	99.9
Bituminous coal fly ash (no BC)	31.4 (1.03)	33.0 (1.08)	33.8 (1.11)	24.9 (0.82)
Subbituminous coal fly ash in tangential-fired boiler (no BC)	40.0 (1.31)	42.7 (1.40)	44.1 (1.45)	31.4 (1.03)
Other coal (no BC)	21.1 (0.69)	21.4 (0.70)	21.5 (0.71)	17.0 (0.56)
Cement kiln (no BC)	6.4 (0.21)	5.6 (0.18)	5.0 (0.16)	5.7 (0.19)
Glass plant (no BC)	4.6 (0.15)	4.5 (0.15)	4.3 (0.14)	3.8 (0.13)
Iron/steel sinter plant dust with mechanical precollector (no BC)	14.0 (0.46)	13.7 (0.45)	13.3 (0.44)	11.6 (0.38)

[a]To convert cm/s to fps, multiply cm/s by 0.0328, but computational procedure uses SI units. To convert cm/s to m/s, multiply cm/s by 0.01. Assumes same particle size as given in full computational procedure. BC = back corona.
[b]All sources assumed at 200°F.

TABLE 4. Flat-Plate ESP Migration Velocities (cm/s [fps])[a]

Particle Source	Design Efficiency, %			
	95	99	99.5	99.9
Bituminous coal fly ash[b] (no BC)	13.2 (0.43)	15.1 (0.50)	18.6 (0.61)	16.0 (0.53)
Subbituminous coal fly ash in tangential-fired boiler[b] (no BC)	28.6 (0.94)	18.2 (0.60)	21.2 (0.70)	17.7 (0.58)
Other coal[b] (no BC)	15.5 (0.51)	11.2 (0.37)	15.1 (0.50)	13.5 (0.44)
Cement kiln[c] (no BC)	2.4 (0.079)	2.3 (0.075)	3.2 (0.11)	3.1 (0.10)
Glass plant[d] (no BC)	1.8 (0.059)	1.9 (0.062)	2.6 (0.085)	2.6 (0.085)
Iron/steel sinter plant dust with mechanical precollector[b] (no BC)	13.4 (0.44)	12.1 (0.40)	13.1 (0.43)	12.4 (0.41)
Kraft-paper-recovery boiler[b] (no BC)	5.0 (0.16)	4.7 (0.15)	6.1 (0.20)	5.8 (0.19)
Incinerator fly ash[e] (no BC)	25.2 (0.83)	16.9 (0.55)	21.1 (0.69)	18.3 (0.60)

[a]To convert cm/s to fps, multiply cm/s by 0.0328, but computational procedure uses SI units. To convert cm/s to m/s, multiply cm/s by 0.01. Assumes same particle size as given in full computational procedure. These values give the grounded collector plate SCA, from which the collector plate area is derived. In flat-plate ESPs, the discharge or high-voltage plate area is typically 40% of the ground-plate area. The flat-plate manufacturer usually counts all the plate area (collector plates plus discharge plates) in meeting an SCA specification, which means that the velocities tabulated above must be divided by 1.4 to be used on the manufacturer's basis. BC = back corona.
[b]At 300°F.
[c]At 600°F.
[d]At 500°F.
[e]At 250°F.

Procedure

1. Compute the design efficiency, $E(\%)$: $= 100 \times (1 -$ outlet load/inlet load).
2. Compute design penetration, p: $= 1 - (E/100)$.
3. Compute or obtain the operating temperature: *Tf, Tc,* and/or *Tk* (°F, °C, K). *Tk* will be needed in the calculations.
4. Determine whether severe back corona is present. Severe back corona usually occurs for dust resistivities above 2×10^{11} or 3×10^{11} ohm-cm. Its presence will greatly increase the size of the ESP required to achieve a certain efficiency.
5. Determine the MMD of the inlet particle distribution MMDi (μm). If this is not known, assume a value from the following table:

Source	MMDi, μm
Bituminous coal	16
Subbituminous coal, tangential boiler	21
Subbituminous coal, other boiler types	10–15
Cement kiln	2–5
Glass plant	1
Wood burning boiler	5
Sinter plant,	50
with mechanical precollector	6
Kraft process recovery	2
Incinerators	15–30
Unknown	1
Copper reverberatory furnace	1
Copper converter	1
Coke plant combustion stack	1

6. Assume values for sneakage, S_N, and rapping reentrainment, *RR,* from the following table:

$S_N = 0.07$	Plate-wire ESPs
$S_N = 0.05$	Wet wall
$S_N = 0.10$	Flat plate
ESP type	*RR*
Coal fly ash, or not known	0.124
Wet wall	0.0
Flat plate with gas velocity >1.5 m/s (not glass or cement)	0.15
Glass or cement	0.10

7. Assume values for the most penetrating size, MMD_p (i.e., the MMD of the size distribution emerging from a very efficient collecting zone), and rapping puff size, MMDr (i.e., the MMD of the size distribution of rapped/reentrained material):
   MMDp $= 2\ \mu$m
   MMDr $= 5\ \mu$m for ash with MMDi $>5\ \mu$m, or
   MMDr $= 3\ \mu$m for ash with MMDi $<5\ \mu$m
8. Use or compute the following factors for pure air:
   $\epsilon 0 = 8.845 \times 10^{-12}$, free space permittivity, F/m
   $\eta = 1.72 \times 10^{-0.5} \times (Tk/273)^{0.71}$, gas viscosity, kg/ms
   $Ebd = 630{,}000 \times (273/Tk)^{0.8}$, electric field at sparking, V/m
   For plate-wire ESPs:
   $Eavg = Ebd/1.75$, average field with no back corona
   or

$Eavg$ = 0.7 × Ebd/1.75, average field with severe back corona

For flat-plate ESPs:

$Eavg$ = Ebd × 5/6.3, average field, no back corona, positive polarity

$Eavg$ = 0.7 × Ebd × 5/6.3, average field, severe back corona, positive polarity

LF = $S_N + RR - S_N \times RR$, loss factor (dimensionless)

9. Assume the smallest number of sections for the ESP, n, such that $LF^n < p$. Suggested values of n are:

E %	n
<96.5	2
<99	3
<99.8	4
<99.9	5
>99.9	6

These values are for an LF of 0.185, corresponding to a coal fly ash precipitator. The values are approximate, but the best results are for the smallest allowable n.

10. Compute the average section penetration, $p_s = p^{1/n}$.
11. Compute the section collection penetration, p_c: = $(p_s - LF)/(1 - LF)$. If the value of n is too small, then this value will be negative and n must be increased.
12. Compute the particle size change factors, which are constants used for computing the change of particle size from section to section, D and MMDrp:

$$D = p_s = S_N + p_c(1 - S_N) + RR(1 - S_N)(1 - p_c)$$

$$\text{MMDrp} = RR(1 - S_N)(1 - p_c)\text{MMDr}/D$$

13. Compute a table of particle sizes for section 1 through n:

Section	MMDs
1	MMD1 = MMDi
2	MMD2 = {MMD1 × S_N + [(1 − p_c) × MMD_p + p_c × MMD1] × p_c}/D + MMDrp
3	MMD3 = { MMD2 × S_N + [(1 − p_c) × MMDp + p_c × MMD2] × p_c}/D + MMDrp
.	
.	
.	
n	MMDn = {MMDn − 1 × S_N + [(1 − p_c) MMDp + p_c × MMDn − 1] × p_c}/D + MMDrp

14. Calculate the SCA for sections 1 through n, using MMDn, η, ϵ, $Eavg$, and p_c:

$SCA1 = -(\eta/\epsilon) \times (1 - S_N) \times \ln (p_c)/(Eavg^2 \times \text{MMD1} \times 1 \times 10^{-6})$

.

.

.

$SCAn = -(\eta/\epsilon) \times (1 - S_N) \times \ln (p_c)/(Eavg^2 \times \text{MMDn} \times 1 \times 10^{-6})$

where the factor 1×10^{-6} converts micrometers to meters. Note that the only quantity changing in these expressions is MMDx; therefore, the following relation can be used:

$SCAn + 1 = SCAn \times \text{MMDn}/\text{MMDn} + 1$

15. Calculate the total SCA and the English SCA:

SCA = $SCA1 + SCA2 + \ldots + SCAn$ (s/m)

$ESCA$ = 5.079 × SCA (ft²/kacfm)

This sizing procedure works best for p_c values less than the value of LF, which means the smallest value of n. Any ESP model is sensitive to the values of particle diameter and electric field. This one shows the same sensitivity, but the expressions for electric field are based on theoretical and experimental values. The SCA should not be strongly affected by the number of sections chosen; if more sections are used, the SCA of each section is reduced.

Specific Collecting Area for Tubular Precipitators

The procedure given above is suitable for large plate-wire or flat-plate ESPs, but must be used with restrictions for tubular ESPs. Values of $S_N = 0.015$ and $RR = 0$ are assumed, and only one section is used.

Table 5 gives migration velocities that can be used with equation 23 to calculate SCAs for several tubular ESP applications.

Flow Velocity

A precipitator collecting a dry particulate material runs a risk of nonrapping (continuous) reentrainment if the gas velocity becomes too large. This effect is independent of SCA and has been learned through experience. For fly ash applications, the maximum acceptable velocity is about 1.5 m/s (5 fps) for plate-wire ESPs and about 1 m/s (3 fps) for flat-plate ESPs. For low-resistivity applications, design velocities of 3 fps or less are common to avoid nonrapping reentrainment. The frontal area of the ESP must be chosen to keep gas velocity low and to accommodate electrical requirements (e.g., wire-to-plate spacing), while also ensuring that total plate area requirements are met. This area can be configured in a variety of ways. The plates can be short in height, long in the direction of flow, with several in parallel (making the width narrow). Or the plates can be tall in height, short in the direction of flow, with many in parallel (making the width large). After selecting a configuration, the gas velocity can be obtained by dividing the volume flow rate, Q, by the frontal area of the ESP:

$$v_{gas} = Q/WH \tag{24}$$

where: v_{gas} = gas velocity, m/s
W = width of ESP entrance, meters
H = height of ESP entrance, meters

When meeting the above restrictions, this value of velocity also ensures that turbulence is not strongly developed, thereby assisting in the capture of particles.

TABLE 5. Tubular ESP Migration Velocities (cm/s [fps])[a]

Particle Source	Design Efficiency, %	
	90	95
Cement kiln (no BC)	2.2–5.4 (0.072–0.177)	2.1–5.1 (0.069–0.167)
(BC)	1.1–2.7 (0.036–0.089)	1.0–2.6 (0.033–0.085)
Glass plant (no BC)	1.4 (0.046)	1.3 (0.043)
(BC)	0.7 (0.023)	0.7 (0.023)
Kraft-paper-recovery boiler (no BC)	4.7 (0.154)	4.4 (0.144)
Incinerator 15 μm MMD, (no BC)	40.8 (1.34)	39 (1.28)
Wet, at 200°F		
MMD, (μm)		
1	3.2 (0.105)	3.1 (0.102)
2	6.4 (0.210)	6.2 (0.203)
5	16.1 (0.528)	15.4 (0.505)
10	32.2 (1.06)	30.8 (1.01)
20	64.5 (2.11)	61.6 (2.02)

[a]BC = back corona. These rates were calculated on the basis of: $S_N = 0.015$, $RR = 0$, one section only. These are in agreement with operating tubular ESPs; extension of results to more than one section is not recommended.

Pressure Drop Calculations

The pressure drop in an ESP is due to four main factors:

- Diffuser plate (usually present) (perforated plate at the inlet)
- Transitions at the ESP inlet and outlet
- Collection plate baffles (stiffeners) or corrugations
- Drag of the flat collection plate

The total pressure drop is the sum of the individual pressure drops, but any one of these sources may dominate all other contributions to the pressure drop. Usually, the pressure drop is not a design-driving factor, but it needs to be maintained at an acceptably low value. Table 6 gives typical pressure drops for the four factors. The ESP pressure drop, usually less than about 0.5 in. H_2O, is much lower than for the associated collection system and ductwork. With the conveying velocities used for dust collected in ESPs, generally 4000 fpm or greater, system pressure drops are usually in the range of 2–10 inches H_2O.

The four main factors contributing to pressure drop are described briefly below.

The diffuser plate is used to equalize the gas flow across the face of the ESP. It typically consists of a flat plate covered with round holes of 5–7 cm in diameter (2–2.5 inches) having an open area of 50–65% of the total.

TABLE 6. Components of ESP Pressure Drop

Component	Typical Pressure Drop, in. H_2O	
	Low	High
Diffuser	0.010	0.09
Inlet transition	0.07	0.14
Outlet transition	0.007	0.015
Baffles	0.0006	0.123
Collection plates	0.0003	0.008
Total	0.09	0.38

Pressure drop is strongly dependent on the percent open area, but is almost independent of hole size.

The pressure drop due to gradual enlargement at the inlet is caused by the combined effects of flow separation and wall friction and is dependent on the shape of the enlargement. At the ESP exit, the pressure drop caused by a short, well-streamlined gradual contraction is small.

Baffles are installed on collection plates to shield the collected dust from the gas flow and to provide a stiffening effect to keep the plates aligned parallel to one another. The pressure drop due to the baffles depends on the number of baffles, their protrusion into the gas stream with respect to electrode-to-plate distance, and the gas velocity in the ESP.

The pressure drop of the flat collection plates is due to friction of the gas dragging along the flat surfaces and is so small compared with other factors that it may usually be neglected in engineering problems.

Particle Characteristics

Several particle characteristics are important for particle collection. It is generally assumed that the particles are spherical or spherical enough to be described by some equivalent spherical diameter. Highly irregular or elongated particles may not behave in ways that can be easily described.

The first important characteristic is the mass of particles in the gas stream, that is, the particle loading. This quantity usually is determined by placing a filter in the gas stream, collecting a known volume of gas, and determining the weight gain of the filter. Because the ESP operates over a wide range of loadings as a constant-efficiency device, the inlet loading will determine the outlet loading directly. If the loading becomes too high, the operation of the ESP will be altered, usually for the worse.

The second characteristic is the size distribution of the particles, often expressed as the cumulative mass less than a given particle size. The size distribution determines how many particles of a given size there are, which is important because ESP efficiency varies with particle size. In practical terms, an ESP will collect all particles larger than 10 μm in diameter better than ones smaller than 10 μm. Only if most of the mass in the particles is concentrated above 10 μm would the actual size distribution above 10 μm be needed.

In lieu of cumulative mass distributions, the size distribution is often described by lognormal parameters. That is, the size distribution appears as a probabilistic normal curve if the logarithm of particle size used is the abscissa. The two parameters needed to describe a lognormal distribution are the mass median (or mean) diameter and the geometric standard deviation.

The MMD is the diameter for which one half of the particulate mass consists of smaller particles and one half is larger (see step 5 of the procedure). If the MMD of a distribution is larger than about 3 μm, the ESP will collect all particles larger than the MMD at least as well as a 3-μm particle, representing one half the mass in the inlet size distribution.

The geometric standard deviation is the equivalent of the standard deviation of the normal distribution: It describes how broad the size distribution is. The geometric standard deviation is computed as the ratio of the diameter corresponding to 84% of the total cumulative mass to the MMD: It is always a number greater than 1. A distribution with particles all of the same size (monodisperse) has a geometric standard deviation of 1. Geometric standard deviations less than 2 represent rather narrow distributions. For combustion sources, the geometric standard deviations range from 3 to 5 and are commonly in the 3.5–4.5 range.

A geometric standard deviation of 4–5, coupled with an MMD of less than 5 μm, means that there is a substantial amount of submicrometer material. This situation may change the electrical conditions in an ESP by the phenomenon known as "space charge quenching," which results in high operating voltages but low currents. It is a sign of inadequate charging and reduces the theoretical efficiency of the ESP. This condition must be evaluated carefully to be sure of adequate design margins.

Gas Characteristics

The gas characteristics most needed for ESP design are the gas volume flow and the gas temperature. The volume flow, multiplied by the design SCA, gives the total plate area required for the ESP. If the volume flow is known at one temperature, it may be estimated at other temperatures by applying the ideal gas law. Temperature and volume uncertainties will outweigh inaccuracies of applying the ideal gas law.

The temperature of the gas directly affects the gas viscosity, which increases with temperature. Gas viscosity is affected to a lesser degree by the gas composition, particularly the water vapor content. In lieu of viscosity values for a particular gas composition, the viscosity for air may be used. Viscosity enters the calculation of SCA directly, as seen in step 14 of the procedure.

The gas temperature and composition can have a strong effect on the resistivity of the collected particulate material. Specifically, moisture and acid-gas components may be chemisorbed on the particles in a sufficient amount to lower the intrinsic resistivity dramatically (orders of magnitude). For other types of materials, there is almost no effect. Although it is not possible to treat resistivity here, the designer should be aware of the potential sensitivity of the size of the ESP to resistivity and the factors influencing it.

The choice of power supplies' size (current capacity and voltage) to be used with a particular application may be influenced by the gas characteristics. Certain applications produce gas whose density may vary significantly from typical combustion sources (density variation may result from temperature, pressure, and composition). Gas density affects corona starting voltages and voltages at which sparking will occur.

Cleaning

Cleaning the collected materials from the plates often is accomplished intermittently or continuously by rapping the plates severely with automatic hammers or pistons, usually along their top edges, except in the case of wet ESPs that use water. Rapping dislodges the material, which then falls down the length of the plate until it lands in a dust hopper. The dust characteristics, rapping intensity, and rapping frequency determine how much of the material is reentrained and how much reaches the hopper permanently.

For wet ESPs, consideration must be given to handling wastewaters. For simple systems with innocuous dusts, water with particles collected by the ESP may be discharged from the ESP system to a solids-removing clarifier (either dedicated to the ESP or part of the plant wastewater treatment system) and then to final disposal. More complex systems may require skimming and sludge removal, clarification in dedicated equipment, pH adjustment, and/or treatment to remove dissolved solids. Spray water from the ESP preconditioner may be treated separately from the water used to flood the ESP collecting plates so that the cleaner of the two treated waters may be returned to the ESP. Recirculation of treated water to the ESP may approach 100%.

The hopper should be designed so that all the material in it slides to the very bottom, where it can be evacuated periodically, as the hopper becomes full. Dust is removed through a valve into a dust-handling system, such as a pneumatic conveyor. Hoppers often are supplied with auxiliary heat to prevent the formation of lumps or cakes and the subsequent blockage of the dust-handling system.

Construction Features

The use of the term "plate-wire geometry" may be somewhat misleading. It could refer to three different types of discharge electrodes: weighted wires hung from a support structure at the top of the ESP, wire frames in which wires are strung tautly in a rigid support frame, or rigid electrodes constructed from a single piece of fabricated metal. In recent years, there has been a trend toward using wire frames or rigid discharge electrodes in place of weighted wire discharge electrodes (particularly in coal-fired boiler applications). This trend has been stimulated by the user's desire for increased ESP reliability. The wire frames and rigid electrodes are less prone to failure by breakage and are readily cleaned by impulse-type cleaning equipment.

Other differences in construction result from the choice of gas passage (flow lane) width or discharge-electrode-to-collecting-electrode spacing. Typically, discharge-to-collecting-electrode spacing varies from 11 to 19 cm (4.3–7.5 inches). Having a large spacing between discharge and collecting electrodes allows higher electric fields to be used, which tends to improve dust collection. To generate larger electric fields, however, power supplies must produce higher operating voltages. Therefore, it is necessary to balance the cost savings achieved with larger electrode spacing against the higher cost of power supplies that produce higher operating voltages.

Most ESPs are constructed of mild steel. The ESP shells are constructed typically of 3/16-inch or 1/4-inch mild steel plate. Collecting electrodes are generally fabricated from lighter-gauge mild steel. A thickness of 18 gauge is common, but it will vary with the size and severity of the application.

Wire discharge electrodes come in various shapes, from round to square or barbed. A diameter of 2.5 mm (0.1 inch) is common for weighted wires, but other shapes used have much larger effective diameters, for example, 64-mm (0.25-inch) square electrodes.

Stainless steel may be used for corrosive applications, but it is uncommon except in wet ESPs. Stainless-steel discharge electrodes have been found to be prone to fatigue failure in dry ESPs with impact-type electrode cleaning systems.[3]

Precipitators used to collect sulfuric acid mist in sulfuric acid plants are constructed of steel, but the surfaces in contact with the acid mist are lead lined. Precipitators used on paper mill black-liquor-recovery boilers are steam jacketed. Of these two, recovery boilers have by far the larger number of ESP applications.

ESTIMATING TOTAL CAPITAL INVESTMENT

Total capital investment (TCI) includes costs for the ESP structure, the internals, rappers, power supply, and auxiliary equipment, and the usual direct and indirect costs associated with installing or erecting new structures. These costs, in second-quarter 1987 dollars, are described in the following subsections.

Equipment Cost

ESP Costs

Five types of ESPs are considered: plate wire, flat plate, wet, tubular, and two stage. Basic costs for the first four are taken from Figure 5, which gives the flange-to-flange, field-erected price based on required plate area and a rigid electrode design. This plate area is calculated from the sizing information given previously for the four types. Adjustments are made for standard options listed in Table 7. Costs for two-stage precipitators are given later.

The costs are based on a number of actual quotes. Least squares lines have been fitted to the quotes, one for sizes between 50,000 and 1,000,000 ft^2, and a second for sizes between 10,000 and 50,000 ft^2. An equation is given for each line. Extrapolation below 10,000 or above 1,000,000 ft^2 should not be used. The reader should not be surprised if quotes are obtained that differ from these curves by as much as ±25%. Significant savings can be had by soliciting multiple quotes. All units include the ESP casing, pyramidal hoppers, rigid electrodes and internal collecting plates, transformer rectifier (TR) sets and microprocessor controls, rappers, and stub supports (legs) for 4-foot clearance below the hopper discharges. The lower curve is the basic unit without the standard options. The upper curve includes all of the standard options (see Table 7) that are normally utilized in a modern system. These options add approximately 45% to the basic cost of the flange-to-flange hardware. Insulation costs are for 3 inches of field-installed glass fiber encased in a metal skin and applied on the outside of all areas in contact with the exhaust gas stream. Insulation for ductwork, fan casings, and stacks must be calculated separately.

Impact of Alternative Electrode Designs

All three designs—rigid electrode, weighted wire, and rigid frame—can be employed in most applications. Any cost differential between designs will depend on the combination of vendor experience and site-specific factors that dictate equipment size factors. The rigid frame design will cost up to 25% more if the mast or plate height is restricted to the same used in other designs. Several vendors can now provide rigid frame collectors with longer plates, and thus the cost differential can approach zero.

The weighted wire design uses narrower plate spacings and more internal discharge electrodes. This design is being employed less; therefore, its cost is increasing and currently is approximately the same as that for the rigid electrode collector. Below about 15,000 ft^2 of plate area, ESPs are of different design and are not normally field erected, and the costs will be significantly different from values extrapolated from Figure 5.

Impact of Materials of Construction: Metal Thickness and Stainless Steel

Corrosive or other adverse operating conditions may suggest the specification of thicker metal sections in the precipitator. Reasonable increases in metal sections result in minimal cost increases. For example, collecting plates are typically constructed of 18-gauge mild steel. Most ESP manufacturers can increase the section thickness by 25% without significant design changes or increases in manufacturing costs of more than a few percent

TABLE 7. Standard Options for Basic Equipment

Item	Cost Adder, %
1. Inlet and outlet nozzles and diffuser plates	8 to 10
2. Hopper auxiliaries/heaters, level detectors	8 to 10
3. Weather enclosure and stair access	8 to 10
4. Structural supports	5
5. Insulation	8 to 10
Total options 1 to 5	1.37 to 1.45 × base

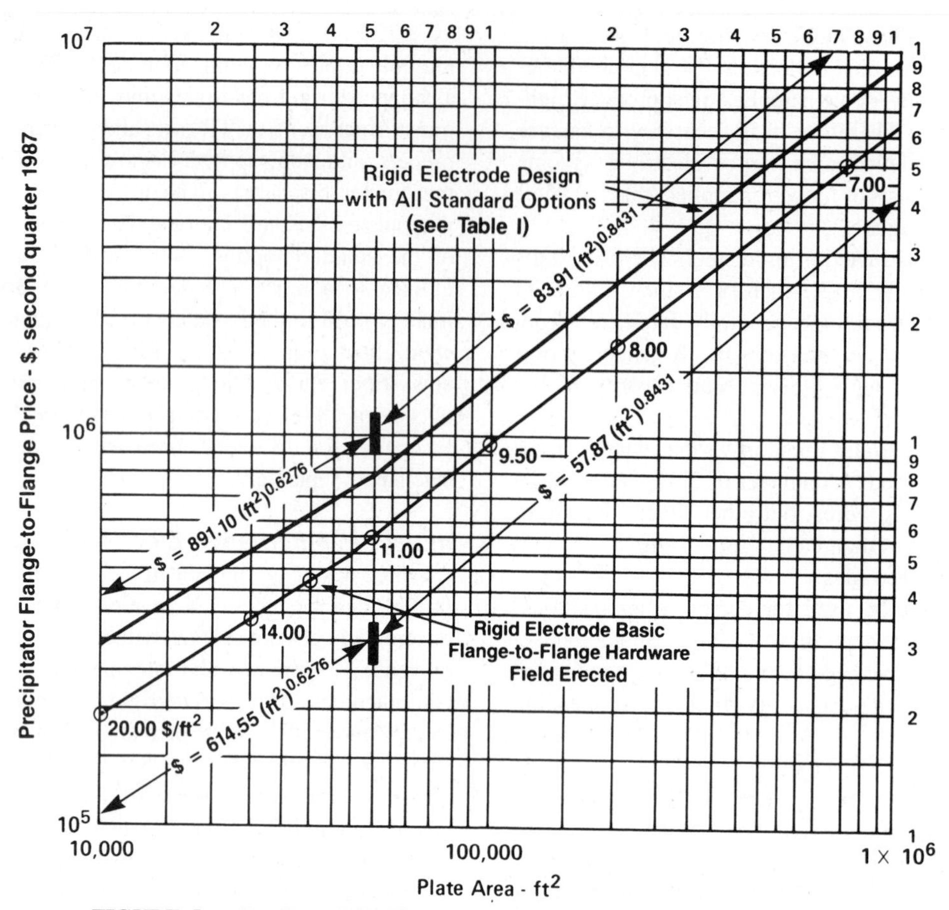

FIGURE 5. Dry-Type ESP Flange-to-Flange Purchase Price versus Plate Area

Changes in type of material can increase the purchase cost of the ESP from about 30% to 50% for type 304 stainless-steel collector plates and precipitator walls, and up to several hundred percent for more expensive materials used for all elements of the ESP. Based on the type 304 stainless-steel cost, the approximate factors given below can be used for other materials:

Material	Factor	Reference(s)
Stainless steel, 316	1.3	3, 4, 5
Carpenter 20 CB-3	1.9	5
Monel-400	2.3	3, 5
Nickel-200	3.2	5
Titanium	4.5	5

See the appendix for more detail on the effects of material thickness and type.

Recent Trends

Most of the market (1987) is in the 50,000 to 200,000 ft² plate area size range. Selling prices of ESPs have increased very little over the past 10 years because of more effective designs, increased competition from European suppliers, and a shrinking utility market.

Design improvements have allowed wider plate spacings that reduce the number of internal components and higher plates and masts that provide additional plate area at a low cost. Microprocessor controls and energy management systems have lowered operating costs.

Few, if any, hot-side ESPs (those used upstream from an air preheater on a combustion source) are being specified for purchase. Recognition that low sodium coals tend to build resistive ash layers on the collection plates, thus reducing ESP efficiency, has almost eliminated sales of these units. Of about 150 existing units, about 75 are candidates for conversion to cold-side units over the next 10 years.

Specific industry application has little impact on either ESP design or cost, with two exceptions: a paper mill ESP and a sulfuric acid mist ESP. Paper mill ESPs use drag conveyor hoppers that add approximately 10% to the base flange-to-flange equipment cost. The sulfuric acid mist application utilizes wet ESPs. Wet ESPs are used for two different applications. The first is the collection of acid mists during sulfuric acid manufacture. These precipitators usually are small, and they use lead for all interior surfaces; hence, they normally cost \$65 to \$95 per square foot installed (mid-1987 dollars). Special installations can be as much as \$120 per square foot. The other primary application, using a wet circular ESP, is for a detarring operation for coke oven off-gases. These precipitators are made using high-alloy stainless steels and typically cost \$90 to \$120 per square foot. Because of the small number of sales, small size of units sold, and dependency on site-specific factors, more definitive costs are not available.

Retrofit Cost Factor

Retrofit installations increase the costs of an ESP because of the common need to remove something to make way for the new ESP. Usually, the ducting is much more expensive. Its path is often constrained by existing structures, additional supports are required, and the confined areas make erection more labor intensive and lengthy. Costs are site-specific; however, for estimating purposes, a retrofit multiplier of 1.3 to 1.5 on the total installed cost can be used. The multiplier should be selected within this range based on the relative difficulty of the installation.

A special case is conversion of hot- to cold-side ESPs for coal-fired boiler applications. The magnitude of the conversion is very site-specific, but most projects will contain the following elements.

- Relocating the air preheater and the ducting to it.
- Resizing the ESP inlet and outlet duct to the new air volume and rerouting it.
- Upgrading the induced-draft fan size or motor to accommodate the higher static pressure and horsepower requirements.
- Adding or modifying foundations for fan and duct supports.
- Assessing the required SCA and either adding collector or installing an SO_3 gas-conditioning system.
- Adding hopper heaters.
- Upgrading the analog electric controls to microprocessor-type controls.
- Increasing the number of collecting plate rappers and perhaps the location of rapping.

In some installations, it may be cost-effective to gut the existing collector totally, utilize only the existing casing and hoppers, and upgrade to modern internals.

The cost of conversion is a multimillion dollar project, typically running at least 25% to 35% of the installed cost of a new unit.

Auxiliary Equipment

The auxiliary equipment depicted in Figure 2 is discussed elsewhere in the *E.A.B. Control Cost Manual*.[18] Because hoods, precoolers, cyclones, fans, motors, and stacks are common to many pollution control systems, they will be given extended treatment in separate sections of the *Manual*. These sections will be added to the *Manual* when they are completed.

Costs for Two-Stage Precipitators

Purchase costs for modular two-stage precipitators, which should be considered separately from large-scale, single-stage ESPs, are given in Figure 6.[7] To be consistent with industry practice, costs are given as a function of flow rate through the system. The lower cost curve is for a two-cell unit without precooler, an installed cell washer, or a fan. The upper curve is for an engineered, package system with the following components: inlet diffuser plenum, prefilter, cooling coils with coating, coil plenums with access, water-flow controls, triple-pass configuration, system exhaust fan with accessories, outlet plenum, and in-place foam cleaning system with semiautomatic controls and programmable controller. All equipment is fully assembled mechanically and electrically, and it is mounted on a steel structural skid.

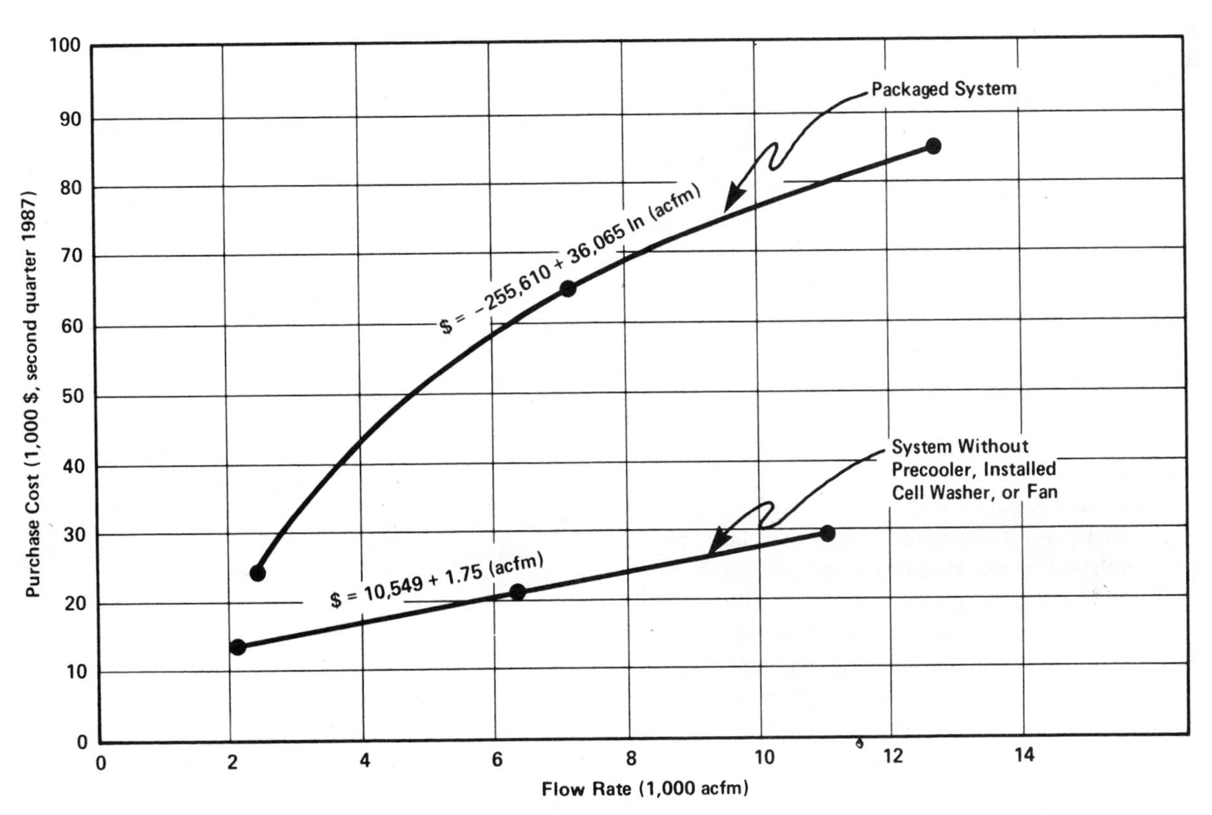

FIGURE 6. Purchase Costs for Two-Stage, Two-Cell Precipitators

TABLE 8. Items That Increase ESP Costs

Item	Factor or Total Cost	Applied to
Rigid frame electrode with restricted plate height	1.0–1.25	ESP base cost
Type 304 stainless-steel collector plates and precipitator walls[a]	1.3–1.5	ESP base cost
All-stainless construction[a]	2–3	ESP base cost
ESP with drag conveyor hoppers (paper mill)	1.1	ESP base cost
Retrofit installations	1.3–1.5	ESP base cost
Wet ESP		
Sulfuric acid mist	$65–$95/ft^2	—
Sulfuric acid mist (special installation)	Up to $120/ft^2	—
Coke oven off-gas	$90–$120/ft^2	—

[a]Other materials' cost factors are given in the section on "Impact of Materials of Construction: Metal Thickness and Stainless Steel."

Total Purchased Cost

The total purchased cost of an ESP system is the sum of the costs of the ESP, options, auxiliary equipment, instruments and controls, taxes, and freight. The last three items generally are taken as percentages of the estimated total cost of the first three items. Typical values are 10% for instruments and controls, 3% for taxes, and 5% for freight.

Costs of standard and other options can vary from 0% to more than 150% of bare ESP cost, depending on site and application requirements. Other factors that can increase ESP costs are given in Table 8.

Total Capital Investment

Total capital investment (TCI) is estimated from a series of factors applied to the purchased equipment cost to obtain direct and indirect costs for installation. The TCI is the sum of these three costs. The required factors are given in Table 9. Because ESPs may vary from small units attached to existing buildings to large, separate structures, specific factors for site preparation or for buildings are not given. However, costs for buildings may be obtained from such references as *Means Square Foot Costs* 1987.[10] Land, working capital, and off-site facilities are excluded from the table because they are not normally required. For very large installations, however, they may be needed and could be estimated on an as-needed basis.

Note that the factors given in Table 9 are for average installation conditions, for example, no unusual problems with site earthwork, access, shipping, or interfering structures. Considerable variation may be seen with other-than-average installation circumstances. For two-stage precipitators purchased as packaged systems, several of the costs in Table 9 would be greatly reduced or eliminated. These include instruments and controls, foundations and supports, erection and handling, painting, and model studies. An installation factor of 0.25 B would be more nearly appropriate for the two-stage ESPs.

TABLE 9. Capital Cost Factors for ESPs[8]

Costs		Factor
Direct costs		
Purchased equipment costs		
ESP	A = Sum of	As estimated
Auxiliary equipment		As estimated
Instruments and controls		0.10 A
Taxes		0.03 A
Freight		0.05A
Total purchased equipment cost	B =	1.18 A
Installation direct costs		
Foundations and supports		0.04 B
Erection and handling		0.50 B
Electrical		0.08 B
Piping		0.01 B
Insulation for ductwork[a]		0.02 B
Painting		0.02 B
Site preparation (SP)		As required
Buildings (Bldg.)		As required
Total installation direct costs		0.67 B + SP + Bldg.
Total direct costs		1.67 B + SP + Bldg.
Indirect costs		
Engineering and supervision		0.20 B
Construction and field expense		0.20 B
Construction fee		0.10 B
Startup fee		0.01 B
Performance test		0.01 B
Model study		0.02 B
Contingencies		0.03 B
Total indirect costs		0.57 B
Total direct and indirect costs = Total capital investment		2.24 B + SP + Bldg.

[a]If ductwork dimensions have been established, cost may be estimated based on $10 to $12 per square feet of surface for field application. Fan housings and stacks may also be insulated.[9]

ESTIMATING TOTAL ANNUAL COSTS

Direct Annual Costs

Direct annual costs include operating and supervisory labor, operating materials, replacement rappers and electrodes, maintenance (labor and materials), utilities, dust disposal, and wastewater treatment for wet ESPs. Most of these costs are discussed individually below. They vary considerably with location and time and, for this reason, should be obtained to suit the specific ESP system being costed. For example, current labor rates may be found in such publications as the *Monthly Labor Review,* published by the U.S. Department of Labor, Bureau of Labor Statistics.

Operating and Supervisory Labor

Proper operation of the ESP is necessary both to meet applicable particulate emission regulations and to ensure minimum costs. An ESP is an expensive piece of equipment. Even well-designed equipment will deteriorate rapidly if improperly maintained and will have to be replaced long before it should be necessary. Not only can proper operation and maintenance save the operator money, such a program can also contribute to good relations with the governing pollution control agency by showing good faith in efforts to comply with air regulations.

Although each plant has its own methods for conducting an operation and maintenance program, experience has shown that plants that assign one individual the responsibility of coordinating all the pieces of the program operate better than those where different departments look after only certain portions of the program. The separate departments have little knowledge of how their portion affects the overall program. In other words, a plant needs one individual to coordinate the operation, maintenance, and troubleshooting components of its ESP program if it expects to have a relatively trouble-free operation. The coordinator typically is an engineer who reports to plant management and interfaces with the maintenance and plant process supervisors, the laboratory, and the purchasing department. For companies with more than one plant, that person would be responsible for all ESPs. The portion of the person's total time that is spent on the ESP then becomes an operating expense for the ESP. This can be expressed as:

$$A.C. = X(LCC) \qquad (25)$$

where: $A.C.$ = annual coordination cost, \$/yr
X = fraction of total time spent on ESP
LCC = individual annual labor cost for ESP coordinator, \$/yr

In addition to coordination costs, typical operating labor requirements are one half to two hours per shift for a wide range of ESP sizes.[8] Small or well-performing units may require less time, and very large or troublesome units may require more time. Supervisory labor is taken as 15% of operating labor.

Operating Materials

Operating materials are generally not required for ESPs. An exception is the use of gas-preconditioning agents for dust resistivity control.

Maintenance

The reader should obtain publication no. EPA/625/1-85/017, *Operating and Maintenance Manual for ESPs,*[11] for suggested maintenance practices. Routine ESP maintenance labor costs can be estimated using data provided by manufacturers. For example, for a 100,000-ft^2 collector, maintenance labor is estimated to require 15 hours per week, 44 weeks a year. At a direct labor cost of \$12.50 an hour (mid-1987 costs), an estimated annual maintenance labor cost of \$8250 or \$0.0825 per square foot of collector area is established. To this must be added the cost of maintenance materials. This relationship can be assumed to be linear above a 50,000-ft^2 collector size and constant at \$4125 below this size. Based on an analysis of vendor information, annual maintenance materials are estimated as 1% of the flange-to-flange precipitator purchase cost:

$$MC = 0.01FCC + \text{labor cost} \qquad (26)$$

where: MC = annual maintenance cost, \$/yr
= $0.01FCC + 0.0825A$ above 50,000 ft^2
= $0.01FCC + 4125$ below 50,000 ft^2
A = ESP plate area, ft^2
FCC = ESP flange-to-flange purchase cost, dollars

Electricity

Power is required to operate system fans, TR sets, and cleaning equipment. Fan power for primary gas movement can be calculated from equation (2-7) of the *E.A.B. Control Cost Manual.*[18] After substituting into this equation a combined fan-motor efficiency of 0.65 and a specific gravity of 1.0, we obtain:

$$F.P. = 0.000181(Q)(\Delta P)(\theta') \qquad (27)$$

where: $F.P.$ = fan power requirement, kWh/yr
Q = system flow rate, acfm
ΔP = system pressure drop, in. H_2O
θ' = annual operating time, h/yr

Pump power for wet ESPs can be calculated from:[8]

$$P.P. = (0.746\ Q_1 Z\ S_g\theta')/(3{,}960\ \eta') \qquad (28)$$

where: $P.P.$ = pump power requirement, kWh/yr
Q_1 = water flow rate, gal/min
Z = fluid head, feet
S_g = specific gravity of water being pumped compared to water at 70°F and 29.92 in. Hg

θ' = annual operating time, h/yr
η' = pump efficiency (fractional)

Energy for TR sets and motor-driven or electromagnetic rapper systems is the sum of the energy consumption for operating both items. Manufacturers' averaged data indicate that the following relationship can be used:

$$O.P. = 1.94 \times 10^{-3} A\theta' \quad (29)$$

where: $O.P.$ = annual ESP operating power, kWh/yr
A = ESP plate area, ft^2
θ' = annual operating time, h/yr

For installations requiring hopper heaters, hopper heater power can be similarly estimated:

$$H.H. = 2H.N.\theta' \quad (30)$$

where: $H.H.$ = annual hopper heater power consumption, kWh/yr
$H.N.$ = number of hoppers
θ' = annual operating time, h/yr.

For two-stage precipitators, power consumption ranges from 25 to 100 W/kacfm, with 40 W/kacfm being typical.

Fuel

If the ESP or associated ductwork is heated to prevent condensation, fuel costs should be calculated as required. These costs can be significant, but they may be difficult to predict. For methods of calculating heat transfer requirements, see Perry.[12]

Water

Cooling process gases for preconditioning can be done by dilution with air, evaporation with water, or heat exchange with normal equipment. Spray cooling requires consumption of plant water (heat exchangers may also require water), although costs are not usually significant. Section 4.4 of the *Manual*[18] provides information on estimating cooling-water costs. Water consumption in wet ESPs is estimated at 5 gal/min kacfm[13] for large units and 16 gal/min kacfm for two-stage precipitators.[14]

Compressed Air

Electrostatic precipitators may use compressed air at pressures of about 60 to 100 psig for operating rappers. Equivalent power cost is included in equation 29 for operation power.

Dust Disposal

If collected dust cannot be recycled or sold, it must be landfilled or disposed of in some other manner. Costs may typically run \$20 per ton or \$30 per ton for nonhazardous wastes exclusive of transportation (see Subsection 2.4 of the *Manual*[18]). Landfilling of hazardous wastes may cost 10 times as much. The disposal costs are highly site-specific and depend on transportation distance to the landfill, handling rates, and disposal unloading (tipping) fees. If these factors are known, they lead to the relationship:

$$D.D. = 4.29 \times 10^{-6}\, G\, \theta' Q\, (T + TM \cdot D) \quad (31)$$

where: $D.D.$ = annual dust disposal costs, \$/yr
G = ESP inlet grain loading or dust concentration, gr/ft^3
θ' = annual operating time, h/yr
Q = flow rate through ESP, acfm
T = tipping fee, \$/ton
TM = mileage rate, \$/ton-mile
D = dust hauling distance, miles

Wastewater Treatment

Water usage for wet ESPs is about 5 gal/min kacfm.[13] Treatment cost of the resulting wastewater may vary from about \$1.30 to \$2.15 per 1000 gallons[15] depending on the complexity of the treatment system. More precise costs can be obtained from Gumerman et al.[16]

Conditioning Costs

Adaptation of information on utility boilers[17] suggests that SO_3 conditioning for a large ESP (2.6×10^6 acfm) costs from about \$1.60 per 10^6 ft^3 of gas processed for a sulfur burner providing 5 ppm of SO_3 to about \$2.30 per 10^6 ft^3 (in first-quarter 1987 dollars) for a liquid SO_2 system providing 20 ppm of SO_3.

Indirect Annual Costs

Capital recovery, property tax, insurance, administrative costs ("G&A"), and overhead are examples of indirect annual costs. The capital recovery cost is based on the equipment lifetime and the annual interest rate employed. (See an engineering economy text, such as *Engineering Economy* by Grant, Ireson, and Leavenworth, for a thorough discussion of the capital recovery cost and the variables that determine it.) For ESPs, the system lifetime varies from five to 40 years, with 20 years being typical. Therefore, when figuring the system capital recovery cost, one should base it on the installed capital cost. In other words:

$$CRC_s = TCI \times CRF_s \quad (32)$$

where: CRC_s = capital recovery cost for ESP system, \$/yr
TCI = total capital investment, dollars
CRF_s = capital recovery factor for ESP system (defined in Section 2 of the *Manual*[18])

For example, for a 20-year system life and a 10% annual interest rate, the CRF_s would be 0.1175.

The suggested factor to use for property taxes, insurance, and administrative charges is 4% of the *TCI*. Overhead is calculated as 60% of the sum of operating, supervisory, and maintenance labor, as well as maintenance materials.

Recovery Credits

For processes that can reuse the dust collected in the ESP or that can sell the dust in a local market, such as fly ash sold as an extender for paving mixes, a credit should be taken. As used below, this credit (*RC*) appears as a negative cost.

Total Annual Cost

The total annual cost for owning and operating an ESP system is the sum of the components listed in the previous three subsections (direct annual costs, indirect annual costs, and recovery credits), that is:

$$TAC = DC + IC - RC \quad (33)$$

where: TAC = total annual cost, dollars
DC = direct annual cost, dollars
IC = indirect annual cost, dollars
RC = recovery credits (annual), dollars

Example Problem

Assume an ESP is required for controlling fly ash emissions from a coal-fired boiler burning bituminous coal. The flue gas stream is 50 kacfm at 325°F and has an inlet ash loading of 4 gr/ft^3. Analysis of the ash shows an MMD of 7 μm and a resistivity of less than 2×10^{11} ohm-cm. Assume that the ESP operates for 8640 hours per year (360 days) and that an efficiency of 99.9% is required.

Design SCA

The SCA can be calculated from equation 23. Assuming that a flat-plate ESP design is chosen, the fly ash migration velocity is 16.0 cm/s (see Table 4). Then:

$$\text{SCA} = -\ln (1 - 0.999)/16.0 = 0.432 \text{ s/cm} = 43.2 \text{ s/m}$$

Converting to English units (see "Procedure," step 15):

$$\text{ESCA} = 5.079 \times 43.2 = 219 \text{ ft}^2\text{/kacfm}$$

Total collector plate area is then:

$$219 \text{ ft}^2\text{/kacfm} \times 50 \text{ kacfm} = 10{,}950 \text{ ft}^2$$

To obtain a more rigorous answer, we can follow the steps of the "Procedure" previously given:

1. Design efficiency is required as 99.9.
2. Design penetration:
   $1 - (99.9/100) = 0.001$
3. Operating temperature in degrees Kelvin:
   $(325°\text{F} - 32°\text{F}) \times 5/9 + 273°\text{C} = 436$ K
4. Because dust resistivity is less than 2×10^{11} ohm-cm, no severe back corona is expected and back corona = 0.
5. The MMD of the fly ash is given as 7 μm.
6. Values for sneakage and rapping reentrainment (from the table presented in step 6, "Procedure," are as follows:
   $S_N = 0.10$
   $RR = 0.124$ (assuming gas velocity <1.5 m/s)
7. The most penetrating particle size, from step 7 of the Procedure is:
   MMDp = 2 μm
   The rapping puff size is:
   MMDr = 5 μm
8. From the Procedure:
   $\epsilon 0 = 8.845 \times 10^{-12}$
   $\eta = 1.72 \times 10^{-5} (436/273)^{0.71}$
   $= 2.40 \times 10^{-5}$
   $E\text{bd} = 6.3 \times 10^5 (273/436)^{0.8}$
   $= 4.33 \times 10^5$ V/m
   $E\text{avg} = E\text{bd} \times 5/6.3 = 3.44 \times 10^5$
   $LF = S_N + RR - S_N \times RR = 0.1 + 0.124 - 0.1 \times 0.124 = 0.212$
9. Choose the number of sections for $LF^n < p$, $p = 0.001$. Try four sections:
   $LF^n = 0.212^4 = 0.002$
   This value is larger than p. Try five sections:
   $LF^n = 0.212^5 = 0.000428$
   This value is smaller than p and is acceptable.
10. Average section penetration is:
    $p_s = p^{1/n} = 0.001^{1/5} = 0.251$
11. Section collection penetration is:
    $p_c = (p_s - LF)/(1 - LF) = (0.251 - 0.212)/(1 - 0.212) = 0.0495$
12. Particle size change factors are:
    $D = p_s = S_N + p_c (1 - S_N) + RR (1 - S_N)(1 - p_c)$
    $= 0.10 + 0.0495 (1 - 0.1) + 0.124 (1 - 0.1) \times (1 - 0.0495)$
    $= 0.251$
    $\text{MMDrp} = RR(1 - S_N) \times (1 - p_c)\, MMD\text{r/D}$
    $= 0.124 (1 - 0.1) \times (1 - 0.0495)(5) \div 0.251 = 2.11$
13. Particle sizes for each section are:

Section	MMD, μm
1	MMD1 = MMDi = 7
2	MMD2 = {MMD1 × S_N + [(1 − p_c) × MMDp + p_c × MMD1] × p_c}/D + MMDrp = {7 × 0.1 + [(1 − 0.0495) × 2 + 0.0495 × 7] × 0.0495}/0.251 + 2.11 = 5.34
3	MMD3 = {5.34 × 0.1 + [(1 − 0.0495) × 2 + 0.0495 × 5.34] × 0.0495}/0.251 + 2.11 = 4.67
4	MMD4 = {4.67 × 0.1 + [(1 − 0.0495) × 2 + 0.0495 × 4.67] × 0.0495}/0.251 + 2.11 = 4.39
5	MMD5 = {4.39 × 0.1 + [(1 − 0.0495) × 2 + 0.0495 × 4.39] × 0.0495}/0.251 + 2.11 = 4.28

14. SCAs for each section are:

Section	SCA, s/m
1	SCA1 = $-(\eta/\epsilon 0) \times (1 - S_N) \times \ln(p_c)/(Eavg^2 \times \text{MMD1} \times 1 \times 10^{-6})$ = $-(2.40 \times 10^{-5}/8.845 \times 10^{-12})(1 - 0.1)\ln 0.0495/[(3.44 \times 10^5)^2 (7 \times 1 \times 10^{-6})] = 8.86$
2	SCA2 = SCA1 × MMD1/MMD2 = 8.86 (7/5.34) = 11.61
3	SCA3 = SCA2 ×MMD2/MMD3 = 11.61 (5.34/4.67) = 13.28
4	SCA4 = SCA3 × MMD3/MMD4 = 13.28 (4.67/4.39) = 14.13
5	SCA5 = SCA4 × MMD4/MMD5 = 14.13 (4.39/4.28) = 14.49

15. Total SCA = 8.86 + 11.61 + 13.28 + 14.13 + 14.49 = 62.37 s/m

English SCA = 5.079 × 62.37 = 317 ft^2/kacfm

Note that the more rigorous procedure calls for an SCA that is considerably higher than the value found by using equation 23. This discrepancy is caused by the considerably smaller particle size used in the example problem than is assumed for Table 4. In this case, the shorter method would lead to an unacceptably low cost estimate.

Total collector plate area is:

$$317 \text{ ft}^2/\text{kcafm} \times 50 \text{ kacfm} = 15{,}850 \text{ ft}^2$$

ESP Cost

From Figure 5, the basic flange-to-flange cost of the rigid electrode ESP is $265,764 (mid-1987 dollars). Assuming all standard options are purchased, the ESP cost rises to $385,358 (mid-1987 dollars).

Costs of Auxiliaries

Assume the following auxiliary costs have been estimated from data in other parts of the *Manual*[18] or from other sources:

Ductwork	$16,000
Fan	16,000
Motor	7,500
Starter	4,000
Dampers	7,200
Pneumatic conveyor	4,000
Stack	8,000
Total	$62,700

Total Capital Investment

Direct costs for the ESP system, based on the factors in Table 9, are given in Table 10. (Again, we assume site preparation and building costs to be negligible.) The TCI is $1,180,000 (rounded, mid-1987 dollars).

Annual Costs—Pressure Drop

Table 11 gives the direct and indirect annual costs, as calculated from the factors given in the subsection on "Estimating Total Annual Costs." Pressure drop (for energy costs) can be taken from Table 6. Using the higher values from the table, pressure drop for the inlet diffuser plate, inlet and outlet transitions, baffles, and plates is:

$$\Delta P = 0.09 + 0.14 + 0.015 + 0.123 + 0.008 = 0.38 \text{ in. } H_2O$$

Assume the ductwork contributes an additional 4.1 in. H_2O. The total pressure drop is, therefore, 4.48 in. H_2O. As is typical, the ductwork pressure drop overwhelms the ESP pressure drop.

Total Annual Cost

The total annual cost, calculated in Table 11, is $424,000 (rounded). Had the particle size being captured been larger, the ESP cost would have been considerably less. Also, for a much larger gas flow rate, the $/acfm treated cost would be

TABLE 10. Example Costs for ESP System

Direct costs	
Purchased equipment costs	
ESP with standard options	$385,358
Auxiliary equipment	62,700
	$448,058 = A
Instruments and controls, 0.1A	44,806
Taxes, 0.03A	13,442
Freight, 0.05A	22,403
Total purchased equipment cost	$528,709 = B
Installation direct costs	
Foundation and supports, 0.04B	21,148
Erection and handling, 0.50B	264,355
Electrical, 0.08B	42,297
Piping, 0.01B	5,287
Insulation for ductwork, 0.02B	10,574
Painting, 0.02B	10,574
Site preparation	—
Facilities and buildings	—
Total installation direct costs	$354,235
Total direct costs	$882,944
Indirect costs	
Engineering and supervision, 0.20B	105,742
Construction and field expense, 0.20B	105,742
Construction fee, 0.10B	52,871
Startup fee, 0.01B	5,287
Performance test, 0.01B	5,287
Model study, 0.02B	10,574
Contingencies, 0.03B	15,861
Total indirect costs	$301,364
Total capital investment	$1,184,308

TABLE 11. Example Annual Costs for ESP System

Item	Cost
Direct annual costs	
Operating labor	
Operator, 3h/day × 360 d/yr × $12/h =	$12,960
Supervisor, 15% of operator =	1,944
Coordinator, ⅓ of operator =	4,320
Operating materials	—
Maintenance	
MC = 0.01 × $528,709 + $4,125	9,412
Utilities	
Electricity for fan power, 0.000181 × 50,000 acfm × 4.48 in. H_2O × 8,640 h/yr × $0.06/kWh =	21,018
Electricity for operating power, 194 × 10^{-3} × 15,850 ft^2 × 8,640 h/yr × $0.06/kWh	15,940
= Waste disposal, at $20/ton tipping fee at two miles and $0.50/ton mile for essentially 100% collection efficiency:	
$4.29 \times 10^{-6} \times 4 \frac{gr}{ft^3} \times \frac{8,640\ h}{yr} \times 50,000\ acfm \times (20 + 0.50 \times 2) =$	155,676
Total direct annual costs	$221,270
Indirect annual costs	
Overhead, 0.6 × (12,960 + 1,944 + 12,044) =	16,169
Property tax, 0.01 × 1,184,308 =	11,843
Insurance, 0.01 × 1,184,308 =	11,843
Administration, 0.02 × 1,184,308 =	23,686
Capital recovery cost, (1,184,308) × 0.1175 =	139,156
Total indirect annual costs	$202,697
Total annual cost	$424,000 (rounded)

more favorable. Reviewing components of the TAC, dust disposal is the largest single item. Care should be taken in determining this cost and the unit disposal cost ($/ton). Finding a market for the dust, for example, as an extender in asphalt or a dressing for fields, even at giveaway prices, will reduce TAC dramatically.

Acknowledgments

The authors gratefully acknowledge C. G. Noll, United McGill Co., for extensive review and the following companies for contributing data to this article: Research-Cottrell; Joy Industrial Equipment Co., Western Precipitation Division; and Environmental Elements.

References

1. H. J. White, *Industrial Electrostatic Precipitation,* Addison-Wesley, Reading, MA, 1963.
2. P. A. Lawless and L. E. Sparks, "A review of mathematical models for ESPs and comparison of their successes," *Proceedings of Second International Conference on Electrostatic Precipitation,* S. Masuda, Ed.; APCA, 1984, pp 513–522.
3. R. L. Bump, "Evolution and design of electrostatic precipitator discharge electrodes," presented at the 75th APCA Annual Meeting, New Orleans, LA, June 1982.
4. Correspondence of Richard Selznick, Baron Blakeslee, Inc., Westfield, NJ, to William M. Vatavuk, U.S. Environmental Protection Agency, Research Triangle Park, NC, April 23, 1986.
5. Correspondence of James Jessup, M&W Industries, Inc., Rural Hall, NC, to William M. Vatavuk, U.S. Environmental Protection Agency, Research Triangle Park, NC, May 16, 1986.
6. *Modern Cost Engineering,* J. Matley, Ed.; McGraw-Hill Co., New York, 1984, p 142.
7. Personal communication from Robert Shipe, Jr., American Air Filter Co., Louisville, KY, and S. A. Sauerland, United Air Specialists, Inc., Cincinnati, OH, to Roger Ellefson, JACA Corp., Fort Washington, PA, June 1987.
8. W. M. Vatavuk and R. B. Neveril, "Estimating costs of air pollution control systems, part II. Factors for estimating capital and operating costs," *Chem. Eng.,* 157–162 (November 3, 1980).
9. Telecon of Gary Greiner, ETS, Inc., Roanoke, VA, to James H. Turner, Research Triangle Institute, Research Triangle Park, NC, October 1986.
10. *Means Square Foot Costs 1987,* R. S. Means Co., Inc., Kingston, MA, 1987.
11. *Operating and Maintenance Manual for ESPs*, PEDCo Environmental, Inc., EPA/625/1-85/017, Office of Research and Development, Air and Energy Engineering Research Lab, Research Triangle Park, NC, September, 1985.
12. R. H. Perry, et al., *Perry's Chemical Engineers' Handbook,* 6th ed., McGraw-Hill, New York, 1984.
13. E. Bakke, "Wet electrostatic precipitators for control of submicron particles," *Proceedings of the Symposium on Electrostatic Precipitators for the Control of Fine Particles,* Pensacola, FL, September 30 to October 2, 1974, EPA-650/2-75-016, U.S. Environmental Protection Agency, Research Triangle Park, NC, 1975.

14. "Poly-stage precipitator for stack and duct emissions," Beltran Associates, Inc., November 1978.
15. W. M. Vatavuk and R. B. Neveril, "Estimating costs of air-pollution control systems, part XVII. Particle emissions control," *Chem. Eng.* (adapted), 97–99. (April 2, 1984).
16. R. C. Gumerman, B. E. Burris, and S. P. Hansen, *Estimation of Small System Water Treatment Costs,* EPA/600/2-84/184a, NTIS No. PB85-161644, U.S. Environmental Protection Agency, Research Triangle Park, NC, 1984.
17. J. P. Gooch, *A Manual on the Use of Flue Gas Conditioning for ESP Performance Enhancement,* Electric Power Research Institute Report No. CS-4145, 1985.
18. *E.A.B. Control Cost Manual,* 3rd ed., EPA-450/5-87-001A, U.S. Environmental Protection Agency, Office of Air Quality Planning and Standards, Economic Analysis Branch, Research Triangle Park, NC, February 1987.

Appendix: Effects of Material Thickness and Type on ESP Costs

The impact of material thickness and composition of collecting plates and the ESP casing can be *estimated* using the following equations and Figure 1.

Plates:

$$I = \frac{\left[\left(\frac{Wt}{2} \times FS\right) - 0.90\right] M + SP}{SP} \tag{A1}$$

Casing:

$$I = \frac{\left[\left(\frac{Wt}{10} \times FS\right) - 0.58\right] M + SP}{SP} \tag{A2}$$

where: I = incremental increase of flange-to-flange selling price, \$/ft^2
Wt = weight of steel, lb/ft^2
FS = fabricated steel selling price, \$/lb (normally assume approximately two times material cost)
M = manufacturer's markup factor of fabricated cost (direct labor, wages, and material cost before applying general and administrative expense and profit) to selling price (normally 2 to 3)
SP = flange-to-flange selling price from Figure 5, \$/ft^2.

Most vendors can produce ESPs with collecting plate material thickness from 16 to 20 gauge and casing material thickness from ⅛ through ¼ inch without affecting the two times material cost = fabricated cost relationship. Thus the impact of increasing the collecting plates from 18 to 16 gauge and the casing from 3⁄16- to ¼-inch plate on a 72,000-ft^2 collector having a selling price of \$10/square foot and assuming a markup factor of 2 is as follows:

Plates:

$$I = \frac{\left[\left(\frac{2.5}{2} \times 0.90\right) - 0.9\right] 2 + 10}{10}$$

$$= 1.045 = 4.5\% \text{ increase}$$

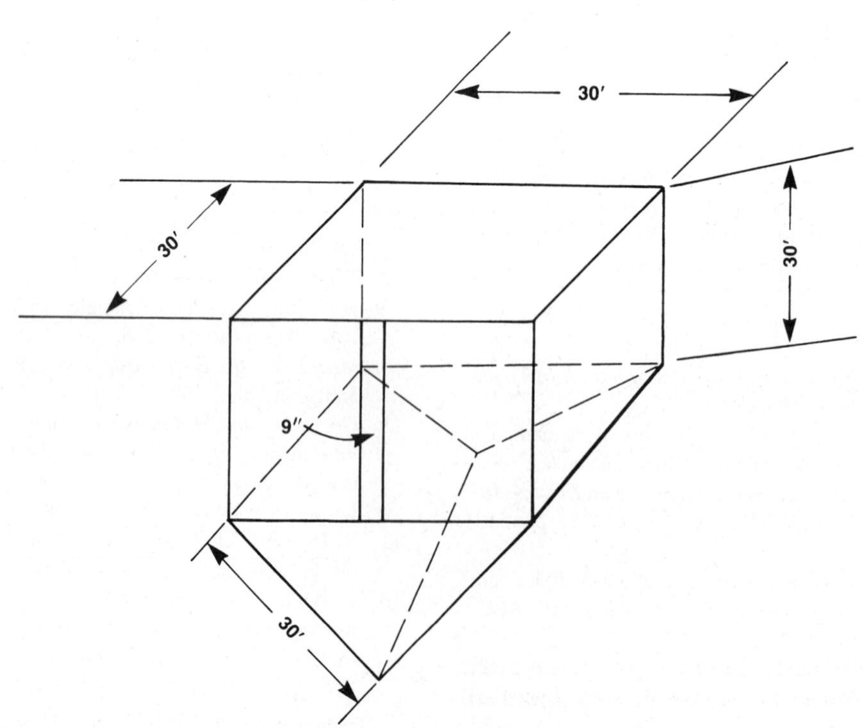

FIGURE 7. ESP Dimensions

Casing:

$$I = \frac{\left[\left(\frac{10.21}{10} \times 0.76\right) - 0.58\right]2 + 10}{10}$$

$$I = 1.039 = 3.9\% \text{ increase}$$

Equations A1 and A2 were developed using the following assumptions:

$$I = \frac{\text{Material selling price increase + Standard ESP selling price}}{\text{Standard ESP selling price}}$$

Because Figure 5 identifies the standard ESP selling price per square foot of collecting area, the material selling price increase = (New material cost – Standard material cost)M. Then it follows that:

$$\text{Material selling price} = \frac{\text{lb steel}}{\text{ft}^2 \text{ collecting area}} \times \text{Fabricated cost in \$/lb} \times M$$

The ESP dimensions given in Figure 7 include:

- Casing area = 30 ft × 30 ft × 8 = 7,200 ft^2 (assume four walls, one top, two hopper sides, two triangular hopper ends ≈ eight equivalent sides)
- Collecting plate area = 30 ft × 30 ft × 2 sides per plate

$$\times \frac{30 \text{ ft}}{s} \text{ plates} = \frac{54,000}{s} \text{ ft}^2$$

$$= 72,000 \text{ ft}^2 \text{ for } s = 0.75 \text{ ft}$$

where s = plate spacing, feet.

Thus, there are:

- $7.50/s$ ft^2 of collecting area per 1 ft^2 of casing and
- 2 ft^2 of collecting area per 1 ft^2 of collecting plate.

Material cost per square foot of collecting area is:

$$\text{Plates} = \frac{\text{lb steel/ft}^2}{2} \times \text{\$/lb}$$

$$\text{Casing} = \frac{\text{lb steel/ft}^2}{7.50/s} \times \text{\$/lb}$$

For the standard ESPs described by Figure 5, 18-gauge collecting plates and 3/16-inch plate casing were specified. Assuming:
Material cost for 18-gauge mild steel = \$0.45/lb
Material cost for 3/16-inch plate mild steel = \$0.38/lb
Material cost to fabricated cost factor = 2

These costs yield fabricated material costs of:

Plates:

$$\frac{2 \text{ lb/ft}^2}{2} \times \$0.45\text{/lb} \times 2 = \$0.90\text{/ft}^2 \text{ of collecting area}$$

Casing:

$$\frac{7.66 \text{ lb/ft}^2}{7.50/s} \times \$0.38 \times 2 = \$0.78\ s\text{/ft}^2 \text{ of collecting area}$$

At a typical 9-inch plate spacing, the casing cost would be \$0.58 per square foot of collecting area.

$$\text{Selling price impact} = \frac{\left(\text{Cost of new material} - \text{Cost of old material}\right)M + \text{Original overall selling price}}{\text{Original overall selling price}}$$

which gives us equations A1 and A2. Note that the value 0.58 will change significantly if a plate spacing other than 9 inches is chosen.

Thus for a less than 5% increase in flange-to-flange cost, all the precipitator-exposed wall sections can be increased by more than 25% to provide increased life under corrosive conditions. Section thickness increases that are greater than those just discussed would probably result in significant cost increases because of both increased material costs and necessary engineering design changes.

The impact of changing from mild steel to 304 stainless steel (assuming material costs of \$1.63 per pound for 18-gauge collecting plates, \$1.38 per pound for the 3/16-inch casing, and a markup factor of 3) is as follows:

Plates:

$$I = \frac{\left[\left(\frac{2}{2} \times 1.63\right) - 0.9\right]3 + 10}{10} = 21.9\% \text{ increase}$$

Casing:

$$I = \frac{\left[\left(\frac{7.66}{10} \times 1.38\right) - 0.58\right]3 + 10}{10} = 14.3\% \text{ increase}$$

To these material costs must be added extra fabrication-labor and procurement costs that will increase the ESP flange-to-flange cost by a factor of 2 to 3. Note that a totally stainless-steel collector would be much more expensive because the discharge electrodes, rappers, hangers, and so on, are also converted to stainless. The preceding equations can be used for other grades of stainless steel or other materials of construction by inserting material costs obtained from local vendors on a dollar-per-pound basis.

FABRIC FILTERS

John D. McKenna and Dale A. Furlong

Fabric filters remove dust from a gas stream by passing the stream through a porous fabric. Dust particles form a more or less porous cake on the surface of the fabric. It is normally this cake that actually does the filtration.

The manner in which the dust is removed from the fabric is a crucial factor in the performance of the fabric filter system. If the dust cake is not adequately removed, the pressure drop across the system will increase to an excessive amount. If too much of the cake is removed, excessive dust leakage will occur while fresh cake develops. The selection of design parameters is crucial to the optimum performance of a fabric filter system.

Fabric filter systems are frequently referred to as baghouses, since the fabric is usually configured in cylindrical "bags." The two most common baghouse designs are the reverse-air and the pulse-jet types. These names describe the cleaning system used with the design.

Reverse-air baghouses operate by directing the dirty flue gas into the inside of the bags; therefore, the collection of dust is on the inside surface of the bags. The bags are cleaned periodically by reversing the flow of air, causing the previously collected dust cake to fall from the bags into a hopper below. Since this cleaning procedure is accomplished at relatively low gas velocity, the fabric is not exposed to violent movement, and so the reverse-air cleaning technique normally results in maximum bag life. In a variation of reverse-air baghouse design and the forerunner of the reverse-air baghouse (i.e., the shaker baghouse), the bags are shaken during the reverse-air cleaning interval.

Pulse-jet baghouses are designed with internal frame structures, called cages, to allow collection of the dust on the outside of the bags. The dust cake is periodically removed by a pulsed jet of compressed air into the bag causing a sudden bag expansion; dust is removed primarily by inertial forces when the bag reaches its maximum expansion. This bag cleaning technique is quite effective, however, the vigorousness of the technique and frequently the bag-to-cage fit tend to limit bag life and also tend to increase dust migration through the fabric, thus decreasing dust collection efficiency.

The selection of the fiber material and fabric construction is important to baghouse performance. The fiber material from which the fabric is made must have adequate strength characteristics at the maximum gas temperature expected and adequate chemical compatibility with both the gas and the collected dust. Felted fabric construction generally gives better removal of fine dust particles as compared with woven fabrics. However, not all fiber materials can be felted into a fabric of adequate strength and hence most filtration fabrics are constructed, at least in part, of filaments and/or fibers that are first twisted into yarns, and then woven or knitted into a fabric.

The published literature for fabric filtration is quite extensive. The reader is referred to *Fabric Filter—Baghouses I*, available from ETS, Inc., for one listing of references and for a more extensive discussion of subjects briefly included here.[1]

CONTROL OF TOXIC AIR POLLUTANTS

Spray adsorption, or dry scrubbing, followed by a fabric filter is a rapidly maturing technology for simultaneously controlling particulate matter and acidic gases emitted from combustion processes. In addition, this combination is used to control emissions of other toxic air contaminants—heavy metal compounds and products of incomplete combustion such as chlorinated hydrocarbons (including dioxins), for example—that may be emitted from solid waste incinerators.

Dry scrubbing followed by a fabric filter is considered the best available technology for controlling emissions from solid waste incinerators for two basic reasons:

1. The dry scrubber both reacts with acidic gases to form solid particulate and causes other toxic vapors, including most heavy metals and chlorinated hydrocarbons, to condense. The condensed substances have been shown to collect preferentially on the very fine dust particles owing to the greater available surface area of the fine particulate.
2. The fabric filter is the best currently available technology for high-efficiency removal of fine particulate.

Baghouses are generally considered to be a superior choice, relative to electrostatic precipitators (ESPs), for fine-particulate control. A multifield precipitator with a very large plate area is required to provide comparable fine-particulate collection performance; therefore, an ESP of comparable performance is usually more expensive. Figure 1 illustrates the greatly improving trend in the control of particulate emissions by both ESPs and fabric filters since 1970. Figure 2 indicates that for typical two-field ESPs, the removal efficiency for fine, submicron particles is considerably less than that achieved in a fabric filter.

As experience with the use of baghouses has increased, their reliability also has increased as a result of the availabil-

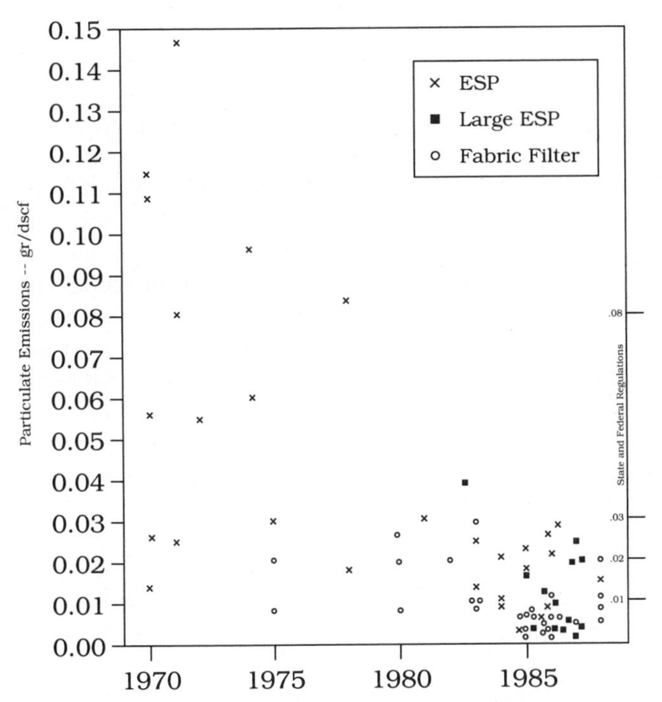

FIGURE 1. Particulate Emissions from MSW Incineration, 1970–1988

ity of different fibers/fabrics and improvements in the design of bag fabrics and in cleaning techniques. These measures have extended bag life to an average of five years, or more in some cases. Well-designed and operated baghouses have been shown to be capable of reducing overall particulate emissions to less than 0.010 gr/dscf, and in a number of cases, to as low as 0.001–0.005 gr/dscf. Based on the potential for greater removal efficiencies overall and in the submicron particle-size range, a number of states, including California, Connecticut, and Michigan, as well as Canada, prefer the use of scrubber/baghouses for solid-waste resource-recovery plants.

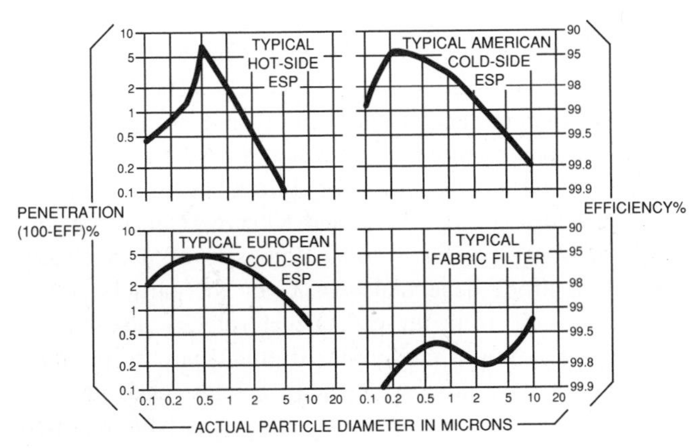

FIGURE 2. Typical Fractional Efficiencies For Existing Collectors (From *Electric Power Research Institute Economics of Fabric Filters vs. Precipitators*, Denver, CO June 1978)

FILTRATION PROCESSES

Mechanisms

Several particle collection mechanisms are normally responsible for filter efficiency. Theoretical equations exist for the capture efficiency of each mechanism based on single particles approaching single fibers. For an operating fabric filter, however, the fabric is covered with a dust cake and the dust cake is of continually varying thickness. Once the dust cake has reformed after each cleaning, sieving is probably the dominant mechanism. As particles approach the porous mass of dust that constitutes the cake, they either will strike one or more surface particles or enter a pore. If the particle is larger than the pore it attempts to enter, it will be sieved out. If the particle is smaller than the pore it enters, it will continue traveling through the pore until it touches the pore wall and adheres (rather than bounding or rolling along the wall); or until the pore narrows to dimensions smaller than the particle, causing the particle to be sieved out; or until the particle passes through the dust pore and a fabric pore and exits on the clean-air side of the air filter. Ordinarily, only one out of 1000, or an even 10,000, particles finds its way through the filter.

One might expect that larger particles would be sieved out with greater efficiency than smaller particles; that is, that the particles leaving the filter would have a smaller median diameter than the particles entering the filter. Experimentation has shown, however, that size distribution across the filter surface changes only slightly. The reason for the lack of change in size distribution stems from the manner in which most particles find their way through the filter. Most particles that transit the filter do so by a leakage process.[2]

The influence of electrically charged particles, a condition that is most common, on oppositely charged fibers of the filter has been shown to enhance attraction to an extent that particle-to-particle agglomeration is increased.

Consider a freshly cleaned bag: As filtration resumes, the cleaned areas of the fabric present pores of various sizes to the oncoming dust cloud. Individual particles strike the edges of pores or attach and begin to form chains or dendrites, a condition encouraged by the attraction produced as a result of opposing charge polarities. Before long, the smaller pores are bridged over by the chains and eventually become completely covered by porous cake. As time passes, more and more pores become covered with cake and the gas velocity through the remaining, uncovered pores becomes higher. Instead of a face velocity of a few feet per minute, we now have a pore velocity of up to several thousand feet per minute. Eventually, the velocity through the few remaining uncovered pores becomes so high and the pores so large that they cannot be bridged. For the remainder of the filtration cycle, these relatively free pores will be leakage points in the filter, and most of the particles passing through the filter will pass through these leakage points.

Billings and Wilder[3] report work by Tomaides suggesting that particles can bridge over a gap in a filter about 10 particle diameters wide. Presumably, the adhesive forces holding the chain together are exceeded by the aerodynamic forces trying to rupture the chain if its length exceeds 10 particle diameters.

Gas-to-Cloth Ratio

Gas-to-cloth ratio, G/C, is a measure of the amount of gas driven through each square foot of fabric in the baghouse, and is given in terms of the number of cubic feet of gas per minute passing through 1 square foot of cloth ($[ft^3/min]/ft^2$ or fpm):

$$G/C = \frac{ft^3 \text{ per minute of gas}}{ft^2 \text{ of fabric}}$$

G/C can correctly be considered a superficial gas velocity. This is not the actual velocity through the openings in the fabric, built rather the apparent velocity of the gas approaching the cloth, although it is often referred to as "the gas velocity" and reported as feet per minute or meters per minute.

As the gas-to-cloth ratio increases, pressure drop (ΔP) also increases. Pressure drop is normally measured and reported as inches of water (in. H_2O, also written as inches water gauge, in. WC), the convention in the United States.

Pressure Drop

During the mid-1800s, Darcy formulated the following law for flow of fluid through a porous bed.

$$\Delta P = \frac{L u_f V}{K}$$

where: ΔP = pressure difference across the bed
L = bed thickness
u_f = fluid viscosity
V = superficial fluid velocity
K = bed permeability

This equation assumes that the fluid is essentially incompressible and steady, the fluid viscosity is Newtonian, and the velocity is low enough that only viscous effects occur. Over the past 100 years, investigators have been trying to find ways to predict K and to refine the Darcy equation.

The basic Darcy equation can be used to predict the pressure drop for an operating fabric filter with dust cake accumulating on the fabric.

$$\Delta P = S_E V + + K_2 C_i V^2 t \qquad (1)$$

where: ΔP = pressure drop, in. H_2O
S_E = effective residual drag, in. H_2O/fpm
V = velocity, fpm
K_2 = specific cake coefficient
C_i = inlet dust concentration, gr/cubic foot
Δt = filtration time, minutes

Energy loss through a fabric filter is composed of two parts. The first, $S_E V$, represents drag or energy expended in pumping system gas through the cleaned equilibrium fabric of the fabric filter. The second part of the equation, $K_2 C_i V^2 t$, represents energy required to pump gas through the filter cake that builds up on the surface of the fabric. Gas velocity appears in both terms, but because it is squared for the cake portion of the equation, it is especially important for describing the energy consumed in pumping gas through the filter cake. Another important part of the equation is K_2, the specific cake coefficient. This term is characteristic of the dust, varies for different dusts, and is a measure of how rapidly pressure drop will build up in a system.

A fabric filter in stable, cyclic operation will normally reach a point of constant drag characteristics. That is, the resistance to gas flow of the freshly cleaned fabric is the same at the beginning of successive filtration cycles. In practice, the value may change as the fabric ages. Residual drag is a measured value. There is no useful predictive equation for residual drag.

Experimental Measurements of K_2

Many researchers have conducted laboratory and pilot-scale fabric filter tests to measure K_2. Billings and Wilder[3] reported an extensive field survey of K_2 as a function of the air–cloth ratio (filtration velocity) and particle size. In this early work, K_2 was determined from the reported values of operating air–cloth ratio (V), dust loading (C_i), filtration time (t), and residual and maximum pressure drops (ΔP_R, ΔP_m). While this earlier work was quantitative, the wide range of dusts, quality of reported data, and configuration of the individual systems (single compartment, multiple compartment, type of cleaning, etc.) led to considerable scatter. The relationship showed order-of-magnitude variations in K_2 at a given particle size.

In more recent tests, obtained under controlled conditions, the relationships among K_2, particle size, and velocity have been shown more clearly.

Data from Dennis et al.[4] and Davis and Kurzynske[5] are shown in Figures 3 and 4, respectively. The solid lines represent each researcher's best fit to the data, where available. The data reported by Dennis et al. were summarized from eight different sources for fly ash, mica, and talc at 2–6 fpm; the data by Davis and Kurzynske were on talc dusts at a velocity of 4 fpm. Both sets of data clearly indicate a strong dependence of K_2 on the particle size.

It is evident from these data that velocity also has an effect on K_2. While this observed effect may be partially attributed to the effect of velocity on dust cake packing and/or Reynold's number, most researchers have reported that K_2 is a function of velocity such that

$$K_2 = kV^x$$

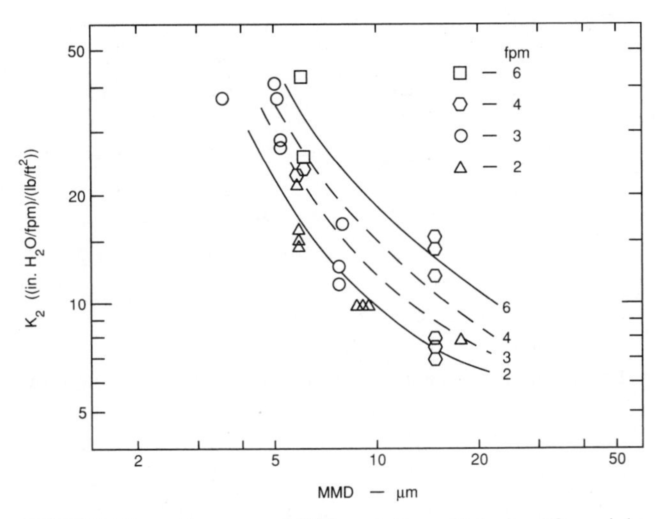

FIGURE 3. K_2 versus *MMD* and Face Velocity (Copyright ASTM. Reprinted with permission.)

Dennis et al.[4] reported that x had a value of 0.5 for fly ash and varied from 0.5 to 1.0. Davis and Frazier,[6] in a series of tests on fly ash using 11 different filter materials, reported an average value of 0.7 for fly ash. The data in Figures 3 and 4, data by Davis and Frazier,[6] and data by Frazier and Davis,[7] were normalized to a velocity of 3 fpm and replotted in Figure 5 assuming an average value for x of 0.6. The normalized data show that there is a well-defined relationship between K_2 and the particle size.

A best-fit equation was determined for the data:

$$K_2 = 118.4\ MMD^{-1.10}$$

where K_2 is measured in the English system and *MMD* is in microns. The best-fit equation predicts the K_2 value within a factor of two. The agreement between various sets of data is excellent considering that measurements obtained under carefully controlled laboratory conditions for a constant particle size distribution have shown a factor-of-two variation within a single laboratory.[6]

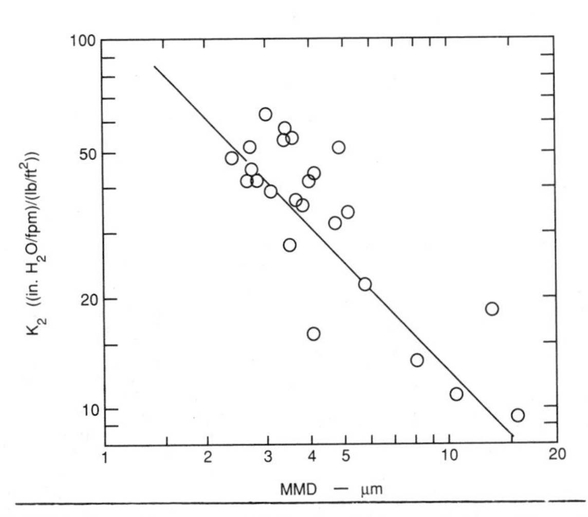

FIGURE 4. K_2 versus *MMD* for Talc at 4 fpm (Copyright ASTM. Reprinted with permission.)

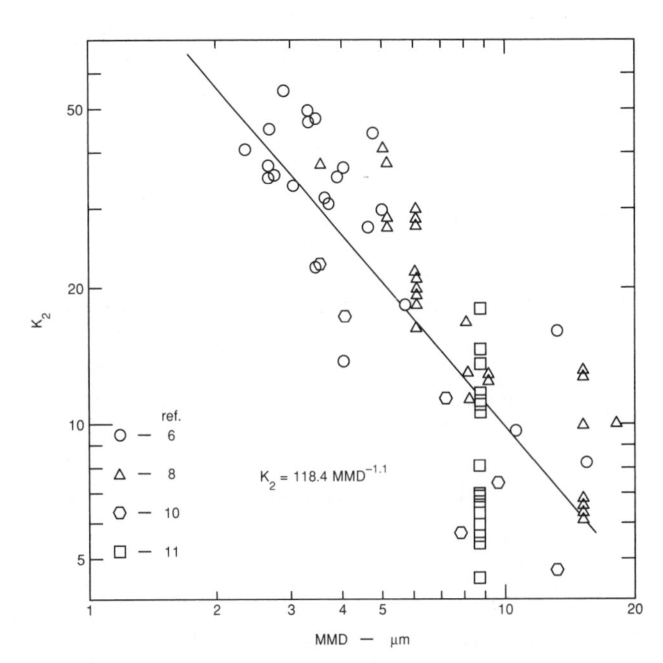

FIGURE 5. K_2 Normalized to 3 fpm versus *MMD* (Copyright ASTM. Reprinted with permission.)

Pressure Drop in Multicompartment Baghouses

Equation 1 describes the instantaneous pressure drop after some definite period of filtration through an area of fabric. When the area is distributed across several compartments operated in parallel, the compartments are generally cleaned sequentially. The total pressure drop across the baghouse, in terms of drag, is analogous to a set of electrical resistances in parallel. Robinson et al.[8] used Hemeon's equation:

$$\frac{N}{S_T} = \frac{1}{S_1} + \frac{1}{S_2} + \cdots + \frac{1}{S_N} \tag{2}$$

where: S_T = total baghouse drag $= \dfrac{\Delta P}{V_T}$

V_T = total gas volume $= V_1 + V_2 \cdots VN$

or

$$S_T = N\left[\sum \frac{1}{S_1}\right]^{-1}$$

N = number of compartments filtering

Subscripts refer to individual compartments

If one knows the drag in each compartment at a given time, the instantaneous drag (and, therefore, pressure drop) could be easily calculated for the entire system. Equation 2 lends itself primarily to computer applications.

Efficiency

Efficiency is a measure of how well a filter separates dust from gas. For an operating baghouse, overall efficiency is calculated from:

$$\text{Efficiency} = \frac{C_i - C_o}{C_i} = n$$

where: C_o = outlet concentration
C_i = inlet concentration

Penetration (1–efficiency) is also used as a measure of performance.

Efficiency and penetration may be measured or calculated for specific particle sizes or size ranges. Inertial impactors are used to measure efficiency by measuring particle concentration over several size ranges on the inlet and outlet streams of a baghouse.

There is no satisfactory set of published equations that allows a designer to calculate efficiency for a prospective baghouse.

FILTRATION FABRICS

Fiber Types

This section discusses fibers and fabric properties that will help the user select and specify fabrics for particular fabric filter applications. Fabrics made of natural fibers such as cotton or wool are still employed for many filter applications; the development of synthetic fibers, however, has greatly extended the possible range of applications for fabric filters. Continuing developments in fiber and fabric technology may be anticipated.

Synthetic fibers are widely used for filtration fabrics because of their low cost, better temperature- and chemical-resistance characteristics, and small fiber diameter. Synthetics used include acetates, acrylics, polyamides, polyesters, polyolefins, and polyvinyl chlorides. Specialty fibers for high-temperature use such as Teflon, Ryton, P84, and carbon fibers have been developed; however, the synthetic fiber most used for high-temperature applications is glass.

The properties of glass fiber, such as good acid resistance, good heat resistance, and high tensile strength, solve many of the problems inherent in baghouses.

Fiberglass has the following characteristics.

- It is noncombustible because it is completely inorganic.
- It has zero moisture absorption; therefore, it is not subject to hydrolysis.
- It has dimensional stability (low coefficient of linear expansion).
- It has very high tensile strength, but poor resistance to flex and abrasion. There are some chemical surface treatments (e.g., silicone, graphite, Teflon B) that improve the flex–abrasion characteristics of glass.
- It has good resistance to acids, but is attacked by hydrofluoric, concentrated sulfuric, and hot phosphoric acids.
- It has poor resistance to alkalis; hot solutions of weak alkalis attack glass.
- It has poor resistance to acid anhydrides and metallic oxides (e.g., fluorides and sulfur oxides). For this reason, glass baghouses should not be operated at or below the dew point.

Table 1 lists the major fiber alternatives for gas filtration and gives some of the important properties of these fibers.

The first, and probably the key, inlet condition that must be identified prior to selecting a filtration medium is the temperature the fabric will experience. Figure 6 compares the recommended operating temperatures for the most often used filtration fabrics. Note that as temperature increases, the fabric choices become fewer and fewer. In 1991, the maximum temperature for which economical filter media were commercially available was about 500°F (260°C), although this may have changed with the advent of ceramic and metallic filters. When confronted with higher temperatures, the usual approach has been to cool the gas down, at least to the vicinity of 500°F (260°C). Once a preliminary selection of filtering fabric has been made, the media supplier can usually provide additional information that should be considered in finalizing the fabric choice.

Important Fiber Characteristics

When selecting a fiber for gas filtration, attention must be paid to the following factors that interact and thus must be considered together.

TABLE 1. Fabric Selection Chart

Fabric	Maximum Temperature, °F	Acid Resistance	Fluoride Resistance	Alkali Resistance	Flex Abrasion Resistance
Cotton	180	Poor	Poor	Good	Very Good
Polypropylene	200	Excellent	Poor	Excellent	Very good
Polyester	275	Good	Poor to fair	Good	Very good
Nomex	400	Poor to fair	Good	Excellent	Excellent
Teflon	450	Excellent	Poor to fair	Excellent	Fair
Fiberglass	500	Fair to good	Poor	Fair to good	Fair

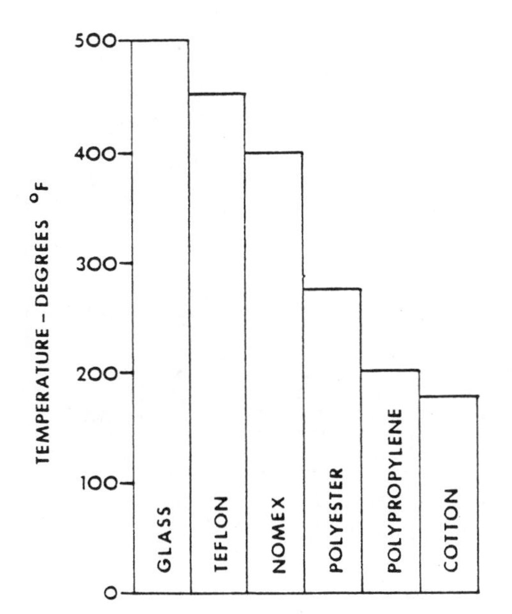

FIGURE 6. Recommended Maximum Operating Temperatures for Fabrics

1. Temperature. The fiber must have a maximum continuous service temperature higher than the normal temperature of the application. If temperature surges above the normal range occur, the ability of the fiber to withstand the expected conditions of surge temperature and duration must be considered.
2. Corrosiveness. The ability of the fiber to resist physical degradation from the expected application levels of acids, alkalies, solvents, or oxidizing agents must be considered.
3. Hydrolysis. Effects of the expected level of humidity must be taken into account.
4. Dimensional stability. If the fiber is expected to shrink or stretch in the application environment, the effects of such a change must be tolerable.
5. Cost. As with any engineering product, the least costly selection that will meet overall requirements is usually the best selection.

Media Selection

Baghouse operating costs are reduced if the baghouse has a high gas-to-cloth ratio, a low pressure drop, and a long life. In each case, the key to operation at minimum cost is the media selected for bag construction. This selection is crucial, but not easy, because many, often conflicting, requirements must be met.

The primary media selection criteria are the compatibility of the selected fiber with the gaseous environment and the physical configuration of the fiber and resulting fabric as it affects filtration performance. Usually, the selection criteria interact so much, or are not well enough understood, that the best selection is not apparent without long-term testing, except for bag-life determination, where experimental bag tests can provide specification data. Sometimes it is possible to equip complete baghouse compartments with various media[9] and observe the results for future selections. In 1985, ETS, Inc., introduced an individual bag flow monitor that is useful in making side-by-side comparisons of alternative media.[10]

The satisfactory performance of a fabric filter in a specific application requires the selection not only of a fiber material that is compatible with the gas–particle environment, but also that is of a fabric design appropriate to the dust collector geometry and collector cleaning requirements. *Fiber, yarn, and fabric parameters all influence the filtration process.*

Fabric strength, stability, and flexibility all are important parameters in determining the ability of the fabric to resist wear caused by abrasion. The term "abrasion" is defined as an eroding of fabric fibers or fiber surface material as a result of moving contact between the fiber and dust particles or between adjacent fibers. The flexibility of a filtration fabric is important for at least two reasons: cleanability and durability. Removal of the dust deposit may be improved by flexing of the fabric substrate, but such flexure may degrade the fabric. Thus fabric flexibility may be both necessary and harmful.

Permeability of the fabric must be considered when selecting a filter media. It must be understood, however, that the permeability of filtration fabrics is so reduced by a residual dust deposit that the permeability of clean fabric appears to have little relationship to permeability during use. The objective in fabric design is to maintain a highly permeable combination of residual dust and fabric while allowing a minimal amount of dust to pass through. To meet this objective, the pores through a fabric must be closely controlled. They must not exceed a certain bridging diameter, but if they are too small, they will either plug or pass too little gas. In an ideal filtration fabric, probably all the fabric pores should be of the same size, that size depending on properties of the dust and gas, fabric characteristics, and so on. Yarn texturizing and the postweave surface treatments contribute greatly to the permeability of clean fabric, and perhaps also to the permeability of residual dust deposit.

The last, but certainly not least, criterion for media selection is the ability of the fabric to *release* collected dust during the cleaning cycle. This ability depends largely on the mode and intensity of cleaning, but also on the adhesive character of the fabric. The way in which fabric construction relates to deposit release has not been completely determined, but it is known that a smooth fabric surface releases dust more readily than does a fuzzy surface. Dust may agglomerate on loose fibers and move away from the surface during cleaning, only to return once filtration resumes. This action, whose outcome is sometimes referred to as "dingleberries," can result in poor cleaning. Some fabric surface treatments are specifically intended to enhance cake release. Dust release also depends on the electrical resistances of selected fibers. Electrical resistance is known to depend on humidity, which itself has a marked effect on filtration fabric performance.[11] An extensive table of electrical properties for a variety of fabrics is given by Frederick.[12] "Electrical effects" is a factor that has rarely been quantified, but is known to influence pressure drop and efficiency. This is one of the factors that can cause major variations in baghouse performance.

In about 90% of the baghouses currently operating on coal-fired boilers, the bags are fabricated from glass fibers. Glass fibers are tiny filaments as small as 0.00015 inches (4 μm) in diameter that are extremely flexible and thus may be woven into fabric. Before 1930, discontinuous "glass wool" fibers were the only form produced commercially. The new technology was first used to develop continuous fibers for high-temperature insulation of fine electrical wires, designated "E"-glass because of its unique electrical insulation properties, and which was industrially practical because an effective lubricating finish protects the filaments. Today, E-glass is used in nearly all glass-fiber applications, ranging from printed circuit boards to boat hulls and filtration fabrics. Other formulations were also developed by altering percentages of the glass composition to obtain special performance characteristics. "C"-glass is particularly resistant to chemical attack; "S"-glass possesses outstanding strength characteristics. Thus far, production constraints and economics have limited glass filtration fabrics to E-glass.

Woven Fabric

Most filtration fabrics are either completely or partially made by weaving. Even felted or so-called "nonwoven" fabrics include a base (scrim) of woven fabric. Baghouses in which the gas flow is from the inside of the bags to the outside, such as reverse-air and shaker-cleaned baghouses, use woven fabrics almost exclusively. These baghouses generally operate at lower gas flow rates where the flow restriction of the fabric is not so significant. Woven fabrics usually have greater flow restriction but greater strength (for a given fabric weight) than do comparable nonwoven fabrics and thus are usually chosen for reverse-air and shaker applications.

Pulse-jet–cleaned baghouse designs offer increased cleaning energy and operate at higher gas flow rates. The tendency of woven fabrics to "bleed," resulting in low filtration efficiency when clean, usually restricts the use of woven fabrics for pulse-cleaned applications. Woven fabrics made with texturized yarn or with membrane films applied to the upstream surface offset the tendency to bleed when cleaned, and these types of fabrics are used in pulse-cleaned applications.

Woven fabric is manufactured by weaving together fibers that previously had been made into yarns. During weaving, longitudinal yarns (the warp) are interlaced at right angles with transverse yarns (the fill) by means of a loom.

Woven fabrics are formed by interlacing yarns at right angles on the loom, after which the raw (greige) fabric may be further treated. While there are many patterns of interlacing, the fabrics most commonly used in gaseous filtration are twill, satin, or plain weaves. These three weave patterns are depicted in Figure 7. Both the twill and the satin weaves have fabric sides where warp or fill yarns predominate, and those sides are referred to as warp face or filling face.

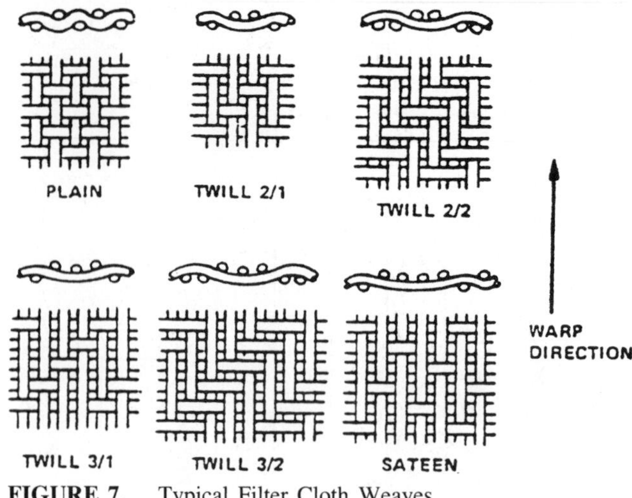

FIGURE 7. Typical Filter Cloth Weaves

In its basic construction, satin weave is similar to twill, but generally uses a pattern of fill yarns going under one, then over four to 12 warp yarns. Satin differs in appearance from twill because the diagonal of satin weave is not visible; it is purposely interrupted in order to contribute to the flat, smooth, lustrous surface desired. There is no visible design on the face of the fabric because the yarns that are to be thrown to the surface are greater in number and finer in count than the yarns that form the reverse of the fabric.

Plain weave is sometimes referred to as tabby, homespun, or taffeta weave. It is the simplest type of construction and consequently the least expensive. Each fill yarn goes alternately under and then over the warp yarns across the width of the fabric. On its return, the yarn alternates the pattern of interlacing. If the yarns are close together, plain weave has a high thread count, often a requirement for suitable efficiency.

Nonwoven Fabrics

Nonwoven fabrics generally are defined as sheet or web structures made by bonding and/or interlocking fibers, yarns, or filaments by mechanical, thermal, chemical, or solvent means.

The International Non-woven and Disposables Association (INDA), a trade association of the nonwoven fabrics industry, has given this definition:

> A non-woven is any fabric made directly from a web of fiber without the yarn preparation required for weaving or knitting. Natural or synthetic fibers can range in length from 0.5–15 cm from crimped staple products, up to continuous filaments for spunbonded products. The fibers may be oriented in one direc-

tion or may be deposited randomly. This web is given structural integrity by 1) mechanical fiber intertangling, 2) thermal or chemically induced fusing of the fibers, or 3) application of any of several adhesives or resins.

Perhaps the most significant feature underlying all nonwoven fabric production, and the one that contributes most to its economic appeal, is the speed at which the fabric is produced. Many nonwoven production units in place today can produce fabric at speeds of 400 fpm (200 cm/s) or more. Some advanced units operate at speeds over 1000 fpm (500 cm/s). These rates may be compared with knitting machine speeds of about 5.0 fpm (2.5 cm/s) and the even slower rates of weaving machines.

A needlepunched fabric is produced by introducing a fibrous web already formed by carding or air-laying into a machine equipped with groups of specially designed, barbed needles (see Figure 8). While the web is trapped between a bed plate and a stripper plate, the needles punch through it and reorient the fibers so that mechanical bonding is achieved among the individual fibers. The needlepunching process is generally used to produce fabrics that have high density and yet retain some bulk. Fabric weights usually range from 1.7 to 10 oz/yd^2 (58–340 g/m^2) and thicknesses from 15 to 160 mils (0.38–4.1 mm).

Needlepunching is often used to combine two or more layers of fiber into a feltlike fabric. Usually one layer is a woven fabric, called a *scrim*, for strength, while the other(s) may consist of fibers of almost any description or combination. Thus considerable control over separate properties of the finished material is possible. For example, the scrim may contribute to desired dimensional stability, while the top, or batting layer, contributes appropriate properties for filtration. Needlepunching is accomplished by punching needles with forward barbs from the batting side into or through the scrim, and the batting fibers thus laced into the scrim remain behind when the needles are withdrawn. Variations on the needling process include changing the needle angle or number of repetitions or using two-sided needling. When a shrinkable scrim is used, the needled material may later be felted in various ways to produce a denser and more uniform material.

To improve the filtration characteristics, primarily the collection efficiency of the fabric surface, the surface may be napped. This is done by teasels (woody, thistlelike parts of weed plants), whose barbs pluck fibers from the surface

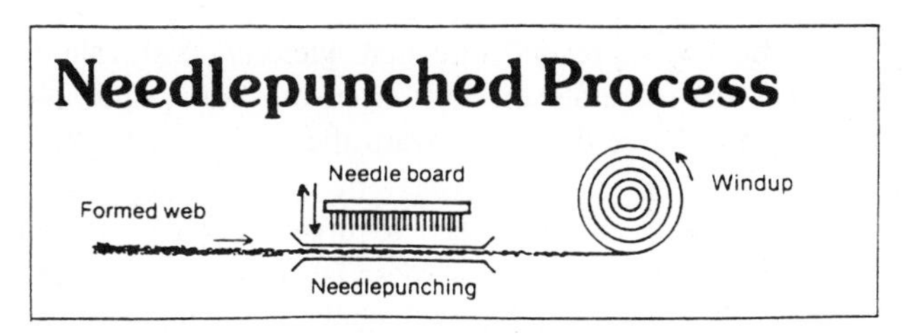

FIGURE 8. Needlepunched Process (From Reference 12)

of the felt. When enough nap has been raised in this manner, it may be singed or otherwise trimmed to the desired thickness.

Needlepunched fabrics, sometimes called needlefelts, are employed in many filtration applications. Modification of the basic structure produced by the needling process determines filtration performance. When punch density is increased during needling, one of the results is an increase in fabric density. Eventually, an optimum punch level is attained above which the density falls off. The maximum density possible from needlepunching alone would be insufficient to produce an efficient filtration medium. Additional processes are required to give the necessary densification and closing-up of the structure. Heat treatments are normally employed at temperatures suited to the thermal characteristics of the particular fiber in use. Suitable processes include the squeezing of the fabric between rollers or a roller and a stationary bed. An increase in fabric density increases filtration efficiency, up to a point.

Successful fabrics are also required to provide good discharge of the dust cake. Good discharge is usually achieved by means of a surface modification in the form of singeing to remove individual fibers protruding from the surface. Good cake release surfaces can also be formed by heating the fabric surface to such an extent that fiber melting and fusion occur, although this will also increase the flow resistance of the fabric.

Relaxation of tensions built into the fabric structure by the needling operation can be brought about by passage through an oven. Such treatment is insurance against dimensional changes occurring when the fabric first encounters operating conditions.

Knitted fabrics, both plain and bulk, are being used (especially the latter) in high-ratio collectors.

Treatments and Finishes

"Finishing" includes those processes that improve the appearance or serviceability of the fabric after it leaves the weaving machine. Greige (unfinished) fabrics intended for use as filtration fabrics are treated and/or finished after weaving to improve their filtration characteristics and cleaning (release) characteristics. Fabric life and strength may also be affected. Treatments are defined as postweaving processes that affect the entire fabric, whereas finishes are postweaving processes that affect only the surface of the fabric.

One fabric treatment of significance for filtration fabrics is heat setting. In the process of heat setting, the fabric is exposed to temperatures exceeding those experienced in service. This treatment is done on a machine known as a pin tenter. The fabric is held under tension in both the warp and fill directions and passed through a heated oven.

Special finishes have been developed for the glass fabrics used in high-temperature filtration. For glass filter bags, organic materials, such as the starch binder that the yarn producer applies and the warp sizing (starch and mineral oil) applied to facilitate weaving, must be removed before a finish is applied to the fabric. This is necessary because these organic lubricants would not be stable at the process temperature, and also because they would interfere with the application of the desired finish. It is desirable that the applied finish have as close a contact as possible with the bare glass filaments, penetrating the yarn bundle and encapsulating individual filaments.

Organic sizings are removed thermally by a process called coronizing, which also sets the crimp in the glass yarns. Crimp is the bend in a yarn caused by interlacing during weaving. The yarn is heated to the softening point and the crimp is set as the yarn cools. It is believed that setting the crimp in the yarn maximizes the flex life of the fabric.

The finish used for glass fabrics must be thermally stable at process temperatures (500–550°F or 260–290°C) and chemically resistant to the gas environments found in fiberglass filter bag applications. The basic purpose of the finish is to protect the glass fibers from abrading themselves, but it can also enhance dust-release characteristics.

In addition to providing lubricity to extend bag life, glass fabric finishes also help to promote dust release from the fabric and offer varying degrees of protection from chemical attack. The success of glass fabric as a viable filter medium depends to a large degree on the quality of the finish. Finish development has occurred in roughly three stages, with finishes making up primarily three groups:

I. Silicones. A glass-to-silicone coupler is the basic prefinish required before subsequent organic finishes are applied. These "couplers" insure the complete individual fiber envelopment needed for effective protection.
II. Silicones and graphite with small amounts of fluorocarbons.
III. Fluorocarbon compounds.

All three groups are still in use, although Group I, the second-stage silicone finish, has largely faded from the scene. Group II finishes are divided into tricomponent and acid-resistant groups.[13]

- Tricomponent
  - Graphite (natural or synthetic)
  - Teflon
  - Silicones
  - Agents to assist application
- Acid-resistant
  - Graphite (natural or synthetic)
  - Polymers
  - Binders
  - Silicone
  - Teflon
  - Agents to assist application

Teflon is used in both Group II and III finishes. Group III finishes consist of Teflon, binders, and agents to assist application.

Most of the experience to date with Group III finishes has been at 10% loading; for example, 10% of the finished fabric weight is Teflon B. Amounts of finish vary greatly, but, in general, the Teflon finish is 7–10% weight added per weight of cloth, the acid-resistant finish is 4–6%, and the tricomponent finish is 1–2%.[14] Some experience indicates that Group II finishes are more resistant to chemical attack, while other experience favors the use of Group III finishes. There has been progress in using Group II finishes to protect Nomex fabrics from acid exposure in coal-fired boiler applications.

A special surface treatment gaining increased use in gaseous filtration is the application of a Gore-Tex™ membrane to the fabric surface.[14] The Gore-Tex membrane is expanded polytetrafluoroethylene (PTFE) deposited as a thin, fibrillated film. Figure 9 shows a photomicrograph of the fibrillated film of PTFE on the fabric surface. A cross section through the fabric is shown in Figure 10. The coarse woven fibers are seen on the right and a thin PTFE film covers the left side. Gore-Tex membranes have been applied to many available backing materials, including woven polyester, Nomex, glass, and Teflon. Usually, the backing fabric is quite porous. Gore-Tex filter bags are constructed for pulse-jet, reverse-air, and shaker collectors. Gore-Tex membrane filter cartridges are also available.

The measured properties of fabric finished with a Gore-Tex membrane show the tensile and burst strengths characteristic of the woven fabric, but permeability is lower. Although lower permeability might suggest higher-than-desirable pressure drops and drags, that often is not the case, because the membrane improves cleanability and reduces residual dust buildup in the fabric.

CLEANING TECHNIQUES

The primary way to categorize a fabric filter is by the method used to clean it. In general, accumulated dust is separated from the fabric by some combination of the following effects.

1. Deflection of the fabric/dust cake, tending to fracture the cake and separate it from the fabric.
2. Acceleration of the fabric/dust cake, yielding separation forces.
3. Gas flow in the reverse direction, yielding aerodynamic forces that separate the dust from the fabric and subsequently move the dust toward the collecting hopper.

Four cleaning methods have evolved, each of which generates some combination of these fabric–dust-cake separation effects. The majority of baghouses currently in use employ one or more of these cleaning methods.

GORE-TEX MEMBRANE vs. POLYESTER FELT STRUCTURE

100x 1000x

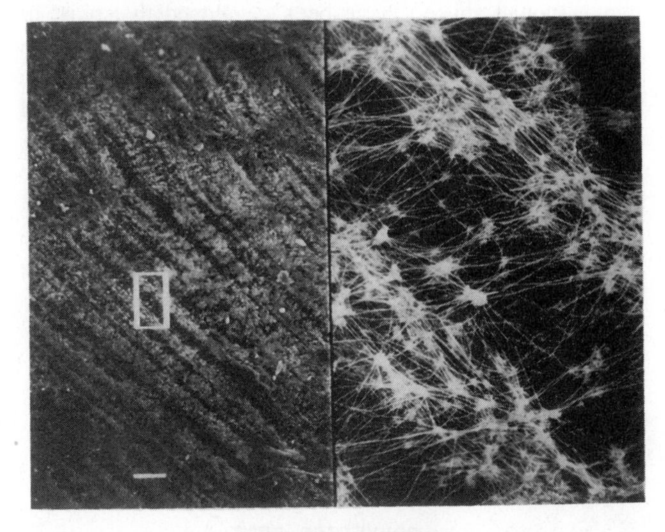

GORE-TEX MEMBRANE

100x 1000x

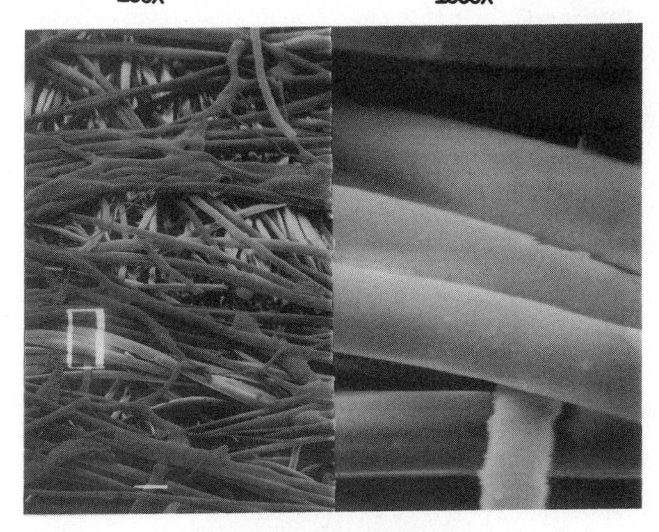

POLYESTER FELT

FIGURE 9. Gore-Tex® Membrane Compared with Polyester Felt

1. Shaker-cleaned baghouses. In shake cleaning, the tops of the bags are shaken, preferably horizontally, resulting in deflections and acceleration forces throughout the bag. Zero or reverse flow is normally combined with the shaking.
2. Reverse-air–cleaned baghouses. In reverse-air cleaning, a combination of bag deflection (inward collapse) and reverse flow is used to remove dust from the fabric. This process, which results in very low stresses on the fabric, was developed specifically for easily damaged fabrics such as fiberglass.
3. Pulse-jet–cleaned baghouses. Pulse-cleaned baghouses use outside–in flow, where the fabric collapses against a wire cage during filtration. During cleaning, a pulse of high-pressure air is directed into the bag (the reverse flow direction), inflating the bag and causing fabric/cake deflection and high inertial forces that separate the dust from the bag. Although reverse-airflow is involved, it is thought to have a minor effect on cleaning.
4. Sonic cleaning. Sonic cleaning, if used, usually augments another cleaning method. Sonic energy is normally introduced into the baghouse by air-powered horns. Although the process is not well understood, the sonic air shock waves apparently generate acceleration forces that tend to separate the dust from the fabric.

The significant parameters of shaker, reverse-air, and pulse-jet cleaning are given in Tables 2, 3, and 4.

Shaker

Although shaker-type baghouses are generally considered to be the oldest known form of fabric filter, they still have a significant place in present-day technology. It is known that in the smelter industry, when bag filters were developed in the mid-1800s and during the early 1900s, filter bags were cleaned by hand shaking. Shake cleaning has subsequently

TABLE 2. Shake Cleaning—Parameters

Frequency	Usually several cycles/second; adjustable
Motion type	Simple harmonic or sinusoidal
Peak acceleration	1–10 *g*
Amplitude	Fraction of an inch to few inches
Mode	Off-stream
Duration	10–100 cycles, 30 seconds to few minutes
Common bag diameters	5, 8, 12 inches

Source: Reference 16

TABLE 3. Reverse-Air Cleaning—Parameters

Frequency	Cleaned one compartment at a time, sequencing one compartment after another; can be continuous or initiated by a maximum-pressure-drop switch
Motion	Gentle collapse of bag (concave inward) upon deflation; slowly repressurize a compartment after completion of a back-flush
Mode	Off-stream
Duration	1–2 minutes, including valve opening and closing and dust settling periods; reverse-air flow itself normally 10–30 seconds.
Common bag diameter	8, 12 inches; length 22, 30 feet
Bag tension	50–75 pounds typical, optimum varies; adjusted after on-stream

Source: Reference 17

TABLE 4. Pulse-Jet Cleaning—Parameters

Frequency	Usually, a row of bags at a time, sequenced one row after another; can sequence such that no adjacent rows clean one after another; initiation of cleaning can be triggered by maximum-pressure-drop switch or may be continuous
Motion	Shock wave passes down bag; bag distends from cage momentarily
Mode	On-stream: in difficult-to-clean applications such as coal-fired boilers, off-stream compartment cleaning being studied
Duration	Compressed-air (100 psi) pulse duration 0.1 second; bag row effectively off-line
Common bag diameter	5–6 inches

progressed through stages, from manually operated racks to today's devices that are automated for either motor or air operation. Many mechanisms have been developed to impart motion to the filter bags to clean them. The motion has been either vertical, horizontal, or some combination of the two, although shakers have been developed that twist or otherwise move the bags. In essence, all of the mechanisms impart energy to the filter fabric in such a way that a change of direction allows inertial forces to remove the collected filter cake from the bags. The important parameters affecting the efficiency of cleaning are frequency, oscillation, and amplitude.

The one constant that must be provided in all shaker-type baghouses, regardless of the type of action imparted to the

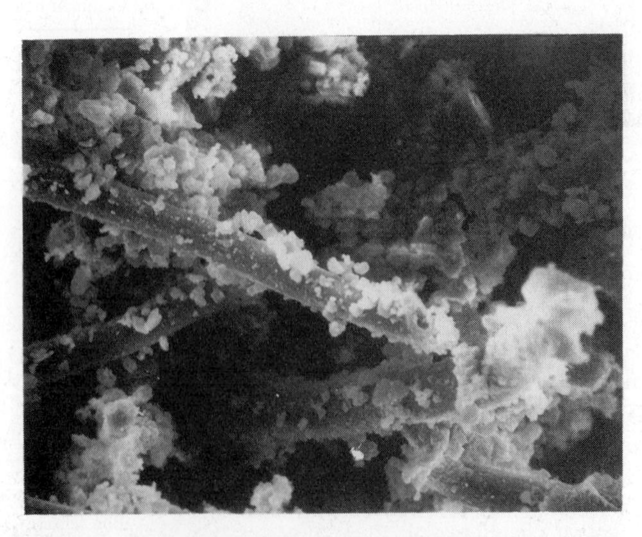

FIGURE 10. Cross Sections of a Gore-Tex® Membrane Filter Bag and a Typical Polyester Felt Filter Bag in Alumina Dust Collection

FIGURE 11. AAF's Shaker Support Logs (Photos Courtesy Snyder General Corp., Manufacturer and Marketer of American Air Filter Products)

fabric, is that flow in the positive direction must be absent during cleaning. Forward differential pressure across the bags of less than 0.05 inch (12 Pa) has been observed to retard bag cleaning significantly.[15] Conversely, a slight reverse flow through the bags during shaking can be beneficial.

Shaking is usually accomplished by the use of a motor driving an eccentric (Figure 11), which, in turn, moves a rod connected to the bags. This bag cleaning technique is currently employed over the full range of baghouse capacities, from very small off-the-shelf units to extremely large structural design units.

The bag is generally open at the bottom and closed at the top, fixed in the tube sheet at the bottom and attached to the shaking mechanism at the top (Figure 12). With this configuration, dust is collected on the inside of the bag. The bag normally contains no rings or cages. The flow of dirty gas to the bags is stopped during the cleaning process.

In the United States, shakers normally employ woven cloth at gas-to-cloth ratios below 4:1. Attempts thus far to use shaker cleaning in combination with domestic felts have led either to ineffective cleaning, and thus high-pressure drop, or to the filter medium being shaken apart.

Reverse Air

Reverse-air cleaning is the gentlest cleaning method. Dust is removed from the bags by back-flushing with low-pressure (a few inches water gauge) reversed flow. In high-temperature applications, the just-cleaned hot gas is employed for back-flush, rather than ambient-temperature air. Woven filter media are generally employed with reverse-air cleaning.

Cleaning airflow is provided by a separate cleaning fan that is normally much smaller than the main system fan, since only a fraction of the total system is cleaned at any one time. In the case of a negative-pressure system, one often can clean without a reverse-flow fan. The flow rate of cleaning gas is normally about equal to that of the dirty gas (see Table 3).

Reverse-flow systems most often include a number of isolatable compartments, or modules. The gas-to-cloth ratios usually employed are less than 4:1. Dust is normally collected on the inside of the bag, the bag being open at the bottom and closed at the top. The bag contains rings to keep it from collapsing completely during flow reversal. Complete collapse would, of course, prohibit cleaning because

FIGURE 12. AAF's Rugged Shaker Mechanism (Photos Courtesy Snyder General Corp., Manufacturer and Marketer of America American Air Filter Products)

the dust particles could not fall down within the bag to the hopper. Cleaning is accomplished both with and without flexing (partial collapsing) of the bag. Reverse flow without flexing can be employed when the dust is very easily dislodged from the bag surface. Figure 13 shows a typical reverse-air cleaning cycle. In the on-stream gas filtering mode, the compartment and outlet dampers are open and dirty gas enters the bag at the bottom. Dust collects on the inside of the bag and the cleaned gas exits the outlet damper. During the bag cleaning cycle, the flow is reversed by closing the outlet plenum and opening a third damper that allows cleaned gas to enter the compartment on the clean side of the bags, thus back-flushing the bags and exiting the compartment through the inlet damper. This now-dirty gas progresses to the balance of the on-stream compartments. It should be noted that this process increases the system gas-to-cloth ratio by adding to the total gas volume the volume of gas employed in the back-flushing process.

One baghouse manufacturer studied the reverse-air cleaning process both analytically and experimentally[18] and concluded that the improved cleaning frequently observed when a thick dust cake is allowed to build up can be attributed to the pistonlike action of a falling plug of dust cake. The manufacturer theorizes that the cascading dust cake plug both scours that cake ahead of it and causes significant evacuation, and hence additional reverse flow behind it. The concept is shown in Figure 14.

Reverse-airflow can serve to flush out loosened particles from fabric interstices and carry the dislodged agglomerates toward the collecting hopper. Based on gravimetric measurements of the filter resistance characteristics, there is, however, little evidence to suggest any significant removal of dust particles by aerodynamic action alone. The above findings are in agreement with those of Larson,[19] who states that air velocities of the order of 200 fpm (102 cm/s) are required to remove a single 20-μm particle from a fiber; and with Zimon,[20] who indicates that air velocities sweeping tangentially over a layer of dust must be in the range of 400 fpm (203 cm/s) before any appreciable dust removal is accomplished. Because significant dust removal is attained in reverse-flow systems, one must conclude that separating forces other than aerodynamic drag are involved. According to the drag theory, the adhesive forces between adjacent particles are actually increased as dirty gas moves

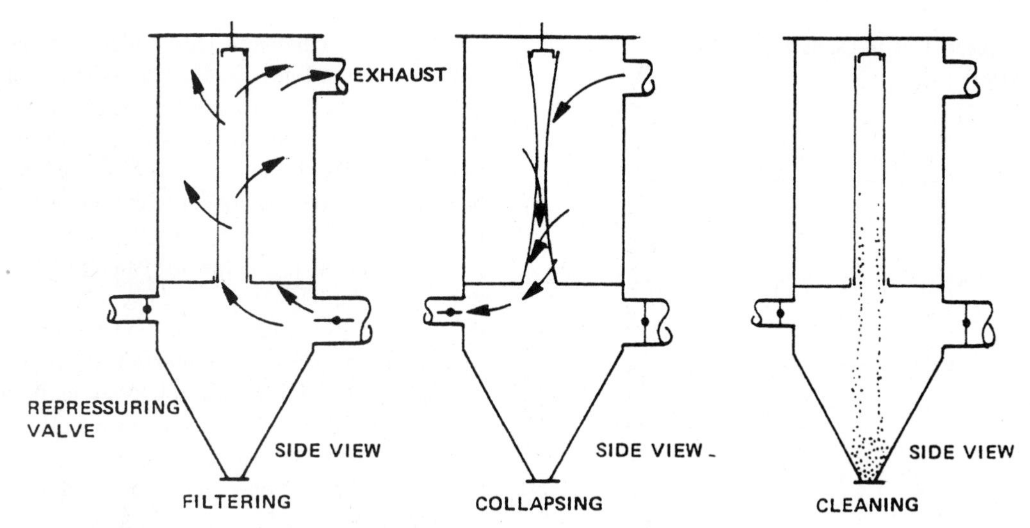

FIGURE 13. Reverse-Air Cleaning (From Reference 17)

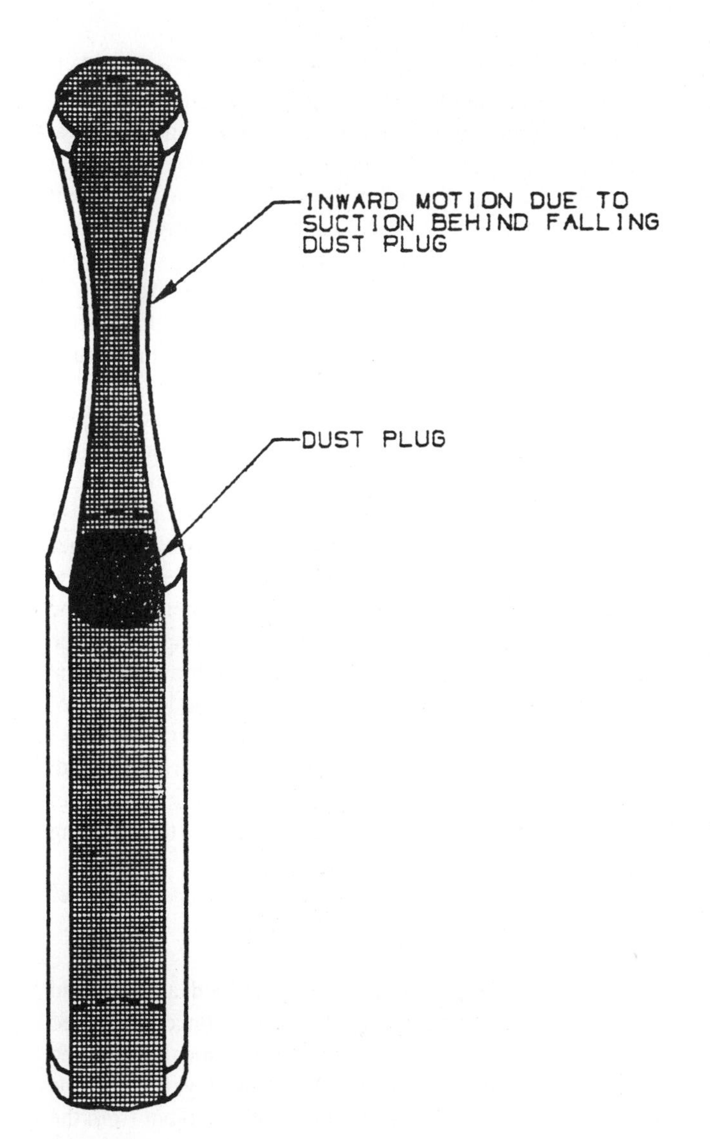

FIGURE 14. Idealization of the Falling Dust Coke upon Cleaning

radially through the dust cake. During reverse-airflow, any dust dislodgement is more likely to follow a spallation process, with the adhesive bond failure probably occurring close to the dirty face of the filter. Aerodynamic drag can be expected to flush out loosened particles.

Another adjunct to dust separation is the flexure produced in the fabric when the flow is reversed. In most systems, sufficient bending of the fabric surface occurs to cause a significant spallation at the dust–fabric interface. This effect is most pronounced for reverse-air–cleaned systems because, with the low gas-to-cloth ratios normally used, a large fraction of the collected dust appears as a superficial layer on the surface of the fabric.

If dust releases readily from the fabric, a reversal of flow alone may suffice for adequate cleaning. Because of its structural depth, and hence greater dust retentivity, a felted fabric is not usually cleaned by reverse flow. When reverse flow is used as the sole method of cleaning, the rate of flexure is probably the controlling factor with respect to fabric (or fiber) failure. Thus a gradual inflation or deflation process is unlikely to cause any serious fabric damage. Increased bag tension and reduced reverse flow rates also minimize the degree of flexure, as well as prevent a complete flattening of the bag. If the bag is flattened, there is no opportunity for loosened dust to fall to the hopper nor is there a pathway for the reverse-flow air. The insertion of restraining rings or a supporting cage eliminates complete bag collapse, but introduces a potential problem of fabric chafing. Sewing the rings to the bag minimizes the problem of chafing or attrition.

Pulse Jet

Pulse-jet cleaning employs high-pressure (60–120-psi) compressed air, with or without a venturi, to back-flush the bags vigorously. This cleaning method creates a shock

TABLE 5. Typical[a] Gas-to-Cloth Ratios for Various Industries

	Shaker/Woven Reverse-Air/Woven	Pulse-Jet/Fest Reverse-Air/Felt
Alumina	2.5	8
Asbestos	3.0	10
Bauxite	2.5	8
Carbon black	1.5	5
Cement	2.0	8
Clay	2.5	9
Coal	2.5	8
Cocoa, chocolate	2.8	12
Cosmetics	1.5	10
Enamel frit	2.5	9
Feeds, grain	3.5	14
Feldspar	2.2	9
Fertilizer	3.0	8
Flour	3.0	12
Fly ash	2.5	5
Graphite	2.0	5
Gypsum	2.0	10
Iron ore	3.0	11
Iron oxide	2.5	7
Iron sulfate	2.0	6
Lead oxide	2.0	6
Leather dust	3.5	12
Lime	2.5	10
Limestone	2.7	8
Mica	2.7	9
Paint pigments	2.5	7
Paper	3.5	10
Plastics	2.5	7
Quartz	2.8	9
Rock dust	3.0	9
Sand	2.5	10
Sawdust (wood)	3.5	12
Silica	2.5	7
Slate	3.5	12
Soap, detergents	2.0	5
Spices	2.7	10
Starch	3.0	8
Sugar	2.0	7
Talc	2.5	10
Tobacco	3.5	13
Zinc oxide	2.0	5

[a]Generally safe design values; application requires consideration of particle size and grain loading.

wave that travels down the bag, knocking dust away from the fabric. Normally, this method is employed in conjunction with felted or bulk knitted filter media and the gas-to-cloth ratio is generally higher than in shake and reverse-air cleaning methods. Typical gas-to-cloth ratios are given in Table 5. The duration of cleaning is shorter than for the other two methods; generally, the pulse lasts only a fraction of a second. The baghouse is often not subdivided when pulse-jet cleaning is employed.

The usual pulse-jet configuration has the bag closed at the bottom and open at the top, as shown in Figure 15. A metal cage is used inside the bag to keep it from collapsing. In the normal mode of operation, dirty gas enters the hopper and proceeds to the bags. Dust collects on the outside of the bags and cleaned gas exits through the top of the bags and baghouse. Usually, a row of bags is cleaned simultaneously by introducing compressed air briefly at the top of each bag. The shock wave created drives the dust off the outside of the bag and down into the hopper. Continuous discharge of dust from the hopper is often employed. Cleaning parameters are given in Table 4. It is noteworthy that this system allows for removal of the bags from the clean side of the house, since they are usually connected only at the top.

The pulsed jets of compressed air commonly used to clean nonwoven fabrics are relatively inefficient at removing deposited dust. Measurements on a pilot-scale pulse-jet fabric filter using fly ash test dust indicate that typically less than 1% of the dust on a bag is removed to the hopper by a cleaning pulse.[18]

The following factors are important to the design and/or operation of a pulse-cleaned baghouse.

1. The location of the pulse-jet nozzle must be considered.
2. Bag material should be flexible, lightweight, and inelastic to obtain maximum acceleration for dust removal during the pulse. The fabric should have sufficient weight (i.e., number of fibers per unit area) to present many targets for dust collection. The pore structure should be as uniform as possible.
3. A large housing and hopper volume on the dirty side of the filter bag will minimize the pressure buildup in this region during the pulse and thus enlarge the magnitude of the pulse differential.
4. The pulse delivered to the bag should begin as abruptly as possible, with sufficient inflating flow to subject the entire bag length to a sudden pressure differential.
5. The back flow of air through the filter that accompanies the pulse assists cleaning in several ways. It flushes from the pore structure agglomerates loosened by the acceleration. It can itself loosen agglomerates if the shock alone was insufficient, although this appears to be a very inefficient use of compressed air. It also accelerates agglomerates that have already left the felt surface, helping to convey them to the hopper.
6. Pulse intensity should be as low as can be tolerated to save on compressed air (and reduce power needs) but sufficiently high to maintain equilibrium in the cleaning process.
7. Pulse duration should ordinarily be as short as possible.
8. Dry, clean (oil-free) air must be used.

Pulse-jet cleaning is sometimes described as the rapid passage of an air bubble through the bag, a concept depicted in Figure 16. During the pulse's descent, localized fabric distension is assumed to loosen the dust layer while the transient reverse flow completes dust dislodgement from the upstream filtering surface. Although there is some truth to this concept, it overlooks the fact that typical pulse durations (of at least 0.06 second) far exceed the time for the

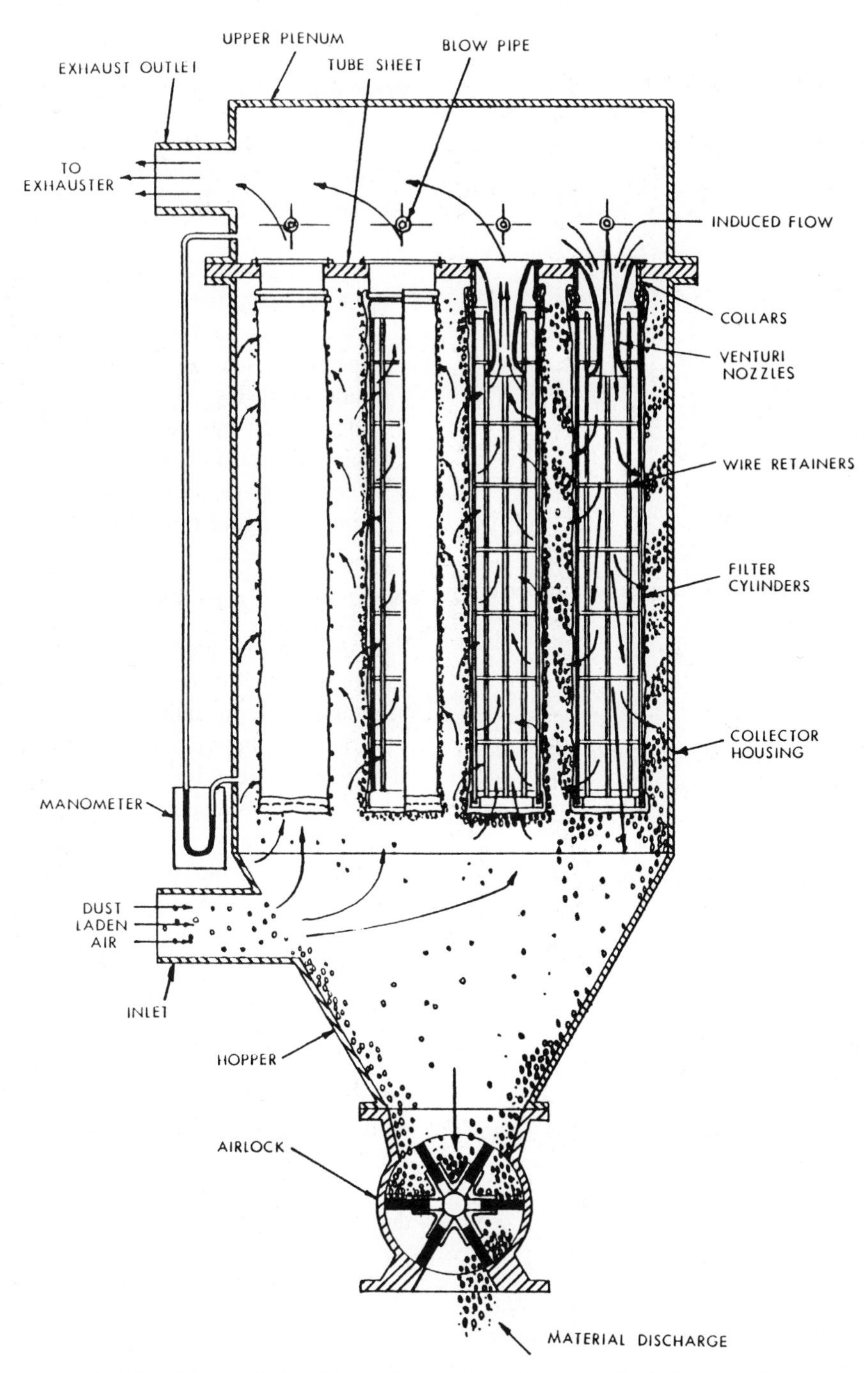

FIGURE 15. Pulse-Air-Jet Type of Bag Filter (From Reference 23)

shock wave to travel to the bottom of a conventional felt bag (roughly 0.01 second for a 10-foot-long bag). Donovan et al.[15] propose that, in reality, while the pressure front is advancing, the continued arrival of pulse air, at successively increasing rates until the solenoid valve is fully open, maintains the inflating pressure behind the front. Because pulse air entry continues well after the bottom of the bag is fully pressurized, the concept of an advancing bubble is questioned. It is suggested that parts (b), (c), and (d) of Figure 16 are more descriptive of the movement of bag fabric during a pulse.

Sonic Horns

Sonic horns are increasingly being used to augment shaker-cleaned and reverse-air–cleaned baghouses. The horns are usually powered by compressed air, and acoustic vibration is introduced by a vibrating metal plate that periodically interrupts the airflow. A cast metal horn bell is normally used for cleaning. Typically, one to four horns are installed in the ceiling of a baghouse compartment containing several hundred bags.

Sonic frequencies of 150–550 Hz have been tested;

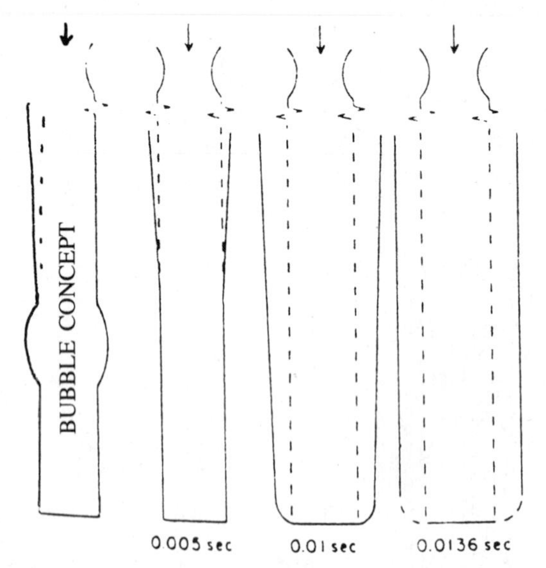

FIGURE 16. Bag Profiles versus Time after Pulse Initiation for a 4-Foot-Long Bag (From Reference 22)

Cushing, Pontius, and Carr[24] got the best results with horns that concentrated most of their energy at the lower frequencies. Relative sound pressure levels of 120–140 dB inside the baghouse are used.

Probably the most significant effect of sonic cleaning is on the weight of residual dust load on the bags. Menard and Richards[25] found that, before the sonic horns were operated, bag weights ranged from 34 to 55 pounds, with an average weight of approximately 46 pounds. After extended use of the sonic horns, bag weights ranged from 12.5 to 25 pounds, with an average weight of approximately 18 pounds. A new bag weighs approximately 9 pounds. The sonic air horns have thus reduced the amount of residual cake on the filtering elements by an average of 76%.

Cushing et al.[24] concluded:

> Sonic horns are an effective method for enhancing fabric filter performance. Their generally low cost and simple construction make them attractive additions to the available methods for cleaning fabric filter bags at coal-fired utility baghouses. Tests of six commercial sonic horns at the EPRI Fabric Filter Pilot Plant have demonstrated that, under appropriate conditions, reverse-air cleaning with sonic assist can be effective in reducing operating pressure losses. The test results have shown that overall sound pressure levels and the output frequency spectrum are important factors in determining whether a particular sonic horn application will enhance baghouse performance. One penalty encountered with sonic assist may be higher particulate penetration due to less residual dust cake on the bag surface. It is possible, however, that less frequent applications of horns will result in low ΔP without a significant increase in emissions.

BAGHOUSE MAINTENANCE

Why Keep Records?

Traditionally, baghouse operators do not keep records of routine baghouse operation. There are several reasons to reconsider this policy, including the following.

1. Records permit the operator to be aware of continuing normal operation.
2. Records permit the operator to be aware of abnormal operation, such as sudden failures or slowly changing parameters (e.g., a slow rise in residual pressure that, if allowed to continue, would limit the ability of the process to reach full load).
3. Records provide a historical record that is useful when troubleshooting problems.

What Records to Keep

Pressure drop is often the only parameter monitored in a baghouse, usually by means of a gauge mounted on the baghouse wall. In multicompartment baghouses, the pressure drops of individual compartments may be monitored along with the total baghouse pressure drop. Unfortunately, over time, the pressure-sensing lines become clogged with dust or the gauge becomes unreadable owing to dust accumulation on the instrument face. An obvious first step in maintenance is to assure that the instrumentation is both functional and accurate.

An instrument reader who routinely records process data should probably also routinely record available baghouse data. A better scheme is to record the data with a strip-chart recorder or as part of a data-logging system. Automated data-recording and parameter alarm systems are becoming more common. If such a system is a part of the process itself, perhaps baghouse data could also be recorded.

Other types of baghouse records, in addition to pressure-drop information, should be considered:

1. Flow rate. Pressure-drop information cannot be interpreted properly unless the flow rate is known. A record of flow rate may be useful in identifying a developing leak in the ducting or in the baghouse itself.
2. Opacity. If a continuous opacity monitor is incorporated into the system, its output should be recorded. If such opacity instrumentation is not available, visual opacity readings recorded manually should still prove useful. The cause of any change in opacity should be pursued and understood.
3. Temperature. The baghouse outlet temperature should certainly be monitored, even if other temperature records are not kept.
4. Dust removal. At least one parameter related to the quantity of dust removed from each baghouse compartment should be monitored and recorded. A change in

dust quantity may be indicative of baghouse failure or of process changes.

The key to baghouse maintenance is frequent and routine inspection. It is essential that a regular program of routine maintenance be established and followed. Records (a log) should be kept of all inspections and maintenance. Inspection intervals will depend on the type of baghouse, the manufacturer's recommendation, and the process on which the unit is installed. The important thing is to be sure that the checks are performed regularly and as frequently as necessary, that no components are neglected, and that all pertinent information is logged for future reference.

When problems are located and isolated during routine or other inspection, it is important that corrections be made as quickly as possible to avoid possible equipment downtime or excess emissions from bypassing the control system. When there is a baghouse failure, the unit is usually shut down and/or bypassed and the malfunction corrected. Plant managers should expect considerable maintenance time to be expended on troubleshooting and correction of baghouse malfunctions. Maintenance personnel must learn to recognize the symptoms that indicate potential problems, and then to determine the cause of the difficulty and remedy it, either by in-plant action or by contact with the manufacturer or some other outside resource. High-pressure drop across the system exemplifies one symptom for which there are many possible causes; for example, difficulties with the bag-cleaning mechanism, low compressed-air pressure, weak shaking action, loose bag tension, or excessive reentrainment of dust. Many other factors can cause excessive pressure drop, and several options are usually available for corrective action. Thus the ability to locate and correct malfunctioning baghouse components requires a thorough understanding of the system. The critical influence of moisture, in the baghouse, in the cleaning air, or with the particulate, should be noted and emphasized.

The operating experience of baghouses in conjunction with electric utility boilers has probably been more extensively documented than any other application. The Electric Power Research Institute (EPRI) currently supports an extensive research program intended to optimize the design, operation, and reliability of baghouse technology for the electric utility industry. Pilot and full-scale evaluations are performed to measure, interpret, and predict baghouse performance. Results show that baghouses routinely achieve a clear stack and particulate matter collection efficiencies well in excess of 99.9%

References

1. J. D. McKenna and J. H. Turner, *Fabric Filter—Baghouses I,* ETS, Inc., Roanoke, VA, 1989.
2. R. Dennis and H. A. Klemm, *Fabric Filter Model Format Change, Vol. I: Detailed Technical Report,* EPA-600/7-79-0432, U.S. Environmental Protection Agency, Research Triangle Park, NC, 1979, National Technical Information Service (NTIS) PB-293-551.
3. C. E. Billings and J. Wilder, *Handbook of Fabric Filter Technology, Vol. I: Fabric Filter Systems Study,* GCA Corp., Bedford, MA.; PB-200-648, NTIS, Springfield, VA, December 1970.
4. R. Dennis, R. W. Cass, D. W. Cooper, et al., *Filtration Model for Coal Flyash with Glass Fabrics,* EPA-600-7-77-084, August 1977.
5. W. T. Davis and R. F. Kurzynske, "The effect of cyclonic precleaners on the pressure drop of fabric filters," *Filtration Separation,* 16(5).
6. W. T. Davis and W. F. Frazier, "A laboratory comparison of the filtration performance of eleven different fabric filter materials filtering resuspended flyash," *Proceedings of the 75th Annual Meeting of the Air Pollution Control Association,* 1982.
7. W. F. Frazier and W. T. Davis, *Effects of Flyash Size Distribution on the Performance of a Fiberglass Filter,* Symposium on the Transfer and Utilization of Particulate Control Technology. Vol. II: *Particulate Control Devices,* EPA-600/9-82-0050c, July 1982, pp 171–180.
8. J. W. Robinson, R. E. Harrington, and P. W. Spaite, "A new method for analysis of multicompartmented fabric filtration," *Atmospheric Environ.,* 1(4):499 (1967).
9. A. A. Reisinger and W. T. Grubb, "Fabric evaluation program at Coyote Unit #1, operating results update," *Proceedings Second EPRI Conference on Fabric Filter Technology for Coal-Fired Power Plants,* Electric Power Research Institute, Palo Alto, CA., EPRI CS03257, November 1983.
10. G. P. Greiner, J. C. Mycock, and D. S. Beachler, "The IBFM, a unique tool for troubleshooting and monitoring baghouses," Paper 85-54.2, Annual Meeting of the Air Pollution Control Association, Detroit, MI, June 1985.
11. J. F. Durham and R. E. Harrington, "Influence of relative humidity on filtration resistance and efficiency of fabric dust filters," *Filtration Separation,* 8:389–393 (July/August 1971).
12. E. R. Frederick, *Electrostatic Effects in Fabric Filtration: Volume II Triboelectric Measurements and Bag Performance* (annotated data), EPA-600/7-78-142b, NTIS PB-287-207, July 1978.
13. International Non-Woven and Disposable Association, New York.
14. C. B. Hotchkiss and L. F. Cox, "Fabric and finish selection, manufacturing techniques and other factors affecting bag life in the coal-fired boiler applications," *Proceedings Second EPRI Conference on Fabric Filter Technology for Coal-Fired Power Plants,* Electric Power Research Institute, Palo Alto, CA, EPRI CS-3257, November 1983.
15. R. P. Donovan, B. E. Daniel, and J. H. Turner, *EPA Fabric Filtration Studies, 3. Performance of Filter—Bags Made From Expended PTFE Laminate,* EPA-600/2-76-168c, U.S. Environmental Protection Agency, Research Triangle Park, NC, December 1976, NTIS PB-263-132.
16. R. L. Adams. "Shaker type filter," *Proceedings APCA Specialty Conference on the User and Fabric Filter Equipment,* Buffalo, NY, October 1973, Air Pollution Control Association, Pittsburgh, PA.

17. *The Fabric Filter Manual,* C. E. Billings, Ed.; McIlvaine Co., Northbrook, IL.
18. M. M. Ketchuk, A. Walsh, O. F. Fortune, et al., "Fundamental strategies for cleaning reverse-air baghouses," *Proceedings Fourth Symposium on the Transfer and Utilization of Particulate Control Technology, Vol. I: Fabric Filtration,* EPA-600/9-84-025a, U.S. Environmental Protection Agency, Research Triangle Park, NC, November 1984.
19. R. I. Larson. "The adhesion and removal of particles attached to air filter surface," *AIHA J.,* 19 (1958).
20. A. D. Zimon, *Adhesion of Dust and Powder,* Plenum Press, New York, 1969, p 112.
21. M J. Ellenbecker and D. Leith, "Dust removal from nonwoven fabrics," Annual Meeting of the Air Pollution Control Association, Montreal, Que., June 22–27, 1980.
22. R. Dennis and L. S. Hovis, "Pulse-jet filtration theory—a state-of-the-art assessment," *Proceedings Fourth Symposium on the Transfer and Utilization of Particulate Control Technology, Vol. I: Fabric Filtration,* EPA-600/9-84-0252, U.S. Environmental Protection Agency, Research Triangle Park, NC, November 1984.
23. C. Orr, *Filtration, Principles, and Particles, Part I,* Marcel Dekker, New York, 1977.
24. K. M. Cushing, D. H. Pontius, and R. C. Carr, "A study of sonic cleaning for enhanced baghouse performance," Annual Meeting of the Air Pollution Control Association, San Francisco, June 1984.
25. A. R. Menard and R. M. Richards, "The use of sonic air horns as an assist to reverse-air cleaning of a fabric filter dust collector," *Proceedings Fourth Symposium on the Transfer and Utilization of Particulate Control Technology, Vol. I: Fabric Filtration,* EPA-600/9-84-025a, U.S. Environmental Protection Agency, Research Traingle Park, NC, November 1984.

4
Fugitive Emissions

John S. Kinsey
Chatten Cowherd, Jr.

Fugitive particulates are emitted by a wide variety of sources in both the industrial and the nonindustrial sectors. Fugitive emissions refer to those air pollutants that enter the atmosphere without first passing through a stack or duct designed to direct or control their flow.

Sources of fugitive particulate emission may be separated into two broad categories: process sources and open dust sources. Process sources of fugitive emissions are those associated with industrial operations that alter the chemical or physical characteristics of a feed material. Examples are emissions from charging and tapping of metallurgical furnaces and emissions from crushing of mineral aggregate. Such emissions normally occur within buildings and, unless captured, are discharged to the atmosphere through forced- or natural-draft ventilation systems. Open dust sources entail the entrainment of solid particles into the atmosphere by the forces of wind or machinery acting on exposed materials. These sources include industrial sources associated with the open transport, storage, and transfer of raw, intermediate, and waste materials and such nonindustrial sources as unpaved and paved public roads and construction activities.

The partially enclosed facilities for the storage and transfer of materials to or from a process operation do not fit well into either of the two categories of fugitive particulate emissions defined above. Examples are partially enclosed conveyor transfer stations and front-end loaders operating within buildings. Nonetheless, partially enclosed materials-handling operations will be classified as open sources.

Unlike ducted sources of particulate emissions, which typically can be characterized as continuously emitting, fugitive emission rates have a high degree of temporal variability. Industrial process sources of fugitive particulate emissions are usually associated with batch operations and emissions fluctuate widely during the process cycle. Open dust sources within industry also exhibit large fluctuations because of the sporadic nature of materials-handling operations and the effects of precipitation and wind on the emissions potential.

In addition, fugitive emissions are characteristically diffuse in nature and are discharged from a wide variety of source configurations. For example, vehicles that entrain surface dust from industrial roads are best represented as individual moving point sources (or as a line source for high traffic density), while process fugitive emissions discharged from building vents are usually depicted as area sources or virtual point sources.

This chapter presents an overview of the generation and control of fugitive particulate emissions. Since the control of emissions from industrial processes is specifically addressed elsewhere in this manual, open dust sources will be emphasized here. Also, since PM_{10} (particulate matter ≤ 10 μm in aerodynamic diameter) is of special concern, this particle size fraction of the emissions will be specifically addressed. The following sections describe the emission characteristics of fugitive sources and available particulate control technology. Additional information can be obtained in the Bibliography.

SOURCE DESCRIPTION

Open Dust Sources

Below is a generic listing of open dust sources,[1] followed by brief descriptions of each component source category.

1. Unpaved travel surfaces
    - Roads
    - Parking lots and staging areas
    - Storage piles
2. Paved travel surfaces
    - Streets and highways
    - Parking lots and staging areas

3. Exposed areas (wind erosion)
   - Storage piles
   - Bare ground areas
4. Materials handling
   - Batch drop (dumping)
   - Continuous drop (conveyor transfer, stacking)
   - Pushing (dozing, grading, scraping)
   - Tilling

Paved Roads

Particulate emissions occur whenever a vehicle travels over a paved surface, such as public and industrial roads and parking lots. These emissions may originate from material previously deposited on the travel surface or from resuspension of material from tires and undercarriages. In general, emissions arise primarily from the surface material loading (measured as mass of material per unit area) and that loading is, in turn, replenished by other sources (e.g., deposition of material from vehicles, deposition from other nearby sources, carryout from surrounding unpaved areas, pavement wear, and litter). Because of the importance of the surface loading, available control techniques attempt either to prevent material from being deposited on the surface or to remove (from the travel lanes) any material that has been deposited.

While the mechanisms of particle deposition and resuspension are largely the same for public and industrial roads, there can be major differences in surface loading characteristics, emission levels, traffic characteristics, and viable control options. For the purpose of estimating particulate emissions and determining control programs, the distinction between public and industrial roads is not a question of ownership, but rather a question of surface loading and traffic characteristics.

Although public roads generally tend to have lower surface loadings than industrial roads, the fact that these roads have far greater traffic volumes may result in a substantial contribution to the measured air quality in certain areas. In addition, many public roads in industrial areas often are heavily loaded and traveled by heavy vehicles. In that instance, better emission estimates would be obtained by treating these roads as industrial roads. In an extreme case, a road or parking lot may have such a high surface loading that the paved surface is essentially covered and it is easily mistaken for an unpaved road. In that event, use of a paved road emission factor may actually result in a higher estimate than that obtained from the unpaved road factor, and the road is better characterized as unpaved in nature rather than paved.[1]

Unpaved Roads

As is the case for paved roads, particulate emissions occur whenever a vehicle travels over an unpaved surface. Unlike paved roads, however, the road itself is the source of the emissions rather than any "surface loading." Within the various categories of open dust sources in industrial settings, unpaved travel surfaces historically have accounted for the greatest share of particulate emissions. For example, unpaved sources were estimated to account for roughly 70% of open dust sources in the iron and steel industry during the 1970s. Recognition of the importance of unpaved roads led naturally to an interest in their control. As a result of these control programs, the portion of total open dust source emissions attributable to unpaved travel surfaces has decreased dramatically over the past five to 10 years. Nevertheless, the need for continued control of these sources is apparent.

Travel surfaces may be unpaved for a variety of reasons. Possibly the most common type of unpaved road is that found in rural regions throughout the country; these roads may experience only sporadic traffic, which, taken with the often considerable road length involved, makes paving impractical.

Other important travel surfaces are found in industrial settings. During the 1980s, industry paved many previously unpaved roads as part of emission control programs. However, some industrial roads are, by their nature, not suitable for paving. These roads may be used by very heavy vehicles or may be subject to considerable spillage from haul trucks. Other roads may have poorly constructed bases that make paving impractical. Because of the additional maintenance costs associated with a paved road in these service environments, emissions from these roads are usually controlled by regular applications of water or chemical dust suppressants.

In addition to roadways, many industries often contain important unpaved travel *areas*. Examples include scraper traffic patterns related to stockpile/reclaim activities in coal yards, compactor traffic in areas proximate to lifts at landfills, and travel related to open storage of finished products (such as coil at steel plants). These areas may often account for a substantial fraction of traffic-generated emissions from individual plants. In addition, these areas tend to be much more difficult to control than stretches of roadway for several reasons. For example, changing traffic patterns make semipermanent controls impractical; increased shear forces from cornering vehicles rapidly deteriorate chemically controlled surfaces; and chemical suppressants may damage raw materials or finished products.

Storage Piles

Inherent in operations that use minerals in aggregate form is the maintenance of outdoor storage piles. Storage piles are usually left uncovered, partly because of the need for frequent material transfer into or out of storage.

Dust emissions occur at several points in the storage cycle: during material loading onto the pile, during disturbances by strong wind currents, and during load-out from the pile. The movement of trucks and loading equipment in the storage pile area is also a substantial source of dust.

Construction and Demolition

Construction and demolition activities are temporary but important open dust sources in urban areas. These activities involve a number of separate dust-generating operations that must be quantified to determine the total emissions from the site and thus their impact on ambient air quality. In turn, the specific type of activities that are conducted on-site will depend on the nature of the construction or demolition project taking place.

In the case of construction, a project may involve the erection of a building or buildings or single- or multifamily homes or the installation of a road right-of-way. Operations commonly found in these types of construction projects consist of land clearing, drilling and blasting, excavation, cut-and-fill operations (i.e., earth moving), materials storage and handling, and truck traffic on associated unpaved surfaces. In addition, secondary impacts associated with construction sites involve mud/dirt carryout onto paved surfaces. The additional loading caused by carryout can substantially increase fugitive dust emissions on city streets over the life of the project.

With regard to demolition, a particular project may involve the razing and removal of entire buildings, a major interior renovation of a structure, or a combination of the two. Dust-producing operations associated with demolition are: mechanical or explosive dismemberment; debris storage, handling, and transport operations; and truck traffic over unpaved surfaces on the site.

Like construction, demolition activities can also create mud/dirt carryout onto paved surfaces with its associated increase in emissions. Also, since building debris is usually being removed from the site, spillage from trucks can be of concern in increasing the amount of surface loading deposited on the paved streets providing access to the site.

Process Sources

Process sources of fugitive emissions are those associated with industrial operations that alter the chemical or physical characteristics of a feed material. Examples are emissions from the charging and tapping of metallurgical furnaces and emissions from the crushing of mineral aggregates. Such emissions normally occur within buildings and, unless captured, are discharged to the atmosphere through forced- or natural-draft ventilation systems. However, a process source of fugitive emissions can occur in the open atmosphere (e.g., scrap metal cutting). The most significant process sources of fugitive particulate emissions are listed by industry in Table 2-1 of Reference 1 and elsewhere in this manual.

AIR EMISSIONS CHARACTERIZATION

Paved Roads

The Environmental Protection Agency's (U.S. EPA) "Compilation of Air Pollutant Emission Factors" (AP-42) indicates that the PM_{10} emission factor for all paved roads may be written in the general form[2]:

$$e_i = a(sL/b)^c \tag{1}$$

where: a, b, c = empirical constants
e_i = PM_{10} emission factor, in units of mass per vehicle-road length traveled
s = surface silt content, fraction of material smaller than 75 μm in physical diameter
L = total surface dust loading, mass per area of road surface

The product sL represents the mass of silt-size dust particles per unit area of the road surface and is usually termed the "silt loading." As is the case for all predictive models, *the use of site-specific values of* sL *is strongly recommended*. If this is not possible, default values have been provided in Table 11.2.6-1 of AP-42.[2]

The appropriate emission factor model (i.e., the constants a, b, and c) for a road segment depends on:

- The mass of loose aggregate material less than 200 mesh per unit area of a paved road surface.
- The average weight of vehicles traveling on the road.

Selection of the appropriate emission factor is summarized in Table 1.

TABLE 1. Selection of Paved Road Emission Factor

Silt Loading (sL)		Range weight (W)		Applicable PM_{10} Emission Factor Model	
g/m²	oz/yd²	Mg	Ton	g/VKT[a]	lb/VMT[a]
$sL < 2$	<0.06	$W > 4$	>4.4	$220\ (sL/12)^{0.3}$[b]	$0.78\ (sL/0.35)^{0.3}$[b]
$sL < 2$	< 0.06	$W < 4$	<4.4	$2.28\ (sL/0.5)^{0.8}$[c]	$0.0081\ (sL/0.015)^{0.8}$[c]
$sL < 2$[d]	>0.06	$W > 6$	>6.6	$220\ (sL/12)^{0.3}$[b]	$0.78\ (sL/0.35)^{0.3}$[b]
$2 < sL < 15$	$0.06 < sL < 0.44$	$W < 6$	<6.6	$220\ (sL/12)^{0.3}$[b]	$0.78\ (sL/0.35)^{0.3}$[b]
$sL > 15$[d]	>0.44	$W < 6$	<6.6	93	0.33

[a]VKT = vehicle kilometers traveled; VMT = vehicle miles traveled.
[b]Commonly referred to as the "industrial" paved road model.[2]
[c]Commonly referred to as the "urban" paved road model.[2]
[d]For heavily loaded surfaces (i.e., $sL > \sim 300$ to 400 g/m² (9 to 12 oz/yd²), it is recommended that the resulting estimate be compared with that from the unpaved road models and the smaller of the two values used.

Unpaved Roads

Unlike paved roads, emission estimates for unpaved surfaces do not require a "decision" process involving surface–vehicle parameters because the AP-42 emission factor equation takes source characteristics into consideration.[2]

$$e = 0.61 \left(\frac{s}{12}\right)\left(\frac{S}{48}\right)\left(\frac{W}{2.7}\right)^{0.7}\left(\frac{w}{4}\right)^{0.5} \frac{(365 - p)}{365} \frac{\text{kg}}{\text{VKT}}$$

$$e = 2.1 \left(\frac{s}{12}\right)\left(\frac{S}{30}\right)\left(\frac{W}{3}\right)^{0.7}\left(\frac{w}{4}\right)^{0.5} \frac{(365 - p)}{365} \frac{\text{lb}}{\text{VMT}} \quad (2)$$

where: e = PM_{10} emission factor, in units stated
s = silt content of road surface material, %
S = mean vehicle speed, km/h (mi/h)
W = mean vehicle weight, mg (ton)
w = mean number of wheels (dimensionless)
p = number of days with ≥ 0.254 mm (0.01 inch) of precipitation (per Figure 11.2.1-1 of AP-42)[2]

Site-specific input parameters are strongly recommended; if this is not feasible, a summary of measured values is presented on page 11.2.1-3 of AP-42.[2]

Bulk Materials Handling

The following equation is recommended for estimating emissions from transfer operations (batch or continuous drop).[2]

$$e = k(0.0016) \frac{(U/2.2)^{1.3}}{(M/2)^{1.4}} \text{ (kg/mg)}$$

$$e = k(0.0016) \frac{(U/5)^{1.3}}{(M/2)^{1.4}} \text{ (lb/ton)} \quad (3)$$

where: e = emission factor, in units stated
k = particle-size multiplier, dimensionless
U = mean wind speed, m/s (mi/h)
M = material moisture content, %

The particle size multiplier k varies with aerodynamic particle diameter as follows:

<30 μm	<15 μm	<10 μm	<5 μm	<2.5 μm
0.74	0.48	0.35	0.20	0.11

Again, source-specific input parameters are recommended for use in equation 3. If such are not available, Table 11.2.3-3 of AP-42 can be consulted.[2]

Storage-Pile Wind Erosion

Dust emissions may be generated by wind erosion of open aggregate storage piles and exposed areas within an industrial facility. These sources typically are characterized by nonhomogeneous surfaces impregnated with nonerodible elements (particles larger than approximately 1 cm in diameter). Field testing of coal piles and other exposed materials using a portable wind tunnel has shown that (1) threshold wind speeds exceed 5 m/s (11 mi/h) at 15 cm above the surface [or 10 m/s (22 mi/h) at 7 meters above the surface], and (2) particulate emission rates tend to decay rapidly (half-life of a few minutes) during an erosion event. In other words, these aggregate material surfaces are characterized by the finite availability of erodible material (mass/area) referred to as the erosion potential. Any natural crusting of the surface binds the erodible material, thereby reducing the erosion potential.

The emission factor for wind-generated particulate emissions from mixtures of erodible and nonerodible surface material subject to disturbance may be expressed in units of g/m²-yr as follows[2]:

$$\text{Emission factor} = k \sum_{i=1}^{N} P_i \quad (4)$$

where: k = particle-size multiplier, dimensionless
N = number of disturbances per year
P_i = erosion potential corresponding to the observed (or probable) fastest mile of wind for the ith period between disturbances, g/m²

The particle-size multiplier (k) for equation 4 varies with aerodynamic particle size, as follows:

<30 μm	<15 μm	<10 μm	<2.5 μm
1.0	0.6	0.5	0.2

This distribution of particle size within the <30-μm fraction is comparable to the distributions reported for other fugitive dust sources where wind speed is a factor. This is illustrated, for example, in the distributions for batch and continuous drop operations encompassing a number of test aggregate materials (see AP-42, Section 11.2.3).[2]

Input to the calculation of wind erosion emissions using equation 4 must be performed in a stepwise manner and is very complex. Therefore, the reader is referred to Section 11.2.3 of AP-42[2] for a detailed description of the required calculations.

For emissions from wind erosion of *active* storage piles, the following total suspended particulate (TSP, particles ≤ ~ 30 μm in aerodynamic diameter) emission factor equation is recommended:

$$E = 1.9 \left(\frac{s}{1.5}\right)\left(\frac{365 - p}{235}\right)\left(\frac{f}{15}\right) \text{ kg/day/hectare)}$$

$$E = 1.7 \left(\frac{s}{1.5}\right)\left(\frac{365 - p}{235}\right)\left(\frac{f}{15}\right) \text{ (lb/day/acre)} \quad (5)$$

TABLE 2. Emissions Increase ($\triangle E$) by Site Traffic Volume[a]

Particle Size Fraction[b]	Sites with >25 Vehicles/Day			Sites with <25 Vehicles/Day		
	Mean	Standard Deviation, σ	Range	Mean	Standard Deviation, σ	Range
$< \sim 30$ μm	52	28	15–80	19	7.8	14–28
<10 μm	13	6.7	4.4–20	5.5	2.3	4.2–8.1
<2.5 μm	5.1	2.6	1.7–7.8	2.2	0.88	1.6–3.2

[a] $\triangle E$ expressed in g/vehicle pass.
[b] Aerodynamic diameter.

TABLE 3. Emissions Increase ($\triangle E$) by Construction Type[a]

Particle Size Fraction[b]	Commercial			Residential		
	Mean	Standard Deviation, σ	Range	Mean	Standard Deviation, σ	Range
$< \sim 30$ μm	65	29	15–110	39	22	10–72
<10 μm	16	9.3	4.2–25	10	5.4	2.8–19
<2.5 μm	6.3	3.6	1.6–9.7	3.9	2.1	1.1–7.3

[a] $\triangle E$ expressed in g/vehicle pass.
[b] Aerodynamic diameter.

where: E = total suspended particulate emission factor, in units indicated
s = silt content of aggregate, %
p = number of days with ≥ 0.25 mm (0.01 inch) of precipitation per year (from Figure 11.2.1-1 of AP-42[2])
f = percentage of time that the unobstructed wind speed exceeds 5.4 m/s (12 mi/h) at the mean pile height

Construction Emissions

At present, the only emission factor available in AP-42[2] is 1.2 tons/acre/mo (related to particles <30 μm Stokes' diameter) for an entire construction site. No factor has been published in AP-42[2] for demolition. However, PM_{10} emission factors have been developed for construction-site preparation using test data from a study conducted in Minnesota for topsoil removal, earth moving (cut-and-fill), and truck haulage operations.[3] For these operations, the PM_{10} emission factors based on the level of vehicle activity (i.e., vehicle kilometers traveled or VKT) occurring on the site are as follows[3]:

- Topsoil removal: 5.7 kg/VKT for pan scrapers
- Earthmoving: 1.2 kg/VKT for pan scrapers
- Truck haulage: 2.8 kg/VKT for haul trucks

Pushing Operations

For pushing (bulldozer) operations, the AP-42[2] emission-factor equation for overburden removal at western surface coal mines can be used. Although this equation actually relates to particulate <15 μm*A*, it would be expected that the PM_{10} emissions from such operations would be generally comparable. The AP-42 dozer equation is as follows:

$$E_p = \frac{0.45(s)^{1.5}}{(M)^{1.4}} \tag{6}$$

where: E_p = PM_{10} emission rate, kg/h
s = silt content of surface material, % (default = 6.9%)
M = moisture content of surface material, % (default = 7.9%)

Mud/Dirt Carryout

Finally, the increase in emissions on paved roads due to mud/dirt carryout has been assessed based on surface loading measurements at eight sites.[4] Tables 2 and 3 provide these emission factors in terms of g/vehicle pass, representing PM_{10} emissions generated over and above the "background" for the paved road sampled. Table 2 expresses the emission factors according to the volume of traffic entering and leaving the site, whereas Table 3 expresses the same data according to type of construction. Either set of factors can be used to estimate the increase in emissions.

Contaminated Emissions

There are, at present, no emission factors that specifically address contaminated particulate matter from fugitive sources. The emission rate R_j of particulate contaminated with a compound j can be estimated as:

$$R_j = \sum_{i=1}^{n} \alpha_{ij} e_i A_i \quad (7)$$

where: R_j = emission rate of contaminant j from source i (mass/unit time)
α_{ij} = the fraction of contaminant j in the particulate emissions from source i (%)
e_i = emission factor for source i (mass/unit source extent)
A_i = source extent (or operating rate) for source i per unit time

Because contamination levels are extremely source-dependent, no general rule of thumb for an appropriate value of α_{ij} should be applied. Consequently, it is *strongly recommended* that material sampling be undertaken at operating facilities, and that those samples be analyzed for compounds of concern.

Information on α_{ij} can be determined by either direct or surrogate measurement. *Direct measurement* of α_{ij} requires the determination of specific contaminants in particulate emission samples collected at the site. A *surrogate measure* of α_{ij} is the fraction of contaminant j in the parent material for source i. An *alternate surrogate measure* would treat α_{ij} as a constant, set equal to the average fraction of contaminant j in all available material samples from the same facility.

AIR POLLUTION CONTROL MEASURES

Introduction

Typically, there are several options for the control of fugitive particulate emissions from any given source. This is clear from the mathematical equation used to calculate the emission rate:

$$R = Me(1 - c) \quad (8)$$

where: R = estimated mass emission rate
M = source extent (e.g., vehicle miles traveled or throughput rate)
e = uncontrolled emission factor, that is, mass of uncontrolled emissions per unit of source extent
c = fractional efficiency of control

To begin with, because the uncontrolled emission rate is the product of the source extent and uncontrolled emission factor, a reduction in either of these two variables produces a proportional reduction in the uncontrolled emission rate.

Although the reduction of source extent results in a highly predictable reduction in the uncontrolled emission rate, such an approach, in effect, usually requires a change in the process operation. Frequently, a reduction in the extent of one source may necessitate an increase in the extent of another, as in the shifting of vehicle traffic from an unpaved road to a paved road. The option of reducing source extent is beyond the scope of this manual and will not be discussed further.

The reduction in the uncontrolled emission factor may be achieved by process modifications (in the case of process sources) or by adjusted work practices (in the case of open sources). The degree of the possible reduction of the uncontrolled emission factor can be estimated from the known dependence of the factor on source conditions that are subject to alteration. For open dust sources, this information is embodied in the predictive emission factor equations for fugitive dust sources as presented in Section 11.2 of the U.S. EPA's "Compilation of Air Pollutant Emission Factors" (AP-42).[2]

The reduction of source extent and the incorporation of process modifications or adjusted work practices are preventive techniques for control of fugitive particulate emissions. In addition, there are a variety of "add-on" measures that can be used for either (1) the prevention of the creation and/or release of particulate matter into the atmosphere or (2) the capture and removal of the particles after they have become airborne.

Selection of suitable control methods depends on the mechanism(s) that generate the particulate emissions and the specific source involved. The methods used to control process sources of fugitive particulate emissions generally take a much different approach from those applied to open dust sources. Differences in source configuration, process requirements, and emissions stream characteristics also affect the selection of specific controls.

The following sections discuss available control technology for sources of fugitive particulate emissions. Generic controls are described first, followed by methods for estimating control efficiency.

Generic Control Technology

Two basic types of controls can be used for fugitive particulate sources—preventive measures and capture/removal methods. Each is described below.

Preventive Measures

Preventive measures include those measures that prevent or substantially reduce the injection of particles into the surrounding air environment. Preventive measures are independent of whether the particulate is emitted directly into the ambient air or into the interior of a building. The main types of preventive measures include passive enclosures (full or partial), wet suppression, stabilization of unpaved surfaces, paved surface cleaning, work practices, and housekeeping.

Descriptions of control techniques within the first four of the above categories are presented below. Work-practice modification and housekeeping are extremely source-specific and thus will not be discussed here.

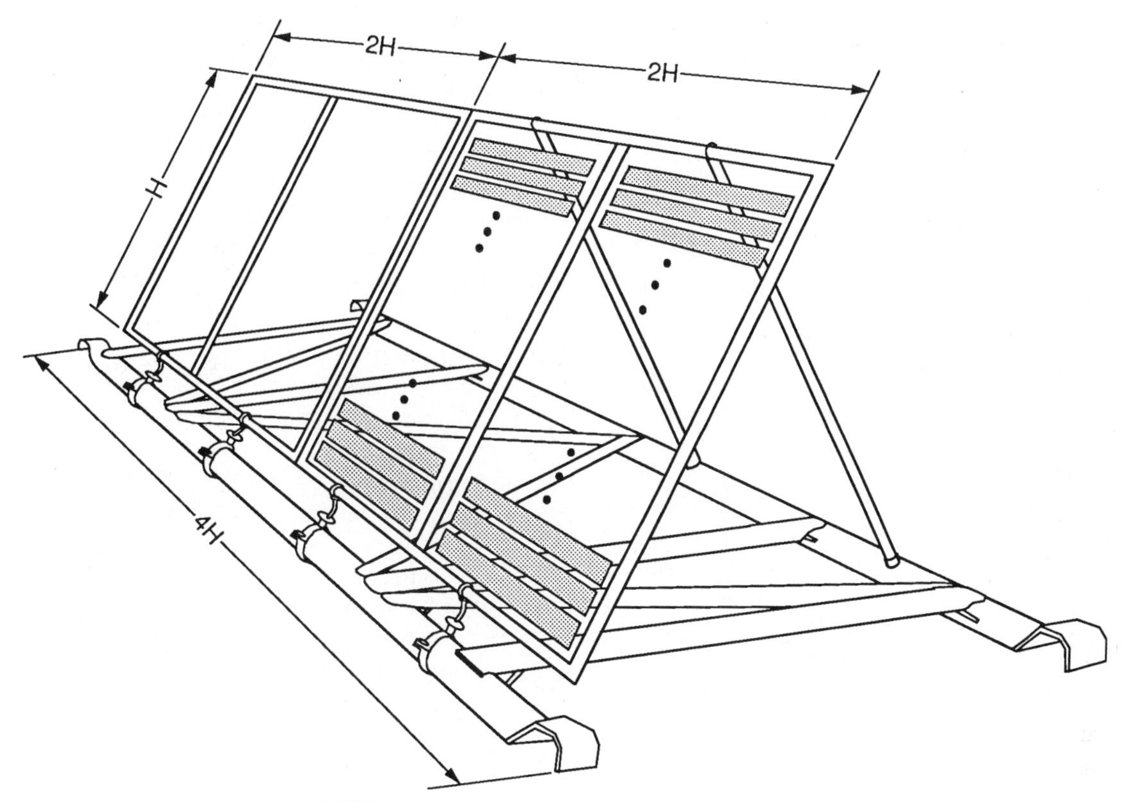

FIGURE 1. Diagram of a Portable Windscreen[1]

Passive Enclosures

A common preventive technique for the control of fugitive particulate emissions is to enclose the source either fully or partially. Enclosures preclude or inhibit particulate matter from becoming airborne as a result of the disturbance created by ambient winds or by mechanical entrainment resulting from the operation of the source itself. Enclosures also help contain those emissions that are generated. They can consist of either some type of permanent structure or a temporary arrangement. The particular type of enclosure used depends on the individual source characteristics and the degree of control required.

A novel variation of the source enclosure method for the control of fugitive particulate emissions involves the application of porous wind fences (also referred to as windscreens). Porous wind fences have been shown to significantly reduce emissions from active storage piles and exposed ground areas. The principle employed by windscreens is to provide a sheltered region behind the fenceline where the mechanical turbulence generated by ambient winds is significantly reduced. The downwind extent of the protected area is many times the physical height of the fence. This sheltered region provides for a reduction in the wind-erosion potential of the exposed surface, in addition to allowing the gravitational settling of the larger particles already airborne. The application of windscreens along the leading edge of active storage piles seems to be one of the few good control options available for this particular source. A diagram of one type of portable windscreen used at a coal-fired power plant is shown in Figure 1.

Wet Suppression

Wet suppression systems apply either water, a water solution of a chemical agent, or a micron-sized foam to the surface of the particulate-generating material. This measure prevents (or suppresses) the fine particles contained in the material from leaving the surface and becoming airborne. If fine water sprays are used to control dust after it has become suspended, this is referred to as plume aftertreatment. Plume aftertreatment (e.g., charged fog) is not a preventive measure, but a capture/removal method, as discussed below.

Chemical agents used in wet suppression systems can be either surfactants or foaming agents for materials handling and processing operations (e.g., crushers, conveyors) or various types of dust palliatives applied to unpaved roads. In either case, the chemical agent acts to agglomerate and bind the fines to the aggregate surface, thus eliminating or reducing its emission potential. A wet suppression system at a crusher discharge point is shown in Figure 2.[1]

One of the more recently developed methods used to augment wet suppression techniques is the use of foam injection to control dust from materials-handling and -processing operations. The foam is generated by adding a proprietary surfactant compound to a relatively small quantity of water, which is then vigorously mixed to produce a

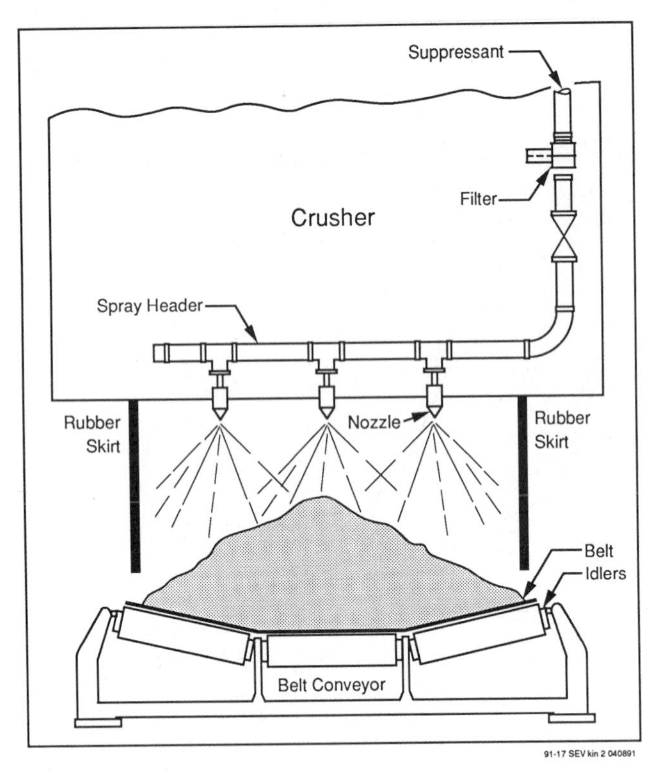

FIGURE 2. Wet Suppression System at a Crusher Discharge Point[1]

small-bubble, high-energy foam in the 100- to 200-μm size range. The foam uses very little liquid volume and, when applied to the surface of a bulk material, wets the fines more effectively than does untreated water. Foam has been used with good success in controlling the emissions from belt transfer points, crushers, and storage pile load-in.

Stabilization of Unpaved Surfaces

Release of particulate from unpaved surfaces can be reduced or prevented by stabilization of those surfaces. Sources that have been controlled in this manner include unpaved roads and parking lots, active and inactive storage piles, and open areas. Stabilizing mechanisms that have been employed successfully include chemical, physical, and vegetative controls. Each of these control types is described below.

The use of chemical dust suppressants for the control of fugitive emissions from unpaved roads has received much attention in the past several years. Chemical suppressants can be classified into six generic categories: (1) salts (i.e., $CaCl_2$ and $MgCl_2$), (2) lignin sulfonate, (3) wetting agents, (4) latexes, (5) plastics, and (6) petroleum derivatives.

Physical stabilization techniques can also be used for the control of fugitive emissions from unpaved surfaces. Physical stabilization includes any measure, such as compaction of fill material at construction and land disposal sites, that physically reduces the emissions potential of a source resulting from either a mechanical disturbance or wind erosion.

Vegetative stabilization involves the use of various species of flora to control wind erosion from exposed surfaces. Vegetative techniques can be used only when the material to be stabilized is inactive and will remain so for an extended period. It is often difficult to establish a vegetative cover over materials other than soil because their physical or chemical characteristics are not conducive to plant growth. Resistant strains that can tolerate the composition of the host material sometimes must be developed.

Paved Surface Cleaning

Other than housekeeping, the only method available to reduce the surface loading of fine particles on paved roads is through some form of street-cleaning practice. Street sweeping does remove some debris from the pavement, thus preventing it from becoming airborne by the action of passing vehicles, but it can also generate significant amounts of finer particulate by the mechanical action used to collect the material.

The three major methods of street cleaning are mechanical (broom) cleaning, vacuum cleaning, and flushing. Mechanical street sweepers utilize large rotating brooms to lift the material from the pavement and discharge it into a hopper for later disposal. Vacuum sweepers remove the material from the street surface by drawing a suction on a pickup head, which entrains the particles in the moving airstream. The debris is then deposited in a hopper, and the air is exhausted to the atmosphere or regenerated back to the pickup head and reused. Street flushers hydraulically remove debris from the surface to the gutter and eventually to the storm sewer system through the use of high-pressure water sprays.

Capture/Removal Methods

The second basic technique for the control of fugitive particulate emissions includes those methods that capture or remove the particles after they have become airborne. Again, this classification is irrespective of whether such emissions are generated inside or outside of a building. The major types of capture/removal processes include capture and collection systems and plume aftertreatment. Methods in both categories are described below.

Capture and Collection Systems

Most industrial process fugitive emissions have traditionally been controlled by capture/collection or industrial ventilation systems. These systems have three primary components: (1) a hood or enclosure to capture emissions that escape from the process, (2) a dust collector that separates entrained particulate from the captured gas stream, and (3) a ducting or ventilation system to transport the gas stream from the hood or enclosure to the air pollution control device.

A wide variety of capture mechanisms, ranging from total enclosure of the source to mobile high-velocity, low-volume (HVLV) hoods to total building evacuation, have been employed. Capture devices (or hoods) generally can be classified as one of three types: enclosure, capture hood, or receiving hood.

Enclosures, partial or complete, surround the source as much as possible without interfering with process operations. Their predominant feature is that they prevent release of particulate to the atmosphere or working environment.

Capture hoods are located in such a manner that the process is external to the hood. Emissions are actually released to the atmosphere or plant environment and subsequently captured by the airstream entering the hood. Capture hoods have also been referred to as exterior hoods by some authors.

In the case of receiving hoods, emissions from the process are also released to the atmosphere or plant environment prior to entering the hood. However, receiving hoods are designed to take advantage of the inherent momentum of some emission streams. This momentum is generally a result of thermal buoyancy, but also may be a result of inertia generated by the process (e.g., a grinding plume).

A variation of the traditional capture/collection concept involves the use of air curtains or jets. Air curtains are usually used in industrial processes that generate a buoyant plume to help isolate it and enhance capture by the emissions control system. One such system is a so-called push/pull arrangement. In such an arrangement, an air curtain consisting of a series of jets is used to contain and direct the plume toward a capture hood.

For capture/collection systems, the controlled emissions are made of (1) that portion of the uncontrolled emissions that is not captured, plus (2) that portion of the uncontrolled emissions that is captured but not collected. This is illustrated in Figure 3 for a canopy hood. Frequently, testing is

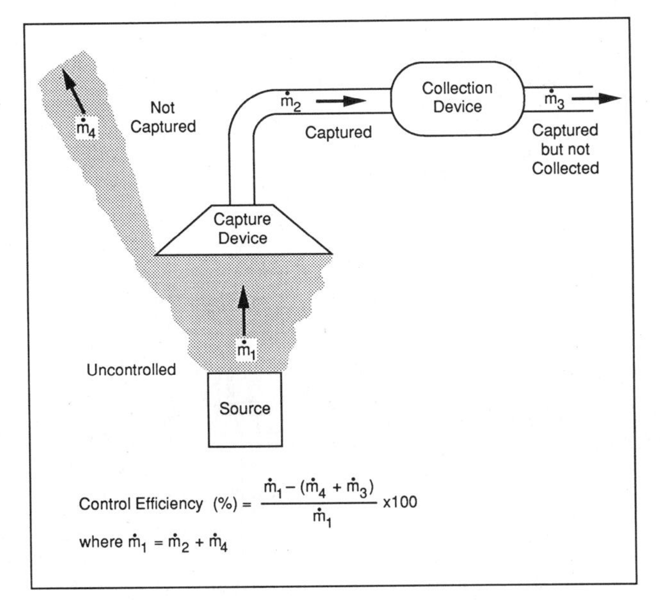

FIGURE 3. Emissions Quantification Requirements for Performance Evaluation of Capture/Collection System

performed at the inlet and outlet of the collection device, but the data are insufficient to determine the overall control efficiency.

Plume Aftertreatment

Plume aftertreatment refers to any system that injects fine water droplets into a dust plume to capture and agglomerate the suspended particles (by impaction and/or electrostatic attraction) to enhance gravitational settling. Plume aftertreatment systems can use water sprays with or without the addition of a chemical surfactant, as well as with or without the application of an electrostatic charge (charged fog).

Finally, Table 4 provides an overall summary of feasible control measures for open dust sources. Because of the wide diversity of process sources, a site-specific evaluation is necessary to determine control feasibility for these operations on a source-by-source basis. Additional information of process source control can also be found in Reference 1 and documents included in the Bibliography.

Estimation of Control Effectiveness

Wet Suppression

The following provides available methods for estimating the control efficiency of wet suppression systems.

Watering of Unpaved Roads

The control efficiency of unpaved road watering depends on (1) the amount of water applied per unit area of road surface, (2) the time between reapplications, (3) traffic volume during that period, and (4) prevailing meteorological conditions during the period. While several investigations have estimated or studied watering efficiencies, few have specified all the factors listed above.

An empirical model for the performance of watering as a control technique has been developed.[5] The supporting database consists of 14 tests performed in four states during five different summer and fall months. The model is as follows:

$$C = 100 - \frac{0.8pdt}{i} \tag{9}$$

where: C = average control efficiency, %
p = potential average hourly daytime evaporation rate, mm/h
d = average hourly daytime traffic rate (h^{-1})
i = application intensity, L/m^2
t = time between applications, hours

Estimates of the potential average hourly daytime evaporation rate may be obtained from:

$$p = 0.0049 \times \text{(value in Figure 4) for annual conditions}$$
$$= 0.0065 \times \text{(value in Figure 4) for summer conditions} \tag{10}$$

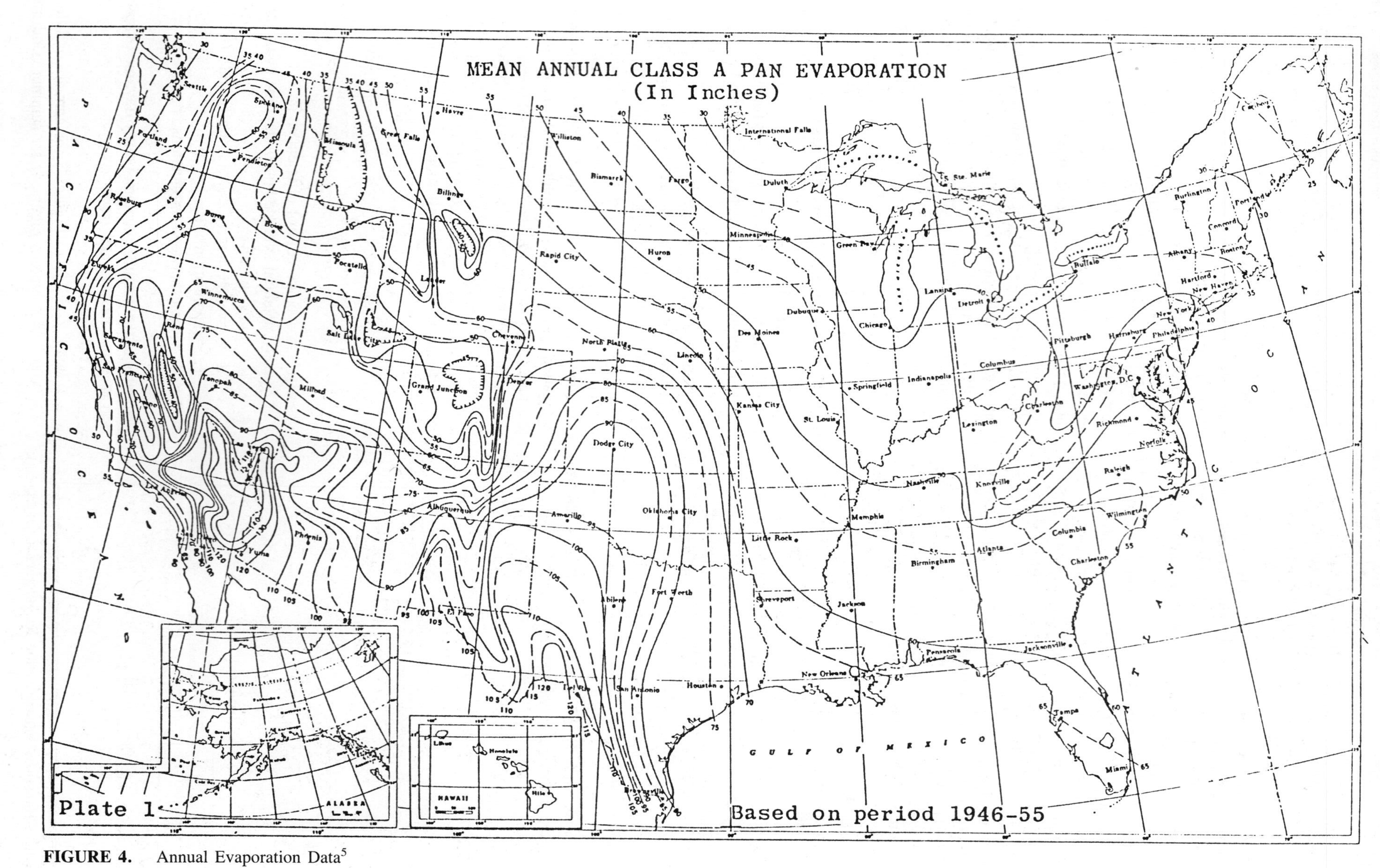

FIGURE 4. Annual Evaporation Data[5]

TABLE 4. Feasible Control Measures for Open Dust Sources

	Fugitive Emission Control Measure						
Source Category	Enclosures[a]	Wet Suppression	Chemical Stabilization	Physical Stabilization	Vegetative Stabilization	Surface Cleaning	Capture/ Removal[b]
Unpaved roads		X	X	X			
Unpaved parking lots and staging areas		X	X	X			
Storage piles	X	X	X	X			
Paved streets and highways						X	
Paved parking lots and staging areas						X	
Exposed areas	X	X	X	X	X		
Batch drop operations[c]	X	X					X
Continuous drop operations[d]	X	X					X
Pushing (dozing, grading, scraping, etc.)		X	X				

[a]Includes full and partial enclosures as well as wind fences.
[b]Includes both capture/collection systems and plume aftertreatment.
[c]Includes operations such as front-end loaders, shovels, etc.
[d]Includes operations such as conveyor transfer, stacking/reclaiming, etc.

An alternative approach is shown in Figure 5. This figure is adapted from 11 field tests conducted at a coal-fired power plant.[5] Measured control efficiencies did not correlate well with either time or vehicle passes after application. However, this is believed to be due to reduced evening evaporation (logistics delayed the start of testing until 3 P.M. and testing continued through the early evening). Surface moisture grab samples were taken throughout the testing period and, not surprisingly, these show a strong correlation with control efficiency.

Figure 5 shows that between the average uncontrolled moisture content and a value of twice that, a small increase in moisture content results in a large increase in control efficiency. Beyond this point, control efficiency grows slowly with increased moisture content. Although it is possible to fit hyperbolas to the data, the relatively simple bilinear relationship shown in the figure provides an adequate description. Furthermore, this relationship is applicable to all particle-size ranges considered:

$$c = 75\,(M - 1), \quad 1 \leq M \leq 2$$
$$62 + 6.7M, \quad 2 \leq M \leq 5 \qquad (11)$$

where: c = instantaneous control efficiency, %
M = ratio of controlled to uncontrolled surface moisture contents

Wet Suppression for Materials Handling

Available control efficiency data for wet dust suppression for materials handling and storage are practically nonexistent. However, certain limited information was compiled by Cowherd and Kinsey[1] that can be used to estimate control efficiencies.

For suppression using plain water, the most applicable efficiency information available is for feeder to belt transfer of coal in mining operations. Control efficiencies of 56% to 81% are reported for respirable particulate (particles <3.5 μmA) at application intensities of 6.7 to 7.1 L/10^6 g (1.6 to 1.7 gal/ton) respectively. Assuming that respirable particulate is essentially equivalent to PM_{10}, the above efficiencies would be representative of controls for similar sources. (The above application intensities were estimated assuming 5 minutes to discharge 7×10^6 gram of coal and 1.4 L/min per spray nozzle.)

In the case of foam suppression, the most appropriate data available are for the transfer of sand from a truck unloading station. Using the respirable particulate control efficiencies at various foam-application intensities (and assuming respirable particulate is equivalent to PM_{10}), the following equation was developed by simple linear regression of the data compiled by Cowherd and Kinsey[1,5]:

$$C = 8.51 + 7.96\,(A) \qquad (12)$$

where: C = PM_{10} control efficiency, %
A = application intensity, cubic feet of foam per ton of material

A coefficient of determination (r^2) of 99.97% was obtained for the above equation based on the three data sets used in its derivation.

An alternative approach involves the use of the materials-handling equation published in AP-42. This equation was presented as equation 3 above. By determining the "uncontrolled" moisture content of the material and again after wet suppression, the control efficiency can be determined by[5]:

$$C = 100(e_u - e_c)/e_u \qquad (13)$$

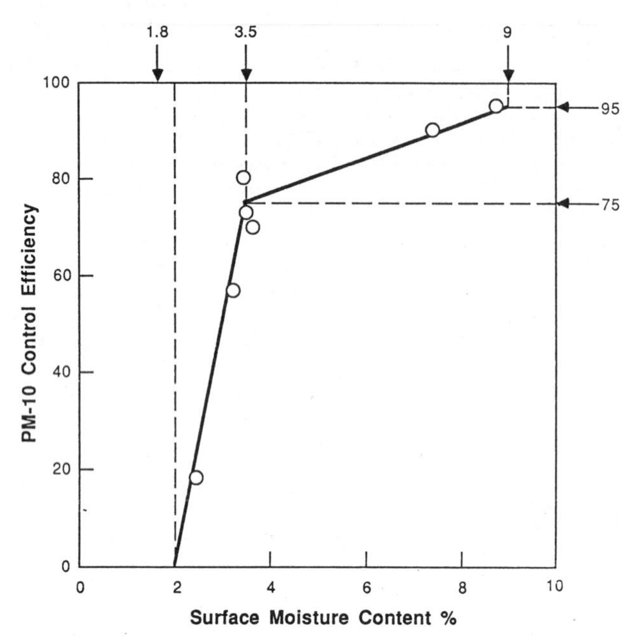

FIGURE 5. Watering Control Effectiveness for Unpaved Travel Surfaces[5]

where: C = PM_{10} control efficiency, %
e_u = "uncontrolled" PM_{10} emission factor, weight per unit source extent
e_c = "controlled" PM_{10} emission factor, weight per unit source extent

Chemical Stabilization of Unpaved Surfaces

Some chemicals (most notably salts) simulate wet suppression by attracting and retaining moisture on the road surface. These methods are often supplemented by some watering. It is recommended that control-efficiency estimates be obtained using Figure 5 for hydroscopic road stabilizing agents.

The more common chemical dust suppressants form a hard cemented surface. It is this type of suppressant that is considered below.

In a recent test report, average performance curves were generated for four chemical dust suppressants: (1) a commercially available petroleum resin, (2) a generic petroleum resin for on-site production of an industrial facility, (3) an acrylic cement, and (4) an asphalt emulsion.[6] (Note that at the time of the testing program, these suppressant types accounted for roughly 85% of the market share in the iron and steel industry.) The results of this program were combined with other test results to develop a model to estimate *time averaged* (rather than instantaneous) PM_{10} control performance. This model is illustrated as Figure 6.

Several items should be noted with regard to Figure 6:

- The term "ground inventory" is a measure of residual effects from previous applications. Ground inventory is found by adding the total volume (per unit area) of concentrate *(not solution)* since the start of the dust control season.

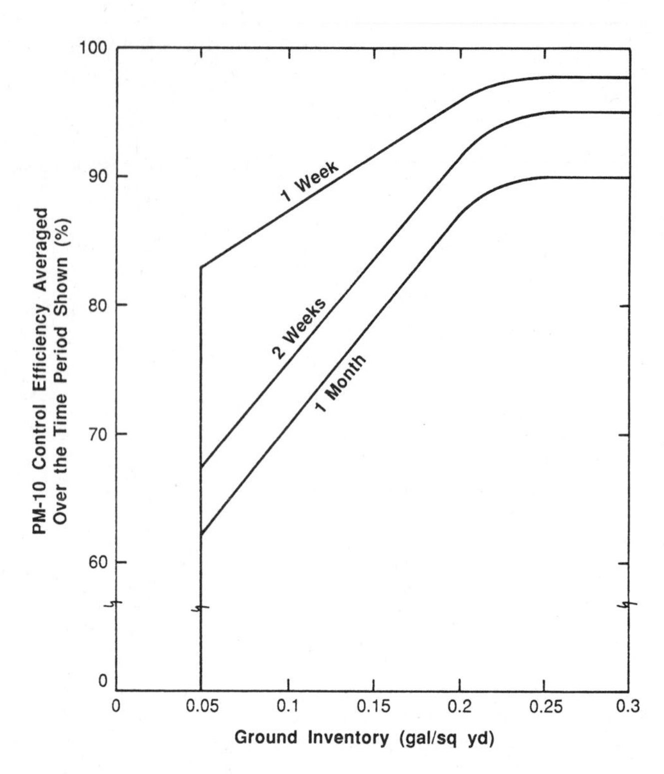

FIGURE 6. Average PM_{10} Control Efficiency for Chemical Suppressants

- Note that no credit for control is assigned until the ground inventory exceeds 0.5 gal/yd^2.
- Because suppressants must be periodically reapplied to unpaved roads, use of the time-averaged values given in the figure is appropriate. Recommended minimum reapplication frequencies are discussed later in this section.
- Figure 6 represents an *average* of the four suppressants given above. The basis of the methodology lies in a similar model for petroleum resins only.[6] However, agreement between the control-efficiency estimates given by Figure 6 and available field measurements is reasonably good.

Because unpaved roads in industry are often used for the movement of materials and are often surrounded by additional unpaved travel areas, spillage and carryout onto the chemically treated road require periodic "housekeeping" activities. In addition, gradual abrasion of the treated surface by traffic will result in loose material on the surface that should be controlled.

It is recommended that at least dilute reapplications be employed every month to control loose surface material, unless paved-road control techniques are used (as described below). More frequent reapplications would be required if spillage and track-on pose particular problems for a road.

Chemical Stabilization of Storage Piles

A portable wind tunnel has been used to measure the control of coal-pile wind-erosion emissions by a 17% solution of

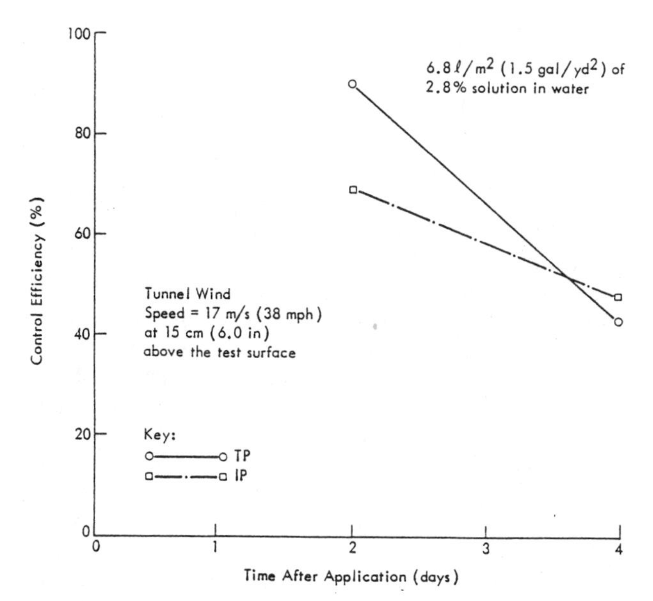

FIGURE 7. Decay in Control Efficiency of Latex Binder Applied to Coal-Storage Piles[5]

Coherex® in water applied at an intensity of 3.4 L/m² (0.74 gal/yd²), and a 2.8% solution of Dow Chemical M-167 Latex Binder in water applied at an average intensity of 6.8 L/m² (1.5 gal/yd²).[5] The control efficiency of Coherex® applied at the above intensity to an undisturbed steam coal surface approximately 60 days before the test, under a wind of 15.0 m/s (33.8 mi/h) at 15.2 cm (6 inches) above the ground, was 89.6% for total particulate (TP) and approximately 62% for inhalable particulate (IP, particles ≤15 μm in aerodynamic diameter) and fine particulate (FP, particles ≤2.5 μm in aerodynamic diameter). The control efficiency of the latex binder on a low-volatility coking coal is shown in Figure 7.

Paved Surface Cleaning

Available control methods for paved roads are largely designed either to prevent deposition of material on the roadway surface or to remove material that has been deposited in the driving lanes. Measurement-based efficiency values for control methods are presented in Table 5. Note that all values in this table are for mitigative measures applied to industrial paved roads.

In terms of public road dust control, only very limited field measurement data are available. One reference was found that could be used indirectly to quantify emission reductions, and this, too, is for mitigative measures. Estimated PM_{10} control efficiencies (Table 6) were developed by applying equation 1 to measurements before and after road cleaning.[6] Note that these estimates should be considered *upper bounds* on efficiencies obtained in practice because no redeposition after cleaning is considered. Note also that these estimated emission-control efficiencies for urban roads compare fairly well with measurements quantifying control efficiency that are available.

TABLE 6. Estimated PM_{10} Control Efficiencies for Public Paved Roads[a]

Method	Estimated PM_{10} Efficiency, %
Vacuum sweeping	34
Improved vacuum sweeping[b]	38

[a]Reference 5. Estimated based on measured initial and residual ≤63-μm loadings on urban paved roads and equation 1. Value reported represents the mean of 13 tests for each method.
[b]Sweeping improvements described in original reference document. See reference 5.

Capture/Collection for Materials Handling

Airflow exhausted from a local capture hood installed on an operation involving material movement serves two purposes: the exhaust must overcome induced airflow created by material motion, and the exhaust must provide sufficient velocity to capture particulate that escapes the confines of the hood. The predominant function depends on hood type.

TABLE 5. Measured Efficiency Values for Industrial Paved Road Controls[a]

Method	Cited Efficiency	Comments
Vacuum sweeping	0–58%	Field emission measurement (PM_{15}) 12,000-cfm blower[b]
	46%	Based on field measurement of 30-μm particulate emissions
Water flushing	69–0.231 $V^{c,d}$	Field measurement of PM_{15} emissions[b]
Water flushing followed by sweeping	96–0.263 $V^{c,d}$	Field measurement of PM_{15} emissions[b]

[a]Reference 5. Broom-sweeping measurements indicate a maximum instantaneous control of 25–30% for industrial and urban paved roads.
[b]PM_{10} control efficiency can be assumed to be the same as that tested.
[c]Water applied at 0.48 gal/yd².
[d]Equation yields efficiency in percent, V = number of vehicle passes since application.

If an enclosure is used, control of induced airstreams is the primary objective. If the operation requires an exterior hood, particulate capture is the primary airflow function.

For those systems that can be controlled by complete or partial enclosure, the airflow at the hood should be sufficient to overcome induced air currents inherent to the process and to provide an inward air velocity through all openings of about 50 to 200 ft/min.[1] The volumes needed to overcome induced air currents associated with specific processes are discussed below.

The flow needed to provide adequate velocities at openings can be calculated by the formula:

$$Q = AV \tag{14}$$

where: Q = required airflow, ft^3/min
A = cross-sectional area of openings, square feet
V = required velocity at openings, ft/min

Material transport creates an induced airflow that must be overcome effectively to control fugitive emissions. Anderson[7] has developed the following equation for calculating induced airflow at transfer points:

$$Q = 10.0\, A_u \sqrt[3]{\frac{RS^2}{D}} \tag{15}$$

where: Q = induced airflow, ft^3/min
A_u = feed opening, square feet
R = rate of material flow, tons/h
S = height of fall, feet
D = average particle diameter, feet

The objective of a capture hood is to provide a capture velocity of 50 to 75 ft/min at the farthest capture point from the hood. The total flow required to achieve this velocity is as follows[1]:

$$Q = V(10X^2 + A) \tag{16}$$

where: Q = required airflow, ft^3/min
V = required capture velocity, ft/min
X = distance from hood to farthest null point, feet
A = cross-sectional area of hood, square feet

Receiving hoods capture particulate as it is directed from the source by thermal or mechanical forces. Examples are canopy hoods for furnace charging and tapping emissions and close capture hoods on grinding equipment. Key design considerations are locating the hood so that the complete exhaust stream is directed to the hood and generating an airflow greater than the induced stream that is directed into the hood. Plume size and cross-draft problems are major concerns in designing receiving hoods. Based on limited information, a capture efficiency of 70% to 98% was measured for a close capture hood used on a metallurgical process.[5]

References

1. C. Cowherd, Jr., and J. S. Kinsey, *Identification, Assessment, and Control of Fugitive Particulate Emissions,* EPA-600/8-86-023, U.S. Environmental Protection Agency, Research Triangle Park, NC, 1986.
2. U.S. Environmental Protection Agency, *Compilation of Air Pollutant Emission Factors,* AP-42, U.S. Environmental Protection Agency, Research Triangle Park, NC, 1988.
3. J. S. Kinsey, P. Englehart and A. L. Jirik, *Study of Construction Related Dust Control,* Contract No. 32200-07976-01, Minnesota Pollution Control Agency, Roseville, MN, 1983.
4. P. J. Englehart and J. S. Kinsey, *Study of Construction Related Mud/Dirt Carryout,* EPA Contract No. 68-02-3177, Assignment 12, U.S. Environmental Protection Agency, Region V, Chicago, July 1983.
5. C. Cowherd, G. E. Muleski and J. S. Kinsey, *Control of Open Fugitive Dust Sources,* EPA-450/3-88-008, U.S. Environmental Protection Agency, Research Triangle Park, NC, September 1988.
6. G. E. Muleski and C. Cowherd, Jr., *Evaluation of the Effectiveness of Chemical Dust Suppressants on Unpaved Roads,* EPA-600/2-87-102, U.S. Environmental Protection Agency, Research Triangle Park, NC, November 1987.
7. D. M. Anderson, "Dust control design by the air induction technique," *Indus. Med. Surg.,* 68–72 (February 1964.).

Bibliography

American Conference of Governmental Industrial Hygienists, *Industrial Ventilation, A Manual of Recommended Practice,* 18th ed., Lansing, MI, 1984 (or latest version).

Climatic Atlas of the United States, U.S. Department of Commerce, Washington, DC, June 1968.

C. Cowherd, et al., *Hazardous Waste TSDF Fugitive Particulate Matter Air Emissions Guidance Document,* EPA-450/3-89-019, U.S. Environmental Protection Agency, Research Triangle Park, NC, May 1989.

Guidelines for Development of Control Strategies in Areas with Fugitive Dust Problems, OAQPS No. 1.2-071, U.S. Environmental Protection Agency, Research Triangle Park, NC, October 1977.

W.C.L. Hemeon, *Plant and Process Ventilation,* Industrial Press, New York, 1963.

E. R. Kashdan, et al., *Technical Manual: Hood System Capture of Process Fugitive Particulate Emissions,* EPA-600/7-86-016, NTIS No. PB86-190444, U.S. Environmental Protection Agency, Research Triangle Park, NC, April 1986.

V. Mody and R. Jakhete, *Dust Control Handbook,* Noyes Data Corp., Park Ridge, NJ, 1988.

Ohio Environmental Protection Agency, *Reasonably Available Control Measures for Fugitive Dust Sources,* Columbus, OH, September 1980.

U.S. Environmental Protection Agency, *Control Techniques for Particulate Emissions from Stationary Sources—Volumes 1 and 2,* EPA-450/3-81-005a and b, Emission Standards and Engineering Division, Research Triangle Park, NC, September 1982.

5

Odors

William H. Prokop, P. E.

SENSORY PROPERTIES OF ODOR

An odor is defined as a sensation resulting from the reception of a stimulus by the olfactory sensory system. The types of human responses to be evaluated depend on the particular sensory property to be measured, including odor intensity, detectability, character, and hedonic tone (pleasantness/unpleasantness). The combined effect of these properties is related to the annoyance that may be caused by an odor.

Intensity

Odor intensity is the strength of the perceived odor sensation and is related to the odorant concentration, which is an entirely different category of measurement. The intensity of an odor is perceived directly without any knowledge of the odorant concentration or of the degree of air dilution of the odorous sample needed to eliminate the odor.

The following equation defines the relationship between the odor intensity (I) and concentration (C) where k is a constant and n is the exponent.

$$I \text{ (perceived)} = k(C)^n$$

or

$$\text{Log } I = \log K + n\log (C)$$

This is known as Stevens' law or the power law. For odors, n ranges from about 0.2 to 0.8, depending on the odorant. For an odorant with n equal to 0.2, a 10-fold reduction in concentration decreases the perceived intensity by a factor of only 1.6; whereas for an odorant with n equal to 0.8, a 10-fold reduction in concentration lowers the perceived intensity by a factor of 6.3. This is an important concept that is related to the basic problem of reducing the odor intensity of a substance by air dilution or other means.

Figure 1 shows data[1] for different chemicals as a log-log plot, where the concentration, in parts per billion, as the abscissa varies with the right-side ordinate shown as relative odor intensity. The spacing of the numbers on the relative-odor-intensity scale is based on data available for methyl sulfide and the four italicized odorants (IIT Research Institute data). The slope of the straight line is equal to n. It is evident that the effect of dilution on the odor intensity of methyl sulfide is much less than that for hydrogen sulfide.

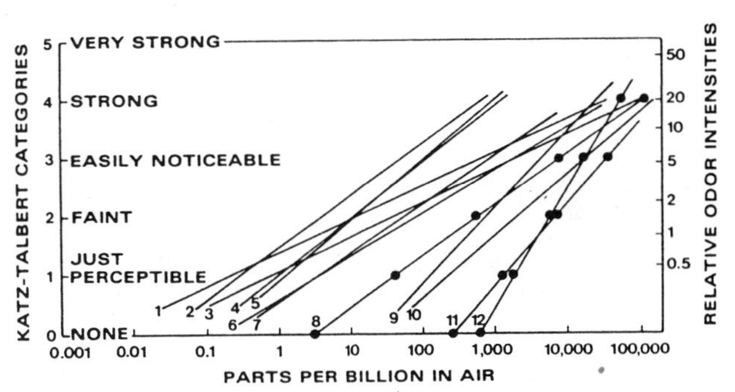

Odor intensity vs. concentration plots for several odorants. Individual points are shown for 8, 11, and 12 to illustrate the linearity. Legend: 1. Thiophenol; 2. Ethyl selenomercaptan; 3. Ethyl sulfide; 4. Phenyl isocyanide; 5. Ethyl selenide; 6. *Methyl sulfide*; 7. *Coumarin*; 8. Methylmercaptan; 9. Hydrogen sulfide; 10. *Pyridine*; 11. *Allyl alcohol*; 12. *Nitrobenzene*.

FIGURE 1. Odor Intensity versus Concentration

Detectability

The detectability or threshold of an odor is a sensory property referring to the minimum concentration that produces an olfactory response or sensation. This threshold usually is determined by an odor panel consisting of a specified number of people and the numerical result is typically expressed as occurring when 50% of the panel correctly detect the odor.

At odor intensity levels at or just above "threshold," odors become difficult to perceive. As a result, the actual values depend on the type of sensory test, the panelist selection, the detectability criterion, and other factors. It can be defined only within the context of these parameters. The odor threshold is not a precisely determined value, as is, for example, vapor pressure. It is highly dependent on the sensitivity of the odor panelists, the method of presenting the odor stimulus to the panelists, including flow rate, and often the purity of the chemical odorant being tested.

An odor detection threshold relates to the minimum odorant concentration required to perceive the existence of the stimulus, whereas an odor recognition threshold relates to the minimum odorant concentration required to identify the stimulus. The detection threshold occurs at a lower concentration than the recognition threshold.

Data in the literature on odor threshold concentrations for any particular compound may differ significantly; in many cases, by 10-fold or more. This is particularly true of the earlier values because of inadequate equipment or methods, too small a panel, or too large a step-wise change in odorant concentration.

Character

Odor character or quality is that property that identifies an odor and differentiates it from another odor of equal intensity. Odors are classified on the basis of odor descriptor terms. Odor character is evaluated by comparison with other odors, either directly or through the use of descriptor words (see Table 1).

The odor character is described by methods known as multidimensional scaling or profiling. In these methods, the odor is characterized by either the degree of its similarity to (or dissimilarity from) a set of reference odors or the degree of applying a scale of various descriptor terms to it. The result is an odor profile.

Hedonic Tone

Hedonic tone is a property of an odor relating to its pleasantness or unpleasantness. A distinction should be made between the acceptability and the hedonic tone of an odor. Acceptability is usually a judgment made by a specific person in the context of a specific situation and with specific expectations. For example, an otherwise pleasant odor may be unacceptable if it persists as part of an air pollution problem in a residential area and originates from a perfume factory instead of from a flower garden.

When an odor is evaluated in the laboratory for its hedonic tone in the neutral context of an olfactometric presentation, the panelist is exposed to a controlled stimulus in terms of intensity and duration. The degree of pleasantness or unpleasantness is determined by each panelist's experience and emotional associations. The responses among panelists may vary depending on odor character; an odor pleasant to many may be declared highly unpleasant by some.

TABLE 1. Odorous Compounds in Industrial Source Emissions

Compound Name	Formula	Molecular Weight	Volatility at 25°C, ppm, v/v	Detection[a] Threshold, ppm, v/v	Recognition[a] Threshold, ppm, v/v	Odor Descriptors
Acetaldehyde	CH_3CHO	44	Gas	0.067	0.21	Pungent, fruity
Allyl mercaptan	$CH_2{:}CHCH_2SH$	74		0.0001	0.0015	Disagreeable, garlic
Ammonia	NH_3	17	Gas	17	37	Pungent, irritating
Amyl mercaptan	$CH_3(CH_2)_4SH$	104		0.0003	—	Unpleasant, putrid
Benzyl mercaptan	$C_6H_5CH_2SH$	124		0.0002	0.0026	Unpleasant, strong
n-Butyl amine	$CH_3(CH_2)NH_2$	73	93,000	0.080	1.8	Sour, ammonia
Chlorine	Cl_2	71	Gas	0.080	0.31	Pungent, suffocating
Dibutyl amine	$(C_4H_9)_2NH$	129	8,000	0.016	—	Fishy
Diisopropyl amine	$(C_3H_7)_2NH$	101		0.13	0.38	Fishy
Dimethyl amine	$(CH_3)_2NH$	45	Gas	0.34	—	Putrid, fishy
Dimethyl sulfide	$(CH_3)_2S$	62	830,000	0.001	0.001	Decayed cabbage
Diphenyl sulfide	$(C_6H_5)_2S$	186	100	0.0001	0.0021	Unpleasant
Ethyl amine	$C_2H_5NH_2$	45	Gas	0.27	1.7	Ammoniacal
Ethyl mercaptan	C_2H_5SH	62	710,000	0.0003	0.001	Decayed cabbage
Hydrogen sulfide	H_2S	34	Gas	0.0005	0.0047	Rotten eggs
Indole	$C_6H_4(CH)_2NH$	117	360	0.0001	—	Fecal, nauseating
Methyl amine	CH_3NH_2	31	Gas	4.7	—	Putrid, fishy
Methyl mercaptan	CH_3SH	48	Gas	0.0005	0.0010	Rotten cabbage
Ozone	O_3	48	Gas	0.5	—	Pungent, irritating
Phenyl mercaptan	C_6H_5SH	110	2,000	0.0003	0.0015	Putrid, garlic
Propyl mercaptan	C_3H_7SH	76	220,000	0.0005	0.020	Unpleasant
Pyridine	C_5H_5N	79	27,000	0.66	0.74	Pungent, irritating
Skatole	C_9H_9N	131	200	0.001	0.050	Fecal, nauseating
Sulfur dioxide	SO_2	64	Gas	2.7	4.4	Pungent, irritating
Thiocresol	$CH_3C_6H_4SH$	124		0.0001	—	Skunk, rancid
Trimethyl amine	$(CH_3)_3N$	59	Gas	0.0004	—	Pungent, fishy

[a]References 2–5.

Adaptation

Adaptation or olfactory fatigue is a phenomenon that occurs when people with a normal sense of smell experience a decrease in perceived intensity of an odor if the stimulus is received continually. Depending on the intensity of the stimulus, this self-adaptation and sensory recovery after removal of the stimulus both take place during a relatively short time. Adaptation to a specific odorant generally does not interfere with the ability of a person to detect other odors. Another phenomenon known as habituation or occupational anosmia occurs when a worker in an industrial situation experiences a long-term exposure and develops a higher threshold tolerance to the odor.

ODOROUS COMPOUNDS

Odorous substances that are emitted from industrial sources include both inorganic and organic gases and particulate. Hydrogen sulfide and ammonia are examples of inorganic gases. Many odorous compounds result from biological activity or are present in emissions from chemical processes. Most of the odorous substances derived from anaerobic decomposition of organic matter contain sulfur and nitrogen.

Table 1 lists some of the odorous compounds emitted from industrial sources and their odor detection and recognition thresholds. Most of the odorous substances are gaseous under normal atmospheric conditions or at least have a significant volatility. The volatility is shown in the table as parts per million (v/v) and is equal to the vapor pressure (mm Hg at 25°C) muliplied by 1316 (1 million ppm per 760 mm Hg). The molecular weights of these substances generally range from 30 to 150. Usually, the lower the molecular weight of a compound, the higher is its vapor pressure and potential for emission to the atmosphere. Substances of high molecular weight are normally less volatile and thus normally have less impact as a cause for odor complaints.

The reduced sulfur compounds, such as the mercaptans and organic sulfides, tend to be the most odorous, based on their relatively low odor threshold concentrations. This also applies to the nitrogen-bearing amines, but to a lesser extent.

ODOR SENSORY MEASUREMENT

An odor emission often consists of a complex mixture of many odorous compounds. As a result, odor sensory methods, instead of instrumental methods, are normally used to measure such odors. Analytical monitoring of individual chemical compounds present in such odors is usually not practical. These odor sensory methods depend on the olfactory response of individuals who serve on panels.

Odor sensory methods are available to monitor odors both from source emissions and in the ambient air. It is important to have both options. For example, an industrial source may be interested in evaluating the performance of odor control equipment by monitoring its odor emission and also may desire to know the downwind odor level in the ambient air beyond its property line to establish compliance with an ambient odor regulatory limit. These two diverse circumstances require different approaches for measuring odor. For example, the collection of odor samples is more easily accomplished for a source emission than for an odor in the ambient air. Also, due to atmospheric dilution, the odor in the ambient air is usually much lower in intensity than it is at the source. Thus the sensitivity of the odor sensory method must be significantly greater for measuring ambient odors than for source odor emissions.

The following discussion on odor dilution to threshold and suprathreshold (intensity scale) measurement methods is based on their treatment by Dravnieks[6,7] and Turk.[8]

Odor Dilution to Threshold

Odor dilution to threshold sensory results are expressed as a dimensionless ratio termed Z.[6,9] If volume v of an odorous sample is diluted to a total volume of V (where V equals the volume of odor diluted to threshold level), the ratio $Z = V/v$. As long as the same volume units are used to measure v and V, the value of Z is independent of the volumetric units.

Numerically, Z is equal to the term "odor units per cubic foot" (or per cubic meter), which has been in use for many years. The use of these units in this context is actually incorrect dimensionally. Other terms that are numerically equivalent to Z include ED_{50} (effective dosage at 50% panel) and D/T (dilution to threshold ratio).

The total emission of odorous substances from a source is then $Q = (Z)(W)$, where W is the volumetric flow rate in either cubic meters per minute or cubic feet per minute. As a result, Q has the same units as W and quantifies the odor emission to the atmosphere. If this emission is not to be detected downwind at ground level, the volumetric emission W must be diluted each minute to a volume either equal to or greater than Q expressed in either cubic meters or cubic feet.

Sampling of Source Odor Emissions

The sampling of source odor emissions for later evaluation by panels to determine odor thresholds requires the proper choice of sampling equipment, collection container, and sampling procedure. As a general rule, a sample should be evaluated by a panel as soon as possible after collection. In some cases, storage tests with various plastic bags of specific classes of odorous compounds at higher concentrations have indicated that a sample does not degrade within 24 hours after collection. In other cases, storage tests have indicated that a much shorter time is required for other classes of compounds, especially at lower concentrations.

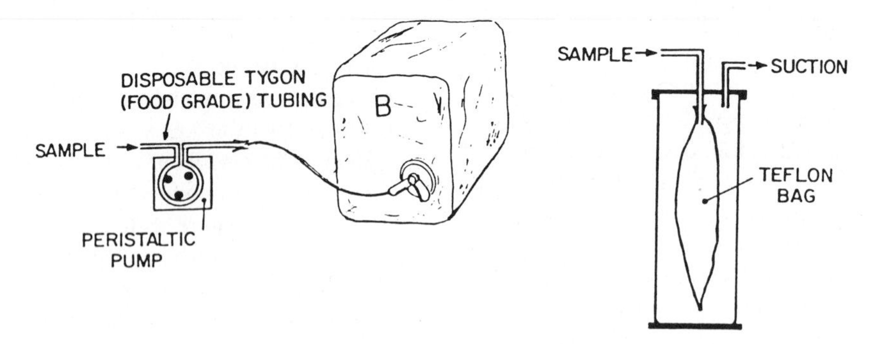

FIGURE 2. Methods for Sampling Source Odor Emission and Ambient Air

Therefore, it is advisable to pretest the storage of a particular type of odorous sample in the proposed bags if previous data are not available.

A peristaltic pump is often used with essentially nonodorous plastic tubing to deliver the sample into a plastic bag. A preflushing technique is used whereby the bag is filled initially and then discharged to allow for the adsorption of the sample to equilibrate with the inside surfaces of the bag and tubing. The peristaltic method allows for a new piece of tubing to be used for each sample without having to clean the pump.

For samples with a lower odor intensity, such as ambient air, it usually is undesirable to have the sample pass through a pump because of adsorption losses. Instead, a bag located in an airtight, stainless-steel cylinder that is evacuated by means of a vacuum pump allows the sample to be drawn into the bag. Again, a preflush is used. Pressure can be applied to the cylinder to discharge the sample from the bag. Both sampling methods are shown in Figure 2.

Plastic bags are the most common containers used for collecting and storing samples of odorous gases. The bags should be free of pores, be mechanically strong, have little background odor, and minimize the loss of odorants by diffusion through the bag walls. For source emission samples, thick-walled collapsible polyethylene bags have been used successfully for various classes of odorous compounds. Also, they are relatively inexpensive and can be discarded after use. For weaker odors, the more expensive Tedlar or heat-sealed FEP Teflon bags are normally used because they contain less background odor. These bags may be reused by repeated flushing with odorless air.

Static Versus Dynamic Methods

Static and dynamic methods of dilution to threshold measurement have significantly different characteristics. Olfactometry consists of receiving the odorous sample, diluting it in various proportions with odorless air, and presenting the diluted samples for evaluation by the panelists.

Earlier methods developed for odor threshold measurement include the American Society for Testing and Materials (ASTM) syringe method,[10] which originally was issued in 1957. The 1978 revision of the ASTM D-1391 syringe method was withdrawn by the ASTM E-18 Committee on March 29, 1985. As a result, no official ASTM syringe method is in effect today. This method is cited in a number of state and local agency odor regulations and is known as a static olfactometric procedure. It consists of diluting the odorous sample in a container (100-mL syringe) and then expelling the entire diluted sample within a few seconds by directing the tip of the syringe toward the nostrils.

In contrast, a dynamic olfactometer provides a continuous and constant diluted odor stimulus by mixing controlled flows of sample and odorless air. With dynamic methods, larger samples are used and dilutions are presented at more reproducible flow rates and for a longer duration for panelists to evaluate.

At relatively low Z values, comparison tests with rendering odors showed that dynamic methods produced results that were numerically three- to fivefold greater than those obtained by the ASTM syringe method.[11] The lower ASTM values are considered to be due to adsorption losses on the glass syringes and the other factors previously discussed. Odor scientists recommend the use of dynamic olfactometric methods instead of static methods.[7]

Factors in dynamic olfactometry that should be considered include the following.

1. Dynamic dilution
2. Delivery of diluted sample at olfactometer–nose interface
3. Scheduled presentation of various dilutions
4. Responses from panelists
5. Calculation of average panel threshold from raw experimental data
6. Number of and selection of panelists

The sample and odorless airflows are mixed and then supplied to the olfactometer–nose interface. The olfactometer should be designed to minimize odor losses due to adsorption on interior surfaces. Also, the design should provide for strong odors to be readily flushed with odorless air or for the sample and diluted flowlines to be easily replaced. It is desirable that the flow rate to the olfactometer–nose interface be constant at all dilutions.

A large variety of methods exist for delivering the odor stimulus to the nose. These include the use of masks, sniffing ports, and nose cones. Each has advantages and disadvantages that relate to the potential for a residual odor to be sensed at the nose interface, and also to minimize any dilution by the surrounding ambient air of the stimulus before reaching the panelist's nose.

The most significant variable in the delivery of the odor stimulus to the nose is the flow rate from the olfactometer. For different olfactometers, it varies from 0.5 to 200 L/min. Higher flow rates produce greater numerical values of dilution to threshold. For example, a 100-fold variation in flow rate can produce a 10-fold variation in the reported dilution ratio or suprathreshold intensity of the same odorant by the same panel. However, lower flow rates consume less sample per test, tending to improve the portability of an olfactometer and to have less impact on producing background odors in the test room for evaluation.

The presentation of odorous sample dilutions to panelists and their responses depend on three sensory effects: judgment criterion, anticipation, and adaptation. The judgment criterion determines how the panelist is to respond when asked whether or not an odor is sensed. This is the case particularly when a single stimulus is presented and a yes or no answer is requested as to the sensation of odor. The anticipation effect is a tendency to expect an odor to occur when odorless or weak samples are consecutively presented. The adaptation effect is a temporary loss of sensitivity after smelling an odor. When a weak odor is detected initially, the same odor may not be detected again after smelling a stronger odor unless the panelist has had sufficient time to recover his or her olfactory sense.

Most of these problems are resolved by adopting a procedure in which a forced-choice multiple stimulus method is combined with a systematic ascending order of odorous sample concentrations (corresponding decrease in dilution). The diluted odor stimulus is presented along with two blanks of odorless air and the panelist is asked to choose the one that contains the stimulus. The ASTM in Method E-679[12] recommends this approach when performing odor threshold measurements.

Data treatment consists of calculating from sets of individual responses an average threshold value representative of the panel as a whole. In most dynamic olfactometer procedures, the sample dilution is changed in discrete steps. For example, the dilution may be changed threefold per step. If a panelist does not detect odor at dilution 3D, but detects it at dilution D (triple the odorant concentration) and at lesser dilutions, the dilution D does not necessarily represent the panelist's threshold. Instead, it might have been in the interval between 3D and D. Statistical considerations indicate that the overall error due to this factor is minimized if it is assumed that the threshold is the geometric mean of the two dilutions, $(3D \cdot D)^{1/2} = 1.7D$.

Individuals differ in their odor sensitivity and panelists who have abnormally low or high sensitivities should be excluded. Panelists should be chosen with the objective in mind that they represent the average sensitivity of the population. Selection of the more sensitive half or third of the prospective panelists produces a data bias toward obtaining higher Z values.

The number of panelists selected for odor sensory evaluation is directly related to the percent confidence level desired to minimize the probability of error occurring by chance. For most odor sensory testing that requires decision making for odor control sytems, a 99% confidence level is usually desired. This requires a panel of nine or more people to obtain a significant number of data points if each panelist evaluates the collected sample only once. If fewer panelists are used, they should evaluate the same sample at least twice.

The forced-choice, triangle olfactometer[13] was designed by the IIT Research Institute (IITRI) to implement the odor evaluation principles described under ASTM Method E-679. The ASTM includes the IITRI method as an example of applying this standard. This particular version of the IITRI olfactometer is used to evaluate source emission odors.

Figure 3 shows this olfactometer with its various component parts coded. In the upper diagram, cup C contains three sniffing ports, one with the diluted odorous sample and the other two with odorless air. The panelist chooses the odorous port by depressing a pushbutton I at the selected port and a light bulb signals the choice at panel box S. The peristaltic pump P delivers the odorous sample to the olfactometer from sample bag B. F and R are flowmeters used to calibrate the sample and dilution airflows. Room air is used for dilution.

There are six dilution levels, ranging typically from 4500× to 15×, or from 80,000× to 450× by use of the attenuator. Adjoining dilution levels differ in concentration by a factor of three. All required dilutions and blanks continuously discharge from the sniffing ports at an essentially uniform flow of 500 mL/min. This flow rate was adopted to prevent an odor buildup in a normally ventilated room.

Another commercially available olfactometer is known as the Hemeon Odor Meter.[14] It uses a forced-choice principle requiring the panelist to decide whether the diluted odor stimulus is (1) stronger than the previous presentation, or (2) has the same strength, or (3) has no odor (blank). The odorous sample is diluted through a series of valves and flowmeters with room air and is delivered at a rate of 150 L/min through each of three sniffing ports to three panelists simultaneously. The test begins with sufficient dilution air to start at an odor concentration below threshold. As the odor concentration is increased sequentially in steps using 50% less dilution air, the panelist rates the odor intensity according to a category scale of 0 to 4. Category 1 relates to detection and category 2 relates to recognition of the odor

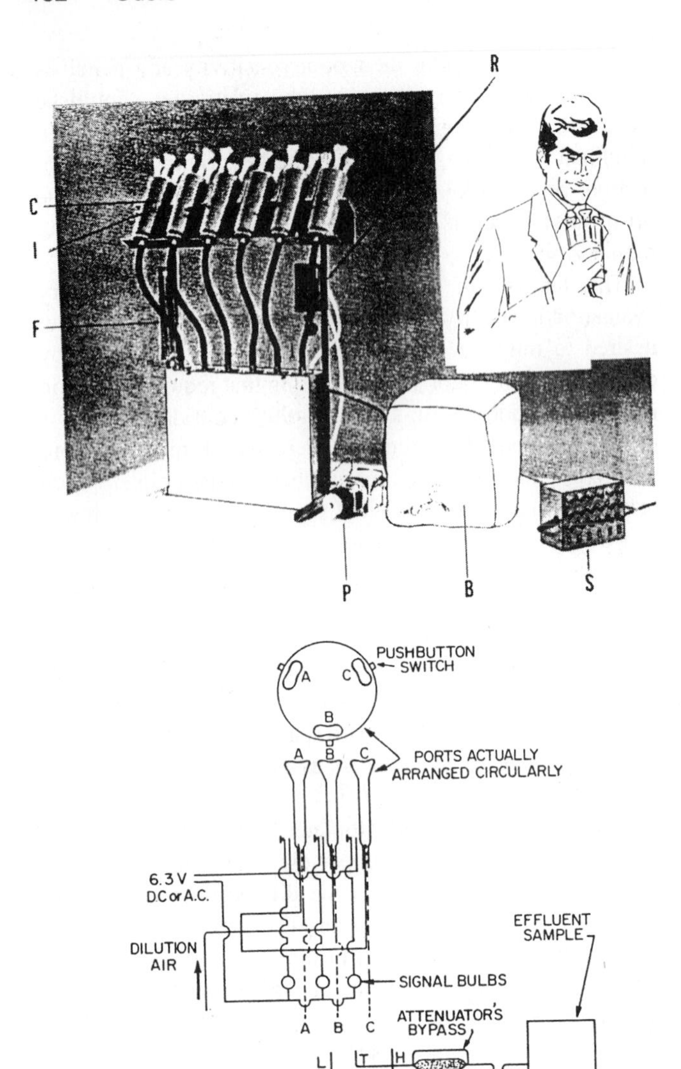

FIGURE 3. Dynamic Forced-Choice Triangle Olfactometer with Six Dilution Levels

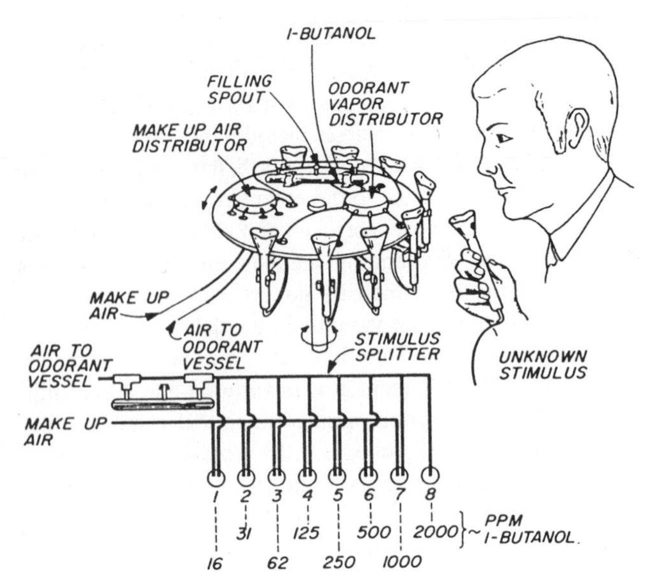

FIGURE 4. Dynamic Odor Intensity Reference Scale Based on 1-Butanol

being presented. Two to four repetitions of the odor dilution series are presented to the panel from the same bag sample. The odor threshold for each panelist is determined as being just below the concentration at which category 1 is followed by category 2.

Suprathreshold Measurements

Suprathreshold (odor intensity) measurements are conducted by the use of scales based on descriptive categories, on estimates of magnitude, and on matching intensity with a reference sample. In all cases, the measurement is performed as a sensory evaluation. Usually, an undiluted odorous sample obtained from ambient air is presented to panelists. The advantages of dynamic olfactometry, as compared with static methods, as described previously for dilution to threshold measurements, apply equally to suprathreshold determinations.

In the reference-sample approach, a series of different concentrations of a selected reference odorant are provided, and each odorous sample to be evaluated is compared with this series to locate the position in the series that most closely matches the odor intensity of the sample. This method has the advantage of avoiding semantic definitions and not requiring any special training.

The ASTM's Method E-544[15] provides an odor intensity reference scale based on 1-butanol (*n*-butyl alcohol). Two procedures are provided: a dynamic-scale method and a static-scale method. The dynamic method uses 1-butanol vapor as the odorant at eight concentrations in the range between approximately 16 and 2000 ppm (v/v) in air. These represent eight different odor intensities. The concentration increase from one intensity to the next is approximately twofold. Figure 4 illustrates the dynamic odor intensity reference scale based on 1-butanol.

The diluted vapor samples are supplied to the panelist as a steady flow from elliptical 20-by 35-mm glass sniffing ports at a rate of 160 mL/min. Panelists use these eight concentrations to match the perceived intensity of the unknown odorous sample, but they can also select positions of intermediate intensity, as well as positions below the weakest or above the strongest intensity.

In the butanol scale method, odor intensities are recorded in terms of concentration, parts per million (v/v), of 1-butanol vapor that has an odor intensity that matches the perceived intensity of the odorous sample. The value of the odor intensity of 200 ppm butanol is not considered to be twice as large as that of 100 ppm butanol. Thus butanol values do not provide information on the relative perceived intensities of odors. However, such information can be obtained from the parts-per-million values for the following experimentally derived equation.

$$I = 0.261(C)^{0.66}$$

where: I = perceived odor intensity of the sample that is matched against butanol
C = concentration in parts per million of 1-butanol (in air emitted at 160 mL/min) that matches the sample's odor intensity

The butanol odor intensity reference scale has the potential for use as a measurement of the odor intensity in the ambient air. Panelists could be used to match the perceived intensity of an ambient odor by comparison with the butanol reference scale. Panelists' judgments could be reinforced by referring to the butanol standards in between measurements of the ambient air for perceived odor intensity.

ODOR CONTROL METHODS

The methods available for controlling odor emissions from industrial sources include the following.

1. Condensation
2. Incineration
3. Wet scrubbing with chemical solutions
4. Activated-carbon adsorption
5. Biofiltration
6. Odor modification
7. Air dilution

Moisture-laden or organic vapor emissions should first be passed through a condenser that is capable of removing a high percentage of the water vapor or condensible volatile organic compounds (VOCs). This minimizes the odorous gas stream to be treated or allows recovery of a costly solvent, providing a more economical solution to solving an odor emission problem. Usually, an indirect condenser, such as a shell and tube or an air-cooled, finned-tube unit, is used to isolate the condensate from the cooling medium.

Thermal incineration of odor emissions is particularly effective at specific temperatures and residence times, normally resulting in the essentially complete removal of odor. The use of separate afterburners and incinerators, even with heat recovery, may be expensive, in terms of both initial investment and fuel operating cost. The use of a boiler firebox for steam generation often is an economical approach to incinerating odor emissions, particularly when the volumetric flow rate is not excessive. Catalytic incineration at lower temperatures is a possible method for controlling odors. However, its use is less prevalent than thermal incineration owing to the potential for fouling the catalyst.

Wet scrubbers with chemical solutions for the absorption of and reaction with odorous compounds are used extensively. Multistage scrubber systems are often used to treat process emissions, whereas single-stage scrubbers are used to treat plant ventilating air. Entrainment separators, venturi scrubbers, and spray-type scrubbers are normally used as the first stage in the multistage system to remove particulate and aerosols from gaseous streams so that the subsequent scrubber stages will be more effective in achieving odor removal. Usually, chemical oxidants, such as sodium hypochlorite and chlorine dioxide, are more effective scrubbing agents than the common bases and acids, such as caustic soda and sulfuric acid. However, the selection of the chemicals to use for scrubbing depends on the odorous compounds present in the source emission.

Activated-carbon adsorption columns are used for the treatment of odor emissions from sewage sources that release highly odorous compounds, such as hydrogen sulfide. Caustic-impregnated activated carbons are used that can be regenerated in-situ. Biofilters, including soil-bed systems, have become prominent in the control of odors from various industrial and municipal sources consisting of chemical (VOCs), wastewater treatment, composting, and rendering facilities.

Odor modification involves the addition of an odorous substance to a given odor emission in order to reduce the odor intensity or to change the odor character to one that is less objectionable. Masking agents change the character of an odor, but also increase its resultant intensity. Counteractants are chemicals that may change the character and also reduce the intensity of an odor emission. Vapor-phase reactions may also occur with certain chemical compounds that reduce the odor intensity by producing less odorous constituents, similar to reactions with chemical oxidants.

Dilution of an odor emission with atmospheric air is a possible method of odor control that may be successfully applied under special conditions of meteorology and surrounding topography. However, odor emissions from point sources are normally treated first by an acceptable control method before being discharged to the atmosphere. An important part of the overall design of an odor control system is related to the stack design to ensure that the height and exit velocity provide the desired dispersion characteristics.

For more details on condensation, incineration, scrubbing and adsorption, refer to Chapters 2 and 3. Also, a comprehensive discussion of odor control methods is available in a recent publication of the Air and Waste Management Association.[16]

References

1. A. Dravnieks, "Odor perception and odorous air pollution," *J. TAPPI*, 55:737–742 (1972).
2. *Odor Thresholds for Chemicals with Established Occupational Health Standards,* American Industrial Hygiene Association, Akron, OH, 1989.
3. J. E. Moore and R. S. O'Neill, "Odor as an aid to chemical safety: Odor thresholds compared with threshold limit values

and volatilities for 214 industrial chemicals in air and water dilution," *J. Appl. Toxicol.*, 3:6 (1983).
4. *Odor Control for Wastewater Facilities*, Manual of Practice No. 22, Water Pollution Control Federation, Washington, DC, 1979.
5. R. J. Sullivan, *Preliminary Air Pollution Survey of Odorous Compounds*, U.S. Department of Health, Education and Welfare Public Health Service, Raleigh, NC, October 1969.
6. A. Dravnieks, "Measurement methods," *Odors from Stationary and Mobile Sources*, National Academy of Sciences, Washington, DC, January 1979, Chapter 4.
7. A. Dravnieks and F. Jarke, "Odor threshold measurement by dynamic olfactometry: Significant operational variables," *J. APCA*, 30(12):1284 (1980).
8. A. Turk, E. D. Switala, and S. H. Thomas, "Suprathreshold odor measurements by dynamic olfactometry—principles and practice," *J. APCA*, 30(12):1289 (1980).
9. A. Turk, "Expressions of gaseous concentration and dilution ratios," *Atmospher. Environ.*, 7:967 (1973).
10. *Standard Test Method for Measurement of Odor in Atmospheres (Dilution Method)*, ASTM D-1391, American Society for Testing and Materials, Philadelphia, PA, 1978.
11. J. P. Wahl, R. A. Duffee and W. A. Marrone, *Evaluation of Odor Measurement Techniques: Vol. 1—Animal Rendering Industry*, EPA 650/2-74-008-a, U.S. Environmental Protection Agency, January 1974.
12. *Standard Practice for the Determination of Odor and Taste Threshold by the Forced-Choice Ascending Concentration Series Method of Limits*, ASTM E-679, American Society for Testing and Materials, Philadelphia, PA, 1979.
13. A. Dravnieks and W. H. Prokop, "Source emission odor measurement by a dynamic forced-choice triangle olfactometer," *J. APCA*, 25(1):28–35 (1975).
14. W. Hemeon, "Technique and apparatus for quantitative measurement of odor emissions," *J. APCA*, 18(3):166–170 (1968).
15. *Standard Recommended Practices for Referencing Suprathreshold Odor Intensity*, ASTM E-544, American Society for Testing and Materials, Philadelphia, PA, 1975.
16. *Transactions of AWMA International Specialty Conference on Recent Developments and Current Practices in Odor Regulations, Controls and Technology*, Air and Waste Management Association, Pittsburgh, PA, 1991.

6
Ancillary Equipment for Local Exhaust Ventilation Systems

H. D. Goodfellow, P. Eng., Ph.D.

FUNDAMENTALS

Introduction

Ventilation is one of the most important techniques available to control contaminant levels in the workplace. As will be discussed later, the control of a potentially hazardous contaminant can be achieved in two ways: by dilution of the concentration of the contaminant before it reaches the workers' breathing zone by mixing it with uncontaminated air (called industrial ventilation or dilution ventilation) or by the removal of the contaminant at or near its source or point of generation to prevent its release into the workplace environment. This latter approach is called local exhaust ventilation (LEV) and its elements, including design and the associated ancillary equipment, are the focus here.

Ventilation systems can be designed for commercial/residential or industrial applications. The design of ventilation systems for the residential/commercial fields is well presented in numerous publications of the American Society of Heating, Refrigerating, and Air Conditioning Engineers.[1] These references also contain valuable information on associated ancillary equipment (e.g., ducts, fans, heating/cooling units) that is of the same generic type as for the industrial ventilation systems. Industrial ventilation systems, however, require unique and specific design features for the ancillary equipment that must be taken into account in order for the design to be successful. The technology for the design and operation of industrial ventilation systems is not as well developed, but significant progress is being made.

A simple definition of ventilation is "the control of the environment with airflow." Ventilation systems must be designed as an integral part of the process. This approach, called the "systems approach," is essential to ensure a successful design of the industrial ventilation system. A textbook entitled *Advanced Design of Ventilation Systems for Contaminant Control*[2] outlines the steps required to implement the systems approach to solving ventilation problems for the processing and manufacturing industries.

The application areas for industrial ventilation systems (both general ventilation and LEV) are as follows:

1. Control of contaminants to acceptable levels
2. Control of heat and humidity for comfort
3. Prevention of fires and explosions

For industrial applications, all of these factors are important considerations at the design stage. In general, the controlling design criterion will be to achieve acceptable contaminant levels in the workplace. The following are four steps to be considered in solving industrial ventilation problems.

1. *Process modifications*. This should always be the first approach to establish whether the process can be modified and the contaminant can be controlled to an acceptable level. For example, a processing plant handling silica using belt conveyors may solve a silica contaminant exposure problem by replacing the belt conveyor with a totally enclosed conveyor. The result of this process modification is that there is no need for a ventilation system.
2. *Local exhaust ventilation*. With the exception of process modifications, LEV systems will almost always be the most cost-effective technology for controlling contaminant levels in the workplace. Contaminants are controlled at the source with low volumes of gas.
3. *Process building or general work area* (i.e., general or dilution ventilation). The concept of general or dilution ventilation is based on supplying or exhausting large volumes of air to an occupied space or the complete processing/manufacturing complex. The ventilation air can be supplied and exhausted by natural ventilation or mechanical ventilation. A subsequent section will de-

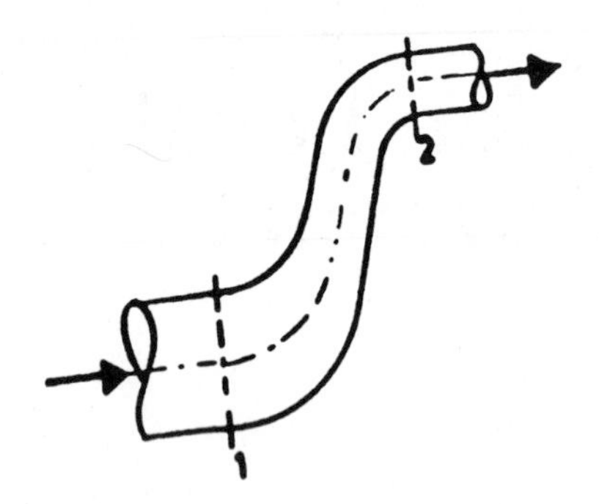

FIGURE 1. A One-Dimensional Fluid System

scribe the features of general ventilation for processing and manufacturing facilities.

4. *Personal protective equipment*. For some specific applications, it may be necessary to use personal protective equipment to reduce the exposure for workers. In general, this is only applicable if process modification or ventilation is not feasible. Personal protective equipment can be considered for nonroutine maintenance work or for processes that do not have feasible engineering control solutions.

Conservation of Mass

A basic principle for airflow consideration is the continuity equation or the conservation of mass. This general physical law states that the mass, m, of a particle of the system remains constant with time, t, which means that the system equation can be written: the rate of change of mass (m) with time (t) must be zero.

$$\frac{dm}{dt} = 0$$

For industrial ventilation applications, gas flow in the ductwork can be treated as one dimensional if the proper averages are used for the variables across the section. A simplified equation can be written for any two points in a fluid system as shown in Figure 1 based on one-dimensional compressible flow with only one entrance and one exit in the control surface. This equation is as follows:

$$\rho_1 A_1 V_1 = \rho_2 A_2 V_2$$

where: ρ = specific density of fluid, lb/ft^3
A = cross-section area of duct, ft^2
V = average gas velocity, fpm

For most industrial ventilation applications, the absolute pressure remains relatively constant and hence the specific density of the gas remains relatively incompressible.

For this case, $\rho_1 = \rho_2$ = constant and the above equation becomes:

$$V_1 A_1 = V_2 A_2$$

or

$$Q_1 = Q_2$$

where: $Q = AV$ = volumetric flow rate, ft^3/min

The above equations must be modified if there is more than one inlet area, A_1, or outlet area, A_2.

Conservation of Energy

Bernoulli developed a general energy equation that relates kinetic energy, flow energy, and potential energy. This general equation can be simplified for the case of a steady flow of a frictionless, incompressible fluid along a streamline to give:

$$\frac{V^2}{2g_c} + \frac{P}{\delta} + Z = \text{constant (different for each streamline)}$$

where: V = velocity, fps
$2g_c$ = gravitational constant, ft/s^2
P = absolute pressure, lb/ft^2
δ = specific density, lb/ft^3
Z = elevation above an arbitrary datum point, feet

Each of the above terms represents a manometric height of fluid (e.g., feet of water).

By taking the above equation and applying it to two points on a streamline and rearranging, the equation becomes:

$$\frac{V_2^2 - V_1^2}{2g_c} + \frac{P_2 - P_1}{\delta} + (Z_2 - Z_1) = 0$$

The equation shows that it is the differences in kinetic energy, flow energy, and potential energy that are significant. For most industrial ventilation applications, the change in potential energy is usually relatively small and is ignored. Also, the change in pressure is only a small percent of the absolute pressure and the gas may be considered incompressible with an average specific weight. In real fluids, the common approach is to add a term to the Bernoulli equation to account for friction losses. For points 1 and 2 in a system, the equation for flow from point 1 to point 2 can be written:

$$\frac{V_1^2}{2g_c} + \frac{P_1}{\delta} + Z_1 = \frac{V_2^2}{2g_c} + \frac{P_2}{\delta} + Z_2 + \text{losses}_{(1-2)}$$

Pressures (Total, Static, Velocity)

Total pressure represents the head required to start and maintain flow for air moving through the duct. Total pres-

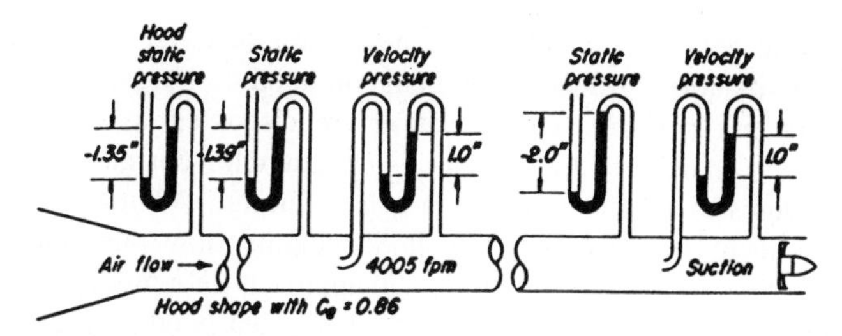

FIGURE 2. Typical Duct under Positive Pressure with a Velocity of 4005 fpm

sure is the algebraic sum of the static and velocity pressure (always positive) terms and is given by:

TP	=	SP	+	VP
Total pressure		Static pressure		Velocity pressure

Static pressure sometimes called "frictional" or "resistance" pressure, can be positive or negative and is the potential pressure exerted in all directions by fluid at rest that tends to expand or compress the fluid. Static pressure exists even if there is no air motion. It produces the initial air velocity and overcomes the frictional resistance of the air against the duct surface, as well as the resistance resulting from any flow restrictions. For most ventilation systems, static pressure is negative in sign upstream from the fan and positive in sign downstream from the fan.

Velocity pressure is the kinetic pressure in the direction of flow necessary to cause a fluid at rest to flow at a given velocity. It exists only when air is in motion, acts in the direction of the air motion, and is always positive. Under certain conditions of fan operations, it is possible for either the velocity or static pressure to be equal to zero, but both can never equal zero at the same time. The relationship between the velocity of air and air pressure can be determined from the general Bernoulli equation and is as follows:

$$V = \sqrt{2g_c h}$$

where: V = velocity, fps
g_c = gravitational constant, 32.12 ft/s^2
h = head of air, feet

For standard dry air at sea level (air density is 0.075 lb/ft^3), velocity can be determined using:

$$V = 4005\sqrt{VP}$$

where: V = velocity, fpm
VP = velocity pressure, in. w.g.

Figure 2 shows the relationship among total pressures, static pressures, and velocity pressures for an open duct under both negative and positive pressure with a velocity of 3000 fpm.

If the velocity through the duct system increases, a part of the available static pressure is converted to additional velocity pressure to accelerate the fluid flowing through the duct. Conversely, if the velocity through the duct system is decreased at some point, a portion of the velocity pressure is converted to static or potential pressure. This conversion is always accompanied by a net loss of total pressure as a result of turbulence and friction losses. For any ducted system, it is important to be able to plot graphically pressures (TP, VP, SP) as a function of position for the ducted system. Figure 3 is a graphical plot of TP, VP, SP for a typical ducted system.

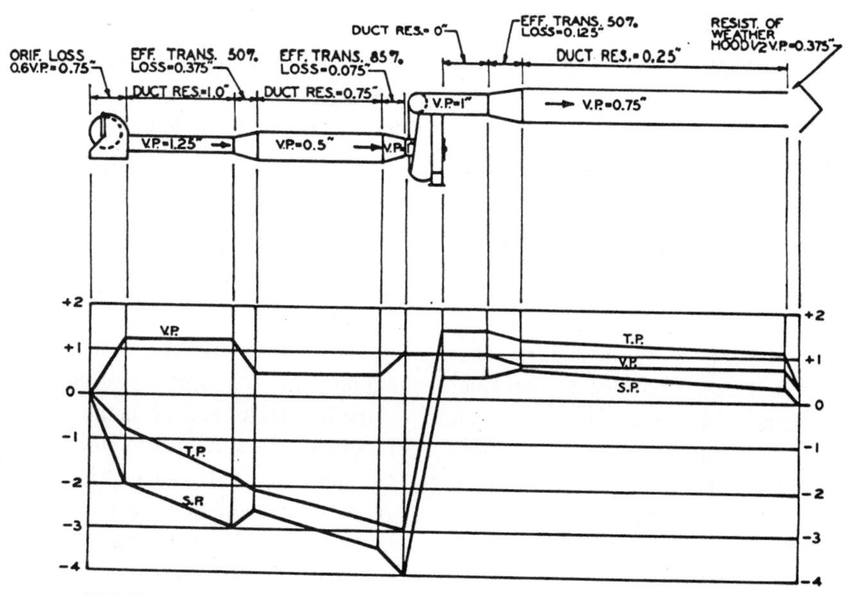

FIGURE 3. Plot of TP, VP, and SP for a Typical Ducted System

Friction and Dynamic Losses

Energy losses (friction plus dynamic) always occur for the flow of gases in industrial ventilation systems. An ideal fluid having zero viscosity and flowing in a straight horizontal pipe of constant diameter would have the same pressures all along the duct. Friction losses occur at the point of movement between the fluid particles and at the boundary layer adjacent to the pipe wall. Dynamic losses occur as a result of turbulence caused by a change in direction or velocity within a duct, such as at a hood entry, elbows, or fittings. The design of a ventilation system requires the establishment of the total energy losses (friction plus dynamic) through the ductwork system for the design flow rate. As will be shown later, the fan static pressure and flow characteristics must be matched to the ductwork system static pressure and flow-rate characteristics. The design flow for the ductwork is established based on the process or equipment requirements and the required level of performance for the ventilation system. Once a system flow rate has been established, a standard design procedure can be developed to determine the friction and static pressures for the ductwork system.

Frictional loss in a duct varies directly as the length, inversely as the diameter, and directly as the square of the velocity of air flowing through the duct. The loss can be calculated from a Moody diagram as shown in Figure 4, which relates friction factor to Reynolds number ($Re = vd/\nu$, where Re = Reynolds number, v = velocity in fps, d = pipe diameter in feet, and ν = kinematic viscosity in ft^2/sec), duct material, and type of construction.

For calculation purposes, Loeffler[3] developed a simplified relationship that calculates friction factors over the range of likely operating conditions for round, galvanized metal ducts.

$$H_f = 0.0307 \frac{V^{0.533}}{Q^{0.612}}$$

$$= \frac{0.4937}{Q^{0.079} D^{1.066}}$$

Standard ventilation references can also be consulted and friction loss values read directly from charts. For gas conditions that are different from standard air, conversion tables exist that allow corrections for relative roughness, gas temperatures, and effects of elevation or gas density.

Dynamic losses occur because of turbulence as a result of a change in velocity or direction within a duct. The resulting pressure drop in a duct system due to dynamic losses increases with the number of elbows and angles and the number of velocity changes within the system. Dynamic losses occur as a gas enters a hood or passes through fittings, such as elbows, expansions and contractions, branch entries, and duct exits. Energy losses at hoods and duct fittings depend on different parameters and the design procedures and equations will be discussed separately.

Hoods are shaped inlets designed to capture contaminated air and conduct it to the exhaust duct system. The classification of hoods and details on hood design parameters and performance evaluation are discussed in detail under "Ancillary Equipment." For example, design procedures to calculate minimum exhaust design volumes are outlined. Based on a given hood volume flow rate, the designer must establish the design approach for a minimum entry loss (dynamic losses) for the selected hood.

For flow into a hood, the cross-sectional area of the flow begins to contract (vena contracta) before expanding again to fill the duct. The energy loss at the vena contracta of the hood is defined as the coefficient of entry (C_e), which is the ratio of the actual flow to the theoretical flow, if there had been no vena contracta. For an ideal hood with no loss, C_e is 1.0.

The hood static pressure (SP_h), measured a short distance downstream from the hood, is a direct measurement of the energy (VP) required to accelerate the fluid from rest to the duct velocity plus the hood entry turbulence losses (h_e), which are a function of the hood shape. The entry loss plus the acceleration energy required to move the air at a given velocity (1 VP) make up the hood static pressure. The hood entry losses can be expressed as a number of velocity pressures using a hood pressure loss factor (K_h).

The SP_h, which can be measured directly at a short distance downstream from the hood entrance, is expressed algebraically as follows:

$$SP_h = (1 + K_h)\ VP$$

The coefficient of entry (C_E) is related to the velocity pressure loss factor K_h by the following equation.

$$K_h = \left(\frac{1}{C_E 2} - 1\right)$$

For compound hoods, it is possible to add pressure losses for each hood. Novel or complex hoods may require experimental tests in a laboratory to establish hood entry losses. Figure 5 provides pressure loss factors for typical hoods.

Fittings in a ducted system will result in additional static pressure losses because of the turbulence produced. The types of fittings include bends, expansions, contractions, dividing flows, combining flows, sucking flows, manifolds, and so on. For design purposes, their losses are expressed as a number of velocity pressures (VP), where K is the loss factor and VP is the velocity pressure in the duct. Figure 6 gives K factors for many common fittings for circular ducting. In other fittings and ducting of rectangular cross section, the necessary design calculations and correction factors can be obtained from the literature.[2,4]

The total static pressure loss for a system is calculated by summing up the friction and dynamic losses. An equation can be written as follows:

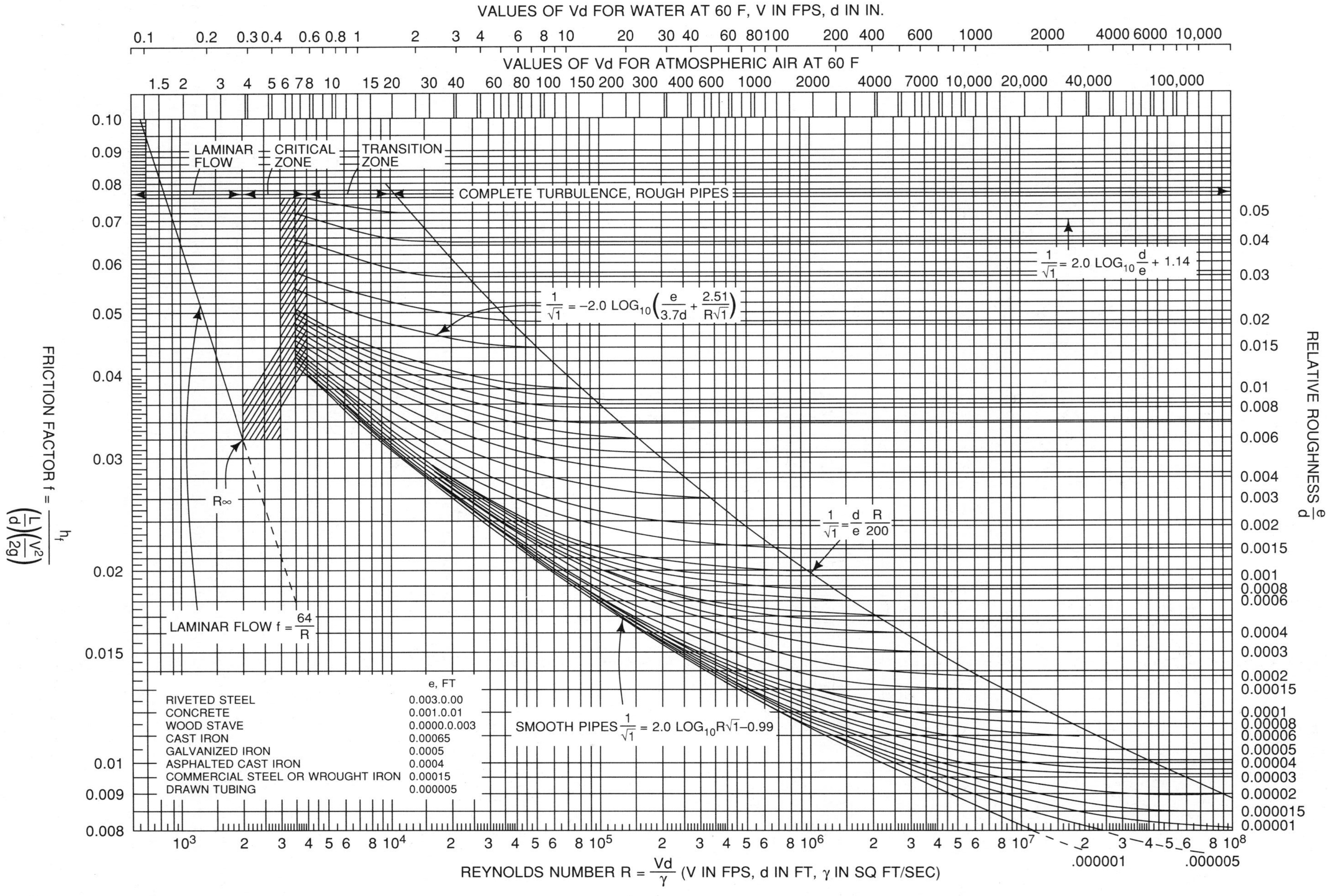

FIGURE 4. Moody Diagram Relating Friction Factor to Reynolds Number

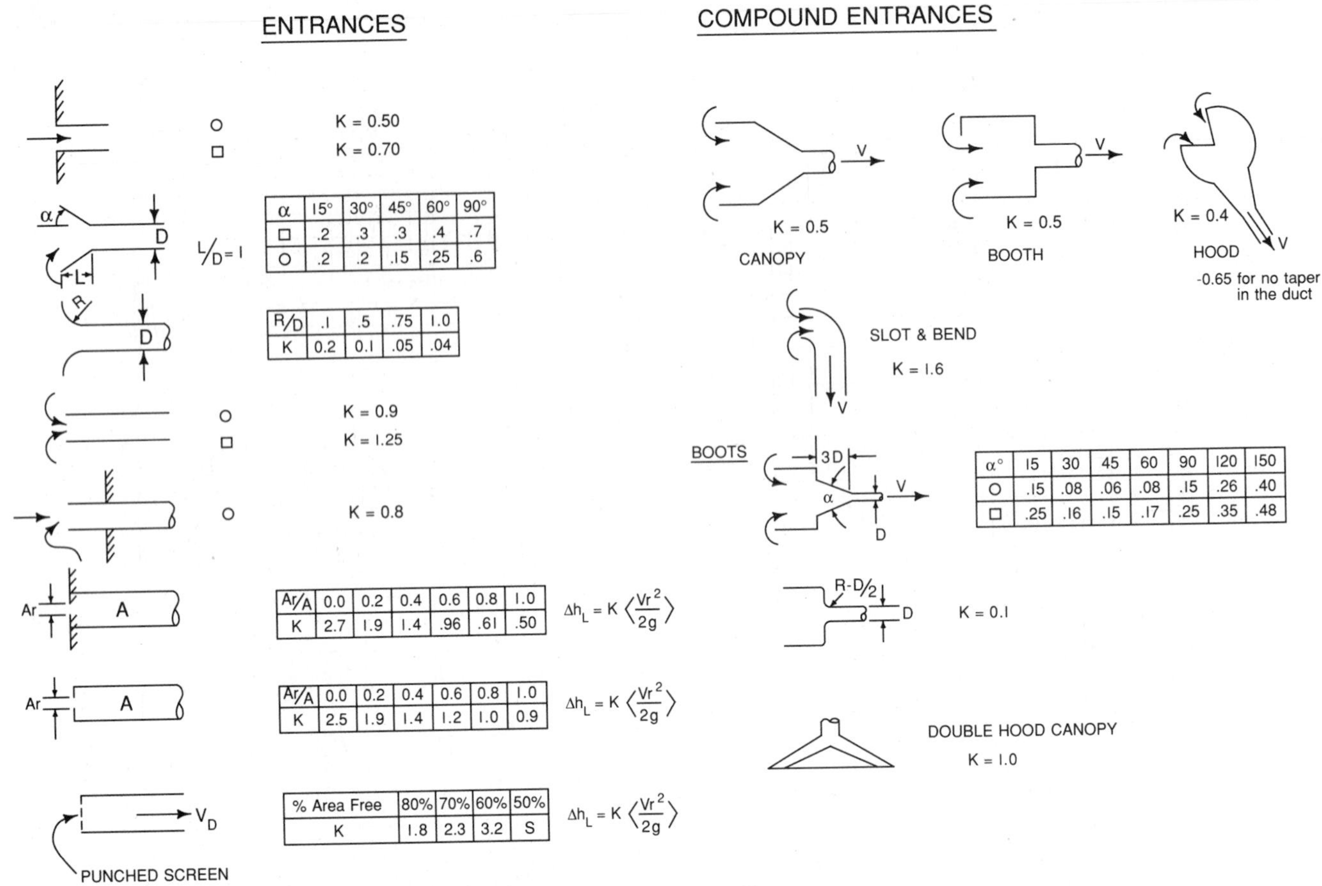

FIGURE 5. Pressure Loss Factors for Typical Hoods (From Reference 2)

$$
\begin{aligned}
\text{Total energy losses} &= \text{friction losses} + \text{dynamic losses} \\
&= (\text{straight duct} + \text{hood} + \text{fittings})\ \text{losses} \\
&= [f(L/D) + (1 + K_h) + \Sigma K]\ VP
\end{aligned}
$$

This general equation is used to calculate the total static pressure losses for an LEV system. This static pressure loss can be combined with the velocity pressure term to give the total fan pressure required. The details of the static pressure loss calculation procedures are presented in a subsequent section.

Measurement Techniques

There are numerous instruments and techniques that can be used to measure flows and pressures for a ventilation system. Detailed descriptions of the different instruments and their applications are available in standard references. An instrument that has wider applications for the measurement of pressures and temperatures is a pitot tube. A brief description of the pitot tube, and its operation and applications, follows.

The pitot tube with an inclined manometer is the standard air velocity meter. It is of rugged design, relatively low in cost, and is available in many different sizes for different duct diameters. Figure 7 shows the proportional dimensions for a standard pitot tube. No calibration curves are required if the tube is carefully made and the velocity pressure readings obtained are considered accurate. Standard pitot tubes are made of Type 304 stainless steel and can be used at gas temperatures up to 1000°F. Above this temperature, special alloy steels or water-cooled pitot tubes are required.

The pitot tube consists of two concentric tubes. The opening of the inner tube is axial to the flow and measures total pressure; the opening of the outer tube is circumferential and measures static pressure. The velocity pressure is the difference between the total pressure and static pressure. Velocity pressure measurements are made with the pitot tube by connecting the static and total pressure tops as shown in Figure 8. By disconnecting one of the hoses, it is possible to perform static or total pressure measurements for a system.

Before taking any measurements with a manometer, the instrument must be calibrated to ensure that the manometer fluid and scales are correct. A simple field check is to use a U-tube or plastic tubing filled with water and the scale reads directly in the pressure units of in. w.g. In general, airflow is not uniform throughout the cross section of the duct. Traverses are used to obtain the average by measuring the velocity pressures at points in a number of equal areas in the cross section, as shown in Figure 9.

To minimize flow distribution measurement problems, it

EXPLANATION

The system resistance is determined by adding the total pressure drops across successive sections of the system of airways conveniently classified under the headings 1-6. This equals the total fan pressure required. Each item of loss, P, is given in terms of the velocity pressure, P_V, and the factor K given in the appropriate section.

$$P = K \times P_V \text{ in. w.g.}$$

P_V is calculated from the average velocity V at the section shown by the arrow in the diagrams, using the formula:

$$P_V = \left(\frac{V}{3970}\right)^3 \times \left(\frac{\text{Actual gas density}}{\text{Standard air density}}\right)$$

$$\text{Total fan pressure} = P = P_1 + P_2 + P_3 + P_4 + P_5 + P_6$$

SYMBOLS

L = Length of duct in feet
D = Diameter of duct in feet
P = Drop in total pressure in w.g.
P_V = Velocity pressure, in w.g.
V = Average velocity, ft./min.
P = Gas density, lb. per cu. ft.
P_2 = 0.0764 lb. per cu. ft.
K = Factor defined in each section

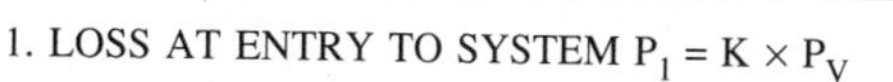

1. LOSS AT ENTRY TO SYSTEM $P_1 = K \times P_V$

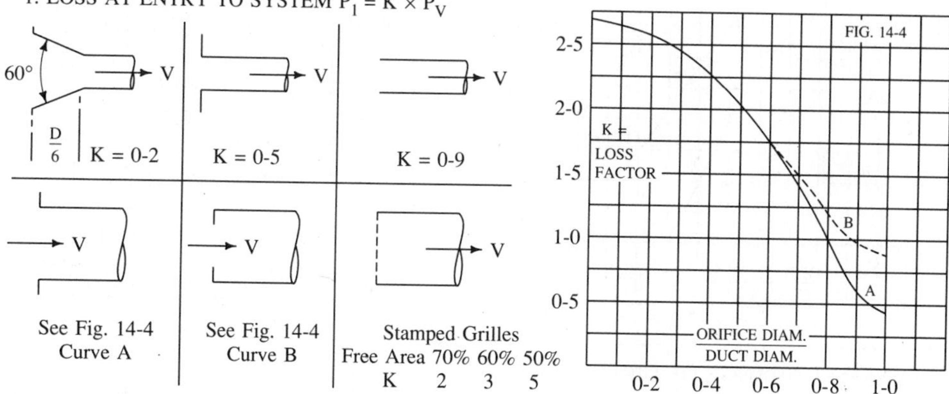

2. LOSSES ALONG STRAIGHT DUCTS $P_2 = K \times P_V$

Approximate loss formulas adequate for short ducts:

Circular Section

$K = 0\text{-}02 \times \frac{L}{D}$

Rectangular Section, A × B ft.

$K = 0\text{-}01 \frac{A + B}{AB} \times L$

Other Sections

$K = 0\text{-}005 \frac{\text{Perimeter ft.}}{\text{Area sq. ft.}} \times L$

Chart based on Friction Factor method for use with sheet metal ducts. Fig. 13-4 is for circular ducts. If duct is rectangular convert to circular equiv. using Table 13-1. Based on e = 5

Friction Factor Method:
For maximum accuracy with long rough walled circular ducts:

$$K = f \times \frac{L}{D}$$

Determine f from Fig. 14-5 estimating surface roughness e (in units of 0-001 in.) using table below as guide:

Typical Surface	e
Drawn Tube	0-1
Rolled Sheet	1
Galvanised Sheet	5
Cement Facing	20
Cast Concrete	80
Heavily riveted	200
Rough Brickwork	500

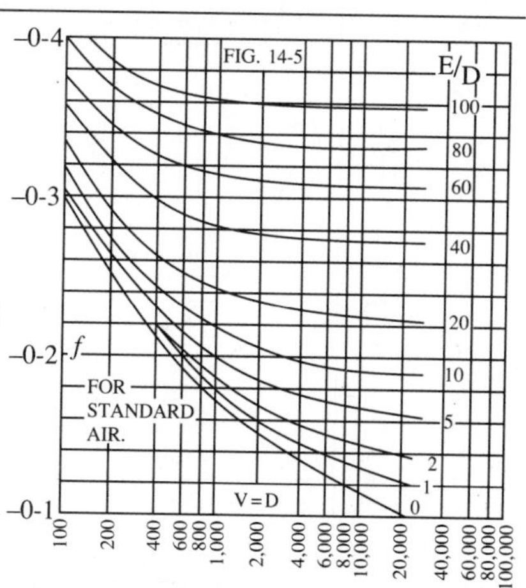

3. LOSSES AT HEATERS, FILTERS, etc. $P_4 = K \times P_V$

Obtain drop in TOTAL PRESSURE across component at required volume rate from manufacturer's date. This equals the drop in STATIC PRESSURE if inlet and outlet areas are equal. Fig 14-6 gives approximate values of K for banks of small gilled tubes. P_V is based on average FACE velocity For banks of plain tubes use formula:

$K = Cn$ where n = number of rows

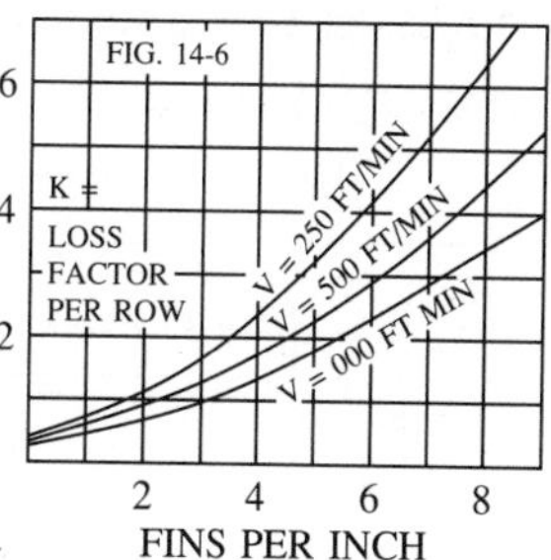

Arrangements of Tubes: d = diameter of tube

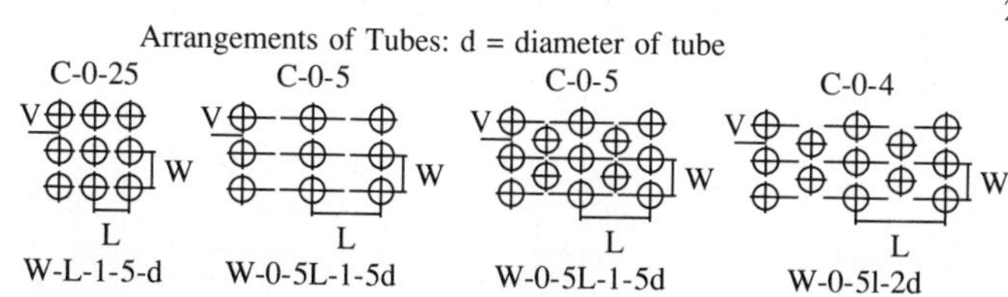

FIGURE 6. K Factors for Common Fittings for Circular Ducting

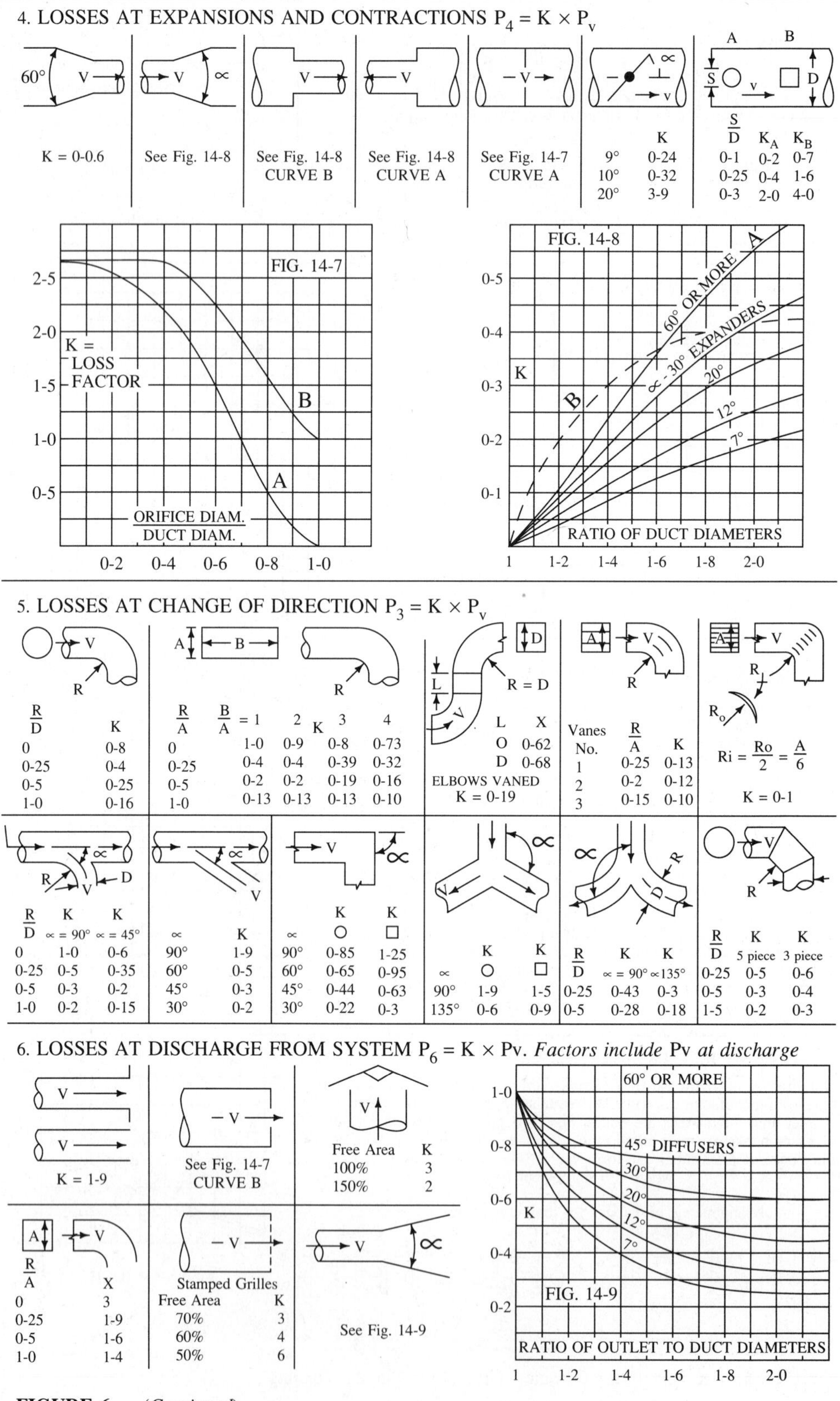

FIGURE 6. (*Continued*)

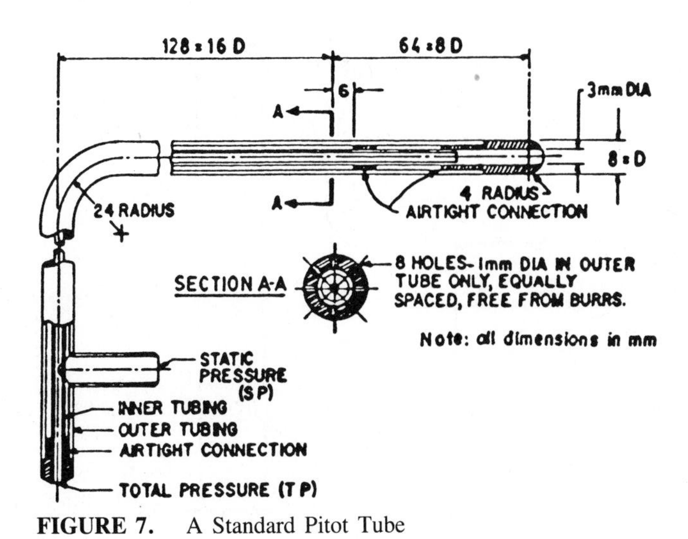

FIGURE 7. A Standard Pitot Tube

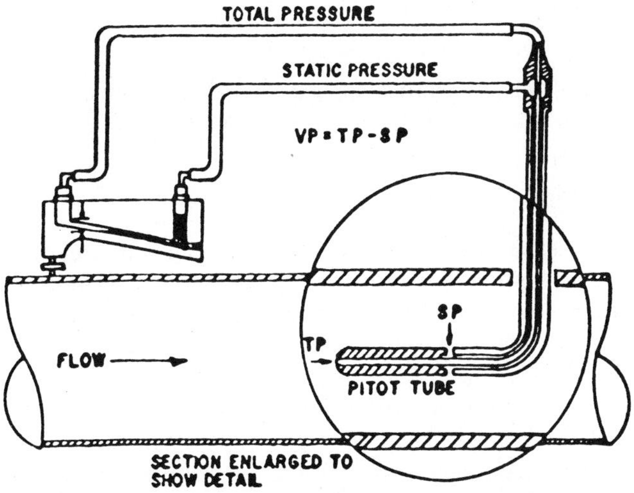

FIGURE 8. Pitot Tube Connected to an Inclined Manometer

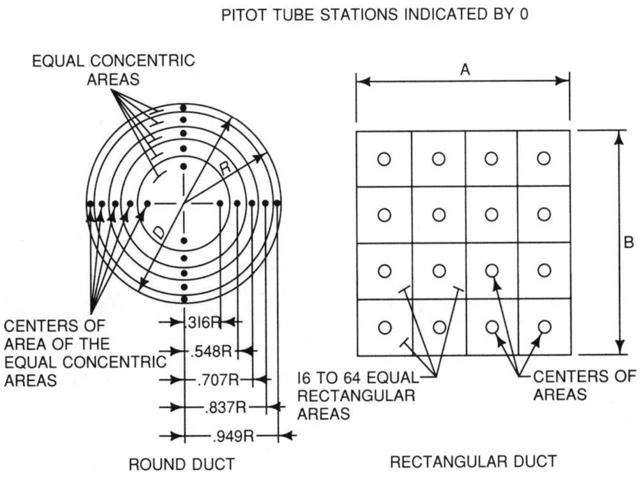

FIGURE 9. Traverse of a Round and Rectangular Duct Area

is recommended that measurements be taken at least 10 diameters downstream and two diameters upstream from any obstructions or flow disturbances.

The design equation developed for a pitot tube from the general Bernoulli equation is as follows:

$$V = 1096 \sqrt{\frac{VP}{(0.075)\ (d_f)}}$$

where: V = velocity, fpm

VP = velocity pressure, in. w.g.

d_f = density factor

$$= \frac{530}{460 + t} \times \frac{B}{29.92} \text{ (dry air)}$$

where: t = air temperature, °F

B = barometric pressure, in. Hg

When the moisture content in the gas is above 0.02 lb/lb of dry air, the density factor should be obtained directly from psychrometric charts. For dry air at standard conditions (70°F and 29.92 in. Hg), the density factor becomes unity and the equation becomes:

$$V = 4005 \sqrt{VP} \text{ (fpm)}$$

To calculate the air volume flow rate from pitot tube readings, it is necessary to calculate velocities for each measured point and then to calculate average velocities. The measured air volume at the temperature in the duct is, then, the average velocity measured by the cross-sectional area of the duct.

LOCAL EXHAUST VENTILATION

Elements of LEV Systems

Figure 10 is a schematic of a typical LEV system used for the capture of a contaminant from an emission source. The major components of an LEV system are the hood, duct, gas cooling, fan, pollution control equipment, and stacks. Chapter 2 describes the pollution control equipment, used for gases (absorption, adsorption, condensation and incineration). Chapter 3 describes the pollution control equipment used for particulates (mechanical collectors, scrubbers, electrostatic precipitators, and fabric filters). Ancillary equipment described here will focus on the other components of an LEV system.

The term "hood" is applied to a shaped inlet that is designed to capture the contaminant from an emission source and conduct it into the exhaust duct system. Hoods can be of any shape or size. A subsequent section will describe a hood classification system, outline hood design methodologies, show schematics of different types of hoods within the groups, and present typical equations and criteria

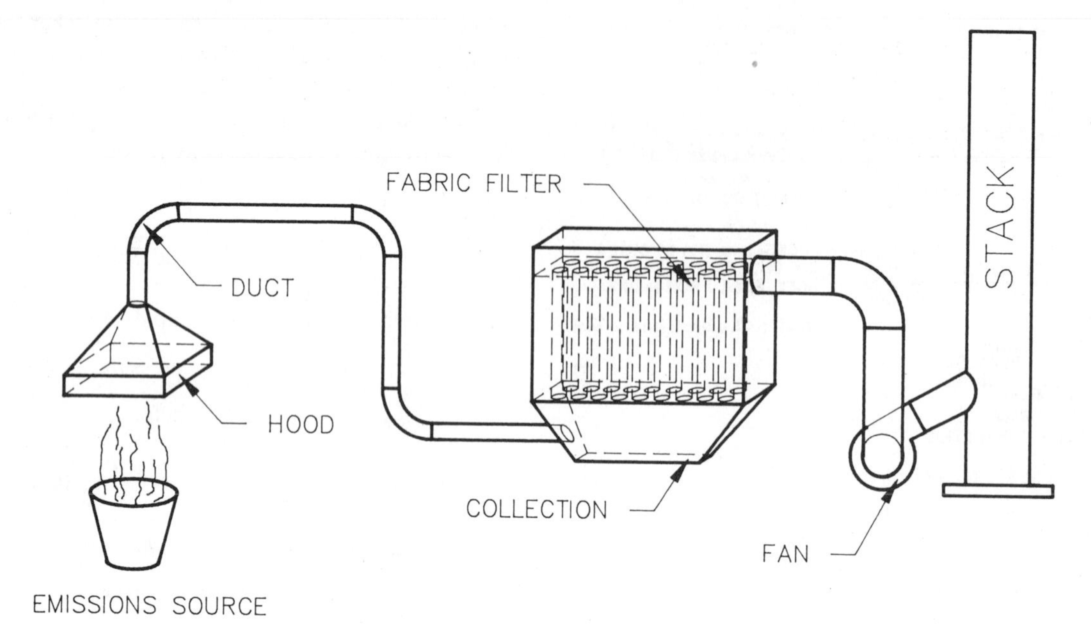

FIGURE 10. Schematic of a Typical LEV System

for hood design. Properly designed hoods should meet the design criteria of maximum efficiency of collection of the contaminant with minimal exhaust volumes and minimal interference with the process and the operators.

The contaminated air is exhausted through the ductwork, which provides a channel for the flow that moves the contaminant from the emission source to be discharged into the atmosphere at the stack. Ducted systems are typically classified as dust control, fume control, and mist/gaseous control. Although these ducted systems have many common features, a successful operation depends on a proper design by an experienced engineer who will incorporate many unique and detailed design features into the specific ducted system. A technical description and design parameters for ducted systems are provided in a later section.

Gas cooling technology applicable to LEV systems includes water-cooled ducting, radiation, forced-draft, evaporative, and air dilution. Each of these gas cooling technologies will be discussed later.

As one of the most important elements to ensure successful operation of an LEV system, fans provide the energy to move the air containing the contaminant through the system. A successful operation requires a fan to deliver the required volumetric flow rate in order to ensure capture of the contaminant at the source and ultimately to discharge the cleaned gas up the stack and into the surrounding atmosphere. A technical description of fans and the pertinent fan law equations for design are presented later in this discussion.

Stacks for LEV systems provide an elevated discharge point for the cleaned gas. The actual height of a stack is determined based on keeping contaminants within a specific contaminant concentration level at the point of impingement. A following section outlines a technical description of stacks and identifies applicable design equations to establish stack diameters and stack heights.

Design Parameters

Local exhaust ventilation is one of the standard "engineering methods of control" used to control contaminant levels in the workplace. Many manufacturing and processing facilities generate contaminants that would present worker exposure problem if the contaminants were not captured. Table 1 identifies the five major types of ducted air systems and shows typical hood velocities, duct velocities, and dust loadings. As also shown in Table 1, design parameters for LEV systems fall in between those for conventional heating, ventilation, and air conditioning (HVAC) systems and pneumatic conveying systems.

TABLE 1. Ducted Air System Categories

	Typical Operating Parameters		
System	Hood Velocities (fpm)	Duct Velocities (fpm)	Dust Loading (grains/ft^3)
1. Heating and ventilation		2000	Insignificant
2. Local exhaust ventilation			
Vapor control	2000	2500	N/A
Fume control	2000–4000	2000–4000	0.05–0.5
3. Local exhaust ventilation			
Dust control	250	4000	5
4. Dedusting or classifying	1250	5000	100
5. Pneumatic conveying		7000	500

Factors to be considered in the design of LEV systems are as follows:

1. Capture of the contaminant at the source (if possible).
2. Provision of an adequate but not excessive capture velocity.
3. Incorporation of containment whenever possible.
4. Use of gravitational force in hood design.
5. Provision of adequate conveying velocity for the contaminant.
6. Use of a design that minimizes pressure losses.
7. Design of the fan so as to accommodate all operating modes.
8. Rigorous application of the systems approach.

The LEV systems have many advantages over general ventilation, some of which are as follows.

1. Level of control of contaminant can be very high (more significant as allowable contaminant levels decrease).
2. Volume of exhaust air is much lower (more economical design).
3. Contaminant level in exhaust gas is more concentrated (more economical design for pollution control equipment).
4. Performance is not grossly affected by cross-drafts.
5. Housekeeping is improved because large particles are collected.
6. Auxiliary equipment can be protected from heat, dust, corrosion, and so on.

The design parameters for LEV systems that are important are the size and shape of the hood; capture velocities and exhaust flow rates for the hood; materials of construction and diameter of ducts; type and size of gas cooling equipment; type of fan and parameters, such as fan speed, design gas flow, and system static pressure; type of pollution control equipment; and diameter and height of stack. Details on the technical specifications and design parameters will be found under "Ancillary Equipment."

Preparation of Technical Specifications

The LEV system is an integral part of the processing/manufacturing facility, and it is essential that the system designer develop the proper design parameters and technical specifications. Goodfellow[2] has described in detail the essential steps and provides a checklist for the preparation of better technical specifications. Typical standard technical specifications for all the major ventilation system components are presented. Components that are included are louvers, roof ventilators, ducted systems (including accessories), gas cooling, fans, gas cleaning, stacks, and dust-disposal systems. The technical specifications should be considered as guidelines or checklists that will assist in the preparation of equipment specifications for particular projects. For successful projects, technical specifications should be prepared only by well-qualified and experienced engineers.

Before preparing detailed technical specifications, a decision must be made as to whether the LEV system will be an "engineered system" or a "turnkey system." The current trend is to purchase a conventional LEV system on a turnkey basis, but any LEV system of a specialized nature will be purchased as an engineered system. The major advantages of an engineered system are that the client can select the best components based on competitive bidding, experienced process specialists can be employed, complex layouts can be integrated into the total project, and major capital expenditures can be deferred. It would appear, then, that emerging technology in the ventilation field will continue to favor the purchase of engineered systems for LEV.

The preparation of technical specifications must focus on communication, criteria, and conciseness. The specification to the bidders must communicate the user's requirement or criteria with clarity and brevity. A checklist of important steps to prepare better technical specifications would include the selection of a qualified engineer, a clear definition of the scope of work, a defined performance requirement, the supplying of all pertinent process and test data, and the attachment of a technical questionnaire. A well-prepared specification will help to ensure that the project will be successful and cost effective.

VENTILATION FOR PROCESSING/MANUFACTURING FACILITIES

Industrial Ventilation

Most industrial ventilation problems for processing and/or manufacturing facilities are complex. The development of cost-effective solutions requires an experienced and qualified ventilation engineer. The design of a ventilation system must be incorporated into the plant design and layout at the earliest conceptual stage of the project. Decisions must be made on the level and requirements for general ventilation, dilution ventilation, and/or LEV. The ventilation engineer must work closely with the project design team at all stages of the project since the ventilation system can have a major impact on the type of process to be used, the plot plan, the building profile, and equipment layout within the plant. Cost-effective ventilation systems can only be developed if the ventilation system design is an integral part of the project planning and design activities from the preliminary conceptual stage.

In many applications, the ideal ventilation scheme may be in conflict with other environmental design criteria, such as noise control and fume control. For example, a building designed for hot processes may be based on the concept of a wide-open building to enhance natural ventilation; this is in

TABLE 2. Design Steps for Ventilation Systems for Greenfield Processing/Manufacturing Facilities

Design Step	Description of Activities
1. Obtain published ventilation design	Literature search to obtain published ventilation design data for similar processing operations
2. Obtain pertinent data on-site condition	Obtain local meteorological data (minimum of three to five years on a monthly basis). Data to include temperature, relative humidity, precipitation, and wind speed and direction
3. Develop a catalog of major emission	For each emission source, establish type of contaminant, gas composition, gas temperature, chemical composition of particulate, particle size of particulate, and dust loading
4. Develop ventilation design objectives	Review list of potential contaminants and prioritize Design must comply with applicable occupational health and safety regulations Client to review and approve design objectives
5. Develop a schedule of building openings	Prepare an isometric drawing Tabulate schedule of building openings
6. Perform analytical work	Heat release calculations Air set in motion Heat and flow balances Preliminary sizing of ventilation equipment
7. Define the design basis for the ventilation system	Use design objectives from above Prepare specific design bases for the ventilation system
8. Develop alternative ventilation system concepts	Preliminary layouts and flow sheets Performance evaluation using computer or fluid dynamic modelling
9. Perform a technical/economic evaluation of the alternatives	Detailed analysis of alternatives Develop a decision matrix (cost/benefits)
10. Prepare an engineering report with recommendations	Detailed description of systems/design calculations Recommended ventilation system
11. Management approval of ventilation systems	Approval of recommended ventilation system
12. Detailed design	Prepare all necessary technical specification for major equipment Detailed design and drawings
13. Installation	Ventilation equipment to be installed in compliance with all detailed drawings, specifications, etc.
14. Start-up and testing	Commissioning in compliance with start-up and operating procedure manuals Testing of ventilation system Documentation of system performance

conflict with noise control considerations that require the plant to be enclosed as much as possible to reduce neighborhood noise. Another example of this conflict is the use of canopies for remote fume capture. Canopies perform best if the building is totally enclosed to prevent cross-drafts from disturbing the rising plume. Totally enclosed buildings reduce the effectiveness of dilution ventilation.

The methodology for the design of a ventilation system for a new facility can be based on implementing a checklist of activities as summarized in Table 2.

Ventilation Design Equations

The design calculations for industrial ventilation systems for processing or manufacturing buildings are based on:

1. Heat release calculations
2. Air set-in-motion calculation
3. Neutral plane calculations

Detailed heat release calculations can be carried out for the different operations within the process building. Heat is released from surfaces by convective forces that cause thermal updrafts, producing the primary ventilation flows. Radiant heat that is not ultimately converted into convective heat does not affect the primary ventilation flow. Both the steady and the intermittent heat release are calculated. For design purposes, it is recommended that an average heat release be established to be somewhat higher than the steady heat release.

For convective heat losses, the heat release from equipment surfaces is calculated by the equation

$$q_c = h_c A(T_s - T_a)$$

where: q_c = convective heat loss, Btu/h
h_c = the natural convective heat transfer coefficient, Btu/h-ft^2 °F
A = the surface area of hot source, ft^2
T_s = the surface temperature, °F
T_a = the surrounding air temperature, °F

The convective heat transfer coefficient h_c can be readily estimated from well-established formulas that can be found in standard heat-transfer reference books.

Air set-in-motion calculations account for major air and fume vertical flows that are produced by the total convective portion of heat release from hot processes. (The convective portion may include some radiant heat loss that is converted to convecting heat.) This hot air is heated by contact with the surfaces of the hot equipment and rises as a result of thermal buoyancy. The size and velocity of this hot air column are a function of the heat release rate and the distance between the source and the roof level.

The heated airstream originating from the surfaces of hot bodies mixes turbulently with the surrounding air as it moves upward. An airstream at a hot body source may have a flow of 1000 cfm. After some vertical travel, this hot airstream may set in motion and become mixed with (i.e., entrained in) an additional flow of 100,000 cfm of surrounding air. To avoid recirculation, roof ventilators should, as a minimum, accommodate this total flow arriving at the roof level.

The plume flow rate can be calculated using

$$Q = 0.166\ (Z)^{5/3}\ (F)^{1/3}$$

where: Q = plume flow rate, cfm
Z = distance from virtual origin, ft
F = buoyancy flux, ft^4/s^3

This equation shows that the plume flow rate depends very strongly on its distance from the source (five-thirds power), but only weakly on the buoyancy or thermal forces (one-third power). Therefore, the accuracy of the calculated plume flow is not highly dependent on the buoyancy flux, which is difficult to determine precisely.

Neutral plane calculations are based on the classical work of Emswiler in 1926.[5] He defined the concept of a neutral zone or plane of neutral pressure as the elevation within a building at which neither the outside air tends to move into the building nor the inside air tends to move out. Figure 11 shows the plane of neutral pressure as defined by Emswiler. Emswiler developed two theorems that apply to industrial ventilation and form the basis for design.

Theorem 1 The sum of flow rates into the building below the neutral zone must equal the sum of the flow rates out of the building above the neutral zone.

$$\sum_{i=1}^{n} A_i v_i + \sum_{j=1}^{n} Q_j = 0$$

where: A_i = area of opening, ft^2
v_i = velocity through the opening, fpm
Q_j = powered ventilation flow rate, cfm

Theorem 2 The driving force at each opening is related to the vertical distance between the opening and the neutral zone and the temperature difference between the opening and the neutral zone.

$$Q_N = A_N\ \sqrt{L_{nz}\ \Delta\ t_{nz}/\ (RT)}$$

where: Q_N = flow rate, cfm
A_N = area of the opening, ft^2
L_{nz} = distance from the centerline of the opening to the neutral zone, feet

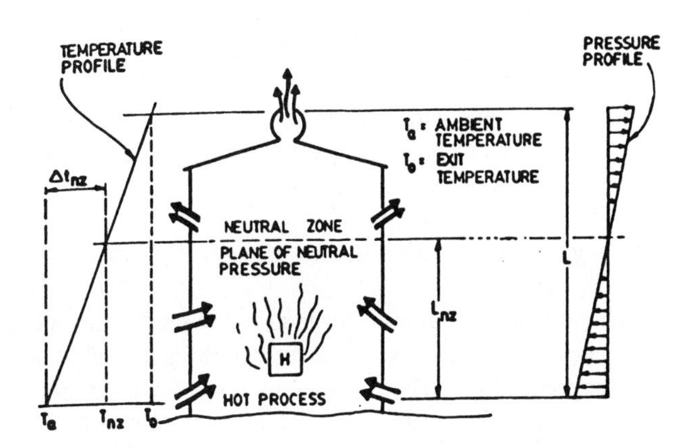

FIGURE 11. Plane of Neutral Pressure as Defined by Emswiler (From Reference 2)

Δt_{nz} = temperature difference between the opening and the neutral zone, °F
R = resistance of the opening
T = absolute temperature of the air at the opening (R)

The elevation of the neutral zone is determined by trial and error, using the preceding equations.

The temperature rise through a hot process building is a function of the amount of heat released in the building and the ventilation flow rate. The heat balance equation can be written as

$$\Delta t = q/\rho C_p Q$$

where: Δt = temperature rise through the building, °F
q = heat release rate, Btu/min
ρ = air density, lb/ft^3
C_p = heat capacity of air, Btu/lb °F
Q = ventilation flow rate, cfm

For design of ventilation systems, the criteria of mass and heat balances must apply. The amount of air carried into the shop per unit of time is equal to the amount of air leaving the shop in the same period. For the heat balance equation, the quantity of heat removed from the shop must equal the sum of heat brought into the shop by the outdoor air plus the surplus heat given off in the shop in the same unit of time.

For natural ventilation, wind has a significant effect on building ventilation rates. Wind on a building manifests itself as an increase in pressure on the building's windward side and as suction on the leeward size. Pressure coefficient is the dimensionless ratio of the pressure at any arbitrary plane on a building to the ambient pressure. Experiments have shown that pressure coefficients remain constant on geometrically similar buildings. Thus pressure coefficients for buildings are found by wind-tunnel tests on geometrically similar models. The wind pressures are obtained by multiplying the pressure coefficient for the points concerned by the velocity pressure of the wind as

$$p = k_w(\rho v^2/2g_c)$$

where: p = wind pressure, lb/ft^2
k_w = wind pressure coefficient (dimensionless)
ρ = air density, lb/ft^3
v = velocity, fps
g_c = acceleration due to gravity, 32.2 ft/sc^2

The amount of ventilation air required to meet allowable contaminant concentrations can be determined using the preceding equations. These equations also allow the areas of the vents in the walls and roof to be calculated. For systems with mechanical ventilation as well as natural ventilation, the quantity of air that is supplied or removed by mechanical ventilation is written into the air balance equation on the inlet or outlet side.

Dilution Ventilation

The terms general ventilation and dilution ventilation are often used interchangeably. In the current context, dilution ventilation refers to the dilution of contaminated air with uncontaminated air for the purpose of health hazard control.

Starting with a fundamental differential material balance, dilution ventilation requirements can be related to the generation and removal rates of a contaminant as

$$\underbrace{V\,dC}_{\text{rate of accumulation}} = \underbrace{G\,dt}_{\text{rate of generation}} - \underbrace{Q'C\,dt}_{\text{rate of removal}}$$

where: V = volume of the room or enclosure, ft^3
C = concentration of gas or vapor at time t, ppm
G = rate of generation of contaminant, cfm
t = time, seconds
Q' = Q/K = effective rate of ventilation corrected for incomplete mixing, cfm
K = design distribution constant or mixing factor, allowing for incomplete mixing

Ventilation of contaminated air in a space would be simple if the outside air could enter the room in a laminar fashion without turbulence and remove the contaminated air in a piston fashion. For this condition, one air change would be required to remove the contaminated air completely (K = 1). In practice, however, the introduction of the fresh air results in turbulent mixing with the contaminated air, and after one air change, the room still contains a dilution mixture of fresh and contaminated air. For most industrial ventilation applications, K varies from 3 to 10.

The material balance equation can be solved for different conditions. Table 3 presents the design equations for three typical conditions.

For any specific contaminant, the design criteria may be based on regulated limits, comfort limits, or odor threshold concentrations. For many industrial solvents, the most stringent design criterion for a dilution ventilation system is based on the odor threshold concentration. Computer models have been developed for calculating contaminant concentrations for different conditions.

Displacement ventilation, a concept developed and tested in laboratories in Norway, is based on ideal unidirectional flow (plug flow). No short-circuiting takes place and the residence time for the room air is exactly the transit time (ideal case). A stratification takes place between the clean supply air and the contaminated air, as shown in Figure 12. Systems can be designed to utilize thermal or density stratification to create a tendency to undirectional flow, that is, to create displacement flow.

The benefits of displacement flow are that it improves

TABLE 3. Dilution Ventilation Equations

Case	Description	Equation	Unit Definition
A	Rate of contaminant concentration buildup ($C_1 = 0$ at $t_1 = 0$)	$t = -\frac{V}{Q'} \ln \left[\frac{(G - Q'C)}{G}\right]$	t = time, minutes V = room volume, ft^3 K = mixing factor Q' = effective ventilation rate, cfm G = rate of generation of contaminant, cfm C = contaminant concentration, ppm
B	Maintaining acceptable concentrations at steady state	$Q = \frac{KG}{C}$	Q = room ventilation rate, cfm
C	Rate of purging	$t_2 = -\frac{V}{Q'} \ln \frac{C_2}{C_1}$	C_2 = concentration at time t_2 C_1 = concentration at time t_1

the air renewal and contaminant removal speed and that it assists in maintaining favorable concentration gradients of the contaminants generated in the room. Calculation procedures based on mathematical two-zone models have been established.

Fluid Dynamic Modeling

The technique of small-scale modeling for fluid flow problems (i.e., fluid dynamic modeling) has been used extensively for a wide variety of industrial ventilation applications.[2]

The fluid dynamic modeling technique is a cost-effective, flexible, and powerful tool for the design of ventilation systems for a greenfield plant and for the elimination of ventilation problems in existing plants. Some of the applications of scale modeling include:

- Finalizing building ventilation flow rates and schemes.
- Examining internal flow patterns and contaminant concentrations at any location.
- Examining external flow patterns, including quantitative measurements of downwash and transport of contaminants to other buildings.
- Establishing the effectiveness of source hoods.

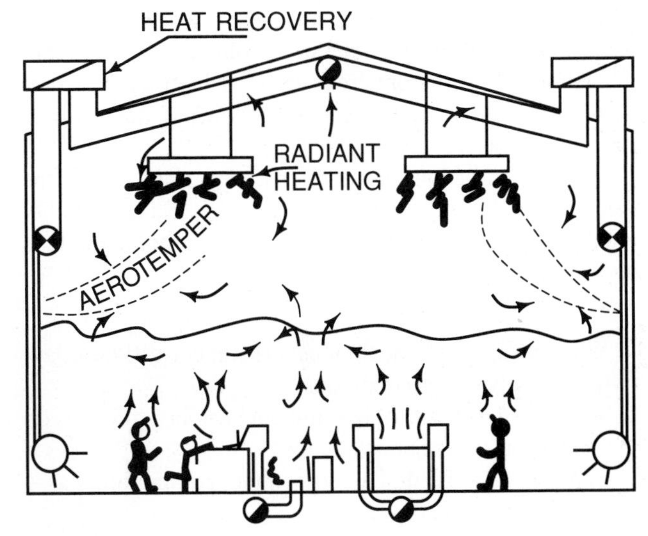

FIGURE 12. Displacement Ventilation in Workshops (From Reference 6)

For new process buildings, the recommended sequence is to use the mathematical models and equations to develop the overall ventilation concepts and architectural constraints. At this stage, the project design team can continue with the design of the structural steel while the small-scale model is being constructed and tested. The results of the small-scale modeling are used to refine the ventilation concept and to finalize all requirements.

For an existing plant, the fluid dynamic model is a valuable tool in modeling the problems and developing alternative cost-effective solutions.

Figure 13 is a typical flowchart of activities required for a ventilation modeling study. The first step is to define the contaminant and the source characteristics. Parameters to be defined are the size of the process building and details of the source flux (i.e., heat and contaminant release rates, etc.). Information is required on the major sources of heat in order to calculate heat balances and volumes of air set in motion. Data are required on the external and internal flow conditions for the prototype. Information may be required on site conditions, such as wind speed, direction, and frequency.

The design of the model system requires an examination

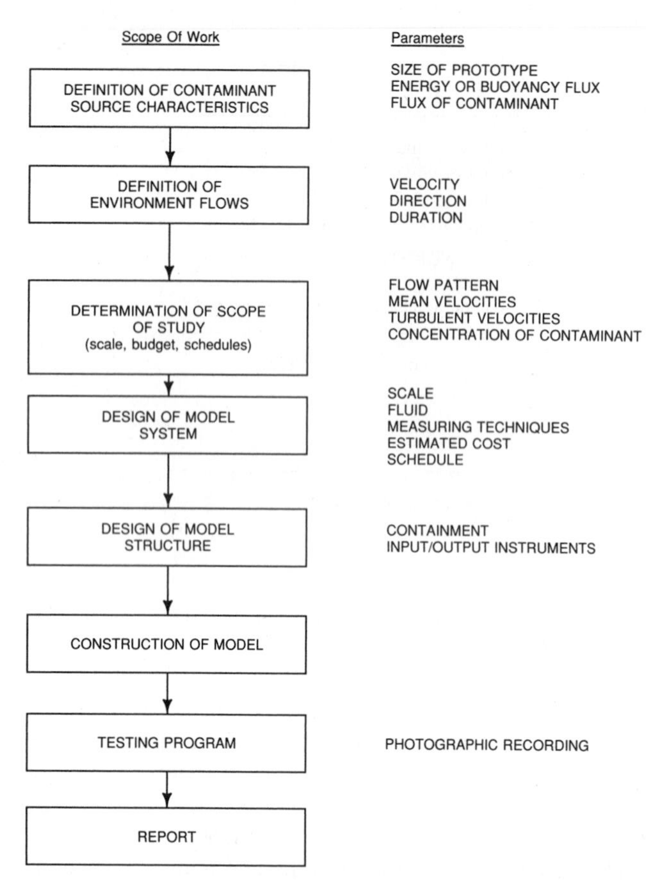

FIGURE 13. Typical Flowchart for Ventilation Modeling Study (Adopted from Reference 2)

of the scaling parameters and the type of fluid medium to be used. The common media used are air and water, but the working or buoyancy-driven fluids for models can vary widely. Fluid systems that have been used include air and heated air, water and saltwater, water and carbon tetrachloride, and mercury and carbon tetrachloride. Air models are usually the simplest and least expensive models to build and test, but usually must be large models to ensure fully turbulent flow. Because water has a smaller kinematic viscosity than air, a small model is required to ensure a high Reynolds number and turbulent flow. Flow visualization is also easier with water-based models because velocities are lower than in air. A detailed analysis of the scaling parameters and possible fluid systems is required in order to select the best modeling technique for the specific ventilation problem. Data measured in the model flow can be related quantitatively to the full-scale prototype flow by establishing dynamic similarity (geometric and kinematic) between the model and the prototype.

The flow required in the prototype is given by

$$Q_p = Q_m(S)^{5/3}(q_p/q_m)^{1/3}$$

where: Q = the volumetric flow rate, cfm
q = heat flow rate, Btu/min
S = model scale (e.g., for a 1:10 scale model, S = 10
p = subscript identifying prototype parameter
m = subscript identifying model parameter

The measuring technique for the contaminant of concern must also be considered since it may have a significant impact on the level of accuracy and the cost of the testing program. For any model testing program, extensive use should be made of photographic and video recording equipment. The photographs and films will be invaluable in analyzing the test results and for subsequent presentation of solutions to management.

Computer Modeling

Available computer programs, based on the application of the design equations in the earlier section, can be applied for solving industrial ventilation problems. Figure 14, a flow diagram of the computer ventilation model, shows a typical input, the initialization and iteration loop, and the output. Figure 15 shows that the overall agreement between predicted and measured flow rates is good.

Once correct modeling of observed ventilation flow rates has been established, the computer program can be used to study the effects of different atmospheric conditions (summer/winter, wind direction and speed, etc.) on the ventilation characteristics for the shop. It is also a very useful tool in the evaluation of proposed ventilation improvement schemes. Different ventilation schemes can be analyzed and cost-effective solutions developed and implemented to solve the ventilation problems.

Computer ventilation models can reliably predict the gross ventilation rates for complex process buildings. The use of high speed computers provides the designer with the capability to examine the impact of architectural changes, wind conditions, or process changes on the performance of the proposed ventilation scheme. Problems such as contamination resulting from cross drafts or high temperatures in the work environment can be identified quickly and corrective measures taken.

Typical ventilation questions that may arise during the planning and design of ventilation systems for a new facility or to improve the environment in an existing facility include the following.

1. What are the internal flow patterns under different layout and operating conditions?
2. What are the effects of intermittent peak heat releases on ventilation flow characteristics?
3. Can the flow fields be represented satisfactorily as a two-dimensional flow?
4. What happens to contaminated plume that misses a hood?

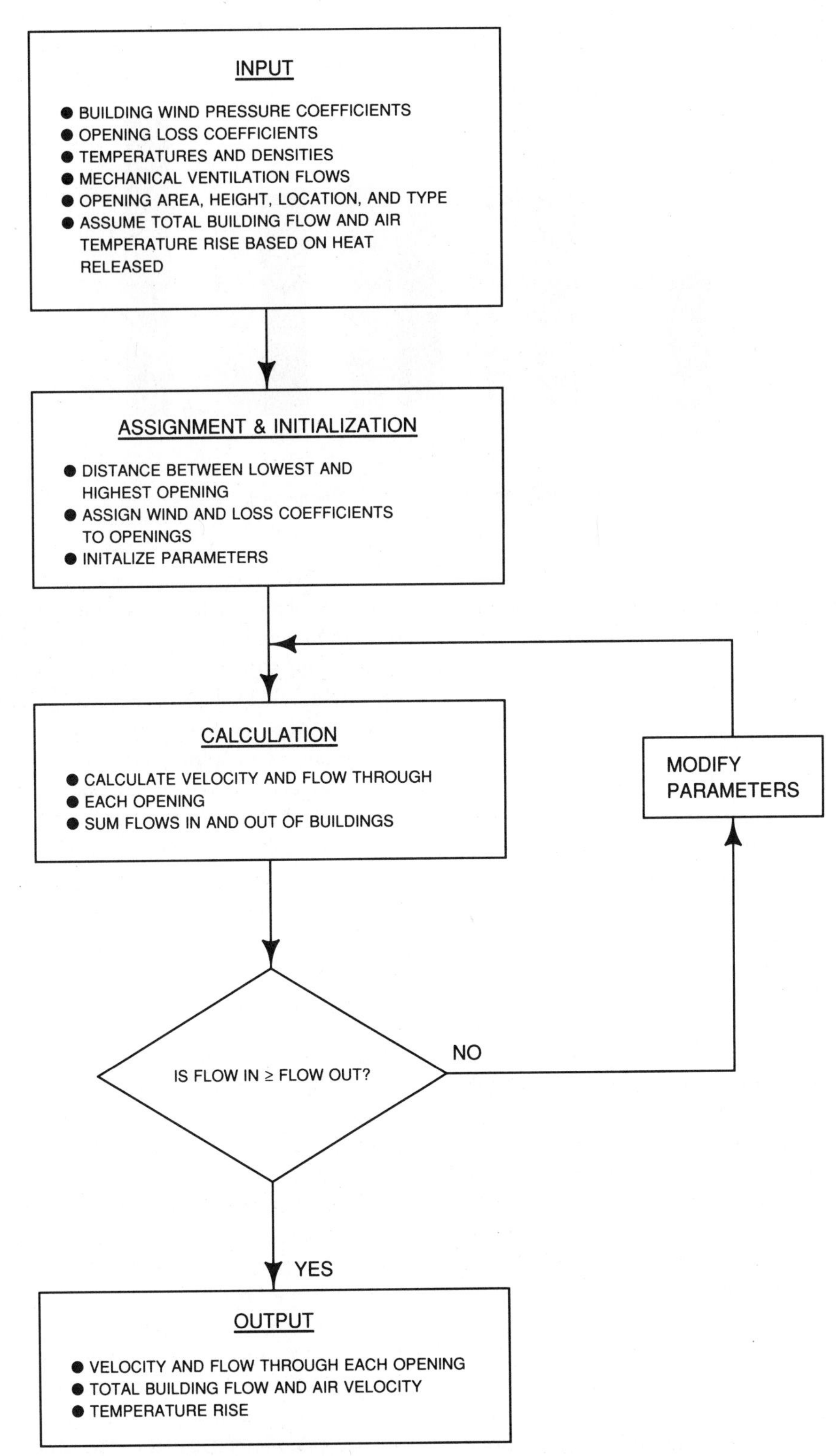

FIGURE 14. Computer Ventilation Model Flow Diagram (From Reference 2)

5. What effect do cross drafts have on the workplace environment?
6. Where does the fresh air enter the building?
7. What are the predicted contaminant concentrations in the breathing zone?
8. What is the source of contamination in a specific area?

Computer programs are available that can operate on a microcomputer and solve three-dimensional complex ventilation problems.[7] Conferences such as Ventilation '88 and RoomVent '90 included many technical papers on new computer programs to calculate contaminant levels as a function of position and time in a workplace environment.

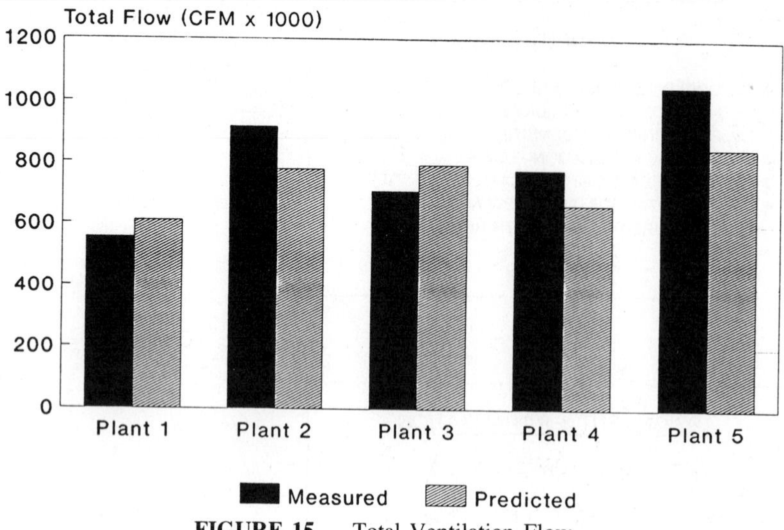

FIGURE 15. Total Ventilation Flow

The practical application of these computer programs represents a significant advancement in the science and technology of ventilation.

Field Testing Methodology

For most complex ventilation problems, it is recommended that both analytical calculations and a field measurement program be carried out in order to develop a proper, cost-effective solution. Experience has shown that the field sampling program must be organized and developed in detail prior to going into the field.

Ventilation flows, design parameters, and field testing protocol are unique for each specific processing or manufacturing operation, but many elements of the field testing program are common to all ventilation problems. Although there may be minor variations for any specific ventilation project, the eight activities listed in Table 4 represent the major activities for a field testing program and the sequence for performing this work. A brief description of the tasks and scope of work required for each activity is included.

As a starting point for the field testing program, all pertinent reports and drawings concerning the plant should be obtained for review and study. Specific information is required on the current and future operating practices, as well as on the nature of the in-plant contaminant concerns (chemical or heat/cold stress). A visit to the site should include a walk-through ventilation survey using a data sheet. The ventilation engineer should develop the questionnaire with experienced process and operating personnel for the specific industry. The prime purpose of the questionnaire is to identify the operating practices (present and future), which will then be used as the design basis for the ventilation system.

It is necessary to develop the details of a field testing program and all necessary field data log sheets prior to the actual field testing. The development of a proper field testing program is the most important step in ensuring a successful ventilation study. This program should be reviewed and approved by plant operating personnel before the actual field testing is carried out. Because the ventilation flow in large process buildings is usually complex, an inexperienced sampling team may collect insufficient data for subsequent analytical calculations or will expend considerable effort collecting field data that are irrelevant. A good approach is to perform preliminary calculations on all sources and then to identify clearly where information gaps exist and what essential data must be collected.

The measurement of air velocities through all the openings must recognize the need to have a "representative" velocity for the survey. It is not acceptable to "be on the safe side" by rounding readings to a higher value. Gusty wind conditions will cause velocity readings to fluctuate. Sufficient time must be allowed to elapse to enable the person monitoring the instrument to arrive at a representative velocity by "integrating" the observed values in his or her mind.

A temperature measurement is required for each velocity measurement. For air entering the building, this would generally be the ambient temperature, which should be recorded on an hourly basis. It will change during the day and could be higher in the wake of the building. The thermometer or temperature probe should not be exposed to sunlight or radiant heat from hot objects. The temperature of air leaving the melt shop will vary considerably. For heat balance calculations, good temperature readings are important.

Temperatures associated with in-plant flow patterns must also be recorded. Sufficient temperature data must be available to allow evaluation of the air density distribution within the shop.

Mean surface temperatures of hot surfaces must be recorded for subsequent heat release and air set-in-motion calculations. The location of the equipment on the floor

TABLE 4. Engineering Activities for a Ventilation Field Testing Program

Activity	Specific Tasks
1. Information gathering	Obtain drawings, reports, operating procedures Review existing data and studies Plant visit Define problem (summer, winter, heat/cold stress, chemical, supply, exhaust)
2. Data on ventilation openings	Isometric drawings: plans, sections Schedule of openings (type, size, location)
3. Develop/complete plant questionnaire	Develop process flow sheet and general layout Identify significant heat sources in building Identify typical operating practice Identify gaps in data to be filled in by field testing program
4. Develop details of field testing program	Building ventilation flow rates In-plant flows Field data log sheets for ventilation measurements, weather conditions, plant operating records, etc.
5. Field testing program	Field measurements (volume, temperature, pressure etc.)
6. Data analysis and calculations	Plant ventilation flow balance In-plant flow patterns Heat balance calculations (total plant/ventilation) Air set-in-motion volumes
7. Computer simulation of ventilation flows (natural ventilation)	Calibrate using field test data Run program for different conditions (summer, winter, etc.)
8. Field testing report	Summarize test conditions and test results Report with recommendations

plan and equipment surface temperature should be recorded on a separate sheet.

Weather data should be measured at the site and obtained from the nearest airport or meteorological station as well. Data to be recorded include ambient temperature, relative humidity, and wind speed and direction. During the test period, this information should be recorded on at least an hourly basis.

A record of plant activities during the testing program is required. Data to be included are the status of operation of all major process and environmental equipment, as well as production levels. Plant records and charts from process operation should be obtained.

An in-plant survey will be required to establish such parameters as levels of contaminants (dust, gases, etc.) and heat stress. These measurements would be concurrent with the ventilation survey. An industrial hygienist should work with the ventilation engineer to establish the scope and extent of the industrial hygiene sampling program.

Ventilation computer models should be used to determine ventilation flow rates for different operating conditions and to estimate contaminant concentrations in the workplace. Data from the computer model calculations can be compared with the measured ventilation and industrial hygiene sampling data.

For any ventilation field testing program, it is essential to prepare a proper engineering report, which includes all the field data, calculations, and test results. Using the results of this field testing program, an experienced ventilation engineer can develop cost-effective solutions for any plant ventilation problem.

ANCILLARY EQUIPMENT FOR LEV SYSTEMS

Hoods

Source Emission Characterization

As described earlier, hoods are the most important element of the LEV system. If the hood fails to capture the contaminant at the source, the performance and efficiency of the rest of the system becomes meaningless. A *Technical Manual: Hood System Capture of Process Fugitive Particulate Emissions*[8] describes design procedures for hoods for numerous industrial processes. The manual shows that each application must be studied for the specific process. A new approach stressed in the development of the hood design methodology is to develop a complete understanding of the source–hood interactions.

Although it is impractical to develop simple standard design procedures for hoods for different plant/process applications, significant progress is being made in the development of technology related to hood design equations. Numerous technical papers have been published at international conferences such as Ventilation '85 and Ventilation '88 that cover both the theoretical and practical design equations for hoods.

Factors known to influence hood performance are as follows:

1. Construction, size, and shape of hood (e.g., degree of enclosure of hood)
2. Suction velocity of exhaust hood
3. Position of hood relative to the source
4. Cross drafts
5. Physical/chemical characteristics of the emission sources
6. Velocity contours for gas and particulate emission sources

Hood design procedures will depend on whether the application is a greenfield or a retrofit plant. In a greenfield plant, the designer has many more options and it is essential that the design be incorporated as an integral part of the development of the plant layout. It is recommended that the hood designer evaluate different alternatives using computer models or fluid dynamic models. For retrofit plants, the layout may already dictate the hood configurations that are acceptable. The opportunity exists to perform field evaluations to collect data on existing hood performance or at least to characterize the emission source.

The emission sources can be characterized by parameters such as the following.

1. Buoyant or nonbuoyant source
2. Continuous or intermittent plume
3. Plume flow rate and geometry
4. Source heat flux and geometry
5. Physical/chemical characteristics of the particulates
6. Gas composition and temperature
7. Layout of the plant

Using the above information, the emission source can be completely characterized as a function of time and position. This becomes the starting point for the establishment of hood design flow rates.

Hood Classification

The four hood classifications[9] are as follows:

Group A Enclosures
Group B Booths
Group C Captor hoods
Group D Receptor hoods

Figure 16 shows typical configurations for the different hood types. Table 5 summarizes characteristics of the four types of hoods, including operating principles, description of types, design considerations, design methodology, and typical industrial applications. Goodfellow[2] has presented further details on the hood classification scheme.

Design Parameters

Hood design methodologies are as follows:

1. Precedent
2. Rule of thumb
3. Analytical methods
4. Diagnosis of an existing site
5. Fluid dynamic modeling

Precedent is based on a similar hood design on an existing application. Rule of thumb is taking data from a standard handbook that gives typical hood volume requirements for similar applications. Table 6 gives recommended control velocities and exhaust rates for enclosures of various materials-handling operations. *Industrial Ventilation*[4] contains more than 100 sketches of hoods for different processing operations. Analytical methods are based on using theoretical or empirical equations developed for different operations. In retrofit plants, it is possible to collect field data and to determine hood performance as a function of flow rate and different hood geometry. A newer approach is based on the use of scale models to simulate the source–hood interactions. This approach has the benefit of permitting hood performance evaluations to be established quantitatively.

Table 7 outlines typical equations and criteria for hood design for contaminant control. (See Table 7A for corresponding nomenclature.)

Hood Performance Evaluation

The evaluation of hood system performance is an important aspect of the technology of the design of LEV systems. Several techniques have been developed for an evaluation of hood system performance. The hood design manual[8] describes the techniques in detail. A brief description of the techniques follows.

Tracer gas techniques[10] are now well established as a procedure to measure hood performance. The technique is to inject a tracer at a known rate at the process source. The quantity captured by the hood source is a measure of the hood performance.

Goodfellow and Bender[11] have described three techniques, which are a more direct means of evaluating hood system performance for the actual process sources. For remote capture of buoyant plumes, plume flow rates can be

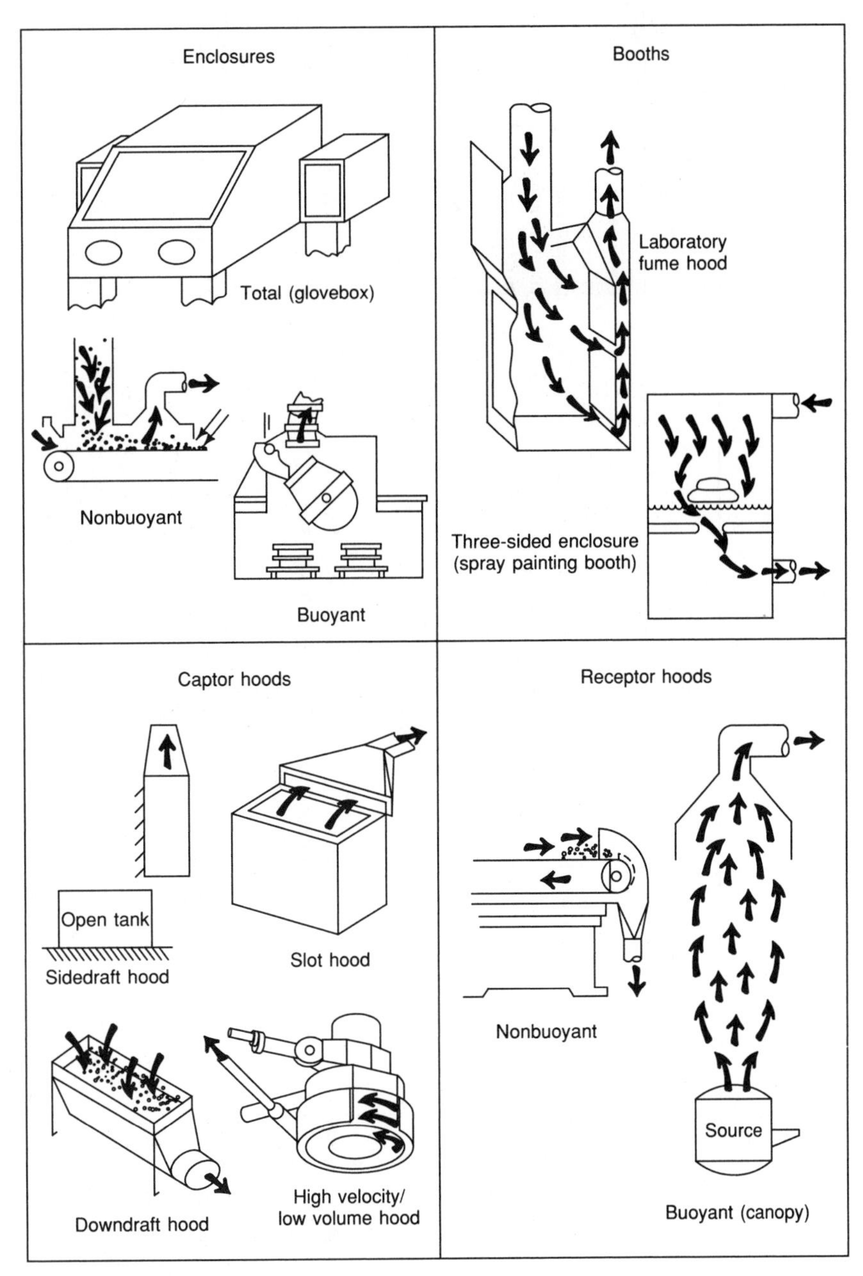

FIGURE 16. Four Different Hood Types (From Reference 2)

measured by movie scaling, stopwatch clocking of the plume, and anemometer measurements at the roof trusses. Using this information on the estimated plume flow rates, it is possible to calculate the hood system performance for a given hood configuration and plant layout. Field measurements can be correlated with opacity measurements to establish performance relative to opacity criteria and overall hood performance evaluation.

Fluid modeling techniques, as described earlier for solving industrial problems, represent a sophisticated and rigorous approach to establishing hood system performance. This type of technique is based on building a scale model of the source–hood configuration in a water tank or air system. For a valid test, the flow in the scale model must be turbulent and the Froude number similarity criteria must be met. Scale modeling permits convenient testing of hood performance for existing hood designs for a proposed modification to an LEV system.

For dust control systems, hood performance can be evaluated using tracer gas techniques, industrial hygiene-type sampling using area monitoring at sources, or a dust lamp. The dust lamp, developed in England, uses a high-intensity light beam utilizing the "Tyndall" effect in which light is scattered forward from the particles in the atmosphere.[2] This technique is extremely sensitive and photographs of the path of the dust clouds can be taken. In addition to its use in research and development applications for new or novel hoods, the dust lamp can be used in conjunction with other measuring equipment to establish hood performance data in the field for existing LEV systems.

TABLE 5. Characteristics of the Four Types of Hoods

	Group A Enclosures (Total or Partial)	Group B Booths	Group C Captor Hoods	Group D Receptor Hoods
1. Operating principles	1. Complete enclosure of emission source. 2. Enclosures should be large enough to avoid direct impingement of particulate cloud on the walls.	1. Emission source inside of enclosure. 2. Enclosures should be large enough to avoid direct impingement of particulate cloud on the walls.	1. Operate at a distance from the source. Contaminant would not naturally enter the hood. 2. Fan supplies the power to establish the airflow patterns at the hood.	1. Positioned to catch the stream of contaminants or contaminated air thrown out in a given direction by an industrial process. 2. Hood exhaust volume is based on the volume of contaminated air that is pumped in by the process.
2. Description of types	*Total* 1. Not used except for radioactive materials. *Partial* 1. Most so-called total enclosures have small openings for entry and egress of materials. 2. Applicable for nonbuoyant sources.	1. Booths are enclosures from which one wall or a large part of one wall has been removed so that articles can be moved inside the enclosure.	*Side-draft hood* 1. Hood located at side as close as possible to emission. *Slot hood* 1. Slot in a narrow opening along the edge of the tank. *Downdraft hood* 1. Hood located under the source of contaminant. *High-velocity, low-volume hood* 1. Use extremely high velocities ($>$ 12,000 fpm) to capture contaminants at the source.	*Nonbuoyant Sources* (local) 1. Local receiving hoods can be designed to capture particles thrown off. *Buoyant sources* 1. Heated air rising vertically from a hot source can be captured remotely using a canopy hood.
3. Design considerations	1. Total enclosures require materials inside to be handled by remote manipulators. 2. Total enclosures must be dust-tight and gastight. 3. For partial enclosures, total surface area of all openings must be small compared with total area of the walls.	1. Booth should be as deep as possible to ensure an even flow at the open face. 2. Operator should never stand between the work and the airstream. 3. Design of offtake may be important to ensure a high capture with an even gas flow at the face.	1. Since air velocities drop off rapidly with increasing distance from the hood, captor hoods must be located as close to the contaminant source as possible. 2. Captor hood performance depends on proper capture velocity and proper distribution of velocity over zone of contamination.	1. Receptor hoods are designed to prevent spillage because of overloading due to high flows or due to random air curtains which cause contaminated air to miss the hood. 2. Remote hoods only practical for buoyant sources.

4. Design methodology	1. Exhaust rate greater than rate at which contaminated air would escape. 2. For airtight design and minimal air disturbance, exhaust inlet airflow at each opening is 100 fpm. 3. Total flow = *v* (design) x *A* (opening). 4. Ventilation rate must be determined for each specific application.	1. Exhaust rate greater than rate at which contaminated air would escape. 2. For minimal air disturbance, exhaust inlet airflow at each opening is 100 fpm. For every small booths, an inlet velocity 400 fpm may be required.	1. Capture velocities range from 100–200 fpm. 2. Summary of design procedures: a. Examine behavior of contaminant. b. Determine capture velocity. c. Establish distance between hood and source. d. Determine flow required. e. Choose a suitable fan.	1. Design of hoods for nonbuoyant sources described in Table 7. 2. Design of hoods (i.e., canopy) for buoyant sources described in Table 7.
5. Typical industrial applications	1. Electric arc furnace enclosure. 2. Pneumatic oxygen vessel enclosure. 3. Copper–nickel converter enclosure. 4. Radioactive glove box. 5. Clamshell unloading enclosure. 6. Belt conveyor transfer point. 7. Screening operation. 8. Bin filling.	1. Spray painting booth. 2. Laboratory fume hood. 3. Portable grinding, polishing, or buffing operations.	*Side-draft hood* 1. Welding hoods. 2. Side shake-out hood. *Slot hood* 1. Narrow open tanks in electroplating. 2. Bench welding. *Down-draft hood* 1. Shake-out hood. 2. Bench soldering 3. Bench torch cutting *High-velocity/low-volume hood* 1. Power hand tools. 2. Manual arc welding.	*Nonbuoyant Sources* 1. Grinding hood. 2. Belt sanders. *Buoyant sources* 1. Electric arc furnace canopy hood.

TABLE 6. Tabulation of Recommended Control Velocities and Exhaust Rates for Materials-Handling Operations

Operation	Velocity fpm	Basis of Control Velocity	Exhaust Rate cfm	Basis of Exhaust Rate
Screens vibrating				
Flat deck	150–200	Openings in enclosure	25–30	Per square foot of screen area
Cylindrical (rotating)	400	Openings in enclosure	100	Per square foot of circular cross-section area of screen
Bag filling				
Paper	100	Openings in enclosure		
Cloth	200	Openings in enclosure		
Nontoxic	<400	Openings in enclosure	400–500	
Toxic	>400	Openings in enclosure	1000–1500	
Belt conveyors				
Transfer points	>150	Per square foot of openings	350	Per foot belt width (speed <200 fpm)
	>200	Per square foot of openings	500	Per foot belt width (speed >200 fpm)
Along length			350	Per foot belt width at 30-foot intervals
Bag emptying	40–100	Openings in insertion enclosures	400–1050	Multiplied by area of openings
Drum and barrel filling	100	Point of origin	100	Per square foot of barrel semitop enclosed
Bins and hoppers	150–200	Openings in enclosures	0.5	Per square foot of bin volume
Bucket elevators			100–200	Per square foot of casing cross-section
Mixers	100–200	Openings in enclosures	2100–3100	Per foot of open area
Crushers	200	Through crusher gap in direction of material flow		

Ducts Plus Accessories

Technical Description

Ducted systems are defined as the actual ducts plus accessories, such as expansion joints, explosion vents, flanges, and manholes, which are required to make a functional LEV system. Standard designs have been developed that have been well proven in the field and are applicable for LEV systems.

Ducted systems should be constructed with materials suitable for the application and installed in a permanent and good workmanlike condition. The interior of all ductwork should be smooth and free of obstructions. Typical materials of construction range from sheet metal and mild steel platework to specialized materials, such as abrasion-resistant steel, high-temperature alloys, plastics, and other corrosion-resistant materials.

For conventional dust control systems, materials of construction are black iron that is welded, flanged, and gasketed or galvanized sheet steel that is riveted and sealed. Corrosive conditions require the selection of plastics, special corrosion-resistant materials, or protective coatings. Galvanized sheet steel is not recommended for gas temperatures above 400°F. Material thickness of at least 16 gauge is recommended for arc welding.

Industrial Ventilation[4] has identified four classifications for ducts for LEV systems for noncorrosive applications.

Class 1: Light duty for nonabrasive applications such as replacement air and general ventilation.

Class 2: Medium duty for applications with moderately abrasive particulates in light concentrations such as woodworking and grain dust.

Class 3: Heavy duty for applications with highly abrasive particulates in low concentrations such as abrasive cleaning and sand handling.

Class 4: Extra heavy duty for applications with highly abrasive particles in high concentrations such as canopying systems in heavy industrial plants.

Round ducts are usually preferred for LEV systems because of their inherent strength and their ability to minimize the settling of airborne dust. The actual metal thickness for round industrial ducts will vary with the classification, static pressure, reinforcement, and span between supports. The Sheet Metal and Air Conditioning Contractors National Association (SMACNA)[12] gives the range of metal thicknesses in Table 8.

If rectangular ducts are necessary, the sections should be as nearly square as possible. All ducts should be constructed in accordance with the "Round or Rectangular Industrial Duct Construction Standards" as published by SMACNA.

Design Parameters

Design parameters to be determined for ducted systems include duct diameter (established from flow and velocity parameters), friction losses, dust filling criteria, materials of construction, duct thickness, structural support of duct, expansion joints, explosion protection, damper requirements, and such ductwork accessories as pitot tube con-

TABLE 7. Typical Equations and Criteria for Hood Design for Contaminant Control

	Total enclosure		Partial enclosure, nonbuoyant[a]		Partial enclosure, buoyant[b]
Enclosures, Group A	$V = 40$ to 60 fpm $Q = VA$ or using a rule-of-thumb: $Q = 50$ cfm per square foot of open area		$Q_1 = 0.043\left(\frac{WH^2 A_s^2}{p_d d}\right)$ $Q_2 = W/p_d$ $Q_3 = AV$; $V = 100$ (well-protected sources[c]) to 200 fpm (vigorous operations[d]) $Q = Q_1 + Q_2 + Q_3$		Volumetric flow rate after dilution $Q = \frac{q}{\rho c_p (T_{amb} - T_s)}$
	Lab fume hood		Three-sided enclosure		
Booths, Group B	50 cfm per square foot for face velocities with good air supply distribution. 150 cfm per square foot for poor air distribution		*Airflow volume required per cross-sectional area* 75 cfm per square foot 100 cfm per square foot 200 cfm per square foot		*Size of booth*[e] > 150 ft^2 < 150 ft^2 < 4 ft^2
	Side draft		Slot	Downdraft	High velocity/low volume
Captor hoods, Group C	$Q = V_u \left(\frac{T_a}{T_u}\right)^{0.5} \left(\frac{x}{y}\right)^{-0.2554} A_c E$ where: $V_u = 0.09V_{max}\,(0.63 + 0.36y)$		$Q = V_u \left(\frac{T_a}{T_u}\right)^{0.5} \left(\frac{x}{y}\right)^{-0.2554} A_c E$ where: $V_u = 0.09V_{max}\,(0.63 + 0.36y)$	$Q = (10X^{0.5} + A)V$ where: $V = 100$ to 200 fpm by rule-of-thumb: or $Q = 150$ to 250 cfm per square foot of bench area	$Q = 25$ to 60 cfm per inch diameter Slot velocity = 30,000 to 39,000 fpm Flexible hose at 1 to 2 inches diameter
	Nonbuoyant[1]		Buoyant[2]		
Receptor hoods, Group D	*Belt width, in.* < 6 in. incl. > 6 to 9 in. incl. > 9 to 14 in. incl. > 14 in.	*Q exhaust* 440 cfm 550 cfm 800 cfm 1100 cfm	$Q = 0.166\,(Z)^{1.67}(F)^{0.33}$ where: $Z = 2D + Y$ $F = \frac{gq}{C_{pm}\,\rho\,T_m}$	Y Z Source D (Applies to Group D, buoyant)	X Y Hood Source (Applies to group C; side-draft, slot, downdraft)

[a]Nonbuoyant, contaminant source at ambient temperature.
[b]Buoyant, contaminant source at elevated temperature.
[c]Protected from significant cross-drafts.
[d]Significant turbulence at the plume source caused by the source or by external factors.
[e]Open area at the front of the booth enclosure.
Source: Reference 2.

nections and clean-out doors. For any application, the need for explosion prevention must be established at the preliminary design stage. Calculation procedures for establishing system pressure losses are outlined in previous sections. Duct thickness and type of reinforcing systems can be determined by using the three-step procedure outlined in the SMACNA duct manual as follows:

Step 1. Duct thickness
Step 2. Size and spacing of reinforcing member and weld specification
Step 3. Flange connection size and bolt requirements

Ducts and their ancillary equipment are key components that must be properly selected to ensure a satisfactory LEV system. Ducted systems should be designed and specified only by experienced engineers and must incorporate designs that will accommodate necessary maintenance features.

Gas Cooling

Technical Description of Types

There are many methods of cooling hot process off-gases. The five common cooling methods for LEV systems are as follows:

TABLE 7A. Nomenclature for Table 7

A	=	control cross-sectional area, ft^2
A_c	=	control surface area calculated using Fletcher's formula and depending on hood configuration, ft^2
A_s	=	cross-sectional area of falling stream, ft^2
b_h	=	plume length scale at hood face, feet
C_p	=	specific heat or air at T_s, Btu/lb/°F
C_{pm}	=	specific heat of air at T_m, Btu/lb/°F
D	=	diameter of plume source, feet
d	=	particle mass median diameter, feet
E	=	1.6 when the vessel is not full and flow separation causes eddy formation
e	=	eccentricity, feet
F	=	buoyancy flux, ft^4/s^3
g	=	gravitational constant, ft/s^2
H	=	drop height, feet
L	=	length, feet
M	=	momentum, ft-lb mass/min^2
q	=	heat transfer rate, Btu/min
Q	=	total exhaust volume, cfm
Q_1	=	volumetric flow rate of induced air, cfm
Q_2	=	flow rate of displaced air, cfm
Q_3	=	flow rate for control velocity through openings, cfm
Q_p	=	peak plume flow rate, cfs
Q_s	=	hood exhaust flow rate, cfs
S	=	scale parameter (ratio of prototpye to model)
T_a	=	ambient fluid temperature, R
T_{amb}	=	ambient dilution air temperature, °F
T_m	=	absolute temperature, R
T_s	=	specified air temperature after dilution, °F
T_u	=	plume fluid temperature, R
t	=	time, minutes
t_d	=	plume surge time, second
U_{max}	=	plume centerline velocity taken at hood face, fpm
V	=	control velocity, fpm
V_{cross}	=	cross-flow velocity, fpm
V_{max}	=	centerline velocity at one source diameter above the hood, fpm
V_u	=	updraft velocity of the plume above the source, fpm
W	=	material flow rate, lb/min
X	=	distance from fume source to hood, feet
Y	=	height to hood face, feet
y	=	vertical distance between source and hood in source diameters
Z	=	effective height from the virtual plume origin to hood face, feet
ρ	=	air density at T_s, lb/ft^3
ρ_d	=	density of dust or other particulate matter, lb/ft^3
α	=	entrainment constant (0.093 for a point plume source)

TABLE 8. Ducts and Gauges

Diameter of Straight Ducts	Range of Metal Thicknesses U.S. Standard Gauge for Steel Duct			
	Class 1	Class 2	Class 3	Class 4
4 to 8 inches	22–20	22–18	16	14
Over 8 to 18 inches	22–12	22–12	16–11	14–11
Over 18 to 30 inches	18–7	16–7	16–6	14–6
Over 30 inches	14–2	14–2	12–2	12–2

[a]Thickness varies with classification, pressure, reinforcement, and span requirements. *Note:* 24- and 26-gauge metal cannot be welded.
Source: Reference 7.

1. Dilution with ambient air
2. Natural convection and radiation from ductwork
3. Forced-draft cooling
4. Quenching with water (evaporative cooling)
5. Use of water-cooled ductwork

Table 9 compares the five different gas cooling techniques based on applicable gas temperature ranges and industrial experience.[2] A detailed technical and economic analysis of the alternative systems is required for any specific application before a selection can be made. A brief description of each system follows.

Design Parameters

Dilution with Ambient Air

Dilution of off-gas with ambient air is a simple method of reducing off-gas temperature.

Dilution cooling does not remove or absorb heat from the off-gas. Its effect can be examined by considering the off-gas volumes produced. Adding dilution air increases the volume of off-gas to be handled by the air pollution equipment. For example, if the air pollution control equipment is a baghouse, the cloth area must be increased to maintain reasonable air-to-cloth ratios. The adequacy of the existing fan must also be evaluated. The large volumes of air introduced when dilution air volumes are increased usually cannot be moved by the existing fan.

The mixed temperature after dilution with ambient air can be calculated from the following equation.

$$T_m = \frac{[Q_g \times c_{p_g} \times (T_g - T_{ref})] + [Q_d \times Cp_d \times (T_d - T_{ref})]}{(Q_m \times Cp_m)} + T_{ref}$$

where:
Q_m = mixed gas volume, scfm
Q_d = dilution air volume, scfm
Q_g = off-gas volume before dilution, scfm
cp_g = $c_{p_g}/387$; c_{p_g} = mean heat capacity of off-gas, Btu/lb mol°F
cp_d = $c_p 1d/387$; $c_p 1d$ = mean heat capacity of off-gas, Btu/lb mol°F
cp_m = $c_p 1m/387$; $c_p 1m$ = mean heat capacity of off-gas, Btu/lb mol°F
T_g = temperature of off-gas, °F
T_d = temperature of dilution air, °F
T_m = temperature of mixed gases, °F
T_{ref} = reference temperature, °F

Natural Convection and Radiation from Ductwork

Gas cooling by natural convection such as radiation from ductwork is a direct function of the outside area of the ductwork, the duct temperature, and ambient conditions, such as wind velocity and air temperatures. In radiation coolers, the off-gas passes through an assembly of U-shaped cooling columns and heat is transferred by radiation

TABLE 9. Gas Cooling Techniques

Cooling Technique	Schematic	Applicable Gas Temperature Range (°C: 50 100 200 500 1000 2000 3000)	Comments
1. Dilution with ambient air	AIR		Applicable over whole temperature range.
2. Natural convection and radiation from ductwork (radiation cooler or U-tube)			Results in smaller gas volumes. No direct operating costs.
3. Forced-draft cooling			Results in smallest gas volumes to be cleaned. Requires energy for cooling.
4. Quenching with water (evaporative cooling)			Potential for heat recovery. Works best for a steady heat flow. Dewpoint considerations important at low end.
5. Water-cooled ductwork			Prime purpose is to protect the ducting from high gas temperatures. High cost for heat removal.

Source: Reference 2.

and free convection. Radiation coolers of mild steel construction are suitable for cooling off-gas from about 1000°F to 400°F. At lower off-gas temperatures, radiation coolers are not efficient. Pressure losses through radiation coolers are high. The ability of an existing system fan to accommodate the pressure requirements of a radiation cooler must be evaluated. Tube blockage and duct erosion are common problems on the bends of radiation coolers. Design must be amenable to regular inspection and ease of cleaning and maintenance.

The temperature reduction of off-gas through a radiation cooler can be estimated from the following equation.

$$T_o = T_{amb} + (T_i - T_{amb}) \exp\left(\frac{-U \times A \times 387}{Q \times c_p \times 60}\right)$$

where: T_o = off-gas temperature out of section of radiation cooler, °F

T_{amb} = ambient air temperature, °F

T_i = off-gas temperature into sections of radiation cooler °F

Q = off-gas flow rate through radiation cooler, scfm

A = heat transfer area of radiation cooler section, ft^2

c_p = specific heat capacity of off-gas, Btu/lb mol °F

U = overall heat transfer coefficient at location along radiation cooler, Btu/h·ft^2 °F

The overall heat transfer coefficient typically varies between 1.5 Btu/h·ft^2°F and 3 Btu/h·ft^2°F, depending mainly on off-gas temperature, off-gas velocity, and ambient conditions.

Forced-Draft Coolers

In a typical forced-draft cooler design, the hot off-gas passes through banks of vertical tubes and axial fans blow ambient air across the outside of the tubes to cool the off-gas. Forced-draft coolers of mild steel construction are suitable for cooling off-gas from about 1000°F to 275°F, as required by polyester filter fabric. Special alloy steel tubes are sometimes used to operate at higher temperatures (e.g., in the ferroalloy industry). Pressure losses through forced-draft coolers can be high, typically 4–6 in. w.g. Typical overall heat transfer coefficients range from 3 to 6 Btu/h·ft^2 °F. In retrofit applications, the ability of the existing system fan to accommodate the pressure requirements introduced by the cooler must be evaluated. The main advantages of forced-draft coolers over radiation coolers are as follows:

1. Reduced surface area with direct cooling to 275°F.
2. Compact design.
3. Potential for heat recovery.
4. Lower maintenance cost because downstream fan and air pollution control equipment operate at a more uniform temperature.

Evaporative Cooling

In evaporative cooling, fine water droplets are injected into the off-gas. The droplets evaporate, absorbing the heat from the gas. Heat is not removed from the system; sensible heat is changed to latent heat, reducing the temperature. Evaporative cooling usually does not increase the volume of off-gas significantly. Unlike forced-draft coolers and radiation coolers, minimal pressure loss is added to the system by evaporative coolers, making their use a good option in a retrofit application. The disadvantage of evaporative cooling is that fairly sophisticated controls may be required to ensure that correct quantities of moisture are injected into the off-gas. Overspraying can lead to condensation of water vapor in the baghouse. Blinding of fabric filter bags will result and condensation on metal surfaces will accelerate corrosion. Also, high off-gas moisture content reduces bag life. The degree of cooling that can be afforded by evaporative coolers is limited by the allowable off-gas moisture levels. However, when evaporative cooling systems are properly engineered, they provide the most cost-effective method of dealing with increased heat loads.

Water-Cooled Ductwork

Water-cooled ductwork provides an effective mechanism for heat removal from the off-gas. Temperature reduction of the off-gas through water-cooled ductwork can be estimated from the following equation.

$$T_o = T_w + (T_i - T_w)\exp - \left(\frac{U \times A \times 387}{Q \times c_p \times 60}\right)$$

where: T_o = off-gas temperature out of section of water-cooled ductwork, °F

T_w = average temperature of cooling water, °F

T_i = off-gas temperature into section of water-cooled ductwork, °F

Q = off-gas flow rate through ductwork, scfm

A = heat transfer area of water-cooled ductwork section, ft^2

c_p = specific heat capacity of off-gas, Btu/lb mol °F

U = overall heat transfer coefficient at location along water-cooled ductwork, Btu/h·ft^2 °F

From the estimated temperature reduction, heat removal from the water-cooled ductwork can be calculated. The equation is valid for heat exchangers in which one of the fluids does not change temperature significantly relative to the temperature change of the other fluid. It provides good results in water-cooled ductwork heat transfer calculations because the temperature change of the water is very small compared with the temperature change of the off-gas.

The overall heat transfer coefficient varies significantly, depending mainly on the temperature of the off-gas. As the temperature of the off-gas decreases, the overall heat transfer coefficient also decreases. As a result, the heat transfer from the water-cooled ductwork decreases significantly at lower off-gas temperatures. Therefore, upstream sections of water-cooled ductwork are more effective in removing heat than are downstream sections of water-cooled ductwork. This should be considered when evaluating the length of ductwork to be installed, because the cost effectiveness of ductwork is reduced as the length of the duct increases. The addition of excess volumes of combustion air will quench the off-gas and will defeat the effectiveness of the ductwork. Water-cooled ductwork is usually sized for maximum velocities of 9000 to 10,000 fpm. High velocities are tolerable because pressure losses through ductwork are less severe at high temperatures.

Fans

Technical Description

Fans have been used in industry for about 100 years. Some of the earlier applications were in metallurgy, mines, and other manufacturing operations. Information covering the design, operation, and maintenance of fans for specific types of systems is widely available.

Fans are the primary moving devices used in industrial applications (such as LEV systems). They are used to convey the air contaminants from the source to a control device, which removes the contaminant from the airstream prior to discharge of the clean air to the atmosphere. The present section will describe the different types of air-moving devices, design equations, fan selection procedures, and fan installations.

Fans can be classified into two main types: centrifugal and axial. A centrifugal fan consists of a wheel or rotor mounted on a shaft that rotates in a scroll-shaped housing. The air enters at the eye of the scroll, makes a right-angle turn, and is forced through the blades of the rotor by centrifugal force into the scroll-shaped housing. These fans have four basic impeller designs: airfoil, backward curved, radial, and forward curved. Figure 17 identifies the common terminology for centrifugal fan components. Figure 18 identifies the common terminology for inlet and outlet fans. Figure 19 identifies the drive arrangements for centrifugal fans.

The performance of a fan is characterized by the volume of gas flow, the pressure at which this flow is produced, the speed of rotation, the power required, and the efficiency.

Testing methods as specified by the Air Movement and

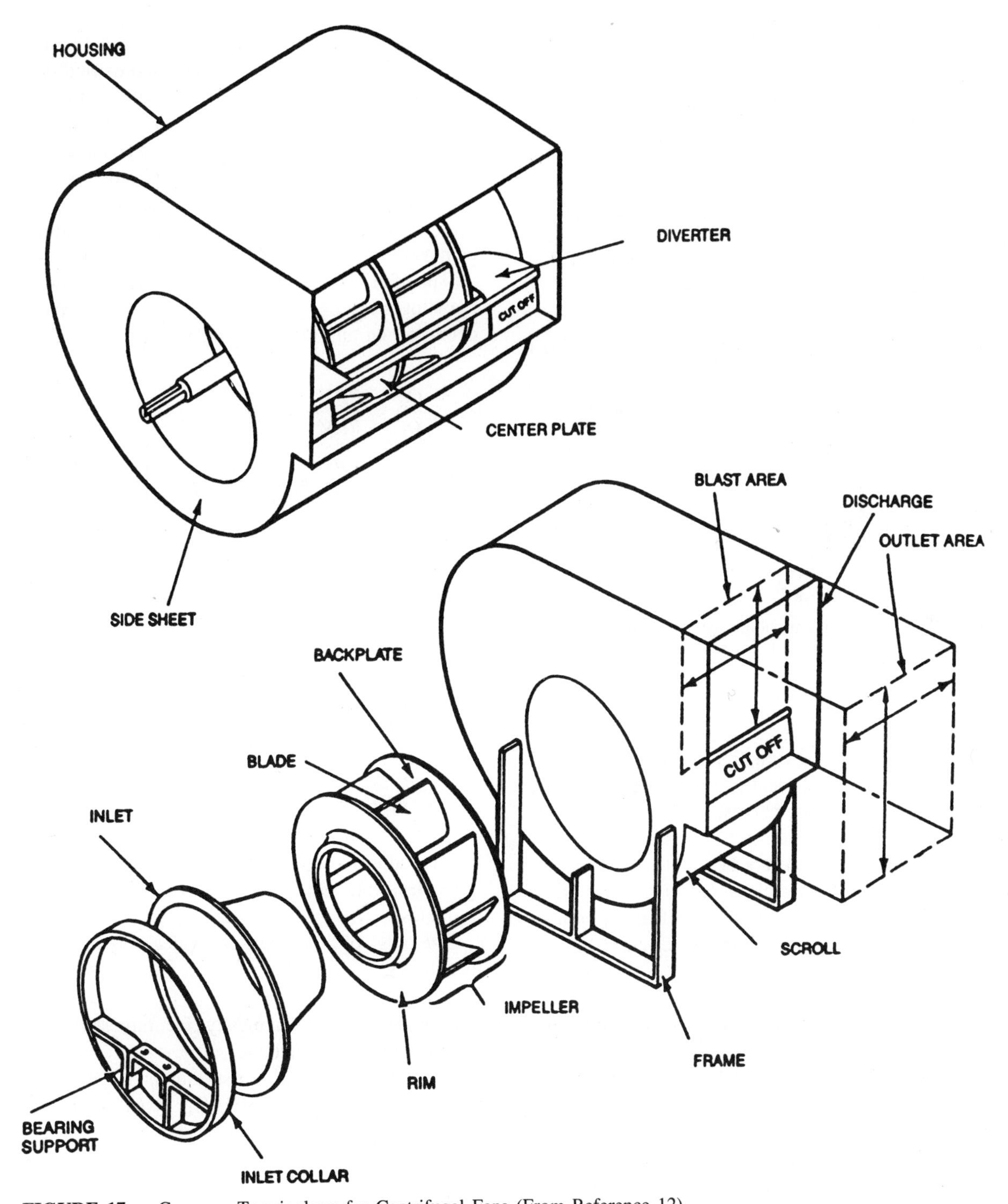

FIGURE 17. Common Terminology for Centrifugal Fans (From Reference 12)

Control Association (AMCA)[13] are used by the fan manufacturers to establish the relationships of these parameters. The test method consists of mounting a fan on the end of a straight length of duct and operating it with various sized orifices (typically about 10) in the duct. For each test, the volume flow is measured, along with pressure, velocity, and power input. Test results are then plotted to obtain the fan characteristic curves. From the volume and pressure, the air horsepower is computed. The efficiency based on total pressure is called mechanical efficiency, while the efficiency based on the friction losses is called static efficiency. Figure 20 identifies the impeller design, housing design, performance curves, performance characteristics, and applications for centrifugal fans.

Axial fans include all those fans in which the air flows through the impeller in a path approximately parallel with the shaft upon which the impeller is mounted. The three basic types of axial fans are propeller, tube axial, and vane

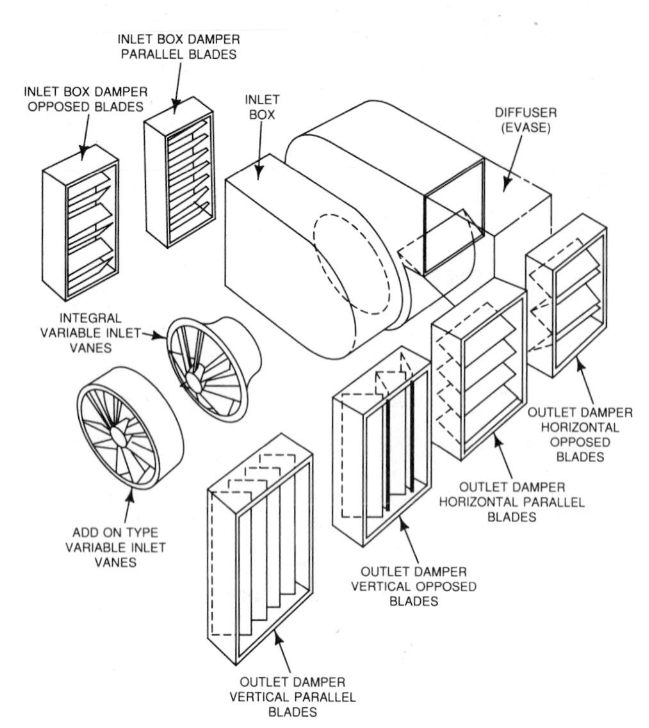

FIGURE 18. Common Terminology for Inlet and Outlet Fans (From Reference 12)

axial. Figure 21 presents the common terminology for axial fan components. Figure 22 identifies the impeller design, housing design, performance curves, performance characteristics, and typical applications for axial fans. It is generally recognized that axial fans are best suited for handling high volumes of a relatively clean gas at a lower static pressure. These fans tend to be of higher efficiency with an essentially flat and self-limiting horsepower curve.

Design Parameters

The starting point for the design of fans is the well-known "fan law" based on the classical theories of fluid mechanics. Simplifying assumptions are made that ignore considerations of Reynolds number, Mach number, kinematic viscosity, dynamic viscosity, surface roughness, impeller blade thickness, relative clearances, and so on. Although the accuracy of the fan law is sufficient for most cases, the AMCA is conducting investigations into the above effects with the aim of developing practical adjustments to the fan law that would give more accurate projections of fan performance. Preliminary results indicate that the correction factors being developed are different for each type of fan and may vary significantly among different manufacturers' designs. Based on these observations, it is recommended that the fan manufacturer be consulted before implementing any major changes in the speed or point of operation of an existing fan installation.

The equations based on the fan law enable the results of fan tests at one speed to be used to calculate performance at other speeds and other gas densities and for a complete series of different-sized, but geometrically similiar (homologous) fans. Each set of equations for changes in speed, size, or density applies only to the same point of rating, and all equations in the set must be used to define the converted condition. "Point of rating" is defined by AMCA as a percentage of wide-open volume flow rate and a percentage of shutoff pressure, or a point of constant efficiency.

The fan law equations are listed below for any homologous series of fans at the same relative point of operation on any two performance curves.

$$\frac{Q_2}{Q_1} = \frac{RPM_2}{RPM_1} \times \left(\frac{D_2}{D_1}\right)^3$$

$$\frac{TP_2}{TP_1} = \frac{VP_2}{VP_1} = \frac{SP_2}{SP_1} = \left(\frac{RPM_2}{RPM_1}\right)^2 \times \left(\frac{D_2}{D_1}\right)^2 \times \frac{\rho_2}{\rho_1}$$

$$\frac{BHP_2}{BHP_1} = \left(\frac{RPM_2}{RPM_1}\right)^3 \left(\frac{D_2}{D_1}\right)^5 \frac{\rho_2}{\rho_1}$$

where: Q = gas flow rate
RPM = revolutions per minute
D = wheel diameter
TP = total pressure
VP = velocity pressure
SP = static pressure
p = gas density
BHP = brake horsepower

Any convenient units may be employed as long as they are consistent, since the equations involve ratios of the variables.

The range of limitations of applicability of the fan law is as follows:

1. Gas is thermodynamically incompressible and flow through fan is adiabatic.
2. T_2/T_1 is in absolute pressure units <1.036.
3. $D_2/D_1 < 3{:}D_2 > D_1{:}\ (D_2/D_1) \times (RPM_2/RPM_1) \leq^3$.
4. $RPM_2/RPM_1 \leq^3$.
5. $(D_2/D_1) \times (RPM_2/RPM_1) \leq^3$.
6. $Re_1 \geq Re_2$.

In practice, the principles expressed by the fan law equations are normally applied to determine the effect of changing only one variable. Table 10 shows equations that can be used to determine the effect of varying speed, changing density, and increasing wheel-diameter size.

The pressure loss in the ducted system is obtained by calculating pressure losses for each of the components of the system. At a fixed volume flow through a given ducted system, a corresponding pressure loss will exist. If the flow changes, the new static pressure (SP) is closely approximated using the following parabolic relationship.

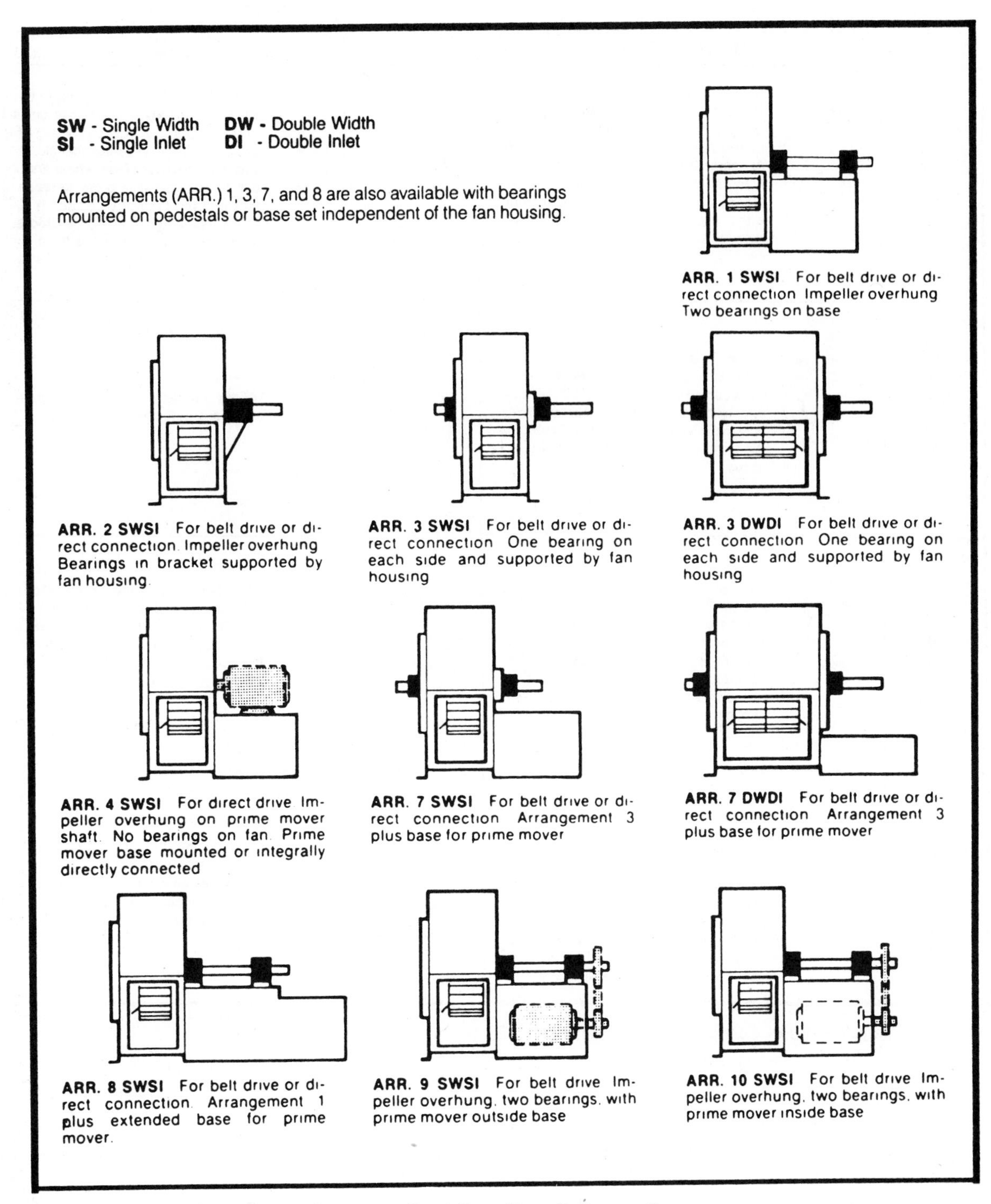

FIGURE 19. Drive Arrangements for Centrifugal Fans (From Reference 4)

$$SP = aQ^2$$

where: a = constant
Q = volume flow rate, scfm

The reason for this relationship is that nearly all losses in the system are function of the velocity squared. Table 11 presents velocity–flow relationships for different types of losses in ducted systems.

Figure 23 illustrates typical plots of the resistance to flow versus volume flow rate for three different and arbitrary fixed-duct systems (A, B, and C). For a fixed system, an increase or decrease in system resistance results from an increase or decrease in the volume flow rate along the given system curve only.

Duct system A has a system design point at 100% volume and 100% duct system resistance. An increase in flow to 120% of the design volume results in an increase

	TYPE	IMPELLER DESIGN	HOUSING DESIGN
CENTRIFUGAL FANS	AIRFOIL	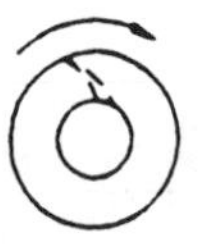 Highest efficiency of all centrifugal fan designs. 9 to 16 blades of airfoil contour curved away from the direction of rotation. Air leaves the impeller at a velocity less than its tip speed and relatively deep blades provide for efficient expansion within the blade passages. For given duty, this will be the highest speed of the centrifugal fan designs.	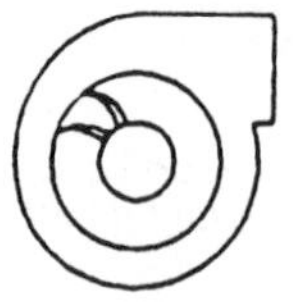 Scroll-type, usually designed to permit efficient conversion of velocity pressure to static pressure, thus permitting a high static efficiency, essential that clearance and alignment between wheel and inlet bell be very close in order to reach the maxiumum efficiency capability. Concentric housings can also be used as in power roof ventilators, since there is efficient pressure conversion in the wheel.
	BACKWARD-INCLINED BACKWARD-CURVED	 Efficiency is only slightly less than that of airfoil fans. Backward-inclined or backward-curved blades are single thickness. 9 to 16 blades curved or inclined away from the direction of rotation. Efficient for the same reasons given for the airfoil fan above.	 Utilizes the same housing configuration as the airfoil design.
	RADIAL	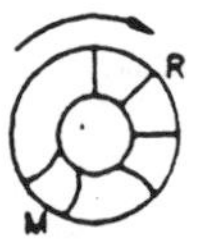 Simplest of all centrifugal fans and least efficient. Has high mechanical strength and the wheel is easily repaired. For a given point of rating, this fan requires medium speed. This classification includes radial blades (R) and modified radial blades (M), usually 6 to 10 in number.	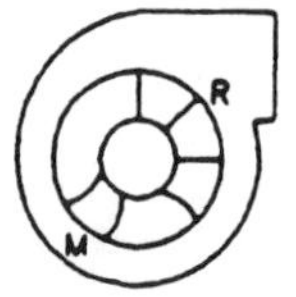 Scroll-type, usually the narrowest design of all centrifugal fan designs described here because of required high velocity discharge. Dimensional requirements of this housing are more critical than for airfoil and backward-inclined blades.
	FORWARD-CURVED	 Efficiency is less than airfoil and backward-curved bladed fans. Usually fabricated of lightweight and low cost construction. Has 24 to 64 shallow blades with both the heel and tip curved forward. Air leaves wheel at velocity greater than wheel. Tip speed and primary energy transferred to the air is by use of high velocity in the wheel. For given duty, wheel is the smallest of all centrifugal types and operates at lowest speed.	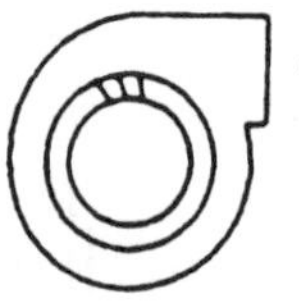 Scroll is similar to other centrifugal-fan designs. The fit between the wheel and inlet is not as critical as on airfoil and backward-inclinded bladed fans. Uses large cut-off sheet in housing.

FIGURE 20. Characteristics of Centrifugal Fans (From Reference 4)

in duct system resistance of 150% from design. A decrease in volume flow to 50% results in a decrease to 30% of the design resistance. Similar relationships also hold for duct system B and duct system C as shown in Figure 23.

The system resistance depends on the density of the gas flowing through the system. The standard gas density is 0.075 lb/ft^3. The dependence of fan performance on gas density is obtained from the fan law equations. The fan pressure and power vary directly as the ratio of the gas density at the fan inlet to standard density. The density ratio must always be considered when selecting fans from manufacturers' catalogs or curves.

The calculation sheets for obtaining system pressure losses are based on static pressures. Most industrial fan rating tables are based on fan static pressures, which cannot be read directly from the calculation sheet. Fan static pressure is defined by AMCA 201 as the total pressure for the fan minus the fan velocity pressure.

$$\text{Fan } SP = \text{Fan } TP - \text{Fan } VP$$
$$= SP_{\text{outlet}} - SP_{\text{inlet}} - VP_{\text{inlet}}$$

Fan power is calculated from the equation

$$P_i = \frac{Q \times \text{Fan}_{tp}}{\eta T}$$

where: P_i = fan input power
Q = volume flow rate
Fan_{Tp} = fan total pressure
η_T = fan total efficiency

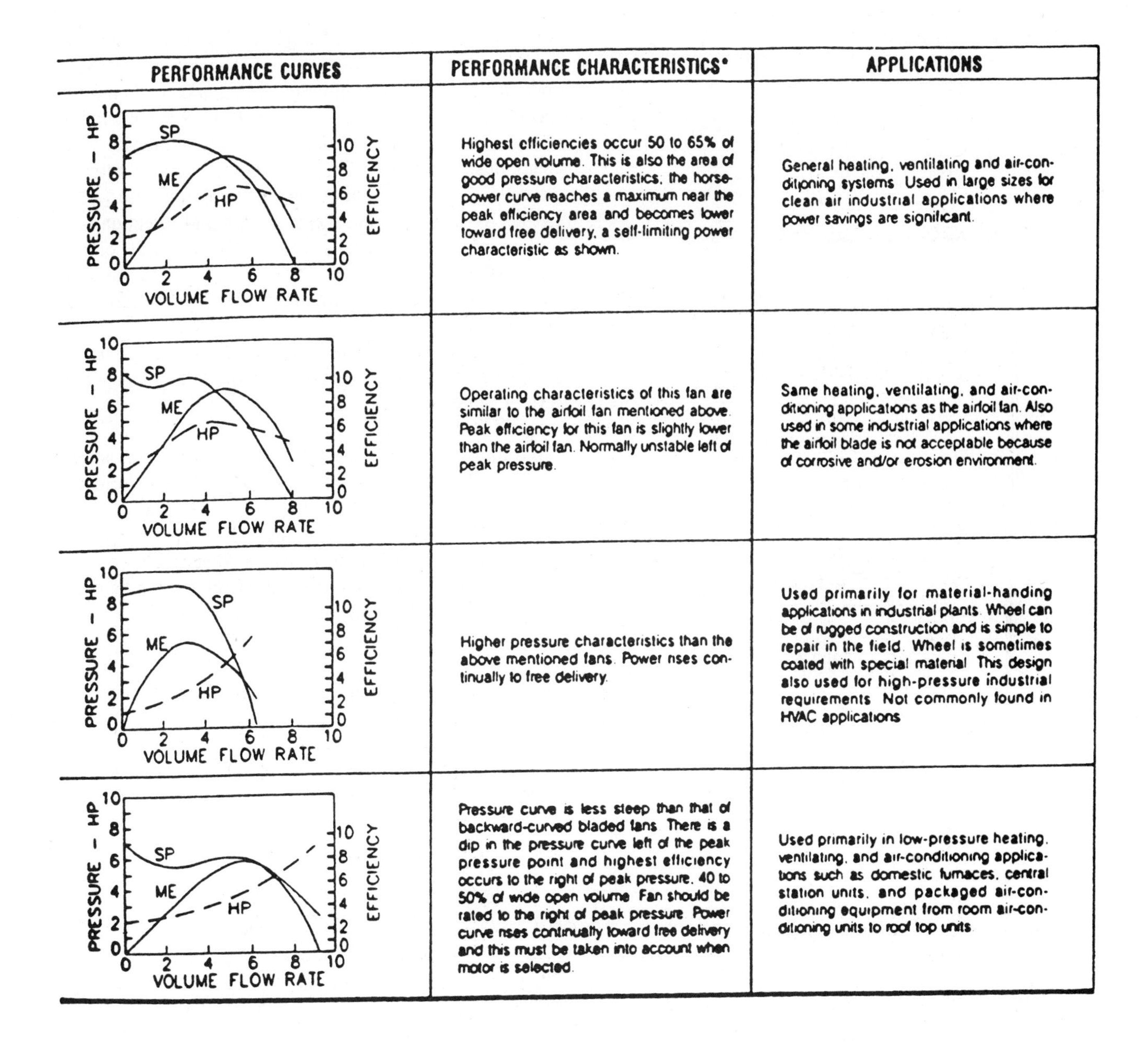

PERFORMANCE CURVES	PERFORMANCE CHARACTERISTICS*	APPLICATIONS
	Highest efficiencies occur 50 to 65% of wide open volume. This is also the area of good pressure characteristics; the horsepower curve reaches a maximum near the peak efficiency area and becomes lower toward free delivery, a self-limiting power characteristic as shown.	General heating, ventilating and air-conditioning systems. Used in large sizes for clean air industrial applications where power savings are significant.
	Operating characteristics of this fan are similar to the airfoil fan mentioned above. Peak efficiency for this fan is slightly lower than the airfoil fan. Normally unstable left of peak pressure.	Same heating, ventilating, and air-conditioning applications as the airfoil fan. Also used in some industrial applications where the airfoil blade is not acceptable because of corrosive and/or erosion environment.
	Higher pressure characteristics than the above mentioned fans. Power rises continually to free delivery.	Used primarily for material-handing applications in industrial plants. Wheel can be of rugged construction and is simple to repair in the field. Wheel is sometimes coated with special material. This design also used for high-pressure industrial requirements. Not commonly found in HVAC applications.
	Pressure curve is less steep than that of backward-curved bladed fans. There is a dip in the pressure curve left of the peak pressure point and highest efficiency occurs to the right of peak pressure, 40 to 50% of wide open volume. Fan should be rated to the right of peak pressure. Power curve rises continually toward free delivery and this must be taken into account when motor is selected.	Used primarily in low-pressure heating, ventilating, and air-conditioning applications such as domestic furnaces, central station units, and packaged air-conditioning equipment from room air-conditioning units to roof top units.

A fan performance curve is a graphical representation of the performance of a fan. Typical fan curves cover the range from no delivery (an airtight system with no air flowing) to free delivery with no obstructions to flow. Information stated on a fan curve should include gas density, impeller size, and speed. Fans selected at nonstandard density require knowledge of the actual volumetric flow rate at the fan inlet, the actual pressure requirement, and the density of the gas at the fan inlet.

Fan Selection

A successful fan installation requires the proper selection of a fan to match the required flows and static pressures for the LEV system. The following is a checklist of factors to be specified or considered.

1. Actual inlet airflow/unit time
2. Fan static pressure
3. Density of gas at the inlet
4. Dust characteristics and loading
5. Barometric pressure
6. Desired fan speed
7. Explosive or inflammable material
8. Whether direct drive or belt drive
9. Noise requirements
10. Fan efficiency
11. Layouts and space limitations
12. Capital and operating cost considerations

Performance is based on fan inlet conditions and the negative pressure developed by the fan must be taken into consideration in flow, pressure, and density calculations.

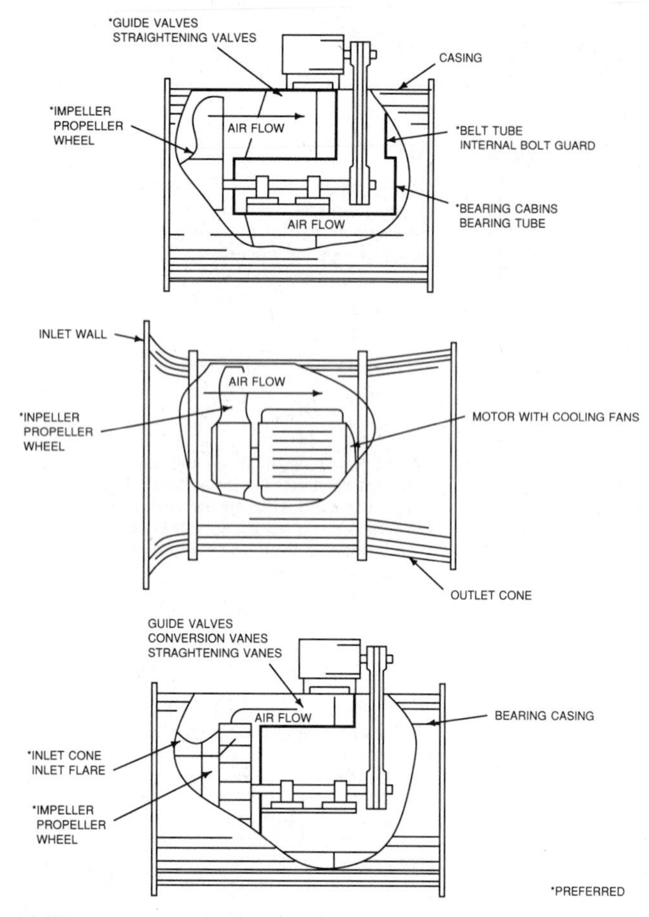

FIGURE 21. Common Terminology for Axial Fans (From Reference 12)

Fan specifications are often based on volume flow units, but sometimes mass flow units are used. The expected range of operating conditions (minimum/maximum/average) must be specified, along with any expected sudden temperature changes, so that the fan design can accommodate those conditions. The type, size, and expected concentration of the entrained dust should be included so that these factors can be considered in the wheel design and selection.

Any information known on acceptable materials of construction for the fan components or the specific environmental conditions of the gas stream should be specified by the user.

The system designer preparing specifications should use as a reference the fan publication manuals[13] prepared by the AMCA entitled:

Publication 201	Fans and Systems (fan testing in laboratories and calculation of performance ratings)
Publication 202	Troubleshooting
Publication 203	Field Performance Measurement
Publication 210	Laboratory Methods of Testing Fans For Rating Purposes

The topics covered by these publications give the designer a realistic estimate of fan-system performance for any specific application.

AMCA prescribes minimum static pressure and air velocity performances for Class I, II, and III fans. There are separate standards for backward-inclined, single- and dou-

TABLE 10. Fan Laws Equations

Fan Parameter	Effect of Varying Speed	Effect of Changed Density	Effect of Increased Size
	$(D_2/D_1 = \rho_2/\rho_1 = 1)$	$(RPM_2/RPM_1 = 1, D_2/D_1 = 1)$	$(RPM_2/RPM_1 = 1, \rho_2/\rho_1 = 1)$
Q	$Q_2 = Q_1 \times \left(\frac{RPM_2}{RPM_1}\right)$	$Q_2 = Q_1$	$Q_2 = Q_1 \times \left(\frac{D_2}{D_1}\right)^3$
TP	$TP_2 = TP_1 \times \left(\frac{RPM_2}{RPM_1}\right)^2$	$TP_2 = TP_1 \times \left(\frac{\rho_2}{\rho_1}\right)$	$TP_2 = TP_1 \times \left(\frac{D_2}{D_1}\right)^2$
VP	$VP_2 = VP_1 \times \left(\frac{RPM_2}{RPM_1}\right)^2$	$VP_2 = VP_1 \times \left(\frac{\rho_2}{\rho_1}\right)$	$VP_2 = VP_1 \times \left(\frac{D_2}{D_1}\right)^2$
SP	$SP_2 = SP_1 \times \left(\frac{RPM_2}{RPM_1}\right)^2$	$SP_2 = SP_1 \times \left(\frac{\rho_2}{\rho_1}\right)$	$SP_2 = SP_1 \times \left(\frac{D_2}{D_1}\right)^2$
BHP	$BHP_2 = BHP_1 \times \left(\frac{RPM_2}{RPM_1}\right)^3$	$BHP_2 = BHP_1 \times \left(\frac{\rho_2}{\rho_1}\right)$	$BHP_2 = BHP_1 \times \left(\frac{D_2}{D_1}\right)^5$

TABLE 11. Velocity and Flow Relationships for Different Types of Losses

Type of Loss	Equation[a]	Relationships: Velocity	Flow[a]
Hood	$h_e = F \times VP$	$h_e = V^2$	$h_e = Q^2$
Fitting	$h_f = K \times VP$	$\text{Loss} = V^2$	$\text{Loss} = Q^2$
Friction	$h_f = K \times Q^{1.9}$	$\text{Loss} = V^{1.9}$	$\text{Loss} = Q^{1.9}$
Baghouse			$\text{Loss} = aQ^2 + bQ^{1.6}$

[a]F, K, a, and b are constants.

ble-width fans; forward-curved, single- and double-width fans; and backward-inclined fans. As an example, the class standard for backward-inclined fans is as follows:

Class I: 5 in. w.g. at 2300 fpm to 2.5 in. w.g. at 3200 fpm
Class II: 8.5 in. w.g. at 3000 fpm to 4.25 in. w.g. at 4150 fpm
Class III: 12.5 in. w.g. at 3850 fpm to 6.5 in. w.g. at 5250 fpm
Class IV: Above Class III minima

The quality of construction is specified based on minimum gauge thickness and not on class.

Fans may be operated in series or in parallel. Figure 24 shows combined characteristic curves for two fans in series and for two fans in parallel. The combined volume–pressure curve for two fans operating in series is obtained by adding the fan pressures at the same air volume flow. For fans in parallel, the combined volume–pressure curve is obtained by adding the air volume flows at the same fan pressures.

The effect of the duct system curve not being at the design point is shown in Figure 25. The design system pressure (point 1) is overestimated from the actual system pressure (point 2). A common mistake made by designers is the belief that an overestimation of system pressure losses is a conservative approach. As shown in Figure 25, an overestimation of the design system pressure loss results in the actual fan power's requirements (point 4) being higher than the design fan power (point 5). The consequence of this situation is that the fan motor may be undersized for the specific application.

The effect of the fan and its inlet and outlet duct connections will often have a significant effect on the flow conditions at the fan inlet and outlet. The result of these disturbances or system effects is that the fan does not perform as expected. Publication 203 of the AMCA has developed an approach to quantifying the system effect losses using a concept called "system effect factor." A family of system effect curves can be established that gives system effect losses as a function of velocity for different configurations. This system effect factor, which must be determined separately for the inlet and the outlet, is added to the estimated total system effect pressure losses to give the system design point of operation. This total estimated system effect pressure loss is used to determine the fan power requirements from the manufacturer's catalog data. Information related to specific types and models of fans may be available directly from the manufacturer.

The following are the four most common causes of system effects.

1. Eccentric flow into fan
2. Spinning flow into the fan
3. Improper outlet ductwork
4. Obstructions at the inlet or outlet

A designer must take system effects into account in order to ensure successful fan and ventilation system performance.

Fan Installation

The proper installation of a fan is critical to ensure a successful start-up and ongoing operation. Some of the important features are as follows:

1. Complete a detailed inspection of fan installation before start-up.
2. Complete the AMCA 202 troubleshooting lists.
3. Make sure that the foundation is sound.
4. Make sure that the fan is firmly secured to the foundation.
5. Check that the impeller is installed correctly.
6. Make sure that inlet sleeves to the wheel are properly located and secure.
7. Check the installation of the damper and its operation.
8. Check the level of the fan shaft and plumb of the casing.
9. Ensure correct bearing installation, including alignment.
10. Ensure proper bearing lubrications.
11. Check the alignment of the fan to the drive for both parallelism and concentricity.
12. Bump the fan for correct rotation.
13. After start-up, check for overall smoothness of operation, along with overall operation of the bearing.

TYPE			IMPELLER DESIGN	HOUSING DESIGN
AXIAL FANS	PROPELLER		 Efficiency is low. Impellers are usually of inexpensive construction and limited to low pressure applications. Impeller is of 2 or more blades, usually of single thickness attached to relatively small hub. Energy transfer is primarily in form of velocity pressure.	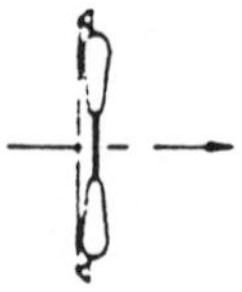 Simple circular ring, orifice plate, or venturi design. Design can substantially influence performance and optimum design is reasonably close to the blade tips and forms a smooth inlet flow contour to the wheel.
AXIAL FANS	TUBEAXIAL		 Somewhat more efficient than propeller fan design and is capable of developing a more useful static pressure range. Number of blades usually from 4 to 8 and hub is usually less than 50% of fan tip diameter. Blades can be of airfoil or single thickness cross section.	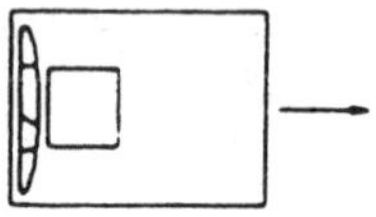 Cylindrical tube formed so that the running clearance between the wheel tip and tube is close. This results in significant improvement over propeller fans.
AXIAL FANS	VANEAXIAL		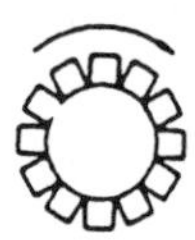 Good design of blades permits medium- to high-pressure capability at good efficiency. The most efficient fans of this type have airfoil blades. Blades are fixed or adjustable pitch types and hub is usually greater than 50% of fan tip diameter.	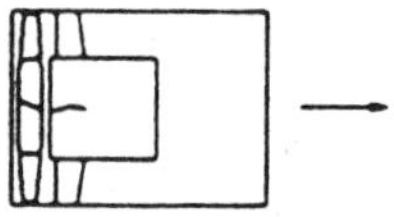 Cylindrical tube closely fitted to the outer diameter of blade tips and fitted with a set of guide vanes. Upstream or downstream from the impeller, guide vanes convert the rotary energy imparted to the air and increase pressure and efficiency of fan.
SPECIAL DESIGNS	TUBULAR	CENTRIFUGAL	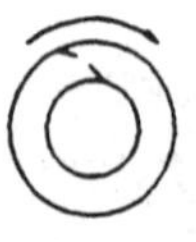 This fan usually has a wheel similar to the airfoil backward-inclined or backward-curved blade as described above. (However, this fan wheel type is of lower efficiency when used in fan of type.) Mixed flow impellers are sometimes used.	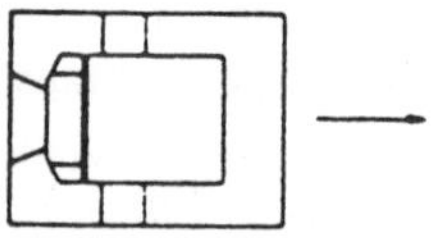 Cylindrical shell similar to a vaneaxial fan housing, except the outer diameter of the wheel does not run close to the housing. Air is discharged radially from the wheel and must change direction by 90 degrees to flow through the guide vane section.
SPECIAL DESIGNS	POWER ROOF VENTILATORS	CENTRIFUGAL	 Many models use airfoil or backward-inclined impeller designs. These have been modified from those mentioned above to produce a low-pressure, high-volume flow rate characteristic. In addition, many special centrifugal impeller designs are used, including mixed-flow design.	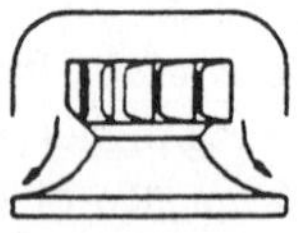 Does not utilize a housing in a normal sense since the air is simply discharged from the impeller in a 360 degree pattern and usually does not include a configuration to recover the velocity pressure component.
SPECIAL DESIGNS	POWER ROOF VENTILATORS	AXIAL	 A great variety of propeller designs are employed with the objective of high-volume flow rate at low pressure.	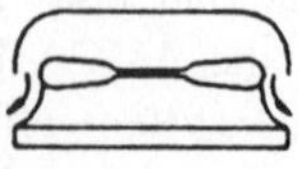 Essentially a propeller fan mounted in a supporting structure with a cover for weather protection and safety considerations. The air is discharged through the annular space around the bottom of the weather hood.

FIGURE 22. Characteristics of Axial Fans (From Reference 4)

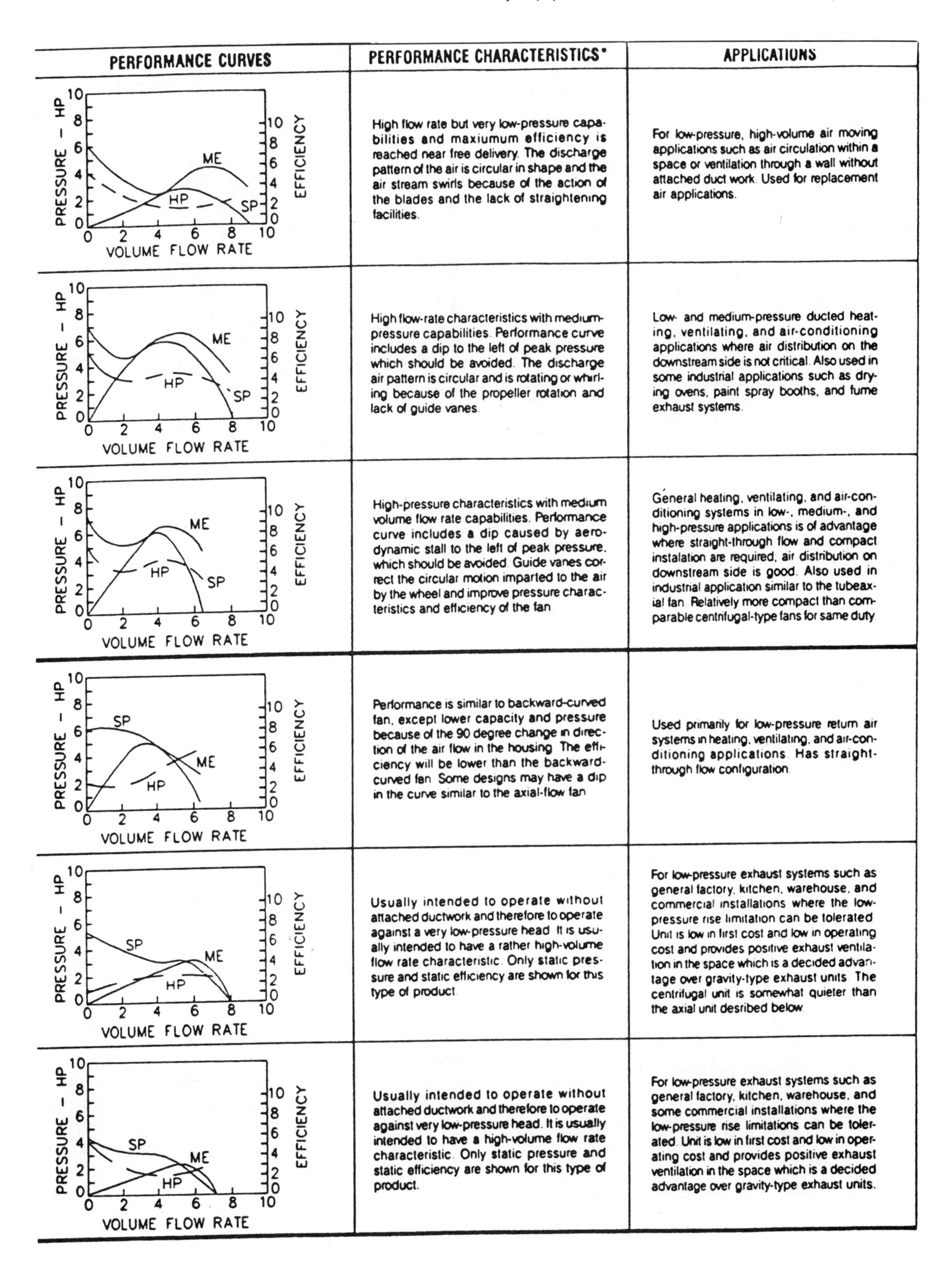

PERFORMANCE CURVES	PERFORMANCE CHARACTERISTICS*	APPLICATIONS
	High flow rate but very low-pressure capabilities and maxiumum efficiency is reached near free delivery. The discharge pattern of the air is circular in shape and the air stream swirls because of the action of the blades and the lack of straightening facilities.	For low-pressure, high-volume air moving applications such as air circulation within a space or ventilation through a wall without attached duct work. Used for replacement air applications.
	High flow-rate characteristics with medium-pressure capabilities. Performance curve includes a dip to the left of peak pressure which should be avoided. The discharge air pattern is circular and is rotating or whirling because of the propeller rotation and lack of guide vanes.	Low- and medium-pressure ducted heating, ventilating, and air-conditioning applications where air distribution on the downstream side is not critical. Also used in some industrial applications such as drying ovens, paint spray booths, and fume exhaust systems.
	High-pressure characteristics with medium volume flow rate capabilities. Performance curve includes a dip caused by aerodynamic stall to the left of peak pressure, which should be avoided. Guide vanes correct the circular motion imparted to the air by the wheel and improve pressure characteristics and efficiency of the fan.	General heating, ventilating, and air-conditioning systems in low-, medium-, and high-pressure applications is of advantage where straight-through flow and compact instalation are required; air distribution on downstream side is good. Also used in industrial application similar to the tubeaxial fan. Relatively more compact than comparable centrifugal-type fans for same duty.
	Performance is similar to backward-curved fan, except lower capacity and pressure because of the 90 degree change in direction of the air flow in the housing. The efficiency will be lower than the backward-curved fan. Some designs may have a dip in the curve similar to the axial-flow fan.	Used primarily for low-pressure return air systems in heating, ventilating, and air-conditioning applications. Has straight-through flow configuration.
	Usually intended to operate without attached ductwork and therefore to operate against a very low-pressure head. It is usually intended to have a rather high-volume flow rate characteristic. Only static pressure and static efficiency are shown for this type of product.	For low-pressure exhaust systems such as general factory, kitchen, warehouse, and commercial installations where the low-pressure rise limitation can be tolerated. Unit is low in first cost and low in operating cost and provides positive exhaust ventilation in the space which is a decided advantage over gravity-type exhaust units. The centrifugal unit is somewhat quieter than the axial unit desribed below.
	Usually intended to operate without attached ductwork and therefore to operate against very low-pressure head. It is usually intended to have a high-volume flow rate characteristic. Only static pressure and static efficiency are shown for this type of product.	For low-pressure exhaust systems such as general factory, kitchen, warehouse, and some commercial installations where the low-pressure rise limitations can be tolerated. Unit is low in first cost and low in operating cost and provides positive exhaust ventilation in the space which is a decided advantage over gravity-type exhaust units.

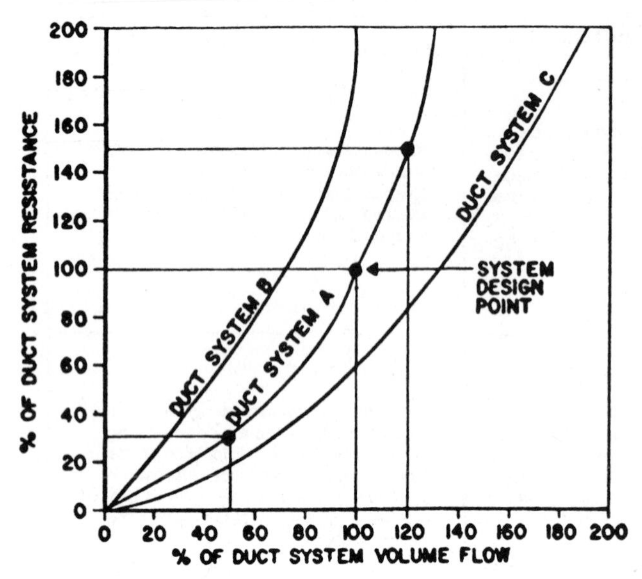

FIGURE 23. Flow System Resistance for Three Different Ducted Systems

An important aspect of fan performance is fan balancing. Technology development in this field is quite recent and is well described in the literature. The techniques for static and dynamic balancing of fan wheels and shaft assemblies, prior to shipment, are described by Myers.[14] The use of vibration measurements and analyses allows the detection and diagnosis of problems in fans before they become serious.

Satisfactory performance of a ventilation system is dependent on matching the fan with the system. It is essential that proper homework and calculations be done when fan specifications are being prepared and duct layouts are being developed. Field modifications to fans to correct system performance problems are always expensive and very often

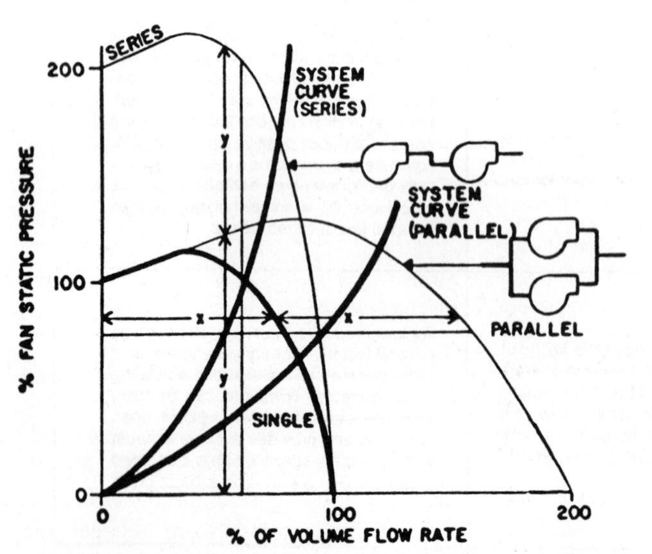

FIGURE 24. Combined Characteristic Curves for Fans in Series or in Parallel

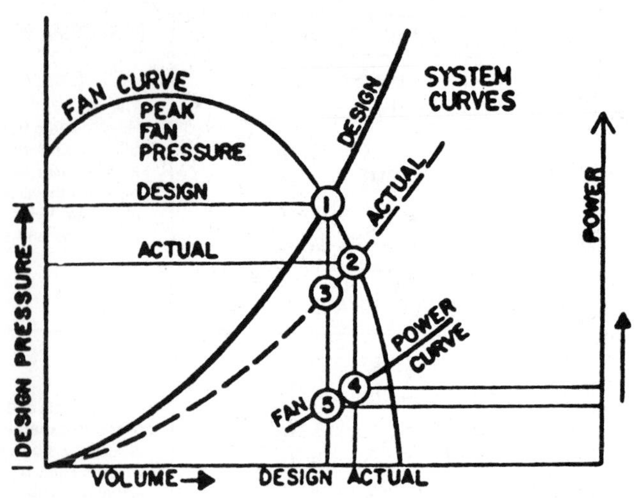

FIGURE 25. Duct System Curve Not at Design Point

unsatisfactory. It is common to have the fan operating away from its point of maximum efficiency and at increased noise levels. The careful selection and specification of the fan and its associated duct system are essential to ensure that the ventilation system operates at its required performance level.

Stacks

Technical Description

In the not so distant past, stacks were used as a method of pollution control, the idea being that the best solution to pollution was "dilution." In other words, the taller the stack, the greater was the dispersion of contaminants and, therefore, the more effective was the "control." Now, as the direct result of changes in environmental legislation and public attitudes, stacks are the final component of sometimes highly complex air pollution control or ventilation systems.

The design of stacks involves rigorous procedures that employ air dispersion modeling techniques to predict downwind contaminant loadings. Stack heights and diameters have to adhere to "good engineering practice" and also to regulatory constraints.

Stacks are used in a multitude of applications and, consequently, their design is invariably specific to the site and process involved. Considering the large capital outlay required, it is important that a proper engineering feasibility study be carried out in order to evaluate all the alternatives so that the most cost-effective solution can be implemented.

There are four important parameters to be considered in specifying a stack.

- Materials of construction
- Liners
- Structural support
- Accessories

Stacks can be made of many different materials, including steel, brick, fiberglass-reinforced plastic (FRP), and reinforced concrete. In order properly to assess which material would best suit the application, information on the physical and chemical properties of the gas stream is very important. Such properties as the corrosiveness and acidity or basicity of the gas must be determined. The anticipated temperature differential between the gas and the outside air is also an important parameter for materials selection.

Stacks can be loosely categorized into two types, lined and unlined. Liners are usually recommended for most designs because they are able to withstand the chemical composition and temperature of the gas. Also, a liner allows the airspace between it and the outer column to act as insulation. Liners can be replaced at much lower cost than can complete stacks, should they be exposed to particularly corrosive gases. Liners exposed to such conditions should be made of stainless steel, brick, or FRP. Where corrosion is not a problem, carbon-steel liners are adequate. For unlined stacks, coatings can be applied to the inside of the stack for protection; however, these tend to be expensive and, in general, do not last as long as liners.

Stack supports can be free standing, intermittently supported, or guyed, although the last is applicable only to short stacks. Regardless of the type of support structure, the designer must design the stacks with wind loads and seismic zones in mind.

Stack accessories include a clean-out or access door, a sampling platform, ladders, lightning protection systems, and aircraft warning lights. The clean-out door is needed to allow the removal of any accumulated materials at the bottom of the stack, and in the case of lined stacks, it should also provide access to the space between the liner and the outer shell. Regulatory authorities require sampling platforms for all stacks for auditing and compliance purposes, and these should be located at least 10 stack diameters downstream of the inlet duct. Access ladders are an important part of sampling and maintenance activities. Lightning protection systems are important to protect a sizable investment. The requirement for aircraft warning lights is based on site-specific aviation requirements.

Since stacks form an integral part of air pollution control systems, complete drawings detailing the above should be developed and included in the required permit application.

Design Parameters

The minimum information required for the proper and successful design of a stack is as follows:

1. Detailed description of the process generating the off-gas, including performance characteristics of the pollution control devices located upstream of the stack.
2. Gas flow rates and inlet temperatures for current, future, and design conditions, including maximum, minimum, and average values.
3. Draft required at the stack opening.
4. Site-specific data, including elevation above sea level and ambient temperature fluctuations.
5. Topographical description and information on potential for seismic activities.
6. Plan and elevation layout of the plant property and buildings.
7. Meteorological data, of which wind data are especially important.

Using this process and plant information, the important stack parameters that need to be calculated and/or specified are as described in the following.

1. *Stack Height.* Stack heights are designed according to good engineering practice (GEP). Accepted GEP is a stack height equal to 2.5 times the height of the nearest building. The U.S. Environmental Protection Agency (EPA) has defined a building within a distance of $5*L$ (where L is the lesser of the building height or diagonal width of the building) to be a building that affects the stack plume and consequently the stack height must be designed according to GEP. The EPA has also defined GEP to be calculated as: GEP = Bldg. height + 1.5 L, or a maximum of 65 meters (213 feet). No credit for stacks of a height greater than 65 meters will be given in modeling calculations.
2. *Effective stack height.* The effective height of a stack is defined as the stack height plus the plume rise. Plume rise is a function of plume velocity and temperature differential between the off-gas and the ambient air. A 1°F differential is expected to produce a corresponding increase in effective stack height of about 9 feet.
3. *Duct entry size.* The diameter or width of the entry duct should not be greater than 50% of the liner diameter or 33% of the outer shell diameter when there is no liner.
4. *Gas exit velocity.* Gas exit velocities are an important parameter in stack design because of their relationship to downwind contaminant concentrations. High stack velocities increase plume height and, therefore, the effective stack height. Low stack velocities would allow the gas to be drawn into the eddy zones created by the wind flowing over the stack and may, in fact, cause a downwash. This, in turn, would cause the off-gas to enter the building cavity and introduce contaminants into the building, as shown in Figure 26. Consequently, the stack exit velocity should be approximately one to five times that of the wind. Exit velocities of between 3000 and 4000 fpm have been found to provide an adequate margin of safety in design. However, in order to be sure, average, maximum, and minimum wind velocities should be obtained for the site.
5. *Stack effect.* The stack effect or stack draft is the draft created by the difference in temperature between the off-gas and the ambient air. Stack draft can be calculated by using the following equation.

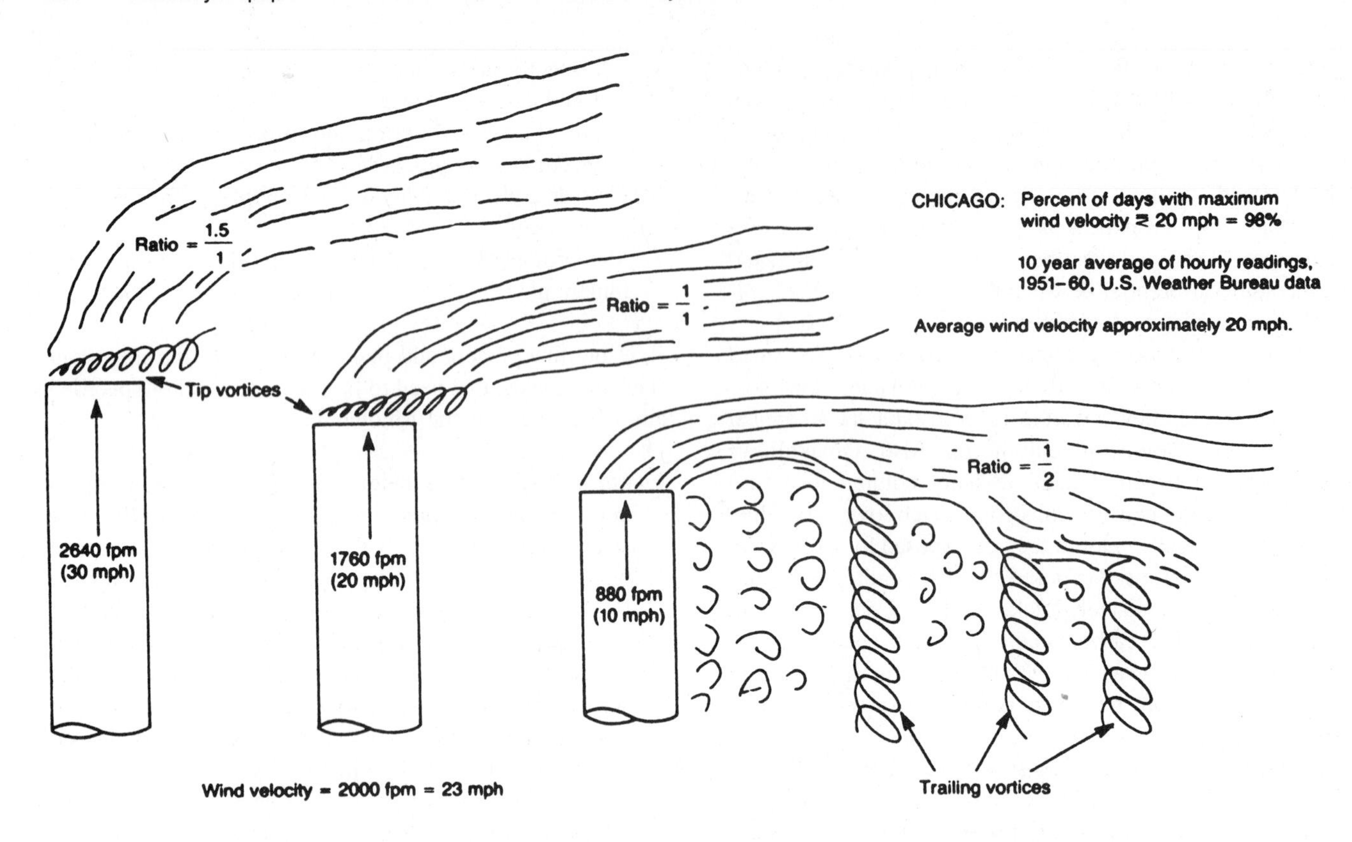

AIR FLOW over stacks is similar to that over buildings. Eddies in the form of tip and trailing vortices will bring the effluent down the stack unless the discharge velocity is high enough. Such downwash may bring the discharge into the building cavity and cause re-entry. Good discharge is obtained with a stack to wind velocity ratio of 1.5:1. Marginal flow occurs with a ratio of 1:1, and very poor flow occurs with a ratio of 1:2. For a 20 mph design wind velocity for Chicago, the stack velocity should be a minimum of 30 mph, or approximately 3000 fpm.

FIGURE 26. Relationship Between Stack Exit Velocity and Wind Velocity (From Reference 11)

$$P_D = 0.034\, L\, P_T \left(\frac{1}{T_{amb}} - \frac{1}{T_{mg}}\right)$$

where: P_D = theoretical stack draft, in. w.g.
L = stack height above breeching, feet
P_T = barometric pressure, in. w.g.
T_{amb} = ambient temperature, °R
T_{mg} = average gas temperature, °R

For stacks with cool gases, the stack effect need not be considered. However, for stacks handling hot gases, the stack effect may be as great as 1 to 2 in. w.g.

Other important design concerns for stacks include:

- Thickness of stack outer shell
- Resonance problems
- Reinforcement of openings in the outer shell
- Design of base plate and anchor bolts
- Design of stack stiffness

The above are described in detail in the *Guide for Steel Stack Design and Construction*[15] and are beyond the scope of this manual.

DESIGN AND OPERATION OF LOCAL EXHAUST VENTILATION SYSTEMS

System Parameters

Design Procedures for Sizing LEV Systems

In the previous sections, details of the design, sizing, and selection of the individual components were discussed (i.e., hoods, ducts, gas cooling equipment, fans, air cleaning devices and stacks). Using this information, the steps required to design an LEV system will be outlined. It is important that the designer accept and use the "systems approach" in order to develop cost-effective solutions. The systems approach requires the integration of the process, the work methods, and the LEV system. The size, cost, and effectiveness of the LEV system applied to any process operation will depend on how well the designer understands the process and how well the LEV system has been integrated into the process.

The preliminary data required for the design of an LEV system are as follows:

1. Process details of the operation (e.g., type of contaminant, level of toxicity, continuous or intermittent).

2. Site conditions (e.g., minimum/maximum/average temperatures, relative humidity, wind speed and directions).
3. Performance level for contaminant control.
4. Layout of operation.
5. Preliminary sketch of each hood.
6. Line sketch, including elevations of the ductwork layout, air cleaner location, fan location, stack location (isometric sketch recommended).

The design of an LEV system is an iterative process. The starting point is defining the parameters or practical ranges from the velocities or flow rates for each hood and through the duct network. The design for the LEV system continuously changes through the course of the project as equipment and plant layouts change. The final design for LEV systems must be based on approved layout drawings and certified drawings of the specific mechanical equipment.

A task to be addressed at the start of the design of an LEV system is the establishment of the nature and extent of the contaminant control problem. As described previously, the design objectives for the LEV system must be clearly defined in terms of expected hood performances or contaminant levels in the workplace environment. The recent development of high-speed computers provides the designer with the capabilities to predict contaminant levels using three-dimensional models for different performance levels of LEV systems in an industrial setting (e.g., RoomVent '90).

Before starting the design, two major decisions are required: the type of system (centralized versus distributed) and the balancing method (balanced system or plenum method). Figure 27 shows a flowchart for the ductwork design. A centralized system is based on the use of a single air-cleaning device using a complex network of properly designed ductwork to convey all the contaminated air from the different sources to a central air-cleaning unit. For some applications, it is advantageous to have a multiplicity of units serving different parts of the plant.

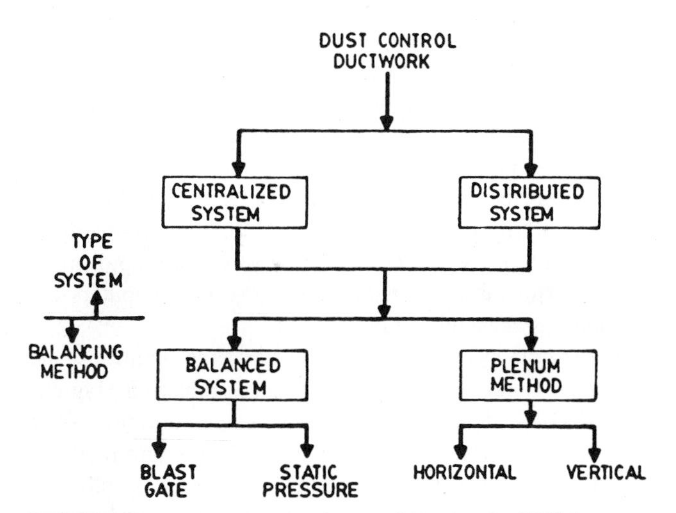

FIGURE 27. Flowchart for Ductwork Design for LEV Systems

The balanced system is the conventional approach and is based on the ducts' being sized for balance and adequate transport velocity. In the plenum approach, the plenum can be either vertical or horizontal, with ducts simply connected to the large plenum, which has a low velocity and a low resistance. For dust or fume control systems, the plenum approach requires that a dust/fume removal system be included in the design.

The procedure for designing an LEV system is as follows:

1. Select and design each exhaust hood.
2. Determine the volume of air that will be required at each hood enclosure in the system in order to ensure adequate contaminant capture.
3. Determine the minimum duct velocity based on the required transport velocity (for systems capturing particulate matter).
4. Determine the branch duct size by dividing the design flow rate by the minimum duct velocity.
5. Determine the design length for each duct segment and the number and type of any special fittings and elbows that are required.
6. Calculate the pressure losses for the entire exhaust system. (Two methods that can be used for these calculations are the velocity pressure method and the equivalent foot method.)
7. Select the fan on the basis of the final volumetric flow rate and the calculated system assistance.
8. Determine makeup air requirements for building air balance.

The procedure for selecting hood type and design exhaust flow rates was outlined in a previous section. The final design exhaust flow rate will be based on the specific hood application and the required level of contaminant control required for the operation.

The calculation of the duct diameter is based on the minimum transport velocity for the given material, as outlined in Table 12. The air velocity in the ductwork must be high enough to ensure that the contaminant does not drop out in the ductwork. A velocity much higher than the minimum transport velocity will result in excessive pressure losses, and for abrasive ducts, it will result in excessive erosion and high maintenance costs. Duct areas, and hence duct diameters, can be determined by dividing the exhaust flow rate by the minimum design velocity.

The general principles for ductwork layout are as follows:

1. Duct layout should be an integral part of the process equipment and plant layout (i.e., simple duct routing).
2. It should minimize horizontal runs.

TABLE 12. Transport Velocities for Different Materials

Material, Operation, or Industry	Minimum Transport Velocity m/s	Minimum Transport Velocity fpm
Abrasive blasting	17.8–20.3	3500–4000
Aluminum dust, coarse	20.3	4000
Asbestos carding	15.2	3000
Bakelite molding powder dust	12.7	2500
Barrel filling or dumping	17.8–20.3	3500–4000
Belt conveyors	17.8	3500
Bins and hoppers	17.8	3500
Brass turnings	20.3	4000
Bucket elevators	17.8	3500
Buffing and polishing		
Dry	15.2–17.8	3000–3500
Sticky	17.8–22.9	3500–4500
Cast iron boring dust	22.9	4000
Ceramics, general		
Glaze spraying	12.7	2500
Brushing	17.8	3500
Fettling	17.8	3500
Dry pan mixing	17.8	3500
Dry press	17.8	3500
Sagger filling	17.8	3500
Clay dust	17.8	3500
Coal (powdered) dust	20.3	4000
Cocoa dust	15.2	3000
Cork (ground) dust	12.7	2500
Cotton dust	15.2	3000
Crushers	15.2	3000 or higher
Flour dust	12.7	2500
Foundry, general	17.8	3500
Sand mixer	17.8–20.3	3500–4000
Shake-out	17.8–20.3	3500–4000
Swing grinding booth exhaust	15.2	3000
Tumbling mills	20.3–25.4	4000–5000
Grain dust	12.7–15.2	2500–3000
Grinding, general	17.8–22.9	3500–4500
Portable hand grinding	17.8	3500
Jute		
Dust	12.7–15.2	2500–3000
Lint	15.2	3000
Dust shaker waste	16.3	3200
Pickerstock	15.2	3000
Lead dust	20.3	4000
with small chips	25.4	5000
Leather dust	17.8	3500
Limestone dust	17.8	3500
Lint	10.2	2000
Magnesium dust, coarse	20.3	4000
Metal turnings	20.3–25.4	4000–5000
Packaging, weighing, etc.	15.2	3000
Downdraft grille	17.8	3500
Pharmaceutical coating pans	15.2	3000
Plastics dust (buffing)	19.3	3800
Plating	10.1	2000
Rubber dust		
Fine	12.7	2500
Coarse	22.9	4000
Screens		
Cylindrical	17.8	3500
Flat deck	17.8	3500
Silica dust	17.8–22.9	3500–4500
Soap dust	15.2	3000
Soapstone dust	17.8	3500
Soldering and tinning	12.7	2500
Spray painting	10.1	2000
Starch dust	15.2	3000
Stone cutting and finishing	17.8	3500
Tobacco dust	17.8	3500
Woodworking		
Wood flour, light dry sawdust and shavings	12.7	2500
Heavy shavings, damp sawdust	17.8	3500
Heavy wood chips, waste, green shavings	20.3	4000
Hog waste	15.2	3000
Wool	15.2	3000
Zinc oxide fume	10.1	2000

Material	Minimum Transport Velocity m/s	Minimum Transport Velocity fpm
Very fine, light dusts	10.2	2000
Fine, dry dusts and powders	15.2	3000
Average industrial dusts	17.8	3500
Coarse dusts	20.3–22.9	4000–4500
Heavy or moist dust loading	22.9 and up	4500 and up

Source: Reference 2.

3. Headers should be located as close to equipment as possible.
4. Lengths of ducts and number of bends should be minimized.
5. Small branches should join the header as close to the fan as possible.
6. Ductwork should provide for ease of clean-out and incorporate necessary maintenance features.

Static Pressures and System Balancing

The required exhaust hood flow rates for an LEV system can be achieved only if the selected fan can provide the required airflow against the static pressure losses for the system. Hence the calculation of the static pressure or resistance losses through the system is a critical element in the design of LEV systems.

The two main procedures for the calculation are the velocity pressure method and the equivalent length method. The velocity pressure method for calculating pressure losses is based on the fact that all frictional losses in ducts and fittings are proportional to the velocity pressure in the duct. The velocity pressure method is generally more rapid, more flexible, and more amenable to the use of computers for calculations than other methods.

For simple systems operating at ambient conditions, the American Conference of Governmental Industrial Hygienists (ACGIH)[4] developed a simplified calculation procedure with design charts, based on the equivalent length method. This technique involves the addition of losses by elbows, entries, fittings, and so forth, expressed as the equivalent

TABLE 13. Relative Advantages and Disadvantages of the Two Balancing Methods

Balance-by-Design Method	Blast Gate Method
1. Volumetric flow rates cannot be changed easily by workers or at the whim of the operator.	1. Volumetric flow rates may be changed relatively easily. Such changes are desirable where pick-up of unnecessary quantities of material may affect the process.
2. There is little flexibility for future equipment changes or additions. The duct is "tailor-made" for the job.	2. Depending on the fan and motor selected, there is somewhat greater flexibility for future changes or additions.
3. The choice of exhaust volumetric flow rates for a new operation may be incorrect. In such cases, some duct revision may be necessary.	3. Correcting improperly estimated exhaust volumetric flow rates is relatively easy within certain ranges.
4. No unusual erosion or accumulation problems will occur.	4. Partially closed blast gates may cause erosion, thereby changing resistance or causing particulate accumulation.
5. Duct will not plug if velocities are chosen wisely.	5. Duct may plug if the blast gate insertion depth has been adjusted.
6. Total volumetric flow rate may be greater than design due to higher air requirements.	6. Balance may be achieved with design volumetric flow rate; however, the net energy required is usually greater than for the balance-by-design method.
7. The system layout must be in complete detail with all obstructions cleared and length of runs accurately determined. Installation must follow layout exactly.	7. Moderate variations in duct layout are possible.

Source: Reference 4.

length of straight duct and added to the centerline length of straight duct. For conditions other than ambient and for different duct roughness factors, the ACGIH supplies correction tables for density and roughness. Standard calculation sheets are used for the design of LEV systems. The *Industrial Ventilation* manual contains many worked examples for conventional applications and the reader should refer to the manual for specific details of design.

For an LEV system that has multiple branches, it is necessary to provide a means of distributing airflow among the branches, since air will always take the path of least resistance. At each junction, the exhaust volumetric flow rate will distribute itself according to the pressure losses of the available flow paths. If the system is not designed correctly, the design flows at each hood may not be achieved. In order to provide the proper airflow distribution, the designer must ensure that all flow paths (ducts) entering a junction will have equal calculated static pressure requirements. Two methods are available to do this: the balance-by-design method and the blast gate method. The advantages and disadvantages of these two methods are shown in Table 13.

In the balance-by-design method, duct sizes are chosen so that the static pressure balance at each junction will achieve the desired air volume in each branch duct and hood. With the blast gate method, blast gates are installed that must be adjusted after installation in order to achieve the desired flow at each hood.

The design procedures for the two balancing approaches are outlined in the following.

Method A: Balanced System Design (Static Pressure Balance Method)

1. Begin in branch of greatest resistance.
2. Based on design data, calculate total loss from exhaust hood to junction of the next branch.
3. At each junction point, the static pressure necessary to achieve desired flow in one air stream must match the static pressure (SP) in the joining airstream.

4. Where differences in SP are greater than 20%, the branch with the lower resistance should be redesigned to increase its pressure drop (e.g., decrease duct size).
5. If SPs are greater than 5%, but less than 20%, balance can be obtained by increasing airflow through the run with lower resistance. Increase in flow can be quickly calculated by:

$$\text{Corrected flow} = \text{Design flow}\sqrt{\frac{\text{SP run with larger SP loss}}{\text{SP run with smaller SP loss}}}$$

6. If SP is within 5%, it is usual to ignore the small error and treat it as if the paths were in complete balance.

Method B: Blast Gate Adjustment (Velocity Method)

1. Begin in branch of greatest resistance.
2. Calculate pressure losses as for method A.
3. At each junction, the desired volume of that branch is added to the main flow. No attempt is made to balance the static pressure in the joining airstreams. Joining branches are merely sized to give the desired minimum duct velocity at the desired flow.
4. Blast gates are adjusted after installation in order to achieve the desired airflow at each hood.

The final step is to select a suitable fan to give the design flow rates and static pressures (i.e., point of operation) at an acceptable efficiency. A preceding section described the different types of fans and their operating characteristics. Both the fans and the system have variable operating characteristics, but the actual "point of operation" will be the one single point at the intersection of the fan curve and the system curve. The designer's responsibility is to match the fan performance and system requirement. For complex LEV systems, the fan selection is the critical component to ensure a satisfactory performance. The experienced designer must work closely with the process operating engineers and the fan manufacturer's technical staff.

It is important to balance a system after the installation in order to ensure that it is performing according to the design specifications. Modifications to the system may be necessary to guarantee adequate contaminant capture. The objectives of checking an exhaust system are as follows:

1. To determine the exhaust volume and capture velocity at each pickup point and evaluate the adequacy of contaminant capture.
2. To determine the total exhaust volume and evaluate the size and performance of the air cleaning device.
3. To determine the system's static pressure and evaluate the fan capacity, speed, and horsepower requirements.
4. To determine the temperature at all points in the system in order to evaluate the materials of construction of the ductwork and the collector.

Testing of LEV Systems

System Evaluations

Testing of LEV systems should be carried out both for new systems, to verify original design data, and for existing systems that have had or need modifications to perform satisfactorily. For more specific details on measurement equipment and testing procedures, the reader is referred to the standard ventilation references,[2,4] which identify the initial tests and periodic field tests, including test data sheets, to be completed. For each LEV system, static pressure and velocity measurements should be made in each branch and the main duct. The LEV ductwork design should include appropriate testing points and suitable access to these points. Static pressures and velocity measurements should be made upstream and downstream of the fan and the air cleaning equipment. Details of the procedures to follow for the performance field testing of fans are given in the next section.

Once the system has been commissioned and properly balanced, it may not be necessary to carry out the rigorous testing procedure described above. Instead, the routine approach should be to measure static pressures at hoods and to compare them with the original or design values. A marked reduction in hood static pressure could be caused by such factors as the following.

1. Decrease in fan performance
2. Accumulation of dust in ductwork
3. Holes in ductwork or hood
4. Addition of further exhaust points to the system
5. Increased pressure drop across gas cleaning unit
6. Modifications to the system or fan dampers

The reduction in hood static pressure would indicate the need for detailed testing and review of the LEV system performance.

Field Performance Testing of Fans

Fans, being the critical component of LEV systems, require rigorous testing for the commissioning of new systems and identification of problems or lack of performance for existing installations.

Publication 203-90 of the AMCA[13] on field performance measurements of fan systems describes fan performance as a function of fan flow rate, fan total or static pressure, and fan power input at a stated fan speed and fan air density. The major difficulty in making reliable field performance tests is selecting suitable locations for measurements. This section provides the general calculation procedures for making such tests and provides guidance in selecting suitable measurement locations. The procedure is based on those outlined in AMCA publication 203-90.

Fan Flow Rate

Publication 203-90 describes the conditions for an acceptable measurement of fan flow rate as follows:

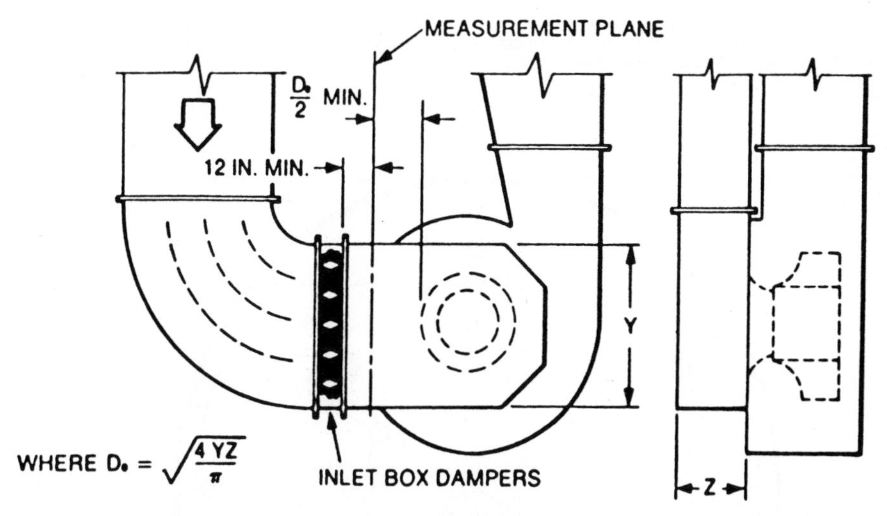

FIGURE 28. Measurement Plane at Fan Inlet Box (From Reference 12)

1. The velocity distribution should be uniform throughout the traverse plane. The uniformity of distribution is considered acceptable when more than 75% of the velocity pressure measurements are greater than 1/10 of the maximum measurement.
2. The flow stream should be at right angles to the plane.
3. The cross-sectional shape of the airway in which the plane is located should not be irregular.
4. The cross-sectional shape of the airway should be uniform throughout the length of the airway in the vicinity of the plane.
5. The plane should be located in such a way as to minimize the effects of leaks in the portion of the system that is located between the plane and the fan.

Figure 28 is a guide for locating the measurement plane within an inlet box.

To determine fan flow rate, the velocity pressure is measured at the traverse location. Figures 29 and 30 show the recommended data collection points on the traverse plane for rectangular and circular ducts.

The fan flow rate is calculated as follows:

$$Q = VA$$

where: Q = flow rate, acfm
A = area of the traverse plane, ft^2
V = average velocity at the traverse plane, fpm

$$= 1096\ (P_v/\rho)^{0.5}.$$

where: P_v = root-mean-square velocity pressure at the traverse plane, in. w.g.
= [Σ(velocity pressure readings)$^{0.5}$/number of readings]^2
ρ = gas density at the traverse plane, lb/ft^3

Fan Static Pressure

Static pressure is measured at the fan inlet and the fan outlet. The requirements for selecting a traverse plane for measurements of fan static pressure are the same as those used for the selection of a traverse plane for flow rate.

The static pressure at a traverse plane *(Ps)* is calculated from:

$$Ps = \frac{\sum Ps_r}{\text{Number of readings}}$$

where Ps_r = static pressure reading, in. w.g.

The fan static pressure (P_s) is given by the equation

$$P_s = P_{s2} - P_{s1} - P_{v1}$$

where: P_{s2} = fan static pressure at the fan outlet, in. w.g.
P_{s1} = fan static pressure at the fan inlet, in. w.g.
P_{v1} = velocity pressure at fan inlet, in. w.g.

P_{s1} and P_{s2} usually cannot be measured directly. In most cases, the measurements will be made a short distance

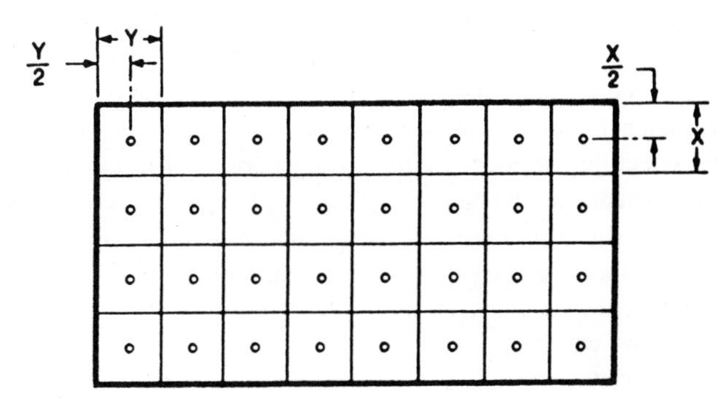

FIGURE 29. Traverse Points for Rectangular Ducts (From Reference 12)

In order to obtain a representative average velocity in a duct, it is necessary to locate each traverse point accurately. It is recommended that the number of traverse points increase with increasing duct size. The distributions of traverse points for circular ducts, as indicated below, are based on log-linear Pitot traverse method.

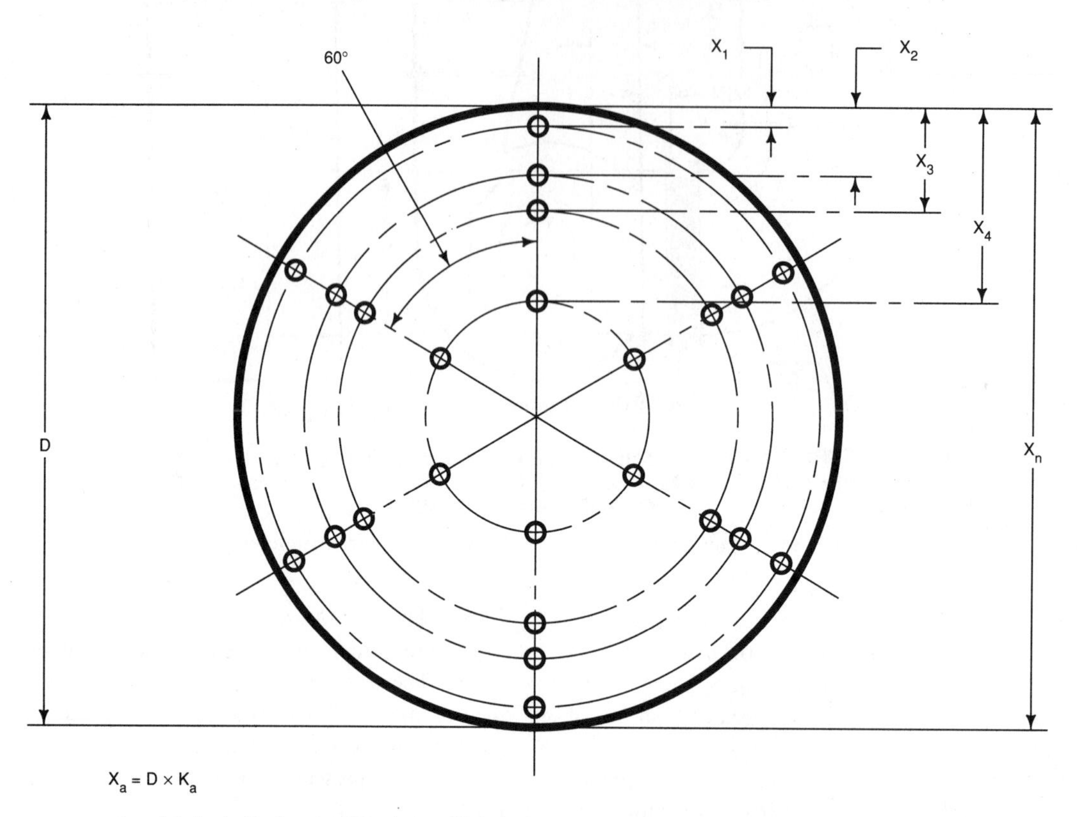

where D is the inside diameter of the duct and K_a is the factor corresponding to the duct size and the traverse point location as indicated in the table below.

INSIDE DIAMETER OF DUCT	NUMBER OF TRAVERSE POINTS IN EACH OF 3 DIAMETERS	K_1	K_2	K_3	K_4	K_5	K_6	K_7	K_8	K_9	K_{10}	K_{11}	K_{12}	K_{13}	K_{14}	K_{15}	K_{16}
LESS THAN 8 ft	8	.021	.177	.184	.345	.655	.816	.883	.979	—	—	—	—	—	—	—	—
8 ft THRU 12 ft	12	.014	.075	.114	.183	.241	.374	.626	.759	.817	.886	.925	.986	—	—	—	—
GREATER THAN 12 ft	16	.010	.055	.082	.128	.166	.225	.276	.391	.609	.724	.775	.834	.872	.918	.945	.990

FIGURE 30. Traverse Points for Circular Ducts (From Reference 12)

upstream of the fan inlet and downstream of the fan outlet.

The downstream static pressure of the fan should be measured a short distance from the outlet because of the turbulence in the vicinity of the outlet. Figure 31 is a guide to selecting the location for static pressure measurements downstream of the fan.

The static pressure measurement upstream of the fan inlet should be measured a minimum of one half of the equivalent diameter from the inlet. Figure 28 is a guide for locations within the inlet box.

The static pressure at the fan inlet (P_{s1}) can be calculated from the following equation.

$$P_{s1} = P_{s4} + P_{v4} - P_{v1} - \Delta P_{4,1}$$

where:
- P_{s4} = static pressure at measurement plane
- P_{v4} = velocity pressure at measurement plane
- P_{v1} = velocity pressure at fan inlet
- $\Delta P_{4,1}$ = pressure losses between inlet and measurement plane from friction, fittings, etc.

The static pressure at the fan outlet (P_{s2}) can be calculated from the following equation.

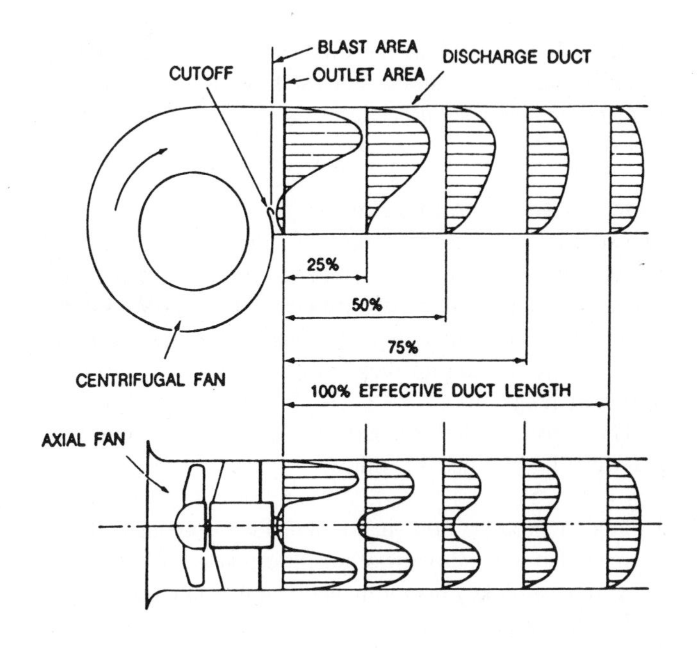

TO CALCULATE 100% EFFECTIVE DUCT LENGTH, ASSUME A MINIMUM OF 2-1/2 DUCT DIAMETERS FOR 2500 FPM OR LESS. ADD 1 DUCT DIAMETER FOR EACH ADDITIONAL 1000 FPM.

EXAMPLE: 5000 FPM = 5 EQUIVALENT DUCT DIAMETERS

IF DUCT IS RECTANGULAR WITH SIDE DIMENSIONS a AND b, THE EQUIVALENT DUCT DIAMETER IS EQUAL TO $(4ab/\pi)^{0.5}$

FIGURE 31. Velocity Profiles in Outlet Duct (From Reference 12)

$$P_{s2} = P_{s5} + P_{v2} + \Delta P_{2,5}$$

where: P_{s5} = static pressure at measurement plane
P_{v5} = velocity pressure at measurement plane
P_{v2} = velocity pressure at fan inlet
$\Delta P_{2,5}$ = pressure losses between outlet and measurement plane from friction, fittings, etc.

The pressure losses $\Delta P_{4,1}$ and $\Delta P_{2,5}$ must be calculated. When the pressure losses between the measurement planes and the planes of interest are negligible, the equations reduce to:

$$P_{s1} = P_{s4} + P_{v4} - P_{v1}$$

$$P_{s2} = P_{s5} + P_{v5} - P_{v2}$$

In addition, when the cross-sectional areas between the measurement planes and the planes are equal, the equations reduce to:

$$P_{s1} = P_{s4}$$

$$P_{s2} = P_{s5}$$

For more accurate fan static pressure calculations, conditions affecting fan performance should be calculated when the fan is installed. For example, fan inlet and outlet restrictions will reduce fan performance. To account for these effects, *system effect factors* must be used in the analysis. Refer to AMCA publications 203-90 and 201-90 for this information.

Fan Power Input

Several measurement methods are available to determine fan power input, including the following.

1. Phase current method
2. Typical motor performance data
3. Calibrated motors methods

In this section, only the phase current method is discussed. This method is applicable to three-phase motors. For a detailed discussion of other methods, refer to AMCA Publication 203-90.

The method requires measurement of the phase currents and voltages supplied to the motor. The closer the phase current is to the full-load current, the greater will be the accuracy. Two equations are used.

Equation A:

$$H_{mo} = NPH\left(\frac{\text{Measured amps}}{FLA}\right)\left(\frac{\text{Measured volts}}{NPV}\right)$$

where: H_{mo} = motor power output, hp
NPH = nameplate horsepower
FLA = full-load amperes
NPV = nameplate volts
Measured volts = average of measured phase volts
Measured amperes = average of measured phase amperes

Equation B:

$$H_{mo} = NPH\left(\frac{\text{Measured amps} - NLA}{FLA - NLA}\right)\left(\frac{\text{Measured volts}}{NPV}\right)$$

where: NLA = average of measured phase values of no-load amperes
NPH = nameplate horsepower
FLA = full-load amperes
NPV = nameplate volts

NLA must be determined with the motor operating and disconnected from the fan shaft.

Equation A is used to estimate the H_{mo} with fan motors greater than 5 hp operating at greater than 90% of *FLA*. The average of equations A and B is used when the fan is operating at lower than 90% *FLA*.

Fan Speed
A minimum of three measurements should be averaged to determine fan speed.

Operation and Maintenance

Introduction
The proper operation and maintenance of LEV air pollution control systems are important. The promulgation of more stringent air pollution legislation has necessitated a much higher performance level from these systems. Regulations in many countries require good operation and maintenance practices (including record keeping) and a comprehensive preventative maintenance program. It is necessary to approach maintenance not only of the actual air pollution control equipment, but also of the entire LEV system, in the same scientific manner that has been adopted for other plant functions.

It is still common to find LEV systems that are badly neglected and receive only minimal maintenance attention. The reasons for this lack of maintenance are as follows:

1. Process equipment has a much higher priority than pollution control equipment.
2. Pollution control equipment is dusty and difficult to maintain.
3. The equipment is often bought on a low-bid basis; attention to maintenance features is not of paramount concern.
4. The owner does not keep an inventory of parts for repairs.
5. No specific individual is responsible for equipment operation.
6. No proper training is conducted for operation and maintenance of equipment.
7. Equipment is complex and not easy to troubleshoot.
8. Equipment is inadequately designed for ease of maintenance.

Successful operation of an installation requires that a proper approach be adopted at the early design stage of the project. Details of the important steps for system design are summarized below.

1. Prepare technical and economic analyses of alternative types of environmental control systems.
2. Develop technical and economic specifications for major equipment for competitive bidders.
3. Prepare technical and economic bid analyses for each bid.
4. Develop detailed design and layouts to accommodate the operation and maintenance requirements.
5. Develop detailed start-up and operating manuals and procedures, including operator training programs.
6. Implement a preventative maintenance program.

Items 1 through 4 of the preceding list are covered in detail in previous sections. For establishing correct operation and maintenance procedures, we will focus on items 5 and 6.

Start-up and Commissioning
The start-up and commissioning of any system are critical. System operation must be established and performance adjusted to obtain design performance. Since LEV systems often work in conjunction with general or dilution ventilation systems, by design they must function in a specified operating range.

The steps involved in a start-up and commissioning program can be outlined as follows:

1. Obtain process flow instrumentation drawings of the overall system. Review of this information with the goal of reviewing key operating and ventilation requirements is necessary.
2. Obtain and review detailed operating and maintenance manuals provided on each specific piece of equipment by vendors. For large and significant components, schedule manufacturer's training sessions prior to start-up.
3. Prepare an overall commissioning and start-up schedule, starting from the completion of installation and proceeding through to system turnover. Ensure that adequate tradespeople are on hand during commissioning to allow immediate corrections to any deficiencies that may become apparent.
4. Establish sufficient measurement points in the system to obtain field flow and static pressure measurements. Ensure that access points for measurements have been installed.
5. Perform mechanical checkout on all moving components to ensure freedom of motion throughout range of travel and verify correct installation of drives and linkages.
6. Perform electrical checkout of individual components of system manually (i.e., isolated from interlocks). This may require overrides on some automated systems. Ensure that each piece of equipment is responding to electrical and instrumentation controls.
7. Calibrate all instruments and instrumentation.
8. Verify that all interlocks and process control automation are functioning correctly.
9. Obtain field flow and static pressure measurements throughout system. Compare with original design values and make adjustments as necessary. Compare final field settings and system flows for operating records.

Preventative Maintenance Program
Operating and maintenance problems may develop with equipment and components in the system. The main equipment components and a brief discussion of the potential problems follows. Further details on air pollution control equipment can be found elsewhere in this manual.

Baghouses

Experience with baghouses has identified high maintenance problems in the following decreasing order of frequency.

1. Filter bags
2. Flow control (blast gases, dampers)
3. Hoppers (screw conveyors, air locks, dust disposal)
4. Cleaning system maintenance

The pressure drop across the bags should be checked frequently during normal operation. This is the most important parameter indicating overall system performance.

TABLE 14. Field Inspection Checklist

Inspection Item	Frequency
1. Dust inspection for dust settlement (and removal if required)	Quarterly, yearly
2. Bag compartment—painting, patching, regasketing, recaulking	Yearly
3. Supporting structure and framing—painting, bolt tightening, weld inspection	Yearly
4. Heating equipment for baghouse—operating temperature sensors and controls, insulation	Daily
5. Fan drives, electrical motors, gear trains—checks for wear, lubrication	Daily, weekly
6. Repressure fan (for reverse air cleaning) check for temperature and vibration	Daily
7. Compartment isolation and bypass dampers—mechanical drive, jamming, leakage	Daily
8. Bag (fabric) inspection, clean side—wear, damage, dust seepage, surface deposits, bag clamps	Monthly, quarterly
9. Lubrication of damper valves, shaker mechanism	Weekly, monthly
10. Access doors—jamming, insulation sealing, hinges, safety locks	Quarterly
11. Hopper—dust level, bridging, transfer valve material balance on dust recovery	Daily, yearly
12. Bag suspensions—frames, drive attachment hoods, tension adjustment system	Quarterly, yearly
13. Baghouse—check for floor deposits, gross leakage	Weekly, quarterly, yearly
14. Control panel—indicator lights, switch position, dial displays	Daily
15. Baffle plate and guide vanes—erosion or dust deposition	Yearly
16. Shaking mechanism—linkage, slack, amplitude or frequency adjustment	Yearly
17. Compressed air supply—pressure, oil and water separators	Daily, weekly
18. Fire detectors, sprinklers, explosion panels	Semiannually
19. Instrumentation—direct display, continuous monitors	
a. Baghouse temperature	Daily
b. Overall baghouse pressure loss	Daily
c. Individual compartment pressure loss	Daily
d. Fan static pressure	Daily
20. Stack emissions—visual estimates, automatic in-stack opacity meters	Daily
21. Fabric cleaning controls—timing circuits, pressure sensors	Daily
22. Manometer—check and record pressure drop readings. Watch for trends	Daily
23. Collector—observe all indicators on control panel and listen to system for properly operating subsystems	Daily
24. Check all bolts and welds. Inspect entire collector thoroughly, clean, and touch up paint where necessary	Yearly

Scrubbers

For scrubbers, high maintenance items are related to the physical and chemical characteristics of the fluids being handled in the ducts, housing, and piping. The highly reactive scrubbing liquids can produce corrosion, erosion, and plugging. The scrubbing liquids must usually be filtered and treated chemically before being reused. A common problem for scrubbers is dust buildup at the wet–dry interface. The solution to this problem may require a redesign of the scrubber inlet ducting or a change in the humidification of the incoming gas. Pipe clogging, buildup, and erosion can occur in the scrubber liquor-handling system. Improperly specified pipes, elbows, reducers, and other fittings can lead to plugging and abrasion. It is also common to find valves misused to throttle flow, which results in high erosion and plugging. Mist eliminators are common sources of plugging and can result in increased system pressure drop and lower system flows.

Electrostatic Precipitators

Experience has shown the following to be typical of users of electrostatic precipitators.

1. The majority of users are satisfied with the precipitator as a functioning piece of equipment.
2. The principal sources of malfunction are the discharge electrodes and design improvements are required.

Fans

The successful operation and maintenance of fans were discussed in earlier sections.

For fume control, the high-temperature gases containing dust-laden materials often present a severe erosion and corrosion problem. Fan balancing is also a common problem and requires careful attention.

Field Inspection Checklist

Troubleshooting to solve problems for existing LEV systems can be organized into the following five steps.

Step 1—Identifying the problem
Step 2—Developing alternative solutions and recommendations
Step 3—Implementation
Step 4—Testing and sampling
Step 5—Doing preventative maintenance

By maintaining thorough, accurate records of system and equipment performance, the system operation can be tracked and changes in performance recorded. Table 14 provides a checklist of field inspection items and a suggested frequency for a LEV system using a baghouse for the air pollution control equipment. A similiar list could be developed for other types of pollution control equipment (e.g., scrubbers, electrostatic precipitators).

Acknowledgment

The Department of Chemical Engineering and Applied Chemistry of the University of Toronto provided financial support for the preparation of this chapter.

References

1. *ASHRAE Handbook—Heating, Ventilating, and Air Conditioning Systems and Application,* American Society of Heating, Refrigerating and Air Conditioning Engineers, Atlanta, GA, 1987.
2. H. D. Goodfellow, *Advanced Design of Ventilation Systems for Contaminant Control,* Elsevier Science Publishing Co., New York, 1985.
3. J. J. Loeffler, "Simplified equations for HVAC duct friction factors," *ASHRAE J.,* 76–79 (January 1980).
4. *Industrial Ventilation,* 20th ed., American Conference of Governmental Industrial Hygienists, Cincinnati, OH, 1988.
5. J. E. Emswiler, "The neutral zone in ventilation," *J. Am. Soc. Heating Ventilating Eng.,* 32(1): 1–16 (January 1926).
6. H. D. Goodfellow, Ed., *Ventilation '85, Proceedings of the 1st International Symposium on Ventilation for Contaminant Control,* Elsevier, New York, 1986.
7. G. C. Cawkwell and H. D. Goodfellow, "Multiple cell ventilation model with time-dependent emission sources," *Proceedings of Engineering Aero and Thermo Dynamics of Ventilated Room 2nd International Conference,* A1-9, Norsk VVS, Oslo, Norway, June 13–15, 1990.
8. E. R. Kashdon et al., *Technical Manual: Hood System Capture of Process Fugitive Particulate Emissions,* EPA 1600/7-86/016, Environmental Protection Agency, Research Triangle Park, NC, April 1986.
9. H. D. Goodfellow, "Hood design for ventilation systems," *Heating, Piping/Air Conditioning,* 60–67 (February 1987).
10. V. Hampl, "Evaluation of local exhaust hood efficiency by a tracer gas technique," *Am. Indus. Hygiene Assoc. J.,* 45(7): 485–490 (July 1984).
11. H. D. Goodfellow and M. Bender, "Design considerations for fume hoods and process plants," *Am. Indus. Hygiene Assoc. J.,* 41: 473–484 (1980).
12. *HVAC Duct Construction Standards, Metal and Flexible,* Sheet Metal and Air Conditioning Contractors National Association, Vienna, VA, 1985.
13. AMCA Publication 201: *AMCA Fan Application Manual—Part 1—Fans and Systems,* 1973; AMCA Publication 202: *AMCA Fan Application Manual—Part 2—Troubleshooting,* 1972; AMCA Publication 203: *AMCA Fan Application Manual—Part 3—Field Performance Measurements,* 1976; AMCA Publication 210 *(AMCA Standard 210-74): Laboratory Methods of Testing Fans for Rating Purposes,* 1974. Air Movement and Control Association, Arlington Heights, IL.
14. R. C. Myers, "Industrial fans—guidelines for a successful installation," *Iron Steel Eng.,* 38–44 (October 1976).
15. *Guide for Steel Stack Design and Construction,* Sheet Metal and Air Conditioning Contractors National Association, Vienna, VA, 1983.

Bibliography

AMCA Publication 203-90, *Field Performance Measurements of Fan Systems,* Air Movement and Control Association, Arlington Heights, IL.

M. O. Amdur, J. R. Anticaglia, E. C. Barnes, et al., *The Industrial Environment—Its Evaluation and Control,* U.S. Department of Health and Human Services, Washington, DC, 1973.

T. Hayashi, R. H. Howell, M. Shibata, et al., *Industrial Ventilation and Air Conditioning,* CRC Press, Boca Raton, FL, 1987.

R. Jorgensen, *Fan Engineering,* 8th ed., Buffalo Forge Co., Buffalo, NY, 1983.

Research and Education Association, *Modern Pollution Control Technology,* Vol 1, New York, 1980.

J. H. Vincent, Ed., *Ventilation '88, The Annals of Occupational Hygiene,* Argamon Press, New York, 1989.

R. A. Wadden and P. A. Scheff, *Engineering Design for the Control of Workplace Hazards,* McGraw-Hill Book Co., New York, 1987.

7 Combustion Sources

Coal
Fuel Oil
Natural Gas
Wayne T. Davis and Arijit Pakrasi
Wood Waste
Anthony J. Buonicore

COMBUSTION PROCESS

Combustion refers to the rapid oxidation of substances (usually referred to as fuels) with the evolution of heat. Most fuels are mixtures of chemical compounds referred to as hydrocarbons. When these burn, they are converted into the final products of carbon dioxide and water. There are three distinctly different components involved in the combustion process—the fuel, the oxidant, and the diluent.

- *Fuel*—substance containing energy-rich carbon–carbon bonds or carbon–hydrogen bonds. During combustion, these bonds are broken up and the chemical energy is released as heat. Fuels are available naturally as solids, liquids, or gases. Examples of fuels are coal, petroleum oil, and natural gas.
- *Oxidant*—substance that aids combustion by breaking the C-C or C-H bonds in fuels and releasing the chemical energy as heat. The most common example of an oxidant is oxygen. Ambient air, containing approximately 21% oxygen (v/v), is the most common source of oxygen.
- *Diluent*—substance that does not take part in the combustion process, but is present during the combustion as a carrier. The major diluent in most combustion processes is nitrogen, which is present in the ambient air at approximately 79% (v/v). While not taking part in the combustion process, nitrogen absorbs a substantial part of the heat released during combustion and is present in the combustion product. A small portion of the nitrogen may be oxidized to NO and/or NO_2, which are precursors of acidic rain and photochemical oxidant formation. Other diluents are carbon dioxide, moisture in the air, and unburned fuel and noncombustibles in the fuel. In cases where excess air is added to the combustion chamber, this excess air also acts as a diluent.

The reaction products of the combustion process are collectively called combustion products. The combustion products are made up of the following.

- Oxidized products of carbon–carbon and carbon–hydrogen bonds in fuels, namely, carbon dioxide (CO_2) and water (H_2O). In the case of incomplete combustion of the fuel, some carbon monoxide (CO) and other partially oxidized hydrocarbons may be formed.
- Oxidized products of other substances (mostly impurities) in the fuel. For example, metals may be oxidized to their oxides. Sulfur and phosphorus in fuel are converted to sulfur oxides (SO_2, SO_3) and phosphorus pentoxide (P_2O_5).
- Diluents, which pass through the system mostly unchanged. A small portion of the major diluent nitrogen is converted to nitrogen oxides.

In the case of fuels containing halides (usually chlorine and fluorine), the combustion products may contain hydrochloric or hydrofluoric acids (HCl, HF) and trace quantities of organohalides such as dioxins and furans.

In general, the combustion process can be represented by the following equation, which describes a stoichiometric combustion equation (one in which the amount of air supplied is just sufficient to react with the fuel to achieve complete combustion).

Fuel + Oxidant + Diluent → Combustion products

$$C_xH_y + bO_2 + (79/21)bN_2 \rightarrow xCO_2 + y/2\ H_2O + (79/21)bN_2$$

where: C_xH_y = the general formula for a HC fuel
b = $x + (y/4)$, the number of moles of oxygen required to combust the fuel, C_xH_y
79/21 = the approximate ratio of the moles of nitrogen to the moles of oxygen found in air

Most combustion systems (exclusive of the gasoline engine used in automobiles) operate with an overall excess of air to ensure complete combustion. A system operating with 50% excess air would require 50% more air (oxidant and diluent) than shown in the equation (i.e., the total air required would be $1.5 \times (bO_2 + 79/21\ bN_2)$. The excess air ($O_2 + N_2$) would then function as an additional diluent and would contribute to the combustion products. Similarly, other residual substances in the fuel (metals, sulfur, chlorine, etc.) would react in the combustion zone, resulting in the emission of gaseous or particulate pollutants.

Combustion is mainly used as a source of heat energy that can then be converted to different forms of energy, including electric power, steam generation, and process heating. In recent years, combustion, in the form of incineration, has been used more frequently for the destruction of toxic substances and for the burning of municipal solid wastes. These two applications are dealt with elsewhere in this manual.

The noncombustible, nonvolatile portion of the fuel results in the formation of a solid residue referred to as ash. The coarser, heavier portion remains within the combustion chamber and is withdrawn as "bottom ash." The finer portion, referred to as "fly ash," becomes airborne and exits with the flue gas.

Based on the physical states in which they are found in nature, fuels can be classified as solid, liquid, or gaseous, with coal, fuel oil, and natural gas being the three most common examples of each. This chapter describes these three fuels, their emission characteristics and the state of the art of air pollution control for each fuel. The chapter also includes sections on wood waste as a fuel and on prescribed burning.

COAL

Wayne T. Davis and Arijit Pakrasi

PHYSICAL/CHEMICAL CHARACTERISTICS

The most common example of solid fuel is coal. Different ranks of coal are found in nature, depending on the conditions under which the coal was formed. Classification by rank is based on the volatile matter, moisture content, and fixed carbon content of the coal as shown in Table 1.[1] The natural transition of coal with time is as follows, with each step in the process enriching the carbon content:

Peat → Lignite → Subbituminous → Bituminous → Anthracite

The composition of coal varies with the rank and also within the rank. Coal compositions are reported in a standard format based on American Society for Testing and Materials (ASTM) standard analyses referred to as "proximate" and "ultimate" analyses. Proximate analyses report coal composition in the broad subgroups of moisture content (MC), volatile matter (VM) content, fixed carbon (FC) content, and ash content (noncombustibles). The volatile matter is that portion of the coal that is driven off (exclusive of H_2O vapor) when it is exposed to a temperature of 950°C for seven minutes (ASTM D-3175). The fixed carbon is the combustible residue that is left after the coal has been subjected to the moisture and volatile matter tests. The noncombustible residue is referred to as the ash.

The ultimate analysis reports the carbon, hydrogen, sulfur, oxygen, nitrogen, and water content of the coal and is used to compute the air requirements and approximate composition of flue gas products. An overall average formula for the coal can be determined from the ultimate analysis of the coal. Table 2 shows typical proximate and ultimate analyses based on the rank of the coal.[2]

Coal contains many noncombustibles collectively termed "ash." The major constituents of ash are various minerals and mineral oxides. During combustion, minerals and oxides partition into either the bottom ash or the fly ash, depending on such factors as combustion temperature, heat-to-fluid flow pattern in combustion, vapor pressure characteristics of the minerals, fusion properties, and the sintering properties of the ash. The fly ash in the flue gas varies in composition and loading (i.e., mass of fly ash per unit volume of flue gas) for each type of coal, combustor, and process operation.

It is possible to conduct a detailed metals analysis of the coal ash using ASTM D-3174. The results are reported as the mass percent for the metal oxides as follows:

$$SiO_2 + Al_2O_3 + Fe_2O_3 + CaO + MgO + Na_2O + K_2O + TiO_2 + P_2O_5 + SO_3 = 100\%$$

While the coal ash properties, as measured by conducting a laboratory ash analysis of the coal sample, may be different from those of the fly ash emitted from the combustion chamber, the coal ash analysis is frequently used to determine slagging characteristics in the boiler and to predict the electrical resistivity of the fly ash. This parameter is a significant one used in the design of electrostatic precipitators (ESPs). A model for the prediction of fly ash resistivity based on coal ash properties was developed by

TABLE 1. Classification of Coals by Rank

Class	Fixed Carbon Limits, %[a]		Volatile Matter Limits, %[a]		Calorific Value Limits, Btu/lb[b]	
	Equal or Greater Than	Less Than	Equal or Greater Than	Less Than	Equal or Greater Than	Less Than
I. Anthracitic						
Meta-anthracite	98	. . .	. . .	2	. . .	. . .
Anthracite	92	98	2	8	. . .	. . .
Semianthracite[c]	86	92	8	14	. . .	. . .
II. Bituminous						
Low-volatile bituminous coal	78	86	14	22	. . .	. . .
Medium-volatile bituminous coal	69	78	22	31	. . .	. . .
High-volatile A bituminous coal	. . .	69	31	. . .	14,000[c]	. . .
High-volatile B bituminous coal	. . .	. . .	. . .	. . .	13,000[c]	14,000
High-volatile C bituminous coal	. . .	. . .	. . .	. . .	11,500	13,000
					10,500	11,500
III. Subbituminous						
Subbituminous A coal	. . .	. . .	. . .	. . .	10,500	11,500
Subbituminous B coal	. . .	. . .	. . .	. . .	9,500	10,500
Subbituminous C coal	. . .	. . .	. . .	. . .	8,300	9,500
IV. Lignitic						
Lignite A	. . .	. . .	. . .	. . .	6,300	8,300
Lignite B	. . .	. . .	. . .	. . .	. . .	6,300

[a]Dry mineral-free basis.
[b]Moist mineral-free basis.
[c]Coals having 69% or more fixed carbon shall be classified by fixed carbon, regardless of calorific value.
Source: Reference 1.

Bickelhaupt.[3] Table 3 shows the broad range of variation in chemical composition of coal ashes.[2]

Although coal ash properties are frequently used to predict the behavior of the ash, the fly-ash properties are also important with regard to the design of particulate control systems. The fly-ash particle-size distribution and surface area directly affect the pressure drop across fabric filter dust collectors and the efficiency of all of the control devices. Vann Bush et al.[4] showed that the resistance to flow through fabric filters (baghouses) increased as the surface area increased and the size of the particles decreased. Table 4 gives a summary of key fly-ash characteristics for 24

TABLE 2. Analyses of Typical U.S. Coals, as Mined

Coal Type	% Proximate Analysis				% Ultimate Analysis						HHV, Btu/lb	Air,[a] lb/10^6 Btu
	H_2O	VM	FC	Ash	H_2O	C	H_2	S	O_2	N_2		
Anthracite	2.5	5.7	83.8	8.0	2.5	83.9	2.9	0.7	0.7	1.3	13.72	787
Bituminous	3.3	20.5	70.0	6.2	3.3	80.7	4.5	1.8	2.4	1.1	14.31	765
Sub-bituminous	23.2	33.3	39.7	3.8	23.2	54.6	3.8	0.4	13.2	1.0	9.42	757
Lignite	34.8	28.2	30.8	6.2	34.8	42.4	2.8	0.7	12.4	0.7	7.21	750

[a]Theoretical air required for combustion under stoichiometric conditions (no excess air).
Source: Adapted from Reference 2.

TABLE 3. Coal Ash Properties

Component	Percentage
SiO_2	10–70
Al_2O_3	8–38
Fe_2O_3	2–50
CaO	0.5–30
MgO	0.3–8
Na_2O	0.1–3
K_2O	0.1–3
TiO_2	0.4–4
SO_3	0.1–30

Source: Reference 2.

coal-fired plants. Specific descriptions of each parameter and its effect on performance were reported. Figure 1 shows typical scanning electron microscope (SEM) photographs of the fly-ash dust cake collected from four different coal-fired power plant ashes, ranging from an ash composed of nearly spherical fly-ash particles with a low surface area of 0.85 m^2/g (Figure 1a) to 2.23, 3.23, and 14.44 m^2/g for Figures 1b, 1c, and 1d respectively. The mass mean diameter of the 24 fly ashes ranged from 3.6 to 9.4 μm.

PROCESS DESCRIPTIONS*

There are two major coal combustion techniques—suspension firing and grate firing. Suspension firing is the

*Taken in part from AP-42.[6]

TABLE 4. Fly Ash Characteristics for 24 Coal-Fired Plants

	Chemical Characteristics of Dust-Cake Ashes, % Wt													
Fabric Filter	Li_2O	Na_2O	K_2O	MgO	CaO	Fe_2O_3	Al_2O_3	SiO_2	TiO_2	P_2O_5	SO_3	LOI	Soluble SO_4[a]	Equilibrium pH
Monticello	0.02	0.49	0.99	2.20	9.70	3.4	20.4	60.5	1.90	0.08	0.44	0.29	0.48	9.81
Scholz HSFP	0.04	0.58	2.83	0.94	2.72	21.1	22.4	46.3	1.42	0.55	0.65	11.10	3.65	4.77
Brunner Island	0.06	0.32	2.70	0.93	2.33	11.7	28.5	49.6	1.53	0.71	0.36	6.10	3.10	4.26
IPP	0.03	2.10	1.46	2.68	10.72	6.1	19.6	53.0	1.10	0.61	1.56	0.64	1.95	10.68
Nixon	0.03	2.60	1.10	1.55	6.00	4.8	28.4	50.2	0.96	1.65	1.02	2.40	2.00	7.15
Shawnee Unit 5	0.06	0.40	2.00	0.84	1.80	4.5	31.5	53.8	2.70	0.19	0.34	6.30	1.83	4.38
Cameo	0.03	0.75	1.20	1.95	6.35	5.0	28.4	50.5	1.70	1.55	1.35	1.73	1.35	8.39
Harrington	0.01	1.55	0.40	6.05	29.00	6.3	18.8	30.5	1.25	1.45	2.60	0.53	3.15	11.05
Escalante	0.02	0.50	1.20	1.60	5.30	6.4	23.5	58.3	0.92	0.07	0.70	0.89	0.90	10.00
Arapahoe Unit 4	0.02	1.60	1.30	1.80	6.30	5.2	25.7	53.4	0.83	1.30	0.76	2.60	2.00	7.80
Arapahoe FFPP	0.02	2.80	1.30	1.60	4.90	4.1	26.4	55.5	1.10	1.00	0.78	1.80	1.50	7.27
Arapahoe FFPP (downstream of multiclone)	0.02	1.90	1.40	2.50	7.60	6.6	25.8	48.0	1.00	2.00	1.30	3.10	2.10	7.40
Arapahoe Eco-laire	0.02	1.40	1.30	2.00	6.20	5.7	26.3	52.0	0.92	1.60	0.71	3.00	1.80	7.40
Crane Unit 1	0.04	1.20	2.10	1.10	4.30	18.6	24.3	44.3	1.20	0.86	1.50	15.50	6.30	4.04
Comanche Ecolaire	0.01	1.20	0.18	6.60	35.90	5.3	18.8	21.8	1.80	2.00	3.30	0.55	4.20	11.40
Cherokee	—	0.57	1.10	1.80	5.60	4.6	28.2	52.1	1.50	1.70	0.95	1.60	1.07	—
Eraring	0.02	0.53	1.70	0.73	1.30	4.4	26.8	61.5	0.92	0.17	0.17	2.00	0.57	5.77
Tallawarra	0.02	0.11	1.20	0.52	0.95	4.9	20.1	70.5	0.84	0.19	0.13	3.00	0.65	5.45
TVA AFBC	0.03	1.33	1.05	1.85	34.67	10.0	11.4	22.4	0.53	0.27	15.78	11.10	16.58	11.47
Plant A CFBC	—	—	—	—	—	—	—	—	—	—	—	—	—	—
Plant B CFBC	0.01	0.28	0.97	1.20	29.20	23.5	7.3	19.2	0.42	0.84	16.30	6.50	18.60	11.32
Nucla CFBC	0.03	0.18	0.93	0.81	16.40	3.8	23.9	45.1	1.00	0.04	5.80	8.80	6.60	11.40
EPRI High Sulfur Test Center	0.61	0.40	0.65	0.70	31.00	4.1	7.4	15.2	0.35	0.24	39.10	5.80	43.60	10.30
Plant C Spray Dryer	0.02	1.70	1.20	1.30	12.20	4.4	18.0	48.4	0.91	0.05	10.60	2.50	12.10	10.40

primary combustion mechanism in pulverized-coal–fired and cyclone-fired units. Grate firing is the primary mechanism in underfeed and overfeed stoker-fired units. Both mechanisms are employed in spreader stokers.

Pulverized-coal furnaces are used primarily in utility and large industrial boilers. In these systems, the coal is pulverized in a mill to the consistency of talcum powder (i.e., at least 70% of the particles will pass through a 200-mesh sieve). The pulverized coal is generally entrained in primary air before being fed through the burners to the combustion chamber, where it is fired in suspension. Pulverized-coal furnaces are classified as either dry or wet bottom, depending on the ash removal technique. Dry bottom furnaces fire coals with high ash fusion temperatures, and dry ash removal techniques are used. In wet bottom (slag tap) furnaces, coal with a low ash fusion temperature is used, and molten ash is drained from the bottom of the furnace. Pulverized-coal furnaces are further classified by the firing position of the burners, that is, single (front or rear) wall, horizontally opposed, vertical, tangential (corner fired), turbo, or arch fired.[7] Figure 2 shows a typical pulverized-coal–fired boiler.[8]

Cyclone furnaces burn coal of low ash fusion temperature crushed to a 4-mesh size. The coal is fed tangentially, with primary air, to a horizontal cylindrical combustion chamber. In this chamber, small coal particles are burned in suspension, while the larger particles are forced against the outer wall. Because of the high temperatures developed in the relatively small furnace volume, and because of the low fusion temperature of the coal ash, much of the ash forms a liquid slag, which is drained from the bottom of the furnace through a slag tap opening. Cyclone furnaces are used mostly in utility and large industrial applications. See Figure 3.[8]

In spreader stokers, a flipping mechanism throws the coal into the furnace and onto a moving fuel bed. Combustion occurs partly in suspension and partly on the grate. Because of significant carbon in the particulate, fly ash reinjected from mechanical collectors is commonly employed to improve boiler efficiency. Ash residue in the fuel

TABLE 4. *(Continued)*

	Physical Characteristics of Dust-Cake Ashes										
	Particle Diameters, μm										
	Bahco MMD	Coulter MMD	Drag Equivalent	Density, g/cm^3	Specific Surface Area, m^2/g	Morphology Factor	Effective Angle of Internal Friction, °	Uncompacted Bulk Porosity	Compacted Bulk Porosity	Dust-Cake Porosity[a]	Relative Gas-Flow Resistance in. H_2O · min · ft/lb
Monticello	7.78	6.0	3.75	2.42	0.85	2.3	38.7	0.65	0.36	0.57	8.1
Scholz HSFP	8.62	7.8	2.37	2.48	3.23	9.9	42.8	0.76	0.61	0.77	2.4
Brunner Island	4.97	5.6	2.15	2.46	1.45	3.5	42.5	0.78	0.56	0.74	4.2
IPP	8.71	7.0	3.54	2.37	1.35	3.9	39.0	—	0.42	0.62	5.7
Nixon	5.79	7.3	2.55	2.23	3.70	10.5	42.8	—	0.50	0.68	5.6
Shawnee Unit 5	—	5.3	1.43	2.55	—	—	—	—	0.63	0.78	5.2
Cameo	5.00	6.3	2.16	2.25	—	—	—	—	0.54	0.72	5.7
Harrington	7.82	5.7	2.20	2.67	1.45	3.8	45.0	—	0.39	0.60	16.0
Escalante	9.39	8.3	3.95	2.23	0.91	2.7	40.0	—	0.44	0.63	4.2
Arapahoe Unit 4	6.22	6.1	2.59	2.25	2.00	5.0	41.5	—	0.50	0.68	5.7
Arapahoe FFPP	7.40	6.7	3.22	2.22	1.70	4.5	42.0	0.73	0.49	0.67	4.2
Arapahoe FFPP (downstream of multiclone)	3.33	3.6	1.25	2.47	3.70	5.9	44.3	—	0.58	0.74	11.4
Arapahoe Ecolaire	6.67	6.4	2.32	2.29	2.10	5.2	42.3	—	0.61	0.73	4.2
Crane Unit 1	7.43	6.6	2.20	2.55	3.80	11.0	45.5	—	0.52	0.69	6.3
Comanche Ecolaire	4.40	5.0	1.85	2.87	1.70	4.0	40.0	—	0.42	0.62	16.3
Cherokee	—	—	—	—	—	—	—	—	—	—	—
Eraring	5.51	5.7	2.84	2.34	1.50	3.5	45.0	—	0.49	0.68	4.6
Tallawarra	4.40	7.0	1.82	2.33	1.80	4.7	46.5	—	0.56	0.73	6.4
TVA AFBC	3.91	4.6	0.84	2.62	14.44	31.5	46.6	0.88	0.68	0.82	9.7
Plant A CFBC	—	4.3	0.84	2.50	21.80	44.0	45.8	—	0.66	0.80	12.0
Plant B CFBC	6.19	4.0	1.19	2.95	4.20	9.1	47.5	—	0.61	0.77	8.0
Nucla CFBC	5.67	3.2	0.91	2.73	10.60	16.7	46.8	0.87	0.68	0.82	7.7
EPRI High-Sulfur Test Center	7.32	7.3	1.07	2.49	6.90	20.0	42.8	—	0.73	0.86	4.0
Plant C Spray Dryer	3.06	7.8	3.31	2.26	2.70	7.6	41.0	—	0.52	0.70	2.8

[a]Some values estimated from compacted bulk porosity data.

Source: Reference 4, p 232.

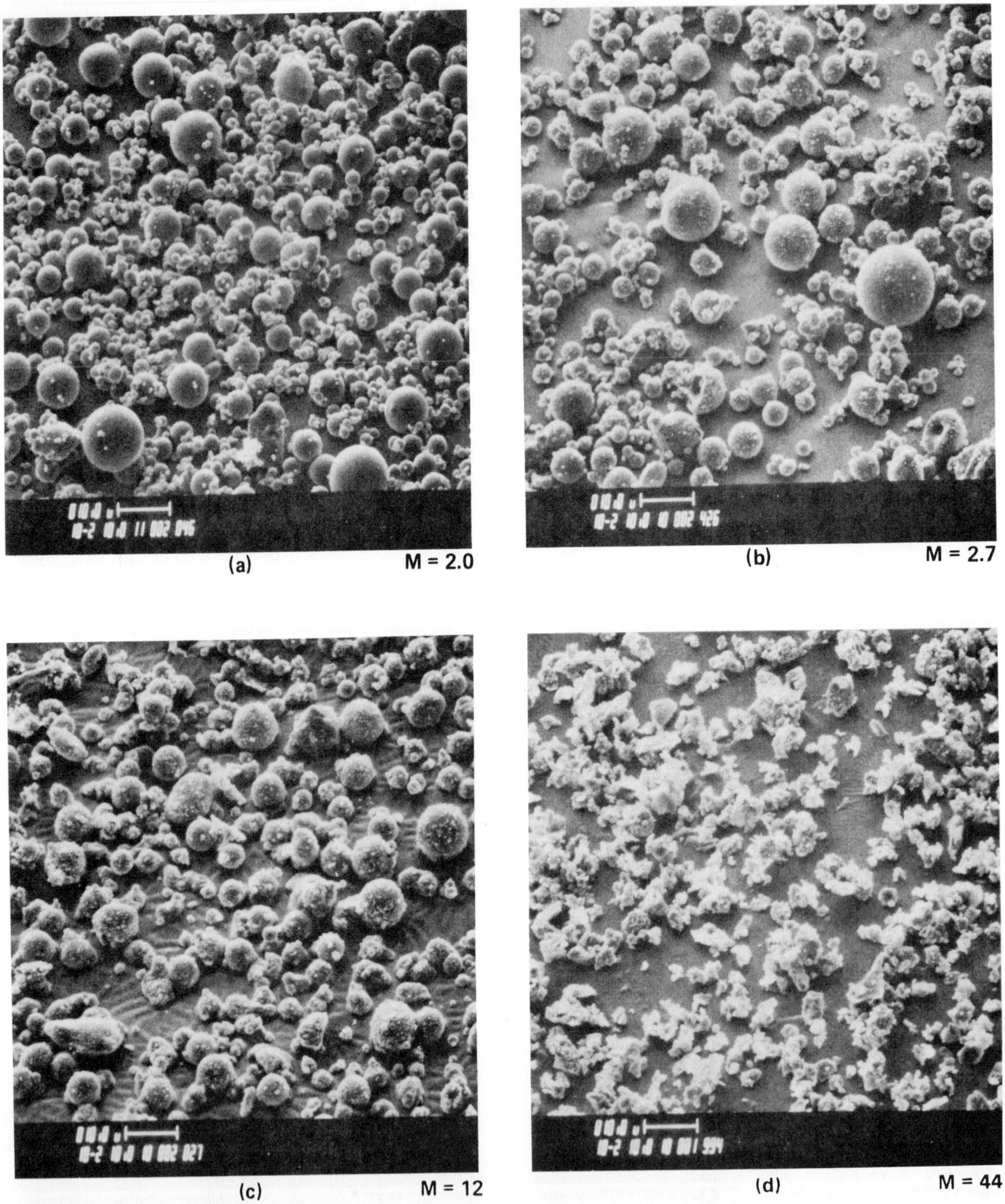

(a) M = 2.0

(b) M = 2.7

(c) M = 12

(d) M = 44

FIGURE 1. SEM Photographs Comparing the Appearance and Morphology Factor, *M*, of Dust-Cake Ashes from (a) Monticello, (b) Escalante, (c) Scholz HSFP, and (d) TVA Atmospheric Fluidized-Bed Combustor (From Reference 4, p 233)

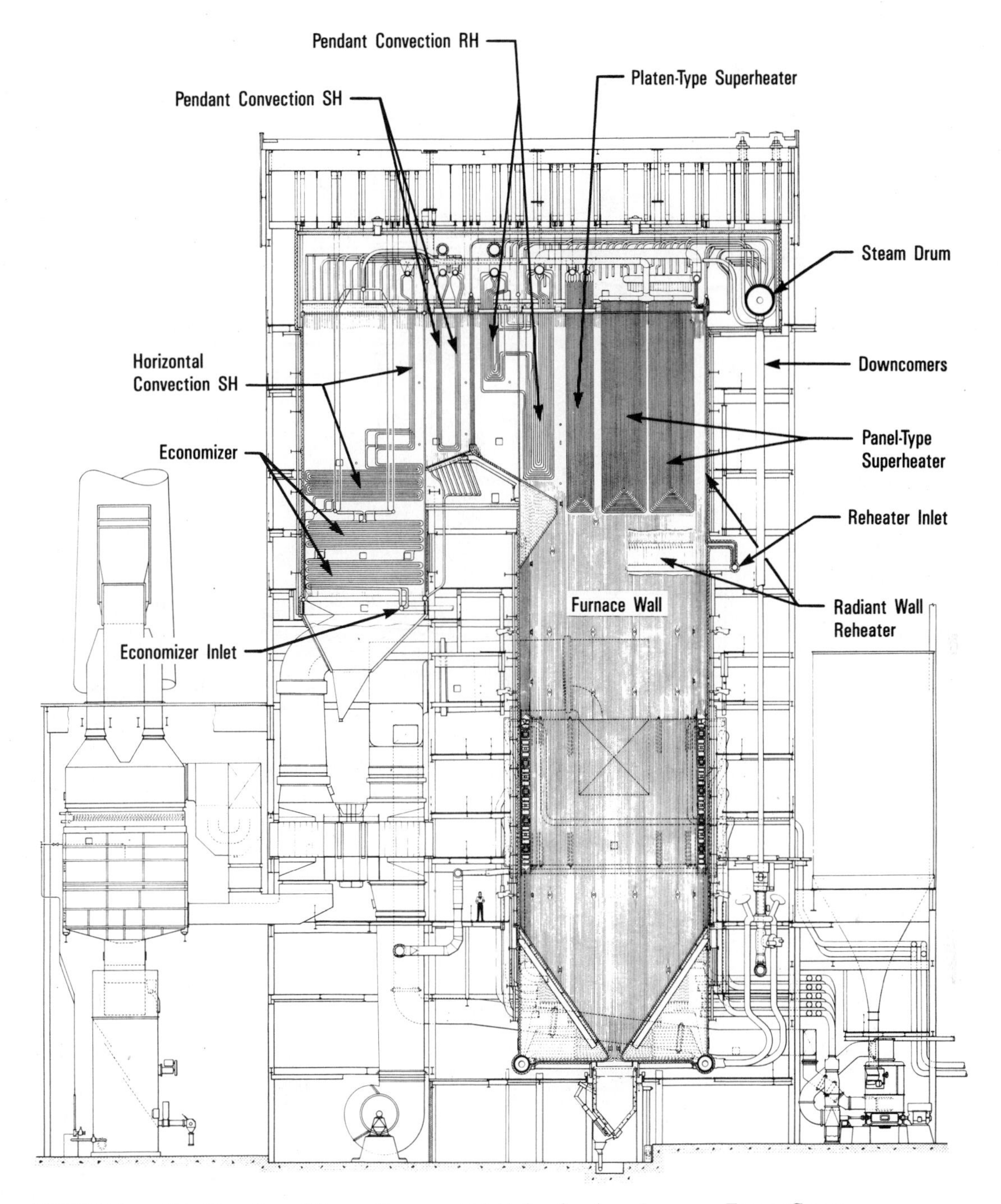

FIGURE 2. Side Elevation of Kansas Power and Light Gas Services, Lawrence Energy Center Unit No. 5 as Originally Built

bed is deposited in a receiving pit at the end of the grate.

In overfeed stokers, coal is fed onto a traveling or vibrating grate, and it burns on the fuel bed as it progresses through the furnace. Ash particles fall into an ash pit at the rear of the stoker. The term "overfeed" applies because the coal is fed onto the moving grate under an adjustable gate. Conversely, in "underfeed" stokers, coal is fed into the firing zone from underneath by mechanical rams or screw conveyers. The coal moves in a channel, known as a retort, from which it is forced upward, spilling over the top of each side to form and to feed the fuel bed. Combustion is completed by the time the bed reaches the side dump grates from which the ash is discharged to shallow pits. Underfeed stokers include single-retort units and multiple-retort units, the latter having several retorts side by side. Figure 4 shows a schematic of a stoker-fired boiler.[9]

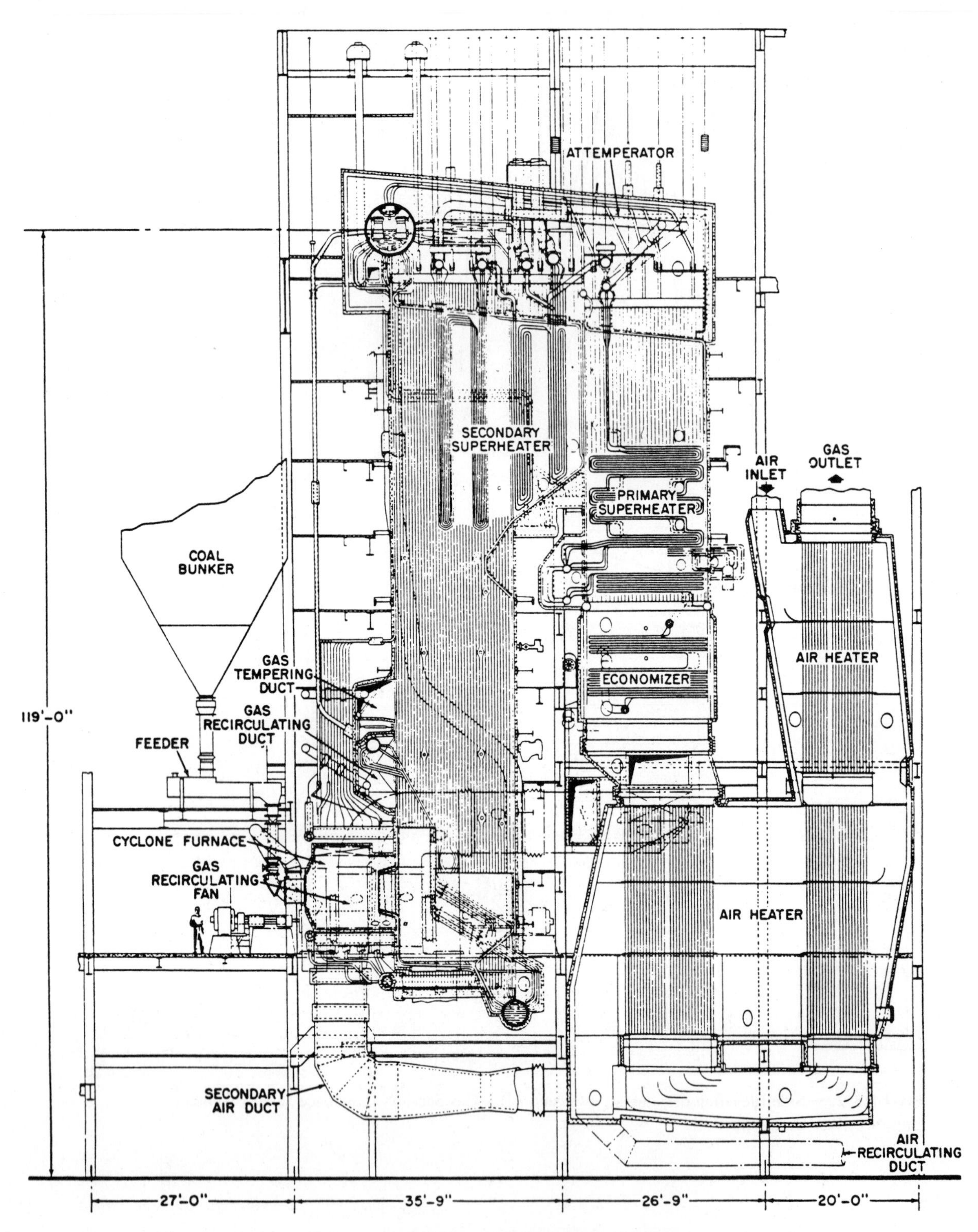

FIGURE 3. Wisconsin Power and Light Company's Nelson Dewey Station Unit 2

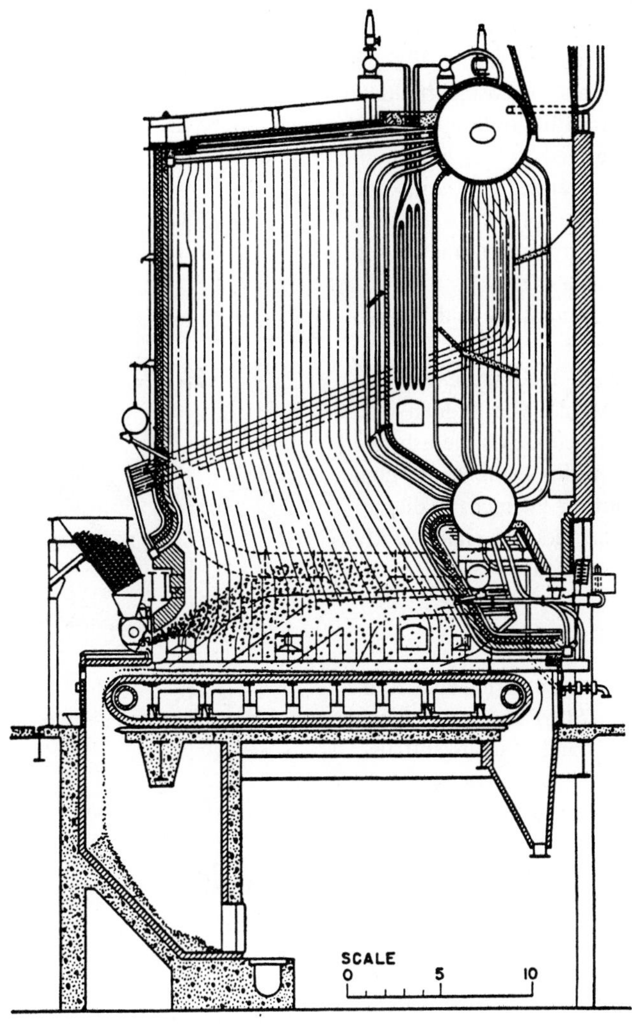

FIGURE 4. A Steam Generating Unit Equipped with Continuous-Discharge Type of Spreader Stoker. Capacity, 150,000 Pounds of Steam per Hour at 450 psi and 750°F When Burning Midwest Bituminous Coal. (From *Combustion Engineering,* 5–25 [1948])

EMISSIONS FROM COAL COMBUSTION*

The major pollutants of concern from external coal combustion are particulate, sulfur oxides, and nitrogen oxides. Some unburned combustibles, including numerous organic compounds and carbon monoxide, are generally emitted even under proper boiler operating conditions.

Particulate Emissions

Particulate composition and emission levels are a complex function of firing configuration, boiler operation, and coal properties. In pulverized-coal systems, combustion is almost complete, and thus the particulate is largely composed of the inorganic ash residue. In wet-bottom, pulverized-coal units and cyclones, the quantity of ash leaving the boiler is less than in dry bottom units, since some of the ash liquifies, collects on the furnace walls, and drains from the furnace bottom as molten slag. To increase the fraction of ash drawn off as wet slag, and thus to reduce the fly-ash disposal problem, fly ash may be reinjected from collection equipment into slag tap systems. Dry bottom unit ash may also be reinjected into wet bottom boilers for the same purpose.

Because a mixture of fine and coarse coal particles is fired in spreader stokers, significant unburned carbon can be present in the particulate. To improve boiler efficiency, fly ash from collection devices (typically multiple cyclones) is sometimes reinjected into spreader stoker furnaces. This practice can dramatically increase the particulate loading at the boiler outlet and, to a lesser extent, at the mechanical

*Taken in part from AP-42.[6]

collector outlet. Fly ash can also be reinjected from the boiler, air heater, and economizer dust hoppers. Fly-ash reinjection from these hoppers does not increase particulate loadings nearly so much as that from multiple cyclones.

Uncontrolled overfeed and underfeed stokers emit considerably less particulate than do pulverized-coal units and spreader stokers, since combustion takes place in a relatively quiescent fuel bed. Fly-ash reinjection is not practiced in these kinds of stokers.

Variables other than firing configuration and fly-ash reinjection also can affect emissions from stokers. Particulate loadings will often increase as load increases (especially as full load is approached) and with sudden load changes. Similarly, particulate can increase as the ash and fines contents increase. ("Fines," in this context, are coal particles smaller than about 1.6 mm, or 1/16 inch, in diameter.) Conversely, particulate can be reduced significantly when overfire air pressures are increased.

Sulfur Oxides Emissions

Gaseous emissions of sulfur oxides from external coal combustion are largely sulfur dioxide (SO_2) and, to a much smaller extent, sulfur trioxide (SO_3) and gaseous sulfates. These compounds form as the organic and pyritic sulfur in the coal is oxidized during the combustion process. On average, 98% of the sulfur present in bituminous coal will be emitted as gaseous sulfur oxides, whereas somewhat less will be emitted when subbituminous coal is fired. The more alkaline nature of the ash in some subbituminous coal causes some of the sulfur to react to form various sulfate salts that are retained in the boiler or in the fly ash. Generally, boiler size, firing configuration, and boiler operations have little effect on the percent conversion of fuel sulfur to sulfur oxides.

Nitrogen Oxides Emissions

Nitrogen oxides (NO_x) are produced in the combustion process by at least two different mechanisms: (1) from the molecular nitrogen in the combustion air (thermal NO_x), and (2) from the nitrogen in the fuel being burned (fuel NO_x). The relative contributions of these two formation pathways depend on the combustion conditions, the type of boiler, and the type of fuel being burned. Figure 5 shows typical uncontrolled NO concentrations in the flue gas for various types of coal-fired boilers.[10,11]

The thermal NO_x formation rate is markedly dependent on gas temperature and becomes rapid at gas temperatures of 3000–3600°F.[11] The currently accepted model for thermal NO_x formation was developed by Zeldovich (1946). The most important reactions in the Zeldovich model are as follows:

$$N_2 + O \rightarrow NO + N$$
$$N + O_2 \rightarrow NO + O$$
$$N + OH \rightarrow NO + H$$

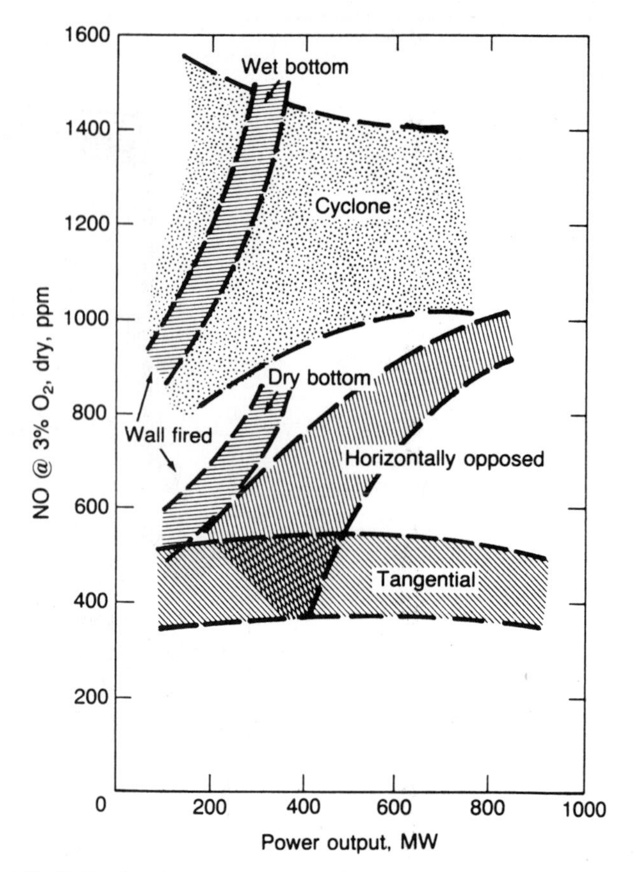

FIGURE 5. Uncontrolled NO Concentrations for Coal-Fired Boilers (From Reference 11, Chapter 15)

It has been observed that the most important operating variables in the formation of thermal NO_x are (1) peak gas temperature, (2) gas residence time in the peak temperature, and (3) fuel–air stoichiometric ratio. In general, a lower operating temperature, gas residence time, and off-stoichiometric fuel–air ratio reduce NO_x formation. Figure 6 shows the simulated effects of these operating variables on the formation of NO_x in a combustion system.[11]

The nitrogen oxides can also be formed in the combustion zone through an alternative mechanism in which molecular nitrogen in the combustion air reacts with free hydrocarbon radicals available from the fuel in the combustion zone, as shown in the following reaction.

$$N_2 + CH \rightarrow HCN + N$$

The HCN then combines with free OH radicals to form CN, which is then oxidized to NO via a number of intermediate steps. The nitrogen oxide formed through this pathway is also called "prompt NO." The contribution of prompt NO to the overall NO_x formation is rather small.[12]

If the fuel contains organically bound nitrogen, as in the case of most coals (0.5–2.0%) and residual fuel oils (0.1–0.5%), reactions with the oxygen in the combustion air produce NO_x. The amount of nitrogen in the fuel is relatively small compared with that in the combustion air. Howev-

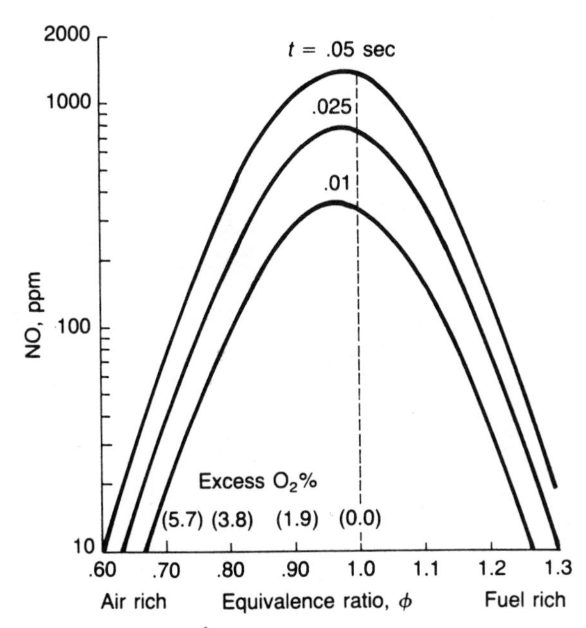

FIGURE 6. NO versus Fuel Equivalence Ratio and Time (Stoichiometric A/F = 16.3, Air Preheat = 650°F) (From Reference 11, Chapter 15, p 460)

er, this nitrogen is highly reactive; depending on combustion conditions, the contribution of fuel nitrogen can be significant.

A schematic representation of the relative contributions of the three mechanisms of NO_x formation to the overall NO_x concentration is shown in Figure 7.[12]

VOC and CO Emissions

Volatile organic compounds (VOCs) and carbon monoxide (CO) are unburned gaseous combustibles that are also emitted from coal-fired boilers, but generally in quite small amounts. However, during start-ups, temporary upsets, or other conditions preventing complete combustion, unburned combustible emissions may increase dramatically.

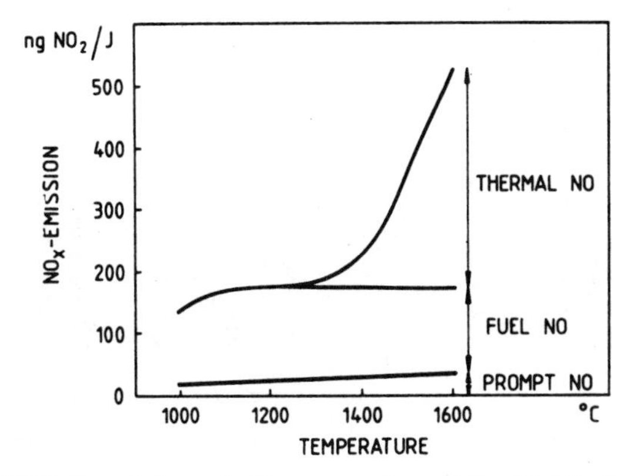

FIGURE 7. Schematic Representation of the Distribution of Formation of NO between Different Formation Mechanisms when Burning Coal, as a Function of Temperature (From Reference 12)

The VOC and CO emissions per unit of fuel fired are normally lower from pulverized-coal or cyclone furnaces than from smaller stokers and hand-fired units where operating conditions are not so well controlled. Measures used for NO_x control can increase CO emissions, so to reduce the risk of explosion, such measures are applied only to the point at which CO in the flue gas reaches a maximum of about 200 ppm. Other than maintaining proper combustion conditions, control measures are not applied to control VOCs and CO.

Table 5 lists the uncontrolled emission factors from the different types of furnaces discussed for particulate, sulfur oxides, nitrogen oxides, carbon monoxide, and VOCs as taken from AP-42.[6]

EMISSION CONTROL

Particulate Control

The primary kinds of particulate control used for coal combustion are (1) multiple cyclones, (2) ESPs, (3) fabric filters, and (4) venturi scrubbers. The application of particular equipment will depend on the type of furnace, coal properties, and operating conditions.

Prior to the Clean Air Act of 1970, multiple cyclones were generally utilized to control fly-ash emissions. These have generally been replaced in recent years by more efficient control systems or retained in the system as a precleaner upstream of the more efficient ESPs or fabric filters, with the intent of reducing the loading of the abrasive fly-ash particles.

Electrostatic precipitators have been used widely for the control of emissions from steam generation for nearly 60 years, principally on coal-fired boilers. For many years, all fly-ash ESPs were installed downstream of the air heaters at temperatures of 270–350°F (referred to as cold-side ESPs). As a result of the increased combustion of low-sulfur coals with higher resistivities, ESPs have also been installed upstream of the air heaters in a temperature range of 600–750°F to take advantage of the lower fly-ash resistivity at higher temperatures. These units are referred to as hot-side ESPs. Figure 8 illustrates a plan view of a typical cold-side ESP.[13] Applications on coal-fired boilers include ESPs that use the weighted wire electrodes (commonly referred to as the "American" design) and the rigid frame-supported electrodes (referred to as the "European" design).

In the early 70s, the ESP was the generally preferred choice for a higher-efficiency particulate control device over fabric filters or wet venturi scrubbers. In 1986, it was reported that ESPs accounted for about 95% of all utility particulate controls in the United States, with a generation capacity of approximately 330 GW.[14] In general, the ESP was a more economical option than the baghouse, particularly on high-sulfur-coal applications where the ESP performance was unaffected by the resistivity of the dust.

The new Clean Air Act of 1990 requires electric utilities

TABLE 5. Emission Factors for External Bituminous and Subbituminous Coal Combustion[a]

	Particulate[b]		Sulfur Oxides[c]		Nitrogen Oxides[d]		Carbon Monoxide[e]		Nonmethane VOC[a,f]		Methane[e]	
Firing Configuration	hg/Mg	lb/ton	hg/Mg	lb/ton	Kg/Mg	lb/ton	Kg/Mg	lb/ton	Kg/Mg	lb/ton	Kg/Mg	lb/ton
Pulverized coal-fired Dry bottom	5A	10A	19.5S(17.5S)	39S(35S)	10.5(7.5)g	21(15)g	0.3	0.6	0.04	0.07	0.015	0.03
Wet bottom	3.5A[h]	7A[h]	19.5S(17.5S)	39S(35S)	17	34	0.3	0.6	0.04	0.07	0.015	0.03
Cyclone furnace	1A[h]	2A[h]	19.5S(17.5S)	39S(35S)	18.3	37	0.3	0.6	0.04	0.07	0.015	0.03
Spreader stoker												
Uncontrolled	30[j]	60[j]	19.5S(17.5S)	39S(35S)	7	14	2.5	5	0.04	0.07	0.015	0.03
After multiple cyclone with fly ash reinjection from multiple cyclone	8.5	17	19.5S(17.5S)	39S(35S)	7	14	2.5	5	0.04	0.07	0.015	0.03
No fly ash reinjection from multiple cyclone	6	12	19.5S(17.5S)	39S(35S)	7	14	2.5	5	0.04	0.07	0.015	0.03
Overfeed stoker[k]												
Uncontrolled	6[a]	16[a]	19.3S(17.5S)	39S(35S)	3.25	7.5	3	6	0.04	0.07	0.015	0.03
After multiple cyclone	4.5[h]	9[h]	19.5S(17.5S)	39S(35S)	3.25	7.5	3	6	0.04	0.07	0.015	0.03

Underfeed stoker												
Uncontrolled	7.5[p]	15[p]	15.5S	31S	4.75	9.5	5.5	11	0.65	1.3	0.4	0.8
After multiple cyclone	5.5[a]	11[a]	15.5S	31S	4.75	9.5	5.5	11	0.65	1.3	0.4	0.8
Handfired units	7.5	15	15.5S	31S	1.5	3	45	90	5	10	4	8

[a]Factors represent uncontrolled emissions unless otherwise specified and should be applied to coal consumption as fired.

[b]Based on EPA Method 5 (front half catch). Where particulate is expressed in terms of coal ash content, *A*, factor is determined by multiplying wt % ash content of coal (as fired) by the numerical value preceding the *A*. For example, if coal having 8% ash is fired in a dry bottom unit, the particulate emission factor would be 5×8, or 40 kg/Mg (80 lb/ton). The condensible-matter collected in back half catch of EPA Method 5 averages <5% of front half, or "filterable", catch for pulverized coal and cyclone furnaces; 10% for spreader stokers; 15% for other stokers; and 50% for hand-fired units.

[c]Expressed as SO_2, including SO_2, SO_3, and gaseous sulfates. Factors in parentheses should be used to estimate gaseous SO_x emissions for subbituminous coal. In all cases, "S" is wt % sulfur content of coal as fired. See footnote b for example calculation. On average for bituminous coal, 97% of fuel sulfur is emitted as SO_2, and only about 0.7% of fuel sulfur is emitted as SO_3 and gaseous sulfate. An equally small percent of fuel sulfur is emitted as particulate sulfate. Small quantities of sulfur are also retained in bottom ash. With subbituminous coal, generally about 10% more fuel sulfur is retained in the bottom ash and particulate because of the more alkaline nature of the coal ash. Conversion to gaseous sulfate appears about the same as for bituminous coal.

[d]Expressed as NO_2. Generally, 95–99 volume % of nitrogen oxides present in combustion exhaust will be in the form of NO, the rest NO_2. To express factors as NO, multiply by factor of 0.66. All factors represent emission at baseline operation (i.e., 60–110% lead and no NO_x control measures.

[e]Nominal values achieveable under normal operating conditions. Values one or two orders of magnitude higher can occur when combustion is not complete.

[f]Nomethane volatile organic compounds (VOC), expressed as C_2 to C_{16} n-alkane equivalents. Because of limited data on NMVOC available to distinguish the effects of firing configuration, all data were averaged collectively to develop a single average for pulverized coal units, cyclones, spreaders, and overfeed stokers.

[g]Parenthetic value is for tangentially fired boilers.

[h]Uncontrolled particulate emissions, when no fly ash reinjection is employed. When control device is installed, and collected fly ash is reinjected to boiler, particulate from boiler reaching control equipment can increase by up to a factor of two.

[j]Accounts for fly ash settling in an economizer, air heater, or breeching upstream of control device or stack. (Particulate directly at boiler outlet typically will be twice this level.) Factor should be applied even when fly ash is reinjected to boiler from boiler, air heater, or economizer dust hoppers.

[k]Includes traveling grate, vibrating grate, and chain grate stokers.

[m]Accounts for fly ash settling in breeching or stack base. Particulate loadings directly at boiler outlet typically can be 50% higher.

[n]Discussion of apparently low multiple cyclone control efficiencies, regarding uncontrolled emissions available in Reference 6.

[p]Accounts for fly ash settling in breeching downstream of boiler outlet.

Source: Reference 6.

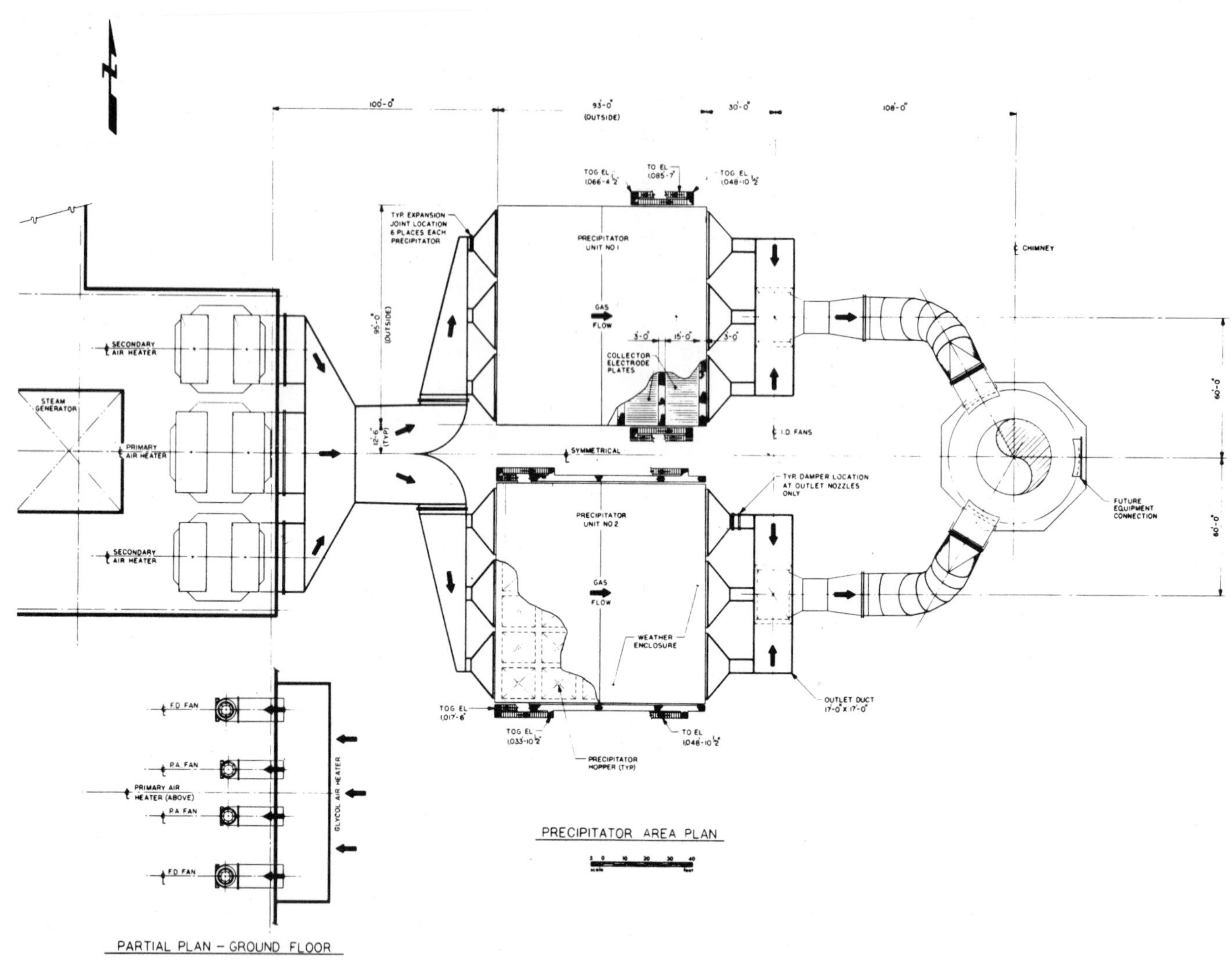

FIGURE 8. European Cold-Side ESP, Plan View (From Reference 13)

to reduce their SO_2 emissions by approximately 10 million tons per year by the year 2000. If utilities choose to do this by using wet scrubbing techniques (downstream of the ESP), the changes may not affect existing ESPs. However, a number of utilities will likely retrofit existing plants with dry SO_2 technologies or switch to low-sulfur coals. Either of these choices will increase dust loadings and change the resistivity of the fly ash, making it more difficult to collect the materials with ESPs.

Offen and Altman[14] concluded in 1991 that:

> The selection of particulate control equipment by U.S. utilities and independent power producers in the 1990s will depend largely on regulatory trends and the technical and commercial success of ongoing particulate control research and development efforts. Well-designed, modern ESPs theoretically can meet emission targets somewhat lower than the current New Source Performance Standards (NSPS). Therefore, if new particulate emission limits are not significantly lower than the current NSPS, utilities will be able to continue to choose either ESPs or baghouses based on considerations such as unit size, fuel type, and boiler type. Requirements for greater particulate reductions or stringent emission limits based on respirable particulate (PM_{10}) may limit the number of sites where an ESP would be acceptable.

Options for upgrading existing ESPs to meet the greater demands are on the horizon and are likely to influence the future application of ESPs.

Electric utilities have made significant progress in recent years in designing and operating fabric filter collectors (baghouses) for the collection of coal fly ash. As a result, baghouses have become an accepted, and sometimes preferred, alternative to ESP. Interest in baghouses will continue to increase because of the increasing stringency of air emission standards, concerns over fine particulates, and the use of dry SO_2 control technologies (duct injection and spray drying). A comparison of capital and levelized costs for ESPs versus baghouses is shown in Figures 9 and 10.[15] In general, the capital costs favored ESPs on high-sulfur coals with 1971 emission standards and favored baghouses in the lower-sulfur applications with 1979 emission standards.

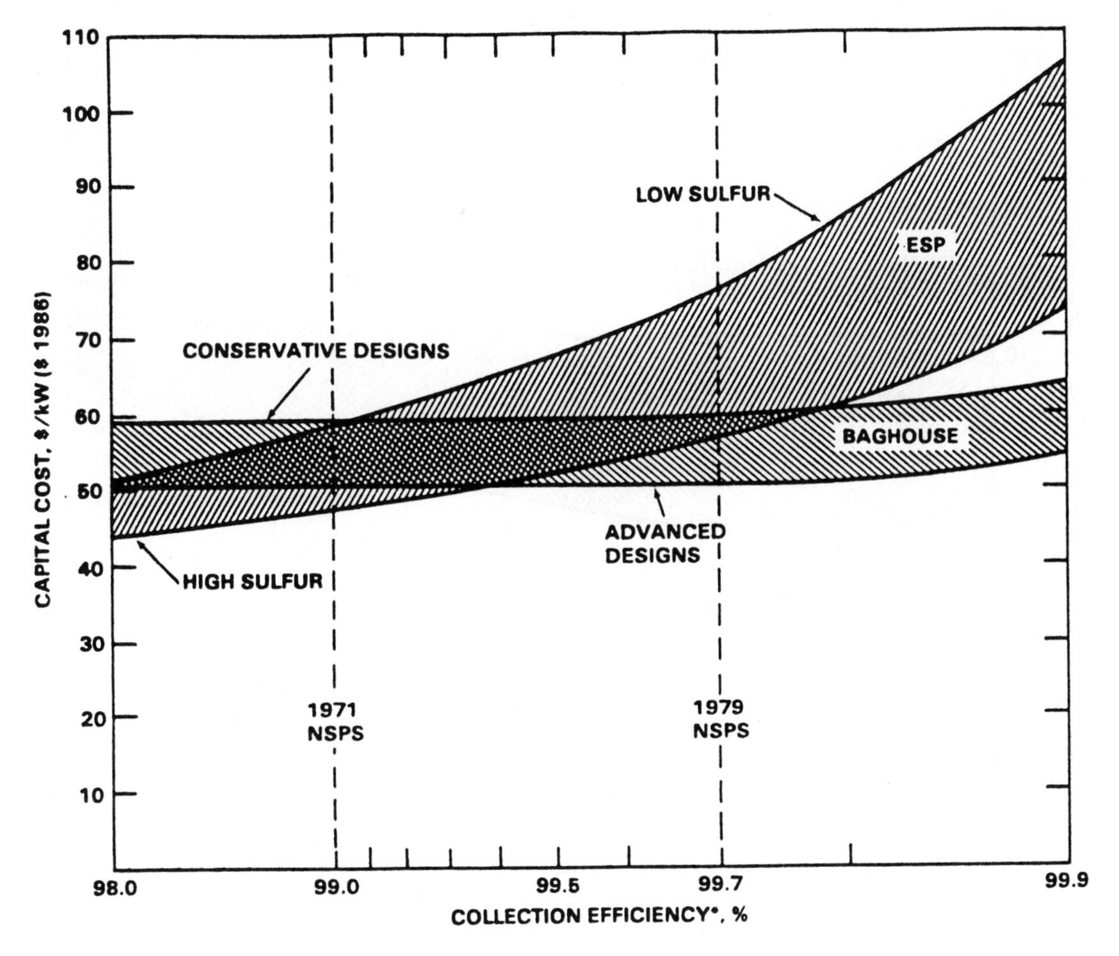

* BASED ON A COAL RATED AT 10,000 Btu/lb AND CONTAINING 10% ASH. THE 1971 NSPS IS 0.1 lb/10^6 Btu, AND THE 1979 NSPS IS 0.03 lb/10^6 Btu.

FIGURE 9. Capital Costs for Baghouses and ESPs, Expressed as a Function of Nominal Collection Efficiency (From Reference 15, p I-5)

Cushing et al.[16] reported that as of December 1989, there were 99 baghouses in operation on utility boilers with 21,359 MW of generation capacity, ranging in size from 6 MW to 860 MW. The largest operating baghouses were two 739-MW units at the Four Corners utility, with 19,872 filter bags in each unit. The oldest unit was reported to be the Sunbury installation, commissioned in 1973. Figure 11 illustrates the rapid growth in the application of baghouses on coal-fired utility boilers since 1972. Table 6 summarizes the primary operating characteristics of the baghouses applied to coal-fired boilers. The majority of the baghouses are cleaned by reverse gas cleaning, and, in some cases, use sonic horns to assist in the cleaning. Shaking and shake/deflate are also used. Figure 12 shows a schematic of a typical reverse-gas–cleaned baghouse.[15] These baghouses have air–cloth ratios typically ranging from 1.5 to 2.0 acfm/ft^2 with operating pressure drops of 3.5–9.0 in. H_2O with an average of 6.4 in. H_2O.

Design efficiencies for the baghouses ranged from 98% to over 99.9% for particulate removal. In general, reported emission levels from the baghouses with reverse air cleaning ranged from 0.005 to 0.03 lb/MBtu; with reverse air with sonic horn assistance, from 0.008 to 0.125 lb/MBtu; and with shake and deflate cleaning, from 0.007 to 0.07 lb/MBtu. All of the baghouses reported opacities of less than 5.0%.

The average bag life was reported to vary from 2.9 years on high-sulfur applications to 4.6 years for low-sulfur applications. Seven of 22 of the systems reported a bag life of more than six years. The majority of the installations used woven fiberglass bags with fabric coatings of Teflon®, silicon graphite, or a proprietary acid-resistant finish.

Figures 13 to 18, taken from AP-42[6], show the extent of particulate emission control achieved by multiple cyclones, ESPs, baghouses, and wet scrubbers for the various types of furnaces. Uncontrolled emission factors are also shown for comparison. As a result of the concern over emissions in various size ranges, the emissions are presented as a function of particle size and are referred to as cumulative size-specific emission factors.

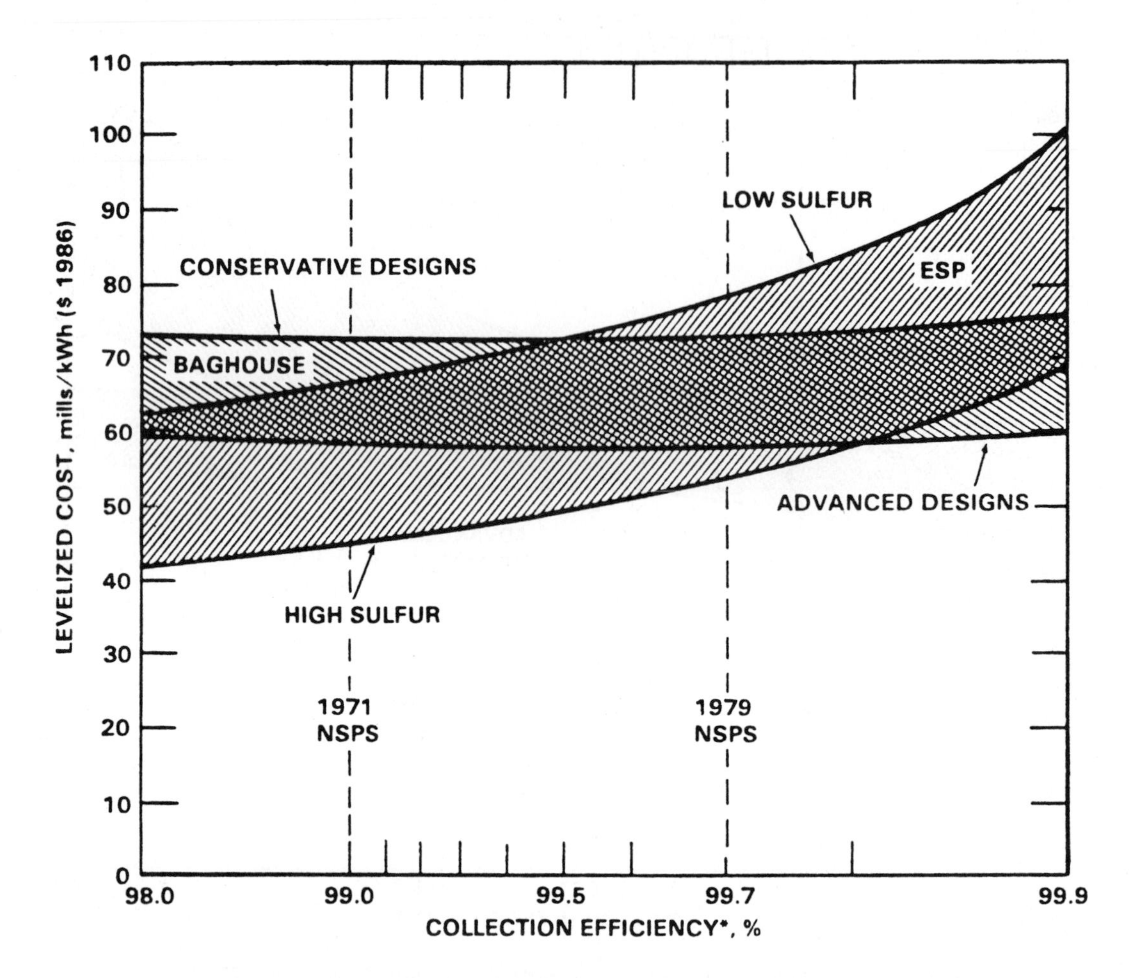

FIGURE 10. Levelized Cost for Baghouses and ESPs, Expressed as a Function of Nominal Collection Efficiency (From Reference 15, p I-6)

Sulfur Oxides Control

Several techniques are used to reduce sulfur oxides from coal combustion. One way is to switch to lower-sulfur coals, since sulfur oxides emissions are proportional to the sulfur content of the coal. This alternative may not be possible where lower-sulfur coal is not readily available or where a different grade of coal cannot be fired satisfactorily. In some cases, various cleaning processes may be employed to reduce the fuel's sulfur content. Physical coal cleaning removes mineral sulfur such as pyrite, but is not effective in removing organic sulfur. Chemical cleaning and solvent refining processes are being developed to remove organic sulfur.

The primary development of sulfur dioxide control technology has occurred as a result of the need to control emissions from the electric utility power industry and is referred to as flue gas desulfurization (FGD). In 1990, the electric utility power industry in the United States had a generating capacity of 731 GW, with the capacity estimated to increase to 773 GW by 1999.[17] Approximately 44% (340 GW) was produced by coal-fired units. (see Table 7).

While coal accounted for 44% of generating capacity in 1989, it actually produced 56% of the actual power since coal is generally used to supply the base load while other fuel sources are often used to meet peak load requirements. In 1989, 21.1% of the coal-fired generating capacity (or 9.3% of total generating capacity) was controlled by FGD systems with an equivalent scrubbed capacity of approximately 63 GW (or 63,000 MW).

Table 8 presents a summary of FGD systems by process type for 1989 (actual capacity) and 2010 (projected capacity).[18] The three dominant processes are all throwaway-product systems, including the two wet scrubbing systems (limestone and lime) with 48.7% and 16.3% of the capacity and the lime spray drying system with 8.8% of the capacity.

The current NSPS for coal-fired utility boilers requires from 70% up to 90%$^+$ for SO_2 removal, depending on the sulfur content of the coal. In 1990, the average design

TABLE 6. Baghouse Performance Data

Plant Generating Capacity, MW	Coal Type[a]	Coal Sulfur Content, %	Bag Cleaning Method[b]	Gas Temp, °F	Flange/Flange Pressure Drop, in. H_2O	Tube-Sheet Pressure Drop, in. H_2O	Gross Air/Cloth Ratio[c]	Dust-Cake Areal Density, lb/ft^2	Emission Rate, lb/MBtu	Stack Opacity, %
Pulverized-coal boilers										
150	WS	0.24	RG	325	7.5–8.0	—	1.95	—	—	—
85	WS	0.37	RG	—	6.0	—	1.77	—	—	—
223	WS	0.37	RG	270	7.0	6.0	1.58	—	0.012	—
223	WS	0.37	RG	282	6.0	—	1.81	—	—	—
405	WS	0.41	RG	267–305	5.0–5.5	4.6–4.8	1.65	0.78	—	0.5–2.0
447	WS	0.43	RG	273–306	3.5–5.0	2.5–3.5	1.72	0.35	0.0045	—
840	WS	0.43	RG	260–280	5.0–6.0	3.8–4.5	1.89	0.24	—	2.0–3.0
245	WS	0.47	RG	320	4.0–5.0	—	1.46	—	0.01	—
24	WB	0.49	RG	360	6.0	—	1.65	—	—	—
110	WS	0.52	RG	290	6.0–7.1	—	1.80	—	0.015	—
150	WS	0.52	RG	283–296	5.0–5.2	4.2–4.7	1.49	0.86	0.013	3.0–4.0
295	WS	0.52	RG	309	4.0–5.0	3.5–4.5	1.97	0.35	—	3.0
30	WS	0.61	RG	290	6.0–7.0	—	1.90(D)	—	—	—
565	WS	0.3	RG/S	275	8.0	5.6–5.8	1.7	0.35	0.03	3.0–5.0
565	WS	0.3	RG/S	275	8.0	—	1.7	0.28	0.023	3.0–5.0
254	WS	0.33	RG/S	230	5.5–6.8	4.5–5.5	1.98(D)	0.29	—	1.0
570	WS	0.45	RG/S	325	5.5–6.5	4.0–5.0	1.91	0.19	0.008	1.0–2.0
44	WS	0.52	RG/S	290	6.5–8.2	—	2.09(D)	—	0.016	—
100	WS	0.52	RG/S	290	4.2–6.0	—	1.98(D)	—	—	—
166	WS	0.6	RG/S	315	5.5–6.5	—	2.0 (D)	—	—	—
44	WS	0.61	RG/S	290	6.0–8.0	—	1.93(D)	—	—	—
739	WS	0.69	RG/S	240–280	4.8–5.5	4.0–4.4	1.50	0.64	0.023	—
185	EB	0.85	RG/S	301	5.0–6.5	2.5–3.5	1.76	0.32	0.029	3.0
185	EB	0.86	RG/S	305	5.7	2.7	1.87	—	0.018	3.0–5.0
185	EB	0.87	RG/S	300	5.0–6.5	2.5–3.5	1.91	—	0.036	3.0–5.0
79	AP	1.79	RG/S	325	6.0	—	1.71	—	—	—
350	EB	1.83	RG/S	—	5.0–9.0	4.0–8.0	1.83(D)	—	—	—

TABLE 6. *Continued*

Plant Generating Capacity, MW	Coal Type[a]	Coal Sulfur Content,%	Bag Cleaning Method[b]	Gas Temp, °F	Flange/ Flange Pressure Drop, in. H_2O	Tube-Sheet Pressure Drop, in. H_2O	Gross Air/Cloth Ratio[c]	Dust-Cake Areal Density, lb/ft^2	Emission Rate, lb/MBtu	Stack Opacity, %
191	EB	2.2	RG/S	304	7.0	—	1.5	—	0.039	—
191	EB	2.4	RG/S	303	7.0	—	1.5	—	0.125	—
87.5	AP	2.6	RG/S	400	3.5	—	1.89(D)	—	0.085	—
87.5	AP	2.7	RG/S	400	3.5	—	1.89(D)	—	0.085	—
384	WS	0.35	S/D	305	9.0	8.0	3.2	0.23	0.03	2.0–4.0
384	WS	0.36	S/D	320	7.5	—	2.8	—	0.051	—
593	TL	0.43	S/D	350	9.0–13.0	7.0–11.5	2.6	—	—	—
593	TL	0.49	S/D	350	9.0–13.0	7.0–11.5	2.6	—	—	—
(79)	AP	1.79	S/D	350	6.0	—	1.9	—	0.01–0.07	—
Pulverized-coal boilers with dry FGD systems										
279	WS	0.31	RG	185	4.0	—	1.58	—	—	—
319	WS	0.36	RG	165	6.0	—	1.60(D)	0.09	—	—
44	WS	0.52	RG	180	6.0	—	1.54(D)	—	0.03	—
860	WS	0.6	RG	165	9.8	—	2.00(D)	—	0.024	—
415	NDL	1.08	S/D	200	4.0–8.0	—	2.24(D)	—	0.018	—
Fluidized-bed combustion boilers										
160	EB	0.33	RG	290	7.2	6.8	1.53	—	<0.03	—
110	WS	0.39	S/D	294	5.0–6.5	3.7–5.2	2.4–2.9	0.23	0.0072	—

[a]WS = western subbituminous, WB = western bituminous, AP = anthracite/petroleum coke, TL = Texas lignite, EB = eastern bituminous, NDL = North Dakota lignite.
[b]RG = reverse gas, RG/S = reverse gas with sonic assistance, S/D = shake/deflate.
[c]D = design air-to-cloth ratio.

Source: Reference 16.

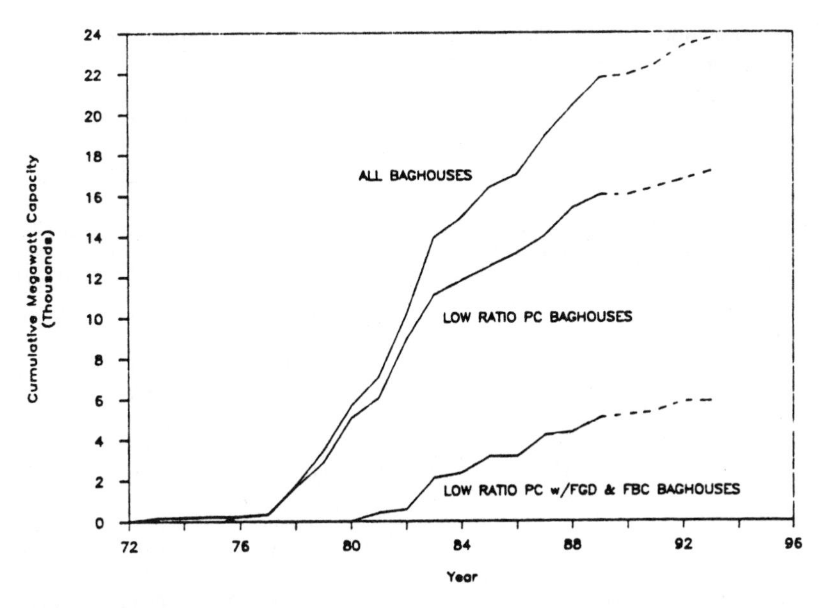

FIGURE 11. Chronological History of Cumulative Megawatt Capacity of Utility Baghouses. Data Are Presented for All Baghouses and Separately for Low-Ratio Baghouses Downstream from Pulverized-Coal Boilers and Low-Ratio Baghouses Downstream from Dry FGD Systems on Pulverized-Coal Boilers and Fluidized-Bed Boilers. Data for Baghouses in Design or Construction Are Shown Beyond 1989 (From Reference 16)

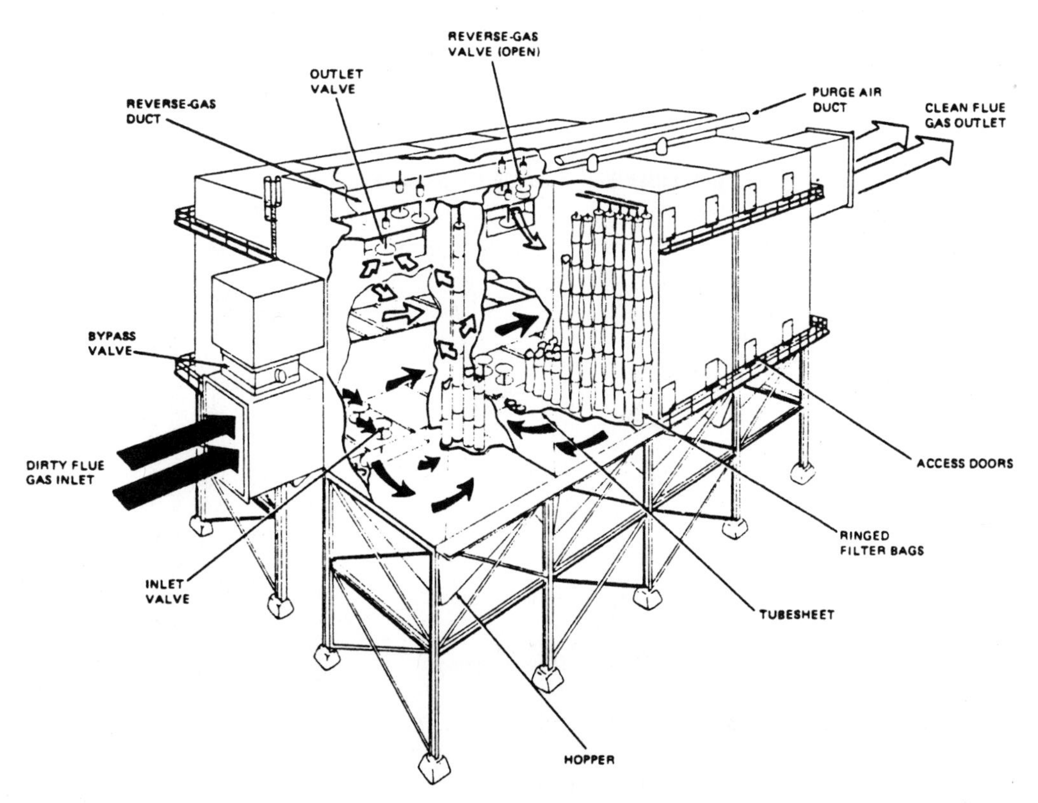

FIGURE 12. Cutaway View of a Typical Reverse-Gas-Cleaned Baghouse; the Compartment in the Foreground Is Filtering Flue Gas While the Adjacent Compartment Is Being Cleaned (From Reference 15, p 2–5)

Cumulative Particle Size Distribution and Size-Specific Emission Factors ***(Emission Factor Rating: C [Uncontrolled], D [Scrubber and ESP Controlled], E [Multiple Cyclone and Baghouse])***

	Cumulative Mass, % ≤ Stated Size					Cumulative Emission Factor[b] [kg/Mg (lb/ton) coal, as fired]				
		Controlled					Controlled[c]			
Particle Size,[a] μm	Uncontrolled	Multiple Cyclone	Scrubber	ESP	Baghouse	Uncontrolled	Multiple Cyclone	Scrubber	ESP	Baghouse
15	32	54	81	79	97	1.6A (3.2A)	0.54A (1.06A)	0.24A (0.48A)	0.032A (0.06A)	0.010A (0.02A)
10	23	29	71	67	92	1.15A (2.3A)	0.29A (0.58A)	0.21A (0.42A)	0.027A (0.05A)	0.009A (0.02A)
6	17	14	62	50	77	0.85A (1.7A)	0.14A (0.28A)	0.19A (0.38A)	0.020A (0.04A)	0.006A (0.02A)
2.5	6	3	51	29	53	0.30A (0.6A)	0.03A (0.06A)	0.15A (0.3A)	0.012A (0.02A)	0.005A (0.01A)
1.25	2	1	35	17	31	0.10A (0.2A)	0.01A (0.02A)	0.11A (0.22A)	0.007A (0.01A)	0.003A (0.006A)
1.00	2	1	31	14	25	0.10A (0.2A)	0.01A (0.02A)	0.09A (0.18A)	0.006A (0.01A)	0.003A (0.006A)
0.625	1	1	20	12	14	0.05A (0.10)	0.01A (0.02A)	0.06A (0.12A)	0.005A (0.01A)	0.001A (0.002A)
Total	100	100	100	100	100	5A (10A)	1A (2A)	0.3A (0.6A)	0.04A (0.08A)	0.01A (0.02A)

[a]Expressed as aerodynamic equivalent diameter.
[b]A = coal ash weight %, as fired.
[c]Estimated control efficiency for multiple cyclone, 80%; scrubber, 94%; ESP, 99.2%; baghouse, 99.8%.

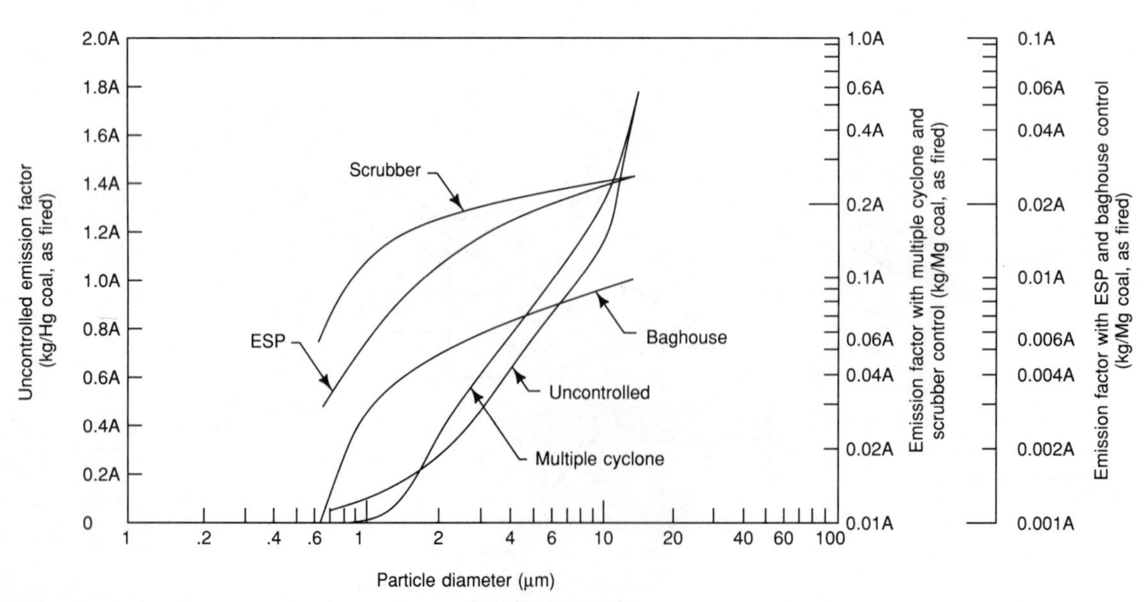

FIGURE 13. Emission Factors for Dry Bottom Boilers Burning Pulverized Bituminous Coal (From Reference 6)

Cumulative Particle Size Distribution and Size-Specific Emission Factors
(Emission Factor Rating: E)

	Cumulative Mass, % ≤ Stated Size			Cumulative Emission Factor[b] [kg/Mg (lb/ton) coal, as fired]		
		Controlled			Controlled[c]	
Particle Size,[a] μm	Uncontrolled	Multiple Cyclone	ESP	Uncontrolled	Multiple Cyclone	ESP
15	40	99	83	1.4A (2.8A)	0.69A (1.38A)	0.023A (0.046A)
10	37	93	75	1.30A (2.6A)	0.65A (1.3A)	0.021A (0.042A)
6	33	84	63	1.16A (2.32A)	0.59A (1.18A)	0.018A (0.036A)
2.5	21	61	40	0.74A (1.48A)	0.43A (0.86A)	0.011A (0.022A)
1.25	6	31	17	0.21A (0.42A)	0.22A (0.44A)	0.005A (0.01A)
1.00	4	19	8	0.14A (0.28A)	0.13A (0.26A)	0.002A (0.004A)
0.625	2	[d]	[d]	0.07A (0.14A)	[d]	[d]
Total	100	100	100	3.5A (7.0A)	0.7A (1.4A)	0.028A (0.056A)

[a]Expressed as aerodynamic equivalent diameter.
[b]A = coal ash weight %, as fired.
[c]Estimated control efficiency for multiple cyclone, 80%; ESP, 99.2%.
[d]Insufficient data.

Cumulative Size-Specific Emission Factors

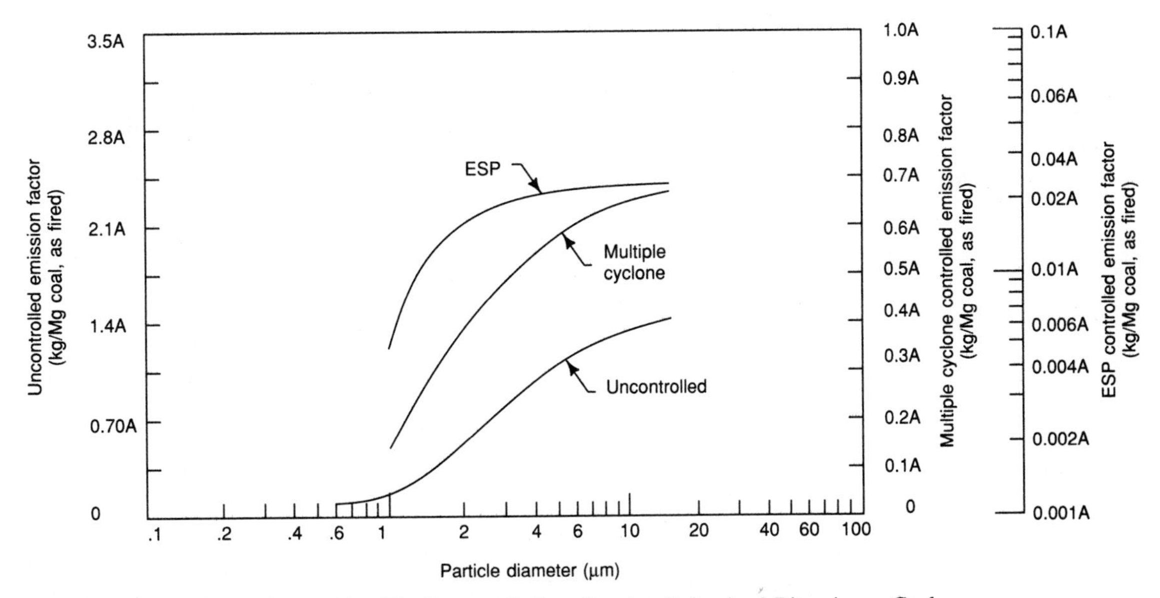

FIGURE 14. Emission Factors for Wet Bottom Boilers Burning Pulverized Bituminous Coal (From Reference 6)

Cumulative Particle Size Distribution and Size-Specific Emission Factors
(Emission Factor Rating: E)

Particle Size,[a] μm	Cumulative Mass, % ≤ Stated Size			Cumulative Emission Factor[b] [kg/Mg (lb/ton) coal, as fired]		
		Controlled			Controlled[d]	
	Uncontrolled	Scrubber	ESP	Uncontrolled	Scrubber	ESP
15	33	95	90	0.33A (0.66A)	0.057A (0.114A)	0.0064A (0.013A)
10	13	94	68	0.13A (0.26A)	0.056A (0.112A)	0.0054A (0.011A)
6	8	93	56	0.08A (0.16A)	0.056A (0.112A)	0.0045A (0.009A)
2.5	0	92	36	0 (0)	0.055A (0.11A)	0.0029A (0.006A)
1.25	0	85	22	0 (0)	0.051A (0.10A)	0.0018A (0.004A)
1.00	0	82	17	0 (0)	0.049A (0.10A)	0.0014A (0.003A)
0.625	0	c	c	0 (0)	c	c
Total	100	100	100	1A (2A)	0.06A (0.12A)	0.008A (0.016A)

[a]Expressed as aerodynamic equivalent diameter.
[b]A = coal ash weight %, as fired.
[c]Insufficient data.
[d]Estimated control efficiency for scrubber, 94%; ESP, 99.2%.

Cumulative Size-Specific Emission Factors

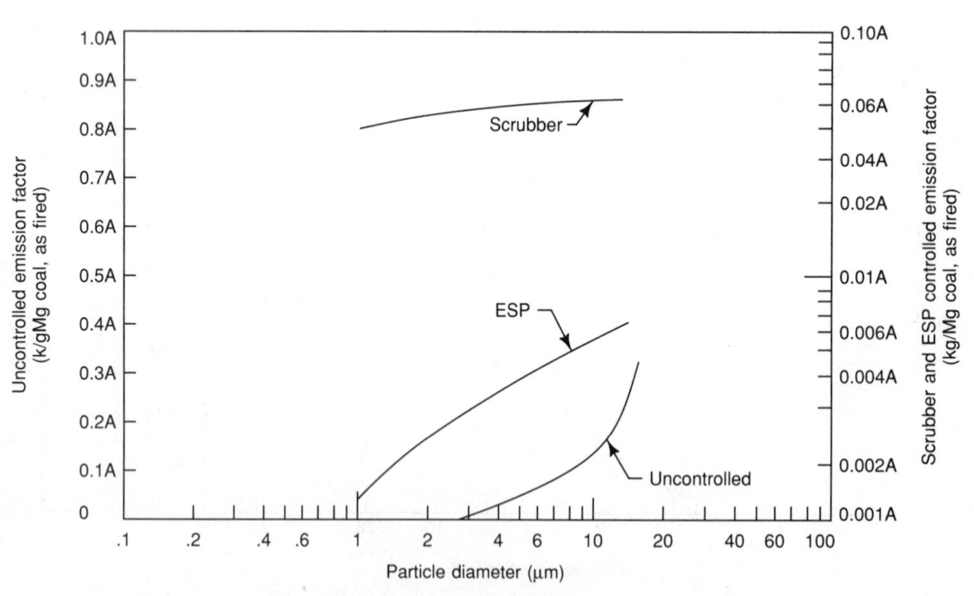

FIGURE 15. Emission Factors for Cyclone Furnaces Burning Bituminous Coal (From Reference 6)

Cumulative Particle Size Distribution and Size-Specific Emission Factors *(Emission Factor Rating: C (uncontrolled and controlled for multiple cyclone without fly-ash reinjection, and with baghouse), E (multiple cyclone controlled with fly-ash reinjection, and ESP controlled))*

	Cumulative Mass, % ≤ Stated Size					Cumulative Emission Factor [kg/Mg (lb/ton) coal, as fired]				
		Controlled					Controlled			
Particle Size,[a] μm	Uncontrolled	Multiple Cyclone[b]	Multiple Cyclone[c]	ESP	Baghouse	Uncontrolled	Multiple Cyclone[c]	Multiple Cyclone[d]	ESP	Baghouse
15	28	86	74	97	72	8.4 (16.8)	7.3 (14.6)	4.4 (8.8)	0.23 (0.46)	0.043 (0.086)
10	20	73	65	90	60	6.0 (12.0)	6.2 (12.4)	3.9 (7.8)	0.22 (0.44)	0.036 (0.072)
6	14	51	52	82	46	4.2 (8.4)	4.3 (8.6)	3.1 (6.2)	0.20 (0.40)	0.028 (0.056)
2.5	7	8	27	61	26	2.1 (4.2)	0.7 (1.4)	1.6 (3.2)	0.15 (0.30)	0.016 (0.032)
1.25	5	2	16	46	18	1.5 (3.0)	0.2 (0.4)	1.0 (2.0)	0.11 (0.22)	0.011 (0.022)
1.00	5	2	14	41	15	1.5 (3.0)	0.2 (0.4)	0.8 (1.6)	0.10 (0.20)	0.009 (0.018)
0.625	4	1	9	[d]	7	1.2 (2.4)	0.1 (0.2)	0.5 (1.0)	[d]	0.004 (0.008)
Total	100	100	100	100	100	30.0 (60.0)	8.5 (17.0)	6.0 (12.0)	0.24[e] (0.48)	0.06[e] (0.12)

[a]Expressed as aerodynamic equivalent diameter.
[b]With fly-ash reinjection.
[c]Without fly-ash reinjection.
[d]Insufficient data.
[e]Estimated control efficiency for ESP. 99.2%; baghouses, 99.8%.

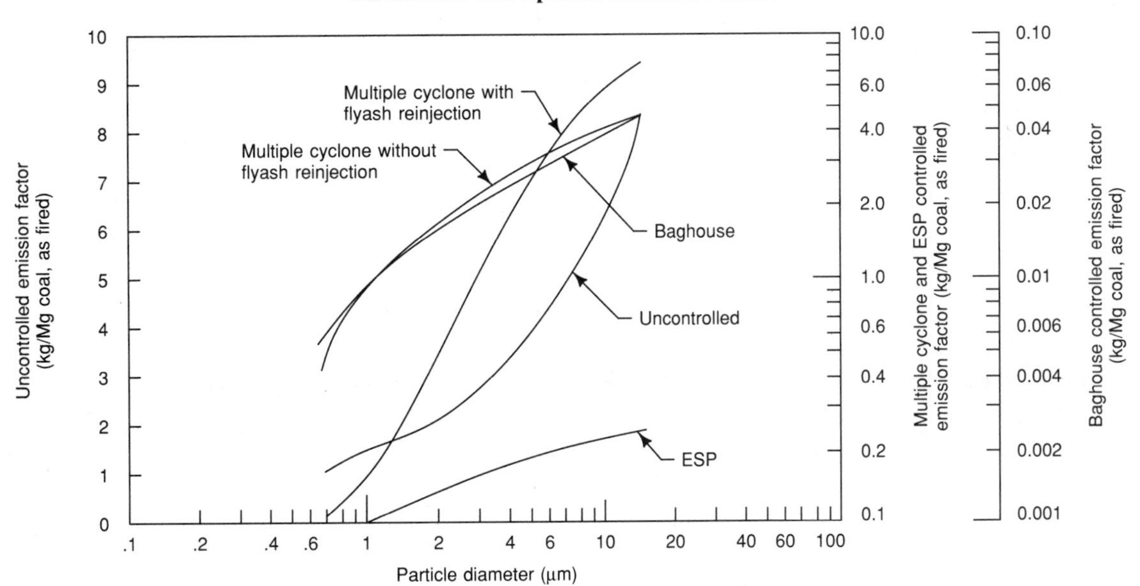

FIGURE 16. Emission Factors for Spreader Stokers Burning Bituminous Coal (From Reference 6)

Cumulative Particle Size Distribution and Size-Specific Emission Factors ***(Emission Factor Rating: C [uncontrolled], E [multiple cyclone controlled])***

	Cumulative Mass, % ≤ Stated Size		Cumulative Emission Factor [kg/Mg (lb/ton) coal, as fired]	
Particle Size,[a] μm	Uncontrolled	Multiple Cyclone Controlled	Uncontrolled	Multiple Cyclone Controlled[c]
15	49	60	3.9 (7.8)	2.7 (5.4)
10	37	55	3.0 (6.0)	2.5 (5.0)
6	24	49	1.9 (3.8)	2.2 (4.4)
2.5	14	43	1.1 (2.2)	1.9 (3.8)
1.25	13	39	1.0 (2.0)	1.8 (3.6)
1.00	12	39	1.0 (2.0)	1.8 (3.6)
0.625	[b]	16	[b]	0.7 (1.4)
Total	100	100	8.0 (16.0)	4.5 (9.0)

[a]Expressed as aerodynamic equivalent diameter.
[b]Insufficient data.
[c]Estimated control efficiency for multiple cyclone, 80%.

Cumulative Particle Size Distribution and Size-Specific Emission Factors

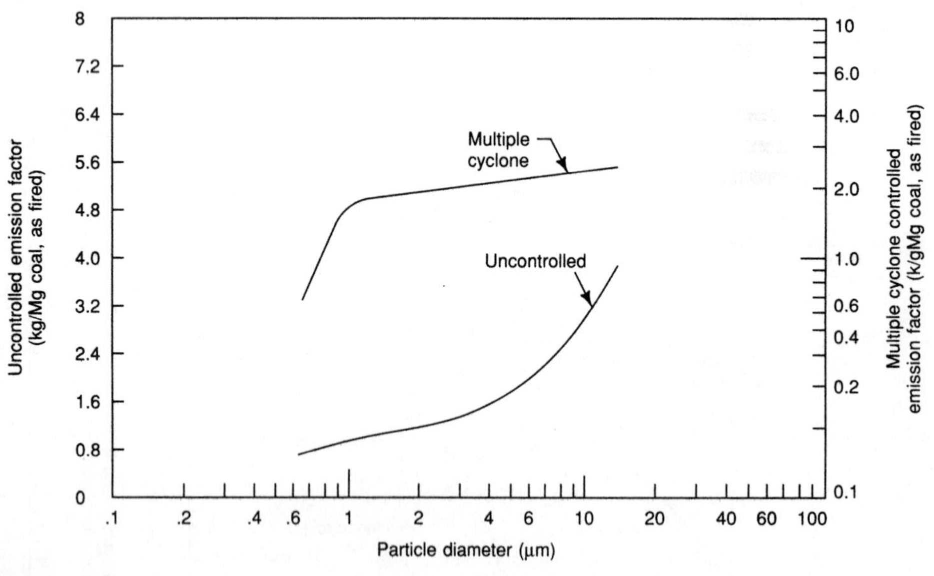

FIGURE 17. Emission Factors for Overfeed Stokers Burning Bituminous Coal (From Reference 6)

Cumulative Particle Size Distribution and Size-Specific Emission Factors *(Emission Factor Rating: C)*

Particle size,[a] μm	Cumulative Mass, % ≤ Stated Size	Uncontrolled Cumulative Emission Factor[b] [kg/Mg (lb/ton) coal, as fired]
15	50	3.8 (7.6)
10	41	3.1 (6.2)
6	32	2.4 (4.8)
2.5	25	1.9 (3.8)
1.25	22	1.7 (3.4)
1.00	21	1.6 (3.2)
0.625	18	1.4 (2.7)
Total	100	7.5 (15.0)

[a]Expressed as aerodynamic equivalent diameter.
[b]May also be used for uncontrolled hand fired units.

Cumulative Size-Specific Emission Factors

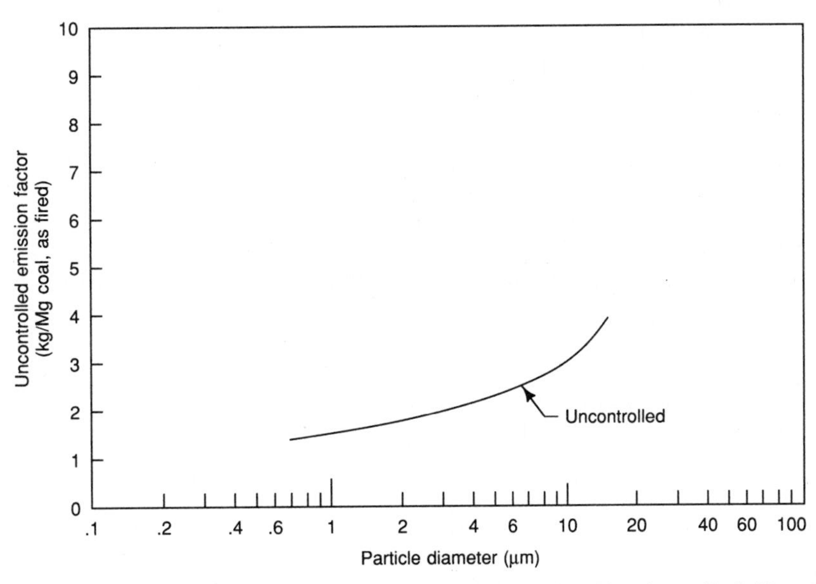

Figure 18. Emission Factors for Underfeed Stokers Burning Bituminous Coal (From Reference 6)

TABLE 7. Power Generation Sources

	Coal	Nuclear	Oil	Hydro	Gas	Other	Total Capacity, GW
1989	44%	14%	12%	12%	17%	1%	731
1999	44%	14%	11%	12%	18%	1%	773

TABLE 8. Summary of FGD Systems by Process (Percentage of Total MW)

Process	By-product	December 1989	December 2010
Throwaway product			
Wet scrubbing			
Dual alkali		3.4	2.3
Lime		16.3	13.5
Lime/alkaline fly ash		7.0	4.9
Limestone		48.2	43.9
Limestone/alkaline fly ash		2.4	1.6
Sodium carbonate		4.0	3.3
Spray drying			
Lime		8.8	7.9
Sodium carbonate		0	0.5
Reagent type not selected		0.7	2.1
Dry injection			
Lime		0.2	0.1
Sodium carbonate		0	1.6
Reagent type not selected		0	0.2
Process not selected		0	2.2
Saleable product			
Wet scrubbing			
Lime	Metals/fly ash/other	<0.1	<0.1
Limestone	Gypsum	4.1	4.6
Magnesium oxide	Sulfuric acid	1.4	1.0
Wellman Lord	Sulfuric acid	3.1	2.1
Spray drying			
Lime	Dry scrubber waste	0	0.3
Process undecided		0	7.8
Total		100.0	100.0

Source: Reference 17.

efficiency for new and retrofit systems was 82% and 76% respectively.[17]

The principal types of absorbers used in the wet scrubbing systems include venturi scrubber absorbers, static packing scrubbers, moving-bed absorbers, tray tower absorbers, and spray tower absorbers.[19] In the case of the venturi scrubber absorber, the venturi (for particulate removal) is also an integral part of the absorber. The particulate collection (using either an ESP or baghouse) precedes the wet scrubber to minimize operational problems. Table 9 provides a summary of typical design parameters for lime scrubber/absorbers. A flowchart for a typical lime FGD system (wet scrubber) is shown in Figure 19.[19] The limestone scrubber is similar except that it does not involve the lime slaking step, but instead includes limestone handling and crushing. See Figure 20.[20]

In applications of spray dryer technology for FGD, the particulate collector is downstream and is considered an integral part of the FGD system. The major equipment items found in a typical spray dryer scrubbing system are the spray dryer absorber, the particulate-collection system, reagent and slurry preparation and handling equipment, solids transfer and recycle equipment, and process control and instrumentation.

Figure 21[21,22] shows a typical utility application that recycles both spray dryer and baghouse (or ESP) products.

TABLE 9. Summary of Scrubber/Absorber Designs

Plant Name	No. of Modules per Unit	% Capacity per Module	Type of Module	L/G, gpm/1000 acfm		Gas Flow per Module, acfm	ΔP in. H_2O	Gas Velocity fps	Materials of Construction		
				Presaturator/ Scrubber	Absorber				Presaturator/ Scrubber	Absorber	Internals
Pleasants 1 and 2	4	25	Venturi/tray tower	14–17	35	600,000	6	10.4	CS, inorganic lined	Cs, rubber lined	316L SS
Four Corners 1, 2, and 3	2	50	Venturi	25	—	407,000 (1&2) 515,000 (3)	12	NR	CS, 316L SS with polyester lining	—	316L SS
R. D. Green 1 and 2	2	50	Spray tower	NR[a]	45	500,000	3	9.2	CS, inorganic lined	CS, inorganic lined	316L SS
Conesville 5 and 6	2	60	Moving bed	0.55	57	500,000	6	10	Carpenter 20	CS, rubber lined	CS, rubber lined
Coal Creek 1 and 2	4	33	Spray tower	—	60	685,000	4.5	10.6	—	316L SS	316L SS
Elrama Station	5 (total)	25	Venturi	—	40	550,000	11	NR	—	CS, polyester lined	316L SS
Phillips Station	5 (total)	20	Venturi	—	40	547,000	16	40	—	CS, polyester lined	316L SS
Hawthorn 3 and 4	2	60	Marble bed	26	—	250,000	12	10	316L SS and CS polyester lined	—	316L SS
Green River Station	1 (total)	100	Venturi/moving bed	34	34	288,000	7/4	14	CS, inorganic lined	CS, inorganic lined	316L SS
Cane Run 4	2	50	Venturi/moving bed	NR	60	368,000	4	11.5	CS, inorganic lined	CS, inorganic lined	316L SS
Mill Creek 3	4	30	Venturi/moving bed	5	60	400,000	11	10	CS, 316L SS	CS, inorganic lined	316L SS
Paddy's Run 6	2	50	Marble bed	—	16.5	175,000	11.5	9	—	CS, polyester lined	316L SS
Clay Boswell 4	4	33	Venturi/spray tower	20	50	640,000	20	12	CS, rubber lined	CS, rubber lined	CS, rubber lined and 316L SS
Milton R. Young 2	2	56	Spray tower	—	80	859,000	8	9	—	CS, polyester lined	—
Colstrip 1 and 2	3	40	Venturi/spray tower	15	18	426,000	17/1	200/8.7	CS, polyester and acid brick lined	CS, polyester lined	316L SS
Bruce Mansfield 1 and 2	6	17	Venturi/venturi	23	20	558,000	18/6	200/100	CS, polyester lined	CS, polyester lined	316L SS
Bruce Mansfield 3	5	25	Horizontal spray chamber	0.7	88	992,000	2.8	22	Inconel 625	CS, polyester lined	316L SS
Hunter 1 and 2	4	25	Spray tower	—	70	330,000	5–6	9.6	—	CS, polyester lined	FRP
Huntington 1	4	25	Spray tower	—	70	330,000	5–6	9.6	—	CS, polyester lined	FRP

[a]NR—Not reported.

Note: gpm/1000 acfm = 0.1337 m^3/km^3, acfm = 0.00047 m^3/s, in. H_2O = 0.2488 kPa, ft/s = 0.3048 m/s.

Source: Reference 19.

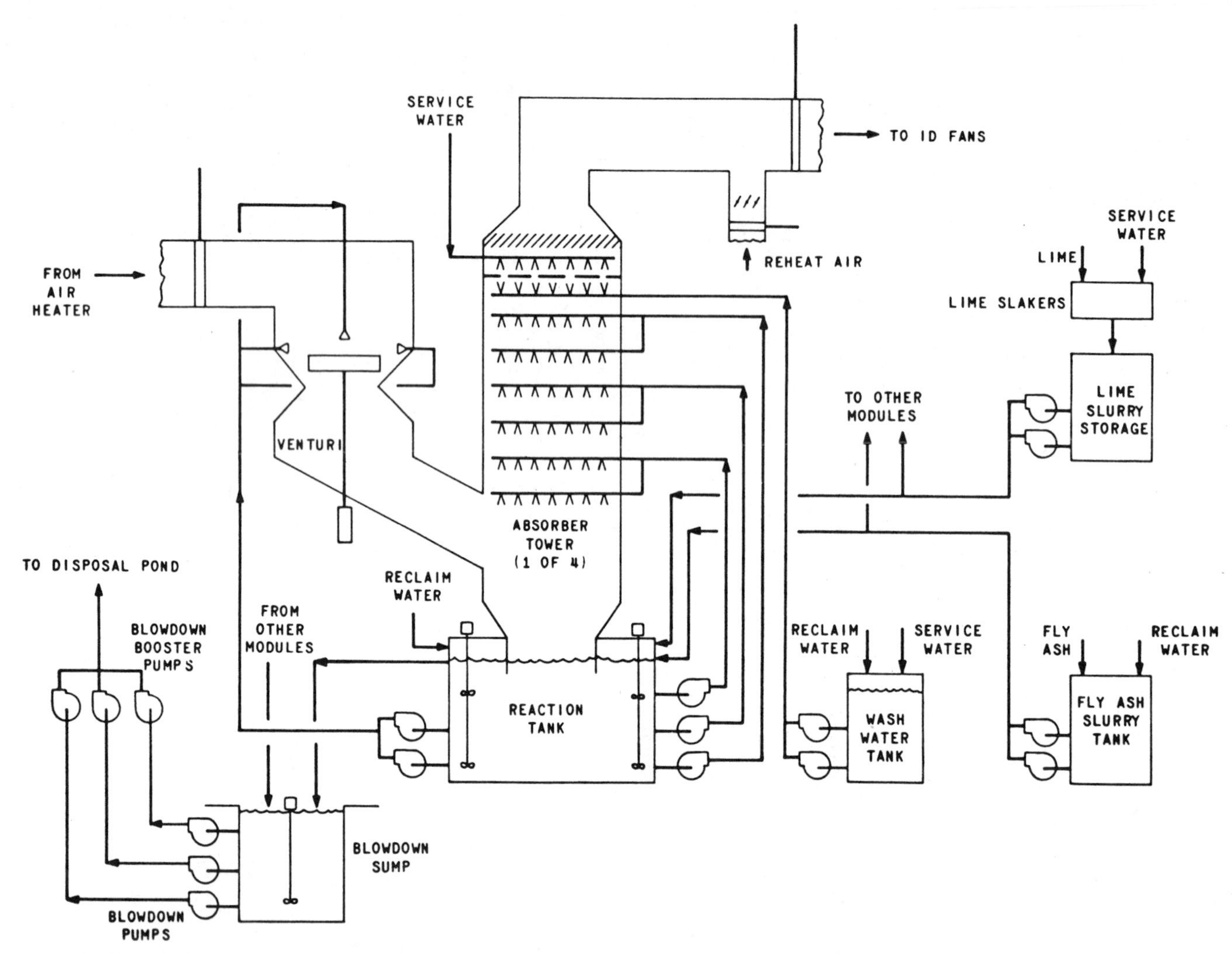

FIGURE 19. Clay Boswell Station Unit 4 Process Flow Diagram (From Reference 19, p 4-1-17)

Applications may or may not use partial recycle of the reacted products. The spray dryer absorber provides the initial contact between the atomized reactive alkali and the acid-gas contaminants. There are two types of atomizers: rotary disks or wheels and dual-fluid pneumatic nozzles. In each case, the slurry is atomized as droplets into the dryer, reacts with the SO_2, and dries to a fine powder, which is then carried over into the ESP or fabric filter and removed. The particulate collector also serves as an additional contact point between the dried reactants and SO_2, providing additional removal.

The chosen atomization method affects the design of the spray dryer absorber vessel, including the physical dimensions. For a rotary-atomizer type of spray dryer, which projects the droplets radially outward and perpendicular to the gas flow, the length-to-diameter ratio of the dryer *(L/D)* is typically 0.8 to 1. Figure 22A illustrates two typical configurations of rotary atomizer spray dryers.[23,24] The droplets (which range in size from 25 to 150 μm) decelerate rapidly owing to the drag forces of the downward-moving flue gas and eventually attain the speed and direction of the flue gas. To avoid wall deposition, the designed radial distance between the atomizer and the dryer wall must be sufficient to allow for adequate drying of the largest droplets. This is accomplished by proper choice of the *L/D*, droplet size, and residence time.

For a two-fluid pneumatic nozzle type of spray dryer, which atomizes the droplets in the direction of the gas flow (downward), the *L/D* is typically 2:1. In this case, sidewall deposition is a minor problem. Two examples of these spray dryers are shown in Figure 22B. Experience has shown that with the above *L/D* ratios, spray dryers on flue-gas systems can be operated with design residence times of 10–12 seconds, based on the actual outlet flow rate conditions (i.e., residence time [minutes] = dryer volume/acfm). Although the designed residence time for most spray dryers is 10–12 seconds, few systems operate at 100% of the design flow rate. The actual operating residence time of most systems is 12–15 seconds.

Table 10 summarizes selected spray dryer applications

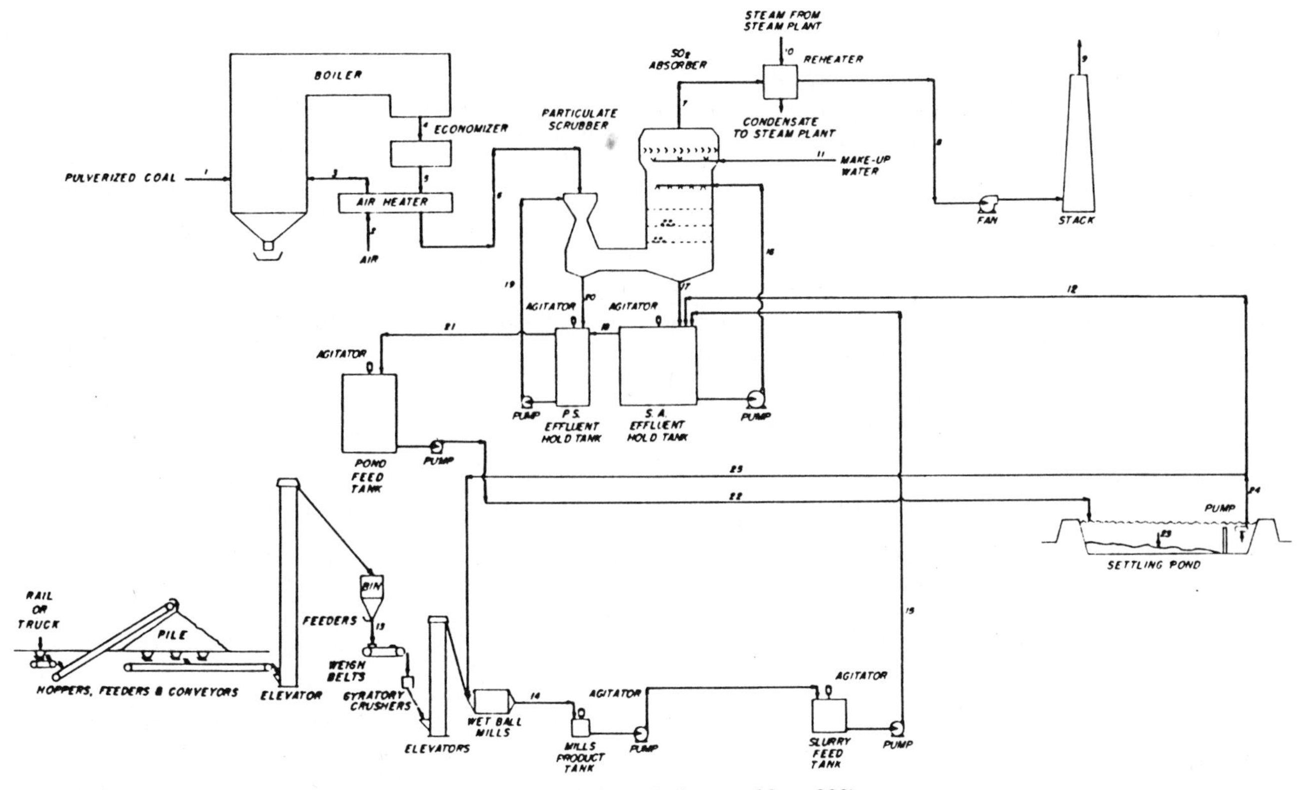

FIGURE 20. Limestone Slurry Process. Flow Diagram (From Reference 20, p 322)

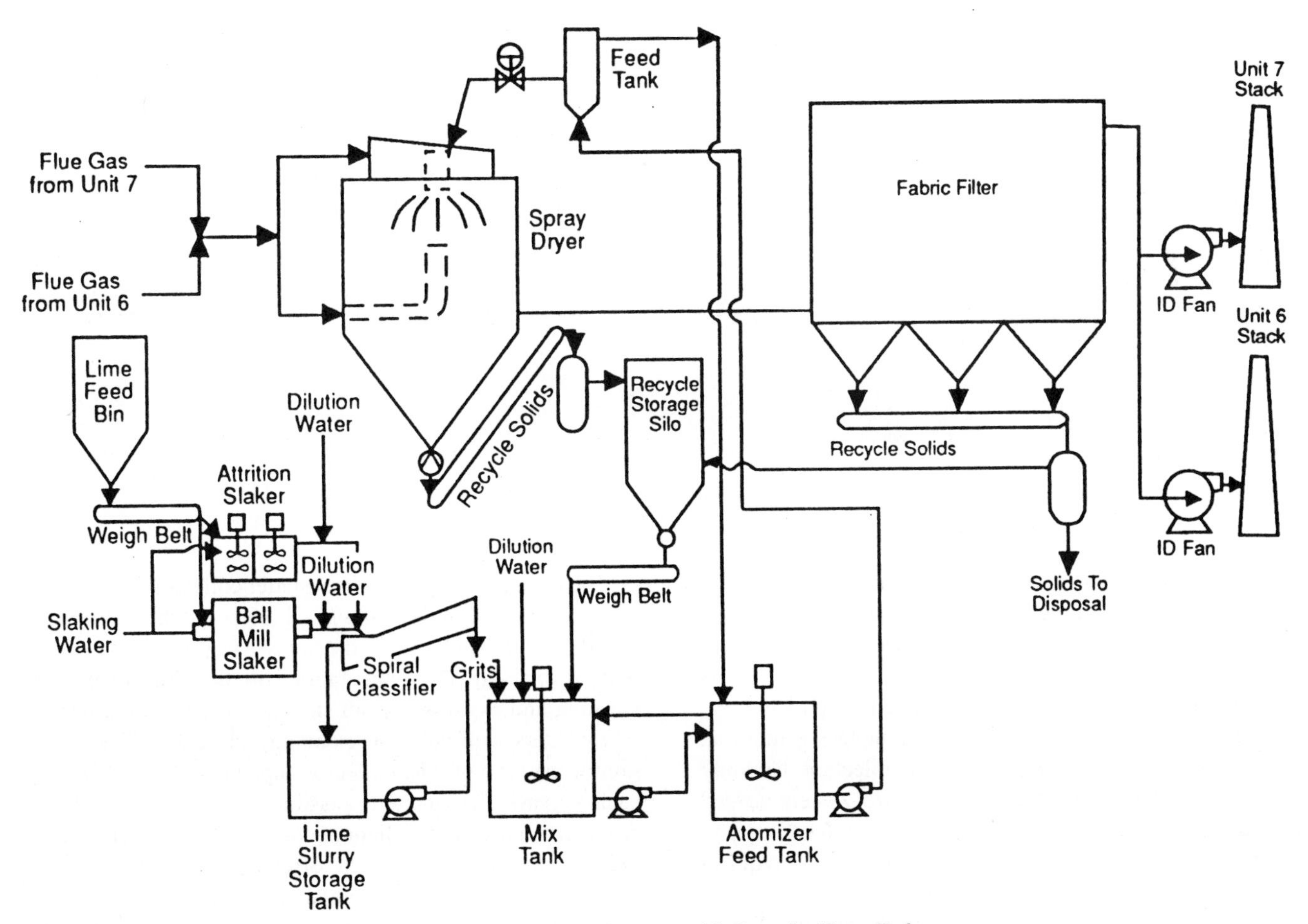

FIGURE 21. Diagram of a Utility Spray Dryer Scrubbing System, with Recycle (From References 21 and 22)

TABLE 10. Summary of Spray Dryer Applications

Type	Size	Unit	Type[a]	Atomizers per Dryer/ Number of Dryers	RT,[b] seconds	T,[c] °F	ΔT,[d] °F	Spray Dryer Diameter (feet)
Utilities (size in MW)								
	440 (each)	Antelope Valley 1,2 Basin Electric	R	1/5	12	310	20	46
	110	Riverside 6,7 Northern States Power	R	1/1	10	350	Var.[e]	46
	60	Stanton 10 United Power	R	3/1	10	323	20	38
	450	Craig 3 Colorado-Ute	N	12/4	NR[f]	NR	25	NR
	280	Rawhide 1 Platte River Power	R	1/3	11	276	NR	46
	320	Holcomb 1 Sunflower Coop.	R	1/3	10.6	249	50	51
	44	Shiras 3 City of Marquette	R	1/1	10	265	25	36
	270	North Valmy Sierra Pacific Power Idaho Power	R	3/3	NR	260–300	30	NR
	570	Laramie River 3 Basin Electric	N	12/4	8	286	23	55
	370	Springerville 1,2 Tucson Electric	R	1/3	12	256	20	46
	575	GRDA	R	3/4	12	310	20	52

and the operating characteristics of each.[21] Typically, industrial boiler spray dryers have diameters of 25–30 feet, whereas utility spray dryers have diameters of 40–50 feet. In most cases, multiple spray dryers are required to handle the large volume of flue gas found in utility boilers.

The spray dryer FGD system is incomplete without an appropriate means of particulate-matter collection. Not only is a well-designed particulate-matter control system needed to meet particulate-matter and opacity emissions requirements, but it can help to meet acid-gas–removal requirements. Acid gases are removed when the flue gas comes into contact with lime-containing particles and encounters the collected particulate matter in the fabric filter or ESP.

The four types of particulate-matter collectors most commonly employed in the removal of the spray-dried products and fly ash downstream of the spray dryer are fabric filters with reverse-air cleaning, fabric filters with reverse-air cleaning and assistance from shaking or from sonic horns, fabric filters with pulse-jet cleaning, and ESPs. The selection of the type of dust collector depends on many factors, such as particulate-matter emission limits, overall acid-gas removal requirements, client and/or vendor preference, and the system flow rate. Each offers advantages and disadvantages that must be evaluated for each site.

The fabric filter has been used in the majority of new installations, whereas ESPs have been used in a few large

TABLE 10. (*Continued*)

Type	Size	Unit	Type[a]	Atomizers per Dryer/ Number of Dryers	RT,[b] seconds	T,[c] °F	ΔT,[d] °F	Spray Dryer Diameter (feet)
Industrial (size in acfm)								
	75,000	Argonne National Lab, Argonne, IL	R	1/1	12	330–340	≈20	25
	90,500	Container Corp. Philadelphia, PA	R	1/1	NR	350	NR	NR
	46,500 (3 units)	Fairchild Air Force Base, Spokane, WA	R	1/1	NR	375	≈25	20
	167,000	General Motors, Buick Div., Flint, MI	R	1/1	NR	300	NR	32
	48,600	Griffis Air Force Base, Rome, NY	R	1/1	NR	400	NR	22
	44,400	Malstrom Air Force Base, Great Falls, MI	R	1/1	NR	325	≈35	20
	40,000	Strathmore Paper Woronoco, MA	N	4/1	NR	NR	NR	NR
	62,000	University of	N	1/1				
	96,000	Minnesota	N	1/1	12	375	20	24
	97,000	Rockwell Int'l Columbus, OH	R	1/1	12	450	30	30
	205,000	M. M. Carbon Long Beach, CA	R	1/1	10	405	90–120	36
	81,710	Ohio State Univ. Columbus, OH	N	NR/1	NR	400	NR	24

[a]R = rotary; N = nozzle.
[b]Residence time.
[c]Flue-gas temperature at entrance.
[d]Approach to saturation at exit.
[e]Varies.
[f]Not reported.
Source: Reference 21.

utility plants and in one retrofit plant. The high percentage of fabric filters found in new installations of dry scrubbers is due to the inherently greater ability of the fabric filter than that of the ESP to remove acid gases. The fabric filter has been observed to remove 15–30% of the SO_2, depending on the operating conditions of the spray dryer and the filter.

Although the characteristics of the fly ash, the presence of spray dryer products, and the flue-gas temperature exiting the spray dryer are considerably different from conditions in a system without a spray dryer, many systems have bypass provisions or may need to be operated without cooling in the spray dryer because of the spray dryer downtime. Therefore, systems are generally designed with dust collector specifications that meet both operating conditions. Otherwise, the entire system (SO_2 control and particulate-matter control) would need to be bypassed if a spray dryer component were to fail. This may or may not be an acceptable practice, depending on the local air pollution regulations and the time involved. In addition, it is not uncommon to start up the system (bypassing the spray dryer) until the system has been operating for some time.

The primary system operating conditions affected by the application of the spray dryer (which, in turn, affect the performance of the particulate-matter collector) are dust loading, particle size distribution, temperature, and gas-flow rate. For boilers burning coal without SO_2 control, the

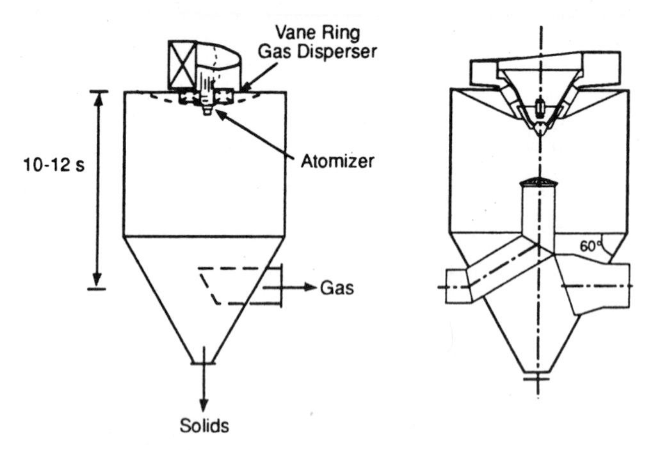

A. Rotary-Atomizer Dryers

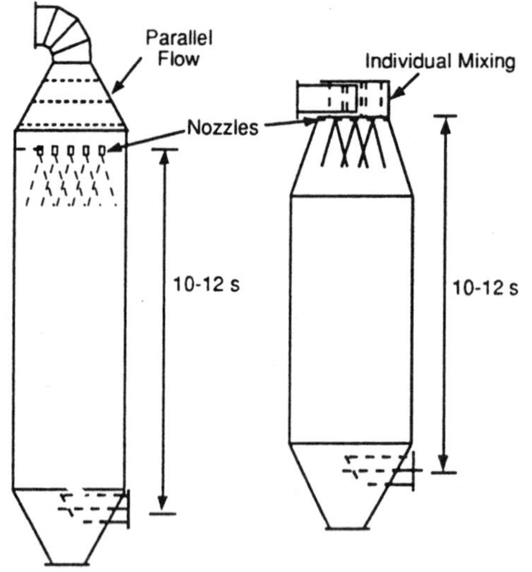

B. Two-Fluid Pneumatic Nozzle Dryers

FIGURE 22. Spray Dryer Arrangements (From References 23 and 24)

fly-ash loading may vary from as little as 0.5 gr/ft^3 to in excess of 4 gr/ft^3, depending on (1) the presence of cyclonic precleaners (commonly found on industrial-size boilers), (2) the type of boiler, and (3) the ash content of the coal. The particle size distribution varies similarly, with mass-mean diameters ranging from 6 to 15 μm. Flue-gas temperatures are typically 275–350°F at the particulate collector.

In contrast to the foregoing conditions, the application of a spray dryer results in (1) flue-gas temperatures of 150–170°F, (2) approximately 20% less gas flow through the particulate-matter collector due to lower gas temperatures, (3) a particle size distribution with a mean diameter of 20–30 μm, and (4) a significantly increased dust loading. The increased particle size distribution and decreased flow rate are beneficial, but the decreased temperature and increased dust loading tend to increase pressure drops across fabric filters and to decrease efficiency across ESPs. The increased dust loading caused by the reacted products ranges from as little as 1.5 gr/acf at 500 ppm of SO_2 (with no recycling) to as high as 14 gr/acf at 3000 ppm (with a 50% recycle) of reacted products. This increased loading, added to the fly-ash loading, results in overall dust loadings of 2–20 gr/acf, depending on the overall conditions. Therefore, increased dust loading must be taken into account in the design of the particulate-matter collector and the associated dust-handling support systems.

As a result of the higher fraction of $Ca(OH)_2$, calcium sulfites, calcium sulfates, increased moisture (5–15%) in the product, and closer approach to saturation, it is important to minimize the heat loss in the particulate collector to prevent cold-end corrosion. Operating experience and related problems have been summarized by a number of vendors.[21,23] Most of the operational problems have been a result of the lower operating temperatures near the water dew point and associated moisture–acid condensation. Five methods that have been employed or recommended to minimize these problems involve the use of:

1. Insulation (usually a minimum of 4 inches).
2. Control of air in-leakage.
3. Hopper heating.
4. Improved operating procedures.
5. Protective coatings on inner surfaces.
6. A crusher (delumper) on the spray dryer hopper.

To date, most spray dryer applications have relied on a fabric filter collector for final particulate-matter control. Table 11 summarizes the baghouse applications at several installations where fabric filter design and operating information was available.[21]

Generally, reverse-air collectors have been used for large utility boilers, while both reverse-air and pulse-cleaned collectors have been used on small industrial boilers. As shown in Table 11, the air-to-cloth ratios have ranged from 1.6 to 2.5 acfm/ft^2 for the reverse-air collectors and from 3 to 5 acfm/ft^2 for the pulse types of collectors.

The filter material used in reverse-air collectors is typically a woven fiberglass with a Teflon B® coating with a filter weight of 9.5–10 oz/yd^2. Although other materials may be better suited for the low-temperature operation downstream of the FGD spray dryer, the bags must also be able to operate at temperatures in excess of 350°F and up to 500°F (i.e., during start-up without the spray dryer). Fiberglass bags with the Teflon® coating have also been used in pulse collectors, although the filter material generally weighs 16–22 oz/yd^2. The heavier weight is required because of the mechanical wear associated with the higher-energy pulsed cleaning, the greater abrasion resulting from the increased air-to-cloth ratios, and the greater potential for abrasion associated with the internal bag cages.

The major advantage of the pulse type of collectors is that they operate at relatively higher filter velocities than the reverse-gas collector (typically 4 fpm, compared with 2 fpm for the reverse-air collectors), thus resulting in a substantially smaller collector. Ironically, the higher velocity tends to reduce bag life, increase outlet emissions, and increase pressure drop. Although these problems can be

TABLE 11. Summary of Baghouse Applications

Type	Size	Unit	A/C, acfm/ft^2	ΔP,[a] in. H_2O	Cloth[b] Material	Bag Life, years	Type of Cleaning	Cleaning Cycle, hours	No. of Compartments
Utilities (size in MW)									
	440	Antelope Valley Basin Electric	2.2	4	FG/TC (10)	2	Reverse gas	2	14 (2 units)
	110	Riverside 6,7 Northern States	1.70	4	FG	4	Reverse gas	1	12
	60	Stanton 10 United Power	1.61	6	FG/SGT (14)	2	Reverse gas	1	NR[c]
	450	Craig 3 Colorado-Ute	2.3	6	NR[c]	2	NR	NR	
	280	Rawhide 1 Platte River Pow.	1.78	3.5	FG/TC (10)	2	Reverse gas	1	12 (2 units)
	320	Holcomb 1 Sunflower Coop.	1.55	3.5	FG/TC (10)	2	Reverse gas	2 (or 4 in.)	14 (2 units)
	44	Shiras 3 City of Marquette	2.0	6.8	FG/AR	2	Reverse gas	8	NR
	270	North Valmy Sierra Pacific Power, Idaho Power	1.8	5.8	FG/TC	NR	Reverse gas	0.5	10
Industrial (size in acfm)									
	75,000	Argonne National Laboratory, Argonne, IL	3.0	3.5	FG/TC (16)	2	Pulse	1.5	4
	90,500	Container Corp. Philadelphia, PA	4.5	5.5	FG/TC	NR	Pulse	NR	8
	46,500	Fairchild Air Force Base, Spokane, WA	2.0	NR	NR	NR	Reverse gas	NR	6
	167,000	General Motors, Buick Div., Flint, MI.	2.0	NR	NR	NR	Reverse gas	NR	8
	48,600	Griffis Air Force Base, Rome, NY	4.0	9.0	FG/TC	NR	Pulse	NR	4
	44,400	Malstrom Air Force Base, Great Falls, MI	2.0	NR	NR	NR	Reverse gas	NR	5
	40,000	Strathmore Paper Woronoco, MA	NR	NR	Acrylic	NR	Pulse	NR	NR
	62,000	University of Minnesota	2.5	2.9	FG/TC	>6	Reverse gas	5	8
	96,000		1.7	3.0	FG/TC	>6	Reverse gas	15	8
	97,000	Rockwell Int'l Columbus, OH	2.1	3.5	FG/TC	2	Reverse gas	30	8
	205,000	M. M. Carbon Long Beach, CA	1.9	4	FG/TC	NR	Reverse gas	24	12
	81,710	Ohio State Univ. Columbus, OH	5.3	8	FG/TC	NR	Pulse	NR	6

[a]Baghouse pressure drop (ΔP).
[b]FG/TC (10) = fiberglass with Teflon® coat (10 oz/yd^2); FG/AR = fiberglass with acid-resistant finish; FG/SGT = fiberglass with silicon-graphite-Teflon® finish.
[c]NR = not reported.
Source: Reference 21.

minimized by the choice of filter material and design changes, there is a cost for these factors that must be considered in evaluating design trade-offs.

Electrostatic precipitators have been used in only a few large utility applications and, in one case, where units were already in existence. The primary disadvantage of the ESP is the decreased sorbent utilization that occurs when the fly ash and reaction products, collected on the plates of the precipitator, are removed from the gas stream. Several units have been installed, including Grand River Dam Authority (575 MW), Pacific Power and Light—Wyodak 1 (300 MW), and Basin Electric's Laramie River Plant (570 MW). The specific collector area (SCA) of the existing ESPs, which have spray dryers upstream, are 685 and 700 $ft^2/10^3$ acfm for Grand River Dam Authority and Wyodak, respectively, based on the actual cubic feet per minute at the outlet of the spray dryers.

Nitrogen Oxides Control

The methods for controlling nitrogen oxides can be divided into (1) combustion control methods, in which operating conditions for combustion are modified to reduce the formation of NO_x; and (2) postcombustion control methods, in which NO_x is removed from the gas stream after formation.

The underlying concepts of the combustion control methods are as follows:

1. To reduce peak temperatures in the combustion zone by operating the primary flame zone under fuel-rich conditions, cooling the flame at a high rate, and decreasing the adiabatic flame temperature by dilution.
2. To reduce the gas residence time in the high-temperature zone.
3. To operate on an off-stoichiometric ratio by using a rich fuel–air ratio in the primary flame zone and lower overall excess air conditions.

The actual method of combustion control used depends on the type of boiler, especially the method of firing the fuel. For example, in the case of pulverized-coal–fired boilers (which accounted for 70% of all utility NO_x emissions in the United States in 1989), low-NO_x burners (LNB) and overfire air (OFA) have been successfully applied to tangentially and wall-fired units, whereas reburning is the only currently available option for cyclone boilers.[25]

The uncontrolled NO_x emissions from tangentially fired and wall-fired units at full load are estimated to range from 0.4 to 1.1 lb/MBtu and 0.5 to 1.4 lb/MBtu respectively.[26] In the United States, Europe, and Japan, most utilities incorporate LNB systems with OFA for NO_x control through combustion modifications. The LNB combustion scheme is shown in Figure 23. After the initial combustion zone, a pyrolitic zone is formed in which the liberation of volatile nitrogenous species takes place. In the next stage, a fuel-rich combustion zone is formed to limit the formation of NO_x. This region is followed by a burnout zone in which completion of combustion is allowed to take place.[27]

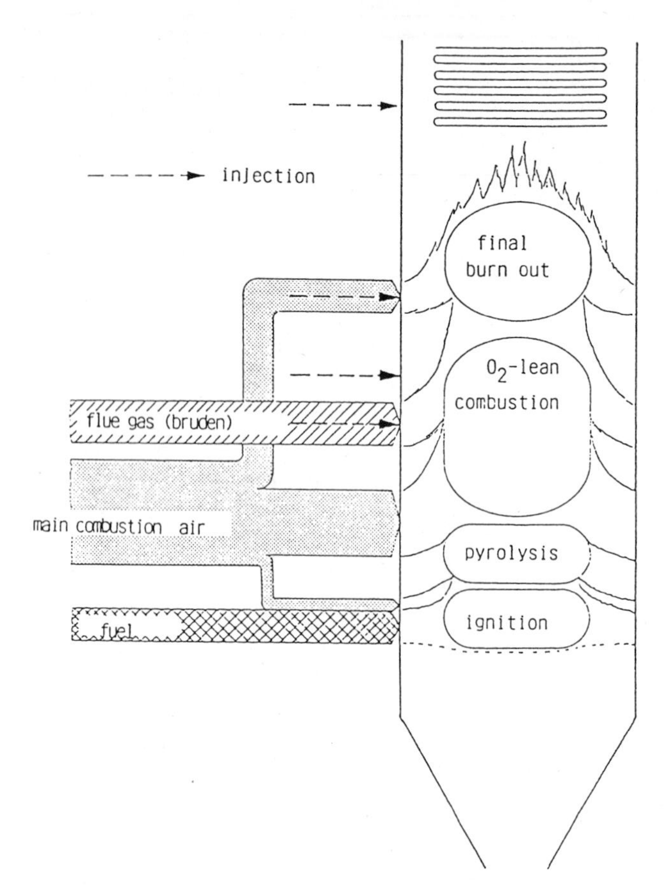

FIGURE 23. Low NO_x-Combustion (Scheme) (From Reference 27)

The actual design of the LNBs differs substantially between the two types of boilers. Figure 24 shows the schematic of an LNB for a wall-fired unit.[26] Figure 25 compares the configurations of a conventional burner and an LNB for a tangentially fired unit.[28]

One of the major findings during recent retrofit LNB

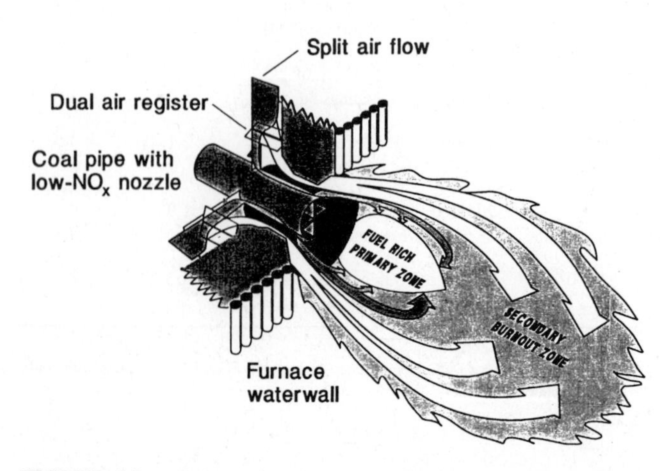

FIGURE 24. Schematic of a Low-NO_x Burner for Wall-Fired Boilers (From Reference 26, p 18)

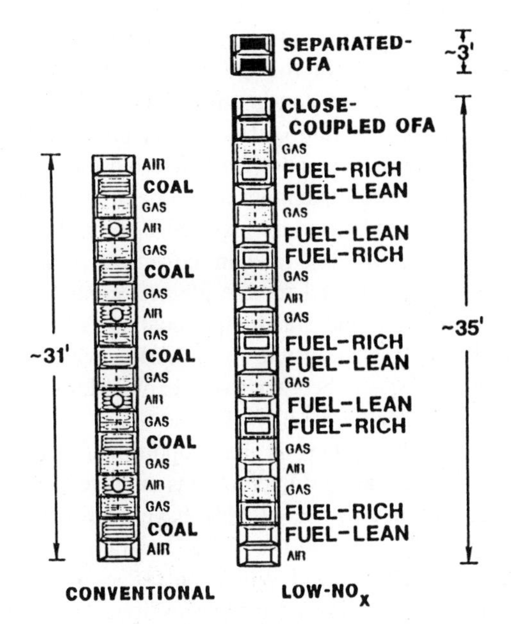

FIGURE 25. Comparison of Conventional and PM Burner Configurations (From Reference 28, p 2-77)

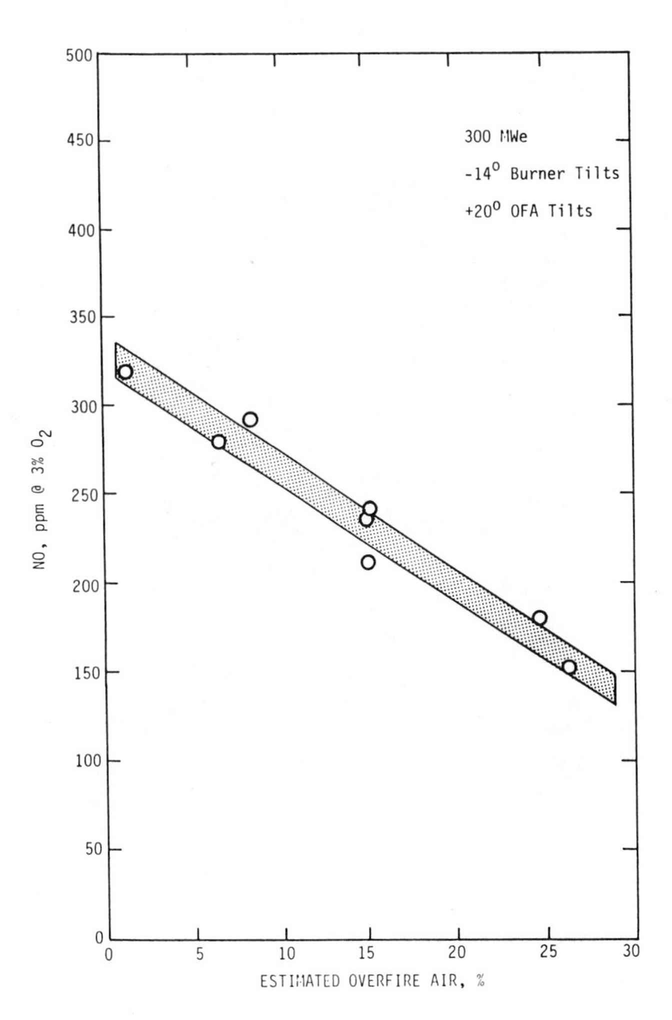

FIGURE 26. NO_x Dependence on Overfire Air Flow (From Reference 28, p 2-82)

demonstrations was the importance of OFA in NO_x reduction. In fact, OFA could be used as a stand-alone modification in some cases to meet the emission standards. Results have indicated that up to 30% NO_x reduction can be achieved by conventional OFA systems that stage 10–20% of the combustion air. Advanced OFA systems, staging 20–30% of the combustion air, have a potential of higher reductions in NO_x.[26] Figure 26 shows the dependence of NO_x emissions on the percentage of combustion air staged in the OFA nozzles and Figure 27 shows a furnace view of an OFA nozzle in a 400-MW wall-fired boiler.[28]

Reductions in NO_x of up to 50% have been demonstrated in LNB systems incorporating OFA. Capital costs for conventional and advanced OFA alone were estimated to be $5 per kilowatt and $10 per kilowatt in 1991. For LNB with OFA, the capital costs were estimated to range between $15 and $25 per kilowatt in the same period. Operating costs in all cases are expected to be low, but have not been estimated.[26]

As of 1989, there were 105 operating cyclone-fired utility boilers with a capacity of 26,000 MW (approximately 14% of the pre-NSPS coal-fired generating capacity). However, these units contributed to approximately 21% of the total NO_x emissions, largely because of their high operating temperatures which are conducive to NO_x formation. Uncontrolled NO_x emissions from cyclone-fired boilers, shown in Figure 5, normally range between 660 ppm and 1400 ppm.[29]

The typical configuration of the cyclone furnaces prevents the successful application of standard LNB technology. The cyclone furnace consists of a cyclone burner connected to a horizontally water-cooled cylindrical chamber, as shown in Figure 28. Fuel (crushed coal) and air are introduced into the chamber by the cyclone burner. The large coal particles are forced onto the molten slag layer formed on the chamber wall by the cyclonic effect of the injection. Combustion of these large particles takes place in this slag layer. The finer particles are held in suspension by the gas stream and are burned in the primary combustion zone within the cylindrical chamber. The flue gases and the remaining ash leave the cyclone and enter the main furnace, which is a separate part of the boiler. Since the combustion occurs outside the main furnace, modification of the furnace, such as with LNBs, is not suitable for these applications.[29] Other conventional NO_x control methods, such as staged combustion, have been shown to create cyclone corrosion problems.[26] Reburning remains the only available promising combustion control method for NO_x control in these types of boilers. At present, this technology is still in the pilot/field-demonstration stage.[29]

The concept of reburning technology is shown in Figure 29.[29] This technology involves injection of a second fuel in the furnace above the main cyclonic combustion zone. This section is maintained at reducing conditions by a controlled supply of combustion air. The reducing atmosphere con-

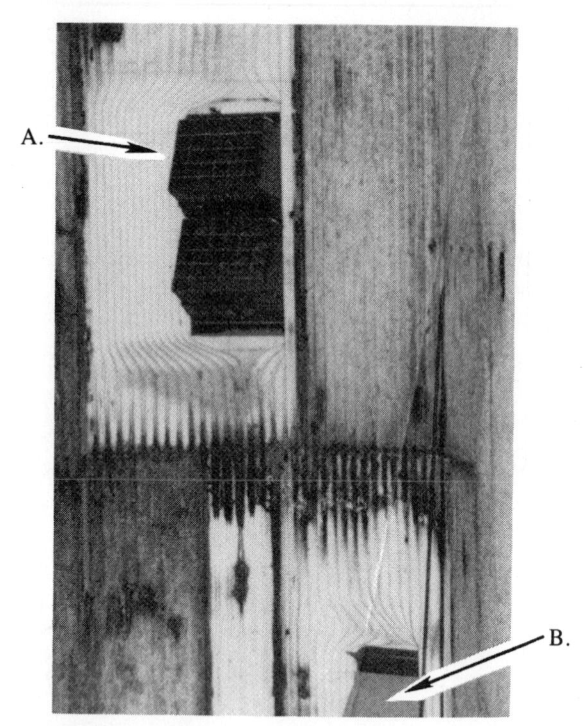

FIGURE 27. Furnace View of Separated OFA Nozzles (Offset) (From Reference 28); A. Separated Overfire Air Nozzles; B. Top of Close Coupled Overfire Air Nozzles

verts NO_x to molecular nitrogen, thus destroying a portion of the NO_x formed in the main combustion zone. This secondary combustion zone is maintained at a lower temperature by recirculation of flue gas to prevent additional NO_x formation in this zone. Complete burnout of the fuel takes place in a subsequent burnout zone in which additional combustion air is used through OFA ports. The fuel–air ratio and the heat release rates in the various zones need to be optimally maintained for the greatest control of NO_x.

From an economic standpoint, coal is favored as the secondary fuel for coal-fired utilities. However, the use of oil and natural gas is being actively explored, especially in light of the better burnability of these fuels. An additional advantage of these fuels is the proportional reduction of SO_x, particulate, and NO_x emissions achieved over coal.

Reductions in NO_x of 40–60% have been measured in pilot-scale and early short-term demonstration tests. Data obtained by Babcock and Wilcox in studies sponsored by EPRI and GRI using a pilot-scale cyclone boiler showed NO_x reductions of 50–75% from a baseline of about 1000 ppm.[29] More data were expected to be available from ongoing studies. From the preliminary studies in the pilot-scale unit mentioned above, it was found that the type of reburn fuel used, the percent of flue gas recirculated in the reburn-

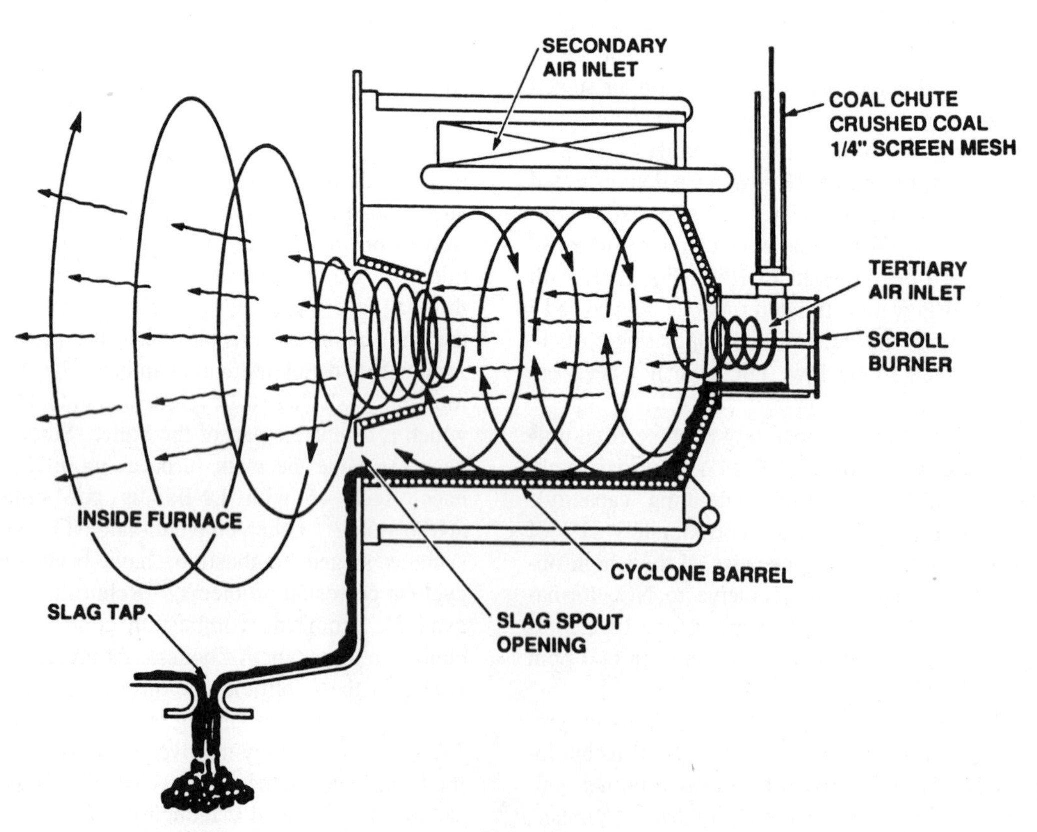

FIGURE 28. Cyclone Furnace (From Reference 29, p 3-16)

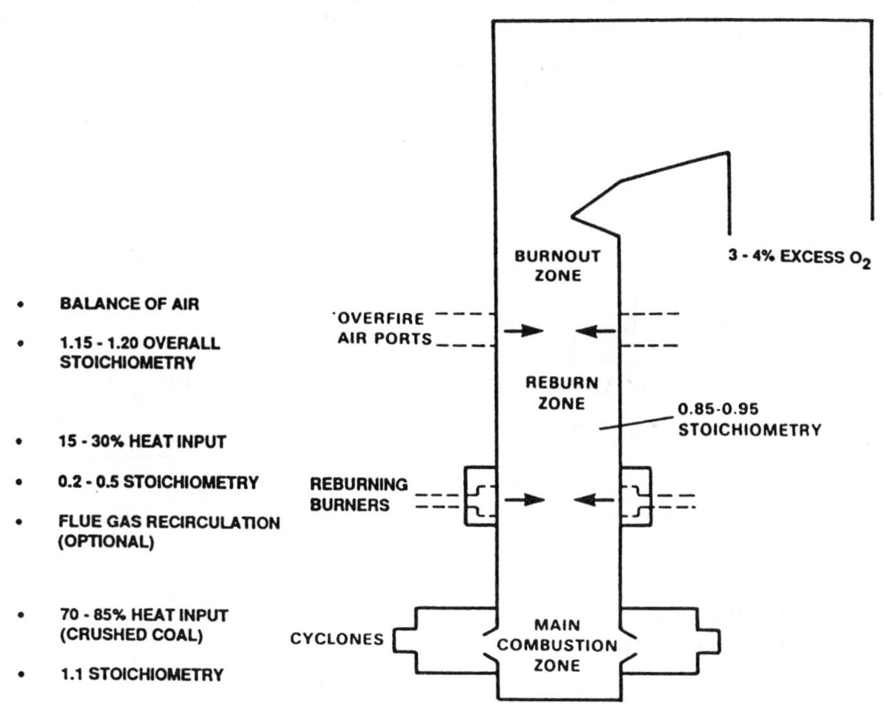

FIGURE 29. Reburning Technology (From Reference 29, p 3-17)

ing zone, and the fuel–air ratio maintained in the cyclone burner zone affected the level of NO_x control achievable with this control method. Depending on the size of the boiler and the type of secondary fuel used in the reburn chamber, the capital costs for installing reburning technology are estimated to be \$35 to \$45 per kilowatt. Operating costs are expected to be 2–6 mills per kilowatt-hour, depending largely on the type of reburn fuel used.[26]

Currently, the most developed and widely applied postcombustion NO_x control technology is selective catalytic reduction (SCR). Selective noncatalytic reduction (SNCR) using ammonia- or urea-based compounds is in the developmental stage and may be available in the future as another postcombustion NO_x control method. In the SCR process, ammonia, used as the reducing agent, is injected into the NO_x-laden flue gas, usually upstream of the air heater. This location provides the temperature window of 550–750°F, which is optimum for the reduction reactions. The NO_x is reduced to molecular nitrogen in a separate reactor vessel containing a catalyst, which is usually a mixture of titanium dioxide, vanadium pentoxide, and tungsten trioxide.[30] The schemes of the reducing reactions are as follows[11]:

$$4NO + 4NH_3 + O_2 \rightarrow 4N_2 + 6H_2O$$

$$2NO_2 + 4NH_3 + O_2 \rightarrow 3N_2 + 6H_2O$$

A schematic of the SCR process is shown in Figure 30.[11] The location of the catalytic converter in the overall equipment layout determines the space velocities and the catalyst geometry used. In "high-dust" SCR systems, in which the converter is located between the boiler and a cold-side ESP, the mesh size of the catalyst has to be kept higher to avoid dust buildup and catalyst fouling. This results in lower catalyst specific surface area. In the "low-dust" systems, in which the converter is located after a hot-side ESP, finer mesh catalysts with higher specific areas can be used. Figure 31 shows the two configurations.[30]

Pressure drop and catalyst space velocities are important considerations in the design of SCR systems. Pressure drops in the catalytic converters of between 5 and 7 mbar[31] have been reported, and depend on the type of catalyst geometry used, that is, plate type (lower pressure drop) or honeycomb type. For the desired NO_x-removal efficiencies, space velocities of 2000 per hour to 2500 per hour are reported to be necessary.[25] Since catalyst costs dominate the SCR system costs, maximum allowable space velocities are desired from an economic standpoint.

Catalyst deactivation and residual NH_3 in the flue gas are the two key operating considerations in an SCR system. Catalyst activity decreases with operating time due to fouling of the catalyst. Thus a low-dust SCR system in which a hot-side ESP precedes the SCR reactor is expected to have a longer life, though the capital costs and operating costs may also be higher.

Unreacted NH_3 passing through the reactor forms ammonium sulfate and causes clogging in downstream equipment such as air heaters. With increasing catalyst lifetime and consequent deactivation, the residual NH_3 in the flue gas or the "NH_3 slip" increases. The particulate ammonium sulfate

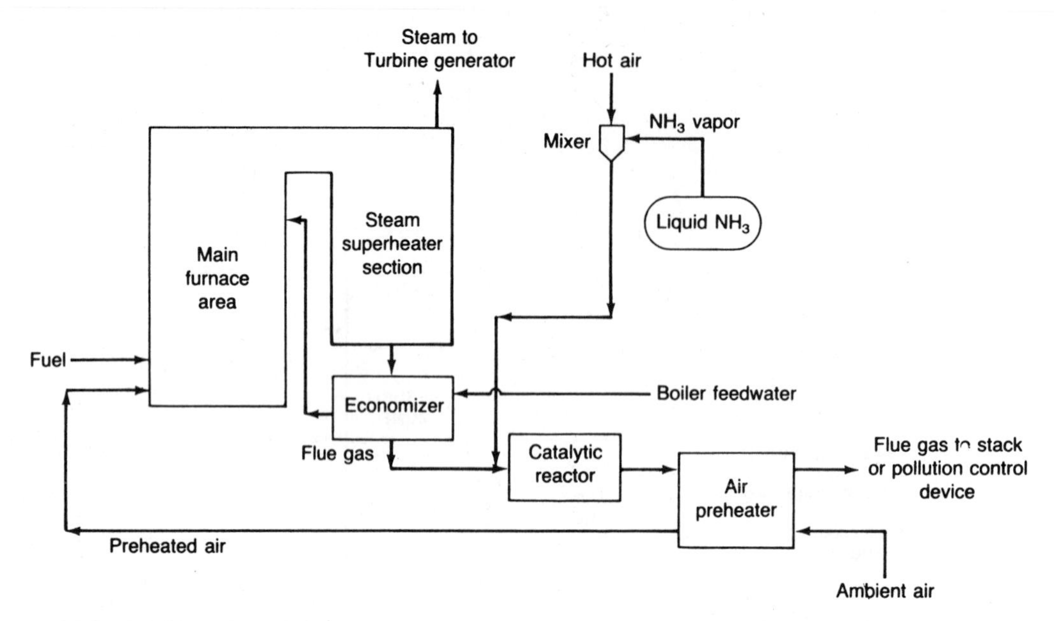

FIGURE 30. Schematic Flow Diagram for the Selective Catalytic Reduction Method of NO_x Control (From Reference 11)

is also known to deposit on the induced-draft fan and to create vibrational problems. From operating experiences in Japanese and European installations, the maximum allowable NH_3 slips seems to be 5 ppmv.

Reductions of 60–90% in NO_x have been reported in full-scale SCR systems internationally and in pilot-scale systems in the United States. Retrofit capital costs are expected to range from \$75 to \$150 per kilowatt, with levelized operating costs ranging between 5 and 9 mills per kilowatt-hour.[26]

Several issues remain to be resolved before SCR can be extensively used in the United States. One of the main issues is whether the data obtained from low-sulfur alkaline-coal–fired boilers in Japan and Europe can be directly extrapolated to the use in the United States of higher-sulfur coals. The other potential concern is that the NH_3 absorbed by the flue gas in this process could lower the resale value of the fly ash by increasing the leachability of metals.[32]

In the SNCR process, urea- or ammonia-based chemicals (either alone or with additives) are injected as the reducing agent to convert NO to molecular nitrogen. However, the reaction is allowed to take place at a higher temperature of 1600–2000°F to obtain a high activation energy and thereby eliminate the use of catalysts. This means that the injection point is usually within the furnace. Based on current pilot-scale and full-scale experiences, 30–70% NO_x reductions are expected for SCNR systems. Capital costs are estimated to range between \$10 and \$20 per kilowatt. Operating costs have not been documented as of 1991.[26]

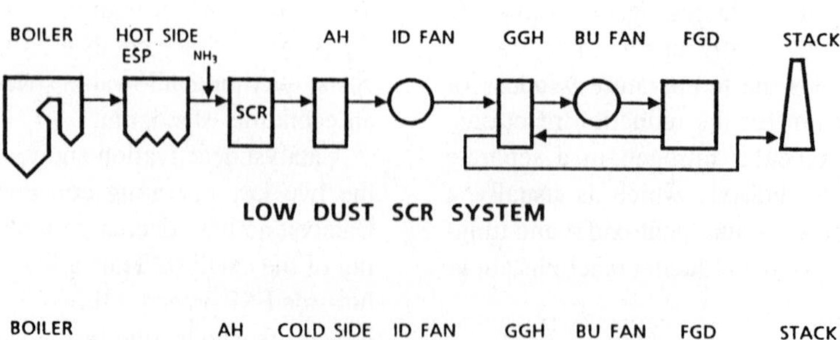

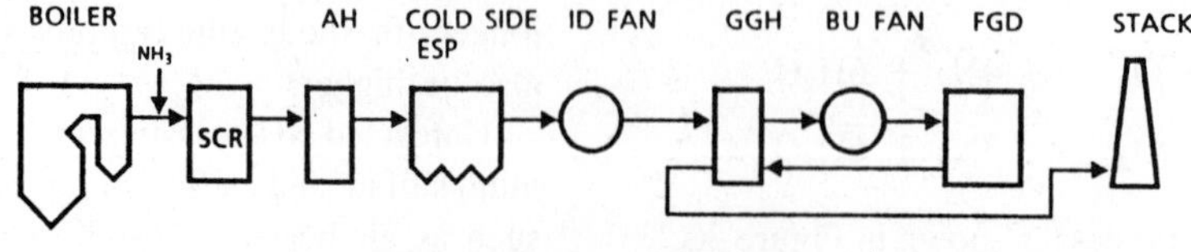

FIGURE 31. Type of SCR System (From Reference 30)

References

1. *Classification of Coals by Rank,* ASTM Standard D-388.
2. J. G. Singer, Ed.; *Combustion—Fossil Power Systems,* Combustion Engineering Inc., 1981.
3. R. E. Bickelhaupt, *A Technique for Predicting Fly Ash Resistivity,* EPA-600/7-79-204, NTIS PB80-102379, U.S. Environmental Protection Agency.
4. P. Vann Bush, T. R. Snyder, and R. L. Chang, "Determination of baghouse performance from coal and ash properties: Part I," *J. APCA,* 39(2): 228 (1989).
5. P. Vann Bush, T. R. Snyder, and R. L. Chang, "Determination of baghouse performance from coal and ash properties: Part II," *J. APCA,* 39(3): 361 (1989).
6. *Compilation of Air Pollutant Emission Factors, Volume I: Stationary Point and Area Sources,* AP-42, U.S. Environmental Protection Agency 1985.
7. *Steam,* 38th ed., Babcock & Wilcox, New York, 1975.
8. H. Farzan, et al., "Pilot evaluation of reburning for cyclone boiler NO_x control," Presented at the 1989 Symposium on Stationary Combustion Nitrogen Oxide Control, Volume 1, EPRI Publication No. GS-6423, 1989, p 2–1.
9. R. D. Lewis, et al., "Retrofit and boiler performance evaluation of its low-NO_x PM firing system at Kansas Power and Light", Presented at the 1989 Symposium on Stationary Combustion Nitrogen Oxide Control, Vol. 1, EPRI Publication No. GS-6423, 1989, p 2–87.
10. *Control Techniques for Nitrogen Oxide Emission from Stationary Sources,* 2nd ed., EPA-450/3-83-002, U.S. Environmental Protection Agency, 1983.
11. C. D. Cooper and F. C. Alley, *Air Pollution Control: A Design Approach,* PWS Engineering, 1988, p 452.
12. M. Hupa, R. Backman, and S. Bostrom, "Nitrogen oxide emission of boilers in Finland," *J. APCA,* 39(11):1496 (1989).
13. *Electrostatic Precipitator Reference Manual,* EPRI Publication No. CS-2809, 1983.
14. G. R. Offen and R. F. Altman, "Issues and trends in electrostatic precipitation technology for U.S. utilities," *J. AWMA,* 41(2):222 (1991).
15. C. Bustard, et al. *Fabric Filters for the Electric Utility Industry, Vol. 1—General Concepts,* EPRI Publication CS-5161, 1988.
16. K. M. Cushing, R. L. Merritt, and R. L. Chang, "Operating history and current status of fabric filters in the utility industry," *J. AWMA,* 40(7):1051 (1990).
17. S. B. Hance and J. L. Kelly, "Status of flue gas desulfurization systems," Paper No. 91-157.3, 84th Annual Meeting of the Air and Waste Management Association, 1991.
18. *Flue Gas Desulfurization Information System (FGDIS), Computerized Database,* U.S. Department of Energy, Energy Information Administration, Forrestal Building, Washington, DC.
19. *Lime FGD Systems Data Book—Second Edition,* EPRI Publication No. CS-2781, Black & Veatch Consulting Engineers, 1983.
20. K. E. Noll and W. T. Davis, *Power Generation: Air Pollution Monitoring and Control,* Ann Arbor Science, Ann Arbor, MI, 1976.
21. *Spray-Dryer Flue-Gas-Cleaning Systems Handbook,* Publication No. ANL/ESD-7, Energy Systems Division, Argonne National Laboratory, U.S. Department of Energy, 1988.
22. G. M. Blythe, et al. *Field Evaluation of a Utility Dry Scrubbing System—Project Summary,* EPA/600/s7-85/020, U.S. Environmental Protection Agency, 1985.
23. J. R. Donnelly, et al., "Spray dryer FGD experience: Joy-Niro installation," *Proceedings of Ninth Symposium on Flue Gas Desulfurization,* U.S. Environmental Protection Agency/EPRI, Cincinnati, OH, 1985.
24. P. J. Kroll and P. Williamson, "Application of dry flue gas scrubbing to hazardous waste incineration," Paper No. 86-10.4, 79th Annual Meeting of the Air Pollution Control Association, 1986.
25. D. Eskinazi, et al., "Stationary combustion NO_x control—a summary of the 1989 symposium," *J. APCA,* 39:1131 (1989).
26. D. Eskinazi, "Retrofit NO_x control strategies for coal fired utility boilers," Paper No. 91-94.5, 84th Annual Meeting of the Air and Waste Management Association, 1991.
27. K. G. Hein, "The application of combustion modifications for NO_x reduction to low rank coal fired boilers," Presented at the 1989 Symposium on Stationary Combustion Nitrogen Oxide Control, Vol. 1, EPRI Publication No. GS-6423, 1989, p 2-1.
28. R. E. Thompson, et al., "NO_x emission results for a low NO_x PM burner retrofit," Presented at the 1989 Symposium on Stationary Combustion Nitrogen Oxide Control, Vol. 1, EPRI Publication No. GS-6423, 1989, p 2-67.
29. H. Farzan, et al., "Pilot evaluation of reburning for cyclone boiler NO_x control," Presented at the 1989 Symposium on Stationary Combustion Nitrogen Oxide Control, Vol. 1, EPRI Publication No. GS-6423, 1989, p 3-1.
30. B. Schonbucher, "Reduction of nitrogen oxides from coal fired power plants by using the SCR process. Experience in the Federal Republic of Germany with pilot and commercial scale deNO_x plants," Presented at the 1989 Symposium on Stationary Combustion Nitrogen Oxide Control, Vol. 1, EPRI Publication No. GS-6423, 1989, p 6A-1.
31. T. Mori and N. Shimuzu, "Operating experience of SCR systems at EPDC's coal fired power station," Presented at the 1989 Symposium on Stationary Combustion Nitrogen Oxide Control, Vol. 1, EPRI Publication No. GS-6423, 1989, p 6A-85.
32. P. Necker, "Experience gained by Neckerwerke from operation of SCR deNO_x units," Presented at the 1989 Symposium on Stationary Combustion Nitrogen Oxide Control, Vol. 1, EPRI Publication No. GS-6423, 1989, p 6A-19.

FUEL OIL

Wayne T. Davis and Arijit Pakrasi

GENERAL DESCRIPTION

The most common liquid fuel used in utilities and industrial processes is fuel oil derived from crude petroleum. Fuel oils are broadly classified into (1) distillates—kerosene used mostly for domestic and small commercial applications and (2) residual fuel oils used in utility and industrial boilers. Fuel oils are classified into six grades, depending on the physical and chemical properties of the oils. Table 1 shows typical analyses and properties for each grade of fuel oil.

Pumpability and flowability are important considerations in using oils as fuel. These properties are characterized by the pour point and temperature–viscosity relationship of the oil. Figure 1 shows a typical relationship between the temperature and viscosity of No. 6 fuel oil. Other important properties are the heating value and the residual carbon content. The residual carbon content is an indicator of the particulate emission expected from the fuel oil.

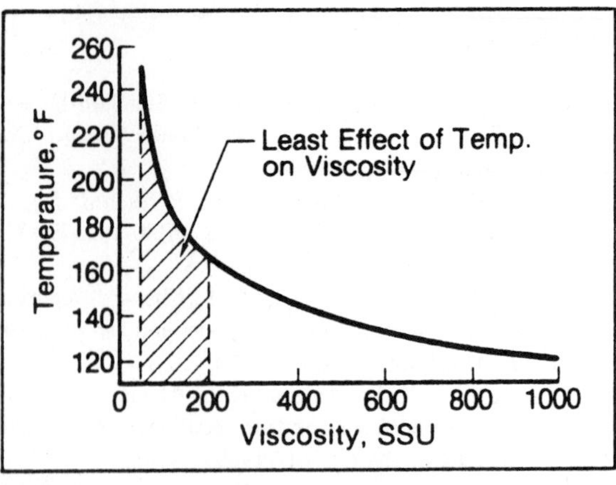

FIGURE 1. Viscosity versus Temperature, No. 6 Fuel Oil (From *Combustion—Fossil Power Systems,* J. G. Singer, Ed.; published by Combustion Engineering Inc.)

AIR EMISSIONS*

Emissions from fuel oil combustion depend on the grade and composition of the oil, the type and size of the boiler, the firing practices used, and the level of equipment maintenance. Table 2 presents emission factors for pollutants from fuel oil combustion. Cumulative size distribution data and size-specific emission factors for particulate emissions from fuel oil combustion and uncontrolled and controlled size-specific emission factors are presented in Figures 2 through 5. Distillate and residual oil categories are given separately, because these produce significantly different particulate, sulfur dioxide (SO_2), and nitrogen oxide (NO_x) emissions.

Particulate Matter

Particulate emissions depend on the grade of oil fired. The lighter distillate oils result in particulate formation significantly lower than with heavier residual oils. Among residual oils, Nos. 4 and 5 usually produce less particulate than does the heavier oil, No. 6.

In boilers firing No. 6 oil, particulate emissions can be described, on the average, as a function of the sulfur content of the oil. As shown in Table 2, particulate emissions can be reduced considerably when low-sulfur No. 6 oil is fired. This is so because low-sulfur No. 6, either refined from naturally low-sulfur crude oil or desulfurized by one of several current processes, exhibits substantially lower viscosity and reduced asphaltene, ash, and sulfur, which results in better atomization and cleaner combustion.

Boiler load can also affect particulate emissions in units firing No. 6 oil. At low-load conditions, particulate emissions may be lowered 30–40% from utility boilers and by as much as 60% from small industrial and commercial units. No significant particulate reductions have been noted at low loads from boilers firing any of the lighter grades. At too low a load condition, proper combustion conditions cannot be maintained and particulate emissions may increase significantly. It should be noted, in this regard, that any condition that prevents proper boiler operation can result in excessive particulate formation.

Trace elements are also emitted from the combustion of fuel oil. The quantity of trace elements emitted depends on the combustion temperature, the fuel feed mechanism, and the composition of the fuel. The temperature determines the degree of volatilization of specific compounds contained in the fuel. The fuel feed mechanism affects the separation of emissions into bottom ash and fly ash.

Sulfur Oxides

Total SO_x emissions are almost entirely dependent on the sulfur content of the fuel and are not affected by boiler size, burner design, or grade of fuel being fired. On the average, more than 95% of the fuel sulfur is emitted as SO_2, about 1–5% as SO_3, and about 1–3% as sulfate particulate. The SO_3 readily reacts with water vapor (in both air and flue gases) to form a sulfuric acid mist.

Nitrogen Oxides

Two mechanisms form NO_x—oxidation of fuel-bound nitrogen and thermal fixation of the nitrogen in combustion air. Fuel NO_x is primarily a function of the nitrogen content of the fuel and the available oxygen. On average, about 45% of the fuel nitrogen is converted to NO_x, but this may

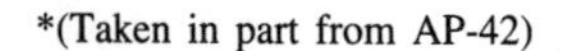

*(Taken in part from AP-42)

TABLE 1. Typical Analyses and Properties of Fuel Oils

Grade	No. 1 Fuel Oil	No. 2 Fuel Oil	No. 4 Fuel Oil	No. 5 Fuel Oil	No. 6 Fuel Oil
Type	Distillate (Kerosene)	Distillate	Very Light Residual	Light Residual	Residual
Color	Light	Amber	Black	Black	Black
API gravity, 60°F	40	32	21	17	12
Specific gravity, 60/60°F	0.8251	0.8654	0.9279	0.9529	0.9861
Pounds/U.S. gallon, 60°F	6.870	7.206	7.727	7.935	8.212
Viscosity, centistokes, 100°F	1.6	2.68	15.0	50.0	360.0
Viscosity, Saybolt univ., 100°F	31	35	77	232	. . .
Viscosity, Saybolt Furol, 122°F	. . .	. . .	. . .	. . .	170
Pour point, °F	Below zero	Below zero	10	30	65
Temperature for pumping, °F	Atmospheric	Atmospheric	15 min.	35 min.	100
Temperature for atomizing, °F	Atmospheric	Atmospheric	25 min.	130	200
Carbon residue, %	Trace	Trace	2.5	5.0	12.0
Sulfur, %	0.1	0.4–0.7	0.4–1.5	2.0 max.	2.8 max.
Oxygen and nitrogen, %	0.2	0.2	0.48	0.70	0.92
Hydrogen, %	13.2	12.7	11.9	11.7	10.5
Carbon, %	86.5	86.4	86.10	85.55	85.70
Sediment and water, %	Trace	Trace	0.5 max.	1.0 max.	2.0 max.
Ash, %	Trace	Trace	0.02	0.05	0.08
Btu/gallon	137,000	141,000	146,000	148,000	150,000

Source: Combustion–Fossil Power Systems, J. G. Singer, Ed.; published by Combustion Engineer Inc.

vary from 20% to 70%. Thermal NO_x is largely a function of peak flame temperature and available oxygen, factors that depend on boiler size, firing configuration, and operating practices.

Fuel nitrogen conversion is the more important NO_x-forming mechanism in residual oil boilers. Except in certain large units having unusually high peak flame temperatures or in units firing a low-nitrogen residual oil, fuel NO_x will generally account for over 50% of the total NO_x generated. Thermal fixation, on the other hand, is the dominant NO_x-forming mechanism in units firing distillate oils, primarily because of the negligible nitrogen content in these lighter oils. Because distillate oil-fired boilers usually have low heat-release rates, however, the quantity of thermal NO_x formed in them is less than that of larger units.

A number of variables influence how much NO_x is formed by these two mechanisms. One important variable is firing configuration. Nitrogen oxide emissions from tangentially (corner) fired boilers are, on the average, less than those from horizontally opposed units. Also important are the firing practices employed during boiler operation.

Other Pollutants

As a rule, only minor amounts of volatile organic compounds (VOCs) and carbon monoxide will be emitted from the combustion of fuel oil. The rate at which VOCs are emitted depends on combustion efficiency. Organic compounds present in the flue gas streams of boilers include aliphatic and aromatic hydrocarbons, esters, ethers, alcohols, carbonyl, carbolic acids, and polycyclic organic matter. The last includes all organic matter having two or more benzene rings.

CONTROL OF EMISSIONS

Since air emissions from oil-fired boilers are substantially lower than those from coal-fired boilers, little or no control of the emissions normally is required, except in large oil-fired boilers. Large oil-fired boilers are often equipped with mechanical collectors to control particulates generated during soot blowing, upset conditions, or when very dirty fuel oil is fired. The particulates emitted during normal operation are too fine to be collected efficiently by mechanical collectors, and in such cases, electrostatic precipitators (ESPs) are used. Up to 90% collection efficiencies have been reported for these ESPs in oil-fired-boiler applications.

Wet scrubbing and spray dryers have been used to reduce sulfur oxide emissions in boilers using high-sulfur oil. Normally, however, sulfur oxide emissions are low and no controls are required.

Nitrogen oxide emissions from oil-fired boilers are usually low because of the low amount of nitrogen present in the fuel. However, since many of these are located in large metropolitan areas, where ozone nonattainment and other health issues related to acid rain are of great concern, these units are also subjected to significant reductions of NO_x. The NO_x-control methods for oil-fired boilers are similar to those for coal-fired boilers. Combustion control methods, including limited excess air firing, flue gas

TABLE 2. Uncontrolled Emission Factors for Fuel Oil Combustion *(Emission Factor Rating: A)*

Boiler Type[a]	Particulate[b] Matter		Sulfur Dioxide[c]		Sulfur Trioxide		Carbon Monoxide[d]		Nitrogen Oxide[e]		Volatile Organics[f]			
											Nonmethane		Methane	
	kg/10^3l	lb/10^3gal	kg/10^3l	lb/10^3gal	kg/10^3l	lb/10^3gal	kg/10^3l	lb/10^3gal	kg/10^3l	lb/10^3gal	kg/10^3l	lb/10^3gal	kg/10^3l	lb/10^3gal
Utility boilers														
Residual oil	g	g	19S	157S	0.34S	2.9S	0.6	5	8.0 (12.6)(5)[h]	67 (105)(42)[h]	0.09	0.76	0.03	0.28
Industrial boilers														
Residual oil	g	g	19S	157S	0.24S	2S	0.6	5	6.6[i]	55[i]	0.034	0.28	0.12	1.0
Distillate oil	0.24	2	17S	142S	0.24S	2S	0.6	5	2.4	20	0.024	0.2	0.006	0.052
Commercial boilers														
Residual oil	g	g	19S	157S	0.24S	2S	0.6	5	6.6	55	0.14	1.13	0.057	0.475
Distillate oil	0.24	2	17S	142S	0.24S	2S	0.6	5	2.4	20	0.04	0.34	0.026	0.216
Residential furnaces														
Distillate oil	0.3	2.5	17S	142S	0.24S	2S	0.6	5	2.2	18	0.085	0.713	0.214	1.78

[a]Boilers can be approximately classified according to their gross (higher) heat rate:
Utility (power plant) boilers: >106 × 10^9 j/h (>100 × 10^6 Btu/h)
Industrial boilers: 10.6 × 10^9 to 106 × 10^9 j/h (10 × 10^6 to 100 × 10^6 Btu/h)
Commercial boilers: 0.5 × 10^9 to 10.6 × 10^9 j/h (0.5 × 10^6 to 10 × 10^6 Btu/h)
Residential furnaces: <0.5 × 10^9 j/h (<0.5 × 10^6 Btu/h)

[b]References 3–7 and 24 and 25. Particulate matter is defined in this section as that material collected by EPA Method 5 (front half catch).

[c]References 1–5. S indicates that the wt % of sulfur in the oil should be multiplied by the value given.

[d]References 3–5 and 8–10. Carbon monoxide emissions may increase by factors of 10 to 100 if the unit is improperly operated or not well maintained.

[e]Expressed as NO_2. Test results indicate that at least 95% by weight of NO_x is NO for all boiler types except residential furnaces, where about 75% is NO.

[f]Volatile organic compound emissions are generally negligible unless boiler is improperly operated or not well maintained, in which case emissions may increase by several orders of magnitude.

[g]Particulate emission factors for residual oil combustion are, on average, a function of fuel oil grade and sulfur content:
Grade 6 oil: 1.25(S) + 0.38 kg/10^3 liter [10(S) + 3 lb/10^3 gal] where S is the wt % of sulfur in the oil. This relationship is based on 81 individual tests and has a correlation coefficient of 0.65.
Grade 5 oil: 1.25 kg/10^3 liter (10 lb/10^3 gal)
Grade 4 oil: 0.88 kg/10^3 liter (7 lb/10^3 gal)

[h]Use 5 kg/10^3 liters (42 lb/10^3 gal) for tangentially fired boilers, 12.6 kg/10^3 liters (105 lb/10^3 gal) for vertical fired boilers, and 8.0 kg/10^3 liters (67 lb/10^3 gal) for all others, at full load and normal (>15%) excess air. Several combustion modifications can be employed for NO_x reduction: (1) limited excess air can reduce NO_x emissions 5–20%; (2) staged combustion, 20–40%; (3) using low NO_x burners, 20–50%; and (4) ammonia injection can reduce NO_x emissions 40–70% but may increase emissions of ammonia. Combinations of these modifications have been employed for further reductions in certain boilers.

[i]Nitrogen oxide emissions from residual oil combustion in industrial and commercial boilers are strongly related to fuel nitrogen content, estimated more accurately by the empirical relationship: kg $NO_2/10^3$ liters = $2.75 + 50(N)^2$ [lb $NO_2/10^3$ gal = $22 + 400(N)^2$] where N is the wt % of nitrogen in the oil. For residual oils having high (>0.5 wt %) nitrogen content, use 15 kg $NO_2/10^3$ liter (120 lb $NO_2/10^3$ gal) as an emission factor.

Source: Reference 31.

Cumulative Particle Size Distribution and Size-Specific Emission Factors *(Emission Factor Rating: C [uncontrolled], E [ESP controlled], D [scrubber controlled])*

Particle size,[a] μm	Cumulative Mass, % ≤ Stated Size			Cumulative Emission Factor[b], kg/10³ 1 (lb/10³ gal)		
		Controlled			Controlled[c]	
	Uncontrolled	ESP	Scrubber	Uncontrolled	ESP	Scrubber
15	80	75	100	0.80A (6.7A)	0.0060A (0.05A)	0.06A (0.50A)
10	71	63	100	0.71A (5.9A)	0.0050A (0.042A)	0.06A (0.50A)
6	58	52	100	0.58A (4.8A)	0.0042A (0.035A)	0.06A (0.50A)
2.5	52	41	97	0.52A (4.3A)	0.0033A (0.028A)	0.058A (0.48A)
1.25	43	31	91	0.43A (3.6A)	0.0025A (0.021A)	0.055A (0.46A)
1.00	39	28	84	0.39A (3.3A)	0.0022A (0.018A)	0.050A (0.42A)
0.625	20	10	64	0.20A (1.7A)	0.0008A (0.007A)	0.038A (0.32A)
Total	100	100	100	1A (8.3A)	0.008A (0.067A)	0.06A (0.50A)

[a]Expressed as aerodynamic equivalent diameter.
[b]Particulate emission factors for residual oil combustion without emission controls are, on average, a function of fuel oil grade and sulfur content:
Grade 6 oil: $A = 1.25(S) + 0.38$ where S is the wt % of sulfur in the oil
Grade 5 oil: $A = 1.25$
Grade 4 oil: $A = 0.88$
[c]Estimated control efficiency for scrubber, 94%; ESP, 99.2%.

Cumulative Size-Specific Emission Factors

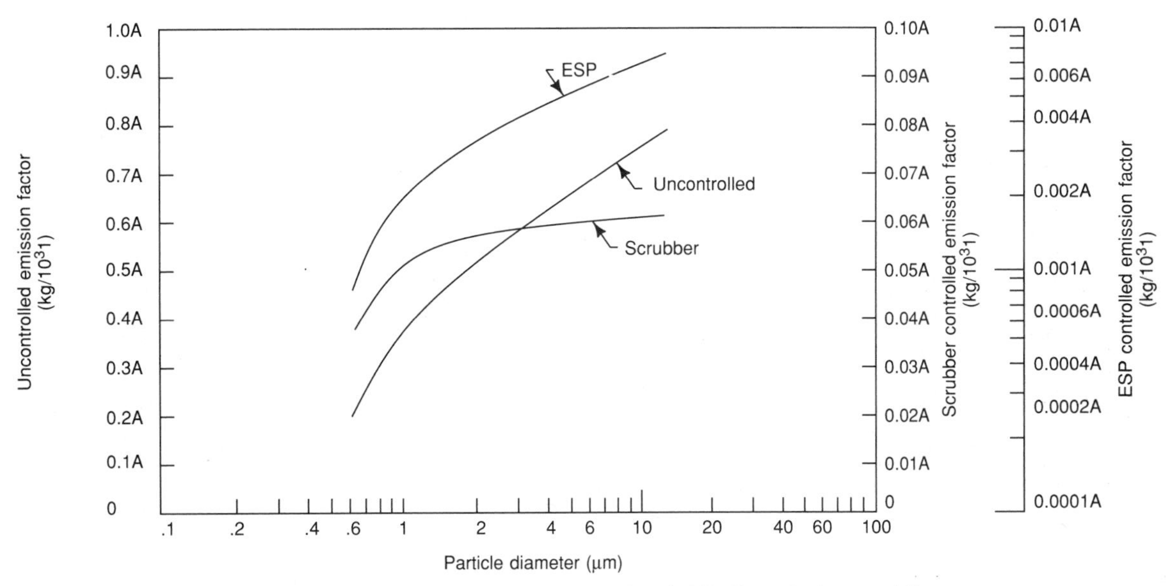

FIGURE 2. Emission Factors for Utility Boilers Firing Residual Oil (From Reference 31)

Cumulative Particle Size Distribution and Size-Specific Emission Factors for Industrial Boilers Firing Residual Oil *(Emission Factor Rating: D [uncontrolled], E [Multiple Cyclone Controlled])*

	Cumulative Mass, % ≤ Stated Size		Cumulative Emission Factor,[b] kg/10³ 1 (lb/10³ gal)	
Particle size,[a] μm	Uncontrolled	Multiple Cyclone Controlled	Uncontrolled	Multiple Cyclone Controlled[d]
15	91	100	0.91A (7.59A)	0.20A (1.67A)
10	86	95	0.86A (7.17A)	0.19A (1.58A)
6	77	72	0.77A (6.42A)	0.14A (1.17A)
2.5	56	22	0.56A (4.67A)	0.04A (0.33A)
1.25	39	21	0.39A (3.25A)	0.04A (0.33A)
1.00	36	21	0.36A (3.00A)	0.04A (0.33A)
0.625	30	c	0.30A (2.50A)	
Total	100	100	1A (8.34A)	0.2A [c](1.67A)

[a]Expressed as aerodynamic equivalent diameter.
[b]Particulate emission factors for residual oil combustion without emission controls are, on average, a function of fuel oil grade and sulfur content:

Grade 6 oil: $A = 1.25(S) + 0.38$ where S is the wt % of sulfur in the oil
Grade 5 oil: $A = 1.25$
Grade 4 oil: $A = 0.88$

[c]Insufficient data.
[d]Estimated control efficiency for multiple cyclone, 80%.

Cumulative Size Specific Emission Factors

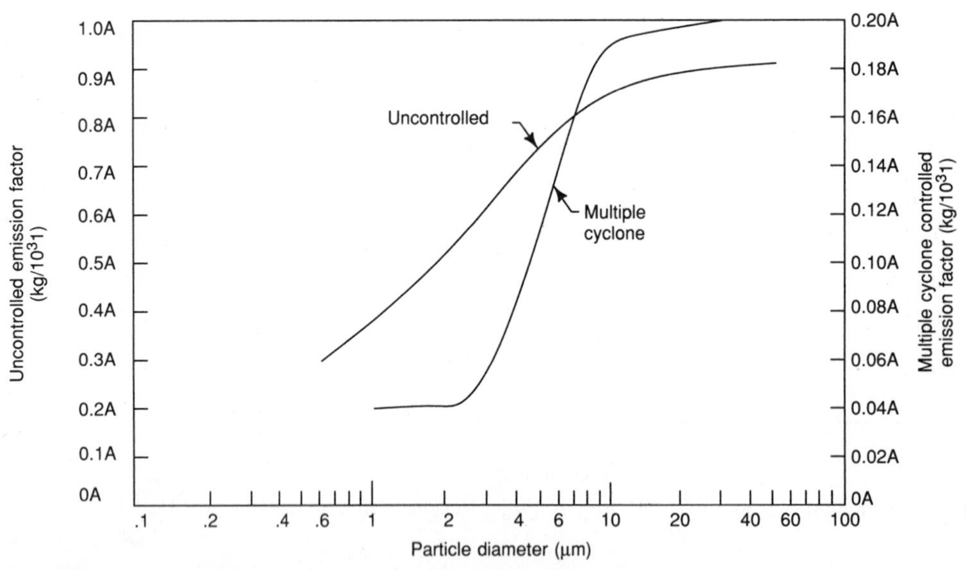

FIGURE 3. Emission Factors for Industrial Boilers Firing Residual Oil (From Reference 31)

Cumulative Particle Size Distribution and Size-Specific Emission Factors for Industrial Boilers Firing Distallate Oil *(Emission Factor Rating: E)*

	Cumulative Mass, % ≤ Stated Size	Cumulative Emission Factor kg/10^3 1 (lb/10^3 gal)
Particle Size,[a] μm	Uncontrolled	Uncontrolled
15	68	0.16 (1.33)
10	50	0.12 (1.00)
6	30	0.07 (0.58)
2.5	12	0.03 (0.25)
1.25	9	0.02 (0.17)
1.00	8	0.02 (0.17)
0.625	2	0.005 (0.04)
Total	100	0.24 (2.00)

[a]Expressed as aerodynamic equivalent diameter.

Cumulative Size-Specific Emission Factors

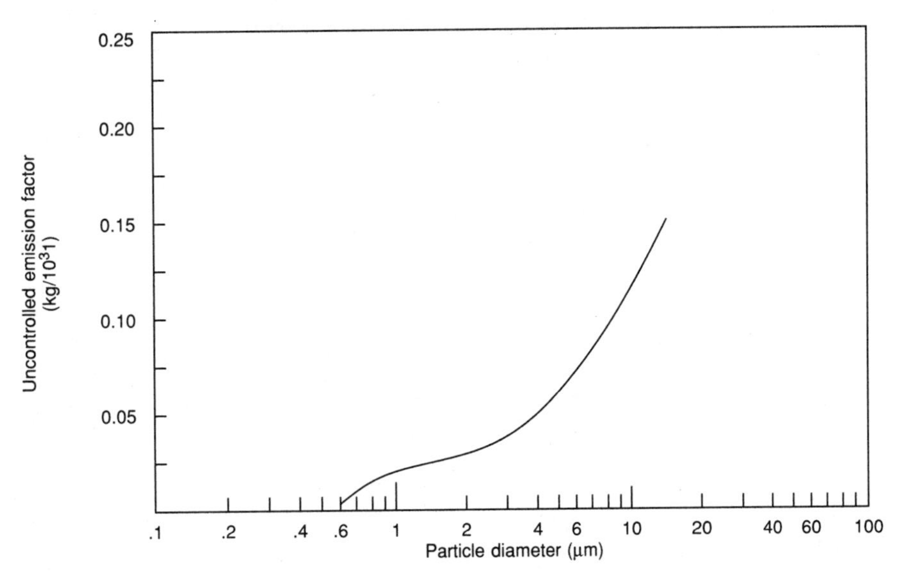

FIGURE 4. Emission Factors for Uncontrolled Industrial Boilers Firing Distillate Oil (From Reference 31)

Cumulative Particle Size Distribution and Size-Specific Emission Factors for Commercial Boilers Burning Residual and Distallate Oil ***(Emission Factor Rating: D)***

	Cumulative Mass, % ≤ Stated Size		Cumulative Emission Factor, $kg/10^3$ 1 ($lb/10^3$ gal)	
Particle Size, [a] μm	Uncontrolled with Residual Oil	Uncontrolled with Distillate Oil[b]	Uncontrolled with Residual Oil	Uncontrolled with Distillate Oil
15	78	60	0.78A (6.50A)	0.14 (1.17)
10	62	55	0.62A (5.17A)	0.13 (1.08)
6	44	49	0.44A (3.67A)	0.12 (1.00)
2.5	23	42	0.23A (1.92A)	0.10 (0.83)
1.25	16	38	0.16A (1.33A)	0.09 (0.75)
1.00	14	37	0.14A (1.17A)	0.09 (0.75)
0.625	13	35	0.13A (1.08A)	0.08 (0.67)
Total	100	100	1A (8.34A)	0.24 (2.00)

[a]Expressed as aerodynamic equivalent diameter.

[b]Particulate emission factors for residual oil combustion without emission controls are, on average, a function of fuel oil grade and sulfur content:

Grade 6 oil: $A = 1.25\ (S) + 0.38$ where S is the weight % of sulfur in the oil

Grade 5 oil: $A = 1.25$

Grade 4 oil: $A = 0.88$

Cumulative Size-Specific Emission Factors

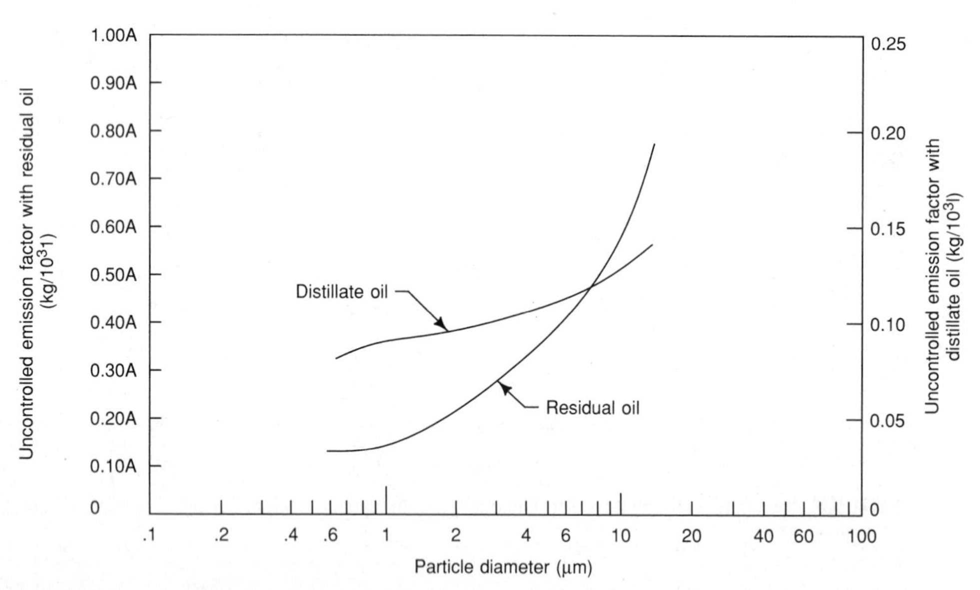

FIGURE 5. Emission Factors for Uncontrolled Commercial Boilers Burning Residual and Distillate Oil (From Reference 31)

recirculation, staged combustion, or some combination thereof, may result in NO_x reductions of 5–60%. In Japan, however, selective catalytic reduction technology is more common for oil-fired-boiler NO_x control.[30]

Load reduction can likewise decrease NO_x production. Nitrogen oxide emissions may be reduced from 0.5% to 1% for each percentage reduction in load from full-load operation. It should be noted that most of these variables, with the exception of excess air, influence the NO_x emissions only of large oil-fired boilers. Limited excess air firing is possible in many small boilers, but the resulting NO_x reductions are not nearly as significant.

One U.S. utility noted, in a study, that the particulate emissions tended to increase with NO_x controls. Further studies have been planned and emphasis is being placed on improving atomizer design and on staging of air to reduce NO_x without increasing the particulate emissions.

Retrofit capital costs for installing LNB and OFA systems in oil-fired boilers are estimated to be $20 to $40 per kilowatt, based on 1989 estimates.[30]

References

1. W. S. Smith, *Atmospheric Emissions from Fuel Oil Combustion: An Inventory Guide,* 999-AP-2, U.S. Environmental Protection Agency, Washington, DC, 1962.
2. *Air Pollution Engineering Manual,* J. A. Danielson, Ed.; 2nd ed., AP-40, U.S. Environmental Protection Agency, Research Triangle Park, NC, 1973 (out of print).
3. A. Levy, et al., "A field investigation of emissions from fuel oil combustion for space heating," *API Bulletin 3099,* Battelle Columbus Laboratories, Columbia, OH, 1971.
4. R. E. Barrett, et al., *Field Investigation of Emissions from Combustion for Space Heating,* EPA-R2-73-084a, U.S. Environmental Protection Agency, Research Triangle Park, NC, 1973.
5. G. A. Cato, et al., *Field Testing: Application of Combustion Modifications to Control Pollutant Emissions from Industrial Boilers—Phase I,* EPA-650/2-74-078a, U.S. Environmental Protection Agency, Washington, DC, 1974.
6. G. A. Cato, et al., *Field Testing: Application of Combustion Modifications to Control Pollutant Emissions from Industrial Boilers—Phase II,* EPA-600/2-76-086a, U.S. Environmental Protection Agency, Washington, DC, 1976.
7. *Particulate Emission Control Systems for Oil Fired Boilers,* EPA-450/3-74-063, U.S. Environmental Protection Agency, Research Triangle Park, NC, 1974.
8. W. Bartok, et al., *Systematic Field Study of NO_x Emission Control Methods for Utility Boilers,* APTD-1163, U.S. Environmental Protection Agency, Research Triangle Park, NC, 1971.
9. A. R. Crawford et al., *Field Testing: Application of Combustion Modifications to Control NO_x Emissions from Utility Boilers,* EPA-650/2-74-066, U.S. Environmental Protection Agency, Washington, DC, 1974.
10. J. F. Deffner et al., "Evaluation of Gulf Econojet equipment with respect to air conservation," Report No. 731RC044, Gulf Research and Development Co., Pittsburgh, PA, December 18, 1972.
11. C. E. Blakeslee and H. E. Burbach, "Controlling NO_x emission from steam generators," *J. APCA,* 23:37–42 (1973).
12. C. W. Siegmund, "Will desulfurized fuel oils help?" *ASHRAE J.,* 11:29 (1969).
13. F. A. Govan et al., "Relationships of particulate emissions versus partial to full load operations for utility-sized boilers," *Proceedings of Third Annual Industrial Air Pollution Control Conference,* Knoxville, TN, 1973.
14. R. E. Hall et al., *A Study of Air Pollutant Emissions from Residential Heating Systems,* EPA-650/2-74-003, U.S. Environmental Protection Agency, Washington, DC, 1974.
15. "Flue gas desulfurization: Installations and operations," PB 257721, National Technical Information Service, Springfield, VA, 1974.
16. *Proceedings: Flue Gas Desulfurization Symposium—1973,* EPA-650/2-73-038, U.S. Environmental Protection Agency, Washington, DC, 1973.
17. R. J. Milligan et al., *Review of NO_x Emission Factors for Stationary Fossil Fuel Combustion Sources,* EPA-450/4-79-021, U.S. Environmental Protection Agency, Research Triangle Park, NC, 1979.
18. N. F. Suprenant et al., *Emissions Assessment of Conventional Stationary Combustion Systems, Vol. I: Gas and Oil Fired Residential Heating Sources,* EPA-600/7-79-029b, U.S. Environmental Protection Agency, Washington, DC, 1979.
19. C. C. Shih et al., *Emissions Assessment of Conventional Stationary Combustion Systems, Vol. II: External Combustion Sources for Electricity Generation,* EPA Contract No. 68-02-2197, TRW, Inc., Redondo Beach, CA, 1980.
20. N. F. Suprenant et al., *Emissions Assessment of Conventional Stationary Combustion Systems, Vol. IV: Commercial Institutional Combustion Sources,* EPA Contract No. 68-02-2197, GCA Corp., Bedford, MA, 1980.
21. N. F. Suprenant et al., *Emissions Assessment of Conventional Stationary Combustion Systems, Vol. V: Industrial Combustion Sources,* EPA Contract No. 68-02-2197, GCA Corp., Bedford, MA, 1980.
22. *Fossil Fuel Fired Industrial Boilers—Background Information for Proposed Standards (Draft EIS),* Office of Air Quality Planning and Standards, U.S. Environmental Protection Agency, Research Triangle Park, NC, 1980.
23. K. J. Lim et al., *Technology Assessment Report for Industrial Boiler Applications: NO_x Combustion Modification,* EPA-600/7-79-178f, U.S. Environmental Protection Agency, Washington, DC, 1979.
24. *Emission Test Report,* Docket No. OAQPS-78-1, Category II-I-257 through 265, Office of Air Quality Planning and Standards, U.S. Environmental Protection Agency, Research Triangle Park, NC, 1972 through 1974.
25. *Primary Sulfate Emission from Coal and Oil Combustion,* EPA Contract No. 68-02-3138, TRW, Inc., Redondo Beach, CA, 1980.
26. C. Leavitt et al., *Environmental Assessment of an Oil Fired Controlled Utility Boiler,* EPA-600/7-80-087, U.S. Environmental Protection Agency, Washington, DC, 1980.
27. W. A. Carter and R. J. Tidona, *Thirty-day Field Tests of Industrial Boilers: Site 2—Residual-Oil-Fired Boiler,* EPA-600/7-80-085b, U.S. Environmental Protection Agency, Washington, DC, 1980.
28. G. R. Offen et al., *Control of Particulate Matter from Oil Burners and Boilers,* EPA-450/3-76-005, U.S. Environmental Protection Agency, Research Triangle Park, NC, 1976.

29. *Inhalable Particulate Source Category Report for External Combustion Sources,* EPA Contract No. 68-02-3156a, Acurex Corp., Mountain View, CA, 1985.
30. D. Eskinazi, J. E. Cichanowicz, W. P. Linak, et al., "Stationary combustion NO_x control—a summary of the 1989 symposium," *J. APCA,* 39:1131 (1989).
31. *Compilation of Air Pollutant Emission Factors, Volume 1: Stationary Point and Area Sources,* AP-42, U.S. EPA, 1985.

NATURAL GAS*

Wayne T. Davis and Arijit Pakrasi

Natural gas is used for power generation, and also for heat/steam generation in the industry. The primary component of natural gas is methane, though varying amounts of ethane and trace compounds such as nitrogen, helium, and carbon dioxide are also present. The average gross heating value of natural gas is approximately 1050 Btu/scf, usually varying between 1000 Btu/scf and 1100 Btu/scf. Typical properties of natural gas are presented in Table 1.

AIR EMISSIONS

Natural gas is considered as a clean fuel with no appreciable particulate or sulfur oxides (SO_x) emissions at normal operating conditions. Typical emissions from natural-gas–fired boilers are presented in Table 2.

* Taken in part from AP-42.

CONTROL OF EMISSIONS

During improper operating conditions of the boilers, such as poor air–fuel mixing, insufficient air, and so on, significant amounts of particulate emissions may result. However, gas-fired boilers are usually not equipped with any particulate collectors.

Sulfur dioxide emissions from gas-fired boilers depend on the sulfur content of the gas. Normally, natural gas is free from sulfur. In the processing stage, however, small amounts of mercaptans are intentionally added to the gas to permit detection; these will produce small quantities of SO_x in the combustion process. Usually, the emissions of SO_x are sufficiently small that a control system is not warranted.

Nitrogen oxides (NO_x) are the major pollutants in natural-gas–fired boilers. The amount of NO_x emission is dependent on the combustion temperature and the rate of cooling of the combustion products. The NO_x control strategies for gas-fired boilers are all based on combustion modifications and are similar to the strategies described for coal-fired boilers in the section on coal combustion. These include staged combustion, off-stoichiometric firing, flue gas recirculation, and low-NO_x burner systems with overfire air. These control strategies can reduce the NO_x emissions by 5% to 50%. Several recent NO_x compliance studies at gas-fired boilers have concluded that emission control strategies for these boilers are very site specific. For example, for tall boilers, flue gas recirculation and staging of the air can be more economical than low-NO_x burners. In the case of short furnaces or where operational flexibility is required, low-NO_x burners seem to be more appropriate.[18]

TABLE 1. Characteristics of Typical Natural Gas

% by Volume								Density,	HHV		A at Zero
CO_2	N_2	H_2S	CH_4	C_2H_6	C_3H_8	C_4H_{10}	C_5H_{12}	lb/ft³	Btu/ft³ [b]	Btu/lb	Excess Air, lb/10⁶ Btu
5.50	...	7.00	77.73	5.56	2.40	1.18	0.63[a]	0.05621	1061	18,880	738
3.51	32.00	0.50	52.54	3.77	2.22	2.02	3.44[a]	0.06610	874	13,220	729
26.2	0.7	...	59.2	13.9	...	...	...	0.06747	849	12,580	732
0.17	87.69	...	10.50	1.64	...	...	...	0.07120	136	1,907	732
0.20	0.60	...	99.20	...	...	...	...	0.04491	1006	22,410	732
...	0.60	...	...	79.40	20.00	...	...	0.08812	1935	21,960	735
...	0.50	...	...	21.80	77.70	...	...	0.11079	2389	21,560	738

[a]All hydrocarbons heavier than C_5H_{12} are assumed to be C_5H_{12}
[b]If gas is saturated with moisture at 60°F and 30.0 in. Hg. reduce by 1.74%.
Source: Combustion—Fossil Power Systems, by J. G. Sunger, Ed.; Combustion Engineering Inc.

TABLE 2. Uncontrolled Emission Factors for Natural Gas Combustion

									Volatile Organics			
	Particulate[b]		Sulfur Dioxide[c]		Nitrogen Oxides[d]		Carbon Monoxide[e]		Nonmethane		Methane	
Furnace Size and Type (10^6 Btu/h heat input)	kg/10^6m^3	lb/10^6 ft^3	kg/10^6m^3	lb/10^6 ft^3	kg/10^6m^3	lb/10^6 ft^3	kg/10^6m^3	lb/10^6 ft^3	kg/10^6m^3	lb/10^6 ft^3	kg/10^6m^3	lb/10^6 ft^3
Utility boilers (>100)	16–80	1–5	9.6	0.6	8800[h]	550[h]	640	40	23	1.4	4.8	0.3
Industrial boilers (10–100)	16–80	1–5	9.6	0.6	2240	140	560	35	44	2.8	48	3
Domestic and commercial boilers (< 10)	16–80	1–5	9.6	0.6	1600	100	320	20	84	5.3	43	2.7

[a]Expressed as weight/volume fuel fired.
[b]References 6–11.
[c]Reference 1. Based on average sulfur content of natural gas, 4600 g/10^6 Nm^3 (2000 gr/10^6 scf).
[d]References 1–5, 7, and 11–13.
[e]Expressed as NO_2. Tests indicate about 95 weight % NO_x is NO_2.
[f]References 1, 3, 4, 9, 11, and 14–17.
[g]References 9 and 11. May increase 10–100 times with improper operation or maintenance.
[h]For tangentially fired units, use 4400 kg/10^6 m^3 (275 lb/10^6 ft^3).

Source: AP-42

References

1. W. Bartok, et al., *Systematic Field Study of NO_x Emission Control Methods for Utility Boilers,* APTD-1163, U.S. Environmental Protection Agency, Research Triangle Park, NC, 1971.
2. F. A. Bagwell, et al., "Oxides of nitrogen emission reduction program for oil and gas fired utility boilers," *Proceedings of the American Power Conference,* 1970, pp 683–693.
3. H. E. Dietzmann, "A study of power plant boiler emissions," Final Report No. AR-837, Southwest Research Institute, San Antonio, TX, 1972.
4. R. E. Barrett, et al., *Field Investigation of Emissions from Combustion Equipment for Space Heating,* EPA-R2-73-084, U.S. Environmental Protection Agency, Research Triangle Park, NC, 1971.
5. C. E. Blakesless and H. E. Burbock, "Controlling NOx emissions from steam generators," JAPCA, 23:37–42 (1979).
6. J. W. Bradstreet and R. J. Fortman, "Status of control techniques for achieving compliance with air pollution regulations by the electric utility industry," presented at the Third Annual Industrial Air Pollution Control Conference, Knoxville, TN, March 1973.
7. *Study of Emissions of NO_x from Natural Gas Fired Steam Electric Power Plants in Texas, Phase II, Vol. II,* Radian Corp., Austin, TX, 1972.
8. N. F. Suprenant, et al., *Emissions Assessment of Conventional Stationary Combustion Systems, Vol. I: Gas and Oil Fired Residential Heating Sources,* EPA-600/7-79-029b, U.S. Environmental Protection Agency, Washington, DC, 1979.
9. C. C. Shih, et al., *Emisson Assessment of Conventional Stationary Combustion Systems, Vol. III: External Combustion Sources for Electricity Generation,* EPA Contract No. 68-02-2197, TRW, Inc., Redondo Beach, CA, 1980.
10. N. F. Suprenant, et al., *Emission Assessment of Conventional Stationary Combustion Sources, Vol. IV: Commercial Institutional Combustion Sources,* EPA Contract No. 68-02-2197, GCA Corp., Bedford, MA, 1980.
11. N. F. Suprenant, et al., *Emissions Assessment of Conventional Stationary Combustion Systems, Vol. V: Industrial Combustion Sources,* EPA Contact No. 68-02-2197, GCA Corp., Bedford, AM, 1980.
12. R. J. Milligan, et al., *Review of NO_x Emission Factors for Stationary Fossil Fuel Combustion Sources,* EPA-450/4-79-021, U.S. Environmental Protection Agency, Research Triangle Park, NC, 1979.
13. W. H. Thrasher and D. W. Dewerth, *Evaluation of the Pollutant Emissions from Gas Fired Forced Air Furnaces,* Research Report No. 1503, American Gas Association, Cleveland, OH, 1975.
14. G. A. Cato, et al., *Field Testing: Application of Combustion Modification to Control Pollutant Emissions from Industrial Boilers, Phase I,* EPA-650/2-74-078a, U.S. Environmental Protection Agency, Washington, DC, 1974.
15. G. A. Cato, et al., *Field Testing: Application of Combustion Modification to Control Pollutant Emissions from Industrial Boilers, Phase II,* EPA-600/2-76-086a, U.S. Environmental Protection Agency, Washington, DC, 1976.
16. W. A. Carter and H. J. Buening, *Thirty-day Field Tests of Industrial Boilers—Site 5,* EPA Contract No. 68-02-2645, KVB Engineering, Inc., Irvine, CA, 1981.
17. W. A. Carter and H. J. Buening, *Thirty-day Field Tests of Industrial Boilers—Site 6,* EPA Contract No. 68-02-2645, KVB Engineering, Inc., Irvine, CA, 1981.
18. ECS Update, Summer 1991, No. 23, published by EPRI, Palo Alto, CA, 1991.

Bibliography

C. E. Blakesless and H. E. Burbock, "Controlling NO_x emissions from steam generators," *J. APCA,* 23:37–42 (1979).

R. L. Chass and R. E. George, "Contaminant emissions from the combustion of fuels," *J. APCA,* 10:34–43 (1980).

Confidential Information, American Gas Association Laboratories, Cleveland, OH, 1970.

ECS Update, Summer 1991, No. 23, published by EPRI, Palo Alto, CA, 1991.

H. H. Hovey, et al., *The Development of Air Contaminant Emission Tables for Non-process Emission,* New York State Department of Health, Albany, NY, 1965.

D. M. Hugh, et al., *Exhaust Gases from Combustion and Industrial Processes,* EPA Contract EHSD 71–36, Engineering Science, Inc., Washington, DC, 1971.

L. K. Jain, et al., *State of the Art for Controlling NO_x Emissions: Part 1, Utility Boilers,* EPA Contract No. 68-02-0241, Catalytic, Inc., Charlotte, NC, 1972.

K. J. Lim, et al., *Technology Assessment Report for Industrial Boiler Applications: NO_x Combustion Modification,* EPA Contract No. 68–02-3101, Acurex Corp., Mountain View, CA, 1979.

J. H. Perry (Ed.), *Chemical Engineers' Handbook,* 4th ed., McGraw-Hill Book Co., New York, 1963.

H. J. Taback, et al., *Fine Particle Emissions from Stationary and Miscellaneous Sources in the South Coast Air Basin,* California Air Resources Board Contract No. A6-191-30, KVB, Inc., Tustin, CA, 1979.

W. H. Thrasher and D. W. Dewerth, *Evaluation of the Pollutant Emissions from Gas Fired Water Heaters,* Research Report No. 1507, American Gas Association, Cleveland, OH, April 1977.

Unpublished data on domestic gas fired units, U.S. Environmental Protection Agency, Cincinnati, OH, 1970.

WOOD WASTE

Anthony J. Buonicore, P.E.

Wood has always been an important boiler fuel in the forest products industry. Each year considerable quantities of timber are harvested and millions of tons of wood products manufactured. In the process, very large quantities of wood "wastes" or residues are generated. Logging operations alone produce billions of cubic feet of residues. In fact, in conventional logging, nearly 40% of the harvested tree is left behind on the forest floor, as well as diseased and rotted trees. Primary processing and manufacture of the logs create additional supplies of residues, including trimmings, edging, sawdust, sander dust, bark, and low-grade chips.

Such residues can represent as much as 50% of the original log that reaches the mill. These wastes also represent a considerable opportunity for energy recovery. Heating values typically range between 7500 and 9000 Btu/lb on a dry basis, with as-fired heating values between 4000 and 8000 Btu/lb.

Local governments also add to the available supply of wood residues. Large volumes of wood "wastes" are developed as communities construct roads, sewers, transmission lines, drainage basins, wastewater treatment plants, and reservoirs. Cleanup operations after storms add to the wood waste supply, as do the normal maintenance of parks and tree-lined streets and local construction and demolition activities. Traditionally, municipalities have discarded these wastes in dumps or landfills. However, as disposal sites become less available, alternatives must be investigated. Wood waste-to-energy projects have been developed and continue to be proposed as one such alternative.

PROCESS DESCRIPTION

Various boiler firing configurations are used in burning wood waste. One common type in smaller operations is the Dutch oven, or extension type of furnace with a flat grate. This unit is widely used because it can burn fuels with very high moisture. Fuel is fed into the oven through apertures atop a firebox and is fired in a cone-shaped pile on a flat grate. The burning is done in two stages: drying and gasification and the combustion of gaseous products. The first stage takes place in a cell separated from the boiler section by a bridge wall. The combustion stage takes place in the main boiler section. Unfortunately, the Dutch oven is not adequately responsive to changes in steam load, and it provides poor combustion control.

In another type, the fuel cell oven, fuel is dropped onto suspended fixed grates and is fired in a pile. Unlike the Dutch oven, the fuel cell also uses combustion air preheating and repositioning of the secondary and tertiary air injection ports to improve boiler efficiency.

In many large operations, more conventional boilers have been modified to burn wood waste. These units may include spreader stokers with traveling grates, vibrating grate stokers, and so on, as well as tangentially-fired or cyclone-fired boilers. The most widely used of these configurations is the spreader stoker. Fuel is dropped in front of an air jet, which casts the fuel out over a moving grate, spreading it in an even, thin blanket. The burning is done in three stages in a single chamber: (1) drying, (2) distillation and burning of volatile matter, and (3) burning of carbon. This type of operation has a fast response to load changes, has improved combustion control, and can be operated with multiple fuels. This is done to maintain constant steam when the wood waste supply fluctuates and/or to provide more steam than is possible from the waste supply alone.

Fluid-bed combusters using a sand-bed medium have also been developed to burn wood residues. These units are able to combust fuels with very high moisture content (up to approximately 65% by weight). The waste fuel is typically hogged to minus 3 inches. Some of the sand in the bed with ash is continuously removed and passed over a vibrating screen. Material suitable for reinjection is returned to the bed. The material removed from the bed is usually fused ash or gravel and rocks initially present in the hogged fuel.

Sander dust is often burned in various boiler types at plywood, particle board, and furniture plants. Sander dust contains fine wood particles with low moisture content (less than 20% by weight). It is fired in a flaming horizontal torch, usually with natural gas as an ignition aid or supplementary fuel.

AIR EMISSIONS CHARACTERIZATION

In wood-fired boiler plants, the air pollutant of primary concern is particulate matter. Removal of sulfur oxides is required only when large amounts of high-sulfur coal or oil, together with wood waste, are burned. Nitrogen oxide emissions are considerably less for wood-fired steam generators than for coal-fired boilers of comparable size, despite the fact that larger quantities of excess air are normally used in burning wood. Generally, nitrogen oxide emissions can be controlled by boiler design; however, in some locations, regulations have required the installation of nitrogen oxide emission control systems.

If demolition debris is part of the wood waste fuel, additional air pollution control equipment may be required, depending on what and how much is burned. In addition to waste wood, demolition debris can contain plastics, paint, creosote-treated wood, glues, synthetics, wire, cable, insulation, and so forth. Such materials can contribute to acid gas emissions (such as HCl), heavy metal emissions (including lead, cadmium, chromium, copper, and zinc), inorganic emissions (including arsenic, cyanides, and asbestos), and organic emissions (including formaldehyde, dioxins, and furans).

Actual emissions will depend on many variables, including:

- Composition of the wood waste fuel and its variability.
- Type and amount of fossil fuel burned in combination with the wood waste (if any).
- Firing method and type of furnace.
- Extent of carbon reinjection.
- Air pollution control system.

The composition of wood waste depends largely on the industry from which it originates. Pulping operations, for example, produce great quantities of bark that may contain more than 70% by weight moisture, along with sand and other noncombustibles. Because of this, bark boilers in pulp mills may emit considerable amounts of particulate matter to the atmosphere unless they are well-controlled. On the other hand, some operations, such as furniture manufacturing, produce a clean dry wood waste, 5–50% by weight moisture, with relatively low particulate emissions when properly burned. Still other operations, such as sawmills, burn a varying mixture of bark and wood waste that results in particulate emissions somewhere between these two extremes.

Furnace design and operating conditions are particularly important when firing wood waste. For example, because of the high moisture content that can be present in this waste, a larger than usual area of refractory surface is often necessary to dry the fuel before combustion. In addition, sufficient secondary air must be supplied over the fuel bed to burn the volatiles that account for most of the combustible material in the waste. When proper drying conditions do not exist or when secondary combustion is incomplete, the combustion temperature is lowered and increased particulate, carbon monoxide, and hydrocarbon emissions may result. The lowering of combustion temperature generally means decreased nitrogen oxide emissions. Also, emissions can fluctuate over short periods if significant variations occur in fuel moisture content.

Fly-ash reinjection, which is common to many large boilers to improve fuel efficiency, has a considerable effect on particulate emissions. Because a fraction of the collected fly ash is reinjected into the boiler, the dust loading from the furnace, and consequently from the collection device, increases significantly per unit of wood waste burned. It is reported that full reinjection can cause a 10-fold increase in the dust loadings of some systems, although an increase of 1.2 to 2 times is more typical for boilers using 50%–100% reinjection. A major factor affecting this dust loading increase is the extent to which the sand and other noncombustibles can be separated from the fly ash before reinjection to the furnace.

Although reinjection increases boiler efficiency from 1% to 4% and reduces emissions of uncombusted carbon, it increases boiler maintenance requirements, decreases average fly-ash particle size, and makes particulate collection more difficult. Properly designed reinjection systems should separate sand and char from the exhaust gases, reinject the larger carbon particles to the furnace and divert the fine sand particles to the ash disposal system.

Uncontrolled particulate emissions of wood waste–fired boilers generally range between 0.5 and 5 gr/dscf or 1.5–6.0 pounds per million Btu heat input, or 10–75 pounds per ton of wood waste burned. Approximately 80–95% of the ash in wood waste becomes fly ash off the boiler. The particle size distribution of this fly ash may range from 10% to 30% by weight less than 10 μm. The particulate emission problem becomes more complex if the wood waste fuel contains salt residue resulting from seawater-floated logs. The salt is discharged as a submicron-sized fume much more difficult to capture in any air pollution control equipment.

Uncontrolled nitrogen oxide emissions are typically less than 200 ppmv, sulfur dioxide emissions less than 10 ppmv, carbon monoxide emissions less than 200 ppmv, and total hydrocarbons less than 10 ppmv. Reasonable emission factors for modern boiler designs include 0.15–0.20 lb/MM Btu for nitrogen oxides and 0.06–0.10 lb/MM Btu for hydrocarbons.[10–12] Sulfur dioxide emissions rarely exceed 0.02 lb/MM Btu with wood waste alone.[1] Emission factors of toxic air pollutants for industrial wood-fired boilers are presented in Table 1.[15]

AIR POLLUTION CONTROL MEASURES

Since the air pollutant of primary concern is particulate matter, air pollution control equipment designed to remove particulates will be the principal focus.

Particulate control equipment used on wood-fired boilers generally falls into one of four categories:

1. Mechanical collectors
2. Wet scrubbers
3. Electrostatic precipitators (ESPs)—dry and wet
4. Filters

Multitube cyclonic mechanical collectors have traditionally been the sole source of particulate control for many existing wood-fired boiler plants. These consist of a series of small-diameter (typically 9-inch inner diameter) cyclonic collector tubes operating in parallel at a total gas pressure drop typically in the 3 in. w.c. range. Unfortunately, mechanical collectors alone are not capable of meeting the current New Source Performance Standard (NSPS) for wood-fired boilers. Consequently, these systems are most often used in series with scrubbers, ESPs, or fabric filters. Large pieces of char and much of the sand can be removed from the gas stream before it flows to these secondary dust collectors. Reinjection of the char by mechanical collectors can increase boiler efficiency by as much as 4%.

The most common type of scrubber used for particulate removal on wood-fired boilers is the venturi scrubber, operating at gas pressure drops ranging between 6 and 25 in. w.c. The venturi scrubber normally is provided with an adjustable throat to maintain constant pressure drop (and, therefore, particulate collection efficiency) at variable boiler load. Liquid-to-gas ratios in the venturi system typically range between 8 and 10 gallons of water per 1000 acfm saturated. Solids buildup in the recirculation loop rarely is allowed to exceed 5% by weight and wastewater treatment facilities may be necessary to treat the scrubber blow-down.

High-carbon ash is not easily collected by an ESP. Hence, the highly conductive (very low resistivity) ash off wood fired boilers often necessitates a very conservative design. Specific collections areas (SCA) in the 300–500 ft^2/1000 acfm area and power requirements at the 150–400 watts per 1000 acfm level would be reasonable, depending on the efficiency required and the nature of the fuel fired.

TABLE 1. Emission Factors of Toxic Air Pollutants (Uncontrolled) for Industrial Wood-Fired Boilers[15]

Pollutant	Emission Factor	Remarks	Reference
Dioxins		Salt-laden wood	16
Heptachlorodibenzo-*p*-dioxins	39 ng/dscm @ 3%O_2		
Hexachlorodibenzo-*p*-dioxins	49 ng/dscm @ 3%O_2		
Octachlorodibenzo-*p*-dioxins	11 ng/dscm @ 3%O_2		
Pentachlorodibenzo-*p*-dioxins	48 ng/dscm @ 3%O_2		
2,3,7,8-Tetrachlorodibenzo-*p*-dioxins	0.28 ng/dscm @ 3%O_2		
Tetrachlorodibenzo-*p*-dioxins	47 ng/dscm @ 3%O_2	Not including 2,3,7,8-isomer	
Furans		Salt-laden wood	16
Heptachlorodibenzo-furans	6.5 ng/dscm @ 3%O_2		
Hexachlorodibenzo-furans	15 ng/dscm @ 3%O_2		
Octachlorodibenzo-furans	0.92 ng/dscm @ 3%O_2		
Pentachlorodibenzo-furans	23 ng/dscm @ 3%O_2		
2,3,7,8-Tetrachlorodibenzo-furan	1.8 ng/dscm @ 3%O_2		
Tetrachlorodibenzo-furans	37 ng/dscm @ 3%O_2	Not including 2,3,7,8-isomer	
Polycyclic organic matter	0.00044–0.0062 lb/ton dry wood	Gaseous and particulate POM	17
Acenaphthene	0.0000022–0.00011 lb/ton	Particulate control with mechanical collector	18
Anthracene	0.0000044 lb/ton	As above	18
Barium	0.007 lb/ton	As above	18
Benzo (k) fluoranthene	0.0000044 lb/ton	As above	18
Chromium	0.0003 lb/ton	As above	18
Chrysene	1 × 10E-9 lb/ton	As above	18
Cobalt	0.0000176–0.0002 lb/ton	As above	18
Copper	0.0004–0.001 lb/ton	As above	18
Fluoranthene	0.0000066–0.000028 lb/ton	As above	18
Lead	0.00022–0.00040 lb/ton	As above	18
Naphthalene	2 × 10E-7 lb/ton	As above	18
Nickel	0.0052–0.0064 lb/ton	As above	18
Phenanthrene	0.000044–0.00032 lb/ton	As above	18
Phenol	0.000022–0.0002 lb/ton	As above	18
Pyrene	0.0000044–0.000013 lb/ton	As above	18
Silver	0.00032–0.034 lb/ton	As above	18

Source: Reference 15.

Superficial gas velocities typically range between 3 and 5 fps. While the use of a wet scrubber eliminates concern about fires, the potential for a fire must be considered in the design if a dry ESP is to be used. For example, glowing char can cause hopper fires if air is allowed to leak in via the hopper discharge port. For this reason, the induced-draft fan is installed between the mechanical collector and precipitator. In this way, the precipitator is slightly pressurized. Figure 1 shows an ESP installed on a wood-fired boiler.

To eliminate the fire concern, ESPs can operate in the wet mode (Figure 2). In such systems, collection plates and internal parts are wetted continuously with water. Because of lower gas volumes, a wet precipitator would be smaller in size than the equivalent dry version. A prequench is typically used to saturate the gas stream and ensure that "condensable" pollutants have condensed into particulate and can be collected in the unit. Water is recirculated around the precipitator, but not completely, since a small blowdown stream is necessary. Wet precipitators on wood-fired boilers are capable of achieving outlet grain loadings similar to dry precipitators, that is, 0.01–0.02 gr/dscf corrected to 12% CO_2 at power levels typically in the 30–150 watts per 1000 acfm range.

Fabric filters have only limited experience on wood-fired boilers, despite the advantage of very high removal efficien-

FIGURE 1. Electrostatic Precipitator on a 300,000-lb/h Wood-Fired Boiler (Courtesy PPC Industries)

cy for fine particulates. The majority of these systems are operating on the West Coast, with more than half on boilers burning bark from logs transported in saltwater. The filters were installed to collect the chloride fumes that resulted from the bark combustion. Operating experience indicates that collected chloride can act as a fire inhibitor. The principal concern with the use of a fabric filter is the potential for fire damage resulting from burning cinders, temperature excursions, and/or operating upsets. Both glass and Nomex bags have been used, with bag life generally not exceeding 18 months to 2 years. Reverse air–cleaned fabric filters generally operate at an air-to-cloth ratio of 1–2 (acfm/ft^2). Pulse jet–cleaned fabric filters generally operate with an air-to-cloth ratio in the range of 4 to 5.

Gravel-bed filters have been developed to eliminate the potential for fire hazard. Instead of a fabric medium, gravel-bed filters have a slowly moving bed of granular "rock" as the filtration medium through which the flue gas must travel. To enhance efficiency, these systems have been electrostatically augmented (10–20 watts/1000 acfm). An electrical conductor configured in the form of a cage is positioned within the medium bed. A high-voltage (50 kV) is applied to this conductor, and the electrical field generated between the conductor and the inlet and outlet louvers enhances particulate collection. Gas pressure drops are typically in the 5–6 in. w.c. range.

Performance experience with particulate removal air pollution control equipment is presented in Table 2. For a major new wood-fired source that must comply with a 0.1 lb/MM Btu particulate emission limitation, the alternatives include a high-energy venturi scrubber, an ESP, an electrostatically augmented moving granular-bed filter, or a fabric filter. The principal disadvantages of the venturi scrubber are the relatively high energy utilization and the need for disposal or utilization of the scrubber blowdown stream. Electrostatic precipitators are relatively expensive, but have a satisfactory operating history on wood-waste–fired boilers. The potential for fires in fabric filters may make it necessary to provide a bypass (for temperature excursions and operating upsets), which may complicate the permitting process.

If nitrogen oxides removal is required and combustion modifications (such as lowering the excess air) for existing units or design techniques for new units are not sufficient, ammonia injection (for example, the Exxon DeNOX process) can be used. Such systems have already been used on woodwaste–fired boilers in California. The process requires the injection of ammonia into the hot flue gases off the boiler at a point where the temperature is in the range of 1700°F to 1800°F. Nitrogen oxide removal efficiencies in the 50–70% range are possible with this technology.

Should acidgas removal equipment be required because of the use of high-sulfur auxiliary fuel or the addition of demolition debris containing plastics or other chloride-containing matter, the following alternatives are possible.

- Use of limestone in a fluid-bed boiler.
- Addition of a spray dryer acid-gas reactor using a lime slurry upstream of the fabric filter or ESP.
- Injection of lime/limestone in the ductwork upstream of the ESP or fabric filter.

TABLE 2. Performance Expectation of Air Pollution Control Equipment on New Wood-Waste–Fired Boilers

Device	Efficiency	Outlet Particulate Loading	
		lb/MM Btu	gr/dscf at 12% CO_2
1. Mechanical—collector multitube, cyclonic	85–92	0.30–0.60	0.15–0.25
2. Two mechanical collectors in series	90–94	0.20–0.30	0.08–0.15
3. Venturi scrubber			
ΔP = 6–8 in. w.c.	95–96	0.10–0.15	0.04–0.06
ΔP = 10–15 in. w.c.	97–98	0.06–0.08	0.025–0.03
ΔP = 20–25 in. w.c.	99	0.03–0.05	0.015–0.02
4. Electrostatic precipitator	99–99.5	0.025–0.05	0.01–0.02
5. Granular-bed (moving) filter	90–95	0.20–0.30	0.10–0.15
6. Electrostatically augmented granular-bed (moving) filter	98–99.2	0.03–0.1	0.01–0.04
7. Fabric filter	99.5–99.9	0.01–0.03	0.004–0.015

FIGURE 2. Wet Electrostatic Precipitator on a Wood-Waste–Fired Boiler (Courtesy United McGill Corp.)

Addition of a packed-tower absorber using a sodium hydroxide (or some other alkaline-based) scrubbing solution downstream of the venturi or other particulate removal device.

CONCLUSION

In summary, air pollution control equipment able to meet any of today's stringent air regulations is readily available for wood waste combustion facilities. The choice of particulate removal equipment will depend, among other things, on the specific emission limitation, economics, the availability of water and wastewater treatment facilities, wastewater discharge limits, whether or not emergency bypassing (for protection) of an air pollution control device is permissible, the risk (of fire hazard) one is willing to assume, space availability, and energy-use considerations.

Whether or not nitrogen oxide removal pollution control equipment is required will depend, among other things, on site location (attainment versus nonattainment area), local emission requirements, boiler type (for example, fluid-bed boilers have demonstrated the capability for lower nitrogen oxide emission levels), and specific design (for example, staged combustion resulting in lower nitrogen oxide emissions).

If the wood waste feed stream contains contaminants that would make it necessary to install acid gas removal equipment, such equipment is readily available. Alternatives include the use of limestone in a fluid-bed boiler, a spray dryer acid-gas removal reactor using a lime slurry, or a lime/limestone injection (in the ductwork) system, or a packed tower absorber using a sodium-based scrubbing solution. The specific choice will be influenced to a large extent by the degree of acid gas reduction required.

References

1. *Atmospheric Emissions from the Pulp and Paper Manufacturing Industry,* EPA-450/1-73-002, U.S. Environmental Protection Agency, Research Triangle Park, NC, September 1973.
2. S. F. Galeano and K. M. Leopold, "A survey of emissions of nitrogen oxides in the pulp mill," *J. Tech. Assoc. Pulp Paper Indus.*, 56(3):74–76 (March 1973).
3. *Control of Particulate Emissions from Wood Fired Boilers,* EPA-340/1-77-026, U.S. Environmental Protection Agency, Research Triangle Park, NC, 1978.
4. *Wood Residue Fired Steam Generator Particulate Matter Control Technology Assessment,* EPA-450/2-78-044, U.S. Environmental Protection Agency, Research Triangle Park, NC, October 1978.
5. H. S. Oglesby and R. O. Blosser, "Information on the sulfur content of bark and its contribution to SO_2 emissions when burned as a fuel," *J. Air Pollution Control Assoc.*, 30(7):769–772 (1980).
6. *A Study of Nitrogen Oxides Emissions from Wood Residue Boilers,* Technical Bulletin No. 102, National Council of the Paper Industry for Air and Stream Improvement, New York, November 1979.
7. A. Nunn, *NO_x Emission Factors for Wood Fired Boilers,* EPA-600/7-79-219, U.S. Environmental Protection Agency, September 1979.
8. *Nonfossil Fueled Boilers—Emission Test Report: Weyerhaeuser Company, Longview, Washington,* EPA-80-WFB-10, Office of Air Quality Planning and Standards, U.S. Environmental Protection Agency, Research Triangle Park, NC, March 1981.
9. *A Study of Wood Residue Fired Power Boiler Total Gaseous Nonmethane Organic Emissions in the Pacific Northwest,* Technical Bulletin No. 109, National Council of the Paper Industry for Air and Stream Improvement (NCASI), New York, September 1980.
10. *Volatile Organic Carbon Emissions from Wood Residue Fired Power Boilers in the Southeast,* NCASI, Technical Bulletin No. 455, New York, April 1985.
11. *A Study of Factors That Affect Nitrogen Oxides and Carbon Monoxide Emissions from Wood-Residue Fired Boilers,* NCASI, Technical Bulletin No. 448, New York, December 1984.
12. *Carbon Monoxide Emissions from Selected Combustion Sources Based on Short-Term Monitoring Records,* NCASI, Technical Bulletin No. 416, New York, January 1984.
13. *Nonfossil Fuel Fired Industrial Boilers—Background Information,* EPA-450/3-82-007, March 1982.
14. W. O. Dameworth, "Application of an ESP for the control of particulate emissions from a wood-residue-fired boiler," *J. TAPPI,* 225 (July 1989).
15. A. Pope, et al., *Toxic Air Pollutant Emission Factors—A Compilation for Selected Air Toxic Compounds and Sources,* 2nd ed., EPA-450/2-90-011, U.S. Environmental Protection Agency, October 1990.
16. *National Dioxin Study: Report to Congress,* EPA-530/SW-87-025, Environmental Protection Agency Office of Solid Waste and Emergency Response, Washington, DC, August 1987.
17. G. W. Brooks, et al., *Locating and Estimating Air Emissions from Sources of Polycyclic Organic Matter (POM),* EPA-450/4-84-007p, U.S. Environmental Protection Agency, Research Triangle Park, NC, May 1988.
18. *Environmental Assessment of a Wood Waste-Fired Industrial Watertube Boiler,* ORD Project Summary, EPA-600/S7-87/012, AEERL, U.S. Environmental Protection Agency, Research Triangle Park, NC, May 1987.

Bibliography

A. J. Buonicore and D. Scerbo, "Air pollution control alternatives for wood waste combustion facilities," Air and Waste Management Association Conference on Air Quality Issues Related to Power Production, Enfield, CT, October 24–25, 1989.

L. Theodore and A. J. Buonicore, *Industrial Air Pollution Control Equipment for Particulates,* CRC Press, Boca Raton, FL, 1976.

8
Waste Incineration Sources

REFUSE

James R. Donnelly

Environmentally sound disposal of municipal and industrial refuse has become a major issue in the past decade. The volume of waste generated has continued to grow annually and traditional disposal methods (landfilling or ocean dumping) are becoming less acceptable because of cost and environmental concerns. Incineration of refuse in modern high-efficiency combustors is being employed for a growing fraction of waste streams to achieve significant reductions in refuse volume, while, in many instances, also achieving energy recovery in the form of steam or electricity. This increase in incineration has been coupled with an increase in the complexity and efficiency of air-pollution controls to limit incinerator emissions. This discussion focuses on commonly employed incineration devices and the air-pollution controls applied to them.

REFUSE INCINERATION

Refuse or municipal solid waste includes nonhazardous waste generated in households, institutions (excluding hospital wastes), and commercial and light industrial facilities, as well as agricultural wastes and sewage sludge. In 1988, the United States generated approximately 180 million tons of refuse.[1] Table 1 lists the major components of refuse. In addition to the major components listed in the table, several toxic trace metals are found in refuse. These include arsenic, cadmium, chromium, lead, and mercury.

The refuse per capita generation rate in the United States is the highest in the world and has shown an annual growth over the past several decades. Because of this continual refuse growth, environmental concerns with landfilling or ocean dumping, and the lack of availability and cost of landfills, the public has begun looking into alternative ways to handle its refuse. This has included composting, recycling, and incineration. Of these three alternatives, incineration has shown the ability to achieve the greatest reduction in refuse volume (70–90%) and can be used in conjunction with the other two alternatives to achieve even greater reductions in disposal volume. Incineration is currently being used to treat about 15% of the refuse stream.

This increase in refuse incineration has led to increased concern over air pollution from these incinerators, which, in turn, has led to the promulgation of air emissions standards and the application of modern air-pollution controls to limit these emissions. The U.S Environmental Protection Agency

TABLE 1. U.S. Municipal Solid Waste Stream *(Annual Generation Rate—180,000 Tons/Year)*

Composition	%
Paper and paperboard	41.0
Glass	8.2
Metals	8.7
Plastics	6.5
Rubber, leather, textiles, and wood	8.1
Food wastes	7.9
Yard wastes	17.9
Miscellaneous inorganic wastes	1.6

Source: Reference 2.

TABLE 2. U.S. Environmental Protection Agency Municipal Waste Combustion Emission Standards

	New Source Performance Standards	Emission Guidelines for Existing Facilities	
Capacity, tons/day	Unit	Unit	Facility
	>250	>250 ≤ 1100	>1100
Particulate matter, gr/dscf	0.015	0.030	0.015
Opacity, %	10	10	10
Organic emissions, ng/dscm			
Total chlorinated PCDD plus PCDF			
Mass burn units	30	125	60
RDF fired units	30	250	60
Acid-gas control			
% Reduction or emissions, ppm			
HCl	95 (25)	50 (25)	90 (25)
SO_2	80 (30)	50 (30)	70 (30)
NO_x	(180)	None	None
Carbon monoxide, ppm	50–150[a]	50–250[a]	50–250[a]

Note: All emissions limits are referenced to dry gas conditions at 7% oxygen concentration.
[a]Range of values reflect differing types of MWCs.
Source: Reference 3.

(EPA), in response to the public's concerns, issued proposed "New Source Performance Standards and Emissions Guidelines for Existing Facilities" in December 1989 and promulgated these standards in February 1991. Table 2 summarizes these standards as they were adopted.

In setting these standards, the EPA recognized differences in facility size, the type of incineration (mass burn fired versus refuse-derived fuel fired), and new sources versus existing sources. The facility capacity refers to the total burn rate for all refuse combustors at a single site. The EPA selected total particulate matter emission limits as the way of controlling trace heavy-metal emission limits. It will add emission limits based on applying maximum achievable control technology (MACT) for mercury, cadmium, and lead emissions in the coming year. The agency has until late 1992 to establish comparable emission standards for smaller combustors, those less than or equal to 250 tons per day per train.

Emissions limits are established for the total emissions of the polychlorinated dibenzyl-dioxins (PCDD) plus polychlorinated dibenzyl-furans (PCDF). These compounds were selected as surrogates for organic emissions because of their potential adverse effects on health. In addition, the EPA has established carbon monoxide (CO) emission limits as a measure of "good combustion practices" that limit the formation of PCDD, PCDF, and their key precursors. The CO emission limits vary from 50 to 150 ppm (1 at 7% O_2 dry gas conditions), depending on the type of combustion.

Acid-gas emissions (HCl and SO_2) are based on either a percent reduction or a maximum stack emission level, whichever is the least stringent. Nitrogen oxides (NO_x) emissions levels are proposed only for large new sources.

In addition to EPA emission standards, many states and local air-pollution control districts have developed their own sets of emission standards for new sources. Many of these standards (or permit conditions) are more stringent than the EPA standards.

INCINERATION SOURCES

Several different technologies are employed in refuse combustion. These include mass burn combustors (modular, traveling grate, and rotary combustors); refuse-derived, fuel-fired combustors; and, to a lesser extent, fluid-bed combustors.

Mass Burn Combustors

Mass burn combustors are the predominant type of incinerators currently being employed for refuse. These are characterized by their accepting refuse that has undergone very limited preprocessing other than removal of oversized items. Mass burn combustors may be of the modular (or starved-air) type, the traveling grate type, or the rotary combustor type.

Modular Combustors

Modular combustors are typically small, two-chamber, starved-air systems with capacities between about 5 and 100 tons/day. Figure 1 shows a cutaway view of a modular incinerator.

Typically, refuse is moved from the tipping floor to a ram feeder, where it is pushed into the primary combustion chamber. Here the refuse is combusted under starved-air conditions, which leads to the formation of combustible gases and ash. The combustible gases pass into the second-

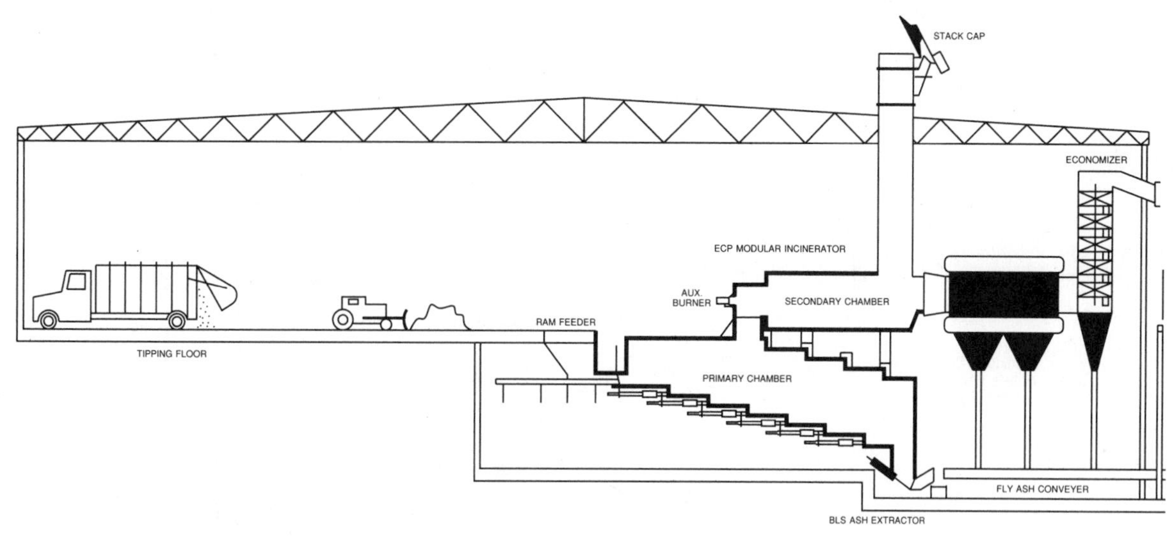

FIGURE 1. Modular Mass Burn Incinerator (Joy Technologies, Inc.)

ary chamber, where they are fired with auxiliary fuel to complete combustion and raise the temperature to above 1800°F. After leaving the secondary chamber, the flue gases often pass through a heat-recovery boiler and economizer, where they are cooled to approximately 450°F before entering the fuel gas cleaning system. The ash formed in the primary chamber is pushed through the chamber and typically exits the end of the chamber via an ash quench system. Two or more modular combustors are typically employed for incinerating refuse at smaller facilities. Modular mass burn incinerators typically involve lower capital costs than do traveling grate or rotary combustors.

Traveling Grate Combustors

Traveling grate mass burn combustors are typically much larger than modular combustors and have capacities ranging from about 150 to 750 tons/day. Figure 2 shows a traveling grate incinerator with water walls, a burnout zone, a heat-recovery boiler, and an economizer.

Typically, refuse is dumped into a tipping bay, where it is mixed and oversized objects are removed by an overhead crane. The overhead crane dumps the mixed refuse into a feed chute, where it is fed into the grate system by a hydraulic ram. The traveling grates move the refuse through the various zones of the combustion chamber in a tumbling

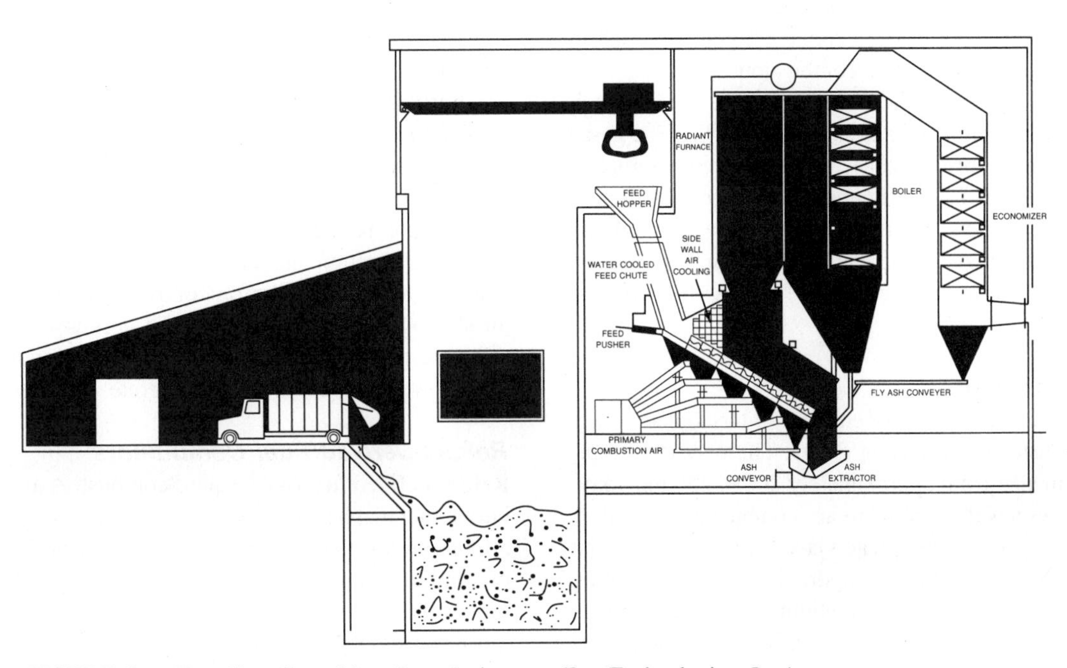

FIGURE 2. Traveling Grate Mass Burn Incinerator (Joy Technologies, Inc.)

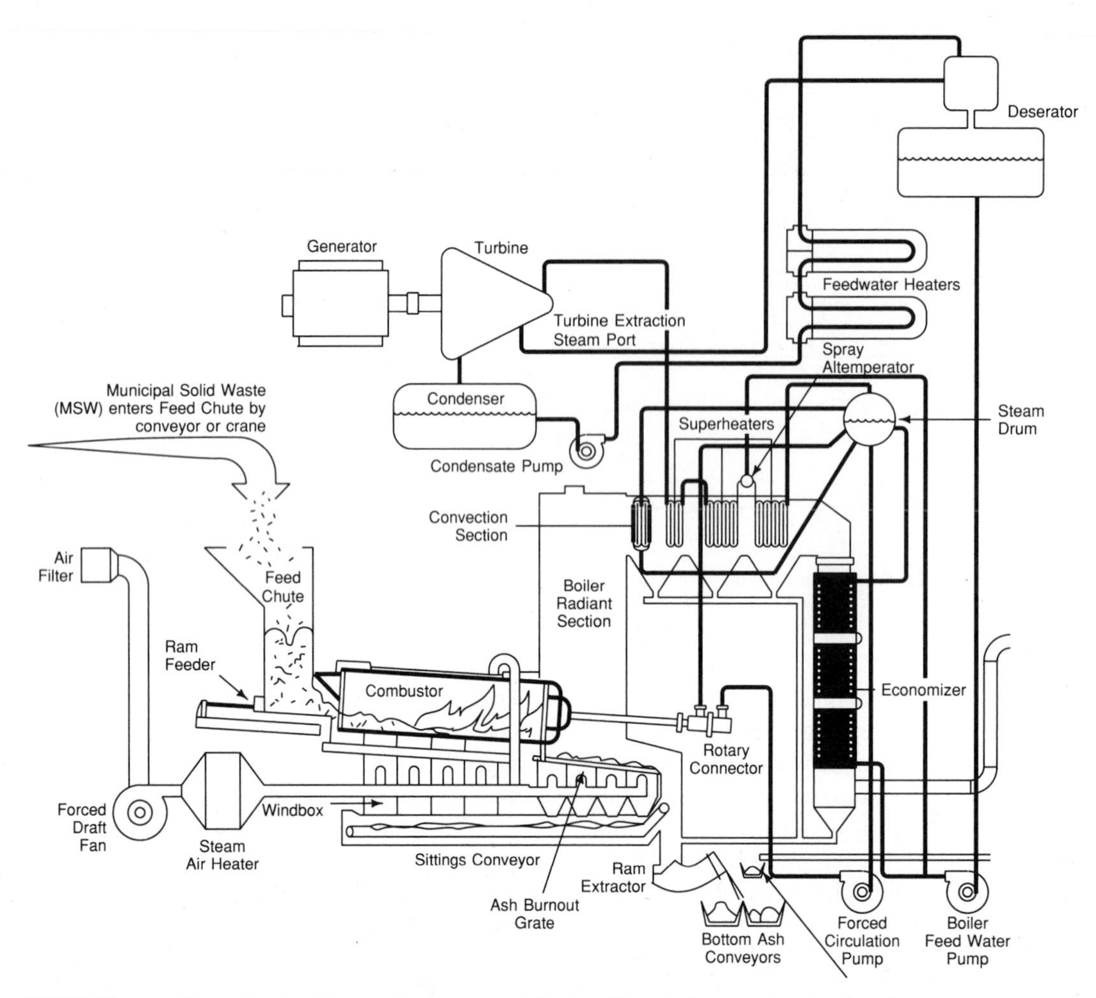

FIGURE 3. Water-Cooled Rotary Combustor and Boiler (Westinghouse Electric Corp.)

motion. Air is added at multiple points through the grate to assist in refuse dryout, combustion, and burnout. Additional combustion air may be added above the grate to assist in flue gas mixing and complete combustion.

Residual ash falls from the grate into a wet quench system and then is removed and sent to a landfill. Flue gases pass upward to a burnout zone, where the temperature is maintained at about 1800°F to ensure complete destruction of organic compounds. The flue gas then passes through the boiler and economizer, where its temperature is reduced to 350–450°F prior to its entering the flue gas cleaning system.

Rotary Combustors

Some mass burn incinerators employ water-cooled rotary combustors instead of traveling grates. These systems typically range in size from approximately 200 to 450 tons/day. Figure 3 shows a water-cooled rotary combustor and boiler.

Refuse is fed to the feed chute via an overhead crane and pushed into the combustor by a hydraulic ram. The rotating combustor barrel is mounted at a slight angle and rotates at 10 to 20 revolutions per hour. Preheated combustion air is fed to the combustor through six air zones (four longitudinal and two transverse) to provide good combustion control. The rotation and angle of the combustor barrel cause the refuse to tumble and slowly move through the combustor onto a water-cooled afterburning grate, where remaining combustibles are consumed.

The residue ash is then discharged via a quench system and sent to a landfill. The flue gases at 1800°F leave the combustor barrel, pass up through a radiant zone where the temperature is reduced to between 900 and 1200°F, and then pass through the boiler and economizer before discharging to the air-pollution control system. Because of the air-staging arrangement and good combustion control for this type of combustor, NO_x generation is typically somewhat lower than for a traveling grate combustor.

Refuse-Derived Fuel Combustors

Refuse-derived fuel (RDF)-fired combustors are designed to burn a fuel that has been preprocessed to produce a fuel with a more uniform size, composition, and heat rate. A variety of combustor designs can be employed with RDF, depending on the degree of preprocessing. The simplest form and most commonly employed method of RDF processing is shredding of refuse followed by magnetic separation to remove ferrous metals, and, in some cases, by air classification to remove ash. This type of fuel is often burned in a spreader stoker combustor or is suspension fired over a stoker.

The RDF may be further processed to produce a densified fuel by pelletizing, a recovery-prepared RDF in which a larger portion of metals and glass is removed, or a fluff RDF for cofiring with coal in suspension-fired combustors.

The RDF combustors range in size from approximately 400 to 1000 tons/day in capacity. Because of the nature of the fuel and firing, particulate matter carryover to the air-pollution control system is generally much higher than for a mass burn combustor.

EMISSIONS CHARACTERIZATION

Refuse incineration has the potential of emitting a wide range of pollutants to the environment. These potential emissions arise from compounds present in the refuse stream, are formed as a part of the normal combustion process, or are formed as a result of incomplete combustion. Table 3 lists principal potential municipal-waste-combustion (MWC) emissions and the prime source for each.

Particulate matter consists primarily of noncombustible inorganic material entrained in the flue gas, and it typically ranges in size from less than 1 μm to about 50 μm. The uncontrolled particulate matter emission rate varies substantially for the different types of MWCs. Modular incinerators produce the lowest levels of uncontrolled emissions, with RDF-fired units having the highest.

Table 4 presents emission factors for a number of pollutants for modular, mass burn, and RDF-fired combustors. The acid gases hydrogen chloride (HCl), sulfur dioxide (SO_2), and hydrogen fluoride (HF) are formed during the combustion of chloride-, sulfur- and, fluoride-containing compounds found in the waste stream. A small fraction (approximately 1% to 5%) of the SO_2 in the flue gas is oxidized to sulfur trioxide (SO_3). These gases, in the presence of water or water vapor, react to form hydrochloric, sulfurous, hydrofluoric, or sulfuric acid. Nitrogen oxides are found predominantly in the form NO and are formed primarily through the conversion of fuel-bound nitrogen, although some nitrogen in the combustion air may also be converted. Carbon monoxide is formed through the incomplete combustion of organic compounds in the waste stream and is used as an indicator of combustion conditions.

TABLE 3. Principal Municipal-Waste-Combustion Emissions and Sources

Pollutant	Principal Source
Particulate matter	Ash in waste stream
Acid gases	
HCl	Chlorinated plastic in waste stream
SO_2	Sulfur compounds in waste stream
SO_3	Oxidation of SO_2 in flue gas
HF	Fluorocarbons in waste stream
NO_x	Air and fuel nitrogen conversion
CO	Incomplete combustion
Heavy metals (arsenic, cadmium, lead, mercury)	Metal compounds in waste stream
Organic compounds (dioxins, furans)	Products of incomplete combustion or contained in waste stream

TABLE 4. Emission Factors for Municipal Waste Combustion ***(All Values Are Pounds per Ton, Uncontrolled)***

	Type of Incinerator		
	Modular Starved Air	Mass Burn	Refuse-Derived Fuel
Particulate matter			
PM_{10}	1.4	14	44
Total	0.12	0.18	0.13
Lead	0.12	0.18	0.13
SO_2	1.7	1.7	1.7
NO_x	4.4	3.6	5.0
CO	3.4	2.2	3.6

Source: Reference 6.

Heavy-metal compounds of concern emitted from MWCs include the oxides and chlorides of arsenic, cadmium, lead, and mercury. These compounds are formed from the combustion of heavy-metal–containing components of the waste stream, such as batteries, plastics, paper products, and metal alloys. A number of these compounds have boiling points or sublime at temperatures below the 1800°F typical of incineration systems and are thus vaporized into the flue gas. As the flue gas temperature cools, they tend to condense out and are concentrated on fine particulate matter in the flue gas. For the compounds of mercury and lead, a significant fraction may remain in the vapor state at typical incinerator-exit flue gas temperatures.

Organic emissions are a result of incomplete combustion of compounds found in the waste stream. The prime organic compounds of concern are PCDD and PCDF. These emissions can arise from incomplete thermal destruction of PCDD- and PCDF-containing materials in the waste stream, from incomplete thermal destruction of other organic compounds that produce PCDD/PCDF precursors, and through chemical reactions that occur at relatively low temperatures downstream of the combustor.

As shown in Table 4, uncontrolled particulate emissions and the fine particulate fractions vary widely for the three major types of combustors. These variations are a function of the turbulence in the combustion zone, the flue gas velocity through the combustor, and the fineness of the fired fuel. The RDF-fired boilers have substantially higher uncontrolled total and fine particulate fractions, reflecting these conditions. The variations of other uncontrolled pol-

TABLE 5. Typical Refuse Incinerator Uncontrolled and Controlled Emissions

Pollutant	Uncontrolled Emissions	Controlled Emissions	Percent Reduction
Particulate matter, gr/dscf	0.5–4.0	0.002–0.015	99.5+
Acid gases, ppmdv			
HCl	400–100	10–50	90–99+
SO_2	150–600	5–50	65–90+
HF	10–0	1–2	90–95+
NO_x	150–300	60–180	30–65[a]
Heavy metals, mg/nm^3			
Arsenic	<0.1–1	<0.01–0.1	90–99+
Cadmium	1–5	<0.01–0.5	90–99+
Lead	20–100	<0.1–1	90–99+
Mercury	<0.1–1	<0.1–0.7	10–90+
Total PCDD/PCDF, ng/nm^3	20–500	<1–10	80–99

Note: Reference conditions: dry gas at 12% CO_2.
[a]Reduction associated with nonselective catalytic reduction.

lutant emissions rates are much less for the various types of combustors and are more a function of waste-stream composition and combustor operating conditions.

Table 5 shows typical uncontrolled and controlled emissions for a number of pollutants of concern from refuse incineration. Percent reduction ranges typical of levels being achieved utilizing the best available control technologies are also shown for each pollutant.

Modern refuse incinerators are achieving very low emissions as a result of the proper application and operation of available air-pollution control systems. The average incinerator emission levels for all pollutants has been decreased substantially over the past five years as more modern installations have been brought into service.

AIR-POLLUTION CONTROL SYSTEMS

Air-pollution control systems for refuse incinerators can be classified by either the pollutant they control or their operating principles. Table 6 presents a list of the common pollutants of concern found in refuse incinerator flue gas and the methods used to control their emissions.

Often more than one control device will be used in series to control a number of pollutants. The most common examples of this are the use of an electrostatic precipitator followed by a wet scrubber or a spray dryer absorption system including an electrostatic precipitator or fabric filter.

Carbon-Monoxide Controls

Carbon-monoxide emissions are controlled by employing "good combustion practices." These practices include operational and incinerator design elements to control the amount and distribution of excess air in the flue gas to ensure that there is enough oxygen present for complete combustion. The design of modern efficient combustors is such that there is adequate turbulence in the flue gas to ensure good mixing, a high-temperature zone (greater than 1800°F) to complete burnout, and a long enough residence time at the high temperature (one to two seconds).

The feed to the combustor is controlled to minimize fuel spikes that lead to fuel-rich firing. The combustor is equipped with adequate instrumentation and combustion air controls to adjust for rapid changes in fuel conditions.

Good combustion practices also limit PCDD/PCDF emissions exiting the incinerator. This is accomplished by maintaining firing conditions that destroy PCDD/PCDFs

TABLE 6. Incinerator Emissions and Controls

Pollutant	Control Device
Carbon monoxide (CO)	Good combustion practices
Nitrogen oxides (NO_x)	Staged combustion selective noncatalytic reduction (SNCR)
Particulate matter	Electrostatic precipitator (ESP)
	Pulse jet fabric filter (PJF)
	Reverse air fabric filter (RAF)
	Wet scrubber (WS)
Acid gases (HCl, SO_2, SO_3, HF)	Dry sorbent injection (DSI)
	Spray dryer absorption (SDA)
	Wet scrubbing
Heavy metals	ESP, PJF, RAF, SDA, WS
PCDD/PCDF	Good combustion practices
	ESP, PJF, RAF, SDA, WS

found in the fuel and by destroying PCDD/PCDF precursors that may be formed from the combustion of other chlorinated organic compounds.

Nitrogen Oxide Controls

Nitrogen oxide emissions are controlled by limiting their formation in the incinerator using staged combustion or applying selective noncatalytic reduction to reduce the NO_x content in the flue gas. Staged combustion is accomplished by splitting up the introduction of combustion air into the combustor so that areas of fuel-rich and fuel-lean firing are established. This will lower the peak flame temperatures and limit the amount of oxygen available to react with nitrogen in the air at the peak temperature. The introduction of additional secondary air downstream in the combustor will ensure complete combustion and minimize CO formation. Generally, staged combustion is effective in reducing NO_x formation due to air–nitrogen conversion, but is not very effective for conversion of fuel-bound nitrogen to NO_x.

The NO_x present in the flue gas can be reduced by employing either a selective catalytic or noncatalytic reduction process. The selective catalytic reduction (SCR) process utilizes ammonia injection upstream of a catalytic reactor, at about 600–650°F, to reduce NO_x to nitrogen. Selective catalytic reduction has been applied to a wide range of combustion sources where 80–85% NO_x reduction has been demonstrated. However, because of the nature of the compounds found in refuse incinerator flue gas, the successful application of SCR requires installation downstream of the acid-gas and particulate control systems with subsequent reheat to the reactor operating temperature. Because of these constraints, only limited SCR applications to refuse incinerator flue gases have been attempted.

Selective noncatalytic reduction (SNCR) reduces flue gas NO_x through a reaction with ammonia in a temperature range of 1700–1900°F. The ammonia may be supplied as anhydrous ammonia, aqueous ammonia, or urea. At flue gas temperatures above 1900°F, the oxidation of ammonia to NO_x increases and SNCR can actually result in an increase in overall NO_x. At temperatures below about 1700°F, NO_x reduction falls off and ammonia breakthrough increases, leading to the potential for a visible ammonium-chloride plume.

Ammonia injection, also known as thermal De-NO_x, has been applied to many different combustion sources, including mass burn refuse incinerators. Reductions in NO_x levels of up to 65% have been demonstrated at an ammonia-to-NO_x ratio of about two, with ammonia breakthrough as low as 5 ppm. This corresponds to a NO_x emission level of approximately 60 ppm. Thermal De-NO_x operates most efficiently under steady-state operating conditions. Changes in fuel feed rate, excess air rate, or incinerator load can significantly change flue gas conditions at the ammonia-injection point, leading to a major change in control efficiency.

Urea injection has been demonstrated full scale on refuse combustors in the United States and Europe. Urea injection offers the advantage of not requiring a hazardous material for operation. At the injection temperatures employed (1600–1900°F), the urea quickly breaks down to form the active reagent. In some cases, reaction enhancers are added to the urea to expand the effective temperature window to as low as 1200°F. Tests with urea injection have achieved greater than 65% NO_x reduction with very low (approximately 5 ppm) ammonia slip.[7]

Particulate Matter Controls

Particulate emissions are primarily controlled by electrostatic precipitators (ESPs) or fabric filters, although wet scrubbers are sometimes used on small incinerators or in series with ESPs for additional control. The ESPs are installed either alone, to control particulate emissions, or after a spray dryer, as a part of an acid-gas cleaning system. Fabric filters are typically installed downstream of a quench tower or spray dryer, where the conditions of increased flue gas moisture and lowered temperature aid in protecting filter bags from hot embers.

Electrostatic Precipitators

Electrostatic precipitators collect particulate matter by introducing a strong electrical field in the flue gas, which imparts a charge to the particulates present. These charged particles are then collected on large plates, which have an opposite charge applied to them. The collected particulate is periodically removed by rapping the collection plates. The agglomerated particles fall to a hopper, where they are removed. Key design parameters for ESPs include particulate composition, density, and resistivity; flue gas temperature and moisture content; inlet particulate loading and collection efficiency; specific collection area (SCA = square feet of collecting surface per 1000 acfm of flue gas) and number of fields; flue gas velocity and collector plate spacing; rapping frequency and intensity; and transformer rectifier power levels.

Table 7 presents sizing parameters typical for ESPs applied for incinerator particulate emissions control. The ranges in parameters shown reflect straight ESP particulate control applications and ESP applications as part of an acid-gas cleaning system. Although the inlet particulate loading to the ESP is much higher as part of an acid-gas cleaning system, the number of fields and specific collecting areas required to achieve a similar outlet emission do not change significantly. This is due to lower ash resistivity values and increased flue gas moisture contents, which improve the ESP's performance. Incinerators that have had spray dryers retrofitted in front of existing ESPs have been able, in most cases, to maintain the same level of particulate emissions (e.g., 0.01 to 0.015 gr/dscf at 12% CO_2).

TABLE 7. Electrostatic Precipitator—Design Parameters

Particulate loading, gr/acf		0.5–9
Required efficiency, %		98–99.9
Number of fields		3–4
SCA, ft^2/1000 acfm		400–550
Average secondary voltage, kV		35–55
Average secondary current, mA/1000ft^3		30–50
Gas velocity, ft/s		3.0–3.5
	Particulate	Acid-Gas Control
Flue gas temperature, °F	350–450	230–300
Flue gas moisture, % vol.	8–16	12–20
Ash resistivity, ohm-cm	10^9–10^{12}	10^8–10^9

Source: Reference 8.

Weighted-wire, rigid-frame, and rigid-electrode types of precipitators are employed for incinerator applications, however, rigid-frame and rigid-electrode types predominate. This is related to the corrosive gas conditions and sticky nature of the fly ash being collected. Electrode failures associated with rigid-frame and rigid-electrode systems are less frequent than for weighted wire. This is especially true where higher rapping forces are needed to dislodge the sticky fly ash. For rigid-frame systems, high-alloy (e.g., Incoloy 825) spring-wound electrodes are also used to minimize electrode corrosion problems.

The insulator compartment ventilation system is designed to minimize the effects of the corrosive nature of the flue gas and fly-ash stickiness. A pressurized ventilation system employing heated air is recommended to maintain clean insulators and reduce potential electrical tracking problems.

Fabric Filters

Both the reverse-air and pulse-jet types of fabric filters are used for particulate emission control on refuse incinerators. Each type offers advantages that should be evaluated on a site-specific basis. Both types are capable of achieving particulate emissions of the order of 0.01–0.015 gr/dscf at 12% CO_2 or lower. Table 8 presents design parameters typical of incinerator fabric filter applications.

The temperature ranges shown represent operation after both a dry quench chamber (350–450°F) and a spray dryer (230–300°F). For these temperature ranges, woven fiberglass is typically used as the bag material, although Nomex fabric is also used. A 10% Teflon B coating is the most commonly specified, with acid-resistant coating also used.

The bag sizes differ substantially for the two types of filters. Reverse-air filters generally employ bags 8 inches in diameter by up to 24 feet long. Pulse-jet bags are usually 6 inches in diameter by 12 to 14 feet long. However, some vendors offer a low-pressure pulse filter with bags up to 24 feet long. The biggest differences in operating parameters

TABLE 8. Fabric Filter—Design Parameters

	Reverse Air	Pulse Jet
Operating temperature, °F	230–450	
Type of fabric	Woven fiberglass	
Fabric coating	10% Teflon B or acid Resistant	
Fabric weight, oz/yd^2	9.5	16 or 22
Bag diameter, inches	8	6
Net air-to-cloth ratio	1.5–2.0:1	3.5–4.0:1
Minimum compartments	6	4
Overall pressure drop, in. w.g.	4–6	8–10
Estimated Bag Life, Years	3–4	1.5–2

are in the air-to-cloth ratio and system pressure drop. Pulse-jet filters generally operate at double the air-to-cloth ratio that reverse-air filters do and at nearly double the pressure drop. This results in more frequent bag cleaning and a substantially shorter bag life.

The main advantages of a pulse-jet fabric filter are a lower capital cost and a smaller footprint. However, because of the shorter bag life and higher pressure drop, the pulse-jet filter generally has a higher total evaluated cost for plants exceeding 15 years of life. A reverse-air filter typically has lower particulate emissions when compared with a pulse-jet filter.

The majority of fabric filter applications are as part of an acid-gas cleaning system and incorporate specific design features for operating after a spray dryer. The flue gas, after a spray dryer has been cooled (240–300°F), has a high moisture content (12–20%), is closer to the dew point (80–160°F), and may have a higher particulate loading. These flue gas conditions can lead to severe corrosion and baghouse plugging.[8]

Corrosion control is accomplished by insulation design, control of air in-leakage into the filter, hopper heating, and, in some instances, coating of the fabric filter internals with an acid-resistant material. Insulation specifications usually require a minimum of 4 inches with double lapping on side panels and with the insulation extending into the hopper crotch areas. Air in-leakage is controlled by good quality control during erection and by minimizing the number of openings into the filter. Hopper heating is used to maintain the hopper skin temperature at the flue gas temperature to prevent cold spots and to aid in maintaining product flowability.

As part of an acid-gas cleaning system, the fabric filter also acts as a reactor to aid in acid-gas absorption, especially for sulfur dioxide. Sulfur dioxide in the flue gas is absorbed by alkaline material in the filter cake on the bags. Therefore, when a bag is freshly cleaned, SO_2 absorption decreases. In order to minimize this impact on overall absorption, the number of bags being cleaned simultaneously should be minimized. This can be accomplished by increasing the number of compartments. A minimum of six

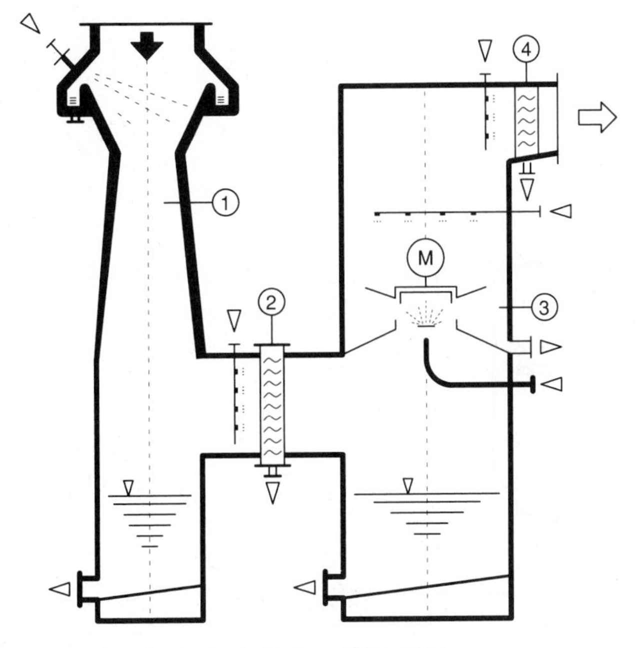

FIGURE 4. Saturator Venturi with Radial-Flow Scrubber (Courtesy Lurgi Corp. Reprinted with permission.)

compartments is generally specified for acid-gas cleaning systems.

Wet Scrubbers

Wet scrubbers are typically employed as part of a two-stage flue gas cleaning system downstream of an ESP. They function as a particulate-removal polishing stage and as an acid-gas absorber. A venturi scrubber followed by a packed or tray tower is commonly used, however, other types of wet scrubbers, such as charged droplet scrubbers, are also used. Figure 4 shows a typical wet scrubber design used for both particulate and acid-gas control.

Typically, water is recycled in the venturi stage to achieve particulate removal. Hydrogen chloride present in the gas would also be removed in this stage. Additional particulate and acid-gas removal can take place in the second scrubbing stage. Absorption of SO_2 is enhanced in this stage by maintaining a recirculating solution pH in the range of about 6.5 to 9 through addition of caustic (sodium hydroxide). A blowdown stream is maintained for each stage to control the recirculating solution solids content. Typical design parameters for refuse-incinerator wet scrubber applications are presented in Table 9.

The venturi section is subjected to severe corrosive conditions due to the low circulating solution pH, the high hydrochloric-acid concentration, and the presence of small amounts of sulfuric, nitric, and hydrofluoric acids. The scrubber inlet temperature may be as high as 450°F, which precludes the use of corrosion-resistant resins, therefore,

TABLE 9. Wet Scrubber—Design Parameters

	Venturi Stage	Absorber Stage
Gas velocity, ft/s	90–150	6–10
Pressure drop, in. w.c.	40–70	4–8
L/G, Gal/Kacfm	10–20	20–40
Scrubbing medium	Water	Caustic
Solution pH	< 1–2	6.5–9
Materials of construction	High-alloy steel (eg., Inconel, Hastelloy)	FRP Lined carbon steel

high-alloy steels are typically specified as the materials of construction. In cases where the inlet flue gas contains high levels of particulate matter, the venturi section may be lined with a corrosion-resistant material such as bricks. The venturi section typically is equipped with a set of emergency quench nozzles to ensure that the flue gas temperature leaving the venturi is maintained at an acceptable level for the absorber-stage materials of construction.

The absorber stage may be a packed tower, a tray tower, or a radial flow tower, as shown in Figure 4. The materials of construction for the absorber are typically fiberglass-reinforced plastics (FRP), although carbon-steel vessels lined with rubber or a corrosion-resistant resin material are also used.

ACID-GAS CONTROLS

Control of refuse incinerator acid-gas (HCl, SO_2, SO_3, and HF) emissions is achieved by dry sorbent injection, spray dryer absorption, or wet scrubbing. Each of these types of technologies has been successfully applied to meet existing emissions regulations, however, as emissions limitations become more stringent, the trend is toward spray dryer absorption and wet scrubbing.

Dry Sorbent Injection

Dry sorbent injection (DSI) involves the addition of an alkaline material—usually hydrated lime, $Ca(OH)_2$, or soda ash, $Na_2(CO_3)$—to the gas stream to react with acid gases present, thus producing a salt that is collected in a particulate-collection device. This very simple process can capture up to 90% of the HCl present in the flue gas and about 50% of the SO_2. However, stoichiometric ratios (equivalents of alkali added per equivalents of acid in the flue gas) are high, typically of the order of 2 to 4. Therefore, simple DSI applications are normally limited to small facilities with moderate emissions control requirements.

The overall acid-gas control efficiency of DSI can be improved and reagent consumption decreased by:

- Increasing flue gas humidity
- Recycling reaction products into the flue gas stream

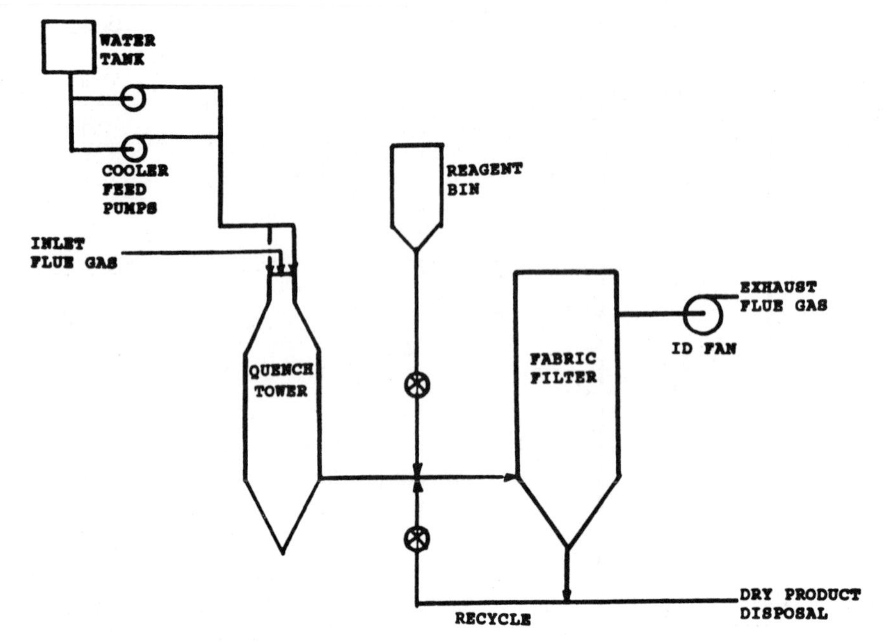

FIGURE 5. Humidification—Dry Sorbent Injection Process

Increasing the flue gas relative humidity can be accomplished by cooling the flue gas using heat exchangers or by quenching the flue gas using water sprays. Both approaches are commercially applied, however, the use of a quench chamber predominates. Figure 5 presents a simplified process flow diagram for a humidification DSI process.

Flue gas from the incinerator enters a three- to five-second retention time cooling tower (or dry quench chamber), where water is sprayed into the gas to lower the temperature. The flue gas temperature leaving the cooling tower is maintained at a temperature high enough to ensure that all water droplets evaporate (300–350°F). Dry reagent is then mixed with the flue gas via pneumatic transport systems or eductor venturis. The reagent reacts with acid gases prior to removal in a dust collector (typically a fabric filter). A portion of the collected reaction products in some cases is reinjected to increase acid-gas removal and decrease reagent consumption. Humidification and reagent-injection steps can also be carried out together in specially designed reactors. This type of process can achieve greater than 95% HCl removal and 90% SO_3 removal at stoichiometric ratios between 1 and 2.

Spray Dryer Absorption

Spray dryer absorption (SDA) has been widely applied for refuse incinerator emissions control and has been specified as best available control technology (BACT) in a number of air permits. The SDA process combines a spray dryer with a dust collector. Reagent addition, flue gas humidification, and some acid-gas absorption take place in the spray dryer. Additional acid-gas absorption and collection of the dry fly-ash reaction products mixture take place in the dust collector. The SDA process is capable of achieving very high removal efficiencies for all acid gases (99+% HCl, 95% SO_2, 99+% SO_3, 95%HF), as well as for the removal of trace metals and organic compounds at stoichiometric ratios between 1 and about 1.8. Figure 6 is a simplified flow diagram for the SDA process.

Incinerator flue gas enters the spray dryer, where it is contacted by a cloud of finely atomized droplets of reagent (typically, a hydrated lime slurry). The flue gas temperature is decreased and the flue gas humidity is increased as the reagent slurry simultaneously reacts with acid gases present and evaporates to dryness. In some systems, a portion of the dried product is removed from the bottom of the spray dryer, while in others, it is carried over to the duct collector. Collected reaction products are sometimes recycled to the feed system to reduce reagent consumption.

Several different spray-dryer design concepts have been employed for incinerator SDA applications. These include single rotary, multiple rotary, and multiple dual fluid nozzle atomization; downflow, upflow, and upflow with a cyclone precollector spray dryers; and single and multiple gas inlets. Flue gas retention times range from 10 to 18 seconds and flue gas temperatures leaving the spray dryer range from 230°F up to 300°F.

Generally, the lower the spray dryer outlet temperature, the more efficient will be the acid-gas absorption. The minimum reliable operating outlet temperature is a function of the spray dryer and dust collector design and the composition of the dry fly-ash reaction product mixture. The spray dryer outlet temperature must be maintained high enough to ensure complete reagent evaporation and the production of a free-flowing product. Low outlet temperature operation requires efficient reagent atomization, good

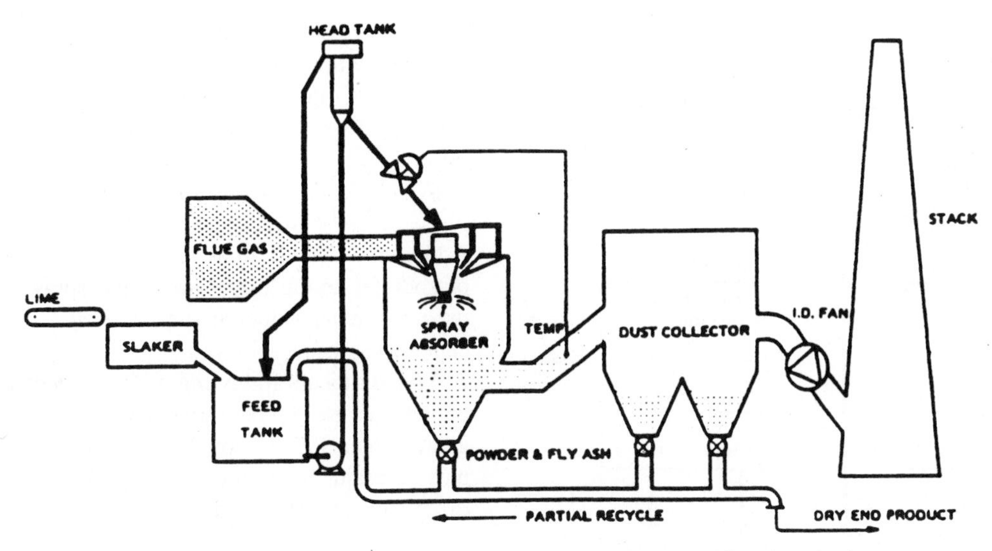

FIGURE 6. Spray Dryer Absorption Process (Courtesy Niro Atomizer)

gas dispersion and mixing, adequate residence time for drying, and design of the dust collector to minimize heat loss and air in-leakage.

The dust collector downstream of the spray dryer may be an ESP, a reverse-air baghouse, or a pulse-jet type of baghouse. The selection of a specific type of dust collector is dependent on such site-specific factors as particulate emission limits, overall acid-gas removal requirements, and project economics. Each of these dust-collection devices offers process advantages and disadvantages that are evaluated on a site-specific basis. Generally, where high acid-gas control is required (95+%HCL, 85+%SO_2), a baghouse is utilized as it is a better reactor than an ESP.

Whether a fabric filter or an ESP is selected as the dust collector, minimization of heat loss from the dust collector to avoid corrosion and increased product stickiness is a prime design consideration. Four methods employed to achieve this are as follows:

- Insulation, to control heat loss
- Control of air in-leakage, to minimize cold spots
- Hopper heating, to maintain product temperature
- Operating procedures to maintain product flowability and minimize cold areas

The end product of the SDA process is a fine hygroscopic material with a significant soluble fraction. It tends to be stickier than MSW fly ash and more difficult to convey and store. Major end-product constituents include:

- Fly ash
- Calcium hydroxide
- Calcium chloride
- Calcium carbonate
- Calcium sulfite
- Calcium sulfate
- Calcium fluoride
- Moisture

The calcium chloride formed at typical spray dryer outlet temperatures is a mixture of mono- and dihydrates ($CaCl_2 \cdot H_2O$ and $CaCl_2 \cdot 2H_2O$) and at lower temperatures will absorb moisture until it reaches the hexahydrate form ($CaCl_2 \cdot 6H_2O$) and melts. Therefore, it is necessary to keep the product from being exposed to cold and/or moist air. This is accomplished by proper design of the product conveying and storage systems.

Wet Scrubbing

Wet scrubbing systems are capable of achieving high acid-gas removal efficiencies and have been applied to a large number of installations in Europe. Typical wet scrubbing applications include two-stage scrubbers located downstream of an ESP. The first stage is used for HCl removal and the second for SO_2 removal. Water is used to capture the HCl and either caustic or hydrated lime is used for SO_2 capture. Figure 4 shows a typical two-stage wet scrubber and Figure 7 shows a process flow diagram for an application of wet scrubbing with fly-ash treatment.

In this process, the HCl stream from the first scrubbing stage is pumped to a fly-ash leaching tank, where it is used to leach out heavy metals from the fly ash collected in the dust collector. After leaching, residual fly-ash solids are either disposed of or used in construction applications. The heavy-metals–bearing HCl stream is then treated alone or with the sodium sulfite–sulfate solution from the second scrubber stage in a neutralization/precipitation stage to concentrate the heavy metals and produce salt-containing wastewater for disposal. When lime is used in the SO_2 absorption section of the scrubber, the calcium sulfite slurry can be oxidized to calcium sulfate (gypsum) for utilization.

Wet scrubbers offer some advantages:

- They are relatively inexpensive to install and require relatively small plot space.

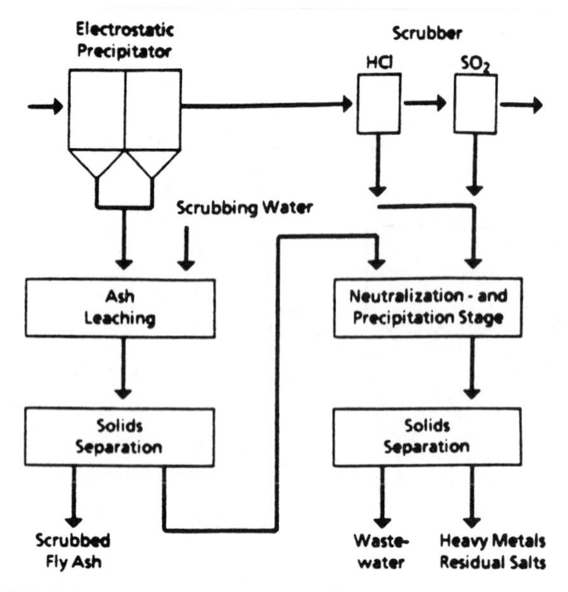

FIGURE 7. Wet Scrubbing with Ash Treatment

- They are capable of achieving very high removal efficiencies for acid gases (99+%HCl, 95+%SO_2).
- They are capable of high removal efficiencies for many volatile trace compounds.
- They require the lowest reagent stoichiometrics (1.0–1.2) of any of the alternatives considered.

Wet scrubbers also have some disadvantages:

- They produce a wet effluent that requires additional treatment with complex effluent treatment systems.
- Economics and space requirements are not as attractive as for the other alternatives.
- Wet scrubbers are more prone to corrosion problems and may require expensive materials of construction.
- Historically, wet scrubbers have experienced more operating problems and higher maintenance requirements than the alternatives.

HEAVY METALS CONTROL

The primary heavy metals of concern from refuse incinerators (arsenic, cadmium, lead, and mercury) are collected in the particulate control device or in the acid-gas control system. Most of these metals exist as solid particulates at incinerator-exit flue gas temperatures and are collected as particulate matter. However, some arsenic, lead, and mercury compounds exist in the vapor state at incinerator flue gas exit temperatures, and these compounds must be collected by condensation through cooling of the flue gas. This can be accomplished with either an SDA or wet scrubbing process.

In the SDA process, the flue gas cooling takes place rapidly in a cloud of finely atomized droplets. These droplets serve as sites on which metals can condense or into which they can be absorbed. The condensed metal is then removed with the reaction products in the downstream dust collector. Collection efficiencies for arsenic and lead at typical SDA system operating temperatures are greater than 90%.

A significant fraction of mercury remains in the vapor phase, even at SDA system outlet temperatures of 250°F. Additions of small amounts of powdered activated carbon or sodium sulfide upstream of the spray dryer have been used to enhance mercury control and greater than 90% capture has been achieved.[9]

Wet scrubbers following a dust collector operate at saturated flue gas temperatures (150–180°F) and can achieve greater than 90% removal of mercury. They can also remove a major fraction of the other metals that may escape the particulate control device.

PCDD/PCDF CONTROL

The PCDD/PCDF emissions are controlled by good combustion practices that inhibit their formations and by particulate and acid-gas controls. Combustion temperatures above 1800°F for more than two seconds are specified as a method of destroying PCDD/PCDF found in the waste stream and their precursors formed from the combustion of other organic and chlorine-containing compounds. However, some PCDD/PCDF compounds may still form downstream of the incinerator on the surface of fly ash at temperatures from 500°F to 700°F.

Control of PCDD/PCDF compounds found in the flue gas leaving the incinerator is achieved by ESPs operating below 450°F or by acid-gas control systems. Acid-gas control systems achieve a higher PCDD/PCDF capture efficiency because of their reduced outlet temperatures and the large droplet surface area available for adsorption to take place. Capture efficiencies of PCDD/PCDF of up to 99% can be achieved and total emissions can be reduced to less than about 10 ng/nm^3.

References

1. *EPA Characterization of Municipal Solid Waste in the United States 1960–2000 (update 1988),* Final Report, U.S. Environmental Protection Agency, Office of Solid Waste and Emergency Responses, Franklin Associates Ltd., March 30, 1988.
2. *Decision-Makers Guide to Solid Waste Management,* U.S. Environmental Protection Agency Office of Solid Waste and Emergency Responses, EPA/530-500-89-072, November 1989.
3. *New Source Performance Standards and Emissions Guidelines for Existing Municipal Waste Combustors,* 40 CFR Part 60 Ca, *Federal Register,* Vol. 56, No. 28, February 11, 1991.
4. Incineration brochures, Joy Energy Systems, Inc., Charlotte, NC.
5. D. G. Jones et al., "Two stage De NO_x process test data from Switzerland's largest incineration plant," International Con-

ference on Municipal Waste Combustions, Hollywood, FL, April 1989.
6. W. Panknins, "Research into activated carbon technology on harmful organic substances, heavy metals and NO_x control," Air and Waste Management Association 83rd Annual Meeting, Pittsburgh, June 1990.
7. B. Brown, J. R. Donnelly, T. D. Tarnok, et al., "Dust collector design considerations for MSW acid gas cleaning systems," EPA/EPRI 7th Particulate Symposium, Nashville, March 1988.
8. J. R. Donnelly, "Design considerations for MSW incinerator APC systems retrofit," Air and Waste Management Association 83rd Annual Meeting, Pittsburgh, June 1990.

Bibliography

T. G. Brna and J. D. Kilgroe, "The impact of particulate emissions control on the control of other MWC air emissions," *J. AWMA,* 40(9):1324–1329 (1988).

M. J. Clarke, "Technologies for minimizing the emission of NO_x from MSW incineration," International Conference on Municipal Waste Combustion, Hollywood, FL, April 1989.

D. J. Collins and P. W. Guletsky, "Operating experiences with spray dry scrubbing systems: Case histories of three industrial plants," IGCI Forum 90, Baltimore, March 1990.

J. R. Donnelly and K. S. Felsvang, "Joy/Niro SDA-FGC systems North American and European operating results," International Conference on Municipal Waste Combustion, Hollywood, FL, April 1989.

R. Emery, R. Gregporini, and A. Berst, "Design and start-up of the Imperial resource recovery spray reactor/fabric filter system," IGCI Forum 90, Baltimore, March 1990.

S. Greene, *Municipal Waste Combustion Study: Report to Congress,* U.S. Environmental Protection Agency, NTIS PB 87-206074, 1987.

K. S. Felsvang, T. S. Holms, and B. Brown, "Control of mercury and dioxin emissions from European MSW incinerator by spray dryer absorption systems," AIChE Summer National Meeting, San Diego, August 1990.

J. H. Hahn, H. P. Von Dem Fange, and D. S. Sofaer, "Recent air emissions data from Ogden Martin Systems, Inc., resource recovery facilities which became operational during 1988," Air and Waste Management Association 83rd Annual Meeting, Pittsburgh, June 1989.

K. Masters, *Spray Drying Handbook,* 3rd ed., George Goodwin Ltd., London, 1979.

Municipal Waste Combustion Study: Sampling and Analysis of Municipal Waste Combustors, U.S. Environmental Protection Agency, NTIS PB 87-206124, 1987.

Municipal Waste Combustion Study: Characterization of the Municipal Waste Combustion Industry, U.S. Environmental Protection Agency, NTIS PB 87-206140, 1987.

L. P. Nelson, P. Shindley, and J. D. Kilgroe, "Development of good combustion practices to minimize air emissions from municipal waste combustors," International Conference on Municipal Waste Combustion, Hollywood, FL, April 1989.

P. Schindler, *Municipal Waste Combustion Study: Emissions Data Base for Municipal Waste Combustors,* U.S. Environmental Protection Agency, NTIS PB 87-206082, 1987.

W. K. Seeker, W. S. Lanier, and M. P. Heap, *Municipal Waste Combustion Study: Combustion Control of MSW Combustors to Minimize Emission of Trace Organics,* U.S. Environmental Protection Agency, NTIS PB 87-206090, 1987.

E. Wheless, "Air emission testing at the Commerce refuse-to-energy facility," Environment Canada MSW Incinerator Workshop, Toronto, October 1987.

HAZARDOUS WASTE

Anthony J. Buonicore, P.E.

One of today's major environmental issues is the proper disposal of hazardous waste. Of all the permanent treatment technologies, properly designed incineration systems are capable of achieving the highest overall degree of destruction and control for the broadest range of hazardous waste streams. Over the past 20 years, significant advances have been made in incineration technology, particularly in the air pollution control systems developed to respond to increasingly more stringent regulation. Today, it is not unusual to find air pollution control equipment representing as much as one third of the total installed cost of the hazardous waste incineration system.

PROCESS DESCRIPTION

A wide variety of incinerator types have been developed to handle wastes. Several of these are adaptable, with minimum modification, for hazardous waste application. The selection of the incinerator type for hazardous waste service is made by considering the amounts, types, and properties of the hazardous waste to be destroyed. Among these are the waste's physical form; whether it is a liquid, sludge, or solid; its constituents, such as water and solids; its heating value; and its chemical composition.

Most hazardous wastes are not similar to fuel oil, natural gas, or coal. Moreover, the incinerator will often be required to perform on a variety of waste streams simultaneously. Fortunately, there is a broad variety of proven incinerator designs, the more common of which are liquid injection, multiple-hearth, rotary kiln, and fluidized-bed incinerators.

Liquid Injection

In order for a waste to be incinerated in a liquid injection incinerator, the waste must be pumpable and atomizable (dispersible into very small droplets). The waste is delivered to the incinerator by a conventional pumping system and passes through a burner into the incineration chamber (see Figure 1). The burner has two components—an atomizing nozzle and a turbulent mixing section wherein atomized waste is mixed with sufficient primary air for complete combustion. The ignitable mixture of atomized waste and air burns. The mixture is then turbulently mixed with addi-

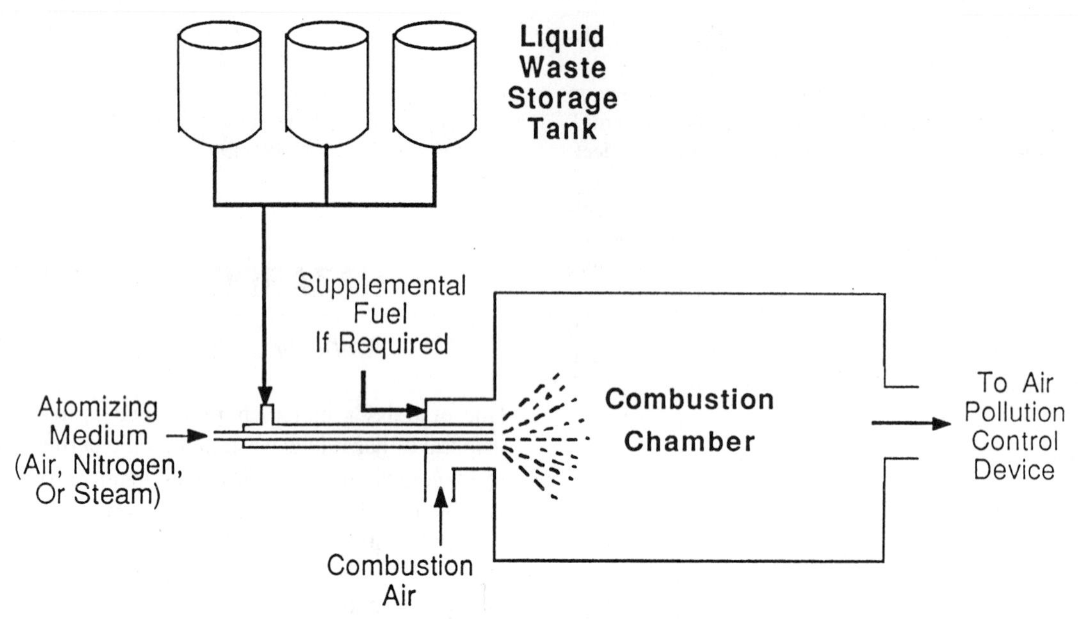

FIGURE 1. Typical Liquid Injection Incinerator

tional air (secondary air) in sufficient quantity to complete the combustion process. Secondary air also helps to moderate bulk temperature so as not to exceed the design limits of the furnace.

The incineration chamber frequently consists of a cylindrical steel shell lined with refractory and mounted in either a horizontal or a vertical position. A boxlike configuration is also common. For hazardous waste, the volume of the chamber is designed to allow the combustion products and excess air, at their bulk temperature, to remain in the chamber for a predetermined time to insure complete destruction of organics.

The most important portion of the liquid injection incinerator is the burner through which the waste is delivered. Once dispersed or atomized, the liquid waste must be thoroughly mixed with sufficient air (primary air) to provide oxygen for combustion. This can be partially accomplished in the atomization process, but it is completed in the turbulent mixing section of the burner where primary air is directed at the atomized liquid through a multiple number of small ports.

A small amount of solid particles in liquid wastes can be tolerated. Liquid feed preparation, for any liquid injection system, should be designed in terms of atomizer nozzle passage-size contraints.

Ignition of the combustible mixture of atomized waste and air is accomplished by its exposure to gases that are already circulating within the burner zone and are above the ignition temperature. It is a regulatory requirement for all types of incinerators that the furnace be brought to a prescribed temperature by conventional fuel before hazardous waste is injected. Instrumentation is provided that will cause the hazardous liquid-waste feed system to shut off automatically if this condition is not met.

Secondary air is provided in sufficient quantity to keep the temperature of the furnace below its maximum design temperature and within a set operating range, for example, from 1600°F to 2200°F. If the waste alone is not sufficient to achieve and maintain required temperatures, auxiliary fuel such as oil or gas is used to achieve the desired temperature.

Liquid injection incinerators are the least versatile of the more common incinerator types; however, liquid injection frequently is part of other incineration systems. For example, liquid injection incinerators can act as secondary combustion chambers on large rotary kiln hazardous-waste incineration systems.

Multiple Hearth

A multiple-hearth incinerator is capable of handling high volumes of waste and may use supplementary fuel. The incineration chamber is almost always a vertical steel cylinder lined with refractory. A number of horizontal platforms are located at various levels in the chamber. The top platform receives, usually from an auger, a continuous charge of waste material that is plowed over by a mechanical arm (rabble arm or rake) and finally moved to a discharge opening, through which the waste falls onto the next lower platform. Any number of platforms or hearths may be designed into the unit. A minimum of six are usually used for combustion of wastes (see Figure 2).

The bottom hearth is usually supplied with overfire air. The other hearths are underfired with hot gases from the hearth(s) below and air. In advanced designs, the combustion air is the fluid that passes through and cools the rabble arms and the shaft to which they are attached. The combustion air is thus preheated.

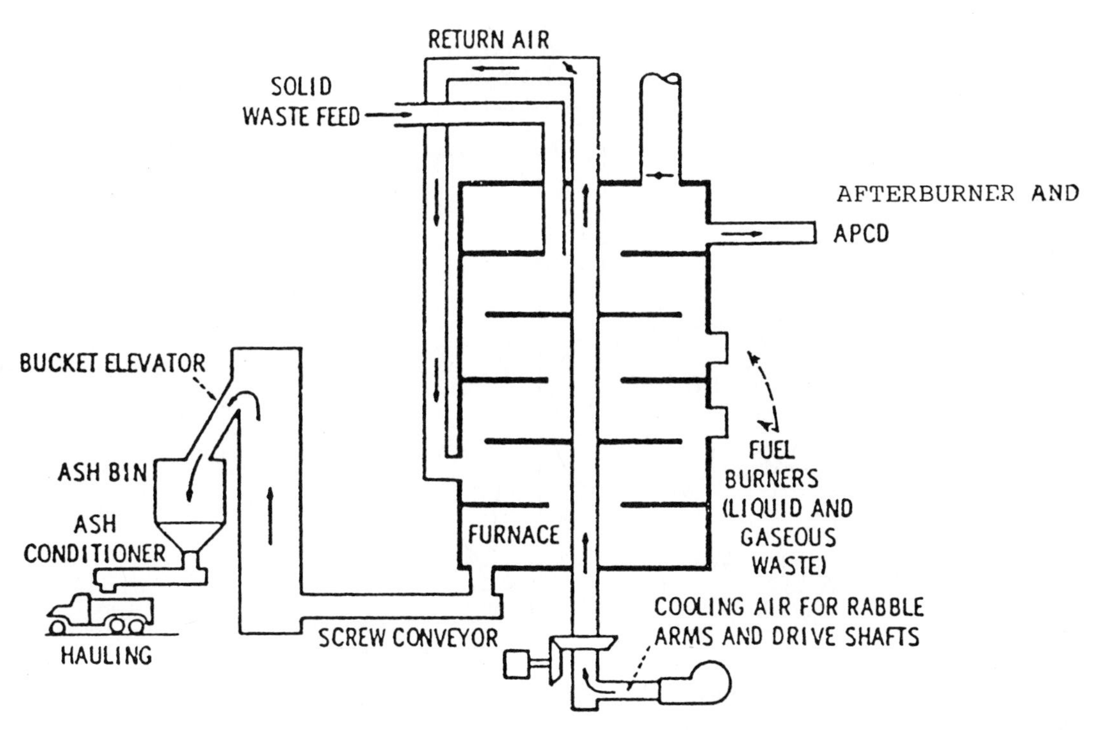

FIGURE 2. Typical Multiple-Hearth Incinerator

A vertical shaft (main shaft) passes up through the center of the chamber and the hearths. This shaft, to which the horizontal rabble arms are attached, is mechanically rotated, causing the rabble arms to sweep across the hearths, plow over the waste, and eventually move the waste to discharge ports. The metal shaft and rabble arms are air cooled.

The multiple-hearth incinerator, because of its inherent features, has three operating zones. The uppermost set of hearths will remove moisture/volatiles at moderate temperatures from top-loaded incoming waste; the middle set of hearths serves as an incineration zone; the lower set of hearths achieves cooling of the ash, which is discharged from the chamber, usually by means of a second auger.

If the waste feed does not contain the required heat content to sustain combustion or achieve a desired temperature, auxiliary fuel is fired into the chamber through side ports. The fuel may be oil, gas, or liquid hazardous waste.

Multiple-hearth incinerators may be designed in various sizes, from 6 to 25 feet in diameter and from 12 to 75 feet high. Upper-zone temperatures, depending on the heat content of the waste and supplemental fuel amount, range from 650°F to 1000°F, mid-zone temperatures from 1500°F to 1800°F, and lower-zone temperatures from 600°F to 1100°F. These incinerators are best suited for incinerating wastes with a high water content and uniform fine size (such as sludge). Bulk solid wastes will not flow downward easily through the hearth openings and could jam the rabble arms.

Although seldom used in conjunction with secondary combustion chambers, a multiple-hearth incinerator operating on hazardous waste may require one. The upper zone, or drying zone, may not be at a temperature sufficiently high to destroy organics completely. Gaseous effluents exit the chamber from this zone and may carry with them volatilized organics.

Rotary Kiln

Generally, the rotary kiln incinerator is considered the most versatile and most durable of the common incinerator types. It can incinerate almost any waste, regardless of type and composition. There are no internal moving parts, maintenance is moderate, and the temperature restrictions on operation are not severe. A basic system is shown schematically in Figure 3.

A rotary kiln is a refractory-lined cylindrical steel shell slightly tilted on its horizontal axis. The shell is usually supported on two or more heavy steel tracks (trundles) that band the shell. These ride on rollers, allowing the kiln to rotate around its horizontal axis. Waste material is "tumbled" through the kiln by gravity as it rotates. The rate of rotation and horizontal angle of tilt determine the amount of time that the waste is held in the kiln (solids residence time).

Rotary kilns can receive solid waste through one end, which is nonrotating, by means of an auger screw or ram feeder. Pumpable, nondispersible waste and sludges may be introduced through a water-cooled tube (wand) and liquid waste may be injected into the kiln through a burner nozzle. As with a liquid injection incinerator, auxiliary fuel can be fired into the kiln chamber. Combustion air (secondary air to the liquid waste and primary air to the sludges and solids)

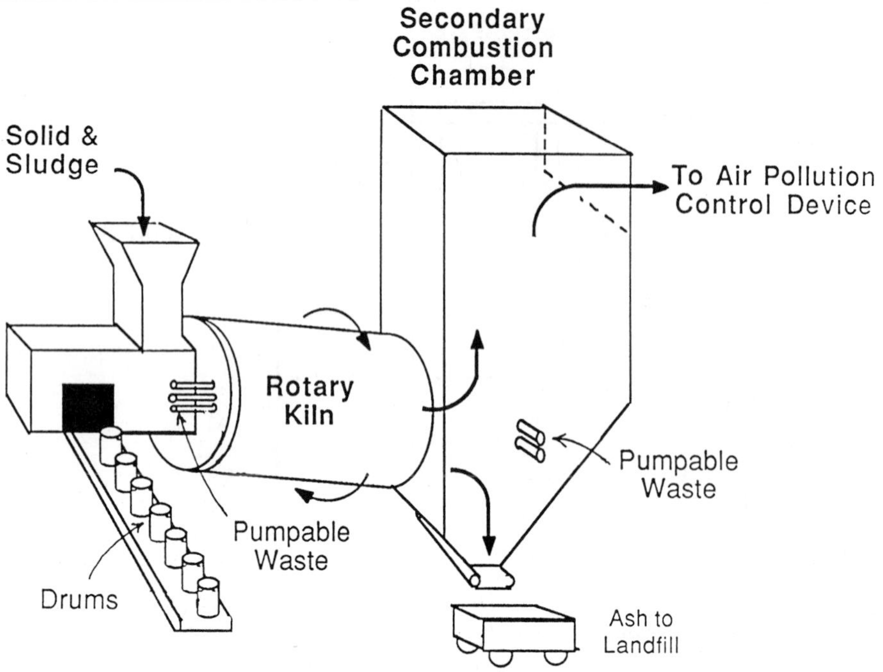

FIGURE 3. Typical Rotary-Kiln Incinerator

can be introduced in a variety of ways to enhance turbulence in the kiln chamber. The rotating motion of the kiln tumbles the sludges and/or solids and allows fresh surfaces to be continually exposed to combustion air.

Length-to-diameter ratios of rotary kilns are typically 2.5–5 to 1. A kiln diameter may range from 4 to 16 feet. The refractory lining of the steel shell of a large rotary kiln is covered with high-temperature fire brick that is resistant to chemical attack and erosion.

Heavy-duty designs can be employed that will allow operation of rotary kilns at temperatures higher than 2600°F; however, alternative designs that are less costly and easier to operate generally restrict operating temperatures to below 2000°F. The high-temperature designs, with sufficient chamber volume, may not require a secondary combustion chamber for incinerating hazardous wastes. The low-temperature primary chamber designs do require a secondary combustion chamber to ensure the destruction of organics.

Neither bulky solid wastes nor fusible ash is seriously detrimental to the operation of a rotary kiln. Usually, if the waste can be fed to the unit, it can be incinerated. Sintered ash typically fractures away from the kiln wall in small fragments and is discharged from the end of the kiln opposite the feed end. At high temperatures, a molten ash discharge is possible.

One design concern with rotary kilns is air leakage around end seals. Because operation is under vacuum, air leakage is into the kiln combustion chamber. The amount of air leaked in has to be considered in the overall design of the combustion air supply rate. Maintenance is required because of the interface between rotating and nonrotating parts and because of thermal shock to the refractory.

Fluidized Bed

The fluidized-bed incinerator has long served the chemical processing industry as a unit operation. Its function has been as a reactor vessel in which intimate contact between gases and particulate solids can be achieved.

The principle of operation is relatively simple. A bed of particulates such as sand is contained in the lower portion of the vertical vessel. Gas is passed upward through the bed at a velocity sufficient to suspend the particles and cause the solids–gas mixture to behave more like a liquid (fluidized). The fluidized particles circulate within the expanded column and mingle with the upward flowing gas. Intimate solid–gas contact is achieved. In chemical process industry applications, the gas usually reacts with the solid.

For hazardous waste incineration, the fluidizing gas is air and the particle bed is an inert material, such as sand, although the waste ash may serve as bed material for high-ash streams (refer to Figure 4). The bed is preheated by overfired or underfired auxiliary fuel. Solid wastes are introduced into the fluidized bed by gravity, conveyors, augers, or pneumatic air transport systems. Sludge or liquid waste is pumped in through a wand. The waste mixes with the bed material and achieves ignition temperature, and the oxygen in the air reacts with the organic constituents of the waste for combustion. Fluidized-bed incinerators can also use calcium as a constituent of the fluidizing material so that acid gas production is reduced.

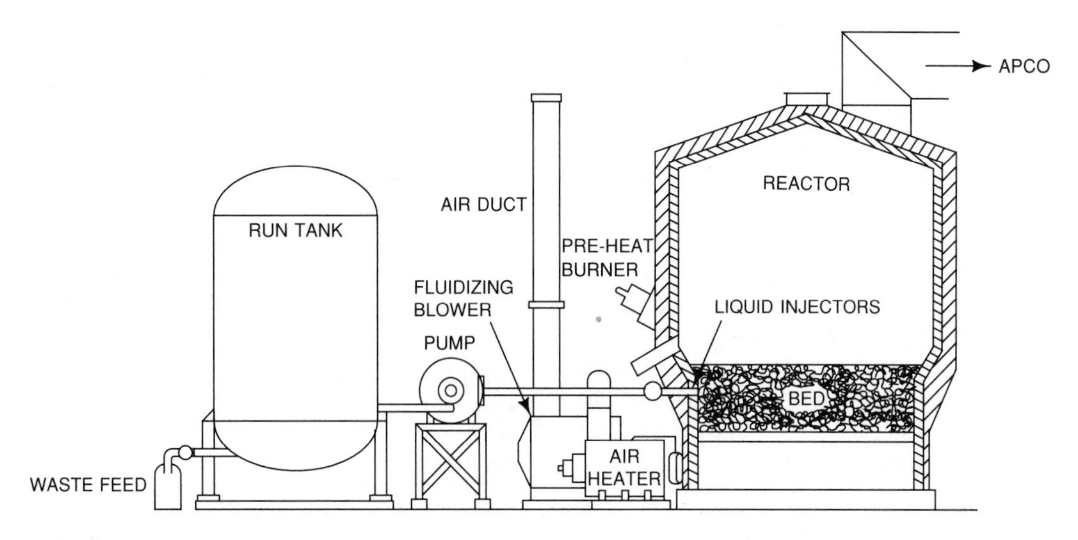

FIGURE 4. Typical Fluidized-Bed Incinerator

In most cases, ash removal from the bed is "over the top" with the flue gas. Because the inert bed material becomes smaller through abrasive action, some smaller particles (usually less than 5% of the ash) will be carried over with the flue gas to downstream air pollution control devices. Bottom ash removal is only applied to wastes with high inert material.

Fluidized-bed incineration chambers usually consist of vertically mounted steel shells lined with either brick or suitable refractories. The chamber typically ranges in size from 4 to 28 feet in diameter. Fluidized-bed depths typically range from 3 to 6 feet.

Fluid beds are not operated above the melting point of the material in the bed or the sintering temperature of the ash. If the materials in the bed form a low-melting-point mixture (eutectic), the temperature constraint may be more severe; however, chemical additives that change the eutectic formed to one with a higher melting point have been demonstrated with certain wastes. Like multiple hearths, fluid beds are also limited in maximum feed-particle size. In this case, the material in the bed must be capable of being fluidized by the air.

The bed portion of the fluidized-bed incinerator typically operates at temperatures between 1200°F and 1700°F. The upper portion of the incinerator, called the freeboard, is usually two to five times the volume of the bed and may operate at slightly higher temperatures, up to 1800°F, because of the combustion of volatized materials.

Fluid-bed incinerators are seldom used with secondary combustion chambers because the volume above the bed acts as an integral afterburner and because fluid beds have inherently longer gas residence times than other incinerator types. A fuel-fired secondary combustion chamber may be used when higher temperatures are required to achieve additional destruction.

Fluid-bed incinerators have few moving parts and there is minimal thermal shock to the refractory, so maintenance requirements may be lower than for other types. However, high-pressure blowers are required to overcome the bed pressure drop. Fluid-bed reactors are a proven chemical processing technology. Adaptation of this technology to the incineration of hazardous waste has been accomplished for some wastes and units are used routinely for the incineration of municipal sewage sludge.

AIR EMISSIONS CHARACTERIZATION

The types and concentrations of contaminants in the flue gases of hazardous waste incinerators depend on incinerator type, the waste being burned, and combustion conditions. Flue gas contaminants can generally be categorized as follows:

- Particulate matter
- Acid gases
- Heavy metals
- Products of incomplete combustion

Particulate matter consists primarily of entrained noncombustible matter in the flue gas, as well as the products of incomplete combustion that exist in solid or aerosol form. Uncontrolled particulate loadings in the flue gas have been found to range from 0.5 to 3.5 grains per dry standard cubic foot.

Acid gases are the flue gas constituents that, when combined with water or water vapor, form acids, including nitric oxides, nitrogen dioxide, sulfur dioxide, sulfur trioxide, hydrogen chloride, and hydrogen fluoride. Hydrogen chloride and sulfur dioxide are often present in concentrations ranging from a few hundred parts per million by volume to several thousand parts per million by volume. The concentrations of nitrogen oxides, hydrogen fluoride, and sulfur trioxide are typically below a few hundred parts per million by volume.

Metallic compounds, such as lead, cadmium, arsenic, nickel, zinc and mercury, are present in the flue gas primarily as oxides and chloride salts. Most of the metallic compounds are in the vapor phase within the incineration system, since these compounds boil or sublime at temperatures around 1800°F. The metallic compounds tend to condense as the flue gas is cooled and become adsorbed onto fine particulate matter (generally submicron in size, i.e., 0.2–0.7 μm). It is possible that a portion of the more volatile metals, such as mercury and lead, may remain in the vapor phase, depending on temperature conditions.

Products of incomplete combustion include carbon monoxide and trace organics. Organic emissions are highly dependent on constituents in the waste feed and combustion conditions. Depending on the temperature, some of the organic constituents may also condense on fine particulates, just as the heavy metals can.

AIR POLLUTION CONTROL SYSTEMS

Air pollution control systems for hazardous waste incinerators generally have been characterized as either wet or dry systems. Wet systems typically utilize either venturi-type scrubbers (conventional, collision type, or ejector type) or wet electrostatic precipitators (ESPs)—such as Ceilcote's Ionizing Wet Scrubber—for particulate control, and packed towers for acid gas control.[1] Representative wet scrubbing facilities are identified in Table 1. Dry systems typically utilize either fabric filters or ESPs for particulate control and spray dryer reactors for acid gas control.[1] Representative dry scrubbing facilities are identified in Table 2. Typical design parameters for air pollution control equipment on hazardous waste incinerators are presented in Table 3.

Representative facilities using either wet scrubbers, fabric filters, or ESPs for particulate control are identified in Tables 4 through 11, along with typical vendor guarantee performance levels for each type of equipment. It should be emphasized that the performance data in the tables are based on what vendors are typically willing to guarantee for their equipment. Actual performance, however, should exceed

TABLE 1. Representative Wet Scrubbing Hazardous Waste Incineration Facilities

Rollins, Deerpark, Tex.; Baton Rouge, La.; Bridgeport, N.J.
CWM, Chicago, Ill.; Port Arthur, Tex.; Sauget, Ill. (unit no. 1)
ENSCO, El Dorado, Ark.
APTUS, Coffeyville, Kans. (in series with dry scrubbing system)
General Electric, Waterford, N.Y.; Pittsfield, Mass.
Chem-Security, Swan Hills, Alberta, Canada
Dow Chemical, Freeport, Tex.; Midland, Mich.; Plaquemine, La.
Robert Ross and Sons, Grafton, Ohio
3M Chemolite, St. Paul, Minn.
McDonald Douglas, St. Louis, Mo.
Kodak, Rochester, N.Y.
L.W.D., Calvert City, Ky.
DuPont, Deepwater, N.J.; LaPlace, La.; Orange, Tex. (wet-dry system)
CIBA-GEIGY, McIntosh, Ala.
BASF, Geismar, La.

TABLE 2. Representative Dry Scrubbing Hazardous Waste Incineration Facilities

Kommunekemi, Nyborg, Denmark
Sakab, Norrtorp, Sweden
Ongelmajate, Riihimake, Finland
Tricil Ltd., Sarnia, Ont., Canada
Calgon Corp., Cattlesburg, Ky.
Trade Waste Incineration, Sauget, Ill. (units no. 2, no. 3, and no. 4)
APTUS, Coffeyville, Kans. (in series with wet scrubbing system)

TABLE 3. Typical Design Parameters for Air Pollution Control Equipment on Hazardous Waste Incinerators

Equipment	Design Parameters
Particulate	
ESPs	SCA = 400–500 ft^2/1000 acfm w = 0.2 fps
Fabric filters	Pulse jet A/C = 3–4:1 Reverse air A/C = 1.5–2:1
Venturi scrubbers	ΔP = 40–70 in. w.c. L/G = 8–15 gal/1000 acfm
Acid gases	
Packed towers	Superficial velocity = 6–10 fps Packing depth = 6–10 feet L/G = 20–40 gal/1000 acfm Caustic scrubbing medium, maintaining pH = 6.5 Stoichiometric ratio = 1.05
Spray dryers	Low temperature: Retention time 15–20 seconds Outlet temperature 250–320°F High temperature: Retention time 25–30 seconds Outlet temperature 350–450°F Stoichiometric ratio (lime) = 2–4

TABLE 4. Representative Hazardous Waste Incineration Facilities with Venturi-Type Wet Scrubbing Systems for Particulate Removal

Conventional
- CWM-Trade Waste, Sauget, Ill., unit no. 1 (40–50 in. w.c. pressure drop venturi)
- Eli Lilly, Ireland (70 in. w.c. pressure drop venturi)
- 3M Chemolite, St. Paul, Minn. (22 in. w.c. pressure drop venturi downstream from a wet ESP)

Collision (Calvert Environmental)
- Rollins, Deer Park, Tex. (45–50 in. w.c. pressure drop venturi)
- Rollins, Bridgeport, N.J. (45–50 in. w.c. pressure drop venturi)

Ejector venturi (Zink Hydro-sonics)
- Ensco, El Dorado, Ark. (steam injection, no fan, 100 in. w.c. pressure drop equivalent)
- Texas Eastman, Longview, Tex. (tandom nozzle, water injection, fan drive, 36 in. w.c. pressure drop on gas side)

TABLE 5. Guarantee Particulate Removal Performance Levels of Venturi-type Wet Scrubbing Systems

Pressure Drop, in. w.c.	Outlet gr/dscf Corrected to 7% Oxygen
30–40	0.06–0.08
40–50	0.03–0.05
50–70	0.02–0.03
70–100	0.015–0.02

TABLE 6. Representative Hazardous Waste Incineration Facilities with Ionizing Wet Scrubbers for Particulate Removal

Rollins, Baton Rouge, La.
APTUS, Coffeyville, Kans. (two stage)
SCA, Chicago, Ill. (two stage)
CWM, Port Arthur, Tex. (four stage)
Oak Ridge Gaseous Diffusion Plant, Oak Ridge, Tenn.
General Electric, Waterford, N.Y. (two stage)
General Electric, Pittsfield, Mass.
Chem-Security, Swan Hills, Alberta, Canada (three stage)
Dow Chemical, Freeport, Tex.
Robert Ross and Sons, Grafton, Ohio

TABLE 7. Guarantee Particulate Removal Performance Levels of Ionizing Wet Scrubbing Systems

No. of Stages	Outlet gr/dscf Corrected to 7% Oxygen
One	0.07–0.08
Two	0.05–0.07
Three	0.03–0.05
Four	0.02–0.03

TABLE 8. Representative Hazardous Waste Incineration Facilities with Electrostatic Precipitators for Particulate Removal

Kommunekemi, Nyborg, Denmark (two fields)
Sakab, Norrtorp, Sweden (three fields)
PPG, Circleville, Ohio (three fields)

TABLE 9. Guarantee Particulate Removal Performance Levels of Electrostatic Precipitators

No. of Fields	Outlet gr/dscf Corrected to 7% Oxygen
Two	0.02–0.04
Three	0.01–0.02
Four	0.008–0.01

TABLE 10. Representative Hazardous Waste Incineration Facilities with Fabric Filters for Particulate Removal

Ongelmajate, Riihimake, Finland
Tricil, Sarnia, Ont., Canada
CWM-Trade Waste, Sauget, Ill., units no. 2, no. 3, and no. 4
Calgon, Big Sandy Plant, Cattlesburg, Ky.
Marine Shale Processors, Amelia, La.
APTUS, Coffeyville, Kans.
DuPont, Sabine River Works, Orange, Tex.

TABLE 11. Guarantee Particulate Removal Performance Levels of Fabric Filters

Bag Type	Outlet gr/dscf Corrected to 7% Oxygen
Fiberglass bags with acid-resistant coating	0.007–0.015
Goretex Bags	0.004–0.01

TABLE 12. Fabric Filters versus Electrostatic Precipitators

Criterion	Fabric Filter	Electrostatic Precipitator
Total particulate removal	Higher	Adequate
Fine particulate removal	Better	Adequate
Relative sensitivity to particulate characteristics	No	Yes
Contribution to acid gas removal capability	Yes (15–20%)	No
On-line maintenance capability	Yes	No
Reliability	Adequate	Higher
Flue gas temperature limitations	Yes	No

guaranteed performance, the extent being dependent upon what precisely is being burned. A comparison of fabric filters and ESPs for particulate removal in the dry systems is presented in Table 12. A comparison of wet versus dry scrubbing systems in hazardous waste incinerators is presented in Table 13.

Recently, hybrid systems (wet–dry combinations) have come onto the market with a goal of achieving the highest possible acid gas removal efficiency and the lowest possible particulate emission rate. A hybrid system, for example, is now operated by APTUS Environmental Services in Coffeyville, Kansas, and another has been proposed for a commercial hazardous waste incinerator in Massachusetts. The APTUS system consists of a spray dryer–fabric-filter system in series with a saturator, a two-stage cross-flow acid gas absorber, and a two-stage ionizing wet scrubber. Recently collected performance data are presented in Table 14.[2] The system proposed in Massachusetts included a spray dryer for evaporation and some gas absorption, a fabric filter for particulate removal (and some additional gas

TABLE 13. Wet versus Dry Scrubbing

Criterion	Wet	Dry
Hydrogen chloride removal	Better	Adequate
Sulfur oxide removal	Better (caustic)	Adequate (lime)
Particulate removal	Adequate	Significantly better
Condensed metal (fine particulate) removal	More condensation; less efficient fine particulate removal	Less condensation; more efficient fine particulate removal
Waste disposal requirements	Wastewater blowdown	Dry solids
Stack plume	Visible	Relatively clear
Flue gas reheat	May be required	Not necessary
Corrosion potential	Higher	Lower
Space Requirements	Lower (assuming no wastewater treatment facility necessary)	Higher

absorption), and a multistage gas scrubber (refer to Figure 5). The multistage gas scrubber consists of a quench section (stage 1), followed by two packed beds (stages 2 and 3) for final acid gas removal and a multiple venturi ring jet fourth stage to remove any remaining particulate, including any that condenses at the lower wet scrubber temperature. The venturi scrubber backup system also provides insurance protection in the event of a bag breaking in the fabric filter. There is minimum liquid blowdown from this hybrid system. Blowdown from the wet scrubber system is sent to the spray dryer for liquid evaporation. This proposed incineration facility would also use ammonia injection for nitrogen oxide control.

In summary, the following conclusions can be drawn about air pollution control equipment on hazardous waste incinerators.

1. Wet systems, dry systems, and hybrid (wet–dry combination) systems can meet current RCRA requirements for particulate and hydrogen chloride removal.
2. Dry systems are capable of achieving the highest particulate removal efficiency.
3. Wet systems are capable of achieving the highest acid gas removal efficiency.
4. Hybrid systems appear capable of achieving the lowest particulate emission rate and highest acid gas removal efficiency.

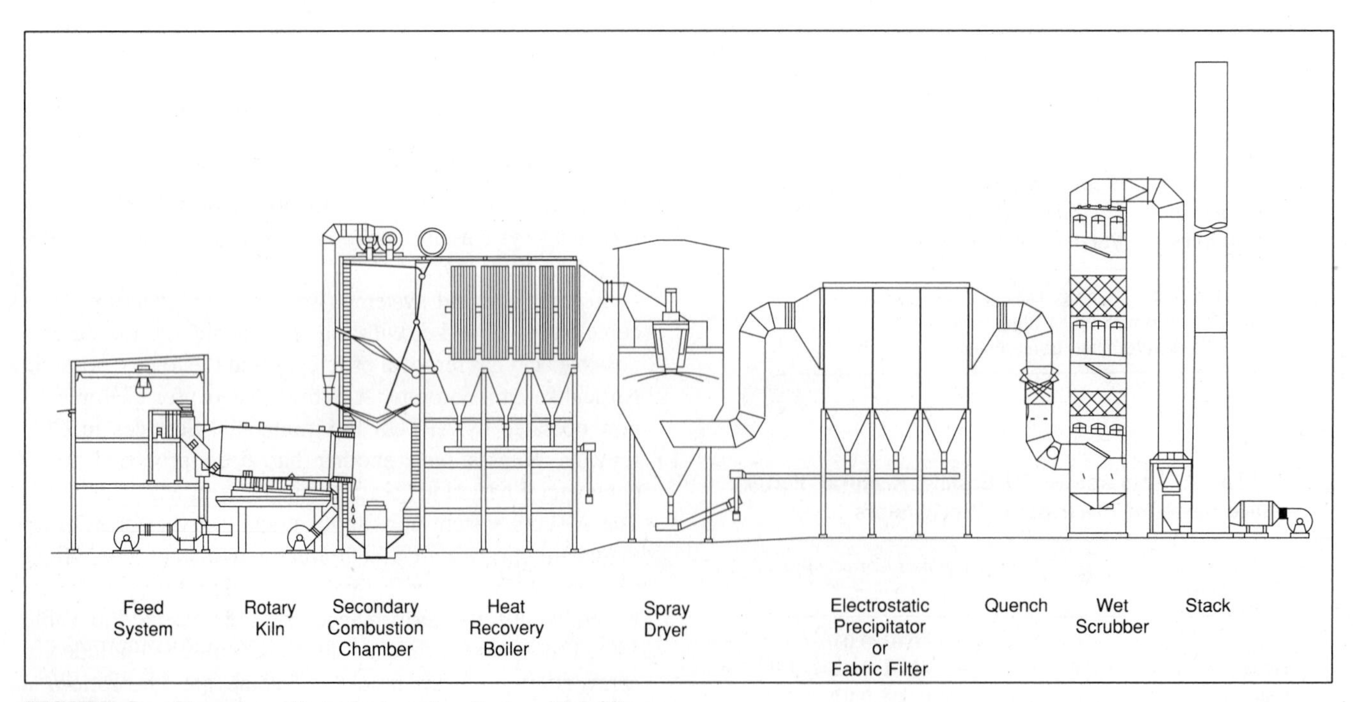

FIGURE 5. Hazardous Waste Incineration System (Hybrid)

TABLE 14. Performance of Hybrid Air Pollution Control System at APTUS Facility in Coffeyville, Kans.

	Condition[a]	Spray Dryer Inlet	Fabric Filter Outlet	Scrubber Outlet
Particulates, gr/dscf at 7% oxygen	1	2.43	0.0087	0.0054
	2	2.08	0.0477	0.0075
HC1, ppmv	1		5,590	1.15
	2		21,200	38.3
Metals, μg/dscf at 7% oxygen				
Arsenic	1	15,009	2.36	1.58
	2	1,763	0.389	0.113
Cadmium	1	357	3.36	2.98
	2	127	5.14	1.74
Chromium	1	2,167	12.1	6.78
	2	921	15.0	4.08
Lead	1	87,675	5.42	4.44
	2	76,253		27.8
Mercury	1	23,622		208
	2	24,075		15.3
Nickel	1	536	9.73	3.54
	2	903	21.1	8.65
Zinc	1	179,932	129	44.7
	2	77,589	340	0.338

[a]Test condition 1 average feed over tests: 3454 lb/h total solids, including 487 lb/hr contaminated soil and 2730 lb/h capacitors; 7302 lb/h total liquids, including 2432 lb/h PCB liquid; 699 lb/h total chlorine in feed.

Test condition 2 average feed over tests: 1853 lb/h total solids, including 682 lb/h contaminated soil and latex paint and 1054 lb/h capacitors; 5993 lb/h total liquids, including 4223 lb/h PCB liquid; 1855 lb/h total chlorine in feed.

References

1. A. J. Buonicore, "Experience with air pollution control equipment and continuous monitoring instrumentation on hazardous waste incinerators," *J. Hazardous Materials,* 22:233–242 (1989).
2. Radian Corp., *Draft Test Report: A Performance Test on a Spray Dryer, Fabric Filter and Wet Scrubber System,* DCN: 89-232-011-034-06, prepared for Office of Solid Waste, U.S. Environmental Protection Agency, Washington, DC, October 1989.

Bibliography

Guidance on Metals and Hydrogen Chloride Controls for Hazardous Waste Incinerators, Volume IV of the Hazardous Waste Incineration Guidance Series, Office of Solid Waste, U.S. Environmental Protection Agency, Washington, DC, March 1989.

Hazardous Waste Incineration: A Resource Document, American Society of Mechanical Engineers, New York, 1988.

E. T. Oppelt, "Incineration of hazardous waste: A critical review," *J. Air Pollution Contl. Assoc.,* V37(5):558–586 (May 1987).

A. Trenholm, et al., *Total Mass Emissions from a Hazardous Waste Incinerator,* EPA/600/S2-87/064, Environmental Protection Agency, November 1987.

A. Trenholm, et al., *Performance Evaluation of Full-Scale Hazardous Waste Incinerators,* EPA/600/S2-84/181, Environmental Protection Agency, May 1984.

MEDICAL WASTE INCINERATION

Anthony J. Buonicore, P.E.

Of the 160 million tons of U.S. waste generated each year, approximately 3.2 million tons are medical waste generated by hospitals. It is estimated that 10–15% of this hospital waste is infectious. Most states have laws that prohibit the disposal of infectious waste in landfills. To qualify for disposal, wastes first must be rendered innocuous. The two most commonly used methods to accomplish this today are autoclaving and incineration, with incineration often the most practical option. Incineration can reduce waste volume by as much as 90%. Hence the volume and cost of ultimate disposal in a landfill can be reduced significantly. An additional benefit of incineration is that the system can be designed with heat recovery to supply a portion of a hospital's steam or hot water requirements. Landfill restrictions, combined with a new "cradle-to-grave" tracking system under the Medical Waste Tracking Act of 1988, are further accelerating the interest in medical waste incineration.

Hospital waste is characteristically heterogeneous, consisting of objects of many different sizes and composed of many different materials. The daily activities and procedures in a hospital can vary dramatically from day to day, thus making it difficult to predict what will be thrown away. Very little data are available on the composition of hospital waste. This may be due to the fact that the amount of sampling and the number of chemical analyses required to generate representative characterization data would be extensive and costly to develop.

PROCESS DESCRIPTION

The primary functions of hospital waste incineration are to render the waste innocuous and to reduce its size and mass. These objectives are accomplished by exposing the waste to high temperatures over a period long enough to destroy threatening organisms and by burning all but the incombustible portion of the waste.

The design of a hospital waste incinerator, like any combustion system, requires consideration of a number of interrelated factors, including residence time, temperature, and turbulence, along with sufficient oxygen. Other factors that can influence combustion performance include fuel feeding patterns, air supply and distribution, heat transfer, and ash disposal. Some of the particular characteristics associated with hospital wastes that must be considered by the incineration system designer and equipment operator include:

- Fuel of nonhomogenous and variable composition. The physical and chemical compositions of hospital waste are highly variable. Furthermore, the waste feed consists of chemically diverse articles of different size and shape. Hospital waste is seldom preprocessed; it is burned in bulk on a mass feed basis. Nonhomogenous and variable compositions must be accounted for in system design and operation to ensure that these factors do not pose problems in feeding, flame stability, particle entrainment, and emission control.
- Variable ash content. Hospital waste contains varying amounts of glass, metals, and ceramics that are not consumed in the combustion process. Fluctuations in ash composition and combustion temperatures can lead to clinker formation, slagging, and fouling in some systems. To avoid these problems, primary combustion chamber temperatures are generally maintained below 1800°F. However, this tends to reduce carbon burnout and the overall energy utilization efficiency.
- Low heating value. Hospital wastes often have low heating values because of their high moisture content. This can cause flame stability problems, and, in some cases, it may become necessary to fire an auxiliary fuel to maintain proper combustion conditions. Alternatively, dry waste batches (especially those with a high plastics content and, therefore, a high heating value) can produce high flame temperatures that can result in overheating of the hearth or other combustion system components. To avoid these problems, the combustion conditions (principally excess air, air distribution, and auxiliary fuel firing rate) must be controlled closely.
- Corrosive materials. Hospital wastes may contain varying amounts of fluorine and chlorine, principally from plastics. The acid gases produced from the incineration of these materials can corrode combustion and air pollution control equipment, especially convective heat-transfer tubes. For this reason, the materials of construction of hospital incineration systems must be carefully selected.

Three major types of incinerators currently are used to incinerate hospital wastes in the United States: retort, controlled air, and rotary kiln. The design and operating principles for each are discussed in the following.

Retort Incinerators

Retort incinerators are variable capacity units that are mostly field fabricated. These units also typify older, existing hospital incinerators. They have been referred to as "pyrolitic," "multiple chamber," and "excess air" incinerators in the literature. The units appear from the outside as a compact cube and have a series of chambers and baffles on the inside. The two principal design configurations for retort incinerators are illustrated in Figures 1 and 2.

In retort incinerators, combustion of the waste begins in the primary, or ignition, chamber. The waste is dried, ignited, and combusted using heat provided by a primary chamber burner, as well as by hot chamber walls heated by flue gases. Moisture and volatile components in the waste feed are vaporized and pass, along with combustion gases, out of the primary chamber and through a flame port connecting the primary chamber to the secondary, or mixing, chamber. Secondary air is added through the flame port and is mixed with the volatile components in the secondary chamber. Burners are also fitted to the secondary chamber to maintain adequate temperatures for combustion of the volatile gases. Incinerators designed to burn general hospital waste operate at total excess air levels of up to 300%; if only pathological wastes (i.e., animal and human remains) are combusted, excess air levels near 100% are more common.

For the in-line incinerator configuration, combustion gases pass in a straight-through fashion from the primary chamber to the secondary chamber and out of the incinerator, with 90-degree flow direction changes only in the vertical direction. The other configuration for retort incinerators causes the combustion gases to follow a more "tortuous" path, with 90-degree flow direction changes in both the horizontal and vertical directions. These flow direction changes, as well as contraction and expansion of the combustion gases, enhance turbulent mixing of the air and gases. In addition, fly ash and other particulate matter drop from the gas stream as a result of the direction and gas velocity changes and collect on the floors. Gases exiting the secondary chamber are directed to the incinerator stack.

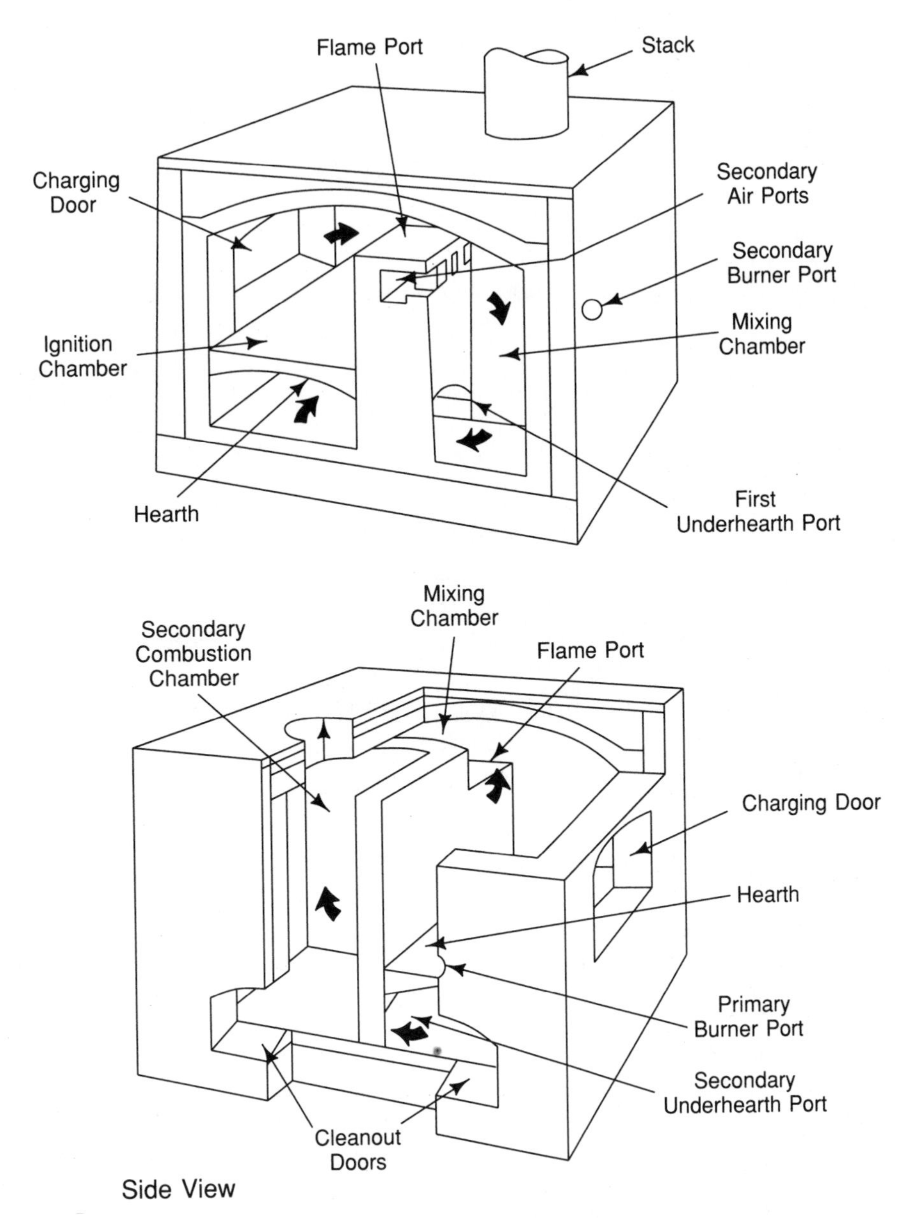

FIGURE 1. Multiple-Chamber Pathological Waste Incinerator

Controlled-Air Incinerators

Controlled-air incineration has become the most widely used hospital waste incineration technology over the past 15 years and dominates the market. This technology is also referred to as "starved air" incineration, "two-stage" incineration, and "modular" combustion. While there are some similarities in operating principles between retort and controlled-air incinerators, the overall equipment design and appearance are quite different, as illustrated in Figure 3.

Like retort incinerators, combustion of waste in controlled-air incinerators occurs in two stages. Waste is fed into the primary, or lower, combustion chamber, which is operated, as the name implies, with less than the full amount of air required for combustion. Under these substoichiometric conditions, the waste is dried, heated, and pyrolized, thereby releasing moisture and volatile components. The nonvolatile, combustible portion of the waste is burned in the primary chamber to provide heat, while the noncombustible portion accumulates as ash. Depending on the heating value of the waste and its moisture content, additional heat may be provided by auxiliary burners to maintain desired temperatures. Combustion air is added to the primary chamber either from below the waste through the floor of the chamber or through the sides of the chamber. The air addition rate is usually 40–70% of stoichiometric requirements. As a result of the low air addition rates in the primary chamber and the corresponding low flue gas velocities and turbulence levels, the amount of

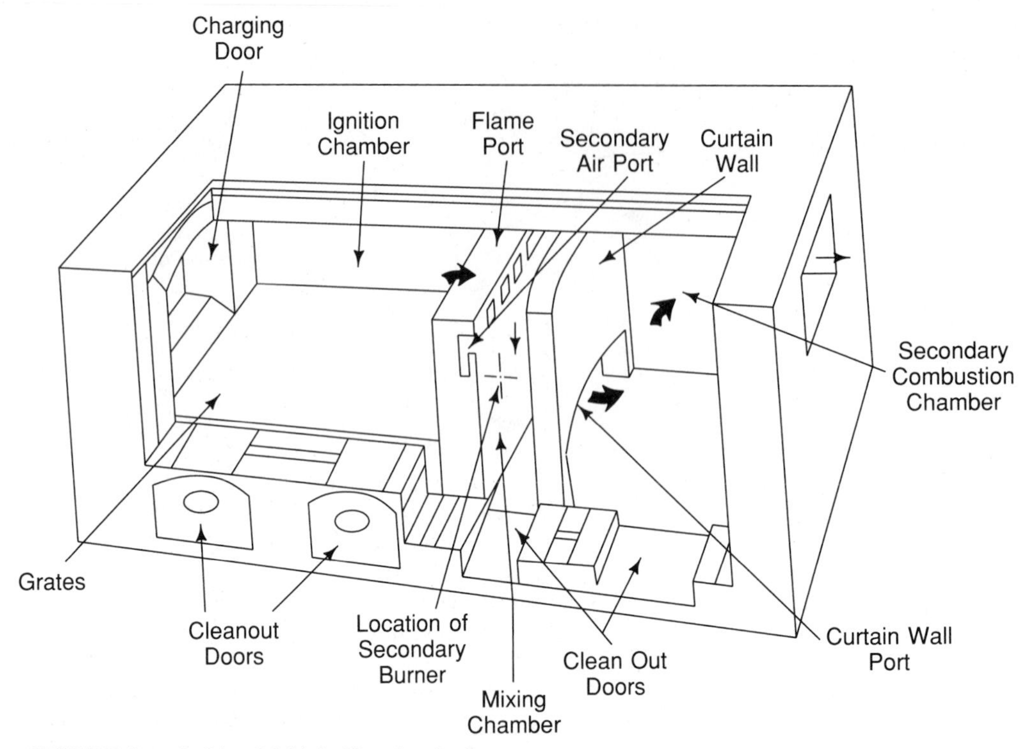

FIGURE 2. In-Line Multiple-Chamber Incinerator

solids entrained in the gases leaving the primary chamber is minimized.

Moisture, volatiles, and combustion gases from the primary chamber flow upward through a connecting section, where they are mixed with air prior to entering the secondary, or upper, combustion chamber. If the primary chamber gases are sufficiently hot, they will self-ignite when mixed with air. A second burner is located near the entrance to the upper chamber, however, to provide additional heat for ignition of the combustible gases and to maintain a flame in the chamber at all times of operation. Mixing of these gases with air is enhanced by the flow direction changes and a contraction–expansion step that the gases undergo as they pass from the lower to upper chambers. The air injection rate in the secondary chamber is generally between 100% and 180% of total stoichiometric requirements (based on the waste feed). From an emissions viewpoint, optimum conditions are generally found between 110% and 140% excess air (i.e., total air added to both chambers of 210–240% of stoichiometric requirements), as illustrated in Figure 4.

The secondary chamber burner is located near the entrance to this section to maximize the residence time of gases at high temperatures in this area. Bulk average gas residence times in the secondary chamber typically range from 0.25 to 2.0 seconds. Design exit gas temperature from the secondary chamber generally ranges from 1600°F to 2200°F. Natural gas or distillate oil is the normal fuel used for both primary and secondary chamber burners. Temperatures in the primary and secondary chambers are monitored by thermocouples and controlled automatically by modulating the airflow to each chamber. Thermocouples are normally located near the exits of these chambers. In the primary (air-starved) chamber, combustion airflow is kept substoichiometric to release volatiles and to keep the gas velocities low to prevent particulates from going out through the secondary combustion chamber. In the secondary combustion chamber, air is generally increased to burn the volatile organics and to create extra turbulence to obtain mixing between the air and volatiles for proper combustion efficiencies. Flue gases exiting the secondary chamber are

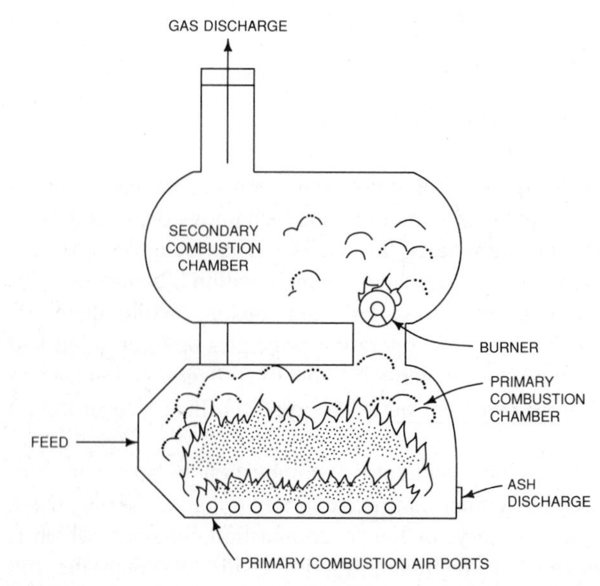

FIGURE 3. Schematic for Controlled-Air Incinerator

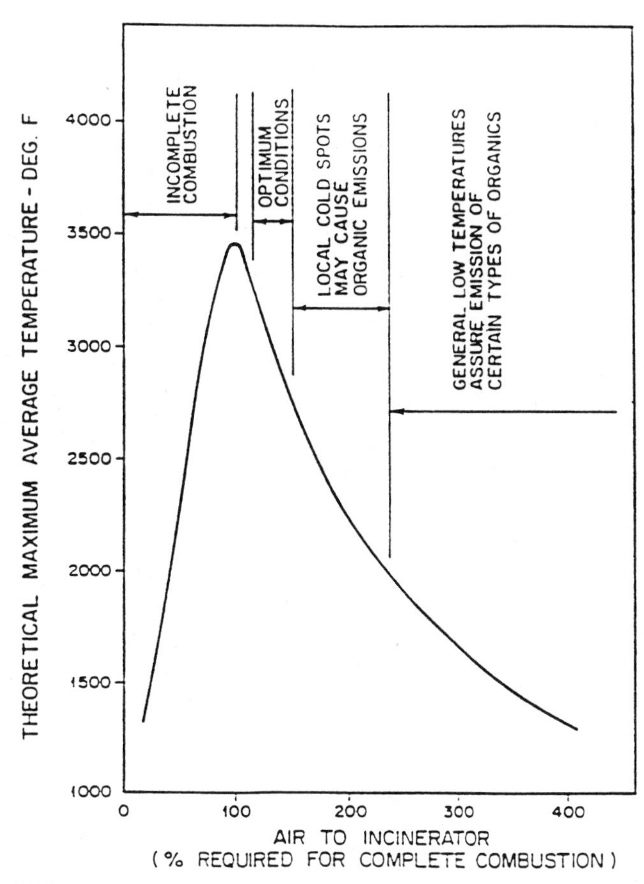

FIGURE 4. Effect of Incinerator Excess Air on Temperature and Emissions

sent either directly to a stack, to air pollution control equipment, or to a waste-heat–recovery boiler.

Both the primary and secondary chambers are usually lined with refractory material. Most chambers are cylindrical; however, some are rectangular. Smaller units (i.e., with waste feed capacities less than 500 lb/h) are usually vertically oriented, with both chambers in a single casing. Larger units generally include two separate horizontal cylinders located one above the other. Some manufacturers offer a third chamber for final air addition to the combustible gases and a fourth chamber for gas conditioning (i.e., gas cooling and condensation of vapors) to minimize effects on downstream heat-recovery equipment or air pollution control equipment.

Waste feed capacities for controlled-air incinerators typically range from about 75 to 6500 lb/h. Capacities for lower-heat-content wastes may be higher since feed capacities are limited by primary chamber heat release rates. Heat release rates for controlled-air incinerators typically range from about 15,000 to 25,000 Btu/h-ft^3.

Rotary Kiln Incinerators

Like other incinerator types, rotary kiln incineration consists of a primary chamber in which waste is heated and volatilized and a secondary chamber in which combustion of the volatile fraction is completed. In this case, however, the primary chamber consists of a horizontal rotating kiln. The kiln is inclined slightly so that the waste material migrates from the waste charging end to the ash discharge end as the kiln rotates. The waste migration, or throughput, rate is controlled by the rate of rotation and the angle of incline, or rake, of the kiln. Air is injected into the primary chamber and mixes with the waste as it rotates through the kiln. A primary chamber burner is generally present for heat-up purposes and to maintain desired temperatures. Both the primary and secondary chambers are usually lined with refractory brick, as shown in the schematic drawing in Figure 5.

Volatiles and combustion gases from the primary chamber pass to the secondary chamber where combustion is completed by the addition of more air and the high temperatures maintained by a second burner. Like other incinerators, the secondary chamber is operated at above-stoichiometric conditions. Due to the turbulent motion of the waste in the lower primary chamber, particle entrainment in the flue gases is higher for rotary kiln incinerators than for the other two chamber incinerator designs previously discussed.

Waste-Feed and Ash-Handling Systems

Feed systems for hospital waste incinerators range from manually operated charging doors to fully automatic systems. Ash-removal systems also range from periodic manual removal of ash by operators to continuous automated quench and removal systems. In general, automated systems are prevalent among large continuously operated incinerators, while manual systems are employed on smaller incinerators or those that operate on an intermittent basis.

For retort incinerators, waste loading is almost always accomplished manually by means of a charging door on the incinerator. The charging door is attached to the primary chamber and may be located either at the end farthest away from the flame port (for burning general wastes) or on the side (for units handling pathological wastes such as large animals or cadavers). As much as 10% of the total air supplied to retort units can be drawn through these charging doors. Ash removal from retort units is accomplished manually with a rake and shovel at the completion of the incinerator cool-down period. Typical operation for a retort incinerator calls for incinerator heat-up and waste charging at the end of the operating day, waste combustion and burnout by the morning, and cool-down and ash clean-out during the following day.

Controlled-air incinerators may be equipped with either manual or mechanical loading devices. For units with capacities of less than 200 lb/h, manual loading through a charging door in the primary chamber typically is the only option. Mechanical loaders, on the other hand, are standard features for incinerators with capacities above 500 lb/h

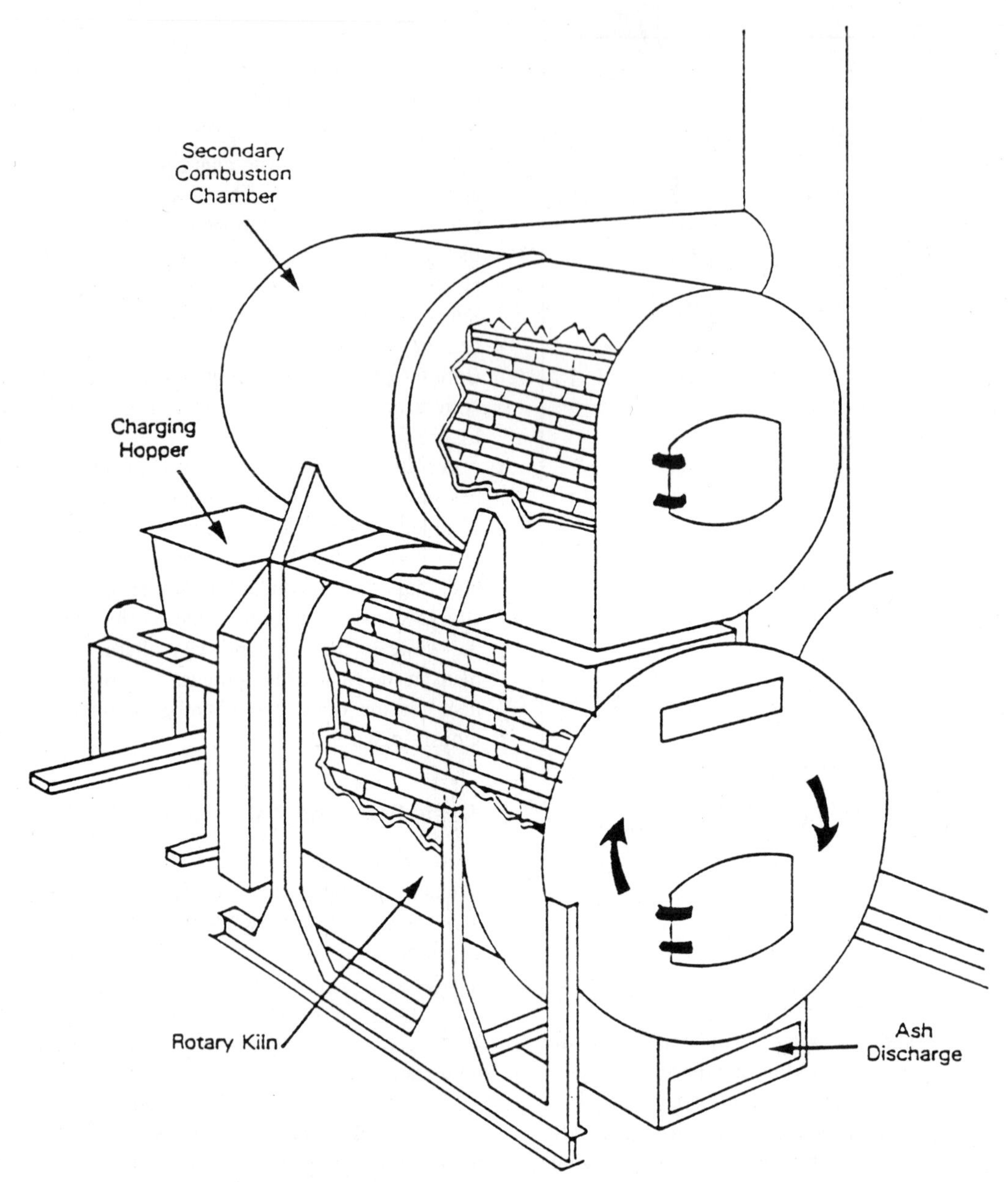

FIGURE 5. Refractory Rotary Kiln System

waste. For units between these size ranges, mechanical feed loaders are usually available as an option. Most mechanical loader designs currently offered employ a hopper-and-ram assembly. In this system, waste is loaded into a charging hopper and the hopper cover is closed. The fire door isolating the hopper from the incinerator opens and a ram comes forward to push the waste into the front section of the incinerator. After reaching the end of its travel, the ram reverses and retracts to the point at which it just clears the fire door. The fire door closes and the ram retracts to its starting position. These operations are normally controlled by an automatic control panel. For smaller incinerators, waste loading into the charging hopper is usually accomplished manually, bag by bag. Larger systems frequently use such waste-loading devices as car dumpers, conveyors, skid-steer tractors, or pneumatic systems.

In addition to improving personnel safety and safety from fire, mechanical loaders limit the amount of ambient air that can leak into the incinerator during waste feeding operations. This is important for controlled-air incinerators since excess air in-leakage can cause lower temperatures, incomplete combustion, and smoking at the stack. Mechanical loaders also permit the feeding of smaller waste batches at more frequent, regular intervals. As the intervals become shorter, this feeding procedure approximates continuous or steady-state operation and helps to dampen fluctuations in combustion conditions.

Ash-removal techniques for controlled-air incinerators

also range from manual to mechanical systems. For smaller units below 500 lb/h capacity (and units constructed before the mid-1970s), operators must rake and shovel ash from the primary combustion chamber into disposal containers. For larger systems, mechanical ash removal may be accomplished by extension of the waste charging ram, augmented by internal transfer rams. The positive displacement action of the rams pushes the ash along the bottom of the primary chamber until it reaches a drop chute. Another mechanical system offered by one manufacturer uses a "pulsed hearth" whereby ash is moved across the chamber floor via pulsation created by end-mounted air cushions. After falling through the drop chute, the ash falls either into a drop cart positioned in an air-sealed enclosure or into a water quench trough. The drop cart is removed manually, generally after spraying the ash with water for dust suppression. In the water trough system, quenched ash is removed by either a drag conveyor or a backhoe trolley system.

When estimating air emissions for controlled-air incinerators with manual ash removal, it is important to recognize that operating, and hence emission, rates will vary over time. A typical operating cycle for such a unit is given by the following.

Operation	*Duration*
Ash clean-out	15–30 minutes
Preheat	15–60 minutes
Waste loading	6–14 hours
Burn-down	2–4 hours
Cool-down	5–8 hours

A waste loading period approaching 14 hours is a maximum value for units with manual ash removal. A more typical value might be five to six hours, since this would be compatible with one-shift-per-day operation. On the other hand, large incinerators with continuous mechanical ash-removal systems may operate on an around-the-clock basis.

Since rotary kiln systems operate in a continuous mode, the waste-feed and ash-removal systems that service these incinerators must also be of a continuous or semicontinuous type. A charging hopper-and-ram system is commonly used to load waste into the kiln. After traveling through the kiln, ash is discharged on a continuous basis into either an ash cart or a water quench system.

Waste Heat Recovery

Waste-heat–recovery operations have generally not been considered for retort incinerators owing to the smaller gas flow rates, lower temperatures, higher particulate matter loadings, and intermittent operation that characterize these systems. For controlled-air and rotary kiln incinerators, however, the higher stack gas temperatures and flow rates can make heat recovery economically attractive in cases where steam or hot water generation rates can be matched with the needs of the hospital. For most systems, heat is recovered by passing hot gases through a waste heat boiler to generate steam or hot water. Boiler equipment can range from a spool piece with heat exchange coil inserted in the stack to a single-drum D-type water-tube waste heat boiler. Most manufacturers, however, use conventional fire-tube boilers because they are low in cost and simple to operate. Options for these boilers include supplemental firing of oil or natural gas and automatic soot-blowing systems. Outlet temperatures from waste heat boilers are generally limited to about 400°F by stack gas dew-point considerations.

AIR EMISSIONS CHARACTERIZATION

Emissions from uncontrolled hospital waste incinerators can generally be characterized as:

- Particulates
- Acid gases
- Trace metals
- Trace organics, including dioxins and furans

Acid gases include hydrogen chloride, sulfur dioxide, and nitrogen oxides. Chlorine, which is chemically bound within the hospital waste in the form of polyvinyl chloride (PVC) or other compounds, will be predominantly converted to hydrogen chloride (HCl), assuming there is sufficient hydrogen available to react with the chlorine. In view of the high hydrogen content of hospital waste, due principally to its paper, plastics and moisture content, there is ample hydrogen available to promote HCl formation. Depending on what is burned, uncontrolled HCl emissions may range from as little as a few hundred to a few thousand parts per million by volume. Uncontrolled sulfur dioxide emissions are generally less than 50 ppmv and nitrogen oxide emissions less than 200 ppmv.

Particulate matter (PM) is emitted as a result of incomplete combustion and by the entrainment of noncombustibles in the flue gas stream. It may exist as a solid or an aerosol, and may contain heavy metals or polycyclic organics. Depending on the method used to measure the PM in the flue gas, lower-boiling point volatile compounds (i.e., boiling point below 100°C) may or may not be included in the measurement.

There are three general sources of PM:

- Inorganic substances contained in the waste feed that are entrained in the flue gas from the combustion process
- Organometallic substances formed by the reactions of precursors in the waste feed
- Uncombusted fuel molecules

Inorganic matter is not destroyed during combustion; most of this material leaves the incinerator as ash. Some, however, becomes entrained in the stack gas as PM.

Organometallic compounds present in the waste stream

that is being incinerated can be volatilized and oxidized under the high temperatures and oxidizing conditions in the incinerator. As a result, inorganic oxides or salts of metals can be formed.

The fuel molecules themselves can also contribute significantly to PM formation. It is known that pyrolitic reactions can lead to the formation of large organic molecules. Inorganics, which may act as nucleation sites, may then further induce growth. The result can be organic particles with inorganic cores.

In general, good combustion conditions, which depend on residence time, temperature, and turbulence, lead to lower PM emissions. As the residence time increases, particle size and the mass of PM tend to decrease. Smaller particle sizes and lower PM emissions are also associated with higher temperatures since, at higher temperatures, oxidation rates are increased so that more of the combustible PM is oxidized to gaseous products.

The amount of trace metals in the flue gas is directly related to the quantity of trace metals contained in the incinerator waste. Some of the trace metal sources in the waste feed include surgical blades, foil wrappers, plastics, and printing inks. Plastic objects made of PVC can contain cadmium heat-stabilizing compounds. In addition, cadmium, chromium, and lead may also be found in inks and paints.

Some metals are selectively deposited on the smaller particulate sizes that are emitted. This is known as fine-particle enrichment. There are three general factors that affect enrichment of trace metals on fine particulate matter:

- Particle size
- Number of particles
- Flue gas temperature

The influence of particle size on trace metal enrichment is thought to be due to specific surface area effects (i.e., the ratio of particle surface area to mass). Particles with large specific surface areas show more enrichment since there is more surface area for condensation per unit mass of PM. The influence of the number of particles is simply due to the increased probability of contact associated with higher particle population. There is some evidence that less enrichment occurs at higher flue gas temperatures. Higher temperatures are thought to lead to increased activity levels, which, in turn, make the metals less likely to condense and bond with PM.

Mercury, because of its high vapor pressure, does not show significant particle enrichment; rather, it is thought to leave largely in the vapor form due to high typical exit gas temperatures. An efficient burning process will result in a high degree of conversion of volatile organics to carbon dioxide and water. Failure to achieve efficient combustion can result in high emission rates of unreacted or partially reacted organics.

Generally, insufficient combustion can occur as a result of charging the incinerator with waste materials in a batch mode. When waste materials are initially charged into the incinerator, the oxygen level in the incinerator is momentarily reduced, resulting in higher emission rates of unburned gaseous and particulate hydrocarbons. The unreacted or partially reacted combustion products include the chlorinated isomers of dibenzo-*p*-dioxin (CDD) and dibenzofuran (CDF), lower-molecular-weight organic compounds, and carbon monoxide (CO).

Emission factors for the various contaminants that can be discharged from uncontrolled hazardous waste incinerators have been developed by the U.S. Environmental Protection Agency (EPA) and are presented in Table 1. The State of California Air Resources Board (CARB) recently com-

TABLE 1. Hospital Waste Incineration Emission Factors (Uncontrolled)

Group/Compound	Emission Factor (lb/ton feed) High	Emission Factor (lb/ton feed) Low
Acid Gases:		
Hydrochloric Acid	99.4	6.6
Sulfur Dioxide	3.01	1.47
Nitrogen Oxides	7.82	4.64
Particulate Matter:		
100 lb/hr feed	26.92	1.69
1,000 lb/hr feed	5.45	1.37
Trace Metals:		
Arsenic	2.14×10^{-4}	7.1×10^{-5}
Cadmium	6.8×10^{-3}	2.48×10^{-3}
Chromium	6.08×10^{-4}	1.02×10^{-4}
Iron	1.83×10^{-2}	3.98×10^{-3}
Manganese	1.14×10^{-3}	1.58×10^{-4}
Nickel	5×10^{-4}	1.08×10^{-4}
Lead	5.6×10^{-2}	3.04×10^{-2}
Dioxins:		
TCDD (tetra)	1.07×10^{-6}	4×10^{-8}
PCDD (penta)	7.6×10^{-7}	1.1×10^{-7}
HxCDD (hexa)	1.52×10^{-6}	2.7×10^{-7}
HeCDD (hepta)	3.84×10^{-6}	3.2×10^{-7}
DCDD (Octa)	5.48×10^{-6}	3.4×10^{-7}
Total PCDD	1.25×10^{-5}	1.43×10^{-6}
Furans:		
TCDF (tetra)	2.08×10^{-6}	5×10^{-7}
PCDF (penta)	3.8×10^{-6}	9×10^{-7}
HxCDF (hexa)	5.64×10^{-6}	8.2×10^{-7}
HeCDF (hepta)	6.46×10^{-6}	5.5×10^{-7}
DCDF (octa)	4.36×10^{-6}	2.1×10^{-7}
Total PCDF	2.18×10^{-5}	3.26×10^{-6}
Low Molecular Weight Organics:		
Ethane	0.003	
Ethylene	0.020	
Propane	0.024	
Propylene	0.022	
Trichlorotrifluoroethylene	8.26×10^{-5}	
Tetrachloromethane	9.92×10^{-5}	
Trichloroethylene	2.4×10^{-5}	
Tetrachloroethylene	2.5×10^{-5}	
Carbon Monoxide	1.7	1.32

Source: Reference 1.

TABLE 2. Uncontrolled Emissions Corrected to 12% CO_2 from Hospital Waste Incinerators Tested in California

Group/Compound	Average Facility Emission Concentrations
Particulates (grams/dry std.cu.m.)	0.15 – 0.57
HCl (grams/dry std.cu.m.)	0.97 – 4.71
Metals (micrograms/dry std.cu.m.)	
Arsenic	N.D. – 18.7
Cadmium	32.3 – 400
Chromium	10.1 – 179
Chromium (+6)	0.003 – 19.5
Iron	76 – 3,250
Lead	404 – 14,600
Manganese	4.71 – 194
Mercury	0.51 – 12,300
Nickel	N.D. – 63.2
Total	542 – 18,500
Dioxins (nanograms/dry std.cu.m.)	
PCDD	106.5 – 2,164.3
PCDF	256 – 4,139
2,3,7,8 TCDD Equiv. (PCDD)	1.26 – 26.7
2,3,7,8 TCDD Equiv. (PCDF)	10.27 – 188.33

Source: Reference 2.

pleted a series of tests on eight different medical waste incinerators in California. A summary of uncontrolled emissions from the various facilities tested is presented in Table 2.

AIR POLLUTION CONTROL MEASURES

In the past, hospital waste incinerators operated largely without requirements for add-on pollution control equipment or special combustion modification techniques. However, this situation is changing rapidly as more and more states are requiring such equipment. There are three broad categories of emission reduction technologies that can be applied to hospital waste incinerators:

1. Source separation
2. Combustion control
3. Flue gas controls (add-on control devices)

Source Separation

Source separation refers to both the segregation of infectious and noninfectious wastes and the removal of specific compounds from the waste stream prior to incineration.

After segregation of infectious and noninfectious wastes, further segregation of the noninfectious portion could be possible. Plastics and metal-containing components of the waste, such as sharps, could be segregated; this could result in lower HCl, polychlorinated dibenzo-*p*-dioxins (PCDDs), polychlorinated dibenzo-*p*-furans (PCDFs), and trace-metal emission rates. However, no data are available on such practices at hospitals. Another approach to possibly lowering HCl and PCDD/PCDF emission rates would be to have hospitals use low-chlorine-content plastics. This could be accomplished if the health-care industry were to use such plastics as polyethylene and polystyrene in place of PVC, which contains over 45 wt % chlorine. Again, no data are available on such practices at hospitals.

Combustion Control

Data presented previously indicate that there is significant variation in the uncontrolled emission rates from hospital incinerators. These variations are partially due to the variability in the chemical and physical properties of hospital wastes, partially to the variations in incinerator design, and partially to the variations in operating practices. This section will review the relationships between combustion processes and emissions of nitrogen oxides (NO_x) and polycyclic organic matter (including dioxins and furans).

Based on NO_x-reduction techniques applied to other combustion processes, at least three control options may be applicable to reduce NO_x emissions from incinerator processes: flue gas recirculation, reburning, and ammonia (NH_3) injection. However, these techniques are not generally being applied to hospital waste incinerators at this time.

Flue gas recirculation introduces a thermal diluent and reduces combustion temperatures. However, lowering of flame and furnace temperatures could be counter to the control of PCDD/PCDF.

Reburning uses a hydrocarbon-type fuel such as natural gas or oil as a reducing agent. Hydrocarbon radicals produced by the reburning fuel react with NO_x to form N_2, H_2O, and CO_2. This control technology is being developed for use in fossil-fuel–fired boilers because only minor modifications are required to the main heat release zone.

A third option available for NO_x reduction on incinerators is NH_3 injection. When injected into combustion gases, NH_3 can react with NO_x to form diatomic nitrogen (N_2) and water. The optimum temperature range for promotion of this NO_x reduction reaction is near 870°C to 1000°C. Thus for application to a controlled-air hospital waste incinerator, NH_3 would likely be injected either into the secondary chamber or into a section of the stack immediately following the secondary chamber.

Available data indicate that the PCDD and PCDF emission rates are closely related to the efficiency of the combustion process. In general, when the flame temperature and combustion efficiency are increased, PCDD and PCDF emission rates decrease. Poor combustion conditions, on the other hand, can lead to polycyclic organic matter

(POM), dioxin, and furan emissions; however, the combustion process may be controlled to minimize these emissions.

Emissions of POM, PCDD, and PCDF are either the products of incomplete combustion or are formed by postcombustion mechanisms. A critical component of the combustion formation process is the presence of fuel-rich pockets of gas. The primary combustion chamber in a controlled air incinerator is operated as a large fuel-rich pocket. To maximize the extent of combustion, the following steps can be taken.

- Control the combustion air supply to the primary chamber to minimize transients in the outlet flow rate and composition.
- Proportion combustion air between the primary and secondary chambers to maintain desired temperatures.
- Promote efficient mixing of air and combustion gases in the secondary chamber.

Each of these combustion parameters is adjustable during the incinerator design and/or as part of the unit operating procedure. The assertion that combustion-generated PCDD and PCDF emissions can be reduced by combustion control is clear, but the types of modifications likely to be effective will depend on the specific design and operating conditions of a given model and size. That is, combustion modifications must be tailored to the specific type of incineration hardware under consideration. Appropriate control strategies for existing facilities must be evaluated on a case-by-case basis and some processes may require extensive hardware modification and/or altered operational procedures.

To the extent that PCDD/PCDF emissions are formed downstream of the combustion chamber, emission reductions will be influenced by the mechanisms involved and practices employed in this area. If emissions are the result of reactions between products of incomplete combustion (PICs), these products must be avoided in the temperature window in which the reactions occur. Possible emission-reduction techniques would be the minimization of PICs in the secondary combustion chamber by operating the chamber at higher temperatures and/or higher gas residence times.

If PCDD/PCDF emissions are the result of *de novo* synthesis on fly ash or from particulate carbon within a given temperature window, then the removal of fly ash and carbon at temperatures above the critical window would reduce flue gas emissions. However, it may still be necessary rapidly to quench the captured particles (e.g., with a scrubbing liquid or air) or to cover their active surface area with an inert material (e.g., powdered lime) in order to minimize subsequent *de novo* synthesis.

Flue Gas Controls

The equipment needed for the control of emissions from hospital waste incinerators must remove effectively both HCl and particulate emissions. The following emission control systems are most often considered in a best available

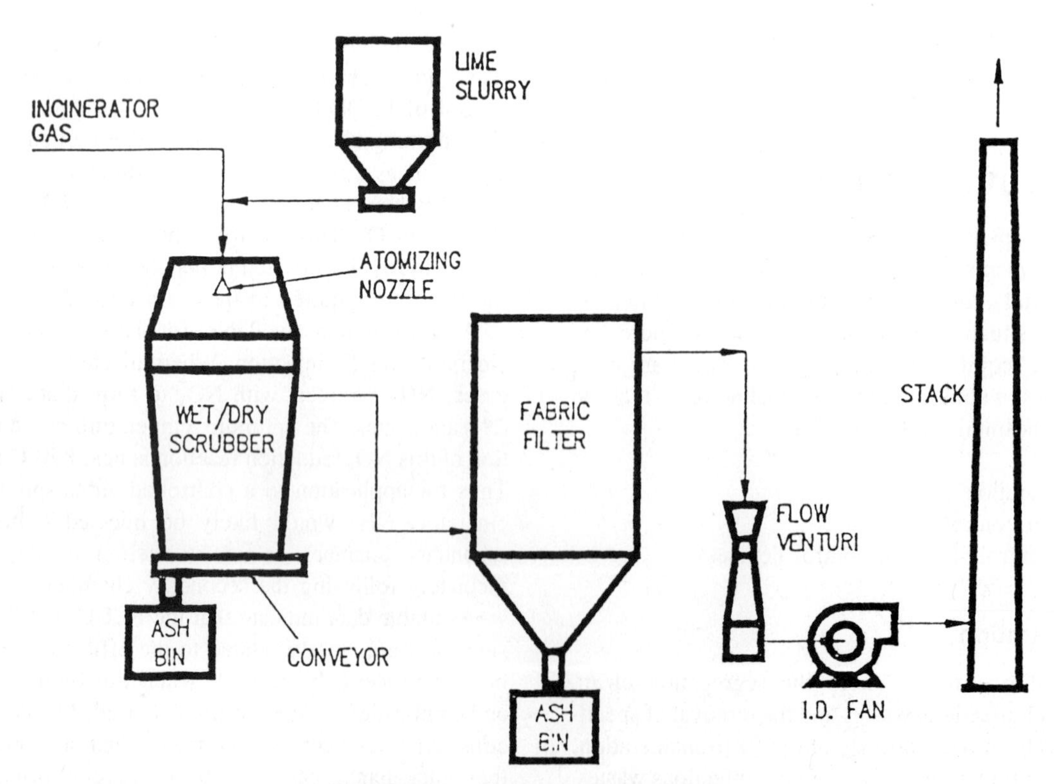

FIGURE 6. Wet/Dry Scrubbing System

control technology (BACT) analysis, as they have demonstrated the ability to comply with regulatory requirements.

- Wet/dry scrubber
- Dry/dry scrubber
- Venturi scrubber with gas absorber

Wet/Dry Scrubber (Lime Slurry Injection Type)

Wet/dry scrubbers combined with fabric filters (or electrostatic precipitators [ESPs]) have received considerable attention recently for use in hospital waste incinerators. In these systems, a lime slurry is injected into a spray dryer, where it contacts the flue gas. The water is evaporated and dry salts result from the reaction of lime with acidic constituents of the flue gas. The salts, unreacted lime, and particulate matter are collected in a fabric filter (or ESP) downstream of the reactor. An advantage of using a fabric filter is that filter cake buildup can provide additional reaction sites for continued neutralization of acid pollutants in the flue gas.

TABLE 3. Performance of Air Pollution Control Systems on Hospital Waste Incinerators

Incineration System Description	Air Pollution Control Device	Performance	Reference
(1) Two chamber, controlled air, 1600–1800°F primary, 1800–2200°F secondary (Thermtec, Model 800, 800 lb/hr design capacity	Emcotek Rotary Atomizing Wet Scrubber, 4 in. w.c. pressure drop, caustic scrubbing reagent	Particulate: 0.013–0.033 gr/dscf corrected to 7% O_2 HCl: 11–14 ppmv dry SO_2: 1–2 ppmv dry	(3)
(2) Two chamber, controlled air, 1600–1800°F primary, 1800–2000°F secondary, Waste Heat Boiler (Ecolaire Model 1500, 1200 lb/hr design capacity)—Cedar Sinai Hospital, California	Micro-pul Fabric Filter Baghouse, pulse-jet	Removal efficiencies: Particulate: 96.5% (0.001 gr/dscf) Cadmium: 99.9% Arsenic: 99.9% + Chrome: 98.9% Lead: 99.8% Total PCDD: 129.5 ng/Nm^3 Total PCDF: 270.5 ng/Nm^3	(1,2)
(3) Two chamber, controlled air, Thermtec EP800 AR, 800 lb/hr design capacity,	Sly Wet Scrubber (Impinjet, single stage, perforated plate with impingement baffle), once through scrubbing using magnesium hydroxide reagent, 1.5 in. w.c. pressure drop	Removal efficiencies: HCl: 85% Particulate: 28.6% (0.15 gr/dscf corrected to 12% (CO_2) Total Dioxin: 65.3% Total Furans: 55.5% Hexavalent Chromium: 46.2% Mercury: 62.5%	(2,4)
(4) Two chamber, controlled air, 1700–2000°F primary, 1900–2100°F secondary, Ecolaire	Ducon Venturi Scrubber using sodium hydroxide scrubbing reagent	HCl: 99.7% efficiency	(2)
(5) Skovde Hospital (Skovde, Sweden) Incinerator with boiler, 2300 lb/hr waste feed rate	Research Cottrell/Teller Dry Venturi with a pulse-jet fabric filter, hydrated lime reagent (3.5 stoichiometric ratio)	Removal efficiency, corrected to 10% CO_2: Particulate: 99.3% (1–2 mg/Nm^3) HCl: 57–99% (12 mg/Nm^3) HF: 87.2–97% (0.23–1 mg/Nm^3) Mercury: 96.9–99.8% (5–7 μg/Nm^3) TCDD (eq.) 44–99.1% (0.2–0.6 ng/Nm^3) Chlorobenzenes: 75–94% (0.2–13 μg/Nm^3 Lead: 6.3–12 μg/Nm^3 Cadmium: 0.24–0.45 μg/Nm^3 Arsenic: <0.15 μg/Nm^3	(5)
(6) Southland Exchange Waste Incineration Facility, Hampton, SC (municipal waste and medical waste); 240 tons/day capacity; waste heat boilers and economizers	Dry Sodium Bicarbonate Sorbent Injection (pulverized), acid gas reactor with 3 second residence time and electrostatic precipitators for particulate removal	Acid gas removal Efficiency: 96% Particulate: 0.04 gr/dscf	(6)

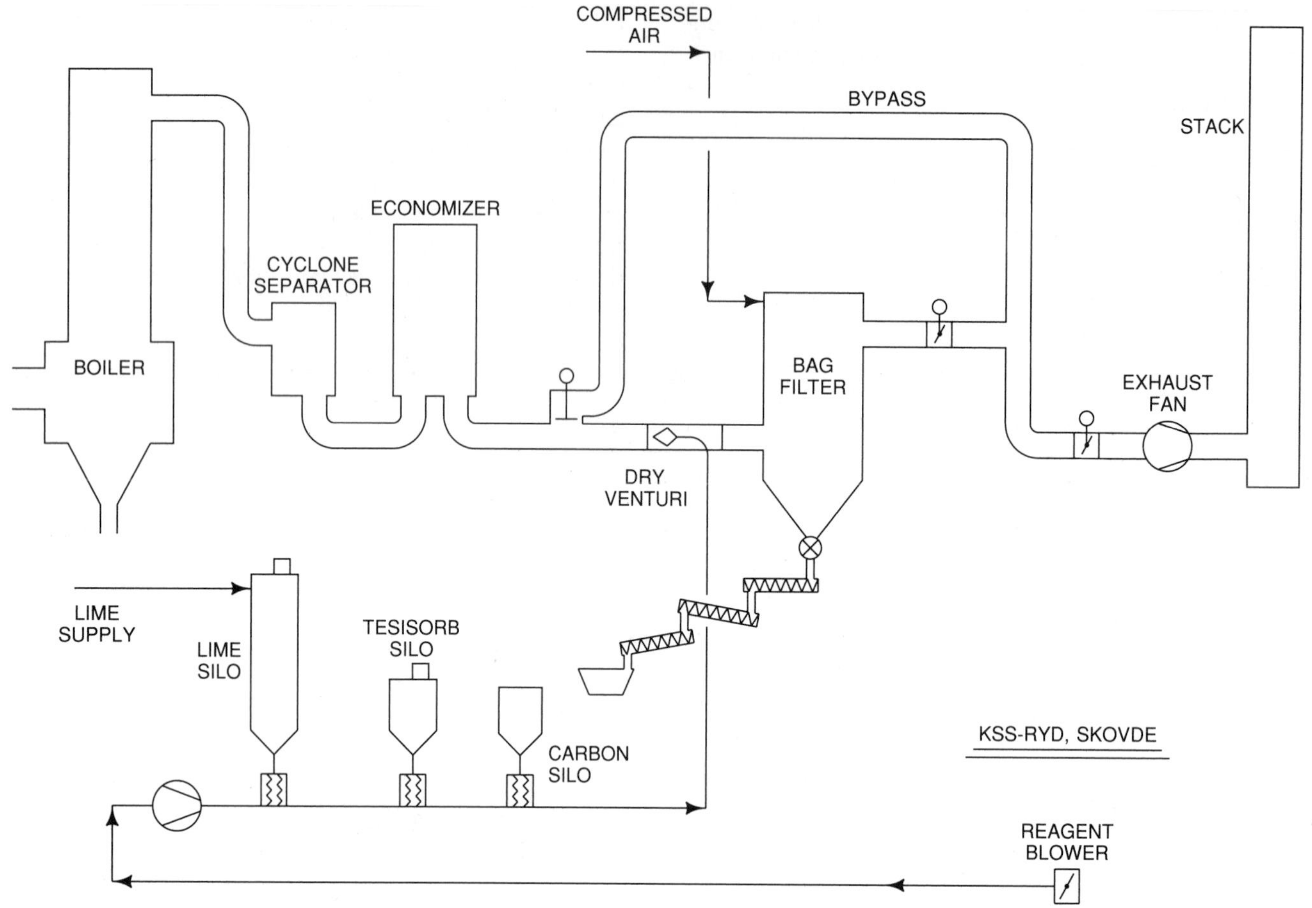

FIGURE 7. Dry/Dry Scrubbing System

Test results on waste incinerators with spraydryerfabric filter systems show enhanced particulate matter emission reductions in all particle size ranges as compared with wet scrubbers operating at comparatively higher pressure drops. These systems can achieve HCl removal efficiencies as high as 99% and outlet particulate loadings of 0.015 gr/dscf or less. Figure 6 shows a typical wet/dry scrubbing system.

Dry/Dry Scrubber (Dry Lime Injection Type)

Both dry lime and sodium bicarbonate injection scrubbing technologies have been applied to hospital waste incineration (refer to Table 3). Dry hydrated lime dust or pulverized sodium bicarbonate is metered directly into the incinerator exhaust, where it absorbs and reacts with the acidic components in the gas stream. A dust condition drum is some-

FIGURE 8. Injection System with a Horizontal Reactor Downstream from a Hospital Incinerator Waste Heat Boiler (Courtesy PPC Industries)

FIGURE 9. Two-Field Electrostatic Precipitator on a Hospital Waste Incinerator (Courtesy PPC Industries)

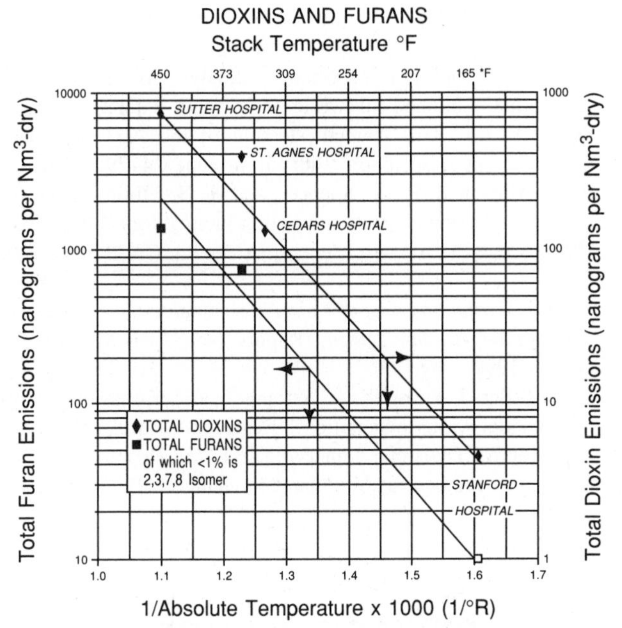

FIGURE 10. Effect of Pollution Control Device Temperature on Hospital Incinerator Emissions (From References 1 and 3)

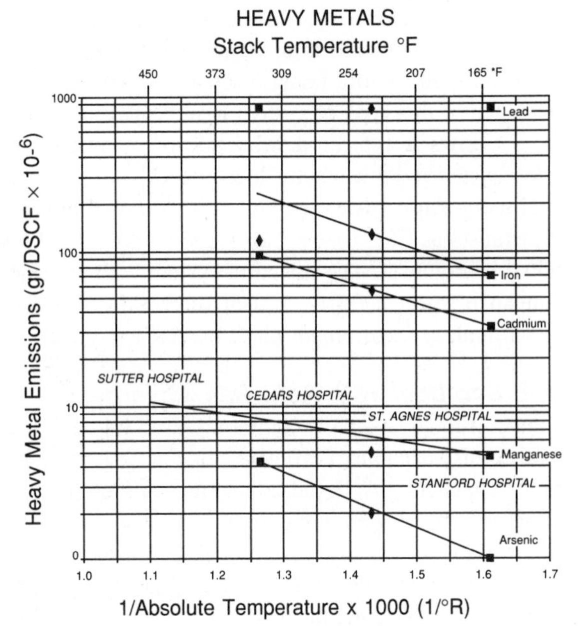

FIGURE 11. Effect of Pollution Control Device Temperature on Hospital Incinerator Emissions (From References 1 and 3)

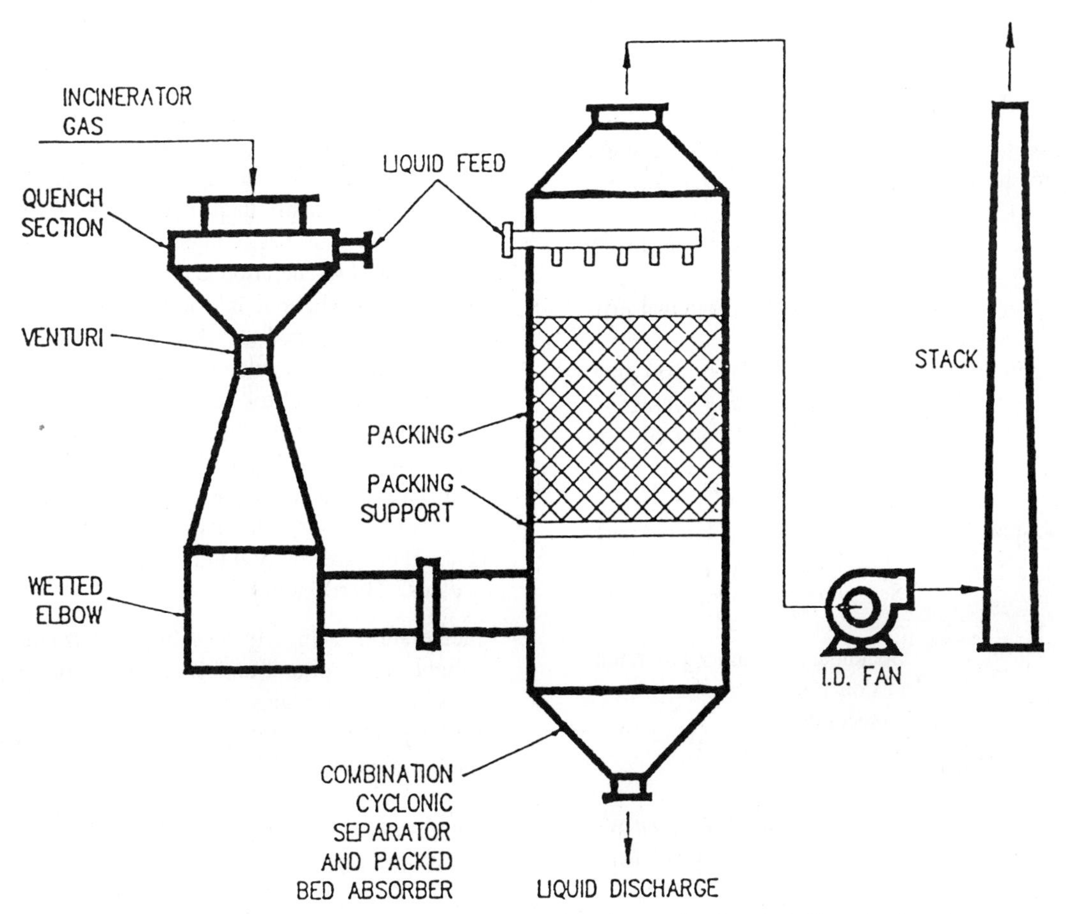

FIGURE 12. Particulate Scrubbing and HCl Absorption System for Hospital Waste Incinerator

times provided to assist in solids–gas contact. The exhaust gas stream is then filtered in a baghouse (although an ESP may also be used) before being discharged to a stack. The performance of this type of system is indicated in Table 3. Figure 7 shows a typical dry/dry scrubber system. The injection system with a reactor located downstream of the hospital incinerator waste heat boiler is presented in Figure 8. The exaust gas from this reactor is fed to a two-field ESP as shown in Figure 9. The injection system with an ESP is guaranteed to meet a 0.015 gr/dscf (corrected to 7% oxygen) particulate requirement and 90% HCl removal.

Venturi Scrubber with Acid Gas Absorber

A venturi scrubber, which is commonly used for particulate removal, is also capable of efficient acid gas absorption as it promotes excellent gas–liquid contacting. Although a venturi scrubber can be used alone for simultaneous particulate and HCl removal, it is often combined with a gas absorber to maximize HCl removal efficiency. The gas absorber typically is made an integral part of the venturi's cyclonic separator and can be either a packed tower or a tray tower. The scrubbing liquid is typically a caustic solution. These systems can achieve 99% HCl removal and outlet particulate loadings of the order of 0.015–0.05 gr/dscf at venturi scrubber pressure drops in the 25–50 in. w.c. range. Since these systems operate at a lower temperature than the dry/dry or wet/dry scrubber systems, more of the volatile metals, such as mercury, will condense out. The impact of pollution control device temperature is presented in Figures 10 and 11. Figure 12 shows a typical system.

Best Available Control Technology

Performance data collected and evaluated thus far suggest that BACT for hospital waste incinerators includes:

Particulates:	0.015 gr/dscf corrected to 12% CO_2
HCl:	50 ppm or 99% removal efficiency
Combustion efficiency:	99.9% minimum or 100 ppmv CO on an hourly average (CO emissions monitored continuously)
Opacity:	Less than 10%
Incinerator design:	Primary chamber temperature ≥1500°F
	Secondary chamber temperature ≥1800°F
	Secondary chamber gas retention time of one to two seconds

References

1. E. F. Aul, P. A. May, and G. E. Wilkins, *Hospital Waste Combustion Study, Data Gathering Phase, Final Report,* EPA-450/3-88-017, U.S. Environmental Protection Agency, December 1988.
2. H. Glasser and D.P.Y. Chang, "Analysis of the State of California's biomedical waste incinerator database," Paper No. 90-27.3, 83rd Annual Meeting of the Air and Waste Management Association, Pittsburgh, June 24–29, 1990.
3. H. L. Marschall, H. W. Spencer, and H. H. Elliott, "Retrofitting air pollution controls to existing incinerators," Paper No. 89-23B.6, 82nd AWMA Meeting, Anaheim, CA, June 25–30, 1989.
4. CARB, *Evaluation Test of Kaiser Permanente Hospital Waste Incinerator in San Diego,* Report No. ARB/ML-90-030, March 1990.
5. A. J. Teller, J. Y. Hsieh, A. Astrand, et al., "Emission control for hospital waste incineration," Research-Cottrell, Air Pollution Control Division, Somerville, NJ, March 1989.
6. "Acid gas removal at Southland Exchange," *Pollution Eng.* (April 1990).

SEWAGE SLUDGE

Calvin R. Brunner, P.E., D.E.E.

Incineration has been utilized for the treatment and disposal of sewage sludge for over 50 years. The Herreshoff multiple-hearth furnace was first used for this application in 1932. The initial use of a fluid-bed furnace for sewage sludge incineration occurred in 1952. Among other types of incineration systems used for sewage sludge are the cyclone furnace, electric (or radiant heat) furnace, and codisposal systems.

Other methods of treatment and disposal include land application, land burial, composting, soil amendments (fertilizers), and ocean dumping. These technologies were found to be less costly than incineration in the past, but as land application is becoming more strictly regulated, the relative cost of incineration is decreasing. Ocean disposal is not an option where it is not now practiced; where it is allowed, this method of disposal is being phased out through restrictive regulatory action.

With the potential for heat recovery and low fuel costs, incineration of sewage sludge is growing in popularity. Its capital cost is high, but with appropriate selection of upstream equipment to obtain dryer, or autogenous burning sludge cake, the annual fuel cost can be significantly reduced. There are many unique advantages to incineration processes. These include:

- Incineration achieves a substantial reduction in the weight, volume, and toxicity of the initial waste stream.
- Less material remains for ultimate disposal.
- Residue is often a sterile ash that has potential for use in road surfacing, building block, metals reclamation, and so on.
- Equipment is available to meet the most stringent air emissions requirements.

- The incineration process is compact and requires nominal land area as compared with land-intensive disposal techniques such as land-filling.
- Incineration is immediate. Months or years are not required as with other treatment or disposal methods, such as burial or land application.
- Incineration takes place at the point of generation of sludge, thus eliminating the need to transport sludge to a distant disposal site. (Often the air emissions resulting from truck hauling of sludge are far greater than those of an incinerator discharge for the same sludge quantity.)
- The potential for recovery of heat exists.
- Incineration systems have a useful life of 10 to 15 years before replacement or major overhaul is necessary. Public acceptance of a continued incinerator presence is easier to obtain than acceptance of necessary searches for new land disposal sites, particularly when a distant area is asked to accept exported sludge.
- Incinerators are stationary, fixed-point dischargers. As such, they can be readily observed and monitored as compared with discharges from nonpoint sources such as trucks or landfills. This feature can provide the public with greater confidence in the assessment of incinerator operation.

The disadvantages of incineration as a sludge treatment method include the following.

- An external source of fuel is always required, if not for maintaining combustion, then to bring the furnace to operating temperature.
- A relatively high investment in capital equipment is necessary.
- Skilled operators are required.
- Metals concentration in the residue (ash) may classify the ash as a "special" or hazardous waste. This could represent a relatively high cost for its ultimate disposal.

THE INCINERATION PROCESS

Sewage sludge is normally fired as a sludge cake produced from dewatering equipment such as a centrifuge, belt filter press, vacuum filter, or plate-and-frame press. It normally contains from 55% to 85% moisture and from 40% to 80% combustibles (as a percent of dry weight). It cannot start to burn until its moisture content has been reduced to less than 30%. The ultimate purpose of incineration is to reduce its moisture content until it will burn, to allow the sludge to burn out to a sterile residue, or ash, and to condition the exhaust gas stream to an innocuous discharge to the atmosphere.

Sludge incineration systems in use today include the following.

- Multiple-hearth incinerator
- Fluid-bed incinerator
- Electric furnace
- Cyclonic furnace
- Codisposal systems

All of these systems rely on the excess air process in which air in excess of the ideal or stoichiometric requirement is provided to the burning sludge to assure that all of the sludge is effectively and efficiently burned. A number of years ago, substoichiometric (starved-air or so-called pyrolysis) processes were attempted in an effort to reduce equipment size and supplemental fuel requirements. However, this technology was not successful anywhere in the United States and these substoichiometric processes were abandoned in favor of conventional burning.

Air requirements for sludge incinerators range from 20% in excess of the stoichiometric requirement for the electric furnace to over 100% of the stoichiometric requirement for multiple-hearth furnaces.

Multiple-Hearth Incinerators

The multiple-hearth furnace is the most common incinerator for the disposal of sewage sludge in this country. It was developed at the turn of the century for ore roasting and has been adapted to carbon regeneration, recalcining and many other industrial applications.

Multiple-hearth furnaces range from 6 to 25 feet (183 to 762 cm) in diameter and from 12 to 65 feet (366 to 1981 cm) in height. The number of hearths depends on the waste feed and processing requirements, but it generally varies between five and 12. Waste retention time is controlled by the rabble tooth pattern and the rotational speed of the center shaft.

Sludge cake is introduced at the top of the furnace (see Figure 1). The furnace interior consists of a series of circular refractory hearths, one above the other. The hearths are arch structures, self-supporting off the refractory-lined cylindrical wall of the furnace. They are numbered no. 1 as the top hearth, no. 2 as the next to top, and so on. Where the top hearth is used as an afterburner section, with sludge deposited on the hearth beneath the top hearth, the top hearth may be referred to as a "zero" hearth. In this case, the next to top hearth would be referred to as hearth no. 1, and so on.

A vertical shaft is positioned in the center of the furnace. Rabble arms are attached to the center shaft above each hearth. The center shaft, along with the rabble arms, rotates relatively slowly, approximately one revolution per minute. Teeth on the rabble arms are positioned to move sludge across the hearth and through the furnace.

Every other hearth has a large annular opening between the hearth and the center shaft. These are termed "in" hearths. The teeth on the rabble arms of these hearths will "rabble" or wipe sludge to the center of the hearth, where the sludge will fall off the edge of the inner refractory ring, landing on the hearth below, an "out" hearth.

There are a series of openings on the outside or periphery

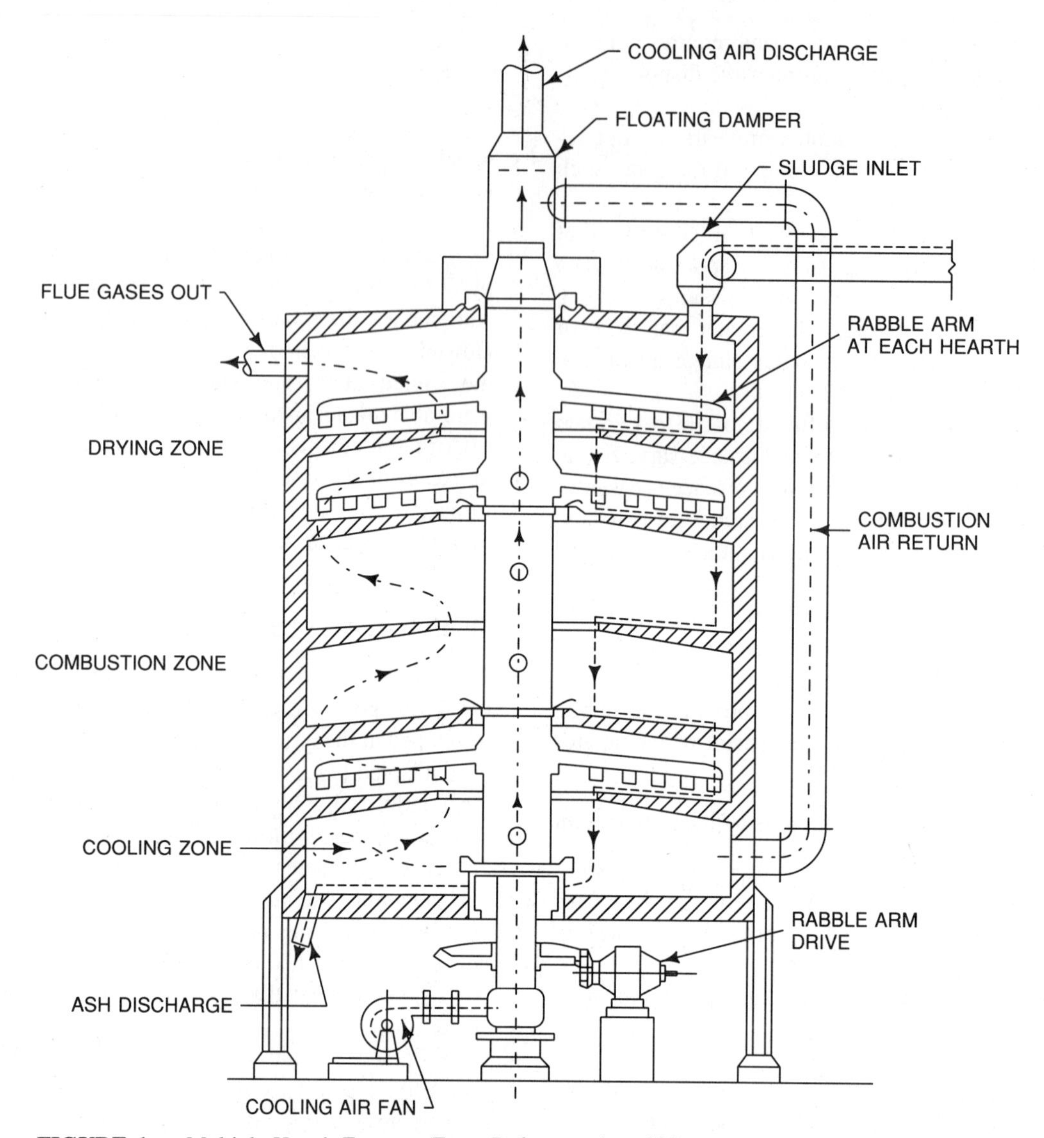

FIGURE 1. Multiple-Hearth Furnace (From Reference 4, p 230)

of the out hearth. The inside of an out hearth is fairly close to the center shaft. A "lute ring cover" is a collar located above the hearth, attached to and moving with the center shaft, preventing sludge from dropping adjacent to the center shaft. Teeth on the rabble arms on the out hearths move sludge to the drop holes on the outside of the hearth, where the sludge drops to the hearth below, an in hearth. This process repeats until sludge, or ash, reaches the bottom hearth, the floor of the furnace, where it discharges from the furnace. Teeth on each hearth agitate the sludge, exposing new surfaces of the sludge to the gas flow within the furnace. As sludge falls from one hearth to another, it again has new surfaces exposed to the hot gas.

The upper hearths of the furnace provide a drying zone where the sludge cake gives up moisture (evaporation) while cooling the hot flue gases. Flue gas exits the top hearth of the furnace at from 800°F to 1200°F (427°C to 649°C). The center hearths make up the burning zone, where gas temperatures can reach 1600°F (871°C). Burnout of sludge to ash occurs in the lower hearths of the furnace.

The center shaft and rabble arms are hollow. Air is passed through the center shaft, which is constructed to distribute this air to each of the rabble arms and discharge it through the top of the shaft. After passing through the rabble arms and center shaft, this cooling airflow reaches temperatures of 300°F to 450°F (149°C to 232°C). This heated air often is recycled into the furnace as preheated sludge combustion air.

The temperature above at least two hearths should be maintained at approximately 1600°F (871°C) at all times when burning sludge cake. The off-gas temperature can range from 800°F to 1400°F (427°C to 760°C). In general, the exit temperature of the multiple-hearth furnace is from 400°F to 600°F (204°C to 316°C) lower than the maximum furnace temperature. This is an important consideration. The maximum allowable temperature anywhere within the furnace is 1800°F (982°C). Above this temperature the sludge ash fusion temperature will be approached and ash

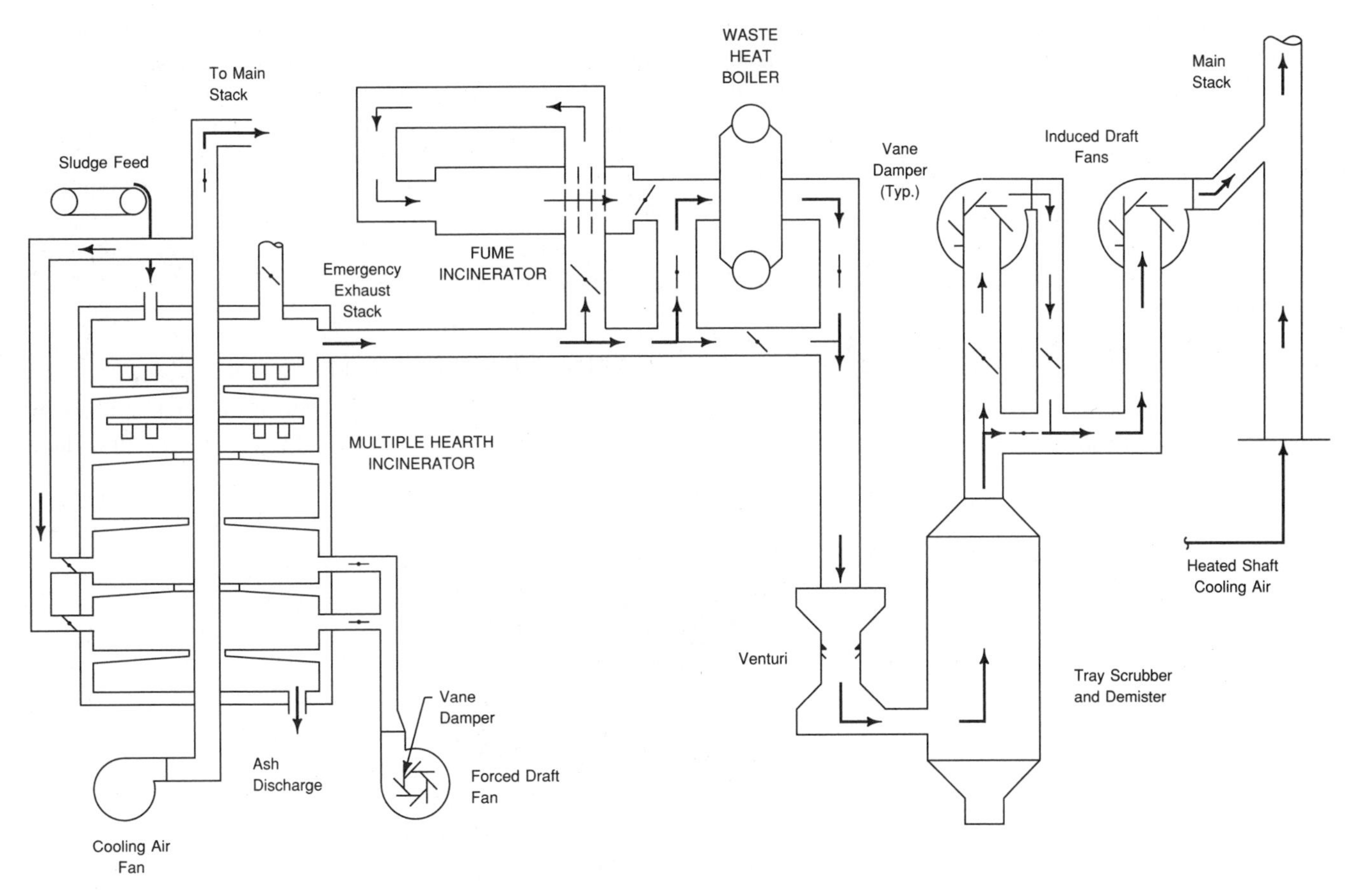

FIGURE 2. Multiple-Hearth Sludge Incinerator System (From Reference 5)

clinkering will occur. The formation of clinker and slag can clog drop holes, interfere with rabble-arm motion, and create other problems within the furnace that can result in shutdown and damage.

Occasionally, odor problems can exist (unburned hydrocarbon emissions), which may require the installation of afterburning equipment. This is of particular concern where there is a significant industrial component to the waste stream, in excess of 20% of the total influent flow to the treatment plant.

Multiple-hearth incinerators are designed to burn materials with a low gross (as received) heating value, such as sewage sludge and other high–moisture-content wastes. Their design includes drying as well as burning sections. Materials with less moisture content, or higher heating values, such as coal or solid waste, will start to burn too high in the furnace. There would be insufficient residence time above these top hearths and the temperatures in this volume of the furnace would be too low for effective burnout. Solid fuel or solid waste is not compatible with a multiple-hearth incinerator system and should not be introduced into it.

Figure 2 is a diagram of a generalized multiple-hearth incinerator system. It includes a fume incinerator (afterburner) separate from the multiple-hearth furnace and a waste-heat boiler. The boiler is placed in the flue gas stream whenever the fume incinerator is in operation. The fume incinerator is designed to bring the temperature of the gases exiting the incinerator from 800°F (427°C) to the range of 1400°F to 1500°F (760°C to 816°C), which represents a significant amount of heat. The waste-heat boiler can reclaim over 50% of the heat input from the supplemental fuel.

Ash exits this system dry, from beneath the furnace. It can be collected and disposed of dry, or it can be dropped into a wet hopper, mixed with water, and pumped to a lagoon for dewatering and ultimate disposal. The multiple-hearth furnace is flexible enough to allow wet or dry ash handling.

Feeding sludge to a multiple-hearth furnace is usually done by gravity; it is dropped onto the top hearth. Sludge also can be deposited on the top hearth or onto a lower hearth by a screw conveyor. Attempts to feed sludge to a multiple-hearth furnace pneumatically have met with failure. Pneumatic systems necessarily inject air into the furnace in strong discrete bursts, which makes the maintenance of constant furnace draft difficult, if not impossible.

Generally, grease (scum) should not be added to sludge feed. If grease is to be incinerated in a multiple-hearth furnace, it should be added at a lower hearth (a burning hearth) through a separate nozzle(s). If it is mixed with or placed on the sludge fed to the furnace, it will tend to volatilize upon entering the furnace at too high a hearth. The volatilized grease would not experience a high enough temperature or sufficient residence time for effective destruction.

The multiple-hearth furnace should always be run at a negative pressure (draft) to prevent external leakage of hot, toxic flue gas. As shown in Figure 2, an induced-draft (ID) fan drives, or pulls, gases through the furnace system. An inlet damper on the ID fan is normally controlled by furnace draft. The position of this damper automatically modulates to maintain constant draft (constant negative pressure) within the furnace. If a separate afterburner is needed, and if an emergency discharge stack is provided, the stack should always be located downstream of the afterburner.

Incinerator instrumentation should include a temperature strip chart that at least records temperatures on each hearth. A chart recording scrubber pressure drop and fraction oxygen in the flue gas also should be included on the control panel. Carbon monoxide indication is a useful tool in the control of sludge combustion air. Opacity meters should not be used downstream of a wet scrubber, as entrained water from the scrubber will appear opaque and the meter will give a false reading.

Fluid-Bed Incineration

The fluid-bed furnace was developed for catalyst recovery by the Exxon Corp. during World War II. The first fluid-

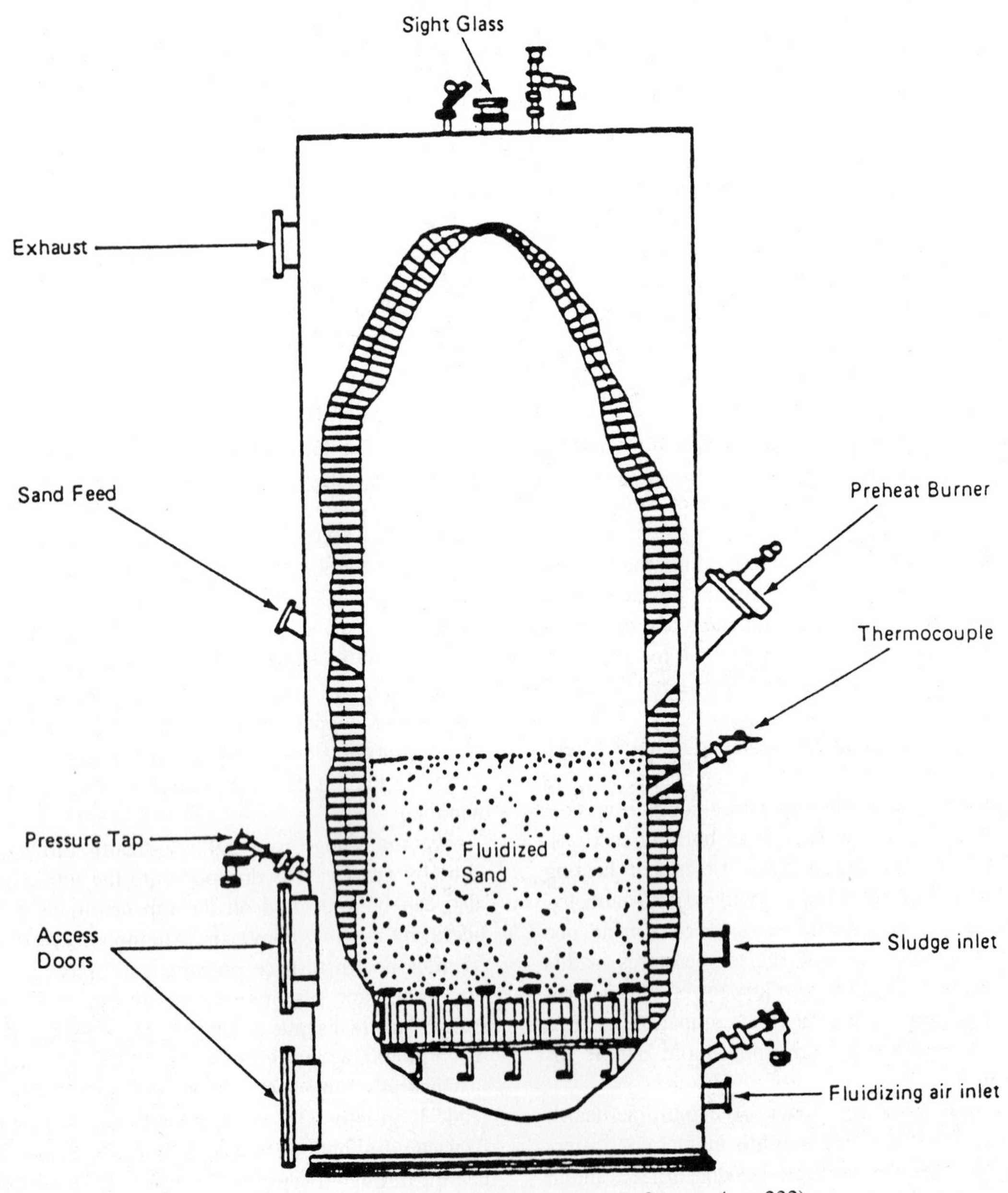

FIGURE 3. Fluid-Bed Furnace (From Reference 4, p 232)

bed furnace used for incineration of sewage sludge was installed in 1962.

As illustrated in Figure 3, the fluid-bed furnace is a cylindrical refractory-lined shell with a grid structure, or bed plate (either a refractory arch or an alloy steel plate), above a wind box to support a sand bed. Air is introduced at the fluidizing air inlet at pressures in the range of 3.5 to 5 psig (72 mb). The air passes through openings (tuyeres) in the bed plate supporting the sand and generates a high degree of turbulence within the sand bed. The sand undulates and has the appearance of a fluid in motion.

Air can be introduced to the wind box cold or, as is usually the case, preheated by the exiting flue gas (see Figure 4). The sand bed is normally maintained at approximately 1300°F to 1400°F (704°C to 760°C). It expands from 30% to 60% in volume when fluidized.

Sludge cake is normally introduced within or just above the fluid bed. Fluidization provides maximum contact of air with the sludge cake particles for optimum burning. The drying process is practically instantaneous. Moisture flashes into steam upon entering the hot bed.

Most of the fluid-bed furnaces used for sludge incineration have one major item of air-moving equipment, the forced-draft fan (or fluidizing air blower). The fan is sized to move (push) air/flue gas through the heat exchanger, the fluid bed, the hot gas side of the heat exchanger, and then through the gas scrubbing system. This requires that the

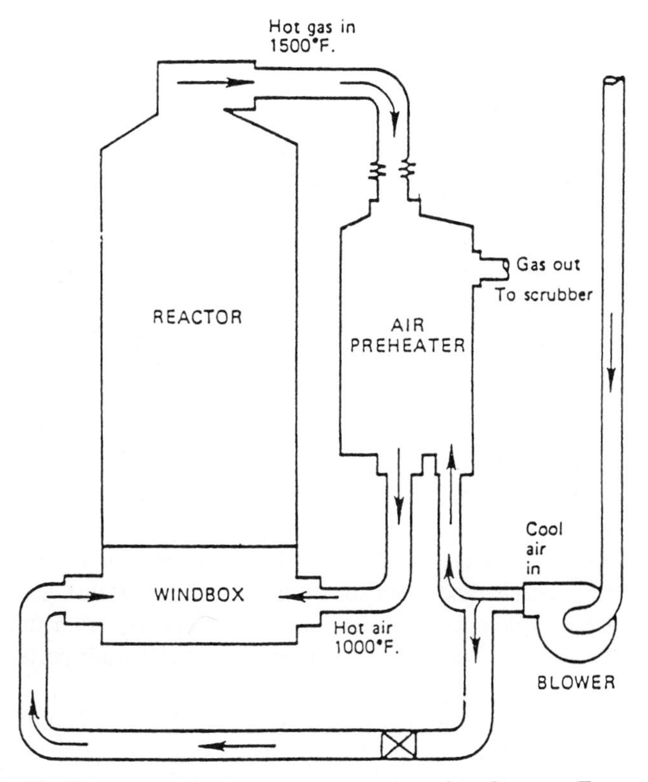

FIGURE 4. Fluid-Bed Furnace Incineration System (From Reference 4, p 233)

reactor be pressurized and that it be airtight to prevent leakage of flue gas from the incinerator.

The large amount of sand within the furnace is a heat sink that provides a significant thermal inertia within the system. This allows the furnace to be shut down with minimal heat loss. It is a relatively tight system; for example, the sand will retain heat to allow quick start-up after an overnight shutdown. A weekend shutdown will normally require only two or three hours of heating on a Monday morning. A multiple-hearth furnace, by comparison, cannot effectively maintain temperature when not burning sludge without the firing of supplemental fuel.

All the sludge ash and some sand become airborne and exit the furnace within the flue gas stream. The gas cleaning system has to be sized for this relatively high particulate loading.

As a result of the intimate mixing of air and sludge in the fluid sand bed, excess air requirements are low, generally 40%. The large volume (termed freeboard) within the furnace above the sand bed is maintained at 1500°F to 1600°F (816°C to 871°C). Residence time of the flue gases at these temperatures is normally sufficient to obtain complete burnout and to eliminate odor.

Fuel is used for start-up, for reheat, and, depending on the net heat content of the sludge, for maintenance of temperature during incineration. It can be injected directly into the bed or sprayed on top of the bed when the bed is fluidized and is at a temperature high enough to insure ignition.

The fluid-bed furnace is sensitive to waste consistency. Wastes containing soluble alkalies or phosphates, lead, or low-melting-point eutectics may cause seizure or fusion of the bed. A major issue associated with a fluid-bed furnace is its ability to handle a wide variety of waste streams. Certain wastes, particularly those containing salts, will tend to slag (seize) the bed, preventing fluidization. Test burns on materials that have not previously been fired in a fluid-bed furnace should be made to determine the bed reaction to that material. If bed seizure is indicated, bed additives may be available to help eliminate this problem by changing the physical properties (eutectic) of the waste feed. If additives are not available, a fluid-bed furnace may not be appropriate.

A related issue is agglomeration. Waste materials may build up on individual sand particles within the bed. With changes in operation (such as continuous bed withdrawal), this can be controlled. If not addressed, agglomeration can result in defluidization. This is of particular concern with sewage sludge generated in shore (coastal) communities where the sludge will have a relatively high salt content. The residual sodium chloride and other sea salts will bring the melting temperature of sludge below 1400°F (760°C), which can cause both agglomeration and seizure of the bed. Note that the multiple-hearth furnace is not as sensitive as the fluid-bed furnace to salts and other low-melting-point materials. In a multiple-hearth furnace, the wastes and their residuals are continually being wiped out of the furnace. By contrast, in a fluid-bed furnace, the wastes are resident within the bed and their high residence time encourages agglomeration and seizure.

An internal water-spray system is often employed in fluid-bed systems to help protect the air preheater from high-temperature excursions. Water normally is automatically injected into the freeboard when the heat exchanger inlet temperature is above a preset figure, generally no higher than 1600°F (871°C).

A fluid-bed furnace requires a minimum amount of air to maintain bed fluidization, regardless of the waste feed. When the bed is operating at a feed rate below the design capacity, approximately the same air quantity is required at the design load to maintain fluidization. This results in a system that may have good fuel consumption at its design point, but a relatively high fuel consumption at lesser feeds.

Ash from a fluid-bed furnace normally exits with the flue gas. Some materials, such as grit, when burned in a fluid-bed furnace, will not be airborne, but will build up within the bed. This requires periodic bed tapping. Provisions should be included in the fluid-bed system to accommodate the possibility of excessive bed buildup, which would include automatic and continous bed-tapping and sand-injection systems.

The air emissions control system must be designed to remove 100% of the sludge ash and elutriated sand from the gas stream. This requires high-energy venturi scrubbing systems with higher pressure requirements than many other incineration systems. Typical fluid-bed systems need scrubbers rated from 30 to 60 in. w.c. (16–31 mb).

Feeding of a fluid-bed furnace requires particular attention. Except for furnaces that use an ID fan where negative pressure is maintained in the freeboard and where waste can be dropped into the furnace by gravity, most fluid-bed furnaces are under positive pressure. Waste must be force-fed into the furnace and gravity feed cannot be utilized. Positive displacement pumps or screws are normally used for feeding fluid-bed furnaces. Sludge feeding is a potential problem area because of the tendency of the sludge within the feed lines to dry during periods when the furnace is maintained hot without sludge feed (hot standby conditions). Hard piping should be used from the pumps to the furnace and a hot-water purge system should be included to help prevent caking and to provide clearing of the piping when not in use.

Electric Furnace

The electric or radiant heat (or infrared) furnace is basically a conveyor belt system passing through a long, rectangular refractory-lined chamber, as shown in Figure 5. An ID fan maintains a negative pressure within the system.

Combustion air is introduced at the discharge end of the belt, shown in Figure 5 as the viewport. Air will pick up heat from the hot burned sludge as sludge and air travel countercurrent to each other. Supplemental heat is provided by electric infrared heating elements within the furnace above the belt. Cooling air is injected into the incinerator chamber to prevent local hot spots in the immediate vicinity of the heaters and is used as secondary combustion air within the furnace.

The conveyor belt is continuous woven wire mesh made of high-temperature steel. The refractory is not brick, but ceramic felt fiber. It does not have a high capacity for holding heat and can be started up from a cold condition relatively quickly (within one to two hours).

Sludge cake is fed by gravity onto the belt, and it is immediately leveled to a depth of approximately 1 inch (2.5 cm). There is no other sludge contact or sludge-handling mechanism. The belt speed and travel are sized to provide burnout of the sludge without agitation. This feature results in a relatively low level of particulate emissions.

Supplemental fuel, in this case, electricity, is required for start-up. Electric power also is used to provide the heat required to maintain combustion temperatures. Unfortunately, the power needed for start-up results in a large connected load.

Electric energy costs approximately four times that of fossil fuel on a heat-value basis. Therefore, unless the sludge will burn autogenously, an electric furnace may not be cost effective—and even with an autogenously burning sludge, the cost of installed kilowatts (demand charge) may be prohibitive.

Cyclonic Furnace

In the cyclonic furnace, the hearth moves and the rabble teeth are stationary (see Figure 6). Sludge is rabbled toward the center of the hearth, where, as ash, it is discharged. This type of incinerator is built in relatively small sizes and has been used in shipboard and shore (Navy base) applications.

Air is introduced at tangential burner ports on the shell of the furnace. The furnace is a refractory-lined cylindrical shell with a domed top. The air, heated with the introduction of supplemental fuel, creates a violent swirling pattern that provides good mixing of air with the sludge feed. The air, later flue gas, swirls up vertically in cyclonic flow through the discharge flue in the center of the domed roof. Sludge is fed into the furnace with a screw feeder and deposited on the periphery of the rotating hearth. A progressive cavity pump also can be used to feed sludge.

A variation of the cyclonic furnace is shown in Figure 7. This is a horizontal skid-mounted cyclonic reactor system. Ash is discharged from the furnace within the flue gas. Sludge is pumped into the furnace tangentially from the furnace wall. Air is introduced at tangential burner ports, creating a cyclonic effect.

There is no hearth, only the furnace shell and refractory. The sludge detention time is no greater than 10 seconds in this furnace. The products of combustion exit the furnace in vortex flow at 1500°F (816°C) and complete combustion is

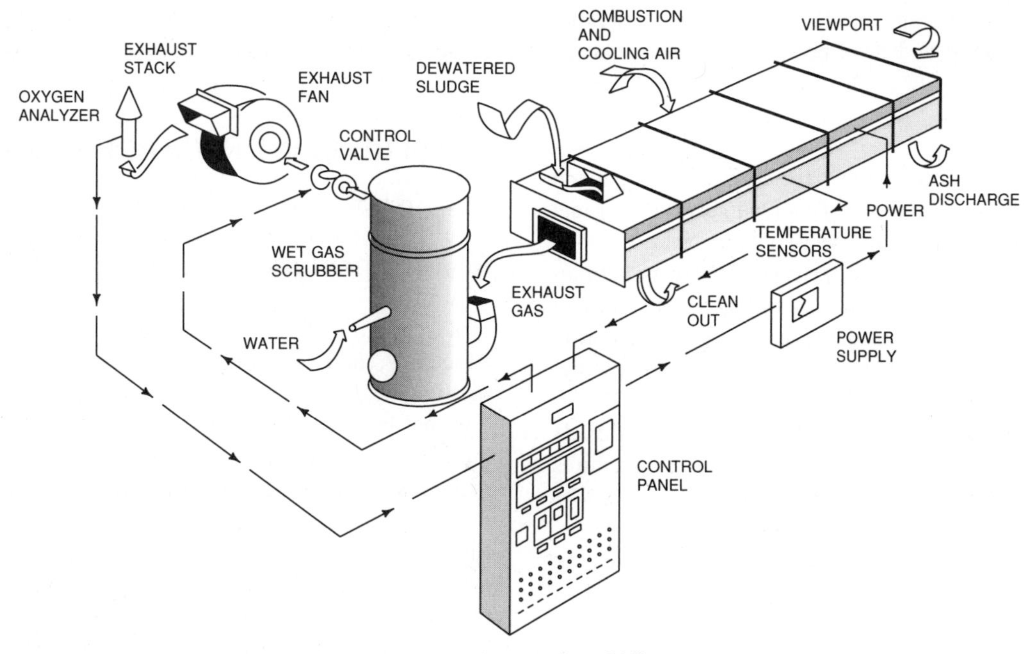

FIGURE 5. Radiant Heat Furnace (From Reference 4, p 235)

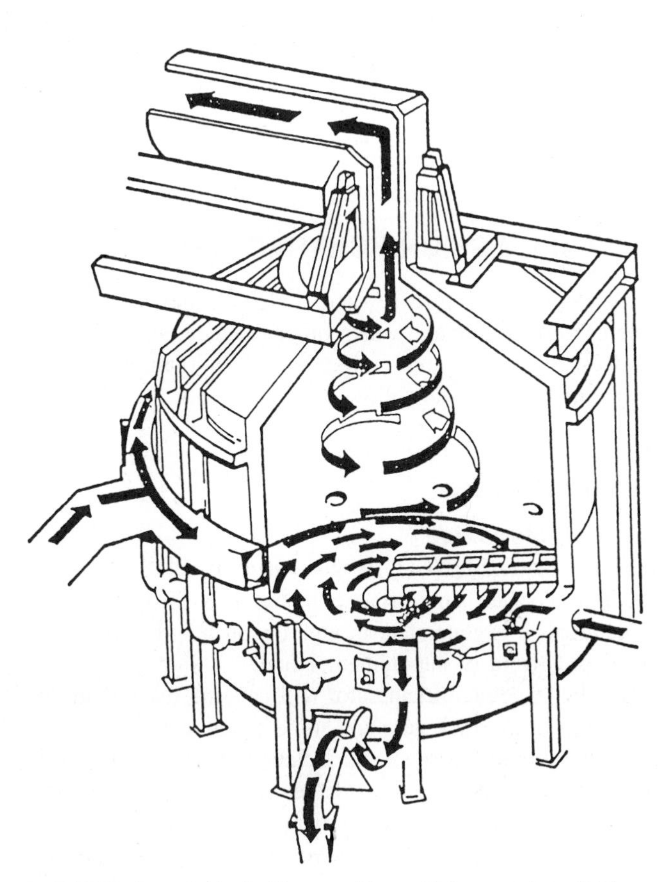

FIGURE 6. Cyclonic Furnace (From Reference 4, p 237)

assured. Temperatures within the furnace are from 1500°F to 1600°F (816°C to 871°C). These furnaces are relatively small and can be placed in operation, at operating temperature, within an hour's time.

Cyclonic furnaces are suited for sludge generated from relatively small influent flows, 2 million gallons per day and less. They are relatively inexpensive, mechanically simple units. These systems can be purchased skid mounted for installation as a complete independent package, requiring only utility and feed connections and a stack. Many

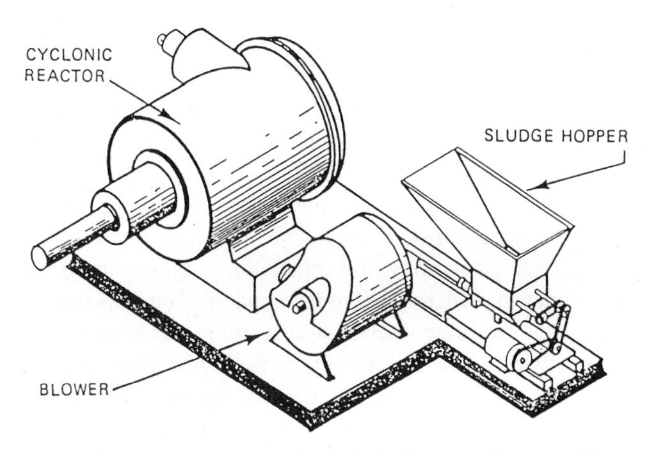

FIGURE 7. Skid-Mounted Cyclonic Reactor System (From Reference 4, p 237)

packaged units like this are sold for on-site sludge and other liquid waste disposal and have been used on shipboard.

Codisposal

A small number of installations have combined sludge and refuse disposal in a single unit. This technique has not met with much success where refuse is charged into a sludge incinerator. Refuse cannot be effectively fired in a multiple-hearth furnace. As noted previously, adding a solid fuel will result in burning too high in the furnace. Feed preparation also is required to size the refuse and to remove low-melting-point materials.

With codisposal in a fluid-bed incinerator, the materials within the refuse stream with low melting temperatures (such as aluminum cans and some glass products) must be removed from the charge prior to injection into the furnace. This is necessary to avoid slagging and subsequent seizure of the bed. The process of preparing the refuse, and of sizing it to less than a 1-inch (2.5-cm) particle size for effective feeding to the bed, has been maintenance intensive, unreliable, and costly.

The charging of sludge into a refuse incinerator has been found to be successful in a limited number of installations. The refuse moisture content is approximately 25%, whereas that of the sludge cake is approximately 75%. To assure that the sludge will burn out as the refuse burns within the incinerator, its moisture content must be reduced to that of the refuse. Drying equipment is used that generally utilizes heat from the gas stream exiting the refuse incinerator to dry the sludge to from 80% to 90% solids content. This high-solids sludge is mixed with sludge cake entering the facility (at 75% moisture content), generally in a pug mill, and the resulting sludge consistency is approximately 75% solids (or 25% moisture). The sludge cake is then introduced into the incinerator, either on the ceiling of the incinerator firing chamber or above the refuse as it enters the incinerator. The moisture-laden stream leaving the drying sludge is directed back to the refuse incinerator, where elutriated organics are destroyed.

The nature of emissions from sludge burning is different from that from refuse burning. In the former, the emissions are predominantly inorganic, whereas unburned organics may be the major emission of concern with the latter. While water scrubbers may be the control system of choice for the sludge incinerator, fabric filters or electrostatic precipitators (ESPs) may be required to capture the small organic constituents of the refuse exhaust stream. These two duties—that is, sludge burning and refuse burning—require emissions control devices that are incompatible. For this reason, the sludge feed rate in codisposal systems has been held to no more than 10% (by weight) of the refuse feed throughput. At this ratio, there has been no noticeable effect of sludge burning on emissions from a refuse incinerator.

The need for drying equipment and associated sludge receiving, storage, and handling facilities adds further operating requirements and costs to codisposal facilities. As a result, very few of these facilities are in operation today.

SLUDGE INCINERATION OVERVIEW

Past experience has indicated that the multiple-hearth furnace is a reliable and predictable sludge incinerator. However, it is relatively complex and requires a high degree of skill and dedication to operate properly. The fluid-bed furnace requires fewer operational decisions to operate satisfactorily, and is a far less complex machine than the multiple-hearth furnace. The fluid-bed furnace, however, does not have universal application. Not all sludges can be incinerated in this furnace. The presence of salts or other materials with low melting temperatures will agglomerate the sand bed and can cause bed seizure. Fuel usage in the fluid-bed furnace is generally lower than in a multiple-hearth furnace at the design point, but at lower feed rates, the fluid-bed furnace becomes less efficient.

Preventive maintenance requires more attention in a multiple-hearth than in a fluid-bed furnace; however, when a failure occurs in a fluid-bed furnace system, it is usually more serious (replacing a recuperator heat exchanger section, for instance).

Other types of equipment discussed here have limited application. The electric power rate structure may eliminate the electric furnace from consideration, and the cyclonic furnace is applicable only to smaller systems and may not be able to meet air emission requirements.

The selection of an appropriate sludge incinerator or thermal treatment system must be evaluated on a case-specific basis.

AIR EMISSIONS CHARACTERIZATION

The emissions from a sewage sludge incinerator, besides the carbon dioxide and water vapor produced from burning processes, will be both organic and inorganic and will generally include the following constituents.

Inorganic Gases

Carbon Monoxide

Carbon monoxide (CO) is generated in relatively small quantities when any organic material is burned. In incinerators, it is generated through the burning of the sludge feed and firing of supplemental fuel.

A well-operated incinerator will have no more than 200 ppm CO in its stack gas, and some states require less than half that amount from sludge incinerators. The CO generation rate will increase with a decrease in air supply and with a decrease in combustion temperature. The lower the CO concentration, the more effective and complete is the combustion process.

Nitrogen Oxides

Nitrogen oxide (NO) is a minor product of combustion. It is considered harmless; however, it will slowly oxidize to nitrogen dioxide (NO_2) and this gas is a problem. NO is the predominant nitrogen oxide from the combustion process; NO_2 will be generated at approximately 5% of the quantity of NO generated. Nitrogen oxides contribute to gross atmospheric effects such as haze and smog.

The generation of nitrogen oxides will increase with increasing temperature and, to a lesser extent, with an increase in the air supply. The sludge incinerator is generally limited to a temperature both within and outside the furnace of 1800°F (982°C). At this temperature, the generation of nitrogen oxides is relatively low and is not a major concern. The burner usually operates at a temperature much higher than the furnace temperature, in excess of 2200°F (1204°C). Often, nitrogen oxides are generated primarily by the supplemental fuel burner.

Another source of nitrogen oxides in a sludge incinerator is in the burning of the sludge itself. Organic nitrogen is present in the sludge and can be as high as 15% of the weight of sludge volatiles. As the sludge volatiles burn, this organic nitrogen will oxidize and generate nitrogen oxides.

Sulfur Oxides

Sulfur is a constituent of sewage sludge, representing up to 2% of the weight of the sludge volatiles. In the burning process, sulfur will oxidize to sulfur dioxide (SO_2) and sulfur trioxide (SO_3). Less than 5% of the sulfur oxides will be SO_3, which is a less desirable compound. It is readily soluble in water and will form sulfuric acid, which is highly corrosive. All sulfur oxides have gross atmospheric effects, including the generation of smog and tree damage.

Hydrogen Chloride

When a chlorine-containing organic compound (organic chlorine) is burned, the chlorine will combine with hydrogen from the water produced by the burning material to generate hydrogen chloride, HCl. The HCl is soluble in water and will readily form hydrochloric acid. As with SO_2, HCl is highly corrosive and does contribute to gross atmospheric effects.

Particulate Matter

Ash

A fraction of the sludge ash will elutriate into the gas stream. In a fluid-bed incinerator, the entire ash load, plus some bed sand, will enter the gas stream, whereas only from 5% to 10% of the sludge ash will enter into the gas stream in an electric furnace. In a multiple-hearth furnace, from 15% to 25% of the ash produced by the burning sludge will enter the gas stream.

The amount of ash elutriated into the exhaust gas is a function of the turbulence the sludge experiences in the burning process. In a fluid-bed furnace, the sludge experiences maximum turbulence. The bed is designed with turbulence in mind, which is the mechanism utilized for effective sludge burning. In the electric furnace, the sludge lies on the conveyor belt and no motion is applied to the sludge at all.

Heavy Metals

Heavy metals will generally concentrate in a sludge. They enter the plant with the influent, and the amount and nature of these metals are related to the industrial profile of the service area of a plant. In general, the more industrial users that feed the plant, the higher will be the heavy metal loading.

Metals may enter the sludge as part of organic compounds, as metals, or as metal oxides. Within the incinerator, they will oxidize and either stay with the ash or exit in the exhaust as a particulate or as a vapor. As the furnace temperature increases, the release of heavy metals to the incinerator hot gas stream increases. One limitation in furnace operation is in the temperature above the burning hearths. At a temperature in excess of 1650°F (899°C), the release of heavy metals will significantly increase. For this reason, the furnace temperature should not be allowed to exceed 1650°F (899°C).

The heavy metals of greatest concern in the sludge incineration process are arsenic, cadmium, chromium, lead, mercury, and nickel. Most of these metals and their oxides (and other compounds) have been found to be toxic when ingested in relatively large quantities.

Metal oxides will either be adsorbed onto other particulate matter or be free within the gas stream. When adsorbed onto particulate, they will be removed from the gas stream as the particulate is removed. As small free oxides, or as metallic oxides in a gaseous form that has condensed through a quench or other low-temperature process, they are generally less than ½ μm in mean particle size. This is too small for many conventional scrubbing systems to remove, and they will likely pass through a wet scrubbing system and discharge through a stack.

The effectiveness of removal of these metals through a water scrubber is noted in Table 1.

TABLE 1. Heavy Metals Removal by Wet Scrubbing

Metal	% Removed
Arsenic	96%
Beryllium	99%
Cadmium	65%
Chromium	96%
Lead	67%
Mercury	0%
Nickel	95%

Source: Reference 1.

Hydrocarbons (Organics)

Volatile organic compounds (VOCs) are undesirable in a furnace off-gas. They can be visible, generating a smoke, and can result in the presence of odor. Some of these organics will be condensible, that is, they will condense out of the gas stream when the stream is reduced in temperature to below 400°F (204°C).

The presence of these organics results from two factors. First, the lighter organic fraction will tend to vaporize as the sludge first enters the incinerator. If the temperature at the sludge entrance is too high, the release of VOCs will increase. If there is insufficient residence time at a temperature in excess of 1400°F (760°C), the temperature required for destruction of the majority of VOCs, these compounds will be present.

The second factor contributing to the generation of these compounds is the lack of sufficient oxygen in the incinerator. The less oxygen there is, the more likely it is that incomplete combustion will result, and incomplete combustion is, almost by definition, a designation of the presence of organics (VOCs) in the furnace exhaust. In a multiple-hearth furnace, the oxygen level in the exhaust should not be allowed to fall below 6%. In a fluid-bed furnace, the exhaust oxygen should always be above 4% (measured on a dry volume basis).

Although not always the case, operation of the multiple-hearth furnace may result in release of a greater quantity of VOCs than will a similarly sized fluid-bed furnace. This can occur as a result of higher temperatures than desired on the top hearth of the furnace. At times, the top-hearth temperatures may be difficult to control. When skimmings are mixed with the sludge, or if industrial wastewater going to the plant generates additional volatiles in the sludge feed, a higher release of VOCs will take place.

An afterburner may be required in a multiple-hearth installation to control VOC destruction. Many multiple-hearth furnaces have an expanded top hearth where residence time is designed for a minimum of one second. This residence time is limited by the geometry of the furnace. Another limitation is the temperature to which the top hearth of the furnace can be brought. At higher than 1400°F (760°C), sludge wiped across, or even dropped through, the top hearth will begin to heat up, and by the time the sludge reaches the burning zone of the furnace, it may be hot enough to slag. In this case, an external afterburner is desirable. With the off-gas from the incinerator at or less than 1200°F (649°C), a separate chamber at the outlet of the top hearth of the incinerator would be designed for a minimum 1400°F (760°C) temperature maintained for two seconds, minimum, to assure the complete destruction of VOCs without affecting sludge burning within the furnace.

In a fluid-bed furnace, VOC release will increase if the temperature within the bed is too low or the bed has insufficient height, or with too high a velocity (excessive fluidizing airflow) through the bed, VOCs are released into the freeboard. Rather than provide additional afterburning equipment, the operation of the fluid-bed furnace can usually be adjusted to minimize the release of VOCs into the exhaust.

Polycyclic organic matter (POM), which includes polychlorinated biphenyls (PCBs), dioxins, and furans, are suspected carcinogens. As techniques for detecting these compounds are becoming more and more sophisticated, they can be found in concentrations of one part per trillion or less. It appears that at these low levels, POM is generated in practically every combustion process, from cigarette smoking to automobile exhaust and from charcoal broiling to incineration.

The generation of POM from a sludge incinerator has not been found to date to result in a significant health risk. The control of these compounds is through efficient combustion. It has been found that when the generation of carbon monoxide is low and the generation of VOCs is low, the POM generation rate is also low.

Opacity

Opacity is a measure of the light interference of an emission. As shown in Figure 8, light interference (a measure of opacity) will vary with particle size. For a given particulate emission rate, the smaller the particle size, the less visible it will be. At the same rate of emissions, for larger particles, fewer particles will be present and their visibility less. In the range of 0.3 μm to 0.7 μm mean particle size, opacity will be greatest. Noting these properties, it is not possible to relate opacity to an emissions quantity without knowing the particle size distribution of that emission. A low emission might be visible if the 0.5-μm particles predominate; however, if the particle size were less than 0.2-μm, the visibility could be nil.

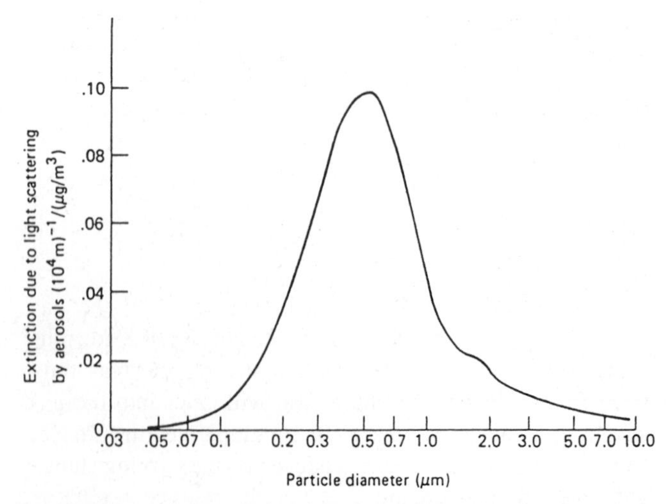

FIGURE 8. Light Scattering as a Function of Particle Diameter (From Reference 4, p 318)

AIR EMISSIONS CONTROL

Air emissions control is critical to an incinerator system. Particulates, including metals emissions, hydrocarbon emissions, and both organic and inorganic gaseous discharges, are of concern, as are opacity and odor.

Control devices can limit particulate, metals, and acid gas emissions from the stack discharge, but the control of other gaseous, condensible and noncondensible hydrocarbon emissions is generally a function of the combustion process and may not be specifically affected by gas scrubbing or conditioning.

Metals and other inorganics (ash), as well as some noncondensible organics, will form into or become attached to particulate matter within the exhaust gas stream. When removing particulate from the gas stream, therefore, most of the metals and many of the organic constituents, as well as most of the elutriated ash, will be removed.

The Hot Gas Stream

The gas stream from an incinerator has a very high moisture content. The sludge moisture, which is approximately 80% of the sludge weight, will eventually be discharged in the incinerator off-gas. The gas exiting the incinerator also will contain organic gases, some of which will be condensible. When the gas is reduced in temperature, these condensibles will drop out of the gas and will produce a highly corrosive liquid.

The high moisture content of the gas stream discourages the use of a fabric filter (baghouse) for emissions control. The effectiveness of a baghouse requires that the accumulated particulate matter on the bags be cleaned from them effectively and regularly. With the high moisture content of the exhaust gas, the particulate will tend to cake on the bags and blind them. It would be extremely difficult, if not impossible, to keep the bags clean and functioning (filtering the gas stream) in this moist and dirty gas environment.

With regard to other dry systems for air emissions control, these systems lack the means to capture and remove condensibles from the gas stream. They must be used after the gas stream is wetted. Such wetting will remove the condensibles from the gas, but also will produce a gas stream saturated with moisture.

For these reasons, the conventional baghouse or the dry ESP is not applicable to sludge incineration systems. In all sewage sludge incinerations systems in this country, wet scrubbing is employed, although additional air emissions control systems are used downstream of the scrubber in some installations.

Particle-Size Analysis

The efficiency of any particulate control device will vary as a function of the particulate size. Particulate matter is normally designated by mean diameter, measured in microns (μm, 1 millionth of a meter).

The particulate size distribution of an emission cannot be calculated, but can only be inferred from prior data based on measurement. As would be expected, data on particle size distribution are rare. It is a function of waste properties, furnace design, and furnace operation, and will vary from test to test. It is virtually impossible to define particle size distribution with any degree of confidence prior to construction and operation of the incinerator in question. As an estimate of the type of distribution that can be expected, note Figures 9, 10, and 11, for multiple-hearth, fluid-bed, and infrared incinerators. These are only estimates, and should not be used as a basis for systems or equipment design.

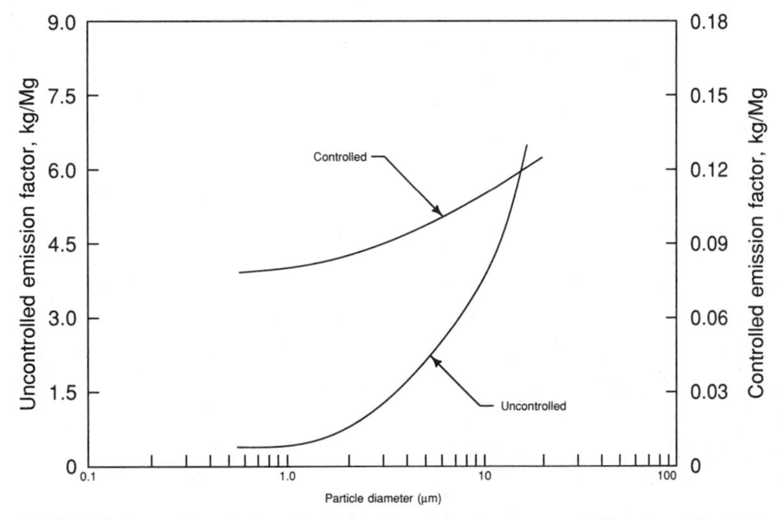

FIGURE 9. Cumulative Particle Size Distribution and Size-Specific Emissions Factors for Multiple-Hearth Incinerators (From Reference 2)

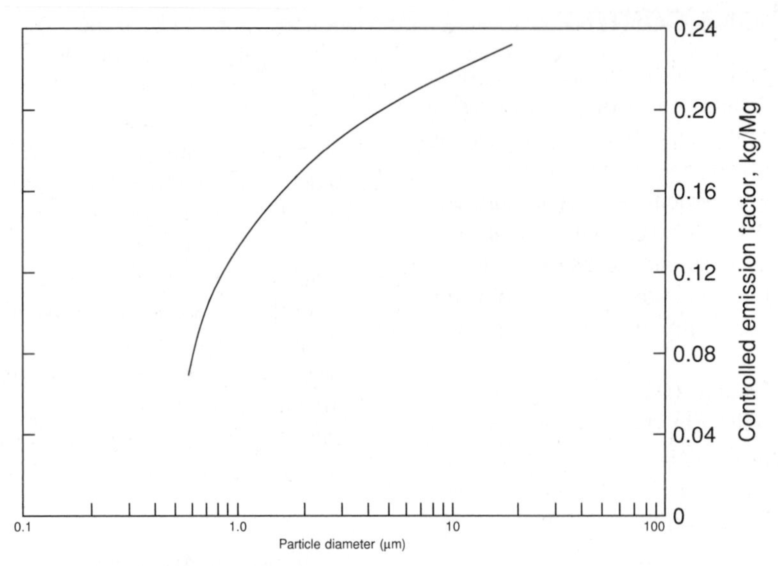

FIGURE 10. Cumulative Particle Size Distribution and Size-Specific Emissions Factors for Fluidized-Bed Incinerators (From Reference 2)

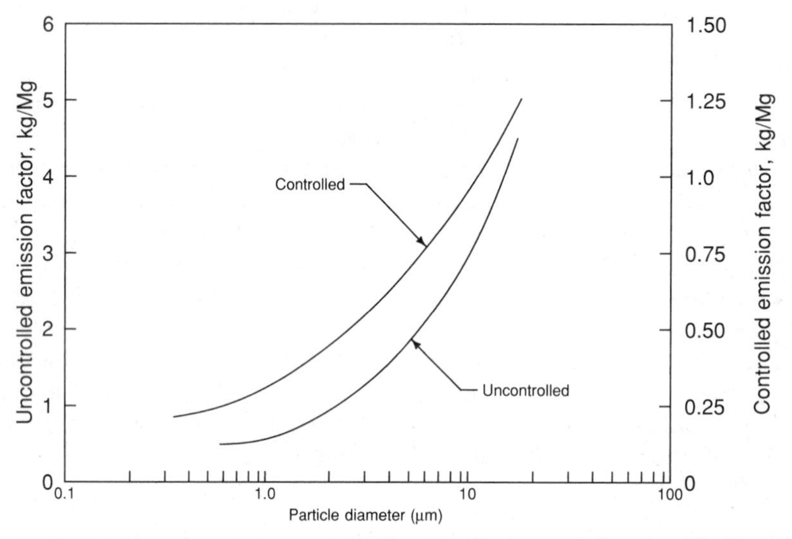

FIGURE 11. Cumulative Particle Size Distribution and Size-Specific Emissions Factors for Electric (Infrared) Incinerators (From Reference 2)

Wet Scrubbing

Wet gas scrubbing with plant effluent is used almost universally for sludge incinerator exhaust gas treatment. Wet scrubber operation is described in the chapter on particulate removal. Table 2 lists typical emissions from sewage sludge incinerators using wet scrubbers.

The effectiveness of a scrubbing system is usually directly related to the pressure drop across the scrubber. The higher the pressure drop, the greater is the turbulence/mixing, and, therefore, the more effective is the scrubbing action. This feature is illustrated by the graph in Figure 12. For a 2-μm-diameter particle, for instance, a pressure differential of less than 3 in. w.c. (1.6 mb) will provide a removal efficiency of 99%, whereas a 20-in. w.c. (10.5-mb) differential is necessary for 99% removal of a 0.8-μm particle from the gas stream.

Scrubbing systems are often described by system pressure drop. A low-energy system is normally defined as one producing less than 12 in. w.c. (6.3 mb) for particulate removal, whereas high-energy systems will have significantly higher pressure differentials, from 20 to over 60 in. w.c. (10 to over 31 mb). Likewise, medium-energy systems operate with a 7- to 20-in. w.c. (3.7- to 10.5-mb) differential.

Gases exiting a sludge incinerator will contain hydrogen chloride and sulfur oxides. These constituents result in acid formation in the gas scrubbing liquid, as do condensible organic residuals. The use of nonmetallic chambers or nonmetallic coated carbon steel is a growing method of

TABLE 2. Emission Factors for Sewage Sludge Incinerators
Values in Pounds per Dry Ton (kg/Mg Dry)

	Multiple Hearth		Fluid Bed		Electric	
Pollutant	Uncontrolled	After Scrubber[a]	Uncontrolled	After Scrubber[a]	Uncontrolled	After Scrubber[a]
Particulate	80 (40)	0.80 (0.40)	NA[b]	6.0 (3.0)	20 (10)	4.0 (2.0)
Lead	0.10 (0.05)	0.06 (0.03)	NA	0.006 (0.003)	NA	NA
Sulfur dioxide	20 (10)	4.0 (2.0)	20 (10)	4.0 (2.0)	20 (10)	4.0 (2.0)
Nitrogen oxides	11 (5.5)	5.0 (2.5)	NA	2.0 (1.0)	8.6 (4.3)	6.0 (3.0)
Carbon monoxide	60 (30)	4.0 (2.0)	NA	4.2 (2.1)	NA	NA
VOC—methane	4.6 (2.3)	4.6 (2.3)	1.6 (0.80)	1.6 (0.80)	NA	NA
—nonmethane	1.7 (0.85)	1.7 (0.85)	NA	NA	NA	NA

[a]Scrubbers are either venturi or venturi plus impingement tray.
[b]NA = not available.
Source: Reference 2.

corrosion control. These materials, such as flake polyester, fiberglass, neoprene, and epoxy, have good anticorrosion properties; however, they are temperature sensitive. In most cases, they cannot withstand temperatures above the range of 160°F to 200°F (71°C to 93°C). Use of such materials must include provisions for emergency quenching to provide cooling if the normal cooling water supply is lost.

Wet gas cleaning equipment floods a gas with scrubbing liquid. Water droplets usually are carried off by the gas stream exiting the scrubber. A mist eliminator is a passive component that removes most of the entrained water droplets from the gas stream. This is desirable to remove collected particulate that is sorbed onto the water particles and to reduce the size of the plume exiting the stack (excess

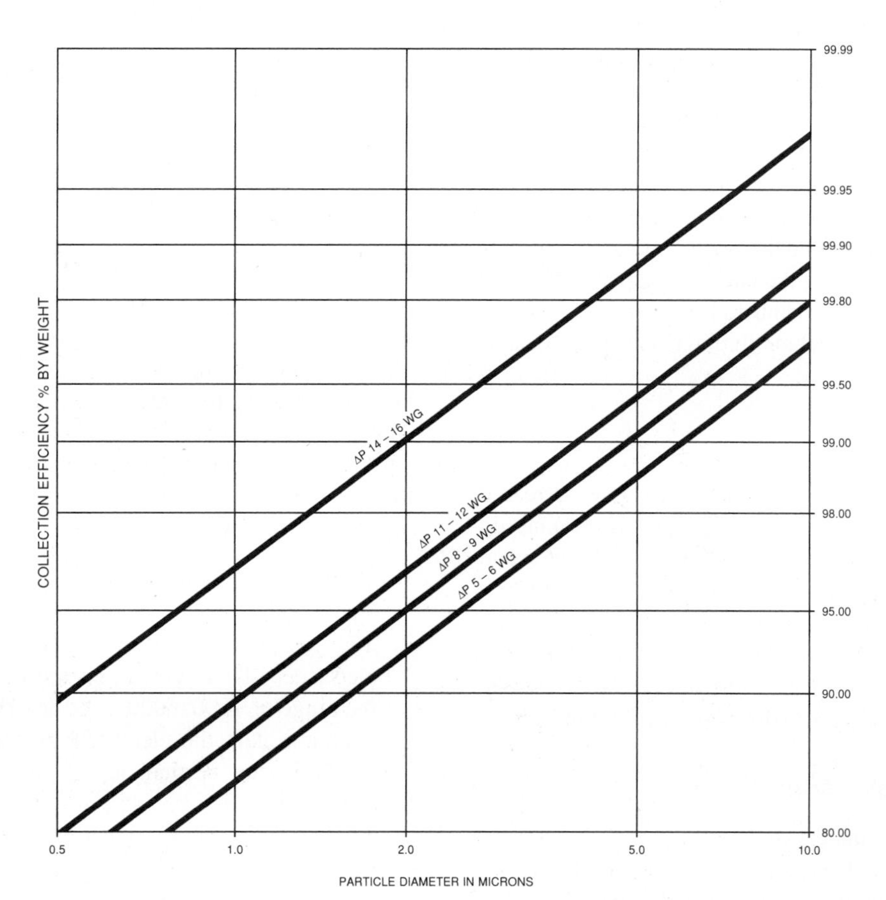

FIGURE 12. Collection Efficiency Versus Particle Size (From Reference 4, p 331)

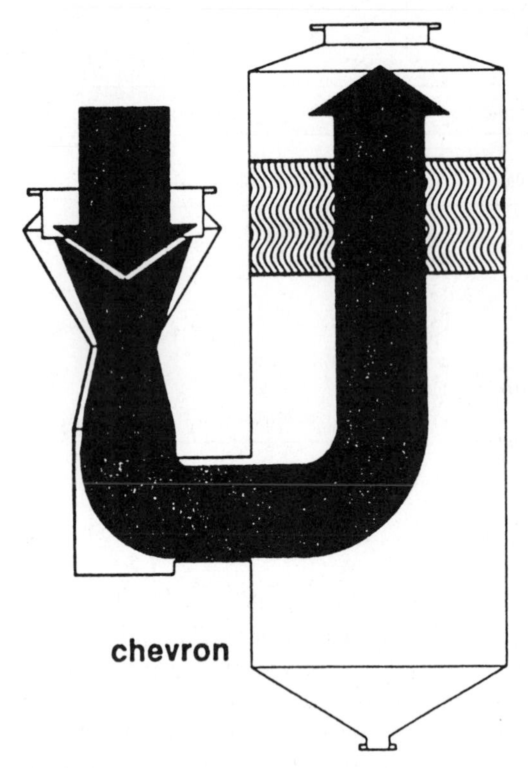

FIGURE 13. Chevron-type Mist Eliminator (From Reference 4, p 328)

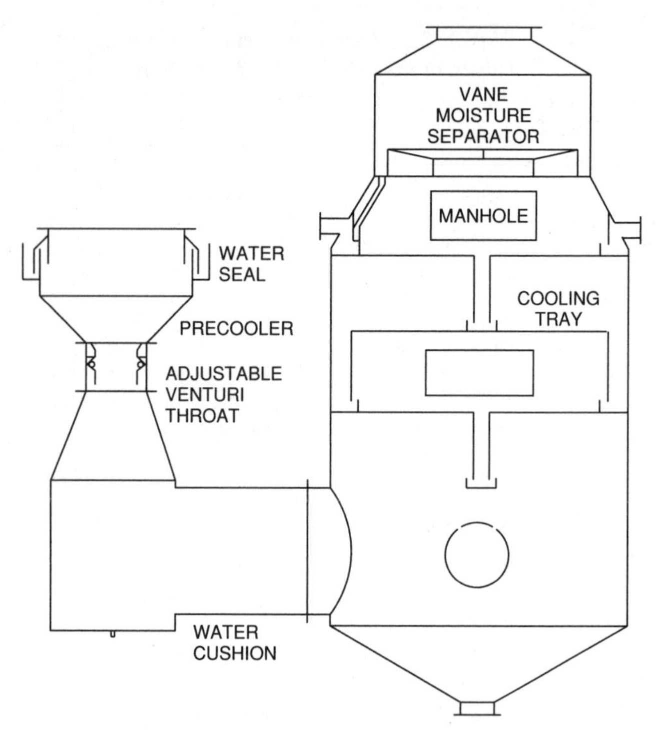

FIGURE 14. Variable Throat Venturi Scrubber (From Reference 4, p 334)

water in off-gas will appear as a white plume at the stack discharge), and will reduce the moisture collected in the ID fan and other downstream equipment. Decreasing this liquid accumulation will also decrease attendant corrosion.

One of the more common mist eliminators in use on sludge incinerators is the chevron type, Figure 13. The moisture-laden gas stream travels through a series of baffles. Gas exits the baffles and the heavier moisture particles, because of their inertia, hit the chevron-shaped baffles and fall to the bottom of the chamber, leaving the gas stream. Other mist eliminators include impingement plate designs and packed towers.

Venturi Scrubber

Venturi scrubbers are widely used, where water is readily available, as high-efficiency, high-energy gas-cleaning devices. The heart of the system is a venturi throat where gases pass through a contracted area, such as that shown in Figure 14, reaching velocities of 200 to 600 fps (6096 to 18,288 cm/s), and then pass through an expansion section. From the expansion section, the gas enters a large chamber for separation of particles or for further scrubbing.

Impingement Tray Scrubbers

Impingement tray scrubbers are essentially perforated plates with target baffles. Tray scrubbers have no large gas-directing baffles, but are simply perforated plates within a tower, usually immediately downstream of a venturi. A water level is maintained above the trays (there are usually two or more trays). The geometrical relationship of the tray thickness, hole diameter, and spacing, as well as the impinger details, results in a high-efficiency device for the removal of small particulate of less than 2 μm in mean diameter.

Wet Electrostatic Precipitator

Although wet ESPs have been used on very few sludge incinerator installations to date, they have been found to be effective devices for the removal of airborne particulate matter. These existing installations utilized wet ESPs after a conventional scrubber as a retrofit to reduce the scrubber particulate and heavy metals emissions discharge. One of these systems is shown schematically in Figure 15 and operates as follows:

1. The hot gas stream is initially wetted either by upstream equipment or by the water sprays within the wet ESP.
2. The gas stream passes through a series of discharge electrodes. These electrodes are negatively charged, in the range of 1000–6000 volts dc. This voltage creates a corona around the electrode. A negative charge is induced in the particulate matter passing through the corona.
3. A grounded surface, or collector electrode, surrounds the discharge electrode. Charged particulate will collect on the grounded surface.

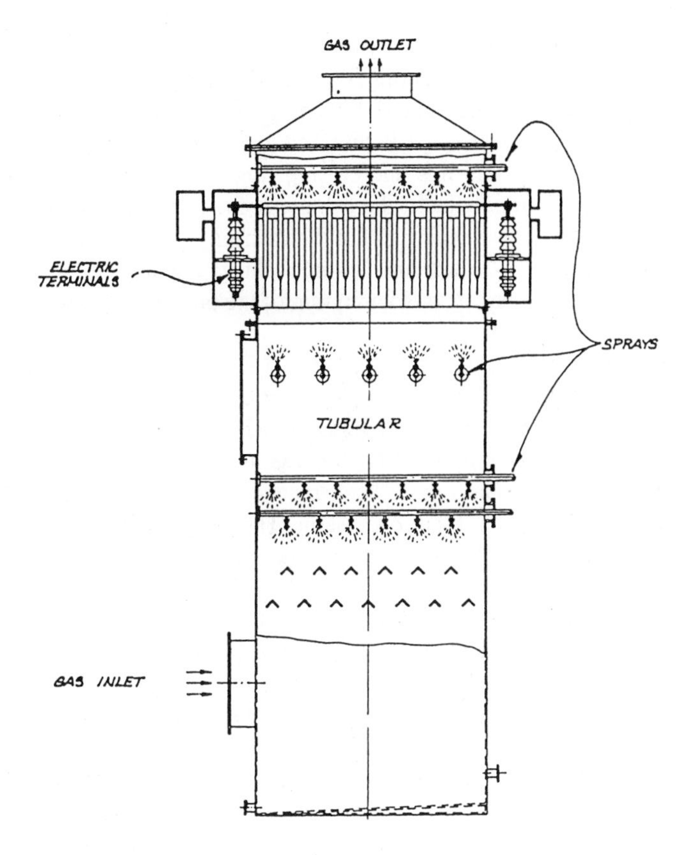

FIGURE 15. Wet Electrostatic Precipitator (Courtesy Beltran Associates Inc., Brooklyn, N.Y.)

TABLE 3. Particulate Discharge After Wet Electrostatic Precipitator *Maryville, Tenn., Wastewater Treatment Plant*

	Particulate Discharge, lb/Dry Ton (kg/Mg)
Test 1	0.49 (0.25)
Test 2	0.23 (0.12)
Test 3	0.21 (0.11)
Average	0.31 (0.16)

Source: Reference 3.

4. Particulate matter will be removed from the collector surface by intermittent water sprays directed within the electrode–grounding-surface assembly.

The wet ESP is very efficient in the collection of small particulate, down to the submicron range. Table 3 lists the results of particulate emissions tests (three replications) measured downstream of a wet ESP. This precipitator is located after a conventional venturi–impingement-tray scrubber on a multiple-hearth sludge incinerator.

References

1. *The Federal Register,* February 6, 1989, p 5890.
2. Draft report. *Emission Factor Documentation for AP-42 Section 2.5—Sewage Sludge Incineration,* Monitoring and Data Analysis Division, Office of Air Quality Planning and Standards, U.S. Environmental Protection Agency, Research Triangle Park, NC, September 1987.
3. *Maryville Wastewater Treatment Plant Sewage Sludge Incinerator Test Report,* EMI Consultants, Knoxville, TN, September 18, 1986.
4. C. R. Brunner, *Incineration Systems: Selection and Design,* Incinerator Consultants Inc., Reston, VA, 1988.
5. C. R. Brunner, "Sewage sludge incineration at the Cleveland Southerly wastewater treatment center," *Proceedings of the 16th Intersociety Energy Conversion Conference,* Atlanta, 1980.

Bibliography

C. R. Brunner, *Design of Sewage Sludge Incineration Systems,* Noyes, Park Ridge, NJ, 1980.

C. R. Brunner, *Hazardous Air Emissions from Incineration,* 2nd ed., Chapman & Hall, New York, 1986.

R. Coulter, "Smoke, dust, fumes closely controlled in electric furnaces," *Iron Age,* 173:107 (1954).

T. D. Ellis, *Industrial Hazardous Waste Incineration Course Notes,* AIChE Short Course, New York, 1987.

M. Kraus, "Baghouses," *Chem. Eng.* (1979).

National Air Pollution Control Administration, *Control Techniques for Particulate Pollutants,* AP-51, 1981.

Water Pollution Control Federation, *Incineration Manual of Practices,* OM-11, 1988.

DRUM RECONDITIONING

Stanley M. Krinov

Steel-drum reclamation (reconditioning) is an important segment of the steel-shipping-container industry. Approximately 50 million 55-gallon drums are reconditioned and reused annually.[1] This compares with approximately 25 million new drums (of the same sizes and gages) manufactured annually.[2] Accordingly, roughly twice as many reconditioned used drums than new drums of 55-gallon size, 20 guage or heavier, are filled and shipped in the United States. This represents a very important savings in resources (1.25 million tons of steel) and a reduction in solid waste (disposal of empty drums).

The U.S. Department of Transportation (DOT) regulates the construction, testing, and marking, as well as the filling and shipping requirements, of drums used for shipment of hazardous materials. The DOT standards are generally parallel with other standards-setting organizations (UN, UFC, NMFC, ANSI).*

The U.S. Environmental Protection Agency (EPA), under the authority of the Resource Conservation and Recovery Act (RCRA), regulates the use of steel drums for the storage and disposal of hazardous wastes and, under the Federal Insecticide, Fungicide, and Rodenticide Act (FIFRA), the procedures for the disposal of empty pesticide containers.

Steel drums essentially are used for every type of liquid and semiliquid material that is shipped in bulk in commerce, including petroleum products, industrial chemicals, paints, varnishes, inks, adhesives, resins, food and agricultural products, cleaning products, pesticides, and industrial wastes.

There are about 250 reconditioning plants in the United States, 50% of which account for more than 90% of the volume.[3] Although steel shipping containers are made in many standard sizes, ranging from 5 to 57 gallons, the reconditioning industry generally restricts itself to the 55-gallon (208-liter) and the 30-gallon (114-liter) sizes.

The two general types are the tight-head and the fully removable head (open-head) drum. About one third of the drums processed by the reconditioning industry are of the open-head type. Both tight-head and open-head new drums can include a removable plastic liner to isolate the contents from the metal to avoid product contamination or corrosion. About half of the open-head drums are lined with a sprayed-on and baked, chemically resistant coating.

Many reconditioning plants "convert" tight-head drums into the open-head type by "deheading and beading." In these operations, one of the drum heads (usually the top head) is cut out, and the cut edge of the body is form-rolled into a bead, similar to the top bead of an open-head drum. Some plants also produce "rebuilt" tight-head drums by flanging instead of beading the cutout drum, then processing the unit through the open-head line. Near the end of the process, a new drum head is seamed onto the body to reconvert it to a tight head, similar to the way a drum head is attached in the manufacturing process for new drums.

PROCESS DESCRIPTION

Tight-head drums are reconditioned by a washing process, whereas open-head drums are reconditioned by a burning (pyrolysis) process. The open-head process steps comprise removal of closing rings and covers (lids), inspection and grading, burning or pyrolyzing, abrasive blasting, straightening and dedenting, leak testing, interior coating (lining) and baking, exterior coating (painting) and paint baking, and, finally, gasketing and assembly (heading up and marking). The lids and closing rings are processed in steps paralleling those of the drums. Some plants clean the closing rings by washing rather than burning. The threaded bung closures (2-inch and ¾-inch plugs) are also either washed or burned. A block diagram of the process is shown in Figure 1.

The pyrolysis step is accomplished in a continuous tunnel furnace, the drum reclamation furnace. A process diagram of the pyrolysis system is shown in Figure 2. The diagram shows not only the furnace and afterburner, but also an integrated waste-heat recovery boiler, a spray quench unit, and a predrain system. The ancillary units are described later. The purpose of the pyrolysis step is to change the previous coatings (i.e., the paint and linings), as well as the adhering and residual contents of the drums, into a dry and friable condition, which allows them to be easily removed in the subsequent abrasive blasting operation.

The drums are conveyed through the furnace in their inverted position by means of a drag-chain conveyor. The lids (also the rings and plugs) are included by placing them on top of the inverted drums or they can be pyrolyzed in a separate furnace. The furnace feed rates range from two or three drums per minute to 10 to 12 drums per minute.

Drum-reclamation furnaces are constructed in a wide variety of sizes and configurations. Many are of "in-house" design. A typical design consists of a refractory-lined tunnel furnace with a number of natural-gas–fired burners mounted in the sides of the tunnel. (A few furnaces are fueled with No. 2 fuel oil.)

The furnace can be viewed as being divided into three zones. After entering the tunnel, the drums pass through the preheat zone, where they are heated by radiation from the burning zone. They then pass through the ignition zone, where combustibles in contact with furnace flames ignite and burn. Last, they pass through the cooling zone, where a small amount of burning continues and the drums are cooled somewhat by the induced air. The exhaust gases from the furnace pass into an afterburner incinerator, where smoke and the unburned or partially burned vapors are completely burned. Figure 3 is a view of an installation looking into the feed end of the furnace, showing the empty drums arriving by conveyor to the operator, who overturns the drums onto the furnace conveyor. The afterburner and the waste-heat recovery boiler are located at the left of the furnace.

---

*U.S. Department of Transportation (DOT), Hazardous Materials Transportation Act (HMTA), 49 CFR Part 178.

United Nations (UN) Economic and Social Council, *Transport of Dangerous Goods*, Recommendations of the Committee of Experts.

Uniform Freight Classification (UFC) Rule 40, Uniform Classification Committee, agency of the railroads.

National Motor Freight Classification (NMFC) Rule 260, National Classification Board, agency of the motor carrier industry.

American National Standards Institute, Inc. (ANSI), *Universal Steel Drums and Pails* (MH2.1–2.6, MH2.11–2.13).

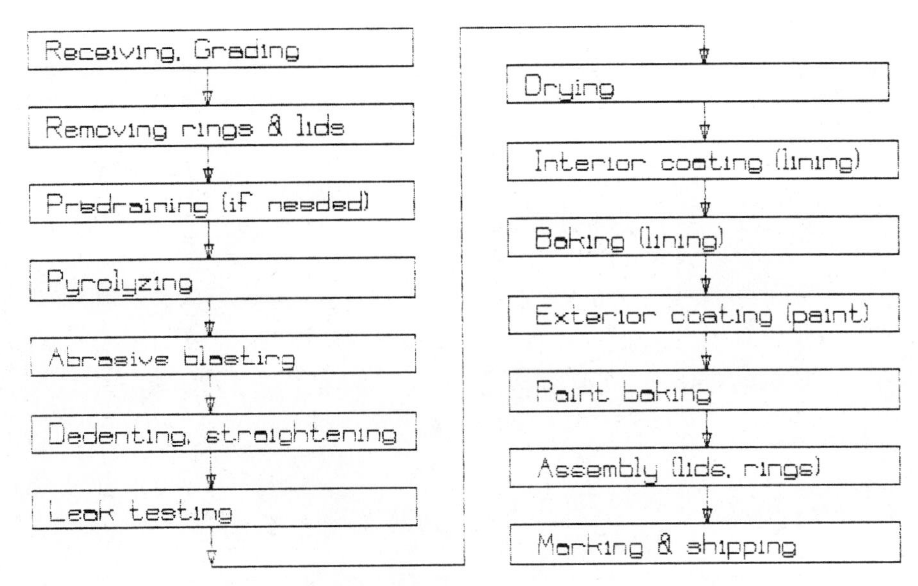

FIGURE 1. Open-Head Drum Reconditioning Process Flow Diagram

Afterburners

The afterburner incinerator, when designed as an integral part of the drum furnace structure, is a rectangular chamber located above the furnace tunnel, the roof of the tunnel being the base of the afterburner chamber. Alternatively, the afterburner can be a separate, free-standing chamber connected to the furnace exhaust outlet by a refractory-lined breeching, as shown in Figure 3.

The controlling mechanism for the combustion reaction, at the temperatures recommended for afterburner incinerators, is the rate of mixing of the fuel (organic vapors) with air (oxygen). Therefore, the design of the afterburner should provide good mixing by means of baffles and abrupt right-angle turns of the flow path or by inducing cyclonic flow patterns.

Waste-Heat–Recovery Boilers

Over the past decade, many of the drum furnaces have been retrofitted with waste-heat–recovery boilers. The boiler installations are designed for either side-stream or integrated (full-flow) operation. In the side-stream design, some of the flue gases from the afterburner are drawn through the boiler by means of an induced-draft (ID) fan and the balance of the flue gases exhaust up the stack by natural draft. In the

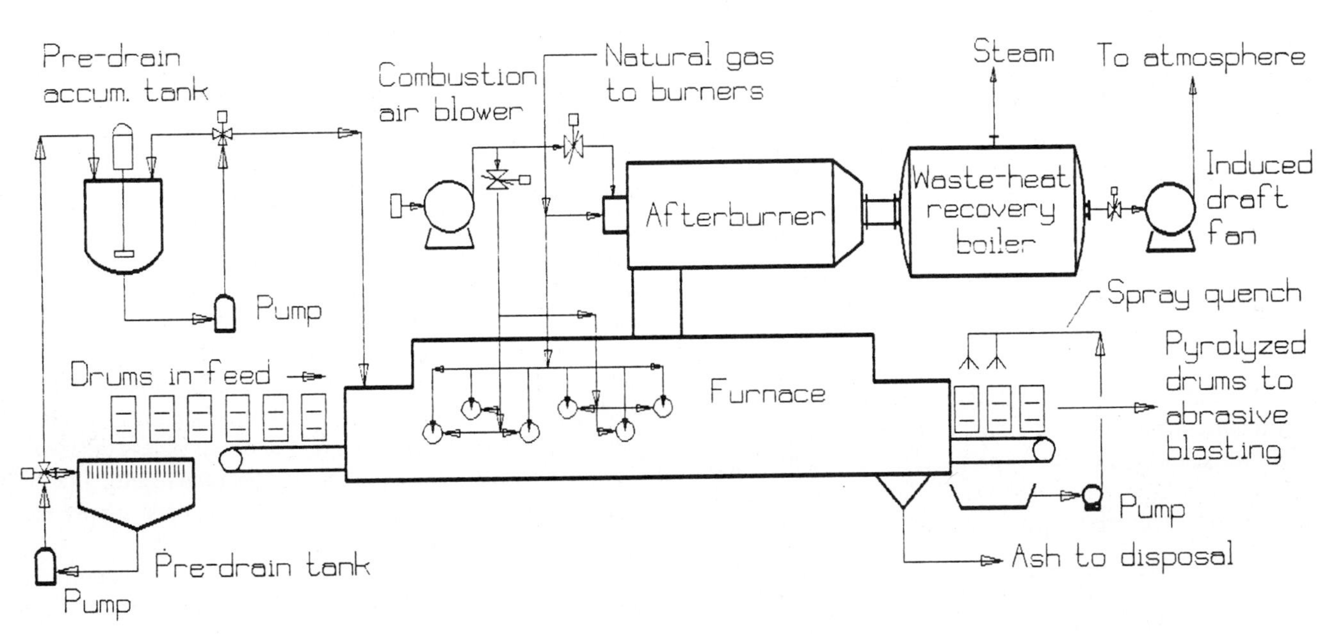

FIGURE 2. Drum Reclamation Furnace System with Afterburner—Waste-Heat–Recovery Boiler, Spray Quench, and Predrain System

FIGURE 3. Furnace System with Cylindrical Afterburner Installed Above Furnace. *Waste-heat recovery boiler and induced draft fan installed on grade* (Spencer Boiler and Engineering, Inc.)

integrated design, the entire flue gas stream is drawn through the boiler by the ID fan. The choice of side-stream versus integral design depends on many factors. In the side-stream design, the rate of withdrawal of hot flue gases is dictated by the facility's steam demand, whereas in the integrated design, all of the recovered heat is converted to steam and any excess steam above demand is vented to the atmosphere.

The chief disadvantage of the side-stream design is that the flow of induced combustion air through the furnace is limited by the size of the stack and, being dependent on natural draft, is not amenable to good control. Consequently, when drums being processed contain relatively small amounts of combustible material, the percent of excess air becomes extremely large and requires the consumption of large amounts of natural gas in order to maintain the required incineration temperatures.

The disadvantage of the integrated design is that the furnace operation is interrupted when the boiler must be shut down for maintenance. Interrupted operation can be avoided, however, by designing the unit to enable bypassing of the boiler. In this case, of course, the stack must be sized for a natural draft capability appropriate to the production requirements during the boiler outage.

Facilities that have little or no need for steam obviously will not install waste-heat–recovery boilers. In these cases, in order to protect the ID fan from the high temperatures of the afterburner gases, the units are designed to induce ambient air for tempering at the fan inlet. An example is shown in Figure 4.

The Combustion Process

The combustion process in a drum furnace system is quite complex. There are at least four sources of air for combustion: the air induced into the entrance and exit openings of the furnace tunnel and the forced air provided by the combustion air blowers for the gas burners in the furnace and the afterburner. There are at least three sources of fuel: the natural gas to the furnace and the afterburner burners and the waste fuel in the drums being fed. There can be additional sources of both air and fuel, such as secondary air to the afterburner and reintroduced fuel from the drum predraining operation. While the fuel rate and the fuel–air ratios in natural-gas combustion are relatively easily controlled by conventional combustion equipment and controls, the induced airflow and the rate of introduction of the fuel-in-drums are more difficult to control.

Conventional practice has been to provide induced air and secondary air in quantities sufficient to ensure complete combustion at all times, based on those periods when the "heaviest" drums are being fed. (Heavy drums are those that contain the maximum amount of combustible residual material that the system is designed to burn at the designed production rate.) This practice is very wasteful of natural gas. During periods when the fuel in drums is less than that needed to pyrolyze the drums and to heat the resulting large amount of excess air, the auxiliary burners must supply the needed additional heat.

A more energy-efficient approach is to predrain all drums that contain more than the design maximum fuel and

FIGURE 4. Furnace System Cylindrical Afterburner Above. *No waste-heat recovery. Tempering air is induced at fan inlet* (Spencer Boiler and Engineering, Inc.)

to reintroduce the accumulated predrain liquids during periods when the drums contain appreciably less than maximum. The reintroduction of liquid fuel is accomplished at a controlled rate and automatic control systems are thereby able to maintain the fuel–air ratio at a reasonably efficient level of excess air.

The adhering residual materials in drums will be completely reduced to dry and friable ash and char when the metal temperature has reached approximately 1000°F. While previously published guidelines and standards[4,5] have recommended that the furnace length be designed to provide a minimum four-minute residence time for proper pyrolysis of the adhering materials, there is no engineering basis for this criterion. Obviously, any furnace length (or residence time) that achieves a drum-metal temperature of 1000°F will have properly pyrolyzed the drum. Accordingly, it is the *rate* at which the drum heats up, from ambient to 1000°F, that determines the required furnace length for a specified production rate and drum spacing. The heat-up rate, in turn, is solely a function of the furnace temperature.

The relationship of furnace length to the production rate, furnace temperature, and drum spacing can be expressed by the equation

$$L = \frac{PI}{720} \ln\left(\frac{T}{T - 1000}\right) \tag{1}$$

where: L = length of furnace heat-up zone, feet
P = production rate, drums/hour
I = drum spacing on furnace conveyor, inches
T = average furnace temperature in the heat-up zone, °F

The equation is derived from the following relationships. The heat-transfer rate is expressed by the formula:

$$q_1/t = UA(\mathrm{lm}\Delta T) \tag{2}$$

where: t = heat-up time
U = the overall heat-transfer coefficient
A = drum surface area exposed to heat absorption
$\mathrm{lm}\Delta T$ = logarithmic mean temperature difference between the furnace and the drum

For typical applications, U = 20 Btu/ft^2·h·°F, A = 18 ft^2/drum, and t = residence time in the furnace heat-up zone in hours.

The value of the overall heat-transfer coefficient is based on furnace temperatures in the range of 1000°F to 1600°F. The mean temperature difference is based on the drum-metal temperature rising from zero to 1000°F. The zero-degree figure is conservatively based on coldest winter periods for most geographical locations in the United States.

The quantity of heat absorbed by the drum is expressed by

$$q_2 = mC_P\Delta T \tag{3}$$

where: m = average weight of a heavy-gage drum
C_P = mean specific heat of steel
ΔT = temperature increase of the drum

Typically, m = 50 pounds per drum, C_P = 0.12 Btu/lb·°F, ΔT = 1000°F.

Care should be taken when discussing furnace capacity. Capacity means different things to different people. For the owner or production manager, the furnace capacity is the capacity for producing a finished product, typically, the number of reconditioned drums per day. When converting this capacity figure to a required furnace feed rate, expressed in drums per hour, it is necessary to take into account the plant operating factor and the hours of operation per day.

For the design engineer and for pollution control considerations, the furnace capacity is the maximum capacity for complete combustion of all fuels introduced into the system.

Fuel in Drums

The most elusive parameters in the design of drum-reclamation furnace systems are the estimates of the quantity, the burning characteristics, and the distribution of residual material in the drums. Many furnace owners and operators have only a general feeling as to the quantity and combustion characteristics of the materials in the drums that are processed in their facilities. Often, these operators will quote a figure that is markedly lower than the true values. The Touhill[3] report reflects this—the available published information (six references) reported amounts ranging from 0.12 gallon to 1.6 gallons (0.75 to 13.0 pounds) per drum. The variable amount of residue remaining in the drums results not only from the nature of the contained material (viscosity, caking tendency, etc.), but also from the unpredictable degree of thoroughness with which the drum is emptied. Hence the residual contents of a drum can range from zero up to 20 pounds or more. If the material is classed as a hazardous material, the RCRA regulations require the drum to be completely empty by normal emptying procedures; in any case, it may not contain more than 1 inch of residual material. If the material is a pesticide or a RCRA-listed "acutely toxic" material, the drum must be triple-rinsed before it can be offered for transport to a reconditioning or disposal facility.

Previously published guidelines[4] have based the furnace design on an assumed maximum of 4 pounds per drum and on the assumption that the average heating value of the material is equivalent to that of residual fuel oil (No. 6), that is, 18,000 Btu/lb. However, there is no sound basis for either of the assumed values. Ideally, the material in representative samples of drums would be removed, weighed, and analyzed. On a practical basis, this might prove to be a very difficult task. In order to establish a representative sample of drums, a large number of them must be included to ensure that the many and varied sources of supply of used drums are represented. The cost of such a sampling program could be prohibitive.

It has been suggested that a representative sample of drums be weighed, then processed through the furnace and the blasting operation, and then weighed again. The loss in weight would then represent the weight of residual material in the sample. This method, however, would provide no information on heating value, water content, ash content, and burning characteristics, nor would it identify the weight of material in the heaviest drums, a fact that must be known in order to establish the maximum combustion capacity of the system.

A new approach (developed by the author) is a method based on the "1-inch rule" of RCRA. This rule defines an empty drum that previously contained hazardous materials as having not more than 1 inch of residual material. Accordingly, a reconditioner could not legally accept drums with more than an inch of hazardous material unless it is a permitted treatment, storage, or disposal facility (TSDF). This legal requirement naturally will result in a greater awareness of the percentage of drums with more than an inch that the reconditioner normally receives into the plant. Presumably, it is easier to judge an inch of liquid than, say, 4 pounds. Estimates of the percentage of drums with more than an inch are assuredly more reliable than estimates of average and maximum *pounds* of material.

Calculation of System Capacity

Using the estimate of the percentage of all drums received that contain less than 1 inch, the distribution and the average weight of materials in drums can be calculated as follows:

1. Determine the common ratio by solving for r in the equation

$$Z = (100 - r^{16}) \tag{4}$$

where: Z = percent of drums containing less than 1 inch
r = the common ratio in a geometric progression
r^{16} = the term in the progression (having 1-pound increments) where the increment is 16 lb/drum

Note: One inch of material in a 55-gallon drum weighs approximately 16 pounds, assuming an average density of 9.3 lb/gal.

2. The total weight of material in all drums is calculated by the formula for a combined arithmetic and geometric progression:

$$\lim n \rightarrow \infty\ S = [1/(1\text{-}r) + rd/(1\text{-}r)^2] \quad (5)$$

where: S = total weight of material
d = the common difference (= 1 pound)
r = the common ratio (established in step 1)

3. The total number of drums is given by the equation for the geometric progression:

$$\lim n \rightarrow \infty\ S = 1/(1\text{-}r),\ (r^2 < 1) \quad (6)$$

4. The design basis for the furnace combustion capacity is the average weight of material per drum and is calculated by dividing the total weight of material (equation 5) by the total number of drums (equation 6). The expressions reduce to:

$$C = 1/(1 - r) \quad (7)$$

where C = the minimum design basis capacity, pounds/per drum

After having established the quantity of fuel in drums, the material and heat balances can be calculated.

Ideally, a representative sample of the material should be analyzed for heating value, char, ash, and free water. If this is impractical or unavailable, the following assumed values may be used.

50% vaporizable hydrocarbon (HC)
15% char
20% free water
15% ash
Heating value (LHV) of HC = 17,000 Btu/lb
Heating value of char (ash-free basis) = 12,000 Btu/lb

Note: The analysis methods for HC and char should duplicate, as nearly as practicable, the conditions expected in the drum furnace.

Material Balances

The material balances for solid and liquid inputs are straightforward, based on the design basis capacity determined above. The weight of steel of the drums and furnace conveyor chain should also be calculated, as these figures are needed in the heat balances. The material balance for gases (natural gas, air, and flue gases) is necessarily a part of the heat-balance calculations.

Heat Balances

The heat balances are considerably more complex. For existing facilities, where the required production rate, furnace length, heat-up distance, and drum spacing are given, the furnace operating temperature is calculated from equation 1. For proposed facilities, the furnace operating temperature may be (arbitrarily) chosen, and the furnace length is calculated from the equation. The optimum choice of furnace length versus operating temperature can be determined by analysis of furnace construction costs, fuel and maintenance costs, and the cost of capital. For higher temperatures, a more expensive refractory must be used. At lower temperatures, a less expensive refractory is acceptable, but the furnace length must be increased proportionately. The drum spacing should be set as close as possible for best efficiency, but will be influenced by the drum feeding and overturning procedures and equipment.

The heat balance is calculated stepwise as follows: First, establish the furnace operating temperature (e.g., from equation 1) and the heating value of the fuels, and then calculate the total heat input at the fuel rate determined from the material balances. The total heat input should include the natural gas supplied to the furnace auxiliary burners at low fire. Next, calculate the heat outputs. These include the heat needed to bring the drums from ambient to 1000°F, plus the sensible heat losses to the discharged ash and char, plus the total heat in the exhaust gases leaving the furnace (entering the afterburner), plus the radiation and convection heat losses from the furnace. The total heat content of the furnace exit gases includes the heat of vaporization of the water content of the fuel, the heat of vaporization of the volatile hydrocarbons, and the sensible heat content of the gases (i.e., the enthalpy increase from ambient to furnace operating temperature). The volume of the exit gases is based on the stoichiometric amount of combustion air needed to satisfy the heat balance. This quantity of air establishes the induced air requirement. Generally, only a portion of the hydrocarbon fuel is burned in the furnace; the balance passes to and is completely burned in the afterburner. The amount of this fuel entering the afterburner determines the stoichiometric air requirement in the afterburner. Finally, calculate the additional air needed to temper the exit gases to maintain the temperature below the maximum allowable inlet temperature to the waste-heat–recovery boiler or, in the case of units without waste-heat recovery, the maximum allowable inlet temperature of the ID fan. The fan capacity requirement is the sum of induced air plus the combustion air for the gas burners in both the furnace and the afterburner at low fire, plus the combustion air (including excess air) supplied to the afterburner.

The temperature-control instrumentation must be carefully designed, taking into account the fact that a change in the flow of secondary combustion air (or tempering air) to the afterburner will result in an equal and opposite change in the flow of induced air, which, in turn, will affect the furnace temperature. The two control loops, afterburner temperature and furnace temperature, are interdependent.

AIR EMISSIONS CHARACTERIZATION

Source Classification

Drum reconditioning furnaces are properly classified as "process furnaces," as defined by the Clean Air Act (CAA) regulations. Unfortunately, many people refer to the furnaces as incinerators. Consequently, regulatory authorities sometimes apply (improperly) CAA incinerator rules in the compliance demonstration and permission procedures. Furthermore, after having classified the unit as an incinerator, some authorities then classify it as a hazardous waste incinerator. The proper approach is to begin with RCRA's definitions of hazardous waste, specifically, the definition of empty containers (see 40 CFR 261.7). Clearly, if the drums being fed to the furnace are empty drums, that is, are not classified as hazardous waste, then the furnace is not a hazardous waste incinerator. In fact, the furnace is not an incinerator in view of its function, which is to prepare the surfaces of used drums for the next process step, abrasive blasting. The surface-preparation operation is done by pyrolysis, not by incineration.

Particulates

PM_{10} is the particulate standard that replaced the total suspended particulate (TSP) matter standard in the federal guidelines, effective July 31, 1987 (40 CFR 50.6). It includes only particles with an aerodynamic diameter of less than 10 μm. Particle-size distribution analyses of the particulate emissions from a typical drum reconditioning furnace indicate that 80–90% of the TSP emissions are in the PM_{10} range.

The chief potential source of air contaminant emissions from drum furnaces is that of smoke (soot and incompletely burned hydrocarbons) resulting from excessive fuel feed rates or the corollary deficiency of combustion air. These emissions are adequately controlled by the proper design and operation of the afterburner. The liquid and solid fuels (residual material in drums) are burned in the furnace in what may be described as "quiet burning"—that is, the fuel is not atomized or tumbled. Rather, the materials burn with "lazy" flames generally spread along the hearth of the furnace. Therefore, there is very little potential for entrainment and emission of fly-ash particulates above the allowable limits. This is demonstrated by the results of a series of stack tests of a typical drum furnace, listed in Table 1.

At a drum production rate of 480 drums per hour, the TSP is 0.006–0.01 pound per drum. The maximum allowable particulate emission rate for this furnace, under the process furnace rules, was 25 lb/h, or 0.052 pounds per drum at 480 drums per hour.

The previously published[7] emission factor for TSP was 0.035 pound per drum (16.8 lb/h at 480 drums per hour production). This is considerably higher than the stack test data of Table 1. The widely differing values are probably due to the fact that the published factor in AP-42[4] is based on data generated several years earlier than the data of the stack tests of the table. Also, the published factors are selected (appropriately) from the high end of the range of available data.

Organometallic Compounds

Organometallics represent a special case where particulate emissions might occur in excessive amounts, but cannot be reduced by the control measures normally effective for other particulate matter, namely, the proper design and operation of the furnace and afterburner systems. Silicone compounds are the most common example. These compounds burn in the vapor state, but the products of combustion contain the oxides of the metal portion of the molecule. These metal oxides invariably exist in the flue gases as finely divided solid particles.

Organics and Carbon Monoxide

Products of incomplete combustion (PICs) and carbon monoxide can appear in the exhaust gases only in instances of improper design or operation of the furnace/afterburner system. For example, fuel rates that are too high, when burned with sufficient combustion air, can result in high flow rates of the gases, such that the residence time in the afterburner is too short to allow complete combustion. Similarly, excessive temperatures can result in reduced residence times as a result of the thermal expansion factor.

Acid Gases

Acid emissions may result from processing drums containing compounds of chlorine or sulfur. The chlorine-containing compounds that are shipped in open-head drums are primarily chlorinated polymeric materials such as poly-

TABLE 1. Air Contaminant Emissions[a] from a Typical Large Drum Furnace

Test No.	TSP	CO	HC (as C)	As	Be	Cd	Pb	Hg
1	3.5	53	13.1	<0.001	<0.002	0.03	1.72	<0.0001
2	5.3	60	7.2	<0.001	<0.002	0.26	2.35	<0.0002
3	2.7	82	5.4	<0.002	<0.002	0.04	0.75	<0.0002

[a]TSP is reported in pounds per hour, carbon monoxide (CO) in ppm corrected to 50% excess air, hydrocarbons (HC) in pounds per million standard cubic feet, and the heavy metals arsenic (As), beryllium (Be), cadmium (Cd), lead (Pb), and mercury (Hg) in mg/m^3.

vinyl chloride (PVC). Chlorinated solvents are not likely to appear because these solvents are usually shipped in tight-head drums. Likewise, relatively few sulfur-containing compounds are shipped in open-head drums.

Toxic Air Contaminants

If a facility is located in an area where ambient air quality standards for air toxics are being exceeded, the facility may be required to reduce its emissions of these contaminants. As is indicated by the data in Table 1, the only toxic contaminants that are likely to appear in measurable concentrations are lead and cadmium. These metals are usually present in paints and maintenance coatings.

Pesticides are not likely to pose an emission problem in view of the triple-rinsing requirements of FIFRA. Similarly, drums that previously contained certain "acutely hazardous" materials must be triple-rinsed in accordance with the RCRA regulations.

Fugitive Emissions

The potential for fugitive emissions is dependent on the design and operation of the furnace system. Three sources of fugitive emissons are possible: flashbacks, puff-outs, and smoldering drag-out.

Flashbacks can occur when flammable or combustible material is allowed to accumulate on the bed of the furnace conveyor from the point where drums are overturned onto the conveyor to the point inside the furnace where the material ignites. The flames of the ignited material travel back to the point outside the furnace where the drums are placed on the conveyor. The "outside-the-furnace" burning is usually somewhat smoky and would constitute open burning, which is generally prohibited.

Puff-outs can occur if a drum contains an appreciable amount of volatile hydrocarbons that is locked in by being caked or skinned over. The material is suddenly released and rapidly vaporized when the drum reaches the ignition zone. The ignition of this vapor can be rapid enough that the outward expansion of the gases in the furnace caused by the rapid rise in temperature is greater than the in-flow velocity of the induced air.

Some materials, such as resins, plastics, and gasket materials, will not be completely burned during the relatively short residence time in a typical drum furnace. In the pyrolysis process, the hydrocarbon portion vaporizes and burns or passes partially burned into the afterburner. However, the char and partially pyrolyzed polymeric materials remain in the furnace and are dragged out by the furnace conveyor. These materials may continue to smolder after leaving the furnace.

Odors

Those facilities that produce "converted" and/or "rebuilt" drums may encounter a problem with odors. In the grading of drums for purposes of scheduling a production order, the experienced operator will recognize, from logos and markings, certain *tight-head* drums as those that cannot be properly cleaned by washing. These are directed to the deheading operation and then processed in the open-head line. If this deheading operation is remote from the furnace, the drums may remain open too long and thus release odors to the atmosphere. Similarly, if the lid-removal operation for open-head drums is remote, the duration of release of odors may be too long. These odors are the characteristic odors of the materials of the previous contents.

AIR-POLLUTION CONTROL MEASURES

Particulates (PM_{10})

The proper design, operation, and maintenance of the furnace system, including the afterburner (also, the predraining and the waste-heat–recovery systems, if used), and all instruments, controls, and alarms will assure that the emission of particulate matter will not exceed the regulatory limits. Facilities that do not install the predrain system described previously should establish and demonstrate a reliable alternative method for ensuring that the furnace system is not overloaded with respect to its combustion capacity. Some reconditioners have adopted a policy of refusing to accept drums that contain any drainable liquid.

Those facilities that exceed emission limits because of unique equipment design factors or operating procedures either should make the necessary design or procedure changes or should install add-on control equipment, such as wet scrubbers, electrostatic filters, or bag filters.

Facilities that process significant numbers of "silicone drums" (or other drums that contain organometallic compounds, but are not triple-rinsed) should either prerinse these drums in a dedicated preflush system or install the add-on controls mentioned above. The considerable costs of these add-on controls are likely to render the enterprise uneconomical or noncompetitive.

Organics and Carbon Monoxide

As in the case of the PM_{10} emissions, the proper design and operation of the furnace system will ensure compliance with regulatory limits. As is indicated by the data in Table 1, CO concentrations will normally be far below allowable limits. The maximum allowable concentration of CO was 200 ppm in that particular ACQR.

Acid Gases

As the relative number of open-head drums containing chlorine or sulfur compounds is very small, the likelihood of exceeding the emission limits for acid gases is minimal. As in the case of silicone drums, the practical choice is to avoid processing these drums. The alternative for the reconditioner whose market in these drums is important is to install wet scrubbers. The facility should recognize that the installation of a wet scrubber inevitably includes pH control and other associated problems of wastewater pretreatment.

Air Toxics

Considering the fact that the paint industry has been reformulating the coatings that contain toxic metal compounds, replacing them with nontoxic substitutes, the concentrations of these contaminants in the exhaust gases are expected eventually to become insignificant. In those cases where a prompt reduction of the air toxics is mandatory, the prudent choice would be to identify the drums most likely to contribute to the problem and either exclude them from the open-head reconditioning process or install a dedicated prerinsing system. Again, the decision to install add-on control equipment and preflush systems is likely to be difficult to justify because of their cost.

Fugitive Emissions

Flashbacks can be prevented by careful design of the furnace conveyor. The construction of the apron section of the conveyor bed should be liquid-tight and the crossbars connecting the two strands of the dual-strand chain conveyor should be positioned so that they continuously scrape the bed of the apron section, thus preventing the buildup of ignitable material. A variation of the type of furnace conveyor, shown in Figure 5, is a wide-link, refuse-style conveyor. Scraper attachments are welded to a link every 6 or 8 feet.

Puff-outs can be prevented operationally by avoiding the feeding of drums that may contain appreciable quantities of trapped-in flammable liquids. Additionally, the potential for puff-outs can be reduced by maintaining a higher rate (velocity) of induced air. Modifying the equipment by lengthening the entrance and exit vestibules would also be helpful in reducing the tendency for puff-outs.

Smoldering drag-out material should be quenched by water to avoid open burning and fugitive emission problems. The quench unit should be located close to the exit end of the furnace. As the drums pass through the quench unit, they are deluged by the water sprays, effectively stopping the smoldering. The quench water can be recirculated to concentrate the solids and then filtered or decanted. The recovered solids can be returned to the furnace to extract additional fuel value.

Odor Control

An acceptable level of odor control can be accomplished by conducting the lock and lid removal and deheading operations in an enclosed room or building located close to the

FIGURE 5. View of Refuse-Type Furnace Conveyor, Showing Scraper Attachments Welded to Chain Link (Spencer Boiler and Engineering, Inc.)

feed end of the furnace. The room should be equipped with a ventilation system that includes odor-removal units, such as wet scrubbers or activated-carbon adsorbers.

References

1. *Introducing the National Barrel and Drum Association,* National Barrel and Drum Association, Washington, DC, 1983.
2. *Current Industrial Reports. Steel Shipping Drums and Pails. Summary for 1983,* Bureau of the Census, MQ34 (83) Report, U.S. Department of Commerce, October 1984.
3. C. J. Touhill and S. C. James, *Barrel and Drum Reconditioning Industry Assessment,* EPA-600/52-81-231, U.S. Environmental Protection Agency, Municipal Environmental Research Laboratory, Cincinnati, OH, January 1982.
4. *Air Pollution Engineering Manual,* 2nd ed., John A, Danielson, Ed.; AP-40, U.S. Environmental Protection Agency, Research Triangle Park, NC, 1973.
5. S. M. Krinov, *Drum Reclamation Furnaces: Standards of Design and Operation,* National Barrel and Drum Association, Washington, DC, December 1979.
6. Private communication with staff at a large reconditioning facility located in the Midwest (1984).
7. T. R. Blackwood et al., *Source Assessment: Rail Tank Car, Tank Truck, and Drum Cleaning, State of the Art,* EPA-600/2-78-004g, U.S. Environmental Protection Agency, Cincinnati, OH, April 1978. (Cited in *Compilation of Air Pollutant Emission Factors,* 3rd ed., AP-42, U.S. Environmental Protection Agency, Research Triangle Park, NC, 1984).

Bibliography

C. J. Touhill and S. C. James, "Environmental assessment of the barrel and drum industry," *Envir. Prog.* 2(4):220 (1983).

9
Evaporative Loss Sources

Dry Cleaning
Stephen V. Capone
Petroleum Storage
Randy J. McDonald
Gasoline Marketing
Stephen A. Shedd
Organic Solvent Cleaning (Degreasing)
Mark B. Turner and Richard V. Crume

DRY CLEANING

Stephen V. Capone

Dry cleaning is the process of cleaning fabrics by washing them in a substantially nonaqueous solvent. Three classes of solvents are currently in use by the dry cleaning industry: petroleum solvents, chlorinated hydrocarbons, and chlorofluorocarbons. Although perchloroethylene is the most widely used solvent in the industry,[1] a dry cleaning facility may use more than one class or type of solvent. The use of solvents and their evaporation make dry cleaning operations an air pollution concern. However, because of the flammability of solvents and their health effects, the industry is also regulated under the National Fire Protection Code and local fire codes and Occupational Safety and Health Administration rules.

The dry cleaning industry can be categorized into three types of facilities: coin operated, commercial, and industrial. The coin-operated facilities are directly available for use by the consumer. Commercial facilities, the most common type, offer the general public dry cleaning services for most types of fabrics. Industrial dry cleaning facilities are the largest plants. They are typically operated in conjunction with a service that rents uniforms or other items to commercial, industrial, or institutional clients.

The process of dry cleaning fabrics, regardless of facility category, is performed in three steps. The first, washing, is conducted by agitating the fabric in a solvent bath. In the next step, extraction, excess solvent is removed by centrifugal force. The third step, drying, is conducted by tumbling the fabric in a stream of warm air to vaporize and remove the solvent from the fabric.

The three steps can be conducted in one or two machines, depending on the age of the equipment or the solvent being used for washing. When older equipment or petroleum solvents are used, the washing and extraction steps are usually performed in one machine and drying in a second machine. These are referred to as transfer operations because the fabrics being cleaned are transferred from the washing/extractor machine to the drying machine. When one machine performs all three steps, it is referred to as a dry-to-dry operation. One machine is often employed at facilities that use nonpetroleum solvents. Coin-operated dry cleaning machines typically involve dry-to-dry operations and are prohibited by the National Fire Protection Code from using petroleum solvents.[2]

PROCESS DESCRIPTION

The key factor determining the equipment employed for dry cleaning is the solvent used. This section begins with information on the solvents employed in the industry. The process steps and the types of equipment that make up a dry cleaning facility are then discussed, followed by a discussion of the sources and amounts of solvent emission.

Solvents

The solvents currently in use by the dry cleaning industry, by type, are as follows:

Petroleum solvents
- Stoddard solvents
- 140-F solvents
- Odorless solvents
- Low-end-point solvents

Chlorinated hydrocarbons
- Perchloroethylene
- 1,1,1-trichloroethane

Chlorofluorocarbons
- Trichloro-trifluoroethane (CFC-113)

TABLE 1. Dry Cleaning Solvent Properties

Property	Perchloroethylene	1,1,1-Trichloroethane	CFC-113	Petroleum
Flash point, °F	None	None	None	100–146
Initial boiling point, °F	248–252	162–165	114–118	300–370
Dry end point, °F	254	190	—	330–416
Density, lb/gal	13.49–13.60	10.97–11.16	13.09–13.16	6.26–6.78
Heat of vaporization, Btu/lb	90	102	63	104–122
Molecular weight	165.8	133.4	187.4	128–166

TABLE 2. Petroleum Solvent Properties

Property	Stoddard Solvents	140-F Solvents	Odorless	Low End Point
Flash point, °F	100–110	140–146	100–128	100
Boiling point range, °F	300–400	350–410	300–415	300–365
Density, lb/gal	6.40–6.60	6.49–6.60	6.10–6.40	6.35–6.45
Heat of vaporization, Btu/lb	110–120	—	—	100–105
Average molecular weight	140–150	—	—	130–140

It is estimated that perchloroethylene is employed in 75–80% of dry cleaning. The petroleum solvents represent 20–23% of dry cleaning solvent usage, while CFC-113 (or F-113) represents 1–2% and 1,1,1-trichloroethane usage is less than 1%.[3]

Many manufacturers produce dry cleaning solvents for the industry. The physical characteristics of a given solvent type may vary among manufacturers, who may add materials to enhance or inhibit a function or characteristic of the solvent. Tables 1 and 2 report solvent properties as ranges of values because specific manufacturers are not identified. Whenever a study of a dry cleaning facility is undertaken, the Material Safety Data Sheet (MSDS) for the specific manufacturer's solvent in use at the facility should be obtained. Table 1 presents general properties for three solvents and the petroleum solvent category. Table 2 provides information on some of the specific petroleum solvent types.

To be satisfactory for dry cleaning use, a solvent should demonstrate good fat, grease, and oil removal and be sufficiently volatile to permit easy drying. In addition to having a low toxicity and being capable of easy purification, an acceptable dry cleaning solvent should not harm or weaken fabrics or cause dyes to bleed. The solvent should have a flash point above 100°F for safety purposes and must be compatible with the metal and seal materials used in the dry cleaning equipment. It is important to note that synthetic chlorinated hydrocarbons manufactured for other industries may not be suitable for dry cleaning.

Numerous manufacturers offer a wide variety of the four types of petroleum dry cleaning solvents. Stoddard solvents are the oldest petroleum solvents in widespread use. As with all the petroleum solvents, they are a blend of hydrocarbons obtained from the distillation of crude oil. Different manufacturers provide slightly different blends and use different additives to vary some of the Stoddard solvent characteristics. The 140-F solvents are similar in specification to Stoddard solvents. However, they have a higher flash point (140°F minimum—hence their name) than Stoddard solvents and their boiling ranges represent the higher end (350–410°F) of the Stoddard solvent ranges. The higher flash point makes these solvents safer then the Stoddards, but the higher boiling range means slower drying. Odorless solvents are Stoddards that have been formulated to exhibit the minimum amount of odor. Low-end-point or fast-dry solvents are formulated to have the dry point end of their boiling point range as low as possible (335–365°F) to reduce the time and energy necessary to complete drying.

As noted, only 1–2% of dry cleaning is currently done with CFC-113. The environmental concerns regarding this compound and the other fluorocarbons are likely to lead to a reduction in its use as a dry cleaning solvent in the future.

Although dry cleaning is essentially the laundering of fabrics in a nonaqueous solvent, it is necessary to add small amounts of detergent to the solvent to enhance the cleaning process. Detergents serve three functions here. First, they assist in the removal of insoluble soils and prevent the redeposition of those soils onto the fabrics being cleaned. Second, they aid in the removal of water-soluble soils. Free water in the dry cleaning solvent could damage the fabrics or affect the dyes. Detergents are used to carry small amounts of water in the solvent for removal of water-soluble soils not removed from fabrics by the solvent. Third, detergents in the dry cleaning solvent provide a wetting action that helps the water to penetrate the fabrics and remove the water-soluble soils. Because moisture enters the dry cleaning process with the fabrics, it is not often necessary to add water to the solvents. In fact, too much water in the solvent without enough detergent will damage fabrics and can inhibit proper cleaning. Detergents are never added to dry cleaning solvents in excess of 2–4% without both an increase in operating costs and the onset of adverse effects such as spotting and streaks.

The Dry Cleaning Process

The equipment used in the dry cleaning process consists of a number of pieces of equipment in addition to the washer, extractor, and dryer. These include tanks and pumps to hold and recirculate the solvent, solvent filters, distillation units and condensers to purify and collect the solvent, muck cookers to recover solvent from filtered solids, and the associated boilers, steam coils, fans, and ducts to provide and move warm air for drying. In addition to these process-related pieces of equipment, facilities using petroleum solvents and perchloroethylene will often employ some type of emission control equipment to prevent release of vapors to the atmosphere. Figure 1 is a schematic of a perchloroethylene dry cleaning plant.

Washers and Extractors

There are two types of washers. "Belly"-type washers, which only wash fabrics in the solvent and require the transfer of the fabrics to a separate extractor for excess solvent removal, are now found only in older facilities using petroleum solvents. In most newer facilities that use either petroleum solvents or perchloroethylene, a unit that performs both washing and extraction in a single piece of equipment is employed. Dry cleaners who use CFC-113 perform washing, extraction, and drying in a single piece of equipment because of the high volatility of that solvent.

The primary part of a washer is the cylinder or wheel into which the fabrics to be cleaned are placed. The cylinder rotates in the solvent during washing. Three types of cylinders currently are in use: open pocket, two pocket, and three pocket. Of these, the open-pocket cylinder design with

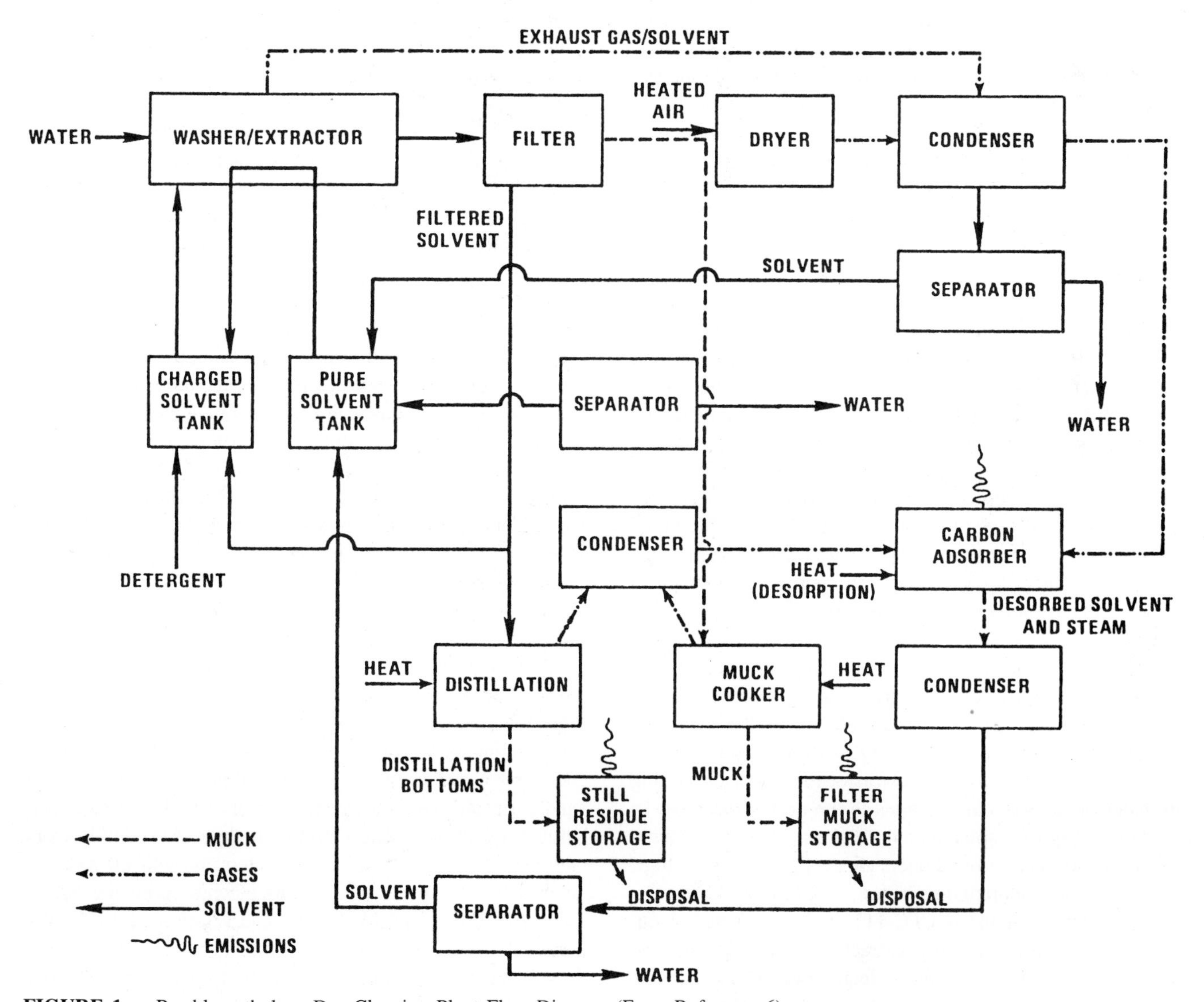

FIGURE 1. Perchloroethylene Dry Cleaning Plant Flow Diagram (From Reference 6).

internal ribs has demonstrated the best cleaning action for the removal of soils, whereas the divided-pocket design has become obsolete and will be found only in older machines.

Regardless of its design, the cylinder is always placed inside a liquid- and vapor-tight enclosure to prevent the escape of the solvent or its vapors. In the small number of plants that continue to use separate extraction units, these extractors are also designed to be vaportight when operating. Machines using petroleum solvents for washing must employ special explosion proof motors, electrical devices, and lighting because of the combustion properties of those solvents.

During the washing step, fabrics are tumbled in solvent for varying amounts of time depending on such factors as their weight, the tightness of their weave, and how soiled they were before washing. During the 15 to 40 minutes of tumbling, solvent is constantly circulated through the cylinder, a button trap, and a filter in a procedure called bath dry cleaning. In most machines using the bath procedure, the solvent passes from the cylinder, through the button trap, and into the base tank, where it is recirculated by a pump through the filter and back into the cylinder.

Variations on this procedure include the batch-type operation, in which the same solvent remains in the cylinder throughout the tumbling cycle, and the batch-bath procedure. In machines designed for the batch-bath procedure, the fabrics are first tumbled for a short period in a batch of solvent that is not recirculated, but is pumped directly to the recovery still. The washer is then refilled with solvent from the base tank, and washing proceeds in the typical bath cleaning manner.

At the completion of the washing cycle, all solvent is drained from the cylinder through the button trap and into the base tank. In machines that perform washing and extraction in the same unit, the cylinder rotation is then increased to approximately 10 times faster than its rotation during the washing cycle. During extraction, the excess solvent removed from the fabrics again drains through the button trap and into the base tank. At a small number of older facilities, the fabrics must be transferred from the washer to a separate extractor, where they are similarly spun at high speed to remove excess solvent that again drains through the button trap into the base tank.

The release of solvent vapors during the washing and extraction steps occurs primarily during removal of the fabrics from the unit at the completion of the cycle. In machines that are not dry-to-dry, solvent vapors escape when the door to the washer/extractor is opened and during transfer of the still damp fabrics to the drying machine. In addition, fugitive solvent vapor release can occur from fittings and seals around the cylinder door, button trap, filter housing, recirculating pump, and base tank.

Facilities that employ CFC-113 as the cleaning solvent always perform the washing, extraction, and drying steps in a single unit (i.e., dry-to-dry). Because of the solvent cost and its high volatility, these machines utilize control equipment to collect the solvent vapors emitted from the cylinder during operation. This equipment, which is an integral part of the process operation, is most often a refrigeration and condensation system. Although emissions of solvent vapor are associated with other parts of CFC-113 dry cleaning operations, there are no solvent vapor releases from well-maintained washing–extraction–drying units at these facilities until the machine door is opened at the end of the dry-to-dry cycle.

Dryers

The next step in the dry cleaning process is the drying of the fabrics that have been washed and from which the excess solvent has been removed by extraction. The machines in which the drying takes place are referred to in the industry as tumblers. The purpose of dryers/tumblers is to evaporate the remaining solvent from the fabrics.

All dryers comprise three major pieces of equipment: the perforated cylinder into which the fabrics are placed, steam or electric heating coils, and a fan to circulate the air through the machine. All dryers also circulate the air through a lint filter. There are slight differences between perchloroethylene dryers and dryers used at facilities that clean with petroleum solvents.

At facilities that employ perchloroethylene as the cleaning solvent, the dryer is always fitted with a reclaimer to capture and recover the solvent removed from the fabrics. These units are referred to as reclaiming tumblers and continually recirculate the airstream during the drying cycle. Air is passed over steam or electric heating coils and through the revolving perforated cylinder. The heated air removes solvent from the fabrics and the resulting vapor-laden air is passed through water-cooled or refrigerant-cooled coils. The cooling coils lower the temperature of the airstream and condense out the solvent and water. Research has shown that the optimum temperatures for efficient solvent reclamation are in the range of 85°F to 90°F.[3] The collected condensed water and solvent flow to a water separator, enter it, and are allowed to settle. Perchloroethylene is heavier than water and will not mix with it. The water rises to the top of the water separator and the recovered solvent sinks to the bottom, where it is drained off and returned to the base tank. The cooled air then passes over the heating coils and begins another circuit.

Reclaim tumblers have an inlet and an outlet damper on the recirculating ductwork. Closed during the drying cycle, these dampers are opened during the cool-down (aeration/deodorize) cycle after drying is complete. When the inlet and outlet dampers are open, air from outside the system is drawn in and passes through the tumbler. Air coming from the tumbler is sent outside the system through the exhaust vent. The exhausted air contains perchloroethylene and is sent to a vapor control device. The aeration cycle usually requires about five minutes.[3]

Temperature during the drying cycle is controlled by monitoring the temperature of the air leaving the tumbler cylinder. This is typically done with a thermometer in the duct in front of the lint bag. Temperatures in the range of 120°F to 140°F are safe for most fabrics and provide for efficient solvent recovery.[3] To achieve this optimum range, the air entering the tumbler should be around 180–190°F.[4] Most recovery tumblers at perchloroethylene facilities run in the dry cycle for 20 minutes. After that time, the amount of solvent being removed and recovered falls off and the drying cycle is considered complete.[3] The actual drying time for a load depends on the weave and weight of the fabrics.

Petroleum solvent dryers/tumblers differ from perchloroethylene units in that the majority of the older units do not have a reclaimer as part of the machine. In these older units, the air is passed across steam-heated coils, circulated through the tumbler, and exhausted to the atmosphere without being recirculated. This is done because of the explosivity of the petroleum solvents and also because at the time these units were designed, the solvents were not expensive.

In standard (nonrecovery) petroleum solvent dryers, the blower pulls ambient air into the tumbler cylinder through steam coils, where it is heated to about 212°F. Temperature is controlled by monitoring the air temperature exiting the tumbler with a thermometer located before the lint filter. The optimum temperature is in the range of 140°F to 150°F. The drying cycle lasts between 15 and 50 minutes. The actual drying time for a load depends on the weave and weight of the fabrics.[2]

The increasing costs of petroleum solvents and environmental regulations have led to the development of a recovery dryer. These units are similar to recovery dryers used at perchloroethylene facilities in that they employ a condenser to remove solvent vapor from the tumbler exhaust. The petroleum-recovery tumbler has a greater airflow through the unit than does the perchloroethylene-recovery tumbler in order to prevent the buildup of dangerous concentrations of petroleum solvent vapors in the cylinder.[3]

The petroleum recovery dryer currently on the market employs a chilled-water condenser to reclaim the solvent. The solvent and water that are condensed flow to a water–solvent separator. Petroleum solvents are lighter than water and are removed from the top of the separator. During the cool-down cycle, inlet and outlet air dampers are opened and ambient air is drawn through the fabrics and exhausted from the unit. When the door to the unit is open, the blower pulls ambient air in through the dryer and exhausts it directly from the unit. This feature prevents the solvent vapor from escaping to the operating area from the fabrics being loaded and unloaded.[2]

The amount of solvent vapors that could be released to the ambient air during the drying cycle depends on the type of dryer. Dry-to-dry machines, which are used only at perchloroethylene or CFC-113 facilities, release solvent vapors only during the cool-down cycle or when the door is opened to remove the cleaned fabrics. Reclaim dryers, used at all perchloroethylene facilities and at newer petroleum solvent facilities, also release solvent vapors only during the cool-down cycle and when the door is opened to remove the cleaned fabrics. Older petroleum dryers release solvent vapors throughout the drying and cool-down cycle, as well as when the door is opened to remove the cleaned fabrics. The dry-cleaned fabrics themselves will continue to release small amounts of the dry cleaning solvent for extended periods.[5]

Filters

Filters are installed at all dry cleaning facilities to remove suspended and dissolved materials from the solvent. The filter medium can be either a tubular or flat plate element of cloth, metal woven fabric, or metal screening. Cartridge filters also use paper as one of the filtering media they employ. All noncartridge filters use some type of filtering aid to enhance their performance.

Powder filters are a commonly used older type of filter. They typically employ diatomateous earth as the filtering medium. In these units, the filtering medium is deposited on the filter medium. These filters also may use an after-precoat of activated carbon or activated clay. The activated carbon is added to remove color from the solvent and the activated clay is used to reduce some types of nonvolatile residue in the solvent. The filtering materials are periodically removed and replaced. Replacement takes place when the buildup of contaminants on the medium causes the pressure across the filter to reach a predetermined level. The removed diatomaceous earth, after-precoats, and accumulated solids, called muck, are cooked to vaporize and recover the solvent for reuse.

Cartridge filters eliminate the need for muck cooking because the entire cartridge is disposed of when the pressure across the filter reaches the predetermined level. Cartridge filters are manufactured in three types: carbon core, all carbon, and all paper.

From an air pollution standpoint, the importance of filters at dry cleaning facilities is the potential for release of solvent vapors during filter changing and when the filter medium and accumulated filter solids (muck) are processed to recover solvent. The development of cartridge filters for perchloroethylene, petroleum solvent, and CFC-113 applications is reducing the use of muck cookers at plants that previously employed powder filters.

Muck Reclaimers

In older petroleum plants, the muck from filters is usually discarded, although a few plants may still be employing a press to squeeze the muck for recovery of some solvent. At perchloroethylene facilities, muck cooking is performed in the same unit that is used for distillation of the solvent. Powder filters are not in use at the other types of synthetic plants (CFC-113 and 1,1,1-trichloroethane), which eliminates the need for a muck reclaimer.

Solvent Distillation Units

Distillation units are used at perchloroethylene and petroleum solvent facilities to regenerate the solvent. Distillation is the most effective way to remove dissolved solvent-soluble oils, waxes and grease, solvent-soluble dyes, and insoluble soils such as carbon. Stills for petroleum solvents are operated under vacuum and are usually continuous units. Perchloroethylene stills are atmospheric and can be either batch or continuous units.

Distillation of petroleum solvents typically takes place under a vacuum of between 22 and 27 in. Hg so as to lower the boiling range for these solvents to 225°–235°F. A petroleum vacuum still has four parts: the boiling chamber, a condenser, a gravimetric separator, and the moisture separator. Solvent from the washer, extractor, settling tank, or filter enters the boiling chamber where steam coils volatilize the solvent and water, leaving the still residue. Still residue is composed of the oils, grease, and dirt removed from the fabrics and some small amount of solvent. The solvent and water vapor from the boiling chamber enter the water-cooled condenser. The resulting liquid solvent and water are piped to a gravimetric separator where they are separated by the differences in their densities. The solvent is then piped to a moisture separator. This unit is a tank containing rags or salt pellets that remove any remaining water.[2]

Perchloroethylene stills have the same four parts as petroleum stills and operate in the same manner, with the exception that they are not under a vacuum. Perchloroethylene stills can operate at atmospheric pressure because the solvent boils at 250°F. However, there is a vent pipe installed in the system to prevent pressure buildup. Stills used for CFC-113 are the same as perchloroethylene stills (i.e., atmospheric), but require less time for distillation and lower temperatures because CFC-113 boils at 118°F.[3]

All stills, whether continuous or batch, require a boil-down on a periodic basis. The need for boil-down is indicated in both types of still by the reduced flow of solvent exiting the condenser. The reduced flow results from the accumulation of high boilers in the boiling chamber that prevents efficient transfer of heat from the steam coils to the liquid. During boil-down, the steam flow to the still is increased to the maximum available and the still residues are allowed to boil for up to 30 minutes.[2] In continuous stills, the flow of solvent must be turned off before boil-down is started. At perchloroethylene facilities, steam or air sweeping or steam injection may be employed at the end of the boil-down operation to increase the recovery of solvent from the still residues.[3]

Releases of solvents from stills occur during disposal of still residues and from the vacuum pump or pressure-relief vents. Releases from perchloroethylene stills can be increased during steam or air sweeping and steam injection if the steam or air is introduced at a rate that the condenser cannot handle.

AIR EMISSIONS CHARACTERIZATION

This section provides information on the emissions from the various dry cleaning processes. Table 3 presents emission factors from various sources within a dry cleaning facility and is excerpted from the *Compilation of Air Pollutant Emission Factors* document that is better known as AP-42.[6] In order to apply an emission factor in Table 3 correctly, a knowledge of the specific facility to which it will be applied is recommended.

Table 4 presents emission factors for commercial and coin-operated facilities. In order to use these emission factors correctly, it is necessary to know the population of the area for which the emissions are to be estimated. Alternatively, for coin-operated facilities, emission rates of 15.9 and 11.3 kg of volatile organic compounds per 100 kg of clothes cleaned are suggested in the literature[8] for uncontrolled and controlled facilities, respectively.

The emission factors in Tables 3 and 4 have an AP-42 emission factor rating of "B." This means that the factors are based on the results of six to nine emission tests and can be considered quite representative of the emissions from the processes described. However, the data in both tables are dated "4/81" in AP-42 and may not include some of the more recent data on emission tests that have been developed to support New Source Performance Standards (NSPS) for the dry cleaning industry.

Additional emission factor information is provided in the background information document for the petroleum dry cleaners' NSPS,[2] the control techniques guideline for petroleum dry cleaners,[7] and the background information document for the perchloroethylene dry cleaners' NSPS.[8]

The emission factors presented and referenced here are based on emission tests that were conducted in the late 1970s and early 1980s. It is again emphasized that a knowledge of the specific equipment at a facility being evaluated should be obtained before assigning an emission factor to the operations. In some cases, the use of newer equipment, maintenance, and operational techniques will result in actual emissions that are significantly less than those reported in the literature.

AIR POLLUTION CONTROL MEASURES

Air pollution control measures currently available include variations in dry cleaning equipment, add-on control devices, and combinations of dry cleaning equipment, control devices, and operating procedures. Also of importance for the control of fugitive emissions from dry cleaning facilities is the proper maintenance of gaskets, seals, valves, and piping connections.

Regardless of the actual dry cleaning solvent used, there are only two types of emission control equipment in widespread use: refrigeration/condensation and carbon adsorption. The extent of emission reduction achieved by the use

TABLE 3. Solvent Loss Emission Factors for Dry Cleaning Operations

Solvent Type (Process Used)	Source	Emission Rate[a] Typical System kg/100 kg (lb/100 lb)	Well-Controlled System kg/100 kg (lb/100 lb)
Petroleum (transfer process)	Washer/dryer[b]	18	2[c]
	Filter disposal		
	Uncooked (drained)	8	
	Centrifuged		0.5–1
	Still residue disposal	1	0.5–1
	Miscellaneous[d]	1	1
Perchloroethylene (transfer process)	Washer/dryer/still/muck cooker	8[e]	0.3[c]
	Filter disposal		
	Uncooked muck	14	
	Cooked muck	1.3	0.5–1.3
	Cartridge filter	1.1	0.5–1.1
	Still residue disposal	1.6	0.5–1.6
	Miscellaneous[d]	1.5	1
Trichlorotrifluoroethane (dry-to-dry process)	Washer/dryer/still[f]	0	0
	Cartridge filter disposal	1	1
	Still residue disposal	0.5	0.5
	Miscellaneous	1–3	1–3

[a]Units are in terms of weight solvent per weight of clothes cleaned (capacity × loads). Emissions also may be estimated by determining the amount of solvent consumed. Assuming that all solvent input is eventually evaporated to the atmosphere, an emission factor of 2000 lb/ton (1000 kg/Mg) of solvent consumed can be applied.
[b]Different material in wash retains a different amount of solvent (synthetics, 10 kg/100 kg; cotton, 20 kg/100 kg; leather, 40 kg/100 kg).
[c]Emissions from washer, dryer, still, and muck cooker are passed collectively through a carbon adsorber.
[d]Miscellaneous sources include fugitive from flanges, pumps, pipes, and storage tanks, and fixed losses such as opening and closing dryers, etc.
[e]Uncontrolled emissions from washer, dryer, still, and muck cooker average about 8 kg/100kg (8 lb/100 lb). About 15% of solvent emitted is from washer, 75% dryer, 5% each from still and muck cooker.
[f]Based on the typical refrigeration system installed in fluorocarbon plants.

of either of these control devices is determined by the number of emission points at a facility that are ducted to the device.

Refrigeration/Condensation

As discussed earlier, the majority of dry cleaning facilities now use solvent-recovery dryers to capture and reuse the cleaning solvent. The recovery system on these dryers is most often a refrigeration or water-cooled condensation unit. These recovery units are not designed to control emissions resulting from solvent releases during washer and dryer loading and unloading or from dryer floor vents, the distillation unit vent, and the dryer aeration cycle.

Refrigeration/condensation units can be designed to control emissions from the aeration cycle. When installed on dry-to-dry perchloroethylene units, refrigeration systems can achieve emission rates comparable to those of a well-operated carbon adsorber equipped facility. However, floor vents cannot be vented to a refrigeration system because the packed bed used as a heat sink in the system would be heated by the ambient air entering the floor vents. The refrigeration system would not operate effectively if the packed bed were not at a sufficiently low temperature.[8]

Carbon Adsorption

Carbon adsorption has been used mostly in perchloroethylene dry cleaning facilities, but can be used in petroleum dry cleaning facilities to control emissions from washer and dryer loading and unloading, or from dryer floor vents, the distillation unit vent, and the dryer aeration cycle.[2,8] Efficiencies of carbon adsorber units have been documented at above 95% in controlling these emission sources.[8]

TABLE 4. Per Capita Solvent Loss Emission Factors for Dry Cleaning Plants[a]

Operation	Emission Factors kg/year/capita (lb/year/capita)	g/day/capita[b] (lb/day/capita)
Commercial	0.6 (1.3)	1.9 (0.004)
Coin operated	0.2 (0.4)	0.6 (0.001)

[a]All nonmethane volatile organic compounds.
[b]Assumes a six-day operating week (313 days per year).

When carbon adsorption is employed, the airstream must be conditioned before entering the adsorber unit. Airstreams containing lint must be passed through a filter so as not to clog the carbon bed. If the airstream is at an elevated temperature or contains moisture, then it must be cooled to prevent damage to the carbon bed.[2]

Azeotropic Vapor Recovery

The azeotropic method of recovery of solvent vapor[3] is used to reclaim solvent left in the garments at the end of the drying cycle. The method is completely different from adsorption and does not require refrigeration.

Azeotropic units take advantage of the fact that when water is combined with some solvents, the resultant mixture will evaporate at a lower temperature than will either the water or the solvent when pure. A typical azeotropic system will operate only during the aeration cycle.

When the reclaimer cycles to aeration, the reclaimer fan stops and the inlet and outlet dampers open. An air pump located in the vapor-recovery unit starts and draws air from the tumbler and bubbles it through a water tank. The moisture-enriched air is then circulated back to the tumbler, where it will form an azeotrope with the solvent in the garments. The solvent vapors can be recovered as the exiting air is passed over condensing coils. The air is then passed through the water again and the cycle is repeated. There is no venting to the atmosphere when this type of recovery is employed.

Fugitive Emissions Control

Fugitive emissions in the form of liquid or vapor losses result from poor housekeeping or maintenance. Liquid losses can be detected by sight; vapor losses, if significant, can be detected by smell.

Liquid leakage areas include hose connections, unions, couplings, and valves; machine door gaskets and seating, filter head gasket and seating, and pump seals; and base tanks and storage containers, distillation units, diverter valves, saturated lint, and cartridge filters. Vapor leakage can arise from the seals on deodorizing and aeration valves, from holes or tears in exhaust ducts, from open button traps and lint baskets, from open containers of solvent, and during removal of articles prior to complete drying.[8]

The only way to eliminate these emissions effectively is with an efficient program of maintenance and training. Equipment operators and maintenance personnel should be taught to inspect and identify solvent liquid and vapor leaks. When identified, leaks should be repaired immediately. Operators should be trained to close containers of solvent when not in use and to keep washer and dryer doors closed whenever loading or unloading is not taking place.[2]

References

1. *Focus on Drycleaning Equipment and Plant Operators Survey*, Vol. 13, No. 1, International Fabricare Institute, March 1989.
2. *Petroleum Dry Cleaners—Background Information for Proposed Standards*, Draft EIS, EPA-450/3-82-012a, U.S. Environmental Protection Agency, Office of Air Quality Planning and Standards, Research Triangle Park, NC, November 1982.
3. *Fundamentals of Drycleaning*, International Fabricare Institute, Silver Spring, MD, revised 1988.
4. *Focus on Drycleaning, An Equipment Handbook*, Vol. 8, No. 3, International Fabricare Institute, July 1984.
5. B. A. Tichenor et al., "Emissions of perchloroethylene from dry cleaned fabrics," *Atmospheric Environment*, 24A(5) (1990).
6. *Compilation of Air Pollutant Emission Factors, Vol. 1, Stationary Point and Area Sources*, 4th ed., U.S. Environmental Protection Agency, Office of Air Quality Planning and Standards, Research Triangle Park, NC, 1985.
7. *Control of Volatile Organic Emissions from Petroleum Dry Cleaners*, Preliminary Draft, U.S. Environmental Protection Agency, Office of Air Quality Planning and Standards, Research Triangle Park, NC, February 1981.
8. *Perchloroethylene Dry Cleaners—Background Information for Proposed Standards*, Draft EIS, EPA-450/3-79-029a, U.S. Environmental Protection Agency, Office of Air Quality Planning and Standards, Research Triangle Park, NC, August 1980.

PETROLEUM STORAGE

Randy J. McDonald

The most common types of tanks for the storage of organic liquids are fixed roof tanks, internal floating roof tanks, external floating roof tanks, and pressure tanks. The selection of tank type and construction materials may depend on such chemical characteristics as volatility, flammability, solubility in water, corrosiveness, and reactivity with air. The selection of tank type for the control of emissions may depend on product conservation or environmental regulations.

PROCESS DESCRIPTION

The fixed roof tank is considered the minimum standard for the storage of organic liquids. It is liquid- and vaportight and is equipped with a pressure-vacuum relief valve. These atmospheric tanks can only maintain internal pressures, or vacuums, of 1–2 oz/in.[2]

Emissions result from breathing losses and working losses. Breathing loss is the expulsion of vapor from the tank due to vapor expansion resulting from diurnal temperature and barometric pressure changes. Working loss occurs

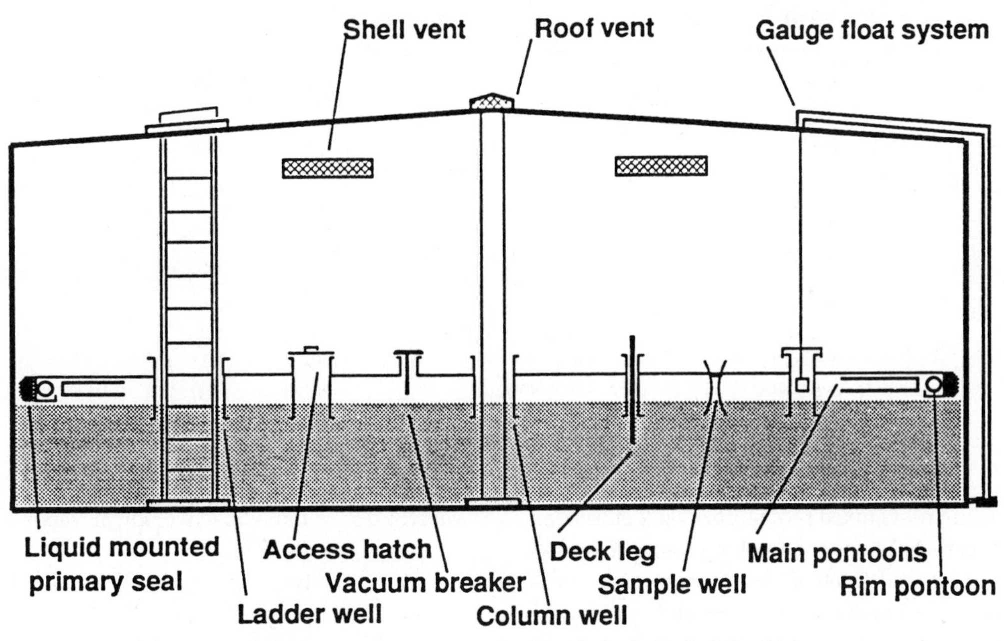

FIGURE 1. Internal Floating Roof Tank (Bolted Deck)

when vapor is displaced during tank loading operations and when air drawn into the tank during unloading operations becomes saturated with vapor and expands.

Internal floating decks can be installed in fixed roof tanks to reduce evaporation from the liquid surface (see Figure 1). Common deck types include (1) welded steel pans, (2) bolted aluminum decks on pontoons, and (3) bolted aluminum buoyant sandwich panels. Deck fittings and seams, the rim seal area, and the wetted tank wall are sources of evaporation losses from internal floating roof tanks.

The external floating roof tank does not have a fixed roof, but incorporates an external floating roof that rests on the liquid surface (see Figure 2). Three basic designs of external floating roofs in operation are pans, pontoon roofs, and double-deck roofs. The pontoon roof is a pan with an added annular ring of pontoons around the outer perimeter of the floating roof. The double-deck roof is composed of pontoon compartments or bulkheads that provide an insulating airspace over the liquid surface. The rim seal area, the deck fittings, and the wetted walls are sources of emissions from external floating roof tanks.

AIR EMISSIONS CHARACTERIZATION

Emissions of organic compounds from storage tanks are a function of the size and type of tank, the vapor pressure of the liquid inside the tank, and atmospheric conditions at the tank location. Three types of tanks are most commonly used to store organic liquids: fixed roof, external floating roof,

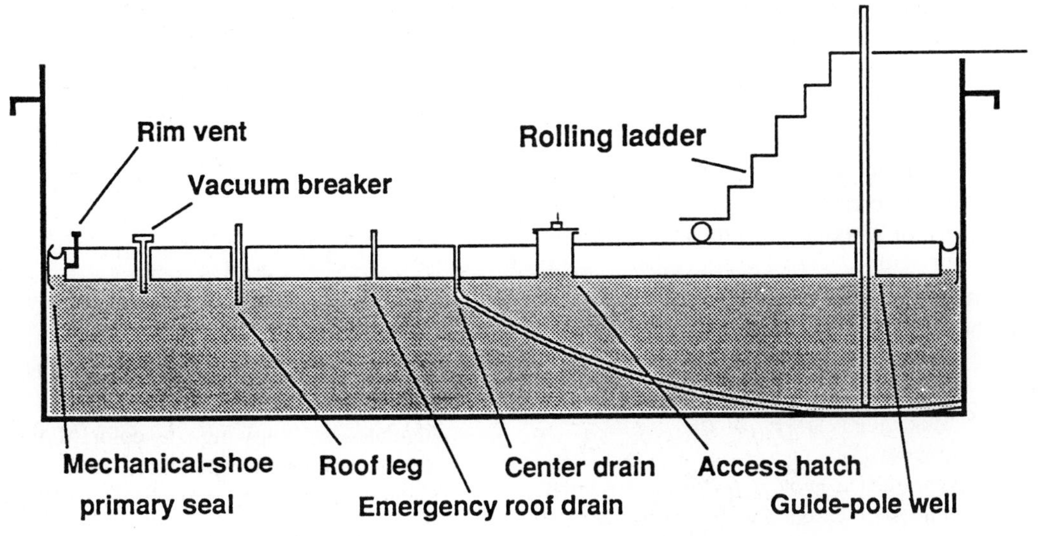

FIGURE 2. External Floating Roof Tank (Double-Deck Roof)

TABLE 1. Paint Factors for Fixed Roof Tanks

Tank Color		Paint Factors *(Fp)* Paint Condition	
Roof	Shell	Good	Poor
White	White	1.00	1.15
Aluminum (specular)	White	1.04	1.18
White	Aluminum (specular)	1.16	1.24
Aluminum (specular)	Aluminum (specular)	1.20	1.29
White	Aluminum (diffuse)	1.30	1.38
Aluminum (diffuse)	Aluminum (diffuse)	1.39	1.46
White	Gray	1.30	1.38
Light gray	Light gray	1.33	1.44[a]
Medium gray	Medium gray	1.40	1.58[a]

[a]Estimated from the ratios of the seven preceding paint factors.

and internal floating roof. Fixed roof storage tank emissions are the sum of breathing losses and working losses. External floating roof and internal floating roof storage tank emissions are the sum of standing storage loss and withdrawal loss. Standing storage loss includes rim seal loss, deck fitting loss, and deck seam loss.

The U.S. Environmental Protection Agency (EPA) publication AP-42 contains equations to estimate emissions of petroleum products and volatile organic liquids from fixed roof and external or internal floating roof tanks.[1] These equations are based on evaporative loss–estimation procedures developed by the Committee on Evaporation Loss Measurement of the American Petroleum Institute.[2,3] This section presents the AP-42 emission factors for fixed roof tanks, external floating roof tanks, and internal floating roof tanks.

Fixed Roof Tanks

The following equations apply to vertical tanks with cylindrical shells and fixed roofs. These tanks are uninsulated and operate approximately at atmospheric pressure.

Total emissions from fixed roof tanks can be estimated as:

$$L_T = L_B + L_W \qquad (1)$$

where: L_T = total loss, lb/yr
L_B = fixed roof breathing loss, lb/yr
L_W = fixed roof working loss, lb/yr

Breathing Loss

Fixed roof tank breathing losses can be estimated as:

$$L_B = 2.26 \times 10^{-2} M_v \left[\frac{P}{P_A - P}\right]^{0.68} \times D^{1.73} H^{0.51} \Delta T^{0.50} F_p C K_c \qquad (2)$$

where: M_V = molecular weight of vapor in storage tank, lb/lb-mol
P = true vapor pressure at bulk liquid conditions, psia, see note 1
P_A = average atmospheric pressure at tank location, psia
D = tank diameter, feet
H = average vapor space height, feet, see note 2
ΔT = average ambient diurnal temperature change, °F
F_P = paint factor, dimensionless, see Table 1
C = adjustment factor for small tanks (dimensionless), see note 3
K_C = product factor (dimensionless), see note 4

Working Loss

Fixed roof tank working losses can be estimated as:

$$L_W = 2.40 \times 10^{-5} M_V P V N K_N K_C \qquad (3)$$

where: V = tank capacity, gallons
N = number of turnovers per year, dimensionless

$$N = \frac{\text{Total throughput per year, gallons}}{\text{Tank capacity, gallons}}$$

K_N = turnover factor, dimensionless:

$$K_N = \frac{180 + N}{6N} \quad \text{for turnovers} > 36$$

$$K_N = 1.0 \quad \text{for turnovers} \leq 36$$

M_V, P, and K_C are as defined for equation 2.

Notes

1. True vapor pressures for organic liquids are determined at the stored liquid temperature, T_S, which may be calculated by knowing the color of the tank and the average ambient temperature. Table 2 shows T_S as a function of ambient temperature and tank color.

*The AP-42 emission factors will be updated in 1992 to include a revised breathing loss equation for fixed roof tanks and a roof fitting loss equation for external floating roof tanks. The fitting loss equation is presented in this section but, due to time constraints, the new breathing loss equation is not.

TABLE 2. Average Annual Storage Temperature as a Function of Tank Paint Color

Tank Color	Average Annual Storage Temperature, T_S
White	$T_A + 0$[a]
Aluminum	$T_A + 2.5$
Gray	$T_A + 3.5$
Black	$T_A + 5.0$

[a]T_A is the average annual ambient temperature in degrees Fahrenheit.

2. If information is not available on the vapor space height, assume that H equals one half of the corrected tank height. To correct for a cone roof, the vapor space in the cone is equal in volume to a cylinder that has the same base diameter as the cone and is one third the height of the cone.
3. The small tank diameter adjustment factor, C, can be calculated using the following equations:

   For diameter ≥ 30 feet, $C = 1.0$

   For diameter <30 feet,
   $$C = 0.0771(D) - 0.0013(D^2) - 0.1334$$

4. For crude oil, $K_C = 0.65$. For all other organic liquids, $K_C = 1.0$

External Floating Roof Tanks

External floating roof tank emissions are the sum of rim seal, withdrawal, and roof fitting losses. External floating roof tanks do not have deck seam losses because the deck has welded sections. The equations are more appropriate for substances with vapor pressures of approximately 1.5 to 14.7 psia. The equations have been developed for tanks having diameters greater than 20 feet and having average wind speed ranging from 2 to 15 mph. The equations are not intended to be used in the following applications: (1) to estimate losses from unstable or boiling stocks, (2) to estimate losses from mixtures of hydrocarbons or other organic liquids for which the vapor pressure is not known or cannot be readily predicted, (3) to estimate losses from tanks in which the materials used in the rim seal system and/or roof fittings are either deteriorated or significantly permeated by the stored liquid. Emissions from external floating roof tanks can be estimated as:

$$L_T = L_R + L_{WD} + L_{RF} \tag{4}$$

where: L_T = total loss, lb/yr
L_R = rim seal loss, lb/yr
L_{WD} = withdrawal loss, lb/yr
L_{RF} = roof fitting loss, lb/yr

Rim Seal Loss

Rim seal loss from floating roof tanks can be estimated by the following equation:

$$L_R = K_s V^n P^* D M_V K_C \tag{5}$$

where: K_S = seal factor for average or tight fit seals, see Table 3
v = average wind speed at tank site, mph, see note 1
n = seal-related wind speed exponent (dimensionless), see Table 3
P^* = vapor pressure function (dimensionless)

$$P^* = \frac{\dfrac{P}{P_A}}{\left[1 + \left[1 - \dfrac{P}{P_A}\right]^{0.5}\right]^2} \tag{6}$$

where: P = true vapor pressure at average actual liquid storage temperature
P_A = average atmospheric pressure at tank location, psia
D = tank diameter, feet
M_V = average vapor molecular weight, lb/lb-mol
K_C = product factor, dimensionless, see note 2

Notes
1. If the wind speed at the tank site is not available, wind speed data from the nearest local weather station may be used as an approximation.
2. For all organic liquids except crude oil, $K_C = 1.0$. For crude oil, $K_C = 0.4$.

Withdrawal Loss

The withdrawal loss from external and internal floating roof storage tanks can be estimated using the following equation.

$$L_{WD} = \frac{(0.943)\, Q\, C_F\, W_L}{D}\left[1 + \frac{N_c F_c}{D}\right] \tag{7}$$

where: Q = annual throughput, bbl/yr
C_F = shell clingage factor, bbl/1000 ft^2, see Table 4
W_L = liquid density, lb/gal
D = tank diameter, feet
N_C = number of columns, dimensionless, see note 1
F_C = effective column diameter, feet (column perimeter/π), see note 2

TABLE 3. Rim Seal Loss Factors[a]

Tank Construction and Rim Seal System	Average-Fitting Seals K_S (lb-mol / [mph]^n ft yr)	Average-Fitting Seals n (Dimensionless)	Tight-Fitting Seals[b] K_S (lb-mol / [mph]^n ft yr)	Tight-Fitting Seals[b] n (Dimensionless)
Welded external floating roof tanks:				
1. Mechanical shoe seal				
a. Primary only	1.2[c]	1.5[c]	0.8	1.6
b. Shoe-mounted secondary	0.8	1.2	0.8	1.1
c. Rim-mounted secondary	0.2	1.0	0.2	0.9
2. Liquid-mounted resilient filled seal				
a. Primary only	1.1	1.0	0.5	1.1
b. Weather shield	0.8	0.9	0.5	1.0
c. Rim-mounted secondary	0.7	0.4	0.5	0.5
3. Vapor-mounted resilient filled seal				
a. Primary only	1.2	2.3	1.0	1.7
b. Weather shield	0.9	2.2	1.1	1.6
c. Rim-mounted secondary	0.2	2.6	0.4	1.5
Riveted external floating roof tanks:				
1. Mechanical shoe seal				
a. Primary only	1.3	1.5	d	
b. Shoe-mounted secondary	1.4	1.2	d	
c. Rim-mounted secondary	0.2	1.6	d	
Internal floating roof tanks[e]:				
1. Liquid mounted resilient seal				
a. Primary seal only	3.0	0	2.6	0
b. With rim-mounted secondary seal	1.6	0	1.2	0
2. Vapor mounted resilient seal				
a. Primary seal only	6.7[f]	0	5.6	0
b. With rim-mounted secondary seal	2.5	0	2.3	0

[a]The rim seal loss factors (K_S, n) may be used for wind speeds from 2 to 15 mph.
[b]No gaps more than one-eighth inch wide between the rim seal and the tank shell. The consistently tight-fitting seal condition is unusual and difficult to verify.
[c]If no specific information is available, a welded tank with an average-fitting mechanical shoe primary seal only can be assumed to represent the most common or typical tank construction and rim seal system in use.
[d]No evaporative loss information is available for riveted tanks with consistently tight-fitting rim seal system.
[e]Based on emissions from tank seal systems in reasonably good working condition, no visible holes, tears, or unusually large gaps between the seals and the tank wall.
[f]If no specific information is available, a vapor-mounted primary seal only can be assumed to represent the most common or typical seal system in use.

Notes

1. For external floating roof tank or a self-supporting fixed roof: $N_C = 0$. For internal floating roof tank with column-supported fixed roof: N_C = use tank-specific information, or see Table 5.
2. For internal floating roof tank with column-supported fixed roof, use tank-specific effective column diameter, or $F_C = 1.1$ for 9-inch by 7-inch built-up columns, 0.7 for 8-inch-diameter pipe columns, and 1.0 if column construction details are not known.

Roof Fitting Loss

Fitting losses from external floating roof tanks can be estimated by the following equation:

TABLE 4. Average Clingage Factors (bbl/1000 ft^2)

Liquid	Shell Condition: Light Rust[a]	Dense Rust	Gunite Lined
Gasoline	0.0015	0.0075	0.15
Single-component stocks	0.0015	0.0075	0.15
Crude oil	0.0060	0.030	0.60

[a]If no specific information is available, these values can be assumed to represent the most common condition of tanks currently in use.

TABLE 5. Typical Number of Columns as a Function of Tank Diameter for Internal Floating Roof Tanks with Column-Supported Fixed Roofs[a]

Tank Diameter Range, D (feet)	Typical Number of Columns, N_C
0 < D F 85	1
85 < D F 100	6
100 < D F 120	7
120 < D F 135	8
135 < D F 150	9
150 < D F 170	16
170 < D F 190	19
190 < D F 220	22
220 < D F 235	31
235 < D F 270	37
270 < D F 275	43
275 < D F 290	49
290 < D F 330	61
330 < D F 360	71
360 < D F 400	81

[a]This table was derived from a survey of users and manufacturers. The actual number of columns in a particular tank may vary greatly with age, fixed roof style, loading specifications, and manufacturing prerogatives. Data in this table should not supersede information on actual tanks.

$$L_{RF} = F_F\, P^*\, M_V\, K_C \tag{8}$$

where: F_F = total roof fitting loss factor, lb-mol/yr

$$F_F = [\,(N_{F_i} + K_{f_i}) + (N_{F_2} K_{F_2}) + \ldots + (N_{F_n} 108\; K_{F_n} 108)\,] \tag{9}$$

where: N_{F_i} = number of deck fittings of a particular type (i = 0,1,2, . . .,n), dimensionless, see Tables 6, 7, and 8

K_{F_i} = roof fitting loss factor for a particular type of fitting (i = 0,1,2 . . .,n), lb-mol/yr

n = total number of different types of fittings, dimensionless

The roof fitting loss factor for individual fitting types can also be estimated from equation 10:

$$K_{F_i} = K_{fa_i} + K_{fb_i}\, V^{m_i} \tag{10}$$

where: K_{fa_i} = roof fitting loss factor for a particular roof fitting type (lb-mol/yr), see Table 6

K_{fb_i} = roof fitting loss factor for a particular roof fitting type [lb-mol/(mph)^m yr], see Table 6

m_i = roof fitting loss factor for a particular roof fitting type (dimensionless), see Table 6

v = average wind speed (mph).

The number of each type of roof fitting can vary significantly from tank to tank and should be determined for each tank under consideration. If specific tank information is not available, see Tables 6, 7, and 8 for typical numbers for common roof fittings.

Internal Floating Roof Tanks

The equations provided in this section are applicable only to freely vented internal floating roof tanks storing liquids with true vapor pressures ranging from approximately 0.01 to less than 14.7 psia. These equations are not intended to be used in the following applications: (1) to estimate losses from closed internal floating roof tanks (tanks vented only through a pressure/vacuum vent), (2) to estimate losses from unstable or boiling liquids, (3) to estimate losses from liquids for which the vapor pressure is not known, or (4) to estimate losses from tanks in which seal materials are either deteriorated or significantly permeated by the stored liquid.

Emissions from internal floating roof tanks may be estimated as:

$$L_T = L_R + L_{WD} + L_F + L_D \tag{11}$$

where: L_T = total loss, lb/yr

L_R = rim seal loss, see equation 5

L_{WD} = withdrawal loss, see equation 7

L_F = deck fitting loss, lb/yr

L_D = deck seam loss, lb/yr

Deck Fitting Losses

Fitting losses from internal floating roof tanks can be estimated by the following equation:

TABLE 6. External Roof Fitting Loss Factors and Typical Number of Roof Fittings[a]

Roof Fitting Type and Construction Details	Roof Fitting Loss Factors K_{fa} (lb-mol/year)	K_{fb} (lb-mol/[mph]^m yr)	m (dimensionless)	Typical Number of Fittings, N_F[b]
1. Access hatch (24-inch-diameter well)				
a. Bolted cover, gasketed	0.0	0.0	0.0[b]	1
b. Unbolted cover, ungasketed	2.7	7.1	1.0	
c. Unbolted cover, gasketed	2.9	0.41	1.0	
2. Guide-pole well (8-inch-diameter unslotted pole, 21-inch-diameter well)				1
a. Ungasketed sliding cover	0	67	0.98[b]	
b. Gasketed sliding cover	0	3.0	1.4	
3. Guide-pole/sample well (8-inch-diameter slotted pole, 21-inch-diameter well)				[c]
a. Ungasketed sliding cover, without float	0	310	1.2	
b. Ungasketed sliding cover, with float	0	29	2.0	
c. Gasketed sliding cover, without float	0	260	1.2	
d. Gasketed sliding cover, with float	0	8.5	2.4	
4. Gauge float well (20-inch-diameter well)				1
a. Unbolted cover, ungasketed	2.3	5.9	1.0[b]	
b. Unbolted cover, gasketed	2.4	0.34	1.0	
c. Bolted cover, gasketed	0	0	0	
5. Gauge hatch/sample well (8-inch-diameter well)				1
a. Weighted mechanical actuation, gasketed	0.95	0.14	1.0[b]	
b. Weighted mechanical actuation, ungasketed	0.91	2.4	1.0	
6. Vacuum breaker (10-inch-diameter well)				See Table 7
a. Weighted mechanical actuation, gasketed	1.2	0.17	1.0[b]	
b. Weighted mechanical actuation, ungasketed	1.1	3.0	1.0	
7. Roof drain (3 inch diameter)				See Table 7
a. Open	0	7.0	1.4[e]	
b. Closed, 90%	0.51	0.81	1.0[e]	
8. Roof leg (3-inch-diameter leg)				See Table 8
a. Adjustable, pontoon area	1.5	0.20	1.0[b]	
b. Adjustable, center area	0.25	0.067	1.0[b]	
c. Adjustable, double-deck roofs	0.25	0.067	1.0	
d. Fixed	0	0	0	
9. Roof leg (2.5-inch diameter)				
a. Adjustable, pontoon area	1.7	0	0	
b. Adjustable, center area	0.41	0	0	
c. Adjustable, double-deck roofs	0.41	0	0	
d. Fixed	0	0	0	
10 Rim vent (6-inch diameter)				1.0[d]
a. Weighted mechanical actuation, gasketed	0.71	0.10	1.0[b]	
b. Weighted mechanical actuation, ungasketed	0.68	1.8	1.0	

[a]The roof fitting loss factors (K_{fa}, K_{fb}, m) may be used only for wind speeds from 2 to 15 mph.
[b]If no specific information is available, this value can be assumed to represent the most common or typical roof fittings currently in use.
[c]Guide-pole/sample well is an optional fitting not typically used.
[d]Rim vents are used only with mechanical shoe primary seals.
[e]Roof drains that drain excess rainwater into the product are not used on pontoon floating roofs. They are, however, used on double-deck floating roofs and are typically left "open."

$$L_F = F_F \, P^* \, M_V \, K_C \quad (12)$$

where: F_F = total deck fitting loss factor, lb-mol/yr
$= [(N_{F_1} K_{F_1}) + (N_{F_2} K_{F_2}) + \cdots + (N_{F_n} K_{F_n})]$

where: N_{F_i} = number of deck fittings of a particular type ($i = 0, 1, 2, \ldots, n$), dimensionless, see Tables 5 and 9

K_{F_i} = deck fitting loss factor for a particular type of fitting (i = 0, 1, 2, . . ., n), lb-mol/yr, see Table 9

n = total number of different types of fittings

The value of F_F may be calculated by using actual tank-specific data for the number of each fitting type (N_F) and then multiplying by the fitting loss factor for each fitting

TABLE 7. Typical Number of External Floating Roof Vacuum Breakers and Drains[a]

Tank Diameter, feet[b]	Vacuum Breakers: Pontoon Roof	Vacuum Breakers: Double-Deck Roof	Roof Drains: Double-Deck Roof[c]
50	1	1	1
100	1	1	1
150	2	2	2
200	3	2	3
250	4	3	5
300	5	3	7
350	6	4	—
400	7	4	—

[a]This table was derived from a survey of users and manufacturers. The actual number of vacuum breakers may vary greatly depending on throughput and manufacturing prerogatives. The actual number of roof drains may also vary greatly depending on the design rainfall and manufacturing prerogatives. For tanks over 300 feet in diameter, actual tank data or the manufacturer's recommendations may be needed for the number of roof drains. This table should not supersede information based on actual tank data.

[b]If the actual diameter is between the diameters listed in this table, use the closest diameter listed. If midway, use the next larger diameter.

[c]Roof drains that drain excess rainwater into the product are not used on pontoon floating roofs. They are, however, used on double-deck floating roofs and are typically left "open."

TABLE 8. Typical Number of External Floating Roof Legs[a]

Tank Diameter, feet[b]	Pontoon Roof: Pontoon Legs	Pontoon Roof: Center Legs	Double-Deck Roof	Tank Diameter, feet[b]	Pontoon Roof: Pontoon Legs	Pontoon Roof: Center Legs	Double-Deck Roof
30	4	2	6	210	31	77	98
40	4	4	7	220	32	83	107
50	6	6	8	230	33	92	115
				240	34	101	127
60	9	7	10	250	35	109	138
70	13	9	13	260	36	118	149
80	15	10	16	270	36	128	162
90	16	12	20	280	37	138	173
100	17	16	25	290	38	148	186
				300	38	156	200
110	18	20	29				
120	19	24	34	310	39	168	213
130	20	28	40	320	39	179	226
140	21	33	46	330	40	190	240
150	23	38	52	340	41	202	255
				350	42	213	270
160	26	42	58	360	44	226	285
170	27	49	66	370	45	238	300
180	28	56	74	380	46	252	315
190	29	62	82	390	47	266	330
200	30	69	90	400	48	281	345

[a]This table was derived from a survey of users and manufacturers. The actual number of roof legs may vary greatly depending on age, floating roof style, loading specifications, and manufacturing prerogatives. This table should not supersede information based on actual tank data.

[b]If the actual diameter is between the diameters listed in this table, use the closest diameter listed. If midway, use the next larger diameter.

TABLE 9. Summary of Internal Floating Deck Fitting Loss Factors and Typical Number of Fittings[a]

Deck Fitting Type	Deck Fitting Loss Factor, K_F (lb-mol/yr)	Typical No. of Fittings, N_F
Access hatch (24-inch diameter)		1
Bolted cover, gasketed	1.6	
Unbolted cover, gasketed	11	
Unbolted cover, ungasketed	25[b]	
Gauge float well (24-inch diameter)		1
Bolted cover, gasketed	5.1	
Unbolted cover, gasketed	15	
Unbolted cover, ungasketed	28[b]	
Column well[c] (24-inch diameter)		(See Table 5)
Built-up column—sliding cover, gasketed	33	
Built-up column—sliding cover, ungasketed	47[b]	
Pipe column—flexible fabric sleeve seal	10	
Pipe column—sliding cover, gasketed	19	
Pipe column—sliding cover, ungasketed	32.1	
Ladder well[c] (36-inch diameter)		1
Sliding cover, gasketed	56	
Sliding cover, ungasketed	76[b]	
Deck leg or hanger well		$\{5 + D/10 + (D^2/600)\}^2$
Adjustable	7.9[b]	
Fixed	0	
Sample pipe or well (24-inch diameter)		1
Slotted pipe—sliding cover, gasketed	44	
Slotted pipe—sliding cover, ungasketed	57	
Sample well—slit fabric seal, 10% open area	12[b]	
Stub drain (1-inch diameter)[d]	1.2	$(D^2/125)$[d]
Vacuum breaker (10-inch diameter)		1
Weighted mechanical actuation, gasketed	0.7[b]	
Weighted mechanical actuation, ungasketed	0.9	

[a]For wind speeds ranging from 2 to 15 mph.
[b]If no specific information is available, this value can be assumed to represent the most common/typical deck fittings currently used.
[c]Not used in welded contact internal floating decks.
[d]D = tank diameter, feet.

(K_F). Values of fitting loss factors and typical number of fittings are presented in Tables 5 and 9.

Deck Seam Loss

Welded internal floating roof tanks do not have deck seam losses. Deck seam losses may be present for tanks with bolted decks. Deck seam loss can be estimated by the following equation:

$$L_D = K_D\, S_D\, D^2\, P^*\, M_V\, K_C \tag{13}$$

where: K_D = deck seam loss per unit seam length factor, lb-mol/ft yr
= 0.0 for welded deck and external floating roof tanks
= 0.34 for bolted deck

TABLE 10. Deck Seam Length Factors for Typical Deck Constructions for Internal Floating Roof Tanks[a]

Deck Construction	Typical Deck Seam Length factor, S_D (ft/ft^2)
Continuous-sheet construction[b]	
5-foot-wide sheets	0.20[c]
6-foot-wide sheets	0.17
7-foot-wide sheets	0.14
Panel construction[d]	
5 by 7.5-foot rectangular	0.33
5 by 12-foot rectangular	0.28

[a]Deck seam loss applies to bolted decks only.
[b]$S_D = 1/W$, where W = sheet width (feet).
[c]If no specific information is available, these factors can be assumed to represent the most common bolted decks currently in use.
[d]$S_D = (L + W)/LW$, where W = panel width (feet) and L = panel length (feet).

S_D = deck seam length factor, ft/ft^2 (equal to the total length of deck seams divided by the area of the deck)

If the total length of the deck seam is not known, Table 10 can be used to determine S_D.

AIR POLLUTION CONTROL MEASURES

The emission factors in the previous section can be used to evaluate how tank type and equipment affects storage tank emissions. Floating roof technology is usually the most economical control technology for storage tanks. The major equipment options that affect emissions from floating roofs include the floating roof seal system and the type of deck fittings. The most effective seal system is a liquid-mounted primary seal and a rim-mounted secondary seal. Deck fitting losses can be controlled by gasketing and sealing the openings in the roof that accommodate the fittings and by substituting lower emitting fitting types that serve the same purpose.

Pressure tanks were not included in the previous section because emission factors are not available. Pressure tanks are a viable control option if the pressure setting is high enough; however, it is a costly method of control.

The use of add-on vapor control or recovery techniques, such as incinerators and refrigerated condensers, is an alternative method for controlling emissions from fixed roof tanks. For example, vapor recovery is used for chlorinated solvents that cause corrosion problems for aluminum internal floating roofs. In addition, inert blanket gases can be used to provide a dry, inert atmosphere above the liquid. In a typical vapor-collection system, vapors remain in the tank until the internal pressure reaches a preset level. A pressure-sensitive device activates blowers that collect and transfer the vapors through a closed-vent system to the control device.

References

1. *Compilation of Air Pollutant Emission Factors*, AP-42, 4th ed., U.S. Environmental Protection Agency, Office of Air Quality Planning and Standards, Research Triangle Park, NC, September 1985.
2. *Evaporation Loss from External Floating Roof Tanks*, Bulletin 2517, 3rd ed., American Petroleum Institute, February 1989.
3. *Evaporation Loss from Internal Floating Roof Tanks*, Bulletin 2519, 3rd ed., American Petroleum Institute, June 1983.

Bibliography

P. Blakey and G. Orlando, "Using inert gases for purging, blanketing, and transfer," *Chem. Eng.*, 91(8):97 (May 1984).

Evaporation Loss From Fixed Roof Tanks, Bulletin 2518, American Petroleum Institute, June 1962.

Evaporation Loss From Fixed Roof Tanks, Bulletin 2518, 2nd ed., American Petroleum Institute, October 1991.

Evaporation Loss From Low-Pressure Tanks, Bulletin 2516, American Petroleum Institute, March 1962.

Standards of Performance for Volatile Organic Liquid Storage Vessels (Including Petroleum Liquid Storage Vessels), Codes of Federal Regulation, Title 40, Part 60, Subpart Kb.

VOC Emissions from Volatile Organic Liquid Storage Tanks—Background Information for Proposed Standards, EPA 450/3-81-003a, U.S. Environmental Protection Agency, Office of Air Quality Planning and Standards, Research Triangle Park, NC, July 1984.

VOC Emissions From Volatile Organic Liquid Storage Tanks—Background Information for Promulgated Standards, EPA-450/3-81-003b, U.S. Environmental Protection Agency, January 1987.

GASOLINE MARKETING

Stephen A. Shedd

The gasoline marketing industry stores gasoline and transfers it from petroleum refineries to the consumer. Gasoline is composed of volatile organic compounds, including hazardous air pollutants, that escape to the atmosphere during storage and each time it is transferred into a container. The air pollution control approach is to suppress or collect the gasoline vapors and recover or destroy them.

PROCESS DESCRIPTION

Figure 1 shows a schematic of the typical arrangement of the gasoline marketing system. Gasoline is transferred from the refinery by ship, barge, or pipeline to large bulk-storage facilities called bulk terminals. Daily gasoline throughput at an average bulk terminal is about 250,000 gallons. Gasoline is then transferred into tank trucks (8000 to 10,000 gallons) for delivery to either bulk plants or service stations.

Bulk plants are intermediate storage facilities that provide gasoline usually to rural or small-volume clients, such as small service stations and farms. Gasoline arriving from the bulk terminal in tank trucks is loaded into fixed-roof, above- or below-ground tanks. Daily gasoline throughput at an average bulk plant is about 5000 gallons. The gasoline is then loaded in smaller tank trucks (1500 to 2900 gallons) called bobtail trucks.

Service stations are facilities that refuel motor vehicles. The term service station as used here includes the familiar neighborhood and off-the-highway gasoline service station, but also convenience stores, parking garages, and fleet service agencies (rental car, private companies, government agencies, etc.) with gasoline-dispensing pumps. The gasoline delivered in tank trucks from bulk terminals and plants is dropped through hoses from the tank truck into the underground storage tank at the service station. It is then pumped from the underground tank through the meter on the dispenser and into a motor vehicle's fuel tank. Daily

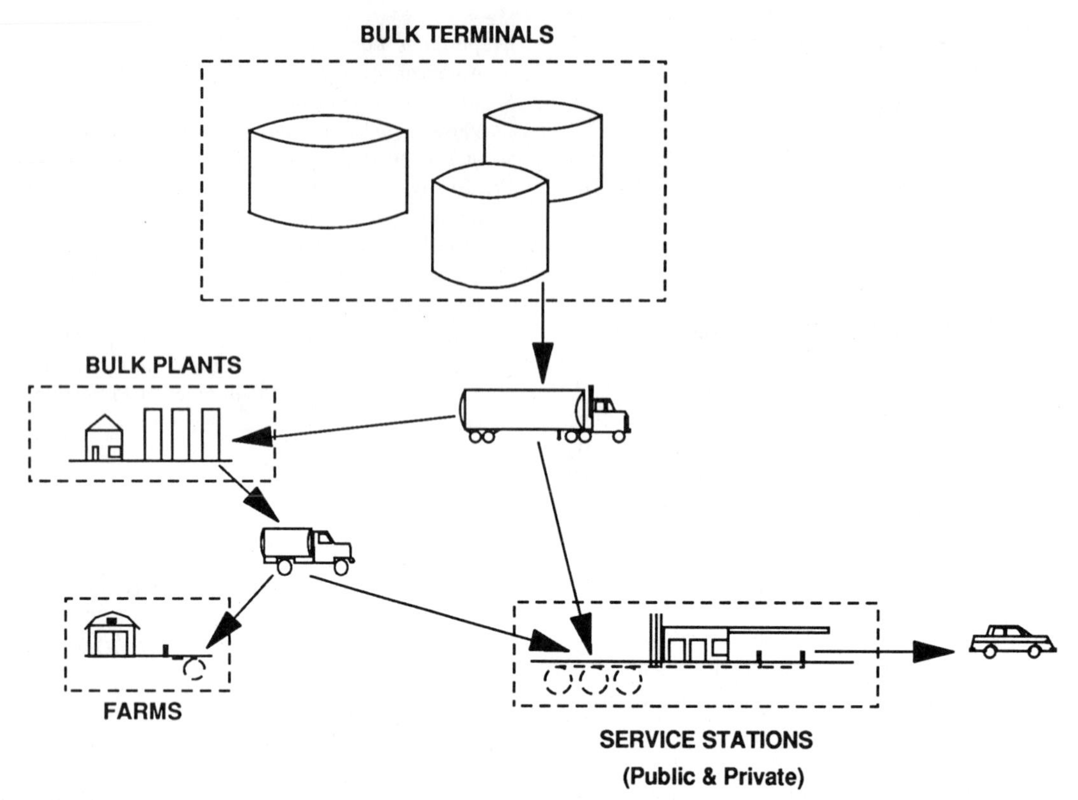

FIGURE 1. Gasoline Marketing System

gasoline throughput at an average service station is about 2000 gallons.

AIR EMISSION CHARACTERIZATION

Gasoline vapors contain volatile organics, many of which are considered volatile organic compounds (VOCs) and hazardous air pollutants (air toxics). In the gasoline marketing industry, emissions ordinarily occur at the time the gasoline is transferred from one container to another (called "displacement emissions or losses") and during storage. Emissions during storage are discussed in the previous section and will not be described here.

Gasoline Vapor Composition

Motor gasolines are a complex mixture of paraffins, naphthenes, olefins, and aromatics originating from complex processes of refining crude oil and blending of additional components. There are two air pollution concerns with gasoline vapors: the amounts of VOCs and of air toxic compounds contained in the vapors. Both vary by (1) the season and area of the country in order to provide gasolines that properly function in the engine under the many seasonal and atmospheric changes across the United States, (2) the many types of refining processes and blending operations providing gasoline to a given area, and (3) the type of crude oil used to produce the gasoline.

Volatility is a measure of the ease with which a compound changes from liquid to vapor at atmospheric pressure and temperature. Therefore, the amount of volatile organic compound emissions is directly proportional to the vapor pressure of the liquid evaporating. Vapor pressure is usually measured for gasoline as Reid vapor pressure (RVP). Gasoline vapor pressure has been increasing in the 1970s and 1980s. However, U.S. Environmental Protection Agency (EPA) recently set vapor pressure limits for the ozone nonattainment areas during the summer months as low as a RVP of 7.8 psia[1] in some cases.

Gasoline vapors contain a number of the air toxics compounds listed in the Clean Air Act Amendments of 1990. For example, an analysis of a limited number of gasoline samples showed that 2% to 22% by weight of gasoline vapors contained benzene, cumene, hexane, ethyl benzene, methyl-tertiary-butyl ether (MTBE), naphthalene, toluene, 2,2,4-trimethylpentane, *m*-xylene, *o*-xylene, and *p*-xylene. Historically, exposure assessments have focused on benzene in gasoline[2]; the other air toxics have not been evaluated. Additionally, unleaded gasoline vapors as a whole have been evaluated and found by the EPA to be a probable human carcinogen.[3] During this decade of the 1990s, the EPA will be investigating the reformulation of gasoline to reduce its hazardous constituents and volatility, as well as the installation of additional air pollution control equipment.

Displacement Losses

Liquid gasoline, flowing into the receiving container, displaces vapors in the vapor space of that container to the atmosphere. Thus emissions caused by displacement occur (1) at bulk terminals when gasoline is loaded into tank trucks; (2) at bulk plants when the gasoline is unloaded from large tank trucks into the storage tanks, and then again when it is loaded from the storage tanks into smaller tank trucks; and (3) at service stations when gasoline is unloaded into the underground storage tanks, and then again when the gasoline is dispensed into vehicles' gas tanks. Additionally,

TABLE 1. Saturation Factors for Calculating Petroleum Liquid Loading Losses

Carrier	Mode of Operation	Saturation Factor
Tank trucks and rail cars	Submerged loading:	
	Clean cargo tank	0.50
	Dedicated normal service	0.60
	Dedicated vapor balance service	1.00
	Splash loading:	
	Clean cargo tank	1.45
	Dedicated normal service	1.45
Marine vessels	Submerged loading:	
	Ships	0.2
	Barges	0.5

Source: Reference 4.

vapors are emitted from leaks and spills during transfers. Emissions from gasoline storage and transfer operations at bulk terminals, plants, and underground storage tanks at service stations are commonly referred to as Stage I emissions. Emissions from the refueling of motor vehicles at service stations are commonly referred to as Stage II emissions.

Displacement losses from loading operations can be estimated using the following expression.[4]

$$D_L = 12.46 \left[\frac{S\,P\,MW}{T}\right]$$

where: D_L = displacement loss, pounds per 10^3 gallons transferred
P = true vapor pressure of liquid loaded, psia (see equation 3 for conversion of RVP to true vapor pressure)
MW = molecular weight of vapors, lb/lb-mol (see equation 2)
T = temperature of transferred liquid, °R or (°F + 459.6)
S = saturation factor (see Table 1)

Equation 1 is derived from the ideal gas law and contains one major assumption, the degree of saturation (S). This saturation factor accounts for the variations observed in emission rates from different unloading and loading methods. Table 1 lists suggested saturation factors.

As shown in Table 1, the degree of saturation is dependent on the mode of operation (loading method employed, the previous load, and the size of the container). With splash loading, there is space between the end of the liquid-fill tube and the liquid surface. This free fall allows for increased turbulence and splashing in the container, providing more liquid surface area for evaporation to occur. Splashing also can cause a mist of fine liquid particles to escape from the open hatchway. In submerged loading, the product enters from the bottom of the tank (bottom loading) or from a long fill pipe extending to within a few inches of the bottom of the tank (submerged fill pipe), thus causing less turbulence and splashing when product is loaded. Additionally, loading into a clean cargo tank, as opposed to a tank that contains vapors from the previous load, reduces the amount of vapor displaced from the container.

Following is an equation for estimating the molecular weight of gasoline vapors for equation 1, given the RVP of the gasoline.[5]

$$MW = 72.833 - 1.3183(P_{RVP}) + 0.15079(P_{RVP})^2 - 0.0087302(P_{RVP})^3 \quad (2)$$

where: MW = molecular weight of vapors, lb/lb-mole
P_{RVP} = Reid vapor pressure, psia

Following is an equation for estimating the true vapor pressure of gasoline given the RVP.[6]

$$P = \exp\left[\left(0.7553 - \frac{413.0}{T}\right) S_D{}^{1/2} \log(P_{RVP}) - \left(1.854 - \frac{1042}{T}\right) S_D{}^{1/2} + \left(\frac{2416}{T} - 2.013\right) \log(P_{RVP}) - \left(\frac{8742}{T}\right) + 15.64\right] \quad (3)$$

where: P = true vapor pressure of liquid loaded, psia
P_{RVP} = Reid vapor pressure, psia
T = stock temperature, °R or (°F + 459.6)
S_D = slope of American Society for Testing and Materials distillation curve at 10% evaporated (for gasoline, $S_D = 3$)
exp[ ] = e^x

Only one emission factor is available from AP-42[4] for vehicle refueling displacement emissions for a given RVP and temperature. Vehicle refueling emission testing was performed by the EPA to develop an expression to take into account different gasoline temperatures and vapor pressures. From this test program, the following expression was developed for estimating vehicle refueling displacement losses.[7]

$$D_{L\ \mathrm{Refuel}} = 2.205\,[-5.909 - 0.0949\,\Delta T + 0.0884\,T_D + 0.485\,P_{RVP}] \quad (4)$$

where: $D_{L\ \mathrm{Refuel}}$ = refueling displacement loss, pounds per 10^3 gallons transferred
P_{RVP} = Reid vapor pressure of dispensed gasoline, psia
T_D = temperature of dispensed gasoline, °F
ΔT = difference in temperature between gasoline in vehicle tank and the dispensed gasoline tank, °F

AIR-POLLUTION CONTROL MEASURES

Gasoline-vapor, air pollution control equipment is being used as of this writing in about half of the United States to control VOC emissions in ozone nonattainment areas. As mentioned earlier, standards have been set on limiting the volatility of gasoline and additional studies and standards on reducing gasoline volatility and hazardous components in gasoline are expected. Following is a discussion of the control equipment installed and operated to reduce displacement emissions from each facility.

Vapor collection and processing equipment has been installed and operated for over 10 years at gasoline-marketing emission sources. The basic strategy has been to collect or suppress vapors emitted at the end of the gasoline marketing chain (service stations and bulk plants) and to transport the vapors back to the beginning of the chain (bulk terminals) for recovery or destruction. Equipment is then needed at each facility, as well as on the tank trucks, to collect and transport the vapors.

Submerged Fill

As discussed earlier, using submerged fill loading into containers (truck or storage tanks), as opposed to splash fill, reduces the amount of vapors generated and potentially escaping to the atmosphere. From the data in Table 1, conversion to submerged loading ($S = 0.60$) from splash fill ($S = 1.45$) reduces the amount of vapors generated by about 60%.

Tank Trucks

To collect and transport displaced vapors during tank-truck and storage-tank loading operations requires the installation of equipment on the tank truck. In addition, older tank trucks may require conversion to another loading method. Without vapor collection, two truck loading methods are used, either top or bottom loading. Top loading is simply inserting a liquid-fill pipe into an open hatch and allowing the displaced vapors to escape from the hatch opening. Bottom loading involves introducing the liquid through sealed pipes and connectors under the truck's tank, with the top hatches closed, and displacing the vapors through a vent on the top of the truck's tank. Bottom loading is shown in Figures 2 and 3. Since top loading requires the hatches to remain open in order to insert the fill pipe, it does not lend itself to vapor collection and recovery.* Most in the gasoline marketing industry now use bottom loading for industry standardization, faster and safer loading, and ease of vapor recovery. However, some older bulk terminals and plants and the tank trucks servicing them still use top open-hatch loading. Those tank trucks and loading facilities not only will have to be equipped with vapor-collection equipment, but also will have to be converted to bottom loading.

To convert the tank truck to vapor collection, the open vent on the top of the truck is enclosed in metal or rubber and is connected to metal vapor piping installed along the top of the tank truck and down the rear outside of the truck tank. Testing by the EPA in the late 1970s at bulk terminals found that on average 30% of the vapor volume displaced through the tank truck and its vapor-collection piping was escaping capture due to vapor leaks. The majority of vapor leaks were found around the seals and pressure and vacuum vents on the truck tank's manhole and hatch assembly on top of the tank truck. By requiring tank trucks to pass an annual pressure and vacuum decay test for leaks on their tank and vapor-collection systems, vapor leakage was reduced, on average, to 10%. In 1978, a reasonably available control technology (RACT) control level was set that required all vapor-collection tank trucks servicing gasoline marketing facilities using vapor processors or balance systems to (1) pass an annual pressure and vacuum decay test, and (2) undergo periodic monitoring for leaks with a portable gas detector.[8] The decay test requires that the truck tank and vapor-collection equipment not sustain a pressure change of more than 3 inches of water in five minutes when pressurized to 18 inches of water or evacuated to 6 inches of water. The same test is also required for tank trucks loading at bulk terminals under the New Source Performance Standards (NSPS) discussed below. In addition, the bulk terminal NSPS provides the standard EPA reference test method (Method 27) for testing truck tanks.

Bulk Terminals

Figure 2 shows a typical arrangement of vapor collection and processing at a bulk terminal. As gasoline is pumped from the storage tank into the truck tank, the air–vapors mixture is displaced through a vapor-collection header system of hoses and pipes to a vapor processor. Bulk terminals currently use vapor-recovery or -destruction units to process vapors. Multistage refrigeration units and double-bed self-regenerating carbon absorbers are the most common vapor-recovery devices installed at bulk terminals. Ground flares or thermal oxidizers are the most common types of destruction devices used at bulk terminals.

Historically, bulk terminals were required to meet 90% control efficiencies. In 1977, a RACT control level[9] was set for new and existing bulk terminals in ozone nonattainment areas to limit vapor processor outlet emissions to no more than 80 mg of VOC per liter of gasoline loaded (about 90% control efficiency). In 1983, the EPA promulgated NSPS[10] for bulk terminals, limiting outlet emission rates to no more than 35 mg/L transferred (about 95%) and promulgated reference test methods for testing bulk terminal control devices and tank trucks.

Bulk Plants

At bulk plants and service stations, the primary method for controlling emissions caused by displacement during load-

*In the 1970s, some bulk terminals and plants used specially designed top loading vapor closure devices that sealed the open hatchway and provided both liquid-fill and vapor-return lines. However, those devices abused the truck tank's hatch rims, resulting in vapor leakage when vapor balancing at other facilities, and lessened the tank integrity in rollover situations. Those devices have been abandoned in favor of bottom loading.

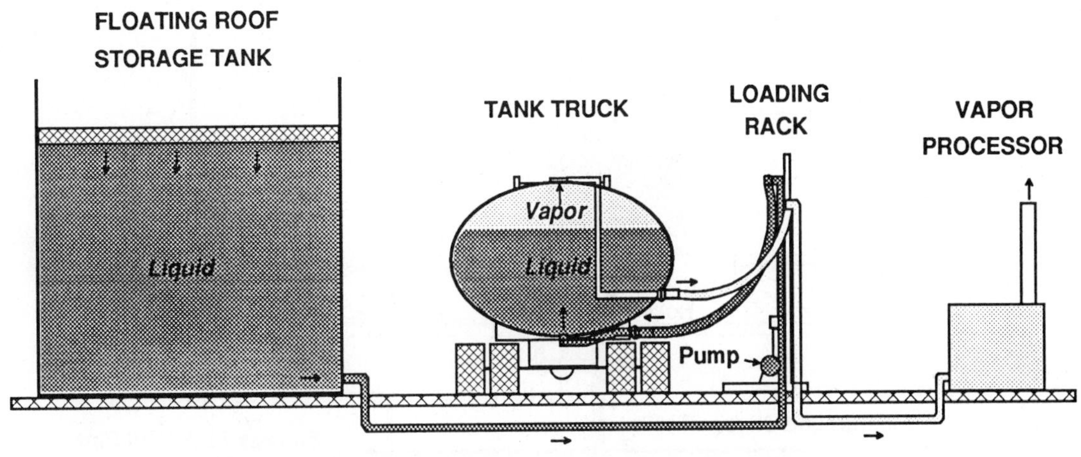

FIGURE 2. Bulk Terminal with Vapor Collection and Processor

ing and unloading is to transfer the vapors being displaced from the receiving container back to the dispensing container during the loading or unloading operation. This method is called vapor balancing. In addition to collecting the transfer or displacement emissions, balancing suppresses the additional evaporation of liquid inside the container. Without vapor balancing, the container ingests air when product is removed, thus providing the opportunity for liquid remaining in the container to evaporate into that air space. Balancing provides a vapor blanket on the surface of the liquid, thus retarding further evaporation. Figure 3 shows an example of vapor balancing at bulk plants. Gasoline unloaded from tank trucks into the bulk plant's storage tank displaces the vapors inside the storage tank through vapor piping to the tank truck's vapor-collection system and then into the truck tank. The collected vapors in the tank truck are transported back to the bulk terminal, where they are collected and processed. In reverse, vapors displaced by liquid loaded into the tank truck is routed though the vapor collection system on the tank truck to the vapor piping at the plant and then into the storage-tank vapor space.

In theory, vapor balancing should transfer 100% of the vapors displaced. In actuality, leaks in the system (connectors, hatch seals, pressure vacuum vents, etc.) at the bulk plant and on the tank truck, as well as the temperature differences in the liquids and vapor–air mixtures being transferred (volume growth or decrease), reduce the effectiveness of vapor balancing. Based on EPA testing, vapor balancing at bulk plants has been shown to achieve an efficiency of greater than a 90% reduction in displacement losses. In 1977, the EPA issued a RACT document recommending vapor balancing to control 90% of displacement losses at bulk plants.[11]

Service Stations

Figures 4 and 5 illustrate vapor balancing during storage-tank loading and vehicle refueling. Vapor balancing the displaced vapors from the underground storage tanks to the tank truck is commonly called Stage I controls. Tests demonstrate that Stage I vapor balance systems achieve reductions of greater than 90% control of underground-tank

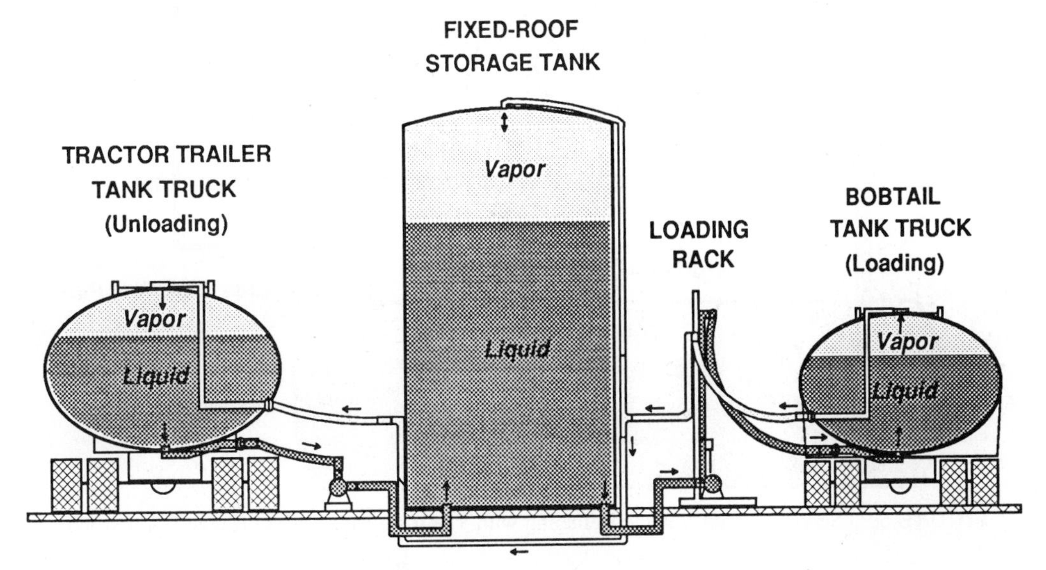

FIGURE 3. Bulk Plant with a Vapor Balance System

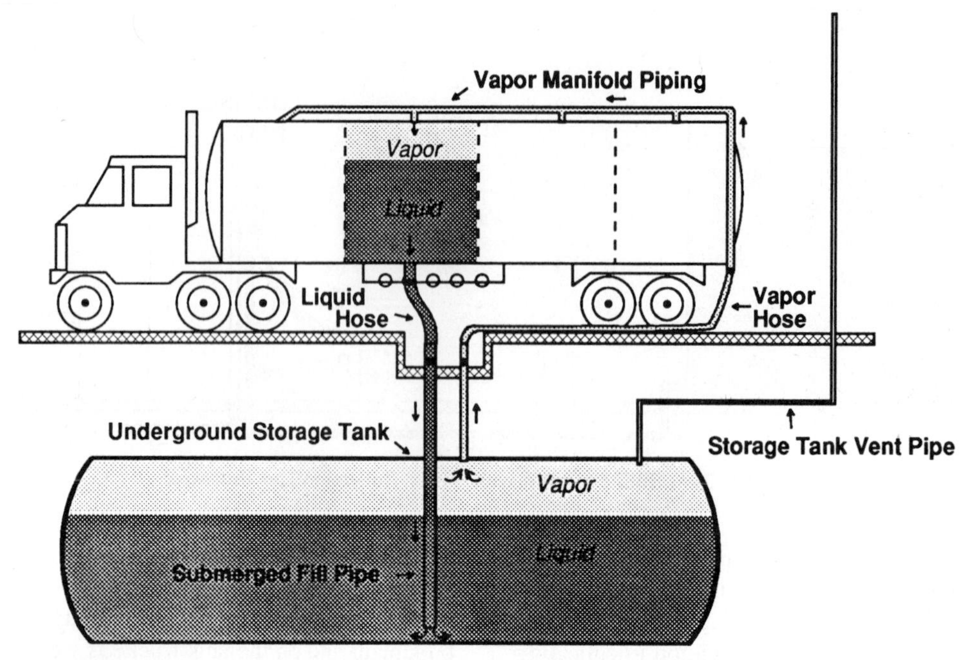

FIGURE 4. Loading of Service Station Underground Storage Tank with Vapor Balance System (Stage I Controls)

loading losses. In 1975, the EPA issued a guidance document providing equipment design criteria for achieving at least 90% control with Stage I vapor balance systems.[12] Again, the effectiveness of these controls is adversely affected by leaks. In addition, a continuing problem has been that some tank truck operators do not connect the vapor hose between the storage-tank connector and the tank truck. Displaced vapors from the filling of vehicle gas tanks at service stations can be controlled by either of two systems, Stage II or onboard controls. Stage II controls are equipment installed at the station and onboard controls are installed on the motor vehicle.

Figure 5 illustrates one type of Stage II system, a vapor balance system. Other Stage II systems look similar, but use vacuum pumps, aspirators, or fluid-powered pumps to enhance the vapor recovery at the nozzle–fill-neck interface. Some of these assist-type systems create excess vapor–air volume that requires an additional vapor processor to process the excess vapors. The integral part of the Stage II system is the vapor-recovery nozzle and hose that efficient-

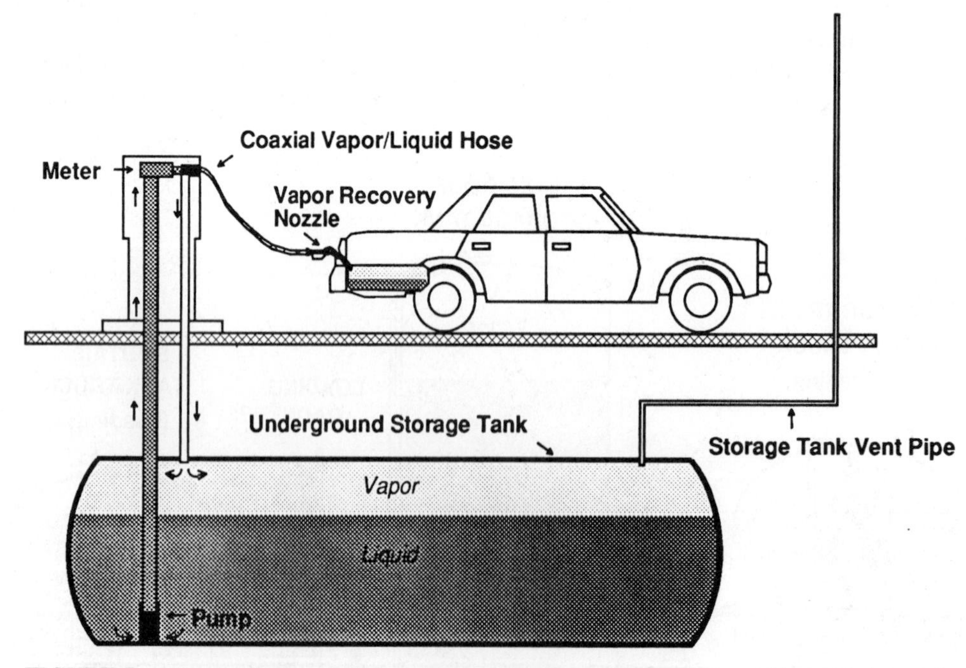

FIGURE 5. Service Station Vehicle Refueling with Vapor Balance System (Stage II Controls)

ly collect the displaced vapors. The special Stage II nozzle is equipped with a rubber boot to collect and then route the vapors through the nozzle into a coaxial vapor–liquid hose and to the dispenser and into the storage tank. Historically, Stage II nozzles were bulky and heavy, and components (especially the rubber boots and vapor hoses) wore out and required periodic replacement or maintenance. Since the 1970s, Stage II nozzles and hoses have become as lightweight as standard nozzles and hoses, boot and hose materials have become more durable and pliable, and coaxial liquid–vapor hoses and high-hang hose arrangements have made the system easier to operate and maintain. In addition, a newer design of Stage II nozzle does not use a rubber boot, but has a perforated pipe surrounding the nozzle's fill tube to collect vapors, and it uses a fluid-powered pump to enhance the vapor recovery at the nozzle–fill-neck opening.

Stage II systems have been tested and some designs have been certified to reduce displaced vapors by at least 95% by the California Air Resources Board. However, inadequate maintenance by station operators and inspections by control agencies have lessened the in-use effectiveness of Stage II systems in some areas. For equipment installed in the early 1980s, the EPA estimated in-use efficiencies to range from 64% for areas with low maintenance and little inspection to 86% with proper maintenance and annual control agency inspections.[2] No information or studies are available to assess in-use efficiencies of the improved Stage II systems marketed today. To date, Stage II systems are in use in four states and the District of Columbia. The Clean Air Act as amended in 1990 requires Stage II to be installed in nearly all ozone nonattainment areas in the 1990s.

The second vehicle-refueling control technique is to use an onboard system. Onboard refueling systems route the displaced vapors to a carbon absorber on the vehicle, instead of out of the fuel tank's fill pipe. When the vehicle is in operation, the carbon is regenerated by a reverse-flow air sweep through the carbon canister. Vapors are then passed into the engine's air intake and burned. Onboard systems have been demonstrated to achieve 94% in-use efficiencies. In 1987, the EPA proposed national regulations requiring onboard refueling systems.[13] The Clean Air Act as amended in 1990 requires onboard rules to be promulgated in the 1990s.

References

1. *Code of Federal Regulations, Protection of Environment,* Title 40, Part 80—Regulation of Fuels and Fuel Additives, U.S. Government Printing Office, Washington, DC.
2. *Draft Regulatory Impact Analysis: Proposed Refueling Emission Regulations for Gasoline-Fueled Motor Vehicles, Vol I: Analysis of Gasoline Marketing Regulatory Strategies,* EPA-450/3-87-001a, U.S. Environmental Protection Agency, Washington, DC, 1987.
3. *Evaluation of the Carcinogenicity of Unleaded Gasoline,* EPA/600/6-87/001, U.S. Environmental Protection Agency, Washington, DC, 1987.
4. *Compilation of Air Pollution Emission Factors, Volume I: Stationary Point and Area Sources,* 4th ed., AP-42, U.S. Environmental Protection Agency, Research Triangle Park, NC, 1985, pp 4.4-1–17.
5. M. R. Beychok, "Calculate tank losses easier," *Hydrocarbon Proc.* (March 1983).
6. *Evaporation Loss from Internal Floating-Roof Tanks,* API Publication 2519, American Petroleum Institute, Washington, DC, 1983, p 18.
7. D. Rothman and R. Johnson, *Refueling Emissions from Uncontrolled Vehicles,* EPA-AA-SDSB-85-6, U.S. Environmental Protection Agency, Ann Arbor, MI, 1985.
8. *Control of Volatile Organic Compound Leaks from Gasoline Tank Trucks and Vapor Collection Systems,* EPA-450/2-78-051, U.S. Environmental Protection Agency, Research Triangle Park, NC, 1978.
9. *Control of Hydrocarbons from Tank Truck Gasoline Loading Terminals,* EPA-450/2-77-026, U.S. Environmental Protection Agency, Research Triangle Park, NC, 1977.
10. *Code of Federal Regulations, Protection of Environment,* Title 40, Part 60, Subpart XX—Standards of Performance for Bulk Gasoline Terminals, U.S. Government Printing Office, Washington, DC.
11. *Control of Volatile Organic Emissions from Bulk Gasoline Plants,* EPA-450/2-77-035, U.S. Environmental Protection Agency, Research Triangle Park, NC, 1977.
12. *Design Criteria for Stage I Vapor Control Systems Gasoline Service Stations,* U.S. Environmental Protection Agency, Research Triangle Park, NC, 1975.
13. "Control of air pollution from new motor vehicles and new motor vehicle engines; Refueling emission regulations for gasoline-fueled light-duty vehicles and trucks and heavy-duty vehicles; Notice of proposed rulemaking," *Federal Register,* Vol. 52, No. 160, U.S. Government Printing Office, Washington, DC, August 19, 1987, pp 31162–31271.

Bibliography

Air Pollution, 3rd ed., Vol. VII, A. C. Stern, Ed.; Academic Press, Orlando, FL, 1986.

Bulk Gasoline Terminals—Background Information for Proposed Standards, EPA-450/3-80-038a & b, U.S. Environmental Protection Agency, Research Triangle Park, NC, 1980.

Evaluation of Air Pollution Regulatory Strategies for Gasoline Marketing Industry, EPA-450/3-84-012a, U.S. Environmental Protection Agency, Washington, DC, 1984.

ORGANIC SOLVENT CLEANING (DEGREASING)

Mark B. Turner and Richard V. Crume

Organic solvent cleaners use organic solvents, solvent blends, or their vapors to remove water-insoluble soils such as grease, oils, waxes, carbon deposits, fluxes, and tars from metal, plastic, fiberglass, printed circuit boards, and other surfaces. Typically, organic solvent cleaning is performed prior to such processes as painting, plating, inspection, repair, assembly, heat treatment, and machining. The three basic types of solvent cleaning equipment are open top vapor cleaners (OTVCs), in-line (cold and vapor) cleaners (also known as conveyorized cleaners), and batch cold cleaners. The vast majority of halogenated solvent use is in vapor cleaning, both open top and in-line. The five commonly used halogenated solvents are methylene chloride (MC), perchloroethylene (PCE), trichloroethylene (TCE), 1,1,1,-trichloroethane (TCA), and trichlorotrifluoroethane (CFC-113). The primary solvents used in batch cold cleaners are mineral spirits, Stoddard solvents, and alcohols. Very little halogenated solvent use has been identified in batch cold cleaning.[1]

Data on the number of organic solvent cleaners in use are scarce. In 1974, the following estimates were available: 1,220,000 cold cleaners, 21,000 OTVCs, 3700 in-line cleaners. In 1987, the following estimates were presented based on those cleaners that use halogenated solvents: 100,000 cold cleaners, 25,000 to 35,000 OTVCs, 2000 to 3000 in-line vapor cleaners, and 500 to 1000 in-line cold cleaners.[2]

PROCESS DESCRIPTION

This section describes in detail the three most commonly used organic solvent cleaners—OTVCs, in-line cleaners, and cold cleaners—including the equipment used, how parts are cleaned, and the types of cleaning applications for which each cleaner is used.

Open Top Vapor Cleaners[3]

Open top vapor cleaners are used primarily in metalworking operations and other manufacturing facilities. They are seldom used for ordinary maintenance cleaning because cold cleaners using petroleum distillate solvents can usually perform this type of cleaning at a lower cost. Exceptions include maintenance cleaning of electronic components, small equipment parts, and aircraft parts, where a high degree of cleanliness is needed.

A basic OTVC, shown in Figure 1, is a tank designed to generate and contain solvent vapor. At least one section of the tank is equipped with a heating system that uses steam, electricity, hot water, or heat pumps to boil liquid solvent. As the solvent boils, dense solvent vapors rise and displace the air inside the tank. The solvent vapors rise to the level of the primary condensing coils. Coolant (such as water) is circulated through the condensing coils to provide continuous condensation of the rising solvent vapors and, thereby, create a controlled vapor zone that prevents vapors from escaping the tank. Condensing coils generally are located around the periphery of the inside walls of the cleaner,

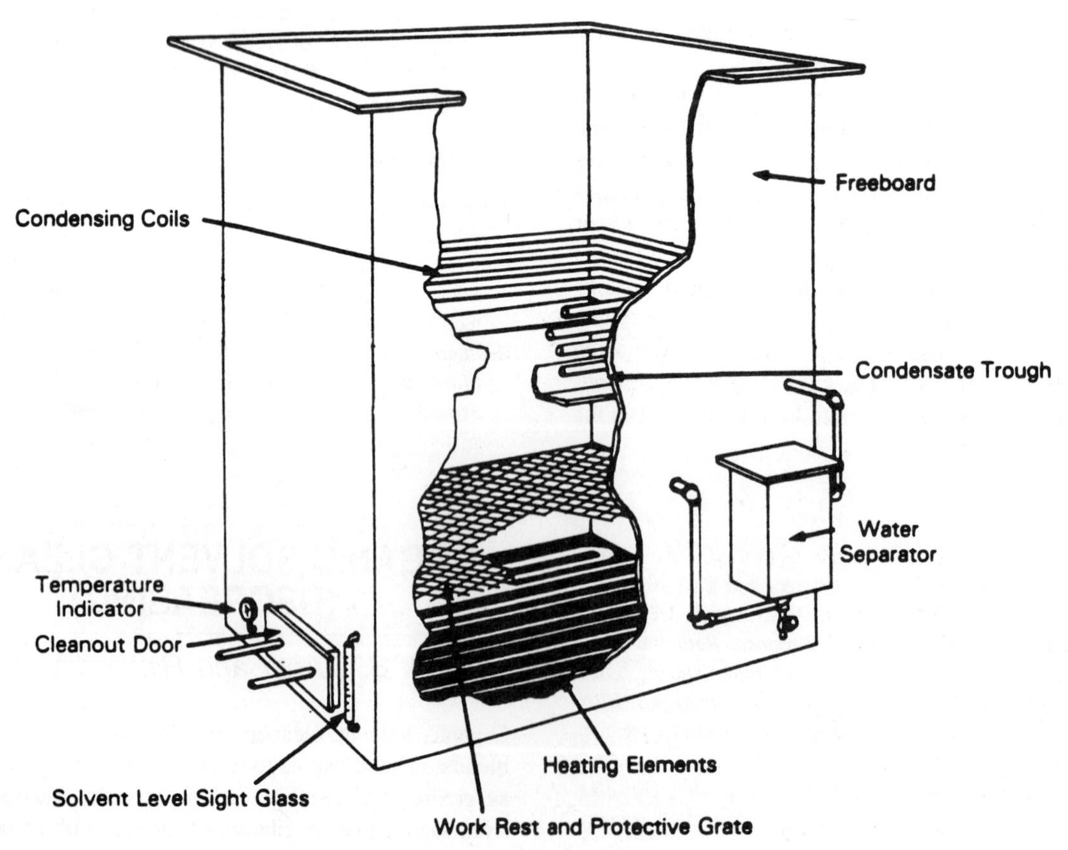

FIGURE 1. Schematic of an Open Top Vapor Cleaner

although in some equipment, they consist of offset coils on one end or side of the cleaner.

All machines have covers of varying design to limit solvent losses and contamination during downtime or idle time. Additional control of the solvent vapor is provided by the freeboard, which is that part of the tank wall extending from the top of the solvent vapor level to the tank lip. The freeboard ratio (FBR), or ratio of freeboard height to machine width (smaller dimension of vapor–air interface area), usually ranges from 0.75 to 1.0, depending on the manufacturer's design. The FBR can be as low as 0.5 on some older machines. Air currents in an OTVC can cause excessive solvent emissions. Increasing the FBR reduces the disturbance of the vapor zone caused by workplace air currents and slows solvent diffusion out of the machine.

Moisture may enter the OTVC on work loads and also can condense from ambient air and solvent vapors on primary cooling coils or freeboard refrigeration coils. If allowed to accumulate, water in an OTVC will lead to higher emissions and may contribute to solvent decomposition and corrosion in the cleaner. Therefore, nearly all vapor cleaners are equipped with a water separator. The water separator is a simple container in which the water phase (the water being essentially immiscible with and less dense than halogenated solvents) separates from the liquid solvent. The water is directed to disposal while the solvent is allowed to return to the cleaner. To reduce water contamination further or to replace the water separator, some manufacturers produce machines using a cannister of desiccant, such as a molecular sieve.

During the vapor cleaning operation, solvent vapors condense on the cooler work load entering the vapor zone. Condensing solvent dissolves some contaminants and flushes both dissolved and undissolved soils from the work load. Condensed solvent and dissolved or entrained contaminants then drain back into the sump below. When the temperature of the work load reaches that of the vapor, condensation ceases and the vapor-phase cleaning process is complete.

Organic impurities (greases, soils, etc.) cleaned from parts will accumulate in the solvent sump. Therefore, contaminated solvent periodically is drained from the machine and replaced with fresh solvent. Alternatively, a still adjacent to the cleaner can be used to extract soils building up in the solvent sump and return clean solvent to the machine.

One OTVC design variation is an immersion–vapor-spray cycle. In this design, the work load is lowered into a warm or boiling immersion compartment, which may be equipped with ultrasonics, for precleaning. In a machine using ultrasonics, high-frequency sound waves are used to produce pressure waves in the liquid solvent. In areas of low pressure within the liquid, minute vapor pockets are formed. These pockets collapse as the pressure in the zone cycles to high pressure. The constant creation and collapse of these vapor pockets (called cavitation) provide a scrubbing action to aid cleaning. Ultrasonically agitated liquids often need to be heated to specific temperatures to achieve optimum cavitation. After this first stage of cleaning has been completed, the work load is cleaned in a vapor section and then sprayed with solvent. The spray nozzle must be below the vapor line, to avoid spraying solvent directly to the atmosphere, and directed downward, to avoid turbulence at the air–solvent-vapor interface.

Lip or slot exhausts are designed to capture solvent vapors escaping from the OTVC and carry them away from operating personnel. These exhaust systems disturb the vapor zone or enhance diffusion, thereby increasing solvent losses. The increased losses can be significant. In properly designed lip exhaust systems, the cover closes below the lip exhaust inlet level.

In-Line Cleaners[4]

In-line cleaners (also called conveyorized cleaners) employ automated load on a continuous basis. Although in-line cleaners can operate in either the vapor or nonvapor phase, the majority of all in-line machines using halogenated solvents are vapor cleaners. A continuous or multiple-batch loading system greatly reduces manual parts handling associated with OTVCs or batch cold cleaners. The same cleaning techniques are used in in-line cleaning, but usually on a larger scale than with open top units.

In-line cleaners are nearly always enclosed, except for parts/conveyor inlet and exit openings, to help control solvent losses from the system. In-line cleaners are most often found in plants where there is a constant stream of parts to be cleaned.

There are five main types of in-line cleaners using the halogenated solvents: cross-rod, monorail, belt, strip, and printed-circuit-board processing equipment (photoresist strippers, flux cleaners, and developers.) While most of these may be used with cold or vaporized solvent, the last two are almost always vapor cleaners. The photoresist strippers are typically cold cleaners.

The cross-rod cleaner (Figure 2) gets its name from the rods from which parts baskets are suspended as they are conveyed through the machine by a pair of power-driven chains. The parts are contained in pendant baskets or, where tumbling of the parts is desired, in perforated or wire mesh cylinders. These cylinders may be rotated within the liquid solvent and/or the vapor zone. This type of equipment lends itself particularly well to handling small parts that need to be immersed in solvent for satisfactory cleaning or that require tumbling to drain solvent from cavities and/or to remove metal chips.

A monorail vapor cleaner (Figure 3) is usually chosen when the parts to be cleaned are being transported between manufacturing operations on a monorail conveyor. The monorail cleaner is well suited to automatic cleaning with solvent spray and vapor. It can be of the straight-through design illustrated or can incorporate a U-turn within the machine so that parts exit through an opening parallel to the

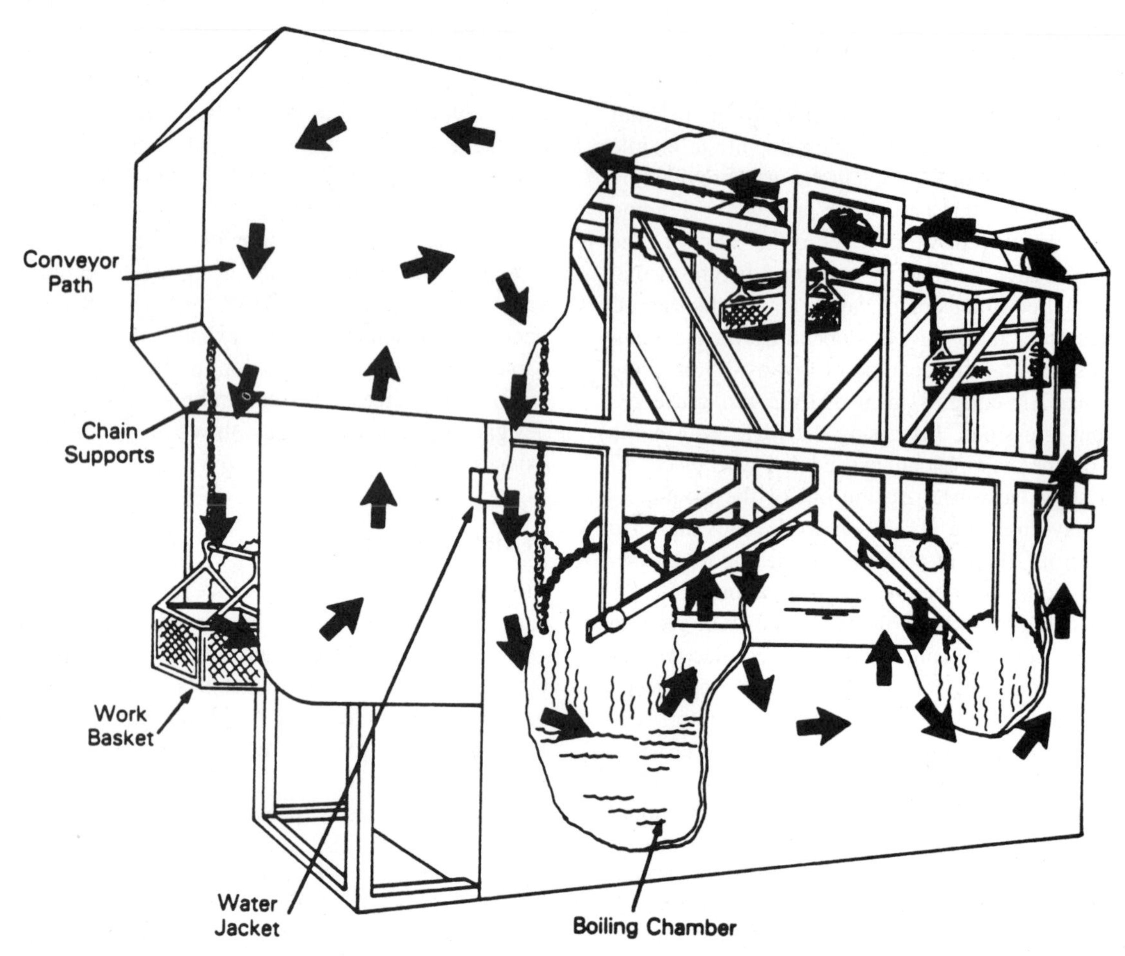

FIGURE 2. Schematic of a Cross-Rod In-Line Cleaner

entrance. The U-turn monorail cleaner benefits from lower vapor loss because the design eliminates the possibility of drafts flowing through the machine.

Both the belt cleaner (Figure 4) and the strip cleaner are designed to allow simple and rapid loading and unloading of parts. A belt cleaner conveys parts through a long, narrow boiling chamber in which the parts are cleaned, either by the condensing vapor or by immersion in the solvent sump. The strip cleaner is similar to the belt cleaner except that the strip itself is the material being cleaned.

Cleaning of printed circuit boards is a common application of a type of mesh belt cleaner (Figure 5). In the production of printed circuit boards, solvent-based photoprocessible resists can be used. The circuit pattern is contained in an artwork film. This pattern is reproduced by projecting ultraviolet rays through the artwork film onto a copper sheet covered with resist. A developer (typically TCA) dissolves the unexposed areas of the resist and thereby reveals the circuit pattern. The resist-covered board is then placed in plating solutions to add more metal to the circuit pattern areas. Next, a photoresist stripper dissolves the remaining resist. The circuit boards are then put in an alkaline etching solution to remove all the copper in the noncircuitry areas. The processing is completed by passing the circuit boards through molten solder.

Because of the nature of the materials being cleaned, photoresist strippers use ambient (room temperature) solvents. Spraying and brushing may be used to enhance cleaning. Methylene chloride is the solvent most often used in photoresist stripping; however, the printed-circuit-board industry has largely converted to aqueous and semi-aqueous materials to replace the use of both TCA and MC.

Circuit board cleaners are used to dissolve and remove flux from the circuit board after the molten soldering step. Unlike photoresist strippers, circuit board cleaners have a heated or boiling sump. However, circuit-board cleaning occurs in the liquid solvent (not vapor) phase, although a vapor phase may be present. Circuit board cleaners commonly use chlorofluorocarbons; however, aqueous fluxes and aqueous flux cleaners are becoming more widely used in the printed-circuit industry as a replacement.

Cold Cleaners[5,6]

The two basic types of cold cleaners are maintenance cleaners and manufacturing cleaners. The maintenance cold cleaners are usually simpler, less expensive, and smaller. The manufacturing cold cleaners usually perform a higher quality of cleaning than do maintenance cleaners and are

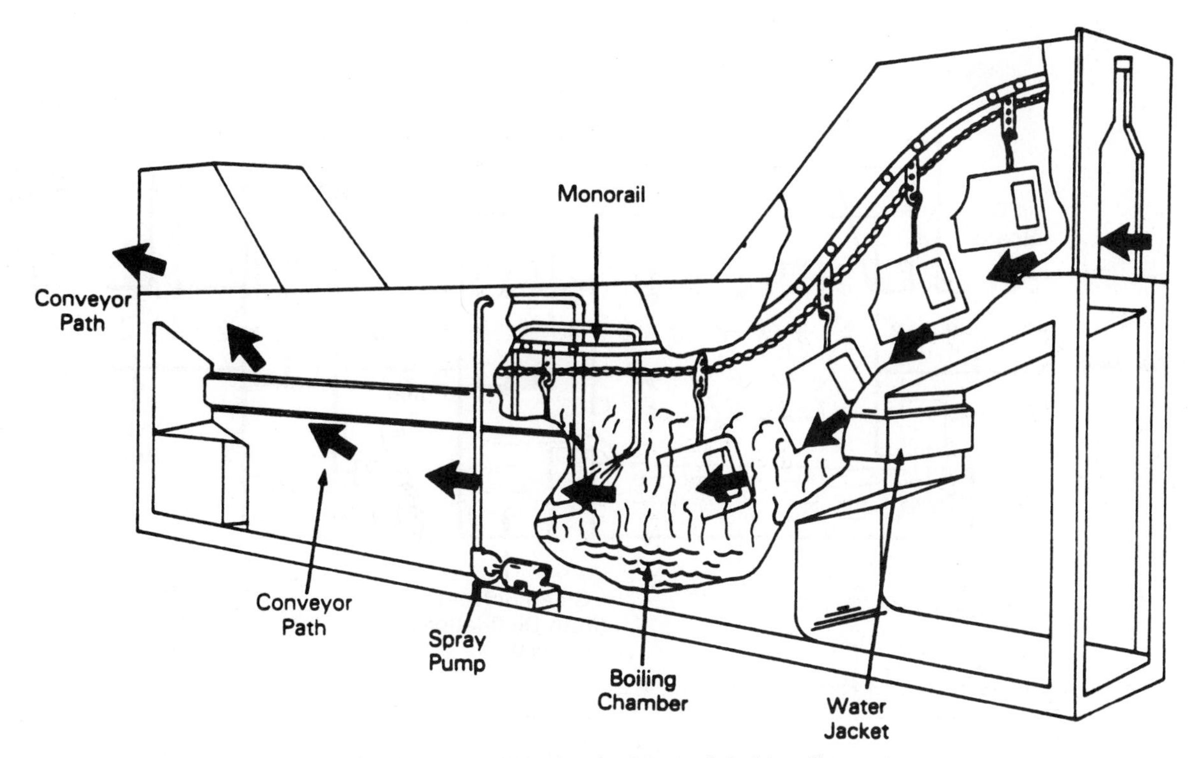

FIGURE 3. Schematic of a Monorail In-Line Cleaner

thus more specialized. While there are many more maintenance cleaners in use than manufacturing cleaners, the manufacturing cleaners tend to emit more solvent per cleaner because of their larger size and work load. Additionally, manufacturing cleaners use a wide variety of solvents, whereas maintenance cleaners use mainly petroleum solvents such as mineral spirits, Stoddard solvents, and alcohols. Cold cleaners that use halogenated solvents are of a type called carburetor cleaners. Figure 6 shows a schematic of a carburetor cleaner. In these cleaners, MC is blended with other solvents and additives to reduce flammability and to increase dissolving power.

The type of cold cleaner to be used for a particular application depends on two main factors: (1) the work load and (2) the required cleaning effectiveness. Work load is a function of tank size, frequency of cleaning, and type of parts. Naturally, the larger work loads require larger degreasers. The more frequently the cold cleaner is used, the greater will be the need to automate and speed up the cleaning process; more efficient materials-handling systems help to automate, while agitation speeds cleaning. Finally, the type of parts to be cleaned is important because more thorough cleaning and draining techniques are necessary for parts that are more complex in shape.

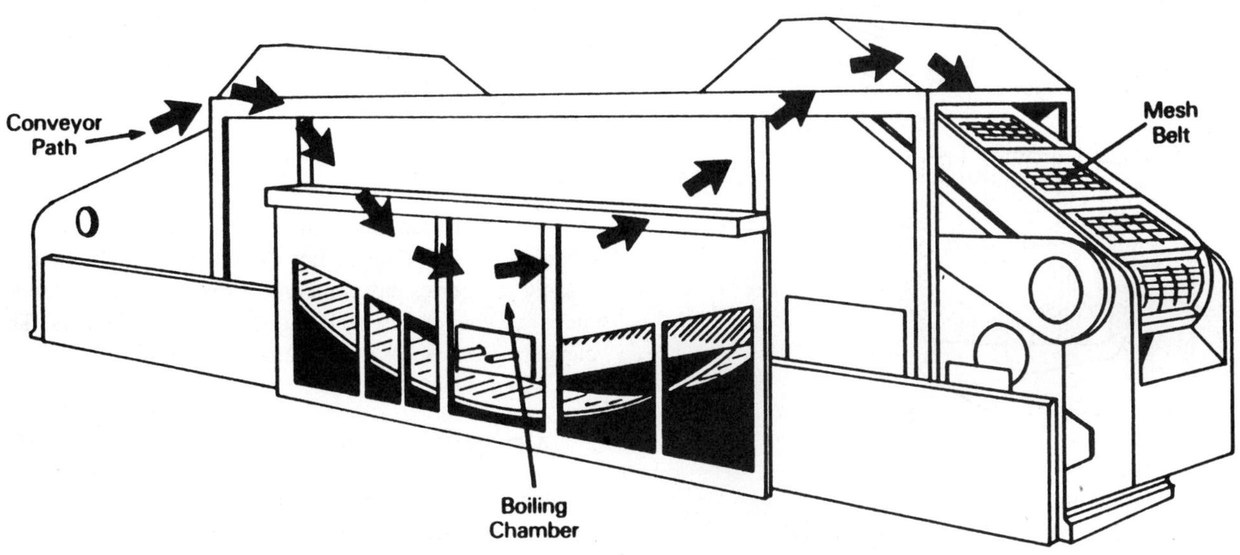

FIGURE 4. Schematic of a Mesh Belt In-Line Cleaner

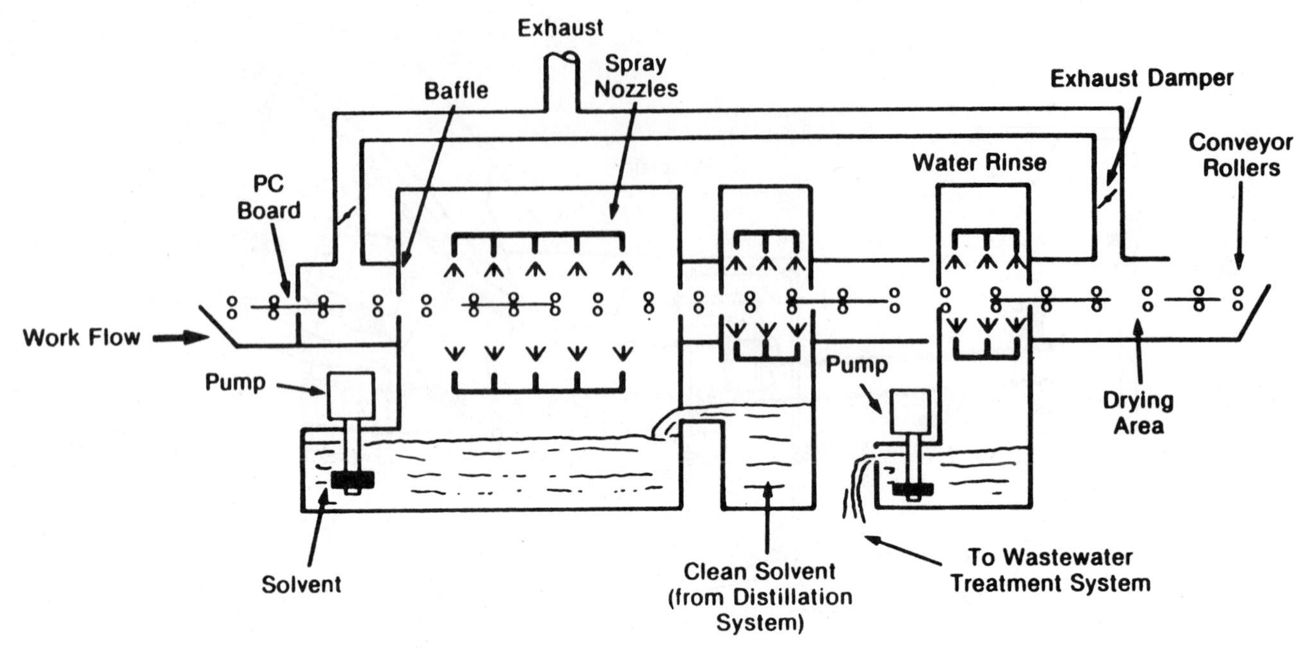

FIGURE 5. Schematic of an In-Line Photoresist Stripping Machine

The required cleaning effectiveness establishes the choice of solvent and the degree of agitation. For greater cleaning effectiveness, more powerful solvents and more vigorous agitation are used. Generally, emissions will increase with agitation and with higher solvency.

The two basic tank designs are the simple spray sink and the drip tank. The simple spray sink is more appropriate for cleaning applications that are not difficult and require only a relatively low degree of cleanliness. The dip tank provides more thorough cleaning through soaking of dirty parts. Dip tanks also can employ agitation, which improves cleaning efficiency.

Agitation is generally accomplished through the use of pumping, compressed air, vertical motion, or ultrasonics. In the pump-agitated cold cleaner, the solvent is rapidly circulated in the soaking tank. Air agitation involves dispersing compressed air from the bottom of the soaking tank, with the air bubbles providing a scrubbing action. In the vertically agitated cold cleaner, dirty parts move up and down while submerged in order to enhance the cleaning process. Finally, in the ultrasonically agitated tank, the solvent is vibrated by high-frequency sound waves.

The materials-handling technique can be important in reducing emissions from cold cleaning. Regardless of the system, the work loads need to be handled so that the solvent has sufficient time to drain from the cleaned parts into an appropriate container.

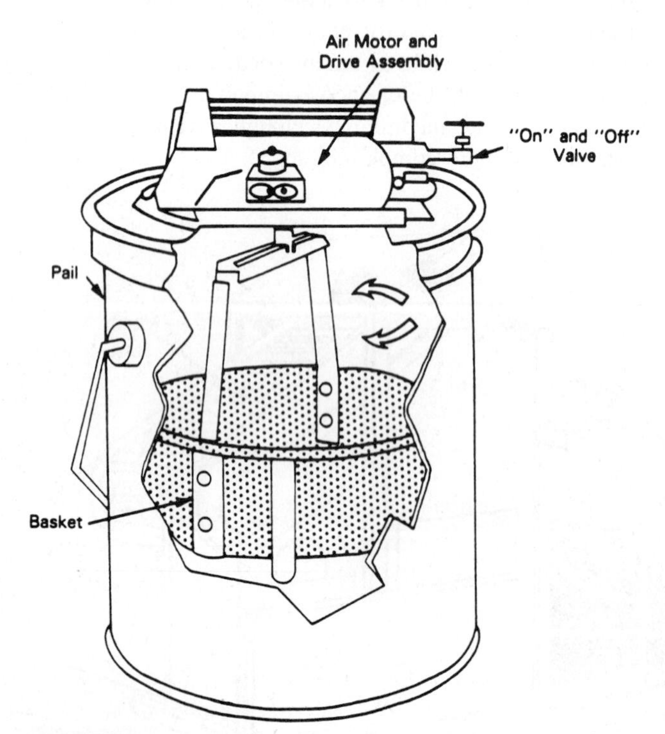

FIGURE 6. Schematic of a Carburetor Cleaner

Hybrid Cleaners

As the solvent cleaning industry has developed, specialized cleaning devices that do not fit into the OTVC or in-line cleaner categories have emerged. Among these cleaners are the vibra, the ferris wheel, and the carousel cleaners. These cleaners are not discussed here.

AIR EMISSIONS CHARACTERIZATION[7]

There are many sources of solvent loss to the atmosphere from an organic solvent cleaner. Two significant sources are air–solvent-vapor interface losses and work-load–related losses. Air–solvent-vapor interface losses during idling consist of solvent vapor diffusion (or evaporation from liquid solvent in a cold cleaner) and solvent vapor convection induced by warm freeboards. Work-load losses are solvent emissions that are liberated during the cleaning process by the introduction to and extraction from the cleaner of parts and by the spraying of parts (if sprays are used). Other potentially significant losses that contribute to the total solvent emissions from a solvent cleaner include losses from filling/draining, wastewater, start-up/shutdown, downtime, and equipment leaks. Air–solvent-vapor interface losses during idling, work-load losses, and other losses are described in the following.

Air–Solvent-Vapor Interface Losses During Idling

Open Top Vapor Cleaners

The main source of idling losses from an OTVC is diffusion, or the movement of solvent vapors from the vapor zone to the ambient air above. Diffusion rates depend on temperature because molecular activity increases at higher temperatures. As solvent vapor molecules diffuse from the high concentration in the vapor zone to the lower concentration in the air, an equilibrium diffusion rate is established.

Solvent vapor convection up the tank walls occurs as heat from the boiling solvent and hot vapor heats the walls. Convective losses can be minimized by cooling the walls, either through the use of a water jacket around the outside of the cleaner or by ensuring that the primary condenser coils in contact with the walls provide adequate cooling.

Room drafts create turbulence at the air–solvent-vapor interface, thereby disturbing the equilibrium diffusion rate and causing increased emissions. Lip (or lateral) exhausts that draw solvent-laden air from around the top perimeter of the cleaner to reduce solvent vapor concentration in the area where operators are working also disturb the equilibrium diffusion rate.

Table 1 shows a summary of the available emission data for idling OTVCs. All of these data were obtained on uncovered machines with no refrigerated freeboard devices or lip exhausts. Emissions rates are lowest in tests where the primary condensing temperature of the cleaner is lowest.

In-Line Cleaners

The primary sources of idling losses from in-line vapor cleaners, convection and diffusion, and their mechanisms are the same as for OTVCs. While no emission test data are available on idling losses from in-line cleaners, the diffusional and convective losses from these cleaners would likely be less per unit of air–solvent-vapor interface area than for an OTVC because the units are almost always enclosed and so are less subject to drafts.

Cold Cleaners

Idling emissions from a cold cleaner occur through both evaporation and subsequent diffusion. The rate of solvent loss is solvent dependent and is affected by room drafts that increase the equilibrium evaporation rates.

Work-Load Losses

Open Top Vapor Cleaners

Increased losses occur when parts and parts baskets enter the solvent cleaner, thereby creating turbulence and increasing diffusional and convective losses at the air–solvent-vapor interface. The amount of loss increases as the speed of the basket increases. Additionally, losses are greater when the parts and parts basket take up a larger percentage of the interface area.

When very cold parts or a large quantity of parts are introduced into the cleaner, more heat will be required to bring the parts up to the temperature of the solvent vapor. When the heat is transferred from the solvent vapor to the parts, the vapor line lowers. As the vapor line rebuilds and rises back to its original level, the air–solvent-vapor mixture above the layer is displaced out of the cleaner.

Spraying of parts with solvent through fixed nozzles or spray wands also can cause turbulence in the air–solvent-vapor interface and vapor line lowering, thereby increasing emissions. Sprays with too high a pressure can cause

TABLE 1. Summary of Available Tests for Idling OTVCs

					Conditions		
Test No.	Solvent	Cleaner Size, m^2	FBR	Cleaner Make	Primary Condenser Temperature, °F	Emission Rate, $lb/ft^2/h$	Reference
1-1	Freon-TF	0.3	1.0	Delta Sonics	55	0.060	8
1-2	1,1,1-TCA	0.9	0.7	Auto-Sonics	50	0.087	9
1-3	1,1,1-TCA	0.9	0.7	Auto-Sonics	70	0.120	9
1-4	1,1,1-TCA	0.9	0.7	Auto-Sonics	85	0.143	9
1-5	CFC-113	0.9	0.7	Auto-Sonics	40	0.062	9
1-6	CFC-113	0.9	0.7	Auto-Sonics	50	0.094	9
1-7	CFC-113	0.9	0.7	Auto-Sonics	70	0.169	9

splashing of solvent against the parts, parts baskets, and tank wall, thereby increasing emissions further.

As parts are removed from the cleaner, the air–solvent-vapor interface again is disturbed. The speed of removal directly affects work-load loss. If parts are extracted too rapidly, solvent vapor will be entrained behind the work load and pulled out of the cleaner.

The final source of solvent loss during work-load removal is solvent drag-out. Drag-out includes solvent pooled in cavities and other surfaces of the parts, as well as the solvent film remaining on all surfaces as they leave the cleaner. If the work load is withdrawn slowly and allowed to dwell in the freeboard area, the solvent film and much of the pooled solvent can evaporate before the work load is withdrawn. A significant portion of the evaporated solvent in the freeboard area will sink back into the vapor layer or be condensed on the coils and returned to the cleaner. If the work load is withdrawn quickly, most liquid solvent will not evaporate from the parts until after they are withdrawn from the cleaner.

Table 2 presents a summary of the available work-load emission rates from OTVCs. The large variability in the data is due to the wide range of operating parameters during the tests. Unlike idling emissions, which are more a function of the machine design, work-load emissions are largely a factor of the operating parameters discussed above. The speed of parts movement in many of the tests is unknown. All of these tests were performed using electric hoists for parts entry and removal. Test results for manually operated machines would be significantly higher because it is difficult for a human operator consistently to achieve the low work-load–related losses exhibited by hoists. Furthermore, these tests included a wide range of room air speeds. Last, these tests did not include tests on OTVCs using a lip exhaust system.

In-Line Cleaners

The principal sources of work-load emissions from in-line cleaners are essentially the same as those from OTVCs. Because in-line systems are automated, the work-load losses are less on a per-part basis than in a manually operated OTVC. However, because of the large volume of parts cleaned in an in-line cleaner, overall emissions typically are higher than from OTVCs.

Cold Cleaners

Work-load losses from cold cleaners are primarily due to solvent drag-out. These losses can be reduced by allowing

TABLE 2. Summary of Emission Tests on Working OTVCs

Test No.	Solvent	Cleaner Size, m^2	Cleaner Make	Air Speed, fpm	Conditions Primary Condenser, °F	FBR	Emission[a] Rate, lb/ft^2/h	Reference Temperature
1	1,1,1-TCA	1.8	Detrex	Calm	—[b]	0.75	0.099	10
2	1,1,1-TCA	1.8	Detrex	130	—[b]	0.75	0.173	10
3	1,1,1-TCA	1.8	Detrex	160	—[b]	0.75	0.233	10
4	1,1,1-TCA	1.4	Auto-Sonics	—[b]	—[b]	—[b]	0.063	11
5	MC	1.2	Crest	—[b]	—[b]	0.83	0.186	12
6	MC	1.2	Crest	—[b]	—[b]	0.75	0.354	12
7	1,1,1-TCA	0.9	Auto-Sonics	—[b]	50	—[b]	0.100	9
8	1,1,1-TCA	0.9	Auto-Sonics	—[b]	70	—[b]	0.140	9
9	1,1,1-TCA	0.9	Auto-Sonics	—[b]	85	—[b]	0.170	9
10	CFC-113	0.9	Auto-Sonics	—[b]	40	—[b]	0.090	9
11	CFC-113	0.9	Auto-Sonics	—[b]	50	—[b]	0.110	9
12	CFC-113	0.9	Auto-Sonics	—[b]	70	—[b]	0.186	9
13[c]	CFC-113		Branson	—[b]	60	1.0	0.775	13
14	MC blend	0.4	Auto-Sonics	30	70	0.75	0.220	14
15	MC	0.4	Auto-Sonics	30	70	0.75	0.180	14
16	CFC-113	0.4	Auto-Sonics	30	70	0.75	0.165	14
17	MC blend	0.4	Auto-Sonics	30	70	0.75	0.125	14
18	1,1,1-TCA	0.4	Auto-Sonics	30	70	0.75	0.112	14
19	TCE	0.4	Auto-Sonics	30	70	0.75	0.080	14
20	MC blend	0.4	Auto-Sonics	30	70	1.0	0.175	14
21	MC	0.4	Auto-Sonics	30	70	1.0	0.145	14
22	CFC-113	0.4	Auto-Sonics	30	70	1.0	0.132	14
23	MC blend	0.4	Auto-Sonics	30	70	1.0	0.100	14
24	1,1,1-TCA	0.4	Auto-Sonics	30	70	1.0	0.092	14
25	TCE	0.4	Auto-Sonics	30	70	1.0	0.065	14

[a] "Working" emissions include diffusion, convection, and work-load losses but not leaks, solvent transfer losses, or downtime losses.
[b] Information unknown or not available.
[c] Constant cycling of parts into and out of machine and use of perforated metal basket that retained significant solvent upon exit from machine account for elevated emission number.

longer drainage time and by tipping parts to drain solvent-filled cavities.

Other sources of solvent loss during cold cleaning are agitation and spraying. Agitation, either through movement of the solvent or through the introduction of compressed air at the bottom of the tank, can increase solvent evaporation by increasing the effective air–solvent interface area. Spraying solvent can increase solvent evaporation by exposing more solvent to the air. As for OTVCs and in-line cleaners, the amount of solvent loss from spraying depends on the spray pressure.

Other Losses

In addition to idling and work-load losses, there are several other loss mechanisms that contribute to overall losses from an organic solvent cleaner. These losses result from downtime, leaks, filling/draining, wastewater, start-up/shutdown, distillation, and solvent decomposition.

Downtime losses are defined as solvent loss when the heat to the sump is turned off (vapor cleaners) and the machine is not in operation. Downtime losses for cold cleaners are identical to idling losses. They result from the evaporation of solvent from the liquid–solvent surface and subsequent diffusion. Tight-fitting covers help to reduce these losses. Equipment vendor estimates of downtime losses range from 0.03 to 0.07 $lb/ft^2/h$.

Loss of solvent through leaks can result from manufacturing defects or from machine use. These leaks are often difficult to detect because the solvent evaporates quickly, leaving no signs of the leak. Solvent loss during filling or draining the solvent cleaner can be significant, especially if the filling or draining of machines is accomplished manually using open buckets or drums. Losses will increase if a large amount of splashing occurs. Solvent loss also takes place when the water in a water–solvent separator is decanted while it contains a small amount of solvent. However, if a separator is properly designed, operated, and maintained, little solvent will be lost.

Start-up losses occur when solvent-laden air within the machine is pushed out after the sump heat has been activated and the solvent vapor layer is being established. Shutdown losses are the result of evaporation of hot liquid solvent from the sump (after the heat has been turned off and the vapor layer has collapsed) and subsequent diffusion of the solvent vapor from the cleaner. These start-up/shutdown losses are associated with vapor cleaners only.

Solvent lost from the on-site distillation of spent solvent results from evaporation during transfer to and from the distillant unit or, if a piping system is used, from leaks in the equipment. Solvent also may evaporate from distillation sludge or spent solvent that is removed for disposal.

Some solvents and blends contain stabilizers that prevent the mixture from turning acidic after reacting with water. Solvent that becomes acidic (through improper monitoring) must be discarded. Dangerous fumes (chlorine gas, hydrochloric acid) can be emitted from solvent decomposition. Emissions can occur during the handling and disposal of the solvent.

AIR POLLUTION CONTROL MEASURES

With the exception of the application of carbon adsorption, almost all of the air pollution control measures that apply to organic solvent cleaners involve equipment design features and the implementation of good operating practices rather than the application of add-on air pollution control devices. Additionally, the application of combinations of control measures can provide significant emission reductions over the application of a single control measure. Finally, alternative cleaning agents may be employed to eliminate emissions of the five common halogenated solvents.

The following sections briefly describe the various air pollution control measures that can be applied to OTVCs, in-line cleaners, and cold cleaners, and the emission reductions that can be achieved.

Open Top Vapor Cleaners[15]

Covers, refrigerated freeboard devices, refrigerated primary condensers, increased FBR, reduced room drafts, enclosed designs, carbon adsorbers, electric or mechanically assisted parts handling/reduced part movement speeds, and selected operation and maintenance practices are described in the following.

A significant amount of emission test data was available for both idling and working OTVCs to characterize the effectiveness of the various control measures. These tests were performed by companies that either manufacture solvent cleaners or sell solvents. No standard test methods were used. The test data and procedures have been reviewed by the U.S. Environmental Protection Agency (EPA) and appear to have given valid, repeatable results. Emission test data on both idling and working OTVCs are presented in Tables 3 and 4, respectively. The test data for working OTVCs are from machines employing automated mechanical systems for parts handling. In most cases, work loads used for these tests can be described as inherently producing low carryout losses. Therefore, emission rates would likely be higher from machines in regular industrial applications.

Covers

Covers are used on OTVCs to eliminate drafts within the freeboard and to reduce diffusion losses. They can be manually operated or powered electrically or penumatically. Manual covers are typically provided as standard equipment and can be flat-hinged, sliding, or roll-top. Hinged covers are not recommended because opening and closing these covers can disturb the air–solvent-vapor interface, thereby unnecessarily exposing the operator to increased emissions. To minimize disturbance of the air–solvent-vapor interface, roll-top and biparting covers that close horizontally can be

TABLE 3. Summary of Available Tests on Idling OTVCs

					Baseline					Controlled						
Test No.	Tested[a] Control	Solvent	Cleaner Size, ft^2	Cleaner Make	Air Speed, fpm	Cover	FBR	Freeboard Refrigeration	Emission, $lb/ft^2/h$	Air Speed, fpm	Cover	FBR	Freeboard Refrigeration	Emission, $lb/ft^2/h$	Control Efficiency[b]	Reference
I-1	AFC (PC@50F)	Freon-TF	3.3	Delta Sonics	30	Open	1.0	Off	0.060	30	Open	1.0	AFC	0.049	18%	8
I-2	BFC (PC@50F)	Freon-TF	3.2	Delta Sonics	30	Open	1.0	Off	0.060	30	Open	1.0	BFC	0.050	17%	8
I-3	BFC (PC@50F)	TCA	9.7	Auto-Sonics	LE off	None	0.7	Off	0.087	LE off	None	0.7	BFC	0.040	54%	9
I-4	BFC (PC@70F)	TCA	9.7	Auto-Sonics	LE off	None	0.7	Off	0.120	LE off	None	0.7	BFC	0.050	58%	9
I-5	BFC (PC@85F)	TCA	9.7	Auto-Sonics	LE off	None	0.7	Off	0.143	LE off	None	0.7	BFC	0.063	56%	9
I-6	BFC (PC@40F)	CFC-113	9.7	Auto-Sonics	LE off	None	0.7	Off	0.062	LE off	None	0.7	BFC	0.055	11%	9
I-7	BFC (PC@50F)	CFC-113	9.7	Auto-Sonics	LE off	None	0.7	Off	0.094	LE off	None	0.7	BFC	0.070	26%	9
I-8	BFC (PC@70F)	CFC-113	9.7	Auto-Sonics	LE off	None	0.7	Off	0.169	LE off	None	0.7	BFC	0.072	57%	9
I-9	PC—70°F to 40°F	CFC-113	9.7	Auto-Sonics	LE off	None	0.7	Off	0.169	LE off	None	0.7	off	0.062	63%	9
I-11	PC—85°F to 50°F	TCA	9.7	Auto-Sonics	LE off	None	0.7	Off	0.143	LE off	None	0.7	off	0.087	39%	9
I-12	PC—85°F to 50°F	TCA	9.7	Auto-Sonics	LE on	None	0.7	Off	0.211	LE on	None	0.7	off	0.171	19%	9
I-13	BFC&LipExh P@50F	TCA	9.7	Auto-Sonics	LE on	None	0.7	Off	0.171	LE off	None	0.7	BFC	0.040	77%	9
I-14	BFC&LipExh P@70F	TCA	9.7	Auto-Sonics	LE on	None	0.7	Off	0.190	LE off	None	0.7	BFC	0.050	74%	9
I-15	BFC&LipExh P@85F	TCA	9.7	Auto-Sonics	LE on	None	0.7	Off	0.211	LE off	None	0.7	BFC	0.063	70%	9
I-16	LIP EXH (PC@50F)	TCA	9.7	Auto-Sonics	LE on	None	0.7	Off	0.171	LE off	None	0.7	off	0.087	49%	9
I-17	LIP EXH (PC@70F)	TCA	9.7	Auto-Sonics	LE on	None	0.7	Off	0.190	LE off	None	0.7	off	0.120	37%	9
I-18	LIP EXH (PC@85F)	TCA	9.7	Auto-Sonics	LE on	None	0.7	Off	0.211	LE off	None	0.7	off	0.143	32%	9
I-19	FBR: 0.75 → 1.0	TCA	8.0	Detrex	Calm	None	0.75	Off	0.051	Off	None	1.0	off	0.054	–6%	12
I-20	FBR: 0.75 → 1.0	TCA	8.0	Detrex	30–100	None	0.75	Off	0.272	Off	None	1.0	off	0.167	39%	12

[a]AFC = above-freezing freeboard refrigeration; BFC = below-freezing freeboard refrigeration; LE = lip exhaust; PC = primary condenser (e.g. PC@50F means the primary condenser temperature was 50° F).
[b]These control efficiency values refer to the percent control of idling emission (i.e., diffusion and convection losses) only.

used. Biparting covers can be made to close around the cables holding parts baskets when the basket is inside the cleaner. The most advanced biparting covers are automated to coordinate the movement of the cover with that of an automated parts-handling system.

Four tests (tests 35, 36, 37, and 38) are available for an automatic biparting, roll-top cover that was closed 79% of the time (275 seconds out of the 350-second OTVC cycle). Based on these tests, working loss emission reductions of 38% (calm air conditions) to 53% (160 fpm room drafts) were achieved.

Refrigerated Freeboard Devices

Freeboard refrigeration devices consist of a second set of cooling coils located above the primary condenser coils of the cleaner. These secondary coils cool the air immediately above the vapor zone, forming a cool air blanket that slows solvent diffusion and creates a temperature inversion zone within the freeboard that reduces the mixing of air and solvent vapors. Above-freezing (AFC) and below-freezing (BFC) freeboard refrigeration devices can be used. The AFC systems operate at a temperature around 5°C (41°F), while the BFC systems operate within a temperature range of –20°C to –30°C (–4°F to –22°F). The BFC units typically are defrosted on a timed cycle to remove the solvent–water ice that may accumulate on the coils.

Twenty-six tests from five sources are available to evaluate the effect of freeboard refrigeration devices on OTVCs under working conditions (20 tests) and idling (six tests) conditions. Four tests (working conditions) evaluated AFC systems (tests 1 through 4). The remaining tests (16 working and six idling conditions) evaluated BFC systems. Under working conditions using AFCs, emission control efficiencies ranged from 18% to 50%, with three of the four tests showing at least a 37% reduction. Under working conditions using BFCs (tests 9 through 20), emission control efficiencies ranged from 28% to 82%. The 82% reduction (test 12) should be considered atypical. Because solvent losses from carryout on cleaned parts are typically greater than diffusional losses, it would be impossible to achieve an 82% reduction from refrigerated freeboard devices that are designed primarily to control diffusional losses.

Efficiencies for BFCs under idling conditions (tests I-3 through I-8) ranged from 11% to 58%. This series of tests shows that as primary condensing temperature decreases, the additional benefit of a BFC decreases. This effect is more pronounced for CFC-113 than it is for TCA.

Refrigerated Primary Condensers

A lower-temperature primary condenser, generally using a refrigerant as opposed to water, will lower diffusional losses. Colder primary condenser temperatures, in addition to condensing solvent vapor, also cool the air above the air–solvent-vapor interface in a way similar to that of freeboard refrigeration devices. The effect of lower primary condenser temperature varies by solvent.

Six emissions tests (tests 30, 31, 33, 34, and I-9 and I-11) are available that show the effect of lowering primary condenser temperatures for both working (four tests) and idling (two tests) conditions using TCA and CFC-113. For TCA, under working conditions, two tests are available: lowering primary condenser temperature from 85°F to 50°F and from 70°F to 50°F reduced emissions by 41% and 29%, respectively. For CFC-113, under working conditions, two tests are available: lowering primary condenser temperature from 70°F to 40°F and from 50°F to 40°F reduced emissions by 52% and 18%, respectively. For TCA, under idling conditions, the only available test shows a 39% emission reduction if the primary condenser temperature is reduced from 85°F to 50°F. For CFC-113, under idling conditions, the only available test shows a 63% emission reduction if the primary condenser temperature is reduced from 70°F to 40°F.

The addition of a freeboard refrigeration device to an OTVC with a refrigerated primary condenser reduces emissions when either TCA or CFC-13 is used. However, the effect of the freeboard refrigeration device on emissions for an OTVC using CFC-113 is not as significant as that for an OTVC using TCA. Using a primary condenser temperature of 50°F and BFC, the working (test 15) and idling (test I-3) emission reductions from an OTVC using TCA are 47% and 54%, respectively. Under the same conditions but using an OTVC with CFC-113, the working (test 19) and idling (test I-7) emission reductions are 27% and 26%, respectively.

Increased Freeboard Ratio

The freeboard zone serves to reduce the air–solvent-vapor interface disturbances caused by room drafts and provides a column through which diffusing solvent molecules must migrate before escaping into the ambient air. Higher freeboards reduce diffusional losses by diminishing the effects of air currents and lengthening the diffusion column. The FBR is used in describing the adequacy of the freeboard height to reduce solvent loss and is defined as the freeboard height divided by the interior width of the cleaner. The FBR is used because as the cleaner width increases, its susceptibility to adverse room drafts also increases, unless the freeboard height is increased proportionally to compensate for the machine width.

Fourteen tests (12 under working conditions, two under idling conditions) are available to evaluate the effect of an increased FBR on emissions. Under working conditions (tests 47–58), the control efficiencies associated with raising the FBR from 0.75 to 1.0 ranged from 19% to 21%; in raising the FBR from 1.0 to 1.25, the control efficiency ranged from 6% to 10%. Therefore, using these data, the emission control efficiency associated with raising the FBR from 0.75 to 1.25 is at least 25%. Under idling conditions (tests I-19, I-20), the control efficiencies associated with raising the FBR from 0.75 to 1.0 were –6% and 39%. The

TABLE 4. Summary of Available Tests on Working OTVCs

Test No.	Tested Control[a]	Solvent	Cleaner Size, m^2	Cleaner Make	Baseline: Air Speed, fpm	Baseline: Cover	Baseline: FBR	Baseline: Secondary Chiller	Baseline: Emission, lb/ft^2/h	Controlled: Air Speed, fpm	Controlled: Cover	Controlled: FBR	Controlled: Secondary Chiller	Controlled: Emission, lb/ft^2/h	Control Efficiency[b]	Reference
1	AFC	TCA	1.8	Detrex	Calm	None	0.75	None	0.099	Calm	None	0.75	AF	0.082	18%	10
2	AFC	TCA	1.8	Detrex	130	None	0.75	None	0.173	130	None	0.75	AF	0.105	39%	10
3	AFC	TCA	1.8	Detrex	160	None	0.75	None	0.233	160	None	0.75	AF	0.116	50%	10
4	AFC	TCA	1.4	Auto-Sonics		None		None	0.063		None		AF	0.040	37%	11
5	AFC	TCE						None	4.30E+06 g/mo				AF	3.60E+06 g/mo	16%	16
6	AFC	TCE						None	6.20E+06 g/mo				AF	3.50E+06 g/mo	44%	16
7	AFC (spray loss)	Freon TF	0.3	Delta Sonics	Calm	None	1.0	None	0.0093 lb/ft^2/cy		None	1.0	AF	0.0079 lb/ft^2/cy	15%	8
8	BFC				Calm			None					BF			17
9	BFC	TCA	1.8	Detrex	30	None	0.75	None	0.099	Calm	None	0.75	BF	0.059	41%	10
10	BFC	TCA	1.8	Detrex	130	None	0.75	None	0.173	130	None	0.75	BF	0.091	47%	10
11	BFC	TCA	1.8	Detrex	160	None	0.75	None	0.233	160	None	0.75	BF	0.150	36%	10
12	BFC	TCA	1.4	Auto-Sonics		None		None	0.063		None		BF	0.011	82%	11
13	BFC	MC	1.2	Crest		Manual	0.83	None	0.186		Manual	0.83	BF	0.112	40%	12
14	BFC	MC	1.2	Crest		None	0.75	None	0.354		None	0.75	BF	0.254	28%	12
15	BFC (P@50°F)	TCA	0.9	Auto-Sonics	LE off	None	0.7	None	0.100	LE off	None	0.7	BF	0.053	47%	9
16	BFC (P@70°F)	TCA	0.9	Auto-Sonics	LE off	None	0.7	None	0.140	LE off	None	0.7	BF	0.070	50%	9
17	BFC (P@85°F)	TCA	0.9	Auto-Sonics	LE off	None	0.7	None	0.170	LE off	None	0.7	BF	0.082	52%	9
18	BFC (P@40°F)	CFC-113	0.9	Auto-Sonics	LE off	None	0.7	None	0.090	LE off	None	0.7	BF	0.075	17%	9
19	BFC (P@50°F)	CFC-113	0.9	Auto-Sonics	LE off	None	0.7	None	0.110	LE off	None	0.7	BF	0.080	27%	9
20	BFC (P@70°F)	CFC-113	0.9	Auto-Sonics	LE off	None	0.7	None	0.186	LE off	None	0.7	BF	0.110	41%	9
21	(BFC&LipExh,P@50°F	TCA	0.9	Auto-Sonics	LE on	None	0.7	None	0.219	LE off	None	0.7	BF	0.053	76%	9
22	(BFC&LipExh,P@70°F	TCA	0.9	Auto-Sonics	LE on	None	0.7	None	0.25	LE off	None	0.7	BF	0.070	72%	9
23	(BFC&LipExh,P@85°F	TCA	0.9	Auto-Sonics	LE on	None	0.7	None	0.277	LE off	None	0.7	BF	0.082	70%	9
24	DWELL TIME	Freon TF	0.3	Delta Sonics	1.0	None	1.0	None	0.014 lb/cy	1.0	None	1.0	None	0.008 lb/cyc	46%	8
25	HOIST: 11–3	Freon TF	0.3	Delta Sonics	1.0	None	1.0	None	0.039 lb/cy	1.0	None	1.0	None	0.008 lb/cyc	81%	8
26	HOIST: 20–10°			Branson	1.0	None	1.0	None	0.775		None	1.0	None	0.555	28%	13
27	(LIP EXH (P@50°F)	TCA	0.9	Auto-Sonics	LE on	None	0.7	None	0.219	LE off	None	0.7	None	0.100	54%	9
28	(LIP EXH (P@70°F)	TCA	0.9	Auto-Sonics	LE on	None	0.7	None	0.25	LE off	None	0.7	None	0.140	44%	9

29	(LIP EXH (P@85°F)	TCA	0.9	Auto-Sonics	LE on	None	0.7	None	0.277	LE off	None	0.7	None	0.160	42%	9
30	PC—70°F to 40°F	CFC-113	0.9	Auto-Sonics	LE off	None	0.7	None	0.186	LE off	None	0.7	None	0.090	52%	9
31	PC—85°F to 50°F	TCA	0.9	Auto-Sonics	LE off	None	0.7	None	0.160	LE off	None	0.7	None	0.100	38%	9
32	PC—85°F to 50°F	TCA	0.9	Auto-Sonics	LE on	None	0.7	None	0.277	LE on	None	0.7	None	0.219	21%	10
33	PC—70°F to 50°F	TCA	0.9	Auto-Sonics	LE off	None	0.7	None	0.140	LE off	None	0.7	None	0.100	29%	10
34	PC—50°F to 40°F	CFC-113	0.9	Auto-Sonics	LE off	None	0.7	None	0.110	LE off	None	0.7	None	0.090	18%	10
35	Biparting cover	TCA	1.8	Detrex	30	None	0.75	None	0.099	30	Bipart	0.75	None	0.061	38%	10
36	Biparting cover	TCA	1.8	Detrex	100	None	0.75	None	0.121	100	Bipart	0.75	None	0.071	41%	10
37	Biparting cover	TCA	1.8	Detrex	130	None	0.75	None	0.173	130	Bipart	0.75	None	0.090	48%	10
38	Biparting cover	TCA	1.8	Detrex	160	None	0.75	None	0.233	160	Bipart	0.75	None	0.109	53%	10
39	Biprtng cvr&AFC	TCA	1.8	Detrex	30	None	0.75	None	0.099	30	Bipart	0.75	AFC	0.054	45%	10
40	Biprtng cvr&AFC	TCA	1.8	Detrex	100	None	0.75	None	0.121	100	Bipart	0.75	AFC	0.070	42%	10
41	Biprtng cvr&AFC	TCA	1.8	Detrex	130	None	0.75	None	0.173	130	Bipart	0.75	AFC	0.083	52%	10
42	Biprtng cvr&AFC	TCA	1.8	Detrex	160	None	0.75	None	0.233	160	Bipart	0.75	AFC	0.105	55%	10
43	Biprtng cvr&BFC	TCA	1.8	Detrex	30	None	0.75	None	0.099	30	Bipart	0.75	BFC	0.055	44%	10
44	Biprtng cvr&BFC	TCA	1.8	Detrex	100	None	0.75	None	0.121	100	Bipart	0.75	BFC	0.064	47%	10
45	Biprtng cvr&BFC	TCA	1.8	Detrex	130	None	0.75	None	0.173	130	Bipart	0.75	BFC	0.080	54%	10
46	Biprtng cvr&BFC	TCA	1.8	Detrex	160	None	0.75	None	0.233	160	Bipart	0.75	BFC	0.078	67%	10
47	FBR: 0.75 → 1.0	MC blend	0.4	Auto-Sonics	30	None	0.75	None	0.220	30	None	1.0	None	0.175	20%	14
48	FBR: 0.75 → 1.0	MC	0.4	Auto-Sonics	30	None	0.75	None	0.180	30	None	1.0	None	0.145	19%	14
49	FBR: 0.75 → 1.0	CFC-113	0.4	Auto-Sonics	30	None	0.75	None	0.165	30	None	1.0	None	0.130	21%	14
50	FBR: 0.75 → 1.0	MC blend	0.4	Auto-Sonics	30	None	0.75	None	0.125	30	None	1.0	None	0.100	20%	14
51	FBR: 0.75 → 1.0	TCA	0.4	Auto-Sonics	30	None	0.75	None	0.112	30	None	1.0	None	0.090	20%	14
52	FBR: 0.75 → 1.0	TCE	0.4	Auto-Sonics	30	None	0.75	None	0.080	30	None	1.0	None	0.065	19%	14
53	FBR: 1.0 → 1.25	MC blend	0.4	Auto-Sonics	30	None	1.0	None	0.175	30	None	1.25	None	0.165	6%	14
54	FBR: 1.0 → 1.25	MC	0.4	Auto-Sonics	30	None	1.0	None	0.145	30	None	1.25	None	0.135	7%	14
55	FBR: 1.0 → 1.25	CFC-113	0.4	Auto-Sonics	30	None	1.0	None	0.132	30	None	1.25	None	0.122	8%	14
56	FBR: 1.0 → 1.25	MC blend	0.4	Auto-Sonics	30	None	1.0	None	0.100	30	None	1.25	None	0.092	8%	14
57	FBR: 1.0 → 1.25	TCA	0.4	Auto-Sonics	30	None	1.0	None	0.092	30	None	1.25	None	0.083	10%	14
58	FBR: 1.0 → 1.25	TCE	0.4	Auto-Sonics	30	None	1.0	None	0.065	30	None	1.25	None	0.059	9%	14
59	Draft 160-calm	TCA	1.8	Detrex	160	None	0.75	None	0.233	30	None	0.75	None	0.099	58%	10
60	Draft 130-calm	TCA	1.8	Detrex	130	None	0.75	None	0.173	30	None	0.75	None	0.099	43%	10

[a]AFC = above-freezing freeboard refrigeration; BFC = below-freezing freeboard refrigeration; LE = lip exhaust; PC = primary condenser (e.g., PC@50°F means the primary condenser temperature was 50°F).

[b]These control efficiency values refer to percent control of working losses (i.e., diffusion/convection losses plus workload related losses). They do not reflect control of other possible emission sources such as: leaks, startup/shutdown losses, sol vent transfer losses, and downtime losses.

[c]The relatively high emission rates were due to the configuration of the parts basket (i.e., a large horizontal surface area) and the constant cycling of parts (i.e., no time was allowed for the parts/basket to reach the temperature of the solvent vapor).

negative efficiency result can likely be attributed to measurement inaccuracies.

Reduced Room Draft/Lip Exhaust Velocities

Air movement over an OTVC affects the solvent emission rate by sweeping away solvent vapors that have diffused into the freeboard area and by creating turbulence, thereby enhancing solvent diffusion, as well as solvent vapor and air mixing. Reducing room drafts to calm conditions (30 fpm or less) can greatly reduce emission rates.

Two tests (tests 59, and 60) are available that show the effect of reduced room draft on emissions under working conditions. Reducing room drafts to calm conditions corresponds to a 43% reduction from working emissions with room drafts of 130 fpm and a 58% reduction from working emissions at 160 fpm.

A lip exhaust affects emissions much like air speed: It increases mixing and diffusion in the vapor layer. Six tests are available that show the effect of turning off a lip exhaust for both working and idling conditions. Based on test data for working conditions (tests 27, 28, and 29) with the lip exhaust turned off, emission reductions ranged from 54% (with primary condenser temperature of 50°F) to 42% (at 85°F). Based on test data for idling conditions (tests I-16, I-17, and I-18) with the lip exhaust turned off, emission reductions ranged from 49% (at 50°F) to 32% (at 85°F).

Enclosed Designs

The enclosed design as a control option for OTVCs involves completely enclosing the cleaner, except for a single opening through which parts enter and leave the enclosure. The enclosed design reduces idling and work-load–related losses by creating a still air environment inside the machine that limits solvent diffusion. These OTVCs may have an opening in the side of the cleaner, above the freeboard, through which parts enter and may be totally enclosed (horizontal port) or may have a sliding cover (vertical port) that closes when the parts are inside the cleaner.

Two data sources are available to evaluate the control efficiency associated with enclosed-design OTVCs. These sources show that uncontrolled OTVC emissions were reduced 42–67% upon conversion to an enclosed design machine.[18,19]

Carbon Adsorption

Carbon adsorption can be employed as an add-on emission control technique in conjunction with a lip exhaust system. These systems are most commonly used on large solvent cleaners where the credit from solvent recovery helps to offset the high capital equipment costs.

The lip exhaust draws the diffusing solvent vapors and, to some extent, solvent evaporating from clean parts, and directs them through an activated-carbon bed. The solvent vapor molecules are adsorbed onto the surface of the activated carbon. At intervals, when the carbon becomes saturated with solvent, the bed is desorbed, usually with steam, to remove the solvent from the carbon. The solvent–steam mixture is then condensed and passed through a water separator and the recovered solvent returned to the cleaner.

One test is available to evaluate the efficiency of carbon adsorbers for controlling solvent emissions.[12] This test indicates that a lip exhaust/carbon adsorber system could control solvent emissions by 65%. However, the test report does not specify whether the baseline emission rate includes lip exhaust. If the baseline OTVC did have a lip exhaust, the 65% emission reduction overstates the achievable reduction for a carbon adsorber and lip exhaust installed on an OTVC without a lip exhaust.

Mechanically Assisted Parts-Handling/ Parts-Movement Speed

Rapid movement of parts through the OTVC cleaning cycle increases solvent emissions because of increased carryout of liquid solvent and entrainment of solvent vapor and increased disturbance at the air–solvent-vapor interface. A human operator is generally unable to move parts at or below the maximum speed of 11 fpm, as required in many state regulations and recommended in EPA guidelines.[20–22] Use of a mechanical parts-handling system can reduce emissions by consistently moving parts into and out of the machine at appropriate rates, thereby eliminating excess losses caused by manual operation.

One test is available that simulates the effect of switching from a human operator to a mechanically assisted system (test 26). This test compared a hoist operated at 20 fpm (to simulate a human operator) with a hoist operated at 10 fpm. The lower speed was found to reduce working losses by 28%. However, because human operator speeds are generally higher than 20 fpm, the reduction attributable to a hoist is likely to be larger than 28%. Another test (test 25) evaluated the effectiveness of reducing hoist speed even further. During the test, a variable-speed, programmable hoist was used to lower the hoist speed to 3 fpm as the parts basket moved through the solvent vapor. Decreasing the hoist speed from 11 fpm to 3 fpm resulted in an 81% decrease in total working losses.

Proper Operation and Maintenance

As with any operating process, proper operation and maintenance (O&M) practices are critical to maintaining low solvent emissions. These practices include the following.

1. Reducing room drafts through the use of baffles or reduced ventilation rates around the cleaner.
2. If using a solvent spray system, spraying within the vapor zone at a downward angle.
3. On start-up, starting the condenser coolant flows prior to starting the solvent sump heater, and the reverse on shutdown.
4. Covering the OTVC during downtime and operating a freeboard refrigeration device, or operating a sump

TABLE 5. Summary of Available Tests—In-Line Cleaners

			Baseline			Controlled			
Tested Control	Solvent	Cleaner Make	Secondary Chiller	Carbon Adsorber	Emission ($lb/ft^2/hr$)	Secondary Chiller	Carbon Adsorber	Emission ($lb/ft^2/hr$)	Control Efficiency
AFC	Gensolv DFX	Allied	off	none	6.2 lb/hr	AFC	none	5.7 lb/hr	8
BFC	Gensolv DFX	Allied	off	none	6.2 lb/hr	BFC	none	1.95 lb/hr	69
BFC	PCE	Detrex	off	none	1.0	BFC	none	0.4	62
CADS	ICE	Blakeslee	none	off	1.2	none	on	0.5	61

AFC = above-freezing freeboard refrigeration device
BFC = below-freezing freeboard refrigeration device
CADS = carbon adsorption system

cooler to reduce solvent vapor pressure, or pumping the solvent to an airtight storage drum.

5. Utilizing a mechanically assisted parts entry/removal system to minimize work-load emissions.
6. Allowing sufficient parts drainage time above the vapor zone.
7. Repairing visible leaks and replacing or repairing cracked gaskets, malfunctioning pumps, water separators, and steam traps promptly.
8. Performing solvent transfer in closed systems using submerged fill piping.
9. Utilizing control safety switches on OTVCs, including vapor level control thermostat (water-cooled machines), sump thermostat, liquid solvent level control, spray pump control switch, and secondary heater switch.

In-Line Cleaners[23]

The following sections describe minimizing the entrance/exit openings, carbon adsorption, freeboard refrigeration devices, drying tunnels, rotating baskets, and hot vapor recycle/superheated vapor systems.

Only four emission tests are available for control techniques on in-line cleaners—three evaluated the effectiveness of a freeboard refrigeration device (two below freezing, one above freezing) and one evaluated the effectiveness of a carbon adsorber. These test data are presented in Table 5.

Minimizing Entrance/Exit Openings

A reduction in the area of entrance and exit openings reduces idling and working losses resulting from diffusion by minimizing air drafts inside the cleaner. In-line machines utilizing U-bend designs eliminate the problem of air currents flowing through the machine. Additionally, in-line cleaners, such as monorail cleaners, can utilize internal baffles to control the effect of airflow through the machine. When the in-line cleaner is not in use, port covers should be used to reduce downtime emissions.

Carbon Adsorption

Carbon adsorption is a major emission control technology for solvent losses from in-line cleaners. The enclosure around in-line cleaners makes it easier to capture and duct emissions to the carbon adsorber. Control of solvent vapor by carbon adsorption was discussed previously.

The available test on carbon adsorbers shows a 61% emissions reduction efficiency when applied to an in-line cleaner (i.e., circuit board stripper).[7]

Freeboard Refrigeration Device

The refrigerated freeboard device on an in-line vapor cleaner functions in the same way as one on an OTVC. The available emission test data indicate that BFC systems can reduce emissions from in-line cleaners by 62–69%, while an AFC system can reduce emissions by 8%.

Drying Tunnels

A drying tunnel is an add-on enclosure that extends the exit area of in-line cleaners. The tunnel allows solvent from cleaned parts to evaporate in an enclosed area where it can be recovered either back into the cleaner or in a carbon adsorption system. No emission test data are available to quantify the effect of drying tunnels.

Rotating Baskets

A rotating basket is a perforated cylinder containing parts to be cleaned that is slowly rotated during the cleaning process. The rotation prevents trapping of liquid solvent on parts, thereby reducing solvent carryout. No emission test data are available to quantify the effect of rotating baskets.

Hot Vapor Recycle/Superheated Vapor

Hot vapor recycle and superheated vapor are promising new technologies that reduce carryout solvent emissions. In both systems, cleaned parts are slowly passed through a superheated solvent vapor zone within the vapor layer where they are warmed and liquid solvent is evaporated before they are removed from the cleaner. Solvent vapor is heated to approximately 1.5 times the solvent boiling point.[21] The vapor recycle process continuously recirculates the solvent vapor from the vapor zone through a heater and back to the vapor zone. The superheated vapor process utilizes heating coils in the vapor zone. The hot vapor recycle process is generally applicable to in-line cleaners because an enclosure is needed for effective recirculation of solvent vapors. Superheated vapor technology can be applied to both in-line and OTVC cleaners.

No emission test data are available to quantify the effect of these control techniques. However, one industry contact claims that a 90% reduction in carryout emissions is possible.[7]

Proper Operation and Maintenance

The O&M practices described previously for OTVCs also apply to in-line cleaners. The only difference is that for in-line cleaners, the conveyor speed should be kept at or below 11 fpm to minimize vapor zone turbulence and solvent carryout.

Cold Cleaners

Carburetor cleaners are the only type of cold cleaner that is currently manufactured for use with a halogenated solvent. These machines typically are well controlled with a water cover. Based on one available test, water covers can reduce evaporation losses by 90%.[24] For all cold cleaners, reducing room drafts, allowing adequate drainage of parts, flushing parts only within the confines of the cleaner, and using close-fitting covers can reduce the quantity of solvent losses from the use of cold cleaners.

References

1. *Alternative Control Technology Document—Halogenated Solvent Cleaners,* EPA-450/3-89-030, U.S. Environmental Protection Agency, Research Triangle Park, NC, August 1989. pp 3-1–3-3.
2. R. F. Pandullo, Radian Corp., memo to D. A. Beck, U.S. Environmental Protection Agency/CPB, February 15, 1989, concerning an estimation of nationwide number of halogenated solvent cleaners and halogenated solvent usage.
3. Reference 1, pp 3-3–3-13.
4. Reference 1, pp 3-13–3-20.
5. Reference 1, p 3–25.
6. *Control of Volatile Organic Emissions from Solvent Metal Cleaning,* EPA-450/2-77-022, U.S. Environmental Protection Agency, Research Triangle Park, NC, November, 1977, pp 2-8–2–22.
7. Reference 1, pp 3-25–3-44.
8. D. A. Beck, U.S. Environmental Protection Agency/CPE, letter and attachments from Delta Sonics, February, 1988, concerning estimation of Freon solvent usage in open-top series of Delta Sonics degreasers.
9. G. C. Hylon and H. F. Osterman, "Cool it to cut degreasing costs," *Amer. Machinist* (November 1982).
10. J. Goodrich, Detrex Corp., memo to L. Schlossberg, Detrex Corp., concerning degreaser emissions control test report.
11. R. Irvin, GCA/Technology Division, trip report submitted to D. A. Beck, U.S. Environmental Protection Agency/CPB, June 14, 1979, summarizing visit to Autosonics, Inc.
12. D. S. Suprenant and D. W. Richards, *Study to Support New Source Performance Standards for Solvent Metal Cleaning Operations,* prepared for U.S. Environmental Protection Agency, Research Triangle Park, NC, Contract no. 68-02-1329, Task Order no. 9, April 1976.
13. R. L. Polhamus, Branson Ultrasonics Corp., letter and attachments to P. A. Cammer, Halogenated Solvents Industry Alliance, February 19, 1988, concerning automated hoist test data.
14. R. F. Pandullo, Radian Corp., trip report submitted to D. A. Beck, U.S. Environmental Protection Agency/CPB, January 1989, summarizing visit to Allied Corp., Buffalo Research Facility.
15. Reference 1, pp 4-1–4-46.
16. D. W. Fiester, Sealed Power Corp., letter to U.S. Environmental Protection Agency Central Docket Section, Washington, DC, commenting on organic solvent cleaner NSPS.
17. E. M. Ryan, to U.S. Environmental Protection Agency Central Docket Section, Washington, DC, commenting on organic solvent cleaner NSPs.
18. R. Rehm, GCA/Technology, report on trip to Finishing Equipment, Inc., St. Paul, MN, to R. George, U.S. Environmental Protection Agency, September 9, 1980.
19. J. Tang, GCA/Technology Division, letter from W. Sabatka, Finishing Equipment, Inc., May 4, 1981, commenting on solvent and labor savings from enclosed degreasers.
20. J. J. Sherman, Scanex, Inc., letter and attachments, December 29, 1988, commenting on issue of solvent release from open top solvent cleaning tanks.
21. S. J. Miller, Radian Corp., trip report submitted to D. A. Beck, U.S. Environmental Protection Agency/CPB, January 1989, summarizing visit to Unique Industries, Sun Valley, California.
22. S. J. Miller, Radian Corp., trip report submitted to D. A. Beck, U.S. Environmental Protection Agency/CPB, January 1989, summarizing visit to Delta Sonics, Paramount, CA.
23. Reference 1, pp 4-47–4-59.
24. J. W. Brown and P. R. Westlin, U.S. Environmental Protection Agency/EMB, memo to D. A. Beck, U.S. Environmental Protection Agency/CPB, May 22, 1981, commenting on effect of water blanket to reduce organic evaporation rates.

10
Surface Coating

Mark B. Turner

Many manufactured items receive surface coatings for decoration and/or protection against damage or corrosion. The coatings used and application methods employed vary with the purpose and the desired properties of the coatings. Surface coating of manufactured items takes place in many industrial categories, including transportation, metal furniture, wood furniture, appliances, cans, and metal coils.

PROCESS DESCRIPTION

Surface Coating Process

In general terms, the surface coating process comprises several distinct steps: surface preparation, application of coatings, and curing of coatings.

Surface Preparation

Surface preparation of parts prior to coatings application typically includes a surface cleaning step (often with an aqueous alkaline cleaner for metal parts). In automobile refinishing and automobile original equipment manufacturing, a solvent wipe is often used after the cleaning step to remove traces of oil and grease.[1,2] Many surface preparation operations include a phosphate treatment or the application of a chromate conversion coating.[3,4] These chemical treatment steps promote good coating adhesion and corrosion resistance.[3,4] Finally, the articles are dried prior to coating application.

Application of Coatings

Following the surface preparation steps, one or more coatings are applied. Typically, a primer is applied first. The primer provides corrosion resistance, fills in surface imperfections, and provides a bonding surface for the top coat.[2] The top coat, often a series of coats, is applied over the primer and determines the final color of the article.

Before describing the various application methods, the concept of transfer efficiency must be defined. Transfer efficiency is the ratio of the amount of coating solids deposited on the surface to the total amount of coating solids used.[5]

The following paragraphs describe these coating application methods: dip coating, flow coating, roller coating, electrodeposition, spray coating (air, airless, and electrostatic), and electrostatic bell and disk coating. Table 1 provides a matrix of these methods and the industries in which they are used.

Dip-coating equipment consists of a large main tank in which the mixed coating is applied. As the coated parts emerge from the tank, they move into an area where excess paint drips off. The excess paint is collected and returned to the main tank.[5] The paint in the main tank is kept at a constant solids concentration by the addition of fresh, properly mixed paint and water or organic solvent to make up for usage and evaporation. This recycling and reuse ensure an overall transfer efficiency of 85%.[5]

In flow-coating operations, a coating is fed through overhead nozzles so as to flow in a steady stream over the article to be coated, which is suspended from a conveyor line.[6] Excess paint drips off the part and back into the holding tank for reuse.[5] This recovery ensures a transfer efficiency of 85%.[5]

Roller coating machines typically have three or more power-driven rollers. One roller runs partially immersed in the coating and transfers the coating to a second, parallel roller. The strip or sheet to be coated is run between the second and third roller and is coated by transfer of coating from the second roller. The quantity of coating applied to the sheet or strip is established by the distance between the rollers.[7]

TABLE 1. Matrix of Surface Coating Application Methods and Industries in Which They Are Used

Surface Coating Method	Coil Coating	Metal Furniture	Auto and Light-Duty Truck	Large Appliance	Can	Auto Refinish	Traffic Marking
Dip		X		X			
Flow		X		X			
Roller	X				X		
Electrodeposition	X		X	X			
Spray							
Air atomized			X	X	X	X	X
Airless				X		X	
Electrostatic		X	X	X		X	
High volume, low pressure			X			X	
Electrostatic bell and disk				X			

In electrodeposition (EDP), a dc voltage is applied between the coating bath (or carbon or stainless-steel electrodes in the bath) and the part to be coated.[8,9] The part, which can act as the cathode or the anode, is dipped into the bath.[8] Coating particles are attracted from the bath to the part because they are oppositely charged, yielding an extremely even coating.[7,8] The coatings used in EDP tanks are waterborne solutions.[8–10] Transfer efficiencies for EDP are commonly above 95%.[11]

The four basic types of coating spray application methods are air-atomized spray, airless spray, electrostatic spray, and high-volume, low-pressure spray. Typically, coatings are sprayed in a spray booth to protect the coated surface from dirt and to provide a well-ventilated area that protects workers from solvent vapors. Air-atomized spray guns use compressed air to atomize the coating into tiny droplets and to spray the coating onto the surface of the article to be coated.[12,13] The transfer efficiency for spray guns varies based on the configuration of the surface being coated, the skill of the operator, and the type of spray gun used.[13] Table 2 shows the transfer efficiency as a function of spraying method and sprayed surface.

In airless spray coating, the coating is atomized without air as it is forced through specially designed nozzles at pressures of 17 to 14MPa (1000–2000 psi). A diagram of an airless spray gun is shown in Figure 1.[12]

There are electrostatic versions of both air-atomized and airless spray guns.[12] Electrostatic spraying involves the use of an electrical transformer capable of delivering up to 60,000+ volts to create an electric potential between the paint particles and the surface to be coated.[14] These charged paint particles are thus electrically attracted to the surface, increasing transfer efficiency over that of the nonelectrostatic air-atomized and airless spray guns.[14]

A relatively new development in coating spray equipment is the high-volume, low-pressure or turbine spray gun. In this system, a turbine is used to generate and deliver atomizing air.[14] The turbine draws in filtered air, which is driven through several stages at up to 10,500 rpm.[15] The result is a high volume of warm, dry, atomizing air that is delivered to the spray gun at less than 7 psi.[15] This low-pressure air gives greater control of the spray, with less

TABLE 2. Transfer Efficiency as a Function of Spraying Method and Sprayed Surface

Method of Spraying	Flat Surface	Table Leg Surface	Bird Cage Surface
Air atomized	50	15	10
Airless	75–80	10	10
Electrostatic			
Disk	95	90–95	90–95
Airless	80	70	70
Air atomized	75	65	65

Source: Adapted from Reference 7, Table 232, which showed overspray percentages.

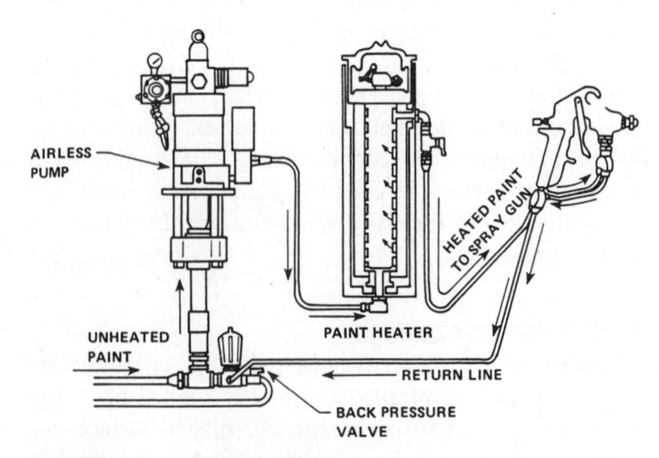

FIGURE 1. Diagram of Airless Spray Gun with an Attached Paint Heater (From Reference 12)

overspray and paint fog because of the absence of the blasting effect common with conventional high-pressure systems.[14]

Other electrostatic methods use bells or disks (see Figures 2 and 3). In these methods, atomization of the coating is caused to a small extent by the centrifugal forces associated with rapid spinning of the bell or disk and to a greater extent by the high voltage applied to repel the particles from the disk or bell and from each other.[11] In addition, the bell or disk housing may reciprocate up and down or back and forth to allow complete coating of the object. The surface of the bell or disk is negatively charged, giving a negative charge to the coating particles passing across it. The particles are then attracted to the positively grounded parts, as controlled by the applied voltage and the centrifugal force of the system.[8,12]

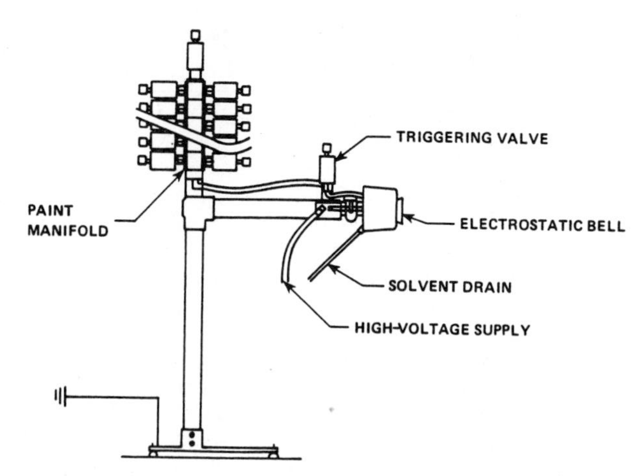

FIGURE 2. Diagram of Stationary Electrostatic Bell (From Reference 12)

Curing of Coatings

Primer and top-coat applications are individually dried and cured in large ovens. Such industries as automobile and light-duty truck manufacturing, coil coating, metal furniture, can manufacturing, and large-appliance manufacturing use multiple-bake ovens, usually after each coating is applied. Bake ovens typically have multiple heat zones that operate at successively higher temperatures.[16,17] Oven temperatures can range from 93°C (200°F) to 260°C

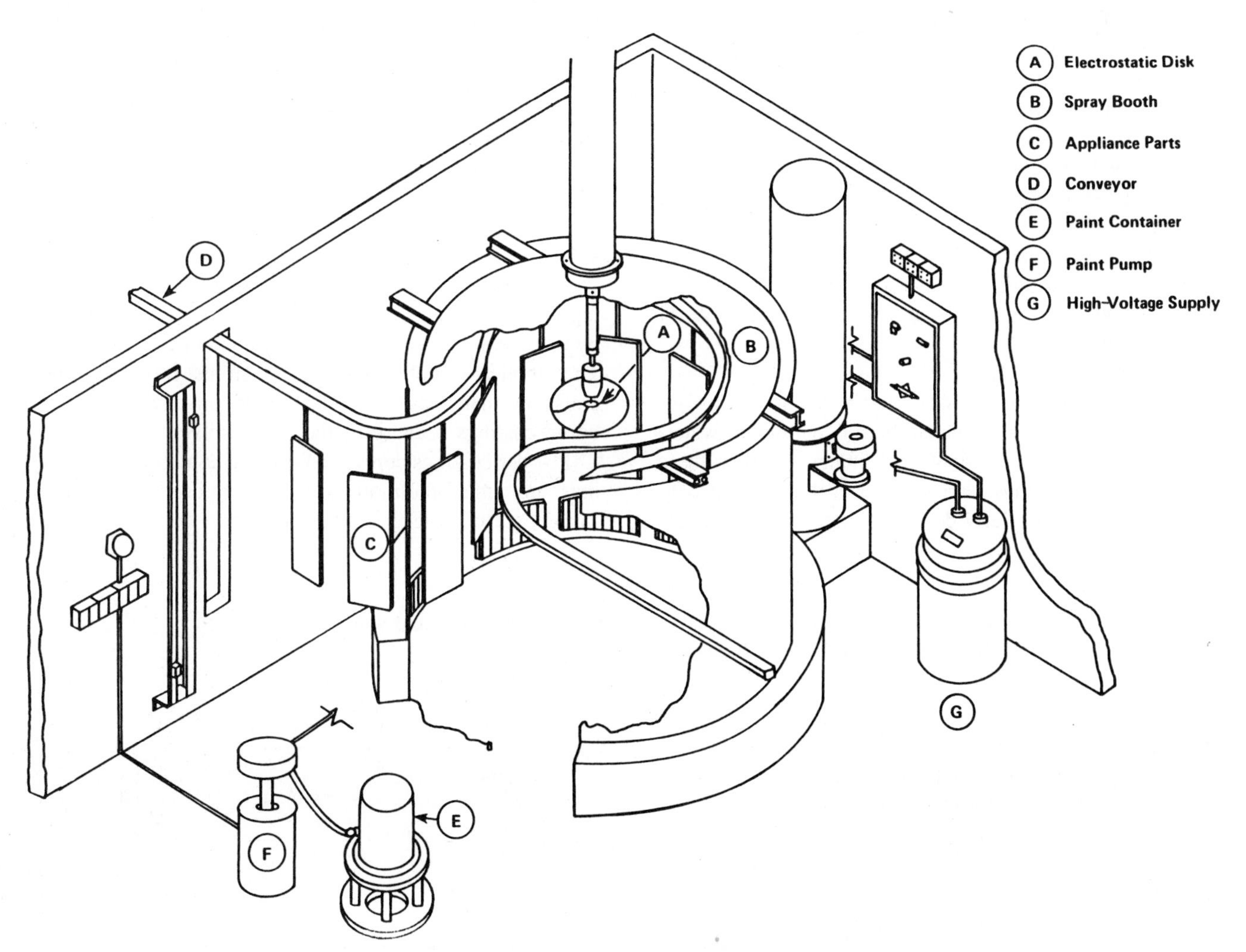

FIGURE 3. Diagram of a Reciprocating Electrostatic Disk and Spray Booth (From Reference 12)

TABLE 3. Coatings Used in Coil Coating

Coatings	Volatile Content, wt %[a]
Acrylics	40–45
Adhesives	70–80
Alkyds	50–70
Epoxies	45–70
Fluorocarbons	55–60
Organosols	15–45
Phenolics	50–75
Plastisols	5–30
Polyesters	45–50
Silicones	35–50
Vinyls	60–75
Zincromet™	35–40
Dacromet™	—

[a]The volatile content by volume is generally 5–10% greater than the volatile content by weight.
Source: Reference 4.

(500°F), depending on the industry, the type of coating used, and the zone.[16,17] During passage through the oven, the solvent is evaporated, and the substrate is heated to a design peak temperature to achieve proper curing of the coating. If bake oven environments are not controlled through exhaust systems, solvent concentrations can increase to an explosive level. Lower explosive limits (LEL) are published by solvent manufacturers in their Material Safety Data Sheets. A bake oven may be operated safely at 25% of the LEL. The automobile industry operates bake ovens at 5–10% of the LEL[18–20] because the ovens are very long with large openings that allow large amounts of air to be drawn into them, thus lowering solvent concentrations.[16]

Surface Coatings Used

The types of coatings used in the surface coatings industries range from conventional solvent-borne coatings to waterborne, high-solids, and powder coatings. Within these categories, the coatings vary significantly in solvent content. The following briefly describes these coating categories and, where possible, identifies those industries in which the coatings are used and the solvent or solids content.

With the exception of powder coatings and radiation-cured coatings, all surface coatings contain a carrier solvent along with the coating solids that include a resin, pigment, and various additives. For many years, various coatings have been used that employ organic solvents as the carrier solvent. Once the coating is applied, the organic solvent evaporates, allowing the coating to harden. Traditionally, these coatings have been used because the solvent evaporates rapidly, allowing the coatings to dry quickly. Conventional solvent-borne coatings range in solvent content up to about 90% by volume. An example of the range of solvent contents in various coatings used in the coil coating industry is shown in Table 3. (The plastisols and organosols listed in Table 3 are examples of high-solids coatings.) Table 4 shows ranges of solvent contents for conventional solvent-borne coatings that traditionally have been used in other industries.

Waterborne coatings use water as the carrier solvent. They also contain a small amount of organic solvent to aid in wetting the pigments, to produce solubility (in the case of partially water-soluble, film-forming components), and to promote good flow and viscosity characteristics in the coating mixtures.[26] The organic solvent content of waterborne compositions varies between 2% and 15% of the total volume of the coating formulation. Because of the lower solvent content of waterborne coatings, significant emission reductions over conventional solvent-borne coatings applications can be realized.

As the name suggests, high-solids coatings have a higher solids content and, therefore, a lower solvent content than conventional solvent-borne coatings. In some industries, high-solids coatings are those with solids contents of 62% by volume or higher (i.e., a solvent content of 38% or less). The application of high-solids coatings produces lower emissions than does the application of conventional solvent-borne coatings because of the lower solvent content.

Powder coatings contain no organic solvent. These coatings are either thermoplastic or thermosetting powders. Thermoplastic powder coatings melt and flow when heat is

TABLE 4. Typical Organic Solvent Contents for Conventional Solvent-Borne Coatings Used in Various Industries

Industry	Solvent-Borne Coating	Typical % Solvent by volume
Metal furniture	Not specified	65
Automobile and light-duty truck	Enamel	67–76
	Lacquer	82–88
Automobile refinishing	Enamel	72–76
	Lacquer	87–91
Large appliance	Not specified	70
Traffic marking	Alkyd	50

Source: References 21–25.

applied, but continue to have the same chemical composition upon cooling and solidifying.[27] Thermosetting powder coatings melt when exposed to heat, flow into a uniform thin layer, and chemically cross-link within themselves or with other reactive components to form a higher-molecular-weight reaction product.[28] Based on the quantities of powder coatings sold, powder coating use in North America is increasing at a rate approaching 20% per year.[29] Powder coatings may be applied electrostatically or by using the fluidized-bed process.

Radiation-cured coating involves photocuring mixtures of low-molecular-weight polymers or oligomers dissolved in low-molecular-weight acrylic monomers. These formulations contain no solvent carriers and can be cured using either electron-beam or ultraviolet-light sources. They are used in a variety of industrial applications, including wood top coatings, floor tile coatings, and metal coatings, and by the paper coating industry on such products as record album covers.[30–32]

AIR EMISSIONS CHARACTERIZATION

Air emissions from surface coating operations result from the evaporation of the organic solvents in the coatings and consist primarily of volatile organic compounds (VOCs). These VOC emissions can occur in a number of places along the production line: during atomization and application of the coating, during initial air drying of the part after it leaves the spray booth (flash-off), and in the bake oven.[30] In the case of automobile refinishing where neither spray booths nor bake ovens may be used, VOC emissions take place during coating application and during drying. Fugitive emissions only result when coatings are mixed and loaded into the application device, during transport of coated parts from the spray booth to the oven (flash-off), and during postcuring.[30] Table 5 shows the percentage of VOC emissions emitted during the various process steps for selected industries.

Estimating emissions from surface coating operations is very site specific because of the wide range of coating formulations and the variety of coating techniques in use. One method of estimating VOC emissions is to gather data on coating usage over a certain period and multiply the actual usage of each coating by the VOC content of that coating to obtain VOC emissions for the time frame for each coating. Summing the VOC emissions from the application of each coating for the given time frame provides an estimate of total VOC emissions. This method can become time consuming if several formulation changes are made during the time studied. Another method is to use the average VOC content for the coatings used and multiply by the total quantity of coatings used.

TABLE 5. Percentage of VOC Emissions Emitted During Surface Coating Operations for Selected Industries

	Percentage of Total VOC Emissions	
Industry	Spray Booth or Application Area and Flash-off	Bake Oven
Metal furniture	70	30
Automobile and light-duty truck	85–90	10–15
Large appliance	80	20
Coil coating[a]	8	90

[a]Remaining VOC emissions (2%) come from the quench section after the bake/curing oven.

Source: References 22, 33, 34.

AIR POLLUTION CONTROL MEASURES

Air pollution control measures that can be applied to surface coating operations generally fall into three distinct categories: reduced-VOC coatings, higher-transfer-efficiency equipment, and add-on air pollution control devices (thermal and catalytic incinerators, carbon adsorbers, and condensers). The following sections provide industry-specific data on the emission reduction efficiency associated with applicable air pollution control measures as applied to the following industries: metal furniture coating, coil coating, automobile and light-duty truck manufacturing, large-appliance manufacturing, automobile refinishing, and the application of traffic markings. These emission reduction efficiencies provide an estimate of the emission reductions achievable from a given baseline case (usually the application of conventional solvent-borne coatings). In many cases, facilities in these industries may be using a different baseline coating as the result of regulatory requirements or for technical reasons. Therefore, the potential emission reduction efficiencies for these facilities will be different from those shown here.

Metal Furniture Coating

Tables 6, 7, and 8 show control efficiencies for the application of powder coatings, of high-solids coatings, and of

TABLE 6. Emission Reduction via Powder Coatings

Coating	Emission Reduction, %	Reference
Thermosetting powders[a]		
Epoxy	97–99	35, 36
Acrylics	99	35, 36
Polyester (urethane)	96–98	35, 36
Thermoplastic powders[b]		
Polyester (others)	99	35, 36
Acrylics	99	35, 36
Polyvinyl chloride and cellulose acetate		
Butyrate	90–95	35, 36

[a]VOC emissions from bake oven only.

[b]VOC emissions from application and cool-down areas.

TABLE 7. Control Efficiencies for High-Solids Coatings

High-Solids Coatings	Emission Reduction by Weight or Volume, (%)[a,b]
60% by volume solids	
Flat surface	62
3Complex surface	61
65% by volume coatings	
Flat surface	69
Complex surface	69
70% by volume coatings	
Flat surface	75
Complex surface	75
80% by volume coatings	
Flat surface	85
Complex surface	85

[a]Approach the same as reported in references 37–40.
[b]Some round-off error (approximately 2%) may exist for the calculations (e.g., 60% solids).

TABLE 8. Control Efficiencies for Water-Based Coatings

Application Technique	Control Efficiency, % by Weight or Volume	Reference
82/18 Waterborne—electrostatic spraying[a,b]		
Flat surface	80–82	41, 42, 43
Complex surface	80–82	44, 45
67/33 Waterborne—electrostatic spraying[b,c]		
Flat surface	67	41, 42, 44, 45
Complex surface	67	
82/18 Waterborne—dip and flow coating[a]	82	41, 43, 44, 45
67/33 Waterborne—dip and flow coating[c]	67	41, 43, 44, 45
82/18 Waterborne—electrodeposition[d]	95	44, 46, 47, 48

[a]For a 35% by volume solids with 82-to-18 water-to-solvent ratio.
[b]Control efficiencies for any electrostatic spraying technique.
[c]For a 35% by volume solids with 67-to-33 water-to-solvent ratio.
[d]Regardless of chemical composition of coating—20% solids.

waterborne coatings from the baseline of the coating of metal furniture with a solvent content of 65%. Table 9 presents control efficiency data for add-on air pollution control equipment.

The control efficiency values shown in Table 6 are based on the percent by weight change in the applied solids coating. The VOC emissions are due to polymerization by-products, mainly from thermo-setting coatings.[36,37] The control efficiency values shown in Table 7 are for the electrostatic application (gun, disk, and bell) of high-solids coatings.[38,39] The control efficiencies given in Table 8 show the effect of applying different waterborne coatings electrostatically and using various application techniques. The control efficiency data presented in Table 9 are for surface coating operations other than the metal furniture industry.[56] It is not clear whether or not these control efficiencies factor in the capture efficiency of the VOC emissions from the sources indicated.

Coil Coating

Control options for reducing emissions from coil coating operations include the use of thermal incineration, catalytic incineration, and waterborne coatings. Coating rooms are

TABLE 9. Control Efficiencies for Add-On Air Pollution Control Equipment

Add-On Control Device	Process Being Controlled	Control Efficiency for Control Device, % by weight	Reference
Carbon adsorber	Spray booth	90	49, 50, 51
Carbon adsorber	Entire coating line	80	52
Thermal incinerator	Bake oven	96	37, 53, 54, 55
Catalytic incinerator	Bake oven	90	37, 53, 54

TABLE 10. Theoretical Emission Reduction Potential Associated with Various New Coating Materials for Use as Automotive Body Coatings

Coating Type and Percent Solids By Volume	Application Method	Transfer Efficiency, %	Solvent/Dry Solids Applied l/l	Potential Percent Emission Reduction When Replacing: Lacquer, 16 vol % Solids	Enamel, 28 vol % Solids
Solvent-based enamel, 28 vol %	Air spray	40	6.4	51	—
Solvent-based lacquer, 16 vol %	Air spray	40	13.1	—	—
Powder coating, 97 vol %	Electrostatic spray	98	0.03	99+	99+
Water-based	Electrodeposition	96	0.12	99	98
Water-based,[a] 25 vol %	Air spray	40	1.3	90	79
Water-based,[b] 25 vol %	Electrostatic spray	87	0.39	97	94

[a]Assumed 82-18 water–organic-solvent ratio by volume.
[b]Assumed 88-12 water–organic-solvent ratio by volume.
Source: Reference 60.

TABLE 11. Common Methods of Reducing VOCs in the Large-Appliance Surface Coating Industry

Control Method	Means of Reduction	% Reduction Over Uncontrolled Conventional Coatings
High-solids coatings (0.340 kg VOC/L)	Lower organic-solvent content	70[a]
Powder coatings	No use of organic solvent	99[a]
EDP (0.040 kg VOC/L)	Lower organic-solvent content	94[a]
Water-based coatings (0.140 kg VOC/L)	Lower organic-solvent content	47[a]
Carbon adsorption[b,c,d]	Adsorption of hydrocarbon emissions on a carbon bed	Top-coat spray booth—33 Top-coat oven—15
Incineration[b,c]	Catalytic or thermal oxidation of hydrocarbon emissions	Top-coat spray booth—33 Top-coat oven—15

[a]Calculated by determining the difference in the organic-solvent content of the coating and a conventional coating containing 0.61 kg VOC per liter (30% solids).
[b]81% overall control efficiency.
[c]Based on 40% of emissions occurring in the spray booth, 40% in the flash-off, and 20% in the curing oven.
[d]In addition to control of VOCs, solvents may be recovered through carbon adsorption.
Source: Reference 63.

TABLE 12. Matrix of VOC Emission Reduction Alternatives for a Medium-Sized Automobile Refinishing Shop and Estimated Emission Reductions

	VOC Emission Reduction Alternatives						
	Solvent Recovery		Transfer Efficiency, %			Emission Reduction	
Coating Alternatives	Yes	No	35	65	Emissions, tons/yr	tons/yr	%
Use of lacquers and enamels (current practice)							
Baseline		X	X		3.63	NA	NA
Alternative 1	X		X		3.08	0.54	15
Alternative 2		X		X	2.46	1.17	32
Alternative 3	X			X	1.91	1.72	47
Replace lacquer and enamel primers with waterborne primers							
Alternative 4		X	X		2.73	0.90	25
Alternative 5	X		X		2.34	1.28	35
Alternative 6		X		X	1.83	1.80	50
Alternative 7	X			X	1.44	2.19	60
Replace lacquer and enamel clears with higher-solids clears							
Alternative 8		X	X		2.82	0.81	22
Alternative 9	X		X		2.28	1.35	37
Alternative 10		X		X	2.02	1.61	44
Alternative 11	X			X	1.48	2.15	59
Replace lacquers with enamels							
Alternative 12		X	X		2.27	1.36	37
Alternative 13	X		X		1.73	1.90	52
Alternative 14		X		X	1.55	2.08	57
Alternative 15	X			X	1.00	2.63	72
Replace lacquers and enamels with urethanes							
Alternative 16		X	X		1.89	1.73	48
Alternative 17	X		X		1.35	2.28	63
Alternative 18		X		X	1.52	2.11	58
Alternative 19	X			X	0.98	2.65	73

Source: Reference 65.

TABLE 13. VOC Emission Reductions for Alternative Traffic Marking Materials from Baseline

Traffic Marking Material	Typical Annual VOC Emissions, lb/mile-yr	Typical Annual VOC Reduction from Baseline, lb/mile-yr	% Reduction from Baseline
Solvent-borne paints	69	NA[a]	NA[a]
Waterborne paints	13	56	81
Thermoplastics	0	69	100
Preformed tapes			
Without adhesive primer	0	69	100
With adhesive primer	58	11	16
Field reacted	0	69	100
Permanent markers	0	69	100

[a]NA = not applicable.
Source: Reference 66.

also used to capture VOC emissions and are vented to a control device.

Emissions test data for thermal incinerators controlling coil coating lines show conversion (destruction) efficiencies of captured VOC emissions that range from 87.6% to 99.9%.[57] For catalytic incinerators, two tests show 92.2% and 99.5% conversion.[58] The use of waterborne coatings in EDP systems for the application of primers was estimated to reduce emissions almost to zero.[59]

Automobile and Light-Duty Truck Manufacturing

Table 10 shows the potential VOC emission reductions associated with the application of powder and waterborne reduced-VOC coatings over the use of lacquer (16% solids) and enamel (28% solids) coatings. If high-solids coatings with a 50–60% solids content replaced the 16 vol % solids lacquer coating, potential emission reductions of over 80% could be realized.[61] Thermal incinerators are used in the automobile and light-duty truck industry to control emissions from bake ovens. Efficiencies of at least 90% are typical.[62]

Large-Appliance Manufacturing

In the large-appliance industry, several control options are available to limit VOC emissions. Table 11 demonstrates the reduction in emissions that these options offer over a simple application of solvent-borne coatings with a 70% solvent content. Assuming 90% capture efficiency, 90% adsorption of captured emissions, 40% flash-off, and 20% emissions in the curing oven, carbon adsorption and incineration will yield a 33% reduction in emissions if used on the spray booth and a 15% reduction in VOC emissions if used on the curing ovens.[64]

Automobile Refinishing

Table 12 shows a matrix of VOC emission reduction alternatives and estimated emission reductions for a medium-size automobile refinishing shop. In this industry, the baseline spray application technique is the use of air-atomized spray guns with a transfer efficiency of 35%. For the medium-size shop, the baseline coatings application is lacquers and enamels. Recovery of solvent used to clean spray guns and high-volume, low-pressure spray equipment has been investigated, along with the various alternative coating application techniques. It is clear that significant overall VOC emission reductions can be achieved with combinations of control options. This information source also developed similar tables for small- and large-volume shops using different baseline assumptions based on the levels of sophistication of the shops.

Traffic Markings

Table 13 shows the VOC emissions and emission reductions for alternative traffic marking materials from the baseline case of solvent-borne paint application. It is clear from Table 13 that almost all of the alternative marking materials show a 100% emission reduction from baseline. Waterborne paint application results in an emission reduction of 81%.

References

1. *Automobile and Light-Duty Truck Surface Coating Operations—Background Information for Proposed Standards,* Draft EIS, EPA-450/3-79-030, U.S. Environmental Protection Agency, Research Triangle Park, NC, September 1979, p 3–16.
2. *Reduction of Volatile Organic Compound Emissions from Automobile Refinishing,* EPA-450/3-88-009, U.S. Environmental Protection Agency, Research Triangle Park, NC, October 1988, p 3-2.
3. *Industrial Surface Coating: Appliances—Background Information for Proposed Standards,* Draft EIS, EPA-450/3-80-037, U.S. Environmental Protection Agency, Research Triangle Park, NC, November 1980, p 3-2.
4. *Metal Coil Surface Coating Industry—Background Information for Proposed Standards,* Draft EIS, EPA-450/3-80-035a, U.S. Environmental Protection Agency, Research Triangle Park, NC, October 1980, pp 3-2, 3-4.

5. Reference 3, p 3-4.
6. *Air Pollution Engineering Manual*. AP-40, 2nd ed., J. A. Danielson, Ed.; Air Pollution Control District of Los Angeles, Environmental Protection Agency, Office of Air and Water Programs, Office of Air Quality Planning and Standards, Research Triangle Park, NC, May 1973, p 858.
7. Reference 6, p 861.
8. Reference 3, p 3-10.
9. Reference 1, p 4-4.
10. Reference 6, p 860.
11. K. A. Daum, Research Triangle Institute, memo to docket, April 12, 1979, meeting with DeVilbiss Co., Toledo, OH.
12. Reference 3, pp 3-6–3-9.
13. Reference 2, p 3-8.
14. Reference 2, p 5-3.
15. K. Marg, Bessam-Aire, Inc., Cleveland, OH questionnaire response to R. Blaszczak, ESD/U.S. Environmental Protection Agency, Research Triangle Park, NC, January 8, 1988.
16. Reference 1, p 3-25.
17. Reference 4, p 3-5.
18. V. H. Sussman, Ford Motor Co., letter to Radian Corp., March 15, 1976, commenting on report, "Evaluation of a Carbon Adsorption/Incineration Control System for Auto Assembly Plants."
19. W. R. Johnson, General Motors Corp., letter to J. A. McCarthy, Environmental Protection Agency, August 13, 1976, commenting on *Guidelines for Control of Volatile Organic Emissions from Existing Stationary Sources*.
20. J. A. McCarthy, Environmental Protection Agency, conversation with F. Porter, Ford Motor Co., September 23, 1976.
21. *Surface Coating of Metal Furniture—Background Information for Proposed Standards*, Draft EIS, EPA-450/3-80-007a, U.S. Environmental Protection Agency, Research Triangle Park, NC, September 1980, p 2-5.
22. Reference 1, p 3-19.
23. Reference 2, p 4-3.
24. Reference 3, p 3-13.
25. *Reduction of Volatile Organic Compound Emissions from the Application of Traffic Markings*, EPA-450/3-88-007, U.S. Environmental Protection Agency, Research Triangle Park, NC, August 1988, p 10.
26. Reference 4, p 4-1.
27. *Powder Coatings Technology Update*, EPA-450/3-89-33, U.S. Environmental Protection Agency, Research Triangle Park, NC, October 1989, p 5.
28. Reference 27, p 6.
29. Reference 27, p 1.
30. S. B. Levinson, "Radiate," *J. Paint Technol.*, 44(571):32–36 (August 1972).
31. A. G. North, "Progress in radiation cured coatings," *Pigment Resin Technol.*, 3(2):3–11 (1974).
32. R. S. Nickerson, "The state of the art in UV coating," *Indus. Finishing*, 50(2):10–14 (1974).
33. Reference 3, p 4-1.
34. Reference 21, pp 5-6–5-11.
35. G. E. Cole, Jr., and D. Scarborough, *Volatile Organic and Safety Considerations for Automotive Powder Finishes*, Society of Automotive Engineers, Dearborn, MI.
36. W. H. Holley, Springborn Labs., Inc., Enfield, CT, teleconference with J. Pfiefer, Pratt & Lambert, Buffalo, NY, and T. Birdsall, Warren, IN, October 18, 1977, concerning emissions of VOCs from cured powder coatings.
37. B. J. Suther and U. Potasku, *Controlling Pollution from the Manufacturing and Coating of Metal Products—Metal Coating Air Pollution Control*, Vol. I, EPA-625/3-77-009, U.S. Environmental Protection Agency, May 1977.
38. W. H. Holley, Springborn Labs., Inc., Enfield, CT, teleconference with J. A. Scharfenberger, Ransburg Corp., Indianapolis, IN, August 29, 1977, concerning high-solids application equipment.
39. A. Mercurio and S. N. Lewis, "High solids coatings for low emission industrial finishing," *J. Paint Technol.*, 47(607):37–44 (August 1975).
40. Question corner, *High Solids Coatings*, 1(1):3–4 (1976).
41. P. H. Goodell, *Economic Justification of Powder Coating*, Association for Finishing Process of SME, Dearborn, MI, FC 76-459, 1976.
42. *Rausburg Electrostatic Swirl-Air*, Equipment Bulletin, Rausburg Electrostatic Equipment, Indianapolis, IN.
43. *Product Profile 18100, 17264 Power Supply System*, Technical Bulletin, Rausburg Electrostatic Equipment, Indianapolis, IN.
44. E. J. Vierling, "Conversion to water-reducible paint," presented at NPCA, Chemical Coatings Conference, Cincinnati, OH, April 23, 1976.
45. "The latest in water-borne coatings technology," *Indust. Finishing*, 51(9):48–53, (1975).
46. S. D. Levinson, "Electrocoat, powder coat, and radiate—which and why?" *J. Paint Technol.* 44(569):41–49 (June 1972).
47. *Electrocoating Twins Offer Two Colors*, Product Bulletin F-38, George Koch Sons, Inc., Evansville, IN.
48. M. T. Oge, Springborn Labs., Enfield, CT, Trip report no. 103, Angel Steel Co., Plainwell, MI, April 5, 1976.
49. Reference 6.
50. M. T. Oge, Springborn Labs., Enfield, CT, Trip report no. 141, Fasson Co., Painesville, OH, July 14, 1976.
51. R. A. McCarthy, Springborn Labs., Enfield, CT, Trip report no. 77, Raybestos-Manhattan, Inc., Mannheim, PA, February 27, 1976.
52. T. Gabris, Springborn Labs., Enfield, CT, Trip report no. 89, American Can Co., Lemonyne, PA, March 11, 1976.
53. J. R. Fisher, Springborn Labs., Enfield, CT, Trip report no. 31, Supracote Svc., Cucamonga, CA, January 16, 1976.
54. G. R. Godowski, A. V. Gimborne, W. J. Green, et al., "An evaluation of emissions and control technologies for the metal decorating process," *J. APCA*, 24(6):579–585 (1974).
55. T. Hurst and R. F. Jongleux, *Stack Emission Sampling at Ford Motor Company, Pico Riveria, California*, TRW (rough draft), Durham, NC, U.S. Environmental Protection Agency, Research Triangle Park, NC, February 13, 1979.
56. Reference 21, p 3-82.
57. Reference 4, pp 4-2, 4-3.
58. M. Wright, Trip report, Kaiser Aluminum, Toledo, OH, Research Triangle Institute, Research Triangle Park, NC, October 31, 1979, attachment A.
59. Reference 4, p 4-11.
60. Reference 1, p 4-36.
61. Reference 1, p 4-39.
62. V. H. Sussman, Ford Motor Co., letter to J. McCarthy, EPA-CTO. March 16, 1976.
63. Reference 3, p 4-4.
64. Reference 3, pp 4-7, 4-10.
65. Reference 2, p 2-9.
66. Reference 25, p 25.

11
Graphic Arts Industry

Nonheatset Web Printing
H. Wilson Cunningham
Heatset Web Offset
Robert M. Birkland, Gerald Bender, and William Schaeffer
Flexography
Warren J. Weaver
Sheetfed Offset Printing
Jerry Bender, Nelson Ho, David Johnson, Frank Little, and William Schaeffer
Gravure Printing
Jerry Bender and Robert Oppenheimer
Screen Printing
Marcia Y. Kinter

INTRODUCTION

The graphic arts industry, often referred to collectively as the printing/publishing/packaging industry(within SIC Code 2700 series), has been evolving and growing rapidly, especially since the introduction of computerized prepress and press operations in the last decade. With nearly 70,000 printing facilities, the graphic arts industry ranks sixth among all U.S. industries. This chapter focuses on the printing operations currently used in the graphic arts industry: flexography, nonheatset web offset, rotogravure, screen printing, sheetfed offset, heatset web offset, and others. Each type of printing operation employs distinctive processes and materials and has its own unique characteristics of air emission and control measures.

The material contained in this chapter may serve as a reference for regulatory agencies, other governmental agencies, universities, and schools that need a greater understanding of the technologies, processes, operations, equipment, air emissions, and emission control measures currently utilized by the graphic arts industry. Six major printing operations are treated separately in the following sections, with emphasis on process description, characteristics of air emission, and emission control technologies. A reference guide for additional, detailed information on technologies and technical terms common to the graphic arts industry is given in Appendix A.

The Environmental Conservation Board of the Graphic Communications Industries (ECB) coordinated and formed an Air Pollution Engineering Manual (AP40) Committee composed of distinguished experts from every field of the graphic arts industry to write this chapter. Many trade associations within the graphic arts industry also contributed tremendously to the making of this chapter. Members of the committee, as well as those associations, are listed in Appendix B. The ECB deeply appreciates the endeavors of those dedicated men, women, companies, and associations that made this chapter possible.

The Environmental Conservation
Board of the Graphic
Communications Industries

Appendix A. Reference Guide for the Graphic Arts Industry

The Lithographers Manual, 8th ed., Ray Blair, Ed.; Graphic Arts Technical Foundation, Pittsburgh, 1988.

A. Glassman, *Printing Fundamentals*, TAPPI Press, Atlanta, 1985.
A. Kosloff, *Screen Printing Techniques*, Signs of the Times Publishing Co., Cincinnati, 1981.
The Printing Ink Manual, 4th ed., R. H. Leach, Ed.; Fishburn, England, 1988.
B. Magee, *Screen Printing Primer*, Graphic Arts Technical Foundation, Pittsburgh, 1985.
A. Peyskens, *The Technical Fundamentals of Screen Making*, SAATI, Como, Italy, 1989.
V. Strauss, *The Printing Industry*, Printing Industry of America, Washington, DC, 1967.
Flexography: Principles and Practices, 3rd ed., Flexographic Technical Association, Long Island, NY, 1980.
Printing Ink Handbook, 5th ed., National Association of Printing Ink Manufacturers, Harrison, NY, 1988.
Rauch Guide to the U.S. Ink Industry, Rauch Associates, Inc., Bridgewater, NJ, 1987.
Rauch Guide to the U.S. Packaging Industry, Rauch Associates, Inc., Bridgewater, NJ, 1987.
Technical Guide for the Gravure Industry, 3rd ed., Gravure Technical Association, New York, 1975.

Appendix B.

The AP-40 Air Pollution Engineering Manual Committee

Jerry Bender
Robert Birkland
Doug Cook
Wilson Cunningham
Frederick Falk
C. Nelson Ho, Coordinator
Lloyd Hoeffner
David Johnson
Marcia Kinter
Frank Little
Mark Nuzzaco
Robert Oppenheimer
Thomas Purcell
William Schaeffer
Kip Smythe
Paul Volpe
Ed Weary
Warren Weaver

Trade Associations

American Newspaper and Publishers Association (ANPA)
Environmental Conservation Board of the Graphic Communications Industries (ECB)
Flexible Packaging Association (FPA)
Flexographic Technical Association (FTA)
Graphic Arts Technical Foundation (GATF)
Gravure Association of America (GAA)
National Association of Printing Ink Manufacturers (NAPIM)
National Association of Printers and Lithographers (NAPL)
National Printing Equipment and Supply Association (NPES)
Printing Industry of America (PIA)
Screen Printing Association International (SPAI)

NONHEATSET WEB PRINTING

H. Wilson Cunningham

In web printing, a rotary press is used to print an image on a continuous web of paper that may exceed 8 miles (12 km) in length. After printing at speeds between 600 and 2100 fpm (180–670 meters per minute), the web is mechanically cut and manipulated into its final form.

The nonheatset or cold-set web printing process can be distinguished from other web printing processes by its unique ink and paper requirements. Nonheatset web inks are semifluid materials whose function depends on the rapid absorption of the liquid component of the ink vehicle by the surface of the paper. The capillary action of the paper surface draws the liquid into the paper surface and "sets" the ink. This process does not require the application of heat to cure the resin or evaporate the liquid component.

Newsprint, which accounts for most of the paper used for nonheatset web printing, is a lightweight paper stock that weighs 44–53 g/m^2. The paper must have enough absorptive capacity to allow rapid penetration of the ink vehicle. This requirement excludes the use of coatings and permits a minimum of filler materials or machine finishing that smooths the surface and decreases the rate of absorption.

Offset lithography and letterpress printing represent the nonheatset web printing processes. The flexographic process, another such process, is covered in another section of this chapter.

PROCESS DESCRIPTION

Letterpress printing uses a relief plate similar to a rubber stamp to transfer ink from its surface directly to the surface of the paper (Figure 1). Newspaper printing accounts for 90% of letterpress production and book and speciality printing represent the remainder.[1] Letterpress news inks are relatively simple inks consisting of an oil vehicle, resinous binder, and solid pigment (Table 1). Letterpress printing, the only mode of newspaper printing up to 1960, represented less than 31% of the daily circulation in 1987.[2] Conversion to offset lithography—and, since 1987, to aqueous flexography—has caused the decrease in newspaper letterpress printing.

Nonheatset web printing represents about 35% of the diverse offset lithography classification of printing.[1] The 1400 daily newspapers that use offset printing[2] account for most of the nonheatset web offset market.[1] Offset lithography differs from letterpress printing in that a planographic printing surface is used that has been chemically modified into a hydrophobic (oleophilic) ink-receptive image area and water-loving nonimage area. To keep the hydrophobic ink film from migrating into the nonimage region of the plate, a second fluid, water, is applied to the plate in conjunction with the ink. A thin film of ink is transferred from the plate to an intermediate blanket cylinder before being transferred to the moving paper web (Figure 1).

The required presence of an immiscible liquid and the amount of ink transferred from a planographic plate have a profound effect on the composition of nonheatset web offset (NHWO) ink (Table 1). An offset print receives about 30% less ink than a corresponding print made from a relief plate.

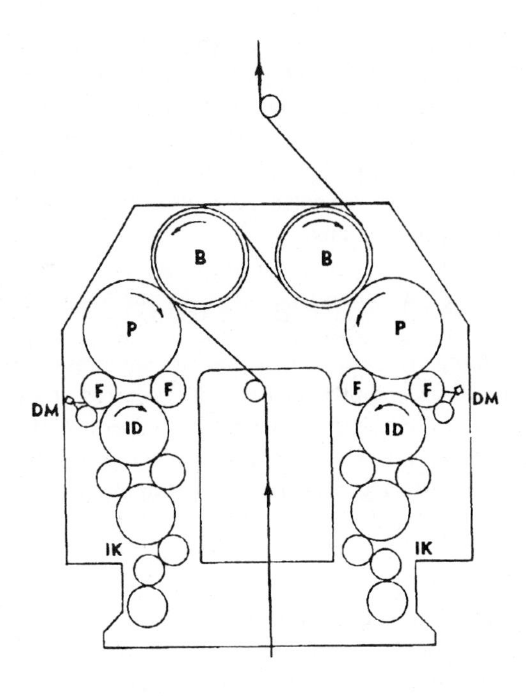

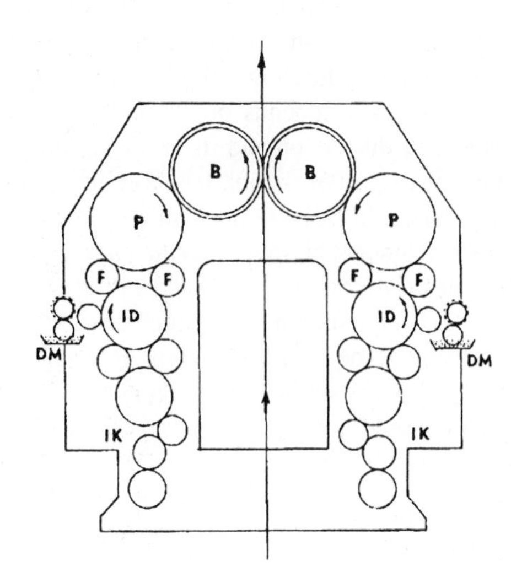

LETTERPRESS UNIT OFFSET UNIT

FIGURE 1. Typical Components of Offset and Letterpress Nonheatset Web Printing Units. Identification of parts: IK = ink train roller system, ID = ink drum roller, F = form roller, P = plate cylinder, B = blanket cylinder for offset and impression cylinder for letterpress, DM = ink train type of dampening system

TABLE 1. Composition of a Typical Nonheatset Web Printing Ink

	Component[a]	Typical[b] Range, %	
		Letterpress	Offset
Pigment	Carbon black	8–15	15–20
	Organic color pigment	6–12	8–15
	Rheological solid[c]	0–1	0–10
Resinous binder[d]	Alkyd resin	0–7	0–20
	Thermoplastic resin	0–7	0–20
Hydrophobic vehicle[e]	High-viscosity oil, 310–550°C bp	0–85	0–80
	Low-viscosity oil, 230–300°C bp	0–10	0–50
	Vegetable oil	0–85	0–80

[a]Typical materials used in formulation. Materials in this category may be used singly or in combinations.
[b]Maximum and minimum amounts of component that may be found in a typical formulation.
[c]Rheological solids are inert materials, such as clays, used to adjust viscosity and tack of ink.
[d]Resinous binders are solid thermoplastic materials solubilized in the hydrophobic liquid to create the ink vehicle.
[e]Hydrophobic liquid is a hydrotreated or severely solvent-refined mineral or vegetable oil. Low- and high-viscosity petroleum oils can be distinguished by boiling-point ranges and viscosity range. High-viscosity oils have viscosities between 750 and 2400 Saybolt universal seconds (SUS). Low-viscosity oil has a 50–100 SUS range.

Source: Reference 3.

This thinner ink film necessitates an increased pigment concentration to provide equal print density. Suspending and properly dispersing more solid pigment within the liquid vehicle require an increase in the amount of resinous binders. Solubilization of the resinous binder affects the types and amounts of hydrophobic liquids in NHWO inks. The choice of the resin depends on its ability to wet the pigment surface for dispersion and its ability to release the liquid component only after the ink has been applied to the surface of the paper. In NHWO inks, this release process, which "sets" the resin and pigment into the paper, cannot be dependent on the evaporation of the liquid by the addition of heat. Finally, the resinous binders cannot dramatically increase the adhesion of the ink film that will pick paper fibers from the uncoated paper surface. The chemical characteristics of the resin should not drastically increase the amount of water emulsified by the ink.

The dampening system is the portion of an offset press that applies water to the surface of the planographic plate. The contact dampening system applies a film of water directly to the plate by some type of roller or misting system. The ink train dampening system applies the water solution to the ink roller train, where an emulsion forms. This ink–water emulsion is then transferred to the plate surface, and the shear of the printing nip causes the emulsion to break into its hydrophobic ink component and nonimage-area–protecting water component. Both types of dampening systems are open to the atmosphere. Refrigeration of the dampening system in NHWO is extremely rare because the use of low-molecular-weight alcohols is not common.

The water component, referred to as fountain solution or etch, used in the lithography printing process is a dilute chemical mixture. The chemicals are added to lower the surface tension of the water and provide a controlled acidity level (Table 2). Fountain solutions used in NHWO printing can be distinguished by the lack of low-molecular-weight alcohols. The low tack and viscosity of the NHWO inks do not require the aggressive wetting power provided by the addition of low-molecular-weight alcohols. Instead, alcohol substitutes, high-boiling-point glycols and glycol ethers, have the advantage of providing lubrication and improved release characteristics at the blanket-plate printing nip. In addition, concerns about workplace exposure to isopropanol have virtually eliminated alcohols from NHWO fountain solution formulations.

TABLE 2. Chemical Additives Listed on Material Safety Data Sheets for Fountain Solution[a] Used in Nonheatset Web Offset Printing

Phosphate salts
Silicates
Colloid materials such as gum arabic or dextrins
Organic surfactants
Alcohol substitutes

[a]Fountain solution is a dilute water solution with specific conductance between 1000 and 2500 mho that may contain some or all of the listed components. Typical fountain solutions are either acidic (pH 3.0–5.5), neutral (pH 6.5–8.5), or alkaline (pH 9.5–11). Phosphate salts are used to maintain the pH of the fountain solution. Alcohol substitutes are high-molecular-weight glycols, polyols, and possibly glycol ethers in concentrations below 0.5%. Use of low-molecular-weight alcohols such as isopropanol is not common.

AIR EMISSIONS CHARACTERIZATION

Ink, fountain solution, platemaking, and press-cleaning operations represent the four potential sources of air emission from the nonheatset web printing process. Unfortunately, actual emission factors are not known at the time of this writing and most data must be extrapolated.

Ink

Potential emissions from the ink used in nonheatset web printing could occur during the printing process. The design of a nonheatset web printing press requires the ink to travel through a series of ink-distribution rollers, known as the ink train, that control the application of ink to the plate. The potential evaporation from the operating press would depend on the temperatures of this roller system and the volatile components of the ink.

In 1989, the American Newspaper Publishers Association (ANPA) used an infrared pyrometer to measure ink-train temperatures for large and small offset and letterpress printing operations. The measurements were taken in February and repeated in July. The maximum temperature observed during the production cycles was 42°C. As it occurred at press stoppage, only a small amount of ink would be exposed to this temperature. Temperature of an operating press was about 3°C lower. No difference was observed between the summer and winter operating temperatures.[4]

Exposure to temperatures that are relatively low in comparison with the boiling point of the liquids used to manufacture nonheatset web inks would suggest that emissions during press operations are minimal. The ANPA, in cooperation with the National Association of Printing Ink Manufacturers (NAPIM), measured the emissions during press operations for three different formulations of NHWO inks (Table 3). Air samples from the ink train and the paper web as it exited the press were taken using standard National Institute for Occupational Safety and Health (NIOSH) procedures for measuring hydrocarbon exposure. Emissions, determined by gas chromotography, were between 0.6 and 5.7 $mg/m^3/h$. The observed differences in the emissions were a result of ink formulation. Extrapolation of these data to a nine-unit press operating for 24 hours predicts an emission of less than 1.3 grams of hydrocarbon into the atmosphere.

These emission data support the theory that all materials transferred to the paper penetrate the surface and are re-

TABLE 3. Volatile Organic Component Emission Rates Measured with Three Different NHWO Inks

Sampling Location	Ink Type		
	Low-Rub[a] Black	No-Rub[b] Black	Petroleum[c] Black
Method 24 VOC	10%	19%	5%
Right ink form roller, mg/m³/h	3.1	5.0	2.6
Left ink form roller, mg/m³/h	3.7	5.2	3.2
Pipe roller, mg/m³/h	0.6	5.7	1.1

[a]Low-rub formulation contained 60% petroleum oil that is a mixture of low-viscosity and high-viscosity oils.
[b]No-rub formulation composed of 45% low-viscosity oil and 10% high-viscosity oil.
[c]Regular black contained about 75% high-viscosity oil.

Source: Reference 4.

tained within the product. Measurement of the loss of any retained materials from the printed sheet is hampered by the low concentration of ink (about 1% to 2% by weight) on a printed page and the very hygroscopic nature of uncoated paper. Even under controlled humidity and temperature, the gain or loss of water exceeds the accuracy of any gravimetric procedure to detect emissions of organic components from the printed product.

Laboratory oven tests such as Method 24 of the American Society for Testing and Materials (ASTM) were developed to accelerate the evaporation of volatile components from products designed to dry by evaporation. The tests were not designed to be applied to products such as NHWO inks where the drying mechanism is not evaporation. However, certain regulatory authorities have chosen to apply Method 24 to nonevaporative systems in an attempt to predict the potential loss of material over its extended lifetime.[5] This type of testing does not take into consideration the reduction of emissions that can occur in a nonheatset web product as a result of the pacification of the liquid component within the "set" resin and pigment complex or the absorption of the material within the cellulose fiber. Until this reduction in emissions can be determined quantitatively, the severe conditions of Method 24 will continue to overestimate the emissions from nonheatset web inks.

Method 24 demonstrates that different nonheatset inks have different potential emissions (Table 4). Most of the differences seen in this table reflect the ratio of low-viscosity to high-viscosity mineral oil required to formulate the performance of a given ink or the substitution of vegetable oils for the petroleum oil. Nonheatset web ink formulated with vegetable oil (e.g., soya oil) has significantly lower emission potential than an ink formulated with petroleum oils. The choice of different types of petroleum oils not traditionally used in ink formulations may have a similar

TABLE 4. Content of Four Classes of Printing Inks Determined by ASTM Method D-2369-81 (Method 24)

Ink Type	Mean, %	Range, %
Color offset[a]	9	2–20
Black offset	9	2–13
Color letterpress	12	10–14
Black letterpress	5	5–10

[a]The lower range of these samples represents samples formulated with soya oil. No soya oil formulations of black offset, color letterpress, or black letterpress were available for testing.

Source: Reference 4

effect of lowering the potential emissions from nonheatset printing inks.

Air Pollution Control Measures

The low volatility of the components used in nonheatset web printing inks limits the control technology options. Applications of drier systems would be expensive and would have little effect on the removal of the high-boiling-point petroleum component from the printed page. The heat from the drier system also would reduce the viscosity of the ink vehicle and would allow the materials to penetrate the paper instead of being isolated near the surface. This penetration would destroy the visual integrity of the product. The operating-press emission data demonstrate that press enclosures and air purification systems would have little benefit in reducing the release of volatile compounds into the environment.

If nonheatset web inks are considered a significant source of volatile materials, the best control technology would be in ink formulation. The data indicate that the inks are not a major source of volatile materials during printing.[4]

Fountain Solutions

Nonheatset letterpress printing does not require a dampening system; therefore, emission from the fountain solution does not apply to this printing process.

Fountain solutions used in NHWO printing are generally purchased in concentrated form and diluted with water to create a "working solution." The working solutions are at least 99.5% water by volume. High-boiling-point glycols and polyols represent the remainder of the working-solution liquid composition. These compounds, in combination with organic surfactants, have effectively replaced isopropanol as the primary wetting agent in the fountain solution.

The mechanical design of the dampening systems on most presses requires that the working solution be open to the atmosphere. Any evaporation from these systems may result in a measurable worker exposure to any organic compound in the fountain solution. Since 1985, ANPA has conducted about 45 chemical air sampling studies in various nonheatset web printing operations. Concentrations of glycols and glycol ethers were below 5 mg/m^3 at all sites tested. These data suggest that the potential emission from the dampening system of an NHWO press operating for 24 hours would be less than 1.2 grams per day.

Air-Pollution Control Measures

The limited amount of information that is available from manufacturers' Material Safety Data Sheets suggests that the NHWO dampening system is not a major source of emission. If the printer adds isopropanol to the dampening system, refrigeration of the system or replacement with less volatile alcohol substitutes may be required.

Platemaking

Nonheatset web letterpress platemaking chemistry is very dependent on the manufacturer's polymer system. The chemistry of these systems is highly proprietary and limited information is available.

The NHWO plates use an aqueous diphenylamine azide coating as the photoactive polymer. After polymerization of the image area by ultraviolet light, the image is developed by an acidified asphalt–water emulsion. Traditionally, this emulsion contained an alcohol-ketone component to stabilize the emulsion. Concerns about wastewater discharges and worker exposure have all but eliminated this volatile component from the plate developer.

Air Pollution Control Measures

All systems are vented, and if the emissions are considered a problem, air purification systems could be installed. Obviously, any exhaust containing significant levels of air contaminants could be processed through air pollution devices such as activated-charcoal filters.

Press-Cleaning Operations

Press cleaning, the removal of the built-up layer of ink and paper dust from rollers and blankets, is a necessary maintenance operation for the nonheatset web press. The cleaning of a nonheatset web press, however, differs from other printing processes in that the inks do not readily dry or oxidize into a solid film. These "soft" films do not require aggressive degreasers for removal. Traditionally, the press-cleaning solvent of choice has been petroleum distillates such as deodorized mineral spirits and kerosenes. These materials sometimes are reduced with water and detergents to form a cleaning solution.

The best available data from the ANPA chemical air sampling in nonheatset web operations would suggest that press cleaning is the source of 90–95% of all emissions from the nonheatset process.[4] The total amount of emissions is hard to estimate because of the different cleaning procedures used in different printing locations.

Air Pollution Control Measures

Viable control technologies to reduce the emission from this periodic action are to reduce the amount of solvent in the cleaning solution required for cleaning or to use less volatile solvents. The water extension is a common means of reducing the volume of solution used. The unique nature of nonheatset web inks may also permit the use of an aqueous-based cleaner. Various manufacturers are currently testing these products, but their functionality has not been proved as yet.

References

1. *The Rauch Guide to the U.S. Ink Industry,* Rauch Associates, Bridgewater, NJ, 1988.
2. *Facts about Newspapers '90,* American Newspapers Publishers Association, Reston, VA, 1990.
3. P. Volpe, National Association of Printing Ink Manufacturers, Harrison, NY, personal communication, 1990.
4. C. S. McLendon and H. W. Cunningham, *Volatile Organic Content of Non-Heatset Newspaper Printing Inks,* American Newspaper Publishers Association, Reston, VA, 1989.
5. J. Berry, Chief of Chemical Applications Section U.S. Environmental Protection Agency, Research Triangle Park, NC, personal communication, 1990.

Bibliography

N. R. Eldred and T. Scarlett, *What the Printer Should Know About Ink,* 2nd ed., Graphic Arts Technical Foundation, Pittsburgh, 1990.

R. H. Leach, *The Printing Ink Manual,* 4th ed., Northwood Books, London, 1988.

HEATSET WEB OFFSET

Robert M. Birkland, Gerald Bender, and William Schaeffer

Heatset web offset lithography is the process that produces over 75% of magazines, catalogs, books, and tabloids. This process is similar to other offset lithographic processes in the type of plates used and the requirement for fountain solution to maintain a printing image. Differences include the web speed, the inks that require energy to dry, energy-application devices (dryers) to dry the ink, and web cooling devices (chill rolls).

PROCESS DESCRIPTION

The heatset web offset printing process is a complex process that varies depending on the type of product to be produced. This process is used to produce a vast array of materials, from complex technical manuals with highly illustrated images to simple one-color text. As a result, the process can vary from the use of one-color printing stations to complex

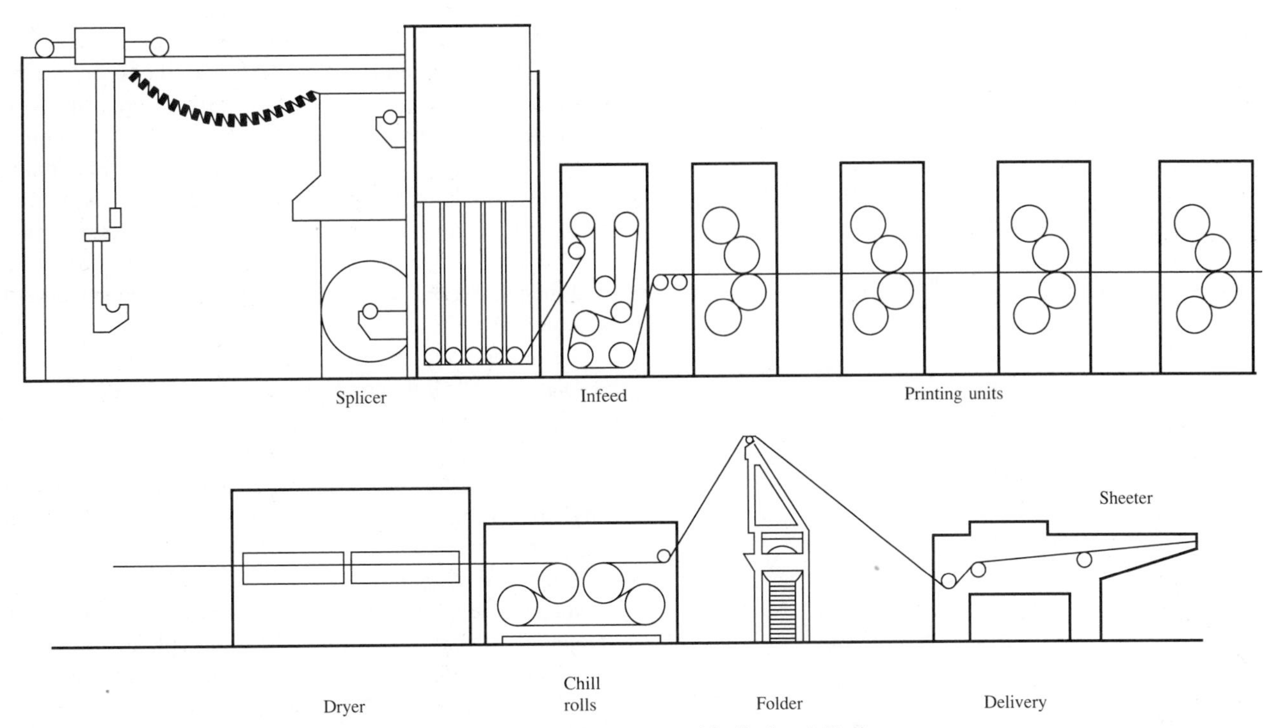

FIGURE 1. A Four-Unit Blanket-to-Blanket Web Offset Press with Optional Delivery to a Folder or Sheeter

multicolor presses utilizing up to eight or more printing units. In its basic form, the web offset process consists of an unwind stand, in-feed, printing units, dryer, chill stand, and folder. A typical web offset printing press is shown in Figure 1.

The reel is the mechanism that allows the continuous unwinding of rolls of paper for subsequent printing in the printing units. This mechanism has the capability of splicing the expiring roll of paper to a new roll of paper without stopping, thus maximizing the operational efficiency of the press. The in-feed tension device keeps a constant tension throughout the press and isolates the press from the tension upsets that occur when a splice is made. Without this device, web offset printing would be extremely difficult to accomplish because the physical appearance of the printed material will vary depending on the tension applied to the paper.

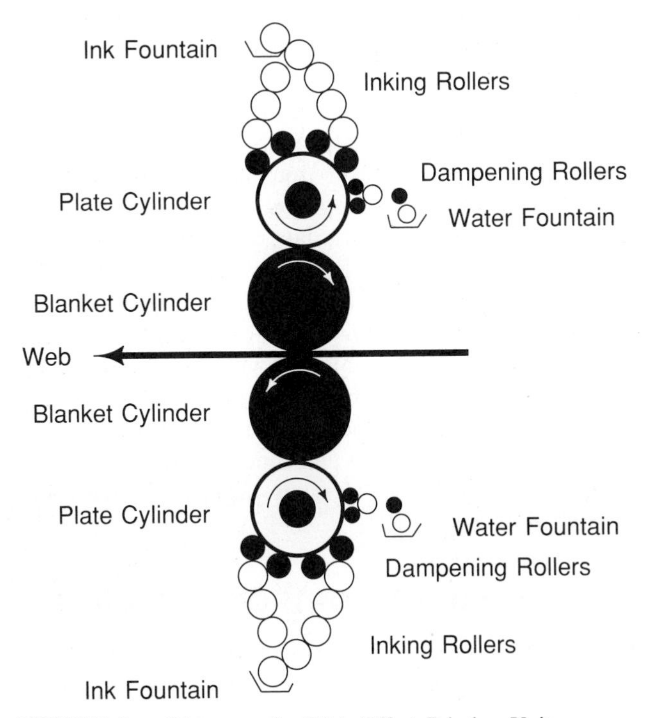

FIGURE 2. Diagram of a Web Offset Printing Unit

After the paper web unwinds on the roll stand or reel, it may pass through a heated web conditioner before entering the first printing unit. Ink and fountain solution are applied to the web, typically on both sides, as shown in Figure 2. Depending on the number of colors being printed, the web will pass either into the dryer or to a number of additional printing units before entering the dryer. The dryer will apply energy to cause the ink oil to evaporate. Because the web has been heated by the energy application, it is passed onto the chill rolls, where it is cooled to about 20 degrees above ambient temperature before folding and cutting.

Dryers for heatset web offset-type inks are typically recirculating hot air systems, although some direct flame impingement and infrared dryers are still in use. Usually, air, flame, or radiation is applied to both sides of the web within the dryer to raise the web temperature to about 275°F ± 50°F. The volatile portions of the ink are exhausted from the dryer, either to an emission control device or to the atmosphere. Evaporation from the hot web exiting the dryer is collected by either a tunnel or a hood over the chill rolls.

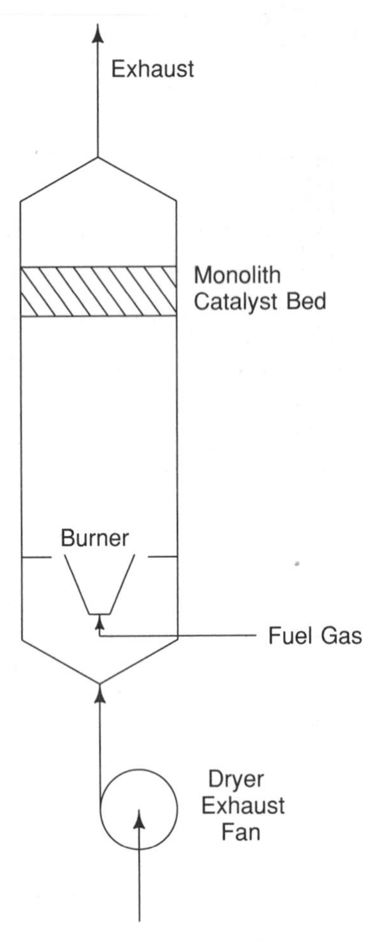

FIGURE 3. Catalytic Afterburner without Heat Recovery

In addition to the inks that dry by evaporation, others cure by chemical reaction among components. Reactive-cure ink systems include three major categories:

1. Ultraviolet-radiation cure (UV)
2. Electron-beam-radiation cure (EB)
3. Thermal catalytic cure

Reactive cure inks contain only a nonvolatile reactive binder or film former and pigment. Dryers for chemical reaction inks are usually very-high-energy UV lamps, electron-beam generators, or thermal ovens. The absorption of this energy within the ink causes the film former to cure and harden. Very few, if any, volatiles are evaporated. The emissions, if any, are associated with the dryer energy sources, not the ink. A low level of ozone may be present during warm-up of UV lamps and decreases as the lamps reach temperature.

AIR EMISSIONS CHARACTERIZATION

Air emissions from heatset web offset lithography are primarily from the three sources—the ink, the fountain solution, and the cleaning solvents. Ink oils used in heatset inks have vapor pressures usually less than 0.1 mm of mercury and when in the vapor state, have a smog-generating reactivity similar to that of ethane. The ink oil is a condensable organic material that, if emitted in high concentrations, will condense into an opaque plum. Higher concentrations may require controls to reduce opacity to acceptable limits.

The fountain solution and the cleaning solvents contain constituents that are defined as volatile organic compounds (VOCs). Fountain solution is essentially water with nonvolatile additives, but may contain isopropyl alcohol or replacements, such as 2-butoxyethanol or ethylene glycol. Isopropyl alcohol ranges from 5% to 20% concentration in the fountain solution, while the alcohol substitutes are present at 2–5% total concentration. Emission factors vary greatly, depending on the additives involved. As much as 75% of the isopropyl alcohol may evaporate into the room, while up to 25% of 2-butoxyethanol may evaporate. Ethylene glycol is even less volatile. The amounts remaining in the dried web are also variable, ranging from essentially zero to as much as 5%.

Manual cleaning solvent emissions are limited to the amount of solvent transferred to blankets, plates, rollers, or paper webs, if the cleaning cloths are placed in sealed containers after use. Cleaning solvents for heatset inks usually consist of mixtures of C_7 to C_{10} hydrocarbons, with some aromatic components. These mixtures have flash points from under 80°F to over 140°F. Cleaning solvents for reactive cure inks usually are alcohol systems rather than aliphatic hydrocarbon types.

The dryer exhaust emissions from heatset inks are primarily a mixture of hydrocarbons of about an average C_{15} chain length. Typical concentrations would range from 2% to 12% of the lower flammable limit. Some heat distortion (oxidation) can occur as a result of contact with either very hot air or flame. Depending on the rate of ink use and type of paper being printed, approximately 40% to 90% of the input ink oil is evaporated in the dryer. The balance of the input ink is retained in the printed paper web. A low ink-input rate usually results in a low percentage of ink oil being evaporated. Other materials present would be either alcohol or alcohol substitute from the fountain solution. Fewer heatset printers use isopropyl alcohol because of cost and environmental concerns, but isopropyl alcohol does provide great operational flexibility. Alcohol substitutes usually include 2-butoxyethanol and/or ethylene glycol. Both are listed as air toxics by SARA III, Section 313.

Presses with automatic blanket washers also will have cleaning solvent vapors in the dryer exhaust during the wash cycle. Relatively small amounts of solvent are used because the wash cycle is very short and concentrations are closely controlled to maintain a 4:1 safety factor below the lower flammable limit.

AIR POLLUTION CONTROL MEASURES

Air pollution control equipment has been installed only to control the emissions from evaporative-style ink dryer exhausts. Three methods of hydrocarbon emission con-

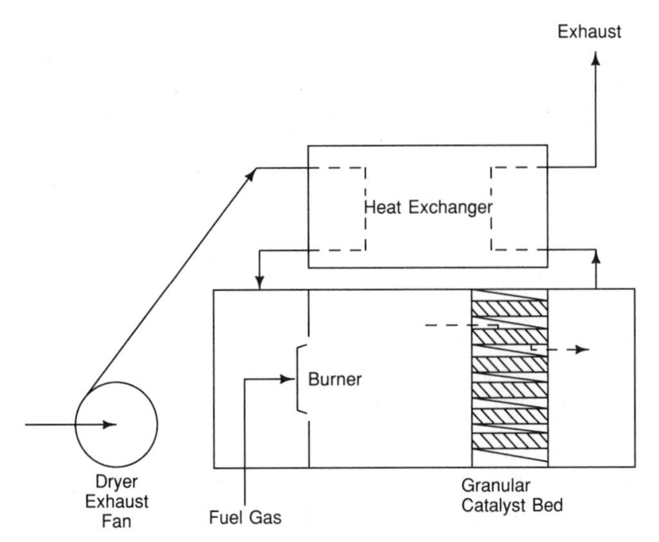

FIGURE 4. Catalytic Afterburner with Heat Recovery

trol—carbon adsorption, incineration, and condensation/removal—theoretically are applicable. Adsorption has been tried, but desorption of ink oil from carbon requires extreme conditions of elevated temperature and low pressure and one-time use of carbon is prohibitively expensive. The primary methods of ink oil emission control are fume incinerators (afterburners) and condenser–droplet-removal systems. Afterburners are the usual method of emission control, with 95–98% of the controls being of this type.

Afterburners are divided into catalytic and thermal types. Both afterburner types often incorporate heat exchanger to reduce fuel consumption if operating hours are sufficient and airflow volume is large enough to provide an economic advantage.

A catalytic afterburner consists of a preheating section, a temperature indicator-controller, a chamber containing the catalyst, safety equipment, and, usually, heat-recovery equipment. Figures 3 and 4 show two arrangements. A catalyst is a substance that changes the rate of a chemical reaction and does not appear to change chemically in doing so. In the case of afterburners, the catalyst functions to promote the oxidation reactions at a somewhat lower temperature than occurs in a thermal afterburner. The catalyst often is platinum combined with other metals and deposited in porous form on an inert substrate. Metallic oxide catalysts are usually homogeneous granules. Catalyst afterburners typically control air temperature into the catalyst at 650–800°F. Typical hydrocarbon destruction efficiencies range from 90% to 98%, depending on the ratio of the volume of catalyst used to the air-flow volume. Catalysts may be supported on granular particles or on rigid structures (monoliths). Catalysts are susceptible to loss of performance as a result of coating (masking) and reaction with contaminants (poisoning). Dryer exhaust gases containing silicones, tars, resins, and dusts will cause masking, and those containing high levels of phosphorus compounds, heavy metals, halogens, or sulfur will cause poisoning. Excessive temperatures can cause crystal growth of the catalyst support and loss of activity. The upper temperature limit appears to be 1400°F for platinum catalysts and lower for metallic oxide catalysts. Granular catalysts are susceptible to abrasion, which may extend the useful life of the catalyst by removing masking and surface poisons. However, it may ultimately cause failure by converting the catalyst into fines, which may eventually pass through the retaining screens. The preheating section may have either electric or gas fuel heaters. Liquid fuels are avoided as inefficient burning will contribute to masking of the catalyst.

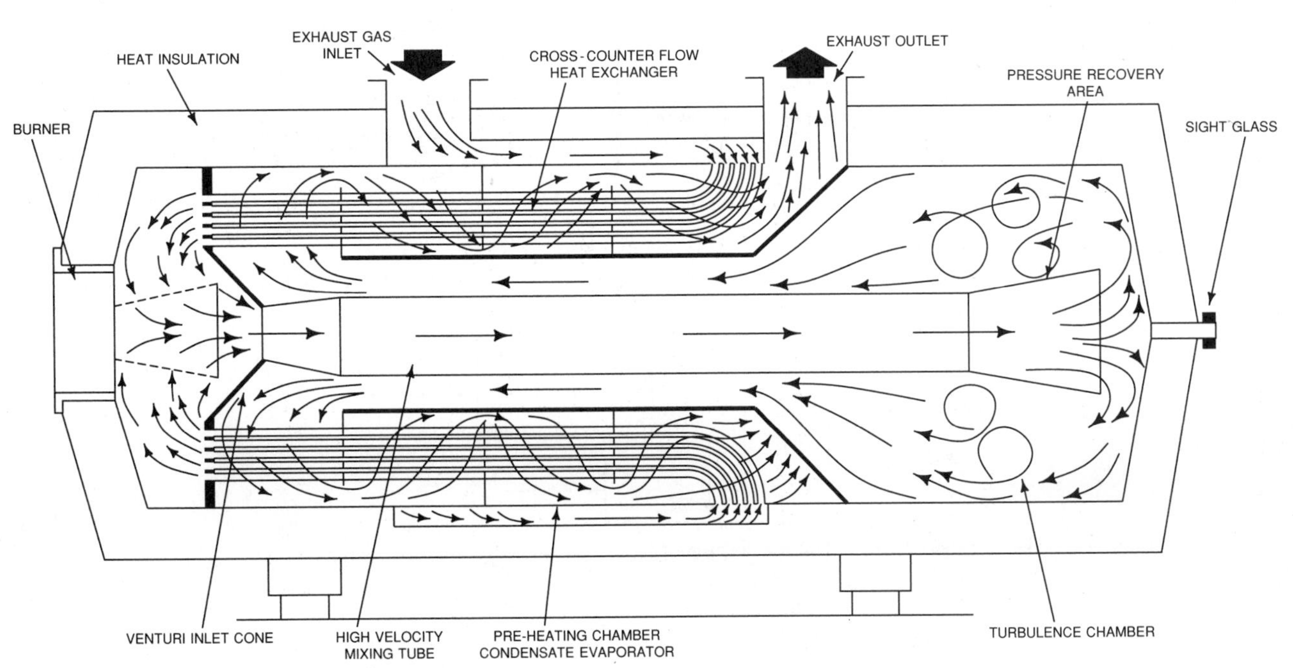

FIGURE 5. Recuperative-Type Thermal Afterburner

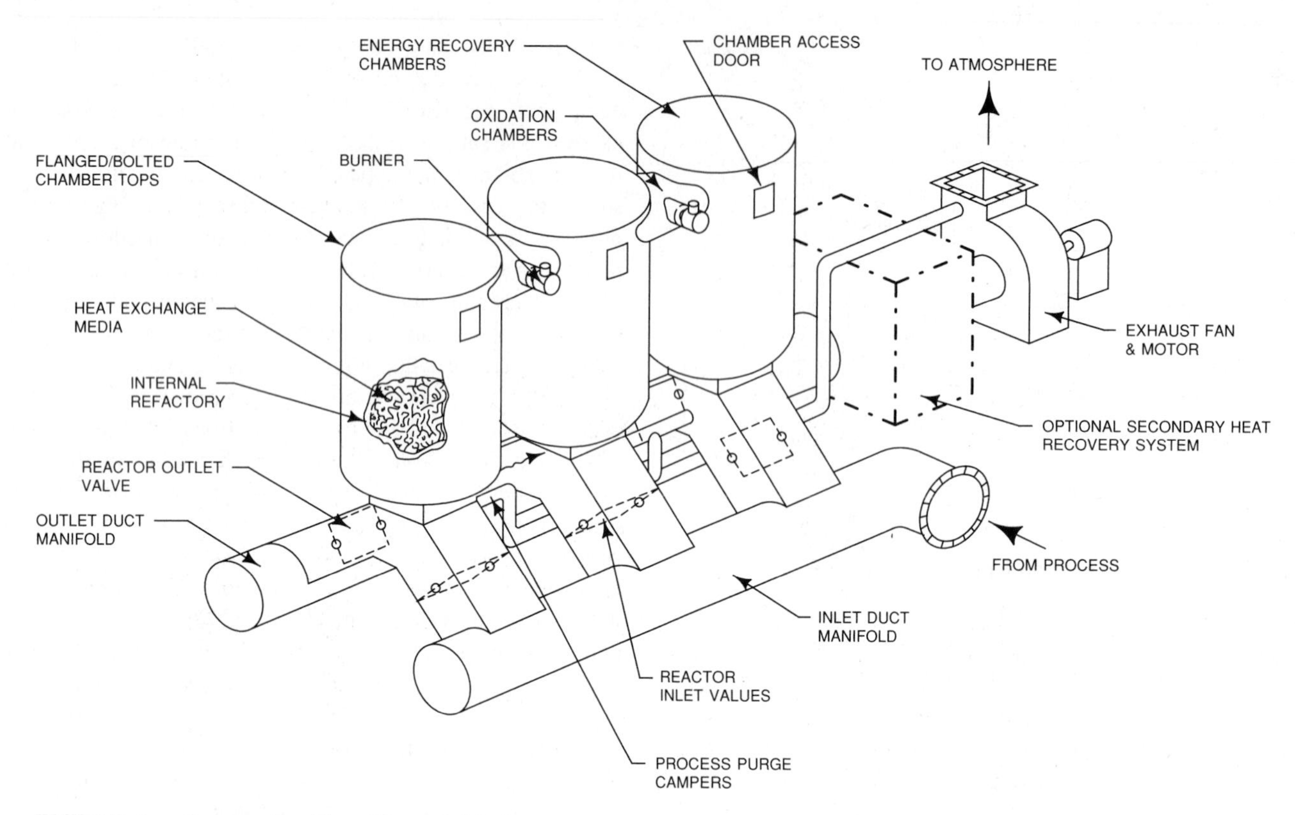

FIGURE 6. Regenerative-Type Thermal Afterburner

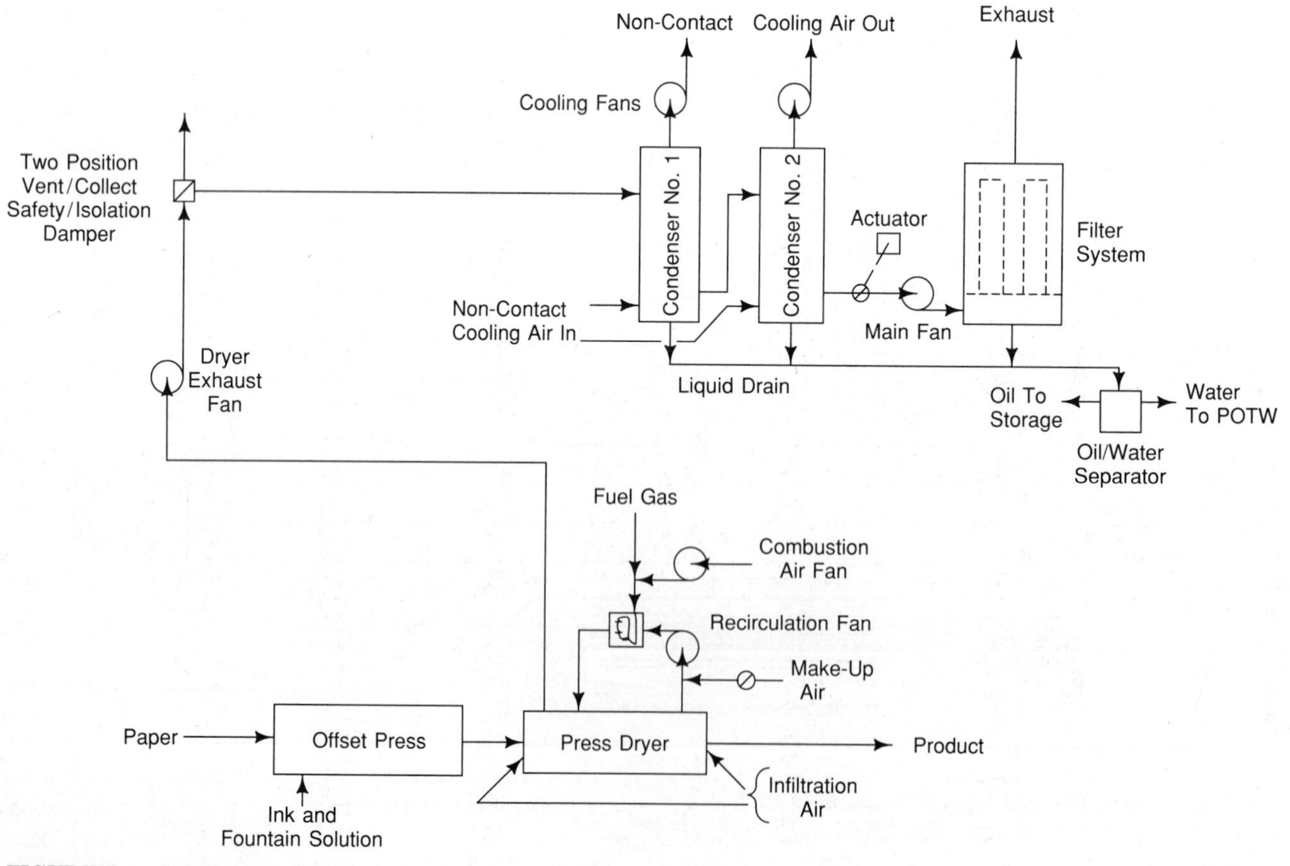

FIGURE 7. Condenser–Droplet-Removal Emission Control

A thermal afterburner has functional components similar to those of a catalytic afterburner, except that the catalyst bed is replaced with a larger combustion chamber, heat recovery is greater and nearly mandatory economically, and construction materials are changed to handle higher temperatures. An effective thermal afterburner design must provide for an adequate combination of (1) a sufficiently high temperature, (2) a high enough residence time (usually above 0.5 second), and (3) adequate mixing or turbulence in the combustion chamber. A small deficiency in any one component may be compensated for by an increase in another. Direct flame contact is not required, although the extremely high temperature in the flame even for a short period is beneficial. Heat recovery is standard owing to the high temperatures involved and has resulted in two types of systems—recuperative and regenerative heat recovery. Recuperative units continually transfer heat from a hot stream to a colder stream in a countercurrent flow arrangement. Figure 5 illustrates a typical shell and tube recuperative unit. Regenerative units cyclically store and transfer heat between streams, and temperatures are always changing. Figure 6 is an example of a regenerative thermal afterburner in which the temperature storage unit consists of ceramic pieces. Operating temperatures of recuperative units range between 1250°F and 1450°F. Regenerative units operate in the range of 1400°F to 1800°F. Hydrocarbon destruction efficiencies range from 97% to 99.8% for the recuperative units and from 95% to 99% for the regenerative units. Thermal recovery effectiveness typically ranges from 45% to 76% for recuperative units and from 80% to 95% for regenerative units. The use of heat exchangers reduces fuel consumption, which results in fewer NO_x compounds.

Condenser–droplet-removal controls consist of heat exchangers followed by either filters or electrostatic precipitators to remove oil droplets. The exchangers are often followed by an oil–water separator. Optional equipment often includes additional mist eliminators to reduce droplet loading or aftertreatment by a carbon filter further to reduce hydrocarbon vapor. The heat exchangers cool the dryer exhaust to a range of 100–120°F. Hydrocarbon oils with concentrations above a dew point of these temperatures will condense. Condensate will be divided between liquid that drains from the heat exchangers and an aerosol fog that must be removed from the airstream. The droplet-removal system usually involves fiberglass filters or electrostatic precipitators, as the droplets are very small in size (0.1 μm or less). Because cooling is usually accomplished using ambient air, lower temperatures are achieved in winter months. Hydrocarbon removal efficiencies may vary from 5% to 95%, depending on the condensing temperature reached and the input hydrocarbon concentration. Condenser–droplet-removal emissions have a relatively constant output concentration and can be a mixture of uncontrolled oil droplets and some uncondensed oil vapor. The condensed liquid is a water and oil mixture, with most of it being water. It usually goes through an oil–water separator where the separated oil is recovered. Heat exchangers, droplet-removal systems, and oil–water separators must be cleaned regularly because of the buildup of residues.

STACK TESTING

Stack testing is often required to determine the emissions entering and leaving a control device. The U.S. Environmental Protection Agency (EPA) Method 25 is not suitable for testing hydrocarbon emissions that consist of a mixture of aerosol droplets and vapor concentrations. It is very difficult to obtain consistent results from operations that are as intermittent as heatset web offset and are further complicated by the installation of multiple press dryers into a single emission-control device. If aerosols are known to be present, the appropriate test is EPA Method 5. For complex emission sources, a continuous readout hydrocarbon monitor is more appropriate. Hydrogen flame ionization detectors provide sufficient sensitivity and heated sampling lines promote sampling accuracy. Frequently, dryer exhaust concentrations of hydrocarbons can be less than 500 ppm carbon. If 90% control is mandated, it is difficult to demonstrate compliance using EPA Method 25, which has not been found to produce accurate or reproducible results below 50 ppm carbon. The alternative methods described above should be considered in these situations.

Bibliography

Air Pollution Control in the Heatset Web Offset Printing Industry, Report 8403, Tech Systems, W. R. Grace Co., DePere, 1989.

D. Crous and R. J. Schneider, *Web Offset Press Operating,* 3rd ed., Graphic Arts Technical Foundation, Pittsburgh, 1989.

The Lithographers Manual, 8th ed., R. Blair, Ed.; Graphic Arts Technical Foundation, Pittsburgh, 1988.

FLEXOGRAPHY

Warren J. Weaver

Flexography is a method of direct rotary printing using raised-image printing plates and fluid inks. The inks are applied to the plate with a metering cylinder called an anilox roll. From the plate, the ink is transferred directly to the substrate. Flexography is similar to letterpress in that both use a raised-image plate. It is similar to rotogravure by virtue of the ink's being transferred directly to the substrate and of the fact that it is an inherently keyless printing process (no ink keys are required to adjust the ink film). It least resembles offset lithography as offset is an indirect printing process and utilizes relatively viscous oil-based inks.

PROCESS DESCRIPTION

The flexographic printing process of today continues to advance and grow at an impressive rate. With its origin rooted in rubber stamping, flexography, when first adapted to a rotary printing system, was called aniline printing. It was originally used for transferring aniline dyes to fabric. Original printing plates were cast or engraved rubber. As the process advanced, rubber plates were formed by pressing the pliable rubber—either natural or synthetic—into a matrix to form a negative version of the image to be printed. This technology has been developed to the point that high-quality process printing is routinely achieved.

More recently, photopolymer technology has been applied to flexography. Organic prepolymers, sensitive to ultraviolet light, are used in the manufacture of the flexo printing plate. To form the plate, prepolymer is coated onto a backing material, usually steel or polyester. When exposed to ultraviolet light, the prepolymer hardens and becomes a tough, elastomeric printing surface. To form the image on the plate, a photographic negative is placed between the plate and the light source and the plate is exposed. Then the unexposed prepolymer is washed away and the plate is cured (postexposed) and mounted on the press for printing.

Photopolymer plate technology, along with precise ink metering systems, has universally transformed flexography from a relatively low-quality linework process to one that is now capable of printing high-quality images and mixing type, linework, tints, reverses, and halftones in process color, spot color, and/or black and white, all in the same image area. Halftone resolutions of 150 lines per inch and finer are possible.

Originally using inks that were exclusively organic solvent based, flexography now uses many varieties of inks, including a number of varieties that are both waterborne and water reducible.

Publication Flexography

It was the combination of process capabilities, environmental friendliness, and simplicity of operation that appealed to the publication printing industry. With their high affinity for cellulose fiber, waterborne inks transfer quite well to newsprint. Utilizing advanced materials and equipment, flexography can be used to print many publications, including newspapers, flyers, direct mailers, newspaper inserts, phone directories, catalogs, and Sunday comics.

Development of publication flexography began in the late 1970s. By the late 1980s, flexography had become a viable printing process for newspapers and a number of commercial printing applications. As the 1990s began, there were more than 1000 printing couples (stations) installed in newspapers and commercial printing plants worldwide.

Packaging and Specialty Flexography

Prior to 1980, flexography was used exclusively for packaging and specialty printing. Today, these applications still constitute more than 90% of all flexography used. A wide range of products are produced within these broad categories, including labels, preprinted linerboard, corrugated board, Christmas wrap, multiwalled bags, bread bags, cartons, cans, tissues, paper towels, and molded containers. Specific combinations of substrate, ink, press, anilox, and operating parameters are necessary to meet the large variety of job requirements in these flexographic applications.

Flexographic Printing Parameters

In both publication and packaging flexography, printing parameters vary considerably from application to application, including the following.

- Press speeds of a few feet per minute to more than 2200 fpm.
- Plates from 0.019 to 0.250 inch thick.
- Substrates from kraft paper to board to cellophane to plastic film to newsprint to filled and coated paper.
- Web widths from 2 inches to more than 100 inches.
- Inks that are hydrocarbon based, alcohol based, or waterborne; pigment based or dye based; that vary from high viscosity to low, from expensive to relatively inexpensive, and from simple to formulate to complex.
- Anilox cylinders with fine engravings, coarse engravings, or no engravings; large cell volumes or small cell volumes; random cell placement or very precise cell placement; and electroplated surfaces or plasma-coated surfaces.
- Ink applicator techniques that vary from a fountain roller (two-roll system) to a single-doctor-bladed system to a hybrid system (two rolls with doctor blade) to a double-doctor-bladed system.
- Print repeat lengths that vary from a few inches to more than 100 inches.
- Ink drying needs that vary from none to extensive (utilizing many technologies).

Flexographic Presses

Flexographic press designs fall into one of the following categories, depending on the application:

- Common or central impression
- In-line
- Stack
- Newspaper unit
- Dedicated four-, five-, or six-color unit.

The common or central impression (CI) design uses one impression cylinder common to all printing stations (called couples in publication printing). The number of stations

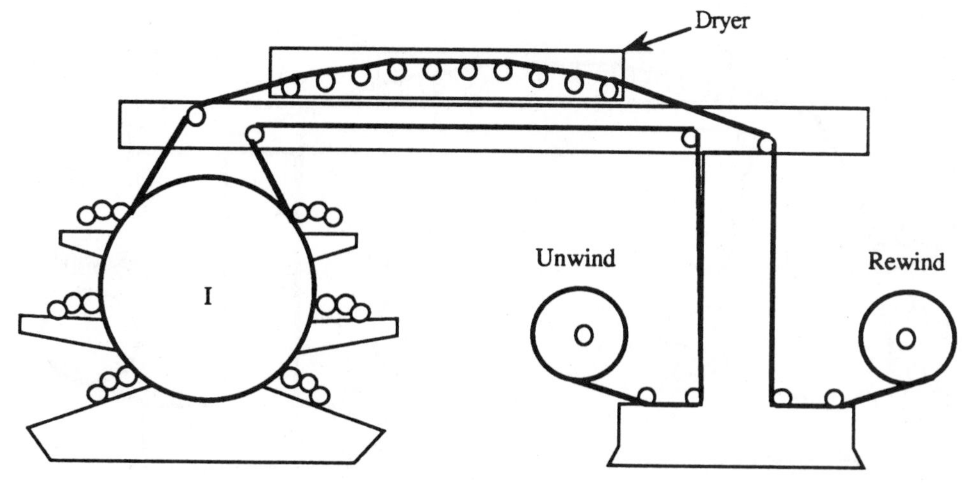

FIGURE 1. Packaging Flexographic Central Impression Press

using the same impression cylinder vary from two through eight. The advantages of a common impression design include compactness, excellent registration of multi-colored images and excellent control of the substrate. Figure 1 shows a typical CI design for a packaging application. Figure 6 shows a CI design (upper couple) used in publication printing.

With the in-line design, the printing stations are separate, discrete units mounted in-line to one another (Figure 2) and are driven by a common driveshaft or by independent, electronically controlled drives in the newer designs. With this design, any number of stations can be supplied and the printing width and repeat length can be selected from a wide range of possibilities. This arrangement has the advantages of complete flexibility as to the number of colors used to print a given job and flexibility to reverse the rotational direction of individual printing stations to select one- or two-sided printing with as many colors as are needed for each (two-sided printing is rarely needed for packaging printing).

In a stack press design (Figure 3), individual color stations are stacked one over another (usually on one or both sides of a main press frame). There are two primary advantages of a stack type-press. First, it is usually possible to reverse the web to allow both sides to be printed during one pass through the press. Second, the printing stations are usually quite accessible, which facilitates easy changeover from one job to another and easy wash-up.

A newspaper printing unit (Figure 4) consists of two dedicated printing couples (stations) located back-to-back in a common pair of frames. This permits the printing of black and white images on both sides of the web using a dedicated lead. Multiple units are arranged into a pressline to print the many pages required of a large newspaper (Figure 5). Color decks (often called half-decks), each consisting of one dedicated printing couple, are positioned above those unit positions where the publisher wants to offer (single) spot color capability. Occasionally, double decks, stacked units, or four-, five-, or six-color units are provided for multiple spot color, spot color on both sides of the web, or process color (four over one) printing on section fronts or center spreads (Figure 6 shows a five-couple process color unit).

Commercial publication flexographic presses utilize dedicated four-color units (Figures 7 and 8). Two units are combined into one press for printing four colors on both sides of the web. The advantages of this design include wide web capability, high operating speed, and compact design. Dryers, usually infrared, are included to ensure adequate drying of the waterborne ink after each side of the web is printed.

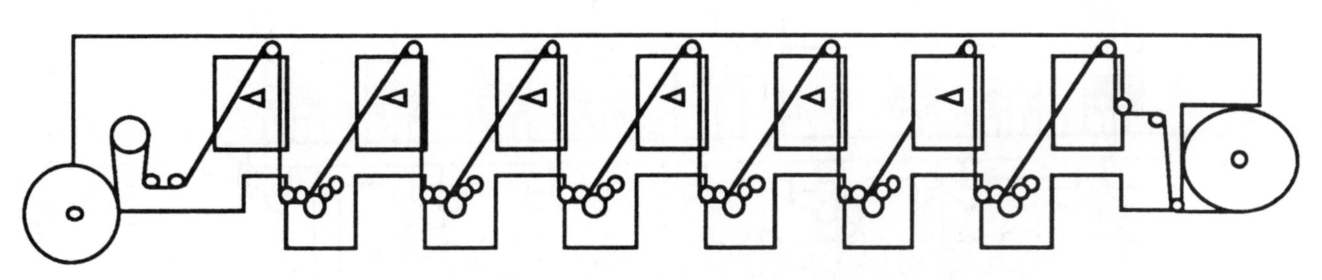

FIGURE 2. Packaging Flexographic-In-Line Press

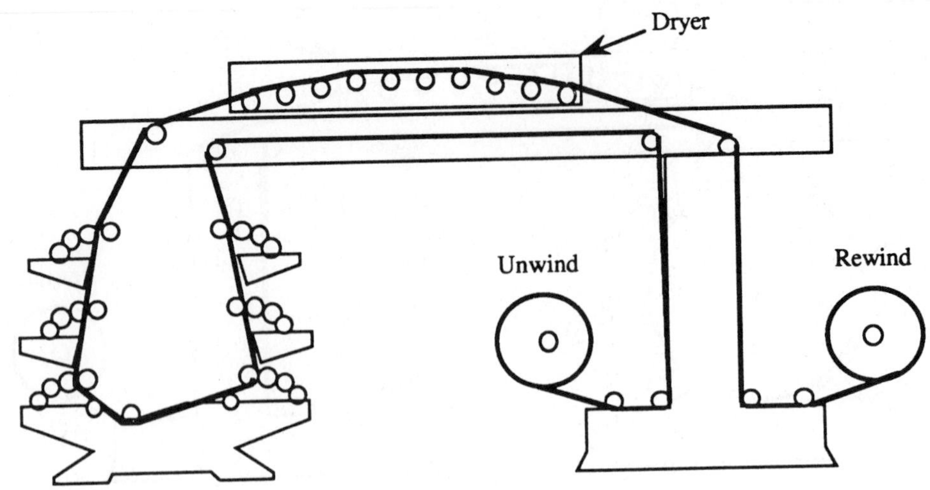

FIGURE 3. Packaging Flexographic Stack Press

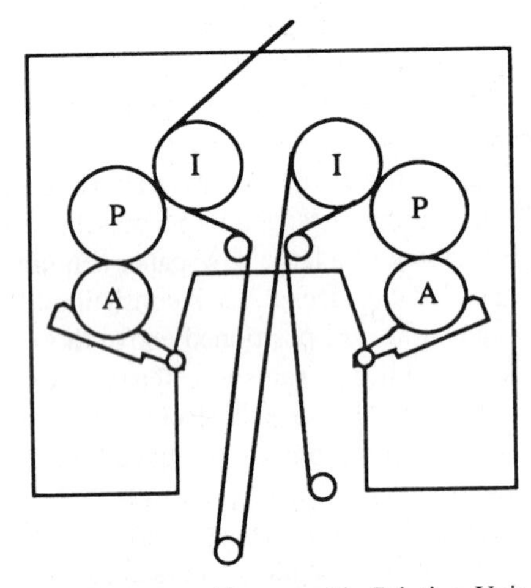

FIGURE 4. Newspaper Flexographic Printing Unit

Inking Systems

There are several inking systems employed in flexographic printing. On some presses, a fountain roller rotating in an ink bath transfers the ink to an anilox cylinder, which, in turn, transfers it to the printing plates and web (see Figure 9). The ink film thickness is controlled by the dynamics of the nip between the rubber-covered fountain roller and the anilox cylinder. In some presses, the fountain roller rotates at a slightly different speed than the anilox (may be operator adjustable) to control ink film thickness.

A second system uses a single reverse-angle doctor blade to scrape or doctor the excess ink off of the anilox surface (see Figure 10). With this system, the anilox rotates in the open ink fountain directly in contact with the ink. A doctor blade is located on the entrance side of the anilox-plate

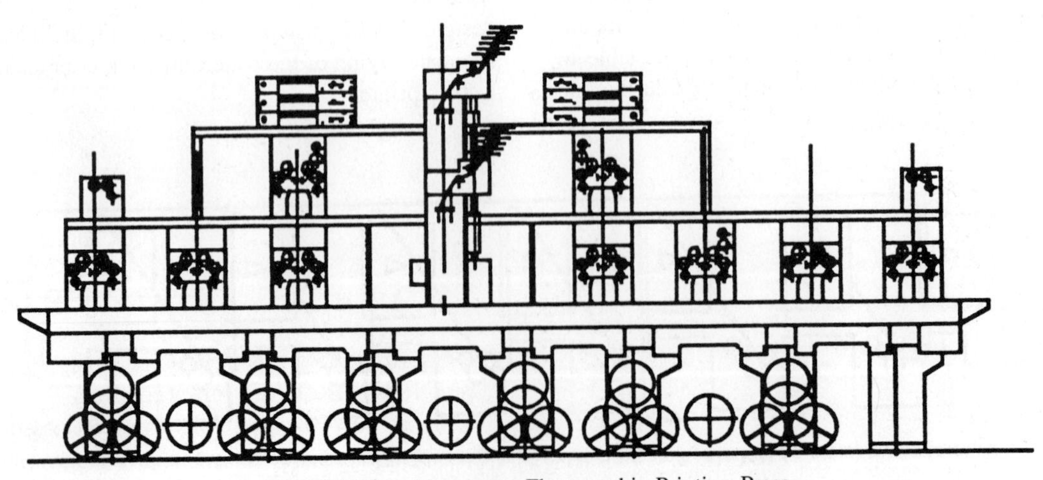

FIGURE 5. Newspaper Flexographic Printing Press

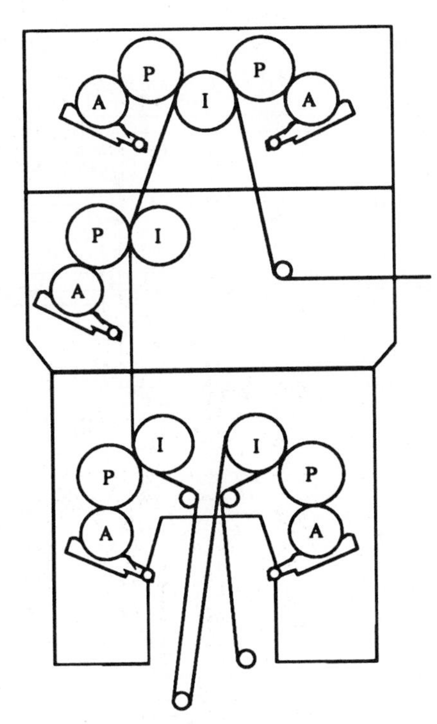

FIGURE 6. Newspaper Five-Couple Flexographic Unit Position

cylinder nip to scrape off the excess ink. This system generally has superior ink-thickness control and is used where print quality is more critical.

The third and most recently introduced system utilizes an anilox cylinder, an enclosed ink chamber, and two doctor blades (see Figure 11). Ink is pumped into the chamber and doctored (wiped) by one of the blades. Ink containment and easy reversibility of the printing station (couple) are two of this system's advantages. The double-doctor-bladed ink chamber is often used where both high speed and high-quality printing are requirements. Since the late 1970s, enclosed inking systems have found wide use in packaging printing. They are used exclusively in publication printing.

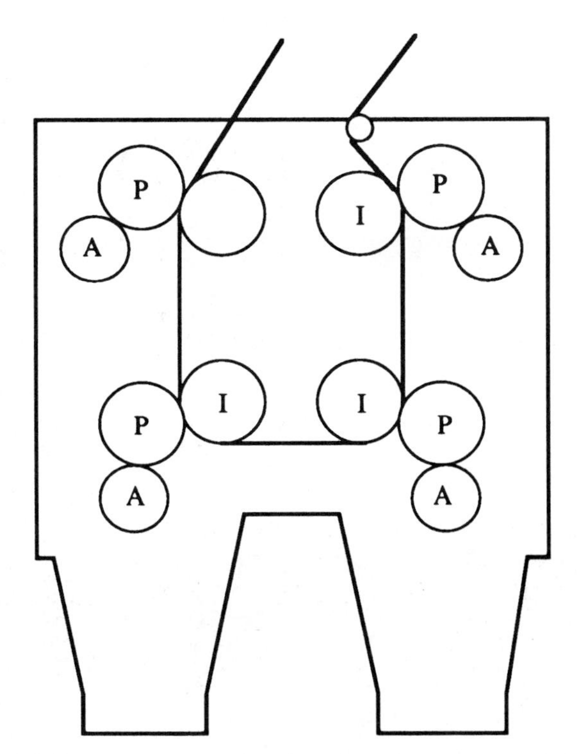

FIGURE 7. Commercial Publication Flexographic Printing Unit

Other Press Components

In addition to printing units, all presses have a number of other components. These components vary depending on the application. They include an unwind or reel—sometimes called a reel/tension/paster (RTP)—that provides a source location for the substrate; a rewind or folder that takes the web away after printing (and in the case of the folder, converts it into a partially or completely finished product); a press drive; and a control system. As with all other types of rotary presses, web tension control devices are utilized on both the infeed and outfeed sides of the flexo unit(s). Flexo presses utilize other systems, including slitters, dryers, chill rollers, manual and/or automatic register controls, wash-up systems, and die cutters, depending on the job requirements.

Dryers are an important part of most packaging flexo presses and some publication presses. They are often necessary to cause the ink to dry fast enough to make the process economically viable. Dryer types are frequently infrared or hot air; other types are rarely applied.

AIR EMISSIONS CHARACTERIZATION

Emissions from flexographic printing arise from organic solvent that may be present in the ink and from solvent that may be used for press cleaning. Emissions are limited in all cases to volatile organic compounds (VOCs). The types of VOCs emitted from a specific operation depend on the specific job requirements being met by the printer. Because of these widely varying job requirements, ink formulators have developed a wide variety of inks to meet them. Consequently, widely diverse resin systems are utilized in the vehicles of flexographic inks. Each resin system requires specific solvent characteristics to maintain solubility and to meet the drying requirements of the application.

Solvents commonly used in flexographic printing include ethanol, isopropanol, *N*-propanol, hexane, toluene, isopropyl acetate, *N*-propyl acetate, glycols, glycol ethers, and water. Waterborne inks were first applied in the 1950s. Their use began to expand in the late 1970s with the development of better inks and in response to environmental pressures. High-solids ink technology was attempted beginning in the late 1970s in response to these pressures, but these efforts were largely unsuccessful.

Organic solvent-based inks are consumed in greater quantities than are waterborne inks. Though waterborne ink technology has come a long way, it remains unsuited to

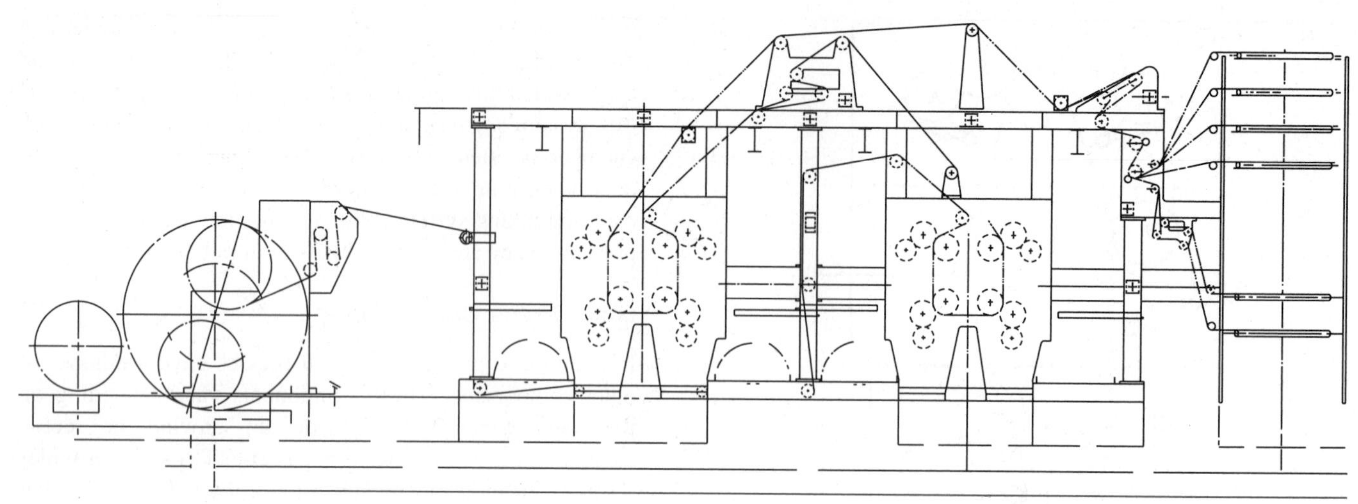

FIGURE 8. Commercial Publication Flexographic Press

many flexo applications. Waterborne inks work best where substrates are hydrophilic and porous. The more hydrophobic the substrate and the higher the press speeds, the less suitable are waterborne inks. Film and foil are two examples of substrates to which waterborne inks are less adaptable. Waterborne ink formulations have been successful in printing some films, but cannot be universally applied.

Newsprint is an excellent substrate for waterborne inks. Thus the products where newsprint is the substrate of choice are the ones with the highest application of water flexo. These products include newspapers, Sunday comics, newspaper inserts, and telephone directories.

Waterborne inks also contain VOCs. These are required to optimize the solubility of the resins in water. The VOCs used in waterborne inks are one or more of the following: amines, alcohols, glycols, and glycol ethers (the latter two may or may not be VOCs depending on the specific compounds and test methods used).

In publication flexo, amines and glycols are the only compounds that meet or could meet the definition of VOC. Alcohols are specifically *not* used. Amines and glycols are utilized in quantities that never exceed 5% of the ink formulation (prior to dilution with water).

In packaging flexo, VOC content may be higher than 5%, particularly if alcohols are used as cosolvents. In such cases, the VOC content is limited so that the ink is exempt from emission controls either by virtue of the specific VOC content or because it meets the definition of a high-solids ink. (A high-solids waterborne ink is one that contains 60% solids and 40% solvent after the water portion has been extracted under Environmental Protection Agency testing guidelines.) In use at the press, the allowable VOCs in the diluted ink may be no more than 25% of the solvent portion.

In packaging flexo, organic solvents are used to clean systems that use solvent-based inks. Where water-based inks are successfully used, nonorganic cleaning techniques are applicable.

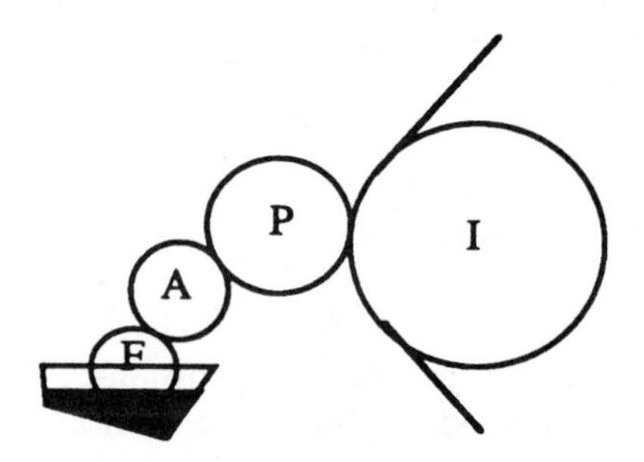

FIGURE 9. Fountain Roller Style of Printing Couple

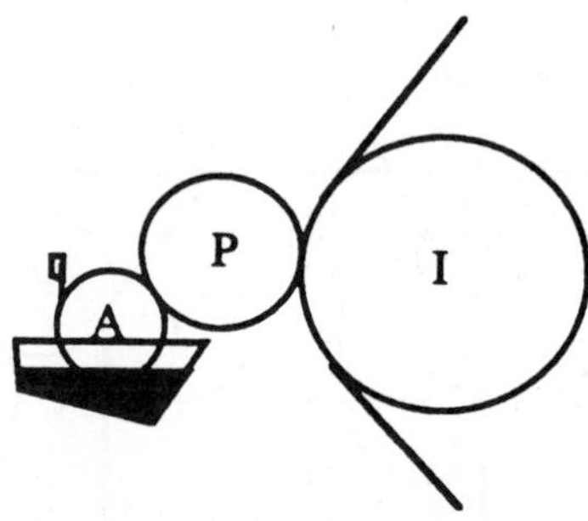

FIGURE 10. Single-Doctor-Bladed Style of Printing Couple

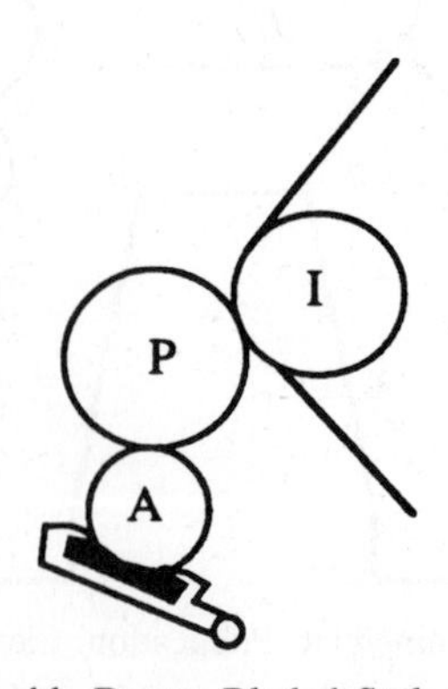

FIGURE 11. Double-Doctor-Bladed Style of Printing Couple

In publication flexo, no organic-based cleaning agents are utilized. If tap water alone does not adequately clean, household cleaners such as 409, Windex, or Fantastic are used. Occasionally, industrial-strength alkaline cleaners such as Zep Spree are used. Insignificant levels of VOCs, if any, are emitted from these compounds.

AIR POLLUTION CONTROL MEASURES

In operations utilizing waterborne inks exclusively, no emission controls are provided nor are they necessary. This specifically includes all publication flexography locations. Locations where high-solids inks have been successfully adapted are also exempted. (As stated earlier, this has so far been limited to waterborne inks with VOC content above the action level.)

In flexo pressrooms utilizing organic-based inks, the applicability of emission controls is related to a number of factors. These include ink consumption rate, statutory emission control requirements, and the degree of success in applying high-solids ink technology (if attempted). The type of emission controls employed is based on other factors, including the type(s) of solvents used in the ink formulations, the need to change inks from press run to press run as job requirements change, the number of inks used in a given pressroom, the statutory requirements, and the relative cost effectiveness of competing alternatives.

Were there no technological stumbling blocks, solvent-recovery strategies using such technologies as adsorption, absorption, and condensation theoretically would be the preferred emission control strategy, as is the case in publication gravure. Since these difficulties do exist, solvent-recovery systems and the subsequent reuse of flexographic solvents have been applied to only a few flexographic printing operations. As many of the flexo solvents are alcohols and acetates, and because many packaging operations do, in fact, utilize multiple inks and solvents in a single pressroom, attempts to recover those solvents for reuse have been mostly or wholly unsuccessful.

The reason for this lack of success is that these solvents are miscible with water and form an azeotropic mixture that is nearly impossible to separate, even with expensive thin-film evaporation techniques. Furthermore, printability and adhesion can only be optimized with inks whose physical characteristics—and, therefore, chemical makeup—are held within a narrow range. Most attempts to reuse mixed solvents collected from a recovery system as diluents for the flexographic ink have met with failure and the others have had only limited success. Thus the used solvents are valuable only for their fuel value.

A second problem with solvent recovery is that many flexo inks contain some high-molecular-weight components. These compounds are sometimes difficult to desorb from the carbon bed. Bed efficiency would drop precipitously from such contamination.

For the above reasons, the add-on emission control strategy most frequently used in solvent-based flexographic printing operations is incineration. The two types of incinerator are catalytic and thermal (see Figures 12 and 13). The catalytic category consists of fluidized-bed catalytic, fixed-bed catalytic, and precious metal catalytic using either metal-on-metal or metal-on-ceramic monolith. The thermal incinerator category can be divided into direct thermal, thermal regenerative, and thermal recuperative.

The thermal incineration process relies on high temperature (1400–1500°F), turbulence, and adequate retention time (>0.5 second) to facilitate the combustion of organic vapors into carbon dioxide and water. Each of the three

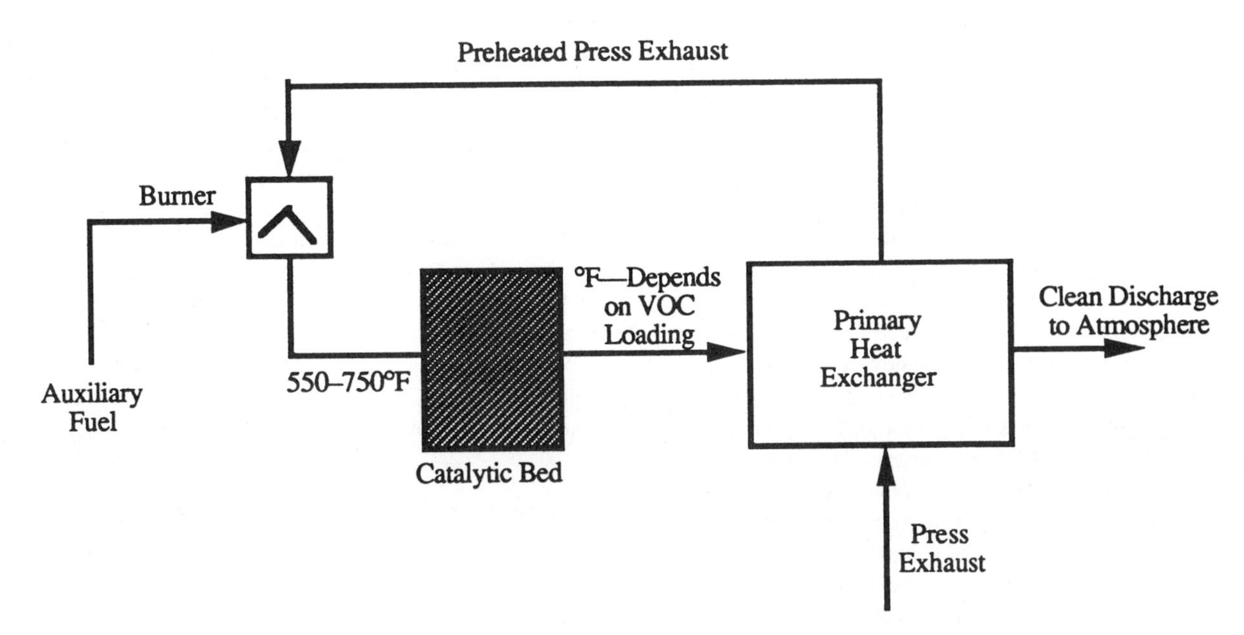

FIGURE 12. Catalytic Incinerator with Heat Exchanger

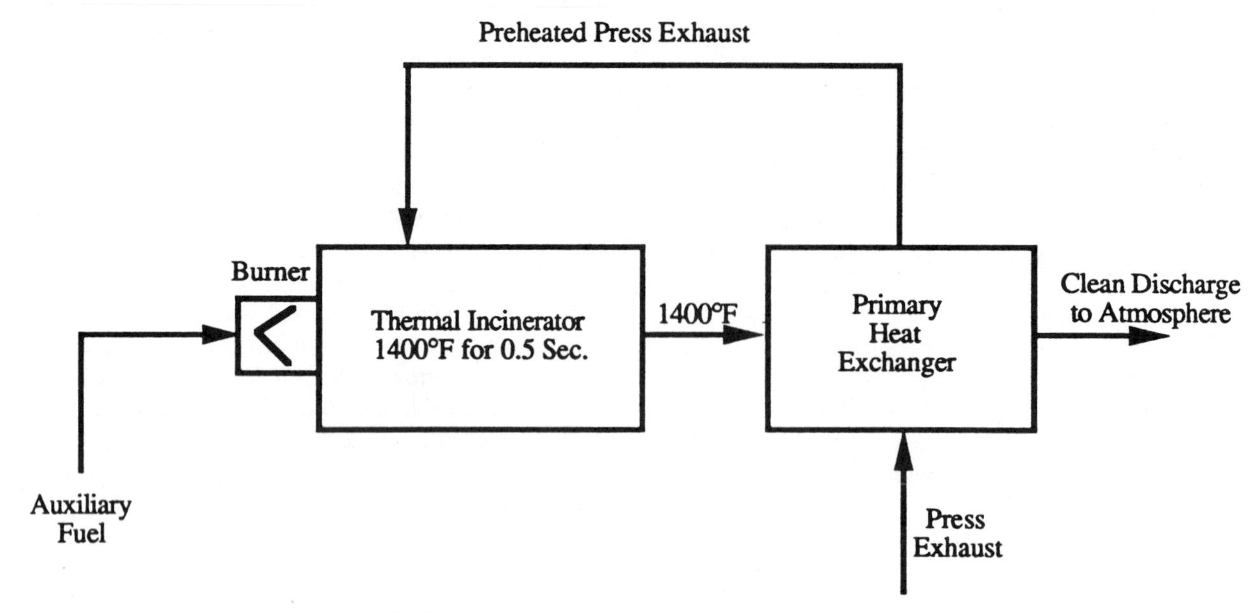

FIGURE 13. Thermal Incinerator with Heat Exchanger

types of thermal incinerators utilizes a different approach to the destruction of the organic vapors. All require a separate fuel, often natural gas, to maintain continuous combustion as the solvent concentration in the waste stream varies depending on press utilization, press speeds, and ink coverage on the substrate.

The catalytic incineration process relies on a catalyst operating in the activation range of 500°F to 750°F to convert the organic vapors to carbon dioxide and water. Uniform flow presentation, uniformity of inlet temperature across the bed, and retention time (referred to as space gas hourly velocity [SGHV]) directly affect not only the destruction efficiency, but the life of the catalyst itself. The specific mechanical and catalyst technology utilized in each type of catalytic incinerator is different.

No system has an inherent advantage that makes it universally superior to another. The decision by a flexographer to purchase one system rather than another (or one type versus another) would depend more on such issues as system design, purchase price, ease of installation, available space, installation costs, operating costs, and source and quantity of emissions to be controlled.

In very general terms, catalytic systems are well suited for processes that have a minimum potential for poisoning or masking the catalyst. Well-designed catalytic units can achieve 90–98% conversion efficiency. Conversion efficiency of a catalytic system depends on properly applying the catalyst to the manufacturing system to be controlled and protecting the catalyst from damage. This is especially critical when striving for high-end conversion efficiencies.

Thermal systems are better suited for operations where poisoning and masking would be a problem with a catalytic system or where large process exhaust flows are involved. Thermal units normally achieve 95–98% conversion efficiency.

Thus, unlike the selection of a particular system within each category, choosing between categories (thermal or catalytic) involves analyzing all factors relevant to a specific printing operation. This initial decision is often dictated by the sum of those factors. Also, reduction in the press exhaust flow rate (by recirculation and/or cascading between stations) can have a major, beneficial effect in reducing both capital and operating costs.

Inherent in an efficient control strategy is the control of fugitive emissions. Proper control here involves well-designed enclosures for ink sumps, fountains, and tanks; efficient, properly enclosed dryers; negative pressroom air pressure; good operating practices, such as keeping doors and tank lids closed; efficient press cleaning using minimal amounts of solvent; and a preventative maintenance program for the control device and ancillary equipment. Periodic hose and damper inspection, keeping lids closed on ink containers, selecting presses that use enclosed ink chambers, and using enclosed cleaning tanks either connected to the control device or that utilize low-evaporation-rate cleaning solvents also minimize fugitive emissions. Improving capture efficiency should be a high priority in planning an installation, as control efficiency is improved and supplemental fuel consumption is often minimized. This is even more important where heat recovery is practiced and used to supplement process and room heating needs. System components such as powered pickups, floor sweeps, and efficient dryers improve capture efficiency.

Acknowledgments

The author would like to thank the following persons for contributions to this effort. Douglas C. Cook of Printpack, Inc., made a substantial contribution relative to packaging flexography including much of the emission source and emission control sections. Joel Shulman and Lloyd Hoeffner of the Flexographic Technical Association also made contributions to the packaging discussion. S. Edward Weary of the Flexible Packaging Association contributed his organization's publication *Incineration as an Add-on VOC Control Technology* as a reference document. Figures 12 and 13 were taken directly from this document.

SHEETFED OFFSET PRINTING

Jerry Bender, Nelson Ho, David Johnson, Frank Little, and William Schaeffer

Sheetfed offset lithography is the process used by most small- to medium-sized printers because of its versatility. Among the establishments that employ the process are quick printers, in-plant printing operations, commercial printers, paperboard converters, and metal decorators, and the products printed range in size from individual envelopes, newsletters, advertising brochures, books, magazines, and folding cartons to billboards. While paper is the dominant substrate, significant volumes of paperboard, tinplate, and plastic are printed with the process. Inasmuch as it uses inks formulated with drying oils, resins, and high-boiling-point, petroleum-based oils that form dry films under ambient conditions through oxidation and polymerization, the air emissions generated are normally very small.

PROCESS DESCRIPTION

The sheetfed lithographic printing press consists of a feeder, a printing unit, and a delivery unit. The feeder supplies sheeted paper, paperboard sheets, or other substrates to the printing unit. A typical printing unit usually consists of the following.

1. An inking system containing multiple metal and polymeric rollers to produce and deliver, at high speed, a very thin, uniform ink film to the printing plate.
2. A dampening system that delivers fountain solution to the printing plate by one of a number of different mechanical systems and from which the nonprinting areas of the printing plate are preferentially wet and maintained during printing runs.
3. A plate cylinder on which the printing plate with its printing images is mounted and from which only the inked images are transferred from the plate to the blanket cylinder.
4. A blanket cylinder on which an elastomeric rubber-covered, multiple-tiered lithographic blanket is mounted, and which receives the inked images and passes them onto the substrate.
5. An impression cylinder that controls the force imposed on the substrate in contact with the lithographic blanket so that the inked images are transferred properly from the blanket to the substrate.

Figure 1 illustrates one configuration of a typical lithographic printing unit. As in all other offset lithographic processes, printing is done by offsetting an inked image from the printing plate onto the blanket cylinder and then transferring the image from the blanket cylinder onto the substrate (paper sheet). Any number of printing units can be installed sequentially, with a six-unit printing press being the most common printing configuration used today. Depending on the product to be printed and the type of ink to be used, the delivery unit may contain various accessories, such as coating devices, radiation sources for ink drying and curing (infrared, ultraviolet [UV], and electron beam), antisetoff agent applicators, and a substrate-handling system for delivering and stacking the printed paper sheets. Figure 2 depicts a modern six-color sheetfed offset lithographic printing press.

AIR EMISSIONS CHARACTERIZATION

A variety of chemicals, including cleaning solvents, fountain solution additives, printing inks, and coating materials, can be used in the sheetfed offset printing process, a number of which may be sources of air emissions. Due to the nature of the printing process, most of the air emissions are fugitive, evaporative losses and are very small in quantity. The emissions can be categorized as described in the following.

Prepress Operations

Most sheetfed lithographers have photographic operations, almost exclusively for processing images on photographic film to develop the desired images for lithographic reproduction. Trace amounts of air emissions may be generated from the use of isopropyl alcohol and other solvents in film developing, color proofing, and some printing-plate processing.

Dampening System

The key to successful sheetfed lithographic printing is the proper maintenance of image boundaries by the application of aqueous fountain solution to the printing plate. This technique is known as dampening and it is done either by applying the fountain solution directly onto the printing plate through a series of rollers or by applying it to the surface of the inkform rollers.

The fountain solution is usually composed of 80–95% water, 5–20% isopropyl alcohol or alcohol substitutes (usually ranges from 2% to 5%), and many other additives in smaller quantities, such as gum and phosphoric acid to maintain the pH of the fountain solution. Air emissions are

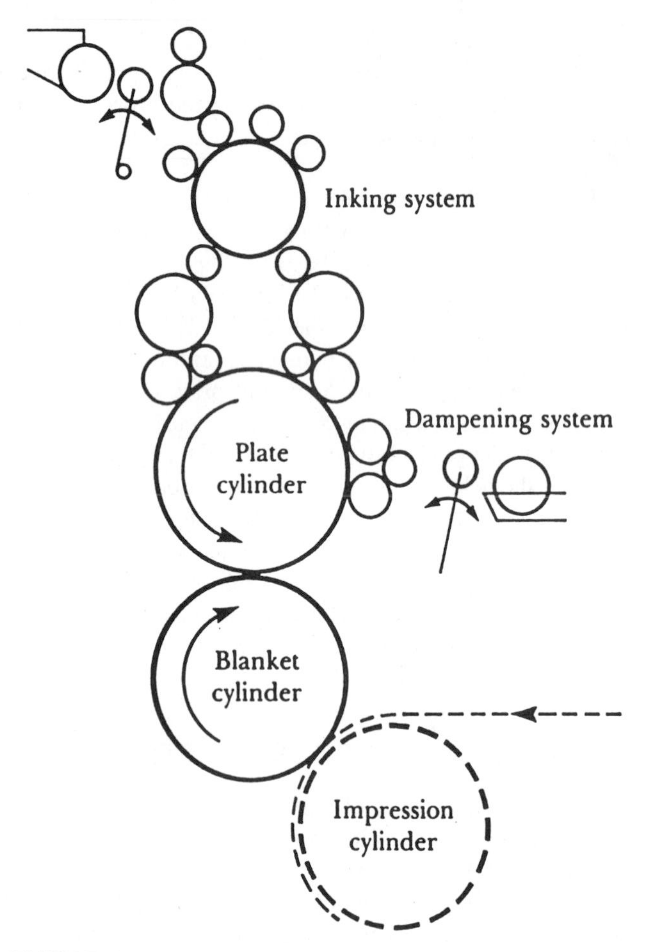

FIGURE 1. Configuration of a Typical Lithographic Printing Unit (From Reference 1, p 14:9)

usually generated during the dampening process from the evaporation of the isopropyl alcohol or its substitutes from the fountains, the dampening rollers, and the printing plates.

A small amount of the fountain solution is transferred to the substrate during printing, where subsequently a minimal amount of alcohol carried over by the fountain solution may evaporate and cause emission. Some minor emission also may occur from the fountain solution transferred to the substrate and emulsified in the ink on the inking roller trains. The amounts of these emissions are very small and of a fugitive type. Some presses use dampening systems that function with fountain solutions that do not require the addition of isopropyl alcohol or alcohol substitutes. These presses, frequently serving the one- or two-color end of the market, do not generate any emissions from their dampening systems.

Inking System

Most sheetfed offset inks used today for general commercial printing are formulated to set quickly, usually within minutes after they are applied. This enables the printer to print the other side of the sheet within a few hours instead of waiting overnight. The main components of a typical sheetfed offset printing ink normally include drying oil (about 30%), such as linseed oil, tung oil, or soy oil; resins (about 30%); pigments (about 15%); petroleum distillates (about 15%); waxes; and other additives.

The ink-setting mechanism involves the penetration of the distillate–resin phase of the ink into the substrate and

FIGURE 2. Miller TP104PLUS Six-Color Press (Roland and Miller Sheetfed Press Systems, Pittsburgh)

deposition of the film-forming drying oil–resin phase with the pigments onto the substrate surface for subsequent oxidation and polymerization.

Minor emissions may result from the evaporation of petroleum distillates from the ink fountain, from the distribution of ink on the ink roller in the inking system, and from the ink in and on the substrates after the ink has been applied. The amount of emission is very small because of the low volatility of the petroleum distillate and the low operating temperature encountered during printing. Inks also may be formulated without petroleum distillates and those organic emissions will not occur.

Ink Roller and Blanket Cylinder Washing

Blankets are cleaned either manually or automatically—during press stops, between runs, and between jobs—with solvent formulations (principally petroleum based) or solvent and water emulsions and mixtures. Ink rollers are normally manually cleaned with cleaning rags or paper wipes soaked with solvent between runs or ink color changes. The emissions are of a fugitive type and often generated while these solvents or solvent mixtures are being applied, and the amounts are relatively small. In some applications, blankets and rollers must be cleaned frequently using lower-boiling-point volatile solvents so that less solvent residue remains on the rollers and blankets. In these situations, considerable quantities of the solvent may become fugitive air emissions. The used cleaning rags and wipes are usually placed in metal containers with lids that curtail the amount of solvent vapor emitted into the press room.

In-Line Coating Systems

Printers frequently utilize in-line coating finishes following printing, either as the last unit on the printing press or as auxiliary equipment. The coating provides a glossy or dull finish for surface protection of the final product. There are three main types of coating compositions: varnish, aqueous coating, and UV-cured coating. Press varnish is essentially a sheetfed offset ink without pigments and may generate very small amounts of ink oil emission. Water-based coatings may also emit a minimal amount of emissions owing to the use of a small amount of alcohol in their formulation. A UV-cured coating normally does not generate any emissions, except for a small amount of ozone from the use of the UV lamps.

Antisetoff Agents (Spray Powders)

Atomizers are used to deposit antisetoff agents of controlled particle sizes (usually 15 to 60 μm) on the surface of the printed sheets to prevent contact of printed sheets in the pile. Antisetoff agents normally consist of chemically modified cornstarch. Excess antisetoff agents usually are collected in the delivery unit by a vacuum device and transported to a cyclone and/or bag filters. Printers often use antisetoff agents only on printing jobs with high ink coverage.

AIR POLLUTION CONTROL MEASURES

Since most air emissions from sheetfed offset printing are fugitive, add-on devices to control the emissions are not practical. The only feasible way to minimize emissions is to reduce evaporative losses during the printing operations. The methods routinely used by sheetfed printers include the following.

1. Refrigeration of the fountain solution. Refrigeration devices to control fountain solution temperature in the 55–60°F range are normally used by sheetfed offset printers. By lowering the temperature of the fountain solution, the evaporation rate of isopropyl alcohol in the dampening systems is considerably reduced. But reducing the temperature of the fountain solutions to below the 55–60°F range is not recommended because random water condensation may form on the printing plate and the condensation may affect print quality.
2. Isopropyl alcohol substitutes in the dampening solution. Dampening agents other than isopropyl alcohol have been employed to reduce emissions. Glycol ethers have been found to be the most acceptable substitutes. These substitutes are considerably less volatile and can be used in much lower concentration (usually about 1–3%). This combination of factors can reduce emissions effectively. The use of alcohol substitutes does require much better control over press operating parameters, as well as a well-balanced fountain solution. Many printers still need to use fountain solutions with isopropyl alcohol for special printing jobs. Nevertheless, substitutes have been gaining acceptance within the printing industry.
3. Use of organic solvents with less volatile organic compound (VOC) content. A wide variety of degreasing, cleaning, and washing organic solvents are used in sheetfed offset printing operation, with most of the organic solvents used to clean press rollers and blankets. The amount of VOC emissions that these cleaning solvents contribute is second to the amount contributed by the isopropyl alcohol in the dampening solution. A large number of printers have successfully tried several different types of cleaning solvents available on the market with less VOC content and have achieved the same cleaning effectiveness. This substitution has helped many printers in several states to maintain compliance with stringent state VOC emission regulations.

References

1. *The Lithographers Manual,* 8th ed., Ray Blair, Ed.; Graphic Arts Technical Foundation, Pittsburgh, 1988.

Bibliography

A. Glassman, *Printing Fundamentals,* TAPPI Press, Atlanta, 1985.

V. Strauss, *The Printing Industry,* Printing Industry of America, Washington, DC, 1967.

GRAVURE PRINTING

Jerry Bender and Robert Oppenheimer

The gravure printing industry consists of three branches, namely, publication gravure, packaging gravure, and product gravure. Publication gravure plants print magazines, catalogs, and free-standing advertising material. Presses in the United States can accommodate paper webs up to 110 inches wide and can run at speeds up to 3000 fpm.

Packaging gravure is used to print cartons, paper wraps and labels, plastic film, and aluminum foil. Some presses operate in-line with laminating or extrusion equipment, producing sophisticated structures of paper, film, and foil with barrier properties for water vapor and oxygen to preserve the freshness of the contents of the package. Press widths range from 12 to 64 inches and press speeds typically range from 300 to 1000 fpm and higher.

Product gravure is used to print gift-wrap paper, wall coverings, floor coverings, wood grains for laminates, some postage stamps, and other specialty products. Press widths and press speeds in product gravure printing vary considerably, with floor-covering presses up to 192 inches in width.

PROCESS DESCRIPTION

Almost all gravure is web fed, that is, the printing substrates, such as paper, paperboard, plastic film, aluminum foil, or laminates of these materials, are supplied in roll form rather than in the form of sheets. During printing, the substrate rolls are unwound in a reel stand. The web may be preconditioned to increase its dimensional stability before entering the printing section of the press.

For each application of primer coatings, printing inks, or top lacquers, a separate printing unit (sometimes called printing station) is used, into which a gravure cylinder has been installed prior to the press run (Figure 1).

In rotogravure printing, the image areas (the areas from which inks or coatings are transferred) consist of small, recessed cells engraved into the surface of the gravure cylinder. The material that is engraved is typically an electroplated layer of copper. After engraving, the copper is protected from wear by a very thin electroplated layer of chromium.

As the cylinder rotates around a horizontal axis, its lower portion is flooded with liquid ink. The rotation carries the ink-covered surface past a doctoring station where a thin doctor blade removes the ink from the nonprinting surface of the cylinder, leaving ink only in the recessed cells. The rotation then carries the inked and doctored surface of the cylinder to the printing nip, where ink transfer onto the substrate is accomplished by pressing the substrate web against the cylinder by means of a rubber-covered impression roll.

Each printing station is followed by a high-volume air dryer, before the next ink or coating is applied. The ink dries by evaporation of its liquid phase at temperatures that typically range from 80°F to 150°F for publication gravure and up to 280°F for packaging gravure (Figure 1).

Gravure inks have to be highly fluid with a viscosity in the 7- to 30-centipoise range to allow for proper flooding of the cylinder by ink, for doctoring, and for transfer of the ink from the cells to the substrate that is printed. The low ink viscosity and fast drying are traditionally achieved by the use of low-boiling-point organic solvents in the ink. The ink in the press fountains can contain as much as 75% solvent by weight.

To maintain ink viscosity during the press run, further solvent is usually added to compensate for solvent that evaporates from the inking system.

In publication gravure, the printing stations are followed by a folder that delivers either cut and folded signatures for bindery use or products that may be trimmed in-line with the press. In packaging and product gravure, the printed web is either rewound, sheeted, or die-cut into carton blanks.

AIR EMISSIONS CHARACTERIZATION

The solvents used in gravure printing inks and coatings are defined by the U.S. Environmental Protection Agency (EPA) as volatile organic compounds (VOCs). Most of the solvents are evaporated in the press dryers, although some solvents (up to 7%) may be retained in the web. Some solvent evaporates from exposed portions of the cylinder, the inking system, and the web, either before or after it has passed through the dryers. Measurements have shown that most of these fugitive vapors are drawn into the dryers, provided there is sufficient dryer exhaust flow, the dryer inlet slots are within about 4 feet of the printing area proper, and the fountains are enclosed. Some solvents used for press cleanup may also evaporate and escape as fugitive emissions.

In addition, VOC emissions may result from breathing and working losses from bulk ink and solvent tanks. These emissions are small and normally uncontrolled, although, in some cases, they may be vented to the pollution control device.

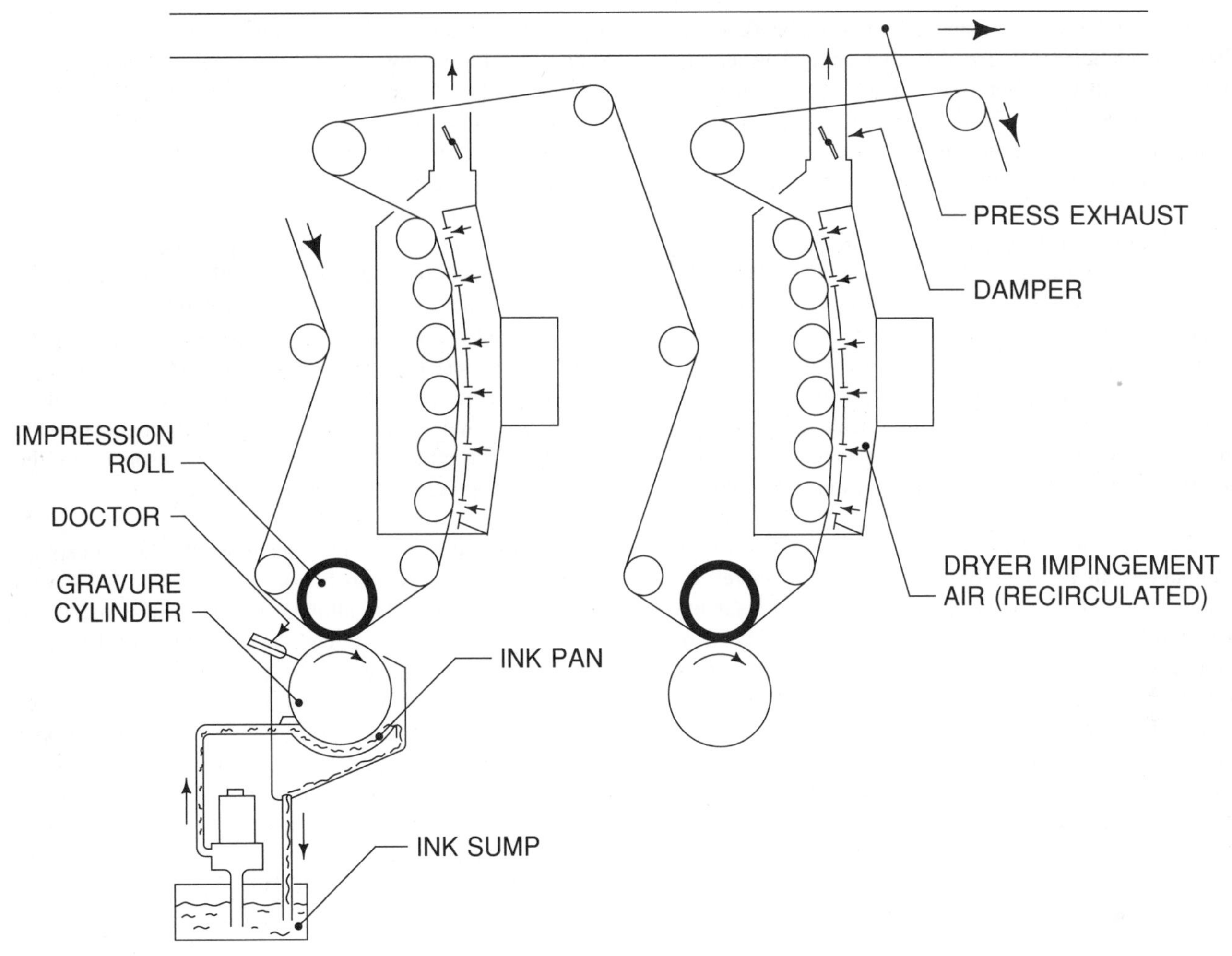

FIGURE 1. Section of a Gravure Press

Ancillary sources of emissions connected with gravure printing include the boilers used to produce steam for steam-heated dryers or, in plants with solvent-recovery systems, to strip the solvent from the carbon beds. The control of boiler emissions is covered elsewhere in this manual.

In plants that operate chrome-plating tanks, the chromic acid mist generated is normally controlled by a demister that, in some cases, is preceded by a wet scrubber. Where paper trimmings and other paper waste are pneumatically transported for collection, the paper is mechanically separated from the airstream using a cyclone. In cases where high loadings of paper and clay dust are present in the airstream, the dust may be removed from the cyclone exhaust be using a baghouse fabric filter. No VOC emissions result from these operations.

AIR POLLUTION CONTROL MEASURES

The following options to control solvent vapor emissions for compliance with regulations arising from the Clean Air Act are recognized:

- Solvent recovery
- Solvent vapor incineration
- Use of water-based inks and coatings
- Use of high-solids inks and coatings

Solvent recovery and incineration require pollution control systems, while the other control options employ low-VOC-content materials to achieve compliance. The high solids compliance option could not be implemented in gravure printing.

Capture, Control Device, and Overall Control Efficiencies

For solvent recovery and incineration, overall emission control efficiency is the product of the capture efficiency and the control device efficiency. In plants with solvent recovery, overall efficiency is typically determined as the ratio of solvent recovered to solvent used.

The amount of solvent used is the sum of the solvent contained in the ink and varnishes that are received from the ink maker plus the solvent added by the printer to reduce the ink viscosity to the level required for printing. The solvent content of the ink can be obtained from the ink makers. Alternatives to determine the solvent level of ink are

Methods 24 and 24A in the appendix to 40 CFR (Code of Federal Regulations) 60, for inks that contain water and inks that contain no water, respectively. As the methods specify temperatures of 110°C and 120°C to evaporate the volatile phase of the inks, these methods can indicate higher values of VOC content than are likely to evaporate in practical situations.

The control device efficiency of a solvent-recovery or incineration system is given by one minus the ratio of the VOC content of the exhaust airstream of the device over the content at the device inlet. Control device efficiencies are typically very high, and hence the overall efficiency of a control system is normally limited by the capture efficiency.

Control devices operate most efficiently at reasonably high vapor concentration levels. All publication and the newer packaging and product gravure presses use two dryer air systems, namely, a recirculating impingement air system and a dryer exhaust system (Figure 1). This cuts down on the amount of air that has to be exhausted and, therefore, increases the exhaust concentration. However, solvent vapor concentration levels are limited by safety considerations, since gravure solvents are flammable. Peak concentrations in dryers and ducts are typically limited to 25% of LFL (lower flammable limit) or 40% to 50% of LFL when continuous monitors are employed to provide automatic warning and equipment shutdown. Dryer exhaust dampers must be set to keep concentrations below these limits at all times.

Solvent use is roughly proportional to press speed. When a press is running slowly, the vapor concentrations in the exhaust tend to be low. In packaging and product gravure, there are wide differences in the ink coverage among different jobs, a factor that leads to lower average solvent concentration levels in the captured air. This, in turn, can lead to reduced control device efficiency.

Capture efficiency is the ratio of solvent in the airstream going to the control device to the solvent used. Methods to measure solvent concentration and airflow volume are also specified in the appendixes to 40 CFR 60. One hundred percent capture efficiency is assumed for plants designed with fully enclosed pressrooms at specified levels of negative air pressure and with all exhaust flow through a control device. Fire and explosion-hazard protection dictates the provisions for emergency venting to atmosphere in the event of equipment malfunction or breakdown.

Low capture efficiencies may result from fugitive losses as well as solvent retention in the substrate. When capture efficiency is measured on presses with gas-heated dryers drawing pressroom air, allowances should be made for solvents that are burned. Depending on the age of the plant, plant layout, and the plant air-handling system design, fugitive losses can be minimized by selective enclosures of parts of the presses, high dryer exhaust rates, the proper location of makeup air inlets, and similar measures. Efficient capture of fumes is important not only in maximizing overall efficiency, but also in ensuring that pressroom solvent fume levels are kept within Occupational Safety and Health Administration exposure levels.

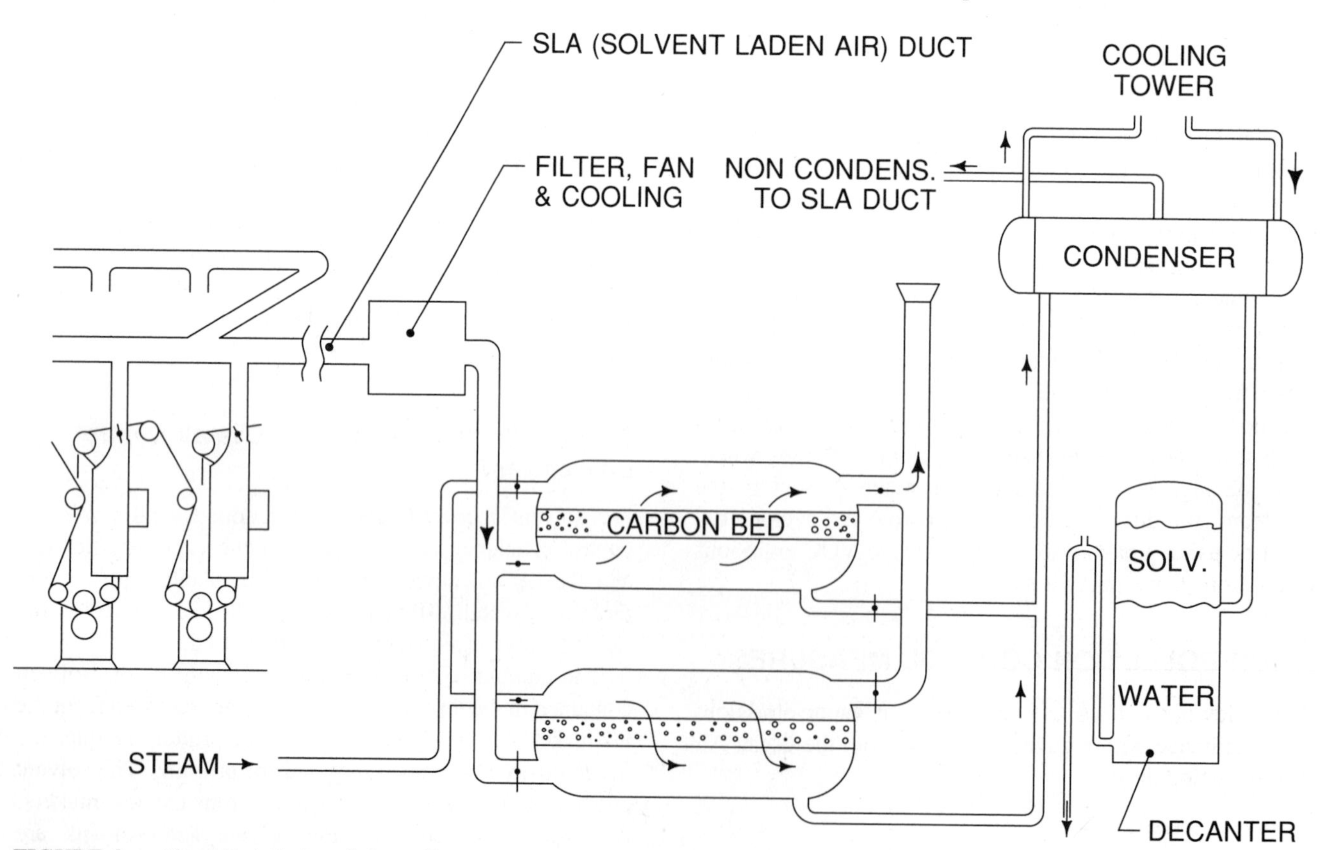

FIGURE 2. Fixed-Bed Carbon Solvent Recovery

Emission Control Options

Solvent Recovery

Solvent recovery is used in all publication gravure printing plants and in some packaging gravure printing plants, where the solvent can be reused without too much further processing.

In the operation of a solvent-recovery system, the exhaust air from the press dryers and, where practical, from other exhausts near the production presses is collected and passed through large beds of activated carbon (Figure 2). The carbon adsorbs all (>95%) but trace amounts of the solvent vapor. When a bed is saturated, the exhaust stream is diverted to a nonsaturated bed and the saturated bed is regenerated by steaming. The resulting steam–solvent-vapor mixture is condensed. The solvent and the condensed water separate by gravity in a decanter vessel. In newer systems, a brief cooling cycle follows steaming. This avoids steam plumes and solvent loss that would occur when solvent-laden air (SLA) is passed over a hot, wet carbon bed and then exhausted.

Solvent-recovery systems are equipped with control instrumentation not shown in Figure 2. For example, a sensitive vapor concentration monitor on the exhaust of the adsorbers will initiate the steaming cycle when a bed is saturated and solvent "breakthrough" occurs. This system is usually backed up by a system that schedules steaming on a time basis.

For efficient operation, it is essential that the air that enters the solvent recovery be properly conditioned. The temperature of the SLA that enters the adsorbers should be below 110°F. The air must be filtered to avoid particulates that can clog the carbon beds. Compounds with high boiling points will stay on the carbon permanently and reduce its working capacity. The use of chlorinated solvents anywhere in the building can lead to corrosion and should be avoided.

Publication gravure solvents are composed of either toluene or mixtures of toluene and similar-boiling-point aliphatic hydrocarbons. These solvents are miscible with water only in trace amounts and can be separated from water in the steam–solvent condensate by gravity in the decanter. The recovered solvents are used within the plant for ink dilution and viscosity control, and some solvent is returned to the ink supplier.

Solvents used for packaging and product gravure are usually more water miscible. These solvents are usually mixtures that contain acetates, ketones, and alcohols. Such recovered solvents may require water removal or redistillation before reuse. In any case, solvent traces may have to be removed from the decant water before it can be reused or discarded.

The ratio of pounds of steam to pounds of recovered solvent, which ranges from 1:1 to 5:1, is a major criterion in operating costs. However, given the relative costs of solvent and steam, plants are usually operated to recover as much solvent as possible.

Solvent Vapor Incineration

Solvent vapor incineration oxidizes the solvent vapors into harmless carbon dioxide and water vapor. By and large,

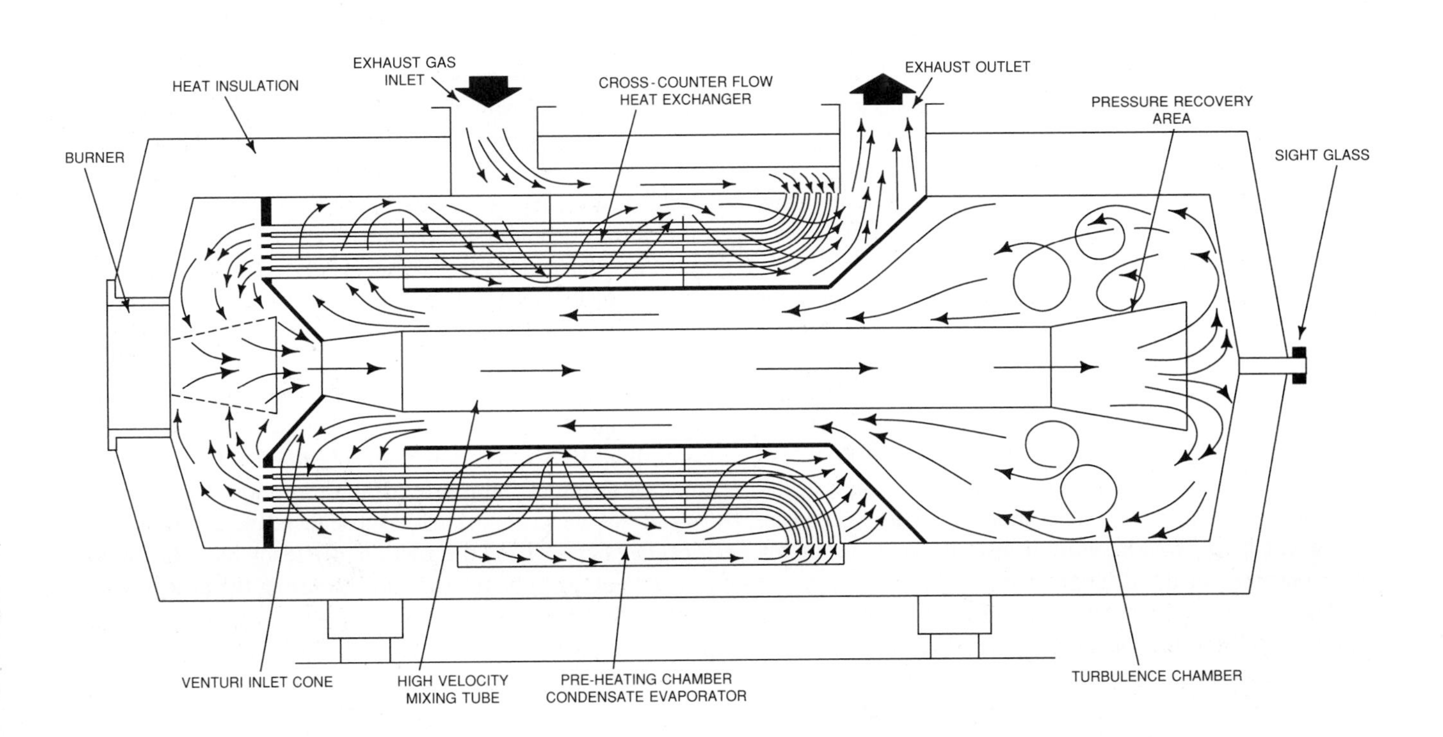

FIGURE 3. Recuperation-Type Thermal Afterburner

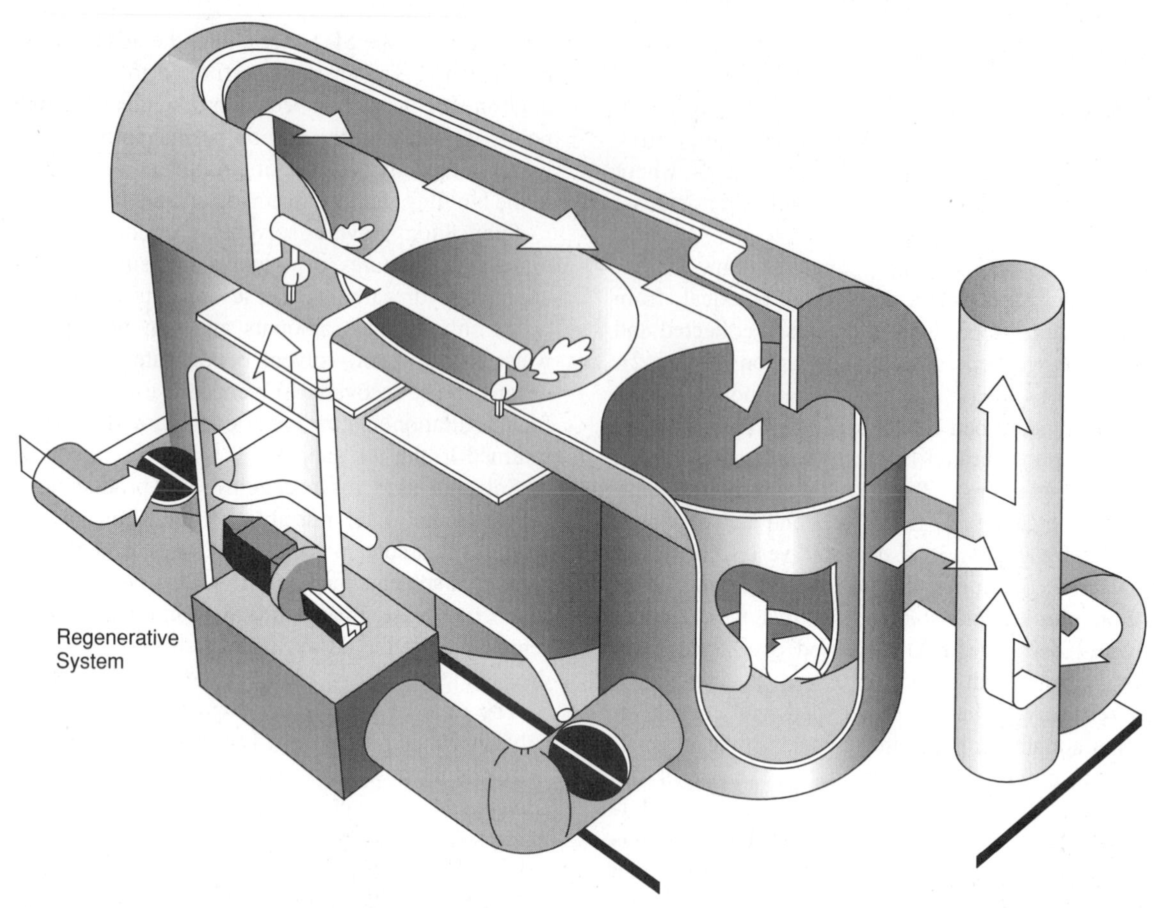

FIGURE 4. Regenerative Afterburner

incinerators are employed where a variety of solvents and water-miscible solvents are used that would require redistillation after solvent recovery. This applies mainly to packaging and product gravure plants where complex end-use requirements mandate the use of special ink resins and solvents such as acetates, alcohols, and ketones, as well as aliphatic and aromatic hydrocarbons. In all solvent incineration systems, high overall VOC destruction requirements will increase energy consumption and may shorten control equipment life.

Thermal oxidation with heat recovery is appropriate for large, steady exhaust streams. Operating at 1325°F to 1600°F, VOC destruction of over 99% of the captured VOCs can be achieved. To reduce energy requirements, heat recovery is usually employed. Recuperative systems incorporate primary and sometimes secondary heat exchangers. Primary heat exchangers are used to preheat the SLA before it enters the combustion chamber. Secondary heat exchangers provide return heat to the building or for other purposes in the plant (Figure 3).

Regenerative systems use a number of chambers filled with a heat-absorbing medium (usually ceramic) to preheat the airstream before it enters the combustion chamber. When needed, a burner further elevates the temperature in the combustion chamber where oxidation of the VOCs occurs. The airstream is redirected every few minutes so that air going to the combustion chamber is passed through chambers that have previously been heated by the high-temperature exhaust air from the combustion chamber (Figure 4). Regenerative systems are more frequently used to control exhaust streams above 15,000 scfm.

Catalytic incinerators operate in the 500°–750°F temperature range, with device efficiencies that range from 90% to 98 (Figure 5). They are usually employed for exhaust streams below 15,000 scfm and where press operations are frequently interrupted by job changes. Due to their lower operating temperatures, catalytic units are better able to withstand the thermal stress placed on the equipment by cycling between standby and active status. Care has to be taken to avoid materials that can poison or mask the catalyst.

Concentrator devices have been introduced to raise afterburner inlet VOC concentrations by a factor of about ten by adsorbing VOCs from low-concentration airstreams by carbon adsorption and subsequent desorption by relatively small volumes of hot air for subsequent incineration. This increases energy and destruction efficiency by reducing the volume of air to be treated and increasing the energy value of the SLA stream.

Water-Based Inks

Water-based ink and coating systems have been under development in the gravure industry for many years. Certain packaging and product gravure operations are successfully printing with water-based systems routinely. Although most current water-based systems contain some VOCs, they

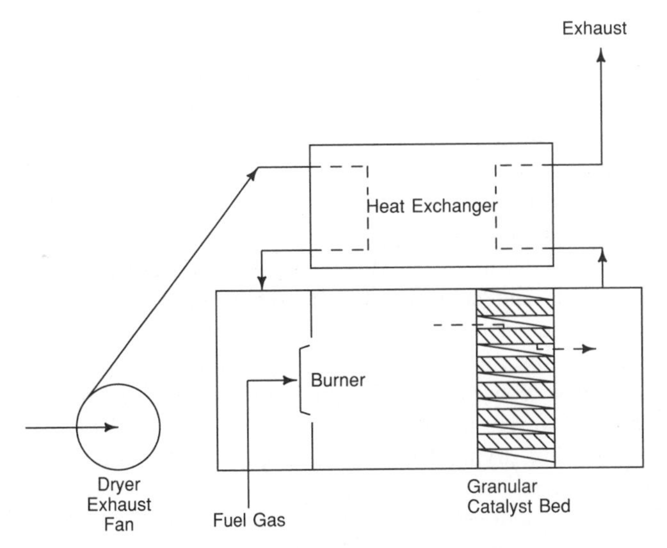

FIGURE 5. Catalytic Afterburner with Heat Recovery

appear to be the alternative to control solvent vapor emissions without the need for capital and energy-intensive control devices. Water-based systems also reduce safety hazards and the complications arising from the need to transport, store, and handle large amounts of flammable solvents. The inherently higher surface tension and the slower drying rate of water are major obstacles. Extensive work on ink formulation, cylinder engraving, press operations, and dryer design has achieved success in some areas.

Control by high-efficiency solvent-recovery systems is employed throughout publication gravure. Water-based ink systems are seldom used in this segment of the industry because of rough printing, paper distortion, and press-speed limitations. However, some limited success has been achieved with uncoated paper on slow presses.

In packaging gravure, some paperboard and paper products are successfully printed at speeds up to 1000 fpm. Water-based usage is expected to increase in this area, although problems still exist at higher press speeds. Water-based systems are used to print a number of aluminum foil products. On plastic film, there has been little success.

Product gravure involves a great variety of printing substrates and end-use requirements, so that no generalizations can be made. Water-based systems have been very successful on medium-weight paper, but on lightweight paper, problems with paper distortion and curl are encountered. Water-based systems are used successfully on some nonabsorbent substrates at low press speeds.

Bibliography

A. Glassman, *Printing Fundamentals,* TAPPI Press, Atlanta, 1985.

V. Strauss, *The Printing Industry,* Printing Industry of America, Washington, DC, 1967

Technical Guide for the Gravure Industry, 3rd ed., Gravure Technical Association, New York, 1975.

SCREEN PRINTING

Marcia Y. Kinter

Although generally classified as "Commercial Printing, Not Elsewhere Classified" by the Office of Management and Budget's Standard Industrial Classification Code System, the process known as screen printing has little in common with lithography, gravure, flexography, and letterpress. The flexibility of the screen printing process allows the controlled deposit of inks onto a substrate in a manner that provides the user with the needed legibility, color opacity, and durability. This method is known for its ability to impart relatively heavy deposits of ink onto practically any type of surface, in a controlled pattern, with few limitations on the size and shape of the object being printed.

Today, there are over 40,000 screen printing operations located throughout the United States. It has been estimated that in 1986, the screen printing industry posted gross sales of $13 billion. This printing process is an integral part of every industry that produces products that need imprinted instructions, identification, or color application. Over the past two decades, the rate of growth of the screen printing industry has exceeded that of any other type of printing process.

The average screen printing company has 15 employees, in both production and management. Nearly half of the screen printing facilities produce imprinted textile garments. This category includes clothing, caps, children's wear, and towels. Other major categories of screen-printed products include banners and billboards for outdoor advertising, point-of-purchase displays, posters (both single and multisheet), electronics such as circuit boards, containers of all types and sizes, and pressure-sensitive decals. End products produced by screen printing include in-store displays, decals, labels, emblems, banners, flags, pennants, binders, transit advertising, nameplates, wallpaper, and product-identification markings.

Unlike the other major printing processes, screen printing is able to utilize a wide variety of substrates. All substrates can be classified as either absorbent or nonabsorbent. Substrates for screen printing include all types of plastics (acrylic, epoxies, vinyl, topcoated and nontopcoated polyester, and polycarbonate); fabric (both natural and synthetic); metals (aluminum, brass, copper, lacquer-coated metals and steel); papers (uncoated, coated, corrugated coated fiberboard, poster, and cardboard); and other substrates, such as leather, masonite, wood, and electronic circuit boards.

This listing of substrates and end products graphically illustrates the differences between screen printing and all other major printing processes. It needs to be noted at this point that the majority of screen printers do not restrict their operations to printing on one substrate or to the production of one end product.

PROCESS DESCRIPTION

In the most simplistic terms, the screen printing process involves the flow of ink through a porous screen mesh to which a stencil has been added to define the image. The ink flows through the imaged screen by virtue of hydraulic pressure that is initiated by the action of a flexible rubber or synthetic blade known as a squeegee. The squeegee blade sweeps across the surface of the pretensioned, preimaged screen, pressing the ink through those areas of the screen not blocked by the stencil and onto the substrate in the pattern defined by the stencil image. The substrate is then either manually placed onto drying racks or mechanically/manually placed onto a conveyor transport system for conveyance into a drying unit.

Mesh

Many different material supplies are necessary for the completion of the screen printing process; however, mesh is a major player when discussing the emission of volatile organic compounds (VOCs). Mesh is available in the following materials: stainless steel, natural silk, monofilament nylon, monofilament polyester, multifilament polyester, metalized monofilament polyester, and calandered (one side flattened) monofilament polyester. All of these fabrics are available in differing numbers of threads per inch and thread diameters, which affect the fabric thickness, the size of the mesh open areas, and subsequently the volume of ink deposited.

The selection of the screen fabric is critical to the screen printing process. The diameter of the mesh thread determines the overall mesh thickness. This, in combination with the number of threads woven per inch, determines the amount of ink deposited onto the substrate. Ink can be deposited onto a substrate in a thickness of 0.04 mil to 10 mils.

Printing Equipment

Most commercial and industrial screen printing is produced on automated printing presses. These presses typically consist of an automatic squeegee, an indexing system to place the screen and printing base in register each time a substrate is loaded, and some type of substrate transport method. Within the industry, there are manually operated, semiautomatic, three-quarters automatic, and fully automatic presses. All types are considered power-assisted presses. The automation refers to the substrate transport method and the mechanical printing device mechanism.

Within the industry, flatbed, cylinder, and web presses are used, with the flatbed press probably the most common. The print bed of such presses is horizontal and parallel to the screen. The cylinder press prints the substrate on a cylinder as the screen, substrate, and cylinder move together.

Also found within the screen printing industry are web presses. On a web press, the substrate unwinds off a roll, moves through various print and dry stations, and finishes by passing through a die cutter or slitter or is rewound onto another spool. Web presses can be of either the flatbed or cylinder types.

The press is typically configured in-line with a conveyorized drying oven. This oven takes the freshly printed substrates and subjects them to the energy level necessary for quick and effective solidification of the printed image. Absorbent substrates are typically printed with all colors consecutively applied prior to the ink-drying step. However, nonabsorbent substrates, such as plastics, require ink drying between each individual color application.

Energy sources used to dry the wet printed ink include infrared (IR), convection, and ultraviolet (UV) light. Each of these methods must be suited to the chemical composition of the ink used. All drying/curing units include adjustable temperature/wattage controls, variable belt speeds, and exhaust systems vented to the outside to remove emissions from the workplace. These vents are viewed as air emission sources. Also, in some cases, the printed product is air dried on racks prior to final packaging.

The most common type of drying system utilized is the hot-air convection oven. The key to a convection oven is the recirculation of hot air. Within this type of drying oven, typically 80% of the air is recirculated to maintain operating efficiencies, while 20% is exhausted. Temperatures are absolutely controllable.

Infrared systems use inductive, as compared with convectional, energy to create heat within the material and effect the drying process. The IR energy meets the substrate and excites the molecules, which creates heat within the material. Heat is produced through electrically conductive alloys with 3- to 7-μm wavelengths being the most useful for the curing process. The IR spectrum can be shifted to suit the absorption characteristics of the ink and/or substrate.

The IR elements are contained in an insulated and heated section that also contains the controls for the elements, airflow, and exhaust devices. The heated section separates the heating elements from the environment and provides for a controllable airflow and exhaust for the system.

The accuracy of the conveyor belt speed is a major contributor to the accuracy of the entire curing process. The amount of energy absorbed by the material depends on the amount of time spent under the IR element. Air is circulated within the heating system; however, very little is exhausted to the outside. The air is swept over the substrate and recirculated within the system.

Ultraviolet curing units provide relatively energetic photons of 200–400 nm. Photons are emitted from mercury vapor lamps within the curing unit. The intensity of commonly used lamps is 200 w/in^2. The lamp is either housed in an extruded aluminum reflector with highly polished walls for maximum UV reflectance or its energy is focused onto the substrate by parabolic reflectors.

The UV energy provided by the lamps is absorbed by the photoinitiators present within the liquid formulation of the ink. Free radicals are produced that attack the double bonds of the resin molecules, increasing the molecular weight as the particles add to themselves and form a cross-linked chain. This chain produces a solid polymer film. This process takes place in a fraction of a second. The UV curing units are 10 to 12 times more energy efficient than the traditional thermal curing units.

Ink Systems

The basic chemistry of ink systems revolves around three major elements: pigments that are the colorants for impact, identification, and contrast; binders that hold the colorants together in a continuous film and to the substrate onto which the film is printed; and solvents that dissolve the resins so that they are put in a fluid state and, together with the pigments, make a pliable and screenable material.

Resins are the important element of any ink system. They are the solids of the ink and bond the pigment to the surface. These resins have to be dissolved by the solvents. Ink manufacturers develop products from a host of resin systems to meet end-product demands for substrate compatibility, opacity, flexibility, abrasion resistance, and durability. For example, to print on a vinyl substrate, an appropriately formulated ink with a vinyl resin system would be required to assure durability. The volatile content of the ink is dependent on the demands of the resin system.

A variety of ink systems are available for use by the screen printing industry. The selection of a given ink system for a particular job is determined in large part by the end use of the product and the substrate being printed.

Within the screen printing ink family, four different major types of ink systems currently are utilized: UV inks (UV curables), water-based inks, plastisols, and the solvent-based ink systems.

Ultraviolet inks consist of pigments, monomers, oligomers, additives, and modifiers. The photoinitiator reacting to the UV-light source produces free radicals that initiate a chain reaction, resulting in polymerization of the monomers and oligomers. Some UV inks are suitable for outdoor usage because they exhibit resistance to abrasion, solvents, chemicals, and other environmental menaces. They do have limitations in the area of flexibility and are not recommended for use on most corrugated surfaces.

Water-based ink systems use water as part or all of their solvent component. These ink systems all contain organic pigments, resins, and additives, such as flow promoters, retarders, and other performance enhancers. Water-based systems require the use of water-soluble resins and contain up to 65–70% solids. These inks can be used on a wide variety of substrates.

A distinction must be made between water-reducible and water-based ink systems. Not all water-reducible systems are water based. Water-reducible systems can contain very little water, as low as 10%, but as long as the system can be reduced or thinned with water, it is considered water reducible. Water-based systems have replaced the traditional solvent with water. This is a major difference between true water-based systems and a water-reducible system that needs to be recognized.

Plastisol inks are generally used in the textile market. This ink system consists of two basic ingredients: polyvinyl chloride (PVC) resins and the plasticizer. This ink system is considered to be 100% solid with little or no VOC emission. Because of their chemical makeup, plastisols fuse rather than dry or cure; they must be exposed to heat in order to fuse. The PVC resin is present as discrete particles dispersed in the plasticizer. Only when the appropriate fusing heat (320°F) is applied to the ink does this condition change.

Conventional solvent-based ink systems are still widely used within the screen printing industry. These ink systems consist of pigments, resins, solvents, and additives. They are dried by the evaporation of the solvent from the system, leaving behind a thin film of resins and pigments. These ink systems are used on virtually all substrates. The amount of volatiles emitted by these different ink systems depends on the resin system utilized and the amount of volatile additive required.

Inks are rarely used as supplied by the ink manufacturers. Because of variations in temperature and humidity, the screen printer will generally use ink additives. These additives perform various functions. Thinners hold the resin in the binding agent and reduce the viscosity of the inks. Retarding agents are used to slow the drying time of the ink, so that the ink will not block the mesh openings of the screen. These additives are similar to thinners in that they hold the solid elements of the ink system soluble.

Generally, solvents are used as ink additives. The three general categories of solvents used in the screen printing industry are aliphatic hydrocarbons, aromatic hydrocarbons, and oxygenated solvents that are miscible in water.

Ultraviolet and plastisol ink additives are not solvent based and are considered nonvolatile materials. As previously mentioned, water-based ink systems are reduced with a combination of water and solvent, depending on the substrate. Solvent-based ink systems are reduced with VOC materials.

Screen/Stencil Reclamation

Depending on the ink system utilized during the printing process, screen reclamation activities can be undertaken with organic solvents or water-based systems. During the screen reclamation process, the ink residue is removed after all excess ink is removed from the screen. Water-soluble ink degradents or the appropriate solvents are used for this purpose. The fabric/stencil must then be degreased. Any chemical or solvent residue left on the stencil will impede the effectiveness of the stencil remover. After the stencil remover has been applied, the emulsion begins to dissolve and is completely removed using a high-pressure washer.

After repeated uses, hardened ink and stencil material removed appear as a ghost image from previous stencil applications. Ghost/haze removers are used to remove this unwanted condition. After using the haze removers, the screen fabric must be degreased again, to prepare it for reuse.

AIR EMISSIONS CHARACTERIZATION

The major emission problem in the screen printing industry remains the emissions of VOCs. Very little information is available concerning the release of air toxic or particulate air emissions.

It has been estimated that 95% of the VOCs emitted in a screen printing facility are fugitive emissions. These emissions result from the ink and solvents used throughout the printing process—two elements that account for most air emission sources in a screen printing facility. Petroleum-based solvents are used in both the ink formulation and screen reclamation operations. The ink systems themselves may contain up to 60% solvent. The previous discussion on the ink systems in the screen printing industry makes it very clear that they differ dramatically from the inks used in other printing processes.

The current method utilized for the calculation of VOCs in inks is Reference Method 24, which has been accepted by the U.S. Environmental Protection Agency (EPA) for the analysis of most printing inks. However, because of the wide variety of ink systems used by the screen printing industry, this test method does not apply for several very important reasons.

First, the test method calculation requires the removal of the water from the product in order to determine the actual VOC content. The water-based ink systems developed for screen printing use have replaced VOCs with water. Although water-reducible systems are used, the industry also uses true water-based ink systems. In utilizing Reference Method 24 to determine the VOC content of these inks, the industry is penalized for technological improvements designed to reduce the amount of VOCs emitted.

Second, the test method calls for the baking of the product to determine the VOC content. Many of the inks in the screen printing industry, most notably the UV curables, are cured, not baked. This requirement seriously distorts the actual VOC content of these ink systems and again penalizes the industry for the utilization of low-VOC emitting systems.

For these reasons, it is advocated that a test method that evaluates the VOC content in grams per liter of applied solids be used by the screen printing industry. Furthermore, if Reference Method 24 is used for the UV curable ink systems, then it is recommended that a cure cycle be inserted before the bake cycle to simulate actual use.

AIR POLLUTION CONTROL MEASURES

Air pollution control equipment with a capture efficiency of 90% and an overall efficiency of 95% currently is not available for the screen printing industry. Both thermal and catalytic afterburners have been installed in several screen printing facilities; however, the average capture–control efficiency level for this equipment ranges between 70% and 80%. The air pollution control equipment either can be retrofitted to existing equipment or new drying units that include the control equipment can be purchased. The cost of an afterburner unit ranges from $12,000 to $2,000,000, depending on the configuration and the size of the facility and whether or not new drying units are purchased. Annual operating costs range as high as $500,000 per year.

The major air pollution control method currently applied by the screen printing industry involves the use of low-VOC-content products. Generally, the VOC content of inks is expressed in either grams per liter, as applied, or pounds per gallon, as applied. Because of the relatively small amounts of solvents used during the screen reclamation process, these VOC products are generally regulated by a vapor pressure figure, that is, 45 mm Hg at 20° C.

Several of the ink systems discussed previously do offer the screen printer the option of utilizing low-VOC-content inks. The UV curable inks have very low emissions of VOCs. True water-based ink systems also are viewed as viable low-VOC-content inks, as are plastisols. However, these low-VOC-content inks are not suitable for use on all substrates or for all end products. Considering the end use of the product, such as outdoor graphics, or the type of substrate, it may still be necessary to use an ink system with a relatively high solvent content.

Substrates for which there are no viable low-VOC inks on the market include cast styrenes, plastic films, paper used to manufacture water-slide decals, rubber, and low-weight, water-sensitive papers. There may be other problem substrates, but their representative volume in the screen printing industry is currently unknown.

There are still certain end uses, regardless of the substrate, that require an ink with a higher solvent content. These situations include ceramic inks, inks used to manufacture ceramic decals, conductive inks used for circuitry printing, and inks for substrates used in the production of durable signs, such as stop signs and other highway signs. Ink manufacturers are still in the process of verifying low-solvent-content inks for durability.

Bibliography

A. Kosloff, *Screen Printing Techniques,* Signs of the Times Publishing Co., Cincinnati, 1981.

B. Magee, *Screen Printing Primer,* Graphic Arts Technical Foundation, Pittsburgh, 1985.

A. Peyskens, *The Technical Fundamentals of Screen Making,* SAATI, Como, Italy, 1989.

Technical Guidebook of the Screen Printing Industry, Vols. I–VI, Screen Printing Association International, Fairfax, VA, 1989.

Manuals

Manual on Determination of Volatile Organic Compounds in Paints, Inks and Related Coatings, J. J. Brezenski, Ed.; American Society for Testing Materials, Philadelphia, 1989.

1986–1987 Industry Profile Study, Summary Report, Screen Printing Association International, Fairfax, VA, 1987.

Screen Printing Fact Book, J. Letuner, Ed.; Screen Printing Network, Denver, 1990–91, pp 33–72.

Articles

H. Bubley and F. Matthews, "A comparison of conveyorized drying systems," *Tech. Guidebook Screen Printing Ind.,* 4:Q1–Q8 (1978).

J. Clark, "Plastisol inks: construction and application," *Screen Printing Mag.,* 58–60 (August 1988).

S. Ducilli, "Making a splash (again): Waterbased textile inks stage a comeback," *Screen Printing Mag.,* 126–131, 133–145 (August 1989).

S. Ducilli, "UV braves the elements, newer formulations find acceptance outdoors," *Screen Printing Mag.,* 63–66, 153 (August 1988).

T. Frescka, "Infrared drying and equipment," *Screen Printing Mag.,* 60–69 (June 1987).

K. Gilleo, "Rheology and surface chemistry for screen printing," *Screen Printing Mag.,* 128–133 (February 1989).

P. R. Herman, "Inks as applied to close tolerance," *Tech. Guidebook Screen Printing Ind.,* 2:H1–H11 (1978).

A. Kosloff, "Ultraviolet curing or drying," *Tech. Guidebook Screen Printing Ind.,* R1–R4 (1978).

C. Moret, "Pigmented UV ink review—an incurable craving of curable color," *Screen Printing Mag.,* 133–135, 154 (August 1988).

C. Moret, "Waiting in the wings," *Screen Printing Mag.,* 124–125, 132, 142–143 (August 1989).

W. J. Ramler, "Ultraviolet curing: For now and for the future," *Screen Printing Mag.,* 58 (September 1989).

Personal Communications

J. Coburn, Screen Printing Association International Technical Services Department, Fairfax, VA, Personal communication, May 1990.

D. Hohl, Screen Printing Technical Foundation, Fairfax, VA, Personal communication, May 1990.

12
Chemical Process Industry

ACRYLONITRILE

Ronald D. Bell and Alan V. Mechtenberg

Acrylonitrile, $CH_2{=}CH{-}C{\equiv}N$, was initially prepared in 1893 by the dehydration of acrylamide or ethylene cyanohydrin using phosphorus pentoxide; however, it did not gain importance until shortly before World War II when it was discovered that it could be used as a copolymer to increase the resistance of synthetic rubbers to oil and solvents. While a number of synthesis methods have been used to produce acrylonitrile (ACN) in the laboratory, relatively few have proven to be of commercial interest. Some of the early production methods included the addition of hydrogen cyanide to acetylene using a cuprous chloride catalyst, a catalytic reaction of propylene with nitric oxide, and a reaction of ethylene oxide with hydrogen cyanide followed by catalytic dehydration of ethylene cyanohydrin.[1] Acrylonitrile has several synonyms and trade names, including propenenitrile, vinyl cyanide, cyanoethylene, Acrylon®, Carbacryl®, Fumigrain®, and Ventox®.

The commercial production of ACN has increased rapidly since the 1960s when technical advances involving the introduction of ammoxidation processes revolutionized the economics of its production. Reduced production costs stimulated widespread research into new applications, and the resulting market development spurred tremendous growth in production capacity worldwide. Production costs have declined steadily since the introduction of ammoxidation processes because of the continuous development and use of improved catalysts and the increasing size of production units. Table 1 lists the ACN production sites in the United States, with corresponding annual production capacities based on 1990 data. A comparison of the total 1990 U.S. ACN capacity of over 3 billion pounds with 1960 U.S. ACN production of less than 200 million pounds[2] provides an indication of the growth that has occurred during this period.

Currently, the major end use of ACN is in the production of acrylic fibers for clothing and textiles, which make up more than half of current ACN monomer demand. The ACN-containing plastics, such as acrylonitrile-butadiene-styrene (ABS) and styrene acrylonitrile (SAN), constitute another major consumer of ACN. Other uses include the application of ACN monomer in the production of nitrile rubbers and nitrile barrier resins, adiponitrile (an intermediate in the manufacture of nylon), and acrylamide.[2]

TABLE 1. U.S. Production Capacity of Acrylonitrile[a]

Manufacturer	Location	Annual Capacity (Millions of Pounds)
American Cyanamid Co.	Avondale, LA	350
BP Chemicals America, Inc.	Greenlake, TX	720
	Lima, OH	380
E.I. du Pont de Nemours and Co., Inc.	Beaumont, TX	375
Monsanto Chemical Co.	Alvin, TX	500
Sterling Chemicals, Inc.	Texas City, TX	700
Total capacity		3025

[a]1990 data from reference 3.

Air emissions from the prevailing processes for ACN monomer production primarily consist of propane, propylene, acetonitrile, acrylonitrile, and hydrogen cyanide. One previous study of a model ACN plant states that approximately 86% of all air emissions consist of propane and propylene.[4] The study also states that ACN production has been identified by preliminary studies as causing the highest emissions of volatile organic compounds (VOCs) of all processes in the synthetic organic chemicals industry. Acrylonitrile is defined in 29 CFR 1910.1200 (d) (4) as a carcinogen.

PROCESS DESCRIPTION

Nearly all of the ACN produced in the world today is produced using the Sohio process for ammoxidation of propylene and ammonia. The overall reaction takes place in the vapor phase in the presence of a catalyst according to the following equation.

$$CH_2{=}CH{-}CH_3 + NH_3 + \tfrac{3}{2}O_2 \rightarrow CH_2{=}CH{-}C{\equiv}N + 3H_2O$$

Reference 1 provides a detailed discussion and illustration of the reaction mechanisms and catalyst interaction involved.

Figure 1 shows a typical simplified process flow diagram for an uncontrolled Sohio process. Streams are designated as main product flow and other flow streams and are described in Table 2. The table separates vent/discharge streams that would result in air emissions from other product and flow streams.

The primary by-products of the process are hydrogen cyanide, acetonitrile, and carbon oxides. The recovery of these by-products depends on such factors as market conditions, plant location, and energy costs. Hydrogen cyanide and acetonitrile, although they carry a market value, are usually incinerated, indicating that the production of these by-products has little effect on the economics of producing ACN.

In the process represented in Figure 1, by-product hydrogen cyanide and acetonitrile are routed to an incinerator. As previously discussed, variations within the Sohio process may provide for purification, storage, and loading facilities for these recoverable by-products. Reference 1 provides further details regarding basic by-product recovery processes. Other variations of the Sohio process include the recovery of ammonium sulfate from the reactor effluent to allow for biological treatment of a wastewater stream and variations in catalysts and reactor conditions.[2]

In the standard Sohio process, air, ammonia, and propylene are introduced into a fluid-bed catalytic reactor operating at 5–30 psig and 750–950°F. Ammonia and air are

TABLE 2. Stream, Vent, and Discharge Descriptions for Figure 1

Stream/Vent/Discharge	Discharge
Stream	
1	Propylene feed
2	Ammonia feed
3	Process air
4	Reactor feed
5	Reactor product
6	Cooled reactor product
7	Sulfuric acid
8	Quenched reactor product
9	Stripping steam
10	Wastewater column volatiles
11	Water recycle
12	Crude acetonitrile
13	Crude acrylonitrile
14	Acetonitrile
15	Hydrogen cyanide
16	Light ends column bottoms
17	Product acrylonitrile
18	Heavy ends
Vent/Discharge	
A	Wastewater column bottoms
B	Absorber vent gas
C	Recovery column purge vent
D	Acetonitrile column bottoms
E	Acrylonitrile plant wastewater
F	Acetonitrile column purge vent
G	Light-ends column purge vent
H	Product column purge vent
I	Column purge waste gas
J	Fugitive emissions
K	Incinerator stack gas
L	Deep well/pond emissions
M	Storage tank emissions
N	Product transport loading facility vent

fed to the reactor in slight excess of stoichiometric proportions because excess ammonia drives the reaction closer to completion and air continually regenerates the catalyst. A significant feature of the process is the high conversion of reactants on a once-through basis with only a few seconds residence time. The heat generated from the exothermic reaction is recovered via a waste-heat–recovery boiler.

The reactor effluent is routed to a water quench tower, where sulfuric acid is introduced to neutralize any unconverted ammonia. The product stream then flows through a countercurrent water absorber-stripper to reject inert gases and recover reaction products. The operation yields a mixture of ACN, acetonitrile, and hydrogen cyanide, which is distilled to remove acetonitrile and water and then is sent to a fractionator to remove hydrogen cyanide. The final two steps involve the drying of the ACN stream and the final distillation to remove heavy ends. The fiber-grade ACN obtained from the process is 99+% pure.[5]

Several fluid-bed catalysts have been used since the inception of the Sohio ammoxidation process. Catalyst 49, which represents the fourth major level of improvement, is currently recommended in the process. Catalyst 49 replaced bismuth-molybdenum Catalyst 41, which was preceded by uranium Catalyst 21 and the older bismuth-molybdenum Catalyst A.

AIR EMISSIONS CHARACTERIZATION

Air emissions of ACN and other VOCs resulting from ACN production via the Sohio process may come from several sources, including the absorber vent, column purge vents, storage tanks, transport and loading facilities, and volatilization from deep wells and ponds. Fugitive emissions also result from valves, compressor seals, pumps, and flanges throughout the process. The ACN itself represents a relatively small fraction of the total VOCs emitted from these sources, with propane and propylene making up the majority of VOC emissions.[6]

Description of Sources

Figure 1 and Table 2 identify 14 waste streams and volatilization points that contribute to emissions from an uncontrolled ACN facility. Seven of these streams/points (vent/discharge B, I, J, K, L, M, and N in Figure 1) are considered the primary air pollutant emission points for the typical ACN plant. Table 3 lists these emission points and provides a checklist of uncontrolled pollutants that may be found in each of these streams. The checklist was compiled using a material balance for a representative ACN plant[7] that did not include estimates of nitrogen oxides (NO_x), carbon monoxide (CO), and carbon dioxide (CO_2) emissions from the by-product incinerator.

Emissions of ACN during start-up are substantially higher than during normal operation. During start-up, the reactor is heated to operating temperature before the reactants (propylene and ammonia) are introduced. Effluent from the reactor during start-up begins as oxygen-rich, then passes through the explosive range before reaching the fuel-rich zone that is maintained during normal plant operation. To prevent explosions in the lines to the absorber, the reactor effluent is vented to the atmosphere until the fuel-rich effluent mixture can be achieved. The ACN emissions resulting from this start-up procedure have been estimated to exceed 10,000 lb/h.[7]

The absorber vent gas (stream B) contains nitrogen and unconverted oxygen from the air fed to the reactor, propane and unconverted propylene from the propylene feed, product ACN, by-product hydrogen cyanide and acetonitrile, other organics not recovered from the absorber, and some water vapor.

The ACN content of the combined column purge vent gases (stream I) is relatively high, about 50 wt% of the total VOCs emitted from the recovery, acetonitrile, light ends, and product columns.[6] The rest of the vent gases consist of noncondensibles that are dissolved in the feed to the columns, the VOCs that are not condensed, and, for the col-

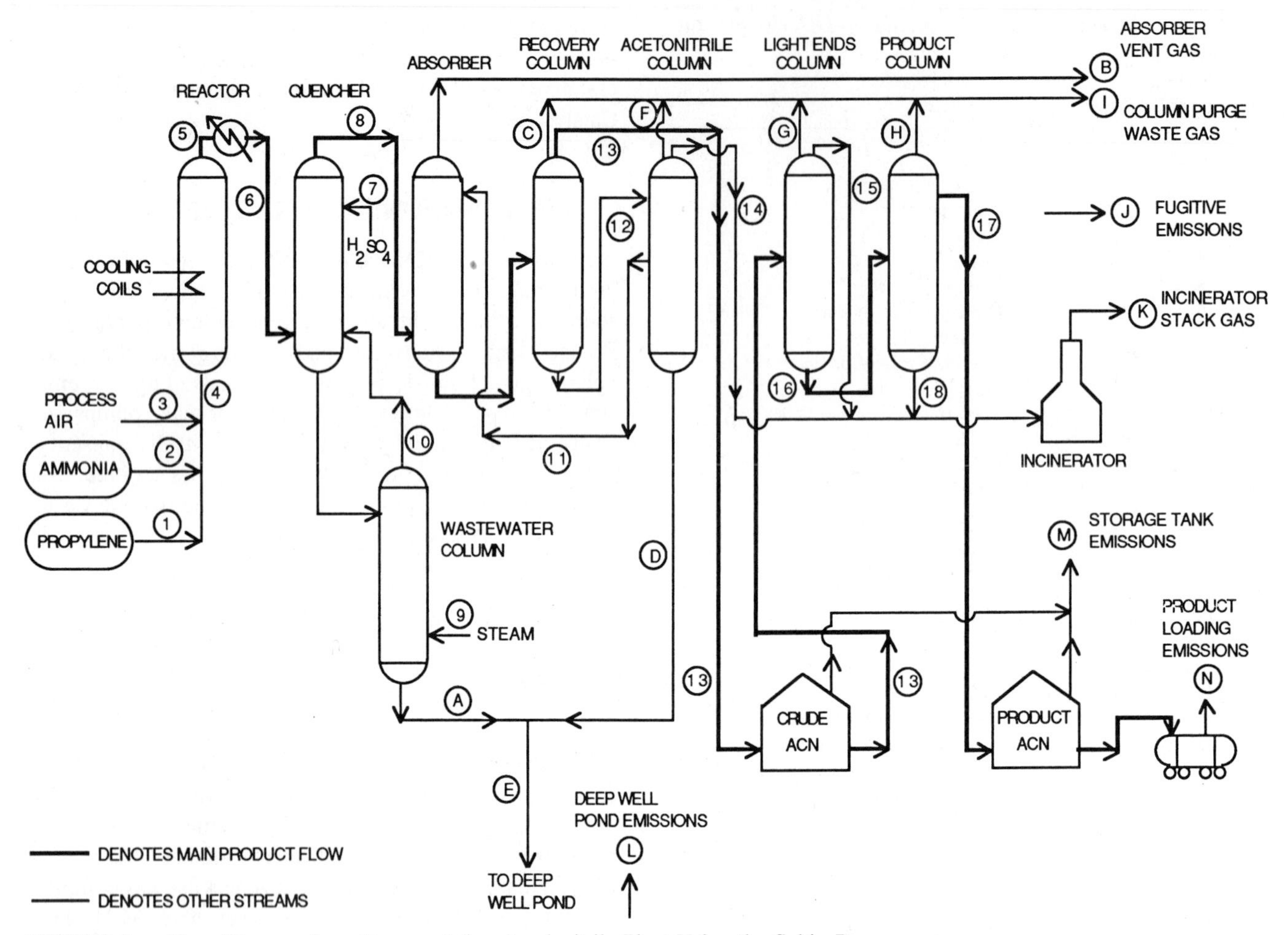

FIGURE 1. Flow Diagram for a Representative Acrylonitrile Plant Using the Sohio Process

TABLE 3. Characterization of Pollutants from Air Emission Points for Acrylonitrile Production Facilities[a]

	Stream Identifier						
	B	I	J	K	L	M	N
	Description						
	Absorber Vent Gas	Column Purge Waste Gas	Fugitive Emissions	Incinerator Stack Gas[b]	Deep Well/Pond Emissions	Storage Tank Emissions	Product Transport Loading Facility Vent
Carbon dioxide	X	X					
Propylene	X	X	X				
Propane	X		X				
Ammonia			X				
Carbon monoxide	X						
Acrylonitrile	X	X	X			X	
Hydrogen cyanide	X	X		X		X	X
Hydrocarbons (general)					X		

[a]Data taken from material balance for a representative ACN plant (reference 7).
[b]The material balance cited did not include combustion products such as CO_2, CO, and NO_x.

TABLE 4. Uncontrolled Emissions of Acrylonitrile and Total Volatile Organic Compounds from Model Plant Using Sohio Process[a,b]

Source	Stream Designation (Figure 1)	Emission Ratio (g/kg)[c]		Emission Rate (kg/hr)	
		Acrylonitrile	Total VOC	Acrylonitrile	TotalVOC
Absorber vent, normal[a]	B	0.1000	100.0000	2.0500	2050.0000
Absorber vent, start-up[d]	B	0.1870	0.2490	3.8400	5.1100
Column vents	I	5.0000	10.0000	103.0000	205.0000
Storage vents					
Crude acrylonitrile	M	0.0480	0.0480	0.9860	0.9860
Acetonitrile			0.0210		0.4300
Hydrogen cyanide	N/A[e]		0.0410		0.8420
Acetonitrile run tanks	N/A[e]	0.1280	.1280	2.6200	2.6200
Acrylonitrile storage	M	0.5310	0.5321	10.9000	10.9000
Handling					
Hydrogen cyanide[f]	N/A[e]		0.0260		0.5400
Acetonitrile[f]	N/A[e]				
Acrylonitrile[f]	N	0.1670	0.1670	3.4400	3.4400
Acrylonitrile[g]	N/A[e]	0.1500	0.1500	2.1600	2.1600
Fugitive[h]	J	0.4630	0.9470	9.5200	19.4700
Secondary					
Incinerator	K		0.3600		7.4000
Deep well/pond	L		13.0000		267.0000
Total		6.7700	126.0000	139.0000	2576.0000

[a]Reference 4.
[b]Model plant has an annual ACN capacity of 180 billion grams. Factors are based on 8760 hours of operation annually.
[c]Grams of emission per kilogram of acrylonitrile produced.
[d]Average rate for entire year, based on eight start-ups per year lasting one hour each.
[e]Streams are not shown on Figure 1.
[f]By tank car.
[g]By barge.
[h]Reference 8.

umns operating under vacuum, the air that leaks into the column and is removed by the vacuum jet systems.

For the ACN process illustrated in Figure 1, by-product hydrogen cyanide and acetonitrile are incinerated, along with product column bottoms. The primary pollutant problem related to the incinerator stack is the formation of NO_x from the fuel nitrogen of the acetonitrile stream (stream 14) and hydrogen cyanide (stream 15). Carbon dioxide and lesser amounts of CO are emitted from the incinerator stack gas.[7]

Other emission sources involve the volatilization of hydrocarbons through process leaks (fugitive emissions) and from the deep well ponds, breathing and working losses from product storage tanks, and losses during product loading operations. The fugitive and deep well/pond emissions primarily consist of propane and propylene, while the storage tank and product loading emissions consist primarily of ACN.

Emission Factors

Emission factors have been developed for ACN and total VOC emissions from a model ACN plant utilizing the Sohio process.[4] The model plant has an ACN capacity of 180 billion grams per year, based on 8760 hours of operation annually. The model plant, unlike that represented in Figure 1, is arranged either to incinerate or to recover and sell by-product hydrogen cyanide and acetonitrile; therefore, emission factors are provided for storage and handling losses of these by-products through tank and loading emission points. Emission factors are based on the assumption that the plant incinerates 8.64 billion grams of the 21.6 billion grams of hydrogen cyanide product and all of the 5 billion grams of the acetonitrile produced. The emission factors resulting from this study are listed in Table 4.

Emissions from the absorber vent (stream B in Figure 1) are listed for normal operation and start-up conditions involving the venting of the reactor effluent to the atmosphere until normal fuel-rich conditions are reached. Start-up emissions are listed as yearly averaged values based on four start-ups per year.

Storage tanks are assumed to be fixed-roof tanks with atmospheric vents for all materials except ammonia and propylene reactants, which are stored in pressure vessels. Details on the number and dimensions of the tanks and other parameters used to determine the listed storage loss factors are available in reference 4.

The fugitive emission factors listed in Table 4 represent the component counts used in the model study; however, the various uncontrolled emission factors were reestablished based on the more recent average uncontrolled fugitive emission factors for the synthetic organic chemical manufacturing industry.[8] Component counts used for the model plant include 50 pumps in light-liquid service, 200 pipeline valves in gas/vapor service, 1000 valves in light-liquid service, 80 relief valves in gas/vapor service, and two compressor seals.

A second study was conducted with emphasis on ACN emissions only.[6] The study utilized the uncontrolled emission factors for ACN from the source referenced in Table 4 and applied hypothetical control devices and assumed efficiencies to develop controlled emission factors for ACN only.

AIR POLLUTION CONTROL MEASURES

Table 3 characterizes the pollutants from air emission points for an ACN facility. Seven point sources are included:

1. Absorber vent gas
2. Column purge waste gas
3. Fugitive emissions
4. Incinerator stack gas
5. Deep well/pond emissions
6. Storage tank emissions
7. Product transport loading facility vent

Absorber Vent Gas

The absorber vent gas stream contains nitrogen, oxygen, unreacted propylene, hydrocarbon impurities from the propylene feed stream, CO, CO_2, water vapor, and small quantities of ACN, acetonitrile, and hydrogen cyanide. Two control methods are used to treat this stream: thermal incineration and catalytic oxidation.

The thermal incineration units have demonstrated VOC destruction efficiencies of 99.9% or greater, while most catalytic units can only achieve destruction efficiencies in the 95–97% range. Destruction efficiencies in the 99% and greater range can be achieved with catalytic oxidizers, but these are not achieved on a long-term basis because of deactivation of the catalyst by a number of causes. The advantage of catalytic oxidation is low fuel usage, but emissions of NO_x formed in the reactors and not destroyed across the catalyst can pose problems.

The thermal units require additional fuel to maintain operating temperature and usually have some form of preheat and energy recovery in the form of steam generated by a heat recovery boiler. In addition to higher VOC destruction efficiency, the thermal incinerators can treat other streams, such as the by-product hydrocyanic acid (HCN) stream and the crude aceto stream. These streams, however, generate high concentrations of NO_x when treated by a standard incinerator. To minimize the formation of NO_x, staging of the combustion air stream and the absorber off-gas stream is practiced. In some units, ammonia is injected to reduce the level of NO_x in the incinerator stack exhaust.

If the incinerator or the catalytic oxidizer has an emergency shutdown, the absorber vent stream is routed to a flare on a temporary basis.

Column Purge Waste Gas

Waste gas releases from the recovery column, light-ends column, product column, and the acetonitrile column are frequently tied together and vented to a flare. The estimated VOC destruction efficiency of the flare is 98–99% for all streams with a heat content of 300 Btu/scf or greater. The use of a flare is ideally suited for streams that are intermittent and have heating values in excess of 300 Btu/scf.

Fugitive Emissions

Fugitive emissions from piping, valves, pumps, and compressors are controlled by periodic monitoring by leak checking with a VOC detector and a directed maintenance program.

Incinerator Stack Gas

Staged combustion and ammonia injection are used to control the emissions of NO_x from the incinerator that treats the absorber off-gas vent, the crude aceto waste liquid stream, and the by-product liquid HCN stream. Staged combustion suppresses the formation of NO_x by operating under fuel-rich conditions (less than stoichiometric air) in the flame zone where most of the NO_x is formed and oxygen-rich conditions downstream at lower temperatures where NO_x is not appreciably formed. The staging is achieved by splitting the air stream and absorber off-gas stream; one portion of each stream is introduced near the flame zone, while the remainder is introduced downstream. Staging the absorber off-gas not only enhances the "fuel-rich/fuel-lean" condition, but also lowers the temperature since this stream is overall endothermic.

Ammonia injection reduces NO_x by selectively reacting ammonia with NO_x. The reaction occurs at temperatures in the range of 1600–1800°F and, as such, the ammonia must be injected in the postflame zone of the combustion chamber. Residence times of 0.5–1.0 second are required for NO_x destruction efficiencies in the range of 80%, which is compatible with the residence time required for VOC destruction. Selective catalytic reduction using ammonia injection can also be used to control NO_x emissions, but no such units are in service at this time. For this control, the catalyst bed and the ammonia injector are placed downstream of the heat recovery boiler, because the operating temperature must be in the 600–700°F range.

Deep Well/Pond Emissions

Emissions of acrolein and other odorous components in vents from wastewater treatment steps are controlled with water scrubbers. In some cases, pond emissions are controlled by adding a layer of a low-vapor-pressure oil on the surface of the pond to limit volatilization.

Storage Tank Emissions

Product storage tank emissions are controlled with double-seal floating roofs or, in some cases, water scrubbers. Field experience indicates that a removal efficiency of 99% can be achieved with water scrubbing.

Product Transport Loading

Product transport loading vents are gathered and sent to a flare or incinerator for VOC control. Destruction efficiencies of 98–99% are achieved using the flare and greater than 99% using incineration.

References

1. W. Löwenbach and J. Schlesinger, *Acrylonitrile Manufacture: Pollutant Prediction and Abatement,* MTR-7752, MITRE Technical Report, McLean, VA, 1978.
2. *Kirk-Othmer Encyclopedia of Chemical Technology,* 3rd ed., Vol. 1, John Wiley & Sons, New York, 1978.
3. *1990 Directory of Chemical Producers United States of America,* SRI International, Menlo Park, CA, 1990.
4. *Organic Chemical Manufacturing, Vol. 10: Selected Processes,* EPA-450/380-028e, U.S. Environmental Protection Agency, Research Triangle Park, NC, 1980.
5. *Hydrocarbon Processing, Petrochemical Handbook '85,* Gulf Publishing Co., Houston, 1985.
6. Radian Corp., *Locating and Estimating Air Emissions from Sources of Acrylonitrile,* U.S. Environmental Protection Agency, Research Triangle Park, NC, 1983.
7. M. T. Anguin and S. Anderson, *Acrylonitrile Plant Air Pollution Control,* EPA-600/2-79-048, U.S. Environmental Protection Agency, Washington, DC, 1979.
8. *Protocols for Generating Unit-Specific Emission Estimates for Equipment Leaks of VOC and VHAP,* EPA-450/3-88-010, U.S. Environmental Protection Agency, Research Triangle Park, NC, 1988.

Bibliography

T. W. Hughes and D. A. Horn, *Source Assessment: Acrylonitrile Manufacture (Air Emissions),* EPA-600/2-77-107, U.S. Environmental Protection Agency, Research Triangle Park, NC, 1977.

Industrial Process Profiles for Environmental Use: Chap. 6: The Industrial Organic Chemicals Industry, EPA-600/2-77-023f, U.S. Environmental Protection Agency, Cincinnati, OH, 1977.

W. A. Schwartz et al., *Engineering and Cost Study of Air Pollution Control for the Petrochemical Industry, Vol. 2: Acrylonitrile Manufacture,* EPA-450/3-73-006-b, U.S. Environmental Protection Agency, Research Triangle Park, NC, 1975.

CARBON BLACK

Barry R. Taylor

Carbon black is a very finely divided form of industrial carbon with median stokes diameters ranging from approximately 30 to 600 nm. The principal uses of carbon black are as a reinforcing agent in rubber compounds (e.g., for tires, belts, and hoses); as a black pigment in printing inks, surface coatings, paper, and plastics; and as a conductive additive for plastics used in the wire and cable industry. There are six major producers of carbon black in the United States operating a total of 21 plants: Cabot Corp. (four plants), Columbian Chemicals Division of Phelps Dodge (five plants), Degussa (three plants), J. M. Huber (three plants), Sid Richardson Carbon & Gasoline Co. (three plants), and Continental Carbon Division of Witco Corp. (three plants). Typical plant production capacities range from 50 to 300 million pounds per year of carbon black, with total U.S. capacity estimated at approximately 3.3 billion pounds annually.

PROCESS DESCRIPTION

Types of Processes

Carbon black is a product of endothermic hydrocarbon pyrolysis. It can be produced by partial combustion processes involving flames or by purely thermal decomposition processes in the absence of flames. Thermal cracking processes, namely, the thermal and acetylene processes, are still used to produce small quantities of special carbon black grades having performance properties that have not been duplicated with the oil furnace process. However, among the major producers, the oil furnace process, which is a partial combustion process, currently accounts for more than 98% of the total production of carbon black in the United States and is the subject of further discussion.

The Oil Furnace Process

Figure 1 is a flow diagram of a typical oil furnace process. The reactor inputs are preheated air, auxiliary fuel (natural gas or oil), and preheated feedstock. The feedstock is preferably a highly aromatic petrochemical or carbochemical heavy oil preheated to 150 to 250°C to achieve a viscosity appropriate for atomization. The feedstock is injected, frequently as an atomized spray, into a highly turbulent high-temperature zone, which is achieved by burning the auxiliary fuel with air. The oxygen in the system is in stoichiometric excess with respect to the auxiliary fuel, but is insufficient for complete combustion of the feedstock. Consequently, a major portion of the feedstock is pyrolyzed to form carbon black.

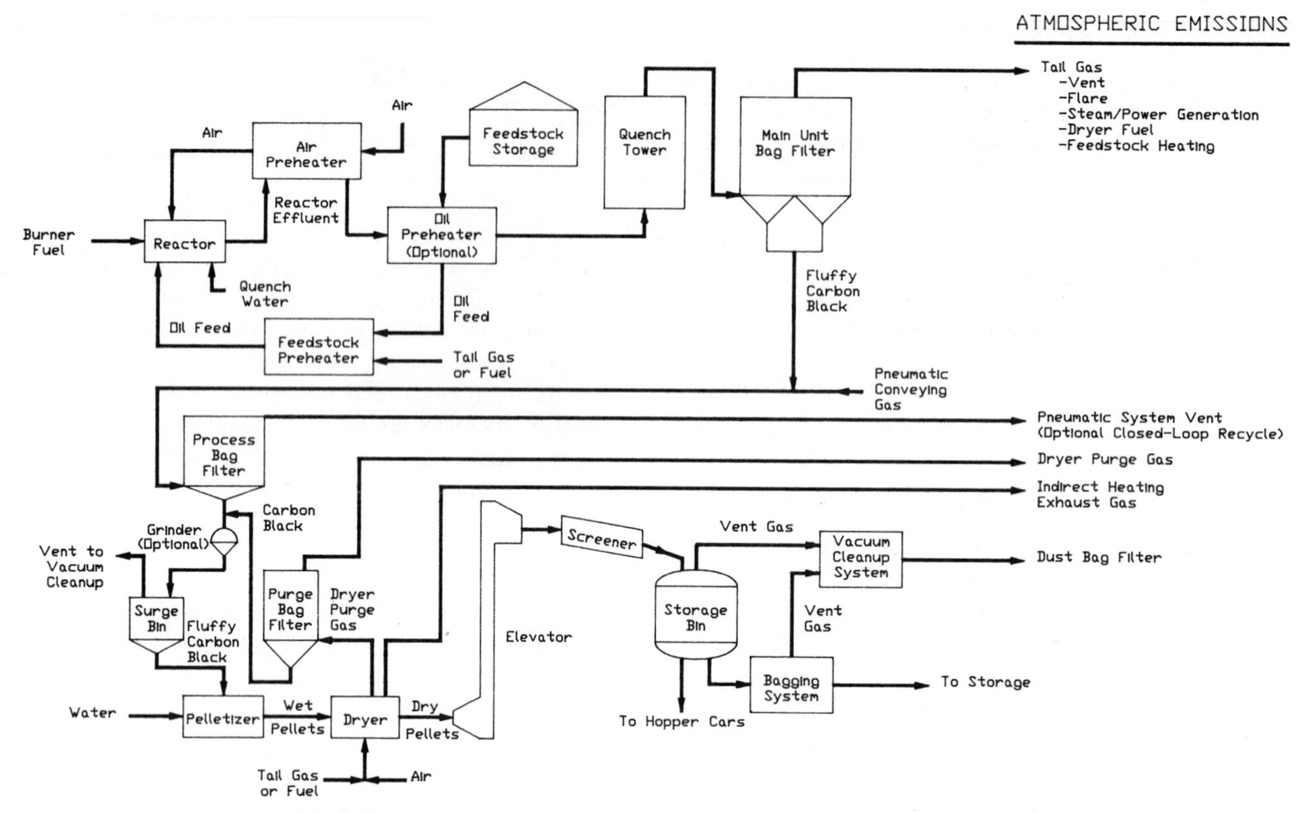

FIGURE 1. Schematic Flow Diagram for Typical Oil Furnace Process

The carbon-containing combustion products are quenched in the reactor with water sprays and pass through heat exchangers that preheat the combustion air and, in some instances, the feedstock oil. The product stream is then cooled further to a temperature in the range of 230°C to 260°C before flowing to the main unit filter, where the carbon black is separated from the gas-phase combustion products (tail gas).

The carbon black, in a finely divided fluffy form, is conveyed pneumatically to a process filter, from which the black drops into a grinder and subsequently into a surge tank. From the surge tank, the carbon black may be packed in fluffy form or may be fed to a pelletizer, where it is compacted into pellets for ease of handling. The majority of rubber blacks are pelletized by a wet process in which carbon black, water, and small quantities of a binder are mixed in pelletizers to form wet pellets. The wet pellets are dried in continuous-flow, rotating drum dryers heated indirectly by burning tail gas or natural gas.

Some grades of carbon black are pelletized by a dry process (not shown in Figure 1) in which the fluffy carbon black is fed to a rotating drum along with a recycle stream of pelletized black. The recycled pellets act as seed material for the formation of new pellets and, in concert with the rotation of the drum, act to densify the new pellets.

Carbon black pellets are transported from the pelletization process by elevators, screw conveyors, and belt conveyors to bulk storage tanks for subsequent loading into railcars and trucks or packing in bags.

AIR EMISSIONS CHARACTERIZATION

Emissions from the manufacture of carbon black include particulate matter, carbon monoxide, hydrocarbons, and sulfur compounds.

Particulate

Carbon black is the main particulate emitted in a carbon black plant. It is a relatively innocuous material with an Occupational Safety and Health Administration permissible exposure limit of 3.5 mg/m^3.[1] As a black, finely divided material that is readily transported by the wind, it can be a nuisance if allowed to disperse. Consequently, sources of emission must be identified and controlled.

Since fluffy carbon black cannot be completely removed from a gas stream, carbon black is emitted when main process gas streams in the manufacturing process are discharged to the atmosphere. The principal process gas stream, reactor tail gas, as well as gases used for dryer purge and open-loop pneumatic conveying, fall into this category. Current emission factors (in pounds of particulate per ton of carbon black produced) used by the U.S. Environmental Protection Agency (EPA) to characterize particulate emissions for carbon black production are 6.53, 0.24, and 0.58 for tail gas, dryer purge gas, and conveying gas, respectively. The carbon black content in tail gas can be reduced when the tail gas is burned; consequently, an emission factor of 2.70 has been established for tail gas that has been flared and of 2.07 for tail gas that has been incinerated.[2]

Intermittent sources are another category of carbon black emissions. During some portions of the reactor warm-up cycle, and in some instances, during operation in a mode to maintain the reactor temperature without making carbon black (referred to as heat load), the oxygen-bearing combustion products that flow through the reactor are vented to the atmosphere and some entrained carbon black will be emitted if not controlled. Intermittent emissions of entrained carbon black also occur during the venting of carbon black storage silos because of displacement of air during filling operations and the expansion of air in the head space upon heating. Carbon black can also be emitted during the bulk loading of railcars and trucks owing to the entrainment of black in the air that is displaced from the vessel being loaded.

Fugitive sources constitute a final category of carbon black emissions. Quantitative information on fugitive emissions is limited. In a survey conducted in 1974, average fugitive emissions of carbon black at one plant were estimated to be 0.1 pound per ton of product.[3] The emission factor currently used for fugitive emissions by the EPA is 0.2 pound per ton of product.[2]

Fugitive emissions of carbon black can occur through imperfect seals on grinders, rotary screens, bucket elevators, pellet conveying systems, and bag packers. Other potential sources of fugitive emissions include ports for equipment access and product sampling, leaks on process equipment, and accidental spills of carbon black that may take place during equipment clean-out operations or from broken packages.

Gas-Phase Emissions

Reactor tail gas is the major source of gas-phase pollutants generated during the production of carbon black. Because carbon black is produced by a fuel-rich combustion process, reactor tail gas contains significant quantities of hydrogen and carbon monoxide, along with small quantities of light hydrocarbons, primarily methane and acetylene. Table 1 shows these species, as well as the other major components and pollutants found in carbon black tail gas. The table also shows tail gas composition (vol%) for a typical grade of carbon black (American Society for Testing and Materials [ASTM] grade N330) and corresponding emission factors (weight of emissions per unit weight of carbon black) for each component. Another column in the table contains data on the range of emission factors measured for tail gas components in surveys of operating plants conducted for the EPA. These data help to show that concentrations of various components in tail gas, as well as the emissions per unit weight of product, are strongly dependent on the grade of carbon black being produced, the particular processing equipment and conditions that are being used to produce it, and the composition of the feedstock used. The final column in the table shows average emission factors currently used by the EPA to characterize emissions of gas-phase pollutants from carbon black tail gas.

Sulfur enters the carbon black manufacturing process via sulfur-bearing feedstocks. Sulfur exiting the process is distributed between the carbon black and the tail gas, with 50–80% of the sulfur ending up in the tail gas, depending on the grade of carbon black being produced, the equipment and the operating conditions used for production, and the

TABLE 1. Tail Gas Emission Information

Component	Data for ASTM N330[a]		Range of Emissions[a]	EPA Emission Factors[b]
	Concentration (vol%)	Emissions (lb/ton Carbon Black)	(lb/ton Carbon Black)	(lb/ton Carbon Black)
Nitrogen	40.	17,700	9,150–26,800	—
Argon	0.48	302	133–456	—
Oxygen	0.95	480	59–1,280	—
Carbon dioxide	2.0	1,400	838–3,440	—
Water	42.	12,000	5,330–25,900	—
Carbon monoxide	6.3	2,800	1,400–4,380	2,800
Hydrogen	7.5	240	104–366	—
Methane	0.20	50	22–120	50
Acetylene	0.22	90	12–258	90
Hydrogen sulfide	0.11	60	Trace–60	60
Carbon disulfide	0.05	60	No analysis	60
Carbonyl sulfide	0.02	20	No analysis	20
Sulfur dioxide	—[c]	0	Trace–24	0

[a]Reference 3.
[b]Reference 2.
[c]Not detectable at 1 ppm.

level of sulfur in the feedstock. As with other tail gas species, the partitioning of sulfur among species in the tail gas depends on the processing equipment and operating conditions employed.

Since the overall environment in the reactor is a reducing one, the sulfur in the tail gas is expected to be in a reduced state. Sulfur is present as hydrogen sulfide (H_2S), carbon disulfide (CS_2), carbonyl sulfide (COS), and, in smaller quantities, sulfur dioxide (SO_2). Measurements of H_2S concentrations have accounted for approximately 40% to more than 99% of the sulfur in the tail gas; generally, more than half of the sulfur in the tail gas is H_2S. Measured fractions of sulfur as CS_2 in the tail gas have ranged from less than 1% to approximately 60%. Carbonyl sulfide has always been detected in the tail gas, but has not been observed to exceed 10% of the total tail gas sulfur. The emission factor for sulfur oxides in vented tail gas currently used by the EPA is zero pounds per ton of carbon black.[2] In published emission surveys of a wide variety of operating conditions, measured SO_2 concentrations have been relatively low, but have varied from undetectable levels (<1 ppmv) to 210 ppmv.[3]

AIR-POLLUTION CONTROL MEASURES

Particulate

High-performance bag filters are used to remove carbon black from the main process gas streams. The collection efficiency of these filters during normal operation is enhanced significantly by the buildup of a filter cake of carbon black on the filter fabric. New bags typically require up to 24 hours of operation to achieve maximum efficiency. Performance testing indicates that effluent carbon black concentrations of 6 mg/nm^3 (wet basis) are at the upper end of capture efficiency capabilities and that 30 mg/nm^3 (wet basis) is an effluent concentration that is consistently achievable for a broad range of carbon black grades with a bag filter system maintained in good operating condition. Using an effluent concentration of 30 mg/nm^3 and the data provided in the Monsanto Research Corp. report[4] on tail gas compositions for various operations, particulate emissions attributable to vented tail gas are calculated to be less than 2.0 pounds of particulate per ton of carbon black. This is significantly less than the emission factor of 6.5 pounds per ton currently used by the EPA.

Vent scrubber systems, which consist of a venturi scrubber, a cyclone, and a spray tower with demisters, can be used effectively to remove the entrained carbon black from reactor effluent gases during intermittent reactor warm-up and heat-load operations. Bin vent filters, which consist of a filter bag connected to the vent, are used to remove entrained carbon black from the vent gases of carbon black storage silos. To control the emission of carbon black during the loading of railcars and trucks, air displaced from the vessels is collected in a plant vent system and discharged to the atmosphere through a bag filter.

Fugitive emissions of carbon black through imperfect seals on process equipment are prevented by connecting the pieces of equipment to a vacuum system and keeping them under reduced pressure. Air collected by the vacuum system is discharged to the atmosphere through a bag filter. Emissions from product sampling ports, process equipment leaks, and accidental spills are minimized through proper maintenance of operating equipment and aggressive housekeeping practices. Housekeeping equipment includes sweeping machines to clean smooth-surfaced portions of the plant and plantwide vacuum systems to enable the cleanup of isolated spills of carbon black before they are dispersed by the wind.

Gas-Phase Emissions

Because of the high concentrations of hydrogen and carbon monoxide in carbon black tail gas, it can be used as a low-Btu fuel. In carbon black plants that have wet pelletization units, some of the tail gas is burned to supply thermal energy to the dryers. In situations where it is not economical to burn the tail gas to generate process steam or electric power, the gas can be flared to convert reduced tail gas species to others more compatible with the environment.[2]

The amount of sulfur released to the atmosphere in North America through the production of carbon black (estimated to be 210,000 tons per year as SO_2 using emission factors of 60 pounds H_2S, 60 pounds CS_2, and 20 pounds COS per ton of carbon black and an annual carbon black production rate of 3.3 billion pounds) is small relative to the estimated total quantity of anthropogenic sulfur emissions (22.8 million tons per year as SO_2).[5] In carbon black plants where the emission of sulfur species to the atmosphere must be reduced, control is accomplished through the use of low-sulfur feedstocks. Tail gas that is not burned in pellet dryers or in power-generating systems can be flared to oxidize reduced sulfur species to less objectionable sulfur oxides.[2] Converting reduced sulfur species to SO_2 during combustion has little impact on sulfur oxides in the environment because species such as H_2S are oxidized to SO_2 in the atmosphere over time.

References

1. Occupational Safety and Health Administration, *OSHA Safety and Health Standards: General Industry Standards* (revised), publication no. 29CFR1910.1000, U.S. Department of Labor, Washington, DC, 1983.
2. Office of Air Quality Planning and Standards, *Compilation of Air Pollutant Emission Factors, Volume I: Stationary Point and Area Sources*, 4th ed., publication AP-42, U.S. Environmental Protection Agency, Research Triangle Park, NC, 1985.
3. Office of Air Quality Planning and Standards, *Engineering and Cost Study of Air Pollution Control for the Petrochemical*

Industry, Vol. 1, report EPA-450/3-73-006-a, U.S. Environmental Protection Agency, Research Triangle Park, NC, 1974.
4. Monsanto Research Corp., *Source Assessment: Carbon Black Manufacture,* NTIS no. PB-273-068, U.S. Department of Commerce, Springfield, VA, 1977.
5. Office of Air Quality Planning and Standards, *National Air Quality and Emissions Trends Report, 1988,* report EPA-450/4-90-002, U.S. Environmental Protection Agency, Research Triangle Park, NC, 1990.

Bibliography

Dannenberg, E. M., "Carbon black," *Encyclopedia of Chemical Technology,* 3rd ed., Wiley-Interscience, New York, 1978.

CHARCOAL/ACTIVATED CARBON

William D. Byers

Charcoal and activated carbon are produced by the pyrolysis (carbonization or destructive distillation) of carbonaceous raw materials. Charcoal is principally formed into briquettes and used as a fuel for outdoor cooking. Activated carbon is used for the decolorizing of sugar and for adsorbing pollutants from water (liquid phase) or air (gas phase). The demand for activated carbon is increasing sharply in response to increasing environmental awareness and tighter regulations. The U.S. consumption of activated carbon in 1987 was reported at 212.6 million pounds and is predicted to be 270 million pounds in 1992.[1]

Almost any carbonaceous material of animal, plant, or mineral origin can be converted to charcoal or activated carbon if properly treated. Charcoal is most commonly prepared from hardwoods such as beech, birch, hard maple, and oak. Lignite and coal are the most commonly used raw materials for liquid-phase carbons, while coconut shells, coal, and petroleum residues are most commonly used for the preparation of gas-phase carbon.

PROCESS DESCRIPTION

Process Overview

The manufacture of charcoal and activated carbon begins with a carbonization operation that converts organic material to primary carbon or "char." In the manufacture of charcoal, the char is mixed with a binder and pressed to form briquettes as the final product. In activated-carbon production, the char is "activated" physically, by contact with hot gases, or chemically, by the addition of certain inorganic chemicals. In the 15 years prior to 1978, there were over 190 patents applicable to the manufacture and reactivation of activated carbon.[2]

Carbonization

During the carbonization operation, organic material is reduced by the slow application of heat in the (relative) absence of oxygen to form a char. Carbonization is accomplished in a four-step operation:

1. Heat is applied to the organic material and, as the temperature approaches 100°C, water and highly volatile organics are driven off.
2. As the moisture is driven off, the temperature rises to approximately 170°C, causing degradation with the evolution of CO, CO_2, and organic (acetic and other) acids.
3. At temperatures of 270–280°C, the process becomes exothermal, with the formation of substantial amounts of hydrocarbon distillates, including methanol, tars, and other by-products.
4. At a temperature of about 350°C, exothermic pyrolysis ends and heat is applied to raise the temperature to the 400–600°C range. At this temperature, some of the less volatile tarry materials are removed from the char.

The overall yield from the carbonization process varies widely depending on the raw material and the actual process conditions. Yields of 25–30% are reported for the production of charcoal from wood.

The carbonization of wood for charcoal production is most commonly accomplished as a continuous process in large, cylindrical, multiple-hearth furnaces where the charge is stirred and moved from one hearth to the next lower one by rotating rabble arms. Figure 1 shows a typical cross section of a multiple-hearth furnace.[3] Batch kilns are also used for the carbonization of wood for charcoal production. The most common style of batch kiln is the "Missouri-type" kiln shown in Figure 2.[3]

Coal and nut (coconut, walnut, etc.) shells are the most common raw materials for activated-carbon manufacture. In commercial plants, the material is carbonized in horizontal rotary kilns or vertical retorts.

Charcoal Manufacture

Briquetting

The fabrication of briquettes from raw charcoal involves hammer-milling or crushing the char, mixing the char with a starch binder, forming the briquettes in a press, and drying the briquettes at approximately 135°C to achieve a product moisture content of about 5%.

Activated-Carbon Manufacture

Activation

Activation is essentially a two-phase process requiring the burn-off of tars formed in carbonization plus the enlarge-

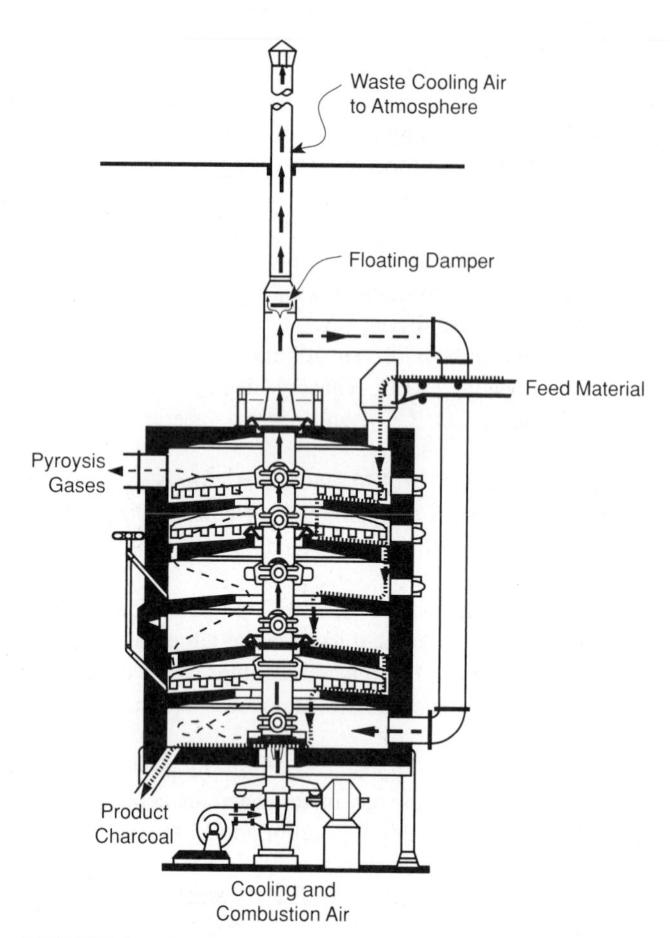

FIGURE 1. Typical Cross Section of Multiple-Hearth Furnace

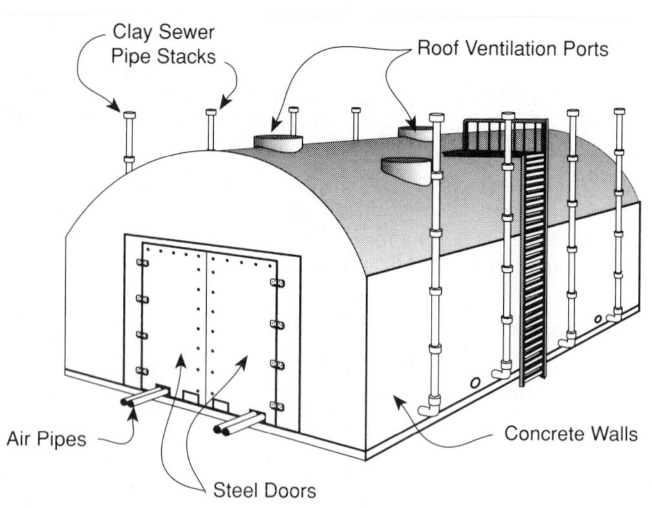

FIGURE 2. The Missouri-Type Charcoal Kiln

ment of pores in the carbonized material. Two types of activation processes are used—chemical and physical. Chemical activation involves the addition of a chemical agent such as phosphoric acid or zinc chloride. Physical activation involves the use of a hot gas such as steam or CO_2 at temperatures of 800°C to 1000°C.

Physical activation is the more commercially important process, as well as the process of greater interest from an air emission standpoint. It is accomplished in continuous internally or externally fired rotary retorts or in multiple-hearth furnaces. The activation process uses sufficient oxygen to burn off residual volatile carbon remaining on the char after carbonization without burning excessive amounts of char. The activation process is exothermic, using the heating value of the volatile carbon plus heat supplied from supplemental fuel.

Reactivation

Carbon that has been used for its intended purpose can be reactivated using essentially the same process as for the original activation. Organics adsorbed on the carbon during its use are burned off in the reactivation process. In addition, a portion of the carbon particles is consumed during reactivation. Reactivation may be performed at the site where the carbon was used (on-site reactivation) or conducted at a commercial manufacturing or reactivation facility. For smaller-scale on-site reactivation operations, fluidized-bed and infrared furnaces are gaining acceptance, in addition to the conventional multiple-hearth furnace.

Furnace types used for activated-carbon reactivation, in the order of their total on-line capacity, are shown in Table 1.[4] Also listed are the approximate number of operating reactivation furnaces of each type.

AIR EMISSIONS CHARACTERIZATION

Five types of products result from the manufacture of charcoal and activated carbon: the solid charcoal or activated carbon; noncondensible gases (carbon monoxide, carbon dioxide, methane, and ethane); pyroacids (primarily acetic acids and methanol); tars and heavy oils; and water. The products and product distribution are varied, depending on the raw materials and the process conditions used.

Particulates

Particulate emissions result from handling the charcoal or activated carbon in transfer, size reduction/screening, briquetting, and related activities. In addition, tars emitted from the process, if not combusted, may solidify to form

TABLE 1. Types of Equipment Used for Activated-Carbon Reactivation

Furnace Type	No. of Units
Multiple hearth	>100
Fluidized bed	<20
Indirect fired rotary kiln	>50
Direct fired rotary kiln	<30
Vertical tube type	<30
Infrared horizontal	<5
Infrared vertical	4

TABLE 2. Uncontrolled Emission Factors for Charcoal Manufacturing[a]

Pollutant	Charcoal Manufacturing		Briquetting	
	kg/Mg	lb/ton	kg/Mg	lb/ton
Particulate[b]	133	266	28	56
Carbon monoxide	172	344	—	—
Nitrogen oxides[c]	12	24	—	—
VOC				
Methane	52	104	—	—
Nonmethane[d]	157	314	—	—

[a]Expressed as weight per unit charcoal produced. Afterburning is estimated to reduce emissions of particulates, carbon monoxide, and VOC >80%. Briquetting operations can control particulate emissions with centrifugal collection (65% control) or fabric filtration (99% control).
[b]Includes tars and heavy oils. Polycyclic organic matter (POM) carried by suspended particulates was determined to average 4.0 mg/kg.
[c]Based on 0.14% wood nitrogen content.
[d]Consists of noncondensibles (ethane, formaldehyde, unsaturated hydrocarbons) and condensibles (methanol, acetic acid, pyroacids).

particulate emissions and pyroacids may form aerosol emissions.

Gases

Noncondensible gases are products of the carbonization process (charcoal and activated-carbon manufacture) and the activation process (activated-carbon manufacture). The quantity and relative concentrations of noncondensible gases depend heavily on the character of the raw material and the operating conditions of the process used.

Uncontrolled emissions from carbonization in batch kilns have been reported as shown in Table 2.[3]

Average noncondensible gas composition ranges from batch kilns are presented in Table 3.[5]

Acetic acid is seen to be 4–7% of the condensible products, methanol is 3–6%, and insoluble tars are 8–13%.[5]

Air Toxics

The uncontrolled emissions from carbonization and activation contain several constituents classified as air toxics (e.g., methanol). Reactivation has an even greater potential for causing air toxics emission, since the carbon being reactivated has often been employed in adsorbing compounds that would be classified as air toxics.

TABLE 3. Composition Range for Noncondensible Products of Charcoal Manufacture

Product	Percent of Noncondensibles
Carbon dioxide	50–60
Carbon monoxide	22–33
Methane	3–18
Hydrogen	1–4
Higher hydrocarbons[a]	1–6

[a] "Higher hydrocarbons" are assumed to be nonmethane, noncondensible hydrocarbons.

Of special interest is the potential for formation of polychlorinated dibenzodioxins (PCDDs) and polychlorinated dibenzofurans (PCDFs) in the high-temperature, low-oxygen environment of the activation (or reactivation) furnace. One test program found no evidence of PCDDs or PCDFs emitted from the reactivation of virgin carbon, but detected the presence of both families of compounds when reactivating spent carbon from water treatment operation. This would tend to indicate that the by-products were formed from the adsorbed organics on the spent carbon rather than from impurities in the virgin carbon.[6]

AIR POLLUTION CONTROL MEASURES

Control of particulates from material sizing, handling, and briquetting operations is typically achieved by the use of centrifugal or fabric filter collectors. Collection efficiencies of 65% for centrifugal collection and 99% for fabric filtration have been reported.[5]

Emissions from carbonization, activation, and reactivation retorts are generally controlled by afterburners followed by water scrubbers.[7] Typical afterburner operating conditions include combustion temperatures of 1800°F (980°C) or greater and residence times in excess of two seconds. There are no data available on destruction-removal efficiency (DRE) for such a control system in this application. The configuration and conditions are similar to those used for controlling hazardous waste incinerator emissions where DREs of 99.99% are expected.

In one application, where data have been published on PCDD and PCDF emissions from reactivation systems using afterburners, a fluidized-bed reactivation system was fitted with an afterburner designed to provide a two-second residence time at a temperature of 2400°F (1316°C). In another application, a horizontal infrared reactivation fur-

TABLE 4. Total Dioxin and Furan Equivalent Emissions as 2,3,7,8-TCDD[a]

Pollutant	2,3,7,8,-TCDD Toxic Equivalence Factor	Average Concentration, mg/dscm	Average 2,3,7,8-TCDD Toxic Equivalent Concentration, ng/dscm
2,3,7,8-TCDD	1.00	ND (0.023)[b]	ND (0.023)[b]
Other TCDD	0.01	0.012	0.00012
Penta-CDD	0.5	ND (0.024)[b]	ND (0.012)[b]
Hexa-CDD	0.04	0.448	0.0179
Hepta-CDD	0.001	0.032	3.2×10^{-5}
Octa-CDD	0.000	0.250	0.000
2,3,7,8-TCDF	0.1	0.068	0.0068
Other TCDF	0.001	0.189	1.89×10^{-4}
Penta-CDF	0.1	0.075	0.0075
Hexa-CDF	0.01	0.03	3×10^{-4}
Hepta-CDF	0.001	0.043	4×10^{-5}
Octa-CDF	0.000	0.033	0.000
Total			0.068

[a]Annual emissions for a 196-dry standard m^3/h combustion gas flow rate, 7000 hours per year facility operation—100 μg per year.
[b]ND—not detected; values in parentheses denote the average detection limit, used in lieu of actual data in the risk assessment.

nace was fitted with an afterburner sized for a 0.3-minute (20-second) residence time at 1850°F (1010°C).

Table 4 shows the measured PCDD and PCDF emissions from the latter installation.[6] The operating rate of the system was 215 lb/h of spent activated carbon.

References

1. S. Irving-Monshaw, "New zip in activated carbon," *Chem. Eng.*, 43–47 (February 1990).
2. A. Yehaskel, *Activated Carbon Manufacture and Regeneration,* Noyes Data Corp., Park Ridge, NJ, 1978.
3. *AP-42,* U.S. Environmental Protection Agency, Research Triangle Park, NC, 1985.
4. W. G. Schuliger and L. G. Knapil, "Reactivation systems," Sunday Seminar, American Water Works Association Annual Conference, June 1990.
5. C. M. Moscowitz, *Source Assessment: Charcoal Manufacturing State of the Art,* EPA-600/2-78-004z, U.S. Environmental Protection Agency, Cincinnati, OH, December 1978.
6. W. E. Koffskey and B. W. Lykins, Jr., "GAC adsorption and infrared reactivation: case study," *J. AWWA,* 48–56 (January 1990).
7. L. Sorrento, AtoChem North America, Philadelphia, personal communication, June 5, 1991.

CHLOR-ALKALI

Kenneth S. Walborn

Carl Wilhelm Scheele, a young Swedish apothecary, is generally recognized as the first person to isolate pure chlorine gas in the year 1774. The major coproduct, caustic soda, has an older but more obscure history. Caustic soda was produced in ancient times by reacting lime and sodium carbonate for use as a cleanser.

The first commercial use for chlorine was in the bleaching of textiles in France in the year 1789 and the bleaching of textiles, paper, and flour remained its major applications until the early 1900s, when it was used to sterilize the Jersey City water supply. The ability of chlorine to destroy lower forms of organic life resulted in its rapid acceptance as a treatment for sewage sludge and drinking water, thus virtually eliminating typhoid fever epidemics. The production history of chlorine is characterized by rapid growth to match its uses in the manufacture of inorganic chemicals, plastics, organic chemicals, solvents, pulp, and paper, and for water treatment.

The growth of caustic soda paralleled that of chlorine and the production volume of caustic soda by the electrolytic chlor-alkali processes passed that of the lime soda process in 1940. The uses of caustic soda have grown from soaps and cleaners to include the manufacture of organic chemicals, detergents, and inorganic chemicals, and its utilization in the pulp and paper, petroleum, aluminum, and water treatment industries.

The growth of the chlor-alkali industry is shown in Figure 1. The units of production are in electrochemical units (ECUs), with 1 ECU equal to 1 ton of chlorine plus 1.1 tons of caustic soda (the ratio of formation).

PROCESS DESCRIPTION

Chlorine and caustic soda are produced concurrently by the electrolysis of salt dissolved in water in either a diaphragm

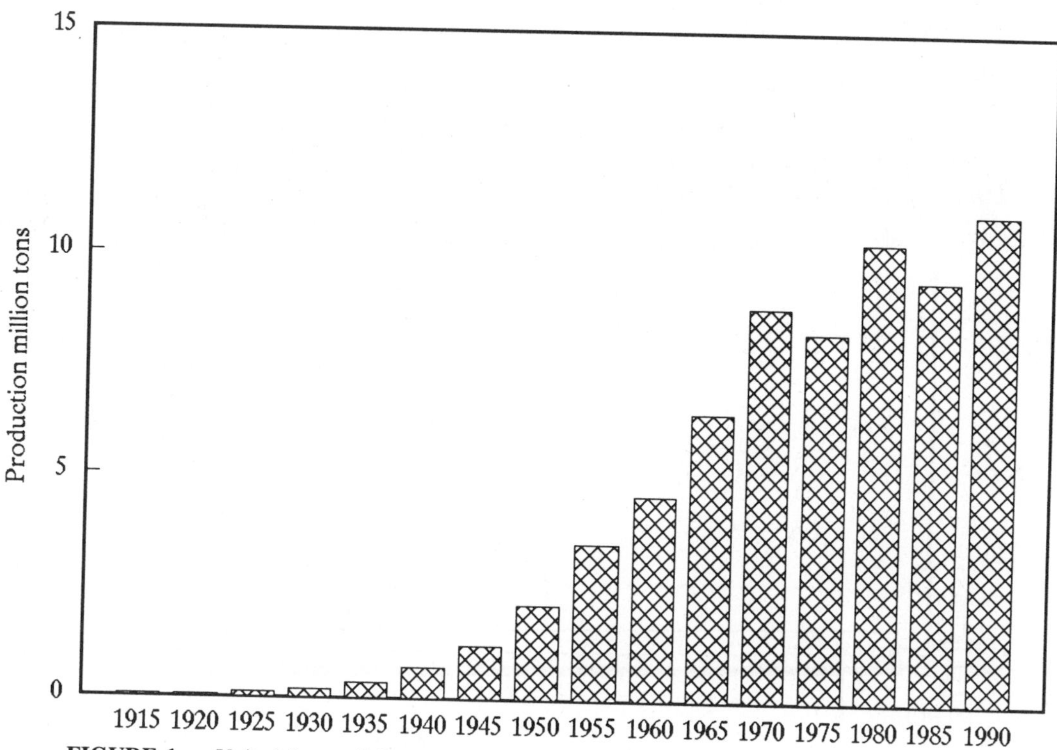

FIGURE 1. United States Chlor-Alkali Production in ECUs for the Years 1915 Through 1990 Projection

cell, a mercury cell, or a membrane cell. In a diaphragm cell (Figure 2), chlorine gas is produced at the positive electrode (anode) and caustic soda and hydrogen are produced at the negative electrode (cathode). The overall reaction can be expressed as:

$$2NaCl + 2H_2O \xrightarrow{\text{Electricity}} Cl_2 + 2NaOH + H_2$$

The chlorine produced at the anode is physically separated from the coproducts by an asbestos diaphragm that allows passage of sodium ions to the cathode, where caustic soda and hydrogen gas are produced.

In the mercury cell, two liquid layers flow through a long cell (Figure 3). The bottom layer is liquid mercury, which serves as the cathode for the cell. The upper layer is a saturated brine solution. Anodes protrude into the brine layer, and when an electric current is passed through the cell, chlorine is produced at the anode and is withdrawn from the cell. The sodium ions are drawn to the cathode where a sodium–mercury amalgam is formed. This amalgam is removed from the cell and reacted in a secondary cell with purified water to form caustic soda and hydrogen.

The membrane cell is an improvement on the diaphragm cell that was made possible with the development of stable membranes in the late 1960s that allowed the elimination of asbestos diaphragms (Figure 4). The passage of chloride through the membrane is reduced, allowing the production of high-quality caustic soda directly from the cell. The brine from the anode side of the cell is depleted of its chloride and sodium values and is recycled for reconcentration. Pure water is added to the cathode side of the cell to produce pure sodium hydroxide.

The choice of cell type or diaphragm material depends on space limitations, salt balance, brine purity, steam and power costs and availability, and caustic-soda strength and purity requirements. The cells are installed in groups operating in electrical series, with production of products in parallel.

The caustic soda from diaphragm cells is low in caustic

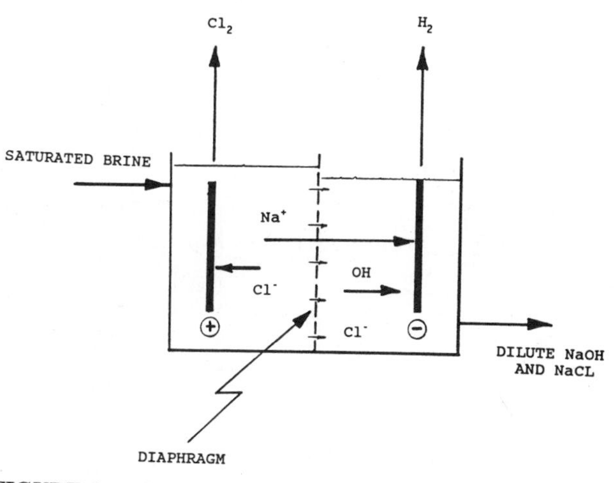

FIGURE 2. Simplified Diaphragm Cell

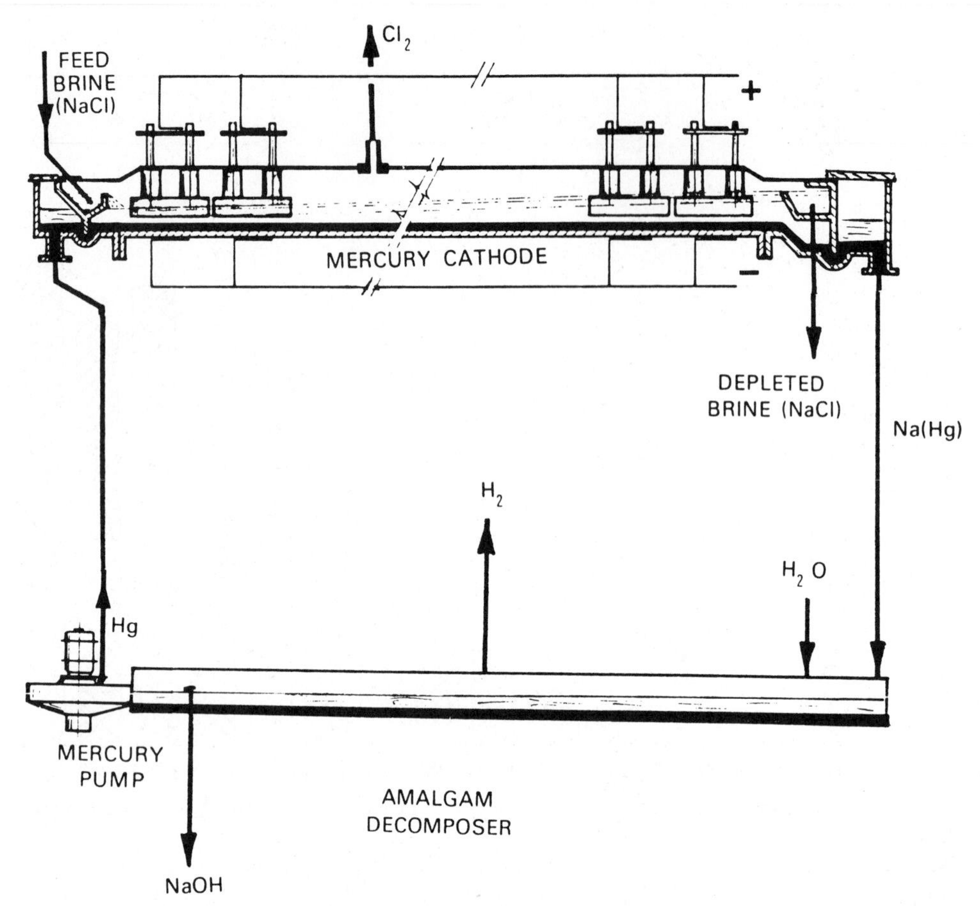

FIGURE 3. Typical Mercury-Cell Construction

concentration and contains unreacted salt, and it must be concentrated in evaporators to increase its strength and to recover salt. The caustic from mercury cells is produced at commercial strength and high purity and requires no further treatment before sale. Membrane cells produce high-purity caustic of somewhat less strength than its mercury cells.

The hydrogen gas produced in the cells is cooled and scrubbed prior to sale or use. Hydrogen from mercury cells must receive further chemical treatment or carbon absorption to remove all traces of mercury before it is used.

The chlorine gas produced in the cells is saturated with water and must be cooled and dried prior to compression. Cooling is normally done in noncontact heat exchangers and the drying is carried out with sulfuric acid in a series of direct-contact scrubbers. The moisture level must be reduced to eliminate corrosion of conventional (steel) equipment that is employed as the construction material in the system from the compressor on. After the chlorine gas is dried, it is compressed and cooled. The high-pressure gas (35–150 psig) is liquified in several stages of refrigeration. The liquified chlorine is stored under pressure in tanks for shipment in tank cars, trucks, and barges. The chlorine gas stream entering the refrigeration units contains small quantities of "noncondensibles," primarily as a result of air in-leakage at the cells. The noncondensible gases leaving the refrigeration units still contain chlorine gas, which must be removed by recovery units or scrubbers or used in processes capable of utilizing low-chlorine-concentration streams. A schematic of a simplified chlorine system is shown in Figure 5.

AIR EMISSIONS CHARACTERIZATION

The air emissions to be controlled from diaphragm-, mercury-, and membrane-cell chlorine plants include chlorine gas, mercury vapors, asbestos, and hydrogen.

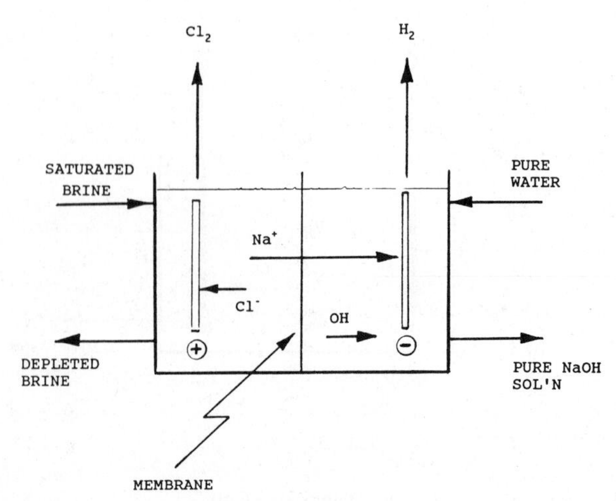

FIGURE 4. Simplified Membrane Cell

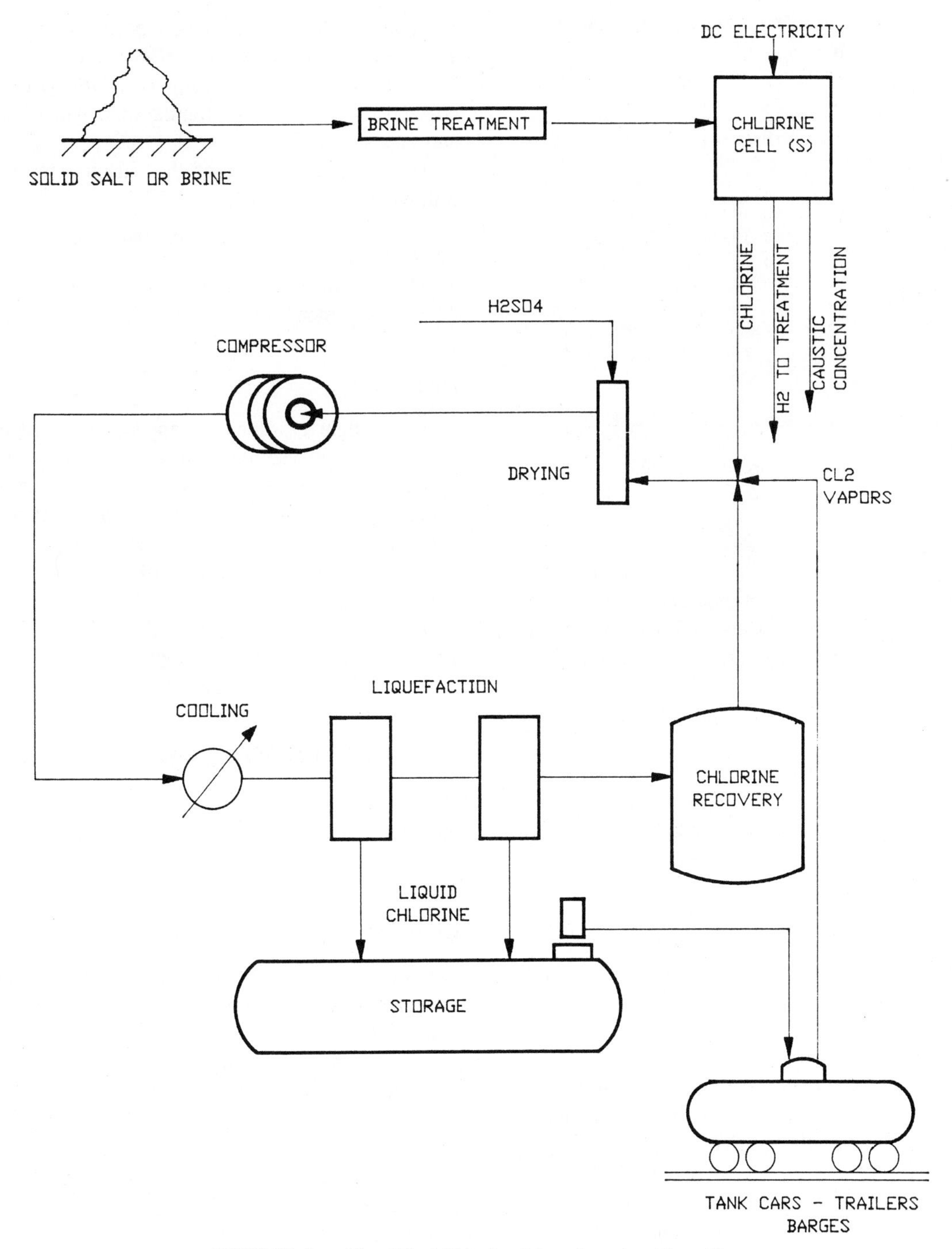

FIGURE 5. Simplified Chlorine Manufacturing Flow Sheet

Chlorine

Chlorine gas is listed as an extremely hazardous substance under SARA Section 302 and any person handling in excess of 100 pounds of the gas must take certain emergency precautions. It is also subject to the release-reporting requirements of SARA Section 304 and noncontinuous releases exceeding 10 pounds in 24 hours must be reported to the National Response Center (NRC), the Local Emergency Planning Commission (LEPC), and the State Emergency Planning Commission (SEPC). Continuous releases that exceed the reportable quantity must also be reported to the same agencies, by both telephone and letter, with a one-year follow-up letter required. Under the requirements of SARA 313, detailed emission data must be reported annually to all groups if a manufacturer produces in excess of 25,000 pounds per year or a consumer uses in excess of 10,000 pounds per year.

The potential sources of the various emissions are as follows.

TABLE 1. Emission Factors for Chlor-Alkali Plants[a]—Emission Factor Rating B

	Chlorine Gas	
Type of Source	lb/100 tons	kg/100 MT
Liquefaction blow gas		
Diaphragm cell	2,000–10,000	1,000–5,000
Mercury cell[b]	4,000–16,000	2,000–8,000
Water absorber[c]	25–1,000	12.5–500
Caustic or lime scrubber	1	0.5
Loading of chlorine		
Tank car vents	450	225
Storage tank vents	1,200	600
Air blowing of mercury-cell brine	500	250

[a]References 1 and 2.
[b]Mercury cells lose about 1.5 pounds of mercury per 100 tons (0.75 kg/100 MT) of chlorine liquified.
[c]Control devices.

1. From the cell flanges, connections, and relief devices
2. From the low-pressure headers during normal operations, start-ups and shutdowns, and emergency operations
3. From chlorine dissolved in the sulfuric acid and then released to the atmosphere
4. From chlorine compressor seals
5. From high-pressure piping and liquefaction equipment
6. Tail gas from recovery units and scrubbers
7. From tanks and equipment being loaded

Chlorine emission factors[1,2] for uncontrolled plant processes and, to some extent, specific control practices are shown in a table of emission factors (Table 1).

Mercury

Mercury liquid or vapor is included on the SARA hazardous chemical list and spills or releases in excess of 1 pound in 24 hours must be reported to the NRC, LEPC, and SEPC. Annual reporting under the terms of Section 313 is also required. Chlor-alkali mercury emissions are also covered by promulgated NESHAP limits, which impose a limit of 2300 grams a day for total emissions.

The potential sources of the various emissions are as follows:

1. From the cells during operation and maintenance
2. In the vent gas from the cells
3. In the hydrogen from the cells
4. From piping leaks

Asbestos

Friable asbestos (dry) is included on the SARA hazardous chemicals list and spills or releases of dry asbestos in excess of 1 pound in 24 hours must be reported to the NRC, LEPC, and SEPC. Annual reporting under the terms of Section 313 is required as well. Handling of friable asbestos is also covered under NESHAP regulations, which set standards and emission limits.

The potential sources of the various emissions are the following.

1. Receiving and handling
2. Diaphragm manufacture
3. Disposal methods

Hydrogen

This very light gas does not appear on the regulatory lists and spill and release reporting is not required. Hydrogen is an extremely flammable gas and extreme care and good equipment design are required in its handling to avoid fires and explosions.

Among the potential sources of the various emissions are the following.

1. From the cells during operation
2. From vent leak-through

AIR POLLUTION CONTROL MEASURES

The control and elimination of fugitive and point source emissions depend on plant design, operating philosophy, and the maintenance program.

Chlorine

Emissions from diaphragm cells and low-pressure headers are controlled or eliminated by the use of corrosion-resistant materials, the use of elastomer gaskets retained by bolts or heavy compression springs, and operation at atmospheric pressure or under a very slight vacuum. All headers and major equipment are vented through safety relief systems to caustic scrubbers during start-ups, shutdowns, or emergency conditions to eliminate releases to the atmosphere. These scrubbers must have sufficient capacity to neutralize full production capacity for the time necessary to start up with inerts or to shut down and evacuate the system.

A control system capable of maintaining a consistent pressure throughout the system under varying rates and stream compositions is extremely important to successful operation. Operation of the low-pressure system above atmospheric pressure can result in the release of fugitive chlorine and operation at deep vacuums can result in the intrusion of atmospheric air, which causes high noncondensible flows in the high-pressure system. Cells, piping headers, and major equipment are evacuated to a vacuum system and purged prior to opening them for repairs.

The design and operation of the sulfuric acid drying

system are extremely important to long-term successful operation. The rate of corrosion of steel used in the high-pressure system is accelerated to unacceptable levels as the moisture level increases. The sulfuric acid used to dry the chlorine will contain dissolved chlorine, which must be recovered and recycled by vacuum stripping or neutralized with sulfides to prevent release to the atmosphere after discharge and neutralization of the spent acid.

Most manufacturers now use multistage centrifugal compressors, although reciprocal and sulfuric acid wetted-lobe compressors have been used. Double seals with a dry air purge between the seals are used to prevent seal leakage. It is important to design all high-pressure piping and equipment with the minimum number of flanges and no leakage is acceptable. Because of the rapid corrosion of steel when it is exposed to wet chlorine, any small leak will rapidly become a large leak. High-pressure equipment is also vented through safety relief valves to a caustic scrubber with sufficient capacity to handle full production rates for the period required to shut down and evacuate the equipment.

The high-pressure chlorine is condensed in liquefaction equipment at temperatures ranging from +75°F to –50°F, depending on the operating pressure of the system and the chlorine removal desired. The tail gas from the system, which consists of the noncondensibles and residual chlorine, can be sent to chlorination processes that can use low-concentration streams or to recovery systems. Recovery systems are capable of complete recovery and recycling of the chlorine in the tail gas. The design of recovery systems must recognize the presence of low levels of hydrogen that can react violently with the chlorine in the system. The hydrogen level must be monitored closely and steps taken when required to insure that it is kept below the explosive range. These systems consist of multiple stages of scrubbing, absorption/reaction, and refrigeration of the stream, with the chlorine being recovered by desorption/reaction from the scrubbing medium. The capability to divert the tail gas to a caustic scrubber should be included in the process to handle upsets or outages.

The liquid chlorine is transferred to scale tanks or spheres for storage prior to shipment. All liquid lines are equipped with expansion chambers between the valves to allow for the expansion of trapped liquid without pipeline rupture. The storage tank vapors are equalized to the drying tower system for recovery. Vapors from transportation equipment (i.e. tank cars, tank trucks, and barges) are recovered by a vacuum system. Pumps are equipped with double seals with a dry air pad to eliminate chlorine leakage at the seals. Valves and cocks are of a special leak-free design with bellows or multiseal construction.

Mercury

The control of mercury emissions is dependent on equipment and facility design, a good preventive maintenance program, and observation of appropriate operations procedures.

The primary cell is operated at a slight vacuum imposed by the chlorine compressor to eliminate emissions. The end boxes, which are the mercury inlet and outlet chambers, should have vaportight covers. In addition, water is added to the end boxes to wash, cool, and control vapor emissions from the mercury. A slight vacuum is pulled on the end boxes by a fume system to further reduce the possibility of mercury release. These vent streams are cooled, scrubbed, and/or passed through treated carbon to insure a mercury-free discharge. Mercury pump tanks, pumps, and seal pots are treated in the same way: vacuum operation, cover water, and tight covers.

Cell maintenance procedures and program quality are extremely important to the prevention of the escape of mercury vapors. Leaks in pipes, flanges, and other equipment must be repaired immediately and a preventive maintenance program should be in place to prevent their occurrence. Any time that the cell cover or top must be removed for maintenance, the cell should be cooled and vented and, when possible, the bottom kept covered with water or brine to reduce emissions. The time that the cell is open and the frequency of such openings should be held to the minimum.

A potentially major source of mercury loss from the system is with the hydrogen generated in the decomposer or denuder cell. The hydrogen from the cells should be cooled at each individual cell and the pipes arranged to return condensed mercury directly to the cell, thus reducing handling of the mercury in later recovery steps. The hydrogen product is further cooled in refrigerated coolers and/or scrubbed with chemical scrubbers in series to recover the maximum quantity of mercury. Liquid mercury recovered is recycled directly to the cells and the mercury recovered in scrubbers can be returned with the brine.

The floors in the cell room area should be smooth with no cracks and all seams sealed, so that no mercury or mercury solutions can be held for later release. Trenches should be sloped to prevent mercury puddles and a layer of water kept on sumps. Where possible, wastewaters are recycled and when this is not possible, they are treated to remove mercury before they are discharged.

A good inspection program involving daily inspections and floor washdown can further limit emissions. Special vacuum cleaners equipped with carbon filters are available for the cleanup of mercury droplets discovered by the inspection program.

Operator training and compliance with operating standards are an important part of the maintenance of minimum emissions. They involve housekeeping, spill control, personal hygiene, and control of operating parameters that affect emission levels.

A waiver from cell room emissions measurement is available if the requirements of Regulation 40CFR61.53(c)(4)—which generally spell out design, maintenance, inspection, and operation requirements for mercury-cell rooms—are met. The Environmental Protection Agency assumes that a facility meeting these requirements will emit less than 1300 grams per day of mercury from the cell room itself. Well-designed systems can reduce emissions to less than 100 grams per day from end-box systems and to less than 10 grams per day from the hydrogen system.

Asbestos

The control of asbestos (friable) emissions starts with the receipt of the asbestos, which arrives in airtight plastic bags, shrink wrapped with plastic, on pallets in boxcars or trailer trucks. The pallet loads are inspected before removal is started. If any tears or spills are noted, they are repaired with tape and the spilled material cleaned up using a vacuum cleaner. Workers wear special clothing and masks during all phases of unloading. The asbestos bags should be stored in a dedicated area with appropriate warning signs. If water is used for the cleanup of spills, care must be taken to limit pressure to insure that fibers are not broadcast into the air.

Periodically, a pallet of asbestos bags is transferred to the cell repair area. Access to this area is limited and appropriate warning signs are posted. When it is necessary to add asbestos to the liquid bath used to produce the asbestos diaphragms, a bag is placed inside a glove box through a hinged Plexiglas door. The box is equipped with rubber gloves hermetically sealed through openings into the box. These gloves are used for all operations inside the glove box. A vacuum is pulled on the box system through the vacuum mix tank, which insures that no fibers escape the box system. The asbestos bags can be opened inside the glove box and the asbestos transferred to the mix tank through a transfer pipe. From this point on, the asbestos is handled in the wet (nonfriable) state, eliminating exposure. The empty asbestos bags are transferred to a clean disposal bag attached to an opening in the glove box. Before adding the next bag, the box can be cleaned with a vacuum wand located inside the box. All spills of asbestos slurry are hosed down as required to prevent any later generation of fibers.

Cell diaphragms eventually must be replaced because of the gradual plugging caused by trace impurities in the brine. At that time, the wet asbestos is washed from the cathode into a sump. It is then recovered in a filter operation and, while still in the wet form, is placed in plastic-lined boxes for secure landfill. Cleanliness is important and any spilled material should be washed to the sump for recovery.

Hydrogen

Hydrogen is a very light flammable gas that does not appear on regulatory lists. Hydrogen from mercury cells can contain mercury, for which the relevant control measures are as discussed under "Mercury." The major emphasis for hydrogen is on control systems to provide safety in the handling of this extremely flammable gas.

Cell and suction header pressures are maintained near atmospheric pressure to reduce losses from flanges. Because of the difficulty in preventing hydrogen from leaking through isolation and vent valves, the use of water seals is recommended where pressure permits. Double-block and bleed arrangements are recommended for higher-pressure service. High-pressure valves can be checked for leak-through by venting through seal tanks that can be filled with water to determine valve tightness.

References

1. *Atmospheric Emissions from Chlor-Alkali Manufacture,* Publication AP-80, U.S. Environmental Protection Agency, Air Pollution Control Office, Research Triangle Park, NC, January 1971.
2. A. L. Duprey, *Compilation of Air Pollutant Emission Factors,* PHS Publication 999-AP-42, U.S. Department of Health, Education, and Welfare, PHS, National Center for Air Pollution Control, Durham, NC, 1968, p 49.

Bibliography

A. E. Dungan, Chlorine Institute, Washington DC, Personal communication, December 1990.

P. M. King, PPG Industries, Inc., Pittsburgh, PA, Personal communication, August 1990.

R. J. Samelson, PPG Industries, Inc., Pittsburgh, PA, Personal communication, August 1990.

ETHYLENE OXIDE

Robert F. Dye

Ethylene oxide (EO) is an important petrochemical intermediate used in the production of a large number of products, such as glycols for polyester fibers or antifreeze, surfactants, ethanolamines, polyethylene glycols, and glycol ethers. It is produced by the direct oxidation of ethylene carried out in the vapor phase, using air or oxygen over a silver-base catalyst. An alternative EO production process utilizing chlorohydrin intermediate no longer is used in this country. As of January 1, 1986, 12 companies, at 13 locations, most in the Gulf Coast area, produced EO in the United States. Nameplate capacity for these facilities totaled about 6.50×10^9 pounds, approximately 37% of the global EO capacity of 17.39×10^9 pounds. Historically, the U.S. industry has operated at about 85% of nameplate capacity and captive use (producers' on-site use) levels have been approximately 90% of actual EO production rates.[1]

Ethylene oxide is a member of the epoxide family of chemicals, named *oxirane* by the International Union of Pure and Applied Chemistry. The Chemical Abstracts Service registry number for the compound is 75-21-8. It is a colorless gas condensing at low temperatures to a mobile liquid. It is miscible in all proportions with water, alcohol,

TABLE 1. Physical Properties of Ethylene Oxide, C_2H_4O

Property	Value
Molecular weight	44.053
Physical state, room temperature	Colorless gas
Freezing point, °C	−112.5
Boiling point, °C at 101.3 kPa (760 mmHg)	10.4
Flash point (open cup), °C	−18
Density at 4°C, kg/m.^{3}	890
Flammable limits in air, %vol	3–100
Autoignition temperature, °C in air at 1 atm	429
Decomposition temperature, °C in absence of air	560
Latent heat of vaporization, kJ/kg	569
Specific heat (liquid), cal/°C-g at 20°C	0.44
Heat of polymerization of liquid, kJ/kg mole	92,100
Heat of decomposition of gas, kJ/kg mol	83,700
Heat of combustion, kJ/kg	29,400
Threshold limit value (1989), ppmv	1
Miscible with water at all concentrations	—

Source: References 2 and 3.

ether, and most organic solvents. Its vapors are flammable and explosive. Some of EO's physical properties are given in Table 1.

PROCESS DESCRIPTION[4]

This section covers (1) a discussion of the process chemistry and design principles of the direct oxidation of ethylene to ethylene oxide, (2) process descriptions of the air- and oxygen-oxidation processes, and (3) brief comments on wastes and emissions associated with ethylene oxide production.

Process Chemistry and Design Principles

In commercial processes, the direct vapor-phase oxidation of ethylene to ethylene oxide is carried out with a silver catalyst at 10–30 atm pressure and 200–300°C. The main oxidation reaction is as follows:

$$CH_2 = CH_2 + 1/2\ O_2 \xrightarrow{Ag} H_2C\overset{O}{—}CH_2 + 106.7\ kJ^* \quad \text{(Ethylene oxide)} \qquad (1)$$

A second reaction is as follows:

$$CH_2 = CH_2 + 3\ O_2 \longrightarrow 2CO_2 + 2H_2O + 1323\ kJ^* \qquad (2)$$

*Calculated at 15 atm and 250°C using JANAF thermochemical tables.

These reactions also produce acetaldehyde at less than 0.1% basis EO product and trace amounts of formaldehyde.

Both ethylene reaction rate and purge losses increase when operated with higher concentrations of ethylene and oxygen in the reactor feed. Thus, depending on the process and the oxidant, the ethylene concentration in the reactor feed will vary between 5 and 40 mol %. Inlet oxygen concentrations will be controlled between 5 and 9 mol %, as set by flammability considerations. Normally, air-oxidation and oxygen-oxidation reactors operate at the low and the high end of this concentration range respectively.

By controlling the appropriate reactor operating conditions, modern reactor systems operate at selectivities (moles EO formed per 100 moles ethylene reacted) in the range of 65% to 75% for air plants and 72% to 82% for oxygen plants. The selectivity range for the air process is lower because of the purge reactors, which, of necessity, operate with low ethylene concentrations and at high conversions per pass. The high reactor efficiencies discussed here are achieved by operating with low ethylene conversions per pass and by adding controlled trace levels of chlorinated inhibitors to moderate the reaction, suppressing reaction 2.

Since the reactor exit gas contains substantial amounts of unconverted ethylene, a large gas recycle back to the reactor is required to achieve a high overall ethylene conversion in both the air- and oxygen-based processes. Before reactor exit gas is recycled, it is first scrubbed, using water to absorb the EO—which, if not removed from the feed, would seriously inhibit the reaction efficiency, as well as result in excessive oxidation losses to carbon dioxide and water.

In the air-based process, a large amount of nitrogen is introduced in the air feed to the reaction. Hence a sizable fraction of the scrubbed reactor gas must be withdrawn to remove this nitrogen. This purge stream is usually mixed with additional air to boost the oxygen concentration and then passed through a secondary, or purge reactor, similar to the primary reactor, to convert most of the contained ethylene. In large air-based plants, a third-stage reactor can often be economically justified to achieve improved yields on ethylene. Ethylene conversions in excess of 95% of the total feed are common for three-stage configurations.

In the oxygen-based process, since high-purity oxygen is used, much less inert gas is introduced into the reaction system. This reduces the required purge and the unconverted ethylene can be more completely recycled. To avoid a buildup of by-product CO_2, a portion of the recycled gas is sent to a CO_2-removal system where the CO_2 is preferentially absorbed and vented, with a minimal loss of ethylene. The scrubbed gas is then returned to the recycle gas system. The reduced purge permits the oxygen-based reactor to operate at higher feed ethylene concentrations without excessive vent losses and allows the use of different diluent gases.

In the air-based process, nitrogen *must* be used as the diluent since any other diluent would be vented together with the nitrogen. The oxygen process, however, can use other diluents, such as methane (most common), which raise the flammable limit and allow the safe use of higher concentrations of ethylene and oxygen. This contributes directly to increased productivity and higher selectivities.

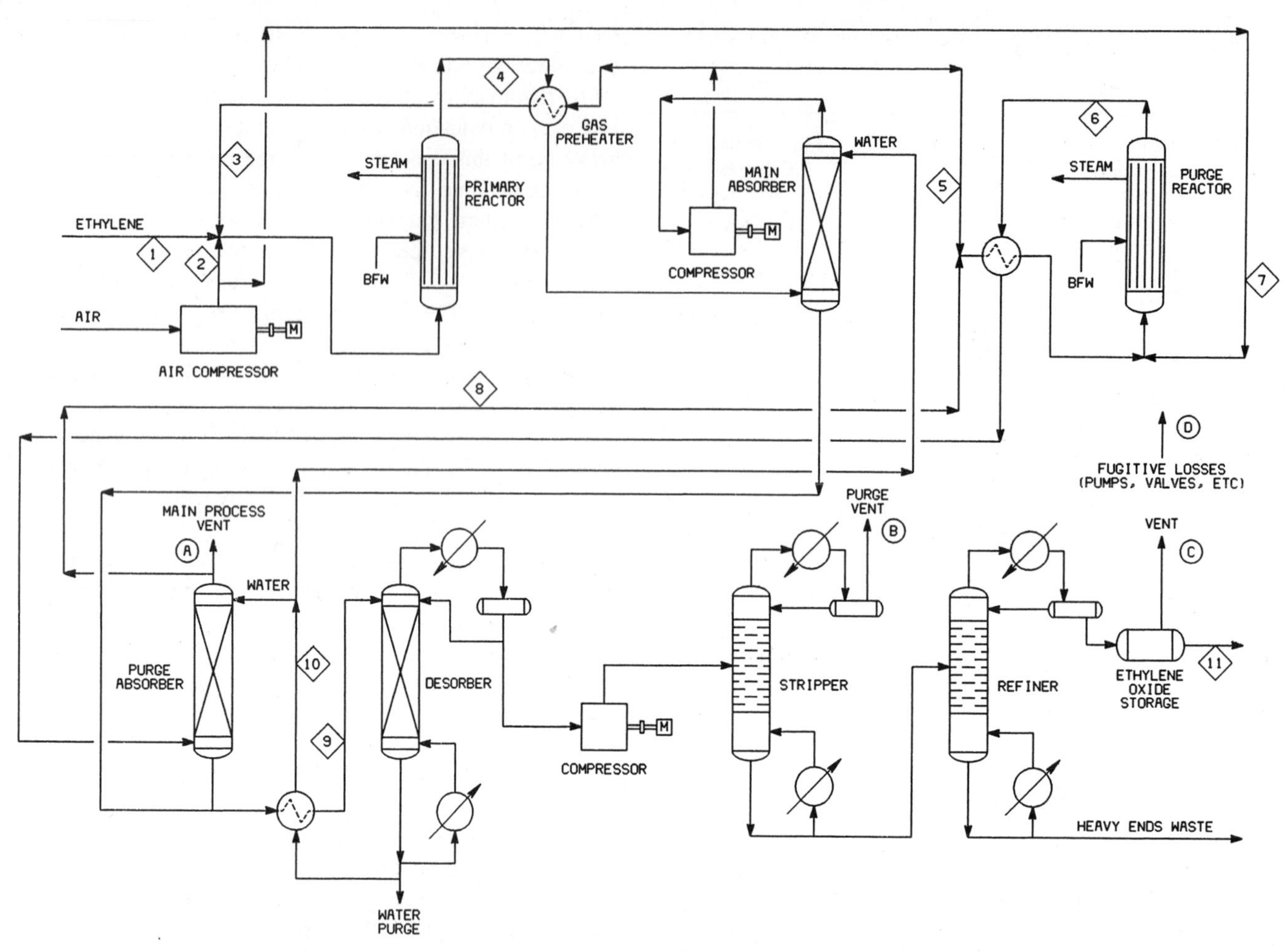

FIGURE 1. Production of Ethylene Oxide by the Air-Oxidation Process

Both the air- and oxygen-based processes use shell and tube reactors, oriented vertically, with the catalyst contained in the tubes and a cooling medium in the shell. Boiling water is now the preferred shell-side coolant for new reactors and is favored over the nonboiling heat-transfer oils and boiling kerosene used in the past.

Air-Oxidation Process

Figure 1 shows the basic operations in a two-stage air-based EO reaction system. Ethylene and compressed air (streams 1 and 2) are mixed with a recycle gas (stream 3) and then fed to one of several multitube catalytic reactors operated in parallel. Reaction temperature is controlled by boiling water on the shell side, generating high-pressure steam that is exported or used in the plant.

The hot effluent gas leaving the primary reactor (stream 4) containing EO is cooled by heat exchange with cold reactor feed gas. The cooled gas then passes into the main absorber where the contained EO is absorbed in water, producing a dilute aqueous solution. The scrubbed EO-free gas from the main absorber is compressed and the major portion is recycled to the primary reactor via the gas preheater. The remainder of the gas (stream 5) is sent to the purge reactor to reject not only the nitrogen introduced in the air feed, but also the CO_2 produced in the reactor.

The purge reactor completes the reaction of most of the ethylene remaining in the purge gas from the primary reactor system. The basic flow scheme for this reactor is essentially the same as for the primary reactor, with additional air (stream 7) being added to provide the required oxygen. In the purge reactor system, the scrubbed gas from the reactor is partly recycled to the purge reactor (stream 8), with the balance vented to purge inerts (vent A). The dilute aqueous solutions of EO, CO_2, and other volatile organic compounds from the absorbers are combined (stream 9) and fed to the desorber, where the EO and dissolved inerts are distilled under reduced pressure. The desorber bottoms stream, virtually free of EO, is recirculated to the absorbers (stream 10). The crude EO from the desorber is sent to a stripper for removal of CO_2 and inert gases and then to a final refining column. (*Note:* In some plants, the EO from the absorbers [stream 9 in Figure 1] may go first to a stripper, then to a light-ends column. The nomenclature is different, but the basic operations are the same.) Light gases separated in the stripper as a tops stream (vent B) are

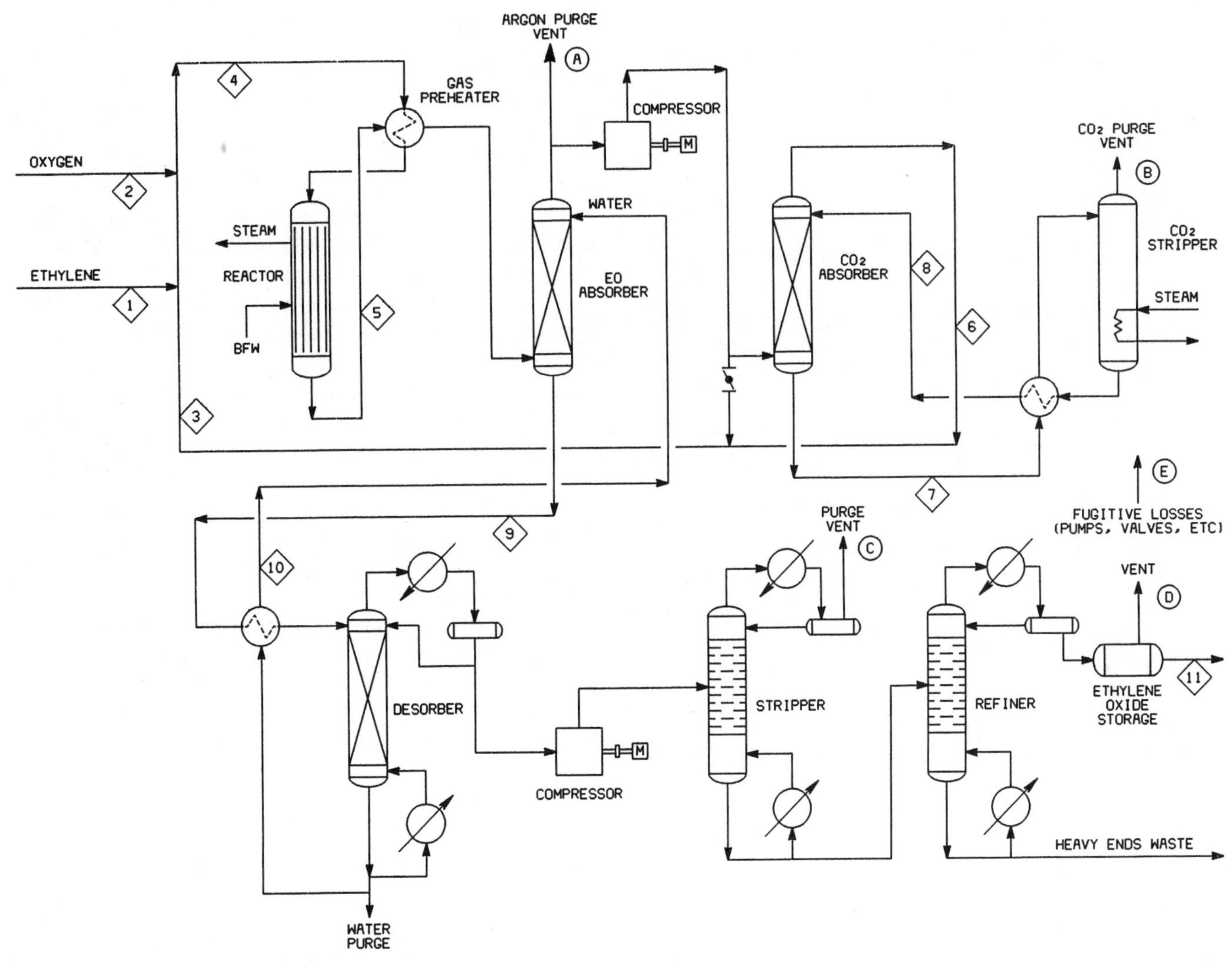

FIGURE 2. Production of Ethylene Oxide by the Oxygen-Oxidation Process

usually sent to a reabsorber for EO recovery prior to passing to atmosphere. The final product (stream 11), 99.5–99.9 mol % EO, is stored under nitrogen pressure in pressurized tanks, typically, at 50–60 psig. In some plants, crude EO is sent directly to a glycol plant without undergoing complete refining.

Oxygen-Oxidation Process

Virtually all the differences between the air-oxidation and oxygen-oxidation processes result from the difference in the oxygen content of the oxidants, namely, 21 mol % for the air-based process versus 98–99.5 mol % for the oxygen-based case. Figure 2 shows the oxygen-oxidation process.

The hot effluent gas from the reactor (stream 5) is cooled by heat exchange and then scrubbed with water in the EO absorber to recover ethylene oxide. The scrubbed gas from the absorber is then compressed and recycled back to the reactor inlet (stream 3). Part of the recycle gas stream is first sent to a CO_2-removal section where CO_2 produced in the EO reactor is chemically absorbed in a recirculated solution of hot potassium carbonate (stream 8). The CO_2-lean gas from the CO_2 absorber (stream 6) is then returned to the recycle gas system. Buildup of by-product CO_2, if not removed, could inhibit catalyst performance, as well as contribute to a rise in recycle loop pressure. To avoid excessive buildup of inerts such as argon, introduced via the oxygen feed, a small purge of recycle gas (vent A) is sent to fuel or is incinerated.

The CO_2-rich carbonate solution is regenerated in the CO_2 stripper, with the desorbed CO_2 being vented to atmosphere from the top of the column (vent B).

Ethylene oxide recovery from the EO absorber bottoms (stream 9) and refining of the crude EO are the same as described for the air-oxidation process.

Vents and Wastes

Both processes can be designed significantly to reduce venting of ethylene and ethylene oxide to the atmosphere.

This is somewhat more difficult and costly for the air-oxidation process because the main cycle gas purge stream (vent A, Figure 1) is large compared with the argon vent (vent A, Figure 2) of the oxygen-oxidation route. The latter is usually incinerated satisfactorily by directing the stream to the plant steam boiler firebox.

Aqueous waste problems are essentially the same for both processes, although many operators of the oxygen-oxidation process have some added problems related to chemicals used in the hot carbonate CO_2 absorption system. The following two sections will discuss vents and air emissions and the process equipment required for containing, controlling, and completely eliminating some of these streams.

AIR EMISSIONS CHARACTERIZATION

Emphasis in this section will be mainly on EO as a component of streams vented or leaked to the atmosphere. Other organics, particularly ethylene, are also present in most vented streams, and measures discussed to contain or control EO will usually serve also to reduce these organics.

If all vents and emissions for the EO processes shown by Figures 1 and 2 were allowed to go to the environment without control measures, most rates and concentrations for EO and other organics would be in excess of those permitted by current air pollution regulations. In reality, not one U.S. facility is known to operate with uncontrolled emissions as indicated in Figures 1 and 2. Moreover, in recent years, much progress has been made by U.S. producers in ongoing efforts to reduce and contain pollutants from EO production. It is important to note that the level of EO emissions from any given plant is a function of such variables as capacity, throughput, and control measures in place. Accordingly, it is the object of this section to characterize EO process vents and emissions as a prelude to discussing control equipment and engineering choices for an abatement program.

Ethylene oxide gas is highly flammable, as well as being a health hazard. A 1983 study concluded that excessive exposure to EO can cause chromosomal rearrangement akin to mutagenic and carcinogenic effects.[5] Subsequently, the U.S. Occupational Safety and Health Administration announced that, as of August 1984, a new standard for workplace exposure to EO would be 1 ppm for an eight-hour time-weighted average, instead of the previous 50-ppm limit.[6] Ethylene oxide released to the environment has an atmospheric lifetime, estimated in days required for the compound to be reduced to $1/e$ (37%) of its original value, of between 200 and 600 days.[7] Half-lives of EO in the aqueous phase have been reported to fall between 200 and 400 hours for a wide variety of types of water, such as sterile distilled water, seawater, fresh water, and sterile and nonsterile river water.[8]

Air-Oxidation Process Emissions

Vent A (Figure 1) is potentially the largest vent source of EO emissions of air-oxidation plants. This stream also contains nitrogen, unreacted oxygen from the air feed, ethane and unreacted ethylene from the ethylene feed, and by-product CO_2. In any given facility, the exact composition of vent A depends on ethylene feed purity, reactor operating conditions, number of purge stages, and absorber operating conditions. Because EO is completely soluble in water, the purge absorber (Figure 1) can be 99.9+% effective in its recovery, and hence the EO content of vent A is quite low. Few, if any, EO production facilities in the United States now vent this stream to the atmosphere. It is normally burned in a thermal or catalytic oxidizer, with about 80% and 100% destruction, respectively, of organics and EO content. Subsequently, vent B from the stripper releases the inert gases and ethylene that were absorbed in the main and purge absorbers. The composition of this stream depends on the solubilities of the gases in the circulating absorbent (water). The quantity is established by the water circulation rate. Ethylene oxide in this stripper vent is normally recovered in a reabsorber or vent scrubber similar to one described later in this discussion. The scrubbed vent stream therefrom is combusted in a boiler firebox, effecting essentially 100% destruction of EO content.

Storage Emissions

Ethylene oxide is a gas at ambient temperatures and hence it is normally stored under nitrogen at approximately 50°F and 60 psig. Emissions from storage, vent C in Figure 1, result from nitrogen displacement during filling operations. Often, Vent C includes displacement nitrogen from a 20,000-gallon railroad tank car, normally loaded directly from plant rundown/storage tanks.

Many plants route vent C to a vent scrubber for recovery of EO. The scrubber tops nitrogen stream goes to incineration.

Oxygen-Oxidation Process Emissions

The volume of vent A (Figure 2) is much smaller than the corresponding vent in the air-oxidation process. Prior to any abatement, the EO content ranges from 0.01 to 0.025 mol % (100–250 ppm), about the same as for the air-oxidation case. This vent also contains argon, nitrogen, and unreacted oxygen from the oxygen feed, along with ethane and unreacted ethylene from the ethylene feed and by-product CO_2. The composition and quantity of vent A depend directly on the purity of the feed oxygen and ethylene. In many plants, methane is a component of this stream. Methane, which is inert in the oxidation reaction, is added as ballast because it raises the flammability limit of reactor feed gas favorably. Vent A is incinerated or sent to boiler

TABLE 2. Oxygen-Oxidation Process Vents, Pounds/Ton Ethylene Oxide Capacity

Point		Ethylene Oxide	Ethylene	Methane
Argon purge	(vent A)	0.022	15.9	14.6
CO_2 purge	(vent B)	0.14	3.65	0.21
Purge	(vent C)	84.3	23.6	6.7
Storage	(vent D)	2.74	—	—

Note: Air-oxidation vents B and C are similar in size and content to vents C and D for oxygen-based operation. Quantities are prior to use of control technology.

fireboxes in virtually all U.S. oxygen-oxidation plants, thus eliminating any organic emissions from this source.

The CO_2 stripper vent, vent B, is, typically, 99.8+ mol % CO_2 and water. Ethylene oxide content varies from facility to facility, depending on whether or not the CO_2 absorber bottoms stream is processed in intermediate operations, such as a flasher or nitrogen stripper, before final CO_2 rejection in the stripper. The ranges of EO and ethylene contents are 10–100 ppm and 1000–2000 ppm respectively. Table 2 gives some emission data for this stream and other vents prior to use of any control technology.[9] Although a number of EO units allow vent B to go to the atmosphere, some producers process this stream to recover CO_2 for sale. Others send this stream to syngas production. Without fuel-gas enrichment, incineration of vent B is not a feasible option.

Vent C, the EO stripper vent, is equivalent to vent B for the air-oxidation case. Sometimes, after a reabsorber wash for EO recovery, the exit gas is recompressed and then injected into the reactor recycle gas stream. In this instance, inerts are purged via the argon vent. Vent D in Figure 2 derives from displacement of storage tank nitrogen in filling, as in the case of vent C for the air-oxidation case.

A process cooling tower (not shown in Figure 2) is used in many oxygen-oxidation plants to augment cooling of the recirculated absorbent water. Cooling is effected by evaporation when the process water and air are contacted. The residual EO content of the desorber bottoms is stripped to the environment in this operation. Because of this, process cooling towers have lost favor with EO producers in recent years and are being phased out. Replacement absorbent cooling systems are closed-loop operations with auxiliary cooling effected by refrigerated chillers. The most common of these are lithium bromide (Li-Br) absorption refrigeration units.

Equipment Leaks

Those emissions that result when a process fluid, either liquid or gaseous, leaks from plant equipment are called "fugitive" emissions. Among the many potential sources for leaks, are pump and compressor seals, in-line process and open-ended valves, flanges, and sampling connections. Emissions from such equipment depend on many variables, including design and maintenance practices.

One early publication on the subject is *VOC Fugitive Emissions in Synthetic Organic Chemicals Manufacturing Industry—Background Information for Proposed Standards,* EPA-450/3-80-033a, November 1980. In a 1985 publication,[10] fugitive emission estimates were made by applying the chemical industry's average emission factors to the number of pump and compressor seals, valves, flanges, and so on, in a typical EO production facility and by adjusting these totals to reflect the EO content in each stream. For a hypothetical model plant, the analysis projects fugitive EO emissions ranging from 148 kg/day to 188 kg/day in oxygen and air facilities respectively. More recently (1988), a joint EPA–industry study[11] specifically examined equipment leaks from 12 of 13 EO production plants. In this study, fugitive emissions ranged from 8.7 kg/day to 40.4 kg/day. This reduced level of emissions represented the application of a range of engineering and work practice controls, but still relies on the basic chemical industry average leak correlations developed around 1980. Application of more stringent leak detection and repair procedures may reduce this level of emissions further. In addition, quantification of emissions could be improved by determining leak rates using procedures involving "bagging" of sources, as illustrated by Figure 3.

Liquid Purge Stream Emissions

The desorber bottoms water purge and EO refiner heavy-ends waste normally contain low levels of EO. Most common treatment used in the industry for the former is biooxidation. However, both are compatible with an on-site glycols process and may be merged into the feed to that operation.

Desorber bottoms may have an EO content of 100–500 ppm. The water purge, even after cooling, could lose a major fraction of its EO content to the atmosphere, depending on the configuration of the downstream biotreatment process. Secondary emissions from this source can be essentially eliminated by in-line hydrolysis prior to admittance to biotreatment.

The refiner heavy-ends stream, containing EO, water, and glycol, is normally sent to incineration, but can also be a candidate for hydrolysis treatment to remove its EO content as described.

AIR POLLUTION CONTROL MEASURES

In EO production facilities, safety is of paramount importance. Listed in Table 3 are measures that can be taken largely for safety reasons that would contribute to reduced EO emissions.

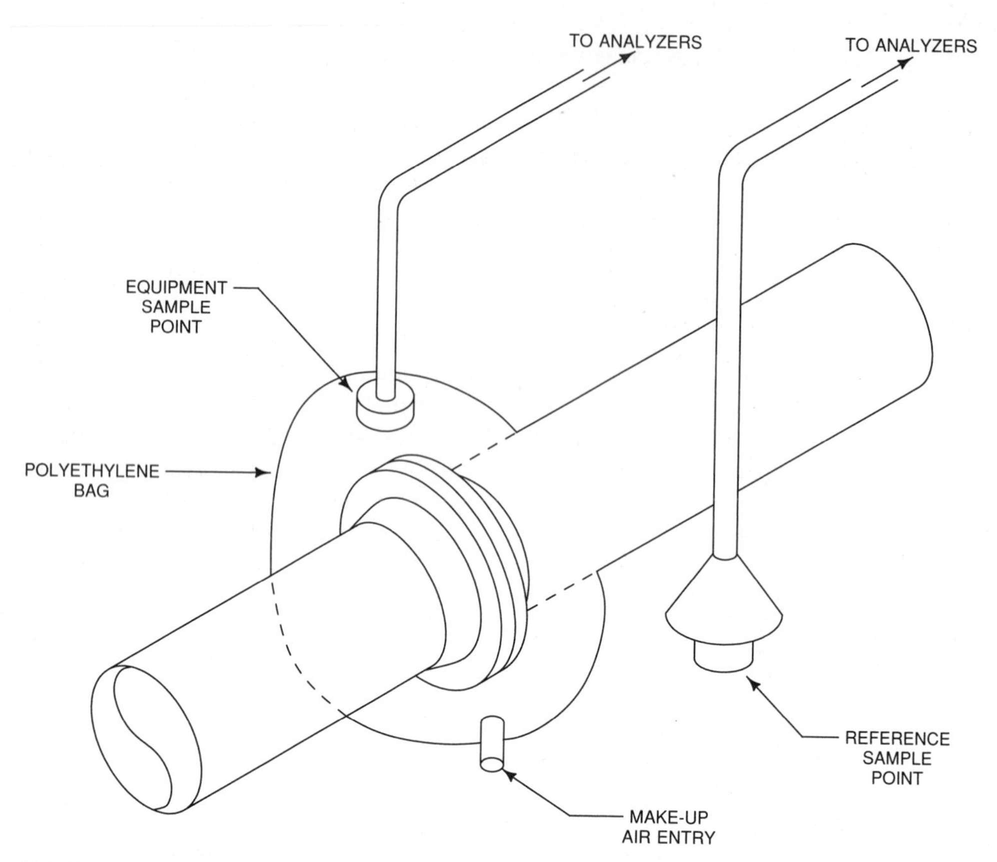

FIGURE 3. Bagged Sample Point for Leakage Measurements

Vent Scrubbers

Figure 4 gives details for a vent scrubber capable of reducing the EO content of a vent gas stream of 1512 lb/h from 50 mol % EO to the 5-ppm range.[12] The premise is that the vent gas feed for the scheme originates from two sources. One is EO refiner reflux accumulator sweep gas. The second is nitrogen gas being displaced from a rundown/storage tank. Each is assumed to be saturated at its originating conditions. The tops stream, 591 lb/h of nitrogen plus water vapor in Figure 4, is sent to the boiler firebox in most plants. It is feasible to merge this stream with other vents in the process served. The "fat" absorbent water containing the EO picked up in the vent scrubber is directed to the main train desorber for EO recovery.

Centrifugal Compressor Shaft Seals[13]

Mechanical shaft seals used for most EO recycle gas compressors have provisions for injection of a compatible buffer gas via a labyrinth seal. A commonly used buffer gas is nitrogen for the air-based process and oxygen-based plants that use nitrogen ballast gas. For oxygen-based plants that use methane ballast gas, methane may be used. In a few

TABLE 3. Ethylene Oxide Plant Safety Features

1. Install EO and flammable gas leak detectors at critical pump seals, flanges, and so on, with regular monitoring and sample analysis.
2. Make routine gasket replacements during planned maintenance turnarounds.
3. Use all-welded construction where feasible to minimize number of flanged joints.
4. Provide upstream rupture disks for safety/relief valves.
5. Use closed-loop sample systems.
6. Use a pressurized buffer gas in labyrinth shaft seals of centrifugal EO compressors (discussed with reference to Figure 5).
7. Equip EO-handling pumps with double mechanical seals of the tandem type (discussed with reference to Figures 6 and 7).
8. Further, provide alarms and automatic pump shutdown switches for high-temperature and seal failure.
9. Have plant personnel make routine inspections for leaks, with immediate leak repair.

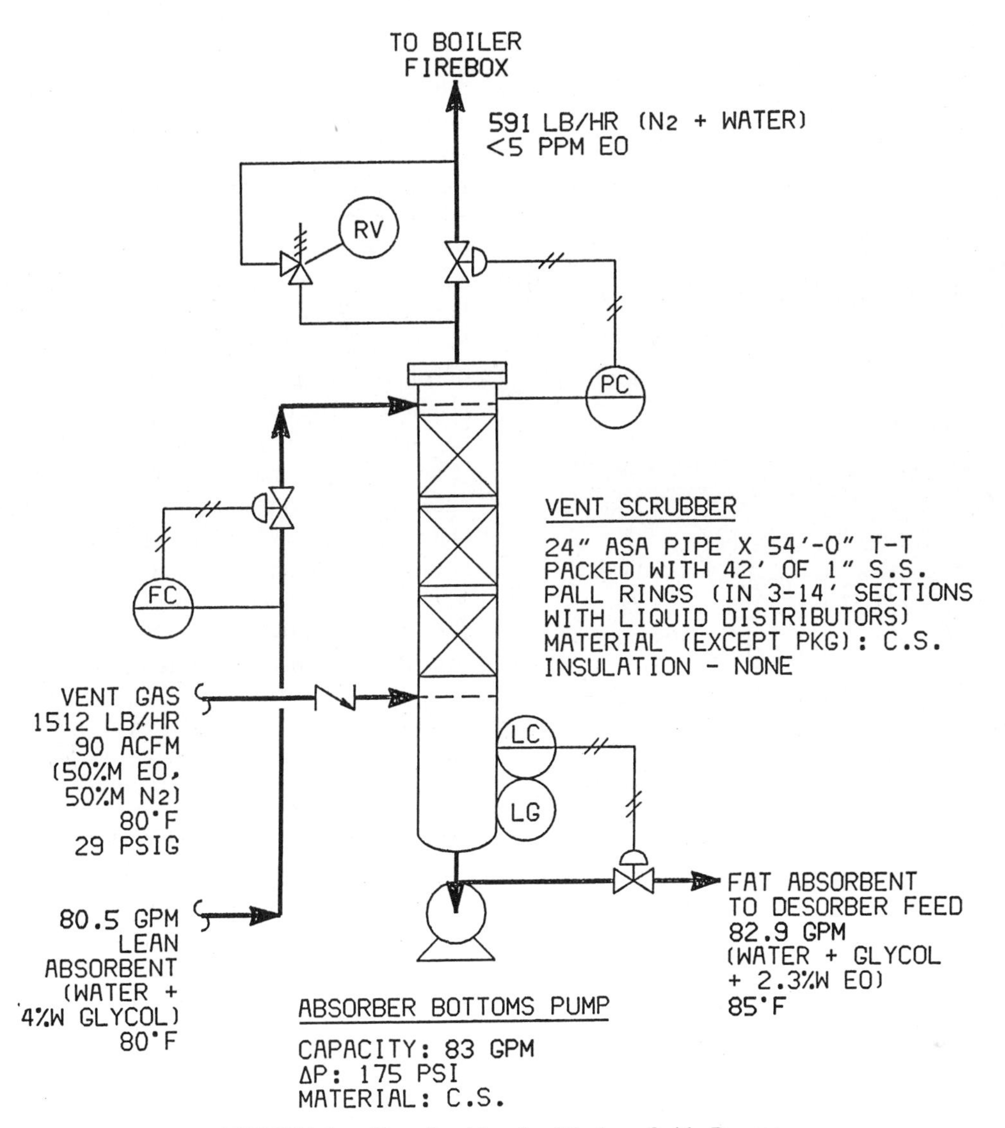

FIGURE 4. Vent Scrubber for Ethylene Oxide Process

instances, steam has been used as the buffer gas. Figure 5 shows special details of a mechanical seal as referred to here for centrifugal compressors. Associated equipment with this type of seal (not shown) is a seal oil circulating system with cooler and vented drain pot. The vented gas from this drain pot should be routed to flare or to incineration.

Centrifugal Pump Shaft Seals[14]

A special double-mechanical-seal configuration, referred to as a tandem seal, is recommended for EO-handling pumps. The rule of thumb that most users follow is to provide this seal for pumps handling 50+% EO streams. No known producers at present use this type of shaft seal for dilute water solutions containing EO, such as the bottoms streams from the main absorber/purge absorber system. Figure 6 is a detailed representation of this seal type. Figure 7 gives additional features of tandem seal application. As noted above for the compressor oil seal drain pot vent, the vent from the tandem seal reservoir should go to flare or incineration. A compatible seal fluid, and most commonly used for this system, is water. Ethylene glycol is also acceptable.

Valves

To eliminate fugitive emissions completely from valve stems, bellows-sealed valves may be applicable in certain services. Figure 8 shows a bellows-sealed manual valve. Similar seals are available for motor-actuated process control valves. Failure of bellows-sealed valves, however, may result in significant emissions until the valve can be isolated.

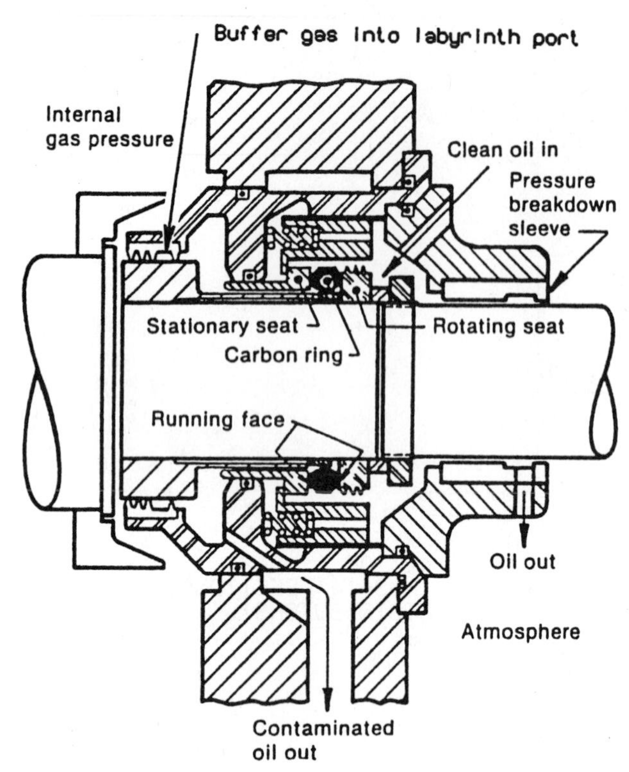

FIGURE 5. Mechanical (Contact) Shaft Seal for Centrifugal Compressors (API Standard 617. *Centrifugal Compressors for General Refinery Service,* Fifth Edition, April 1988. Reprinted courtesy of the American Petroleum Institute.)

Recent Development

Although pumps without seals have been in use by the chemical process industry for about 30 years, only in the past few years have EO producers given consideration to magnetic-drive centrifugal pumps as an alternative for reducing emissions. This type of pump is designed to be leakfree by eliminating seals altogether. For EO service, use of double mechanical seals as described above have been the standard. But double mechanical seals, although highly reliable with the prospect for the simultaneous failure of both seals remote, *have failed* and serious plant fires have resulted. Today, only one EO producer known to this writer has had in use—for about three years—a number of magnetic-drive pumps, chiefly in EO transfer service. This producer has been so pleased with sealless pump performance that a large number have been placed on order to be used in refiner column reflux, bottoms, and other EO-handling service. Figure 9 shows some of the features of one vendor's magnetic-drive pump. This particular vendor offers an American National Standards Institute version.

Costs

Table 4 gives some June 1990 costs that may be of interest.

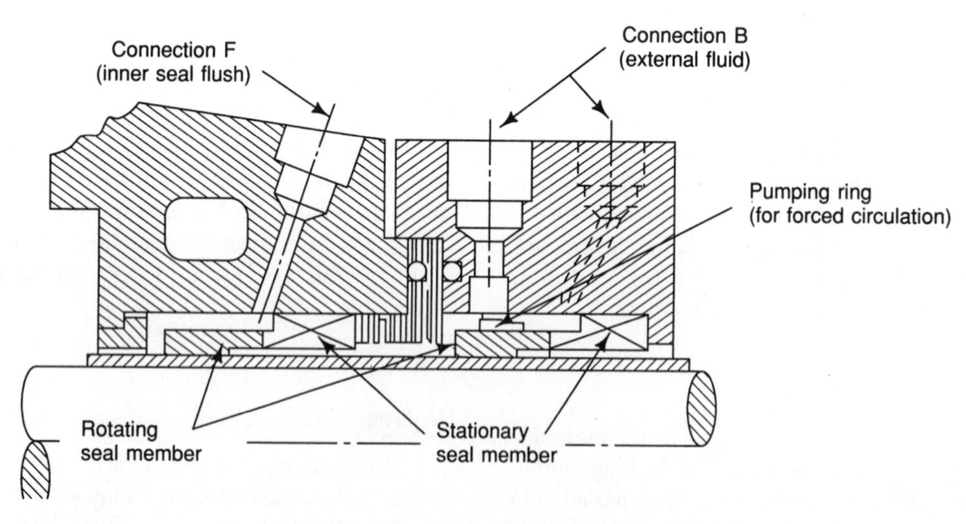

FIGURE 6. Tandem Seals for Centrifugal Pumps. (API Standard 610, *Centrifugal Pumps for General Refinery Service,* Seventh Edition, February 1989. Reprinted courtesy of the American Petroleum Institute.)

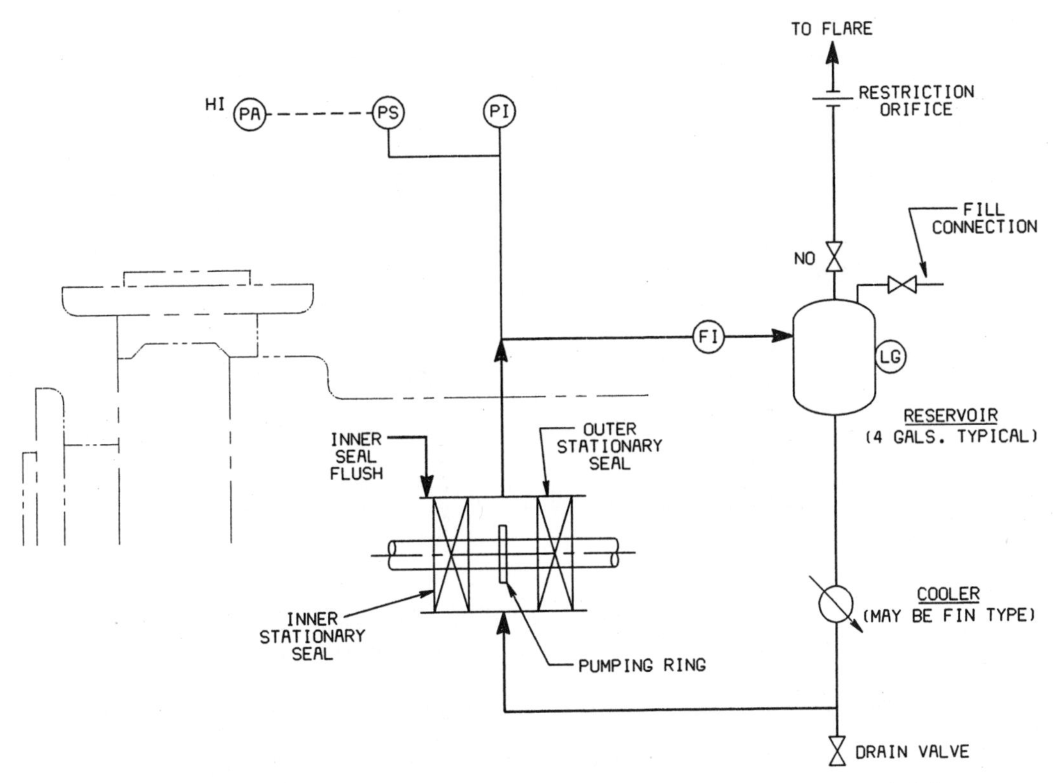

FIGURE 7. Tandem Seal Flush Plan for Centrifugal Pumps (Plan 52, API Standard 610, *Centrifugal Pumps for General Refinery Service,* Seventh Edition, February 1989. Reprinted courtesy of the American Petroleum Institute.)

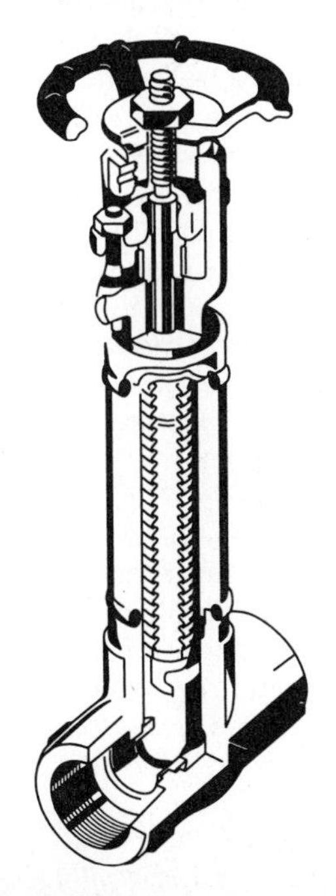

METALLIC BELLOWS IS WELDED AT THE TOP TO THE BONNET ASSEMBLY AND AT THE BOTTOM TO THE STEM, A CONNECTOR TUBE IS THEN WELDED TO THE BODY AND THE BONNET TO PROVIDE A COMPLETELY SEALED UNIT. A CONVENTIONAL GLAND IS STILL EMPLOYED TO PREVENT THE INGRESS OF MOISTURE OR DEBRIS INTO THE INSIDE OF THE BELLOWS AND TO PROVIDE A BACK-UP TO THE BELLOWS IN THE EVENT OF FAILURE.

FIGURE 8. Bellows-Sealed Valve (Hattersley Heaton)

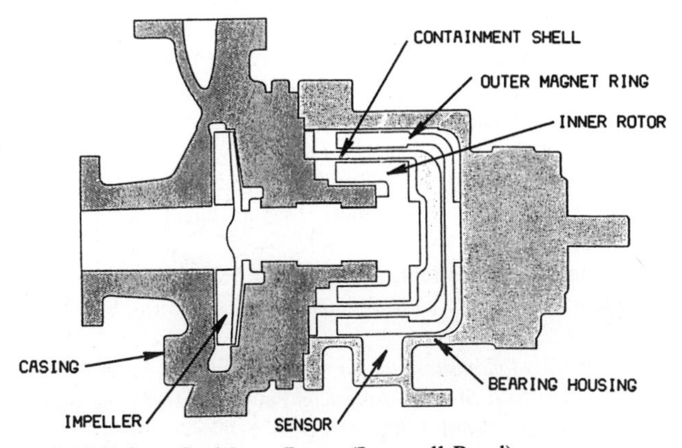

FIGURE 9. Seal-Less Pump (Ingersoll-Rand)

TABLE 4. Emissions Containment Equipment Costs

Item	Cost ($1000)	Remarks
Vent scrubber (Figure 4)	159	Raw installed cost, including allowances at 15%, before engineering and contractors' charges
Tandem pump seal versus single seal	5.9	Delta (difference) cost, including allowances at 15%, factory-installed basis
Manual bellows valve (6 inch, 150 psi) versus standard nonbellows	1.2	Delta cost
Motor-actuated bellows valve (6 inch) versus standard nonbellows	2.8	Delta cost, 150-psi valve
	4.2	Delta cost, 300-psi valve

Source: Reference 15.

References

1. S. A. Cogswell, "Ethylene oxide," *Chemical Economics Handbook,* SRI International, Menlo Park, CA, May 1986, pp 654.5031A–654.5032H.
2. *Guidelines for Bulk Handling of Ethylene Oxide,* Chemical Industries Association, London, 1983.
3. *Threshold Limit Values for 1989–1990,* American Conference of Governmental Industrial Hygienists, Cincinnati, OH, 1989.
4. B. R. Ozero and J. V. Procelli, "Can developments keep ethylene oxide viable?", *Hydrocarbon Proc.,* 55–58 (March 1984).
5. "Calling for tighter controls over ethylene oxide," *Chem. Week,* 132(11):35 (March 16, 1983).
6. "Ethylene oxide rule published by OSHA to cut exposure," *Chem. Marketing Reporter,* 225(26):7–20 (June 25, 1984).
7. L. T. Cupitt, *Atmospheric Persistence of Eight Air Toxics,* EPA-600/3-87-004, U.S. Environmental Protection Agency, Research Triangle Park, NC, January 1987.
8. R. A. Conway et al., "Environmental fate and effects of ethylene oxide," *Environ. Sci. Technol.,* 17(2):107–112 (February 1983).
9. R. F. Dye, Shell Oil Co., personal communications with five U.S. Gulf Coast area EO unit engineers, June 1990.
10. D. W. Markwordt, *Sources of Ethylene Oxide Emissions,* EPA-450/3-85-014, U.S. Environmental Protection Agency, Research Triangle Park, NC, April 1985.
11. *Fugitive Emissions from the Ethylene Oxide Production Industry: A Field Study,* Radian Corp., Austin, TX, April 1989 (prepared for the Chemical Manufacturers Association, Washington, DC).
12. Design developed June 1990 via PROCESS® Simulation Program, Version 4.01 L3, Simulation Sciences Inc., Fullerton, CA.
13. *Centrifugal Compressors for General Refinery Service,* API Standard 617, 5th ed., American Petroleum Institute, Washington, DC, April 1988.
14. *Centrifugal Pumps for General Refinery Service,* API Standard 610, 7th ed., American Petroleum Institute, Washington, DC, February 1989.
15. Costs developed June 1990 via COST® System, Version 18.1A, ICARUS Corp., Rockville, MD.

Bibliography

D. A. Bogyo et al. *Investigation of Selected Potential Environmental Contaminants: Epoxides,* Syracuse Research Corp., Syracuse, NY, March 1980 (prepared for the U. S. Environmental Protection Agency, Washington, DC).

L. T. Cupitt, *Fate of Toxic and Hazardous Materials in the Air Environment,* EPA-600/3-80-084, U.S. Environmental Protection Agency, Research Triangle Park, NC, August 1980.

Evaluation and Selection of Models for Estimating Air Emissions from Hazardous Waste Treatment, Storage, and Disposal Facilities, EPA-450/3-84-020, U.S. Environmental Protection Agency, Research Triangle Park, NC, December 1984.

D. E. Field et al., *Engineering and Cost Study of Air Pollution Control for the Petrochemical Industry—Vol. 6: Ethylene Oxide Manufacturing by Direct Oxidation of Ethylene,* Air Products and Chemicals, Inc., Marcus Hook, PA, June, 1975 (prepared for the U. S. Environmental Protection Agency, Research Triangle Park, NC).

Guideline Series—Control of VOC Equipment Leaks from Synthetic Organic Chemical and Polymer Manufacturing Plants, EPA-450/3-83-006, U.S. Environmental Protection Agency, Research Triangle Park, NC, March 1984.

R. K. June and R. F. Dye, "Explosive decomposition of ethylene oxide," *Plant/Operations Prog.,* 9(2):61–74 (April 1990).

V. Kalcevic and J. F. Lawson, *Organic Chemical Manufacturing, Vol. 9: Selected Processes, Report 4,* EPA-450/3-80-028d, U.S. Environmental Protection Agency, Research Triangle Park, NC, December 1980.

Kirk-Othmer, *Encyclopedia of Chemical Technology,* 3rd ed., Vol. 9, *Ethylene Oxide,* John Wiley and Sons, New York, 1984, pp 432–471.

Locating and Estimating Air Emissions From Sources of Ethylene Oxide, EPA-450/4-84-0071, U.S. Environmental Protection Agency, Research Triangle Park, NC, September 1986.

NIOSH Current Intelligence Bulletin 35, *Ethylene Oxide (EtO),* U.S. Department of Health and Human Services, Cincinnati, OH, May 22, 1981.

EXPLOSIVES

Institute of Makers of Explosives

Explosives were introduced into commerce when Marco Polo brought the art of making black powder back to Italy from China. Explosives are designed and manufactured to detonate and today's safe manufacturing processes and practices are the result of experience gained from over 500 years of manufacturing. These practices include siting explosive manufacturing facilities in accordance with the American Table of Distances, intraplant or interline separation distances, or other quantity/distance tables. Buildings are often of frangible construction protected, by barricades, and posted with personnel and product limitations. History has dictated that the manufacturing of explosives is one of minimization. They are manufactured in the smallest possible quantity to reduce the hazards to the minimum with practicality.

The commercial explosives industry is involved in manufacturing a wide variety of explosive products, which the industry generally defines as "explosive materials." Commercial explosives are typically used for blasting dirt or rock, including excavating for road building or construction, removing overburden at a strip mine, fragmenting ore in an underground mine, and quarrying. Some situations require the use of specialty products, such as oil well charges (which are used for perforating oil well casings and cladding metal parts together), seismic activities, and demolition. Because there are numerous applications for commercial explosives, there are many types of explosives, each with its own manufacturing process.

Commercial explosives can be divided into the following groups: (1) dynamite, (2) emulsion, (3) ammonium nitrate fuel oil mixtures, (4) cast boosters, (5) detonators, (6) detonating cord, (7) initiating explosives, (8) black powder, (9) water gels and slurries, (10) specialty explosives. Each of these processes uses different raw materials, methods, and equipment to manufacture the explosives. Most production plants might produce one or more of these products, but there are no plants that produce all of them at one location.

Only the manufacturing processes in the commercial explosives industry will be covered here. The discussion will not include explosives that are manufactured under the control of the Department of Defense, and also will not cover workplace air contaminants that are subject to Occupational Safety and Health Administration regulation.

DYNAMITE

Process Description

Dynamites are combinations of nitrate esters, nitrate salts such as ammonium nitrate, and wood pulp or similar absorbent materials. Nitrocellulose is also used as a plasticizer or gelatinizer, particularly in semigelatin and gelatin dynamites. The dynamite is then packaged in paper, plastic, or cardboard shells and tubes of varying diameters, lengths, and weights.

The first step in the manufacture of dynamites is the nitration of organic alcohols to produce nitrate esters that sensitize the dynamite, such as nitroglycerin (NG), ethylene glycol dinitrate (EGDN), diethylene glycol dinitrate (DEGDN), and metriol trinitrate (MTN). The nitration process involves reacting the organic substrate (usually glycerin and ethylene glycol) with a mixture of nitric and sulfuric (oleum) acids. Nitration can be conducted in batches or in a continuous process. The by-products of the nitration are nitrate esters, spent acids, and water. There are no significant air emissions from the nitration process.

Manufacture of NG

$$H_2SO_4/HNO_3 + C_3H_8C_3 \rightarrow C_3H_5N_3O_9 + HNO_3 + H_2SO_4 + H_2O$$

Manufacture of NG/EGDN

$$H_2SO_4/HNO_3 + C_3H_8O_3 + C_2H_6O_2 \rightarrow C_3H_5N_3O_9 + C_2H_4N_2O_6 + HNO_3 + H_2SO_4 + H_2O$$

Dry carbonaceous materials such as corn flour, rice hulls, and sawdust are blended together and are called "dopes" in the industry. These dopes are used to absorb the NG/EGDN. All dry ingredients are received in containers and stored in separate hoppers in the "dope house." The various formulae are mixed on a batch basis by metering portions of each ingredient from its respective storage hopper into transport containers. Particulates may be trapped within the material-handling systems or dope-house air emissions may be controlled by the use of exhaust fans with dry cyclone dust collectors, baghouses, or both.

The dopes and nitrated esters are then blended together in the "mix house" to make dynamite, which has either a dry, damp, or paste consistency. The dynamite is then transferred to pack houses, where it is loaded into the various packages. Dust generated may be contained within the process equipment and the building. If it is not, mix- and pack-house air emissions are controlled with conventional particulate-control technology. The loaded sticks or tubes are placed in fiberboard cases either at the pack house or at a casing house. No air emissions result from the packaging process.

Air Emissions Characterization

The spent acid generated during the nitration process contains weak nitric acid, weak sulfuric acid, water, and some trace amounts of NG/EGDN. To remove the trace amounts of NG/EGDN, the spent acids are passed through an absorption tower to separate the two acids from the water and to destroy the trace amounts of NG/EGDN. In the commercial industry, this is called a "denitrating tower." The weak acids are reclaimed and are cycled back to the acid plants, where they are concentrated into their proper strengths. Nitrogen oxide fumes generated during this process are recovered with high efficiency in the absorption tower as weak nitric acid. The other by-product of the denitration process is steam, which is condensed into water.

Denitration of NG/EGDN

$$H_2SO_4 + HNO_3 + C_3H_4N_2O_6 + C_3H_5N_3O_9 \rightarrow HNO_3 + H_2SO_4 + H_2O + NO_x + CO + CO_2$$

Air Pollution Control Measures

Dynamite plants have extensive engineering controls installed to correct the dust and fumes in the workplace environment. Ventilation of buildings for industrial hygiene purposes produces minimal air pollution. Air pollution control is sometimes provided to control particulate dust emissions into the environment. In these cases, the pollution control technology utilized is conventional wet scrubbers, cyclones, or filters.

EMULSIONS

Process Description

Commercial explosive emulsions are typically water-in-oil emulsions of a concentrated aqueous solution of one or more oxidizer salts (typically ammonium nitrate, calcium nitrate, and sodium nitrate) and about 6% of a water-insoluble organic fuel mixture (e.g., mineral oil, fuel oil, or waxes), which are emulsified to form a product with a soft to firm consistency. These emulsions are less sensitive than dynamites and their processed densities cannot be detonated by a blasting cap. In order to make them cap sensitive, it is common to add sensitizers. These compounds may be explosives, aluminum, solid fuel, and/or such density-control materials as hollow glass, ceramic, or plastic beads.

The fuels and oils are usually heated to about 70–110°C in their separate tanks. A solution of nitrate salts is prepared in another heated tank and kept at 70–105°C. The formulated amounts of the fuels and oils are then either batched or continuously forced through a high-sheer mixer to form a thick, high-viscosity emulsion. This is either packaged in paper or plastic cartridges or prepared for use in bulk applications.

Air Emissions Characterization

Prior to blending, the oils, waxes, solutions, fuels, emulsifiers, and oxidizer salts are usually stored in metal or plastic tanks, which have submerged fill lines for the liquids. Fugitive emissions from these bulk tanks are minimal and are not controlled because of the low volatility of these liquids. Once the process has started, it is common to have all of the ingredients in a closed-loop system. The pumping, heating, blending, and packaging of the explosive emulsion are usually done in the same room of the plant. This allows for the collection of dust, scrap, and waste materials within the plant. The processed emulsions are packaged in plastic film or paper cartridges or produced for shipment in bulk quantities. Particulate emissions from the handling of aluminum and density-control materials are minimal. Handling takes place in buildings and, in some cases, ventilation may be required for industrial hygiene purposes.

Air Pollution Control Measures

Since most of the process systems are closed loops and air emissions from tanks and buildings are minimal, air pollution control systems are not warranted. Submersible fill pipes in fuel oil tanks provide emission control. Lids and control valves are used on ammonium nitrate bins.

AMMONIUM NITRATE–FUEL OIL MIXTURES

Process Description

Ammonium nitrate–fuel oil mixtures are classified as blasting agents by the Department of Transportation. These blasting agents are usually mixtures of industrial-grade ammonium nitrate and No. 2 fuel oil. Aluminum, bagasse, nitroparaffins, or plastic spheres are sometimes added to afford more energy or to control the density of the packaged product. Often a red dye is added to the fuel for quality-control purposes. Most blasting agent plants are modest installations and are the most common explosive plants in the United States.

Prilled ammonium nitrate is usually received in bulk by truck or railcar and unloaded into storage bins. The fuel oil is stored in tanks and would have the typical fugitive emissions of No. 2 fuel oil in storage tanks. The explosive is produced by metering the raw materials into a mixer/blender in the proper proportions, where, after mixing, it is either packaged or put into bulk bins for delivery directly to the consuming site.

Air Emissions Characterization

The ammonium nitrate prill and fuel oil are usually stored in large tanks and bulk bins outside the plants. The fuel oil would have the same emissions characteristics as would any common above-ground storage tank with a submerged fill pipe and which is vented to the atmosphere. Ammonium nitrate prill bins use containment lids and valves to limit the generation of dust. The mixing equipment usually is enclosed and this limits air emissions. Fugitive emissions from production are usually contained within the building and do not escape to the atmosphere.

Air Pollution Control Measures

Emission control measures include submersible fill pipes in the fuel oil tanks, lids and flow control valves on the ammonium nitrate prill bins, and enclosed mixing systems in the production buildings. These measures limit the small amount of fugitive air emissions from storage and handling. Fugitive emissions from production are minimal and typically stay within the buildings.

CAST BOOSTERS

Process Description

Cast boosters are formulated by adding pentaerythritol tetranitrate (PETN), trimethylentrinitramin (RDX), or other, similar high-brisance explosives with trinitrotoluene (TNT) and casting the mix into small charges. The TNT is melted in a steam-jacketed kettle that has an agitator to blend the materials. Various proportions of TNT, PETN, and/or RDX are used, depending on market conditions and the physical performance required. The molten explosives are cast into a variety of forms and shapes, but they are typically cylindrical with one or more channels through the axis of the cylinder to accept either a blasting cap, a detonating cord, or both. The weights of these units range from about 50 grams to 2500 grams and they are usually packaged in fiberboard boxes.

Air Emission Characterization

The drying of the PETN and RDX may generate a minimum amount of dust that is typically contained within the dryer system. No chemical by-products are produced in the blending of these materials. The explosives are very sensitive to heat, friction, impact, and shock, and thus safety considerations are of paramount importance in the collection and retention of this dust.

Air Pollution Control Measures

The fumes generated by the mix kettles and the cooling operations are typically exhausted through a water scrubber that is contained in or is a part of the product building. The cleaning of the exterior of the cast booster and packaging results in the generation of small amounts of dust and chips, which are contained within the building.

DETONATORS

Process Description

Detonators (also known as blasting caps) are copper or aluminum tubes containing a very small quantity of initiating explosive(s). They are used to provide an energetic impulse capable of initiating a detonation wave in larger charges of high explosives. The two most prevalent types of detonators are electric and nonelectric. Both types are fairly similar in their methods of construction and differ substantially only in the type of energy input and the method of ignition involved.

An explosive base charge of granular dry PETN is press-loaded remotely into the metal cap shell. A flame-sensitive priming charge of lead azide, diazodinitrophenol, or hexanitromannitol (or mixtures thereof) is remotely loaded atop the base charge. One or more delay elements are usually next introduced on top of the priming charge. These provide a specific time delay to the detonator and contain various proprietary delay powders (fuel/oxidizer mixtures) encased in a steel or lead sheath.

Electric detonators typically utilize either of two configurations above the delay column: a bridge wire surrounded by a proprietary flash powder or an electric match of proprietary composition. Both configurations are connected to two plastic-jacketed wires (either copper or iron) that exit the top of the shell and are used to deliver the energy input to ignite the bridgewire or match head.

Nonelectric detonators incorporate a hollow plastic tube above the delay column. This plastic tube contains a low dose of proprietary energetic material and provides the nonelectric alternative energy input to the delay column. Another form of nonelectric detonator is the fuse-cap type, which contains only the base charge and the priming charge and uses the flame from a safety fuse to provide the energy input.

Air Emissions Characterization

The materials utilized in the manufacturing process are combinations of dry powders. These raw materials remain process captive during the drying, blending, and loading operations and do not give rise to fugitive air emissions. Mixing of fuel/oxidizer raw materials for detonator manufacture may involve the use of a liquid mix vehicle (i.e., isopropanol), which is discharged to air as the mixtures are dried.

Air Pollution Control Measures

There are minimal fugitive emissions and, therefore, no pollution controls are employed.

DETONATING CORD

Process Description

Detonating cord is a flexible linear product containing a center core of high explosive and is used to initiate other explosives or other lines of detonating cord. The explosive core loading is usually PETN, but other explosives such as RDX and cyclotetramethylene tetranitramine (HMX) can also be used. The detonating cord is made by either a dry or wet loading process. The dry loading process is generally more popular because of the lower production costs for the ordinary types of cord.

Dry process manufacturing requires that the PETN be dried to a very low moisture content. Additives such as flow enhancers, water-repellent compounds, or desensitizers may also be added if required. The PETN is placed in a hopper leading to an orifice on a spinning machine. A continuous strip of plastic is folded around the core of PETN as the plastic is drawn through a loading die. The tube thus formed, which contains relatively loose PETN, is overspun with several layers of yarns and has a layer of plastic extruded over it. Additional yarns may be overspun on the cord and wax may be applied to hold these yarns in place. The finished cord is usually coiled onto spools and packaged in fiberboard boxes.

The wet-process detonating-cord manufacturing system uses a fine-granulation explosive such as PETN. The explosive is mixed to a fluid paste in water with a suspending agent. This paste is then pressure loaded into a core surrounded by a braid of textile fiber and subsequently dried. The process is finished by countering with various yarns, overextruding with a plastic jacket, and final overspinning and waxing, if desired. The slow, limited speed of the braiding process makes this method of manufacturing more expensive than the dry method.

Air Pollution Characterization

Both methods of manufacturing employ explosive powder as feedstock. Only a very small amount of dust is generated during drying, and it is contained within the dryer system. Any dust generated during the braiding or spinning process is process captive within the building.

Air Pollution Control Measures

There are no fugitive or point source emissions from this manufacturing process and, therefore, no pollution controls are employed.

INITIATING EXPLOSIVES

Process Description

Initiating explosives, as the name implies, are used to provide a localized point-source impulse to trigger the detonation of other explosives. An example of this function of an initiating explosive is found in a detonator (blasting cap). The principal initiating explosives used commercially are lead azide, diazodinitrophenol (DDNP), and hexanitromannite (HNM). Pentaerythritol tetranitrate, the explosive of choice for use in detonating cord, is also an initiating explosive.

Lead azide, $Pb(N_3)_2$, is synthesized by combining aqueous solutions of sodium azide and lead nitrate (or lead acetate). Lead azide precipitates as a white solid from the combined solutions. The liquid is filtered away from the crystalline product and the lead azide is washed with water. The process uses no solvents nor are there any gaseous by-products of the reaction. There are no fugitive airborne emissions from this process. The wash water is collected, treated, and released.

Pentaerythritol tetranitrate (PETN) is manufactured by either a batch or a continuous nitration process. Pentaerythritol, $C(CH_2OH)_4$, a white, granular solid, is added to concentrated (98%) nitric acid. The nitration conversion to the tetranitrate is rapid. This produces a crude PETN that is separated from the spent nitric acid by filtration while being washed with water. The crude PETN can be used in its present state or it can be recrystallized further after being dissolved in acetone in the next downstream process.

Air Emissions Characterization

The principal airborne process-related fugitive emissions are oxides of nitrogen (NO_x) and acetone vapor. Sources of NO_x fumes include nitration reactors, drown tanks (specific to batch nitrators), spent acid filters, spent acid storage tanks, and strong nitric acid (98%) storage tanks. Wet scrubbing systems are employed to capture NO_x vapors from these various sources. Many scrubber designs can be used to accomplish this goal. The difficulty in the removal of oxides of nitrogen is a combination of low solubility of nitric oxide and the conversion of nitrogen dioxide (NO_2) into nitric oxide (NO). When NO_2 is absorbed, it converts to nitrous acid (HNO_2) and NO. This is unstable and breaks down into nitric acid (HNO_3) and NO. As a result, without any other consideration, one third of the NO_2 is lost as NO because of the low solubility of NO. One chemical that will interfere with this quick conversion is sodium sulfide and it is utilized in some scrubbers. Here, the NO is converted to NO_2 in the first of two towers. The NO_2 is reduced in tower 2 by the presence of sodium sulfide to nitrogen gas (N_2). The reaction of sodium sulfide is fast enough to interfere with the normal conversion of part of the NO_2 to NO.

As noted above, two different sets of reactions take place in the scrubber columns. The following equations show which reactions take place in each tower.

Tower 1

$$2NO + NaClO_2 \rightarrow 2NO_2 + NaCl$$

Tower 1

$$3NO_2 + H_2O \rightarrow 2HNO_3 + NO$$

Towers 1 and 2

$$HNO_3 + NaOH \rightarrow NaNO_3 + H_2O$$

Towers 1 and 2

$$2NO_2 + 2Na_2S + H_2O \rightarrow N_2 + Na_2S_2O + 2NaOH$$

Tower 2

$$2NO_2 + Na_2S_2O + 2NaOH \rightarrow N_2 + 2Na_2SO_4 + H_2O$$

The scrubber chemicals are used in two packed towers in series. The exact liquid rates and the residence time help to ensure that the NO_x removal is accomplished. The acetone vapors generated at the dissolvers, crystallizers, and final filtration locations are commonly recondensed by shell and tube condensers using chilled (40°F) water as the condensing medium.

Nitromannite (HNM) is produced by the mixed-acid (sulfuric and nitric) nitration of mannitol ($C_6H_{14}O_6$). The powdered solid mannitol is fed into agitated mixed acid in a nitrator. After the nitration phase is completed, the liquid mixture, composed essentially of suspended HNM and spent acids, is discharged into drowning tanks containing water. The resultant suspension is then separated in a centrifuge. The retained solid is washed to remove residual surface acids. Solid HNM is then dissolved in acetone, at which point the residual acid is neutralized. The solution separates into layers and the water and acid are drawn off; the acetone layer is diluted with water in a continuous precipitator to form a slurry mixture and then filtered and washed. Both the acetone–water mixture and the filtrate wash water are collected for acetone recovery by distillation.

Diazodinitrophenol (DDNP) is produced by the diazotization of picramic acid. Sodium picramate and a DDNP seed mixture are reacted with sodium nitrate and nitric acid over a four-hour period in a cooled agitated reactor. After brief settling, the liquor and the very fine waste DDNP are pumped off, leaving the product in the reactor. Wash water is added, the DDNP is resuspended with agitation, and the fine-grain-size DDNP is washed and pumped to a vacuum filter. The coarse fraction of DDNP is again suspended in water and filtered. The liquor, waste DDNP fines, DDNP wash water, and building wastewater are discharged into a digestion tank. This waste is treated with hydrogen peroxide, ferrous sulfate, and sodium carbonate to oxidize all phenol compounds to water and carbon dioxide. The by-product fugitive emissions are water and carbon dioxide and, therefore, no controls are required for entrapment or recovery of gases or vapors.

Air Pollution Control Measures

The air pollution controls for the initiating explosives are outlined in the process description or the "Air Emission Characterization" section.

BLACK POWDER

Process Description

Black powder is a deflagrating or low explosive of an intimate mixture of potassium (or sodium) nitrate, charcoal, and sulfur. The raw materials are stored in either bins or containers prior to usage. The potassium or sodium nitrate is crushed and batch weighted to formulation. The sulfur and charcoal are batch weighted by formula and pulverized to a very fine powder. These ingredients are remotely incorporated, plasticized with a small amount of water, and mechanically kneaded to produce a relatively loose, low-density and moist mass known as "wheel cake." The wheel cake is pressed to the desired density and then chipped into the maximum dimension of choice for the grade to be produced. The chips are granulated through a system of crushing rollers and screens. The rate of feed, the spacing of the rolls, and the screen mesh produce the desired sizes of "green grains." The rough edges of the green grains are rounded through the friction process within the glazing cycle. They are then dried and graphited to add polish and luster. Some black powder is not glazed by choice.

Air Emission Characterization

The production buildings are designed to contain most of the dust generated during the different stages of the process.

Air Pollution Control Measures

As there are minimal fugitive emissions from this manufacturing process, no air pollution controls are employed.

WATER GELS AND SLURRIES

Process Description

Water gels and slurries compromise a wide variety of explosive materials used for blasting. As manufactured, they have varying degrees of sensitivity to initiation. They usually contain a substantial proportion of ammonium nitrate, some of which is in a water solution. These products may be sensitized with metals such as aluminum, fuels, or molecular explosives. A metal cross-linker, typically antimony or chromium compound, is responsible for controlling the viscosity and selling consistency of the materials.

The process is usually started by making a heated solution of ammonium nitrate and water. This solution can then be further processed by either a continuous or batch method. Both systems use the same general methods. The solution is added to the process stream in measured amounts with fuel and sensitizer metals, if required. The addition of gelling agents, which are activated by the heat of the product, causes the material to form a gelled mass that can be from a loose to a very firm form. The consistency is chosen to match the method of packaging or bulk delivery desired.

Air Emission Characterization

The manufacturing process of water gels and slurries is a closed loop and very few air emissions are generated. Liquid raw materials and the products are nonvolatile. Minimal fugitive air emissions result from handling of aluminum and dry oxidizer ingredients.

Air Pollution Control Measures

The fuel oils are stored in large above-ground storage tanks that usually have submerged fill pipes. The solid ammonium nitrate is usually stored in large hopper bins that have lids covering the openings.

SPECIALTY EXPLOSIVES

Process Description

Specialty explosives are those that are used in a specific application and usually are in small quantities. There are two general categories of these products: sheet explosives and oil-field specialty explosives. There may be other classifications of commercial explosives, but they would be so small that they are not categorized by the industry.

Sheet explosives' commercial applications lie principally in the area of shock-hardening of metals. These explosives are pliable, flat material most commonly manufactured in thin sheets, although they can also be made in various other shapes, such as cylindrical. The process is typically composed of three steps: mixing of components, homogenization, and extrusion into finished form.

The mixing operation combines the explosive (usually PETN or RDX) with a matrix material and plasticizer (nitrocellulose and acetyl tributyl citrate for PETN-based explosives) in an aqueous slurry in a mixer. The process water is then removed from the mixture, usually by a stationary vacuum filter.

The mixture generally is further homogenized in an intensive mixer, at which point residual process water from the mixing step is removed. The final step is processing the uniformly mixed explosive and matrix system through an extruder to form the final desired shapes.

Oil-field specialty explosives are a wide variety of products manufactured for commercial oil-well-completion applications. The perforating charges make up the principal product line and they consist of a charge case, a main charge, a high explosive primer, and a cone-shaped charge liner. The charge is usually a hollow, machined steel cup. It is the receptacle into which the explosive components are loaded. The most common explosives used are RDX, PETN, and HMX.

A primer charge (a pure form of the main charge explosive) is first added to the case and usually accounts for 5% of the net explosive weight. The main charge is a pressed granular powder (usually RDX, PETN, or HMX with a waxlike coating around the individual particles).

The conically shaped charge liner is a hollow cone previously formed by pressing a powdered copper and lead mixture that has been pressed into the explosive-loaded charge case.

Air Pollution Characterization

The materials utilized in the manufacturing process are combinations of dry powders. These raw materials remain process captive during the blending, loading, and pressing operations and do not give rise to fugitive air emissions.

Air Pollution Control Measures

There are no fugitive emissions and, therefore, no pollution controls are required.

Bibliography

American National Standards Institute, *Safety Requirement for the Transportation, Storage, Handling and Use of Commercial Explosives and Blasting Agents in the Construction Industry*, Standard A10.7, American National Standards Institute, New York, 1970.

Atlas Powder Co., *Handbook of Electric Blasting*, 1985 ed., Atlas Powder Co., Dallas, 1985, 74 pp.

A. Bailey and S. G. Murray, *Explosives, Propellants, and Pyrotechnics*, Brassey's, United Kingdom, 1989.

D. G. Borg, R. F. Chiapetta, R. C. Morhard, et al., *Explosives and Rock Blasting*, Atlas Powder Co., Dallas, 1987, 662 pp.

G. H. Damon, C. M. Mason, N. E. Hanna, et al., *Safety Recommendations for Ammonium Nitrate-Based Blasting Agents*, Information Circular 8746, U.S. Bureau of Mines, Washington, DC.

Design Considerations for Toxic Chemicals and Explosives Facilities, R. A. Scott and L. J. Doemeny, Eds.; ACS Symposium Series #345, American Chemical Society, Washington, DC, 1987.

R. A. Dick, L. R. Fletcher, and D. V. D'Andrea, *Explosives and Blasting Procedures Manual*, Information Circular 8925, U.S. Bureau of Mines, Washington, DC, 1983.

R. A. Dick, *Factors in Selecting and Applying Commercial Explosives and Blasting Agents*, Information Circular 8405, U.S. Bureau of Mines, Washington, DC.

Encyclopedia of Explosives and Related Items, B. Federoff and S. M. Kaye, Eds.; PATR 2700, U.S. Army ARRADCOM, Dover, NJ, 1960–1983 (10 vols.).

Engineering Design Handbook: Explosives Series—Properties of Explosives of Military Interest, ACMP 706-177, U.S. Army Material Command, Washington, DC, 1971.

Explosive Materials Regulations, Title 27, Code of Federal Regulations, Parts 55 and 181, U.S. Bureau of Alcohol, Tobacco and Firearms, Washington, DC.

Explosives Technologies International, *Blaster's Handbook*, 16th ed., Explosives Technologies International, Wilmington, DE, 1980.

H. D. Fair and R. F. Walker, *Energetic Materials*, Plenum Press, New York, 1977.

S. Fordam, *High Explosives and Propellants*, Macmillan Co., New York, 1980.

C. E. Gregory, *Explosives for North American Engineers*, Trans Tech Publications, Rockport, MA, 1979.

Hazardous Materials Regulations, Title 49, Code of Federal Regulations, Part 174.16, U.S. Department of Transportation, Washington, DC.

C. H. Johanssen and P. A. Persson, *Detonics of High Explosives*, Academic Press, London, 1970, 326 pp.

R. Meyer, *Explosives*, 3rd ed., Verlag Chemie, Weinheim/New York, 1987.

T. Urbanski, *Chemistry and Technology of Explosives*, Macmillan Co., New York, 1984 (four vols.).

PHOSPHORIC ACID MANUFACTURING

Gordon F. Palm

In the past 10 years, wet process phosphoric acid has been the source of perhaps 100% of the new phosphate production. This is expected to continue in the future.

There are two basic types of processes for the production of phosphoric acid: wet processes and furnace processes. Furnace processes include the blast furnace process and the electric furnace process. The blast furnace is no longer used, however, the electric furnace process is used extensively to make elemental phosphorus, most of which is converted to phosphoric acid for nonfertilizer uses. As it is unlikely that the electric furnace process will become competitive for producing phosphoric acid for fertilizer use, it will not be described here. For details of this process, refer to AP-40, February 1980, or to reference 8.

Wet processes may be classified according to the acid used to decompose the rock.[1] Sulfuric, nitric, and hydrochloric acid have been used in commercial processes. However, processes involving nitric acid are no longer used in the United States and those using hydrochloric acid are not competitive for fertilizer production. Therefore, the discussion will be limited to the sulfuric acid processes.

PROCESS DESCRIPTION

Wet process phosphoric acid is produced, as shown in Figure 1, by digesting ground phosphate rock, a fluorapatite, with an amount of sulfuric acid approximately stoichiometrically equal to the calcium oxide in the rock.[2] About 97% of the sulfate is precipitated as calcium sulfate. Some 2% to 3% appears as unreacted sulfate (free acid) in the phosphoric acid. For efficient operation the free acid content must be closely controlled. The concentration of phosphorus pentoxide (P_2O_5) in the product and the solids concentration in the reaction slurry are both controlled by the quantity and concentration of recycle acid returned from the filter. Gypsum from the filter is slurried with water and pumped to a gypsum stack for storage. The water is recycled through a surge-cooling pond to the phosphoric acid process.

Commercial wet processes may be classified according to the hydrate form in which the calcium sulfate crystallizes.[3]

Anhydrite:	$CaSO_4$
Hemihydrate:	$CaSO_4 \cdot \frac{1}{2}H_2O$
Dihydrate:	$CaSO_4 \cdot 2H_2O$

At present there is no commercial use of the anhydrite process, mainly because the required reaction temperature is high enough to cause severe corrosion difficulties. Processes in commercial use are listed in Table 1.

Straight dihydrate processes are the most popular today. Recently, some have been converted to a one-step hemihydrate process that has the advantage of producing phosphoric acid of a relatively high concentration and with lower impurity levels. Hemihydrate-dihydrate processes without intermediate filtration and hemihydrate-dihydrate processes with two separation steps are not used in the United States.

CHEMISTRY

Digestion is considered to take place in several stages.[2] First, tricalcium phosphate in the rock is attacked by phosphoric acid to form monocalcium phosphate. The monocalcium phosphate then reacts with sulfuric acid to yield additional phosphoric acid and gypsum. The overall reaction is as follows:

$$\underset{\text{Fluorapatite}}{Ca_{10}(PO_4)_6F_2CaCO_3} + \underset{\text{Sulfuric acid}}{11H_2SO_4}$$

$$= \underset{\text{Phosphoric acid}}{6H_3PO_4} + \underset{\text{Gypsum}}{11CaSO_4 \cdot nH_2O} + \underset{\text{Hydrogen fluoride}}{2HF}$$

$$+ \underset{\text{Carbon dioxide}}{CO_2} + \underset{\text{Water}}{H_2O} \qquad (1)$$

Simplified versions of these reactions are as follows:

$$\underset{\text{Tricalcium phosphate}}{Ca_3(PO_4)_2} + \underset{\text{Phosphoric acid}}{4H_3PO_4} = \underset{\text{Monocalcium phosphate}}{3CaH_4(PO_4)_2} \qquad (2)$$

$$\underset{\text{Monocalcium phosphate}}{3CaH_4(PO_4)_2} + \underset{\text{Sulfuric acid}}{3H_2SO_4} + \underset{\text{Water}}{6H_2O}$$

$$= \underset{\text{Gypsum}}{3CaSO_4 \cdot 2H_2O} + \underset{\text{Phosphoric acid}}{6H_3PO_4} \qquad (3)$$

Depending on the temperature and phosphoric acid strength maintained during digestion, dihydrate, hemihydrate or even the anhydrite of calcium sulfate may result.[3]

A number of minor reactions occur between fluorides, silica, and water in the scrubbing of gaseous fluorides from the digester. These reactions are discussed under "Air Pollution Control Measures."

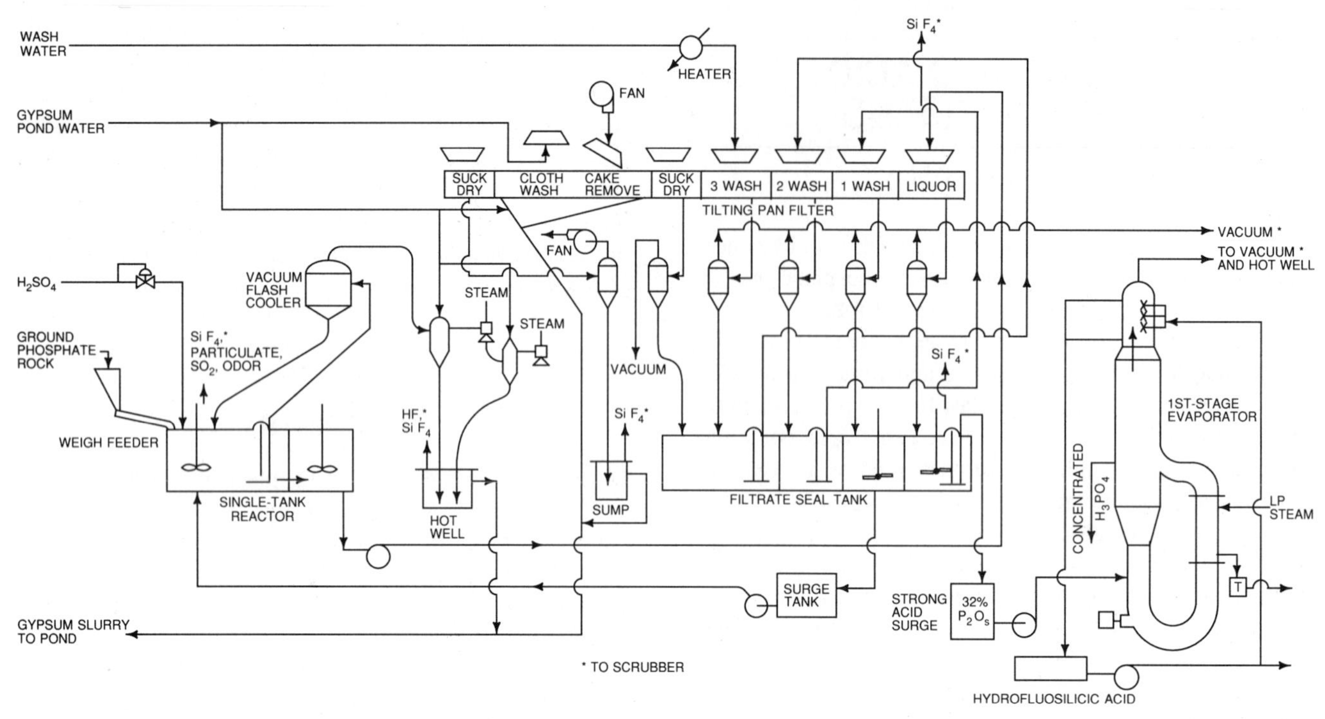

FIGURE 1. Diagram of Wet Process Phosphoric Acid Plant (AP-42, *Compilation of Air Pollution Factors* U.S. Environmental Protection Agency, revised February 1980, p 5. 41-1.)

AIR EMISSION CHARACTERISTICS

Fluorides

In the digester, about 3–7% of the fluoride in the phosphate rock is evolved.[2,4,5] At the higher P_2O_5 concentrations and temperature of the hemihydrate and hemihydrate-dihydrate processes, more silicofluoride fumes are emitted from the attack tank than in the dihydrate process.[5] The digesters are vented to wet scrubbers for fluoride removal.

The relative quantities of fluorides in the filter acid and gypsum depend on the type of rock and the operating conditions.[6] The main factors appear to be the following.

1. The quantity of sodium or potassium salts present, which will precipitate insoluble fluorine compounds.
2. The reaction temperature at which digestion occurs (increasing temperature results in increasing gaseous fluorides).
3. The concentration of the product phosphoric acid.

Filter acid contains about 67% of the fluoride in the phosphate rock,[7] much of which is volatilized during evaporation. The gaseous fluorides are removed by scrubbing to prevent pollution. In some cases, salable products are produced[8] (fluosilicic acid, fluosilicates, cryolite, or aluminum fluoride). The amount of fluoride evolved during concentration from 30% to 54% P_2O_5 may be 70–80%[8] of that originally present in the acid. This is equivalent to 50% of the fluoride in the phosphate rock. In the production of superphosphoric acid (69–72% P_2O_5) by concentrating wet process acid, most of the fluoride is volatilized[8] so the acid contains only 0.2–0.3% as F. The addition of reactive silica

TABLE 1. Commercial Wet Processes in Use

Crystal Form(s)	Number of Separation Steps[a]	Usual Concentration of Acid, % P_2O_5	Usual Temperature, °C Reactor	Usual Temperature, °C Recrystallizer
Dihydrate	1	26–32	70–85	
Hemihydrate	1	40–50	85–100	
Hemihydrate-dihydrate	1	26–30	90–100	50–60
Hemihydrate-dihydrate	2	40–50	90–100	50–65
Dihydrate-hemihydrate	2	35–38	65–70	90–100

[a]Filtration of centrifuging steps.

during evaporation enhances silicon tetrafluoride volatization and reduces fluoride content to about 0.1% as F.

The evaporators used to concentrate phosphoric acid operate under a vacuum. The steam from the evaporated acid and the evolved fluorides goes to a total condenser where they are directly contacted with cool process water. The steam is condensed and the fluorides are absorbed in a large quantity of process water. Gaseous fluoride emissions from the condenser waters are negligible.

Wet process phosphoric acid[9] is defluorinated for use in animal feed products. The stripping of the fluoride from the acid is more easily attained when the fluoride is present in the silicon tetrafluoride form,[10] which is more volatile than hydrogen fluoride. Silicon dioxide is added as needed to the phosphoric acid to allow the chemical reactions necessary for producing silicon tetrafluoride. The various methods of defluorinating are as follows:

1. Air stripping by blowing air through hot 54% P_2O_5 acid (the exit air must be scrubbed).
2. Steam stripping 54% P_2O_5 phosphoric acid by circulating through a stripping column with steam blown countercurrent to the acid (exit steam must be condensed to remove fluorides).
3. Evaporation method by diluting 54% P_2O_5 acid and then reconcentrating the acid to 54%; uses conventional vacuum evaporation equipment.
4. Concentrating to superacid under vacuum will give a P/F (weight ratio) of 80 at 70% P_2O_5 acid. Reactive silica, such as diatomaceous earth, when added to the acid, allows it to be stripped of fluorides in the evaporator to a P/F ratio of 100 or higher.

Acid used in animal feed products must have a minimum P/F of 100. If there is fluoride in the limestone used for feed-grade dicalcium phosphate, the acid P/F must be about 125. Fifty-four percent P_2O_5 acid has a P/F of about 15; 30% P_2O_5 acid is hard to defluorinate.

Particulates

In the older plants, dry ground phosphate rock is generally used. It is transported to the phosphoric acid plant digester by conveyor belt, screw conveyor, and/or airveying systems. The handling of the dry ground rock results in the generation of fugitive dust which is collected by ventilation ductwork and removed by bag collectors which discharge the dust to the rock feed bin. In addition to the gaseous fluoride in the digester vent, there is usually a small amount of rock dust generated by the mechanical handling of the rock.[11] This dust is removed in the wet scrubber designed for the removal of fluorides.

In newer plants, there are no particulate emissions since wet rock grinding is used.[12] The rock is ground as a slurry in a ball mill to about 65% solids. The advantages of wet rock grinding are a 30–40% reduction in horsepower in the grinding area, elimination of the rock dryer, elimination of atmospheric pollution by dust, and a savings in the cost of fuel for drying. Many older plants have been converted to wet rock grinding to realize these significant environmental and cost benefits.

Gypsum Stacks and Process Water Ponds

The gypsum stacks and process water ponds are an integral part of a typical phosphoric acid plant.[13] The gypsum stacks serve primarily as storage for stacking of the by-product gypsum. Adjoining these stacks are process water ponds that provide surge volume for wet and dry weather conditions to avoid water treatment and water discharge and to conserve on fresh water makeup. The water in the ponds, commonly called "pond water," is recirculated for process cooling, recovery of P_2O_5 and sulfate values, control of effluents, scrubbing of process emissions, process makeup water, and hydraulic transport of by-product gypsum. The tops of the gypsum stacks are frequently used for supplemental cooling of the pond water. During the manufacturing process, the pond water comes in direct contact with raw materials, intermediate products, by-products, waste products, or finished products. The cooling ponds, including the cooling water on top of the gypsum stacks, also serve as surge equalization ponds for the rainfall runoff from the gypsum stacks.

Pond water is used mainly in the phosphoric acid processes to wash the filter cake, in scrubbers, in barometric condensers, and for slurrying waste gypsum. After settling of gypsum solids in the gypsum stack, the pond water is allowed to cool by natural evaporation and is recirculated. Pond water is acidic, has a pH of about 1.5–2.0, and contains weak sulfuric, phosphoric, and fluosilicic anion components. Phosphate analysis, reported as P_2O_5, varies from about 1% to 2% P_2O_5. Fluoride analysis, reported as percent of fluorine, varies from about 1% to 1.8% F.

Hydrogen fluoride (HF) and silicon fluoride (SiF_4) are prevented from escaping from the facility into the atmosphere by the use of wet scrubbers. These recycle pond water as the scrubbing medium.[13] In the scrubbers, HF and SiF_4 are removed from the air, forming H_2SiF_6.

Low-level fluoride emissions occur from the pond water on top of the gypsum stack and pond water cooling ponds. Measurements of these levels are extremely difficult to correlate because of changing air velocity and direction and the concentration variation of the fluorides along the elevation over the pond water surface. Estimations based on fluoride vapor pressure are also difficult to make because of the wide variations in pond water composition and the lack of available data.[14]

The U.S. Environmental Protection Agency, in discussing the verification of predicted fluoride emissions, made the following statement[13]: "Based on our findings concerning the emissions of fluoride from gypsum ponds, it was

TABLE 2. Emission Factors for Wet Process Phosphoric Acid Production

	Fluorides	
Source	lb/ton[a]	kg/MT
Wet process, uncontrolled		
Reactor, dihydrate	13	6.5
Reactor, hemihydrate	27–54	14–27
Evaporation, 30–54% P_2O_5	100	50
Evaporation, 54–70% P_2O_5	41	21
Gypsum settling and cooling ponds	[b]	[b]
Typical controlled emissions[c]	0.02–0.07	0.01–0.04

[a]References 7 and 8. Pounds of fluoride (as gaseous fluoride) per ton of P_2O_5 produced.
[b]Site specific. Acres of cooling pond required: ranges from 0.10 acre per daily ton P_2O_5 produced in the summer in the southeastern United States to zero in the colder locations in the winter months when the cooling ponds are frozen. Also, EPA states: "Based on our findings concerning the emissions of fluoride from gypsum ponds, it was concluded that no investigator had as yet established experimentally the fluoride emission from gypsum ponds."[13]
[c]AP-42, Compilation of Emission Factors, U.S. Environmental Protection Agency, Table 5.11-1, February 1980.

concluded that no investigator had as yet established the fluoride emission rate from gypsum ponds."

Cooling towers are mentioned for cooling condenser pond water[15] rather than a pond water cooling pond. However, if a cooling tower is used, surge equalization provisions must also be considered to avoid excessive treatment and the discharge of treated water. Cooling towers are used for cooling gypsum pond water in four U.S. locations instead of pond water cooling ponds. Fluoride emissions from the cooling towers are a function of the fluoride vapor pressure of the pond water. The fluoride emissions are low, but must meet limits of the state regulatory authorities.

EMISSION FACTORS

Emission factors for wet phosphoric acid production are shown in Table 2. These were obtained from Table 5.11-1 of AP-42 dated February 1980 with modifications as shown.

AIR POLLUTION CONTROL MEASURES

Fluorides

Gaseous fluoride compounds are removed from the vent air of the digester, filter, and various tanks. Wet scrubbers have been used exclusively for this service in the United States.

Wet scrubbing combines the ability to remove particulates (rock dust) from gas streams by impaction of the particulates on the surface of liquid droplets and the ability to absorb gaseous constituents into the liquid phase.[16] Both of these functions are limited by the characteristics of the scrubbing liquor, the properties of the materials to be removed, and sometimes by the two in combination. In the phosphoric acid plant wet scrubbers, the pond water and the fluoride-containing gases from the wet process phosphoric acid plant reactor can produce a gelatinous silica precipitate. This plugs the packing in the scrubbers and limits the types of scrubbing equipment that can be used.

The basic chemistry of the compounds of fluoride, silica, and water must be considered to characterize the application. In the reactor, the fluoride contained in the fluorapatite or fluorspar goes into solution according to the following reactions.[16]

$$\underset{\text{Calcium fluoride}}{CaF_2} + \underset{\text{Sulfuric acid}}{H_2SO_4} = \underset{\text{Calcium sulfate}}{CaSO_4} + \underset{\text{Hydrogen fluoride}}{2HF} \tag{4}$$

or

$$\underset{\text{Hydrogen fluoride}}{2HF} + \underset{\text{Silicon tetrafluoride}}{SiF_4} = \underset{\text{Fluosilicic acid}}{H_2SiF_6} \tag{5}$$

A vapor equilibrium is set up between the reactants in equation 5, which may be regarded as the reverse of the following equation.

$$\underset{\text{Fluosilicic acid}}{H_2SiF_6} = \underset{\text{Hydrogen fluoride}}{2HF} + \underset{\text{Silicon tetrafluoride}}{SiF_4} \tag{6}$$

High temperature drives the reaction to the right, increasing the vapor pressure of both the HF and SiF_4, and increases the relative significance of SiF_4 as the fluorine-containing species.[16] These vapor pressures set the lower limit of fluoride concentration in the gas phase leaving the scrubber.

In addition to the reactions given, hydrolysis of SiF_4 occurs when the concentration of this component is higher than the equilibrium values, according to the following reaction.

$$\underset{\text{Silicon tetrafluoride}}{3SiF_4} + \underset{\text{Water}}{4H_2O} = \underset{\text{Silicic acid}}{Si(OH)_4} + \underset{\text{Fluosilicic acid}}{2H_2SiF_6} \tag{7}$$

Reaction 7 occurs as the temperature of a gas stream is reduced in the presence of water and leads to the formation of gelatinous deposits of polymeric silica, which plug scrubber packings. This problem limits the use of conventional packed countercurrent absorbers in this service, as well as other contacting devices that have small gas passages that might plug up.

Transfer Unit Concept

Removal of fluorine compounds from effluent gases is achieved by absorption in water. With the decrease of allowable emissions in the past few years, scrubber efficiencies in the range of 99+% are required.[17] Scrubber emissions are now expressed as pounds per hour of fluorides emitted. If once-through or neutralized water can be used, at least 4.6 transfer units are required to achieve this efficiency. If recycled pond water is used, the number of transfer units required for the same efficiency has been as many as eight units in several installations. This depends on the partial pressure of fluorides in the incoming water. The number of transfer units required is established for these lean systems as shown in the following.

It is first assumed that enough absorbing liquid is used to prevent the concentration of fluoride in the liquid from varying significantly as a result of the absorption. Then,[17]

$$N_{og} = \ln[(y_1 - a)/(y_2 - a)]$$

where: N_{og} = number of transfer units required
y_1 = inlet concentration of fluorine-containing compound in gas phase
y_2 = outlet concentration of fluorine-containing compound in gas phase
a = concentration of fluorine-containing compounds in gas phase in equilibrium with scrubbing liquid

All concentrations are expressed in the same units, which may be volume percent, mole percent, or partial pressure. For very dilute systems, little accuracy is lost by expressing concentrations in weight percent or weight per unit volume.

In those cases where the vapor pressure of fluorides in the scrubbing liquid, inlet and outlet, is negligible, then

$$N_{og} = \ln[y_1/y_2]$$

Scrubbers

Scrubbing systems used in phosphoric acid plants are venturis, wet cyclonic, and semi–cross-flow[18] scrubbers.

Figure 2 is an artist's sketch of a cyclonic scrubber with outside wall-mounted spray boxes. This design feature allows the spray nozzles, which are mounted in a number of boxes, to be cleaned or replaced while the scrubber is in operation. Since the nozzles tend to plug with silica, this feature improves the operating time of the scrubber. Figure 3 shows a two-stage cyclonic scrubber recently installed on a phosphoric acid reactor.

In the current design of semi–cross-flow scrubbers, pond water is sprayed countercurrent to the gas stream in several

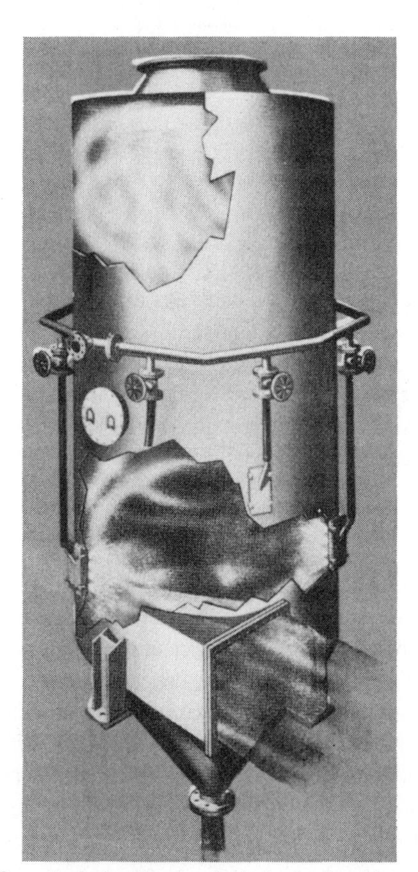

FIGURE 2. Artist's Sketch of Cyclonic Scrubber with Wall-Mounted Spray Boxes Showing Flows of Internal Gases and Scrubbing Liquid (used with permission Hydronics Engineering Corp., Midland Park, NJ)

FIGURE 3. Two-stage Cyclonic Scrubber Installed on a Wet Process Phosphoric Acid Reactor (Hydronics Engineering Corp., Midland Park, NJ.)

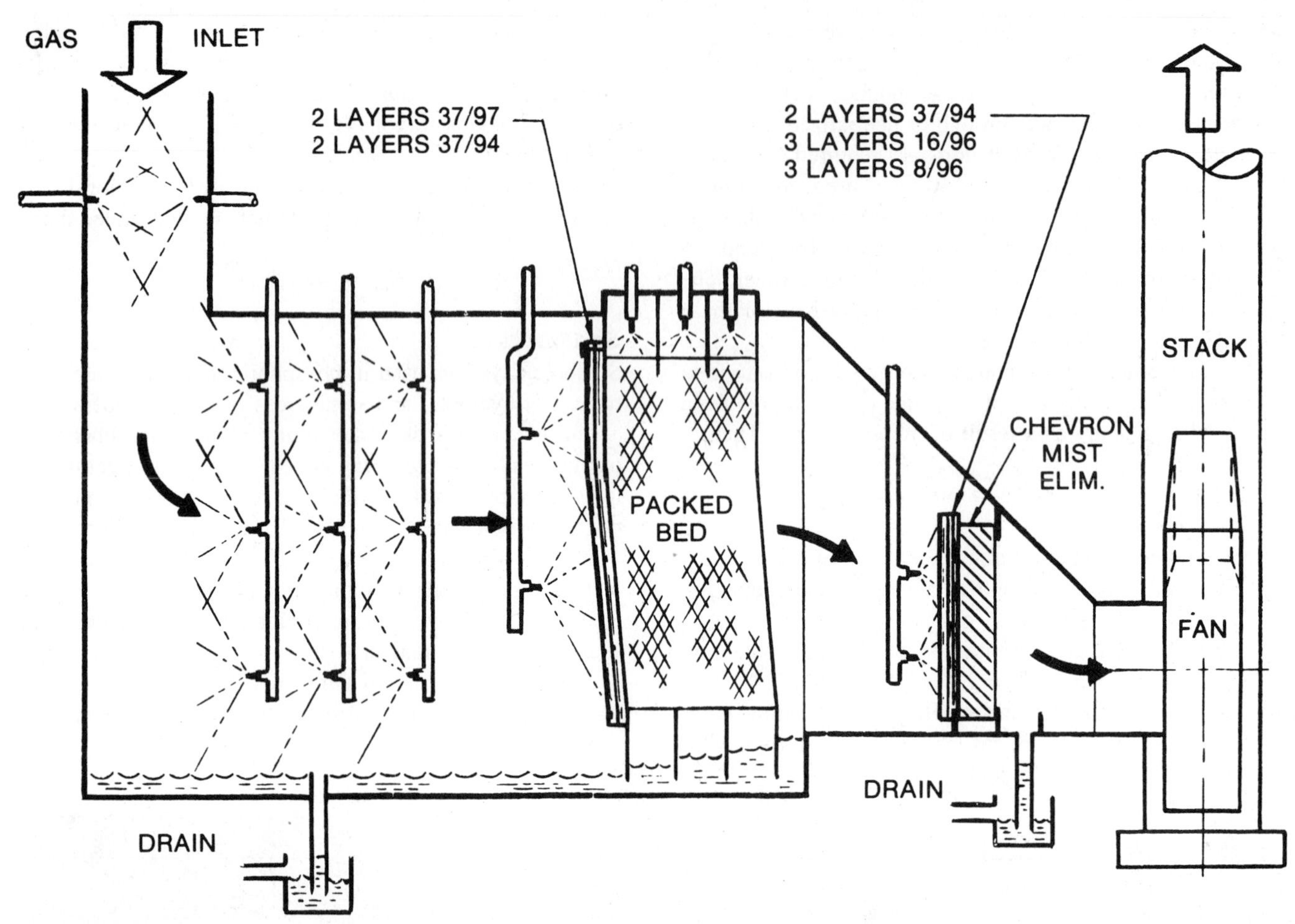

FIGURE 4. Phosphoric Acid Fluoride Scrubber after Changes (Kimre, Inc., Perrine, FL.)

rows of spray headers in the front of the scrubber. Then spray headers are used to spray cocurrent with the gas stream onto two or more Kimre tower packing sections (multiple woven pads) in series followed by a dry Kimre tower packing for drift elimination. Several older spray cross-flow packed scrubbers have been modified because of plugging of the initial type of tower packing used. In some plants, the packing was removed and the scrubber converted by installing Kimre tower packing. In other plants, the Kimre tower packing was installed prior to the initial tower packing and a dry Kimre tower packing installed before the fan suction for drift elimination. Figure 4 presents a sketch of a phosphoric acid scrubber showing these changes.

REGULATIONS

Fluorides

Emission standards for fluorides in the states with operating phosphoric acid plants are summarized in Table 3. Ambient air and vegetation standards are shown in Table 4.

Florida

Florida rules[19] state that discharges to the atmosphere from wet process phosphoric acid plants shall not contain total fluorides in excess of 10 grams per metric ton (0.020 lb/ton) of equivalent P_2O_5 feed and 5 grams per metric ton (0.010 lb/ton) for superphosphoric acid plants. These were adopted by reference to the federal EPA rules.[20] Florida has no ambient air or vegetation fluoride rules.

Texas

Texas fluoride rules[21] limit inorganic fluoride emissions, calculated as HF to 6 ppb by volume average ambient air difference between upwind level and downwind level for the property (measured at the property lines). Allowable fluoride emissions for point sources must be calculated by Sutton's equation for dispersion from stacks. The equation has been modified to consider the critical wind speed and to correspond to a three-hour air sample. The equation contains terms for stack velocity, exit stack diameter, and stack height.

Texas also has ambient air fluoride standards for 12-hour, 24-hour, 7-day, and 30-day periods, including background, as shown in Table 4. Vegetation fluoride levels

TABLE 3. Fluoride Emission Limits

Regulator	Facility	Emission Limits
Federal EPA	Wet process phosphoric acid[a]	10.0 g/metric ton P_2O_5[c] (0.02 lb/ton)
	Superphosphoric acid[b]	5.0 g/metric ton P_2O_5[d] (0.01 lb/ton)
Florida	Wet process phosphoric acid	Same as EPA
	Superphosphoric acid	Same as EPA
Wyoming	Wet process phosphoric acid	Same as EPA
	Superphosphoric acid	Same as EPA
Idaho	Total plant	0.30 lb F/ton P_2O_5[e]
Louisiana	Wet process phosphoric acid	Same as EPA
	Superphosphoric acid	Same as EPA
Texas	Individual stacks	Calculated from Sutton See Rule Total property limited to 6 ppb fluoride (as HF) difference upwind and downwind concentration[f]
North Carolina	Wet process phosphoric acid	Same as EPA
	Superphosphoric acid	Same as EPA

[a]40 CFR 60.200 Subpart T.
[b]40 CFR 60.000 Subpart U.
[c]Total fluorides from reactors, filters, evaporators, hot wells.
[d]Total fluorides from evaporators, hot wells, acid sumps, cooling tanks.
[e]Total fluoride emissions in gaseous and particulate form, expressed as (F), per ton of P_2O_5 input, to the calciner operations, calculated at the maximum feed rate.
[f]Total inorganic fluorides from three consecutive hours, by volume.

TABLE 4. Fluoride Ambient Air and Vegetation Limits

Regulator	Ambient Air Limits		Vegetation Limits	
Texas	Time period	ppb F (1)	Time period	ppm F (2)
	12 hours	4.5	12 months	40
	24 hours	3.5	3 months	60
	7 days	2.0	2 months	80
	30 days	1.0		
Wyoming	Regional Standard			
	Time Period	μg/cu. meter (3)	Time Period	ppm (4)
	12 hours	10.0	1 year	30
	24 hours	4.0	60 days	60
	7 days	1.8	30 days	80
	30 days	1.2		
Idaho	None	None		
Louisiana	None	None		
Florida	None	None		
North Carolina	None	None		

1. Average gaseous fluorides (calculated as HF); by volume.
2. Average inorganic fluoride, by wt in forage based on samples once a month for consecutive calendar months. Total of absorbed and deposited fluoride.
3. Calculated as HF, maximum allowable for averaging time. See rule for specific areas.
4. In forage for animal consumption. Maximum allowable, measured as fluoride, dry basis.

have been set for samples taken once a month for various consecutive monthly periods, as shown in Table 3.

Wyoming

Wyoming has also adopted the EPA fluoride emissions standards by reference[22] for wet process phosphoric acid plants and superphosphoric plants. Ambient air standards for Wyoming provide statewide values that vary with sampling times. More restrictive standards are used for the areas that contain phosphoric acid and superphosphoric acid plants (Table 3). Vegetation standards vary with sampling time, as shown in Table 4.

Idaho

Idaho fluoride emissions are based on the entire fertilizer complex[23] and are limited to 0.30 lb/ton P_2O_5 fed to the phosphate rock calciner. There are no standards for ambient air or vegetation.

Louisiana

Louisiana has also adopted by reference[24] the EPA rules for emissions of fluorides from wet process phosphoric acid plants and superphosphoric acid plants. Louisiana has no ambient air or vegetation fluoride standards.

North Carolina

North Carolina has also adopted by reference[25] the EPA rules for emissions of fluorides from wet process phosphoric acid plants and superphosphoric acid plants. North Carolina has no ambient air or vegetation fluoride standards.

References

1. International Fertilizer Development Center, *Fertilizer Manual,* Muscle Shoals, AL, 1981, p 163.
2. A. V. Slack, *Phosphoric Acid,* Vol. 1, Marcel Dekker, New York, 1969, p 183.
3. International Fertilizer Development Center, *Fertilizer Manual,* Muscle Shoals, AL, 1981, p 164.
4. A. V. Slack, *Phosphoric Acid,* Vol. 1, Marcel Dekker, New York, 1969, p 197.
5. J. Gobbitt, "Hemihydrate phosphoric acid plant retrofits at Geismar and Chinhae," AIChE Central Florida Section, Annual Clearwater Meeting, 1990, p 5.
6. A. V. Slack, *Phosphoric Acid,* Vol. 2, Marcel Dekker, New York, 1969, p 745.
7. A. V. Slack, *Phosphoric Acid,* Vol. 2, Marcel Dekker, New York, 1969, p 765.
8. International Fertilizer Development Center, *Fertilizer Manual,* Muscle Shoals, AL, 1981, pp 170–171.
9. W. E. Rushton, "Defluorination of wet process acid," *Chem. Eng. Prog.,* 52 (November 1978).
10. A. C. Perkins, "Acid defluorination for animal feeds," AIChE Central Florida Section, Annual Clearwater Meeting, 1988, p 2.
11. P. A. Boys, *Air Pollution Control Technology and Costs in Seven Selected Areas,* EPA-450/3-73-010, U.S. Environmental Protection Agency, Research Triangle Park, NC, 1973, p 38.
12. S. V. Houghtaling, "Wet grinding of phosphate rock holds down dollars, dust and fuel," *Eng. Mining J.,* 94–96 (January 1975).
13. A. A. Linero and R. A. Baker, *Evaluation of Emissions and Control Techniques for Reducing Fluoride Emissions from Gypsum Ponds in the Phosphoric Acid Industry,* EPA-600/2-78-124, U.S. Environmental Protection Agency, Research Triangle Park, NC, 1978, pp 1–219.
14. P. Beker, *Phosphates and Phosphoric Acid,* Marcel Dekker, New York, 1983, p 486.
15. P. Tiberghien and F. L. Prado, "The Rhone-Poulenc phosphoric acid process and the environment," AIChE Central Florida Section Annual Clearwater Meeting, 1986, pp 2–5.
16. P. A. Boys, *Air Pollution Control Technology and Costs in Seven Selected Areas,* EPA-450/3-73-010, U.S. Environmental Protection Agency, Research Triangle Park, NC, 1973, pp 38–47.
17. A. V. Slack, *Phosphoric Acid,* Vol. 2, Marcel Dekker, New York, 1969, pp 753–754.
18. G. C. Pederson, "Design procedure for fluorine scrubbers in the fertilizer industry," AIChE Central Florida Section Meeting, September 8, 1987, p 15.
19. Rules of the Department of Environmental Regulation. Florida Administrative Code, Title 17, Chapter 17-2, Air Quality, *17-2.660 Standards of Performance for New Stationary Sources (NPS),* September 17, 1980.
20. 40 CFR 60.200, *Subpart T & U—Standards for Fluorides,* U.S. Environmental Protection Agency, Research Triangle Park, NC.
21. Texas Regulation III: Control of Air Pollution from Toxic Materials, Subchapter A. Inorganic Fluoride Compounds and Beryllium, Para. (a)(1), *Emission limits for inorganic fluoride compounds,* Texas Air Control Board, last amendment October 1, 1987.
22. Wyoming Air Quality Standards and Regulations, *Section 11, Fluorides, Ambient Air and Forage Standards for Fluorides,* Department of Environmental Quality, Division of Air Quality, April 17, 1986.
23. Idaho Department of Health and Welfare Rules and Regulations, Sections 1.1401 through 1.1450, *Rules for Control of Fluoride Emissions,* June 23, 1980.
24. Louisiana Administrative Code, Volume 11, Environmental Quality, *Subchapter T Standards Of Performance For The Phosphate Fertilizer Industry: Wet-Process Phosphoric Acid Plants (Subpart T) & Subchapter U: Superphosphoric Acid Plants (Subpart U),* December 1987, pp 426–432.
25. North Carolina Air Pollution Control Regulations, Subchapter 2D—Air Pollution Control Requirements, para. 0534, *Fluoride Emissions From Phosphate Fertilizer Industry,* Bureau of National Affairs, Washington, D.C., August 24, 1990, p 183.

Bibliography

Kirk-Othmer, *Encyclopedia of Chemical Technology,* Vol. 9, 1968, pp 87, ff.

V. Sauchelli, *Chemistry and Technology of Fertilizers,* Rheinhold, New York, 1960, pp 192–250.

R. N. Shreve, *Chemical Process Industries,* 3rd ed., McGraw-Hill Book Co., New York, 1956, pp 273–276.

A. W. Waggaman, *Phosphoric Acid, Phosphates and Phosphatic Fertilizers,* 2nd ed., Rheinhold, New York, 1952, pp 174–209.

PHTHALIC ANHYDRIDE

Herbert P. Dengler

The demand for phthalic anhydride (PAN) in the United States hovered around a billion pounds per year in the decade of the 1980s. A record peak of 1050 million pounds per year was reached in 1988 compared with 950 million pounds per year in 1989. Demand growth rates in the early 1990s are expected to accelerate up to 2.5% per year. Over the past two decades, the number of producers has dwindled to six, all with capacities in the range of 150–250 million pounds a year. Total capacity was estimated at 1120 million pounds per year as of mid-1989.[1]

All current production is obtained by catalytic air oxidation in fixed-bed units. Five of the six U.S. plants use *o*-xylene as feedstock. The sixth uses naphthalene derived from coal tar for at least a portion of its feedstock. As of mid-1989, the price of *o*-xylene was about 21 cents per pound and molten PAN was selling in the range of 40–44 cents per pound.[2] The price of flaked and bagged PAN in the United States is typically 4 cents per pound higher than for molten PAN.

The major end use for PAN is in the manufacture of plasticizers. For the last quarter of a century, domestic end uses in the U.S. have been 50% to plasticizers, 25% to polyester resins, 20% to alkyd resins, and the remaining 5% to a myriad of uses such as insecticides and dyestuffs. Exports have ranged to as high as 5% of the production in some years.

PROCESS DESCRIPTION

Commercially, PAN is made by the catalytic air oxidation of either *o*-xylene or naphthalene (Figure 1). *O*-xylene is obtained from petroleum sources and typically is about 98% pure, with the major impurity being *p*-xylene. The naphthalene is obtained from coal tar sources and the purity typically exceeds 95%.[3] The catalyst is a vanadium-pentoxide–titanium-dioxide mixture containing small quantities of promoters.[4] This active material is surface coated on a support that is inert at the operating temperature levels. The reaction is strongly exothermic. To facilitate heat removal in commercial plants, the reaction is carried out in nominal 1-inch-diameter vertical tubes in a reactor with circulating molten heat-transfer salt in the shell.[5]

The major reactions are as follows:

$$o\text{-Xylene} + 3O_2 \rightarrow \text{PAN} + 3H_2O + \sim 265 \text{ kcal/g-mol}$$

$$\text{Naphthalene} + 4.5O_2 \rightarrow \text{PAN} + 2H_2O + 2CO_2 + \sim 427 \text{ kcal/g-mol}$$

Temperatures of molten salt in the reactor are in the 630–730°F range. Process temperatures are limited to less than 900°F to avoid damage to the catalyst, especially because surface temperatures can be considerably higher than bulk temperatures.

The air oxidation process is carried out at pressures only slightly above atmospheric pressure. Process air is compressed to approximately 10 psig and preheated. Preheated *o*-xylene or naphthalene is mixed with the air in carefully controlled concentrations. Typically, the *o*-xylene concentration is in the range of 60–70 grams of *o*-xylene feed per cubic meter of wet air and is lower for naphthalene. The mixture is less than 1.5 vol% *o*-xylene, but it is flammable. In the event of an ignition, the equipment is protected by rupture disks vented to the atmosphere. Because ignitions are very infrequent and the concentrations of hydrocarbons are low, atmospheric emissions from this source are insignificant.

The residence time in the reactor is less than one second. Essentially complete conversion of the feed occurs. The molar selectivity for the conversion of *o*-xylene to PAN has improved steadily over the years and is now about 80% of theoretical, while that of naphthalene is in the 90–95% range. The corresponding weight yield for *o*-xylene is 1.1 pound of PAN per pound of 100% *o*-xylene. Naphthalene weight yields are about 10% lower. Theoretical weight yields are 1.4 for *o*-xylene and 1.16 for naphthalene (pounds of PAN per pound of 100% pure feedstock).

The effluent from the reactor is primarily nitrogen and oxygen. Reactor effluent oxygen concentrations are in the 12–15 vol% range because less than half of the oxygen in the air is consumed in the reaction. In descending order of volumetric concentration, the effluent contains nitrogen, oxygen, water, carbon dioxide, PAN, carbon monoxide, maleic anhydride, and other oxygenated by-products. The other oxygenated by-products are about 1 wt% of the hydrocarbon feed when using *o*-xylene and slightly higher when naphthalene is the feed, depending on feed purity.

The by-products with lower boiling points than that of PAN are primarily maleic anhydride and benzoic acid. With naphthalene feedstocks, naphthaquinone production can reach 0.1–0.5 wt% on feed. The naphthaquinone must be removed from the final product by distillation and/or chemical treatment to much lower levels than those of benzoic acid and maleic anhydride because it has a detrimental effect on product color. The by-products with higher boiling points include phthalide, trimellitic anhydride, and a variety of carboxylic acids, together with a small quantity of unidentified tars and carbon particles.

After exiting the reactor, the process stream is cooled and ≥99% of the PAN is removed as a solid by desublimation in switch condensers. These condensers alternately are switched onstream to a cooling/desublimation service and isolated from the process stream for heating up and melting out. During melting, the crude PAN is recovered as a liquid.

The crude PAN is then subjected to a heat treatment of

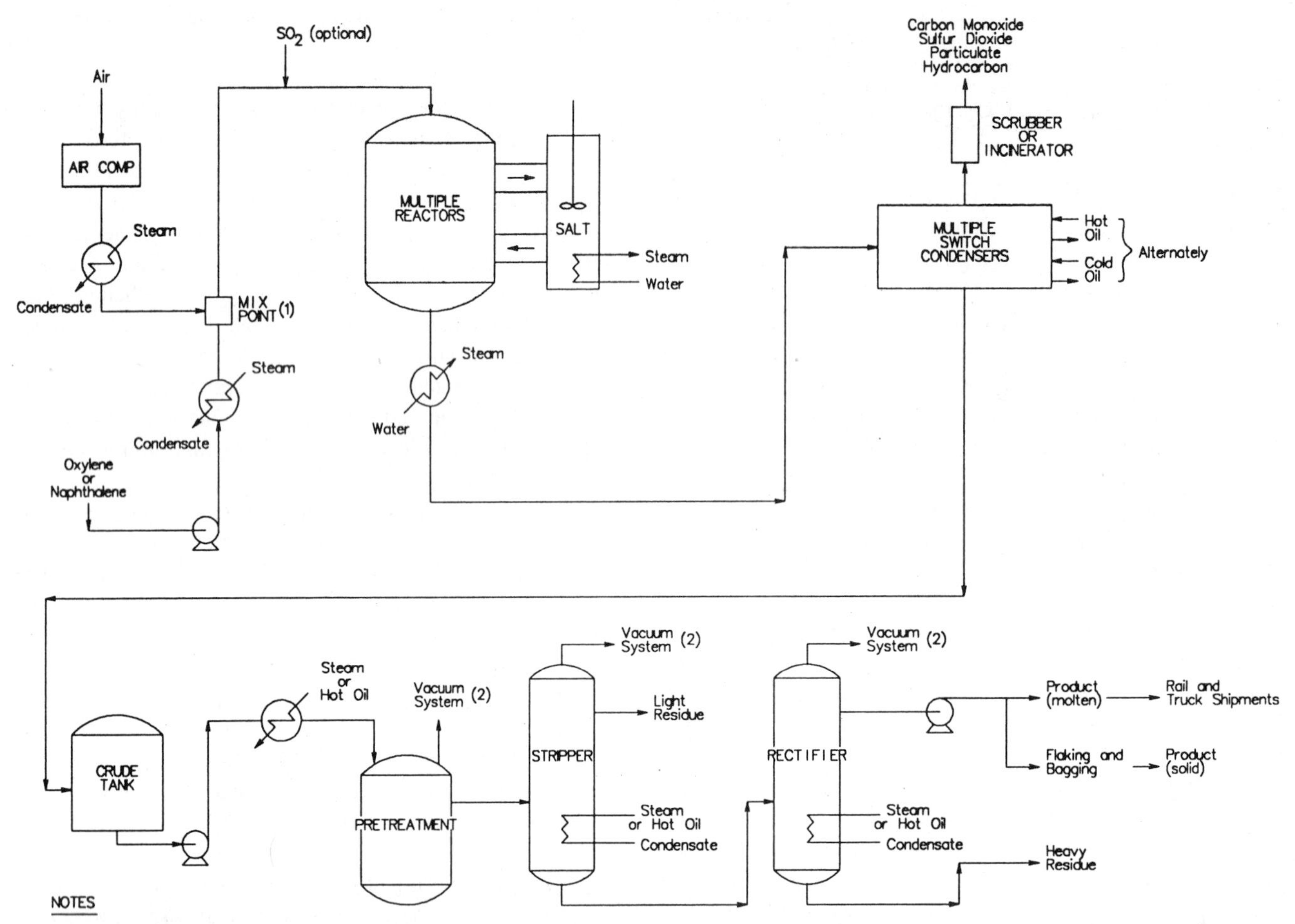

FIGURE 1. Phthalic Anhydride Process

12–24 hours at 450–550°F, which improves product quality. The pressure level ranges from about 10 mmHg vacuum to slightly above atmospheric pressure. In the latter case, a vacuum jet would not be required. The pretreatment step is frequently called decomposition because one of the reactions that occurs is the decomposition of phthalic acid to PAN and water vapor. The phthalic acid concentration in the crude can be as high as 1 wt%, depending on the operating conditions in the switch condensers. Chemicals such as sodium carbonate are sometimes added during the decomposition step to improve product quality further, and, in the case of naphthalene feeds, especially to reduce naphthaquinone levels in the product.

After the heat treatment, the light ends are removed from the crude product by vacuum distillation. The pure product is then taken overhead in a second vacuum distillation tower, leaving behind the higher-boiling-point materials. The light and heavy cuts are about equal in volume, depending on the purity of the feedstock. The PAN content of these cuts is about 50%, but can be lower if additional distillation facilities are provided.

The light and heavy residues are usually mixed together and typically are used as fuel to generate steam or to heat oil in a high-temperature heat-transfer oil circuit. The gases leaving the switch condensers contain light ends, unrecovered PAN, carbon oxides, and water. This stream is either incinerated or water scrubbed to remove the organic materials.

The molten PAN product is a water-white liquid with fluid properties similar to those of water. Most of the PAN used in the United States is transferred in its molten state. This requires keeping it above its freezing point of 268°F. Shipments are made in insulated railcars or tank trucks, most of which are fitted with steam coils to melt cargo that has partly solidified during transit. The material that is flaked and bagged is used for long-term storage or exported; this is the only method currently available for overseas shipments.

AIR EMISSIONS CHARACTERIZATION

Sources

Air emission sources from the process include the following:

1. Incinerator or scrubber effluent
   a. Oxygenated hydrocarbons
   b. Particulates (*Note:* Most are organic solids that eventually vaporize on exposure to air at ambient conditions.)
      (1) Primarily due to PAN, which desublimates when cooled below its freezing point of 268°F.
      (2) Minor quantities of ash, primarily from corrosion of carbon steel heat-transfer equipment.
      (3) In the case of scrubbers, liquid particles from entrainment of the scrubber solution.
      (4) Residue particulates if residue is incinerated.
   c. Sulfur dioxide (SO_2) is optionally injected into the reactor as a catalyst promoter. It passes essentially unchanged to the incinerator or scrubber. Also, SO_2/SO_3 is produced by combustion of sulfur-containing fuels used in process heaters and thermal incinerators.
   d. Carbon monoxide (CO) is produced in the process reaction, as well as in thermal incinerators and process heaters.
   e. Nitrogen oxides (NO_x) are not produced in the process, but are produced in the thermal incineration and in process furnaces from the combustion of the fuel used to provide the heat.
2. Vacuum system vents
   a. Steam jets with barometric condensers—similar to scrubber effluent, but with very little CO and maleic anhydride present.
   b. Vents discharged to a furnace—part of furnace emissions.
   c. Recycle to switch condensers—part of switch condenser effluent.
   d. If the pretreatment is operated above atmospheric pressure, a vacuum system is not required to vent it to, for example, the switch condenser effluent circuit.
3. Storage tanks
   a. Crude and product storage and transfers—particulates (PAN).
   b. Feed tankage—*o*-xylene or naphthalene hydrocarbon emissions.
   c. Heat transfer oils—weathering of light ends if degradation occurs.
4. Fugitive emissions valves, flanges, and so on. Very small quantities are involved because most of the process operates close to atmospheric pressure or under a vacuum. Values are based on heavy liquid because of the low vapor pressure of PAN. Hydrocarbon emissions also occur from the *o*-xylene–naphthalene feed system up to the reactor catalyst bed.
5. Emissions from flaking operations—particulates (PAN).

Emission Factors

Table 1 is a revision of Table 5.12-1 from AP-42[6] updated on the following basis.

1. The portion of Table 5.12-1 entitled "oxidation of naphthalene," which was largely based on the older fluid-bed processes, was deleted because PAN is currently produced only in fixed-bed units. Because the fixed-bed units are similar, whether *o*-xylene or naphthalene is used as the feedstock, only one set of values for both feedstocks is shown.
2. Since the completion of the 1975 engineering study, which set the base values for AP-42, technological changes in the industry have permitted operations at higher-weight concentrations of feedstock in the process air.[5] Considerable energy savings have resulted because it has been possible to increase production at constant air capacity. Typical concentration increases were from 40 grams of *o*-xylene per cubic meter of air to about 66. While the main process stream concentrations of SO_2 and CO are increased by the capacity ratio of 66/40, this is offset by the increased plant production ratio of 66/40. Thus the SO_2 and CO values reported in Table 5.12-1 are unchanged because they were based on a unit weight of product. Although it is unlikely that all the plants are running at the 66 g/m^3 air level, this value was chosen to reflect an average of current and near-future trends in the industry.
3. For particulates, the recovery systems have been improved with the increased production so that losses per ton have remained constant. Incineration and scrubbing technology has been improved so that 98% instead of 95% destruction of the particulates is achieved, as shown in the table.
4. Only incineration technology is shown because it is the best of current technology. Combinations of scrubbers and incinerators were deleted because it is currently more likely that an incinerator would be used to replace a scrubber than be added to a scrubber.
5. NO_x emissions have been added. The process produces nil NO_x but NO_x is produced in the thermal incineration and in any process furnaces. NO_x can also be produced in catalytic incineration prior to start-up. Values shown are based on fuel gas firing in a thermal incinerator.

AIR POLLUTION CONTROL MEASURES

Incinerator

The incinerator or scrubber is an integral part of the process because of the large volumes of air used. Although about

TABLE 1. Emission Factors for Phthalic Anhydride[a,b] *(Emission Factor Rating: B)*

	Particulate[c] (Revised)		NO_X (Added)	
	g/kg	lb/ton	g/kg	lb/ton
Main process stream[d]				
Uncontrolled	69	138.0	0.0	0.0
With incinerator[e]	1.4	2.8	0.5	1[f]
Pretreatment (vacuum system)				
Uncontrolled	6.4	13.0	0.0	0.0
With incinerator[e]	0.1	0.3	0.0	0.0
Distillation (vacuum system)				
Uncontrolled	45	89.0	0.0	0.0
With incinerator[e]	0.9	1.70	0.0	0.0

The following factors are unchanged from AP-42

	SO_x		Nonmethane VOC[g]		CO	
	g/kg	lb/ton	g/kg	lb/ton	g/kg	lb/ton
Main stream						
Uncontrolled	4.7[h]	9.4[h]	0.0	0.0	151.0	301.0
With incinerator[e]	4.7	9.4	0.0	0.0	3.0	6.0
Pretreatment						
Uncontrolled	0.0	0.0	0.0	0.0	0.0	0.0
With incinerator[e]	0.0	0.0	0.0	0.0	0.0	0.0
Distillation						
Uncontrolled	0.0	0.0	1.2[c,e]	2.4[c,e]	0.0	0.0
With incinerator[e]	0.0	0.0	<0.1	0.1	0.0	0.0

[a]Revised Table 5.12-1 in reference 7, which was based on data in reference 6. Factors are in grams of pollutant per kilogram of PAN produced (or in pounds per ton).
[b]Control devices listed are those currently being used by PAN plants.
[c]Consists of PAN, maleic anhydride, benzoic acid, and for naphthalene feedstocks, naphthaquinone.
[d]Main process stream includes reactor and multiple switch condensers as vented through condenser unit.
[e]Normally a vapor, but can be present as a particulate at low temperature.
[f]Typical value for a thermal incinerator operating on a fuel gas that does not contain nitrogen compounds. Varies considerably with design, operating temperature, and fuel composition.
[g]Emissions contain no methane.
[h]Value shown is for a relatively fresh catalyst that requires an SO_2 promoter. The requirement can change with age and can be 9.5–13 kg/mg (19–25 lb/ton) for aged catalysts. Catalysts that do not require SO_2 are available, but with debits of shorter life and lower yield of product.

99% of the phthalic is recovered in the switch condensers, the remaining phthalic and by-products have to be removed. Water scrubbing recovers the materials for possible sales or for combustion as a liquid solution. Scrubbing has practical limitations in the recovery of the oxygenated hydrocarbons and does not convert CO to carbon dioxide. Therefore, the preferred choice, currently, is either catalytic or thermal incineration.

In thermal incinerators, the effluent is heated to about 1500°F by burning additional fuel. Residence times are about one second at this temperature, after which the gases are cooled, usually by generating steam.[8] Catalytic incinerators operate in the 500–800°F temperature range and, depending on the amount of organics in the waste gas, may not require auxiliary fuel, except for start-up and end-of-run operations.[9]

Both types achieve a destruction efficiency of the hydrocarbons and CO of 98%. Increased destruction efficiency can be obtained in a thermal incinerator by increasing residence time and/or temperature. Similar results can be obtained in a catalytic system by increasing either inlet temperature or the amount of catalyst (i.e., residence time in the catalyst bed). Higher temperature in a thermal incinerator increases NO_x formation and requires more steam generation to remain efficient. Catalytic incinerators are limited on catalyst bed outlet temperatures because of potential catalyst damage at elevated temperatures.

Vacuum Systems

Because PAN solidifies at 268°F and reacts with water to form solid phthalic acid, vacuum pumps with water seals normally are not used to provide a vacuum. Instead, vacuum jets are used. The effluent from the jets can be treated in one of three ways: scrubbed with water and the resultant solution treated in a biox system, burned as fuel, or recycled back to the process. In the last case, the noncondensibles in the stream are eventually treated in the switch condenser effluent incinerator or scrubber. Alternatively, condensers can be installed ahead of the jets to reduce the organic levels to the point where further treatment of the jet effluent is not required.

Tank Vents and Transportation

Vent control consists of two types: desublimators or a vacuum system using jets (as described above and including any vents at elevated pressure that can bypass the jets and go directly to the downstream treatment steps). The simplest type of desublimator is a large sheet-metal box that provides sufficient surface area to cool the vented gases. On cooling, the PAN desublimes from the vapor and collects in the box. Suitable baffles are provided to prevent the vapor from exiting the box until cooled and for periodic removal and recycling of the solid PAN crystals. The vent gas composition corresponds to the vapor pressure of PAN at the exit temperature. Because PAN has a low vapor pressure (0.0035 mm Hg at 100°F), the concentration is low. The total amount of PAN vented will depend on tank filling rates and the rate of the gas blanketing on the tank.

Desublimators mounted on the top of the tank and equipped with heating and cooling facilities also are used. These can be operated to lower outlet temperatures depending on the cooling-medium temperature and do not require manual transfer of the recovered solid phthalic.

References

1. *Chemical Marketing Reporter,* July 24, 1989, p 58.
2. *Chemical Marketing Reporter,* July 24, 1989, p 15.
3. H. P. Dengler, *Phthalic Anhydride. Vol. 36, Encyclopedia of Chemical Processing and Design,* J. J. McKetta and W. A. Cunningham (Eds); Marcel Dekker, New York, 1990 p 33.
4. R. Y. Saleh and I. E. Wachs U. S. Patent 4,582,912, April 15, 1986 (to Exxon Research & Engineering Co.).
5. W. Gierer and O. Weideman, "Phthalic anhydride made with less energy," *Chem. Eng.* 86(3):26 (1979).
6. *Compilation of Air Pollutant Emission Factors,* AP-42, U.S. Environmental Protection Agency, Research Triangle Park, NC, 1985, pp 5.12–5.
7. *Engineering and Cost Study of Air Pollution Control for the Petrochemical Industry, Vol. 7, Phthalic Anhydride Manufacture from Ortho-xylene,* EPA-450/3-73-006g, U.S. Environmental Protection Agency, Research Triangle Park, NC, July 1975.
8. T. F. McGowan and R. D. Ross, "Incineration: flexible and efficient," *Chem. Proc.,* 53(11):79 (1990).
9. R. G. McInnes, S. Jelinek, and V. Putsche, "Cutting toxic organics," *Chem. Eng.,* 97(9):108 (1990).

PRINTING INKS

Stephen W. Paine and Stephen K. Harvey

There are four major classes of printing inks: letterpress and lithographic, which are oil or paste inks, and flexographic and rotogravure, which are fluid inks (solvent inks). Printing inks are used for the printing of publications, packaging, and certain specialty products. While letterpress and offset inks are used for almost every type of publication and packaging, gravure inks are used principally for magazines, advertising circulars, packaging, and related products. Flexographic ink is used almost exclusively for package printing, although there is a trend toward increased use of the process to print newspapers. Products such as gift wrap, wall coverings, and floor coverings are direct printed using the gravure process, with lesser amounts being printed using the flexographic process. Other printing processes (which, combined, consume less ink on an annual basis than any of the four major processes) include screen printing, intaglio, ink jet printing, and xerography. Figure 1 provides a breakdown of ink manufacturing in the United States.

All inks consist of a pigment, which is the colorant, dispersed in a vehicle, which is the component that binds the pigment to the receiving surface. Receiving surfaces include paper, paperboard, foil, and plastic films. Generally, while the pigments used in all four types of printing ink are similar, the vehicles are quite different. Letterpress and lithographic processes use paste inks that incorporate a thick, viscous vehicle, while gravure and flexographic inks are fluid and utilize a much lower viscosity vehicle. The letterpress inks and lithographic inks dry by oxidation of the vehicle or vehicle absorption into the printing surface, while fluid inks dry by solvent evaporation.

The nature of oil and paste inks is such that they are less

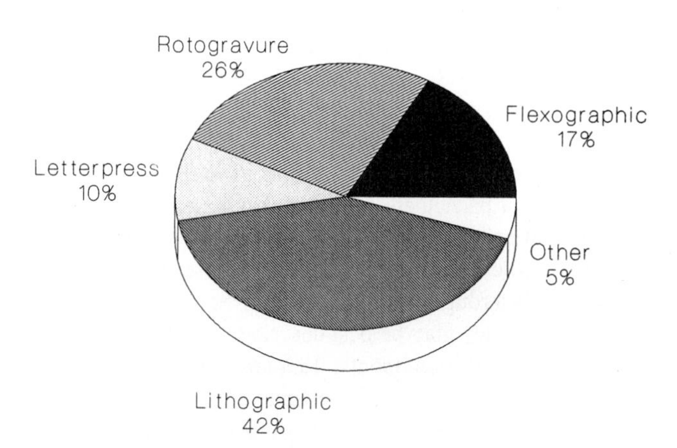

FIGURE 1. Ink Production in the United States—Percent of Total Production by Weight

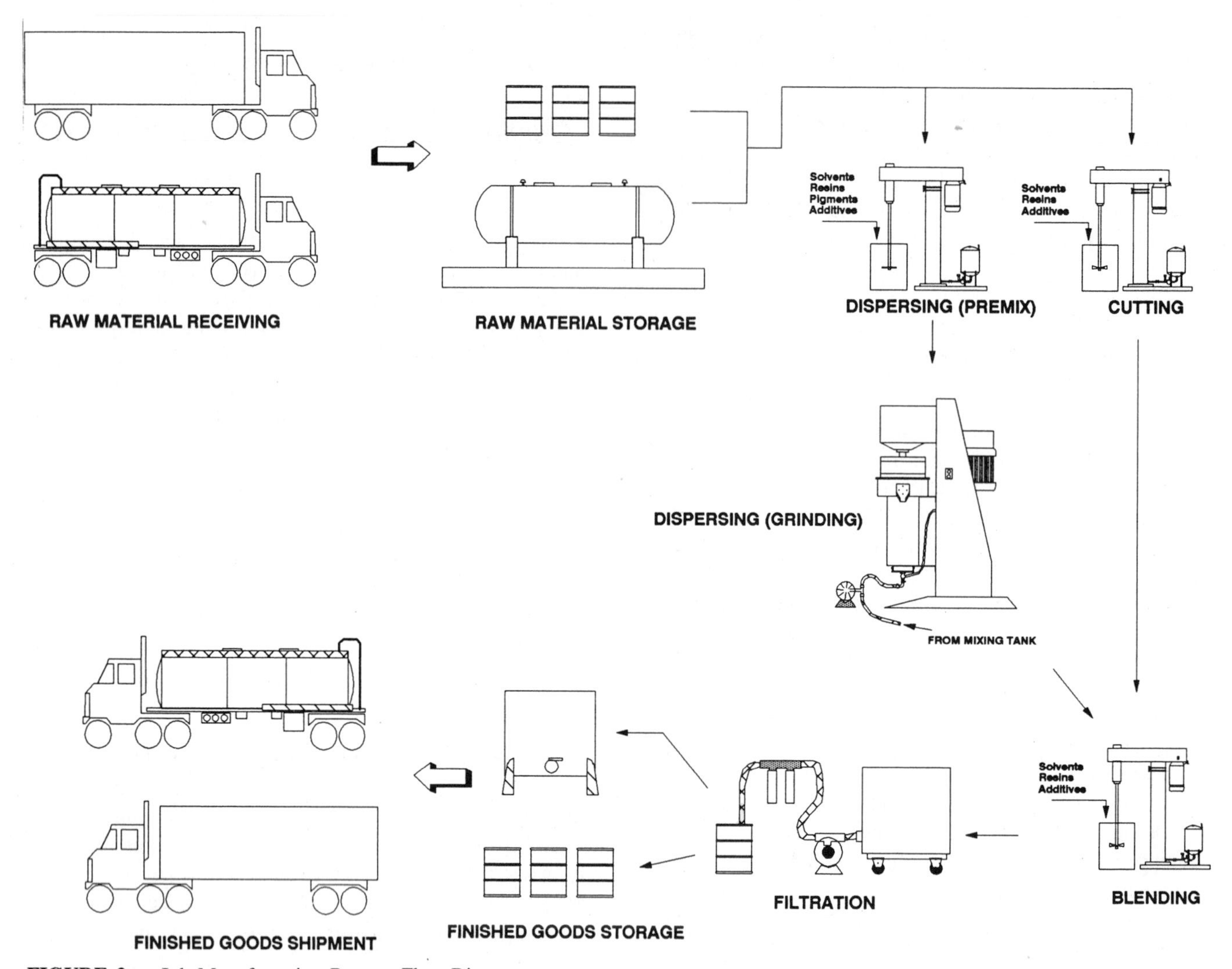

FIGURE 2. Ink Manufacturing Process Flow Diagram

important emissions sources. This is so because the processes which utilize them, lithography and letterpress printing, do not normally use volatile organic compounds (VOCs).[1] One exception, a significant source, occurs where vehicle (varnish) is cooked on-site. This is, where practiced, the primary source of VOC emissions from lithography and letterpress ink manufacturing.[2] While many companies purchase the vehicle after cooking, emissions estimates are included under "Air Emissions Characterization", for use when applicable.

Flexographic and rotogravure inks have many properties in common: a particulate (pigment) dispersed in a vehicle, low viscosities, and drying after application by the evaporation of volatile solvents and/or water. These inks are manufactured via a roughly similar process sequence (see Figure 2). This sequence may be best described as a simple blending of prepared components. Preparation of components consists of wet grinding (dispersing) pigments in vehicles and dissolving (cutting) of resins into solvent mixtures.

The principal difference between flexographic and rotogravure inks relates directly to their respective end uses. Flexographic inks are used in a printing process where the design is transferred from a rubber or synthetic relief cylinder onto the printing surface. A solvent that does not degrade the surface of the relief cylinder must be used. The principal solvents used in the flexographic process are alcohols or mixtures containing alcohols. The gravure printing process differs from the flexographic process in that the design to be printed is etched or engraved into the surface of a metal printing cylinder. Gravure ink formulations include solvents that would degrade flexible printing cylinders. Various hydrocarbon-based solvents such as toluene and other solvents such as isopropyl acetate can be used within gravure formulations.

Since the mid-1970s, much of the flexographic ink for printing of corrugated containers and folding cartons has been formulated with water-soluble materials. The *Census of Manufacturers* reported a total of 90 million pounds of water-based flexographic ink produced in 1982. The emissions from the manufacture of these inks contain few, if any, VOCs. In the 1980s, water-based gravure inks for packaging increased in usage, reaching an estimated 30 million pounds annually at the end of the decade. Future increases in the production and use of water-based fluid ink is limited by the high surface tension and slower evaporation rate of water as compared with VOC solvents. Ink manufacturers continue to attempt to overcome these problems, which would result in the increased manufacture of water-based inks and a corresponding decrease in the production of solvent-based ink.

Both flexographic and rotogravure inks for packaging printing are typically formulated for specific customer applications and to match a myriad of different colors. Variations in color, adhesion, drying, printing equipment, substrate, operator preference, and regulatory requirements all can result in complex formulation changes. As a result, the flexographic and rotogravure ink manufacturing processes for packaging printing can be best described as batch operations that must be highly adaptable to customer demands. Gravure inks used for publication printing can be produced in more or less continuous operations because only four standard colors normally are manufactured.

PROCESS DESCRIPTION

Raw-Materials Receiving

Most raw materials are purchased from outside sources rather than being manufactured by the ink company. However, some paste-ink manufacturers do manufacture their own resins from oils and other materials. Raw materials are received via tank truck, portable tanks, drums, and bags. Materials include resins, dry pigments, solvents (such as ketones, alcohols, aromatics, and acetates), and other processing chemicals. Except for accidental spills and leaks, the only emissions from material receiving occur during tank truck unloading, when vapors are displaced to the atmosphere as the storage tanks are filled (working losses).

Storage

Storage of solvents in tanks may result in breathing losses. Since breathing losses result from diurnal temperature variation of the stored material, there are negligible emissions of this type from underground tanks; only aboveground tanks located outside of buildings are relevant. Other materials stored in portable tanks, drums, and bags emit air pollutants only while undergoing transfers into process containers.

Vehicle Preparation

Vehicles for paste inks are prepared by mixing oils such as petroleum oils and rosin oils with various resins to impart suitable tack and flow characteristics. Some paste inks incorporate reactive oils, such as linseed, soybean, or china wood oil, that allow the ink to dry by oxidation. By chemical modification of these oils, alkyd vehicles are produced that are much superior in drying and certain other properties. Many of these vehicles must be heated or cooked during preparation. Most paste-ink manufacturers purchase these vehicles from unaffiliated suppliers, but some companies have a vehicle cooking capability on the same site as that of ink manufacturing. Primary emissions are the oils and solvents that evaporate during the cooking process. Because most of the cooking vessels are closed, few vapors escape.

Vehicles for fluid inks are prepared by the cutting operation described below.

Cutting

Cutting, or dissolving, is the primary production step for vehicles for fluid inks. These solutions may be used as intermediates, vehicles, and extenders or as finished products such as varnishes. Processing these clear solutions involves pumping solvents into a receiving vessel and then manually adding dry resins and/or other chemicals. Mixing is accomplished by a mechanically driven medium shear blade equipped with slotted teeth. This mixing speeds the dissolving of the resin in the solvent. During mixing, some frictional heating usually occurs.

Emissions are a function of the frictional heating, mix time, the extent to which the vessel is closed, and losses associated with liquid transfers (displaced vapor). The degree of containment of air emissions can vary by ingredient, available equipment, and operator technique.

Dispersing

Dispersing is the primary production step in making an intermediate, usually referred to as a base. A base is a colored concentrate used to produce a finished ink. The first step in dispersing is the preparation of a "premix." A premix involves the addition of vehicles (either manually or automatically), dry pigments (manually), and other chemicals into a mixing container. The batch is then mixed mechanically by the use of a special slotted blade in the case of fluid inks, or by the use of a slower-speed, higher-torque mixing blade in the case of paste inks. Frictional heating may occur, depending on the duration and angular velocity of the mixing blade and the viscosity of the premix. Following the premix stage, wet grinding is accomplished by any of several technologies (see Figure 2). These technologies utilize spherical media or the abrasiveness of the mixture itself to reduce the agglomerated pigment to the micron or submicron size.

Common technologies operate as follows:

1. *Sand and shot mills*. The premix is pumped through a vertical chamber containing sand, steel shot, or other small spherical media. Mechanically driven disks within the chamber churn the media, dispersing the pigment

particles in the vehicle by breaking apart agglomerates. Most mills have self-contained screens through which product is strained and sand or media retained. A water jacket is usually included with this unit to dissipate resulting frictional heating. Air emissions (evaporative losses) can occur at the screen assembly, though in some mills the screen is totally enclosed.

2. *Horizontal mills (media mills).* This process is very similar to the sand mill operation except that the chamber is horizontal and the media are retained by an internal screen or a slot. In this case, premix material is pumped through a horizontal chamber filled with grinding media (ceramic or steel). A rotor within this chamber churns this mixture of premix and media. Dispersion is thus accomplished within a closed system. The media are retained in the mill with an internal screen or very narrow slot. A water jacket is provided to dissipate resulting frictional heating.
3. *Ball mills.* Ball mills consist of a water-jacketed steel cylinder filled with ceramic or steel balls. In this process, a premix may be used or the raw materials added directly into the ball mill. Raw materials and balls are tumbled within this rotating cylinder, mixing and grinding the ingredients for periods up to 24 hours. After the milling cycle has been completed, the base mixture is strained into containers, thus retaining the shot within the mill. Air emissions result when vapors within the cylinder are displaced to atmosphere while the unit is being charged.
4. *Rotor stator devices.* These are devices in which a high-speed turbine moves inside a stator and disperses pigment by forcing the pigment and vehicle through the close clearances between the rotating rotor and stationary stator. These devices may be fixed inside a large vessel or mounted on a lift and lowered inside portable vessels. Usually there is no premix step and the pigment is added directly to the vehicle in the vessel while the rotor is turning. Thus premix and dispersion are achieved simultaneously. These devices generate considerable frictional heat. However, since they are usually used with covers, emissions to air result only from the displacement of vapors when loading and from losses transfer during discharge of the product.
5. *Dispersion via mixing.* For certain applications and easily dispersed pigments, the premix step, which occurs in a large mixing vessel, may be continued until dispersion is accomplished. Again, emissions result from the displacement.
6. *Double-arm mixers.* Unlike other processes, dispersion in a double-arm mixer requires ingredients to be added directly to the unit. This process requires the premix to be an extremely thick pastelike material. Within this mixer, two counterrotating arms force the pastelike material against itself, causing dispersion and grinding. No separate media are required in this process. This vessel is completely closed and little loss occurs except for transfer losses.
7. *Two-roller mills and three-roller mills.* Two- and three-roller mills are composed of two or three horizontal, closely spaced steel rollers that revolve in opposite directions at different speeds. The premix is fed to the rear roller from a tub by pouring or by hand. As the ink moves through the rollers, the pigment is dispersed by the shearing action that occurs at the nips between the rollers. The ink or dispersion is removed from the last roller with a scraper blade. These are very open devices and the potential for evaporative losses is great. For this reason, they are used for dispersions that incorporate only nonvolatile materials or very slowly evaporating solvents.

Once the base solution has been reviewed for quality, it may be used directly or placed into intermediate storage. This intermediate storage can be in drums or in bulk. Air emissions can result during routine transfer operations (working losses and breathing losses).

Further Steps

Blending In the blending process, intermediates and other raw materials (which do not require previous processing) are mixed in specific quantities to produce a specific finished ink. The blending process sequence and ingredients often have subtle variations that impart specific qualities to the final ink product. These are usually considered trade secrets. Emissions include transfer losses during material transfer and evaporative losses during mechanical mixing.

Filtering Because of customer and quality-related concerns, most inks are filtered prior to or during packaging. This is accomplished via cloth or bag filters, vibrating screens, or porous resin cartridge filter systems. Filtration is an extra handling step that may lead to some transfer and evaporative losses.

Packaging After filtration the finished product is pumped or emptied into the final shipping container (tins, pails, drums, portable tanks, or tank trucks). Emissions occur during transfer operations.

Shipment Finished ink products are shipped in tins, pails, drums, portable tanks, or tank trucks. Under normal conditions, no significant emissions result during this stage.

Cleaning Blending vessels, mills, filter screens, and other equipment are routinely cleaned with solvent after the production of an ink formulation is complete. An exception is that in a few large plants, only one color or a limited number of colors are produced, and dedication of each piece of equipment to a single color eliminates the need to clean between colors. Where cleaning is done, the final rinse utilizes a cleaning agent, typically a solvent, that is compatible with the next ink formulation to be produced in the equipment. Air emissions result from evaporative and displacement losses during and after cleaning.

AIR EMISSIONS CHARACTERIZATION

Emissions from printing ink manufacturing consist of solvent vapors and dusts from the addition of dry pigment and other fine solids to mixing vessels.

Because of the small size of most ink manufacturing plants, few detailed emissions surveys exist.[2,3] Theoretically, solvent emissions can be calculated for sequential process steps, starting with product receiving and storage and carrying through to product packaging. Storage tank breathing and losses can be estimated using algorithms published by the U.S. Environmental Protection Agency (EPA). Emissions from processing steps can be estimated using engineering principles. Often, many simplifying assumptions must be made and final estimates may be off by a considerable margin.

Stack testing can be used to supplement engineering estimates, but this is difficult to implement because of the batch nature of ink manufacturing, the wide array of solvents being used in the ink products, and the often complex and diverse nature of the emissions capture and collection systems employed. In short, accurately estimating emissions from an ink manufacturing facility is a challenging task requiring considerable planning.

For typical printing-ink manufacturing facilities, solvent losses to air as a percentage of total usage are between 0.3% and 1.6% for newer facilities and can exceed 2% for older facilities or for batch operations requiring frequent cleaning because of equipment changeover. These ranges reflect differences in facility age, equipment, operating and maintenance methods, and the degree to which water-based production capacity has replaced solvent-based production capacity. Flexographic ink producers with a high percentage of water-based products may tend to be at the low end of the range, or even slightly below it at individual facilities.

Dry pigment varies in particle size. This characteristic directly affects potential emissions, along with the following parameters: method of addition to the mixing vessel (manual versus automated), product mix (e.g., clear varnishes use no pigments), ventilation, and operating practices. This includes the use of covers on mixers, the size of pigment bags, how empty drums or bags are handled, and even how bags are split open for emptying, lengthwise versus at one end.

To represent a variety of conditions, an emission factor of 2 pounds per ton of pigment mixed has been developed.[4] It is felt that this represents the upper end of actual emissions and that good practice, such as the use of mixer covers while emptying bags, splitting bags endwise, and emptying them thoroughly before disposal, would reduce emissions further.

Cooking an ink vehicle is a special case. This step is more prevalent in lithographic and letterpress ink manufacturing than in rotogravure and flexographic ink manufacturing facilities. Vehicle cooking is not applicable to facilities that purchase a varnish or other clear solution that has already been heated, or that make varnish or other clear solutions without heating them at all. Vehicle cooking involves heating a vessel containing resins, solvents, and other additives at temperatures up to 340°C (650°F) for an average of eight to 12 hours.[5]

Where it is used, this process is the primary source of VOC emissions at a facility.[2] Emissions result from decomposition or evaporation of varnish ingredients at peak temperatures or from purging the reactor vessel with air or inert gas. The quantity, composition, and rate of emissions from the varnish cooking step are a function of cooking temperature and time, the ingredients, the way additives are introduced, the degree of stirring, and the extent of air/inert gas purging. Uncontrolled VOC releases from this operation is estimated at from 40 to 160 pounds per ton for various varnish types.[4]

AIR POLLUTION CONTROL MEASURES

Air emissions from the manufacturing of printing ink consist of dusts and solvent vapors. Controlling their source is cost effective and goes a long way toward minimizing air pollution.

Source Control

For pigments, good manufacturing practice would include splitting bags at their ends, minimizing the vertical drop of the pigment to the extent practicable, and emptying bags and drums thoroughly to reduce fugitive emissions after the fact.

For vehicle cooking, where applicable, the use of retractable hoods, enclosed reactors, and/or add-on controls will substantially reduce emissions. Besides condenser systems, potentially applicable add-on controls are reported to include thermal and catalytic incineration and carbon absorption. However, this is based on a study involving vehicles working in an allegedly similar operation in the paint industry.[5]

For solvents, equipping vessels with well-fitting covers and paying attention to housekeeping help to minimize solvent losses. "Proper" manufacturing techniques are a function of each facility, reflecting its age, layout, type of equipment, and personnel experience and training. However, in general, best manufacturing practice would be considered to include, for solvents, use of well-fitting covers on mixing vessels; keeping vessel transfers to a practical minimum; closure of mixer cover openings immediately after addition of ingredients; cleaning of mixers and/or tanks by use of a combined washer/cover unit, where applicable; and good housekeeping in general. Improvements in initial plant design or via retrofits can also reduce VOCs. For example, equipping aboveground storage tank vent lines with conservation vents (breathing losses) and equipping bulk tanks with vapor return lines can reduce VOCs. Through proper plant design and manufacturing techniques, emissions can be kept to a practical minimum.

End-of-Line Controls

Dusts, usually particulate associated with the transfer of resins and pigments, are collectible by fabric or cartridge filters (baghouses). Depending on pigment particle size, this control equipment may achieve up to the 98–99% collection efficiencies observed for other processes. In theory, collected material may be reused within the process. However, because of cross-contamination and customer specifications, most material is ultimately disposed of as waste.

A unique source of solvent vapors is resin cooking; while performed by some ink manufacturers, it should be considered a chemical process because chemical reactions take place. Emissions from this high-temperature process are reduced by the use of vessel covers and by the use of condensers to return vapors to the process as condensate.

Other solvent vapors, usually a mixture of solvents depending on product mix, present a much more difficult problem. Three factors make it difficult to control the solvent vapors in the exhaust from ink plants:

1. The concentration of VOCs is quite low (5–50 ppm is typical).
2. The VOC concentration is highly variable because of the nature of batch operations.
3. The VOC in the exhaust is a highly variable mixture of chemicals, ranging from polar to nonpolar species and high-boiling-point to low-boiling-point compounds.

Because of these factors, a number of control technologies can be eliminated as nonfeasible on technical grounds. Vapor refrigeration is generally not reasonable owing to the dilute concentration of VOCs and the resulting low operating temperature required in the condenser. Because of the wide range in physical properties of the chemicals in the exhaust, carbon absorption may have limited effectiveness and scrubber systems may not be feasible.

Oxidizing systems (thermal or catalytic) are technologically feasible control processes. Ink manufacturing tends to be a relatively small-scale operation and any control technology found to be technologically feasible must be evaluated for economic viability. It is quite common for all control options to prove prohibitively expensive in terms of cost per ton of VOC removed. In these instances, control of emissions at the source must be emphasized.

References

1. S. V. Capone and M. Petruccia, *Guidance to State and Local Agencies in Preparing Regulations to Control Volatile Organic Compounds from Ten Stationary Source Categories*, EPA-450/2-79-004, U.S. Environmental Protection Agency, Research Triangle Park, NC, 1989, p 155.
2. R. Blaszczak, "A preliminary review of 19 source categories of VOC emissions," unpublished document, U.S. Environmental Protection Agency, Research Triangle Park, NC, May 1988, p 8-2.
3. *1982 Census of Manufacturers, Industry Series*, "Miscellaneous chemical products," U.S. Department of Commerce. Washington, DC.
4. *Compilation of Air Pollutant Emission Factors*, Vol. I, 4th ed. AP-42. U.S. Environmental Protection Agency. May 1983.
5. *Air Pollution Control Engineering and Cost Study of the Paint and Varnish Industry* EPA-450/3-74-031, U.S. Environmental Protection Agency, June 1974.
6. *GAA Solvent Manual;* Gravure Association of America, New York, 1989.
7. *Improving Air Quality: Guidance for Estimating Fugitive Emissions from Equipment*, Chemical Manufacturers Association, Washington, DC, 1989.
8. *Fugitive Emission Sources of Organic Compounds—Additional Information on Emissions, Emission Predictions, and Costs*. EPA-450-3-82-010, U.S. Environmental Protection Agency Office of Air Quality Planning and Standards, Research Triangle Park, NC, 1982.
9. Center for Environmental Assurance, *SARA Title III, Section 313 Guidance Manual*, 2nd ed., Arthur D. Little Co., 1988, p 88.

SOAPS AND DETERGENTS

Richard C. Scherr

The term "synthetic detergent products" applies broadly to cleaning and laundering compounds containing surfactants along with other ingredients formulated for use in aqueous solutions. These products are marketed as heavy- or light-duty granules or liquids, cleansers, and laundry or toilet bars. The heavy-duty granules and liquids represent the major portion of all products manufactured, with the light-duty liquid and light-duty granules produced to a much lesser degree. Heavy-duty powders and liquids for home and commercial laundry, laundry detergent, make up 60% to 65% of the U.S. soap and detergent market and were estimated at 2.6 million metric tons in 1990.[1] The manufacture of all detergent products incorporates equipment and processes similar to those used for manufacturing soap products. The manufacture of the granular products via spray drying is of paramount interest, with more severe air pollution problems than those encountered with soap granule spray drying. The manufacture of liquid detergent and bar products is of lesser importance, with little or no difference from similar soap products in either process equipment or air pollution potential, and will not be discussed.

Laws requiring zero- or low-phosphate content in detergents and the emergence of heavy-duty liquid detergents have had a significant impact on the production of granule laundry detergents. Until the early 1970s, almost all laundry detergents sold in the United States were heavy-duty powders. Public concerns regarding the role of phosphates in the eutrophication of lakes resulted in the banning of phosphates in detergents in some parts of the United States and Canada in the early 1970s. Nitrilotriacetic acid (NTA), a leading substitute builder for phosphates, was eliminated for use in the United States because of health concerns. Manufacturers, forced to formulate granule laundry products without an adequate builder, used sodium carbonate and zeolites. This significantly decreased cleaning performance and also resulted in a product that was difficult to dissolve. The introduction of heavy-duty laundry liquids utilizing sodium citrate and sodium silicate offered the consumer superior cleaning performance and improved solubility at only a slightly increased cost. While the performance of zero-phosphate granules has improved in recent years, heavy-duty liquids now account for 40% of the laundry detergents sold in the United States, up from 15% in 1978.[2] This growth in liquid share and consequent reduction in the demand for granule production, combined with improvements in manufacturing techniques, resulted in the shutdown of over half of the spray-drying facilities for laundry granule production since 1970. Consumption of laundry detergents has grown approximately 4% per year since 1980, reaching 7 billion pounds in 1989. Future growth is forecasted at between 2% and 3%.[1]

RAW MATERIALS

Components of detergents fall into three major categories: surfactants, to remove dirt and other unwanted materials from clothes; builders, to treat the water to improve the efficiency of the surfactants by softening the water; and additives, to improve cleaning or physical properties. Additives may include bleaches, bleach activators, antiredeposition agents, optical brighteners, antistatic agents, fabric softeners, and fillers. Each particular formulation depends on the ultimate design for consumer use. Table 1 illustrates typical detergent formulations for the United States, Europe, and Japan.[2]

Surfactants

The surfactants used in formulating synthetic detergent products are either anionic, cationic, nonionic, or amphoteric. Anionic surfactants derive their detergency from the large anion part of the molecule. Anionic surfactants are alkyates of benzene and propylene trimer or tetramer. Typical detergent surfactants are linear alkylbenzene sulfonates (LAS), alcohol sulfates, alcohol ether sulfates, and alkylglyceryl ether sulfonates. Cationic surfactants derive their detergency from the cation. Most cationic surfactants are derivatives of amines. Nonionic surfactants are hydrophobic molecules made water soluble by the addition of ethylene oxide groups. Typical hydrophobes include fatty acids, alcohols and acids, amines, and amides. Typical detergent nonionics include alkyl polyethylene glycol ethers and nonylphenyl polyethylene glycol ethers. Amphoterics are surfactants that contain both a cationic and an anionic end group.

Detergent surfactants include alkyl sulfonics, alkyl sulfates, and alcohol sulfates, discussed above, and almost the entire range of anionic and nonionic surfactants, including soap. Anionics and nonionics are the predominant surfactants used in laundry detergents. Cationics and amphoterics are generally used in specialty products because of their poor cleaning efficiency and high cost respectively.[2] The most widely used surfactant for laundry detergents is LAS, followed closely by the alcohol sulfonates. Some manufacturers combine these surfactants, with the balance between the two being determined by their relative cost.[1]

Anionic surfactant materials such as LAS and alcohol sulfonates are sometimes produced at detergent plants by sulfonation and sulfation or, as with LAS, received as liquid sulfonic acids. In either case, the "acid mix" is neutralized with 30% sodium hydroxide in an exothermic reaction to form the final surfactant. Plants manufacturing their own sulfonic and sulfate surfactants also use other surfactants in some or all of their products. The detergents are received and handled mostly as liquid solutions of varying strength, but some surfactants are received as flake or powder. The exact method and processing vary for each surfactant, but anionic surfactants are principally mixed with the slurry before drying, while some nonionic surfactants can be sprayed on the finished granule.

Amides of various types can be used as supplementary surfactants in many formulations. They improve the detergency of the sulfonic and sulfate surfactants and act as foam boosters or stabilizers. Amides used include the higher fatty amides (e.g., cocomonethanolamide), ethanolamides, dialkyl and alkylol (hydroxyalkyl) amides, morpholides, and nitriles, as well as the lower acyl derivatives of higher fatty amines. These materials are handled as liquids and received in tank cars or barrels. In granule manufacture, they can be incorporated in the slurry before drying, but are usually blended with the detergent granules after drying to avoid emission problems from the spray dryer.

Builders

Sodium tripolyphosphate (STP) or tetrasodium pyrophosphate (TSPP) is incorporated in granular formulations as a "builder" or sequestering agent where phosphates are not prohibited. These agents serve to eliminate interference with the detergent action by the calcium and magnesium ions (hardness) in the water used in the wash solution. Both

TABLE 1. Household Detergent Formulations in the United States, Western Europe, and Japan

		Composition %					
		United States		Western Europe		Japan	
Component	Examples	With builders	Without builders	With builders	Without builders	With builders	Without builders
HEAVY-DUTY POWDERED DETERGENTS							
Anionic surfactants	Alkylbenzene sulfonates	0–15%	0–20%	5–10%	5–10%	5–15%	5–15%
	Fatty alcohol sulfates	—	—	1–3	—	0–10	0–10
	Fatty alcohol ether sulfates	0–12	0–10	—	—	—	—
	Alpha-olefin sulfonates	—	—	—	—	0–15	0–15
Nonionic surfactants	Alkyl and nonylphenyl poly(ethylene glycol) ethers	0–17	0–17	3–11	3–6	0–2	0–2
Suds-controlling agents	Soaps, silicon oils, paraffins	0–1.0	0–0.6	0.1–3.5	0.1–3.5	1–3	1–3
Foam boosters	Fatty acid monoethanol amides	—	—	0–2	—	—	—
Chelators (builders)	Sodium tripolyphosphate	23–55	—	20–40	—	10–20	—
Ion exchangers	Zeolite 4A, poly(acrylic acids)	—	0–45	2–20	20–30	0–2	10–20
Alkalies	Sodium carbonate	3–22	10–35	0–15	5–10	5–20	5–20
Cobuilders	Sodium citrate, sodium nitrilotriacetate	—	—	0–4	0–4	—	—
Bleaching agents	Sodium perborate	0–5	0–5	10–25	20–25	0–5	0–5
Bleach activators	Tetraacetylethylenediamine	—	—	0–5	0–2	—	—
Bleach stabilizers	Ethylenediaminetetraacetate	—	—	0.2–0.5	0.2–0.5	—	—
Fabric softeners	Quaternary ammonium compounds	0–5	0–5	—	—	—	0–5
Antiredeposition agents	Cellulose ethers	0–0.5	0–0.5	0.5–1.5	0.5–1.5	0–2	0–2
Enzymes	Proteases, amylases	0–2.5	0–2.5	0.3–0.8	0.3–0.8	0–0.5	0–0.5
Optical brighteners	Stilbene derivatives	0.05–0.25	0.05–0.25	0.1–0.3	0.1–0.3	0.1–0.8	0.1–0.8
Anticorrosion agents	Sodium silicate	1–10	0–25	2–6	2–6	5–15	5–15
Fragrances		a	a	a	a	a	a
Dyes and blueing agents		a	a	a	a	a	a
Formulation aids		0–1.0	0–1.0	—	—	—	—
Fillers and water	Sodium sulfate	Balance	Balance	Balance	Balance	Balance	Balance
HEAVY-DUTY LIQUID DETERGENTS							
Anionic surfactants	Alkylbenzene sulfonates	5–17	0–10	5–7	10–15	5–15	—
	Fatty alcohol ether sulfates	0–15	0–12	—	—	5–10	15–25
	Soaps	0–14	—	—	10–15	10–20	—
Nonionic surfactants	Alkyl poly(ethylene glycol) ethers	5–11	15–35	2–5	10–15	4–10	10–35
Suds-controlling agents	Soaps	—	—	1–2	3–5	—	—
Foam boosters	Fatty acid alkanolamides	—	—	0–2	—	—	—
Enzymes	Proteases	0–1.6	0–2.3	0.3–0.5	0.6–0.8	0.1–0.5	0.2–0.8
Builders	Potassium diphosphate, sodium tripolyphosphate	—	—	20–25	—	—	—
	Sodium citrate, sodium silicate	6–12	—	—	0–3	3–7	—
Formulation aids	Xylene sulfonates, ethanol, propylene glycol	7–14	5–12	3–6	6–12	10–15	5–15
Optical brighteners	Stilbene derivatives	0.1–0.25	0.1–0.25	0.15–0.25	0.15–0.25	0.1–0.3	0.1–0.3
Stabilizers	Triethanolamine	—	—	—	1–3	1–3	1–5
Fabric softeners	Quaternary ammonium salts	0–2	0	—	—	—	—
Fragrances		a	a	a	a	a	a
Dyes		a	a	a	a	a	a
Water		Balance	Balance	Balance	Balance	Balance	Balance

[a]Component may be present in very small concentrations.

STP and TSPP may be used in powder, prill, or granule form and are received in car lots. These ingredients are most often blended into the slurry before spray drying.

Nitrilotriacetic acid and its sodium salts are used in Canada in zero- and low-phosphate detergents. The acid is a crystalline powder and the salts (disodium or trisodium) are powders. They are received in car lots. The NTA is added to the slurry mix before drying.

Zero- and low-phosphate detergents in the United States use sodium carbonate (soda ash), sodium silicate, and calcium carbonate, or a combination of these compounds, with sodium aluminosilicates (zeolites) as builders. Sodium carbonate and calcium carbonate are powders received in bulk. Sodium silicates and zeolites are generally received as liquids. Like other builders, these materials are added to the slurry before drying.

Other Ingredients

Fillers, usually sodium sulfate or sodium carbonate, are incorporated in granule products. They are either powders or crystalline powders and are added in bulk form to the slurry before drying.

Trisodium phosphate (TSP) is used in detergent granule formulations such as dishwasher compounds and wall cleaners that are designed to clean hard surfaces. Considered functionally as an alkali rather than a sequestering agent, TSP is usually handled as a crystalline powder and is received in car lots, drums, or bags.

Carboxy methylcellulose (CMC; sodium cellulose glycolate) usually is added to heavy-duty granules and serves to prevent redeposition onto the fabric of the dirt removed by the detergent. This chemical is received in bags or drums as a powder or granule. It is added to the slurry mixture before drying.

In most synthetic detergent formulations, sodium silicate is added to inhibit the surfactant's tendency to corrode metal. It also is used to overcome production and packaging problems encountered with detergent granules. It is functionally used as a primary detergent alkali in compounds designed for hard-surface washing, (e.g., machine dishwashing compounds). It can serve to retain uniform viscosity in the mixing and pumping of the slurry before it is dried, and it reduces the "tackiness" of the dried granules, facilitating their handling and reducing caking of the product after packaging. The sodium silicates are received in tank cars as water solutions.

Optical brighteners are added to many formulations. These are usually fluorescent dyes that absorb ultraviolet rays and reflect them as visible light. The dyes are received as powders in bags or as liquids in drums, and they are usually blended in the slurry before drying. Also, perfumes are added to almost all detergent products to overcome unpleasant odors and impart a pleasing scent to laundered fabrics. The perfumes are added by spraying onto the dried granules or by mixing with the liquid detergents. They are handled as liquids in small containers or drums.

Bleaches of various kinds are frequently incorporated in heavy-duty detergents. Sodium perborate, along with magnesium silicate as a stabilizer, is commonly employed. They are received as powders or crystals in boxes or bags and are added to the granules after drying. Because of lower washing temperatures, perborate activators are being introduced in some U.S. heavy-duty laundry products. These compounds are more commonly used in European laundry products. Tetraacetylethylenediamine (TAED) is the predominant bleach activator used.

In Europe and Japan, 90% of the detergents contain enzymes. Enzymes are limited to 45–50% of the heavy-duty laundry detergents in the United States because of the high use of chlorine bleaches that destroy enzymes; however, lower washing temperatures and increasing use of perborate bleaches by consumers are leading to increased enzyme usage.[1] Enzymes assist in the removal of protein-based stains from fabrics. The enzymes, which are received as prills (coated powders) in bags or drums, are heat sensitive and are destroyed if heated to 212°F. Most manufacturers blend the enzymes into the detergent granules after drying.

Many other compounds may be incorporated in various products. Preservatives, antioxidants, foam suppressors, and other types of additives are used. The scouring cleansers are composed principally of finely pulverized silica, active detergent, small amounts of phosphates, and frequently a bleach.

PROCESSES

The only manufacturing process discussed here will be the production of detergent granule formulations incorporating spray-drying processes. All other products are produced in processes, such as drum drying, similar to soap production, discussed elsewhere.

The manufacture of detergent granules incorporates three separate steps: slurry preparation, spray drying, and granule handling (including cooling, additive blending, and packaging). Figure 1 illustrates the various operations.

SLURRY PREPARATION

Process Description

The formulation of slurry for detergent granules requires the intimate mixing of various liquid, powdered, and granulated materials. The soap crutcher is almost universally used for this mixing operation. Premixing of various minor ingredients is performed in a variety of equipment prior to charging to the crutcher or final mixer. The slurry, mixed in batch operations, is then held in surge vessels for continuous pumping to the spray dryer.

Air Emissions Characterization

The receiving, storage, and batching of the various dry ingredients create dust emissions. Pneumatic conveying of fine materials causes dust emissions when conveying air is separated from the bulk solids. Many detergent products require raw materials with a high percentage of fines. Typical specifications for some raw materials include the following percentage of fine materials passing a 200-mesh screen: sodium sulfate, 12%; TSPP, 74%; STP, 53%.

The storage and handling of the liquid ingredients, including the sulfonic acids, sulfonic salts, and sulfates, do not cause emission problems other than mild odors. In the batching and mixing of the fine dry ingredients to form slurry, dust emissions are generated at scale hoppers, mixers, and the crutcher. Liquid-ingredient addition to the slurry creates no visible emissions, but may cause odors.

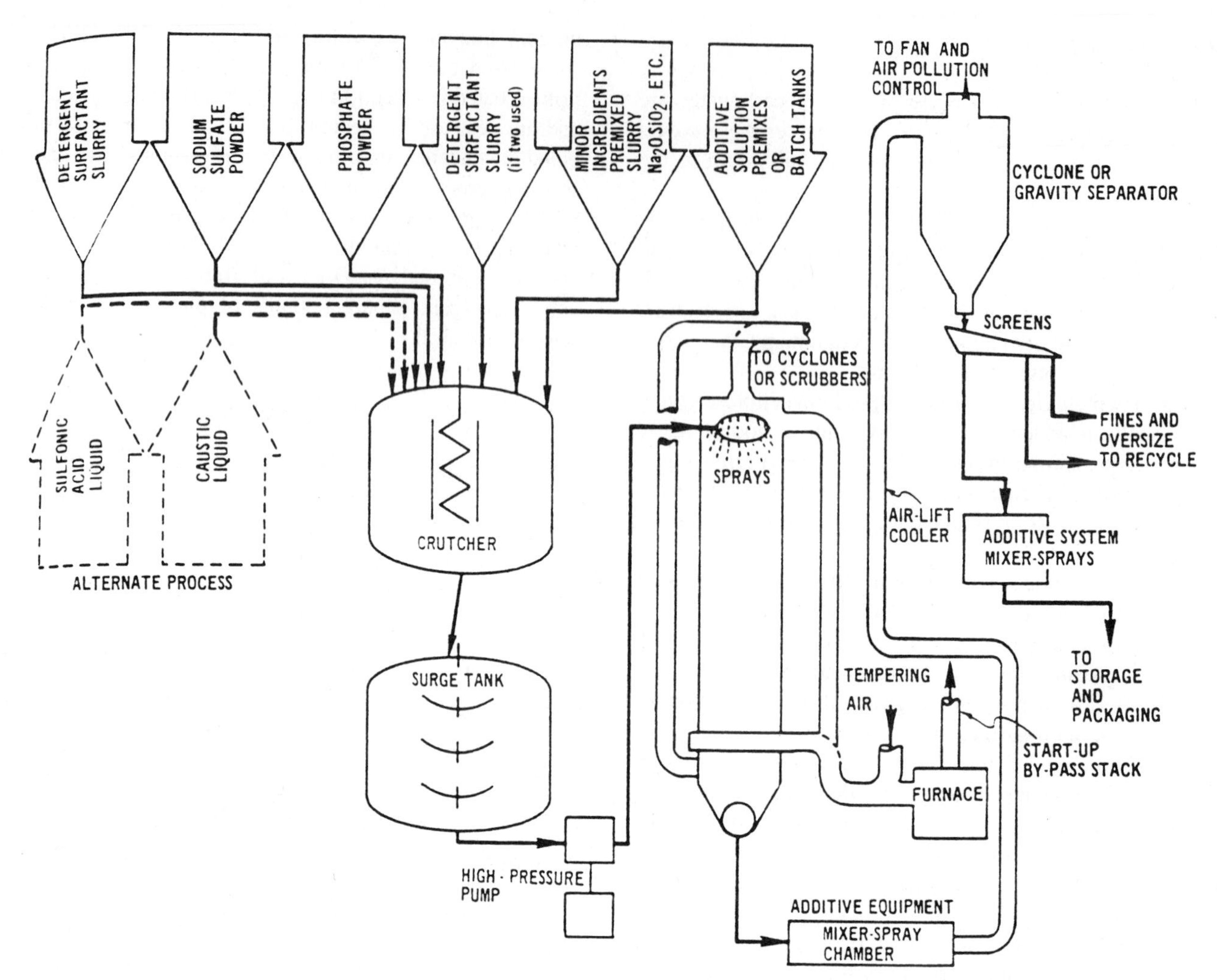

FIGURE 1. Detergent Spray Drying with Tower Equipped for Either Concurrent or Countercurrent Operation (Dampers in Countercurrent Mode of Operation)

Air Pollution Control Measures

Control of dusts generated from pneumatic or mechanical conveying or from discharge of fine materials into bins or vessels is described in Chapter 3. No unique problems occur in hooding or exhaust systems for controlling dust emissions from conveying and slurry preparation.

Baghouses are employed not only to reduce and eliminate the dust emissions, but also for the salvage of raw materials. None of the dusts cause any serious corrosion problems. Filter fabrics should be selected that have good resistance to alkalis. Filter ratios for baghouses with intermittent shaking cleaning mechanisms should be under 3 cfm/ft^2.

SPRAY DRYING

Process Description

All spray-drying equipment designed for detergent granule production incorporates the following components: spray-drying tower, air heating and supply system, slurry atomizing equipment, slurry pumping equipment, product cooling equipment, and conveying equipment. The towers are cylindrical with cone bottoms and range in size from 12 to 24 feet in diameter and 40 to 125 feet in height. Single towers may vary in diameter, being larger at the top and smaller at the bottom. Air is supplied to the towers from direct-heated furnaces fired with either natural gas or fuel oil. The products of combustion are tempered with outside air to lower temperatures and then blown to the dryer under forced draft. The towers are usually maintained under slightly negative pressure, between 0.05 and 1.5 inches of water column, with exhaust blowers adjusted to provide this balance. Most towers designed for detergent production are of the countercurrent type, with the slurry introduced at the top and the heated air introduced at the bottom. A few towers of the concurrent type are used for detergent spray drying, with both hot air and slurry introduced at the top. Some towers are equipped for either mode of operation, as illustrated in Figure 1.

In most towers today, the slurry is atomized by spraying through a number of nozzles rather than by centrifugal action. The slurry is sprayed at pressures of 600–1000 psi in single-fluid nozzles and at pressures of 50–100 psi in two-fluid nozzles. Steam or air is used as the atomizing fluid in the two-fluid nozzles.

Tower operations vary widely among manufacturers and among products. Heated air supplied to the tower varies from 350°F to 750°F. Temperatures of air supplied to countercurrent towers are generally lower and most often range from 500°F to 650°F. Concurrent tower temperatures are somewhat higher. Solids content of slurries for detergent spray drying varies from 50% to 65% by weight, with some operations as high as 70%. Moisture content of the dried product varies from 10% to 17%. Towers are designed for specific airflow rates, and these rates are maintained throughout all phases of operation. Slurry temperatures may vary, but in most formulations they do not exceed 160°F and are frequently as low as 80°F. Exit gas temperatures range from 150°F to 250°F, with wet-bulb temperatures of 120°F to 150°F. Air velocities in concurrent towers are usually higher than velocities in countercurrent towers. The concurrent towers produce granules that are mostly hollow beads of light specific gravity (0.05 to 0.20). Countercurrent towers produce multicellular, irregularly shaped granules that have higher specific gravities ranging from 0.25 to 0.45.

In countercurrent towers, with lower air velocities and droplets descending against a rising column of air, most of the dried granules fall into the cone at the bottom of the tower. They are discharged through a star valve, or regulated opening, while still hot. Cooling of the granules is discussed below with other granule processing procedures. Unlike in other product spray-drying operations, (e.g., powdered milk), the desired detergent granule product is comparatively large. The specifications for some well-known granular products require 50% by weight or more to be retained on a 28-mesh screen. A certain amount of the product is dried to a comparatively small size. This amount depends on tower feed rates, the liquid droplet size in slurry atomization, the paste viscosity, the particular product, and other variables. Usually, the exhaust air entrains 7–10% of that portion of the granular product that is too fine to settle at the base of the tower.

Concurrent towers operate with higher air velocities than do countercurrent towers. The air is vented just above the bottom of these towers through a baffle that causes violent changes of direction to the exhaust air dynamically to separate the dried granules, which then fall to the cone bottom for discharge. Concurrent towers producing very-low-gravity granules vent air still conveying the product to auxiliary equipment for separation. The loss of detergent fines entrained in the exhaust air stream will be somewhat higher from concurrent towers than from countercurrent towers.

Air Emissions Characterization

The exhaust air from detergent spray drying towers contains two types of air contaminants. One is the fine detergent particles entrained in the exhaust air discussed above; the second consists of organic materials vaporized in the higher-temperature zones of the tower.

The detergent particles entrained in the exhaust air are relatively large in size. Over 50% by weight of these particles are over 40 μm. These particles constitute over 95% of the total weight of air contaminants in the exhaust air.[3] They consist principally of detergent compounds, although some of the particles are uncombined phosphates, sulfates, and other mineral compounds.

The second type of air contaminant, organic compounds, originates primarily from the surfactants included in the slurry. Various organic components in the slurry vaporize in the tower. The amount vaporized depends on many variables, such as tower temperatures and volatility of the organics in the slurry mixture. Volatility of the organics is a function of the chain length of surfactant compounds and the reaction completeness in the sulfonation and sulfation of anionic surfactants. The vaporized organic materials condense upon cooling in the tower exhaust air stream into micron- and submicron-size droplets or particles.

The variety of possible detergent compounds is almost infinite and manufacturers are continually introducing new formulations or reformulating older ones. It is not always possible to predict how certain organic compounds in a slurry mixture will affect stack emission. Work in the early 1970s indicated that if amides are present in the slurry in amounts greater than 0.5% by weight, emission problems will occur. Source tests of exhaust air from an air-pollution control scrubber with amides present in the slurry being dried in the spray tower revealed 0.08 grain of organic particulate per standard cubic foot of exhaust gas. The presence of this relatively low concentration of submicron-size aerosols causes water vapor plumes to persist for long distances. Following the break or end of the water vapor plume, a highly visible contaminant plume persists for even greater distances. The amide emission rate increases as tower operating temperatures rise. Many tower operating variables affect air contaminant emissions. The most significant variables are the formulation's final granule temperature and moisture, the dryer inlet air temperature, and temperature profiles in the dryer.

Slurry formulations containing alcohol ethoxylate surfactants or alcohol sulfates with high levels of unsulfated alcohols cause similar aerosol emissions. Source tests of the aerosol leaving the scrubber indicate a particle size range of 0.2 to 1 μm where ethoxylated alcohol surfactants were added to the crutcher. It is believed that very small amounts of unreacted alcohols coat water droplets like amides, inhibiting the evaporation of the droplets and resulting in gray water vapor plumes that persist for long distances.

Air Pollution Control Measures

The collection of air contaminants not only provides for the economic return of detergent fines to the process, but it also provides for control of submicron particles and aerosols to ensure compliance with air pollution prohibitory rules.

Manufacturers producing detergent granules have developed several separate approaches for capturing the detergent fines in the spray dryer effluent for return to process. Dry cyclones and cyclonic impingement scrubbers are the primary approaches used. Cyclonic impingement scrubbers return a slurry to the crutcher, while dry cyclones return detergent fines to the crutcher. The dry cyclone separators can remove 90% or more by weight of the detergent product fines in the tower exhaust air. The detergent dust remaining in the effluent vented from the cyclones consists of particles 95% by weight less than 10 μm.[4] Particulate concentrations vary from 0.1 to 1.0 gr/scf. The cyclones are designed for relatively high efficiencies and operate at pressure drops of from 8 to 10 inches of water column.[3]

Secondary collection equipment is used to collect fine dust and organic aerosols that pass the primary collectors. Mist eliminators are used after cyclonic impingement scrubbers. Cyclones are generally followed by scrubbers, scrubber/precipitators, or fabric filters. Several types of scrubbers can be used following the cyclone collectors. Venturi scrubbers operating at 8 to 10 inches of water pressure drop (using water at 8 to 10 psig distributed through nozzles in the throat, throat velocities of 8500 fpm, and water supplied to the throat at a ratio of 4.5 to 5.0 gallons per 1000 ft^3 of effluent) have been used following the cyclones. Venturi scrubbers have been replaced largely by packed bed scrubbers operating at 0.5 to 2.0 inches of water pressure drop and water rates of 1 to 3 gallon per 1000 acfm. Packed bed scrubbers are usually followed by wet-pipe-type electrostatic precipitators constructed immediately above the packed bed in the same vessel. Fabric filters are sometimes used after cyclones, but their application is limited to spray dryers having low drying loads and drying products with low surfactant concentrations. On efficient spray dryers with exhaust conditions near the dew point drying products with more than 10–15% surfactant, condensing water vapor and condensing organic aerosols will bind the filter fabrics.

Particulate emissions, as measured by the Environmental Protection Agency's Method 5, vary slightly with formula and operating conditions. Typical control efficiencies and emission factors for various control options are shown in Table 2.

As previously mentioned, small amounts of organic, especially paraffin alcohols and amides, can result in a highly visible plume that can persist after the condensed water vapor plume has dissipated. The precipitator/scrubber appears to provide the best control for these compounds, although no conclusive data exist. Differences in formulation among manufacturers and differences in drying systems make comparison impossible. Fabric filters operating at higher temperatures will not collect organic aerosols, some of which will be caught on the Method 5 filter.

Opacity and the organics emitted are influenced by granule moisture and temperature at the end of drying, temperature profiles in the dryer, and formulation of the slurry. An alternative method for controlling visible emissions from the dryer caused by volatile organics in the slurry is to reformulate the slurry to eliminate these offending organic compounds, Sometimes the purity of raw materials such as alcohol ethoxylates can be improved by stripping short-chain molecules from the raw material. When amide compounds were identified as causing the emission problems, some manufacturers developed other formulations or methods for adding the amides to the spray-dried granules after the dryer to achieve a comparable product.

When reformulation is not possible, the tower production rate may be reduced, permitting operation at lower air inlet temperatures and lower exhaust gas temperatures. When tower temperatures are reduced, lower temperatures may result in lower granule temperature and higher granule moisture, thus reducing organic emissions.

GRANULE HANDLING

Process Description

Many manufacturers discharge hot granules from the spray tower into mixers where dry or liquid ingredients are added. The granules are usually mechanically conveyed away from the tower or mixer discharge and then are air-conveyed to storage and packaging. Air conveying cools the granules and elevates them for gravity flow through further processing equipment to storage and packaging. Air conveying of low-density granules usually is designed for 50–75 scfm air per pound of granules conveyed. At the end of the conveyor, gravity or centrifugal separators remove granule product from the conveying air. Some manufacturers mechanically lift the granules from the spray tower to aeration bins where they are cooled or aged by injecting air at the bottom of the bin. This air percolates upward through the scrubber.

The cooled granules are screened to deagglomerate the large granules and to remove undersize- or oversize-particles. Further mixing or blending may be performed to add heat-sensitive compounds such as enzymes to the detergent products. Many manufacturers do not store the finished granules, but convey them directly to packaging equipment. Some detergent products are held in storage, in either large fixed bins or small-wheeled buggy bins, and then are charged to packaging equipment. Packaging is done with either scale or volumetric filling machines.

TABLE 2. Particulate Emmission Factors for Detergent Spray Drying[10]

Control Device	Efficiency, %	Emission Factor, kg/Mg of Product	Emission Factor, lb/ton of Product
Uncontrolled		45	90
Cyclone	85	7	14
Cyclone			
With spray chamber	92	3.5	7
With packed scrubber	95	2.5	
With venturi scrubber	97	1.5	
With wet scrubber	99	0.544	1.08
With scrubber/ESP[a]	99.9	0.023	0.046
Fabric filter	99	0.544	0.1

[a]ESP = electrostatic precipitator.

Air Emissions Characterization

Conveying, mixing, packaging, and other equipment used for granules can cause dust emissions. The granule particles, which are hollow beads, are crushed during mixing and conveying and generate fine dusts. With continuous exposure dusts emitted from screens, mixers, bins, mechanical conveying equipment, and air-conveying equipment are quite irritating to eyes and nostrils. Some additive materials, such as enzymes, bleaches, bleach activators, and brighteners, also can cause health problems. Equipment involving enzymes requires very efficient ventilation in addition to proper dust collection. Dust emissions in most cases represent a significant product loss, and their collection and return to process (usually as an ingredient of the slurry to be spray dried) is necessary for cost-efficient plant operation, as well as for air pollution control.

Air Pollution Control Measures

Dust generated by granule processing, conveying, and storage equipment does not create unique air pollution control problems. Usually, baghouses provide the best control. Collection efficiencies for baghouses are high; efficiencies can exceed 99%. No extreme conditions of temperature or humidity have to be met, but filter fabrics selected must show good resistance to alkaline materials. Baghouses utilizing intermittent shaking mechanisms should not have filtering velocities exceeding 3 fpm. Baghouses with continuous cleaning mechanisms may have filtering velocities as high as 6 fpm.

CURRENT DEVELOPMENTS

The recent introduction of superconcentrated detergents and a new push for more environmentally friendly products and packaging may cause a change equal to or greater than the shift to liquids sparked by phosphate bans. Manufacturers are developing more biodegradable surfactants from natural oils. The introduction of polymers for builder systems offers improved performance for zero- and low-phosphate detergents. Superconcentrated powders offer the environmental benefit of reducing packaging material and shipping energy by eliminating fillers and compressing the detergent to achieve two-to-one and three-to-one volume reductions. These factors may push the market back toward powder detergents.[1] While this may increase the demand for spray-drying capacity, newer formulations might also allow other manufacturing processes with less pollution potential. Only two detergent plants have been built in the United States since 1970, and both were completed after 1988. One of those plants was the typical spray-drying plant described here; the other utilized a low-temperature fluid-bed dryer to dry the detergent with a fabric filter to control emissions. While the second plant is still limited as to the type and amount of surfactant used in the formulation, changes in types of surfactant and surfactant sulfation technology may shift significant granule production away from spray-drying towers.

References

1. "Soaps and detergents, new opportunities in a mature business," *Chem. Wk.*, 20–64 (January 31, 1990).
2. B. F. Greek and P. L. Layman, "Higher costs spur new detergent formulations," *Chem. Eng News*, 67(4):29 (1989).
3. A. H. Phelps, "Air pollution aspects of soap and detergent manufacture," *J. Air Poll. Cont. Assoc.* 17(8):505 (1967).
4. *Compilation of Air Pollution Emission Factors, Volume 1. Stationary Point and Area Sources*, 4th ed., Supplement B, PB89-128631, U.S. Environmental Protection Agency, Research Triangle Park, NC, 1988.

SODIUM CARBONATE

Michael J. Barboza, P.E. and Neil M. Haymes M.S.P.H.

Sodium carbonate, or soda ash, is a white, crystalline, hygroscopic powder whose major use is in the production of glass. Other major users of sodium carbonate include the

TABLE 1. U.S. Sodium Carbonate Plants

Process Type	Owner	Location
Direct carbonation	Kerr-McGee	Trona, CA
	Kerr-McGee	Trona, CA
Monohydrate	Allied Chemical	Green River, WY
	FMC Corp.	Green River, WY
	Stauffer Chemical	Green River, WY
	Texasgulf Inc.	Green River, WY
Sesquicarbonate	FMC Corp.	Green River, WY
Solvay	Allied Chemical	Syracuse, NY

chemical, pulp and paper, cleaning agent, and water treatment industries.

Four different processes have been used domestically to produce sodium carbonate. Three of these processes—monohydrate, sesquicarbonate, and direct carbonation—are classified as *natural* processes. The fourth, the Solvay process, is classified as a *synthetic* process. Since the mid-1960s, use of this last process has declined significantly, probably as the result of increasing fuel costs that have had a direct impact on this fuel-intensive process.

As of March 1979, there were eight sodium carbonate plants in the United States, with a total capacity of approximately 8.5 million Mg/yr or 9.4 million tons per year. Table 1 presents a list of these plants and indicates that the Solvay and sesquicarbonate processes are each used in only one of these plants. Thus the following sections will describe the monohydrate and direct carbonate processes in detail and only briefly discuss the Solvay and sesquicarbonate processes, as their respective contributions to the total industrial production and the resulting atmospheric emissions are relatively small.

INDUSTRIAL PROCESSES

The monohydrate process, as shown in Figure 1, produces sodium carbonate from the mining and processing of trona ore. Some producers use a single crushing/screening step, while others use two stages of crushing. The crushed and sized ore is heated to approximately 200°C (392°F) in the calciner. Carbon dioxide and water vapor are removed, leaving crude sodium carbonate. The crude sodium carbonate, which also contains insoluble impurities, is fed into leach tanks, or dissolvers, where the sodium carbonate dissolves. The liquor is passed to a clarifier, where suspended solids are allowed to settle.

Multiple-effect evaporators are used to crystallize sodium carbonate monohydrate from the clear liquor. The mechanism of crystallization involves the increase in the concentration of dissolved sodium carbonate monohydrate until the liquor becomes supersaturated and crystallization is initiated. The liquor is returned to the process and the sodium carbonate monohydrate crystals are transferred to product dryers. In the product dryers, both free and chemically bound moisture is evaporated from the sodium carbonate monohydrate at approximately 120–180°C (248–350°F). At some facilities, air clarifiers or rotary tubes with external cooling water are used to cool the product. The product is conveyed to intermediate storage silos and then to loading facilities.

The processing steps in the sesquicarbonate process are very similar to those in the monohydrate process, but the order in which they occur is different. Here the raw trona ore is purified before rather than after calcining, as is done in the monohydrate process.

Sodium carbonate is produced from brine-containing sodium sesquicarbonate, sodium carbonate, and other salts in the direct carbonation process. A flow diagram of this process is presented in Figure 2. Initially, carbon dioxide is used to carbonate the brine. Further carbonation of the brine occurs in primary and secondary carbonation towers. This carbonation converts the sodium carbonate to sodium bicarbonate. Vacuum crystallizers are used to recover sodium bicarbonate from the brine. The sodium bicarbonate filter cake enters steam-heated predryers, where some of the moisture is evaporated. The temperature in these predryers is kept below approximately 50°C (122°F) so that no carbon dioxide is evolved. The partially dried sodium bicarbonate is then further heated in a steam-heated calciner. Carbon dioxide and the remaining water vapor are removed, forming impure sodium carbonate, which is bleached with sodium nitrate to burn off discoloring materials. The light sodium carbonate from the bleacher is recrystallized to sodium carbonate monohydrate. The monohydrate crystals

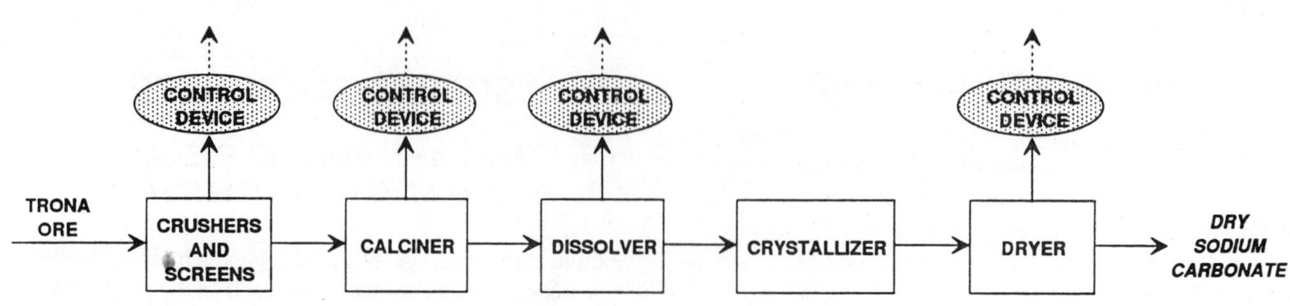

FIGURE 1. Sodium Carbonate Production by Monohydrate Process

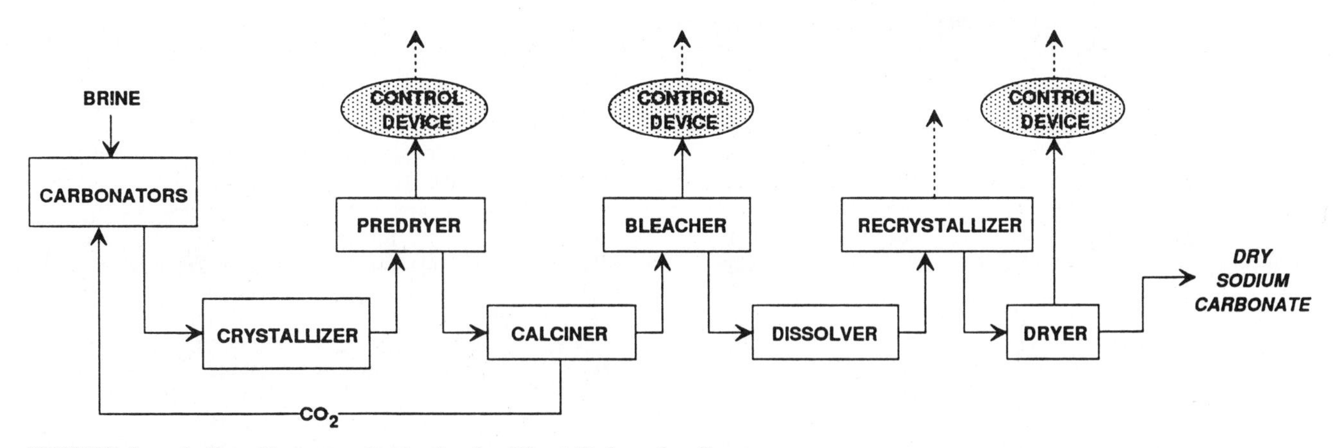

FIGURE 2. Sodium Carbonate Production by Direct Carbonation Process

are transferred to product dryers, where free and bound moisture is evaporated.

Sodium carbonate is made by carbonating a sodium chloride brine in the Solvay process. Ammonia is used as a catalyst for this reaction, as shown in Figure 3.

PROCESS EMISSIONS[1,2]

There are a number of emission sources within the natural sodium carbonate industry. Those judged most significant include calciners, dryers, bleachers, and predryers. These are process emission sources that emit large quantities of particulate matter.

Calciners and Bleachers

Calciners are the largest source of particulate emissions from plants using the monohydrate process. These particulates consist of sodium carbonate and inerts. The exit gas from coal-fired calciners also contains fly ash. Particulate emissions from calciners are affected by the gas velocity and the particle size distribution of the ore feed. As the gas velocity increases, the rate of increase in the particulate emissions steadily increases. Thus coal-fired calciners may have higher particulate emissions than gas-fired calciners because of higher gas flow rates. Additionally, small particles are more easily entrained in a moving stream of gas than are larger particles. Estimated uncontrolled particulate emission rates, particulate concentrations, and exit gas flow rates extrapolated from U.S. Environmental Protection Agency (EPA) test data are presented in Table 2 for small, medium, and large gas- and coal-fired calciners.

Sulfur oxides are produced from fuel combustion. The quantities produced depend on the sulfur content of the fuel. Organics are also emitted from calciners, some of which are present in the feed in the form of oil shale. At the calcina-

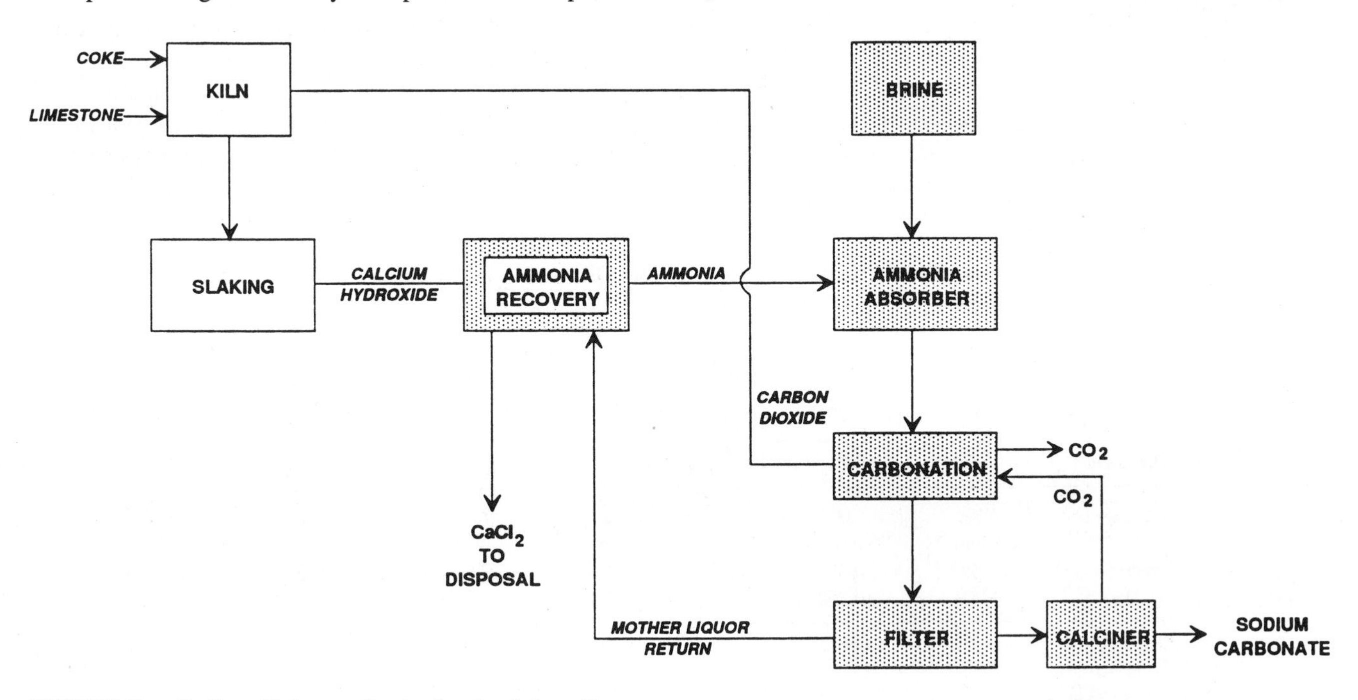

FIGURE 3. Sodium Carbonate Production by Solvay Process

TABLE 2. Uncontrolled Particulate Emissions from Calciners in the Monohydrate Process

Size	Ore Feed Rate, Mg/h (tons/h)	Fuel	Particulate Concentration g/dNm^3 (gr/dscf)	Particulate Emission Rate,[a] 1000 kg/h (1000 lb/h)	Exit Gas Flow Rate[a], 1000 dNm^3/min (1000 dscf/min)	Particulate Emission Factor, kg/Mg (lb/ton)
Small	40	Coal[a]	110	8.5–8.8	1.2–1.3	213–222
	(44)		(52)	(19–18)	(41–44)	(425–442)
		Gas[a]	167–263	4.6–7.8	0.45–0.52	115–195
			(73–115)	(10–17)	(16–18)	(230–389)
Medium	118	Coal	110	25–26	3.4–3.7	213–222
	(130)		(52)	(55–57)	(120–130)	(425–442)
		Gas	167–263	14–23	1.3–1.5	115–195
			(73–115)	(30–51)	(47–54)	(230–389)
Large	200	Coal[a]	110	43–44	5.8–6.3	213–222
	(220)		(52)	(94–97)	(210–220)	(425–442)
		Gas	178–238	23–35	2.2–2.4	115–195
			(78–104)	(51–77)	(77–86)	(230–389)

[a]Particulate concentration is the same as that measured for a medium-size calciner. Reported particulate emission rate and exit gas flow rate are based on values for a medium-size calciner weighted for the different ore feed rate.

Source: Reference 1.

tion temperatures, they may vaporize or be partially combusted. In addition, some organics may result from partial or incomplete combustion of the fuel.

Emissions from bleachers consist mainly of particulates of sodium carbonate. Small amounts of compounds formed from the reactions of sodium nitrate may also be present in the particulates. Estimated uncontrolled particulate emission rates and exit gas flow rates extrapolated from source test data are presented in Table 3 for a bleacher with a feed rate of 82 megagrams per hour (Mg/h) (90 tons per year).

Dryers and Predryers

Three types of dryers are used for product drying in the monohydrate and direct carbonation processes: rotary steam tube, rotary gas fired, and fluid-bed steam tube. Sodium carbonate fines are emitted from each of these dryers. Estimated uncontrolled particulate emission rates, particulate concentrations, and exit gas flow rates for small and medium-sized rotary dryers and for medium and large fluid-bed dryers extrapolated from EPA test data are presented in Table 4. No data on uncontrolled particulate emission rates for gas-fired dryers are available.

Particulate emissions from dryers are affected by the gas velocity of the feed. As the gas velocity increases, the rate of increase in the total emission rate of particulates steadily increases. Therefore, because of higher gas flow rates and higher gas velocities, fluid-bed steam tube dryers and rotary gas-fired dryers have higher emission rates than do rotary steam tube dryers.

In the direct carbonation process, rotary steam-heated predryers are used to lower the water content of wet sodium bicarbonate crystals before they are calcined. Particulates of sodium bicarbonate are the primary type of emissions from predryers. Estimated uncontrolled particulate emission rates and exit gas flow rates extrapolated from EPA test data are

TABLE 3. Uncontrolled Particulate Emissions from Bleachers and Predryers in the Direct Carbonation Process

Operation	Feed Rate, Mg/h (tons/h)[a]	Particulate Emission Factor, kg/Mg Feed (lb/ton Feed)[a]	Particulate Emission Rate, kg/h (lb/h)	Particulate Concentration, g/dNm^3 (gr/dscf)	Exit Gas Flow Rate, dNm^3/min (dscf/min)
Bleacher	82	45.5–228	3,700–19,000	105–380	590–820
	(90)	(90.1–455)	(8,100–41,000)	(46–166)	(21,000–29,000)
Predryer	59	0.377–3.21	22.2–189	0.261–1.49	1,400–2,400
	(65)	(0.754–6.42)	(49.0–417)	(0.114–0.653)	(46,000–78,000)

[a]Reported as dry impure sodium carbonate feed to the bleachers. The impurity content ranges from 0 to 15% for bleachers and 0 to 10% for predryers.

Source: Reference 1.

TABLE 4. Uncontrolled Particulate Emissions from Dryers in the Monohydrate and Direct Carbonation Processes

Steam Tube Dryer Type	Size	Production Rate Dry Product, Mg/h (ton/h)	Particulate Emission Factor, kg/Mg Feed (lb/ton Feed)	Particulate Emission Rate, 1000 kg/h (1000 lb/h)	Particulate Concentration g/dNm^3 (gr/dscf)	Exit Gas Flow Rate,[a] 1000 dNm^3/min (1000 dscf/min)
Rotary	Small	23	25.5–33.9	0.581–0.767	69–77	0.125–0.186
		(25)	(51.1–67.7)	(1.28–1.69)	(30–34)	(4.42–6.58)
Rotary	Medium	63	25.5–33.9	1.62–2.15	69–77	0.351–0.521
		(70)	(51.1–67.7)	(3.5–4.74)	(30–34)	(12.4–18.4)
Fluid bed	Medium	63	51.5–116	3.27–7.35	42–91	~1.42
		(70)	(103–231)	(7.21–16.2)	(18–40)	(~50.2)
Fluid bed	Large	113	51.5–116	6.08–13.6	42–91	~2.6
		(130)	(103–231)	(13.4–30.0)	(18–40)	(~94.5)

[a]Particulate emission rate and exit gas flow rate were calculated by ratioing values measured in source tests according to production rate.

Source: Reference 1.

presented in Table 3 for a predryer with a dry feed rate of 59 Mg/h (64 tons per hour).

Table 5 presents a summary of particle size distributions and emission factors for natural process sodium carbonate plants.[3]

EMISSION CONTROLS[1]

Particulate emission control techniques applicable to sources in sodium carbonate plants include the following:

- Centrifugal separation
- Wet scrubbing
- Electrostatic precipitation
- Fabric filtration

Centrifugal separators, or cyclones, rely on centrifugal forces to effect particulate separation from the gas stream. Cyclones are frequently used upstream of a scrubber or electrostatic precipitator (ESP).

Scrubbers rely mainly on the inertial impaction of particles with water droplets to effect particulate separation from the gas stream. Particles are contacted with a wetted surface or atomized liquid droplets. Although gas streams will diverge to pass such obstructions, the inertia of particles in the gas stream will carry the particles into the water droplets or wetted surface. The particulate-laden liquid is then separated from the gas stream and either recycled to the production process or discharged as waste.

Electrostatic precipitators generate an electrical field by applying a high voltage to a discharge electrode system consisting of rows of vertical wires. The strength of the field depends in part on the gas composition. The subsequent migration of the charged particles to the collected plates depends on the particle size, resistivity, gas velocity and distribution, rapping, and field strength. The collecting electrodes are rigid plates that are baffled. Electromagnetic or pneumatic hammers are used to rap the electrodes, dislodging the collected dust, which then falls into hoppers. Baffling on the collecting electrodes provides shielded air pockets that reduce reentrainment of particles after rapping.

Fabric filtration is a process where dust particles in a gas stream are filtered out and collected.

Calciners and Bleachers

Calciners and bleachers are typically controlled by cyclones in series with ESPs. Venturi scrubbers, also used to control emissions from calciners, achieve lower removal efficiencies than do ESPs. Higher removal efficiencies may be achieved with higher scrubber pressure drops. Based on the removal efficiency achieved in an EPA test on a gas-fired calciner with a scrubber pressure drop of about 85 cm (33.5 inches) of water, it appears that a pressure drop of 154 cm (60 inches) of water may be required to achieve a removal efficiency comparable to that achieved with a four-stage ESP.

Fabric filters have not been reported to be in use to control emissions from calciners or bleachers. The sticky, hygroscopic nature of sodium carbonate could lead to problems with bag blinding or caking. Baghouses are used to control particulate emissions from other sources in sodium carbonate plants, such as conveyor transfer points, crushing, and product sizing.

Dryers and Predryers

Venturi scrubbers are the devices used to control emissions from rotary steam tube dryers. Cyclones in series with venturi scrubbers are used to control emissions from fluid-bed steam tube dryers and rotary steam-heated predryers. Both venturi scrubbers and ESPs have been used to control emissions from gas-fired dryers.

The exhaust gas, from both rotary and fluid-bed steam tube dryers and predryers, is well suited to control by wet

TABLE 5. Particle Size Distribution and Emission Factors for Natural Process Sodium Carbonate Plants

Operation/Particle Size, μm[a]	Particle Size Distribution,[b] % <			Emission Factors[c] Size Specific,[d] kg/Mg			Emission Factors[c] Total Particulates, kg/Mg	Emission Factors Rating[e]
	2.5	6	10	2.5	6	10		
Rotary dryer								
Uncontrolled	2.8	4.2	5.2	0.04	0.065	0.08	1.55	C
After cyclone/scrubber	46.0	51.0	52.5	—	—	—	—	D
Calciner								
Gas fired, uncontrolled	2.8	5.2	6.7	5.2	9.6	12.3	184	C
Gas fired, after cyclone/ESP	64.5	79.0	86.0	—	—	—	—	E
Gas fired, after cyclone/scrubber	60.0	69.5	71.0	—	—	—	—	E
Gas fired, uncontrolled	2.0	6.5	9.5	3.9	12.7	18.5	195	E
Rotary gas-fired bleacher								
Uncontrolled	0.6	1.5	2.5	0.9	2.3	3.9	155	C
After cyclone/ESP	8.0	22.0	35.0	—	—	—	—	D
Product dryers								
Fluid-bed steam tube, uncontrolled	6.5	12.5	13.0	4.7	9.1	9.5	73	E
Rotary steam tube, uncontrolled	20.0	20.5	21.0	6.6	6.8	6.9	33	E

[a]Particle size is aerodynamic particle diameter in microns.
[b]Cumulative weight % < given particle size.
[c]For predryers, calciners, and bleachers, emission factors are kilograms of particulates/megagrams of feed to process unit. For product dryers, factors are kilograms of particulates/megagrams of product.
[d]Size-specific emission factor = Total particulate emission factor × (Particle size distribution/100).
[e]Rating of emission factor reliability and accuracy (A through E, with A the best).
—indicates no data available.

Source: Reference 3.

scrubbing. The removed sodium carbonate particles are quite soluble and hygroscopic. These characteristics enhance the removal of sodium carbonate particles in wet scrubbers. However, these characteristics, coupled with the high water content of the dryer exit gas, also can result in operating problems for ESPs or baghouses. Moisture in the exit gas condenses in the ESP or baghouse. Wet, sticky dust adheres to the electrodes and hoppers of the ESP or blinds and cakes the bags in the baghouse. Electrostatic precipitators have been used to control emissions from rotary gas-fired dryers, but the exit gas from these dryers is at a higher temperature and lower relative humidity than is gas from steam tube dryers.

RECENT AND FUTURE INDUSTRY TRENDS[1]

Sodium carbonate facilities are major sources of particulate emissions. These emissions can be effectively controlled through the proper use and maintenance of conventional add-on particulate control techniques. Tests conducted at sodium carbonate plants, along with industry data, contributed to the selection of ESPs as the best system of emission reduction for calciners and bleachers and of venturi scrubbers as the best system for dryers and predryers.

Small amounts of sulfur oxides and organics are also emitted from direct-fired calciners. However, source tests have indicated that these emissions are very low relative to uncontrolled particulate emissions.

Environmental issues have contributed to the decline in the production of sodium carbonate by the Solvay process. For example, substantial quantities of aqueous waste, containing high concentrations of calcium chloride, are produced in this process. Another reason for the declining Solvay production has been increasing fuel costs. The Solvay process is more fuel intensive than any of the natural processes.

New sodium carbonate plants in the United States are likely to use the monohydrate process, direct carbonation process, or anhydrous process. The anhydrous process involves the same unit operations as the monohydrate process, but the operating conditions of the crystallizer are such that anhydrous sodium carbonate rather than sodium carbonate monohydrate is produced in the crystallizers. Because of the limited availability of natural gas, future plants are expected to make greater use of coal than do existing plants.

References

1. *Sodium Carbonate Industry—Background Information for Proposed Standards—Draft EIS,* EPA-450/3-80-029a, U.S. Environmental Protection Agency, Office of Air Quality Planning and Standards, Emission Standards Division, Research Triangle Park, NC, August 1980.

2. *Sodium Carbonate Emission Test Report Kerr-McGee Chemical Corporation, Trona, California,* EMB Report 79-SOD-3, U.S Environmental Protection Agency, Office of Air Quality Planning and Standards, Research Triangle Park, NC, March 1980.
3. *Compilation of Air Pollution Emission Factors. Vol. I: Stationary Point and Area Sources,* AP-42, 4th ed., U.S. Environmental Protection Agency, Office of Air Quality Planning and Standards, Research Triangle Park, NC, September 1985.

Bibliography

Sodium Carbonate Emission Test Report F.M.C. Green River, Wyoming. EMB Report 79-SOD-2, U.S. Environmental Protection Agency, Office of Air Quality Planning and Standards, Research Triangle Park, NC, May 1979.

SULFURIC ACID

Thomas L. Muller

Sulfuric acid is a basic raw material in an extremely wide range of industrial processes and manufacturing operations. Because of its widespread use and relatively low production cost as compared with shipping cost, sulfuric acid plants are scattered throughout the nation near every industrial complex.

The production of sulfuric acid involves the generation of sulfur dioxide (SO_2), its oxidation to sulfur trioxide (SO_3), and the reaction of SO_3 with water to form sulfuric acid. In practice, the reaction of SO_3 with water is accomplished by first absorbing the SO_3 in strong sulfuric acid and then adding water.

The combustion of elemental sulfur is the predominant source of SO_2. The combustion of hydrogen sulfide (H_2S) from waste gases, the thermal decomposition of spent sulfuric acid or other sulfur-containing materials, and the roasting of pyrites are also used as sources of SO_2. In recent years, primarily for environmental reasons, many nonferrous metal producers have built sulfuric acid plants to recover the large amounts of SO_2 generated in the smelting process.

PROCESS DESCRIPTION

The two main processes for sulfuric acid manufacture are the chamber process and the contact process. The chamber process uses the reduction of nitrogen dioxide (NO_2) to nitric oxide (NO) as the oxidizing mechanism to convert SO_2 to SO_3. The contact process uses a vanadium-based catalyst to oxidize SO_2 to SO_3. During the first part of this century, the contact process gradually replaced the chamber process. Continued improvements have made it highly efficient from the standpoint of conversion efficiency (SO_2 to SO_3) and energy recovery. As essentially all sulfuric acid today is manufactured by the contact process, further discussion will be restricted to that process.

Contact Process

Contact plants may be divided into two broad classifications—"hot gas" plants and "cold gas" plants. In hot gas plants, the SO_2-containing gas, generally from burning sulfur, need only be cooled before entering the catalytic converter. Cold gas plants have dirty gas streams (from spent acid regeneration, pyrite roasting, smelting operations, etc.) as their SO_2 source. These gas streams must be cooled and cleaned before entering the contact plant. So-called "wet gas" plants generally use H_2S as the sulfur source and are a special type of hot gas plant. The combustion gas in wet gas plants contains appreciable water, but is sent to the converter without drying.

A flow diagram of a typical single-absorption, sulfur-burning contact sulfuric acid plant is shown in Figure 1. Combustion air is typically provided by a single-stage centrifugal blower. The blower is usually driven by a steam turbine, but electric drives are not uncommon, especially in enhanced energy recovery and cogeneration plants. The pressure at the blower discharge may be as high as 6 psig, or 180 inches water.

Combustion air passes through a drying tower to remove moisture before entering the sulfur burner. In the drying tower, moisture is removed from the air by countercurrent scrubbing with sulfuric acid. Drying tower acid strength is usually maintained at 93% or 98–99%, depending on product requirements. Since the acid in the drying tower picks up moisture from the air, some drying tower acid is continuously removed and replaced with stronger acid from the absorbing tower. Drying towers can operate successfully over a wide range of acid temperatures, depending on selected acid concentration; however, 40–70°C is most common.

Drying towers are usually brick-lined steel vessels filled with ceramic packing. To prevent mechanically generated spray and mist from carrying over into the rest of the process, drying towers usually have internal spray catchers or mist eliminators built into the top of the tower. The two most common types of mist and spray eliminators are the mesh pad and the packed fiber "candle" or tubular mist eliminator. These will be discussed later.

Molten sulfur is pumped to the burner, where it is burned with the dry combustion air to form SO_2, typically 7–11% SO_2 by volume. Some SO_3 also may be produced in the sulfur burner.[1,2] If not accounted for, the SO_3 may be high enough to cause calculated SO_2 conversion to be low and emissions (lb_m SO_2 emitted per ton of acid produced) to be erroneously high. The combustion reaction is highly exothermic, and in the absence of reducing substances, is irreversible. The SO_2 and excess air leave the sulfur burner

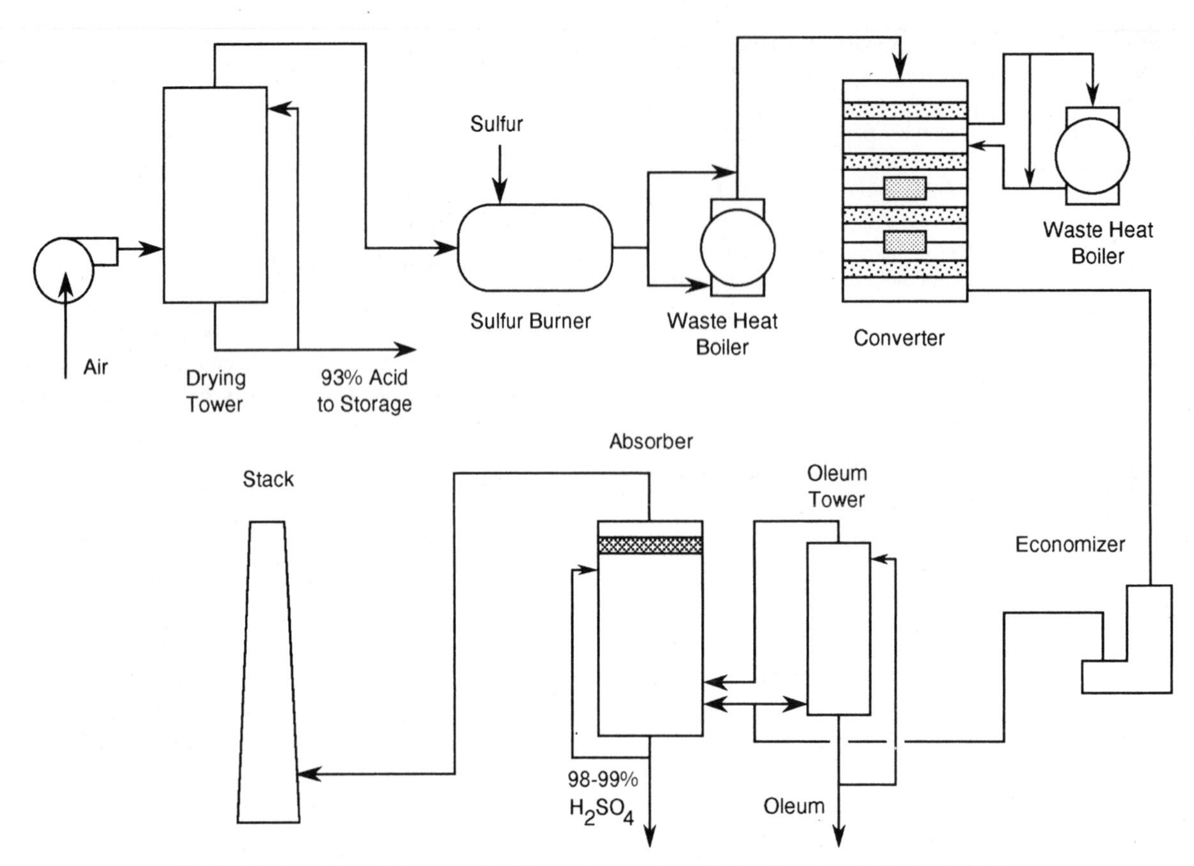

FIGURE 1. Flow Diagram of a Single-Absorption, Sulfur-Burning Sulfuric Acid Plant

at 700–1000°C and must be cooled, typically in a waste-heat boiler, to about 425°C before entering the converter.

Sulfur dioxide is converted to SO_3 in a multipass catalytic converter with two to five (most often four) catalyst beds. Process gas enters the first pass of the converter to begin the oxidation of SO_2 to SO_3.

$$SO_2(g) + \frac{1}{2}O_2(g) \leftrightarrow SO_3(g)$$
$$\Delta H \approx -41{,}400 \text{ Btu/lb-mol}$$

This reaction, too, is highly exothermic, but unlike the combustion reaction, it is an equilibrium reaction (i.e., it is reversible). High temperature favors the reverse reaction, decomposition of SO_3 to SO_2. Therefore, as the oxidation reaction proceeds and the temperature and amount of SO_3 both increase, the reverse reaction rate approaches the forward reaction rate. When the forward and reverse reaction rates are equal, the overall reaction is said to be in equilibrium. No additional conversion of SO_2 to SO_3 is possible unless the process gas is cooled or SO_3 is removed. Gas leaving the first pass is typically cooled in a waste-heat boiler or steam superheater. Steam superheaters and boiler feedwater economizers are commonly used to cool the gas leaving the second and third passes.

Three primary factors influencing overall converter performance are as follows:

- Equilibrium considerations (gas strength and temperature)
- Minimum temperatures required for catalyst to operate
- The need to keep catalyst masses to a practical size

Together, these factors limit overall conversion in a single-absorption plant to about 98%. Table 1 shows typical gas temperatures and conversions leaving each pass of a four-pass, single-absorption converter.

Gas leaving the converter is cooled to 185–250°C in an economizer. This increases overall plant energy recovery and improves SO_3 absorption by lowering the process gas temperature entering the absorption tower.

From the economizer, the process gas enters the bottom of the absorbing tower, where SO_3 is absorbed in a recirculating stream of acid. To maintain acid strength as SO_3 is absorbed, water or weak acid is added to the absorbing tower. Acid from the drying tower also may be used for this purpose, although additional water probably will still be required.

Acid concentration entering the absorbing tower should be 98–99% sulfuric acid (H_2SO_4). This is near the azeotrope where the partial pressures of SO_3, H_2O and H_2SO_4 are lowest. Below 98%, the partial pressure of water is high enough that it reacts with SO_3 in the gas to form H_2SO_4 mist, before the SO_3 can be absorbed in the acid. The partial

TABLE 1. Temperatures and Conversions in a Single-Absorption Sulfur-Burning Contact Plant at 9.0% SO_2.

Location of Gas	Temperature °C	Temperature °F	Δ Conversion, %
First-pass inlet	410	770	
First-pass exit	602	1115	
Temperature rise	192	345	74.0
Second-pass inlet	438	820	
Second-pass exit	485	906	
Temperature rise	47	86	18.4
Third-pass inlet	432	810	
Third-pass exit	443	830	
Temperature rise	11	20	4.3
Fourth-pass inlet	427	800	
Fourth-pass exit	430	806	
Temperature rise	3	6	1.3
Total rise	253	457	98.0

pressure of SO_3 increases at strengths much above 99%, allowing SO_3 to leave with the tail gas.

The temperature of acid entering the tower is typically 60–80°C, but may be significantly higher or lower. In some newer towers designed for high-energy recovery, temperatures may be as high as 200°C. Construction of the absorbing tower is similar to that of the drying tower. Process gas, with the SO_3 removed, passes through mist-removal equipment and is discharged to the atmosphere through the stack.

Figure 2 is a flow diagram of a spent acid regeneration plant. In this plant, the sulfur source may be waste sulfuric acid, H_2S, other sulfur-containing compounds, or a mixture of these. At temperatures of 1000–1050°C, these materials decompose to produce SO_2. Carbon dioxide (CO_2) and water are the other two main products. With the exception of H_2S, these materials usually provide little heat of combustion, so natural gas or fuel oil is added to the spent furnace to provide the necessary heat. Elemental sulfur also may be burned in the spent furnace. Decomposition efficiency is aided by the near stoichiometric ratio of oxygen to combustibles.

The gases leaving the spent furnace contain excess water and, depending on the spent materials burned, ash. Various equipment combinations, including waste-heat boilers, humidification towers, reverse jet scrubbers, packed gas cooling towers, impingement tray columns, Karbate® coolers, and electrostatic precipitators, are used to cool and clean the gas before it is sent to the drying tower.

The balance of the spent plant is the same as the sulfur-burning plant, except that the gas leaving the drying tower must be heated to the reaction temperature (about 425°C) before it enters the converter. This is done using the heat of reaction in the converter as the heat source, with gas-to-gas heat exchangers replacing waste-heat boilers and superheaters.

Metallurgical acid plants (i.e. those receiving wet, dirty SO_2 from burning or smelting sulfide ores) have essentially the same flow diagram as the spent acid regeneration plant shown in Figure 2. The only major difference is that the smelter or ore roaster replaces the combustion chamber. In operation, they are similar, with metallurgical plants having continuously varying feed rates and concentrations, causing difficulty in achieving high conversions (of SO_2 to SO_3), that is, low SO_2 emissions.

Figure 3 is a flow diagram of a dual-absorption, sulfur-burning plant (also called double absorption, interpass absorption, or double catalysis). This type of plant was developed in the early 1960s as a way of increasing conversion of SO_2 to SO_3. The oxidation of SO_2 to SO_3,

$$SO_2 + \tfrac{1}{2} O_2 \leftrightarrow SO_3$$

being an equilibrium reaction, can be shifted strongly to the right by removing the SO_3 from the process gas. Process gas leaves the primary converter, and is cooled and sent to an "interpass" absorbing tower, where the SO_3 is removed. The SO_3-free gas leaves the interpass absorber, is reheated, and enters the final converter. Gas leaving the final converter is cooled before it enters a final absorbing tower, where the remaining SO_3 is absorbed. Overall conversion greater than 99.7% is possible in a dual-absorption plant. Virtually all new plants built in the United States since the implementation of New Source Performance Standards (NSPS) have been dual-absorption plants.

There are many variations on the plants described. Where oleum is desired, a separate oleum tower is used ahead of the primary absorbing tower. In cold gas and dual-absorption plants, gas-to-gas heat exchangers are used instead of boilers and superheaters to cool the converter gas. In some older converters, dried air is injected into the process gas following the second and third passes to provide cooling. Especially when older single-absorption plants are converted to dual absorption, other combinations of equipment can be found.

AIR EMISSIONS CHARACTERIZATION

The only significant source of air emissions from a contact sulfuric plant is the tail gas leaving the final absorbing tower. This gas contains small amounts of SO_2 and even smaller amounts of SO_3, sulfuric vapor, and sulfuric acid mist.

Sulfur Dioxide Emissions

Sulfur dioxide emissions and limits are reported either as pounds of SO_2 per ton of 100% H_2SO_4 produced or percent SO_2 converted to SO_3, the two being related by a simple

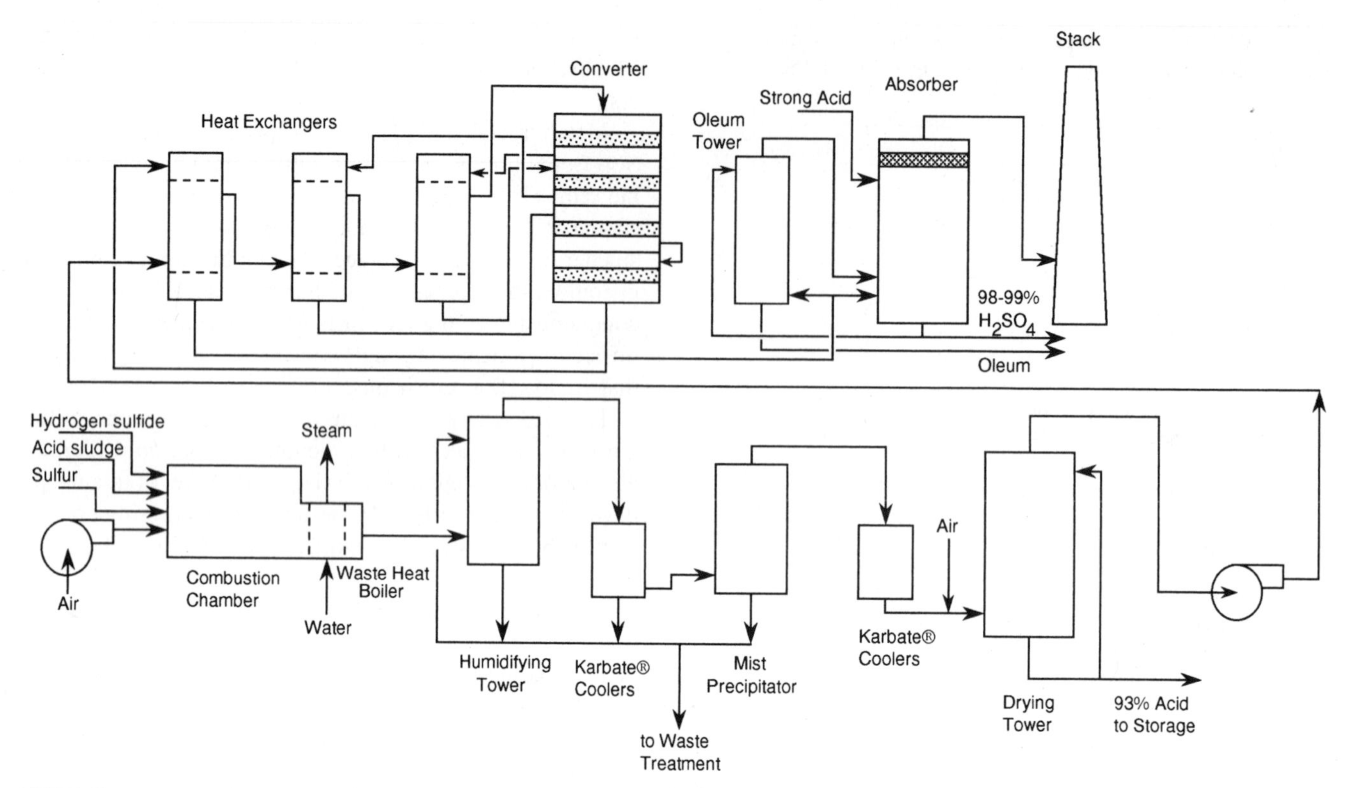

FIGURE 2. Flow Diagram of a Single-Absorption, Spent Acid Regeneration Sulfuric Acid Plant

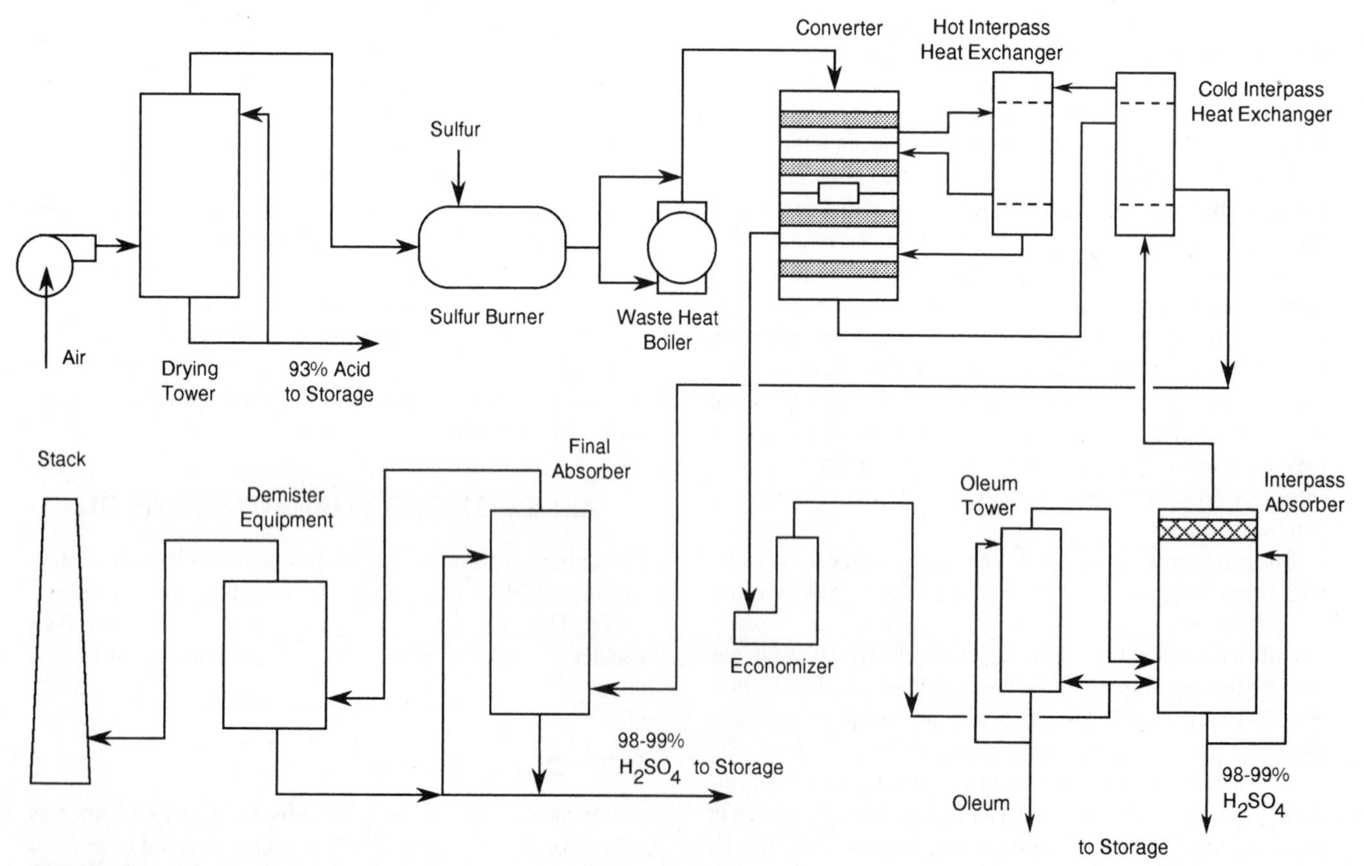

FIGURE 3. Flow Diagram of a Dual-Absorption, Sulfur-Burning Sulfuric Acid Plant

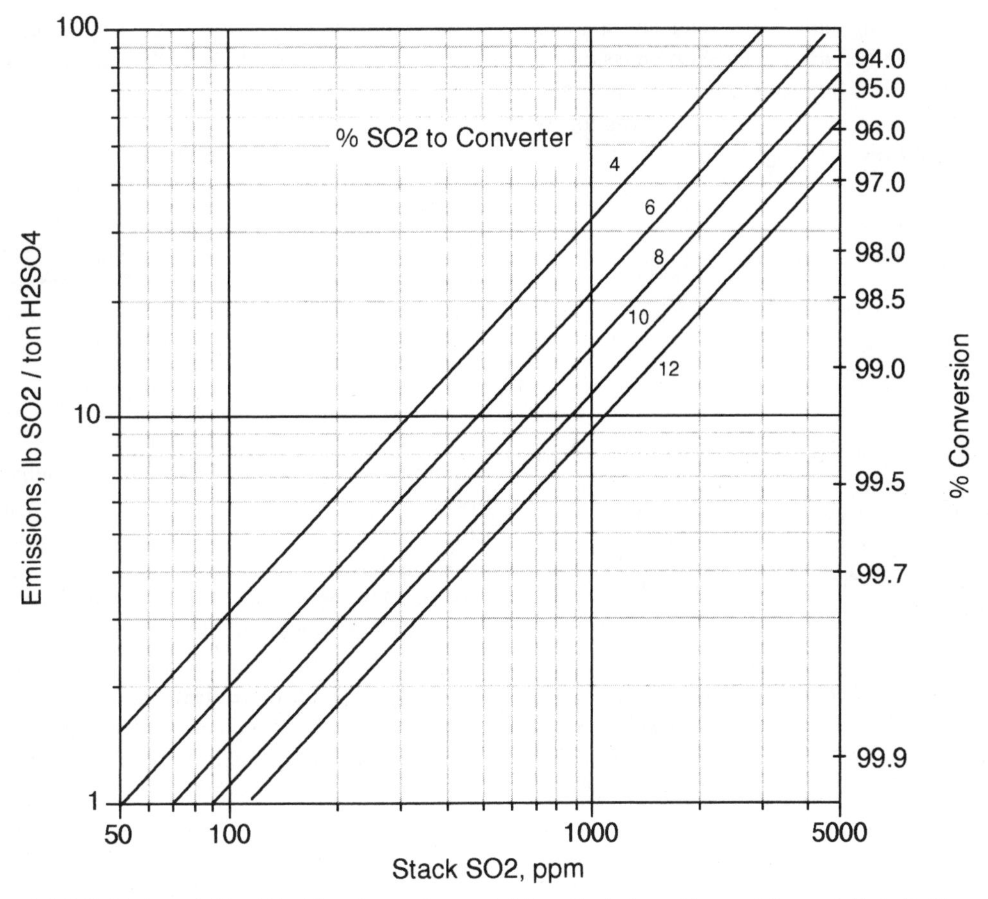

FIGURE 4. Sulfur Dioxide Conversion and Mass Emissions Versus Inlet and Exit Sulfur Dioxide Concentrations

formula. (Some old "grandfathered" plants have only "in-stack" SO_2 limits. These typically limit emissions to 2000 ppm, or sometimes 3500 ppm for spent acid regeneration plants.)

A minor source of air emissions from spent acid regeneration and metallurgical acid plants is the stack of a fuel-fired heater. These are typically fired with natural gas or light oils and have characteristic emissions for those burners. These heaters are used to add heat indirectly to the process and may be used only at start-up, or continuously in some cases.

Single-absorption, sulfur-burning plants are generally capable of meeting emission standards of 27 lb_m SO_2 per ton of 100% H_2SO_4, equivalent to 97.97% conversion. Dual-absorption plants can meet NSPS requirements of 4 lb_m SO_2 per ton of 100% H_2SO_4 produced, or 99.7% conversion. Dual absorption is well demonstrated and is the technology most frequently used today to reduce SO_2 emissions, although various scrubbing technologies are also available. These will be briefly discussed below.

Conversion and SO_2 emissions are generally calculated from SO_2 entering the converter and SO_2 leaving the stack using the following formulas. These formulas are plotted in Figure 4.

$$C = \frac{A - B}{A - 1.5AB}$$

$$SO_2 \text{ emissions} = \frac{1306}{C} - 1306$$

where: C = conversion, fraction
A = inlet SO_2, fraction
B = stack SO_2, fraction
SO_2 emissions = lb_m SO_2 per ton 100% H_2SO_4

The conversion equation has been used for years and is based on simple material balance. Fairlie[3] and Miles[4] both give its derivation. The emissions equation is simple algebra based on the fact that it takes 1306 lb_m of SO_2 to make a ton of sulfuric acid at 100% conversion efficiency.

The above formulas are valid for hot or cold gas plants, but do not apply to plants that use air injection for cooling in the lower converter passes unless a correction is made to include the amount of air added. These formulas assume that the conversion in the gas entering the converter is zero, that is, no SO_3 is present. If significant SO_3 is present, as can happen in the combustion of sulfur, the measured inlet SO_2 will be low and the calculated SO_2 emissions will be high.

For hot gas plants using only sulfur and air, a more precise way of measuring conversion and SO_2 emissions is based on SO_2 and oxygen (O_2) in the stack gas. Conversion is calculated from the following formula: (The basis for its derivation is given by Miles.[5])

$$C = \frac{0.209 - Y - B}{0.209 - Y + 0.186B}$$

where: C = conversion, fraction
B = stack SO_2, fraction
Y = stack O_2, fraction

This formula is also applicable and especially useful to air injection plants if their only SO_2 source is sulfur and air. Trace amounts of H_2S and hydrocarbons in sulfur cause slight errors in this calculation since the hydrogen and carbon consume oxygen during combustion. These errors are usually negligible.

Sulfur Trioxide and Sulfuric Acid Mist

The gas leaving the stack contains small amounts of SO_3, H_2SO_4 vapors and H_2SO_4 mist. Sulfur trioxide is normally combined with mist when determining acid plant emission factors because, upon leaving the stack, SO_3 rapidly reacts with atmospheric moisture to form H_2SO_4 mist. Hence the resultant pollution problem is the same whether the initial cause is SO_3 or H_2SO_4 mist. However, the sources of SO_3 and mist emissions and their control are different. An increase in either SO_3 or mist emissions is usually first seen as increased stack opacity. Except in extreme cases, however, it is nearly impossible to tell the source of the problem by observing the stack plume. Hence understanding the difference between SO_3 and mist is necessary to properly diagnose and control these emissions. Mist and SO_3 emission factors can be determined only by testing.

When reporting the emission factor for SO_3 and mist, SO_3 is calculated as 100% H_2SO_4 and added to the mist value. The NSPS requirement for combined SO_3 and mist is 0.15 lb_m/ton of acid. Some "grandfathered" plants have limits of 0.5 lb_m/ton. (The very small amount of H_2SO_4 vapor in the gas stream will be measured as SO_3 by existing test methods such as the Environmental Protection Agency's Method 8.)

Sulfur trioxide in the stack gas comes from a poorly operating final absorbing tower. Significant variables for proper tower operation are acid strength, acid temperature, circulation rate, and acid distribution over the top of the tower. The least important to good SO_3 absorption is temperature, although temperature can significantly affect internal mist generation. (See later discussion.) The SO_3 concentration leaving a properly operating tower will normally be < 0.1 mg/scf (68°F, 1 atm.).

Mist is formed within the process by mechanical means (splashing/entrainment) and physicochemical means (reaction/condensation). Mechanically generated mist is very coarse and is easily collected. Reaction/condensation-generated mist forms from water in the gas stream. Any water vapor in the gas stream will react with SO_3 to form H_2SO_4. If the gas temperature falls below the dew point, this H_2SO_4 will condense into mist particles. The dew point is a function solely of the H_2SO_4 concentration. Therefore, two ways to prevent mist condensation are to keep the gas temperature elevated and to keep the H_2SO_4 concentration in the gas low, that is, keep water out of the system.

Water vapor may come from poor drying tower performance, boiler tube or other steam leaks, or hydrocarbon impurities in the sulfur. Even with efficient gas drying and no steam leaking into the system, mist formation is impossible to prevent completely. Sufficient water vapor is always present in the process gas so that if the gas stream is cooled too quickly, before the SO_3 can be absorbed, a mist will be formed. This is particularly a problem in plants with oleum towers ahead of the final absorber, with bypass oleum towers producing more mist than full flow oleum towers. Acid mist formed by condensation is very difficult to collect. Because of its small particle size, sufficient mist will pass through the absorbing tower and, if unabated, will create a highly visible stack plume. Typical uncontrolled mist levels leaving the final absorber are 30–50 mg/scf for dry gas plants. Wet gas plants may have uncontrolled mist levels of several thousand milligrams per standard cubic foot.

The size of mist particles ranges from about 0.1 μm to greater than 10 μm. Mist particles greater than 5–10 μm are probably from mechanical entrainment. Plume visibility is a strong function of particle size. Acid mist composed of particles less than 10 μm can be visible even at concentrations less than 1 mg/scf. As the particle size decreases, the plume becomes more opaque because of the greater light-scattering effect of smaller particles. Maximum light scattering occurs when the particle size approximates the wavelength of light, about 0.3 μm. Thus particles about 0.3 μm in diameter are the most visible. In practice, these are also the most difficult particles to collect.

Sulfuric acid plants must meet maximum opacity limits on stack emissions, however, opacity is not an accurate indication of mass emissions. In one study of 24 variables affecting plume visibility, the angle of the sun caused a greater change in apparent opacity than did any other variable.[6] Other significant factors listed included stack (plume) diameter, color of the sky, distance of observer from the stack, wind direction, particle density, particle index of refraction, and plume color. The last three variables are normally fixed for H_2SO_4 mists. However, these may change significantly if the use of tail gas scrubbers changes the chemical nature of the emissions (such as with ammonia scrubbers).

AIR POLLUTION CONTROL MEASURES

Sulfur Dioxide

Dual Absorption

Dual absorption has been generally accepted as Best Available Control Technology (BACT) for meeting NSPS emission limits. Conceptually, dual absorption is the addition of another converter and absorbing tower to the tail end of a single-absorption plant (with appropriate heating and cooling of the gas stream), so no new technology is involved. Only sulfuric acid is produced in the dual-absorption equipment. There are no by-products or waste scrubbing materials. This typically has the least economic impact of abatement methods available to the acid producer. (See "Process Description" section for detailed discussion of dual absorption technology.)

Scrubbing Technologies

Various tail gas scrubbing processes are also available for SO_2 abatement. Tail gas scrubbers may be of the "throwaway" type, where the scrubbing chemicals are used on a once-through basis and disposed of as by-products or discarded, or of the "regenerative" type, where the SO_2 is removed from the gas and the active scrubbing solution is regenerated. Even with a regenerative process, some waste materials are produced, which must be purged from the system. For example, in the Wellman-Lord process, part of the active sulfite ingredient is converted to sulfate and must be discarded. The economics of processes that produce significant by-products is usually determined by the market for the by-product or the cost of its disposal.

Three scrubber processes will be briefly discussed: sodium sulfite, hydrogen peroxide, and ammonia-based scrubbers.

Sodium Sulfite (Wellman-Lord Process)

This process is chiefly used for desulfurizing flue gases from power stations and tail gases from Claus sulfur plants, but it has also been used for reducing SO_2 emissions from single-absorption sulfuric acid plants. The typical emission rate is around 200 ppm, which is equivalent to about 3–4 lb_m/ton, depending on the inlet gas strength.

Gas leaving the absorbing tower is scrubbed with a saturated aqueous solution of sodium sulfite and sodium bisulfite. The reaction is as follows:

$$SO_2 + Na_2SO_3 + H_2O \rightarrow 2NaHSO_3$$

Sulfur dioxide is regenerated from the bisulfite-rich scrubbing solution in a forced-circulation evaporator. Wet sulfur-dioxide–containing gas is recycled to the acid plant. About 10–20% of the recovered sulfur values end up as sodium sulfate as a result of the oxidation of the sulfite in solution and must be removed as a by-product.

Peroxide Scrubbing

Du Pont developed a hydrogen peroxide (H_2O_2) scrubbing process to reduce SO_2 emissions in the early 1970s. Dilute sulfuric acid and H_2O_2 are circulated over a packed-bed countercurrent to the stream of SO_2-containing gas. Sulfur dioxide is absorbed in the solution, where a rapid, high-yield reaction takes place to produce sulfuric acid. The acid produced in the scrubber becomes part of the plant's total production by blending with high-strength acid in the drying or absorbing towers. Thus there is no by-product or purge stream to dispose of with this process. The basic reaction is as follows:

$$SO_2 + H_2O_2 \rightarrow H_2SO_4$$

The Dupont process has been successfully applied to single-absorption plants.

Lurgi and Süd-Chemie developed the similar Peracidox® process in Europe.[7] This process uses either hydrogen peroxide or electrolytically produced peroxymonosulfuric acid (H_2SO_5) to oxidize SO_2 to sulfuric acid according to the following reactions

$$SO_2 + H_2O_2 \rightarrow H_2SO_4$$
$$SO_2 + H_2SO_5 + H_2O \rightarrow 2H_2SO_4$$

Both versions of the Peracidox process have been successfully demonstrated.

Ammonia Scrubbing

Ammonia scrubbing is capable of achieving the same level of performance as other tail gas scrubbers. Although some ammonia scrubbing processes include some SO_2 regeneration, ammonia scrubbing is essentially a once-through process and produces a considerable amount of ammonium sulfate that must be disposed of. This process may be especially favorable if the sulfuric plant is part of a fertilizer complex, as this represents an outlet for the ammonium sulfate, as well as a source of reasonably priced ammonia.

Sulfur Trioxide and Acid Mist Removal

Acid mist removal from sulfuric acid plant tail gases is accomplished almost exclusively with packed-fiber mist eliminators or demister pads. Although a small portion of the SO_3 that leaves the final absorber will be absorbed in fiber mist eliminators and demister pads, SO_3 emission control depends primarily on proper plant operation.

A successful packed-fiber tubular mist eliminator using treated glass fibers was developed by the Monsanto Chemical Co. in 1959. Figure 5 shows a diagram of this style of mist eliminator. The original name, Brink mist eliminator, is still sometimes used. These units are available in two styles, the hanging (shown) and stand-up varieties. These devices capture particles using a combination of three dif-

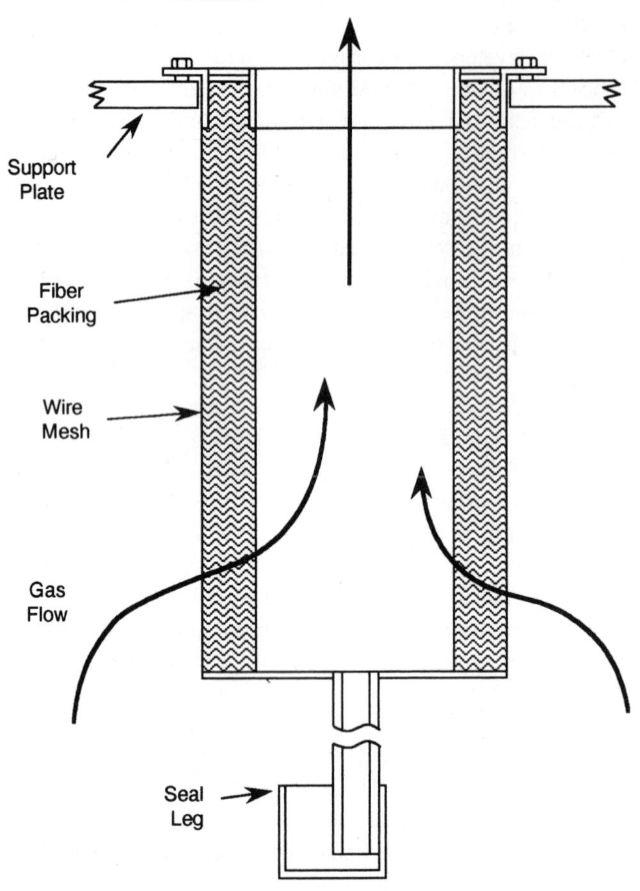

FIGURE 5. Packed-Fiber Mist Eliminator

ferent mechanisms: interception, impaction, and Brownian motion. Each mechanism operates most efficiently for a particular particle size. Together, they provide overall collection efficiencies that can exceed 99%, depending on the inlet mist loading.

Packed-fiber mist eliminators must be properly sized for the gas flow to provide optimum collection efficiency. Collection by the impaction mechanism suffers if gas velocity is too low. Reentrainment becomes a problem if gas velocity is too high. Fiber mist eliminators typically operate with 6–8 inches H_2O pressure drop. While high-pressure drop can be an indication of fouling, there is no direct correlation between pressure drop and collection efficiency or mass emissions. The most frequent causes of poor mist eliminator performance are the following.

- Poor sealing at the mist eliminator flanges that allows gas bypassing
- Loss of liquid in the seal leg (see Figure 5), which allows gas bypassing
- Fouling, which causes increased local velocities through the mist eliminators

Packed-fiber mist eliminators are capable of meeting NSPS requirements of 0.15 lb_m/ton of acid.

Demister pads are mesh pads designed to capture larger mist particles by the interception and impaction mechanisms. Sometimes a coalescing pad is used ahead of the demister pad to provide higher collection efficiency.

Coalescer pads use high velocity to coalesce smaller particles into larger particles via the impaction mechanism. Because of the high gas velocity, the larger coalesced particles are almost always reentrained into the gas stream leaving the coalescer pad. These reentrained particles are then collected in the larger, low-velocity demister pad.

Demister pads, with or without a coalescer section, are not able to collect submicron particles as efficiently as will packed-fiber demisters. Coalescer pads also operate at much higher pressure drop than packed-fiber demisters, up to 20 inches water. Nevertheless, they require significantly less investment and many plants have used them to achieve nearly invisible stack emissions. The successful use of demister pads requires careful control of plant operating parameters to minimize internal mist formation.

References

1. F. D. Miles, *The Manufacture of Sulphuric Acid,* D. Van Nostrand Co., New York, 1925, p 29.
2. T. J. P. Pearce, "Sulphuric acid: physico-chemical aspects of manufacture", in *Inorganic Sulphur Chemistry,* G. Nickless, Ed.; Elsevier Publishing Co., New York, 1968, p 538.
3. A. M. Fairlie, *Sulfuric Acid Manufacture,* Reinhold Publishing Corp., New York, 1936, pp 496–498.
4. F. D. Miles, op. cit., pp 187–189.
5. Ibid., p 193.
6. A. Weir, Jr., D. G., Jones, L. T. Papay, et al., "Measurement of particle size and other factors influencing plume opacity", "International Conference on Environmental Sensing and Assessment, Las Vegas, NV, September 14–19, 1975.
7. U.H.F. Sander, H. Fischer, U. Rothe, et al., *Sulphur, Sulphur, Dioxide and Sulphuric Acid,* Verlag Chemie International Inc., English edition, 1984, pp 338–340.

Bibliography

J. R. Donovan, and J. M. Salamone, "Sulfuric acid and sulfur trioxide," in *Kirk-Othmer Encyclopedia of Chemical Technology,* 3rd ed., John Wiley & Sons, New York, Vol. 22, 1983, pp 190–232.

The Manufacture of Sulfuric Acid, W. W. Duecker and J. R. West, Eds.; Van Nostrand Reinhold Co., 1959, reprinted by Robert E. Krieger Publishing Co., Huntington, NY, 1977.

SULFUR

Bruce Scott

Elemental sulfur is recovered as a by-product of the processing of crude oil and natural gas. This source of recovered sulfur, over 6.5 million tonnes in 1989, represents about 63% of the total elemental sulfur market in the United States, the remainder being derived from Frasch mining or imports.[1] The Frasch mining process has little direct impact on air emissions and will not be discussed. Elemental sulfur has a wide variety of industrial uses, the chief being the manufacture of sulfuric acid. From the standpoint of air pollution, the sulfur-recovery plants operated by petroleum refineries and natural gas plants dramatically reduce the emissions of sulfur compounds into the atmosphere from those facilities.

All crude oils and most natural gases contain sulfur compounds in widely varying amounts. Crude oils may contain from as little as 0.1% sulfur to as much as 5.0%. Natural gas is sometimes sulfur-free or may contain 15–20% or more of hydrogen sulfide (H_2S) or other sulfur compounds. These sulfur compounds are not tolerable in most finished petroleum products, and can prove to be corrosive in refinery processes used to make salable products.

While the sulfur in natural gas is usually H_2S, which can be directly removed by the sulfur-recovery plants described here, the removal of sulfur from raw crude oil is much more complex. In most modern refineries, this is usually accomplished by treating distillate cuts from crude oil with hydrogen in the presence of a catalyst at elevated temperature and pressure. This treatment converts the various sulfur compounds into H_2S. Some residual cuts are also desulfurized by this same means to produce low-sulfur fuel oil or to provide feedstocks for further processing. Heavy nonsalable cuts are usually cracked into lighter, salable fractions. The process of cracking normally converts organic sulfur compounds into H_2S. Being a gas at normal refinery conditions, H_2S tends to concentrate in the off-gas streams used for plant fuel. If this gas were to be burned as is, it would frequently result in unacceptable levels of atmospheric sulfur dioxide (SO_2) pollution.

PROCESS DESCRIPTION

The sulfur-recovery facilities used in most refineries and gas plants to prevent this pollution consist of three distinct steps: gas scrubbing to remove the H_2S, sulfur recovery to convert H_2S to elemental sulfur, and tail gas treating to clean the effluent from the sulfur-recovery plant to levels appropriate for release to the atmosphere. The refinery or gas plant operator has considerable flexibility in the choice of specific processes to be used to accomplish each of the three steps, as discussed in the following.

Gas Scrubbing

The processes described here are used primarily for removing H_2S from gas streams, but are also used to scrub H_2S from liquid streams such as propane and butane. Application to liquids differs from gas scrubbing only in the engineering details and will not be described further.

Removal of H_2S from gas streams is accomplished by scrubbing with a water solution of an organic amine in a packed or trayed tower (scrubber or absorber). The amine solution is alkaline and the weakly acidic H_2S in the gas stream readily dissolves in it. The acid and base react to form a salt, as illustrated in the following reaction.

$$R_2NH + H_2S \leftrightarrow R_2NH_2HS + \text{Heat} \qquad (1)$$

It should be noted that carbon dioxide (CO_2) is also a weak acid and will react by the same mechanism as reaction 1. Hydrocarbon gases, being insoluble in water and neutral rather than acidic, are not affected by the amine solution and so pass out of the scrubbing tower. As indicated by reaction 1, the reaction of H_2S or CO_2 with an amine is exothermic. Reaction 1 is shifted to the right by low temperatures and to the left by high temperatures.

In commercial plants, the rich amine solution, loaded with H_2S and CO_2 (acid gases), is sent to a stripping tower (regenerator), where it is boiled and the acid gases stripped out. The amine solution, now containing very little acid gas, is cooled and returned to the scrubbing tower for reuse. Heat exchangers, pumps, filters, and a flash drum complete the equipment list for a typical plant. Frequently, one regenerator serves several absorbers. The acid gases from the stripping tower are cooled and sent to the sulfur-recovery plant for conversion to elemental sulfur. Figure 1 illustrates a typical gas scrubbing plant.

A number of amines are in commercial use in gas scrubbing plants. The most common is diethanolamine (DEA), a secondary amine as illustrated in reaction 1. Monoethanolamine (MEA), a primary amine, has been used extensively and is still widely used where its higher reactivity is needed, particularly for CO_2 removal. Methyldiethanolamine (MDEA), a tertiary amine, is coming into greater favor with refiners because of its ability selectively to absorb H_2S and reject most of the CO_2 in the gas stream. There are other amines in use, and the above amines can be blended or otherwise promoted to enhance their performance in specialized applications.

Typical operating conditions in an amine plant are approximately as follows, although individual plants may vary greatly. Absorbers in refineries are operated at about 90 to 120°F and at pressures ranging from 80 psig to 150 psig in most towers to over 2000 psig in certain applications. Natural gas plant absorbers are operated at the same temperatures, but normally at higher pressures, typically 400 psig to 1000 psig. Regenerators operate at low pressure, usually 5–25 psig, and at temperatures in the range of 245°F to 265°F.

In some cases, usually where there is very low H_2S in the sour gas, other treatment processes may be used in lieu of amine treatment. These generally use a solid or liquid scrubbing agent that reacts chemically with H_2S, removing it from the gas stream. The spent scrubbing agent frequently is sent directly to disposal, although some can be regenerated and reused. There are a number of these scrubbing agents on the market and a detailed discussion will not be attempted here.

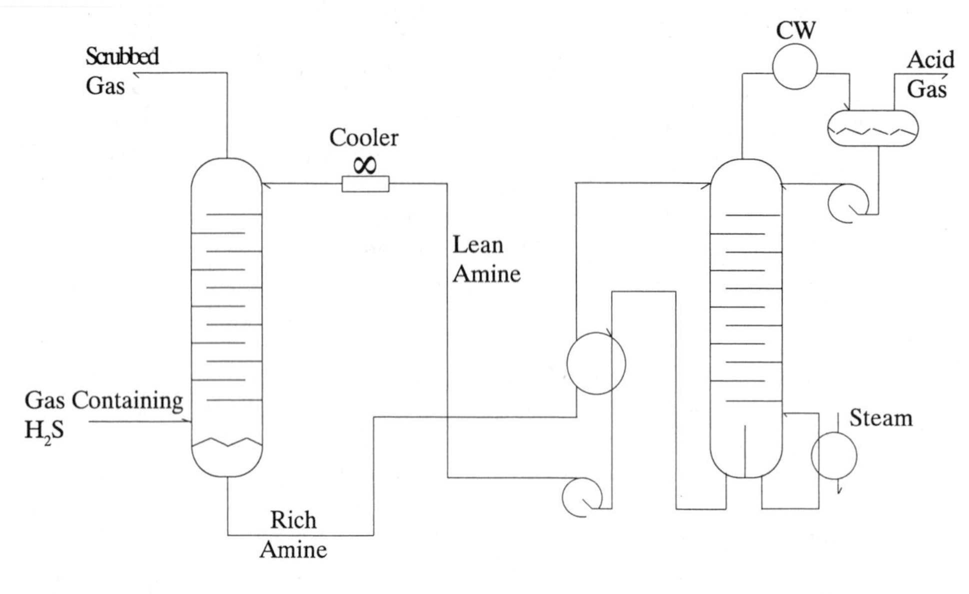

FIGURE 1. Typical Gas Scrubbing Plant

Sour gases are also sometimes scrubbed directly with liquid redox solutions. The H_2S is removed and converted directly to elemental sulfur, which can be recovered and the scrubbing solution regenerated for further use. These processes are more often used as tail gas cleanup processes and will be discussed in more detail later.

Sulfur Recovery

Acid gases from the amine plant are processed in a sulfur-recovery plant, utilizing the Claus process in the vast majority of cases. In the Claus process, one third of the H_2S in the acid gas stream is burned with air to form SO_2:

$$H_2S + 3/2O_2 \rightarrow SO_2 + H_2O + \text{Heat} \tag{2}$$

The SO_2 reacts with the remaining H_2S to form sulfur, as follows:

$$2H_2S + SO_2 \leftrightarrow 3S + 2H_2O + \text{Heat} \tag{3}$$

Reaction 3 takes place in the high-temperature combustion part of the plant and accounts for about half of the total conversion. The remaining sulfur is formed in lower-temperature catalytic parts of the plant. Because equation 3 represents an equilibrium chemical reaction, it is not possible for a Claus plant to convert all the incoming sulfur compounds to elemental sulfur. Each catalytic stage can recover half to two thirds of the sulfur entering it. Some H_2S, SO_2, sulfur vapor, and traces of other sulfur compounds formed in the combustion section escape with the inert gases from the tail end of the plant (tail gas). Thus it is frequently necessary to follow the Claus plant with a tail gas cleanup plant.

Since the 2:1 ratio of $H_2S:SO_2$ indicated by equation 3 is essential to efficient operation of the plant, an extensive control system is usually provided to insure the correct injection of air into the combustion chamber.[2] Many of the engineering differences among Claus plants are centered on the combustion chamber and the need for proper combustion conditions for a wide range of compositions and flammabilities of the acid gas.[3] Hot gas from the combustion chamber is quenched by passing through a waste-heat boiler, in which high- or medium-pressure steam is generated, before entering a sulfur condenser. Since reaction 3 is an equilibrium reaction, conversion to sulfur is improved by removing that product.

This step is followed by a catalytic stage, consisting of a gas reheater, a catalyst chamber, and a condenser to remove the product sulfur as a liquid. This catalytic stage can be repeated one to four times, depending on the level of conversion desired. Most plants are now built with two stages (some jurisdictions require three), since the need for a tail gas cleanup plant makes the third and fourth stages uneconomical. From the condenser of the final catalytic stage, the process stream passes to some form of tail gas treatment process.

The liquid sulfur from the condensers runs through a seal leg into a covered pit, from which it is pumped to trucks or railcars for shipment to end users. Some plants, remote from end users, form molten sulfur into solid pellets or other shapes for shipment. Figure 2 illustrates a simple, two-stage Claus plant.

Claus plants operate at low pressures, rarely higher than 10 psig. Temperatures range from 1800–2800°F in the combustion chamber to 400–600°F in the catalytic reactors. Tail gas from the last condenser is at about 270°F. The catalyst

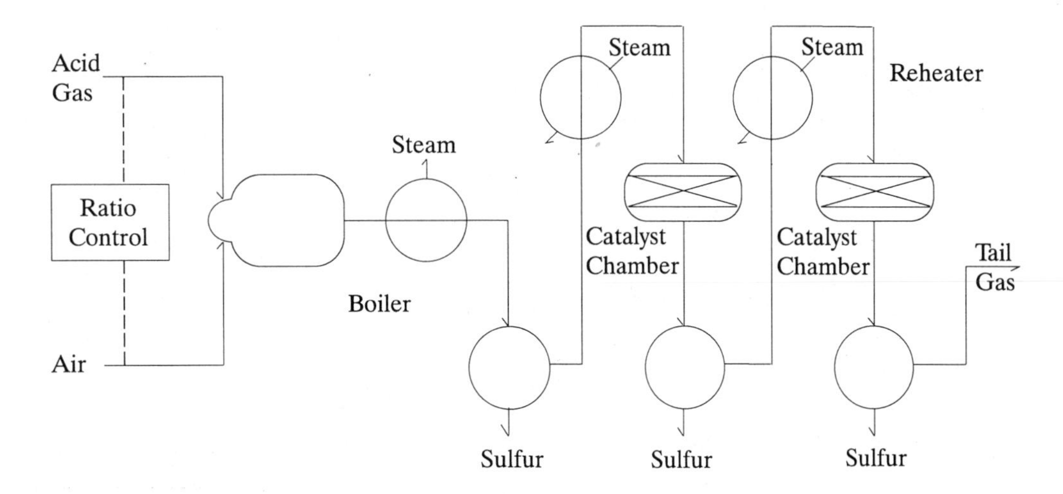

FIGURE 2. Typical Claus Plant—Two Bed

used is alumina, sometimes promoted and sometimes in the form of bauxite.

Claus plants in refineries are frequently used to process gas from the treatment of refinery sour waters. This gas may contain significant quantities of ammonia, as well as H_2S. The ammonia can be efficiently converted to nitrogen and water in the Claus plant, if proper attention is given to maintaining operating conditions in the right range.

Tail Gas Treatment

Gas from the Claus plant still contains 0.8–1.5% of sulfur compounds, which must be removed before the gases can be safely vented to the atmosphere. This is done in a tail gas treatment, or cleanup plant. The number and variety of tail gas cleanup options are beyond our scope here and only the two most effective types will be described. These are the amine scrubbing and the liquid-redox scrubbing processes.

Amine Scrubbing Process

Tail gas from the Claus plant is heated to about 550°F, hydrogen is added (or generated in the heater, if necessary), and the gas is passed through a catalyst bed, where virtually all the sulfur compounds are converted to H_2S; for example:

$$SO_2 + 3H_2 \rightarrow H_2S + 2H_2O + \text{heat} \qquad (4)$$

Other sulfur compounds are similarly hydrogenated. The hot gas is cooled in a waste-heat boiler and passes to a quench column to remove excess water. Figure 3 shows the above reaction section. The cool dry gas stream flows to the bottom of an amine absorber, where the H_2S is removed in the same manner as indicated in the amine section. Acid gas is stripped from the amine, cooled, and recycled to the front of the Claus plant to be converted to sulfur. In its general layout, this amine plane is quite similar to that described above, although many engineering details are different. Gas from the amine absorber is routed to an incinerator (see below) before being vented to the atmosphere. Newer plants, where proper amines are selected and details engineered, may have low enough sulfur contents to be vented directly.

Liquid-Redox Scrubbing Process

Tail gas from the Claus plant is treated essentially as indicated above to convert all sulfur compounds to H_2S and to cool and dry the gas stream and make it ready for final scrubbing. The gas is contacted in a packed tower with a highly alkaline water solution containing an oxidized redox catalyst. Hydrogen sulfide is readily absorbed into the alkaline solution and the cleaned tail gas passes from the absorber directly to the atmosphere.

In the scrubbing solution, the H_2S is oxidized directly to elemental sulfur by the redox catalyst, which is reduced. The sulfur is in the form of tiny particles that must be removed from the solution by settling, centrifuging, or filtering. The redox catalyst is reoxidized by sparging air into the scrubbing solution in an oxidation vessel. The regenerated solution is then recirculated to the absorber for further reaction with H_2S. The chemistry of this process can be summarized as follows:

$$\text{Absorption:} \quad H_2S\ (\text{gas}) \longrightarrow 2H^+ + S^= \qquad (5)$$

$$\text{Oxidation:} \quad S^= + 2V^{+5} \longrightarrow S^O + 2V^{+4} \qquad (6)$$

$$\text{Regeneration:} \quad 2V^{+4} + 2H^+ + 1/2O_2 \longrightarrow 2V^{+5} + H_2O \qquad (7)$$

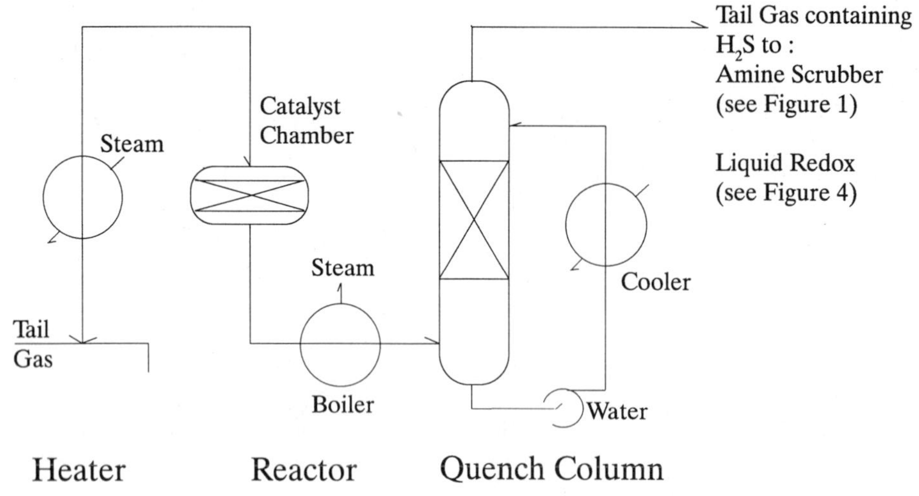

FIGURE 3. Typical Tail Gas Treater, Reaction Section

The redox catalyst is shown as vanadium in equations 6 and 7, but iron is gaining considerable acceptance because of its lower toxicity and lower cost. Figure 4 shows, in simplified form, a typical liquid redox scrubber.

Other Processes

A number of other processes that are used commercially are not discussed here because they would not normally meet the criterion of best available control technology. These include processes that continue equation 3 at temperatures below the sulfur dew point to achieve better equilibrium and oxidation processes in which sulfur compounds are oxidized to SO_2, and recycled to the claus plant.

While not strictly pollution control devices, most sulfur-recovery facilities end by routing the tail gas through a thermal or catalytic incinerator. These incinerators convert the small amount of H_2S remaining in the cleaned tail gas to SO_2.

AIR EMISSIONS CHARACTERIZATION

Air emissions are affected at both ends of the sulfur recovery train. The amine plant sets the sulfur level in the refinery fuel gas system, which controls stack emissions from furnaces. The tail gas treatment plant controls sulfur emissions from the sulfur-recovery complex itself.

Typical refinery amine plants are designed to limit the H_2S content of fuel gases to below 160 ppm, the current Environmental Protection Agency (EPA) limit, and generally are able to clean to well below that level. Uncontrolled emissions from refineries and gas plants could be as high as several thousand parts per million, depending on the raw materials and processing equipment used.

Natural gas plants are able to treat gas to below 4 ppm H_2S, the normal pipeline specification. This is better than accomplished at refineries because of the higher treating pressure, which favors the scrubbing reaction, and the nor-

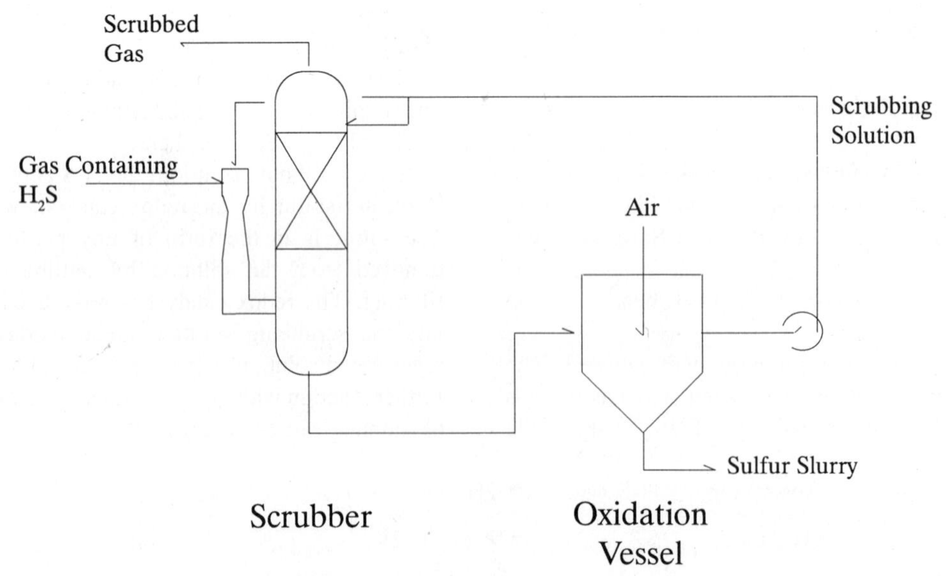

FIGURE 4. Typical Liquid Redox Scrubber

mally lower content of other trace sulfur compounds. The other pollution issue in amine treating plants involves occasional losses of amines into the plant effluent water system.

Older, or very small (less than 20 LTPD sulfur), Claus plants without tail gas cleanup plants influence air pollution directly by their sulfur-recovery efficiency. This is a function of the number of catalytic stages in the plant and the concentration of H_2S in the acid gas fed to the plant. The higher the acid gas concentration and the more stages, the higher will be the recovery of elemental sulfur. Contaminants (e.g., carbonyl sulfide, hydrocarbons, water) in the acid gas stream also reduce the expected recovery efficiency. While it is difficult to generalize, one would expect a two-catalytic-bed Claus plant to recover 94–96% of the inlet sulfur as liquid. The range covers acid gas H_2S concentrations from 35% to 90%. The equivalent recoveries from a three-bed plant are 96–97.5% and from a four-bed plant, 97–98.5%.[4] The liquid sulfur produced from a Claus plant is saturated with H_2S and, unless degassed, some low parts-per-million levels of H_2S and elemental sulfur will be released from pit and storage tank vents.

Tail gas treatment plants are rated by the concentration of sulfur compounds in the treated vent gas, corrected for moisture and air dilution. Amine scrubbing plants normally release 50–250 ppm of sulfur compounds, but new, specifically designed plants may be expected to release no more than 20–25 ppm, and of this, less than 15 ppm is H_2S. Liquid-redox-type plants should release below 40 ppm, although older plants may be designed for less efficient cleanup. Current EPA limits are 250 ppmv.

Both amine and liquid-redox processes are subject to leakage of scrubbing solution into effluent water systems.

AIR POLLUTION CONTROL MEASURES

Measures taken by plant operators to maintain the pollution control efficiency of their sulfur recovery facilities include the following.

Gas Scrubbing

The principal cause of excess emissions from an amine plant is foaming of the amine solution in the absorber. Foaming can be caused by excessive particulate contamination, the presence of liquid hydrocarbons in a gas scrubber, and contamination by surface-active chemicals. Thus the operator's first concern must be for the cleanliness of the amine solution. Strongly acidic compounds in the gas stream, or oxygen that will form acidic compounds, react with the amine to form salts that will not decompose in the stripper. These are called heat-stable salts and will gradually cause the performance of the amine to deteriorate. At levels above 4–5% in the amine solution, heat-stable salts promote corrosion, which can lead to foaming and equipment failure. Amines (and H_2S) are known to cause cracking in the welds of carbon-steel vessels. Care must be exercised in the fabrication and repair of steel equipment for this service and stress relief of vessel welds is required.

Sulfur Recovery

Careful control of the air injection rate into the Claus plant is the chief factor affecting the recovery efficiency. In addition to the front-end control system, most plants have a "tail gas ratio" analyzer to measure the concentrations of H_2S and SO_2 in the gas leaving the last condenser. Poor combustion control can cause the catalyst to "sulfate" and lose activity. Periodic rejuvenations can control this phenomenon temporarily. Because of the toxic nature of H_2S, Claus plants must be equipped with an interlock system to put the plant automatically into a safe condition in case of a critical failure. Some plants remove (degas) the H_2S dissolved in the liquid sulfur produced. This may reduce the emissions from the complex slightly, but its main value is in transportation safety. It is possible to reach explosive levels of H_2S in the vapor space of railcars with undegassed sulfur.

Tail Gas Treatment

The pollution control concerns in the two principal processes discussed include the maintaining of proper air-to-fuel ratios in the heater, adequate hydrogen to accomplish the necessary hydrogenation, and control of the pH of the quench tower water. Concerns in the amine process parallel those in the amine gas scrubbing process, with the extra need to insure adequate stripping of the amine. Amine stripping sets the H_2S leak rate through the absorber.

Emissions from the liquid-redox plants are controlled primarily by maintaining the proper alkalinity and flow rate of the scrubbing solution. Attention must be paid to the cleanliness of the solution as well. The buildup of sulfur particles, catalyst degradation products, and salts can cause mechanical problems, which can interfere with efficient scrubbing. Biological infestations also can cause foaming problems. Most plants purge a small stream of scrubbing solution from the plant to keep the salt levels under control.

Incinerators must operate at a temperature of 1200°F or higher if all the H_2S is to be combusted. Proper air-to-fuel ratios are needed to eliminate pluming from the incinerator stack. Stacks should be equipped with analyzers to monitor the SO_2 level.

References

1. "The United States sulphur industry," *Sulphur,* 209:17 (1990).
2. "Control: The key to successful Claus sulphur recovery," *Sulphur,* 202:24 (1989).
3. J. B. Hyne, "Optimum furnace configuration for sulphur recovery units," *Sulphur,* 198:24 (1988).
4. H. G. Paskall, *Capability of the Modified Claus Process,* Western Research, Calgary, 1979, p 172.

Bibliography

A. L. Kohl and F. C. Riesenfeld, *Gas Purification*, 4th ed., Gulf Publishing Co., Houston, 1988.
U. H. F. Sander, H. Fischer, et al., *Sulphur, Sulphur Dioxide and Sulphuric Acid*, British Sulphur Corp. Ltd., London, 1984.
"Sweetening of natural gas: I," *Sulphur*, 192:30 (1987).
"Sweetening of natural gas: II," *Sulphur*, 193:30 (1987).

SYNTHETIC RUBBER

Excerpted from AP-42, U.S. Environmental Protection Agency

Two types of polymerization reaction are used to produce styrene butadiene copolymers, the emulsion type and the solution type. This discussion addresses volatile organic compound (VOC) emissions from the manufacture of copolymers of styrene and butadiene made by emulsion polymerization processes. The emulsion products can be sold in either a granular solid form, known as crumb, or in a liquid form, known as latex.

Copolymers of styrene and butadiene can be made with properties ranging from those of a rubbery material to those of a very resilient plastic. Copolymers containing less than 45 wt % styrene are known as styrene-butadiene rubber (SBR). As the styrene content is increased over 45 wt %, the product becomes increasingly more plastic.

PROCESS DESCRIPTION

Emulsion Crumb Process

As shown in Figure 1, fresh styrene and butadiene are piped separately to the manufacturing plant from the storage area. Polymerization of styrene and butadiene proceeds continuously through a train of reactors, with a residence time in each reactor of approximately one hour. The reaction product formed in the emulsion phase of the reaction mixture is a milky white emulsion called latex. The overall polymerization reaction is not carried out beyond a 60% conversion of monomers to polymer, because the reaction rate falls off considerably beyond this point and product quality begins to deteriorate.

Because recovery of the unreacted monomers and their subsequent purification are essential to economical operation, unreacted butadiene and styrene from the emulsion crumb polymerization process normally are recovered. The latex emulsion is introduced to flash tanks where, using vacuum flashing, the unreacted butadiene is removed. The butadiene is then compressed, condensed, and pumped back to the tank-farm storage area for subsequent reuse. The condenser tail gases and noncondensibles pass through a butadiene adsorber/desorber unit, where more butadiene is recovered. Some noncondensibles and VOC vapors pass to the atmosphere or, at some plants, to a flare system. The latex stream from the butadiene-recovery area is then sent to the styrene-recovery process, which usually takes place in perforated-plate, steam-stripping columns. From the styrene stripper, the latex is stored in blend tanks.

From this point in the manufacturing process, latex is processed continuously. The latex is pumped from the blend tanks to coagulation vessels, where dilute sulfuric acid (H_2SO_4 of pH 4 to 4.5) and sodium chloride solution are added. The acid and brine mixture causes the emulsion to break, releasing the styrene-butadiene copolymer as crumb product. The coagulation vessels are open to the atmosphere.

Leaving the coagulation process, the crumb and brine-acid slurry is separated by screens into solid and liquid. The crumb product is processed in rotary presses that squeeze out most of the entrained water. The liquid (brine/acid)

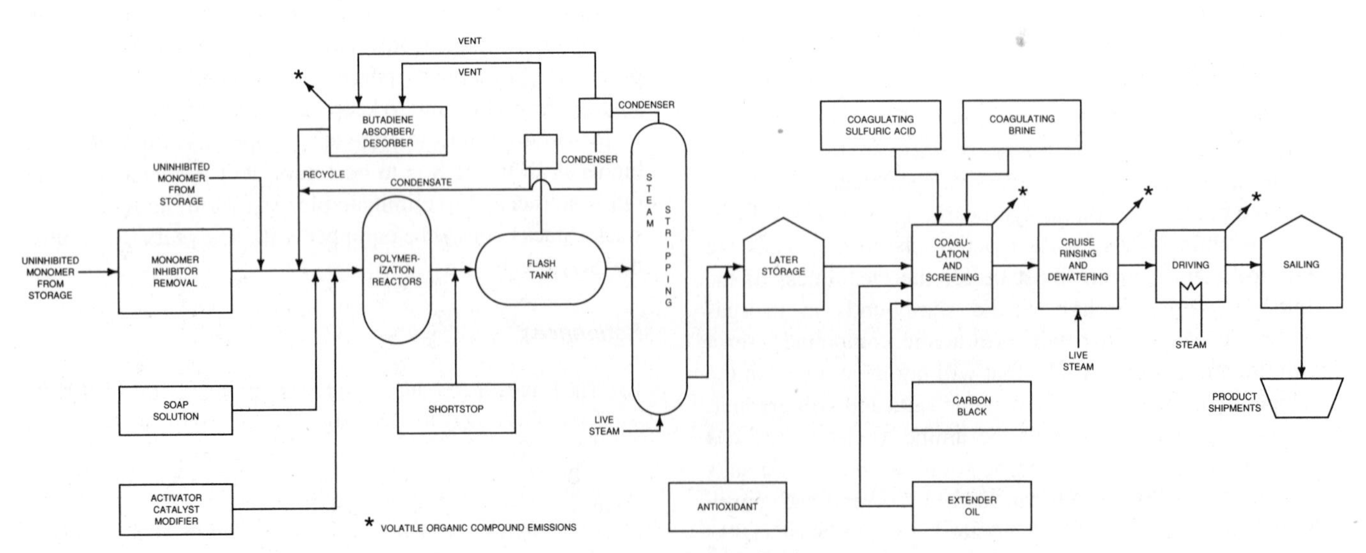

FIGURE 1. Typical Process for Crumb Production by Emulsion Polymerization

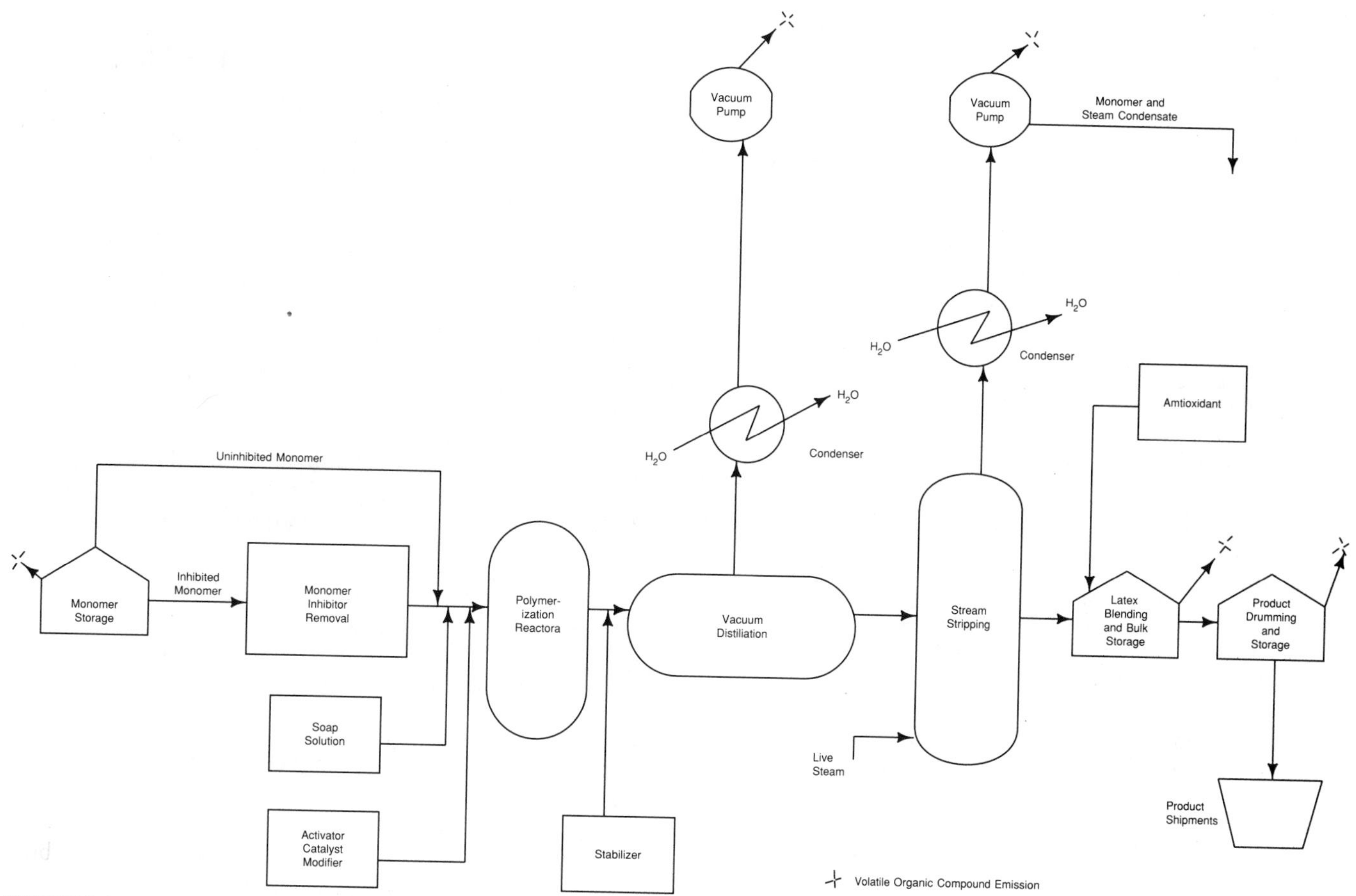

FIGURE 2. Typical Process for Latex Production by Emulsion Polymerization

from the screening area and the rotary presses is cycled to the coagulation area for reuse.

The partially dried crumb is then processed in a continuous belt dryer that blows hot air at approximately 93°C (200°F) across the crumb to complete the drying of the product. Some plants have installed single-pass dryers, where space permits, but most plants still use the triple-pass dryers that were installed as original equipment in the 1940s. The dried product is baled and weighed before shipment.

Emulsion Latex Process

Emulsion polymerization can also be used to produce latex products. These latex products have a wider range of properties and uses than do the crumb products, but the plants are usually much smaller. Latex production, shown in Figure 2, follows the same basic processing steps as emulsion crumb polymerization, with the exception of final product processing.

As in emulsion crumb polymerization, the monomers are piped to the processing plant from the storage area. The polymerization reaction is taken to near completion (98–99% conversion), and the recovery of unreacted monomers, therefore, is uneconomical. Process economy is directed toward maximum conversion of the monomers in one process trip.

Because most emulsion latex polymerization is done in a batch process, the number of reactors used for latex production is usually smaller than for crumb production. The latex is sent to a blow-down tank where, under vacuum, any unreacted butadiene and some unreacted styrene are removed from the latex. If the unreacted styrene content of the latex has not been reduced sufficiently in the blow-down step to meet product specifications, the latex is introduced to a series of steam-stripping steps to reduce the content further. Any steam and styrene vapor from these stripping steps are taken overhead and sent to a water-cooled condenser. Any uncondensibles leaving the condenser are vented to the atmosphere.

After discharge from the blow-down tank or the styrene stripper, the latex is stored in process tanks. Stripped latex is passed through a series of screen filters to remove unwanted solids and is stored in blending tanks, where antioxidants are added and mixed. Finally, latex is pumped from the blending tanks to be packaged into drums or to be bulk loaded into railcars or tank trucks.

TABLE 1. Emission Factors for Emulsion Styrene-Butadiene Copolymer Production[a]

Process	Volatile Organic Emissions[b]	
	g/kg	lb/ton
Emulsion crumb		
Monomer recovery, uncontrolled[c]	2.6	5.2
Absorber vent	0.26	0.52
Blend/coagulation tank, uncontrolled[d]	0.42	0.84
Dryers[e]	2.51	5.02
Emulsion latex Monomer removal		
Condenser vent[f]	8.45	16.9
Blend tanks		
Uncontrolled[f]	0.1	0.2

[a]Nonmethane VOC, mainly styrene and butadiene. For emulsion crumb and emulsion latex processes only.
[b]Expressed as units per unit of copolymer produced.
[c]Average of three industry-supplied stack tests.
[d]Average of one industry stack test and two industry-supplied emission estimates.
[e]No controls available. Average of three industry-supplied stack tests and one industry estimate.
[f]EPA estimates from industry-supplied data, confirmed by industry.

AIR EMISSIONS CHARACTERIZATION

Emission factors for emulsion styrene-butadiene copolymer production processes are presented in Table 1.

AIR POLLUTION CONTROL MEASURES

In the emulsion crumb process, uncontrolled noncondensed tail gases (VOCs) pass through a butadiene absorber control device, which is 90% efficient, to the atmosphere or, in some plants, to a flare stack.

No controls are employed at present for the blend tank and/or coagulation tank areas on either crumb or latex facilities. Emissions from dryers in the crumb process and the monomer-removal part of the latex process are not currently subject to control devices.

Individual plant emissions may vary from the average values listed in Table 1 with facility age, size, and plant modification factors.

Bibliography

Control Techniques Guideline (Draft), U.S. Environmental Protection Agency Contract No. 68-02-3168, GCA, Inc., Chapel Hill, NC, April 1981.

Emulsion Styrene-Butadiene Copolymers: Background Document, U.S. Environmental Protection Agency Contract No. 68-02-3063, TRW Inc., Research Triangle Park, NC, May 1981.

C. Fabian, U.S. Environmental Protection Agency, Research Triangle Park, NC, confidential written communication to Styrene-Butadiene Rubber File (76/15B), July 16, 1981.

NATURAL FIBER TEXTILE INDUSTRY

Anthony J. Buonicore, P.E.

Textiles are any product derived from the processing of natural fibers such as wool and cotton, and/or from the processing of fibers synthesized and extruded from petrochemicals and modified wood pulp, such as polyester, nylon, rayon, and acetate. The processing that the raw fibers receive functions to remove natural impurities (in the case of cotton and wool) and to impart particular qualities of appearance, touch, and durability. The textile industry consists of companies that process natural and/or synthetic fibers into yarns, fabrics, and certain consumer products. The production of synthetic fibers, however, is not included in this discussion as the nature of such plants is more closely identified with that of organic chemical plants.

The activities of the natural fiber textile industry (excluding fiber production) typically include:

- The production of yarns by spinning
- The production of fabrics by weaving, knitting, and other processes
- The dyeing and finishing of fibers, yarns, and fabrics
- The production of clothing, household linens, curtains, carpets, and other consumer products

The textile "complex" is shown schematically in Figure 1.[1] The industry uses many different raw materials and chemicals to produce a great variety of products. Many of the chemicals are not retained in the final textile product and are either recovered for reuse, discarded in the plant effluent, or driven off into the atmosphere.

PROCESS DESCRIPTION

The classical method of categorizing the industry involves grouping the manufacturing plants according to the fiber being processed, that is, cotton, wool, or synthetics. The modern approach to textile industry categorization, however, involves grouping the manufacturing plants according to their particular operation. Two broad categories can be identified: low-water-use processing and wet processing. Low-water-use processing includes the manufacture of yarns and unfinished fabrics ("greige" goods). Wet processing includes several categories according to the processes and raw materials employed in the production of the finished product. For example, such categories might include:

- Wool scouring
- Wool finishing
- Woven fabric dyeing and finishing
- Knit fabric dyeing and finishing
- Carpet finishing

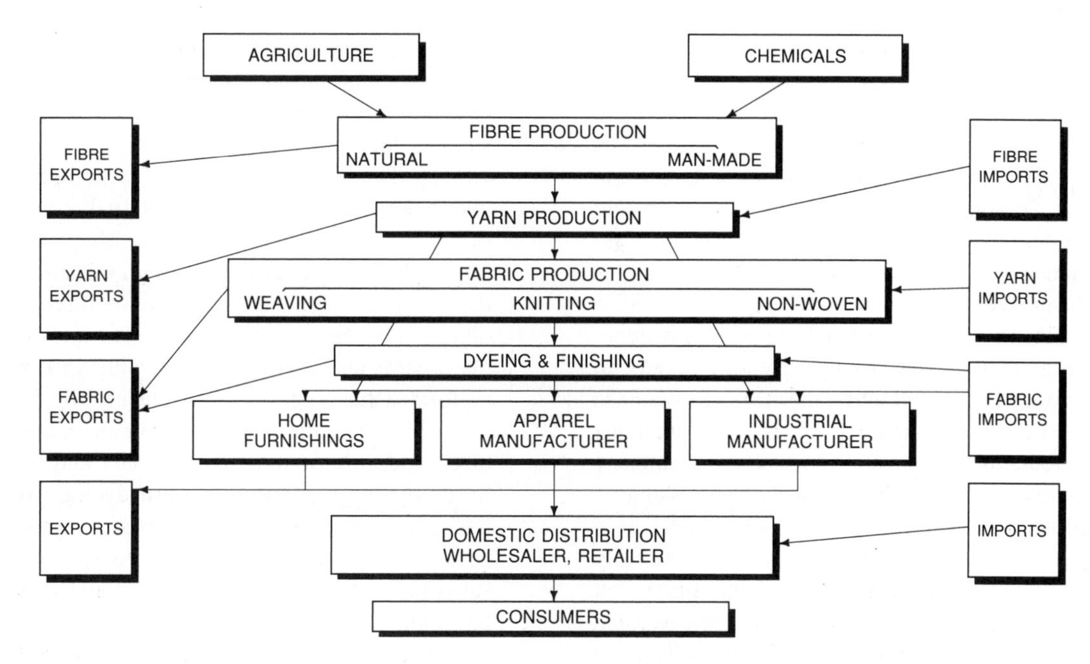

FIGURE 1. Schematic Diagram of the Textile Complex (Courtesy Canadian Textile Institute)

- Stock and yarn finishing
- Nonwoven manufacturing
- Felted fabric finishing

The low-water-use process unit operations consist of carding, spinning, slashing (sizing), weaving, knitting, and tufting. The wet process unit operations include singeing, desizing, scouring, wool carbonizing, wool fulling, cotton mercerizing, bleaching, dyeing/printing, and finishing. These processes are briefly described below.[1]

Low-Water-Use Processing Unit Operations

a. **Carding** is the preliminary process in spun yarn manufacture. The fibers are separated, distributed, equalized, and formed into a thin web and condensed into a continuous, untwisted strand of fibers called a sliver. This process also removes impurities and a certain amount of short, broken, or immature fibers. The operation is performed on a card.
b. **Spinning** is the process of making yarn from fibers by a combined drawing out and twisting operation or from filament tow by the combination of cutting/breaking with drafting and twisting in a single series of operations.
c. **Slashing** (sizing) is the process of sizing warp yarns on a slasher to protect the yarns against injury during weaving. It also gives temporary added strength and abrasion resistance to the yarns, enabling them to be processed successfully for weaving. The entire warp is coated by immersion in the sizing material that may contain starch, polyvinyl alcohol (PVA), wax, gelatin oils, and/or other materials, and is dried before winding on the beam of the loom prior to weaving.
d. **Weaving** is the process of interlacing two yarns of similar materials so that they cross each other at right angles to produce woven fabric. The three basic weaving patterns are plain, twill, and satin.
e. **Knitting** is the process of constructing fabric by an interlocking series of loops of one or more yarns. The three classes of knitted fabrics are circular knit, flat knit, and warp knit.
f. **Tufting** is the process of making carpets and involves a wide multiple-machine needle process that sews pile yarns to a broad fabric backing.

Wet Processing Unit Operations

a. **Singeing** removes protruding surface fibers to reduce pilling. The fabric passes over an impinging flame, which burns the fibers from the fabric surface.
b. **Desizing** removes the sizing compounds applied to yarns to impart tensile strength. The starch sizing compounds are solubilized with alkali, acid or enzyme, and the fabric is washed thoroughly. Alkaline desizing utilizes a weak alkaline solution to facilitate size removal, while acid desizing employs a dilute acid solution to hydrolyze the size and render it water soluble. Enzyme desizing utilizes vegetable or animal enzymes to decompose size (usually only starch size) to a water-soluble form. After solubilizing the size, the fabric is rinsed clean.
c. **Scouring** removes natural and acquired impurities from fibers and fabric. Synthetic fibers require less scouring than does cotton or light wool. Scouring agents include detergents, soaps, and various assisting agents, such as alkalis, wetting agents, foamers/defoamers, and lubri-

cants. After scouring, the goods are thoroughly rinsed (or washed) to remove excess agents.

d. **Wool carbonizing** removes burrs and other vegetable matter from loose wool or woven fabric goods. The process consists of acid impregnation, baking, and mechanical agitation. A dilute solution of sulfuric acid is used to degrade cellulosic impurities to hydrocellulose without damaging the wool. The excess acid is squeezed from the wool and the wool is baked to oxidize the contaminants to gases and a solid carbon residue. The material then passes through pressure rollers to crush the solid residue and into a mechanical agitator to shake loose the crushed material. The acid content in the material remains high after agitation, requiring neutralization and rinsing before further processing.
e. **Wool fulling** gives woven woolen cloth a thick, compact feel, finish, or appearance. Cloth is mechanically worked in fulling machines to make the fibers felt together, causing shrinkage, area weight increase, and concealment of the woven threads of the cloth. The two common methods of fulling are acid and alkali. In acid fulling, an aqueous solution of sulfuric acid, hydrogen peroxide, and small amounts of metallic catalyst are used. In alkali fulling, soap or detergent, sodium carbonate, and a sequestering agent are added to provide the required lubrication. Acid fulling is usually followed by alkali fulling.
f. **Mercerizing** is usually used in cotton processing to increase the tensile strength, luster, dye affinity, and abrasion resistance of the goods. In this operation, the cotton, usually in fabric form, is impregnated with a cold sodium hydroxide solution. The solution causes the cotton fibers (cellulose) to swell. After the desired contact period, the alkali is thoroughly washed out, sometimes with the use of a dilute acid bath to ensure neutralization. It has been normal practice to recover the caustic solution for concentration and reuse in scouring or mercerization.
g. **Bleaching** is a common process used to whiten cotton, wool, and some synthetic fibers by removing the natural coloring. It is usually performed after scouring and prior to dyeing or printing. Bleaching chemicals include sodium hypochlorite, hydrogen peroxide, and sodium perborate, as well as optical brighteners. Batch bleaching is done in kiers (continuous processes use J-boxes) where fabric is stacked for a given period to allow the chemical to work before goods are withdrawn from the bottom of the box. Bleaching is followed by thorough rinsing.
h. **Dyeing** can be performed in the stock, yarn, or fabric state, and single- or multiple-fiber types can be dyed. Multiple-fiber types may require multiple or sequential steps.

   Stock dyeing is performed before the fiber is converted to the yarn state and can be a batch or continuous process. The batch process consists of placing the stock in a vat or pressure kettle, applying a sufficient amount of dye solution, providing optimum environmental conditions, allowing time for dye fixation, and rinsing. The stock is then processed into yarn.

   Yarn dyeing is performed on yarns used for woven goods, knit goods, and carpets. Usual methods include skein, package, and space dyeing. Skein dyeing is performed by placing turns of yarn on a frame and putting the frame in a dye bath under the required dyeing conditions. Package dyeing is the most common yarn dyeing process and is effected by placing yarn wound onto perforated tubes on a frame, placing the frame into a pressure vessel, circulating dye liquor in and out of the tubes under the required conditions, and then rinsing. Space dyeing is a specialty yarn dyeing process. Dye liquor is applied to yarns at a repeat or random interval by a roller-type dye pad. The dyed yarn then enters a subsequent step(s) for development and fixation of the color and is rinsed.

   Fabric dyeing is the most common method in use today because it can be continuous or semicontinuous, as well as a batch process. Methods employed include beck (winch), jet, jig, beam, and continuous range. Beck dyeing is accomplished with the fabric in rope form in either atmospheric or pressure vessels. The fabric is connected end-to-end and rotated over a large drum through the dye liquor. The length of the fabric is such that the fabric lies in a heap at the bottom of the vessel for a short time. As with all dyeing methods, the proper conditions of time and chemicals must be provided.

   Jet dyeing is also accomplished with the fabric in the rope form. Jet machines are similar to pressure becks, except that the cloth passes through a venturi tube, which helps transport the fabric. A pump circulates the dye liquor through the tubes and suction at the venturi causes the fabric to rotate. Jet machines have improved on certain deficiencies of beck dyeing by providing faster dyeing times, lower liquor-to-cloth ratios, reduced tangling, more uniform temperatures, and reduced elongation of fabric due to tension.

   Jig dyeing is performed with the fabric in an open-width configuration. Both atmospheric and pressure equipment are available. The fabric is wound onto rollers that pass through a shallow trough containing the dye liquor. The rollers rotate in clockwise and counterclockwise directions alternately and guide rolls ensure complete immersion of the cloth. Only a few meters of cloth are immersed at a time and it is possible to work with a low liquor-to-cloth ratio.

   Continuous dyeing also is performed with the fabric in an open width on what are termed ranges. The ranges consist of a number of dip troughs, from which dyes and chemicals are applied; drying and fixation sections; wash boxes to remove excess dye liquor; steam chambers; and a final drying step.

Beam dyeing is a batch process involving the use of a kettle where open-width fabric is wound onto a perforated drum with dye solution circulated through the material in a manner similar to package yarn dyeing. This is unlike the other fabric dyeing processes described, in which the fabric is circulated through the dye solution.

Thermosol dyeing is a continuous process normally used for dyeing polyester and polyester–cotton blends. Dye is applied to the fabric in the pigment form and dried, causing a film of pigment to adhere to the fabric. The fabric is then heated to 350–400°F (near the melting point of polyester), causing the synthetic fibers to swell, and as the dispersed dye sublimes, it penetrates the polyester. The cotton portion is dyed by passing the fabric through an additional chemical pad and steamer. The fabric is then washed and dried.

i. **Printing** is similar to dyeing, except that print color is applied to specific areas of the cloth to achieve a planned design instead of the whole cloth being colored. Dyes and dyeing assists are similar to those used in fabric dyeing; however, the color application techniques are quite different. Textiles are usually wet-printed by roller, rotary screen, or flatbed screen methods.

Roller printing is accomplished by applying print paste to an etched or engraved roller and transferring the design to the fabric by contact with the rollers. After printing, the pattern is fixed by steaming, aging, or some other treatment.

Screen printing differs from roller printing in that the print paste is transferred to the fabric through openings in specially designed screens. The process can be manual, semiautomatic, or completely automatic. Automatic screen printing can be either flatbed or rotary, while manual and semiautomatic are flatbed processes only.

Screens are made by manually or photographically transferring the design onto the screen material. The area outside the design is specially treated so that the print paste is retained. The screens are securely stretched over a frame for correct positioning. A separate screen is required for each color in the design. Modern techniques for rotary screen production include the use of laser engravers, which etch the cotton in a coated screen.

In manual screen printing, the fabric is placed on a long table, the screen frames correctly positioned for the pattern, and the selected print paste forced through the screen mesh onto the fabric by squeegee. The fabric is then dried and given other finishing treatments. In the semiautomatic process, the fabric travels and the screens remain in place. Handling of the screens and application of the print paste are performed manually.

Automatic flatbed screen printing takes place on a machine that electronically performs and controls each step. The process is continuous in that the fabric moves along a table, the screens are positioned, and print paste is applied and squeezed through the screen onto the fabric. The fabric moves forward one frame between each screen, and when all color has been applied, it moves into a steaming and/or drying box.

In rotary screen printing, the color is applied to the fabric through lightweight metal-foil screens that resemble the cylinder rollers of the roller printing process. The fabric moves continuously under the cylinder screens and print paste is forced from the inside of the screens through and onto the fabric. A separate screen is required for each color in the design.

j. **Finishing** operations are included as wet processes because the majority of treatments are applied in the liquid state. Wet finishing involves the application of a wide range of chemicals to add properties to a fabric. Finishes can be applied, for example, to make a fabric wrinkle resistant, crease retentive, water repellant, flame resistant, mothproof, mildew resistant, bacteriostatic and/or stain resistant. Application is usually by means of a pad that contains immersion rollers in a pan with finish and a set of nip rolls that squeeze the finish excess and assure an even finish application. The fabric is then dried to cure the finish onto the fabric.

Synthetic resins are most often used to achieve wrinkle resistance and crease retention (permanent press). The resins are adhesive and are permanently cross-linked with the fibers by curing with heat and a catalyst. Silicones and other synthetic materials are used to provide water repellency. The silicones successfully repel oily fluids as well. Flame-resistant finishes are applied to cellulosic fabrics (cotton) to prevent them from supporting combustion.

Mothproofing finishes are typically applied to wool and other animal hair fibers to make them unfit as food for moth larva. The growth of mildew, mold, rot, and fungus is inhibited by the application of compounds that usually contain chlorinated phenols, metallic salts or zinc, copper, or mercury. Selected hygienic additives can also be used to inhibit the growth of bacteria.

Soil-release finishes usually contain organosilicone compounds, fluoro compounds, or oxazoline derivatives. These compounds allow stains to be removed by ordinary washing.

Cotton Processing

Cotton processing consists of two basic steps—weaving (greige manufacturing) and finishing (wet processing) (refer to Figure 2).

Greige manufacturing entails mostly dry operations, as follows:

1. Trash and foreign matter are manually or mechanically removed from the raw fibers (opening, cleaning, carding, combing).

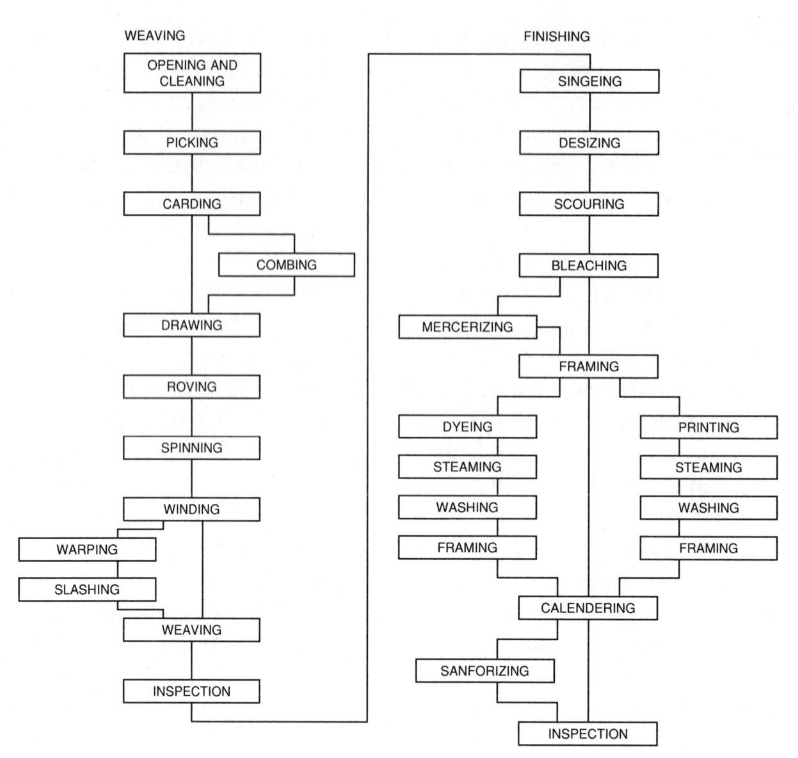

FIGURE 2. Cotton Processing Flow Diagram

2. The fibers are joined, straightened, drawn into yarn, and wound on spools (drawing, roving, spinning, winding).
3. The yarn is run through a starch solution (sizing or slashing) and then dried so that it has the strength and stiffness required to withstand the abrasion and friction generated in the weaving operation. Yarn for knitting is waxed to provide lubrication for the knitting machine and reduce yarn-to-metal friction.
4. The strengthened or waxed yarn is woven into cloth (greige goods).

The finishing plant receives the greige goods and must process them to satisfy various client demands. Finishing is essentially wholly of the wet type, wherein:

1. Natural impurities (wax, pectins, alcohols, etc.) and acquired impurities (unremoved size agent, dirt, oil, grease, etc.) must be removed to make the cloth suitable for chemical processing.
2. Wet chemical processing is performed to impart to the cloth the desired properties and appearance.

Wool Processing

As in the case of cotton, wool must undergo both weaving (greige manufacturing) and finishing operations; the former is necessary in order to develop a continuous cloth from the bits of raw fiber, and the latter in order to impart the desired characteristics to the cloth.

Greige manufacturing entails numerous mechanical and manual dry operations (see Figure 3). Only two operations are of the wet type (scouring and slashing).

Greige operations include sorting and blending of raw fibers, scouring (to remove natural grease and foreign matter), drying, carding (opening and paralleling of fibers), rewashing, oiling (for reducing static electricity), gilling (paralleling of fibers), combing (to remove trash), roving (or drawing of fibers into rope form), spinning (or further drawing of fibers into yarns), slashing (starch sizing of wool yarn to protect against abrasion and friction generated in the weaving operation), weaving (yarns are woven into cloth), and mending. Scouring is performed by detergents or solvents.

Dyes and Chemicals

Dyes used by the textile industry are classified according to the application and include:

a. **Acid dyes** are sodium salts, usually of sulfonic acids or carboxylic acids, and are the main class of dyes used in wool dyeing and for dyeing nylon.
b. **Direct dyes** resemble acid dyes in that they are sodium salts of sulfonic acids and are often azo compounds. They are used primarily for the dyeing of cellulosic fibers (e.g., cotton).
c. **Basic dyes** are usually hydrochlorides or salts of organic bases. The pigment is found in the cation and the dyes are often referred to as cationic dyes. They are mainly used on acrylic fibers.

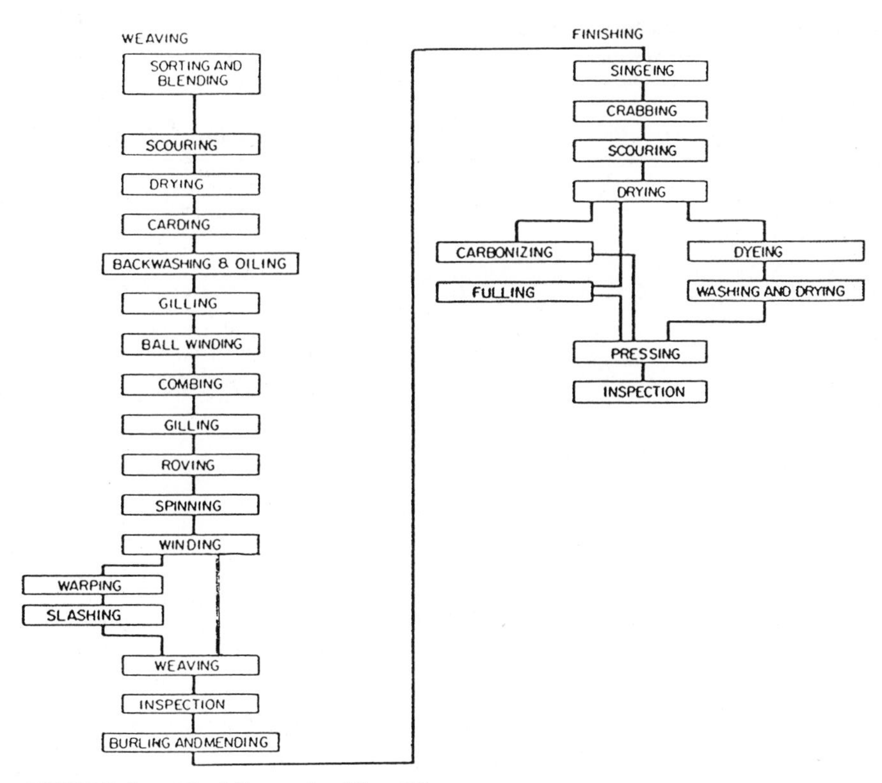

FIGURE 3. Wool Processing Flow Diagram

d. **Disperse dyes** are suspensions of finely divided organic compounds with very slight aqueous solubility. These dyes were developed to dye hydrophobic fibers such as cellulose acetate and the synthetic fibers.

e. **Mordant dyes** include natural and synthetic dyes. They have no natural affinity for textile fibers, but can be applied if the fiber has been mordanted with a metallic oxide. The most commonly used mordant is chrome and these dyes can be referred to as chrome dyes. The dyes are most commonly applied in a boiling acid dye bath after application of the chrome mordant to the wool. Alternatively, the order of application can be reversed.

f. **Premetallized dyes** are a variation on mordant dyes with the metallic oxide, again usually chrome, already included in the formulation. This eliminates the need for addition of the dichromate as a second step in the process.

g. **Reactive dyes** are generally referred to as fiber-reactive dyes. They react chemically with functional groups on the fiber, and because of this feature, possess good wash-fastness properties. They are primarily applied to natural and some synthetic fiber types (e.g., cotton, rayon, linen, wool, and silk). Fiber-reactive dyes are applied either by exhaust or continuous methods. A common method, utilizing a continuous system, is to impregnate the fiber with the dye and an alkali followed by steaming to promote the reaction between fiber and dye. Most fiber-reactive dyes are used to dye cellulose.

h. **Sulfur dyes** are complex organic compounds that contain sulfur linkages within the molecules. They are usually insoluble in water, but will form soluble components in the presence of a reducing agent such as sodium sulfide. The soluble components have an affinity for cellulose (cotton), and once on the fiber, are oxidized back to the original insoluble form. These dyes have good resistance to washing and moderate resistance to sunlight, and are not very bright.

i. **Vat dyes** are made from indigo, anthraquinone, and carbazol, and are used mainly on cotton, linen, and rayon. They are insoluble in water, but when treated with reducing agents, they are converted into soluble compounds in the presence of alkali. The soluble compounds have an affinity for cellulose and reoxidize to the insoluble pigment within the fiber when exposed to air.

Auxiliaries typically recommended for satisfactory dyeing are listed in Table 1.[1] Other chemicals used in different textile processes are listed in Table 2.[1]

AIR EMISSIONS CHARACTERIZATION

Possible emissions from textile processing include:

a. Oil mists and organic emissions produced when textile materials containing knitting and lubricating oils, plasticizers, and other materials that can volatilize or be

TABLE 1. Auxiliaries Required for Satisfactory Dyeing

Chemical Auxiliary	Type of Dye								
	Acid	Direct	Basic	Disperse	Mordant	Premetallized	Reactive	Sulfur	Vat
Acetic acid	X		X	X	X	X	X	(or alternative acids)	X
Ammonium acetate	X								
Ammonium phosphate	X								
Ammonium sulfate	X				X	X			
Aromatic amines		X							
Buffer							X		
Defoamer				X					
Dispersing agents				X					
Formic acid	X		X		X	X			
Gelatin									X
Hydrochloric acid		X							
Hydrogen peroxide								X	X
Leveling or retarder agents	X		X	X					X
Oxalic acid			X	X					
Penetrating agents					X	X			
Potassium dichromate					X	X			
Sequestering agents		X		X					
Sodium acetate			X						
Sodium carbonate		X					X	X	
Sodium chloride		X					X	X	
Sodium dichromate					X	X		X	
Sodium hydrosulfite									X
Sodium hydroxide							X		X
Sodium nitrate		X							
Sodium sulfate (Glauber's salt)	X		X		X	X			
Sodium sulfide								X	
Soluble oil									X
Sulfuric acid	X				X	X			
Urea							X		
Wetting agent		X							

thermally degraded into volatile substances, are subjected to heat. Processes that can be the sources of oil mists include tentering, calendaring, heat setting, drying, and curing.

b. Acid mists produced during the carbonizing of wool.
c. Solvent vapors released during and after solvent processing operations such as dry cleaning.
d. Dust and lint produced by the processing of natural fibers and synthetic staple prior to and during spinning, as well as by napping and carpet shearing.

AIR POLLUTION CONTROL MEASURES

Emissions from finishing operations, including drying, curing, and heat setting, have been and continue to be one of the more significant air pollution problems in the textile industry. The blue haze and odor that are characteristic of these emissions have posed both technical and economic problems for large and small facilities alike. The amount of air pollution depends on the finishing processes used. Various kinds and amounts of oils and finishing resins can vaporize from the cloth in tenter frames (commonly used for drying, curing, and heat setting) operating at temperatures typically in the 300–400°F range. Some exhausts also can carry a significant amount of lint.

There are process modifications that can reduce some of this air pollution. For example, most of the oils can be removed by effectively prescouring the cloth. Alternatively, the selection of "less polluting" finishing chemicals may help, but cloth finishing quality and economics often limit the ability of such changes to correct the problem. Over the longer term, new technology, such as solvent-free, radiation-curable coatings, may prove to be a factor in overcoming the air pollution problem.[2] Such technology not only would eliminate the problem, but would result in significant energy savings, higher line speeds, and greatly reduced floor space. Unfortunately, lack of availability of radiation-curable coatings and high cost have been the major limitations to commercializing this technology.

Historically, the solution to the characteristic blue haze and odor problem has been the use of air pollution control equipment.[3–6] Wet (two-stage) electrostatic precipitators, fabric-type (glass fiber–wool) mist eliminators, and fume incinerators have traditionally been the equipment of

TABLE 2. Other Chemicals Used in Different Textile Processes

Textile Process	Chemicals: Major	Chemicals: Minor
Printing	Dyestuffs	Ammonium
	Pigments	Sodium alginate
	Gums/thickeners	Soda ash
	Solvents	Hypochlorite
	Alcohols	Soaps and detergents
Bleaching	Hydrogen peroxide	Sodium phosphates
	Sodium hypochlorite	Sodium carbonate
	Sodium chlorite	Sulfur dioxide
	Sodium silicate	Sodium bisulfite
	Sodium hydroxide	Epsom salt
	Organic stabilizers	Calcium carbonate
	Surfactants, detergents	Sodium nitrite
	Chelating agents	
Slashing	Cornstarch	
	Polyvinyl alcohol	
	Carboxymethyl cellulose	
	Soaps	
Spinning	Wool oil	
Carpet finishing	Latex	
	Calcium carbonate	
	Aluminum hydrate	
	Thickeners	
	Styrene-butadiene rubber	
Desizing	Enzymes	Sodium persulfate
	Sodium hydroxide	Sodium bromite
	Hydrogen peroxide	Potassium perphosphate
	Surfactants, detergents	Glauber salt
	Chelating agents	Sodium chloride
	Sodium carbonate	Potassium chloride
	Epsom salt	Calcium chloride
Scouring	Sodium hydroxide	Sodium silicate
	Surfactants, detergents	Solvents
	Chelating agents	Salt
	Sodium phosphates	
	Sodium carbonate	
Fiber preparation	Oleic acid	
Special finishing	Softening agents (oils, waxes, soaps)	
	Stiffeners (PVA, ethyl cellulose)	
	Permanent press (resins)	
	Shrinkproofing (cross-linking resins)	
	Oil repellents (perfluorofatty acids and acrylates)	
	Water repellents (silicone resins)	
	Antistatic finishes	
	Fire retardants	
	Bactericides (trichlorophenol)	
	Mothproofer (dieldrin)	

choice. Wet precipitators and mist eliminators have been plagued by continuous and costly operation and maintenance problems. In general, their performance has been strongly dependent on the effectiveness of gas stream cooling prior to treatment in the control device. Experience has shown that temperatures generally should not exceed 120°F. Fume incinerators operating at 1200–1400°F with approximately a 0.3-second gas residence time have been effective in eliminating blue haze and odor problems; however, capital and operating costs may have a significant impact on the facility.

Air pollution problems resulting from greige processing

of cotton consist predominantly of dust generation in the high-speed mechanical operations, such as spinning, drawing, carding, and twisting. Dry dust collectors of the vacuum type are well suited to maintain a dust-free working environment.

Finely divided lint is present in the air in the vicinity of various operations associated with spinning the yarn and weaving the cloth. Most modern mills have air tunnels built under the floor to effect a downward flow of air around the machines that produce the largest volumes of waste lint (carding, roving, spinning, twisting, weaving, etc.). In the opening and picking rooms in a cotton plant, stock is usually handled by overhead ductwork. Lint and dust are collected by the vacuum system as close as possible to their point of origin. Automatic traveling vacuum cleaning systems are frequently used in conjunction with the underfloor system. The suspended lint and dust are conveyed by air movement through ducts to a centrally located filter system for removal. Automatic dry-type traveling disposable air filters are extensively used for this purpose. Since any filtering medium tends to hold the lint fibers tenaciously, a low-cost, disposable paper medium is often used.

The fumes from wool carbonizing pose another problem that must be dealt with. This process generates very fine carbon particles that appear as smoke, as well as some fumes and odors. These fumes probably include residual sulfur oxides (if sulfuric acid has been used for carbonizing), in addition to organic decomposition products, and are generally very corrosive. Corrosion-resistant stainless steels and/or plastics have been used in the collection systems, which usually exhaust the fumes to the atmosphere. Since a significant quantity of particulate matter is in the submicron range, a visible emission is usually apparent unless there is a very efficient air pollution control system. The severity of the opacity problem will depend to a large extent on location, topography, and local meteorological conditions. Incinerators, wet scrubbers, and fabric filters have been used to resolve the problem. Wet scrubbing systems have the added advantage of reducing residual gases such as sulfur oxides, as well as particulate emissions.

References

1. E. C. Chen, *Environmental Assessment of the Canadian Textile Industry,* Environment Canada Report EPS 5/TX/1, June 1989.
2. *Symposium Proceedings: Textile Industry Technology* (December 1978, Williamsburg, VA), F. A. Ayers, Ed.; prepared for U.S. Environmental Protection Agency, Report No. EPA-600/2-79-104, May 1979.
3. P. Miller and J. L. Wilki, "Removing visible oil fumes from tenter frame exhausts," *Mod. Textiles* (February 1974).
4. M. R. Beltran, "How to recover heat from your oven exhausts," *Am. Dyestuff Reporter* (April 1974).
5. W. Hebrank and L. Nelson, "Heat recovery system saves 45% fuel at Cannon Mills," *Textile Indust.* (February 1980).
6. Textile finishing application literature from the following companies: wet electrostatic precipitators—Beltran (Brooklyn, NY), Trion (Sanford, NC), and United Air Specialists (Cincinnati, OH); fume filters—CVM (Wilmington, DE) and Monsanto's Brink's mist eliminator (St. Louis, MO); incinerators—Hirt Combustion (Montebello, CA), AER (Ramsey, NJ), and Smith Environmental (Duarte, CA).

TEREPHTHALIC ACID

Ronald D. Bell, Alan J. Mechtenberg, and Kevin McQuigg

Terephthalic acid (TPA) produced in a purified form (P-TPA) is used almost exclusively in the manufacture of polyethylene terephthalate (polyester) fibers. A small percentage of P-TPA is used for the manufacture of polyester films, polybutylene terephthalate resins, and barrier resins for carbonated-beverage bottles. Table 1 provides a listing of the U.S. producers of P-TPA, the locations of their facilities, and the annual P-TPA capacities of these facilities. Production capacity in the United States has grown to 4.8 billion pounds from a reported 2.9 billion pounds in 1978.[2]

Beginning in the United Kingdom in 1949 and the United States in 1953, polyester production initially involved the use of TPA in a technical grade as an intermediate to dimethyl terephthalate. The acid was produced by a dilute nitric acid oxidation of *p*-xylene, and a number of subsequent variations were made to the process that reduced the consumption of nitric acid by introducing air into the reaction. Neither of these nitric acid processes is currently in commercial application.

Prior to 1963, monomer produced from the direct esterification of TPA did not meet required standards for purity. This was primarily due to the difficulty in purifying TPA itself, since it is extremely insoluble in most organic solvents, as well as in water. Dimethyl terephthalate, on the other hand, was susceptible to most common methods of purification available at the time and was the source for transesterification to all monomer.

TABLE 1. U.S. Production Capacity of Purified Terephthalic Acid[a]

Manufacturer	Location	Annual Capacity (millions of pounds)
Amoco Corp.	Charleston, NC	1600
	Decatur, AL	1800
Cape Industries	Wilmington, NC	380
E.I. du Pont de Nemours & Co., Inc.	Wilmington, NC	598
Eastman Kodak Co.	Columbia, SC	400
Total Capacity		4778

[a]1990 data from reference 1.

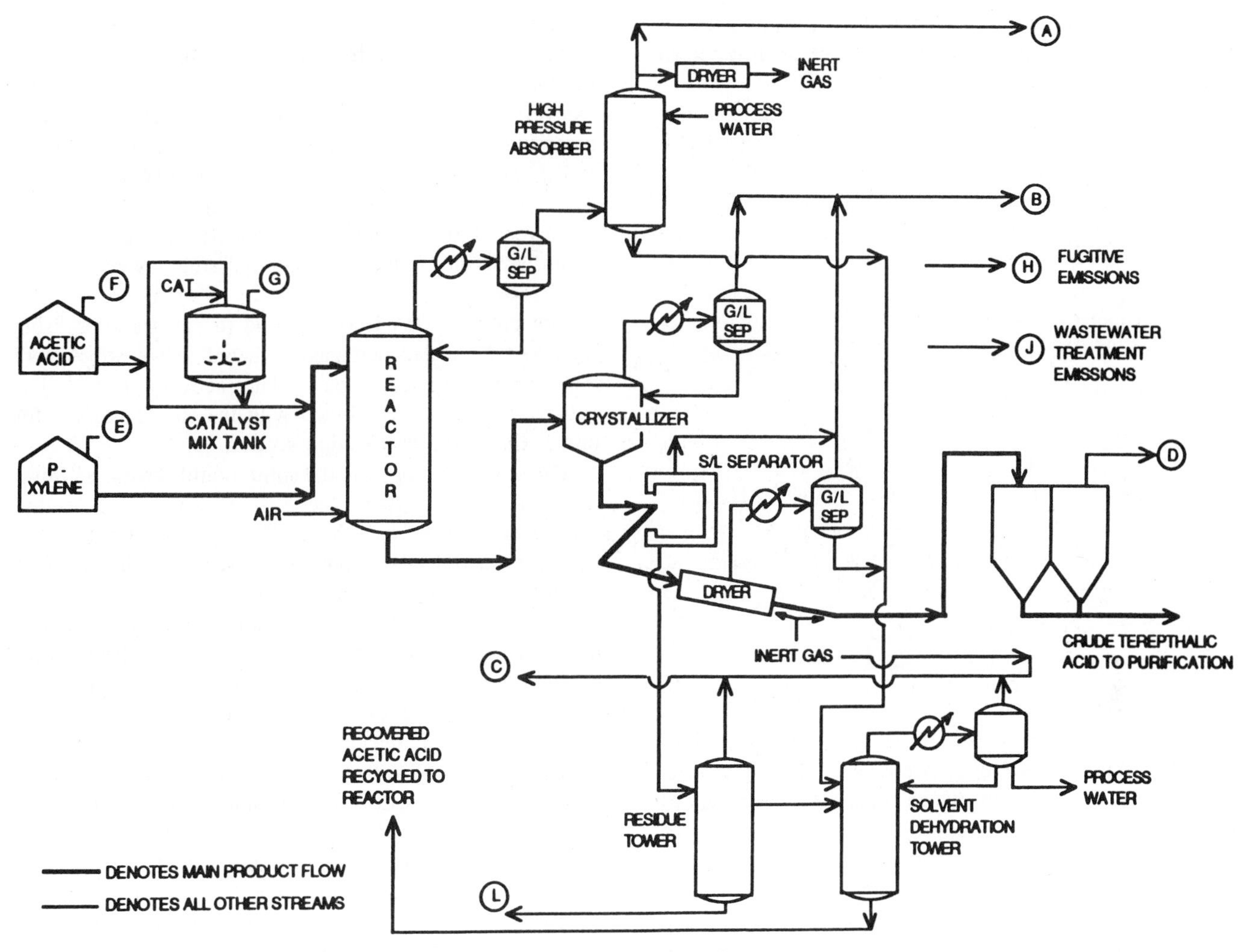

FIGURE 1. Flow Diagram for Crude TPA Production Using the Amoco Process

New methods of producing and purifying TPA from the liquid-phase oxidation of *p*-xylene were commercialized in 1963. Thus P-TPA is currently used interchangeably with dimethyl terephthalate in polyester production in which 1.17 grams of dimethyl terephthalate as a reactant is equivalent to 1 gram of P-TPA. While production of P-TPA requires two separate processes to synthesize and purify crude TPA, the use of P-TPA in polyester production via the direct esterification process offers distinct cost advantages to the producer. Specifically, the use of dimethyl terephthalate requires recovery of methanol, a by-product of the transesterification. Also, P-TPA requires less ethylene glycol for polyesterification, thereby improving final polymer quality.

Air emissions from the synthesis of P-TPA vary with respect to solvent-recovery portions of the process because of variations from facility to facility. Emissions of concern include *p*-xylene, acetic acid, carbon dioxide (CO_2), carbon monoxide (CO), methyl acetate, and TPA in particulate and vapor form. One information source describes the emissions from the industry as a whole as "moderate."[3] This study also indicates that, compared with other commercial P-TPA processes, the currently dominant liquid-phase air oxidation process is one of the lowest polluters.

PROCESS DESCRIPTION

The predominant process for technical-grade TPA manufacture is the Amoco process for liquid-phase air oxidation of *p*-xylene. Of the domestic P-TPA producers listed in Table 1, the Eastman Kodak Co. and Cape Industries utilize different methods that account for 780 million pounds of the 4.8 billion pounds of P-TPA production capacity. The E.I. du Pont de Nemours & Co. and Eastman Kodak Co. facilities also produce crude TPA, which is subsequently esterified with methanol to produce dimethyl terephthalate. The purification processes for the production of P-TPA also vary within the industry; however, the Amoco process for catalytic hydrogenation of impurities is dominant in the United States.

A simplified process flow diagram of the Amoco process for production of crude TPA is depicted in Figure 1. The process represented is essentially that practiced by Amoco at the Charleston and Decatur facilities. Streams are designated as main product flow and other flow streams. Streams with letter designations are those that potentially result in air pollutant emissions.

The overall reaction in the Amoco process takes place in the liquid phase in the presence of a catalyst, according to the following reaction. The parallel reaction that results in the formation of carbon oxides is considered a minor reaction.

$$\underset{\text{Acetic acid}}{CH_3COOH} + \underset{p\text{-Xylene}}{CH_3-C_6H_4-CH_3}$$

$$\xrightarrow{CAT} \underset{\text{Terephthalic acid}}{CH-\overset{O}{\overset{\|}{C}}-C_6H_4-\overset{O}{\overset{\|}{C}}-CH} + \underset{\text{Water}}{2\,H_2O}$$

$$\longrightarrow \underset{\text{Carbon Monoxide}}{CO} + \underset{\text{Carbon Dioxide}}{CO_2} + \underset{\text{Water}}{H_2O}$$

The primary by-products of the reaction system are products of partial oxidation of *p*-xylene, such as *p*-toluic acid and *p*-formyl benzoic acid, which appear to some extent as impurities in the TPA product. Methyl acetate is also formed in significant amounts in the reaction.

Fresh and recycled acetic acid, *p*-xylene, and a heavy-metal catalyst (normally manganese or cobalt acetate and sodium bromide) are continuously fed to the reactor, which is maintained at about 175–230°C and 220–435 psi. The reaction temperature is controlled by regulating the pressure at which the reaction mixture is permitted to boil and form the vapor stream leaving the reactor. Air is added in greater than stoichiometric amounts to minimize the formation of by-products. The residence time in the reactor ranges from 30 minutes to three hours, depending on process conditions. Reportedly, the conversion of *p*-xylene exceeds 95% and the yield of TPA is at least 90 mol%.[4]

Inert gases, excess oxygen, CO, CO_2, and volatile organic compounds (VOCs) are routed through a gas–liquid separator to a high-pressure absorber where the gas stream is scrubbed with process water to reduce the VOC content. Part of the stream is dried and used as a source of inert gas and the remainder is vented to the atmosphere (vent A in Figure 1). The scrubbing liquor from the absorber is routed to the solvent dehydration tower for acetic acid recovery.

The effluent from the reactor is a slurry, since TPA is somewhat soluble in acetic acid. The slurry flows to a series of crystallizers where the pressure is relieved and the liquid is cooled by the vaporization and return of condensed VOCs and water. Noncondensible inert gases containing some VOCs are vented from the condenser (vent B).

The impurities from the partial oxidation of *p*-xylene are somewhat soluble in acetic acid and remain in the mother liquor as the effluent from the crystallizers is routed to solid–liquid separators (normally centrifuges). The TPA recovered as wet cake is sent to a rotary dryer where inert gases and heat are used to drive off the remaining water and VOCs. The hot VOC-laden vapors are routed to a condenser where the noncondensibles containing some VOCs are again vented to the atmosphere (vent B) and the condensed liquids are routed to the solvent dehydration tower for acetic acid recovery.

Dry product TPA is transported to storage silos using inert gas; there the inert gas is vented to the atmosphere after passing through a particulate-recovery device. The collected solids are combined with the stored product for processing in the purification section.

The mother liquor from the solid–liquid separators flows to a residue still where acetic acid, water, and methyl acetate are recovered overhead. The bottoms from the residue still contain tars, products of partial oxidation of *p*-xylene, catalyst residue, and some acetic acid.

The overhead from the residue still, along with the condensed vapors from the dryer and the scrubbing liquor from the high-pressure absorber, is routed to the solvent dehydration tower that produces a relatively pure acetic acid, which is recycled to the reactor. A reflux accumulator at the column is used to recover condensed water and methyl acetate, which is used as the scrubbing liquor in the high-pressure absorber. Noncondensibles from the reflux accumulator and the residue column are vented to the atmosphere (vent C), along with some VOCs and inert gas, which is used for blanketing and instrument purging.

The purpose of the purification process is to produce a product (P-TPA) that is suitable for polyester production. In the Amoco purification process presented in Figure 2, this is accomplished by converting certain impurities to a more water-soluble form via catalytic hydrogenation in an aqueous medium. The converted impurities remain in solution upon subsequent recrystallization of TPA.

The primary impurity formed by the partial oxidation of *p*-xylene in the reactor, *p*-formyl benzoic acid, is converted to the more water-soluble *p*-toluic acid, according to the following reaction.

$$\underset{p\text{-Formyl benzoic acid}}{CH-\overset{O}{\overset{\|}{C}}-C_6H_4-\overset{O}{\overset{\|}{C}}-OH} + \underset{\text{Hydrogen}}{2\,H_2}$$

$$\xrightarrow{CAT} \underset{p\text{-Toluic acid}}{H_3C-C_6H_4-\overset{O}{\overset{\|}{C}}-OH} + \underset{\text{Water}}{H_2O}$$

Crude TPA is sent to the feed slurry tanks and combined with hot water, resulting in the release of gases that had been trapped in the TPA crystals (vent A in Figure 2). The

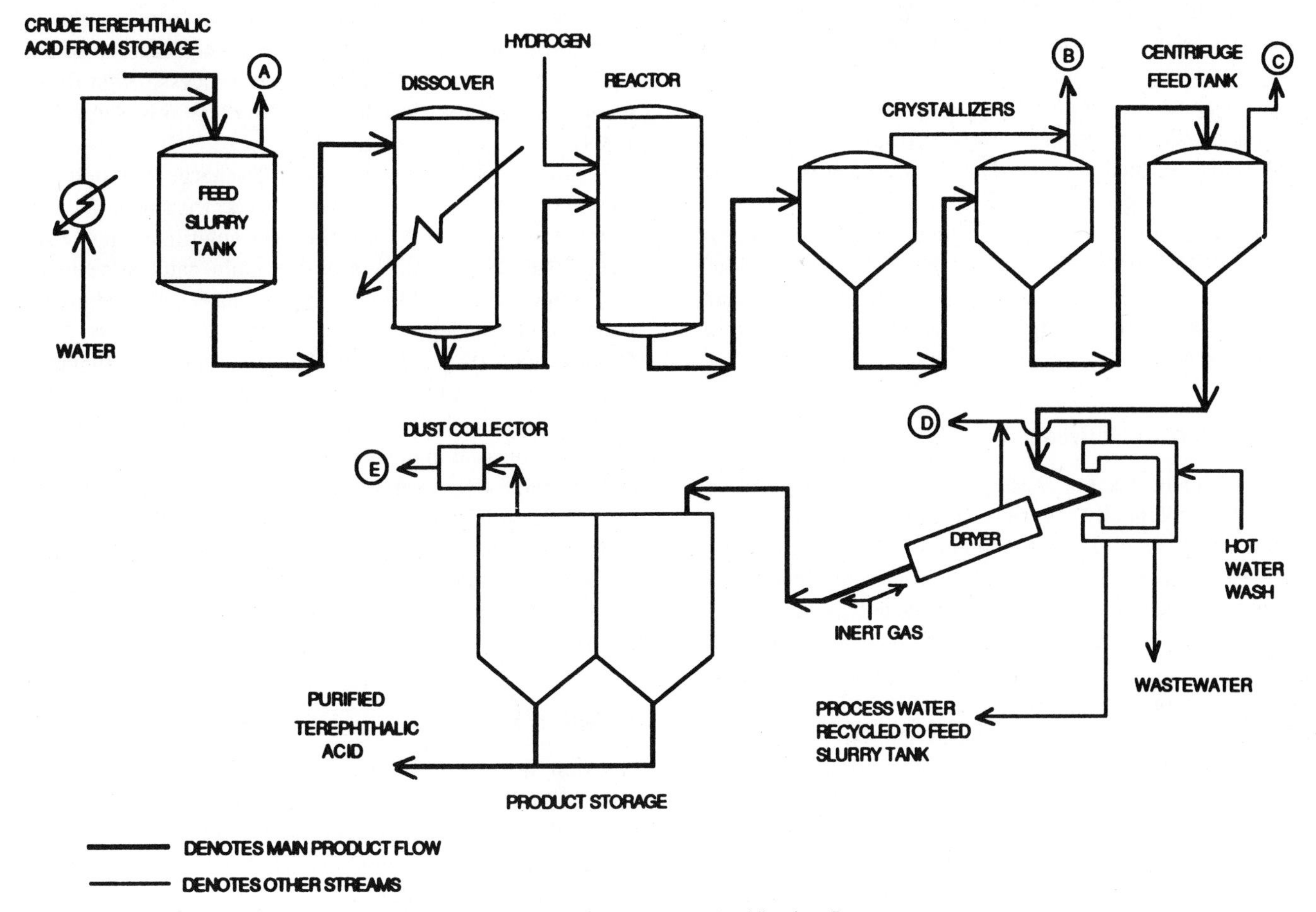

FIGURE 2. Flow Diagram for P-PTA Production Using the Amoco Purification Process

slurry is then routed to a dissolver, which is maintained under slight positive pressure and approximately 250°C, to achieve dissolution of TPA.

The dissolved TPA slurry and hydrogen in the amount of about 0.004 gram per gram of TPA, which represents a significant stoichiometric excess, are fed to the reactor, which is maintained at the same temperature as the dissolver. The reactor contains a noble-metal catalyst on a carbon support.

The reactor effluent flows to the crystallizers where the temperature is reduced by allowing the water to boil at a controlled rate. Minor amounts of TPA in the vapor phase are released with the water vapor (vent B). Since TPA will sublime rather than melt, the released TPA vapors are cooled in the atmosphere and fall as particles.

The slurry from the crystallizers is sent to a solid–liquid separator (normally a centrifuge) and the mother liquor is sent to wastewater treatment. The wet cake in the centrifuge is then washed with hot water and the water is recyled to the feed slurry tanks. The wet cake is dried in a rotary drier and transported to product storage silos with inert gas; this gas is vented to the atmosphere after passing through a particulate-control device. The final concentration of *p*-formyl benzoic acid is less than 20 ppm and the yield from the purification process is reportedly 98%.[4]

Several variations of the Amoco processes for crude and purified TPA production are used in the United States. The Cape Industries facility listed in Table 1 obtains P-TPA directly by hydrolyzing dimethyl terephthalate. The Eastman Kodak facility utilizes acetaldehyde to replace acetic acid solvent losses and a somewhat different catalyst in the crude TPA process. Eastman Kodak also employs further steps in the solvent-recovery sections of the process in which an azeotroping agent is applied and methyl acetate is recovered further to reduce VOC content in the wastewater from the solvent-recovery operations. The additional solvent-recovery steps are typical of those carried out in dimethyl terephthalate technology.[5] Variations within the purification operations do not result in a significant difference in air emissions from those resulting from the Amoco process.

AIR EMISSONS CHARACTERIZATION

Air emissions resulting from crude TPA production may arise from several sources, including the reactor vent, the crystallization/separation/drying vent, the solvent-recovery vent, the product-transfer vent, material storage and handling vessel vents, and the wastewater treatment operations. Fugitive emissions from valves, compressor seals, pumps, and flanges throughout the process also result. In addition, TPA is also emitted in particulate form, primarily from the crude TPA storage areas.

Air emissions from the purification process consist primarily of TPA vapor and particulate matter from various processing vents. No VOCs are introduced or generated in the purification process; therefore, no VOCs are emitted to the atmosphere.

Description of Sources

Figure 1 identifies the waste streams and volatilization points that potentially contribute to air emissions from an uncontrolled crude TPA production process. Letter designations for each of these streams are referenced with the following discussion of emission characterization.

The reactor vent gas (vent A) consists of the substances within the vapor-phase reactor effluent that are recovered via the high-pressure absorber. These substances include nitrogen from air oxidation, unreacted oxygen and *p*-xylene, CO, and CO_2. Acetic acid and methyl acetate are also present in certain amounts, depending on the removal efficiency of the absorber.

The crystallization/separation/drying vent (vent B) consists primarily of the noncondensible off-gases from the crystallization and drying portions of the process. Carbon dioxide and CO are not present in an appreciable amount; however, some VOCs carried by the noncondensibles are emitted through this vent.

The residue and solvent-recovery column vent (vent C) contains the noncondensible vapors from these operations, including some CO_2 and CO. Inert gases that are introduced for blanketing and instrument purging are released at this point and VOCs are carried with the vent gas stream. The VOC content within the discharge stream is reportedly very low.[2]

The product storage vent (vent D) releases the inert gases used to transport solid TPA product to the storage silos and contains the TPA particulate matter carried with these gases. The VOCs that are carried with recovered inert gases from the high-pressure absorber may also be present in this stream, where the recovered inert gases are used for product transfer.

The emissions from storage and handling vessels (vents E, F, and G) consist of *p*-xylene and acetic acid from the vessels in which they are present. Fugitive emissions throughout the process consist primarily of VOCs that are introduced and generated in the various process areas.

Wastewater treatment emissions (point J) consist of volatilized pollutants from the treatment area. The sources of wastewater from facilities that use the Amoco process as described include the blowdown from the high-pressure absorber (scrubbing liquor) and the effluent from equipment wash-down and process drains. Where the process does not employ a methyl acetate recovery process, significant amounts of methyl acetate are present in this scrubbing liquor.

Producers of P-TPA may also discharge the bottoms from the residue still to a wastewater treatment facility where emissions of VOCs through volatilization would occur. The residue still bottoms contain catalyst residue, tars, some acetic acid, and the products of incomplete oxidation of *p*-xylene.

The crude TPA purification process as shown in Figure 2 does not include VOCs in significant amounts since VOCs are not introduced or generated. Very small amounts of VOCs are released in the dissolution process from the crude TPA crystals. The venting of excess hydrogen and water from the hydrogenation reactor carries some TPA in vapor form, which is cooled in the atmosphere and converted to TPA particulate matter. Additionally, some particulate P-TPA product is released from the product storage facility.

Emission Factors

Emission factors have been developed in terms of total VOC and CO emissions from an uncontrolled model plant that produces crude TPA.[2] The model plant has a crude TPA capacity of 507 million pounds per year based on 8760 hours of annual operation.

The model plant is a combination of the Amoco process shown in Figure 1 and the solvent-recovery portion representative of dimethyl terephthalate technology as practiced at the Eastman Kodak facility. The model plant's solvent-recovery operation involves the use of *n*-propyl acetate as an azeotroping agent in the solvent dehydration tower and additional separation steps for methyl acetate and plant wastewater. The *n*-propyl acetate and methyl acetate emissions resulting from storage of these materials are, therefore, included in the factors. The recovered methyl acetate, along with the bottoms from the residue tower, is incinerated in the model plant; therefore, incinerator stack gas emission factors are listed. The emission factors resulting from the study are given in Table 2.

Storage tank emission factors were developed assuming two raw-material tanks and one in-process *p*-xylene tank, two storage tanks and one catalyst mix tank containing acetic acid, and four crude TPA tanks. The tanks are typical-size, fixed-roof tanks. Details regarding additional parameters used to determine the storage loss factors are available in reference 2.

The fugitive emission factors listed in Table 2 represent the component counts used in the model study; however, the various uncontrolled factors were reestablished based on the more recent average uncontrolled fugitive emission factors for the synthetic organic chemical manufacturing industry.[6] Component counts used for the model plant include

TABLE 2. Uncontrolled Emissions of VOC and CO from Model Plant for Crude TPA Production[a,b,c]

Source	Stream Designation (Figure 1)	Emission Ratio (g/kg)[d]		Emission Rate (kg/h)	
		VOC	CO	VOC	CO
Reactor vent	A	14.6	17	383.3	446
Crystallization, separation, and drying vent	B	1.9		49.9	
Distillation and recovery vent[e]	C	1.14		29.9	
Product transfer vent[f]	D	1.78	2	46.7	53
Raw material storage[g]	E,F	0.112		2.94	
Other storage[h]	G	0.006		0.017	
Fugitive emissions	H	0.50		13.02	
Incinerator stack[i]	N/A	0.00605		0.1583	
Wastewater treatment[j]		<0.004		<0.1	
Total		20.05	19	526.04	499

N/A = Not applicable.
[a]Reference 2.
[b]Model plant has an annual crude TPA capacity of 507 million pounds. Factors are based on 8760 hours of annual operation.
[c]Factors are uncontrolled factors with the exception of the incinerator stack as noted in this table.
[d]Grams of emission per kilograms of crude TPA produced.
[e]Emission factors include methyl acetate and wastewater towers in the model plant, which are not shown in Figure 1.
[f]VOC and CO emissions originate from inert gases that are used to transport product and are recovered from the high-pressure absorber.
[g]Emission factors include emissions from *n*-propyl acetate storage in the model plant, which is not shown in Figure 1.
[h]Emission factors include emissions from methyl acetate storage in the model plant, which is not shown in Figure 1.
[i]Incinerator stack emission factors are based on the model plant in which residue still bottoms and recovered methyl acetate are incinerated.
[j]Wastewater treatment emissions are based on the model plant in which methyl acetate is removed from plant wastewater.

50 pumps, 900 process valves, and 40 pressure-relief devices in light-liquid VOC service.

The emission factors in Table 2 are considered uncontrolled factors, with the exception of the incinerator stack emissions, since the incinerator itself is considered a control device. The factor for vent A in Figure 1 is an uncontrolled factor since the high-pressure absorber is considered a part of the production process.

The study[2] also resulted in the development of solid TPA emission factors for the Amoco purification process represented in Figure 2, which were based on a screening study of particulate emissions from the purification process.[7] The factors are considered controlled emission factors since particulate emissions from each of the vents (vents A–D in Figure 2) were measured following the application of a water scrubber for particulate control. The emissions from the product storage vent (vent E in Figure 2) were measured following the passage of the gas stream through a dust collector. Controlled particulate TPA emission factors resulting from the screening study are listed in Table 3.

AIR-POLLUTION CONTROL MEASURES

The types and quantities of pollutants from air emission sources within the TPA and P-TPA processes are shown in Table 2. Nine point sources are included as follows:

- Reactor vent
- Crystallization, separation, and drying vent
- Distillation and recovery vent
- Product transfer vent
- Storage and handling vents
- Fugitive
- Wastewater treatment
- Residue tower bottoms
- TPA purification

Reactor Vent and Product Transfer Vent

The reactor vent gas normally contains nitrogen, unreacted oxygen, unreacted *p*-xylene, acetic acid, CO, CO_2, methyl acetate, and water. The quantity of VOCs in the reactor vent varies depending on the absorber operating pressure and temperature. The product transfer vent gas has the same composition as the reactor vent gas, except that it includes water and TPA particulate matter. Particulate matter in the product transfer vent gas is typically removed by bag filters downstream of the product silos. Three technologies for control of VOCs are applicable to these vent streams: aqueous absorbers, carbon adsorption, and thermal oxidation.

In general, VOCs in the reactor vent and product transfer vent gases are controlled by the use of aqueous absorbers. The liquid streams from the absorbers are separated and returned to the process. The vent gas from the absorbers, which still contains from 3% to 10% of the original VOC and an unchanged amount of CO, is vented to the atmosphere.

TABLE 3. Controlled TPA Particulate Emissions from Model Plant for TPA Purification Using the Amoco Purification Process[a,b,c]

Source	Stream Designation (Figure 2)	Emission Ratio (g/kg)[d]	Emission Rate (kg/h)
Feed slurry tank vent	A	0.088	5.08
Crystallizer vent	B	0.098	5.69
Atmospheric centrifuge feed tank	C	0.023	1.32
Dryer vent	D	0.0012	0.07
Silo dust collector vent	E	0.0017	0.10

[a]Reference 2.
[b]Model plant has an annual P-TPA capacity of 507 million pounds. Factors are based on 8760 hours of annual operation.
[c]Control of source streams A–D includes a water scrubber.
[d]Grams of emission per kilogram of P-TPA produced.

Carbon adsorption can be expected to provide approximately 97% control of the VOCs in the vent streams. However, no change in the level of CO results. For continuous operation, a typical carbon-adsorption system utilizes two carbon beds—one in operation and one undergoing regeneration. Regeneration is generally achieved by steam stripping the VOCs from the carbon bed. The VOC-laden stripping steam is condensed and decanted. The organic layer is returned to the reactor section and the aqueous layer is sent to distillation for recovery of water-soluble VOCs.

Aqueous absorbers and carbon adsorption may prove to be inadequate as regulations regarding VOC and CO emissions become more and more stringent. Thermal oxidation can reduce VOC and CO levels in the vent gases by 99% to 99.9%. Depending on the VOC and CO content of the stream, considerable supplemental fuel may be required to achieve the operating temperature required (about 875°C for 99% control to 1000°C for 99.9% control) to obtain this level of control efficiency. Supplemental fuel cost can be offset by adding a waste-heat boiler to provide steam for the TPA production process.

Crystallization, Separation, and Drying Vent

Noncondensible gases carrying VOCs are released during crystallization of the TPA and separation of crystallized solids from the solvent. Different methods incorporated into the process affect the amount of noncondensible gases and accompanying VOCs. Control of the VOCs can be obtained by carbon adsorption, thermal oxidation, or, as is typically practiced, by the use of aqueous absorbers. The use of aqueous absorbers to remove acetic acid will reduce VOC emissions from this vent stream by approximately 98%.

Distillation and Recovery Vent

This small discharge consists of the gases that are dissolved in the feed stream sent to distillation; the inert gases used for blanketing, instrument purging, and pressure control; and the VOCs carried by the inert gases. This vent gas is typically incorporated with gas from the crystallizer, separation, and drying vent, and sent to aqueous absorbers. Removal of acetic acid by the aqueous absorbers reduces VOC emissions from this vent stream by approximately 96%. Alternatively, the vent gas can be sent to a carbon-adsorption system or to a thermal oxidizer to achieve VOC control.

Storage and Handling Vents

Some emissions result from the storage of *p*-xylene and acetic acid. Most emissions occur only during tank filling, especially for the *p*-xylene storage tanks that are maintained at a constant temperature and are equipped with conservation vents. Emissions from acetic acid storage tank vents are typically controlled using aqueous absorbers. The resultant aqueous stream is returned to the process. Alternatively, these vents could be routed to a carbon-adsorption system or thermal oxidizer.

Fugitive Emissions

Fugitive emissions from piping, valves, pumps, and compressors are controlled by periodic monitoring by leak checking with a VOC detector and a directed maintenance program.

Wastewater Treatment

Currently, the small amount of VOC emissions from the various wastewater treatment steps is not controlled. If control of these emissions becomes necessary, aqueous absorbers would be the most easily applied technology. Other treatment alternatives might include carbon adsorption or air stripping of the wastewater. Although these technologies seem applicable, lack of experience would require some development to determine the effect of inorganic species in the wastewater on carbon adsorption and the efficiency of air stripping on the organic compounds in the wastewater stream.

Residue Tower Bottoms

The residue tower bottoms stream contains products of partial oxidation, tars, catalyst residue, and some acetic acid. If not treated, volatilization of light ends will result in VOC emissions. This stream is most effectively treated by sending it to a liquid-waste incinerator. The incinerator is capable of an efficiency of 99.99% in the destruction and removal of the organic compounds. Bromine salts and other inorganic solids contained in the still bottoms must be controlled by removal prior to incineration (usually very difficult) or by the treatment of the incinerator flue gas with some type of particulate-control device, such as a bag filter, electrostatic precipitator, or venturi scrubber.

Terephthalic Acid Purification

There are no VOC emissions associated with the purification of TPA, since the purification process does not involve the use of any VOC. Emissions of TPA, not considered a VOC, as a vapor result from the venting of excess hydrogen and water vapor at elevated temperatures. The TPA vapors sublime in the atmosphere, thereby creating particulate emissions. Water scrubbers are currently used to provide effective control of the TPA in the emissions from the purification process.

References

1. *1990 Directory of Chemical Producers United States of America,* SRI International, Menlo Park, CA, 1990.
2. *Organic Chemical Manufacturing, Vol. 7: Selected Processes,* EPA-450/3-80-028b, U.S. Environmental Protection Agency, Research Triangle Park, NC, 1980.
3. J. W. Pervier et al., *Survey Reports on Atmospheric Emissions from the Petrochemical Industry, Vol. II,* EPA-450/3-73-005-6, U.S. Environmental Protection Agency, Research Triangle Park, NC, 1974.
4. *Kirk-Othmer Encyclopedia of Chemical Technology,* 3rd ed., Vol. 17, John Wiley & Sons, New York, 1978.
5. *Supplement B to Compilation of Air Pollutant Emission Factors, Vol. I: Stationary Point and Area Sources,* U.S. Environmental Protection Agency, Research Triangle Park, NC, 1988.
6. *Protocols for Generating Unit-Specific Emission Estimates for Equipment Leaks of VOC and VHAP,* EPA-450/3-88-010, U.S. Environmental Protection Agency, Research Triangle Park, NC, 1988.
7. L. M. Elkin, "Terephthalic acid and dimethyl terephthalate," pp 49–55 in *Report No. 9, A Private Report by the Process Economics Program,* Stanford Research Institute, Menlo Park, CA, 1966.

Bibliography

D. F. Durocher et al., *Screening Study to Determine Need for Standards of Performance for New Sources of Dimethyl Terephthalate and Terephthalic Acid Manufacturing,* EPA Contract No. 68-02-1316, Task Order no. 18, 1976, p 4.

THERMOPLASTIC RESINS

John A. Fey and J. Mitchell Jenkins III

Thermoplastic resins are materials that soften and melt upon heating, with no chemical change and with no permanent change in physical properties. These resins are readily extruded or molded into film, fibers, bottles, and many types of molded articles. In the following sections, two of the major thermoplastic resins are discussed: polyethylene and polypropylene.

POLYETHYLENE

The earliest polyethylene product was low-density polyethylene (LDPE), which was first produced in the United States during World War II for use as radar cable insulation. It is made from ethylene by a polymerization process that operates at pressures as high as 50,000 psi.

During the 1950s, catalysts were discovered that allowed production of polyethylene at much lower pressures, a few hundred pounds per square inch. The polyethylene was found to have somewhat different physical properties than the LDPE produced by high-pressure processes. Because of its higher density, the material was called high-density polyethylene (HDPE).

Further catalyst development has allowed the copolymerization of ethylene with propylene, 1-butene, and other higher α-olefins. Incorporation of these higher α-olefins resulted in a decrease in the density of the polyethylene, along with other physical property changes. It was found, however, that this "new" material had properties similar to, but not identical with, those of the LDPE produced by the high-pressure processes. This new low-density polyethylene was called linear low-density polyethylene (LLDPE).

A large percentage of the HDPE and LLDPE produced worldwide is made by gas-phase fluidized-bed processes. The most widely used process of this type is the UNIPOL™ polyethylene process, which was developed by Union Carbide Chemicals and Plastics Co. Inc. and has been licensed for use by polyethylene producers worldwide.

Process Description

Following is a brief description of facilities in each part of a typical gas-phase fluidized-bed polyethylene plant. References such as R-1 indicate process streams as shown on the block flow diagram Figure 1. Nomenclature for Figure 1 is identified in Table 1.

*UNIPOL™ is a trademark of Union Carbide Chemicals and Plastics Technology Corp.

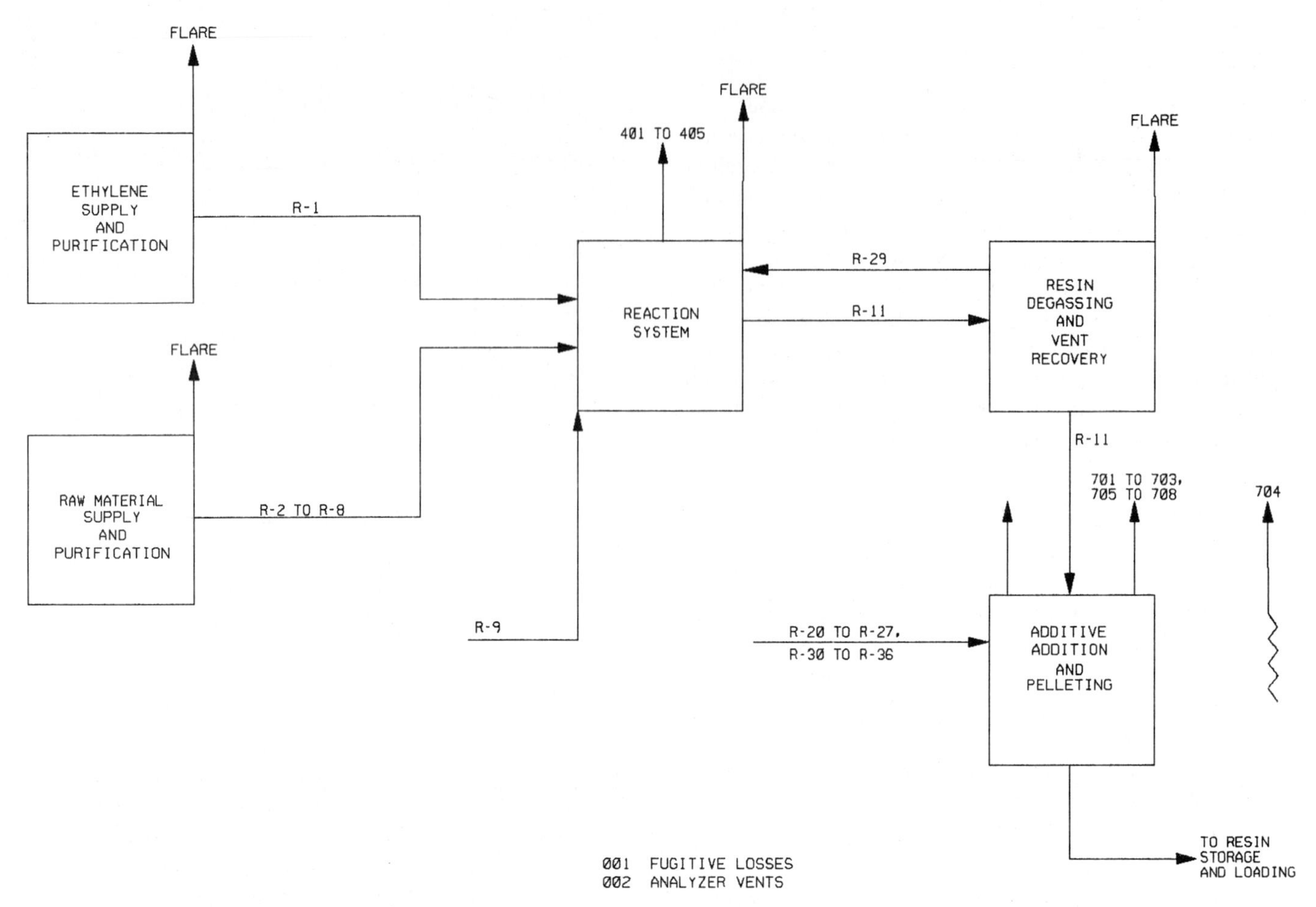

FIGURE 1. Polyethylene Plant Using Gas-Phase Fluidized-Bed Process

Ethylene Supply and Purification

The ethylene feed to the plant (R-1) is compressed or pumped to the required feed pressure, after which it is passed through a series of purifiers where trace quantities of impurities are removed. The purifiers are vented to a flare before and during regeneration. The fresh ethylene is combined with recycled ethylene and is injected into the reactor.

Raw-Materials Supply and Purification

Other reactor feed materials are identified as R-2 through R-8. R-2, R-3, and R-7 are passed through degassing columns and temporarily stored. R-2 (or alternatively R-7) and R-3 are passed through purifiers to remove impurities and are injected into the reactor. The purifiers are vented to the flare before and during regeneration. R-4 is passed through a series of purifiers, where trace quantities of impurities are removed, and then injected into the reactor. R-8 also is passed through a series of purifiers, where trace quantities of impurities are removed. A portion of R-8 is compressed and then routed to the reactor. The remaining R-8 is used throughout the plant. R-5 and R-6 are routed directly to the reactor.

Reaction System

The reaction system consists of a fluidized-bed reactor, a cycle gas compressor and cooler, and product discharge tanks. Ethylene (R-1), R-2, R-3, and/or R-7, R-4, R-5, R-6, and catalyst (R-9) are fed continuously to the reactor. The unreacted ethylene is recycled. Polyethylene resin (R-11) is removed from the reactor and separated from the small amount of gas accompanying it in the discharge tank. The separated gas either is pumped back to the reactor as part of the recycle stream or is flared.

Resin Degassing and Vent Recovery

Resin from the reactor (R-11) is conveyed to a purger where the dissolved hydrocarbons are stripped from the resin and

TABLE 1. Nomenclature for Figure 1

001	Hydrocarbons (fugitive losses)
002	Hydrocarbons (from analyzers)
401	Cycle gas compressor seal and lube oil reservoir vent
402, 403	Catalyst feeder vents
404	Reactor seed bed vent
405	Catalyst vent filter
701	Dust collector exhaust
702	Pelleter dryer exhaust
703	Polyethylene fines while conveying to storage
704	Hydrocarbon evolution from storage
705	Resin surge bin vent
706	Additive dump hopper vent
707	Resin cooler vent
708	Seed bed resin rotary feeder vent

returned to the process (R-29). Any excess gas not used in the process is vented to the flare. Resin from the purger flows to the additive system.

Additive System

Resin (R-11) and solid additives (R-21 to R-27, R-30 to R-36) are metered by gravimetric feeders and flow by gravity to the pelleting system.

Pelleting

The additives and resin are thoroughly mixed, melted, and pelleted in the pelleting system. The pellets are dried, cooled, and conveyed to product storage and packaging or shipping.

POLYPROPYLENE

Polypropylene was first produced commercially in the late 1950s. In the early plants, polypropylene was polymerized as a slurry in a hydrocarbon diluent. Later, processes were developed in which the polypropylene was polymerized as a slurry in liquid propylene. Mechanically stirred gas-phase processes also were developed. In the early 1980s, Union Carbide Chemicals and Plastics Co. Inc. and the Shell Chemical Co. (a division of Shell Oil Co. [USA]) jointly developed the UNIPOL polypropylene process, a gas-phase fluidized-bed process. This process has been licensed for use by polypropylene producers worldwide.

Polypropylene products can generally be divided into three classes: homopolymers, random copolymers, and impact copolymers. Polypropylene homopolymers are crystalline polymers containing only propylene. Random copolymers are copolymers of propylene with small amounts of ethylene (or other α-olefins). Both homopolymers and random copolymers are made in a single reactor, and the same plant can produce both types of polymers as market demand may require.

Impact copolymers are a mixture of polypropylene and ethylene-propylene copolymer produced by two reactors operating in series. Catalyst is fed to the first reactor where the polypropylene homopolymer is produced. The homopolymer, containing still-active catalyst, is transferred to a second reactor, where the ethylene-propylene copolymer is produced. Most UNIPOL™ polypropylene plants are built with two reactors so that the full range of homopolymer, random copolymer, and impact copolymer products can be produced.

Process Description

Following is a brief description of facilities in each part of a typical gas-phase fluidized-bed polypropylene plant. References such as R-1 indicate process streams as shown on the block flow diagram, Figure 2. Nomenclature for the block flow diagram is identified in Table 2.

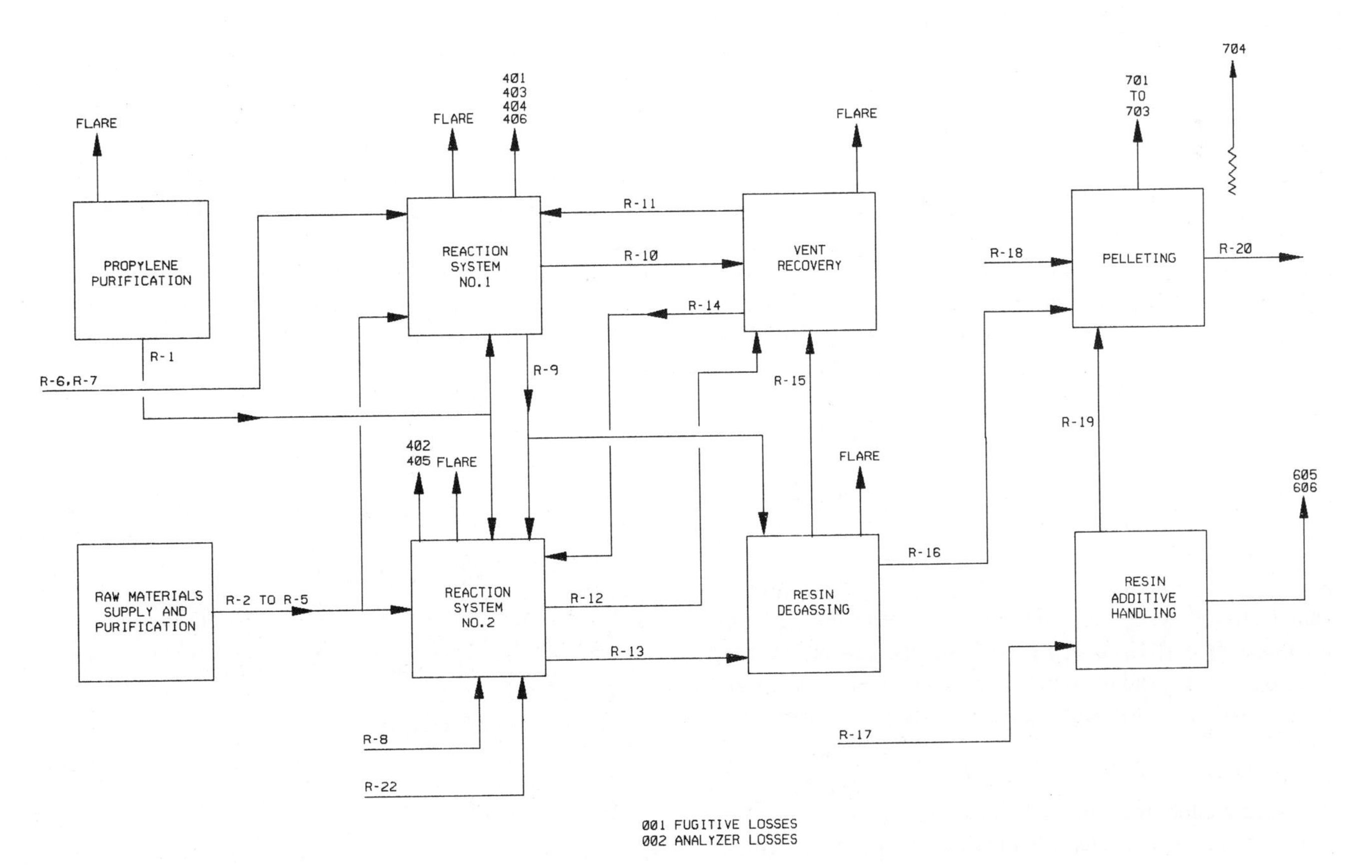

FIGURE 2. Polypropylene Plant Using Gas-Phase Fluidized-Bed Process

TABLE 3. Emissions Characterization for Polyethylene[a]

UNIPOL™ Polyethylene emissions	
Hydrocarbon	
Process	0.20 kg/ton
Fugitive	0.03 kg/ton
Particulate (resin)	0.02 kg/ton
Aqueous	None
Nonaqueous liquids	< 0.1 ton/yr
Solid wastes	
Catalysts and treatment beds	< 0.15 kg/ton
Polymer spills	Normally recovered and sold

[a]Data provided by Union Carbide Chemicals and Plastics Co. Inc.

Propylene Supply and Purification

Fresh propylene feed to the unit is passed through a degassing column where trace quantities of impurities are removed. The column is vented to a flare. This propylene (R-1) is pumped through a purifier, where additional impurities are removed. The purifier is vented to a flare before and during regeneration. R-1 is then routed to one or both reactors.

Raw-Materials Supply and Purification

The R-2, R-3, and R-4 are passed through separate series of purifiers where trace quantities of impurities are removed. Each is supplied as required to various parts of the unit. All except one of the purifiers are vented to a flare before and during regeneration. R-5 is routed directly to the reactors.

Reaction System No. 1

Reaction system no. 1 consists of a fluidized-bed reactor, a cycle gas compressor and cooler, and product discharge tanks. Streams R-1 to R-6, catalyst (R-7), and R-11 from the vent recovery system are fed continuously to the reactor. Polypropylene (R-9) is removed from the reactor by the discharge tanks and is sent either to reaction system no. 2 or resin degassing. A purge (R-10) is sent to the vent recovery system.

Reaction System No. 2

Reaction system no. 2 also consists of a fluidized-bed reactor, a cycle gas compressor and cooler, and product discharge tanks. Streams R-1 to R-5, additives (R-8 and R-22), R-9 from reaction system no. 1, and R-14 from the vent recovery system are fed continuously to the reactor. Polypropylene R-13 is removed from the reactor by the discharge tanks and is sent to resin degassing. A purge (R-12) is sent to the vent recovery system.

Resin Degassing

Resin from either reaction system (R-9 or R-13) is conveyed to a receiver, where interstitial and some dissolved hydrocarbons are stripped from the resin and are sent to the vent recovery system (R-15). The resin flows to a purger, where additional hydrocarbons are removed. The purged resin (R-16) is sent to the pelleting system. The gas from the purger is flared.

Vent Recovery

The vent recovery system recovers as much hydrocarbon as possible from the streams sent to it (R-10, R-12, and R-15). The recovered materials are returned directly to both reaction systems (R-11 and R-14). Any gas not returned to the process is vented to the flare.

Resin Additive Handling

Solid additives (R-17) are metered and sent to the pelleting system (R-19).

Pelleting

The resin (R-16), solid additives (R-19), and liquid additive (R-18) are mixed, melted, and pelleted in the pelleting system. The pellets (R-20) are dried, cooled, and sent to product storage and packaging or shipping.

Air Emissions Characterization

Table 3 characterizes the emissions from a UNIPOL polyethylene plant and Table 4 characterizes the emissions from a UNIPOL polypropylene plant.

AIR POLLUTION CONTROL MEASURES

The gas-phase fluidized-bed processes for producing polyethylene and polypropylene are low-pressure, solventless processes with a low inventory of in-process hydrocarbons. This provides for very low hydrocarbon emissions to the atmosphere. The low operating pressure and virtual absence of polluting effluents enable this type of process to meet current government regulations for air, water, and solid wastes. Low operating pressure, a minimum of process steps, and absence of solvent combine to insure compliance with hydrocarbon emission standards.

Hydrocarbon discharges from occasional routine de-

TABLE 3. Emissions Characterization for Polyethylene[a]

UNIPOL™ Polyethylene emissions	
Hydrocarbon	
Process	0.20 kg/ton
Fugitive	0.03 kg/ton
Particulate (resin)	0.02 kg/ton
Aqueous	None
Nonaqueous liquids	< 0.1 ton/yr
Solid wastes	
Catalysts and treatment beds	< 0.15 kg/ton
Polymer spills	Normally recovered and sold

[a]Data provided by Union Carbide Chemicals and Plastics Co. Inc.

TABLE 4. Emissions Characterization for Polypropylene[b]

UNIPOL™ Polypropylene emissions	
Hydrocarbon	
Process	0.16 kg/ton
Fugitive	0.03 kg/ton
Particulate (resin)	0.02 kg/ton
Aqueous	None
Solid wastes	
Catalysts and treatment beds	< 0.1 kg/ton
Polymer spills	Normally recovered and sold

[b]Data provided by Union Carbide Chemicals and Plastics Co. Inc.

pressurization of the reactor and from purifier regeneration for polyethylene and sieve regenerations for polypropylene are routed to a high-efficiency flare. All but the last traces, less than 50 ppm, of hydrocarbons removed from the resin granules are recovered or flared. Polymer particulate emissions to the air are very low, with dust controlled by filters on gas exhausted from conveying stations. Surface drainage is collected and screened to remove polymer spills. Solid waste consists of a minimal amount of polymer scrap and spills that can be recovered and sold or incinerated. There are also small amounts of inactive catalyst and treatment beds that can be sold or disposed of. There are no aqueous process effluents whatsoever from the gas-phase process.

The simplicity of the gas-phase process allows for the minimum use of valves, pumps, and compressors, which are the primary source of fugitive emissions. Fugitive emissions, therefore, are minimal.

For polyethylene, less than 0.15 kg per ton and for polypropylene less than 0.1 kg per ton of waste treatment beds and, in both cases, 1.1 kg per ton of spilled and scrap polymer are generated each year. The spilled polymer is washed down to a collection pit, filtered from the water, and sold. Spilled polymer, therefore, does not normally contribute to the solid waste load.

As a result of the exceptionally low emissions, the low-pressure gas-phase fluidized-bed process has been considered "best available technology" by regulatory agencies. Government permits have been obtained for UNIPOL polyethylene and polypropylene plants in Texas, Louisiana, and Illinois, and in over a dozen countries around the world.

Bibliography

Anonymous, "New route to low-density polyethylene," *Chem. Eng.*, 86:80 (December 3, 1979).

I. D. Burdett, "Polypropylene: Lowest cost process," *Hydrocarbon Proc.*, 75 (November 1986).

H. K. Ficker, G. L. Goeke, and G. W. Powers, "Gas-phase process for PP surpasses other methods," *Plast. Eng.*, 29 (February 1987).

F. D. Hussein and T. L. Nemzek, *UNIPOL™ PP-Innovation Through Combined Technologies,* AIChE 1989 Spring National Meeting, Houston, TX, April 2–6, 1989.

J. M. Jenkins III, "Polyethylene: Continuing technological advancement," in *Hydrocarbon Technology International,* P. Harrison, Ed.; Sterling Publications, London, 1987, pp 93–95.

F. J. Karol, "Catalysis and the polyethylene revolution," in *History of Polyolefins,* R. B. Seymour and T. Cheng, Eds.; D. Reidel Publishing Co., Dordrecht, Holland, 1986, pp 193–211.

F. J. Karol and F. I. Jacobson, "Catalysis and the UNIPOL™ process," *Stud. Surf. Sci. Catal.*, 25:323 (1986).

S. Royse, "UNIPOL™ Process brings UC continued success," *Europ. Chem. News* 28 (May 28, 1990).

THERMOSETTING RESINS

John A. Gannon

Thermosetting synthetic resins constitute a variety of manufactured polymers characterized by conversion to an insoluble, infusible state brought about by a second-stage chemical reaction, usually employing heat. The process is often referred to as curing or hardening, and the result is a cross-linked, three-dimensional structure. Prior to curing, the resinous material can be processed, along with the curing or hardening agent, in a variety of applications, including molding, casting, powder coating, and embedding, during which the composition is cured to develop the physical and chemical properties required for the specific application.

In contrast to thermoplastics, the thermosetting synthetic resins depend on the curing reaction to provide the useful properties they possess. Prominent thermosetting resins are varieties of polyesters, including both unsaturated and saturated types with linear and branched varieties sometimes containing fatty acids (alkyds). A family of polyethers with oxirane groups represents the commonly known epoxy resins that can exist in many forms, but principally are based on epichlorohydrin and bisphenol-A.

Polyurethanes composed of diisocyanate reactants with various diols or polyols make up another prominent category of synthetic resins. An important family of thermosetting resins is also derived from the condensations of phenols, ureas, and melamine derivatives with formaldehyde to form useful commercial products upon conversion to a three-dimensional state or as curing agents for other synthetic resins.

This discussion will feature synthetic resin processes used in manufacturing epoxy resins.

Epoxy resins derived from epichlorohydrin and bisphenol-A exist in two major forms: liquid epoxy resins (base liquid resin [BLR]) and solid epoxy resins derived from the liquid resin by further reaction with bisphenol-A ("advancement process"). Both the liquid and solid resins are thermoset or cured by a variety of reagents that react with the terminal epoxy groups of the resin and/or the pendant hydroxyl groups characteristic of advanced products.

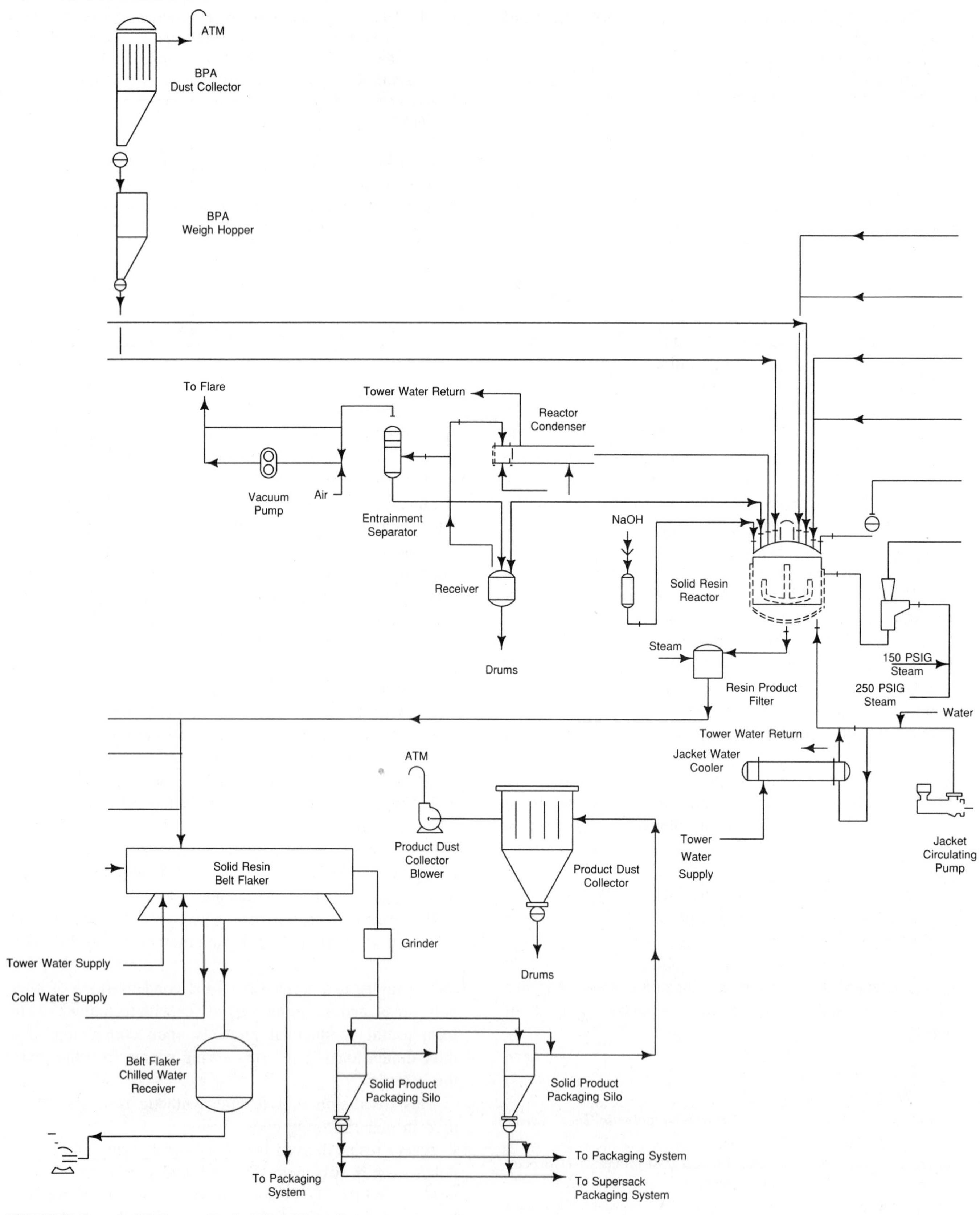

FIGURE 1. Solid Epoxy Resin Plant Flow Sheet

Important multifunctional epoxy resins are prepared by glycidylation (epoxidation with epichlorohydrin) of polyfunctional amines or amino phenols with subsequent use of sodium hydroxide as a dehydrohalogenating agent. Such resins are *tri-* or *tetra*-functional in epoxy content and upon curing with aromatic amines, for example, exhibit enhanced chemical resistance and thermal properties as a consequence of the higher cross-linking realizable with these resins.

Epoxy cresol novolak resins are derived from epoxidation of polyfunctional cresol-novolaks, which are, in turn, derived from *o*-cresol and formaldehyde in various ratios using an acid catalyst for the condensation reaction. A range of epoxy cresol novolak resins is manufactured that includes resins of different molecular weights, functionalities, and softening points.

Curing of the multifunctional epoxy cresol novolak resins again leads to products with enhanced electrical, chemical, and thermal properties. Similar resins are based on glycidylation of phenol novolaks derived from the condensation of phenol and formaldehyde in the presence of acidic catalysts.

PROCESS DESCRIPTION

The manufacture of solid epoxy resins and related solid epoxy resin solutions is accomplished by the so-called advancement process wherein a liquid epoxy resin is further reacted in a polymerization process in the presence of a catalyst to form a higher-molecular-weight solid resin possessing terminal epoxy groups and a secondary hydroxyl group in each repeat unit of the polymer chain. A series of solid resins and solid resin solutions is manufactured to various epoxy values (epoxy equivalent weights) and resin viscosities (melt or solution viscosities at various temperatures) depending on the charge ratios of liquid epoxy resin to bisphenol-A, with the former employed in excess to

TABLE 1. Emission Source Characteristics

Description	Pollutant	Emission Rate, g/s	Height, meters	Exit Gas Temperature, K	Exit Gas Velocity, m/s	Stack Diameter, meters
Packaging dust collector	Product dust	0.0210	9.1	289	23.2	0.508
BPA[a] dust collector	BPA[a]	0.161	38.1	294	14.0	0.254
BPA[a] dust collector	BPA[a]	0.161	35.1	294	14.0	0.254
Product blend or term.[b] bag	Term.[b]	0.0328	38.1	294	10.7	0.289
Charge station	Product dust	0.0328	38.1	294	10.7	0.289
Celite[c] and term.[b] bag	Term.[b]	0.0328	30.5	294	10.7	0.289
Charge station	Celite[c]	0.0328	30.5	294	10.7	0.289
Term.[b] and celite[c] bag	Term.[b]	0.0328	38.1	294	10.7	0.289
Charge station						
Vent gas combustor	Dowanol PM[d]	0.00756	48.8	1,144	27.95	0.254
	Dowanol PMA[e]	0.00171	48.8	1,144	27.95	0.254
	MAK[f]	0.00237	48.8	1,144	27.95	0.254
	MIBK[g]	0.0151	48.8	1,144	27.95	0.254
	Cyclosol-53[h]	0.00529	48.8	1,144	27.95	0.254
	N-Butanol	0.00277	48.8	1,144	27.95	0.254
	Isopropanol	0.107	48.8	1,144		
	Toluene	0.360	48.8	1,144	27.95	0.254
	Methanol	0.0464	48.8	1,144	27.95	0.254
	Acetone	0.0628	48.8	1,144	27.95	0.254
	MEK[i]	0.0788	48.8	1,144	27.95	0.254
	Xylene	0.00738	48.8	1,144	27.95	0.254

[a]Bisphenol-A.
[b]Chain terminator.
[c]Solid filter aid.
[d]Monomethyl ether of propylene glycol (Dow Chemical Co.).
[e]Monomethyl ether of propylene glycol acetate (Dow Chemical Co.).
[f]Methyl amyl ketone.
[g]Methyl isobutyl ketone.
[h]Aromatic hydrocarbon.
[i]Methyl ethyl ketone.

provide epoxy terminal groups. In some instances, a monofunctional phenol (chain terminator) is also used in minor amounts relative to the bisphenol-A charge in order to exert better control of the product viscosity.

Isolation of the solid advanced resin from the reaction vessel is performed by transferring the bag-filtered molten mass to a flaker feed tank and then to a chilled-water–cooled belt flaker. The flakes are sized and dropped into a jacketed blending hopper, from which they are fed to a packaging system. The dust generated from the sizing, flaking, and packaging is collected in a dust collector and blended into subsequent batches. See Figure 1.

The solid resin solutions are prepared by quenching (in reactors equipped with a suitable cooled reflux condenser) the molten advanced resin with solvent(s) from the solvent weigh tank via the tank farm or from drums. The resin solution, after analysis, is charged with a solid filter aid through a charging hopper and then filtered, after which it is transferred through a bag filter to a storage tank, tank truck, railcar, or drums.

Possible air emission sources in solid resin manufacture include the bisphenol-A and chain terminator storage and charging bins, as well as vents from vessels used in processing/purification of the resinous products and corresponding resin solutions.

AIR EMISSIONS CHARACTERIZATION

Table 1 details the air emission source characteristics for the various pollutants that are part of the solid epoxy resin advancement systems and the epoxy resin solution manufacture. As is evident from the table, solid bisphenol-A and chain terminator reactants are handled with multiple dust collectors or in charging stations with built-in dust collectors and the solvents used in the processing are vented to vent gas combustors from reactors and storage/processing vessels.[1]

Table 2 details the emissions inventory for solid resins and resin solutions, whereby control of total suspended particulates (TSP) and volatile organic compound (VOC) pollutant is effected.[2,3]

Table 3 summarizes the data obtained on fugitive VOC emissions, including emission factors at an epoxy resin plant.[4–7]

AIR POLLUTION CONTROL MEASURES

Two major emission control techniques are utilized in the epoxy resin manufacturing plant described here—recovery via baghouses and combustion via vent gas combustors. Display of the efficiency of these techniques is detailed in

TABLE 2. Emissions Inventory

				Pollutant Emissions with Controls		
Emission Point Description	Pollutant	Compound	Control Efficiency	Peak lb/h	Average lb/h	Annual tons/year
Packaging dust collector	TSP	Product dust	99.%	0.167	0.167	0.74
BPA[a] dust collector	TSP	BPA[a]	99.%	1.28	0.2256	0.988
BPA[a] dust collector	TSP	BPA[a]	99.%	1.28	0.024	0.104
BPA[a] dust collector	TSP	BPA[a]	99.%	1.28	0.1095	0.475
Product blend or term.[b] bag	TSP	Term.[b]	99.%	0.26	0.001	0.0046
Charge station	TSP	Product dust	99.%	0.26	0.0034	0.0148
Term.[b] and celite[c] bag Charge station	TSP	Term.[b]	99.%	0.26	0.0014	0.0125
Term.[b] and Celite[c] bag	TSP	Term.[b]	99.%	0.26	0.0014	0.0125
Celite[c] and term.[b] bag	TSP	Term.[b]	99.%	0.26	0.0015	0.0066
Charge station	TSP	Celite[c]	99.%	0.26	0.00028	0.0012
Vent gas combustor	VOC	Isopropanol	98.%	0.8466	0.064	0.2803
	VOC	Toluene	98.%	2.8555	0.296	1.2964
	VOC	Methanol	98.%	0.3683	0.052	0.2277
	VOC	Acetone	98.%	0.4983	0.0238	0.1042
	VOC	MEK[d]	98.%	0.625	0.0108	0.0473
	VOC	Xylene	98.%	0.0586	0.049	0.0214

[a]Bisphenol-A.
[b]Chain terminator.
[c]Solid filter aid.
[d]Methyl ethyl ketone.

TABLE 3. Fugitive VOC Emission Inventory

Source	Number	Emission Factor, lb/h/source	VOC Emissions without Controls, tons/year	Control Efficiency, %	VOC Emissions with Controls, tons/year
Pumps					
Light liquid (mech. seal)	53	0.109	25.3	61[a]	9.9
Heavy liquid (mech. seal)	25	0.0472	5.2	61[a]	2.0
Light liquid (sealless)	40	0.109	19.1	100[b]	0
Valves					
Light liquid	2235	0.016	162.9	59[a]	66.8
Heavy liquid	625	0.00051	1.4	—	1.4
Compressors	0	0.503	0	—	0
Relief valves	6	0.229	6.0	50[a]	3.0
Flanges	4720	0.0018	37.2	—	37.2
Open-ended lines	148	0.0037	2.4	100[c]	0
Sampling connections	37	0.0331	5.4	100[d]	0
Totals			264.9		120.3

[a]Leak detection and repair program credit.
[b]Using sealless pumps as an equipment alternative results in no leaks.
[c]Using caps on open-ended lines as an equipment alternative results in no leaks.
[d]Using closed purge sampling systems as an equipment alternative results in no leaks.

Table 2, and it can be seen that both control measures are effective in a plant devoted to solid epoxy resin and resin solution manufacture.

References

1. Central Engineering Dept. CIBA-GEIGY Corp.
2. *Rules and Regulations,* Alabama Air Pollution Control Commission, revised September 21, 1989.
3. *Compilation of Air Pollutant Emission Factors,* EPA AP-42, 4th ed., U. S. Environmental Protection Agency, September 1985.
4. "Method 21—Determination of volatile organic compounds leaks", *Code of Federal Regulations,* Title 40, Part 60, Appendix A, U. S. Environmental Protection Agency, July 1, 1986.
5. *Fugutive Emission Sources of Organic Compounds—Additional Information on Emissions, Emission Reductions, and Costs,* EPA-450/2-82-010, U.S. Environmental Protection Agency, April 1982.
6. "Standards of performance for new stationary sources: Volatile organic liquid storage vessels (including petroleum liquid storage vessels)," *Federal Register,* Vol. 52, No. 67, April 8, 1987.
7. "Subpart VV—Standards of performance for equipment leaks of VOC in the synthetic organic chemicals manufacturing industry," *Code of Federal Regulations,* Title 40, Part 60, July 1, 1986.

Bibliography

"Performance tests," *Code of Federal Regulations,* Title 40, Part 60 Section 8, July 1, 1986.

Guidelines on Air Quality Models (revised), EPA-450/2-78-027R, U. S. Environmental Protection Agency, July 1986.

P. C. Siebert, K. R. Meardon, and J. C. Serne, "Emission controls in polymer production, *Chem. Eng. Prog.,* 80(1):68 (1984).

Industrial Source Complex (ISC) Dispersion Model User's Guide, 2nd ed., Vol. I, EPA-450/4-86-005a, U. S. Environmental Protection Agency, June 1986.

Addendum to Regional Workshops on Air Quality Modeling: A Summary Report, EPA-450/4-82-015, U. S. Environmental Protection Agency, revised October 1983.

Users Manual for Single Source (CRSTER) Model, EPA-450/2-77-013, U. S. Environmental Protection Agency.

13
Food and Agricultural Industry

BREAD BAKING

Patrick J. Cafferty

In recent years, bread baking has come under increasing regulatory scrutiny by air pollution control agencies considering the control of ethanol emissions from bakery ovens.[1] Ethanol is generated from the fermentation process that takes place in yeast-leavened bakery products such as bread and bread-type rolls. Chemically leavened bakery products, such as cakes, cookies, cake doughnuts, and biscuits, do not generate significant ethanol emissions and, therefore, are not currently being considered.

The amount of ethanol generated during the bread-baking process varies depending on the type of dough process employed, with those processes that use a higher percentage of yeast and having a longer fermentation time tending to generate the largest amount of ethanol. Several different types of dough processes are employed in large commercial bakeries, including sponge dough, straight dough, liquid and flour brews, and continuous-mix or "no time" doughs. Sponge dough processes typically will have the highest percentage of yeast and longest fermentation time and, therefore, will usually generate the greatest amount of ethanol.[2] But while the sponge dough process is commonly used in the formulation of white breads, the popularity of and new demand for other bread varieties have resulted in the broad use of virtually all dough processes. As a result, no single dough process is prevalent today.

Bread ovens are predominantly directly fired by natural gas. In such ovens, the ethanol vapors driven from the bread during baking and the combustion product gases are removed from the oven through the same exhaust stacks.

Unlike industrial drying ovens where controls can be readily installed on exhaust stacks, controlling bakery oven exhaust emissions, particularly from existing uncontrolled ovens, presents difficult problems because of the sensitivity of the baking process. Indeed, bread baking, even on the largest commercial scale, is still both an art and a science, with the oven playing an intricate role in the transformation of a living dough into bread. To accomplish this result, ovens are critically balanced to allow the baking of a finished bread product conforming to federal food safety, moisture, and weight requirements, while achieving proper crust formation, loaf volume, internal texture, flavor, color, and shelf life. Any interference with this critical balance can result in the baking of "unsalable" bread.

PROCESS DESCRIPTION

Bread baking at large commercial bakeries is a highly mechanized process consisting of high-speed production lines with ovens capable of baking 20,000 pounds or more of bread per hour. The process starts with the mixing of flour, water, sugar, and yeast to form dough, thereby initiating a long series of complex biochemical changes that ends in the oven where the bread is baked.

There are four basic types of dough mixing processes: sponge dough, straight dough, brew, and continuous-mix ("no-time"). These processes vary in the manner in which the various dough ingredients are mixed, which determines the amount of fermentation time available. Fermentation time can vary from five hours or more in the sponge dough process, where a "sponge" is formed when two thirds of the flour, part of the water, and the yeast are initially mixed and allowed to ferment before the remaining ingredients are added, to 20 minutes or less for the continuous-mix or no-time process, where all of the dough ingredients are mixed at the same time and fermentation time is minimized through the use of processing agents and higher temperatures.

The baking process actually occurs in the oven itself, which causes expansion of the loaf to final volume, crust formation, yeast and enzymatic activity inactivation, coagulation of dough proteins, partial gelatinization of starch, and reduction of loaf moisture, all of which are necessary to produce high-quality, salable bread products. To accomplish all of these product and process effects in the proper sequence, commercial bread ovens have from three to eight temperature gradient zones, which are maintained in critical balance. Oven rise, which determines the final loaf volume and internal texture, occurs during the first five to six minutes of baking. Thermal death of the yeast occurs as the internal bread temperature reaches 140–145°F, there-

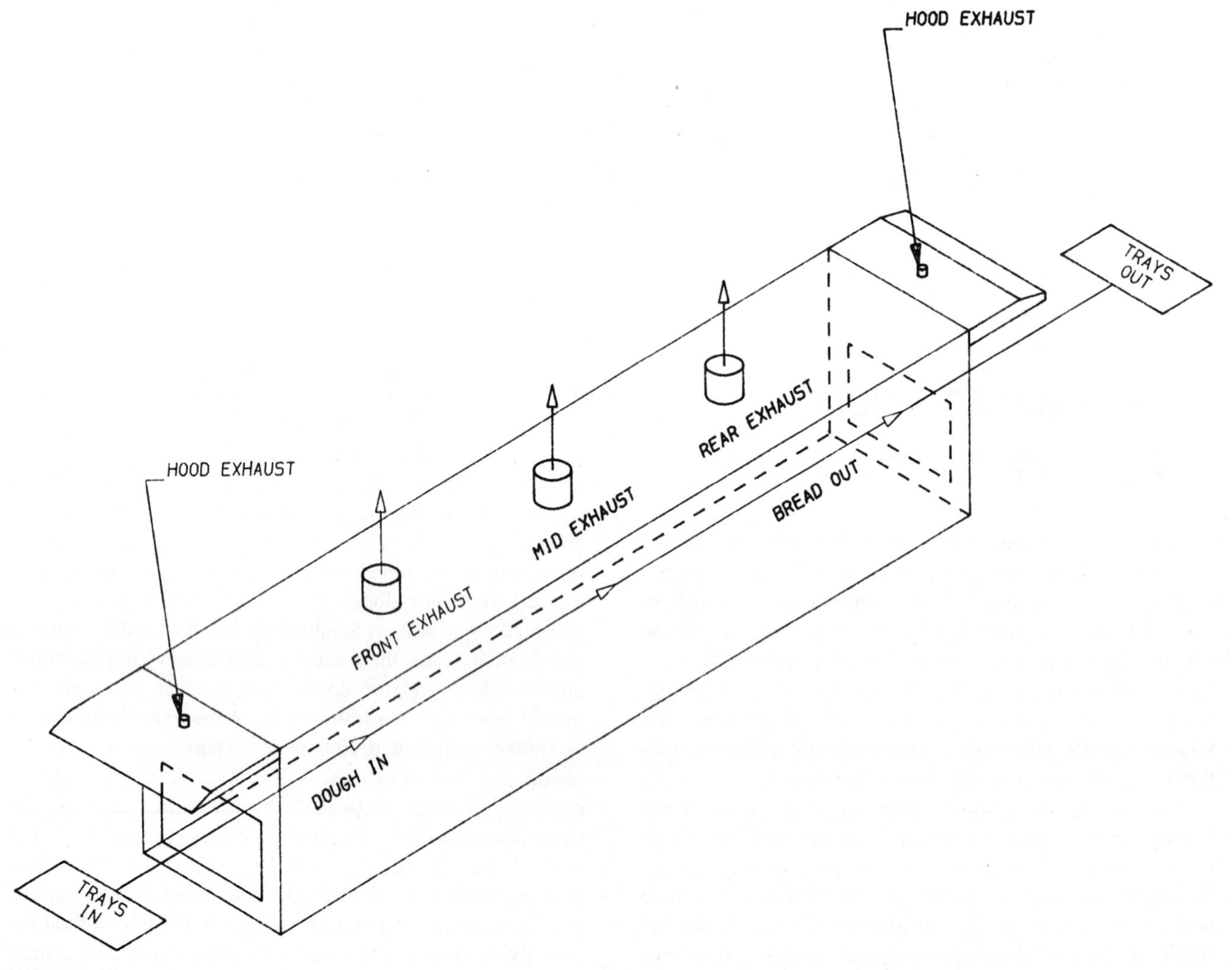

FIGURE 1. Tunnel Oven

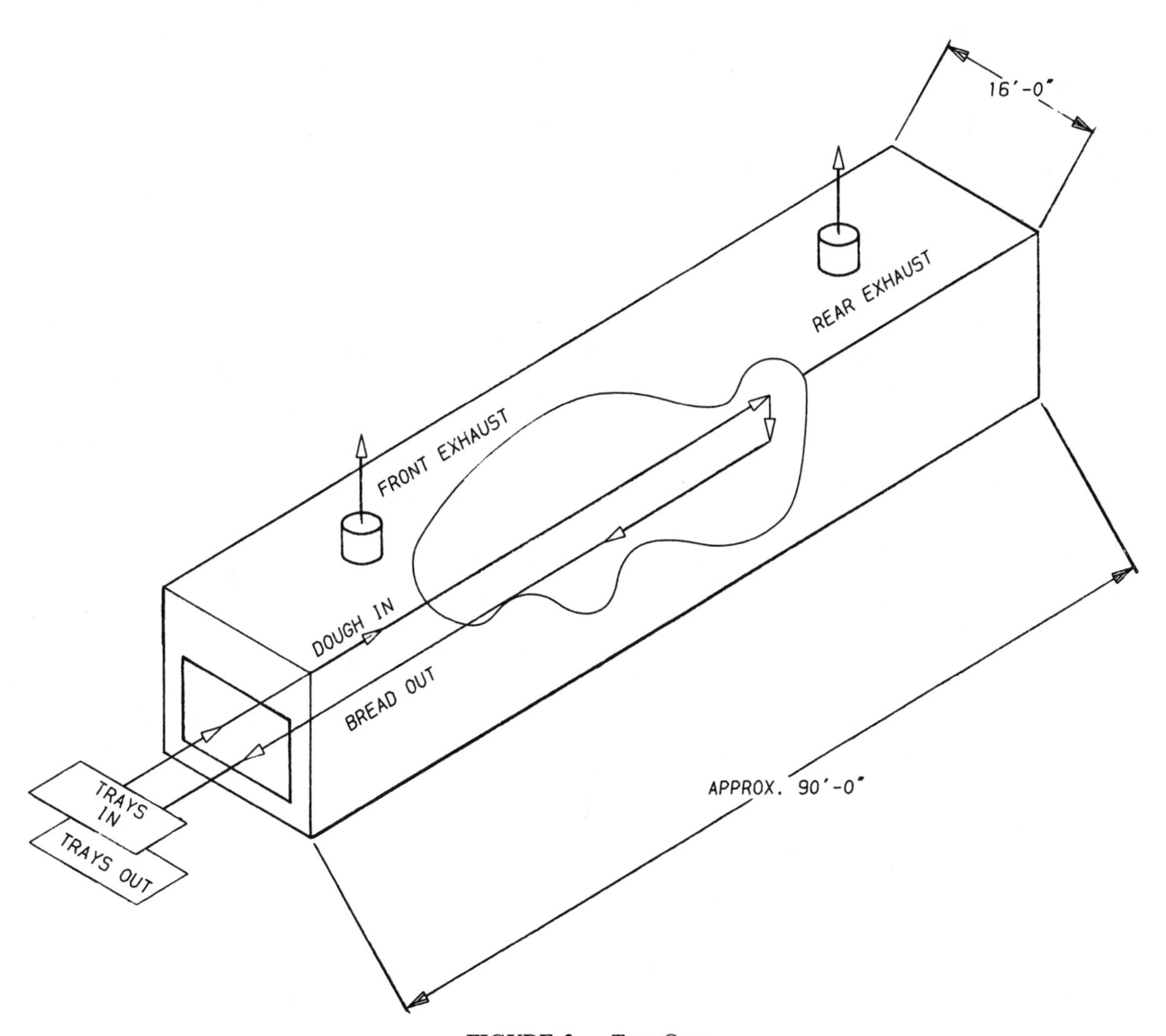

FIGURE 2. Tray Oven

by stopping the fermentation process. Protein is denatured at between 140°F and 180°F. At the end of the process, browning and crust color develop, while ethanol and moisture are evaporated to cool the loaf and prevent the internal temperature from reaching the boiling point of water.[3]

There are three fundamental oven types: tunnel, tray, and spiral (Figures 1–3). Tunnel ovens are long horizontal ovens in which dough enters at one end and is conveyed to the opposite end, where it exits as bread. Tray ovens are also horizontal, but the dough entering the oven exits on the same side as that on which it entered, after being conveyed the length of the oven, lowered to a second level, and conveyed to the exit near where it entered. In spiral ovens, dough enters at the top corner of the oven and is conveyed in a downward spiral to the bottom corner of the oven, where it exits through an opening diagonally below where it entered the oven. Tunnel and tray ovens typically contain from three to five exhaust stacks, with one stack typically used for purging the oven of natural gas during ignition and the remaining stacks in use during normal baking operations. In contrast, spiral ovens usually contain just one stack, which is used during both purging and normal operations.[4]

AIR EMISSIONS CHARACTERIZATION

Although combustion gases from direct-fired ovens are also emitted through oven exhaust stacks, stack tests performed by the Bay Area Air Quality Management District (BAAQMD) demonstrated that ethanol from fermentation constitutes more than 99% of the volatile organic compound (VOC) emissions from bread ovens.[5] Regulatory concern has focused on ethanol because it is considered to be a reactive hydrocarbon; that is, capable of forming smog. However, several scientific studies have demonstrated that ethanol is actually much less reactive than hydrocarbons routinely emitted by other mobile and stationary sources (as little as one quarter as reactive).[6] Based on this information and given the relatively small amount of ethanol emitted by even the largest bread ovens, air quality modeling performed by the San Francisco Bay area has demonstrated that

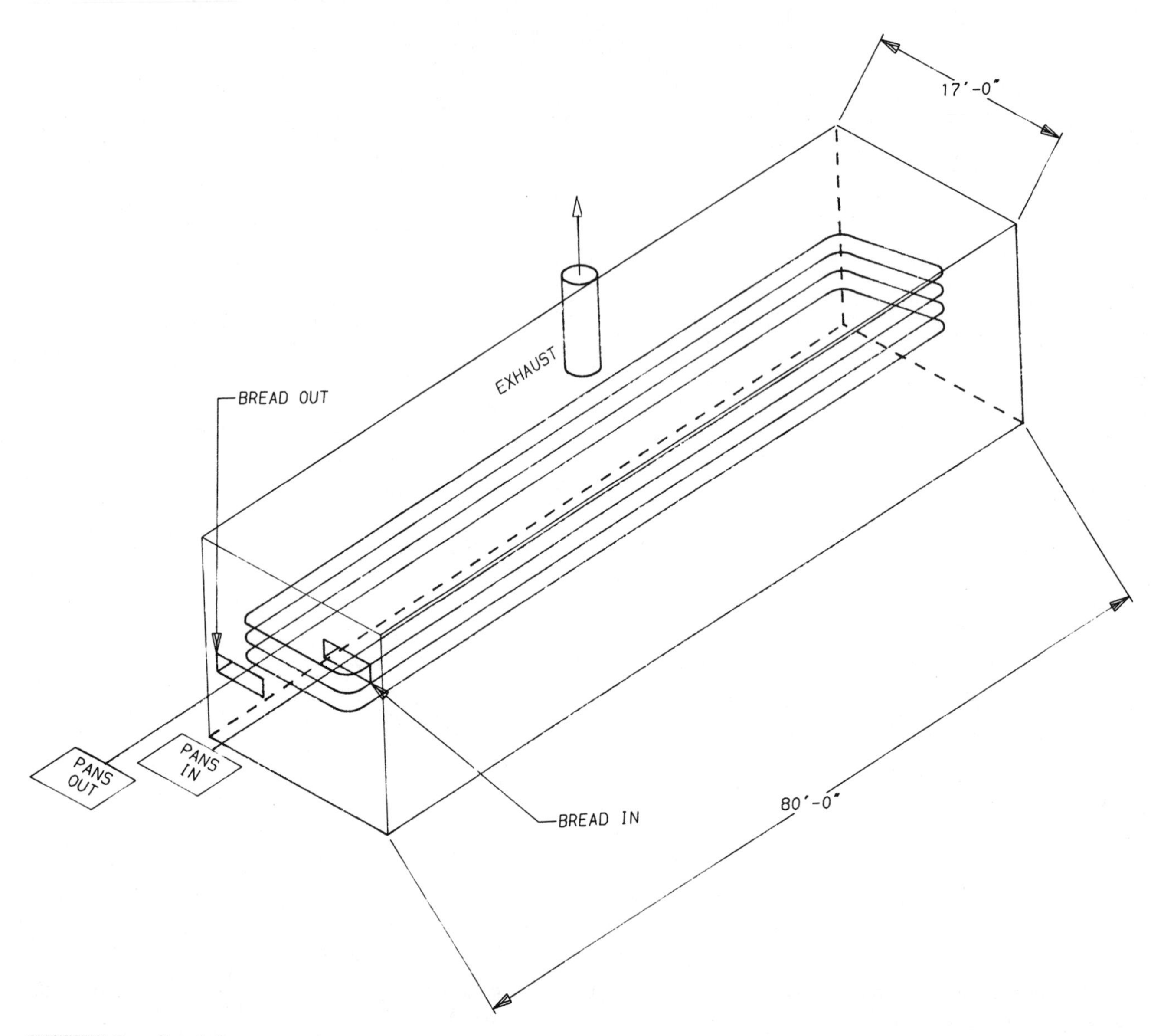

FIGURE 3. Spiral Oven

ethanol emissions from large commercial bakeries in the area have a virtually immeasurable effect on ozone levels.[7]

The ethanol is generated as a result of the fermentation process, which causes the sugars and starches to be converted to ethanol, carbon dioxide, and water. Fermentation begins immediately following the initial mixing of ingredients and continues until the yeast is killed in the oven. During the initial fermentation period, a skin forms on the top of the dough. The skin keeps the ethanol and carbon dioxide inside the dough, thereby allowing the dough to rise and minimizing fugitive emissions. As a result, most of the ethanol generated by fermentation is released in the oven during baking. This is confirmed by the stack testing, which has demonstrated that ethanol emissions from bread bakery vents and stacks other than oven exhaust stacks are very low, if not negligible.[8] In multiple-stack tunnel and tray ovens, the BAAQMD stack tests have shown that 60–70% or more of the ethanol typically exits from one "principal" stack in the oven, although the exact percentages appear to vary from oven to oven and by bread type.[9]

The BAAQMD stack test results were also analyzed by the American Institute of Baking (AIB) to develop appropriate ethanol emission factors for bread baking.[10] Based on that evaluation, as well as a separate chemical engineering evaluation, the AIB has developed a formula for calculating ethanol emission factors based on the percent of yeast in the dough and the total fermentation time. The AIB emission factor formula is set forth in Table 1.

AIR POLLUTION CONTROL MEASURES

Given the sensitivity of the bread-baking process, the application of air pollution controls to bakery oven exhaust stacks, particularly retrofit applications, presents a difficult

TABLE 1. Ethanol Emissions (AIB Formula)

$[(y \cdot t)\ 0.444585] + 0.40425$ = ethanol factor, pounds per ton
where: y = total bakers percent yeast in the formula*
t = total of all proofing, floor, and fermentation times expressed in hours to the closest tenth of an hour
*Bakers percent yeast = yeast added in pounds/100 pounds flour in bread formula

Note: The AIB formula should be used to predict ethanol emissions only for yeast-raised products.

technical feasibility problem and is generally more expensive than other applications. In fact, the BAAQMD, which carefully studied the feasibility of controlling bakery oven emissions, concluded that only incinerator control devices are feasible for bakery ovens. More important, the BAAQMD also concluded that the serious technical feasibility obstacles warranted an extended period for retrofitting existing ovens and made it reasonable to control only the single highest-emitting stack on a multistack oven.[4,9]

Available Controls

The BAAQMD evaluated several types of controls commonly applied to reduce hydrocarbon emissions, including incineration, activated-carbon absorption, wet scrubbing, and refrigerator condensation, to assess their potential applicability to bakery ovens. With the exception of incineration (either thermal or catalytic), the BAAQMD concluded that none of the control systems could be applied to a bakery oven. Specifically, the BAAQMD found that high water and energy requirements precluded the use of wet scrubbing and refrigeration/condensation systems and that carbon regeneration and waste disposal problems precluded the use of an activated absorption system. Similar conclusions were reached by Frederiksen Engineering, which also examined these issues.[11]

With either direct or catalytic incineration, a combustion process is used to oxidize the organic compounds in the bakery oven exhaust to primarily carbon dioxide and water vapor. Auxiliary heat required to raise the oven exhaust to combustion temperatures is provided by burning natural gas or some other gaseous fuel. Temperatures up to 1450°F are required for conversion in direct incineration and up to 1000°F for catalytic incineration. In both cases, fuel consumption can be reduced by primary heat exchange of inlet and outlet gases. Catalytic incineration has been successfully applied to the installation of a new single-stack spiral oven in the San Francisco area. This unit is achieving a 90% reduction in organic emissions on a consistent basis.

Incinerators are most effective at controlling exhaust streams with relatively high concentrations of organics. Ethanol concentrations in the oven exhaust gas can substantially decrease during gaps in production. The incinerator may not be able to achieve 90% reduction during these periods.

Catalytic incinerators usually require a very clean exhaust stream to avoid fouling the catalyst bed. There is some concern that oils and other trace compounds in bakery oven exhaust may foul the catalyst over time and reduce the destruction efficiency. The one San Francisco area application has not yet experienced these problems.

Control Constraints

While the installation of controls on a new bakery oven presents difficulties not found with other oven applications where the process is not as sensitive as the baking process, the application of controls to the new single-stack oven in San Francisco has demonstrated that a new oven designed to accept controls can be controlled successfully. However, the retrofitting of existing bakery ovens, particularly multiple-stack ovens, raises technological feasibility issues concerning the risk of interference with the baking process, increased risk of bacterial contamination, constructibility problems based on space and sanitation constraints, and bread production demands.[4,12]

The most significant unresolved question concerning the retrofitting of existing bakery ovens is the ability to maintain the airflow balance if a single control unit is used for multiple stacks. To assure a consistent quality of bake, the exhaust flows in each oven zone must be steady and the operator must be able to adjust each zone independently of the other zones. Adjusting the exhaust damper in one stack may affect the airflow through another stack. These exhaust dampers may require small incremental adjustments, relying on the skill of the oven operator to assure an even bake. Turbulent airflows within the oven result in unstable ribbon burner flames, uneven lateral heat distribution, incorrect zone temperatures, and, ultimately, improperly baked bread and unsalable product.

While damper controls theoretically might solve the airflow balance problem, they have not been used on bakery ovens and their application is technologically uncertain. Independent operation of the dampers could be achieved by having separate control devices for each stack. However, a multiple-control system would be significantly more expensive than a single unit.

Another potential problem concerns the possible buildup of grease or biological growth in long runs of horizontal ductwork. Contamination can be minimized by installing insulated ducting, with the ductwork sloped toward the control device. This would reduce the potential for condensation and cause any condensate to flow toward the incinerator instead of the oven. Traps could be installed to collect the condensate and to allow for easier cleaning.

Maintenance of tight sanitation standards is also of concern during construction because such standards must be observed even when the operating oven is being retrofitted with controls. Special construction techniques must be used to avoid bread contamination during retrofit, particularly since space constraints typically require installation of the incinerator controls on the roof.

Bakeries normally operate on a schedule that will permit the shutdown of equipment two days per week for sanitary purposes. But since bakeries must supply fresh bread for the marketplace, the ovens cannot be taken out of service for any period longer than that allowed by a bakery's normal schedule without interrupting customer supplies. Disruption of the operating schedule for any reason could severely hamper the baker's ability to meet market demands, resulting in the possible loss of revenue and market selling space. The combination of market-supply risks posed by the technological uncertainties of retrofit and the high cost of controls has caused the Illinois Environmental Protection Agency to conclude that controls cannot be cost-effectively applied to bakery ovens, while the BAAQMD has concluded that it is reasonable to control only the single highest-emitting stack on a multiple-stack oven. Therefore, to reduce the risk of business interruption, retrofit controls should be limited to a single stack on multistack bread ovens.

References

1. See Bay Area Air Quality Management District ("BAAQMD") Rule 8-42 (regulating ethanol emissions from bread ovens) and Illinois Environmental Protection Agency Rule 86-18 (deciding not to regulate ethanol emissions from bread ovens).
2. E. J. Pyler, *Baking Science and Technology,* 3rd ed., Sosland Publishing Co., Merriam, KS, 1988, p. 589.
3. J. W. Stitley, *Baking Technology, Oven Emissions and Control Devices,* American Institute of Baking, Manhattan, KS, 1986.
4. BAAQMD Staff Report supporting adoption of proposed Rule 8-42, July 1988.
5. BAAQMD bakery oven source test results, 1985 and 1986.
6. G. Whitten et al., *Potential Impact of Ethanol Emissions on Urban Smog: A Preliminary Modelling and Smog Chamber Study,* Systems Applications, Inc., San Rafael, CA, 1986.
7. G. Whitten et al., *Assessment of the Impact of Baking Emissions on Air Quality, Impact of Ethanol Emissions on Ozone Formation in the San Francisco Bay Area,* Systems Applications, Inc., San Rafael, CA, 1988.
8. H. Weiss, Chief Environmental Services, Pennsylvania Department of Environmental Resources, 1989 field test results.
9. BAAQMD Supplemental Staff Report supporting adoption of Rule 8-42, September 1988.
10. J. W. Stitley et al., *Bakery Oven Ethanol Emissions Experimental and Plant Survey Results,* American Institute of Baking Technical Bulleting IX, 12, Manhattan, KS, 1987.
11. J. Fox et al., *Preliminary Assessment of Cost of Ethanol Control Strategies, Safeway Stores, Inc., Bakery Division, Richmond, California,* Frederiksen Engineering, Oakland, CA, 1986.
12. *Draft Control of Ethanol Emissions From Ovens: Bakery Demonstration Program,* American Bakers Association, Washington, DC, 1988.

COFFEE PROCESSING

*Ronald G. Ostendorf, Editor**

Most coffee is grown in Central and South America. After harvesting and drying at or near the coffee plantation, most "green" coffee beans are exported and further processed before sale to the consumer. Coffee processing in the United States consists essentially of cleaning, roasting, grinding, and packaging.

Roasting is the key operation and produces most of the air contaminants associated with the industry. Roasting reduces the sugar and moisture contents of green coffee and also renders the bulk density of the beans about 50% lighter. An apparently desired result is the production of water-soluble degradation products that impart most of the flavor to the brewed coffee. Roasting also causes the beans to expand and split into halves, releasing small quantities of chaff.

BATCH ROASTING

The oldest and simplest coffee roasters are direct-fired (usually by natural gas), rotary, cylindrical chambers. These units are designed to handle from 200 to 500 pounds of green beans per 15- to 20-minute cycle and are normally operated at about 400°F. A calculated quantity of water is

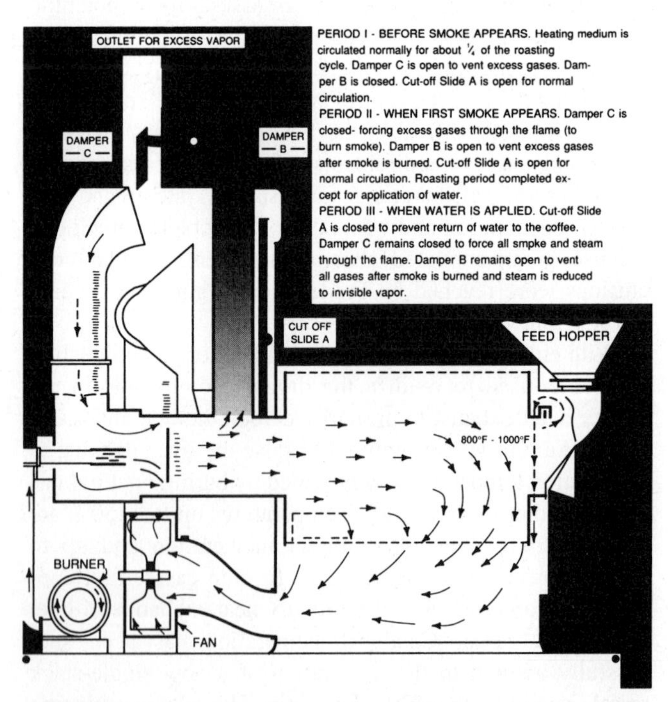

FIGURE 1. A Recirculating-Batch Coffee Roaster (Jabez Burns—Gump Division, Blaw-Knox Co., New York, N.Y.)

*Source: AP-40 *Air Pollution Engineering Manual,* Second Edition, U.S. Environmental Protection Agency

added at the completion of the roast to quench the beans before discharge from the roaster. After they are dumped, the beans are further cooled with air and run through a "stoner" air classifier to remove metal and other heavy objects before the grinding and packaging. The roaster and cooler and all air-cleaning devices are normally equipped with cyclone separators to remove dust and chaff from exhaust gases. Most present-day coffee roasters are of batch design, though the newer and larger installations tend to favor continuous roasters.

In the batch roaster shown in Figure 1, some of the gases are recirculated. A portion of the gases is bled off at a point between the burner and the roaster. Thus the burner incinerates combustible contaminants and becomes both an air pollution control device and a heat source for the roaster.

AN INTEGRATED COFFEE PLANT

A process flow diagram of a typical large integrated coffee plant is shown in Figure 2. Green beans are first run through mechanical cleaning equipment to remove any remaining hulls and foreign matter before the roasting. This system includes a dump tank, scalper, weigh hopper, mixer, and several bins, elevators, and conveyors. Cleaning systems such as this commonly include one or more centrifugal separators from which process air is exhausted.

The direct, gas-fired roasters depicted in Figures 2 and 3 are of continuous rather than batch design. Temperatures of 400°F to 500°F are maintained in the roaster, and the residence time is adjusted by controlling the drum speed. Roaster exhaust products are drawn off through a cyclone separator and afterburner, with some recirculation from the cyclone to the roaster. Chaff and other particulates from the cyclone are fed to a chaff collection system. Hot beans are continuously conveyed through the air cooler and stoner sections. Both the cooler and the stoner are equipped with cyclones to collect particulates.

The equipment following the stoner is used only to blend, grind, and package roasted coffee. Normally, there are no points in these systems where process air is emitted to the atmosphere. At the plant shown on the flow diagram, chaff is collected from several points and run to a holding bin from which it is fed at a uniform rate to an incinerator. Conveyors in the chaff system may be of almost any type, though pneumatic conveyors are most common.

THE AIR POLLUTION PROBLEM

Dust, chaff, VOCs, coffee bean oils (as mists), visible emission, and odors are the principal air emissions issues with coffee processing. In addition, combustion contaminants are discharged if chaff is incinerated. Dust is ex-

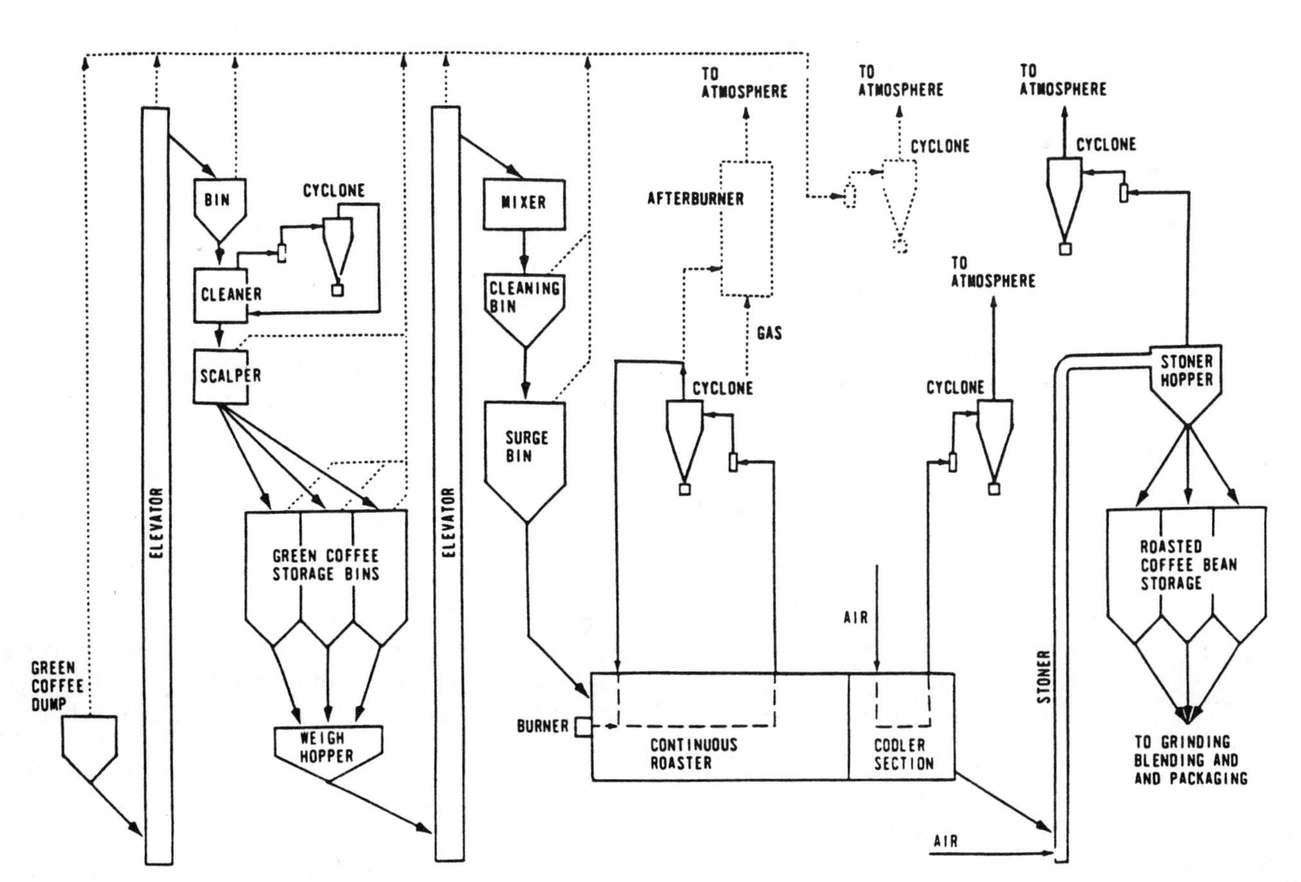

FIGURE 2. Typical Flow Sheet for a Coffee-Roasting Plant

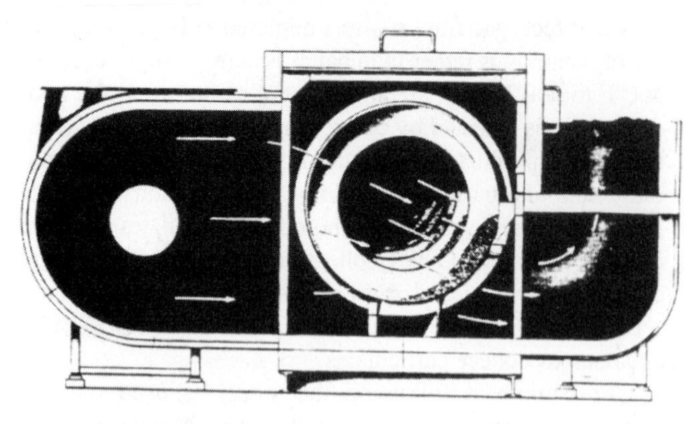

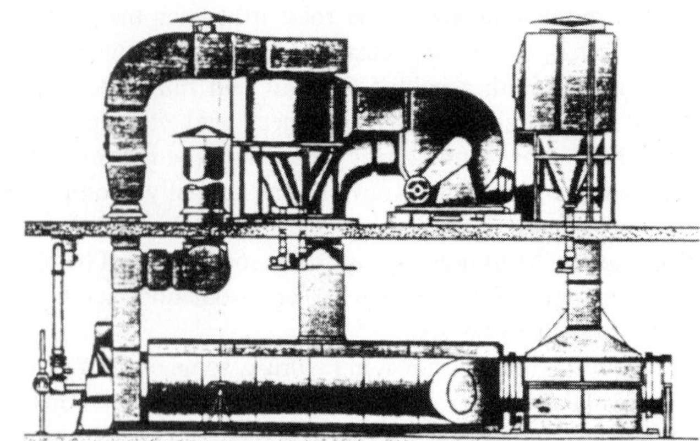

FIGURE 3. Continuous Coffee Roaster Showing Course of the Heated Gases as They Are Drawn Through the Coffee Beans in the Perforated, Helical-Flanged Cylinder and then into the Recirculation System. (Right) Left-Side Elevation of Continuous Coaster, Showing Relationship of Recirculating and Cooler Fans and the Respective Collectors on the Roof (Jabez Burns—Gump Division, Blaw-Knox Company, New York, N.Y.)

hausted from several points in the process, while smoke and odors are confined to the roaster, chaff incinerator, and, in some cases, the cooler.

Coffee chaff is the main source of particulates, but green beans, as received, also contain appreciable quantities of sand and miscellaneous dirt. The major portion of this dirt is removed by air washing in the green coffee-cleaning system. Some chaff (about 1% of the green weight) is released from the bean on roasting and is removed with roaster exhaust gases. A small amount of chaff carries through to the cooler and stoner. After the roasting, coffee chaff is light and flaky, particle sizes usually exceeding 100 μm. As shown in Table 1, particulate-matter emissions from coffee processing are well below the limits permitted by typical dust and fume prohibitions.

Coffee roaster VOC odors are attributed to alcohols, aldehydes, organic acids, and nitrogen and sulfur compounds, which are all probably breakdown products of sugars and oils. Roasted coffee odors are considered pleasant by many people, and they may often be pleasant under certain conditions. Nevertheless, continual exposure to uncontrolled roaster exhaust gases usually elicits widespread complaints from adjacent residents. The pleasant aroma of a short sniff apparently develops into an annoyance upon long exposure.

TABLE 1. Analysis of Coffee Roaster Exhaust Gases

	Contaminant Concentration		
	Continuous Roaster		
	Roaster	Cooler	Batch Roaster
Particulate matter, gr/scf	0.189	0.006	0.160
Aldehydes (as formaldehyde), ppm	139	—	42
Organic acids (as acetic acid), ppm	223	—	175
Oxides of nitrogen (as NO_2), ppm	26.8	—	21.4

Visible bluish-white emissions from coffee roasters are caused by distilled oils and organic breakdown products. The moisture content of green coffee is only 6% to 14%, and thus there is not sufficient water vapor in the 400–500°F exhaust gases to form a visible steam plume. From uncontrolled, continuous roasters, the opacity of exhaust gases exceeds 40% almost continuously. From batch roasters, exhaust opacities normally exceed 40% only during the last 10 to 15 minutes of a 20-minute roast. Plume opacity appears to be a function of the oil content, with the oilier coffee producing the heavier smoke. The water quenching of batch-roasted coffee causes visible steam emissions that seldom persist longer than 30 seconds per batch.

HOODING AND VENTILATION REQUIREMENTS

Exhaust volumes from coffee-processing systems do not vary greatly from one plant to another as far as roasting, cooling, and stoning are concerned. Roasters equipped with gas recirculation systems exhaust about 24 scf per pound of finished coffee. Volumes from nonrecirculation roasters average about 40 scf per pound. A 10,000-lb/h, continuous roaster with a recirculation system exhausts about 4000 scfm. A 500-pound-per-batch, nonrecirculation roaster exhausts about 1000 scfm. Each batch cycle lasts about 20 minutes.

Coolers of the continuous type exhaust about 120 scf per pound of coffee. Batch-type coolers are operated at ratios of about 10 scfm per pound. The time required for batch cooling varies somewhat with the operator. Batch-cooling requirements are inversely related to the degree of water quenching employed.

Continuous-type stoners use about 40 scf air per pound of coffee. Batch-stoning processes require from 4 to 10 scfm per pound, depending on ductwork size and batch time.

AIR POLLUTION CONTROL EQUIPMENT

Air contaminants from coffee-processing plants have been successfully controlled with afterburners and cyclone separators, and combinations thereof. Incineration is necessary only with roaster exhaust gases. There is little smoke in other coffee plant exit gas streams where only dust collectors are required to comply with air pollution control regulations.

Separate afterburners are preferable to the combination heater-incinerator of the batch roaster shown in Figure 1. When the afterburner serves as the roaster's heat source, its maximum operating temperature is limited to about 1000°F. A temperature of 1200°F or greater is necessary to provide good particulate incineration and odor removal.

A roaster afterburner should always be preceded by an efficient cyclone separator in which most of the particulates are removed. A residence time of 0.3 second is sufficient to incinerate most vapors and small-diameter particles at 1200°F. Higher temperatures and longer residences are, however, required to burn large-diameter, solid particles.

Properly designed centrifugal separators are required on essentially all process airstreams up to and including the stoner and chaff collection system. With the plant shown, cyclones are required at the roaster, cooler, stoner, chaff storage bin, and chaff incinerator. In addition, the scalper is a centrifugal classifier venting process air. Some plants also vent the green coffee dump tank and several conveyors and elevators to centrifugal dust collectors.

The inorganic ash content of the chaff, at approximately 5% by weight, is considerably greater than that of most combustible refuse fed to incinerators. Provisions should be made in the incinerator design so that this material does not become entrained in the exhaust gases. If most of the noncombustible material is discharged with products of combustion from the incinerator, the combustion contaminants then exceed 0.3 gr/ft^3 calculated to 12% carbon dioxide.

GRAIN HANDLING AND PROCESSING

Dennis Wallace

Grain handling and processing facilities move grain (wheat, corn, soybeans, rice, oats, etc.) from the farm, through storage and transfer locations (grain elevators), to the mills and plants that generate a variety of products (flour, starch, oil, animal feed, etc.).[1] A key characteristic of the industry is its diversity, which reflects the geographic dispersion of the raw-materials sources (farms), the variety of raw materials (different types and qualities of grain), and the wide range of final products. Consequently, the industry comprises a large number of facilities (as many as 9,000 grain elevators,[2] about 500 grain mills,[3] and 15,000 feed manufacturing plants[4]) spread throughout the United States in areas with vastly different population densities. Individual facilities vary in size and use different unit processes. As a consequence of their diversity of location and operating characteristics, each facility has different emission levels, emissions control needs, and technical and economic feasibility of emissions control. This discussion includes a brief overview of the emission characteristics and control alternatives, but because the industry is so diverse, the information should be used very cautiously in assessing emission problems or control options at an individual facility or group of facilities.

In the broadest sense, the grain handling and processing industry can be defined as including all operations, from the point of production (the farm) to the points where final products are prepared for consumer use. This review will focus on a narrower segment of the industry in which facilities have similar air pollution problems. Specific segments that will be considered are grain elevators, grain milling (including flour mills and wet corn and rice milling), and animal feed manufacturing, segments that have many common unit processes. The remainder of this presentation provides a general description of each of these industry segments. The operating characteristics of the different unit processes and the mechanisms by which they generate emissions are described later.

Grain elevators transfer, condition, and store grains and soybeans as they are passed from the farmer to the grain processor or exporter. The industry divides elevators into two classes—country elevators and terminal elevators—as a function of their size, source of grain, and destination of shipments. Country elevators are those facilities that receive the bulk of their grain directly from the farm. Terminal elevators, those that ship grain directly to a processor or export grain, are often further divided into port terminals and inland terminals (or subterminals). The subterminals receive the bulk of their grain from country elevators, but also receive a small amount directly from farmers; they ship grain to port terminals or directly to processors. Although these different elevator classes vary greatly in size, all generally perform the same receiving, conditioning, internal handling and transfer, and storage and shipping operations.

A schematic of a traditional country elevator is shown in Figure 1.[5] Newer country elevators differ from the one shown in that they tend to have free-standing metal legs and metal storage bins. The typical subterminal or terminal elevator is much more complex than the facility shown. However, the figure illustrates the operations found at all grain elevators.

Grain is received at the elevator by truck, rail, or barge

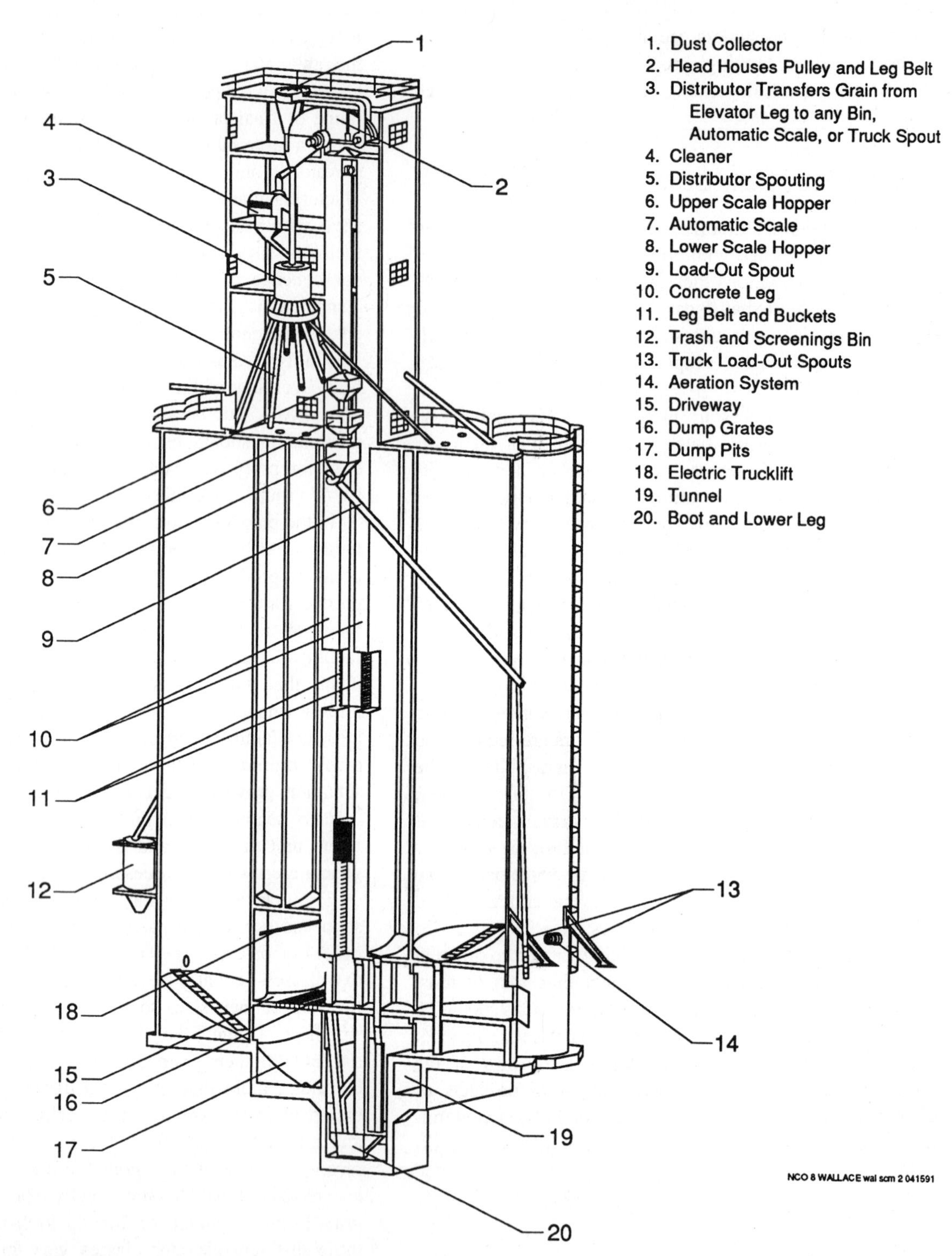

FIGURE 1. Schematic of a Traditional Country Elevator

and is unloaded by one of a variety of receiving systems, which include gravitational processes, power shovels, and augers. As grain is unloaded, it is transferred by gravity to the heart of the elevator operations—an enclosed bucket elevator called the "leg." Almost all grain transfer within the elevator is accomplished by a combination of gravity flow, belt conveyor, and the elevator leg. Grain flows into the bottom, or "boot," of the leg and is picked up by the bucket elevator and carried up to the top of the leg—"the headhouse." Here, the grain is discharged to the distributor, which can discharge it directly to a load-out spout, to a holding bin for load-out, to the gallery belt for transfer to a storage bin, or to an ancillary operation such as weighing, cleaning, scalping, or drying. Grain moves by conveyor through an enclosed area called the gallery to the appropriate bin, where it is discharged from the belt into the bin by a tripper or is transferred directly from the distributor through a metal spout to a detached bin. When stored grain is to be shipped or conditioned further, it is discharged by gravity from the bottom of the bin onto the tunnel belt, a belt conveyor that transfers the grain to the leg for subsequent distribution.

Grain processing facilities have receiving, handling, and transfer operations comparable to those found in grain elevators. Each processing facility also has unit processes that are specific to its final products. Detailed process descriptions for different types of grain mills are presented in reference 6; Figure 2 illustrates one type of mill, flour milling from wheat.[6] Although specific operations vary from plant to plant, grain mills generally use four major types of operation: a series of cleaning steps to remove light impurities (hulls, straw, dust, and metallic objects); tampering or conditioning to adjust the moisture content of the grain; milling operations such as grinding, sifting, rolling, aspirating, and drying; and packaging or bagging as illustrated in Figure 2.

A detailed description of the animal feed manufacturing industry and the production processes used in the industry is presented in reference 7. A simplified diagram of a typical smaller plant is shown in Figure 3.[8] Again, these facilities use receiving and storage operations that are comparable to those used in grain elevators. As grain is received, it goes through scalpers to remove coarse materials such as sticks, husks, and straw. Then the grain is stored or transferred directly to the grinding area. The ground materials are mixed, generally in batch units. The material from the mixer is a meal, or mash, and may be bagged and marketed as is.[9] However, the material is often formed into pellets in the pellet mill. Further size reduction, if needed, is provided by the crumbler or granulator. Finally, screens are used to remove undersized and oversized materials, and the final product is transferred to the bagging area, to storage, or to bulk load-out.

PROCESS DESCRIPTION

Although each grain handling and processing facility is unique, many different facilities use the same unit operations. Further, different unit operations are similar in that they are either materials handling operations or operations such as grinding, sizing, and mixing that mechanically change grain characteristics. Consequently, common mechanisms generate emissions from these operations. Each process is a potential source of process fugitive emissions because of the dusty nature of grain and the dirt that is mixed with the grain during harvest and transport. Even processes involving "clean" grain are subject to emissions because additional grain dust is generated by abrasion and breakage as grain is handled and processed. Particulate emissions are generated by a combination of mechanical energy imparted to particles in handling and processing operations and particle entrainment by drag forces in the locally turbulent airflows that accompany these operations. These entrained particles are then emitted from the facility via natural airflows or process or building ventilation systems. The following briefly describes the more common grain handling and processing operations and identifies how emissions are generated from these processes.

Grain Receiving

Grain handling and processing facilities receive grain by truck, railroad hopper car, railroad boxcar, barge, and ship. Unloading procedures depend on the type of vehicle delivering the grain, and multiple options are available for unloading from each type of vehicle. However, most receiving procedures generate emissions by the same mechanisms. The two principal factors that affect emission levels during bulk unloading are local wind currents in the receiving area and the quantity of dust generated when the falling stream of grain strikes the receiving pit. When falling grain strikes an immovable object (e.g., the base of a pit), the energy expended causes extreme air turbulence. A violent generation of dust occurs from a combination of splashing of dust particles upon impact and entrainment of dust in the turbulent airstream.[10]

Trucks are generally unloaded at some type of dumping platform or dump pit. The area may be unenclosed, partially enclosed, or fully enclosed. Typically, the front of the truck is elevated and grain flows from the back of the truck through a steel grate into an underground pit. Emission levels are affected by drop height and wind currents in the pit area. Hopper cars are unloaded into a pit similar to the one used for truck unloading, but grain is unloaded through the bottom of the car. The dust emissions from this operation can be reduced by allowing the grain to flow at a velocity high enough that a cone forms around the hopper discharge at the receiving pit (i.e., choke unloading) and by limiting wind velocities in the pit area.

Two methods are commonly used for boxcar unloading. In the first, the inside grain door is broken, allowing a surge of grain to flow from the car into the receiving pit. After this initial surge stops, the remaining grain is unloaded into the pit with a power shovel or bobcat. Alternatively, some large facilities use a mechanical car dump, which clamps the car to a movable section of track and rotates and tilts it to dump grain from the car door into the receiving pit. Again, the primary factors that affect emissions are the drop pattern and wind currents in the receiving area. Barges and ships typically are unloaded by either a bucket elevator that is lowered into the ship's hold or a pneumatic discharge system.[10]

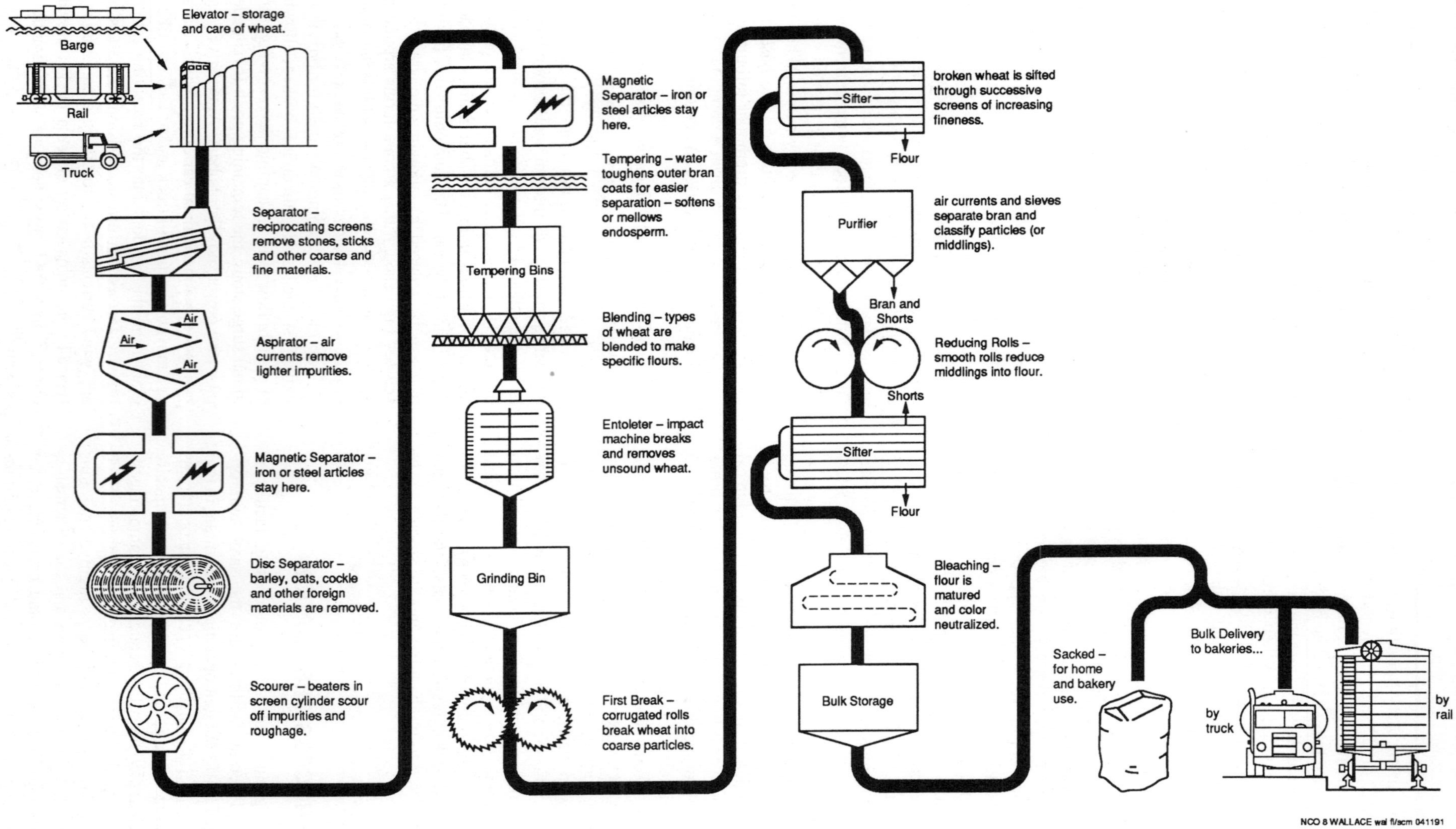

FIGURE 2. Grain Milling Process

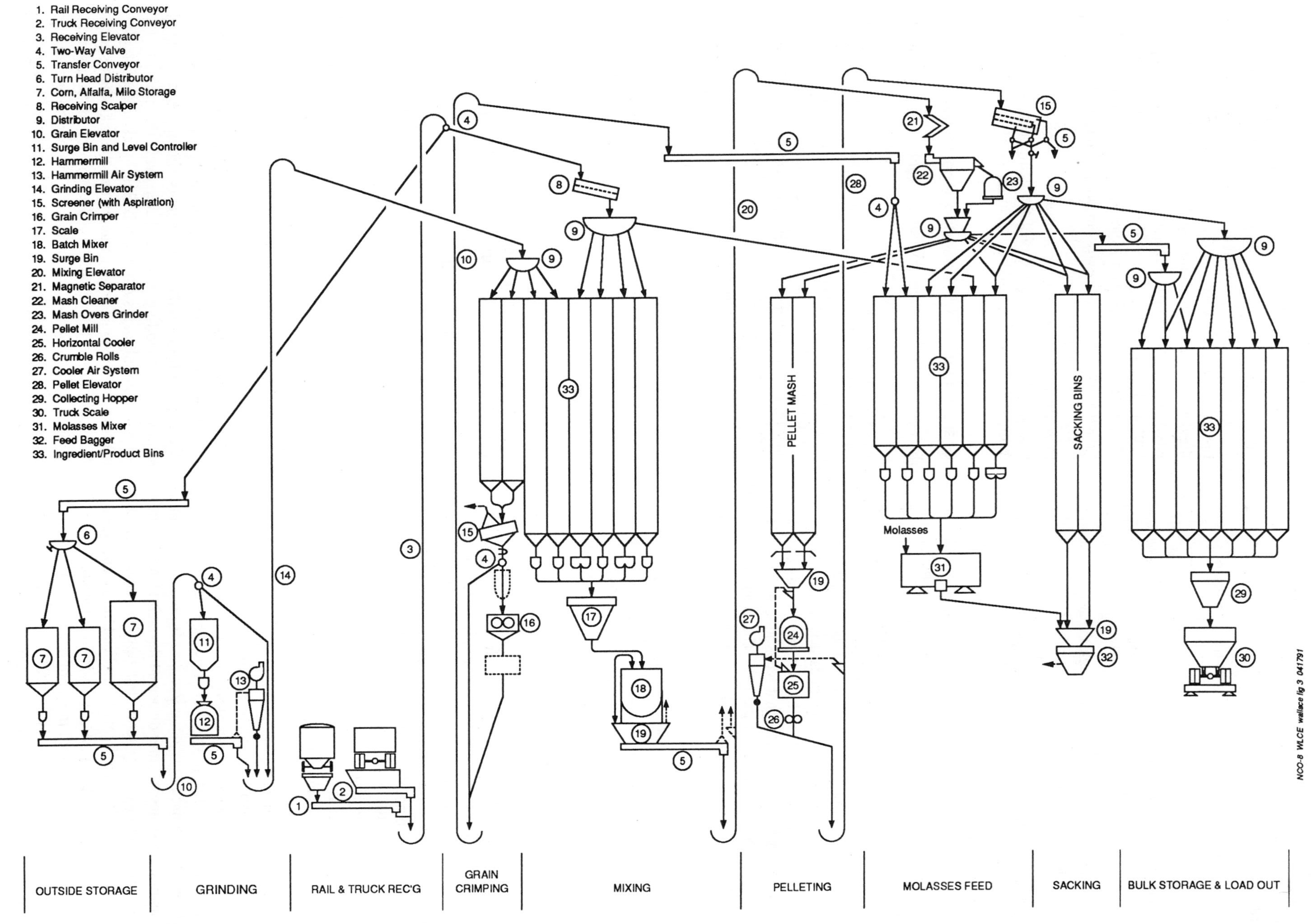

FIGURE 3. Flow Diagram for a Small Feed Manufacturing Process

Grain Transfer

The three primary operations used to transfer grain in both grain elevators and grain processing plants are by the bucket elevator (or leg), by belt conveyors, and by gravity flow through spouts. The moving pulley system and attached buckets in the legs pick up grain in the boot and transport it to the head of the leg. Dust is generated as grain is dumped into the boot, as buckets move through the boot to pick up grain, and as grain is discharged from the elevators at the head. Typically, the legs are aspirated at the top, and possibly at the boot, to relieve air pressure and remove dust generated by the moving elevator, but some systems are operated without venting.

Belt conveyors are used for the horizontal transfer of grain within grain elevators and grain processing plants. Typically, grain is transferred to the belt from a spout by gravity feed. Grain can be removed from the belt at a dump station at the tail pulley or by a tripper that removes the grain at an intermediate point in the belt. The primary source of emissions is falling grain as it is transferred onto the belt and discharged from the belt. Emissions are also generated when dust particles that have adhered to the belt drop off of the return side of the belt. These particles may be entrained immediately in local air currents or may fall to an exposed floor or equipment surface and subsequently be reentrained. Frequently, transfer of the grain to storage and load-out is accomplished via gravity flow through spouts. The primary source of emissions is the "splash" of dusty material at the spout discharge. Such emissions are most pronounced from barges, ships, and load-out operations.

Grain Cleaning

Both terminal grain elevators and grain processing plants employ various cleaning devices to remove impurities from the grain. These devices rely on differences in size, shape, specific gravity, and material between the grain and the impurities to promote separation. All of the different cleaning devices involve grain movement, potential for abrasion, and substantial airflows. Consequently, cleaning operation can emit significant levels of particulate matter (PM).

The primary cleaning operations are scalping, air aspiration, metals removal, and scouring. A preliminary step widely used at both elevators and processing plants is called scalping. In the scalper, grain is run through a coarse mesh shaker or reel screen to remove straw, sticks, stones, strings, or other coarse material. Low-velocity vibrating screens are also used to remove fine dust from the grain. In air aspiration systems, air is dispersed through falling grain in a cross-current or countercurrent flow to remove field dust, fibers, chaff, and light trash. Feed mills and feed manufacturing plants use magnetic drum separators to remove bolts and screws from the grain or from milled grain dust. Many mills also employ scouring as the final step in the cleaning room. A scourer is a machine in which beaters attached to a central shaft throw grain violently against a surrounding drum. This violent action buffs each kernel and breaks off loose material. Air currents are used to separate the detached dust and bran from the grain.[11,12] The exhaust gases from aspiration systems and scourers contain significant quantities of dust that will be emitted if these streams are not controlled. If grain velocities are kept low and the systems are tightly enclosed, the scalper and the screening operations will have essentially no emissions.[13] However, these operations can generate substantial fugitive emissions if grain flows are excessive or if the equipment is open and grain streams are exposed to local air currents.

Grain Milling and Processing Operations

In addition to the handling and cleaning steps defined above, grain mills and feed manufacturing plants use a series of milling and processing steps to convert the raw grain to a usable final product. All facilities use some type of milling operation to reduce the size of the grain material. Options for milling include hammer mills, attrition mills, burr mills, and roller mills. Other operations conducted at many facilities are proportioning and mixing, pelletizing and subsequent sizing, and grain drying.

Hammer mills are commonly used at grain mills and feed manufacturing plants to grind ingredients. These units accomplish both grinding and air conveyance. Grain is fed into the center of a high-speed rotor, where the pivoted hammers move it to the perimeter screen. The grain is forced through the screen, usually with the aid of supplemental air. The ground product is pneumatically conveyed to a cyclone separator or filter receiver that separates the air from the ground ingredient.[11]

Attrition and burr mills are used in feed plants for preliminary size reduction. Attrition mills and burr mills comprise two disks, or plates. In the burr mill, one plate is stationary, the other rotating. The plates have rough surfaces, and different types of plate surfaces can be used for different materials or to produce different particle sizes. The material is fed between the plates and reduced by cutting, crushing, and shearing. The "double-runner" attrition mill is a heavy-duty precision plate mill, and each plate rotates at high speeds.[14]

Proportioning and mixing operations are the heart of any formula-feed manufacturing plant. Each ingredient has a single identity until it is mixed with one or more ingredients to form a formula feed. Key operations are moving both sacked and bulk ingredients from storage to the mixing center; proportioning these ingredients; mixing, conveying, scalping, and blending them; and conveying the mixed feed to processing bins, such as pelleting, sacking, or bulk load-out bins.[15] Dust is generated when the ingredients are dumped into the hopper scale and when dry materials are dumped from the scale into the mixer. The mixing operation itself produces minimal emissions if the system is tightly closed.

Pelleting, extruding, and blocking are methods of producing moist agglomerated particles; they generate little or no dust emissions. However, following pelleting and extruding, cooling or drying processes use high airflow rates that can generate dust. However, minimal dust is typically emitted because high-efficiency collection systems are used to prevent product loss. The pelleted or extruded product passes through a shaker to size particles before packaging, bulk storage, or bulk shipping. Some dust may be generated during this operation, depending on product moisture; therefore, a dust-tight enclosure is required to prevent product loss.[11]

Many grain elevators, grain mills, and feed manufacturing plants use grain dryers to reduce the moisture content of the grain. Almost all grain dryers can be classified as either batch or continuous-flow dryers. In batch dryers, the grain is essentially stationary during the drying and cooling process. The dryer does not operate during loading and unloading.[16]

In continuous-flow dryers, grain flows continuously downward through the dryer. Air temperature and grain flow rates are usually not adjustable. Continuous-flow dryers can be further classified by how air moves in relation to the grain column. If air flows across the column, the dryer is a cross-flow type. Cross-flow dryers are the easiest and simplest to design and are most common. If air moves in the same direction as the grain, the dryer is called a concurrent flow model. This design allows the use of higher air temperature and produces better-quality grain, but it is more susceptible to fire and more complicated to build. If air moves opposite to the direction of grain flow, the dryer is a counterflow type. This version accomplishes the most drying in the least amount of space, but it is less efficient, more difficult to build, and produces the lowest quality of grain of the three types of dryers.[16]

All types of grain dryers present a substantial air pollution problem. The emission stream from grain dryers is characterized by large air volumes, a large cross-sectional area of the exhaust stream, low PM specific gravity, and high moisture content. The PM from dryers generally comprises particles of relative large diameter, but the individual particles are light and have proven difficult to collect. "Bees' wings," a light, flaky material that breaks off from the corn kernel during drying, is particularly difficult to control.

AIR EMISSIONS CHARACTERIZATION

Each of the handling and processing operations in this industry has the potential to release PM emissions. A recent study for the U.S. Environmental Protection Agency provided a comprehensive review of all available PM data.[17,18] Tables 1 and 2 present emission factors that were generated by that survey. The estimates in Table 1 represent average uncontrolled emissions based on mass of grain handled by an operation. Note that for a particular grain elevator, the amount of grain handled in relation to the amount of grain received or shipped varies from operation to operation.

TABLE 1. Total Particulate Emission Factors for Uncontrolled Grain Elevators[a]
Emission Factor Rating: B

Type of Operation	Total Particulate lb/ton
Country elevators	
Unloading (receiving)	0.6
Loading (shipping)	0.3
Removal from bins (tunnel belt)	1.0
Drying[b]	0.7
Cleaning[c]	3.0
Headhouse (legs)	1.5
Inland terminal elevators	
Unloading (receiving)	1.0
Loading (shipping)	0.3
Removal from bins (tunnel belt)	1.4
Drying[b]	1.1
Cleaning[c]	3.0
Headhouse (legs)	1.5
Tripper (gallery belt)	1.0
Export elevators	
Unloading (receiving)	1.0
Loading (shipping)	1.0
Removal from bins (tunnel belt)	1.4
Drying[b]	1.1
Cleaning[c]	3.0
Headhouse (legs)	1.5
Tripper (gallery belt)	1.0

[a]Expressed as weight of dust emitted per unit weight of grain by each operation.
[b]Based on 1.8 lb/ton for uncontrolled rack dryers and 0.30 lb/ton for uncontrolled column dryers, prorated on the basis of the distribution of these two types of dryers.
[c]Average of values, from <0.6 lb/ton for wheat to 6.0 lb/ton for corn.

While these estimates are reliable industry-wide averages, they do not provide reliable emission estimates for a particular facility. Earlier studies indicate that emissions exhibit large variability, both between facilities and within a given facility, over time.[20] Factors that contribute to this variability include the type of grain handled, quality or grade of the grain, moisture content of the grain (usually 10–30%), amount of foreign material in the grain (usually 5% or less), moisture content of the grain at harvest (hardness), amount of dirt harvested with the grain, degree of enclosure at loading and unloading areas, and type of cleaning and conveying.[20] Of particular note are the studies from a subterminal elevator that show a wide variability of emissions as a function of grain type with emissions from soybean handling consistently two or four times higher than emissions from corn, wheat, and milo handling.[19]

The emission estimates in Table 2 are generally based on a process unit of mass of grain entering the plant. Again, these estimates represent industry-wide averages and are not likely to provide reliable estimates for a particular operation because of the factors defined above. Also, the author

TABLE 2. Total Particulate Emission Factors for Uncontrolled Grain Processing Operations[a] ***Emission Factor Rating: D***

Type of Operation	Emission factor, lb/ton	Type of Operation	Emission factor, lb/ton
Feed milling		Soybean milling	
Receiving	2.5	Receiving	1.6
Shipping	1.0	Handling	5.0
Handling	5.5	Cleaning	—
Grinding		Drying[g]	7.2
Hammermilling[b]	0.2[c,d]	Cracking and dehulling	3.3
Flaking[b]	0.2[d]	Hull grinding	2.0
Cracking[b]	0.02[c,d]	Bean conditioning	0.1
Pellet cooler[b]	0.4[c]	Flaking	0.57
		Meal drying	1.5
Wheat milling		Meal cooling	1.8
Receiving	1.0	Bulk loading	0.27
Precleaning and handling	5.0		
Cleaning house	—	Dry corn milling	
Mill house	70.0	Receiving	1.0
		Drying[g]	0.5
Durum milling		Precleaning and handling	5.0
Receiving	1.0	Cleaning house	6.0
Precleaning and handling	5.0	Degerming and milling	—
Cleaning house	—		
Mill house	—	Wet corn milling	
		Receiving	1.0
Rye milling		Handling	5.0
Receiving	1.0	Cleaning	6.0
Precleaning and handling	5.0	Drying[h]	0.48
Cleaning house	—	Bulk loading	—
Mill house	70.0		
Oat milling[e]	2.5		
Rice milling			
Receiving	0.64		
Precleaning and handling	5.0		
Drying[f]	0.30		
Cleaning and mill house	—		

[a]Most emission factors are expressed as weight of dust emitted per unit weight of grain entering the plant, not necessarily the same as the amount of material processed by each operation. Dash = no data.
[b]Expressed as weight of dust emitted per unit weight of grain processed.
[c]With cyclones.
[d]Measured on corn processing operations at feed mills.
[e]Represents several sources at one plant, some controlled with cyclones and others with fabric filters.
[f]Average for uncontrolled column dryers.
[g]Drier types unknown.
[h]For rotary steam tube dryers.

indicates that the emission factors for grain processing have limited overall reliability.[18]

While a reasonable quantity of information is available on total PM emissions from grain handling and processing, data on the particle size distribution of these emissions and on the fraction of emissions that might be toxic air pollutants are quite limited. The earlier version of this manual indicated that 70–98% of the control device catch from shipping and receiving operations was 10 μm in diameter or larger.[11] The recent survey shows results that are generally in agreement with these earlier results. For uncontrolled rice dryers, about 90% of the emissions are greater than 10 μm in diameter, with less than 1% being 2.5 μm in diameter or smaller. For uncontrolled ship loading, about 60% of the emissions are 10 μm diameter or larger and only 10% are 2.5 μm or smaller.[18] Hence the available data, although quite limited, do suggest that PM emissions from grain handling and processing operations tend to have relatively small fractions of fine PM.

No quantitative or qualitative information is available on toxic air pollutant emissions from grain handling and processing. Generally, the potential appears to be relatively small. The only two exceptions appear to be contamination of entrained grain dust with pesticides or herbicides applied in the fields and the use of fumigants for rodent and fungus control in grain storage facilities. No data are available on the air toxic emissions potential of these processes. However, limited information is available on adverse health effects associated with grain dust emissions.

Data indicate that dust from grain handling operations have adverse effects on the skin, eyes, and respiratory systems of humans. Allergic skin reactions and eye irritation have occurred in workers exposed to high concentrations of grain dust, but the most severe effects are on the respiratory system. The two main hypotheses on mechanisms by which grain dusts affect the respiratory system are (1) that the fine particles in the grain dust may act as a mechanical irritant and (2) that the organic, or perhaps the inorganic, part of the dust may produce an antigenic reaction, with asthmatic reactions in the case of the former and fibrotic changes in the latter.[20] Detailed descriptions of the effects of grain dust on workers' respiratory systems are presented in references 21–24.

Data also indicate that these same respiratory effects can be found in some persons exposed to airborne dusts emitted from grain elevators. In particular, a study done at the University of Minnesota over a five-year period supported a hypothesis of asthma attacks induced by respiratory irritation from atmospheric particles generated by grain handling operations. Evidence also suggests that asthmatic attacks in the vicinity of a New Orleans elevator are partially attributable to airborne effluent from the large grain elevator. Generally, no evidence exists for adverse effects on healthy people from grain emissions at concentrations less than 100 $\mu g/m^3$. However, people with preexisting respiratory disorders may be affected by rapid increases above the seasonal mean concentration of particulate grain dust.[20]

AIR POLLUTION CONTROL MEASURES

The two general types of measures that are available to reduce emissions from grain handling and processing operations are process modifications that are designed to prevent or inhibit emissions and capture/collection systems. Detailed descriptions of both process controls and capture systems for specific operations are available in the literature.[25–29] These detailed descriptions are not repeated here, but Table 3 identifies the types of controls available for each source and directs the reader to specific documents for more detailed information. The following describes the general approaches to process controls and capture systems. The characteristics of the collection systems most frequently applied to grain handling and processing plants (mechanical collectors [or cyclones] and fabric filters) are then described, and common operation and maintenance problems found in the industry are discussed.

Because PM emissions from grain handling operations are fugitive emissions generated as a consequence of

TABLE 3. Process Control and Exhaust Systems for Grain Handling and Processing[a]

		Source of Control System Information				
Grain Handling and Processing Operation	Control Mechanism	Dust Control[b]	Feed Manufacturing[c]	Environmental Controls[d]	Emission Control[e]	Industrial Ventilation[f]
Receiving operations	Grain flow control	35–38				
	Exhaust	41–52	113–114	12–20	297–300	
Belt conveyors	Enclosure	17–19				
	Flow control	90				
	Hooding/exhaust	98–108	115–116	30–31		5–39
Elevator legs	Exhaust	80–86	115	29–30		5–38
Distributors	Exhaust					5–128
Cleaners	Enclosure/exhaust		117	26–27		5–40
Scales	Enclosure/exhaust	109–113		21–24		5–128
Grain driers	Screens	203–262			287–297	
Hammer mills	Exhaust		117–118	41–46		
Roller mills	Exhaust			48		
Mixers	Exhaust		118	56–61		5–128
Truck/rail load-out	Dust suppression	127–142				
	Exhaust	143–160			300–302	
Barge/ship load-out	Dust suppression	99				
	Exhaust	166–173		301–314		

[a]Numbers in table are page references in these documents. Dashes in columns 3 to 6 represent page ranges, while those in column 7 represent chapter page
[b]Reference 29.
[c]Reference 27.
[d]Reference 26.
[e]Reference 25.
[f]Reference 30.

mechanical energy imparted to the dust by the operations themselves and of local air currents in the vicinity of the operations, an obvious control strategy is to modify the process or facility to limit the effects of those factors that generate emissions. The primary preventive measures that facilities have used are construction and sealing practices that limit the effect of air currents, and minimizing grain free-fall distances and grain velocities during handling and transfer. Some recommended practices of this type are to enclose the receiving area to the degree practicable, preferably with doors at both ends of a receiving shed; to specify dust-tight cleaning and processing equipment; to use lip-type shaft seals at bearings on conveyor and other equipment housings; to use flanged inlets and outlets on all spouting, transitions, and miscellaneous hoppers; and to fully enclose and seal all areas in contact with products handled.[30]

A substantial reduction in emissions from receiving, shipping, and handling and transfer areas can be achieved by reducing grain free-fall distances and grain velocities. Figure 4 illustrates a choke unloading procedure used to reduce free-fall distance during hopper car unloading. The same principle can be used to control emissions from grain transfer onto conveyor belts and from load-out operations. An example of a mechanism that reduces grain velocities is a dead box spout, which is used in load-out operations. The dead box spout slows down the flow of grain and stops the grain in an enclosed area (Figure 5). The dead box is mounted on a telescoping spout to keep it close to the grain pile during operation. In principle, the grain free-falls down the spout to an enclosed impact dead box, with grain velocity going to zero. It then falls about 12 inches to the grain pile. Typically, the entrained air and dust liberated at the dead box are aspirated back up the spout to a dust collection system.[31]

While the preventive measures described can minimize emissions, most facilities also require ventilation, or cap-

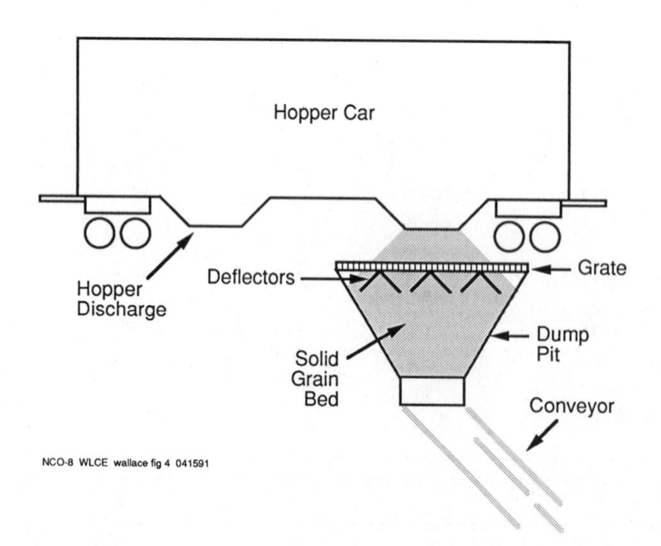

FIGURE 4. Choke Unloading for a Grain Receiving Process

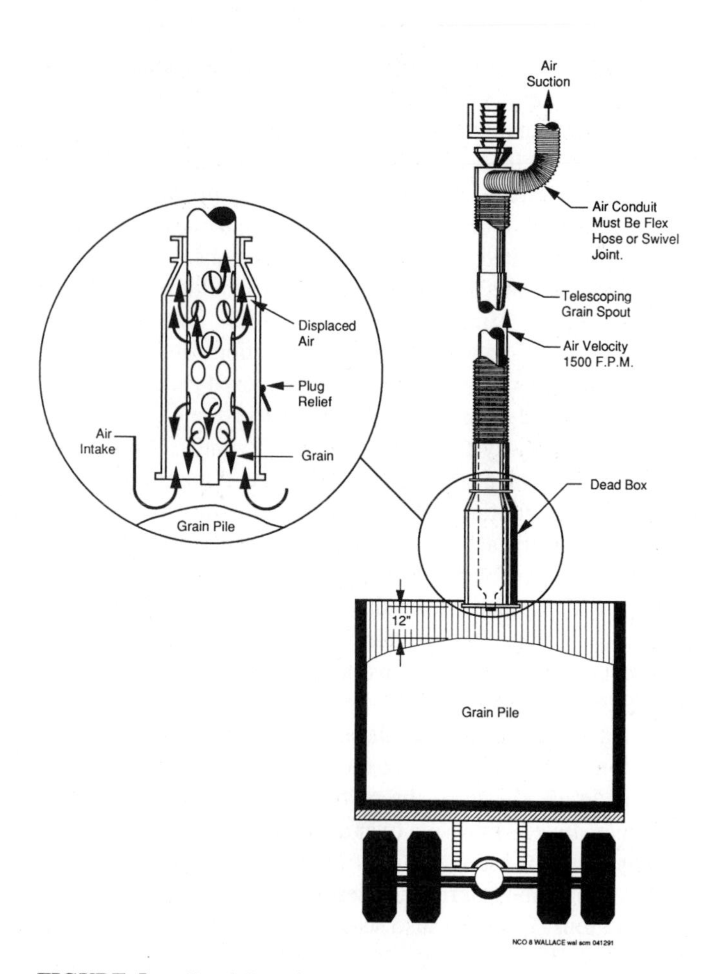

FIGURE 5. Dead Box for Reducing Load-Out Emissions

ture, systems to reduce emissions to acceptable levels. In fact, air aspiration (ventilation) is a part of the dead box system. Almost all grain handling and processing facilities, except relatively small grain elevators, use ventilation systems on the receiving pits, cleaning operations, and elevator legs. Generally, milling and pelletizing operations at processing plants are ventilated, and some facilities use hooding systems on all handling and transfer operations. An example of a ventilation system at a truck receiving station is illustrated in Figure 6. Details on capture systems can be found in the references noted in Table 3.

The PM control devices typically used in the grain handling and processing industry are cyclones (or mechanical collectors) and fabric filters. Cyclones are used only on country elevators and small processing plants located in sparsely populated areas. Terminal elevators and processing plants in densely populated areas, as well as some country elevators and small processing plants, use fabric filters for PM control. Both of these systems can achieve acceptable levels of PM control for many grain handling and processing sources, as noted in Table 4.

Although cyclone collectors can achieve acceptable performance in some scenarios, and fabric filters are highly efficient, both devices are subject to failure if they are not properly operated and maintained. Also, malfunction of the

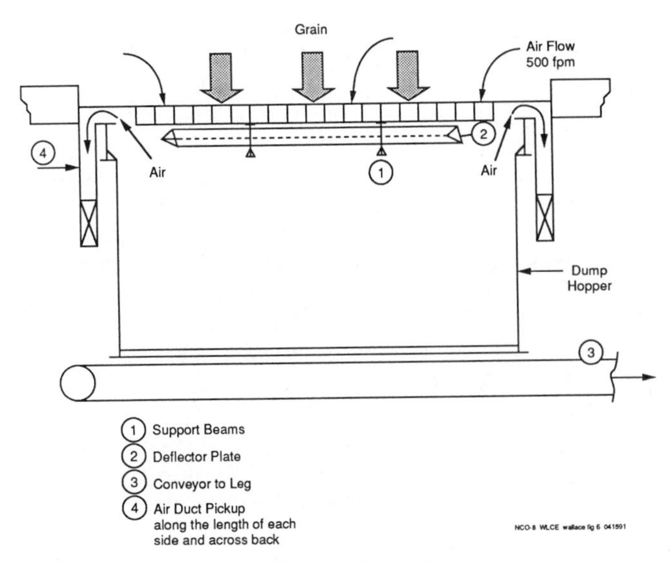

FIGURE 6. Receiving Pit Ventilation System

ventilation system can lead to increased emissions at the source. Some typical malfunction problems at grain handling and processing facilities are described in the following.

Limited emission test data indicate that cyclone malfunctions can cause elevated emissions at country elevators. The efficiency of a cyclone depends on its design parameters and the velocity of the gas stream through it. Reduced gas velocities also result in decreased efficiency. Factors that cause an increased pressure drop through the control system, such as plugging in the cyclone, improper damper settings in the ductwork, or plugging in the ductwork, lead to decreased flow through the cyclone and a drop in cyclone efficiency. Periodic checks of cyclone pressure drop, exhaust fan current, the condition of cyclone walls and fan blades, and the exhaust rate at pickup points provide indications of any problems that might reduce efficiency. One other problem with cyclone operation that often occurs is leakage of air into the cyclone from the dust discharge. Therefore, dust discharge mechanisms should be checked regularly to control leakage.

A major contributor to excess emissions from well-controlled grain elevators is the malfunction of fabric filters. Typical causes are torn bags that allow excess emissions from the control device outlet and plugging of the bags, which creates excess emissions at the source. Industry personnel, environmental control officials, and equipment vendors indicate that there is some evidence that operating problems may result in significant periods during which the control equipment is shut down or operating inefficiently.[32] Substantial bag blinding problems in areas of high humidity have been reported. Blinding problems also have been reported in cooler climates when the exhaust stream collected from handling operations for relatively warm, moist grain, which has been stored inside the elevator, is ducted to a control device located in a cooler outdoor atmosphere. Using baghouse heating systems and cleaning the bags thoroughly prior to shutdown can reduce these problems. Dust bridging in the baghouse hopper can be a major problem. Potential causes are poor design of hoppers and improper shutdown procedures. Problems can be reduced if systems are designed so that elevator personnel have easy access to the hopper for inspection and cleaning. Finally, malfunction of the bag cleaning mechanism can result in dust buildup on the bags and eventually in total plugging of the systems. At some facilities, the pulse-jet control panel is located on top of an outdoor baghouse and not in a weath-

TABLE 4. Cyclone and Fabric Filter Performance

	Cyclone Performance			Fabric filter Performance		
		Outlet Levels			Outlet Levels	
Process Source	Efficiency,%	gr/dscf	lb/ton	Efficiency, %	gr/dscf	lb/ton
Grain receiving						
Truck			0.06[a]		0.002–0.01[a]	0.005[a]
Rail			0.1[a]		<0.002[a]	0.0002[a]
Not specified	85–95[b]	0.02–0.09[b]		99+[b]	0.002–0.006[b]	
Handling operations (Legs, conveyors)			0.1[a]		<0-002[a]	0.0002[a]
Cleaning						
Elevators			0.3–0.6[a]		0.002–0.004[a]	0.003–0.014[a]
Grain mills	88–99[b]	0.03–0.1[b]		99+[b]	0.005–0.02[b]	
Grain milling	96–99[b]	0.04–0.2[b]		99+[b]	0.005–0.01[b]	
Pellet cooling	88–99[b]	0.02–0.1[b]				
Grain load-out						
Ship			0.03[a]		<0.002[a]	0.001[a]
Rail			0.06[a]		0.003–0.01[a]	0.001[a]

[a]Reference 33.
[a]Reference 34.

ertight enclosure. Accumulation of dust and moisture inside the control panel can cause short circuits in the solenoid valves for the pulse jet system, ultimately leading to system failure. To avoid such problems, general baghouse operation and maintenance practices should be routinely applied at grain handling and processing facilities.

Generally, emissions from multiple operations at grain handling and processing facilities are controlled by a single air pollution control system. If the ventilation system that connects these sources is not properly designed and operated, excess fugitive emissions are generated at the source. References 35 and 36 provide excellent information on designing, operating, and maintaining balanced ventilation systems.

References

1. G. LaFlam, *Documentation for AP-42 Emission Factors: Section 6.4, Grain Elevators and Processing Plants,* EPA Contract No. 68-02-3887, WA54, U.S. Environmental Protection Agency, Research Triangle Park, NC, 1987, p. 2.
2. D. Wallace and V. Ramananthan, *Review of Compliance Monitoring Programs with Respect to Grain Elevators,* EPA Contract No. 68-01-4139, WA 14, U.S. Environmental Protection Agency, Washington, DC, 1980, p. 44.
3. U.S. Department of Commerce, *1987 Census of Manufacturers—Grain Mill Products,* U.S. Department of Commerce, Washington, DC, 1990, p 200–7.
4. R. R. McElhiney, *Feed Manufacturing Technology III,* American Feed Manufacturing Association, Arlington, VA, 1985, p 9.
5. L. J. Shannon, R. W. Girstel, P. G. Gorman, et al., *Emissions Control in the Grain and Feed Industry, Vol. I—Engineering and Cost Study,* EPA-450-/3-73-003a, U.S. Environmental Protection Agency, Research Triangle Park, NC, 1973, p 112.
6. *Ibid.*, pp 203–269.
7. Reference 4, pp 1–216.
8. *Ibid.*, pp 36–37.
9. Reference, 5 p 159.
10. *Ibid.*, pp 120–129.
11. J. A. Danielson, *Air Pollution Engineering Manual, AP-40,* U.S. Environmental Protection Agency, Research Triangle Park, 1973, NC, pp 352–361.
12. Reference 5, p 205.
13. Reference 4, p 436.
14. *Ibid.*, p 143.
15. *Ibid.*, p 151.
16. *Ibid.*, p 16.
17. Reference 1, 19 pp.
18. U.S. Environmental Protection Agency, *Compilation of Air Pollutant Emission Factors,* U.S. Environmental Protection Agency, Research Triangle Park, NC, 1988, pp 6.4-1–6.4-15.
19. P. G. Gorman, *Potential Dust Emissions from a Grain Elevator in Kansas City, Missouri,* EPA Contract No. 68-02-0228, U.S. Environmental Protection Agency, Research Triangle Park, NC, 1974.
20. Reference 2, pp 56–77.
21. N. Williams, A. Skoulas, and N. Merriman, "Exposure to grain dust: I. Survey of effects," *J. Occupational Med.*, 6(8):319 (1964).
22. A. Skoulas, N. Williams, and J. Merriman, "Exposure to grain dust: II. A clinical study of the effects," *J. Occupational Med.*, 6(9):359 (1964).
23. R. Gordon, *Dust Control for Grain Elevators,* National Grain and Feed Association, Washington, DC, 1981, pp 358–375.
24. T. Hurst and J. Dosman, "Characterization of health effects of grain dust exposures," *Am. J. Ind. Medi.*, 17(1):27 (1990).
25. Reference 5, 544 pp.
26. W. Briggs, M. Shiver, and T. Stivers, *Environmental Controls for Feed Manufacturing and Grain Handling,* American Feed Manufacturer's Association, Chicago, 1971, 184 pp.
27. Reference 4, 608 pp.
28. Committee on Industrial Ventilation, *Industrial Ventilation, a Manual of Recommended Practice, 19th ed,* American Conference of Governmental Industrial Hygienists, Lansing, MI, 1986, 392 pp.
29. Reference 23, 466 pp.
30. Reference 26, pp 8–9
31. Reference 23, p 149.
32. Reference 2, pp 15, 16.
33. Reference 5, pp 293–294.
34. *Standards Support and Environmental Impact Statement, Vol. I: Standards of Performance for Grain Elevator Industry,* U.S. Environmental Protection Agency, Research Triangle Park, NC, 1977, pp 4-1–4-30, 5-2.
35. Reference 23, pp 274–306.
36. Reference 28, pp 6-1–6-55.

FERMENTATION

Joseph A. Mulloney, Jr., P.E.

Fermentation as an industrial process is currently employed primarily in the manufacture of beer, alcoholic spirits, wine, and fuel-grade ethanol (ethyl alcohol). Minor applications include a wide variety of food (including enzyme and amino acid), pharmaceutical, and industrial processes. In the past, fermentation processes were used to produce a variety of chemicals that are currently derived from petroleum feedstocks. Future petroleum prices or availabilities may spur renewed interest in these processes. Additionally, the potential of biotechnology may result in both improvements to current processes and totally new processes and products from industrial fermentations.

With respect to characterizing the air emissions from distilleries, breweries, and wineries, a distinction is necessary between a distillery that produces concentrated volatile organic streams (e.g., ethanol and fusel oil) similar to the fuel-grade ethanol facility described below and breweries and wineries where the volatile compounds are always present in solution or associated with water vapor. While ethanol is the major VOC, there are many other VOCs present, such as isoamyl alcohol, ethyl acetate, isopropyl alcohol, and *n*-propyl alcohol, that contribute to beverage bouquet and taste but constitute only 1% or less of the amount of ethanol. Emissions of beverage alcohol are primarily from spillage and breakage in packaging operations and secondarily from processing operations. A typical brewery will lose 3% of liquid volume after fermentation (where the VOCs are formed) as processing and packaging losses that primarily run to sewers, but also partially leave the premises as evaporate.

In 1989, approximately 800 million gallons of fuel-grade ethanol were produced, as well as 200 million barrels of beer (200 million gallons ethanol), 1200 million tax gallons of distilled spirits (600 million gallons ethanol), and 475 million gallons of still wines (50 million gallons ethanol).[1]

PROCESS DESCRIPTION

The base case alcohol fermentation plant is designed to produce 50 million gallons per year of 99.5 vol % (199 proof) fuel-grade ethanol from corn. In addition, it will produce 177,111 tons per year of Distillers' Dark Grain (also known as Distillers' Dried Grain with Solubles or DDGS), a commercial animal feed. The plant is assumed to be located in central Illinois, close to a source of Illinois No. 6 coal, which is used as fuel.

The alcohol plant, designed by Raphael Katzen Associates,[2] Cincinnati, Ohio, generally uses existing process technology currently employed in grain alcohol plants. A drawing of the facility is presented in Figure 1. The plant operates as a continuous-flow process, except for the fermentation and fungal amylase sections, which are operated in a continuous batch mode. The distillation system employs a two-pressure concept currently utilized in industrial and beverage alcohol production and in other chemical processing fields. The process also utilizes several heat economy measures that result in a total steam usage of 31.7 lb/gal of ethanol. The distillation system uses 21.4 lb/gal of steam, of which 2.8 lb/gal is obtained as flash vapors from mash cooking. Feedstock to the plant will consist of shelled corn at a rate of 58,900 bushels per day. No distress corn (e.g., corn contaminated with aflatoxins, pesticides, etc.) is contemplated for use in this facility; No. 2 shelled corn (less than 15.5% moisture content) will be used.

All the utility requirements, with the exception of electricity, are produced within the boundaries of the plant. Water is obtained from a well field located close to the plant. The boiler burns relatively low-cost, high-sulfur coal. The plant is designed effectively to utilize most of the waste streams, with final disposal in an environmentally acceptable manner. Flue gas from the boiler is used to dry the stillage residue to yield Distillers' Dark Grain as a byproduct. Wastewater is treated in a two-stage, activated sludge treatment facility. The resultant sludge is dewatered and fed to the boiler. Cooling water from the various con-

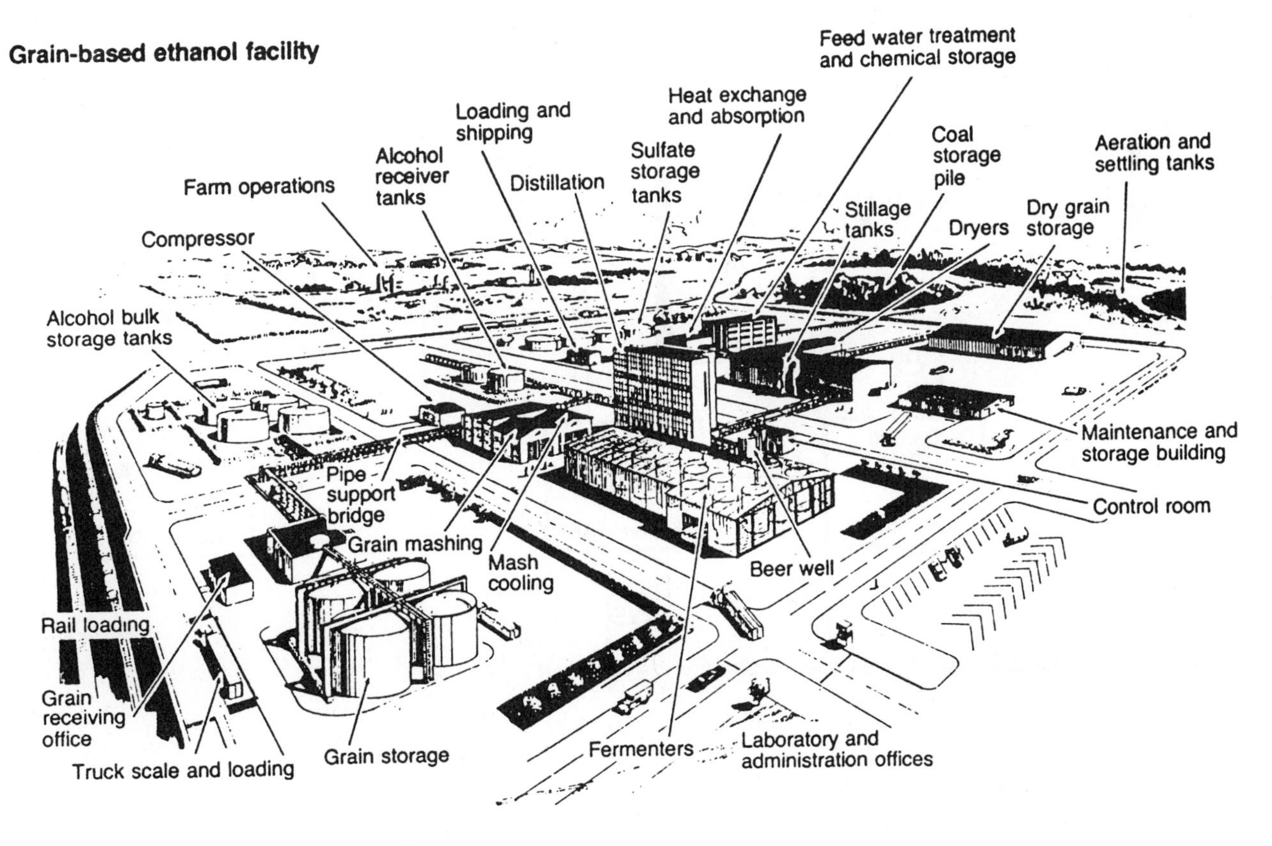

FIGURE 1. Plan of 50-Million-Gallon-per-Year Grain-Based Ethanol Facility

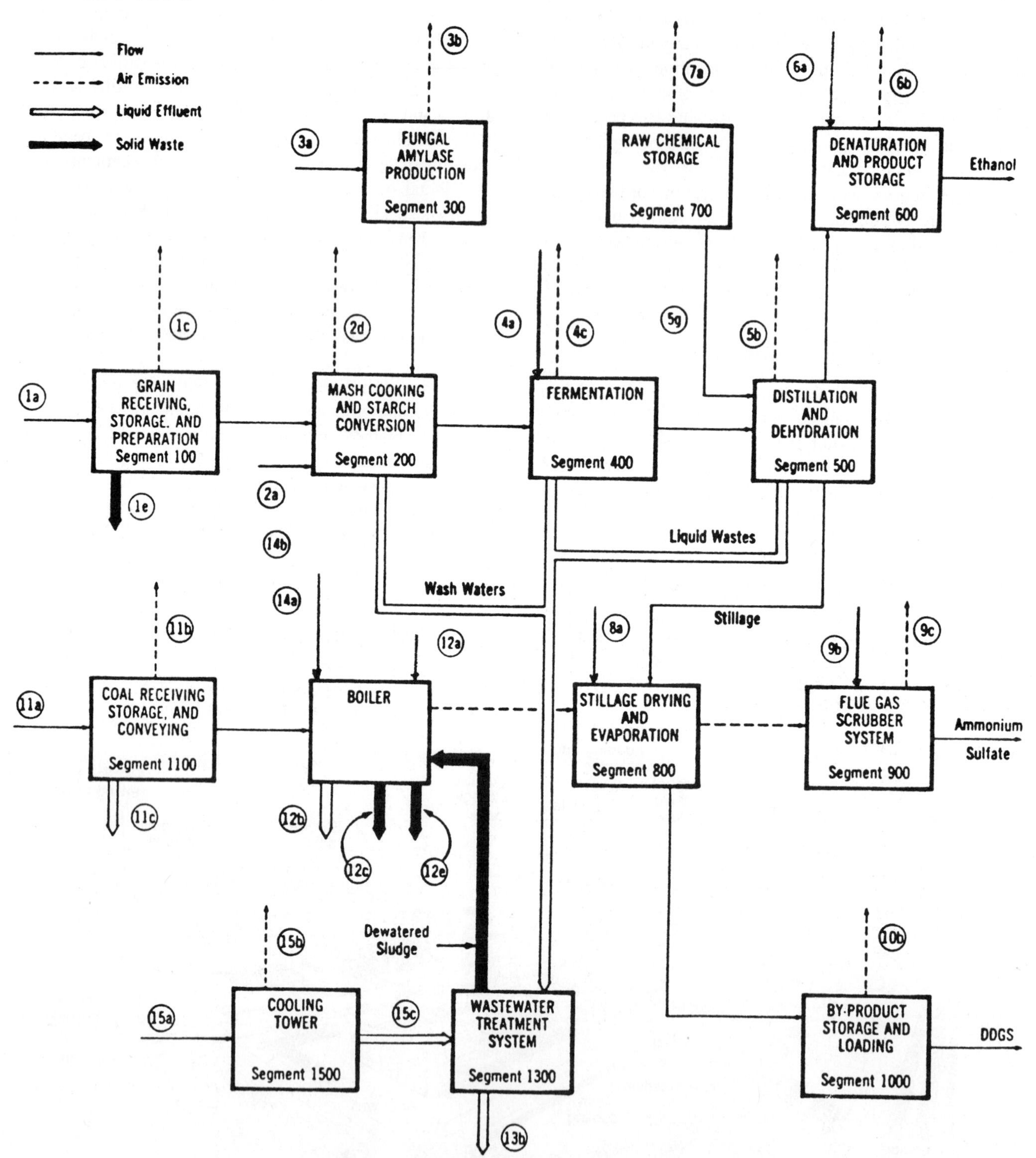

FIGURE 2. Grain-Based Ethanol Fermentation Facility Process Flowchart

densers is recycled through a two-cell cooling tower. A flue gas scrubbing system employing ammonia is utilized to remove particulates and sulfur dioxide emissions, producing ammonium sulfate. The latter could be utilized as a fertilizer.

AIR EMISSIONS CHARACTERIZATION

Air emissions from an ethanol plant arise principally from three sources:

- Combustion of conventional and unconventional fuels
- Feedstock preparation and by-product processing operations
- Overall process schemes employed, such as the distillation/dehydration systems, flash coolers, evaporators, and cooling towers.

Figure 2[3] presents the process flowchart for the facility and should be reviewed in conjunction with Table 1,[3] which provides data on the resources used and products produced. Table 2,[3] also to be reviewed in conjunction with Figure 2, presents the facility's annual releases of air and water pollutants and solid wastes.

Most of the plant air emissions are associated with the combustion process used to supply steam and electricity to the plant. The type of fuel used and the degree of combustion will dictate the nature of these emissions. For example, uncontrolled emissions from coal- or biomass-fired boilers will be greater than those from facilities using natural gas or residual oil. The degree of local impact of emissions from facilities using solar energy or process waste, such as bagasse, would be considerably different from that of a conventional fuel source. These air emissions, therefore, are not inherently coupled to the biomass-to-ethanol process.

Particulate emissions, sulfur oxide (SO_x) emissions, and, to a lesser extent, nitrogen oxide (NO_x) emissions associated with coal combustion are likely to constitute the primary air-related environmental problems for most facilities. Polycyclic organic matter (POM) emissions from some of these sources are also significant. In a test performed by

TABLE 1. Resources Used and Products Shipped

Code No.[a]	Resource	Annual Usage[b]
	Feed materials	
1a	Corn	544 × 10^3 tons
	Fuel	
11a	Coal	97.9 × 10^3 tons
	Water	
2a	Process water	330 × 10^6 gallons
14b	Raw water makeup	83.1 × 10^6 gallons
5a	Cooling tower makeup	280 × 10^6 gallons
	Processing materials	
3a	Air input to amylase production	223 × 10^3 tons
4a	Yeast	396 tons
	Iodine sterilizing solution	7.92 × 10^3 gallons
5g	Hydrocarbon solvent	9.03 × 10^3 gallons
6a	Denaturant	1.00 × 10^6 gallons
8a	Air input to dryer	652 × 10^3 tons
9b	Anhydrous ammonia	3.04 × 10^3 tons
12a	Air input to boiler	1.10 × 10^6 tons
14a	Water treatment chemicals	
	Sodium chloride	396 tons
	Lime	792 tons
	Sludge polymer	7.9 tons
	Land	50.0 acres
	Personnel	
	Operation	159 workers

Product	Annual Production
Primary	
Ethanol (199 proof)	50 × 10^6 gallons
By-products	
Distillers' Dark Grains	177 × 10^3 tons
Dry ammonium sulfate	10.4 × 10^3 tons
Fusel oils[c]	760 tons

[a]See Figure 2.
[b]Assumes a 90 percent capacity factor.
[c]Fusel oils are usually mixed with the grain ethanol product prior to blending with gasoline.

TABLE 2. Annual Releases of Air and Water Pollutants and Solid Wastes

Code No.[a]	Environmental Residuals	Annual Quantities Released
	Air Pollutants	
1c	Particulates from grain cleaning	163 tons
2d	Emissions from starch conversion	Negligible
3b	Emissions from enzyme production	Negligible
4c	Emissions from fermentation:	
	Ethanol	1.80×10^3 tons
	Water vapor	2.48×10^3 tons
	Carbon dioxide	170×10^3 tons
5b	Distillation/dehydration pollutants:	
	Carbon dioxide	$2.,81 \times 10^3$ tons
	Water vapor	41.2 tons
6b	Emissions from product handling	Negligible
7a	Emissions from raw chemical storage	Negligible
9c	Scrubber emissions:	
	Sulfur dioxide	947 tons
	Particulates	104 tons
	Nitrogen oxides	726 tons
	Water vapor	120×10^3 tons
10b	Exhaust from DDGS handling	5.31 tons
11b	Fugitive dust from coal handling	49.0 tons
15b	Cooling tower evaporative drift losses	242×10^6 gallons
	Water pollutants	
11c	Water runoff from coal storage	Not quantifiable
12b	Boiler blowdown	7.92×10^6 gallons
13b	Wastewater effluent to river	363×10^6 gallons
15c	Cooling tower blowdown	38×10^6 gallons
	Solid wastes	
1e	Grain cleaning rejects	54.4 tons
12c	Boiler bottom ash	3×10^3 tons
12e	Boiler fly ash	4.75×10^3 tons

[a]See Figure 2.

TRW Environmental Engineering Division (Redondo Beach, Calif.), the POM emission factors for coal and, especially, for wood were found to be extremely high, approximately 13.8 mg/kg and 484 mg/kg respectively.[4] In the study, dibenz[a,h]anthracene, a carcinogen, was identified, and the presence of other carcinogens, such as benzo[a]pyrene and benzo[g,h,i]perylene, was also indicated. For this reason, particulate emissions, especially respirable particulates, and associated POMs from wood and wood residue combustion are of concern.[5]

Stack emissions from burning corn stalks or bagasse are primarily in the form of particulates and NO_x since there is very little sulfur present. Little analysis of the chemical composition of these emissions has been done. On the basis of experience with burning bagasse in the sugar industry, the particulates can be expected to be lightweight and high in unburned carbon content.[6] The moisture content of these residues will determine the feasibility of their use as fuel sources. At 50% moisture content, most of the agricultural residues are good fuel sources. At higher percentages, however, the moisture causes some problems, limiting combustion.

Several types of trace elements also will be present in the boiler flue gas from conventional coal and oil combustion. Besides chlorine, the elements of greatest concern appear to be aluminum, barium, beryllium, chromium, lithium, nickel, phosphorus, and silicon.[3] Again, the amount of these emissions will depend largely on the type and grade of fuel and on the type of boiler and stack emission controls.

Other sources of emissions are the fermentation and enzyme-producing and distillation and dehydration sections, as well as ethanol denaturing, storage, and handling operations. The principal pollutants of concern from these operations are VOC emissions. Vents from the fermentation vats, flash coolers, enzyme-producing reactors, and distillation and dehydration columns produce the highest levels of VOCs. The VOC emissions generated during chemical storage are relatively small.

Fermentation facilities generally produce large amounts of carbon dioxide (CO_2). For example, in the ethanol facility reviewed above, for each molecule of sugar fermented, two molecules of ethanol and two molecules of CO_2 are produced. Carbon dioxide from annually renewable feedstocks is not generally considered a net contributor to atmospheric CO_2. The CO_2 is often recovered and sold as liquified CO_2 where the local market conditions are favorable.

AIR POLLUTION CONTROL MEASURES

The emissions most likely to require controls are SO_x and particulates. Control of NO_x is possible with boiler design and operating parameters; where NO_x emissions are not regulated, no control is likely to be employed. Generally, the controlled emissions of air pollutants from fuel combustion are expected to be within regulated limits in most states.[7]

Grain-handling or feedstock-preparation operations within an ethanol plant generate emissions similar in amount and characteristics to those from other grain-handling operations, such as grain elevators and milling activities. Particulate emissions are the main pollutant from grain-processing facilities; they are generated in an ethanol plant by grain receiving and unloading, cleaning, and conveying, as well as by storage and milling operations. The majority of these emissions arise principally from cleaning and milling operations. Stillage (residual mash remaining after distillation) drying or by-product processing will also generate particulates; their amounts may be greater than those from grain-milling operations. Although these emissions are fugitive, control by conventional techniques is both possible and feasible.

References

1. U.S. Bureau of Alcohol, Tax and Firearms, various publications and statistical abstracts.
2. Raphael Katzen Associates, *Grain Motor Fuel Alcohol Technical and Economic Assessment Study*, for U.S. Department of Energy, Report no. HCP/J6639-01, Washington, DC, June 1979.
3. Mueller Associates, Inc., *Alcohol Fermentation Plant—Environmental Characterization Information Report*, for U.S. Department of Energy, Washington, DC, September 1981.
4. C. C. Shih and A. M. Takata, *Emissions Assessment of Conventional Stationary Combustion Systems—Summary Report*, TRW Environmental Engineering Division, for U.S. Environmental Protection Agency, Research Triangle Park, NC, July 1981, NTIS no. P882-109414.
5. Mueller Associates, Inc., *Wood Combustion: State-of-Knowledge Survey of Environmental Health and Safety Aspects*, for U.S. Department of Energy, Washington, DC, October 1981.
6. Argonne National Laboratory, *Draft Report on Environmental Concerns of Ethanol from Corn*, Energy and Environmental Systems Division, for U.S. Department of Energy, Washington, DC, January 1980.
7. R. M. Scarberry, M. P. Papai, and M. A. Braun, *Source Test and Evaluation Report: Alcohol Facility for Gasohol Production*, Radian Corp., for U.S. Environmental Protection Agency, Cincinnati, OH, November 1979, NTIS no. PB82-237041.

Bibliography

Commercial Biotechnology: An International Analysis, U.S. Congress, Office of Technology Assessment, Washington, DC, OTA-BA-218, January 1984.

"Industrial Microbiology" (issue theme), *Sci. Am.*, 245(8):66 (1954).

FISH, MEAT, AND POULTRY PROCESSING

William H. Prokop, P.E.

This chapter discusses the application of odor control technology to various agricultural operations which consist of the production and processing of fish, meat and poultry. For a comprehensive discussion of odor sensory measurement and a brief discussion of odor control methods, refer to Chapter 5 on odors.

This chapter is divided into four sections. These subjects are of sufficient importance to merit individual discussions of the pollution problems and the control technology associated with these agricultural activities.

LIVESTOCK PRODUCTION AND MEAT PROCESSING

Livestock production is a major activity in the United States. The raising of cattle, hogs, and sheep for meat production is an important food source. Feedlots for beef cattle and hog production have become major operations where large herds of animals are concentrated in a single location. For example, more than half of the beef cattle raised in the United States are located in feedlots containing more than 10,000 head. As a result, the collection, storage, transport, treatment, and disposal of manure has resulted in major odor problems. These emissions and their control are discussed in detail later in this chapter.

Cattle and hog slaughter operations are becoming more concentrated into fewer and larger companies, with only three major companies currently accounting for more than 60% of the annual total beef slaughter, estimated at approximately 35 million head. Less than five years ago, six or seven companies accounted for the same percentage of the total.

A typical beef slaughter operation includes receiving cattle in holding pens, stunning the animals and draining their blood at the kill floor, removing their hides, and evisceration and trimming. Each animal's carcass is separated into edible parts for human consumption and inedible by-products, which are processed in rendering plants. Choice fatty parts from the cutting operations are processed into edible fats by a special rendering process. Manure is collected from the holding pens and paunch manure is

TABLE 1. Livestock Slaughter and Meat Production[a]

Category	Cattle	Calves	Hogs	Sheep and Lambs
Slaughter,				
1000 head	33,220	1,807	85,120	5,316
pounds/head	1,139	280	250	126
Slaughter,				
million pounds	37,840	506	21,280	670
Dressed weight				
million pounds	22,620	321	15,290	360
Rendering,				
million pounds	15,220	185	5,990	310

[a]*Livestock and Poultry Situation and Outlook Report*, USDA Economic Research Service, February 1991.

TABLE 2. Poultry Slaughter and Production[a]

Category	Chickens	Turkeys
Slaughter,		
million birds	5,837	271
pounds/bird	4.37	21.3
Slaughter,		
million pounds	25,510	5,772
Dressed weight,		
million pounds	18,570	4,560
Rendering,		
million pounds	6,940	1,212

[a]*Livestock and Poultry Situation and Outlook Report*, USDA Economic Research Service, February 1991.

separated from the viscera and inedible materials removed by the trimming and cutting operations. Both the edible and inedible rendering processes are discussed in detail later.

The dressed beef and pork are refrigerated before shipment for human consumption. Dressed meat may be further processed to manufacture such specialty products as smoked meat. This type of product results from a special type of operation known as a "smokehouse." The air emissions from meat smokehouses and their control are described in detail later.

Slaughterhouse operations are conducted under the supervision of the U.S. Department of Agriculture, Food Safety and Inspection Service (USDA/FSIS). As a result, special sanitation and processing conditions are required to be met, which normally are conducive toward preventing the emission of objectionable odors. Also, these operations, except for the holding pens, are normally completely enclosed and are ventilated. Odor control equipment may be provided if the slaughterhouse is near residential areas.

Table 1 provides a summary of the livestock slaughter, dressed meat production, and raw material available for rendering during 1990. The total dressed weight of red meat for human consumption was 39,425 million pounds in 1989 and 38,590 million pounds in 1990. The total raw material available for rendering from the livestock slaughter was 22,330 million pounds in 1989 and 21,700 million pounds in 1990.

POULTRY PRODUCTION AND PROCESSING

Poultry production also is a major agricultural activity in the United States and consists of broilers (young chickens) and turkeys. Currently, broilers represent approximately 83% of the total poultry production in the United States. Poultry hatcheries are usually large integrated operations owned or leased by the companies that sell the retail poultry products to the consumer. The poultry feed ingredients supplied to hatcheries are often produced in poultry by-product rendering plants owned by these same companies. Poultry hatcheries have odor problems similar to those of feedlots, but on a smaller scale, particularly as to the quantities of manure to be collected and disposed of.

Poultry hatcheries and processing plants are totally integrated within a few large companies, which process a large percentage of the total production. Only three major companies currently account for approximately 45% of the total slaughter of broilers in the United States, estimated at 100 million birds per week. Less than five years ago, five or six companies accounted for the same percentage of the total.

Poultry processing plants receive the live birds in cages. They are unloaded and their feet are shackled to a continuous overhead conveyor for the kill and bleeding operations to take place. Scalding of the carcass removes the feathers and subsequent operations include removal of the head, feet, and viscera, followed by separate removal of the heart and liver as a part of the edible product. The carcasses are held in chilled-water coolers containing chlorine-treated water for bacteria control before subsequent refrigerating, packing, and shipping.

Inspectors from the USDA/FSIS are present in the processing plant to ensure that conditions are sanitary. These plants are usually totally enclosed and ventilated. Odor control equipment may be provided, depending on the proximity of residential neighbors.

The heads and feet, together with the viscera are referred to as offal. The offal and feathers often are conveyed by water through a special channel to the nearby inedible rendering plant that is integrated with the poultry processing operation. The inedible rendering process as it applies to poultry offal and feathers is discussed in detail later.

Table 2 provides a summary of the poultry slaughter, dressed poultry production, and raw material available for rendering during 1990. The total dressed weight of poultry for human consumption was 21,510 million pounds in 1989 and 23,130 million pounds in 1990. The total raw material available for rendering from the poultry slaughter was 7640 million pounds in 1989 and 8152 million pounds in 1990.

FISH PROCESSING

William H. Prokop, P.E.

The processing of fish is divided into two main categories. Freshly caught fish such as salmon and tuna are either frozen or canned for human consumption. The by-product material from these operations is processed by "fish meal plants," which are similar to the rendering plants for meat and poultry processing. The products from these plants are used for pet food and animal feed supplements.

At fish freezing plants, the fresh fish is filleted and the pieces trimmed to size and packaged for freezing, either in the raw state or after cooking. Fish canneries use one of the two basic processes: wet-fish and precooked fish.[1]

The wet-fish method is used to preserve salmon, mackerel, sardines, and similar species that are collected locally and are brought to the cannery quickly. The distinctive feature of the wet-fish process is the complete removal of heads, tails, and offal before cooking. The raw, trimmed fish are cooked directly in unsealed cans. These cans are drained of any liquid, and then sealed and pressure cooked before shipment.

The precooked process is used for larger species of fish such as tuna. Whole eviscerated fish are cooked in live-steam–heated chambers operated at about 5 psig pressure. After cooking, the flesh is cooled so that it becomes firm before further handling. Only about one third of the raw tuna weight is canned for humans and pets. The remaining skin, bone, offal, and other waste parts are processed in fish meal plants.

Since the odor emissions from plants that process fresh fish for human consumption and pet food are considerably less than those associated with fish meal plants, the remaining discussion is devoted to the fish meal process and associated odor problems.

PROCESS DESCRIPTION

The raw material for fish meal processing consists of the by-product waste material from fresh fish processing and also whole fish, such as menhaden, that are specifically caught for conversion to fish meal. In the fish meal process, three main product streams result from the raw material: fish solids known as fish meal, marine oil, and stickwater solubles concentrate. It requires slightly over 3 pounds of raw fish to produce 1 pound of product, including the meal, oil, and stickwater solubles.

Table 1 illustrates the production of fish meal, marine oil, and stickwater solubles in the United States.[2] The predominant fish being processed is menhaden, which accounts for nearly 80% of the total. Shellfish account for less than 5%.

TABLE 1. U.S. Production[a] of Fish Meal

Year	Fish Meal	Marine Oil	Stickwater Solubles
1986	680	267	219
1987	666	249	174
1988	644	225	224
1989	618	225	233
1990	577	282	186

[a]In million pounds.

Source: Reference 2.

Figure 1 provides a flow diagram of the wet reduction (rendering) process for a fish meal plant. As shown, the total operation includes the unloading of whole fish from fishing vessels into storage tanks or silos for a short time before the various processing stages of cooking and pressing take place.

Unloading of whole fish from boats is divided into dry and wet methods. Dry unloading methods include the use of a bucket elevator and conveyor, an air suction method that moves fish in a current of air, and a vacuum method employing a vacuum pump and a rotary valve to maintain a proper seal. Direct pumping without water is also possible to convey fish in a sealed pipeline. The wet unloading method uses a vacuum pump and water as the transporting medium to convey the fish from the boat to the plant. The water is separated at the plant and recycled back to the vessel for reuse.

The fish meal process typically is a continuous operating system. The following description is divided into two process phases, solids flow and liquid flow.[3,4]

Solids Flow

Cooking

Raw fish enter the cookers, which are horizontal cylindrical vessels equipped with steam-heated jackets and steam-heated screws with hollow flights. The cooking step coagulates the fish protein so that liquids and solids can be separated mechanically. Fat cells are also ruptured, releasing the oil into the liquid phase. A typical cooking process heats the fish for 10 to 15 minutes at 90°C.

Pressing

The cooked fish are conveyed to the press by an inclined screw conveyor with perforations for dewatering the cooked material. The purpose of pressing is to separate the liquid phase from the remaining solids in the cooked fish. Both continuous single- and twin-screw presses are used in the fish meal industry. The single-screw press is designed with a taper and exerts an increasing pressure on the cooked fish by reducing its volume as it passes through the press. A problem may be experienced with a single-screw press on soft fish, which can slip through the press without being

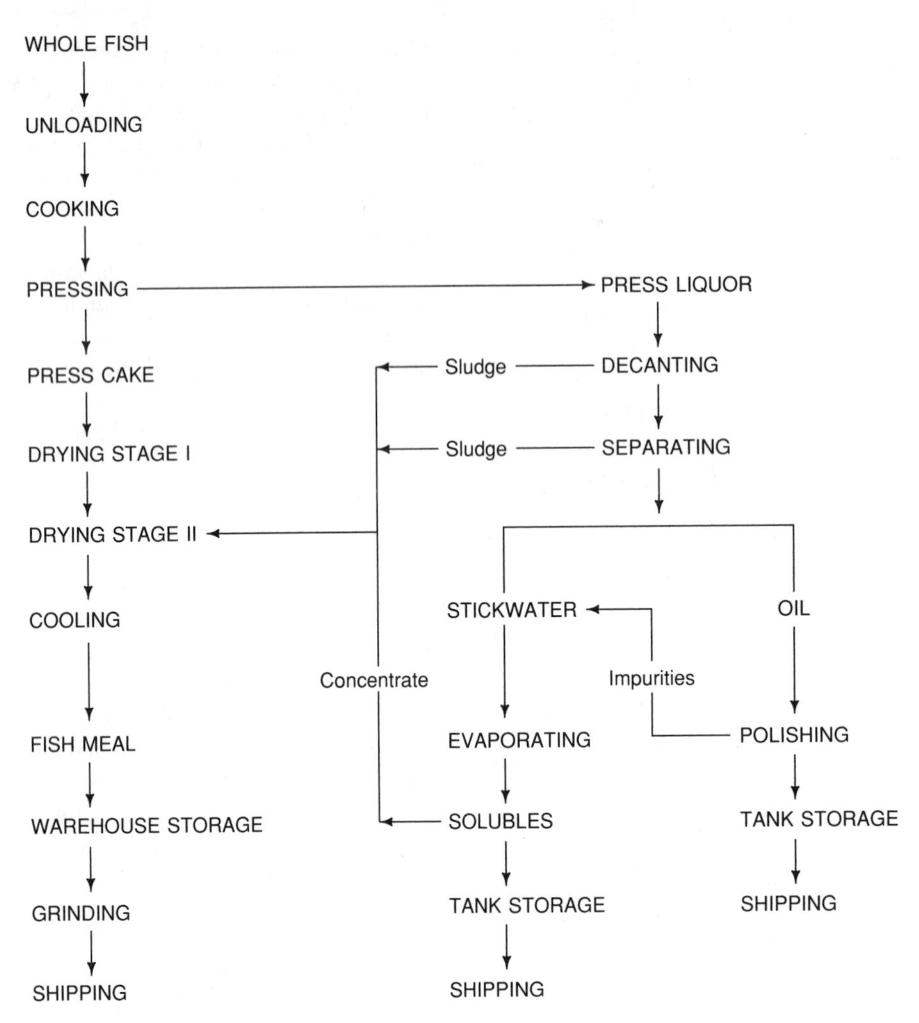

FIGURE 1. Typical Fish Meal Process

processed. This difficulty may be minimized by the use of a twin-screw press.

Drying

The pressed fish solids are then dried to produce fish meal. Drying reduces the moisture content in order to prevent microbial decomposition. The press cake fed to the dryer normally averages about 50% moisture, and the dried fish meal has a moisture content varying from 5% to 10%. Two main types of dryers are used: direct-fired rotary units and indirect rotary dryers.

The conventional direct-fired or flame-contact rotary dryer mixes the hot gases of combustion from the dryer firebox with a larger volume of fresh air before contact with the wet meal inside the rotating section. The exhaust air from this dryer usually passes through a cyclone to recover the entrained fish meal product. In this drying process, known as convection drying, the drying air and moisture-laden solids flow cocurrently, coming into direct contact with each other. The drying air inlet temperature is maintained below 600°C. The residence time of the solids in the dryer is about 15 minutes. This type of dryer is shown in Figure 2.

The indirect rotary dryer relies on heated surfaces that provide conduction drying for moisture removal. These heated surfaces consist of a series of parallel hollow disks or coils mounted vertically on a rotating hollow shaft through which passes either hot air or steam under 75–100 psig pressure. A small volume of fresh air passes countercurrent to the solids flow for moisture removal. The steam temperature generally ranges from 160°C to 170°C and the temperature of the exiting product is usually 85°C to 95°C. The retention time in this dryer is about 30 minutes. It is shown in Figure 3.

Another type of indirectly heated dryer combines certain features of both the direct-fired and the indirect rotary dryers. In the dryer, the combustion gases from the furnace pass through a heat exchanger to heat another stream of air that subsequently enters the rotary dryer, coming into direct contact with the solid particles, and removes moisture by convection. This type of indirect dryer can be referred to as an indirect rotary convection dryer to differentiate it from

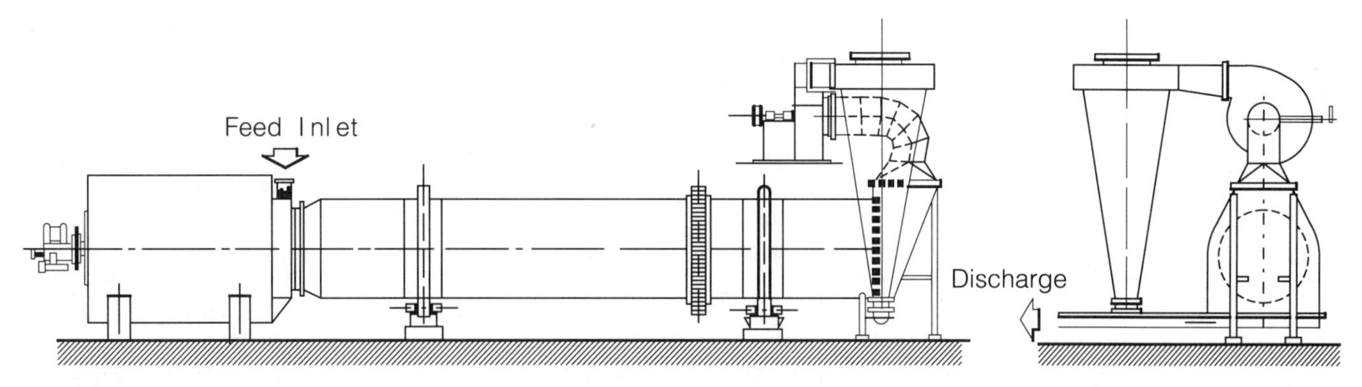

FIGURE 2. Direct-Fired Rotary Dryer (Reprinted with permission from the International Association of Fish Meal Manufacturers.)

the indirect rotary conduction dryer previously described.

Low-temperature, gentle drying systems have become more prevalent in the fish meal industry during the past five years. These improved drying methods include the increased use of indirect hot-air, vacuum, and fluid-bed dryers.[5] For example, the vacuum dryer, which is steam heated, is operated under a reduced pressure, resulting in correspondingly lower product temperatures of 50°C to 60°C.

Sand and Burt[5] compare the direct-fired dryer with various types of indirect dryers (steam, hot air, and vacuum) regarding certain criteria, including product quality, capital investment and operating costs, odor control, energy consumption, process control, maintenance, and space requirements. Wiedswang[6] compares the theoretical drying requirements for various types of dryers: direct-fired rotary, indirect rotary conduction, indirect rotary convection, and indirect vacuum.

Two stages of drying (I and II) are shown in Figure 1. A current trend exists to utilize both the direct-fired and the indirect hot-air rotary dryer in two stages. The direct-fired unit is used in the first stage to reduce the moisture in the press cake to 20–25% moisture. The indirect hot-air unit in the second stage completes the moisture removal. The stickwater concentrate and the sludge removed by the decanter centrifuges (in the liquid-flow part of the process) are added to the feed inlet to the second-stage dryer, since this additional moisture load is better assimilated at this point than in the feed inlet to the first-stage dryer.

Cooling, Storage, and Shipping

The dried meal is normally cooled in a rotary unit through which ambient air is passed countercurrent to the meal flow. The meal is then ground to pass 100% through a U.S. No. 7 Standard Screen. Fish meal must be stored in weatherproof, well-ventilated areas. Currently, the meal is being stored in bulk in sheds of single- or multi-unit construction with concrete floors and walls. Besides bulk storage, fish meal may be stored and shipped in multiwall paper, burlap, or woven plastic bags. A new type of bulk bag is capable of holding about 1 ton of product.

Liquid Flow

During the pressing operation, two intermediate products result: press cake and press liquor. As described under "Solids Flow," the press cake is dried to produce fish meal. The press liquor squeezed from the cooked fish contains coarse particles of fish and bone that must be removed before the liquor is centrifuged. These solids are removed by passing the liquor over a vibrating screen with 5–6-mm perforations. The recovered solids are sent back with the press cake to be dried.

Separation of the oil–water mixture is normally carried

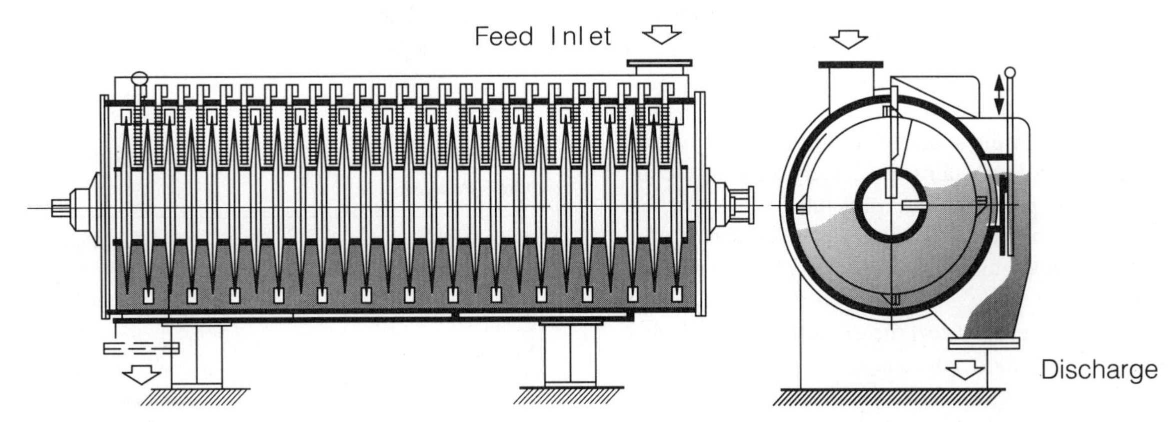

FIGURE 3. Rotary Disk Dryer (Reprinted with permission from the International Association of Fish Meal Manufacturers.)

out in three separate stages, or types of centrifuges: (1) a decanter to remove suspended solids; (2) a separator to recover oil from the press liquid, which becomes the stickwater; and (3) a polisher to remove water and other impurities from the oil concentrate. The stickwater is evaporated in a multiple-effect evaporator to produce a stickwater concentrate containing 50% dry substance. This concentrate is recycled back to the dryer for further moisture removal or is sold as a high-protein concentrate for animal feed.

Decanting (Solids Removal)

The decanter centrifuge consists of a horizontal cylindrical bowl that rotates and contains a rotating screw conveyor or scroll. The press liquor is fed into the bowl and the solids are forced to the outer wall. They are scraped off and conveyed out of the centrifuge by the scroll, which rotates at a slower speed than the bowl. Before discharge, the solids pass through a dewatering zone that tends to thicken or concentrate the solids. This sludge is returned to the dryers.

Separation

The liquid discharge from the decanter passes through a heat exchanger that raises its temperature to about 95°C before it enters the separator centrifuge. The separator is a three-phase centrifuge in which the feed liquor is separated into oil, stickwater, and a solids phase. The centrifuge consists of a series of cone-shaped disks rotating about a vertical axis and enclosed in a circular bowl. The feed liquor enters at the top center of the bowl and a sensor in the bowl is able to control the separation cycle by automatically discharging the accumulated solids. Since the main purpose of the separator is to maximize the recovery of oil, its operation is adjusted accordingly.

Polishing

The design of the polisher centrifuge is similar to that of the separator. However, its purpose is to polish or purify (ensure maximum removal of impurities from oil). As a result, the operation of the polisher is different from that of the separator. In order to obtain the desired purity, hot water equal to 10% of the oil volume is added to the feed liquor. The water and solids removed by polishing are added to the stickwater.

Evaporation of Stickwater

The water phase from the separator centrifuge is known as stickwater and usually contains 6–12% dry substance consisting of soluble proteins, fat, minerals, and vitamins. These stickwater solubles are concentrated by means of a multiple-effect evaporator that may consist of two, three, or four stages.

For a three-stage evaporator, the dilute stickwater enters the first stage, with a more concentrated solution passing on to the second stage and a final stickwater solubles concentrate of 50% dry substance discharging from the third stage. Steam under pressure is injected into the jacket at the first stage. The vapor from evaporation in the first stage passes on to the second-stage jacket and the vapor from the second stage passes on to the third-stage jacket. The vapor from the third stage is condensed and a vacuum is applied to the third-stage evaporator, which reduces the boiling points of evaporation in stages 2 and 3.

Most recent multiple-effect evaporators operate according to the falling film principle. Because the inside of the vertical evaporator tubes becomes gradually fouled due to deposits of proteins and calcium phosphate, it is necessary to circulate a caustic soda solution through the evaporator periodically in order to maintain production.

Since the raw material processed in fish meal plants contains 70–80% moisture, considerable fuel energy is required to evaporate this water. The recent trend in European and many U.S. fish meal plants has been to provide systems that require less energy.[7,8] This is accomplished by utilizing the exhaust vapors at atmospheric pressure from the indirect steam-heated or hot-air-heated dryers for moisture removal in the multiple-effect evaporator, which is operated under vacuum.

Fish meal has been used traditionally as a nutritious, high-protein feed ingredient in the diets of poultry and pigs. Currently, more fish meal is produced under improved drying methods and process control. This type of product is used to feed salmon, trout, shrimp, and mink. Fish oil is a normal constituent of many animal feeds. It is used as an industrial drying oil, such as linseed oil and tall oil, and also in margarine, shortening, and bakery fats.

AIR EMISSIONS CHARACTERIZATION

The largest odor emission source in fish meal plants normally is at the dryer, especially direct-fired dryers that require significant volumes of air for moisture removal. Lesser emissions occur at the cookers, evaporators, and other parts of the process. Table 2 provides an estimate of the odor emissions from various stages of the fish meal process.[9]

An extremely important consideration in the handling and storage of raw fish aboard a vessel concerns its temperature. The deterioration of raw fish begins rather quickly after the catch because of the bacteria and enzymes in the

TABLE 2. Odor Emissions from the Fish Meal Process

Stage of Process	Total Emission, %
Dryer	60–80
Cookers	10–20
Raw-material conveying	10–20
Pneumatic meal conveying	2–5

Source: Reference 9.

digestive tract. It has been determined for raw fish that the formation of volatile amines doubles with an increase in temperature of 6°C, and the odor unit profile increases tenfold.[7] It is essential to keep the raw fish at a low enough temperature to prevent this deterioration. As a result, fishing vessels are equipped with ice to be mixed with the fish or with refrigerated seawater circulated through their holds to maintain the temperature of the fish as low as is possible.

The storage of the raw material before processing is a definite source of odor and can vary from 1,000 to 100,000 odor units per cubic foot ASTM syringe method in the head space above the material, depending on its degree of deterioration.[10] These odor emissions from open areas, unless they are enclosed, are virtually impossible to control.

Where fugitive odor emissions from fish plants may affect the surrounding community more severely, a plant ventilating air system with subsequent odor control equipment may be necessary. Other potential sources of fugitive emissions include the leakage from certain process equipment, such as presses and open screens, and at the transfer point from one stage of the process to the next.

The odor emissions from the cookers may vary from 5,000 to 100,000 odor units per cubic foot (ASTM syringe method), depending on the state of decomposition of the raw material. The cooker is normally vented to odor control equipment at a volumetric emission rate varying from 100 to 1,000 ft^3/min.[1] Although the multiple-effect evaporator system may remove as much as 50% of the water evaporated from the total process, its operation normally is under vacuum and the odor emission is slight.

Table 3 illustrates the odor emissions from several rotary direct-fired dryers without odor control measures being applied.[11] It should be noted for dryer B that the odor units per standard cubic foot increased from 1500 to 4000 when the dryer discharge temperature increased from 240°F to 300°F. This illustrates the importance of controlling process conditions adequately in order to minimize odor generation. The exhaust from direct-fired dryers normally averages about 200°F and its moisture content ranges between 15% and 25% by volume.

Exhaust from indirect steam dryers is 30–45% lower in volume than from comparable direct-fired units. Steam dryers typically have a 25% moisture content in their exhaust, compared with about 15% moisture from a direct-fired unit processing the same material. Odor concentrations from steam dryers are generally in the same range as those from direct-fired units when fresh fish waste material is processed under proper operating conditions. Steam dryers are less likely to overheat the meal and result in excessive odor concentrations.

In many European countries, direct-fired dryers that use relatively large volumes of air and higher temperatures have been replaced by indirect steam-heated dryers because a lesser volume of fresh air is circulated through this type of dryer.[10] As a result, the operation of drying equipment that reduces odor volumetric emissions from the process allows the odor control systems to be more effective.

The composition of the odorous compounds present in the fish meal process is variable due to the various species of fish to be considered, as well as the degree of spoilage that has occurred and the actual process operating conditions being used. The concentration of gases in the head space above the storage of raw materials has been found to be quite high.[12] For example, inorganic gases such as hydrogen sulfide and ammonia were reported at levels up to 2000 ppm and 1000 ppm, respectively. The threshold limiting values (TLVs) for H_2S and NH_3 are specified at 10 ppm and 50 ppm, whereas their odor detection thresholds are reported at 0.00021 ppm and 21.4 ppm, respectively.[13] Also present were organic amines such as trimethyl amine, with an odor detection threshold of 0.00021 ppm.[13] Because of these low thresholds, even small volumes of these gases escaping to the surrounding atmosphere are capable of causing complaints.

With regard to dryer exhaust emissions, the use of gas chromatography and mass spectrometry[14] indicated the presence of sulfur compounds, such as hydrogen sulfide, carbon disulfide, carbonyl sulfide, and methyl and *n*-propyl mercaptans. In addition to ammonia, the only amine present was trimethyl amine. Analyses of the cooker exhaust in-

TABLE 3. Typical Odor Emissions from Rotary Fish Meal Dryers

Dryer	Feed Rate, tons/hour	Type of Fish	Temperature at Dryer Discharge, °F	Exhaust Gas Volume, scf/min[a]	Odor Units[b] per scf
A	10	Tuna	220	18,500	1500
A	15	Mackerel	220	18,500	1500
B	70	Tuna	220	9,000	700
B	10	Tuna	240	10,000	1500
B	14	Tuna	300	8,000	4000
C	9	Tuna	200	17,000	2500

[a]Standard cubic feet per minute.

[b]Odor units per standard cubic foot (70°F and 14.7 psia), ASTM syringe method.

Source: Reference 11.

TABLE 4. Odor-Removal Ability of Direct Condensers

System	Odor Units/scf[a]		Percent Removal	Air Temperature, °C	
	Inlet	Outlet		Inlet	Outlet
1	138,000	19,000	86		
	18,000	3,600	80	92	35
2	52,000	18,000	65	79	28

[a]Odor units per standard cubic foot, ASTM syringe method.

Source: Reference 10.

dicated that a number of sulfur compounds were present, but only one amine, trimethyl amine.

Dust emissions from the fish meal process are usually limited to the direct-fired dryers and to the grinding and pneumatic conveying of the dried fish meal. There are relatively few fines in ground fish meal. A sample of the dried product was collected from a pneumatic conveyor transporting ground fish meal. Only 0.6% by weight of the sample was less than 5 μm in diameter and 1.4% was less than 10 μm in diameter.[1] The cyclone collecting the pneumatically conveyed fish meal was found to be better than 99.9% efficient with an exhaust dust concentration of less than 0.01 gr/scf.

AIR POLLUTION CONTROL MEASURES

Since the process emissions from the cookers and dryers contain considerable moisture at temperatures of about 200°F (94°C), the necessary means should be provided to remove most of this moisture and to cool the process exhaust air before further odor control treatment. Also, there may be dust particles in the dryer cyclone exhaust that should removed before effective odor control measures can be applied. This is normally accomplished by either direct-contact or indirect water-cooled condensers. The indirect type includes the shell and tube condenser. The direct contact type includes cocurrent-flow venturi scrubbers and countercurrent flow spray-type scrubbers. They are often used where large volumes of seawater are available.

In addition to condensing water vapor and cooling the gas, a specific degree of odor reduction is obtained with direct-contact condensers since certain odorous compounds, such as NH_3, H_2S and $N(CH_3)_3$, are quite water soluble. Table 4 shows the ability of direct-contact condensers to reduce the inlet odor level.[10]

Two basic methods of odor control have been applied to the process emissions from cookers and dryers after condensing the water vapor: boiler incineration by direct-flame oxidation and wet scrubbing by chemical oxidation or the use of other scrubbing agents.

Thermal oxidation of the odorous process air is readily achieved in a boiler firebox operated at a temperature ranging from 650°C to 800°C (1200°F to 1475°F). However, the amount of odorous air to be treated must be compatible with the combustion air required for the boiler.

Where indirect steam-heated dryers are used, it may be feasible to use the dryer exhaust as combustion air after condensing the water vapor. However, if a direct-fired dryer is used, it normally is more feasible to use wet scrubbing. The fish meal industry in Europe uses the process air as a source of secondary combustion air in the boiler.[10] Table 5 illustrates the odor reduction obtained by boiler incineration.

As fuel costs have increased significantly, particularly in Europe, wet scrubbing systems have become more prevalent. Also, if fugitive odor emissions from the raw material storage area and the processing area are significant and cannot be captured by the exhaust duct system, a plant ventilating air system with adequate wet scrubbing equipment is required. Two wet scrubbing systems located in Norway and Denmark are described by Hansen.[15]

Depending upon the type of dryer (steam-heated disk units versus direct-fired units), the wet scrubbing system treats 500–1500 m^3 of dry air at standard conditions per ton of raw material processed. If plant ventilating air is required, the capacity of the scrubber system may increase to 3000–4000 m^3 of dry air at standard conditions per metric ton of raw material processed.

Seawater scrubbers are used to condense water vapor and cool the incoming gas stream. These may be venturi, packed-tower, or spray-type scrubbers and are normally

TABLE 5. Odor-Removal Ability of Boiler Incineration

System	Odor Units/scf[a]		Percent Removal
	Inlet	Outlet	
1	144,000	5,600	96
	18,000	840	95
2	4,750	850	82
	3,600	600	83
3	18,000	1,200	93
	9,000	1,000	89

[a]Odor units per standard cubic foot, ASTM syringe method.

Source: Reference 10.

TABLE 6. Odor Removal with Two Stages of Wet Scrubbing

Category of Scrubber	D/T[a]
Seawater Scrubber Inlet	200,000
Seawater Scrubber Outlet	20,000
Chemical Scrubber Outlet	1,000

[a]Odor dilution to threshold by IITRI dynamic olfactometer.
Source: Reference 15.

used as the first stage of a two-stage wet scrubbing system. The exhaust air from this first-stage scrubber passes on to the second-stage chemical oxidation scrubber, which normally contains packing. In the past, the packing consisted of 1-inch or 2-inch Rasching rings. Currently, more efficient types of packing are available. The use of sodium hypochlorite solution as a wet scrubbing agent in the second stage is preferred over potassium permanganate and hydrogen peroxide. Table 6 illustrates the odor reduction results obtained with two stages of scrubbing.[15]

Another fish meal plant in Norway is described by Onarheim and Utvik.[16] It uses two separate odor control systems—one exclusively for the dryer exhaust and the other for treating the remaining process odor emissions and the plant ventilating air. This plant processes 700 tons of raw fish (mackerel, herring, fish waste, etc.) every 24 hours.

The exhaust air from three Rotadisc driers passes to a packed tower scrubber operating at a linear velocity of 2 m/s (6½ fps) with 6 m^3/min (1600 U.S. gpm) of seawater. The scrubber outlet passes through an indirect heat exchanger, which uses the steam vapors from the dryers as a heating medium to preheat the scrubber exhaust. Most of this air is recirculated back through the dryers and the remainder is incinerated in the boiler firebox.

The second odor control system, which treats the remaining process emissions and plant ventilating air, consists of two stages of scrubbers. A packed-tower scrubber through which seawater is circulated precedes a second packed tower through which sodium hypochlorite solution is recirculated. The first-stage scrubber consumption of seawater is 2½ m^3/min (660 U.S. gpm). In the second stage, the consumption of NaOCl is estimated at 13 to 19 pounds per 1000 metric tons of raw material. Amines and ammonia could not be detected analytically in the second-stage exhaust.

In summary, fish meal plants in Europe and the United States have developed an odor control capability to treat their emissions and avoid becoming local nuisances. Process emissions are controlled by a combination of wet scrubbing with seawater and boiler incineration. Plant ventilating air is treated by wet scrubbing.

Biofilter technology developed during the past 10 years in Europe and elsewhere has been applied to the treatment of odor emissions from fish meal plants.[17] Biofilters consist of large beds of porous media that are capable of adsorbing odorous gaseous compounds and reducing these by aerobic microbial action to nonodorous components. This technology is now being applied to fish meal plants in the United States.

The Torry Research Station[18] in Scotland has developed a set of recommended practices to be used to reduce odor emissions during the production of fish meal. These practices include the delivery and handling of raw material, processing, handling and storage of fish meal, plant and equipment design, plant management, and maintenance.

Acknowledgment

This section was reviewed and comments were furnished by Anthony P. Bimbo, Director of Applied Development for Zapata Haynie Corp., Reedville, Va.

References

1. "Fish canneries and fish reduction plants," *Air Pollution Engineering Manual,* 2nd ed., AP-40, U.S. Environmental Protection Agency, Research Triangle Park, NC, May 1973, pp 760–770.
2. *Fish Meal and Oil Quarterly Report,* NOAA National Marine Fisheries Services, U.S. Department of Commerce, No. 2162-3 in the *American Statistics Index.*
3. A. P. Bimbo, "Production of fish oil," *Fish Oils in Nutrition,* M. E. Stansby, Ed.; Van Nostrand Reinhold, New York, 1990, p 155.
4. A. P. Bimbo, "Fish meal and oil," *The Seafood Industry, A Self-Study Guide,* Sea Grant Extension Division, Virginia Tech University, Blacksburg, VA, 1987, pp 1–45.
5. G. Sand and J. Burt, "Which kind of dryer is best," *Processing Bulletin No. 2,* International Association of Fish Meal Manufacturers, Herts, England, July 1987, pp 1–11.
6. H. Wiedswang, "Low temperature dryers," paper presented at the 29th Annual Conference of the International Association of Fish Meal Manufacturers, Lima, Peru, November 1988.
7. P. Fosbol, "Production of fish meal and fish oil," INFOFISH International, May 1988, pp 26–32.
8. A. O. Utvik, "Modern fish oil and meal processing," 75th Annual Meeting of American Oil Chemical Society, Dallas, TX, 1984.
9. W. Schmidtsdorff, Symposium, International Exhibition, Jonkoping, Denmark, 1974.
10. J. Wingall, "Odor pollution in the fish meal and oil industry," *News Summary No. 44, Special Issues on Processing,* International Association of Fish Meal Manufacturers, Herts, England, September 1978, pp 44–65.
11. J. L. Mills, J. A. Danielson, and L. K. Smith, "Control of odors from inedible rendering and fish meal reduction in Los Angeles County," 60th Annual Meeting of APCA, Cleveland, OH, June 1967.

12. J. B. Dalgaard, F. Denker, and B. Fallentin, *Br. J. Indus. Med.*, 29:307 (1972).
13. G. Leonardos, D. Kendall, and H. Barnard, "Odor thresholds for 53 commercial chemicals," *J. Air Poll. Control Assoc.*, 19(2):91 (1969).
14. Y. Hoshika, S. Kadowaki, I. Kojima, et al., *Oil Chemistry* (Japan), 24(4):233 (1975).
15. K. W. Hansen, "Odor control in the fish meal industry," *News Summary No. 44, Special Issues on Processing,* International Association of Fish Meal Manufacturers, Herts, England, September 1978, pp 66–74.
16. R. Onarheim, and A. O. Utvik, "A system of complete effluent control," *News Summary No. 44, Special Issues on Processing,* International Association of Fish Meal Manufacturers, Herts, England, September 1978, pp 75–80.
17. G. Leson and A. M. Winer, "Biofiltration: an innovative air pollution control technology for VOC emissions," *J. Air Waste Manage. Assoc.*, 41(8):1048 (1991).
18. "Reducing odour in fish meal production," Torry Advisory Note No. 72, Torry Research Station, Aberdeen, Scotland, May 1979, pp 1–7.

LIVESTOCK FEEDLOTS

John M. Sweeten, Ph.D., P.E.

The purposes of this discussion are to identify and assess (1) the relationship of livestock production to air pollution emissions and (2) technology and management practices to reduce potential for air contamination from livestock and poultry operations.

MANURE PRODUCTION AND DISTRIBUTION IN THE UNITED STATES

Total U.S. manure production (dry basis) for all livestock and poultry species were calculated by Van Dyne and Gilbertson.[1] They found that total annual manure production was 112 million dry tons and the manure contained 8.2 billion pounds of nitrogen, 2.1 billion pounds of phosphorus, and 4.8 billion pounds of potassium. Volatilization, leaching, and runoff losses reduced the dry weight by 10%, total nitrogen by 36%, phosphorus by 5%, and potassium by 4%. Recoverable (collectable) manure and nutrients were estimated by subtracting the portion of materials voided on pasture areas. Collectable (economically recoverable) dry manure was estimated at 52 million tons, and the fractions from various animal species were dairy cattle, 39%; feeder cattle, 31%; hogs, 11%; laying hens, 6%; broilers, 5%; sheep, 3%; and turkeys, 2%.

The estimates[1] were based on an engineering standard adopted by the American Society of Agricultural Engineers[2] based on constituent production per unit weight of live animal. These standard values were recently updated to reflect current research data.[3] In most cases, average values of dry manure and nutrients (pounds per day per 1000 pounds live weight) were revised upward, and standard deviations were calculated to reflect the degree of variability.

PROCESS DESCRIPTION

Intensive Animal Production Systems

Major types of livestock and poultry production facilities, design information, and associated manure management systems have been described in other reports.[4–7] Roofed or total confinement facilities are common for poultry and swine and, to a lesser extent, dairy and beef production.[8] Open feedlots (nonroofed) are the type of intensive confinement facilities most commonly used for beef cattle. They are also widely used for dairy, swine, and sheep production in the southwestern United States.

Intensive livestock production systems are regarded as "animal feeding operations," which are defined (for purposes of water pollution control) in U.S. Environmental Protection Agency (EPA) regulations for the feedlots point source category as areas where animals are "stabled or confined and fed or maintained for a total of 45 days or more in any 12 month period, and . . . crops, vegetation, forage growth or post-harvest residues are not sustained in the normal growing season over any portion of the lot or facility."[9] The EPA definition is not specific as to animal species, type of confinement facility, or animal density, but essentially integrates these factors (along with climate and soils) into a single, visually determined criterion—absence of vegetation—which develops where manure production and/or animal traffic are sufficient to prevent germination or growth of forages.

Cattle Feedlots

The United States has 9.4 million beef cattle in feedlots and ranging in live weight from 450 to 1200 pounds per head, typically averaging 850 pounds. Each animal fed in a normal 130- to 150-day fattening period results in approximately 1 dry ton of collectable manure solids, or about 2 dry tons collected manure per year per head of feedlot capacity. The animal spacing per head varies according to rainfall and temperature, slope, and other factors from typically 100 to 125 ft^2 per head in the desert southwest (less than 10 inches annual rainfall) to 175–200 ft^2 per head in the southern central Great Plains (15 to 25 inches per year), which contains the largest cattle feeding states, and to 300 to 400 ft^2 per head in the eastern and northern Great Plains (25 to 35 inches per year).

Most of the manure deposited on the feedlot surface is compacted by cattle into a relatively moist manure pack of 35–50% moisture content (wet basis). At higher moisture contents, odors can develop, especially in warm weather.

The surface manure may become pulverized by cattle hooves during prolonged drying conditions to only 10–25% moisture wet basis. When surfaces are excessively dry, there is a potential for dust problems, and these problems have been most severe in the arid cattle-producing areas of Arizona, California, and Texas. Dust from cattle feedlot surfaces, alleys, and roads can annoy neighbors, irritate feedlot employees, possibly impair cattle performance, and create a traffic hazard on adjacent highways.[10] Total particulate emissions are affected by feedlot area, cattle density in pens, wind speed, and precipitation and evaporation patterns.[11]

Odors from Livestock Feeding Operations

Odors from livestock facilities are sometimes an annoyance and affect the well-being of downwind neighbors. However, odorous gases are not toxic at concentrations found downwind of livestock feeding facilities. Public or private nuisance lawsuits can threaten the survival of an operation, and livestock producers need to control the evolution of odorous compounds.

These odorous gases arise from feed materials (food-processing wastes and fermented feeds), fresh manure, and stored or decomposing manure. Research has been directed toward the control of volatile compounds released by the storage and treatment of manure. Odor is generally less objectionable from fresh manure than from anaerobically decomposing manure. Fresh manure evolves large quantities of ammonia, but it is generally not accompanied by other decomposition products that contribute the most objectionable characteristics. Odorous compounds evolved from manure treatment facilities are a function of the material as excreted, the biological reactions occurring in the material, and the configuration of the storage or treatment unit.

Roofed confinement facilities typically have high odor production potential, owing to the high animal density involved, the large inventory of manure frequently in storage, and limited rate of air exchange.[8] Manure-covered surfaces (e.g., building floors and animals), manure storage tanks beneath slotted floors, and anaerobic lagoons often used for manure storage and treatment are important odor sources.

When open feedlot surfaces become wet, particularly in warm weather, they support anaerobic decomposition with an associated large surface area for the evolution of odorous gases. Feedlot odor problems are most frequent in warm, humid areas and in feedlots constructed in areas of inadequate drainage or poor drying conditions.

AIR EMISSIONS CHARACTERIZATION

Particulate Emission Standards

In 1971, the EPA[12] promulgated primary and secondary national ambient air-quality standards for total suspended particulate matter (TSP). The primary standards were set at 260 $\mu g/m^3$ for 24-hour average, not to be exceeded more than once per year, and an annual geometric mean of 75 $\mu g/m^3$. Secondary standards were set at 150 $\mu g/m^3$ for a 24-hour sampling period, not to be exceeded more than once per year.

Effective July 31, 1987, the EPA[12] replaced TSP as the indicator for particulate matter for the ambient standards in favor of a new indicator that includes only those particulates with an aerodynamic particle diameter less than or equal to a nominal 10 μm (PM-10). The new standard replaced the 24-hour primary TSP standard with a PM-10 standard of 150 $\mu g/m^3$, with no more than one expected exceedance per year allowed; replaced the annual primary standard for TSP with a PM-10 standard of 50 $\mu g/m^3$ expected annual arithmetic mean; and replaced the secondary TSP standard with 24-hour and annual PM-10 standards that are identical to the primary standards.

Dust Emissions in Cattle Feedlots

Elam et al.[13] collected feedlot dust samples inside of 65 pens at 10 California feedlots using a Staplex high-volume air sampler operated in one- to three-hour increments during 24-hour continuous sampling periods. Peak particulate concentrations, collected between 1900 and 2200 Pacific Daylight Time (PDT), ranged from 1946 to 35,536 $\mu g/m^3$ and averaged 14,200 $\mu g/m^3$ ($\pm$11,814 $\mu g/m^3$ standard deviation). Lowest concentrations occurred in early morning when concentrations were only 130 to 250 $\mu g/m^3$ in some feedlots.

Algeo et al.[14] measured total suspended particulates using 24-hour samplings both upwind and downwind of 25 California feedlots. Net particulate concentrations (downwind minus upwind) for 24-hour continuous sampling time ranged from 54 to 1268 $\mu g/m^3$. The average value for all 25 feedlots was 654 $\pm$ 376 $\mu g/m^3$. Upwind concentrations averaged 25% of the downwind concentrations. Both upwind and downwind particulate levels usually exceeded the EPA ambient air-quality standards for TSP. Most feedlot particulates will settle out rapidly, however.

Peters and Blackwood[11] cited as major limitations of the results of Algeo et al.[14] the facts that:

1. All sampling was performed in the dry season.
2. Details such as feedlot size, cattle number, distances from samplers to feedpens, and climate conditions were not reported.

Nevertheless, using the California data from Algeo et al.,[14] *worst-case projections* were developed for cattle feedlots.[11] A mass particulate emission rate of 0.036 $\pm$ 0.022 g/s/m of feedlot length was assumed. The projections would place feedyards with more than 2000 head, at an average animal spacing of 150 ft^2 per head, above a particulate emission level of 100 tons per year, not including the feedmill.

Based on the Peters and Blackwood[11] treatment of the California data, the EPA[15] published so-called emission factors (AP-42) for cattle feedlots and termed them "crude estimates" at best. These estimates of particulates were indexed according to cattle feedlot size and cattle throughput as follows[15]:

1. Feedlot capacity basis (280 pounds particulates per day per 1000 head).
2. Feedlot throughput basis (27 tons particulates per 1000 head fed).

Other emissions factors were similarly written for ammonia, amines, and total sulfur compounds.[15]

The EPA emission factors[15] ignore the major climatic differences that exist between southern California and the larger cattle regions of the Great Plains and the Midwest, in which both total rainfall and seasonality of rainfall are different. California represents only 5% of the U.S. cattle on feed as compared with Texas, Nebraska, Iowa, and Kansas, which combined had 68% of the 9.9 million head on feed as of January 1, 1990.

To obtain a broader database, dust emissions were measured at three Texas cattle feedlots ranging from 17,000 to 45,000 head on 15 occasions in 1987 to determine concentrations of feedlot dust measured both as TSP and dust below 10-μm aerodynamic particle size (PM-10).[16] Net feedlot dust concentrations (downwind minus upwind) ranged from 16 to 1700 $\mu g/m^3$ and averaged 412 $\pm$ 271 $\mu g/m^3$ (which is 37% less than the earlier California data). Feedlot dust concentrations were generally highest in early evening and lowest in early morning, and upwind concentrations averaged 22% of downwind concentrations.

Using two types of PM-10 sampler (Wedding and Andersen-321A), the PM-10 dust concentrations were 19% and 40%, respectively, of mean TSP concentrations in direct comparisons.[16] There was good correlation between PM-10 and TSP concentrations with $r^2 = 0.634$ and 0.858 for Wedding and Andersen 321-A samplers respectively.[16] Mean particle sizes of feedlot dust were 8.5 to 12.2 μm on a population basis, while respirable dust (below 2 μm) represented only 2.0–4.4% of total dust on a particle volume basis.[17]

When the Wedding sampler was used for PM-10 measurements, feedlots were below the new EPA standard and peak concentrations did not coincide with expected early-evening peaks caused by cattle activity. Hence comparatively little of the actual feedlot manure dust may have been captured in the Wedding PM-10 instruments.

Analysis with a Coulter counter resulted in aerodynamic particle size distribution curves for TSP and PM-10 samplers (Figure 1)[18]. Mass median diameters (MMD) of dust particles captured on high-volume samplers averaged 14.2 μm downwind as compared with values of 12.3 μm upwind of feedlots.[18] The PM-10 sampler oversampled particles larger than 10 μm.

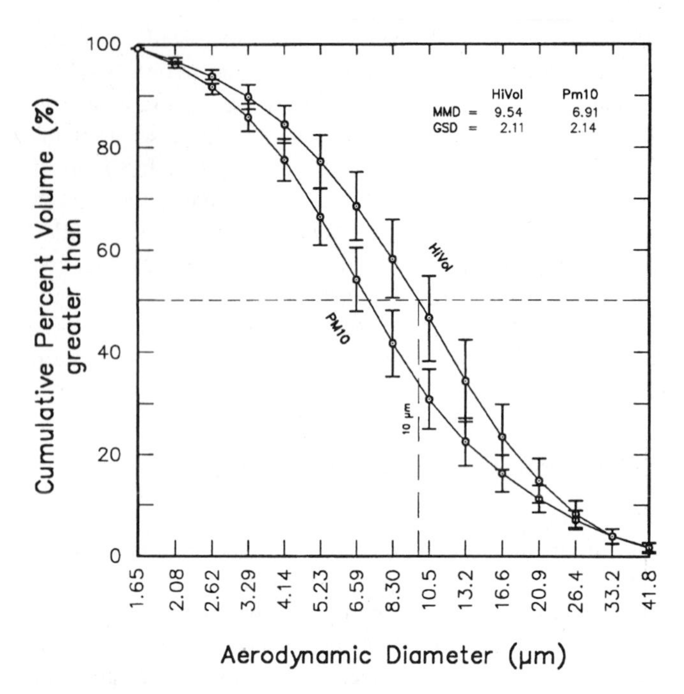

FIGURE 1. Cumulative Volume Fraction of Feedlot Dust Particles of Given Size Captured on Filters of High-Volume and PM10 Samplers; Downwind Samplers at Feedlots A, C, and B (Experiments 11, 14, and 16) (From Reference 18)

Odor from Livestock Facilities

Animal manure odor is composed of gaseous compounds that are intermediate and final products of biodegradation and includes these groups (Table 1): ammonia and amines, sulfides, volatile fatty acids, alcohols, aldehydes, mercaptans, esters, and carbonyls.[8,19,20,21] Although concentrations of these compounds at downwind locations are low, some may exceed olfactory threshold values and create nuisance conditions.

There is nearly universal acceptance of sensory approaches, involving the use of trained human panelists, for measurement of odor. However, instruments and techniques to facilitate sensory odor measurement may vary. Odor measurement technology applicable to livestock operations includes determining:

1. Concentrations of specific compounds (ammonia, hydrogen sulfide, volatile organic acids, etc.)
2. Dilutions-to-threshold with a dynamic forced-choice olfactometer or scentometer.
3. Equivalent concentration of butanol vapor that matches the ambient odor intensity using a butanol olfactometer.

Several states and municipalities have property-line odor standards based on these and other measurement methods.[22]

Ammonia and propionic acid concentrations were correlated with odor intensity.[23] Correlations between odor intensity and gas concentrations were developed in two- and three-component mixtures of hydrogen sulfide, ammonia, and methyl amine.[24]

TABLE 1. Compounds Resulting from the Anaerobic Decomposition of Livestock and Poultry Manure

Alcohols	Amines
	Methylamine
Acids	Ethylamine
Butyric	Trimethylamine
Acetic	Diethylamine
Propionic	
Isobutyric	Esters
Isovaleric	
	Fixed gases
Carbonyls	Carbon dioxide (odorless)
	Methane (odorless)
Sulfur compounds	Ammonia
Hydrogen sulfide	
Dimethyl sulfide	Nitrogen heterocycles
Diethyl sulfide	Indole
Methylmercaptan	
Disulfides	Skatole

Odor caused by anaerobic decomposition of swine manure were determined by Meyer and Converse,[25] who found that H_2S and NH_3 production were, respectively, 218% and 118% higher at 73°F as compared with 60°F during a 23-day storage period. In European research,[26] the odor emission rate from swine houses with anaerobically stored manure increased 20-fold for each 10°F rise in manure temperature and, including ventilation rate influences, was over four times greater in summer than in winter. Emissions were 73% greater for fully slotted floors than for partially slotted floors.

Odor-intensity observations were made with a scentometer during eight experiments at three cattle feedlots at upwind and downwind locations alongside the dust samplers.[16] Upwind odor intensities were typically in the range of 0 to 2 dilutions to threshold (DT), while downwind concentrations were 7, 31, or 170 DT. Arithmetic average downwind odor intensities ranged from 13 to 49 DT.

AIR POLLUTION CONTROL MEASURES

Dust Emissions from Cattle Feedlots

Guidelines on feedlot dust control[10,27,28] usually involve water sprinkling at strategic times and in proper amounts. Carroll et al.[29] compared two feedlots, one unsprinkled and the other sprinkled at the rate of two hours on, two and one-half hours off, one and one-half hours on each day, and reported that sprinkling reduced dust emissions by at least half.

Elam et al.[13] reported that feedlot manure moisture content of 20% to 30% was needed for dust control. Particulate concentrations (24-hour averages) increased from 3,150 to 23,300 $\mu g/m^3$ when daily water sprinkling was terminated for seven days.

Sweeten et al.[16] found that feedlot dust concentrations decreased with increasing moisture content in the top 1 inch of feedlot surface. Regression equations indicated that the manure moisture needs to be 26–31% (wet basis) in the loose surface manure and 35–41% at 0- to 1-inch depths in order to control feedlot dust to TSP limits of 150 and 260 $\mu g/m^3$. However, odor intensity (DT) increased as net dust concentrations decreased, perhaps because of moisture influences.

Odor Control Methods

Specific measures have been devised to reduce odors from livestock manure,[19,20] including the development of a standard engineering practice.[21] These methods fall into three broad categories[22]:

1. Treatment of manure—aeration, anaerobic digestion, biochemical treatment.
2. Capture and treatment of odorous gases—containment, covered storage pits or lagoons, wet scrubbing, packed-bed adsorption, soil incorporation, soil filter fields, or absorption beds.
3. Odor dispersion—site selection, separation distance, elevated source, topography, frequency of wind direction, and atmospheric stability.

Research involving several of these methods will be discussed in the following.

Manure Treatment

Controlled-rate anaerobic digestion of liquid swine manure at 32°C for methane gas production reduced the odor emission rate from land-applied digested slurry by 90% as compared with pit-stored slurry.[26] Anaerobic digestion also reduced the time for odor dissipation (in terms of 50% panelist detection threshold) from 72 hours to 24 hours.

Anaerobic lagoons must have adequate capacity (i.e., low loading rate) to obtain relatively low odor production. Design criteria have been developed based principally on the volatile solids loading rate, which is proportional to the volume per pound of live weight.[30–33]

Mechanical aeration of liquid manure in oxidation ditches or lagoons has been determined to be an effective odor-control method.[34] Aerating only the top third or half of swine lagoon contents proved successful and reduced power requirements as compared with complete mixing.[34] Converse et al.[35] used limited aeration of liquid swine manure without a measurable dissolved oxygen residual and reduced odor as compared with nonaerated storage. Phillips et al.[36] rapidly reduced hydrogen sulfide and methanol emissions from swine manure by aeration, but less volatile and less offensive compounds such as phenols persisted. Aeration just prior to land spreading could reduce odors from field application.

Frequent manure collection by flushing, cable scraping, or pit drainage/recharge is useful for absorbing odorous gases and eliminating anaerobic storage conditions in confinement buildings.[25,37,38]

Biochemicals for odor control include masking agents, counteractants, digestive deodorants, chemical deodorants, adsorbents, and feed additives.[39] Digestive deodorants are the most prevalent, and necessitate frequent additions to allow selected bacteria to become predominant. Potassium permanganate (100–500 ppm), hydrogen peroxide (100–125 ppm), and chlorine are oxidizing chemicals capable of controlling hydrogen sulfide emissions.

Warburton et al.[40] significantly reduced odors from anaerobic swine manure slurry with four treatments: aeration, chlorination, and two biochemical formulations. Numerous other chemical, bacterial, or enzyme products that were tested were not effective for either odor control or solids reduction. Lindvall et al.[41] reduced odors of liquid swine manure with ammonia persulfate, and Miner and Stroh[42] determined that zeolites (clinoptilolite and erionite) were somewhat effective for odor reduction from a dirt-surfaced cattle feedlot.

Odor Capture and Treatment

Installing a cover on an outside manure storage pit, tank, or lagoon is an effective means of odor control because it effectively reduces the ventilation rate and hence the rate of odor emission. However, rigid covers are expensive and flexible membrane covers over large surfaces are subject to photodegradation and wind damage.

Wet scrubbers that involve spraying exhaust air with water or oxidizing chemicals are widely used for industrial and food processing plant odors, and some researchers have adapted them to livestock confinement buildings. Van Geelen and Van Der Hoek[43] obtained a 88% reduction in odor concentration with wet scrubbing of exhaust from a swine house, although captured dust formed a sludge that caused difficulty in recirculating the scrubbing water. Schirz[44] cited problems with clogging of spray nozzles when scrubbing with recycled water and biological treatment was required. Licht and Miner[45] built a horizontal cross-flow packed-bed wet scrubber for swine confinement building exhaust air and achieved 50% and 90% removal of particulates larger than 1 and 5 μm, respectively; an ammonia reduction of 8–38%; and 82% reduction of odor intensity.

A packed-bed dry scrubber filled with a zeolite (clinoptilolite) reduced ammonia emissions from a poultry house by 45% initially, but efficiency dropped to only 15% in 18 days.[46]

The soil is an excellent odor-scrubbing medium because of chemical absorption, oxidation, and aerobic biodegradation of organic gases.[47] Lindvall et al.[41] determined that soil injection reduced odor emissions (measured as DT) from liquid swine manure by 90–99% as compared with surface spreading. Odor from soil-injected manure was about equal to that from a non-manured soil surface. Disk harrowing or plowing of surface spread manure reduced odor threshold levels by 67% to 95%.

Soil filters with perforated pipe in a shallow soil bed have proved effective for scrubbing odors from a process or building exhaust air. Kowalewsky[48] removed 52–78% of the ammonia and 46% of the organic constituents from ventilation air from a swine-confinement building using a soil filter system. Prokop and Bohn[49] reported 99.9% odor reduction (ED_{50}) from a soil filter used to treat high-intensity odors in exhaust from rendering plant cookers. Soil filters required a moderately fine-textured soil, sufficient moisture, and pH control of 7 to 8.5. A typical land area requirement is 2,500 to 4,600 ft^2 per 1000 cfm, depending on the airflow rate.[49] Sweeten et al.[50] measured 95–99% reduction in ammonia emissions and 30–82% reduction in odor intensity (matching butanol concentrations) using a 0.25-acre sand filter field to scrub air from a poultry manure composting operation.

Odor Dispersion

Odorous gases are diluted with distance downwind of the source, depending on atmospheric turbulence and odorant reactions. Atmospheric dispersion models are sometimes used to predict the travel of odor emissions[51] and their impact on communities. The use of dispersion models is limited to short distances and to nonreactive odorous gases.[8] One or more versions of the Gaussian diffusion model is used in most regulatory applications. The prediction models require that atmospheric stability, wind speed, and odor emission rates be known.

Odors, ammonia, and propionic acid traveled farther downwind from land-spreading liquid manure than from above-ground manure storage tanks or confinement buildings.[23] An odor panel observed a 90% reduction in odor intensity, as determined by a matching butanol olfactometer,[52] over a distance 0.5 mile downwind of a cattle feedlot in Texas.[53]

Klarenbeek[26] reported an ordinance in the Netherlands that stipulates a "required minimum distance" of swine finishing houses from residences, for example, for a 1500-hog finishing operation, a 1000-foot distance to residential areas or 550 feet to an individual nonagricultural residence. The Queensland Department of Primary Industries[54] developed guidelines for feedlot separation distance from residences on populated areas based on size, design, and management factors.

References

1. D. L. Van Dyne, and C. B. Gilbertson, *Estimating U.S. Livestock and Poultry Manure and Nutrient Production,* ES-CS-12, Economics, Statistics and Cooperative Service, U.S. Department of Agriculture, Washington DC, 1978, 150 pp.
2. *Manure Production and Characteristics,* ASAE Data D384, American Society of Agricultural Engineers, St. Joseph, MI, 1976, 1 p.

3. *Manure Production and Characteristics,* ASAE Data D384.1, American Society of Agricultural Engineers, St. Joseph, MI, 1988, 4 pp.
4. *Development Document for Proposed Effluent Limitations Guidelines and New Source Performance Standards for the Feedlots Point Source Category,* EPA-440/1-73/004, Washington, DC, 1973, pp 59–64.
5. R. K. White, and D. L. Forster, *A Manual on Evaluation and Economic Analysis of Livestock Waste Management Systems.* EPA 600/2-78-102, U.S. Environmental Protection Agency, Robert S. Kerr Environmental Research Laboratory, Ada, OK, 1978, 302 pp.
6. *Beef Housing and Equipment Handbook,* MWPS-6, Midwest Plan Service, Iowa State University, Ames, IA, 1987.
7. J. Foster, and V. Mayrose, *Pork Industry Handbook.* Cooperative Extension Service, Purdue University, West Lafayette, IN, 1987.
8. National Research Council, *Odors from Stationary and Mobile Sources,* National Academy of Sciences, Washington, DC, 1979.
9. U.S. Environmental Protection Agency "State program elements necessary for participation in the national pollutant discharge elimination system—concentrated animal feeding operations," 40 CFR 124.82, *Federal Register,* March 18, 1976, p 11460. (See also 40 CFR 122.23 including Appendix B thereof.)
10. J. M. Sweeten, *Feedlot Dust Control,* L-1340, Texas Agricultural Extension Service, Texas A&M University System, College Station, TX, 1982.
11. J. A. Peters and T. R. Blackwood, *Source Assessment: Beef Cattle Feedlots,* Montsanto Research Corp., EPA-600/2-77-107, U.S. Environmental Protection Agency, Industrial Environmental Research Laboratory, Research Triangle Park, NC, 1977.
12. U.S. Environmental Protection Agency, 40CFR50, *Revisions to the National Ambient Air Quality Standards for Particulate Matter* and Appendix J—*Reference Method for the Determination of Particulate Matter as PM-10 in the Atmosphere, Federal Register,* 52(126):24634-24669, 1987.
13. C. J. Elam, J. W. Algeo, T. Westing, et al., "Measurement and control of feedlot particulate matter," Bulletin C, *How to Control Feedlot Pollution,* California Cattle Feeders Association, Bakersville, CA, January 1971.
14. J. W. Alego, C. J. Elam, A. Martinez, and T. Westing, "Feedlot air, water and soil analysis," Bulletin D, *How to Control Feedlot Pollution,* California Cattle Feeders Association, Bakersville, CA, June 1972, 75 pp.
15. Supplement A to *Compilation of Air Pollution Emission Factors,* Section 6.15, *Beef Cattle Feedlots* (*Stationary Point and Area Sources,* Vol 1). AP-42, U.S. Environmental Protection Agency. Office of Air Quality Planning and Standards, Research Triangle Park, NC, 1986.
16. J. M. Sweeten, C. B. Parnell, R. S. Etheredge, and D. Osborne, "Dust emissions in cattle feedlots," Veterinary Clinics in North America, *Food Animal Practice,* 4 (3):557–578 (November 1988).
17. D. J. Heber, and C. B. Parnell, "Comparison of PM-10 and high-volume air samplers using a Coulter counter particle size analyzer," Paper no. SWR 88-109, presented at 1988 Southwest Region Meeting of ASAE, Lubbock, TX, 1988.
18. J. M. Sweeten, and C. B. Parnell, "Particle size distribution of cattle feedlot dust emissions," ASAE Paper no. 89-4076, International Summer Meeting of American Society of Agricultural Engineers, Quebec, Canada, June 25–28. 1989, 20 pp.
19. J. R. Miner, "Management of odors associated with livestock production, in *Managing Livestock Wastes, Proceedings of Third International Symposium on Livestock Wastes,* American Society of Agricultural Engineers, St. Joseph, MI, 1975, pp 378–380.
20. C. L. Barth, L. F. Elliot, and S. W. Melvin, "Using odor control technology to support animal agriculture," *Trans. ASAE,* 27:859–684 (1984).
21. "Control of manure odors," ASAE EP-379, *Agricultural Engineers Yearbook of Standards,* American Society of Agricultural Engineers, St. Joseph, MI, 1987, pp 405–406.
22. J. M. Sweeten, "Odor measurement and control for the swine industry," *J. Envir. Health,* 50(5):286 (1988).
23. H. H. Kowalewsky, R. Scheu, and H. Vetter, "Measurement of odor emissions and imissions," in *Effluents from Livestock,* (J.K.R. Gasser, Ed.); Applied Science Publishers, London, 1979, pp 609–625.
24. D. T. Hill, and C. L. Barth, "Quantitative prediction of odor intensity," *Trans. ASAE,* 19:939–944 (1976).
25. D. J. Meyer, and J. C. Converse, "Gas production vs. storage time on swine nursery manure," Paper no. 81-4512, American Society of Agricultural Engineers, St. Joseph, MI, 1981.
26. J. V. Klarenbeek, "Odour emissions of Dutch agriculture," *Agricultural Waste Utilization and Management, Proceedings of the Fifth International Symposium on Agricultural Wastes,* American Society of Agricultural Engineers, St. Joseph, MI, 1985, pp 439–445.
27. P. D. Andre, "Sprinklers solved this feedlot dust problem," *Beef* 70–72, 74, 79–81 (February 1985).
28. F. M. Simpson, "The CCFA control of feedlot pollution plan," Bulletin A. *How to Control Feedlot Pollution,* California Cattle Feeders Association, Bakersville, May 28, 1970.
29. J. J. Carroll, J. R. Dunbar, R. L. Givens, et al. 1984. "Sprinkling for dust supression in a cattle feedlot," *Calif. Agr.* 12–13 (March 1984).
30. C. L. Barth, "A rational design standard for anaerobic livestock waste lagoons," *Agricultural Waste: Utilization and Management, Proceedings of the Fifth International Symposium on Agricultural Wastes,* American Society of Agricultural Engineers, St. Joseph, MI, pp 638–647.
31. F. J. Humenik, and M. R. Overcash, *Design Criteria for Swine Waste Treatment Systems,* EPA-600/2-76-233, U.S. Environmental Protection Agency, Ada, OK, 1976, 291 pp.
32. J. M. Sweeten, C. L. Barth, R. E. Hermanson et al., "Lagoon systems for swine waste treatment," PIH-62, *National Pork Industry Handbook,* Cooperative Extension Service, Purdue University, West Lafayette, IN, 1979, 6 pp.
33. "Design of anaerobic lagoons for animal waste management," Engineering Practice EP403.1, *Standards 1990,* American Society of Agricultural Engineers, St. Joseph, MI, 1990, 1 p.
34. F. J. Humenik, R. E. Sneed, M. R. Overcash, et al., "Total waste management for a large swine production facility," *Managing Livestock Wastes, Proceedings of the Third International Symposium on Livestock Wastes,* American Society of Agricultural Engineers, St. Joseph, MI, 1975, pp 168–171.

35. J. C. Converse, D. L. Day, J. T. Pfeffer, et al., "Aeration with ORP control to supress odors emitted from liquid swine manure system," *Livestock Waste Management and Pollution Abatement*. Proceedings of International Symposium on Livestock Wastes, American Society of Agricultural Engineers, St. Joseph, MI. 1971, pp 267–271.
36. D. Phillips, M. Fattori, and N. R. Bulley, "Swine manure odors: Sensory and physico-chemical analysis," Paper No. 79-4074, American Society of Agricultural Engineers, St. Joseph, MI, 19 pp.
37. W. Korsmeyer, M. D. Hall, and T. H. Chen, "Odor control for a farrow-to-finish swine farm—a case study," *Livestock Waste: A Renewable Resource, Proceedings of the Fourth International Symposium on Agricultural Wastes*, American Society of Agricultural Engineers, St. Joseph, MI, 1981, pp 193–197, 200.
38. S. J. Raabe, J. M. Sweeten, B. R. Stewart, et al., "Evaluation of manure flush systems at caged layer operations," *Trans. ASAE*, 27:852–858 (1984).
39. W. F. Ritter, "Chemical odor control of livestock wastes," Paper no. 80-4059, American Society of Agricultural Engineers, St. Joseph, MI, 1980, 16 pp.
40. D. J. Warburton, J. N. Scarbrough, D. L. Day, et al., "Evaluation of commercial products for odor control and solids reduction of liquid swine manure," *Livestock Waste: A Renewable Resource, Proceedings of the Fourth International Symposium on Livestock Wastes*, American Society of Agricultural Engineers, St. Joseph, MI, 1981, pp 309–313.
41. T. Lindvall, O. Noren, and L. Thyselius, "Odor reduction for liquid manure systems," *Trans. ASAE*, 17:508–512 (1974).
42. J. R. Miner, and R. C. Stroh, "Controlling feedlot surface odor emission rates by application of commercial products," *Trans. ASAE*, 19:533–538 (1976).
43. M. A. Van Geelan, and K. W. Van Der Hoek, "Odor control with biological air washers," *Agr. Envir*. 3:217–222 (1972).
44. S. Schirz, "Odour removal from the exhaust of animal shelters," *Agr. Envir.*, 3:223–228 (1977).
45. L. A. Licht, and J. R. Miner, "A scrubber to reduce livestock confinement building odors," Paper No. PN-78-203, American Society of Agricultural Engineers, St. Joseph, MI, 1978, 12 pp.
46. J. K. Koelliker, J. R. Miner, M. L. Hellickson, et al., "A zeolite packed air scrubber to improve poultry house environment," *Trans. ASAE*, 23:157–161 (1980).
47. H. Bohn, "Soil absorption of air pollutants," *J. Environ. Quality*, 1:372–377 (1972).
48. H. H. Kowalewsky, "Odor abatement through earth filters," *Landtechnik*, 36(1):8–10 (1981).
49. W. H. Prokop, and H. L. Bohn, "Soil bed system for control of rendering plant odors," Paper 85-79.6, 78th Annual Meeting, Air Pollution Control Association, Pittsburgh, 1985, 17 pp.
50. J. M. Sweeten, R. E. Childers, and J. S. Cochran, "Odor control from poultry manure composting plant using a soil filter," ASAE Paper No. 88-4050, International Summer Meeting, American Society of Agricultural Engineers, Rapid City, SD, June 26–29, 1988, 40 pp.
51. K. A. Janni, "Modeling dispersion of odorous gases from agricultural sources," *Trans. ASAE.*, 25:1721–1723 (1982).
52. J. E. Sorel, R. O. Gauntt, J. M. Sweeten, et al., "Design of a 1-butanol scale dynamic olfactometer for ambient odor measurements," *Trans. ASAE*, 26:1201–1206 (1983).
53. J. M. Sweeten, D. L. Reddell, A. R. McFarland, et al., "Field measurement of ambient odors with a butanol olfactometer," *Trans. ASAE*, 26:1206–1216 (1983).
54. Queensland Department of Primary Industries, *Queensland Government Guidelines for Establishment and Operation of Cattle Feedlots*, Brisbane, Queensland, Australia, September 1, 1989, 16 pp.

MEAT SMOKEHOUSES

John R. Blandford

The smoking of food, especially meat and fish, has been practiced for thousands of years. The original purpose of smoking was to preserve food and add flavor to it. Modern smoking operations are very similar to the ancient methods, except for the level of technology used. Smoking is now used to impart flavor, color, aroma, and sensory appeal to food. Smoking food also provides antioxidant and antimicrobial properties. Refrigeration systems significantly reduce the need for smoking for preservation.

Pork and beef meat products are the primary smoked foods, although increasing amounts of turkey, chicken, and fish are also smoked as these products enjoy greater consumer popularity. Some vegetables and nuts are also smoked as gourmet or snack foods. Because meat is the prevalent food smoked, the remainder of this discussion is devoted to meat smokehouses.

PROCESS DESCRIPTION

Smokehouses

The meat production process in a smokehouse involves four individual operations: tempering or drying, smoking, cooking, and chilling. There are endless variations of these individual operations throughout the industry. One or more of the individual steps may be deleted for any given product. The process is varied by adjusting such parameters as time, temperature, humidity, smoke density, and product type, shape, and density.

Virtually all modern meat processing systems use either conventional batch-type smokehouses or continuous-processing ovens. Atmospheric, nonrecirculating smokehouses are no longer used commercially. In both the batch and continuous-meat-processing systems, heated air with controlled temperature, humidity, and smoke density (when used) is circulated uniformly over the surface of the meat. Figure 1 depicts a typical batch smokehouse in which meat is placed on racks or "trees," which remain stationary in the

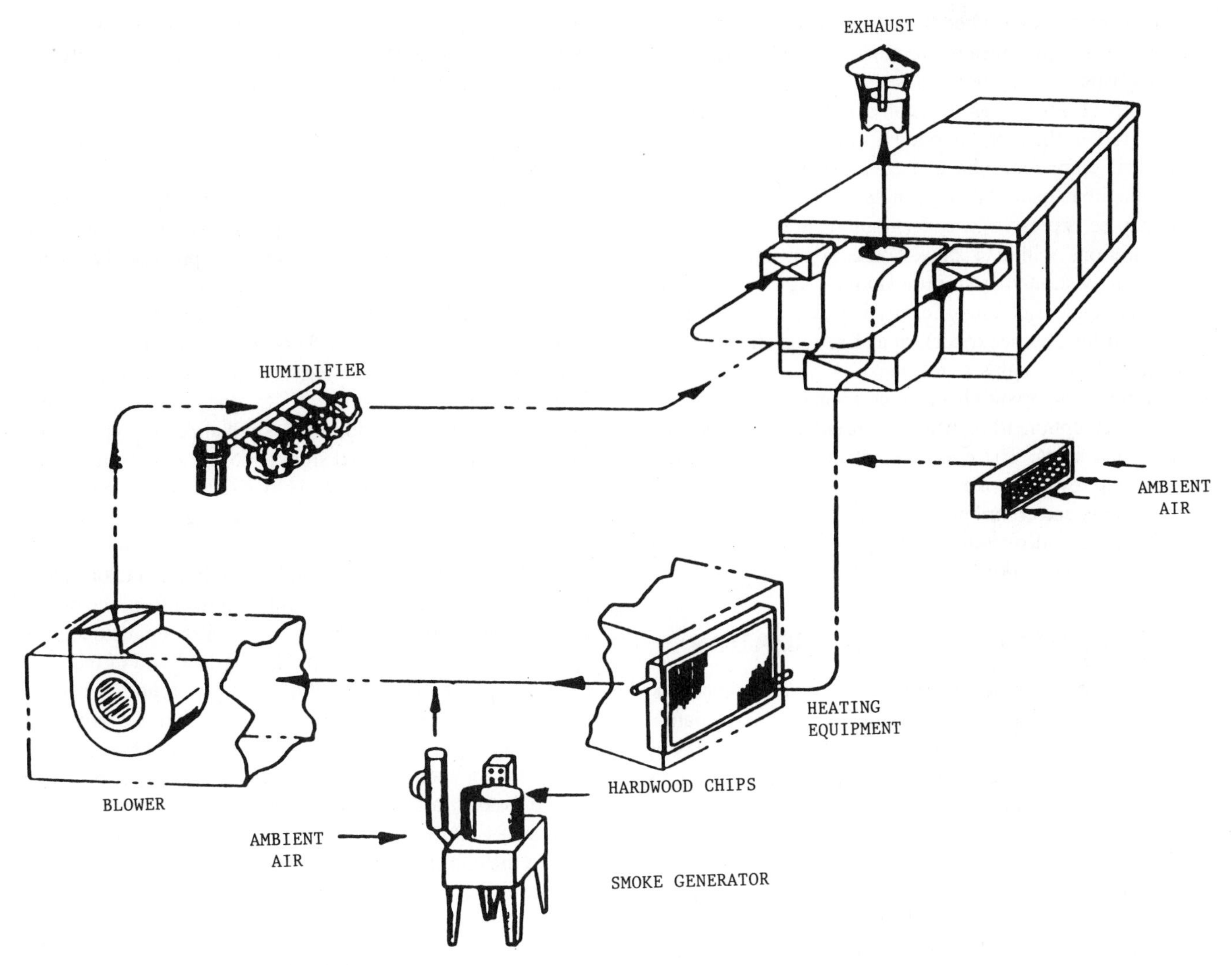

FIGURE 1. Conventional Smokehouse

smokehouse throughout the production cycle. Figure 2 is a schematic of a continuous system where meat is hung on sticks or hangers and is conveyed from zone to zone within the oven.

There are a variety of mechanisms for creating smoke, the most common of which is to pyrolyze hardwood chips (hickory, oak, maple, and beech are the most frequently used) in an external smoke generator. The chips are fed onto a gas-fired or electrically heated metal surface. Typically, this surface is heated to 350–400°C. Volatile components are carried away by a controlled air stream to a smoke tube leading into the air recirculation system. Smoke produced by this or similar methods is often referred to as natural smoke.

Another form of smoke flavor application that is increasingly popular in the industry is the use of liquid smoke, frequently called artificial smoke. Liquid smoke is produced from natural smoke through a smoke washing and concentration process. The concentrate is dissolved in water or oils and is sold commercially to meat processors. Liquid smoke is atomized and dispersed into smokehouses through the air recirculation system as a fine aerosol.

Smoke Composition

Natural smoke consists of an extremely fine particulate phase and a gaseous phase in equilibrium. The smoke mixes in the smokehouse with moisture evaporated from the meat to form a diffused aerosol. The aerosol from sawdust smoke

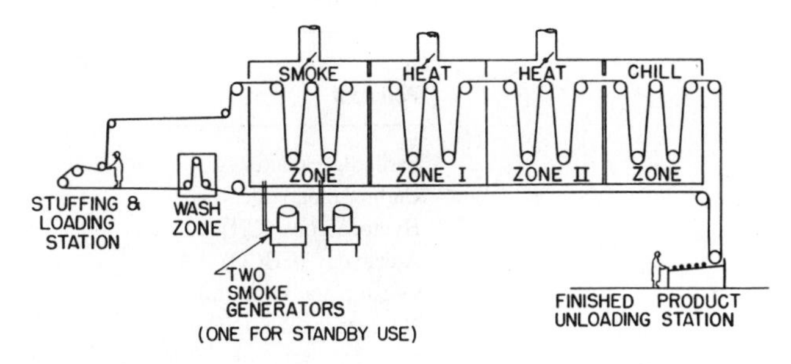

FIGURE 2. Continuous Meat Processing Oven

or liquid smoke is absorbed through the surface of the meat and reacts to give meats, or other smoked foods, their characteristic sensory properties.

Smoke research has generally focused on chemical groups rather than on individual compounds. Over 400 individual compounds have been identified qualitatively, and sometimes quantitatively, in smoke.[1] The quantitative data may be expressed as concentrations in pure smoke, as concentrations within a smokehouse or an exhaust stack (significantly diluted from pure smoke), as concentrations in smoke condensate (significantly higher than pure smoke because of the absence of air), or as fractions relative to the quantity of wood chips pyrolyzed.

The most important chemical components of smoke for sensory perception have been identified as phenols, carbonyls, and acids. These are present in both the solid and gaseous phases of smoke.[2–4] Phenols are mainly responsible for aroma and contribute to flavor. Carbonyls are important for color and overall sensory properties. Acids act as surface sterilants and also contribute to flavor.[5]

AIR EMISSIONS CHARACTERIZATION

Just as smoke researchers have generally focused on chemical groups rather than individual compounds, the regulation of smokehouse emissions has generally focused on classes of pollutants. Most local and state air pollution regulations set limits for particulate matter opacity, concentration, and/or mass emission rate.[6] Certain large urban areas also regulate smokehouses to control organics or odors. This generic regulation of smokehouses is reflected in emission factors published by the Environmental Protection Agency (EPA), that identify pollutants as particulate matter, carbon monoxide, hydrocarbons, aldehydes (all as formaldehyde), and organic acids (all as acetic acids).

Particulate Matter

The size of particulate matter within smoke aerosols is extremely fine. A particle-size determination conducted on the smoke zone exhaust of a continuous oven using sawdust smoke yielded a mass median diameter of 0.9 μm. Particles smaller than 2.0 μm represented over 90% by weight of all particles.[8] Particles this size have maximum light-scattering properties while having minimum mass. As a result, opacity limitations rather than concentration or mass emission limits have normally dictated the extent of emission control required.

Another factor contributing to generally low mass emission rates of pollutants from smokehouses is the small exhaust gas volumes during the smoking process. The minimum airflow rate is established by the volume of air entering the smokehouse through the smoke tube connection from the smoke generator. This volume is approximately 300 scfm. Actual exhaust gas volume is dependent on the humidity in the makeup air to the smokehouse and the desired humidity within the smokehouse. Typical airflow rates for conventional smokehouses range from 500 to 1500 scfm during the smoke cycle. Continuous oven smoke zones exhaust at a rate of 750 to 2000 scfm. The exhaust temperature is normally 35°C to 50°C.

Table 1 gives the long-standing emission factors from the EPA and the uncontrolled emission data from the previous edition of this manual.[7,9] Concentration data correspond well with more recent published data[6] and of industry experience.[8] The only exception to this is found in certain high-product-volume, low-residence-time smoke zones in continuous ovens. Smoke concentration may be as high as 0.5 gr/scf for such systems since the product is exposed to smoke for only a few minutes. Opacity from such a system approaches 100%.

The uncontrolled emission factor of 0.3 pound of particulate matter per ton of meat (lb PM/ton) is based on 1 pound of wood chips consumed per 110 pounds of meat. Similar emission factors from recent industry experience give a range of 0.2 to 0.75 lb PM/ton for the lightest to the heaviest smoked meats. This is also corrected to 1 pound of wood chips per 110 pounds of meat. This range suggests that the emission factor from the EPA is perhaps low as an overall average.

One reason for this low factor may be that current emission tests often measure particulate matter in the front and back sections of the EPA Method 5 sampling equipment, whereas the original data are probably based on front-half

TABLE 1. Meat Smokehouse Emission Factors and Concentration Data

Pollutant	Uncontrolled Emission Factor (pounds/ton of meat)	Concentration
Particulate matter	0.3	0.14 gr/scf
Carbon monoxide	0.6	—
Hydrocarbons (CH_4)	0.07	—
Aldehydes (CHCO)	0.08	40 ppm
Organic acids (acetic)	0.2	87 ppm
Nitrogen oxides (NO_2)	—	3.9 ppm

Source: References 7 and 9.

results only. Recent industry tests show the back-half particulate matter (condensible gases and fine aerosols, identified before as present in smoke) account for 10–30% of all particulate matter measured.

Organics

Research on organics in smoke has focused primarily on identifying the ways in which different classes of compounds affect the many aspects of sensory perception. Until recently, when concern about the presence of air toxics such as formaldehyde and individual polycyclic aromatic hydrocarbons (PAHs) has increased, quantification of organics in smoke was directed at classes of compounds rather than individual substances.

Research in sensory perception has focused on phenols, carbonyls, and acids. For each 100 grams of dry wood used to generate smoke, phenols are expected to total 0.12 to 0.3 grams (as phenol), carbonyls 1.1 to 9.0 grams (as acetone), and acids 3.6 to 5.2 grams (as acetic acid).[10] The amount of these groups present is a function of the temperature at which the wood is pyrolyzed, as seen in Table 2.[11] Note that both of these citations apparently refer to concentrations in pure smoke or in smoke condensate.

The amounts of organic acids and aldehydes have been measured in smokehouse exhausts by regulators in the Los Angeles area. In tests at a number of sources, organic acids ranged from 0.82 to 1.85 pounds per hour (pph), averaging 1.42 pph.[12] Aldehydes ranged from 0.13 to 0.83 pph, averaging 0.50 pph.[12] The ratio of aldehydes to acids in these tests is 0.35 to 1.0, which is very near the ratio of 0.4 to 1.0 in the EPA emission factors in Table 1.

The dominant compound within the organic acids group is acetic acid. Table 3 shows the contributions of formic, acetic, propionic, butyric, and higher acids to overall acid quantities present in smoke. Note that the formulation for liquid smoke is quite similar to the two wood smokes used in meat processing. The other wood smoke analysis is from seafood processing.

Formaldehyde is of special interest to present day regulators as an air toxin and possible carcinogen. While the EPA reports all smokehouse aldehydes as formaldehyde in smokehouse emission factors, formaldehyde is not the predominant aldehyde present. An early researcher reported formaldehyde in wood smoke at 20–40 ppm, whereas higher aldehydes were at 140–180 ppm (cited in the second edition of this publication).[14] More recent data show formaldehyde and acetaldehyde composition in smoke from six different types of wood. Formaldehyde content averaged 17% and acetaldehyde averaged 83% for these two components. Higher aldehydes were not listed.[15,1] A recent agency-witnessed emission test on a continuous-processing system gave these results: formaldehyde = 0.06 pph, acetaldehyde = 0.34 pph, and total particulate matter = 4.8 pph. The reported wood chips use was 90–120 pph.[16] Note that formaldehyde is 15% of total aldehydes reported.

TABLE 2. Smoke Condensate Composition (mg/100 g Sawdust) as a Function of Temperature

Temperature°C	Total Phenols	Total Carbonyls	Total Acids
380	998	9996	2506
600	4858	14952	6370
760	2632	7574	2996

Source: Reference 11.

TABLE 3. Individual Organic Acids in Smoke (Percent of Total)

Acid	Meat Smoke A[2]	Meat Smoke B[3]	Seafood Smoke[13]	Liquid Smoke[4]
Formic	13.2	19.1	51.3	5.0
Acetic	74.5	69.3	7.0	75.6
Propionic	6.0	6.4	21.4	11.8
Butyric	1.8	1.3	3.4	1.3
Others	4.5	3.9	16.9	6.3

Extensive research has been conducted in recent years to determine the presence and effect of PAHs in meat and other food products. Many individual PAH compounds have been identified, but at extremely low concentrations. One study found 31 individual PAHs in mesquite smoke at a total concentration of 1280 μg/kg of wood. The same study detected only 22 compounds in hickory smoke at a total concentration of 688 μg/kg of wood.[17] An earlier study found eight compounds in whole smoke and the vapor phase (particulate removed by electrostatic precipitation) from maple, but only at a total of less than 100 μg/4.5 kg of wood.[18] Table 4 lists the compounds found in both studies. The data for hickory smoke PAHs were recently applied to a large meat processing facility in the Midwest (four continuous-processing ovens, 33 conventional smokehouses). The potential uncontrolled PAH emissions were predicted to be 10.5 lb/yr for all PAHs listed[17] and 1.9 lb/yr for the PAHs regulated by the state environmental agency.[19]

As Table 2 shows, total phenols are a relatively minor constituent of smoke when compared with carbonyls and acids. Early research showed that phenols were present in smoke at 20 to 30 ppm compared with aldehydes at 150 to 200 ppm and acids at 550 to 625 ppm.[14] More recent research identified "the four major phenols" as phenol, guaiacol, 4-methylguaiacol, and syringol. Their relative concentrations in whole-smoke condensate and the vapor phase are given in Table 5.[20]

AIR POLLUTION CONTROL MEASURES

Smokehouse emissions have several characteristics that affect decisions concerning control equipment. Essentially all the emissions come from smoking rather than drying,

TABLE 4. PAH Composition of Different Wood Smokes

Compound	Smoke Mesquite[17]	Hickory[17]	Maple–Whole[18]	Maple–Vapor[18]
	(μg/kg wood)		(ug/4.5 kg wood)	
Phenanthrene	204	114	51.5	28.4
Benzo(a)anthracene	60	38	7.0	4.3
Fluoranthene	162	94	5.7	4.2
Pyrene	155	104	5.5	4.1
Anthracene	47	31	3.8	1.9
Chrysene	72[a]	49[a]	2.6	0.3
Benzo(a)pyrene	74	41	1.2	0.4
Benzo(e)pyrene	36	20	0.9	Trace

[a]Chrysene and triphenylene

cooking, or chilling. All of the particulate matter is classified as "fine" with virtually no inorganics present. Smoke particles are hydrophobic, making water scrubbing difficult. Condensed wood smoke is viscous, sticky because of the tars and resins present, and mildly acidic. The major organic compound groups are present in both the gaseous and solid phases. In high concentrations, smoke is irritating to the eyes and is considered odorous. At low concentrations, smoke is generally considered pleasant.

A conventional smokehouse control device must be operated only during the smoke cycle and perhaps the first moments of the cook cycle as residual smoke is vented. The control device may be bypassed at other times, utilizing water dampers or mechanical dampers to direct exhaust gases appropriately. Damper operators are generally interlocked with smokehouse or smoke generator controls. Continuous-processing ovens may be designed to vent the individual zones together or separately. A combined exhaust may be considered if opacity is the only parameter of concern, provided the combined exhaust does not interfere with the recirculating air balance in each processing zone. If regulatory limits other than just opacity are applicable, the smoke zone should be separately vented and the control device sized accordingly.

Control devices that have proved acceptable for smokehouse control include high-pressure filters, certain scrubbers, modular electrostatic precipitators, and afterburners. Each of these devices has advantages and disadvantages. All are generally effective for controlling particulate matter when applied and operated appropriately. As the data in Table 4 and 5 imply, all can be somewhat effective in removing organic compounds traditionally considered to be gaseous.

TABLE 5. Major Phenol Concentrations in Smoke Concentrate and Vapor Phase

Compound	Whole Smoke (mg/l)	Vapor Phase (mg/l)
Phenol	59	5.9
Guaiacol	417	32.1
4-Methylguaiacol	333	13.6
Syringol	392	6.5

Source: Reference 20

High-pressure filters utilize cleanable or disposable filter media at a cross-filter pressure drop of 20 inches to 30 inches water gauge. At this pressure differential, the fiberglass filaments composing the filter medium collapse together to promote particle growth and aerosol condensation. Particulate matter removal efficiencies in excess of 95% have been measured, yielding essentially clear stacks. High-pressure filters must be protected from adverse weather. Gas-contacting surfaces must be manufactured from stainless steel and must be cleaned periodically with smokehouse cleaners to prevent the buildup of smoke tars and resins. Because these filters function best when exhaust rates do not fluctuate much, they are more suitable for continuous ovens or groups of conventional smokehouses.

Next to afterburners, scrubbers have been utilized in the meat industry more than any other control device. Venturi scrubbers are the most common type used because they are more efficient for particulate matter removal. Scrubbers are more effective if a small amount of surfactant or wetting agent is used in the water sprays to offset the hydrophobic nature of smoke particles. A highly efficient mist separator is mandatory to prevent smoke-water-droplet carryover. The dilute smoke–water condensate can discolor building walls and roofs. The scrubber blow-down can be discharged into the plant process wastewater system as there is no known instance of adverse impacts of smoke condensate on biological treatment plants.

Modular electrostatic precipitators (ESPs) have been utilized on smokehouses in southern California since the 1950s. Multistage units are required to achieve desired particulate matter removal efficiencies. In combination with a wet centrifugal collector upstream, a two-stage ESP had a 65% particulate removal efficiency and 35–50% efficiency for aldehydes and organic acids.[9] More recent tests show that a three-stage ESP can achieve over 95% particulate removal and a virtually clear stack. However, these units must have clean-in-place systems to remove smoke condensate and must have a heat source to keep ionizing ele-

TABLE 6. Smokehouse Afterburner Efficiency as Function of Temperature

Pollutant	800°F	1000°F	1200°F	1400°F
Particulate matter	70	90–93	96	97
Organic acids	60–80	75–90	90–98	95–100
Aldehydes	60–80	70–90	87–92	95
Total organics[a]	NA	68	88	77
Carbon monoxide	NA	99+	99+	99+
Nitrogen oxides[b]	NA	5	12	12

[a]Includes particulate matter and organic gases together. 1400°F cited as unexplained test anomaly.
[b]Reported as concentration (ppm) rather than removal efficiency.
Source: Reference 12.

ments dry. Otherwise, arcing will occur and efficiencies will suffer significantly.[6] A promising technology for smokehouse emission control is the combination of a low-pressure-drop venturi scrubber with a modular wetted-wall ESP. Such a combination, properly developed, should offer high particulate matter efficiency, good control of organic gases, and continuous disposal of smoke condensate in a dilute wastewater stream.

Until the energy crises of the 1970s, afterburners had long been viewed as the preferred control device for smokehouses. Their low capital cost; high efficiencies for particulate matter, organic gas, and carbon monoxide control; low maintenance requirements; and high flexibility make afterburners attractive. For limited operating hours on conventional smokehouses, their cost of operation is reasonable when compared with amortized capital costs for other, more expensive technologies. Extensive studies in southern California document high afterburner efficiencies.[12] Table 6 lists efficiencies from several stack tests as a function of afterburner temperature. In consideration of such information, an afterburner temperature of 1200°F with a residence time of 0.5 second was accepted as best available control technology (BACT) for all pollutants of concern in southern California in 1989.

Another important factor for BACT is the temperature of the smoke generator, which should be minimized. Table 2 shows how organic gases increase as the smoke generator temperature increases. It has also been recommended that the smoke generator temperature be kept below 400°C to limit the formation of polycyclic aromatic hydrocarbons.[21]

References

1. J. A. Maga, *Smoke in Food Processing,* CRC Press, Boca Raton, FL 1988, pp 49–68.
2. S. A. Husaini and G. E. Cooper, "Fractionation of wood smoke and the comparison of chemical composition of sawdust and friction smokes," *Food Technol.* 11(10):499 (1957).
3. R. W. Porter, L. J. Bratzler, and A. M. Pearson, "Fractionation and study of compounds in wood smoke," *J. Food Sci.,* 30:615 (1965).
4. W. Baltes, R. Wittkowski, I. Sochtig, et al., "Ingredients of smoke and smoke flavor preparations," in *The Quality of Food and Beverages,* Vol. 2, G. Charalambous and G. Inglett, Eds., Academic Press, New York, 1981, Chapter 1.
5. R. W. Porter, *Fractionation and Identification of Some Compounds in Wood Smoke,* Ph.D. thesis, Michigan State University, 1963.
6. J. R. Blandford, "Industrial experience in meat smokehouse emission control," 72nd Annual Meeting of the Air Pollution Control Association, No. 79–44.2, Cincinnati, 1979.
7. *Compilation of Air Pollutant Emission Factors, Vol. 1,* 4th ed., USEPA Publication AP-42, p 6.7-1, 1985.
8. T. Roberts, W. Krill, and J. Blandford, Miscellaneous smokehouse emission tests, Oscar Mayer Foods Corp., 1972–1985.
9. *Air Pollution Engineering Manual,* 2nd ed. Publication AP-40, U.S. Environmental Protection Agency, 1973, pp 794–799.
10. D. J. Tilgner and H. Daun, "Polycyclic aromatic hydrocarbons (polynuclears) in smoked foods," *Residue Rev.,* 27:19 (1969).
11. R. Hamm and K. Potthast, "Einflub verschiendener techniken des raucherns und der anwendung von rauchermittein auf den gehalt von fleischwaren an cancerogenen kohlenwasserstoffen, phenolen und anderen rauchbestandtellen." Final report, research projects Ha 517/6, Ha 517/11, and Ha 517/14 of the Deutsche Forschungsgemeinschaft (1976).
12. M. Hickman, South Coast Air Quality Management District, Los Angeles, personal communication, 1987.
13. K. Kasahara and K. Nishibori, "Volatile carbonyls and acids of smoked sardines," *Bull. Jap. Soc. Scien. Fisheries,* 48(5):691 (1982).
14. L. B. Jensen, *Microbiology of Meats,* 2nd ed., Garrand Press, Champaign, IL, 1945.
15. L. Toth and K. Potthast, "Chemical aspects of the smoking of meat and meat products," *Adv. Food Res.,* 29:87 (1984).
16. Control agency-witnessed stack test results, private communication, 1990.
17. J. A. Maga, "Polycyclic aromatic hydrocarbon (PAH) composition of mesquite (prosopis fuliflora) smoke and grilled beef," *J. Agric. Food Chem.,* 34:249 (1986).
18. K. S. Rhee and L. J. Bratzler, "Polycyclic hydrocarbon composition of wood smoke," *J. Food Sci.,* 33:626 (1968).
19. "Oscar Mayer Foods Corporation Air Toxics Reports—Madison, WI," Dames and Moore, 1989.
20. M. R. Kornreich and P. Issenberg, "Determination of phenolic wood smoke components as trimethylsilyl ethers," *J. Agric. Food Chem.,* 20(6):1109 (1972).
21. K. Potthast, "Smoking methods and their effect on the content of 3,4-benzopyrene and other constitutents of smoke in smoked meat products," *Die Fleischwirtschaft,* 3:371 (1978).

RENDERING PLANTS

William H. Prokop, P.E.

Rendering plants process or "recycle" animal and poultry by-product materials in order to produce tallow, grease, and protein meals. The rendering industry is divided into two basic groups: integrated and independent. The integrated rendering plants are operated in conjunction with animal slaughterhouses or poultry processing plants. This type of plant is an integral part of the animal slaughterhouse or meat packing operation and processes all of the residual material from these operations. This also applies to those rendering plants that process the offal and feathers from poultry processing plants.

Independent rendering plants collect their raw materials from a variety of sources that are off-site or separate from the plant itself. These renderers send out special-body route trucks to collect discarded fat and bone trimmings, meat scraps, restaurant grease, blood, feathers, offal, and entire animal carcasses from a variety of sources: butcher shops, supermarkets, restaurants, fast-food chains, poultry processors, slaughterhouses, farms, ranches, feedlots, and animal shelters. Independent renderers collect raw materials primarily in metropolitan areas. However, a number of these renderers collect raw material from a variety of regional areas and require the use of a transfer station, which receives the collected material for further transport to a rendering plant.

Animal rendering systems are divided into two classes: edible rendering of animal fatty tissue into edible fats and proteins for human consumption, and inedible rendering of animal by-product materials into fats and proteins for animal feed and nonedible applications. Edible rendering plants are normally operated in conjunction with meat packing plants under the inspection and processing standards established by the U.S. Department of Agriculture, Food Safety and Inspection Service (USDA/FSIS). Edible tallow or lard is produced from beef or swine fatty tissue.

Inedible rendering plants are operated by independent renderers to produce inedible tallow and grease. These are used in the feed rations for livestock and poultry, for soap production, and for fatty-acid manufacture. A number of different grades of inedible tallow and grease are produced for these different categories of use. For example, "yellow grease" or feed-grade animal fat is produced from waste cooking fats received from restaurants and deep fat fryers. This product is used as an ingredient in animal and poultry feeds. Protein meal includes animal and poultry meals that are used as feed supplements for livestock and poultry. Rendering plants also produce blood meal and feather meal. These products usually have their protein content specified.

The meat packing industry during the past 15 years has converted its meat production facilities almost totally from an animal carcass to a "boxed beef" and "boxed pork" operation.[1] This has resulted in more fat and bone being trimmed at the slaughterhouse and less raw material being available to the independent renderer. Also, there has been a recent downward trend in tallow and grease prices.

During the past 10 years or more, the number of independent rendering plants has been reduced significantly. Currently, an estimated 150 independent rendering plants are operating in the United States.[2] This consolidation also applies to integrated rendering, with an estimated 75 meat packing plants and 25 poultry processing plants also including rendering operations.

PROCESS DESCRIPTION

Current edible and inedible rendering systems are described in detail by Prokop.[3] These include the batch cooker, Duke continuous and Stord waste heat dewatering systems for inedible rendering. The Sharples Trim-R process is described for edible rendering.

Edible Rendering

The current edible rendering process is continuous and consists of two stages of centrifugal separation. A typical feedstock of beef fat trimmings from USDA inspected meat processing plants consists of 14–16% fat, 60–64% moisture, and 22–24% protein solids.

The Sharples Trim-R Edible Fat Process is shown in Figure 1. Fat trimmings are ground through a Weiler grinder and belt conveyed to a melt tank equipped with an agitator and steam-heated jacket. The melted fatty tissue at 110°F is pumped to a Reitz disintegrator to rupture the fat cells. A Sharples Super-D-Canter Centrifuge separates the proteinaceous solids from the melted fat and moisture containing a small percentage of solids or fines.

A second-stage centrifuge is required to "polish" the edible fat, which is first heated to 200°F by a shell-and-tube heat exchanger with steam. The Westfalia De-Sludging Separator makes a two-phase separation where the polished edible fat discharges from the top and the water fraction containing the protein fines is discharged as sludge from the bottom. The edible fat is pumped to storage, whereas the sludge from the centrifuge is either transported to an inedible rendering plant or passed through a primary treatment system for wastewater.

Since no cooking vapors are emitted from current edible rendering processes and heat contact with the edible fat is minimal, odor emissions from these rendering plants are perceived as having little impact. This also is due to the freshness of the raw material being processed and the plant sanitation and housekeeping practices that have been established by the USDA/FSIS. As a result, the remainder of this section is devoted to inedible rendering processes and the control of their odor emissions.

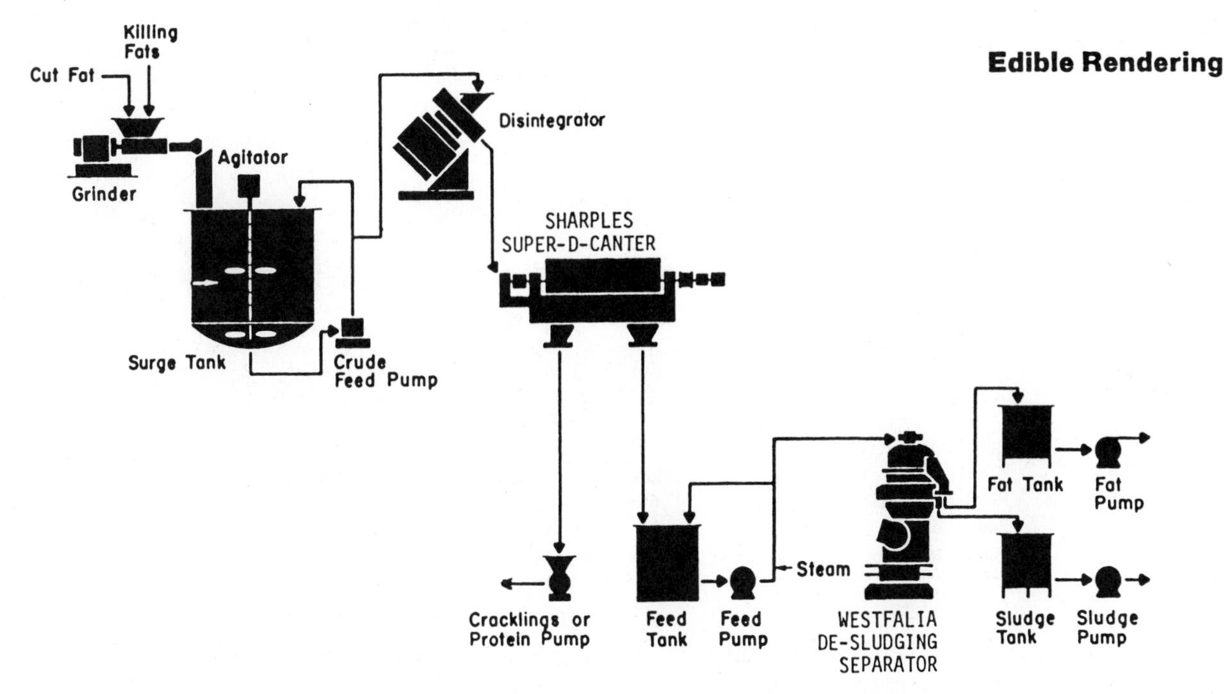

FIGURE 1. Edible Rendering System (From Reference 3)

Wet Rendering Versus Dry Rendering

Wet rendering is defined as a process of separating fat from raw material by boiling with water. This is normally accomplished by boiling the raw material in a tank of water. The products of wet rendering are fat, stick water containing glue, and wet tankage (protein solids). The wet rendering process involves the addition of water to the raw material and the use of live steam to cook the raw material and accomplish separation of the fat.

The wet rendering process no longer is used in the United States because of the high costs of energy and of an adverse effect on fat quality. The water added to the raw material must be evaporated, which increases fuel costs to generate additional steam for moisture removal. Also, contact of the fat with excess water under a boiling temperature tends to increase the free fatty acid content of the fat.

The wet rendering process has been exclusively replaced by dry rendering, which is defined as a process for releasing fat by dehydrating raw material in a batch or continuous cooker. After moisture removal in the cooker, the melted fat is separated from the protein solids. No excess water or live steam is added to the raw material in this process.

Raw Materials for Rendering

The integrated rendering plant for an animal slaughterhouse or poultry processor normally receives only one type of raw material. This simplifies the control of the processing conditions, which usually require only minor adjustments. Also, the raw material is relatively fresh, undergoing little or no noticeable deterioration. Conversely, the independent renderer often handles a variety of raw materials that require either the operation of multiple rendering systems in parallel or significant changes in the operating conditions for a single system to process variable raw material.

Table 1 provides specific raw-material yield data and the split between the fat and protein solids content of various categories of raw materials for inedible rendering.[1]

In comparing Table 1 with a similar table (Table 221) in the previous edition of this manual, we see that the fat content of steers and hogs has increased significantly as a result of improved breeding and the nutritional feeding of fat to livestock. However, the fat content of butcher shop fat and bone has decreased due to the production of "boxed beef/pork," since more of the available fat is trimmed at the packing house.

Basic Rendering Process—Batch Cooker System

Figure 2 from reference 3 illustrates the basic rendering process where batch cookers are used. These are multiple units arranged in a row or series of rows, depending on the size and arrangement of the rendering plant. Each cooker consists of a horizontal, steam-jacketed cylindrical vessel equipped with an agitator. This vessel is known as a batch cooker because it follows a repetitive cycle: the cooker is charged with the proper amount of raw material, the cook is made under controlled conditions, and, finally, the cooked material is discharged.

The raw material from the receiving bin is screw conveyed to a crusher or similar device for size reduction. For batch cookers, the raw material is reduced in size to 1 or 2 inches to provide efficient cooking, which normally re-

TABLE 1. Composition of Raw Materials for Inedible Rendering

Source	Tallow/Grease, Wt %	Protein Solids, Wt %	Moisture, Wt %
Packing house offal and bone			
Steers	30–35	15–20	45–55
Cows	10–20	20–30	50–70
Calves	10–15	15–20	65–75
Sheep	25–30	20–25	45–55
Hogs	25–30	10–15	55–65
Dead stock (whole animals)			
Cattle	12	25	63
Calves	10	22	68
Sheep	22	25	53
Hogs	30	28	42
Butcher shop fat and bone	31	32	37
Blood	None	16–18	82–84
Restaurant grease	65	10	25
Poultry offal	10	25	65
Poultry feathers	None	33	67

Source: Reference 1.

quires 1½ to 2½ hours. The raw material is quite variable, depending on the source, and adjustments in the cooking time and temperature may be required to process it. The final temperature of the cooked material ranges from 250°F to 275°F, depending on the type of raw material.

After the cooking process is completed, the cooked material is discharged to the percolator drain pan, which contains a perforated screen that allows the free-run fat to drain and be separated from the protein solids, which are known as "tankage." After one or two hours of drainage, the protein solids still contain about 25% fat and are conveyed to the screw press, which completes the separation of

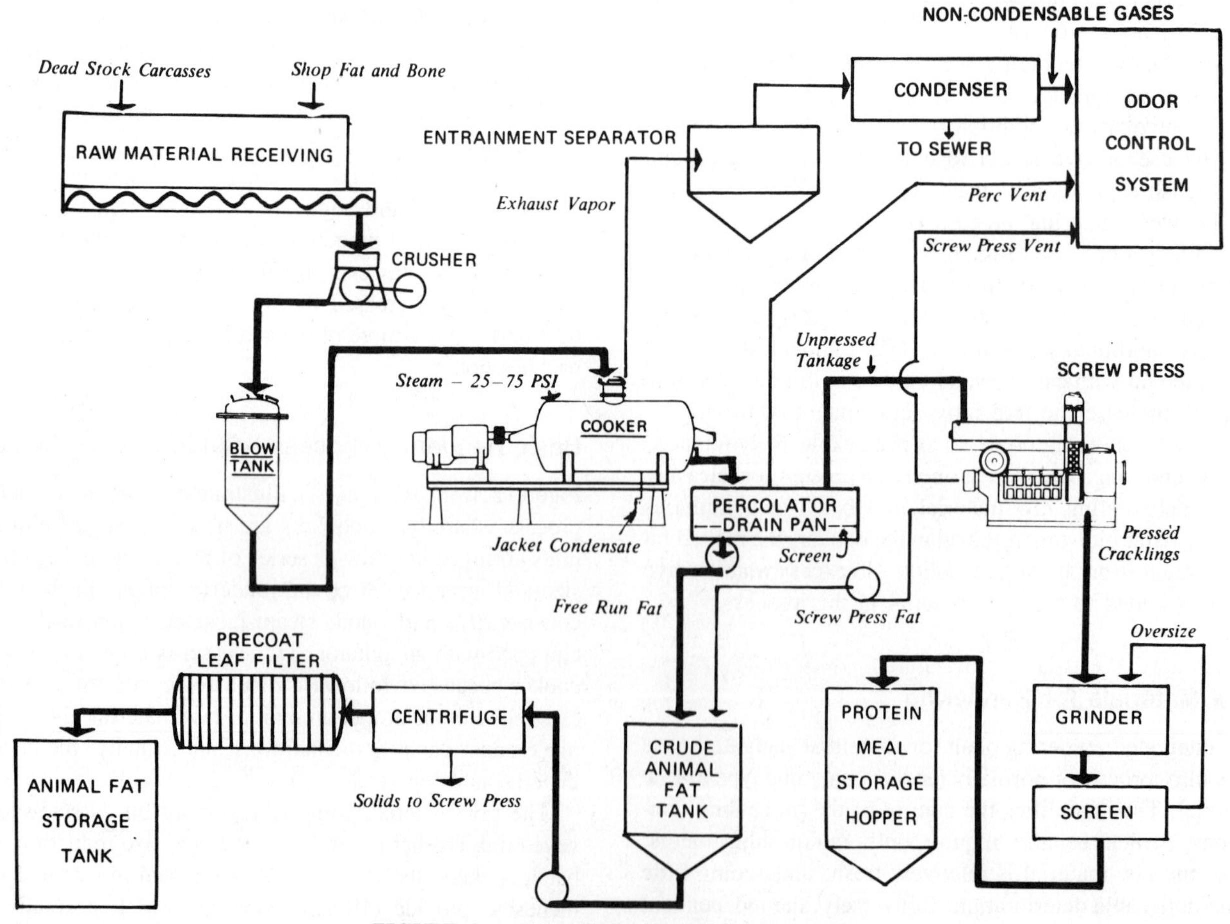

FIGURE 2. Batch Cooker Rendering Process

fat from solids. The final protein solids have a residual fat content of 10%.

The solid protein material discharged from the screw press is known as "cracklings." It is normally screened and ground with a hammer mill to produce protein meal that essentially passes a 12-mesh screen. The fat discharged from the screw press usually contains fine solid particles that are removed by either centrifuging or filtration.

Continuous Rendering Systems

Since the 1960s, a variety of continuous rendering systems have been installed to replace the batch cooker systems. Continuous rendering is synonymous with continuous cooking. The raw material is fed continuously to the cooker, and the cooked material is likewise discharged at a constant rate.

A continuous rendering system normally consists of a single continuous cooker, whereas the batch cooker system consists of multiple cooker units. A continuous system usually has a higher capacity than the batch cooker system it replaces. This increased capacity provides for more efficient processing of the raw material by processing more material in less time.

Continuous rendering also has a number of other inherent advantages over the batch system. Since a continuous process requires less cooking time or exposure to heat, improved product quality normally results. Further, the continuous system occupies considerably less space than a batch cooker system with equivalent capacity, thus saving building construction costs. Finally, a single-cooker unit is inherently more efficient than multiple-cooker units in terms of steam consumption and achieves a significant savings in fuel usage by the boilers. Likewise, less electric power is consumed for agitation in the single continuous cooker unit.

The Duke continuous rendering system is manufactured by the Dupps Co., Germantown, Ohio. This rendering system is shown in Figure 3.[3] The Duke system is designed to provide a method of cooker operation similar to that of the batch cooker. The Equacooker is a horizontal steam-jacketed cylindrical vessel equipped with a rotating shaft to which are attached paddles that lift and move the material horizontally through the cooker. Steam also is injected into the hollow shaft to provide increased heat transfer.

The feed rate to the Equacooker is controlled by adjusting the speed of the variable-speed drive for the feed screw, which establishes the production rate for the system. The discharge rate for the Equacooker is controlled by the speed at which the control wheel rotates. The control wheel contains buckets, similar to those used in a bucket elevator, that pick up the cooked material from the Equacooker and discharge it to the drainer.

The drainer performs the same function as the percolator drain pan in the batch cooker process. It is an enclosed screw conveyor that contains a section of perforated trough

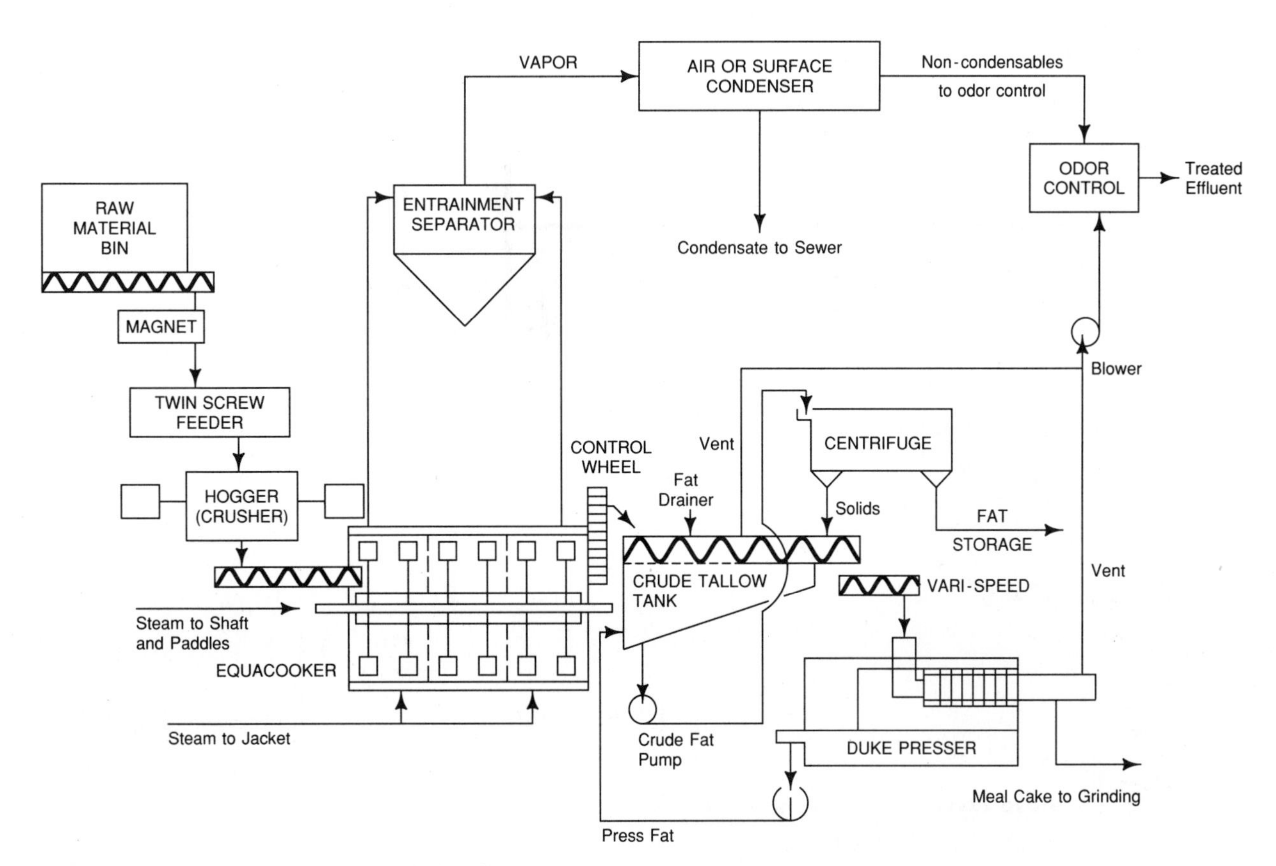

FIGURE 3. Duke Continuous Rendering System

for the free melted fat to drain through to the crude tallow tank. The protein solids containing residual fat are then conveyed to the pressers for the additional separation of fat. The pressers and other components of the Duke continuous system are similar to those used for the batch cooker system. A central control panel is provided for the operator to consolidate the instruments required for process control of these continuous systems.

New Continuous Systems with Reduced Energy Usage

During the early 1980s, considerable effort was spent on developing new rendering systems that utilize the cooking vapors from either a batch cooker system or a continuous rendering system to obtain further moisture removal. Most of this new rendering technology has evolved in Europe, where energy costs have been significantly higher than in the United States.

The Stord waste-heat dewatering (WHD) system, for example, is manufactured by Stord, Inc., in Bergen, Norway. It is illustrated by Figure 4 from reference 3. The Stord WHD system consists of a preheater, twin-screw press, and evaporating system. It usually is installed in conjunction with an existing rendering system.

In this system, raw material is screw conveyed as usual from the raw material bin over an electromagnet and fed to a prebreaker for coarse grinding. This ground material passes through the preheater, which is a horizontal, steam-jacketed, cylindrical vessel with a rotating shaft and agitator to move the material through the vessel continuously and to improve heat transfer. The temperature of the raw material in the preheater is controlled and ranges from 65°C to 85°C (150°F to 190°F), depending on the type of raw material

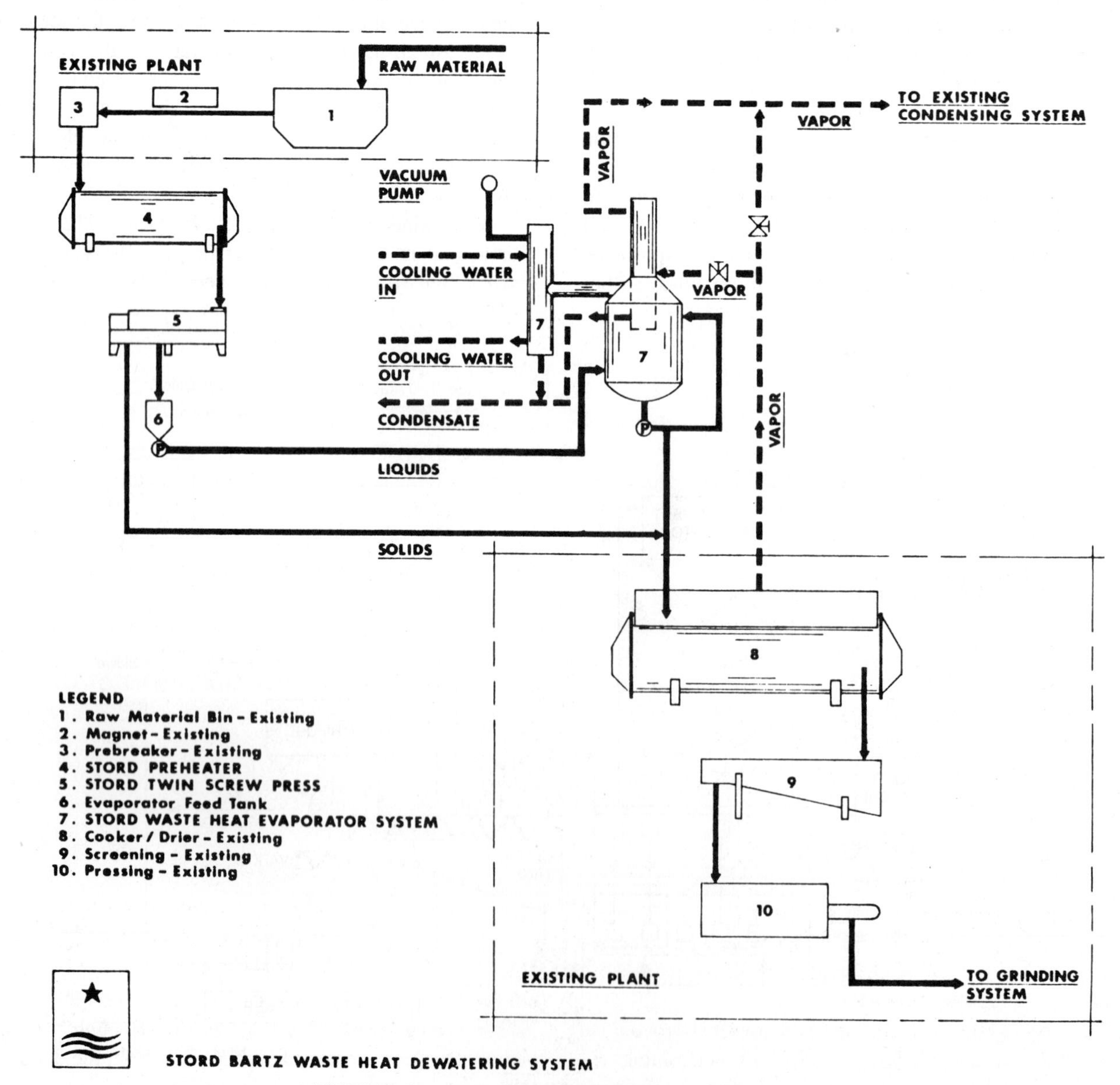

FIGURE 4. Stord Waste-Heat Dewatering System

being processed. This heating step is necessary to melt the fat and condition the animal fibrous tissue properly for the subsequent pressing operation.

The twin-screw press consists of intermeshing, counter-rotating, twin screws moving inside a press cage assembly. It includes a perforated screen through which the liquid is pressed and a series of vertical support plates to secure the perforated screen in place. The perforated screen with heavy backing plate is constructed to follow the contour of the rotating flights of the twin screws. The feed material fills the free space between the screws and the press cage. At the feed end, the twin screws are constructed with a lower-diameter shaft and deeper flights to provide a larger volume of space. As the press screws rotate, this space decreases and the material is subjected to a steadily increasing pressure to squeeze out the liquid through the perforated screen.

The twin-screw press separates the heated and ground raw material into two phases, a press cake of solids containing fat and moisture and a liquid containing mostly the melted fat and water. The solids are screw conveyed to the existing cooker or dryer, which is steam heated to remove the moisture. Final separation of the fat from the solids is completed with a screw press.

Liquid from the twin-screw press is pumped from the feed tank to the evaporator, which consists of a tubular heat exchanger mounted vertically and integrally with the vapor chamber. The vapors from the existing cooker or dryer provide the heating medium for evaporation. The liquid pumped to the evaporator enters at the top of the heat exchanger and flows by gravity downward through the tubes discharging into the vapor chamber, which is maintained under a vacuum of 24 to 26 in. Hg provided by a vacuum pump. The temperature of the liquid ranges from 70°C to 90°C (160°F to 200°F) at which point the moisture is evaporated. The water vapor from the vapor chamber is condensed with a shell-and-tube condenser through which cooling water is circulated.

The basic concept of the Stord WHD system is to use the waste heat in the vapors from the existing cooker to evaporate the moisture from the liquid removed by pressing the raw material, thus converting the existing plant into a two-stage evaporation system. It is essential to balance the operation of this system so that sufficient vapors are available from the existing cooker to evaporate the moisture from the pressed liquid. Operation of the preheater and twin-screw press requires adjustment for different types of raw material with varying moisture content in order to achieve proper balance of this system.

In addition to reducing fuel costs by 30–40%, the operation of the Stord WHD system, in conjunction with the existing cooker system, is capable of increasing production throughput by 75%.

These new systems[4] incorporate the use of microcomputer control concepts that are capable of performing essentially all start–stop sequences, monitoring specific process elements and recording process data to provide trend or deviation outputs. If a particular malfunction occurs, the control system automatically shuts down key operating elements that otherwise could cause serious damage or loss of production.

Blood Processing and Drying

Whole blood received from animal slaughterhouses contains 16–18% total protein solids. Of this amount, approximately 70% of the total protein is recovered as blood meal after steam coagulation and drying. The soluble protein fraction remains in the serum water.

In the past, batch cookers were used to coagulate and dry the blood. However, the dried blood becomes quite dusty and is easily entrained in the exit gases from the cooker. The batch cookers have been replaced by continuous drying processes that are more energy efficient and provide a product with improved quality. Blood meal is highly valued as an animal feed ingredient because of its relatively high lysine (amino acid) content. The continuous drying processes maintain the lysine content of blood nearly at the maximum available, whereas the batch process reduces this lysine availability by 30–50%.[5]

In the continuous blood process, whole blood passes through an inclined tubular vessel into which steam is injected to coagulate the blood solids. This slurry is pumped to a horizontal, solid bowl centrifuge, which separates the blood solids from the blood serum water. The blood solids at 50–55% moisture are fed into a continuous drying system: either a gas-fired, direct-contact ring dryer or steam tube, rotary dryer.

The Dupps Ring Dryer is manufactured by the Dupps Co. in Germantown, Ohio. The coagulated blood solids enter the dryer through a hammer-type mill and are air conveyed through the "ring" duct at a temperature of 200°F to the manifold. At this point, the moist product is separated from the dried product and is recirculated back to the ring duct for further drying. Heated air at 600°F from the direct-fired furnace passes through the hammer mill and provides the heat necessary to dry the moist product. This heated air consists of 60% recycled exhaust air and 40% makeup air. The dried product is separated from the exhaust air by twin cyclone collectors with rotary valves. The exhaust air normally passes through a venturi scrubber and packed tower scrubber in series before discharging to atmosphere. This dryer is furnished in four different models with varying evaporative capacities, ranging from 1000 to 4000 pounds of water evaporated per hour.

The Anderson 72 Tube Rotary Steam Dryer is manufactured by Anderson International Corp. in Cleveland, Ohio. The moist feed material cascades over the rotating, steam-heated tubes. Fresh air passes through the dryer across the steam-heated tubes and exits from the dryer with the moisture released from the solids. The 72-tube unit is the only model available and has an evaporative capacity estimated to be 650 pounds of water evaporated per hour.

Poultry Feathers and Hog Hair Processing and Drying

The feathers and hog hair consist mostly of keratin, which is a long-chain, highly cross-linked, relatively indigestible protein. The rendering process converts the keratin by chemical hydrolysis, combining with water at elevated temperatures (280–300°F), into shorter-chain, more digestible amino acids. This hydrolyzation is accomplished by processing the feathers or hair in a batch cooker with an internal cooker pressure of 40–50 psig maintained for 30 to 45 minutes. The moisture content after hydrolyzation is approximately 50%. The feather hydrolysis process is discussed,[6] including the effect of the processing conditions on the quality of the feather meal product. The feathers or hair may be dried in the batch cooker. However, the drying operation normally is conducted with either the Dupps Ring Dryer or Anderson Rotary Steam Tube Dryer, both of which were described under "Blood Processing and Drying."

Grease Processing

The recent growth of the restaurant business, and of fast-food chains in particular, has made the recycling of restaurant grease an important part of the rendering industry. In the past, 55-gallon drums of grease were picked up at restaurants and unloaded manually by plant employees for processing at the grease plant. Currently, much of this grease is bulk loaded into specially designed and constructed vehicles, which transport and discharge it directly to the grease processing system without any manual labor.

The melted grease is screened to remove coarse solids, then heated to 200°F in vertical processing tanks, and stored for 36–48 hours to separate the grease from the water and fine solids by gravity. Four phases of separation normally occur: (1) solids, (2) water, (3) emulsion layer, and (4) grease product. The solids settle to the bottom of the processing tank and are separated from the water layer above. The emulsion is processed through two stages of centrifuges: a horizontal, solid bowl type to remove solids and a vertical disk type to remove water and fines. The grease product is skimmed off the top. No air is introduced into the processing tanks for "blowing" or drying of the grease.

AIR EMISSIONS CHARACTERIZATION

Odor is the primary air pollutant emitted from the rendering process. There is a potential dust problem in a few instances, such as the exhaust from the Dupps Ring Dryer when processing blood or feathers. However, the provision of proper control equipment, such as cyclones and venturi scrubbers, is normally adequate to abate this type of emission.

Points of Emission

For the batch cooker rendering process, the primary sources of high intensity odors include the noncondensibles from the cooker exhaust and emissions from the screw press, since the material in both cases is heated to temperatures of about 220°F. The processing of blood or poultry feathers normally results in high odor levels. Other sources of high intensity odors include dryers, centrifuges, tallow processing tanks, and the perc pans that are open to the plant atmosphere and receive the discharge from the batch cookers. The hot, cooked material from the batch cooker not only releases odor, but also fat particles, which tend to become airborne and are deposited on equipment and building surfaces within the plant.

The raw material is another source of odor, but it normally is not significant when processed without delay.

TABLE 2. Odor Concentration of Emissions from Inedible Batch Cookers during Discharge to Perc Pan

Type of Raw Material Cooked	Emission During Cooker Discharge odor units/cf[a]	Emission Five Minutes After Discharge odor units/cf[a]
Poultry feathers	200	20
Poultry and turkey offal	2,000	500
Slaughterhouse viscera and bones	150	150
Fresh meat and bone trimmings from beef slaughterhouse	100	70
Meat & bone trimmings with	25,000	3,000
high percent rancid restaurant grease	40,000	200
Mixture of dead cats and dogs, fish scrap, poultry offal, etc.	1,000	1,500

[a]Odor units per cubic foot by ASTM syringe method.

Source: Reference 7.

However, the age of the raw material is important because older material that has deteriorated will result in substantially higher odors being generated during the cooking and pressing operations. Also, the type of raw material being cooked is a significant factor. For example, dead stock will tend to result in the emission of higher-intensity odors during the rendering process.

Table 2 illustrates the effect of raw material on the odor emission from an inedible batch cooker when discharged to the perc pan.[7]

An important trend during the 1970s and 1980s involved the replacement of batch cooker systems with continuous rendering systems that are essentially enclosed and are capable of confining the odors and fat aerosol particles within the equipment. By providing proper equipment seals and locating suction pickup vents at strategic points, a major percentage of the odor generated from the continuous rendering process can be confined and treated by a low-volume scrubber system or by boiler incineration.

Odor control measures have been applied successfully to the Duke continuous rendering system because of its ability to confine the process odors within the system. Recent innovations in rendering equipment technology have resulted in the replacement of screw conveyors delivering raw material to and cooked material from the Equacooker with specially designed pumping systems. This has provided a better seal at the Equacooker, preventing the leakage of high-intensity odors into the plant operating area.

The primary sources of high-intensity odors from the Duke continuous process are similar to those for batch cooker systems: the cooker exhaust noncondensibles and the press vents. In the past, only these two odor emissions from a Duke system were normally vented to an afterburner to incinerate them. However, experience has proven that additional pickup vents are required at the drainer and at the centrifuge to prevent leakage of high-intensity odor to the plant atmosphere. Depending on the leakage occurring at the feed end of the Equacooker, an odor pickup may be located at the inlet chute.

The degree of tightness of an enclosure at each pickup point is an important factor to consider in designing an exhaust duct system. A definite but regulated excess of air is needed at each pickup point to minimize leakage of odor from the rendering process. Often in the past, the exhaust system for an afterburner was undersized because it was desired to incinerate the least volume of air and so minimize fuel costs. A damper is provided at each pickup point to adjust and maintain each flow at the desired rate and also to balance the exhaust duct system.

Composition of Rendering Odors

Odor emissions from rendering plants are relatively complex mixtures of organic compounds. Samples of rendering plant odors have been analyzed by a combination of gas chromatograph and mass spectrometric methods. A total of 30 or more odorous compounds were identified.[8]

Further research work identified the odorous compounds present in rendering plant emissions. Improved analytical techniques provided a more complete list of components. The major compounds included organic sulfides, disulfides, C-4 to C-7 aldehydes, trimethylamine, and various C-4 amines, quinoline, dimethyl pyrazine and other pyrazines, C-3 to C-6 organic acids. Compounds of lesser significance included C-4 to C-7 alcohols, ketones, aliphatic hydrocarbons and aromatic compounds. Odor panel tests were also conducted to relate odor intensity with the various important peaks identified by gas chromatography.[9]

Table 3 lists odor-detection and recognition threshold values for certain odorous compounds known to be present in emissions from rendering plants. Some of the compounds have extremely low odor threshold values and can be detected at concentrations as low as 1 ppb or less.

Odor Emission Data

It is essential that accurate odor emission data be available in order to design an odor control system that successfully abates the emission. If such data are not available, it may be necessary to conduct appropriate odor sensory measurements and obtain accurate gas-flow data. In this section, odor dilution to threshold values are provided for various high-intensity odor emissions from the rendering processes, together with their corresponding volumetric emission rates expressed in cubic feet per minute.

Odor emissions from batch rendering cookers are discussed in detail.[7] In a batch cooker, the rate of moisture removal rises initially, reaching a peak usually within one hour, then decreases rapidly until the end of the cooking cycle. The cooking time for a batch cycle normally ranges from 1½ to 2½ hours, depending on the initial moisture content and type of raw material being processed.

The average steam rate during the batch cooking cycle will normally vary from 450 to 900 ft^3/min. Shell-and-tube condensers through which cooling water is circulated or air-cooled, finned-tube condensers are used to condense the steam vapors and cool the condensate to normally below 120°F. The batch cooker noncondensibles range in odor intensity from 5000 to 1 million odor units/standard cubic feet by the ASTM syringe method, depending on the age and type of raw material. The volumetric emission rate of the noncondensibles may vary from 25 to 75 ft^3/min, depending on the tightness of the batch cooker top cover/discharge door openings and shaft seals.

Table 4 shows odor sensory data obtained[13] during a batch cooker cycle, with polyethylene bag samples being collected at half-hour intervals during the cook cycle. These samples were evaluated using the IITRI dynamic olfactometer. Readings were also obtained with an instrument measuring the concentration of total organic compounds (as methane) in parts per million.

TABLE 3. Odorous Compounds in Rendering Plant Emissions

Compound Name	Formula	Molecular Weight	Detection[a] Threshold (ppm, v/v)	Recognition Threshold (ppm, v/v)
Acetaldehyde	CH_3CHO	44	0.067	0.21
Ammonia	NH_3	17	17	37
Butyric acid	C_3H_7COOH	88	0.0005	0.001
Dimethyl amine	$(CH_3)_2NH$	45	0.34	—
Dimethyl sulfide	$(CH_3)_2S$	62	0.001	0.001
Dimethyl disulfide	CH_3SSCH_3	94	0.008	0.008
Ethyl amine	$C_2H_5NH_2$	45	0.27	1.7
Ethyl mercaptan	C_2H_5SH	62	0.0003	0.001
Hydrogen sulfide	H_2S	34	0.0005	0.0047
Indole	$C_6H_4(CH)_2NH$	117	0.0001	—
Methyl amine	CH_3NH_2	31	4.7	—
Methyl mercaptan	CH_3SH	48	0.0005	0.0010
Skatole	C_9H_9N	131	0.001	0.050
Trimethyl amine	$(CH_3)_3N$	59	0.0004	—

Source: References 7, 10, 11, and 12.

TABLE 4. Batch Cooker Noncondensible Odor Emissions

Cook Cycle, hours	Cooker Temperature, °F	Total Organics, ppm	Odor Dilution[a] to Threshold
¼	150		
½	220	180	40,000
1	245	1,000	45,000
1½	245	700	97,000
2	245	400	75,000
2½	245	260	93,000
3	245	200	127,000

[a]IITRI dynamic olfactometer

Source: Reference 13.

Another series of six samples were taken simultaneously on a different day approximately three hours after the batch cooker cycle began. The cooker temperature was recorded at 265°F and the noncondensibles temperature at 90°F, and the total organics varied from 230 to 330 ppm during the sampling. The odor dilution to threshold values (IITRI dynamic olfactometer) varied from 184,000 to 276,000. The batch rendering plant was known to process raw material such as dead stock, which emits unusually high-intensity odors during the cooking process.

The steam vapor emission rate from continuous rendering processes is relatively constant and can be calculated for a specific moisture content and tonnage rate of raw material being processed. The capacity of cookers and dryers usually is expressed in terms of evaporative capacity as pounds of water evaporated. For example, the Series 1800 Duke Equacooker with regular shaft has a rated evaporative capacity of 12,000 pounds of water evaporated per hour based on a steam pressure of 100 psig.

During the past five years, a number of Duke Series 1800 continuous systems have been evaluated to quantify the high-intensity odors emitted from the process. These emissions include the noncondensibles from the Equacooker, the drainer, centrifuge, and other sources of odor. Samples were taken of the noncondensibles alone and also of the total odor emissions from the Duke system or other rendering plant operations. All samples were taken during the late spring or summer months and were evaluated using the IITRI dynamic olfactometer. Also, these evaluations were conducted with varying types of raw material, including that from beef slaughterhouse, shop fat and bone, and restaurant grease and fish. Table 5 from reference 14 summarizes the data, which include the volumetric emission rates.

There has been a tendency in the past to have inadequate odor pickup flow rates for certain emissions from Duke continuous systems. Table 6 provides a guide for establishing the volumetric rate of pickup for various odor emissions from a Series 1800 Duke continuous system. These should not be considered as rigid requirements, but instead as guidelines.

No odor sensory data are available for the new con-

TABLE 5. Odor Emissions From Continuous Rendering Processes

Plant	Category of Renderer and Material	Rendering Process and Type of Emission	Odor Dilution[a] to Threshold	Emission Rate ft³/min
A	Integrated beef slaughterhouse	Duke—noncond.	20,000–50,000	450
		Duke—total[b]	39,100–43,200	1,500
		Blood ring dryer[c]	300– 1,000	2,000
B	Independent beef slaughterhouse and restaurant grease	Duke—noncond.	24,400– 62,700	—
		Duke—total[d]	56,000–138,000	—
B	Shop fat and bone	Duke—noncond.	11,000– 16,000	665
		Duke—total[d]	7,600– 13,200	1,200
C	Independent beef slaughterhouse and restaurant grease	Duke—noncond.	36,100– 39,800	800
		Duke—total[e]	21,600– 73,700	1,700
D	Independent herring (fish)	Duke—noncond.	39,500	600
		Duke—total[f]	23,800	2,600
D	Meat scraps and beef slaughterhouse	Duke—noncond.	59,400–93,800	600
		Duke—total[f]	28,100–59,200	2,600

[a]IIRTI dynamic olfactometer
[b]Includes noncondensibles, two presses, drainer, centrifuge, and blood coagulator and centrifuge. Tests conducted in mid-April in Texas.
[c]Venturi scrubber discharge after dryer exhaust.
[d]Both tests include noncondensibles, presses, drainer, and centrifuge. First test conducted during late July and second test conducted during late September in New England.
[e]Includes two presses, two tallow tanks, two centrifuges, drainer, and grease vapor noncondensibles. Test conducted in mid-August in New York.
[f]Includes two presses, centrifuge, drainer, meal product conveyor, and storage bins. Tests conducted in late May on West Coast.

Source: Reference 14.

tinuous systems with reduced energy usage. However, it has been observed in these plants that the odor emissions from the rendering process are definitely lower in intensity. This no doubt can be attributed to the lower processing temperatures that are used, particularly in the evaporator where the liquid from the twin-screw press is under vacuum for moisture removal.

AIR POLLUTION CONTROL MEASURES

Rendering plant operation and maintenance considerations are of basic importance in developing odor control measures to abate plant emissions. Raw material received at the plant should be processed with a minimum of delay. Cooking and pressing operations should be conducted to prevent overheating and burning of the processed material. As an example, a Duke Equacooker normally operated at a temperature of 270°F experienced a process upset, causing an increase to 320°F. This resulted in a corresponding increase from 14,000 to 102,000 odor units per standard cubic foot (ASTM syringe method).[15]

Start-up and shutdown operating procedures should ensure that all odor control equipment is operating properly while any raw material is being processed through the rendering system. Process equipment leaks to the floor or to the plant atmosphere should be corrected in a timely manner. Daily plant cleanup should normally follow shutdown of the rendering plant. Sanitation practices are crucial. A substantial amount of odor may be generated from within a plant building whose walls and ceiling have become permeated with fat aerosol odor emissions.

The basic purpose of providing odor control in a rendering plant is to reduce the odor emissions from the plant to a level that will result in the surrounding ambient air *not* containing odors that are a source of valid nuisance complaints. In designing a new control system or revising an old one, each individual plant situation must be evaluated separately based on a variety of factors, including proximity of neighbors to the plant, categories of neighbors present,

TABLE 6. Suggested Odor Pickup Flow Rates for Duke Process Emissions

Category of Odor Emission Pickup	Duct ϕ, inches	Flow Rate ft³/min
Equacooker inlet chute (if not sealed)	6	500
Equacooker noncondensibles	6	500
Drainer section	6	500
Equacooker unloading Elevator	4	250
Duke presser (10-inch or 12-inch)	6	500
Centrifuge (24 inches × 60 or 24 inches × 38 inches)	4	250
Tallow processing tank	4	250

surrounding topography, prevailing winds, plant building features, ability of rendering process to confine odors, residual plant odors, type of raw material, and seasonal climatic conditions.

The fundamental question often to be resolved is whether to treat the high-intensity odors only or also to treat a large volume of air that would be used to ventilate the operating area within the plant. A decision to treat only the high-intensity odors is usually predicated on the ability of the rendering process to confine these odors within the equipment. As discussed before, continuous rendering systems usually have this capability. Boiler incineration or multistage, low-volume scrubbing of the high-intensity odors is particularly compatible with this type of system.

Boiler Incineration of Process Odors

The installation and operation of afterburners or incinerators solely for pollution control is relatively uncommon due to the capital investment and fuel costs required. Currently, boiler incineration of the high-tensity process odors is a regular practice throughout the United States since all rendering plants require the generation of steam for the cooking and drying processes.

Two basic choices are available for the odorous air to be introduced into the boiler: primary combustion air (that mixed with fuel before ignition) or secondary combustion air (that mixed with the burner flame to complete combustion). The following factors should be considered for boiler incineration of high intensity odors.

1. The volume of odorous air to be handled should be minimized. Ensure that odor pickup points in the rendering process are not pulling excessive quantities of air.
2. Maximum fuel economy is achieved by using the odorous stream as primary combustion air, whereas its use as secondary combustion air probably requires additional fuel. When used as primary combustion air, particular care must be taken to see that the air stream is essentially free of moisture and particulate, which can interfere with the operation of the burner and controls.
3. Cooker noncondensibles can be successfully used as primary combustion air for incineration, provided that proper precautions are taken to "clean" it up. This is accomplished with a combination scrubber and entrainment separator of proper design. A water spray is provided to cool the odorous air and condense out the moisture. Likewise, the solid and fat aerosol particles should be removed.
4. Any high-intensity odors used as secondary combustion air should also be "cleaned up" in a manner similar to that described. When used as secondary combustion air, it is essential that the odorous stream come into intimate contact with the burner flame to accomplish efficient odor removal by incineration. Merely ducting the odorous air to the firebox without achieving contact with the flame may result in unsatisfactory odor removal.
5. The boiler size and burner capacity should be compatible with the amount of odorous air to be incinerated. If multiple boilers are used, the odorous air can be split among the various boilers or a smaller single unit can be used to incinerate the odors. The boiler should be equipped with suitable burner controls to ensure that the minimum firing rate is sufficient to incinerate the volume of odorous air passing through the firebox, regardless of the steam demand. A temperature of 1200°F or more is usually obtained in the firebox at the minimum firing rate. The residence time in the boiler firebox at maximum fuel rate is normally more than one second.
6. If an existing boiler is to be used, a thorough analysis should be made to establish that the combustion, control and safety requirements are satisfied for incinerating odors. The boiler manufacturer should be consulted regarding any details to modify the unit. Likewise, the insurance company should be contacted to receive their approval of any such revisions.

Figure 5 illustrates the boiler incineration of process odors used as primary combustion air. A two-stage spray scrubber with tangential inlet and entrainment separator is shown.

Previously, Table 5 summarized odor emission data for high-intensity odors from the Duke continuous process for Plants B, C, and D. Table 7 illustrates the odor removal efficiency achieved by boiler incineration for these same plants. The odor dilution to threshold values shown in this table were obtained with the IITRI dynamic olfactometer.

These results clearly show that boiler incineration is a very efficient method of odor control for treating the high-intensity odors from the rendering process. Two different incineration conditions are shown for Plant B to illustrate the higher stack exhaust odor levels resulting from the use of high-sulfur No. 6 fuel (2% sulfur) oil compared with those for natural gas. A strong, pungent, sulfur dioxide smell was observed in the exhaust stack emission. When natural gas or low-sulfur fuel oil is used as boiler fuel, the odor character of the stack exhaust is that of combustion gases only. In both cases, a rendering odor was not detected.

Wet Scrubbing of Process Odors

Multistage scrubber systems for treating the high-intensity odors provide an alternative to boiler incineration when the latter approach may not be feasible because of the boiler plant's being remotely located or for other reasons. These multistage systems have been successfully applied to rendering plant emissions since the early 1970s.

An evaluation[9] was made of various scrubbing agents to determine their comparative effectiveness to absorb and

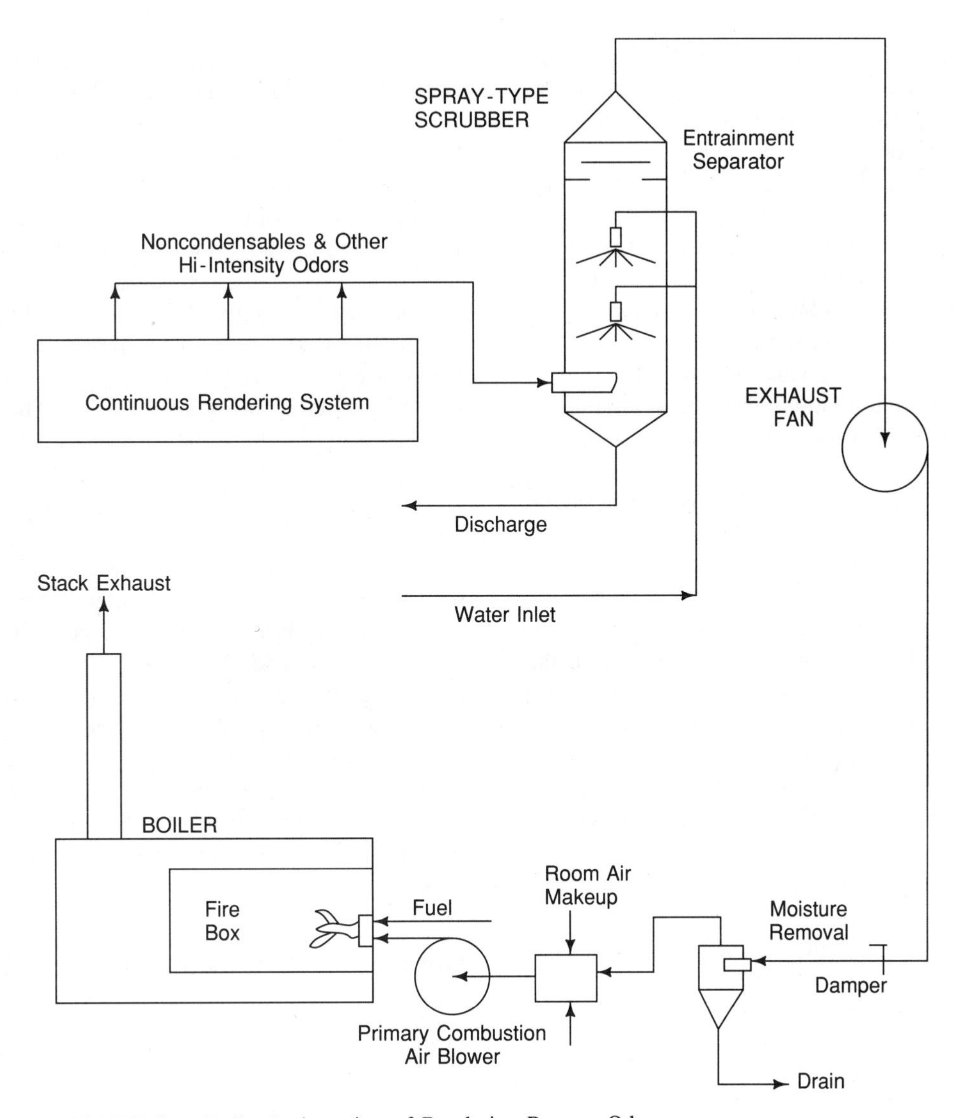

FIGURE 5. Boiler Incineration of Rendering Process Odors

neutralize selected odorous compounds in an experimental packed column. These odorous compounds were synthetically generated in the laboratory. Their selection included two sulfides, two amines, an organic acid, a ketone, an aldehyde, and an alcohol. Included among the various chemicals studied as scrubbing agents were water, soda ash, caustic soda, hydrochloric acid, sodium bisulfite, hydrogen peroxide, potassium permanganate, and sodium hypochlorite. The chemical oxidizing agents were found to be the most effective. From an overall viewpoint of treating a variety of odorants present in rendering plant emissions, sodium hypochlorite was considered to be the most effective agent.

Sodium hypochlorite or the addition of chlorine gas to a

TABLE 7. Boiler Incineration of Rendering Process Odors

Plant	Fuel Used	Fire Box Temperature, °F	Odor Dilution to Threshold Values[a]		Odor Removal %
			Boiler Inlet	Stack Exhaust	
B	No. 6 oil	—	56,000–138,000	234–650	99.5
B	Natural gas	—	7,600– 13,200	88–128	99.0
C	No. 6 oil	1,400	21,600– 73,000	76–157	99.6
D	Natural gas	1,250	28,100– 59,200	202–356	99.3

[a]IITRI dynamic olfactometer

Source: Reference 14.

caustic soda solution has been used since the early 1970s. More recently, chlorine dioxide generated on-site by the addition of chlorine gas to sodium chlorite solution has been used in a number of these scrubber systems. Chlorine dioxide is known to have a higher oxidizing potential than sodium hypochlorite.[16]

It is important to establish the design and operational criteria for a wet scrubbing system that uses a chemical oxidant solution, not only to obtain effective odor control, but also to achieve practical operation that is economical. This subject is discussed by Lundgren and others.[17]

The chemical oxidant solutions used in wet scrubbers normally are recirculated to conserve water and minimize chemical and wastewater treatment costs. The concept of balancing the oxidant chemical addition rate with the chemical use rate is important to achieve optimum usage and minimum cost. It is advisable to treat gas streams containing solid and fat aerosol particles with a preconditioning device, such as a low- to medium-pressure-drop venturi scrubber, to remove this particulate before passing to the chemical scrubber. Since the mass of the particulate matter suspended in the gas stream may exceed significantly that of the gaseous odor components, the rate of chemical oxidant consumption could rise quickly because the presence of the particles in the recirculating solution will tend to use up most of the oxidizing agent.

A two-stage scrubber system consisting of a venturi and a packed tower is described by Prokop.[18] This scrubber system treats high-intensity odors, including those from the screw press vents, blood dryer exhaust, raw feather receiving, feather noncondensibles, feather cooker discharge, and feather dryer exhaust. The scrubber system capacity is 32,000 ft^3/min. The scrubbing solutions consist of water circulating through the venturi at 110 gpm and sodium hypochlorite solution circulating through the packed tower at 350 gpm. Chemical consumption consists of 2 gph of 12% NaOCl solution.

A three-stage scrubber system consisting of a venturi and two packed towers in series is described by Prokop.[15] This scrubber system treats high-intensity odors, including those from the Equacooker noncondensibles, drainer, presser vents, centrifuge, and the steaming of grease barrels. The scrubber system capacity is 7500 ft^3/min. The scrubbing solutions consist of a weak solution of trisodium phosphate for circulation through the venturi at 22 gpm, a solution of phosphoric acid at 2–3 pH and recirculated through the first packed tower at 80 gpm, and a sodium hypochlorite solution at 10 pH and recirculated through the second packed tower at 80 gpm. The NaOCl solution is generated by an electrolytic cell from sodium chloride. The NaOH solution is added to control the pH of the NaOCl solution.

Figure 6 illustrates the operation of this multistage scrubber system. The high-intensity odors enter the throat of the low-energy venturi scrubber, which removes the particulate matter and cools and saturates the air with water vapor. A demister removes entrained droplets from the air before it passes upward through the first-stage counterflow packed-bed scrubber. Acid solution is recirculated through this stage to remove the amine-type odors and NaOCl solution is recirculated throughout the second stage to remove the sulfide-type odors.

Table 8 summarizes the odor sensory data obtained for these two multiple-stage scrubber systems. The odor removal efficiency of these two systems is approximately 99%.

Wet Scrubbing of Plant Ventilating Air

This approach provides a more complete solution to an overall plant odor problem. Fugitive odors within the plant can be captured and treated in a uniform manner with this type of scrubber. It is particularly suited for rendering plants located near sensitive population areas, such as residential or commercial. For this application, it is essential to have adequate distribution and flow of air throughout the plant in order to pick up and capture in the ventilating air those fat

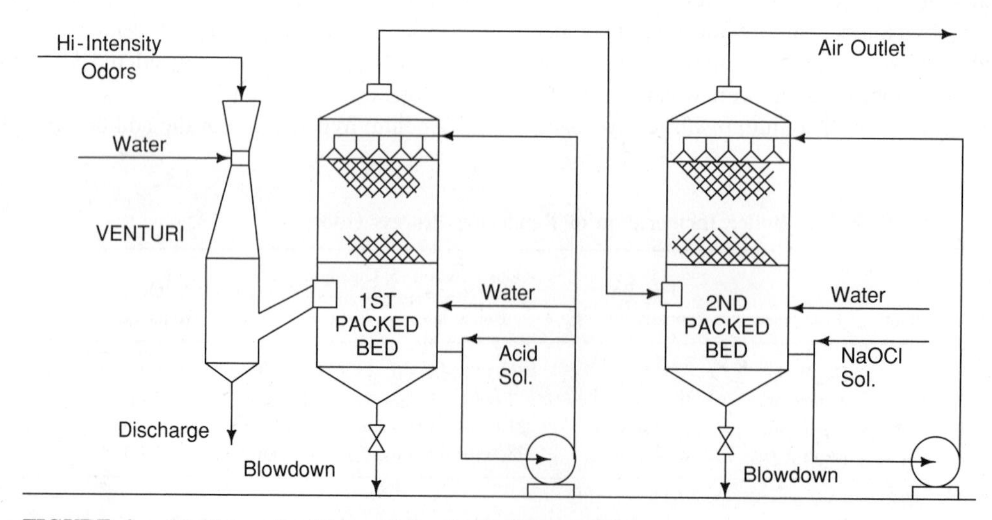

FIGURE 6. Multistage Scrubbing of Rendering Process Odors

TABLE 8. Multistage Scrubbing of Rendering Process Odors

Scrubber Category	Exhaust Flow, ft^3/min	Scrubber Solutions			Odors Units/scf[a]	
		First	Second	Third	Inlet	Outlet
Venturi and one packed tower	32,000	Water	NaOCl		5000–20,000	50–100
Venturi and two packed towers	7,500	Na_3PO_4	H_3PO_4	NaOCl	14,000	185

[a]ASTM syringe method.

Source: References 15, 18

aerosols and other particles that are emitted from the rendering process.

In addition to accomplishing effective odor control, these scrubbers provide proper ventilation of the plant operating areas, thereby maintaining satisfactory working conditions for the employees and improved compliance with the Occupational Safety and Health Administration standards. It is important to recognize that sufficient ventilating air must pass through the operating area during the summer months. Otherwise, doors and windows will be opened excessively, allowing the plant odors to escape from the building instead of being treated by the scrubber system. Ideally, a slight negative pressure should be maintained within the rendering plant.

Certain factors should be considered for the installation of a plant ventilating air scrubber.[19] These include the following.

1. Different types of operations should be separated. For example, protein meal grinding and storage should be isolated from rendering processes and raw material receiving in order to prevent bacterial contamination of the protein meal.
2. Ventilation air should flow from low odor sources to high odor sources. Thus, air from the finished product storage area should flow *toward* the process operating area where the scrubber is normally located. This also applies to the raw material receiving area.
3. Building openings need to be controlled in some positive manner. The overhead truck doors need to be kept closed whenever possible. A building fitted with gravity intake louvers that automatically close when a door is opened may be able to maintain a constant air volume exhausted from the building.

An important consideration in designing a plant ventilating air scrubber system concerns the number of room changes per hour for various categories of rendering plant operations. Table 9 provides a listing of these ventilation rates. A range of values is provided to allow sufficient flexibility in selecting an appropriate ventilation rate that takes into consideration the presence of an operator(s), emission of vapors or odors into the room atmosphere, ambient temperature (summer conditions), and raw material being processed.

A typical plant ventilation air scrubber is shown in Figure 7 as a single packed tower through which NaOCl solution is circulated. These scrubbers have capacities normally ranging from 40,000 to 80,000 ft^3/min.

The addition of NaOCl and caustic soda is controlled by means of oxidation–reduction potential (ORP) and pH measurements. This type of scrubber is capable of achieving 95% or more of odor removal based on inlet values of 2000 to 5000 and exhaust values of 100 to 200 (IITRI dynamic olfactometer).

Other types of wet scrubbers that have been successfully applied to treating the plant ventilating air in rendering plants include the following.

1. Cross flow, packed-bed scrubber where the air flows horizontally and is contacted by the scrubbing solution as it flows downward by gravity through the packing. The capacity for this type of scrubber ranges above 100,000 cfm and the scrubbing solution is recirculated at a rate equivalent to 8 gallons per 1000 ft^3 of odorous air. This type of scrubber is described by Frega and Prokop.[20]
2. Cross-flow, coarse-spray scrubber where the air flows horizontally past a series of baffles countercurrent to the spray pattern. The capacity for this type of scrubber ranges from 38,000 to 150,000 cfm and the scrubber solution is recirculated through each stage of scrubbing at 7½ gallons per 1000 ft^3 of odorous air. Either one or two stages are furnished where different scrubbing solu-

TABLE 9. Ventilation Guidelines for Rendering Plant Processing Areas

Processing Area Category	Number of Room Changes per Hour
Batch rendering system operating area	20–40
Continuous rendering system (suitably enclosed)	10–20
Grease melting and processing	10–20
Raw-material storage and handling	5–15
Fat processing (filtering and bleaching)	5–10
Protein meal milling and conveying	5–10
Fat storage	3–5
Protein meal storage	3–5

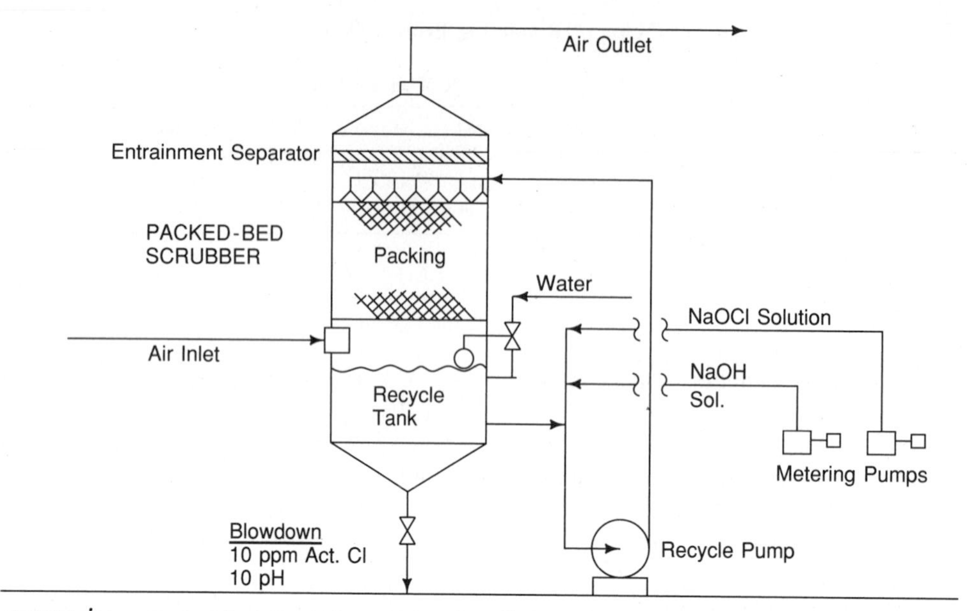

FIGURE 7. Packed-Bed Scrubbing of Plant Ventilating Air

tions are normally used in both stages. A two-stage scrubber of this type is described by Prokop.[18]

3. Mist spray, cocurrent flow scrubbers where very fine droplets are formed by atomizing nozzles normally supplied with compressed air. The mist spray at the top of the scrubber is directed downward in the same direction as the flow of odorous air. The capacity for this type of scrubber is available up to 75,000 cfm, with the once-through scrubbing solution delivered at one-half gallon per 10,000 ft^3 of odorous air.

Current practice for designing wet scrubbing systems to treat both the high-intensity process odor and plant ventilating air is to combine both types of scrubbers into a single system. A three-stage scrubber system is described[20] that consists of the following.

1. A venturi scrubber in series with a vertical countercurrent packed tower to treat the high-intensity odors, which include the Rotadisc noncondensibles, the drainer screw, and two Duke pressors. The scrubbing solutions consist of fresh water circulated through the venturi at 70 gpm and sulfuric acid solution at 2 pH recirculated through the packed tower at 75 gpm. This two-stage scrubber system treats 6300 cfm of odorous air.
2. A horizontal, cross-flow, packed-bed scrubber that receives the exhaust from the two-stage scrubber system described above and the plant ventilating air. Sodium hypochlorite solution at 9 pH is recirculated through this packed-bed scrubber at 575 gpm and its capacity is 70,000 cfm.

Table 10 summarizes the odor sensory data obtained with this three-stage scrubber system.

The overall reduction of the high-intensity odors by three stages of scrubbing (venturi, small packed tower, and large packed bed) exceeded 99%. The plant ventilating air scrubber had an odor removal efficiency ranging from 90% to 95%, depending on the inlet odor level.

Other Methods of Odor Control

Biofilter technology developed during the past 10 years in Europe and elsewhere has been applied to the treatment of rendering odors.[21,22] Biofilters consist of large beds of porous media that are capable of adsorbing odorous gaseous compounds and reducing these by aerobic microbial action to nonodorous components.

Two basic types of biofilter media influence flow rate. Compost, peat moss, heather, and other fibrous media, because of their greater porosity, allow high airflow rates ranging from 35 to 180 m^3/hm^2 (2 to 10 cfm/ft^2) of bed surface without excessive pressure drop resulting. This category of biofilter is analogous to the plant ventilating air wet scrubbers described earlier. Rendering plants currently

TABLE 10. Multiple-Stage Scrubbing of Process Odor and Plant Ventilating Air

	Odor Dilution to Threshold[a]	
	Average	Range
High-intensity odor scrubber		
Venturi scrubber inlet	112,000	42,900–175,000
Plant ventilating air scrubber		
Inlet	3,810	2,180–6,180
Outlet	332	165–628

[a]IITRI dynamic olfactometer.

Source: Reference 20.

TABLE 11. Soil-Bed Treatment of Cooker Noncondensibles

Date	4/11/84	4/12/84	6/26/85	6/27/85
Noncondensibles[a]	52,500	29,000	93,000	209,000
Soil-bed exhaust[a]	43	23	46	120
Odor removal, %	99.9	99.9	99.9	99.9

[a]Odor dilution to threshold values (IITRI dynamic olfactometer)

Source: Reference 23.

are operating such biofilters in Canada, Denmark, France, Holland, Germany, and New Zealand. Also, rendering plants in the United States are now installing these biofilters.

Soil-bed systems are another type of biofilter. Less porous media such as soil result in lower airflow rates, ranging from 2 to 10 $m^3/h/m^2$ (0.1 to 0.5 cfm/ft^2) of bed surface. Such a system is described[23] for a soil bed with a surface area of 4500 ft^2 that treats 650 cfm of Duke Equacooker noncondensibles. A bed depth of 60 cm (24 inches) was provided and the pressure drop across the bed was measured at 5 cm (2 inches) of water column. Two series of odor sensory tests were conducted approximately 14 months apart. Composite samples of soil-bed exhaust were evaluated, along with samples of the inlet noncondensibles. Table 11 summarizes the results, which are expressed as odor dilution to threshold (IITRI dynamic olfactometer). These odor sensory performance results clearly indicate the potential for soil-bed treatment of high-intensity process odors of relatively low volume. Compared with wet scrubbing systems, soil beds require considerably less initial investment and have lower operating costs.

Other odor control methods include activated-carbon adsorption and catalytic oxidation. No actual odor sensory performance tests related to rendering plant emissions have been reported for these methods since the publication of the second edition of this manual.

A very important part of the overall design of an odor control system is related to the emission from the control system into the surrounding atmosphere. Specific precautions should be taken to ensure that the stack design takes into account the height and exit velocity needed to provide the desired dispersion characteristics.

In performing atmospheric dispersion calculations to predict downwind ground-level concentrations, it should be recognized that estimates based on averaging times of 10 to 60 minutes typical of traditional models may be in error. Objectionable odors are detected in much shorter time intervals and peak concentrations for these short times can be much higher than the long-term concentrations for traditional models. Jann and Cha[24] discuss the importance of the odor dispersion averaging times being expressed in seconds in order to be able to predict the maximum odor concentration at ground level.

References

1. *U.S. Production, Consumption, Trade and Stocks of Tallow/Grease, Lard and Animal Protein Meals; Estimates and Projections;* P-318 (April 1980), prepared by Development Planning & Research Associates, Manhattan, KS, for the USDA Foreign Agricultural Service, Washington, DC.
2. *Listing of Member and Non-member Plants,* National Renderers Association, Washington, DC, 1990.
3. W. H. Prokop, "Rendering systems for processing animal by-product materials," *J. Am Oil Chem. Soc.*, 62(4):805–811 (April 1985).
4. W. H. Prokop, "Low temperature rendering by ATLAS system," *Meat Proc.*, p. 24 (April 1986).
5. W. R. Boehme, "New drying techniques preserve lysine in blood meals," *Director's Digest No. 116,* Fats and Proteins Research Foundation, Fort Myers Beach, FL, February 1974.
6. *Feather Meal,* National Renderers Association, Washington, DC, revised March 1985, 11 pp.
7. R. T. Walsh, "Reduction of inedible rendering matter," *Air Pollution Engineering Manual,* 2nd ed., J. A. Danielson (Ed.); U.S. Environmental Protection Agency, Research Triangle Park, NC. Publication No. AP-40, May 1973, p 823.
8. T. A. Burgwald, "Identification of chemical constituents in rendering industry odor emissions," Project No. C8172, IIT Research Institute, Chicago, IL. January, 1971, p 39. (supported by Fats and Proteins Research Foundation).
9. R. H. Snow and H. G. Reilich, "Investigation of odor control in the rendering industry," Project No. C8210, IIT Research Institute, Chicago, August 1972, p 117 (performed under Contract No. 68-02-0260 with U.S. Environmental Protection Agency and with the support of Fats and Proteins Research Foundation).
10. *Odor Thresholds for Chemicals with Established Occupational Health Standards,* American Industrial Hygiene Association, Akron, OH, 1989.
11. J. E. Moore and R. S. O'Neill, "Odor as an aid to chemical safety: Odor thresholds compared with threshold limit values and volatilities for 214 industrial chemicals in air and water dilution," *J. Appl. Toxicol.*, 3:6, 1983.
12. G. Leonardos, D. Kendall, and N. Bernard, "Odor thresholds for 53 commercial chemicals," *J. AWMA,* 19(2):91, 1969.
13. F. Jarke, "Investigation of high-intensity odors," Project C8563, IIT Research Institute. Chicago, June 1981 (supported by Fats and Proteins Research Foundation).
14. W. H. Prokop, "Control methods for treating odor emissions from inedible rendering plants," Paper 91-146.8 at Annual Meeting of AWMA, Vancouver, B.C., Canada, 1991.
15. W. H. Prokop, "Wet scrubbing of high intensity odors from rendering plants," *Proceedings AWMA Specialty Conference on Odor Control Technology II,* Pittsburgh, PA, March 1977, pp 153–166.
16. G. C. White, *Handbook of Chlorination,* Van Nostrand Reinhold Co., New York, 1972, p 600.
17. D. A. Lundgren, L. W. Rees, and L. D. Lehmann, "Odor control by chemical oxidation: cost, efficiency and basis for selection," Paper 72-116 at Annual Meeting of AWMA, Miami Beach, FL, 1972.
18. W. H. Prokop, "Wet scrubbing of inedible rendering plant odors," *Proceedings AWMA Specialty Conference on Odor Control Technology I,* Pittsburgh, PA, 1974, pp 132–150.

19. J. M. Sweeten, G. I. Wallin, and J. A. Chittenden, "Control of rendering plant odors: State of the art," Paper 80-6004 at Annual Meeting of American Society of Agricultural Engineers, 1980.
20. V. Frega and W. H. Prokop, "Wet scrubbing of multiple source odors in a rendering plant," Paper 81-39.2 at Annual Meeting of APCA, Philadelphia, PA, 1981.
21. M. B. Rands, D. E. Cooper, C. P. Woo, et al., "Compost filters for H_2S removal from anaerobic digestion and rendering exhausts," *J. WPCF,* 53:185 (1981).
22. W. Koch, H. G. Liebe, and B. Striefler, "Betriebserfahrungen mit Biofiltern zur Renduzierung geruchsin-tensiver Emissionen," *Staub-Reinhaultung der Luft,* 43:488 (1982).
23. W. H. Prokop and H. L. Bohn, "Soil bed system for control of rendering plant odors," *J. AWMA,* 35(12):1332–1338 (1985).
24. P. R. Jann and S. S. Cha, "Atmospheric dispersion and odor modeling," *Odor Control for Wastewater Facilities,* Manual of Practice No. 22, Water Pollution Control Federation, Washington, DC, 1992.

NITRATE FERTILIZERS*

Horace C. Mann†

Ammonium nitrate (NH_4NO_3) is produced by neutralizing nitric acid with ammonia. The reaction can be carried out at atmospheric pressure or at pressures up to 410 kPa (45 psig) and at temperatures between 405K and 458K (270°–365°F). An 83 wt % solution of ammonium nitrate is produced when concentrated nitric acid (56–60 wt %) is combined with gaseous ammonia in a ratio of from 3.55 to 3.71 to 1, by weight. When solidified, ammonium nitrate is a hygroscopic white solid that contains 35% nitrogen.

In 1989, approximately 7,557,000 tons of ammonium nitrate solution were produced.[1] It is estimated that 15–20% of this amount was used for explosives and other purposes, while the balance was used as fertilizer. It is used either as a straight fertilizer material or in mixtures (with calcium carbonate, limestone, or dolomite) called calcium ammonium nitrate (CAN), ammonium nitrate-limestone (ANL), or various trade names, and in compound fertilizers, including nitrophosphates. It is also a principal ingredient of most nitrogen solutions.

Ammonium nitrate is used for blasting purposes in conjunction with fuel oil; relatively small amounts are consumed by the brewing and chemical industries. The earlier "grained" type of ammonium nitrate, made by rolling the semimolten salt in an open pan and coating with resins or waxes, has been largely superseded by prilled, granular, and crystalline end products.

The main disadvantages of ammonium nitrate are (1) it is quite hygroscopic, (2) there is some risk of fire or even explosion unless suitable precautions are taken, (3) it is reported to be less effective for flooded rice than urea or ammoniacal nitrogen fertilizers, and (4) it is more prone to leaching than ammoniacal products.

Some countries forbid the sale of straight ammonium nitrate as fertilizer because it can be used as an explosive when mixed with fuel oil or other synthesizers. In these countries, the mixture of ammonium nitrate with calcium carbonate (CAN) is permitted. Formerly, CAN contained 20.5% nitrogen (N), corresponding to about 60% ammonium nitrate (AN); at present, the most common grade is 26% N (75% AN).

Ammonium nitrate is generally regarded as posing no unacceptable hazard when suitable precautions are taken and is commonly used as a fertilizer with strict regulations. "Fertilizer-grade" ammonium nitrate cannot be exploded by impact. There are no records of explosions resulting from heat and fire alone.

Some compound fertilizers containing ammonium nitrate and chloride, such as potassium chloride, are subject to propagated decomposition or "cigar burning" when ignited. Once initiated, the decomposition propagates through the mass of material at a rate that usually ranges from 5 to 50 cm/h. The ignition temperature is about 200°C, and temperature in the decomposition zone is usually 300–500°C, but it may be lowered by certain sensitizing agents such as copper salts. As little as 4% potassium chloride (about 1.9% chlorine) is sufficient to make some mixtures susceptible to cigar burning. The reaction is inhibited by ammonium phosphate; therefore, many NPK compositions containing ammonium nitrate and potassium chloride are free from this hazard.

The exact nature of the reaction is not entirely clear, but it results in complete destruction of the ammonium nitrate and evolution of some of the chloride. Noxious red, white, yellow, or brown fumes are given off that contain NH_4Cl, HCl, Cl_2, NO_2, and other oxides of nitrogen, N_2, and H_2O. The fumes are toxic and have resulted in several fatalities in some incidents.

Because the reaction does not require oxygen, other than that present in ammonium nitrate, the fire cannot be extinguished by smothering. It can only be stopped by flooding with water. If a localized area of decomposition in a bin or pile is discovered early enough, the decomposing material may be removed from the building by a power shovel, for example, and extinguished by water, thereby saving the remainder of the material.

Perbal[2] concluded that compound fertilizers in general may be regarded as safe from explosion hazard if they contain less than 70% NH_4NO_3, unless there is a high

*Portions of this discussion were edited from AP-42, the U.S. Environmental Protection Agency's *Emission Factors Handbook,* Vol 1, Chapter 6.8, and *International Fertilizer Development Center Reference Manual,* IFDC-R.1, Chapter VIII.

†With assistance from authorities at Arcadian Corp. and TVA's NFERC.

percentage of $(NH_4)_2SO_4$ or other reducible material in the mixture, in which case the material should be tested.

PROCESS DESCRIPTION

Several proprietary processes for ammonium nitrate manufacture are available, using various combinations of different neutralization, evaporation, and drying and finishing methods. Solid ammonium nitrate is produced in the form of prills, crystals, and granules, either alone or in combination with other materials. Large tonnages of ammonium nitrate also are made in the form of solutions having concentrations in the range of 80–90% for use in granular compound fertilizers. The solution also is used to prepare nitrogen solutions that usually also contain ammonia or urea for use in liquid fertilizer.

The process for manufacturing ammonium nitrate can contain up to seven major unit operations. These operating steps are solution formation or synthesis, solution concentration, solids formation, solids finishing, solids screening, solids coating, and bagging and/or bulk shipping. In some cases, solutions may be blended for marketing as liquid fertilizers. A schematic flow diagram of the various possible ways of processing ammonium nitrate solution into either solid or fluid products is shown in Figure 1.

The number of operating steps employed is determined by the desired end product. For example, plants producing ammonium nitrate solutions alone use only the solution formation, solution blending, and bulk shipping operations. Plants producing a solid ammonium nitrate product can employ all of the operations.

All ammonium nitrate plants produce an aqueous ammonium nitrate solution through the reaction of ammonia and nitric acid in a neutralizer. To produce a solid product, the ammonium nitrate solution is concentrated in an evaporator or concentrator to drive off water. A melt is produced containing from 95% to 99.8% ammonium nitrate at approximately 422K (300°F). This melt is then used to make solid ammonium nitrate products.

Finishing Processes

In the past, several finishing processes were used, including graining, flaking, granulation, crystallization, and "low-density" prilling. In low-density prilling, the ammonium nitrate solution is fed to the prill tower at about 95% concentration and the resulting prills are dried and cooled. The prills are somewhat porous and may have an apparent specific gravity of 1.29 compared with 1.65 for high-density prills. Some of these methods are still in use, particularly for ammonium nitrate used as a blasting agent. A porous prill or granule that will absorb oil is preferred for this use.

For fertilizer use, the high-density prilling process, using $99^{+}\%$ solution concentration, has been used in most new plants that make straight ammonium nitrate. Quite recently, however, there has been a trend toward greater use of granulation processes that also use $99^{+}\%$ solution.

Prilled or granulated ammonium nitrate materials are often coated with a powdery conditioning agent such as china clay, *kieselguhr,* or calcined fullers earth, in amounts ranging from 1% to 3%. However, the use of additives in the ammonium nitrate melt before prilling may preclude the use of coatings. Conditioned ammonium nitrates usually have a guaranteed N content of 33.5–34.0%. In some climates, coating to prevent caking is not considered necessary, especially with stabilized, high-density prills of very low moisture content. In this case, the guaranteed nitrogen content is 34.0–34.5%. Stabilizers that prevent change in crystal form at 32°C have been developed and used. These stabilizers include magnesium nitrate (produced by dissolving magnesite in the nitric acid) and "Permalene," a combination of ammonium sulfate and diammonium phosphate.

Ammonium nitrate may be stored in bulk, although in most climates, controlled-humidity storage buildings are advisable. In most countries, the product is distributed in bags that should be "moistureproof"; at least one ply should be impermeable to moisture. Plastic film bags or bags with plastic liners are suitable if properly constructed. In the United States, bulk shipment is common using covered, hopper-bottom rail cars.

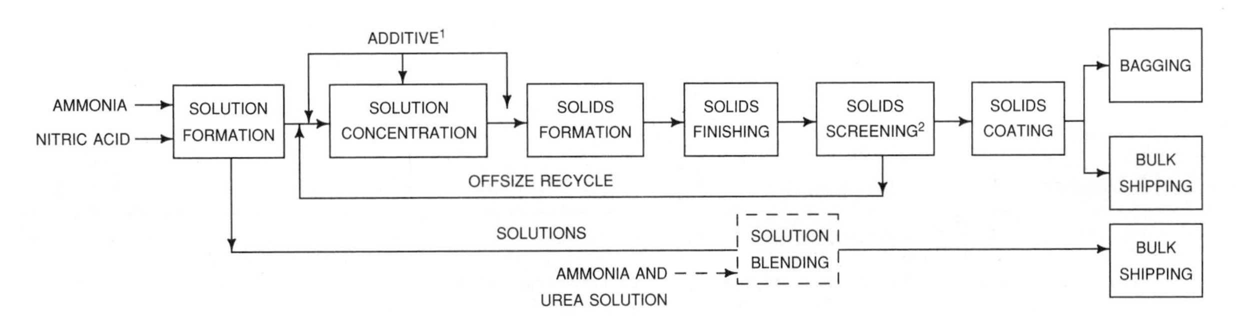

FIGURE 1. Ammonium Nitrate Manufacturing Operations

AIR EMISSION CHARACTERIZATION

Emissions from ammonium nitrate production plants are particulate matter (ammonium nitrate and coating materials), ammonia, and nitric acid. Ammonia and nitric acid are emitted primarily from solution formation and concentration processes, with ammonia also being emitted from prill towers and granulators. Particulate matter (largely as ammonium nitrate) is emitted from most of the process operations and is the primary emission addressed here.

The emission sources in solution formation and concentration processes are neutralizers and evaporators, primarily emitting nitric acid and ammonia. Specific plant operating characteristics, however, make these emissions vary, depending on the use of excess ammonia or acid in the neutralizer. Since the neutralization operation can dictate the quantity of these emissions, a range of emission factors is presented in Table 1. Particulate emissions from these operations tend to be smaller in quantity than those from solids production and handling processes and generally are recycled back to the process.

Emissions from solids formation processes are ammonium nitrate particulate matter and ammonia. Primary sources are prill towers (for high-density and low-density prills) and granulators (rotary drum and pan). Emissions from prill towers result from carryover of fine particles and fumes by the prill cooling air flowing through the tower. These fine particles are from microprill formation, attrition of prills colliding with the tower or one another, and rapid transition of ammonium nitrate between crystal states. The uncontrolled particulate emissions from prill towers, therefore, are affected by tower airflow, spray melt temperature, condition and type of melt spray device, air temperature, and crystal state changes of the solid prills. The amount of microprill mass that can be entrained in the prill tower exhaust is determined by the tower air velocity. Increasing spray melt temperature causes an increase in the amount of gas-phase ammonium nitrate generated. These gaseous ammonium nitrate fumes, on cooling, may form ammonium nitrate particulates in the atmosphere. Fuming is much more severe in high-density prilling because the ammonium nitrate melt to be prilled must be at a high temperature (about 180°C) to keep it from freezing. At this temperature, there is an appreciable vapor pressure of $NH_3 + HNO_3$ resulting from dissociation of ammonium nitrate, according to the equation:

TABLE 1. Emission Factors for Processes in Ammonium Nitrate Manufacturing Plants[a]

	Particulate Matter		Ammonia	Nitric Acid	
Process	Uncontrolled	Controlled[b]	Uncontrolled[c]		Emission Factor Rating
Neutralizer	0.045–4.3 (0.09–8.6)	0.002–0.22 (0.004–0.43)	0.43–18.0 (0.86–36.0)	0.042–1[d] (0.084–2)[d]	B
Evaporation/concentration operations	0.26 (0.52)	—	0.27–16.7 (0.54–33.4)	—	A
Solids formation operations					A
High-density prill towers	1.59 (3.18)	0.60 (1.20)	28.6 (57.2)	—	
Low-density prill towers	0.46 (0.92)	0.26 (0.52)	0.13 (0.26)	—	
Rotary drum granulators	146 (292)	0.22 (0.44)	29.7 (59.4)	—	
Pan granulators	1.34 (2.68)	0.02 (0.04)	0.07 (0.14)	—	
Coolers and dryers					A
High-density prill coolers[e]	0.8 (1.6)	0.01 (0.02)	0.02 (0.04)	—	
Low-density prill coolers[e]	25.8 (51.6)	0.26 (0.52)	0.15 (0.30)	—	
Low-density prill dryers[e]	57.2 (114.4)	0.57 (1.14)	0–1.59 (0–3.18)	—	
Rotary drum granulator coolers[e]	8.1 (16.2)	0.08 (0.16)	0.59 (1.118)	—	
Pan granulator coolers[e]	18.3 (36.6)	0.18 (0.36)	0 (0)	—	
Coating operations	≤2.0 (≤4.0)	≤0.02 (≤0.04)	—	—	B
Bulk loading operations[f]	≤0.01 (≤0.02)	—	—	—	B

[a]Factors are g/kg (kg/mg) and lb/ton of ammonium nitrate fertilizer produced. Some ammonium nitrate emission factors are based on data gathered using a modification to EPA Method 5 (see title footnote). Dash—no data.

[b]Based on the following control efficiencies for wet scrubbers, applied to uncontrolled emissions: neutralizers, 95%; high-density prill towers, 62%; low-density prill towers, 43%; rotary drum granulators, 99.9%; pan granulators, 98.5%; coolers, dryers, and coaters, 99%.

[c]Given as ranges because of variation in data and plant operations. Factors for controlled emissions not presented due to conflicting results on control efficiency.

[d]Based on 95% recovery in a granulator recycle scrubber.

[e]Factors for coolers represent combined precooler and cooler emissions and factors for dryers represent combined predryer and dry emissions.

[f]Fugitive particulate emissions arise from coating and bulk loading operations.

$$NH_4NO_3 \rightarrow NH_3 + HNO_3$$

The dissociation products recombine in the cooler air to form a blue haze consisting of ammonium nitrate particles of submicron size. Particles of this size are difficult to collect, and they present a highly visible and stable haze or fog. The problem is much less serious with low-density prilling because of lower ammonium nitrate solution temperatures. It is less serious in granulation processes because of much smaller volumes of air in contact with hot solution.

Microprill formation resulting from partially plugged orifices of melt spray devices can increase fine dust loading and emissions. Certain designs (spinning buckets) and practices (vibration of spray plates) help reduce microprill formation. High ambient air temperatures can cause increased emissions because of entrainment as a result of the higher airflow required to cool prills and because of increased fumes formation at the higher temperatures.

The granulation process in general provides a larger degree of control in product formation than does prilling. Granulation produces a solid ammonium nitrate product that, relative to prills, is larger and has greater abrasion resistance and crushing strength. The airflow in granulation processes is lower than that in prilling operations. Granulators, however, cannot produce low-density ammonium nitrate economically with current technology. The design and operating parameters of granulators may affect emission rates. For example, the recycle rate of seed ammonium nitrate particles affects the bed temperature in the granulator. An increase in bed temperature resulting from the decreased recycling of seed particles may cause an increase in dust emissions from granule disintegration.

Cooling and drying are usually conducted in rotary drums. As with granulators, the design and operating parameters of the rotary drums may affect the quantity of emissions. In addition to design parameters, prill and granule temperature control is necessary to control emissions from the disintegration of solids caused by changes in crystal state.

Emissions from screening operations are generated by the attrition of the ammonium nitrate solids against the screens and against one another. Almost all screening operations used in the ammonium nitrate manufacturing industry are enclosed or have a cover over the uppermost screen. Screening equipment is located inside a building, and emissions are ducted from the process for recovery reuse.

Prills and granules are typically coated in a rotary drum. The rotating action produces a uniformly coated product. The mixing action also causes some of the coating material to be suspended, creating particulate emissions. Rotary drums used to coat solid product are typically kept at a slight negative pressure, and emissions are vented to a particulate control device. Any dust captured is usually recycled to the coating storage bins.

Bagging and bulk loading operations are a source of particulate emissions. Dust is emitted from each type of bagging process during final filling, when dust-laden air is displaced from the bag by the ammonium nitrate. The potential for emissions during bagging is greater for coated than for uncoated material. It is expected that emissions from bagging operations are primarily the kaolin, talc, or diatomaceous earth coating matter. About 90% of solid ammonium nitrate produced domestically is bulk loaded. While particulate emissions from bulk loading are not generally controlled, visible emissions are within typical state regulatory requirements (below 20% opacity, but varies with the states).

Table 1 summarizes emission factors for various processes involved in the manufacture of ammonium nitrate. Uncontrolled emissions of particulate matter, ammonia, and nitric acid are given in the table. Emissions of ammonia and nitric acid depend on specific operating practices, so ranges of factors are given for some emission sources.

TABLE 2. Particle Size Distribution Data for Uncontrolled Emissions from Ammonium Nitrate Manufacturing Facilities

	Cumulative Weight %		
	≤2.5 μm	≤5 μm	≤10 μm
Solids formation operations			
Low-density prill tower	56	73	83
Rotary drum granulator	0.07	0.3	2
Coolers and dryers			
Low-density prill cooler	0.03	0.09	0.4
Low-density prill predryer	0.03	0.06	0.2
Low-density prill dryer	0.04	0.04	0.15
Rotary drum granulator cooler	0.06	0.5	3
Pan granulator precooler	0.3	0.3	1.5

AIR POLLUTION CONTROL MEASURES

Emission factors for controlled particulate emissions are also shown in Table 1, reflecting wet scrubbing particulate control techniques. The particle size distribution data presented in Table 2 indicate the applicability of wet scrubbing to control ammonium nitrate particulate emissions. In addition, wet scrubbing is used as a control technique because the solution containing the recovered ammonium nitrate can be sent to the solution concentration process for reuse in the production of ammonium nitrate, rather than to waste disposal facilities.

A solution to the air pollution problem when producing high-density prills has been developed by the Cooperative Farm Chemicals Association (CFCA) and is in use at its plants in Lawrence, Kansas and at least 14 other plants in North America.[3] The fume abatement system consists of a bell-shaped shroud installed around the spray head in the upper part of the prill tower to collect fume-laden air from that part of the tower where fumes are formed as a result of contact of the air with hot ammonium nitrate solution or prills in the process of solidification. The airflow through this shroud is only about 25% of the total airflow through the tower; the remaining 75% is practically free of dust and fume and is discharged directly to the atmosphere. The air from the shroud is drawn through a scrubber and Brink high-efficiency mist eliminators. Fume and vapors from the neutralizer and evaporators are treated in the same scrubbing system. The scrubber solution is recirculated to build up its concentration and eventually recycled to the solution preparation step. The system recovers 3–7 kg of ammonium nitrate per ton of product from all sources (neutralizer, evaporator, and prill tower). From a pollution abatement viewpoint, the system has met applicable standards; atmospheric emissions of less than 0.5 kg per ton of product and opacity of less than 10% have been attained.[2]

References

1. *Chemical and Engineering News,* April 9, 1990, p 12.
2. G. Perbal, "The thermal stability of fertilizers containing ammonium nitrate," *Proc. Fertilizer Soc.* (London), no. 124 (1971).
3. "The control of fume from ammonium nitrate prilling towers," *Nitrogen* 107:34–39 (1977).

AMMONIUM PHOSPHATES*

Horace C. Mann†

Ammonium phosphates are produced mainly as either monoammonium phosphate (MAP) or diammonium phosphate (DAP), with most of the ammonium phosphate produced being used in fertilizers. Ammonium phosphates are also used as fire retardants and in animal feeds. Because of its high analysis and good physical properties, DAP is the most popular phosphate fertilizer on a worldwide basis. The composition of the pure MAP salt is 12.2% nitrogen and 61.7% P_2O_5; the composition of DAP is 21.2% nitrogen and 53.7% P_2O_5. These pure salts are usually made from electric-furnace or thermal phosphoric acid that contains essentially no impurities. Fertilizer-grade or wet-process phosphoric acid does have a number of impurities, such as iron, aluminum, magnesium, and fluorine, and grades made from acid of average impurity content are 11-55-0 for MAP and 18-46-0 for DAP. (*Note*: Grade in fertilizer terminology indicates % N–% P_2O_5–% K_2O by weight in the product.) Processes are available for removing the impurities and color from wet-process acid so that essentially pure salts can also be made from fertilizer-grade acids. The grades of the various ammonium phosphates will vary depending on the amount of impurities in the wet-process phosphoric acid. In 1989, the reported fertilizer production in the United States of DAP was about 13.5 million tons and of MAP was about 2 million tons.[1]

PROCESS DESCRIPTION

Either solid or fluid ammonium phosphate fertilizers can be produced. Ammonium phosphates are made by reacting phosphoric acid with anhydrous ammonia. Commercial grades of phosphoric acid usually contain about 30% P_2O_5 (filter grade), 54% P_2O_5 (merchant grade), or 70% P_2O_5 (superphosphoric acid). After ammoniation, the molten material is either dissolved in aqua ammonia to make a fluid fertilizer or solidified to make solid fertilizer by a granulation process involving either a pug-mill mixer or a rotary drum granulator.

Approximately 95% of the solid ammonium phosphates made in the United States are produced using a rotary drum granulation process developed and patented by the Tennessee Valley Authority (TVA). In the TVA process, phosphoric acid is mixed in an acid surge tank with 93–98% sulfuric acid (used for product analysis control) and with recycle and acid from wet scrubbers (see Figure 1). The mixed acids are then partially neutralized with liquid or

*Portions of this discussion were edited from AP-42, U.S. Environmental Protection Agency *Emission Factors Handbook,* Vol 1, Chapter 6.10.3, and International Fertilizer Development Center *Reference Manual IFDC-R.1,* Chapter XIV.

†With assistance from authorities at the Tennessee Valley Authority's National Fertilizer and Environmental Research Center.

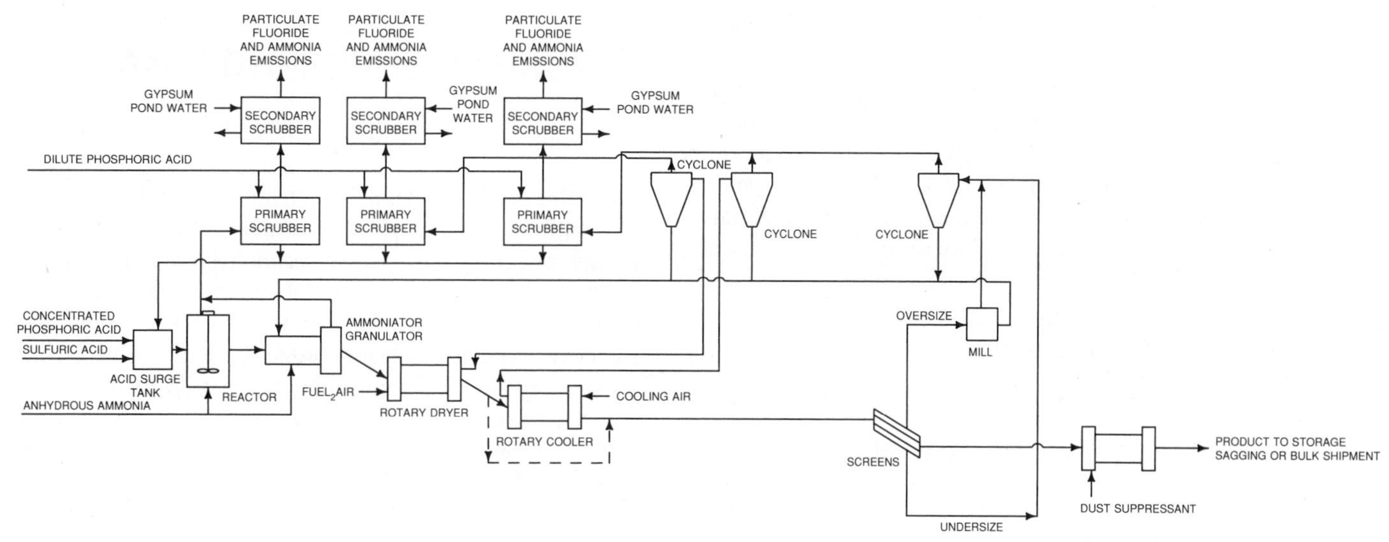

FIGURE 1. Ammonium Phosphate Process Flow Diagram

gaseous anhydrous ammonia in either a tank-type reactor or a pipe reactor. Ammonia-rich off-gases from the reactor and/or ammoniator granulator are wet scrubbed before exhausting to the atmosphere. Primary scrubbers use raw-materials–mixed acid as scrubbing liquor and secondary scrubbers use gypsum pond water.

The basic rotary-drum ammoniator granulator consists of a slightly inclined open-end rotary cylinder with retaining rings at each end and a scraper or cutter mounted inside the drum shell. A rolling bed of recycled solids is maintained in the unit. Slurry from the reactor or pipe reactor is distributed on or in the bed, and the remaining ammonia (approximately 30%) is sparged underneath the bed. Granulation, by agglomeration and by coating particles with slurry, takes place in the rotating drum. Ammonia-rich off-gases pass through a wet scrubber before exhausting to the atmosphere. Moist ammonium phosphate granules are transferred to a rotary, generally cocurrent, dryer and then to a cooler. Before exhausting to the atmosphere, these off-gases pass through cyclones and wet scrubbers.

The granules can be cooled either before or after screening, usually in a double-deck screen in which oversized and undersized particles are separated from product-size particles. The oversized material is usually crushed and then mixed with the finer particles and recycled back to the drum granulator. Most of the product is shipped to the consumer in bulk via railcars or barges; however, a small proportion of MAP or DAP is bagged.

AIR EMISSIONS CHARACTERIZATION

Air emissions from the production of ammonium phosphate fertilizers by ammoniation of phosphoric acid and granulation of the resulting melt occur from five different process operations. The reactor and ammoniator granulator produce emissions of gaseous ammonia, gaseous fluorides (HF and SiF_4), and particulate ammonium phosphates. These two exhaust streams generally are combined and passed through primary and secondary scrubbers. Exhaust gases from the dryer and cooler also contain ammonia, fluorides, and particulates, and these streams commonly are combined and passed through cyclones and primary and secondary scrubbers. Particulate emissions and low levels of ammonia and fluorides from product-sizing and material-transfer operations are controlled in the same way.

Emission factors for ammonium phosphate production are summarized in Table 1. These emission factors are

TABLE 1. Average Controlled Emission Factors for the Production of Ammonium Phosphates[a]

Emission Point	Controlled Emission Factors	
	lb/ton P_2O_5	kg/MT P_2O_5
Reactor/ammoniator granulator		
Fluoride (as F)	0.05	0.02
Particulates	1.52	0.76
Ammonia	b	b
Dryer/cooler		
Fluoride (as F)	0.03	0.02
Particulates	1.50	0.75
Ammonia	b	b
Product sizing and material transfer		
Fluoride (as F)[c]	0.01	0.01
Particulates[c]	0.06	0.03
Ammonia	b	b
Total plant emissions		
Fluoride (as F)[d]	0.08	0.04
Particulates[e]	0.30	0.15
Ammonia	0.14	0.07

[a]See Reference 1.
[b]No information available. Although ammonia is emitted from these unit operations, it is reported as a total plant emission.
[c]Represents only one sample.
[d]The EPA has promulgated a fluoride emission guideline of 0.03 g/kg P_2O_5 input.
[e]Based on limited data from only two plants.

averages based on source test data from controlled phosphate fertilizer plants in Florida.

Exhaust streams from the reactor and ammoniator granulator pass through a primary scrubber in which phosphoric acid recovers ammonia and particulate. Exhaust gases from the dryer, cooler, and screen go first to cyclones for particulate recovery and from there to primary scrubbers. Materials collected in the cyclone and primary scrubbers are returned to the process. The exhaust is sent to secondary scrubbers, where recycled gypsum pond water is used as a scrubbing liquid to control fluoride emissions. The scrubber effluent is returned to the gypsum pond.

AIR POLLUTION CONTROL MEASURES

Primary scrubbing equipment commonly includes venturi and cyclonic spray towers, while cyclonic spray towers, impingement scrubbers, and spray–cross-flow or vertical packed-bed scrubbers are used as secondary controls. Primary scrubbers generally use phosphoric acid of 20–30% P_2O_5 as scrubbing liquor, principally to recover ammonia. Secondary scrubbers generally use gypsum pond water for fluoride control.

Throughout the industry, however, there are many combinations and variations. Some plants use reactor-feed concentration phosphoric acid (40% P_2O_5) in both primary and secondary scrubbers and others use phosphoric acid near the dilute end of the 20–30% P_2O_5 range in only a single scrubber. Existing plants are equipped with ammonia-recovery scrubbers on the reactor, ammoniator granulator, and dryer, and particulate controls on the dryer and cooler. Additional scrubbers for fluoride removal are common, but not typical. Only 15–20% of installations contacted in an Environmental Protection Agency (EPA) survey were equipped with spray–cross-flow packed-bed scrubbers or their equivalent for fluoride removal.

Emission control efficiencies for ammonium phosphate plant control equipment have been reported as 94–99% for ammonia, 75–99.8% for particulates, and 74–94% for fluorides.

References

1. "Production by the U.S. chemical industry," *Chem. Eng. News* (June 18, 1990).
2. *Emission Factors Handbook,* AP-42, Vol. 1, U.S. Environmental Protection Agency, p 4.

NORMAL SUPERPHOSPHATE*

Horace C. Mann†

The solid product obtained by treating natural phosphates with sulfuric acid alone is often designated simply as superphosphate. However, precise technology requires its differentiation from other materials in the superphosphate category. Definitions used for this purpose include the adjectives regular, standard, ordinary, single, and normal superphosphate. The term normal superphosphate is now generally used in statistical publications of the U.S. government. As defined by the Bureau of the Census, normal superphosphate (prepared with the aid of sulfuric acid alone) contains not more than 22% of available P_2O_5. In the United States, the superphosphate manufacturer needs to achieve a minimum of about 15% P_2O_5‡ in a marketed product. There are currently only about eight fertilizer facilities producing normal superphosphates in the United States, with an estimated total production of about 300,000 tons per year.

The raw materials generally used for the production of normal superphosphate are ground phosphate rock and sulfuric acid. The phosphate rock now comes mainly from either the Florida or western phosphate fields; the rock can be ground either at the mining site or at the manufacturing site. Rock for superphosphate manufacture is, for the most part, produced by beneficiation of mined ore.

Important characteristics of rock introduced into the superphosphate manufacturing cycle are those that govern its performance in processing. The most important index of this quality is grade, usually expressed as bone phosphate of lime (BPL), where BPL $\times$ 0.4576 = % P_2O_5. The most skilled operator can use phosphate rock that runs no lower than 68–70% BPL; in a reasonably efficient operation, rock running 75–76% BPL will yield a superphosphate containing 20% available P_2O_5 or more. A factor closely related to grade is the content of iron and aluminum in the phosphate rock. Amounts of Al_2O_3 plus Fe_2O_3 greater than about 5% cannot be tolerated because of the extreme stickiness imparted to the superphosphate. Another widely recognized index of quality is fineness of the phosphate rock. The finer the rock, the more rapid is its reaction with the sulfuric acid.

The two general types of sulfuric acid used in superphosphate manufacture are virgin and spent acid. Virgin acid is produced from elemental sulfur, pyrites, and industrial gases. Spent acid is derived as a waste product in sundry

*Portions of this discussion were edited from AP-42, The Environmental Protection Agency's *Emission Factors Handbook,* Vol 1, Chapter 6.10.1, and U.S. Department of Agriculture and Tennessee Valley Authority, *Superphosphate: Its History, Chemistry, and Manufacture,* December 1964.

†With assistance from authorities at International Fertilizer Development Center, Texasgulf, Inc., and the Tennessee Valley Authority's National Fertilizer and Environmental Research Center.

‡North Carolina requires 18% P_2O_5 minimum; the *Official Journal of the European Communities* lists 16% P_2O_5 as the minimum for normal superphosphate.

industries that use large quantities of sulfuric acid in processing a great variety of products. Problems that arise when using spent acid include unusual color, unfamiliar odor, and toxicity to crops in the normal superphosphate made using this acid.

The major chemical compounds present in normal superphosphate are (1) calcium sulfate ($CaSO_4$) with 0, 0.5, or 2 moles of hydration water; and (2) monocalcium phosphate, monohydrate ($CaH_4P_2O_8 \cdot H_2O$). Minor components are dicalcium phosphate, calcium iron phosphate, calcium aluminum phosphate, and such inert materials as silicon, fluosilicate salts, unreacted rock, organic matter, and phosphates of other metals present in the rock. Apatite is the major constituent of the phosphate rock that reacts with the sulfuric acid to form these chemical compounds.

As in any manufacturing process, the main operating objective is to make an acceptable product at as low a cost as possible. The principal factors affecting the acceptability of the product in superphosphate manufacture are as follows:

1. Physical condition, which is affected adversely by high moisture and free acid content.
2. Content of available phosphate when the product is sold on a guaranteed minimum analysis basis. It is desirable to have as high a phosphate content as possible, and it is helpful to the user to have a fairly uniform available P_2O_5 content from shipment to shipment.
3. Ammoniating characteristics. The product should be sufficiently porous and reactive to allow good ammoniation when it is used in making mixed fertilizer.
4. Graininess. There appears to be a growing demand, in the United States and abroad, for a grainy rather than a powdery product. Aside from the improvement in physical condition, this makes the product more acceptable for the various uses to which nongranular superphosphate is put.

Considerable emphasis is placed on the efficient use of raw materials, especially with regard to getting a high conversion of phosphate to the available (citrate soluble) form. Finding the ratio of acid to rock that gives a good product at the lowest overall raw-material cost is an important operating objective. The main process variables in the manufacture of superphosphate are acid temperature and concentration and acid–rock ratio. The particle size of the

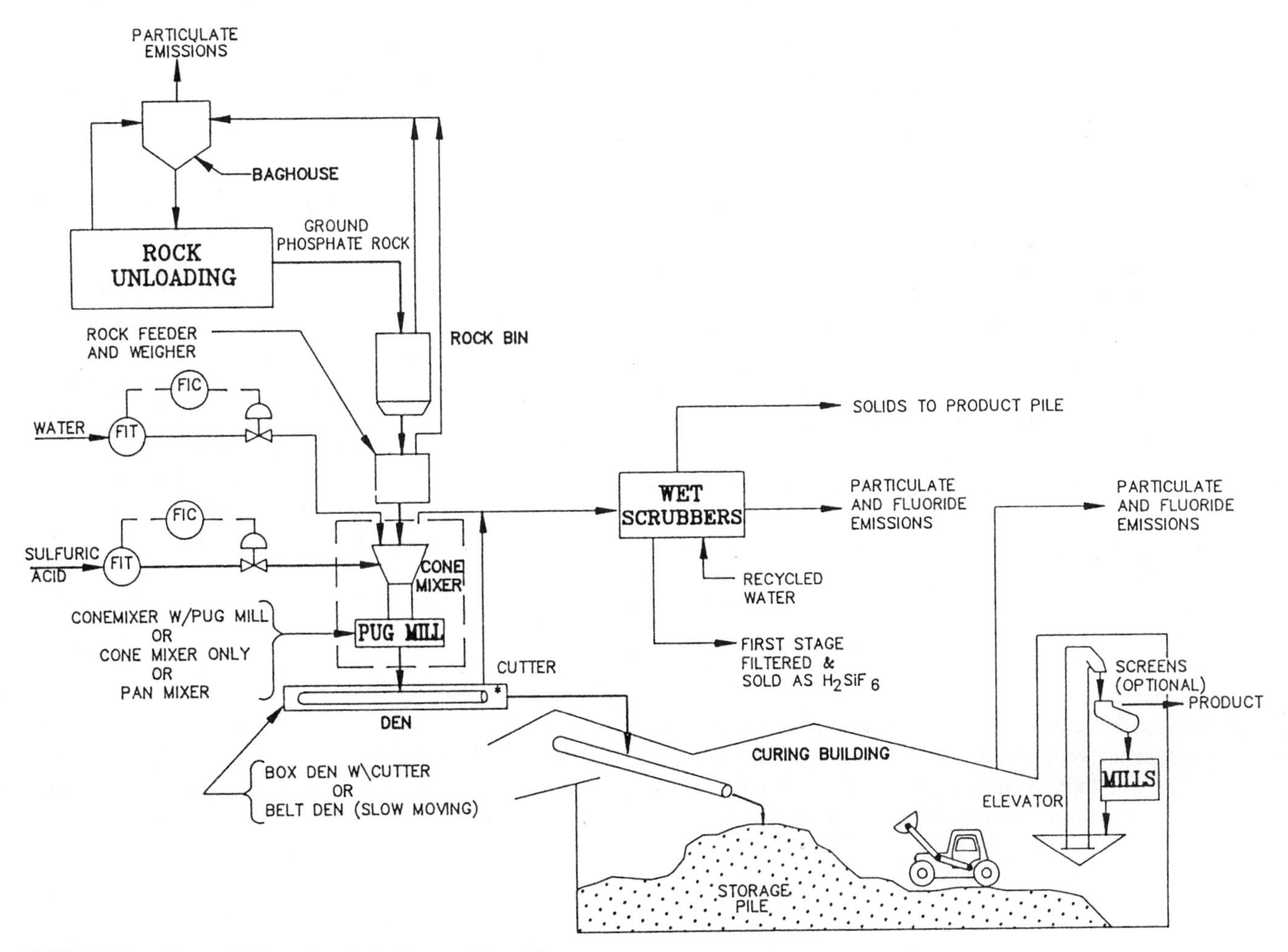

FIGURE 1. Normal Superphosphate Process Flow Diagram

rock is important also, as are the grade and type of rock and acid used.

PROCESS DESCRIPTION

The term normal superphosphate is used to designate a fertilizer material containing 15–21% P_2O_5. It is prepared by reacting ground phosphate rock with 65–75% sulfuric acid. Rock and acid are mixed in a reaction vessel, held in an enclosed area (den) while the reaction goes to partial completion (approximately 30 minutes), and transferred to a storage pile, where the reaction continues. Following storage for "curing," the product is most often used as a raw material in the production of granular fertilizers. It can also be granulated for sale as granular superphosphate. To produce granular normal superphosphate, cured superphosphate is fed through a clod breaker and sent to a rotary drum granulator, where steam, water, and acid may be added to aid in granulation. The material then passes through a rotary dryer and a rotary cooler, is screened to specification, and is sent to storage for sale as bulk or bagged product. A generalized flow diagram of the process for the production of normal phosphate is shown in Figure 1.

AIR EMISSIONS CHARACTERIZATION

Sources of emissions at a normal superphosphate plant include rock unloading and feeding operations, the mixer (reactor), the den, the curing building, and during fertilizer-handling operations. Rock unloading, handling, and feeding generate particulate emissions of phosphate rock dust. The mixer, den, and curing building emit gaseous fluorides (HF and SiF_4) and particulates composed of fluoride and phosphate material. Fertilizer-handling operations release fertilizer dust. Emission factors for the production of normal superphosphate are presented in Table 1. These emission factors are averages based on source test data from controlled phosphate fertilizer plants in Florida.

AIR POLLUTION CONTROL MEASURES

At a typical normal superphosphate plant, the emissions from the rock unloading, handling, and feeding operations are controlled by a baghouse. The mixer and den emissions are controlled by a wet scrubber. The emissions from the curing building and fertilizer handling operations normally are not controlled.

Particulate emissions from ground rock unloading, storage, and transfer systems are controlled by baghouse collectors. These cloth filters have reported efficiencies of over 99%. Collected solids are recycled to the process.

TABLE 1. Emission Factors for the Production of Normal Superphosphate[a]

Emission point	Pollutant	Emission Factor	
		lb/ton P_2O_5	kg/MT P_2O_5
Rock unloading[b]	Particulate	0.56	0.28
Rock feeding[b]	Particulate	0.11	0.06
Mixer and den[c]	Particulate	0.52	0.26
	Fluoride	0.20	0.10
Curing building[d]	Particulate	7.20	3.60
	Fluoride	3.80	1.90

[a]See reference 1.
[b]Factors are for emissions from baghouse with an estimated collection efficiency of 99%.
[c]Factors are for emissions from wet scrubbers with a reported 97% control efficiency.
[d]Uncontrolled.

Silicon tetrafluoride and hydrogen fluoride emissions and particulate from the mixer, den, and curing building are controlled by scrubbing the off-gases with recycled water. Gaseous silicon tetrafluoride in the presence of moisture reacts to form gelatinous silica, which has the tendency to plug scrubber packings. The use of conventional packed countercurrent scrubbers and other contacting devices with small gas passages for emissions control therefore is limited. Scrubber types that can be used are cyclonic, venturi, impingement, jet ejector, and spray cross-flow packed scrubbers. Spray towers also find use as precontactors for fluorine removal at relatively high concentration levels (greater than 3000 ppm, or 4.67 g/m^3).

Air pollution control techniques vary with particular plant designs. The effectiveness of abatement systems in the removal of fluoride and particulate also varies from plant to plant, depending on a number of factors. The effectiveness of fluorine abatement is determined by (1) inlet fluorine concentration, (2) outlet or saturated gas temperature, (3) composition and temperature of the scrubbing liquid, (4) scrubber type and transfer units, and (5) effectiveness of entrainment separation. Control efficiency is enhanced by increasing the number of scrubbing stages in series and by using a fresh-water scrub in the final stage. Reported efficiencies for fluoride control range from less than 90% to over 99%, depending on inlet fluoride concentrations and the system employed. The scrubbed fluorides are collected as hydrofluosilicic (H_2SiF_6) acid and are sold for municipal water fluoridation. An efficiency of 98% for particulate control is achievable.

Reference

1. *Emission Factors Handbook,* Vol. 1, U.S. Environmental Protection Agency, p 6.

TRIPLE SUPERPHOSPHATE*

Horace C. Mann†

Triple superphosphate, also known in the trade as double, treble, or concentrated superphosphate, is produced by reacting phosphate rock‡ with phosphoric acid or with a mixture of phosphoric and sulfuric acids in which phosphoric acid preponderates. The phosphoric pentoxide (P_2O_5) content of the acid ranges up to about 55%; the available P_2O_5 in the superphosphate runs 40% to 49%, with the bulk of production in the range 45% to 47%. Wet-process phosphoric acid, which is produced by treating phosphate rock with an excess of sulfuric acid, is commonly used and usually contains 1–5% of sulfuric acid.

The major chemical component of concentrated superphosphate is monocalcium phosphate monohydrate ($CaH_4P_2O_8 \cdot H_2O$). However, there are other constituents, as shown in the following analyses, of commercial concentrated superphosphate made from wet-process phosphoric acid.

$Ca(H_2PO_4)_2 \cdot H_2O$ or $CaH_4P_2O_8 \cdot H_2O$, 63–73%
$CaSO_4$, 3–6%
$CaHPO_4$ and Al phosphates, 13–18%
Silica, fluosilicates, unreacted rock, organic matter, 5–10%
Free moisture, 3–6%

The chemical composition of the reactants, rock and acid, as well as the physical characteristics of the rock, affect the type and amounts of compounds formed; for example, if furnace-type phosphoric acid is used, sulfate is practically eliminated and the content of the other impurities is reduced. Other factors influencing the quality of the product and the proportions of components in the product include the ratio of reactants, the reaction temperature, the degree of mixing, and the curing temperature.

As of 1990, there were only six fertilizer facilities in the United States with the capability to produce triple superphosphate. One of the plants is located in North Carolina, four in Florida, and one in Idaho. All use phosphate rock mined and beneficiated near their facilities, except for the plant in Idaho, which uses limestone instead of phosphate rock to minimize fluorine evolution.[1] In 1989, an estimated 3.5 million tons of triple superphosphate were produced.[2]

PROCESS DESCRIPTION

The two principal types of triple superphosphate are run-of-the-pile (ROP) and granular; at this time, no facilities are producing ROP triple, but a description of the process is presented for information purposes. The ROP type is essentially a pulverized mass of variable particle sizes produced in a manner similar to that for producing normal superphosphate. Phosphoric acid (50% P_2O_5) is reacted with ground phosphate rock in a cone mixer. The resultant slurry begins to solidify on a slowly moving conveyor (den) en route to the curing area. At the point of discharge from the den, the material passes through a rotary mechanical cutter that breaks up the solid mass. Coarse product is sent to a storage pile and cured for three to five weeks. The final product is mined from the "pile" in the curing shed and then crushed, screened, and shipped.

Granular triple superphosphate yields larger, more uniform particles with improved storage and handling properties. In this process, shown in Figure 1, ground phosphate rock (67–73% BPL, 85%–200 mesh) or limestone (–50 mesh) is reacted with phosphoric acid in one or two reactors in series. The phosphoric acid used in this process is appreciably lower in concentration (40% P_2O_5) than that used to manufacture ROP product. The lower-strength acid helps to maintain the slurry in a fluid state during a mixing period that ranges from 10–15 minutes to up to two hours. A thin slurry is continuously removed and distributed onto dried, recycled fines in the granulator, where it coats the granule surfaces and builds up size.

Pug mills and rotating drum granulators have been used in the granulation process. Currently, there is only one pug mill operating in the United States. The remainder of the producers use drum granulators. The basic rotary drum granulator consists of an open-ended, slightly inclined rotary cylinder, with retaining rings at each end and a scraper or cutter mounted inside the drum shell. A rolling bed of dry material is maintained in the unit while the slurry is introduced through distributor pipes or spray nozzles above the bed. Slurry-wetted granules then discharge to a rotary dryer, where excess water is evaporated. Evaporation of water essentially stops the chemical reaction. Dried granules are then sized on vibrating screens. Oversized particles are crushed and recirculated to the screen; undersized particles are recycled to the granulator. Product size granules are cooled and sent to a storage pile. After a curing period of

*Portions of this chapter were edited from AP-42, U.S. EPA's Emission Factors Handbook, Vol 1, Chapter 6.10.2 and U.S. Department of Agriculture and Tennessee Valley Authority, *Superphosphate: Its History, Chemistry, and Manufacture*, December 1964.

†With assistance from authorities at IMC Fertilizer, J. R. Simplot Co., Texasgulf, Inc., and the Tennessee Valley Authority's National Fertilizer and Environmental Research Center

‡Bone phosphate of lime (BPL); BPL × 0.4576 = % P_2O_5.

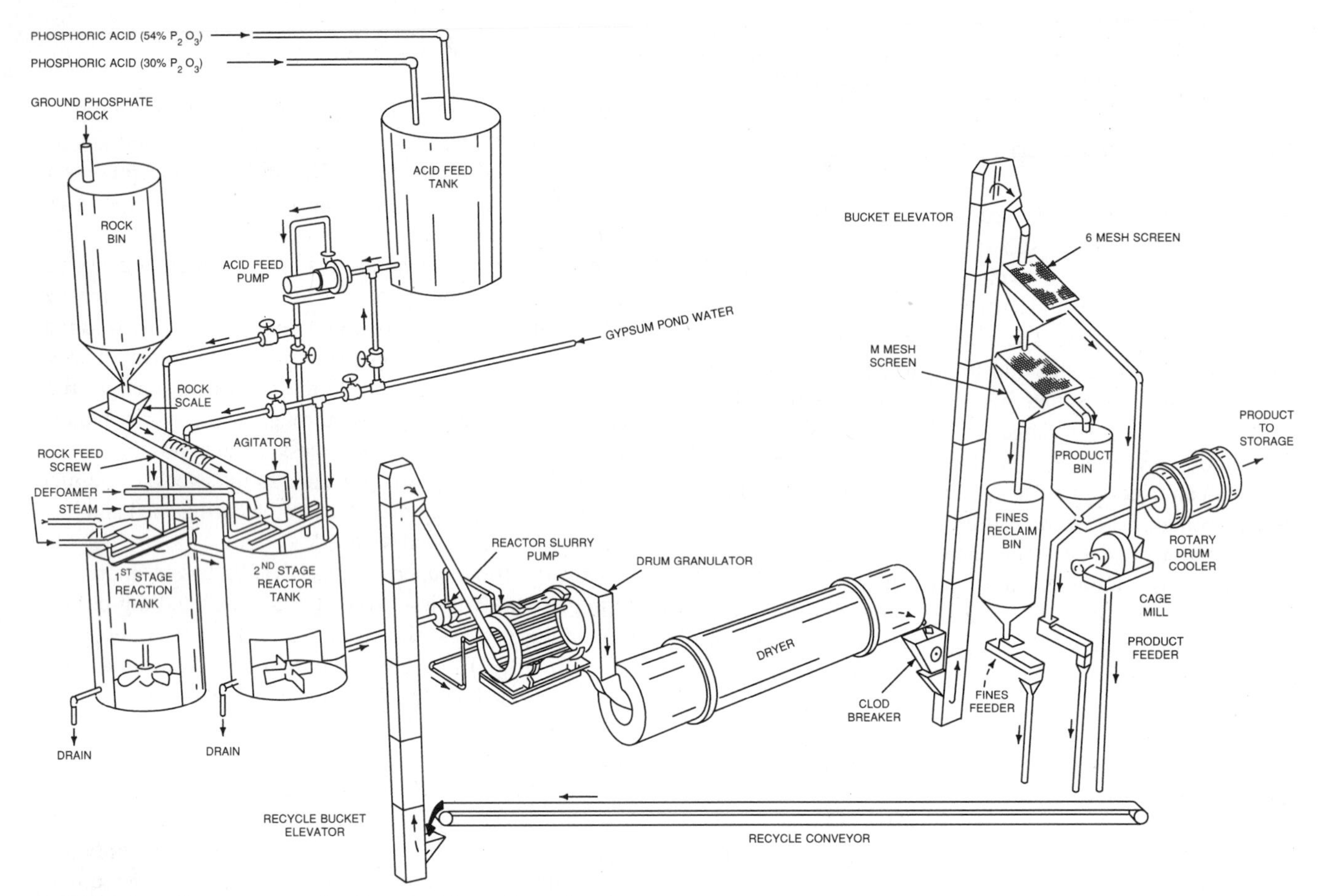

FIGURE 1. Slurry Process for Granular Triple Superphosphate

three to five days, granules are removed from storage, screened, and shipped. The granular product contains about 45–47% P_2O_5 in an available form. Production rates from the various facilities range from about 25 to 100 tons per hour.

AIR EMISSION CHARACTERISTICS

Sources of particulate emissions include the reactor, granulator, dryer, screens, cooler, mills, and transfer conveyors. Additional emissions of particulate result from the unloading, grinding, storage, and transfer of ground phosphate rock. The limestone is received already ground and no further milling is necessary.

Emissions of fluorine compounds and dust particles occur during the production of granular triple superphosphate. Silicon tetrafluoride and hydrogen fluoride are released by the acidulation reaction from the reactors, den, granulator, and dryer. Evolution of fluoride is essentially finished in the dryer and little fluoride evolves from the storage pile in the curing building.

AIR POLLUTION CONTROL MEASURES

At a typical plant, baghouses are used to control the fine rock particles generated by the rock grinding and handling activities. Emissions from the reactor, den, and granulator are controlled by scrubbing the effluent gas with recycled gypsum pond water in cyclonic scrubbers. Emissions from the dryer, screens, mills, product-transfer systems, and storage buildings are sent to a cyclone separator for removal of a portion of the dust before going to wet scrubbers to remove fluorides. In some instances, where little particulate is present, cyclonic separators are used to remove both the particulate and fluorides.

Emission factors for the production of granular triple superphosphate are given in Table 1. The fluoride emission factors are the maximum allowable emissions based on federal regulations; in Florida and Idaho, more stringent regulations have been adopted and these are shown also. There currently are no federal emission limits for particulates, but state regulations are given for the three states that have triple manufacturing facilities.

Baghouse collectors (cloth filters) have reported efficiencies of over 99%. Collected solids are recycled to the process. Emissions of silicon tetrafluoride, hydrogen fluoride, and particulate from the production area and curing building are controlled by scrubbing the off-gases with recycled water. Exhausts from the dryer, screens, mills, and curing building are sent first to a cyclone separator and then to a wet scrubber. Tail-gas wet scrubbers are installed to handle final cleanup of the plant off-gases.

TABLE 1. Allowable Emissions from Triple Superphosphate Production Facilities

Process	Emission Point	Maximum Allowable Emissions	
		Fluoride	Particulates
Granular triple superphosphate from phosphate rock and H_3PO_4	Manufacturing facilities (reactors, granulators, dryers, screens, and mills)	Federal regulations[a] 0.2 lb/ton P_2O_5 fed 0.1 kg/MT P_2O_5 fed	None
		North Carolina regulations same as federal regulations	Varies with production rate, i.e., at 50 tons/hour = 31 lb/h; at 100 tons/h = 38 lb/h
		Florida regulations 0.15 lb/ton P_2O_5 fed	E = 17.3 Po.16 when P (production rate, tons/hour > 30 tons/hour E = emissions, lb/h
	Storage facilities (storage for curing piles, conveyors, elevators, screens, and mills)	Federal regulations[a] 0.0005 lb/h ton P_2O_5 stored 0.00025 kg/MT P_2O_5 stored	None
		North Carolina regulations Same as federal regulations	Included in manufacturing facility rate
		Florida regulations 0.05 lb/ton P_2O_5 fed	Included in manufacturing facilities rate
Granular triple superphosphate from limestone and H_3PO_4	Manufacturing facilities (reactors, granulators, dryers, screens, and mills)	Idaho regulations 1.7 lb/h	Site specific
	Storage facilities (storage for curing piles, conveyors, elevators, screens, and mills)	Idaho regulations 0.036 lb/h	Site specific

[a]*Source*: Federal regulations (for facilities that began construction or modifications after 10/22/74) 40FR33155 August 6, 1975.

Gaseous silicon tetrafluoride in the presence of moisture reacts to form gelatinous silica, which has the tendency to plug scrubber packings. The use of conventional packed countercurrent scrubbers and other contracting devices with small gas passages for emissions control is, therefore, limited. Scrubber types that can be used are (1) spray tower, (2) cyclonic, (3) venturi, (4) impingement, (5) jet ejector, and (6) spray–cross-flow packed.

Air pollution control equipment varies with particular plant designs. The effectiveness of abatement systems for the removal of fluoride and particulate also varies from plant to plant, depending on a number of factors, including (1) inlet fluorine concentration, (2) outlet or saturated gas temperature, (3) composition and temperature of the scrubbing liquid, (4) scrubber type and transfer units, and (5) effectiveness of entrainment separation. Control efficiency is enhanced by increasing the number of scrubbing stages in series and by using a freshwater scrub in the final stage. Reported efficiencies for fluoride control range from less than 90% to over 99%, depending on inlet fluoride concentrations and the system employed. An efficiency of 98% for particulate control is achievable.

In routine operation, all of the facilities are reported to meet or produce less emission than is allowable by the appropriate federal or state regulations.

References

1. Larry Bierman and Bary Long, U.S. Patent No. 4,101,63, July 18, 1978.
2. North American Fertilizer Capacity Data, TVA/NFDC 89/1, Circular Z-243.

UREA

Horace C. Mann

Urea ($CO[NH_2]_2$) is also known as carbamide or carbonyl diamide. All commercial production of urea is from carbon dioxide and ammonia[1] The carbon dioxide is obtained as a by-product from ammonia production. The reaction proceeds in two steps: (1) formation of ammonium carbamate and (2) dehydration of ammonium carbamate.

$$2NH_3 + CO_2 \rightarrow NH_2CO_2NH_4$$

$$NH_2CO_2NH_4 \rightarrow CO(NH_2)_2 + H_2O$$

The urea solution produced by this synthesis reaction usually contains about 75% urea. It is then processed further into either a solid or a fluid that is stable at room temperature; a urea ammonium nitrate solution made using either the urea solution or solid urea usually contains about an equal amount of nitrogen derived from ammonium nitrate to improve its low-temperature storage properties. Solid urea (prills, granules, and crystals) contains 46% total nitrogen, while the commonly used nitrogen solutions contain 28–32% total nitrogen.

Biuret (NH_2-CO-NH-CO-NH_2) is formed during urea synthesis and in the processing of solutions containing urea following the synthesis. It is toxic to citrus plants and some other crops when applied as a foliar spray. For foliar application on citrus crops, urea containing less than 0.25% biuret is preferred. Urea of very low biuret content (less than 0.25%) can be obtained by vacuum crystallization of urea solution. For other crops, sensitivity to biuret in foliar sprays varies widely; solutions made from urea containing 1.5% biuret were considered acceptable for foliar application to maize or soybeans. For most other urea fertilizer uses, biuret content up to 2% is of no consequence; it decomposes in the soil and its nitrogen content becomes available to plants. Biuret is preferred to urea for use in cattle feed as a protein substitute.

On a worldwide basis, urea is now the most popular solid nitrogen fertilizer, and its use is growing more rapidly than that of other materials. About 6 million tons of urea (calculated as solid urea, but composed of urea solution and solid urea) were produced in the United States in 1989. About 85% was used in fertilizers (either in solid or solution form), 3% in animal feed supplements, and the remaining 12% in plastics and for other uses. About 35 plants producing from 25 to 1000 tons per day of urea were in operation in 1986.[2] Urea, usually solid, is also imported into the United States.

PROCESS DESCRIPTION

The process for manufacturing urea involves a combination of up to seven major unit operations. These operations, illustrated by the flow diagram of Figure 1, are solution synthesis, solution concentration, solids formation, solids cooling, solids screening, solids coating, and bagging and/or bulk shipping.

The combination of processing steps is determined by the desired end products. For example, plants producing urea solutions use only the solution formation and bulk shipping operations. Facilities producing solid urea employ these two operations and various combinations of the remaining five operations, depending on the specific end product being produced.

In the solution synthesis operation, ammonia and carbon dioxide are reacted to form ammonium carbamate. Three basic types of synthesis operations are marketed by a number of companies that involve various procedures for handling the unreacted ammonia and carbon dioxide. The methods are (1) once-through, (2) partial recycle, and (3) total recycle of the unreacted NH_3 and CO_2. Most new plants now use total-recycle processes. However, once-through or partial-recycle processes have been popular in some countries and, in some cases, still may be preferred.

Once-Through and Partial-Recycle Processes

The once-through method is the simplest and least expensive (in both capital investment and operating costs) of the three basic types of processes. A typical unit flowchart (synthesis section only) is given in Figure 2. Liquid ammonia and gaseous carbon dioxide are pumped into the urea reactor at about 200 atm. The reactor temperature is maintained at about 185°C by regulating the amount of excess ammonia; about 100% excess NH_3 is required, and about 35% of the total NH is converted to urea (75% of the CO_2 is converted). The reactor effluent solution contains about 80% urea after carbamate stripping. The unconverted NH_3 and CO_2 are driven off at moderate pressures by steam heating the effluent solution in the carbamate strippers.

While this process is the simplest of the urea processes, it is the least flexible and cannot be operated unless some provision is made to utilize the large amount of off-gas ammonia. It is thus tied to the coproduction of some other material, such as ammonium sulfate, ammonium nitrate, nitric acid, or ammonium phosphate, for which the ammonia can be used.

In the partial-recycle process, Figure 3, part of the off-gas ammonia and carbon dioxide from the carbamate strippers is recycled to the urea reactor. Recycling is accomplished by absorbing the stripper gases in a recycle stream of partially stripped urea effluent, in process-steam condensate, or in mother liquor from a crystallization finishing process. In this manner, the amount of NH_3 in the off-gas is reduced. Any proportion of the unreacted ammonia can be recycled; typically, the amount of ammonia that must be used in some other process is reduced to about 15% of that from a comparable once-through unit.

In a typical process, liquid NH_3 and gaseous CO_2 are pumped to the urea reactor at 200 atm. The temperature of the reactor is maintained at about 185°C by proper balance of excess NH_3 and carbamate solution recycle feed. About 100–110% excess NH_3 is used; about 70% of the NH_3 and 87% of the CO_2 are converted to urea. The remaining 30% of the NH_3 must be used in some other process. The reactor effluent contains about 80% urea.

Unreacted NH_3 and CO_2 are separated from the urea solution in the high-pressure separator and in two to three steam-heated carbamate strippers at successively lower pressures. The off-gas from the separator and the first-stage

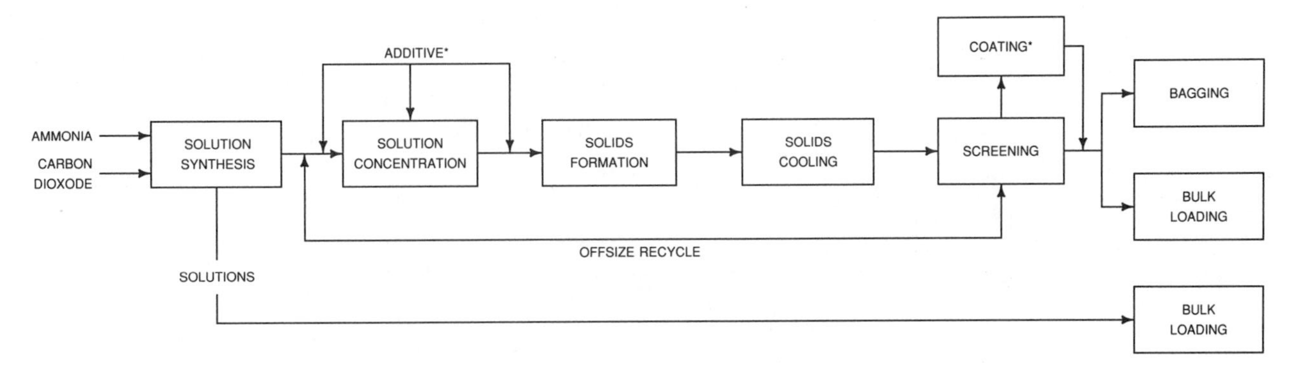

FIGURE 1. Major Urea Manufacturing Operations

stripper is absorbed in the high-pressure absorber by a side stream of partially stripped reactor effluent from the high-pressure separator. Heat evolved in the absorber reaction is removed (to increase absorption capacity) by the addition and expansion of part of the liquid ammonia feed at this point. Pure gaseous ammonia from the top of the absorber is also recycled to the urea reactor after being condensed.

Since the amount of ammonium carbamate that can be absorbed in the absorber solution described above is limited by its solubility in the system H_2O–urea—NH_3, part of the ammonia and carbon dioxide cannot be recycled and must be used in the production of a coproduct nitrogen material. As in the once-through process, the operation of the urea plant still must coincide with that of the coproduct plant.

Other partial-recycle processes differ in detail, but accomplish similar results. While the investment cost is somewhat lower than that for total recycle, this advantage apparently does not compensate for the inflexibility arising from the necessity to operate a coproduct plant with mutual interdependency problems.

Total-Recycle Processes

In total-recycle processes, all the unconverted ammonia–carbon dioxide mixture is recycled to the urea reactor (conversion is about 99%) and no nitrogen coproduct is necessary. This is the most flexible of the urea processes because it depends only on the CO_2 and NH_3 supply from its supporting ammonia plant for operation. However, it is also the most expensive in terms of investment and operating costs. Therefore, if the production of other materials requiring ammonia is planned, an integrated once-through or partial-recycle unit would have lower investment and perhaps lower operating costs. The disadvantages are decreased reliability arising from mutual dependence of two plants, inflexibility in the proportions of coproducts, and difficulties in synchronizing the operation of two plants. Because of these difficulties, most manufacturers prefer a total-recycle process, even when a second nitrogen product is desired.

Total-recycle processes can be classified in five groups according to the recycle principle: (1) hot-gas mixture recycle, (2) separated-gas recycle, (3) slurry recycle, (4) carbamate-solution recycle, and (5) stripping. The first four groups use carbamate decomposition steps basically similar to those of the once-through and partial-recycle processes, whereas the last one differs even in this respect.

Some years ago, there was considerable variation in the design features of the various solution-recycle processes; these differences have gradually disappeared. Today, most of the "conventional" processes are much the same. All use similar reactor conditions (temperature about 185°C and pressure about 200 atm), maintain an $NH_3:CO_2$ mole ratio of about 4:1 in the synthesis loop, and get about the same conversion (65–67%) of CO_2 to urea for each pass through the synthesis reactor. Overall conversion of NH_3 to urea is

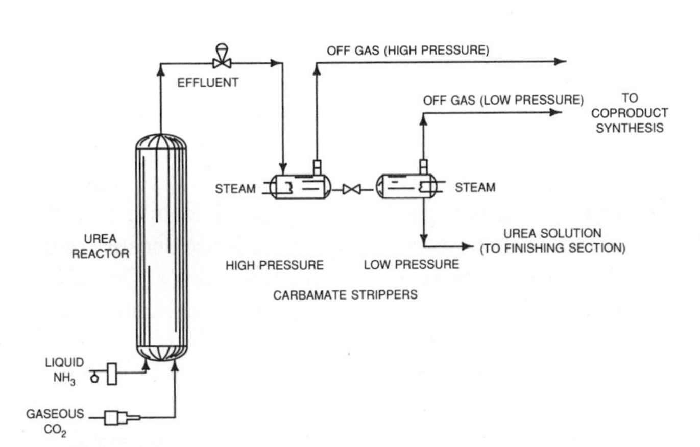

FIGURE 2. Typical Once-Through Urea Process

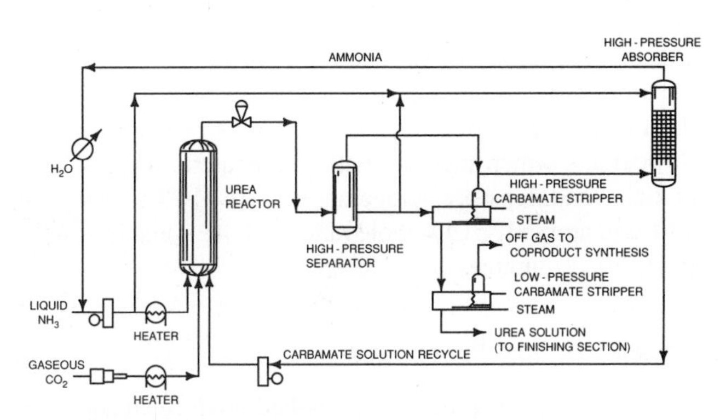

FIGURE 3. Typical Partial-Recycle Process

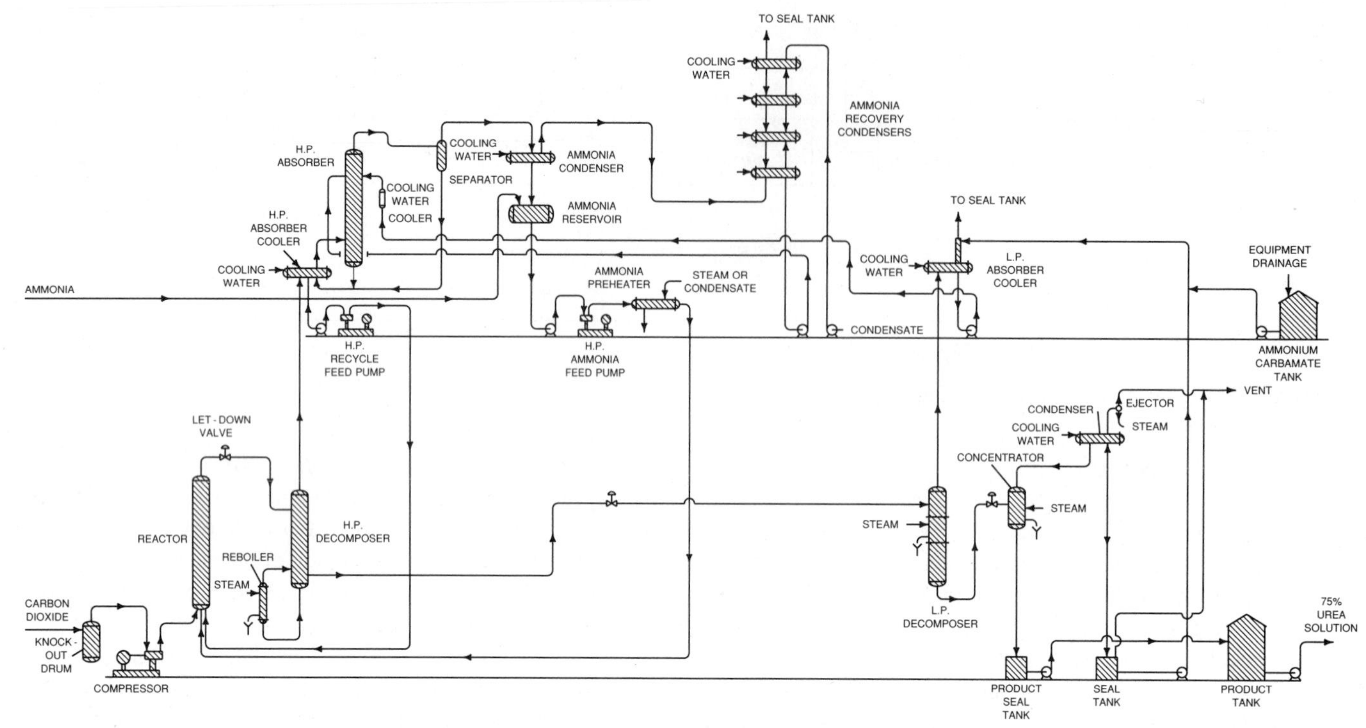

FIGURE 4. Typical Total-Recycle Urea Process

99% or more. All reduce the reactor effluent pressure to an intermediate level and then pass the solution through two or three stages of decomposition (by heating) at successively lower pressure levels. In each stage, the evolved gas mixture is condensed (or absorbed in weak solution condensed in a later stage), and the resulting solutions are worked back through the system to the reactor. The excess ammonia (from the excess used in the initial reactor feed) passes through the absorbers, is condensed, and is fed back to the reactor.

Because of the large number of processes offered and frequent changes in flowcharts, it is not feasible to show flow diagrams for all processes. Figure 4 shows a flow diagram for one total-recycle process. No implication that this process is necessarily superior to others is intended.

Optimum Conditions

It is generally not economical to maximize the percentage urea conversion in the reactor since this would require an excessive retention time. The aim, therefore, is to attain maximum quantity of urea production per unit of time with due regard to the cost of increased reactor size and corrosion difficulties, which increase with temperature. Typical operating conditions are temperature, 180–210°C; pressure, 140–250 atm; $NH_3:CO_2$ mole ratio, 3:1–4:1; and retention time, 20–30 minutes.

Urea Finishing Processes

The urea synthesis processes described produce an aqueous solution containing about 75% urea. The solution can be used directly to prepare nitrogen fertilizer solutions. It also can be used to prepare granular compound fertilizers, although further concentration usually is desirable for this purpose.

Methods of producing solid urea include flaking, prilling, granulation, crystallization, and a combination of crystallization followed by melting and prilling or granulation.

Flaking

Flaking is done by evaporating the solution to a melt (less than 1% H_2O) and solidifying the melt on a cooled metallic surface such as a Sandvik conveyor. The solidified melt is broken up into flakes. This is a convenient means for producing a solid material for shipment when the product is to be used in industrial processes or in solutions. Flaked urea is not used directly for fertilizer.

Crystallization

In the vacuum crystallization process, the crystals are separated from the mother liquor by centrifuging and then dried, usually in a rotary fuel-fired dryer. In some countries, crystalline urea is bagged and distributed for use as fertilizer, but it is not very satisfactory for this use because of the small size of the crystals, which leads to caking. Crystalline urea also may be used for preparing fertilizer solutions for foliar application or for nonfertilizer purposes.

Prilling

Until recently, nearly all straight urea was prepared for fertilizer use by prilling, and this is still the most widely used process. The 75% urea solution is evaporated to a melt and prilled using one of two drop-forming devices: (1) a "shower-head" spray consisting of a number of pipes with holes drilled in them or (2) a rotating perforated bucket, usually conical in shape. Alternatively, the urea melt may be formed by melting crystalline urea.

The droplets formed by the prilling devices cool and solidify as they fall through an ascending airstream. As urea has a lower melting point than other fertilizer materials that are prilled, its prills in general are smaller than the usual granular fertilizers. Prilled urea is also weaker than granular urea, in both crushing strength and resistance to abrasion.

Granulation

Granulation of urea is carried out commercially in either a pan or rotary drum granulator or in a spouted-bed granulator. All processes involve spraying molten urea on a moving bed of fines. The product-size urea is then removed, cooled, and sent to storage, while the fines and crushed oversized granules are returned to the granulator.

In addition to improved strength, granular urea has the advantage of greater flexibility in particle size; any desired size can be made, at least in the range of 1.5 to 15.0 mm. For instance, a size-matching granular diammonium phosphate or KC1 (1.5–3.3 mm or 2.5–4.0 mm) can be produced for bulk blending or a larger size (6–10 mm) can be produced for aerial forest fertilization or for other special uses, such as deep placement in flooded rice fields. Another advantage of granulation over prilling is the greater ease of control of fume and dust.

The advantages of granulation as compared with prilling are sufficiently important that most new urea plants built in the United States and Canada use granulation, and granulation facilities have been added to older plants that previously used prilling.

Effect of Finishing Process on Biuret Content

Several investigators have found that biuret is not formed in detectable amounts during the prilling or granulation operation itself, but during the preparation of the melt, whether by evaporation of solution or by melting crystalline urea. Also, biuret continues to form in the melt so long as it remains in molten form. The practice of recycling recovered urea dust and fume to the evaporation step tends to increase the biuret content. Granular urea produced in U.S. plants averages about 1.5% biuret. The biuret content of prilled urea ranges from 1.0 to 1.6%.

Low-biuret product can be produced by vacuum crystallization of urea, followed by drying of the crystals and then remelting them for prilling or granulation.

Conditioning

Prilled or granular urea is usually treated with a conditioner to increase its physical strength and resistance to caking. The most popular conditioning treatment at present is the addition of 0.3–0.4% formaldehyde (either as a 37% aqueous solution or as UF Concentrate 85) to the concentrated solution just before prilling or granulating. It is believed that formaldehyde reacts to form mainly methylene diurea—no formaldehyde has been reported to have been found in the solid urea. Several manufacturers have recently begun using a calcium lignosulfonate in place of formaldehyde; urea containing calcium lignosulfonate[3] is tan in color and produces a somewhat stronger product.

Coating of prills and granules also is used as a conditioning treatment, although this use is decreasing because the formaldehyde treatment is generally preferred. Coating materials include powdery materials such as kieselguhr, china clay, and talc. A combination of about 1% clay with 0.5% oil has been used; the oil promotes adherence of the clay to the urea prills or granules. Some types of powdery coating materials will adhere without oil.

A coating of medium-viscosity oil (without clay) has been developed by the Dutch State Mines for prilled urea to retard moisture absorption in warm, humid climates. In such climates, the free-flowing properties of prilled urea can deteriorate very rapidly after the bag is opened and before the urea is used. The treatment is reported to be effective in preventing rapid deterioration.

The majority of solid urea products are bulk shipped in trucks, enclosed railroad cars, or barges, but approximately 10% is bagged.

AIR EMISSONS CHARACTERIZATION

Emissions from urea manufacture include ammonia and particulate matter. Ammonia is emitted during the solution-synthesis and solids-production processes. Particulate matter is emitted during the solution-concentration, solids-production, and product-handling steps. Formaldehyde emissions from the process are thought to be nil, since it reacts with the hot urea to form methylenediurea.

In the synthesis process, emission control is inherent in the recycle process where carbamate gases and/or liquids are recovered and recycled. Typical emission sources from the solution synthesis process are noncondensible vent streams from ammonium carbamate decomposers and separators. Emissions from synthesis processes generally are combined with emissions from the solution concentration process and are vented through a common stack. Combined particulate emissions from urea synthesis and concentration are much less than particulate emissions from a typical solids-producing urea plant. The synthesis and concentration operations are usually uncontrolled, except for recycle provisions to recover ammonia.

The total of mass emissions per unit is usually lower for feed-grade prill production than for agricultural-grade prills, because of the lower airflows. Uncontrolled particulate emission rates for fluidized-bed prill towers are higher than those for nonfluidized-bed prill towers making agricultural-grade prills and are approximately equal to those for nonfluidized-bed feed-grade prills. Ambient air con-

TABLE 1. Emission for Urea Production[a]

	Particulates				Ammonia			
	Uncontrolled		Controlled		Uncontrolled		Exiting Control Device	
Operation	kg/Mg	lb/ton	kg/Mg	lb/ton	kg/Mg	lb/ton	kg/Mg	lb/ton
Solution formation and concentration	0.0105	0.021	—	—	9.12	18.24	—	—
Solids formation								
Nonfluidized-bed prilling								
Agricultural-grade	1.9	3.8	0.032	0.064	0.43	0.87	—	—
Feed grade	1.8	3.6	NA	NA	NA	NA	NA	NA
Fluidized-bed prilling								
Agricultural grade	3.1	6.2	0.39	0.78	1.46	2.91	—	—
Feed grade	1.8	3.6	0.24	0.48	2.07	4.14	1.04	2.08
Drum granulation	120	241	0.115	0.234	1.07	2.15	—	—
Rotary drum cooler	3.72	7.45	0.10	0.20	0.0256	0.051	NA	NA
Bagging	0.095	0.19	NA	NA	NA	NA	NA	NA

[a]Data from Table 6:14-1, reference 4.

ditions can affect prill tower emissions. Available data indicate that colder temperatures promote the formation of smaller particles in the prill tower exhaust. Since smaller particles are more difficult to remove, the efficiency of prill tower control devices tends to decrease with ambient temperatures. This can lead to higher emission levels for prill towers operated during cold weather. Ambient humidity also can affect prill tower emissions. Airflow rates must be increased with high humidity, and higher airflow rates usually cause higher emissions.

Drum granulators have an advantage over prill towers in that they are capable of producing very large particles without difficulty. Granulators also require less air for operation than do prill towers. A disadvantage of some granulators is their inability to produce smaller feed-grade granules economically. To produce small granules, the drum must be operated at a higher seed-particle recycle rate. It has been reported that, although the increase in seed material results in a lower bed temperature, the corresponding increase in fines in the granulator causes a higher emission rate. Cooling air passing through the drum granulator entrains approximately 10–20% of the product. This airstream is controlled with a wet scrubber that is standard process equipment in drum granulators.

In the solids screening process, dust is generated by abrasion of urea particles and the vibration of the screening mechanisms. Therefore, almost all screening operations used in the urea manufacturing industry are enclosed or have covers over the uppermost screen. As this operation is a smaller source of emissions, particulate emissions from solids screening are not treated here.

Emissions attributable to coating include entrained clay dust from loading and in-plant transfer and leaks from the seals of the coater. No emissions data are available to quantify this fugitive dust source.

Bagging operations are a source of particulate emissions. Dust is emitted from each bagging method during the final stages of filling, when dust-laden air is displaced from the bag by urea. Bagging operations are conducted inside warehouses that are usually vented to keep dust out of the workroom area, according to Occupational Safety and Health Administration regulations. Most vents are controlled with baghouses. Nationwide, approximately 90% of urea produced is bulk loaded. Few plants control their bulk loading operations, but the generation of visible fugitive particles is slight.

TABLE 2. Uncontrolled Particle Size Data for Urea Production[a]

Operation	≤ 10 μm	≤ 5 μm	≤ 2.5 μm
Solution formation and concentration	NA	NA	NA
Solids formation			
Nonfluidized-bed prilling			
Agricultural grade	90	84	79
Feed grade	85	74	50
Fluidized-bed prilling			
Agricultural grade	60	52	43
Feed grade	24	18	14
Drum granulation	b	b	b
Rotary drum cooler	0.70	0.15	0.04
Bagging	NA	NA	NA
Bulk loading	NA	NA	NA

[a]Data from Table 6:14-2, reference 4.

AIR POLLUTION CONTROL MEASURES

The synthesis and concentration operations are usually uncontrolled, except for recycle provisions to recover ammonia.

Urea manufacturers currently control particulate emissions from prill towers, coolers, granulators, and bagging operations. With the exception of bagging operations, urea emission sources usually are controlled with wet scrubbers. The preference for scrubber systems over dry collection systems is due primarily to the easy recycling of dissolved urea collected in the device. Scrubber liquors are recycled to the solution concentration process to eliminate waste disposal problems and to recover the urea collected.

Fabric filter (baghouses) are used to control fugitive dust from bagging operations, where humidities are low and blinding of the bags is not a problem. However, many bagging operations are uncontrolled.

Table 1 summarizes the uncontrolled and controlled emission, by processes, for urea manufacture. Table 2 summarizes particle sizes for these emissions.

References

1. *Fertilizer Manual,* Reference Manual IFDC-1, International Fertilizer Development Center, Muscle Shoals, AL, 1979.
2. Tennessee Valley Authority bulletin Y-195.
3. U.S. Patent 4, 587, 358, May 6, 1986.
4. *Compliation of Air Pollutant Emission Factors,* Vol. 1, AP-42, U.S. Environmental Protection Agency, 1985.

14
Metallurgical Industry

PRIMARY ALUMINUM INDUSTRY

Maurice W. Wei

The process used to make aluminum was discovered in 1886 almost simultaneously by Charles Martin Hall of Oberlin, Ohio, and Paul-Louis-Toussaint Héroult of France. Working independently, Hall and Héroult discovered that alumina (Al_2O_3) dissolved in cryolite (Na_3AlF_6) will undergo electrolysis and produce aluminum at the carbon cathode. The Hall process was commercialized in 1888 when the first smelter to use this technology was built in Pittsburgh, Pennsylvania.

Aluminum is the second most common metallic element in the earth's crust, and economically recoverable quantities of aluminum oxide exist principally as boehmite and gibbsite in bauxite deposits found in many areas of the world. Major deposits are found in Australia, West Africa, Brazil, Jamaica, Surinam, and Venezuela. Aluminum oxide is recovered from bauxite, usually by the Bayer process. The world production of primary aluminum in 1990 was 17,832,000 tonnes (metric tons), of which 4,048,000 tonnes were produced in the United States.

Because the production of aluminum requires a significant amount of electricity, primary production facilities are generally located in countries with relatively low-cost energy and/or a good supply of bauxite. Thus most aluminum plants were originally built in North America, Latin America, and certain areas of Europe. With the discovery of bauxite in Australia and the availability of coal and gas there, the Australian aluminum industry has grown rapidly in the past few years. The U.S.S.R. and the Peoples Republic of China are also major producers of primary aluminum.

Pure aluminum is a relatively soft metal and has low tensile strength. But by alloying aluminum with other metals, such as copper, magnesium, manganese, and silicon, several series of aluminum alloys have been developed with properties tailored to their specific applications. Some properties exhibited by these alloys are a high strength-to-weight ratio, corrosion resistance, and the ability to be extruded, formed, machined, and to accept certain finishes.

In the United States, the major uses for aluminum are in containers and packaging (28.3%), such as beverage and food cans, followed by transportation (17.3%), building and construction (15.9%), electrical (7.8%), consumer durables, machinery, and equipment (12.4%), and others (3.4%). Over 14% of the U.S. primary production was exported.

Aluminum scrap retains high residual value, resulting in a thriving secondary aluminum industry. In 1990, 2,393,000 tonnes of aluminum were recovered from scrap. Reclamation from aluminum cans accounted for about 877,300 tonnes. Over 63% of the estimated 83 billion cans produced in this country were recycled, and this percentage has been going up every year. The target for 1995 is 75%. As energy and other costs go up, and the public's concern over the recyclability of consumer and commercial products increases, the incentive for recycling increases. Recycling aluminum scrap saves 95% of the energy required to make new aluminum from bauxite.

PROCESS DESCRIPTION

Alumina Production

Aluminum is produced by the electrolysis of alumina, and the latter is produced from bauxite by the Bayer process. In the Bayer process, aluminum hydroxides or hydrates are selectively separated from the other components by extracting them with sodium hydroxide to form sodium aluminate. Alumina is then precipitated from the sodium aluminate solution, washed, and calcined (Figure 1).

Bauxite Preparation

To prepare bauxite for processing, mined ore is crushed and ground in ball mills to a finely divided state. Most bauxite, as mined, has a relatively low moisture content. However, in certain parts of the world, the moisture content is very high, and the bauxite must be dried in rotary kilns before it is sent to storage. The drying operation emits significant amounts of bauxite dust requiring dust control equipment. The types of control equipment used are usually mechanical collectors such as a cyclone followed by either a venturi scrubber or an electrostatic precipitator (ESP).

Digestion of Bauxite

To produce high-purity alumina suitable for aluminum reduction cells, finely divided bauxite containing 30–70% alumina is slurried with sodium hydroxide solution and reacted at a high temperature and pressure in reactors called digesters. Sodium aluminate is formed according to the general equation

$$Al_2O_3 . x(H_2O) + 2NaOH = 2NaAlO_2 + (1+x)H_2O$$

while leaving behind substantially all of the silicon, iron, titanium, and calcium oxides as insoluble components in the solid waste residue.

Clarification and Precipitation

The hot slurry is usually cooled by flash evaporation, producing steam that is used to preheat incoming slurry. Next, the slurry is processed in a series of clarification steps designed not only to separate the solid residue from the liquor, but also to rid the solution of the impurities that would lower the quality of the aluminum when the alumina later is electrolyzed. These impurities are mainly oxides of silicon and iron.

If the bauxite contains coarse material (mainly sand), it

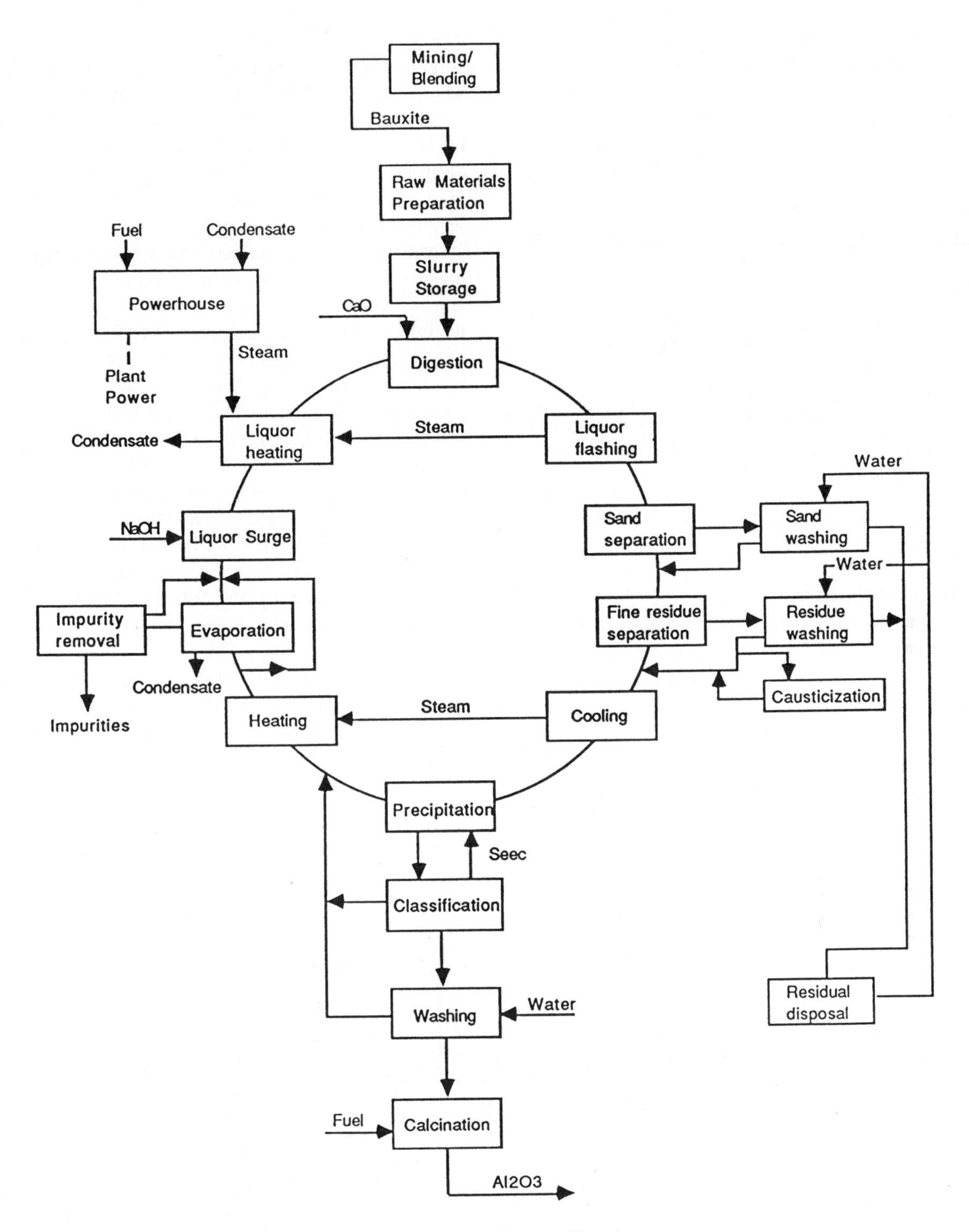

FIGURE 1. Bayer Process Flowsheet

is first removed from the slurry in hydrocyclones, and starch or some settling agent is added to the overflow before it enters a clarifier. Starch promotes agglomeration and settling of the fine particles, and the relatively clear overflow from the clarifier is filtered in either filter presses or sand filters to remove any remaining suspended solids. The clarifier underflow is then washed in thickeners to remove as much caustic as possible before it is pumped to a bauxite residue pond. The filtrate containing sodium aluminate, now at 120°F to 160°F, is processed in precipitators where aluminum hydroxide crystallizes and precipitates as shown in the following.

$$2NaAlO_2 + 4H_2O = 2Al(OH)_3 + 2NaOH$$

After precipitation, the slurry containing trihydrate crystals is sent to classification, where the product is separated according to crystal size. The coarse product is washed to remove excess sodium hydroxide, dewatered on vacuum filters, and sent to calcination. The finer crystals are recycled to the precipitation process as seed, thereby providing nucleation sites for precipitation to occur. Spent liquor from the precipitators is recycled to the digesters.

Calcination

The coarse alumina is calcined in rotary kilns or fluid-bed calciners at about 1800°F. During calcination, water (mechanically and chemically bound) is driven off. Calciners produce hot flue gases containing alumina and water vapor, and the usual types of control equipment consist of cyclonic separators followed by ESPs. The control equipment not only minimizes air pollution, but also recovers valuable product as well. Calcined alumina is stored for shipment to the smelter as feedstock for the aluminum electrolytic cells.

Aluminum Production

Electrolytic Reduction

Aluminum production is carried out in a semibatch manner in large electrolytic cells called pots with a dc input of up to 280,000 amperes at about 5 volts. The pot, a rectangular steel shell ranging in size from 30 to 50 feet long, 9 to 12 feet wide, and 3 to 4 feet high, is lined with a refractory insulating shell on which carbon blocks are placed to form the cathode. Figure 2 shows a cross section of the commonly used center-worked prebaked cell. Steel collector bars are inserted in the bottom blocks to carry current from the pot. The insulation is designed to allow enough heat loss at the vertical walls that a protective layer of frozen electrolyte will form, but not on the bottom where electrical continuity must be maintained between molten aluminum and the cathode block. Molten cryolite is placed in the cavity formed by the cathode blocks, and anodes, also of baked carbon, are immersed in the cryolite, completing the electrical path. There are usually two rows of 20 to 30 closely spaced anodes hung on busbars along the full length of the pot that supplies the current.

Alumina is fed to the pots either from a crane-mounted hopper or from a hopper running the full length of the pot attached to the pot superstructure. A hole is first punched in the frozen cryolite crust on the pot by a crust breaker on the overhead crane or by breakers on the superstructure. Alumina is then metered to the opening in the crust. The anodes can be lowered or raised simultaneously or individually. As the anodes are consumed by combining with oxygen from the alumina, they are lowered into the bath and are replaced when most of the carbon is gone. The carbon butt from the spent anode is allowed to cool and the bath adhering is later removed, usually by shot blasting. The anode butt is then crushed and recycled to make more anodes. Operation is semicontinuous because the pots are fed periodically and tapped about once a day as the aluminum reaches a certain level.

In the Soderberg cells, continuous self-baking anodes are used instead of prebaked anodes (Figure 3). A paste consisting of coal tar pitch and petroleum coke is fed to the top of a steel casing and is baked into an anode by heat from the cell and also by heat produced by the current passing through the anode. Electricity is introduced by the spikes inserted into the anode either vertically or at an angle from the sides. As the lower portion of the anode is consumed, the mass of baked material moves downward through the casing. More

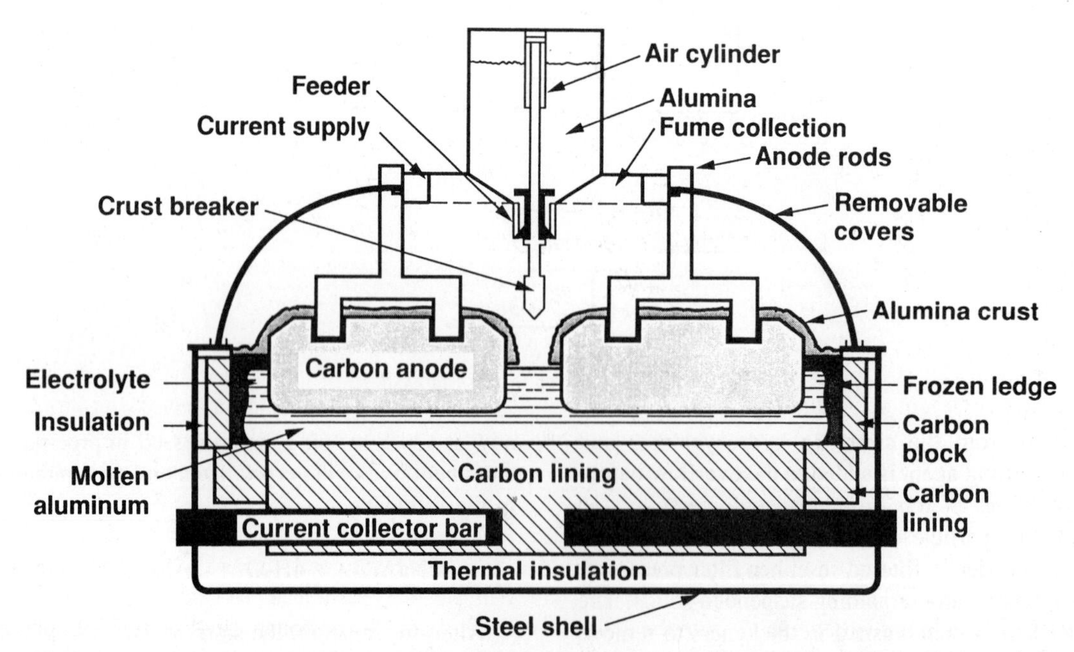

FIGURE 2. Aluminum Electrolyzing Cell—Prebaked Anode (*Kirk Othmer Encyl. of Chem. Tech.*, vol. 2, 4th ed.)

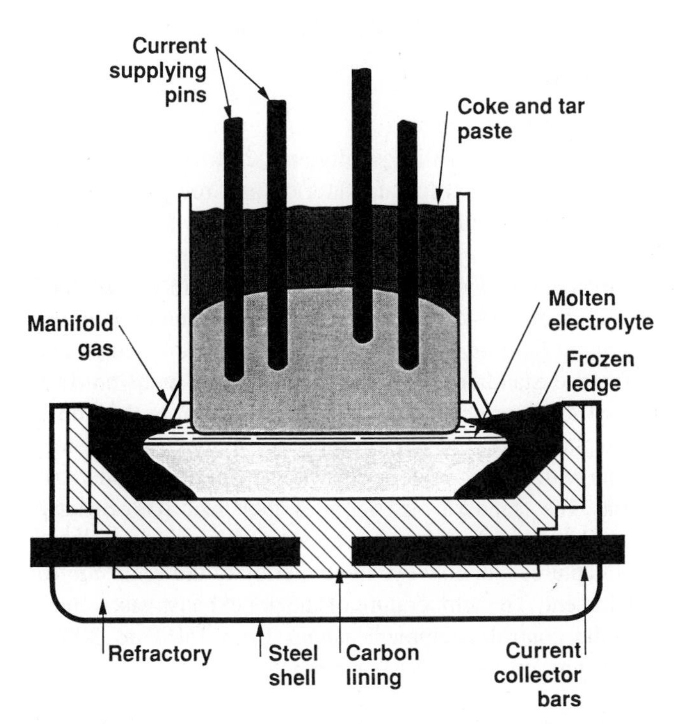

FIGURE 3. Aluminum Electrolytic Cell—Soderberg (*Kirk Othmer Encycl. of Chem. Tech.*, vol 2, 4th ed.)

paste is added to the top, and the spikes must be pulled and reset at a higher level.

Electrolysis of the molten bath involves several complex electrode reactions. However, in the overall reaction, molten aluminum is produced at the cathode and oxygen, released at the anode, reacts with carbon to form carbon dioxide (CO_2).

$$Al_2O_3 + 3/2C = 2Al + 3/2CO_2$$

About 0.4–0.5 pound of carbon are consumed for each pound of aluminum produced.

The major gaseous emissions besides CO_2 are hydrogen fluoride (HF) gas, sulfur dioxide (SO_2), and carbon monoxide (CO). Particulate emissions are particulate fluoride, carbon dust, and alumina. The HF gas is produced by the hydrolysis of fluoride compounds due to adsorbed water, to water of hydration associated with the alumina, and from hydrogen in the anodes. Particulate fluoride results from the vaporization of fluoride compounds from the molten bath and subsequent condensation in the cooler zone above the pot. Sulfur dioxide comes from the sulfur present in the coke and coal tar pitch binder in the anodes. The transportation and handling of the materials used in the process, such as alumina, coke, and pitch, are also sources of emissions.

In the Soderberg cell, unlike the prebake cell, there is continuous evolution of hydrocarbon vapors from the anode paste that not only creates a less desirable working environment, but also makes collection and treatment of the cell gases more difficult. On the cells with vertical studs, gas burners are sometimes installed at the gas collection skirt to destroy some of the pitch fumes prior to the pollution control system. Although Soderberg cells avoid costly anode baking facilities, the majority of new smelters are designed with prebake cells because of their higher current efficiency and more manageable emission problems.

Anode Production

Green Mill

Anodes for prebake pots are prepared in two stages. The first step is the production of green anodes where pitch and petroleum coke are mixed and heated to 140–180°F in either batch or continuous mixers. In the batch mixer, liquid or pencil pitch and coke are added batch-wise, and each batch is heated and blended separately until the mixture is uniformly mixed and at the proper temperature. The batch of anode mix is then discharged and transported in a covered conveyor belt to the anode forming machine. Coke dust and hydrocarbon vapors from pitch evolve during the mixing, discharging, and conveying of the mix to the anode former.

To produce good high-density anodes, an important parameter is the proportion of coarse and fine coke particles used in the mix. To produce different size fractions, the coke is sized in classifiers and transported to different tanks according to particle size. Coke dust is emitted in this operation.

Included with the coke used in this operation are other carbonaceous materials obtained from anode butts. These butts are first mechanically cleaned to remove most of the adhering bath material and then crushed to the required size fractions. Good butt cleaning to remove the fluoride compounds will be discussed in the next section since it affects both fluoride emission and furnace performance.

With the continuous mixer, pitch and coke are fed continuously to the inlet of the mixer, and as the material travels through the mixer, it is heated, blended, and then transported by conveyor belt to the anode forming area. At the anode former, the mix is metered into molds and compressed to form green anodes. Hydrocarbons are also emitted from this operation.

Anode Baking

The second step in anode production is the baking operation. Green anodes are placed in a baking furnace consisting of refractory-lined pits with flues on both sides of each pit. The anodes are then covered with coke and fired for 30 hours or less at temperatures up to 2400°F. Hot flue gas from the firing section is used to preheat a section of packed green anodes located ahead of the fired section. Simultaneously, combustion air for the burners is drawn through the flues of the previously fired section that preheats the air and also cools the baked anodes.

Thus at any given moment, one section is being cooled

while another is being fired and a third section is undergoing preheat. After firing is completed, the fired section is cooled, the preheat section is fired, and a new section is preheated. A movable waste-gas manifold is placed after the preheat section to collect the flue gases and direct them to the waste gas main, which connects to the fume control equipment. In a baking furnace operation, several fires are usually in progress at a given moment. Because firing of the anodes proceeds around the building in a circular manner, the furnace is usually referred to as a ring furnace.

Negative pressure created in the flues by the waste gas fans pulls air through the packing over the anodes into the flues, where it blends with products of combustion. Flue gas from the baking furnace contains HF, SO_2, CO_2, CO, and hydrocarbons from pitch. Particulate emissions are carbon dust and particulate fluoride. Gaseous and particulate fluorides are produced from residual bath material in the anode butts.

Peak hydrocarbon emission occurs when the firing frame containing the burners and the waste-gas manifold are moved. At this point, firing of the preheat section commences and green anodes in the preheat section are exposed to heat for the first time. The temperature of the flue gas will normally range from 300°F to 450°F, with the lower temperature usually occurring when a fire is moved.

As mentioned above, bath material adhering to the anode butts is removed before the butts are crushed and recycled. Thorough removal of the bath is important since it not only will reduce the fluoride emissions, but will provide several other benefits. The HF acid aggressively attacks silica in the furnace refractory; therefore, with lower HF in the flue gases, the furnace flues will require less maintenance. Furnaces in good condition will permit better temperature control and higher finishing temperatures and thus higher-quality anodes. Higher flue gas temperatures also promote better incineration of hydrocarbons in the flue gas, which not only provides heat to the process, but also results in lower emissions.

In contrast, furnaces with poor flue conditions and high air in-leakage invariably produce lower-quality anodes and higher hydrocarbon emissions. This is a good example of the principle that a well-controlled process will usually produce a better product and less pollution.

There is another type of furnace such as the Riedhammer furnace, which by design will produce a tar by-product. The furnace is closed on top to prevent air in-leakage and to create a non-oxidizing environment. Tar in the off-gases is condensed, collected, and sold or reused as binder in green anode production. The environmental control system, therefore, must be capable of collecting tar as well as fluorides. Such a system is described later in the section on baking furnace emission control systems. The design of this furnace is more complex than the open ring furnace, but a more uniform temperature distribution is claimed for the closed furnace.

ENVIRONMENTAL

Potroom

Potroom Emissions

An aluminum pot will typically emit 20 to 35 kg per tonne of gaseous and particulate fluoride and roughly an equal amount of particulate matter. The emissions factors for the various operations in primary aluminum production are summarized in Tables 1 to 4 reproduced from Tables 7.1-2 to 7.1-5 of AP-42 Supplement A to *Compilation of Air Pollutant Emission Factors*.[1] U.S. federal New Source Performance Standards (NSPS) require emissions of no more than 1 kg fluoride per tonne of aluminum from the Soderberg plant and 0.95 kg fluoride per tonne from prebake plants with 0.05 kg fluoride per tonne from the anode bake plant. The standard also requires less than 10% opacity. Each cell is equipped with a hooding and manifold system that captures and conveys the emission to the fume control equipment. The temperature of the pot exhaust gases entering the control equipment ranges from 150°F to 300°F, depending on the ambient temperature.

Pot Hooding

Pot hooding is an integral part of the smelter fume control system. The purpose of the hooding is to enclose the pot as tightly as possible to collect the cell off-gases and, at the same time, to allow the pot attendant to change anodes and perform other functions. This enclosure may consist of individual panels in close contact forming a continuous covering or large continuous covers on each side of the pot. Each panel is removable to allow access to the cell. Cell covers must perform many functions well. They must seal tightly as they lie side by side on the cell, and they should be rugged, but also fairly light, so that they can be easily removed by the operator.

From this hooding, the cell off-gases rise and enter a rectangular enclosure above the cell superstructure from which they enter an exhaust duct. The duct from each cell is connected to a manifold running the length of the potroom that connects to the fume control system. In a well-designed system, the off-gases are exhausted uniformly from the whole cell, and draft from each pot is adjusted so that there is equal volume from each cell.

In theory, hooding efficiency is the amount of off-gas produced by the electrolytic process that is collected by the hooding system as a percentage of the total off-gas generated by the cell.

$$\text{Hooding efficiency (\%)} = \frac{\text{Off-gas collected by hoods} \times 100}{\text{Off-gas produced by cells}}$$

In actual practice, hooding efficiency is calculated as follows:

$$\text{Hooding efficiency (\%)} = \frac{A \times 100}{A + B}$$

TABLE 1. Emission Factors for Primary Aluminum Production Processes[a,b] *(Emission Factor Rating: A)*

	Total Particulate[c]		Gaseous Fluoride		Particulate Fluoride	
Operation	kg/Mg	lb/ton	kg/Mg	lb/ton	kg/Mg	lb/ton
Bauxite grinding						
Uncontrolled	3.0	6.0	Neg	NA		
Spray tower	0.9	1.8	Neg	NA		
Floating-bed scrubber	0.85	1.7	Neg	NA		
Quench tower and spray screen	0.5	1.0	Neg	NA		
Aluminum hydroxide calcining						
Uncontrolled[d]	100.0	200.0	Neg	NA		
Spray tower	30.0	60.0	Neg	NA		
Floating-bed scrubber	28.0	56.0	Neg	NA		
Quench tower	17.0	34.0	Neg	NA		
ESP	2.0	4.0	Neg	NA		
Anode baking furnace						
Uncontrolled	1.5	3.0	0.45	0.9	0.05	0.1
Fugitive	NA	NA	NA	NA	NA	NA
Spray tower	0.375	0.75	0.02	0.04	0.015	0.03
ESP	0.375	0.75	0.02	0.04	0.015	0.03
Dry alumina scrubber	0.03	0.06	0.0045	0.009	0.001	0.002
Prebake cell						
Uncontrolled	47.0	94.0	12.0	24.0	10.0	20.0
Fugitive	2.5	5.0	0.6	1.2	0.5	1.0
Emissions to collector	44.5	89.0	11.4	22.8	9.5	19.0
Multiple cyclones	9.8	19.6	11.4	22.8	2.1	4.2
Dry alumina scrubber	0.9	1.8	0.1	0.2	0.2	0.4
Dry ESP plus spray tower	2.25	4.5	0.7	1.4	1.7	3.4
Spray tower	8.9	17.8	0.7	1.4	1.9	3.8
Floating-bed scrubber	8.9	17.8	0.25	0.5	1.9	3.8
Coated bag filter dry scrubber	0.9	1.8	1.7	3.4	0.2	0.4
Cross-flow packed bed	13.15	26.3	3.25	6.7	2.8	5.6
Dry plus secondary scrubber	0.35	0.7	0.2	0.4	0.15	0.3
Vertical Soderberg stud cell						
Uncontrolled	39.0	78.0	16.5	33.0	5.5	11.0
Fugitive	6.0	12.0	2.45	4.9	0.85	1.7
Emissions to collector	33.0	66.0	14.05	28.1	4.65	9.3
Spray tower	8.25	16.5	0.15	0.3	1.15	2.3
Venturi scrubber	1.3	2.6	0.15	0.3	0.2	0.4
Multiple cyclones	16.5	33.0	14.05	28.1	2.35	4.7
Dry alumina scrubber	0.65	1.3	0.15	0.3	0.1	0.2
Scrubber plus ESP plus spray screen and scrubber	3.85	7.7	0.75	1.5	0.65	1.3
Horizontal Soderberg stud cell						
Uncontrolled	49.0	98.0	11.0	22.0	6.0	12.0
Fugitive	5.0	10.0	1.1	2.2	0.6	1.2
Emissions to collector	44.0	88.0	9.9	19.8	5.4	10.8
Spray tower	11.0	22.0	3.75	7.5	1.35	2.7
Floating bed scrubber	9.7	19.4	0.2	0.4	1.2	2.4
Scrubber plus wet ESP	0.9	1.8	0.1	0.2	0.1	0.2
Wet ESP	0.9	1.8	0.5	1.0	0.1	0.2
Dry alumina scrubber	0.9	1.8	0.2	0.4	0.1	0.2

[a]For bauxite grinding, expressed as kg/Mg (lb/ton) of bauxite processed. For aluminum hydroxide calcining, expressed as kg/Mg (lb/ton) of alumina produced. All other factors are per Mg (ton) of molten aluminum product. NA = not available. Neg = negligible.

[b]Sulfur oxides may be estimated, with an emission factor rating of C, by the following calculations.

Anode baking furnace, uncontrolled SO_2 emissions (excluding furnace fuel combustion emissions):

$20(C)(S)(1-0.01K)$ kg/Mg $[40(C)(S)(1-0.01\ K)$ lb/ton$]$

Prebake (reduction) cell, uncontrolled SO_2 emissions:

$0.2(C)(S)(K)$ kg/Mg $[0.4(C)(S)(K)$ lb/ton$]$

where: C = anode consumption* during electrolysis, lb anode consumed/lb Al produced

S = % sulfur in anode before baking

K = % of total SO_2 emitted by prebake (reduction) cells.

*Anode consumption weight is weight of anode paste (coke + pitch) before baking.

[c]Includes particulate fluorides.

[d]After multicyclone.

Source: Reference 3

where: A = measured total fluoride to the fume treatment system
B = measured total fluoride escaping from the roof monitor

A good pot hooding system should be at least 95% efficient on a center-worked prebaked line.

Much effort has gone into the design of pot hood systems over the years, and new designs are being tested constantly in an effort to improve their performance and durability. It should be mentioned that a primary source of the fluoride emission from a smelter is from around the pot enclosure, either from leaks or when the covers are removed. Significant amounts of fluoride are also emitted to the environment when an anode is removed from the pot and placed in the aisle. The hot carbon continues to air burn for some time while fluorides evolve from the adhering bath material.

Finally, it should be pointed out that fume control systems are expensive, costing $35 to $50 per actual cubic foot per minute of gas treated. Minimizing the exhaust volume from each pot can produce considerable savings in the cost of the control equipment. A well-designed and well-maintained pot hooding will require less exhaust volume per pot and, therefore, a smaller, less expensive fume treatment system.

Data on potroom and anode baking operations, environmental control systems, regulations, and capital cost are summarized in the appendix.

Control Systems

Dry Systems

Most smelters constructed within the last 20 years use alumina to scrub the gaseous fluoride emission and the alumina containing fluoride is then fed to the cells. The two types of dry scrubbing systems in use today differ primarily in the manner in which the potroom gases are brought into contact with alumina. One system capitalizes on the ability of a fluid-bed reactor to promote intimate mixing between solids and fluids. In this process, potroom gases fluidize a bed of alumina. In the other system, the alumina is injected into a zone of high gas velocity and turbulence. Alumina is entrained and turbulent mixing promotes a reaction between HF and alumina.

These dry scrubbing systems are capable of removing over 99% of the HF gas and particulate matter. The alumina containing HF is then used to make aluminum. With the high recovery of fluorides and their subsequent reuse, consumption of fluoride has been reduced significantly in smelters with dry scrubbing systems. Because these dry scrubber systems are so efficient in capturing particulate and gaseous fluorides, the emissions from their stacks contribute very little to the overall plant fluoride emission.

Wet Systems

A few plants still use the older control system consisting of electrostatic precipitators (ESPs) to collect particulate emis-

TABLE 2. Uncontrolled Emission Factors and Particle Size Distribution for Roof Monitor Fugitive Emissions from Prebake Aluminum Cells *(Emission Factor Rating: C)*

Particle size,[a] μm	Cumulative Mass, % ≤ Stated Size	Cumulative Emission Factor	
		kg/Mg Al	lb/ton Al
15	65	1.62	3.23
10	58	1.45	2.90
5	43	1.08	2.15
2.5	28	0.70	1.40
1.25	18	0.46	0.92
0.625	13	0.33	0.67
Total	100	2.5	5.0

[a]Expressed as equivalent aerodynamic particle diameter.
Source: Reference 3.

TABLE 3. Uncontrolled Emission Factors and Particle Size Distribution for Roof Monitor Fugitive Emissions from HSS Aluminum Cells *(Emission Factor Rating: D)*

Particle size,[a] μm	Cumulative Mass, % ≤ Stated Size	Cumulative Emission Factors	
		kg/Mg Al	lb/ton Al
15	39	1.95	3.9
10	31	1.55	3.1
5	23	1.15	2.3
2.5	17	0.85	1.7
1.25	13	0.65	1.3
0.625	8	0.40	0.8
Total	100	5.0	10.0

[a]Expressed as equivalent aerodynamic particle diameter.
Source: Reference 3.

TABLE 4. Uncontrolled Emission Factors and Particle Size Distribution for Primary Emissions from HSS Reduction Cells *(Emission Factor Rating: D)*

Particle size,[a] μm	Cumulative Mass, % ≤ Stated Size	Cumulative Emission Factors	
		kg/Mg Al	lb/ton Al
15	63	30.9	61.7
10	58	28.4	56.8
5	50	24.5	49.0
2.5	40	19.6	39.2
1.25	32	15.7	31.4
0.625	26	12.7	25.5
Total	100	49.0	98.0

[a]Expressed as equivalent aerodynamic particle diameter.
Source: Reference 3.

sions, followed by spray towers to scrub the gaseous fluoride. A small amount of particulate matter is also removed in the scrubber. The scrubber effluent is then neutralized, usually by the addition of lime to form calcium fluoride, and the water is recirculated to the scrubber. Wet systems have many disadvantages, such as corrosion by hydrofluoric acid, scaling and the expense involved in recovering the fluoride values from calcium fluoride. The only benefit of a wet scrubber system is that it removes some of the sulfur oxide in the cell off-gases. Since new smelters no longer install wet systems, only dry scrubber systems will be covered here.

Dry Scrubber Chemistry

In potroom dry scrubbers, gaseous HF reacts rapidly with alumina through chemisorption to form a monolayer of HF gas on the Al_2O_3. The HF is then converted to AlF_3 when exposed to heat in the cell. To achieve high HF-removal efficiencies, it is necessary to use alumina with a high surface area. If the surface area is too low, HF will be adsorbed in excess of a monolayer and could later be desorbed in the pot. In this case, the HF concentration in the exhaust gas stream will build up as more HF is added to the gas stream than is removed by the alumina, and high fluoride emission will result from the scrubber system.

A surface area of 45 m^2/g or greater as measured by the BET method is generally considered to be satisfactory. For example, an alumina with a surface area of 50 m^2/g will adsorb 1.79% of its weight as HF. In some earlier dry scrubbing systems, to obtain high surface area, kiln-activated alumina (KAH) was ground and then used in injection systems. For the most part, these earlier systems have been converted to the fluid-bed process.

Dry Scrubbers

Fluid-Bed Scrubber

The heart of this system is a fluid-bed reactor roughly 12 feet wide by 50 feet long and 8 feet high with a fabric filter located on top of the reactor (Figure 4). Metal-grade alumina is fed into one end, usually with a rotary vane feeder, and exits at the other end of the reactor through a seal leg. Potroom gases, introduced at the bottom of the reactor through dribble plates, fluidize the alumina while, simultaneously, HF gas reacts with the alumina. The alumina bed in the nonfluidized state is usually 4 to 8 inches deep. The reaction between HF and alumina is very rapid and HF removal efficiency is 99% or more. The pressure drop through the system ranges from 12 to 20 in. w.c.

The fabric filter on top of the reactor captures and returns the entrained alumina particles to the reactor below, as well as particulate fluoride and carbon particles produced by the cells. The fume control system (Figure 5) consists of a bank of reactors (only one reactor shown) fed by alumina supplied from storage tanks. Alumina leaving the reactors is stored in adjacent tanks that supply the cells. In a typical smelter, most of the alumina is used in the dry scrubber system before it goes to the potroom, but in some plants, a portion of the fresh ore may be segregated and used in cells where higher-purity aluminum is desired because reacted ore contains higher levels of trace impurities such as iron and silica. The anode baking operation could also use up to 10% of the fresh alumina in the dry scrubbing system.

The choice of fabric material for the filter is dictated by the gas temperature and resistance to HF and sulfur oxide gases. Polyester and acrylic fabric are both suitable, except where the cell exhaust gas temperature exceeds 275°F. Nomex is usually selected for higher temperatures. The

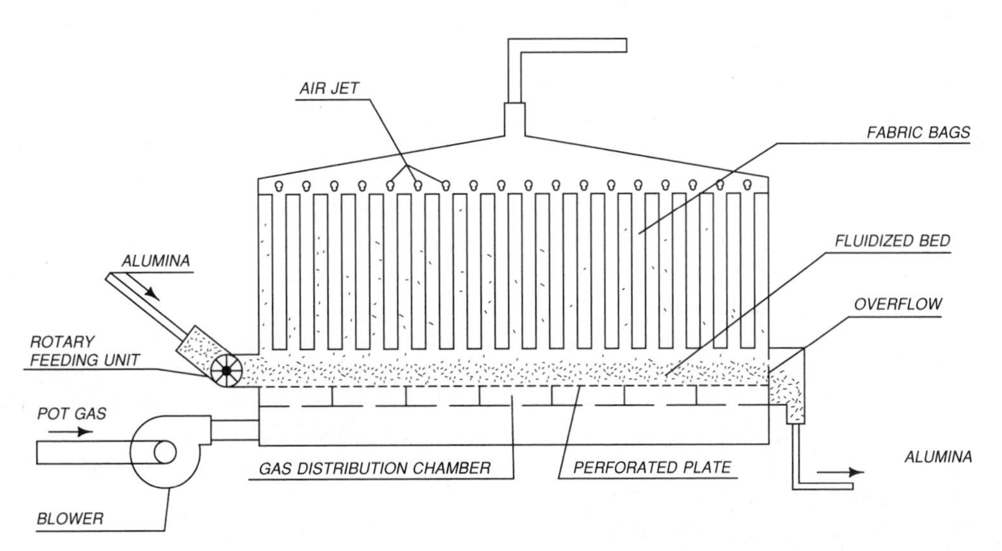

FIGURE 4. Fluid-Bed Reactor for Potroom Fumes (Aluminum Company of America, Pittsburgh, PA)

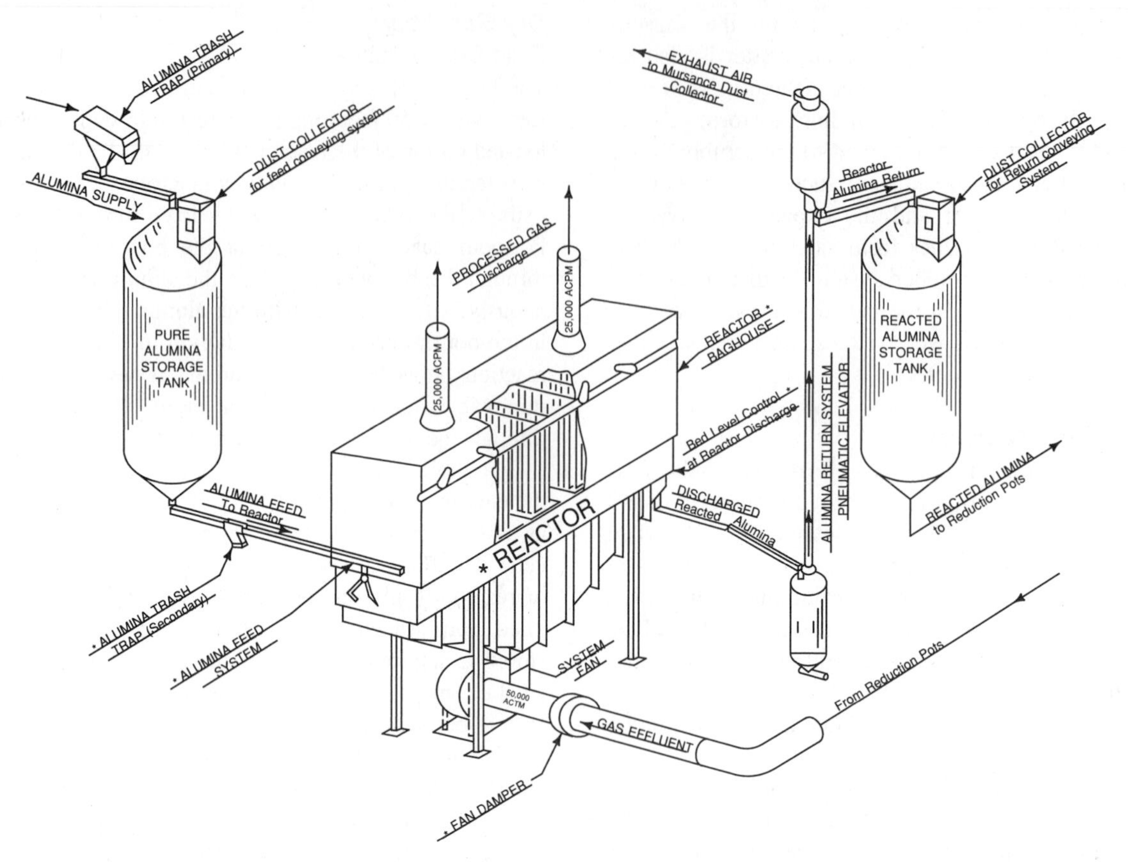

FIGURE 5. Fluid-Bed Scrubber System for Potroom Fumes (Aluminum Company of America, Pittsburgh, PA)

filter bags in the baghouse must be cleaned periodically as the pressure drop builds up across the reactor. To clean the bags, the reactor–baghouse unit is isolated by shutting off the potroom gases, causing the bags to deflate. They are then cleaned by a shaker mechanism. After bag cleaning, the reactor is started up, and another reactor unit is taken off-line for cleaning. One or more spare reactor units thus are needed to pick up the load during bag cleaning, as well as when one is down for bag changing. The cleaning and reactor start-up cycle is usually performed automatically.

Most fluid-bed installations on prebake cells use a shaker-type fabric filter, but in a few installations, reverse-pulse-type filters are used. The shaker collector, using woven filter bags, is usually designed with an air-to-cloth ratio of 2 : 1 while the reverse-pulse collectors with felt bags have air-to-cloth ratios ranging from 4.0 : 1 to 5.5 : 1. A set of woven filter bags in the shaker collector will provide up to five years' service before replacement is required, and felt bags in the reverse-pulse collector will provide a bag life of about two years.

The working environment for bag changing in a shaker collector is usually better than in the reverse-pulse unit, since the dirty side is on the inside surface of the bags, while with pulse collector bags, filtering occurs on the bag's outside surface.

The exhaust volume from a pot ranges from 2000 to 5000 acfm, depending on its size. With up to 200 pots per line, dry scrubber systems are designed to handle from 500,000 to over 1,000,000 cfm of exhaust gases. Since a reactor can handle up to 50,000 cfm, a large installation will consist of up to 25 reactor–baghouse units.

Injection System

In the injection system (Figure 6), alumina is injected into the gas stream where turbulent contact occurs between the alumina and the potroom gas, resulting in the adsorption of HF. Alumina particles are entrained in the gas stream and recovered in a fabric filter. There are variations in the method of injecting alumina into the gas stream. In some units, alumina is injected in a venturi throat, while others introduce the alumina through a specially designed pipe into a plenum. The main objectives of these designs are to create good mixing and to distribute the alumina evenly into the gas stream. To increase utilization of the alumina, a portion of it is usually removed from the fabric filter hopper and reinjected into the gas stream. The balance of the alumina is conveyed to the potroom storage feed tank.

The system is designed with two large multicompartment bag filters side by side. A reverse-pulse collector is the usual choice, although shaker collecters can be used. Compartmentalization of the baghouse allows maintenance of the collector and bag changes while the system is operating.

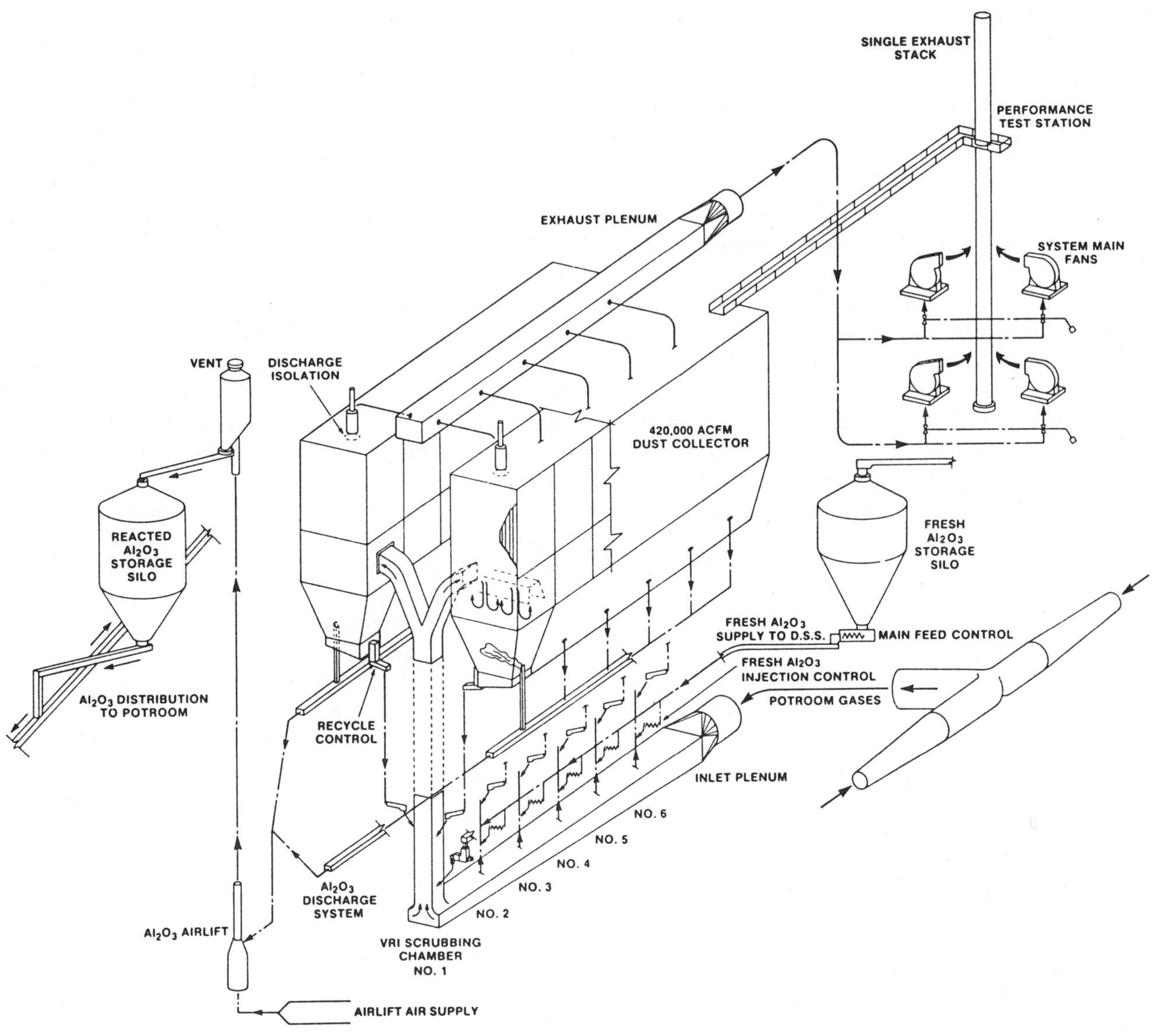

FIGURE 6. Dry Scrubber Injection System—Potroom Fumes (Busch International, Pittsburgh, PA)

With the reverse-pulse collector, which uses felt bags, it is possible to use a higher air-to-cloth ratio of 3.0:1 to 3.5:1, resulting in fewer bags and a smaller baghouse; thus, it is a more economical design from a capital cost standpoint. However, the pulse collector uses a large amount of compressed air for bag cleaning. Shaker-type collectors, on the other hand, with woven bags require a lower air-to-cloth ratio of about 2.0:1, and so are larger and more costly. However, bag life is greater for the shaker because the cleaning method is gentler than with pulsing.

Baking Furnace

Baking Furnace Emissions

The amount of emissions produced by baking furnaces depends on many factors, such as the type of furnace, the condition of the flues, firing practice, and type of fuel used, that is, heavy oil or gas. Also, as mentioned earlier, the fluoride levels in the emissions are a function of the amount of fluorides retained with the recycled anode butts. Uncontrolled fluoride emissions will range from 0.25 to 1.5 kg per tonne alumina with similar amounts of coke dust and hydrocarbon vapors from the pitch binder. New Source Performance Standards (NSPS) for fluoride emissions from baking furnaces are 0.05 kg per tonne alumina and 10% maximum opacity.

Control Systems

Wet Systems

As with potrooms, wet scrubbers were originally used to reduce baking furnace emissions. These scrubbers ranged from simple cyclonic to high-energy venturi scrubbers. While cyclonic scrubbers were fairly effective in removing gaseous fluorides, they removed only some of the pitch fumes and the escaping matter consisted of very fine condensed tar particles, which could cause unacceptably high stack opacity. The high-energy venturi scrubber is a more efficient collector, but both scrubber systems generate large amounts of low-pH, high-fluoride water containing suspended tar, polynuclear aromatic compounds, and particulate matter. Under the best circumstances, this effluent is difficult to treat and expensive to dispose of.

Wet ESPs have also been used, and if properly maintained, will provide high collection efficiency on tar. The tar vapors must first be condensed in a quench chamber since the ESP will collect only particulate matter. The maintenance cost of wet ESPs is quite high because the collector plates must be cleaned regularly to maintain high collection efficiency and the water treatment problems are the same as with scrubbers. Wet ESPs are used where the emission from the closed-type furnace is basically a tar product and must be collected before the fluorides can be removed in the dry scrubbers. However, for the closed furnace emission, a dry system will be described in a later section for collecting tar vapor and fluorides.

Dry Scrubbers

Dry scrubbers, which are widely used today on anode baking furnaces, are similar to those used on potrooms, but quite different in their mission. Dry scrubbers for anode furnaces were originally developed to collect, primarily, the hydrocarbons that caused high stack opacity. As federal regulations evolved, however, the NSPS for aluminum smelters also limited fluoride emissions from the baking furnace; consequently, fluoride removal is now as important as hydrocarbon control.

The major difference between the baking furnace and potroom control systems is that the former is designed to cool the furnace gas to condense the tar vapors so that they can be collected on the alumina. Cooling also increases fabric filter bag life. The first dry scrubber installation used a fluid-bed reactor, but later the injection system used on smelters was adapted to the baking furnace by adding a gas cooler to the system.

The fluoride collection efficiency of these systems is similar to that obtained for potroom emissions—99% or greater. The collection efficiency for hydrocarbon emissions is about 90%, but this value can vary depending on the manner in which the baking furnace is operated, the amount of tar in the flue gas, and the operating temperature of the dry scrubber.

In some baking furnaces, the hydrocarbon level in the flue gas is so low that the scrubber system is used mainly to control fluoride emissions. In other plants where local regulations do not require hydrocarbon control and hydrocarbon levels are relatively low, the gas is cooled by convective cooling in long runs of ductwork, thus avoiding the evaporative coolers.

The cost of dry scrubber systems for anode furnace emissions ranges from $60 to $75 per actual cubic foot per minute. The higher cost as compared with potroom scrubbers reflects the added cost for cooling the gas and the smaller gas volumes to be treated. The large potroom's treatment systems can take advantage of the economy of scale.

Fluid-Bed System

The fluid-bed reactor is quite different from that used on potroom emissions. The baking furnace flue gas cannot be used to fluidize the alumina because the tars would plug the small holes in the distribution or dribble plates. Therefore, the alumina is fluidized, as shown in Figure 7, by a small amount of ambient air, and the flue gas enters the reactor through a row of pipes above the dribble plate. Another difference, mentioned earlier, is the need to cool the flue gases to condense and capture the hydrocarbons. This is accomplished by spraying water into the fluidized bed. Fluidization provides efficient mixing among the water droplets, alumina, and pollutants, thus promoting good mass and heat transfer. The water droplets evaporate as they contact the hot alumina, causing the bed to cool, and condensation of the tar fume takes place on the cooled alumina particles. Fresh alumina is introduced at one end of the reactor bed, which is 12 inches to 15 inches deep, and is removed at the other end. The entrained alumina and other solids are separated in a bag filter located on the reactor, and the scrubbed flue gas exits from stacks on the baghouse.

The reactors are designed to process 35,000–50,000 acfm of flue gas, with an alumina feed rate of 600 to 1000 lb/h per reactor. The bed is cooled to 150°F to 180°F, requiring from 10 to 15 gallons per minute of water, depending on the flue gas temperature. The material handling system is quite similar to that used in the potroom fluid-bed scrubber system except on a smaller scale, and reacted alumina is also sent to the potrooms.

In some installations, the reacted alumina containing coke and tar is heated to destroy the combustibles before the alumina is used in the pots. In this case, the alumina is fed to a regenerator, where it is fluidized and heated to 1200°F. The bed is about 55 inches deep and is fluidized by ambient air, which provides oxygen for combustion. Radiant tube gas-fired burners immersed in the bed supply the heat. Hot alumina emerging from the regenerator is cooled to about 200°F in a water-cooled heat exchanger before being transferred to storage. A typical system with the regenerator is shown in Figure 8.

Injection System

The injection system, as pointed out earlier, is similar to that used to control potroom emissions, with the addition of an evaporative cooler up front to cool the flue gas (Figure 9). The cooler will range in size from 12 to 15 feet in diameter and will be about 100 feet tall, depending on the gas volume to be treated and the desired residence time. Flue gas enters the cooler at the top, where a bank of nozzles spray water mist into the gas stream. The entrained

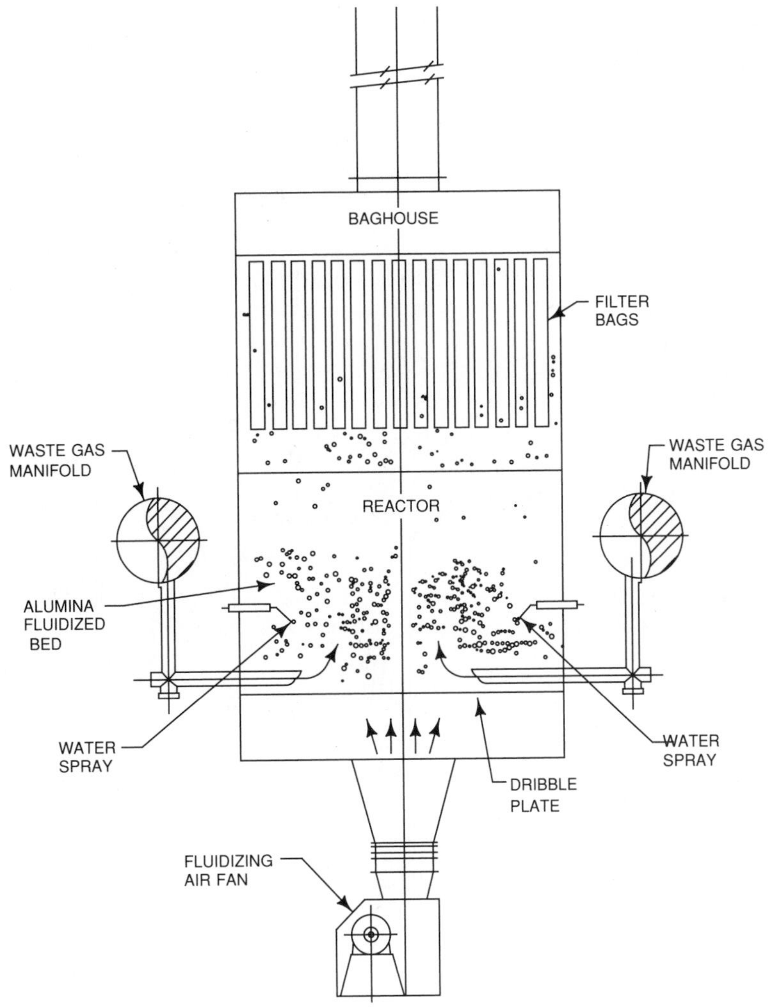

FIGURE 7. Dry Scrubber Fluid-Bed Reactor—Baking Furnace (Aluminum Co. of America, Pittsburgh, PA)

water droplets travel co-currently with flue gas down the tower. Cooled gas exits the cooler bottom and enters the injection system.

The water spray is produced from two-fluid nozzles using compressed air to create the fine mist particles ranging in size from 50 to 100 μm. The tower is designed to provide several seconds of residence time in order to ensure complete evaporation of the water droplets. The operating temperature of the cooler is usually 200°F to 210°F, which is a practical limit. Attempts to lower the temperature can result in wetting of the inside wall, as well as carryover of water into the bag filter. Perhaps lower gas temperatures could be achieved with spray nozzles that produce smaller water droplets. Coke particles and condensed hydrocarbons settle out in the cooler and are removed continuously from the bottom.

After the cooler, alumina is injected into the gas stream in a contact chamber or in the venturi section in the ductwork, and adsorption of HF gas and tar deposition occurs on the alumina. The gases then enter the compartmentalized baghouse where entrained alumina, together with coke particles and condensed hydrocarbons, are removed. Cleaned flue gas exits to a blower and out of the stack. As in the potroom system, a portion of the alumina from the bag filter hopper is recycled to the contact chamber to increase alumina utilization and the balance sent to the potroom.

The size of a system will vary with the volume of the flue gas to be treated and will range from about 50,000 to 150,000 cfm. As mentioned, the system is similar in design to the potroom dry scrubber. It has a multicompartment baghouse reverse-pulse bag cleaning or a shaker system that permits each compartment to be isolated for maintenance and bag change.

Tar Collection and Dry Scrubbing

As mentioned previously, there are closed baking furnaces that produce a flue gas low in oxygen and high in tar content. The control scheme for some of the earlier installations consisted of cooling the flue gas to 150°F to 160°F in an evaporative cooler to condense the tar into aerosol particles, which were then removed in an ESP. To achieve this relatively low temperature, it was necessary to

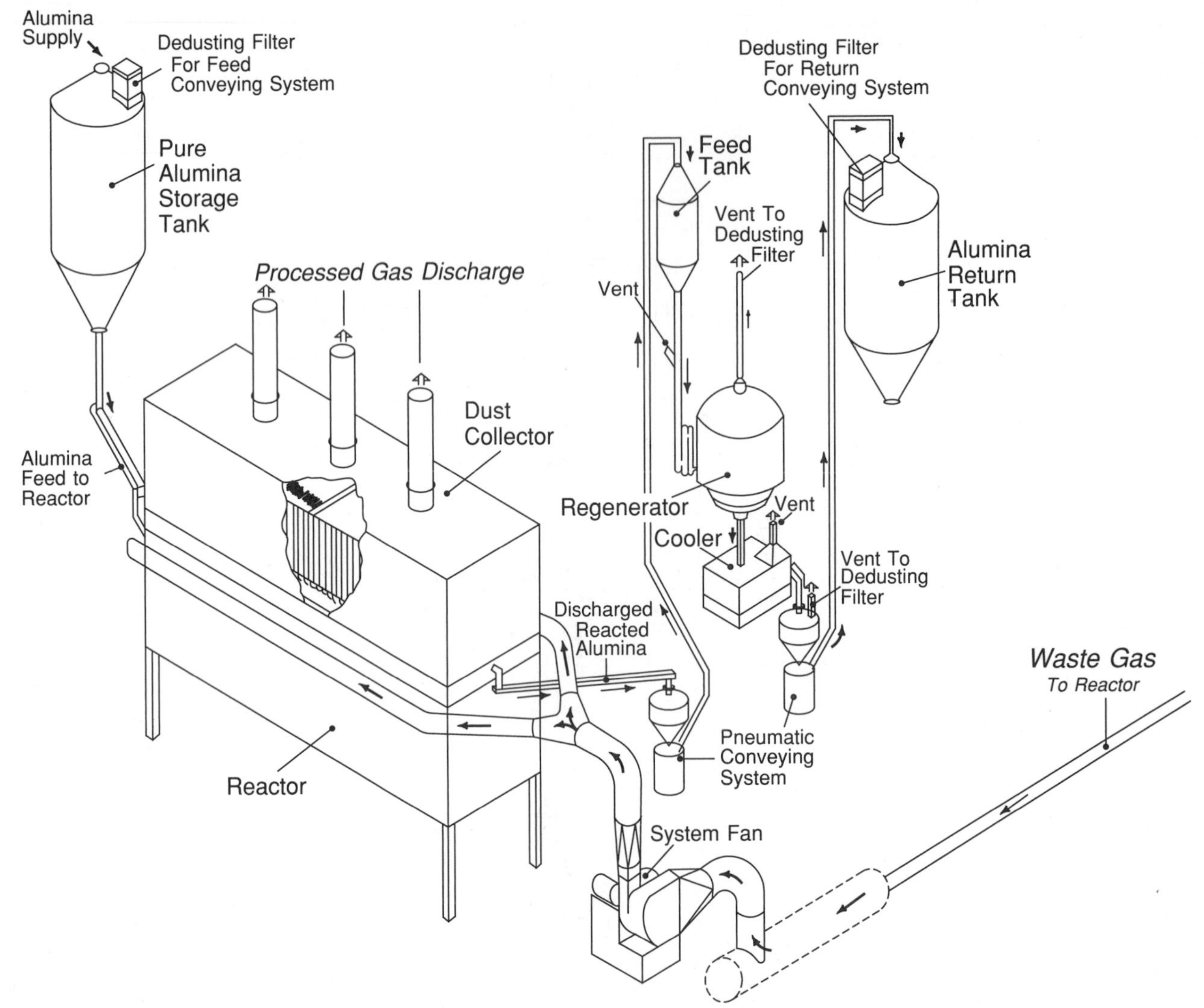

FIGURE 8. Fluid-Bed Reactor System—Baking Furnace (Aluminum Co. of America, Pittsburgh, PA)

operate the cooler with a wet bottom; that is, with excess water flowing from the cooler bottom. This effluent contained dissolved HF and suspended tar and required treatment before it could be disposed of.

To avoid the evaporative cooler and water treatment, a system was recently developed in which the flue gases are cooled in shell-and-tube water-cooled heat exchangers (Figure 10), with flue gas on the shell side and cooling water on the tube side. Through careful control of the water temperature in each cooling zone, the flue gas is cooled to 160°F without plugging the heat exchanger. Tar that condenses on the tubes drips to a heated sump below. This particular temperature was chosen because it is low enough to cause most of the tar vapor to condense, but warm enough for the tar to flow. The cooled gases then go to a plate-type ESP, where the remaining fine aerosol tar particles are captured on the plates and tar drips off to a heated hopper below. Finally, the gases are treated in a dry alumina injection system to remove fluoride.

In these systems, the detarring equipment must be insulated and heat traced to prevent tar from freezing in cold weather, and a fire protection system is also required.

Green Anode

Green Mill Control

As mentioned earlier, emissions produced in the various steps in green anode production are coke dust from the coke classifiers and hydrocarbon vapors from the mixing, conveying and anode forming equipment. Coke dust from the classifiers and conveying equipment is usually controlled by fabric filters.

Control of hydrocarbon emissions is more difficult. Since the gases are at relatively low temperature—150°F or less—tar vapors readily condense in the ductwork and, together with the coke dust, cause plugging. Heat tracing the ductwork will help to minimize the problem, but will not completely prevent it. Gas velocity in the ductwork

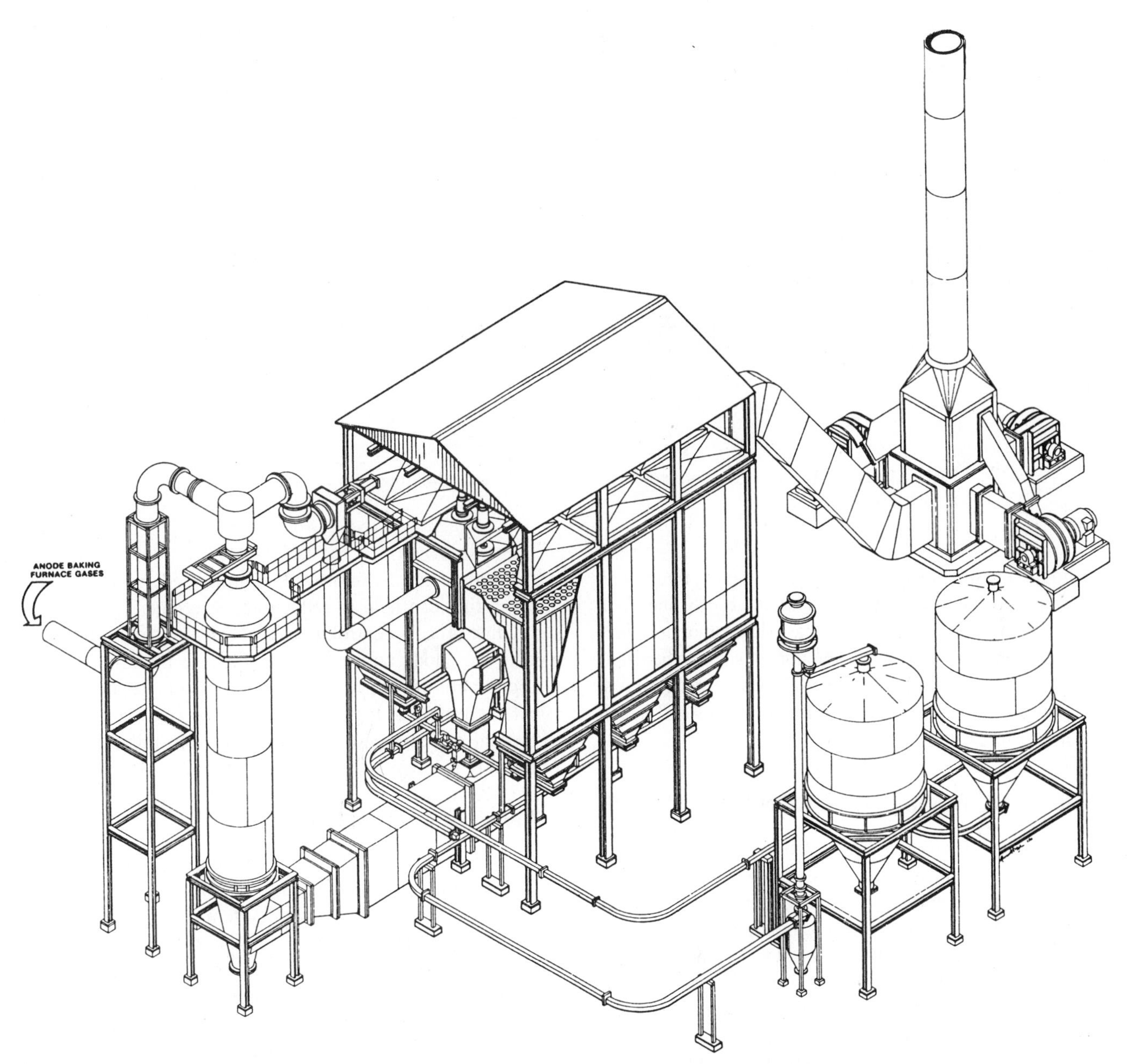

FIGURE 9. Dry Scrubber Injection System—Baking Furnace (Busch International, Pittsburgh, PA)

should be at least 3000 fpm to prevent coke particles from dropping out and cleanout ports in the ductwork will facilitate cleaning.

An effective control technology for green mill emission is a precoat bag filter using coke dust as the precoat material, which not only absorbs/adsorbs the hydrocarbon vapors, but also prevents the tar from blinding the filter bags. Coke dust can be obtained from coke pneumatic conveying systems, coke classifiers, and other exhaust streams. The bags are precoated by injecting the coke dust into the ductwork prior to the bag house. Since the gas is relatively cool, the vapors are almost totally condensed at this stage and exist as finely condensed particles of tar together with a small amount of vapor. Some vapor adsorption probably occurs as well.

The amount of coke required will depend on the concentration of hydrocarbon vapors in the gas stream, but in a shaker collector, a minimum thickness of about ⅛ inch of precoat material is necessary for tar absorption and to protect the filter bags from blinding. Another operating parameter is to provide enough coke dust so as not to exceed 4–5% tar content on the coke dust. As a general rule, tar-removal efficiency will increase with the amount of coke used since more coke will produce more available surface, as well as a thicker precoat layer for increased filtration efficiency.

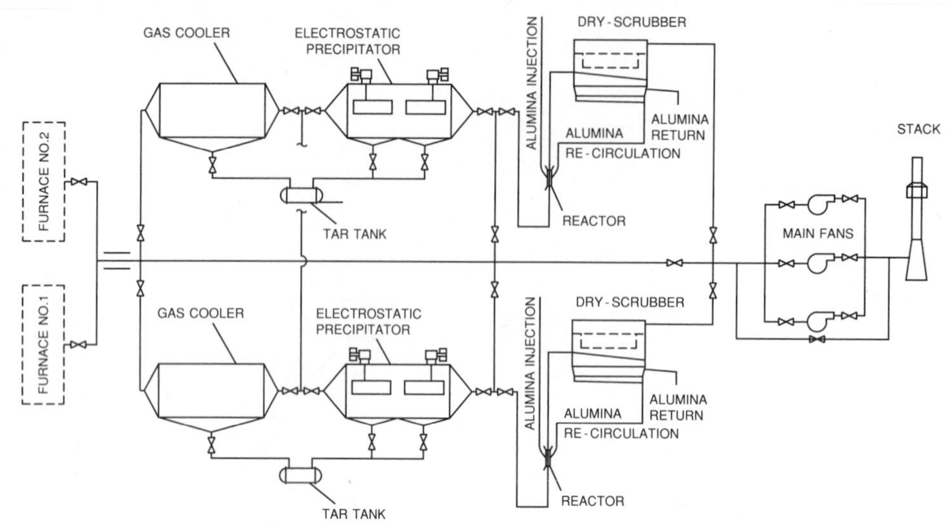

FIGURE 10. Tar Precipitation and Dry Scrubber System (Mikropul Env. Systems, Morris Plains, NJ)

Another possible method for controlling pitch fumes is to scrub the gases with a heavy oil. The oil absorbs the pitch fumes, and since the scrubbing oil has a low vapor pressure, it does not emit a significant amount of vapor when it is exposed to the gases being scrubbed. The scrubbing oil will eventually become saturated with hydrocarbons and it will have to be replaced.

OTHER ENVIRONMENTAL CONSIDERATIONS

The foregoing discussion has dealt largely with control systems used to abate air emissions from the primary aluminum industry. Historically, as environmental problems were identified, suitable control technology was applied or developed to correct the problems. This is the so-called end-of-the-pipe treatment approach. However, with mounting environmental costs and an increased awareness of the global effects of pollution, there is an emerging consensus that a broader approach must be taken to address these problems not only in the aluminum industry, but in other industries as well. Therefore, it is appropriate to mention some of the other programs and steps—long and short range—that will find wider acceptance as industry takes a more holistic view toward addressing these environmental issues.

Process Control

Good process control in baking furnace operation to reduce emissions and minimization of exhaust volume from pots through a well-designed hood system were mentioned earlier. The use of modern computer control systems has led to improved process control in the potrooms. For example, computerized process control can reduce emissions by minimizing the "anode effect." An anode effect is caused by a deficiency of alumina feed to a pot, resulting in high cell resistance, high temperature, and higher fluoride emissions. Computer controls, improved alumina feeders, and process monitoring can help to reduce anode effects.

Other process control items that can affect environmental costs are anode burn-off and premature potlining failure. Anode burn-off can be caused by poor-quality anodes or a bad connection at the carbon–anode rod interface. And the disposal of spent potlining removed from failed linings is currently receiving a great deal of attention in the industry.

Choice of Control Technology

The proper choice of control technology is also important. The example most often cited to illustrate a poor choice of technology is one where a wet system instead of a dry one was chosen because of lower capital cost, overlooking the high cost of waste treatment. Thus, an air pollution problem was merely exchanged for a worse one.

In many respects, the aluminum industry is fortunate to have developed dry environmental control systems for potroom and baking furnace emissions, thus replacing the wet systems. However, it should be mentioned that solving the major air emission problems resulted in solid and liquid waste problems with which to contend. The large number of used filter bags, some of which are impregnated with hydrocarbons from the anode baking operation, presents one problem, and the tar collection system used on the closed anode furnace generates tar contaminated with acids.

Equipment Operation and Maintenance

Good operation and maintenance of pollution control and process equipment cannot be overstressed. Poorly maintained pot hooding, fabric filters, and pneumatic conveyors, for instance, can cause major emissions. Poor flue maintenance in the baking furnace, and its relationship to good product quality, has been pointed out already. In the long run, well-maintained environmental control equipment not only emits less, but also provides longer service life.

Environmental control systems should also be integrated into the production process and not operated as an appendage. The fume control system for the baking furnace is a good example. It is an important part of the baking operation, providing draft for combustion, and must work in unison with the process. However, the two systems are sometimes operated and maintained as separate systems by different groups. As a result, the baking operation is often listed as the cause of poor performance of the control system, and vice versa. Because of this separation, the relationship between good process control and good performance of the fume control system often goes unrecognized.

Process Modification, TQM, Early Involvement, and Waste Minimization

Process modification, if it can be carried out without major expenditures, can be an inexpensive method for reducing and perhaps even eliminating an emission. To take advantage of process modification as a control scheme, continuous evaluation of the process and input from operators, supervisors, and others familiar with the operation are necessary.

The approach of total quality management (TQM) could also be applied to environmental control to improve the operation of the systems in terms of reliability, cost of operation and maintenance, and reduction in emissions. This concept, which also embodies a goal of continuous improvement, is becoming more accepted in this country as an industrial management tool. It is a system that uses quality control tools, such as data gathering and statistical controls, and applies them in every step of an *operation* to produce the best-quality *product* in the most efficient manner. The terms operation and product are used in the broadest sense and can range from a highly sophisticated production process to the most mundane task.

Another long-range program is one of early involvement during process development. By identifying environmental problems in the early stages of development, they can be resolved before the process is in place and expensive end-of-the-pipe treatment has to be installed.

Finally, a program of waste minimization has been shown by many companies to be quite beneficial in terms of resource conservation, pollution reduction, and good payback. Like process modification and TQM, waste minimization is a never-ending program requiring plant-wide support, environmental awareness, and a clear definition of the objectives of waste minimization.

These approaches are proactive in nature, and in order to succeed, will require some training, teamwork, and the commitment of an entire organization. However, they represent a more logical and integrated management of environmental issues, and in the long run, a combination of these programs with the best end-of-the-pipe solutions will prove to be less stressful to the environment and far more cost effective.

Appendix. Summary of Smelter Environmental Data

Potroom

• NSPS regulations	1 kg F/tonne Al (Soderberg pot) 0.95 F/tonne Al (prebake pot) 10% opacity
• Exhaust volume per pot	2000–5000 cfm
• Hooding efficiency	95% (approximately)
• HF removal efficiency—dry scrubbers	99%
• Minimum required surface area of alumina for dry scrubber	45 m^2/g (BET method)
• Carbon consumption	0.4–0.5 lb C/lb Al
• Capital cost—dry scrubber system	$35–$50/cfm

Carbon Baking Furnace

• NSPS regulations	0.05 kg F/tonne Al 10% opacity
• Emission factors	
—Fluoride	1–5 lb/tonne Al
—Hydrocarbons + particulate	1–5 lb/tonne Al
• Dry scrubber efficiency	
—Fluoride	99%
—Hydrocarbon	90% or less[a]
—Particulate	99%
• Capital cost—dry scrubber system	$60–$75/cfm

[a]Hydrocarbon removal efficiency will depend on many factors, such as gas temperature, alumina feed rate, and gas composition.

References

1. *AP-42 Supplement A to Compilation of Air Pollutant Emission Factors, October 1986. Vol. 1: Stationary Point and Area Sources.* U.S. Environmental Protection Agency, Office of Air and Radiation, Research Triangle Park, NC.

Bibliography

Air Pollution Control in the Primary Aluminum Industry, EPA-450/33-73-004A and EPA-450/33-73-004B, Singmaster and Breyer, New York, 1973.

Aluminum Statistical Review for 1990, Aluminum Association, Washington, DC.

P. R. Atkins and K. P. Karsten, "In plant practices for job related health hazards control: aluminum," Chap. 4, Vol. 1, *Production Processes,* John Wiley & Sons, New York, 1989.

W. W. M. Boon and R. de Fluiter, "Leaning of flue gas from anode bake furnaces," *Light Metals,* Metallurgical Society of AIME, 1980, pp 721–733.

C. N. Cochran, W. C. Sleppy, and W. B. Frank. "Fumes in aluminum smelting: chemistry of evolution and recovery," *J. Metals* (September 1970).

C. N. Cochran, "Recovery of hydrogen fluoride fumes on alumina in aluminum smelting," *Environ. Sci. Technol.*, 8(1) (1974).
P. A. Comella and R. W. Rittmeyer, "Waste minimization/pollution control," *Pollution Eng.* (April 1990).
C. C. Cook, G. R. Swany, and J. W. Colpitts, "Operating experience with the Alcoa 398 process for fluoride recovery," *J. APCA,* 21(8) (August 1971).
C. C. Cook and G. R. Swany, "Evolution of fluoride recovery processes Alcoa smelters," *Light Metals,* Metallurgical Society of AIME, Vol. II, 1971, pp 465–477.
Development Document for Effluent Guidelines and New Source Performance Standards—Primary Aluminum Smelting, EPA-440/1-74-019-d. U.S. Environmental Protection Agency, Washington, DC.
J. Edwards, *The Immortal Woodshed,* Dodd, Mead & Co., New York, 1955.
J. C. Files, "Treatment of bake oven exhaust gases using unique system of alumina dry scrubbing," *Light Metals,* Metallurgical Society of AIME, 1981, pp 931–939.
L. K. Hudson, *Alumina Production,* Alcoa Research Laboratories, Alcoa Center, PA, 1982.
Primary Aluminum: Guidelines for Control of Fluoride Emissions from Existing Primary Aluminum Plants, EPA-450/2-78-049b, Office of Air Quality Planning Standards, Research Triangle Park, NC, 1979.
Review of New Source Performance Standards for Primary Aluminum Reduction Plants, EPA-450/3-86-010, U.S. Environmental Protection Agency, Research Triangle Park, NC.
P. D. Stobart. *Centenary of the Hall & Heroult Processes 1886–1986,* International Primary Aluminum Institute, London, 1986.
T. J. Robare and M. W. Wei, "Control of pitch volatiles in green electrode plant," *Light Metals* Metallurgical Society of AIME, Vol. II, 1984, pp 1521–1529.
D. Rush, J. C. Russel, and R. E. Iverson, "Air pollution abatement on primary aluminum potlines: effectiveness and cost." *J. APCA,* 23(2), (1973).
Ullman's Encyclopedia of Industrial Chemistry. 5th ed., Vol. A., Verlagsgesellschaft mbH, Weinheim, Germany, 1985.
M. W. Wei, "The Alcoa 446 process for the control of anode baking furnace fumes." *Light Metals,* Metallurgical Society of AIME, Vol. II, 1975, pp 261–268.

METALLURGICAL COKE

Thomas W. Easterly, P.E., Stefan P. Shoup, and Dennis P. Kaegi

Metallurgical coke is manufactured by heating high-grade bituminous coal (low sulfur and low ash) to around 1050°C (1925°F) in an enclosed oven chamber without oxygen. The resulting solid material consists of elemental carbon and any minerals (ash) that were present in the coal blend that did not volatilize during the process.

PURPOSE OF METALLURGICAL COKE

There are two major categories of metallurgical coke: furnace coke and foundry coke. Furnace coke is used in steel mill blast furnaces and foundry coke is used in foundry cupolas. In general, foundry coke is made from coals with a lower volatility, is heated longer, and is larger than furnace coke. Both types of coke can be made with the same equipment by changing the coals and the operating practices. In the United States in the early 1990s, over 90% of coke produced is furnace coke for use in steel mill blast furnaces.

Furnace coke is essential to the blast furnace method of iron production where it serves three major functions. First, the coke provides the porous support for the other materials in the blast furnace burden (principally iron ore pellets and limestone), allowing the hot blast to blow through the furnace. Second, coke provides fuel to react with the blast air, which, in turn, provides the heat and reducing atmosphere required to refine the ore into iron. Third, the coke acts as a chemical reductant, removing oxides from the ore.

Because of the costs and environmental concerns associated with coke manufacturing, the steel industry has been working on processes to reduce the need for coke in blast furnaces. To date, there has been success in substituting oil, natural gas, and coal for some of the fuel value provided by the coke, but large amounts of coke are still needed both to provide a porous support to the blast furnace burden and to act as a chemical reductant. As less coke is used in the blast furnace per ton of hot metal produced, it is expected that the quality of coke will have to be improved, particularly its strength or stability.

TYPES OF BATTERIES

Two types of coke oven batteries are currently employed to produce metallurgical coke. The most common is the slot oven by-product type, which was used to produce over 99% of the metallurgical coke in 1990. In this type of battery, the volatiles driven off in the process are refined in the coal chemical plant to produce clean coke oven gas, tar, sulfur, ammonium sulfate, and light oil. Some 33% to 40% of the coke oven gas produced is used to heat the battery, while the rest can be utilized as a fuel at other facilities. Figures 1 and 2 show the general layout and sources of emissions of a typical by-product coke oven battery.

The other type is the nonrecovery battery, which is currently used to produce coke at only one location in the United States and is seriously being considered by a number of steel companies as an alternative to the slot oven by-product battery when present coke plants need to be reconstructed. A nonrecovery coke battery is one in which the coal volatiles driven off in the process are immediately combusted within and around the oven to provide the heat required for carbonization. This eliminates the need for associated by-products plants. Emission control for exhaust

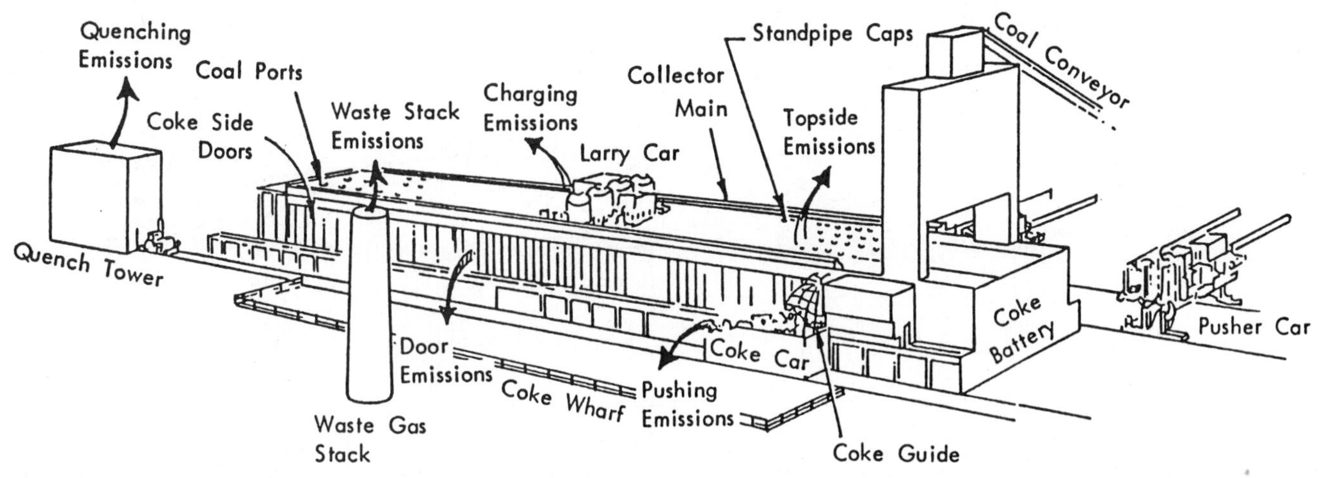

FIGURE 1. General Layout of By-Product Coke Oven Battery

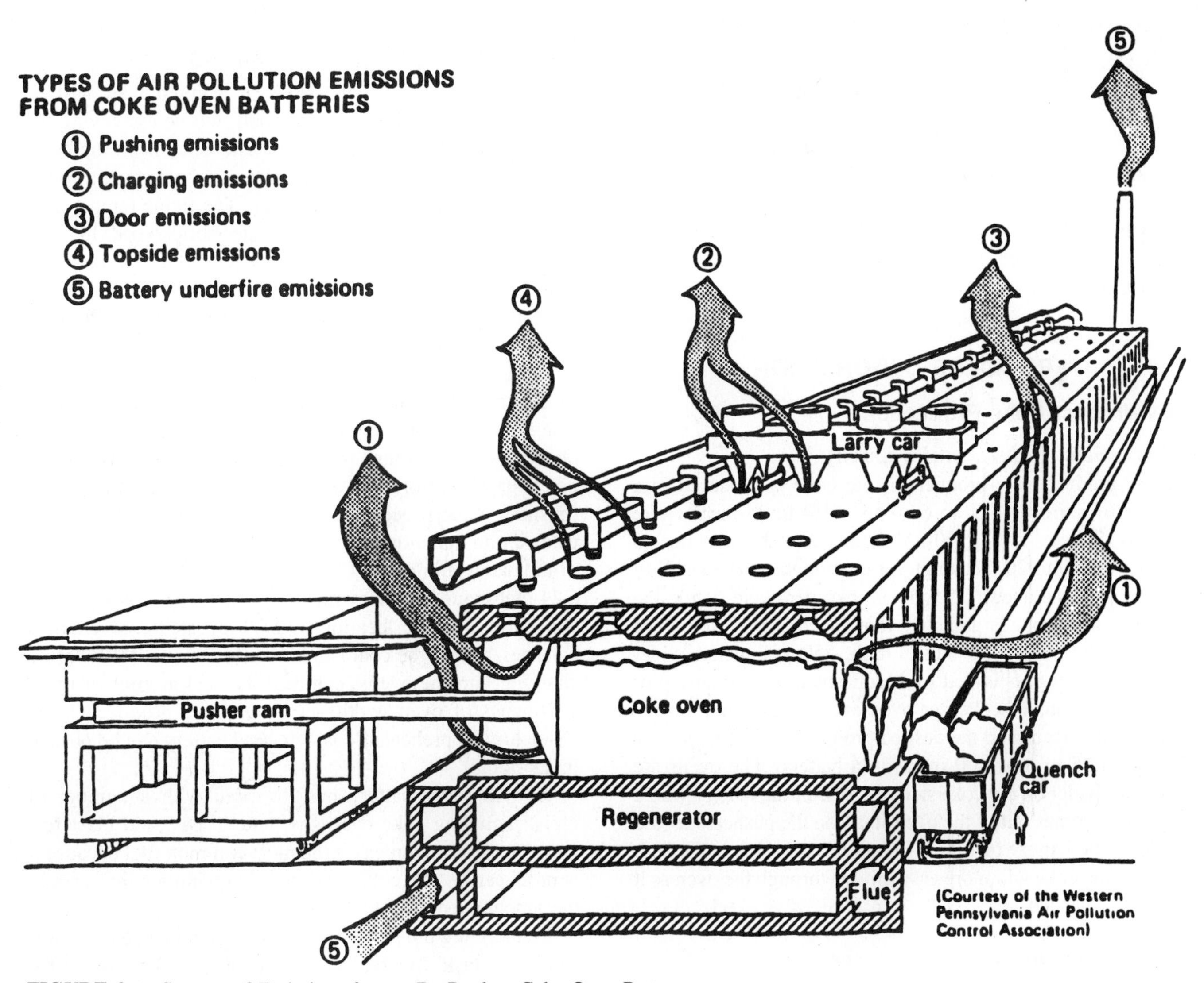

FIGURE 2. Sources of Emissions from a By-Product Coke Oven Battery

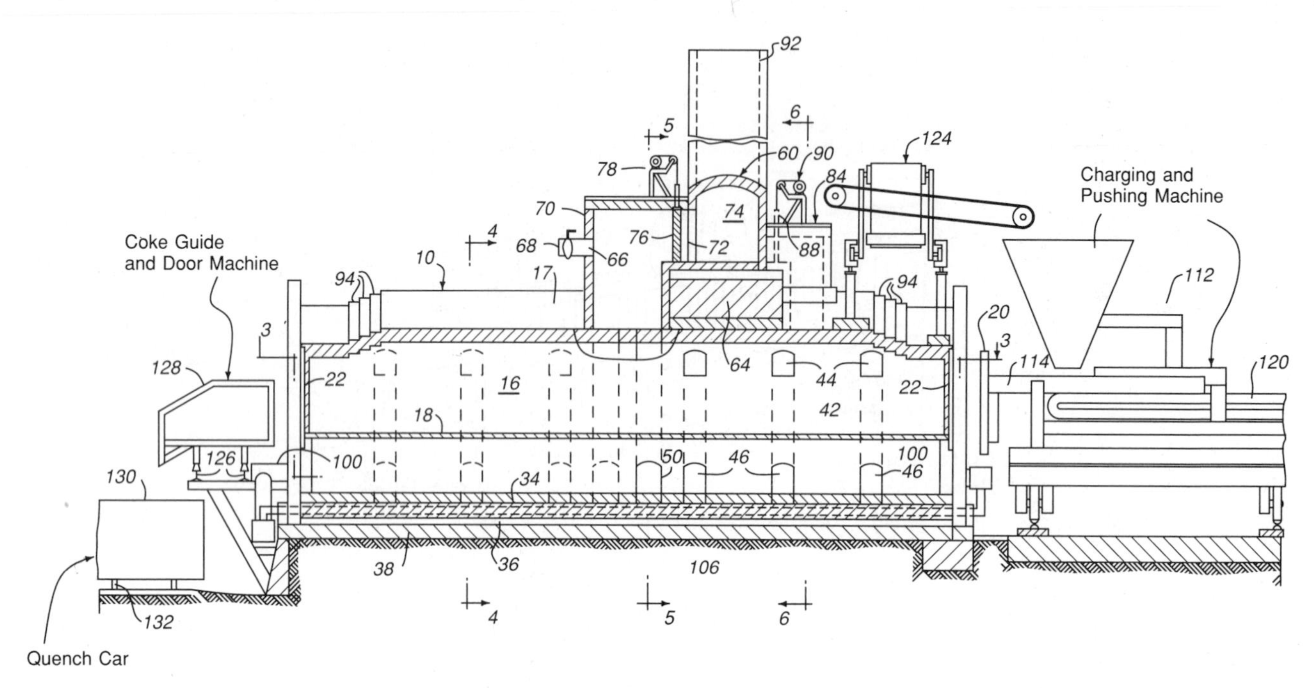

FIGURE 3. Side View of an Oven on a Nonrecovery Coke Battery

gas generally would consist of removal of particulate matter and sulfur dioxide prior to discharge through the waste gas stack. Figures 3 and 4 show a typical individual oven and layout of a nonrecovery battery facility.

An older type of coke oven that is no longer in use is the beehive oven. It was an environmentally uncontrolled and manually operated forerunner of the current design of nonrecovery ovens.

PROCESS DESCRIPTION

By-product Coke Production

Oven Size, Shape, and Capacity

In by-product coke production, the coke is produced in a series of narrow, 400- to 600-mm- (16- to 24-inch) wide, 12- to 18-meter- (40- to 60-foot) long ovens that may be 3 to 6 meters (10 to 20 feet) high, or even higher. The smallest ovens produce about 7.5 tons of coke per cycle, while the larger ones produce about 39 tons of coke per cycle. The new jumbo ovens (7.5 meters high, 550 millimeters wide, 18 meters long) operated by Mannesman in Germany produce 50.6 tons of coke per oven cycle. Depending on the size of the ovens and the desired production rate, there may be from 35 to over 100 ovens in a battery. The ovens are slightly wider on the coke side (the side of the battery where the coke comes out of the oven) than on the pusher side (the side of the battery on which the pusher machine is located) so that the coke will more easily move through the oven as it is being pushed.

Oven Heating

In a battery, the wall separating adjacent ovens, as well as each end wall, is made up of a series of heating flues. At any one time, half of the flues in a given wall will be combusting gas, while the other half will be conveying the heated products of combustion (waste heat) from the combustion flues to the combustion or waste-heat stack through the "checker brick" heat exchanger. The flame temperature is above the melting point of the battery brickwork. In order to avoid melting the battery brickwork, and to provide more uniform heating to the coke mass, every 20 to 30 minutes the battery "reverses." The former waste-heat flues become combustion flues, and the former combustion flues become waste-heat flues.

The maximum design heating rates of a slot-oven type (long, narrow, and high ovens) of coke battery, which charges wet coal, range from 1 to 1.25 inches of oven width per hour. Therefore, a 16-inch-wide oven may be pushed (pushing is the process of forcing the coke out of the oven using a ram that moves from the pusher side to the coke side of the battery) in about 13 hours after being charged, while a 24-inch-wide oven may require 24 hours per coking cycle, depending on the heating rate. Preheated-coal–charged batteries, where the coal is preheated to about 400°F before charging into the ovens, can heat the coal at rates of up to 1.5 inches of oven width per hour. Coke produced in an 18-inch-wide preheated-coal–charged battery can be pushed in 12 hours.

A by-product battery may be fired with a number of fuels, including coke oven gas, natural gas, and blast furnace gas. Coke oven gas is the most common fuel because it is produced by the coking process in quantities that exceed the amount required to heat the battery.

Maintaining the proper heat distribution in a coke oven is a complex task. In a typical battery, there will be over 2000 individual flues with from one to four burners per flue. Each burner has a nozzle that must be of the correct size and must

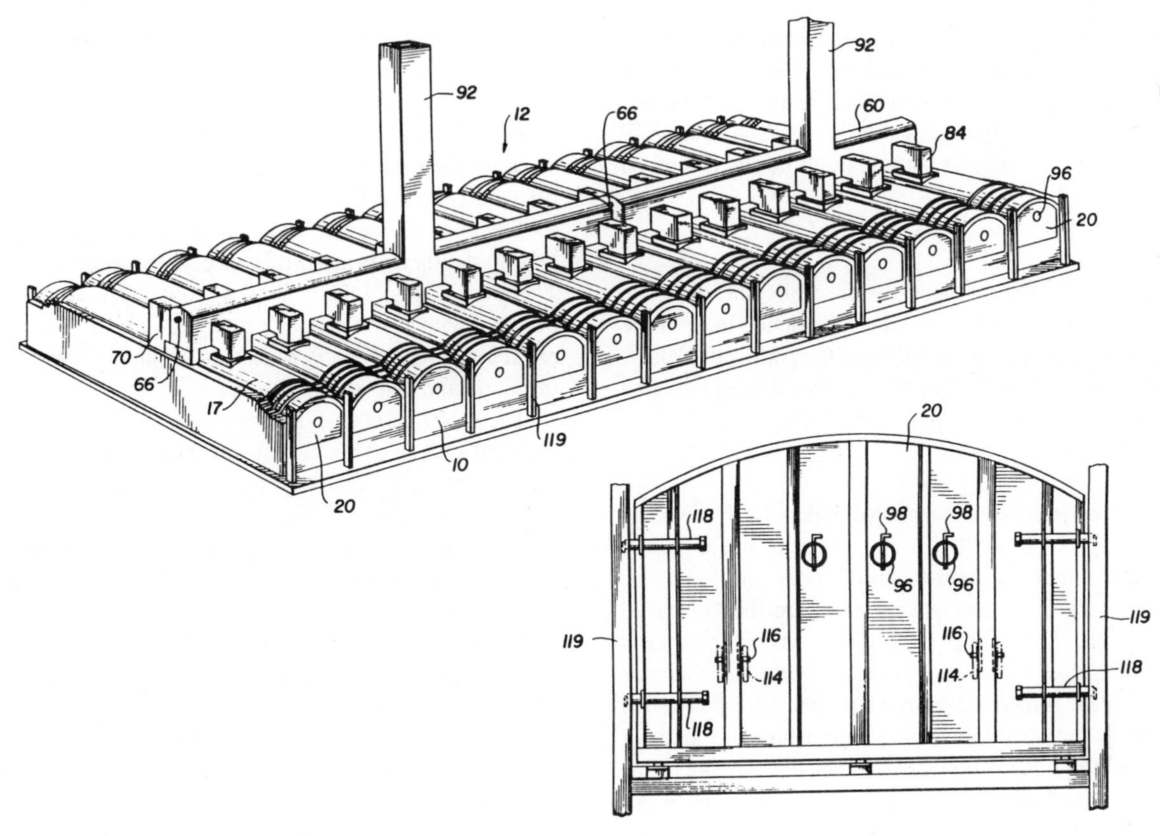

FIGURE 4. General Arrangement of a Nonrecovery Coke Battery (Also an End View of a Door)

be kept clean. A number of factors affect battery heating, including the types of coal used, the moisture content of the coal, the bulk density of the coal, the level and volume of the coal charged, variations in the coke production schedule, the fact that the coke side of the oven is wider than the pusher side, and the order in which the ovens are charged and pushed.

Coke Oven Gas Control

During the coking process, about one quarter of the mass of the coal charged is volatilized and removed as gas. Collecting, cleaning, and distributing this gas properly is a major component of the emission control effort in the coke production process. Each oven is connected via offtake piping assemblies to one or two collector mains, which run the length of the battery. The gas reaching the collector main(s) is evacuated through the by-product recovery plant, where it is processed into clean gas and a number of by-products, including tar, light oil, naphthalene, ammonia, sulfur, and many other components.

The Coke Production Operation

Coal Handling

Most coke-making operations use a blend of from two to seven distinct coals selected to produce coke with the desired qualities of size, strength, reactivity, and so on. To reduce variations in the quality of the coke and the required coking times, it is desirable for the coal to be of consistent blend, moisture content, and bulk density at all times. Over a million tons of coal are processed at a typical coke plant in a year. Therefore, the coal grinding, blending, moisture control, and bulk density control steps of coal handling are critical to the successful operation of the coke plant. While they will not be discussed here, the coal mining, drying and cleaning, and quality assurance steps done at the mine are also critical to the successful operation of the coke plant.

Charging

With the oven doors in place and properly sealed, a larry car that contains the proper blend of pulverized coal is placed (spotted) over the oven to be charged. With the oven under a slight negative pressure via steam or liquor aspirated into the off-take piping, the charging hole ports or lids on top of the oven are removed and the coal is discharged into the oven until it is full. Near the end of the charging sequence, the leveler (chuck) door within and near the top of the pusher side door is opened and a bar about as long as the oven is inserted to level the charge and aid in filling the oven. When all of the coal is in the oven and leveled, the leveler door is closed, the charging port lids are replaced and sealed, the steam or liquor aspiration is turned off, and the off-take piping assembly is adjusted so that the gas evolved from the coke reaches the collector main.

Note: A number of batteries have been built that utilize a pipeline, rather than a larry car, for charging the oven. Pipeline-charged batteries have proved to be very difficult to operate and the last remaining such battery (Inland no. 11) is likely to be changed to larry-car–type charging. Therefore, pipeline-charged batteries will not be discussed here.

Coking

The coal in the oven is heated by the walls and the coke mass increases in temperature from the outside in until the volatile matter has been removed from the entire coke mass. During this time, all gas evolved is removed from the oven via the off-take piping to the collector main.

Pushing

The oven to be pushed is isolated (dampered off) from the collector main so that gas in the collector main is not discharged through the oven when the doors are removed. Both doors (coke side and pusher side) are then removed from the oven. The pusher ram, which has a face that is taller than the coke mass and is almost as wide as the narrow end of the oven, is inserted from the pusher side and pushed through the oven toward the coke side, pushing the coke out of the oven. The coke is typically collected in a quench car, which is mounted on rails and is used to transport the hot coke to the quench tower.

Quenching

The coke pushed from the oven is very hot (about 1050°C) and will burn when it contacts oxygen from the air. To keep the coke from burning, it is normally quenched with water. The quench car, which receives the hot coke pushed out of the oven, is driven to the quench tower, where the coke is sprayed with water to reduce its temperature to below the ignition point of the coke and below the temperature at which the wharf conveyor belts will burn. Ideally, the temperature of the quenched coke on the wharf will vary from 120°F to 400°F.

Nonrecovery Coke Production

Oven Size, Shape, and Capacity

The individual ovens are of a horizontal design with batteries of 30 to 60 ovens being typical. Each oven has two doors, one on each end. The individual ovens are generally between 30 and 40 feet long and between 6 and 10 feet wide with little or no taper in their width along the length. The internal oven chambers are usually semicylindrical in shape, with the apex of the arch 5 to 12 feet above the oven floor.

The Coke Production Operation

Coal is charged either through lidded holes located above the oven chamber (as with by-product ovens) or through the oven doorway via conveyor. The combustion products of the coal volatiles driven from the coal charge are burned in the chamber above the coal charge and in the gas pathway through the walls, beneath the oven (the sole flues), and out a stack. Each oven chamber has two to six off-take ports (in each oven wall) and the sole flue may be subdivided into separate flues supplied by individual or combinations of the off-take ports.

The oven doors are fabricated from steel structural units that hold either a bricked face or cast monolithic refractory. The outer edges of the doors that contact the oven ends can be fitted with a heat-resistant rope material or be sealed with refractory grout to prevent excessive air infiltration around the door and into the oven.

The sole flue is designed to supply heat to the bottom of the coal charge. This conductive heat flow, coupled with the radiant and convective heat flow from above the coal charge, is responsible for distillation of the coal. As with by-product coke making, coal used in nonrecovery coke batteries is normally washed and sized at the mines to allow for the production of coke of suitable chemical and physical properties. Further size reduction of the coal is performed at the coke plant, with coal particle size normally being 75–90% less than ⅛ inch when it is charged. The coal charge can consist of a blend of coals or a single coal.

When coal is charged into the oven from topside larry cars (as in by-product ovens), it heaps into mounds in the oven and must be leveled through use of a leveling bar inserted into the oven through an opening in the oven door (usually on the pusher side of the oven). Owing to the width of the nonrecovery ovens as compared with by-product ovens, the leveling bar is fitted with a spring-loaded expanding sweep that extends to almost the full width of the oven after insertion into the oven. For conveyor-charged ovens, no such leveling is required.

Upon charging of an oven, carbonization begins as a result of the hot oven brickwork from the previous charge. Combustion air is allowed to enter the oven through vents in the doors and into the sole flue and stacks. All gas flow emits from the stack via natural draft and, as a result, a slight negative pressure is maintained in the oven at all times. This negative pressure eliminates door, refractories, and charging lid leakages associated with by-product coke ovens.

Upon completion of the coking cycle, both doors are removed from the oven and the coke is pushed into a "hot coke railcar." The pusher is normally a steel beam structure with a steel plate face nearly the width of the oven and slightly taller than the coal charge depth. The door mechanism can be powered either pneumatically or through cable drive. The hot car then moves to a coke quench station essentially identical in design and function to those of by-product coke plants. Subsequently, the quenched coke is transported via conveyor to a screening plant in which it is subdivided into various size fractions for its many uses.

The oven dimensions of the nonrecovery battery offer great flexibility in operation in terms of coke cycle. Various amounts of coal can be charged into the oven over a wide range of charge depths. The amount of coal charged determines the total time required for carbonization to be completed. It is not uncommon for charge height to range from 24 to 76 inches in order to control coking times between one and four days overall. Thus holiday scheduling, maintenance, and production changes can be accommodated.

AIR EMISSIONS CHARACTERIZATION

Byproduct Coke Batteries

Coke Oven Doors

Door leaks occur when there are gaps between the coke door and jamb or between the door jamb and the surrounding brickwork. Coke ovens are kept at a slight positive pressure (relative to atmospheric) during the coking cycle. Therefore, if there is an imperfection in the seal between the door and jamb or between the jamb and the brickwork, the material volatilized during the coking process is likely to leak from the door or jamb.

TABLE 1. Typical Emissions from Coke Oven Door Leaks

Material	Typical Wt % Composition	Expected Emissions (g/h/leak)
Acetylene	0.11	0.01
Ammonia	1.90	0.16
Benzene	4.05	0.34
Butane	0.20	0.02
Butylene	0.85	0.07
Carbon dioxide	5.59	0.47
Carbon monoxide	18.28	1.53
Ethane	3.42	0.29
Ethylene	5.97	0.50
Hydrogen	9.58	0.80
Hydrogen cyanide	0.45	0.04
Hydrogen sulfide	2.90	0.24
Methane	40.92	3.43
Nitrogen	3.04	0.25
Oxygen	0.70	0.06
Propane	0.42	0.04
Propylene	1.02	0.09
Toluene	0.38	0.03
Xylene	0.22	0.02

Emissions

The emissions from door leaks are referred to by the U.S. Environmental Protection Agency (EPA) as "coke oven emissions" and are listed as a hazardous air pollutant under Section 112 of the Clean Air Act. These emissions are essentially uncleaned coke oven gas. There have been few reported measurements of coke oven door emission rates or composition. What has been reported is that estimates of coke oven door emission rates vary by a factor of ten to 100 or more.

Particulates Since coke oven door emissions are actually uncleaned coke oven gas, the particulate emissions are the fractions of raw coke oven gas that condense out at the filter temperature (248°F or 120°C during a standard EPA Method 5 test). This condensed gaseous material is virtually all "benzene soluble organics" (BSO).

There is little information on the rate of emission generation. In AP-42, the EPA reports 0.54 pound of particulate emissions per ton of coal charged for "uncontrolled" door leaks. However, in the EPA's *Coke Oven Emissions from Wet-Coal Charged By-product Coke Oven Batteries—Background Information for Proposed Standards,*[1] door leak emissions for a model battery with 10% leaking doors are reported as 0.01 to 0.12 kg BSO per hour per leak. The EPA's data are based on two sets of measurements of emissions from leaking doors under a coke side shed. One of these measurements was taken on a battery with about 70% leaking coke side doors, and the other on a battery with about 30% leaking coke side doors. During the tests reported by the EPA, the amount of BSO per leak per hour from the battery with 70% door leaks was significantly lower than that from the battery with 30% leaking doors.

Recent industry measurements have generally confirmed the range of the EPA's reported door leak emissions and have provided further insight into the reason for the range. The numbers at the low end of the range are associated with the light wisp type of leaks now commonly found on most batteries, while values near the high end of the EPA's range are associated with the very large leaks that are now relatively rare. Most large leaks now observed have mass emissions closer to the midpoint of the EPA's reported range of emissions. There currently is no reported information on the size distribution of door leak particulates. However, since they are condensed gaseous material, the size of the particles for these emissions are expected to be very small (i.e., submicron).

Gases Because door leaks are essentially raw coke oven gas, the gaseous portion of the emission is the lighter portion of the coke oven gas and can be approximated by the typical composition shown in Table 1.

Based on recent industry measurements of the condensible portion of an EPA Method 5 particulate emission test procedure, the emission rate for gaseous benzene from a leak appears to be essentially independent of the particulate (BSO) emission rate from a leak; it is about 0.34 gram per hour per leak. The other gases are expected to be present in the same relative proportions in which they are found in coke oven gas.

Toxics As stated, the EPA has listed "coke oven emissions" as a hazardous air pollutant under the Clean Air Act. In addition, the gaseous portions of the door emissions contain the following chemicals that are identified as toxics under SARA: ammonia, benzene, ethylene, propylene, toluene, xylene.

Coke Oven Lids and Off-takes

Coke oven lids are normally closed and sealed except when the oven is being charged or pushed. During these periods,

TABLE 2. EPA Emission Factors for Wet-Coal Larry Car Charging

	Uncontrolled		Stage Charging		Scrubber Control	
Pollutant	kg/Mg	lb/ton	kg/Mg	lb/ton	kg/Mg	lb/ton
Particulates	0.24	0.48	0.008	0.016	0.007	0.014
Sulfur dioxide	0.01	0.02				
Carbon monoxide	0.3	0.6				
Volatile organics	1.25	2.5				
Nitrogen oxide	0.015	0.03				
Ammonia	0.01	0.02				

the oven is either at atmospheric pressure or at a negative pressure so there should be no emissions. Similarly, coke oven off-take piping assemblies are normally sealed closed, except when the oven is being decarbonized or pushed so that they should not leak. However, both lids and off-takes are under pressure during the coking cycle and can leak if sealing methods fail.

Emissions

Emissions from lids and off-takes are the same hazardous air pollutant that the EPA calls "coke oven emissions." The characteristics of these emissions are the same as those of emissions from door leaks.

The EPA's AP-42 does not list any emission rate for this source. *The background information document for the coke oven emissions for the National Emission Standards for Hazardous Air Pollutants, (NESHAP)*, does estimate benzene-soluble particulate emissions from lid leaks as ranging from 0.0033 to 0.021 kg per leak per hour. Based on industry measurements of door leak emissions, it is expected that a typical lid leak has benzene-soluble particulate emissions of 0.006 kg/h and a normal off-take leak has benzene-soluble particulate emissions of 0.01 kg/h. A major off-take leak may have benzene-soluble particulate emissions of up to 0.1 kg/h.

The gaseous and air toxics emissions from these sources are expected to be similar to coke oven door emissions.

Coke Oven Charging

Charging is the process of introducing coal into the oven through the charging ports on top of the oven.

Emissions

Emissions from charging are similar in character to those from door, lid, and offtake leaks. All of these emissions are listed by EPA as the hazardous air pollutant "coke oven emissions." However, in addition to raw coke oven gas, charging emissions may contain coal dust. The EPA's AP-42 lists the emissions from wet coal larry car charging shown in Table 2.

Unfortunately, AP-42 does not relate emissions to the typical charging standards of seconds of visible emissions for a series of charges. Based on measurements of door emissions, it would be more accurate to relate charging emissions to seconds of light, moderate, and significant emissions during the charge. The emissions of benzene-soluble organic particulate emissions would be about 0.1 g/s for light visible emissions, 0.8 g/s for moderate visible emissions, and 1.7 g/s for heavy visible emissions. The difference between particulate emissions determined by this method and those predicted by other emission factors are likely to be coal-dust emissions of a fairly large particle size. The ratio of other gaseous emissions to benzene-soluble organic emissions would be the same as for door emissions.

In AP-42, the EPA has also estimated particle size distributions for stage charging operations. These estimates are given in Table 3.

Coke Oven Pushing

The process of pushing hot coke out of the oven is often one of the most spectacular and hard-to-control processes in the coke-making process. The hot coke breaks up as it falls out of the coke guide on its way to the quench car. As a result of this fall, the coke, which has newly exposed surfaces to the air and may contain volatile matter not removed during the coking cycle, often ignites violently. This ignition and the mechanical impact of falling coke hitting the quench car often release coke particles into the atmosphere. On the other hand, perfectly coked material can be pushed with

TABLE 3. EPA Emission Factor and Particulate Size Distribution for Wet-Coal Larry Car Stage Charging

Particle Size, μm	Cumulative Mass % Less Than or Equal to Stated Size	Cumulative Mass Emission Factors	
		kg/Mg	lb/ton
0.5	13.5	0.001	0.002
1.0	25.2	0.002	0.004
2.0	33.6	0.003	0.005
2.5	39.1	0.003	0.006
5.0	45.8	0.004	0.007
10.0	48.9	0.004	0.008
15.0	49.0	0.004	0.008
30.0	100.0	0.008	0.016

virtually no flames and little smoke or other emissions. The variation in uncontrolled pushing emissions from battery to battery and from oven to oven on the same battery makes reliable control difficult.

Emissions

Coke pushing emissions are primarily carbon particulate. The EPA has published the particulate emission factors by particle size for uncontrolled coke pushing as shown in Table 4.

There is very little information on gaseous emissions from pushing. The EPA does report gaseous emissions from uncontrolled coke pushing (see Table 5). However, these emission factors are based on a 1968 United Nations report and are not well documented. For example, a recent industry measurement indicates that uncontrolled coke pushing volatile organic compound emissions may be 0.005 kg/mg or 0.01 lb/ton, which is less than one tenth of that reported in AP-42.

Coke pushing is generally not considered a source of air toxics. The emissions from coke pushing are *not* included in the listed hazardous air pollutant "coke oven emissions."

TABLE 4. EPA Emission Factors and Particulate Size Distribution for Uncontrolled Coke Pushing Operations

Particle Size, μm	Cumulative Mass % Less Than or Equal to Stated Size	Cumulative Mass Emission Factors	
		kg/Mg	lb/ton
0.5	3.1	0.02	0.04
1.0	7.7	0.04	0.09
2.0	14.8	0.09	0.17
2.5	16.7	0.10	0.19
5.0	26.6	0.15	0.30
10.0	43.3	0.25	0.50
15.0	50.0	0.29	0.58
30.0	100	0.58	1.15

TABLE 5. EPA Gaseous Emission Factors for Uncontrolled Coke Pushing Operations

Material	Emission Factor	
	kg/Mg	lb/ton
Carbon monoxide	0.035	0.07
Volatile organics	0.1	0.2
Ammonia	0.05	0.1

Coke Oven Underfire Stack

The underfire (or combustion) stack's main purpose is to provide natural draft to the battery combustion system and to convey and emit the combustion products from heating the battery.

Emissions

The emissions from the underfire stack come from two sources. Every underfire stack conveys the emissions from fuel combustion of the gas used to heat the battery. In addition, there are often some unburned coal fines that make their way to the combustion stack via oven-to-flue leakage. These coal fines can often be traced to a recently charged oven and thus stack opacity provides useful feedback to the operator concerning oven wall damage needing repair.

Coke battery stack emissions are no more toxic than emissions from other fuel-burning sources and are *not* included in the listed hazardous air pollutant "coke oven emissions." The effect of fuel type on particulate emissions is indicated by the data from the EPA's AP-42 uncontrolled emission factors shown in Table 6.

In AP-42, EPA also gives the particle size estimates for uncontrolled emissions when using undesulfurized coke oven gas (see Table 7). While this emission information is from a reputable source, it should be remembered that coke battery underfire particulate emissions are highly dependent on the condition of the battery walls and on whether the underfire gas has been desulfurized. It is not uncommon for batteries fired on similar fuels to have particulate emissions that vary by a factor of 4 or 5. Similarly, the sulfur dioxide emission factor for underfiring with undesulfurized coke oven gas is based upon a number of assumptions. It is much

TABLE 6. EPA Emission Factors for Coke Battery Combustion Stacks

Fuel Type	Particulate Emission Factor		Sulfur Dioxide Emission Factor	
	kg/Mg	lb/ton	kg/Mg	lb/ton
Coke oven gas (undesulfurized)	0.234	0.47	2.0	4.0
Blast furnace gas	0.085	0.17		

TABLE 7. EPA Particulate Emission Factors and Size Distribution for Combustion Stack of Coke Batteries Fired with Undesulfurized Coke Oven Gas

Particle Size, μm	Cumulative Mass % Less Than Stated Size	Cumulative Mass Emission Factor	
		kg/Mg	lb/ton
1.0	77.4	0.18	0.36
2.0	85.7	0.20	0.40
2.5	93.5	0.22	0.44
5.0	95.8	0.22	0.45
10.0	95.9	0.22	0.45
15.0	96	0.22	0.45
30.0	100	0.23	0.47

TABLE 8. EPA Emission Factors and Size Distribution for Coke Quenching Operations

Particle Size, μm	Cumulative Mass % Less Than or Equal to Stated Size				Cumulative Mass Emission Factors (lb/ton of Coal Charged to Oven) Divide by 2 to get kg/Mg			
	Clean Water		Dirty Water		Clean Water		Dirty Water	
	No Baffles	Baffles	No Baffles	Baffles	No Baffles	Baffles	No Baffles	Baffles
1.0	4.0	1.2	13.8	8.5	0.05	0.006	0.72	0.11
2.5	11.1	6.0	19.3	20.4	0.13	0.03	1.01	0.27
5.0	19.1	7.0	21.4	24.8	0.22	0.04	1.12	0.32
10.0	30.1	9.8	22.8	32.3	0.34	0.05	1.19	0.42
15.0	37.4	15.1	26.4	49.8	0.42	0.08	1.38	0.65
30.0	100	100	100	100	1.13	0.54	5.24	1.30

more reliable to use the actual gas sulfur content and the quantity of fuel gas consumed to determine battery stack emissions.

Coke Quenching

Quenching is the process of spraying water on hot coke to reduce its temperature so that it will not burn.

Emissions

Emissions from quenching are typically larger-sized particulates created by the breakup of the hot coke when it is hit with water. This particulate is carried up the quench tower by the velocity of the steam plume. Quenching is normally conducted with relatively clean water (less than 1500 mg/L total dissolved solids); however, at some facilities, water containing over 5000 mg/L of total dissolved solids may be used (this is often called dirty water quenching). The typical control for conventional quenching is to install baffles in the quench tower. The EPA has published the size-specific emission factors shown in Table 8 for the various combinations of coke quenching practices.

Very little data are available about other emissions from quench towers. Some researchers have reported various volatile organic emissions from quenching, while some industry tests have been unable to duplicate these findings when measuring emissions.

Coke quenching emissions are *not* included in the listed hazardous air pollutant "coke oven emissions."

Battery Venting

When the exhausters have insufficient suction to remove the gas being generated by the battery, emergency relief vents on the battery will open and the battery will vent raw coke oven gas.

Emissions

These emissions are essentially raw coke oven gas and are hazardous air pollutant "coke oven emissions." The composition of the emission is expected to be similar to that reported for coke oven door leaks. The quantity of emissions depends on the amount of gas not evacuated by the exhausters.

Coke By-Product Recovery Plants

As their name indicates, the original purpose of coke oven by-product–recovery plants was to recover by-products such as tar, benzene, sulfur, and ammonium sulfate for sale. Because of current economics in North America, most by-products plants now function primarily to clean the coke oven gas for downstream users. The "by-products" are now often produced as a necessary part of cleaning the gas. This has resulted in the more recently constructed coal chemical plants being designed to destroy, rather than produce, many of the by-products historically produced. Older plants, which have the capability to refine the tar and light oil into many constituents, are now often operated to produce tar and light oil for sale as feedstocks to chemical manufacturers.

Because it is essentially a chemical plant, the coke oven by-products emission sources are similar to those found at other chemical manufacturing facilities. These sources are product storage tanks, decanters, reactors, and heat exchangers. Figure 5 shows the simplified flows at a typical coal chemical plant. Notice that flushing liquor, circulating liquor, and weak (or excess) ammonia liquor are similar. Similarly, light oil and BTX are essentially the same material.

Emissions

Product storage tanks for tar, light oil (or BTX), and excess ammonia liquor have emissions related to the vapor pressures of the materials and the operation of the tanks. Decanters and sumps also can have emissions related to the material they contain, flow rates, temperature, and decanter design. Cooling towers and process tanks for materials (such as final cooler cooling water and sulfuric acid) that come into contact with the coke oven gas can emit materials that have been adsorbed from the gas.

There are normally no particulate emissions from the coal chemical plant. In addition, the only sulfur emissions

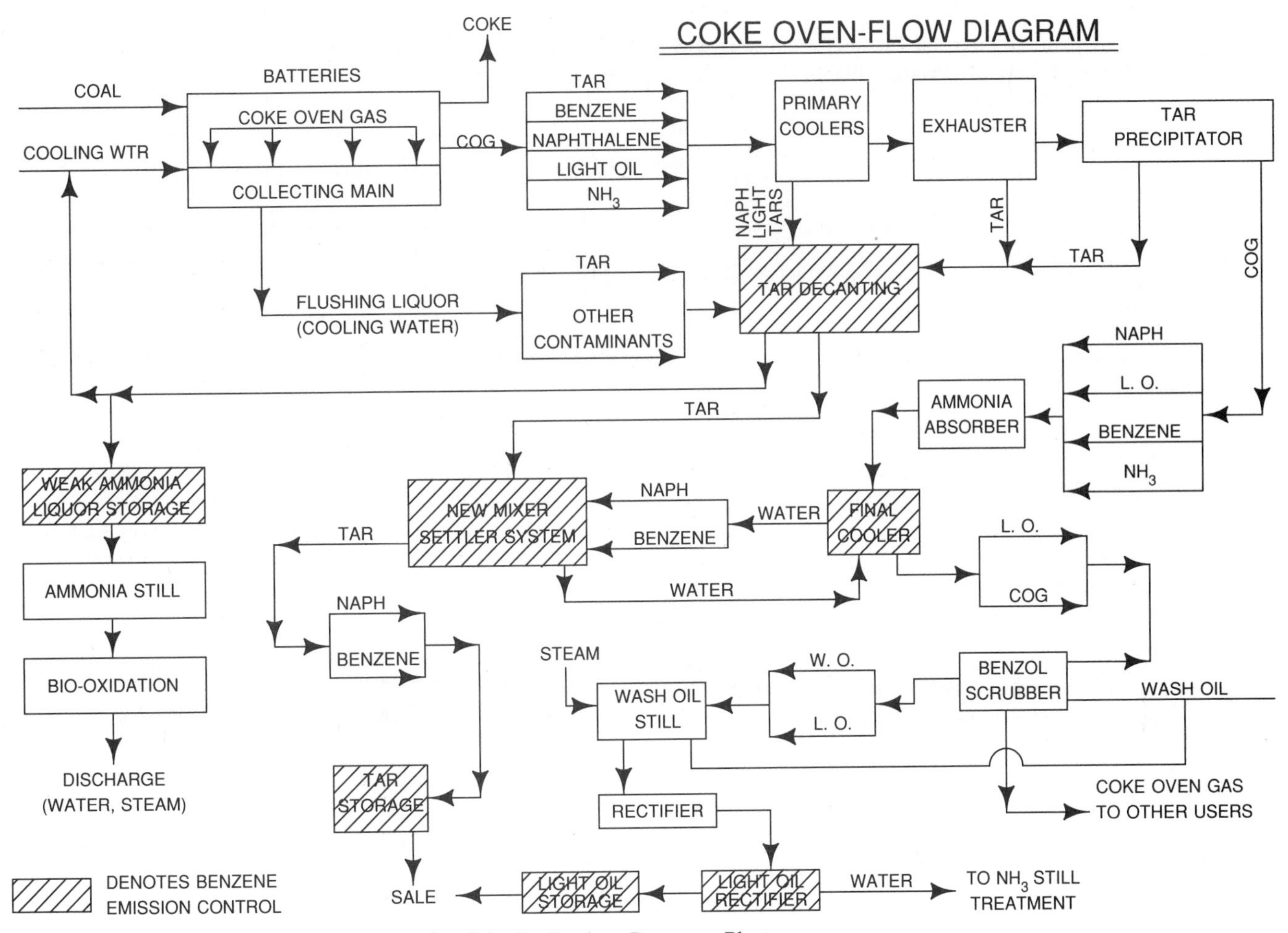

FIGURE 5. Simplified Flow Diagram of a Coke By-Product–Recovery Plant

are likely to occur from the "tail gas" of any sulfur-recovery plant that is used to convert hydrogen sulfide into elemental sulfur. The only nitrogen dioxide or carbon monoxide emissions will be from combustion sources such as those associated with the sulfur-recovery operation.

The most common emissions from coke oven by-product–recovery plants are the various species of volatile organic compounds (VOCs), including benzene and other air toxics. Chemicals that may be found in typical coke oven gases and are listed as toxics under SARA Title III include ammonia, anthracene, benzene, biphenyl, dibenzofuran, ethylbenzene, ethylene, naphthalene, phenol, propylene, quinoline, styrene, toluene, 1,2,4-trimethylbenzene, and xylene.

Most of the effort in estimating emission rates from by-product–recovery plants has focused on benzene emissions. Some work has also been done on estimating total VOC emissions. The data in Table 9 from the EPA's *Benzene Emissions from Coke By-product Recovery Plant—Background Information for Proposed Standards*[2] indicate the estimated range of uncontrolled benzene emissions from both furnace and foundry coke plants.

While the emission estimates in Table 9 are the best commonly available, they should be used with caution. Not all by-product plants have all of these components. In addition, some of the emission estimates do not make sense. For example, the emission estimate for the storage of tar, which has a relatively small amount of benzene, is twice as high as the emission estimate for the storage of benzene (where all emissions are benzene). It is best to use these factors as a screening tool to determine the likely sources of benzene emissions at a facility and then to use facility-specific information to determine actual facility benzene emissions.

The other source of emissions at a by-product–recovery plant is equipment leaks, primarily from equipment in the light-oil–recovery plant carrying liquids that contain significant concentrations of benzene. The EPA estimates of the emissions from such sources are given in Table 10.

Nonrecovery Battery Sources

Coke Oven Doors, Lids, and Off-takes

Emissions

Because of the oven design and operating practices (negative oven pressure), emissions from doors, lids, and off-takes do not exist.

TABLE 9. EPA Benzene Emission Factors for Uncontrolled Furnace and Foundry Coke By-Product Plants (Grams of Benzene per Megagram of Coke)

Source	Furnace Plant, g/Mg	Foundry Plant, g/Mg
Direct-water final cooler cooling tower	270	197
Tar-bottom final cooler cooling tower	70	51
Naphthalene separation and processing	107	79
Light oil condenser vent	89	48
Tar intercepting sump	90	45
Tar dewatering	21	9.9
Tar decanter	77	36
Tar storage	12	5.6
Light-oil sump	15	8.1
Light oil, BTX, or benzene storage	5.8	3.1
Flushing liquor circulation tank	9	6.6
Excess ammonia liquor circulation tank	9	6.6
Wash-oil decanter	3.8	2.1
Wash-oil circulation tank	3.8	2.1

TABLE 10. EPA Benzene Emission Factors for Equipment Leaks (Kilograms of Benzene per Leak per Day)

Source	VOC Emissions	Furnace Plant		Foundry Plant	
		LO[a]	Benzene[b]	LO[a]	Benzene[b]
Valves	0.26	0.18	0.22	0.16	0.20
Pumps	2.7	1.9	2.3	1.7	2.1
Exhausters	1.2	0.28	0.28	0.25	0.25
Pressure-relief devices	3.9	2.7	3.4	2.5	3.1
Sampling connections	0.36	0.25	0.31	0.23	0.28
Open-ended lines	0.055	0.038	0.047	0.035	0.043

[a]"LO" indicates that the plant recovers light oil. The amount of benzene in the light oil is assumed to be 70% at furnace plants and 63.5% at foundry plants.
[b]"Benzene" indicates that the plant recovers refined benzene. The amount of benzene averaged over the light oil and benzene portions of the plant is assumed to be 86% at furnace plants and 78% at foundry plants.

Coke Oven Charging
Emissions

Limited testing at a point above the door during 3.5-minute charges indicates the emission factors shown in Table 11.

Coke Oven Pushing and Quenching
Emissions

Pushing and quenching operations are similar to conventional by-product oven operations. Therefore, emissions will also be similar in terms of quantity and constituents. Limited testing inside a hot-car shed has provided some qualitative results. It was not possible to quantify exhaust gas volume during pushing. However, assuming the AP-42 factor of 1.15 pounds per dry ton of coal, the factors shown in Table 12 were indicated.

Combustion Stack
Emissions

Stack emissions are, to a large extent, a function of the coal mix used in the coking process. Some testing has been done for specific coal mixes. Emission factors indicated were as given in Table 13.

None of the polycyclic aromatic hydrocarbons (PAHs) typically associated with coke oven gas emissions were detected (e.g., benzo-α-pyrene, anthracene, phenanthrene, chrysene, pyrene). Among the volatile and semivolatile organic compounds, only four were consistently present. They were benzene, bromomethane, methyl ethyl keytone (2-butanone), and carbon disulfide.

TABLE 11. Charging Emissions for Nonrecovery Batteries

Pollutant	Emission Factors Uncontrolled, Pounds per Dry Ton Coal Charged
TSP (30% less than 10 μm)	0.007–0.025
Benzene-soluble organics (BSO)	ND[a]
Benzo-α-pyrene (BaP)	2.23×10^{-8}

[a]BSO detection limit = 0.48 mg per sample (average of 0.15% of total particulate).

TABLE 12. Pushing Emissions for Nonrecovery Batteries

Pollutant	Emission Factor Uncontrolled Pounds per Dry Ton Coal Charged
TSP	1.15 (AP-42)
PM-10 fraction (0–30 feet from plume), 37%	
PM-10 fraction (45–60 feet from plume), 90%	
Benzene-soluble organics (BSO)	0.0011
Benzo-α-pyrene (BaP)	ND

TABLE 13. Combustion Stack Emissions for Nonrecovery Batteries

Pollutant	Emission Factor, Uncontrolled Pounds per Dry Coal Ton Charged
TSP	3–4
SO_2	10–20
NO_X	0.5–1.0
CO	0.08–0.16
Naphthalene	6.0×10^{-4}
Di-*n*-butylphthalate	2.4×10^{-4}
Benzene	3.4×10^{-4}
Bromomethane	2.5×10^{-5}

AIR POLLUTION CONTROL MEASURES

By-product Coke Batteries

Coke Oven Doors

Equipment

Door and Sealing Designs In the past, when batteries were shorter, it was typical to utilize rigid doors and jambs and to hand lute (seal) the opening between the door and jamb with a claylike material. More recently, door leakage control has relied on maintaining very close tolerances between the door sealing edge and the door jamb. The tars contained in the raw coke oven gas escaping from small leaks remaining between the door sealing edge and the door jamb will condense and seal the opening—and thus these doors are called self-sealing.

The task of minimizing gaps between the door knife (or sealing) edge and the door jamb is complicated by the tendency of the door jambs and door frames to warp over time. This tendency for warpage is greater on taller ovens that have been constructed since the early 1970s. A number of techniques, in addition to the continual replacement of doors and jambs that are not perfectly straight, have been employed to reduce leaks from coke oven doors. These include adding adjustment to the door knife (or sealing) edge to allow it to reconform to a jamb that is no longer straight and plumb and using sealants such as sodium silicate to plug gaps remaining after the door is seated.

In addition to simply increasing the adjustability of the door knife edge with various spring arrangements, a number of manufacturers are now offering "flexible" doors that are designed to bend. This allows them to seal with jambs that have many inches of warpage. These doors have not yet been fully tested in long-term operation. The initial results after replacing all doors and jambs on a battery with flexible doors are typically positive; that is, there is a significant decrease in leakage. However, based on experience to date and depending on battery and door operation and maintenance practices, the improvement has not always proved to be long lasting. When maintained similarly to conventional doors with adequate adjustability, the new flexible-door designs appear to have about the same leak performance.

The use of sodium silicate has been shown to dramatically reduce door leaks for short periods. However, over the longer term, use of this material tends to mask the underlying problem of the door and jamb's not sealing correctly. Typically, after less than a year's use, the gaps between the door sealing edges and the jambs become so large that even the sodium silicate no longer can seal the opening. After that time, the battery typically will have significant door leakage until the core problems are addressed. The use of door sealants such as sodium silicate also creates major door and jamb cleaning problems, which make proper door maintenance difficult.

Auxiliary Capture Devices An alternative method of controlling emissions from door leaks is to capture the emissions and duct them to an appropriate control device. Two methods have been used to capture door leak emissions after they have been generated: door hoods and coke side sheds.

A few batteries have been constructed with evacuation hoods positioned over the doors to capture the door leak emissions as they rise above the battery. Reports on at least one of these batteries indicate that the hood did not have a significant impact on the control of door emissions. This was so because typical door leaks are so small in relation to the distance from the leak to the hood that they are often affected by winds and are blown away from the battery before they reach the hood.

There are also a number of batteries with coke side sheds. A properly designed coke side shed will capture virtually all leaks from coke side doors (which are typically the most difficult doors to keep from leaking). If the control device is capable of cleaning the door leak emissions, there will be no impact on the ambient air regardless of the amount of leakage. A number of the control devices used on the exhaust gases from coke side sheds have the potential to remove the benzene-soluble particulate emissions from the shed exhaust.

Work Practices

Work practices play a very significant part in door emission control. With the exception of the coke side shed, all of the current methods of door emission control are highly dependent on the work practices discussed in the following.

Jamb and Brickwork Maintenance The surface between the door jamb and the battery brickwork must be reasonably level and properly packed to prevent leakage there. The jamb also must be properly supported by the underlying brickwork or it will soon warp, making proper sealing difficult. In addition, if the oven walls are too close together or have excessive carbon buildup, the door plug will not fit fully into the oven. This will prevent the door sealing edge from fully mating with the door jamb, thus virtually ensuring that the door will leak.

There are significant differences in battery designs. On some batteries, changing door jambs is relatively routine, whereas on others, doing so requires a major effort, with the possibility that significant brickwork damage will occur that will necessitate major wall repairs coincident with the door jamb change. Thus while maintaining true and flat door jambs is an important part of some door leakage programs, other programs rely on being able to mate the doors to the jambs properly as they currently exist. In these programs, changing door jambs is a last resort.

Door Maintenance In order for the door to seal properly, the sealing knife edge must not have any nicks or other imperfections and it must mate with the door jamb. This means that the door sealing edge and the springs that hold it in place must be properly maintained. There are many other parts of a door assembly that affect its leak performance, such as the door latches and door plug. To maintain these items in proper condition, regular rebuilding of each coke oven door is essential. In addition, doors that leak before their next scheduled rebuild, because they either are damaged or have some component failure, often need additional maintenance. The schedule for regular rebuilds varies depending on door designs, the age of the doors, and battery operational practices. To date, there is no consensus on an ideal routine door-rebuild frequency. (In the United Kingdom, the standard frequency is reported to be one door repaired or rebuilt per day for a two-battery operation or about once a year for each door.)

Door Adjusting The sealing edges of most modern coke oven doors have a significant range of adjustment that can be used to ensure that the door mates with the jamb properly. In addition, each door has stop blocks that prevent the door from being pressed against the jamb so hard that the sealing edge is permanently deformed. Proper adjustment of the door to its jamb is essential to control door leaks. However, there is significant disagreement on the advisability of adjusting the door sealing edge after the door has been in service on a jamb. Some operators believe that door leaks can be controlled by routinely adjusting the sealing edge to stop observed leaks. Others believe that once the door is adjusted to the jamb, routine adjustments of the sealing edge are likely to cause additional leaks and so it is better to determine the cause of the leak and make repairs as needed or even to rebuild the door completely if it starts leaking. There are no reported controlled long-term studies to determine which method of operation results in less door leakage.

Door Cleaning In normal battery operation, the door seal gas channel, the door plug, both the outside and inside areas of the jamb, and the leveler door should be cleaned of tar and carbon buildup prior to every charge. This cleaning may be done manually (typically on short batteries) or by machine. However, this cleaning, particularly manual cleaning, is generally erratic in frequency and in quality, resulting in hard carbon buildup on the seal, plug, and jamb areas. Therefore, in most operations, each coke oven door is cleaned by sand blasting or high-pressure water blasting on a routine basis between door rebuilds. As with door rebuilds, there is no consensus on the ideal type of cleaning or the ideal timing of the cleaning. What seems to work at one facility often increases leaks when tried at another, similar facility. However, because clean doors and jambs are essential to good sealing performance, all new batteries are equipped with mechanical cleaners and some older batteries are being retrofitted with mechanical cleaners on the door and pusher machines.

Sodium Silicate Sealing As indicated above, sodium silicate can be used to seal a door leak when the opening causing the leak is too large to self-seal. However, in addition to the practical problems of properly applying the sealant to portions of the door that are more than 20 feet above the nearest supporting surface, use of this material often leads to more significant leakage later. This happens because sodium silicate will mask the underlying cause of the door leakage until the gap causing the leak becomes so large that even this sealant cannot stop the emission successfully. It becomes evident whenever there is a loss of gas pressure control on the battery and the higher pressure in the oven breaks all of the sodium silicate seals. At that time, there is often a major leak instead of the minor wisp leak that was originally controlled with the sealant. Thus while the limited application of sodium silicate, along with proper door maintenance, may be used to improve door leak performance, reliance on sodium silicate to the extent that the need for proper door maintenance is hidden will simply result in more significant leaks in the future.

Battery Back-Pressure Optimization The collector main is maintained at a positive pressure designed to ensure that when the coal has completed coking and is no longer generating gas, the pressure at the bottom of an oven will not be negative. This is to ensure that air does not leak into the battery. Air leaking into the battery can cause the coke oven gas to burn, resulting in permanent damage to battery brick- and ironwork that will result in greatly increased emissions. In addition, too much air can cause the gas mixture in the collector main to become explosive, and possibly to explode, resulting in personal injury or major damage to the by-product plant, as well as very large emissions.

The rate of gas evolution from each single oven is not constant during the coking cycle. In addition, the gas flow through the collector main is affected by the oven charging process. Therefore, a damper system called an askania valve is used to regulate the battery back pressure. If this valve allows significant variations in back pressure, the tar that forms the final seal in a self-sealing door will be dislodged and the door will resume leaking. In addition, if the back pressure is set too high, the pressure on the door seals will be harder to contain within the oven. Similarly, if the back pressure is set too low, the fires associated with air infiltration will cause significant damage to the battery brickwork and door jamb, resulting in additional leakage.

Door Removal and Installation, Spotting, and Latching The sealing edges of coke oven doors are easily damaged by contacting adjacent battery surfaces during their removal or installation. Frequently, the seals are damaged by hitting the latch hooks while the doors are being inserted into the oven. The removal and replacement of a door is a complex process involving unlatching, lifting, and twisting the door, and then reversing the process. If the machine performing this task is not properly adjusted or is not properly located in relation to the oven, the door sealing edge may contact the fixed portion of the door latches or buckstays and be damaged.

In addition to routine maintenance to ensure that the machines are properly adjusted, there must be a reliable method of "spotting" or properly locating the machine in front of each oven. Spotting methods range from the simple (such as aligning a pointer on the machine with a pointer on the battery) to sophisticated laser-alignment methods. All methods can be made to work. While the laser systems are more precise, all methods rely on the target on the battery being in the proper location. These targets are often located on, or are attached to, the battery buckstays. Over time, the buckstays move and warp. Therefore, no matter what spotting system is used, there must be routine realignment to ensure that the targets are properly located.

Door latching is important to ensure that the door sealing edge is held securely against the door jamb, but is not under so much pressure that it is permanently deformed. As discussed above, the door stop blocks are often used as an indication that the door is properly secured. In addition, door latching pressure is routinely measured as an indication of proper door placement and latching.

Coke Oven Lids and Off-takes

Equipment Design

Standpipe Caps The standpipe cap on the off-take piping is normally opened prior to each push. This means that the cap must be sealed to the standpipe in such a way that it can be opened, but that it must seal reliably during each coking cycle. The three most common methods of sealing standpipe caps are water seals, luting, and a sealing knife edge.

In a water seal cap, a lip on the cap is submerged in a water-filled trough. Major concerns with this design are keeping the standpipes horizontal so that the lip on the cap is submerged in the proper amount of water to maintain the seal and keeping the water inputs and drains on the troughs clean so that the troughs are filled with water, but do not overflow.

Many caps are designed to seal on a slant and must be sealed with a luting compound each time they are closed. The major concern with this type of system is ensuring that the luting compound is properly mixed and applied.

Caps with knife edge seals are similar in operation to the sealing edges of flexible coke oven doors. These caps require adequate levels of cleaning and maintenance to seal reliably.

Standpipe Slip Joints Standpipes extend from the oven brickwork to the gooseneck, which is, in turn, connected to the collector main. In many batteries, there is a slip joint where the standpipe emerges from the battery top paving (in other batteries, the slip joint may be between the standpipe and the gooseneck or between the gooseneck and the collector main). These slip joints must be kept gastight to prevent leaks, but must slip to compensate for normal battery movement. They are often sealed with a packing material. However, especially when the joint is located where the standpipe exits from the battery top paving, currently available packing materials often burn out. Another sealing method is to use tar in the slip joint.

Charging Port Sealing Charging port lids are normally luted whenever they are replaced after a charge. On some batteries, charging port lids are designed to seal without the need for luting.

Work Practices

Control of off-take and charging port lid leaks is mainly a function of paying attention to the details of cleaning, luting, and maintenance. There are occasions when castings need to be replaced and the refractory linings of the off-takes need to be repaired or replaced.

The operating practices of dampering the oven off from one of the collector mains partway through the coking cycle on a dual-collector main battery and of decarbonizing properly can reduce the need for casting and refractory repairs.

Coke Oven Charging

Equipment

Stage Charging Virtually all larry car charging emission control systems now rely on stage charging. Stage charging is a process that uses steam aspiration at the gooseneck assembly of the off-take piping to create a vacuum in the oven being charged. The material that would otherwise be emitted is evacuated to the collector main and cleaned in the by-product plant.

During the stage charging process, prior to leveling, coal will tend to pile and extend to the roof of the oven, effectively blocking the gas channel. Therefore, it is important that there be evacuation from both ends of the oven being charged. In a dual-collector main battery, steam is aspirated into each off-take assembly so that emissions can be evacuated from each end of the oven. In a single-collector main battery, steam is aspirated into the off-take assembly of a nearby oven referred to as a "jumper oven." A jumper pipe, which may either be integral with the larry car or a separate assembly, is used to connect the charging port furthest from the collector main of the oven being charged to the jumper oven.

In addition to proper evacuation of the oven, stage charging relies on proper seals between the larry car drop sleeves and the charging ports, as well as a "smoke seal" on the leveler bar, to reduce the infiltration of clean air that would overwhelm the evacuation capabilities of the steam aspiration system. Stage charging also relies on proper placement of the coal in the oven to ensure that there are no empty spots in the oven and that the gas channel is not blocked in more than one place at any one time. In stage charging, generally, the coal is dropped out of the end larry car hoppers first and then out of the center hoppers after the end hoppers are empty. On many batteries, the charging hole lids are removed only when the coal is ready to be dropped and are replaced whenever the hopper is empty. When this method is used, no more than two charging hole lids are off the oven at any one time in order to maximize the emission control effect of the steam aspiration. Since batteries are of different sizes and have different numbers of charging ports and different methods of removing and replacing the charging hole lids, the coal charging portion of the stage charging process is custom designed for each battery.

Scrubber-Equipped Larry Cars Early attempts to control charging emissions produced many types of controls, including the scrubber-equipped larry car. This car functioned by evacuating the emissions through a scrubber on the larry car. Because of weight limitations, the fans and scrubbers on these cars were often not sufficiently powerful to properly evacuate and clean the emissions from the charging process. Therefore, virtually all batteries now use stage charging as the means of controlling charging emissions. However, there are still a few scrubber-equipped larry cars that are used as backups when the cars equipped for stage charging are out of service for maintenance or repair.

Evacuation to a Land-Based Control Device In an attempt to overcome the problems with a mobile scrubber, some designs evacuated the charging emissions through a series of ducts to a fixed, land-based fan and control device. Because stage charging has proved to be simpler and more workable, only one land-based charging emission control system is currently in use in North America.

Pipeline Charging Pipeline charging provided another means of controlling emissions from the charging process. This system had both production-related problems (e.g., inability to fill ovens, shortened battery life) and emission-related problems (e.g., sometimes lids were blown off the oven being charged owing to off-take plugging and often created large door leaks during charging). Therefore, all pipeline-charged batteries, except for Inland Battery No. 11, have been shut down or converted to larry-car–type charging. Inland reports that it is considering conversion of its No. 11 Battery to a wet charge battery using stage charging.

Work Practices

Stage Charging Stage charging, which is as much a work practice as it is an equipment-related control, is described in detail above. For stage charging to control emissions effectively, the drop sleeves on the larry car must mate effectively with the charging ports on the battery. In addition, the steam nozzles must be clean, properly sized, properly oriented, and supplied with adequate steam pressure and volume; the smoke boot must be effective in reducing the infiltration of air around the leveler bar; and the gas channels from the oven to the collector main must be open.

Carbon and Tar Control Proper evacuation of the ovens depends on the presence of a clear gas channel for the material being evacuated from the oven. This channel can be clogged by excessive oven roof carbon, the buildup of carbon in the standpipe, the buildup of carbon or tar in the gooseneck, and the buildup of tar in the collector main.

Carbon is built up as coke oven gas passes over hot coke oven surfaces and is cracked (i.e., broken into its simple components) into carbon and hydrogen. The carbon attaches to the hot surface and can grow to many inches in depth, blocking the gas channel. The growth of roof carbon is controlled by operating practices, by using carbon cutters and decarbonizing air on the pusher machine ram, and by letting the oven sit with its charging ports and standpipes open either just before or just after the push. Standpipes and goosenecks are typically cleaned before each charging operation to ensure that they have adequate gas-handling capacity. In addition, periodic water blasting or other cleaning methods are used to keep this equipment clean.

Coke Oven Pushing

Equipment

Devices Connected to Land-Based Pollution Controls The following types of capture devices are ducted to fixed control systems located on land. The control devices in use include baghouses, venturi scrubbers, and wet electrostatic precipitators. As long as the design of the system protects the control device from possible failures (e.g., baghouses from high temperature or flame), all types of control devices can function acceptably for pushing emission control.

Sheds Coke side sheds are typically large steel structures that are at least twice the height of the battery and extend the entire length of the battery from its coke side face across the quench car tracks. These sheds are constantly evacuated. Depending on the size of the battery and the design of the shed, the evacuation rate can be from 154,000 to 600,000 acfm. The fume rises to the top of the shed, where it is stored in a plenum until it is completely evacuated to the control device. Sheds can be very effective in capturing all pushing and travel emissions, as well as door leak emissions.

Traveling Hoods A traveling hood is connected to a duct extending the length of the battery in such a way that the hood can be evacuated while it is moving. A typical design is the use of a hood with a flexible belt as the top member. The belt is lifted by, and moves on, rollers through the hood as the hood travels. Because of hood travel, these systems can control emissions during both coke pushing and quench car travel to the quench tower.

Typical designs for hood systems use from 6000 to 9000 acfm per ton of coke pushed. The fan is dampered, except during the pushing and travel cycle, resulting in less energy use than in a shed system. A major challenge in hood system design is capturing the emissions at the coke guide-door jamb interface. Fume generated at this point has a high thermal drive in the vertical direction. Weight and space restrictions on the hood, when combined with the horizontal distance from the point of fume generation to the source of suction at the pollution control duct, make it extremely difficult to capture these emissions. A related concern with hood systems is controlling heating so that the emissions generated during the push do not overwhelm the system and spill from the hood, thereby escaping capture.

Indexing Hoods Indexing hoods are similar to traveling hoods except that they are connected to the pollution control duct by inserting one to three connecting ducts from the hood into openings in the pollution control duct. Because they can only provide evacuation when mated with the pollution control duct at fixed locations, they cannot travel with the coke to the quench tower. An advantage of the indexing hood is that, because there is no airflow through the hood except when it is mated with the pollution duct, it is practical to maintain a "spare" hood that can be used to control emissions when the other hood is being repaired.

Coke Transfer Machines A coke transfer machine is an integrated coke guide, door machine, and pushing emission control hood and is connected to a pollution control duct that extends the length of the battery. Because of the integration of all pushing-emissions–related equipment into a single unit, close tolerances are maintained and these devices provide excellent emission control at relatively low air flows (3000 to 7000 acfm per ton of coke pushed). Because it integrates three functions into one device, a coke transfer machine is relatively complex and heavy, and, therefore, it is costly. Concerns over the ability of existing battery foundations to accept the weight of these machines, along with the relatively high cost of providing a second machine as a backup for use when the primary machine is being maintained, have resulted in their limited use in North America.

Mobile Systems

Mobile Scrubber Cars There are a number of proprietary-design pushing emission control cars that combine a mobile scrubber control device with the quench car in a single unit. These devices are generally used on batteries that are no larger than 4 meters high. Mobile scrubber cars are typically used to control emissions from existing, rather than new, batteries. An advantage of these units is that, with sufficient spare units, all pushes can be controlled without the need to construct a fixed system around an operating battery. Mobile scrubber cars can reduce pushing emissions to the same levels as those attained by land-based systems.

These devices put a scrubber, its energy source, and the water supply on wheels. They are heavy, use a lot of energy, and require a lot of maintenance. Owing to the close fit required between the machine and the battery to collect the pushing emissions successfully and the weight of the machines, substantial quench car track maintenance is typically required when these cars are used. Because of the high maintenance requirements associated with these cars, a number of them are being replaced with land-based systems.

Mobile Fume Suppression Systems An attempt to overcome the maintenance and energy consumption problems associated with mobile scrubber cars has resulted in the development of the simpler mobile fume suppression car. These cars typically include pumps, water spray nozzles, and a water tank. They are much lighter and more reliable than mobile scrubber cars.

These systems work by spraying water on the coke as it is pushed. This can stop combustion, knock down particulates, and agglomerate particles into larger sizes that will fall out near the battery. These cars are typically used on small batteries with longer than normal coking times. Tests indicate that they can reliably reduce particulate emissions by about 50% over uncontrolled pushes.

Kress Indirect Dry Cooling (KIDC) This proprietary system works to prevent the generation of the emissions so that capture and control are not needed. The system consists of a number of steel boxes that are roughly the size of the ovens being pushed, a cooling rack, and mobile equipment. In operation, the hot coke is pushed into a box that mates directly to the door jamb. The oxygen in the box is quickly consumed, and since the coke moves as an oven-shaped mass that does not break up, emissions are not generated. Once the coke is in the box, the box is sealed and taken to the cooling rack, where it is cooled from the outside. When the coke in the box has cooled, it can be handled as dry, cool coke. There are essentially no pushing or quenching emissions from this system.

The KIDC system has operated as a prototype at National Steel's Granite City facility and Bethlehem Steel's Sparrows Point facility. The system has significant pollution control benefits; however, it has not yet been proven in production operation. The first full-production KIDC system will be tried at the Sparrows Point facility in 1991. If it is successful, the system will be extended to an adjacent battery at this facility.

Work Practices

Heating Control The amount of volatile material left in the coke when it is pushed significantly affects the amount of uncontrolled pushing emissions. Control of the heating process to ensure that volatile matter and emission generation are minimized is most important when controls such as hoods, mobile scrubber cars, and mobile fume suppression are used. The control of coke volatile matter is also important to the blast furnace that uses the coke and where the presence of excess volatile material has an adverse impact on the coke's properties.

In order to heat the coal properly, the coal mixtures, coal bulk density, coal moisture, and oven charge levels must be as uniform as possible. Variations in these parameters will change the amount of heat needed to ensure proper coking. Because the coke battery is a massive structure that holds a lot of heat, it is virtually impossible to adjust coke heating to respond to short-term variations in the coal inputs to the process.

Ram Speed Within narrow limits, reducing the ram speed may reduce the rate of emission generation.

Coke Oven Underfire Stack

Equipment (Add-on Controls)

As stated above, the major function of a coke battery stack is to provide sufficient natural draft for optimum battery combustion operation. Since all of the add-on controls interfere with the natural draft of the battery, they are employed on a small number of batteries where all other means of emission control have failed.

For the control of opacity caused by coal leaking from the oven to the flue, electrostatic precipitators have been used most often. In 1990, Inland Steel operated electrostatic precipitators on four of its Indiana Harbor Works coke batteries. Electrostatic precipitators can operate at the temperatures commonly encountered at the base of a stack and thus do not require the gas to be cooled. Precipitators are not effective against sulfate particulate because the sulfate condenses at a temperature lower than those found at the base of the stack. The main operating problems associated with precipitators include fires in the dust hoppers and the shutdown of the induced-draft fans during power failures. At least one battery is believed to have had its life substantially shortened because failures of both the induced-draft fan and the bypass dampers resulted in explosions in the heating system.

Other controls, such as baghouses, can effectively clean the gas with low-pressure drop, but would normally require the gas to be cooled before it reaches the baghouse. Scrubbers should also be effective in controlling emissions from coke underfire stacks, but would have the disadvantages of high energy consumption and the need for subsequent wastewater treatment. Both of these control devices would also have the same induced-draft problems that have been associated with precipitators.

Work Practices

Because a well-operated battery in good condition does not usually have excessive particulate emissions, the focus on battery stack emission problems is normally on operating and maintenance practices.

Carbon Control While excessive carbon growth can cause charging emission problems, insufficient wall carbon leads to increased oven-to-flue leakage. Wall carbon is an effective means of sealing minor cracks between the oven and flue. Operating the battery to keep the proper amount of wall carbon is an important part of stack emission control.

Oven Dusting Silica dusting—which involves blowing a dust, under pressure, into a hot empty oven before it is charged—can seal cracks that are too large to be controlled by normal carbon growth. Typical oven dusting programs involve dusting a certain number of ovens each week so that each oven is dusted on a routine recurring basis.

Battery Maintenance Practices

The most important battery work practices for the control of stack emissions are those relating to battery brickwork maintenance. If the brickwork's condition is adequate to prevent excessive oven-to-flue leakage, control of any remaining stack emission problems is a function of ensuring proper combustion.

Oven Spraying A number of companies offer proprietary refractory spray products that are used to make effective, but temporary, repairs to small oven-wall imperfections. When used on a routine basis, these products can effectively control oven-to-flue leakage that does not require more extensive repairs.

Panel Patching Panel patching involves the use of castable zero-expansion brick in place of conventional brickwork to replace damaged sections of battery brickwork. The advantages of this system include quicker repair time, fewer joints with the potential to leak, easier tie-in to the remaining brickwork, and no problems with expansion during heat-up of the repaired area. As this practice is still fairly new, the life of these repairs is not yet known, but it is over four years.

End Flue Repairs Normal battery brickwork damage is usually confined to a few flues at each end of each oven. The general practice is to tear out the damaged flues and replace them with new bricks that are identical to the original bricks. When done properly, these repairs are effective in returning the battery to its original condition.

Through-Wall Repairs If the damage extends further than the end flues, it is often possible to replace the entire oven wall with brickwork similar to that originally used to construct the battery. This type of repair is also effective in returning the battery walls to their original condition.

Combustion Control

Combustion Air Control The combustion air to the battery is controlled by a series of finger-box adjustments for each oven flue wall. Although it is rare, it is possible for these adjustments to be sufficiently far off that they cause incomplete combustion and the emission of unburned carbon particulate.

Cleaning of Underfire Fuels The combustion of properly cleaned fuels (e.g., properly treated coke oven gas) affects emissions in a number of ways. Combustion of undesulfurized gas often results in a sulfate plume. The sulfate also reports as particulate during the standard EPA Method 5 stack test.

Adequate removal of materials such as naphthalene from the coke oven gas it is necessary to maintain the cleanliness of the nozzles that control the fuel flow to the heating flues. When the nozzles are not clean, it is difficult to maintain proper combustion conditions.

Coke Quenching

Equipment

Baffles in Quench Tower According to the EPA's emission factors, the installation of baffles in the quench tower can reduce quenching emissions by at least 50%. This is the only control device for quenching emissions that is currently in widespread use.

Kress Indirect Dry Cooling This proprietary pushing and dry quenching system (described under "Coke Oven Pushing") will completely eliminate quenching emissions. It has not yet been used in actual production, but is scheduled for a full production trial in late 1991.

Dry Quenching Quenching of coke with inert gas under controlled conditions instead of water can also eliminate quenching emissions and recover heat energy from the hot coke. While this process has been used in Europe and Japan, it has not been tried in North America.

Work Practices

Quench Water Quality Using relatively clean water for quenching can reduce quenching emissions. Most North American coke facilities use clean water for quenching.

Battery Venting

Equipment (Add-on Controls)

Flares When a battery is venting, the only effective control measure is to combust the emissions with a flare. There will still be sulfur dioxide emissions from the combustion of the undesulfurized gas.

Work Practices

Exhauster Redundancy and Maintenance Battery venting incidents, which indicate a breakdown in the exhauster system, are best prevented. While prevention is never perfect, maintaining multiple exhausters on different power systems (e.g., some on steam, some on electricity) and using power systems that are as independent of each other as possible can minimize battery venting incidents.

Change of Battery Operation During Venting Incident The gas being released during venting incidents is generated from the heating of coal. When a venting incident is occurring, it is prudent to stop all coal charging. In addition, if the venting incident is expected to last for some time, reducing the heat input to the battery may be prudent (because the battery has a large thermal mass, this action will have no effect on emissions during short venting incidents).

Coke By-product–Recovery Plants

Process Changes to Eliminate Sources

In by-product–recovery plants, major reductions in plant benzene (and VOC) emissions can be obtained by replacing the direct-water final cooler cooling tower and naphthalene separation and processing units with either a wash-oil final cooler or a tar-bottom mixer settler with indirect cooling. In many cases, this one process change will reduce facility emissions by over 90%.

The replacement of sumps with pipes and tanks is often a relatively simple means of reducing benzene (and VOC) emissions. The goal is to reduce the amount of material exposed to the air.

Gas Blanketing

Once process changes have been made, the major remaining sources of emissions are likely to be from storage and process tank vents. In late 1989, the EPA promulgated regulations requiring that all openings in such tanks be sealed and that they be vented to the coke oven gas main (or a control device). To allow the draw-down of product storage tanks connected to such systems, some provision must be made to replace the volume of the product with another material. The materials commonly chosen for this are nitrogen, coke oven gas, or natural gas.

There has been very little operating experience with these systems. Among concerns that have been expressed are tank rupture resulting from the plugging of gas lines by condensibles, transmission of fires and explosions from one tank to another through the gas manifold, and adverse effects on the quality of the coke oven gas (that accepts the material vented from the tanks). The EPA regulations require that these systems were to be in operation by September 1991. Since that time, all operators should have been gaining experience to determine which, if any, of the potential problems with the systems occur in operation.

Leak Detection and Repair

Once the two emission-reduction steps listed above have been taken, the remaining source of emissions is leaks from devices in liquid service that contain benzene or other VOCs and from the gas blanketing system and various tanks. The control of these leaks can be accomplished by routine monitoring and repair. The EPA has promulgated regulations requiring that the exterior of all valves, pumps, exhausters, and gas blanketing systems be sampled on a routine basis for VOC concentration levels above background. The agency has currently established 500 ppmv above background as the level of "no detectable emissions" and 10,000 ppmv as the level of a "leak." Once a leak is detected, there are time limits set for the first attempt at repair and in which to accomplish the final repair.

Nonrecovery Coke Batteries

Coke Oven Charging

Equipment

Charging of nonrecovery ovens can be accomplished either via topside charging holes or via horizontal conveyors through the oven door. In the latter case, only the bottom part of the door need be removed. Because there is suction on the oven, fugitive charging emissions are minimal. Jumper pipes can be used to improve suction. Sheds can be installed, but are not generally deemed necessary.

Work Practices

Minimizing emissions is not difficult since the draft is into the oven. Charging takes about 3.5 minutes on average using a retractable coal conveyor.

Coke Oven Pushing and Quenching

Equipment and Work Practices

These are similar to the equipment and work practices for by-product coke ovens.

Combustion Stack

Equipment

Air toxic compounds have been shown either not to be present or to be occasionally present in trace amounts. Therefore, off-gas controls for only SO_2 and particulate matter are indicated in the form of scrubbers, baghouses, or electrostatic precipitators.

Work Practices

Conventional work practices to maintain any installed pollution control equipment and ducting on the waste gas stream are all that are generally required.

References

1. *Coke Oven Emissions from Wet-Coal Charged By-product Coke Oven Batteries—Background Information for Proposed Standards,* EPA-450/3-85-028a, U.S. Environmental Protection Agency, Research Triangle Park, NC, April 1987.
2. *Benzene Emission from Coke By-product Recovery Plants—Background Information for Revised Proposed Standards,* EPA-450/3-83-016b, U.S. Environmental Protection Agency, Research Triangle Park, NC, June 1988.
3. *Compilation of Air Pollution Emission Factors,* AP-42 Supplement A, U.S. Environmental Protection Agency, Research Triangle Park, NC, October 1986, pp 7.2-1–7.2-22.

COPPER SMELTING

*Jacques Moulins, Editor**

In the United States, copper is produced from sulfide ore concentrates, principally by pyrometallurgical smelting methods. Because the ores usually contain less than 1% copper, they must be concentrated before transport to smelters. Concentrations of 15–35% copper are accomplished at the mine site by crushing, grinding, and flotation. Sulfur content of the concentrate ranges from 25% to 35% and most of the remainder is iron (25%) and water (10%). Some concentrates also contain significant quantities of arsenic, cadmium, lead, antimony, and other heavy metals.

PROCESS DESCRIPTION[1-3]

A conventional pyrometallurgical copper smelting process is illustrated in Figure 1. The process includes roasting of ore concentrates to produce calcine, smelting of roasted (calcine feed) or unroasted (green feed) ore concentrates to produce matte, and converting of the matte to yield blister copper product (about 99% pure). Typically, the blister copper is fire refined in an anode furnace, cast into "anodes," and sent to an electrolytic refinery for further impurity elimination.

In roasting, charge material of copper concentrate mixed with a siliceous flux (often a low-grade ore) is heated in air to about 650°C (1200°F), eliminating 20–50% of the sulfur as sulfur dioxide (SO_2). Portions of such impurities as antimony, arsenic, and lead are driven off, and some iron is converted to oxide. The roasted product, calcine, serves as a dried and heated charge for the smelting furnace. Either multiple-hearth or fluidized-bed roasters are used for roasting copper concentrate. Multiple-hearth roasters accept moist concentrate, whereas fluid-bed roasters are fed finely ground material (60% minus 200 mesh). With both of these

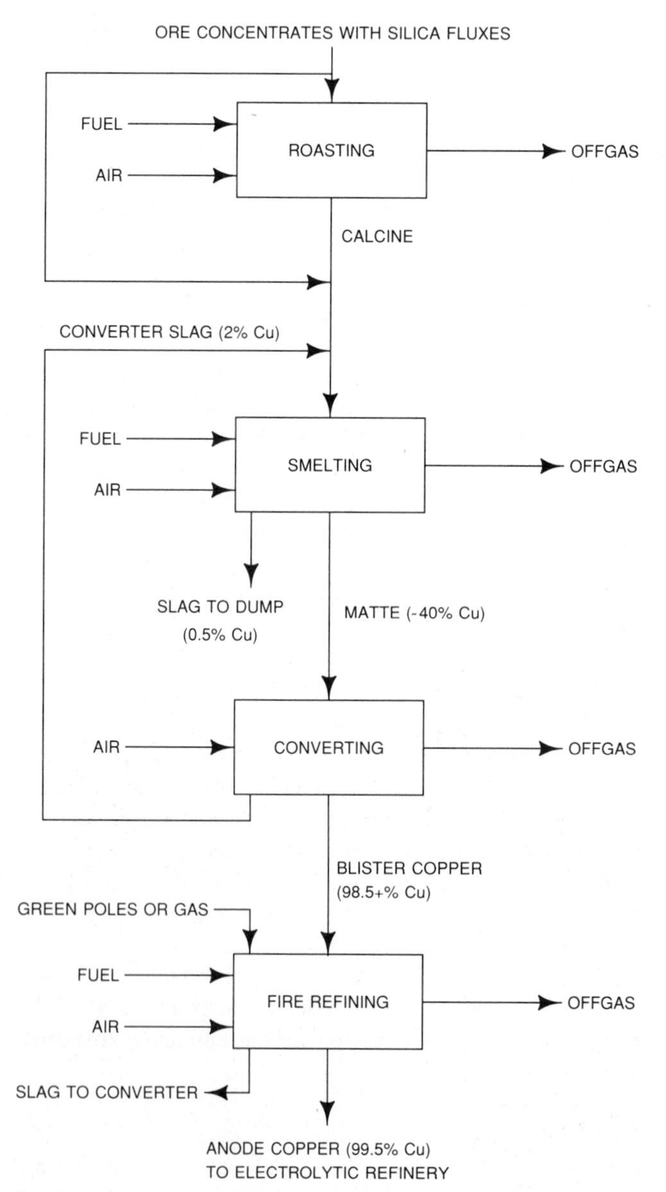

FIGURE 1. Typical Primary Copper Smelter Process

*Taken from AP-42, *Compilation of Air Pollutant Emission Factors,* U.S. Environmental Protection Agency

types, the roasting is autogenous. Because there is less air dilution, higher SO_2 concentrations are present in fluidized-bed roaster gases than in multiple-hearth roaster gases.

In the smelting process, either hot calcines from the roaster or raw unroasted concentrate is melted with siliceous flux in a smelting furnace to produce copper matte, a molten mixture of cuprous sulfide (Cu_2S), ferrous sulfide (FeS), and some heavy metals. The required heat comes from the partial oxidation of the sulfide charge and from the burning of external fuel. Most of the iron and some of the impurities in the charge oxidize with the fluxes to form a slag atop the molten bath; this slag is periodically removed and discarded. Copper matte remains in the furnace until tapped. Mattes produced by the domestic industry range from 35% to 65% copper, with 45% the most common. The copper content percentage is referred to as the matte grade. Currently, five smelting furnace technologies are used in the United States—reverberatory, electric, Noranda, Outokumpu (flash), and Inco (flash).

Reverberatory furnace operation is a continuous process, with frequent charging of input materials and periodic tapping of matte and skimming of slag.

For smelting in electric arc furnaces, heat is generated by the flow of an electric current in carbon electrodes lowered through the furnace roof and submerged in the slag layer of the molten bath. The feed generally consists of dried concentrates or calcines, and charging wet concentrates is avoided. The chemical and physical changes occurring in the molten bath are similar to those occurring in the molten bath of a reverberatory furnace. Also, the matte and slag tapping practices are similar at both furnaces. Electric furnaces do not produce fuel combustion gases, so flow rates are lower and SO_2 concentrations higher in the effluent gas than in that of reverberatory furnaces.

Flash furnace smelting combines the operations of roasting and smelting to produce a high-grade copper matte from concentrates and flux. In flash smelting, dried ore concentrates and finely ground fluxes are injected, together with oxygen, preheated air, or a mixture of both, into a furnace of special design, where the temperature is maintained at approximately 1000°C (1830°F). Flash furnaces, in contrast to reverberatory and electric furnaces, use the heat generated from the partial oxidation of their sulfide charge to provide much or all of the energy (heat) required for smelting. They also produce off-gas streams containing high concentrations of SO_2.

Slag produced by flash furnace operations contains significantly higher amounts of copper than does that from reverberatory or electric furnace operations. As a result, the flash furnace and converter slags are treated in a slag cleaning furnace to recover the copper. Slag cleaning furnaces usually are small electric furnaces. The flash furnace and converter slags are charged to a slag cleaning furnace and are allowed to settle under reducing conditions, with the addition of coke or iron sulfide. The copper, which is in oxide form in the slag, is converted to copper sulfide, and is subsequently removed from the furnace and charged to a converter with regular matte. If the slag's copper content is low, the slag is discarded.

The Noranda process, as originally designed, allowed the continuous production of blister copper in a single vessel by effectively combining roasting, smelting, and converting into one operation. However, for the operation of these reactors, the copper matte mode is preferred. As in flash smelting, the Noranda process takes advantage of the heat energy available from the copper ore. The remaining thermal energy required is supplied by natural gas burners or by coal mixed with the ore concentrates.

The final step in the production of blister copper is converting, with the purposes of eliminating the remaining iron and sulfur present in the matte and leaving molten "blister" copper. All but one U.S. smelter use Pierce-Smith converters, which are refractory-lined, cylindrical, steel shells mounted on trunnions at either end and rotated about the major axis for charging and pouring. An opening in the center of the converter functions as a mouth through which molten matte, siliceous flux, and scrap copper are charged and gaseous products are vented. Air or oxygen-rich air is blown through the molten matte. Iron sulfide is oxidized to iron oxide (FeO) and SO_2, and the FeO blowing and slag skimming are repeated until an adequate amount of relatively pure Cu_2S, called "white metal," accumulates in the bottom of the converter. A renewed air blast oxidizes the copper sulfide sulfur to SO_2, leaving blister copper in the converter. The blister copper is subsequently removed and transferred to refining facilities. This segment of converter operation is termed the finish blow. The SO_2 produced throughout the operation is vented to pollution control devices.

One domestic smelter uses Hoboken converters, the primary advantage of which lies in emission control. The Hoboken converter is essentially like a conventional Pierce-Smith converter, except that this vessel is fitted with a side flue at one end shaped as an inverted U. This flue arrangement permits siphoning of gases from the interior of the converter directly to the off-gas collection system, leaving the converter mouth under a slight vacuum.

Blister copper usually contains from 98.5% to 99.5% pure copper. Impurities may include gold, silver, antimony, arsenic, bismuth, iron, lead, nickel, selenium, sulfur, tellurium, and zinc. To purify blister copper further, fire refining and electrolytic refining are used. In fire refining, blister copper is placed in a fire refining furnace, a flux is usually added, and air is blown through the molten mixture to oxidize remaining impurities, which are removed as a slag. The remaining metal bath is subjected to a reducing atmosphere to reconvert cuprous oxide to copper. The temperature in the furnace is around 1100°C (2010°F). The fire-refined copper is cast into anodes, after which further electrolytic refining separates copper from impurities by electrolysis in a solution containing copper sulfate and sulfuric acid. Metallic impurities precipitate from the solution and form a sludge, which is removed and treated to recover precious metals. Copper is dissolved from the anode and deposited at the cathode. Cathode copper is remelted and made into bars, ingots, or slabs for marketing purposes. The copper produced is 99.95% to 99.97% pure.

AIR EMISSIONS CHARACTERIZATION

Particulate matter and SO_2 are the principal air contaminants emitted by primary copper smelters. These emissions are generated directly from the processes involved, as in the liberation of SO_2 from copper concentrate during roasting or in the volatilization of trace elements as oxide fumes. Fugitive emissions are generated by leaks from major equipment during material-handling operations.

Roasters, smelting furnaces, and converters are sources of both particulate matter and sulfur oxides. Copper and iron oxides are the primary constituents of the particulate matter, but other oxides, such as arsenic, antimony, cadmium, lead, mercury, and zinc, may also be present, with metallic sulfates and sulfuric acid mist. Fuel combustion products also contribute to the particulate emissions from multiple-hearth roasters and reverberatory furnaces.

Fugitive Emissions

The process sources of particulate matter and SO_2 emission are also the potential fugitive sources of these emissions: roasting, smelting, converting, fire refining, and slag cleaning. Table 1 presents the potential fugitive emission factors for these sources, while Tables 2 through 4 and Figures 2–7 present cumulative size-specific particulate emission factors for fugitive emissions from reverberatory furnace matte, slag tapping, converter slag, and copper blow operations. The actual quantities of emissions from these sources depend on the type and condition of the equipment and on the smelter operating techniques. Although emissions from many of these sources are released inside a building, ultimately they are discharged to the atmosphere.

Fugitive emissions are generated during the discharge and transfer of hot calcine from multiple-hearth roasters, with negligible amounts possible from the charging of these roasters. Fluid-bed roasting, a closed-loop operation, has negligible fugitive emissions.

Matte tapping and slag skimming operations are sources of fugitive emissions from smelting furnaces. Fugitive emissions can also result from the charging of a smelting furnace or from leaks, depending on the furnace type and condition. A typical single matte tapping operation lasts from five to 10 minutes and a single slag skimming operation lasts from 10 to 20 minutes. Tapping frequencies vary with furnace capacity and type. In an eight-hour shift, matte is tapped five to 20 times and slag is skimmed 10 to 25 times.

Each of the various stages of converter operation—charging, blowing, slag skimming, blister pouring, and holding—is a potential source of fugitive emissions. During blowing, the converter mouth is in stack (i.e., a close-fitting primary hood is over the mouth to capture off-gases). Fugitive emissions escape from the hoods. During charging, skimming, and pouring operations, the converter mouth is out of stack (i.e., the converter mouth is rolled out of its vertical position and the primary hood is isolated). Fugitive emissions are discharged during rollout.

At times during normal smelting operations, slag or blister copper cannot be transferred immediately from or to the converters. This condition, a holding stage, may occur for several reasons, including insufficient matte in the smelting furnace or the unavailability of a crane. Under these conditions, the converter is rolled out of its vertical position and remains in a holding position and fugitive emissions may result.

TABLE 1. Fugitive Emission Factors for Primary Copper Smelters[a] *(Emission Factor Rating: B)*

	Particulate		Sulfur Dioxide	
Source of Emission	kg/Mg	lb/ton	kg/Mg	lb/ton
Roaster calcine discharge	1.3	2.6	0.5	1
Smelting furnace[b]	0.2	0.4	2	4
Converter	2.2	4.4	65	130
Converter slag return	NA	NA	0.05	0.1
Anode furnace	0.25	0.5	0.05	0.1
Slag cleaning surface[c]	4	8	3	6

[a]References 16, 22, 25–32. Expressed as mass units per unit weight of concentrated ore processed by the smelter. Approximately four unit weights of concentrate are required to produce one unit weight of copper metal. Factors for flash furnace smelters and Noranda furnace smelters may be lower than reported values. NA = not available.

[b]Includes fugitive emissions from matte tapping and slag skimming operations. About 50% of fugitive particulate emissions and about 90% of total SO_2 emissions are from matte tapping operations, with the remainder from slag skimming.

[c]Used to treat slags from smelting furnaces and converters at the flash furnace smelter.

TABLE 2. Uncontrolled Particle Size and Size-Specific Emission Factors for Fugitive Emissions from Reverberatory Furnace Matte Tapping Operations[a] *(Emission Factor Rating: D)*

Particle Size,[b] μm	Cumulative Mass, % ≤ Stated Size	Cumulative Emission Factors	
		kg/Mg	lb/ton
15	76	0.076	0.152
10	74	0.074	0.148
5	72	0.072	0.144
2.5	69	0.069	0.138
1.25	67	0.067	0.134
0.625	65	0.065	0.130
Total	100	0.100	0.200

[a]Reference 25. Expressed as units per unit weight of concentrated ore processed by the smelter.
[b]Expressed as aerodynamic equivalent diameter.

TABLE 3. Particle Size Distribution and Size-Specific Emission Factors for Fugitive Emissions from Reverberatory Furnace Slag Tapping Operations[a] *(Emission Factor Rating: D)*

Particle Size,[b] μm	Cumulative Mass, % ≤ Stated Size	Cumulative Emission Factors	
		kg/Mg	lb/ton
15	33	0.033	0.066
10	28	0.028	0.056
5	25	0.025	0.050
2.5	22	0.022	0.044
1.25	20	0.020	0.040
0.625	17	0.017	0.034
Total	100	0.100	0.200

[a]Reference 25. Expressed as units per unit weight of concentrated ore processed by the smelter.
[b]Expressed as aerodynamic equivalent diameter.

TABLE 4. Particle Size Distribution and Size-Specific Emission Factors for Fugitive Emissions from Converter Slag and Copper Blow Operations[a] *(Emission Factor Rating: D)*

Particle Size,[b] μm	Cumulative Mass, % ≤ Stated Size	Cumulative Emission Factors	
		kg/Mg	lb/ton
15	98	2.2	4.3
10	96	2.1	4.2
5	87	1.9	3.8
2.5	60	1.3	2.6
1.25	47	1.0	2.1
0.625	38	0.8	1.7
Total	100	2.2	4.4

[a]Reference 25. Expressed as units per unit weight of concentrated ore processed by the smelter.
[b]Expressed as aerodynamic equivalent diameter.

TABLE 5. Lead Emission Factors for Primary Copper Smelters[a] *(Emission Factor Rating: C)*

Operation	Emission Factor[b]	
	kg/Mg	lb/ton
Roasting	0.075	0.15
Smelting	0.036	0.072
Converting	0.13	0.27
Refining	NA	NA

[a]Reference 33. Expressed as units per unit weight of concentrated ore processed by smelter. Approximately four unit weights of concentrate are required to produce one unit weight of copper metal. Based on test data for several smelters with 0.1–0.4% lead in feed throughput. NA = not available.
[b]For process and fugitive emissions totals.

Lead Emissions

At primary copper smelters, both process emissions and fugitive particulate from various pieces of equipment contain oxides of many inorganic elements, including lead. The lead content of particulate emissions depends on both the lead content of the smelter feed and the process off-gas temperature. Lead emissions are effectively removed in particulate control systems operating at low temperatures, about 120°C (250°F).

Table 5 presents process and fugitive lead emission factors for various operations of primary copper smelters. Fugitive emissions from primary copper smelters are cap-

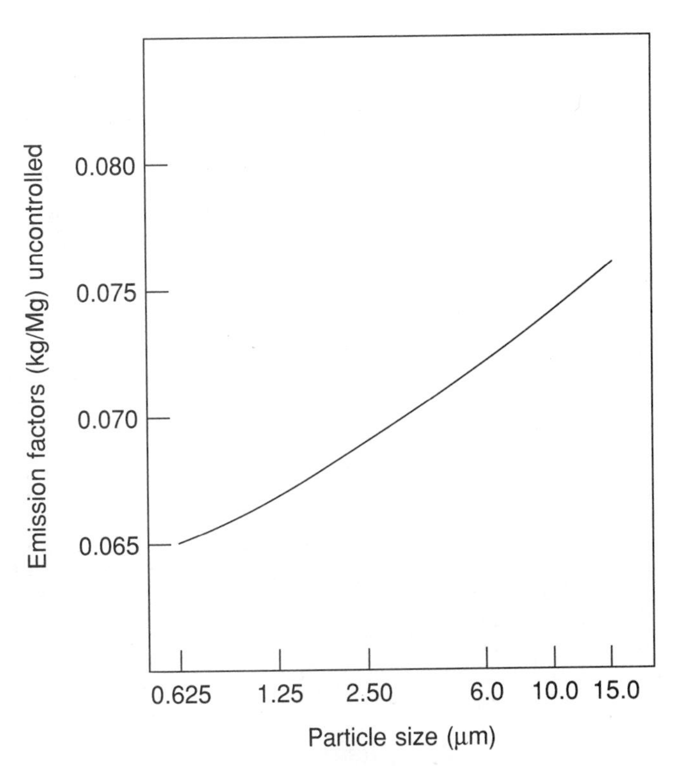

FIGURE 2. Size-Specific Fugitive Emission Factors for Reverberatory Furnace Matte Tapping Operations

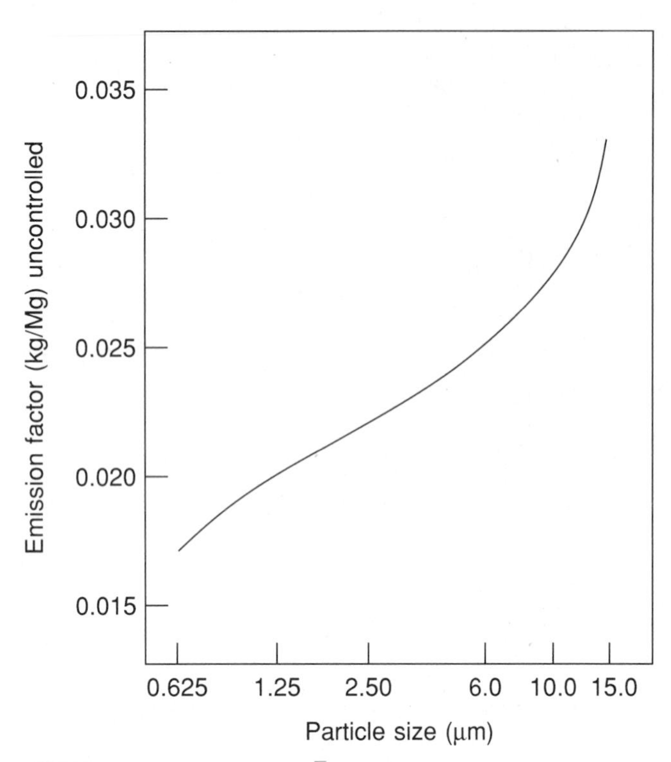

FIGURE 3. Size-Specific Fugitive Emission Factors for Reverberatory Furnace Slag Tapping Operations

tured by applying either local ventilation or general ventilation techniques. Once captured, emissions may be vented directly to a collection device or be combined with process off-gases before collection. Close-fitting exhaust hood capture systems are used for multiple-hearth roasters and hood ventilation systems for smelt matte tapping and slag skimming operations. For converters, secondary hood systems or building evacuation systems are used.

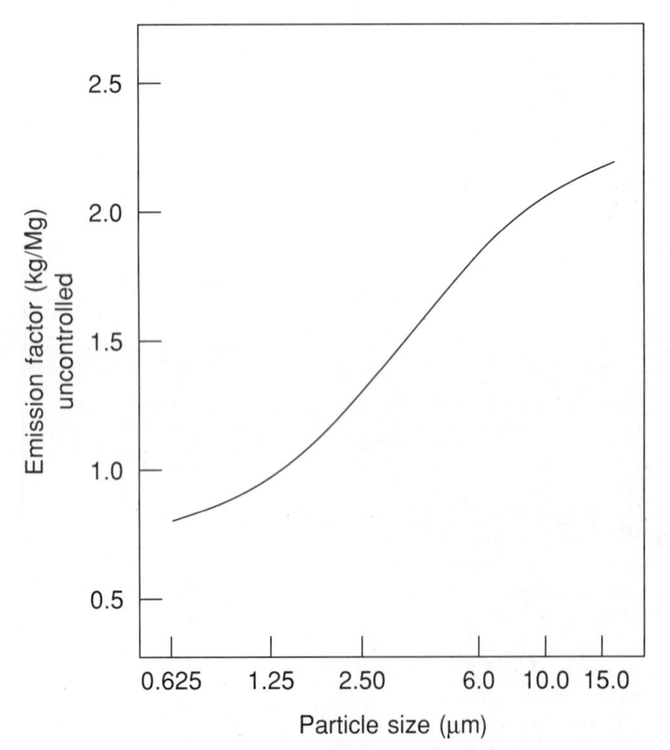

FIGURE 4. Size-Specific Fugitive Emission Factors for Converter Slag and Copper Blow Operations

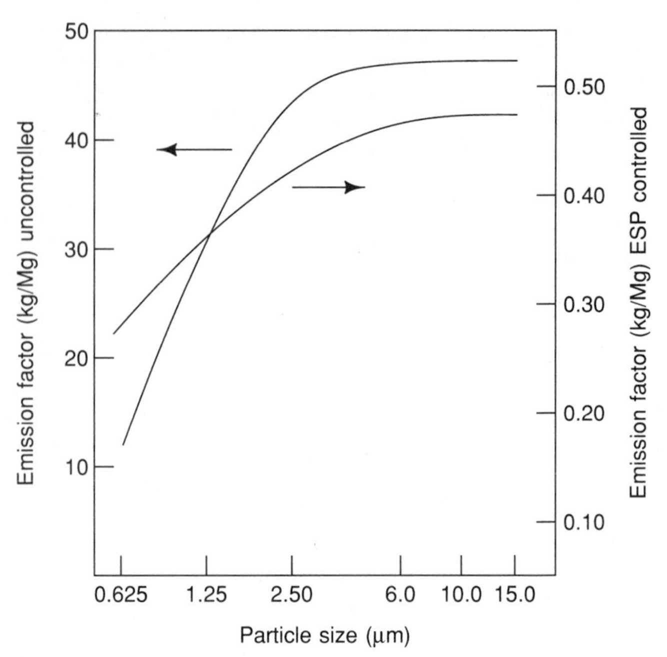

FIGURE 5. Size-Specific Emission Factors for Multiple-Hearth Roaster and Reverberatory Smelter

AIR POLLUTION CONTROL MEASURES

Single-stage electrostatic precipitators (ESPs) are widely used in the primary copper industry to control particulate emissions from roasters, smelting furnaces, and converters. Many of the existing ESPs are operated at elevated temperatures, usually from 200°C to 340°C (400°F to 650°F) and are termed "hot ESPs." If properly designed and operated, these ESPs remove 99% or more of the condensed particulate matter present in gaseous effluents. However, at these elevated temperatures, a significant amount of volatile emissions, such as arsenic trioxide (As_2O_3) and sulfuric acid mist, is present as vapor in the gaseous effluent and thus cannot be collected by the particulate control device at elevated temperatures. At these temperatures, the arsenic trioxide in the vapor state will pass through an ESP. Therefore, the gas stream to be treated must be cooled sufficiently to assure that most of the arsenic present is condensed before entering the control device for collection. At some smelters, the gas effluents are cooled to about 120°C (250°F) temperature before entering a particulate control system, usually an ordinary ("cold") ESP. A spray chamber or air infiltration is used for gas cooling. Fabric filters can also be used for particulate matter collection.

Gas effluents from roasters usually are sent to an ESP or spray chamber/ESP system or are combined with smelter furnace gas effluents before particulate collection. Overall, the hot ESPs remove only 20% to 80% of the total particulate (condensed and vapor) present in the gas. Cold ESPs may remove more than 95% of the total particulate present

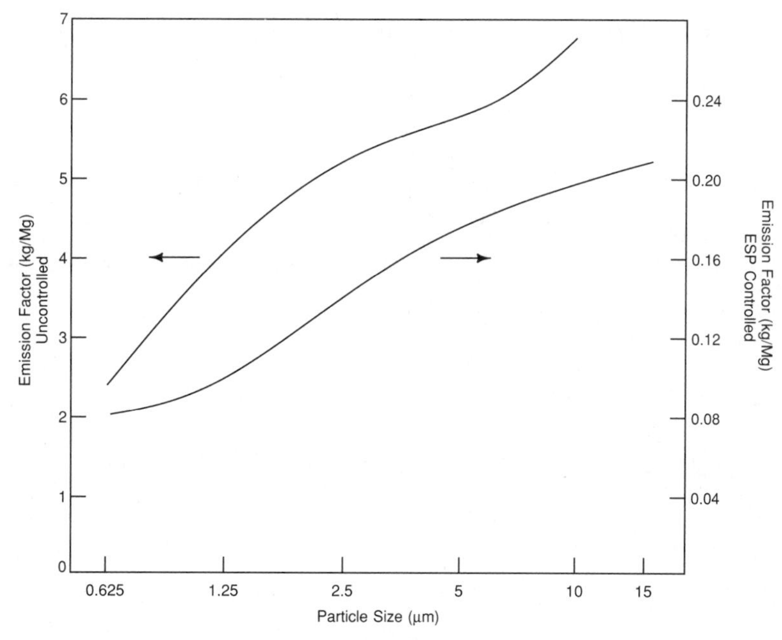

FIGURE 6. Size-Specific Emission Factors for Reverberatory Smelting

in the gas. Particulate collection systems for smelting furnaces are similar to those for roasters. Reverberatory furnace off-gases are usually routed through waste-heat boilers and low-velocity balloon flues to recover large particles and heat, and then are routed through an ESP or spray chamber/ ESP system.

In the standard Pierce-Smith converter, flue gases are captured during the blowing phase by the primary hood over the converter mouth. To prevent the hood's binding to the converter with splashing molten metal, there is a gap between the hood and the vessel. During charging and pouring operations, significant fugitives may be emitted when the hood is removed to allow crane access. Converter off-gases are treated in ESPs to remove particulate matter and in sulfuric acid plants to remove SO_2.

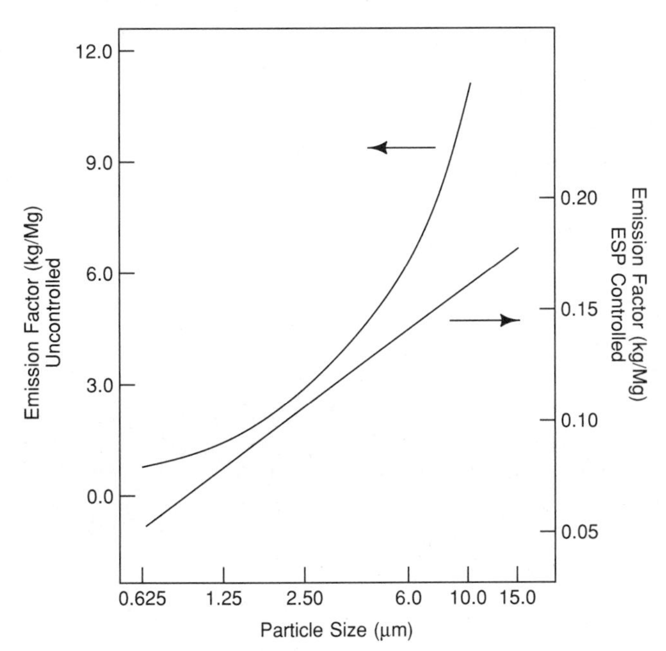

FIGURE 7. Size-Specific Emission Factors for Copper Converting

The remaining smelter processes handle material that contains very little sulfur, and hence SO_2 emissions from these processes are relatively insignificant. Particulate emissions from fire refining operations, however, may be of concern. Electrolytic refining does not produce emissions unless the associated sulfuric acid tanks are open to the atmosphere. Crushing and grinding systems used in ore, flux, and slag processing also contribute to fugitive dust problems.

The control of SO_2 emissions from smelter sources is most commonly performed in a single- or double-contact sulfuric acid plant. Use of a sulfuric acid plant to treat copper smelter effluent gas streams requires that gas be free from particulate matter and that a certain minimum SO_2

TABLE 6. Typical Sulfur Dioxide Concentrations in Off-Gases from Primary Copper Smelting Sources

Unit	SO_2 Concentration, Vol %
Multiple-hearth roaster	1.5–3
Fluidized-bed roaster	10–12
Reverberatory furnace	0.5–1.5
Electric arc furnace	4–8
Flash smelting furnace	10–70
Continuous smelting furnace	5–15
Pierce-Smith converter	4–7
Hoboken converter	8
Single-contact H_2SO_4 plant	0.2–0.26
Double-contact H_2SO_4 plant	0.05

TABLE 7. Emission Factors for Primary Copper Smelters[a,b] *(Emission Factor Rating: B)*

Configuration[c]	By Unit	Particulate kg/Mg	Particulate lb/ton	Sulfur Dioxide[d] kg/Mg	Sulfur Dioxide[d] lb/ton	References
Reverberatory furnace (RF)	RF	25	50	160	320	4–10,
followed by converters (C)	C	18	36	370	740	9,11–15
Multiple-hearth roaster (MHR)	MHR	22	45	140	280	4,5,16,17
followed by reverberatory	RF	25	50	90	180	4–9,18,19
furnace (RF) and converters (C)	C	18	36	300	600	8,11–13
Fluid-bed roaster (FBR) followed	FBR	NA	NA	180	360	20
by reverberatory furnace (RF)	RF	25	50	90	160	e
and converters (C)	C	18	36	270	540	e
Concentrate dryer (CD) followed	CD	5	10	0.5	1	21,22
by electric furnace (EF) and	EF	50	100	120	240	15
converters (C)	C	18	36	410	820	8,11–13,15
Fluid-bed roaster (FBR) followed	FBR	NA	NA	180	360	20
by electric furnace (EF) and	EF	50	100	45	90	15,23
converters (C)	C	18	36	300	600	e
Concentrate dryer (DC) followed	CD	5	10	0.5	1	21,22
by flash furnace (FF),	FF	70	140	410	820	24
cleaning furnace (SS), and	SS[f]	5	10	0.5	1	22
converters (C)	C[e]	NA[g]	NA[g]	120	240	22
Concentrate dryer (CD) followed	CD	5	10	0.5	1	21,22
by Noranda reactors (NR) and	NR	NA	NA	NA	NA	
converters (C)	C	NA	NA	NA	NA	

[a]Expressed as units per unit weight of concentrated ore processed by the smelter. Approximately four unit weights of concentrate are required to produce one unit weight of blister copper. NA = not available.
[b]For particulate matter removal, gaseous effluents from roasters, smelting furnaces, and converters usually are treated in hot ESPs at 200–340°C (400–650°F) or in cold ESPs with gases cooled to about 120°C (250°F) before the ESP. Particulate emissions from copper smelters contain volatile metallic oxides that remain in vapor form at higher temperatures (120°C or 250°F). Therefore, overall particulate removal in hot ESPs may range from 20% to 80% and in cold ESPs may be 99%. Converter gas effluents and, at some smelters, roaster gas effluents are treated in single-contact acid plants (SCAP) or double-contact acid plants (DCAP) for SO_2 removal. Typical SCAPs are about 96% efficient and DCAPs are up to 99.8% efficient in SO_2 removal. They also remove over 99% of particulate matter. Noranda and flash furnace off-gases are also processed through acid plants and are subject to the same collection efficiencies as cited for converters and some roasters.
[c]In addition to sources indicated, each smelter configuration contains fire-refining anode furnaces after the converters. Anode furnaces emit negligible SO_2. No particulate emission data are available for anode furnaces.
[d]Factors for all configurations except a reverberatory furnace followed by converters have been developed by normalizing test data for several smelters to represent 30% sulfur content in concentrated ore.
[e]Based on the test data for the configuration of multiple-hearth roaster followed by reverberatory furnace and converters.
[f]Used to recover copper from furnace slag and converter slag.
[g]Since converters at flash furnace and Noranda furnace smelters treat high-copper-content matte, converter particulate emissions from flash furnace smelters are expected to be lower than those from conventional smelters with multiple-hearth roasters, reverberatory furnace, and converters.

TABLE 8. Particle Size Distribution and Size-Specific Emission Factors for Multiple-Hearth Roaster and Reverberatory Smelter Operations[a] *(Emission Factor Rating: D)*

Particle Size,[b] μm	Cumulative Mass, % ≤ Stated Size: Uncontrolled	Cumulative Mass, % ≤ Stated Size: ESP Controlled	Cumulative Emission Factors: Uncontrolled kg/Mg	Uncontrolled lb/ton	ESP Controlled[c] kg/Mg	ESP Controlled[c] lb/ton
15	100	100	47	95	0.47	0.95
10	100	99	47	94	0.47	0.94
5	100	98	47	93	0.46	0.93
2.5	97	84	46	80	0.40	0.80
1.25	66	76	31	72	0.36	0.72
0.625	25	62	12	59	0.29	0.59
Total	100	100	47	95	0.47	0.95

[a]Reference 25. Expressed as units per unit weight of concentrated ore processed by the smelter.
[b]Expressed as aerodynamic equivalent diameter.
[c]Nominal particulate removal efficiency is 99%.

TABLE 9. Particle Size Distribution and Size-Specific Emission Factors for Reverberatory Smelter Operations[a] ***(Emission Factor Rating: E)***

	Cumulative Mass, % ≤ Stated Size		Cumulative Emission Factors			
			Uncontrolled		ESP Controlled[c]	
Particle Size,[b] μm	Uncontrolled	ESP Controlled	kg/Mg	lb/ton	kg/Mg	lb/ton
15	NR	83	NR	NR	0.21	0.42
10	27	78	6.8	13.6	0.20	0.40
5	23	69	5.8	11.6	0.18	0.36
2.5	21	56	5.3	10.6	0.14	0.28
1.25	16	40	4.0	8.0	0.10	0.20
0.625	9	32	2.3	4.6	0.08	0.16
Total	100	100	25	50	0.25	0.50

[a]Reference 25. Expressed as units per unit weight of concentrated ore processed by the smelter. NR = not reported because of excessive extrapolation.
[b]Expressed as aerodynamic equivalent diameter.
[c]Nominal particulate removal efficiency is 99%.

TABLE 10. Particle Size Distribution and Size-Specific Emission Factors for Copper Converter Operations[a] ***(Emission Factor Rating: E)***

	Cumulative Mass, % ≤ Stated Size		Cumulative Emission Factors			
			Uncontrolled		ESP Controlled[c]	
Particle Size,[b] μm	Uncontrolled	ESP Controlled	kg/Mg	lb/ton	kg/Mg	lb/ton
15	NR	100	NR	NR	0.18	0.36
10	59	99	10.6	21.2	0.17	0.36
5	32	72	5.8	11.5	0.13	0.26
2.5	12	56	2.2	4.3	0.10	0.20
1.25	3	42	0.5	1.1	0.08	0.15
0.625	1	30	0.2	0.4	0.05	0.11
Total	100	100	18	36	0.18	0.36

[a]Reference 25. Expressed as units per unit weight of concentrated ore processed by the smelter. NR = not reported because of excessive extrapolation.
[b]Expressed as aerodynamic equivalent diameter.
[c]Nominal particulate removal efficiency is 99%.

concentration be maintained. Practical limitations have usually restricted sulfuric acid plant application to gas streams that contain at least 3% SO_2. Table 6 shows typical average SO_2 concentrations for the various smelter unit off-gases. Currently, converter gas effluents at most smelters are treated for SO_2 control in sulfuric acid plants. Gas effluents of some multiple-hearth roaster operations and of all fluid-bed roaster operations also are treated in sulfuric acid plants. The weak-SO_2-content gas effluents from reverberatory furnace operations are usually released to the atmosphere with no reduction of SO_2. The gas effluents from the other types of smelter furnaces, because of their higher contents of SO_2, are treated in sulfuric acid plants before being vented. Typically, single-contact acid plants achieve 92.5% to 98% conversion of SO_2 to acid, with approximately 2000 ppm SO_2 remaining in the acid plant effluent gas. Double-contact acid plants collect from 98% to more than 99% of the SO_2 and emit about 500 ppm SO_2. Absorption of the SO_2 in dimethylaniline (DMA) solution has also been used in U.S. smelters to produce liquid SO_2.

Emissions from hydrometallurgical smelting plants generally are small in quantity and are easily controlled. In the Arbiter process, ammonia gas escapes from the leach reactors, mixer/settlers, thickeners, and tanks. For control, all of these units are covered and are vented to a packed-tower scrubber to recover and recycle the ammonia.

Actual emissions from a particular smelter unit depend on the configuration of the equipment in that smelting plant and its operating parameters. Table 7 gives the emission factors for various smelter configurations, and Tables 5, 8, 9, and 10 give size-specific emission factors for those copper production processes where information is available.

References

1. *Background Information for New Source Performance Standards: Primary Copper, Zinc and Lead Smelters, Volume I, Proposed Standards,* EPA-450/2-74-002a, U.S. Environmental Protection Agency, Research Triangle Park, NC, October 1974.
2. *Arsenic Emissions from Primary Copper Smelters—Background Information for Proposed Standards, Preliminary Draft,* EPA Contract No. 68-02-3060, Pacific Environmental Services, Durham, NC, February 1981.
3. *Background Information Document for Revision of New Source Performance Standards for Primary Copper Smelters,* EPA Contract No. 68-02-3056, Research Triangle Institute, Research Triangle Park, NC, March 31, 1982.
4. *Air Pollution Emission Test: Asarco Copper Smelter, El Paso, TX,* EMB-77-CUS-6, Office of Air Quality Planning and Standards, U.S. Environmental Protection Agency, Research Triangle Park, NC, June 1977.
5. W. F. Cummins, Inc., El Paso, TX, written communications to A. E. Vervaert, U.S. Environmental Protection Agency, Research Triangle Park, NC, June 1977.
6. AP-42 Background Files, Office of Air Quality Planning and Standards, U.S. Environmental Protection Agency, Research Triangle Park, NC, March 1978.
7. *Source Emissions Survey of Kennecott Copper Corporation, Copper Smelter Converter Stack Inlet and Outlet and Reverberatory Electrostatic Precipitator Inlet and Outlet, Hurley, NM,* EA-735-09, Ecology Audits, Inc., Dallas, TX, April 1973.
8. *Trace Element Study at a Primary Copper Smelter,* EPA-600/2-78-065a and 065b, U.S. Environmental Protection Agency, Research Triangle Park, NC, March 1978.
9. *Systems Study for Control of Emissions, Primary Nonferrous Smelting Industry, Vol. II: Appendices A and B,* PB 184885, National Technical Information Service, Springfield, VA, June 1969.
10. *Design and Operating Parameters for Emission Control Studies: White Pine Copper Smelter,* EPA-600/2-76-036a, U.S. Environmental Protection Agency, Washington, DC, February 1976.
11. R. M. Statnick, *Measurements of Sulfur Dioxide, Particulate and Trace Elements in Copper Smelter Converter and Roaster/Reverberatory Gas Streams,* PB 238095, National Technical Information Service, Springfield, VA, October 1974.
12. AP-42 Background Files, Office of Air Quality Planning and Standards, U.S. Environmental Protection Agency, Research Triangle Park, NC.
13. *Design and Operating Parameters for Emission Control Studies, Kennecott-McGill Copper Smelter,* EPA-600/2-76-036c, U.S. Environmental Protection Agency, Washington, DC, February 1976.
14. *Emission Test Report (Acid Plant) of Phelps Dodge Copper Smelter, Ajo, AZ,* EMB-78-CUS-11, Office of Air Quality Planning and Standards, Research Triangle Park, NC, March 1979.
15. S. Dayton, "Inspiration's design for clean air," *Eng. Mining J.* 175:6 (June 1974).
16. *Emission Testing of Asarco Copper Smelter, Tacoma, WA,* EMB-78-CUS-12, Office of Air Quality Planning and Standards, U.S. Environmental Protection Agency, Research Triangle Park, NC, April 1979.
17. Written Communication from A. L. Labbe, Asarco, Inc., Tacoma, WA, to S. T. Cuffe, U.S. Environmental Protection Agency, Research Triangle Park, NC, November 20, 1978.
18. *Design and Operating Parameters for Emission Control Studies: Asarco-Hayden Copper Smelter,* EPA-600/2-76-036j, U.S. Environmental Protection Agency, Washington, DC, February 1976.
19. *Design and Operating Parameters for Emission Control Studies: Kennecott, Hayden Copper Smelter,* EPA-600/2-76-036b, U.S. Environmental Protection Agency, Washington, DC, February 1976.
20. R. Larkin, *Arsenic Emissions at Kennecott Copper Corporation, Hayden, AZ,* EPA-76-NFS-1, U.S. Environmental Protection Agency, Research Triangle Park, NC, May 1977.
21. *Emission Compliance Status, Inspiration Consolidated Copper Company, Inspiration, AZ,* U.S. Environmental Protection Agency, San Francisco, CA, 1980.
22. M. P. Scanlon, Phelps Dodge Corp., Hidalgo, AZ, written communication to D. R. Goodwin, U.S. Environmental Protection Agency, Research Triangle Park, NC, October 18, 1978.
23. G. M. McArthur, Anaconda Co., written communication to D. R. Goodwin, U.S. Environmental Protection Agency, Research Triangle Park, NC, June 2, 1977.
24. V. Katari, Pacific Environmental Services, Durham, NC, telephone communication to R. Winslow, Hidalgo Smelter, Phelps Dodge Corp., Hidalgo, AZ, April 1, 1982.
25. *Inhalable Particulate Source Category Report for the Nonferrous Industry,* Contract 68-02-3159, Acurex Corp., Mountain View, CA, August 1986.
26. *Emission Test Report, Phelps Dodge Copper Smelter, Douglas, AZ,* EMB-78-CUS-8, Office of Air Quality Planning and Standards, U.S. Environmental Protection Agency, Research Triangle Park, NC, February 1979.
27. *Emission Testing of Kennecott Copper Smelter, Magna, UT,* EMB-78-CUS-13, Office of Air Quality Planning and Standards, U.S. Environmental Protection Agency, Research Triangle Park, NC, April 1979.
28. *Emission Test Report, Phelps Dodge Copper Smelter, Ajo, AZ,* EMB-78-CUS-9, Office of Air Quality Planning and Standards, U.S. Environmental Protection Agency, Research Triangle Park, NC, February 1979.
29. R. D. Putnam, Asarco, Inc., written communication to M. O. Varner, Asarco, Inc., Salt Lake City, UT, May 12, 1980.
30. *Emission Test Report, Phelps Dodge Copper Smelter, Playas, NM,* EMB-78-CUS-10, Office of Air Quality Planning and Standards, U.S. Environmental Protection Agency, Research Triangle Park, NC, March 1979.
31. *Asarco Copper Smelter, El Paso, TX,* EMB-78-CUS-7, Office of Air Quality Planning and Standards, U.S. Environmental Protection Agency, Research Triangle Park, NC, April 25, 1978.
32. A. D. Church et al., "Measurement of fugitive particulate and sulfur dioxide emissions at Inco's Copper Cliff smelter," Paper A-79-51, Metallurgical Society, American Institute of Mining, Metallurgical and Petroleum Engineers, New York.
33. *Copper Smelters, Emission Test Report—Lead Emissions,* EMB-79-CUS-14, Office of Air Quality Planning and Standards, U.S. Environmental Protection Agency, Research Triangle Park, NC, September 1979.

FERROALLOY INDUSTRY PARTICULATE EMISSIONS

*Steve Stasko, Editor**

A ferroalloy is an alloy of iron and one or more other elements, such as silicon, manganese, or chromium. Ferroalloys are used as additives to impart unique properties to steel and cast iron. The iron and steel industry consumes approximately 95% of the ferroalloy produced in the United States. The remaining 5% is used in the production of nonferrous alloys, including cast aluminum, nickel–cobalt base alloys, and titanium alloys, and in making other ferroalloys.

Three major groups—ferrosilicon, ferromanganese, and ferrochrome—constitute approximately 85% of domestic production. Subgroups of these alloys include silicomanganese, silicon metal, and ferrochromium. The various grades manufactured are distinguished primarily by their carbonol silicon content. The remaining 15% of ferroalloy production is specialty alloys, typically produced in small amounts and containing such elements as vanadium, columbium, molybdenum, nickel, boron, aluminum, and tungsten.

Ferroalloy facilities in the United States vary greatly in size. Many facilities have only one furnace and require less than 25 mw of electricity. Others consist of 16 furnaces, produce six different types of ferroalloys, and require over 75 mw.

PROCESS DESCRIPTION

A typical ferroalloy plant is illustrated in Figure 1. A variety of furnace types produce ferroalloys, including submerged electric arc furnaces, induction furnaces, vacuum furnaces, exothermic reaction furnaces, and electrolytic cells. Furnace descriptions and their ferroalloy products are given in Table 1. Ninety-five percent of all ferroalloys, including all bulk ferroalloys, are produced in submerged electric arc furnaces and it is the furnace type that is principally discussed here.

The basic design of submerged electric arc furnaces is generally the same throughout the ferroalloy industry in the United States. This furnace is composed of a cylindrical steel shell with a concave hearth. The interior of the shell is lined with two or more layers of carbon blocks. Raw materials are charged through feed chutes from above the furnace. The molten metal and slag are removed through one or more tapholes that extend through the furnace shell at the hearth level. Three carbon electrodes, arranged in a delta formation, extend downward through the charge material to a depth of 3 to 5 feet.

Submerged electric arc furnaces are of two basic types, open and covered, with about 80% of these furnaces in the United States of the open type. Open furnaces have a fume collection hood at least 3 feet above the top of the furnace. Movable panels or screens sometimes are used to reduce the open area between the furnace and hood to improve emissions capture efficiency. Covered furnaces have a water-cooled steel cover to seal the top, with holes through it for the electrodes. The degree of emission containment provided by the covers is quite variable. Air infiltration some-

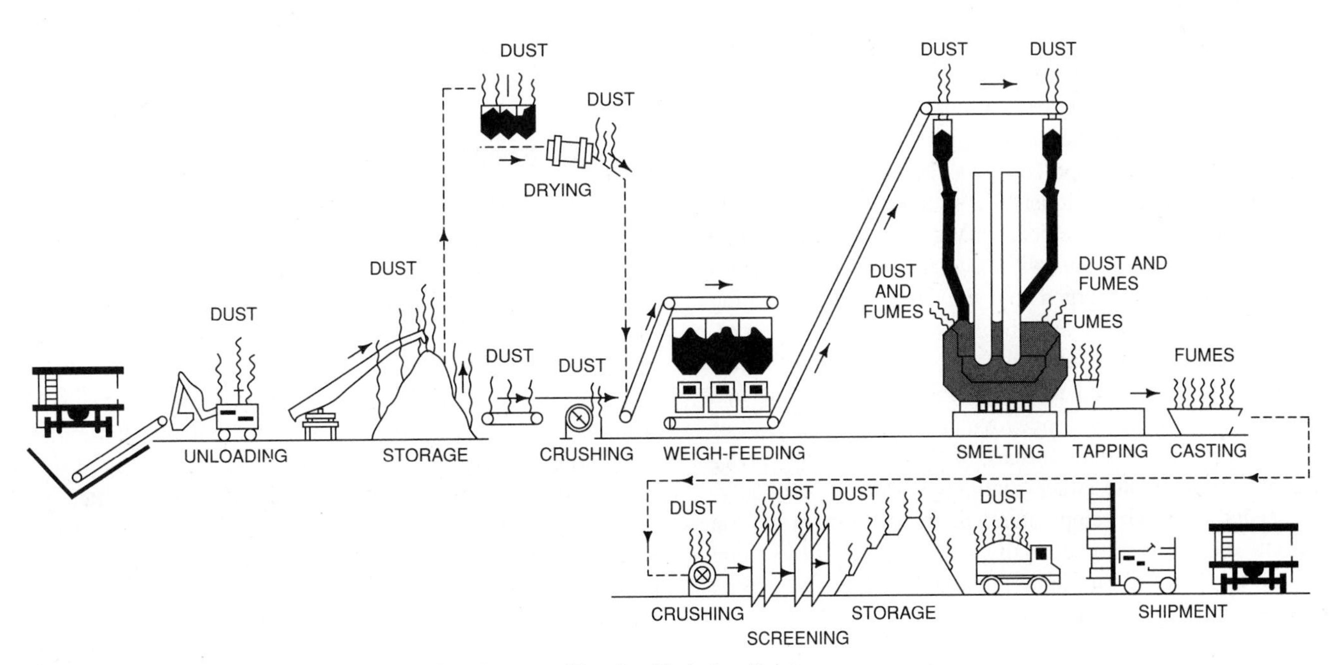

FIGURE 1. Typical Ferroalloy Production Process, Showing Emission Points

Taken from AP-42, *Compilation of Air Pollutant Emission Factors*, U.S. Environmental Protection Agency

TABLE 1. Ferroalloy Processes and Respective Product Groups

Process	Product
Submerged arc furnace[a]	Silvery iron (15–22% Si) Ferrosilicon(50% Si) Ferrosilicon (65–75% Si) Silicon metal Silicon/manganese/zirconium (SMZ) High carbon (HC) ferromanganese Siliconmanganese HC ferrochrome Ferrochrome/silicon FeSi (90% Si)
Exothermic[b]	
Silicon reduction	Low-carbon ferrochrome, low-carbon ferromanganese, medium-carbon ferromanganese
Aluminum reduction	Chromium metal, ferrotitanium, ferrocolumbium, ferrovanadium
Mixed aluminothermal/ silicothermal	Ferromolybdenum, ferrotungsten
Electrolytic[c]	Chromium metal, manganese metal
Vacuum furnace[d]	Low-carbon ferrochrome
Induction furnace[e]	Ferrotitanium

[a]Process by which metal is smelted in a refractory-lined, cup-shaped steel shell by three submerged graphite electrodes.
[b]Process by which molten charge material is reduced, in exothermic reaction, by addition of silicon, aluminum, or a combination of the two.
[c]Process by which simple ions of a metal, usually chromium or manganese in an electrolyte, are plated on cathodes by direct low-voltage current.
[d]Process by which carbon is removed from solid-state, high-carbon ferrochrome in vacuum furnaces maintained at temperature near the melting point of alloy.
[e]Process that converts electric energy without electrodes, into heat to melt metal charge in a cup or drum-shaped vessel.

times is reduced by placing charge material around the electrode holes. This type is called a mix seal or semi-enclosed furnace. Another type is a sealed or totally closed furnace with mechanical seals around the electrodes and a sealing compound packed around the cover edges.

The submerged arc process is a reduction smelting operation. The reactants consist of metallic ores and quartz (ferrous oxides, silicon oxides, manganese oxides, chrome oxides, etc.). Carbon, usually as coke, low-volatility coal, or wood chips, is charged into the furnace as a reducing agent. Limestone also may be added as a flux material. After crushing, sizing, and, in some cases, drying, the raw materials are conveyed to a mix house for weighing and blending, and then by conveyors, buckets, skip hoists, or cars to hoppers above the furnace. The mix is fed by gravity through a feed chute either continuously or intermittently, as needed. At high temperatures in the reaction zone, the carbon sources react chemically with oxygen in the metal oxides to form carbon monoxide and to reduce the ores to base metal. A typical reaction illustrating 50% ferrosilicon production is as follows

$$Fe_2O_3 + 2SiO_2 + 7C \rightarrow 2FeSi + 7CO$$

Smelting in an electric arc furnace is accomplished by conversion of electric energy to heat. An alternating current applied to the electrodes causes a current to flow through the charge between the electrode tips. This provides a reaction zone of temperatures up to 2000°C (3632°F). The tip of each electrode changes polarity continuously as the alternating current flows between the tips. To maintain a uniform electric load, electrode depth is continuously varied automatically by mechanical or hydraulic means, as required. Furnace power requirements vary from 7 mw to over 50 mw, depending on the furnace size and the product being made. The average is 17.2 mw.

The molten alloy and slag that accumulate on the furnace hearth are removed at one- to five-hour intervals through the taphole. Tapping typically lasts 10 to 15 minutes. Tapholes are opened with a pellet shot from a gun, by drilling, or by oxygen lancing. The molten metal and slag flow from the taphole into a carbon-lined trough, then into a carbon-lined runner that directs the metal and slag into a reaction ladle, ingot molds, or chills. Chills are low, flat iron or steel pans that provide rapid cooling of the molten metal. Tapping is terminated and the furnace resealed by inserting a carbon paste plug into the taphole.

When chemistry adjustments are necessary to produce a specified product, a reaction ladle is used. Ladle treatment reactions are batch processes and may include chlorination, oxidation, gas mixing, and slag metal reactions.

During tapping and/or in the reaction ladle, slag is skimmed from the surface of the molten metal. It can be disposed of as a raw material in a furnace or reaction ladle to produce a chemically related ferroalloy product.

After cooling and solidifying, the large ferroalloy castings are broken with drop weights or hammers. The broken ferroalloy pieces are then crushed, screened (sized), and stored in bins until shipment.

AIR EMISSIONS CHARACTERIZATION

Particulate is generated from several activities at a ferroalloy facility, including raw-material handling, smelting, and product handling. The furnaces are the largest potential sources of particulate emissions. The emission factors in Tables 2 and 3 reflect emissions from ferroalloy smelting furnaces.

Electric arc furnaces emit particulate in the form of fume, accounting for an estimated 94% of the particulate emissions in the ferroalloy industry. Large amounts of carbon monoxide and organic materials also are emitted by submerged electric arc furnaces. Carbon monoxide is formed as a by-product of the chemical reaction between oxygen in the metal oxides of the charge and carbon contained in the reducing agent (coke, coal, etc.). Reduction gases containing organic compounds and carbon monoxide continuously rise from the high-temperature reaction zone, entraining fine particles and fume precursors. The mass

TABLE 2. Emission Factors for Particulate from Submerged ARC Ferroalloy Furnaces[a]

Product[b]	Furnace Type	Particulate Emission Factors Uncontrolled[c]: kg/Mg (lb/ton) Alloy	Particulate Emission Factors Uncontrolled[c]: kg(lb)/Mwh	Size Data	Notes	Emission Factor Rating	Control Device[d]	Particulate Emission Factors Controlled[c]: kg/Mg (lb/ton) alloy	Particulate Emission Factors Controlled[c]: kg (lb)/MWh	Size Data	Notes	Emission Factor Rating
FeSi (50%)	Open	35 (70)	7.4 (16.3)	Yes	e,f,g	B	Baghouse	0.9 (1.8)	0.2 (0.44)	Yes	e,f	B
	Covered	46 (92)	9.3 (20.5)		h	E	Scrubber					
							High energy	0.24 (0.48)	0.05 (0.11)		h,j	E
							Low energy	4.5 (9.0)	0.77 (1.7)		h,j	E
FeSi (75%)	Open	158 (316)	16 (35)		k	E	Scrubber					
	Covered	103 (206)	13 (29)		h,j	E	Low energy	4.0 (8.0)	0.5 (1.1)		h,j	E
FeSi (90%)	Open	282 (564)	24 (53)	Yes	m	E						
Si metal (98%)	Open	436 (872)	33 (73)	Yes	n,p	B						
FeMn (80%)	Open	14 (28)	4.8 (11)	Yes	q,r	B	Baghouse	16 (32)	1.2 (2.6)	Yes	n,p	B
							Baghouse	0.24 (0.48)	0.078 (0.172)	Yes	q,r	B
							Scrubber					
FeMn (1% Si)	Covered	6 (12)	2.4 (5.3)		h,t	E	High energy	0.8 (1.6)	0.34 (0.75)		h,s	E
	Sealed	37 (74)	17 (37)		u,v	E	High energy	0.25 (0.5)	0.10 (0.22)		h,s,w	C
FeCr (high carbon)	Open	78 (157)	15 (33)	Yes	x,y	C	Electrostatic precipitator	1.2 (2.3)	0.23 (0.5)	Yes	x,y	C
							Scrubber	2.1 (4.2)	0.44 (1.0)	Yes	aa,bb	C
SiMn	Open	96 (192)	20 (44)	Yes	z,aa	C	High energy	0.15 (0.30)	0.016 (0.035)		v,w	E
	Sealed	— (–)	— (–)									

[a]Factors are for main furnace dust collection system before and after control device. Where other emissions, such as leaks or tapping, are included or quantified separately, such is noted. Particulate sources not included: raw-material handling, storage, preparation; product crushing, screening, handling, packaging.
[b]Percentages are of the main alloying element in product.
[c]In most source testing, fugitive emissions not measured or collected. Where tapping emissions are controlled by primary system, their contribution to total emissions could not be determined. Fugitive emissions may vary greatly among sources, with furnace and collection system design and operating practices.
[d]Low-energy scrubbers are those with $\Delta P < 20$ inches H_2O; high energy, with $\Delta P > 20$ inches H_2O.
[e]Includes fumes captured by tapping hood (efficiency estimated near 100%).
[f]References 4, 5, 15.
[g]Factor is average of three sources, fugitive emissions not included. Fugitive emissions at one source measured an additional 10.5 kg/Mg alloy or 2.7 kg/MWh.
[h]References 4, 5.
[j]Does not include emissions from tapping or mix seal leaks.
[k]References 18, 19.
[m]Reference 16.
[n]Estimated 60% of tapping emissions captured by control system (escaped fugitive emissions not included in factor).
[p]References 5, 7.
[q]Estimated 50% of tapping emissions captured by control system (escaped fugitive emissions not included in factor).
[r]References 4, 5, 6.
[s]Includes fume only from primary control system.
[t]Includes tapping fumes and mix seal leak fugitive emissions. Fugitive emissions measured at 33% of total uncontrolled emissions.
[u]Assumes tapping fumes not included in emission factor.
[v]Reference 8. Dash—No data.
[w]Does not include tapping or fugitive emissions.
[x]Tapping emissions included. Factor developed from two test series performed on the same furnace seven years apart. Measured emissions in latter test were 36% less than in former.
[y]References 2, 9–11.
[z]Factor is average of two test series. Tests at one source included fugitive emissions (3.4% of total uncontrolled emissions). Second test insufficient to determine whether fugitive emissions were included in total.
[aa]References 2, 12, 13.
[bb]Factors developed from two scrubber controlled sources, one operated at $\Delta P = 47–57$ inches H_2O, the other at unspecified ΔP. Uncontrolled tapping operations emissions are 2.1 kg/Mg alloy.

TABLE 3. Size-Specific Emission Factors for Submerged Arc Ferroalloy Furnaces

Product	Control Device	Particle size,[a] μm	Cumulative Mass% ≤ Stated Size	Cumulative Mass Emission Factor kg/ton Alloy	(lb/ton) Alloy	Emission Factor Rating
50% FeSi						
Open furnace	None[b,c]	0.63	45	16	(32)	B
		1.00	50	18	(35)	
		1.25	53	19	(37)	
		2.50	57	20	(40)	
		6.00	61	21	(43)	
		10.00	63	22	(44)	
		15.00	66	23	(46)	
		20.00	69	24	(48)	
		d	100	35	(70)	
	Baghouse	0.63	31	0.28	(0.56)	B
		1.00	39	0.35	(0.70)	
		1.25	44	0.40	(0.80)	
		2.50	54	0.49	(1.0)	
		6.00	63	0.57	(1.1)	
		10.00	72	0.65	(1.3)	
		15.00	80	0.72	(1.4)	
		20.00	85	0.77	(1.5)	
			100	0.90	(1.8)	
80% FeMn						
Open furnace	None[e,f]	0.63	30	4	(8)	B
		1.00	46	7	(13)	
		1.25	52	8	(15)	
		2.50	62	9	(17)	
		6.00	72	10	(20)	
		10.00	86	12	(24)	
		15.00	96	13	(26)	
		20.00	97	14	(27)	
		d	100	14	(28)	
80% FeMn						
Open furnace	Baghouse[e]	0.63	20	0.048	(0.10)	B
		1.00	30	0.070	(0.14)	
		1.25	35	0.085	(0.17)	
		2.50	49	0.120	(0.24)	
		6.00	67	0.160	(0.32)	
		10.00	83	0.200	(0.40)	
		15.00	92	0.220	(0.44)	
		20.00	97	0.235	(0.47)	
		d	100	0.240	(0.48)	
Si Metal[h]						
Open furnace	None[g]	0.63	57	249	(497)	B
		1.00	67	292	(584)	
		1.25	70	305	(610)	
		2.50	75	327	(654)	
		6.00	80	349	(698)	
		10.00	86	375	(750)	
		15.00	91	397	(794)	
		20.00	95	414	(828)	
		d	100	436	(872)	
	Baghouse	1.00	49	7.8	(15.7)	B
		1.25	53	8.5	(17.0)	
		2.50	64	10.2	(20.5)	
		6.00	76	12.2	(24.3)	
		10.00	87	13.9	(28.0)	
		15.00	96	15.4	(31.0)	
		20.00	99	15.8	(31.7)	
			100	16.0	(32.0)	

TABLE 3. (*Continued*)

Product	Control Device	Particle size,[a] μm	Cumulative Mass% ≤ Stated Size	Cumulative Mass Emission Factor kg/ton Alloy	(lb/ton)	Emission Factor Rating
FeCr (HC)						
Open furnace	None[b,j]	0.5	19	15	(30)	C
		1.0	36	28	(57)	
		2.0	60	47	(94)	
		2.5	63[k]	49	(99)	
		4.0	76	59	(119)	
		6.0	88[k]	67	(138)	
		10.0	91	71	(143)	
		d	100	78	(157)	
	Electrostatic precipitator					C
		0.5	33	0.40	(0.76)	
		1.0	47	0.56	(1.08)	
		2.5	67	0.80	(1.54)	
		5.0	80	0.96	(1.84)	
		6.0	86	1.03	(1.98)	
		10.0	90	1.08	(2.07)	
		d	100	1.20	(2.30)	
SiMn						
Open furnace	None[b,m]	0.5	28	27	(54)	C
		1.0	44	42	(84)	
		2.0	60	58	(115)	
		2.5	65	62	(125)	
		4.0	76	73	(146)	
		6.0	85	82	(163)	
		10.0	96[k]	92[k]	(177)[k]	
		d	100	96	(192)	
	Scrubber[m,n]	0.5	56	1.18	(2.36)	
		1.0	80	1.68	(3.44)	
		2.5	96	2.02	(4.13)	
		5.0	99	2.08	(4.26)	
		6.0	99.5	2.09	(4.28)	
		10.0	99.9[k]	2.10[k]	(4.30)[k]	
			100	2.10	(4.30)	

[a]Aerodynamic diameter, based on Task Group on Lung Dynamics definition. Particle density = 1 g/cm^3.
[b]Includes tapping emissions.
[c]References 4, 5, 15.
[d]Total particulate, based on Method 5 total catch.
[e]Includes tapping fume (capture efficiency 50%).
[f]References 4, 5, 6.
[g]Includes tapping fume (estimated capture efficiency 60%).
[h]References 5, 7
[j]References 1, 9–11.
[k]Interpolated data.
[m]References 2, 12, 13.
[n]Primary emission control system only, without tapping emissions.

weight of carbon monoxide produced sometimes exceeds that of the metallic product. The chemical constituents of the heat-induced fume consist of oxides of the products being produced, carbon from the reducing agent, and enrichment by SiO_2, CaO, and MnO, if present in the charge.[14]

AIR POLLUTION CONTROL MEASURES

In an open electric arc furnace, all carbon monoxide burns with induced air at the furnace top. The remaining fume, captured by hooding about 3 feet above the furnace, is directed to a gas cleaning device. Baghouses are used to control emissions from 85% of the open furnaces in the United States. Scrubbers are used on 13% of the furnaces and electrostatic precipitators on 2%. Control efficiencies for well-designed and operated control systems [i.e., baghouses with air-to-cloth ratios of 1:1 to 2:1 ft^3/ft^2 and scrubbers with a pressure drop from 14 to 24 kPa (55 to 96 inches H_2O)], have been reported to be in excess of 99%.[4] Air-to-cloth ratio is the ratio of the volumetric airflow through the filter medium to the medium area.

Two emission capture systems, not usually connected to the same gas cleaning device, are necessary for covered furnaces. A primary capture system withdraws gases from beneath the furnace cover. A secondary system captures fume released around the electrode seals and during tapping. Scrubbers are used almost exclusively to control exhaust gases from sealed furnaces. The gas from sealed and mix sealed furnaces is usually flared at the exhaust of the scrubber. The carbon monoxide–rich gas has an estimated heating value of 300 Btu/ft^3 and is sometimes used as a fuel in kilns and sintering machines. The efficiency of flares for the control of carbon monoxide and the reduction of organic emission has been estimated to be greater than 98% for steam-assisted flares with a velocity of less than 60 fps and a gas heating value of 300 Btu/scf.[17] For unassisted flares, the reduction of organic and carbon monoxide emissions is 98% with a velocity of less than 60 fps and a gas heating value greater than 200 Btu/scf.

Tapping operations also generate fumes. Tapping is intermittent and is usually conducted during 10%–20% of the furnace operating time. Some fumes originate from the carbon lip liner, but most are a result of induced heat transfer from the molten metal or slag as it contacts the runners, ladles, casting beds, and ambient air. Some plants capture these emissions to varying degrees with a main canopy hood. Other plants employ separate tapping hoods ducted to either the furnace emission control device or a separate control device. Emission factors for tapping emissions are unavailable because of a lack of data.

A reaction ladle may be involved to adjust the metallurgy after furnace tapping, by chlorination, oxidation, gas mixing, and slag metal reactions. Ladle reactions are an intermittent process and emissions have not been quantified. Reaction ladle emissions often are captured by the tapping emissions control system.

Available data are insufficient to provide emission factors for raw-material handling, pretreatment, and product handling. Dust particulate is emitted from raw-material handling, storage, and preparation activities (see Figure 1), from such specific activities as unloading of raw materials from delivery vehicles (ship, railcar, or truck), storage of raw materials in piles, loading of raw materials from storage piles into trucks or gondola cars, and crushing and screening of raw materials. Raw materials may be dried before charging in rotary or other types of dryers, and these dryers can generate significant particulate emissions. Dust may also be generated by heavy vehicles used for loading, unloading, and transferring material. Crushing, screening, and storage of the ferroalloy product emit particulate in the form of dust. The properties of particulate emitted as dust are similar to the natural properties of the ores or alloys from which they originated, ranging in size from 3 to 100 μm.

Approximately half of ferroalloy facilities have some type of control for dust emissions. Dust generated from raw-material storage may be controlled in several ways, including sheltering storage piles from the wind with block walls, snow fences, or plastic covers. Occasionally, piles are sprayed with water to prevent airborne dust. Emissions generated by heavy vehicle traffic may be reduced by using a wetting agent or paving the plant yard.[3] Moisture in the raw materials, which may be as high as 20%, helps to limit dust emissions from raw-material unloading and loading. Dust generated by crushing, sizing, drying, or other pretreatment activities is sometimes controlled by dust collection equipment such as scrubbers, cyclones, or baghouses. Ferroalloy product crushing and sizing usually require a baghouse. The raw-material emission collection equipment may be connected to the furnace emission control system.

References

1. F. J. Schottman, "Ferroalloys," *1980 Mineral Facts and Problems,* Bureau of Mines, U.S. Department of the Interior, Washington, DC, 1980.
2. J. O. Dealy and A. M. Killin, *Engineering and Cost Study of the Ferroalloy Industry,* EPA-450/2-74-008, U.S. Environmental Protection Agency, Research Triangle Park, NC, May 1974.
3. *Background Information on Standards of Performance: Electric Submerged Arc Furnaces for Production of Ferroalloys, Volume I: Proposed Standards,* EPA-450/2-74-018a, U.S. Environmental Protection Agency, Research Triangle Park, NC, October 1974.
4. C. W. Westbrook and D. P. Dougherty, *Level I Environmental Assessment of Electric Submerged Arc Furnaces Producing Ferroalloys,* EPA-600/2-81-038, U.S. Environmental Protection Agency, Washington, DC, March 1981.
5. C. W. Westbrook, *Multimedia Environmental Assessment of Electric Submerged Arc Furnaces Producing Ferroalloys,* EPA-600/2-83-092, U.S. Environmental Protection Agency, Washington, DC, September 1983.
6. T. Epstein et al., *Ferroalloy Furnace Emission Factor Development, Roane Limited, Rockwood, Tennessee,* EPA-600/X-85-325, U.S. Environmental Protection Agency, Washington, DC, June 1981.
7. S. Beaton et al., *Ferroalloy Furnace Emission Factor Development, Interlake Inc., Alabama Metallurgical Corp., Selma, Alabama,* EPA-600/X-85-324, U.S. Environmental Protection Agency, Washington, DC, May 1981.
8. J. L. Rudolph et al., *Ferroalloy Process Emissions Measurement,* EPA-600/2-79-045, U.S. Environmental Protection Agency, Washington, DC, February 1979.
9. Joseph F. Eyrich, Macalloy Corp., Charleston, SC, written communication to GCA Corp., Bedford, MA, February 10, 1982, citing Airco Alloys and Carbide test R-07-7774-000-1, Gilbert Commonwealth, Reading, PA, 1978.

10. *Source Test, Airco Alloys and Carbide, Charleston, SC,* EMB-71-PC-16(FEA), U.S. Environmental Protection Agency, Research Triangle Park, NC, 1971.
11. Joseph F. Eyrich, Macalloy Corp., Charleston, SC, telephone communication with Evelyn J. Limberakais, GCA Corp., Bedford, MA, February 23, 1982.
12. Source test, *Chromium Mining and Smelting Corporation, Memphis, TN,* EMB-72-PC-05 (FEA), U.S. Environmental Protection Agency, Research Triangle Park, NC, June 1972.
13. *Source Test, Union Carbide Corporation, Ferroalloys Division, Marietta, Ohio,* EMB-71-PC-12 (FEA), U.S. Environmental Protection Agency, Research Triangle Park, NC, 1971.
14. R. A. Person, "Control of emissions from ferroalloy furnace processing," *J. Metals, 23*(4):17–29 (April 1971).
15. S. Gronberg, *Ferroalloy Furnace Emission Factor Development, Foote Minerals, Graham, W. Virginia,* EPA-600/X-85-327, U.S. Environmental Protection Agency, Washington, DC, July 1981.
16. *Air Pollutant Emission Factors, Final Report,* APTD-0923, U.S. Environmental Protection Agency, Research Triangle Park, NC, April 1970.
17. Leslie B. Evans, Office of Air Quality Planning and Standards, U.S. Environmental Protection Agency, Research Triangle Park, NC, telephone communication with Richard Vacherot, GCA Corp., Bedford, MA, October 18, 1984.
18. R. Ferrari, "Experiences in developing an effective pollution control system for a submerged arc ferroalloy furnace operation," *J. Metals,* 95–104 (April 1968).
19. Fredriksen and Nestaas, *Pollution Problems by Electric Furnace Ferroalloy Production,* UN Economic Commission for Europe, September 1968.
20. A. E. Vandergrift et al., *Particulate Pollutant System Study—Mass Emissions,* PB-203-128, PB-203-522 and P-203-521, National Technical Information Service, Springfield VA, May 1971.
21. *Control Techniques for Lead Air Emissions,* EPA-450/2-77-012, U.S. Environmental Protection Agency, Research Triangle Park, NC, December 1977.
22. W. E. Davis, *Emissions Study of Industrial Sources of Lead Air Pollutants,* 1970, EPA-APTD-1543, W. E. Davis and Associates, Leawood, KS, April 1973.
23. *Source Test, Foote Mineral Company, Vancoram Operations, Steubenville, Ohio,* EMB-71-PC-08 (FEA), U.S. Environmental Protection Agency, Research Triangle Park, NC, August 1971.

Bibliography

1. F. J. Schottman, "Ferroalloys," *Minerals Yearbook, Vol. I: Metals and Minerals,* Bureau of Mines, U.S. Department of the Interior, Washington, DC, 1980
2. S. Beaton and H. Klemm, *Inhalable Particulate Field Sampling Program for the Ferroalloy Industry,* TR-80-115-G, GCA Corporation, Bedford, MA, November 1980.
3. G. W. Westbrook and D. P. Dougherty, *Environmental Impact of Ferroalloy Production Interim Report: Assessment of Current Data,* U.S. Environmental Protection Agency, Research Triangle Institute, Research Triangle Park, NC, November 1978.
4. K. Wark and C. F. Warner, *Air Pollution: Its Origin and Control,* Harper & Row, New York, 1981.
5. M. Szabo and R. Gerstle, *Operations and Maintenance of Particulate Control Devices on Selected Steel and Ferroalloy Processes,* EPA-600/2-78-037, U.S. Environmental Protection Agency, Washington, DC, March 1978.
6. S. Gronberg et al., *Ferroalloy Industry Particulate Emissions: Source Category Report,* EPA-600/7-86-039, U.S. Environmental Protection Agency, Cincinnati, OH, November 1986.
7. R. W. Gerstle et al., *Review of Standards of Performance for New Stationary Air Sources—Ferroalloy Production Facility,* EPA-450/3-80-041, U.S. Environmental Protection Agency, Research Triangle Park, NC, December 1980.

STEEL INDUSTRY

Bruce A. Steiner

In the years that have passed since the first publication of the *Air Pollution Engineering Manual* (AP-40) in 1967, the iron and steel industry has changed dramatically in many respects.

The number of steel-making facilities and the domestic capacity have been greatly reduced over the past quarter of a century. As an indication, steel industry employment has dropped from a high of over 500,000 persons to about 160,000 today. One reason is that increased imports have cut into domestic production. In 1967, imports represented only 13% of steel shipments, but in recent years, they have been as high as 26%. Yields have also increased substantially. For example, in 1967, the U.S. industry produced 127 million tons of raw steel and shipped 88 million tons. In 1990, the industry shipped nearly the same amount (85 million tons), but required only 98 million tons of raw steel. The increased ratio of shipped tons to raw steel production reflects substantial increases in productivity, accomplished in large part by the increased installation of continuous casting facilities, which now account for about 67% of the raw steel output. Moreover, the development of stronger and lighter steels and coated grades allow steel to be more efficiently utilized than in past years, and the tailoring of steel properties and shapes to the needs of consuming industries results in less scrap and greater yield of finished product.

The changing face of the U.S. iron and steel industry has also brought about many process changes. In 1967, the dominant method of making steel was the open hearth furnace. Since then, however, most open hearth shops have been closed, and the basic oxygen process as a percentage of the total production has climbed from 33% to 60% and electric furnace production has increased from 12% to 37%. Over that period, annual domestic coke consumption has fallen to less than half, due in part to the lower raw steel production, but also to improved blast furnace operations that require lower coke rates. Improved yields, improved processing technologies, and energy conservation programs have allowed the industry to reduce its energy consumption per ton of shipped steel by about 40% since 1975. New processing techniques also continue to emerge. Among the technologies now commonly employed, but rarely used or unknown in 1967, are hot metal desulfurization, argon–oxygen decarburization, electrogalvanizing, vacuum degassing, ladle metallurgy, and slag skimming.

With these numerous changes come new challenges for air pollution control technology. Emphasis cannot be placed solely on the selection and efficiency of the control device for criteria pollutants, but we must also consider the full range of air pollutants that may be emitted, and we must include other factors, such as the capture efficiency of the entire system, and the energy efficiency, cost effectiveness, and ease of maintenance and reliability of the control system.

MISCELLANEOUS FUGITIVE EMISSION SOURCES

Robert E. Sistek and Frank Pendleton

Fugitive emissions can be defined as those emissions that enter the atmosphere from other than a stack, chimney, or similar device, and are commonly classified as either process or nonprocess emissions. Such emissions occur when pollutants escape capture by a control device or are generated without employing control measures. Many of the significant sources of fugitive emissions in the iron and steel industry are identified, along with appropriate emission factors, in Sections 7.5 and 11.2 of the U.S. Environmental Protection Agency (EPA) publication AP-42.[1]

SOURCES OF FUGITIVE EMISSIONS

Emissions of particulate matter, the most common fugitive emissions in the iron and steel industry, are generated by mechanical operations, process reactions, and fuel combustion. Gaseous pollutants are generated during certain processes, as well as during fuel combustion. Process emissions, which result from the chemical or physical alteration of a raw material, are normally generated inside a building or enclosure. Certain process emissions, such as oxygen–fuel cutting or lancing, can also occur as open source emissions. Potential sources of fugitive emissions from iron and steel mill operations are indicated in Figures 1 and 2 and identified in Table 1. Sources for which emission data are included in AP-42 are listed in Table 2. Several additional sources that are less significant or occur less frequently are addressed in the following.

Oxygen–Fuel Cutting and Lancing

Emissions from cutting and lancing are assumed to be similar to those from uncontrolled scarfing.[2] An emission factor of 0.0015 pound of particulate matter per pound of metal removed has been derived from the AP-42 emission

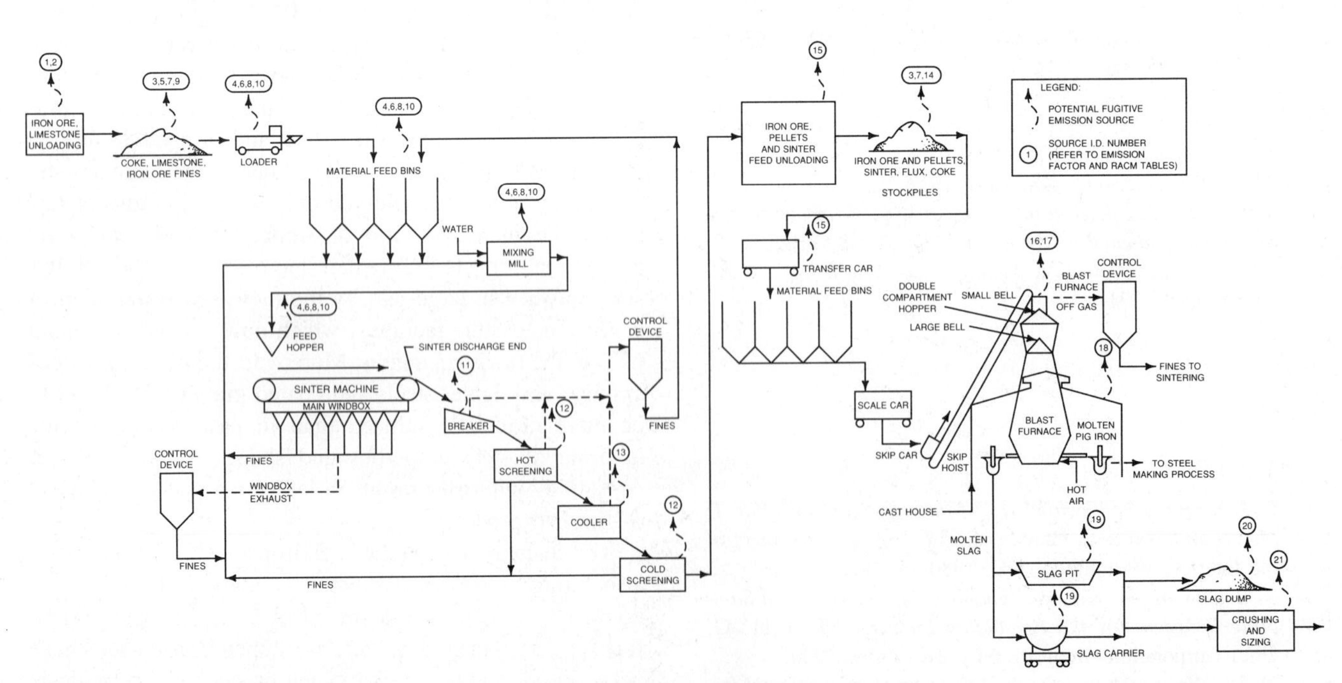

FIGURE 1. Process Flow Diagram of a Typical Iron Production Facility (From Ohio EPA, *Reasonably Available Control Measures for Fugitive Dust Sources,* September, 1980)

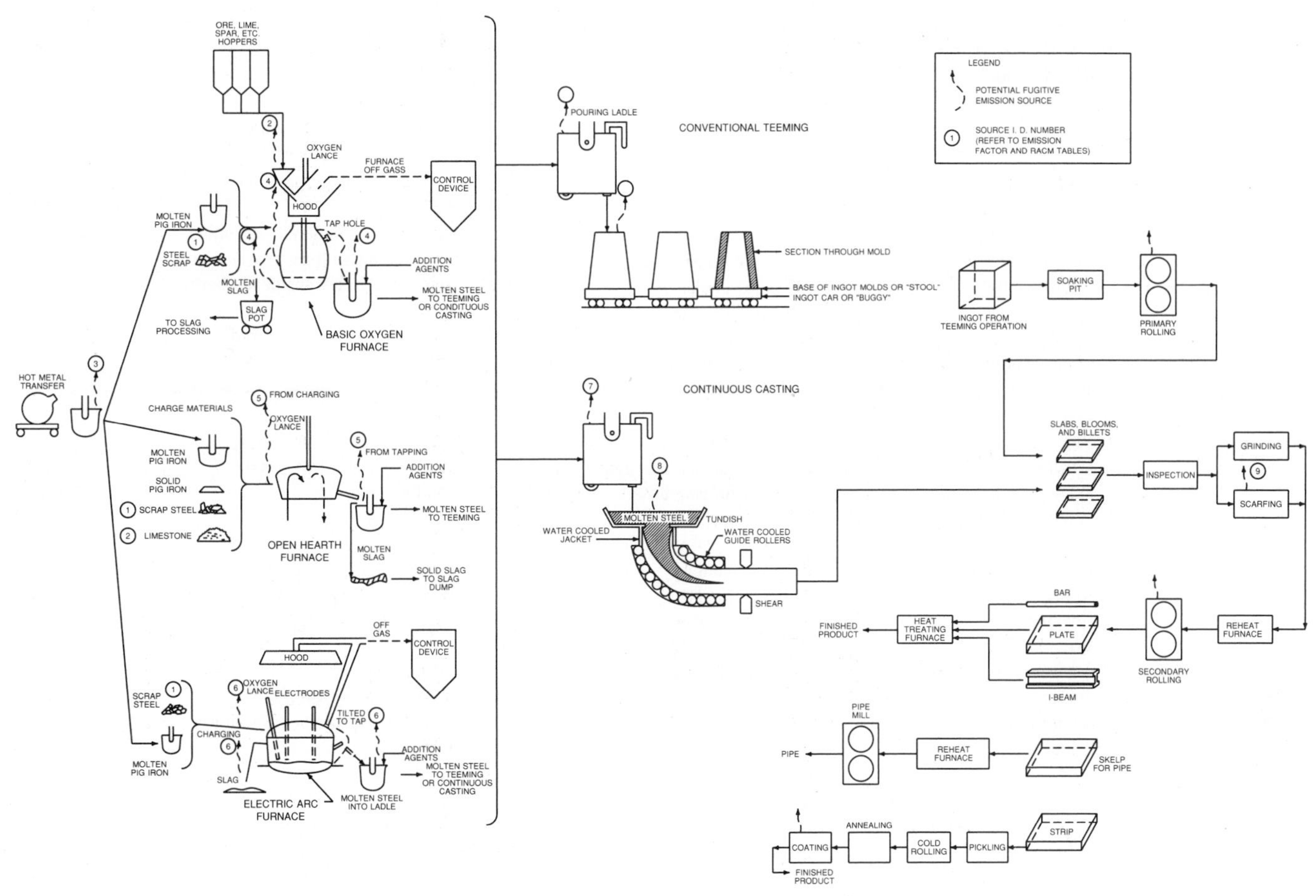

FIGURE 2. Process Flow Diagram of a Typical Steel Production Facility (From Ohio EPA, *Reasonably Available Control Measures for Fugitive Dust Sources,* September, 1980)

factor for uncontrolled scarfing.[3] A torch operator typically removes 200 pounds of metal in one hour and generates about 0.3 pound of particulate matter emissions.[4] Characteristics of the emissions are expected to be similar to those from scarfing. Emissions can be controlled by enclosing the operation and exhausting the emissions to a control device. Some degree of control can be achieved by the enclosure alone and can be as much as 70% for certain process emissions (see Table 7.5-1 of reference 1). Visible emissions from the operation can be reduced by injection of CO_2 into the oxygen–fuel mixture. Whether CO_2 injection actually reduces the mass emissions or simply prolongs the emission rate by extending cutting times is uncertain.

Kish Collection and Disposal

As the solubility of carbon in molten iron decreases during cooling, carbon precipitates out of solution as a fine graphite particle commonly referred to as kish. Although kish is generated during several operations, most is produced during the desulfurization process, generally utilizing an efficient emission control system. Following desulfurization, the kish is skimmed from the surface of the ladle, along with other slag components, and transported to a processing site. Emissions generated during skimming can be controlled by an appropriately placed hood, which exhausts to a control device, such as a baghouse. Control of fugitive dust emissions from the dumping and subsequent processing of kish, as well as other slag components, can best be achieved by moving the entire process indoors. Emissions from uncontrolled disposal of kish can be estimated using the material-handling equation in Section 11.2.3 of AP-42. Because of the hot, dry nature of the material, a moisture content of 0.25% should be used, which reflects the lowest value included in the data set used in developing the equation. A typical two-vessel basic oxygen furnace (BOF) shop can be expected to produce about 480 tons of kish per day.[5]

Ladle and Tun-Dish Repairs[5]

Emissions from ladle and tundish repairs are made up of dust and fumes from the removal and installation of refractory products, as well as combustion products from curing and preheating the repaired unit. Typically, a two-vessel BOF shop repairs an average of nine tundishes and 35 ladles each day. Removal of the metallic residue (skulls) requires about 15 minutes of lancing or oxygen–fuel cutting per unit.

TABLE 1. Identification of Potential Sources of Fugitive Emissions from Iron and Steel Production

Iron Production Sources		Steel Production Sources

1 Iron ore unloading
2 Limestone unloading
3 Iron ore storage
  Loading onto pile
  Vehicular traffic
  Loading out
  Wind erosion
4 Iron ore handling and transfer
5 Limestone storage
  Loading onto pile
  Vehicular traffic
  Loading out
  Wind erosion
6 Limestone handling and transfer
7 Coke storage
  Loading onto pile
  Vehicular traffic
  Loading out
  Wind erosion
8 Coke handling and transfer
9 Blast furnace flue dust storage
10 Blast furnace flue dust handling and transfer
	11 Sinter machine wind box discharge
	12 Sinter machine discharge (breaker and screen)
	13 Sinter cooler
	14 Sinter storage
	  Loading onto pile
	  Vehicular traffic
	  Loading out
	  Wind erosion
	15 Sinter handling and transfer
	16 Blast furnace charging
	17 Blast furnace upsets (slips)
	18 Blast furnace tapping—iron and slag
	19 Slag handling
	20 Slag storage
	  Loading onto pile
	  Vehicular traffic
	  Loading out
	  Wind erosion
	21 Slag crushing
		1 Scrap steel unloading, transfer, and storage
		2 Ore and flux material unloading, transfer, and storage
		3 Molten pig iron transfer to charge ladles
		4 Basic oxygen furnace—roof monitor
		  Charging
		  Refining
		  Tapping (steel and slag)
		5 Open hearth furnace—roof monitor
		  Charging
		  Refining
		  Tapping
		6 Electric arc furnace—roof monitor
		  Charging
		  Refining
		  Tapping
		7 Molten steel reladling
		8 Continuous casting/conventional teeming
		9 Scarfing

Note: Emission factors or predictive equations for the above sources can be found in Sections 7.5 and 11.2 of the EPA publication AP-42.

TABLE 2. Iron and Steel Mill Fugitive Emission Sources Included in AP-42, Section 7.5

		Data Available	
Source	Operation	TSP	Particle Size
Sintering	Wind box	Yes	Yes
	Discharge	Yes	No[a]
Blast furnace	Slip	Yes	No
	Cast-house	Yes	Yes
Basic oxygen furnace	Charging, tapping	Yes	Yes
	Hot metal transfer	Yes	No[a]
Electric arc furnaces	Charging, tapping, slagging	Yes	No[a]
Open hearth furnaces	Roof monitor	Yes	Yes
Teeming	—	Yes	No[a]
Machine scarfing	—	Yes	No
Conveyor transfer	Sinter	Yes	Yes[b]
Pile stacking	Lump ore	Yes	Yes[b]
	Coal	Yes	Yes[b]
Batch drop	Slag	Yes	Yes[b]
Paved roads	—	Yes	Yes[b]
Unpaved roads	—	Yes	Yes[b]

[a]Particle size data can be estimated using data for similar sources or equations provided in AP-42, Section 11.2.

[b]Use of predictive equations in AP-42, Section 11.2, are preferred over the stated emission factors.

Note: Appropriate sections of AP-42 are included in the appendix.

Damaged refractory is either patched or replaced by spray methods. Tundish spraying is often conducted utilizing local hoods to control overspray. Curing and preheating typically require about 4100 ft^3 of natural gas per tundish and 1000 ft^3 per ladle. Repairs to tundishes and ladles are normally conducted indoors. Dumping of the old refractory usually produces little, if any, dust emission. However, certain types of tundish construction can produce considerable dust when being removed. In such cases, a hood with an appropriate control device can be utilized. Emissions from lancing and natural gas combustion can be estimated using the factor for lancing contained here and the AP-42 factors for fuel combustion.

Ground Casting[5]

Ground casting (pouring molten metal on the ground) results from misalignment of the torpedo or subcars at the blast furnace cast house or burnout of a ladle. It is normally considered a nonroutine operation or a malfunction. The practice is, however, employed at times to allow for continued iron production during BOF maintenance outages. Estimates of ground cast metal can vary from about 600 tons per year when used only for emergency situations to as high as 190,000 tons per year when used in conjunction with BOF outages. An estimate of uncontrolled fugitive dust emissions can be made using the AP-42 factor for hot metal transfer. However, because of the much greater emissions associated with ground casting, it is recommended that the factor be doubled. The most suitable method of controlling emissions from ground casting is to conduct the operation indoors.

Conditioning Slag Pots[5]

Slag from the BOF process is transported in thimble-shaped pots to a slag processing site. To aid in removal of the slag, a release agent (commonly hot solid slag) is placed in the pot prior to its use. Typically, about 2–3 cubic yards of this material are charged in each slag pot after it is emptied. Some facilities, however, use as much as 6 cubic yards per slag pot. A two-vessel BOF shop will use an estimated 24 slag pots per day. Emissions from this conditioning process can be estimated using the material-handling equation from Section 11.2.3 in AP-42. Because of the hot, dry nature of the material, a moisture content of 0.25% should be used, which reflects the lowest value included in the data set used in developing the equation. Fugitive dust emissions from this practice can be minimized by controlling the amount of dry slag being charged in the pots and by moving the entire process indoors.

References

1. *Compilation of Air Pollutant Emission Factors,* AP-42, 4th ed., *Vol. I—Stationary Point and Area Sources,* U.S. Environmental Protection Agency. September 1988.
2. *Fugitive Emissions from Integrated Iron and Steel Plants,* EPA-600/2-78-050, Midwest Research Institute, March 1978, p 2–1.
3. F. Pendleton and R. Sistek, "PM_{10} characterization of iron and steel sources not previously included in fugitive emission inventories," Presented at the 83rd Annual Meeting and Exhibition of AWMA, Pittsburgh, June 1990.
4. W. Hass, Heckett Corp., East Chicago, IN, personal communication, January 1990.
5. *Characterization of Sources Not Previously Included in LTV's PM_{10} Inventory for the Indiana Harbor Works,* Project 9415-L, Midwest Research Institute, June 20, 1990.

Appendix A. AP-42 Emission Factors for Use with Iron and Steel Industry Emissions

In using the emission factors, attention should be given to the letter rating assigned to the specific factor. The ratings, which indicate A as being best and E as the lowest rating, reflect the quality of the data base used to determine the emission factor. A rating of A indicates that the factor is based on several high-quality tests from different sites. A single test of lesser quality, or data extrapolated from another, similar source, would likely result in a D or E rating.

AP-42 SECTION 7.5: IRON AND STEEL PRODUCTION

AP-42 TABLE 7.5-1. Particulate Emission Factors for Iron and Steel Mills[a]

Source	Units	Emission Factor	Emission Factor Rating	Particle Size Data
Sintering wind box	kg/Mg (lb/ton) finished sinter			
Uncontrolled				
Leaving grate		5.56 (11.1)	B	Yes
After coarse particulate removal		4.35 (8.7)	A	
Controlled by dry ESP		0.8 (1.6)	B	
Controlled by wet ESP		0.085 (0.17)	B	Yes
Controlled by venturi scrubber		0.235 (0.47)	B	Yes
Controlled by cyclone		0.5 (1.0)	B	Yes

(Continued)

AP-42 TABLE 7.5-1. *(Continued)*

Source	Units	Emission Factor	Emission Factor Rating	Particle Size Data
Sinter discharge (breaker and hot screens)				
	kg/Mg (lb/ton) finished sinter			
Uncontrolled		3.4 (6.8)	B	
Controlled by baghouse		0.05 (0.1)	B	Yes
Controlled by venturi scrubber		0.295 (0.59)	A	
Wind box and discharge	kg/Mg (lb/ton) finished sinter			
Controlled by baghouse		0.15 (0.3)	A	
Blast furnace				
Slip	kg/Mg (lb/ton) slip	39.5 (87.0)	D	
Uncontrolled cast house	kg/Mg (lb/ton) hot metal			
Roof monitor[b]		0.3 (0.6)	B	Yes
Furnace with local evacuation[c]		0.65 (1.3)	B	Yes
Taphole and trough only (not runners)		0.15 (0.3)	B	
Hot metal desulfurization				
Uncontrolled[d]	kg/Mg (lb/ton) hot metal	0.55 (1.09)	D	Yes
Controlled by baghouse		0.0045 (0.009)	D	Yes
Basic oxygen furnace (BOF)				
Top blown furnace melting and refining	(kg/Mg (lb/ton) steel			
Uncontrolled		14.25 (28.5)	B	
Controlled by open hood vented to:				
ESP		0.065 (0.13)	A	
Scrubber		0.045 (0.09)	B	
Controlled by closed hood vented to:				
Scrubber		0.0034 (0.0068)	A	Yes
BOF charging	kg/Mg (lb/ton) hot metal			
At source		0.3 (0.6)	D	Yes
At building monitor		0.071 (0.142)	B	
Controlled by baghouse		0.0003 (0.0006)	B	Yes
BOF tapping	kg/Mg (lb/ton) steel			
At source		0.46 (0.92)	D	Yes
At building monitor		0.145 (0.29)	B	
Controlled by baghouse		0.0013 (0.0026)	B	Yes
Hot metal transfer	kg/Mg (lb/ton) hot metal			
At source		0.095 (0.19)	A	
At building monitor		0.028 (0.056)	B	
BOF monitor (all sources)	kg/Mg (lb/ton) steel	0.25 (0.5)	B	
Q-BOP melting and refining	kg/Mg (lb/ton) steel			
Controlled by scrubber		0.028 (0.056)	B	Yes
Electric arc furnace				
Melting and refining	kg/Mg (lb/ton) steel			
Uncontrolled carbon steel		19.0 (38.0)	C	Yes
Charging, tapping, and slagging	kg/Mg (lb/ton) steel			
Uncontrolled emissions escaping monitor		0.7 (1.4)	C	
Melting, refining, charging, tapping, and slagging	kg/Mg (lb/ton) steel			
Uncontrolled				
Alloy steel		5.65 (11.3)	A	
Carbon steel		25.0 (50.0)	C	
Controlled by:[e]				
Building evacuation to baghouse for alloy steel		0.15 (0.3)	A	
Direct shell evacuation (plus charging hood) vented to common baghouse for carbon steel		0.0215 (0.043)	E	Yes
Open hearth furnace				
Melting and refining	kg/Mg (lb/ton) steel			
Uncontrolled		10.55 (21.1)	D	Yes
Controlled by ESP		0.14 (0.28)	D	Yes
Roof monitor		0.084 (0.168)	C	

AP-42 TABLE 7.5-1. (*Continued*)

Source	Units	Emission Factor	Emission Factor Rating	Particle Size Data
Teeming				
Leaded steel	kg/Mg (lb/ton) steel			
Uncontrolled (measured at source)		0.405 (0.81)	A	
Controlled by side draft hood vented to baghouse		0.0019 (0.0038)	A	
Unleaded steel				
Uncontrolled (measured at source)		0.035 (0.07)	A	
Controlled by side draft hood vented to baghouse		0.0008 (0.0016)	A	
Machine scarfing				
Uncontrolled	kg/Mg (lb/ton) metal through scarfer	0.05 (0.1)	B	
Controlled by ESP		0.0115 (0.023)	A	
Miscellaneous combustion sources[f]		f f		
Boiler, soaking pit and slab reheat	$kg/10^9$ J ($lb/10^6$ Btu)			
Blast furnace gas[g]		0.015 (0.035)	D	
Coke oven gas[g]		0.0052 (0.012)	D	

[a]Reference 3, except as noted.
[b]Typical of older furnaces with no controls or for canopy hoods or total cast-house evacuation.
[c]Typical of large, new furnaces with local hoods and covered evacuated runners. Emissions are higher than without capture systems because they are not diluted by outside environment.
[d]Emission factor of 0.55 kg/Mg (1.09 lb/ton) represents one torpedo car; 1.26 kg/Mg (2.53 lb/ton) for two torpedo cars, and 1.37 kg/Mg (2.74 lb/ton) for three torpedo cars.
[e]Building evacuation collects all process emissions and direct shell evacuation collects only melting and refining emissions.
[f]For various fuels, use the emission factors in Chapter 1 of this document. The emission factor rating, for these fuels in boilers is A and in soaking pits and slab reheat furnaces it is D.
[g]Based on methane content and cleaned particulate loading.

AP-42 TABLE 7.5-2. Size-Specific Emission Factors

Source	Emission Factor Rating	Particle Size, μm[a]	Cumulative Mass % ≤ Stated Size	Cumulative Mass Emission Factor	
				kg/Mg	(lb/ton)
Sintering					
Wind box					
Uncontrolled					
Leaving grate	D	0.5	4[b]	0.22	(0.44)
		1.0	4	0.22	(0.44)
		2.5	5	0.28	(0.56)
		5.0	9	0.50	(1.00)
		10	15	0.83	(1.67)
		15	20[c]	1.11	(2.22)
		d	100	5.56	(11.1)
Controlled by wet ESP	C	0.5	18[b]	0.015	(0.03)
		1.0	25	0.021	(0.04)
		2.5	33	0.028	(0.06)
		5.0	48	0.041	(0.08)
		10	59[b]–	0.050	(0.10)
		15	69	0.059	(0.12)
		d	100	0.085	(0.17)

(*Continued*)

AP-42 TABLE 7.5-2. *(Continued)*

Source	Emission Factor Rating	Particle Size, μm[a]	Cumulative Mass % ≤ Stated Size	Cumulative Mass Emission Factor kg/Mg	(lb/ton)
Controlled by venturi scrubber	C	0.5	55	0.129	(0.26)
		1.0	75	0.176	(0.35)
		2.5	89	0.209	(0.42)
		5.0	93	0.219	(0.44)
		10	96	0.226	(0.45)
		15	98	0.230	(0.46)
		d	100	0.235	(0.47)
Controlled by cyclone[e]	C	0.5	25[c]	0.13	(0.25)
		1.0	37[b]	0.19	(0.37)
		2.5	52	0.26	(0.52)
		5.0	64	0.32	(0.64)
		10	74	0.37	(0.74)
		15	80	0.40	(0.80)
		d	100	0.5	(1.0)
Controlled by baghouse	C	0.5	3.0	0.005	(0.009)
		1.0	9.0	0.014	(0.027)
		2.5	27.0	0.041	(0.081)
		5.0	47.0	0.071	(0.141)
		10.0	69.0–	0.104	(0.207)
		15.0	79.0	0.119	(0.237)
		d	100.0	0.15	(0.3)
Sinter discharge (breaker and hot screens) controlled by baghouse	C	0.5	2[b]	0.001	(0.002)
		1.0	4	0.002	(0.004)
		2.5	11	0.006	(0.011)
		5.0	20	0.010	(0.020)
		10	32[b]	0.016	(0.032)
		15	42[b]	0.021	(0.042)
		d	100	0.05	(0.1)
Blast furnace					
Uncontrolled cast-house emissions, roof monitor[f]	C	0.5	4	0.01	(0.02)
		1.0	15	0.05	(0.09)
		2.5	23	0.07	(0.14)
		5.0	35	0.11	(0.21)
		10	51	0.15	(0.31)
		15	61	0.18	(0.37)
		d	100	0.3	(0.6)
Furnace with local evacuation[g]	C	0.5	7[c]	0.04	(0.09)
		1.0	9	0.06	(0.12)
		2.5	15	0.10	(0.20)
		5.0	20	0.13	(0.26)
		10	24	0.16	(0.31)
		15	26	0.17	(0.34)
		d	100	0.65	(1.3)
Hot metal desulfurization[h]		0.5	j		
Uncontrolled	E	1.0	2[c]	0.01	(0.02)
		2.5	11	0.06	(0.12)
		5.0	19	0.10	(0.22)
		10	19	0.10	(0.22)
		15	21	0.12	(0.23)
		d	100	0.55	(1.09)
Hot metal desulfurization[h] Uncontrolled baghouse	D	0.5	8	0.0004	(0.0007)
		1.0	18	0.0009	(0.0016)
		2.5	42	0.0019	(0.0038)
		5.0	62	0.0028	(0.0056)
		10	74	0.0033	(0.0067)
		15	78	0.0035	(0.0070)
		d	100	0.0045	(0.009)

AP-42 TABLE 7.5-2. (*Continued*)

Source	Emission Factor Rating	Particle Size, μm[a]	Cumulative Mass % ≤ Stated Size	Cumulative Mass Emission Factor kg/Mg	(lb/ton)
Basic oxygen furnace					
Top blown furnace melting and refining controlled by closed hood and vented to scrubber	C	0.5	34	0.0012	(0.0023)
		1.0	55	0.0019	(0.0037)
		2.5	65	0.0022	(0.0044)
		5.0	66	0.0022	(0.0045)
		10	67	0.0023	(0.0046)
		15	72[c]	0.0024	(0.0049)
		d	100	0.0034	(0.0068)
BOF charging					
At source[k]	E	0.5	8[c]	0.02	(0.05)
		1.0	12	0.04	(0.07)
		2.5	22	0.07	(0.13)
		5.0	35	0.10	(0.21)
		10	46	0.14	(0.28)
		15	56	0.17	(0.34)
		d	100	0.3	(0.6)
Controlled by baghouse	D	0.5	3	9.0×10^{-6}	1.8×10^{-5}
		1.0	10	3.0×10^{-5}	6.0×10^{-5}
		2.5	22	6.6×10^{-5}	(0.0001)
		5.0	31	9.3×10^{-5}	(0.0002)
		10	45	0.0001	(0.0003)
		15	60	0.0002	(0.0004)
		d	100	0.0003	(0.0006)
BOF tapping at source[k]	E	0.5	j	j	j
		1.0	11	0.05	(0.10)
		2.5	37	0.17	(0.34)
		5.0	43	0.20	(0.40)
		10	45	0.21	(0.41)
		15	50	0.23	(0.46)
		d	100	0.46	(0.92)
BOF tapping					
controlled by baghouse	D	0.5	4	5.2×10^{-5}	(0.0001)
		1.0	7	0.0001	(0.0002)
		2.5	16	0.0002	(0.0004)
		5.0	22	0.0003	(0.0006)
		10	30	0.0004	(0.0008)
		15	40	0.0005	(0.0010)
		d	100	0.0013	(0.0026)
Q-BOP melting and refining					
Controlled by scrubber	D	0.5	45	0.013	(0.025)
		1.0	52	0.015	(0.029)
		2.5	56	0.016	(0.031)
		5.0	58	0.016	(0.032)
		10	68	0.019	(0.038)
		15	85[c]	0.024	(0.048)
		d	100	0.028	(0.056)
Electric arc furnace melting and refining carbon steel					
Uncontrolled[m]	D	0.5	8	1.52	(3.04)
		1.0	23	4.37	(8.74)
		2.5	43	8.17	(16.34)
		5.0	53	10.07	(20.14)
		10	58	11.02	(22.04)
		15	61	11.59	(23.18)
		d	100	19.0	(38.0)
Electric arc furnace melting, refining, charging, tapping, and slagging					

(*Continued*)

AP-42 TABLE 7.5-2. (*Continued*)

Source	Emission Factor Rating	Particle Size, μm[a]	Cumulative Mass % ≤ Stated Size	Cumulative Mass Emission Factor	
				kg/Mg	(lb/ton)
Controlled by direct shell evacuation (plus charging hood) vented to common baghouse for carbon steel[n]					
	E	0.5	74[b]	0.0159	(0.0318)
		1.0	74	0.0159	(0.0318)
		2.5	74	0.0159	(0.0318)
		5.0	74	0.0159	(0.0318)
		10	76	0.0163	(0.0327)
		15	80	0.0172	(0.0344)
		d	100	0.0215	(0.043)
Open hearth furnace melting and refining					
Uncontrolled	E	0.5	1[b]	0.11	(0.21)
		1.0	21	2.22	(4.43)
		2.5	60	6.33	(12.66)
		5.0	79	8.33	(16.67)
		10	83	8.76	(17.51)
		15	85[c]	8.97	(17.94)
		d	100	10.55	(21.1)
Open hearth furnaces					
Controlled by ESP[p]	E	0.5	10[b]	0.01	(0.02)
		1.0	21	0.03	(0.06)
		2.5	39	0.05	(0.10)
		5.0	47	0.07	(0.13)
		10	53[b]	0.07	(0.15)
		15	56[b]	0.08	(0.16)
		d	100	0.14	(0.28)

[a]Particle aerodynamic diameter micrometers (μm) as defined by Task Group on Lung Dynamics. (Particle density = 1 gr/cm^3).
[b]Interpolated data used to develop size distribution.
[c]Extrapolated, using engineering estimates.
[d]Total particulate based on Method 5 total catch. See Table 7.5-1.
[e]Average of various cyclone efficiencies.
[f]Total cast-house evacuation control system.
[g]Evacuation runner covers and local hood over taphole, typical of new state-of-the-art blast furnace technology.
[h]Torpedo ladle desulfurization with CaC_2 and $CaCO_3$.
[j]Unable to extrapolate because of insufficient data and/or curve exceeding limits.
[k]Doghouse-type furnace enclosure using front and back sliding doors, totally enclosing the furnace, with emissions vented to hoods.
[m]Full-cycle emissions captured by canopy and side draft hoods.
[n]Information on control system not available.
[p]May not be representative. Test outlet size distribution was larger than inlet and may indicate reentrainment problem.

AP-42 SECTION 11.2.3: AGGREGATE HANDLING AND STORAGE PILES

AP-42 TABLE 11.2.3-1. Typical Silt and Moisture Content Values of Materials at Various Industries

Industry	Material	Silt (%)			Moisture, %		
		No. of Test Samplers	Range	Mean	No. of Test Samplers	Range	Mean
Iron and steel production[a]	Pellet ore	10	1.4–13	4.9	8	0.64–3.5	2.1
	Lump ore	9	2.8–19	9.5	6	1.6–8.1	5.4
	Coal	7	2–7.7	5	6	2.8–11	4.8
	Slag	3	3–7.3	5.3	3	0.25–2.2	0.92
	Flue dust	2	14–23	18.0	0	NA	NA
	Coke breeze	1		5.4	1		6.4
	Blended ore	1		15.0	1		6.6
	Sinter	1		0.7	0	NA	NA
	Limestone	1		0.4	0	NA	NA

AP-42 TABLE 11.2.3-1. *(Continued)*

		Silt (%)			Moisture, %		
Industry	Material	No. of Test Samplers	Range	Mean	No. of Test Samplers	Range	Mean
Stone quarrying and processing[b]	Crushed limestone	2	1.3–1.9	1.6	2	0.3–1.1	0.7
Taconite mining and processing[c]	Pellets	9	2.2–5.4	3.4	7	0.05–2.3	0.9
	Tailings	2	NA	11.0	1		0.35
Western surface coal mining[d]	Coal	15	3.4–16	6.2	7	2.8–20	6.9
	Overburden	15	3.8–15	7.5	0	NA	NA
	Exposed ground	3	5.1–21	15.0	3	0.8–6.4	3.4
Coal-fired power generation[e]	Coal	60	0.6–4.8	2.2	59	2.7–7.4	4.5

[a]References 2–5. NA = not applicable.
[b]Reference 1.
[c]Reference 6.
[d]Reference 7.
[e]Reference 8. Values reflect "as received" conditions of a single power plant.

The quantity of particulate emissions generated by either type of drop operation, per ton of material transferred, may be estimated, with a rating of A, using the following empirical expression[2]:

$$E = k(0.0016)\,\frac{\left(\frac{U}{2.2}\right)^{1.3}}{\left(\frac{M}{2}\right)^{1.4}}\ (\text{kg/Mg})$$

$$E = k(0.0032)\,\frac{\left(\frac{U}{5}\right)^{1.3}}{\left(\frac{M}{2}\right)^{1.4}}\ (\text{lb/ton})$$

where: E = emission factor
k = particle size multiplier (dimensionless)
U = mean wind speed, m/s (mph)
M = material moisture content (%)

The particle size multiplier, k, varies with aerodynamic particle diameter, as shown in Table 11.2.3-2.

AP-42 TABLE 11.2.3-2. Aerodynamic Particle Size Multiplier *(k)*

<30 μm	<15 μm	<10 μm	<5 μm	<2.5 μm
0.74	0.48	0.35	0.20	0.11

The equation retains the assigned quality rating if applied within the ranges of source conditions that were tested in developing the equation, as given in Table 11.2.3-3. Note that silt content is included in Table 11.2.3-3, even though silt content does not appear as a correction parameter in the equation. While it is reasonable to expect that silt content and emission factors are interrelated, no significant correlation between the two was found during the derivation of the equation, probably because most tests with high silt contents were conducted under lower winds, and vice versa. It is recommended that estimates from the equation be reduced one quality rating level, if the silt content used in a particular application falls outside the range given in Table 11.2.3-3.

AP-42 TABLE 11.2.3-3. Ranges of Source Conditions for Equation 1

Silt Content	Moisture Content	Wind Speed	
		m/s	mph
0.44–19	0.25–4.8	0.6–6.7	1.3–15

References for AP-42 Section 11.2.3

1. C. Cowherd, Jr., *et al.*, *Development Of Emission Factors For Fugitive Dust Sources,* EPA-450/3-74-037, U.S. Environmental Protection Agency, Research Triangle Park, NC, June 1974.
2. R. Bohn, *et al.*, *Fugitive Emissions From Integrated Iron And Steel Plants,* EPA-600/2-78-050, U.S. Environmental Protection Agency, Cincinnati, OH, March 1978.
3. C. Cowherd, Jr., *et al.*, *Iron And Steel Plant Open Dust Source Fugitive Emission Evaluation,* EPA-600/2-79-103, U.S. Environmental Protection Agency, Cincinnati, OH, May 1979.
4. R. Bohn, *Evaluation Of Open Dust Sources In The Vicinity Of Buffalo, New York,* EPA Contract No. 68-02-2545, Midwest Research Institute, Kansas City, MO, March 1979.
5. C. Cowherd, Jr., and T. Cuscino, Jr., *Fugitive Emissions Evaluation,* MRI-4343-L, Midwest Research Institute, Kansas City, MO, February 1977.
6. T. Cuscino, *et al.*, *Taconite Mining Fugitive Emissions Study,* Minnesota Pollution Control Agency, Roseville, MN, June 1979.
7. K. Axetell and C. Cowherd, Jr., *Improved Emission Factors For Fugitive Dust From Western Surface Coal Mining Sources,* 2 Volumes, EPA Contract No. 68-03-2924, PEI, Inc., Kansas City, MO, July 1981.
8. E. T. Brookman, *et al.*, *Determination of Fugitive Coal Dust Emissions From Rotary Railcar Dumping,* 1956-L81-00, TRC, Hartford, CT, May 1984.
9. G. A. Jutze, *et al.*, *Investigation Of Fugitive Dust Sources Emissions And Control,* EPA-450/3-74-036a, U.S. Environmental Protection Agency, Research Park, NC, June 1974.

BLAST FURNACE

Marek S. Klag, B.E.

The blast furnace is at the present time the most efficient method of producing hot metal (molten iron). Depending on the size, modern furnaces can produce up to 10,000 tons of hot metal daily. It is possible to achieve uninterrupted furnace operation of more than 10 years. The blast furnace process control systems have been developed to optimize operational efficiency, increase productivity, and protect the environment.

PROCESS DESCRIPTION

A blast furnace as shown in Figure 1 consists of a refractory-lined steel shaft in which the charge is fed into the top through a gas seal. Air heated from 871°C to 1100°C is blown through tuyeres into the lower part of the furnace. The combustion of coke provides the carbon monoxide (CO) to reduce the iron oxides to iron and provides additional heat to melt the iron and impurities. Auxiliary fuels such as coal, oil, natural gas, or tar may also be injected through the tuyeres. As the burden moves downward through the furnace, it is heated by the countercurrent upward flow of gases that exit at the top of the furnace. Molten iron and slag accumulate in the hearth of the furnace and a taphole is drilled into the hearth to drain the slag and iron accumulation into a trough. The trough is equipped with a skimmer and dam at the outlet end, allowing the iron to be drawn off separately from the slag. A typical charge required to produce a ton of hot metal consists of 1.55 tons of iron ore, 0.5 ton of coke, and 0.05–0.15 ton of limestone and dolomite. Higher productivity has been achieved through the use of high-purity iron units, sized raw materials, higher blast temperatures, blast humidity control, higher top pressures, and fuel and oxygen injection.

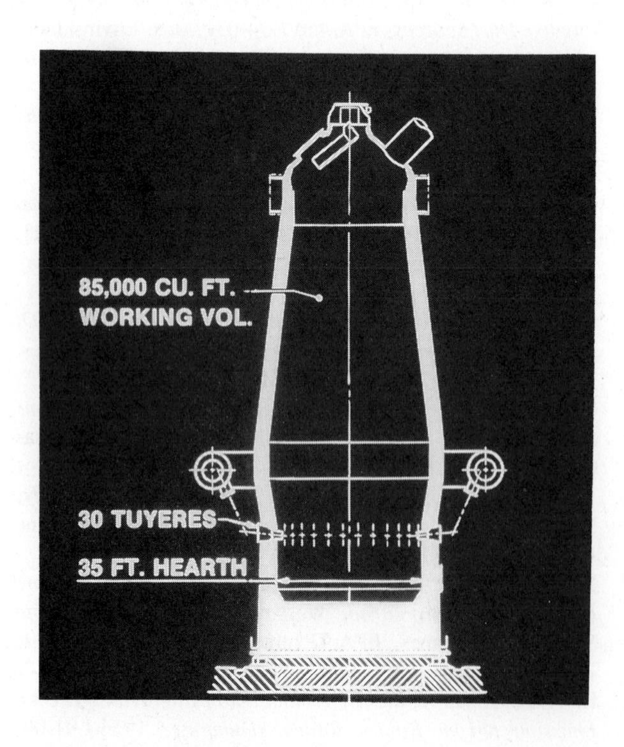

FIGURE 1. Modern Blast Furnace

Pollution Control Equipment

Pollution control in the iron-making industry focuses on primary and secondary emissions. The blast furnace top gas and emissions problems associated with the iron production activities have been essentially solved. The difficulty in capturing dust often requires the extraction of a large volume of dilution air; particularly for secondary processes. Emissions contained in blast furnace top gas are generally handled through wet collection systems. These include dust-catchers, venturi scrubbers, and precipitators (Figure 2). The collection efficiency of wet systems can be related to the energy losses resulting from the pressure drop of gas through the units. Disadvantages include high energy requirements and large volumes of scrubbing liquid to be treated. Collection efficiency is in the range of 99% for particulate larger than 1 μm.[1] Secondary emission collection (e.g., cast-house operation) is accomplished using a baghouse. The dirty gases are forced through filter bags that trap the particulate. Dust particulates are separated from the bags by gravity. The material from which the bags are woven depends on particle size, the gas temperature, acidity, alkalinity, and the desired life of the filter element. The collection efficiency for fabric filtration is greater than 99%. Reference 2 shows particulate emission factors and particulate size for iron and steel mills.

AIR EMISSIONS CHARACTERIZATION

Nature of Emissions

Top Gas

The top gas contains CO and particulate levels. The blast furnace gas has a relatively low heating value in the range of 3.7 MJ/m^3.[3,4] The top gas can be used as a fuel in steel plants. Submicron particles are formed when vapour from low-boiling-point materials such as sodium (Na), potassium (K), cerium (Ce), and zinc (Zn) compounds vaporize in the bottom of the furnace, where temperatures rise above 1650°C and condense in the upper section of the furnace. The gas leaving the furnace contains trace quantities of hydrogen sulfide (H_2S) and sulfur dioxide (SO_2) due to the efficiency of the iron oxide and flux in removing sulphurous gases. See Table 1.

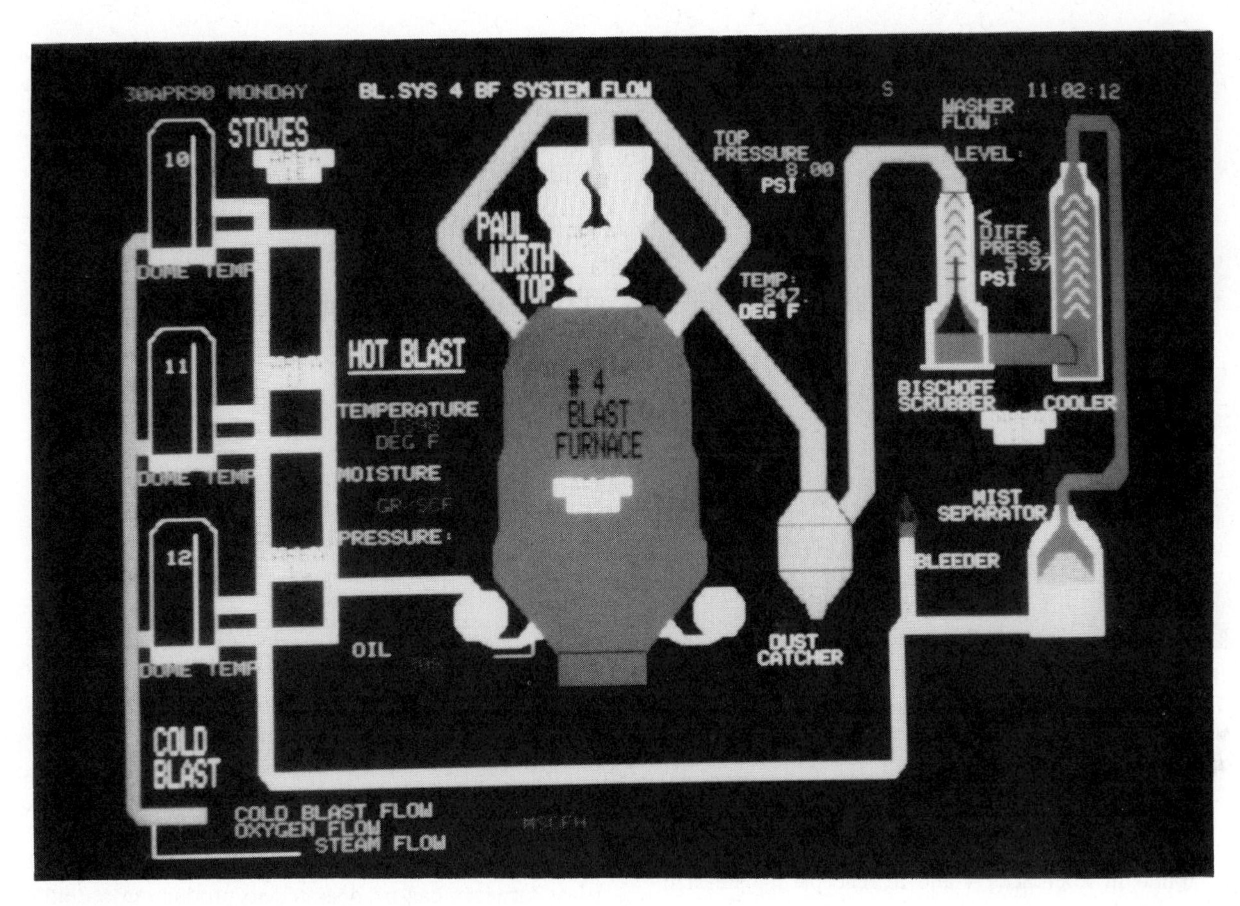

FIGURE 2. Blast Furnace System Flow

Tapping (Casting) Liquid Iron

Every 45 to 90 minutes, the taphole in the hearth of the furnace is drilled open. The slag metal mixture exits at a temperature of from 1300°C to 1500°C. The slag floats to the top of the trough and a dam system separates the hot metal and slag into two separate streams. The hot metal is drained through a runner system into a torpedo car mounted on railroad wheels. Hot metal cars typically hold 80 to 300 tons of liquid iron for delivery to oxygen steel-making facilities. The slag is drained into a pit or slag pots by slag runners. Casting emissions generated in the cast-house are shown in Table 2.

Iron oxides emissions are generated during casting either by the direct vaporization of compounds or the partial pressure of CO bursting bubbles at the metal atmosphere interface. The ejection of hot metal into the air quickly oxidizes to produce iron oxide fume (see Figure 3). If vaporization were the major mechanism, then enrichment of the manganese (Mn) level in the collected dust should occur. Since the Mn in the dust collected by the blast furnace cast-house baghouse is only 1.6 times the Mn

TABLE 1. Particulate, Energy Content, Gas Flow, Temperature, and Analysis of Blast Furnace Top Gas

		Chemical Analysis of Top Gas (wt %)	
Temperature	175–250°C	Fe	47
Gas quantity	1000–2000 nm^3/ton of iron	C	12
Gas analysis	CO: 23–40%	SiO_2	12
	CO_2: 15–22%	Ca	2.9
	H_2: 1.5–6.0%	Mg	1.2
	N_2: remainder	Mn	0.38
Energy content	3.7 MJ/m^3		
Dust content	Up to 30g/nm^3		

TABLE 2. Tapping Emissions and Particle Size Distribution

Tapping Emission, wt%		Particulate Size Distribution, μm[a]		Emission, kg/ton of Hot Metal	
Fe	51	0.25	0.026	Taphole	0.06–0.11
C	33	0.50	0.083	Runner	0.04
SiO_2	4.5	0.75	0.110	Ladle spot	0.06–0.11
CaO	4.0	1.0	0.177	Slag	0.01
MgO	1.5				
MnO	0.5				

[a]Blast furnace cast-house total building ventilation.

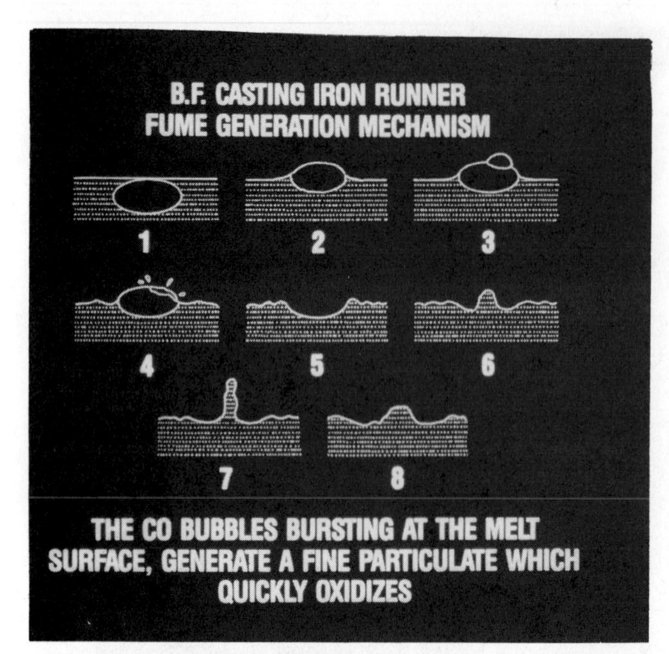

FIGURE 3. Proposed Mechanism of Fume Generation

TABLE 3. Hydrogen Sulfide Concentration at Source

Slag Processing	Average of 30-minute Average, ppm
Pit quenching	22
Expending (palletizing)	10

TABLE 4. Chemical Analysis of Calcium Carbide Desulfurization; Particulate Size Distribution

Desulfurization using CaC_2 wt%		Particulate Size Distribution, μm^a
Fe	11	0.015
C	33	0.025
SiO_2	4	0.03
Ca	46	—
Mg	<1.0	—
Mn	<1.0	—

[a]Fraction smaller than.

analysis in the hot metal, very little evaporation takes place.[5] Droplet ejection must be the major mechanism for fume generation.

Further discussion on the bursting-bubble phenomenon can be found in references 4 and 5. The particulate size analysis collected from a total building ventilation system is shown in Table 2. Sulfur dioxide and CO, collected at a local hood mounted over a tilting iron runner, amounted to 0.018 kg per ton and 0.025 kg per ton respectively. Particulate collected at the same location amounted to 0.14 kg per ton. Total particulate collected by a local collection system was reported at 0.65 kg per ton with 0.14 kg per ton less than 15 μm.[6]

Slag Processing

A common practice is to drain the molten slag, after separation from the hot metal, into a pit or pot adjacent to the furnace. The slag emissions make a contribution to the dust generated by oxidizing sulfur in the slag to SO_2 during a casting. The light-colored and low-mass slag emissions are estimated to contribute approximately 10% of the total emissions. Studies have correlated slag composition and slag emissions.[7] There is a direct relationship with lime content, lime–silica ratio, and slag basicity. An inverse relationship has been found with silica and potassium oxide content. The authors propose a mechanism that relates the behavior of potassium in the furnace to the slag emissions.

Additional emissions occur during processing of the slag. Water may be used as a coolant onto the slag to increase the cooling rate in the pit or produce granulated slag. The sulfur in the slag cast from the furnace is present as calcium sulfide (CaS).[8] The use of water in the cooling process with the hot slag results in production of H_2S emissions. Hydrogen sulfide concentrations measured over a section of the slag pit being quenched peaked at 250 ppm H_2S by volume.[9] The emission factors shown in Table 3 are based on actual measurements.[10]

Hot Metal Desulfurization

Blast furnaces can be operated to produce iron containing from 0.02 to 0.03% sulfur, which can be utilized directly in steel-making operations. Certain plants have increased blast furnace productivity and lowered operating costs by producing iron containing from 0.04% to 0.1% sulfur, which is externally desulfurized prior to use in steel making by injecting calcium carbide (CaC_2), lime (CaO), magnesium (Mg), or soda-based reagents. Desulfurization station particulate emission characteristics are shown in Table 4. The particulate emission factor for a system injecting 5.2 kg (CaC_2) per ton of metal was 0.8 kg per ton.

FACTORS AFFECTING EMISSION

Top Gas

Most of the raw materials charged into a blast furnace are screened. The quality and size play the most decisive role in influencing the gas distribution. Blast furnace operation is influenced by the reactions taking place:

- At the burden surface at the top of the furnace
- In the granular zone in the middle of the furnace
- In the cohesive zone in the lower shaft belly and bosh
- In the zone in front of the tuyeres
- In the hearth (see Figure 4)

These influences govern the counterflow of gas to solids or liquids. As the burden descends in the furnace, it is heated by the ascending reducing gases and is prereduced. The iron ores begin to soften and melt and the ore layers

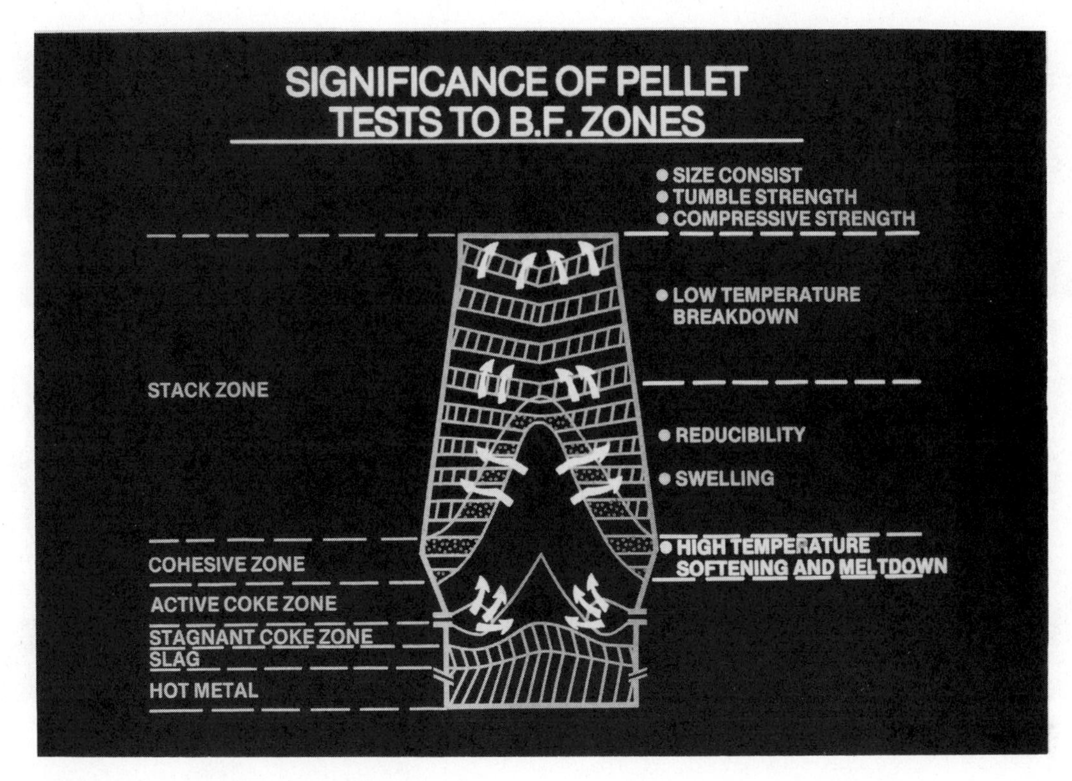

FIGURE 4. Significance of Pellets

become impermeable to gas. The gas flows through the "coke channels" between the ore layers. Buildups or secretions may occur on the walls of the furnace. This condition causes extremely slow, erratic, or stopped stock-line (raw materials) movement. When the burden movement exhibits a sudden large drop, there is a resulting increase in a furnace top pressure. The excess gas is vented out the safety release bleeder valve. Estimated emissions from slips can be found in references 11 and 12. The shape of the cohesive zone is largely responsible for the behavior of the furnace and productivity (see Figure 5).

Tapping (Casting) Liquid Iron

Tapping temperature of hot metal is in the range of 1400°C. The hot metal is saturated with carbon. During furnace casting, turbulent conditions break up the slag layer, increasing oxygen transport, and reducing the temperature by exposing more metal surface to the air. With a drop in temperature, carbon is ejected from solution in the form of graphite flakes (kish). After slag separation, the iron becomes exposed to oxygen from the air, resulting in oxidation of molten iron (dust). One of the major problems with blast furnace cast-house fume emissions is the long length of runners over which the fumes are generated. Trozzo et al,[13] estimated iron runner emissions at 0.04 kg per ton and slag runner emissions at 0.01 kg per ton of hot metal. A number of factors increase emissions generated during casting:

- The turbulence at the trough, which is a function of the angle of the drill and the size of the hole (Figure 6).
- Hot metal casting temperature and sulfur content.
- Runner width-to-depth ratio.

The blast furnace refractory material for runners, trough, and mud gun clay may use organic coal-tar binders, resulting in emissions during casting or closing the taphole. After completion of the cast, the liquid metal reservoir has been drained away. This will allow high-pressure gases inside the furnace to channel out through the taphole, generating large quantities of fume.

Slag Processing

There are two common blast furnace slag-handling processes:

- Discharge slag into a pit/pot
- Granulate/pelletize slag

The blast furnace slag discharge temperature is in the range of 1500°C. A typical cooling rate at 1.4 meters below the surface is 2.8°C per hour.[9] The outer layer of slag cools down between casts, but inner layers can remain molten. Therefore, the slag must be water cooled to allow the pit to be excavated safely. This practice of water quenching generates high levels of H_2S (Table 3).

The method available for granulation of liquid slag utilizes the centrifugal force imparted by a rotating drum or plate and water jets. Hydrogen sulfide gas and fibers are generated.

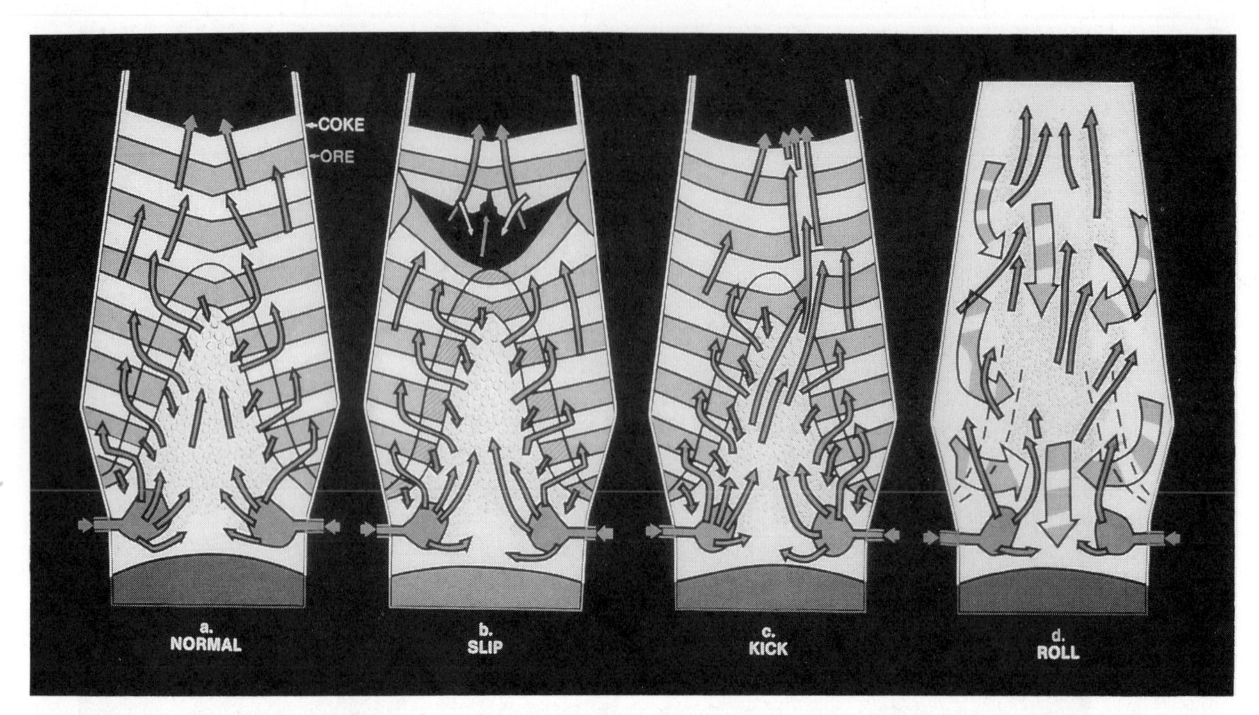

FIGURE 5. Blast Furnace Zones

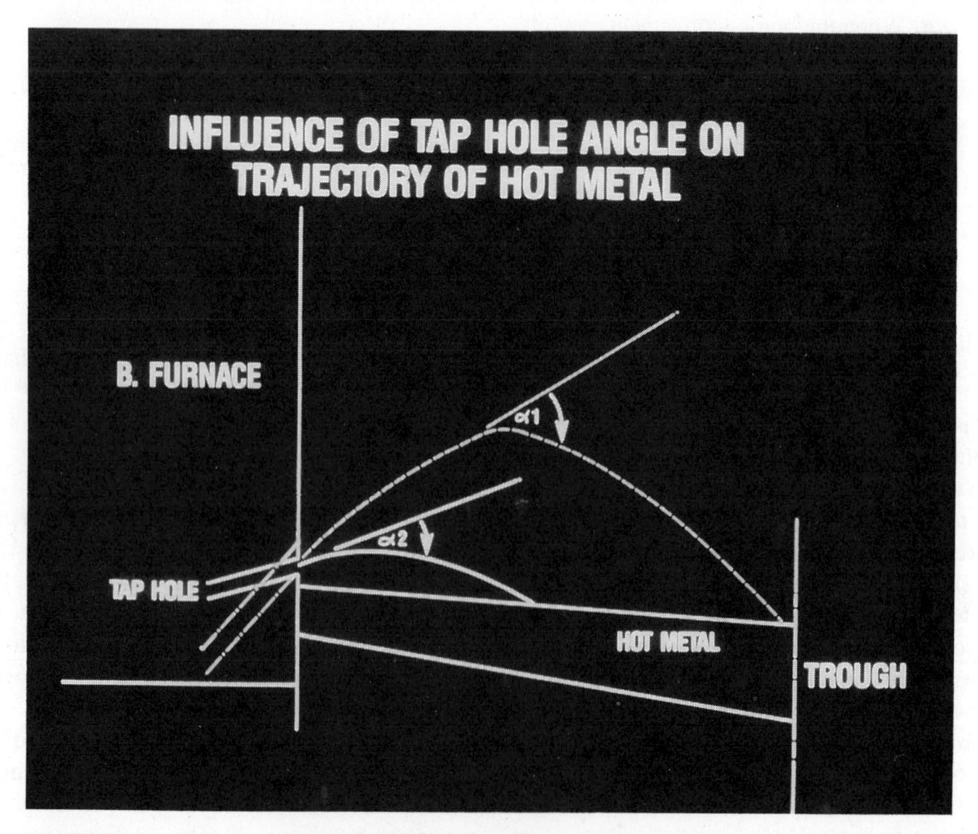

FIGURE 6. Influence of Taphole Angle on Trajectory of Hot Metal

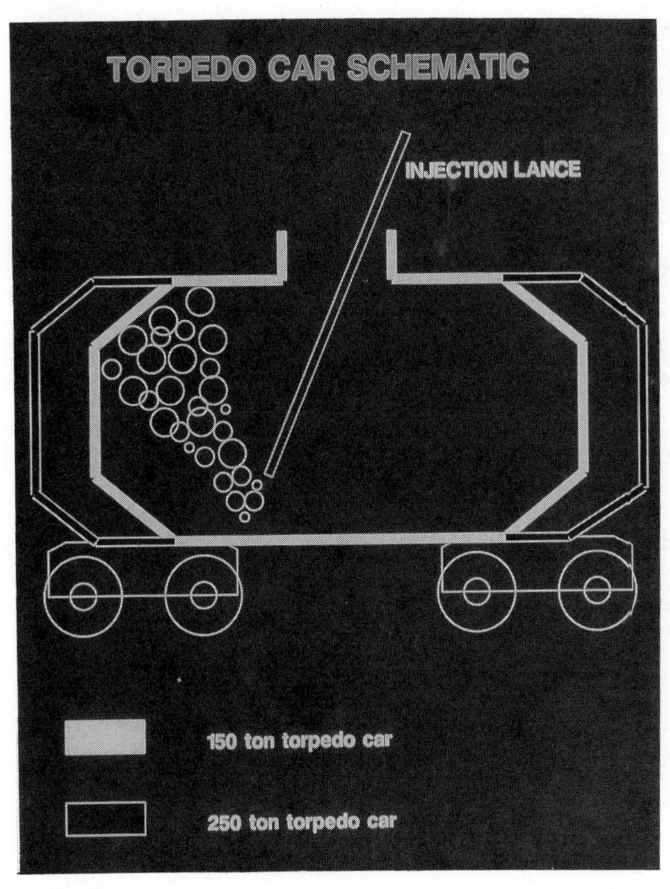

FIGURE 7. Torpedo Car Schematic

Hot Metal Desulfurization

Sulfur is one of the most notorious impurities in steel. The hot metal may be desulfurized at the blast furnace, at a separate desulfurization facility, or at the steel-making operation. A common method of desulfurization is to inject a desulfurization reagent through a lance into the hot metal using a carrier gas (Figure 7). The bubbles from the carrier gas and the reacting agent cause rapid stirring, breaking the surface and creating emissions. The turbulence might also cause molten metals to be ejected from the ladle, creating excess emissions. Insufficient metal level in the ladle, which limits the required lance submergence, may also result in increased emissions.

The constituents of this reagent are typically CaC_2, which desulfurizes the hot metal, and calcium carbonate ($CaCO_3$), which provides the dioxide (CO_2) gas required to mix the hot metal and the desulfurizing powder. Another common injection material is salt-coated Mg granules containing approximately 85% Mg. Approximately 80% of the injected reagent is captured by the slag and the remaining salt escapes as a fume.[14] During desulfurization, Mg salt goes throughout the steps of boiling dissolution. The Mg vapor is extremely volatile. The particle size of salt-coated material would be that of a vaporized fume due to the evaporation of the alkali salts and the evaporation and oxidation of nonreacted metallic Mg.

BLAST FURNACE BEST AVAILABLE POLLUTION CONTROL TECHNOLOGY

Top Gas

The productivity of a furnace is governed by the aerodynamics of gas flow from the tuyere to the top of the furnace. Due to the energy content, blast furnace top gas is used as fuel in blast furnaces, stoves, reheat furnaces, boilers, and so on. The practice of achieving low-maintenance burners and high-efficiency cleaning of the gas is widespread.

The gas permeability in the granular zone and the lower part of the furnace are very important. A modern charging installation (Paul Worth top) makes it possible to distribute coke and burden materials in separate layers or in any mixtures on the surface, compared with two bell-top systems that have batching charges (Figure 8). The permeability in the burden zone depends on the quality, size of raw materials, their fines content, the reduction degradation of the iron-bearing materials, and the burden distribution. For large blast furnaces, the installation of an energy-recovery turbine can lead to a contribution to the works' electricity needs.[15] The installation of a gas expansion turbine, utilizing the excess pressure, therefore provides a useful method of recovering both pressure energy and some sensible heat energy from the gas (Figure 9).

Tapping (Casting) Hot Metal Liquid Iron

The modern blast furnace may be equipped with the following to control casting emissions.

- Local collections at the taphole, skimmer, and tilting runners.
- Each hood with its own control system to adjust air flow independently.
- Local hoods that require the evacuation rates indicated in Table 5.

Tilting runners are used to reduce the runner length, and, therefore, to reduce surface iron oxide fume emissions. Dofasco's No. 4 Blast Furnace was originally fitted with runner length of over 60 feet with four ladle spots per cast-house. For production and environmental reasons, the runner length was reduced to 25 feet, with two ladle spots and tilting runners (Figure 10). The iron runner emissions were reduced from an estimated 0.04 kg per ton to 0.017 kg per ton. The iron temperature losses were reduced from 27.2°C to 4.4°C.[16,17]

Covering the runners is an effective way to reduce emissions (Figure 11). The evacuated runners generate emissions that are captured by the control device. A favored practice is tightly covered runners without evacuation. The use of covered runners also reduces the heat loss from iron in runners. The hot metal temperature delivered to the steel-making operations increases to 12.7°C if covers are used (see Table 6).[18] The emission temperature during

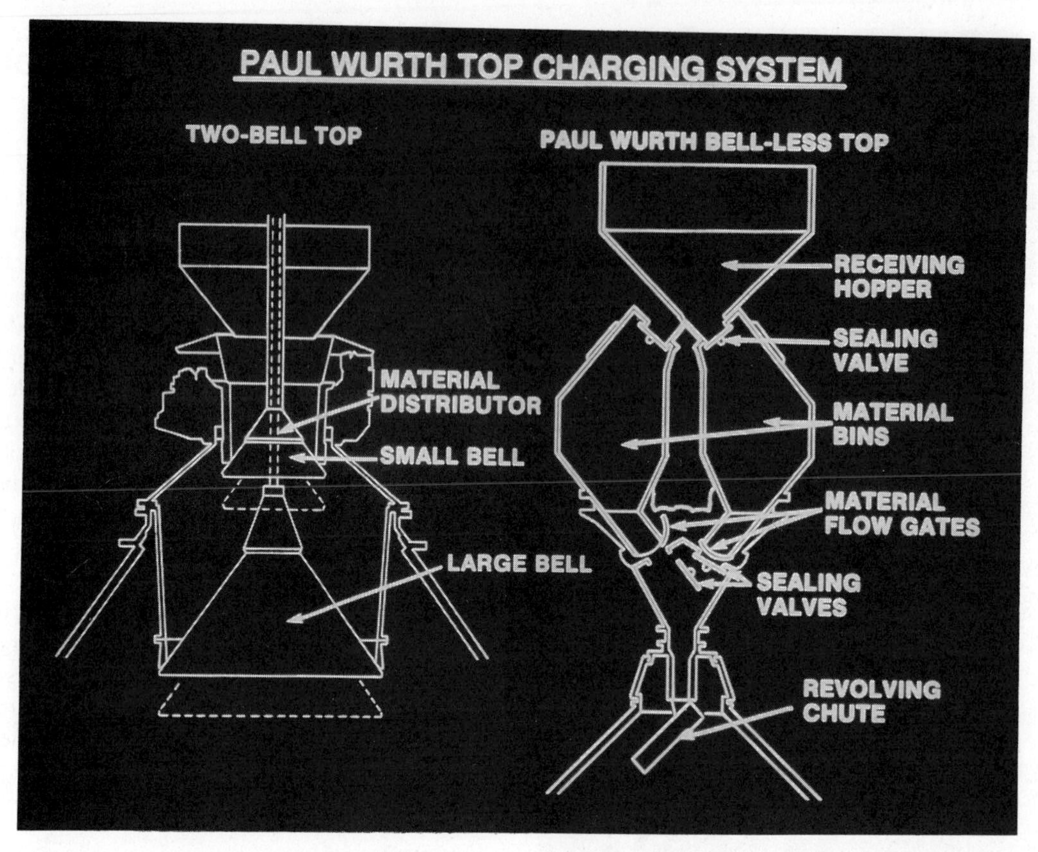

FIGURE 8. Paul Wurth Top Charging System

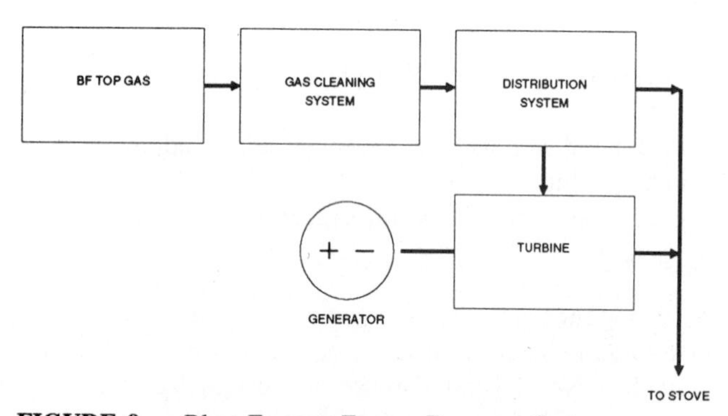

FIGURE 9. Blast Furnace Energy Recovery System

casting does not usually exceed 90°C. Satisfactory gas cleaning installations for secondary emissions are bag-houses.

The casting emissions can also be controlled using suppressant gases like methane (CH_4) and nitrogen (N_2). Certain North American companies such as USX and LTV have attained considerable success with gas suppression at the ladle spot, trough, and runner.[19] The principle of suppression is to eliminate oxygen from contact with liquid iron during casting. Figure 12 shows the arrangement for fume suppression during a trial period at Dofasco's blast furnaces.

TABLE 5. Local Hoods Evacuation Rate[a]

Location	Rate
Taphole	3450 nm^3/min
Dam skimmer	1100 nm^3/min
Tilting iron runner	3450 nm^3/min

[a]Dofasco no. 4 Blast Furnace cast-house local hoods rate.

TABLE 6. Hot Metal Temperature Loss with and without Runner Covers

	Sample Number	Average Temperature	Standard Deviation
Without covers	47	27.1°C	6.04
With covers	43	14.37°C	11.51

Slag Processing

Using water as a coolant for slag quenching results in H_2S emissions. A reduction in H_2S generation can be achieved by increasing the air cooling time as long as possible before water is used. Depending on the pit size and water cooling system, the process of slag quenching can allow effective control of H_2S emissions. Based on the adopted practice at

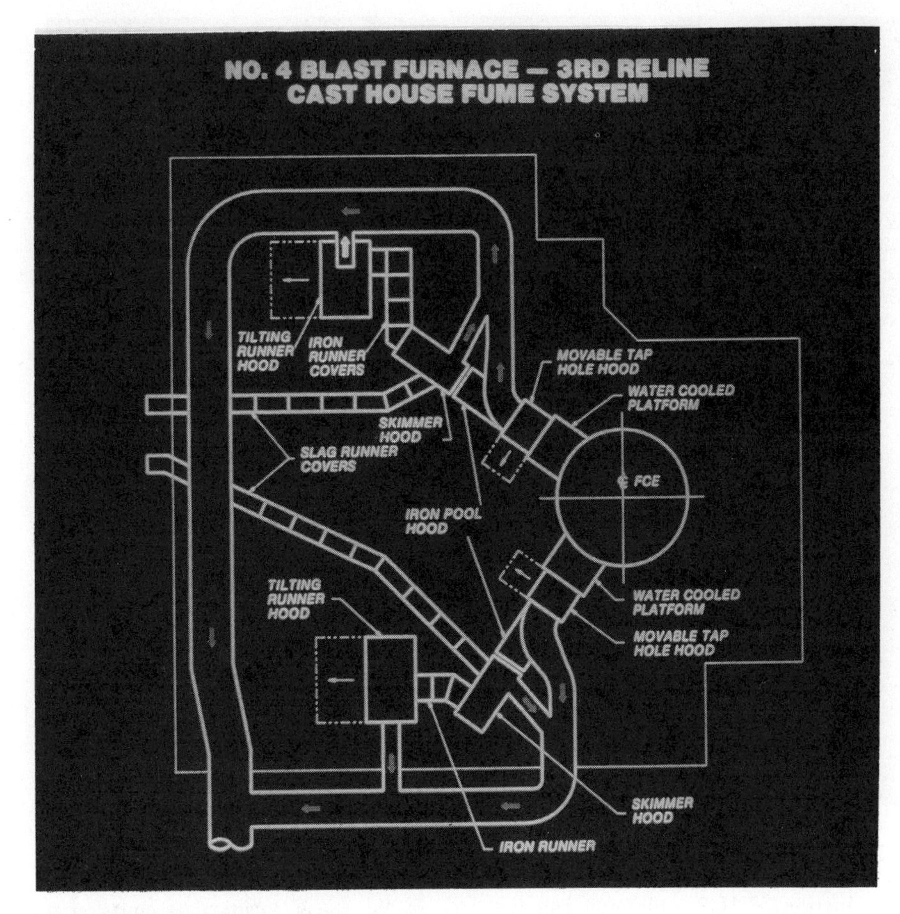

FIGURE 10. Casthouse Fume System

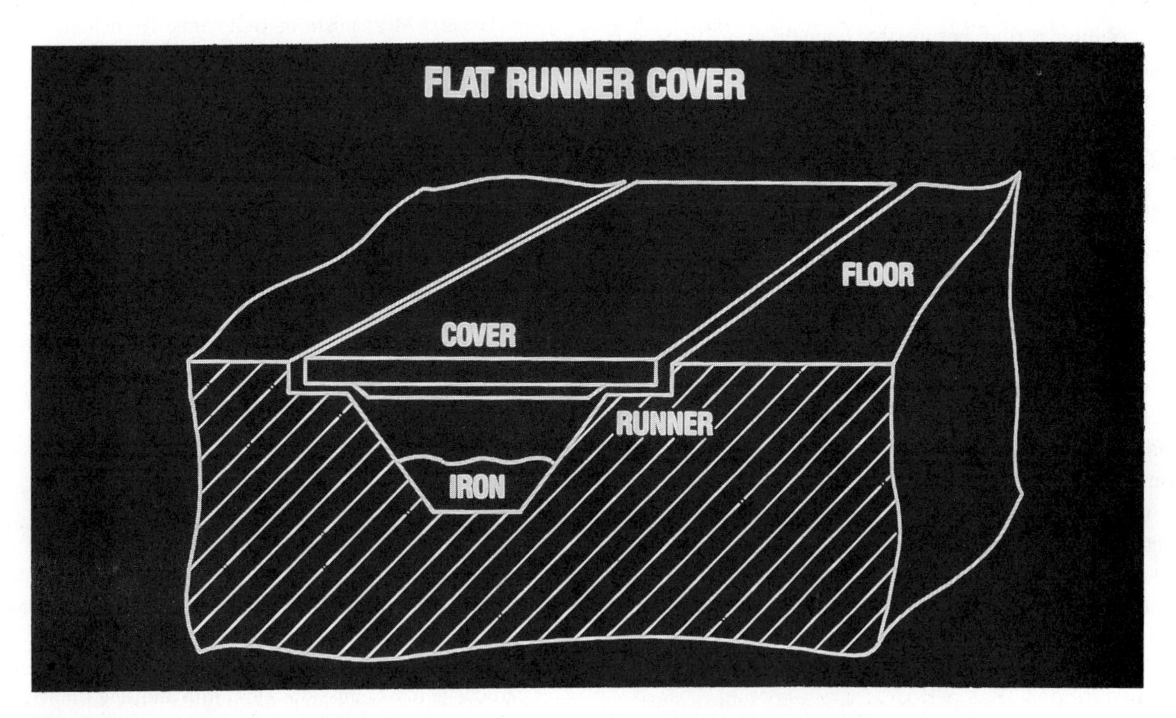

FIGURE 11. Flat Cover Runners

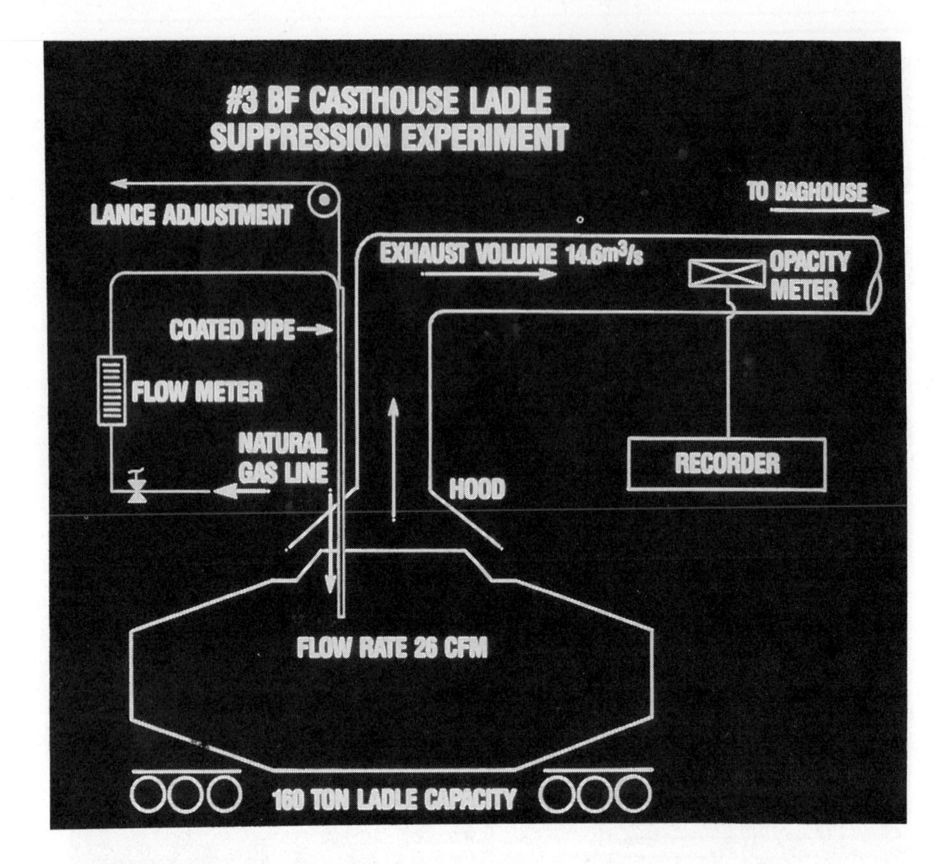

FIGURE 12. Suppression Experiment

Dofasco, up to 85% reduction of H_2S emissions can be anticipated. The pits with high surface area and shallow depth are favoured.

A reduction in H_2S can also be attained by adding caustic or potassium permanganate ($KMnO_4$) to the quench water. A total reduction of 88% was attained by a combination of a three-day cooling period before the quench as well as a 100-ppm addition of $KMnO_4$ to the quench water.[3] For slag granulating process, a water-free system may be considered by individual plants (see references 20 and 21).

Hot Metal Desulfurization

A collection volume of 1000 nm^3/min for a hood located within 0.3 meters of the ladle car will satisfactorily collect emissions from the injection of 80 kg/min of a CaC_2–$CaCO_3$ mixture.[22] The average temperature of gases would be approximately 80°C and could peak much higher if a lance break occurred. Scrubbers, baghouses, and WESPs are satisfactory gas cleaning devices. Provisions for the safe collection of unreacted CaC_2 or Mg reagents are required. Calcium carbide and salt-coated magnesium are the most prominent of the external treatment process agents. The treatment can reduce the sulfur level to less than 0.005%. The selection of the desulfurization techniques is governed by the characteristics and objective of each plant.[14]

References

1. J. H. Seinfeld. *Air Pollution, Physical and Commercial Fundamentals,* McGraw-Hill Book Co., New York, 1975.
2. W. Davis, "Particulate emission factor for iron and steel mills," Tables 7.5-1 and 7.5-2 of AP-42.
3. S. Calvert and H. M. Englund, *Handbook of Air Pollution Technology,* John Wiley & Sons, New York, 1984.
4. R. D. Pehlke et al., *BOF Steelmaking,* Theory Process Technology Division, Iron and Steel Society, AIME.
5. F. Goetz, "Mechanism of BOF fume formation," thesis for McMaster University, Hamilton Ont., April 1984.
6. T. J. Maslany, "Blast furnace control technology update," Symposium of Iron and Steel Pollution Abatement Technology, Chicago, 1981.
7. A. W. Simon, and J. F. Kelly, *Characterization of slag Emissions at a Blast Furnace Casthouse,* U.S. Steel Corp., Research Laboratory, Monroeville, PA.
8. R. A. Stoehr and J. P. Pezze, "Effect of oxidizing and reducing conditions on the reaction of water with sulfur bearing blast furnace slag," *J. APCA,* 25 (II) (1975).
9. F. H. Remmus et al., "Control of H_2S emissions during slag quenching," *J. APCA.* 23 (October 1973).
10. M. Klag and S. Iaboni, "Determination and control of H_2S emissions from slag processing operations," Dofasco internal report., November 1987.
11. *Pollution Effects of Abnormal Operations in Iron and Steelmaking. Vol. III. Blast Furnace Iron Making Manual of Practice,* U.S. Environmental Protection Agency.
12. *Blast Furnace Slips and Accompanying Emissions as an Air Pollution Source,* U.S. Environmental Protection Agency.
13. D. L. Trozzo, and C. F. Hoffmann, "Method for determining particulate mass emissions from iron and slag runners," EPA Conference, June 1980.
14. *Blast Furnace Iron Making—Intensive Course,* Vols. II, III, McMaster University, Hamilton, Ont., May 1989.

15. D. Young, "A project to recover energy from the blast furnace," *Steel Technol. Int.*, (1989).
16. "Minimizing energy costs in the pollution control system," *J. APCA* (1985).
17. P. Spawn and R. Craig, "Status of casthouse control technology in the USA, Canada and West Germany in 1980," GCA Technology Division, Second Symposium on Iron and Steel Pollution Abastement Technology, Bedford, MA., November 1980.
18. A. M. McClure, *Hot Metal Temperature Loss with/without Runner Covers,* Dofasco Research, January 25, 1983.
19. "Molten metal fume suppressions," U.S. Patent 445,883, July 10, 1984.
20. M. Yoshinaga, K. Fujii, T. Shigematsu, et al., "Dry granulation and solidification of molten blast furnace slag" *Tetsu-To-Hagane,* 67:917 (1981); (English version), *Iron Steel Inst., Jpn* 22:823–829. (1982).
21. S. J. Pickering, N. Hay, F. Roylance, et al., *New Process for Dry Granulation and Heat Recovery from Molten Blast Furnace Slag*. Institute of Metals, 1985.
22. C. Borgianni and A. Praitoni, "Injection of powdered reagents into the melt," *Steel Technol. Int'l.* (1988).

BASIC OXYGEN FURNACE SHOPS

S. S. Felton, E. Cocchiarella, and M. S. Greenfield

Basic oxygen steel making was developed in Linz-Donawitz, Austria, in the 1950s and is a variation of the older Bessemer process. The basic oxygen process now accounts for most steel-making capacity worldwide and accounted for 58.4% of U.S. steel production in 1988.[1]

PROCESS DESCRIPTION

A basic oxygen process furnace (BOPF) is a large open-mouthed vessel lined with a basic refractory material.[2] The furnace (Figure 1) is mounted on trunnions to permit full rotation. A typical vessel can be 12 to 14 feet across and 20 to 30 feet high. Furnace capacities range from 15 to 400 tons. A BOPF receives a charge composed of approximately 30% scrap and 70% molten iron and converts it to molten steel. To do this, it utilizes a jet of high-purity oxygen at rates up to 30,000 cfm, which oxidizes the carbon and the silicon in the molten iron, removes these products, and provides heat for melting the scrap. After the oxygen blow is started, lime, usually in the form of pebble lime, is added to the top of the bath to provide a slag of the desired basicity. An entire heat cycle may take only 30 to 50 minutes.

Three types of furnaces are currently in use. The most common type is the top-blown furnace in which oxygen is blown into the vessel through a water-cooled lance suspended above the mouth of the furnace. The second type of furnace is the bottom-blown Q-BOPF. In this furnace, oxygen is introduced into the vessel through tuyeres (openings) in the furnace bottom. The third type is the combined-blown K-OBM furnace where 30% of the oxygen is blown through the bottom.[3] Powdered lime, natural gas, nitrogen, and argon may also be injected through the bottom tuyeres.

Other process variations involve different techniques for hooding and combustion of the gases evolved from the basic oxygen process. Hoods can be open combustion hoods, in which excess air is introduced in quantities from 10% to 300%, or they can be "suppressed combustion" or closed hoods, into which only 5% to 70% of the air theoretically required for combustion is allowed to infiltrate.[4] With open hoods, there is a gap between the hood and the furnace top into which air can be induced. With closed hoods, a movable skirt seals as tightly as feasible to the furnace top to discourage air inflow. Closed hoods have the advantages of reduced gas volume and reduced heat in the hood, reduced fume generation, and possible savings through increased metallic yield or offsetting credits for recovered carbon monoxide. Several partial (suppressed) combustion systems have also been developed independently and differ only slightly in the method of controlling infiltration and combustion levels at the interface of the vessel mouth and hood. These include the OG, IRSID-CAFL, Baumco, and Krupp systems.[4]

There are several ancillary operations associated with the basic oxygen process of making steel. The first is the handling of scrap. The next is the transfer of molten iron from the torpedo car to the charging ladle and from the charging ladle to the furnace itself. The handling of molten iron may include the operation of mechanically skimming slag from the top of the bath of iron. After the blowing period, there are emissions from tapping the furnace into teeming ladles. A fourth operation is the teeming of the finished steel into ingot molds or into continuous casting machines. Finally, there are the handling and treating of molten slag. After the slag solidifies, it is conveyed in trucks, railroad cars, or special slag pot carriers to recycle facilities. Screening and magnetic separation take place, providing recyclable or salable materials.

AIR EMISSIONS CHARACTERIZATION

The operations in the BOPF shop are directly responsible for two principal types of air pollution. The first is the direct result of the steel-making process itself (oxygen blow) and consists of dense emissions of fumes from the mouth of the basic oxygen vessel. These are commonly referred to as "primary" emissions. The fumes are mostly metallic oxides that result from the reaction between the jet of oxygen and the molten bath. Also included in these fumes are particles of slag. Carbon monoxide produced by the reaction of both carbon and oxygen is also emitted.

The gases that leave the mouth of the furnace vessel, in addition to being dusty, are extremely hot. In a closed hood system, temperatures are in the neighborhood of 1650°C

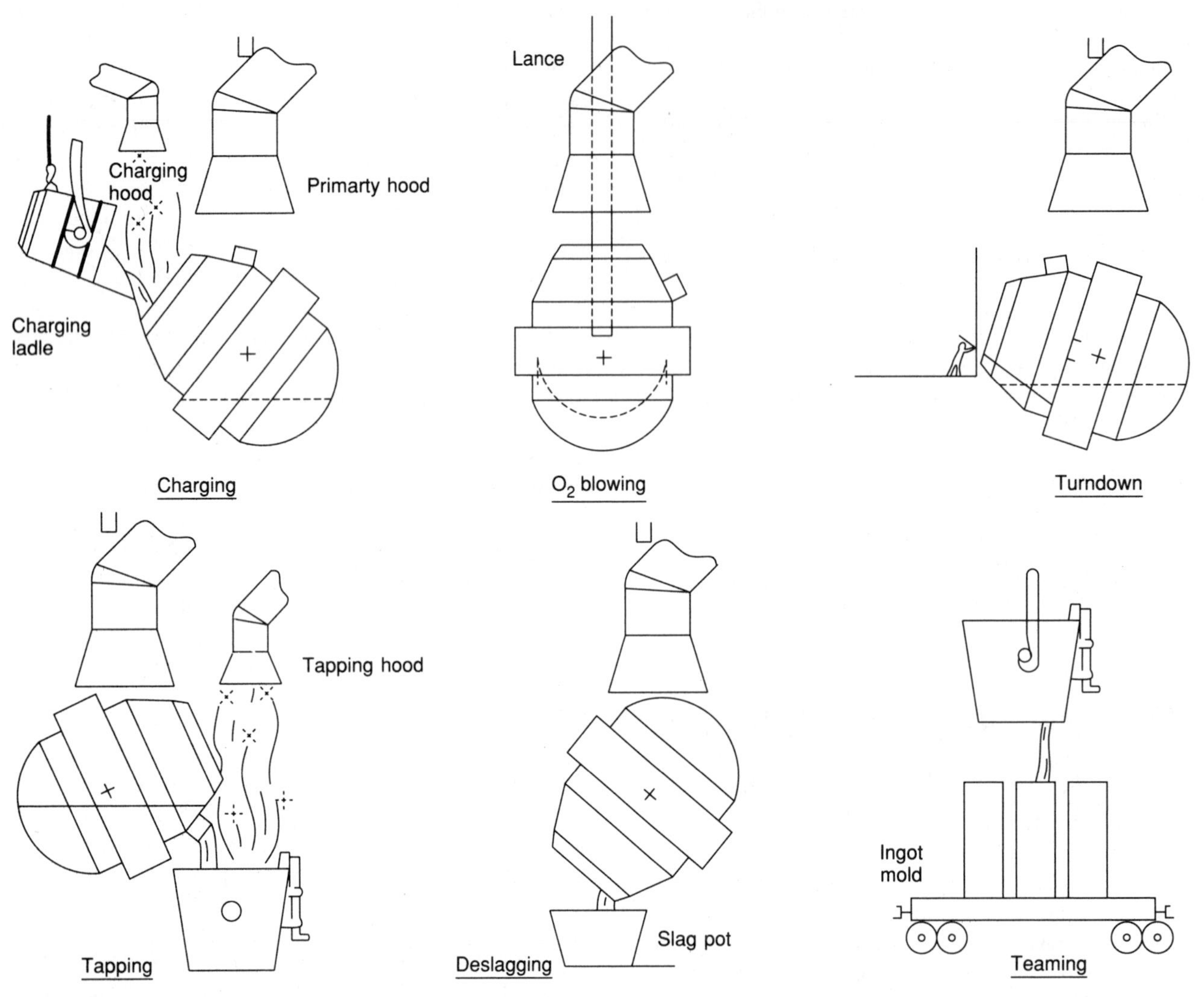

FIGURE 1. Steps for Making Steel by the Basic Oxygen Process. (From reference 2)

(3000°F). In an open hood system, carbon monoxide combustion takes place at the entrance of the hood, raising the temperature perhaps another 540°C (1000°F). Before the gases can be cleaned of their particulate matter, they must be cooled.

The second type of air pollution source comprises a variety of operations that are generally classified as "fugitive" or "secondary" emissions. A description of these sources follows.

- *Reladling or "hot metal transfer"* of molten iron from the torpedo railroad car to the charging ladle is accompanied by emissions of kish, a flat particle usually composed of a mixture of fine iron oxide particulates together with larger graphite particles.
- *Desulfurizing of molten iron* may be accomplished by means of various reagents, such as calcium carbide, soda ash, lime, and magnesium. Injection of the reagents into the molten iron is done pneumatically, with either dry air or nitrogen. Desulfurizing may take place at various locations within the iron- and steel-making facility; however, if the location is the BOPF shop, then it is often carried out at the reladling station to take advantage of the fume collection system at that location.
- *Skimming of slag* from the ladle of molten iron keeps this source of high sulfur out of the steel-making process. Skimming results in the emissions of kish particles described above.
- *Charging of scrap and molten iron* into the BOPF vessel can result in a large flame and a dense cloud of emissions. Emissions from the charging of scrap are particularly severe if the scrap over which it has been charged is dirty or contaminated or has been preheated.
- *Tapping of the molten steel* from the BOPF vessel into the ladle results in iron oxide fumes. The quantity of fumes is substantially increased by additions into the ladle of alloying materials, such as silicon and manganese. The pouring rate and tilt must be carefully controlled to ensure that local hooding captures most of the emissions.
- *Turndown of the vessel* for the purpose of taking samples

or for pouring out the slag results in emissions. These emissions are particularly dense in the case of the Q-BOPF and K-OBM because when the vessel is turned down, a flow of nitrogen must be maintained in the tuyeres in the bottom of the vessel in order to keep out the molten metal and slag.

- *Slag handling* may consist of transporting the ladle of molten slag from the shop to a remote dump area or dumping the molten slag on the ground at the end of the shop and allowing it to cool there. The dumping of slag and its subsequent removal by bulldozer is a very dusty operation that is generally uncontrolled.
- *Teeming of steel* from the ladle to the ingot mold or continuous caster results in emissions that are normally uncontrolled. With teeming into a continuous caster, emissions are usually much lower than with teeming into ingots because of the sophisticated shrouding techniques employed and the reduced air contact.
- *Flux material handling* is accomplished with a sophisticated system of receiving hoppers for accepting deliveries from trucks, railroad cars, a belt conveyor, large overhead bins, weigh hoppers, or special feeders. Particulate emissions result when the bulk materials fall into the vessel, as well as at all transfer points in the conveyor and bin system.
- *Skull burning and ladle maintenance* relate to the fact that the molten steel that remains in the ladle after teeming may cool and solidify between successive uses. After accumulating for some time, these skulls may interfere with proper ladle operation. To prevent this, they are burned out with oxygen lances. This ladle-lancing procedure results in the emission of iron oxide fumes. Ladles must also be relined at intervals to protect the steel shell. The ladles are turned upside down to dump loose material onto the shop floor. This generates fugitive dust.

As previously mentioned, particulate emissions generated from BOPFs are usually classified as either primary or secondary in nature and are produced mainly by the condensation of vaporized metal oxides and coagulation of these particles to form agglomerates.

Generally, more than 50% of the primary particulate emissions generated during the oxygen blow phase is less than 1 μm in physical diameter. Particulates from bottom-

TABLE 1. Particulate Emission Factors for Basic Oxygen Furnace Shops

Source	Units	Emission Factors	Emission Factor Rating	Size Data
Basic oxygen furnaces				
BOF melting and refining	kg/Mg (lb/ton) steel			
Uncontrolled		14.25 (28.05)	B	
Controlled by open hood vented to:				
ESP		0.065 (0.13)	A	
scrubber		0.045 (0.09)	B	
Controlled by closed hood vented to:				
scrubber		0.0034 (0.0068)	A	Yes
BOF charging	kg/Mg (lb/ton) hot metal			
At source		0.3 (0.6)	D	
At building monitor		0.071 (0.142)	B	
Controlled by baghouse		0.0003 (0.0006)	D	Yes
BOF tapping	kg/Mg (lb/ton) steel			
At source		0.46 (0.92)	D	
At building monitor		0.145 (0.29)	B	
Controlled by baghouse		0.0013 (0.0026)	D	Yes
BOF metal transfer	kg/Mg (lb/ton) hot metal			
At source		0.095 (0.19)	A	
At building monitor		0.028 (0.056)	B	
BOF monitor (all sources)	kg/Mg (lb/ton) steel	0.25 (0.5)	B	
Q-BOP melting and refining	kg/Mg (lb/ton) steel			
Controlled by scrubber		0.028 (0.056)	B	Yes
Hot metal desulfurization emissions	kg/Mg (lb/ton hot metal)			
Uncontrolled		0.55 (1.09)	D	Yes
Controlled by baghouse		0.0045 (0.009)	D	Yes

Source: Reference 5.

TABLE 2. Size-Specific Emission Factors for Basic Oxygen Furnaces

Source	Emission Factor Rating, A–E	Particle Size, μm	Cumulative Mass % Less Than Stated Size	Cumulative Mass Emission Factor kg/Mg Alloy	(lb/ton Alloy)
Basic oxygen furnaces					
BOF melting and refining controlled by closed hood and vented to:					
Scrubber	C	0.5	34	0.0012	(0.0023)
		1.0	55	0.0019	(0.0037)
		2.5	65	0.0022	(0.0044)
		5.0	66	0.0022	(0.0045)
		10	67	0.0023	(0.0046)
		15	72	0.0024	(0.0049)
		*	100	0.0034	(0.0068)
BOF charging at source	E	0.5	8	0.02	(0.05)
		1.0	12	0.04	(0.07)
		2.5	22	0.07	(0.13)
		5.0	35	0.10	(0.21)
		10	46	0.14	(0.28)
		15	56	0.17	(0.34)
		*	100	0.3	(0.6)
BOF charging controlled by baghouse	D	0.5	3	7.5×10^{-6}	(1.5×10^{-5})
		1.0	10	3.0×10^{-5}	(6.0×10^{-5})
		2.5	22	6.6×10^{-5}	(0.0001)
		5.0	31	9.3×10^{-5}	(0.0002)
		10	45	0.0001	(0.0003)
		15	60	0.0002	(0.0004)
		*	100	0.0003	(0.0006)
BOF tapping at source	E	0.5	N/A	N/A	N/A
		1.0	11	0.05	(0.10)
		2.5	37	0.17	(0.34)
		5.0	43	0.20	(0.40)
		10	45	0.21	(0.41)
		15	50	0.23	(0.46)
		*	100	0.46	(0.92)
BOF tapping Controlled by baghouse	D	0.5	4	5.210^{-5}	(0.0001)
		1.0	7	0.0001	(0.0002)
		2.5	16	0.0002	(0.0004)
		5.0	22	0.0003	(0.0006)
		10	30	0.0004	(0.0008)
		15	40	0.0005	(0.0010)
		*	100	0.0013	(0.0026)
Q-BOP melting and refining Controlled by scrubber	D	0.5	45	0.013	(0.025)
		1.0	52	0.015	(0.029)
		2.5	56	0.016	(0.031)
		5.0	58	0.016	(0.032)
		10	68	0.019	(0.038)
		15	85	0.024	(0.048)
		*	100	0.028	(0.056)

TABLE 2. *(Continued)*

Source	Emission Factor Rating, A–E	Particle Size, μm	Cumulative Mass % Less Than Stated Size	Cumulative Mass Emission Factor kg/Mg Alloy	(lb/ton Alloy)
Hot metal desulfurization					
Controlled baghouse	D	0.5	8	0.0004	(0.0007)
		1.0	18	0.0008	(0.0016)
		2.5	42	0.0019	(0.0038)
		5.0	62	0.0028	(0.0056)
		10	74	0.0033	(0.0067)
		15	78	0.0035	(0.0070)
		*	100	0.0045	(0.009)
Uncontrolled	E	0.5	N/A		
		1.0	2	0.01	(0.02)
		2.5	11	0.06	(0.12)
		5.0	19	0.10	(0.22)
		10	19	0.10	(0.22)
		15	21	0.12	(0.23)
		*	100	0.55	(1.09)

Source: Reference 5.

blown BOPFs (Q-BOPFs) are smaller and generally estimated to have a mean diameter of about 0.1 μm.

A significant change in particle-size distribution appears to occur when BOPF primary emissions are collected in closed hooded gas collection systems as compared with open hooded gas collection systems. The closed hooded exhaust gas particles are larger than those typically obtained from an open hooded furnace.

The particulate emission rate in the basic oxygen process depends on several factors, including oxygen blow rate, carbon content of iron, percentage of scrap charged, quality of scrap charged, rate of additions, operational practices, and the condition of the refractory lining of the vessel. During the production cycle, the gas evolution rate and gas temperature vary considerably. Because of the resultant variations in the concentration of particulate matter and gas temperature and volume in the inlet gas stream, emissions are greater at the beginning of the blowing period than during the remainder of the oxygen blow and the rest of the cycle. About 30 pounds of particulates are produced per ton of raw steel during the entire oxygen blow segment of the BOPF cycle.

The sources of secondary emissions within the BOPF shop are hot metal transfer, desulfurization, skimming, charging, turndown, tapping, deslagging, teeming, ladle maintenance, flux handling, and slag handling. AP-42 emission factors, particle size distribution, and trace metals analysis have been developed for most steps within the BOPF cycle. This information is summarized in Tables 1, 2, and 3.[5]

AIR POLLUTION CONTROL MEASURES

Just as there are primary and secondary sources of particulate emissions from BOPFs, control systems are also designated as either primary or secondary.

BOPF Primary Emission Collection Systems

Two types of primary emission collection systems are in common use. For open hood systems, the gases that evolve during the oxidation process and the combustion of carbon monoxide are captured by the hood and enter a hood cooling section where heat is extracted. The gas then passes through a conditioning chamber, where it is cooled to the required temperature for the gas cleaning equipment. The gas cleaning system usually consists of a precipitator or wet scrubber, fans, dust handling equipment, and a stack for carrying away the cleaned gases. For closed hood systems, the gases are collected in an uncombusted state, their volume is reduced as compared with those in the open hood, and the yield of the process is increased. Because the gases remain combustible in the closed hood system, gas cleaning is generally performed by means of a wet scrubber, since precipitators present a potential source of explosions. The cleaned gas is then vented to a stack where any combustible gases are flared.

Although both the closed and open hood systems are capable of achieving 99+% particulate removal efficiencies, closed systems are becoming more popular for several reasons. First, they are required to clean only the process

TABLE 3. Selected Metals Analysis in BOF and Q-BOP Fume

	Concentration in gas, mg/nm³	
Element	BOF	Q-BOP
Aluminum	MC[a]	0.43
Antimony	0.006	<0.001
Arsenic	≤0.05	<0.02
Barium	>0.11	0.02
Bismuth	0.003	<0.0007
Cadmium	0.077	0.002
Calcium	MC[a]	64
Chromium	0.84	0.26
Copper	0.18	0.1
Iron	>2.7	85.3
Lead	>0.02	0.41
Magnesium	>0.55	2.3
Manganese	>0.11	3.8
Mercury	0.0008	>0.0031
Nickel	0.31	0.18
Phosphorus	MC[a]	0.53
Selenium	0.087	<0.033
Silicon	>1.1	4.2
Strontium	0.016	<0.056
Sulfur	>0.06	7.9
Zinc	MC[a]	0.14

[a]MC = major component
Source: Reference 5.

gases; therefore, a lesser quantity of gas requires cleaning and equipment can be smaller. Process off-gas temperatures are somewhat lower and, so, are not as damaging to the control system components. Also, less water is required for quenching and gas scrubbing.

There is a much higher percentage of larger particles in the closed system exhaust gas. This is advantageous in that it is easier and less expensive to remove large particles than fine particles. Thus the closed hood system also achieves a lower mass emission rate than the open hood system. Electricity consumption in a closed system is typically about 60% of that required in an open system. With the presence of flammable carbon monoxide in the closed hood off-gas, there is a potential energy recovery for use within or outside the plant.

The amount of gases exhausted from the furnace varies according to the type of fume collection system. An open hood system with an electrostatic precipitator (ESP) utilizes the greatest volume of gas, approximately 2000 scfm per ton of steel. A supplementary benefit of the high exhaust volume is that it facilitates the capture of emissions from the mouth of the vessel when it is tilted away from the hood at various steps in the BOPF cycle.

An open hood system with a wet scrubber generally requires a much lower flow of gases than the ESP, with the amount usually in the vicinity of 900 scfm per ton of steel. The reduced volume results from the need to conserve energy in a scrubber system with a pressure drop in the range of 50 to 70 inches of water.

Most closed hood systems operate in conjunction with a wet scrubber in order to reduce the hazard of igniting the carbon monoxide–laden exhaust gas. The volumetric flow through this closed wet system is considerably lower than in the above two systems at approximately 360 scfm per ton of steel.

Early performance data on the effectiveness of these two primary control systems rank them in terms of particulate control from lowest emission rate to highest as follows: the closed hood with a scrubber, the open hood with a scrubber, and the open hood with a precipitator.

BOPF Secondary Emission Collection Systems

Steps in the BOPF cycle that require the vessel to be tilted out from under the hood are scrap charging, hot metal charging, sampling, tapping, and deslagging.

A large part of these secondary furnace emissions is captured by means of one or more of the following special techniques.

- Use of furnace enclosures
- Use of local hoods
- Building evacuation (full or partial)

A summary of their use and success to date is presented in the following.

Furnace Enclosures

Furnace enclosures may partially (on at least two sides) or fully (on four sides plus the top) enclose a furnace vessel. Most of the BOPFs brought on-line in the United States since 1973 have been enclosed. Partial enclosures can shield the BOPF from most drafts, thus permitting hoods within or adjacent to the enclosure to be more effective at lower airflow rates. A partial enclosure is less expensive, easier to retrofit, and less likely to impede furnace operations. However, the trend in recent years has been toward "total" enclosure. Enclosures are capable of controlling other furnace emissions, such as puffing, turndown, hot metal charging, and skimming.

Local Hoods

The use of localized hoods has become a popular method for capturing particulates from scattered sources throughout the BOPF plant; however, they present design problems as a result of complicated cross-drafts within the building, or they may get in the way of crane operations. Local hoods exhausting to a secondary system make up only a part of the secondary emission control system.

Building Evacuation

Another approach to capturing secondary emissions is to evacuate all air from within the building; however, this can be quite costly and has several disadvantages.

BOPF Secondary Gas Cleaning Systems

Whereas high-performance wet scrubber and ESPs have traditionally been used for primary furnace emissions, baghouses are typically used for secondary furnace emissions. Sometimes, through the use of enclosures, secondary emissions are collected through the primary gas cleaning system.

Additional BOPF Gas Cleaning

Baghouses also have been traditionally used to control emissions from hot metal transfer, desulfurization, slag skimming, and flux material-handling operations. Occasionally, mechanical collectors have been used for these applications, but with limited success.

References

1. "World steel statistics—1989," *Association of Iron and Steel Engineers Year Book—1989,* 416 (1989).
2. J. H. Lucas, *An Evaluation of Basic Oxygen Process Furnace Air Pollution Control Technology,* Mellon Institute, April 1984.
3. Edward Cocchiarella, DOFASCO, Inc., written communication to S. S. Felton, Armco Steel Co., October 10, 1990.
4. B. A. Steiner, "Air pollution control in the iron and steel industry," *Intl. Metals Rev.* 21: 171 (1976).
5. *Compilation of Air Pollutant Emission Factors,* AP-42, U.S. Environmental Protection Agency, October 1986.

Bibliography

1. A. G. Nicola, "Fugitive emission control in the steel industry," *Iron Steel Eng.,* 53(7):25 (1976).
2. J. Pearce, "Q-BOP steelmaking developments," *Iron Steel Eng.,* 52(2):29 (1975).
3. C. M. Parker, "BOP air cleaning experiences," Page no. 66-100, presented at Annual Meeting of the Air Pollution Control Association, San Francisco, CA, June 1966.
4 B. A. Steiner, "Air pollution control in the iron and steel industry," *Int. Metals Rev.,* 21:171 (1976).
5. B. A. Steiner, "Ferrous metallurgical operations," Chapter 21 in *Air Pollution,* Vol. IV, A. C. Stern, Ed.; Academic Press, New York, 1977.
6. *Background Information for Proposed New Source Performance Standards: Iron and Steel Plants. Vol. 1, Main Text,* Publication APTD-1352a, U.S. Environmental Protection Agency, Research Triangle Park, NC, June 1973.
7. *The Making, Shaping and Treating of Steel.* 9th ed., H. E. McGannon, Ed.; U.S. Steel Corp., 1971.
8. *Background Information for Proposed New Source Performance Standards: Iron and Steel Plants Vol. 2, Appendix: Summaries of Test Data,* Publication APTD-1352b, U.S. Environmental Protection Agency, Research Triangle Park, NC, June 1973.
9. *Iron and Steel Industry Particulate Emissions: Source Category Report,* EPA-600/7-86-036., U.S. Environmental Protection Agency, October 1986.
10. *Background Information for an Opacity Standard of Performance for Basic Oxygen Process Furnaces in Iron and Steel Plants,* U.S. Environmental Protection Agency, Research Triangle Park, NC, February 1976.
11. "Standards of performance for new stationary sources—basic oxygen process furnaces: Opacity standard," *Federal Register,* 43:72, 15600 (April 13, 1978).
12. D. W. Coy et al., *Pollution Effects of Abnormal Operations in Iron and Steel Making: Vol. VI. Basic Oxygen Process, Manual of Practice,* EPA-600/2-78-118f, U.S. Environmental Protection Agency, Research Triangle Park, NC, June 1978.
13. *Revised Standards for Basic Oxygen Process Furnaces—Background Information for Proposed Standards. Preliminary Draft,* U.S. Environmental Protection Agency, Research Triangle Park, NC, November 1980.
14. L. J. Goldman et al., "Performance of BOF emission control systems," presented at USEPA Symposium on Iron and Steel Pollution Abatement Technology, Chicago, October 6–8, 1981.
15. *Revised Standards for Basic Oxygen Process Furnaces—Background Information for Promulgated Standards. Final EIS,* EPA-450/3-82-005b, Office of Air Quality Planning and Standards, U.S. Environmental Protection Agency, December 1985.
16. M. Drabin and R. Helfand, *A Review of Standards of Performance for New Stationary Sources—Iron and Steel Plants/Basic Oxygen Furnaces,* EPA-450/3-78-116 (NTIS PB 289877), Metrek Division, Mitre Corp., November 1978.
17. C. W. Westbrook, *Hot Metal Desulfurization, BOF Charging, and Oxygen Blowing: Level 1 Environmental Assessment,* EPA-600/2-81-036 (NTIS PB 179251), December 1981.
18. T. A. Cuscino, Jr. *Particulate Emission Factors Applicable to the Iron and Steel Industry,* EPA-450/4-79-028 (NTIS PB81-145914), September 1971.
19. J. Steiner and L. F. Kertcher, "Fugitive particulate emission factors for BOP operations," *Proceedings: First Symposium on Iron and Steel Pollution Abatement Technology,* Chicago, EPA-600/9-80-012 (NTIS PB80-176258)., February 1980.
20. C. W. Westbrook, *Level 1 Assessment of Uncontrolled Q-BOP Emissions.* EPA-600/2-79-190 (NTIS PB80-100399), September 1979.
21. S. Gronberg and S. Piper. *Characterization of Inhalable Particulate Matter Emissions from a Q-BOP Furnace,* Vols. I and II, EPA-600/X-85-329a,-329b, GCA/Technology Division, August 1982.
22. C. W. Westbrook, "BOF and Q-BOP hotel metal charging emission comparison," *Proceedings Symposium on Iron and Steel Pollution Abatement Technology for 1980,* EPA-600/9-81-017 (NTIS PB81-244808), March 1981.

CASTERS

Manfred Bender, P. E. and Michael S. Peters

Casters are steel plant equipment used to process liquid steel into continuous lengths of solid steel of various cross sections. The most common sections are less than 8 inches square (billets); larger squares are blooms and rectangles averaging 48 inches in width (slabs). Special sections are rounds and so-called "near net shapes," such as beam blanks and slabs or sheets, are as thin as 1 inch.

PROCESS DESCRIPTION

Molten steel is tapped from the furnace into a refractory-lined ladle. The molten steel stream emerges from the slide gate in the ladle bottom and fills a tublike distribution vessel (tundish).

The number of drain holes in the tundish corresponds to the number of strands to be cast simultaneously. The molten steel streams emerging from the tundish bottom fill oscillating water-cooled copper molds in which the molten steel solidifies as it passes through the mold and emerges from its bottom. The molds provide immediate cooling of the outer skin of the steel. As the steel emerges from the mold, the center is still molten. After complete solidification, the continuous strip is cut into convenient lengths using torch machines or shears.

AIR EMISSIONS CHARACTERIZATION

Ladle emissions are negligible. The molten steel is sometimes given a cover of inert material such as rice hulls to provide thermal insulation. In addition, many modern casting facilities employ refractory-lined lids.

The molten steel stream from the ladle bottom generates almost no emissions, particularly when it is surrounded by a ceramic submersion tube. Some emissions are seen to emerge from the molds. The emissions are caused by mold powders and mold lubricating oils. These emissions are mostly a white–blue haze generated by vaporized oil condensate or mold powder combustion products. The emissions are greater if the molten steel stream is nitrogen shrouded because the nitrogen prevents the burn-off of combustibles.

The caster cut-off torches may also generate minor emissions because the red-hot steel section that emerges from the casting machine is cut by a traveling torch with an oxidizing flame. Other emissions are generated during the maintenance of tundishes and casting ladles. These are mostly dust and iron oxide fumes from dumping and oxygen lancing sculls.

AIR POLLUTION CONTROL MEASURES

Casting emissions are generally too insignificant to be controlled at the source. In special circumstances, canopy hoods in the roof trusses above the casting machines capture these emissions. The exhaust is usually in the 100,000-acfm range. The emissions are then ducted to the electric furnace emission control system.

Cut-off machine torches generate a strong vertical downward blast into a water flume below the caster run-out. Most of the heavy particulate is washed away in the flume water. Some slight amount of fine iron oxide particulate is generated. It is rapidly dispersed and is hardly visible, even in the immediate vicinity of the machine.

Ladle and tundish lancing emissions are generally allowed to rise into the plant's fugitive emission control system. In cases where this is not possible and where visible emissions from the process building do occur, custom-designed hooding systems and fabric filters are employed to control the emissions.

ELECTRIC ARC FURNACES/ ARGON–OXYGEN DECARBURIZATION PROCESS

Manfred Bender, P. E. and Michael S. Peters

In 1990, electric arc furnaces (EAFs) generated approximately 40% of the raw steel produced in the United States. The market share of EAFs is increasing because of the continued rapid development of EAF steelmaking technology.

Electric arc furnaces are the prime means of recycling steel scrap into liquid steel. They are, however, also increasingly being used to produce liquid steel from iron sources such as direct reduced iron (DRI) pellets or briquettes and blast furnace liquid molten steel or pig iron.

The EAF steel is batch produced. Each batch is called a "heat". The furnaces generally range in capacity from 50 tons to over 200 tons. Smaller furnaces (5-ton capacity and up) are used to produce batches of special steels. The economical furnace size is limited by the power transformer size and the power the electric supply grid can provide. The largest transformers are rated at approximately 135 MVA. The most powerful furnaces approach a power-to-tapped-steel-weight ratio of 1 MVA per ton.

Because of electric power supply limits, electric power costs, and furnace efficiency considerations, oxygen, natural gas, and coal are increasingly used to augment the melting process. An EAF is only 55% energy efficient, with 45% of the total furnace energy input going into slag (10%), water-cooled panels (12%), the off-gas system (21%), and miscellaneous other losses (2%).

Most EAFs in use today supply ac power to the arc using three roof electrodes. Some furnaces now supply dc power through two electrodes (one through the furnace roof and one in the furnace hearth center) and through three electrodes. These furnaces may lead to new continuously charged, higher-efficiency designs.

A recent trend is to use EAFs as melting machines. The furnace is tapped once the desired percentage of carbon in the melt is established. The steel is tapped at a temperature lower than that required for refining and casting. Alloying is done in the ladle during tapping, in ladle metallurgy sta-

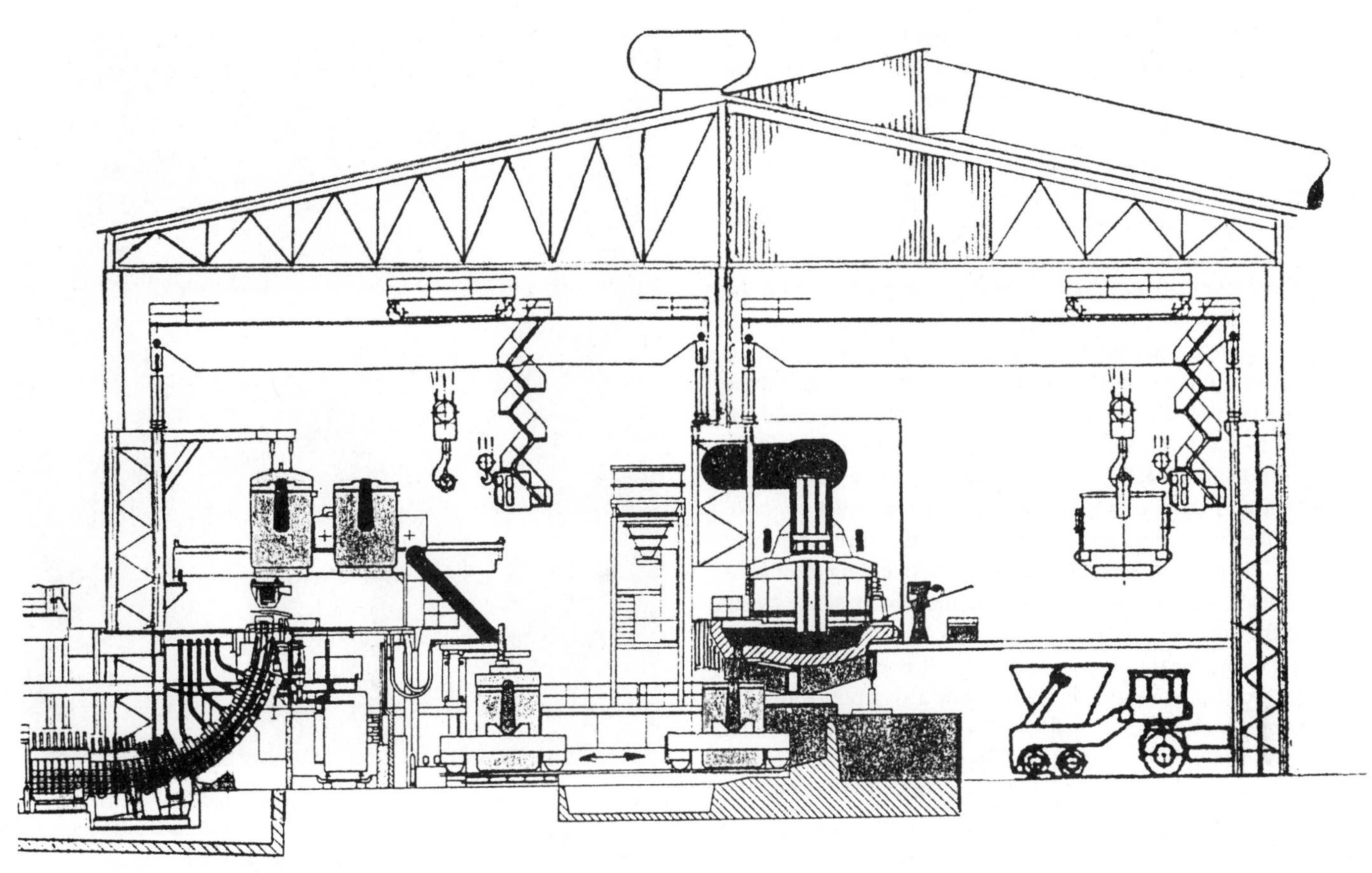

FIGURE 1. Section through a Modern Electric Arc Furnace Melt Shop

tions, or in ladle furnaces. Ladle furnaces are small EAFs that allow the temperature of the molten steel in the ladle to be raised to the temperature required for casting.

An introduction to EAF steel making can be found in reference 1. A critical look at today's EAF steel-making frontier is presented in reference 2.

PROCESS DESCRIPTION

Equipment

Figure 1 shows a cross section through a modern EAF melt shop. Multifurnace melt shops usually consist of a furnace or charging aisle and a tapping or casting aisle. The process flows for each furnace are generally kept totally independent of each other. In single-furnace shops, the furnace and caster are generally in the same aisle. The melt-shop layout has a significant influence on the design of the furnace secondary emission control system.

The furnace consists of three parts: the hearth, the sidewalls, and the roof. The hearth is refractory lined and is sized to contain the molten steel capacity of the furnace. The sidewalls contain the scrap volume. The sidewalls have traditionally been refractory lined as well, but now are almost exclusively water-cooled. The same holds for the roof, except that the roof center, which accommodates the three electrode ports, is generally made from refractory.

The furnace rests on heavy concrete foundations on which the furnace can rock forward and back. Furnaces have the ability to rock 15 degrees backward for slagging or 20 to 40 degrees forward, depending on whether the furnace is a bottom tapping or spout furnace.

Most furnaces have such accessories as burners, limestone injectors, carbon injectors, and oxygen lances. These accessories are attached through ports in the side wall of the furnace. The side opposite to the taphole has a slag door through which slag is allowed to flow out. The tapping spout directs the molten steel into the ladle. More and more furnaces are turned into eccentric bottom tapping (EBT) or offset bottom tapping (OBT) furnaces that discharge the molten steel vertically into a ladle placed underneath the furnace by a transfer car or a ladle turret.

The Process

The steelmaking process (the heat) begins with the first addition of scrap. The furnace roof swings open and the overhead crane operator opens the clam shell bottom of the scrap bucket and lets the scrap drop into the furnace. The roof closes and power is applied. Most furnaces now use oxy-fuel burners. At least three oxy-fuel burners are turned on during the first 5 to 10 minutes of the heat in order to accelerate melting and add heat to the cold spots of the furnaces. Oxygen lances are used to make the scrap collapse into the melt and to burn combustibles in the furnace, thereby accelerating the scrap heating and melting process. Carbon and/or lime is often added with the scrap charge. The coke or coal provides the proper reducing environment for the process. The limestone removes impurities from the steel through the formation of slag.

Subsequent additions of scrap are made in a similar manner until the furnace capacity is achieved. Two or three scrap bucket loads are needed to make a full charge. As the molten steel temperature rises, impurities are removed by the production of slag. After the burners are turned off, carbon and oxygen are added to the slag in order to make the slag foam. Foamy slag improves the furnace power input, lowers hearth refractory wear by shielding arc radiation, and improves heat transfer to the cooler scrap zones near the furnace walls.

The floating slag rises as it foams. It is poured from the bath through the slag door into a pit or a slag pot below the furnace. Metallurgical tests indicate the progress of the process according to established goals for temperature, residual metals, carbon, sulfur, phosphorus, and so on.

When these goals are achieved, a preheated ladle is brought to the furnace and the furnace is tapped. Two general methods of tapping furnaces are employed. A traditional spout furnace is tapped through a liquid metal runner (spout) attached to the furnace wall near the liquid steel elevation. The furnace is tilted to about 40 degrees to tap the hot metal into the ladle. The ladle is supported by either a tapping crane, a ladle stand, a turret, or a transfer car. The spout furnace ladle may become encrusted with slag/solid steel. The most recent innovation in tapping EAFs is the EBT or OBT furnace. Both types of furnaces tap through a hole in the bottom of the furnace. The advantage of this method is rapid, nearly slag-free tapping, molten steel temperature retention, and less taphole and ladle maintenance.

Alloy additions, if needed, can be provided either directly to the furnace or into the ladle. Many melt shops homogenize the melt for temperature and chemistry by injecting inert gas into the ladle. Gas is bubbled through the ladle by either a lance or a porous bottom plug.

AIR EMISSIONS CHARACTERIZATION

The EAF emissions are generally characterized in three forms: charging, melting, and tapping. Each of these will be individually discussed.

Charging

Before the scrap bucket arrives to charge the furnace, the furnace roof swings open. Large volumes of hot fume-laden air rise into the melt-shop roof trusses. Both the heat and the fume contained in the rising plume have to be exhausted from the melt shop. The evacuation system should be able instantaneously to extract these emissions as they are dispersed by the arriving scrap bucket and crane. Most plants keep oxy-fuel burners operational during this period, thereby adding more heat and fume to the rising plume.

FIGURE 2. The Mushroom Cloud Generated by Electric Arc Furnace Charging Spreads around the Scrap Bucket and the Charging Crane on Its Way to the Canopy Fume Hood

The crane operator opens the clam shell bottom of the scrap bucket and lets the scrap drop into the furnace. The drop generates a mushroom cloud and flames, which often engulf the crane (see Figure 2). Peak flow rates in these plumes on medium-sized furnaces have been estimated to exceed 1.5 million acfm.

The second or third bucket of scrap is charged in the same manner, except that the mushroom cloud is usually larger because of the presence of molten steel in the furnace into which the scrap drops. The volume flow rate and emission level in the mushroom cloud are also increased if the steel maker places additives such as coal and lime into the scrap bucket.

Melting

With each charge, as the electrodes are lowered onto the scrap through each of the three ports, the arc is struck and the burners are turned to full power. An immense amount of heat is generated in the furnace. This heat melts the scrap, but it also produces much waste heat, gases, and fume.

These are extracted from the furnace through a hole in the furnace roof (fourth hole). Some smaller furnaces extract the emissions from enlargements of the electrode ports using partly water-cooled or stainless-steel side-draft hoods or a hybrid of the side-draft hood and the fourth-hole concept.

Emissions generated by the EAF while the power is on and/or the roof is closed are called furnace primary emissions. The off-gas volume flow rate from furnaces is relatively low. A modern ultrahigh-power, 100-ton furnace with oxy-fuel burners using the foamy slag practice generates only 50,000 scfm of fourth hole off-gas, but can produce a peak waste heat flow rate equivalent to the electric power input.

The off-gas calculations not only require experience in process engineering, but also an appreciation of the large variety and ever-changing furnace operating practices.[3]

The amount of heat generated depends on many factors:

- The electric power input (up to 1 MVA per ton of molten steel).
- The percentage of combustibles in the scrap (normally 1.5% measured as oil and likely to be higher if the scrap contains more than 5% turnings).
- The amount of air infiltrating the furnace as a function of the furnace air infiltration areas and the pressure maintained by the exhaust system under the furnace roof. All nitrogen entering the furnace as part of infiltration air robs the furnace of valuable melting heat.
- The amount of scrap cutting oxygen injected (15–25 scfm per ton of molten steel).
- The amount of slag foaming oxygen injected (usually 15–20 scfm per ton of molten steel), together with the amount of foaming carbon added (generally 0.9 pounds/min of coal or coke per ton of molten steel).
- The amount of decarburization oxygen injected during refining (generally 25–35 scfm per ton of molten steel).
- The total oxy-fuel burner power input (generally about 15% of the furnace power).
- The electrode consumption rate (4–8 pounds per ton of molten steel).
- The percentage of DRI fines exhausted by the fume system.
- The percentage of oxygen in oxygen carriers such as iron ore pellets, mill scale, recycled baghouse dust, and extra-rusty scrap (such as high-manganese turnings) charged.

The heat in the off-gas from an EAF greatly depends on the furnace operating practice. The off-gas is generally assumed to be air, but in off-gas calculations and emission control equipment selection consideration is given to the individual gaseous components and particulate in the gas stream.

Carbon monoxide (CO) can be generated in the production of steel in EAFs. Carbon containing compounds in the additives, scrap contamination, and particularly the foamy slag practice are the source of these emissions.

TABLE 1. Approximate Size Distribution and Composition of Baghouse Dust from an EAF Producing Carbon Steel

Particle Size		Composition[a]	
% less than	μm	Component	%
15	1	Fe	35.1
55	2	Fe(II)	1.7
80	5	Zn	15.4
90	10	Cd	0.028
99.5	50	Pb	1.5
		Mn	3.0
		Cr	0.38
		Ni	0.07
		SiO_2	5.3
		Al_2O_3	0.37
		CaO	4.8
		Na_2O	0.83
		C	0.44
		S	0.36
		F	0.9
		Cl	1.0
		Plus oxides	

[a]Reference 4

As the furnace contents are heated to approximately 2950°F, any metals that volatilize below this temperature will be carried away by the furnace off-gases. The volatized metals recondense on particulate matter in the off-gases as the temperature is reduced. When coke, coal, or limestone is injected into the furnace, fine particulates of these commodities may be drawn into the off-gas system.

In EAF steelmaking, a fair amount of heavy solid particulate is injected into the off-gas. The quantities of heavy particulate are difficult to estimate because they can vary greatly from furnace to furnace. A furnace using the foamy slag practice can expect to collect 26 pounds of dust per ton of molten steel,[4] but one could expect to collect more with unfavorable oxygen injection practices or too small a fourth hole.

Assuming that the heavy particulate has been properly separated from the particulate carried to the baghouse, the gas contains particulate of the approximate size distribution and composition given in Table 1.

Tapping

Furnace spout and furnace bottom tapping produce similar emissions. The emissions are mostly iron oxide and slag particulate. However, almost all EAF steelmaking processes add alloying elements to the ladle while tapping. This procedure can significantly increase tapping fume evolution. The emissions, therefore, also contain particulate consisting of oxides of these additives.

Ladle Metallurgy

After tapping, the molten steel ladle is moved to the ladle furnace or other ladle metallurgy stations where the liquid steel temperature and composition are adjusted to meet the standard of the specific grade to be produced.

These processes all require emission control. The type of emission capture process used is affected by metallurgical considerations and the main emission control concept employed for the melt shop. The concept, therefore, can vary greatly from case to case.

The ladle then moves to the casting area. For a description of casting processes see preceeding section.

Postproduction Emission Sources

After most of the scrap is melted, enough slag has accumulated in the furnace to cause it to flow over the slag door sill into the slag pit or into slag pots. Without the use of slag pots, the slag is usually removed by a bulldozer from the pit in a solid red-hot state in order to minimize dust generation and heating of the melt-shop air by the cooling slag.

Before charging the first scrap load for the next heat, the furnace is inspected and the hearth is maintained. Hearth maintenance consists of coating the refractory at the slag line. In some meltshops, furnace melting additives are added. All this work is generally performed with the furnace roof open. Emissions from these sources consist of particulate matter and fumes.

AIR POLLUTION CONTROL MEASURES

The peak heat flow rate (Btu/minute) through the fourth hole during the melting cycle usually determines the most critical operation for the design of the direct evacuation control (DEC) system.

A major factor affecting the combustion of furnace off-gas components is the amount of combustion air supplied to the gas leaving the furnace. The off-gas should receive at least 100% excess combustion air based on the period giving the largest heat release from the furnace. This combustion air is introduced for furnace DEC systems at the break flange between the furnace off-gas elbow and the entrance to the water-cooled duct and for side-draft hood systems with the gas cooling air around the electrodes above the furnace roof.

Furnace Primary Emission Control

Direct Evacuation Control Systems

The concept uses a hole in the furnace roof (a fourth hole) and a fourth hole duct elbow to direct the off-gas into the fixed water-cooled duct. The elbow preferably rests on the furnace roof ring and is attached temporarily to the furnace roof lifting beams for roof maintenance.

The inlet to the fixed duct is usually enlarged in order to ensure discharge of the elbow gas into the fixed duct as the furnace tilts forward and back within reasonable limits. The elbow is equipped with a flange that prevents excessive air from leaking into the enlarged duct.

Good furnace operation is achieved when the furnace discharges a nearly constant amount of fume (not flames!) from around the electrodes into the melt shop. The lack of electrode emissions is a sign of excessive infiltration of air into the furnace. This causes increased electrode consumption and excessive heat loss to the fume system.

The fourth hole gas flow rate is roughly proportional to the off-gas heat flow rate because during most phases of the power-on period, the off-gas temperature varies from 2400°F to 3000°F. This is a result of the draft control system limiting the volume flow rate of cool air drawn into the furnace. As another consequence, the combustion air infiltraton rate at the break flange varies nearly in proportion to the off-gas volume flow rate.

The concept provides good combustion of the furnace off-gas with an average CO concentration during the power-on period of less than 800 ppm. Rapid changes in the furnace off-gas volume flow rate caused by scrap cave-ins, coal injection, oxygen injection, the oxy-fuel burner operation, and varying furnace power levels make it difficult for the draft control system to maintain the desired constant furnace pressure at all times. Carbon monoxide peaks of over 7000 ppm (0.7%) are, therefore, not uncommon EAF process total CO emissions are usually less than 2 lb/ton of tapped steel, corresponding to approximately 100 ppm in the bag-house exhaust when combined with canopy hood exhaust.

The heavy particulate in furnace off-gas is mostly slag and limestone dust. It is beneficial to shape the furnace water-cooled elbow and the subsequent fixed duct to prevent obstruction of the gas flow by slag settling in these components. At the same time, the water-cooled duct next to the furnace is most appropriately designed to contain a dropout box, where this material can be collected. In a properly designed water-cooled duct and dropout, the collected material is free from the scrap and frozen slag lumps that preclude the use of conveying equipment for the continuous removal of the dropout box catch.

Furnace draft control is accomplished by either a damper in the DEC duct system (which can sometimes be the fan inlet damper) or a damper in the canopy hood duct.

A dilution air damper at the end of the water-cooled duct protects the uncooled duct from temperatures in excess of 1200°F (1400°F for peaks lasting less than 30 seconds). This concept applies if insufficient water-cooled ducting is provided to cool the gas to 1200°F. If excursions above 1200°F do not occur for more than about five minutes at a time, air-atomized water spray nozzles may be used to inject water into the duct for the purpose of peak shaving occasional temperature spikes.

Side-draft Hood Systems

A few EAFs, and mostly EAFs of less than 50 tons in capacity, still use side-draft hood systems. This system consists of a hood, portions of which may be water-cooled, surrounding the electrodes and capturing emissions from the enlarged electrode openings. The concept does not maintain a negative pressure in the furnace.

Since temperatures downstream of the hoods are usually less than 1200°F, water-cooled ducting is not needed, but the duct system and gas flow rates become very large. Carbon monoxide combustion occurs readily because of the high ratio of ambient air to furnace gases.

In some cases, a fourth hole exhaust is integrated into the side-draft hood.

Secondary Emission Control

Canopy Hood Systems

Secondary emission control in EAFs includes the control of emissions generated by operations not associated with melting. The most significant of these emissions are those from furnace charging and tapping.

Charging and tapping emissions are generally captured by canopy hoods in the melt-shop roof trusses. Tapping emissions are sometimes captured by low level tapping hoods when the ladle is not crane held. Canopy hood shapes and hood exhaust flow rates are determined by experience and from plume rise formulas modified for the presence of obstructions such as cranes.

Because potential emission control investments are usually large, secondary emission capture concepts, including canopy hoods and scavenging duct off-take systems, are often developed in a fluid dynamic model of the melt shop that takes into consideration the particular distribution of heat, fume, and dust sources prevailing in the plant.[5–7]

Charging canopy hoods are not designed to capture the furnace charging mushroom cloud instantaneously. Rather, the hood off-take flow rate is determined by the fume flow rate from the open furnace before and after charging. The hood is either built deep and wide enough to store the charging plume surge temporarily or to spill the mushroom cloud into a space surrounding the primary hood. The spilled fume is drawn from the space by so-called scavenging ducts or allowed to drift back into the hood.

When the DEC duct branch has a higher pressure drop than the secondary emission control (canopy hood) branch, separate hot-gas fans are needed to move the gas to the canopy duct. In this case, it is advisable to provide a coarse particulate separator, such as a low-pressure drop cyclone, ahead of the hot gas fans in order to protect the fans against excessive erosion from abrasive particulate. Since the hot gas fans cannot usually withstand temperatures in excess of 700°F, it becomes advisable to lay the system out in a way that permits drawing dilution air from the canopy system into the DEC system.

Depending on the space constraints and the duct configuration selected, it is often possible to select a simpler system that artificially increases the pressure drop in the canopy branch in order to match it to the pressure drop of the DEC branch. In this case, the canopy branch exhaust fans are selected to also handle the gas from the DEC system, and it is best to design the DEC system to have a low pressure drop and the canopy duct to have a high gas velocity.

In naturally ventilated (open roof) melt shops, a design consideration must be the prevention of the loss of secondary emissions through the roof ventilators. In totally closed roof melt shops, the only means of heat removal is the air pollution control system. The system, therefore, will have to be considerably larger in order to prevent the lower level of the pool of fume in the plant roof from being so low that in-plant visibility is impaired or that fume causes emissions to the environment from low level plant openings. It thus is very important that in closed roof melt shops, the input of clean heat into the building is kept as low as possible. A plant air temperature rise between the ambient air and the air under the melt shop roof of 30–35°F is a good target to use.

In addition to the above, generally applicable plant ventilation criteria should be met in order to prevent emissions from reaching a low level or even escaping through low-level plant openings. When considering these aspects of EAF secondary emission control, special attention must be paid to the nonferrous metallic oxides and respirable particulate because of the applicable Occupational Safety and Health Administration regulated workplace air threshold limit values.

The desire to keep the melt shop fume level high and the plant cool is motivating steel makers to remove ladle preheat and caster billet or slab heat from the melt shop through forced- or natural-draft exhaust systems separated from the meltshop by sheeting. Spout EAFs use tapping canopy hoods to capture emissions from tapping. These hoods are sometimes combined with the charging canopy hood.

The EBT furnaces are sometimes equipped with local tapping hoods that are evacuated by the DEC system. For these furnaces, canopy hoods are less cost-effective because they have to capture emissions dispersed by the furnace tilting platform. The same concept may be applied to short spout furnaces and side tapping furnaces that do not utilize crane-held tapping ladles. Sometimes, however, the tapping process generates so little fume that the canopy hood concept can be employed just as well.

Charging/tapping canopy off-take volume flow rates for well-designed hoods usually range from 250,000 to 500,000 acfm at 150°F, depending on the furnace size and the factors discussed above. Each scavenging off-take may have a volume flow rate ranging from 25,000 to 100,000 acfm at 150°F. Local tapping hoods have off-take volume flow rates from 50,000 to 100,000 acfm at 400°F. Most main fume fans handling unfiltered emissions have a standby fan and operate at speeds of 900 rpm.

Furnace Enclosures

Furnace enclosures are enclosures around the furnace just large enough to accommodate the furnace and its ancillary equipment. The enclosures have movable doors to permit scrap bucket access, tapping ladle access if necessary, access to the slag door, and access for maintenance.

Enclosures are generally used for EAF melting noise control in densely populated neighborhoods. They are also more popular for alloy- and stainless-steel–producing EAFs because the longer tap-to-tap times result in the enclosure being less of an interference with access to the furnace.

Many carbon steel producing furnaces are now too fast for this type of emission capture concept to be practical. Their off-gas heat release is so high that the enclosure does not eliminate the need for a furnace DEC system. It has also become evident that, in most cases, a furnace enclosure does not make canopy hoods or meltshop roof fume offtakes unnecessary. This eliminates the major advantage of the enclosure concept.

A common problem with enclosures is that they have too many open areas around the bottom perimeter. As a result, a negative pressure cannot be maintained in the enclosure and emissions escape around doors near the top. Another problem that is sometimes encountered, because of the confined volume of the enclosure, is the occurrence of high baghouse temperatures during furnace charging. High-temperature bags or larger system capacities are usual solutions to these problems.

Air Pollution Control Equipment

Most EAFs have fabric filters (baghouses) for fume cleaning and collection. Total system exhaust rates of 300,000 SCF/ton of liquid steel are common. Both positive- and negative-pressure fabric filters are used. Cleaning methods of all types—reverse air, shaker, and pulse jet—are employed. The reverse air filter is the most commonly used type of filter.

The advantages of negative-pressure filters include easier filter maintenance because the exterior of the filter usually remains cleaner and the use of higher efficiency and lower cost fume fans without the need for standby fans. The advantage of positive-pressure filters is lower filter cost. However, the price of negative-pressure filters is slowly dropping as a result of filter designs that use less steel for the 25 inches w.c. negative pressure.

Both positive- and negative-pressure fabric filters are generally of the reverse air type. For reverse air and shaker filters, the air-to-cloth ratio (filtration velocity) with one compartment down for cleaning (the net ratio) is usually selected to be 3 ft/min. Well-maintained reverse air filters with bags selected for the specific steelmaking operation have operated at net filtration velocities of up to 3.5 ft/min. Inlet-to-outlet fabric filter pressure drops of 6–8 inches w.c. are normal.

Pulse-jet baghouses are sometimes used for EAF fume cleaning. The air-to-cloth ratio for these filters ranges from 4.5 to 7 ft/min, depending on the dust load and the pulsing method employed. The selection of pulse-jet fabric filters for EAF fume cleaning requires considerable expertise, but with care, often-encountered problems such as high filter pressure drops and high dust penetration can be avoided.

Emission control systems with separate ducts, baghouses, and fans for DEC and secondary emission control often experience high DEC system baghouse temperatures. These temperatures are either reduced with air-to-air heat exchangers called forced draft coolers or are accepted by the fabric filter through the installation of high temperature bags. These bags can handle up to 500°F as compared with the normally employed polyester fabrics, which have a continuous operation limit of 275°F.

Most EAF fume systems carry furnace sparks to the baghouse where they tend to cause bag spark damage. Well-designed baghouse hopper inlets have the ability to drop these sparks out in the hopper before they have a chance to travel onto the bag surface. In many cases, spark arresters in the DEC ducts are used to remove sparks from the gas stream. The units are settling chambers or low-pressure drop cyclones using gravitational or inertial forces to capture glowing particulate. Sometimes the spark arrester includes the canopy duct. In this case, the arrester also acts as a mixing chamber for the DEC and canopy gas.

Collected dust should be removed from the baghouse hoppers as soon as possible in order to prevent dust reentrainment and bridging of dust in baghouse hoppers.

As an alternative to the baghouse (dry) emission control system, scrubbers are also in use today. These systems use water sprays to trap to the particulate matter. The water is collected and recirculated to the scrubber. Sludge is collected in a clarifier. This sludge must be dewatered prior to further handling.

ARGON–OXYGEN DECARBURIZATION VESSELS

Argon–oxygen decarburization (AOD) vessels are used as metallurgical processing units for the production of specialty steels. They are charged with molten steel from EAFs. The molten steel is usually high in carbon and chromium as a result of charging low cost, high carbon ferrochromium into the EAF. The AOD is a converter style vessel that is equipped with bottom gas injection ports (tuyeres). The process involves the injection of oxygen together with an inert gas (usually argon, but sometimes also nitrogen and air) for the purpose of decarburizing the heat without oxidizing the alloying components.[8,9]

Primary AOD emissions are generated when the process gases are blown through the vessel bottom into the molten steel. These emissions are captured in a local movable hood and ducted to the EAF fume system. Sometimes, AOD vessels are equipped with movable accelerator hoods that serve to prevent the growth and dispersion of the AOD off-gas plume on its way to a roof-mounted canopy hood. The concept requires large air pollution control systems because the gas is cooled using large volumes of dilution air.

An AOD primary off-gas hood can require excessive maintenance if it is not designed to withstand impingement by the vessel-mouth flame. In the initial section of AOD off-gas systems, the vessel off-gas, which consists of CO and the inert injection gas, is fully burned and cooled. Therefore, the hood and subsequent ducting are sometimes water cooled.

Canopy hoods are used to capture vessel charging and tapping emissions. These canopy hoods have design considerations similar to those of EAF canopy hoods. Again, it is recommended that the best canopy hood concept be developed in a fluid dynamic model of the facility if the potential emission control investment is large, if plume dispersion above the vessel is affected by a crane and unusually warm air under the plant roof, or if the plant roof structure is not deep enough to accommodate a normal canopy hood design.

For a more detailed investigation of AOD emission control concerns, see reference 10.

References

1. R. J. Fruehan, "A nontechnical introduction to electric furnace steelmaking" in the series "Keeping current," *Iron Steelmaker,* 65–66 (March 1989); 43–44 (April 1989); 58–59 (May 1989); 40–41 (June 1989); 36–37 (July 1989); 52–53 (August 1989).
2. K-H Klein and G. Paul, "Reflections on the possibilities and limitations of cost savings in electric arc furnace steel production," *Iron Steelmaker,* 25–34 (January 1989).
3. M. Bender and L. F. Rostik, "Emission control aspects of modern EAF steelmaking," *Iron Steel Eng.* 64 (9):22–26 (1987).
4. D. R. MacRae, "Electric arc furnace dust; Disposal, recycle and recovery," Report 85–01, *Center for Metals Production,* April 1985.
5. M. Bender, "Fume hoods, open canopy type. Their ability to capture pollutants in various environments," *Am. Ind. Hyg. Assoc. J.,* 40:118–127 (1979).
6. D. L. Harmon, *Technical Manual: Hood System Capture of Process Fugitive Emissions,* Office of Research and Development, U.S. Environmental Protection Agency, January 1985.
7. M. Bender, G. Paul, and R. V. McCabe, "Meltshop ventilation optimization using fluid dynamic models," *Electric Furnace Conference Proceedings,* Vol. 45 1987, pp 161–166.
8. GHH Sterkrade, "Production of special steels in AOD converters with plant description," *Steel Times Int.,* 13–20 (March 1980).
9. A. G. Williams and I. Ludlam, "The AOD process and equipment design," in a series of case histories entitled: "Quo vadis AOD," *Steel Times Int.,* 50–72 (December 1978).
10. R. Iverse, *Background Information Document for the Review of the New Source Performance Standard for Electric Arc Furnaces and Argon Oxygen Decarburization Vessels in the Steel Industry,* U.S. Environmental Protection Agency, February 1982.

LADLE METALLURGY VACUUM DEGASSING

Edward Cocchiarella, P. Eng.

Ladle metallurgy facilities (LMFs) are used primarily to adjust the composition and temperature of a steel heat from a primary melting furnace, such as an electric arc furnace (EAF) or basic oxygen furnace (BOF). They include a series of processes to treat molten steel in the ladle before continuous casting or ingot teeming. The LMF is also a buffer between the primary melting furnace and a continuous casting machine.[1]

Ladle metallurgy, also known as secondary metallurgy, is an increasingly important operation in both electric arc and basic oxygen steel production. The series of processes includes raking or skimming of slag; injection or addition of fluxes, nonferrous metals, and gases; degassing by application of vacuum; and electric arc reheating. These were developed to make cleaner steels in response to more stringent standards and difficult applications.[2,3] The combination of processes used at a particular plant depends on the products made, the quality desired, and the requirements of the melting furnace and continuous caster. At Dofasco Steel in Hamilton, Ont., Canada, for instance, the No. 2 Melt Shop has two deslagging stations, one ladle furnace, one vacuum degasser tank, two strong stir–powder-injection stations, and one wire injection station. The typical sequence of ladle metallurgy processes used between the K-OBM (a variation of the BOF steel-making process) and the continuous caster is shown in Figure 1.

PROCESS DESCRIPTION

In steel-making shops with LMFs, the EAF or BOF becomes primarily a melting tool. The heat is then refined at the LMF. This results in shorter process times, which may increase the shop capacity by some 20%. However, the additional operations produce new sources of emissions. Particulate emissions are generally collected by baghouses, except for vacuum degassing systems, which have steam ejectors and condensers.

The LMFs may have several stations connected by a track, a turntable, or a crane ladle transfer system. The stations are usually centered around a ladle furnace. Other stations may include slag raking, powder or wire injection, vacuum degassing, inert gas or magnetic stirring, alloy addition, and synthetic-slag additions. The different stations

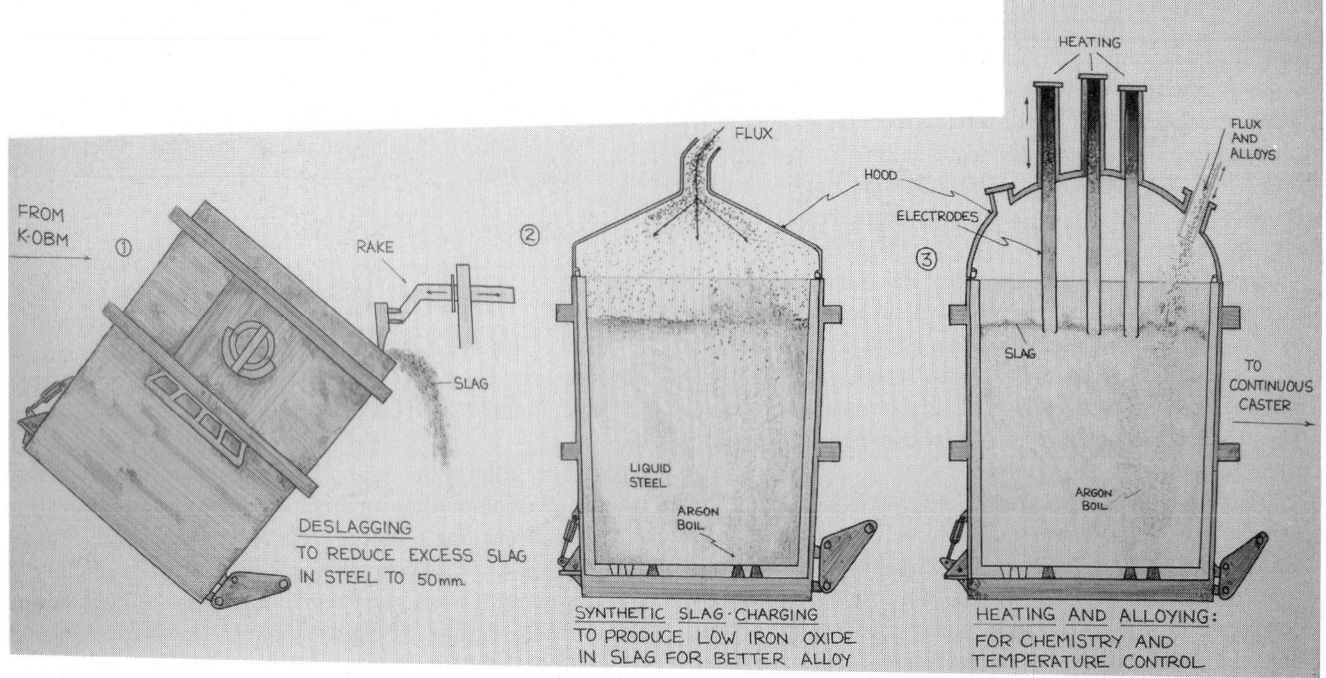

FIGURE 1. Typical Sequence of Ladle Metallurgy Processes

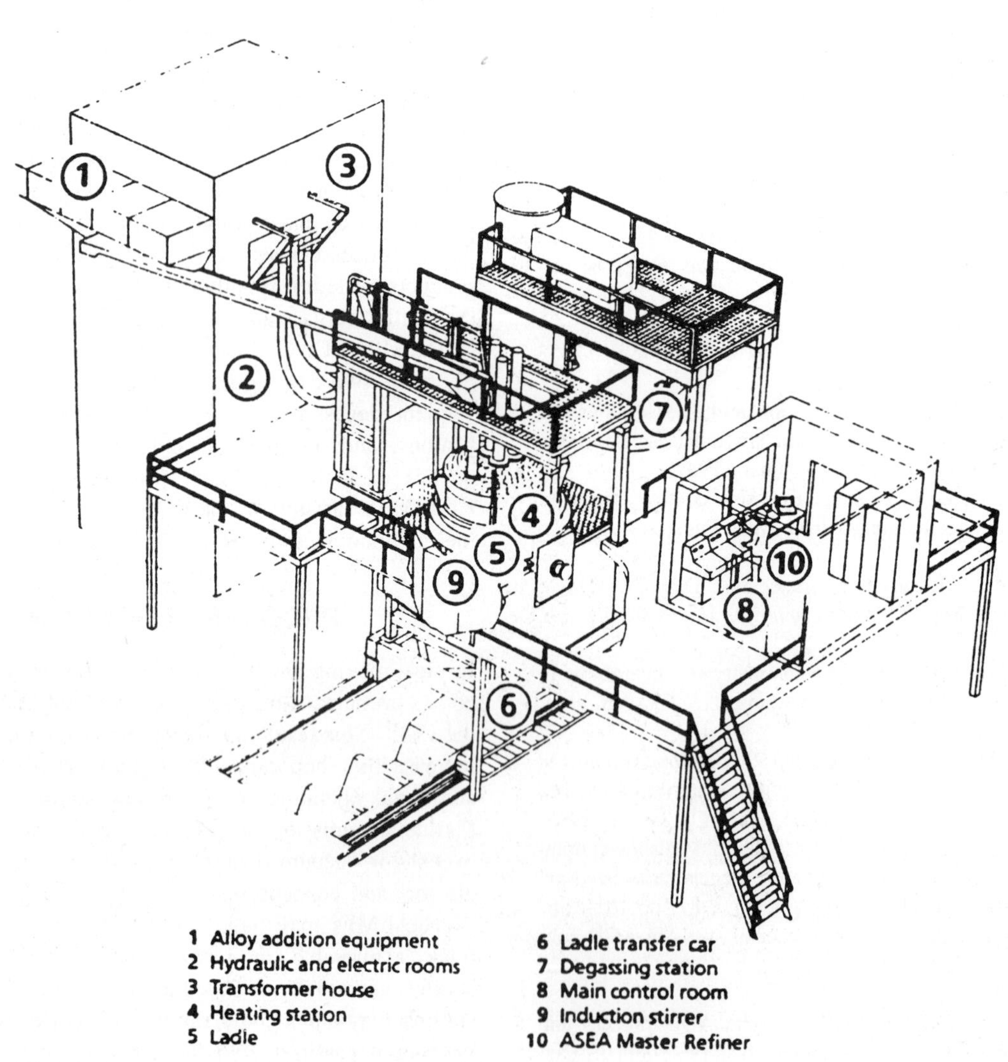

FIGURE 2. Layout of an ASEA Refining System

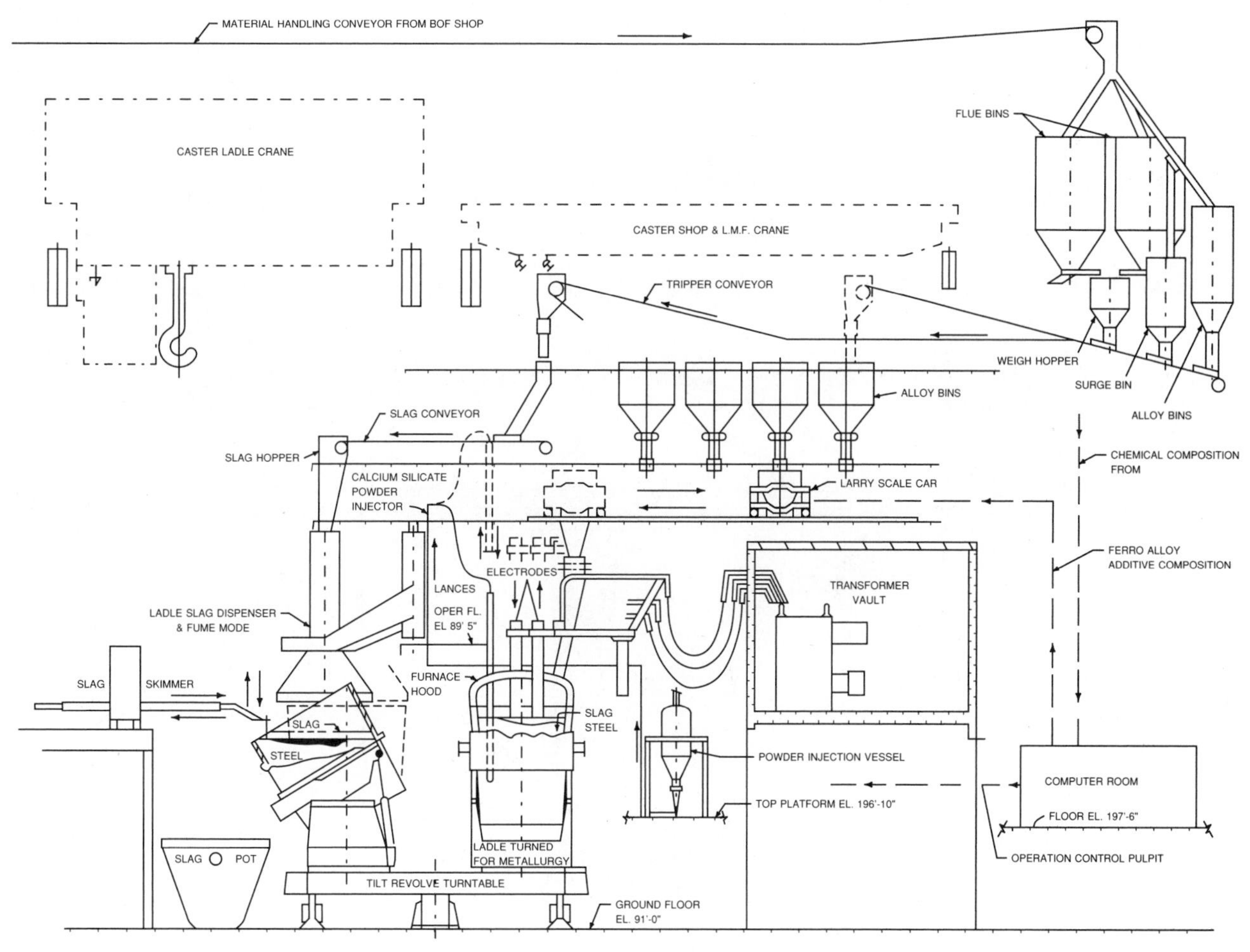

FIGURE 3. Schematic of a Ladle Metallurgy Facility (From reference 5)

may incorporate fixed or movable hoods to capture emissions. Figures 2 and 3 show two examples of LMFs.[4,5]

Vacuum degassing is achieved by three main processes. The simplest is a tank degasser, where the entire ladle is placed in a tank, as in Figure 4. The tank is sealed and vacuum is applied. The steel is stirred by argon bubbling. The second is the Ruhrstahl Heraeus (RH) process, in which a vacuum vessel with two snorkels is lowered into the ladle, as shown in Figure 5. Argon bubbling is used to circulate the molten steel through the vacuum vessel. The RH degassers are prone to the buildup of solidified steel and slag, known as skulling, on the walls of the vessel from steel splashing and low refractory temperatures. The skulls are melted down by gas–oxygen burners.[6] The final process, ladle furnace/vacuum degassing (LF/VD), involves the application of a tight-fitting vacuum hood to a ladle furnace. Several other vacuum technologies are used for special steel applications. These include vacuum arc degassing (VAD), vacuum oxygen decarburization (VOD), vacuum oxygen heating (VOH), and vacuum induction degassing (VID).[1]

The vacuum for degassing processes is usually generated by multiple-stage steam ejectors. Condensers trap particulate and carbon monoxide in the cooling water. The carbon monoxide is released in the condenser sump and cooling tower. The sump vent and cooling tower are minor sources of carbon monoxide emissions.

Secondary sources of particulate emissions arise during the pneumatic or conveyor transport of alloys and fluxes. Emissions from transfer points, hoppers, and transport gases are collected by baghouses.

AIR EMISSIONS CHARACTERIZATION

Ladle emissions can be either particulate or gaseous. Emission factors for ladle furnaces or other ladle metallurgy processes are not included in the U.S. Environmental Protection Agency's *Emission Factor Manual* (AP-42). Particulate emissions factors from two steel plants are listed in Table 1, with AP-42 emission factors for EAF operations shown for comparison.[7]

Particulate emissions include dusts from fluxes, slag, and various additives, and fumes from the oxidation of metals. Table 2 shows a typical LMF baghouse dust analysis for a shop with a ladle furnace, slag raking, and synthetic-slag facilities.[8] The analysis indicates that flux–slag compounds are the main components of the dust, since fine particulate is drawn into the hood as the granular fluxes break down. Handling, inert gas stirring, and electric arc

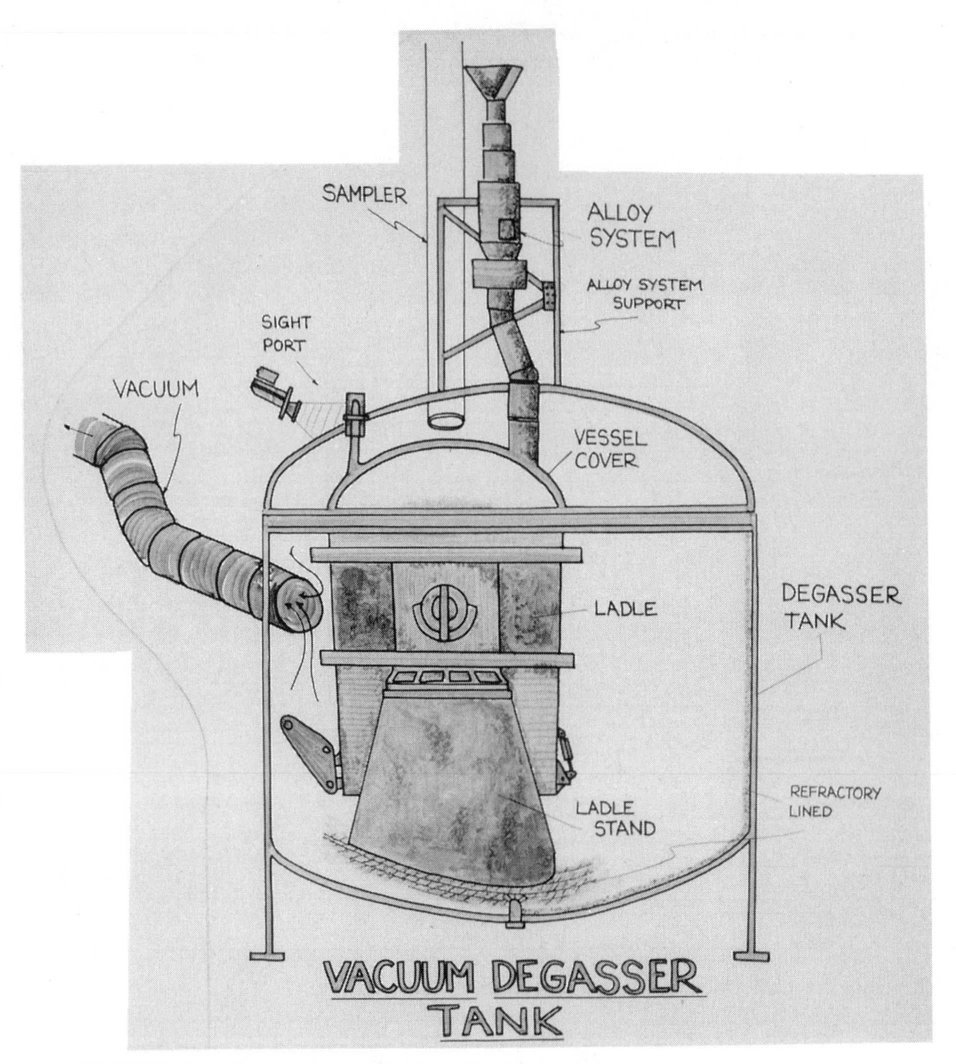

FIGURE 4. Schematic of a Vacuum Tank Degassing Process

heating contribute to the formation of fine flux–slag particulate. The metal oxide fume is mainly formed when the molten steel vaporizes, oxidizes, and finally condenses. Operations that expose molten steel to air or oxygen promote the formation of metal oxide fume, as during slag raking, inert gas stirring, electromagnetic stirring, oxygen lancing, electric arc heating, and pouring.

Another method of fume formation is by bubble bursting. This takes place when an entrained gas bubble erupts at the surface of the steel. Metal and slag droplets are ejected, which partially oxidize.[9] Operations that generate this type of fume include inert gas stirring, oxygen lancing, vacuum degassing, and the injection of certain materials that react vigorously, such as, calcium silicate.

Gases emitted include carbon monoxide and carbon dioxide, which are formed during final decarburization of the steel by oxygen lancing or vacuum degassing at the LMF. Since only residual carbon is removed at the LMF, the generation of carbon monoxide is low.

TABLE 1. Emission Factors for Ladle Metallurgy Facilities

Source	Emission Factor, lb/ton Steel	Reference No.
Uncontrolled LMF		
Ca-Si heats		
Plant A	0.25	12, 13
Plant B	0.23	12, 13
Non–Ca-Si heats		
Plant A	0.012	12, 13
Plant B	0.016	
Controlled by baghouse	0.0077	12, 13
Fugitives at 99% capture efficiency	0.0016	12, 13
Uncontrolled EAF melting/refining	38.0	7
Controlled EAF with direct shell evacuation and baghouse	0.043	7

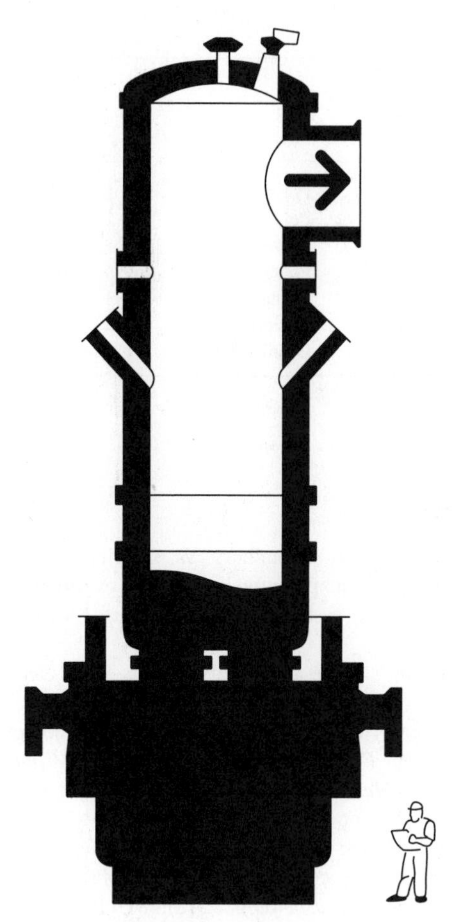

FIGURE 5. Schematic of the RH-OB Process (From reference 6)

TABLE 2. Typical Analysis of LMF Baghouse Dust

Parameter	Composition, %
C	4.1
SiO_2	2.8
Al_2O_3	13.9
CaO	39.3
MgO	10.2
Fe_2O_3	9.3
K_2O	0.60
MnO	10.0
TiO_2	0.47
P_2O_3	0.10
Loss on ignition (1000°C)	4.4

Source: Reference 8.

AIR POLLUTION CONTROL MEASURES

The most effective measure is control at source. Contact of the molten steel with air should be minimized. Maintaining a slag layer, shrouding, and using ladle covers are examples of this type of control and reduce the formation of particulate emissions.

The degree of inert gas stirring may increase the fumes generated. High gas rates lead to vigorous agitation of the steel heat, exposing molten steel to the air. Figure 6 shows

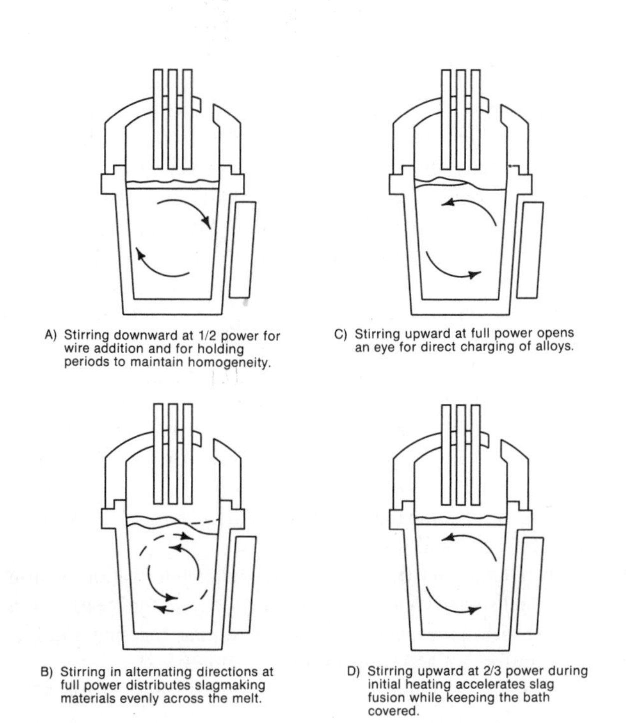

FIGURE 6. Electromagnetic Stirring Patterns (From reference 10, p 338, Figure 5)

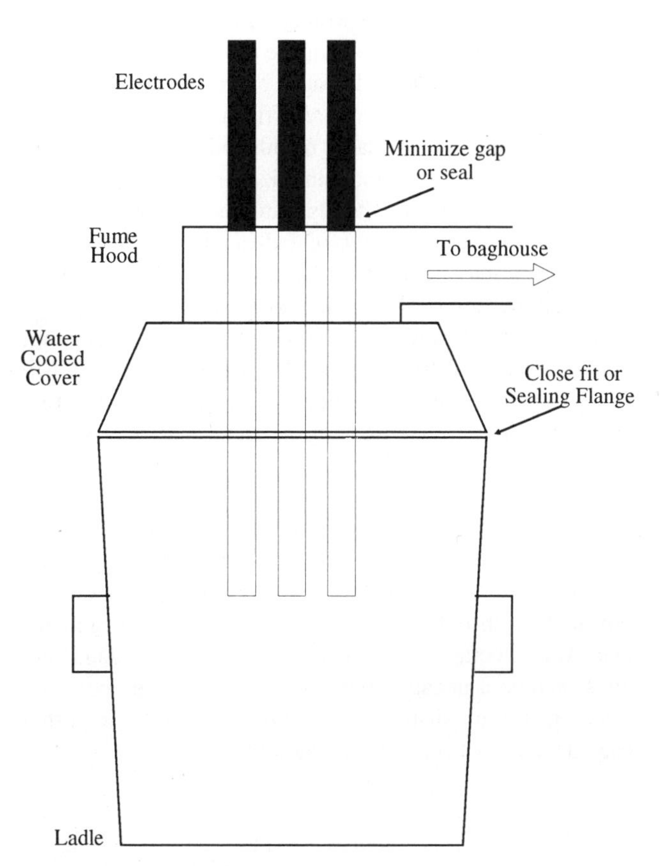

FIGURE 7. Ladle Furnace Cover

TABLE 3. Features of LMF Fume Collection Systems

Plant	Positive/ Negative System	Volume Flow, acfm	Air-to-Cloth Ratio	Average Temperature, °F	Cleaning Method	Bag Fabric	Bag Size, inches
A	Negative	165,000	2.8:1	150	Shaker	Polyester	5 × 171
B	Negative	N/A	N/A	N/A	Pulse jet	Polyester	5 × 168
C	Negative	60,000	4.2:1	>375	Pulse jet	Nomex	5.75 × 144
D	Negative	50,000	5.5:1	145	Pulse jet	Polyester	5.75 × 147
E	Positive	65,000	6.0:1	330	Pulse jet	Nomex	4.63 × 145
F	Negative	85,000	4.5:1	130	Pulse jet	Polyester	5.75 × 146
G	Negative	100,000	4.9:1	est. 150	Pulse jet	Dacron felt	6 × 144

that certain practices of electromagnetic stirring can expose the steel surface.[10] Certain wire or powder additions, such as calcium silicate, greatly increase fume emissions owing to their high vapor pressures and high reactivity. When injected into the ladle, vigorous reactions result in the release of a white fume.

Direct evacuation systems used in vacuum degassing have good particulate collection. Either the entire ladle is enclosed, as in tank degassing, or the exposed molten-steel surface is sealed from air contact by a snorkel or vacuum hood.

Close capture hood systems are employed at various stations. Ladle furnace hoods are mainly water-cooled membrane designs similar to those used in EAFs. Refractory-lined hoods tend to experience more operational problems. Steel hoods can be damaged by heat cycling, causing poor fit and a loss of capture efficiency. Good features include a close fit to the ladle mouth and sealing flanges to maximize fume capture and minimize air infiltration (Figure 7).[11] Some close capture hoods feature swing away designs to allow for crane transfer of ladles. This allows lower exhaust volumes than would be required for side-draft or large canopy hoods above the crane rails.

The different stations, except for vacuum degassing, may be exhausted to a common baghouse. Also, the LMF may share a common baghouse with an EAF or other secondary emission sources in the shop. The duct system may have dampers to control flow and minimize baghouse size.

Basic information on a number of LMF fume collection systems is given in Table 3. The table shows that the baghouses are predominantly negative-pressure (before the fan) units with pulse jet cleaning and air-to-cloth ratios of over 4:1. Operating experience has shown that good features include a negative-pressure design, off-line pulse jet cleaning, and an air-to-cloth ratio of 4:1 or less. Reported bag life ranges from 7 to 18 months.[14]

References

1. G. Bruckmann, A. Choudhury, W. Dietrich, et al., "Vacuum metallurgy and its potential—a general overview of secondary metallurgy systems," *Metall. Plant Technol.,* 12(2):8 (1989).
2. W. Hogan, "Ladle metallurgy," *Iron Steel Eng.,* (11):38 (1989).
3. H. Muller, "Secondary metallurgy plant technology and development," *Proceedings of the International Conference on Secondary Metallurgy,* ICS, Stahleisen, Aachen, September 21–23, 1987, pp 85–94.
4. K. Larsudd and S. Olund, "ASEA master refiner—ladle furnace computer control system," *Fachberichte Huttenpraxis Metallweiterverarbeitung,* 23(8):584 (1985).
5. G. McQuillis, J. Pruett, and J. Bell, "Ladle metallurgy experience at the Granite City Division of National Steel," *Steelmaking Conference Proceedings,* 1988, pp 317–322.
6. D. Kittenbrink, W. Krajcik, and M. Szatkowski, "The startup of the first RH-OB vacuum degasser in North America," *Steelmaking Conference Proceedings,* 1988, pp 299–310.
7. *Compilation of Air Pollution Emission Factors, Vol. 1, Stationary Point and Area Sources,* AP-42, 4th ed., U.S. Environmental Protection Agency, September 1988.
8. "Analysis report—baghouse dust from ladle metallurgy system," internal report, Dofasco, Inc., March 1988.
9. F. Goetz, "Mechanism of BOF fume formation," master's thesis, McMaster University, Hamilton, Ont., April 1984.
10. M. Alavanja and M. Pacillas, "Inland's ladle metallurgy facility—equipment and processing," *AISE Yearbook, 1987,* pp 335–339.
11. H. Legrand and M. Amblard, "Technical evolution of ladle furnaces," *Proceedings of the International Conference on Secondary Metallurgy,* ICS, Stahleisen, Aachen, September 21–23, 1987, pp 449–460.
12. R. Sistek, "Comparative LMF data for two LTV plants," unpublished data, 1987–1988.
13. F. Pendleton and R. Sistek, "PM10 characterization of iron and steel sources not previously included in fugitive emission inventories," presented at the 83rd Annual Meeting and Exhibition of AWMA, Pittsburgh, June 1990.
14. B. Byrd and J. Jenkins, BHA Group Inc., Kansas City, MO, personal communication, March 26, 1991.

FLUXED IRON ORE PELLET PRODUCTION

Gus R. Josephson

The following is an overall description of the fluxed iron ore pellet production process from the initial ground excavation to the final stage of shipping.

PROCESS DESCRIPTION

The production of fluxed iron ore pellets in a traveling grate taconite plant is accomplished by a series of interrelated operations. These operations are as follows:

1. Mining of the ore
2. Ore crushing
3. Ore concentrating
4. Fluxstone processing
5. Pelletizing
6. Shipping

The interrelation of these operations is depicted in Figure 1.

Mining

The mining process involves the stripping of overburden material to expose the ore body by using power shovels to load into the mine haulage trucks. The crude ore is then drilled using rotary drills and blasted so that it can be loaded into mine haulage trucks by power shovels for transport to the primary crusher.

Crushing

The crushing process involves reducing the size of the crude ore to less than 9⁄16 inch through a series of gyratory and cone crushers.

Ore Concentrating

Water is added to the minus-9⁄16-inch crude ore, and it is further ground in a rod mill to minus-eight mesh and is discharged to rotating magnetic drums (magnetic separators), where the good magnetic ore is separated from the waste tailings. The good magnetic ore is then discharged to a ball mill, where it is further ground to 90% minus-325 mesh. The ore concentrate and water mixture forms a slurry that is again magnetically separated over two more sets of magnetic drums to remove silica and other impurities. The ore concentrate slurry is then pumped to a storage tank.

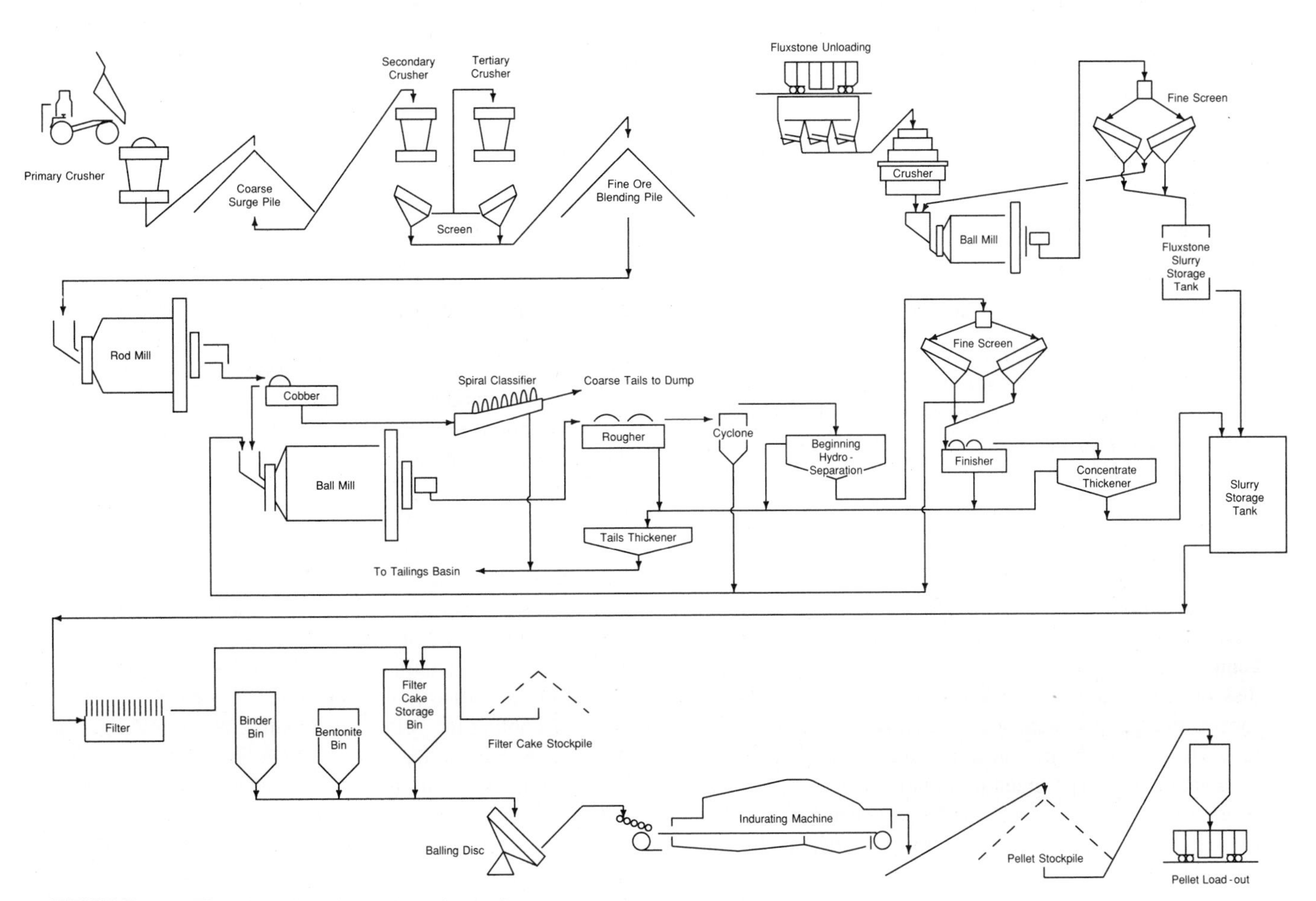

FIGURE 1. Flowchart for Fluxed Iron Ore Pellets

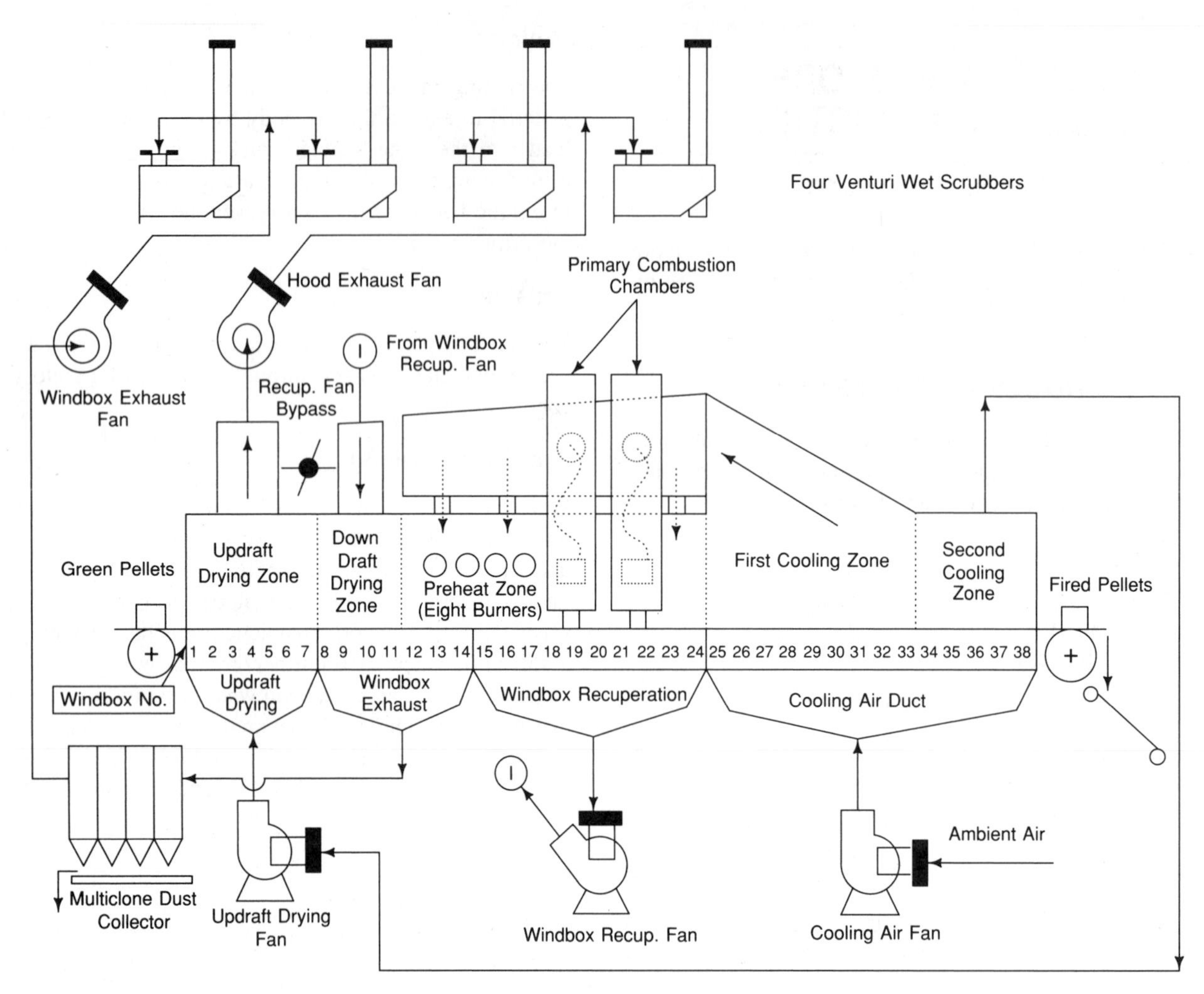

FIGURE 2. Fluxed Pellet Configuration for Traveling Grate Indurating Furnace

Fluxstone Crushing and Grinding

Fluxstone (dolomite and limestone) of less than 1½ inches in size is crushed, milled, and screened to 50% minus 500. Water is added to the fluxstone to form a slurry, and the slurry is added to the iron ore concentrate slurry in the concentrate storage tank.

Pelletizing

The water is removed from the fluxed concentrate slurry with vacuum filters to about 9% moisture content. The fluxed concentrate is then mixed with a binder (organic or bentonite) and formed into ⅜–½-inch pellets in a balling disk or drum. In the rotating disk or drum, the pellets are formed in much the same way as a snowball rolling down a hill. They are then fired in a traveling grate furnace fired with natural gas. The green or unhardened pellets enter the furnace on top of a layer of hardened pellets and move on the traveling grate through the furnace, where they are indurated or heat hardened.

The indurating machine is divided into main drying zones consisting of wind-box sections as viewed in Figure 2. The furnace has two primary and preheat burners that provide the thermal energy to indurate the pellets.

The furnace's air distribution system is designed to preheat, indurate, and cool the taconite pellets. The induration and calcination processes occur in the combustion zone of the furnace. The above-bed air temperature reaches 2200°F in the preheat zone and 2400°F in the primary combustion zone. The combustion gases are next routed through the down-draft drying zone that is used to preheat the pellets. After passing through the down-draft drying zone, the combustion gasses enter the wind-box exhaust duct, followed by a primary dust collector and process gas scrubber. After the completion of the indurating process, the taconite pellets are cooled in the furnace's cooling zones as seen in Figure 2.

After cooling the pellets, the cooling air is split into the first and second cooling zone. The cooling air from the first cooling zone is distributed to the preheat and primary combustion zones. The air from the second cooling zone is recirculated to the up-draft drying zone, where the green pellets are preheated. The air is then routed to the hood exhaust fan, which exhausts the flow to the process gas scrubbers.

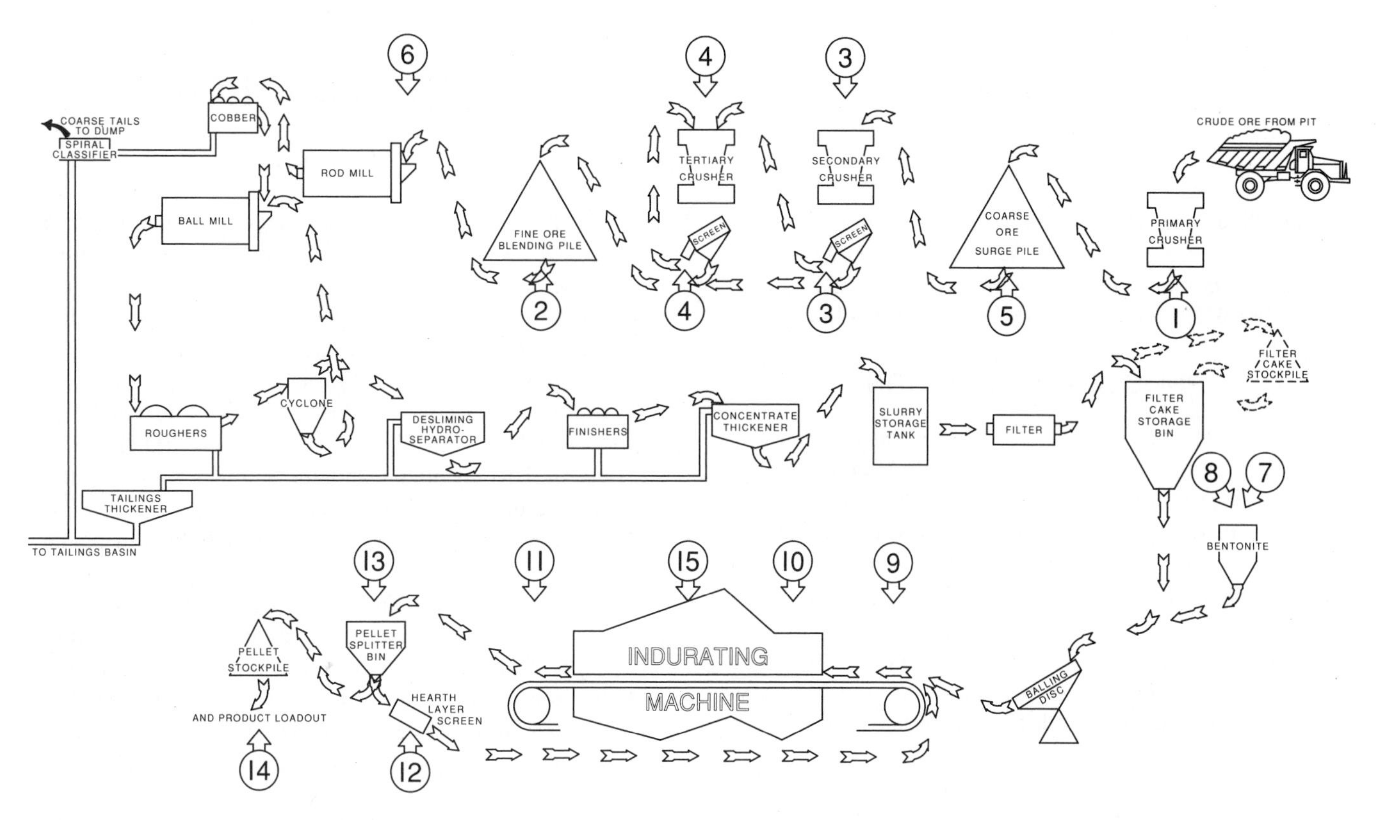

FIGURE 3. Taconite Simplified Flowchart, with Emission Control Points Indicated

Shipping

The fired pellets are shipped by rail and boat to the steel mill.

AIR EMISSIONS CHARACTERIZATION AND CONTROL MEASURES

Mining

Emissions from the mining operation are generated by drilling, blasting, and haulage. Fugitive particulates are the main emission. Dust is controlled in the drilling process by using wet drilling. Blasting emissions are minimized by utilizing controlled blasting techniques that monitor wind direction and temperature inversions. Haulage emissions are controlled by using dust suppression (water, lignosulfanate, chloride) on haulage roads. See Figure 3.

Crushing

Emissions from the crushing operation are generated from the crushing itself and the ore transfer (conveying). Emissions from the crushers, feeders, and transfer points on conveyors are captured by local hoods and ducted to a baghouse.

Concentrating

Emissions from grinding and concentrating occur as fugitive dust where the ore is transferred to the rod mills. Emissions are controlled at these transfer points by local hoods venting to a wet scrubber.

Fluxstone Processing

Emissions from fluxstone processing occur primarily in crushing and conveying as fugitive dust. Emissions are controlled by hooded pickup points vented to a wet scrubber.

Pelletizing

Emissions from the pelletizing operation are nitrogen oxide, sulfur dioxide, carbon monoxide, and carbon dioxide particulates. Particulates are controlled by a dust collector and gaseous emissions are controlled by process gas scrubbers.

Shipping

Emissions from loading and shipping are from fugitive dust. Emissions are controlled through dust suppressant application and water sprays.

Bibliography

Fluxed Pellet Conversion Operating Manual, Dravo Corp., Pittsburgh, PA, July 1987.

PICKLING

R. M. Hudson

During the hot rolling of steel in air, an oxide scale forms that must be removed before subsequent forming or processing operations. The process of removing scale by contacting steel with acid solutions is termed pickling.

PROCESS DESCRIPTION

Bars, rods, and tubing are usually immersed in batches in tanks containing solutions of sulfuric acid. After a sufficient time has elapsed for scale to be removed, the work is lifted out of the pickling tank, rinsed with fresh water, and dried. For alloy steels, scale removal may require the use of other acids, such as nitric acid or hydrofluoric acid. Scale is removed from hot band by continuously passing the strip through several successive tanks that contain solutions of either sulfuric acid or hydrochloric acid. After pickling, the strip is rinsed with water, dried, and usually oiled before recoiling.

Because of the shorter immersion times available during continuous strip pickling, higher temperatures and acid concentrations are required than in batch pickling to assure complete scale removal. To maintain satisfactory pickling rates, fresh acid additions are made to offset the lowered acid concentrations that result when scale dissolves to form iron salts and water. Inhibitors are generally added to the acid solutions to minimize acid attack on the base metal with formation of iron salts and hydrogen gas. Control of the pickling operation requires a knowledge of solution concentrations and temperatures.

AIR EMISSIONS CHARACTERIZATION

A spray comprising droplets of acid solution is formed during hydrogen evolution and the spray and water vapor are present in the air above the surface of pickling solutions. When hydrochloric acid is used, hydrogen chloride vapor is also present in concentrations that increase with solution temperature and with increases in the concentration of hydrochloric and/or ferrous chloride in the solution.

AIR POLLUTION CONTROL MEASURES

An air exhaust system and scrubber are used to recover acid constituents present; the scrubber water can be used for makeup additions to the pickling tanks. Waste acid solutions can be subjected to a regeneration process or treated to prepare iron salts for industrial applications.

Bibliography

ASM Metals Handbook, Vol. 5: Surface Cleaning, Finishing, and Coating, 9th ed., 1982, pp 68–82.

DIRECT REDUCTION

Murray S. Greenfield, P.E.

The blast furnace is the dominant method of smelting iron from oxides. Coke or charcoal is required to form a porous bed in the hearth of the furnace for the passage of the reducing gases. North American coke oven batteries are not being replaced when they are phased out. Steel producers using scrap-based electric arc furnaces are moving upscale rolling products that require lower levels of residual elements available from direct reduced iron (DRI). The amounts of DRI available and the number of hot metal facilities are predicted to grow substantially.

According to Table 1, in 1990, direct reduction plants produced 17,880,000 metric tons per year or 2–3% of the world's steel production, and an additional 10 million tons[1] per year capacity is under construction. Direct reduction covers the nonblast furnace techniques for producing metallic iron from oxides. Some processes produce hot metal (molten iron), while others produce a metallized lump that is suitable as a scrap substitute for an electric arc furnace.

The direct reduction techniques can be categorized into four main types:

1. The vertical shaft type using reformed gas as the reductant. Commercial variations in this category include Midrex, supplied by the Midrex Corp., and HYL, supplied by Hysla. In 1990, Midrex and HYL accounted for about 90% of direct reduction production.
2. The rotary kiln or rotary hearth type using coal as the reductant (for example, SLRN from Lurgi, CODIR

TABLE 1. 1990 Production and Market Share by Process of Steel-Making–Grade DRI[a]

Process	Fuel	Production, tonnes per year	Total World Production, %
Midrex	Gas	10.84	60.6
Hyl I	Gas	3.69	20.6
Hyl III	Gas	1.55	8.6
SL/RN	Coal	0.84	4.7
Fior	Gas	0.39	2.2
Codir	Coal	0.19	1.1
Accar	Coal	0.09	0.5
DRC	Coal	0.14	0.8
Tisco	Coal	0.05	0.3
Purofer	Gas	0.05	0.3
DAV	Coal	0.03	0.2
K-M	Coal	0.02	0.1
Total		17.88	100.0

[a]Preliminary figures as of January 31, 1991.

from Mannesman Demag, Fastmet from the Midrex Corp., Inmetco from Inco).

3. The melter gasifier using coal as the reductant. Partially reduced iron oxides are smelted with noncoking coals in a melter gasifier. (Corex from Deutsche Voest-Alpine is of this type.)
4. Hot metal bath (HMB) using coal as the reductant. The bath may be contained by a hearth or converter. (This type includes HIsmelt from the HIsmelt Corp., NKK, DIOS, Nippon Steel, AISI etc.)

TABLE 2. Analysis of Reducing Gas

Reformed Gas	
Component	Percent
H_2	73
CO	17
CO_2	7
CH_4	3

PROCESS DESCRIPTION

Vertical Shaft Using Reformed Gas

The vertical shaft using reformed gas produces over 86% of the world's DRI. As shown in Figure 1, heated reformed reducing gas at temperatures of up to 930°C is injected into the reduction zone of the vertical shaft through a refractory annular duct countercurrent to the descending pellet ore charge. An analysis of the reducing gas is shown in Table 2. The hydrogen (H_2) content of the reducing gas produces a specific output of 3.5–5.0 tons of direct reduced material per hour per square meter of hearth area.

The system is economic only if low-priced fuels such as natural gas, naphtha, or coke oven gas are available. About 2.3 to 2.5 Gcal of natural gas is required per ton of DRI product. The furnace is divided into an upper reducing zone and a lower cooling zone. Heavy oil and coal require gasifiers to produce the gas for reforming. The reducing gas from the reformer is introduced into the furnace at about 800–870°C through numerous openings around the periphery of the furnace. The spent reducing or top gas is passed through a scrubber and cooled to about 40°C. About one third of the cleaned gas must be purged from the circuit to remove carbon dioxide (CO_2) and nitrogen (N_2). The other two thirds of the gas is recompressed, mixed with fresh reformed gas, and recirculated back to the bottom of the reactor, where it cools the descending hot material. The pressure inside a Midrex and HYL reactor is 1 bar and 5.5 bar gauge respectively.

Rotary Kiln Using Coal

As shown in Figure 2, the rotary kiln is another technique used to process pellets or lump ore. Flux, ore, or pellets are fed into the furnace countercurrent to the gas flow. The charge must be exposed to temperatures exceeding 900°C for five to seven hours to achieve 92% metallization. The coal can be fed either with the pellets or close to the discharge from the kiln. Coal fed closer to the discharge[2] reduces the coal consumption. Carbon monoxide (CO) is

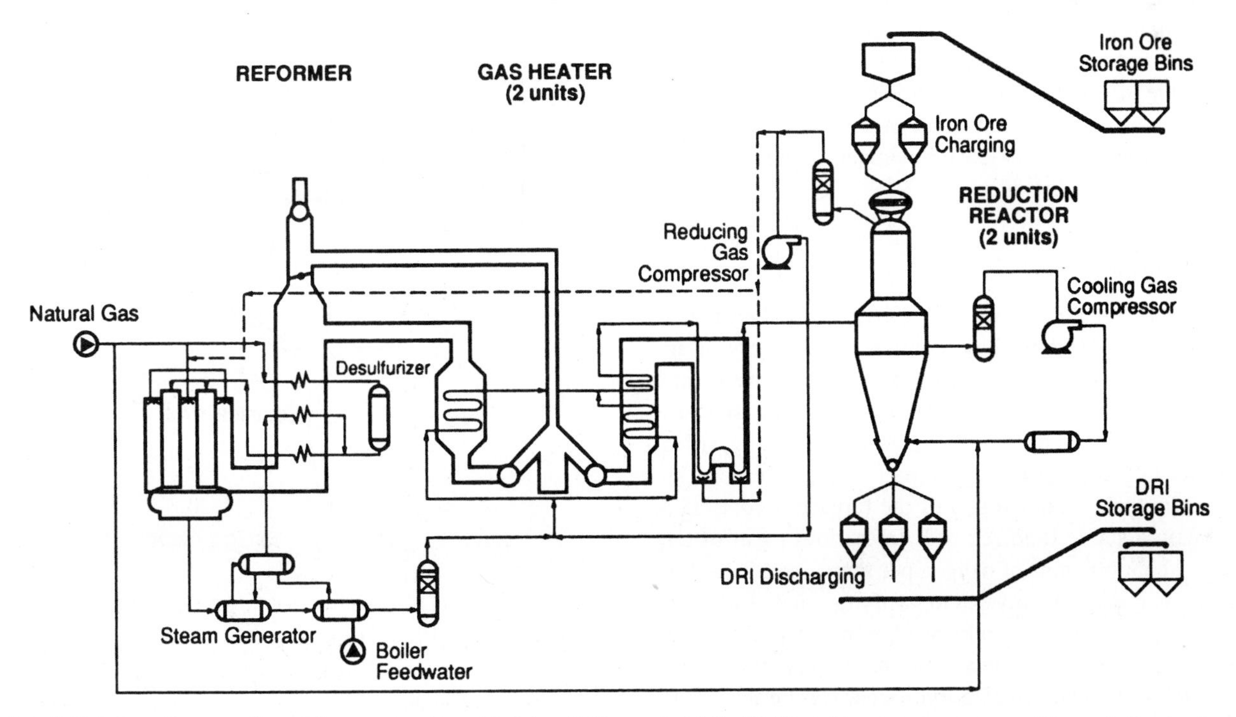

FIGURE 1. Process Flow Diagram for One Module of Sicartsa's HYL III Plant (Reprinted courtesy of *Iron & Steelmaker*.)

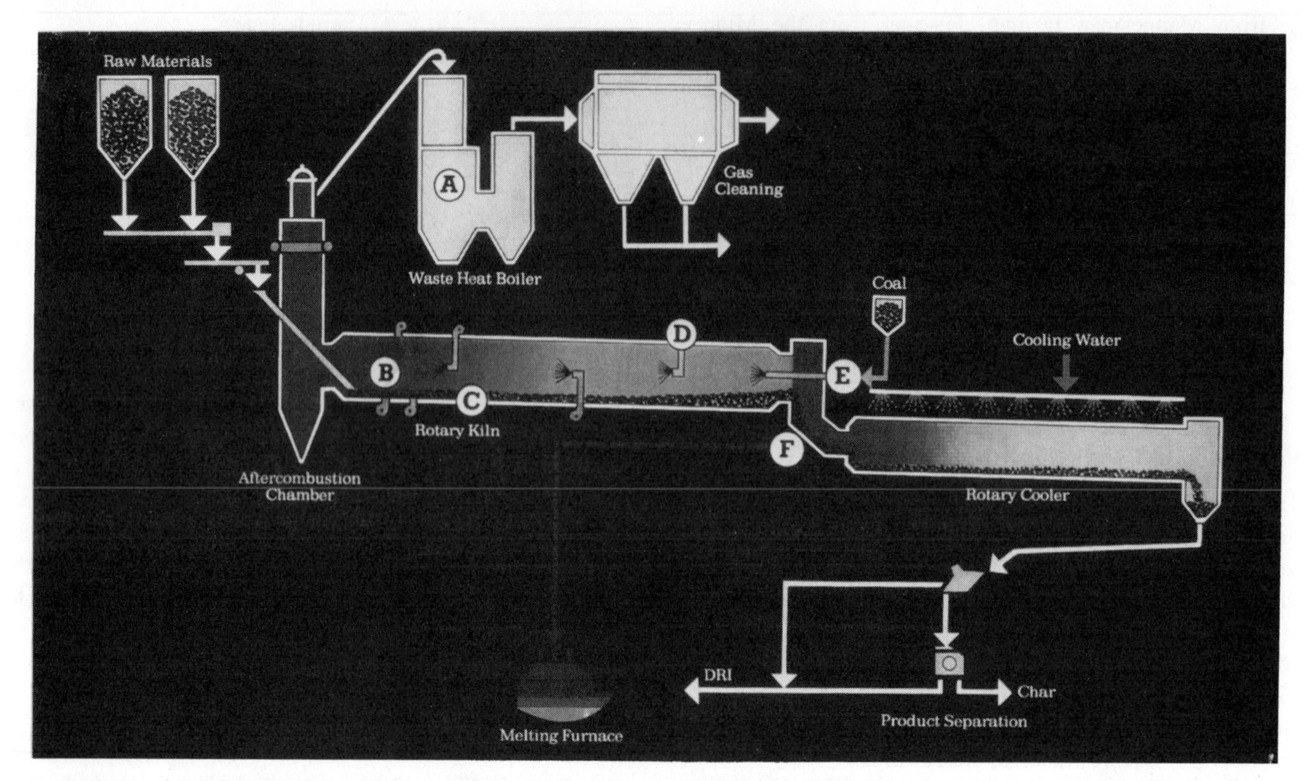

FIGURE 2. SL/RN Coal-Based Direct Reduction (Reprinted courtesy Verlag Stahleisen.)

formed from the char/ore bed at the bottom of the hearth from the fixed carbon (C) as well as the volatile fraction of the coal. There is some direct reduction from the contact of the carbon with the ore. Sufficient flux is also fed into the kiln to trap the sulfur (S) from the coal. After cooling, the direct reduction material is separated magnetically from the residual char and ash. The ash and char are separated and the char is fed back with the feed to the kiln.

Costs are lowest for coal-based technologies,[2] but those savings are offset by the difficult process control, higher energy consumption, poor reactor availability, and the presence of char in the product.

Melter Gasifier

As shown in Figure 3, iron ore is partially reduced in a reduction furnace followed by smelting in a melter gasifier. The coal and iron units are fed by separate lock hoppers into the top of the gasifier melter kept at a pressure of 2 to 5 bar. About 1100 kg of coal per ton of iron is dried and gasified during its fall through the upper part of the furnace. The flow rate in the cylindrical part of the furnace is controlled to maintain a stable fluidized bed. In the lower part of the furnace, about 600 nm^3 of oxygen per tonne (metric ton) of hot metal is injected by tuyeres to gasify the coal. Temperatures at the top of the cylindrical section and in the dome produce a reducing gas containing approximately 65–73% CO, 3–4% CO_2, and 20–25% hydrogen. Before the gas is passed into the prereduction section, it is cooled to 850°C and passed through cyclones. It then serves as a reducing gas in the reduction shaft furnace. The flow from the prereduction section is about 1500 m^3/h of gas per ton of DRI containing 48% CO, 34% CO_2, and 16% hydrogen. An additional volume of gas surplus to the needs of the pre-

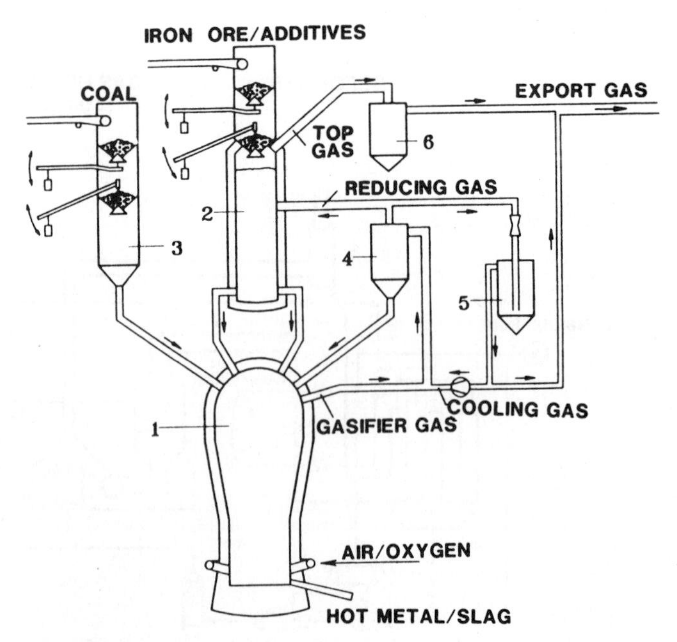

FIGURE 3. Corex Process: Melter Gasifier (1), Reduction Shaft Furnace (2), Coal Feed System (3), Hot-Dust Cyclone (4), Cooling Gas Scrubber (5), Top Gas Scrubber (6). (Reprinted courtesy *Iron and Steel Engineer.*)

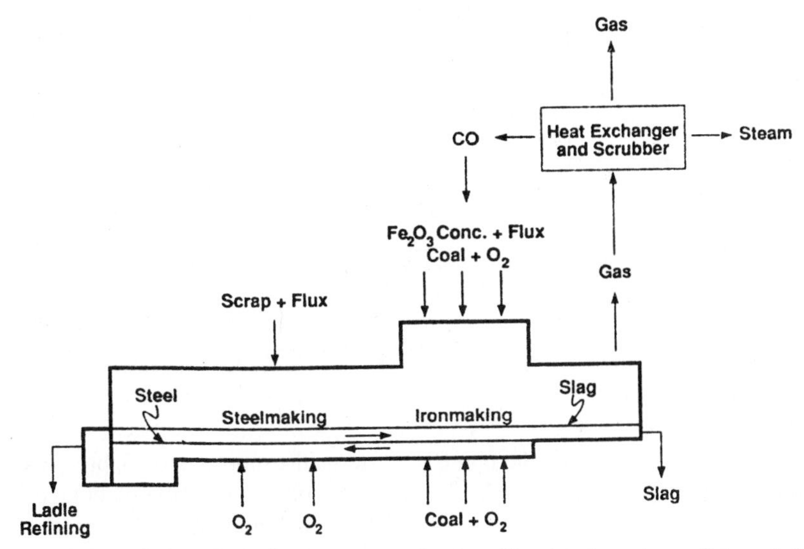

FIGURE 4. Horizontal Two-Zone Countercurrent Smelter (Reprinted courtesy of *Iron & Steelmaker.*)

reduction system is available, depending on the degree of metallization desired.

Hot Metal Bath

The fourth variant is the iron bath type. A schematic of a horizontal two-zone countercurrent smelter is shown in Figure 4.[13] In 1992, AISI will experiment with such a furnace to produce 10 tons per hour of hot metal.

The converter or hearth is first filled with liquid iron, or a heel of liquid from a previous heat is left in. The coal that is fed into the iron bath is gasified by the temperature of the molten iron along with the oxygen that is also injected into the bath. The smelting occurs in the thick, foamy slag layer on top of the hot metal. The oxygen injection provides sufficient agitation to achieve high smelting rates. Either ore or, preferably, hot prereduced wustite-rich ore concentrate is fed into the bath. The reduction and oxidation zones are kept separate by a physical barrier.

Part of the CO evolved is postcombusted above the bath in the hearth and the radiant heat is absorbed by the slag. The off-gases may be directed to a prereduction shaft furnace or used for combustion in boilers and reheat furnaces.

AIR EMISSION CHARACTERIZATION

Vertical Shaft Using Reformed Gas as the Reductant

Air emissions from the shaft furnace occur from three major areas. The first is the material-handling system. The emissions from the pellet handling are strongly dependent on the quality of the pellet and the moisture content. Pellets tend to dust at each transfer location. Another possible emission location is the feed to the shaft furnace. If the devices sealing the lock hopper are a problem, then the pressurized gases inside the shaft may escape.

Fuel (usually natural gas) is required to produce the feed to make the carbon monoxide–hydrogen reformed gas in the reformer, as well as the combustion energy to heat the reformer catalyst tubes externally. Some of the purged process gas is used as the combustion fuel. Air emissions[3,4] from the Midrex and HYL processes are illustrated in Tables 3 and 4 respectively.

The oxides of nitrogen (NO_x) emissions[3] from a Midrex plant are much lower than would be expected considering the flame temperature and air-to-fuel ratios used. Ammonia is formed as a trace by-product in the reformer furnace by the reaction of nitrogen and hydrogen over the catalyst. The ammonia reduces the NO_x. The quantity of purged gas[5] may be reduced by installing a scrubber to remove the CO_2 from the recirculated top gas. The amount of hydrogen and CO removed from the circuit to the combustion section is reduced.

If more than 1 ppm of sulfur is present in the fuel to be reformed, it must be removed to avoid poisoning the reformer catalyst. The process gas will remove sulfur from the ores in the reduction zone and deposit it on the reduced iron in the cooling zone. A limited amount of sulfur from

TABLE 3. Typical Emission Levels

	Kilograms per Ton of Product (Concentration, mg/m^3)		
	Particulate	SO_2	NO_x
Reformer	0.012 (14)	0.035 (40)	0.172 (196)
Shaft furnace	<0.001 (25)	<0.001 (40)	<0.002 (196)
Dust collection[a]	0.016	0.000	<0.001

[a]A Midrex facility typically contains three dust-collection systems.

TABLE 4. Emissions to the Air

	Kilograms per Tonne of DRI			
	Material Handling	Combustion Systems	Dust Collection	Total
Particulates	0.001	0.006	0.032	0.039
SO_2	—	0.0002	—	0.0002
NO_x	—	0.050	—	0.050
CO	—	0.004	—	0.004
HC	—	0.001	—	0.001
Total	0.001	0.0612	0.032	0.0942

the ores will be emitted to the atmosphere when the purge gas is burned as a combustion fuel.

Rotary Kiln/Hearth

Air emissions from the rotary units occur during raw-materials handling and from the process gas exhaust and the product screening and carbon char return to the kiln. The total dust collected[6] from the raw materials and the product-handling system for a kiln amount to from 200 to 300 kg per tonne of DRI produced. The waste gas from the kilns generates 600 kg of dust per tonne of DRI handled. The volume of gas from a kiln is partially dependent on the volatile content of the coal used and the efficiency of utilization of the coal. An optimum fill[7] must be achieved and the bed must be kept in a rolling mode, to minimize the segregation of iron ore and reductant so that coal consumption is minimized.

Changing the coal feed to the discharge end[7] reduced the energy consumption from 20 GJ per ton to 15 GJ per ton (coal use was reduced from 690 to 530 kg of coal per ton of DRI). The waste gas contains CO, hydrogen, and hydrocarbons. The excess air–oxygen conditions in the kiln will influence whether the sulfur is present as oxides or hydrogen sulfide (H_2S). Twenty-seven percent of the sulfur in the feed[7] reverts to the gas as sulfur dioxide (SO_2). The remainder is tied up in the product and the waste flux. The gas volume from a kiln[8] varies from 3300 to 6100 nm^3 per ton of DRI. The waste gas from the kiln contains from 20%[2] to 65%[8] of the energy input (3 GJ per ton of DRI at the 20% level). The retention time of gases[9] in a kiln is from five to nine seconds at temperatures varying from 1200°C at the burner end to 500°C at the exit end, favoring destruction of the polycyclic aromatic hydrocarbons (PAHs). Coal fed closest to the product discharge may also minimize PAH emissions. The medium-heat-content gas from the kiln may be passed through an afterburner or combusted in a boiler.

Melter Gasifier

The melter gasifier generates about 1500 m^3 per ton of DRI or gas per ton of hot metal. The calorific value of the gas (termed export gas) is approximately 7600 kJ/nm^3. Therefore, this gas is similar to coke oven gas and would be used as a fuel for reheating furnaces or boilers. Much of the sulfur from the coal would have an affinity for the slag. Part of the sulfur would leave the smelter as H_2S, most of which would be removed by the oxides in the prereduction shaft as its level[10] in the export gas is less than 50 ppm.

Hot Metal Bath

The emissions from the HMB systems are similar to those from basic oxygen furnaces (BOFs). The major difference is the influence of the coal on the volatile and nonvolatile organic emissions from the furnaces.

In a case[11] where the ratio of the slag cavity formed by the top oxygen jet (L_s) to the foaming slag thickness (L_{so}) exceeded 1, the ferrous content in the dust increased to 100 g/nm^3. At ratios of less than 1 and low mixing intensities, the ferrous content in the dust decreased to less than 10 g/nm^3. Control of the slag *V* ratio[11] and the suspension of coal char or coke in the slag is important for the control of foaming. The size range of particulates from the HMB would be expected to be similar to that from a top blown or combined blown oxygen furnace. Uniform coal feeding rates should result in the destruction of volatile matter fed to the furnace.

Processes[12] that operate at extremely high temperatures with gases that contain nitrogen from infiltration air and partly from the nitrogen content of the oxygen used for postcombustion have a high probability of producing high NO_x concentrations. Postcombustion using low-nitrogen-containing oxygen and limiting the ingress of air into the hoods or ducts above the furnace will limit NO_x emissions. Most of the sulfur from the coal and the metallic feed would stay with the slag and metal.

AIR POLLUTION CONTROL MEASURES

See the following table.

	Contaminant Source		
	Process Gas	Raw-Material Feed	Product Discharge
Vertical shaft, reformed gas	Purged gas from reformer can be cleaned using a high-energy scrubber or (ESP). Moisture is too high for fabric filters. Any sulfur removed from the reformer circuit may be recovered or vented to the atmosphere.	Pellet transfer points must be collected and scrubbed.	Product is hot and dusty. Tight hoods are required at product transfer locations.
Rotary kiln	The medium-heat-content kiln exhaust gas can be combusted in an afterburner or boiler. Use of a boiler would allow use of an ESP for particulate control. An afterburner would require a high-energy scrubber for particulate. Adequate control of hydrocarbons could be attained with close control of the CO in the combustion exhaust. Medium-heat-content gases do not produce high NO_x. The SO_x, and NO_x control technology would be the same as with standard utility boilers.	Lime, coal, and oxide transfer points require collection and cleaning. Fabric filters would be preferred.	Product is hot and dusty. Same control as above.
Hot metal bath	The particulate emissions from the HMB system would be similar to those from a BOF or combined blown steel-making vessel. PAH emissions may not be destroyed as the temperature at the mouth of suppressed combustion hoods may only reach 840°C. The medium-heat-content gas from the bath would be used for prereduction or combusted in a boiler. Much of the SO_x would be retained by the oxides in the prereduction section. Further SO_x control after combustion could be attained using standard utility boiler technologies.	Same as for BOF.	The product is molten hot metal or steel. Control techniques applicable to BOF tapping are applicable.
Melter gasifier	A low- to medium-heat-content gas is produced from the gasifier. The gases would either be used in a prereduction stage or be used for combustion in a boiler. Standard utility boiler cleanup systems could be used for NO_x and SO_x control.	Same as for blast furnace.	The product is hot metal. Control techniques are as discussed under blast furnaces.

References

1. A. Chatterjee, "State of direct reduction ironmaking in India," *Steel Technol. Int.*, 49 (1990–91).
2. W-D. Hausler, "Improvements in the Codir direct reduction process," *Steel Technol. Int.* (1990–91).
3. T. W. Hoffman, "The MIDREX direct reduction process: An environmentally sound route to steel making," *UNEP Industry Environ.* (1984).
4. J. M. Pena, "Environmental aspects of the HYL III direct reduction process," *UNEP Industry Environ.*, 9 (1984).
5. P. Heinrich, K. Knop, and R. Madlinger, "Status report and potentials for development of HYL-III direct reduction technology," *Metallurg. Plant Technol. Int.* (2) (1990).
6. S. Das Gupta, "Environmental aspects of electric arc furnace and direct reduction plants in India," *UNEP Industry Environ.* (1984).
7. S. P. Mehrotra and J. K. Brimacombe, "Kinetics of gasification of coal char with CO_2 in a rotary reactor," *Ironmaking Steelmaking* 17(2):93 (1990).
8. L. Formanek, H. Eichberger, and H. Serbent, "Energy utilisation in direct reduction using rotary kilns," *Metallurg. Plant Technol.* (2):9 (1988).
9. B. R. Nijhawan, "Environmental aspects of the direct reduc-

tion route to steel making," *UNEP Industry Environ.* 25 (1984).
10. R. Steffen, "The Corex process: First operating results in hot-metal making," *Metallurg. Plant Technol. Int.*, (2): 20 (1990).
11. T. Ibaraki, M. Kanemoto, S. Ogata, et al., "Development of smelting reduction of iron ore—an approach to commercial ironmaking," *Iron Steelmaking,* 30 (1990).
12. J. C. Agarwal, "Strategic considerations in direct steelmaking," *Iron Steelmaking,* 18(3):30 (1991).
13. P. E. Queneau, "Direct steelmaking—quo modo," *Iron Steelmaking,* 76 (1990).

ROLLING

Michael T. Unger

The rolling of steel involves the working of metal to attain permanent deformation by applying mechanical forces to a metal's surface. The objective may be to produce a desired shape or size (mechanical shaping) or to improve the physical properties of the metal (mechanical treating).

Rolling operations are usually subdivided into hot working and cold working. Forces required to deform metal in hot rolling are sensitive to the rate of applied loads and to temperature variations. The resultant strength of the metal after deformation is changed very little. During cold rolling, the deformation forces are relatively insensitive to load application and to temperature variations, but the basic strength of the cold-worked metal is increased.

Semifinished forms (slabs, blooms, and billets) are made by the ingot process or the newer continuous casting process. In the ingot method, the finishing stages of steel making begin when ingots are lowered into furnaces called soaking pits that reheat them to an even temperature for rolling. The ingot is rolled in a primary mill into semifinished slabs or into rectangular shapes called blooms. In the newer continuous casting process, steel is poured directly into a tundish, which feeds a curved mold where the steel is solidified directly into semifinished products. The end product can be slabs, blooms, or billets, depending on the final proportions.

Shaping by rolling consists of passing the steel between two rolls that revolve at the same speed, but in opposite directions, as shown in Figure 1. If spaced so that the distance between the rolls is less than the height of the section entering them, the rolls will grip the piece of metal and deliver it reduced in cross-sectional area and increased in length.

PROCESS DESCRIPTION

During hot rolling, steel heated to 2300°C is placed onto the end of the rolling line. For example, a slab's metallurgical

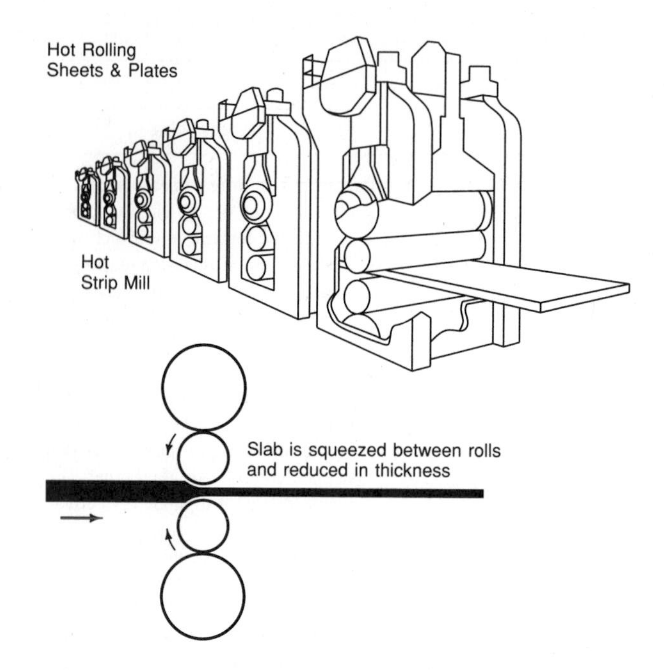

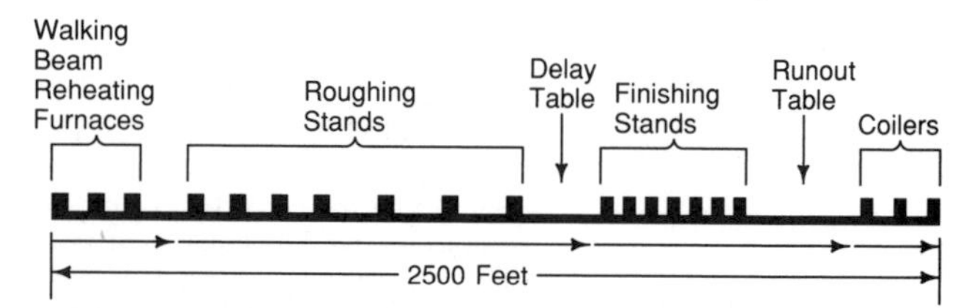

FIGURE 1. Hot Rolling Schematic

analysis, size, weight, and intended end use are fed into the mill's computer. Sensing devices continuously record the slab's thickness, width, and temperature, and the squeezing force of the rolls as the slab moves through the roughing and finishing stands, as shown in Figure 1. The computer instantaneously resets controls or changes the speed of rolling stands to ensure a properly finished product. Steel rolled on the hot strip mill can be sold in this form or further processed into cold-rolled, galvanized, or aluminized steel.

Air emissions from hot-rolling mills generally consist of water vapor from scale breaking or metallurgical water sprays and are typically uncontrolled. However, there are potential emissions from rolling solution aerosols that are localized close to the rolling stands, as well as small pieces of scale that may become airborne. Emissions from combustion sources for the soaking pits and reheat ovens are frequently controlled.

Hot-band or steel rolled on the hot strip mill can be fed into a multistand cold rolling mill. Without reheating the steel, the cold mill rolls steel under great pressure to precision thickness and produces specific surfaces and develops controlled mechanical properties. A schematic of a cold rolling mill started in 1990 is shown in Figure 2. This mill exhibits best available control technology (BACT) and is expected to release minimal quantities of air pollutants because of the high level of pollution control equipment.

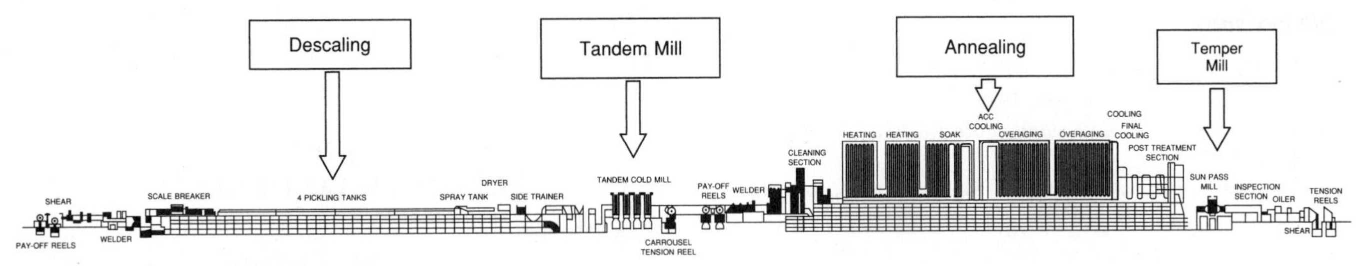

FIGURE 2. Cold Rolling Schematic

AIR EMISSIONS CHARACTERIZATION AND CONTROL MEASURES

During state-of-the-art cold rolling, steel is introduced onto the line and passes through the major processing steps shown in Table 1. Uncontrolled emission estimates, calculated controlled emissions, the types of air pollution control utilized, and the techniques' expected control efficiencies are summarized in Table 1.

In both hot and cold rolling, one can expect nitrogen oxides (NO_x) emissions resulting from combustion in reheat furnaces, in annealing furnaces, or from boilers. The primary means of controlling NO_x emissions is through combustion modification or selective catalytic reduction. Nitrogen oxides formed from combustion air constituents are referred to as thermal NO_x. The general technique to control thermal NO_x is to suppress the natural gas combustion temperature to below 2600°F. Above this temperature, NO_x formation is exponential, while below this temperature, it is linear at a very limited rate. Methods of controlling NO_x emissions by reduced flame temperatures include:

1. *A low heat release furnace*, which rapidly absorbs energy from the flame.
2. *Low-turbulence, laminar burners*, which radiate heat rapidly, resulting in low flame temperatures.
3. *Flue gas recirculation* from the boiler outlet into the wind box, which absorbs heat and inhibits combustion, both of which will lower flame temperature.
4. *Low combustion air temperature*, which will lower combustion efficiency, but can reduce NO_x generation.
5. *Low excess air firing*, which limits the oxygen available and in general will decrease nitrogen oxide formation.

Selective catalytic reduction (SCR) is an expensive postcombustion NO_x-control technique used in Japan,[1,2] but not on industrial combustors in the United States. Ammonia is used as a reducing agent in a gas-phase reaction with oxides of nitrogen in the presence of a catalyst to form nitrogen and water.

References

1. *Prevention of Significant Deterioration Permit Application*, United Engineers and Constructors, 1987.
2. *BACT/LAER Clearinghouse—A Compilation of Control Technology Determinations*, EPA-450/3-85-016B, U.S. Environmental Protection Agency, Research Triangle Park, NC 1985.

TABLE 1. Continuous Cold Mill with BACT Controls

Source	Purpose	Emissions Type	Expected Uncontrolled Emissions, mg/nm^3	Control Technology	Calculated Controlled Emissions, (mg/nm^3)	Percent Control
Pinch roll leveler	Flattens steel	Particulate matter	1000	Fabric filter	10	99
Flash butt welder	Connects coils	Particulate matter	360	Fabric filter	3.6	99
Tension leveler	Removes oxidation	Particulate matter	1000	Fabric filter	10	99
Descaling (pickling)	Cleans surface	HC1 vapor	3250	Countercurrent packed-tower scrubber with mist eliminator	6.5	99
Tandem cold rolling	Reduces thickness of strip	Roll coolant spray (water and oil)	100	Baffle plate collision mist eliminator	20	80
Electrolytic cleaning	Removes oil from strip	Alkali mist	100	Horizontal air washer	5	95
Posttreatment pickling	Removes light oxide	HC1 vapor	840	Gas washing tower	1.7	99
Temper mill	Tempers/restores flatness	Negligible		None		
Electrostatic oiler	Applies protective oil coat	Oil aerosol	13	Mist eliminator	1.3	90

Bibliography

AISE, *The Making, Shaping and Treating of Steel,* 10th ed., Herbich & Held, Pittsburgh, 1985.

W. L. Roberts. *The Cold Rolling of Steel,* Marcel Dekker, New York. 1978.

W. L. Roberts. *Hot Rolling of Steel,* Marcel Dekker, New York, 1983.

SCARFING

S. J. Manganello

Scarfing in the steel mill[1,2] consists of surface removal by use of an oxy–fuel-gas flame process. The flame is used rapidly to melt the steel surface thermochemically while the high-pressure oxygen stream propels the oxidized product from the surface. The process is usually carried out on hot steel between stages of rolling by hot-scarfing machines. Subsequent scarfing, if required, is done by hand torches in the conditioning area.

The tip of a hand-scarfing torch is somewhat similar to the oxygen–gas torch used to cut steel. The oxygen not only promotes the chemical reaction that forms the melted iron oxide, but its kinetic energy must force the liquid, oxidized metal from the path of the torch, thereby removing surface defects from the semifinished steel. In addition to oxygen, fuel gas is necessary to the functioning of the hand-scarfing torch. The gas (commonly natural gas) serves as a preheating agent to elevate a spot on the surface to such a temperature that the oxygen and the steel will begin to combine chemically. The starting rod, a small steel rod that extends into the flame and is heated until a small drop of melted rod falls onto the steel surface, is often used to kindle the reaction quickly.

PROCESS DESCRIPTION[1,2]

The mechanical scarfer, installed directly in-line on the rolling mill, is composed of a number of (usually fixed) scarfing units across the steel, designed in such a way that they form a pass on the mill. The hot-scarfing machine is usually placed before the slab/bloom shears or after the roughing or first billet mill. The high temperature of the product being rolled is enough to obtain quick starting action. The machine scarfer can remove defects up to 4.5 mm (0.18 inch)—which is usually sufficient to remove such defects as rolled seams, light scabs, and checks. General practice is to desurface each steel grade to the depth that strikes the right economic balance between loss in yield from metal removal and processing costs, on the one hand, and savings in lessened rejection loss, on the other. This depth varies with different materials, but the metal loss is generally 1.5–2.5% of the product. Four-sided scarfing machines are now in common use, as are automated spot-scarfing machines. Water jets are employed to help control the slag.

AIR EMISSIONS AND POLLUTION CONTROL[2–4]

Hand Scarfing

Scarfing generates small particles, mainly iron oxides, that may remain suspended in the atmosphere for a considerable time. The concentration in a closed area can build up over time, and particles will eventually settle on the walls, floor, and equipment. The size of the enclosure, the size of the scarfing operation, and the ventilation affect the concentration. Adequate ventilation is the key, whether it be natural, mechanical, or respirator ventilation, and there are safety criteria for each type. General ventilation of the work site is often adequate, but local ventilation may also be required.

Machine Scarfing

Exhaust hoods are commonly used for machine scarfers. They must have sufficient airflow and velocity to keep contaminants at or below permissible limits to control air pollution, which are recommended/stated by the manufacturer.

The scarfing of steel results in the generation of appreciable quantities of smoke, the quantity and density depending on such variables as steel analysis, scarfing oxygen pressure, and efficiency of the slag water jets. It is necessary to collect this smoke and discharge it to the atmosphere outside the building by means of a suitable exhaust system, consisting of a hood over the roll table, a duct system, a fan, and a stack. Systems are normally designed for maximum gas velocities of 3000 fpm, based on free air movement at average ambient mill temperatures.

Scarfing-machine smoke contains some solids in addition to steam and gases. The solids are mainly oxides of iron with traces of the alloying elements present in the scarfed steel. Much of the solid matter in the smoke is submicron in size, with occasional particles up to the maximum that can be airborne at the exhaust velocities. An average four-sided scarfing machine equipped with cross-fire water jets and an efficient smoke-collection system will produce approximately 36 grains (7000 grains equal 1 pound) of smoke solids per pound of metal removed under normal scarfing conditions. The large number of submicron particles give the smoke a yellowish-brown color. Smoke gases on a typical installation have been analyzed to contain 0.22% CO_2, 23.0% O_2, 0.12% H_2, and 76.6% N_2. The smoke is a nuisance and can irritate the eyes. Also, the iron in the smoke may cause short circuits if allowed to accumulate on electrical apparatus. Thus consideration should be given to the most suitable location for discharge to the atmosphere. The smoke can be cleaned so that the stack discharge has no visible color by passing it through an electrostatic precipitator. The construction materials used for the smoke-removal system should be corrosion and abrasion resistant because of the nature of the smoke, which may vary with each installation in accordance with the composition of the steel being scarfed. Bag-type units have not been used because of the moisture content of the air.

References

1. *The Making, Shaping, and Treating of Steel,* 10th ed., 1985, pp 734–737.
2. Private correspondence with R. C. Smith, L-TEC, Florence, SC, March 19, 1990.
3. *Welding Handbook,* Vol. I, *Welding Technology,* 8th ed., 1987, Chapter 16.
4. L-TEC *Scarfing Machine Bulletin:* SMB-04.

SINTER PLANTS

S. S. Felton and L. M. Stuart

Sintering is one of a number of iron ore agglomeration processes used in the steel industry. It was developed in the early 1900s to utilize the large amount of blast furnace dusts being generated each year and the millons of tons of hermatite ore fines that were being accumulated in the Mesabi range. The sintering process fuses iron-bearing fines into a hard clinker suitable for use in blast furnaces. It is capable of converting a wide variety of iron ore fines (iron-bearing blast furnace dusts and sludges, mill scales, and steelmaking dusts and iron-bearing dusts and sludges collected from air and water pollution control processes) into a high-quality blast furnace burden material.

PROCESS DESCRIPTION

The sintering process includes a material storage, blending, and feeding system for the burden materials; a traveling-grate hearth called the sinter strand to produce the sinter; a sinter breaking, screening, and cooling system to size and cool the sinter properly and to recycle undersized sinter to become part of the feed burden materials; and a variety of environmental control facilities. Figure 1 provides a flow diagram of a modern sinter plant.[1]

Storage, Blending and Feeding System

In modern sinter plants, the bedding and blending system takes the ores and other materials from storage bins and transfers them in predetermined amounts to a traveling stacker, which deposits them in thin layers down the length of a large bedding pile. As many as 500 layers of materials may be deposited in a single bedding pile, and a pile may contain enough material to operate a sinter plant for 6 to 12 days. The blended feed material is reclaimed from the pile to large feed storage bins at the sinter plant. Sinter plants without a bedding system meter burden materials from large storage bins onto conveyor belts going directly to the main sinter feed conveyor. Careful blending of the raw feed materials into a homogeneous blend produces a sinter product that is consistently uniform in chemical and physical properties.

Feeding Materials to the Sinter Strand

Conversion of the raw material mix into a sinter product takes place on a sinter strand. The feed to the sinter strand consists of the raw materials from the blending pile, fluxes, solid fuel (usually coke breeze), and recycled sinter fines from the sinter screening process. Each of these materials is added in closely controlled amounts to a mixing–reroll drum for further blending and control of moisture content. The final mix is then transferred onto the sintering strand in a uniform depth by a roll-feeder–mix-level control system.

Ignition

The sinter feed mix travels under an ignition hood where the hot combustion gases ignite the solid fuel in the sinter mix to start the sintering process. The fuels used in the ignition hood are usually gaseous fuels, such as natural gas or coke oven gas. The combustion products from the ignition zone are drawn through the sinter mix into the windbox gas collection main located under the sinter strand.

Sintering

Once the fuel in the sinter mix is ignited, the sintering process has begun. As the sinter strand moves out of the ignition zone, air is drawn downward through the bed into the windboxes under the traveling grate. Air movement is provided by large fans located downstream of the windboxes. As the air is drawn through the sinter bed, the solid fuel in the mix burns at temperatures sufficient to fuse the raw materials into a hard clinker material. The flame front of burning fuel in the mix continues to move downward through the mix as the strand moves forward. Completion or burn-through is controlled to occur in the first section of the last windbox at the end of the strand. All of the combustion products, excess air, and other gaseous materials exiting the sinter bed during the sinter process are captured in the windboxes under the sinter strand.

Sinter Sizing and Cooling

The sinter leaves the sinter strand in large, hot chunks that must be properly sized and cooled before they can be utilized in the blast furnace. This is accomplished by a series of breakers, screens, and crushers. The screening

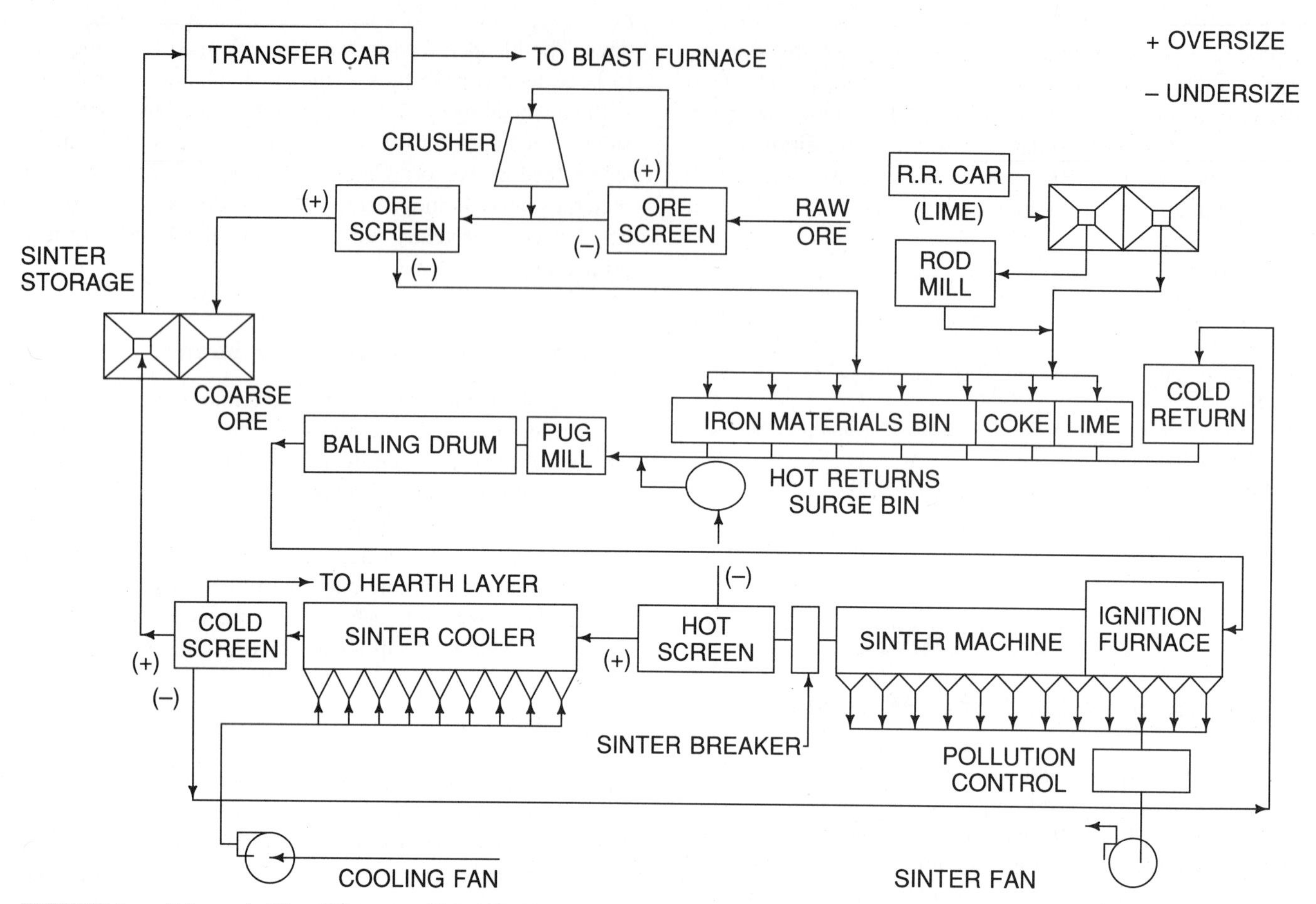

FIGURE 1. Schematic Flow Diagram of Modern Sinter Plant (from Reference 1)

processes remove the less than ¼-inch sinter and recycle it into the raw feed mix ahead of the mix–reroll drum.

Types of Sinter

Three general types of sinter are produced by sinter plants: acid sinter, self-fluxed sinter, and superfluxed sinter.[2] The term basicity, which is the ratio of $(CaO+MgO)/(SiO_2+AL_2O_3)$ found in the sinter, is used to distinguish one type of sinter from another. Acid sinter normally is produced with a basicity of less than 1.0 and is produced primarily from iron-bearing materials. With this type of sinter, all of the fluxes needed by the blast furnace must be added in the blast furnace burden. The use of this type of sinter is declining and most sinter plants are producing either self-fluxing or superfluxed sinters. Self-fluxed sinter is produced by adding sufficient fluxing materials to adjust the basicity to 1.0 to 1.5. This type of sinter provides only the flux needed by the sinter. The fluxes needed for the other blast furnace burden materials must also be added to the blast furnace. Superfluxed sinter has a basicity of 2.0 or higher and is produced with sufficient additional flux materials so that no additional flux is needed in the blast furnace burden. All of the flux needed by the blast furnace is provided by the sinter.

AIR EMISSIONS CHARACTERIZATION

Sinter-plant windbox emissions contain particulate emissions and combustion gases generated during the sintering process. Tables 1 and 2 show particulate emissions factors and size data for sinter plants based on data collected by the U.S. Environmental Protection Agency (EPA), mostly in the 1970s. It is important to note that many of these data may no longer be applicable since there have been major changes in sintering practices and gas cleaning system technology. Tables 3 and 4 present sinter plant emission data collected in the late 1970s at sinter plants equipped with wet scrubber treatment of windbox gases.

Particulate Matter

Chemical and mineralogical examinations of the inorganic fractions of many particulate catches from sinter plant windboxes show that the principal component, after the cyclones, is potassium chloride (KCl).

Hydrocarbon Content

The amount of hydrocarbon particulate generated in the windbox exhaust is a function of the amount of hydrocarbon

TABLE 1. Particulate Emission Factors for Sinter Plants

Source	Units	Emission Factors		Emission Factor Rating[a,b]	Size Data
Sintering					
Windbox emissions	kg/Mg (lb/ton) finished sinter				
Uncontrolled					
Leaving grate		5.56	(11.1)	B	Yes
After coarse particulate removal		4.35	(8.7)	A	
Controlled by dry ESP[c]		0.8	(1.6)	B	
Controlled by wet ESP[c]		0.085	(0.17)	B	Yes
Controlled by venturi scrubber		0.235	(0.47)	B	Yes
Controlled by cyclone		0.5	(1.0)	B	Yes
Sinter discharge (breaker and hot screens)	kg/Mg (lb/ton) finished sinter				
Uncontrolled		3.4	(6.8)	B	
Controlled by baghouse		0.05	(0.1)	B	Yes
Controlled by venturi scrubber		0.295	(0.59)	A	
Windbox and discharge	kg/Mg (lb/ton) finished sinter				
Controlled by baghouse		0.15	(0.3)	A	

[a]The following ratings were applied to each emission factor.

A = *excellent*. Developed from A-rated test data taken from many randomly chosen facilities in the industry population. The source category is specific enough to minimize variability within the source category population.

B = *above average*. Developed only from A-rated test data from a reasonable number of facilities. Although no specific bias is evident, it is not clear whether the facilities tested represent a random sample of the industries. As with the A rating, the source category is specific enough to minimize variability within the source category population.

C = *average*. Developed only from A- and B-rated test data from a reasonable number of facilities. Although no specific bias is evident, it is not clear whether the facilities tested represent a random sample of the industry. As with the A rating, the source category is specific enough to minimize variability within the source category population.

D = *below average*. The emission factor was developed from A- and B-rated test data, and there may be reason to suspect that these facilities do not represent a random sample of the industry. There also may be evidence of variability within the source category population. Limitations on the use of the emission factor are footnoted in the emission factor table.

E = *poor*. The emission factor was developed from C- and D-rated test data, and there may be reason to suspect that the facilities tested do not represent a random sample of the industry. There also may be evidence of variability within the source category population. Limitations on the use of these factors are always footnoted.

[b]The data were rated according to the following criteria.

A = Tests performed by a sound methodology and reported in enough detail for adequate validation. These tests are not necessarily EPA reference method tests, although such reference methods are used as a guide.

B = Tests that are performed by a generally sound methodology, but lack enough documentation for adequate validation.

C = Tests that are based on an unproven methodology or lack a significant amount of documentation.

D = Tests that are based on a generally unacceptable method, but may provide an order-of-magnitude value for the source.

[c]Electrostatic precipitator

Source: Reference 3.

contained in the sinter burden. Most of this material evolves from the mill scale as a part of the burden. Mill scales are generated during rolling of hot steel and consist of a mixture of iron and iron oxides. It is a premium burden material since it helps to produce a hard, dense sinter. However, mill scales contain various types of oils and greases. During the sintering process, most of these oils and greases are burned, but some of them vaporize into the windbox gases ahead of the high-temperature advancing flame front during sintering and then condense at the cooler windbox temperatures. Visible emissions from the windbox collection device containing significant hydrocarbons tend to be bluish.

Gaseous Constituents

Table 5 shows the concentration of oxygen, sulfur dioxide, combustibles (principally carbon monoxide), and nitrogen oxides in the stack gases from a large sinter plant equipped with wet scrubbers. The sulfur content of the stack gases will vary greatly, depending on the sulfur content of the sinter burden materials and the fuels used.

AIR POLLUTION CONTROL MEASURES

Particulate and gaseous emissions are generated at all phases of the sintering process. Each operation has different

TABLE 2. Size Specific Emission Factors for Sinter Plants

Source	Emission Factor Rating (A–E)[a]	Particle Size, μm	Cumulative Mass, % Less Than Stated Size	Cumulative Mass *Emission Factor* kg/Mg Alloy	(lb/ton Alloy)
Sintering					
Windbox emissions					
Uncontrolled leaving grate	D	0.5	4	0.22	(0.44)
		1.0	4	0.22	(0.44)
		2.5	5	0.28	(0.56)
		5.0	9	0.50	(1.00)
		10	15	0.83	(1.67)
		15	20	1.11	(2.22)
		*	100	5.56	(11.1)
Controlled by wet electrostatic precipitator	C	0.5	18	0.015	(0.03)
		1.0	25	0.021	(0.04)
		2.5	33	0.028	(0.06)
		5.0	48	0.041	(0.08)
		10	59	0.050	(0.10)
		15	69	0.059	(0.12)
		*	100	0.085	(0.17)
Controlled by venturi scrubber	C	0.5	55	0.129	(0.26)
		1.0	75	0.176	(0.35)
		2.5	89	0.209	(0.42)
		5.0	93	0.219	(0.44)
		10	96	0.226	(0.45)
		15	98	0.230	(0.46)
		*	100	0.235	(0.47)
Controlled by cyclone	C	0.5	25	0.13	(0.25)
		1.0	37	0.19	(0.37)
		2.5	52	0.26	(0.52)
		5.0	64	0.32	(0.64)
		10	74	0.37	(0.74)
		15	80	0.40	(0.80)
		*	100	0.5	(1.0)
Controlled by baghouse	D	0.5	3	0.005	(0.009)
		1.0	9	0.014	(0.027)
		2.5	27	0.041	(0.081)
		5.0	47	0.071	(0.141)
		10	69	0.104	(0.207)
		15	79	0.119	(0.237)
		*	100	0.150	(0.30)
Sinter discharge (breaker and hot screens) controlled by baghouse	C	0.5	2	0.001	(0.002)
		1.0	4	0.002	(0.004)
		2.5	11	0.006	(0.011)
		5.0	20	0.010	(0.020)
		10	32	0.016	(0.032)
		15	42	0.021	(0.042)
		*	100	0.05	(0.1)

[a]See Table 1, footnotes a and b.

Source: Reference 3.

TABLE 3. Performance of Sinter Plant Windbox Gas Collection Systems

	Sinter Plant No. 1	2
Cyclones—pounds per net ton sinter		
Inlets	4.6	2.03
Outlets	2.4	1.4
Percent removal	48	33
Wet scrubbers—pounds per net ton sinter		
Inlets	2.7	1.1
Outlets	0.14	0.29
Percent removal	95	74

Source: Reference 2.

TABLE 4. Particle Size Distribution of Sinter Windbox Particulates

Particle Size Distribution	Scrubber Inlet	Scrubber Outlet	Percent Removal
Percent less than:			
1 μm	67	90	64
2 μm	70	95	64
3 μm	72	97	65
5 μm	75	98	—
40 μm	85	100	—
			All sizes 74

Source: Reference 2.

types of particulate and gaseous emissions and requires a different type of emission control technology.

Feed Preparation Process

Handling of the large quantity and variety of feed materials in storage bins, feeders, screens, and transfer chutes generates dusts and requires that dust emission control hoods and exhaust systems be installed at several points in the feed preparation process. The emissions from these operations consist primarily of particulate matter. Collection of these emissions is generally accomplished by baghouses or low-energy wet scrubbers.

Sinter Sizing and Cooling

The hot, dirty gases collected during the sizing and cooling of hot sinter also contain primarily particulate matter. These are generally fairly coarse particles and can be removed by cyclones or baghouses. The collected dust is normally recycled directly to the sintering process.

The dusts generated at the cold screening operations and conveyor transfer points, and during sinter crushing are generally collected via a baghouse collector, although low-energy wet scrubbers have also been used.

Sinter Windbox Gases

The cleaning of the sintering process gases is the most difficult gas cleaning problem in the sinter plant and one of the most difficult and costly gas cleaning problems in the steel industry. Many sinter plants are equipped with cyclone collectors followed by the primary emission control equipment. The primary emission control equipment is located after the induced-draft fans and includes either (1) electrostatic precipitators, (2) wet scrubbers, or (3) baghouses.

Cyclone Separators

The gases leaving the sintering grate contain a significant amount of coarse particles. Most sinter plants are equipped with a series of gravity separators or dropout chambers followed by mechanical collectors, such as cyclones or multiclones. The cyclone collectors are installed ahead of the induced-draft fans and are designed to remove particles with diameters of 25 μm or greater. This is done to protect the induced-draft fans from severe wear by these abrasive particles.

Electrostatic Precipitators

In the 1960s, most sinter plants were equipped with dry electrostatic precipitators that operated with removal efficiencies of up to 99% of the windbox emissions produced during the production of "acid" sinters. In recent years, there has been a trend toward the manufacture of "fluxed" sinters—which has been detrimental to the performance of electrostatic precipitators since the dust generated has increased resistivity.

With the increased use of oil-bearing mill scales in the sinter feed materials, there are increased amounts of condensed hydrocarbon particulates in the windbox gases. Dry electrostatic precipitators are not efficient collection devices for hydrocarbons. Also, the presence of hydrocarbons in precipitators can present explosion and fire hazards.

Baghouses

Baghouse installations have been applied to windbox emission controls in several sinter plants where high-basicity sinter is made. Their efficient operation is sensitive to condensation of moisture or hydrocarbon vapors. Condensed moisture and sulfur oxides in the gas stream will react with lime in the dust or in the precoat material on the bags to form calcium sulfites and sulfates on the surface of the bags and thus blind them.

Scrubbers

High-energy wet scrubber devices followed by efficient mist eliminators are very efficient in treating sinter plant windbox gases. The wet scrubber device must be designed

TABLE 5. Chemical Composition of Sinter Plant Windbox Stack Gases, After Scrubber

Fuels Used, % Btu	Sinter Plant 1 Coke Oven Gas 10% Coke Breeze 90%	Sinter Plant 2 Coke Oven Gas 19% Coke Breeze 81%
Stack gas constituents		
Oxygen, %	15.3	16.6
Combustibles, %	0.8	1.9
Sulfur dioxide, ppm	72	—
NO_x, ppm	88	95
Pounds per ton sinter	0.71	0.66
Pounds per million Btu	0.56	0.56
Temperature, °F	—	131

Source: Reference 2.

to remove very fine particulate matter with particle diameters of 1 μm or less. This usually requires operating the wet scrubber at pressure drops across the scrubber of 30 to 70 inches of water. Since there are large volumes of sinter plant windbox gases, the electric power requirements for these gas cleaning systems are very high. Wet scrubbers are also capable of capturing a significant portion of the condensible hydrocarbons and other fine particulates in the windbox gases.

Much of the water used in the wet scrubbers is recirculated to reduce makeup water requirements and the amount of water that must be blown down from the system. However, the percent recirculation rate will vary from 30% to 80%, depending on the mineral scaling potential of the waters in a given system. Treatment of the scrubber waters requires neutralizing of the acidic waters from the scrubber and adding chemical coagulants to enhance the clarification of the scrubber discharge waters prior to recirculation of the waters to the scrubber.

References

1. W. R. Mura, *An Evaluation of Sinter Plant Air Pollution Control Technology,* Mellon Institute, Pittsburgh, March 1982.
2. L. M. Stuart, Bethlehem Steel Corp., written communication to J. A. Grantz, Armco Steel Co., May 11, 1990.
3. *Compilation of Air Pollutant Emission Factors,* U.S. Environmental Protection Agency, AP-42 October 1986.

Bibliography

D. F. Ball, A. F. Bradley, and A. Grieve, "Environmental control in iron ore sintering," Paper 7, presented at Symposium on Minerals and the Environment, Institution of Mining and Metallurgy, London, England, June 4–7, 1974.

B. H. Carpenter et al., *Pollution Effects of Abnormal Operations in Iron and Steel Making*. Vol. II. *Sintering, Manual of Practice,* EPA-600/2-78-118b, U.S. Environmental Protection Agency, Research Triangle Park, NC, June 1978.

T. A. Cuscino, Jr. *Particulate Emission Factors Applicable to the Iron and Steel Industry,* EPA-450/4-79-028 (NTIS PB81-145914), September 1971.

R. G. Genton, "Steel mill sinter plant," Paper No. 72-81, presented at Annual Meeting of the Air Pollution Control Association, Miami Beach, FL, June 18–22, 1972.

Iron and Steel Industry Particulate Emissions: Source Category Report. EPA-600/7-86-036, October 1984. U.S. Environmental Protection Agency.

The Making, Shaping and Treating of Steel, 9th ed., H. B. McGannon, Ed., U.S. Steel Corp., 1971.

D. A. Pengidore, *Sinter Plant Windbox Gas Recirculation System Demonstration—Phase 1. Engineering and Design,* EPA-600/2-75-014 U.S. Environmental Protection Agency, Research Triangle Park, NC, August 1975.

B. A. Steiner, "Air pollution control in the iron and steel industry," *Int. Metals Rev.*, 21:171 (1976).

B. A. Steiner, "Ferrous metallurgical operations, Chapter 21 in *Air Pollution,* Vol. IV, A. C. Stern, Ed.; Academic Press, New York, 1977.

33 Metal Producing, Vol. 27, No. 5, May 1989.

J. Varga, *Control of Reclamation (Sinter) Plant Emissions Using Electrostatc Precipitators,* EPA-600/2-76-002, U.S. Environmental Protection Agency, Research Triangle Park, NC, January 1976.

PRIMARY LEAD SMELTING

*Paul Deveau, Editor**

Lead is usually found naturally as a sulfide ore containing small amounts of copper, iron, zinc, and other trace elements. It is usually concentrated at the mine from an ore of 3–8% lead to a concentrate of 55–70% lead that can contain up to 30% free and uncombined sulfur. Processing involves three major steps—sintering, reduction, and refining.

Source: AP-42, *Compilation of Air Pollutant Emission Factors,* U.S. Environmental Protection Agency

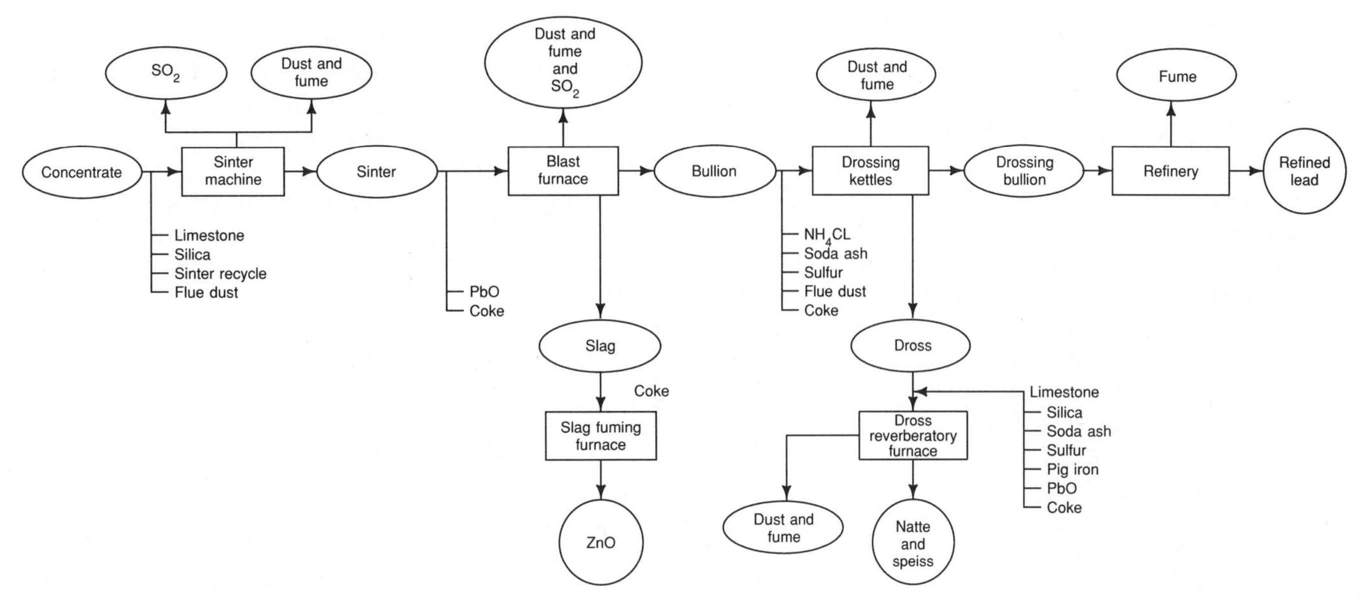

FIGURE 1. Typical Primary Lead Processing Scheme

PROCESS DESCRIPTION

A typical diagram of the production of lead metal from ore concentrate, with particle and gaseous emission sources indicated, is shown in Figure 1.

Sintering

Sinter is produced by a sinter machine, a continuous steel pallet conveyor belt moved by gears and sprockets. Each pallet consists of perforated or slotted grates, beneath which are wind boxes connected to fans to provide a draft, either up or down, through the moving sinter charge. Except for draft direction, all machines are similar in design, construction, and operation.

The primary reactions occurring during the sintering process are autogenous, taking place at approximately 1000°C (1800°F):

$$2PbS + 3O_2 \rightarrow 2PbO + 2SO_2 \quad (1)$$

$$PbS + 2O_2 \rightarrow PbSO_4 \quad (2)$$

Operating experience has shown that system operation and product quality are optimum when the sulfur content of the sinter charge is from 5 to 7 wt %. To maintain this desired sulfur content, sulfide-free fluxes such as silica and limestone, plus large amounts of recycled sinter and smelter residues, are added to the mix. The quality of the product sinter is usually determined by its Ritter Index hardness, which is inversely proportional to the sulfur content. Hard-quality sinter (low sulfur content) is preferred, because it resists crushing during discharge from the sinter machine. Undersized sinter, usually from insufficient desulfurization, is recycled for further processing.

Of the two kinds of sintering machines, the updraft design is superior for many reasons. First, the sinter bed is more permeable (and hence can be larger), thereby permitting a higher production rate than with a down-draft machine of similar dimensions. Second, the small amounts of elemental lead that form during sintering will solidify at their point of formation in updraft machines, but in down-draft operation, they flow down and collect on the grates or at the bottom of the sinter charge, thus causing increased pressure drop and attendant reduced blower capacity. The updraft system also can produce sinter of higher lead content, and it requires less maintenance than the down-draft machine. Finally, and most important from an air pollution control standpoint, updraft sintering can produce a single strong sulfur dioxide (SO_2) effluent stream from the operation by the use of weak gas recirculation. This permits more efficient and more economical use of control methods, such as sulfuric-acid–recovery devices.

Reduction

Lead reduction is carried out in a blast furnace, which basically is a water-jacketed shaft furnace supported by a refractory base. Tuyeres, through which combustion air is admitted under pressure, are located near the bottom and are evenly spaced on either side of the furnace.

The furnace is charged with a mixture of sinter (80–90% of charge), metallurgical coke (8–14% of charge), and various recycled and cleanup materials. In the furnace, the sinter is reduced to lead bullion by reactions 3 through 7.

$$C + O_2 \rightarrow CO_2 \quad (3)$$

$$C + CO_2 \rightarrow 2CO \quad (4)$$

$$PbO + CO \rightarrow Pb + CO_2 \quad (5)$$

$$2PbO + PbS \rightarrow 3Pb + SO_2 \quad (6)$$

$$PbSO_4 + PbS \rightarrow 2Pb + 2SO_2 \quad (7)$$

The carbon monoxide and heat required for reduction are supplied by the combustion of coke. Most of the impurities are eliminated in the slag. Solid products from the blast furnace generally separate into four layers: speiss (the lightest material, basically arsenic and antimony), matte (copper sulfide and other metal sulfides), slag (primarily silicates), and lead bullion. The first three layers are called slag, which is continually collected from the furnace and is either processed at the smelter for its metal content or shipped to treatment facilities.

Sulfur oxides are also generated in blast furnaces from small quantities of residual lead sulfide and lead sulfates in the sinter feed. The quantity of these emissions is a function not only of the sinter's residual sulfur content, but also of the sulfur captured by copper and other impurities in the slag.

Rough lead bullion from the blast furnace usually requires preliminary treatment (drossing) in kettles before undergoing refining operations. First, the bullion is cooled to 370–430°C (700–800°F). Copper and small amounts of sulfur, arsenic, antimony, and nickel collect on the surface as a dross and are removed from the solution. This dross, in turn, is treated in a reverberatory furnace to concentrate the copper and other metal impurities before being routed to copper smelters for their eventual recovery. To enhance copper removal, drossed lead bullion is treated by adding sulfur-bearing material, zinc, and/or aluminum, lowering the copper content to approximately 0.01%.

Refining

The third and final phase in smelting, the refining of the bullion in cast iron kettles, occurs in five steps:

- Removal of antimony, tin, and arsenic
- Removal of precious metals by Parke's process, in which zinc combines with gold and silver to form an insoluble intermetallic at operating temperatures
- Vacuum removal of zinc
- Removal of bismuth by the Betterson process, which is the addition of calcium and magnesium to form an insoluble compound with the bismuth that is skimmed from the kettle
- Removal of remaining traces of metal impurities by the addition of NaOH and $NaNO_3$

The final refined lead, commonly from 99.990% to 99.999% pure, is then cast into 45-kg (10-pound) pigs for shipment.

AIR EMISSIONS CHARACTERIZATION

Each of the three major lead smelting process steps generates substantial quantities of SO_2 and/or particulate. Nearly 85% of the sulfur present in the lead ore concentrate is eliminated in the sintering operation. In handling process off-gases, either a single weak stream is taken from the machine hood at less than 2% SO_2, or two streams are taken, a strong stream (5–7% SO_2) from the feed end of the machine and a weak stream (less than 0.5% SO_2) from the discharge end. Single-stream operation has been used where there is little or no market for recovered sulfur, so that the uncontrolled weak SO_2 stream is emitted to the atmosphere. When sulfur removal is required, however, dual-stream operation is preferred. The strong stream is sent to a sulfuric acid plant and the weak stream is vented to the atmosphere after the removal of particulate.

When dual-gas-stream operation is used with updraft sinter machines, the weak gas stream can be recirculated through the bed to mix with the strong gas stream, resulting in a single stream with an SO_2 concentration of about 6%. This technique decreases machine production capacity, but it does permit a more convenient and economical recovery

TABLE 1. Uncontrolled Emission Factors for Primary Lead Smelting[a]
(Emission Factor Rating: B)

	Particulate		Sulfur Dioxide	
Process	kg/Mg	lb/ton	kg/Mg	lb/ton
Ore crushing[b]	1.0	2.0	—	—
Sintering (updraft)[c]	106.5	213.0	275.0	550.0
Blast furnace[d]	180.5	361.0	22.5	45.0
Dross reverberatory furnace[e]	10.0	20.0	Neg	Neg
Materials handling[f]	2.5	5.0	—	—

[a]Based on quantity of lead produced. Dash = no data. Neg = negligible.
[b]Reference 2. Based on quantity of ore crushed. Estimated from similar nonferrous metals processing.
[c]References 1, 5–7.
[d]References 1, 2, 8.
[e]Reference 2.
[f]Reference 2. Based on quantity of materials handled.

of the SO_2 by sulfuric acid plants and other control methods.

Without weak gas recirculation, the end portion of the sinter machine acts as a cooling zone for the sinter and, consequently, assists in the reduction of dust formation during product discharge and screening. However, when recirculation is used, sinter is usually discharged at 400°C to 500°C (745°F to 950°F), with an attendant increase in particulate. Methods to reduce these dust quantities include recirculating off-gases through the sinter bed (to use the bed as a filter) or ducting gases from the sinter machine discharge through a particulate collection device and then to the atmosphere. Because reaction activity has ceased in the discharge area, these gases contain little SO_2.

Particulate emissions from sinter machines range from 5% to 20% of the concentrated ore feed. In terms of product weight, a typical emission is estimated to be 106.5 kg/Mg (213 pounds per ton) of lead produced. This value, and other particulate and SO_2 factors, appear in Table 1.

Typical material balances from domestic lead smelters indicate that about 15% of the sulfur in the ore concentrate fed to the sinter machine is eliminated in the blast furnace. However, only half of this amount, about 7% of the total sulfur in the ore, is emitted as SO_2. The remainder is captured by the slag. The concentration of this SO_2 stream can vary from 1.4 to 7.2 g/m^3 (500 to 2500 ppm) by volume, depending on the amount of dilution air injected to oxidize the carbon monoxide and to cool the stream before baghouse particulate removal.

Particulate emissions from blast furnaces contain many different kinds of material, including a range of lead oxides, quartz, limestone, iron pyrites, iron–lime–silicate slag, arsenic, and other metallic compounds associated with lead ores. These particles readily agglomerate and are primarily submicron in size, difficult to wet, and cohesive. They will bridge and arch in hoppers. On average, this dust loading is quite substantial, as is shown in Table 1. Minor quantities of particulates are generated by ore-crushing and materials-handling operations, and these emission factors are also presented in Table 1.

Table 2 and Figure 2 present size-specific emission factors for the controlled emissions from a primary lead blast furnace. No other size distribution data can be located for point sources within a primary lead processing plant. Tables 3 through 7 and Figures 3 through 7 present size-specific emission factors for the fugitive emissions generated at a primary lead processing plant. The size distribution of fugitive emissions at a primary lead processing plant is fairly uniform, with approximately 79% of these emissions at less than 2.5 μm. Fugitive emissions less than 0.625 μm in size make up approximately half of all fugitive emissions, except from the sinter machine, where they constitute about 73%.

Emission factors for total fugitive particulate from primary lead smelting processes are presented in Table 8. The factors are based on a combination of engineering estimates, test data from plants currently operating, and test data from plants no longer operating. The values should be used with caution, because of the reported difficulty in accurately measuring the source emission rates.

TABLE 2. Lead Emission Factors and Particle Size Distribution for Baghouse-Controlled Blast Furnace Flue Gases *(Emission Factor Rating: C)*

Particle Size,[a] μm	Cumulative Mass, % ≤ Stated Size	Cumulative Emission Factors	
		kg/Mg	lb/ton
15	98	1.17	2.34
10	86.3	1.03	2.06
6	71.8	0.86	1.72
2.5	56.7	0.68	1.36
1.25	54.1	0.65	1.29
1.00	53.6	0.64	1.28
0.625	52.9	0.63	1.27
Total	100.0	1.20	2.39

[a]Expressed as aerodynamic equivalent diameter.
Source: Reference 9.

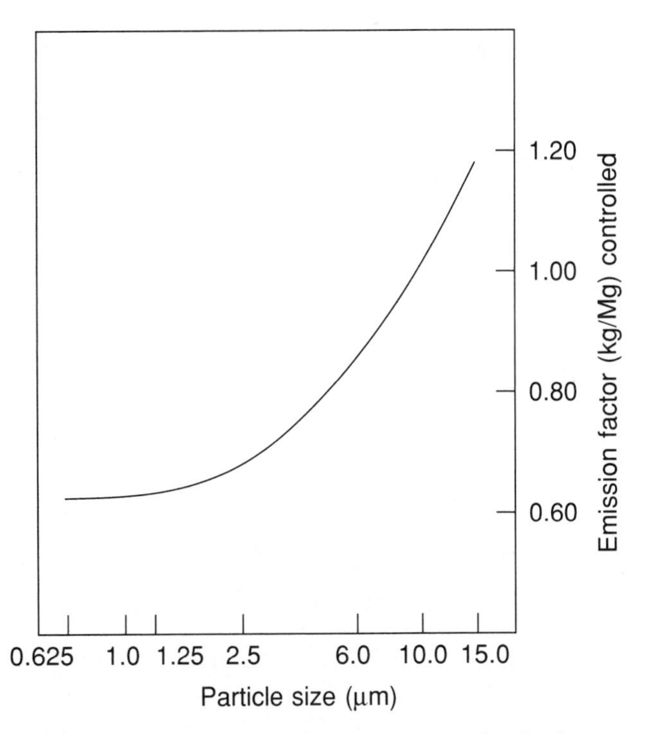

FIGURE 2. Size-Specific Emission Factors for Baghouse-Controlled Blast Furnace

TABLE 3. Uncontrolled Fugitive Emission Factors and Particle Size Distribution for Lead Ore Storage *(Emission Factor Rating: D)*

Particle Size,[a] μm	Cumulative Mass, % ≤ Stated Size	Cumulative Emission Factors kg/Mg	lb/ton
15	91	0.011	0.023
10	86	0.010	0.021
6	80.5	0.010	0.020
2.5	69.0	0.009	0.017
1.25	61.0	0.008	0.015
1.00	59.0	0.007	0.015
0.625	54.5	0.007	0.013
Total	100.0	0.012	0.025

[a]Expressed as aerodynamic equivalent diameter.
Source: Reference 10.

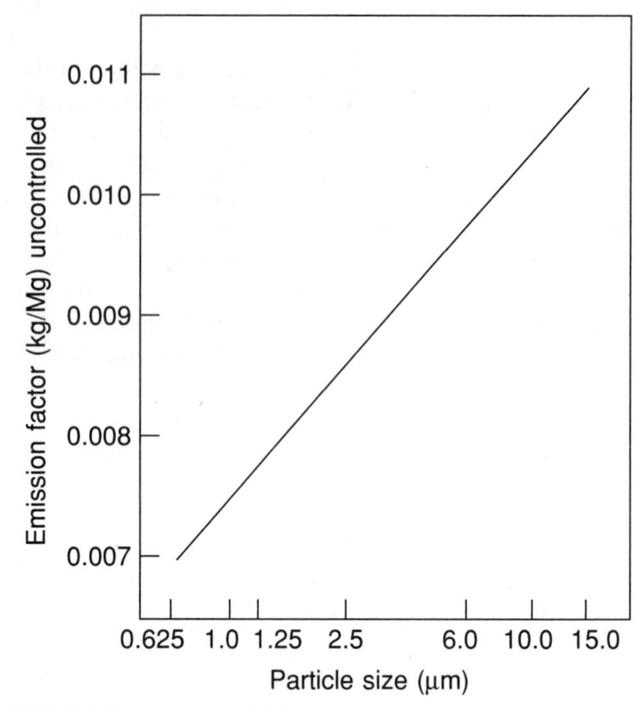

FIGURE 3. Size-Specific Uncontrolled Fugitive Emission Factors for Lead Ore Storage

TABLE 4. Uncontrolled Lead Fugitive Emission Factors and Particle Size Distribution for Sinter Machine *(Emission Factor Rating: D)*

Particle Size,[a] μm	Cumulative Mass, % ≤ Stated Size	Cumulative Emission Factors kg/Mg	lb/ton
15	99	0.10	0.19
10	98	0.10	0.19
6	94.1	0.09	0.17
2.5	87.3	0.08	0.16
1.25	81.1	0.07	0.15
1.00	78.4	0.07	0.15
0.625	73.2	0.07	0.14
Total	100.0	0.10	0.19

[a]Expressed as aerodynamic equivalent diameter.
Source: Reference 10.

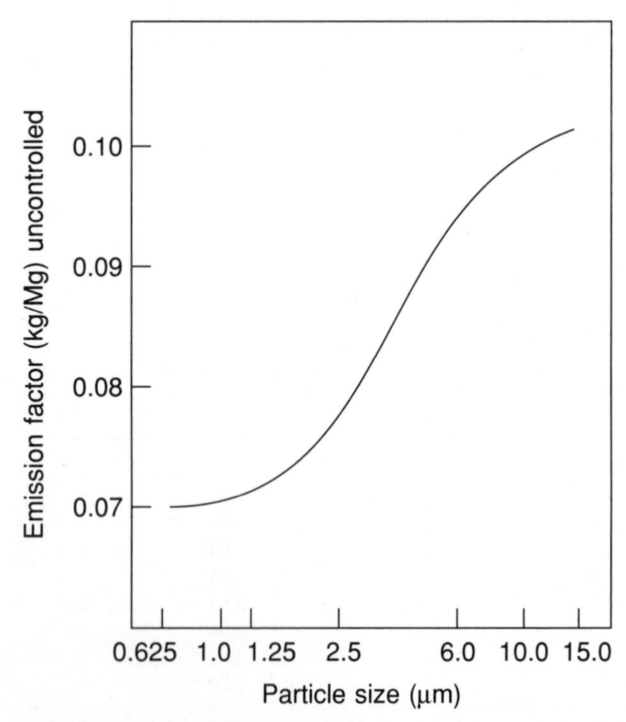

FIGURE 4. Size-Specific Fugitive Emission Factors for Uncontrolled Sinter Machine

TABLE 5. Uncontrolled Lead Fugitive Emission Factors and Particle Size Distribution for Blast Furnace *(Emission Factor rating: D)*

Particle Size,[a] μm	Cumulative Mass, % ≤ Stated Size	Cumulative Emission Factors kg/Mg	lb/ton
15	94	0.11	0.23
10	89	0.11	0.21
6	83.5	0.10	0.20
2.5	73.8	0.09	0.17
1.25	65.0	0.08	0.15
1.00	61.8	0.07	0.15
0.625	54.4	0.06	0.13
Total	100.0	0.12	0.24

[a]Expressed as aerodynamic equivalent diameter.
Source: Reference 10.

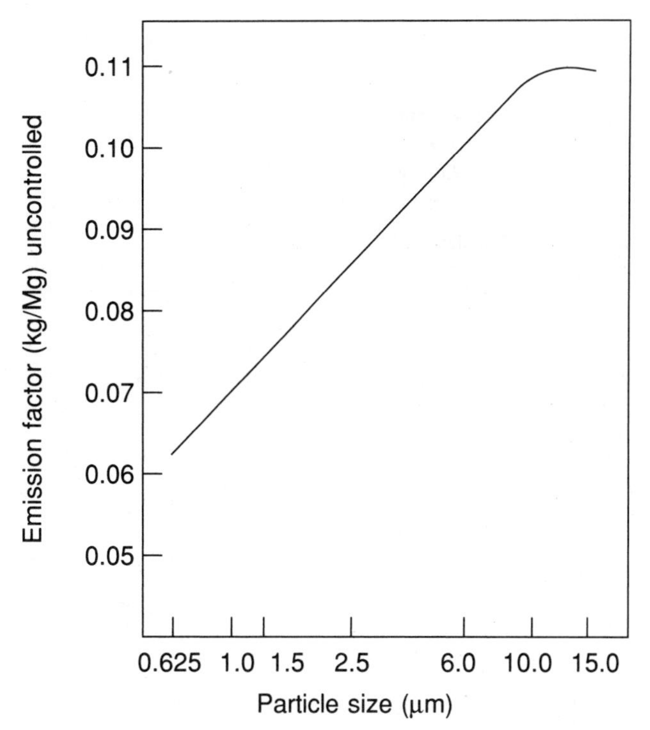

FIGURE 5. Size-Specific Lead Fugitive Emission Factors for Uncontrolled Blast Furnace

TABLE 6. Uncontrolled Lead Fugitive Emission Factors and Particle Size Distribution for Dross Kettle *(Emission Factor Rating: D)*

Particle Size,[a] μm	Cumulative Mass, % ≤ Stated Size	Cumulative Emission Factors kg/Mg	lb/ton
15	99	0.18	0.36
10	98	0.18	0.35
6	92.5	0.17	0.33
2.5	83.3	0.15	0.30
1.25	71.3	0.13	0.26
1.00	66.0	0.12	0.24
0.625	51.0	0.09	0.18
Total	100.0	0.18	0.36

[a]Expressed as aerodynamic equivalent diameter.
Source: Reference 10.

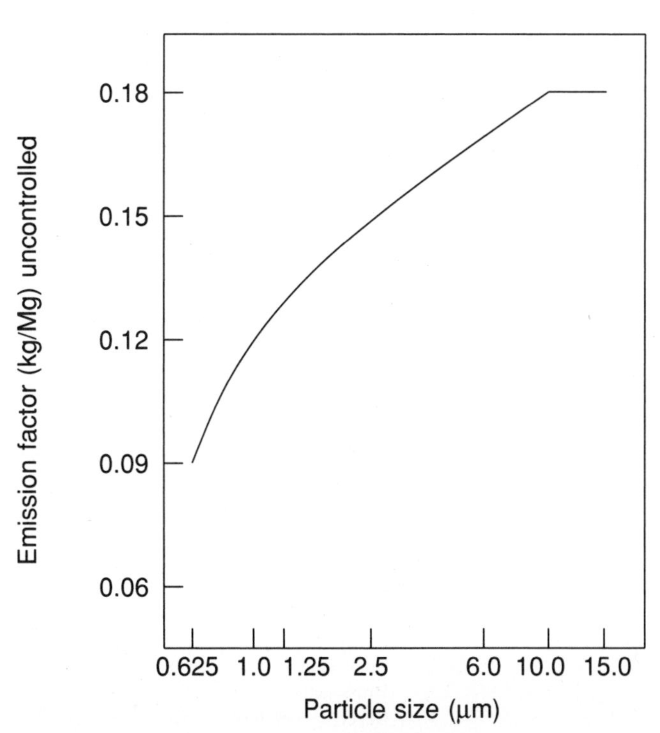

FIGURE 6. Size-Specific Lead Fugitive Emission Factors for Uncontrolled Dross Kettle

TABLE 7. Uncontrolled Lead Fugitive Emission Factors and Particle Size Distribution for Reverberating Furnace ***(Emission Factor Rating: D)***

Particle Size,[a] μm	Cumulative Mass, % ≤ Stated Size	Cumulative Emission Factors kg/Mg	lb/ton
15	99	0.24	0.49
10	98	0.24	0.48
6	92.3	0.22	0.45
2.5	80.8	0.20	0.39
1.25	67.5	0.16	0.33
1.00	61.8	0.15	0.30
0.625	49.3	0.12	0.24
Total	100.0	0.24	0.49

[a]Expressed as aerodynamic equivalent diameter.

Source: Reference 10.

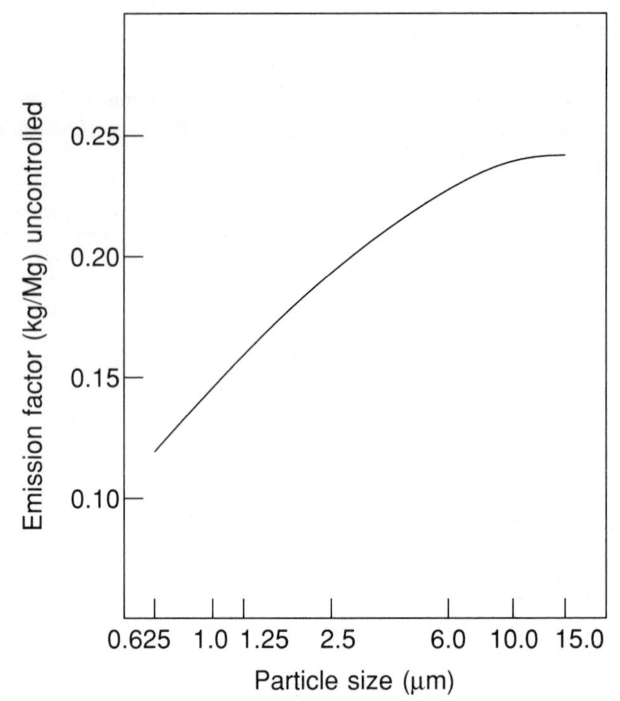

FIGURE 7. Size-Specific Lead Fugitive Emission Factors for Uncontrolled Reverberating Furnace

TABLE 8. Uncontrolled Fugitive Emission Factors for Primary Lead Smelting Processes[a]

Emission Points	Particulate kg/Mg	Particulate lb/ton	Emission Factor Rating
Ore storage[b]	0.012	0.025	D
Ore mixing and pelletizing (crushing)	1.13	2.26	E
Car charging (conveyor loading, transfer) of sinter	0.25	0.50	E
Sinter machine			
Machine leakage[c]	0.34	0.68	E
Sinter return handling	4.50	9.00	E
Machine discharge, sinter crushing, screening[c]	0.75	1.50	E
Sinter transfer to dump area	0.10	0.20	E
Sinter product dump area	0.005	0.01	E
Total building[b]	0.10	0.19	D
Blast furnace			
Lead pouring to ladle, transferring, slag pouring[c]	0.47	0.93	D
Slag cooling[d]	0.24	0.47	E
Zinc fuming furnace vents	2.30	4.60	E
Dross kettle[b]	0.24	0.48	D
Reverberatory furnace leakage[b]	1.50	3.00	D
Silver retort building	0.90	1.80	E
Lead casting	0.44	0.87	E

[a]Expressed in units per end product lead produced, except sinter operations, which are units per sinter handled, transferred, charged.
[b]Reference 10.
[c]References 12, 13. Engineering judgment, using steel sinter machine leakage emission factor.
[d]Reference 2. Engineering judgment, estimated to be half the magnitude of lead pouring and ladling operations.

TABLE 9. Typical Control Device Efficiencies in Primary Lead Smelting Operations

Control Method	Efficiency Range, %	
	Particulate	Sulfur Dioxide
Centrifugal collector[a]	80–90	NA
Electrostatic precipitator[a]	95–99	NA
Fabric filter[a]	95–99	NA
Tubular cooler (associated with waste-heat boiler)[a]	70–80	NA
Sulfuric acid plant (single contact)[b,c]	99.5–99.9	96–97
Sulfuric acid plant (dual contact)[b,d]	99.5–99.9	96–99.9
Elemental-sulfur-recovery plant[b,e]	NA	90
DMA absorption process[b,c]	NA	95–99
Ammonia absorption process[b,f]	NA	92–95

[a]Reference 2. NA = not available.
[b]Reference 1.
[c]High particulate-control efficiency from action of acid plant gas-cleaning systems. With SO_2 inlet concentrations 5–7%, typical outlet emission levels are 5.7 g/m^3 (2000 ppm) for single contact, 1.4 g/m^3 (500 ppm) for dual contact.
[d]Collection efficiency for a two-stage uncontrolled Claus-type plant.
[e]With SO_2 inlet concentrations 4–6%, typical outlet emission levels are from 1.4 to 8.6 g/m^3 (500–3000 ppm).
[f]With SO_2 inlet concentrations of 1.5–2.5%, typical outlet emission level is 3.4 g/m^3 (1200 ppm).

AIR POLLUTION CONTROL MEASURES

Emission controls on lead smelter operations are for particulate and SO_2. The most commonly employed high-efficiency particulate control devices are fabric filters, electrostatic precipitators, which often follow centrifugal collectors and tubular coolers (pseudogravity collectors), and wet scrubbers in limited use for very moist sinter charges.

Three of the six lead smelters currently operating in the United States use single absorption sulfuric acid plants to control SO_2 emissions from sinter machines and, occasionally, from blast furnaces. Single-stage plants can attain sulfur oxide levels of 5.7 g/m^3 (2000 ppm) and dual-stage plants can attain levels of 1.6 g/m^3 (550 ppm). Typical efficiencies of dual-stage sulfuric acid plants in removing sulfur oxides can exceed 99%. Other technically feasible SO_2-control methods are elemental-sulfur-recovery plants and dimethylaniline (DMA) and ammonia absorption processes. These methods and their representative control efficiencies are given in Table 9.

References

1. C. Darvin and F. Porter, *Background Information for New Source Performance Standards: Primary Copper, Zinc and Lead Smelters, Volume I,* EPA-450/2-74-002a, U.S. Environmental Protection Agency, Research Triangle Park, NC, October 1974.
2. A. E. Vandergrift et al., *Particulate Pollutant System Study, Vol. I: Mass Emissions,* APTD-0743, U.S. Environmental Protection Agency, Research Triangle Park, NC, May 1971.
3. A. Worcester and D. H. Beilstein, "The state of the art: Lead recovery," presented at the 10th Annual Meeting of the Metallurgical Society, AIME, New York, March 1971.
4. *Environmental Assessment of the Domestic Primary Copper, Lead and Zinc Industries (Prepublication),* EPA Contract No. 68-03-2537, Pedco Environmental, Cincinnati, OH, October 1978.
5. T. J. Jacobs, visit to St. Joe Minerals Corp. Lead Smelter, Herculaneum, MO, Office of Air Quality Planning and Standards, U.S. Environmental Protection Agency, Research Triangle Park, NC, October 21, 1971.
6. T. J. Jacobs, visit to Amax Lead Company, Boss, MO, Office of Air Quality Planning, and Standards, U.S. Environmental Protection Agency, Research Triangle Park, NC, October 28, 1971.
7. R. B. Paul, American Smelting and Refining Co., Glover, MO, written communication to Regional Administrator, U.S. Environmental Protection Agency, Kansas City, MO, April 3, 1973.
8. *Emission Test No. 72-MM-14,* Office of Air Quality Planning and Standards, U.S. Environmental Protection Agency, Research Triangle Park, NC, May 1972.
9. *Source Sampling Report: Emissions from Lead Smelter at American Smelting and Refining Company, Glover, MO, July 1973,* EMB-73-PLD-1, Office of Air Quality Planning and Standards, U.S. Environmental Protection Agency, Research Triangle Park, NC, August 1974.
10. *Sample Fugitive Lead Emissions from Two Primary Lead Smelters,* EPA-450/3-77-031, U.S. Environmental Protection Agency, Research Triangle Park, NC, October 1977.
11. *Silver Valley/Bunker Hill Smelter Environmental Investigation (Interim Report),* Contract No. 68-02-1343, Pedco Environmental, Durham, NC, February 1975.
12. R. E. Iverson, Meeting with U.S. Environmental Protection Agency and AISI on Steel Facility Emission Factors, Office of Air Quality Planning and Standards, U.S. Environmental Protection Agency, Research Triangle Park, NC, June 1976.
13. G. E. Spreight, "Best practicable means in the iron and steel industry," *Chem. Eng.,* 271: 132–139 (March 1973).
14. *Control Techniques for Lead Air Emissions,* EPA-450/2-77-012, U.S. Environmental Protection Agency, Research Triangle Park, NC, January 1978.

ZINC SMELTING

*Philippe Krick, Editor**

Zinc is found primarily as the sulfide ore sphalerite (ZnS). Its common coproduct ores are lead and copper. Metal impurities commonly associated with ZnS are cadmium (up to 2%) and minor quantities of germanium, gallium, indium, and thalium. Zinc ores typically contain from 3% to 11% zinc. Some ores containing as little as 2% are recovered. Concentration at the mine brings this to 45–60% zinc, with approximately 31% and 10% combined sulfur and iron respectively.

PROCESS DESCRIPTION

Zinc ores are processed into metallic slab zinc by two basic processes—the electrolytic process and a pyrometallurgical smelting process typical of the primary nonferrous smelting industry. A general diagram of the industry is presented in Figure 1.

Electrolytic processing involves four major steps: roasting, leaching, purification, and electrolysis, details of which follow. Pyrometallurgical processing involves three major steps: roasting (as above), sintering, and retorting.

Roasting

Roasting is a process common to both electrolytic and pyrometallurgical processing. Calcine is produced by the roasting reactions in any one of the three different types of roasters—multiple hearth, suspension, or fluidized bed. Multiple-hearth roasters are the oldest type used in the United States, while fluidized-bed roasters are the most modern. The primary zinc roasting reaction occurs between 850°C and 1000°C (1300°F and 1800°F), depending on the type of roaster used, and is as follows:

$$2ZnS + 3O_2 \rightarrow 2ZnO + 2SO_2 \quad (1)$$

In multiple-hearth roaster, the concentrate is blown through a series of nine or more hearths stacked inside a brick-lined cylindrical column. As the feed concentrate drops through the furnace, it is first dried by the hot gases passing through the hearths and then oxidized to produce calcine. The reactions are slow and can only be sustained by the addition of fuel.

In a suspension roaster, the feed is blown into a combustion chamber very similar to that of a pulverized coal furnace. Additional grinding, beyond that required for a multiple-hearth furnace, is normally needed to assure that heat transfer to the material is sufficiently rapid for the desulfurization and oxidation reactions to occur in the furnace chamber. Hearths at the bottom of the roaster capture the larger particles, which require additional time within the furnace to complete the desulfurization reaction.

In a fluidized-bed roaster, finely ground sulfide concentrates are suspended and oxidized within a pneumatically supported feedstock bed. This achieves the lowest sulfide sulfur content calcine of the three roaster designs, or about 0.3%.

Suspension and fluidized-bed roasters are superior to the multiple hearth for several reasons. Although they emit more controlled particulate, their reaction rates are much

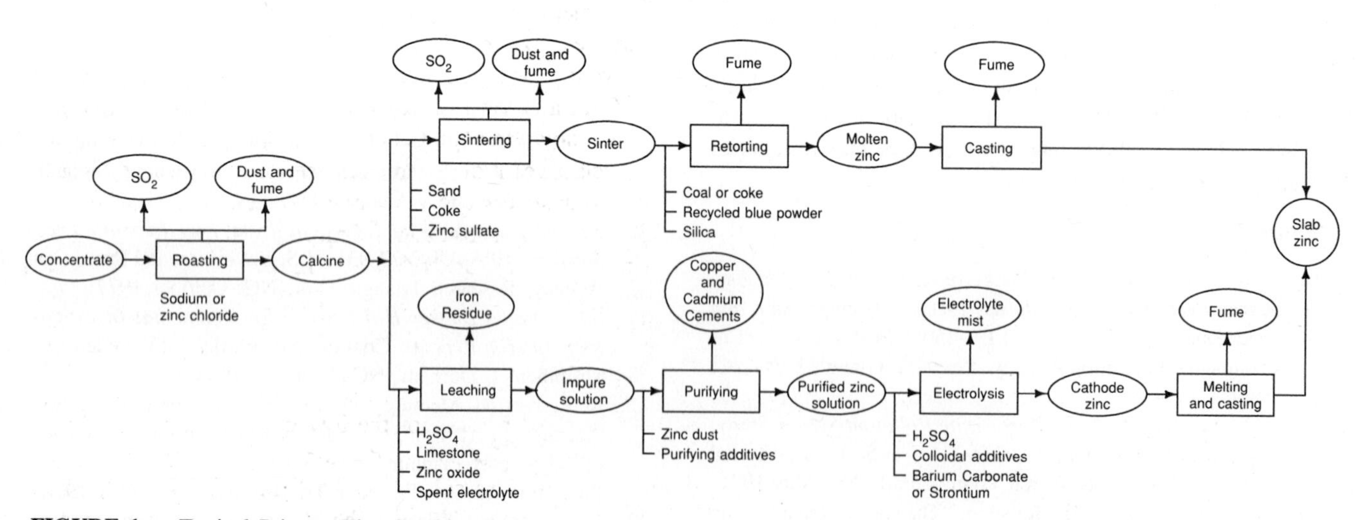

FIGURE 1. Typical Primary Zinc Smelting Process

*Source: AP-42, *Compilation of Air Pollutant Emission Factors*, U.S. Environmental Protection Agency

faster, allowing greater process rates. Also, the sulfur dioxide (SO_2) content of the effluent streams of these two types of roasters is significantly higher, thus permitting more efficient and economical use of acid plants to control SO_2 emissions.

Leaching

Leaching is the first step of the electrolytic reduction, in which the zinc oxide reacts to form aqueous zinc sulfate in an electrolyte solution containing sulfuric acid.

$$ZnO + H_2SO_4 \rightarrow Zn^{+2}(aq) + SO_4^{-2}(aq) + H_2O \quad (2)$$

Single- and double-leach methods can be used, although the former exhibits excessive sulfuric acid losses and zinc recoveries below 90%. In double leaching, the calcine is first leached in a slightly acidic solution. The readily soluble sulfates from the calcine dissolve, but only a portion of the zinc oxide enters the solution. The calcine is then leached in the acidic electrolysis recycle electrolyte. The zinc oxide is dissolved through reaction 2, as are many of the impurities, especially iron, combined as zinc ferrite. The electrolyte is neutralized by this process, and it serves as the leach solution for the first stage of the calcine leaching. This recycling also serves as the first stage of purification, since much of the dissolved iron precipitates out of the solution as either goetite, jarosite, or hematite. Variations on this basic procedure include the use of progressively stronger and hotter acid baths to bring as much of the zinc as possible into solution.

Purification

Purification is a process in which zinc dust is added to the zinc-laden solution to force impurities to precipitate. The solid precipitates are separated from the solution by filtration.

$$\underset{\text{(Zinc dust)}}{Zn^{o} + Cd^{+2}} \xrightarrow{\text{at } 70^{\circ}C} \underset{\text{(Cement)}}{Zn^{+2} + Cd^{o}} \quad (3)$$

The techniques used are among the most advanced industrial applications of inorganic solution chemistry. Processes vary from smelter to smelter, and the details are proprietary and often patented. Metallic impurities, such as arsenic, antimony, cobalt, germanium, nickel, and thallium, interfere severely with the electrolyte deposition of zinc, and their final concentrations are limited to less than 0.05 mg/L (4×10^{-7} lb/gal). Cadmium is recovered as metal and copper as a 60% copper cake.

Electrolysis

Electrolysis takes place in cells containing a number of closely spaced rectangular metal plates acting as anodes (made of lead with 0.75–1.0% silver) and as cathodes (made of pure aluminum). A series of three major reactions occurs within the electrolysis cells:

$$2H_2O \xrightarrow[\text{Anode}]{H_2SO_4} 4H^{+}(aq) + 4e^{-} + O_2 \quad (4)$$

$$2Zn^{+2} + 4e^{-} \xrightarrow{\text{Cathode}} 2Zn \quad (5)$$

$$4H^{+}(aq) + 2SO_4^{-2}(aq) \longrightarrow 2H_2SO_4 \quad (6)$$

Oxygen gas is released at the anode, metallic zinc is deposited at the cathode, and sulfuric acid is regenerated within the electrolyte. The current efficiency is 92% and the specific energy consumption some 3000 kWh per ton.

Electrolytic zinc smelters contain a large number of cells, often several hundred. A portion of the electric energy released in these cells dissipates as heat. The electrolyte is continuously circulated through cooling towers, both to lower its temperature and to evaporate water. Periodically, each cell is shut down and the zinc is removed from the plates.

The final stage of electrolytic zinc smelting is the melting and casting of the cathode zinc into small slabs, 27 kg (60 pounds), or large slabs, 640–1100 kg (1400 to 2400 pounds).

Sintering

Sintering is the first stage of the pyrometallurgical reduction of zinc oxide to slab zinc. Sintering removes lead and cadmium impurities by volatilization and produces an agglomerated permeable mass suitable for feed to retorting furnaces. Down-draft sintering machines of the Dwight-Lloyd type are used in the industry. Grate pallets are joined to form a continuous conveyor system. Combustion air is drawn down through the grate pallets and is exhausted to a particulate control system. The feed is a mixture of calcine, recycled sinter, and coke or coal fuel. The low-boiling-point oxides of lead and cadmium are volatilized from the sinter bed and are recovered in the particulate control system.

Retorting

In retorting, because of the low boiling point of metallic zinc, 906°C (1663°F), reduction and purification of zinc bearing minerals can be accomplished to a greater extent than with most minerals. The sintered zinc oxide feed is brought into a high-temperature reducing atmosphere of 900–1500°C (1650–2600°F). Under these conditions, the zinc oxide is simultaneously reduced and volatilized to gaseous zinc:

$$ZnO + CO \rightarrow Zn(vapor) + CO_2 \quad (7)$$

Carbon monoxide regeneration also occurs.

$$CO_2 + C \rightarrow 2CO \quad (8)$$

The zinc vapor and carbon monoxide produced pass from the main furnace to a condenser, for zinc recovery by bubbling through a molten zinc bath.

Retorting furnaces can be heated either externally by combustion flames or internally by electric resistance heating. The latter approach, electrothermic reduction, has greater thermal efficiency than do external heating methods. In a retort furnace, preheated coke and sinter, silica, and miscellaneous zinc-bearing materials are fed continuously into the top of the furnace. Feed coke serves as the principal electrical conductor, producing heat, and it also provides the carbon monoxide required for zinc oxide reduction. Further purification steps can be performed on the molten metal collected in the condenser. The molten zinc finally is cast into small slabs, 27 kg (60 pounds), or large slabs, 640–1000 kg (1400 to 2400 pounds).

AIR EMISSIONS CHARACTERIZATION

Each of the two zinc smelting processes generates emissions along the various process steps. Although the electrolytic reduction process emits less particulate than does pyrometallurgical reduction, small quantities of acid mist are generated during electrolysis that are exhausted directly to the atmosphere or scrubbed out as in Vielle-Montagne–type cell houses.

Nearly 99% of the potential SO_2 generated from zinc ores is released in roasters. Concentrations of SO_2 in the exhaust gases vary with the roaster type, but they are sufficiently high to allow recovery in an acid plant. Typical SO_2 concentrations for multiple-hearth, suspension, and fluidized-bed roasters are 4.5–6.5%, 10–13%, and 7–12% respectively. Additional SO_2 is emitted from the sinter plant, the quantity depending on the sulfur content of the calcine feedstock. The SO_2 concentration of sinter plant exhaust gases ranges from 0.1% to 2.4%. No sulfur controls are used on this exhaust stream. Extensive desulfurization before electrothermic retorting results in practically no SO_2 emissions from these devices.

The majority of particulate emissions in the primary zinc smelting industry are generated in the roaster area. Depending on the type of roaster used, emissions range from 3.6% to 70% of the concentrate feed. When expressed in terms of zinc production, emissions are estimated to be 133 kg/Mg (266 pounds per ton) for a multiple-hearth roaster and 1000 kg/Mg (2000 pounds per ton) for a fluidized-bed roaster, expressed in terms of zinc production. Particulate emission controls are required for the economical operation of a roaster, with cyclones and electrostatic precipitators (ESP) the primary methods used. No data are available for controlled particulate emissions from a roasting plant.

Controlled and uncontrolled emission factors for point sources within a zinc smelting plant appear in Table 1.

TABLE 1. Particulate Emission Factors for Primary Slab Zinc Processing[a]

Process	Uncontrolled		Emission Factor	Controlled		Emission Factor
	kg/Mg	lb/ton	Rating	kg/Mg	lb/ton	Rating
Roasting						
Multiple hearth[b]	113	227	E	—	—	
Suspension[c]	1000	2000	E	4	8	E
Fluidized bed[d]	1083	2167	E	—	—	
Sinter plant						
Uncontrolled[e]	62.5	125	E	—	—	
With cyclone[f]	NA	NA		24.1	48.2	D
With cyclone and ESP[f]	NA	NA		8.25	16.5	D
Vertical retort[g]	7.15	14.3	D	—	—	
Electric retort[h]	10.0	20.0	E	—	—	
Electrolytic process[j]	3.3	6.6	E	—	—	

[a]Based on quantity of slab zinc produced. NA = not applicable. Dash = no data.
[b]References 3–5. Averaged from an estimated 10% of feed released as particulate emissions, zinc production rate at 60% of roaster feed rate, and other estimates.
[c]References 3–5. Based on an average 60% of feed released as particulate emission and a zinc production rate at 60% of roaster feed rate. Controlled emissions based on 20% dropout in waste-heat boiler and 99.5% dropout in cyclone and ESP.
[d]References 3,6. Based on an average 65% of feed released as particulate emissions and a zinc production rate of 60% of roaster feed rate.
[e]Reference 3. Based on unspecified industrial source data.
[f]Reference 7. Data not necessarily compatible with uncontrolled emissions.
[g]Reference 7.
[h]Reference 2. Based on unspecified industrial source data.
[j]Reference 13.

TABLE 2. Uncontrolled Fugitive Particulate Emission Factors for Primary Slab Zinc Processing[a] ***(Emission Factor Rating: E)***

Process	Emission Factor[b]	
	kg/Mg	lb/ton
Roasting	Negligible	Negligible
Sinter plant[c]		
Wind box	0.12–0.55	0.24–1.10
Discharge and screens	0.28–1.22	0.56–2.44
Retort building[d]	1.0–2.0	2.0–4.0
Casting[e]	1.26	2.52

[a]Based on quantity of slab zinc produced, except as noted.
[b]Reference 8.
[c]From steel industry operations for which there are emission factors. Based on quantity of sinter produced.
[d]From lead industry operations.
[e]From copper industry operations.

Sinter plant emission factors should be applied carefully, because the data source is different from the only plant currently in operation in the United States, although the technology is identical. Additional data have been obtained for a vertical retort, although no examples of this type of plant are operating in the United States. Particulate factors also have been developed for uncontrolled emissions from an electric retort and the electrolytic process.

Fugitive emission factors have been estimated for the zinc smelting industry and are presented in Table 2. These emission factors are based on similar operations in the steel, lead, and copper industries.

AIR POLLUTION CONTROL MEASURES

Control measures for zinc smelting operations are similar to those used in copper and lead smelting.

References

1. V. A. Cammerota, Jr., "Mineral facts and problems: 1980," *Zinc,* Bureau of Mines, U.S. Department of Interior, Washington, DC, 1980.
2. *Environmental Assessment of the Domestic Primary Copper, Lead and Zinc Industries,* EPA-600/2-82-066, U.S. Environmental Protection Agency, Cincinnati, OH, October 1978.
3. *Particulate Pollutant System Study,* Vol. I: *Mass Emissions,* APTD-0743, U.S. Environmental Protection Agency, Research Triangle Park, NC, May 1971.
4. G. Sallee, personal communication anent Reference 3, Midwest Research Institute, Kansas City, MO, June 1970.
5. *Systems Study for Control of Emissions in the Primary Nonferrous Smelting Industry,* Vol. I, APTD-1280, U.S. Environmental Protection Agency, Research Triangle Park, NC, June 1969.
6. *Encyclopedia of Chemical Technology,* John Wiley & Sons, New York, 1967.
7. R. B. Jacko and D. W. Nevendorf, "Trace metal emission test results from a number of industrial and municipal point sources," *J. APCA,* 27(10):989–994 (1977).
8. *Technical Guidance for Control of Industrial Process Fugitive Particulate Emissions,* EPA-450/3-77-010, U.S. Environmental Protection Agency, Research Triangle Park, NC, March 1977.
9. L. J. Duncan and E. L. Keitz, "Hazardous particulate pollution from typical operations in the primary non-ferrous smelting industry," presented at the 67th Annual Meeting of the Air Pollution Control Association, Denver, CO, June 9–13, 1974.
10. *Environmental Assessment Data Systems,* FPEIS Test Series No. 3, U.S. Environmental Protection Agency, Research Triangle Park, NC.
11. *Environmental Assessment Data Systems,* FPEIS Test Series No. 44, U.S. Environmental Protection Agency, Research Triangle Park, NC.
12. R. E. Lund, et al., "Josephtown electrothermic zinc smelter of St. Joe Minerals Corporation," *Proceedings AIME Symposium on Lead and Zinc,* Vol. II, 1970.
13. *Background Information for New Source Performance Standards: Primary Copper, Lead and Zinc Smelters,* EPA-450/2-74-002a, U.S. Environmental Protection Agency, Research Triangle Park, NC, October 1974.

SECONDARY ALUMINUM

Charles A. Licht, P.E.

The reclaiming and recycling of aluminum to a usable form and alloy are termed secondary aluminum processing.

PROCESS DESCRIPTION

Previously, secondary aluminum operations revolved around the sweating of aluminum in scrap yards where scrap was melted down, separated from high-melting-point materials, and poured into ingot or sows. This type of operation still goes on. Foundry-type alloys have been manufactured in facilities that utilize both the sows and ingot generated in scrap yards and also scrap coming from a great number of other sources, including obsolete or postconsumer scrap and new or manufactured scrap. Additionally, over the years, techniques have been developed for the processing of the slags and skimmings (drosses) from various melting operations, which include skimmings from primary facilities, foundries and die casters, and those generated in secondary aluminum facilities. (see Figure 1.)

For many years, the term secondary aluminum had pejorative overtones, fostered in part by the primary aluminum producers who had great surpluses of primary capacity and "talked down" the recycling of aluminum in order to keep their potlines going. These issues are now historic since not only is aluminum remelted by the casting alloy manufacturers, but the casting alloy producers must now vie with primary companies in the purchase of scrap materials, which are reprocessed into sheet and extrusion form by

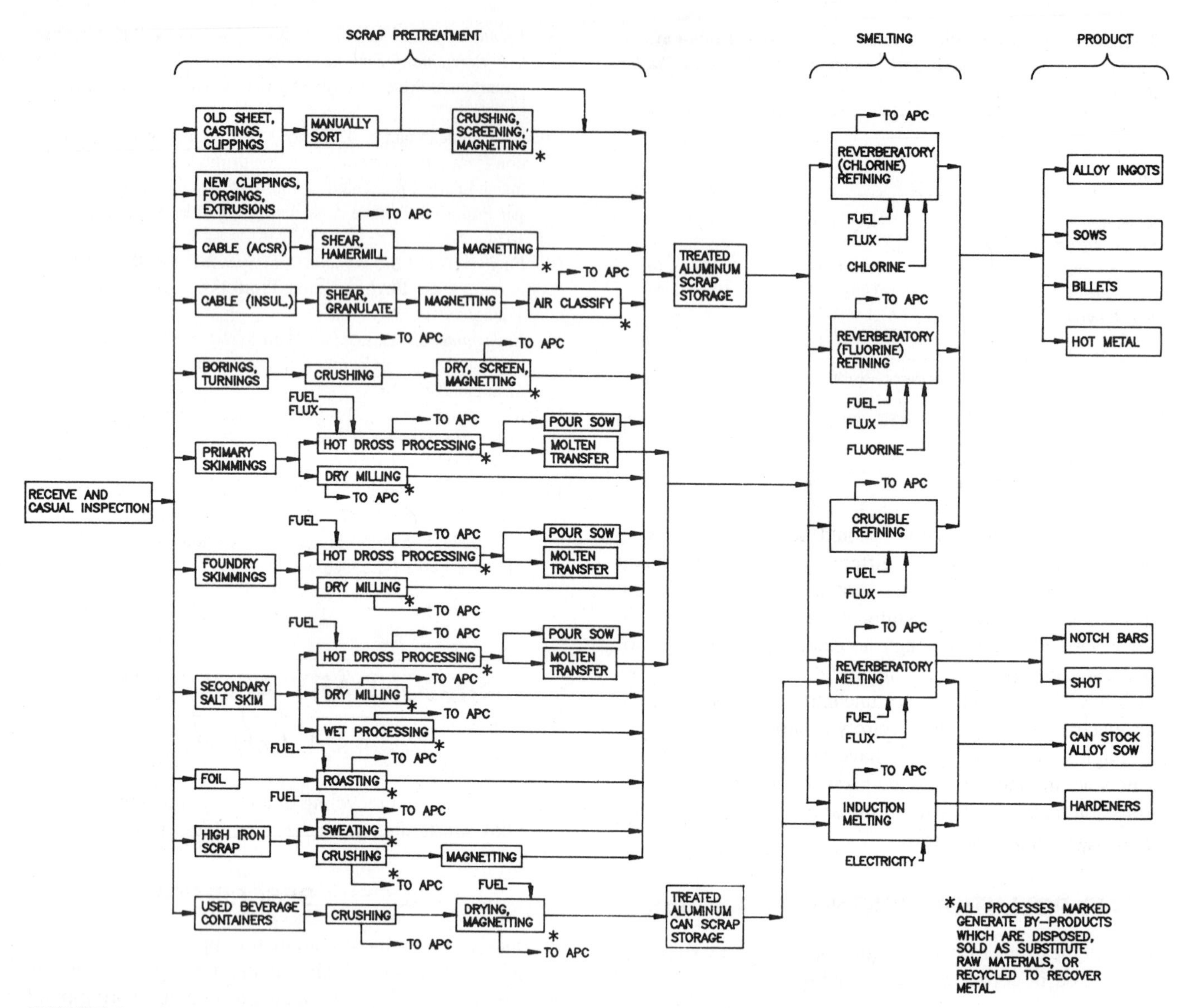

FIGURE 1. Secondary Aluminum Processing

independent sheet rollers and extruders and the primary producers.

The foundry industry takes a vast percentage of the alloys produced by the secondary aluminum smelters. The secondary aluminum smelters, by virtue of both market forces and availability, generally take lower-purity and dirtier materials to recycle into casting alloys. In part, this is due to the broader specifications available in the casting alloy spectrum. The presence of iron, silicon, and other elements makes certain materials impossible to extrude or roll. They do not create the same problem when making castings. Metals, such as silicon and iron, tend to make the materials brittle. Materials that are going to be extruded or rolled cannot have brittleness, but must be ductile and tough.

A number of terms must be defined. These include:

1. *Alloying*. Alloying is a process wherein a molten bath is modified by the addition of alloying elements to bring the alloy into specification. In a typical foundry alloy, the addition of silicon, copper, and certain other minor elements is usually required.
2. *Degassing*. Degassing is a process wherein various gases and/or fluxes are utilized to remove absorbed gases from the molten metal. This is required since the alloyed molten metal will contain much more gas than is possible to be held when the metal solidifies; thus if the gases are not removed from the molten metal as they solidify, holes will develop. This porosity leads to inferior castings and makes it impossible to roll or extrude salable products.
3. *Demagging*. Essentially one of the very few techniques that is available for refining aluminum, demagging al-

lows the removal of magnesium from the molten bath by reacting the bath with either a fluoride salt or chlorine.

4. *Fluxing*. This term covers a very broad array of processes. In melting aluminum scrap in open charging well furnaces (to be described later), a salt–potash–fluoride flux is used on the open surface of the bath to eliminate the formation of aluminum oxides and to cause the agglomeration of small beads of aluminum that will form in the oxide particles, thus allowing these particles to develop sufficient mass to flow into the bath. Additionally, certain fluxes are used in reverberatory furnaces that do not have charging wells. Basically, these fluxes are used to remove the buildup of oxides from the walls or to eliminate and/or reduce oxidation. Fluxes are also used in the degassing and demagging processes.

RAW MATERIALS

Among the major sources of aluminum materials available to the secondary casting alloy producer are manufactured scrap in the form of chips and turnings.

Also, to a lesser extent, clippings or skeletons from stamping operations are available, as well as cutoffs from heavier sections from extrusion operation. The availability of this portion of the scrap market to the secondary alloyers is minimal because of the high cost of these materials, which are coopted by primary and secondary sheet rollers and extruders.

Scrap is obtained from the automobile shredding process after the nonferrous portions are separated from the ferrous portions. Ultimately, the nonferrous portions themselves are split so that fairly high-quality aluminum scrap can be obtained.

Another source of aluminum scrap is aluminum copper radiators used in many heat exchanger applications, which are a source not only of aluminum, but of inexpensive copper. Another very large source is obsolete scrap. This can be in the form of old sheet and siding, painted and unpainted. Occasionally, old sheet and siding carry insulating materials. Old castings from a variety of consumer products, such as automatic transmission housings and engines from automobiles, lawn chairs, lawn mower housings and motors, outboard motors, aluminum doors and windows, and aluminum screens, are also available.

Often, die casters and foundries will have a secondary smelter process for their gates and risers and will return specification alloy ingot to the foundry. Many times, this type of material is sold outright to the smelters.

Used beverage containers (UBCs) represent a very significant volume of metallics. Because the material is essentially one family of alloy, the optimum utilization of this material is for it to be recycled into can sheet stock. The process for handling the cans is similar to other processes in the aluminum cycle in that the cans are generally shredded, run through a delacquering system, which is equivalent to a chip dryer, and then processed in either rotary furnaces or specially equipped reverberatory furnaces. Some can scrap is processed as received and melted in rotary furnaces.

The benefit of direct melting of can scrap in the rotary furnace lies in the fact that the residual coatings are generally burned in the furnace. Aside from that, rotary salt furnaces generally have adequate air pollution controls so that if even partially burned carbonaceous material is generated, it is collected in the baghouse. The use of delacquered cans allows for open-well or induction furnace operation without the need for substantial air pollution controls.

One area that is usually ignored lies in the various chlorine-bearing compounds that the interior coatings and the sealants contain. There is hydrochloric gas emission potential from melting can scrap, which often is ignored, but will eventually require control as toxic air regulations come into place.

FURNACES

Typically, one does not see small crucible or pot furnaces used in the secondary aluminum industry. These furnaces are primarily used today in small sand and permanent mold foundries and to a degree as holding units at die casting machines. The major units used to melt aluminum scrap in secondary alloy facilities are open-well charging furnaces. These furnaces typically have capacities of 30,000 to 300,000 pounds. They are set up to operate so that melting occurs for the most part in the open charging well, while the closed combustion chamber functions to introduce heat. These furnaces can range up to 15 feet to 20 feet wide by 15 feet to 30 feet long, in addition to which a charging well, 5 feet to 6 feet wide by the length of the furnace, is utilized.

This type of furnace allows for easy degassing, demagging, fluxing, and so on, since all of these activities can occur in the outside well, which can be appropriately hooded for the removal of any fumes that are generated.

Reverberatory Furnace

Most of the primary companies that are remelting scrap (both purchased and home) and most of the independent extruders and sheet rollers are utilizing furnaces into which the scrap materials are charged directly into the combustion zone. With the typical clean scrap that these companies must utilize, because of alloy considerations, the materials can be melted without excessive oxidation. This type of furnace, however, is not amenable to the charging of such material as auto shredder scrap, turnings, and chips or light-gauge materials, unless in the form of tightly rolled coils.

Typically, these furnaces are gas fired, although some oil firing occurs because of the lack of gas availability. The benefits of gas firing lie in the fact that the particulate and

SO_x products of combustion of the gas firing are considerably less than those of oil firing, with there being essentially no ash or sulfur from the gas-fired units.

The typical process starts with heating of the furnace to an appropriate temperature (about 1400°F) and the charging into the combustion chamber of a substantial amount of heavy section scrap. This can be sow, ingot, heavy castings, or plate. Generally, this will represent 10–20% of the furnace capacity. When this material is molten, additional scrap will be charged into the charging well of the open-well furnaces with puddling and/or circulation by means of various types of equipment so that the hot molten bath inside the furnace combustion zone can be circulated into the charging well to melt the scrap being charged.

The ideal technique for charging scrap into the charging well is by means of a mechanism such as a vibrating feeder or conveyor so that the scrap is charged at a steady rate. By utilizing mechanical charging techniques, it is possible for the air pollution control hoods to be fitted closely to the furnace to allow minimal potential loss of smoke and fume from the charging well. This type of furnace also has the benefit of not having any "dirty" material introduced into the combustion chamber, and thus the emissions from the combustion chamber will primarily be products of combustion (POCs) of fuels with only minor traces of particulate from the bath. In the case of direct-charged reverberatory furnaces, substantially more scrap will be charged initially and will be allowed to melt down. When molten, the furnace may be skimmed and an additional charge placed in the furnace. Depending on the equipment available for doing the charging and the size of the furnace, charges of 10,000 to 60,000 pounds can be made at a time. It must be noted that the direct-charge furnace will be more thermally efficient than the charging well furnace. This is due in part to the omission of the use of salt fluxes. Substantial amounts of flux are used in the melting process on the charging well that require added thermal input to heat and melt.

While the fluxes require energy, because they are very effective in removing the aluminum content of the oxide formations, the amount of aluminum that is skimmed from the furnace is substantially reduced as compared with that when the skimming is done in a direct-charge reverberatory furnace without fluxing.

After a proportion of the furnace bath is melted, the bath is stirred and sampled. The analysis is compared with the precalculated analysis at that stage. If the analysis is within reasonable limits of the precalculations, then the previously allotted scrap will be added to the furnace. If it is not, an opportunity exists rapidly to recalculate and change the analysis of the material to be charged. By doing this, the chances of having a full furnace come in out of "spec" are dramatically reduced.

In order to assist in the melting process, a variety of techniques are utilized. Pumps made of extremely high temperature components are used to circulate the bath. These pumps can also be used with appropriate fitting to do degassing and/or demagging with chlorine. Additionally, mechanical puddling devices are commercially available that can be used both to puddle the scrap in the open charging wells and to "work out" the skimmings and mechanically skim them from the furnace.

The open charging well furnace, because the charging well is controlled by an air pollution control system, can take advantage of considerably "dirtier" scrap, that is, dirtier in the sense of having combustible materials on the surface such as paint, oil, or grease. The "dirt" from this material, if charged into a direct-charge furnace, would be partially combusted in the combustion zone. However, substantial amounts of unburned material and fine particulate would be exhausted, thus requiring air pollution control on the high-temperature exhaust stream, which is expensive as compared with the relatively low temperatures that can be utilized in controlling the smoke and fume from a charging well furnace.

As the batch proceeds, metallic silicon and/or copper are added as required to meet the final specifications. If a pump is not used, then, toward the end of the batch, chlorine and/or aluminum fluoride are introduced to the bath. If a pump is used, chlorine can be introduced at a low input level over the last half of the heat cycle. This can be done only in an open charging well furnace, which is one of the reasons why the secondary casting alloy producers utilize this type of furnace almost exclusively. Magnesium is removed by reaction of the magnesium with chlorine. This process is very efficient in equipment such as the pump devices mentioned above, and also in the chlorination bell systems that are used in a few plants. When the alloy is in specification, it can be handled in several ways. It can be poured into large 15,000- to 40,000-pound ladles, which can be trucked to foundries and die casters. Or it can be tapped into transfer ladles to be taken to another location for pouring into sow or ingot or tapped via troughs into sow and/or ingot pouring systems adjacent to the furnace.

Rotary Furnaces

Rotary furnaces are used to melt drosses and can scrap (UBC). They are usually directly fired with gas or oil. The fumes and POC are vented into baghouses to control fumes. Salt or salt/potash flux is usually used. The exhausts generally contain salt fumes, oxides, and carbonaceous material. If cans are melted, the exhaust may also contain hydrochloric acid gas.

European technology uses rotary furnaces for making alloy. Typical European rotary furnaces are substantially bigger than U.S. equipment.

FLUXES

Generally fluxes fall into four categories, as follows:

1. *Cover fluxes*. These fluxes prevent oxidation of the molten bath and cause the agglomeration of metal droplets to form larger pieces, which then sink back into the bath. These are used on open well furnaces only.
2. *Solvent fluxes*. These fluxes generally cause suspended particles and oxides to float to the top of the bath to form a dross. This dross can be moved out into the charging well for fluxing down with a cover flux prior to skimming, or in the case of direct charge reverbatory furnaces, the oxide flux mixture can be skimmed from the furnace and cooled.
3. *Degassing fluxes*. Generally, these fluxes are used to remove soluble gases, such as hydrogen, from the molten aluminum. If the bath is to be demagged using chlorine, this is generally sufficient to do an adequate job, although other fluxes, including a number of proprietary mixtures, are used for degassing purposes.
4. *Magnesium reducing fluxes*. Basically, there are two materials that are used for the removal of magnesium—chlorine and aluminum fluoride. The melting point of aluminum fluoride is substantially higher than that of the bath, and thus reactions can be difficult to achieve. When aluminum fluoride is used to remove magnesium from a bath, it is generally mixed with the "smelter's flux" (47 ½% calcium chloride, 47½% sodium chloride, 5% fluoride salt such as cryolite) to reduce the melting point.

Although the literature is replete with descriptions of a great variety of flux materials used both for cover, solvent, and degassing, most secondary aluminum alloy producers use "smelter's" flux for their cover and depend on chlorine or chlorine and nitrogen mixtures for their solvent and degassing fluxes.

Uncontrolled utilization of chlorine can generate enormous volumes of visible air pollutants. Chlorine will react preferentially with magnesium to form magnesium chloride, which, at bath operating temperatures, is a barely molten salt. The chlorine will also react with aluminum to form aluminum chloride, which is a gas above 368°F (187°C). When cooled, it will react with atmospheric moisture to form aluminum oxide and hydrochloric acid gas, causing an acrid visible plume.

With the use of demagging metal pumps, the flow of chlorine into the system can be sustained at a level low enough that the reaction process takes place without the formation of visible aluminum chloride. The use of a chlorination bell restrains the aluminum chloride under the bell and allows it to vent through a scrubbing section so that there is no loss of aluminum chloride to the environment.

A patented (Derham) process for magnesium removal makes use of a pump that pumps metal from the charging well into a secondary chamber, where a thick molten-salt flux bath is maintained with molten aluminum below it. Chlorine is introduced into the chamber well below the bath surface. The magnesium chloride and aluminum chloride are absorbed in the salt bed above the aluminum. The aluminum underflows and returns to the furnace, where it is reheated and recirculated until magnesium specification levels are obtained.

AIR EMISSIONS CHARACTERIZATION

In the secondary alloy producer's operations particularly, but even, to some extent, in sheet rolling and extrusion facilities, scrap being remelted has surface coatings. The obvious paint, dirt, oil, and grease are present, but even when melting new die casting or home scrap, parting agents can be on the surface of the casting. In the case of extrusions and sheet rolling, lubricants and coolants remain on the surface. When remelted, the surface coating will cause the formation of visible emissions on the charging well.

TABLE 1. Particulate Emission Factors for Secondary Aluminum Operations[a]

	Uncontrolled		Baghouse		Electrostatic Precipitator		Emission Factor
Operation	kg/Mg	lb/ton	kg/Mg	lb/ton	kg/Mg	lb/ton	Rating
Sweating furnace[b]	7.25	14.5	1.65	3.3	—	—	C
Smelting							
Crucible furnace[b]	0.95	1.9	—	—	—	—	C
Reverberatory furnace[c]	2.15	4.3	0.65[e]	1.3[e]	0.65	1.3	B
Chlorine demagging[d]	500	1000	25	50	—	—	B

[a]Emission factors for sweating and smelting furnaces expressed as units per unit weight of metal processed. For chlorine demagging, emission factor is kg/Mg (lb/ton) of chlorine used.

[b]Based on averages of two source tests.

[c]Uncontrolled, based on averages of ten source tests. Standard deviation of uncontrolled emission factor is 1.75 kg/Mg (3.5 lb/ton), that of controlled factor is 0.15 kg/Mg (0.3 lb/ton).

[d]Based on average of 10 source tests. Standard deviation of uncontrolled emission factor is 215 kg/Mg (430 lb/ton); of controlled factor, 18 kg/Mg (36 lb/ton).

[e]This factor may be lower if a coated baghouse is used.

Source: Reference 1

TABLE 2. Particle Size Distributions and Size-Specific Emission Factors for Uncontrolled Reverberatory Furnaces in Secondary Aluminum Operations *(Size-Specific Emission Factor Rating: D)*

Aerodynamic Particle Diameter, μm	Particle Size Distribution[a]		Size-Specific Emission Factor,[b] kg/Mg	
	Chlorine Demagging	Refining	Chlorine Demagging	Refining
2.5	19.8	50.0	99.5	1.08
6.0	36.9	53.4	184.5	1.15
10.0	53.2	60.0	266.0	1.30

[a]Cumulative wt % < aerodynamic particle diameter, μm.

[b]Size-specific emission factor = total particulate emission factor × particle size distribution, %/100. From Table 1, total particulate emission factor for chlorine demagging is 500 kg/Mg chlorine used, and for refining, 2.15 kg/Mg aluminum processed.

Source: Reference 1.

Table 1, taken from AP-42, the U.S. Environmental Protection Agency's *Compilation of Air Pollutant Emission Factors,* presents emission factors for the principal emission sources in secondary aluminum operations. Table 2, also from AP-42, presents particle size distributions and corresponding emission factors for uncontrolled chlorine demagging and metal refining in secondary aluminum reverberatory furnaces.

AIR POLLUTION CONTROL MEASURES

The direct-charge reverberatory furnace is generally charged only with clean scrap, sows, T-bar, or pig, resulting in the exhaust flue being clear and generally well within the allowable process weight rate and opacity limitations. On the other hand, the scrap charged into an open charging well, since lower temperatures will exist and the scrap is generally considerably "dirtier," will generate volumes of smoke and fume. By proper design and application of hoods with appropriate closures to the furnaces, the use of appropriate doors, and the use of charging devices allowing for relatively small continuous additions in order to minimize "puffs" of smoke that are difficult to control, control of large charging wells can be sustained with relatively small amounts of air. There are reverberatory furnace operations, with essentially the same furnace (150,000-pound) capacity, that are controlled with airflows that range from about 22,000 acfm to 45,000 acfm. Through adequate design of the hoods and proper control of the charge, the fumes are successfully controlled.

The use of precoating on the dust collector bags utilizing lime, bentonite, and so on, has been known technology for many decades. The use of continuous additions of appropriate additives is a technology that has worked effectively for the past 20 years. The additives react with active chemicals in the gas stream and collect in the control system. Appropriate reactions between active chlorides with an alkaline can result in the reduction of hydrochloric gas to an innocuous form of particulate that can be gathered in the baghouse. Typically, the particulate collected in secondary aluminum casting alloy facilities that employ continuous additive additions will result in opacities ranging from 0 to 10%. The dusts collected are generally nonhazardous, passing Toxicity Characteristics Leaching Procedure (TCLP) tests.

Prior to the melting step, there are a number of processes in secondary aluminum facilities that have the potential to produce pollution.

CHIP DRYING

The aluminum chips and turnings discussed earlier are a manufactured product generated in the turning, milling, boring, and machining of aluminum. These chips generally contain varying amounts of coolants or lubricants, ranging from 2% to 20% of the inbound weight. Machining operations usually utilize miscible compounds for cooling and lubricating, and thus a substantial proportion of the coolant adhering to the chips is water. However, the oil content can create problems. Additionally, charging water-wet material into a molten bath is to be avoided at any cost.

Because of this potential for problems, many secondary aluminum and some primary aluminum facilities have chip-drying equipment. These systems are generally set up so that a small crusher (hammer mill or ring mill) breaks the particles down to uniform size below ¾ to 1-¼ inches. The materials are then fed into a storage bin and withdrawn from that storage bin by a metering conveyor so that uniform feed can be achieved. The chips are fed through air locks into the drying system, where the chips are heated and the oil and moisture driven off. The chips then leave the system via air locks and cross magnetic separators to remove "free" iron. They are then available for melting. Obviously, the optimum situation is when these hot chips are taken to the furnaces directly.

The fumes from the process are ducted to a mechanical separator, such as a cyclone, a multiple cyclone, or a "dropout box." The fumes are then ducted into fume incinerators, where the combustible components are removed.

There are several dryer designs that are commercially available, although many "standard" designs are custom engineered. The standard dryer has an open flame within the rotary drum. While this would appear to cause oxidation, in fact, only trivial oxidation is noted. This is so because the drum is sealed so that leakage air is minimized and the oxygen content in the drum is only sufficient to sustain a small flame.

Another technique that is used is to heat the drum externally. This has the advantage of requiring substantially smaller pollution control systems since the combustion gases do not need to be controlled. Only those gases generated by the evaporation and/or partial burning of the "moisture" are then drawn off into mechanical separation and fume-burning equipment. The direct-fired unit has a greater throughput for physical size, thus offsetting the somewhat higher cost of operating the pollution control system.

Another system utilizes a "black" drum wherein no flame occurs. The hot gases are recirculated through mechanical separators and an afterburner. A portion of the "burned-out" gas is exhausted. The remaining gas is recirculated as the dryer heat source. This patented concept is used extensively in can delacquering.

SCRAP SHREDDING

Aluminum scrap material, such as painted siding, bus bodies, insulated pieces of all sorts, and a great variety of materials that were processed by "sweating" (to be discussed later), are now being processed by large shredding systems. These systems generally utilize hammer mills and/or ring mills, usually in the range of 60 inches in diameter by 60 inches wide, although a few larger units are in service. The scrap is fed into the system after having been picked over to eliminate oversized materials and to remove nonaluminum components such as zinc, magnesium, stainless steel, and other components.

Depending on the installation, scrap may simply drop into the shredding device or may be fed with an appropriate feeder that controls the feed rate based on motor amperage. Systems range from 400 hp to over 1000 hp, with throughput ranges from 5000 pounds an hour to over 50,000 pounds an hour, depending on the grade of scrap, the amount of picking that has to be done, and so forth. The scrap coming from this process has been dramatically cleaned so that the amount of smoke and fume generated in the charging well is substantially reduced. Additionally, free iron and combustibles that can cause oxidation on the charging well are removed.

The hammer mill "windage" is ventilated to a mechanical separation system, which removes large amounts of particulate loading. In some instances, it is necessary to go to secondary pollution controls to deal with extremely fine particulate from the mill.

SWEAT FURNACE

Mentioned earlier was the "sweating" furnace. These direct-fired furnaces are generally small. They are utilized to separate low-melting-point materials from higher-melting-point materials. In the case of aluminum, sweating is frequently done on such material as lawn mowers and other materials that contain steel components. Occasionally, one sees automatic transmissions with the steel shafts and gearing sweated, although typically these are disassembled. Specific materials often are sweated because of the nature of their configuration. Pistons may be sweated because certain ferrous components are cast into the head of the piston. Running the piston through a hammer mill simply does not break these components free.

DROSS PROCESSING

Aluminum skimmings and drosses produced by primary operations, foundries, and secondary smelters can be processed in several different ways.

A technique that was popular earlier, particularly with the high-salt-content secondary smelter skimmings, was to wash the drosses in large volumes of water. Typically, this was carried on near a large river so that many hundreds of gallons per minute of water could be sluiced through a long drum, generally raising the salt content by 100 to 200 ppm of the water flow going through the system and raising the stream levels substantially less than that. Exiting from the end of the system would be a melange of aluminum oxide and aluminum particles. Some systems screened these materials at 20 mesh, salvaging most of the aluminum particles, which were then dried and melted, while the oxides were allowed to go into settling ponds where they settled out, with the overflow returning to the stream. These ponds had to be dredged periodically and the oxides taken off to landfill.

In the middle 1950s, improvements were developed in dry milling dross to the point where some facilities were capable of processing in excess of a million pounds per day of skimmings and drosses. This process basically involved the primary breaking of large chunks to smaller sizes and then running the chunks through a series of hammer mills and screens, separating out the coarser materials that contained the higher metallic components and melting the concentrates. The finer materials that were below the levels that were readily melted were separated by various air-classification techniques and developed into a particle size adequate for the ferrous exothermic market. The extremely fine materials were sold to cement companies for their alumina content or discarded to landfill.

Processes were installed to wash the fine residual materials to develop a more adequate exothermic product. This was carried out by at least one facility in the middle 1960s, generating a product that was widely sought after by the ferrous exothermic manufacturers. Since 1975, a number of companies have started up that melt skimmings and drosses in rotary furnaces, adding appropriate salt fluxes and separating out the metals. The metals are melted and then decanted. The salt cake is poured into molds and, after cooling, is taken to landfill.

At the present time, a few systems are in place that "wet process" salt cake. The systems wash the salt-bearing materials. The brines are separated, cleaned, and processed to form salt solids for reuse as flux, while the oxides are clean enough to be used for exothermic purposes, if sufficient metal content is present, or can be utilized in other processes because of the low chloride content. What metal values are left in the drosses after washing (generally plus 20 mesh), are remelted, in either rotary or reverbatory furnaces.

Reference

1. *Compilation of Air Pollutant Emission Factors, Volume 1: Stationary Point and Area Sources,* AP-42, U.S. EPA, 1985.

Bibliography

C. L. Brooks, "Melt loss and metal conservation," LM 78-42, Metallurgical Society of AIME, New York, 1978.

Air Pollution Engineering Manual, AP-40, J. A. Danielson, Ed.; U.S. Environmental Protection Agency, Research Triangle Park, NC, 1973, pp 283–292.

R. F. Jones, T. H. Ginsburg, and C. A. Licht, "Secondary aluminum plant particulate control," 76-24.6, presented at 69th Annual Meeting of the Air Pollution Control Association, Portland, 1976.

C. A. Licht, *"Demagging" Aluminum Alloys Using an Innovative Technology,* A 77-76, Metallurgical Society of AIME, New York, 1977.

C. A. Licht, "Energy conservation in a large aluminum melter," *Proceedings of the Third International Aluminum Extrusion Technology Seminary,* Vol. 2, Aluminum Association, 1984, pp 291–297.

A. Pultz, and C. A. Licht, " 'New source' pollution controls at a secondary aluminum plant," 78-61.6, presented at 71st Annual Meeting of Air Pollution Control Association, Houston, TX, 1978.

E. L. Rooy, and J.H.L. VanLinden, "Recycling of aluminum," *Metals Handbook,* Vol. 2, 10th ed., ASM International, 1990, pp 1205–1213.

D. L. Siebert, *Impact of Technology on the Commercial Secondary Aluminum Industry,* IC 8445, U.S. Bureau of Mines, Washington, DC, 1970.

SECONDARY BRASS AND BRONZE MELTING PROCESSES

Charles A. Licht, P.E.

Zinc-alloyed copper is termed brass. Tin-alloyed copper is termed bronze. Other copper alloys are identified by the alloying metals, such as aluminum bronze and silicon bronze. However, some alloys called bronze are actually brass. Manganese bronze is a high-zinc brass. Because of their high strength, workability, corrosion resistance, color, and other desirable physical characteristics, the copper-base alloys have found wide use in hardware, radiator cores, condensers, jewelry, musical instruments, plumbing fittings, electric equipment, ship propellers, and many other applications.

The remelting of nearly pure copper or bronze causes minor air pollution as a result of the high boiling points of the materials involved. With good melting practice, total emissions to the air should not exceed 0.5% of the process weight. The brasses containing 15–40% zinc, however, are poured at temperatures above the boiling points of zinc, so vaporization and combustion of desirable elements, particularly zinc, occur. Emissions into the air may vary from less than 0.5% to 6% or more of the total metal charge. A substantial percentage of the zinc content may be lost through fuming, depending on the composition of the alloy, the type of furnace used, and the melting practice.

PROCESS DESCRIPTION

Furnace Types

Brass and bronze shapes for working, such as slabs and billets, are usually produced in large gas- or oil-fired reverberatory furnaces. Most secondary smelters use stationary or rotary reverberatory furnaces for reclaiming and refining scrap metal, casting the alloyed metal into pigs. Brasses and bronzes used to make commercial castings are often melted in induction furnaces in the larger foundries and in crucible-type, fuel-fired furnaces in the smaller job foundries.

AIR EMISSIONS CHARACTERIZATION

Air contaminants emitted from brass furnaces consist of products of combustion from the fuel and particulate matter in the form of dusts, metallics, and oxide fumes. The particulate matter composing the dust and fume load varies according to the fuel, alloy composition, melting temperature, type of furnace, and other operating factors. The particulate matter is made up of fly ash, carbon, and mechanically produced dust. The furnace emissions also contain fumes resulting from condensation and oxidation of the more volatile elements, including zinc and lead.

Air pollution resulting from the volatilization of metals during the melting of nearly pure copper and bronze is not too serious because of the high-boiling-point temperatures of copper, tin, nickel, aluminum, and even lead commonly used in these alloys. Alloys containing zinc ranging up to 7% can be successfully processed with a minimum of fume emission when an inert slag cover is used. This nominal figure is subject to some variation, depending on the composition of the alloy, temperatures, operation procedures, and other factors.

Copper-base alloys containing 20–40% zinc have low boiling points of approximately 2100°F (1150°C) and melting temperatures of approximately 1700°F (925°C) to 1900°F (1040°C). These zinc-rich alloys are poured at approximately 1900°F (1040°C) to 2000°F (1100°C), which is only slightly below their boiling points. Pure zinc melts at 787°F (420°C) and boils at 1663°F (906°C). Even within the pouring range, therefore, fractions of high-zinc alloys usually boil and flash to zinc oxide. The zinc oxide formed is submicron in size. Its escape to the atmosphere can be prevented only by collecting the fumes and venting through efficient air pollution control equipment.

The difficulty of controlling metallic fumes from brass furnaces lies in the physical characteristics of these fumes. The particle sizes of zinc oxide fumes vary from 0.03 to 0.3

μm. Lead oxide fumes are in this same range of particle sizes. The collection of these very small particles requires high-efficiency control devices. These metallic fumes also produce very opaque emissions, since particles of 0.2 to 0.6 μm in diameter produce maximum light scattering.

In copper-base alloy foundries, as much as 98% of the particulate matter contained in furnace stack gases may be zinc oxide and lead oxide, depending on the composition of the alloy. Constituents of fumes include zinc, lead, tin, copper, cadmium, silicon, and carbon. They are present in varying amounts, depending on the composition of the alloy and the foundry practice.

Investigations prove conclusively that the most troublesome fumes consist of particles of zinc and lead compounds submicron in size, and that air pollution control equipment capable of collecting particulate matter from 1.0 down to about 0.03 μm is required.

Factors Causing Large Concentrations of Zinc Fumes

According to the *Air Pollution Engineering Manual,* 1973 edition,[1] four principal factors cause relatively large concentrations of zinc fumes in brass furnace gases:

1. Alloy composition. The rate of loss of zinc is approximately proportional to the zinc percentage in the alloy.
2. Pouring temperature. For a given percentage of zinc, an increase of 100°F (50°C) increases the rate of loss of zinc about three times.
3. Type of furnace. Direct-fired furnaces produce larger fume concentrations than the crucible type does, other conditions being constant. Operation of direct-fired furnaces will require emmision control equipment.
4. Poor foundry practice. Excessive emissions result from improper combustion, overheating of the charge, addition of zinc at maximum furnace temperature, flame impingement on the metal charged, heating of the metal charged, heating the metal too fast, and insufficient flux cover. Excessive superheating of the molten metal is to be avoided for metallurgical and economic reasons, as well as pollution control reasons. From an air pollution viewpoint, the early addition of zinc is preferable to gross additions at maximum furnace temperatures.

In any fuel-fired furnace, the internal atmosphere is of prime importance since there exists a constant flow of combustion gases through the melting chamber in contact with the metal. A reducing atmosphere is undesirable from both the metallurgical and air pollution viewpoints. With too little oxygen, the metal is exposed to a reducing atmosphere of unburned fuel and water vapor, which usually results in gassy metal. Incomplete combustion, especially with oil firing, produces carbon soot. To prevent these difficulties, the atmosphere should be slightly oxidizing. Excess oxygen content should be greater than 0.1%, otherwise castings will be affected by gas porosity. Excess oxygen content must be less than 0.5% to prevent excessive metal oxidation. The need for close control of the internal furnace atmosphere requires careful regulation of the fuel and air input and frequent checking of the combustion gases.

Crucible Furnace—Pit and Tilt Type

Indirect-fired crucible furnaces are used extensively in foundries requiring small- and medium-sized melts. The lift-out type of crucible is frequently employed in small furnaces. According to LAAPC data, tests have demonstrated that, with careful practice and use of slag covers, the crucible furnace is capable of low-fume operation on red brasses but not on yellow brass.

The slag cover, consisting mainly of silicates, is not used as a refining flux, but as an inert, cohesive slag of sufficient thickness to keep the molten metal covered. If the quantity of slag is carefully controlled, minimum emissions result. Before any metal is added to the crucible, the flux should be added so that, as melting takes place, a cover is formed to keep the molten metal separated from the atmosphere.

Electric Furnace—Low-Frequency Induction Type

The induction-type furnace has a number of desirable characteristics for melting brasses. The heating is rapid and uniform and the metal temperature can be accurately controlled. Contamination from combustion gases is eliminated. High-frequency induction furnaces are well adapted to copper- and nickel-rich alloys, but are not widely used for zinc-rich alloys. During melting of clean metal, use of a suitable flux cover over the metal prevents excessive fuming, except during back-charging and pouring phases of the heat. The usual flux covers—borax, soda ash, and others—are destructive to furnace walls. Charcoal is used with satisfactory results.

Cupola Furnace

The cupola furnace is used for reduction of copper-base alloy slag and residues. The residues charged have a recoverable metallic content of 25% to 30%. The balance of the unrecoverable material consists of nonvolatile gangue, mainly silicates. In addition to the residues, coke and flux are charged to the furnace. Periodically, the recovered metal is tapped from the furnace. The slag produced in the cupola is eliminated through a slag tap located slightly above the metal tap.

In addition to the usual metallic fumes, the cupola also discharges carbon particles and fly ash. Collection of these emissions is required at the cupola stack, the charge door, and the metal tap spout. With no control equipment, emissions of 60–100% opacity can be expected from the charge door and stack. The opacity of the fumes emitted at the tap varies from 60% to 80%.

The slag discharged from the cupola is rich in zinc oxide. It does not volatilize since its boiling point is 3590°F (1975°C). The discharge slag is usually cooled by water. The emissions from the slag-tapping operation have low opacity.

HOODING AND VENTILATION REQUIREMENTS

Regardless of the efficiency of the control device, air pollution control is not complete unless all the fumes generated by the furnace are captured. Since different problems are encountered with the various types of furnaces, each will be discussed separately.

Reverberatory Furnace

Stationary Type

In an open-hearth reverberatory furnace, the products of combustion and metallic fumes are normally vented directly from the furnace through a cooling device to a baghouse. Auxiliary hoods are required over the charge door, rabble (or slag) door, and taphole. These may be vented to the baghouse serving the furnace and help cool the hot combustion gases by dilution.

The furnace burners should be turned down or off during periods when the furnace is open for charging, rabbling, lancing, removing slag, or adding or pouring metal. Otherwise, the exhaust system may not have sufficient capacity to handle the products of combustion and the additional air required to capture the fumes. Since no two of these operations occur simultaneously, the required air volume for collection may be reduced by the use of properly placed dampers within the exhaust system.

Well-designed hoods, properly located, with an indraft velocity of up to 200 fpm, adequately capture furnace emissions. If the hood is placed too high for complete capture or is improperly shaped and poorly fitted, higher indraft velocities are required.

The rabble or slag door permits (1) mixing the charge, (2) removing slag from the metal surface, and (3) lancing to refine the metal. Emissions from the furnace may be of 50–90% opacity during these operations, even with the burners partially throttled. Again, up to 200 fpm indraft velocity is recommended for properly designed hoods.

Generally, after the slag has been removed, metal, usually zinc, must be added to bring the brass within specifications. The furnace metal is at a temperature well above the boiling point of zinc and is no longer covered by the tenacious slag cover. Hence, voluminous emissions of zinc oxide result. The addition of slab zinc can produce 100% opaque fumes in great quantity, while a brass addition may generate fumes of 50% opacity. A well-designed hood is required over the charge door or rabble door through which the metal is charged.

Perhaps the most critical operation from the standpoint of air pollution occurs when the furnace is tapped. Nearly continuous emissions of 90–100% opacity may be expected. Much planning is required to design a hood that completely captures the emissions and yet permits sufficient working room and visibility of the molten metal. Again, the burners should be turned off or throttled as much as possible to reduce the quantity of fumes emitted.

The fluxes used in reverberatory furnaces normally present no air pollution problems. Generally, only nonvolatile fluxes such as borax, soda ash, and iron oxide mill scale are used.

Rotary Type

Cylindrical-type reverberatory furnaces present all the collection problems of the open-hearth type with the additional complication of furnace rotation. The cylindrical furnace may be rotated up to 90° for charging, slag removing, and metal tapping. With hoods installed in a fixed position, the source of emissions may be several feet from the hood, and thus no fumes would be collected. Either a hood attached to the furnace and venting to the control device through flexible ductwork or an oversized close-fitting hood covering all possible locations of the emission source is required. A close-fitting hood and high indraft velocities are often necessary.

A cylindrical furnace rotates on its longitudinal axis. A tight breeching is mandatory at the gas discharge end of the furnace. Adequate indraft velocity must be maintained through the breeching connection to prevent the escape of fumes.

The exhaust system for the cylindrical furnace, as well as for all types of reverberatory furnaces, must be designed to handle the products of combustion at the maximum fuel rate. Any lesser capacity results in a positive pressure within the furnace during periods of maximum firing, with resultant emissions from all furnace openings.

Tilting Type

The tilting-type furnace differs from the reverberatory furnaces previously discussed in that the exhaust stack is an integral part of the furnace and rotates with the furnace during charging, skimming, and pouring. One type of tilting furnace is charged through the stack, and skimming and pouring are accomplished through a small taphole in the side of the furnace. Another type has a closable charge door and a small port through which the furnace gases escape. These two furnace openings may describe a full 180° of arc during the various phases of a heat.

Rotary Tilting Type

The rotary-tilting-type furnace not only tilts for charging and pouring, but rotates during the melting period to improve heat transfer. Two types are common. One is charged through the burner end and is poured from the exhaust port of the furnace, opposite the burner. The other has a side charge door at the center of the furnace through which charging, slagging, and pouring operations are conducted.

Because of the various movements of this type of furnace, direct connection to the control device is not feasible. The furnace is under positive pressure throughout the heat, and fumes are emitted through all furnace openings.

Hooding a rotary-tilting-type reverberatory furnace for complete capture of fumes is difficult, and complete collection is seldom achieved. These furnaces are undoubtedly the most difficult type of brass furnace to control. To hood them effectively requires a comprehensive design. The major source of emissions occurs at the furnace discharge. Capture of fumes is accomplished by a hood closely fitted.

Hooding is sometimes installed at the burner end of a furnace to capture emissions that may escape from openings during melting or particularly during the time the furnace is tilted to pour. Because both ends of the furnace are open, a venting action is created during the pour, causing fume emissions to be discharged from the elevated end of the furnace. Close hooding is not practicable because the operator must observe the conditions within the furnace through the open ends. An overhead canopy hood is usually installed.

Additional heavy emissions may be expected during charging, alloying, and slagging. High overhead canopy hoods are generally used. These overhead hoods are, however, unsatisfactory unless they cover a large area, and a high indraft velocity is provided along with very large control air volumes.

The need for numerous hoods and large air volumes, with the resultant larger control device, makes the tilting-type, open-flame furnace expensive to control. This type of furnace is being gradually replaced by more easily controlled equipment.

Crucible-Type Furnaces

Tilting-type crucible furnaces can be controlled by an exhaust system to control emissions during pouring. An exhaust system may vent several furnaces to a baghouse. The hooding collects all the fumes during pouring without interfering with the furnace operation in any way. The hood can be equipped with a damper that is closed when the furnace is in the normal firing position. A linkage system can open the damper when pouring begins. After the furnace is partially tilted, the damper is fully opened, remaining open for the rest of the pour. It swings shut automatically when the furnace is returned to the firing position. The ductwork leading from the hood pivots when the furnace is tilted. The entire hood is fixed to the furnace with a few bolts, which permit rapid removal for periodic repairs to the furnace lining and crucible.

Emissions resulting from the pouring of molten metal from a ladle into molds can be controlled by one of three

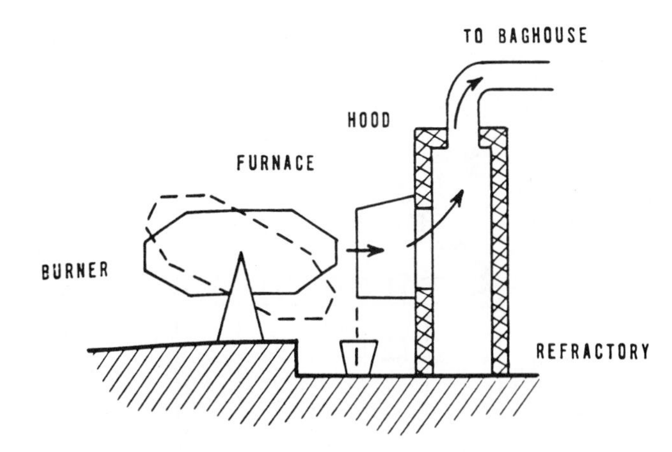

FIGURE 1. Rotary-Tilting-Type Brass Melting Furnace

devices. The first is a fixed pouring station that is hooded so that the emissions from the ladle and molds are captured during the pouring (Figure 1). A second solution, for smaller foundries, is a hood attached to the pouring ladle and vented to the control system through flexible ductwork. One variation places a small baghouse on the crane holding the ladle so that the ladle is vented over the whole pouring floor. Another variation is a stationary baghouse and sufficient flexible ductwork to allow the hood to travel from mold to mold.

A third solution employs a hood attached to the bale, a short section of flexible ductwork, and two ducts—one mounted on the bridge crane and one attached to the stationary crane rails. The ducts are interconnected with a transfer box, which is sealed by a continuous loop of rubberized belt. This allows complete freedom of movement for the ladle within the area served by the overhead crane.

Low-Frequency Induction Furnace

The control of the emissions from an induction furnace is much more expensive and difficult if only turnings are charged to the furnace. In addition to the fumes common to brass melting, great clouds of 100% opacity black smoke are generated when the oily shavings contact the molten heel within the furnace. Adequate hooding enclosing the furnace is required, and a large volume of air is necessary to capture the smoke and fumes.

Cupola Furnace

An exhaust system to control a cupola must have sufficient capacity to remove the products of combustion, collect the emissions from the metal tap spout, and provide a minimum indraft velocity of 250 fpm through the charge door. In addition, side curtains may be required around the charge door to shield adverse crosscurrents. A canopy hood is recommended for the metal tap spout. The air requirement for this hood is a function of its size and proximity to the source of emissions.

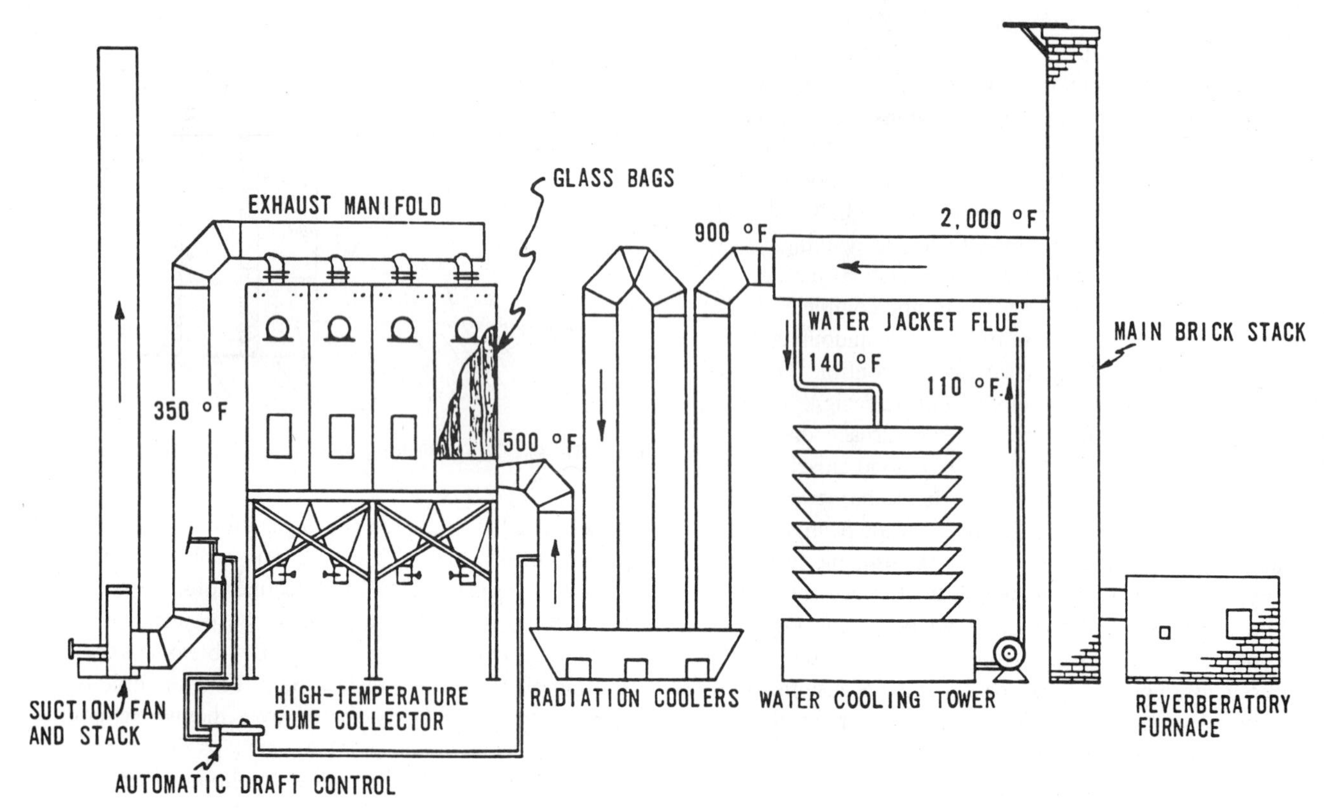

FIGURE 2. Sketch of Small Baghouse for Zinc Fume (Allen et al., 1952).

AIR POLLUTION CONTROL MEASURES

Baghouses

Baghouses with tubular bag filters are used to control the emissions from brass furnaces. This type of collector is available in many useful and effective forms. Wool, cotton, and synthetic filter media effectively separate submicron-sized particulate matter from gases because of the filtering action of the "mat" of particles previously collected.

The gases leaving a reverberatory furnace may be 100°F to 200°F hotter than the molten metal and must be cooled before reaching the filter cloth. Direct cooling, by spraying water into the hot combustion gases, is practiced, with maximum water content being controlled to less than 10% by weight. Some plants prefer not to use water because (1) there is increased corrosion of the ductwork and collection equipment; (2) the water vapor increases the exhaust gas volume, necessitating a correspondingly larger baghouse; and (3) the temperature of the gases in the baghouse must be kept above the dew point to prevent condensation of water on the bags. The exhaust gases may be cooled by dilution with cold air, but this increases the size of the control equipment and the operating costs of the exhaust system.

One type of cooling system employed consists of a water-jacketed cooler followed by air-cooled radiation–convection columns, as shown in Figure 2. The radiation–convection coolers reduce the temperature to protect the fabric of the filter medium. Figure 2 depicts an installation showing the cooling columns and baghouse.

Treated polyester or Nomex$^{(R)}$ has replaced glass cloth as the most favored high-temperature fabric. Although glass bags withstand higher temperatures, the periodic shaking of the bags gradually breaks the glass fibers and results in higher maintenance costs.

Probably the most critical design factor for a tubular baghouse is the filtering velocity. A filtering velocity of 2.5 fpm is recommended for collecting the fumes from brass furnaces with relatively small concentrations of fume. Larger concentrations of fume require a lower filtering velocity. A higher filtering velocity requires more frequent shaking to maintain a pressure drop through the baghouse within reasonable limits. Excessive bag wear results from frequent shaking and higher filtering velocities. A pressure drop of 2 to 5 inches of water column is normal, and high-pressure differentials across the bags are to be avoided.

The baghouse should be completely enclosed to protect the bags from inclement weather and water condensation. The exhaust fan should be placed downstream from the baghouse to prevent blade abrasion and imbalance. Broken bags are easily detected when the exhaust system discharges to the atmosphere through one opening.

Electrostatic Precipitators

Generally, electrostatic precipitators are extremely effective collectors for many substances in any size range from 200 mesh (74 μm) to perhaps 0.001 μm, wet or dry, ambient or up to 1200°F (650°C). This equipment has not, however, proved satisfactory on lead and zinc fumes. High-voltage precipitators have not been available in the small units suitable for small nonferrous foundry use, and the first cost is usually prohibitive.

Scrubbers

Scrubbers or mechanical washers have proved in some applications to be effective from 10 to 1 μm. However, in addition to being ineffective in the submicron range, they have the disadvantage of high power consumption and mechanical wear and usually require separation of the metallic fumes and other particulate matter from the circulating water.

References

1. *Air Pollution Engineering Manual, AP-40;* J. A. Danielson, Ed.; U.S. Environmental Protection Agency, Research Triangle Park, NC, 1973, pp 269–283.

Bibliography

D. V. Neff and R. F. Schmidt, "Recycling of copper," *Metals Handbook,* Vol. 2, 10th ed., ASM International, 1990, pp 1213–1216.

IRON FOUNDRIES

*American Foundrymen's Society Air Quality Committee (10-E)**

Control of the air pollution that results from iron foundries may be considered according to the methods of melting, the handling of sand, the types of molten metals and other materials, and the cleaning of finished castings. The air pollutant characteristics are affected by a number of factors, including the type of melting unit, material-handling and hooding systems, and emissions control systems. The air pollution problem then becomes one of capturing the smoke, dust, and fumes at the furnace and other sources and transporting these contaminants to suitable control devices.

PROCESS DESCRIPTION

Mold and Core Production

Molds are forms used to shape the exterior of castings. The green sand mold, the most common type, uses moist sand mixed with 3–20% clay and 2–5% water, depending on the process. Additives to prevent casting defects include organic material, such as seacoal (a pulverized high-volatility, low-sulfur bituminous coal), wood or corn flour, oat hulls, or similar organic matter.

Cores are molded sand shapes used to form the internal voids in castings. They are made by mixing sand with various binders, shaping it into a core, and curing the core with a variety of processes.

The Melting Process

Electric Furnace (General)

In the electric furnace, the basic process operations are (1) furnace charging, in which metal, scrap, alloys, carbon, and flux are added to the furnace; (2) melting, during which the furnace remains closed; (3) back-charging, which involves the addition of more metal and alloys; (4) refining and treating, during which the chemical composition is adjusted to meet product specifications; (5) slag removal; and (6) tapping molten metal in a ladle or directly into molds.

Induction Furnaces

Electric induction furnaces are either horizontal or vertical, cylindrical, refractory-lined vessels. Heating and melting occur when the charge is energized with a low-, medium- or high-frequency alternating current. Induction furnaces also may be used for holding and superheating.

Electric induction furnaces generally have lower emissions per ton of metal melted than the other furnace types. As a result, in spite of a generally lower unit capacity, induction furnaces have supplanted cupolas in many foundries.

Electric Arc Furnaces

Electric arc melting furnaces are large, welded-steel cylindrical vessels equipped with a removable roof through which three carbon electrodes are inserted. The electrodes are energized by three-phase alternating current, creating arcs that melt the metallic charge material. Additional heat is generated by the electrical resistance of the metal to the current between the arc paths. The most common method of charging an arc furnace is by removing the roof and introducing the charge material directly. Alternatives include charging through a roof chute or side charging door. Once the melting cycle is complete, the metal is tapped by tilting the furnace and pouring the metal into a ladle.

Cupola

The cupola is a vertical, cylindrical shaft furnace and may use pig iron, scrap iron, scrap steel, and coke as the charge components. The mechanism by which melting is accom-

*Charles E. Brown, *Chairman;* Charles H. Borcherding, CIH, CSP, *Vice Chairman;* William A. Baker, *Secretary;* J. Peter Aldred, Charles M. Davis, LeRoy E. Euvrard, Jr., JD, Tom Godbey, Jeff Haworth, William B. Huelsen, Walter M. Kiplinger, Fredrick H. Kohloff, Barry Kornegay, Gerry Lanham, P.E., Gary E. Mosher, CIH, George W. Nicholas, Gene Odenreider, Jack C. Shih, Craig M. Wehr.

plished in the cupola is heat release through the combustion of coke—the reaction between oxygen in the air and carbon in the fuel—that is in direct contact with the metallic portion of the charge and the fluxes.

One of the advantageous features of such a furnace is that counterflow preheating of the charge material is an inherent part of the melting process. The upward flowing hot gases come into close contact with the descending burden, allowing direct and efficient heat exchange to take place. The running or charge coke is also preheated, which aids in the combustion process as it reaches the combustion zone to replenish fuel consumed.

Greater understanding of these features accounts, in part, for the continued popularity of the cupola as a melting unit. However, recent design improvements, such as cokeless, plasma-fired types that alter emission characteristics are now encountered.

Casting, Cooling, and Finishing

After melting, molten metal is tapped from the furnace and poured into a ladle or directly into molds. If poured into a ladle, the molten iron may be treated with a variety of alloying agents predetermined by the desired metallurgical properties. It then is ladled into molds, where it solidifies and is allowed to cool further before separation of the casting from the mold (shakeout).

In larger, more mechanized foundries, the molds are conveyed automatically through a cooling tunnel before they are placed on a vibrating grid to shake the mold and core sand loose from the casting. In some foundries, molds are placed in an open floor space and molten iron is poured into the molds and allowed to cool. Molding and core sand are separated from the casting(s) either manually or mechanically.

Used sand from casting shakeout is usually returned to the sand preparation area and cleaned, screened, and processed to make new molds. Because of process losses and potential contamination, additional makeup sand may be required.

When castings have cooled, any unwanted appendages, such as sprues, gates, and risers, are removed by an oxygen torch, abrasive saw, friction cutting tool, or hand hammer. The castings then may be subjected to abrasive blast clean-

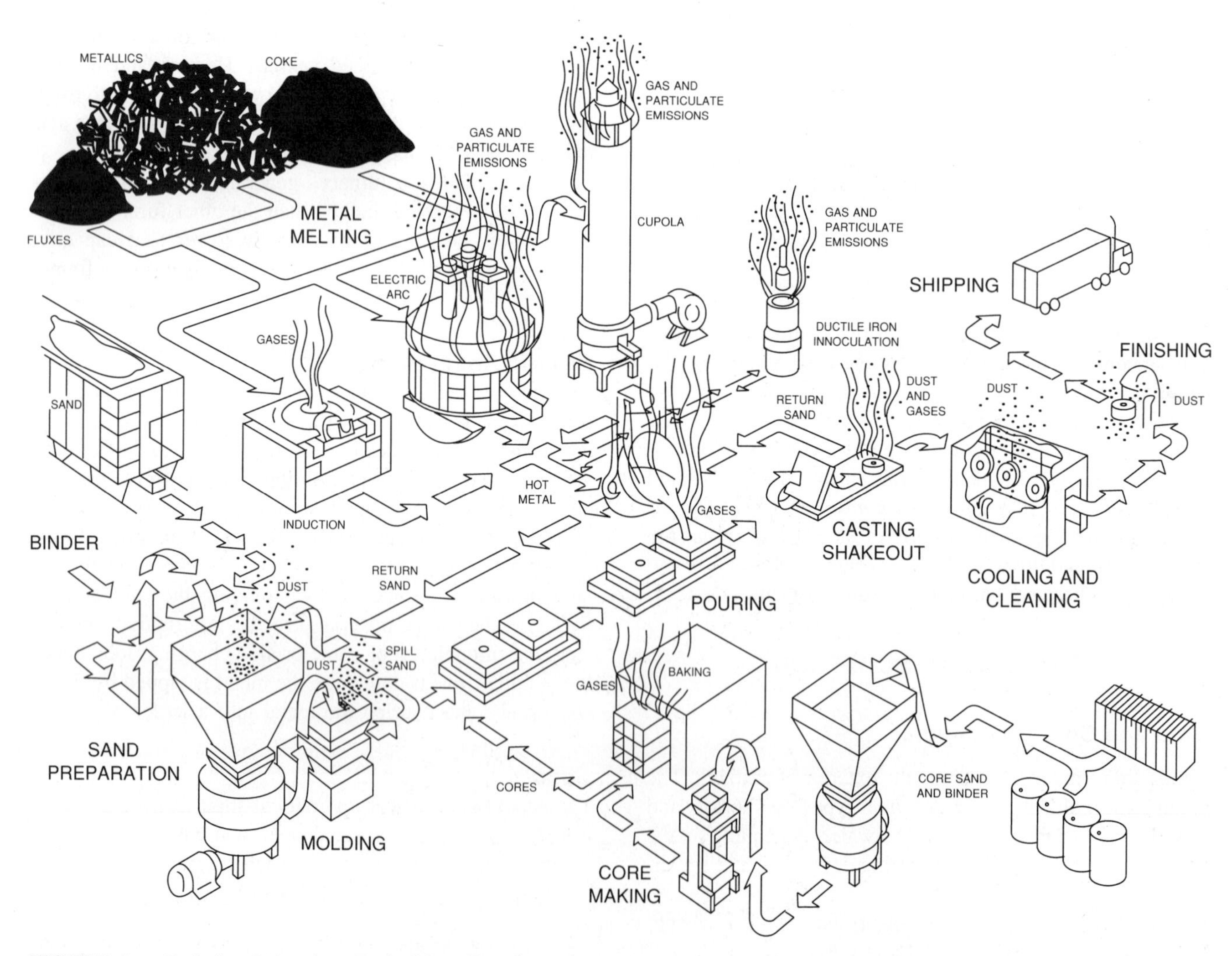

FIGURE 1. Emission Points in a Typical Iron Foundry

ing and/or tumbling to remove any remaining mold sand or scale.

AIR EMISSION CHARACTERIZATION

Mold and Core Production

The major pollutants emitted in mold and core production operations are particulates from sand preparation, mold core forming, and curing. In addition, volatile organic compounds (VOCs), carbon monoxide, and particulates may be emitted during core and mold curing or drying. (See Figure 1.)

Melting

The melting process begins with the handling of charge materials going into the melting furnace. Emissions from the raw materials handling are fugitive particulates generated from the receiving, unloading, storage, and conveying operations. Scrap preparation and preheating may emit one or more of the following: fumes, organic compounds, carbon monoxide, or coarse particulates. Scrap preparation with solvent degreasers may emit VOCs.

Induction and Arc Melting

The highest concentrations of furnace emissions occur during charging, back-charging, alloying, slag removal, and tapping operations. These emissions primarily include particulate (metal oxides) and possibly organics, depending on the scrap quality and pretreatment.

Typical dust loading from electric arc furnaces can range from 10 to 15 pounds per ton melted. However, electric induction furnaces may emit particulates at one tenth of that value.

Cupola Melting

The quantity and composition of particulate emissions vary among cupolas, and even at intervals in the same cupola. Causes include changes in iron-to-coke ratios, air volumes per ton melted, stack velocity, and the quality of the scrap melted.

Where oily scrap is charged, the raw emissions potentially not only will be greater in quantity, but can be much more visible. The American Foundrymen's Society compiled a survey of cupola emissions and found that an average emission from an uncontrolled cupola was approximately 13 to 17 pounds of particulate per ton melted. Eighty-five percent of the emissions may be greater than 10 μm in size.

Each cupola will have varying airflows at different phases in the melt process. This will affect the grains per standard cubic foot in emitted stack gases if all other factors are equal.

Dust composition and amounts vary from cupola to cupola. The source of the raw charge materials also will have a significant impact on dust composition and quantity. The dust could include some or all of the following materials.

- Iron oxide
- Magnesium oxide
- Manganese oxide
- Zinc oxide
- Silicon dioxide
- Calcium oxide
- Lead
- Cadmium

In addition, other gases and organic compounds may be emitted as part of the melting process. These include carbon monoxide, sulfur oxides, lead, and organic emissions. Both sulfur and organic emissions may be affected by the amount of oil or grease on the scrap.

The quantity of sulfur oxides also is large enough to be a definite consideration in the corrosion of air pollution control equipment. There are a number of instances of rapid deterioration of dust collectors on cupolas where corrosion protection was not considered.

Where fluorspar is used as an additive, the fluorine driven off can cause a corrosion problem with dust collection equipment. Fluorine also has the potential to dissolve glass bags. The carbonic acid formed when carbon dioxide reacts with water vapor may cause corrosion problems as well.

Pouring, Casting, Cooling, and Finishing

Particulate emissions can be generated during the treatment and inoculation of molten iron before pouring. For example, the addition of magnesium to molten metal to produce ductile iron causes a very violent reaction. This reaction is accompanied by various emissions of magnesium oxides and metallic fumes, depending on the method of treatment. Some methods, such as the tundish method, result in significantly lower emissions than others. Emissions from pouring consist of metal fumes, carbon monoxide, organic compounds, and particulates evolved when the mold and core materials are contacted by the molten iron. Emissions continue as the molds cool and during the shakeout operation, although at a much lower rate.

Finishing operations emit particulates during the removal of burrs, risers, and gates and during shotblast cleaning. These particulates consist primarily of iron, iron oxide, and abrasive media. The painting of castings also can lead to a variety of VOC emissions.

AIR POLLUTION CONTROL MEASURES

There are two primary collection methods for foundry particulates—wet and dry. Wet scrubbers include low- and high-energy types. Dry collection includes baghouses, mechanical collectors, and electrostatic precipitators.

In addition, the control of organic compounds may require incineration or afterburners. Air toxics merit special consideration, requiring careful selection of the collection method.

Wet Scrubbers

For particulate collection, the mechanisms used in a wet-type collector are inertial impaction and direct interception. These are used either separately or in combination.

In studying wet collector performance, independent investigators developed the contact power theory, which states that for a well-designed wet-scrubber collection, efficiency is a function of the energy consumed in the air-to-water contact process and is independent of the collector design. On this basis, well-designed collectors operating at or near the same pressure drop can be expected to exhibit comparable performance.

All wet collectors have a fractional efficiency characteristic—that is, their cleaning efficiency varies directly with the size of the particle being collected. In general, collectors operating at a very low pressure loss will remove only medium to coarse particles. High-efficiency collection of fine particles requires increased energy input, which will be reflected in higher collector pressure loss.

Some high-energy scrubbers are designed to use as little as 8 gal/min of scrubbing water per 1000 cfm of saturated gas at the throat. The water is usually recirculated and a small amount, ½ to 1 gal/min per 1000 cfm, is bled from the system to remove collected dust. This stream is subject to effluent guidelines for metal molding and casting.

Some sand system scrubbers may operate with as little as 8 to 15 inches w.c. pressure drop. However, when encountering extremely fine particulates, such as emissions from melting operations, a high-energy scrubber is required. For example, a typical cupola wet collector operation could require 30 to 100 inches w.c. pressure drop.

In addition, gas scrubbers may be used to control odors and toxic and sulfur dioxide emissions. In this case, acids, bases, or oxidizing agents may have to be added to the scrubbing liquid. This stream also would be subject to effluent guidelines for metal molding and casting.

Dry Collectors

The most frequently encountered equipment for the removal of solid particulate matter from an air stream or gas stream is the fabric dust collector or baghouse. With a mass median size of 0.5 μm, a collection efficiency of 98–99+% can be expected. As the filter medium becomes coated in a fabric collector, the collection efficiency rises. However, as material continues to build on the bag surface, higher pressure drops occur, which could result in a significant reduction in airflow. To maintain design flows, the bags must be cleaned periodically by mechanical shaking or with pulsed air.

Filter media are now available for hot corrosive atmospheres, such as furnace emissions. Operating inlet temperatures up to 500°F (260°C) are not uncommon. High humidity can be a problem if no provision is included for the condensation of free moisture. Free moisture and acid dew point are the worst enemies of all fabric collectors. It is important to have the following design information in order to select the proper fabric and the quantity of bags required:

- Gas flow rate
- Temperature and dew point
- Acid dew point
- Particle size and distribution
- Concentration of solids
- Chemical and physical properties of solids

Teflon-coated woven-glass-fiber bags have been used on a large majority of cupola installations because of their high temperature resistance. If fluorspar is used, Nomex bags are generally installed. The temperature of the gases entering the baghouse is then reduced to a maximum of 400°F (204°C). Use of these lower-temperature bags creates a potential corrosion hazard because of the acid dew point problem. For reverse-air and mechanical shake collectors, air-to-cloth ratios vary between 1.5 and 2.5 to 1.

Pulse-jet and cartridge collectors also can be used to collect pollutants from sand systems and casting cleaning operations. With either type of unit, care must be taken to select the proper air cloth ratio (maximums of 25:1 with pulse jet and 1.5:1 with cartridge). In general, these types of collectors will have only marginal results with furnace and inoculation emissions. If considered, they should be employed at a very low air cloth ratio. In addition, moisture introduced with compressed air may be significant and cause system failure.

Incineration

Afterburners may be used in some processes to control emissions, particularly when oily scrap or hydrocarbons in any form are charged into the furnaces or scrap preheat systems. Afterburning is required for below-the-door takeoff cupola emission systems. If afterburners are not used, carbon monoxide and oil vapors may be emitted through the discharge stack of the air pollution equipment. In order to achieve the required incineration, sufficient retention time (a minimum of 0.6 second) and ignition temperatures must be maintained.

In general, in the selection of collection devices for all processes, moisture, temperature, and the presence of corrosive materials must be considered. The temptation to operate at higher air cloth ratios in baghouses must be avoided. Similarly, claims that lower pressure drops in scrubbers create high efficiencies have been proved to be false.

Absorption

Charcoal absorption has been used in conjunction with other control devices for VOC control.

Bibliography

1. V. H. Baldwin, Jr., *Environmental Assessment of Iron Casting,* EPA-600/2-80-021, Research Triangle Institute, Research Triangle Park, NC, 1980.
2. R. Bethea, *Air Pollution Control Technology,* Van Nostrand Reinhold Environmental Engineering Series.
3. G. S. Cole, *Odor Pollution Control,* Scientific Research Staff, Ford Motor Co., American Foundrymen's Society Air Quality Committee (10-E), 1976.
4. *Compilation of Air Pollutant Emission Factory, Vol. 1, Stationery Point and Area Sources,* 4th ed., AP-42, Supplement A, U.S. Environmental Protection Agency, Research Triangle Park, NC, October 1986.
5. *Control of External Air Pollution,* American Foundrymen's Society Air Quality Committee (10-E), 1976.
6. *Cupola Handbook,* 5th ed., American Foundrymen's Society, Des Plaines, IL, 1984.
7. G. Engles, *The Nature and Characteristics of Cupola Effluents,* Verein Deutscher Giessereifachleute, Dusseldorf, Germany, 1969.
8. *General Principles of Foundry Ventilation,* American Foundrymen's Society Air Quality Committee (10-E), 1972.
9. *Legal Aspects of Air Pollution Control,* American Foundrymen's Society Air Quality Committee (10-E), 1976.
10. *Systems Analysis of Emissions and Emissions Control in the Iron Foundry Industry,* A. T. Kearney and Co., Chicago, IL, 1971.

SECONDARY LEAD SMELTING

*Charles A. Licht, P.E., Editor**

Control of the air pollution resulting from the secondary smelting and reclaiming of lead scrap may be conveniently considered according to the type of furnace employed. The reverberatory, blast, and pot furnaces are the three types most commonly used.

Various grades of lead metal, along with the oxides, are produced by the lead industry. The grade of product desired determines the type of equipment selected for its manufacture. The most common grades of lead produced are soft, semisoft, and hard. By starting with one of these grades and using accepted refining and alloying techniques, any special grade of lead or lead alloy can be made.

Soft lead may be designated as corroding, chemical, acid copper, or common desilverized lead. These four types are high-purity leads. Their chemical requirements are presented in Table 1. These leads are the products of the pot furnace after a considerable amount of refining has been done.

Semisoft lead is the product of the reverberatory-type furnace and usually contains from 0.3% to 0.4% antimony and up to 0.05% copper.

Hard lead is made in the blast furnace. A typical composition of hard lead is 5–12% antimony, 0.2–0.6% arsenic, 0.5–1.2% tin, 0.05–0.15% copper, and 0.001–0.01% nickel.

REVERBERATORY FURNACES

Process Description

Sweating operations are usually conducted in a reverberatory-type furnace or tube. The reverberatory furnace is also used to reclaim lead from oxides and drosses. Very often, materials for both sweating and reducing, such as lead scrap, battery plates, oxides, drosses, and lead residues, are charged to a reverberatory furnace. The charges are made up of a mixture of these materials and put into the furnace in such a manner as to keep a very small mound of unmelted material on top of the bath. As the mound becomes molten, more material is charged.

This type of furnace may be gas fired or oil fired or a combination of both. The temperature is maintained at approximately 2300°F. Only sufficient draft is pulled to remove the smoke and fumes and still allow the retention of as much heat as possible over the hearth. The molten metal is tapped off at intervals as a semisoft lead as the level of the metal rises. This operation is continuous, and recovery is generally about 10 to 12 pounds of metal per hour per square foot of hearth area.

Air Emissions Characterization

A fairly high percentage of sulfur is usually present in various forms in the charge to the reverberatory furnace. The temperature maintained is sufficiently high to "kill" the sulfides and results in the formation of sulfur dioxide and sulfur trioxide in the exit gases. Also present in the smoke and fumes produced are oxides, sulfides, and sulfates of lead, tin, arsenic, copper, and antimony. An overall material balance shows on the product side approximately 47% recovery of metal, 46% recovery of slag, sometimes called "litharge," and 7% of smoke and fumes.

The unagglomerated particulate matter emitted from secondary lead-smelting operations has been found to have a particle size range from 0.07 to 0.4 μm with a mean of about 0.3 μm.[1] The particles are nearly spherical and have a distinct tendency to agglomerate. The concentration of particulate matter in stack gases ranges from 1.4 to 4.5 gr/ft^3.

*Source: AP-40, *The Air Pollution Engineering Manual,* Second Edition, U.S. Environmental Protection Agency.

TABLE 1. Chemical Requirements for Lead

	Corroding Lead[a]	Chemical Lead[b]	Acid-Copper Lead[c]	Common Desilverized Lead[d]
Silver, max. %	0.0015	0.020	0.002	0.002
Silver, min. %		0.002		
Copper, max. %	0.0015	0.080	0.080	0.0025
Copper, min. %		0.040	0.040	
Silver and copper together, max. %	0.0025		0.040	
Arsenic, antimony, and tin together, max. %	0.002	0.002	0.002	0.005
Zinc, max. %	0.001	0.001	0.001	0.002
Iron, max. %	0.002	0.002	0.002	0.002
Bismuth, max. %	0.050	0.005	0.025	0.150
Lead (by difference), min. %	99.94	99.90	99.90	99.85

[a]Corroding lead is a designation used in the trade for many years to describe lead refined to a high degree of purity.
[b]Chemical lead is a term used in the trade to describe the undesilverized lead produced from southeastern Missouri ores.
[c]Acid-copper lead is made by adding copper to fully refined lead.
[d]Common desilverized lead is a designation used to describe fully refined desilverized lead.
Source: ASTM Standards, Part 2 1958.

Hooding and Ventilation Requirements

All the smoke and fumes produced by the reverberatory furnace must be collected and, since they are combined with the products of combustion, the entire volume emitted from the furnace must pass through the collector. It is not desirable to draw cool air into these furnaces through the charge doors, inspection ports, or other openings to keep air contaminants from escaping from them; therefore, external hoods are used to capture these emissions. The ventilating air for these hoods, as well as for the hoods venting slag stations, must also pass through the collector. In large furnaces, this represents a considerable volume of gases at fairly high temperatures.

Air Pollution Control Measures

The only control systems found to operate satisfactorily in Los Angeles County have been those employing a baghouse as a final collector. These systems also include auxiliary items such as gas-cooling devices and settling chambers.

A pull-through type of baghouse with compartments that can be shut off one at a time is very satisfactory. This allows atmospheric air to enter one compartment and relieve any flow. The bags may then be cleaned by a standard mechanical shaking mechanism.

Provision should be made to prevent sparks and burning materials from contacting the filter cloth, and temperature must be controlled by preceding the baghouse with radiant cooling ducts, water jacketed cooling ducts, or other suitable devices in order that the type of cloth used will have a reasonable life. The type of cloth selected depends on such parameters as the temperature and corrosivity of the entering gases and the permeability and abrasion- or stress-resisting characteristics of the cloth. Dacron bags are being used successfully in this service. The filtering velocity should not exceed 2 fpm. Test results of secondary lead-smelting furnaces venting to a baghouse control device are shown in Table 2.

LEAD BLAST FURNACES

Process Description

The lead blast furnace or cupola is constructed similarly to those used in the ferrous industry. The materials forming the usual charge for the blast furnace and a typical percentage composition are 4.5% rerun slag, 4.5% scrap castiron, 3% limestone, 5.5% coke, and 82.5% drosses, oxides, and reverberatory slags. The rerun slag is the highly silicated slag from previous blast furnace runs. The drosses are miscellaneous drosses consisting of copper drosses, caustic drosses, and dry drosses obtained from refining processes in the pot furnaces. The coke is used as a source of heat, and combustion air is introduced near the bottom of the furnace through tuyeres at a gauge pressure of about 8 to 12 oz/in.^{2}. Hard lead is charged into the cupola at the start of the operation to provide molten metal to fill the crucible. Normal charges, as outlined previously, are then added as the material melts down. The limestone and iron form the flux that floats on top of the molten lead and retards its oxidation.

As the level of molten material rises, the slag is tapped at intervals while the molten lead flows from the furnace at a more or less continuous rate. The lead product is "hard" or "antimonial." Approximately 70% of the molten material is tapped off as hard lead and the remaining 30% as slag. About 5% of the slag is retained for rerun later.

TABLE 2. Dust and Fume Emissions from a Secondary Lead-Smelting Furnace

	Test No. 1	Test No. 2
Furnace data		
Type	Reverberatory	Blast
Fuel used	Natural gas	Coke
Material charged	Battery groups	Battery groups, dross, slag
Process weight, lb/h	2,500	2,670
Control equipment data		
Type	Sectioned tubular baghouse[a]	Sectioned tubular baghouse[a]
Filter material	Dacron	Dacron
Filter area, ft^2	16,000	16,000
Filter velocity, fpm at 327°F	0.98	0.98
Dust and fume data		
Gas flow rate, scfm		
Furnace outlet	3,060	2,170
Baghouse outlet	10,400[b]	13,000[b]
Gas temperature, °F		
Furnace outlet	951	500
Baghouse outlet	327	175
Concentration, gr/scf		
Furnace outlet	4.98	12.3
Baghouse outlet	0.013	0.035
Dust and fume emission, lb/h		
Furnace outlet	130.5	229
Baghouse outlet	1.2	3.9
Baghouse efficiency, %	99.1	98.3
Baghouse catch, wt %		
Particle size, μm		
0 to 1	13.3	13.3
1 to 2	45.2	45.2
2 to 3	19.1	19.1
3 to 4	14.0	14.0
4 to 16	8.4	8.4
Sulfur compounds as SO_2, vol %		
Baghouse outlet	0.104	0.03

[a]The same baghouse alternately serves the reverberatory furnace and the blast furnace.
[b]Dilution air admitted to cool gas stream.

Air Emissions Characterization

Combustion air from the tuyeres passing vertically upward through the charge in a blast furnace conveys oxides, smoke, bits of coke fuel, and other particulates present in the charge. A typical material balance based on the charge to a blast furnace in which battery groups are being processed is 70% recovery of lead, 8% slag, 10% matte (sulfur compounds formed with slag), 5% water (moisture contained in charge), and 7% dust (lead oxide and other particulates discharged from stack of furnace with gaseous products of combustion). Particulate matter loading in blast furnace gases is exceedingly heavy, up to 4 gr/ft^3.

Blast furnace stack gas temperatures range from 1200°F to 1350°F. In addition to the particulate matter, which consists of smoke, oil vapor, fume, and dust, the blast furnace stack gases contain carbon monoxide. An afterburner is necessary to control the gaseous, liquid, and solid combustible material in the effluent.

Hooding and Ventilation Requirements

The only practical way to capture the contaminants discharged from a lead blast furnace is to seal the furnace and vent all the gases to a control system. The hooding and ventilation requirements are very similar to those for gray iron foundries discussed elsewhere in this manual.

Air Pollution Control Measures

The control system for a lead blast furnace is similar to that employed for gray iron cupola furnaces, except that electrical precipitators are not used for economic reasons. Moreover, difficulties are encountered in conditioning the particles to give them resistivity characteristics in the range that will allow efficient collection.

Afterburners

An afterburner should be designed with heat capacity to raise the temperature of the combustibles, inspirated air, and cupola gases to at least 1200°F. The geometry of the secondary combustion zone should be such that the products to be incinerated have a retention time of at least ¼ second. A luminous flame burner is desirable, since it presents more flame exposure. Enough turbulence must be created in the gas stream for thorough mixing of combustibles and air. In large-diameter cupola furnaces, stratification of the gas stream may make this a major problem. One device, proved successful in promoting mixing in large-diameter cupolas, is the inverted cone. The combustion air is inspirated through the charging door and, if necessary, may also be inspirated through openings strategically located in the cupola circumference, above the charging opening. The rapid ignition of the combustible effluent by the afterburner frequently results in a pulsating or puffing emission discharge from the charging door. This can be eliminated by the installation of an ignition burner below the level of the charging door, which ignites and partially burns the combustible effluent.

A cupola afterburner need not be operated through the entire furnace cycle. Even without an afterburner, an active flame can be maintained in the upper portion of the cupola. This requires control of the materials charged, and also, control of combustion air and mixing. The afterburner however, must be in operation during the furnace light-off procedure.

Baghouse Dust Collectors

The temperature of the gas stream discharged from the top of a cupola may be as high as 2200°F. If a baghouse is used as a control device, these gases must be cooled to prevent burning or scorching of the cloth bags. Maximum temperatures allowed vary from 180°F for cotton bags to 500°F for glass fabric bags. Cooling can be effected by radiant cooling columns, by evaporative water coolers, or by dilution with ambient air.

For satisfactory baghouse operation, when metallurgical fumes are to be collected, filtering velocity should not exceed 2½ fpm. Provisions for cleaning collected material from the bags usually require compartmentation of the baghouse so that one section may be isolated and the bags shaken while the remainder of the system is in operation. The gas temperature through the baghouse should not be allowed to fall below the dew point, because condensation within the baghouse may cause the particles on the bag surfaces to agglomerate, deteriorate the cloth, and corrode the baghouse enclosure. A bypass control must also be installed. If the cooling system fails, the bypass is opened, which discharges the effluent gas stream to the atmosphere and thus prevents damage to the bags from excessive temperatures. Properly designed and maintained baghouses normally can be expected to have efficiencies ranging upward from 95%.

POT-TYPE FURNACES

Process Description

Pot-type furnaces are used for remelting, alloying, and refining processes. Remelting is usually done in small pot furnaces, and the materials charged are usually alloys in the ingot form that do not require any further processing except to be melted for casting operations.

The pots used in the secondary smelters range from the smallest practical size of 1-ton capacity up to 50 tons. These furnaces are usually gas fired. Various refining and alloying operations are carried on in these pots. Alloying usually begins with a metal lower in the percentage of alloying materials than desired. The percent desired is calculated and that amount is then added. Antimony, tin, arsenic, copper, and nickel are the most common alloying elements used.

The refining processes most commonly employed are those for the removal of copper and antimony to produce soft lead, and those for the removal of arsenic, copper, and nickel to produce hard lead. For copper removal, the temperature of the molten lead is allowed to drop to 620°F and sulfur is added. The mixture is agitated and copper sulfide is skimmed off as dross. This is known as "copper dross" and is charged into the blast furnace.

When aluminum is added to molten lead, it reacts preferentially with copper, antimony, and nickel to form complex compounds that can be skimmed from the surface of the metal. The antimony content can also be reduced to about 0.02% by bubbling air through the molten lead. It can be further reduced by adding a mixture of sodium nitrate and sodium hydroxide and skimming the resulting dross from the surface of the metal.

Another common refining procedure, "dry drossing," consists of introducing sawdust into the agitated mass of molten metal. This forms carbon, which aids in separating the globules of lead suspended in the dross and reduces some of the lead oxide to elemental lead.

Air Emissions Characterization

Although the quantity of air contaminants discharged from pot furnaces as a result of remelting, alloying, and refining is much less than that from reverberatory or blast furnaces, the capture and control of these contaminants are equally important in order to prevent periodic violations of air pollution regulations and to protect the health of the employees.

Problems of industrial hygiene are inherent in this industry. People working with this equipment frequently inhale and ingest lead oxide fumes, which are cumulative, systemic poisons. Frequent medical examinations are necessary for all employees, and a mandatory compliance with OSHA regulation 40 CFR 1910. 1025 is required.

Hooding and Ventilation Requirements

The canopy-type hood is readily adaptable to capturing the fumes from a pot furnace. Recommended in-draft velocities vary depending on the hood, furnace geometry, cross-drafts, and temperatures involved.

Air Pollution Control Measures

The control systems for pot furnaces, as with the other lead furnaces, require the use of a baghouse as the final collector. The temperature of the gases is, however, generally much lower than that of gases from the other furnaces; therefore, the gas-cooling devices, if needed, will be much smaller. Afterburners generally are not required.

Reference

1. G. L. Allen, F. H. Viets and L. C. McCabe, "Control of Metallurgical and Mineral Dusts and Fumes in Los Angeles County, California," Bureau of Mines Information Circular 7627, U.S. Department of Interior, Washington, D.C. (April, 1952).

SECONDARY ZINC

Charles A. Licht, P.E.

Secondary zinc scrap reclamation operations are carried out using a number of techniques. Among the end products that are generated are zinc powders (for paints and precious metal precipitation), zinc oxide (for use in tire manufacturing, coated-paper manufacturing, paints, etc.), zinc alloy (for manufacturing castings and galvanizing), and various low-grade zincs for off-specification noncritical castings and zinc compounds (used as additives in a variety of products, including fertilizers and animal feeds).

The raw materials are generally scrap die castings from postconsumer products. Also, gates and risers from foundry operations, as well as drosses, are consumed. Flash and trim scraps are used, as are edge trim and sheet scrap, reject battery cases, old zinc roof sheet, and galvanizers' bottom dross.

Iron or contaminated zinc scrap is generally processed in sweating furnaces. The typical sweating furnace is an unlined rotary furnace, although refractory-lined units are also in use. The rotary furnace has the advantage of keeping the material moving, exposing all surfaces of the scrap to heat, and allowing for a separation so that the molten zinc is able to run free to the discharge point. Properly operated zinc rotary sweat furnaces are used to separate zinc from aluminum since there is about a 400°F (200°C) temperature differential between the two melting points.

Molten zinc generated from a rotary sweat furnace is poured into a cast-iron holding pot where the alloy is checked and modified or refined prior to pouring into a usable end product. Typically, the zinc is poured into blocks or slabs. If the material is alloyed, it is poured in either 50-pound (22-kg) slabs or 20-pound (9-kg) bars.

In the manufacture of zinc powder, zinc scrap and drosses and carbon-content material can be placed in retort furnaces, which are heated externally. Zinc oxide reacts with carbon to form zinc metal. The zinc fume is caused to boil over into a condenser. Air is kept out of the system by tight seals. This type of system can create air pollution problems from retort seal failure or when a batch is completed and the seal must be broken to remove the condenser and empty the retort.

Another end product of retort operations is zinc powder, which is generated by condensing zinc vapors in a large condenser. Zinc oxide is also manufactured utilizing retort technology. In this case, zinc is allowed to boil and exit the retort to burn in air. It is then collected in an air pollution control system. Here, the pollution control device collects a salable product.

High-purity zinc is made from scrap by utilizing retorts. The zinc metal from the retort goes into a second retort, where it is condensed to form a molten bath that is tapped after all the metal in the first retort has been boiled off.

Another technique employed in the manufacture of zinc oxide is the use of a muffle furnace. In this unit, the heating is also done indirectly. The furnace is built so that molten zinc can be added to the boiling bath continuously or periodically. The bath is operated at temperatures in excess of the boiling point (1668°F [910°C]) of zinc, thus allowing the zinc vapors to leave the system, where they are burned in air to form zinc oxide particles. The zinc oxide is air conveyed to a baghouse system whose primary purpose is to collect a salable product, but which also does an excellent job of controlling potential emissions. (See Figure 1.)

Zinc skimmings and drosses are melted in refractory-lined rotary furnaces, which are filled, heated, refilled, and reheated. The oxide materials are allowed to overflow into coolers while the metal fills the furnace. When the furnace is filled with metal, the metal is tapped into blocks for resale to galvanizers or alloyers. The oxides generated in the dross-reclamation operations are generally used as micronutrients in fertilizers and animal feed.

A number of secondary operations exist that are secondary only in the sense that they involve remelting. The manufacture of zinc casting alloys is frequently carried out in iron pots, although some companies use refractory-lined reverberatory furnaces. In the process of alloy manufacture, special high-grade (99.99% pure) zinc metal is melted, and minor additions of aluminum, magnesium, copper, and so on are made to meet alloy specifications. The metal is poured off as slabs or 20-pound (9-kg) bars, occasionally in the form of blocks. Other processes include the melting of special high-grade zinc and pouring, without alloying, into special configurations, such as balls used in plating.

The processes utilized in the zinc metal remelt operations

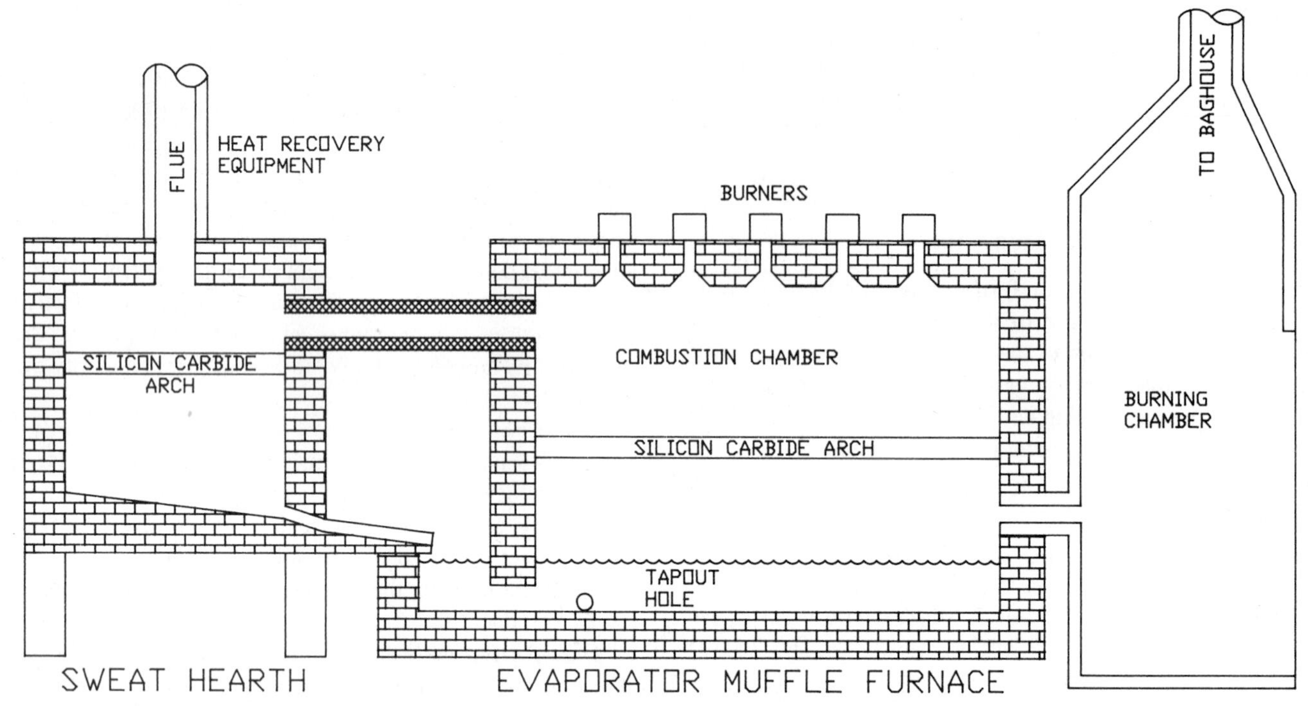

FIGURE 1. Zinc Muffle Furnace with Integral Sweat Furnace

generate almost no fumes. On the other hand, the "wet" and "dry" rotary sweating furnaces generate substantial amounts of fumes, as do the scrap remelting furnaces. In these cases, the furnaces must be controlled by adequate air pollution control systems, such as shaker-type baghouses. The products collected in these baghouses, while a lower-grade material than the manufactured-for-sale zinc oxides, are nonetheless salable.

Fluxing is done in some of the processes, particularly in the reclamation of zinc scrap in iron pots. In this process, zinc chloride is utilized for cleaning the metal. This generally requires that the zinc pots be ventilated to baghouses because of the periodic high fume generation.

Muffle furnaces have much greater vaporizing capacity than any other process available. They are operated for many months at a time. The heat for vaporization, which is supplied by fuel-fired burners, is generally forced through a silicon carbide arch that separates the zinc vapors from the products of combustion. Molten zinc from a melting pot or sweat furnace is charged through a feed well, which is also the air lock. The zinc vapors are carried to a condenser where the purified zinc is collected or vented and burned in air to generate oxides. The zinc fume is collected in a baghouse.

Where scrap is sweated to generate the molten metal necessary for the muffle furnace, the sweating furnace will create fumes. Often, sweat furnaces are constructed as muffle furnaces. The products of combustion used to boil zinc are utilized to heat the scrap in the sweating furnace and then are vented through a heat exchanger system that allows the use of preheated combustion air in the primary chamber. The only contaminated vent is on the sweating furnace at the charge door. The charge door is enclosed by a hood that controls the fumes to an air pollution control system. This secondary fume will not meet product specifications and is used in lower-grade applications, such as fertilizer and animal feed.

AIR POLLUTION CONTROL EQUIPMENT

Generally, the zinc processing furnaces are controlled at the point of fume generation by hoods that encompass the source and act as the pickup point to convey the zinc oxide fumes into baghouses. Typical baghouses are multicompartment, shaker-type units operating at temperatures below 275°F (135°C), thus making it possible for polyester bags to be utilized rather than more expensive fabrics. A few applications use reverse-jet baghouses. The volume of zinc oxide collected from muffle furnaces will range to more than 3000 pounds (1360 kg) an hour. Retorts operate at a much lower rate. The baghouses are typically set up to shut down compartment by compartment for cleaning on a cyclical basis. In some plants, pulsed-air cleaning dust collectors using either bags or cartridges are used. The product generated in this process is usually withdrawn continuously via a screw conveyor to storage bins, where further processing and/or packaging is accomplished.

REDUCTION RETORT FURNACES

Reduction in Belgian Retorts

The Belgian retort furnace is one of several horizontal retort furnaces that had been the most common device for the

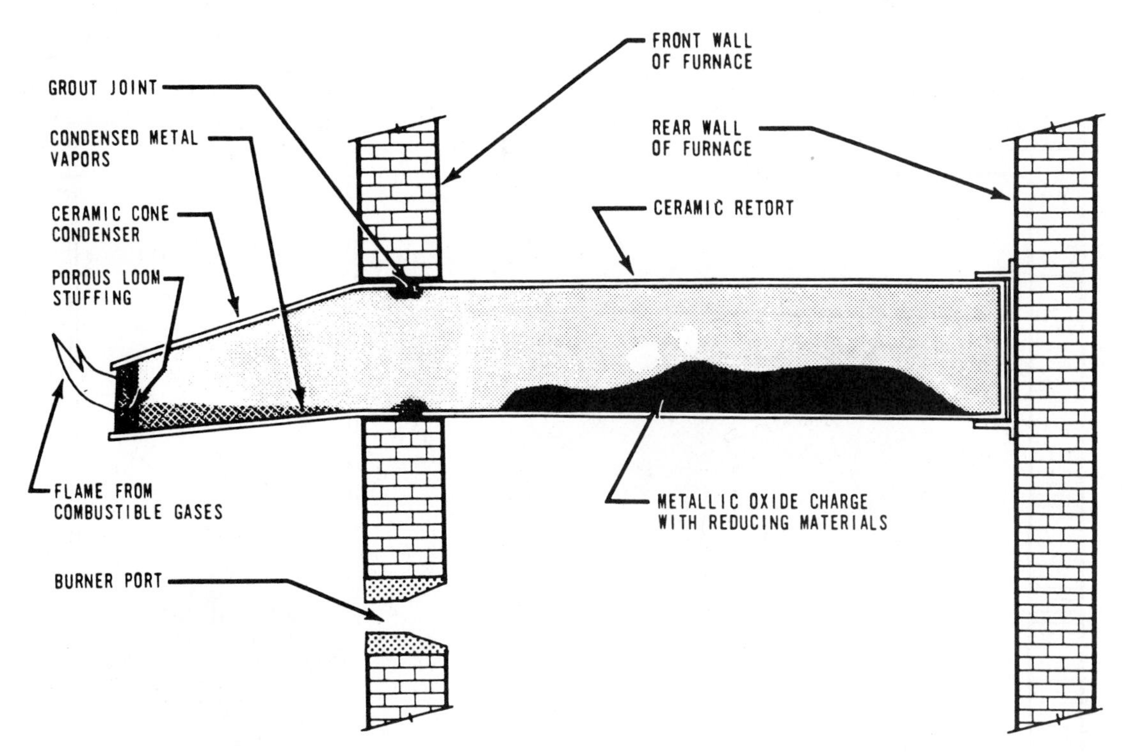

FIGURE 2. Diagram Showing One Bank of a Belgian Retort Furnace

reduction of zinc. The horizontal retort process is now being replaced by other methods capable of handling larger volumes of metal per retort and by the electrolytic process for the reduction of zinc ore. The reduction process is used to reclaim zinc from the dross formed in zinc-melting operations, the zinc oxide collected by air pollution control systems serving zinc-alloy–melting operations, and the contaminated zinc oxide from the zinc oxide plants.

A typical Belgian retort (Figure 2) is about 8 inches (20 cm) in internal diameter and from 48 to 60 inches (120–150 cm) long. One end is closed. A conical clay condenser from 18 to 24 inches (45–60 cm) long is attached to the open end. The retorts are arranged in banks with rows four to seven high and as many retorts in a row as are needed to obtain the desired production. The retorts are generally gas fired today, although previously many units were coal fired.

The retorts are charged with a mixture of zinc scrap, zinc oxide, and powdered coke. If these materials are powdered, water is added to facilitate charging and to allow the mixture to be packed tightly into the retort. From three to four times more carbon is used than is needed for the reduction reaction.

After the charging, the condensers are replaced and their mouths stuffed with a porous material. A small hole is left through the stuffing to allow moisture and unwanted volatile materials to escape. About three hours are needed to expel all the undesirable volatile materials from the retort. About six hours after charging is completed, zinc vapors appear. The charge in the retort is brought up to 1832°F (1000°C) to 2012°F (1100°C) for about eight hours, after which it may rise slowly to a maximum of 2280°F (1250°C). The temperature on the outside of the retorts ranges from 2375°F (1300°C) to 2550°F (1400°C). The condensers are operated at from 780°F (415°C) to 1020°F (550°C), a temperature range above the melting point of zinc, but one at which the vapor pressure is so low that a minimum of zinc vapor is lost.

The reduction reaction of zinc oxide can be summarized by the following reaction.

$$ZnO + C = Zn + CO \quad (1)$$

Very little, if any, zinc oxide is actually reduced by the solid carbon in the retort. A series of reactions results in an atmosphere rich in carbon monoxide, which does the actual reducing. The reactions are reversible, but by the use of an excess of carbon, they are forced toward the right. The reactions probably get started by the oxidation of a small portion of the coke by the oxygen in the residual air in the retort. The oxygen is quickly used, but the carbon dioxide formed reacts with the carbon to form carbon monoxide, according to the following equation.

$$CO_2 + C = 2CO \quad (2)$$

The carbon monoxide in turn reacts with zinc oxide to produce zinc and carbon dioxide:

$$CO + ZnO = Zn + CO_2 \quad (3)$$

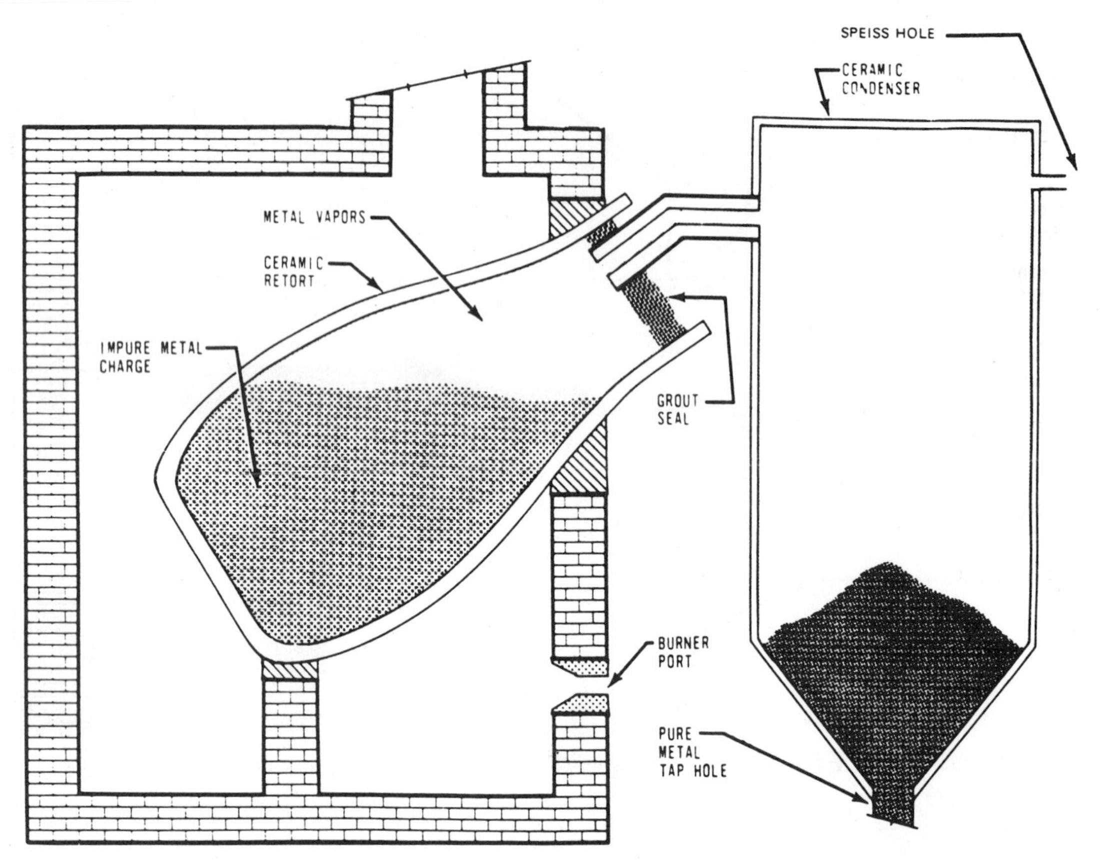

FIGURE 3. Diagram of a Distillation-Type Retort Furnace

Carbon monoxide is regenerated by use of equation 2, and the reduction of the zinc oxide proceeds.

About eight hours after the first zinc begins to be discharged, the heat needed to maintain production begins to increase and the amount of zinc produced begins to decrease. Although zinc can still be produced, the amount of heat absorbed by the reduction reaction decreases and the temperature of the retort and its contents increases. Care must be taken not to damage the retort or fuse its charge. As a result, a 24-hour cycle is usually found to be an economical operation. The zinc values still in the spent charge are recovered by recycling with the fresh charges. A single-pass recovery yields 65–70% of the zinc charged, but, by recycling, an overall recovery of 95% may be obtained.

Air Emissions Characterization

The air contaminants emitted vary in composition and concentration during the operating cycle of Belgian retorts. During charging operation, very low concentrations are emitted. The feed is moist and, therefore, not dusty. As the retorts are heated, steam is emitted. After zinc begins to form, both carbon monoxide and zinc vapors are discharged. These emissions burn to form gaseous carbon dioxide and solid zinc oxide. During the heating cycle, zinc is poured from the condensers about three times at six- to seven-hour intervals. The amount of zinc vapors discharged increases during the tapping operation. Before the spent charge is removed from the retorts, the temperature of the retorts is lowered, but zinc fumes and dust from the spent charge are discharged to the atmosphere.

Hooding and Ventilation Requirements

Air contaminants are discharged from each retort. In one installation reported in Los Angeles, a furnace has 240 retorts arranged in five horizontal rows with 48 retorts per row. The face of the furnace measures 70 feet (21.3 meters) long by 8 feet (2.4 meters) high; therefore, the air contaminants are discharged from 240 separate openings and over an area of 560 square feet (15.7 square meters). A hood 2 feet (0.6 meter) wide by 70 feet (21.3 meters) long positioned immediately above the front of the furnace is used to collect the air contaminants. The hood indraft is 175 fpm (53 meters per minute).

DISTILLATION RETORT FURNACES

The distillation retort furnace (Figure 3) consists of a pear-shaped graphite retort, which may be 5 feet (1.5 meters) long by 2 feet (0.6 meter) in diameter at the closed end by 1½ feet (0.45 meter) in diameter at the open end and 3 feet (1 meter) in diameter at its widest cross section. Normally, the retort is encased in a brick furnace with only the open end protruding and it is heated externally with gas- or oil-fired burners. The retorts are charged with molten impure zinc through the open end, and a condenser is attached to the opening to receive and condense the zinc vapors. After the distillation is completed, the condenser is moved away, the residue is removed from the retort, and a new batch is started.

The vaporized zinc is either conducted to a condenser or discharged through an orifice into a stream of air. Two types of condensers are used—a brick-lined steel condenser operated at from 780°F to 1010°F (415–545°C) to condense the vapor to liquid zinc, or a larger, unlined steel condenser that cools the vapor to solid zinc. The latter condenser is used to manufacture powered zinc. The condensers must be operated at a slight positive pressure to keep air from entering them and oxidizing the zinc. To ensure that there is a positive pressure, a small hole, called a "speiss" hole, is provided through which a small amount of zinc vapor is allowed to escape continuously into the atmosphere. The vapor burns with a bright flame, indicating that there is a pressure in the condenser. If the flame gets too large, the pressure is too high. If it goes out, the pressure is too low. In either case, the proper adjustments are made to obtain the desired condenser pressure.

When it is desired to make zinc oxide, the vapor from a retort is discharged through an orifice into a stream of air where zinc oxide is formed inside a refractory-lined chamber. The combustion gases and air, which bear the oxide particles, are then carried to a baghouse collector where the powdered oxide is collected.

Air Emissions Characterization

During the 24-hour cycle of the distillation retorts, zinc vapors escape from the retort (1) when the residue from the preceding batch is removed from the retort and a new batch is charged, and (2) when the second charge is added to the retort. As the zinc vapors mix with air, they oxidize and form a dense cloud of zinc oxide fumes. Air contaminants are discharged for about one hour each time the charging hole is open. When the zinc is actually being distilled, no fumes escape from the retort; however, a small amount of zinc oxide escapes from the speiss hole in the condenser. Although the emission rate is low, air contaminants are discharged for about 20 hours per day.

Hooding and Ventilation Requirements

To capture the emissions from a distillation retort furnace, simple canopy hoods placed close to and directly over the sources of emissions are sufficient.

The retorts are fuel fired and the products of combustion do not mix with the emissions from the retort or the condenser. The exhausted gases are heated slightly by the combustion of zinc and from radiation and convection losses from the retort, but the amount of heating is so low that no cooling is necessary.

MUFFLE FURNACES

Muffle furnaces (Figure 4) are continuously fed retort furnaces. They generally have a much greater vaporizing capacity than do either Belgian retorts or bottle retorts, and they are operated continuously for several months at a time. Heat for vaporization is supplied by gas- or oil-fired burners by conduction and radiation through a silicon carbide arch that separates the zinc vapors and the products of combustion. Molten zinc from either a melting pot or a sweat furnace is charged through a feed well that also acts as an air lock. The zinc vapors are conducted to a condenser, where purified liquid zinc is collected, or the condenser is bypassed and the vapors are discharged through an orifice into a stream of air, where zinc oxide is formed.

The furnace, including the feed well and sweating cham-

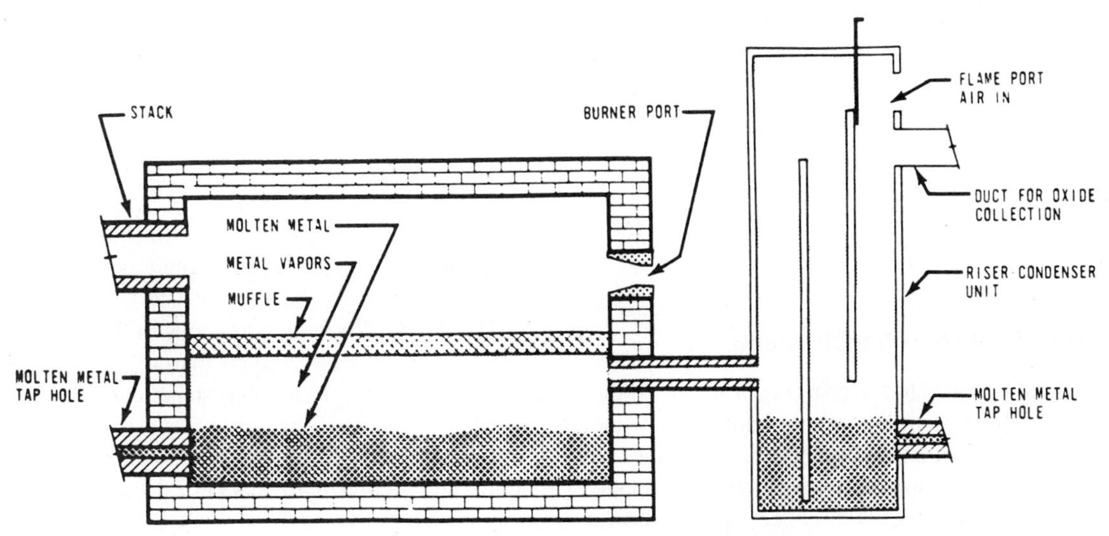

FIGURE 4. Diagram of a Muffle Furnace and Condenser

ber, is heated indirectly with a combination gas- or oil-fired burner. The combustion chamber, located directly over the vaporizing chamber, is heated to about 2400°F (1315°C). On leaving the combustion chamber, the products of combustion are conducted through the sweating chamber to supply the heat needed for melting the zinc alloys from the scrap charged and for heating the zinc prior to discharge into the feed well at 900°F (480°C).

Zinc vapors can be conducted from the vaporizing section into a multiple-chamber condenser. When zinc oxide is the desired product, condensers are not used. The vapors are allowed to escape through an orifice into a burning chamber.

Air Emissions Characterization

Dust and fumes are created by the sweating operation. Scrap is charged into the sweating chamber through a door. After the zinc alloys have been melted, the residue is pushed or pulled out of the chamber into refractory-lined pots. The pots are then taken to a dross stirrer to separate metal from oxide. Excessive dust and fumes are created at the rake-out point and at the dross stirrer, requiring air pollution controls at both locations.

The zinc alloys charged into the vaporizing section contain copper, aluminum, iron, lead, and other impurities. As zinc is distilled from the metals, the concentration of the impurities increases until continued distillation becomes impractical. After periods of operation, the residue, containing 10–50% zinc, must be removed. When tapped, the temperature of the residue is about 1900°F (1055°C), hot enough to release zinc oxide fumes. The molds collecting the residue metal are often arranged so that the metal overflows from one mold to another; however, the metal cools so rapidly that fumes are released only from the pouring spout and the first two or three molds. The fumes, almost entirely zinc oxide, are 100% opaque from the pouring spout and the first mold. At the third mold, the opacity decreases to 10%.

Any discharge of the zinc vapor from the condenser forms zinc oxide of product purity; therefore, the condenser vents into the intake hood of a product-collecting exhaust system. Since some zinc oxide is always produced, even when the condenser is set to produce a maximum of liquid zinc, the product-collecting exhaust system is always in operation to prevent air contaminants from escaping from the condenser to the atmosphere.

TABLE 1. Uncontrolled Particulate Emission Factors for Secondary Zinc Smelting[a] ***(Emission Factor Rating: C)***

Operation	Emissions	
	kg/Mg	lb/ton
Reverberatory sweating		
Clean metallic scrap	Negligible	Negligible
General metallic scrap	6.5	13
Residual scrap	16	32
Rotary sweating	5.5–12.5	11–25
Muffle sweating	5.4–16	10.8–32
Kettle sweating		
Clean metallic scrap	Negligible	Negligible
General metallic scrap	5.5	11
Residual scrap	12.5	25
Electric resistance sweating	<5	<10
Crushing/screening	0.5–3.8	1.0–7.5
Sodium carbonate leaching		
Crushing/screening	0.5–3.8	1.0–7.5
Calcining	44.5	89
Kettle (pot) melting	0.05	0.1
Crucible melting	DNA	DNA
Reverberatory melting	DNA	DNA
Electric induction melting	DNA	DNA
Alloying	DNA	DNA
Retort and muffle distillation		
Pouring	0.2–0.4	0.4–0.8
Casting	0.1–0.2	0.2–0.4
Muffle distillation	22.5	45
Graphite rod distillation	Negligible	Negligible
Retort distillation/oxidation[b]	10–20	20–40
Muffle distillation/oxidation[b]	10–20	20–40
Retort reduction	23.5	47
Galvanizing	2.5	5

[a]Expressed as units per unit weight of feed material processed for crushing/screening, skimming/residues processed; for kettle (pot) melting and retort and muffle distillation operations, metal product. Galvanizing factor expressed in units per unit weight of zinc used. DNA: Data not available.
[b]Factor units per unit weight of zinc oxide produced. The product zinc oxide dust is totally carried over in the exhaust gas from the furnace and is recovered with 98–99% efficiency.

Source: Reference 1

Hooding and Ventilation Requirements

The dust and fumes created by the charging of scrap and the sweating of zinc alloys from the scrap originate inside the sweat chamber. The thermal drafts cause the emissions to escape from the upper portion of the sweat chamber doors. Hoods are placed over the doors to collect the emissions. The charging door hood extends to the furnace wall and covers a little more than the width of the door. When possible, hood sides are installed to the plant floor.

A hood enclosing the rake-out door can be controlled with 5500 cfm (155 m^3/min). An inlet velocity of 250 fpm (76 m/min) is sufficient to capture all of the emissions escaping from both the furnace and the screen.

The ductwork joining the hoods to the control devices is manifolded and dampered so that any or all hoods can be opened or closed. Exhaust systems can provide sufficient ventilation to control the fumes created by multiple furnaces in operation at the same time. Factors for uncontrolled point source and fugitive particulate emissions are listed in Tables 1 and 2 respectively.

TABLE 2. Fugitive Particulate Uncontrolled Emission Factors for Secondary Zinc Smelting ***(Emission Factor Rating: E)***

	Particulate	
Operation	kg/Mg	lb/ton
Reverberatory sweating[b]	0.63	1.30
Rotary sweating[b]	0.45	0.90
Muffle sweating[b]	0.54	1.07
Kettle (pot) sweating[b]	0.28	0.56
Electric resistance sweating[b]	0.25	0.50
Crushing/screening[c]	2.13	4.25
Sodium carbonate leaching	DNA	DNA
Kettle (pot) melting furnace[b]	0.0025	0.005
Crucible melting furnace[d]	0.0025	0.005
Reverberatory melting furnace[b]	0.0025	0.005
Electric induction melting[b]	0.0025	0.005
Alloying retort distillation	DNA	DNA
Retort and muffle distillation	1.18	2.36
Casting[b]	0.0075	0.015
Graphite rod distillation	DNA	DNA
Retort distillation/oxidation	DNA	DNA
Muffle distillation/oxidation	DNA	DNA
Retort reduction	DNA	DNA

[a]Expressed as units per end product, except factors for crushing/screening and electric resistance furnaces, which are expressed as units per unit of scrap processed. DNA: Data not available.
[b]Estimate based on stack emission factor given in reference 1, assuming fugitive emissions to be equal to 5% of stack emissions.
[c]Reference 1. Average of reported emission factors.
[d]Engineering judgment, assuming fugitive emissions from crucible melting furnace to be equal to fugitive emissions from kettle (pot) melting furnace.
Source: Reference 1.

AIR POLLUTION CONTROL MEASURES

For all of these (that is, reduction retort furnaces, distillation retort furnaces, and muffle furnaces), air pollution control is achieved with a baghouse. In some installations on a muffle furnace, a low-efficiency cyclone and a baghouse are used to control the emissions from the sweating chambers and residue pouring operations of the three muffle furnaces. Although the cyclone has a low collection efficiency, it does collect from 5% to 10% of the dust load and it is still used. The cyclone was in existence before the baghouse was installed. More recent systems often do not have cyclones, but only baghouses. Filtration velocity is usually about 1.5 fpm (0.5 m/min), although some older systems are operating successfully at 3 fpm (1 m/min).

Dust collectors for other zinc-melting and zinc-vaporizing furnaces are very similar to the ones already described. Glass bags have been found adequate when gas temperatures exceed the limits of synthetics, although they are seldom used today. Filtering velocities of 3 fpm (1 m/min) have been employed and found adequate, but modern installations realize better bag life and product recovery at a lower filtering velocity.

Reference

1. *Compilation of Air Pollutant Emission Factors, Volume 1: Stationary Point and Area Sources,* AP-42, U.S. EPA, 1985.

Bibliography

1. *Air Pollution Engineering Manual, AP40;* J. A. Danielson, Ed.; U.S. Environmental Protection Agency, Research Triangle Park, NC, 1973, pp 293–299.
2. M. Bess, "Recycling of zinc," *Metals Handbook,* Vol. 2, 10th ed., ASM International, 1990, pp 1223–1224.
3. H. E. Brown, *Zinc Chemicals: Applications,* International Lead Zinc Research Organization, Research Triangle Park, NC, June 1990.
4. B. C. Hafford, W. E. Pepper, and T. B. Lloyd, *Zinc Dust and Zinc Powder: Their Production, Properties, and Applications,* International Lead Zinc Research Organization, Research Triangle Park, NC, 1982.
5. B. Baum and D. S. Carr, *Zinc Oxide: A Weathering Stabilizer for Plastics,* International Lead Zinc Research Organization, and Zinc Institute, New York, rev. ed. June 1984.
6. H. E. Brown, *Zinc Oxide, Properties and Applications,* International Lead Zinc Research Organization, New York, 1976, rev. ed. 1986.
7. *Zinc Oxide and Rubber,* Zinc Institute Inc., New York, 1984.

STEEL FOUNDRIES

Ezra L. Kotzin

OVERVIEW OF THE INDUSTRY

In the production of steel castings (low-carbon, mild-alloy, or high-alloy steel), the technology of mold and core preparation, in a general way, is similar to that used to produce castings from metals other than steel (Figure 1). Although the melting and pouring of steel in the foundry for the production of steel castings are also similar to the melting and pouring of steel into ingots that are subsequently forged or rolled, there are major differences between the melting practice in a steel foundry and that in a steel mill. One difference is the higher tapping and pouring temperatures used for foundry melting to obtain increased fluidity of the molten steel. The producer of ingots for rolling is less concerned with fluidity since it is much simpler to pour steel into ingot molds than into sand molds in order to produce some of the intricate and complex geometries of castings with relatively thin metal sections. It is then easy to understand that at these higher tapping temperatures of approximately 2950°F (1620°C) to 3000–3200°F (1649–1760°C), the gasses, fume, and particulate emissions will change materially in their amount as well as their characteristics.

Today, in steel foundry production, melting is primarily

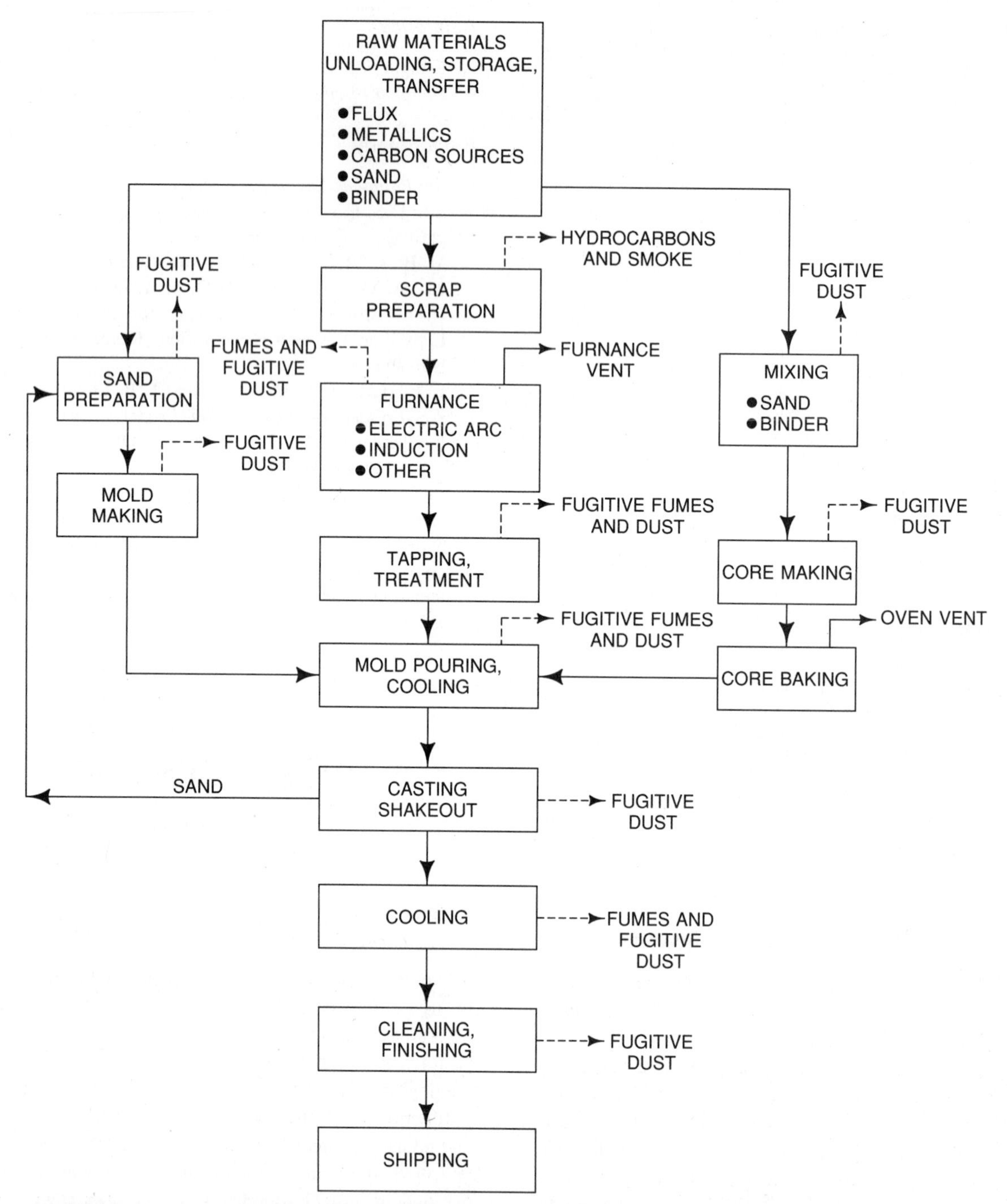

FIGURE 1. Typical Flow Diagram of a Steel Foundry and Sources of Emission Factors

accomplished with three types of electric furnaces—direct arc, indirect arc, and induction. Of these, the indirect arc melter is the lowest tonnage producer, with the direct arc and the coreless induction furnace accounting for most tonnage being produced. Figures 2, 3, and 4 show typical schematics of each of these melting units. Although as we advance into the 1990s, the more recent technological innovations—the argon oxygen decarburization (AOD) process and ultrahigh-power (UHP) electric arc melting—are gaining in popularity because of the steel casting industry's drive for increased productivity and cleaner steel, the fundamental melting units mentioned above are still depended on to furnish the major amount of molten steel for steel castings.

The Indirect Arc Furnace

The indirect arc furnace (Figure 2) may not be a large tonnage contributor, but in the foundries where it is the major melting unit, its melting quality is capable of satisfying today's stringent customer demands. The indirect arc furnace is a highly mechanized high-speed unit that requires

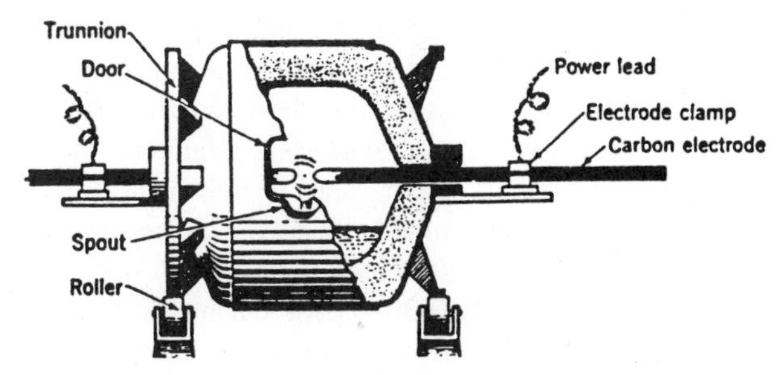

FIGURE 2. Indirect Arc Furnace "Rocking Arc"

efficient supervisory control. A steel cylinder, lined with granular insulation and a preformed refractory lining, rests on power-driven rollers that rock the cylinder. Two graphite electrodes, supported by port blocks, enter the center of the shell, one from each end of the furnace. Current from the transformer passes through the electrodes, which strike an arc in the center of the melting chamber. The molten metal is not part of the electrical circuit, as is the case in the induction furnace. The indirect arc furnace is equipped with one stationary and one movable electrode and cooling water is circulated through the electrode holders. By use of an automatic electrode adjustment control, an input of a fixed amount of power for a known alloy will produce molten metal of uniform pouring temperature. A carefully maintained lining is essential for the production of quality metal. One spare shell, completely relined, is always ready to replace a furnace at the end of its refractory life.

The working conditions around the indirect arc furnaces are excellent, although the noise of the electric arc may be objectionable to some people. If the furnace is not provided with an automatic electrode control, the varying noise of the arc aids the operating personnel in determining when to adjust the electrodes without having to stand constant watch over the furnace during the melting cycle. Metal melted in the indirect arc furnace receives heat from two sources. Radiant heat is obtained from the electric arc, while the secondary source of heat is the refractory lining, which is heated by the arc. The furnace is rocked at predetermined intervals, exposing more surface area of the lining to the heat of the arc, which delivers additional heat to the metal. The rocking permits the furnace to melt the materials faster, producing a more homogeneous melt and increasing the life of the furnace lining. Care must be taken to avoid the presence of oil or water in the metal charge and the refractory lining must be free of moisture. The rocking control is adjusted so that the degree of rock will increase uniformly, at a rate sufficient to ensure thorough mixing of the metal, but not so rapid as to cause electrode breakage during the meltdown.

The Direct Arc Furnace

In the direct arc furnace (Figure 3), the body of the furnace is enclosed in a refractory-lined cylindrical steel shell pivoted or mounted on rockers so that it may be tipped forward and backward. It is provided with three vertical electrodes, each connected to one phase of a three-phase alternating-current circuit. The furnace roof is built into a separate steel ring and rests on the side walls of the furnace shell. The electrodes are suspended from horizontal arms that are attached to structural steel masts and extend through openings in the furnace roof, which is provided with water-cooled rings that fit them quite closely. The electrodes are made of either graphite or amorphous carbon. Because their electrical conductivity is higher, graphite electrodes are smaller for a given current-carrying capacity than are those of amorphous carbon and they also are somewhat stronger, but more expensive. The electrodes have tapered threaded recesses at each end into which graphite or carbon connecting plugs are screwed. As an electrode becomes shorter, a new one is attached above it, and, in this manner, the suspended end of the old electrode can be used almost completely.

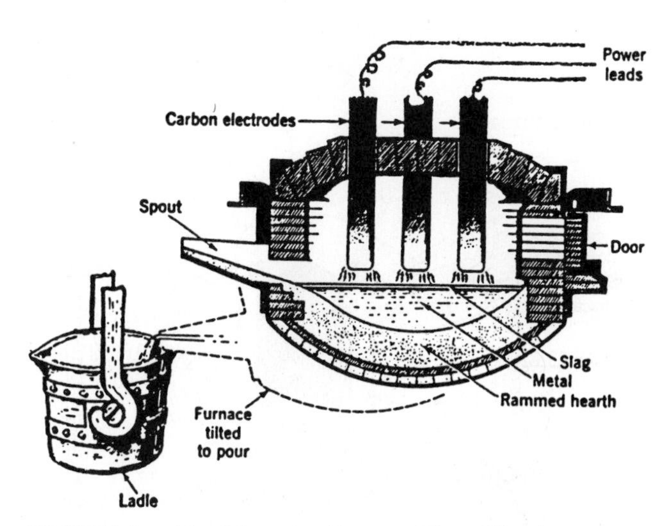

FIGURE 3. The Direct Arc Furnace (Three Phase)

The electric arc furnace usually has a working door on each side and a taphole and spout at the front. In most steel foundries today, the roof of the furnace can quickly be swung off the furnace's body and to one side to allow the charge to be placed in the furnace. The entire charge, contained in a drop-bottom bucket, is lowered into position with an overhead crane. In this way, the furnace can be charged, the roof replaced, and the current turned on in two or three minutes, whereas it previously required 15 to 30 minutes or more to charge the metal through the furnace doors. The metal is then melted and heated by arcs struck between the ends of the graphite or amorphous carbon electrodes and the charge of steel. Depending on the nature of the refractory brick used as the lining material for the electric arc furnace, the practice will identify itself as "basic" or "acid." Melting, slag practice, and deoxidation will vary according to the characteristics of the refractory lining and the alloy being melted.

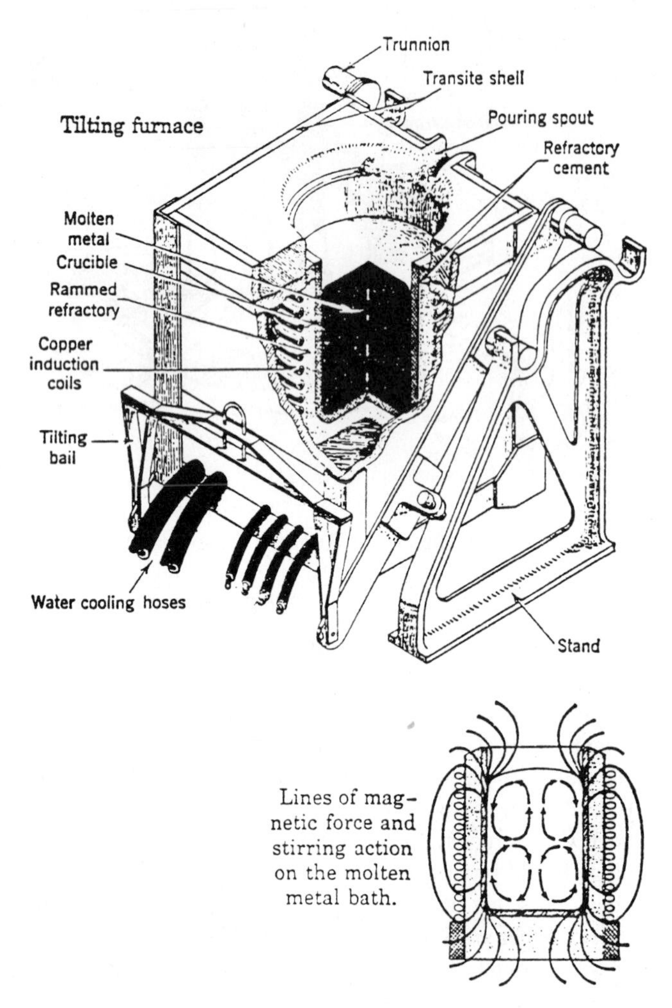

FIGURE 4. Coreless Induction Furnace

ACID PRACTICE

In acid electric practice, the furnace hearth is composed of silica sand or ganister rammed into place. The furnace is charged with selected scrap that is low in phosphorus and sulfur content, because the acid process is not able to eliminate these elements. Approximately 40% of the metallic charge is usually made up of foundry returns (gates and risers). The general practice is to charge small pieces first, in order to form a compact mass in the furnace, thus aiding electrical conductivity. The heavier portions of the charge are placed over the smaller pieces, followed by the lightest section of the charge. The power is turned on and the charge is melted down as quickly as possible. Small amounts of sand and limestone are added to the bath from time to time during the melting period to form a protective slag cover of proper consistency.

As soon as melting is complete, a good grade of iron ore is added or oxygen is used as an oxidizing agent to produce excess oxygen. Iron oxide in the slag reacts with the silicon and the manganese in the metal bath and produces oxides and silicates in the slag. After most of the silicon and manganese have been oxidized, the bath begins to "boil." The boil is evidence of carbon elimination, which results from the interaction of dissolved carbon and free oxygen, giving rise to a carbon monoxide (CO) bubble. In this oxidizing stage of melting, the slag is rich in iron oxide (FeO). As the carbon is eliminated from the molten iron, the FeO content of the slag decreases. The carbon content is reduced to approximately 0.20–0.25% during the vigorous carbon boil. The boil is stopped by the addition of carbon in the form of a carburizing iron of low sulfur and phosphorus content. Then further deoxidizers (ferromanganese and ferrosilicon) are added to the molten metal bath. The metal is then tapped into the pouring ladle. As a final deoxidizer, a small amount of aluminum is added to the ladle as the metal is tapped.

BASIC PRACTICE

For basic electric arc furnace melting, the furnace lining is a basic refractory such as magnesite or dolomite. The metallic charge is usually composed of purchased scrap steel and foundry returns. During the melting period, small quantities of lime are added from time to time to form a protective slag cover over the molten metal. The slag cover becomes highly oxidizing and in the correct condition to take up phosphorus from the metal. Shortly after all of the steel has melted, this first slag is taken off (if a two-slag process is to be used) and a new slag composed of lime and other slag-making agents is added. As soon as this second slag has melted, the current is reduced and, at intervals, pulverized coke, carbon, or ferrosilicon, or a combination of these, is spread over the surface of the molten bath. This period of furnace operation is known as the refining period, and its purpose is to reduce the oxides of iron and manganese in the slag and to form a calcium carbide slag, which is essential for the removal of sulfur from the metal. Adjustments are then made in the carbon content of the bath. After the proper bath temperature is obtained, ferromanganese and ferrosilicon are added to the bath and the furnace is tapped into the pouring ladle. Aluminum generally is added to the ladle during the tap as a final deoxidizer.

The basic electric arc furnace is indispensable in the manufacture of high-alloy steels. Alloy steels containing easily oxidized elements, such as chromium and manganese, can be remelted in this furnace without loss of these elements. For these applications, a single reducing slag is usually used to minimize oxidation loss.

The Induction Furnace

Two types of induction furnaces are available—the core type, or channel furnace, and the coreless type, or crucible furnace. Although both employ the same basic theory of inducing current into the metal to be melted, the number of coil designs, power sources, and frequencies used have shown that some furnaces work better than others in a given foundry situation, depending on the various alloys being melted.

The coreless high-frequency induction furnace (Figure 4) is essentially an air transformer in which the primary is a coil of water-cooled copper tubing and the secondary is the metal charge. A circular winding of copper tubing is placed inside the shell. Firebrick is placed on the bottom of the shell and the space between that and the coil is rammed with grain refractory. The furnace chamber may be a refractory crucible or it may consist of a rammed and sintered lining.

The process consists of charging the furnace with steel scrap and then passing a high-frequency current through the primary coil, thus inducing a much heavier secondary current in the charge, which results in heating it to the desired temperature. As soon as a pool of liquid metal is formed, a very pronounced stirring action takes place in the molten metal, which helps to accelerate melting. In this process, melting is rapid and there is a slight loss of the easily oxidized elements. If a capacity melt is required, steel scrap is added continually during the melting-down period. As soon as melting is complete, the desired superheat temperature is obtained and the metal is deoxidized and tapped into the pouring ladle, although some metallurgists and melters prefer to maximize the recovery of the deoxidizer alloys by adding the deoxidizers to the metal stream during tapping of the heat.

In most induction furnaces, no attempt is made to melt under a slag cover, because the stirring action of the bath makes it difficult to maintain a slag blanket on the metal. However, slag is not required because oxidization is slight. The induction furnace is especially valuable because of its flexibility in operation, particularly in the production of small lots of alloy steel castings. Induction melting is also well adapted to the melting of low-carbon steels because no carbon is picked up from the electrodes, as it is in an electric arc furnace. The induction furnace has continued to gain prominence in quality casting production. Its greatest disadvantage is that it is primarily a melter and not a metal refiner, which is the major advantage of the direct arc furnace.

PROCESS DESCRIPTION

The major processing operations of the typical steel foundry are raw-materials handling, metal melting, mold and core production, and casting and finishing (see Figure 1).

Raw-Materials Handling

The raw-materials handling operations include the receiving, unloading, storage, and conveying of all raw materials for the foundry. Some of the raw materials used by steel foundries are pig iron, iron and steel scrap, foundry returns, metal turnings, ferroalloys, carbon additives, fluxes (limestone, soda ash, fluorspar), sand, sand additives, and binders. These raw materials are received in ships, railcars, trucks, or containers and are transferred by trucks, loaders, and conveyors to both open piles and enclosed storage areas. The materials are then transferred by similar means from storage to the subsequent processing areas.

Metal Melting

Generally, the first step in the metal melting operation is metallic charge preparation. Since scrap metal for melting is normally purchased in the proper size for the particular furnace in which it is to be melted, preparation primarily consists of metallic charge degreasing. This is very important for electric induction furnace use, since organics contained on the metallic charge to be melted may be of an explosive nature. Metal scrap for remelting in the induction furnace should be preheated before use. It must be clean and dry (free of any moisture or concrete-contained encasements). A proper induction furnace melting facility should include a scrap preheater. Preheating of the charge ensures clean and dry scrap and also conserves a considerable amount of energy, since the preheated charge material has now been brought up to a temperature of approximately 1200–1600°F (649–871°C).

After preparation, the materials to be melted are charged into the induction furnace and the power is turned on to begin the melting process. When the total charge is melted, the bath surface is skimmed free of slag and the heat is tapped into the pouring ladle. At this time, the molten metal may be treated by adding alloys or deoxidizers, depending on the type of alloy being melted and the melting procedure that has been established. When the castings have solidified and the molds are partially cooled, the castings are placed on a vibrating grid and the sand of the mold and core is broken away from the casting. The sand is recycled to the sand-preparation center and then to the molding center, where a repeat of the total mold/core procedure for pouring is again carried out.

AIR EMISSIONS CHARACTERIZATION

Emissions from the raw-materials handling operations are fugitive particulates that are generated from the receiving, unloading, storage, and conveying of all raw materials for the foundry (see Figure 1). These emissions are controlled by enclosing the major emission points and routing the air from the enclosures through fabric filters. Emissions from scrap preparation consist of hydrocarbons, if solvent degreasing is used, and of smoke, organics, and CO, if heating is used. Catalytic incinerators and high-efficiency afterburners for CO and organics can be applied to these sources.

Emissions from melting furnaces are particulates, CO, organics, sulfur dioxide, nitrogen oxides, and small quantities of chlorides and fluorides (Table 1). The particulates, chlorides, and fluorides are usually generated by the flux and the carbon additives, dirt, and scale from the metallic

TABLE 1. Typical Emission Factors for Electric Arc and Electric Induction Melting for Steel Casting Production

Process	Particulates[a] lb/ton	Particulates[a] kg/Mg	Nitrogen Oxides lb/ton	Nitrogen Oxides kg/Mg
Melting				
Electric arc[b]	13 (4–40)	6.5 (2–20)	0.2	0.1
Electric induction	0.1	0.05	—	—

[a]Expressed as units per unit weight of metal processed. If the scrap metal is very dirty or oily, or if increased oxygen lancing is employed, the emission factor should be chosen from the high side of the factor range.

[b]Electrostatic precipitator, 92–98% control efficiency; baghouse (fabric filter), 98–99% control efficiency; venturi scrubber, 94–98% control efficiency.

TABLE 2. Chemical Analysis of Particulate Emissions (in wt %) during Various Melting Stages of a Typical 70-Ton Electric Arc Furnace Melting Cycle

Component	Melting Down	Ore Boil	Oxygen Injection	Refining
SiO_2	9.77	0.76	2.42	Traces
CaO	3.39	6.30	3.10	35.22
MgO	0.46	0.67	1.83	2.72
Fe_2O_3[a]	56.75	66.00	65.37	26.60
Al_2O_3	0.31	0.17	0.14	0.45
MnO	10.15	5.81	9.17	6.70
Cr_2O_3	1.32	1.32	0.86	0.53
SO_3	2.08	6.00	1.84	7.55
P_2O_5	0.60	0.59	0.76	0.55

[a]Calculated Fe_2O_3 equivalent of total iron content.

charge material. Organics on the scrap and the carbon additives affect CO emissions. The highest concentrations of furnace emissions occur during charging, backcharging, alloying, oxygen lancing, slag removal, and tapping operations when the furnace covers and doors are opened (Table 2). Traditionally, these emissions have escaped into the furnace building and have been vented through roof vents.

In the casting operations, large quantities of particulates can be generated in the steps prior to pouring. Emissions from pouring consist of fumes, CO, volatile organic compounds (VOCs), and particulates from the mold and core materials, when they are in contact with the molten steel. As the mold cools, emissions continue. A significant quantity of particulate emissions is also generated during the casting shakeout operation. The particulate emissions from the shakeout operations can be controlled by either high-efficiency cyclones or bag filters. Emissions from finishing operations consist of large particulates from the removal of burrs, risers, and gates, as well as during shot blasting and abrasive cleaning of the castings. Particulates from finishing operations typically are large and are generally controlled by cyclone collectors.

AIR POLLUTION CONTROL MEASURES

Controls for emissions during the melting and refining operations focus on venting the furnace gasses and fumes directly to an emission collection duct and control system. Controls for fugitive furnace emissions involve the use of either building roof hoods or special exhaust hoods near the furnace doors to collect emissions and route them to emission-control central gathering points. Emission control systems commonly used to control particulate emissions from the electric arc and induction furnaces are normally bag filters, electrostatic precipitators, or venturi scrubbers. The capture efficiencies of the various collection systems, presented in Table 1, range from 92% to 99%.

Emission control techniques involve an emission capture system and a gas cleaning system. Five emission capture systems used in the industry are fourth-hole (direct shell) evacuation, side-draft hood, combination hood, canopy hood, and furnace enclosures. Direct shell evacuation consists of ductwork attached to a separate or fourth hole in the furnace roof that draws emissions to the control device. The fourth-hole system works only when the electric arc furnace is upright, with the roof in place. Side-draft hoods collect furnace gasses from around the electrode openings and the work doors after the gasses leave the furnace. The combination hood incorporates elements from the side-draft and fourth-hole ventilation systems. Emissions are collected both from the fourth hole and from around the electrodes. An air gap in the ducting introduces secondary air for combustion of CO in the exhaust gas. The combination hood requires careful regulation of furnace internal pressure. The canopy hood is the least efficient of the first four ventilation systems listed, but it does capture emissions during charging and tapping operations. Many of the present-day electric arc furnaces incorporate the canopy hood with one of the other three systems. In the fifth system, an enclosure completely surrounds the furnace and emissions are evacuated through the hooding at the top of the enclosure.

Bibliography

Air Pollutant Control Techniques for Electric Arc Furnaces in the Iron and Steel Foundry Industry, EPA450/2-78-024, U.S. Environmental Protection Agency, Research Triangle Park, NC, 1978.

C. W. Briggs, *Metallurgy of Steel Castings,* McGraw-Hill Book Co., New York, 1946.

E. P. Degarmo, *Materials and Processes in Manufacturing,* 4th ed., Macmillan Publishing Co., 1974.

E. L. Kotzin, *Metalcaster's Reference and Guide,* 2nd ed., American Foundrymen's Society, Des Plaines, IL, 1989.

Steel Casting Handbook, 5th ed., P. Weiser, Ed.; Steel Founder's Society of America, Des Plaines, IL, 1980.

15
Mineral Products Industry

HOT MIX ASPHALT MIXING FACILITIES

Kathryn O'C. Gunkel

Hot mix asphalt (HMA) paving material is a scientifically proportioned mixture of graded aggregates and asphalt cement. The aggregates, which include stone, sand, and mineral dust, and can include reclaimed asphalt pavement (RAP), make up about 92–96% of the total mixture by weight. In addition to serving as paving materials for roadways, parking lots, race tracks, etc., HMA can also be used as liners for reservoirs, landfills, and other containment purposes. It is a unique paving material in that when it is removed from the roadway or other surface, it can be recycled back into new HMA paving materials, providing a pavement as good as one produced from all virgin materials.

The process of producing HMA involves drying and heating the aggregates to prepare them for the asphalt cement coating. The asphalt cement is typically stored at 300°F. Two types of drying and heating methods are available, both of which use a direct-fired, rotating drum. One method is counter-flow drying, in which the aggregates move opposite to the flow of the exhaust gases. This is the drying process used in the original process equipment and in recent design modifications to the drum mix process. The other drying process is a parallel-flow process in which the aggregates and exhaust gas move through the drum in the same direction.

There are two ways to coat the aggregates with asphalt cement. The original method, which still predominates, is to apply the coating in batches. A development in the mid-1970s introduced the asphalt cement into the lower end of the rotating drum of the parallel-flow drying process. This method of coating is called drum mixing. Recent design modifications now allow drum mixing to be performed with the counter-flow drying process.

PROCESS DESCRIPTION

The aggregates are dried in a rotating, slightly inclined, direct-fired drum. They are introduced into the end of the drum at the top of the incline. The interior of the drum is equipped with flights, which are shaped pieces of iron ("iron" is a term used to refer to the equipment materials of fabrication, even though the drum and its interior are typically fabricated from mild steel) that carry the aggregates up through the rotation of the drum while allowing them to shower through the hot exhaust gas. This process is called veiling. The design and installation of the flights have the most significant impact on the drying efficiency of the equipment. Dwell time, which is a function of the slope of the drum, its rotational speed, and also the flight design, is the second most important factor affecting drying efficiency. After drying, the aggregates typically are heated to temperatures ranging from 275°F to 325°F prior to coating with asphalt cement. The temperature of the aggregates is extremely important relative to the bonding of the asphalt cement.

Producing HMA Recycle Mixes

When RAP is used to produce recycle mixes, it must depend primarily on conductive heat transfer from the aggregates to reach temperature because it is difficult to process RAP through the combustion and heating zones. In the presence of very hot gases, RAP can smoke and its constituent asphalt cement is likely to be degraded to the point of affecting the quality of the final product. The amount of RAP used in a recycle mix can range up to 50% by weight. The existing process methods limit the amount of RAP that can be used because of the dependence on conductive heat transfer. The virgin aggregates must be "superheated," that is, heated beyond the desired mix temperature to provide the heat required by the RAP. Therefore, the total amount of virgin aggregates fed to the process can be reduced only so much before there is not enough aggregate, regardless of temperature, to provide sufficient heat for the RAP to reach temperature.

The Batch Mixing Process

In the batch mixing process (Figure 1), the aggregates will leave the drum at about 20–25°F higher than the desired mix temperature as measured in the transport vehicle. The exhaust gas temperature at its exit point should be lower than the aggregate temperature at its exit point. In terms of heat transfer and fuel efficiencies, the lowest possible exhaust gas temperature is desired. However, the type of particulate control equipment used will dictate how low the exhaust gas temperature can be (if a baghouse is used, exhaust gas temperatures at or near the dew point of the gas can result in condensation of moisture in the baghouse). The aggregates are carried to the top of the mixing tower in an elevator, typically referred to as the "hot elevator." At the top of the mixing tower is a set of screens that vibrate to classify the dried aggregates into the various sizes. Each screen transfers the aggregate it collects to a separate storage bin immediately below the screen deck. The effective diameter in the screen cloths decreases with each screen, from the top screen to the bottom screen.

The storage bins in the mixing tower, typically called hot bins, are insulated and/or heated. The individual sizes of aggregate are transferred from the hot bins to a weigh bucket in the proportions specified by the customer. From the weigh bucket, the aggregates are dropped into the mixing chamber, called the pugmill, which is equipped with two mixing shafts. The aggregates are "dry-mixed" for a few seconds to distribute the various sizes uniformly throughout the pug mill before the asphalt cement is added. Meanwhile, the asphalt cement is transferred to its weigh bucket, prior to adding it to the mixed aggregate. "Wet-mixing," whereby the aggregates and asphalt cement are combined, takes place for a few seconds more in the pugmill. The combined dry and wet mixing times will rarely exceed 42 seconds.

When recycling is performed in the batch mixing process, the RAP is added to the hot aggregates *after* they have left the drum, either in the boot of the hot elevator or directly into the pugmill. This can result in a significant fugitive emission problem if the fugitive-dust air handling system on the mixing tower has not been properly modified. The problem is a sudden, rapid release of steam resulting from evaporation of the moisture in the RAP upon mixing it into the superheated (often above 400°F) aggregates in the pugmill. In recent years, one equipment manufacturer has developed process equipment to preheat the RAP prior to introduction to the pugmill, which eliminates the release of steam. The maximum amount of RAP generally recommended when recycling in the batch mixing process is about 20–25% by weight of the total mix. If higher amounts of RAP are going to be used on a regular basis, it is usually necessary to go to one of the drum mixing processes.

The Parallel-Flow Drum Mixing Process

In this process (Figure 2), the aggregates are proportioned through a cold feed system prior to introduction to the drying process. The asphalt cement is introduced into approximately the lower third of the drum. The aggregates are coated with asphalt cement as they veil to the end of the drum. The mix is discharged at a little more than the desired mix temperature as measured in the transport vehicle. The RAP is introduced at some point along the length of the drum, generally as far away from the combustion zone as possible, but with enough drum length remaining to dry and heat the material adequately before it reaches the coating zone. The amounts of virgin aggregates and virgin asphalt cement fed to the process are adjusted according to the amount of RAP used and the condition of the asphalt cement in the RAP relative to the specified mix design being produced.

The Counter-flow Drum Mixing Process

When this process is used, the aggregates are proportioned through a cold feed system prior to introduction to the

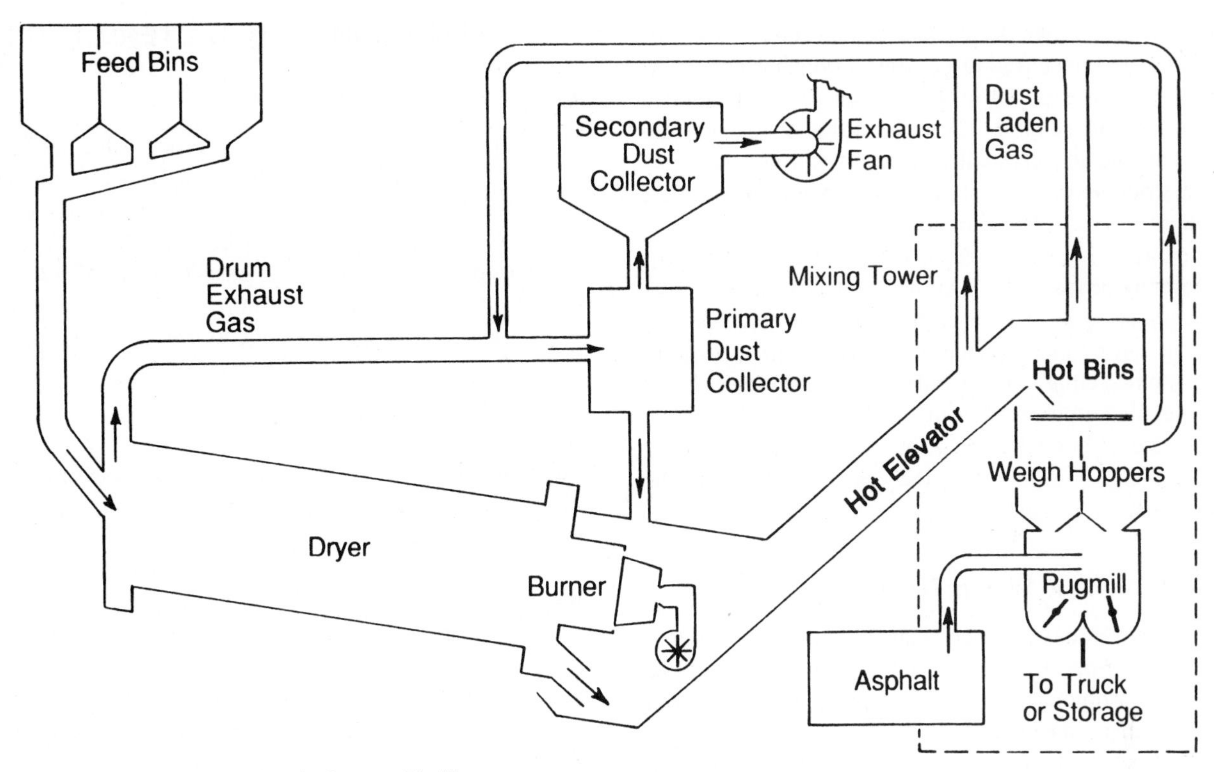

FIGURE 1. The Batch Mix Process Facility

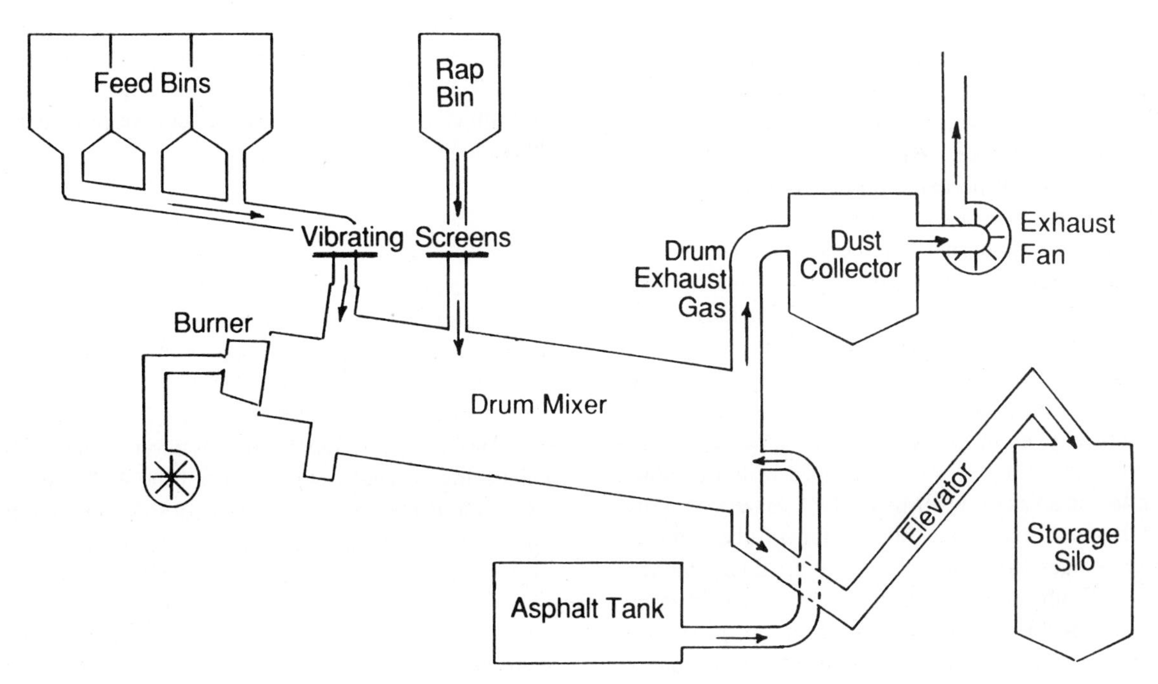

FIGURE 2. The Parallel-Flow Drum Mix Process

drying process. After drying and heating take place, the aggregates are transferred to a chamber, which is part of the drum and is not exposed to the exhaust gas, to be coated with the asphalt cement. There are different, distinct designs available to accomplish this, the purpose being to prevent stripping of the asphalt cement by the exhaust gas, as well as the resulting emissions and odors.

When recycle mixes are produced in a counter-flow drum mixing process, the RAP is introduced, usually, into the coating chamber. However, this limits the percentage of RAP that can be incorporated into a recycle mix, as compared with recycling in the parallel-flow drum mixer. In the event that higher percentages of RAP are required, in some designs, the additional RAP is introduced into the heating section of the drum to utilize the heat available from the exhaust gas.

Burners for HMA Mixing Facilities

There are two types of burners related to the introduction of combustion air into the process and two types related tc the use of combustion chambers. In combination, four types of burner systems can be found at HMA facilities that can have a direct effect on the formation of gaseous emissions from combustion. The burners can be designed to handle almost any type of fuel—natural gas, light fuel oils, liquified petroleum gas, heavy fuel oils, waste fuel oils, and coal.

The original type of burner design, and the most prevalent, is generally referred to as a 70/30 burner, or induced draft burner, by the HMA Industry. The terminology means that 70% of the total combustion air required (stoichiometric and excess air) is induced into the system around the burner by the exhaust fan and 30% is introduced into the system through the burner and used for atomization purposes. While they are generally called 70/30 burners, the actual ratio could be 65/35, 60/40, 55/45, etc. Frequently, these burners are the targets of noise complaints and so are enclosed with sound attenuators.

In the other type of burner, called a 100% air burner or forced draft burner, 100% of the combustion air passes (or is forced) through the burner into the system. This type of burner is an inherently quieter burner than the induced draft burner. Additionally, it is usually more fuel efficient under proper operating and maintenance conditions.

Originally, burners were designed and installed with refractory-lined combustion chambers; however, today's burners generally do not have these chambers. When a chamberless burner replaces one equipped with a combustion chamber on an existing HMA facility, it is essential that flighting be installed at the burner end of the drum to prevent aggregates from interfering with the flame, thereby maximizing combustion efficiency. Drum length is an essential design factor in parallel-flow drum mixers because of the need to provide a properly sized combustion zone, a drying and heating zone, and a coating section.

AIR EMISSIONS CHARACTERIZATION

Process Particulate Emissions

The air pollutant resulting from this process that has been of primary concern and is regulated by a U.S. Environmental Protection Agency (EPA) New Source Performance Standard (NSPS) as particulate matter. The NSPS for HMA facilities prohibits emissions of particulate in excess of 0.04 grains per standard cubic foot dry (gr/dscf) of exhaust gas. It also restricts visible emissions to less than 20% opacity.

Fugitive Particulate Emissions

Fugitive emissions can be a problem at the site of an HMA mixing facility, but can be controlled. There are several potential sources of fugitive emissions—unpaved driving surfaces, aggregate stockpiles, and, in the case of a batch mixing facility, the mixing tower. On batch mixing facilities that were constructed as NSPS facilities, the mixing tower is generally controlled with an auxiliary air handling system that discharges to the main particulate control system or to a separate particulate control system. The former setup is the type most commonly found.

The most frequent sources of particulate fugitive emissions at an HMA facility are unpaved driving surfaces. In past years, it was common to maintain unpaved driving surfaces primarily by utilizing wet-down techniques when the dust stirred up by vehicular traffic became excessive. In recent years, there has been a trend toward paving the driving surfaces to eliminate this problem.

Fugitive emissions from the aggregate stockpiles may be a problem in strong winds. Crusting agents have served fairly well to mitigate this problem. Watering of the stockpiles is not used, because of the burden it puts on the process that is designed to remove moisture from the aggregates. It is rare that fugitive emissions exceed visible emission limits set by some states for fugitive particulate emissions.

Gaseous Emissions

Because combustion is utilized in the HMA process, it is not unreasonable to expect gaseous air pollutant emissions. The possible pollutants are carbon monoxide, hydrocarbons as unburned fuel, oxides of nitrogen, and, if sulfur is present in the fuel, oxides of sulfur. Unfortunately, since there has not been much emphasis on documenting and quantifying gaseous emissions from the HMA mixing process, little data exist in the public domain relative to these emissions. There has been some attempt on the part of equipment manufacturers to identify the source or cause of these emissions. As a result, some manufacturers have modified or developed (or are currently doing so) equipment designed to minimize the cause of most of these gaseous pollutants.

Oxides of Sulfur Emissions

A study conducted by E. I. du Pont de Nemours & Co., Inc.,[1] in assessing gas stream compositions relative to baghouse fabric choices found that there was up to a 50% reduction of sulfur in the stack gases as compared with the amount anticipated on the basis of fuel analyses. It has been believed in the HMA industry that the aggregates adsorb the sulfur compounds from the exhaust gas, and that the higher the lime content of the aggregate, the better will be the removal of sulfur compounds.

Carbon Monoxide and Hydrocarbons (Unburned Fuel)

Carbon monoxide and hydrocarbons can result primarily for one reason. If the burner is of the type that does not utilize a refractory combustion chamber, but instead uses the end of the drum to which it is connected, improper flighting in the first few feet of the drum that allows aggregates to veil through the flame causes quenching. A secondary cause of these gaseous pollutants may be excess air entering the combustion process, particularly in the case of an induced-draft burner. In addition, the moisture content of the aggregate may contribute to the formation of carbon monoxide and unburned fuel emissions.

A two-year study conducted by the HMA industry in New Jersey in cooperation with the New Jersey Department of Environmental Protection found that these design elements appear to be crucial to the control of carbon monoxide formation and minimization of unburned fuel emissions. In general, HMA facilities with proper combustion zone design and/or air handling equipment had relatively low concentrations of carbon monoxide and hydrocarbons. Facilities that had high concentrations of these two pollutants in the first year were modified for the second year of stack sampling and exhibited reductions of from 37% to 71% from the first-year results.* Table 1 lists the results of this stack sampling program.[2] This study was but a small sampling of HMA facilities relative to such factors as fuels burned, equipment configuration, aggregate conditions, process temperatures and other operating parameters, process rates with respect to the rated process capacity, and mix design. Therefore, these data cannot be considered conclusive at this point in time.

The EPA conducted a study of volatile organic compound (VOC) emissions from parallel-flow drum mixers and published the results in 1981 (EPA-600/2-81-026) using Reference Method (RM) 25. The results of this study are not discussed here because, since that time, the EPA determined that when the product of the concentration of moisture as measured in the stack gas multiplied by the concentration of carbon dioxide as measured in the stack gas is 100 or greater, a positive bias in the results occurs. Since it is rare that the product of the stack gas moisture content and the carbon dioxide content is less than 100 in the stack gas of a typical HMA mixing process, except under conditions of excessive dilution that is frequently caused by air leaks into the system, the results of EPA-600/2-81-026 could be considered questionable. The EPA conducted side-by-side testing of RM 25 and a modification of the method that was expected to eliminate the bias with the New Jersey stack sampling program in 1987. The results indicated that the modification did not work.

TABLE 1. Carbon Monoxide and Hydrocarbon Emissions Results from New Jersey Stack Sampling Program

	CO		THC	
Sample No.	ppm	lb/h	ppm	lb/h
Facility 1 (drum mixer)				
WM-1/87	3,265	324.2	158	9.0
WM-2/87	2,401	235.1	227	12.7
WM-3/87	2,651	272.1	705	41.5
Average 1987	2,772	277.2	363	21.1
WM-1/88	76	8.2	114	7.1
WM-2/88	62	6.2	165	9.4
WM-3/88	59	6.2	91	5.6
WM-4/88	57	5.7	142	8.1
WM-5/88	64	5.9	68	3.6
Average 1988	63	6.5	116	6.8
Average 1988[a]	61	6.1	116	6.9
% Change	–97.7%	–97.7%	–68.1%	–67.8%
% Change[b]	–97.8%	–97.8%	–68.2%	–67.3%
Facility 2 (batch mixer)				
RL-1/87	3,287	313.8	358	19.6
RL-2/87	938	115.0	195	13.7
RL-3/87	2,973	261.3	257	12.9
Average 1987	2,399	230.0	270	15.4
RL-1/88	2,427	318.4	147	11.0
RL-2/88	636	59.6	90	4.8
RL-3/88	3,469	705.5	470	54.7
RL-4/88	69	7.2	67	4.0
RL-5/88	45	5.9	58	4.4
RL-6/88	178	18.7	82	4.9
Average 1988	1,137	185.9	152	14.0
Average 1988[a]	827	100.7	97	6.2
% Change	–52.6%	–19.2%	–43.6%	–9.2%
% Change[b]	–65.5%	–56.2%	–64.3%	–59.7%

*These modifications generally involved redesigning the combustion zone and/or the air handling equipment. In the case of liquified petroleum gas, the vaporization process was adjusted and fine tuned, generally focusing on the vaporization pressure.

TABLE 1. *(Continued)*

Sample No.	CO ppm	CO lb/h	THC ppm	THC lb/h
Facility 3 (batch mixer)				
WE-1/87	170	24.2	19	1.5
WE-2/87	114	15.7	13	1.0
WE-3/87	169	24.5	9	0.7
Average 1987	151	21.5	14	1.1
WE-1/88	51	6.1	19	1.3
WE-2/88	66	15.3	0	0.0
WE-3/88	27	4.3	5	0.5
Average 1988	48	8.6	8	0.6
% Change	–68.3%	–60.0%	–41.5%	–46.5%
Average of 1987 samples	1,776	176.2	216	12.5
Average of 1988 samples	520	83.8	108	7.8
Average of all samples	1,011	120.0	150	9.7
% Change from 1987 to 1988	–70.7%	–52.4%	–50.0%	–37.6%

[a]These averages do not include the highest result and the lowest result of the various samples collected and reported.

[b]These values reflect the percent change using the averages that do not include the high result and the low result, as indicated above.

Process Hydrocarbon Emissions

When asphalt cement is heated, fumes are released. Additionally, the industry has determined that light ends stripped from the asphalt cement by the exhaust gas can be an air pollution problem, primarily as a visible emission. The emission is seen as a detached plume that is blue to grayish-white in color. The extent to which the visible emission occurs is the direct effect of process temperatures—the higher the exhaust gas temperature, the worse is the visible emission. In recent years, equipment manufacturers have developed equipment modifications that minimize this problem. Odor is another emission problem with this process and is traced directly to contact between the exhaust gas and the asphalt cement.

There has been concern that the introduction of RAP into the drum in contact with the exhaust gas contributes to the visible emissions and possibly to the odors. Under adverse operating conditions, these concerns are valid. However, it has not been confirmed that RAP contributes to these visible emissions under well-controlled operating conditions in a properly designed HMA mixing facility.

The State of Maryland Air Management Administration conducted a study to determine whether these light ends stripped from the asphalt cement were entirely VOC emissions, and if not, to what extent they were VOCs.[3] The study found that only a small percentage by weight of the hydrocarbon emissions were VOCs. These results were presented at the 78th Annual Meeting of the Air Pollution Control Association in June 1975. The VOC emission factors that were developed from this study are listed in Table 2, and are based on asphalt cement consumption, as opposed to product production.

Fugitive Hydrocarbon Emissions

The potential sources of fugitive hydrocarbon emissions are the vents in HMA storage silos, the vents in asphalt cement storage tanks, and the transport vehicle loading area. To what extent these emissions contain VOCs has not been determined. However, given the means by which asphalt is derived, it is unlikely that there are any significant levels of VOCs in these fugitive emissions. The fugitive emissions from storage silos and the loading area can frequently occur as visible emissions. The fugitive emissions from asphalt cement storage generally take place during charging of the tank with asphalt cement and may be visible emissions.

AP-42 Emission Factors

The EPA's *Compilation of Air Pollution Emission Factors* (AP-42) provides emission factors for various gaseous pollutants. Unfortunately, the emission data come from stack sampling at only one HMA facility, and AP-42 gives the factors the lowest confidence rating possible. The HMA industry, through its national organization, the National Asphalt Pavement Association (NAPA), is developing a program to gather and evaluate existing, unpublished data for gaseous air pollutant emissions resulting from stack sampling of HMA facilities. NAPA, working with the EPA, is also organizing a stack sampling program to gather the necessary data to develop new emission factors for gaseous air pollutant emissions from the HMA mixing process. The emission factors provided in AP-42 are listed in Table 3.

Toxic Air Pollution Emissions

Toxic emissions resulting from the process of an HMA mixing facility have not been fully studied under various conditions. Because toxic air pollutants continue to be an issue, primarily in communities around HMA facility operations, since a number of states have enacted or proposed regulations governing these emissions and the Clean Air Act Amendments of 1990 have addressed them (as hazardous air pollutants [HAPs]), NAPA has included in the protocol of its stack sampling program sampling requirements for possible toxic air pollutants.

AIR POLLUTION CONTROL MEASURES

Particulate Control Equipment

Two types of control equipment are used for the control of particulate emissions from the HMA mixing process. The most common is the baghouse, which is essentially a dry scrubber used to remove particulate entrained in the exhaust

TABLE 2. VOC Emission Factors Developed by Maryland Study

	Pounds VOC/ton Liquid Asphalt			
Sample No.	Plant A	Plant B	Plant C	Plant D
MM-5-1[a,b]	0.658	0.232	0.259	0.309
MM-5-2	1.279	0.258	0.278	0.151
Average	0.969	0.245	0.269	0.230
V-1-1[c]	0.417	0.381	0.188	0.818
V-1-2	0.209	0.228	0.286	0.943
V-1-3	0.228	0.257	0.305	0.818
Average	0.283	0.289	0.260	0.860
V-2-1	0.188	0.193	0.313	0.485
V-2-2	1.072	0.224	0.261	0.581
V-2-3	0.444	0.295	0.583	0.680
Average	0.568	0.237	0.386	0.561
Air pollution control equipment type	Baghouse	Low pressure Venturi	Low pressure Venuri	Baghouse

[a]Average of factors (using 5% of MM-5 results): 0.604 pound VOC per ton asphalt cement.
[b]MM-5 refers to Modified Method 5 sampling method. Mass spectrophotometry was used to analyze the sample. It was estimated that approximately 5%, by weight, of the MM-5 sample was C-9 to C-13 chains, which are VOCs according to the EPA's vapor pressure definition for VOCs.
[c]V-1, etc., refers to the Volatile Organic Sampling Train (VOST) sampling method, which appeared to capture up to C-9 chains. These samples were also analyzed by mass spectrophotometry.

gas of the process prior to discharge into the environment. Some producers of the parallel-flow drum mixing process, where asphalt cement is in contact with the exhaust gas, claim that a reduced dust loading into the control equipment results from impingement of fines entrained in the exhaust gas into the asphalt cement, thereby allowing a higher air-to-cloth ratio in the baghouse for these processes. Condensation of hydrocarbons on the baghouses is always a possibility and poses a serious fire hazard. For the most part, the industry has learned to operate this equipment to minimize this hazard, and the number of fires and explosions occurring at these operations has decreased significantly over the past 10 years.

The other type of particulate control equipment is the

TABLE 3. Emission Factors for Selected Gaseous Pollutants from a Conventional Asphaltic Concrete Plant Stack

Material Emitted	Emission Factor Rating[a]	Emission Factor g/Mg	Emission Factor lb/ton
Sulfur oxides (as SO_2)	C	$146 \times S^b$	$0.292 \times S^b$
Nitrogen oxides (as NO_2)	D	18.000	0.036
Volatile organic compounds	D	14.000	0.028
Carbon monoxide	D	19.000	0.038
Polycyclic organic material	D	0.013	0.000026
Aldehydes	D	10.000	0.02
Formaldehyde	D	0.075	0.00015
2-Methylpropanal (isobutyraldehyde)	D	0.650	0.0013
1-Butanal (*n*-butyraldehyde)	D	1.200	0.0024
3-Methylbutanal (isovaleraldehyde)	D	8.000	0.016

[a]Emission factors are rated from A to D, A being the most reliable data, D being the least reliable.
[b]S = sulfur content of fuel by weight

Source: From the U.S. Environmental Protection Agency's *Compilation of Air Pollution Emission Factors* (AP-42), Section 8.1—Asphaltic Concrete Plants, October 1986 edition, Table 8.1-5.

venturi wet scrubber. It is not uncommon to find wet scrubbers that are not venturi in design, but these types are usually found on HMA mixing facilities that are not covered by the NSPS for HMA facilities. Generally, a high-pressure (20 inches, water gauge) venturi scrubber is required to meet the NSPS requirements. In addition to controlling particulate emissions, the venturi scrubber is likely to remove some of the process hydrocarbon emissions from the exhaust gas before discharge into the environment. However, owing to the high power requirements and water-discharge permitting issues, and also as a result of pressure exerted by some state air quality agencies, the wet scrubber is not as frequently installed as is the baghouse on HMA facilities.

The baghouse has proven to be very reliable with regard to meeting the NSPS requirements for particulate loading of the exhaust gas discharged to the environment. The high-pressure venturi scrubber is also reliable, but it requires considerable attention and daily and weekly maintenance to maintain a high degree of particulate removal efficiency.

Gaseous and Toxic Air Pollution Control Measures

Since no pressure has been exerted on the HMA industry with regard to the control of gaseous and/or toxic air pollutant emissions, add-on air pollution control equipment has not been evaluated for this process. However, the results of the New Jersey stack testing program for carbon monoxide and hydrocarbons indicate that operating management practices and proper process equipment design can significantly reduce the release of these two gaseous air pollutants into the environment.

References

1. H. H. Forsten, "Importance of gas stream analysis in selecting baghouse fabric," DuPont Fibers Department, Wilmington, DE, November 1990.
2. Summary of New Jersey's Asphalt Paving Association and Department of Environmental Protection joint stack testing program, 1987 to 1989, *Evaluation of Hot Mix Asphalt Plants Hydrocarbon and Carbon Monoxide Emissions,* submitted to New Jersey Department of Environmental Protection, 1989.
3. K. O'C. Gunkel and A. C. Bowles, "Drum mix asphalt plants—Maryland's experience," presented at the Air Pollution Control Association's 78th Annual Meeting, Detroit, MI, June 1975.

Bibliography

The Fundamentals of the Operation and Maintenance of the Exhaust Gas System in a Hot Mix Asphalt Facility, K. O'C. Gunkel, Ed.; National Asphalt Pavement Association, Information Series 52 (IS-52), Lanham, MD, 1987.

PORTLAND CEMENT

Walter L. Greer, Michael D. Johnson, Edward L. Morton, Errol C. Raught, Hans E. Steuch, and Claude B. Trusty, Jr.

Portland cement is a fine powder, usually gray in color, that consists of a mixture of the hydraulic cement minerals, dicalcium silicate, tricalcium silicate, tricalcium aluminate, and tetracalcium aluminoferrite, to which one or more forms of calcium sulfate have been added. Portland cement accounts for about 95% of the cement production in the United States. Masonry cement represents most of the balance of domestic cement production and is also produced in portland cement plants.

Portland cement and masonry cement are produced in several different types or formulations for specific purposes or properties. Chemical and physical specifications for the types of portland and masonry cements are written by several agencies, of which the most widely used are those provided by the American Society for Testing and Materials (ASTM). The most common types of portland cement are designated by the Roman numerals I through V. Types of masonry cement are designated by the letters N, S, and M.

The production of portland cement is a four-step process: (1) acquisition of raw materials, (2) preparation of the raw materials for pyroprocessing, (3) pyroprocessing of the raw materials to form portland cement clinker, and (4) grinding of the clinker to portland cement. In terms of productive capacity, a properly designed portland cement plant has the pyroprocessing operation as the limiting factor. Figure 1 is a basic flow diagram of the portland cement process. Figure 2 presents a layout of the cement plant most recently built in the United States and Figure 3 shows another modern cement plant. While the various unit operations and unit processes in portland cement plants accomplish the same end result, no flow diagram can fully represent all plants. Each plant is unique in layout and appearance owing to variations in climate, location, topography, raw materials, fuels, and preferences of equipment vendors and owners. These plants are capital intensive. In 1990, there were 111 plants in the United States producing approximately 80 million tons of portland cement.[1] Portland cement plants can run 24 hours per day for extended periods—six months or more with only minor downtime for maintenance is not unusual.

Raw materials are selected, crushed, ground, and proportioned so that the resulting mixture has the desired fineness and chemical composition for delivery to the pyroprocessing system. The major chemical constituents of portland cement are calcium, silicon, aluminum, iron, and oxygen. Carbon is a major constituent of the cement raw mix, but that element is eliminated during processing. Minor constituents, generally in a total amount of less than

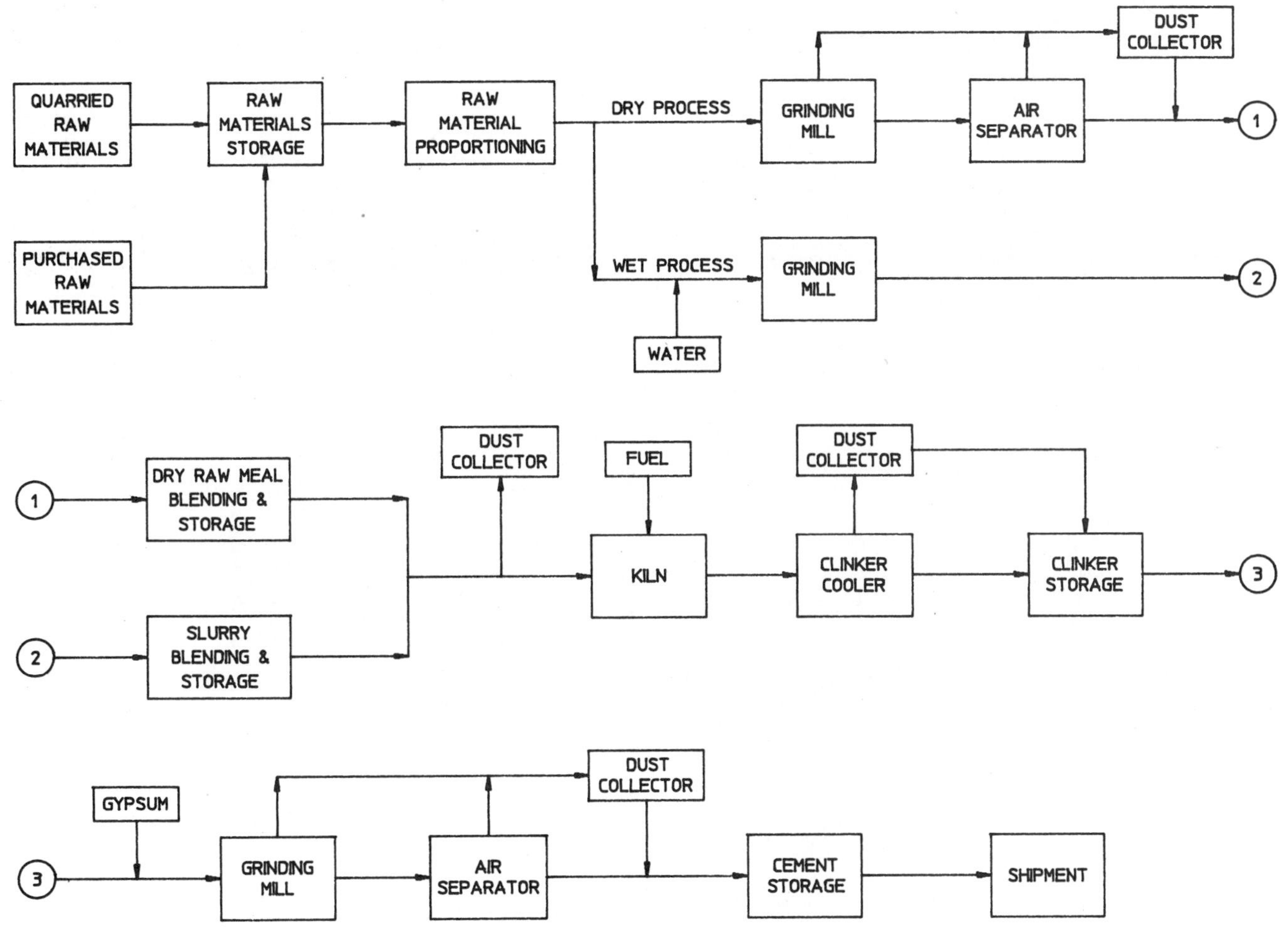

FIGURE 1. Basic Flow Diagram of the Portland Cement Manufacturing Process

5% by weight of the mixture, include magnesium, sulfur, sodium, and potassium. And since raw materials for portland cement usually come from the earth's crust, a wide variety of trace elements can be found in the cement, although these generally total less than 1% by weight of the mixture. Some of these naturally occurring trace elements can affect the performance of portland cement and/or appear in particulate emissions and process residues from cement plants. Most often, however, they harmlessly substitute for the four major metals in the crystalline matrix of the portland cement.

The more than 30 raw materials that are known to be used in the manufacture of portland cement can be divided into four categories: lime (calcareous), silica (siliceous), alumina (argillaceous), and iron (ferriferous). A limestone or other form of calcium carbonate will predominate in the mixture of raw materials. One or more quarries are usually associated with a portland cement plant. The terms slurry, raw meal, raw mix, and kiln feed are somewhat synonymous in naming the prepared raw materials or product of the raw mill department. At least 1575 kg (3388 pounds) of dry raw materials are required to produce 1000 kg (2200 pounds) of cement clinker. This ratio of feed to product can increase by several pounds due to the raw mix composition and dust removal. Most of the weight loss represents carbon dioxide, which is calcined from the calcium carbonate and emitted to the atmosphere during pyroprocessing.

Standard industry practice is to report the chemical analyses of raw materials, process intermediates, by-products, and portland cement as metal oxides, even though the constituents are rarely present in that form. If desired, the theoretical quantities of minerals in the cement matrix are calculated from the oxide analysis using specified formulas. Actual quantities of minerals may be determined by X-ray diffraction.

There are wet-process and dry-process portland cement plants. In the wet process, the ground raw materials are suspended in sufficient water to form a pumpable slurry. In the dry process, they are dried to a flowable powder. A variation of the dry process, the semidry process, uses a moist nodule or pellet to prepare the raw materials for pyroprocessing. Newer portland cement plants in the United States have almost exclusively used the dry process because of its lower thermal energy requirement. The Portland Cement Association estimated in 1988 that the average thermal energy used to produce a ton of cement in the United States

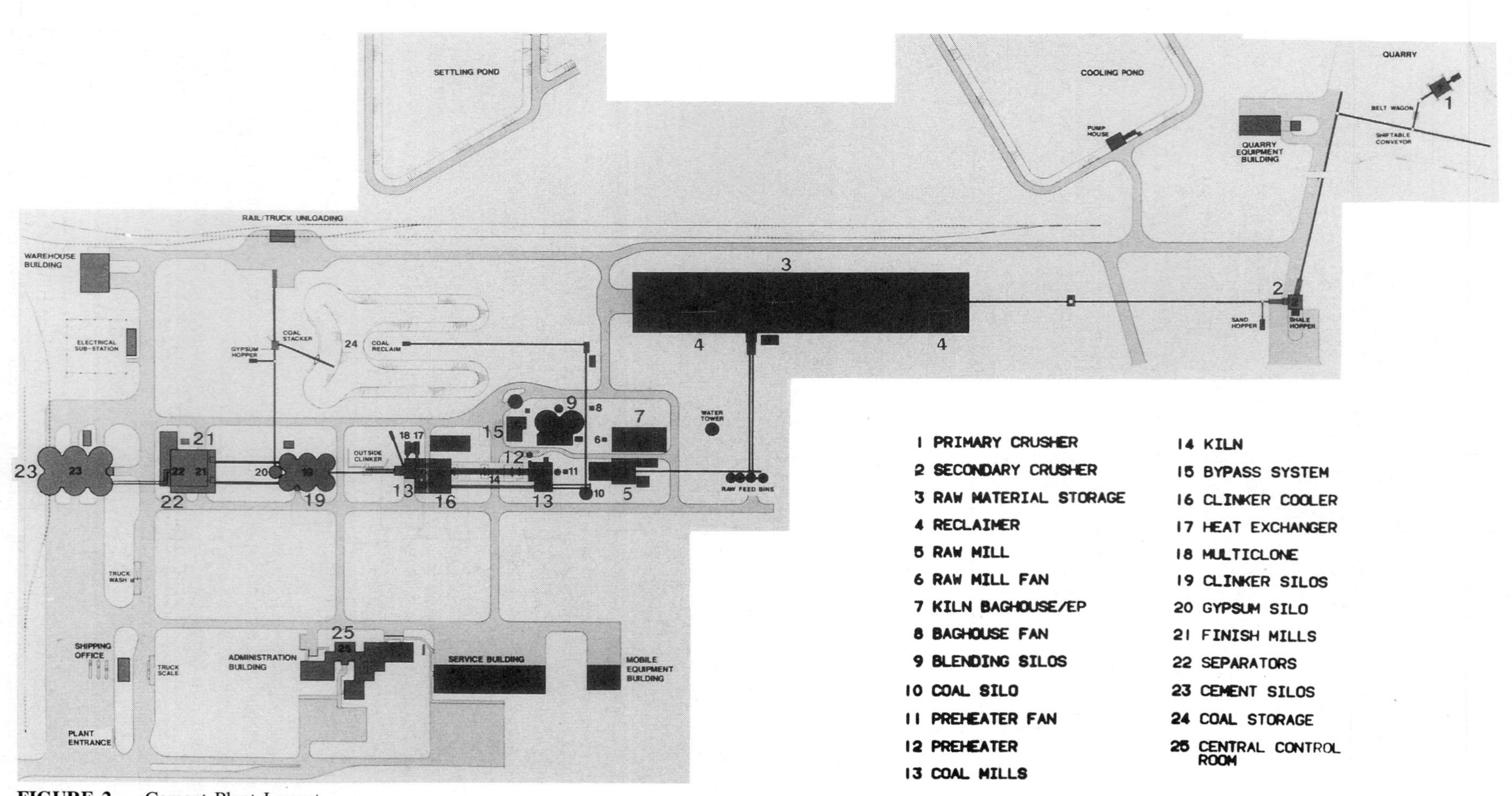

FIGURE 2. Cement Plant Layout

FIGURE 3. Modern Cement Plant

was about 4.4 million Btu. Thermal energy consumption ranges from about 3.0 to 7.0 million Btu per ton, depending on the age and design of the plant. Average electric energy consumption is about 0.5 million Btu (147 kWh) per ton of cement.

The prepared raw materials are fed to one of several pyroprocessing systems in the kiln or burning department. Each system accomplishes the same result via the following basic steps: evaporation of free water, evolution of combined water, calcination of the carbonate constituents (decarbonization), and formation of the portland cement minerals (clinkerization). The wet process uses rotary kilns exclusively. The semidry process uses a moving grate on which the moist nodules are dried and partially calcined by hot kiln exhaust gases before being fed to a rotary kiln. The dry process can also employ simple rotary kilns. Thermal efficiency can be improved, however, through the use of one or more cyclone-type preheater vessels that are arranged vertically, in series, ahead of the rotary kiln in the material flow. It can be further improved by diverting up to 60% of the thermal energy (i.e., fuel) required by the pyroprocessing system to a special calciner vessel located under the preheater vessels.

The rotary kiln is the heart of the portland cement process since the several and complex chemical reactions necessary to produce portland cement minerals take place there. The portland cement kiln is a slightly inclined, slowly rotating steel tube that is lined with appropriate refractory materials. The rotation of the kiln causes the solid materials to be slowly transported downhill from the feed end. Fuel is supplied at the lower or discharge end of the kiln. Many fuels can be used in the kiln, but coal has predominated in the United States since the mid-1970s. The choice of fuel is based on economics and availability. The hot, gaseous combustion products move countercurrent to the material flow, thereby transferring heat to the solids in the kiln load. Flame temperatures in excess of 3400°F result in the material temperatures of 2700–2800°F that are required to produce the hydraulic calcium and aluminum silicates.

The product of the rotary kiln is known as clinker. Heat from this clinker is recuperated in one of three types of clinker cooling devices and returned to the kiln by preheating combustion air.

The cooled clinker is mixed with a form of calcium sulfate, usually gypsum, and ground in ball or tube mills in the finish mill department to produce portland cement. Masonry cement is similarly produced from portland cement clinker, gypsum, and one or more calcareous materials.

Portland cements are shipped from the packhouse or shipping department in bulk or in paper bags by truck, rail, barge, or ship. Masonry cements are shipped in paper bags.

Except for the quarry and rock crushing operation, the New Source Performance Standards (NSPS) that apply to a new or modified portland cement plant constructed after August 17, 1971, are contained in 40 CFR 60, Subpart F, *Standards of Performance for Portland Cement Plants*.[2]

Emission factors for portland cement plants are contained in Section 8.6 of the October 1986 Supplement to the U.S. Environmental Protection Agency (EPA) publication, *Compilation of Air Pollutant Emission Factors*, (AP-42).[3] These emission factors have limited usefulness because of the diversity of cement plant design and operation, as well as the old and limited data on which the factors are based. For these reasons, the AP-42 emission factors have not been included here.

An explanation of the operation of the common dust collection devices used in the cement industry, such as multiclones, fabric filters, and electrostatic precipitators (ESPs), is beyond our scope here but can be found elsewhere in this manual.

ACQUISITION OF RAW MATERIALS

Process Description

The initial step in the manufacture of portland cement is the acquisition of raw materials. The industry is considered an extractive industry since nearly all of the raw materials required are obtained from the earth's crust by mining or quarrying. Most cement plants are located near a source of calcium carbonate, which is most often limestone. Since about one third of the weight of the limestone is lost as carbon dioxide during pyroprocessing, process economics dictate that this lost weight be transported as short a distance as possible. Those plants that are not immediately associated with a limestone quarry often have a source of limestone or other form of calcium carbonate (e.g., aragonite) that is available by less expensive water transportation. However, there are a few exceptions to these generalizations on plant location.

Calcium is the metallic element of highest concentration in portland cement. The calcareous raw materials include limestone, chalk, marl, seashells, aragonite, and an impure limestone known in the industry as natural cement rock. Limestone, chalk, and cement rock are most often extracted from open-face quarries, but underground mining can be employed. Dredging and underwater mining techniques are used to develop deposits of calcareous raw materials in the ocean or below the water table. Gypsum and/or natural anhydrite (i.e., forms of calcium sulfate) from quarries or mines are calcium-bearing constituents of portland cement that are introduced as part of the final stage of its manufacture, finish grinding. It is rare for a cement plant to have a captive source of gypsum or anhydrite and these materials are usually purchased.

Silicon, aluminum, and iron are the next most prevalent metallic elements in normal portland cement and are listed in descending order of concentration. These metals are found in various siliceous, argillaceous, and ferriferous ores and minerals, such as sand, shale, clay, and iron ore. Although usually extracted in open-face quarries or pits, these raw materials can be dredged or excavated from underwater deposits. They can be obtained from captive sources adjacent to or away from the portland cement plant; however, it is often necessary or economical for the cement manufacturer to purchase them from outside sources.

The wastes and by-products of other industries are successfully employed as portland cement raw materials. Such materials include, but are not limited to, power plant fly ash, mill scale, and metal smelting slags.

The cement manufacturing process and the performance of portland cement are sometimes affected by trace elements that are found in virgin raw materials or wastes. Care must be exercised in selecting these raw materials to assure that trace elements will not be present in high enough concentrations to cause problems in the plant or product.

Air Emissions Characterization

Quarries at cement plants are similar to other stone quarries. The necessary operations include rock drilling, blasting, excavation, loading, hauling, crushing, screening, materials handling, stockpiling, and storing. There are many different operating methods, types of equipment, and equipment brands that are used to accomplish these tasks. Particulate matter is the primary air pollutant associated with quarry operations. In some locations, exhaust emissions from mobile equipment may be of concern. There are usually no atmospheric air pollution problems at underground mines or underwater operations.

The NSPS that apply to quarry and crushing operations at portland cement plants are contained in 40 CFR 60, Subpart 000, *Standards of Performance for Nonmetallic Mineral Processing*.[4] These standards are applicable to those affected facilities that commenced construction, reconstruction, or modification after August 31, 1983.

Raw materials can also be the source of some environmentally undesirable emissions from the kiln stack later in the process. If the raw materials contain naturally occurring hydrocarbons, such as petroleum or kerogens, these materials can evaporate in the relatively cooler portions of the pyroprocessing system and appear at the stack exit as a "blue haze." Sulfur and chlorine from raw materials can participate in reactions with the small amount of ammonia in fossil fuel combustion products to form a "detached plume" of ammonium sulfate or ammonium chloride. Nitrogenous constituents of the raw materials can possibly contribute to NO_x emissions that are unrelated to combustion. Sulfides in raw materials have been identified as contributors to sulfur dioxide (SO_2) emissions under some operating conditions.

Air Pollution Control Measures

Control measures for particulate emissions in quarries include water sprays with and without surfactants, foams, chemical dust suppressants, wind screens, equipment enclosures, paving, mechanical collectors and fabric filters on operating equipment, and material storage buildings, enclosures, bins, and silos with and without exhaust venting to fabric filters. Collected dust is returned to the process.

Typical fabric filters found in the quarry are pulse jet in newer plants and reverse air or shaker types in the older plants. Table 1 presents typical data.

Purchased raw materials, including coal or petroleum coke for fuel, can also generate particulate emissions as a result of vehicle loading and unloading, material handling, stockpiling, and haulage. The particulate emission control measures for purchased materials are the same as those listed for quarries.

RAW MILLING

Process Description

The second step in the manufacture of portland cement is the preparation of the raw materials for pyroprocessing. This operation in the raw mill department combines the blending of appropriate raw materials for proper chemical composition with particle-size reduction through grinding.

Grinding is required to achieve optimum fuel efficiency in the cement kiln and to develop maximum strength potential and durability in portland cement concrete. Typically, the raw material in the kiln feed is ground to about 85% passing a 200-mesh (74-μm) sieve or 90% passing a 170-mesh (88-μm) sieve. Usually less than 1% of the material is retained on the 50-mesh (297-μm) sieve. The actual fineness that is required depends on the reactivity of the raw material components. Material that is too finely ground wastes energy and reduces the productive capacity of the raw mill. Unnecessary grinding can also result in excessive dusting in the cement kiln. This dust reduces available draft in the kiln and upsets the combustion process, thereby potentially lowering product quality and kiln production rates. Raw milling processes are either wet or dry, depending on the type of pyroprocessing system(s) at the plant.

When raw materials are dried before grinding or when the physical properties of the moist materials permit handling, the raw materials are usually proportioned with weigh feeder systems located in the process flow ahead of a mill feed bin or the raw mill itself. If required and justified by process economics, raw materials can also be proportioned and blended in large (e.g., 1000 feet long) linear or circular stacker-reclaimer systems, which are sometimes located in closed buildings.

Cement raw materials are received in the raw mill department with a moisture content varying from 2% to 35%. In the dry process, this moisture is usually reduced to less than 1% before or during grinding. Drying prior to grinding is accomplished in impact dryers, drum dryers, paddle-equipped rapid dryers, air separators, or autogenous mills. Drying can also be carried out during grinding of the raw mix in ball and tube mills or roller mills. Thermal energy for drying can be supplied by separate, direct-fired coal, oil, or gas burners that heat the airstream that passes through the drying apparatus or mill. The most efficient and popular source of heat for drying is the hot exit gases from the pyroprocessing system. These gases can come from the kiln, the clinker cooler, the alkali bypass system, or a combination of these sources. Unless the hot gases are supplied from the clinker cooler, the gases passing through dryers and raw mills will contain products of combustion, as well as solid particles. The selection of the drying method depends on the physical properties of the raw materials, the type of pyroprocessing system in the plant, the availability and cost of energy, and the preferences of owners, managers, and vendors.

TABLE 1. Fabric Filters in Quarries

	Pulse Jet	Reverse Air/Shaker
acfm	5,000–25,000	5,000–25,000
Fabric type	Polyester	Polyester
Temperature range, °F	<275	<275
A/C ratio[a]	6:1	2.5:1
Inlet loading, gr/acf	5–40	5–40
Expected outlet emissions, gr/acf	0.02	0.02
Particle size out, μm	>1.0	>1.0

[a]Air-to-cloth ratio (acf/ft^2).

Ball and tube mills (i.e., long ball mills) are rotating, horizontal steel tubes that contain steel balls and are used to provide comminution or grinding of the raw materals. Air separators are frequently used in conjunction with these mills in the dry process to separate materials of adequate fineness from the coarse particles that must be returned to the grinding mill for additional work (i.e., closed-circuit grinding). The design and operation of these mills and air separators are discussed in the following description of finish milling. The hot gases required for simultaneous drying and grinding in a ball mill system can enter the feed end of the mill and flow concurrently with the raw materials. Otherwise, the unground raw materials and the hot gases are introduced simultaneously into the air separator. Some operators feel that this latter procedure provides for more efficient drying, easier operation of the mill circuit, and early removal from the mill system of those materials that are already sufficiently ground. The separator does, however, experience additional wear.

Vertical roller mills are very popular in new, dry-process portland cement plants because of their relative simplicity and high efficiency. The principle of operation of these mills is similar to that of a mortar and pestle. In this case, the pestle (rolls) is stationary and the mortar (table) rotates. Raw materials are dropped on the rotating table to be crushed and ground between the rolls and the table. Hot gases enter the mill through an annular duct at table height. As ground material is forced off the table into the hot gas stream, it is entrained in the gases, dried, and transported upward to internal separators from which coarse material is returned to the mill by gravity. Figure 4 is a process diagram of a typical vertical roller mill raw milling circuit; Figure 5 shows an installed roller mill.

Materials are transported to, within, and away from dry raw milling systems by a variety of mechanisms, including screw conveyors, belt conveyors, drag conveyors, bucket elevators, air slide conveyors, and pneumatic conveying systems.

The dry raw mix is pneumatically blended and stored in specially constructed silos until it is fed to the pyroprocessing system.

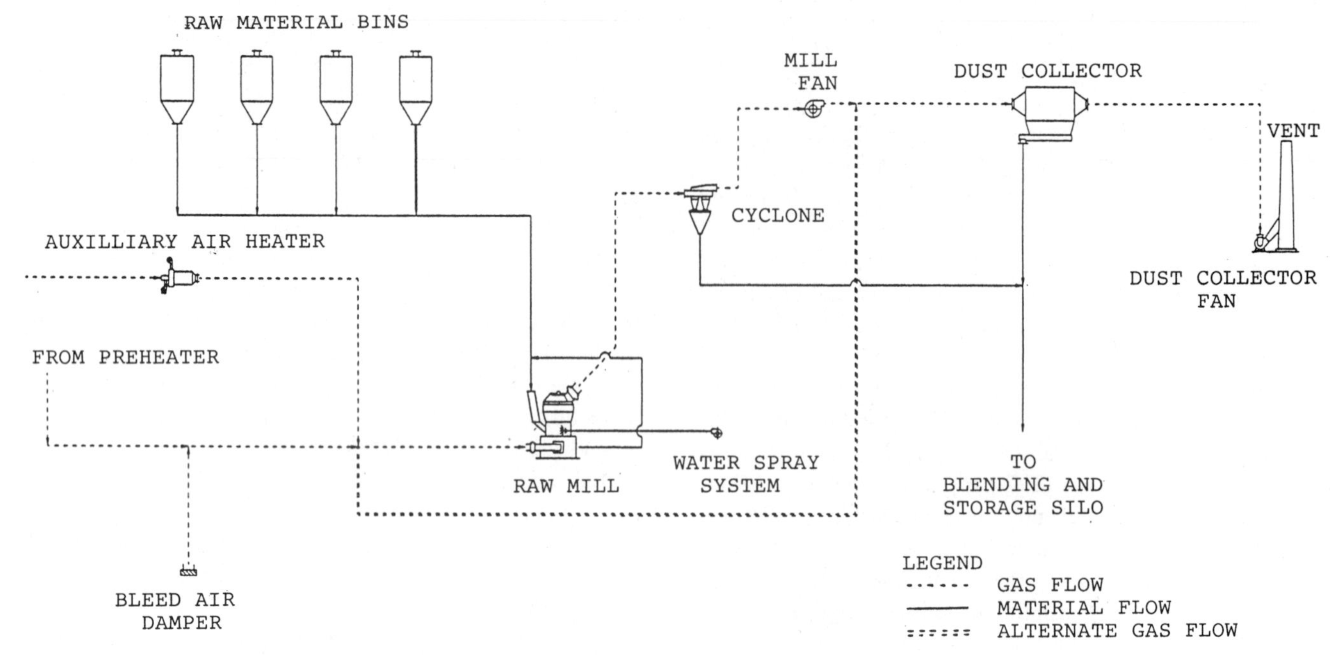

FIGURE 4. Raw Mill Schematic Diagram

In the wet process, water is added to the raw mill during the grinding of the raw materials in ball or tube mills, thereby producing a pumpable slip or slurry of approximately 65% solids. The slurry is agitated, blended, and stored in various kinds and sizes of cylindrical tanks or slurry basins until it is fed to the pyroprocessing system. Until recently, the advantage of the wet process was that the chemical composition of the kiln feed could be controlled more closely since slurries blend more easily than powders. Modern equipment can now blend raw meal powders satisfactorily.

In the semidry process, the dry raw mix is transformed into pellets by the addition of water in a pelletizing device such as a pan pelletizer. These pellets are then directly fed to a moving grate preheater at the start of the pyroprocessing system. In a little-used European system described as the semiwet process, slurry is filtered in filter presses to reduce the moisture content. Nodules of the filter cake are formed in kneading mills and pans before the raw mix is fed to a grate preheater kiln system.

Air Emissions Characterization

The raw material feeders, stackers, blenders, and reclaimers can produce fugitive dust emissions. Transfer points on belt conveyor systems and bucket elevators that serve to transport raw materials from storage to the raw mill department can also generate fugitive dust emissions.

The dry raw mills and the auxiliary equipment are all designed to run under negative pressure to suppress particulate emissions. Nevertheless, poorly designed or maintained seals and closures throughout the system can result in fugitive dust emissions. If these systems experience positive pressure through a fan failure or other cause, short-term particulate emissions can be expected until the system can be shut down.

During colder weather, the vents from dryers, raw mills, and air separators may exhibit a steam plume that is sometimes confused with particulate emissions. The condensate will dissipate within a few feet of the emission point. Fabric filters in the vent circuits for dryers, raw mills, and air separators must be insulated to prevent internal moisture condensation and the resultant blinding of bags.

There are no particulate emissions from the wet grinding process, except for the materials-handling systems ahead of the mills.

Air Pollution Control Measures

Dust collecting devices in the raw mill and raw mix storage areas include mechanical cyclones, fabric filters, and, rarely, ESPs. Mechanical collectors are usually used in series with one of the other devices. The collected dust is returned to the mill system or raw mix stream.

Typical fabric filters found in the raw mill area are pulse jets in the newer or upgraded plants and reverse air or shaker types in the older plants. Cartridge-type filters can be found on materials-handling equipment. Table 2 presents typical data for independent raw mill systems. Vertical mills are most often associated with the pyroprocessing system. The air pollution control measures for these mills will be found in the following section on pyroprocessing.

FIGURE 5. Vertical Roller Mill

PYROPROCESSING

Process Description

The third step in the manufacture of portland cement is the pyroprocessing of the raw mix into portland cement clinker. Clinkers are gray-colored, glass-hard, spherical-shaped nodules that generally range from ⅛ inch to 2 inches in diameter. The clinkers are predominantly composed of the cement minerals, dicalcium silicate, tricalcium silicate, calcium aluminate, and tetracalcium aluminoferrite, that result from chemical reactions between the cement raw materials that are completed at the temperature of incipient fusion. The chemical reactions and physical changes that describe the transformation are very complex. A simplified version of the major sequential events is as follows:

TABLE 2. Fabric Filters for Raw Mill Systems

	Pulse Jet	Reverse Air/Shaker	Cartridge
acfm	10,000–50,000	10,000–50,000	2,000–10,000
Fabric type	Polyester	Polyester	Paper, polyester
Temperature range, °F	<275	<275	<180, 275
A/C ratio	5:1	2.5:1	<2:1, N/A
Inlet loading, gr/acf	5–20	5–20	5–20
Expected outlet emissions, gr/acf	0.02	0.02	0.02
Particle size out, μm	>0.5	>0.5	>0.5

- Evaporation of free water.
- Evolution of combined water in the argillaceous components.
- Calcination of the calcium carbonate to calcium oxide.
- Reaction of calcium oxide with silica to form dicalcium silicate.
- Reaction of calcium oxide with the aluminum and iron-bearing constituents to form the liquid phase.
- Formation of the clinker nodules.
- Evaporation of volatile constituents (e.g., sodium, potassium, chlorides, and sulfates).
- Reaction of excess calcium oxide with dicalcium silicate to form tricalcium silicate.

The pyroprocessing system is generally described as containing three steps or zones: (1) drying or preheating, (2) calcining, and (3) burning or sintering. The pyroprocessing is accomplished in the burning or kiln department. The word "burning" is jargon that is used in the cement industry to describe the intense heat in the kilns. None of the constitutents of the cement raw mix actually combust during pyroprocessing.

The raw mix is fed to the pyroprocessing system as a slurry in the wet process, as a powder in the dry process, and as moist pellets in the semidry process. A rotary kiln is the common element in all pyroprocessing systems, and it will always contain the burning zone and all or part of the calcining zone. All of the pyroprocessing steps occur in the rotary kiln in wet-process and long, dry-process (i.e., no preheater) systems. The application of chemical engineering principles to cement pyroprocessing has resulted in equipment additions to the rotary kiln system that can accomplish preheating and most of the calcining more quickly and efficiently outside of the kiln. In the semidry process, the drying or preheating step and a small degree of calcination are accomplished outside the kiln on a system of moving grates through which hot kiln gases are passed.

Rotary kilns are rotating, cylindrical steel tubes with length-to-diameter ratios in the approximate range of 15:1 to 35:1. The kiln size and relative proportions are determined by the process type and productive capacity of the pyroprocessing system. Wet-process kilns of over 700 feet in length and up to 23 feet in diameter are in operation. However, many wet- and all dry-process kilns are shorter—and dry-process kilns that are equipped with preheaters are shorter yet. The kiln rotates about the longitudinal axis, which is slightly inclined to the horizontal, at a speed of from 1 to 3.5 rpm. Refractory material lines the kiln to protect the steel shell from the intense heat and to retain heat within the kiln. The inclination and rotation of the tube result in the transport of solid materials from the upper or feed end to the lower or discharge end. The solids (i.e., load) occupy no more than 15–20% of the internal volume of the rotary kiln inside the refractory. There will be hundreds of tons of material within the kiln at any particular time. Material transit time is measured in hours. Heat energy is supplied to the discharge end of the kiln through combustion of a variety of fuels. The flow of hot, gaseous combustion products is, therefore, countercurrent to the material flow. Heat is transferred from the flame and hot gases to provide the driving force for the required chemical reactions. The solid material is heated to more than 2700°F by flame temperatures in excess of 3400°F.

Wet-process and long, dry-process pyroprocessing systems consist solely of the simple rotary kiln. Usually, a system of chains is provided at the feed end of the kiln in the drying or preheat zones to improve heat transfer from the hot gases to the solid materials. These chains are attached to the inside of the kiln shell in various patterns. As the kiln rotates, the chains are raised and exposed to the hot gases. Further kiln rotation causes the hot chains to fall into the cooler materials at the bottom of the kiln, thereby transferring the heat to the load.

Dry-process pyroprocessing systems have been improved in thermal efficiency and productive capacity through the addition of one or more cyclone-type preheater vessels in the gas stream after the rotary kiln. The vessels are arranged vertically, in series, and are supported by a structure known as the preheater tower. Hot exhaust gases from the rotary kiln pass countercurrent through the downward-moving raw materials in the preheater vessels. Compared with the simple rotary kiln, the heat transfer rate is significantly increased, the degree of heat utilization is more complete, and the process time is markedly reduced owing to the intimate contact of the solid particles with the hot gases. The required length of the rotary kiln is thereby reduced.

The hot gases from the preheater tower are often used as a source of heat for the drying of raw materials in the raw mill. The mechanical collectors, fabric filters, and/or ESPs that follow the raw mill are production machines as well as pollution control devices.

Additional thermal efficiencies and productivity gains have been achieved by diverting some fuel to a calciner vessel at the base of the preheater tower. At least 40% of the thermal energy is required in the rotary kiln. The amount of fuel that is introduced to the calciner is determined by the availability and source of the oxygen for combustion in the calciner. If available and allowed by environmental regulations, calciner systems can use lower-quality fuels (e.g., less volatile matter) and thereby further improve process economics.

In preheater and calciner kiln systems, it is often necessary to remove the undesirable volatile constituents through a bypass system located between the feed end of the rotary kiln and the preheater tower. Otherwise, the volatile constituents would condense somewhere in the preheater tower and subsequently recirculate to the kiln. Buildups of these condensed materials can also restrict process flows by blocking gas and material passages. In a bypass system, a portion of the kiln exit gas stream is withdrawn and quickly cooled by air or water to condense the volatile constituents to fine particles. The solid particles are removed from the gas stream by fabric filters or ESPs.

Figure 6 is a flow diagram of a four-stage preheater with calciner pyroprocessing system that is equipped with an alkali bypass and a reciprocating grate clinker cooler. Figure 7 shows a four-stage preheater kiln system next to the traditional rotary kiln that it replaced.

Air Emissions Characterization

In simple rotary kiln systems, some finely divided particles of raw mix, calcined kiln feed, clinker dust, and volatile constituents (e.g., potassium sulfate) are entrained in the exiting gas stream. These particles are almost entirely removed from the gas stream before the combustion products are vented to the atmosphere. Affected pyroprocessing systems always meet or exceed the NSPS for particulate emissions from portland cement plants. Even those plants built prior to 1971 that are not subject to NSPS usually meet these standards for particulate emissions.

The powder that is collected from the kiln exhaust gases is known as cement kiln dust (CKD). Some plants return all or a portion of the CKD to the process; others completely remove it from the process. The chemical composition and physical state of the CKD depend on the type of pyroprocessing system, the chemical composition of the raw materials and fuel, and the state of the process at any given time. The chemical composition of the CKD that is caught in the last field of an ESP is very similar to that of the particulate emissions from the kiln stack. The same generalization cannot be made about CKD caught in a fabric filter. Specifications for portland cement often contain limitations on the quantity of sodium and potassium. Since the volatile oxides and salts of these metals tend to migrate

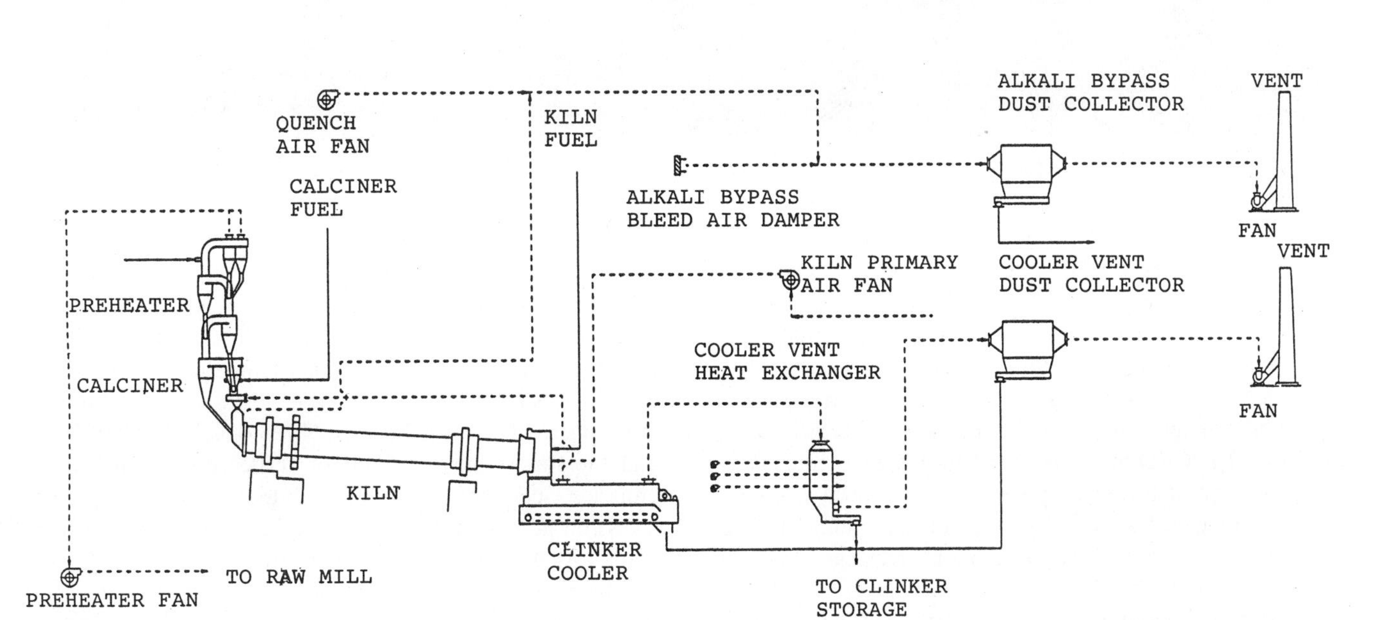

FIGURE 6. Pyroprocessing System Schematic Diagram

FIGURE 7. Preheater and Rotary Kilns

or partition to the CKD, some or all CKD is frequently removed from the pyroprocessing system to meet product quality standards. Bypass CKD is rich in sodium and potassium and is usually not returned to the process. The CKD from a preheater tower has the same general chemical and physical composition as the kiln feed and is returned to the process. The CKD that is removed from the system is used for a variety of beneficial purposes (e.g., waste stabilization) or managed at the cement plant in a monofill. The handling, storage, and deposition of CKD can result in fugitive dust emissions.

The bypass gases may be used in the raw mill, vented through a separate stack, or combined with kiln gases in the main kiln stack after particulate removal. The preheater gases may be vented to the atmosphere after particulate removal or used in the raw mill. Regardless of the treatment or use of combustion products and tempering air from the pyroprocessing system, when these gases are vented to the atmosphere, they must meet the NSPS opacity limit of 20% and the mass emission limit of 0.015 kg of particulate emissions per metric ton (0.30 lb/ton) of kiln feed (dry basis) on a combined basis from all emission points if the plant was built or modified after August 17, 1971.

The principal gaseous emissions from the pyroprocessing system in a typical descending order of volume are nitrogen, carbon dioxide, water, oxygen, nitrogen oxides, sulfur oxides, carbon monoxide, and hydrocarbons. The volumetric composition range is from about 73% to less than 10 ppm. The last four gases are the primary constituents of environmental concern, with the latter two usually of environmental interest only when waste-derived fuel (WDF) is being burned.

In 1982, a Portland Cement Association survey showed that the average SO_2 and NO_x emissions for approximately 50 reporting kilns were 8.41 and 4.62 pounds per ton of clinker respectively. The standard deviation of the survey results for each constituent was nearly equal to the mean value. The frequency distribution revealed a wide range of values.[5] It is impossible, therefore, to characterize the industry for gaseous emissions of SO_2 and NO_x with a single number or narrow numerical range. Each individual pyroprocessing system has its own emission characteristics and the SO_2 and NO_x emissions from proposed or untested pyroprocessing systems are very difficult to predict accurately. Extensive continuous testing of a few cement plants has shown that SO_2 and NO_x emissions from a single source will vary with time over a rather large range for a variety of reasons (e.g., 70–700 lb/h of SO_2). Short-term tests, such as EPA Methods 6 and 7, can lead to very erroneous conclusions regarding SO_2 and NO_x emissions since these methods represent nearly instantaneous process conditions.

Sulfur input to a pyroprocessing system is only from feed and fuel. The relative amounts of sulfur in the feed and fuel, the system design, the chemical form of the input sulfur, and the process conditions, such as the presence of an oxidizing or reducing atmosphere in the kiln, are the variables that determine the quantity of SO_2 emissions at any given time. Oxides of nitrogen result primarily from the combustion of fuel, although nitrogenous constituents in the raw mix may make a contribution to NO_x emissions. The two basic sources of nitrogen oxides from fuel combustion are known as fuel and thermal NO_x. Nitric oxide (NO) predominates among the oxides of nitrogen that are emitted from cement pyroprocessing systems.

The NSPS for cement plants recognize the current uncertainty about SO_2 and NO_x emission rates and the lack of economically feasible control technology through the absence of any emission standards for these pollutants. Regulators, however, often find it necessary to include air pollution permit limitations on SO_2 and NO_x emissions to meet prevention of significant deterioration (PSD) regulatory requirements.

Air Pollution Control Measures

Air pollution control equipment on the kiln system includes reverse air fabric filters and ESPs. Acoustic horns are sometimes used in both devices to assist in cleaning. Table 3 presents typical data.

Cement kiln systems have highly alkaline internal environments that can absorb up to 95% of potential SO_2 emissions. Exceptions to the generalization are found in systems that have sulfide sulfur (pyrites) in the kiln feed. Without unique design considerations or changes in raw materials, the sulfur absorption rate may be as low as 50%. The cement kiln system itself has been determined to be best available control technology (BACT) for SO_2 emissions. Various reports have appeared in the literature that fabric filters on cement kilns absorb SO_2. Generally, this allegation is not true. There must be an absorbing reagent (e.g., calcium oxide) in the filter cake for SO_2 capture to occur. Without the presence of water, which is undesirable in the operation of a fabric filter, calcium carbonate is not an absorbing reagent. It has been observed that as much as 50% of the SO_2 can be removed from the pyroprocessing system exhaust gases when this gas stream is used in a raw mill for heat recovery and drying. In this case, moisture and calcium carbonate are simultaneously present for sufficient time to accomplish the chemical reaction with SO_2.

Energy-efficient pyroprocessing systems have the potential to emit less SO_2 than inefficient systems because of the lower sulfur input from the fuel. Similarly, raw materials with the lowest content of sulfide sulfur usually result in the lowest SO_2 emissions. Selective quarrying or a change in raw materials can lower the input of sulfur to the pyroprocessing system.

A mechanism for the control of NO_x emissions from cement kilns has not been established, although several possibilities exist. Stable kiln operation, such as is found in a successful precalciner system, appears to reduce cumulative, long-term NO_x emissions. Short-term spikes of NO_x emissions during process upsets are currently unavoidable since a higher than normal input of heat from the combustion source is required to restore the process to equilibrium. Several equipment vendors sell burner configurations for the rotary kiln that are alleged to reduce NO_x. These burners have met with varying degrees of success in reducing NO_x emissions. A form of staged combustion can be used on precalciner kilns to reduce NO_x. Fuel is burned under reducing conditions in the riser duct from the rotary kiln to the calciner to generate carbon monoxide. This carbon monoxide chemically reduces the NO_x generated in the kiln to elemental nitrogen. The oxygen-deficient gases thereby generated are then supplied to the calciner further to reduce

TABLE 3. Kiln System Dust Collectors

	Reverse Air	Precipitator
acfm	50,000–300,000	50,000–300,000
Fabric type	Fiberglass	N/A
Temperature range, °F	350–500	350–650
A/C ratio	1.5 : 1 net	SCA: 350–500[a]
Inlet loading, gr/acf	4–18	4–18
Expected outlet emissions, gr/acf	0.02	0.02
Particle size out, μm	>0.5	>0.5

[a]Specific collecting area ($ft^2/1000\ ft^3$).

NO_x generation in that low-temperature combustion source. It is theoretically possible to inject ammonia or urea into a preheater tower at a point where gas temperatures are about 1800°F to achieve a beneficial reaction between ammonia and NO_x. To date, this technology has not been demonstrated in the United States. Ammonia injection is not possible in pyroprocessing systems with only a rotary kiln since the point of optimum temperature is not accessible through the kiln shell. Other possibilities for NO_x emissions reduction exist in the recirculation of flue gas as oxygen-deficient primary air in the rotary kiln and alternative or low-nitrogen fuels and/or raw materials.

CLINKER COOLING

Process Description

The clinker produced in a rotary kiln is cooled in a device called a clinker cooler. This process step recoups up to 30% of the heat input to the kiln system, locks in desirable product qualities by freezing mineralogy, and makes it possible to handle the cooled clinker with conventional conveying equipment.

The more common types of clinker coolers are (1) reciprocating grate, (2) planetary, and (3) rotary. In these coolers, depicted in Figure 8, the clinker is cooled from about 2000°F to 350°F by ambient air that passes through the clinker and into the rotary kiln, where it nourishes the combustion of fuel. In the reciprocating grate cooler, lower clinker discharge temperatures are achieved by passing additional airstreams through the clinker. This air cannot be utilized in the kiln for efficient combustion so it is vented to the atmosphere, used for drying coal or raw materials, or used as a source of heated combustion air in a precalciner. Water sprays are sometimes used to lower clinker discharge temperatures from planetary and rotary coolers.

The reciprocating grate cooler consists of a horizontal box of rectangular cross section that houses horizontal rows of fixed and movable grate plates that bisect the cross section. A grate plate has holes for passage of air and is about 1 foot square. A row may consist of 6 to 12 grate plates, depending on the cooler capacity. The clinker cooler is normally oriented so that the clinker continues its flow parallel to the longitudinal axis of the kiln as it is passed along the top of the grates. The reciprocating movement of every second row of grates forces the clinker through the cooler. The ambient air is forced through the grates and the bed of clinker from the chamber below by a series of fans along the length of the cooler.

Planetary coolers are attached to the kiln shell and rotate with it. Typically, 10 coolers are attached to a single rotary kiln. The clinker drops into the coolers through holes in the kiln shell and is cooled by ambient air that is drawn into the kiln through the cooler tubes. The rotation of the tubes and the internal lifting mechanisms create a cascading of the clinker through the cooling air. The inclination of the kiln and planetary tubes ensures transport of the clinker toward the outlet of the tubes.

A rotary cooler is an independently rotating tube that receives hot clinker by gravity from the rotary kiln discharge point. The clinker is cooled in essentially the same manner as in a planetary cooler with ambient air being drawn through cascading streams of clinker.

Planetary and rotary coolers are vented exclusively into the kiln with an amount of air equal to the combustion air requirements, thereby eliminating the need for a cooler excess air vent. The induced-draft fan for the kiln creates a suction of a few tenths of an inch of pressure, water gauge, at the cooler inlet, which is sufficient to supply the required combustion air.

The reciprocating grate cooler will deliver an identical amount of combustion air to the kiln as the other coolers; however, since this cooler provides for better cooling of the clinker as a result of more airflow, an excess airstream is created. This gas stream must be cleaned of clinker dust before it is vented to the atmosphere.

Air Emissions Characterization

The collected dust from clinker coolers is fairly coarse, with only about 0–15% of it finer than 10 μm. This abrasive dust consists solely of cement minerals and is returned to the process.

The quantity of air used for cooling is about 1 to 2 pounds per pound of clinker, depending on the efficiency of cooling and the desired temperature of the clinker and vent gas. If some of the gases are used for the drying of coal or for other purposes, then the volume of gas to be cleaned at the cooler vent may be reduced by 10–100%.

The dust content of the cooler exhaust gases is affected by the granular distribution of the clinker, the degree of burning of the clinker, the bulk density of the clinker (i.e., liter weight), and the flow rate of the cooling air. Frequently, a clinker breaker (i.e., hammermill) is located at the discharge of the cooler and may increase the dust burden.

If applicable, NSPS set an allowable mass emission limit of 0.050 kg of particulates per metric ton (0.10 lb/ton) of kiln feed (dry basis) from the cooler vent stack. An opacity limit of 10% also applies to the cooler stack. If the cooler gases are used for drying in the raw mill, the same mass emission and opacity limits apply to the raw mill vent.

Air Pollution Control Measures

Upsets in the kiln can rapidly increase the vent gas temperature to 1000°F and the dust load to 13–50 gr/ft^3. In older plants, there may be bypass arrangements to vent these gases directly to the atmosphere until the upset is over. These particulate emission excursions are not permitted in newer plants. Gas temperatures are controlled to protect the dust collector through the use of tempering bleed air, water

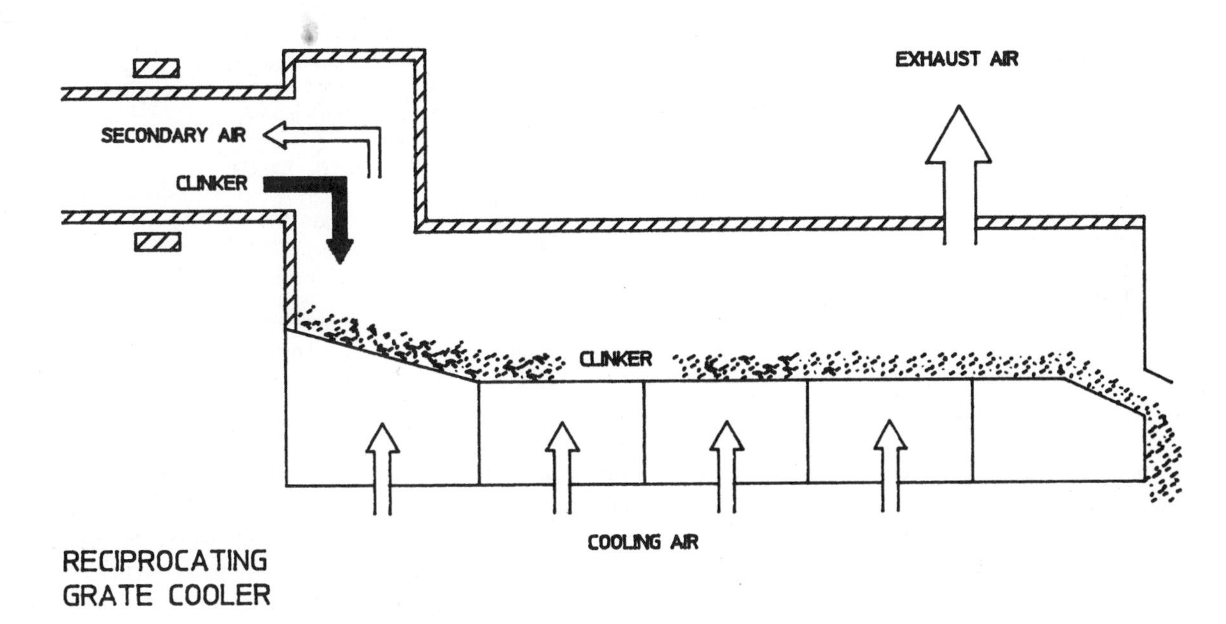

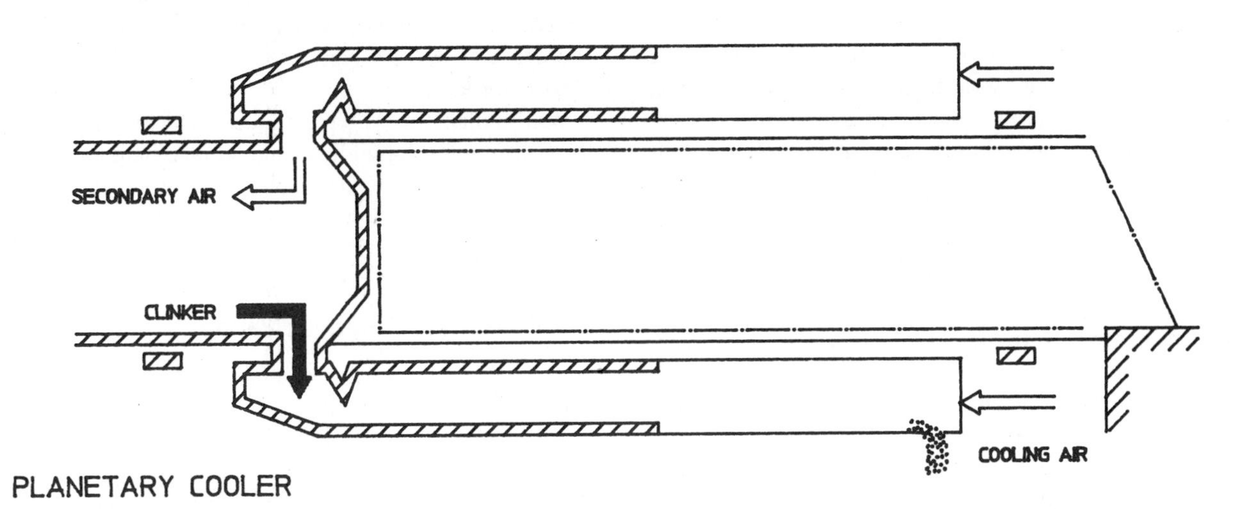

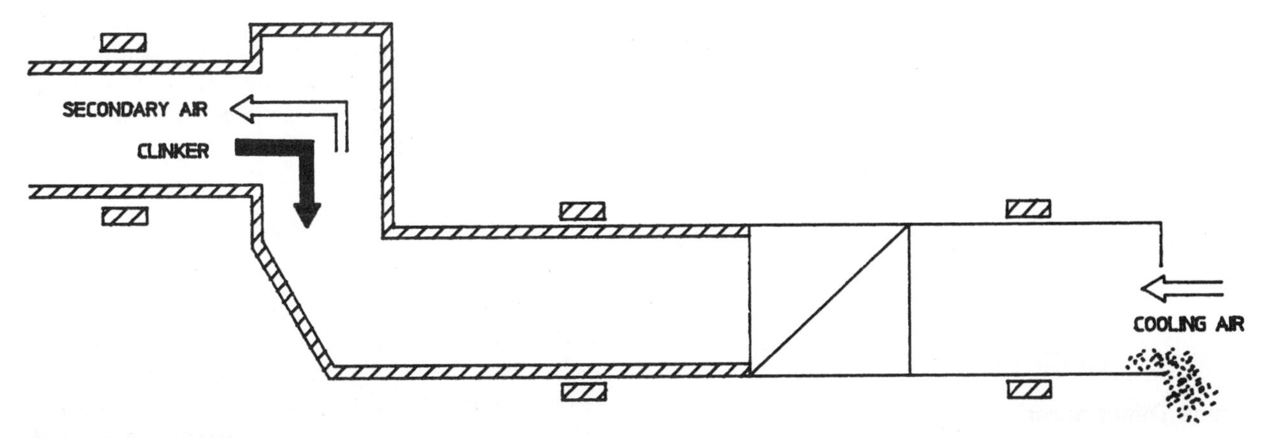

FIGURE 8. Types of Clinker Coolers

TABLE 4. Fabric Filters on Clinker Coolers

	Pulsed Plenum/Pulse Jet	Reverse Air	Precipitator
acfm	20,000–100,000	20,000–100,000	20,000–100,000
Fabric type	Nomex, polyester	Nomex, fiberglass	N/A
Temperature range, °F	<425, 275	<500	350–600
A/C ratio	5 : 1 net	2 : 1 net	SCA: 350–500
Inlet loading, gr/acf	5–10	5–10	5–10
Expected outlet emissions, gr/acf	0.02	0.02	0.02
Particle size, μm	>0.5	>0.5	>0.5

sprays, or an air-to-air heat exchanger. Alternatively, a gravel bed filter, which can tolerate the high temperatures, can be used. All of these methods have costs and limitations and there is no clear universal solution. In a few plants with air-to-air heat exchangers, the cooled air is recirculated to the cooler, thereby eliminating the need for a clinker cooler vent stack.

The dust collectors used on reciprocating grate clinker coolers are most often fabric filters, but ESPs and gravel bed filters are also used. Sometimes these collectors are preceded by a mechanical cyclone or multiclone dust collector. Typical fabric filters on clinker coolers are pulse jets or pulsed plenums in the newer plants and reverse air types in the older plants. In the older plants, the cooler dust collector may be a smaller version of the kiln fabric filter. Table 4 presents typical data.

Gravel bed filters are somewhat unique to the cement industry and may not be described elsewhere in this manual. The gravel bed is contained in several cylindrical compartments and consists of quartz granules of about 2–5 mm in diameter lying on a wire mesh. The dirty gases pass through the bed of quartz. The circuitous path of the gas through the bed causes the dust to drop out of the gas stream and remain in the bed. The beds are individually cleaned at regular intervals by reverse airflow and agitation of the gravel with an internal rake system. The advantage of a gravel bed filter is its ability to tolerate high-temperature excursions without permanent damage. The gravel bed filter is somewhat sensitive to flow volume changes that tend to result in particulate emissions that are higher than normal. Normal particulate emissions from a gravel bed filter are comparable to other dust collection devices on clinker coolers.[6]

CLINKER STORAGE

Process Description

To allow for necessary operational flexibility, a cement plant usually is able to store from 5% to 25% of its annual clinker production capacity. The storage requirement largely depends on market conditions.

The material-handling equipment used to transport clinker from the clinker coolers to storage and then to the finish mill department is similar to that used to transport raw materials. Belt conveyors, screw conveyors, deep bucket conveyors, and bucket elevators are popular. Where possible, drag chains are used because they are less sensitive to abrasion and high temperatures during upset conditions. Gravity drops and transfer points in the conveying and storage systems are normally enclosed and connected to dust collectors.

Older plants were typically designed to store clinker in partially enclosed buildings and storage halls or in outside piles. Newer and modernized plants store at least some clinker in fully enclosed structures or cylindrical, vertical silos, but may also use the other types of storage facilities.

Air Emissions Characterization

Dust in the clinker has a tendency to become airborne during handling. The character of the dust varies by plant and existing process conditions. Dust caught in the clinker cooler exhaust dust collector is usually returned to the clinker stream and can result in reentrainment of this material in air during subsequent handling. Usually, clinker dust is a small proportion of clinker production and is relatively coarse, but some kilns normally produce dusty clinker. During process upsets when the kiln falls below clinkering temperatures and runs "raw," material that is discharged from the kiln is said to be "unburned" (i.e., not fused into clinker) and very dusty.

Air Pollution Control Measures

The air pollution control measures and equipment used in clinker handling systems are similar to those described for raw milling.

The free fall of clinker onto storage piles usually creates visible, fugitive particulate emissions. This dust generation can be reduced by discharging the clinker to piles through a simple device known as a rock ladder or by using variable-height, automatic, stacker belt conveyor systems.

Fugitive dust emissions from open storage piles are mitigated by rain and snow, which cause a crust to form on the piles. Wind breaks and pile covers (e.g., tarpaulins) have also been used to minimize fugitive clinker dust with mixed success. Clinker in open piles is usually reclaimed with mobile equipment, such as front-end loaders. Clinker in storage halls is frequently handled with overhead bucket cranes. Some fugitive clinker dust from operations around open storage piles is usually observed and very difficult to control.

FINISH MILLING

Process Description

The final step in the manufacture of portland cement is the grinding of portland cement clinker to a fine powder. Up to 5% by weight of gypsum and/or natural anhydrite is added to the clinker during grinding to control the setting time of the cement. In the industry, this step is called finish grinding or finish milling and is accomplished in the finish mill department. Small amounts of various other chemicals may be added to the cement during finish grinding to function as processing additions (e.g., grinding aids) or to impart special properties to the cement or resulting concrete (e.g., flowability, air entrainment). Small amounts of water are often sprayed into a cement finish mill to aid in cooling the mill and the cement. Other specification and nonspecification cements with unique properties and constituents can be prepared in the finish mill department. For example, pozzolans or blast furnace slag can be mixed with portland cement clinker and gypsum during the finish grinding process to produce blended cements. In the United States, the most often produced speciality cement derived from portland cement clinker is masonry cement. Typically, a masonry cement is composed of equal portions of portland cement clinker and limestone to which 2–4% by weight of gypsum is added. In addition, chemicals that impart the properties of air entrainment, plasticity, and water repellency to a mortar are added. Each manufacturer of masonry cement has a proprietary formula for its product.

Finish milling is almost exclusively accomplished in ball or tube mills. These mills are rotating, horizontal steel cylinders containing slightly less than half their volume in steel alloy balls, which are called grinding media. These balls can range in size from 4 inches to ½ inch in diameter. Clinkers and gypsum are fed into one end of the mill (feed end) and partially ground portland cement exits from the other end (discharge end). A finish mill might be divided into two or more internal compartments in which the grinding media are segregated by size. The larger balls are at the feed end of the mill. The compartments of the mills are formed by slotted division heads that cover the entire cross section of the mill. The slots are small enough to retain the grinding media in the proper compartment, but large enough to allow the partially ground cement to flow toward the discharge end. At the discharge end of the mill, there is a similar slotted barrier called a discharge grate, which serves to keep the balls in the mill while allowing partially ground cement to exit. A given particle of cement remains in the mill for three to seven minutes. The ends (heads) and sides (shells) of the mills are lined with replaceable alloy steel plates or castings that undergo the wear and abrasion of the grinding process. Mill shell linings are sometimes designed so that the balls are segregated by size during mill operation, thereby eliminating the need for division heads.

Cement is usually ground in a closed circuit with an air separator. This continuously operating device is used to separate particles of cement of acceptable size in the material discharged from the mill from those particles that have not been fully ground. The large particles (i.e., tailings) are returned to the mill and reintroduced to the feed end along with new feed. A figure that is 100 times the ratio of the weight of the returned tailings to the weight of new feed is called the circulating load and is expressed in percent. Circulating loads in the range of 200–500% are typical, but higher and lower circulating loads are found in acceptable mill circuits. Air separators are mechanical devices that use centrifugal force, gravity, and an ascending air current to separate the cement particles. Older air separators have a poor separation efficiency. Equipment manufacturers recently developed high-efficiency separators that are included in most new finish mill projects and are popular retrofit items because of the increased efficiency, lower operating costs, and sometimes improved product performance.

Another device of increasing popularity in the finish mill department is the roll crusher. This device accomplishes the initial size reduction of the clinker and the gypsum outside of and prior to the ball mill. The efficiency and/or the productive capacity of a given ball mill is thereby increased at potentially lower grinding temperatures.

For a variety of reasons, cement customers usually demand cement that is at temperatures of less than 100–150°F when delivered. Cement grinding temperatures can reach 350°F. Water-supplied cement coolers (i.e., heat exchangers) are often installed in the material flow path following the finish mill to reduce the cement temperatures prior to product storage. The high-efficiency separator circuits, with their associated high volumes of mill vent air, provide better cooling of the cement than conventional mill circuits. No air pollution problems are associated with the closed heat exchangers.

Figure 9 is a process flow sheet for a finish mill circuit that includes a roll crusher and a high-efficiency air separator. Figure 10 shows a two-compartment ball mill in finish mill service as viewed from the feed end of the mill.

Air Emissions Characterization

Particulate emissions from mill vents, air separator vents, and material-handling system vents constitute the air pollution concerns in the finish mill department.

About 30–40% of the particles of ordinary Type I portland cement are finer than 10 μm. For Type III, high-early-

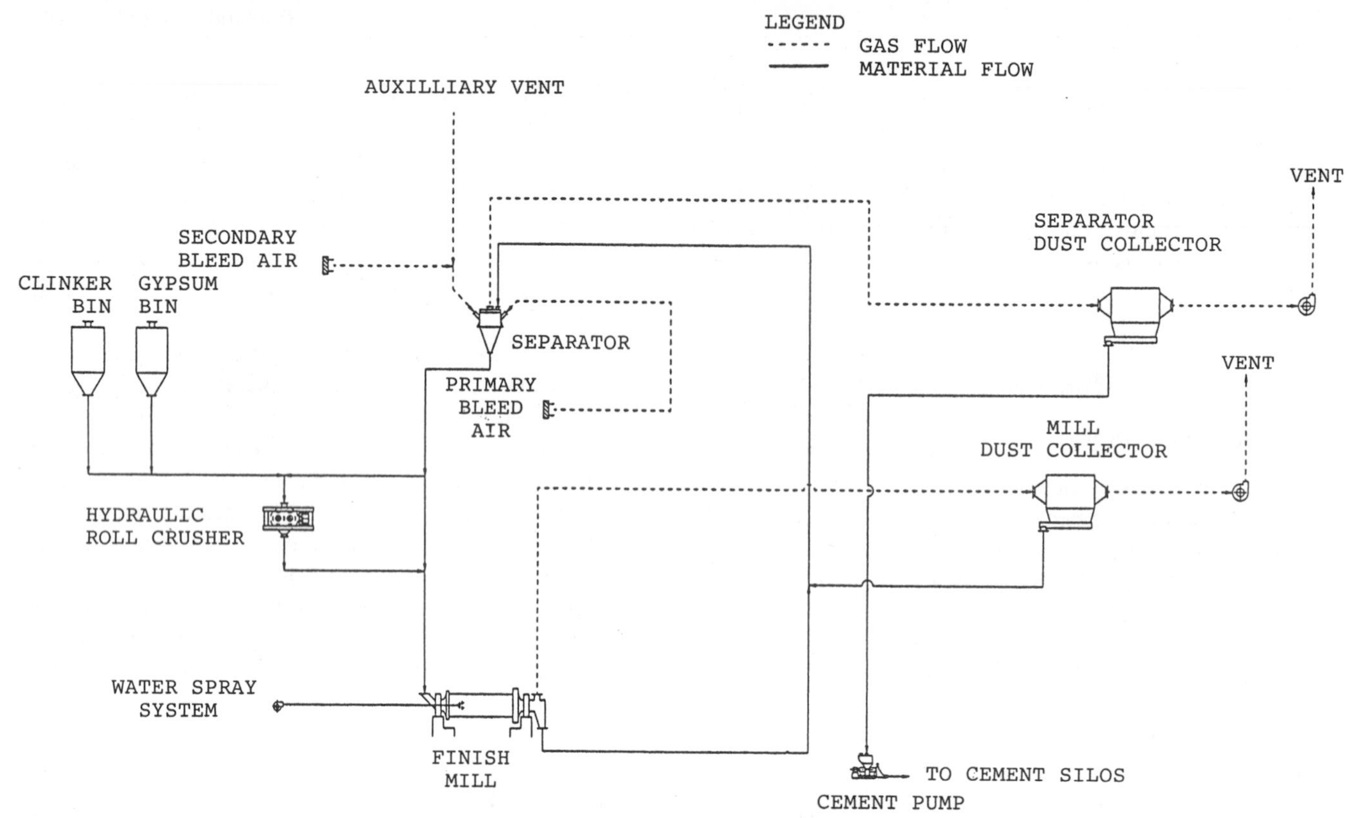

FIGURE 9. Finish Mill Schematic Diagram

FIGURE 10. Finish Mill

TABLE 5. Fabric Filters for Finish Mill Systems

	Reverse Air/Shaker	Pulse Jet	Pulsed Plenum
acfm	10,000–30,000	10,000–30,000	10,000–30,000
Fabric type	Polyester	Polyester	Polyester
Temperature range, °F	<275	<275	<275
A/C ratio	2.5:1	4:1	5:1
Inlet loading, gr/acf	5–20	5–100	5–100
Expected outlet emissions, gr/acf	0.02	0.02	0.02
Particle size out, μm	>0.5	>0.5	>0.5

strength portland cement, the percentage of particles finer than 10 μm increases to the 45–65% range. Typically, about 90% of portland cement will pass a 325-mesh (44-μm) sieve. The potential air pollution problems associated with the manufacture, handling, and transportation of portland cement have their origin in the number of very fine particles in the product.

Air Pollution Control Measures

Emissions from finish mills are adequately controlled by fabric filters. The fabric filters most often found on new or upgraded plants are the pulse-jet and pulsed-plenum types. Reverse air/shaker fabric filters are typically found in older plants. In almost all cases, pulse-jet or pulsed-plenum fabric filters are installed in conjunction with high-efficiency separators. Tables 5 and 6 present typical details.

The cement dust caught in a fabric filter is returned to the process. In colder weather, the water that is used for internal mill cooling can produce a steam plume at the mill baghouse vent. This condensate plume is sometimes confused with excessive particulate emissions, but it will dissipate within a few feet of the vent opening. Fabric filters on finish mill systems that use cooling water must be well insulated to prevent condensation within the baghouse and subsequent blinding of the bags.

PACKING AND LOADING

Process Description

Portland cement is pneumatically conveyed from the finish mill department to large, vertical, cylindrical concrete storage silos in the packhouse or shipping department. Mechanical transfer systems, such as bucket elevators, belt conveyors, screw conveyors, and air slide conveyors, supplement the pneumatic system.

The number and capacity of the storage silos depend on the capacity of the plant, the number (i.e., types) of cements in the product mix, the marketing strategy of the company, and the weather-driven shipping pattern.

Portland cement is withdrawn from the storage silos by a variety of feeding devices and conveyed to loading stations in the plant or directly to transport vehicles using the same kinds of material-transfer systems that were used to put the cement into the silos. Most of the portland cement is shipped from the plant in bulk by rail or truck transport. Those plants located adjacent to water transportation routes usually serve some customers or distribution terminals by barge or ship.

Portland cement is also shipped in multiwall paper bags with a capacity of 94 pounds. These bags are filled on automatic or semiautomatic packing machines. During filling, each bag is vented to allow the escape of displaced air. The filled bags are then manually or mechanically palletized for shipment. A few customers may still require loads of unpalletized bags of cement.

Masonry cement is almost totally shipped in multiwall paper bags. Bag weights range from 70 to 80 pounds, depending on the type of masonry cement in the bag. The packing, palletizing, and dust suppression operations are identical to those used for portland cement.

There are remote distribution terminals associated with some cement plants. Bulk or packaged cement is shipped from the plant to the terminal for storage and subsequent timely distribution to customers. Shipments to terminals are most often accomplished by rail or barge, although trucks and ships are sometimes used. The handling and loading of bulk portland cement at distribution terminals are carried out by the same kinds of pneumatic and mechanical conveying systems as are used at the plant.

TABLE 6. Fabric Filters for High-Efficiency Separators

	Pulse Jet	Pulsed Plenum
acfm	40,000–60,000	40,000–60,000
Fabric type	Polyester	Polyester
Temperature range, °F	<275	<275
A/C ratio	4:1	5:1
Inlet loading, gr/acf	150–300	150–300
Expected outlet emissions, gr/acf	0.02	0.02
Particle size out, μm	>0.5	>0.5

Air Emissions Characterization

Particulate emissions from the silo openings, cement-handling equipment, bulk and package loading operations and the fabric filters constitute the air pollution problems in the shipping department.

Air Pollution Control Measures

Active and passive fabric filters are used to remove dust from the exhaust airstreams from the silos and transport systems. The cement dust is returned to the product.

The dust generated during the loading of trucks, railcars, barges, and ships is controlled by venting the transport vessel to a fabric filter. The collected dust is returned to the shipment of cement. Flexible loading spouts with concentric pipes are among the devices that are successfully used for dust-free loading. In a loading spout, the cement flows to the transport vessel by gravity through a central pipe, while exhaust air is drawn through an annular space.

Dust is controlled at distribution terminals through the venting of silos, bins, and transfer points to fabric filters. The captured cement dust is returned to the product.

The typical fabric filters used in the packing and loading departments of newer plants are of the pulse-jet type. Reverse-air or shaker-type fabric filters are found in older plants. Occasionally, a cartridge-type fabric filter will be employed. Table 7 presents typical data.

SUPPLEMENTAL FUELS AND RAW MATERIALS

The recycling of wastes in portland cement kilns as fuel and raw material substitutes is a reliable and proven technology. This technology offers a cost-effective, safe, and environmentally sound method of resource recovery for some hazardous and nonhazardous waste materials. Following appropriate preparation, the energy and chemical values of selected wastes can be beneficially recovered, thereby enabling a portland cement manufacturer to operate more competitively.

The energy-bearing, ignitable wastes that are currently used in the portland cement industry as fossil fuel substitutes are primarily waste oils and spent organic solvents, sludges, and solids from the paint and coatings, auto and truck assembly, and petroleum industries. Smaller amounts of other waste streams are also being successfully recycled into cement kilns as fuel substitutes. Some waste streams require pretreatment so that they can be effectively introduced into the kiln. For example, liquids with high or variable chlorine levels are blended to a somewhat consistent chlorine concentration to minimize the impact on kiln operations. Sludges are liquefied, solidified, or encapsulated to provide better material-handling properties. Solids may be ground to facilitate blending and sampling. Materials such as petroleum coke and sawdust can be handled with the same equipment as coal and are readily consumed in a cement kiln. Other high-energy waste streams, such as rubber tires, can present materials-handling problems that are overcome with new or modified fuel-delivery equipment.

The cement-making process offers many unique opportunities for the utilization of nonhazardous wastes. Silicon, aluminum, and iron are needed to react chemically with the calcium in the cement raw mix. Such materials as spent cracking catalyst, diatomaceous earth filter material, foundry sand, and contaminated soils have high concentrations of these elements and are used to replace the conventional siliceous, argillaceous, and ferriferous components of the raw mix.

The greatest economic and societal benefits from utilizing wastes in cement manufacturing are derived from the replacement of fossil fuel. Coal provides about three fourths of the thermal energy for manufacturing cement in the United States. If the domestic cement industry had replaced only 10% of its conventional fuels with waste substitutes in 1987, 40 trillion Btu of nonrenewable energy would have been conserved.

Older, less fuel-efficient cement plants often derive the greatest benefit from waste fuel substitution because of an improved competitive position. These older facilities successfully compete with more modern facilities as a result of lower operating costs and improved cash flow.

Cement plants are often located near industrial and metropolitan areas. As these same areas produce most of the waste materials, transportation of the waste can be mini-

TABLE 7. Fabric Filters in Packing and Loading Departments

	Pulse Jet	Reverse Air/Shaker	Cartridge
acfm	3,000–10,000	3,000–10,000	2,000–10,000
Fabric type	Polyester	Polyester	Paper, polyester
Temperature range, °F	<275	<275	<275
A/C ratio	6:1	2.5:1	<2:1
Inlet loading, gr/acf	5–40	5–40	5–40
Expected outlet emissions, gr/acf	0.02	0.02	0.02
Particle size out, μm	>0.5	>0.5	>0.5

mized by recycling appropriate materials into cement kilns rather than transporting them to a remote disposal facility.

Cement kilns have several important advantages that contribute to the effective destruction of waste materials. The gas residence time in the burning zone of the kiln is in excess of 3000°F for a period of approximately three seconds. Temperatures in excess of 2000°F exist for as long as six seconds. Test burns repeatedly demonstrate destruction and removal efficiencies (DREs) of 99.99 to 99.9999% for the most stable organic compounds. The alkaline environment of a portland cement kiln can absorb the hydrogen chloride that may result from the combustion of chlorinated hydrocarbons. Ash resulting from incombustible material, such as metals in the waste, either becomes chemically incorporated in the clinker crystal matrix or is caught with the CKD in the air pollution control device prior to the kiln stack. The more volatile metals, such as lead, primarily migrate to the CKD, while refractory metals, such as chromium, are mostly found in the clinker. Since cement raw materials come from the earth's crust, these and other naturally occurring trace metals are normally found in cement and CKD. The use of waste fuels and waste raw materials usually increases the heavy metal content of cement and CKD only by small and insignificant increments. No significant increases in metals emissions from cement kilns have been observed during numerous test burns.

The CKD is not a listed or a characteristic hazardous waste under 40 CFR 261. In no instance has there been a report of CKD associated with the burning of WDF failing the EP toxicity or TCLP tests for the eight specified metals.

The most significant operating problems with WDF are associated with its chlorine content. Excessive chlorine levels can contribute to material buildups within a kiln system (e.g., kiln rings), deterioration of ESP performance, and excessive corrosion. The cement manufacturer quickly learns the maximum chlorine level that a pyroprocessing system will tolerate and limits chlorine input to below that amount. Each system has a different chlorine tolerance, but chlorine in WDF is usually held to below 5% by weight in the absence of permit limits. The chemical and physical specifications for portland cement also make it necessary for a cement manufacturer carefully to control the performance of the pyroprocessing system during waste-fuel firing and to monitor the input of several trace elements that can be found in WDFs or raw materials. Regulations for the burning of WDF in cement kilns were promulgated under RCRA by the EPA on February 21, 1991. The air emissions of pollutants of concern (i.e., carbon monoxide, hydrocarbons, hydrogen chloride, and heavy metals) during waste burning for energy recovery are regulated.

PROCESS AND QUALITY CONTROL

Modern cement plants are exclusively controlled by digital computers from central control rooms. Process variables in all manufacturing departments are continuously monitored. Control actions are usually initiated by the process control computer, but manual intervention is possible during process upsets, equipment malfunctions, or emergency conditions. Older cement plants use analog control systems in either central control rooms or departmental control stations.

Modern cement plants are usually equipped with continuous opacity monitors on the kiln and clinker cooler stacks. At some plants, gaseous emissions of oxygen, carbon monoxide, nitrogen oxides, and sulfur dioxide from the kiln stacks are also continuously monitored. In the past few years, these monitoring devices have become more reliable and less maintenance intensive. Nevertheless, equipment redundancy may be required if there are to be minimal data gaps in a compliance monitoring scheme. The location for these devices in the plant is often hot and dirty, thereby complicating the monitoring task.

The portland cement process involves rather complex chemistry and close process control. Plant laboratories are staffed around the clock. Frequent chemical and physical tests are made on raw materials, raw mix, clinker, and cement. The procedures may range from elementary wet chemistry to more sophisticated testing by X-ray fluorescence. Some of the newest cement plants are equipped with automatic sampling and analytical systems. The operation of the pyroprocessing system receives particularly close attention since product quality is largely determined in the kiln. If proper process conditions and kiln temperatures are not maintained, the complex chemical reactions that take place in the kiln are incomplete and the clinker is unacceptable.

References

1. W. Greer, personal communications, with R. Crolius, Portland Cement Association staff, June–September 1990.
2. *40 Code of Federal Regulations, Part 60, Subpart F,* U.S. Government Printing Office, Washington, DC. Last revised by 53 FR 50363, December 14, 1988.
3. *Compilation of Air Pollutant Factors* (AP-42), Section 8.6, 4th ed., U.S. Environmental Protection Agency, Research Triangle Park, NC, September 1985. Supplement dated 10/86.
4. *40 Code of Federal Regulations, Part 60, Subpart 000,* U.S. Government Printing Office, Washington, DC. Added by 53 FR 31337, August 1, 1985.
5. Price Waterhouse, *Portland Cement Association Survey of NOx and SOx Emissions,* June 1, 1983.
6. R. E. Shumway, D. W. Janocik, and C. U. Pierson, *The Application of Gravel Bed Filters in Clinker Cooler Vents,* Portland Cement Association, Chicago, 1978.
7. W. Greer, "The calcining zone of the kiln and other considerations" Paper no. 11, Portland Cement Association, Kiln Optimization Short Course, Chicago, 1980.
8. W. Greer, "SO_2/NO_x control, compliance with environmental regulations," *IEEE Trans. Industry Applications,* Vol. 25 (3) (May/June 1989).
9. C. Schneeberger, Portland Cement Association, letter to cement industry executives August 18, 1983, *Kiln Gaseous Emissions Survey.*

Bibliography

R. H. Bogue, *The Chemistry of Portland Cement,* 2nd ed., Reinhold Publishing Corp., New York, 1955.

W. H. Duda, *Cement Data Book,* 3rd ed., Bauverlag GmbH, Weisbaden and Berlin, 1985.

O. Labahn and B. Kohlhaas, *Cement Engineers Handbook,* 4th Ed., Bauverlag GmbH, Wiesbaden and Berlin, 1983.

K. E. Peray, *Cement Manufacturers Handbook,* Chemical Publishing Co., New York.

K. E. Peray, and J. J. Waddell, *The Rotary Cement Kiln,* Chemical Publishing Co., New York, 1972.

VDZ Congress '85, *Process Technology of Cement Manufacturing,* Bauverlag GmbH, Wiesbaden and Berlin, 1987.

FIBERGLASS OPERATIONS

Aaron J. Teller and Joseph Y. Hsieh

The production of fiberglass consists of two different forms of product—continuous-filament fiberglass or textile products and fiberglass blown wool or insulation products.

The general-purpose textile fiberglass, which is moisture and alkali resistant with good electrical and physical properties, is also called E- (electrical) Glass. The major applications of the general-purpose textile fiberglass are in the production of fireproof cloth, fiberglass-reinforced plastics (FRP), and composites. The insulation product lends itself for the application of thermal and acoustical insulation because the small cell of air entrained in the wool or blanket prevents the movement of the air and sound wave. The manufacturing process consists of (Figure 1): materials blending and transport, melting and refining, and fiberforming and textile operation.

As in the manufacturing of glass, the major air emission problem where acid-gas recovery in addition to particulate is required is related to the melting and refining furnace operation. The fiberforming and textile operations produce, primarily, particulate emissions and some volatile organic compounds (VOCs). The textile product manufacture presents a more complex emissions control problem than does insulation manufacture because of the presence of boron and fluorine in the most flexible E-Glass product.

PROCESS DESCRIPTION

The composition of textile and insulation fiberglass, although varied as a function of the producer, has the general composition[1] shown in Table 1.

The raw materials are unloaded from freight cars or trucks and transported to specific silos in the batch house. The materials are then withdrawn to automatic weight machines and blended. The mix is transported by air conveying to the holding vessel at the melter and then fed as a batch to the furnace or melter.

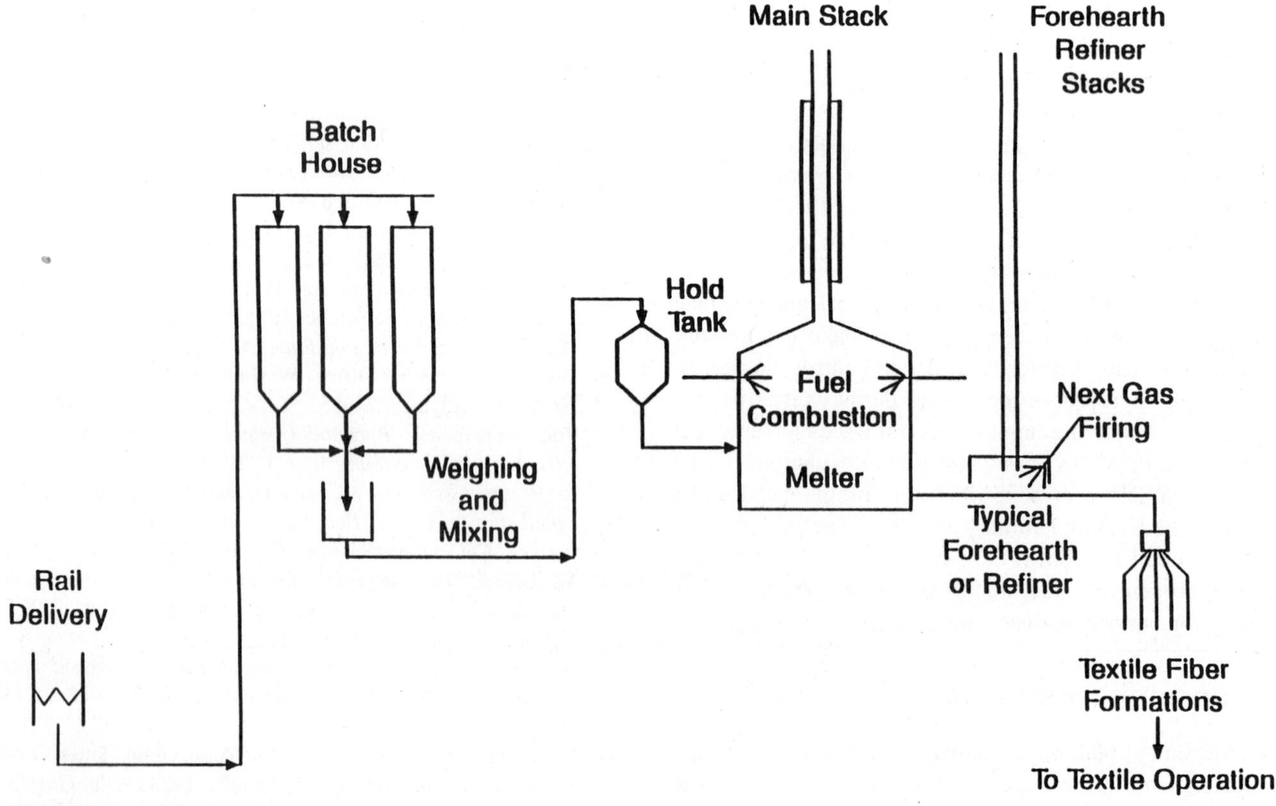

FIGURE 1. Fiberglass Manufacturing Process

TABLE 1. Composition of Textile and Insulation Fiberglass

Component	Textile, % Weight	Insulation, % Weight
SiO_2	54–55	59–65
Al_2O_3	13–14	4–8
Na_2O	0.5	15–18
CaO	22–24	4.5–9
MgO	0.3–0.4	0–4
B_2O_3	5.5–6.8	0–2
Fe_2O_3	0.3	0.1–5
F	0.7	
K_2O		0.3
MnO		0–4
BaO		0–5
SO_3		≤0.6

The melters are of the reverberatory type with gas firing, although oil may be used. There is a significant difference from the container or float glass furnace. Thermal recovery is achieved by a recuperator in a continuous manner rather than cyclical. A part of the reasoning is that the fiberglass (especially textile) melt is extremely sensitive to very small pressure variations and that boron emissions can clog the brick-lattice regenerator.

Because of the time–temperature sensitivity of the final product, liquid glass flows through channels with zones of refiners and forehearths, in addition to the melter. These forehearths and refiners are generally fired with natural gas. The flue gas from the melter flows upward from the combustion zone through the stack containing an annular zone heat exchanger. Combustion air flows in the annular zone of the recuperator for preheating. The flue gas at 500–800°C is emitted, generally by passing through an air damper, to the stack. The forehearth and refiner flue gases are emitted through individual stacks. These flue gases are often combined with the melter flue gas.

AIR EMISSIONS CHARACTERIZATION

The air emissions for the fiberglass operation can be characterized according to the three phases of manufacturing processes. Emission factors for the fiberglass operations are given in Tables 2 and 3.

TABLE 2. Emission Factors for Glass Fiber Manufacturing without Controls[a]
Emission Factor Rating: B

	Particulates		SO_x		CO		NO_x		VOC[b]		Flourides	
	lb/ton	kg/Mg	lb/ton	kg/Mg	lb/ton	kg/Mg	lb/ton	kg/Mg	lb/ton	kg/Mg	lb/ton	kg/Mg
Unloading and conveying[c]	3.0	1.5	d	d	d	d	d	d	d	d	d	d
Storage bins[c]	0.2	0.1	d	d	d	d	d	d	d	d	d	d
Mixing and weighing[c]	0.6	0.3	d	d	d	d	d	d	d	d	d	d
Crushing and batch charging[c]	Neg	Neg	d	d	d	d	d	d	d	d	d	d
Glass furnace—wool												
Electric	0.5	0.25	0.04	0.02	0.05	0.025	0.27	0.14	e	e	0.002	0.001
Gas—regenerative	22	11	10	5	0.25	0.13	5	2.5	e	e	0.12	0.06
Gas—recuperative	25–30	13–15	10	5	0.25	0.13	1.7	0.85	e	e	0.11	0.06
Gas—unit melter	9	4.5	0.6	0.3	0.25	0.13	0.3	0.15	e	e	0.12	0.06
Glass furnace—textile												
Recuperative	2	1	3	1.5	0.5	0.25	20	10	d	d	2	1
Regenerative	16	8	30	15	1	0.5	20	10	d	d	2	1
Unit melter	6	3	e	e	0.9	0.45	20	10	d	d	2	1
Forming—wool												
Flame attenuation	2	1	d	d	d	d	d	d	0.3	0.15	e	e
Forming—textile	1	0.5	d	d	d	d	d	d	Neg	Neg	d	d
Oven curing—wool												
Flame attenuation	6	3	e	e	3.5	1.8	2	1	7	3.5	e	e
Oven curing and cooling—textile	1.2	0.6	d	d	1.5	0.75	2.6	1.3	Neg	Neg	d	d

[a]Expressed as units per unit weight of raw material processed. Neg = negligible.
[b]Includes primarily phenols and aldehydes, and to a lesser degree, methane.
[c]Reference 10.
[d]Not applicable.
[e]No data are available.

TABLE 3. Uncontrolled Emission Factors for Rotary Spin Wool Glass Fiber Manufacturing[a]

	Particulate			Organic Compounds[b]		
Products	Front Half	Back Half	Total	Phenolics[c]	Phenol	Formaldehyde
R-19	17.81 (36.21)	4.25 (8.52)	22.36 (44.72)	3.21 (6.92)	0.96 (1.92)	0.75 (1.50)
R-11	19.61 (39.21)	3.19 (6.37)	22.79 (45.59)	6.21 (12.41)	0.92 (1.84)	1.23 (2.46)
Ductboard	27.72 (55.42)	8.55 (17.08)	36.26 (72.50)	10.66 (21.31)	3.84 (7.68)	1.80 (3.61)
Heavy density	4.91 (9.81)	1.16 (2.33)	6.07 (12.14)	0.88 (1.74)	0.53 (1.04)	0.43 (0.85)

[a]Reference 7. Expressed kg/Mg (lb/ton) of finished product. Gas stream did not pass through any added primary control device (wet electrostatic precipitator, venturi scrubber, etc.)
[b]Included in total particulate catch. These organics are collected as condensible particulate matter and do not necessarily represent the entire organics present in the exhaust gas stream.
[c]Includes phenol.

Materials Blending and Transport

The primary air pollution problem in this phase of the operation is that of containment and fugitive dust handling inasmuch as fine particulates are not involved. The raw materials are generally in the size range of 20–500 μm.

Melting and Refining

The emissions from the melting–refining furnace during the textile fiberglass operation contain:

1. Fine particulates, including calcium carbonate, sodium fluoride, sodium fluosilicate, silica, calcium fluoride, aluminum silicate, sodium sulfate, boron oxides—boric acid.
2. Gases, including fluorides—hydrogen fluoride, silicon tetrafluoride, sulfur oxides, nitrogen oxides, boric acid, carbon dioxide, water vapor.

The solid particulates emitted in the melter flue gas are in the class of fine particulates with most of the particle size below 10 μm. Another class of particulates is formed downstream of the stack for the untreated emissions. This consists primarily of sublimed boric acid. It can form very opaque, long-lasting white plume after cooling resulting from sublimation. This "formed particulate" is created in the emissions from the melter, forehearths, and refiners. About 70% of the boron and fluorine emissions are emitted by the melter.

The concentration of the particulates and condensible vapor boric acid in the efflunt gases are as follows:

Solid particulates	200–1500 mg/nm^3
Boron oxide, solid plus equivalent boric acid	100–1000 mg/nm^3

The gaseous components of the emissions are hydrogen fluoride, silicon tetrafluoride, sulfur oxides, and the boric acid considered in the discussion of particulates. They are present in concentrations ranging as follows:

HF, SiF_4(equivalent HF)	20–160 ppm
SO_x	40–80 ppm

The emissions from the melting–refining furnace of the insulation fiberglass operation contain only small quantities of boron and no fluoride. The emissions from the fiberforming and textile operations include particulate and VOC emissions. The VOC emissions are lower for textile fiberglass than for insulation fiberglass operation because of the small quantities of lubricant and VOC used in the textile fiberglass operation.

AIR POLLUTION CONTROL MEASURES

Materials Blending and Transport

In the materials storage and transport area, the railroad discharge is generally partially sleeved and under either positive or negative pressure. The silos are provided with bin vent filters for collecting the fugitive dust. The blending vessels are vented with airflow directed to bag filters. These are generally pulse jets operating at an air-to-cloth ratio in the range of 4–7 fpm. The solid discharge is recycled to the blending zone.

Melting and Refining

A major consideration in the control of emissions from E-Glass furnaces is the suppression by capture of the boron compounds. If this is not achieved, then the prospect of opaque plumes resulting from sublimation of boric acid and

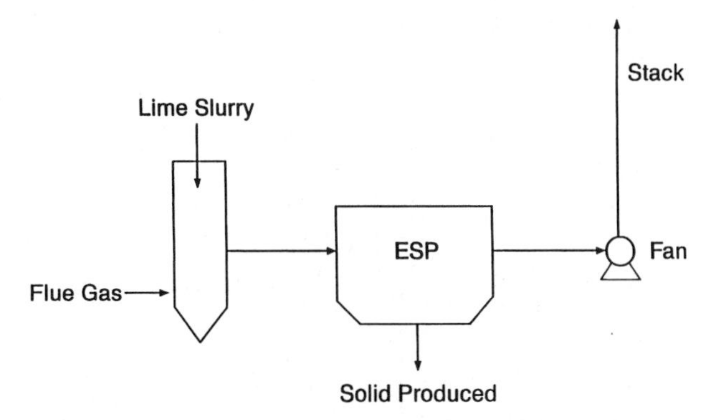

FIGURE 2. Spray Dryer—Electrostatic Precipitator System

the potential for forming a complex of HF and H_3BO_3 such as HBF_4 both exist.

A significant factor in the design of the air pollution control system is the maintenance of the melter pressure, with the permissible variation, as requested by furnace operators, ±0.01 inch wc. Both wet and dry scrubbing methods have been utilized in controlling the emissions from the melting and refining operation.

Wet Scrubbing

Either the venturi-packed bed scrubber or a nucleation cross-flow scrubber can be used. The venturi system is a modification of the conventional packed-tower–venturi–cyclone arrangement for recovering the particulate and acid gases. The treatment of blow-down stream required appears to have limited its application.

A low-energy (less than 15 inches wc) nucleation cross-flow scrubber was operated in a Canadian facility[8]. The system used a cooling tower to circumvent the steam plume appearance. The major problems encountered were the dual alkali transfer for (Na→ Ca), resulting in excess solubility of calcium ions. This often resulted in plugging of the scrubbers by deposition of insoluble calcium salts.

Dry Scrubbing

Two types of systems have been installed in the United States for E-Glass production.

1. Spray drier–electrostatic precipitator (ESP) (Figure 2)
2. Quench reactor–dry venturi–Baghouse (Figure 3).

The latter type has also been installed in the U.S.S.R. and Taiwan.

Spray Dryer–Electrostatic Precipitator

The gases are cooled by a water-lime slurry quench to about 100°C and then flow to the ESP, the system fan, and out the stack. Although gaseous fluoride was reduced to the 5-ppm range, there was evidence of significant plume opacity,

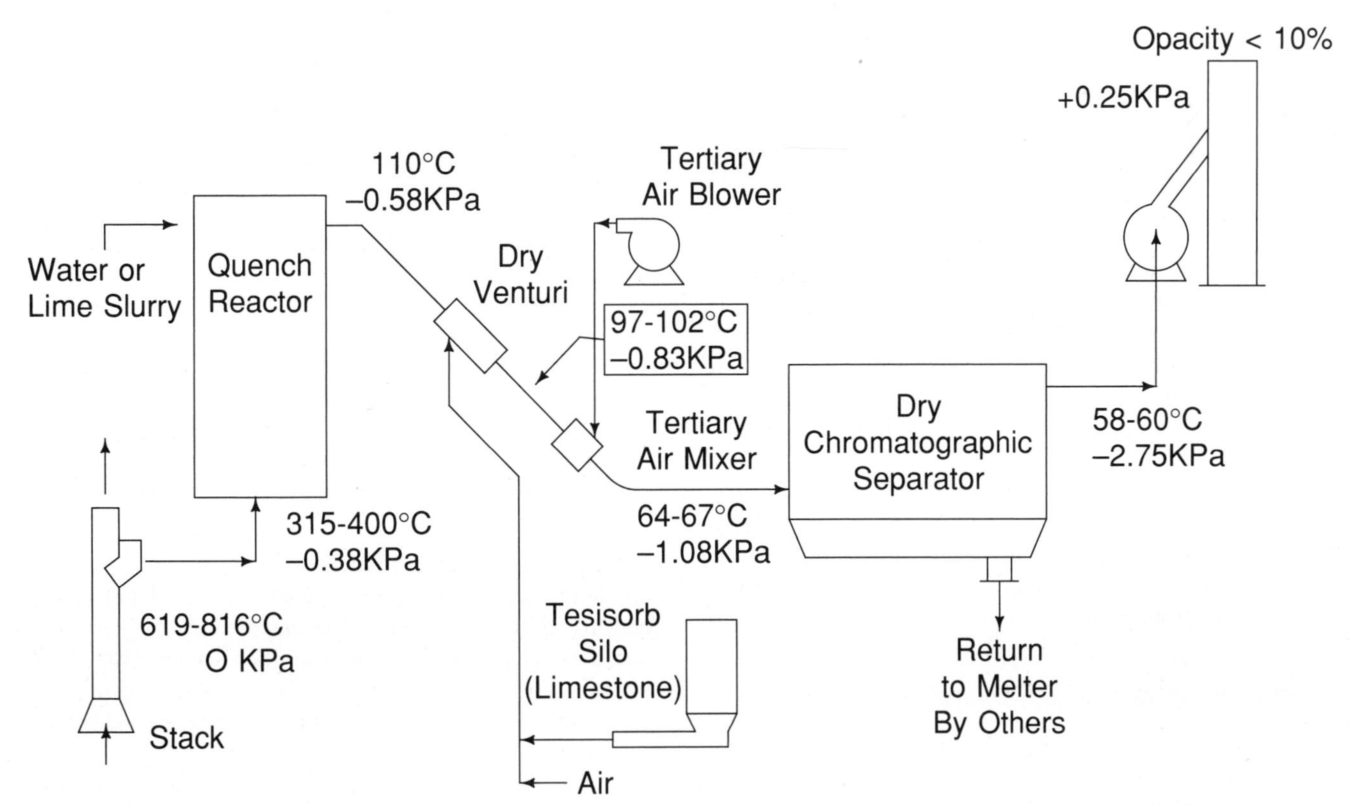

FIGURE 3. Typical Teller Dry Chromatographic System for the Fiberglass Industry—HF, SO_x, and Particulate

TABLE 4. Performance of SemiDry System Fiberglass Emission Control

		Outlet Composition					
		Fluoride			Boron		
Dry Gas Flow, sdcfm	Particulate, mg/nm^3	Solid, lb/hr	Gas, lb/hr	ppm	Solid, lb/hr	Gas, lb/hr	mg/nm^3
18,500	1.5	0.015	0.11	2.12	0.03	0.92	13.7
14,600	2.3	0.004	0.009	0.28	0.030	0.475	9.2
15,500	2.1	0.005	0.017	0.45	0.032	0.568	10.3
14,800	4.5	0.003	0.007	0.21	0.019	0.208	4.1
15,100	1.8	0.003	0.002	0.10	0.03	0.52	9.8

Source: Reference 9.

reflecting the presence of boron compounds in the emitted gases.

Problems were reported regarding condensation and adhesion melting of particulate on the ESP electrodes.[2]

Quench Reactor–Dry Venturi–Baghouse[3,4]

This system is operated with either water quench or lime slurry quench as a function of requirements for reduction of SO_2 and HF. Where the only SO_x reduction requirement is related to opacity, and SO_3 only is to be removed, water quench in the quench reactor is utilized.

The quenched gas, at about 100°C proceeds to the dry venturi, where limestone is introduced into the gas stream as both reagent and cake modifier.[3] The gas is then further cooled by air dilution to the range of 60–70°C. The limestone acts to neutralize the fluorides and aid in the formation of a low-compressibility layer in the baghouse. A significant part of the neutralization reaction occurs in equivalent renewable "fixed bed" cake on the bags. The cleaning cycle is approximately six hours.

The reduction of temperature to 60–70°C results in the sublimation of the boric acid. This boric acid is deposited on the surface of the baghouse cake along with the crystalline reagents. Filtration in a temperature region only 20°F greater than the dew point is maintained with a system pressure drop in the range of 8 inches wg. The performance of the system, with greater than 98% on-line reliability, is given in Table 4.

For the insulation fiberglass operation, inasmuch as the emissions from the melter contains no fluoride and only small quantities of boron, the emission control can be achieved by a system such as a spray dryer–ESP (Figure 2) or spray dryer–baghouse.

Fiberforming and Textile Operation

Fiber formation does not present emission control difficulties. However, the spraying of the mat with lubricants and/or binders to minimize erosion of the fibers can create either VOC or inorganic particulate emissions.

The predominantly inorganic discharges are treated in a cross-flow scrubber with particulate emissions reduced to less than 25 mg/nm^3.[6] The organic emissions are treated by condensation processes for high-molecular-weight emissions and, if necessary, adsorption for low-molecular-weight compounds.

References

1. Duffy, "Glass technology," *Chem. Tech. Rev.,* 184 (1981).
2. Noll, *Glass Tech.,* 25(2); 91–97 (1984).
3. A. J. Teller, U.S. Patent 2,018,157, 1972.
4. A. J. Teller, U.S. Patent 4,319,890, 1982.
5. A. J. Teller, *Glass Indus.* 18–26 (June 1982).
6. Private communications, Ceilcote
7. *Wool Fiberglass Insulation Manufacturing Industry: Background Information for Proposed Standards,* EPA-450/3-83-022a, U.S. Environmental Protection Agency, Research Triangle Park, NC, December 1983.
8. A. J. Teller, *Glass Indus.,* 15–19 (February 1976).
9. Private communications, PPG Industries
10. J. R. Schorr, et al., *Source Assessment: Pressed and Blown Glass Manufacturing Plants,* EPA-600/2-77-005, U.S. Environmental Protection Agency, Research Triangle Park, NC, January 1977.

GLASS MANUFACTURING

Aaron J. Teller and Joseph Y. Hsieh

The production of glass products, although based on an old art, has emerged into sophisticated automated processes, both batch and continuous. The major forms of glass produced are container glass and flat glass, about 75% of the total production. Lesser amounts produced as pressed and blown glasses include lead and borosilicate glass and frit for ceramic coatings.

A typical flow diagram for the production of container glass (Figure 1) establishes three major components of the manufacturing process, raw-material blending and trans-

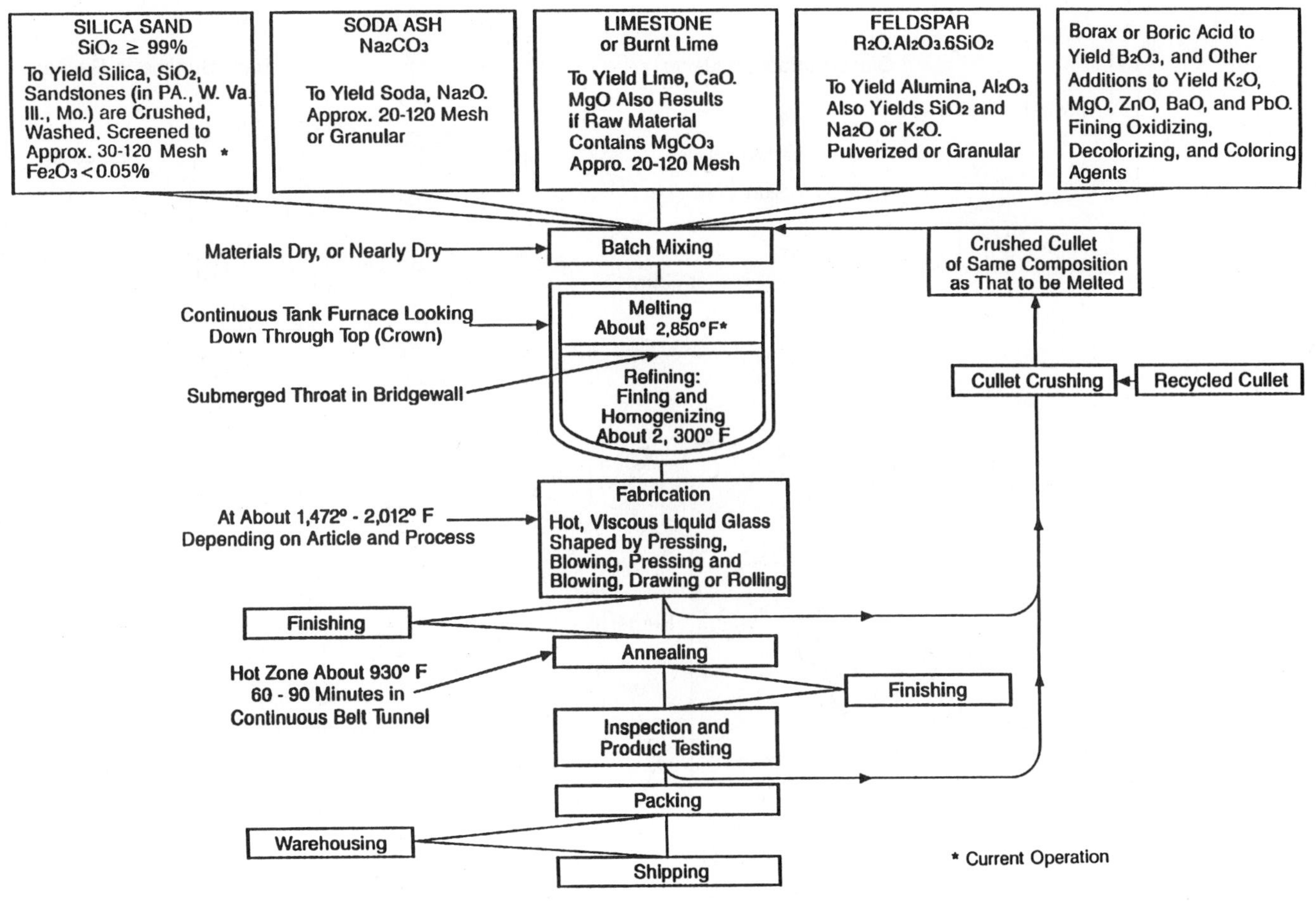

FIGURE 1. Flow Diagram for Soda-Lime Glass Manufacture (Kirk and Othmer, 1947)

port, melting, and forming and finishing. The major difference between container glass and flat glass is in the forming operation. Container glass is formed by blow molding. Flat glass floats the molten glass continuously on a liquid alloy surface with simultaneous cooling to sheet glass. Thus flat glass is sometimes referred to as float glass. The lead and borosilicate glasses and frit may be produced in continuous reverberatory furnaces, but because of small size production, are often conducted in batch, rotary (including tilting), crucible, and hearth furnaces.

The air emissions from glass manufacturing operation are in three zones:

1. Raw-material blending and transport
2. Melting
3. Forming and finishing

The majority of air emissions are in the melting furnace operation.

One of the major concerns in the glass industry was that the addition of an emission control system would adversely affect the pressure stability in the melter and that parallel operation of melters with an emission control system would destablize the operation. Several years of operation with an emission control system on glass furnaces[8] has dispelled this concern and has resulted in even greater melter stability.

PROCESS DESCRIPTION

The major materials used in the production of glass are cullet (recycled glass), silica sand, limestone, and soda ash.

Additional components used in minor quantities for both container and flat glass, and for specialty glasses are salt cake; alumina; barium oxide; boron compounds for borosilicate glass; lead compounds for both lead glass and borosilicate glass; cryolite or fluorospar for frit glass; opacifiers (tin oxide, sodium antimonate, etc.)—frit; and color materials (iron oxide, nickel oxide, copper oxide, cobalt oxide, and so on). These materials are stored in batch house silos and are automatically weighed and blended in batches and transported to the holding zone at the furnace, where they are added to the melt. These materials are in the size range of 10–500 μm with a median in the 300-μm range.

Melting for container and flat glass is generally conducted in a continuous reverberatory furnace (Figures 2 and 3) fired by natural gas or fuel oil. Electric boost furnaces have been introduced in some operations to minimize flue gas emissions (Figure 4).

The reverberatory melters, both end-port and side-port

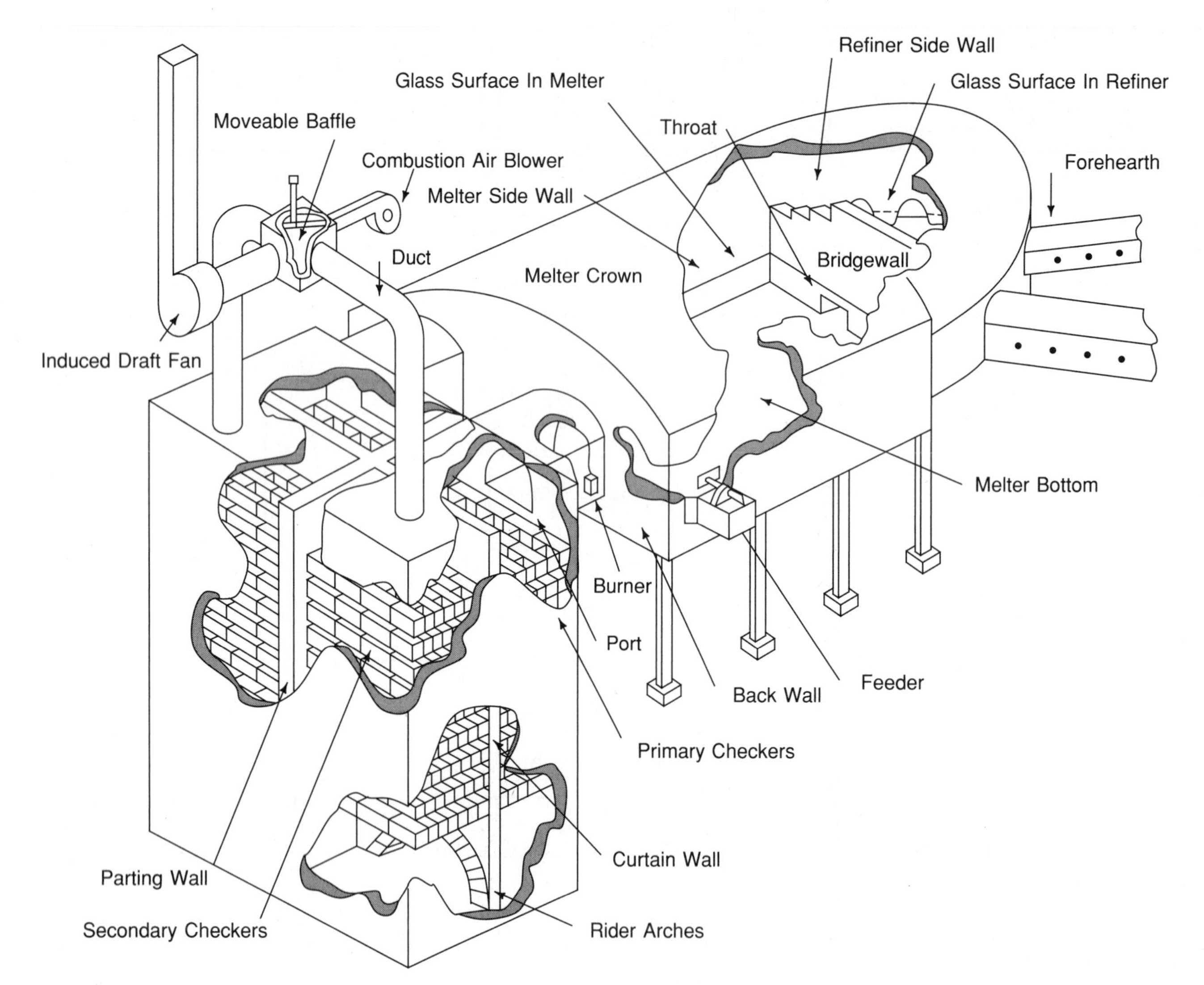

FIGURE 2. Regenerative and Port Glass-Melting Furnace

designs, utilize regenerative heat recovery. These are commonly constructed of checker brick in two separate arrays. The flue gas and combustion gases are alternated between the two checker arrangements, with the flue gas heating one checker while the combustion air flows through the other checker, recovering a portion of the thermal energy of the flue gas.

The diversion of the gases causes spikes in the system pressure and temperature at the time of reversal, the durational cycles is 15 to 25 minutes.[8] The furnace temperature is maintained in the range of 1200°–1575°C. The flue gas, after leaving the regenerative checkers, exhibits a temperature range of 450–500°C.

AIR EMISSIONS CHARACTERIZATION

Raw Material Blending and Transport

Fugitive dust is produced by the preparation transport process. Modern plants contain the fugitive dust in an enclosed system and use air transport of the batch feed. Thus conventional fabric filters are used on all silos and the transport system to confine the particulate emissions.

The emissions consist of particulates in the size ranges of 10, 50, and 300 μm. They are characterized as fugitive dust emissions.

Melting

Two sources contribute to the air emissions from melting furnace operation, namely, the combustion of fuel and the evaporation or dissociation of raw materials.

The emissions from glass melting, in the flue gas, consist of particulates and sulfur oxides, in the case of container and flat glass manufacture. They may contain boron oxides and fluorides in the frit and some specialty glass product manufacture. Almost all the particulates emitted from the melting furnace are less than one μm in size (Figure 5).

The characteristics of uncontrolled emissions from glass manufacture are presented in Table 1.

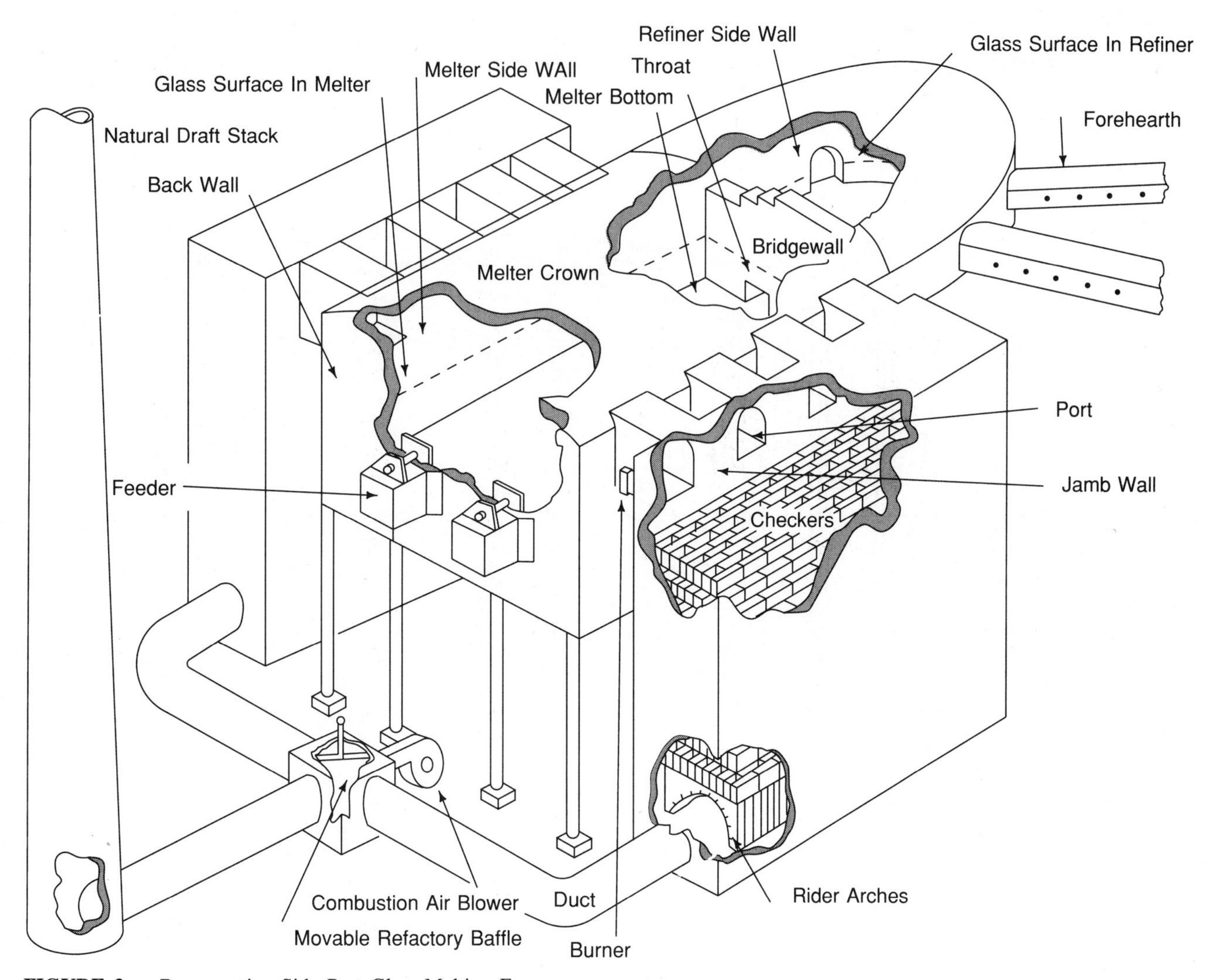

FIGURE 3. Regenerative Side Port Glass-Melting Furnace

TABLE 1. Characteristics of Emissions from Glass Manufacturers

	Container	Float	Opal (Borosilicate)	Frit
Opacity, %	10–50	20–50	20–100	10–60
Particulate, gr/dscf	0.02–0.35	0.02–0.35	0.02–0.35	0.02–0.35
Particulate composition (primary composition)	Na_2SO_4, SO_3	Na_2SO_4, SO_3 NaF, Na_2SiF_6	B_2O_3 P_4O_{10}	Na_2SO_4, NaF
Particulate size (Figure 1), μm	< 5	< 5	< 5	< 5
Gaseous emissions	SO_x, NO_x	SO_x, NO_x	HF, NO_x	SO_x, HF,NO_x
Gas emissions—gas fired, ppm				
SO_x	50–300	150–400		100–300
HF			400–1000	100–300
$NO_x{}^2$	100–600	200–600	200–600	200–600
SO_3 % of SO_x	2–15	2–15		2–15
SO_x when oil-fired, ppm	400–800	400–800	350–500	400–800

Source: Reference 7.

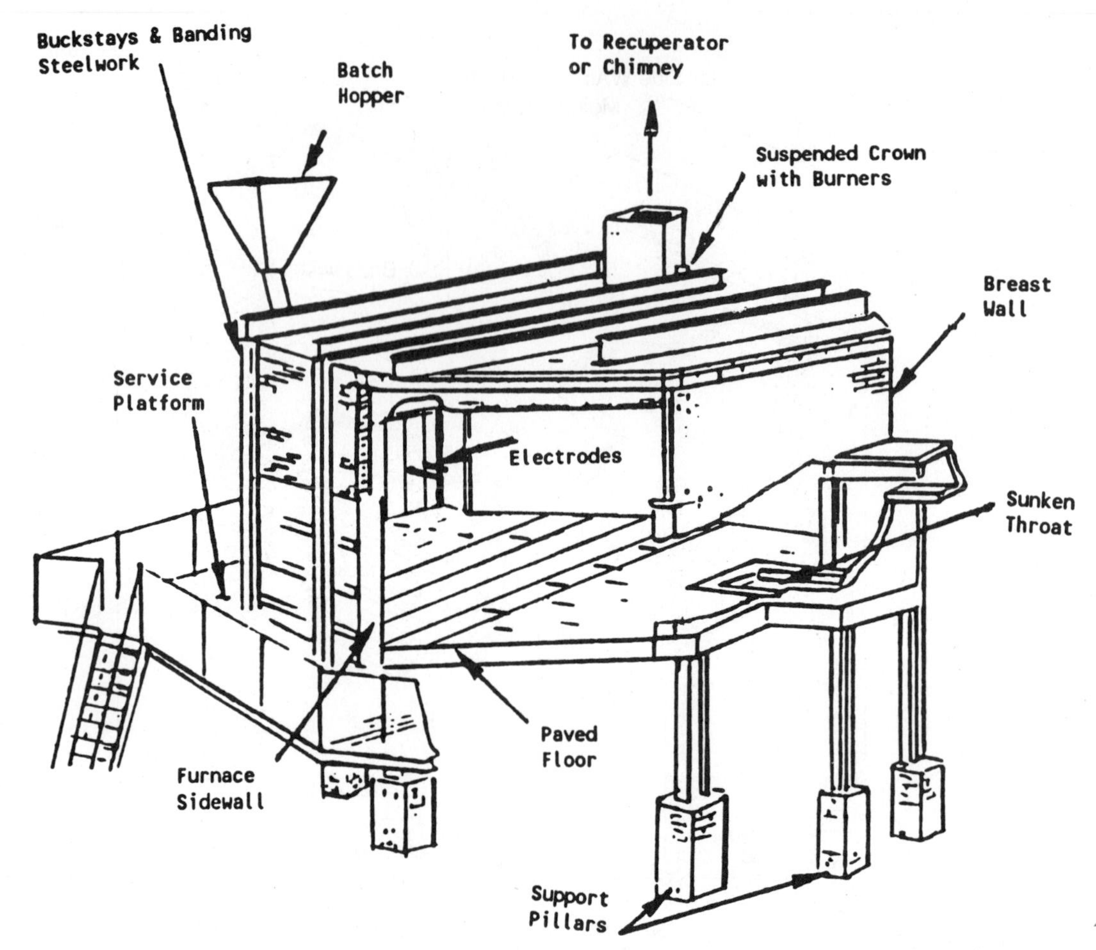

FIGURE 4. Electric Boost Furnace

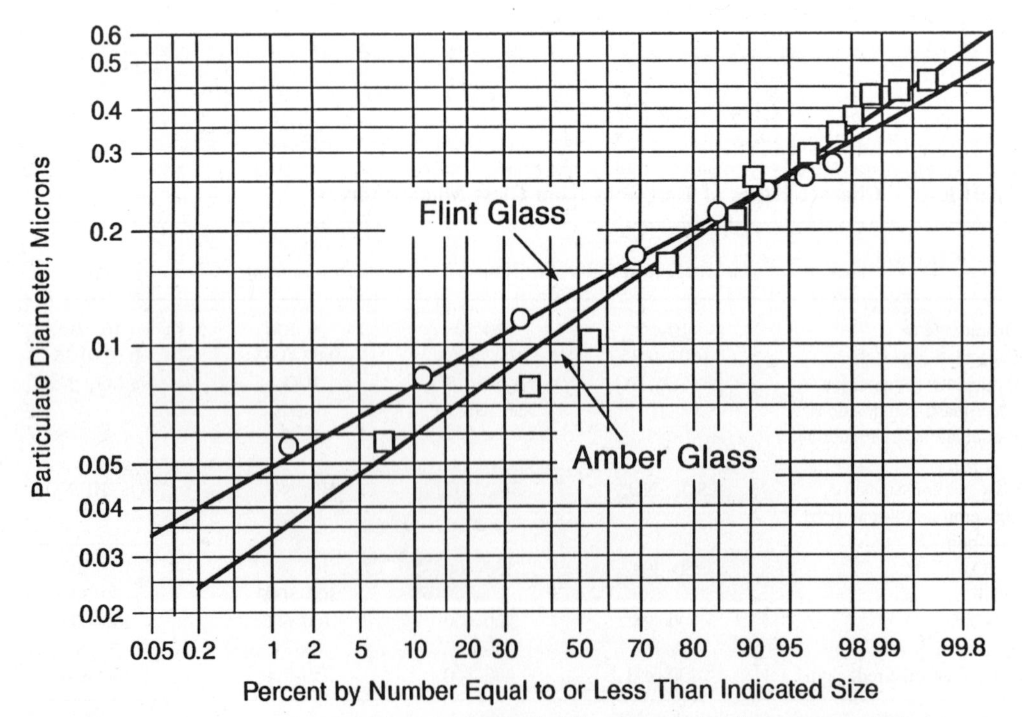

FIGURE 5. Log–Probability Distribution of Particle Sizes Present in Glass Furnace Effluent

TABLE 2. System Comparison-150 TPD-Container Glass

	Wet Scrubbing	Baghouse	ESP	Dry Scrubbing	SemiDry Scrubbing	Electric Boost
Outlet						
Opacity, %	Steam	<10	<10	<5	<5	<20
Particulate, mg/nm^3	50	20	22	20	14	22–40
SO_x removal, %	>95	0	0	20	>95	0
Product disposal	Treatment req.	Recycle	Recycle	Recycle	Recycle	Loss
Power consumption						
Installed horsepower	250	90	25	60	70	0
Power cons kWh/ton	30	12	3	8	9	0
Electrical use, kWh/ton	0	0	3	0	0	950
Total power use, kWh/ton	30	12	6	8	9	950

Forming and Finishing

The emissions from forming and finishing result from thermal decomposition of lubricants for the gob shears and gob delivery systems.

AIR POLLUTION CONTROL MEASURES

Raw-Material Blending and Transport

Bin vent filters on the storage silos are standard units supplied by the silo vendors for batch house operation. Fabric filters are used for collection of particulates emitted at the weigh-blending station. These are primarily pulse jet baghouses operating at an air–cloth ratio of 4–7 fpm.

Typical practice provides hooding at the blending and receiving zones. The railroad hopper car discharge is connected by sleeves to the air transport system. The ventilation rate on hooding is of the order of 8–15 cfm/ft.^{3} of mixer capacity.

Melting

Particulate and Acid Gases

The control of emissions from glass furnaces has utilized the following systems.

Wet scrubbing—venturi and packed bed
Baghouse
Electrostatic precipitator (ESP)
Electric boost
Dry scrubbing
Semidry scrubbing

These have been specifically related to the removal of particulate and acid gases. No NO_x reduction has been reported for any of these systems.

A comparison of the reported information regarding performance and operating cost is given in Table 2.

Wet Scrubbing

The system consisting of a venturi and packed-bed scrubber with a pressure drop of 20–30 inches wc was earlier built to control the emission from container and opal glass operation. The high energy consumption and the complexity of treatment of liquid discharge required have appeared to limit its application.

Baghouse

Both reverse-air cleaning and shaker-type baghouses with an air-to-cloth ratio of 2–4 fpm have been utilized for the particulate control. The high-temperature baghouse, utilizing fiberglass bags operating in the range of 200°C, functions only as a particulate remover. Because of the fine particulate, a rapid rise in pressure drop is encountered, resulting in short cleaning cycles. A rise in the "clean bag" pressure drop increases over time coupled with an operating pressure drop of 10–15 inches wc. The recovered product can be recycled to the production process.

Electrostatic Precipitator

The ESP has been the predominant air pollution control system in the glass industry. It functions only as a particulate separator and, in that type of service, has provided reliable performance for container glass operators. The particulate emissions are low, in the range of 20 mg/nm^3.

It has been reported that adhesion problems of the collected particulate to the electrodes have occurred in borosilicate glass applications.[4] The material recovered by the ESP can be recycled to the glass-making process.

Dry Scrubbing

Dry scrubbing (Figure 6) is operational in the glass industry primarily for the control of both particulate and SO_3, while maintaining low pressure drop and long cleaning cycles in the baghouse. The system consists of the injection of nepheline syenite, with or without limestone addition, into the dry venturi preceding the baghouse. The system is generally operated in the range of 150°C. These Tesisorbs™ permit the formation of a thick cake in the baghouse with low pressure drop, thus enhancing absorption and fine particulate removal. At an operating pressure drop of 4 inches wc in the baghouse, with a four-hour cleaning cycle, and an

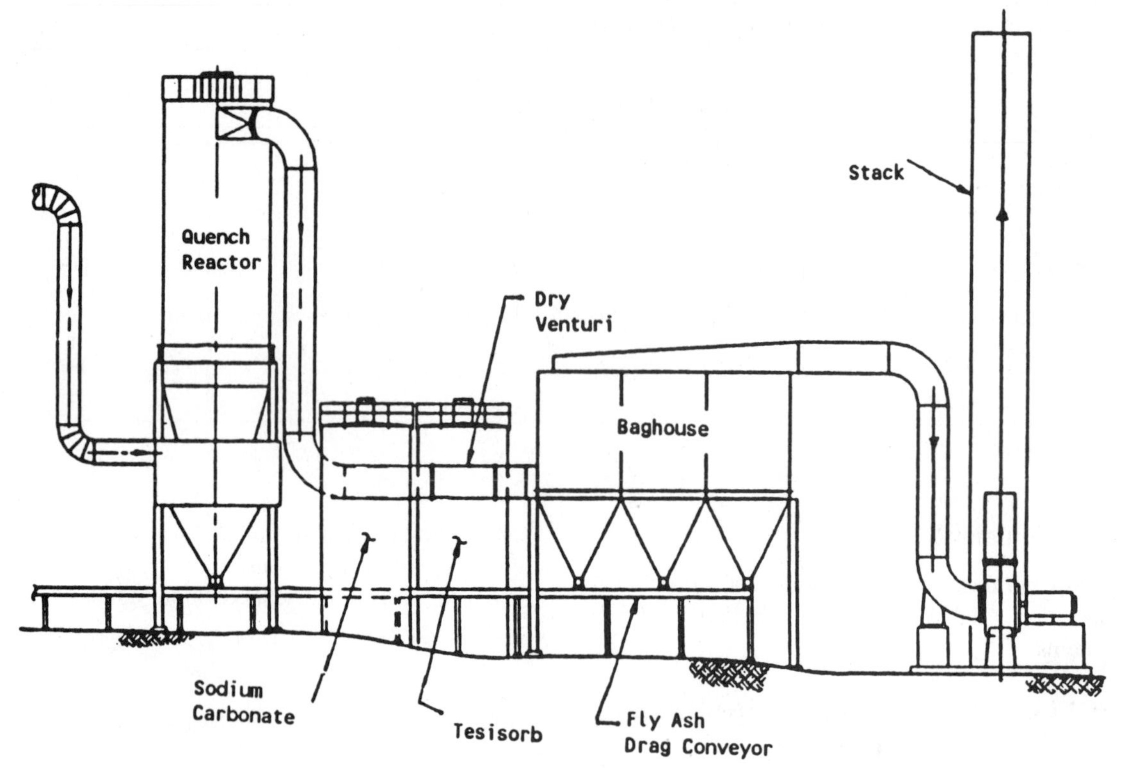

FIGURE 6. Glass Emission Control

air-to-cloth ratio of 2 fpm, the SO_3 concentration was reduced from 25–30 ppm to 0.2–2 ppm, in addition to providing particulate emissions in the range of 10–20 mg/nm.[3]

Semidry Scrubbing

This system, operating in the glass industry, is installed in two configurations: quench reactor–dry venturi–baghouse (QR-DV-BH) (Figure 6), and spray dryer–ESP (SD-ESP). These systems control particulate, SO_2, and SO_3 emissions and permit the use of fuel oil as the combustion fuel in the melter.

The performance reported for the SD-ESP type of system showed a reduction of SO_x by greater than 90%, while no reduction in conventional ESP performance was indicated.[5] The performance for the QR-DV-BH type of system[8] for the control of three furnaces in parallel, with: sodium carbonate solution as the reagent in the quench reactor and limestone as the Tesisorb reagent in the dry venturi resulted in the following reduction in SO_x.

Stoichimetric Ratio	*SO_x Removal, %*
0.39	55.8
0.52	89.1
1.0	>98

Particulate emission was 9.2–20.6 mg/nm^3.

No "brown plume" was noticed during operation, even though the major neutralization was achieved by reaction with Na_2CO_3 solution. The entire baghouse catch is recycled to the batch house inasmuch as the reagents are merely diverted from the melter batch feed. The sulfite–sulfates formed replace the salt cake normally purchased.

Electric Boost

The utilization of the electric boost process maintains a cool surface in the melt, thus minimizing the emissions of the disassociated products, Na_2O, SO_2, and SO_3 and that of particulate emissions from both disassociation and entrainment (Figure 4). The consumption of energy, however, is quite high (Table 2), thus affecting the economics of glass manufactured.

The reported data on emissions[6] of 1050°C from an electric boost furnace indicated fluoride emissions less than 0.5 mg/nm^3. Contrary to expectations, it was reported that replacement of electric boost by oil firing in an opal furnace reduced metal emissions As, Se, Cd, Sb, and Pb.[3]

Nitrogen Oxides

The reduction of NO_x emissions from glass manufacturing processes is an emerging technology. The systems now being offered, planned, or in the process of installation are as follows:

1. Homogeneous catalytic reaction
2. Selective catalytic reduction (SCR)

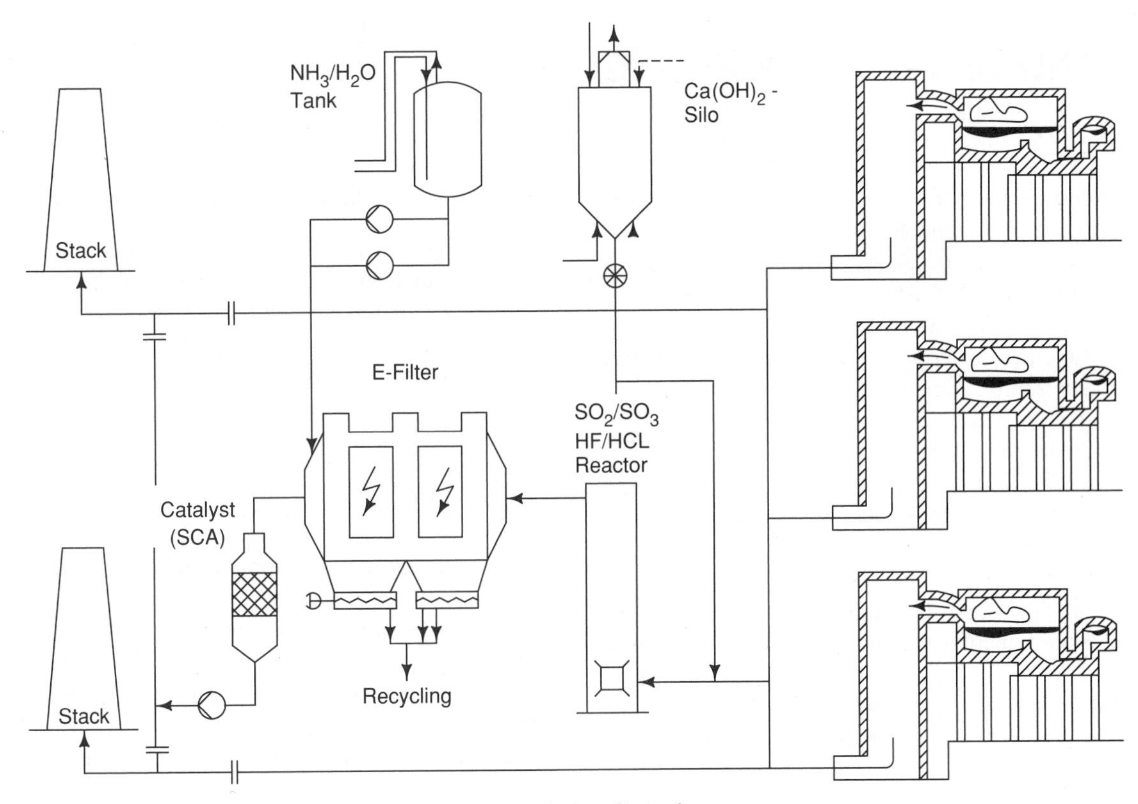

FIGURE 7. Glass Manufacturing Facility with Emission Control

The homogeneous catalytic reaction is based on the introduction of ammonia into the flue gas stream in the temperature range of 700–800°C (known as Exxon De NO_x process), or the introduction of ethyl alcohol, benzene, or kerosene into the flue gas stream at temperatures exceeding 700°C.[9]

The ammonia injection generally achieves a 50–75% reduction in NO_x before excessive slip of ammonia occurs in other industrial applications. In glass furnace emission control, the concentration of the "slip" becomes critical because of the potential for white particulate NH_4HSO_3 formation downstream of the stack. The NO_x reduction by the EtOH injection process is claimed to be from 200 ppm to 30–50 ppm.

Selective catalytic reduction for NO_x control requires a clean stream to protect the catalyst. Thus it must be installed downstream of the particulate control unit. It requires a temperature of 300–500°C to achieve a 75–80% reduction in NO_x with minimum ammonia slip for a single-stage SCR. The catalysts used are generally noble or combinations of metal oxides on a ceramic surface.

A report on a molecular sieve catalyst[2] installed in a glass plant indicates that subsequent to acid-gas removal (Figure 7), an 80% reduction in NO_x was achieved with minimum ammonia slip (10–50 ppm) at an operating temperature of 340–360°C. It should be noted, although not indicated in the flowsheet, that removal of acid gases generally occurs below 200°C and that reheat of the gases must precede SCR treatment.

Forming and Finishing

The development of newer lubricants, including silicone–water emulsions, has decreased the emissions owing to the thermal decomposition of the lubricants. The solution appears to be related to continual improvement of the lubricants. If necessary, a system utilizing a nonplugging condensation-adsorption system can be installed. Such a system has been tested for oily condensibles and volatile organic compound (VOC) recovery in a proprietary dry scrubber system in the cement industry.

References

1. Chikamoto et al., *Kyoto-f-u Eisel Koqai Kenleyushoi Nempo*, 23:124–126 (1978).
2. Grove et al., 49th Conference on Glass Problems, American Ceramic Society 1988.
3. Mamuru et al., *Ann. Rep. Radiat. Cent. Osaka Prof.*, 20:29–35 (1979).
4. Noll, *Glass Technol.*, 24(2):91–97 (1984).
5. Spirina et al., *Steklo Kaam*, (6):6–7 (1979).
6. Tarapore et al., *Glass* 59(12):470–473 (1982).
7. Teller, *Glass Indus.* 15–23 (February 1976).
8. Teller, Hsieh, and van Saun, *Proceedings 49th Conference on Glass Problems*, American Ceramic Society, 1988, pp 196–208.
9. Uetsuki et al., *Osaka Gijutsu Shikensho Kitio*, 36(3):162–169 (1985)
10. Private communication, R-C Co.
11. Teller, *Proceedings 42nd Conference on Glass Problems*, American Ceramic Society, November 1981, pp 1–8.

SAND AND GRAVEL PROCESSING

John E. Yocom, P.E., C.I.H.

The processing of sand and gravel is one of the mineral products industries and generally starts with the removal of material from naturally occurring deposits of unconsolidated granular material that usually consists of siliceous and calcareous materials. The primary operation taking place is separation of the raw material into size classes, which may include some washing of material to remove silt. ("Silt" is usually defined as particulate matter less than 200 mesh [75 μm].) In the quarrying and processing of stone described in the following section, "Stone and Quarrying Processing" the raw material is rock, which must be crushed before use. While some size reduction is sometimes required in sand and gravel operations, it will not be discussed here, but rather will be covered in the next section since the same basic size-reduction methods are used in both processes.

Sand and gravel deposits are located all over the United States, usually close to urban areas or large construction sites to minimize transportation costs. Such deposits are commonly found adjacent to or in river courses or in areas with glaciated or weathered rock and they often contain the fine alluvial silt that is the primary source of process and fugitive dust from sand and gravel operations. Figure 1 shows a sand and gravel operation adjacent to a river course. Even though some deposits of sand and gravel may be quite wet and generate little fugitive dust during removal of the material from the deposit, later in the process, when the material becomes dry, large amounts of fugitive dust may be emitted unless wetting and other dust control measures are utilized. Such dust is nontoxic and is primarily of nuisance concern. Furthermore, such particles, even though small enough to be airborne, usually make minor contributions to PM_{10} in the ambient atmosphere.

The U.S. Environmental Protection Agency (EPA) summarized the annual production quantities for sand and gravel as 716,100 Gg/yr (787,700 tons per year) from 5500 plants during 1975.[1] It estimated an annual growth rate of 1%, which would translate into a 1992 production rate of an estimated 848,100 Gg/yr (932,900 tons per year).

FIGURE 1. Typical Sand and Gravel Operation

PROCESS DESCRIPTION

Depending on the location and characteristics of the deposit, sand and gravel are removed by any one of a number of methods, including power shovels, front-end loaders, draglines, and suction dredge pumps. Occasionally, blasting with light charges may be used to loosen the deposit. Dust emissions from sand and gravel operations vary widely. If the deposit is dry and the material and overburden have a high silt content, dust emissions from various parts of the process may be significant. If the deposit is wet or is removed by dredging, dust emissions tend to be negligible as long as a high moisture content is maintained in the material.

Figure 2 is a flow diagram of a typical sand and gravel operation. A process schematic for a sand and gravel operation is similar, and may even be identical to that for a stone quarrying and process described in the following section. The reader should refer to Figure 1 in that section for such a diagram.

In general, the process starts with extraction of the sand and gravel from the deposit, often preceded by the removal of overburden by draglines, shovels, or mobile equipment, such as bulldozers, scrapers, or front-end loaders. If sand

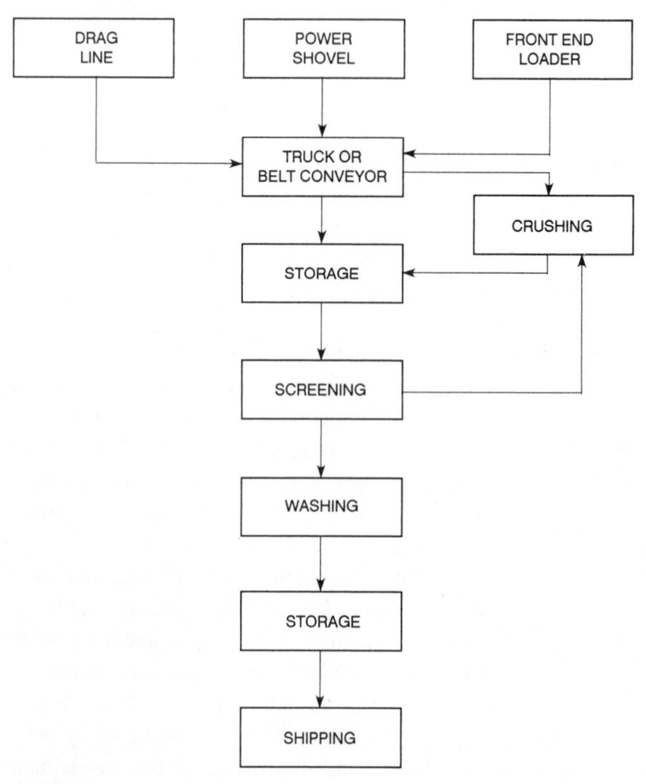

FIGURE 2. Flow Diagram for Typical Sand and Gravel Processing Operation

and gravel operations are part of river dredging, suction pumps may be used. Removal of dry overburden may generate large quantities of dust because such overburden may have a high silt content. Raw sand and gravel are usually stockpiled at the processing area, which may be adjacent to the deposit or at least relatively close to it. Delivery to the stockpile may be by means of conveyor belt, truck, or front-end loader. In any case, if the material is dry, dust may be generated during these transfer processes. In addition, truck and front-end loader traffic on haul roads will generate dust, and dust may be released from open conveyor belts during high winds. Placement on storage piles is usually by means of conveyor belts. Considerable amounts of dust may be generated from the fall of material onto the pile, unless fabric chutes or stone towers are used to suppress such emissions. Note in Figure 1 that a stone ladder is used on a storage pile.

At this point, the material receives preliminary screening with a grizzly and oversized material may be sent to a crusher. In screening, large amounts of dust may be produced if it is carried out dry and if hoods and air pollution control devices are not used. If wet screening is used to produce a washed gravel product, negligible amounts of dust are produced, but effluent water must be clarified by settling before reuse or discharge.

AIR EMISSION CHARACTERIZATION

The only air pollutant of consequence emitted from sand and gravel operations is particulate matter. Emissions from internal combustion engines on vehicles and other equipment will not be discussed here. The particulate matter emitted from sand and gravel is made up principally of inert crustal material (e.g., soil and rock particles), which is of little health concern. Furthermore, particles generated from such operations tend to be of rather large size. However, because of the wide range of particle sizes generated by various types of sand and gravel operations, it is not possible to present a meaningful particle-size distribution representative of the industry.

In sand and gravel operations, dust may be emitted from a variety of locations in the form of fugitive dust. Typical release points are listed as follows:

- Dust during removal of overburden and the sand and gravel from the deposit
- Wind-blown dust from storage piles

TABLE 1. Uncontrolled Particulate Emission Factors for Sand and Gravel Processing Plants[a]

Uncontrolled Options	Emissions by Particle Size Range (Aerodynamic Diameter)[b]			Units	Emission Factor Rating
	Total Particulate	TSP ($\leq$30 μm)	PM_{10} ($\leq$10 μm)		
Process sources[c]					
Primary or secondary crushing (wet)	NA	0.009 (0.018)	NA	kg/Mg (lb/ton)	D
Open dust sources[c]					
Screening[d]					
Flat screens (dry product)	NA	0.08 (0.16)	0.06 (0.12)	kg/Mg (lb/ton)	C
Continuous drop[c]					
Transfer station	0.014 (0.029)	NA	NA	kg/Mg (lb/ton)	E
File formation—stacker	NA	0.065 (0.13)	0.03 (0.06)[e]	kg/Mg (lb/ton)	E
Batch drop[c]					
Bulk loading	0.12 (0.024)	0.028 (0.056)[f]	0.0012 (0.0024)[f]	kg/Mg (lb/ton)	E
Active storage piles[g]				kg/hectare/day[h]	D
Active day	NA	14.8 (13.2)	7.1 (6.3)[e]	(lb/acre/day)	
Inactive day (wind erosion only)	NA	3.9 (3.5)	1.9 (1.7)[e]	kg/hectare/day[h] (lb/acre/day)	D
Unpaved haul roads					
Wet materials	i	i	i		D

[a]NA = not available. TSP = total suspended particulate. Predictive emission factor equations, which generally provide more accurate estimates of emissions under specific conditions, are presented in Chapter 11 of reference 2. Factors for open dust sources are not necessarily representative of the entire industry or of a "typical" situation.

[b]Total particulate is airborne particles of all sizes in the source plume; TSP is what is measured by a standard high-volume sampler (see Section 11.2 in reference 2).

[c]References 4–8.

[d]References 3 and 4. For completely wet operations, emissions are likely to be negligible.

[e]Extrapolation of data, using *K* factors for appropriate operation from Chapter 11 in reference 2.

[f]For physical, not aerodynamic, diameter.

[g]Reference 5. Includes the following distinct source operations in the storage cycle: (1) loading of aggregate onto storage piles (batch or continuous drop operations), (2) equipment traffic in storage areas, (3) with erosion of pile (batch or continuous drop operations). Assume 8 to 12 hours of activity per 24 hours.

[h]Kilograms per hectare (pounds per acre) of storage per day (includes areas among piles).

[i]See Section 11.2 of reference 2 for empirical equations.

Source: Reference 2.

- Dust from traffic on haul roads
- Dust from open conveyors exposed to the wind
- Dust during material dumping from trucks, front-end loaders, and conveyors
- Dust from screening
- Dust from transfer points in conveyor systems

Table 1 presents uncontrolled emission factors for particulate matter from sand and gravel processing plants as presented in AP-42.[2] In AP-42, the emission factors for fugitive dust sources are rated on a separate scale (in comparison with point sources) from A through E, with A being the most reliable and E the least reliable. All data in this table are rated C, D, or E. Note that emissions from haul roads must be calculated from empirical equations presented in this same publication. If one can believe that the particle size distribution on which these data are based is typical, the data show that for a batch drop (bulk loading), the PM_{10} fraction ($\leq 10\ \mu m$) is only 1% of the total particulate matter and only 4% of total suspended particulate (TSP) measured by the Hi Vol sampler. Therefore, the bulk of dust emissions from a sand and gravel operation will tend to settle out close to the source. Such dust can deposit on horizontal surfaces and create a nuisance at points near sand and gravel processing facilities.

Even though the fine particulate matter from sand and gravel operations represents a small fraction of the total dust released, this quantity may have significant effects on downwind air quality. Air pollution control agencies invariably use opacity standards and measured ambient concentrations as a means of assessing the potential impact of sand and gravel operations and other sources of fugitive dust.

AIR POLLUTION CONTROL MEASURES

The methods of controlling dust emissions from sand and gravel operations include using water sprays to keep materials and roads wet, maintaining good housekeeping, limiting drop heights of materials, covering trucks and conveyors, using enclosures or hooding material at transfer points and screening operations, and exhausting air from these points to air pollution control systems. Since these measures are basically the same as used in various parts of a stone quarrying and processing facility, a detailed discussion of these air pollution control systems will not be given here, but is given in the following section. As stated before, the basic difference in the requirements for control of emissions from sand and gravel and rock quarrying and processing is that the quarrying and crushing of rock usually produce larger quantities of dry dust and at more points in the process than do sand and gravel operations. On the other hand, sand and gravel operations commonly rely on relatively wet material that may require little or no crushing. Water sprays alone may be an adequate air pollution control measure for many sand and gravel operations, while rock quarrying and crushing require not only the use of water sprays, but also of industrial ventilation systems, including high-efficiency dust collectors such as baghouses.

References

1. *Control Techniques for Particulate Emissions from Stationary Sources—Vol. 2,* EPA-450/3-81-005b, U.S. Environmental Protection Agency, Research Triangle Park, NC, 1982.
2. *Compilation of Air Pollutant Emission Factors,* AP-42, U.S. Environmental Protection Agency, Research Triangle Park, NC, 1988.
3. *Review Emissions Data Base and Develop Emission Factors for the Construction Aggregate Industry,* Engineering-Science, Inc., Arcadia, CA, September 1984.
4. "Crushed rock screening source test reports on tests performed at Conrock Corp., Irwindale and Sun Valley, CA, plants," Engineering-Science, Inc., Arcadia, CA, August 1984.
5. C. Cowherd, Jr., et al., *Development of Emission Factors for Fugitive Dust Sources,* EPA-450/3-74-037, U.S. Environmental Protection Agency, Research Triangle Park, NC, June 1974.
6. R. Bohn et al., *Fugitive Emissions from Integrated Iron and Steel Plants,* EPA-600/2-78-050, U.S. Environmental Protection Agency, Washington, DC, March 1978.
7. G. A. Jutze and K. Axetell, *Investigation of Fugitive Dust, Vol. I: Sources, Emissions and Control,* EPA-450/3-74-036a, U.S. Environmental Protection Agency, Research Triangle Park, NC, June 1974.
8. *Fugitive Dust Assessment at Rock and Sand Facilities in the South Coast Air Basin,* Southern California Rock Products Association and Southern California Ready Mix Concrete Association, P.E.S., Santa Monica, CA, November 1979.

STONE AND QUARRYING PROCESSING

John E. Yocom, P.E., C.I.H.

The source of crushed stone is usually a deposit of relatively solid rock that may be one of a variety of geological types, such as limestone, dolomite, traprock, granite, sandstone, or quartz. In order to loosen the rock in the quarry, drilling and blasting are usually required. After transfer to the processing plant by various means, the material is subjected to various size reduction, size classification, and transfer operations, all of which have the potential to emit process and fugitive dust. Quarried rock is dry, and unless it is kept wet or industrial ventilation and air pollution control systems are used, large quantities of fugitive dust may be emitted any time the material is processed or otherwise handled. Furthermore, fugitive dust may be emitted from storage piles, roadways, and open conveyors. Figure 1 is a generalized schematic diagram of a typical crushed-stone processing plant. (This diagram also could apply in many

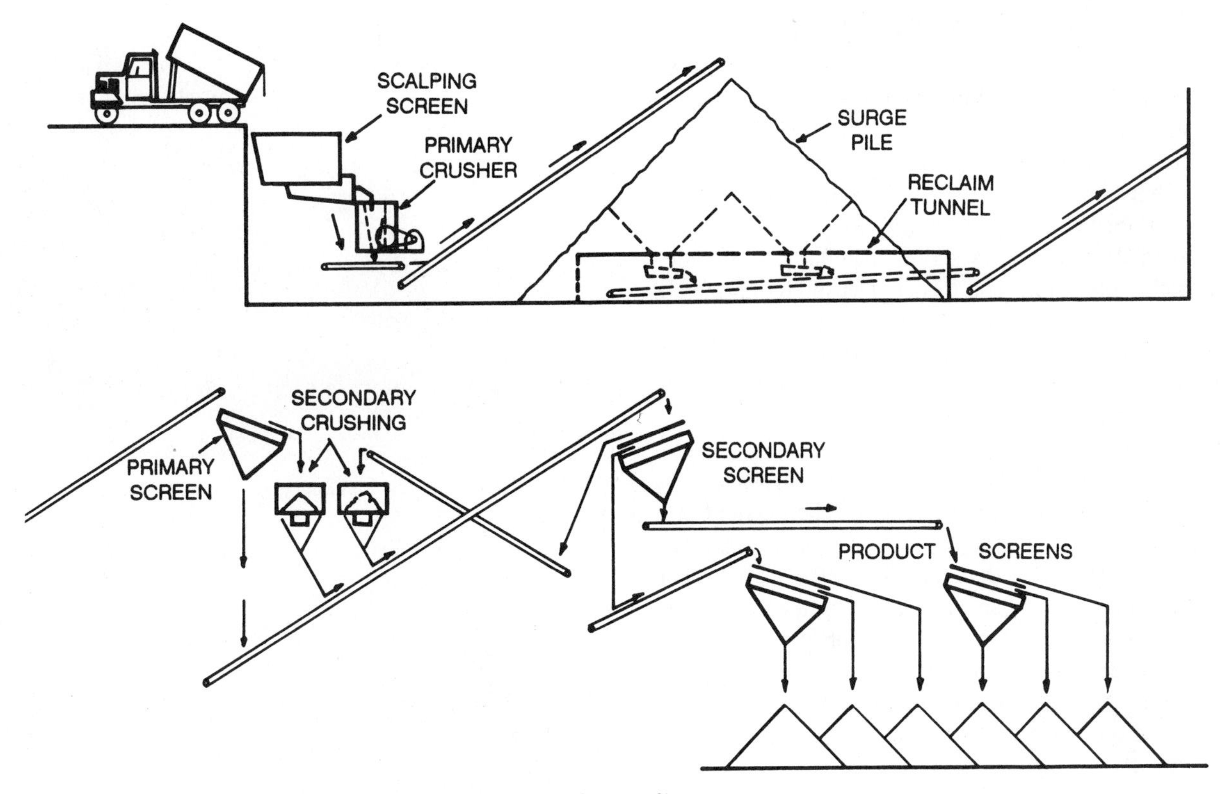

FIGURE 1. Typical Stone Processing Plant (From Reference 2)

respects to a sand and gravel processing plant.) Figure 2 is a generalized view of a rock quarrying and processing plant and Figure 3 shows some of the processing and material handling equipment at a crushed-rock processing plant.

Crushed rock is basically a construction material (e.g., road bed, paving materials, and concrete mixing), and distances between the quarry and processing plant and its point of use must be kept to a minimum to avoid high transportation costs. Thus many crushed-stone operations tend to be located relatively near populated areas. Furthermore, since the life of a quarry will be a number of decades, many such operations that were originally located in undeveloped or rural areas may eventually be surrounded by residential developments, with homeowners complaining about dust settling on their property. Such operations, therefore, must meet not only federal and state standards such as process-weight-based limits and opacity standards, but also local standards that may be developed based on the stringent requirements imposed by citizens' legal actions.

The quantities of crushed stone used in the United States are large. The U.S. Environmental Protection Agency (EPA) estimated the 1975 production level at 819,020 Gg/yr (901,000 tons per year) from 4800 quarries.[1] The EPA further estimated that this production level was growing at a rate of 4% per year, which would yield an estimate for 1992 of 1,755,100 Gg/yr (1,930,600 tons per year).

FIGURE 2. Generalized View of a Rock Quarrying and Processing Plant

FIGURE 3. Processing and Material-Handling Equipment at a Crushed-Rock Processing Facility

PROCESS DESCRIPTION

The process starts with the loosening of the rock by drilling and blasting. Fugitive dust is released when the blast occurs and when the wall of loosened rock falls to the quarry floor. Figure 4 shows the quarry face after blasting. The loosened rock is then loaded into heavy trucks by power shovels or front-end loaders. Figure 5 shows material being loaded into a truck with a front-end loader. Depending on the silt content of the rock and the height of the drop of material between loader and truck, fugitive dust may be released. The quarried material is then transported to the processing plant. During transport, dust from the rock in the truck may be released and road dust may be reentrained. At the processing plant, the material receives primary crushing in a gyratory or jaw crusher. Figure 6 shows the opening of a large gyratory primary crusher. In this plant, oversized pieces of rock are removed with a power claw to be broken into smaller pieces that can be handled by the crusher. In some plants, quarried material is first dumped onto a grizzly or primary screen to scalp large boulders to reduce the load on the primary crusher. Unless this and other screens are enclosed and ventilated through dust collectors or the material is kept wet, copious amounts of dust may be released. The primary crusher product will vary in size, but generally will be somewhere in the range of approximately 1½- to 12-inch material, depending on the plant and its products. The primary crushed material or grizzly throughs (undersized material) are discharged onto a belt conveyor and transported to a secondary crusher or are stockpiled to await further processing. The transfer of crushed rock by belt conveyor may produce fugitive dust emissions if the material is dry, is high in silt content, and is exposed to the wind. Dust may also be emitted at transfer points unless enclosures, industrial ventilation, and air pollution control systems are provided or the material is kept wet.

FIGURE 4. Quarry Face After Blasting

FIGURE 5. Truck Loading of Rock in a Quarry

FIGURE 6. Quarried Rock Entering a Large Gyratory Crusher

At this point, the material is usually screened into one of three sizes (oversize, throughs, and undersize). The sizes of these three materials depend on the plant and the uses to which various products will be put. Oversized material is discharged or transferred to a secondary crusher for size reduction. Secondary crushers are usually of the gyratory, cone or impact type. The discharge of the secondary chambers is transferred by conveyor belt to secondary screens, with the oversized material being conveyed back to the secondary chamber. If tertiary crushing is required, cone crushers or hammer mills are usually employed. Size classification at this point is usually accomplished by mechanically vibrated screens or air classifiers. Throughout this entire process, crushed rock of various sizes and size ranges is stockpiled, usually in the open prior to shipment or awaiting further processing. Figure 7 shows sized crushed rock being dumped onto a stockpile.

AIR EMISSIONS CHARACTERIZATION

As is the case with sand and gravel operations, the only pollutant emission of concern from stone and quarrying

FIGURE 7. Sized Crushed Rock Being Dumped onto a Storage Pile

processing is particulate matter. Depending on the type of rock being quarried and processed, the dust generated tends to be more of a nuisance than a significant health hazard. An exception to this might be from the quarrying of silica rock, which raises concern as an occupational hazard, since airborne concentrations can be high near dust-producing operations. However, it is usually of no great concern in this regard to those outside the quarry because of the low concentrations of dust in the respirable size range (<3 μm). Particulate matter produced in rock quarrying and processing, as is the case with sand and gravel operations, is usually of relatively large particle size. In the case of rock quarrying and processing, the chemical composition of the dust tends to be relatively homogeneous since its ancestry is the rock formation from which the rock deposit was taken. Dust from sand and gravel operations is more often related to the overburden and silt associated with the sand and gravel. Nevertheless, the relative quantities of potential dust released from the quarrying and processing of a given amount of rock tend to be greater than for dust released from producing the same amount of sand and gravel because of the greater number of size-reduction and other dust-producing steps required in the former process.

Emission points for dust release from rock quarrying and processing typically include the following.

- Drilling for blast charges
- Blasting
- Loading of trucks
- Truck travel on dusty roads
- Fugitive dust loss from trucks
- Dumping into primary crusher
- Primary, secondary, and tertiary crushing
- Screening
- Transfer points on conveyor systems
- Loading onto storage piles from conveyors
- The wind blowing dust from storage piles and open conveyors

Table 1 presents uncontrolled particulate emission factors for rock crushing operations as published in the EPA's AP-42.[2] These data are based on extremely limited information. In AP-42, the emission factors for fugitive dust sources are rated on a separate scale (in comparison with point sources) from A through E, with A being the most reliable and E the least. All data in this table are rated D or E.

Table 2 shows uncontrolled emission factors for particulate matter from open sources at rock quarrying and processing plants as presented in AP-42.[2] Note that emission estimates for unpaved haul roads depend on a variety of site-specific factors and must be calculated from empirical equations provided in Section 11 of AP-42.[2]

Relatively large particles (>10 μm) account for most of the weight rate of emissions from rock quarrying and processing plants. These emissions, therefore, tend to be involved more in complaints about settled dust than in possible health effects related to inhalable particles.

AIR POLLUTION CONTROL MEASURES

Control measures for reducing or eliminating dust emissions from quarrying and processing plants include the following.

- Wetting of material or surfaces with water with or without surfactants or foaming agents.
- Covering open operations to prevent dust entrainment by the wind.
- Reducing the drop height of dusty material.
- Using hooding, industrial ventilation systems, and dust collectors (e.g., baghouses) on dusty processes amenable to enclosure.

Table 3 lists typical operations in a rock quarrying and processing plant and shows candidate air pollution control measures applicable to each. Note that wet dust suppression is a common method of controlling dust emissions. Such suppression can often be enhanced by wetting agents. In addition, the use of foaming agents can provide additional dust control capability. The use of such dust suppression techniques has been reviewed by Kester.[9] Typical control measures for several of the processes listed in the previous section are discussed in the following.

Drilling for Blast Charges

Dust emissions from this process can be controlled by water spray or by hooding followed by an air pollution control device. Figure 8 shows a system consisting of a box-shaped hood through which the drill passes that can be lowered to the surface being drilled. Air is drawn from the enclosure

TABLE 1. Uncontrolled Particulate Emission Factors for Crushing Operations[a]

Type of Crushing[b]	Particulate ≤30 μm kg/Mg	Particulate ≤30 μm (lb/ton)	Particulate ≤10 μm kg/Mg	Particulate ≤10 μm (lb/ton)	Emission Factor Rating
Primary or secondary					
Dry material	0.14	(0.28)	0.0085	(0.017)	D
Wet material[c]	0.009	(0.018)	—	—	D
Tertiary dry material[d]	0.93	(1.85)	—	—	E

[a]Based on actual feed rate of raw material entering the particular operation. Emissions will vary by rock type, but data available are insufficient to characterize these phenomena. Dash = no data.
[b]References 5 and 6. Typical control efficiencies for cyclone, 70–80%; fabric filter, 99%; wet spray systems, 70–90%.
[c]References 6 and 7. Refers to crushing of rock, either naturally wet or moistened to 1.5–4 wt % with wet suppression techniques.
[d]Range of values used to calculate emission factor is 0.0008–1.38 kg/Mg. Extremely small database.
Source: Reference 2.

TABLE 2. Uncontrolled Particulate Emission Factors for Open Dust Sources at Crushed-Stone Plants

Operation	Material	Emissions by Particle Size Range (Aerodynamic Diameter)[a] TSP (≤ 30μm)	PM$_{10}$ (≤10 μm)	Units[b]	Emission Factor Rating
Wet quarry					
drilling	Unfractured stone[c]	0.4 (0.0008)	0.04 (0.001)	g/Mg (lb/ton)	E
Batch drop					
Truck unloading	Fractured stone[c]	0.17 (0.0003)	0.008 (0.00002)	g/Mg (lb/ton)	D
Truck loading					
Conveyor	Crushed stone[d]	0.17 (0.0003)	0.05 (0.0001)	g/Mg (lb/ton)	E
Front-end loader	Crushed stone[e]	29.0 (0.06)	NA	g/Mg (lb/ton)	E
Conveying					
Tunnel belt	Crushed stone[c]	1.7 (0.0034)	0.11 (0.0002)	g/Mg (lb/ton)	E
Unpaved haul roads		f	f		
Blasting	Quarried stone	g	g		

[a]Total suspended particulate (TSP) is that measured by a standard high-volume sampler (see section 11.2 in reference 2). Use of empirical equations in Chapter 11 of reference 2 is preferred to single-value factors in this table. Factors in this table are provided for convenience in quick approximations and/or for occasions when equation variables cannot be reasonably estimated. NA = not available.
[b]Expressed as g/Mg (lb/ton) of material through primary crusher, except for front-end loading, which is g/Mg (lb/ton) of material transferred.
[c]Reference 3.
[d]Reference 4.
[e]Reference 7.
[f]See section 11.2 in reference 2 for empirical equations.
[g]Not presented because of sparsity and unreliability of test data.
Source: Reference 2.

and is passed through a cyclone. Collected dust drops through a cloth sleeve to the ground, where it is periodically picked up for disposal. In this case, the dust generated is of sufficiently large particle size that a cyclone provides an adequately high collection efficiency. It should be pointed out here that rock dust recovered from air pollution control systems is a valuable product in road building and other construction operations.

Truck Dumping

Figure 9 shows the dumping of quarried rock into a primary crusher. Because of the scale of the operation and the need for accessibility to the crusher opening by large vehicles, enclosure and venting to a control device are impractical. In this case, water is used to wet the material and reduce dust emissions.

TABLE 3. Particulate Emission Sources and Typical Control Measures for Rock Quarrying and Processing

Operation or Source	Control Options
Drilling	Liquid injection (water or water plus a wetting agent)
	Capturing and venting emissions to a control device
Blasting	Adoption of good blasting practices
Loading (at mine)	Water wetting
Hauling	Water wetting of haulage roads
	Treatment of haulage roads with surface agents
	Soil stabilization
	Paving
	Traffic control
Crushing	Wet dust suppression systems
	Capturing and venting emissions to a control device
Screening	Same as crushing
Conveying (transfer points)	Same as crushing
Stockpiling	Stone ladders
	Stacker conveyors
	Water sprays at conveyor discharge
	Pugmill
Storage bins	Capturing and venting to a control device
Conveying (other than transfer points)	Covering
	Wet dust suppression
Windblown dust from stockpiles	Water wetting
	Surface-active agents
	Covering (i.e., silos, bins)
	Windbreaks
Windblown dust from roads and plant yard	Water wetting
	Oiling
	Surface-active agents
	Soil stabilization
	Paving
	Sweeping
Loading (product into railroad cars, trucks, ships)	Wetting
	Capturing and venting to control device

Source: Reference 1.

FIGURE 8. Air Pollution Control System for Rock Drilling

Secondary Crushing

Control of secondary crushers can usually be achieved by enclosure and ventilation through a control device. Figure 10 shows a secondary crusher with a baghouse controlling emissions from both the transfer into the crusher and from the crusher itself.

Screening and Material Transfer

Emissions from screens and transfer points are usually amenable to enclosure and venting through a baghouse. Figure 11 shows a baghouse serving a screening operation, together with the transfer point onto the screen.

Stockpiling

Dust emissions in stockpiling are not usually amenable to enclosure; therefore, wetting of the material is most commonly used. Figure 12 shows material being dumped onto a material stockpile. Note that the discharge sleeve is wet, indicating that the material has a high moisture content,

FIGURE 9. Water Spray for Controlling Dust Emission from Dumping into a Primary Crusher

FIGURE 10. Secondary Crusher and Transfer Point with a Baghouse Dust Control

FIGURE 11. Baghouse Dust Control on Rock Screening

FIGURE 12. Dust Control for Stockpiling by Maintaining High Moisture Content in the Material

which accounts for the lack of any visible dust release from this transfer operation. If dry material containing silt is dumped on a storage pile, large quantities of dust may be emitted. Furthermore, fugitive dust may be emitted from dry-material storage piles during windy periods. Another method for reducing dust emissions from stockpiling involves reducing the fall height or using a stone ladder. Figure 1 in the section on Sand and Gravel Processing shows a stone ladder on a gravel pile.

References

1. *Control Techniques for Particulate Emissions from Stationary Sources—Vol. 2,* EPA-450/3-81-005b, U.S. Environmental Protection Agency, Research Triangle Park, NC, 1982.
2. *Compilation of Air Pollutant Emission Factors,* AP-42, U.S. Environmental Protection Agency, Research Triangle Park, NC, 1988.
3. P. K. Chalekode et al., *Emissions from the Crushed Granite Industry: State of the Art,* EPA-600/2-78-021, U.S. Environmental Protection Agency, Washington, DC, February 1978.
4. T. R. Blackwood et al., *Source Assessment: Crushed Stone,* EPA-600/2-78-004L, U.S. Environmental Protection Agency, Washington, DC, May 1978.
5. F. Record and W. T. Harnett, *Particulate Emission Factors for the Construction Aggregate Industry,* Draft Report, GCA-TR-CH-83-02, EPA Contract No. 68-02-3510, GCA Corp., Chapel Hill, NC, February 1983.
6. *Review Emissions Data Base and Develop Emission Factors for the Construction Aggregate Industry,* Engineering-Science, Inc., Arcadia, CA, September 1984.
7. C. Cowherd, Jr., et al., *Development of Emission Factors for Fugitive Dust Sources,* EPA-450/3-74-037, U.S. Environmental Protection Agency, Research Triangle Park, NC, June 1974.
8. R. Bohn et al., *Fugitive Emissions from Integrated Iron and Steel Plants,* EPA-600/2-78-050, U.S. Environmental Protection Agency, Washington, DC, March 1978.
9. M. Kester, "Using suppressants to control dust emissions," Parts I and II, *Powder and Bulk Engineering,* February 1989 and March 1989.

LIME MANUFACTURING

*National Lime Association, Editor**

Lime is the product of the high-temperature calcination of limestone. There are two kinds of lime: high-calcium lime (CaO) and dolomitic lime (CaO-MgO). Lime is manufactured in various kinds of kilns by one of the following reactions.

$$CaCO_3 + \text{Heat} \rightarrow CO_2 + CaO \text{ (high-calcium lime)}$$
$$CaCO_3 \cdot MgCO_3 + \text{Heat} \rightarrow CO_2 + CaO \cdot MgO \text{ (dolomitic lime)}$$

In many of the lime plants, some of the resulting lime is reacted (slaked) with water to form hydrated lime.

PROCESS DESCRIPTION

The basic processes in the production of lime are (1) quarrying the raw limestone, (2) preparing the limestone for the kilns by crushing and sizing, (3) calcining the limestone to quicklime (CaO), (4) processing the quicklime further by hydrating to calcium hydroxide, and (5) miscellaneous transfer, storage, and handling operations. A generalized material flow diagram for a lime manufacturing plant is given in Figure 1. Note that some of the operations shown may not be performed in all plants.

The heart of a lime plant is the kiln. The most prevalent type of kiln is the rotary kiln, which accounts for about 90% of all lime production in the United States. This kiln is a long, cylindrical, slightly inclined, refractory-lined furnace through which the limestone and hot combustion gases pass countercurrently. Coal, oil, and natural gas may all be used as fuel in rotary kilns. Product coolers and kiln-feed preheaters of various types are commonly employed to recover heat from the hot lime product and hot exhaust gases respectively.

The next most popular type of kiln in the United States is the vertical, or shaft, kiln. This kiln can be described as an upright, heavy steel cylinder lined with refractory material. The limestone is charged at the top and calcined as it descends slowly to the bottom of the kiln, where it is discharged. A primary advantage of vertical kilns over rotary kilns is their higher average fuel efficiency. The primary disadvantages of vertical kilns are their low to moderate production rates and the fact that coal generally cannot be used without degrading the quality of the lime produced. Although still prevalent in Europe, there have been few recent vertical kiln installations in the United States because of the high production requirements of domestic manufacturers.

Other, much less common, kiln types include rotary hearth and fluidized-bed kilns. The rotary hearth kiln, or "calcimatic" kiln, is a circular kiln with a slowly revolving doughnut-shaped hearth. In fluidized-bed kilns, finely divided limestone is brought into direct contact with hot combustion air in a turbulent zone, usually above a perforated grate. Dust collection equipment must be installed on fluidized-bed kilns for process economics because of the high lime carryover into the exhaust gases. Both kiln types can achieve larger production rates by utilizing gas or oil firing (the calcimatic kiln can also be used with pulverized coal).

About 15% of all lime produced is converted to hydrated (slaked) lime. There are two kinds of hydrators: atmospher-

*Source: AP-42, *Compilation of Air Pollutant Emission Factors,* U.S. Environmental Protection Agency

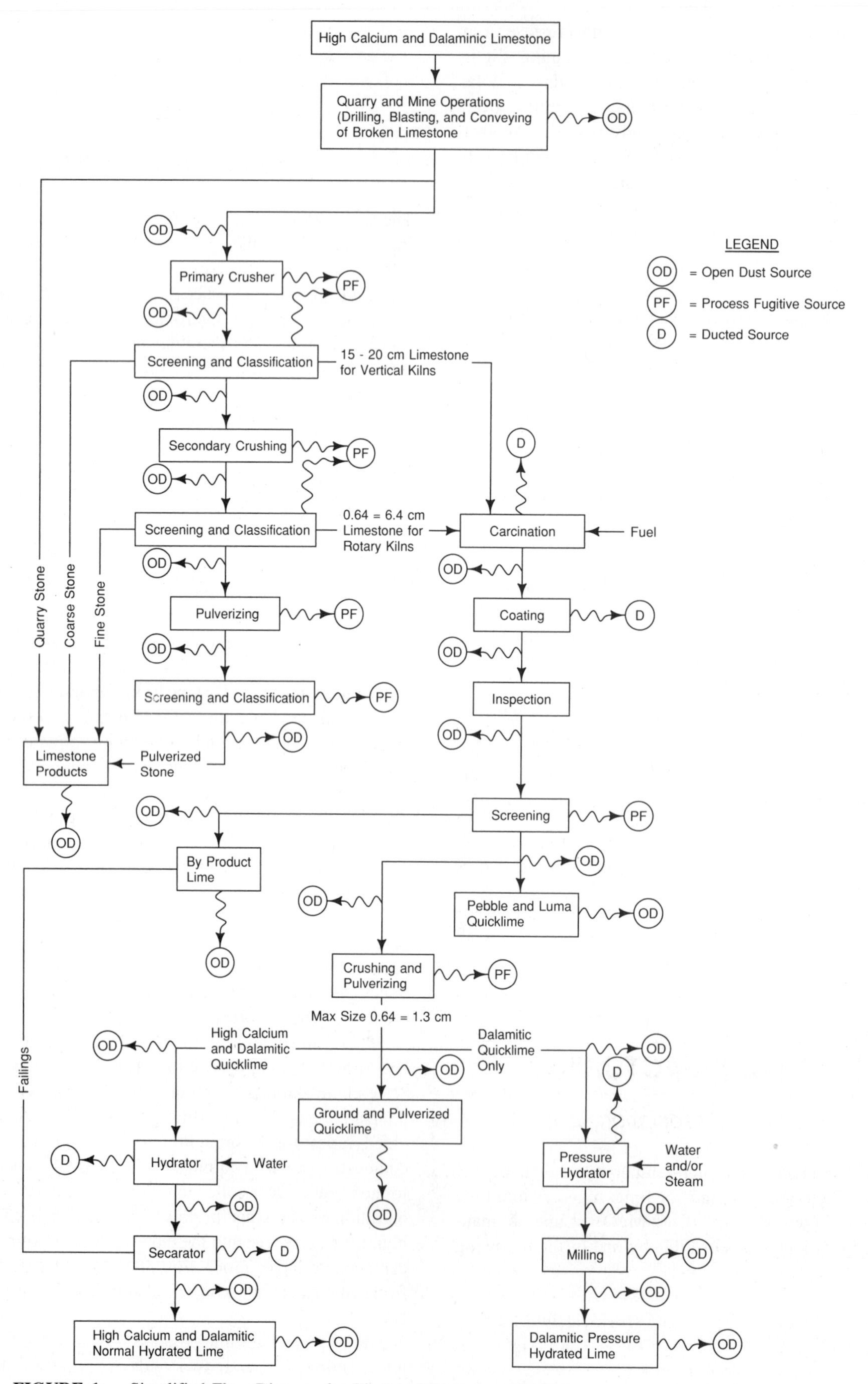

FIGURE 1. Simplified Flow Diagram for Lime and Limestone Products

ic and pressure. Atmospheric hydrators, the most common kind, are used to produce high-calcium and normal dolomitic hydrates. Pressure hydrators, on the other hand, are only employed when a completely hydrated dolomitic lime is needed. Both hydrators operate continuously. Generally, water sprays or wet scrubbers are employed as an integral part of the hydrating process to prevent product losses. Following hydration, the resulting product may be milled and conveyed to air separators for further drying and for removal of the coarse fractions.

In the United States, the major uses of lime are metallurgical (steel, copper, gold, aluminum, and silver), environmental (flue gas desulfurization, water softening and pH control, sewage-sludge stabilization, hazardous waste treat-

TABLE 1. Emission Factors for Lime Manufacturing[a] *(Emission Factor Rating: B)*

	Particulate[b]		Nitrogen Oxides		Carbon Monoxide		Sulfur Dioxide	
Source	kg/Mg	lb/ton	kg/Mg	lb/ton	kg/Mg	lb/ton	kg/Mg	lb/ton
Crushers, screens, conveyors, storage piles, unpaved roads, etc.	c	c	Neg	Neg	Neg	Neg	Neg	Neg
Rotary kilns[d]								
Uncontrolled[e]	180	350	1.4	2.8	1	2	f	f
Large-diameter cyclone	81	160	1.4	2.8	1	2	f	f
Multiple cyclone	42	83	1.4	2.8	1	2	f	f
Electrostatic precipitator[g]	2.4	4.8	1.4	2.8	1	2	h	h
Venturi scrubber	2.4	4.8	1.4	2.8	1	2	h	h
Gravel-bed filter[g]	0.53[i]	1.1[i]	1.4	2.8	1	2	h	h
Multiclone and venturi scrubber[g]	0.44	0.87	1.4	2.8	1	2	h	h
Baghouse	0.45[j]	0.89[j]	1.4	2.8	1	2	h	h
Cyclone and baghouse	0.055	0.11	1.4	2.8	1	2	h	h
Vertical kilns								
Uncontrolled	4	8	NA	NA	NA	NA	NA	NA
Calcimatic kilns[k]								
Uncontrolled	25	50	0.1	0.2	NA	NA	NA	NA
Multiple cyclone	3	6	0.1	0.2	NA	NA	NA	NA
Secondary dust collection[l]	NA	NA	0.1	0.2	NA	NA	NA	NA
Fluidized-bed kilns	m	m	NA	NA	NA	NA	NA	NA
Product coolers								
Uncontrolled	20[n]	40[n]	Neg	Neg	Neg	Neg	Neg	Neg
Hydrators (atmospheric)[p]								
Wet scrubber	0.05	0.1	Neg	Neg	Neg	Neg	Neg	Neg
Crusher, screen, hammermill								
Baghouse	0.0005	0.001	Neg	Neg	Neg	Neg	Neg	Neg
Final screen								
Baghouse	0.0004	0.0008	Neg	Neg	Neg	Neg	Neg	Neg
Uncontrolled truck loading								
Limestone								
Open truck	0.75	1.5	Neg	Neg	Neg	Neg	Neg	Neg
Closed truck	0.38	0.76	Neg	Neg	Neg	Neg	Neg	Neg
Lime—closed truck	0.15[i]	0.30[i]	Neg	Neg	Neg	Neg	Neg	Neg

[a]References 1–4. Factors for kilns and coolers are per unit of lime produced. Divide by two to obtain factors per unit of limestone feed to the kiln. Factors for hydrators are per unit of hydrated lime produced. Multiply by 1.25 to obtain factors per unit of lime feed to the hydrator. Neg = negligible. NA = not available.
[b]Emission factor rating = D.
[c]Factors for these operations are presented in Sections 8.19 and 11.2 of this document.
[d]For coal-fired rotary kilns only.
[e]No particulate control except for settling that may occur in stack breeching and chimney base.
[f]Sulfur dioxide may be estimated by a material balance using fuel sulfur content.
[g]Combination coal/gas-fired rotary kilns only.
[h]When scrubbers are used, < 5% of the fuel sulfur will be emitted as SO_2 even with high-sulfur coal. When other secondary collection devices are used, about 20% of the fuel sulfur will be emitted as SO_2 with high-sulfur fuels and < 10% with low-sulfur fuels.
[i]Emission factor rating = E.
[j]Emission factor rating = C.
[k]Calcimatic kilns generally have stone preheaters. Factors are for emissions after the kiln exhaust passes through a preheater.
[l]Fabric filters and venturi scrubbers have been used on calcimatic kilns. No data are available on particulate emissions after secondary control.
[m]Fluidized-bed kilns must have sophisticated dust collection equipment for process economics, hence particulate emissions will depend on efficiency of the control equipment installed.
[n]Some or all cooler exhaust typically is used in kiln as combustion air. Emissions will result only from that fraction not recycled to kiln.
[p]Typical particulate loading for atmospheric hydrators following water sprays or wet scrubbers. Limited data suggest particulate emissions from pressure hydrators may be approximately 1 kg/Mg (2 lb/ton) of hydrate produced, after wet collectors.

ment, and acid neutralization), and construction (soil stabilization, asphalt additive, and masonry lime). Lime has hundreds of applications and has been an essential part of environmental cleanup activities.

AIR EMISSION CHARACTERIZATION

Potential air pollutant emitting points in lime manufacturing plants are shown in Figure 1. Particulate is the only pollutant of concern from most of the operations; however, gaseous pollutants are also emitted from kilns.

The most important source of particulate is the kiln. Of the various kiln types in use, fluidized-bed kilns have the highest uncontrolled particulate emissions. This is due to the very small feed size combined with the high airflow through these kilns. Fluidized-bed kilns are well controlled for maximum product recovery. The rotary kiln is second to the fluidized-bed kiln in uncontrolled particulate emissions. This is attributed to the small feed size and relatively high air velocities and dust entrainment caused by the rotating chamber. The rotary hearth, or "calcimatic," kiln ranks third in dust production, because of the larger feed size and the stationary position of the limestone to the hearth. The vertical kiln has the lowest uncontrolled dust emissions owing to the large lump-size feed and the relatively slow air velocities and slow movement of material through the kiln.

AIR POLLUTION CONTROL MEASURES

Particulate emission control is generally employed on most kilns. Rudimentary fallout chambers and cyclone separators are commonly used for control of the larger particles. Gravel-bed filters, wet scrubbers, electrostatic precipitators, and fabric filters have all been successfully used in the lime industry for control of particulates.

Nitrogen oxides, carbon monoxide, and sulfur dioxide (SO_2) are all produced in lime kilns. Although SO_2 is the only gaseous pollutant emitted in significant quantities, the vast majority of the sulfur in the kiln fuel is not emitted. The SO_2) reacts with the calcium oxides in the kiln. The SO_2 emissions may be further reduced in the pollution equipment used for particulate control if CaO and SO_2 come into intimate contact.

Particulate emissions from hydrators are low because water sprays or wet scrubbers are usually installed for economic reasons to prevent product loss in the exhaust gases. Emissions from pressure hydrators may be higher than from the more common atmospheric hydrators because the exhaust gases are released intermittently over short time intervals, making control more difficult.

Product coolers are emission sources only when some of their exhaust gases are not recycled through the kiln for use as combustion air to maximize fuel-use efficiencies. The trend is away from the venting of product cooler exhaust to atmosphere. Cyclones, baghouses, and web scrubbers have been employed on coolers for particulate control.

Other particulate sources in lime plants include primary and secondary crushers, mills, screens, mechanical and pneumatic transfer operations, storage piles, and paved and unpaved roads. If quarrying is a part of the lime plant operation, particulate emissions may also result from drilling and blasting.

Emission factors for lime manufacturing are present in Table 1.

References

1. *Standards Support And Environmental Impact Statement, Volume I: Proposed Standards Of Performance For Lime Manufacturing Plants,* EPA-450/2-7-007a, U.S. Environmental Protection Agency, Research Triangle Park, NC, April 1977.
2. Source test data on lime plants, Office Of Air Quality Planning And Standards, U.S. Environmental Protection Agency, Research Triangle Park, NC, 1976.
3. *Air Pollutant Emission Factors,* APTD-0923, U.S. Environmental Protection Agency, Research Triangle Park, NC, April 1970.
4. J. S. Kinsey, *Lime And Cement Industry-Source Category Report, Volume I: Lime Industry,* EPA-600/7-86-031, U.S. Environmental Protection Agency, Cincinnati, OH, September 1986.

Bibliography

C. J. Lewis and B. B. Crocker, "The Lime Industry's Problem Of Airborne Dust", *Journal Of The Air Pollution Control Association, 19*(1):31–39, January 1969.

Kirk-Othmer Encyclopedia Of Chemical Technology, 2d Edition, John Wiley And Sons, New York, 1967.

Screening Study For Emissions Characterization From Lime Manufacture, EPA Contract No. 68-02-0299, Vulcan-Cincinnati, Inc., Cincinnati, OH, August 1974.

COAL PROCESSING

Larry L. Simmons, P.E., and Lisa E. Lambert

This discussion focuses on the emissions from sources at coal processing plants, which are considered to include all coal-preparation plants and coal-handling facilities.

A coal-preparation plant is any facility (excluding underground mining operations) that prepares coal by one or more of the following processes: breaking, crushing, screening, wet or dry cleaning, thermal drying. Generally, coal-preparation plants can be classified into three types[1]:

- Those that clean both coarse and fine coal.
- Those that clean only coarse coal.
- Those that crush coal to a specific size.

Coal-handling facilities include any facility (excluding those associated with mining) that processes coal solely by

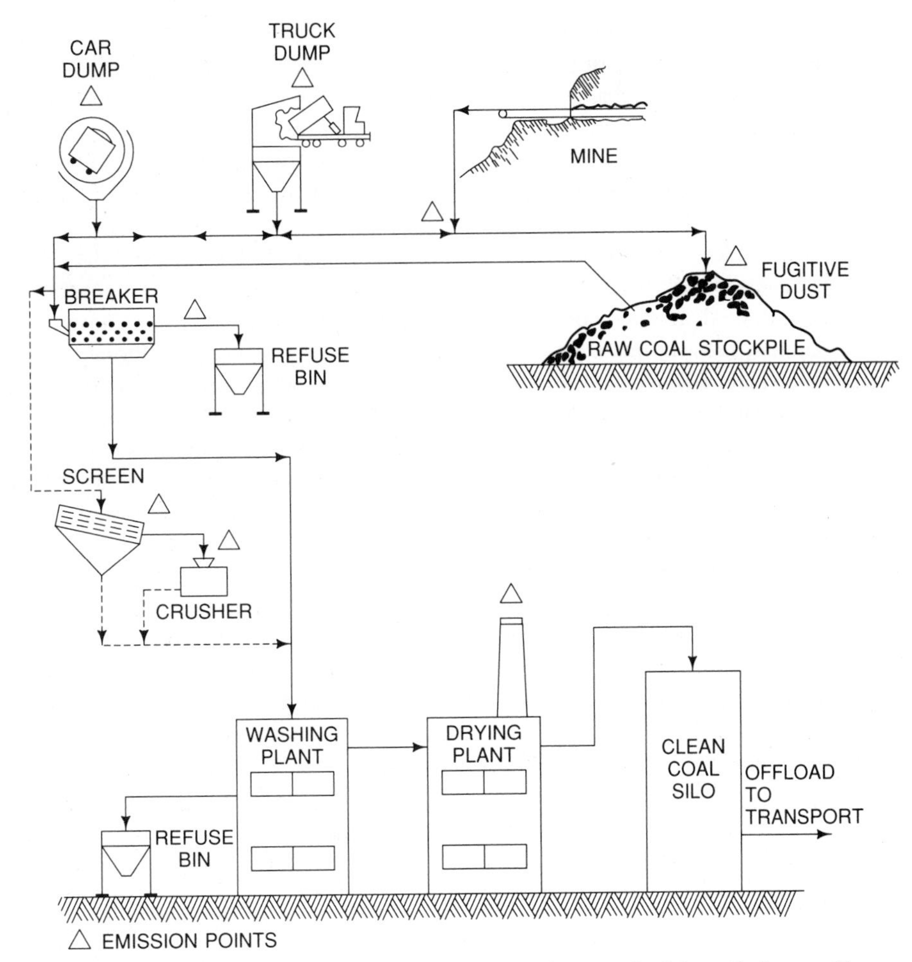

FIGURE 1. Typical Coal-Processing Plant Schematic (Revised from Reference 2)

use of one or more of the following operations: transferring, conveying, loading, unloading, storing.

PROCESS DESCRIPTION

Coal-processing plants can vary according to size, complexity, and purpose. For example, the type of coal-processing plant can range from a simple coal-loading station that handles only a few tons of coal per hour to a complex coal-washing plant processing over 1000 tons of coal per hour. The type of coal processing performed depends on the requirements of the end user. Figure 1 illustrates a typical coal-processing plant.

As shown in Figure 1, this description of particulate matter emissions from coal-processing plants begins at the point where coal is brought to the processing plant via some mode of transportation (truck, railcar, conveyor) and is unloaded. It concludes at the point where the processed coal is loaded into transportation equipment (truck, railcar, conveyor, barge) for shipment to the end user.

At the processing plant, the coal is unloaded from trucks or railcars by dumping into a receiving hopper that discharges to a primary crusher, a tipple, or a feeder. A feeder subsequently empties onto a conveyor for transfer to an open storage pile or enclosed silo. Coal that is carried directly to the processing plant by conveyors is unloaded by dropping onto open storage piles or into enclosed silos.

Coal is normally stored either on open storage piles or in silos to allow for optimum scheduling of processing and transportation equipment. The coal from open storage piles is transferred to either crushing, screening, or loading operations by either front-end loaders or self-feeding tunnel conveyors. The coal that is stored in silos is transferred to processing equipment through the use of belt conveyors or is loaded into trucks or railroad cars by gravity. Because of the large amount of material involved, belt conveyors that have large carrying capacities are the most common method of transporting material at a coal-processing plant.

Coal at any preparation plant undergoes at least a crushing operation. At many smaller plants, the coal that is processed by crusher is loaded directly into either a truck, a railcar, or a barge for shipment to the user. Most of the larger mining operations will also provide secondary crushing, screening, wet or dry cleaning, and drying of the crushed coal. The remainder of this process description will discuss such processing at a typical coal-preparation plant.

The cleaning or beneficiation of coal at preparation plants is performed for a number of reasons. One such reason is to improve the coal quality by removing undesirable impurities. This increases the heating value of the coal and provides a better fuel for the user. In fact, coal cleaning is often necessary to order to market coal, since mined coal may contain up to 60% of reject material.[3]

Another reason for the cleaning of coal is that air pollution control requirements on the user often dictate the partial removal of pyrites with the ash in order to reduce the sulfur content of the coal. Also, ash content must often be monitored and reduced to levels stipulated in sales contracts. However, a minimum ash content must be maintained to ensure optimum combustion characteristics.[4]

Last, substantial savings in freight costs for shipping coal may be achieved by the removal of impurities before loading. Moreover, it is generally easier to dispose of the impurities at the mining site than at the combustion site.

At coal-preparation plants, the initial process operations consist of "tramp iron" removal and size reduction (Figure 1).[5] Coal is first exposed to a high-intensity magnet, which is usually suspended over the incoming belt conveyor, and the iron impurities are extracted from the coal. This high-intensity magnet may also be located after the breaker, but it is always located before the screen and crusher.

The coal is next conveyed to a breaker, which consists of a cylindrical shell with perforated holes (2 to 8 inches in diameter) and interior lifting blades. The perforated shell allows the smaller coal to pass through. The breaker rotates on a horizontal axis and breaks the tumbling coal, which is fed into the breaker at one end. The soft material (coal) is broken in the breaker to a size small enough to pass through the shell, while the hard, larger, unbroken material (reject) passes out through the other end of the breaker and into a refuse bin. The coal (usually less than 4 inches in size) that passes through the breaker shell is then transferred to the cleaning plant.[5]

Instead of entering the breaker, some coal is diverted to a scapling screen. From the scalping screen, the oversized material (>4 inches) falls into a crusher and is reduced in size to less than 4 inches. This material is then combined with the screenings from the scalping screen and transferred to the cleaning plant. This alternative flow is used more often than the breaker circuit despite the disadvantage of exposing the crusher to large pieces of material. A heavy-duty, single-roll crusher with tramp iron protection is most often used for this process.[6]

As in the case of crushing equipment, coal-preparation plants also use a variety of screening equipment for coal sizing. The types generally used are the grizzly, shaker, and vibrating screens. Grizzly screens, which size coal by gravity only, are usually used on coal preceding a crusher or belt conveyor loading operation.[7] Shaker screens are used infrequently and rarely provide a separation of less than 2 inches. Vibrating screens are the most common type of separating device. They are used in both dry and wet processing plants.[8]

After crushing and screening, the raw coal at a preparation plant is usually stored on an open pile or in a silo prior to washing to allow for the smooth, efficient operation of the cleaning plant.[9] At the cleaning plant, the type of mechanical cleaning equipment used depends on the size range of the coal entering the plant and the desired coal size. For example, coal that is larger than 8 inches is usually crushed; however, if lump coal is desired, the large fraction above 8 inches is cleaned by slate pickers.

Coal of a size range of less than ⅜ inch is often cleaned by using pneumatic cleaning devices or air tables. These devices have a perforated bottom plate over which a layer of coal passes. A current of air is passed upward through the bed, which removes the finer particles. The fines are eventually removed from the airstream by cyclones and fabric filters. Generally, air tables are not very efficient with respect to their ability to remove ash from coal. One source reports that their efficiency of ash removal is limited to 2–3%.[10]

Jig-table washing plants use jigs to clean coals greater than ¼ inch in size and Diester tables to clean the ¼-inch to 28-mesh size range. This equipment is often used with Froth cells and/or thermal dryers. In a jig-table wash plant, the raw coal (<8 mesh) is first separated on a wet screen (usually ¼-inch mesh). The larger coal enters the jig, while the remaining coal is transferred to another, separate cleaning circuit. The coal exiting the jig is dewatered on screens and in centrifuges, crushed, and loaded or stored.

The finer coal (<¼ inch) is mixed with water and then poured into the tables, where the refuse is separated from the coal. Water is removed from the refuse by screening and the refuse is deposited into a bin for storage until it may be hauled to a disposal area. The washing coal is then dewatered by using a stationary gravity screen or "sieve bend," where the fines are removed and sent to a centrifuge for dewatering and extraction of the fines. Finally, the washed coal is loaded or conveyed to a thermal dryer.[11]

A Diester table has a flat, riffled surface (about 12 square feet in area) that oscillates perpendicularly to the riffles toward the flow of coal. The heavy reject material falls off one end of the table, while the light coal is discharged off the opposite end. The remaining material is distributed in between.

A heavy-media wash plant performs the cleaning of coal by flotation in a medium with a selected specific gravity, in which a dispersion of finely ground magnetite (Fe_3O_4) is maintained. In this type of plant, the raw coal is first separated at ¼ inch on an inclined screen. The oversized fraction is transferred to a flat, wet screen, where the finer particles are sprayed off the >¼-inch coal. The oversized material from this wet screen is then discharged into a heavy-media bath, where the refuse is separated from the coal. The refuse is then dewatered by discharging to a screen. The medium, which is removed from the bath, is divided into two parts, one returning to circulation via the heavy-medium sump and the other being pumped to the magnetite recovery system. The refuse is discharged from the screen for disposal. The coal is next removed from the washer to a rinse screen, where the coal is dewatered.

The thermal drying of coal is occasionally used at coal-processing facilities employing fine-coal washing equipment. Generally, coal is dried via a thermal dryer for one or more of the following reasons.[12,31]

- To avoid freezing difficulties and to facilitate handling during shipment, storage, and transfer
- To maintain high pulverizer capacity
- To improve the quality of coal used for coking
- To decrease transportation costs

Thermal dryers are generally used to dry the ¼- by 0-inch fraction. However, sometimes the plus-¼-inch portion is dried for ease of screening.

Thermal dryers are simply contacting devices where hot exhaust gases from the combustion process in the dryer are used to heat the wet coal and to evaporate the surface moisture. Seven basic types of thermal dryers are currently used: rotary, screen, cascade, continuous carrier, flash or suspension, multilouver, and fluidized bed.

The most prevalent type of dryer in use today is the fluidized-bed dryer. (See Figure 2.) This dryer contains a perforated plate in a negative-pressure fluidizing chamber above which coal is suspended by a rising column of hot gases. The dried coal exits from the dryer at an overflow weir. A flash or suspension dryer is the second most used type of dryer. In this dryer, the hot gases generated in the combustion furnace of the dryer transport the wet coal up a riser. The turbulence created in the riser provides an excellent drying environment. This type of dryer is used for extremely fine coal, with the top size not exceeding ⅜ inch.

After washing and drying, mines using unit train shipment usually store enough clean coal to fill a train. Silos are often used for this purpose. Furthermore, silo storage prevents the accumulation of moisture and exposure to wind.

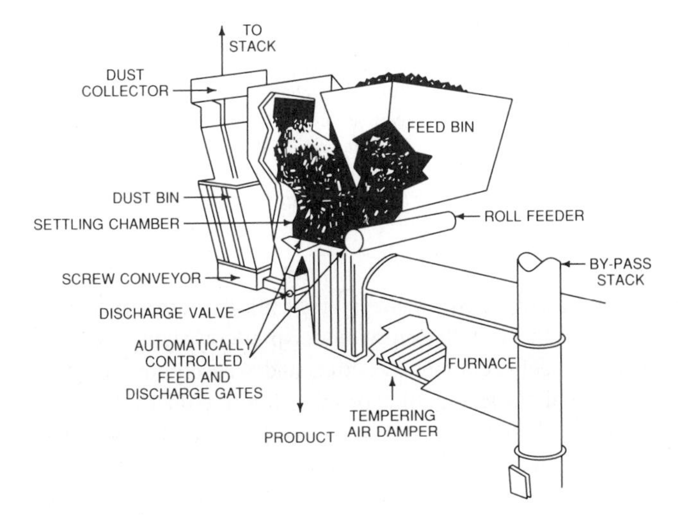

FIGURE 2. Fluidized-Bed Thermal Coal Dryer (Revised from Reference 15)

TABLE 1. Fugitive Dust Emission Factors for Coal-Processing Plants

Unloading	
Truck	0.02 lb/ton unloaded[a]
Railcar	0.40 lb/ton unloaded[b]
Primary crushing	0.02 lb/ton crushed[c]
Secondary crushing/screening	0.16 lb/ton crushed/screened[c]
Transfer and conveying	0.20 lb/ton transferred or conveyed[d]
Cleaning	Negligible[e]
Storage	
Loading onto pile	0.08 lb/ton loaded[f]
Vehicular traffic	0.16 lb/ton stored[f]
Loading out	0.10 lb/ton stored[f]
Wind erosion	0.09 lb/ton stored[f]
Loading	
Truck	0.02 lb/ton loaded[a]
Railcar	0.40 lb/ton loaded[b]
Barge	0.40 lb/ton loaded[b]

[a]Reference 16.
[b]References 17 and 18.
[c]Reference 19.
[d]Reference 20.
[e]Reference 21 (see Table 2 for dryer emissions).
[f]Reference 22.

Some mines employ open storage, using conveyors for loading. At some mines, railroad cars or barges are loaded directly as the coal is processed.

AIR EMISSION CHARACTERIZATION

The estimated emission factors for coal-processing-plant fugitive dust sources are summarized in Table 1. Table 2 presents the emission factors for the coal-cleaning portion of the coal-processing plant. No air toxic emissions were noted as related to coal-processing plants.[32] Coal combustion for thermal drying has the general characteristics of any coal combustion process described for air toxic emission factors.

AIR POLLUTION CONTROL MEASURES

This section presents the fugitive dust control methods that may be used on a case-be-case basis by the coal-processing industry.[17,22] Application of any of these control measures would be to meet the National Ambient Air Quality Standards or to reduce the nuisance potential beyond the property line of the coal-processing facility.

The fugitive dust emissions from a truck or railcar may be controlled by one or a combination of the following.

- Total enclosure with ventilation to a fabric filter
- A partial enclosure
- A water spray system
- A wet suppression system using water and chemical wetting agents or foams

TABLE 2. Emission Factors for Coal Cleaning[a] *(Emission Factor Rating: B)*

	Operation			
	Fluidized Bed[b]		Flash	
Pollutant	lb/ton	kg/MT	lb/ton	kg/MT
Particulates				
Before cyclone	20[c]	10[c]	16[c]	8[c]
After cyclone[d]	12[e]	6[e]	10[f]	5[f]
After scrubber	0.09[e]	0.05[e]	0.04[f]	0.02[f]
SO_2[g]				
After cyclone	0.43[h]	0.22[h]	—[i]	—
After scrubber	0.25	0.13	—	—
NO_x[j]				
After scrubber	0.14	0.07	—	—
VOC[k]				
After scrubber	0.10	0.05	—	—

[a]Emission factors expressed as units per weight of coal dried.
[b]See Figure 2.
[c]References 23 and 24.
[d]Cyclones are standard pieces of process equipment for product collection.
[e]References 25–29.
[f]Reference 30.
[g]References 26 and 27. The control efficiency of venturi scrubbers on SO_2 emissions depends on the inlet SO_2 loading, ranging from 70% to 80% removal for low-sulfur coals (0.7% sulfur) down to 40–50% removal for high-sulfur coals (3% sulfur).
[h]References 26–28.
[i]Not available.
[j]Reference 27. The control efficiency of venturi scrubbers on NO_x emissions is approximately 10–25%.
[k]Volatile organic compounds as pounds of carbon/per ton of coal dried.

For primary and secondary crushing and screening operations, control methods consist of enclosures with ventilation to a fabric filter and wet suppression systems utilizing a chemical wetting agent or foams.

For fugitive dust emissions from conveying operations, the control methods generally used are partial (top) enclosure, total enclosure, or wet suppression. Also, fugitive dust emissions created by the droppings from the return belt conveyors may be controlled through the use of dribble pans.

Fugitive dust emissions from transfer points may be controlled through the use of total enclosure, enclosure with ventilation to a fibric filter, or wet suppression systems using chemical wetting agents.

The fugitive dust emissions from storage piles consist of four sources: load-in, wind disturbance, vehicular traffic, and load-out. The control methods available for those sources vary with the type of source. For load-in at storage piles, control methods include (1) enclosures, (2) wind guards, (3) telescopic chutes, (4) wet suppression, and (5) operating practices.

Fugitive dust emissions from wind disturbances at storage piles may be controlled through the use of wet suppression and the application of surface crusting agents. Also, operating practices such as orienting the storage piles perpendicularly to prevailing winds to reduce the exposed surface are helpful in reducing wind erosion.

TABLE 3. Summary of Control Alternatives and Their Control Efficiencies

Fugitive Dust Sources	Control Alternatives	CE,[a] %
Unloading coal		
Railcar	Enclosure, vent to fabric filter	99
	Enclosure	70
	Wet suppression with chemicals	80
	Watering	50
Truck	Enclosure vent to fabric filter	99
	Enclosure	70
	Wet suppression with chemicals	80
	Watering	50
Primary crushing	Enclosure, vent to fabric filter	99
	Wet suppression with chemicals	90
Secondary crushing/ screening	Enclosure, vent to fabric filter	99
	Wet suppression with chemicals	90
Transfer and conveying	Enclosure of conveyors and transfer points, vent to fabric filter	99
	Enclosure of conveyors and transfer points	70
	Wet suppression with chemicals	90
Cleaning	Enclosure	100
Storage		
Loading onto piles	Enclosure	80
	Telescopic chutes	75
	Wet supression with chemicals	75
	Wind guards	50
Wind erosion	Enclosure	100
	Wet suppression with chemicals	99
Loading out	Under-pile conveyor	80
	Wet suppression with chemicals	95
	Bucket wheel reclaimer	80
Loading		
Railcar/barge	Wet suppression with chemicals	80
	Telescopic chutes	75
Truck	Wet suppression with chemicals	80
	Telescopic chutes	75

[a]CE = Control efficiency.
Source: Adapted from references 17 and 22.

For loading out from coal storage piles, control measures consist of the installation of under-pile conveyor systems, bucket wheel reclaimer systems, and wet suppression.

For coal storage in silos or bins, fugitive dust emissions are normally controlled by covering the conveyors that transport and dump the coal into the silos or bins. The fugitive dust in the displaced air from silos and bins may be controlled by the addition of either bin vent filters or exhaust fabric filters.

Control methods for loading operations consist of telescopic chutes and wet suppression. For operations that use wet suppression in subsequent processes, generally only telescopic chutes are required for dust control if the coal has been adequately treated with a wetting agent. A summary of the control measures and their respective efficiencies is presented in Table 3.

References

1. M. Szabo, *Environmental Assessment of Coal Transportation. PEDCo Environmental, Inc.*, EPA-600/7-78-081, U.S. Environmental Protection Agency, Industrial Environmental Research Laboratory, Office of Research and Development. Cincinnati, OH, Contract No. 68-02-1321, May 1978, p 59.
2. *1977 Division of Mines Report,* State of Ohio, Department of Industrial Relations, p 4–6.
3. *Inspection Manual for the Enforcement of New Source Performance Standards: Coal Preparation Plants, PEDCo Environmental, Inc.,* EPA-340/1-77-022, U.S. Environmental Protection Agency, Division of Stationary Source Enforcement, Washington, DC, Contract No. 68-01-3150, November 1977, p 3–2.
4. Reference 3, p 3–3.
5. Reference 3, p 4–5.
6. Reference 3, p 4–8.
7. Reference 3, p 6–7.
8. M. Weisburg, *Field Operations and Enforcement Manual for Air Pollution Control*. Vol. III: *Inspection Procedures for Specific Industries,* System Development Corp., prepared for U.S. Environmental Protection Agency, Office of Air Programs, Research Triangle Park, NC, Control No. CPA 70-122, APTD-1102, August 1972 pp 6–8
9. Reference 3, p 4–8.
10. Reference 3, p 4–9.
11. Reference 3, p 4–12.
12. P. N. Formica, *Controlled and Uncontrolled Emission Rates and Applicable Limitations for Eighty Processes,* prepared for U.S. Environmental Protection Agency, Research Triangle Park, NC, Control No. 68-02-1382, Task Order No. 12, Report No. EPA-450/3-77-016 (PB 266 978), September 1976, p VIII–20.
13. Reference 12, p VIII–21.
14. Reference 3, p 4–25.
15. Reference 3, p 4–23.
16. G. Jutze, K. Axetell, and R. Amick, *Evaluation of Fugitive Dust Emissions from Mining,* PEDCo Environmental, Inc., prepared for U.S. Environmental Protection Agency, Industrial Environmental Research Laboratory, Office of Research and Development, Cincinnati, OH, Contract No. 68-02-1321, Task No. 36. EPA-600/9-76-001, June 1976, pp 38, 40.
17. *Reasonably Available Control Measures for Fugitive Dust,* Environmental Protection Agency, Ohio, p 2–175.
18. Reference 1, p 63.
19. *Technical Guidance for Control of Industrial Process Fugitive Particulate Emissions,* PEDCo Environmental, Inc., prepared for U.S. Environmental Protection Agency, Office of Air and Waste Management, Office of Air Quality Planning and Standards, Research Triangle Park, NC, Contract No. 68-02-1375, Task No. 33, Project No. 3155-GG, EPA-450/3-77-010, March 1977. p 2–241.
20. Reference 16, p 46.
21. Reference 17, p 2–237.
22. C. Cowherd, G. E. Muleski, and J. S. Kinsey, *Control of Open Dust Sources,* EPA-450/3-88-008, U.S. Environmental Protection Agency, September 1988.
23. "Stack test results on thermal coal dryers" (unpublished), Bureau of Air Pollution Control, Pennsylvania Department of Health, Harrisburg, PA.
24. "Amherst's answer to air pollution laws," *Coal Mining Proc.*, 7(2):26–29 (1970).
25. E. Northcott, "Dust abatement at Bird Coal," *Mining Congr. J.*, 53:26–29 (November 1967).
26. R. W. Kling, *Emissions from the Island Creek Coal Company Coal Processing Plant,* York Research Corp., Stamford, CT, February 14, 1972.
27. *Coal Preparation Plant Emission Tests, Consolidation Coal Company, Bishop, West Virginia,* EPA Control No. 68-02-0233, Scott Research Lab, Inc., Plumsteadville, PA, November 1972.
28. *Coal Preparation Plant Emission Tests, Westmoreland Coal Company, Wentz Plant,* EPA Control No. 68-02-0233, Scott Research Lab, Inc., Plumsteadville, PA, April 1972.
29. *Background Information for Standards of Performance: Coal Preparation Plants,* Vol. 2: *Test Data Summary,* EPA-450/2-74-021b, U.S. Environmental Protection Agency, Research Triangle Park, NC, October 1974.
30. *Background Information for Establishment of National Standards of Performance for New Sources: Coal Cleaning Industry,* Environmental Engineering, Inc., Gainesville, FL, EPA Contract No. CPA-70-142, July 1971.
31. *1986 Keystone Coal Industry Manual,* McGraw-Hill, New York.
32. *Toxic Air Pollutant Emission Factors: A Compilation for Selected Air Toxic Compounds and Sources,* 2nd ed., EPA-450/2-90-011, U.S. Environmental Protection Agency, October 1990.

16
Pharmaceutical Industry

Richard V. Crume and Jeffrey W. Portzer

This discussion addresses air emissions and controls in the pharmaceutical industry. The first section provides a characterization of the industry. The remaining sections furnish more detailed information on the process description, air emissions characteristics, and air pollution control measures.

INDUSTRY CHARACTERIZATION

The pharmaceutical industry consists of the manufacture, packaging, and sales of chemicals used as medication for humans and animals. This includes both "ethical" (i.e., prescription) and "proprietary" (i.e., nonprescription) drugs. Additionally, the industry manufactures related chemicals for nonpharmaceutical purposes, including preventive medicine and health-enhancing products, medicated and nonmedicated cosmetics, and food additives. Some facilities also manufacture medical apparatus, such as surgical and medical instruments and appliances, dental equipment and supplies, and various products used in diagnostic laboratories.

The predominant characteristic of the industry is diversity. Not only are a wide variety of products manufactured, but a significant number of distinct chemical and manufacturing processes are used in producing these products. Consequently, no two manufacturing facilities are alike and any attempt specifically to define the industry in terms of products and processes is a complicated matter. The industry definition is further complicated by the manufacture of (1) nonpharmaceutical products, by some pharmaceutical facilities and (2) pharmaceutical products, reagents, and intermediates by certain specialty synthetic chemical manufacturers whose operations would not ordinarily be classified as pharmaceutical. Finally, because some pharmaceutical products have multiple end uses or are used in conjunction with nonpharmaceutical products, the question of what characteristics best define the pharmaceutical industry is not easily answered.

In preparing this discussion, a more traditional definition of pharmaceutical manufacturing is used. This definition is limited to chemicals intended for use as medications (including both ethical and proprietary drugs) and as preventive medicine and health-enhancing agents. Using this definition, it is possible to classify pharmaceutical manufacturing processes into several clearly defined categories for the purpose of air emission control.

Nature of Operations

Because of the importance of product purity and the presence of toxic and biologically active constituents, pharmaceutical manufacturing operations tend to be very clean and closely controlled. Controlling air emissions is considered important to avoid the inadvertent contamination of products and to prevent occupational and community exposure to harmful substances. Using emission controls also is considered important in recovering high-priced solvents. For these reasons, the air emissions from many pharmaceutical manufacturing facilities are well controlled and serious air pollution problems are relatively rare.

Nevertheless, emissions problems sometimes do occur and are most frequently associated with (1) odorous emissions caused by solvent extraction and fermentation operations and (2) volatile organic compound (VOC) emissions related to general chemical synthesis operations. Additional emission sources include formulation and packaging operations, materials handling and storage, wastewater treatment, spent solvent treatment for recycling and the regeneration of spent adsorber carbon. Emissions include both VOCs (including toxic materials) and particulates. Controls are available for most pharmaceutical operations, although applying controls is not always cost effective owing to the high-volume, low-concentration airstreams that must be handled. Additionally, because many batch operations are involved in manufacturing pharmaceuticals, emission controls also must be operated in a batch mode or else operated continuously while handling an intermittent, variable feed stream. These methods of operation can reduce the effectiveness of control systems, limit the availability of control system options, and increase control system operating costs.

The batch nature of pharmaceutical operations lends itself to decentralized emissions control, where control systems are applied to individual process units rather than to entire production lines. Consequently, regulatory actions tend to focus on controlling specific process units (e.g., reactors, distillation units, crystallizers, centrifuges, dryers, filters, and storage tanks) rather than on limiting emissions from an entire facility. Nevertheless, economical emission reduction for an entire facility is sometimes achieved by using a collection manifold and central control device.

Facilities

A U.S. Environmental Protection Agency (EPA) study identified 464 pharmaceutical manufacturing facilities that accounted for the vast majority of pharmaceutical production in the United States.[1] Of these facilities, about 80% were located in the eastern United States and nearly 10% were located in Puerto Rico. The number of employees per facility ranged from several to several thousand, with about 37% having fewer than 100 employees, 52% employing between 100 and 1000 personnel, and about 11% having more than 1000 employees.

In addition to these 464 facilities, there are probably many more facilities producing pharmaceutical products. However, most of these additional facilities are very small operations, accounting for a nearly insignificant fraction of nationwide sales. Many of these additional facilities are difficult to identify because of their small size. Also, as some of the additional facilities manufacture products other than pharmaceuticals, they often are not classified as pharmaceutical manufacturing facilities, thus making their identification that much more difficult.

PROCESS DESCRIPTION

This section describes the processes and operations that are common in the pharmaceutical industry. Information is presented on the four process categories, the representative design and operation of manufacturing facilities, and the significant sources of emissions.

Process Categories

It is convenient to divide the pharmaceutical industry into four process categories: chemical synthesis, fermentation, extraction, and formulation.[2] Although a majority of U.S. pharmaceutical facilities fall into the formulation category, chemical synthesis operations are reported to account for almost two thirds of air emissions from the industry in the United States.[1,3]

A breakdown of U.S. pharmaceutical manufacturing facilities by category is presented below.[1] It is interesting to note that 25% of all facilities incorporate a combination of process categories. The most common combination is chemical synthesis and formulation.

Manufacturing Category	% of Total U.S. Facilities
Chemical synthesis	10
Fermentation	1
Extraction	5
Formulation	59
Combination	25

The four pharmaceutical process categories are described in the following.

Chemical Synthesis

Chemical synthesis processes use a variety of batch operations to produce high-purity, pharmaceutical-grade chemicals. Common production equipment includes reactors, dryers, distillation units, filters, centrifuges, and crystallizers. Unless a very-high-volume product is manufactured, the equipment is generally campaigned. In this case, the equipment may be used to produce a number of different products over the course of a year and may be dedicated to a specific product for only several weeks at a time.

Organic solvents (including chlorinated compounds) are commonly used in reactors and crystallizers. For example, in synthesis reactions, the solvent serves to dissolve the reactants, to provide dilution to control reaction rate, and to control reaction temperature by removing the heat of the reaction through refluxing (returning condensed solvent vapors to the reaction mixture). Similarly, solvents are required for product purification.

Fermentation

Fermentation involves the combination of various species or strains of fungus or bacterium with selected raw materials and the reliance on biological processes to produce the desired compound. The principal pharmaceuticals produced by fermentation are antibiotics, vitamins, and steroids. The fermentation process takes place in three steps, as described in the following.

Inoculum and Seed Preparation

Spores from a master stock are activated with water, nutrients, and heat. As the mass grows, it is transferred to a seed tank with a capacity of 350 to 7500 liters (100 to 2000 gallons), where further growth occurs.

Fermentation

Seeds from the foregoing step are combined with raw materials in a fermentation vessel, typically of 18,000 to 380,000 liters (5000 to 100,000 gallons) in capacity. Air is sparged through the vessel to help sustain growth, and the action of the microorganisms on the raw materials results in the desired product. At the end of the fermentation step, filtration may be used to separate the microorganisms from the product. Fermentation typically requires a period of 12 hours to one week for completion.

Product Recovery

Product recovery and purification are usually achieved through solvent extraction, direct precipitation, ion exchange, or adsorption.

Extraction

The pharmaceutical extraction process is unique in its low ratio of product weight to raw material weight (often less than 1%). Solvents are used to extract desired compounds from a variety of materials, including roots, leaves, animal glands, parasitic fungi, and human blood. In extracting these compounds, solvents also are used to remove and separate fats and oils. After extracting several different types of separation devices (e.g., centrifuges, filters, evaporators, or dryers) can then be used to isolate the desired compounds. Numerous pharmaceuticals are produced by extraction, including insulin, morphine, digitalis derivatives, and allergy medications.

Formulation

Formulation processes convert bulk chemicals into tablets, capsules, liquids, and ointments (i.e., the final dose forms). Typical operations include drying, blending, coating, polishing, mixing, granulating, tablet pressing, capsule filling, sorting, and packaging. Solvents are used to assist with mixing and coating operations and to clean the equipment. Most of the dust generated by formulation operations is captured and sometimes returned to the process.

Representative Design and Operation

While pharmaceutical facilities differ greatly in design and operation, the pharmaceutical manufacturing process can be illustrated by a description of the manufacture of aspirin. A representative process flow diagram for an aspirin manufacturing facility is shown in Figure 1.[2]

Two of the raw materials used in aspirin manufacturing are acetic anhydride and salicylic acid. These materials are charged into a reaction vessel and heat is applied for up to two to six hours. The resulting solution is filtered and then crystallized at about 0°C (32°F). The crystals are separated from the mother liquor by centrifuge or vacuum filter. Next they are washed, dried, mixed with starch and other fillers, and compressed into tablets. A still may be used to remove excess acetic acid and anhydride, and other recovery operations may be used to reclaim any solvents used in the process. Solvents such as glacial acetic acid, xylene, methanol, carbon tetrachloride, benzene, and toluene may be used during the reaction process or to wash crystals following separation.

Aspirin production is just one of a variety of processes at pharmaceutical manufacturing facilities. However, many production lines resemble aspirin manufacturing in the use of a series of batch operations, such as in reactors, filters, crystallizers, centrifuges, and dryers. Aspirin manufacture differs from some other pharmaceutical production processes in that its equipment tends to be dedicated and the processes may be highly automated. In contrast, other production processes may use campaigned equipment, which is used in the production of a number of different products over the course of a year and where a high degree of automation is not feasible.

Sources of Emissions

Sources of emissions at pharmaceutical plants vary widely from facility to facility, depending on the products manufactured. However, pharmaceutical manufacturing typically consists of several of the following batch processes: in reactors, distillation units, extractors, centrifuges, filters, crystallizers, dryers, and storage and transfer. These operations and associated emissions points are discussed in the following.

Reactors

Many manufacturing processes are initiated with the addition of reactants to glass-lined or stainless-steel stirred-tank reactors. These are closed, pressurized vessels typically equipped with agitation capability, a heating–cooling jacket, a vapor vent with reflux condenser, a pressure-relief (safety) valve, a manway, a sight glass, and multiple nozzles to add reactant and/or remove product. Such a vessel may also function as a mixer, preheater, holding tank, crystallizer, evaporator, or distillation kettle. Vessels of this type range in capacity from as little as 10 liters (3 gallons) for specialty products to 10,000 liters (2500 gallons) or more for high-volume production. Automatic controls range from simple temperature control to fully automated computer batch process control. Operating pressures range from full vacuum to several hundred thousand kilograms per cubic meter (several hundred pounds per square inch).

Emissions from reactors may occur by several mechanisms, including (1) displacement of air during the charging of reactants, (2) inert gas flows during the reaction cycle, (3) opening of the reactor during operation to collect samples or charge reactants, (4) venting of emissions during refluxing, (5) solvent vaporization during cleanup operations, and (6) the relieving of pressure between cycles (where a reactor is operated under pressure). Emissions are maximized when reactions involve highly volatile chemicals and elevated temperatures. In contrast, emissions are

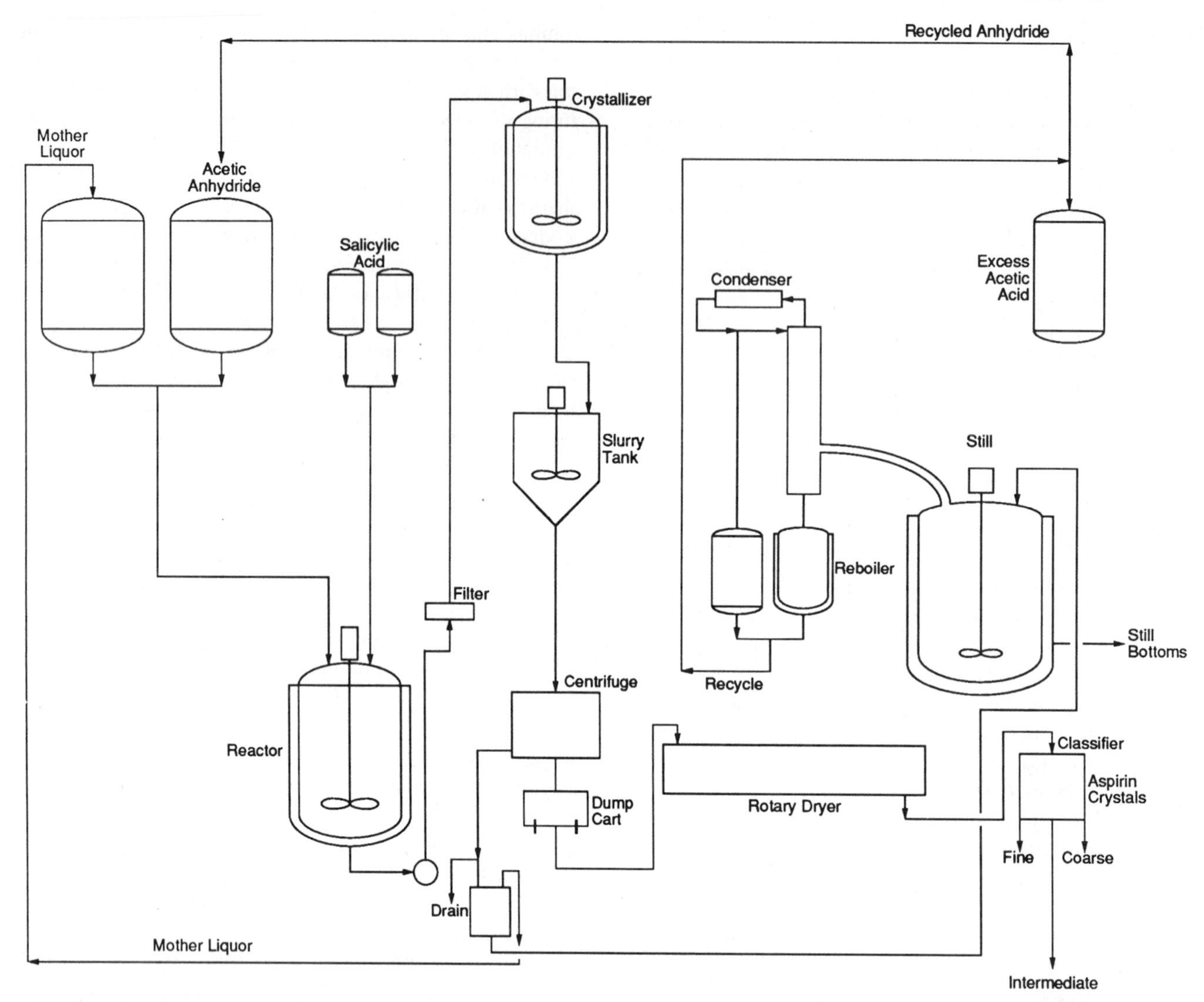

FIGURE 1. Representative Process Flow Diagram for an Aspirin Manufacturing Facility

minimized with less volatile chemicals, low temperatures, and/or sometimes subatmospheric pressures and when water-based reactions are involved.

Distillation Units

Batch reactors, small stills, and large columns may all be used for distillation operations. However, pharmaceutical facilities tend to utilize small equipment relative to that employed in petrochemical operations and refineries. Two methods of distillation are in use: (1) boiling the liquid mixture and condensing the vapors, without returning any liquid to the still (essentially evaporation); and (2) the same process, but with part of the condensate (called the reflux) returned to the still and brought into contact with the vapors. The largest distillation equipment in pharmaceutical facilities handles no more than about 3000 kg/h (6600 lb/h) of feed material. Emissions occur from the condensers used to recover the evaporated liquids and from the vents as a result of vapor displacement during startup.

Extractors

Extraction is used to separate the components of liquid mixtures. It relies on differences in the solubilities of these components and on the use of a solvent that preferentially combines with one of the components. The resulting mixture consists of an extract (containing the preferentially combined material) and a raffinate (containing the residual phase). Extraction may take place in an agitated reaction vessel (essentially a mixer-settler), in a vertical cylinder (where the solvent flows upward or downward through the liquid mixture), or in a column with internals to enhance mechanically the contact between the two liquid phases.

Emissions may occur by (1) the displacement of air while charging, emptying, and cleaning the extractor; (2)

the agitation of liquids during the extraction process; and (3) the filling and emptying of associated surge tanks.

Centrifuges

Where centrifuging is desired for a solid–liquid separation, the pharmaceutical industry typically uses vertical basket centrifuges. In these devices, the solids are retained on a filter cloth, which is supported by a perforated metal basket. Filtration can usually be improved by establishing a "pre-coat" of solids on the cloth at the start of the centrifuge cycle or leaving a "heel" after removing the filter cake at the end of the cycle. Especially when flammable solvents are used, the centrifuge will be blanketed or purged with an inert gas such as nitrogen.

A significant source of emissions is the vaporization of organic compounds from the "wet" solids as they are removed from the centrifuge and transported to the next operation. In addition, emissions can occur as the centrifuge is opened, as the filtrate is discharged into a holding tank, and as inert gases are vented. Some older facilities may have open-top centrifuges, where direct vaporization to the ambient air can occur.

Filters

A variety of types of filters are in use in the pharmaceutical industry, including pressure, cartridge, plate and frame, atmospheric, and vacuum filters. The objective in using a filter is the same as that of the centrifuge, to remove solids from a liquid. These solids can be impurities or contaminants (e.g., catalysts or carbon particles) or they can be process intermediates or the final product. Emissions typically occur when a filter is opened to remove the collected solids or when the filter is purged with steam or inert gas prior to cleaning.

Crystallizers

Products of very high purity can be manufactured via a crystallization process. Crystallization is achieved by creating a supersaturated solution in which the desired compounds form crystals. Supersaturation can result from cooling the solution, especially where solubility is a strong function of temperature. Supersaturation can also result from evaporating a portion of the solvent or by reducing solubility through the addition of a third component.

Emissions from crystallization operations are usually not significant, unless supersaturation is achieved through solvent evaporation. These emissions may be vented directly to the atmosphere or through a capture device (e.g., a condenser).

Dryers

Drying is often necessary to remove excess solvents remaining in products following centrifuging, filtering, and related operations. There are two types of dryers, defined on the basis of their principle of operation—convection and conduction. In a convective dryer, the heat for drying is provided by a large flow of dry air (or inert gas), which also acts to transport the volatilized solvent out of the dryer vent. In a conductive dryer, the heat for drying is supplied by the walls or trays of the dryer and is thus transferred to the wet product by conduction. The flow rate of gas, plus vaporized solvents from the vent of a conductive dryer, is much smaller than that from convective dryers.

The mechanical configuration of dryers varies greatly, with batch tray, rotary tray, rotary drum, moving belt, and fluidized bed designs among the possible types. Some types, such as the rotary tray dryer, can be designed for use as either a convective or a conductive dryer. Vacuum dryers are typically conductive dryers.

In the pharmaceutical industry, drying is the unit operation with generally the highest VOC emissions. With a convective dryer, the exhaust gas is a high-flow, relatively dilute stream, which traditionally has often been vented directly to the atmosphere without emission control. With a conductive dryer, the vent gas is much more concentrated, so the flow rate is low. In the case of vacuum drying, VOC emissions occur downstream from the vacuum pump, where the air in-leakage is exhausted from the system.

Storage and Transfer

Volatile compounds are often stored in tank farms, process storage tanks, and storage drums. These storage devices are all susceptible to breathing and working losses, the magnitude of which depends on the type of compound stored, the size and design of the tank, ambient temperature and diurnal temperature changes, and tank throughput. Emissions may also be associated with manual material transfer operations within the facility and with the transfer of liquids from tankcars and railcars.

Miscellaneous Operations

Power and steam generation, waste incineration, refrigeration equipment maintenance solvent recovery operations, and wastewater treatment can all contribute to emissions from a pharmaceutical manufacturing facility. The need for these operations varies from plant to plant. In addition, they are not by definition part of the pharmaceutical manufacturing process (although they are certainly a necessary component of some of the larger facilities).

AIR EMISSIONS CHARACTERIZATION

This section describes the characteristics of emissions associated with operating a pharmaceutical facility. Additionally, it provides guidance on estimating emission rates. Three subsections are presented: types of emissions, emission estimates, and estimation techniques.

Types of Emissions

Pharmaceutical manufacturing facilities are similar to many chemical manufacturing operations in that both gaseous and particulate emissions can be significant. Additionally, odors can be a problem, especially where fermentation takes place. While odors can create a community nuisance, they generally do not represent a serious air pollution problem.

Another class of gaseous emissions is acid gases; these are principally sulfur dioxide, nitrous oxides, and hydrochloric acid. The oxides of sulfur and nitrogen are typically associated with the large-scale combustion of fossil fuels such as that utilized for power generation or steam production by large chemical facilities. Because of the generally smaller-scale operations of the pharmaceutical industry, acid gases are usually not an issue. The exception may be halogen acids (e.g., hydrochloric), which are formed as by-products of chemical synthesis or as products of the combustion of halogenated compounds.

Emissions of particulates also may be of concern. One reason is that particulate emissions, especially those associated with formulation and packaging operations, may include biologically active ingredients. Additionally, some of these emissions may be in the respirable size range. Within this range, a significant fraction of the particulates may be inhaled directly into the lungs, thereby enhancing the likelihood of being absorbed into the body and damaging lung tissues. Fortunately, because of concern about worker health and product contamination, particulate emissions are often closely controlled, especially where toxic compounds are in use. Consequently, particulate emissions at most pharmaceutical manufacturing facilities are insignificant and are not addressed further here.

Usually of greatest concern at pharmaceutical facilities are emissions of VOCs. A partial list of VOCs used in the pharmaceutical industry is presented in Table 1. Many of these compounds are photochemically reactive and contribute to tropospheric (lower level) ozone formation (a concern for U.S. facilities located in ozone nonattainment areas), while others may deplete the stratospheric (upper level) ozone layer. An added concern is that some of the compounds listed in Table 1 are air toxics, categorized by the EPA as possible or probable human carcinogens, thereby making long-term human exposure undesirable. Finally, some compounds used in the industry may be acute toxicants, where short-term exposure, including emergency releases, can lead to adverse health effects. The primary focus of the remainder of this discussion is the control of VOC emissions.

Emission Estimates

It is very difficult to characterize specific pharmaceutical operations in terms of representative emission rates. This difficulty is due to the batch nature of the industry, where process units vary widely in the following characteristics.

- Size
- Annual and daily throughput
- Number of cycles per year and operational duration
- Types of raw materials used and products manufactured, including the production of multiple products over the course of a year

An additional characteristic of batch operations that makes estimating emissions difficult is that emissions are not steady, but tend to vary throughout a cycle. An example of how VOC emissions can vary during a batch drying cycle is illustrated in Figure 2.[3]

Despite these problems, it can generally be stated that emissions for pharmaceutical manufacturing process units with standard solvent-recovery control systems tend to be in the range of near-zero to 10 kg/h (22 lb/h), with values occasionally as high as 50 kg/h (110 lb/h).

The EPA has ranked pharmaceutical process units according to their VOC emission potentials as indicated in the following (in decreasing order).[4]

Emission Potential	Process Unit
1	Dryer
2	Reactor
3	Distillation
4	Storage and transfer
5	Filter
6	Extractor
7	Centrifuge
8	Crystallizer

The first four units generally account for the majority of emissions at many facilities. However, because of the wide variation in process unit design and operation from facility to facility, the emissions associated with each process unit must be evaluated on a case-by-case basis.

It is also possible to rank the four basic manufacturing categories (see category definitions under "Process Description") according to their VOC emission potential as in Table 2. Note that while chemical synthesis and fermentation account for the vast majority of emissions, the two categories represent a small fraction of all U.S. pharmaceutical manufacturing facilities (see discussion under "Process Description").

Estimation Techniques

As noted above, pharmaceutical manufacturing unit processes vary extensively in size, throughput, operation, and product. This makes determining realistic emission factors virtually impossible. Short cf emission testing, which can be difficult and expensive for batch processes, the only alternatives to determining emission rates are (1) material balances and (2) theoretical calculations.

TABLE 1. Typical Solvents and Their Ultimate Disposition[a]

Solvent	Annual Purchase, metric tons	Ultimate Disposition, %				
		Air Emissions	Sewer	Incineration	Solid Waste	Product
Acetic acid	930	1	82	—	—	17
Acetic anhydride	1,265	1	57	—	—	42
Acetone	12,040	14	22	38	7	19
Acetonitrile	35	83	17	—	—	—
Amyl acetate	285	42	58	—	—	—
Amyl alcohol	1,430	99	—	—	—	1
Benzene	1,010	29	37	16	8	10
Blendan (AMOCO)	530	—	—	—	—	100
Butanol	320	24	8	1	36	31
Carbon tetrachloride	1,850	11	7	82	—	—
Chloroform	500	57	5	—	38	—
Cyclohexylamine	3,930	—	—	—	—	100
o-Dichlorobenzene	60	2	98	—	—	—
Diethylamine	50	94	6	—	—	—
Diethyl carbonate	30	4	71	—	—	25
Dimethyl acetamide	95	7	—	—	93	—
Dimethyl formamide	1,630	71	3	20	6	—
Dimethylsulfoxide	750	1	28	71	—	—
1,4-dioxane	43	5	—	—	95	—
Ethanol	13,230	10	6	7	1	76
Ethyl acetate	2,380	30	47	20	3	—
Ethyl bromide	45	—	100	—	—	—
Ethylene glycol	60	—	100	—	—	—
Ethyl ether	280	85	4	—	11	—
Formaldehyde[b]	30	19	77	—	—	4
Formamide	440	—	67	—	26	7
Freons[c]	7,150	0.1	—	—	—	99.9
Hexane	530	17	—	15	68	—
Isobutylaldehyde	85	50	50	—	—	—
Isopropanol	3,850	14	17	17	7	45
Isopropyl acetate	480	28	11	61	—	—
Isopropyl ether	25	50	50	—	—	—
Methanol	7,960	31	45	14	6	4
Methyl cellosolve	195	47	53	—	—	—
Methylene chloride	10,000	53	5	20	22	—
Methyl ethyl ketone	260	65	12	23	—	—
Methyl formate	415	—	74	—	12	14
Methyl isobutyl ketone	260	80	—	—	—	20
Polyethylene glycol 600	3	—	—	—	—	100
Pyridine	3	—	100	—	—	—
Skelly Solvent B (hexanes)	1,410	29	2	69	—	—
Tetrahydrofuran	4	—	—	100	—	—
Toluene	6,010	31	14	26	29	—
Trichloroethane	135	100	—	—	—	—
Xylene	3,090	6	19	70	5	—

[a]These data were reported by 26 member companies of the Pharmaceutical Manufacturers Association, accounting for 53% of pharmaceutical sales in 1975. (See reference 4.)
[b]Sold as aqueous solutions containing 37–50% formaldehyde by weight.
[c]Some freons are gases and others are liquids weighing 12–14 lb/gal.

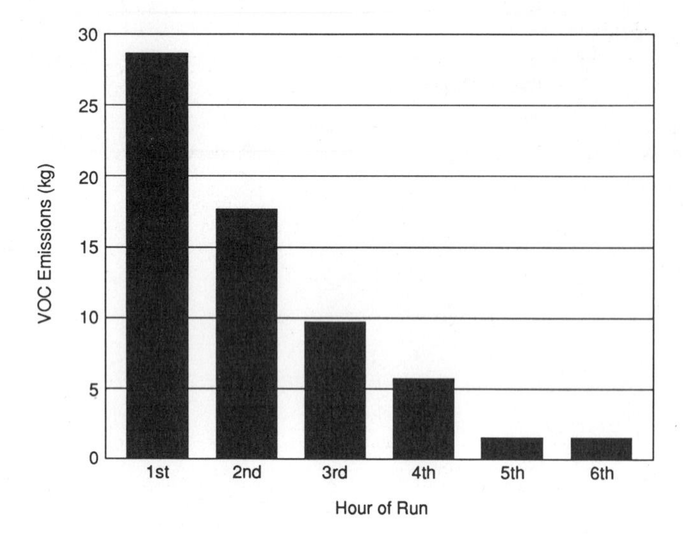

FIGURE 2. Illustration of Variation in VOC Emissions During a Typical Batch Drying Cycle

Material Balances

A material balance can be performed by obtaining records on total solvent purchases for a particular facility and assuming that these purchases represent replacements for solvents lost through evaporation (and thus emitted to the atmosphere). However, other routes by which solvent can be lost also must be taken into account. These routes include discharges to the sewer, incineration, residuals within solid waste, and incorporation into product.

Table 1 presents information on solvent disposition that has been summarized and averaged for 26 large pharmaceutical producers.[4] This information may be useful in designing a material balance study, although this summary information should not be applied directly to a specific facility.

The information in Table 1 is further summarized below, where all columns in the table have been averaged to illustrate the percentage of solvent use according to disposition category.

Solvent Category	Solvent Disposition, % of Total
Remains in product	32
Air emissions	22
Incineration	20
Discharges to sewer	17
Solid waste	9

According to the figures, about 22% of solvent consumption is lost through air emissions.

Theoretical Calculations

It is possible to estimate emission rates from pharmaceutical manufacturing processes using theoretical calculations, based on such factors as vapor pressure, ideal gas behavior, and various simplifying assumptions. The EPA has published a series of VOC emission calculation procedures for charging, evacuation, nitrogen or air sweep, heating, gas evolution, vacuum distillation, and drying operations.[5] When using these and similar calculations to determine hourly, daily, or annual emission rates, correction factors that account for the actual duration of the process must be incorporated into the calculations.

It is important to bear in mind when applying theoretical calculations that the simplifying assumptions inherent in calculations of this type lead to inaccuracies. These inaccuracies usually overstate emissions and can lead to emission estimates that are from several percent to as much as an order of magnitude higher than the actual value. Thus caution is required in interpreting the results of these calculations.

TABLE 2. Emission Potential According to Basic Manufacturing Category

Emission Potential[a]	Category	% Total Industry Emissions (U.S.)
1	Chemical synthesis	64
2	Fermentation	19
3	Extraction[b]	7
4	Formulation	5
5	Other[c]	5

[a]Decreasing order.
[b]Listed in reference 4 as botanicals.
[c]Includes research and development, animal sources, and biological products.

AIR POLLUTION CONTROL MEASURES

This section presents information on air pollution control measures that apply to pharmaceutical processes. This information is divided into the following three subsections: applications, design considerations, and specific control systems.

Applications

Emission control systems are commonly used to reduce emissions of VOCs (including odors) and particulates. Because some VOCs and particulates may contain toxic components, these control systems are also effective in reducing emissions of air toxics. Indeed, air-toxics control practices have been followed in the industry for many years as a result of the biologically active and high-value materials handled by some operations. As discussed under "Air Emissions Characterization," VOC emissions are of primary concern at pharmaceutical facilities and are thus our focus here.

A variety of control systems are available for use in the pharmaceutical industry, depending on the particular application and on process operating parameters. A summary of the types of controls that are in use or available for use at pharmaceutical facilities is presented in the following.

Control Category	Available Control Systems
Add-on controls	Condensers
	Absorbers
	Adsorption units
	Thermal destruction
	Vapor containment
Process design	VOC minimization
	Solvent substitution
Operational practices	Cleanup operations

Design Considerations

Because of the batch nature of many pharmaceutical operations, it is common for critical emission parameters to vary throughout a process operating cycle. These parameters include temperature, pressure, flow rate, VOC composition and concentration, and particulate loading. To control these variable emissions streams effectively, emission controls must be designed to handle both peak and nonpeak flow conditions. Additionally, the controls must be designed to be started up and stopped numerous times over a year and to be effective over short operational cycles. Consequently, it is very important that control equipment be sized correctly and designed to handle a range of operating parameters and conditions.

Specific Control Systems

Specific information is presented in the following on the use of these pollution control concepts: add-on controls, process design, and operational practices.

Add-on Controls

Add-on controls available for use at pharmaceutical manufacturing facilities include condensation, absorption, carbon adsorption, thermal destruction, and vapor containment.

Condensation

The principle of operation of a condenser in emission control service is that a gas stream containing a VOC can be cooled to below the saturation temperature ("dew point"), forming a second phase consisting of the VOC liquid. Separating the liquid from the gas (e.g., by gravity in the condenser combined with a demister pad to trap entrainment) removes a fraction of the VOC from the emission stream. The amount of VOC that remains in the gas stream as a vapor is a function of the temperature and the vapor–liquid equilibrium of the VOC species. In general, the lower the temperature, the lower will be the VOC content in the exit gas stream. In theory, any desired level of VOC removal (control efficiency) approaching 100% is possible with a condenser operating at a low enough temperature. There are some very real practical limitations, however.

Low condensation temperatures are required to control high-volatility VOCs or low concentrations of VOCs with any significant degree of volatility. In the latter case, most of the cooling duty of the condenser is to cool the noncondensible gas to the saturation temperature. Because the heat-transfer coefficient is low under these circumstances, the required heat-transfer area is large, increasing the cost of the condenser.

Producing low temperatures requires energy to drive the refrigeration system. Lower temperatures have progressively higher energy consumption (and thus cost) per unit quantity of heat removed. One scheme to raise the required condensation temperature is to compress the vent stream to a higher pressure so that the dew-point temperature will be raised. This approach is particularly useful where high control efficiencies of low-boiling-point materials (e.g., methylene chloride) are required. It is sometimes possible to raise the dew-point temperature sufficiently so that the problem with water vapor described below can be avoided.

The presence of water vapor in the emission stream along with the VOC is common in actual practice. If room air sweeps or air purges are used, as is often the case in pharmaceutical manufacture, normal humidity will be present. If the process contains water as well as organic solvents, then even isolated vent streams will contain water vapor. Even conditioned "plant air" has traces of water, since the design operation of typical plant air dryers is to lower the water content to a dew point of only about −40°C (−40°F). If a gas stream containing water passes through a condenser that is operating below the freezing point, frost (rime) will form on the tubes, thereby effectively decreasing the capacity of the condenser by blocking efficient heat transfer.

Low-temperature condensation of volatiles from a gas stream containing water can be accomplished with a series of two or more condensers with decreasing temperatures. The first condenser(s) removes the water vapor, while successive, lower-temperature units remove the organic compounds. A concern with this scheme is that the condensate containing the water may also have a substantial organic content (by cocondensation) and thus may pose a water pollution control problem.

Because a vent condenser can be a modest cost control device, it is often recommended. Based on the foregoing discussion, there are clearly some situations that are well suited to condensation. For example, when there is little or no water vapor in the stream and a reasonable concentration of a medium- or high-boiling-point VOC (10% or more), the use of a condenser is particularly appropriate. When the VOC is a low-boiling-point material, when there is water vapor present, or when the VOC concentration is low (less than 1%), the applicability of condensers is marginal, and multiple units operated at low temperatures may be required. When the VOC concentration is quite low (at the fractional percent level), condensers are probably not appropriate, as the achievable level of control will be unacceptably low.

Absorption

Absorption is the selective transfer of one or more components of a gas mixture (solute) into a solvent liquid. For any given solvent, solute, and set of operating conditions (pressure and temperature), there is an equilibrium ratio of solute concentration in the gas mixture to solute concentration in the solvent. If the solute concentration is higher than the equilibrium value, there is a "driving force" for mass transfer of that solute into the solvent, leading to its removal from the gas stream. If the equilibrium concentration in the gas is low compared with the concentration in the liquid solvent, the solute is said to have high solubility. Conversely, if a high concentration of the solute in the gas phase is in equilibrium with the liquid, the solute is said to have low solubility.

The apparent solubility of a solute species may be enhanced if there is a chemical reaction between it and components of the solvent solution. This reaction has the practical effect of minimizing the concentration of the actual solute species in the liquid by changing the solute species into another, more soluble species.

In emission control service for VOC removal, the solvents are chosen for their high organic solubility. These solvents often include water, high-boiling-point mineral oils, and sometimes aqueous solutions of sodium carbonate and sodium hydroxide.

Emission control devices that are based on the principle of absorption are commonly called "scrubbers" and include spray towers, venturi scrubbers, packed columns, and trayed (plate) columns. The basic design concept of all these devices is to provide the most economical means of achieving a high degree of gas–liquid contact in order to promote the desired mass transfer. The degree of control achieved (i.e., the fraction of the solute that is transferred from the gas to the liquid) is a function of residence time, interfacial area, and physical and thermodynamic properties of the VOC species involved.

Spray towers are inherently simple devices consisting of an open column up which the gas stream flows and a group of atomization nozzles to distribute the solvent as a spray of fine drops. Spray towers are versatile in that particulate matter also can be removed, but they have low mass transfer coefficients and typically are used only with high-solubility gases.

Venturi scrubbers, which are slightly more complex mechanically than spray towers, have a high degree of gas–liquid mixing and high particulate removal efficiency, but have relatively short residence times that limit the mass transfer efficiency. They are typically used only with high-solubility gases.

For controlling organic compound emissions, packed and trayed columns are generally preferred. Packed columns are usually more economical than trayed columns for small-diameter (less than 0.6-meter) units and can be readily designed for low-gas-phase pressure drop and for handling corrosive materials and liquids that tend to foam or plug. Trayed columns are preferred for larger-diameter units, where internal cooling is required or where low liquid flow rates would be insufficient to wet the packing.

Scrubbers have been applied particularly to control water-soluble inorganic gases (e.g., sulfur dioxide, hydrogen chloride, hydrogen sulfide, and ammonia) in airstreams using water as the solvent. In the case of acid gases, using an alkaline scrubbing medium increases the effectiveness even more.

In the pharmaceutical industry, scrubbers are also used to remove VOCs. Solvent selection, equipment design parameters, and operating conditions all determine the control efficiency. To absorb organic compounds that have relatively high water solubility (e.g., most alcohols, organic acids, aldehydes, ketones, amines, and glycols), water is the preferred solvent. For organic compounds with low water solubility, another organic liquid (usually one with low vapor pressure) may be chosen. However, in the case of low solubility, water may still be selected, and the equipment will be designed to handle the high flow rate required.

Increasing the depth of the packing or the number of trays increases the control efficiency by increasing the amount of mass transfer area. Decreasing the operating temperature may result in more favorable gas–liquid equilibrium and achieve higher control efficiency with a given piece of equipment.

Absorption can be used effectively with a wide range of concentrations, ranging from several percent to as low as 200 or 300 ppm. Control efficiencies typically range from 60% to 95% or better. However, because of the hydraulic behavior of packing and trays, a relatively steady gas flow is required to maintain good efficiency. Turndown ratios of approximately 50% are standard, thus limiting the ability of absorption to handle the intermittent flow of vents from batch processes.

Adsorption

The principle of operation for adsorption and the application of adsorption to VOC removal from gas streams are discussed in the following.

Principle of Operation Gas adsorption is the selective transfer of one or more components (solutes) of a gas mixture onto the surfaces of the pores of a microcrystalline solid sorbent. The selectivity is most pronounced in a monomolecular layer next to the solid surface, although selectivity may persist to a height of three or four molecules. Both natural and synthetic materials are suitable for use as sorbents if they have a microcrystalline structure. The large number of pores in these materials results in extremely high surface areas (e.g., more than a square kilometer per kilogram, or 0.2 square mile per pound, of solid in some instances). Adsorbents in large-scale commercial use include carbons, silicates, aluminas, and aluminosilicates (molecular sieves).

Generally, adsorption is reversible: The effective adsorption capacity (performance) of a sorbent for a particular solute increases with the concentration of the solute in the gas stream and decreases with increasing temperature. This means that the sorbent may be repeatedly "regenerated"; that is to say, the solute can be desorbed by passing a lower-concentration or higher-temperature gas stream over the bed. Reversible adsorption is desired for VOC removal because it allows for reuse of the adsorbent. However, in some situations, the adsorption of a solute from a gas is irreversible. The capacity of a solid sorbent is nearly constant and independent of the gas stream composition, but because the solute is so tightly bonded to the sorbent, it cannot be reused.

Activated carbon is the preferred adsorbent material for removing VOCs from gas streams. It has a high affinity for nonpolar compounds, with an even greater capacity for high-molecular-weight materials. The high-molecular-weight compounds can pose some problems, as they tend to be irreversibly adsorbed and effectively decrease the capacity of the carbon in a reversible system. A further problem is that high-molecular-weight compounds can displace lower-molecular-weight materials already adsorbed.

The performance of a particular type of activated carbon for a particular solute species must be determined from equilibrium behavior. Where only one species of solute is present in the gas stream, the equilibrium behavior is conveniently represented by a simple curve that plots solute concentration in the solid phase as a function of solute concentration in the gas phase. These curves are usually only valid at a single temperature and are known as "isotherms." Isotherms for common organic vapors being adsorbed on activated carbon are available in the open literature and from carbon suppliers. They may represent strictly experimental data or, more commonly, experimental data fitted to established algebraic formulas (e.g., Langmuir or Freundlich isotherms). In some rare cases, they may represent purely theoretical calculations of equilibrium based on the molecular statistics of the solute–surface interactions.

Application to VOC Removal from Gas Streams In designing a carbon adsorption pollution control device in which the inlet VOC concentration is known and a certain outlet concentration is desired, it is possible to determine from the isotherm and material balance calculations how much carbon will be required. The process design is finalized by applying experience factors in terms of "aging" of the carbon beds (i.e., loss of capacity with time as a result of pore plugging or some other problem) and safety factors. The desired outlet concentration of the solute in the gas phase is used to determine the concentration of the solute on the activated carbon. When the entire mass of carbon reaches the "saturation" level, the solute no longer will be adsorbed from the gas stream to the desired level and "breakthrough" is said to have occurred. The carbon then must be either regenerated (in the case of reversible adsorption) or disposed of (in the case of irreversible adsorption).

To allow vent streams or other gas streams containing VOCs to come into contact with the activated carbon, the carbon granules are usually arranged in a fixed-bed arrangement in either a vertical or horizontal cylindrical vessel. Small units are manufactured with the carbon already in place (e.g., a cannister) with the intent that the entire unit will be replaced when the capacity limit of the carbon is reached. The unit may be regenerated by the supplier, or it may be disposed of. Larger units are constructed so that the carbon granules are loaded after installation of the vessel. In these units, the carbon may be periodically regenerated in place, or it can be removed and shipped off to a central location for regeneration on a contract basis.

Fixed-bed units are operated "batch-wise"; that is, the carbon is gradually saturated with the VOC and must be periodically regenerated, usually in place. This regeneration in place is done by having two or more beds installed and manifolded together so that one bed is in use while the other bed is undergoing regeneration. For very large adsorption operations, "continuous" processes have been designed in which the carbon bed is continuously moved (by a fluidized bed, conveyor belt, or other mechanical means) from an adsorption zone through a regeneration zone and back to the adsorption zone. Continuous adsorption generally does not apply to the smaller-scale operations at pharmaceutical plants.

Activated carbon beds in VOC removal service usually are regenerated using steam. The steam serves as a gas stream with zero concentration of the VOC solute (so that the solute tends to be desorbed) and acts to raise the temperature of the bed so that the equilibrium for desorption will be favored. A much smaller volume of steam is required for regeneration compared with the volume of gas that is treated during the adsorption portion of the cycle. Following desorption, the steam is condensed (further reducing its volume) and sent to a wastewater treatment system to remove the organic content. In some cases, depending on the species and loading on the carbon bed, the VOC concentration in the water is sufficient such that two liquid layers form with the steam condensate. The second layer, with the higher organic compound concentration, can be readily decanted and either recycled or disposed of in a more efficient manner than wastewater.

In the pharmaceutical industry, because of the final product purity requirements, there are limited opportunities for in-plant recycling of recovered solvents. However, where practiced, recovery of a separate organic stream is desirable, since there may be opportunities to sell recovered solvents to another industry or to incinerate them for in-plant heat recovery.

Activated carbon adsorption is currently in use in pharmaceutical plants to remove VOCs from vent streams, back up condensers, and control odors. Removal efficiencies of

95% are common, with values as high as 98 to 99% seen in selected cases. In odor-control applications, the malodorous organic compounds are often fairly large molecules and thus carbon adsorption may be substantially irreversible. In these cases, the use of disposable carbon devices (such as cannisters) may be favored. A related issue is the problem of displacement: if the carbon bed is used to adsorb the larger molecules, it is unlikely that it also will be very effective at simultaneously removing the lighter-molecular-weight compounds. Thus the removal efficiency of the lighter-weight compounds will be poor.

It should be noted that potential fire hazards are associated with using carbon adsorption on air vents or vent streams containing oxygen and flammable hydrocarbons. Whereas the gas streams may be sufficiently dilute in flammables as to be well below the lower flammability limit, the carbon bed concentrates the flammable materials, so that a potentially flammable mixture may be formed inside the bed. Furthermore, the carbon itself, with a high surface area, is combustible. For these reasons, special precautions for early fire detection or bed cooling may be required in some instances or an alternative to carbon adsorption may need to be used.

Thermal Destruction

There are four main types of thermal destruction systems in general use: flares, boilers and process heaters, thermal incinerators, and catalytic incinerators.

Flares Flares burn flammable or combustible vapors with an open flame and can be either tower- or ground-mounted. Nozzle design, steam injection, and air injection are some of the design aspects used to enhance destruction efficiency. Flares are well suited for use with gas streams having a significant fuel value (more than 7.45 MJ/nm^3 or 200 Btu/scf) and are particularly well suited for handling a large, intermittent flow such as might occur from an emergency vent header. They have only limited capability for handling chlorine-containing compounds owing to corrosion of the burner elements and the lack of postcombustion emission control devices (for acid gas removal). The use of flares in the pharmaceutical industry for emission control is very limited; more typical applications are in petroleum refining or large chemical complexes.

Boilers and Process Heaters Boilers and process heaters represent an opportunity to burn a waste stream while simultaneously recovering the stream's heating value. Typical applications are combusting a high-fuel-value liquid waste stream (uncommon in pharmaceutical operations) or using a pollutant-laden airstream as a source of combustion air. A fairly high-fuel-value vent gas stream is required for the boiler or process heater to function without added fuel. As with flares, boilers and process heaters operating with waste streams more typically are found in petroleum refining or large chemical complexes.

Thermal Incinerators Thermal incinerators are controlled combustion devices where fuel and air are added to a combustion chamber to maintain a high minimum operating temperature. Because of these high temperatures, thermal incinerators are able to handle dilute vent streams with high destruction efficiency. Gases with heating values below about 1.86 MJ/nm^3 (50 Btu/scf) can be handled with the addition of fuel. Combustion chamber temperatures range from 700°C to 1300°C (1300°F to 2400°F). Packaged, single-unit thermal incinerators are available in a wide range of sizes and are able to handle flow rates ranging from 0.1 nm^3/s (200 scfm) to about 24 nm^3/s (50,000 scfm).

The control efficiency achieved by thermal incineration is typically 98% destruction or an exit gas concentration of 20 ppm by volume, whichever is less stringent. Thus an inlet stream with a VOC concentration of 2000 ppm by volume or higher (which, with the typical 1:1 air dilution, becomes 1000 ppm by volume at the inlet to the combustion chamber) will have 98% of the incinerator inlet VOC content removed. Inlet streams with lower VOC concentrations will result in outlet gas concentrations of 20 ppm by volume of unburned organics, but with lower destruction efficiencies. Thermal incinerators can handle chlorine- and sulfur-containing compounds, but may require a scrubber on the outlet to control acid gases.

Due to the wide range of sizes available and the ability to handle a wide range of compounds in vent gas streams, thermal incineration is well suited for emission control in the pharmaceutical industry. It is, however, a capital- and energy-intensive approach.

Catalytic Incinerators Catalytic incineration is a variation of thermal incineration. A catalyst is used to promote oxidation of the inlet gas stream at lower temperatures than are required in standard thermal combustion. Catalytic units usually operate over a range of 320°C to 650°C (600°F to 1200°F), and these lower operating temperatures reduce energy requirements. Catalytic incineration units may be smaller than standard thermal incineration units, but the cost of the catalyst may tend to offset any potential savings in the capital investment. Two additional design constraints are that (1) high organic concentrations may cause catalyst failure and (2) the catalyst may be poisoned by sulfur-containing compounds, heavy metals, and halogens.

The control efficiency of catalytic incineration is equivalent to that of thermal incineration, and a wide range of packaged unit sizes is available. As with thermal incineration, catalytic incineration is well suited for emission control in the pharmaceutical industry.

Vapor Containment

Transferring volatile organic liquids from delivery tank trucks and tank cars to storage tanks and from storage tanks to process vessels displaces gas from the headspaces that contain some fraction of VOC vapor. While these intermittent vent streams may be treated by one of the add-on control devices described above, vapor containment may in some cases be the preferred option. In this technology, additional piping is provided so that as the liquid is transferred from the supply tank to the receiving tank, the displaced gas and vapor from the receiving tank are returned in a separate line to the supply tank. With a properly designed system, very little gas/vapor escapes into the atmosphere and VOC emissions are minimized. This concept can be extended to filling reactors from drums by using a drum pump and vapor return line rather than picking up the drum and pouring.

Other unit operations that can be improved in terms of vapor containment are those designed to operate at atmospheric pressure. Because of the relatively unrestricted mixing of the process vapors with the atmosphere, reactor and condenser vents open to the atmosphere are difficult-to-control sources of emissions. An option is to operate them as closed systems, at a slight pressure or vacuum, so that the flow of noncondensible gas (air or inert gas) is restricted. This approach is likely to decrease emissions and the restricted flow rate stream that occurs is easier to control.

Bulk quantities of volatile organic liquids typically are stored in atmospheric pressure tanks with vents to relieve excessive pressure and to break any vacuum that may form. However, diurnal temperature fluctations will result in "breathing losses" where the headspace vapor, approximately saturated with volatile organic vapors, is expelled when the temperature rises. When the temperature drops, fresh air (or an inert blanketing gas) is drawn into the tank, from which it may later be expelled with some degree of saturation. Installing "conservation vents" raises, by a small amount, the pressure or vacuum threshold at which this event occurs, thereby reducing the emission stream volume. Isolated, especially high-pressure, rises or unusually low vacuums will still open the conservation vents, but normal daily fluctuation will not, so that the total volume vented to the atmosphere is substantially reduced.

Process Design

To use add-on controls, including vapor containment, a certain amount of process design must be undertaken. However, in the case of new products or revamps of existing processes, it may be possible to use a different pollution control concept—designing a process that has an inherently lower potential for air emissions. One strategy is to minimize the use of VOCs, to substitute lower-volatility (and/or less toxic) compounds, and to keep these compounds contained.

For example, in the case of drying emissions in pharmaceutical manufacture, a modified process design could use and transfer intermediates as solutions or slurries, rather than as dry products. Where solvents are required, a lower-vapor-pressure solvent would tend to minimize evaporative losses and increase recovery efficiency when using condensers. With some processes, a reaction or formulation step traditionally conducted in a solvent possibly could be carried out in an aqueous medium. One ultimate goal may be the practice of a "solventless" synthesis, where the reaction is conducted "neat." In this case, a customized reactor design may be required to handle potential heat transfer and rheological problems associated with the resulting more concentrated reaction mixture.

Although there are often physical or chemical reasons why a process design must incorporate a particular volatile compound, there are many other instances where, with ingenuity, VOC emissions can be minimized by eliminating the offending material altogether.

Another strategy in a modified process design is to minimize the transfer of materials from process vessels by having a particular unit perform more than one unit operation. An example is a filter–dryer combination. These units, increasingly being used in the pharmaceutical industry, are batch pressure filters that, instead of discharging a wet filter cake, remain closed and convectively dry the cake with a recirculating gas stream.

For maximum effectiveness with some add-on controls, process redesign may be required. An example is the use of closed-loop drying systems. Acknowledging that a convective dryer is potentially a large source of emissions, a condenser, or a condenser combined with gas compression, is often a good choice. Instead of venting the drying gas, a further refinement of this approach is to recirculate the gas from the discharge of the condenser through the heater and back to the dryer. During the drying cycle there are, other than fugitives, essentially no emissions from the closed loop.

Operational Practices

This last control concept deals with the potential for changing workplace practices in such a way as to minimize emissions. Much of the everyday operation of a pharmaceutical synthesis facility is governed by standard operating procedures (SOPs). With proper engineering and management review, the SOPs can sometimes be revised and modified to reduce emissions. One area where emission reductions are particularly possible is in the use of solvents in cleanup practices. Typical procedures may call for the filling and subsequent emptying and air drying of process vessels. The air-drying step results in significant emissions regardless of the vapor pressure of the solvent involved. As an improved SOP, it may be feasible to eliminate the drying step, leaving residual solvent present before the start of the next cycle. Alternatively, a low-VOC aqueous cleaner might be substituted. A similar emission source involves flushing and blowing transfer lines (piping). Again, it may be possible to change the SOPs so that the lines are left filled with solvent rather than blown dry to the atmosphere.

References

1. *Development Document for Proposed Effluent Limitations Guidelines, New Source Performance Standards, and Pretreatment Standards for the Pharmaceutical Manufacturing Point Source Category,* EPA-440/1-82-084, U.S. Environmental Protection Agency, Washington, DC, 1982, pp 6–62.

2. R. V. Crume, *Air Pollution,* Vol. 7, A. C. Stern, Ed.; Academic Press, Orlando, FL, 1986, pp 478–486.
3. L. Ramm and M. Karell, "Technical, economic, and regulatory evaluation of tray dryer solvent emission control alternatives," *Env. Prog.* 9(2):73 (1990).
4. *Control of Volatile Organic Emissions from Manufacture of Synthesized Pharmaceutical Products,* EPA-450/2-78-029, U.S. Environmental Protection Agency, Research Triangle Park, NC, 1979, App. B., p 2–5.
5. *Compilation of Air Pollutant Emission Factors, Vol. I: Stationary Point and Area Sources,* EPA Publication No. AP-42, U.S. Environmental Protection Agency, Research Triangle Park, NC, 1980, p 5.23-3.

Bibliography

Alternative Control Technology Document—Organic Waste Process Vents, Review Draft, U.S. Environmental Protection Agency, Research Triangle Park, NC, 1990, Chapter 3.

Chemical Process Industries, 4th ed., R. N. Shreve and J. A. Brink, Jr., Eds.; McGraw-Hill Book Co., New York, 1977, pp 753–788.

G. A. Herr, *Industrial Odor Technology Assessment,* P. N. Cheremisinoff and R. A. Young, Eds.; Ann Arbor Science Publishers, Ann Arbor, MI, 1975, pp 175–187.

B. S. Lane, *Industrial Pollution Control Handbook,* H. F. Lund, Ed.; McGraw-Hill Book Co., New York, 1971, pp 17-1–17-34.

17
The Petroleum Industry

J. Eldon Rucker and Robert P. Strieter

OVERVIEW OF OPERATIONS

The petroleum industry is organized into the following four broad segments.

1. Exploration and production
2. Transportation
3. Refining
4. Marketing

These segments encompass the operations that produce crude oil and natural gas (petroleum industry raw materials), transform these raw materials into refined products, transport both the raw materials and refined products from one place to another, and market some of the refined products (such as gasoline) to consumers. Although the industry segments are interrelated, each segment has unique functions involving different operations and equipment. Figure 1 shows schematically the interrelationships between the four industry segments and the flow of petroleum industry products.

Petroleum industry operations begin with exploration to locate new sources of crude oil and natural gas. When potential sources are located, wells are drilled to confirm the presence of oil or gas and to determine whether the reserves are sufficient to support production.

During production, crude oil and/or natural gas is recovered from wells and prepared for transportation from the field. Domestic crude oil is transported from the field to refineries by a complex network of pipelines. Alaskan and imported crude oil are transported to refineries by tanker. Natural gas, which may be produced alone or in combination with crude oil, often must be processed at a gas plant to make it suitable for consumer use.

At the refinery, crude oil is converted into a large variety of products, such as hydrocarbon fuels and feedstocks for the petrochemical industry. Refining operations include physical separation of the components in the crude oil and chemical conversion and treating processes to generate the desired products. The refined petroleum products are transported by pipeline, rail, marine vessel, and tank truck to distribution outlets and sometimes directly to the users. The distribution outlets, referred to as bulk terminals and bulk plants, are part of the marketing segment. From these distribution, or marketing, outlets, tank trucks transport the petroleum refinery products directly to large commercial users or, in the case of gasoline, to service stations.

Because the four petroleum industry segments differ significantly in function, they will be discussed separately.

EXPLORATION AND PRODUCTION

Introduction

The first segment of the petroleum industry, exploration and production (E&P), involves the recovery of natural gas and crude oil from underground reservoirs. This industry segment encompasses the following major activities: (1) exploration and well-site preparation, (2) drilling, (3) crude oil and gas processing, and (4) enhanced recovery. There are an estimated 575,000 crude oil wells, 252,000 natural gas wells, and 724 gas plants in the United States.[1] The leading oil- and gas-producing states are Alaska, Texas, Louisiana, California, Oklahoma, New Mexico, and Kansas.[2,3]

Process Description

Seismic and other geophysical methods are used to locate subterranean formations that signal the potential presence of oil and gas reservoirs. When a likely formation is located, drilling is the only way to confirm that oil and gas are present.

Since exploration and well-site preparation activities do not contribute significantly to air emissions, these activities are not discussed in further detail. Instead, this section will

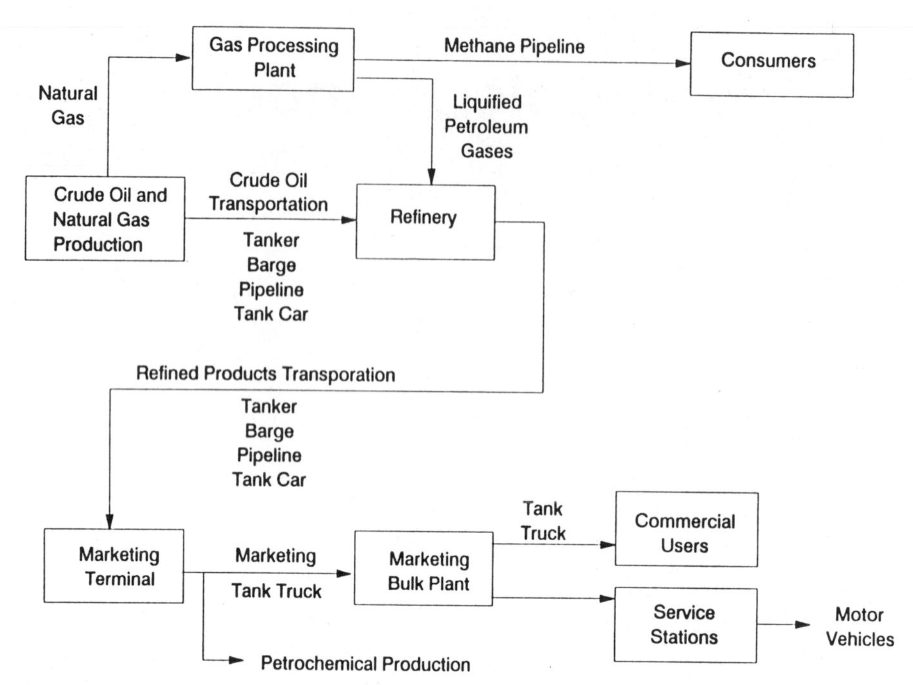

FIGURE 1. Flowsheet of Petroleum Industry Production, Transportation, Refining, and Marketing

concentrate on drilling and production operations, which are more significant sources of air pollutant emissions.

Drilling Operations

Drilling operations include the activities necessary to bore through the earth's crust to access crude oil and natural gas resources. During drilling operations, specially formulated muds are circulated through the hole to remove cuttings from around the drill bit, to provide lubrication for the drill string, to protect the walls of the hole, and to control down-hole pressure. Internal-combustion (IC) engines, often driven by diesel and natural gas, provide energy to rotate the drill bit, hoist the drill string, and circulate the mud, and these engines are sources of hydrocarbon air emissions and criteria pollutants (carbon monoxide [CO], particulate matter, nitrogen oxides [NO_x], and sulfur dioxide [SO_2]).

Cuttings are separated from the mud at the well surface as the mud is passed through shale shakers, desanders, desilters, and degassers.[4] The mud flows to a tank for recycling, and the cuttings, which may be contaminated with hydrocarbons, are pumped to a waste pit for disposal. Because the waste pits may be open to the atmosphere, they are a potential source of hydrocarbon emissions. One other potential air emission source during drilling is blowouts. However, in the 10-year period from 1960 to 1970, blowouts occurred at only 106 of 273,000 wells that were drilled in eight states.[5]

Once a well has been completed and is producing crude oil or natural gas, an arrangement of high-pressure valves termed a "christmas tree" is installed to control production. As the well ages, an artificial lift device may be needed to help bring product to the well surface. Figure 2 shows

FIGURE 2. Typical Piping and Valve Arrangement at Oil and Gas Production Facility

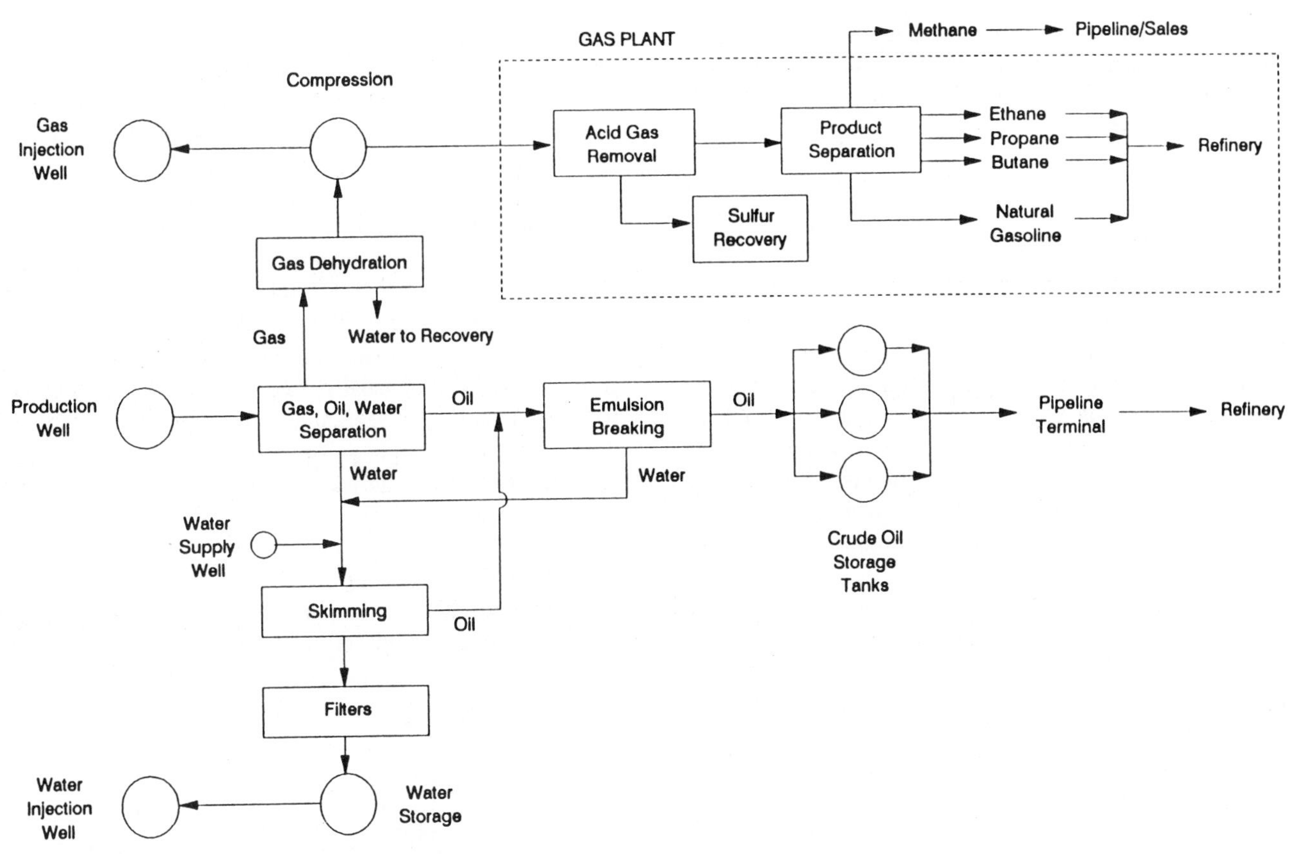

FIGURE 3. Oil and Gas Production Facility

examples of a "christmas tree" and a pumping unit for artificial lift. Pumping units are typically powered by electric motors, but are sometimes powered by IC engines. Gas-fired turbines are another power source typically associated with steam generation or gas processing.

Oil and Gas Production Facilities

The overall objective of the production facility is to prepare the oil and gas for pipeline transport. The oil is usually transferred to a pipeline terminal and eventually to a refinery, while the gas may be sold to a pipeline company for further sale or transported directly to a gas plant for additional treatment. A variety of types of process equipment may be required at a production facility, depending on the quality of the well-head product and whether or not both oil and gas are produced by the well. Figure 3 shows a production facility that is producing and processing both crude oil and natural gas. Not all of the processing steps shown in Figure 3 are needed at all wells. Also, depending on local conditions, the equipment may be used in different combinations or additional equipment not shown may be necessary.

The first processing step employed at many production facilities involves separating the oil, gas, and water produced by the well. The separators provide only one stage of separation, and, in many cases, additional water separation from the oil and gas streams may be required. Water in the oil can form an emulsion. This emulsion is broken using heat in heater treaters or electric energy in devices such as electrostatic coalescers. Cleaned oil flows from the emulsion breakers to crude oil storage tanks, which store the crude oil before it is transferred to the pipeline. The water that is recovered during emulsion breaking is often recycled through skimmers to recover remaining oil, is filtered, and then is stored in water tanks prior to underground injection or other discharge.

Natural gas exiting the first-stage separators is composed primarily of methane, with smaller quantities of ethane, propane, butane, and some natural gasoline. It may also contain such impurities as nitrogen, helium, carbon dioxide (CO_2), hydrogen sulfide (H_2S), mercaptans, and water. Condensible hydrocarbons are recovered from the gas using absorption, refrigeration, or gas expansion processes.[6] Water is removed by passing the gas through equipment such as glycol dehydrators, where the water is absorbed by the glycol. This water is ultimately vented as steam from the glycol reboiler.

The gas-sweetening and product-separation steps shown in Figure 3 are often conducted at a gas plant that is not located at the oil and gas production facility. To sweeten "sour" natural gas, H_2S and CO_2 are usually scrubbed from the gas using an amine solution in an absorption tower.[7] The acid gas stream may be flared, incinerated, or processed further in a sulfur-recovery facility, and the scrubbed gas stream is further processed to separate products. The majority of gas plants use one of the following four processes to separate the hydrocarbon constituents contained in the natural gas stream.[8]

- Absorption
- Refrigerated absorption
- Refrigeration
- Compression

Refrigerated absorption is the process in widest use. In this process, all hydrocarbons except methane are absorbed by an oil operating at low temperature. The methane is transferred to a gas pipeline company, and the heavier hydrocarbon components are separated from the absorber oil and each other in fractionating columns.[9] After separation, these hydrocarbons are transferred to a refinery for additional processing or, in the case of propane and butane, sold directly to users.

Air emissions may occur from a variety of sources at oil and gas production facilities and gas plants. At oil wells, "casing-head gas" may be vented directly to the atmosphere. Hydrocarbons may also be lost with water vapor that is vented from glycol dehydrators and with amines vented from absorbers at gas plants. Crude oil and water are stored in tanks at production facilities and air emissions occur as a result of working and breathing losses from these tanks. Also, at gas plants, natural gasoline may be stored in tanks at atmospheric pressure, and emissions are expected from these storage tanks. The other hydrocarbon products separated at gas plants are stored in pressure vessels, but fugitive emissions may arise from relief valves on these vessels. Fugitive emissions also may occur from valves, flanges, pumps, and compressors located at both gas plants and production facilities.

In addition to hydrocarbon emissions, criteria pollutants are emitted at gas plants and production facilities where IC engines are used to drive pumps, compressors, and other prime movers. Fired vessels such as heater treaters, which often burn natural gas, are another source of criteria pollutants, as are steam generators. A small number of production facilities and all gas plants operate flares, and these devices are an additional source of criteria air pollutants (e.g., SO_2 emissions during acid gas flaring).

Enhanced Recovery

A number of techniques are used to increase the ultimate production of oil and gas from a reservoir. They include water injection, gas injection with CO, thermal processes, and acid treatment and fracturing. Water injection is the most widely applied of these enhanced recovery techniques. Water recovered with oil and gas from production wells is combined with water pumped from water supply wells and then injected into the oil and gas reservoir. This technique forces oil and gas from the reservoir toward the producing wells. Similarly, gas may be injected into a reservoir to create or strengthen a gas cap, which increases the pressure energy in this type of reservoir.[10]

Thermal processes include the injection of steam or heated water, as well as the in situ burning of a portion of the oil in the reservoir.[11] This technique, which is more common in California than in other states, reduces the viscosity of the oil and improves flow characteristics.

Acid treatment and fracturing are two separate techniques, but both increase production by increasing the porosity or permeability of the underground formations.[12] Acid treatment is effective for carbonate rock formations. Hydrochloric acid is usually used to dissolve the carbonate rock. In fracturing, a sand and fluid suspension is pumped down the well at high pressure to crack open sandstone formations, enabling oil and gas to flow more freely to the producing well.

Pumps are used to inject water and other fluids during water flooding, fracturing, and acid treatment. Air emissions will result if IC engines are used to power these pumps. Also, compressors are used during gas injection, and some of these compressors may be driven by IC engines.

Air Emissions Characterization

Volatile organic compound (VOC) air emissions have been characterized for sources in the E&P segment using the 1985 National Acid Precipitation Assessment Program (NAPAP) emissions inventory of the U.S. Environmental Protection Agency (EPA). The principal constituents of VOC emissions include benzene, toluene, xylene(s), 1,3-butadiene, and trimethylbenzene(s). Unfortunately, estimates of emission rates are not well developed in any database and national emission levels cannot be accurately estimated.

Emission estimates have been developed for specific criteria pollutant emissions associated with the E&P industry segment. Nationwide SO_2 emissions were estimated to be 550,000 tons per year in 1979.[13] These estimates were based on data available for 82 sulfur-recovery facilities performing gas-sweetening operations on "sour" gases (i.e., gas containing H_2S). With respect to NO_x emissions, it was estimated that 111,760 tons of NO_x per year were emitted during drilling operations in 1975.[14] These emissions occurred from IC engines that were powering 1580 mobile drilling rigs during the drilling of 40,248 onshore wells. Total NO_x emissions from oil and gas production during 1975 totaled 618,640 tons per year; major sources of these emissions were IC engines (67%), steam generators (29%), and heaters (4%).[15]

Air Pollution Control Measures

Principal sources of emissions from the E&P segment include well casing vents, fired vessels, IC engines, oil and water storage tanks, glycol venting, and fugitive (equipment leak) sources. Control measures for each of these sources of emissions are discussed in the following.

Vent Collection System

Collection systems are used to capture gases discharged from well casings during oil production. The most common disposition of collected gases is to a gas plant or dehydration facility. The collected gases also may be compressed for underground reinjection, used as a fuel for prime movers, or fed to a flare for destruction. Collection systems usually include a knockout pot or other device to remove condensible liquids. The control efficiency of the collection system depends on the capture efficiency and how the collected gas is used or destroyed. For example, flares operate at a destruction efficiency of about 98%.[16]

Air–Fuel Controllers

Criteria and hazardous air pollutant emissions from combustion sources can be reduced by the use of these controls. Proper air-to-fuel adjustment and control can reduce emissions of VOCs and particulate matter (PM) through more efficient combustion. Precombustion chambers vary the air-to-fuel ratio in stages to reduce NO_x emissions. In addition, the low NO_x burners are also used to reduce NO_x formation in combustion. The efficiency of air-to-fuel-ratio controllers is estimated to be about 50%.

Carbon Filter

Carbon adsorption can be used to control organic emissions from vent streams, such as from glycol vents. Organic constituents are adsorbed on the active surface sites of the carbon particles. When the carbon capacity to adsorb organics is exhausted, the carbon must be disposed of or regenerated. Disposable cannisters are sometimes used for low-organic-loading vent streams.

Refrigeration

Low-temperature condensers can be used to recover organic constitutents from vent streams. The temperature of the vent stream is reduced to below the dew point of the condensibles. Chilled water, brine, and low-boiling-point gases can be used as coolants. Condenser efficiencies range between 85% and 95%.[17]

Fixed and Floating Roofs

A fixed roof tank consists of a cyclindrical steel shell with a permanently affixed roof that may vary from cone or dome in shape to flat. Fixed roof tanks commonly operate at a slight internal pressure or slight vacuum. Emissions occur from these tanks as a result of working and breathing losses. As a tank is filled, the increasing liquid level forces saturated vapors to vent from the tank, and these losses are referred to as working losses. Breathing losses are caused by diurnal temperature changes that influence the vapor pressure of the liquid in the tank.

Fixed roof tanks may be equipped with conservation vents that relieve pressure to prevent tank damage during filling. However, a vent recovery system can be installed to collect the vent emissions for recovery or destruction. The control efficiency of this type of system will depend on the ultimate recovery or destruction technique, but usually ranges up to 85% or more, depending on the tank size and application.

An open topped (external) floating roof consists of a cylindrical steel shell equipped with a roof that floats on the liquid surface as it rises and falls (see p. 4.3-2 of reference 18).[18] A seal, which is attached to the roof, contacts the tank wall at the annular space between the roof and the wall. As with internal floating roofs, air emissions occur from external floating roofs as a result of liquid that leaks through seals and stands on top of the roof. However, with a properly designed seal system, the efficiency of an external floating roof ranges from about 95% to 99%.[19]

Inspection and Maintenance

Fugitive emissions due to equipment leaks from pump and compressor seals, valves, flanges, and other equipment connections can be reduced by an active inspection and maintenance program. Leaking equipment is identified during periodic inspections with a vapor-detection device. The leaking equipment is logged on a maintenance schedule and mechanical adjustments (e.g., tightening valve stems, packing) are made to repair the leaks. The efficiency of this control procedure is affected by how often the inspections are conducted. In addition, leakless equipment has been developed to reduce fugitive emission losses from such equipment as valves and pump seals.

TRANSPORTATION

Introduction

This segment of the petroleum industry encompasses the transportation operations used to move raw materials and refined products among the various other interrelated segments. It includes the transportation of oil and gas from well-field production facilities to refineries and gas plants and the transportation of refined products by means other than tank trucks (tank truck transport is covered in the marketing section of this discussion).

Process Description

Figure 4 presents a schematic of the transportation network showing the different modes used to transport petroleum products among segments of the industry. Most domestic crude oil is transported from well-field production facilities to refineries through a complex network of pipelines. The crude oil is introduced into these pipelines by pumping stations that are powered by electric motors, reciprocating

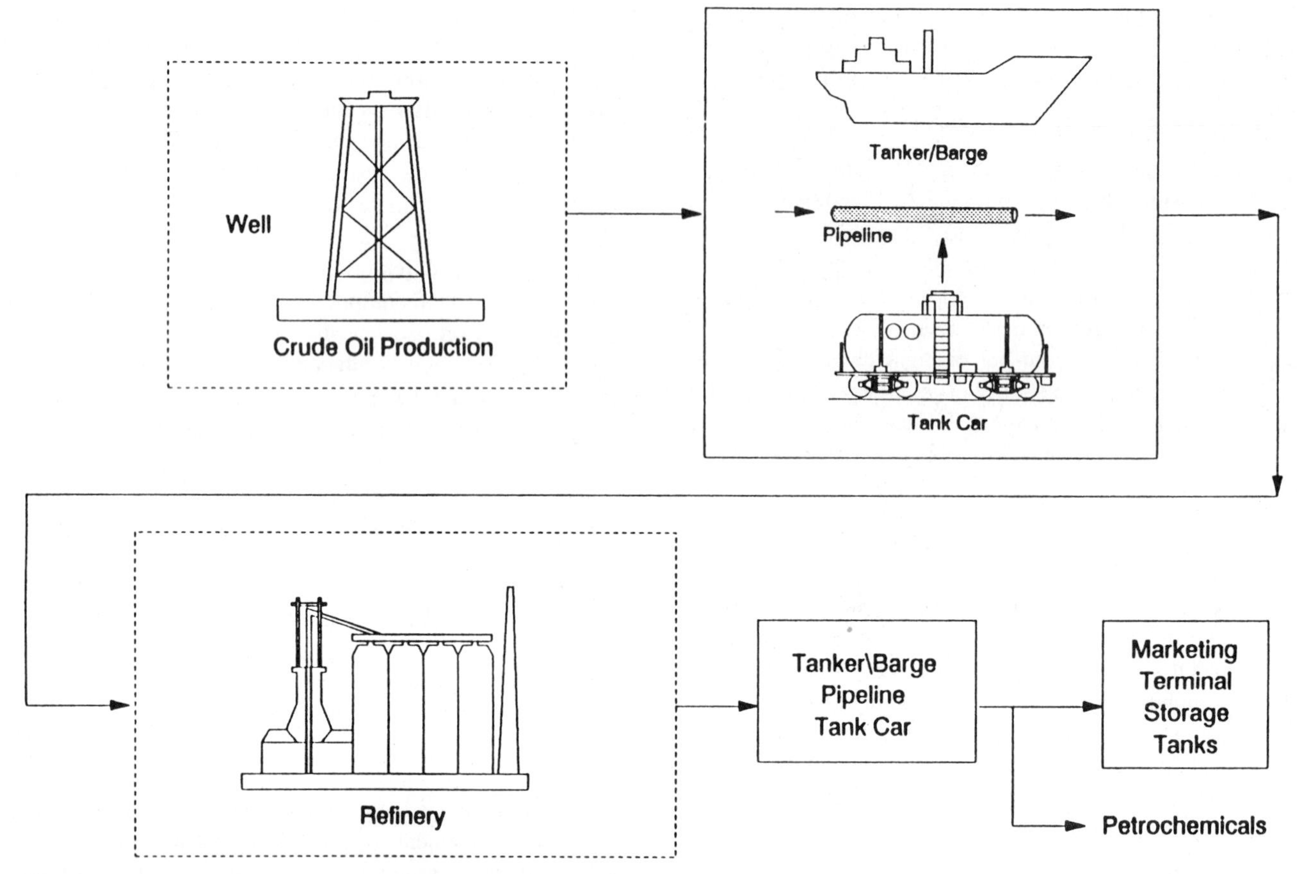

FIGURE 4. Petroleum Transportation and Distribution Systems

engines, or gas turbines. There are approximately 3400 pump stations nationwide. Crude oil is also transported from some production facilities by barges, railroad tank cars, and tank trucks. Unlike domestic crude oil, imported and Alaskan crude oil enter the continental United States by tanker through 52 U.S. marine terminals. Once the crude oil has been processed at refineries, the refined products are transported from the refineries to marketing terminals by barges, railroad tank cars, and tank trucks.

The transportation segment involves many distinct operations that are potential air emission sources. Transportation vessels and vehicles are sources of evaporative hydrocarbon losses from crude and refined products. These losses occur during loading, while in transit, and from marine vessel ballasting operations. Also, pipeline pump stations are sources of fugitive emissions from equipment leaks and storage tanks can be sources of VOC emissions as a result of working and breathing losses (Figure 5).

Air Emissions Characterization

Volatile organic compound air emissions occur from a variety of sources during the transportation of petroleum liquids. There are emission losses of VOCs and criteria pollutants from transportation vehicles, from storage tanks, and from leaks in equipment used in the transportation segment.

TABLE 1. Transportation Segment Emission Sources of VOC Emissions

Source	Number of Sources
Loading (marine terminal)	52
Barges (product/crude)	3,800
Carrier (crude)	125
Carrier (product)	150
Equipment leaks at pump stations	
Pumps	8,500
Valves/flanges	76,800
Pump station storage tanks	3,523

Source: Reference 1.

Table 1 presents estimates for the 1985 annual VOC emissions from each of these sources and provides estimates for the number of sources in each category. The emission estimates shown in Table 1 were derived from the EPA NAPAP emissions inventory.

Emissions from Loading, Transit, and Ballasting Operations

Air emissions from railroad tank cars, tank trucks, and marine vessels are of three types: loading, transit, and

FIGURE 5. View of Trans-Alaska Pipeline

marine vessel ballasting losses. Table 2 gives emission factors for such losses for various petroleum products.

Loading losses occur when organic vapors in cargo tanks are displaced by liquids being loaded into tanks for transport. These vapors are vented to the atmosphere or to a vapor-recovery system. The rate of vapor generation during loading is affected by the type of loading operation—splash loading or submerged loading. In splash loading, the fill pipe is lowered only partway into the cargo tank and the product is introduced above the liquid level in the tank. Turbulence and vapor–liquid contact cause high levels of vapor generation and loss. In submerged loading, the product is introduced into the tank being filled, with the fill pipe extended almost to the bottom of the cargo tank. In bottom loading, the fill pipe is attached to the cargo tank bottom. In both submerged fill and bottom loading, turbulence is controlled and vapor generation and loss are significantly lower than in splash loading.

Transit losses are emissions that take place while the crude or product is being transported from one location to another by a mobile carrier. For gasoline, transit emissions depend predominantly on the extent of venting from the cargo tank during transit and are not as influenced by time in transit.

Ballasting losses occur during the unloading of petroleum liquids from ships and barges at marine terminals. After the petroleum liquid has been unloaded, operators load several cargo tank compartments with seawater. This seawater is called ballast. As the ballast water is added, it displaces organic vapors from the cargo tank to the atmosphere or to a cargo tank being unloaded simultaneously. The ballast water must be unloaded before the cargo tank can be refilled with petroleum liquids.

Storage Tank Emissions

Storage tanks at pipeline pump stations are sources of VOC emissions. Storage tank design, emissions, and emission controls are discussed in the E&P section. In general, storage tank emissions can be attributed to breathing and working losses. Breathing losses result from vapor volume and pressure variations caused by diurnal temperature changes. Working losses result from tank-filling operations where hydrocarbon vapors are displaced from the tank because of the rising liquid surface.

TABLE 2. Transportation Segment Total Organic Emission Factors for Petroleum Marine Vessel Sources[a]

Emission Source	Gasoline[b]	Crude Oil[c]	Jet Naphtha (JP-4)	Jet Kerosene	Distillate Oil No. 2	Residual Oil No. 6
Loading operations						
Ships/ocean barges						
mg/L transferred	NA[d]	73	60	0.63	0.55	0.004
lb/10^3 gal transferred	NA	0.61	0.50	0.005	0.005	0.00004
Barges						
mg/L transferred	NA	120	150	1.60	1.40	0.011
lb/10^3 gal transferred	NA	1.0	1.2	0.013	0.012	0.00009
Tanker ballasting						
mg/L ballast water	100	NA	NA	NA	NA	NA
lb/10^3 gal ballast water	0.8	NA	NA	NA	NA	NA
Transit						
mg/week-L transported	320	150	84	0.60	0.54	0.003
lb/week—10^3 gal transported	2.7	1.3	0.7	0.005	0.005	3×10^{-5}

[a]Emission factors are calculated for a dispensed product temperature of 60°F.
[b]Factors shown for gasoline represent nonmethane-nonethane VOC emissions. The example gasoline has an RVP of 10 psia.
[c]Nonmethane-nonethane VOC emission factors for a typical crude oil are 15% lower than the total organic factors shown. The example crude oil has an RVP of 5 psia.
[d]NA = not available.

Source: Reference 18.

Fugitive Emissions

Pump station pumps, seals, valves, and flanges are sources of VOC fugitive emissions, which are discussed in the refinery section.

Fuel Combustion Emissions

Pumps, tankers, and barges are often powered by IC engines. These engines emit VOC, NO_x, SO_2, CO, and particulate as a result of fuel combustion.

Air Pollution Control Measures

Table 3 presents transportation-segment emission sources and control options as further discussed with the following. Unfortunately, there is a wide range of estimates and uncertainties for these emission control measures that precludes accurate estimation of control efficiencies at this time.

Submerged Loading

In splash filling, the petroleum product is introduced into a tank at the top and above the liquid level. Submerged fill reduces vapors generated during product loading into tanks at marine terminals and into barges. During submerged fill, the product is introduced into the tank being filled, with the transfer line outlet situated below the liquid surface. Submerged filling minimizes droplet entrainment, evaporation, and turbulence, and it can reduce vapor loss by up to 58% over splash filling.[20]

TABLE 3. Transportation Segment Emission Sources and Emission Controls

Emission Source	Emission Control
Loading (marine terminal)	Submerged loading Refrigeration Compression absorption Thermal oxidation
Barges (product/crude)	Submerged loading Refrigeration
Carrier (crude)	
Ballasting losses	Vapor displacement into cargo tank
Transit losses	Refrigeration
Carrier (product)	
Ballasting losses	Vapor displacement into cargo tank
Transit losses	Refrigeration
Equipment leaks (pumps, flanges, valves)	Quarterly inspection and maintenance Monthly inspection and maintenance
Pump station storage tanks	Fixed roofs with conservation vents Internal floating roofs External floating roofs Refrigeration

Sources: References 18, 20, 28–30.

Vapor Balance Systems

During filling operations, vapors displaced by the rising liquid level in the tank being filled can be routed to the tank being emptied. This system is called vapor balance. Vapor balance prevents the compression and expansion of vapor spaces that would otherwise take place during filling. Vapor balance control efficiencies will vary with temperature, barometric pressure, and the type of product being transferred.

Refrigeration

Vapor recovery by refrigeration recovers hydrocarbon vapors by condensing at cryogenic temperatures and atmospheric pressure. Product vapors are first cooled in a dehydrator by direct contact with cooling coils, where most of the vapors are condensed and recovered. From the dehydrator, vapors are further cooled and condensed. The vapor-recovery efficiency of the refrigeration unit depends on the composition and concentration of the hydrocarbon vapors and the operating temperature of the condenser. Control efficiencies range from 85%[5] to 95%.[20]

Absorption

An absorption vapor-recovery unit recovers hydrocarbons by absorption into a lean oil stream. Vapors are compressed and fed to a packed-bed absorber where the hydrocarbons in the vapor are absorbed by the lean oil. The lean oil, which is fed countercurrently to the vapor stream, is cooled by refrigeration to improve efficiency. The removal efficiencies of absorbers depend on the ratio of the lean oil flow rate to the vapor flow rate. For gasoline loading vapors, the efficiency of vapor removal can be as high as 95%.[20]

Thermal Oxidation

Incineration operates by combusting hydrocarbon vapors to CO_2 and water. Incineration can be used to control vapors from product loading at marine terminals, but these systems require sophisticated instrumentation and can be very expensive to install and operate. The efficiency of incineration for controlling vapors at marine terminals is 99%.[20]

Compression

Compression–refrigeration–condensation units recover hydrocarbon vapors by condensation at low temperature and moderate pressure. Such units can pose a significant safety risk in marine loading of gasoline.

Inspection and Maintenance—Leakless Equipment

Pipeline pump station valves, pumps, and flanges are sources of emission losses. These losses can be controlled by regularly scheduled inspection and maintenance. Equipment is inspected for leaks with a vapor-detection device and leaking equipment is repaired. Also, new valves have been developed that significantly reduce emissions from valve stem packing.

Pump Station Storage Tank Controls

Pump station storage tanks are sources of VOC emissions. Storage tank emission controls are discussed in detail in the E&P section. In general, replacing fixed roofs on product storage tanks with internal floating roofs can reduce vapor losses by 60–99%, depending on the type of roof and seals installed and on the type of organic liquid in the storage tank.

REFINING

Introduction

The petroleum refining industry (SIC Code 29) involves numerous processes that convert crude oil into more than 2500 products, including liquified petroleum gas, gasoline, kerosene, jet fuel, diesel fuel, other fuel oils, lubricating oils, and feedstock for the petrochemical industry.[18] Petroleum refinery activities include the storage of crude at the refinery, petroleum handling and refining operations, and storage of the refined products prior to shipment (Figure 6).

As of January 1990, there were 189 operating refineries in the United States with a total crude capacity of 15.4 million barrels per calendar day.[21] Refineries are generally located in coastal areas, with the majority of the capacity (54%) located in Texas, Louisiana, and California. Refineries also are concentrated in the Chicago, Philadelphia, and Puget Sound areas. A geographical distribution of the U.S. refineries and crude capacities is presented in Table 4.

Process Description

Refining operations consist of separation processes, conversion processes, treating processes, feedstock and product handling, and associated auxiliary operations. The flow scheme at a particular refinery is determined by the composition of the crude oil and the chosen slate of products. An example of a complex refinery flow scheme is presented in Figure 7.

FIGURE 6. View of Typical Refinery

TABLE 4. U.S. Refinery Capacity by State as of January 1990

State	Number of Plants	Crude Capacity (barrels per calendar day)
Texas	30	3,872,750
Louisiana	17	2,272,415
California	30	2,222,248
Illinois	6	920,600
Pennsylvania	9	856,000
New Jersey	6	494,250
Washington	7	492,675
Ohio	4	482,650
Indiana	4	426,900
Oklahoma	6	391,500
Mississippi	5	358,600
Kansas	8	348,125
Minnesota	2	285,600
Alaska	6	231,000
Kentucky	2	218,900
Wyoming	5	164,600
Utah	6	154,500
Delaware	1	140,000
Alabama	3	139,250
Montana	4	138,900
Hawaii	2	129,800
Michigan	4	123,700
Colorado	3	91,200
New Mexico	3	80,300
Tennessee	1	60,000
Arkansas	3	58,570
North Dakota	1	58,000
Virginia	1	53,000
New York	1	42,500
Georgia	2	35,500
Wisconsin	1	32,000
West Virginia	2	29,680
Oregon	1	15,000
Arizona	2	13,710
Nevada	1	4,500
Total	189	15,438,923

Source: Reference 21.

The operations associated with petroleum refining are described more fully in the following. Refining operations (separation, conversion, and treating) are discussed first, followed by a brief discussion of auxiliary operations, including wastewater treatment and cooling towers. Handling operations are discussed last and include equipment (tanks, pipes, pumps, and valves) employed throughout the refinery. Potential emission sources from each segment of refining are discussed and are summarized in Table 5. In general, these emission sources are either those resulting from the petroleum products (namely, VOC emissions) or those resulting from combustion sources at the refinery.

Volatile organic compound emissions from refinery operations can be characterized as of two types: process point source emissions and fugitive emissions. Process point source emissions are those emissions directly associated with or generated by a process unit. Process vents are an example of a point source emission. Fugitive emission

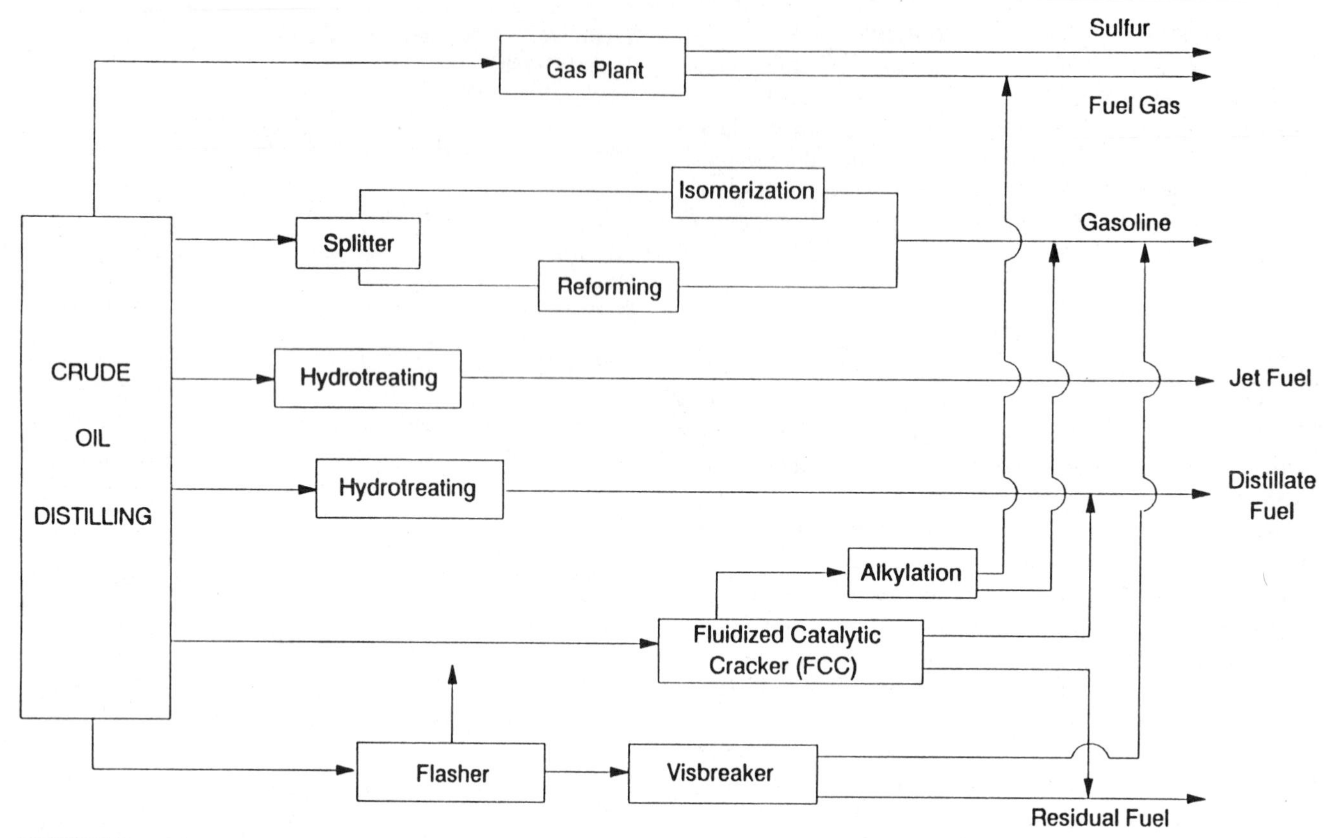

FIGURE 7. Flow Scheme for a Complex Petroleum Refinery

sources are VOC emission sources not specifically generated by a particular process unit. Such emission sources are found throughout a refinery and may or may not be associated with a process unit. They include valves, flanges, pump and compressor seals, cooling towers, storage tanks, transfer operations, and wastewater treatment systems. Fugitive emissions also result from the evaporation of leaked or spilled hydrocarbon liquid and gases. Combustion sources at refineries result in emissions of SO_2, NO_x, and PM.

Refining Operations

Separation Processes

Crude oil is a mixture of paraffinic, naphthenic, and aromatic hydrocarbon compounds with small amounts of impurities, including sulfur, nitrogen, and metals. Refinery separation processes separate the crude into common boiling-point fractions. The three petroleum separation processes are atmospheric distillation, vacuum distillation, and light ends recovery.

Refineries operate some process equipment under a

TABLE 5. Emission Sources for Petroleum Refining

	Potential Emission Sources[a]		
Source Category	Process Point	Process Fugitive	Area Fugitive
Crude separation	G,J,L	F,H,M,N	I
Light hydrocarbon processing	O,G	F,H	I
Middle and heavy distillate processing	G,O,P,R	F,H	Q
Residual hydrocarbon processing	B,G,K,O,R	H	I
Auxiliary processes	G	F,H	I

[a]Key to emission sources: B—visbreaker furnace; F—wastewater disposal (process drain, blow-down, cooling water); G—flare, incinerator process heater, boiler; H—storage, transfer, and handling; I—pumps, valves, compressors, fittings, etc.; J—absorber; K—process vent; L—distillation/fractionation; M—hot wells; N—steam ejectors; O—catalyst regeneration; P—evaporation; Q—catalytic cracker; R—stripper.

Source: Reference 20.

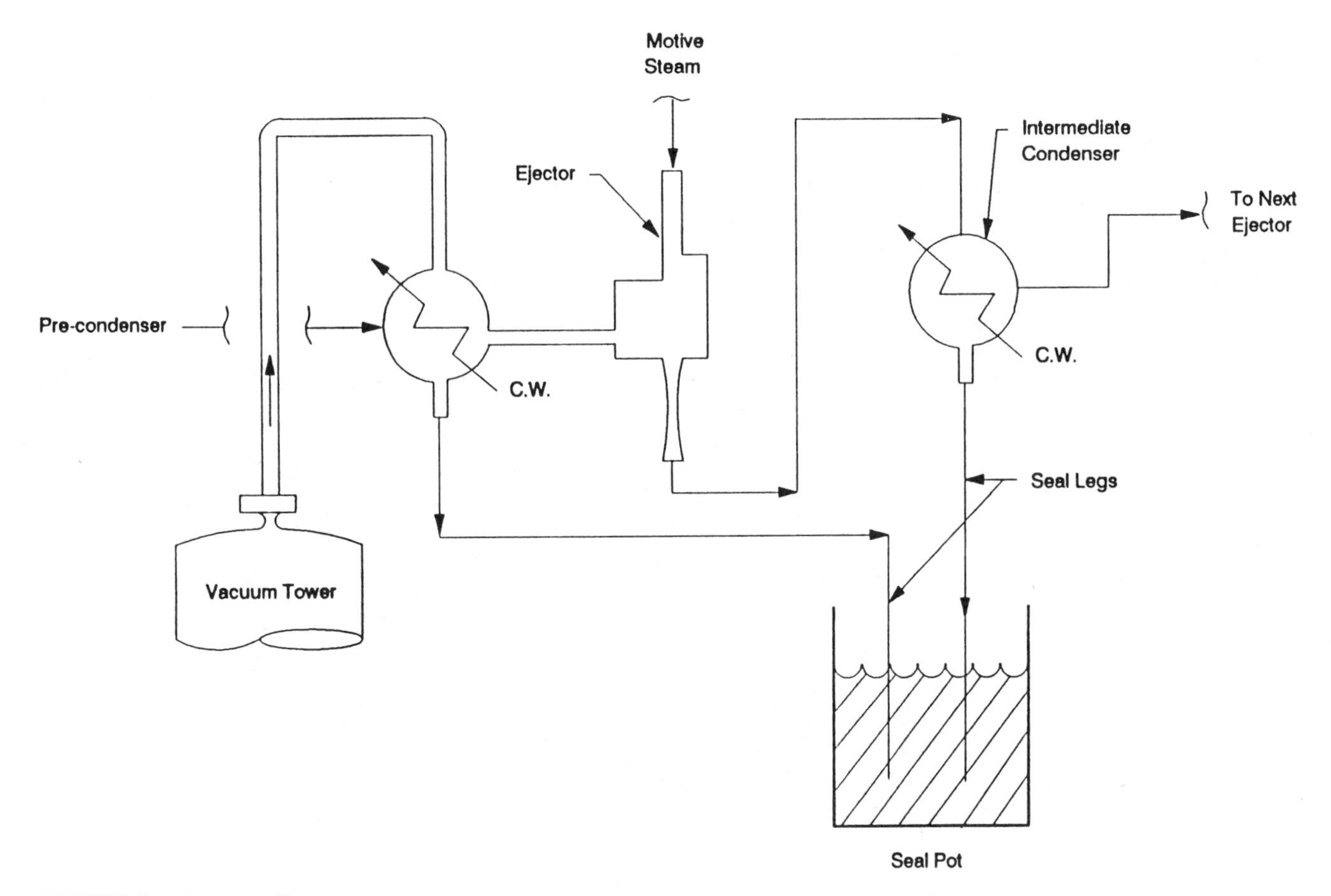

FIGURE 8. Vacuum System

vacuum, as in vacuum distillation, where heavy feedstocks (usually the bottom fraction from atmospheric distillation) are distilled under high temperatures. A steam-driven vacuum jet, sometimes coupled with a barometric condenser, is frequently used to produce and maintain a vacuum (see Figure 8). Light volatile compounds such as H_2S, can be formed during vacuum distillation owing to the high temperatures used in the process. Those compounds that do not condense in the barometric condenser are discharged with the exhaust stream.

Conversion Processes

These processes are used to meet the specifications for high-octane gasoline, jet fuel, and diesel fuel, and to upgrade components such as residual oils and light ends to gasoline. Cracking, coking, and visbreaking processes are used to break large molecules into smaller molecules. Isomerization and reforming processes rearrange the structure of molecules to produce higher-value molecules of a similar molecular size. Polymerization and alkylation processes are used to combine small molecules into larger ones.

Coking is a low-pressure thermal cracking process that produces petroleum coke and gas oil from crude oil residue, decanted oils, and/or tar pitch products. Delayed coking is the most widely used process. In delayed coking, the process stream is initially fed through a fractionator where the lighter material is flashed off. The bottoms subsequently are heated to 900–1100°F and the vaporized material is sent to a coke drum. In the drum, some of the process stream condenses into coke, while more volatile components are returned to the fractionator for separation. Once the coke drum is filled with coke, the removal process is initiated. First the drum is cooled, and then the coke is cut from the drum using a high-pressure water jet, after which it is physically removed from the drum. Significant VOC and PM air emissions can occur during coke removal. Substances expected to be present in these emissions include benzene, toluene, xylene, H_2S, metals, and polycyclic aromatic hydrocarbons (PAH).

Catalytic cracking is commonly used to convert a gas oil feed stream into fuel gas, liquified petroleum gas (LPG), high-octane gasoline, and distillate fuel (see Figure 9). The fluid catalytic cracking (FCC) process uses a catalyst in the form of very fine particles that act as a fluid when charged with a vapor. Fresh feed is preheated in a process heater and introduced into the bottom of a vertical transfer line or riser as the catalyst vaporizes the feed, bringing both to the desired reaction temperature, 880–980°F. The high activity of modern catalysts causes most of the cracking reactions to take place in the riser as the catalyst and oil mixture flows upward into the reactor. The hydrocarbon vapors are sepa-

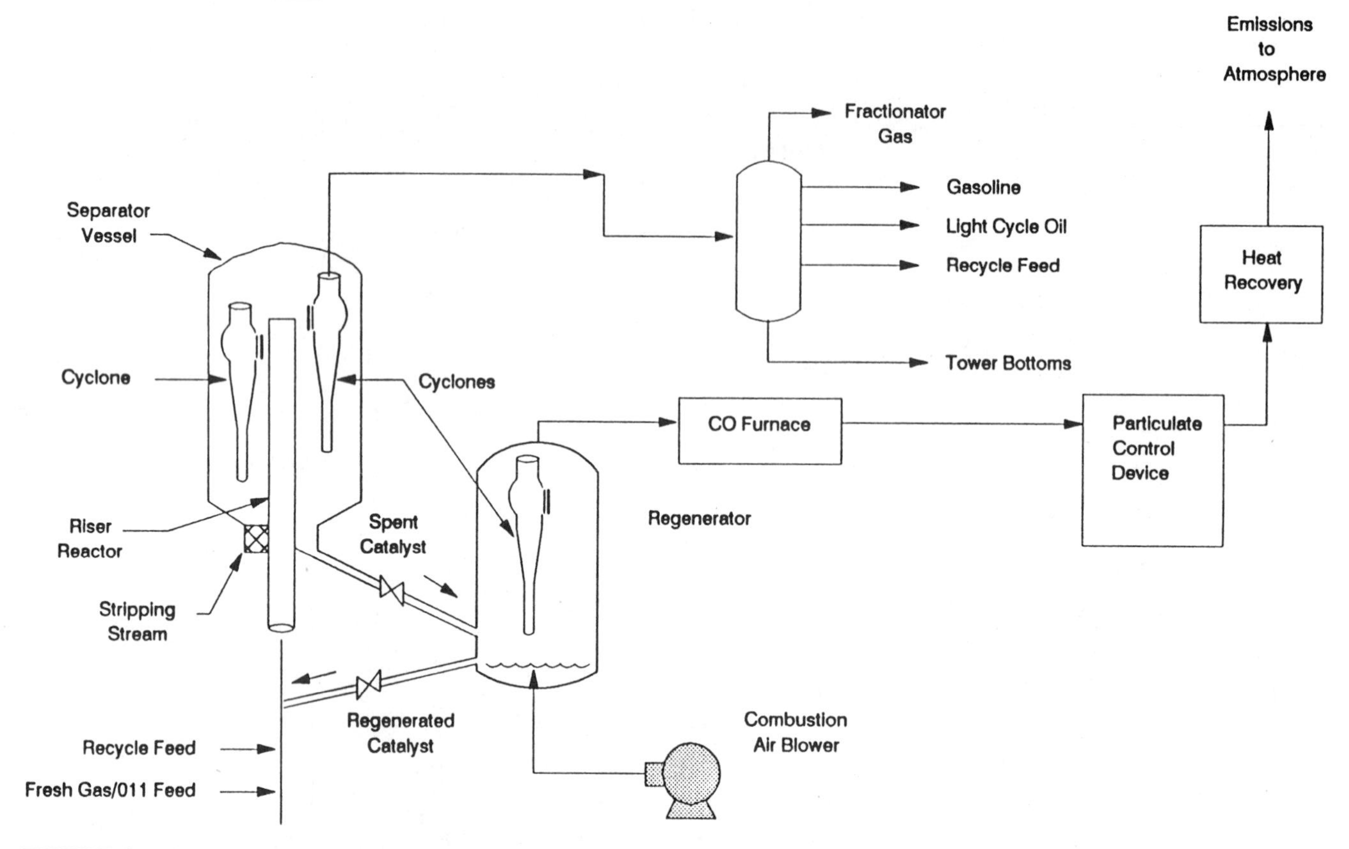

FIGURE 9. Schematic Diagram of a Fluid Catalytic Cracking Unit

rated from the catalyst particles by cyclones in the reactor. The reaction products are sent to a fractionator for separation. The spent catalyst falls to the bottom of the reactor and is steam stripped to remove adsorbed hydrocarbons as it exists the reactor bottom. The spent catalyst is then conveyed to a regenerator, where coke deposited on the catalyst during cracking reactions is burned off in a controlled combustion process with preheated air. Regeneration temperature is usually 1100°F to 1250°F. The catalyst is then recycled to be mixed with fresh hydrocarbon feed.

There are two sources of air emissions from catalytic cracking: process heater and catalyst regenerator. Emissions from the catalyst regenerator include hydrocarbons, oxides of sulfur, ammonia, aldehydes, NO_x, cyanides, CO, metals, and particulates.

Catalytic reformers utilize catalysts to cause the rearrangement of existing molecules into higher-value molecules. Reformers are classified by method and frequency of catalyst regeneration. In the *cyclic* method, regeneration frequencies of one to two reactors per day are typical. In *semiregeneration,* all reactors are regenerated simultaneously at the end of a run, which may range from three to twenty-four months in length. In *continuous* regeneration, the catalyst is not regenerated in the reactor itself. A portion of the catalyst is continually routed to a separate regeneration system and returned to the top of the reactor. Regeneration is done at 200–500 psig and 900–1050°F. Bimetallic catalysts are the most commonly used, with platinum-rhenium on alumina the most popular combination. Catalysts used in reforming become contaminated with carbon, sulfur, nitrogen, water, and metals and must be regenerated to restore activity. During regeneration, the carbon is oxidized to mixed carbon oxides (CO, CO_2, COS) and hydrocarbons are combusted. Compounds that may be emitted during regeneration include ammonia, carbon tetrachloride, dibenzofurans, dioxins, metals, chlorides, and cyanides. Catalyst fines also may be emitted in small quantities, but this is generally minimized owing to the high cost of platinum catalyst. Continuous reforming has the highest potential for emissions.

Treating Processes

These processes are used to stabilize and upgrade petroleum products by separating them from less desirable products and by removing impurities. Deasphalting is a treating process used primarily for the separation of asphaltic and resinous materials from petroleum products. Treatment of products to remove other undesirable elements, such as sulfur, nitrogen, and oxygen, is achieved by hydrodesulfurization, hydrotreating, chemical sweetening, and acid gas removal. Desalting processes remove salt, minerals, grit, and water from crude oil prior to distillation. Asphalt blowing is used for polymerizing and stabilizing asphalt.

Hydrotreating is a commonly used treatment process that

refers to a variety of processes. It is used for viscosity index improvement, desulfurization, denitrification, demetallization, removal of gum-forming compounds, and color improvement. In the hydrotreating process, the oil feed mixed with hydrogen is charged to a fixed-bed catalyst reactor. The catalyst generally has a cobalt or nickel-molybdenum/alumina base. When the catalyst is exhausted, it is regenerated at 700–1200°F using air with steam or flue gas. Common emissions are ammonia, H_2S, nickel, and VOCs. Catalyst handling operations produce negligible particulate emissions of coke fines and catalyst dust.

Sulfur-recovery processes are used to minimize SO_x emissions from refinery fuels and products. Elemental sulfur is produced as a by-product and is often sold commercially. In the widely used Claus process, H_2S is oxidized to form SO_2 and water:

$$H_2S + 1.5O_2 \rightarrow SO_2 + H_2O$$

Additional H_2S reacts with this SO_2 to form elemental sulfur and water:

$$2H_2S + SO_2 \rightarrow 3S + 2H_2O$$

Tail gas from a Claus sulfur-recovery unit contains a variety of pollutants, including SO_2, H_2S, other reduced sulfur compounds (COS and CS_2), CO, and VOCs.

Asphalt blowing refers to the blowing of air through vacuum distillation residuum to polymerize asphalt by oxidation and increase the hardness and melting point. Ferric chloride or phosphorus pentoxide sometimes is added to increase the reaction rate and to create a very-high-melting-point product. The feed is typically preheated to 400–600°F. The oils, containing a large quantity of polycyclic aromatic compounds, are oxidized by blowing heated air through a batch mixture or, in a continuous process, by passing hot air countercurrent to the airflow. The reaction is exothermic and quench steam is sometimes used for temperature control. Air emissions from asphalt blowing are primarily hydrocarbon vapors, including acetaldehyde, benzene, formaldehyde, metals, PAH, toluene, and sulfur compounds.

Distillate sweetening is accomplished by the conversion of mercaptans to alkyl disulfides in the presence of a catalyst. Sulfur is added to the sour distillate with a small amount of caustic and air. The mixture is then passed upward through a fixed-bed catalyst counter to the flow of caustic. Conversion may be followed by disulfide removal in an extraction process. In the conversion and extraction process, the sour distillate is washed with caustic and is then contacted in the extractor with a solution of catalyst and caustic. The extracted distillate is contacted with air to convert mercaptans to disulfides. After oxidation, the distillate is gravity settled, inhibitors are added, and the distillate is sent to storage. The major emissions are VOCs resulting from the contact between the distillate product and air in the "air blowing" step.

Auxiliary Operations

A wide assortment of auxiliary operations not directly involved in the refining of crude oil are used in functions vital to the operation of the refinery. Examples include boilers, wastewater treatment facilities, cooling towers, and combustion devices.

Wastewater Treatment

Process wastewater containing H_2S, ammonia, and oil is termed sour water. It is stripped with steam to remove H_2S, ammonia, and light gases, which are then recovered or incinerated. Because of the low heating value of the process stream to be incinerated, auxiliary fuel is required to maintain the incineration combustion process. The most common emissions from the process are H_2S and ammonia.

All refineries employ some form of wastewater treatment prior to discharge to the environment or reuse in the refinery. The design of wastewater treatment systems is dictated by the diversity of refinery pollutants, including oil, phenols, sulfides, dissolved solids, and toxic chemicals. Although the exact configurations of wastewater treatment processes employed by refineries vary greatly, drain systems, oil–water separators, and air flotation systems are generally included. Drain systems consist of individual process drains, where oily water from various sources is collected, and junction boxes, which receive the oily water from several drains. Oil–water separators generally represent the first step in the treatment of refinery wastewater. The separation and removal of oil from the water are accomplished through density differences that cause the oil to rise to the top and enable it to be skimmed off. Air flotation usually follows the oil–water separator and is used to remove remaining oil (free and emulsified) and solids (colloidal and suspended) by introducing air bubbles into the wastewater by mechanical means. The majority of the air emissions, including VOCs and dissolved gases, evolved by the wastewater treatment are the result of evaporation from the surfaces of wastewater in open drains, separators, and aerated basins. The factors influencing emissions from these systems are wastewater composition, equipment design, and climatic factors.

Cooling Towers

Refineries use large quantities of water for cooling throughout the refining process. The cooling water absorbs heat from process streams in noncontact heat exchangers. Before the water can be reused, it must be cooled. Cooling towers are used to transfer waste heat from the cooling water to the atmosphere by allowing the water to cascade through a series of decks and slat-type grids. Water that enters the tower may contain hydrocarbons from leaking heat exchangers. Atmospheric emissions from the cooling tower can consist of VOCs and gases stripped from the cooling water as the air and water come into contact. They also may contain cooling water treatment chemicals such as chromium and chlorine, although the use of chromium is largely being phased out.

Combustion Sources

There are a number of different types of combustion sources in a typical refinery. In general, these sources can be categorized on the basis of the combustion device and the fuel. Refinery combustion sources include process heaters, boilers, gas turbines, and engines. Refinery fuels are typically either residual oil, distillate oil, refinery gas, or natural gas. All of the criteria pollutants, as well as VOCs and metals, are emitted from combustion sources. The quantity of these emissions is a function of the fuel type, as well as of the combustor type and size.

Boilers and process heaters are used throughout refineries to produce steam and to raise the temperature of feed materials to meet reaction or distillation requirements, respectively. In addition to the typical refinery fuels, CO-rich regenerator flue gas also may be used as a fuel. Gas compressors often are driven by IC engines and usually are fueled by natural or refinery gas. Gas turbines can fire a variety of fuels and typically are used to produce electricity. They also are used as cogeneration units that produce electricity and steam for process needs.

Handling Operations

The refinery feedstock and product handling operations consist of unloading, storage, blending, and loading activities. The types of equipment used in these operations include tanks, unloading and transfer equipment, valves, flanges, pumps, and compressors. Valves, flanges, pumps, and compressors also are used throughout the refining operations.

All refineries have a feedstock and product storage area, termed a "tank farm," which provides surge storage capacity to assure smooth, uninterrupted refinery operations. Individual storage tank capacities range from less than 1000 to more than 500,000 barrels.[18] Storage tank, emissions and emission control technologies are discussed under "Exploration and Production."

Valves and Flanges

Valves and flanges are found throughout a refinery. Under the influence of heat, pressure, vibration, friction, and corrosion, valves and flanges may develop leaks. Valves are subject to product leakage from the valve stem. These leaks may be liquid, vapor, or both, depending on the product carried and on the temperature.

A single pipeline connected through a manifold to a number of feeder lines may be used to carry several different products. To avoid the possibility of product contamination, it is customary to blind off the unused feeder lines. Blinding involves loosening a flanged pipe connection, inserting a flat, solid metal plate between the flanges, and retightening the connection. The installation or removal of blinds results in the spillage of some product. Depending on product volatility, temperature, drainage, and flushing facilities, a certain amount of the spilled product evaporates.

Transfer Operations

Although most refinery feedstocks and products are transported by pipeline, some are transported by truck, railcars, and marine vessels. The emissions from transfer operations and applicable control technologies are discussed under "Marketing" and "Transportation."

Pumps and Compressors

Liquid and gases in the refinery are moved about via pumps and compressors. These pieces of equipment can leak product through their seals, the point of contact between the moving shaft and the stationary casing.

Sampling

The operation of process units is periodically checked throughout the refinery by routine analyses of feedstocks and products. To obtain representative samples for these analyses, sampling lines must be purged, resulting in possible hydrocarbon vapor emissions.

Pressure-Relief Valves

Pressure-relief valves are used to prevent excessive pressures from developing in process vessels and lines. Leaks may develop because of the corrosive nature of the product or the valve may reseal improperly after blowoff. As routine maintenance and observation are often difficult because of a lack of accessibility and the operational difficulties involved with repair, relief valve installations can leak substantially before repair is completed.

The blow-down system allows for the safe disposal of liquid and gaseous hydrocarbon discharged from pressure-relief devices. Refinery equipment subject to planned and unplanned hydrocarbon discharges is manifolded into a collection unit. The blow-down is separated into liquid and vapor, with the liquid being recycled into the refinery and the vapor either recycled or flared. Uncontrolled blow-down emissions primarily consist of hydrocarbons, but also can include any of the other criteria pollutants.

Turnarounds and Tank Cleaning

The periodic maintenance and repair of process equipment and storage tanks are essential to refinery operations. To provide for safe vessel entry, all hydrocarbon liquid and vapor must be removed. Once the liquid has been cooled and pumped to storage facilities, the vessel is depressurized and/or purged with steam, nitrogen, or water to remove any remaining hydrocarbon vapors and odorous material.

Air Emissions Characterization

This section will quantify, where possible, refinery air emissions. The following are 1987 estimates of criteria pollutant emissions based on the 1970 level of control for the entire refining industry.[22] The use of 1970 emission control assumptions undoubtedly results in an overestimation of actual emission rates by perhaps an order of magnitude or more for some pollutants, especially VOCs. More up-to-date emission rate data are not available because even recent estimates rely largely on outdated emission factors to estimate releases.

Pollutant	Thousands of Tons per Year
PM	771
SO_2	1322
NO_x	220
VOC	991
CO	2533

Potential hazardous air pollutant emission sources from refining operations are presented in Table 6. Table 7 summarizes estimated benzene, toluene, xylene, butadiene, and trimethylbenzene emissions from refineries, as reported under SARA Title III community right-to-know requirements.

Air Pollution Control Measures

Control of refinery air emissions can be accomplished by process change, equipment change, procedure change, installation of control equipment, improved housekeeping, increased monitoring, and better equipment maintenance. Specific control devices are addressed in the following for the refining sector. Some combination of these often proves to be the most effective solution. Several of these controls also result in some form of savings.

Vapor Control Systems

Vapor control systems involve the collection of process or fugitive vapor emissions and their recycling or recovery for hydrocarbon content or fuel value or their destruction. In some instances, where economically feasible, the emission stream itself can be recycled to refinery fuel gas production or the fuel value of the emission stream can be recovered via combustion. Hydrocarbons in vapors can be destroyed via flares or incinerators. The specific configuration of vapor control systems depends on the specific configuration and economics of the refinery.

Flares

Flares are commonly used for the disposal of waste gases during process upsets (e.g., start-up, shutdown) and emergencies. They are basically safety devices that also are

TABLE 6. Potential Hazardous Air Pollutants (HAPs) for Petroleum Refining

	Potential HAPs			
	Organic		Inorganic	
Process	Vapor	Particulate	Vapor	Particulate
Refining operations[a]				
Crude separation	a,b,d,e,f,g,h,i,j,k,l,m,o,A,B,C,D,E,F,J	o	c,m,t,u,v,x,y,L	p,I,Q,R
Light hydrocarbon processing	g,h,i,n,N,O,P	R	t,v	G,H,Q
Middle and heavy distillate processing	a,d,e,f,g,h,i,j,k,l,F,J,K,O,P,S,T	o,R	m,t,u,v,x,y,L	p,q,G,H,I,Q,U
Residual hydrocarbon processing	a,d,e,f,g,h,i,j,k,l,n,F,J,M,N,P,S,T	o,R	m,s,t,u,v,x,y,L	p,q,G,H,I,Q,U
Auxiliary processes	a,b,d,e,f,g,h,i,j,k,l,n,A,B,C,D,J,K,M,T	o,R	c,m,s,u,y,L	p,q,r,z,I
Combustion processes[b]				
Oil combustion	14	19	13,17,27	1,2,5,6,8,9, 10,11,15,16,18, 20,22,24
Natural gas combustion	14	19		
Gasoline combustion	12,14	12,19	17	15
Diesel combustion	12	12,19		6,18

[a]Pollutant key: a—maleic anhydride, b—benzoic acid, c—chlorides, d—ketones, e—aldehydes, f—heterocyclic compounds (e.g., pyridines) g—benzene, h—toluene, i—xylene, j—phenols, k—organic compounds containing sulfur (sulfonates, sulfones), l—cresol, m—inorganic sulfides, n—mercaptans, o—polynuclear compounds (benzo pyrene, anthracene, etc.), p—vanadium, q—nickel, r—lead, s—sulfuric acid, t—hydrogen sulfide, u—ammonia, v—carbon disulfide, x—carbonyl sulfide, y—cyanides, z—chromates, A—acetic acid, B—formic acid, C—methylethylamine, D—diethylamine, E—thiosulfide, F—methyl mercaptan, G—cobalt, H—molybdenum, I—zinc, J—cresylic acid, K—xylenols, L—thiophenes, M—thiophenol, N—nickel carbonyl, O—tetraethyl lead, P—cobalt carbonyl, Q—catalyst fines, R—coke fines, S—formaldehyde, T—aromatic amines, U—copper.

[b]Pollutant key: 1—arsenic, 2—antimony, 5—barium, 6—beryllium, 8—cadmium, 9—chromium, 10—cobalt, 11—copper, 12—dioxin, 13—fluoride, 14—formaldehyde, 15—lead, 16—manganese, 17—mercury, 18—nickel, 19—polycyclic organic matter (POM), 20—vanadium, 22—radionuclides, 24—zinc, 27—chlorine.

Source: Reference 20.

TABLE 7. Estimated Emissions from Petroleum Refinery

Chemical	Point, tons/year	Fugitive, tons/year	Total, tons/year
Benzene	114	29	142
Toluene	437	111	548
Xylene (total)	31	1751	1782
Butadiene	3	1	4
Trimethyl-benzene (1,2,4)	310	141	452

Source: Reference 32.

used to destroy organic constituents in waste emission streams. Flares can be used for controlling almost any nonhalogenated VOC emission stream. They are designed and operated to handle large fluctuations in flow rate and VOC content. There are several different types of flares, but the prevalent refinery flare, illustrated in Figure 10, is an elevated steam-assisted flare. Flaring generally is considered a control option when the heating value of the emission stream cannot be recovered because of uncertain or intermittent flow, as in process upsets or emergencies. If the waste gas to be flared does not have sufficient heating value to sustain combustion, auxiliary fuel may be required. According to studies conducted by the EPA, 98% destruction efficiency can be achieved by steam-assisted flares when controlling emission streams with heat contents greater than 300 Btu/scf.

Incinerators

There are two types of incinerators—thermal and catalytic. Thermal incinerators are used to control a wide variety of continuous VOC emission streams. Thermal incineration is preferable to flaring when halogenated or sulfur-bearing compounds are present owing to the corrosive properties of these compounds. Destruction efficiencies up to 98–99+% are achievable with thermal incineration. Although they accommodate minor fluctuations in flow, thermal incinerators are not well suited to streams with highly variable flow because of reduced residence time and poor mixing during increased flow conditions. Thermal incineration is typically applied to emission streams that are dilute mixtures of VOCs and air. In such cases, due to safety con-

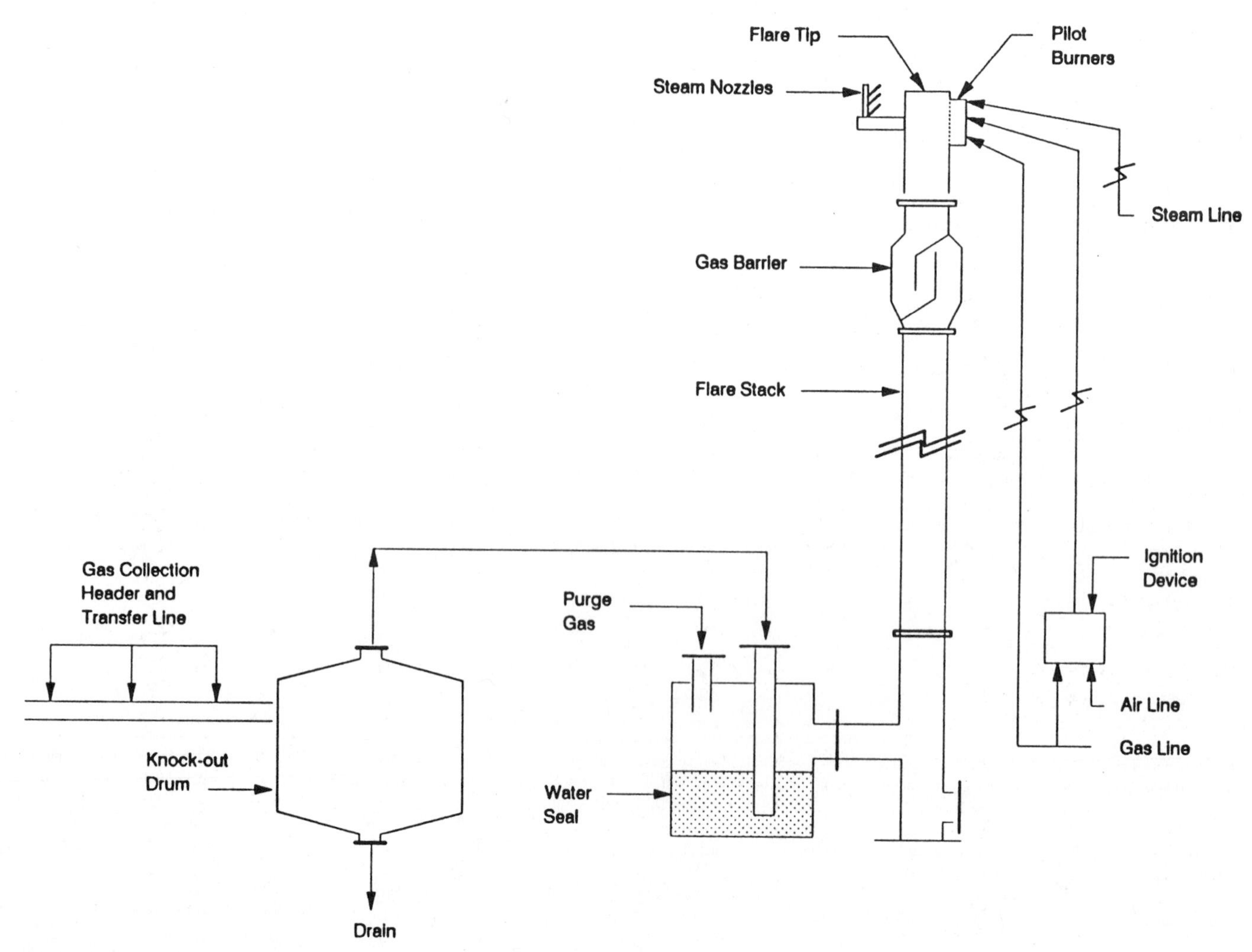

FIGURE 10. Steam-Assisted Elevated Flare System

siderations, the concentration of the VOCs is limited in some instances to 25% of the lower explosive limit (LEL). Thus if the VOC concentration is high, dilution may be required. When emission streams controlled by thermal incineration are dilute (i.e., low heat content), supplementary fuel is required to maintain the desired combustion temperature. Fuel requirements may be reduced by recovering the energy contained in the hot flue gases from the incinerator.

Catalytic incinerators are similar to thermal incinerators in design and operation, but they employ a catalyst to increase the reaction rate. Since the catalyst allows the reaction to take place at lower temperatures, significant fuel savings are possible. Destruction efficiencies of up to 95% are generally achieved with catalytic incineration. Catalytic incineration is not as broadly applicable as thermal incineration because of catalyst sensitivity to pollutants and process conditions.

Refinery Fuel Gas

Existing boilers or process heaters can be used to control emission streams by recovering the fuel value while destroying the VOCs. This is accomplished by diverting the streams to the refinery fuel gas system or directly into the firebox. When used as emission control devices, boilers or process heaters can provide destruction efficiencies of greater than 98% at a small capital cost. Operating costs are reduced as the recovery of the emission stream heat content reduces fuel requirements. Typically, emission streams are controlled in boilers or process heaters and used as supplemental fuel only if they have sufficient heating value (greater than 150 Btu/scf). In some instances, emission streams with high heat content may be the main fuel to the process heater or boiler. Note that emission streams with low heat content can also be burned in boilers or process heaters when the flow rate of the emission stream is small compared with that of the fuel–air mixture. There are some limitations

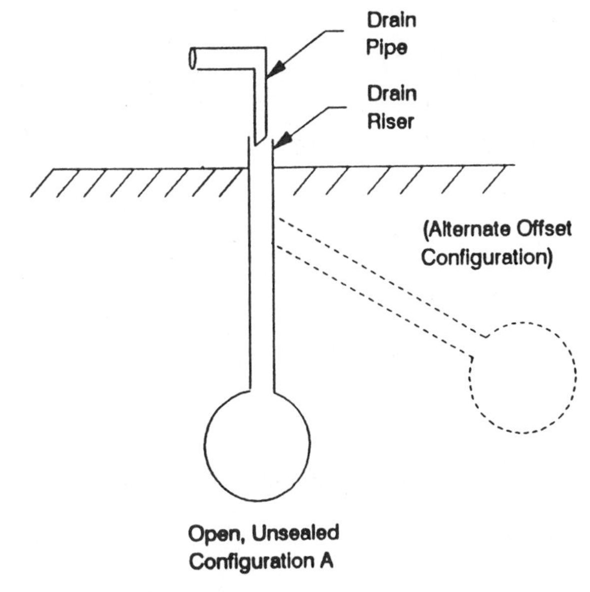

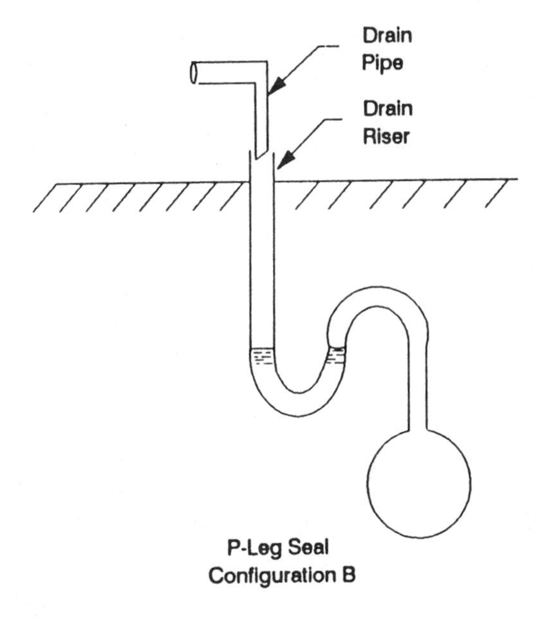

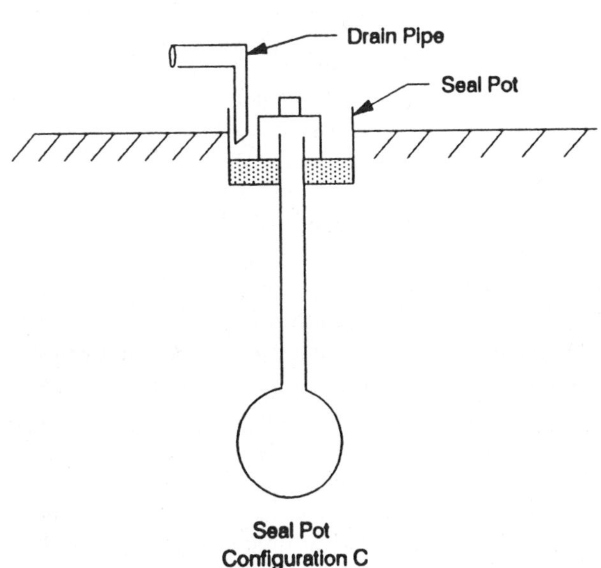

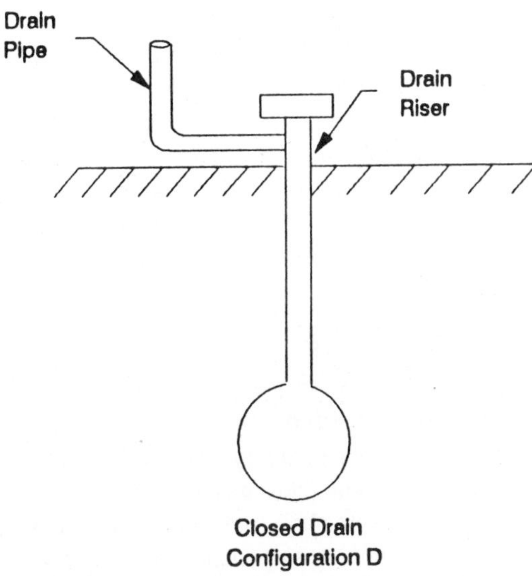

FIGURE 11. Types of Individual Refinery Drains for Oily Wastewater

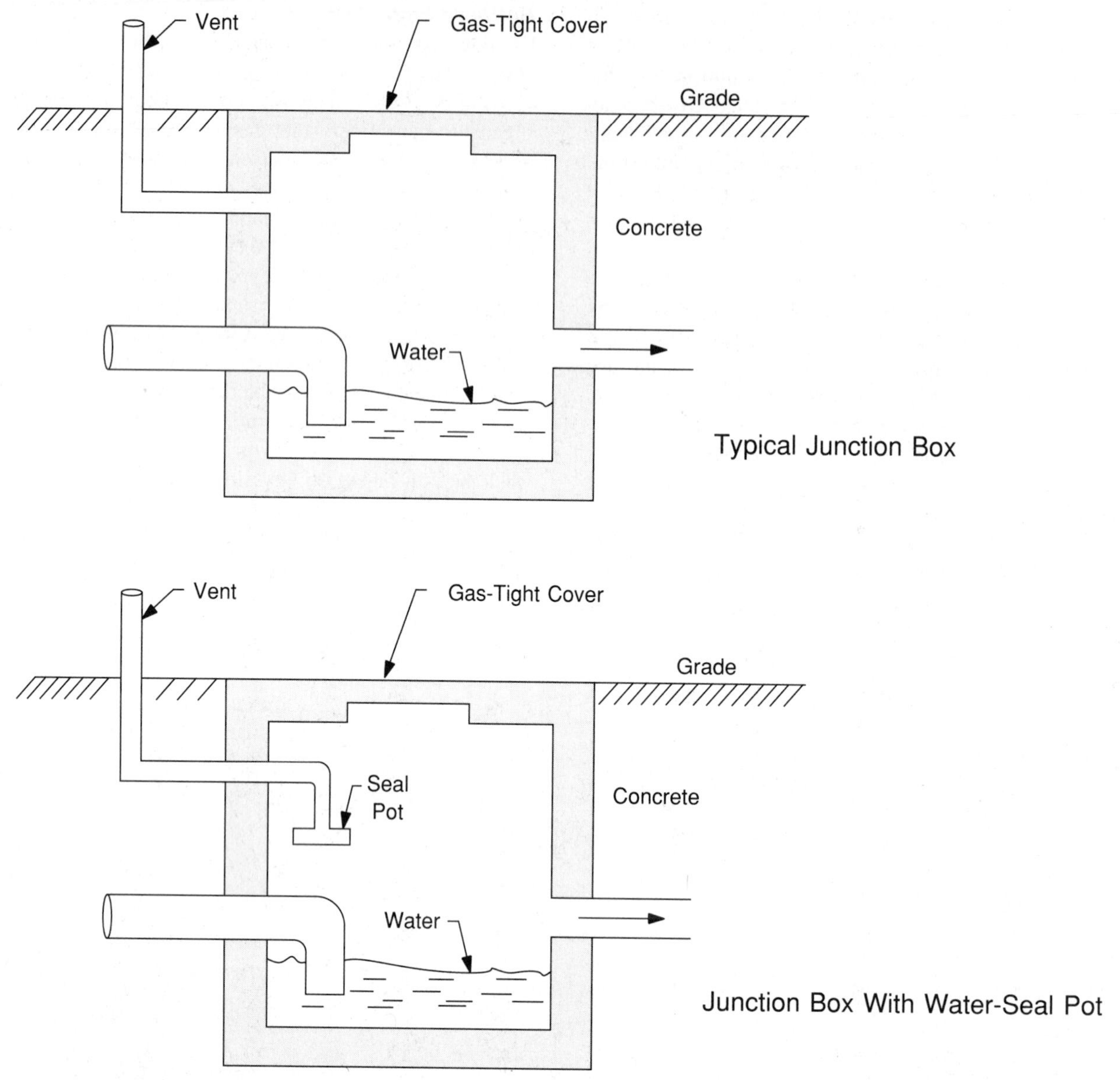

FIGURE 12. Refinery Drain System Junction Boxes

on the application of boilers or process heaters as emission control devices. Because these combustion devices are essential to the operation of the refinery, only those emission streams that will not reduce their performance or reliability can be controlled. Streams not suitable for control include those with varying flow rate and/or heating value, high-volume/low-heating-value streams, and streams with corrosive compounds.

Hydrocarbon Reuse

The ultimate vapor-recovery process recycles the emission stream to the light ends recovery unit (gas processing), where it is recovered as a salable product, usable feedstock, or gasoline blending component. This requires compressing and separating, by distillation, the stream into its respective components.

Covers

Covering a liquid surface or the vapor space above a liquid surface suppresses VOC emissions to the atmosphere. Examples of control by covering are floating roof tanks (external and internal) and wastewater systems. Tank applications are discussed in detail under "Exploration and Production." An equipment-based New Source Performance Standard was promulgated in November 1988 for refinery wastewater systems. The control of wastewater treatment systems requires covering sources where emission generation is greatest, namely, process drains, junction boxes, and oil–water separators. By suppressing emissions through the separator, more hydrocarbon can be recovered as a liquid and recycled back to the refinery. Covering a drain (Figure 11) involves either a physical cover at ground level or a liquid seal in the drain pipe. Emission reductions of 40–50%[23] are achievable by drain seals. Junction boxes (Figure 12) require venting to prevent siphoning and vapor locks.

Vents on junction boxes should be at least 4 inches in diameter and 3 feet in length. To minimize VOC emissions, the vent pipe must also have a water seal. Control efficiencies can be assumed to be equal to those of drain seals.

Any VOC emissions from oil–water separators are controlled by installing a fixed or floating roof. A vapor space under a fixed roof may constitute an explosion or fire hazard. In order to eliminate this problem, the vapor space can be blanketed with either plant gas or an inert gas such as nitrogen. Floating roofs eliminate much of the vapor space, thus greatly reducing the potential for volatilization from the oil layer. Emission reduction estimates based on qualitative information range from 90% to 98%.[23] Data developed by Litchfield indicate that 85% is representative of the emission reduction achievable by a simple fixed or floating roof.[23] The most obvious factor affecting performance is the degree of maintenance.

In addition to covering components in the wastewater system, the system can be completely closed and vented to a combustion device, such as a flare. Emission reductions depend on the efficiency of the control device; for instance, flares would result in 98% reduction.

Inspection and Maintenance

Improved maintenance—including scheduled inspection and monitoring, improved housekeeping, and improved employee training—is a very practical method for reducing hydrocarbon emissions and alleviating odor problems. Moreover, it is often the only control method for some sources, such as valves, relief valves, seals, cooling towers (heat exchanger leaks), and sampling operations. For nearly all sources, especially process drains, wastewater separators, treating units, blind changing, and loading operations, employee awareness of the problem will reduce emissions.

In-Line Sampler

In-line sampling allows the collection of a representative process sample without having to flush any hydrocarbon to the process drain. Typically, small piping/tubing is run parallel to the process piping. A three-way valve is used for sample collection, thus eliminating any "dead legs."

Rupture Disks

Rupture disks as a control device are used to protect relief valves from the corrosive process environment. They are typically thin metal disks located on the pressure side of the relief valve. They are designed to burst at the relief valve setting. Owing to their "one-time" use, rupture disks are applicable for relief valves that are expected to be vented in emergencies only. They also can be used in place of relief valves in certain applications.

New Valve Technology

New valve technology has been developed that can significantly reduce pollutant emissions from the valve stem packing. It includes new types of packing material for valve stems and new seal technologies. New valve designs have been developed for some applications to prevent leaks by eliminating packing material leak sources around valve stems or by encasing potential leak sources in a "bellows" structure to trap emissions. The new "leakless" valve designs do have limitations of applicability. Some designs cannot withstand the high temperatures and pressures of many refinery process streams. Other designs are limited to valve sizes of 6 to 8 inches in diameter and, therefore, are applicable only to smaller line sizes in refineries. For refineries, the most widely used control technique to minimize leakage from valves is an effective schedule of inspection and preventative maintenance.

Mechanical Seals

Mechanical seals are used to control product emissions from centrifugal pump and compressor glands. A simple mechanical seal consists of two rings with wearing surfaces at right angles to the shaft. One ring is stationary, while the other is attached to the shaft and rotates with it. A spring and the action of the fluid pressure keep the two faces in contact. Lubrication of the wearing faces is effected by a thin film of the material being pumped. The wearing faces are precisely finished to ensure perfectly flat surfaces.

A pressure seal can further reduce emissions. A liquid that is less volatile or dangerous than the product being pumped is introduced between a dual set of seals. Since this liquid is maintained at a higher pressure than the product, some of it passes by the seal and into the product. The pressure differential prevents the product from leaking outward and the sealing liquid provides lubrication. As some of the sealing liquid passes the outer seal (hence the need for low volatility), a means should be provided for its disposal.

Quick Change Blind/Manifold Design

Several devices have been developed to reduce spillage, such as Hamer and Greenwood blinds. These "line" blinds do not require a complete break of the flange connection as "slip" blinds do, but rely on a gear mechanism to release the plate. Combinations of these devices in conjunction with gate valves allow changing of the line blind while the line is under pressure from either direction.

In addition, during design stages, the manifold can be designed to minimize the need for blinding, as well as the quantity of liquid that will be spilled. With highly volatile products, water or nitrogen can be used to displace the product, precluding any product spillage during the blind change.

Wetting

Wetting down the coke is used as a PM control in the decoking process.

New Gaskets, Bolts, and Welding

Replacement of leaking gaskets on existing pipe and valve flanges and minimizing the number of flanges with the use of welded pipe are methods of reducing VOC emissions from flanges.

Steam Stripping

Steam stripping is an effective control for removing VOCs from process streams. Two examples within a refinery include FCC catalyst stripping and sour-water treating. Spent FCC catalyst is steam stripped to remove absorbed hydrocarbons at the reactor exit. Process wastewater is stripped with steam to remove H_2S, ammonia, and light gases. The process is generally conducted in the sour-water stripper. The sour water and steam are fed into the column. The stripped water and condensed steam are routed to the wastewater system and the undesirable constituents are incinerated.

Carbon Monoxide Boiler

Depending on the refinery, the catalytic cracker may or may not have a CO boiler. Carbon monoxide boilers are used to recover the energy contained in the catalyst regeneration off-gases. The recovered energy is used as process heat for various refinery processes. The fuel used in the CO boiler consists of the process gas from the catalyst regenerator and an auxiliary fuel source. The process gas may contain up to 5–10% CO.[24] Combusting the gas in the boiler produces heat and reduces emissions of CO and VOCs to negligible levels.

Cyclone

Cyclone separators are used for catalyst dust collection in the upper section of both FCC unit reactors and regenerators. The cyclones are employed as a single unit or in multiple two- or three-stage series units. In general, high-efficiency cyclones have dust collection efficiencies of over 90% for particle sizes of more than 15 μm.[25] The efficiency drops off rapidly for particles of less than 10 μm.

Electrostatic Precipitators

Electrostatic precipitator (ESP) particle removal is accomplished by charging the particles, collecting the particles, and transporting the collected particles into a hopper. An ESP is very sensitive to the aerosol density and the electrical resistivity of the material, but is less sensitive to particle size. The electrical resistivity of the particles influences the drift velocity or the attraction between the particles and the collecting plate. A high resistivity will cause a low drift velocity, which will decrease the overall collection efficiency. Used to control PM emissions on FCC units, ESPs achieve efficiencies as high as 80–85%.[25]

Scrubbers

Scrubbers are used in refineries to separate and purify gaseous streams containing high concentrations of VOCs, SO_2, and PM. Examples of scrubbing within a refinery include applications on the asphalt blowing airstream and on the FCC regenerator off-gas. A common system on FCC regenerators is a caustic scrubber, where particulate removal and sulfur oxides absorption take place in a venturi scrubber. Particulate is removed by inertial impaction of the scrubbing liquid with the entrained particles. The sulfur oxides absorption reactions that take place are as follows:

$$2NaOH + SO_2 \rightarrow Na_2SO_3 + H_2O$$

Sulfur dioxide and PM removal efficiencies of 93–98% have been recorded.

Combustion Controls

All the criteria pollutants and several potential hazardous air pollutants (HAPs) are emitted from combustion sources within refineries. Sulfur oxides can be controlled by fuel selection, fuel desulfurization, or flue gas treatment. Carbon monoxide, hydrocarbons, and PM can be minimized by more efficient combustion. Nitrous oxides can be controlled by combustion modification or by flue gas treatment. Controls specific to each refinery combustion source are described below.

Internal-Combustion Engine

The VOC, CO, and NO_x emissions can be controlled by a three-way catalyst (similar to a catalytic converter in a motor vehicle) on a rich-burn engine (fuel-rich combustion). The NO_x emissions can be controlled by selective catalytic reduction (SCR) on a lean-burn engine. Catalyst systems generally are designed for 80% reduction efficiencies. Precombustion chambers, which control the air-to-fuel ratio in stages, are another NO_x-reduction technique for gas-fired engines. Levels of NO_x as low as 1.5 g/bhp-h have been reported. Injection timing retard also reduces NO_x production. The magnitude of the reduction may vary considerably among engine types. In general, reductions of 20–34%, with corresponding 1–4% fuel consumption penalty and slight increases of VOCs and CO, are obtained.

Combustion Turbines

Wet injection using either water or steam is the most prevalent NO_x-reduction technique for turbines. Depending on turbine type and size, NO_x emission levels of 25–50 ppm at 15% oxygen are obtainable. Selective catalytic reduction is also applicable to turbines for NO_x reductions of generally 80%. Coupling SCR with wet injection can result in NO_x emission levels as low as 5 ppm.

Heaters and Boilers

In addition to three-way catalysts and SCR, selective noncatalytic reduction processes can be applied to process heater and boiler flue gas for NO_x control. The process using ammonia can achieve 40–60% reduction and that using urea can achieve 20–80% reduction, depending on temperature. Low-NO_x burners, which stage the mixing of air and fuel to reduce flame temperature, result in NO_x reductions of 30–40% over conventional burners.

MARKETING

Introduction

The marketing of petroleum products at wholesale and retail facilities involves an extensive distribution network. Of the more than 2500 products produced by the petroleum industry, the major products distributed in bulk include motor gasoline and distillates such as diesel fuel, jet fuel, and heating oil.

Gasoline and other large-volume transportation fuels are distributed primarily by pipeline, ship, and barge to bulk terminals. From bulk terminals, tank trucks generally bring the product to a bulk plant, service station, or commercial customer. Specialty products, such as solvents, thinners, and lubricating oils, as well as some fuels, may be transported by railroad tank cars directly from the refinery. In addition, many products, such as lube oils and solvents, may be put in drums and shipped by truck to warehouses, dealers, or customers.

Process Description

Gasoline and distillate products are made at refineries and sent by pipeline, ship, and barge to bulk storage terminals. The products are stored in large tanks before being dispensed into tank trucks for delivery to customers. Facilities handling more than 20,000 gallons of gasoline per day are considered bulk terminals, but the largest facilities can throughput more than 1 million gallons of petroleum products per day. There are an estimated 637 bulk terminals in the United States.[26]

Some products loaded at bulk terminals are delivered by truck to bulk plants, which are small storage facilities located close to their customers. Bulk plants have limited storage capacity and are equipped with a small loading rack for dispensing products into local delivery trucks. They deliver fuel to such customers as service stations, commercial accounts, and farmers. There are an estimated 9391 bulk plants[26] nationwide, with a typical daily throughput of 5000 gallons per day.

Tank trucks deliver motor fuels to service stations. The fuels are stored in underground tanks and then pumped directly into motor vehicle fuel tanks using a metering dispenser equipped with a flexible hose and nozzle. There are an estimated 160,000 service stations[27] in the United States.

The general distribution network in the petroleum marketing sector is outlined in Figure 13. During operations at marketing facilities, emissions of VOCs can occur as a result of product transfer (e.g., loading), during product storage, and from pump, valve, and flange seals and product spillage. Combustion sources in the marketing segment include vehicles transporting products that emit NO_x, SO_2, CO, and particulates.

Air Emission Characterization

In this section, air emissions of VOCs and criteria pollutants are discussed separately. Little data are available to speciate VOC emissions into components of toxic air pollutants that may be emitted.

Volatile Organic Compounds

Emission of VOCs during marketing operations can be broken down into three major categories: storage tanks, truck loading, and service stations.

Storage Tanks

Storage tank design, emissions, and emission controls are discussed in detail in the E&P section. The greatest potential for VOC emissions comes from motor gasoline storage. Diesel and heating fuels have relatively low vapor pressures and are not as significant sources of releases from storage tanks.

Truck Loading Operations

Vapors containing a mixture of VOCs and air are emitted when tank trucks are loaded with motor gasoline or if a truck contains gasoline vapors from a previous load. As product is loaded, it displaces the vapors in the tank truck, and these vapors are vented to the atmosphere or to a vapor processing system.

Gasoline tank trucks are normally divided into compartments, with a loading dome hatchway at the top of each compartment. The method of loading affects the amount of emissions generated. Loading can be by top splash or submerged fill through the dome using a fill tube or by bottom filling. Vapor collection equipment for tank trucks includes piping to capture vapors from the tank being filled back into the tank truck. The vapors are then discharged into the vapor collection systems at bulk terminals when the tank truck returns for product loading.

Service Station Operations

There are emissions from three sources at service stations: during the filling of underground tanks, underground tank breathing and emptying losses, and vehicle refueling losses. Splash filling losses and uncontrolled displacement losses during vehicle refueling are the largest emission sources at service station operations.

Tank trucks load gasoline into underground storage tanks at service stations. This is called Stage I refueling. Typically, splash or submerged filling methods are used. Vapor balance is utilized to control emissions by returning displaced vapors to the tank truck rather than allowing them to be emitted to the atmosphere.

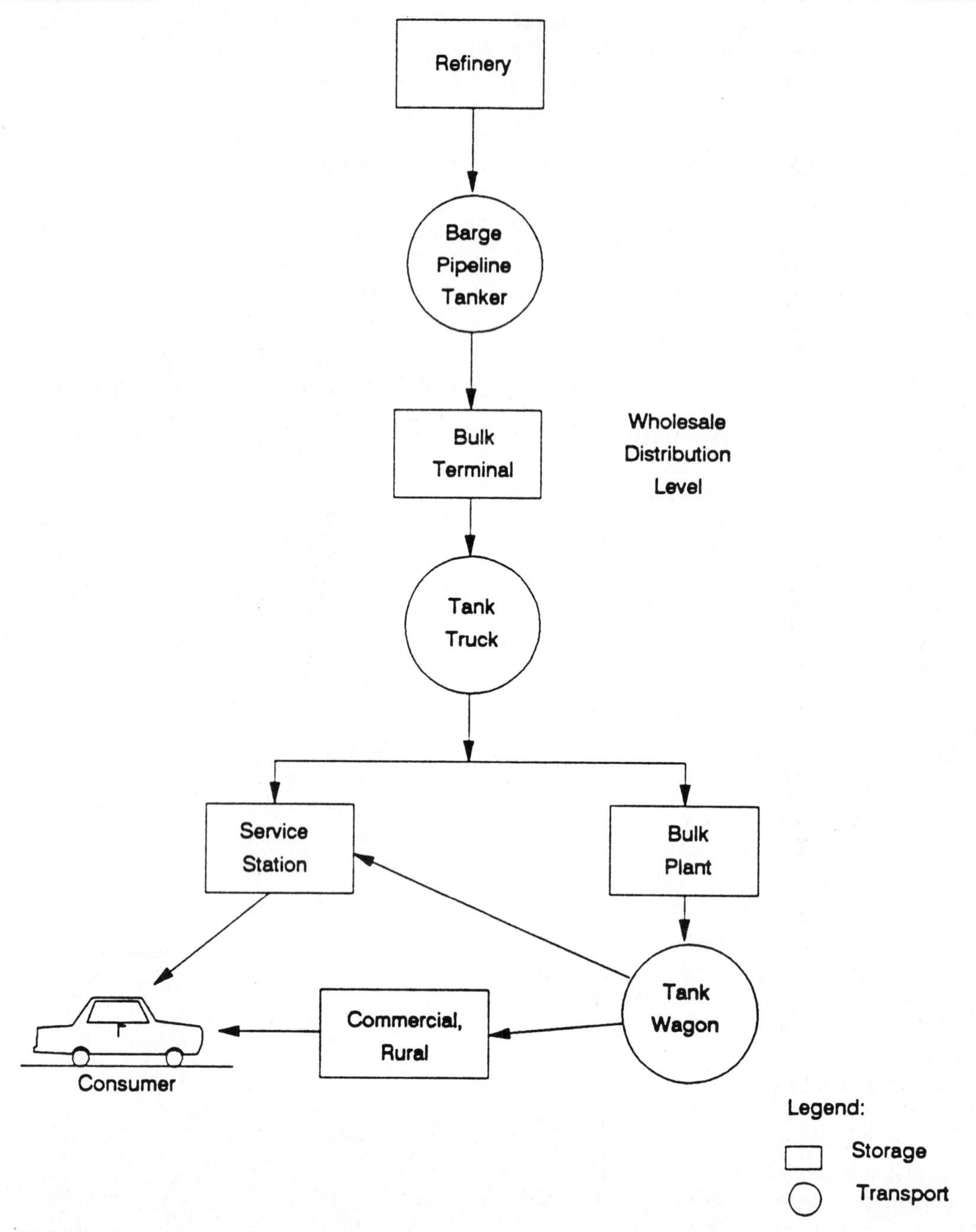

FIGURE 13. Gasoline Marketing in the United States

A second source of VOC emissions from gasoline service stations is breathing losses from underground tanks. Breathing losses vary by the frequency with which gasoline is withdrawn from the tank and also with barometric pressure. Stage II refueling is the process of loading fuel into vehicles. Vehicle refueling produces evaporative emissions (i.e., venting of vapors) when vapors are displaced from the automobile tank by dispensed liquid gasoline and from spillage.

Air Toxics Emissions

The EPA has identified potential hazardous air pollutants that it is reviewing for possible emissions regulations. The agency compiled emissions profiles of these chemicals from various emissions sources by speciating the VOC emissions from these sources. A synopsis of the distribution of these emissions is presented in Table 8. Currently, little information is available on potential hazardous air pollutants from the marketing sector.

Criteria Pollutant Emissions

Criteria pollutant emissions (NO_x, SO_x, and CO) are from combustion sources related to motor vehicle transport in the marketing sector. Motor vehicles transporting product to bulk plants, bulk terminals, or service stations are sources of emissions from IC engines.

TABLE 8. Marketing Sector Emission Sources and Emission Controls

Emission Source	Emission Control
Storage tanks (bulk plants/ terminals)	Fixed roof with conservation vent
	Internal floating roof
	External floating roof
Loading racks (bulk plants/ terminals)	Submerged loading
	Refrigeration
	Adsorption and/or compression
	Thermal oxidation
	Vapor balance system
Service station tank truck unloading	Vapor balance system/vapor recovery
Service station refueling	Onboard controls
	Stage II vapor-recovery controls
	Vapor balance
	Hybrid
	Vacuum assist

Source: References 18, 20, 28–30.

Air Pollution Control Measures

Table 8 presents marketing segment emissions sources and their control options. Several controls are available and in use to control VOC emissions from marketing segment sources.

Storage Tank Controls

Covering a liquid surface or the vapor space above a liquid surface suppresses VOC emissions to the atmosphere. Both internal and external floating roofs provide emission control for storage tanks. Floating roofs rise and fall with the liquid level, preventing formation of a large vapor space. Fixed roof tanks use pressure-vacuum vents to control breathing losses, and some tanks are equipped with vapor-recovery systems to control filling and emptying losses. Tank emission controls were discussed in more detail in the E&P section.

Loading Controls

Submerged Loading

In splash loading, the gasoline is introduced into a tank at the top and above the liquid level in the tank. Submerged fill methods eliminate splash-induced vaporization and reduce vapors generated during the loading of gasoline into tank trucks and storage tanks. During submerged fill, the gasoline is introduced into the tank being filled, with the transfer line outlet situated below the liquid surface. Submerged filling minimizes droplet entrainment, evaporation, and turbulence, and can significantly reduce vapor loss over splash filling.

Vapor Balance Systems

During filling operations, vapor balance systems can be used to transfer vapor displaced by liquid in the tank being filled into the tank being emptied. Vapor balance prevents the compression and expansion of vapor spaces that would otherwise occur during filling. Vapor balance systems can reduce storage tank filling losses by more than 95% and tank truck loading losses by more than 90%.[28]

Refrigeration–Absorption–Adsorption

Refrigeration, absorption, and adsorption are means of recovering vapors as product. Refrigeration recovers hydrocarbon vapors by condensing vapors at cryogenic temperatures and atmospheric pressure. It can achieve reductions of 85–95%.[28] Absorption vapor-recovery units recover hydrocarbon from gasoline vapors by absorption into a lean oil stream. Carbon adsorption units can be used to recover hydrocarbon vapors by passing the gasoline loading vapors through a carbon bed, where the hydrocarbon constituents are adsorbed onto the carbon.

Service Station Operations

Gasoline transfer operations at service stations are sources of hydrocarbon emissions. The transfer takes place in two stages, referred to as Stage I and Stage II. Stage I refers to loading gasoline into underground storage tanks at the station, while Stage II refers to vehicle refueling.

To control emissions, Stage I vapor balance is usually used between the tank truck and the underground storage tank at the service station. Vapor balance systems were described above. Stage II controls can be vapor balanced between the service station underground tank and the vehicle gasoline tank. Vacuum-assist systems differ from vapor balance systems in that they use a vacuum pump to provide extra negative pressure at the nozzle–fill-neck interface to increase vapor recovery. They require some venting of excess vapors to the atmosphere. Hybrid systems combine vapor balance and vacuum assist. They enhance vapor recovery at the nozzle–fill neck by vacuum, while keeping vacuum low enough to minimize vapor returned to the vapor-recovery system. The other control technique for vehicle refueling is the use of onboard controls. Here, a vapor seal in the vehicle fill neck forces vapors being displaced from the tank into a carbon cannister on the vehicle, where they are adsorbed.[29–31]

References

1. *An Analysis of Petroleum Industry Costs Associated with Air Toxic Amendments to the Clean Air Act,* Interim Final Report, Delta Management Group, 1989, p 13.
2. *SO_2 Emissions in Natural Gas Production Industry—Background Information for Proposed Standards,* EPA-450/3-82-023a, U.S. Environmental Protection Agency, Research Triangle Park, NC, 1983, p 9-3.
3. *Industrial Process Profiles for Environmental Use: Chapter 2. Oil and Gas Production Industry,* EPA-600/2-77-023b, U.S. Environmental Protection Agency, Cincinnati, OH, 1977, p 1.
4. *Industrial Process Profiles for Environmental Use,* p 16.
5. *Industrial Process Profiles for Environmental Use,* p 14.
6. *Industrial Process Profiles for Environmental Use,* p 34.

7. *SO_2 Emissions in Natural Gas Production Industry,* p 3-3.
8. *NO_x Emissions from Petroleum Industry Operations,* API Publication 4311, American Petroleum Institute, Washington, DC, 1979, p 13.
9. *Industrial Process Profiles for Environmental Use,* p 43.
10. *Introduction to Oil and Gas Production,* Book 1 of the Vocational Training Series, American Petroleum Institute, Dallas, 1983, p 51.
11. *Introduction to Oil and Gas Production,* p 52.
12. *Industrial Process Profiles for Environmental Use,* pp 53, 55.
13. *SO_2 Emissions in Natural Gas Production Industry,* p 3-9.
14. *NO_x Emissions from Petroleum Industry Operations,* pp 57–58.
15. *NO_x Emissions from Petroleum Industry Operations,* p 66.
16. *Evaluation of Control Technologies for Hazardous Air Pollutants,* Final Report, Radian Corp., Research Triangle Park, NC, 1985, p 3-9.
17. *Evaluation of Control Technologies for Hazardous Air Pollutants,* p 3-14.
18. *Compilation of Air Pollutant Emission Factors* (September 1985), AP-42, 4th ed., and *AP-42, Supplement B to Compilation of Air Pollutant Emission Factors* (September 1988).
19. *Volatile Organic Compound (VOC) Emissions from Volatile Organic Liquid Storage Tanks—Background Information for Proposed Standards,* EPA-450/3-81-003a, U.S. Environmental Protection Agency, Research Triangle Park, NC, 1984, pp 4-17–4-20.
20. *Background Information on Hydrocarbon Emissions from Marine Terminal Operations,* EPA-450/3-76-038a and 038-b, U.S. Environmental Protection Agency, Research Triangle Park, NC, 1976.
21. L.A. Thrash, *Oil Gas J.,* 88(13): 78 (1990).
22. *National Air Pollutant Emission Estimates 1940–1987,* EPA-450/4-88-022, U.S. Environmental Protection Agency, Office of Air Quality Planning and Standards, Research Triangle Park, NC, March 1989.
23. *VOC Emissions from Petroleum Refinery Wastewater Systems—Background Information for Proposed Standards,* EPA-450/3-85-001a, U.S. Environmental Protection Agency, Research Triangle Park, NC, 1985.
24. *Emission Estimation Techniques for Petroleum Refineries and Bulk Terminals,* Radian Corp., Research Triangle Park, NC, 1989.
25. *Air Pollution Engineering Manual,* 2nd ed., U.S. Environmental Protection Agency, Washington, DC, 1973.
26. *Census of Wholesale Trade Geographic Area Series U.S., 1987,* WC87-A-52, Bureau of Census, U.S.D.O.C., 1989, US-14.
27. *Gasoline Marketing in the 1980's: Structure, Practice and Public Policy,* Temple, Barker, and Sloane, Inc., 1988.
28. *Evaluation of Air Pollution Regulatory Strategies for Gasoline Marketing Industry,* EPA-450/3-84-012a, U.S. Environmental Protection Agency, Washington, D.C., 1984.
29. *Fugitive Emission Sources of Organic Compounds—Additional Information on Emissions, Emission Reductions, and Costs,* EPA-405/3-82-010, U.S. Environmental Protection Agency, Research Triangle Park, NC, 1982.
30. *Draft Regulatory Impact Analysis: Proposed Refueling Emission Regulations for Gasoline-Fueled Motor Vehicles. Vol. I—Analysis of Gasoline Marketing Regulatory Strategies,* EPA-450/3-87-001a, U.S. Environmental Protection Agency, Washington, DC, 1987.
31. *Petroleum Supply Annual 1988: Vol. 1,* DOE/EIA-0340(88)1, U.S. Department of Energy, Washington, DC, 1988, p 35.
32. *The Toxics-Release Inventory: A National Perspective, 1987,* U.S. Environmental Protection Agency, Office of Toxic Substances, Economics and Technology Division, Washington, DC, June 1989.

18
Wood Processing Industry

Chemical Wood Pulping
Mechanical Pulping
Paper and Paperboard Manufacture
Arun V. Someshwar and John E. Pinkerton

The processing of wood to yield such varied products as paper, paperboard, lumber, plywood, and reconstituted building board ranks among the ten largest industrial activities in the United States. To produce paper or paperboard, the wood is first pulped. Pulps are made from wood chips, whole tree chips, sawmill residues, or logs. Pulps can be produced through chemical or mechanical means or by a combination of both. The pulp may then be bleached to various degrees of brightness. Finally, bleached or unbleached pulp is processed into thick sheets of paperboard or paper. The manufacture of building boards, with a few exceptions, involves processing the wood itself into sheets or boards, with the aid of adhesives.

The following material is intended to provide a broad overview of selected segments of the wood processing industry. The production process, atmospheric emissions, and emission control technologies employed are described for (1) several chemical wood pulping processes, (2) mechanical wood pulping processes, and (3) paper and paperboard manufacture. It should be recognized that this manual only provides information on the most common manufacturing processes and control technologies. There are a large number of variations of the basic processes and control systems; thus the descriptions here are necessarily generic in nature. Also, this presentation does not cover solid wood products manufacturing operations.

Chemical Wood Pulping

Arun V. Someshwar and John E. Pinkerton

Chemical wood pulping involves cooking wood chips or sawdust in an aqueous solution of pulping chemicals, resulting in the extraction of cellulose from the wood by dissolving the lignin that binds the cellulose fibers together. The pulping chemicals may be alkaline, acidic, or neutral. Three principal chemical wood pulping processes currently in use are (1) kraft, (2) acid sulfite, and (3) neutral sulfite semichemical. Of these three, kraft pulping accounts for nearly 80% of the chemical pulp produced in the United States. In 1988, 126 kraft, 18 sulfite, and 15 other chemical pulp mills were in operation in this country.[1]

KRAFT PULPING

The dominant wood pulping process today is the kraft process. The term kraft or sulfate has been in use since 1879, when sodium sulfate replaced sodium carbonate (soda process) as the makeup chemical. For reasons that include the comparative simplicity and rapidity of the process, its insensitivity to variations in wood condition, and its applicability to all wood species, as well as the valuable properties of the pulp produced, the kraft process is expected to continue as the dominant chemical wood pulping process.[2]

Process Description

The production of pulp by the kraft process can be divided into three areas, namely, (1) the making of pulp, (2) the recovery of cooking chemicals, and (3) the bleaching of pulp. Figure 1 provides a schematic representation of the kraft pulping and chemical-recovery process.

The Making of Pulp

In the kraft process, wood chips are cooked at an elevated temperature (340–360°F) and pressure (100–135 psig) in an aqueous solution of sodium hydroxide and sodium sulfide (also called white liquor). The sodium sulfide in the cook-

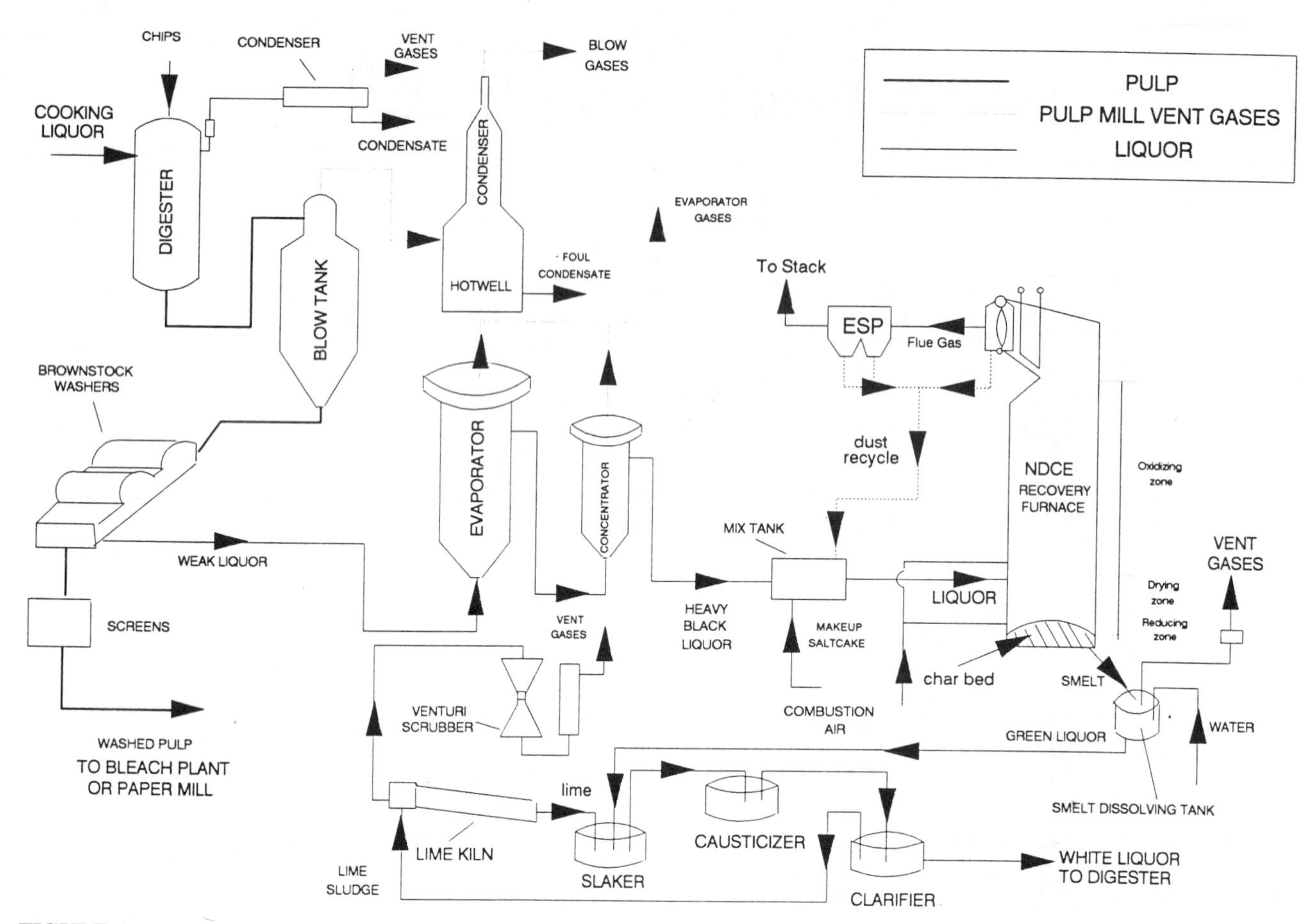

FIGURE 1. Schematic Diagram of the Kraft Pulping and Recovery Process

ing liquor serves to buffer and sustain the cooking reaction, while the sodium hydroxide is consumed by reaction with the lignin and carbohydrates in the wood. Once the cooking is complete, the wood is broken down into two phases: a soluble phase containing the lignin and alkali-soluble hemicellulose and an insoluble phase containing the alpha cellulose or pulp. The wood chips are cooked in large vertical vessels called digesters, which may be operated in either the "batch" or the "continuous" mode.

In the case of batch digesters, air trapped with the chips and gases formed during digestion are relieved intermittently during cooking. In softwood pulping, the relief gases are condensed for turpentine recovery before venting. The cooking period ranges from two to six hours. At the end of the cook, the digester contents are transferred to an atmospheric tank, called the blow tank. Gases leaving the blow tanks pass through a condenser to remove moisture, and the uncondensed gases are incinerated in a combustion device.

In continuous digesters, which are extremely tall and large cooking vessels (see Figure 2), an uninterrupted flow of wood chips and cooking liquor enters from the top. Pulp is withdrawn continuously from the bottom into a blow tank, while the spent liquor is drawn off and transferred to a flash tank. Steam from the flash tank is used to presteam the wood chips after which it passes through a condenser.

FIGURE 2. Continuous Digester

FIGURE 3. Brown-Stock Washers (Vacuum Drum Type)

Noncondensible gases (NCGs) are vented, and in the case of softwoods, turpentine is decanted from the condensate.

Newer digesters may utilize a "cold blow" as compared with the conventional "hot blow." In the cold blow process, the pulp leaves the digester at a much lower temperature, resulting in a significant reduction in blow gas emissions.

From the blow tank, the pulp and spent cooking liquor are diluted and pumped to a series of brown-stock washers, where the spent liquor is separated from the pulp, usually by countercurrent washing. Brown-stock washers are typically of the rotary-drum vacuum filter type (see Figure 3), although the newer diffusion-type washers are becoming increasingly common, especially with continuous digesters. The washed pulp, still brown in color, may then be subjected to a bleaching sequence, before being pressed and dried to yield the finished product.

Recovery of Cooking Chemicals

The spent liquor (called black liquor) is extracted from the washers in a dilute phase (12–18% dissolved solids) and the rest of the chemical-recovery process is concerned with recovering and regenerating the cooking chemicals dissolved in this liquor. The weak black liquor is concentrated in multiple-effect evaporators to about 55% solids. Further concentration to about 65% solids is accomplished in two distinct ways, depending on the configuration of the recovery furnace in which the 65% solids liquor is burned. In the older recovery furnaces, the hot combustion gases are utilized to concentrate the liquor in a "direct-contact evaporator" (DCE). Since the hot, acidic flue gases (primarily CO_2) directly contacting the black liquor react with its reduced sulfur content and cause odorous total reduced sulfur (TRS) compounds to be stripped, the liquor is usually subject to oxidation prior to concentration in the DCE. Weak or strong black liquor oxidation by either air or molecular oxygen is carried out in gas–liquid contactors. This results in oxidation of the Na_2S species to $Na_2S_2O_3$ or other polysulfides with a higher oxidation state for sulfur, which, in turn, leads to significantly reduced TRS emissions from the DCE. Most furnaces built since the early 1970s and many of the older furnaces that have since been modified are designed without a DCE (also called noncontact furnaces or NDCE). A concentrator is used in such instances to concentrate the black liquor from about 55% to over 65% solids prior to burning in the furnace.

Kraft recovery furnaces are much larger and nearly three times as expensive as fossil-fuel–fired boilers of comparable heat input capacity (see Figures 4 and 5). Concentrated black liquor is sprayed into the furnace and the organics in the black liquor, which are derived from pulping the wood, are combusted. This generates sufficient energy to produce steam and, more important, to reduce the sodium sulfate in the liquor to sodium sulfide, a cooking chemical. The bulk of the inorganic molten smelt that forms and collects in the furnace bottom consists of sodium carbonate and sodium sulfide in about a 3:1 weight ratio. The smelt is continuously withdrawn through smelt spouts into a smelt-dissolving tank. The alkali-fume-laden gas streams from both the DCE and NDCE furnaces pass through a particulate control device, usually an electrostatic precipitator (ESP). The ESP serves not only to control particulate emissions, but also to recover and recycle the predominantly sodium sulfate particulate catch.

In the smelt-dissolving tank, jets of water are used to quench the molten smelt to form green liquor, which consists of an aqueous solution of Na_2CO_3 and Na_2S. The

FIGURE 4. Kraft Recovery Furnace

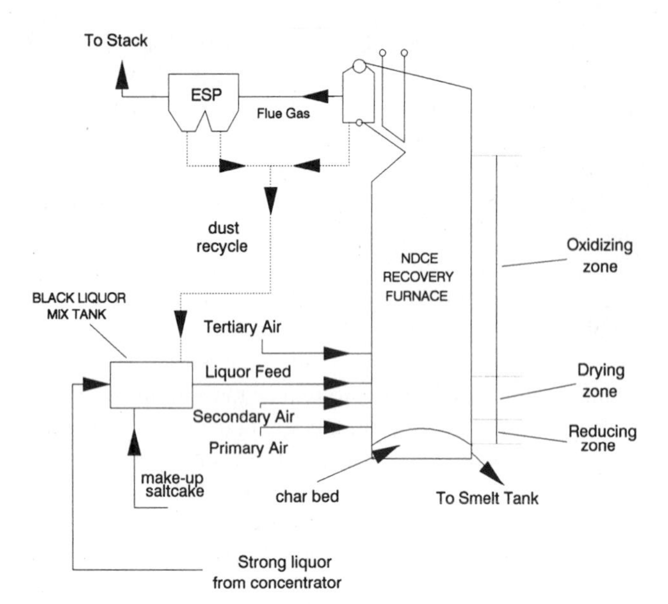

FIGURE 5. Kraft Recovery Furnace Schematic

quenching of molten smelt results in large quantities of steam leaving the tank, carrying with it small amounts of particulate matter, containing mainly Na_2CO_3 and Na_2S. The green liquor is then transferred to a causticizing tank where quicklime (calcium oxide) is added to convert the Na_2CO_3 to NaOH. This results in a white liquor solution containing NaOH, Na_2S, and lime mud precipitate (mainly $CaCO_3$). The white liquor is then recycled to the digesters and the entire process repeated.

The lime mud from the causticizing tank is washed, dried (60–80% solids), and then calcined in a lime kiln to regenerate quicklime. Large rotary kilns are typically used in the kraft pulp industry, although a few fluidized-bed calciners are also being used. Fossil fuels, mainly natural gas or residual fuel oil, are used to provide the energy required for calcining the limestone. Particulate emissions in the lime kiln exhaust gases are controlled by venturi scrubbers or ESPs. The wet or dry particulate catch is recovered and returned to the system for calcining.

The two principal commercial by-products of the kraft pulping industry are crude sulfate turpentine and crude tall oil. Further refining of these products is normally done elsewhere.

Pulp Bleaching

Pulp bleaching imparts whiteness or brightness to the pulp, in addition to yielding certain desirable physical and chemical properties. Nearly 60% of all chemical pulps are bleached. Kraft and other chemical pulps are usually subjected to multistage "lignin-removing" methods as compared with single- or two-stage "lignin-bleaching" methods used for mechanical and some semichemical pulps.[2] Chlorination in one or two stages brings about the degradation of lignin. An alkali stage usually follows chlorination to dissolve and extract the degradation products. The bleaching process is typically finished by one or more oxidative stages with intermediate alkali extraction. Oxidation by chlorine dioxide, hypochlorite, or peroxide is essential to remove the final discoloration and obtain full high brightness.[2] Significant changes in bleaching sequences, bleaching conditions, and use of bleaching chemicals have occurred in recent years, mostly in response to environmental concerns resulting from formation of certain bleaching by-products. These by-products include polychorinated dibenzodioxins and -furans (in the aqueous phase) and chloroform. The substitution of chlorine dioxide for chlorine and the elimination or reduction in the use of hypochlorite are among a number of bleach plant modifications already implemented or planned for the near future. Relative to air emissions from bleach plant vents, chlorine, chlorine dioxide, and the by-product chloroform are the principal compounds.

Air Emissions Characterization

Historically, odor and visible particulate emissions from kraft pulp mills have received considerable attention. Initial state and federal emission regulations addressed TRS and particulate emissions. The characteristic odor of a kraft pulp mill results from TRS compound emissions, which are the by-products of reactions between the wood lignin and sulfide ions in kraft liquors. Hydrogen sulfide (H_2S) and three organic sulfur compounds—methyl mercaptan (CH_3SH), dimethyl sulfide ($(CH_3)_2S$), and dimethyl disulfide ($(CH_3)_2S_2$)—constitute the major TRS compounds. Besides TRS emissions, other gaseous by-products from the pulping process, such as methanol and acetone, are present in NCGs formed in the pulping and evaporation area. TRS, acetone, and methanol are present in digester relief gases or vent gases from turpentine condensers that treat digester relief gases, in noncondensible blow gases after the condenser, in gases vented from brown-stock washers, in seal tank and pulp knotter vents, and in NCGs from the evaporator–concentrator system. Mills that collect their foul condensates and either steam or air-strip them generate an additional gaseous stream containing these chemical by-products.

Table 1 presents estimates of uncontrolled emission factors for TRS, acetone, and methanol in pulp mill and evaporator NCGs. For purposes of uniformity, the emission factors in Table 1 and all succeeding tables are expressed in units of pounds per air-dried ton of pulp. It should be noted, however, that mills differ greatly with respect both to the pulping processes and to type of wood pulped. Thus based on conversion factors alone, some variability in emission factors among mills is to be expected.

Mills operating recovery furnaces with DCEs often practice black liquor oxidation (BLO). This is typically accomplished by employing air-sparging reactors in order to provide the necessary air–liquid contact. Estimates of uncontrolled emissions of TRS compounds and volatile or-

TABLE 1. Uncontrolled Noncondensible Gas Emissions from Kraft Pulping

Source	Type[a] of Digester	Acetone Emissions, lb/ton			Methanol Emissions lb/ton			TRS[b] Emissions lb/ton as sulfur		
		Range	Average	Reference	Range	Average	Reference	Range	Average	Reference
Relief gases or turpentine condenser vent gases	B	0.08–0.25	0.16	5	0.10–0.55	0.30	5	0.01–1.33	0.50	3
Blow gases	B	0.47–1.15	0.90	5	0.80–1.40	1.20	5	0.29–0.97	0.64	3
Condenser vent gases, hardwood[c]	C	0.01–0.04	0.03	7	0.02–0.10	0.05	7	0.20–0.84	0.47	7
Condenser vent gases, softwood	C	0.30–0.55	0.42	5	NA	NA		NA	NA	
Brown-stock washers[d]										
Fresh water	—	0.03–0.07	0.05	7	0.05–0.35	0.18	7	0.01–0.79	0.16[e]	3
Condensate	—	NA	NA		NA	NA		0.13–0.85	0.43	3
Evaporator gases	—	0.01–0.02	0.02	5	NA	0.30	5	0.03–5.93	1.09	3
Miscellaneous vent gases[f]	B	0.03–0.21	0.05	7	0.33–0.90	0.48	7	NA	0.18	3
Condensate stipping system[g]	—	NA	NA		NA	NA		0.30–5.20	1.10	14

All factors are in lb/(air-dried ton pulp) and are valid for untreated gases only. For gases that are collected and burned in a combustion device (lime kiln, power boiler, incinerator, etc.), almost all the methanol, acetone, and TRS will be destroyed. NA—not available.
[a]B for batch and C for continuous digesters.
[b]TRS (includes H_2S, methyl mercaptan, dimethyl sulfide, and dimethyl disulfide).
[c]Estimates are for continuous digesters without diffusion washers only.
[d]Includes gases to roof vents and from undervents (vacuum pump exhausts and filtrate seal tanks).
[e]TRS emissions for vacuum-drum-type washing systems; TRS for newer diffusion washers are <0.001 lb/ADTP.[6]
[f]Estimates for fresh water use only; includes washer seal tanks and pulp knotter vents.
[g]Untreated stripped gases.

TABLE 2. Emissions from Black Liquor Oxidation Tower Vents

Black Liquor Oxidation	VOC Emissions, lb/ton as carbon			TRS Emissions, lb/ton as sulfur		
	Range	Average	Reference	Range	Average	Reference
Weak liquor	0.42–1.94	0.80[a]	7	0.02–0.21	0.10	15
Strong liquor	0.23–0.44	0.34	8	0.01–0.17	0.08	15

All factors are in lb/(air-dried ton pulp). VOC and TRS emissions will be destroyed if vent gases are sent to a combustion device.
[a]Includes methanol, acetone, and alpha pinene only.

ganic compounds (VOCs) from BLO tower vents are given in Table 2.

Major emission sources in the chemical-recovery area include recovery furnaces, smelt-dissolving tanks, and lime kilns. Besides TRS compounds, these sources emit particulates, sulfur dioxide (SO_2), carbon monoxide (CO), and nitrogen oxides (NO_x). The particulate matter emitted from kraft recovery furnaces is mainly Na_2SO_4 (about 80%), with smaller amounts of K_2SO_4, Na_2CO_3, and NaCl. Significant alkali fuming action in the lower furnace causes over 10% of the sodium input to the furnace to vaporize. Sodium vapors are most likely to react rapidly with oxygen and carbon dioxide to form submicron-sized Na_2CO_3 fume particles, which, in turn, scavenge the sulfur dioxide resulting from black liquor combustion. The net result is that typically less than 5% of the total sulfur entering the furnace via the kraft black liquor escapes as SO_2 and over 85% of the particulate catch that is recycled is made up of alkali sulfates.

Finely divided smelt (Na_2CO_3 and Na_2S) entrained in water vapor accounts for the particulate emissions from smelt-dissolving tanks. Lime kiln particulate emissions are mainly sodium salts, calcium carbonate, and calcium oxide, with uncontrolled emissions containing mainly calcium compounds and controlled emissions comprising mainly sodium salts. Sodium salts result from the residual sodium sulfide in the lime mud after washing.

Table 3 gives estimates of uncontrolled and controlled particulate emissions from these three kraft mill sources in units of pounds per ton of pulp. Particulate emission limits for kraft sources subject to New Source Performance Standards (NSPS) are expressed in units other than pounds per ton of pulp (see Table 8). However, for purposes of comparison, the NSPS values, after using typical conversion factors, correspond to about 2.0, 0.30, 0.40 and 0.80 lb/ton pulp for kraft recovery furnaces, smelt-dissolving tanks, and lime kilns firing gaseous and liquid fuels respectively. Table 4 gives the cumulative mass percent less than or equal to 10 μm (PM_{10}) and 2.5 μm ($PM_{2.5}$) in particulate emissions from uncontrolled and controlled recovery furnaces,

TABLE 3. Particulate Emissions from Recovery Sources in Kraft Pulping

Source	Type of Control	lb/ton Range	lb/ton Average	Units Tested	Reference
Recovery furnace with direct-contact evaporator (DCE)	Untreated	140–313	206	10	19
	Venturi scrubber	14–115	48	10	3
	ESP[a]	0.25–11.4	2.22	20	18
	ESP[b]	0.66–1.56	0.74	10	19
Noncontact recovery furnace without DCE	Untreated	204–743	448	19	21
	ESP[c]	0.62–4.33	1.98	13	20
	ESP[d]	0.50–2.66	1.67	19	21
Smelt-dissolving tank	Untreated	0.19–23.7	7.0	10	3
	Mesh pad	0.05–2.3	1.0	6	3
	Scrubber	0.07–0.29	0.2	10	4
Lime kiln	Untreated	41.5–71.3	56	2	3
	Venturi scrubber[e]	0.23–1.1	0.5	5	4

All factors are in lb/(air-dried ton pulp). Particulate emissions from ESPs on recovery furnaces were converted from units of gr/acf and gr/scf to lb/ADTP using conversion factors of 400 acfm/TPD (range, 370 to 510) and 301 scfm/TPD (range, 252 to 344) respectively.
[a]1971 survey of ESPs on DCE furnaces started between 1966 and 1970.[18]
[b]1979 survey of ESPs on DCE furnaces started between 1973 and 1977.[19]
[c]1974 survey of ESPs started between 1969 and 1973.[20]
[d]1979 survey of ESPs started between 1974 and 1978.[21]
[e]Data from five kilns during the NSPS review and development program.[4]

smelt-dissolving tanks, and lime kilns. It is to be noted that the estimates given in Table 4 are based on extremely limited data.[4] It can be seen from Table 4 that nearly 75% of the particulate emissions from ESPs on DCE and NDCE furnaces are ≤ 10 μm in size, whereas this fraction increases to 98% and over 89% for controlled emissions from lime kilns and smelt-dissolving tanks respectively.

The presence of sulfides in kraft liquors invariably leads to some TRS compound emissions from these three sources. TRS emission factors for kraft recovery furnaces, smelt-dissolving tanks, and lime kilns are shown in Table 5. Over

TABLE 4. PM_{10} and $PM_{2.5}$ Emissions from Kraft Mill Sources

Source	Type of Control	Cumulative Mass % ≤ Stated Size, μm: 10 μm	2.5 μm
DCE furnace	Uncontrolled	93.5	83.5
	ESP	75.0	53.8
NDCE furnace	Uncontrolled	NA	78.0
	ESP	74.8	67.3
Lime kiln	Uncontrolled	16.8	10.5
	Venturi scrubber	98.3	96.0
	ESP	88.5	83.0
Smelt-dissolving tank	Uncontrolled	88.5	73.0
	Packed tower	95.3	85.2
	Venturi scrubber	89.5	81.3

Source: Reference 4.

the past 20 years, there have been major reductions in kraft pulp mill TRS emissions as a result of equipment upgrades and replacements, federal and state emission regulations, and a desire to minimize odor complaints in mill communities.

The combustion of sulfur-containing black liquor is expected to result in sulfur dioxide emissions. These emissions from kraft recovery furnaces are a complex function of liquor sulfidity, liquor characteristics, furnace design and loading, combustion air flow and distribution, and so on. Sulfur dioxide emissions from most well-designed furnaces of the DCE and NDCE type are below 500 ppm, with several operating consistently at levels below 100 ppm. However, for reasons not totally understood, on a given day the SO_2 emissions can vary considerably. The emissions of SO_2 from smelt-dissolving tanks and lime kilns are generally insignificant. Moderate to low quantities of NO_x, VOCs, and CO are also released from recovery furnaces and lime kilns. Available emission factors for SO_2, NO_x, VOC and CO from kraft recovery furnaces, smelt dissolving tanks and lime kilns are also presented in Table 5. Other kraft recovery furnace emissions that are occasionally tested for, but for which insufficient data exist at present, include PCDD/Fs (dioxins and furans), certain organic gases, and trace metals. Preliminary test results indicate that the PCDD/F levels range from extremely low to nondetectable.[25]

If pulp bleaching is practiced at a kraft mill, uncontrolled bleach plant vent gases may contain chlorine (Cl_2), chlorine dioxide (ClO_2), and chloroform ($CHCl_3$). Emission factors for uncontrolled Cl_2 and ClO_2 in bleach plant vent gases are

TABLE 5. SO_2, NO_x, CO, VOC, and TRS Emissions from Kraft Recovery Sources

	SO_2, lb SO_2/ton	NO_x, lb NO_2/ton	CO, lb CO/ton	VOC, lb CH_4/ton	TRS, lb S/ton
Recovery furnace DCE with BLO[a]					
Range	2.5–5.2	0.9–3.3	0.4–42	1.8–2.1	0.01–0.15[d]
Average	3.5	1.8[b]	10.6[c]	1.95	NA
Number of furnaces	3	10	5	2	NA
Reference	16	11	9	12	22
Recovery furnace NDCE					
Range	0.2–14.7	0.9–3.3	0.4–42	0.7–1.1	0.07–0.14
Average	4.2	1.8[b]	10.6[c]	0.83	0.11
Number of furnaces	13	10	5	3	4
Reference	16	11	9	12	4
Smelt-dissolving tank					
Range	0.005–0.11[e]	NA	NA	NA	0.01–0.05[f]
Average	0.04	NA	NA	NA	NA
Number of SDTs	6	NA	NA	NA	NA
Reference	24	—	—	—	23
Lime kiln					
Range	0.007–0.13[e]	0.2–3.7	0.04–0.12	0.0–0.75	0.01–0.10[d]
Average	0.05	1.2	0.07	0.22	NA
Number of kilns	11	6	4	3	—
Reference	24	10	9	17	22

All factors are in lb/(air-dried ton pulp). NA—not available.
[a]Values are for "new design" furnaces; "old design" DCE furnaces (built before 1965) have TRS emission limit guidelines of 0.6 lb/ADTP[22] and emission factors for NO_x, CO, and VOC are unavailable.
[b]Average from 10 units tested, five NDCE and five DCE.[11] For all 10 units, the percent solids in black liquor were below 70%. More recent furnaces firing higher solids liquor could have higher NO_x emission levels.
[c]Average value for five units, two NDCE and three DCE[9]
[d]Based on TRS emission guidelines for existing kraft pulp mills[22]
[e]Based on test data obtained from different mills.
[f]Based on NSPS for SDTs of 0.016 g/Kg bls as H_2S[23]

TABLE 6. Chlorine and Chlorine Dioxide Emissions from Pulp Bleaching

Source	Type of Control	Chlorine, lb/ton			Chlorine Dioxide, lb/ton		
		Range	Average	Reference	Range	Average	Reference
Bleach plant	Uncontrolled	0.01–10.1	0.70	5	0.00–2.90	0.50	5
	scrubbers	a	a	5	b	b	5
ClO_2 generators[c]	Absorbers	0.80–14.0	—	5	0.23–6.10	2.30	5

Factors are in lb/(air-dried ton pulp) for bleach plant sources and in lb/ton ClO_2 generated for ClO_2 generators.
[a]Cl_2 removal efficiencies by various scrubbing fluids range from 75% to 99%.[5]
[b]ClO_2 removal efficiencies by various scrubbing fluids range from 50% to 99%.[5]
[c]Range applies to various ClO_2 generating processes with absorbers that are followed by a caustic scrubber if required[5]; the pound ClO_2 used per ton pulp varies considerably, ranging from 0 to as much as 60 (three separate ClO_2 stages).

presented in Table 6 and those for $CHCl_3$ are given in Table 7. Bleach plant emissions of Cl_2, ClO_2, and $CHCl_3$ are extremely site specific and variable. Ranges and median values of available data should be used with caution. The emission of these gases depends on the application rates of Cl_2, ClO_2, and hypochlorite during bleaching, and also on the bleaching sequences and bleaching conditions employed. Chlorination, oxidation, and alkali extraction may each make up a separate bleaching stage. Figure 6 gives a much-simplified diagram of a typical bleaching stage. Gas-phase emissions may occur from each stage's tower, washer

TABLE 7. Chloroform Emissions from Pulp Bleaching

Bleaching Sequence Hypochlorite Use	Chloroform Emissions, lb/ton		
	Range	Average	Reference
0.1–<0.5%	0.06–0.67	0.27	5
0.5– 2.0%	0.11–1.10	0.53	5
>2.0%	0.22–2.01	0.80	5

All factors are in lb/(air-dried ton pulp).
Table 7 only applies to bleaching sequences that use hypochlorite in excess of 0.1%, where percent usage is expressed as (pounds hypochlorite used per pound oven dry brown-stock pulp) × 100.

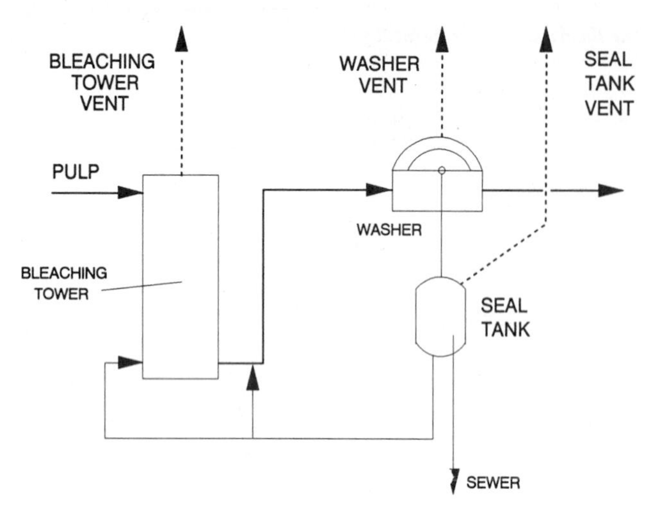

FIGURE 6. Typical Venting from a Bleaching Stage

hood, and seal tank vent. The various emission points may or may not be combined into one or more common vents. Scrubbers utilizing various scrubbing fluids, such as bleach-plant extraction-stage filtrate, sodium hydroxide solution, sodium bisulfite solution, alkaline wash water from causticizing operations (weak wash), white liquor, and chilled water, have been installed for the control of Cl_2 and ClO_2 air emissions. Chloroform formed during bleaching can also be released to the environment in the wastewater treatment area. A study[31] by the National Council of the Paper Industry For Air and Stream Improvement showed that between 10% and 94% (average 48%) of $CHCl_3$ that is formed as a by-product is released to the wastewater treatment area, where most of it is stripped or volatilized.

The production of steam in kraft recovery furnaces is usually sufficient to meet the steam demand for the pulping and evaporation functions at a mill. Additional steam is required for several other operations within a kraft mill, including bleaching, pulp drying, paper making, and stripping. Power boilers firing either wood residues, coal, oil, or natural gas are utilized for the purpose of generating steam and power. Air emissions from such sources have already been dealt with elsewhere in this text.

Air Pollution Control Measures

Kraft pulp mills that began construction or modification after September 24, 1976, are subject to NSPS for particulate and TRS compound emissions as shown in Table 8.[23] In 1979, the U.S. Environmental Protection Agency (EPA) issued retrofit emission guidelines for the control of TRS emissions at existing facilities not subject to NSPS, and these are also shown in Table 8.[22] In addition to particulate and TRS, many recent preconstruction air quality permits[26] for kraft pulp mill sources contain limits for PM_{10}, SO_2, NO_x, CO, and VOCs.

Control of TRS Emissions

The collection and treatment of NCGs from kraft pulp mill sources have been practiced for over 40 years. While most early systems were designed to control TRS emissions from the digester and multiple-effect evaporator NCGs only, during the past 20 years, efforts to include such sources as the brown-stock washer hood vents, condensate stripper system vents, turpentine decanter vents, and a number of other minor sources have been made.[27] Incineration in existing combustion devices, namely, lime kilns and power boilers, is the treatment most commonly used. Incinerators dedicated to this purpose, while used to a lesser degree, are becoming more prevalent. However, the oxidation of TRS results in SO_2 formation and a caustic scrubber is often installed following the incinerator for SO_2 removal. Other combustion devices, such as recovery furnaces and flares, are also used. Alternative disposal practices employed include venting with the bleach-plant chlorination or chlorine

TABLE 8. NSPS and Emission Guidelines for Kraft Pulp Mills

Source	Particulate Matter	Total Reduced Sulfur	
		NSPS	Guidelines[a]
Recovery furnace			
New design	0.044 gr/dscf @ 8% O_2	5 ppmdv @ 8% O_2	5 ppmdv @ 8% O_2
Old design	—	—	20 ppmdv @ 8% O_2
Cross recovery	0.044 gr/dscf @ 8% O_2	25 ppmdv @ 8% O_2	25 ppmdv @ 8% O_2
Smelt-dissolving tank	0.20 lb/ton bls	0.032 lb/ton bls	—
Lime kiln			
Gas	0.067 gr/dscf @ 10% O_2	8 ppmdv @ 10% O_2	20 ppmdv @ 10% O_2
Liquid	0.130 gr/dscf @ 10% O_2	8 ppmdv @ 10% O_2	20 ppmdv @ 10% O_2
Digester system	—	5 ppmdv @ 10% O_2	No control
BSW system	—	5 ppmdv @ 10% O_2	No control
BLO system	—	No control	No control
MEE system	—	5 ppmdv @ 10% O_2	5 ppmdv @ 10% O_2
Condensate stripper system	—	5 ppmdv @ 10% O_2	5 ppmdv @ 10% O_2

[a]EPA guidelines for TRS emission from existing kraft pulp mills not subject to NSPS[22]; ppmdv-ppm dry on volume basis.

dioxide stage washer vent gases for chemical oxidation of the sulfur compounds, and, during process upsets, discharge of selected streams to stacks for improved dispersion, thus minimizing the potential for worker exposure to high concentrations of TRS.[27]

The non-condensible gases in the kraft pulping process are of two types—low-volume, high-concentration (LVHC) and high-volume, low-concentration (HVLC). The LVHC gases include those from the digester area, condensate strippers, evaporators, and so on, while HVLC gases include those from brown-stock washers, pulp knotters, and washer seal tanks. The LVHC gases are typically burned in lime kilns, boilers, or incinerators, while HVLC gases are burned in boilers that can accept such large gas volumes.

Vent gases from BLO systems are not required to be controlled under NSPS due to the prohibitive cost-effectiveness of control and the declining use of BLO.[13] The TRS emissions from the newer diffusion brown-stock washers are extremely small as compared with the older vacuum-drum-type washing systems.[6] The use of fresh water or stripped condensate as a washing medium in brown-stock washers (as opposed to dirty condensate) also leads to reduced TRS emissions. The TRS emissions from smelt-dissolving tanks are similarly most effectively controlled by the choice of a water (both for smelt dissolving and particulate control) containing minimal amounts of reduced sulfur compounds.

The TRS emissions from kraft recovery furnaces are most efficiently controlled by maintaining sufficient oxygen, residence time, and turbulence, and avoiding overloading. Furnaces with DCEs rely additionally on BLO for TRS emission control. By oxidizing the Na_2S in the liquor to $Na_2S_2O_3$ before it enters the DCE, the reactions between the combustion gases and black liquor in a DCE that generate H_2S are inhibited.

Lime kiln TRS control mainly involves achieving a high degree of lime mud washing. Lime mud is the precipitate resulting from the causticizing reaction when Na_2CO_3 in the green liquor is converted to NaOH and lime is converted to $CaCO_3$. Sodium sulfide entrained in the lime mud reacts with CO_2 in the cold end of the kiln, giving rise to H_2S emissions. Proper operation of the lime mud filter to prevent Na_2S from entering the lime kiln and allow for oxidation of residual Na_2S to $Na_2S_2O_3$ is usually sufficient to keep H_2S levels below about 8 ppm. The use of sulfide-free streams such as fresh water or clean condensate for makeup in the scrubber also prevents H_2S formation by contact with kiln gases.

Control of Particulate Emissions

Particulate control on furnaces with DCEs and on NDCE furnaces is achieved predominantly by ESPs. Electrostatic-precipitator particulate-removal efficiencies range from about 90% in older installations to well over 99% in newer units. The range and average particulate emission factors for furnaces with ESPs are shown in Table 3. Direct-contact evaporators used to concentrate black liquor also serve to scrub the particulates leaving the furnace, removing from 20% to 50% of the particulate load prior to the ESP. A few scrubbers have been installed following older ESPs to obtain satisfactory levels of particulate removal.

Particulate control of smelt-dissolving tank vent gases is accomplished by installing demister pads, packed towers, or venturi scrubbers. Most lime kilns are controlled by venturi scrubbers, with pressure drops ranging from 17 to 34 inches of water, although ESPs are increasingly being used in new installations. Typical emission factors for smelt-dissolving tank vents and lime kilns with control are also shown in Table 3.

Fugitive emissions from sources or areas of importance in a kraft mill include coal piles, paved and unpaved roads, bulk materials handling (lime, limestone, starch, etc.), and wood handling. Control strategies include wetting; the use of chemical agents, building enclosures, and windscreens; paving or wetting roads; and modifying handling equipment.[29]

Control of SO_2, NO_x, CO, and VOC Emissions

Besides power boilers in which sulfur-containing fuels are fired, SO_2 emissions from a kraft mill occur principally through kraft-recovery-furnace flue gases. Unlike power boilers firing fossil fuel, the combustion of black liquor in a kraft recovery furnace results in SO_2 emissions that are extremely variable and depend on a variety of factors, which include (1) liquor properties such as sulfidity (or sulfur-to-sodium ratio), heat value, and solids content; (2) combustion air and liquor firing patterns; (3) furnace design features; and (4) other operational parameters.[16] Table 5 gives SO_2 emission factors for DCE and NDCE furnaces developed from averaging recent, long-term, continuous emission monitoring SO_2 data.[16] Strategies to (1) lower liquor sulfidity and (2) optimize liquor and combustion air properties and firing patterns so as to yield maximum and uniform temperatures in the lower furnace have been used to minimize kraft-recovery-furnace SO_2 emissions. Flue gas desulfurization is capital and energy intensive and its efficacy is uncertain, considering the generally low concentrations and rapidly fluctuating levels of SO_2 in the furnace flue gases. Sulfur dioxide may be formed in the lime kiln when fuel oil is combusted. The regenerated quicklime in the kiln acts as an in-situ scrubbing agent and the venturi scrubber that normally follows the kiln augments the SO_2 removal process. Limited SO_2 measurements following ESPs on fuel-oil-burning kilns suggest that over 90% of the SO_2 is captured by the time the flue gas exits the ESP.

The NO_x emissions from recovery furnaces and lime kilns result from black liquor and fossil fuel combustion respectively. The emissions from recovery furnaces are mainly attributed to "fuel NO_x," resulting from partial oxidation of the black liquor nitrogen content. Kraft recovery furnaces operate with a reducing zone in the lower part of the furnace (for reduction of Na_2SO_4 to Na_2S) and an oxidizing zone farther up in the region of the liquor spray guns and secondary air. This staged combustion is a natural deterrent to excessive NO_x formation, with the result that most existing kraft recovery furnaces emit less than about 100 ppm NO_x. However, the current trend for new units or existing units undergoing upgrading is to burn liquors with increasingly higher solids content (>70% solids). Available emission factors for NO_x from kraft recovery furnaces (shown in Table 5) are based on tests on older units burning liquors with <70% solids. These emission estimates are being reviewed, considering the somewhat higher NO_x formation that could result from more intense burning of the higher-solids liquor. Coincidentally, higher temperatures in the lower furnace have been documented to yield significant other environmental (such as lower SO_2 and TRS) and furnace operational (such as smelt-bed stability) advantages.[16] Higher temperatures in the lower furnace zone can result from several factors, including firing higher-solids liquor, firing liquor with higher heat content, and operating the furnace at higher than the design load.

The VOC and CO emissions from recovery furnaces and lime kilns result from incomplete combustion of the organic matter in the fuel. Lime kiln CO and VOC emissions are generally small, as seen from Table 5. Recovery-furnace CO and VOC emissions are a function of the level of excess air used and the degree of mixing achieved within the furnace. Control strategies to minimize CO and VOC emissions involve increasing the residence time, oxygen content, temperature, and level of turbulence in the furnace combustion zone. Unfortunately, an increase in excess air, residence time, and temperature has the opposite effect on NO_x formation.

Control of Bleach Plant Emissions

As shown in Table 6, the uncontrolled Cl_2 and ClO_2 emission from bleach plants exhibit a very broad range. These emissions are expected from bleach towers, washer hood vents, and seal tank vents of the chlorination and chlorine dioxide stages of bleaching respectively. Emissions of Cl_2 and ClO_2 are also expected from the ClO_2 generator absorbers.

Studies have indicated that maintaining low-bleaching chemical residuals in the pulp leaving the bleaching towers results in minimal Cl_2 and ClO_2 emissions from washer hoods and seal tanks.[30] Application rates for Cl_2 and ClO_2 govern the Cl_2 and ClO_2 emission rates from their respective tower vents. Packed-tower scrubbers are used to control Cl_2 and ClO_2 emissions from washer hoods and seal tanks. Smaller scrubbers are used to control emissions from tower vents, especially those of an upflow–downflow design.[30] These smaller scrubbers are designed with chemical recovery in mind. Some mills use the larger packed scrubbers to scrub combined vent gases from tower vents, washer hoods, and seal tanks. Estimates for the removal efficiencies of Cl_2 and ClO_2 emissions from bleach plant vent gases after alkali scrubbing are given in Table 6.

The formation of by-product chloroform ($CHCl_3$) during pulp bleaching is a much more complex phenomenon. Laboratory and field studies have shown that $CHCl_3$ may be formed in the chlorination (C), extraction (E), and hypochlorite (H) stages of bleaching.[31] The use of hypochlorite in the bleaching sequence seems to have the most impact on $CHCl_3$ formation. Elimination of hypochlorite from the bleaching sequence has been used to reduce $CHCl_3$ generation[31] at a number of bleach plants. A reduction in the chlorine factor (defined as percent chlorine applied/kappa number, where kappa number is a measure of the residual lignin content) also helps in minimizing chloroform by-product formation. On an average, slightly over one half of the $CHCl_3$ formed in the bleach plant is expected to be released through the bleach plant vents.[31] Table 7 gives the $CHCl_3$ air emission factors for bleaching sequences with different levels of hypochlorite usage. Emission estimates for bleaching sequences using <0.1% hypochlorite are more complex to determine and are given elsewhere.[5] The feasibility of gas-phase scrubbing of $CHCl_3$ from bleach plant vents has not yet been demonstrated. A reduction in $CHCl_3$ by-product formation would also lead to decreased emissions from the wastewater treatment area.

ACID SULFITE PULPING

In the early 1900s, the sulfite process was predominant as it yielded the brightest unbleached pulp and the most easily bleached one. However, over the years, owing to its sensitivity to the wood raw material and the difficulty of recovering cooking chemicals and utilizing process waste products, the sulfite pulping process has been steadily on the decline.

The production of pulp by the acid sulfite process is carried out in a manner similar to that of kraft pulp, except that the cooking chemicals NaOH and Na_2S are replaced by sulfurous acid. To buffer the cooking solution, a bisulfite solution of one of four bases (ammonium, calcium, magnesium, or sodium) is used. The liquor (pH 1 to 6) is prepared by reacting SO_2 with the base solution in one or more absorption devices.

Process Description

Wood chips are cooked in the acidic cooking solution in batch or continuous digesters at high pressures (90–100 psi) and temperatures (260–320°F). At the end of the cooking period (somewhat longer than kraft cooks), the digester contents are either discharged under pressure into a blow tank or pumped at lower pressures to a dump tank. The spent sulfite liquor (also called red liquor) drains through the bottom of the tank and is either processed to recover certain organic materials, concentrated and incinerated, or sent to a recovery area where the cooking chemicals and

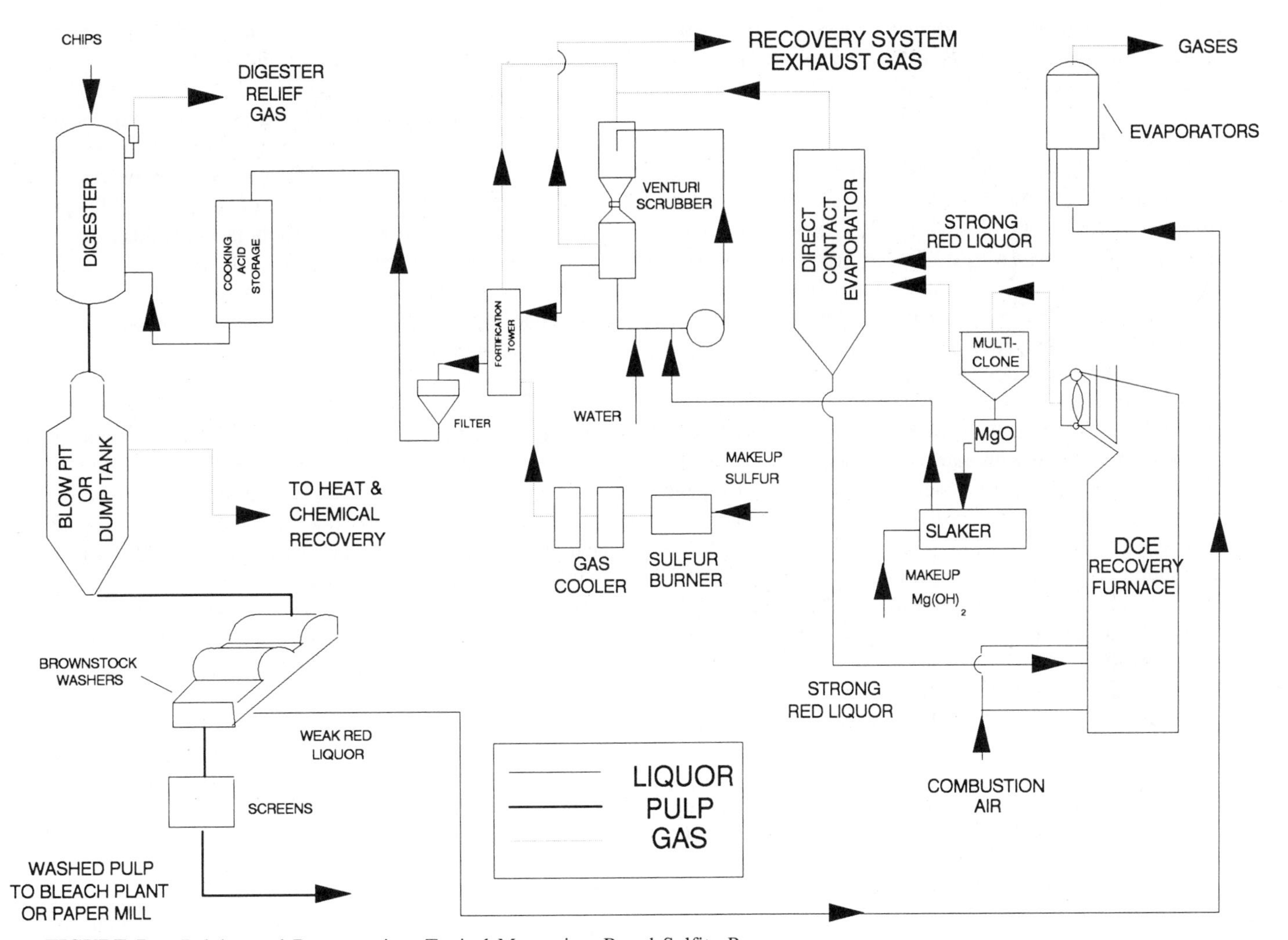

FIGURE 7. Pulping and Recovery in a Typical Magnesium-Based Sulfite Process

liquor heat content are recovered. The pulp is washed, screened, and centrifuged (to remove knots). It may subsequently be bleached, pressed, and dried. The pulp is used in a variety of fine papers or specialty paper products or is converted to numerous nonpaper end uses.

The magnesium-based process is currently the most widely used form of sulfite pulping. A typical magnesium-based sulfite cooking process is shown schematically in Figure 7. Older sulfite mills typically used calcium-based chemicals for cooking. As a result of maintenance problems associated with scale deposition on equipment and, more important, the impracticality of recovering the cooking chemicals, there has been a gradual trend in the industry away from calcium-based sulfite pulping. Existing calcium-based mills usually either process the spent liquor to recover selected organics or sewer the spent liquor or incinerate it, using the ash for by-products. In NH_3-based cooking, the spent liquor is concentrated in evaporators and burned in recovery furnaces. Heat is recovered, but the ammonium base itself is consumed during combustion. However, the sulfur is recombined with aqueous ammonia and recycled to cooking liquor. In sodium- and magnesium-based pulping, it is feasible to recover heat, sulfur, and base economically.

For acid sulfite mills, a sulfur burner is the usual source of SO_2. Sulfur is burned in rotary or spray burners and the gases are cooled by heat exchangers and a water spray and then absorbed in a variety of different scrubbers into solutions of $CaCO_3$ or one of the other base chemicals. The amount of sulfur burned in a mill depends on whether chemical recovery is practiced and also on the efficiency of the SO_2 emission control equipment.

If chemical recovery is practiced, the spent red liquor is concentrated in multiple-effect evaporators and a DCE to between 55% and 60% solids. This strong liquor is then sprayed into a boiler or furnace, where the organic content in the liquor is burned, producing steam to operate the digesters, evaporators, and so on. The inorganic content undergoes different fates, depending on the base chemical.

When magnesium-based liquor is burned, a flue gas containing magnesium oxide (MgO) and SO_2 is produced, and the MgO is recovered in a multiple cyclone as a fine white powder. The MgO is then water slaked and used as the circulating liquor in a series of venturi scrubbers, designed to absorb the SO_2 from the recovery-furnace flue gases leaving the DCE. The bisulfite solution formed is fortified in the acid plant by the SO_2 formed by burning makeup sulfur.

When a sodium-based liquor is burned, the inorganic compounds in the liquor are recovered as a molten smelt containing Na_2S and Na_2CO_3 (just as in kraft recovery furnaces). This smelt may be further processed to absorb SO_2 from the flue gas and sulfur burner. In some sodium-based mills, the smelt may be sold to a nearby kraft mill as raw material for preparing green liquor.

When NH_3-based liquor is burned, the small amount of inorganics is removed as furnace slag. The ammonia is decomposed to nitrogen and water. The SO_2 is absorbed in a heat-recovery unit with aqueous ammonia. The resulting bisulfite solution is further acidified with SO_2 from the burning of elemental sulfur to make the cooking liquor.

There are several recovery processes for magnesium- and sodium-based liquors, including the B&W magnefite process, the Copeland process, the STORA process, the SCA-Billerud process, the Tampella process, and the CE Sevola process.

Air Emissions Characterization and Control Measures

Sulfur dioxide is considered the major pollutant of concern from sulfite pulping. The characteristic odor of kraft pulping is generally absent, as TRS compounds are not formed during the lignin–bisulfite reactions. However, sodium-based sulfite pulping operations practicing recovery may experience some TRS emissions. Particulate, NO_x, and CO emissions from all recovery furnaces constitute the other pollutant emissions. Emissions from bleaching operations may be similar to those from kraft pulp bleaching, but are very dependent on the bleach sequence, the wood source, and the amount of residual lignin entering the bleach stage.

The digester and blow tank areas are a major source of SO_2. The relief gases from the digester and NCG gases from the presteaming vessel and flash evaporators are normally all returned to the acid-preparation system for SO_2 recovery. Considerable quantities of SO_2 (ranging from 10 to 70 pounds SO_2 per ton of pulp[32]) could be flashed during a hot blow, with uncontrolled emissions comprising nearly 95% water vapor, 3% SO_2, and 2% CO_2.[33] The quantity of SO_2 actually evolved and emitted in the blow gases is usually much less, depending on the pH of the cooking liquor and the pressure at which the digester contents are discharged. Raising the pH of the digester contents before blow lowers the free SO_2 in solution. Lowering the digester pressure before blow (to as low as 3 psi) and pumping instead of blowing out the contents are added measures for reducing SO_2 emissions. The SO_2 released to the atmosphere, however, depends on the effectiveness of the heat-recovery and SO_2-absorption systems employed for SO_2 recovery. The SO_2 in blow gases can be scrubbed with an alkaline solution of the base and returned to acid preparation, with recoveries as high as 97%.[33] This method is viable with sodium and NH_3 bases. However, magnesium and calcium bases require slurry scrubbers, which are less practical.[33] The SO_2 released during a cold blow from a blow tank is of a lesser magnitude (4–20 pounds SO_2 per ton of pulp[35]), which makes scrubbing impractical.

Many sulfite mills currently utilize displacement or diffusion washers, quite similar to those in kraft mills. Uncontrolled emissions from acid bisulfite washers and screens can be as high as 16 pounds SO_2 per ton of pulp,[33] although typical losses range from 1 to 4 lb/ton.[34] Scrubbing of SO_2 from the washers is generally accomplished by hooding and directing the collected gases to a direct-contact scrubber.

Spent sulfite liquor (SSL) is evaporated in multiple-effect evaporators similar to kraft liquors. The SO_2 generated during evaporation is pH dependent, ranging from 40 to 60 pounds SO_2 per ton of pulp for acid bisulfite cooking (pH <2) to less than 2 pounds SO_2 per ton of pulp for bisulfite (pH 2 to 6) and neutral sulfite (pH 6 to 9) cooking.[35] Sulfur dioxide from acid bisulfite liquors is usually treated by absorption in an alkaline solution of the base or sent to the acid plant for SO_2 recovery. Some mills practice weak SSL neutralization (usually magnesium-based process), which practically eliminates SO_2 emissions during evaporation.[33]

There exist a number of SSL recovery processes for bisulfite liquors using magnesium, sodium, and NH_3 bases. Either boilers or fluidized-bed reactors are used for SSL combustion. Efficient recovery of SO_2 from the combustion gases is critical for economical operation, as concentrations exceeding 1% (10,000 ppm) SO_2 can result in gases from SSL combustion. Recovery systems at most mills are closed and include the recovery furnace, DCE, multiple-effect evaporator, acid fortification tower, and SO_2-absorption scrubbers. Generally, there exists only one emission point for this recovery system. For magnesium-based liquors, multicyclones and venturi scrubbers are used to control recovery-system SO_2 emissions to levels below about 9 pounds SO_2 per ton of pulp.[32] Gases from NH_3-based liquor combustion are treated in an aqueous ammonia-absorption–heat-recovery tower. Glass-fiber filters for particulate and mist elimination are used before final emission to the atmosphere. The SO_2 emissions from NH_3-based recovery systems amount to less than 7 pounds SO_2 per ton of pulp.[32] The gases from sodium-based liquor-recovery furnaces are scrubbed in a sodium carbonate scrubber and recovery-system SO_2 emissions are <2 lb/ton pulp.[32]

Particulate emissions in the sulfite process result only from SSL combustion. Typical emission factors from recovery systems with additional scrubbing or absorption units for magnesium-, NH_3-, and sodium-based liquors are 2, 0.7, and 4 lb/ton pulp[32] respectively. The lower factor for NH_3-based liquors results from the NH_3 breaking down to N_2 and H_2O and becoming nonrecoverable. However, the sulfur content is emitted as SO_2 and recycled by absorption in aqueous ammonia to generate ammonium bisulfite solution. This is then used to make cooking liquor. A single set of tests on the uncontrolled exhaust gases from a sulfite-recovery furnace showed that 98% and 78% of the particulate emissions were less than 10 μm (PM_{10}) and less than 2.5 μm ($PM_{2.5}$) in size respectively.[36]

NEUTRAL SULFITE SEMICHEMICAL PULPING

Semichemical pulping is a process for obtaining high yields of pulps with characteristics suitable for certain end uses. Conventional chemical pulping achieves yields of 50% to 55%, whereas semichemical pulping can achieve from 60% to 80% yields. The yield is defined as the fraction of wood that results in unbleached pulp, both on a dry basis. The lignin is removed only partially during the cook and a second stage involving mechanical disintegration is necessary. Various types of semichemical pulps are produced by the acid sulfite, neutral sulfite, kraft, soda, and cold soda pulping processes. The major process difference from conventional chemical pulping lies in the use of lower temperatures, more dilute cooking liquor or shorter cooking time, and mechanical disintegration.[37]

Process Description

The neutral sulfite semichemical (NSSC) process is the most widely used semichemical pulping process today. In this method, wood chips are cooked in a neutral solution of sodium sulfite and sodium carbonate. Sulfite ions react with the wood lignin while the carbonate buffers the reaction, neutralizing the organic acid formed and maintaining a neutral pH of about 7. After cooking, the contents are discharged into a blow tank and excess liquor is separated by draining, pressing, or washing. Next, the softened chips are reduced to pulp by mechanical treatment in such equipment as rod mills or rotating-disk refiners.[37] The pulp is then washed in multistage drum filters and the weak liquor separated. Weak black liquor is either disposed of, recovered in furnaces, or blended with kraft liquors to "cross-recover" Na_2S and Na_2CO_3 from the liquor.

Air Emissions Characterization

Because of the milder pulping conditions, NSSC processes are expected to generate lesser amounts of pollutants than kraft or other full chemical pulping processes. Particulate emissions are a potential problem only when recovery furnaces are involved and are similar to those for kraft recovery furnaces. Fluidized-bed reactors are also utilized to burn NSSC spent liquor when a kraft furnace is not available. When recovered in a furnace, the combustion of sulfur containing NSSC liquor can result in TRS and SO_2 emissions that are somewhat higher than those from kraft black liquors. Lower NSSC liquor heat content resulting from lesser amounts of organic matter extracted from the wood and higher ratios of sulfur to sodium are believed to be responsible. With proper instrumentation and control of the furnace operation, Galeano et al.[38] have shown that NSSC recovery furnaces should be able to operate just as efficiently (with respect to air emissions) as comparable kraft recovery furnaces.

Preparation of the pulping chemical is carried out in a manner similar to the acid sulfite process. Sulfur dioxide from sulfur burners is absorbed in the green liquor containing Na_2CO_3 and Na_2S in scrubbing towers. Significant quantities of H_2S are also produced during this process, and they need to be further oxidized to SO_2 and scrubbed by Na_2CO_3 or some other base. Potential emission points of SO_2 are absorbing towers, digester/blow tank systems, and recovery furnaces. Owing to the great variations in the type of NSSC processes practiced and the paucity of measured emission data, emission factors are not readily available.

References

1. *Lockwood Post's Directory of the Pulp, Paper and Allied Trades, 1988,* Miller Freeman Publications, San Francisco, 1988.
2. S. A. Rydholm, *Pulping Processes,* 1st ed., Interscience Publishers, New York, 1965.
3. *Atmospheric Emissions from the Pulp and Paper Manufacturing Industry—Report of NCASI-EPA Cooperative Study Project,* Technical Bulletin No. 69, National Council of the Paper Industry for Air and Stream Improvement, New York, 1974.
4. H. Modetz and M. Murtiff, *Kraft Pulp Industry Particulate Emissions: Source Category Report,* EPA/600/7-87/006, U.S. Environmental Protection Agency, Research Triangle Park, NC, 1987.
5. *NCASI Handbook of Chemical Specific Information for SARA Section 313 Form R Reporting,* National Council of the Paper Industry for Air and Stream Improvement, New York, April 1990.
6. *Emission of Reduced Sulfur Compounds from Kraft Process Brownstock Diffusion Washer Vents,* Technical Bulletin No. 406, National Council of the Paper Industry for Air and Stream Improvement, New York, 1983.
7. J. C. Walther and H. R. Amberg, "A positive air quality control program at a new kraft mill," *J Air Pollut. Contr. Assoc.* 20(1):9 (1970).
8. *TGNMO Emission Potential from Kraft Process Heavy Black Liquor Oxidizers Operated on Liquors from Western Wood Species,* Technical Bulletin No. 371, National Council of the Paper Industry for Air and Stream Improvement, New York, 1982.
9. *Carbon Monoxide Emissions from Selected Combustion Sources Based on Short-term Monitoring Records,* Technical Bulletin No. 416, National Council of the Paper Industry for Air and Stream Improvement, New York, 1984.
10. *A Study of Nitrogen Oxides Emissions for Lime Kilns,* Technical Bulletin No. 107, National Council of the Paper Industry for Air and Stream Improvement, New York, 1980.

11. *A Study of Nitrogen Oxides Emissions from Large Kraft Recovery Furnaces,* Technical Bulletin No. 111, National Council of the Paper Industry for Air and Stream Improvement, New York, 1981.
12. *A Study of Kraft Recovery Furnace Total Gaseous Non-Methane Organic Emissions,* Technical Bulletin No. 112, National Council of the Paper Industry for Air and Stream Improvement, New York, 1981.
13. *Review of New Source Performance Standards for Kraft Pulp Mills,* EPA-450/3-83-017, U.S. Environmental Protection Agency, Research Triangle Park, NC, 1983.
14. K. E. McCance and H. G. Burke, "Contaminated condensate stripping—an industry survey," *Pulp Paper Canada* 81(11):78 (1980).
15. *Factors Affecting Emission of Odorous Reduced Sulfur Compounds from Miscellaneous Kraft Process Sources,* Technical Bulletin No. 60, National Council of the Paper Industry for Air and Stream Improvement, New York, 1972.
16. *Summary of Long-term CEMS SO_2 Data and Review of Factors Affecting Sulfur Dioxide Emissions from Kraft Recovery Furnaces,* Technical Bulletin No. 604, National Council of the Paper Industry for Air and Stream Improvement, New York, 1991.
17. *A Study of Kraft Process Lime Kiln Total Gaseous Non-Methane Organic Emissions,* Technical Bulletin No. 358, National Council of the Paper Industry for Air and Stream Improvement, New York, 1981.
18. J. S. Henderson and J. E. Roberson, "1971 precipitator survey," *TAPPI J.,* 56(4):91 (1973).
19. J. S. Henderson, "Final survey results for direct contact evaporator recovery boiler electrostatic precipitators," *TAPPI J.,* 65(3):91 (1982).
20. J. S. Henderson, "Precipitator survey on non-contact recovery boilers," *TAPPI J.,* 58(5):86 (1975).
21. J. S. Henderson, "Final survey results for non-contact recovery boiler electrostatic precipitators," *TAPPI J.,* 63(12):71 (1980).
22. *Kraft Pulping—Control of TRS Emissions from Existing Mills,* EPA-450/2-78-003b, U.S. Environmental Protection Agency, Research Triangle Park, 1979.
23. *Standards of Performance for New Stationary Sources: Kraft Pulp Mills,* U.S. Environmental Protection Agency [*Federal Register 51* 18544], May 20, 1986.
24. Data File Information, National Council of the Paper Industry for Air and Stream Improvement, New York, 1990.
25. *National Dioxin Study Tier 4—Combustion Sources, Project Summary Report,* EPA-450/4-84-014h, 1987.
26. *BACT-LAER Clearinghouse—A Compilation of Control Technology Determinations (4th Supplement to 1985 Edition),* U.S. Environmental Protection Agency, Office of Air Quality Planning and Standards, June 1989.
27. *Collection and Burning of Kraft Non-Condensible Gases—Current Practices, Operating Experience and Important Aspects of Design and Operation,* Technical Bulletin No. 469, National Council of the Paper Industry for Air and Stream Improvement, New York, 1985.
28. *A Review of Preconstruction Air Quality Permits Issued for Pulp Mill Emission Sources,* Special Report No. 90-03, National Council of the Paper Industry for Air and Stream Improvement, New York, May 1990.
29. *Fugitive Dust Emission Factors and Control Methods Important to Forest Products Industry Manufacturing Operations,* Technical Bulletin No. 424, National Council of the Paper Industry for Air and Stream Improvement, New York, March 1984.
30. A. K. Jain and V. J. Dallons, "Control of chlorine and chlorine dioxide emissions from bleach plants," *Proceedings of the 1989 TAPPI Environmental Conference,* TAPPI, 1989, pp 507–512.
31. *Results of Field Measurements of Chloroform Formation and Release from Pulp Bleaching,* Technical Bulletin No. 558, National Council of the Paper Industry for Air and Stream Improvement, New York, 1988.
32. *Compilation of Air Pollutant Emission Factors, Volume I: Stationary Point and Area Sources (AP-42),* U.S. Environmental Protection Agency, Research Triangle Park, NC, 1986, Chapter 10.1.
33. H. Edde, *Environmental Control for Pulp and Paper Mills, Pollution Technology Review No. 108,* Noyes Publication, Park Ridge, NJ, 1984, pp 148–156.
34. A. L. Caron, "Practices used by the sulphite pulping industry in the handling and treatment of sulfur dioxide from miscellaneous sources," *Proceedings of the 1976 NCASI Central-Lake States Regional Meeting,* Special Report No. 77-02, National Council of the Paper Industry for Air and Stream Improvement, New York, 1977.
35. *Environmental Pollution Control, Pulp and Paper Industry, Part I: Air,* EPA-625/7-76-001, U.S. Environmental Protection Agency, Washington, DC, October 1976.
36. *Air Emissions Species Manual, Vol. II: Particulate Matter Species Profiles,* U.S. Environmental Protection Agency, Research Triangle Park, NC, 1988, p 174.
37. M. Benjamin, I. B. Douglas, G. A. Hansen, et al., "A general description of commercial wood pulping and bleaching processes," *J. Air Pollut. Contr. Assoc.* 19(3):155–161 (1969).
38. S. F. Galeano, D. C. Kahn, and R. A. Mack, "Air pollution: Controlled operation of a NSSC recovery furnace," *TAPPI J.,* 54(5):741 (1971).

MECHANICAL PULPING

Arun V. Someshwar and John E. Pinkerton

Mechanical pulping, as the name implies, relies mainly on mechanical energy to convert wood to pulp. Mechanical pulping dates back to 1840, and the invention of the pulpwood grinder[1] and the stone-ground wood (SGW) process. In addition to the SGW process, current mechanical pulp manufacturing processes comprise several high-energy refining systems for the production of pulp from chips. These include the refiner mechanical pulping (RMP) process, the thermomechanical pulping (TMP) process, the chemimechanical pulping (CMP) process, and the chemithermomechanical pulping (CTMP) process.

PROCESS DESCRIPTION

In the RMP process, chips are refined directly at atmospheric pressure. Chemicals are sometimes added at the various stages of refining. In the TMP process, the chips are usually steamed under a pressure of 20–40 psi for two to four minutes prior to refining. In some modifications of this process, the refiners are operated under pressure. In the mid-1970s, chemically modified versions of RMP and TMP were introduced. Up until the present, the chemical treatment used has been almost exclusively sulfonation.[1] This may be carried out in several ways, including the treatment of wood chips prior to refining, the treatment of coarse pulp between refining stages, the treatment of completely refined pulp, and the treatment of long fiber. Chips are usually cooked (softened) in a sodium sulfite solution (pH between 4 and 9) for about 30 minutes at temperatures between 270°F and 320°F. In the CTMP process, the chips are treated with chemicals for softening and refined under pressure. The usual level of addition of sodium sulfite is kept between 1% and 4% Na_2SO_3 on a bone-dry pulp basis.[2]

The CMP and CTMP processes reduce power consumption at the refiners. The RMP and TMP processes are known for their high consumption of electric energy. Pulp yields of 95% or higher are common with the TMP and RMP processes, whereas the CMP and CTMP processes exhibit yields of about 90%.

AIR EMISSIONS CHARACTERIZATION AND CONTROL MEASURES

Large quantities of steam are generated from mechanical and chemimechanical pulping processes. Moisture emission rates are reported to range from 4000 to 7500 lb/ton of pulp for systems not equipped with heat recovery.[3] As wood contains a fair amount of organic material, volatile organic compounds (VOCs) are likely to be released along with the steam during the cooking and refining process.

Heat recovery from steam emissions is extensively practiced. A study[3] by the National Council of the Paper Industry for Air and Steam Improvement on estimating VOC emissions from TMP processes showed the following: (1) Emissions of VOCs from the TMP process operated on western white wood species ranged from 1.09 to 1.73 and averaged 1.4 pounds of carbon per ton of pulp; from western pine species, it ranged from 0.83 to 3.84 and averaged 1.9 pounds of carbon per ton; and from southern pine species, it ranged from 2.2 to 7.6 pounds of carbon per ton. (2) Emission rates for VOCs were proportional to moisture emission rates, indicating that those heat-recovery systems that drop the exhaust gas temperature well below the boiling point of water would also be expected to reduce VOC emissions by an unknown amount. No emissions other than VOCs are expected from mechanical pulping processes.

References

1. D. Atack, "Mechanical pulping—some highlights of the second 75 years," *Pulp Paper Canada,* 90:10 (1989).
2. D. M. Mackie and J. S. Taylor, "Review of the production and properties of alphabet pulps," *Pulp Paper Canada,* 89(2):58 1988).
3. *TGNMO Emissions from the Thermomechanical Pulping Process,* Technical Bulletin No. 410, National Council of the Paper Industry for Air and Stream Improvement, New York, 1983.

PAPER AND PAPERBOARD MANUFACTURE

Arun V. Someshwar and John E. Pinkerton

Pulping operations using chemical or mechanical processes were discussed in earlier sections of this chapter. Following the washing operation (and bleaching, if performed), the pulp or stock is pumped to high-density storage tanks. From the pulp slurry, thin sheets (paper) or mats (paperboard) are manufactured in the paper mill area on a paper machine (see Figure 1). Individual pulp mills may or may not have paper machines, and mills with stand-alone operations may purchase baled pulp for paper and paperboard making. To obtain paper or board sheet from a pulp suspension, three operations are performed: stock preparation, sheet formation, and drying. Further operations, finishing and converting, yield a final product. Converting is often carried out in stand-alone plants located in the consuming districts.

PROCESS DESCRIPTION

Stock preparation involves mixing of various types of pulps and additives and beating of the pulp fibers. Pulps that are

FIGURE 1. A Typical Paper Machine

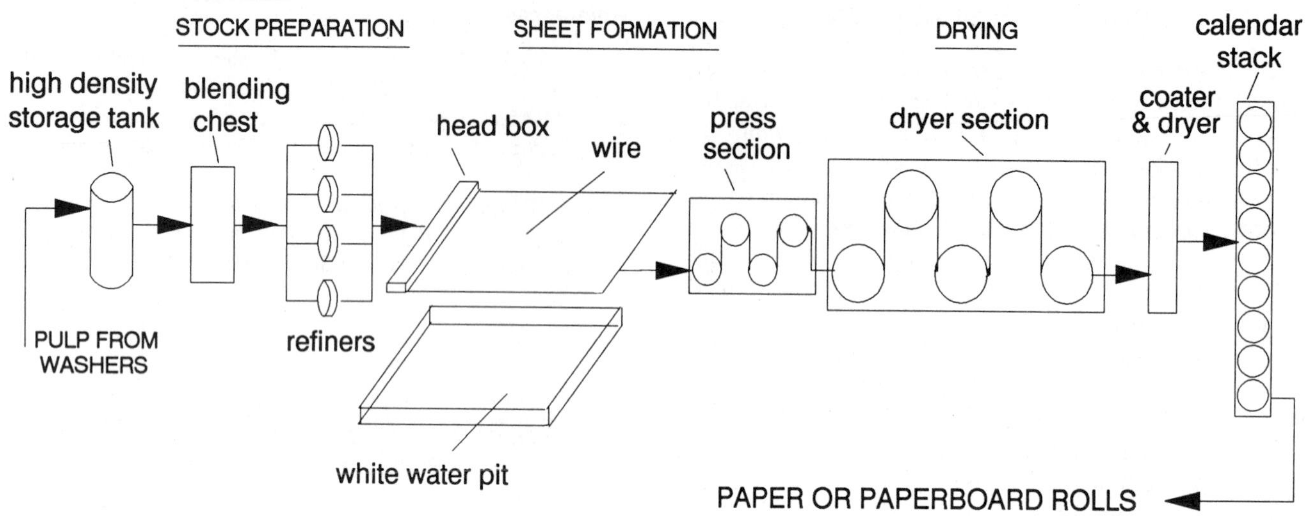

FIGURE 2. Typical Layout of a Paper Mill

derived from virgin wood are termed "virgin" pulp. In recent times, there has been a large increase in the use of "secondary" fiber, derived from recycling paper, for pulp and paper products. Post- or pre-consumer recycled paper is commonly grouped into five categories: deinked, mixed paper, pulp substitutes, newspaper, and corrugated.[1] Depending on the category, one or more of the following processes—asphalt dispersion, cleaning, deinking, pulping, and screening—are used for pulp production. Pulping chemicals used in recycled paper deinking may include caustic as a difibering agent, sodium silicate as a stabilizer, other deinking chemicals as dispersants and collectors, calcium chloride for water conditioning, and hydrogen peroxide for bleaching and preventing yellowing of groundwood.[1] Process chemicals include defoamers and acid for pH adjustment. Bleaching can range from minimal (involving H_2O_2 during pulping and hydrosulfite at end of the deinking process) to extensive (such as a chlorination–hypochlorite or chlorination–extraction–hypochlorite sequence).[1]

Figure 2 shows a typical layout of a paper and paperboard mill. Stock from the high-density storage tanks is mixed in blending chests to control consistency and the mixed pulp is sent to a system of "beaters" or "refiners," which further treat the fibers to obtain the desired finished-product characteristics, such as paper strength. After stock preparation, the stock is pumped to the head boxes of the paper machine prior to which it is diluted to about 0.5% consistency. From the head boxes, the dilute stock is uniformly distributed across the "wire" of the paper machine. About 98% of the water is removed by gravity and vacuum while the stock is on the wire. The newly formed sheet, still very soft and wet, then passes through the press section of the paper machine, where the sheet is smoothed and additional moisture is removed. The sheet then enters the "dryer" section, where steam-heated drying cylinders are used to evaporate the rest of the moisture in the sheet. After drying, many printing grades of paper and paperboard are surface coated with an aqueous suspension of pigments (such as clay) in adhesives (such as starch) in order to improve surface properties. After coating or sizing, the solvent, usually water, is removed from the coating by evaporative drying, first by air impingement or radiant heating and then over steam-heated drums. Finally, the coated sheet is "ironed" in the calendar stack and then wound in large rolls, ready for shipment.

AIR EMISSIONS CHARACTERIZATION

Emissions from the paper machine consist mainly of water vapor; little or no particulate matter is emitted from the dryers.[2] Many papers and paperboards contain noncellulosic additives that improve the use properties of the final product. Additives may be put into the paper stock either before the sheet is formed or later at the size press, the calendar stack, or in a subsequent converting operation.[3] The beater or wet-end method of incorporating additives is generally preferred. Thus only small fractions of these additives are expected to be retained in the sheet before they enter the dryer section. Additionally, because of their high boiling points, very small quantities are likely to volatilize and result in air emissions. Chemicals that may volatilize during the making of paper or paperboard include formaldehyde and phenol present in resins; ammonia added as a coating or for pH adjustment; chemicals such as hexane, xylene, petroleum naphtha, and toluene that may be used periodically in cleaning operations; and minute quantities of the free, unreacted monomeric constituents of the various additives. Because of the variety of chemicals used and conditions of usage, paper machine vent emission estimates have not been developed. Ammonia and formaldehyde are two compounds that have been identified as being present in paper machine vents where ammonia- or formaledhyde-based additives are used in the papermaking process. Estimates of air emissions during the recovery of secondary fiber are not available. However, these are expected to be minimal.

AIR POLLUTION CONTROL MEASURES

Control techniques for paper machine vents are considered impractical because of the high moisture content and high volume of the vent exhaust gases and the minimal pollutant concentrations. Waste minimization techniques, such as reduced chemical usage and modifications in the mode and location of chemical addition, may be considered on a site-specific basis where reductions are desirable.

References

1. L. A. Broeren, "New technology, economic benefits give boost to secondary fiber use," *Pulp Paper,* 69 (November 1989).
2. *Compilation of Air Pollutant Emission Factors, Volume I: Stationary Point and Area Sources (AP-42),* U.S. Environmental Protection Agency, Research Triangle Park, NC, 1986, Chapter 10.2.
3. *Pulp and Paper Science and Technology, Vol II. Paper;* C. E. Libby, Ed.; McGraw-Hill Book Co., New York, 1962, p 113.

19
Treatment and Land Disposal

TREATMENT AND LAND DISPOSAL

T. T. Shen, C. E. Schmidt, and T. P. Nelson

Approximately 30,000 potentially contaminated land disposal sites have been identified in the United States. Among them, 1170 have been listed on the National Priorities List for cleanup.[1] Improper land treatment and/or disposal is one of the most pressing environmental problems facing us today. The environmental consequences of improper land treatment and disposal have resulted in the contamination not only of local groundwater and surface water, but also of the land, air, and food and forage crops.

Air emissions from land treatment and disposal processes include particulate matter (PM) and gases, primarily volatile organic compounds (VOCs). The U.S. Environmental Protection Agency's (EPA's) air emission control strategies for land treatment and disposal are to reduce VOC emissions from various sources to protect public health and the environment. Each land treatment or disposal site is given a specific control level for reducing VOC emissions, restricting the cross-media effects of waste treatment and disposal practices. Control of VOC emissions will ultimately help reduce ozone formation and cancer risks.

This discussion begins with an overview of various land treatment and disposal processes and their sources of VOC emissions. Emphasis will be placed on air emissions characterization and source emission control options.

PROCESS DESCRIPTION

Land treatment and disposal processes may include landfill, surface impoundment, land treatment, and deep well injection. For deep well injection, the major environmental concern is groundwater contamination. Air emissions from well injection appear to be insignificant and thus will not be covered.

Land disposal waste materials are subject to varioustransport processes that may lead to environmental pollution. Figure 1 presents an overall view of initial transport processes at waste disposal sites. The environmental effects arising from the pollution of land disposal sites can be localized or widespread, direct or indirect, short term or long term. Figure 2 illustrates the flow of land-disposed waste contaminants through the environment.[2]

A landfill is defined as a disposal facility or part of a facility where waste is placed in or on land and which is not a land treatment facility, a surface impoundment, or an injection well. Landfilling is the burial of bulk or containerized waste in excavated trenches or cells. Figure 1 also shows VOC emissions from several emission points of a landfill. There are two typical process stages in landfill processing, each of which is a potential VOC emission source. The landfill surface, whether open or covered with soil daily, is an emission source. Waste piles are similar to landfills and the same emission sources can be found there; they are, in essence, temporary landfills. Landfill design and operation are generally determined by local topography, depth of groundwater, and availability of natural clay formations that act as liners. In addition to site topography, the method of filling depends on waste characteristics. Dry, immediately workable waste may be spread in horizontal layers of 6 inches to 2 feet in thickness, with successive layers occurring in heights of 10 to 20 feet. These layers are commonly referred to as lifts. Containerized wastes disposed of in such fashion require careful placement and covering. Void spaces between the containers must be filled to prevent uneven settling of the cover and premature rupture of containers. Special requirements for containerized wastes and incompatible wastes are outlined in 40 CFR 264.

Air emission rates during the early stages of waste disposal are a function of the concentration of each chemical

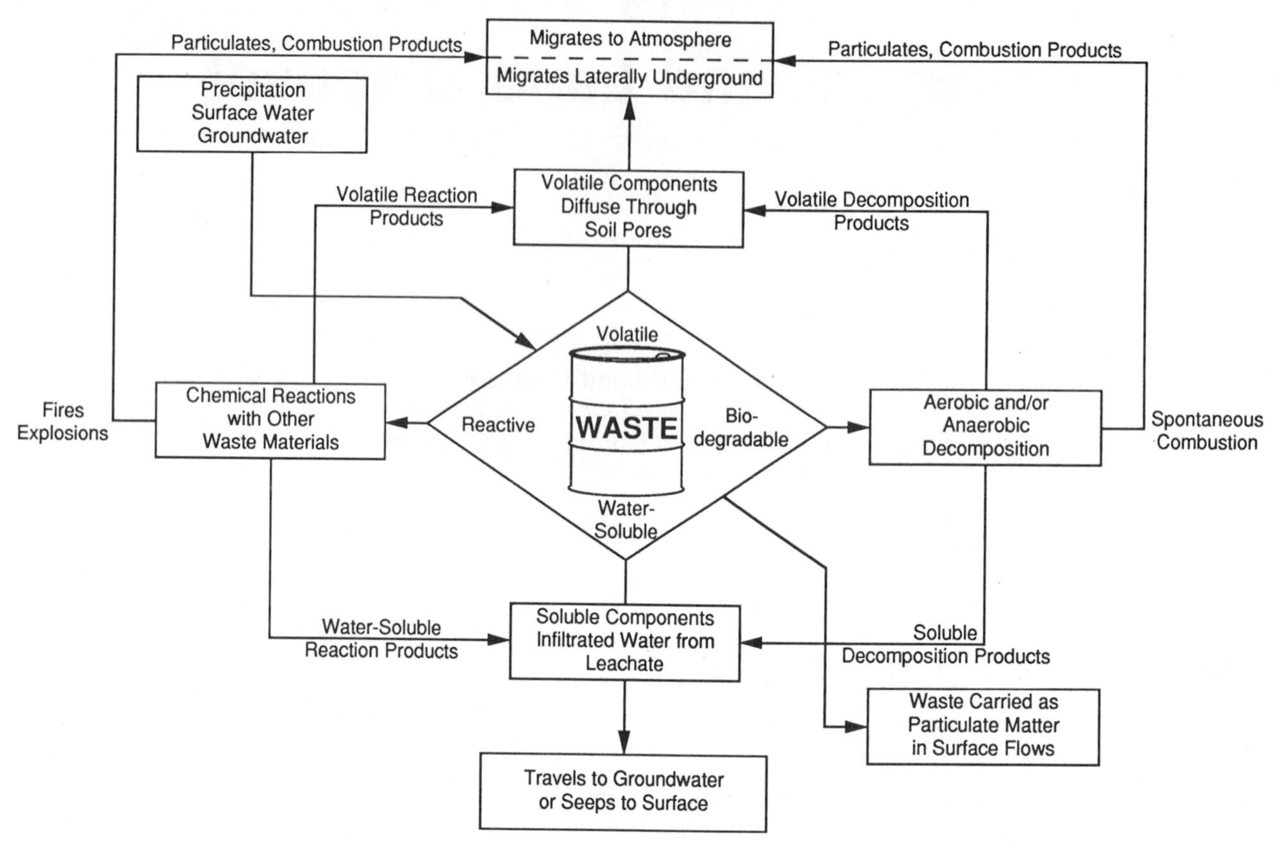

FIGURE 1. Initial Transport Processes at Waste Disposal Sites (From Reference 2)

constituent of concern in the waste, in addition to its vapor pressure and diffusivity. Emission rates after the waste material has been covered are a function of the type of covering material (soil); its moisture content, porosity, and thickness; and the effective diffusivity of the cover material. The state of California recently instituted a "Landfill Gas Testing Program" that is providing significant information on bulk and toxic gases from landfills.[3]

Codisposal (i.e., combining hazardous waste with municipal refuse) can potentially increase air emission rates from a landfill. The anaerobic decomposition of municipal refuse generates such gases as methane and carbon dioxide. The convection effects (upward motion) and bulk flow created by these gases convey entrained volatile chemical vapors to the landfill surface and sometimes laterally away from the landfill. This generation of landfill gas (carbon dioxide and methane) is the dominant force for moving VOCs out of these waste disposal sites. Research results indicate, however, that changes in atmospheric pressure affect volatilization rates from landfills. Computer simulation studies have shown that the barometric pumping effect can increase emission rates by as much as 15%. With diffusion, in limited situations, like non-codisposal hazardous waste sites, the effect of atmospheric pressure is more significant.

A surface impoundment is defined as a facility or part of a facility that is a natural topographic depression, a man-made excavation, or a diked area formed primarily of earth materials (although it may be lined with manufactured materials) that is designed to hold an accumulation of liquid wastes or wastes containing free liquids. Examples of surface impoundments are holding, storage, settling, and aeration pits (46 CFR 27476, May 20, 1981). They may range in depth from 2 feet to as much as 30 feet below the land surface and generally are built above the water table.

Industrial and municipal wastes processed by surface impoundments are highly variable in chemical composition and quantity. The major source of VOC emissions from surface impoundments is the uncovered liquid surface exposed to the air. Emissions of VOCs occur as a result of the volatilization of organic constituents from the liquid mixture. The primary environmental factors that influence the emission rate potential from surface impoundments are temperature and wind velocity. Emissions tend to increase with an increase in surface turbulence caused by the wind or mechanical agitation.

Land treatment of a waste can be described as the application of waste onto land and/or incorporation into surface soil, sometimes with the addition of a fertilizer or soil conditioner. Land treatment may include land farming, land application, land cultivation, land irrigation, land spreading, soil farming, and soil incorporation. Wastes added to the soil environment are subject to decomposition, leaching, volatilization, and assimilation. The main purpose of land treatment is to employ the microbiological activity

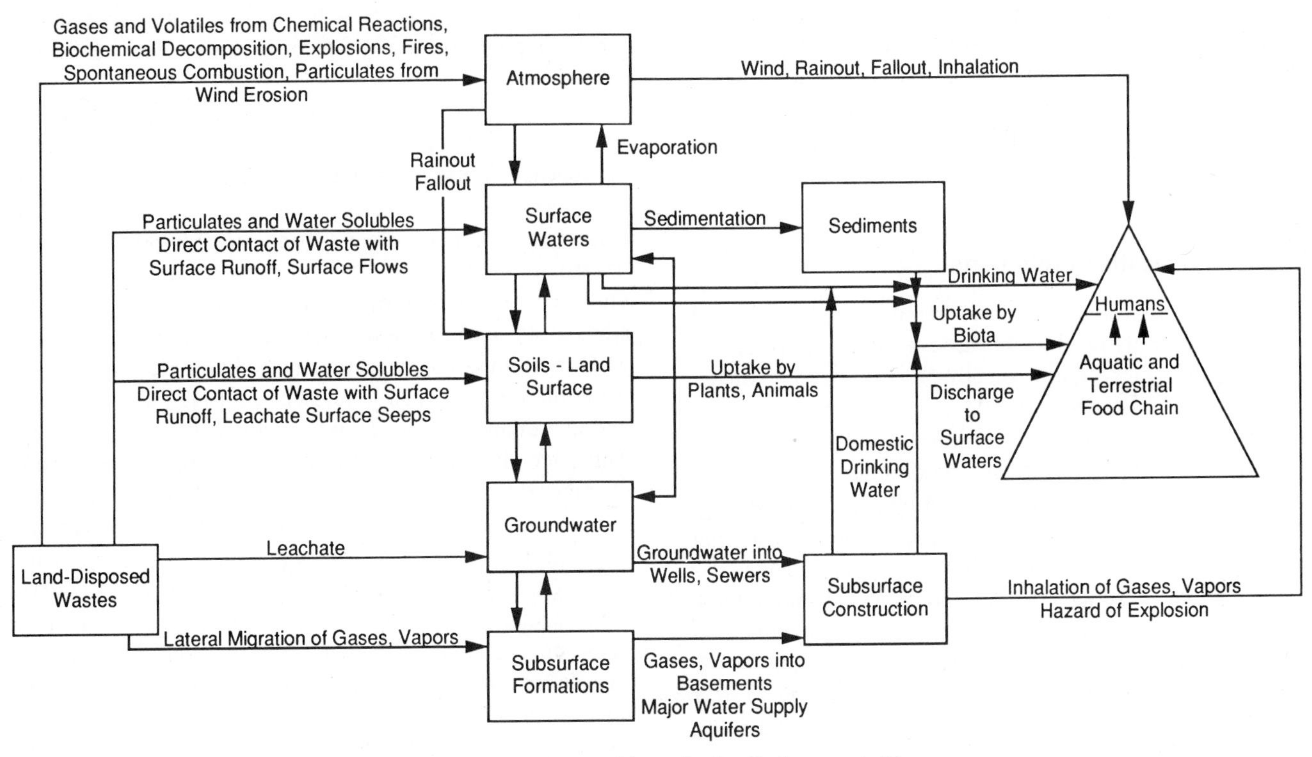

FIGURE 2. Flow of Land-Disposed Waste Contaminants Through the Environment (From Reference 2)

of the upper layers of the soil to decompose organic waste constituents into elemental forms. Since the upper soil layers contain the largest microbial population, land treatment is generally confined to the soil plow zone, that is, the first 6 to 8 inches of soil.

Waste oils and sludges are generally applied to the soil by surface spreading or subsurface injection. Wastes can be effectively mixed into the soil plow zone using a disk or rototiller. The number and frequency of waste–soil cultivations are site dependent. Some operators let fluid waste dry out prior to blending, while others cultivate waste into soil immediately after application. Air emissions from land treatment processes are a function of waste characteristics, soil type, temperature, loading rate, and mode of application. The modes of application include surface spreading and surface injection. Surface spreading has more potential for volatilization than does subsurface injection.

Among other activities that are likely to occur in relation to land treatment processes are transfer, storage, handling, and dewatering of the wastes to be land treated. Examples would include loading and unloading of wastes in vacuum trucks and dewatering of wastes using one of the various types of available filtration devices.

Each of the land treatment or disposal process steps is a potential emission source associated with waste transfer, storage, and handling operations. Loading operations contribute to overall emissions, especially splash loading of waste as opposed to submerged loading. Spills also occur during waste transfer and handling, and for liquid wastes that are pumped, emissions may arise from fugitive sources such as pumps and valves or at open-ended lines, pressure-relief valves, and sampling connections. Emissions of VOCs can result from drums, dumpsters, and tanks. Other sources also contribute to VOC emissions, including drum cleaning or the crushing and landfilling of empty drums containing waste residues. The improper handling of drum residues can lead to VOC emissions, along with lost waste residues.

AIR EMISSIONS CHARACTERIZATION

The air emissions from land treatment and/or disposal practices include gases, VOCs, and PM. They are released from point sources and area sources; however, the area or fugitive sources are the major sources of air emissions. Point sources include vents from landfills and from covered and ducted processes, such as covered solidification operations, liquid–solid transfer operations, and pretreatment activities. These point sources are generally low-temperature and low-pressure gas-phase emission sources. They can easily be characterized using standard point source testing technologies, which involve sampling the exhaust stream, measuring velocity and duct diameter, and calculating emission rates. Since point sources are not the dominant types of emission sources from land disposal practices, they will not be discussed further.

Fugitive air emission sources from land disposal practices are variable, heterogenous, and generally difficult to characterize in comparison with point sources. They are usually related to the surface area of waste exposed, the composition of the waste, and the age of the waste. Certainly, these sources are difficult to generalize by the type of land treatment or disposal practice because differences in operation, design, and, most important, the type of waste material disposed are usually site specific and somewhat unique.

Particulate Matter Emissions

Fugitive PM emissions from land treatment and/or disposal practices generally result from waste-handling activities and include inorganic and sometimes toxic compounds. Although the generation of PM occurs at the surface of landfills, land treatment facilities, and surface impoundments from wind effects, the dominant mechanisms include dumping, grading, compaction, covering, tilling, mixing, and aeration activities. All physical waste handling and disturbance have the potential for generating PM emissions, but the leading related sources are (1) landfills—dumping, grading, vehicle traffic; (2) land treatment—tilling, vehicle traffic; and (3) surface impoundments—aeration. The site distribution and concentration of PM generated depend on the waste matrix and physical activity causing the disturbance. Likewise, the chemical composition of the PM depends on the waste material.

Technologies for characterizing PM emissions from disposal practices are somewhat limited and involve measuring the resulting downwind PM concentration and estimating PM emission rates using models.[4] Typically, an array of PM samplers is positioned upwind (one or two stations) and downwind (three to eight stations) of the fugitive PM source, and upwind and downwind PM concentrations are measured by collecting samples on a filter while monitoring meteorological conditions. Filters are weighed before and after sampling to obtain PM loading and the PM sample is then generally analyzed to determine chemical composition. Size fraction sampling can also be conducted to determine the size distribution of PM, the coarse and fine fractions, or the respirable size fraction (less than 10 μm). Emission rates are then estimated, knowing the dispersion conditions during the test, using a dispersion model.

Gases and Volatile Organic Compounds

Gas-phase fugitive emissions from land treatment and disposal practices originate from a variety of waste surfaces, including solids, sludges, and liquids. In addition, gas-phase emissions can migrate through soil covers or laterally into the natural soil formations or into man-made conduits and then to the atmosphere. Gas-phase emission rates can be significant, resulting in part-per-billion to part-per-million levels of gas species downward of the facility. This is especially true with activities, such as waste aeration, that generate high unit emission rate factors. In addition, facilities that have low unit emission rates can also generate high downwind concentrations if the facility has a large surface area, since fugitive emissions are directly related to surface area.

There are four approaches to characterizing area or fugitive source emissions, three of which are measurement approaches. These approaches are described in volume II of the guidance document series developed by the EPA to assist site managers in conducting air-pathway analyses at hazardous waste sites.[5] The technologies described in this guidance series are directly applicable to assessing emissions from land disposal practices.

Direct emission measurement technologies involve the isolation of representative surfaces emitting gaseous species and quantitating the air emission rate of the surface tested. These technologies, given in Table 1, employ a device or chamber to isolate these surfaces, generally add sweep air to the device, mix the contents, and collect and analyze the resulting exhaust gas stream. Knowing the test parameters (exposed surface area and sweep air flow rate) and the concentration of species in the exhaust gas stream, the emission rate is then calculated from measured parameters. The direct emission measurement approach has the advantage of being a measurement approach and does not involve modeling to estimate emissions. It can also provide emission rate data with very good detection limits and data free of upwind interferences. There is, however, the requirement for representative testing, which involves characterizing each similar type of emissions source. The other major concern when using this approach is that the measurement activity may somehow affect the emissions event, resulting in a biased estimate of the emission rate. The EPA has provided an operations manual for constructing and using a recommended surface isolation emissions flux chamber.[6] This technology has been used successfully on solid, sludge, and liquid surfaces to develop fugitive emission rate data.

One other popular approach to measuring emission rates from fugitive sources is the indirect measurement approach. Here, the effect of the fugitive source is measured downwind of the source (ambient concentrations) and the emission rate is modeled using technology-specific models (see Table 1). Field data (wind speed, direction, temperature, ambient concentration) are collected following model-specific directions so that these data qualify as input to these models. Ambient concentrations can be obtained using a variety of sampling techniques, including fixed station or point sampling (sorbent sampling, whole air sampling), mobile monitoring, and line monitoring (remote sensing). The advantage of using the indirect measurement approach is that the emission measurement can be performed for any land disposal facility or practice regardless of waste homogeneity. The disadvantages of indirect measurement technologies are that they all require modeling to estimate emission rates, rely on dispersion and transport phenomena, and are subject to upwind interferences.

The third approach to measuring emission rates is known

TABLE 1. Application of Emission Rate Measurement Sampling Approaches to Area Emission Sources

Area Source Emission Assessment	Area Source	Limitations/Comments
Surface isolation flux chamber[a]	Active landfills	Most effective small cells with uniform waste composition.
	Inactive landfills	Can be used on surface and for vents at inactive landfills.
	Surface impoundments	Must float equipment.
	Land treatment	Subject to treatment cycle variabilities.
Headspace samplers[a]	Applications similar to the surface isolation flux chamber	Typical use is for concentration measurements; data used for relative comparison purposes.
Wind tunnels[a,b]	Inactive controlled and uncontrolled landfills	Used to estimate emissions under simulated wind flow.
	Surface impoundments	Can be difficult to perform sampling because of air supply needs.
	Waste piles	Provides estimates of particulate matter emissions.
Subsurface direct emission measurement technologies	Inactive controlled and uncontrolled landfills	Used to measure soil gas concentration or emission rates subsurface.
	Subsurface contamination	Typically used to identify and map subsurface contaminants via soil gas concentration. Can be used to estimate emissions from disturbed waste conditions.
Concentration-profile technique[a]	Surface impoundments, land treatment	Requires complex equipment. Meteorological conditions must meet criteria; not suited for small impoundments or land-treatment plots.
Transect technique[a,b]	Active landfills, surface impoundments, land reatment, drum storage	Meteorological conditions must meet criteria; requires minimal interferences from other upwind sources.
Upwind/downwind technique[a,b]	All TSDF facilities and uncontrolled waste sites	Emission estimate limited, technique typically used as survey technique in the development of a program to represent emissions more accurately.
Mass balance technique[a]	Most TSDF facilities, process units	Must identify and be capable of measuring all streams.
Air monitoring/modeling technologies[a,b]	TSDF or hazardous waste site fence line	Meteorological conditions, terrain, and upwind interferences will affect utility; analytical sensitivity is usually a limiting factor.
Predictive modeling[a,b]	Most TSDFs, uncontrolled landfills/ lagoons	Models require site-specific input data to be representative.

[a]Volatile species emissions.
[b]Particulate matter emissions.

as fence-line monitoring and modeling. This approach is like the indirect emission measurement approach, except that the air concentration measurements are performed downwind and a less rigorous sampling design is generally employed. Standard dispersion models rather than technology-specific dispersion models are used to estimate fugitive emission rates. This approach is generally used when facility fence-line monitoring is conducted as part of standard operating practice. The disadvantages of this approach are similar to those identified for the indirect measurement approach, except that further downwind of the source, ambient concentrations are typically lower and more vari-

TABLE 2. Listing of Predictive Models for Estimating Emissions for Area Sources[a]

Predictive Model	Application	Comment References
Farmer original	Covered landfill	Developed for hexachlorobenzene waste[12]
Farmer modified	Covered landfill	Modification of Farmer model to include porosity term[12,13]
Shen (Farmer)	Covered landfill	Simplification of Farmer, some oversimplification[13–17]
Hwang (Farmer)	Covered landfill	Modifications of Shen model to correct oversimplification[18]
Farmer as modified by Shen and Hwang	Covered landfill with codisposal	Includes factor for biogas generation at codisposal sites[19,20]
Thibodeaux	Covered landfill	Two-resistance theory model[13]
	Covered landfill with codisposal	Accounts for biogas flow through soil[13]
	Covered landfill with codisposal and barometric pumping	Include effects of changes in barometric pressure[13]
RTI closed landfill	Covered landfill	Based on Farmer model. Accounts for barometric pumping and decline in emission rate with time.[21]
RTI land treatment	Open landfill, waste pile, land treatment	Applicable to waste mixed with soil[21]
Arnold modified by Ziegler	Open landfill	Includes wind speed variable.[15–17]
RTI open dump	Open landfill	Derived from Arnold model for spills. Applicable to uncovered waste.[21]
MacKay-Leinonen	Nonaerated surface impound	Non-steady-state model based on two-film resistance theory.[20,22–24]
Thibodeaux, Parker, and Heck	Aerated and nonaerated surface impoundment	Based on two-film resistance theory.[19,20]
Smith, Bomberger, and Haynes	Nonaerated surface impoundment	For highly volatile species only.[20]
Surface Impoundment Modeling System (SIMS)	Aerated and nonaerated surface impoundment	For all species; includes biodegradation factors.[26]

[a]This table is not inclusive of all available predictive emission models.

TABLE 3. Typical Emission Ranges for Uncontrolled TSDFs

Facility	Pollutant	Emission Range
Landfills	VOC	4.2×10^{-6}–1.1×10^{-4} kg/m^3-day
Surface impoundments	VOC	6.2×10^{-6} kg/m^3-day
Waste piles	PM	1.6×10^{-6}–6.3×10^{-6} kg/m^3-day
Soil handling	PM	27–170 kg/day

Source: Reference 27.

able, resulting in less sensitivity of the emissions estimation technology and higher imprecision.

Emission Factors

The remaining approach for characterizing emission rates from fugitive sources for both PM and gas-phase species involves predictive modeling and/or the use of empirically derived emission factors. This approach uses site-specific input parameters and descriptive models or emission factors to estimate emissions. Predictive models that can be used for land treatment and disposal practices are listed in Table 2 and empirically derived emission ranges for land disposal activities (waste handling) are given in Table 3. When used with representative model input parameters, emission factors or predictive modeling can provide representative characterization of emission rates from land disposal facilities.

AIR POLLUTION CONTROL MEASURES

Numerous control options have been suggested for the control of VOCs, toxic chemicals, and fine particulates from land treatment and disposal facilities. The options first center on the reduction of emissions through containment techniques. Then, if possible, the emissions may be routed to a control device for recovery or destruction. Finally, preventive measures have been recommended to minimize the formation of air pollutant emissions and thus eliminate the need for containment or control. The following sections address the current technologies available in each of these areas.

It should be pointed out that the EPA has promoted and continues to promote the development of new technologies through the SITE (Superfund Innovative Technology Evaluation) program. Therefore, new technologies are continuously being evaluated for handling and disposing of hazardous wastes with land disposal techniques. In these evaluations, the EPA considers air pollution and controls and innovative measures for reducing emissions.

Treatment Systems

Air pollutant emissions can be directly controlled through the use of treatment systems. These systems can be generally classified as active and inactive controls. With active controls, operators of the treatment, storage, disposal facility (TSDF) will be required to operate equipment designed to reduce PM or VOC emissions. In the case of inactive controls, the technology includes the installation of the treatment system and no further activities by the operator. Most of the treatment technologies focus on the enclosure of the TSDF activities.

For landfills, enclosures such as inflatable domes have been suggested for reducing emissions around active areas. Within the dome, ventilation air will have to be provided to keep ambient levels below the level specified by health standards for the safety of the workers. Also, dust emissions and VOC emissions may be further reduced through the addition of suppressants such as foams, chemical sprays, or water. The inflatable dome will have a discharge point for air within the dome. The need for controls for this source will depend on the quantities of emissions and the regulatory needs for the area. With a high stack discharge point, the emissions from the landfill activities will provide added dispersion and relatively good possibilities that no add-on controls will be required. If add-on controls are needed, they will be large and costly (add-on controls are discussed in a later section).

Solidification buildings at landfills should also be enclosed and have localized ventilation and controls for particularly high PM or VOC emissions sources. These areas include locations where materials are being dumped or where mixing of dusty or highly volatile material is occurring. The local capture systems could be specially designed hooding systems.[7] The exhaust air from the solidification building may need an add-on control for dusts and possibly for VOCs.

Once the landfill has been completed, the emissions from the landfill area are reduced by a relatively impervious covering material such as clay. Particulate matter is controlled by the addition of a vegetation cover over the clay cap. The VOCs contained in the landfill can be controlled by a ventilation system consisting of gathering wells and a VOC control device such as a flare or an adsorber. The spacing of the recovery well depends on the waste properties and the well depth should equal three-fourths of the depth of the waste. A vacuum system is used to remove the vapors from the landfill.

For surface impoundments, VOC treatment systems also include inflatable enclosures or the equivalent.[8] In cases were VOC emissions are felt to be low and the surface is quiescent, simple covering materials such as floating rafts or spheres may be used to reduce the surface area of the impoundment and further reduce emissions. In the situations where the inflatable dome is used, emissions can be captured and discharged through the ventilation system. Many surface impoundments are small. Inside the dome with an aerated system, the air will be very humid and will possibly contain a high concentration of organics. In these cases, access to the inside of the dome is more difficult and will require full protection clothing for worker and maintenance people. This is an added cost for this treatment system.

Open tanks and containers should have covers if they are small and fixed roof tanks with double-sealed internal floating roofs if they are large (greater than 25,000 gallons). If nitrogen blanketing is used, the nitrogen off-gas should be routed to a flare.

Fugitive emissions of PM and VOCs are also a concern at TSDFs. Fugitive PM generally arises from vehicular traffic on dusty roadways. Emissions can be reduced up to 90% or more through the application of asphalt or concrete surfaces on all roadways. Chemical or water sprays can be used for unpaved surfaces. For VOCs, there are uncontrolled gaseous emissions from any operation or equipment, such as leaking valves, pump fittings, flanges, storage tanks, and sampling and instrumentation connections. Such fugitive emissions can be effectively reduced through equipment selection and the establishment of adequate inspection and maintenance programs.[9] Emissions reduction achievable with such programs is limited primarily by the number of components used in the operation, the frequency of monitoring, the ability to repair leaks, and the volatility of the organic waste. About a 60–75% reduction of fugitive emissions associated with covered or closed processes is possible with a quarterly inspection program. Higher reductions are possible through the use of "leak-free" fittings and more frequent inspections. Fugitive emission controls and test methodology for equipment leaks have been established under Sections 111 and 112 of the Clean Air Act.

Land treatment facilities have peak VOC emissions during the sludge or waste application and tilling operations. Control strategies include subsurface application of the sludges and longer periods between tillings. Proper soil conditioning is important to enhance biodegradation activity and to prevent anaerobic activity in the treatment plot. Enclosures may be possible for small land-treatment facilities.

Control Devices

Emission control devices can be used to reduce PM and VOC emissions from enclosures as described above and from process vents. Detailed discussions of add-on control devices are provided in other parts of this manual. Because these control devices are reducing emissions of toxic air pollutants, the control efficiency requirements are generally high, often higher than for a criteria pollutant. In the regulatory environment, the control level will usually be determined through dispersion modeling of the individual toxic species and then assessment of the ground level impacts around the TSDF facility. With compounds that are considered highly toxic, the impact screening levels can be very low.

Dust emission can be effectively controlled with fabric filter devices or wet scrubbers. A properly sized fabric filter operating under dry conditions and within the temperature limitations of the fabric can provide extremely efficient collection (greater than 99%) over a wide range of particle sizes. The dust is removed from the filter by a shaking or air blowing process. The collected PM can be disposed of in a secure landfill. The wet scrubber can be used in conditions that are not suitable for the fabric filter. Most wet scrubbers can operate efficiently in collecting large particles ($>2\ \mu m$); some are also efficient for very small particulates ($<0.2\ \mu m$). Therefore, for many wet scrubbers, there is a particle size range for which removal is less efficient. Particulate collection generally increases as energy input increases.[10]

Thermal incineration for VOC control is a common hazardous-waste control device. It is applicable to a wide variety of organic compounds and is not very sensitive to hazardous compound characteristics or waste-stream conditions. Incineration of emission streams containing organic vapors with halogen or sulfur compounds will generally create additional control requirements. Depending on the concentration of these compounds in the exhaust gas and the applicable regulations, scrubbers may be required to reduce the concentration of the halogen or sulfur species. Packaged single-unit thermal incinerators are available in many sizes to control VOC emissions with flow rates from a few hundred standard cubic feet per minute to about 50,000 scfm. In larger sizes, the units will be erected in the field.

Flares use open flames for destroying VOCs in waste gases during both normal operation and emergencies. They are typically used when the heat value of the waste gas cannot be recovered economically because of intermittent or uncertain flow or when process upsets occur. There are several types of flares, the most common of which are steam-assisted, air-assisted, and pressure-head flares. Regulatory requirements call for 98% destruction efficiency for flares in hazardous VOC service.[11]

Activated-carbon systems can also be used to control VOC emissions from a TSDF. If the carbon is virgin material, the VOC reduction efficiency will be high. This is generally the case for capture of VOCs in low-concentration streams from tankage, vents, and so on. However, once the carbon is spent, it will have to be replaced by additional carbon, and the spent carbon will have to be disposed of as hazardous waste. This is very costly. Activated carbon can also be used in a regenerative system; however, the VOC concentration in the exhaust stream should be of the order of 200 ppmv or higher. Also, with steam-regenerated systems, a wastewater stream is formed that will require treatment and the VOCs recovered as liquids will have to be incinerated in a hazardous-waste incinerator. Again, these are costly options.

Preventive Measures

Preventive measures are ideal means to reduce air pollutant emissions from waste sources or land treatment and disposal operations. These VOC emissions can be recovered or reduced through physical or procedural means before they are ultimately disposed of in the land disposal site. Table 4 lists some of the preventive measures that could be considered for land-disposed materials.

Sludges and organic wastes that readily generate gases and leachate are known to contribute significantly to VOC pollutant emissions. Thus liquids containing organic compounds should be controlled because of their potential for contamination not only of ambient air, but also of groundwater.

CONCLUSIONS

Atmospheric emissions from land treatment and disposal practices are now recognized as significant emission sources when local health issues and regional hydrocarbon loadings are considered. Historically, these area or fugitive sources have been ignored since emission rates from these sources were considered small or not measurable. But the recent development, organization, and publication of fugitive source emission assessment technologies, as well as health concerns, have brought these air emissions into the foreground.

Fugitive emission sources from land treatment and disposal practices can, like classic point sources, be measured or estimated and controlled. The assessment technologies presented can be used to assess emissions, evaluate control technologies, and help maintain acceptable emissions from these facilities. By using treatment systems, control devices, and preventive measures, fugitive emissions from land treatment and disposal practices can be regulated to acceptable levels.

TABLE 4. Preventive Measures for Reducing VOCs at TSDFs

Type of Preventive Measure	Comments
VOC source reduction	Reduce VOC in the waste by recovery, recycling, and reuse at the sources.
Waste devolatilization	Process the waste before disposal to reduce the VOC content. Includes sorption, biodegradation, or other techniques.
Design handling equipment leak-free	Use leak-free components and weld all joints rather than using threaded joints.
Segregation of wastes	Separate wastes according to volatility and handle appropriately.
Regulatory requirements	Manifest and permit low VOC waste measures.

References

1. *Environmental Progress and Challenges,* EPA-230-07-88-033, U.S. Environmental Protection Agency, August 1988.
2. *Handbook of Remedial Action at Waste Disposal Sites,* EPA-625/6-82-006, U.S. Environmental Protection Agency, June 1982.
3. L. Baker, R. Capouya, R. Crooks, et al., *The Landfill Testing Program: Data Analysis and Evaluation Guidelines,* State of California Air Resources Board, Stationary Source Division, Sacramento, CA, August 1990.
4. C. Cowherd, P. Englehart, G. E. Muleski, et al., *Hazardous Waste TSDF (Treatment, Storage, and Disposal Facilities): Fugitive Particulate Matter Air Emissions Guidance Document,* EPA/450/3-89/019, Midwest Research Institute for Environmental Protection Agency, Research Triangle Park, NC, 1989.
5. *Air/Superfund National Technical Guidance Study Series, Volume 2, Estimation of Baseline Air Emission at Superfund Sites,* EPA/450-1-89/002, Radian Corp. for U.S. Environmental Protection Agency, Research Triangle Park, NC, 1989.
6. M. R. Keinbusch, *Measurement of Gaseous Emission Rates from Land Surfaces Using an Emissions Flux Chamber: Users Guide,* EPA Contract 68-02-3889, Radian Corp., Austin, TX, 1986.
7. *Industrial Ventilation: A Manual of Recommended Practice,* 20th ed. American Conference of Governmental Industrial Hygienists, Cincinnati, OH, 1989.
8. T. P. Nelson, B. M. Eklund, and R. G. Wetherold, *Field Assessment of Surface Impoundment Air Emissions and Their Control Using an Inflated Dome and Carbon Adsorption System,* Radian Corp. for U.S. Environmental Protection Agency, Cincinnati, OH, 1985.
9. G. J. Langley, and R. G. Wetherold, *Evaluation of Maintenance for Fugitive VOC Emissions Control,* EPA-600/52-81-080, Radian Corp. for U.S. Environmental Protection Agency, Research Triangle Park, NC, 1981.
10. *Control Techniques for Particulate Emissions from Stationary Sources—Vol. 1 and 2,* EPA-450/3-81-005a/b, U.S. Environmental Protection Agency, Office of Air Quality Planning and Standards, Research Triangle Park, NC, 1982.
11. *Evaluation of the Efficiency of Industrial Flares: Background—Experimental Design Facility,* EPA 600/2-83-070, U.S. Environmental Protection Agency, 1983.
12. W. J. Farmer, M. S. Yang, J. Letey, et al., *Land Disposal of Hexachlorobenzene Wastes: Controlling Vapor Movement in Soil,* EPA-600/280-119, U.S. Environmental Protection Agency, Office of Research and Development, Municipal Environmental Research Laboratory, Cincinnati, OH, 1980.
13. W. Farino, P. Spawn, M. Jasinski, et al., "Review of landfill AERR models," in *Evaluation and Selection of Models for Estimating Air Emissions from Hazardous Waste Treatment, Storage, and Disposal Facilities,* Revised Draft Final Report, Contract No. 68-02-3168, U.S. Environmental Agency, Office of Solid Waste, Land Disposal Branch, 1983, pp 5-1–5-13.
14. T. T. Shen, "Estimating hazardous air emissions from disposal sites," *Pollution Eng.,* 13(8):31–34 (1981).
15. T. T. Shen and G. H. Sewell, "Air pollution problems of uncontrolled hazardous waste sites," *Civil Eng. for Practicing Design Eng.,* 3(3):241–252 (1984).
16. T. T. Shen, "Air pollution assessment of toxic emissions from hazardous waste lagoons and landfills," *Environ. Int.,* 11(1):71–76 (1985).
17. T. T. Shen, "Air quality assessment for land disposal of industrial wastes," *Environ. Mgmt.,* 6(4):297–305 (1982).
18. S. T. Hwang, "Estimating and field-validating hazardous air emissions from land disposal facilities," *Proceedings of the Third Pacific Chemical Engineering Conference,* Seoul, Korea, 1983, pp 338–343.
19. G. B. DeWolf and R. G. Wetherold, *Protocols for Calculating VOC Emissions from Land Applications Using Emission Models,* Radian Corp., Austin, TX, EPA Contract No. 68-02-3850, U.S. Environmental Protection Agency, Research Triangle Park, NC, 1984.
20. R. G. Wetherold and D. A. Dubose, *A Review of Selected Theoretical Models for Estimating and Describing Atmospheric Emissions from Waste Disposal Operations,* EPA Contract 68-03-3038, U.S. Environmental Protection Agency, Office of Research and Development, Industrial Environmental Research Laboratory, Cincinnati, OH, 1982.
21. *Hazardous Waste Treatment, Storage, and Disposal Facilities (TSDF)-Air Emission Models* (Draft Report) Research Triangle Institute for the U.S. Environmental Protection Agency, Office of Air Quality Planning and Standards, Research Triangle Park, NC, 1987.
22. G. D. DeWolf and R. G. Wetherold, *Protocols for Calculating VOC Emissions from Surface Impoundments Using Emissions Models: Technical Note,* Radian Corp., Austin, TX,

EPA Contract No. 68-02-3850, U.S. Environmental Protection Agency, Research Triangle Park, NC, 1984.

23. D. Mackay and P. J. Leinonen, "Rate of evaporation of low-solubility contaminants from water bodies to atmosphere," *Environ. Sci. Technol.*, 17(4):211–217 (1983).
24. D. Mackay and A. T. K. Yeun, "Mass transfer coefficient correlations for volatilization of organic solutes from water," *Environ. Sci. Technol.*, 17(4):211–217 (1983).
25. S. Watkins, *Background Document for the Surface Impoundment Modeling System (SIMS)*, EPA/SW/DK-EPA/450/4; 90/009B-89/013B, Radian Corp. for U.S. Environmental Protection Agency, Research Triangle Park, NC, 1989.
26. D. C. Misenheimer, *Surface Impoundment Modeling System (SIMS), Version 1.0 (for Microcomputers)*, EPA-/SW/DK-90/009, U.S. Environmental Protection Agency, Research Triangle Park, NC, 1989.
27. U.S. Environmental Protection Agency draft document, *Proceedings for Conducting Air Pathway Analysis for Superfund Applications*, Vol. I, prepared by NUS Corp., December 1988.

Bibliography

L. C. Adkins, S. H. Nacht, J. J. Doria, et al., *Methods for Assessing Exposure to Chemical Substances, Volume 3*, EPA/560/5-85/003, Versar, Inc., for U.S. Environmental Protection Agency, Washington, DC, 1985.

C. C. Allen, "Prediction of air emissions from surface impoundments," *Proceedings of the American Institute of Chemical Engineers 1986 Summer National Meeting*, New York, 1986.

R. April and J. Vorbach, *Best Demonstrated Available Technology (BDAT) Background Document for K043*, EPA/530/SW-89/048L, Versar, Inc., for U.S. Environmental Protection Agency, Washington, DC, 1989.

Army Engineer Waterways Experiment Station, *Guide to the Disposal of Chemically Stabilized and Solidified Waste*, EPA/530/SW-82/872, Army Engineer Waterways Experiment Station for U.S. Environmental Protection Agency, Washington, DC, 1982.

K. W. Brown and D. E. Daniel, "Potential groundwater implications of land disposal of toxic substances," *Proceedings of the Water Resources Symposium (12)*, San Antonio, TX, 1984.

T. G. Brna and C. B. Sedman, *Waste Incineration and Emission Control Technologies*, EPA/600/D-87/147, EPA/600/D-87/147-S, U.S. Environmental Protection Agency, Research Triangle Park, NC, 1987.

L. Cahill, *Assessment of Incineration as a Treatment Method for Liquid Organic Hazardous Wastes*, EPA/230/02-86/005, U.S. Environmental Protection Agency, Washington, DC, 1985.

P. N. Cheremisinoff, "Hazardous waste treatment and recovery systems," *Pollution Eng.* 20(2):52 (1988).

R. R. DuPont, "Measurement of volatile hazardous organic emissions," *JAPCA* 37(2):168 (1987).

R. DuPont, T. B. Hardy, and J. A. Reineman-Coover, "Evaluating the uncertainty of estimates of hazardous air emissions from land treatment systems using the Thibodeaux-Hwang AERR model," *Proceedings of the APCA Annual Meeting and Exhibition*, Dallas, TX, 1988.

R. R. DuPont, "Measurement of volatile hazardous organic emissions," *JAPCA* 37(2):168 (1987).

B. M. Eklund, T. P. Nelson, and R. G. Wetherold, *Field Assessment of Air Emissions and Their Control at a Refinery Land Treatment Facility, Volume 1*, EPA/600/2-86/086A, Radian Corp. for U.S. Environmental Protection Agency, Cincinnati, OH, 1987.

ENVIRON Corp., *Documentation for the Development of Toxicity and Volume Scores for the Purpose of Scheduling Hazardous Waste*, ENVIRON Corp. for U.S. Environmental Protection Agency, Washington, DC, 1985.

D. J. Fingleton, H. Oezkaynak, B. Burbank, et al., "Assessing exposure to toxic substances from land disposal of hazardous wastes," *Proceedings of the 1986 Air Pollution Control Association Annual Meeting and Exhibition*, CONF-860606-17, Minneapolis, MN, 1986.

M. Ghassemi, K. Crawford, and M. Haro, *Leachate Collection and Gas Migration and Emission Problems at Landfills and Surface Impoundments*, EPA/600/2-86/017, Multidisciplinary Energy and Environmental Systems and Applications for U.S. Environmental Protection Agency, Cincinnati, OH, 1986.

A. R. Gohlson, J. R. Albritton, and R. K. M. Jayanty, *Evaluation of the Flux Chamber Method for Measuring Volatile Organic Emissions from Surface Impoundments*, EPA/600/3-89/008, Research Triangle Institute for U.S. Environmental Protection Agency, Research Triangle Park, NC, 1989.

J. R. Gronow, A. N. Schofield, and R. K. Jain, *Land Disposal of Hazardous Waste: Engineering and Environmental Issues*, R/D-5401-EN-03, Cambridge University for Army Research Development and Standardization Group, Cambridge, England, 1988.

R. C. Hanisch and M. A. McDevitt, "Protocols for sampling and analysis of surface impoundments and land treatment/disposal sites for VOCs," *Proceedings of the Annual Meeting—Air Pollution Control Association*, Pittsburgh, PA, 1985.

S. Q. Hassan and J. P. Herrin, "Steam stripping and batch distillation for the removal/recovery of volatile organic compounds," *Proceedings of the HAZMAT Central '89 Conference*, EPA/600/D-89/009, U.S. Environmental Protection Agency, Cincinnati, OH, 1989.

International Council of Scientific Unions, *Disposal of Toxic Wastes*, GEN/4/8.27/87, Paris, France, 1987.

L. Jones and J. R. Berlow, *Best Demonstrated Available Technology (BDAT) Background Document for K015*, EPA/530/SW-88/031A, U.S. Environmental Protection Agency, Washington, DC, 1988.

R. G. Lewis, B. E. Martin, D. L. Sgontz, et al., *Measurement of Fugitive Atmospheric Emissions of Polychlorinated Biphenyls from Hazardous Waste Landfills*, EPA/600/J-85/243, Battelle Columbus Labs for U.S. Environmental Protection Agency, Research Triangle Park, NC, 1985.

I. J. Licis, *High Temperature Destruction of Hazardous Waste Practice, Performance, Prospects*, EPA/600/D-86/046, U.S. Environmental Protection Agency, Cincinnati, OH, 1986.

R. C. Loehr and J. F. Malina, "Land treatment: A hazardous waste management alternative," *Proceedings of a Water Resources Symposium (13th)*, EPA/600/9-86/011, Austin, TX, 1986.

R. J. Lutton, *Evaluating Cover Systems for Solid and Hazardous Waste*, EPA/530/SW-82/867, Army Engineer Waterways Experiment Station for U.S. Environmental Protection Agency, Washington, DC, 1982.

R. McDonald and D. Janes, *Hazardous Waste Treatment, Storage, and Disposal Facilities (TSDF)—Air Emission Models*

(For Microcomputers), EPA/SW/DK-88/046, Research Triangle Park, NC, 1987.

C. M. Northeim, C. C. Allen, and B. A. Westfall, *Thin-Film Evaporation as a Pretreatment Technique for Removing Volatile Organics from Petroleum Refinery Wastes,* EPA/600/D-88/063, Research Triangle Institute for U.S. Environmental Protection Agency, Cincinnati, OH, 1988.

NUS Corp., *No Migration Variances to the Hazardous Waste Land Disposal Prohibitions: A Guidance Manual for Petitioners,* EPA/530-SW-89/032, NUS Corp. for Earth Technology Corp., Alexandria VA, and U.S. Environmental Protection Agency, Washington, DC, 1989.

E. T. Oppelt, *Pretreatment of Hazardous Waste,* EPA/60/D-87/047, U.S. Environmental Protection Agency, Cincinnati, OH, 1987.

K. S. Park, D. L. Sorensen, J. L. Sims, et al., "Volatilization of wastewater trace organics in slow rate land treatment systems," *Hazardous Waste Hazardous Mat.,* 5(3):219 (1988).

S. H. Poe, K. T. Valsaraj, L. J. Thibodeaux, et al., "Equilibrium vapor phase adsorption of volatile organic chemicals," *J. Hazardous Mat.* 19(1):17 (1988).

T. T. Shen and G. H. Sewell, "Control of VOC emissions from waste management facilities," *J. Environ. Eng. ASCE,* 114(6):1392 (1988).

C. Springer, K. T. Valsaraj, and L. J. Thibodeaux, *In-Situ Methods to Control Emissions from Surface Impoundments and Landfills,* EPA/600/2-85/124, Arkansas University for Louisiana State University and U.S. Environmental Protection Agency, Cincinnati, OH, 1985.

K. Sumino and E. K. S. Hum, "Toxic wastes: Prevention better than cure," *Proceedings of the Asia-Pacific Symposium on Environmental and Occupational Toxicology,* Singapore, 1987.

U. S. Environmental Protection Agency, *Assessment of Fluidized-Bed Combustion Solid Wastes for Land Disposal,* Washington, DC, 1985.

U.S. Environmental Protection Agency, *Assessment of Incineration as a Treatment Method for Liquid Organic Hazardous Wastes,* EPA/230/02-86/002, Washington, DC, 1985.

U.S. Environmental Protection Agency, *Best Demonstrated and Available Technology (BDAT) Background Document for K016, K019, K020, K030,* EPA/530/SW-88/031B, Washington, DC, 1988.

U.S. Environmental Protection Agency, *Evaluation Guidelines for Toxic Air Emissions from Land Disposal Facilities: Technical Resource Document for Air Emissions, Monitoring, and Control,* Washington, DC, 1981.

U.S. Environmental Protection Agency, *Guidance Document for Subpart F Air Emission Monitoring—Land Disposal Toxic Air Emissions Evaluation Guideline,* Washington, DC, 1981.

U.S. Environmental Protection Agency, *Handbook: Control Technologies for Hazardous Air Pollutants,* EPA/625/6-8/014, Research Triangle Park, NC, and Cincinnati, OH, 1986.

U.S. Environmental Protection Agency, *Hazardous Waste Treatment, Storage, and Disposal Facilities (TSDF)—Air Emission Models, Documentation,* EPA/450/3-87/026, EPA/SW/DK-88/046A, Research Triangle Park, NC, 1987.

U.S. Environmental Protection Agency, "Incineration and treatment of hazardous waste," *Proceedings of the 1985 Annual Research Symposium,* EPA/600/985-028, Cincinnati, OH, 1985.

U.S. Environmental Protection Agency, *Test Methods for Evaluating Solid Waste, Volume 1A through 1C, and Volume 2. Field Manual Physical/Chemical Methods (3rd Edition),* EPA/SW-846, Washington, DC, 1986.

K. T. Valsaraj and L. J. Thibodeaux, "Equilibrium adsorption of chemical vapors on surface soils, landfills, and landfarms—a review," *J. Hazardous Mat.* 19(1):79 (1988).

W. G. Vogt and J. J. Walsh, "Volatile organic compounds in gases from landfill simulators," *Proceedings of the Annual Meeting—Air Pollution Control Association,* Pittsburgh, PA, 1985.

L. K. Wang and M. H. S. Wang, *Guidelines for Disposal of Solid Wastes and Hazardous Wastes, Volume 1–5,* LIR/02-87/230, Lenox Institute for Research, Inc., for U.S. Environmental Protection Agency, Washington, DC, 1987.

R. G. Wetherold, B. M. Eklund, B. L. Blaney, et al., "Assessment of volatile organic emissions from a petroleum refinery land treatment site," *Proceedings of the National Conference on Hazardous Wastes and Hazardous Materials,* EPA/600/D-86/074, Atlanta, GA, 1986.

R. G. Wetherold and W. D. Balfour, "Volatile emissions from land treatment systems," *Proceedings of the Water Resources Symposium (13),* Austin, TX, 1986.

MUNICIPAL SOLID WASTE LANDFILL GAS EMISSIONS

Michael J. Barboza, P.E.

Landfilling of municipal solid waste (MSW) has been a generally acceptable means of disposal for many years. MSW landfills receive primarily household and/or commercial waste. Current practice is to spread the waste in layers, compacting and covering it with soil. The compacted layers compose the landfill building blocks called cells. The buried waste decomposes biologically and chemically to produce solid, liquid, and gaseous products.

MSW landfills are potential sources of emissions of gas mixtures generated from the natural decomposition of organic wastes and vapors from volatile compounds present in the wastes. The concerns associated with landfill gas commonly involve odors, combustion/explosion hazards, and possible toxic effects. Landfill operators must consider subsurface gas migration, gas collection, control and recovery systems, and ambient air quality impacts, including odors. Studies of landfill gas emissions have been implemented to support landfill permits and landfill closures, and to assess impacts and site acceptability for alternative uses. Figure 1 illustrates the activities, emission sources, and control devices at an MSW landfill.

PROCESS DESCRIPTION[1]

There are an estimated 6000 active MSW landfills in the United States receiving around 209 million megagrams

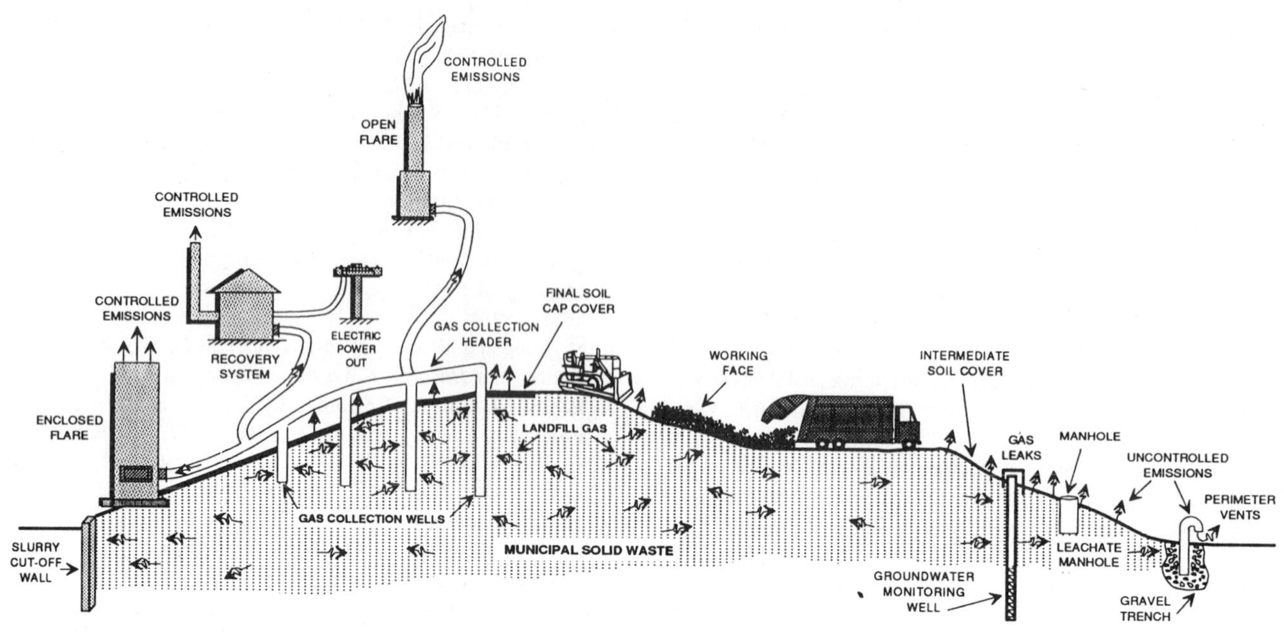

FIGURE 1. Municipal Solid Waste Landfill Gas Emissions

(Mg) (230 million tons) of waste annually. Most (around 93%) active landfills have a design capacity of 5 million Mg (5.5 tons) or less. Many landfills with capacities of 1 million to 20 million Mg (1.1 million to 22 million tons) are around 44–29% filled respectively. Average acceptance rates of landfills range from 50 Mg per day (55 tons per day) for smaller landfills (less than 1 Mg design capacity) to around 4000 Mg per day (4405 tons per day) for larger landfills (greater than 20 Mg design capacity). The average waste acceptance rate is around 31,780 Mg per year or 122 Mg per day (35,000 tons per year or 135 tons per day). The median acceptance rate, however, is 2724 Mg a year or 10.4 Mg a day (3000 tons per year or 11.5 tons per day) because of a small number of large landfills.

Cells are usually around 8 feet high. The working face can extend across the width of the cell. Usually the refuse is covered at the end of each day with at least 6 inches of soil. One foot of intermediate and 2 feet of final soil cover are commonly used. Liners, either soil or a combination of soil and synthetic, are often used beneath the landfill to contain leachate (the liquid produced from waste decomposition) to prevent groundwater contamination.

Composition of Municipal Solid Waste[1]

Municipal waste composition[1] can be described as containing household wastes (71.97%), commercial wastes (17.19%), construction demolition wastes (5.83%), industrial process wastes (2.73%), sewage sludge (0.51%), and lesser amounts of incinerator ash, and a small quantity of generator hazardous wastes, infectious wastes, asbestos-containing waste materials, and other wastes.

Landfill Gas Composition

Landfill gas can be generated at both active and inactive landfills. Natural biological processes occurring in landfills transform the waste's constituents, producing leachate and gas (sometimes referred to as biogas). Initially, decomposition is aerobic until the oxygen supply is exhausted. Anaerobic decomposition of buried refuse produces relatively high concentrations of methane and carbon dioxide. The major components of landfill gas and various phases of gas generation are shown in Figure 2.[1] Decomposition of waste can reach the anaerobic steady methanogenic phase in around two to four years.[2] Landfill gas consists of approximately 50% methane and 50% carbon dioxide (by volume). Other constituents of landfill gas can include ammonia, hydrogen sulfide, nitrogen, hydrogen chloride, carbon monoxide, and a variety of volatile organic compounds (VOCs).

Methane, the major constituent of landfill gas, is a colorless, odorless gas that is only slightly soluble in water and burns readily in air. It is generally very stable; however, when mixed with air at concentrations of around 5–15% by volume, it is highly explosive. Methane is considered a simple asphyxiant and thus is of lesser concern from a toxicity point of view, except in situations where it may displace oxygen (such as confined spaces). With the exception of methane, the other gases are usually present in trace amounts and not of concern with regard to explosion hazard.

The heating value of landfill gas is derived mostly from its methane content. The lower heating value of methane is 991 Btu/ft^3, making the heating value of landfill gas around 4000 KC/m^3 (400–500 Btu/ft^3), about half that of natural

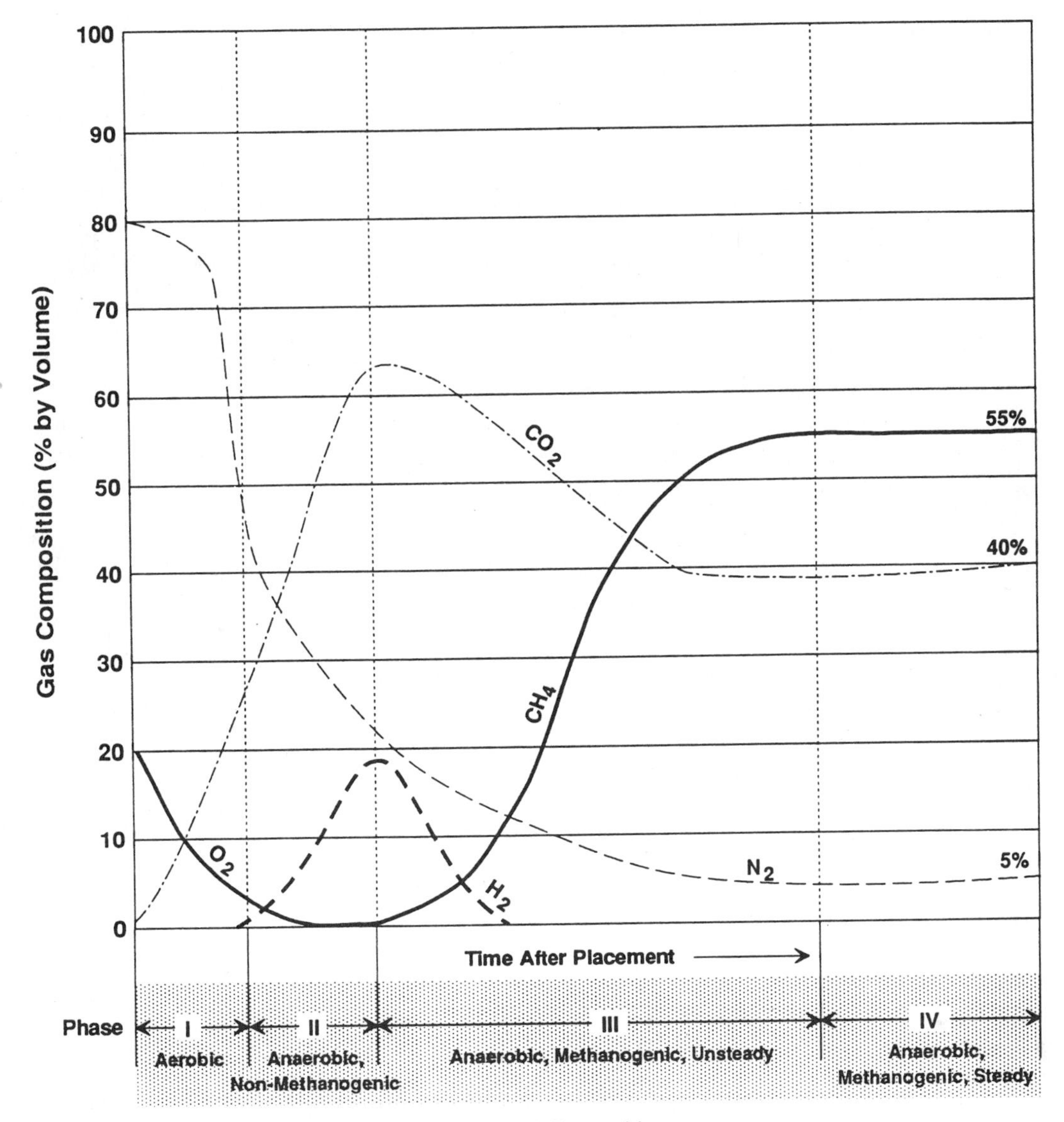

FIGURE 2. Typical Landfill Gas Generation and Composition

gas. This heating value makes disposal of landfill gas by burning practical and efficient. It also can be recovered for use as a fuel for combustion in engines or boilers for the generation of electricity.

Typically less than 1% (by volume) of landfill gas is nonmethane organic compounds (NMOCs).[1] A number of different NMOCs have been detected in landfill gas. Table 1 lists compounds that have been found in landfill gas, are of regulatory concern, and are of toxicological or other significance.[3] The number of times quantitated and the highest and average concentrations found in a U.S. Environmental Protection Agency (EPA) study of landfills are presented.[1] Some of the compounds are carcinogenic and available unit risk factors are listed. Also listed are compounds regulated by the states of California[4] and New Jersey.[5]

Gas Generation

The MSW deposited in landfills decays chemically and biologically to produce solid, liquid, and gaseous products. The rate of landfill gas generation varies and is affected by a number of factors, such as:

- Type and composition of waste
- Amount of biodegradable materials
- Age of waste
- Moisture content and pH
- Temperature

The decay process is accelerated with warm temperatures and moisture. Anaerobic decomposition can produce heat with internal landfill temperatures that range from 29°C

TABLE 1. Air Toxics Compounds

Compound	Regulated Compounds		Potential Carcinogen Inhalation Unit Risk Factor, $(\mu g/m^3)^{-1}$	Max Detected in Landfill Gas at a Landfill,[c] ppm	Typical Landfill Gas Concentration[d]		
	N.J.[a]	Calif.[b]			Ave., ppm	High, ppm	Times Quantified
Acetone				23	5.94	32	26
Benzene	X	X	8.3E-06	5.2	3.6	52.2	45
Benzyl chloride			NA[e]	0.42	—[f]	—	—
Butanone-2				41	8.17	57.5	27
Carbon tetrachloride	X	X	1.5E-05	ND[g]	1.85	68.3	37
Chlorobenzene				0.14	0.38	10	29
Chloroethane				1.6	2.03	9.2	29
Chloroform	X	X	2.3E-05	ND	0.08	1.56	36
Dichlorobenzene-1,4			NA	4.8	<.009	<0.9	28
Dichloroethane-1,1				7.5	3.51	19.5	33
Dichloroethane-1,2	X	X	2.6E-05	13	1.3	30.1	37
Dichloroethene-1,1			5.0E-05	9	0.23	3.1	32
Dimethyl disulfide				1	0.02	0.55	2
Dimethyl sulfide				5.8	0.55	1	2
Dioxane-1,4	X		NA	14	—	—	—
Ethylbenzene				29	21.73	428	31
Ethylene dibromide	X	X	2.2E-04	ND	<0.9	<0.9	2
Ethyleneimine	X		NA	ND	—	—	—
Ethyl toluene-4				14	—	—	—
Freon 114				0.25	—	—	—
Hexanone-2				2.3	—	—	—
Methyl mercaptan				0.26	1.87	3.3	3
Methylene chloride		X	4.1E-06	38	24.5	174	37
Methyl-2-Pentanone-4				7.7	89	89	26
Tetrachloroethane-1,1,2,2	X		5.8E-05	ND	0.1	2.35	28
Tetrachloroethene	X	X	9.5E-07	26	8.43	77	39
Toluene				120	59.34	758	40
Trichloroethane-1,1,1		X			0.84	9	38
Trichloroethane-1,1,2	X		1.6E-05	ND	<0.1	0.1	28
Trichloroethene	X	X	1.7E-06	5.7	3.98	34	44
Trimethylbenzene-1,2,4				14	—	—	—
Trimethylbenzene-1,3,5				6.2	—	—	—
Vinyl chloride		X	4.2E-05	6.5	7.71	48.1	42
Xylenes				77	17.11	70.9	27

[a]New Jersey Subchapter 17 list.[5]
[b]California Calderon compounds.[4]
[c]Reference 3.
[d]Reference 1.
[e]NA = not available.
[f]— = not listed.
[g]ND = not detected.

to 37°C (84°F to 98.6°F). The moisture content of waste can vary, with the highest gas production occurring at a moisture content of around 60–78%.[6]

Landfill gas generation projections can be used to estimate emissions and for designing gas collection and control systems, such as identifying the size and number of collection wells and flare capacities. A number of different landfill gas-flow models have been used, some of which are based on stoichiometric chemically limiting formulas and empirical relationships. A review of gas generation models indicates a fairly broad range of gas production rates. Because of the variability and uncertainties in the factors affecting gas production, projections are usually made for average or typical conditions and worst-case or maximum anticipated gas generation. The available models include Scholl Canyon, Paulas Verde, and Lockman Associates.[1,7]

A number of landfill gas-flow models have been used to estimate gas generation/production from MSW landfills. One such landfill gas-flow model is based on an average rate of production of 100–300 ft^3 of gas per ton of in-place refuse per year, with an average of 180. Gas-flow models used to make gas generation rate productions can be used to estimate maximum gas production over the life of a landfill for various years of operation and postclosure.

AIR EMISSIONS CHARACTERIZATION

Emissions for MSW landfills are commonly associated with direct releases of landfill gas. Releases of gas to the atmosphere can be from a number of different points at a landfill. The major gas emissions are from surface releases, venting from various on-site structures, cracks in soil cover surfaces, and venting from gas collection and venting systems.

The emissions of individual gas constituents can be estimated based on the concentrations of constituents in the gas and using gas generation projections and gas collection system efficiencies. Other methods include estimating emissions from the various source categories with a number of different techniques, such as diffusion equations and leak estimation.[3,8,9]

Gas generation often results in pressure differential with higher pressures within the landfill. This pressure gradient acts as a driving force for gas movement either laterally or upward. Additionally, normal variations in atmospheric pressure result in cyclical changes in gas pressure within landfills that tend to lag changes in atmospheric pressure. These changes in differential pressure within landfills can cause atmospheric pumping action, resulting in increases in gas releases.

An emission factor for total VOCs from landfills ranges from 13.6 to 35.8 tons of VOCs per 10^6 tons of waste per year for dry states and those with greater precipitation, respectively.[1]

Odors are frequently associated with air emissions from MSW landfills.

Landfill Surfaces

Methane is lighter (around 0.544 times) than air and tends to rise within the landfill cell, where it may be released from vents and cracks or openings in the landfill surface and may migrate through the soil cover. There is a tendency for settling to occur at landfills, which can cause cracks and openings in the surface cover and thus result in gas releases. Ninety percent of the settlement usually takes place during the first five years.[1]

Three mechanisms contribute to gas-phase transport of chemical species to the air from landfills via cover soils:

- Molecular diffusion
- Biogas generation
- Atmospheric presure pumping

Molecular diffusion is the dominant mechanism when no biogas is present within the landfill. Biogas can enhance molecular diffusion transport by three to ten times. Time-varying atmospheric pressure fluxuations can result in pumping action, which enhances the vapor-phase diffusion process. Fluxations in atmospheric pressure influence cell pressure in a nonlinear fashion, as can the biogas generation rate. Atmospheric pressure fluxuations have been found to enhance molecular diffusion by around 13%. It is suggested that the most critical parameter for controlling vapor emission rates is the cover (cap) porosity.[9]

Structures

MSW landfills often contain many structures, including leachate collection manholes, bedrock wells, groundwater monitoring and withdrawal wells, piezometers, inclinometers, gas headers, and landfill gas collection wells that are installed through the landfill cover and thus create a potential for gas to escape to the atmosphere. There can be large numbers (over 100 on some sites) of these structures on a given landfill site. Gas can accumulate in structures and may be released to the atmosphere from openings or leaks. The accumulation of gas in some structures may pose explosion hazards.

Other Emissions

Other sources of air emissions at landfills include emissions from heavy-duty vehicles on-site and sometimes fugitive dust. Additionally, there are emissions from trucks delivering waste and soil cover material to the landfill. These sources are usually of minor concern. Emissions from vehicles are controlled by equipment design, manufacture, and periodic equipment maintenance. Fugitive dust may be generated during landfilling activities, although concentrations may be easily controlled with dust suppression techniques (i.e., water spraying of roads). Dust is usually of concern as a nuisance.

Controlled Emissions

Combustion is the primary method of controlling landfill gas emissions. Uncontrolled emissions of constituents may be reduced with combustion controls, but this results in emissions of combustion products such as nitrogen oxides, carbon dioxide, carbon monoxide, and unburned NMOCs. Emissions of sulfur dioxide (SO_2) may result from the burning of sulfur compounds present in the gas, such as hydrogen sulfide (H_2S) and mercaptans.

AIR POLLUTION CONTROL MEASURES

Effective control of MSW landfill gas emissions requires a combination of gas containment, sometimes venting, collection, and effective destruction of organics in collected gas by flaring or recovery systems.

Gas Containment

Containment of landfill gas is most often used in combination with venting and collection systems. The use of low-permeability soil for cover and slurry walls can impede landfill gas movement. Gas containment and control of surface emissions can be accomplished by using a cover soil with low permeability and low porosity—a process referred to as capping. Capping is commonly used to reduce the infiltration of water from precipitation, but can also reduce gas emissions from the landfill surface cover. Slurry walls are commonly used for the containment of contaminated groundwater at landfills. These slurry walls can serve as barriers to lateral subsurface gas migration. To remain effective, the clay material in slurry walls and caps must be kept moist and not allowed to dry out.

Gas Venting

A number of passive and active venting systems have been used to control lateral subsurface migration of gas beyond landfill boundaries. Methods sometimes used in combination with venting systems include the utilization of bentonite slurry walls, concrete grout curtains, and plastic membranes. However, their effectiveness is often limited since they may not be totally impermeable. Methods of passive venting include the use of open trenches and gravel-filled trenches.

Active venting systems can include trenches with collection piping systems and gas extraction wells with mechanical blowers. These systems are more effective, since they provide a positive means of gas extraction under conditions where natural venting alone may not be sufficient, depending on physical and site-specific factors.

Gas Collection

Two basic types of landfill gas collection systems are active and passive. Passive collection systems are typically employed for atmospheric venting of landfill gas to prevent lateral migration or reduce potential explosion hazards. These systems emit uncontrolled landfill gas. Active systems are generally employed to collect landfill gas for control by combustion or for energy recovery. Active systems use mechanical blowers or compressors to create a pressure gradient and extract the landfill gas.

Gas collection systems consist of a series of vertical or horizontal recovery wells that withdraw gas from within the landfill and convey it via a piping header system to control devices, such as flares or recovery systems. These collection systems are used to reduce the amount of uncontrolled landfill gas that can escape to the atmosphere. Gas collection systems are installed along the perimeters of landfills to prevent lateral subsurface gas migration, while others are placed in the landfill refuse for gas venting, control, or recovery. A header system (commonly of polyvinyl chlorine or high-density polyethylene) conveys collected landfill gas from the wells or trenches to vents, flares, or recovery systems, with blowers or compressors. The size and the type of blower or compressor depend on total gas flow rate, total system pressure drop, and vacuum required. Centrifugal blowers allow easy throttling throughout their operational range. Blowers can accommodate system pressure drops of up to 50 inches of water and flow rates of 100–100,000 cfm. For lower flow rates and higher pressures, regenerative (combination of axial and centrifugal) blowers can be used.[1]

The efficiency of landfill gas collection systems can vary widely from system to system and landfill to landfill. Gas collection efficiencies for active landfills can range from around 40% to 60% of gas production, and for closed landfills with gas collection and disposal using state-of-the-art well and piping systems, from 85% to 90%. The South Coast Air Quality Management District (SCAQMD), a district with a large population of landfills with gas controls, estimates greater than 70% gas collection efficiency for both operational and post closure periods. Waste Management Inc., a large single operator of controlled landfill facilities, estimated on the order of 80% on average over both periods considering portions of landfills where waste volumes afforded rather complete enclosure of gas collection systems.[7] In summary, typical gas collection efficiency can be 70% to 80% during operation and 90% following closure. Gas collection efficiencies are related to the density of gas collection wells, withdrawal rates, and differential pressures. Excessive vacuum and gas withdrawal can lead to air intrusion, resulting in reduced biological activity (waste decomposition), reduction of methane content (gas combustibility), and the creation of potential fire or explosion hazards.

Gas Combustion

The high percentage of methane in landfill gas and the resultant heat content make it suitable for disposal by combustion in flares, often without auxiliary fuel, and also for use in energy-recovery systems as fuel.

Flares

Flares are devices used to dispose of large quantities of unwanted flammable gases and vapors, such as gas streams that contain potentially odorous or harmful constituents, by combustion. Flares ideally burn waste gas completely and smokelessly. Thus emissions of landfill gas and its combustible constituents (including VOCs) can be controlled. Use of combustion techniques requires the removal of water from landfill gas by condensate traps, knockout drums, or filters.

Two types of flares used at landfills for gas disposal are open (elevated) flares and enclosed (ground) flares. The major components of a flare are the gas burner, stack, water seal/liquid trap, controls, pilot burner, and ignition system. Some are equipped with automatic pilot ignition systems, temperature sensors, and air and combustion controls.

Elevated flares used for landfill gas have a flare tip with no obstruction to flow. The flare tip is of the same diameter as the stack. Open flares resemble large Bunsen burners with candlelike flames. Combustion and mixing of air and

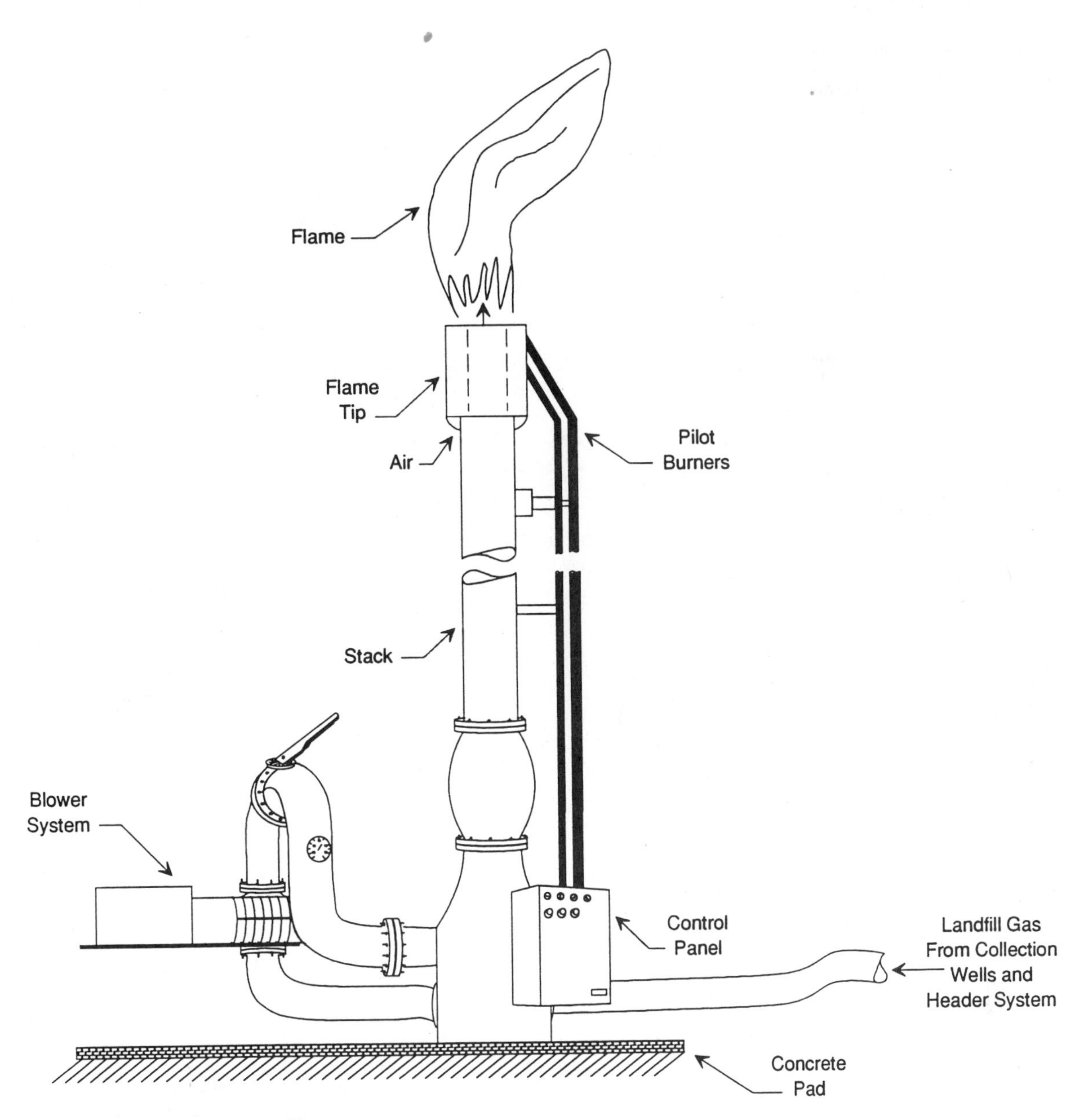

FIGURE 3. Typical Open Flare

gas take place above the flare with the flame in the atmosphere, as shown in Figure 3.

Enclosed flares are composed of multiple gas burner heads placed at ground level in an enclosure that is usually refractory lined. A diagram of a typical enclosed flare is presented in Figure 4. Some enclosed ground flares are equipped with automatic damper controls that regulate supply air depending on combustion temperature, which is monitored within the flare stack around 3 feet below the outlet. Air pollution control regulatory agencies are tending to require the use of enclosed flares because it is easier to perform emission tests on them and they have better combustion control than open flares. Requirements can include limits on carbon monoxide emissions (i.e., less than 100 ppm in exhaust gas) and residence time and combustion temperature of at least 0.3 second and 1000°F.

Flare combustion efficiency is related to flame temperature, residence time of gases in the combustion zone, turbulent mixing of the combustion zone, and amount of oxygen available for combustion. Flares can be quite effective in the destruction of hydrocarbons and provide approximately 98% or better control of hydrocarbons, depending on design and operating parameters.[1,10]

Emissions from flares are related to landfill gas-flow rates, combustibility of gas (use of auxiliary fuel), temperature and retention time, and the composition of the gas feed. The emissions from flares can be estimated from combustion calculations used to estimate the volume and con-

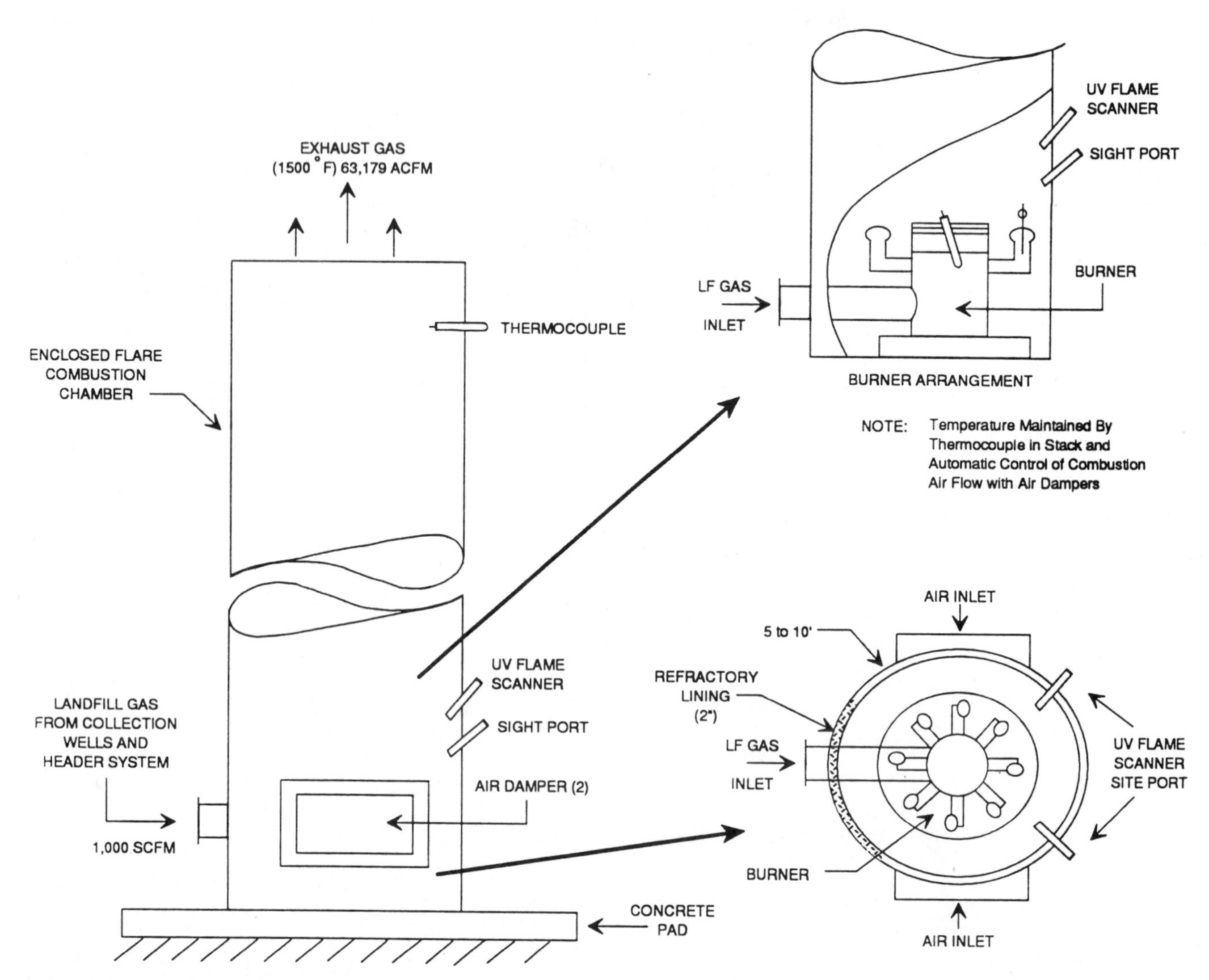

FIGURE 4. Typical Enclosed Flare

stituents of exhaust gases based on the makeup of the feed gas, flare operating parameters, and combustion efficiency.

The temperature of exhaust gases from flares can range from 1000°F to 2000°F. These high temperatures can cause relatively high plume rise, and good dispersion, and coupled with high combustion efficiency, can result in relatively low concentration impacts, depending on the gas constituents. The plume rise associated with open flares is enhanced because of the added height of the open flame. This, however, can be offset by the lack of control of combustion air and the quenching effect of and radiant heat loss from the atmosphere and the open flame. Consideration should be given during placement to potential hazards associated with radiant heat from open flares. This is of less concern with enclosed flares, since the flames are in a shell with refractory lining and have better temperature, excess air, and combustion control. It also is easier to perform emission tests on them.

Recovery Systems

Gas-recovery systems involve combustion techniques that destroy organics and recover energy from the combustion process. The most common energy-recovery systems used at MSW landfills include internal combustion engines, boilers with steam turbines, and gas turbines. They are often used to produce electricity, although boilers may be used to

TABLE 2. Air Emissions from Control of MSW Landfill Gas[a,b]

	Secondary Air Emissions (lb/mm scf LFG)					
Control Technique	NO_x[c]	CO[d]	HCl[e]	CO_2	PM	SO_2[f]
Enclosed flare	4.9	58	12	60,000	Neg.[g]	3.0
Boiler	70	17	12	50,000	Neg.	3.0
Gas turbine	26.4	12.5	12	60,000	37	3.0
Internal-combustion engine	111	259	12	60,000	Neg.	3.0

[a]Reference 1.
[b]Based on control of 21,840 lb/mm scf landfill gas of methane (assuming 50% methane) and a range of NMOCs of 56 to 3,395 lb/mm scf landfill gas (convention ranges from 237 ppm to 14,294 ppm NMOC, assumed molecular weight of NMOC equal to hexane).
[c]NO_x emissions for boilers based on emission factors for natural gas and converted to lb/mm scf landfill gas assuming 500 Btu/set.
[d]CO emissions for boilers based on emission factors for natural gas and converted to lb/mm scf landfill gas assuming 500 Btu/set.
[e]Secondary HCl emissions are based on NMOC compositions assuming all chlorine is converted to HCl.
[f]Secondary SO_2 emissions are based on assuming all sulfur is converted to SO_2.
[g]Neg. = negligible.

produce steam for heating. The factors needed to make landfill gas-recovery systems economically feasible are as follows[1]:

- Greater than 2 million tons (1.8 million Mg) of refuse
- Greater than a 35-foot depth of waste
- Greater than 35 acres of landfill area
- Refuse conducive to gas production
- Continued landfill operation
- Short time lapse after closing for closed landfills

The relative combustion efficiencies (approximate or typical destruction efficiencies) of these recovery systems are as follows[1]:

Internal combustion engines	78.96%
Boilers	99.7%
Gas turbines	99.8%

Emissions from internal combustion engines have the highest by-product emissions (e.g., NO_x CO, SO_x, etc.) when compared with other recovery techniques (boilers, gas turbines). Table 2 presents a summary of emissions from flares, boilers, gas turbines, and internal combustion engines burning landfill gas.[1]

Practices

A number of practices may be used to insure that landfill gas control systems are working properly to control gas emissions and minimize odor complaints. An effective maintenance, repair, and monitoring program includes:

- A crack-sealing program to maintain the effectiveness of the cover soil.
- Ongoing saturation of natural soil barriers such as clay in caps and slurry walls to maintain high efficiency.
- Repair and sealing of gas leaks in structures and maintenance of gas leaks in collection systems.
- Use of good cover material for containment of landfill gas.
- Monitoring of migration control systems such as with combustible gas meters and probes placed around the landfill perimeter.
- Monitoring for explosion hazards.

References

1. *Air Emissions from Municipal Solid Waste Landfills—Background Information for Proposed Standards and Guidelines,* Draft EIS (preliminary draft), U.S. Environmental Protection Agency Office of Air Quality Planning and Standards, Emission Standards Division, Research Triangle Park, NC, March 12, 1990.
2. D. J. Hagerty, J. L. Pavoni, and J. E. Heers, *Solid Waste Management,* Van Nostrand Reinhold, New York, 1973.
3. M. J. Barboza, "An integrated study of air toxics emissions from an MSW landfill, American Society of Civil Engineers, Proceedings of the 1991 National Conference on Environmental Engineering, Reno, NV, July 8–10, 1991.
4. Air Resources Board, State of California, *The Landfill Gas Testing Program: A Report to the Legislature,* prepared by Stationary Source Division Toxic Air Contaminants Identification Branch, 1988.
5. Title 7, New Jersey Administrative Code, Subchapter 17, Control and Prohibition of Air Pollution by Toxic Substances.
6. *ADA Landfill Gas Control at Military Installations,* U.S. Army ADA 140–190 January 1984.
7. A. Eschenroeder, D. Burmaster, S. Wolff, et al., *Health Risk Assessment of a Proposed Landfill for MSW in Douglas, Massachusetts,* for Douglas Environmental Associates, Inc., Allanova, Lincoln, MA, June 1990.
8. L. W. Baker, and K. P. Mackay, "Screening models for estimating toxic air pollution near a hazardous waste landfill," *J. APCA* (November 1985).
9. L. J. Thibodeaux, C. Springer, and L. M. Riley, "Models of mechanisms for the vapor phase emission of hazardous chemicals from landfills," *J. Hazardous Mat.,* 7:63–74 (1982).

10. *Control Technologies for Hazardous Air Pollutants,* EPA/625/6-86/014, U.S. Environmental Protection Agency Air and Energy Engineering Research Laboratory, Research Triangle Park, NC, September 1986.

Bibliography

D. R. Brunner, and D. J. Keller, *Sanitary Landfill Design and Operation,* U.S. Environmental Protection Agency, 1972.

C. Ceni and C. Emerson, *The Landfill Gas Testing Program: A Second Report to the California Legislature,* State of California Air Resources Board, Stationary Source Division, Toxics Program Support Section, June 9, 1989.

Engineering Science, Austin, TX, *A Report on a Flare Efficiency Study,* Vol. I, Chemical Manufacturers Association, Washington, DC, March 1983.

Landfill Methane Recovery, M. M. Schunacher, Ed.; Noyes Data Corp., Park Ridge, NJ, 1983.

John G. Pacey, "Controlling landfill gas," *Waste Age,* 32–36 (March 1981).

20
Groundwater and Soil Treatment Processes

Lori P. Andrews, P.E.

Cleanup activities from Superfund sites, RCRA corrective actions, facility closures, hazardous material emergencies, property transfers, and underground storage tank excavations can result in the release of emissions to the atmosphere. Remediation activities may involve the cleanup of contaminated soil and contaminated groundwater. The most common groundwater cleanup technologies include air stripping, carbon adsorption and bioremediation. Other technologies, such as chemical oxidation, chemical precipitation, reverse osmosis, and ion exchange, are much more site specific.

The most common soil cleanup technologies include excavation/disposal (such as in a landfill), soil vapor extraction, stabilization/fixation, bioremediation, thermal stripping, soil washing, incineration, and vitrification. Emissions from landfills and hazardous waste incinerators were discussed elsewhere in this manual. This discussion will focus on emissions from air stripping, soil vapor extraction, and thermal stripping processes.

PROCESS DESCRIPTION

Historically, air stripping has been the most popular process to remove volatile organic compounds (VOCs) from contaminated groundwater. In a typical air stripper, contaminated groundwater is countercurrently contacted with air in a packed tower. As the groundwater comes into intimate contact with air, stripping of VOCs occurs as a result of the volatilization of the dissolved organics into the gas phase to reach equilibrium. Air, now containing VOCs, exits the stripper and, if left uncontrolled, may pose an air pollution problem.

Soil vapor extraction has proved to be effective for the removal of VOC and light petroleum hydrocarbons from subsurface soils. It is performed by applying a vacuum to the soils to induce volatilization of soil contaminants. The extracted air is usually treated for VOC removal prior to discharge to the ambient air.

Thermal stripping of volatiles in contaminated soils has also proven effective for the treatment of subsurface soils. The process removes volatiles by indirectly heating (at temperatures of 400–800°F) the soils and solids to temperatures sufficient to vaporize the hazardous components. Vapors desorbed must be treated before their release to the atmosphere, where they would pose an air pollution problem.

AIR EMISSIONS CHARACTERIZATION

Contamination in soil and groundwater is highly variable. As such, air emissions from contaminated soil and groundwater treatment processes will be highly variable. Table 1 presents typical air emission values for cleanup processes used at Superfund sites.[1]

AIR POLLUTION CONTROL MEASURES

Emissions from air strippers are typically controlled by activated-carbon adsorbers or incinerators. Activated-carbon adsorber efficiencies typically range between 75% and 95% at inlet VOC concentrations ranging from a few hundred to a few thousand parts per million by volume (ppmv). The VOC emissions leaving the adsorber are usually less than 100 ppmv. The vapor-phase carbon adsorption process is very sensitive to water and, therefore, to the relative humidity of the airstream. Since the effluent from the air stripper is saturated with water, it is necessary to reduce the relative humidity of the airstream, preferably to 50% or less.

Both thermal and catalytic incinerators have been used to control emissions from air strippers. Thermal incineration (at temperatures in the range of 1200–1500°F) may be

TABLE 1. Typical Air Emission Values

Remedial Option	Typical Operation Rate	Uncontrolled VOC Emissions
Air stripping	3500 L/min	5–50 kg/day[a]
Soil vapor Extraction	0.15–0.85 m^3/min[b]	1–110 kg/day

[a]Assumed 1–10 mg/L pollutant.
[b]Exhaust gas rate per recovery well.
Source: Reference 1.

expensive since it requires a substantial energy input to destroy dilute gas-phase contaminants. Catalytic incineration (at temperatures in the 700°F range) operates at much lower temperatures and can offer substantial energy savings. Typical VOC removal efficiencies in the systems range between 90% and 98%. The combustion of VOCs in incinerators may lead to additional pollutants such as hydrogen chloride, which will then have to be factored into the evaluation.

Incineration and carbon adsorption are also the emission control technologies of choice for soil vapor extraction systems.

Thermal stripping systems are generally proprietary processes with their own individual air pollution control systems. Weston's low-temperature thermal treatment (LT^3) system includes a fabric filter for particulate emissions control, a condenser (to reduce the volume of water vapor), and an afterburner to destroy the organics.[2] The afterburner is designed for 1800°F operation and a one-second gas residence time. If the waste feed contains sulfides and/or chlorides, the LT^3 system is configured to include a caustic wet scrubber following the afterburner.

IT's thermal desorption system includes an emission control system consisting of a cyclone, wet scrubber, refrigerated condenser, Brink's Demister, HEPA filter, and carbon adsorber.[3]

Deutsche Babcock's system has an indirectly heated rotary kiln to clean up contaminated soils.[4] The emission control system includes an afterburner for pyrolysis gas destruction, a flue gas cooler, a dry lime injection system to neutralize acid gases, and a fabric filter for particulate control.

References

1. B. Eklund and J. Summerhays, "Procedure for estimating emissions from the cleanup of Superfund sites," *J. A&WMA,* 40(1):17–23 (1990).
2. R. M. Leuser, et al., "Low temperature thermal treatment of contaminated soil," Paper No. 89-2.2, 82nd Annual Meeting of the Air and Waste Management Association, Anaheim, CA, June 1989.
3. R. D. Fox, et al., "Thermal treatment for the removal of PCBs and other organics from soil," *Environ. Prog.,* 10(1):40–44 (1991).
4. D. Schneider and B. D. Beckstrom, "Cleanup of contaminated soils by pyrolysis in an indirectly heated rotary kiln," *Environ. Prog.,* 9(3):165–168 (1990).

Index